Test Yourself

Take a quiz at *The Living World* Online Learning Center to gauge your mastery of chapter content. Each chapter quiz is specially constructed to test your comprehension of key concepts. Immediate feedback on your responses explains why an answer is correct or incorrect. You can even e-mail your quiz results to your professor!

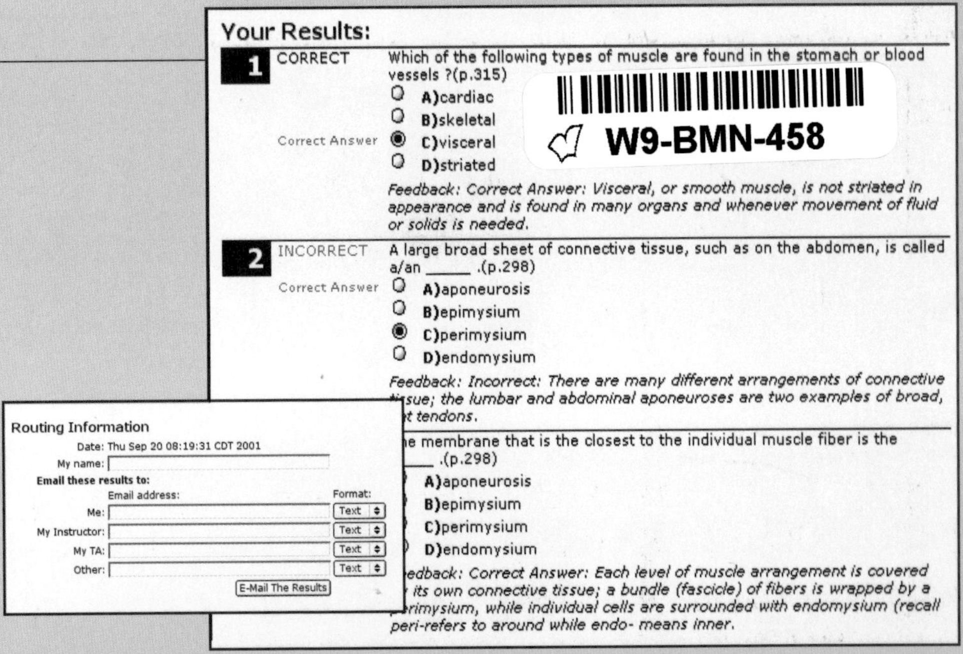

Your Results:
1 CORRECT
Which of the following types of muscle are found in the stomach or blood vessels ?(p.315)
A)cardiac
B)skeletal
Correct Answer C)visceral
D)striated
Feedback: Correct Answer: Visceral, or smooth muscle, is not striated in appearance and is found in many organs and whenever movement of fluid or solids is needed.
2 INCORRECT
A large broad sheet of connective tissue, such as on the abdomen, is called a/an _____ .(p.298)
Correct Answer A)aponeurosis
B)epimysium
C)perimysium
D)endomysium
Routing Information
Date: Thu Sep 20 08:19:31 CDT 2001
My name:
Email these results to:
Email address:
Format:
Me:
My Instructor:
My TA:
Other:
Text
E-Mail The Results
A)aponeurosis
B)epimysium
C)perimysium
D)endomysium

Virtual Labs

The Virtual Lab that accompanies each chapter provides you with an open-ended experience of scientific inquiry. By going to the eBRIDGE and clicking on *Virtual Lab*, you will be able to read about an experiment done by today's leading scientists and then have an opportunity to recreate the lab online. Real data sets and guided follow-up help you gain confidence in working with data and drawing conclusions.

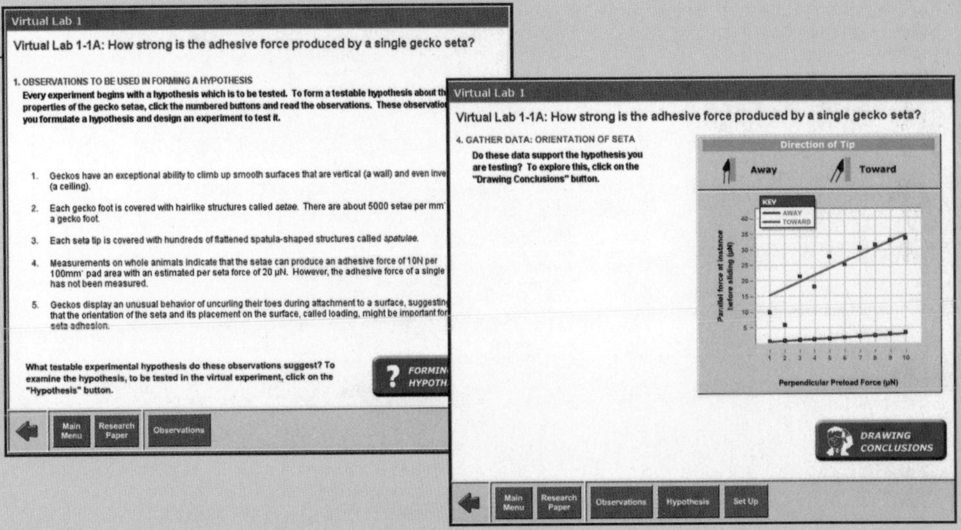

Virtual Lab 1
Virtual Lab 1-1A: How strong is the adhesive force produced by a single gecko seta?
1. OBSERVATIONS TO BE USED IN FORMING A HYPOTHESIS
What testable experimental hypothesis do these observations suggest? To examine the hypothesis, to be tested in the virtual experiment, click on the "Hypothesis" button.
Main Menu
Research Paper
Observations
Virtual Lab 1
Virtual Lab 1-1A: How strong is the adhesive force produced by a single gecko seta?
4. GATHER DATA: ORIENTATION OF SETA
Do these data support the hypothesis you are testing? To explore this, click on the "Drawing Conclusions" button.
Direction of Tip
Away
Toward
Perpendicular Preload Force (μN)
DRAWING CONCLUSIONS
Main Menu
Research Paper
Observations
Hypothesis
Set Up

Access to Premium Learning Materials

The Living World Online Learning Center is your portal to exclusive interactive study tools like McGraw-Hill's Essential Study Partner, BioCourse.com, PowerWeb: Biology, and Tutoring Service.

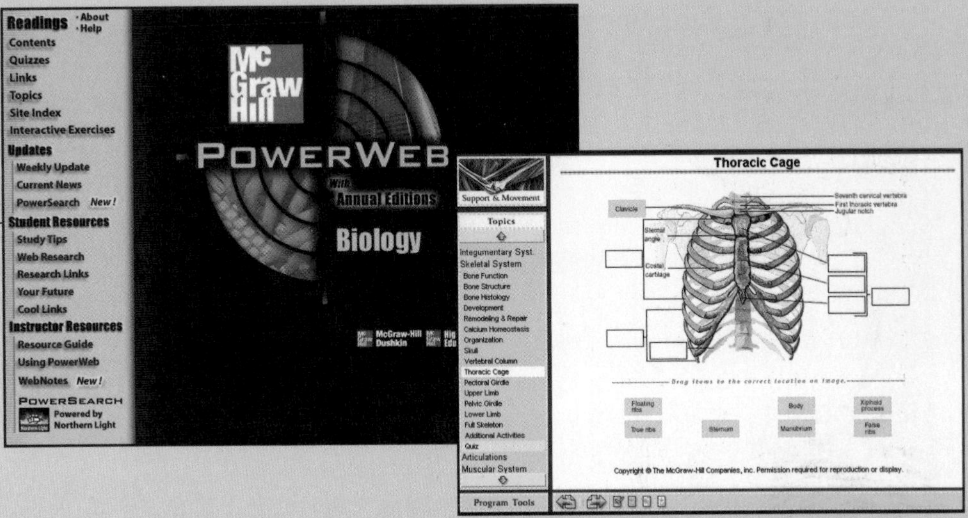

Readings
About
Help
Contents
Quizzes
Links
Topics
Site Index
Interactive Exercises
Updates
Weekly Update
Current News
PowerSearch New!
Student Resources
Study Tips
Web Research
Research Links
Your Future
Cool Links
Instructor Resources
Resource Guide
Using PowerWeb
WebNotes New!
POWERSEARCH
Powered by Northern Light
Mc Graw Hill
POWERWEB
Annual Editions
Biology
Thoracic Cage
Topics
Integumentary Syst.
Skeletal System
Thoracic Cage
Muscular System
Program Tools

Visit: www.mhhe.com/tlw3

THIRD EDITION

The Living World

GEORGE B. JOHNSON

Washington University
St. Louis, Missouri

McGraw Hill

Boston Burr Ridge, IL Dubuque, IA Madison, WI New York San Francisco St. Louis
Bangkok Bogotá Caracas Kuala Lumpur Lisbon London Madrid Mexico City
Milan Montreal New Delhi Santiago Seoul Singapore Sydney Taipei Toronto

McGraw-Hill Higher Education
A Division of The ***McGraw-Hill*** *Companies*

THE LIVING WORLD, THIRD EDITION

Published by McGraw-Hill, a business unit of The McGraw-Hill Companies, Inc., 1221 Avenue of the Americas, New York, NY 10020.

This book is printed on recycled, acid-free paper containing 10% postconsumer waste.

International 1 2 3 4 5 6 7 8 9 0 VNH/VNH 0 9 8 7 6 5 4 3 2 1
Domestic 2 3 4 5 6 7 8 9 0 VNH/VNH 0 9 8 7 6 5 4 3 2

ISBN 0–07–234720–1
ISBN 0–07–119936–5 (ISE)

Publisher: *Martin J. Lange*
Senior sponsoring editor: *Patrick E. Reidy*
Developmental editor: *Anne L. Melde*
Senior development manager: *Kristine Tibbetts*
Off-site developmental editors: *Megan Jackman/Elizabeth Sievers*
Executive marketing manager: *Lisa Gottschalk*
Lead project manager: *Peggy J. Selle*
Production supervisor: *Kara Kudronowicz*
Design manager: *Stuart D. Paterson*
Cover/interior designer: *Christopher Reese*
Cover image: *Charles Lynn Bragg/Raging Art Unlimited*
Senior photo research coordinator: *Lori Hancock*
Photo research: *Meyers Photo-Art*
Lead supplement producer: *Audrey A. Reiter*
Media technology producer: *Janna Martin*
Compositor: *Carlisle Communications, Ltd.*
Typeface: *10.4/12 Times Roman*
Printer: *Von Hoffmann Press, Inc.*

The credits section for this book begins on page 781 and is considered an extension of the copyright page.

Library of Congress Cataloging-in-Publication Data

Johnson, George B. (George Brooks), 1942–
The living world / George Johnson ; illustrations done by Bill Ober, Claire Garrison. —3rd. ed.
p. cm.
Includes bibliographical references (p.) and index.
ISBN 0–07–234720–1
1. Biology. I. Title.

QH308.2 .J62 2003
570—dc21 2001052186
CIP

INTERNATIONAL EDITION ISBN 0–07–119936–5

www.mhhe.com

BRIEF CONTENTS

PART ONE
THE STUDY OF LIFE

PART TWO
THE LIVING CELL

PART THREE
THE CONTINUITY OF LIFE

PART FOUR
THE EVOLUTION AND DIVERSITY OF LIFE

PART FIVE
PLANT LIFE

PART SIX
EVOLUTION OF ANIMAL LIFE

PART SEVEN
ANIMAL LIFE

PART EIGHT
THE LIVING ENVIRONMENT

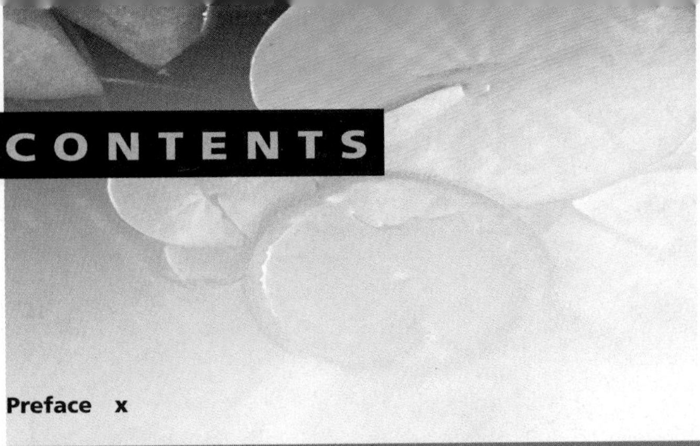

CONTENTS

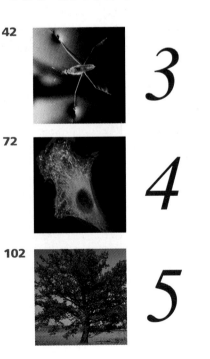

PART THREE
THE CONTINUITY OF LIFE

PART FOUR
THE EVOLUTION AND DIVERSITY OF LIFE

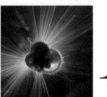

PART SEVEN
ANIMAL LIFE

496 22

The Animal Body and How It Moves

The vertebrate body has the architecture of a crane, with jointed limbs attached to a central skeleton, within which is a protected cavity that contains the body's organs.

The Animal Body Plan 498
Tissues of the Vertebrate Body 506
The Skeletal System 512
The Muscular System 516

524 23

Circulation and Respiration

Vertebrates circulate blood through the body by pumping it with the heart. The blood picks up O_2 in the lungs and delivers it to the tissues; from the tissues it collects CO_2 and carries it back to the lungs.

Circulation 526
Evolution of Vertebrate Circulatory Systems 534
Respiration 540
Lung Cancer and Smoking 548

554 24

The Path of Food Through the Animal Body

The digestive tract consists of a series of chambers that form a continuous tube. As food passes through this tube, large molecules in the diet are broken down into smaller bits the body can use.

Food Energy and Essential Nutrients 556
Digestion 560
Maintaining the Internal Environment 572

586 25

How the Animal Body Defends Itself

Several types of white blood cells protect the body from infection and eliminate cancer cells. These cells screen the surface of body cells for foreign proteins, and eliminate any cells that contain "nonself" proteins.

Three Lines of Defense 588
The Immune Response 596
Defeat of the Immune System 604

610 26

The Nervous System

The nervous system regulates the body's activities and its internal condition, using special signaling cells called neurons to transmit information about the body's condition to the brain, and relay commands back.

Neurons and How They Work 612
The Central Nervous System 620
The Peripheral Nervous System 626
The Sensory Nervous System 628

642 27

Chemical Signaling Within the Animal Body

The body uses chemical signals called hormones to control a wide variety of organs and processes. Many of these hormones are secreted by glands under the direct control of the brain.

The Neuroendocrine System 644
The Major Endocrine Glands 648
Other Hormone-Producing Glands 656

660 28

Reproduction and Development

Vertebrate reproduction involves the fertilization of a female egg by a male sperm. Hormones regulate the development of the female eggs, and prepare her body to nourish the developing embryo.

Vertebrate Reproduction 662
The Human Reproductive System 668
The Female Reproductive Cycle 672
The Course of Development 674
Birth Control 680

PART EIGHT
THE LIVING ENVIRONMENT

684 29

Ecosystems

Ecology is the study of the way organisms interact with one another and with their physical environment. A community of organisms, and the habitat in which they live, is called an ecological system, or ecosystem.

The Energy in Ecosystems 686
Materials Cycle Within Ecosystems 691
How Weather Shapes Ecosystems 696
Major Kinds of Ecosystems 700

712 30

Living in Ecosystems

The species living in an ecosystem have evolved many accommodations to living together. This coevolution has made the members of ecosystems interdependent in many ways.

Population Dynamics 714
How Competition Shapes Communities 720
How Coevolution Shapes Communities 724
Predator and Prey Interactions 728
Community Stability 734

742 31

Planet Under Stress

Our planet faces numerous problems that threaten its future existence, including chemical pollution on a global scale and nonsustainable use of the earth's nonrenewable resources.

Global Change 744
Saving Our Environment 748
Solving Environmental Problems 754

Appendix A: Classification of Organisms 758

Appendix B: Answers to Concept Reviews 763

Glossary: Key Terms and Concepts 766

Credits 781

Index 785

PREFACE

Writing *The Living World* has been the most enjoyable of my academic pursuits. I wrote it to create a text that would be easy for today's students to learn from—a book that focused on concepts rather than information. More than most subjects, biology is at its core a set of ideas, and if students can master these basic ideas, the rest comes easy.

Unfortunately, while most of today's students are very interested in biology, they are put off by the terminology. When you don't know what the words mean, it's easy to slip into thinking that the matter is difficult, when actually the ideas are simple, easy to grasp, and fun to consider. It's the terms that get in the way, that stand as a wall between students and science. With this text I have tried to turn those walls into windows, so that readers can peer in and join the fun.

Analogies have been my tool. In writing *The Living World* I have searched for simple analogies that relate the matter at hand to things we all know. As science, analogies are not exact, but I do not count myself compromised. Analogies trade precision for clarity. If I do my job right, the key idea is not compromised by the analogy I use to explain it, but rather revealed.

A second barrier stands between students and biology, and that is the mass of information typically presented in an introductory biology text. The fun of learning biology becomes swamped by a sea of information. To make the ideas of biology more accessible to students, I have trimmed away a lot of detail traditionally taught in freshman biology courses.

My first step was to attack the traditional table of contents (usually a formidable list of chapters covering a broad range of topics). The number of chapters in biology textbooks has grown over the years, until today the most widely used short text has 44 chapters! I have cut back ruthlessly on this overwhelming amount of information, reducing the number of chapters in this edition of *The Living World* to 31. I think this matches more closely what is actually being taught in classrooms, and, as you will see, all that is really important is preserved.

I have deliberately combined photosynthesis and cellular respiration into a single chapter in *The Living World*, not because metabolism is unimportant, but because the basic principles a student needs to understand are simple and easy to explain. The metabolic activities of organisms are most easily grasped when the many similarities between photosynthesis and cellular respiration reveal their underlying unity.

There is no way to avoid the fact, however, that some of the important ideas of biology are complex. No student encountering photosynthesis for the first time gets it all on the first pass. To aid in learning the more difficult material, I have given special attention to key processes like photosynthesis and osmosis, the ones that form the core of biology. The key processes of biology are not optional learning. A student must come to understand every one of them if he or she is to master biology as a science. A student's learning goal should not be simply to memorize a list of terms, but rather to be able to visualize and understand what's going on. With this goal in mind, I have prepared special "This is how it works" process boxes for some four dozen important processes that students encounter in introductory biology. Each of these process boxes walks the student through a complex process, one step at a time, so that the central idea is not lost in the details.

It is no accident that *The Living World* begins with a chapter on evolution and ecology. These ideas, central to biology, provide the student a framework within which to explore the world of the cell and gene which occupy the initial third of the text. Biology at the gene and cellular level is every bit as much an evolutionary accomplishment as are the animal phyla encountered later in the text. Students learn about cells and genes much more readily when they are presented in an evolutionary context, as biology rather than as molecular machinery.

In organizing *The Living World*, I set out to present the concepts of biology—as much as my writing skills would allow—as a story. I teach that way, and students learn more easily that way. Evolution and diversity are no longer treated in separate sections of the text, for example, but rather are combined into one continuous narrative. Traditionally, students are exposed to weeks of evolution before tackling animal diversity, struggling past the Hardy-Weinberg equilibrium and population growth equations (microevolution) and on through Darwin's discoveries (macroevolution). Then, when all that is done, they are dragged through a detailed tour of the animal phyla, followed by a long excursion into botany. In large measure, the three areas are presented as if unrelated to each other. In *The Living World* I have chosen instead to combine all three of these areas into one treatment, presenting biological diversity as an evolutionary journey. It is a lot more fun to teach this way, and students learn a great deal more, too.

New This Edition: Content Enhancement

Deep into the task of preparing this third edition of *The Living World,* I was challenged by my daughter Caitlin, who was resenting my absence from family: "If your book is so good," she asked, "why do you need to work so much on its revision?" Good question. The answer, of course, is that biology has changed a lot in the few brief years since the last edition.

Genomics

Consider, for example, the Human Genome Project (chapter 10, Genomics). To gain some idea of why the explosion of interest in the human genome, consider the following. If the DNA molecule in one of your cells were to be stretched out straight, it would extend about six feet—very nearly the height of a human. How much of that DNA do you suppose is devoted to genes—to sequences encoding proteins? About an inch. That's right, less than 2% of your DNA is devoted to genes! Over half of the human genome is composed of independently replicating "transposable elements." This astonishing result goes right to the heart of what it means to be human.

Stem Cells

As a second item, consider stem cells. Barely mentioned in the previous edition, stem cells occupy the front pages of today's newspapers. The desirability of federal funding of stem cell research has become one of the major political issues of the day. An early human embryo, prior to implantation at six days, is composed of an outer layer of protective cells, and an inner cell mass of some 200 so-called embryonic stem cells. Each of these stem cells, as yet undeveloped, is capable of becoming any tissue in the body. In mice, these cells, if transplanted, can replace damaged heart muscle lost in heart attacks, neurons from severed spines, brain cells whose loss leads to Parkinson's, or insulin-producing pancreatic cells.

Why the controversy? The great promise of stem cell regenerative medicine is balanced by the fact that embryonic stem cell lines can only be obtained by harvesting embryonic stem cells from human embryos. This raises many ethical questions. Researchers point out that infertile couples using in vitro fertilization to conceive provide the chief source of human embryos—many more embryos are produced than are needed to conceive. These excess embryos would be destroyed if not used to obtain stem cells, researchers claim, mitigating any ethical concerns. Not so, respond critics, who believe that human life begins at conception, and that destroying a human embryo, for whatever purpose, is simply murder. Few issues in science so polarize public opinion. The enhancement chapter, "The Revolution in Cell Technology," provides an in-depth look at this controversial issue.

Cancer

Yet another area of major recent progress that affects every American is the search for a cure for cancer. Great progress has been made in the last few years, as researchers learn more about how cancer "happens." It turns out that everyone who gets cancer has accumulated mutations that accelerate cell proliferation, and other mutations that disable the brakes that cells normally apply when cell division starts to accelerate. To block cancer, researchers are inventing ways to inhibit the out-of-kilter accelerating step, and ways to reestablish brakes on the process. New progress is announced practically every month.

Gene Engineering

Few areas of biology have engendered as much sustained controversy among the general public as the prospect of using genetic engineering to produce so-called genetically modified food (GM food). Over the last two years much of the complexion of the argument has changed. Panic at the rapid pace of change has been replaced with a grudging acceptance, as the very real benefits of modifications have become more apparent. One clear example is provided by so-called "golden rice." A significant fraction of the world's people use rice as their staple food, but because rice is deficient in iron and vitamin A, these people often experience iron deficiency and poor vision. Addressing the problem head on, gene engineers added a battery of genes to rice to correct the deficiencies. As a result of these gene modifications, rice can be a far superior human food.

Bioterrorism

The anthrax attack on America in 2001 removes any doubt that the threat of bioterrorism is real. While a detailed treatment of infectious disease is usually far beyond the scope of an undergraduate nonmajor's text, this issue cries out to be addressed. The enhancement chapter "Infectious Disease and Bioterrorism" is intended to provide the information and background necessary to understand this important topic.

Ribosomes

Not all important progress in biology in the last few years has been reported on the evening news. One extremely important advance occurred in what may seem a prosaic area, ribosomes. Ribosomes are very complex organelles within cells that carry out protein synthesis. Each ribosome is made up of over 50 different proteins and several RNA molecules. It used to be thought that the catalysis of protein synthesis was carried out by the proteins, arrayed on an RNA framework. We have now learned that exactly the opposite is true. RNA molecules catalyze the assembly of protein chains from amino acids, with proteins stabilizing the relative positions of the individual RNA molecules.

Throughout the text, *The Living World,* Third Edition, has been updated to reflect the many changes that have occurred in biology in these last very active years.

New This Edition: The eBRIDGE

The single greatest change that has occurred in biology in the few years since the last edition of *The Living World* has been the blossoming of the Internet as a teaching resource. No student wants a 10-pound textbook, so in the past there have been serious constraints on how much "end-of-chapter" material could be crammed into a text. The Internet has now lifted that limitation. Because the Internet takes up no space in a textbook, I have been free to develop a battery of new tools to facilitate student learning. In this new edition of *The Living World* the Internet serves as an electronic bridge to a wealth of materials that drill, test, explore, and enhance a student's learning. I have called this electronic bridge between text and Internet resources the "eBRIDGE." No other text presents anything remotely like it.

How do you use the eBRIDGE? When you purchased *The Living World,* Third Edition, you received a free 6-month subscription to *The Living World's* Online Learning Center. When you want to use the eBRIDGE, go to *The Living World's* Online Learning Center, www.mhhe.com/tlw3. The first time you go there you will be asked to register by entering the passcode you received in your textbook and creating your individual user name and password. After you have registered, go to "student center" and click on "eBRIDGE." Select the chapter you want, say chapter 5, and a screen will appear that looks exactly like the eBRIDGE pages at the back of chapter 5 of the text—except that on your computer screen version all the underlined items are live. To explore any item, just click on the underlined name of that item, and you will immediately cross the eBRIDGE and enter the virtual space where that item resides.

For each chapter of *The Living World,* Third Edition, four sorts of resources can be reached via the eBRIDGE. On the left page of the eBRIDGE (illustrated above right), you will find Reinforcing Key Points, and Electronic Learning. On the right page of the eBRIDGE, discussed on page xiii, you will find video streaming lectures delivered by me in the Virtual Classroom, and open-ended laboratory investigations in the Virtual Lab.

Reinforcing Key Points

Every chapter is organized as a series of numbered one-page or two-page modules. The Reinforcing Key Points portion of the eBRIDGE is a within-chapter search engine devoted to helping a student explore all the resources of the Online Learning Center that apply to that particular numbered module. This saves a lot of running around looking for things.

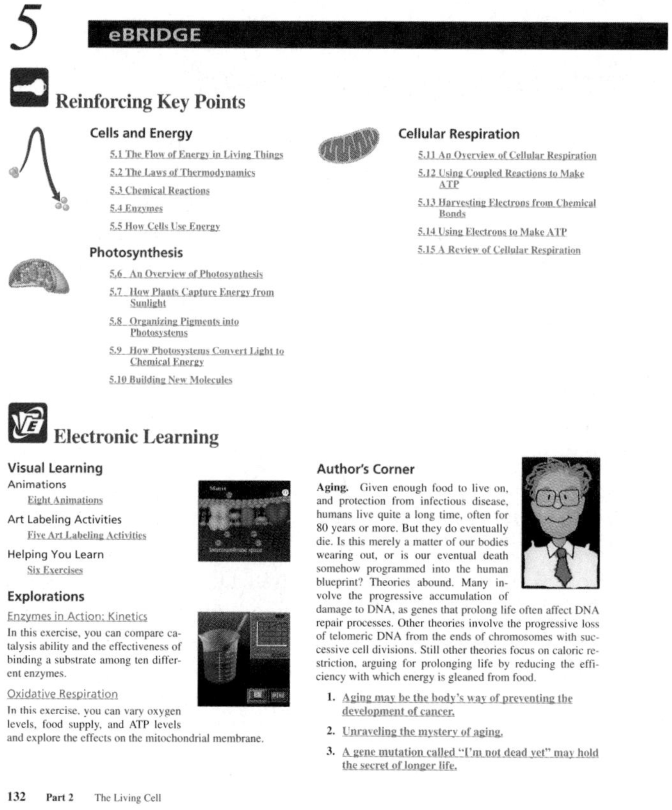
5 eBRIDGE

Reinforcing Key Points

Cells and Energy
5.1 The Flow of Energy in Living Things
5.2 The Laws of Thermodynamics
5.3 Chemical Reactions
5.4 Enzymes
5.5 How Cells Use Energy

Photosynthesis
5.6 An Overview of Photosynthesis
5.7 How Plants Capture Energy from Sunlight
5.8 Organizing Pigments into Photosystems
5.9 How Photosystems Convert Light to Chemical Energy
5.10 Building New Molecules

Cellular Respiration
5.11 An Overview of Cellular Respiration
5.12 Using Coupled Reactions to Make ATP
5.13 Harvesting Electrons from Chemical Bonds
5.14 Using Electrons to Make ATP
5.15 A Review of Cellular Respiration

Electronic Learning

Visual Learning
Animations
Eight Animations
Art Labeling Activities
Five Art Labeling Activities
Helping You Learn
Six Exercises

Explorations
Enzymes in Action: Kinetics
In this exercise, you can compare catalysis ability and the effectiveness of binding a substrate among ten different enzymes.
Oxidative Respiration
In this exercise, you can vary oxygen levels, food supply, and ATP levels and explore the effects on the mitochondrial membrane.

Author's Corner
Aging. Given enough food to live on, and protection from infectious disease, humans live quite a long time, often for 80 years or more. But they do eventually die. Is this merely a matter of our bodies wearing out, or is our eventual death somehow programmed into the human blueprint? Theories abound. Many involve the progressive accumulation of damage to DNA, as genes that prolong life often affect DNA repair processes. Other theories involve the progressive loss of telomeric DNA from the ends of chromosomes with successive cell divisions. Still other theories focus on caloric restriction, arguing for prolonging life by reducing the efficiency with which energy is gleaned from food.
1. Aging may be the body's way of preventing the development of cancer.
2. Unraveling the mystery of aging.
3. A gene mutation called "I'm not dead yet" may hold the secret of longer life.

132 Part 2 The Living Cell

Electronic Learning

The eBRIDGE links the student to a rich array of electronic learning resources.

Visual Learning

The eBRIDGE provides a rich assortment of animations, art labeling activities, and "helping you learn" drills. These visual resources provide a powerful learning tool, particularly for students who learn better visually.

Explorations

Explorations are fully interactive exercises that delve into interesting points covered in the chapter. One exploration allows you to analyze enzyme kinetics, another to construct a gene map from the results of a three-point cross, yet another to use DNA fingerprinting to examine real courtroom cases. While a lot of fun, these explorations are not simply games or simulations. Based on actual lab data, they allow students to gather and analyze data much as they might in a real lab.

Author's Corner

The Author's Corner takes the student to a collection of short "On Science" articles written by me on a topic intended to amplify and enrich some aspect of the chapter. The articles stress issues of current interest such as cloning and stem cells, forging a link between what students are learning and the world in which they live.

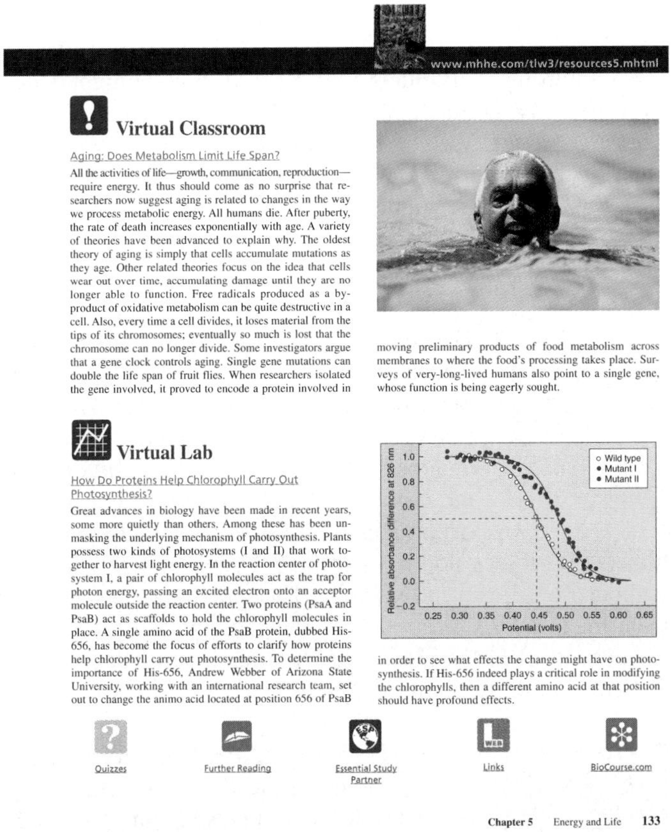
www.mhhe.com/tlw3/resources5.mhtml

Virtual Classroom

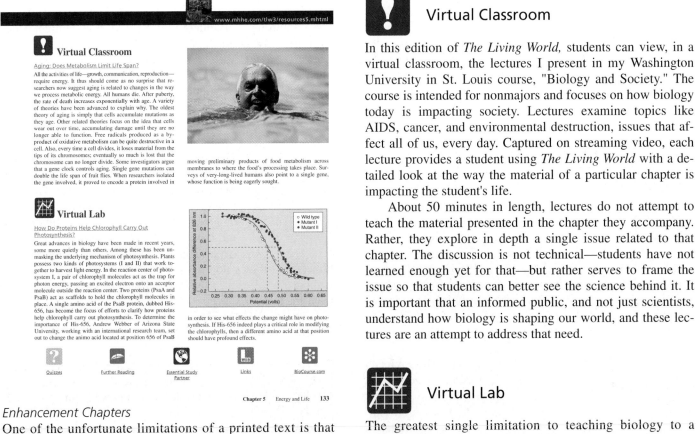

Aging: Does Metabolism Limit Life Span?

All the activities of life—growth, communication, reproduction—require energy. It thus should come as no surprise that researchers now suggest aging is related to changes in the way we process metabolic energy. All humans die. After puberty, the rate of death increases exponentially with age. A variety of theories have been advanced to explain why. The oldest theory of aging is simply that cells accumulate mutations as they age. Other related theories focus on the idea that cells wear out over time, accumulating damage until they are no longer able to function. Free radicals produced as a by-product of oxidative metabolism can be quite destructive in a cell. Also, every time a cell divides, it loses material from the tips of its chromosomes; eventually so much is lost that the chromosome can no longer divide. Some investigators argue that a gene clock controls aging. Single gene mutations can double the life span of fruit flies. When researchers isolated the gene involved, it proved to encode a protein involved in moving preliminary products of food metabolism across membranes to where the food's processing takes place. Surveys of very-long-lived humans also point to a single gene, whose function is being eagerly sought.

Virtual Lab

How Do Proteins Help Chlorophyll Carry Out Photosynthesis?

Great advances in biology have been made in recent years, some more quietly than others. Among these has been unmasking the underlying mechanism of photosynthesis. Plants possess two kinds of photosystems (I and II) that work together to harvest light energy. In the reaction center of photosystem I, a pair of chlorophyll molecules act as the trap for photon energy, passing an excited electron onto an acceptor molecule outside the reaction center. Two proteins (PsaA and PsaB) act as scaffolds to hold the chlorophyll molecules in place. A single amino acid of the PsaB protein, dubbed His-656, has become the focus of efforts to clarify how proteins help chlorophyll carry out photosynthesis. To determine the importance of His-656, Andrew Webber of Arizona State University, working with an international research team, set out to change the animo acid located at position 656 of PsaB in order to see what effects the change might have on photosynthesis. If His-656 indeed plays a critical role in modifying the chlorophylls, then a different amino acid at that position should have profound effects.

Relative absorbance difference at 826 nm
1.0
0.8
0.6
0.4
0.2
0.0
−0.2
0.25 0.30 0.35 0.40 0.45 0.50 0.55 0.60 0.65
Potential (volts)
○ Wild type
● Mutant I
● Mutant II

Quizzes — Further Reading — Essential Study Partner — Links — BioCourse.com

Chapter 5 Energy and Life 133

Enhancement Chapters

One of the unfortunate limitations of a printed text is that it cannot present detailed treatments of everything that a student might enjoy exploring, topics like dinosaurs and stem cells. The eBRIDGE provides a ready solution to this dilemma, as there is no length limitation to material accessed via the Internet. In this edition of *The Living World* you will find four "enhancement chapters," each a complete chapter written by the author devoted to presenting a topic of wide interest, beyond the scope of the printed text but well worth exploring:

The Revolution in Cell Technology. (eBRIDGE, Chapter 9) Stem cells and therapeutic cloning are both medically exciting and ethically controversial.

Infectious Disease and Bioterrorism. (eBRIDGE, Chapter 13) The anthrax attack on America leaves no doubt about the threat.

Dinosaurs. (eBRIDGE, Chapter 20) Dinosaurs dominated life on land for 150 million years, the many kinds presenting a long parade of evolutionary change.

Conservation Biology. (eBRIDGE, Chapter 31) Among the greatest challenges facing the biosphere in the new century is the accelerating rate of species extinction.

Virtual Classroom

In this edition of *The Living World,* students can view, in a virtual classroom, the lectures I present in my Washington University in St. Louis course, "Biology and Society." The course is intended for nonmajors and focuses on how biology today is impacting society. Lectures examine topics like AIDS, cancer, and environmental destruction, issues that affect all of us, every day. Captured on streaming video, each lecture provides a student using *The Living World* with a detailed look at the way the material of a particular chapter is impacting the student's life.

About 50 minutes in length, lectures do not attempt to teach the material presented in the chapter they accompany. Rather, they explore in depth a single issue related to that chapter. The discussion is not technical—students have not learned enough yet for that—but rather serves to frame the issue so that students can better see the science behind it. It is important that an informed public, and not just scientists, understand how biology is shaping our world, and these lectures are an attempt to address that need.

Virtual Lab

The greatest single limitation to teaching biology to a large freshman class is the inability to expose students to open-ended laboratory investigation. There is no substitute for this sort of hands-on experience. However, the interactive nature of the internet provides an opportunity for students to experience the intellectual challenge of scientific inquiry. The Virtual Labs that accompany each chapter of *The Living World,* Third Edition are open-ended investigations of real scientific problems. They require the student to think like a scientist, examining an issue, phrasing a question, forming a testable hypothesis, devising a way to test it, carrying out the experiment and gathering data, analyzing the data, and assessing whether or not the data support the student's hypothesis. Challenging and fun, the Virtual Lab experiments provide a student experience with open-ended inquiry, the intellectual process that real scientists go through every day in research.

The Living World, Third Edition contains 31 Virtual Labs, addressing topics as varied as how gecko lizards can walk on ceilings, to how hormones protect seed development in peas. The experiments in each case are real ones, involving actual data presented in a published research paper. No two replicas of an experiment yield the same data points, as the student experiences the same experimental error the investigator reports. Taken as a whole, the Virtual Labs are a powerful resource for experiencing how science is done, for learning how a scientist thinks.

Virtual Lab: A Closer Look

The Virtual Lab that accompanies each chapter of *The Living World,* Third Edition, provides students with an open-ended experience of scientific inquiry. As an example, consider the Virtual Lab accompanying chapter 31, an experiment attempting to gain a better understanding of why many amphibian populations today are exhibiting decreasing numbers and numerous individuals with severe developmental deformities. By going to the eBRIDGE for chapter 31 and clicking on the Virtual Lab devoted to this experiment, "Identifying the Environmental Culprit Harming Amphibians," a student can undertake an in-depth exploration of this experiment.

EXPLORE THE ISSUE BEING INVESTIGATED provides a detailed look at the experimental issue of amphibian decline, a problem of great concern to environmental scientists today. Frogs and other amphibians have been around since before the dinosaurs. If something in the environment is causing their abrupt decline, we need to know what it is. This initial discussion provides a conceptual framework for the student's examination of Andrew Blaustein's experiment, outlining the extent of the problem and reviewing the sorts of theories that have been advanced to explain the decline.

READ THE ORIGINAL RESEARCH PAPER allows the student to read the scientific paper Blaustein published to report his work, Blaustein, Andrew R. et al., "Ambient UV-B radiation causes deformities in amphibian embryos," *Proc. Nat. Acad. Sci. USA* 1997 (vol. 94):13735–13737, and a related paper, Blaustein, Andrew et al., "UV repair and resistance to solar UV-B in amphibian eggs: A link to population declines?" *Proc. Nat. Acad. Sci. USA* 1994 (vol. 9):1791–1795. There is no better introduction to the reality of an experiment than reading the actual research paper that reports it. While the paper might seem indigestible by itself, read in the context of the supporting materials of the Virtual Lab, it is quite approachable, and adds concreteness to the student's research experience.

MEET THE INVESTIGATOR lets the student into Blaustein's thinking about this experiment. In a personal interview, he describes why he was drawn to this particular hypothesis, why he set up his experiment the way he did, what controls he felt were important, and what he would do different if he could go back in time and do the experiment over again. The interview does not introduce Blaustein, so much as his experiment.

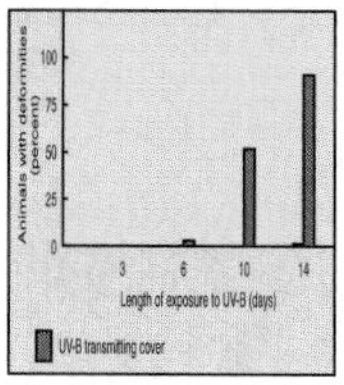
Animals with deformities (percent)
100
75
50
25
0
3
6
10
14
Length of exposure to UV-B (days)
UV-B transmitting cover

GAIN AN OVERVIEW OF THE EXPERIMENT provides a brief summary of what Blaustein actually did. The overview first describes the experiment that Blaustein and his coworkers carried out to investigate the issue of amphibian disappearance. His experimental design involved allowing fertilized eggs to develop in their natural environment with and without a UV-B protective shield. The experimental procedure is outlined, with a discussion of necessary controls, followed by a report of his results—what he found, and what he concluded from these findings.

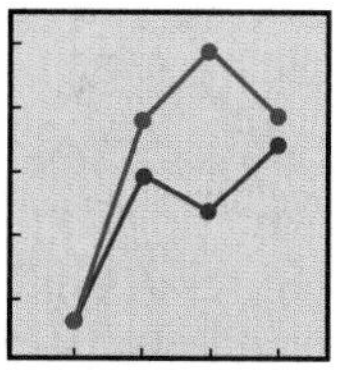

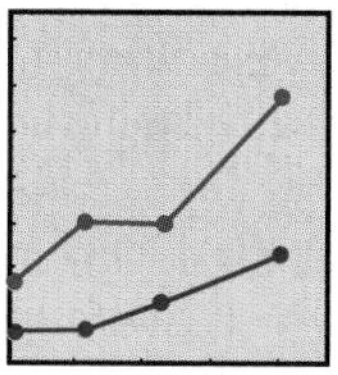

RUN VIRTUAL EXPERIMENTS allows a student to take Blaustein's place, and carry out his or her own investigation. No hands get dirty in this experiment, but all the thought processes of creative scientific investigation are here. The student proposes alternative hypotheses about the cause of amphibian disappearance, devises ways to test the hypotheses, carries out the experiment (virtually), and collects relevant data. Real data are obtained, based on Blaustein's results, with his experimental errors used to introduce variability into the data set much as it was encountered by Blaustein (thus doing the same procedure twice does not yield exactly the same data, but rather similar points, as alike as experimental error would produce). Analyzing the data obtained, the student evaluates the validity of the hypothesis being tested, and comes to a conclusion.

READINGS AND ADDITIONAL RESOURCES provides the student with references to related papers, and to websites of interest. It is important for students encountering research for the first time to realize that experiments like these are not an endpoint, but rather a beginning. If a student's experience in the Virtual Lab is successful, it will open doors to other lines of interest and inquiry.

Real People Doing Real Science

In selecting experiments for the Virtual Lab, I felt it important that the student experience science the way it is actually carried out in most labs. Not every good experiment wins a Nobel Prize or makes the newspapers. In laboratories all over the country, researchers are doing good experiments that most students never read about. With this in mind, I sought to select experiments for the Virtual Labs from the world of real people doing real science—the nuts-and-bolts research upon which scientific progress depends. There is no better way to appreciate how scientific progress occurs than to get down in the trenches with the researchers doing the work.

Chapter 1. John Endler (University of California, Santa Barbara) and **David Reznick** (University of California, Riverside)—*Catching Evolution in Action.*

Chapter 2. Mark Boyce (University of Alberta, Edmonton)—*Why Do Tropical Songbirds Lay Fewer Eggs?*

Chapter 3. Kellar Autumn (Lewis & Clark College) and **Robert Full** (University of California, Berkeley)—*Unraveling the Mystery of How Geckos Defy Gravity.*

Chapter 4. Richard Cyr (Pennsylvania State University)—*How Do the Cells of a Growing Plant Know in Which Direction to Elongate?*

Chapter 5. Andrew Webber (Arizona State University)—*How Do Proteins Help Chlorophyll Carry Out Photosynthesis?*

Chapter 6. Randall Johnson (University of California, San Diego)—*Can Cancer Tumors Be Starved to Death?*

Chapter 7. Simon Rhodes (Indiana University–Purdue University, Indianapolis)—*How Regulatory Genes Direct Vertebrate Development.*

Chapter 8. James Golden (Texas A&M)—*Cyanobacteria Control Heterocyst Pattern Formation /Through Intracellular Signaling.*

Chapter 9. Hamid Habibi and **Maurice Moloney** (University of Calgary)—*Trading Hormones Among Fishes: Gene Technology Lets Us Watch What Happens.*

Chapter 10. John Schiefelbein (University of Michigan)—*The Control of Patterning in Plant Root Development.*

Chapter 11. Julian Adams (University of Michigan)—*Do Some Genes Maintain More Than One Common Allele in a Population?*

Chapter 12. Todd Barkman (Western Michigan University) and **Claude de Pamphilis** (Pennsylvania State University)—*Unearthing the Root of Flowering Plant Phylogeny.*

Chapter 13. Vojo Deretic (University of New Mexico) and **Donald Rowen** (University of Nebraska, Omaha)—*How* Pseudomonas *"Sugar-Coats" Itself to Cause Chronic Lung Infections.*

Chapter 14. Michael McKay (Bowling Green State University)—*Tracking Iron Stress in Diatoms*

Chapter 15. David Drubin (University of California, Berkeley)—*How Actin-Binding Proteins Interact with the Cytoskeleton to Determine the Morphology of Yeasts.*

Chapter 16. Robert Boyd (Auburn University) and **Scott Martens** (University of California, Davis)—*Why Do Some Plants Accumulate Toxic Levels of Metals?*

Chapter 17. James Bidlack (University of Central Oklahoma)—*Which Pest Control Method Is Best for Basil?*

Chapter 18. Jocelyn Ozga (University of Alberta, Edmonton)—*How Hormones Protect Seed Development in Peas.*

Chapter 19. Nels Troelstrup, Jr. (South Dakota State University)—*In Pursuit of Preserving Freshwater Mussels.*

Chapter 20. Christopher Barnhart (Southwest Missouri State University)—*Amphibian Eggs Hatching in Shallow Ponds Thirst for Oxygen.*

Chapter 21. Larry Gilbert (University of Texas, Austin)—*Plotting an Aerial Attack on Maurading Fire Ants.*

Chapter 22. Jon Harrison (Arizona State University)—*How Honeybees Keep Their Cool.*

Chapter 23. Elizabeth Brainerd (University of Massachusetts, Amherst)—*Why Some Lizards Take a Deep Breath.*

Chapter 24. Michael Houghton (Chiron)—*Discovering the Virus Responsible for Hepatitis C.*

Chapter 25. John Dankert (University of Louisiana at Lafayette)—*In Search of New Antibiotics: How Salamander Skin Secretions Combat Microbial Infections.*

Chapter 26. Paul Hamilton (University of Central Arkansas)—*How Snails "See" an Invisible Trail.*

Chapter 27. Deborah Clark (Middle Tennessee State University)—*Pheromones Affect Sexual Selection in Cockroaches.*

Chapter 28. Louis Guillette (University of Florida)—*Are Pollutants Affecting the Sexual Development of Florida's Alligators?*

Chapter 29. Kevin Carman, John Fleeger, and **Steven Pomarico** (Louisiana State University at Baton Rouge)—*Why Does Contamination of a Coastal Salt Marsh with Diesel Fuel Lead to* Increased *Microalgal Biomass?*

Chapter 30. Jerry Wolff (University of Memphis)—*Factors Limiting the Home Range of Male Voles.*

Chapter 31. Andrew Blaustein (Oregon State University)—*Identifying the Environmental Culprit Harming Amphibians.*

SUPPLEMENTS FOR THE INSTRUCTOR AND STUDENT

The third edition of *The Living World* is chapter-by-chapter, full-color customized to better fit the needs of your course. McGraw-Hill also offers various tools and technology products to support this textbook.

For the Instructor

Digital Content Manager—a multimedia tool that enables the user to easily create customized presentations. This CD-ROM is made up of easy to use folders containing the following content:

Active Art Library—files that allow the instructor to manipulate art and adapt figures to meet the needs of the lecture environment.
Animations Library—animations created from figures from the textbook.
Art Libraries—contain all the images in the book in alternate formats (labeled, unlabeled, grayscale). These images are also placed in a PowerPoint presentation for ease of use.
Photo Libraries—contain images from the textbook.
PowerPoint Lectures—outlines for instructors to follow the structure of the text; can be manipulated to add your own topics.
Tables Library—every table found in the text is provided in electronic form.

Online Learning Center—provides a wealth of opportunities for the instructor. It can be found at www.mhhe.com/tlw3. All the libraries found in the Digital Content Manager can be found within the Online Learning Center as well as the following:

BioCourse.com—an electronic meeting place for students and instructors. It provides a comprehensive set of resources in one easy place that is up-to-date and easy to navigate.
Course Integration Guide—helps professors correlate all the ancillary materials to the chapters in the book.
Instructor's Manual—provides the following instructional aides for each chapter: lecture outlines, learning objectives, key terms, lecture suggestions, critical thinking questions, and films/media suggestions.
BioLabs—give instructors and students the opportunity to run online lab simulations to enhance or supplement the wet lab experience. The labs can provide a lab experience when wet labs are impractical due to time constraints, costs, or other factors.
PageOut—McGraw-Hill's exclusive tool for creating your own website for your biology course. It requires no knowledge of coding and is hosted by McGraw-Hill.
PowerWeb—an online supplement with access to the following: course-specific, current articles refereed by content experts; course-specific, real-time news; weekly course updates; refereed and updated research links; daily news; and access to the Northernlight.com Special Collection™ of journals and articles.

Additional features include lecture suggestions, web links, case studies, author's bookshelf, and essays on science.

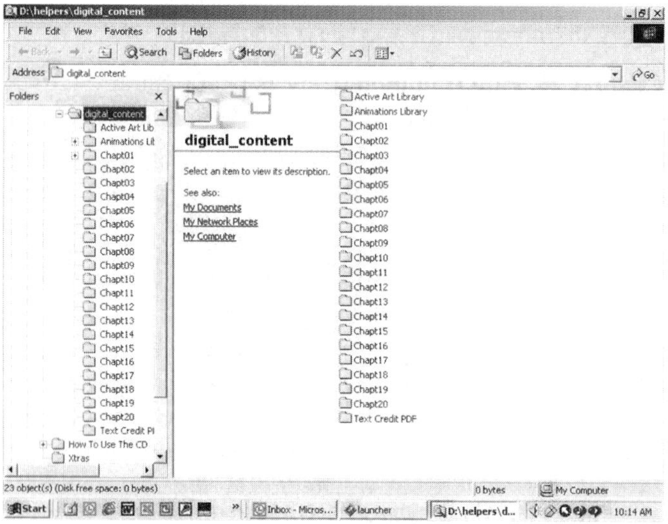
D:\helpers\digital_content
digital_content
Select an item to view its description.
My Documents
My Network Places
My Computer

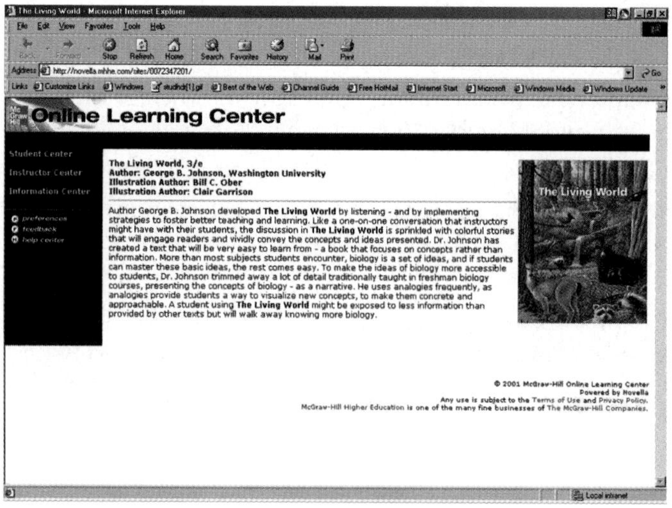
Online Learning Center
Student Center
Instructor Center
Information Center
The Living World

Transparencies—every piece of line art in the textbook is included with better visibility and contrast than ever before. Labels are large and bold for clear projection.

Computerized Test Bank—available on CD-ROM in both Mac and Windows platforms. These questions are the same as those included in the Test Item File of the Instructor's Manual.

Life Science Animations Library CD-ROM—this CD-ROM contains over 400 animations in an easy to use program that enables users to quickly view the animations and import the animations into PowerPoint to create multimedia presentations.

For the Student

Online Learning Center—offers an extensive array of learning tools for the student. The site includes chapter-specific quizzing, end-of-chapter activities, flashcards, crossword puzzles, case studies, and links to related websites. Additional features to the Online Learning Center include:

BioCourse.com—the student portion of this site allows students to search for information specific to the course area they are studying. Information is also available on tips for studying and test taking, surviving the first year of college, and job searches.

Essential Study Partner—contains over 120 animations and more than 800 learning activities to help students grasp complex concepts.

Explorations—interactive modules that cover key concepts in biology.

BioLabs—give students the opportunity to run online lab simulations to enhance or supplement the wet lab experience. BioLabs help students gain understanding of the scientific method as they improve their data gathering and data handling skills.

PowerWeb—an online supplement with access to the following: course-specific, current articles refereed by content experts; course-specific, real-time news; weekly course updates; refereed and updated research links; daily news; and access to the Northernlight.com Special Collection™ of journals and articles.

Student Study Guide—contains chapter reviews, practice quizzes, art exercises and web references for each chapter.

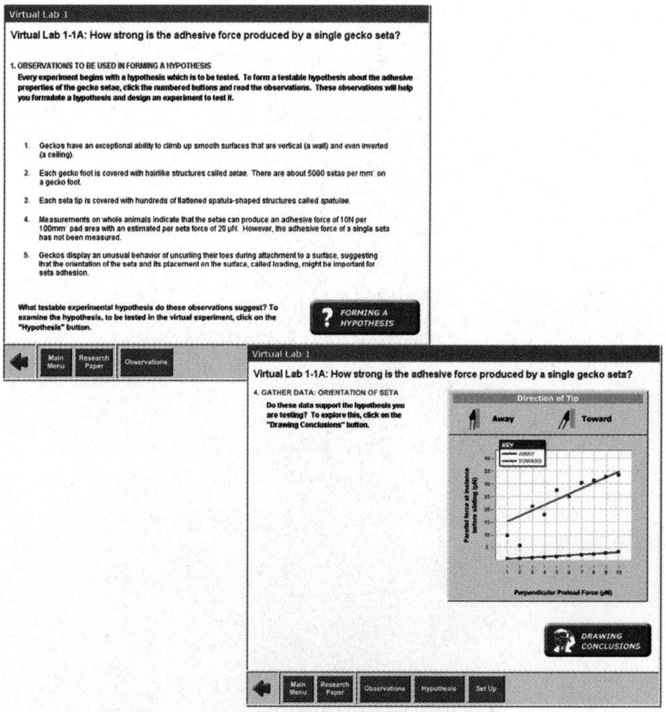

Virtual Lab 1
Virtual Lab 1-1A: How strong is the adhesive force produced by a single gecko seta?
1. OBSERVATIONS TO BE USED IN FORMING A HYPOTHESIS
FORMING A HYPOTHESIS
Main Menu
Research Paper
Observations
4. GATHER DATA: ORIENTATION OF SETA
Direction of Tip
Away
Toward
DRAWING CONCLUSIONS
Hypothesis
Set Up

Readings
Contents
Quizzes
Links
Topics
Site Index
Interactive Exercises
Updates
Current News
PowerSearch New!
Student Resources
Study Tips
Web Research
Research Links
Your Future
Cool Links
Instructor Resources
Resource Guide
Using PowerWeb
WebNotes New!
POWERSEARCH
Powered by Northern Light
Mc Graw Hill
POWERWEB
Biology

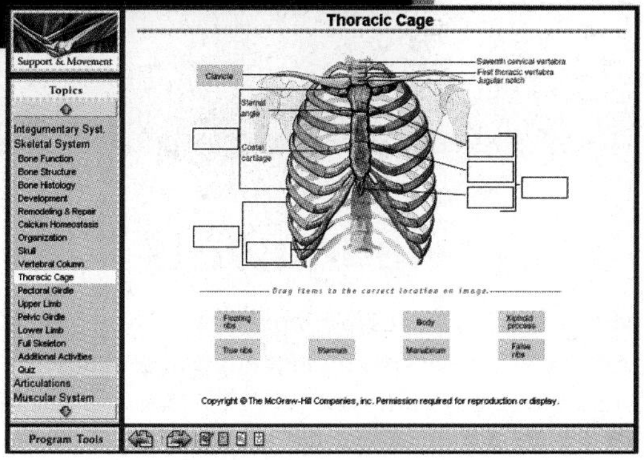

Thoracic Cage
Support & Movement
Topics
Integumentary Syst.
Skeletal System
Articulations
Muscular System
Program Tools
Clavicle
Copyright © The McGraw-Hill Companies, Inc. Permission required for reproduction or display.

BIOLOGY

Acknowledgments

My goal for *The Living World* has always been to present the science in an interesting and engaging manner while maintaining a clear and authoritative text. This is a lofty goal considering the mountains of information and research I must go through just to update the text from one edition to the next. Too lofty for me to accomplish by myself. This third edition would not have been possible without the contributions of many, on the shoulders of whose efforts I have labored. The visuals are critically important in a biology textbook. Many of the superb illustrations were conceived and rendered by Bill Ober and Claire Garrison. I would also like to thank Donald Murie of Meyers Photo-Art for his excellent research of new photographs for this and past editions. Of course I am also indebted to my colleagues from across the country and around the globe who have provided numerous suggestions on how to improve the third edition. Every one of you has my thanks.

A major feature of *The Living World* continues to be the presentation of the information in conceptual modules. It is no small feat to take the information I write, along with my suggestions for figures and tables, and combine them into a conceptual module. This formidable task would not have been possible without the efforts of Megan Jackman, my longtime off-site developmental editor. Her intelligence and perseverance continue to play a major role in the high quality of this book. Liz Sievers, my second off-site developmental editor and other right arm, played an invaluable role in helping organize and produce the Virtual Labs. Their quality directly reflects her effort.

As any author knows, a textbook is made not by a writer but by a publishing team, a group of people that guide the raw book written by the authors through a year-long process of reviewing, editing, fine-tuning, and production. This edition was particularly fortunate in its book team, led by Patrick Reidy, sponsoring editor and supporter; Michael Lange, friend, publisher, and tough critic; Kris Tibbetts, developmental editor and reliable anchor; Peggy Selle, dextrous project manager committed to getting the best possible book; Stuart Paterson, creative and patient design manager; Lisa Gottschalk, tireless marketing manager; and many, many more people behind the scenes.

As in earlier editions, the side-splitting "The Far Side" cartoons of Gary Larson grace each chapter opener, and I want to explicitly thank Gary Larson and Toni Carmichael for letting *The Living World* continue to use so many of their cartoons.

For the third time the powerful and intriguing art of Charles Bragg graces *The Living World* with an arresting cover. Covers have always seemed important to me, the first sniff of what awaits within, and Charles Bragg's pictures speak volumes about the fun and mystery of biology.

This is the seventeenth time I have thanked my family in the preface of one of my books, the twentieth year of a long detour into text writing. I looked for the first time at my first child the same night I held the first edition of my first book in my hands. Since then, as I have written, my family has grown around me. My three girls, Nikki (18), Caitlin (16), and Susie (14), are a far richer reward than any book. They have become accustomed to the many hours this book draws me away from them, a hidden price of textbook writing of which they are fully aware. My wife Barbara, giver of this rich bounty, and in my absence bearer of much of the stress and bother of raising three girls, has provided support without which I could not have written any book, much less seventeen.

Acknowledgments would not be complete without thanking the generations of students and teachers who have used the many editions of my texts. No one is born able to write a textbook of introductory biology. The knowledge and judgment needed to sift through mountains of information, trying always to understand not only the details of what is going on in a particular process but also how it relates to the broader picture of what biology should be to a beginning student—this knowledge and judgment are gifts an author is given by a long parade of teachers and students.

I have been gifted indeed in my teachers. I went to Dartmouth College in 1960 fully intending to be a writer—but of fiction. The change in my career path was a course in biology I took to fulfill a distribution requirement. The course was taught by a new biology faculty member, David Dennison, and it changed my life. His lectures were a model of clarity, intellectually exciting to a young open mind. For the first time, in Dennison's lectures, I saw science as process rather than information, as a give-and-take of inquiry and investigation. I would not have embarked on a career in biology had Dave Dennison not done such a superlative job as a teacher. His example always serves to remind me of the importance of what we do as teachers, that every single student matters, that every lecture we give is important.

An appreciation of what makes a successful experiment lies at the heart of the education of every biologist. In my first year of graduate school at Stanford, I was in the laboratory of a prominent molecular geneticist named Charles Yanofsky. Every week or so the graduate students, postdocs, and faculty of this and a few other labs with related interests got together for lunch and "journal club," and one person described and evaluated a current experiment recently reported in a scientific journal. Faculty and students all took their turns, and were expected to spend weeks preparing. There was no mercy shown to the presenter during the discussion that followed if he or she had not clearly and accurately analyzed the experiment, its results, and its relation to other findings. The free-for-all discussion might involve Paul Berg (now a Nobel laureate), or any of dozens of other sharp minds, and students were expected to hold their own, to justify their opinions, and to argue for what they thought was right. No experience in my life has done more to shape my appreciation of the nature of scientific inquiry than the shattering experience of preparing for these journal club presentations. To this day I can recount the experiments I presented over 30 years ago. I have taught undergraduates biology for 29 years, and I have increasingly come to believe that Charlie Yanofsky had it right—that the best way to understand science in general is to study science in particular. Whatever scientific judgment I have been able to bring to bear in writing this text, I owe in large measure to Charlie.

Finally, I need to thank my reviewers. Every text owes a great deal to those faculty across the country who review it. Serving as sensitive antennae for errors and sounding boards for new approaches, reviewers are among the most valuable tools at an author's disposal. Representing a very diverse array of institutions and interests, they have provided me with invaluable feedback. Many new features and improvements in this edition are the direct result of their suggestions. Every one of them has my sincere thanks.

George Johnson
St. Louis, MO
2002

Felix Akojie
Paducah Community College

Sylvester Allred
Northern Arizona University

William Anyonge
UCLA

Gail F. Baker
La Guardia Community College

Gregorio B. Begonia
Jackson State University

Keith E. Belcher
Austin Peay State University

Arlene G. Billock
University of Louisiana—Lafayette

Lorena V. Blinn
Michigan State University

Randall M. Brand
Southern Union State Community College

Mimi Bres
Prince George's Community College

Young D. Choi
Purdue University—Calumet

Michael A. Davis
Central Connecticut State University

John Dickerman
Northern Illinois University

Jean Dickey
Clemson University

Gary N. Donnermeyer
Kirkwood Community College

Lynn A. Ebersole
Northern Kentucky University

David J. Eisenhour
Morehead State University

Stephen I. N. Ekunwe
Jackson State University

Carl Estrella
Merced College

Ibrahim O. Farah
Jackson State University—Mississippi

Susan Finazzo
Broward Community College

James Fitch
Jones Community College

Malcolm P. Frisbie
Eastern Kentucky University

Suzanne Frucht
Northwest Missouri State University

Farooka Gauhari
University of Nebraska—Omaha

Edwin Ginés-Candelaria
Miami-Dade Community College

Michael Golden
Grossmont College

Brian D. Greene
Southwest Missouri State University

David J. Grisé
Southwest Texas State University

Peggy J. Guthrie
University of Central Oklahoma

Madeline M. Hall
Cleveland State University

Blanche C. Haning
North Carolina State University

Laszlo Hanzely
Northern Illinois University

Joseph F. Hawkins
College of Southern Idaho

Barbara Hetrick
University of Northern Iowa

Diane Hilker
Mercer County Community College

Ellen Porter Holtman
Virginia Western Community College

Jane Aloi Horlings
Saddleback College

Charles D. Howes
Ashland Community College

Pat Hilliard Johnson
Palm Beach Community College

O. Ray Jordan
Tennessee Technological University

Martin A. Kapper
Central Connecticut State University

Arnold J. Karpoff
University of Louisville

Frances G. R. Kennedy
State University of West Georgia

D. T. Kidwell
Southeast Community College

Joanne M. Kilpatrick
Auburn University—Montgomery

Paul N. Kotila
Franklin Pierce College

Beth A. Krueger
Monroe Community College

Geneen Lannom
University of Central Oklahoma

Siu-Lam Lee
University of Massachusetts—Lowell

Jani E. Lewis
University of Toledo

Richard Londraville
University of Akron

David Loring
Johnson County Community College

Doug Lyng
Indiana University—Purdue University

Kenneth A. Mason
Purdue University

Leroy R. McClenaghan, Jr.
San Diego State University

Regina McClinton
University of Louisiana—Lafayette

Heike I. McConnell
Western Illinois University

Michael J. McLeod
Belmont Abbey College

Mary Lou McReynolds
Austin Peay State University

Susan T. Meiers
Louisiana State University—Baton Rouge

Janet Mihuc
Plattsburgh State University

Angela Montel
Bluffton College

Royden Nakamura
California Polytechnic State University—SLO

Nathan O. Okia
Auburn University—Montgomery

Marcy P. Osgood
University of Michigan

John C. Osterman
University of Nebraska—Lincoln

Gregory Paulson
Shippensburg University

Patricia C. Paulson
Bethel College

Kathleen Pelkki
Saginaw Valley State University

Kenneth E. Petit
Springfield Technical Community College

David H. Pistole
Indiana University of Pennsylvania

Barbara Pleasants
Iowa State University

John M. Pleasants
Iowa State University

Calvin A. Porter
Texas Tech University

Kumkum Prabhakar
Nassau Community College

Lansing Prescott
Augustana College

Regina Rector
William Rainey Harper College

Michael H. Renfroe
James Madison University

Robert B. Sanders
University of Kansas

Douglas P. Schelhaas
University of Mary

Stefan O. Schiff
The George Washington University

Margit Schmidt
East Carolina University

Brian W. Schwartz
Columbus State University

Linda Sigismondi
University of Rio Grande

Phillip D. Simpson
Shepherd College

David A. Smith
Lock Haven University of Pennsylvania

Linda D. Smith-Staton
Pellissippi State Technical Community College

David W. Stewart
SUNY Canton

Judith L. Stewart
Community College of Southern Nevada

Andrew W. Turner
Clarion University of Pennsylvania

Anthony Udeogalanya
Medgar Evers College/CUNY

Leslie Vandermolen
Humboldt State University

R. Warwick
Coastline College

Cheryl L. Watson
Central Connecticut State University

W. G. Weaver
Miami-Dade Community College—North Campus

Mary E. White
Southeastern Louisiana University

J. D. Wilhide
Arkansas State University

Thurman E. Wilson
Prairie State College

William L. Wissinger
St. Bonaventure University

Michael Woller
University of Wisconsin—Whitewater

Carol L. Wymer
Morehead State University

James R. Yates
University of South Carolina—Aiken

Calvin Young
Fullerton College

David D. Zeigler
University of North Carolina—Pembroke

The Living World

CHAPTER 1

THE SCIENCE OF BIOLOGY

THE FAR SIDE® By GARY LARSON
© 1984 FarWorks, Inc. All Rights Reserved/Dist. by Creators Syndicate
= 0
Larson

"No doubt about it, Ellington—we've mathematically expressed the purpose of the universe. Gad, how I love the thrill of scientific discovery!"

CHAPTER 1

The Science of Biology

CHAPTER OVERVIEW

Biology and the Living World

- All living things share eight fundamental properties:
  complexity
  movement
  response to stimulation
  cellular organization
  metabolism
  homeostasis
  reproduction
  heredity
- There are many ways to study biology. Five general themes often used to organize the study of biology are
  evolution
  the flow of energy
  cooperation
  structure determines function
  homeostasis

The Scientific Process

- The discovery of how CFCs are reducing levels of ozone in the atmosphere is a good example of science in action.
- The scientific process is founded on careful observation.
- In a control experiment, only one variable is allowed to change.
- Scientific progress is made by rejecting hypotheses that are inconsistent with observation.

Using Science to Make Decisions

- The acceptance of a hypothesis is always provisional.
- Well-tested hypotheses are often combined into general statements called theories.
- There is no surefire way to do science and no foolproof "method."
- One of the most creative aspects of scientific investigation is the formulation of novel hypotheses.

1.1 The Diversity of Life

In its broadest sense, biology is the study of living things—the science of life. The living world teems with a breathtaking variety of creatures—whales, butterflies, mushrooms, and mosquitoes—all of which can be categorized into six groups, or **kingdoms,** of organisms (figure 1.1).

Biologists study the diversity of life in many different ways. They live with gorillas, collect fossils, and listen to whales. They isolate viruses, grow mushrooms, and examine the structure of fruit flies. They read the messages encoded in the long molecules of heredity and count how many times a hummingbird's wings beat each second. In the midst of all this diversity it is easy to lose sight of the key lesson of biology, which is that all living things have much in common.

1.1 The living world is very diverse, but all living things share many key properties.

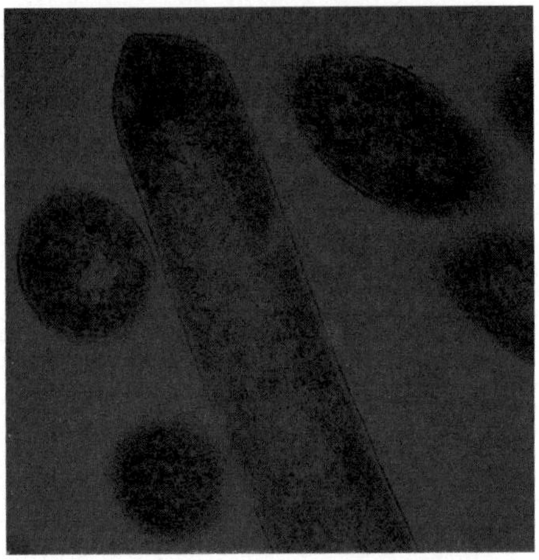

Archaebacteria. This kingdom includes bacteria such as this methanogenic bacterium, which manufactures methane as a result of its metabolic activity.

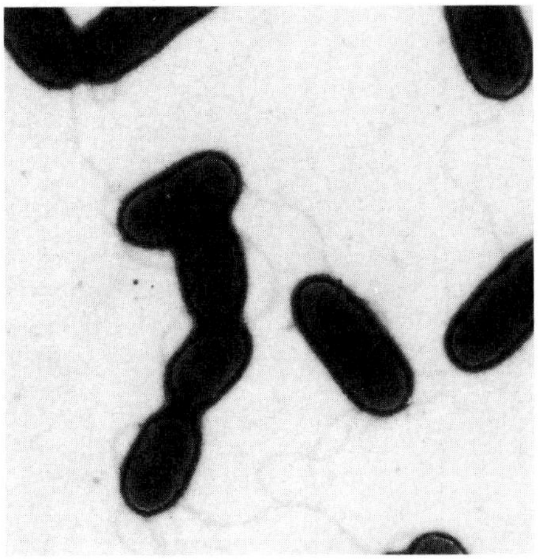

Eubacteria. This group is the second of the two bacterial kingdoms. Shown here is a soil bacterium that is responsible for many plant diseases.

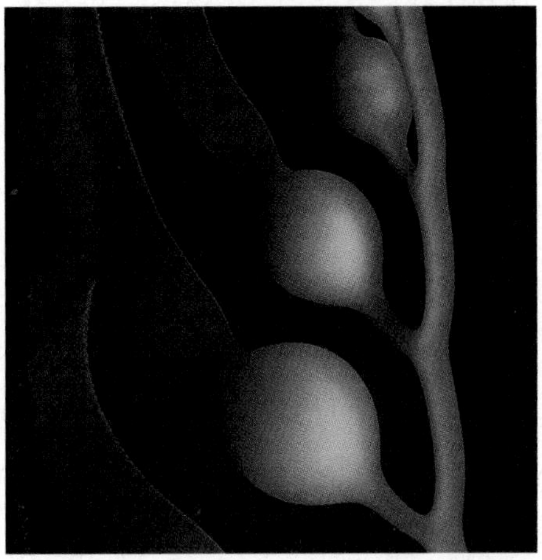

Protista. The unicellular eukaryotes (those whose cells contain a nucleus) are grouped into this kingdom, and so are the algae pictured here.

Fungi. This kingdom contains nonphotosynthetic multicellular organisms that digest their food externally, such as this mushroom.

Plantae. This kingdom contains photosynthetic multicellular organisms that are terrestrial, such as the flowering plant pictured here.

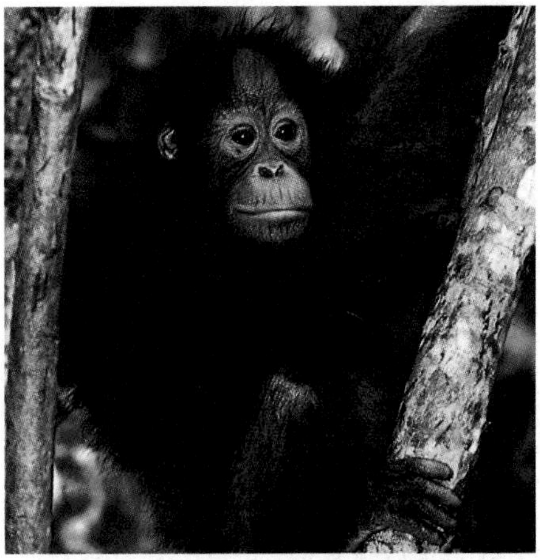

Animalia. Organisms in this kingdom are nonphotosynthetic multicellular organisms that digest their food internally, such as this primate.

Figure 1.1 The six kingdoms of life.

Biologists categorize all living things into six major categories called *kingdoms.* Each kingdom is profoundly different from the others.

1.2 Properties of Life

Biology is the study of life—but what does it mean to be alive? What are the properties that define a living organism? This is not as simple a question as it seems because some of the most obvious properties of living organisms are also properties of many nonliving things. Three of the most important of these are *complexity* (a computer is complex), *movement* (clouds move in the sky), and *response to stimulation* (a soap bubble pops if you touch it). To appreciate why these three properties, so common among living things, do not help us to define life, imagine a mushroom standing next to a television: the television seems more complex than the mushroom, the picture on the television screen is moving while the mushroom just stands there, and the television responds to a remote control device while the mushroom continues to just stand there—yet it is the mushroom that is alive.

All living things share five more basic properties, passed down over billions of years from the first organisms to evolve on earth: *cellular organization, metabolism, homeostasis, reproduction,* and *heredity*.

1. **Cellular organization.** All living things are composed of one or more cells. A cell is a tiny compartment with a thin covering called a *membrane*. Some cells have simple interiors, while others are complexly organized, but all are able to grow and reproduce. Many organisms possess only a single cell (figure 1.2); your body contains about 100 trillion—that's how many centimeters long a string would be wrapped around the world 1,600 times!
2. **Metabolism.** All living things use energy. Moving, growing, thinking—everything you do requires energy. Where does all this energy come from? It is captured from sunlight by plants and algae. To get the energy that powers our lives, we extract it from plants or from plant-eating animals in a process called *metabolism* (figure 1.3). All organisms use energy to grow, and all organisms transport this energy from one place to another within cells using special energy-carrying molecules called ATP molecules.
3. **Homeostasis.** All living things maintain stable internal conditions. While the environment often varies a lot, organisms act to keep their interior conditions relatively constant, a process called *homeostasis*. Your body acts to maintain an internal temperature of 37°C (98.5°F), however hot or cold the weather might be.
4. **Reproduction.** All living things reproduce. Bacteria simply split in two, as often as every 15 minutes, while many more complex organisms reproduce sexually (some as rarely as every thousand years).
5. **Heredity.** All organisms possess a genetic system that is based on the replication and duplication of a long molecule called *DNA* (*deoxyribonucleic acid*). The information that determines what an individual organism will be like is contained in a code that is dictated by the order of the subunits making up the DNA molecule, just as the order of letters on this page determines the sense of what you are reading. Each set of instructions within the DNA is called a *gene*. Together, the genes determine what the organism will be like. Because DNA is faithfully copied from one generation to the next, any change in a gene is also preserved and passed on to future generations. The transmission of characteristics from parent to offspring is a process called *heredity*.

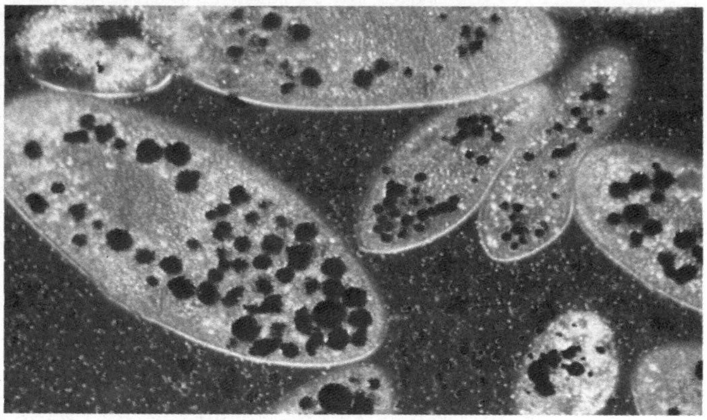

Figure 1.2 Cellular organization.

These paramecia are complex single-celled protists that have just ingested several yeast cells. Like these paramecia, many organisms consist of just a single cell, while others are composed of trillions of cells.

Figure 1.3 Metabolism.

These cedar waxwing chicks obtain the energy they need to grow and develop by eating plants. They metabolize this food using chemical processes that occur within cells.

1.2 All living things possess cells that carry out metabolism, maintain stable internal conditions, reproduce themselves, and use DNA to transmit hereditary information to offspring.

1.3 The Organization of Life

The organisms of the living world function and interact with each other at many levels (figure 1.4). A key factor in organizing these interactions is the degree of complexity. There is a hierarchy of increasing complexity within cells, from the **molecular level** of DNA, at which the chemistry of life occurs, to the **organelle level,** at which cellular activities are organized, to the **cell,** the smallest level of organization that can be considered alive.

There is a further hierarchy of increasing complexity within multicellular organisms. At the cell level, different cells within the body are specialized to do different things (neurons to conduct signals, for example, and muscle cells to contract). Cells with a similar structure and function are grouped together into **tissues** (for example, muscle is a tissue composed of many muscle cells working together). Different tissues are combined into **organs,** which are biological machines that carry out particular jobs (the heart is an organ composed of muscle, nerve, and other tissues that works as a pump). The various organs that carry out major body functions make up **organ systems** (your heart, blood vessels, and the blood within them, for example, together make up your circulatory system).

There is yet another hierarchy of increasing complexity among different organisms. Individuals of the same type of organism living together are called a **population,** and all the populations of a particular kind of organism are members of the same **species.** All the different species that live in a place are called a **community** (a forest community, for example, contains trees and deer and woodpeckers and fungi and many other creatures). A community and the physical environment in which it lives is called an ecological system, or **ecosystem.**

1.3 Cells, multicellular organisms, and ecological systems each are organized in a hierarchy of increasing complexity.

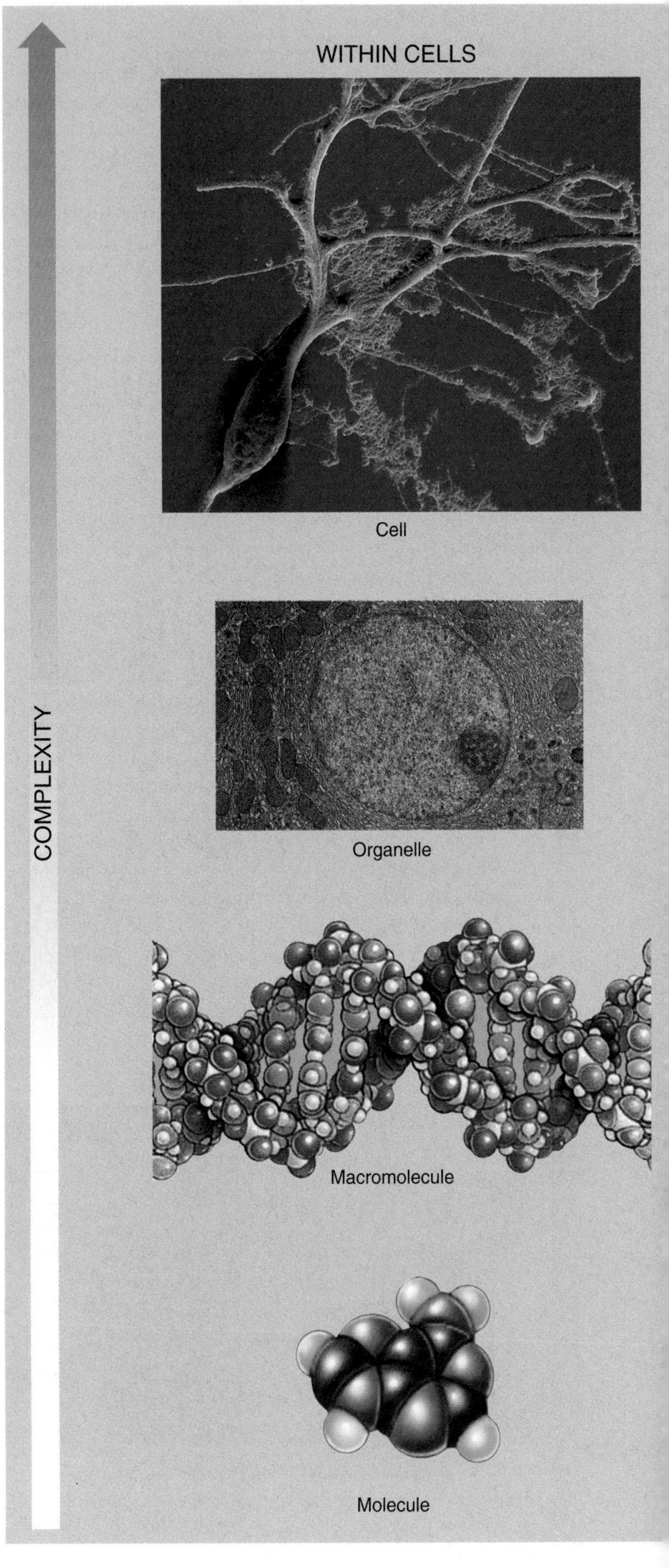

WITHIN CELLS
COMPLEXITY
Cell
Organelle
Macromolecule
Molecule

Figure 1.4 Levels of organization.

A traditional and very useful way to sort through the many ways in which the organisms of the living world interact is to organize them in terms of levels of organization, proceeding from the very small and simple to the very large and complex. Here we examine levels of organization within cells, within multicellular organisms, and among organisms.

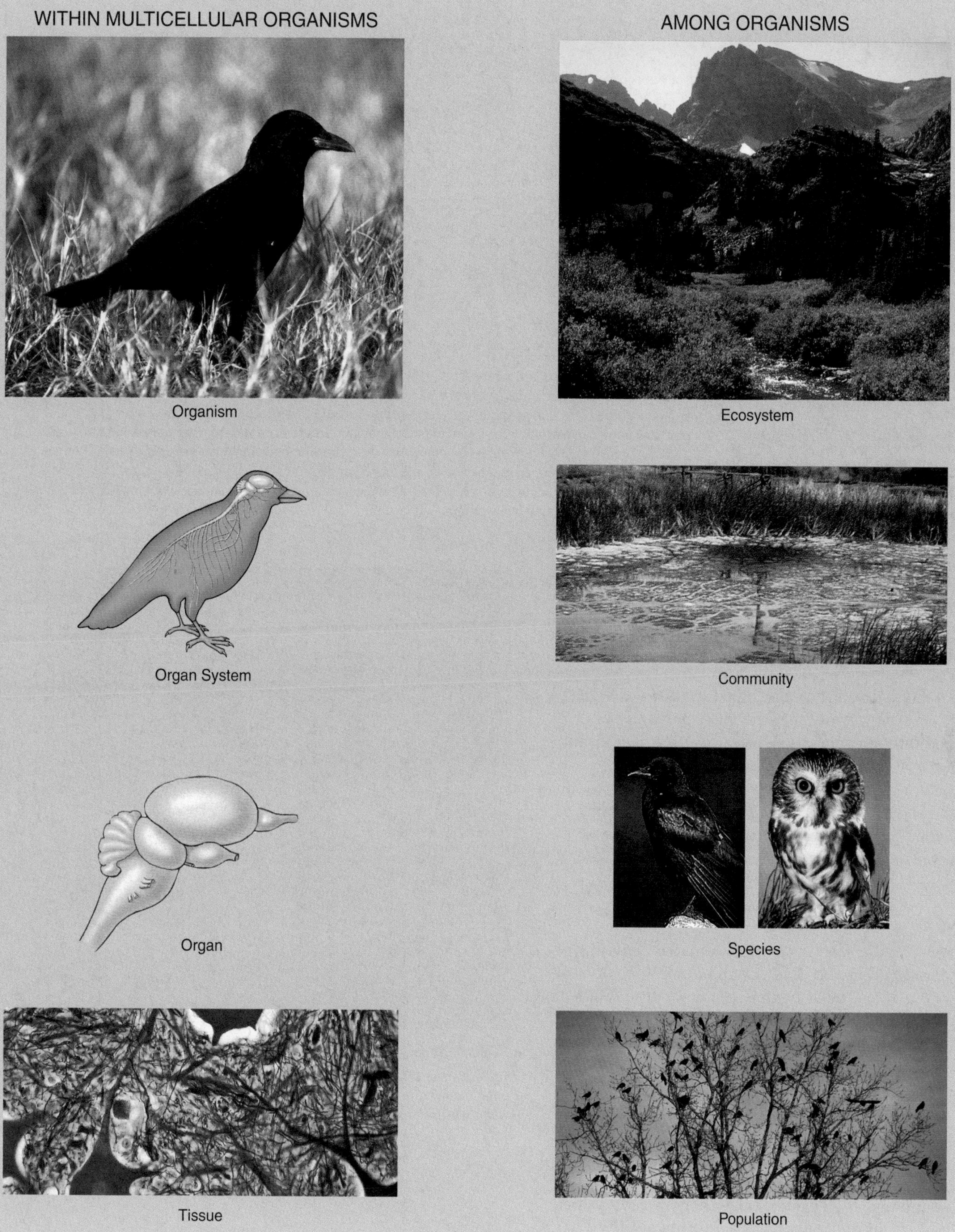
WITHIN MULTICELLULAR ORGANISMS
AMONG ORGANISMS
Organism
Ecosystem
Organ System
Community
Organ
Species
Tissue
Population

1.4 Biological Themes

Just as every house is organized into thematic areas such as bedroom, kitchen, and bathroom, so the living world is organized by major *themes,* such as how energy flows within the living world from one part to another. As you study biology in this text, five general themes will emerge repeatedly, themes that serve to both unify and explain biology as a science:

1. evolution;
2. the flow of energy;
3. cooperation;
4. structure determines function;
5. homeostasis.

Evolution

Evolution is the change in species over time. Charles Darwin was an English naturalist who, in 1859, proposed the idea that this change is a result of a process called **natural selection.** Simply stated, those organisms better able to successfully respond to the challenges of their environment become more common. Darwin was thoroughly familiar with variation in domesticated animals (in addition to many nondomesticated organisms), and he knew that varieties of pigeons could be selected by breeders to exhibit exaggerated characteristics, a process called **artificial selection** (figure 1.5). We now know that the characteristics selected are passed on through generations because DNA is transmitted from parent to offspring. Darwin then visualized how selection in nature could be similar to that which had produced the different varieties of pigeons. Thus, the many forms of life we see about us on earth today, and the way we ourselves are constructed and function, reflect a long history of natural selection.

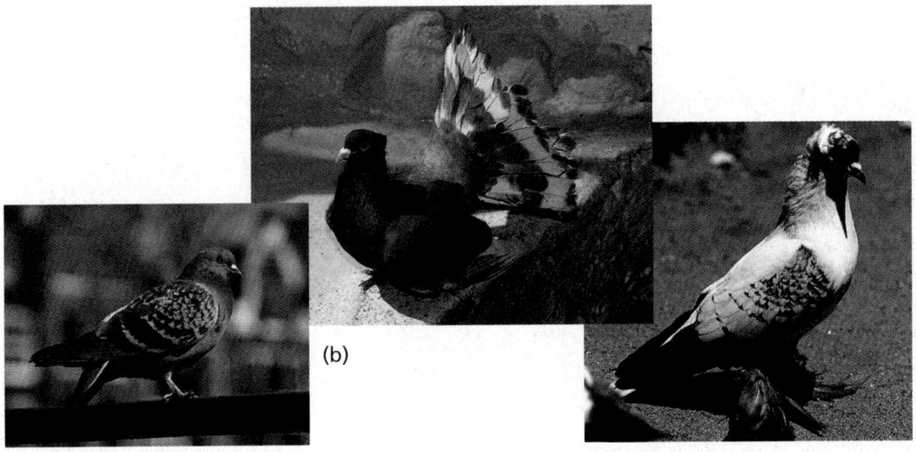

(a) (b) (c)

Figure 1.5 Evolution.

Charles Darwin's studies of artificial selection in pigeons provided key evidence that selection could produce the sorts of changes predicted by his theory of evolution. In *On the Origin of Species,* Darwin wrote about his attempts to produce differences in domestic pigeons using artificial selection. He wrote that the kinds of pigeons he had produced were so different that "if shown to an ornithologist, and he were told that they were wild birds, would certainly, I think, be ranked by him as well-defined species." The differences that have been obtained by artificial selection of the wild European rock pigeon (*a*) and such domestic races as the red fantail (*b*) and the fairy swallow (*c*), with its fantastic tufts of feathers around its feet, are indeed so great that the birds probably would, if wild, be classified in entirely different major groups.

The Flow of Energy

All organisms require energy to carry out the activities of living—to build bodies and do work and think thoughts. All of the energy used by most organisms comes from the sun and is gradually used up as it passes in one direction through ecosystems. The simplest way to understand the flow of energy through the living world is to look at who uses it. The first stage of energy's journey is its capture by green plants and algae in photosynthesis. Plants then serve as a source of life-driving energy for animals that eat them. Other animals may then eat the plant eaters (figure 1.6). At each stage, some energy is used, some is transferred, and much is lost. The flow of energy is a key factor in shaping ecosystems, affecting how many and what kinds of animals live in a community.

Figure 1.6 The flow of energy.

This bald eagle swooping down on its prey is a carnivore, an organism that feeds on other animals. Energy passes from plants to plant-eating animals to animal-eating animals, such as this eagle.

Cooperation

Cooperation between different kinds of organisms (figure 1.7) has played a critical role in the evolution of life on earth. For example, animal cells possess organelles that are the descendants of symbiotic bacteria, and symbiotic fungi helped plants first invade land from the sea. The coevolution of flowering plants and insects has been responsible for much of life's great diversity.

Structure Determines Function

One of the most obvious lessons of biology is that biological structures are very well suited to their functions. You will see this at every level of organization: Within cells, the shape of the proteins called enzymes that cells use to carry out chemical reactions are precisely suited to match the chemicals the enzymes must manipulate. Within the many kinds of organisms in the living world, body structures seem carefully designed to carry out their functions—the long tongue with which a moth sucks nectar from a deep flower is one example (figure 1.8). The superb fit of structure to function in the living world is no accident. Life has existed on earth for over 3 billion years, a long time for evolution to favor changes that better suit organisms to meet the challenges of living. It should come as no surprise to you that after all this honing and adjustment, biological structures carry out their functions well.

Homeostasis

The high degree of specialization we see among complex organisms is only possible because these organisms act to maintain a relatively stable internal environment, a process called homeostasis (figure 1.9). Without this constancy, many of the complex interactions that need to take place within organisms would be impossible, just as a city cannot function without rules and order. Maintaining homeostasis in a body as complex as yours requires a great deal of signaling back-and-forth between cells.

As already stated, you will encounter these biological themes repeatedly in this text. But just as a budding architect must learn more than the parts of buildings, so your study of biology should teach you more than a list of themes, concepts, and parts of organisms. Biology is a dynamic science that will affect your life in many ways, and that lesson is one of the most important you will learn. It is also an awful lot of fun.

1.4 The five general themes of biology are (1) evolution, (2) the flow of energy, (3) cooperation, (4) structure determines function, and (5) homeostasis.

Figure 1.7 Cooperation.

Some animals live in unexpected places. These barnacles live on the back of a gray whale. The whale carries them from place to place so that they have continuous access to fresh sources of the small, free-floating organisms on which they feed.

Figure 1.8 Structure determines function.

With its long tongue, the hummingbird clear-wing moth is able to reach the nectar deep within these flowers.

Figure 1.9 Homeostasis.

Homeostasis often involves water balance. All complex organisms need water—some, like this hippo, luxuriate in it. Others, like the kangaroo rat, never drink. Maintaining a proper water balance is part of the homeostasis necessary for life.

1.5 The Nature of Science

Deductive Reasoning

Science is a particular way of investigating the world. Not all investigations are scientific. For example, when you want to know how to get to Chicago from St. Louis, you do not conduct a scientific investigation—instead, you look at a map to determine a route. Making individual decisions by applying a "map" of accepted general principles is called **deductive reasoning.** Deductive reasoning is the reasoning of mathematics, philosophy, politics, and ethics; deductive reasoning is also the way a computer works. All of us rely on deductive reasoning to make everyday decisions. We use general principles as the basis for examining and evaluating these decisions.

Inductive Reasoning

Where do general principles come from? Religious and ethical principles often have a religious foundation; political principles reflect social systems. Some general principles, however, are not derived from religion or politics but from observation of the physical world around us. If you drop an apple, it will fall, whether or not you wish it to and despite any laws you may pass forbidding it to do so. Science is devoted to discovering the general principles that govern the operation of the physical world.

How do scientists discover such general principles? Scientists are, above all, observers: they look at the world to understand how it works. It is from their observations that scientists determine the general principles that govern our physical world.

This way of discovering general principles by careful examination of specific cases is called **inductive reasoning** (figure 1.10). Inductive reasoning first became popular about 400 years ago, when Isaac Newton, Francis Bacon, and others began to conduct experiments and from the results infer general principles about how the world operates. The experiments were sometimes quite simple. Newton's consisted simply of releasing an apple from his hand and watching it fall to the ground. This simple observation is the stuff of science. From a host of particular observations, each no more complicated than the falling of an apple, Newton inferred a general principle—that all objects fall toward the center of the earth. This principle was a possible explanation, or **hypothesis,** about how the world works. Like Newton, scientists today formulate hypotheses, and observations are the materials on which they build them.

1.5 Science uses inductive reasoning to infer general principles from detailed observation.

DEDUCTIVE REASONING
Because lions are dangerous, they are kept behind locked doors. Therefore, deduce that the lion is behind the right (locked) door. Open the left door.

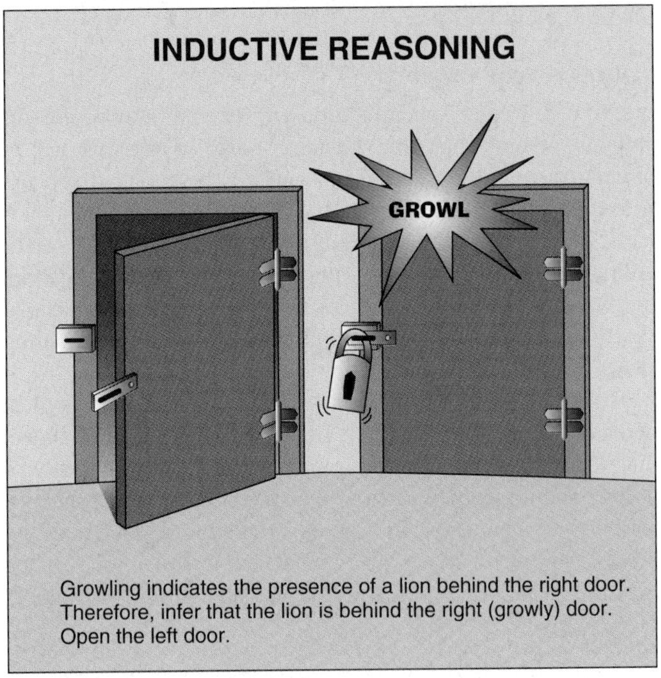
INDUCTIVE REASONING
GROWL
Growling indicates the presence of a lion behind the right door. Therefore, infer that the lion is behind the right (growly) door. Open the left door.

Figure 1.10 Deductive and inductive reasoning.
A deduction is a conclusion drawn from general principles. An inference is a conclusion drawn from specific observations. In this hypothetical example, a gladiator is forced to choose between two doors in a coliseum. Behind one of the doors is a deadly lion; behind the other door is freedom. How can the gladiator make the choice? He can use either deductive or inductive reasoning.

1.6 Science in Action: A Case Study

In 1985 Joseph Farman, a British earth scientist working in Antarctica, made an alarming discovery. Scanning the Antarctic sky, he found less ozone (O_3, a form of oxygen gas) than should be there—not a slight depletion but a 30% drop from a reading recorded five years earlier in the Antarctic!

At first it was argued that this "ozone hole" was an as-yet-unexplained weather phenomenon. Evidence soon mounted, however, implicating synthetic chemicals as the culprit. Detailed analysis of chemicals in the Antarctic atmosphere revealed a surprisingly high concentration of chlorine, a chemical known to destroy ozone. The source of the chlorine was a class of chemicals called **chlorofluorocarbons (CFCs).** CFCs have been manufactured in large amounts since they were invented in the 1920s, largely for use as coolants in air conditioners, propellants in aerosols, and foaming agents in making Styrofoam. CFCs were widely regarded as harmless because they were chemically unreactive under normal conditions. But in the thin atmosphere over Antarctica, CFCs condense onto tiny ice crystals; warmed by the sun in the spring, they attack and destroy ozone without being used up (figure 1.11).

The thinning of the ozone layer in the upper atmosphere 25 to 40 kilometers above the surface of the earth is a serious matter. The ozone layer protects life from the harmful ultraviolet (UV) rays from the sun that bombard the earth continuously. Like invisible sunglasses, the ozone layer filters out these dangerous rays. When UV rays damage the DNA in skin cells, it can lead to skin cancer. Every 1% drop in the atmospheric ozone concentration is estimated to lead to a 6% increase in skin cancers. The drop of approximately 3% that has already occurred worldwide, therefore, is estimated to have led to as much as a 20% increase in skin cancers.

The world currently produces about 1 million tons of CFCs annually, three-fourths of it in the United States and Europe. As scientific observations have become widely known, governments have rushed to correct the situation. By 1990, worldwide agreements to phase out production of CFCs by the end of the century had been signed. Nonetheless, most of the CFCs manufactured since they were invented are still in use in air conditioners and aerosols and have not yet reached the atmosphere. As these CFCs, as well as CFCs still being manufactured, move slowly upward through the atmosphere, the problem can be expected to grow worse. Ozone depletion has now been reported over the North Pole as well, and there is serious concern that the Arctic ozone hole will soon extend over densely populated northern Europe and the northeastern United States. Elevated levels of chlorine were reported over northern Europe in 1992, a warning of ozone destruction to come.

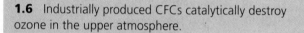

1.6 Industrially produced CFCs catalytically destroy ozone in the upper atmosphere.

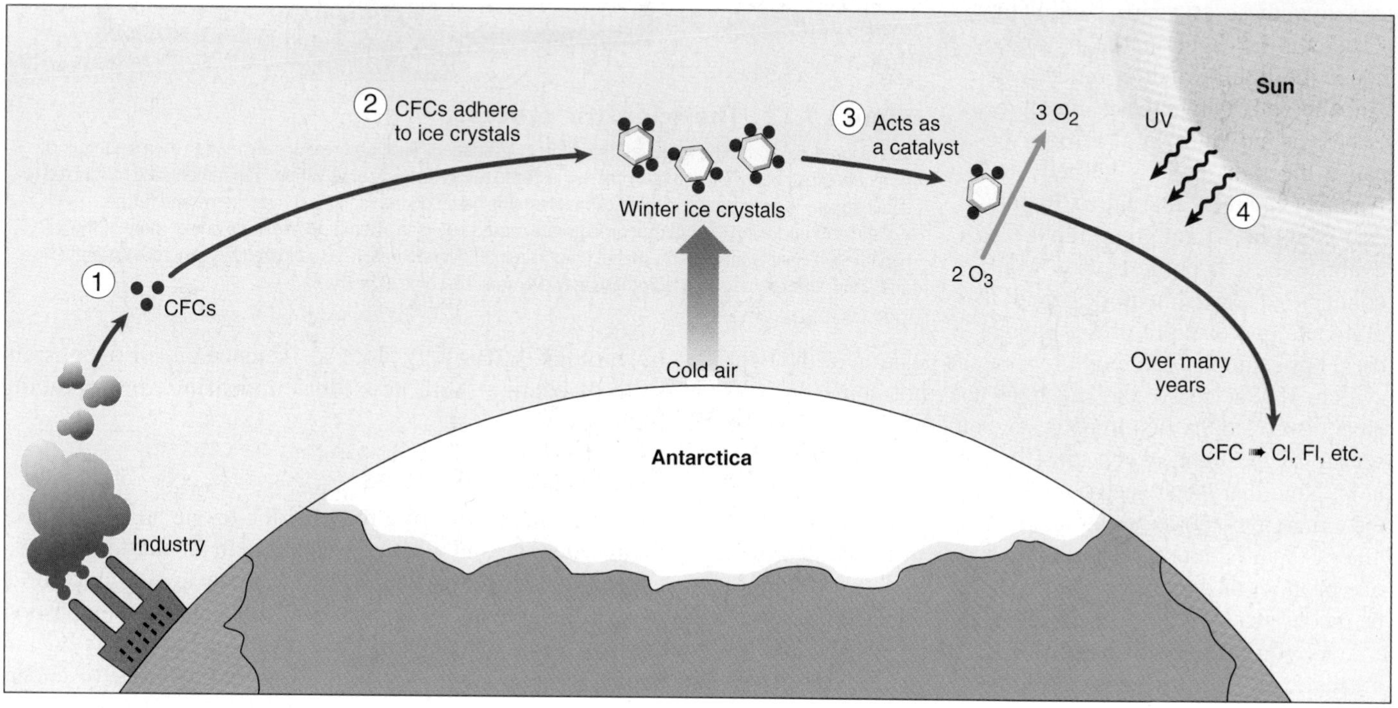
2 CFCs adhere to ice crystals
3 Acts as a catalyst
3 O2
UV
Sun
4
Winter ice crystals
2 O3
1
CFCs
Cold air
Over many years
Antarctica
CFC → Cl, Fl, etc.
Industry

Figure 1.11 How CFCs attack and destroy ozone.
CFCs are stable chemicals that accumulate in the atmosphere as a by-product of industrial society (*1*). In the intense cold of the Antarctic, these CFCs adhere to tiny ice crystals in the upper atmosphere (*2*), where they catalytically destroy ozone (*3*). As a result, more harmful UV radiation reaches the earth's surface (*4*).

1.7 Stages of a Scientific Investigation

How Science Is Done

How do scientists like Joseph Farman establish which general principles are true from among the many that might be? They do this by systematically testing alternative proposals. If these proposals prove inconsistent with experimental observations, they are rejected as untrue. After making careful observations concerning a particular area of science, scientists construct a hypothesis, which is a suggested explanation that accounts for those observations. A hypothesis is a proposition that might be true. Those hypotheses that have not yet been disproved are retained. They are useful because they fit the known facts, but they are always subject to future rejection if, in the light of new information, they are found to be incorrect.

We call the test of a hypothesis an experiment. Suppose that a room appears dark to you. To understand why it appears dark, you propose several hypotheses. The first might be, "There is no light in the room because the light switch is turned off." An alternative hypothesis might be, "There is no light in the room because the light bulb is burned out." And yet another alternative hypothesis might be, "I am going blind." To evaluate these hypotheses, you would conduct an experiment designed to eliminate one or more of the hypotheses. For example, you might reverse the position of the light switch. If you do so and the light does not come on, you have disproved the first hypothesis. Something other than the setting of the light switch must be the reason for the darkness. Note that a test such as this does not prove that any of the other hypotheses are true; it merely demonstrates that one of them is not. A successful experiment is one in which one or more of the alternative hypotheses is demonstrated to be inconsistent with the results and is thus rejected.

As you proceed through this text, you will encounter a great deal of information, often accompanied by explanations. These explanations are hypotheses that have withstood the test of experiment. Many will continue to do so; others will be revised as new observations are made by biologists. Biology, like all science, is in a constant state of change, with new ideas appearing and replacing old ones.

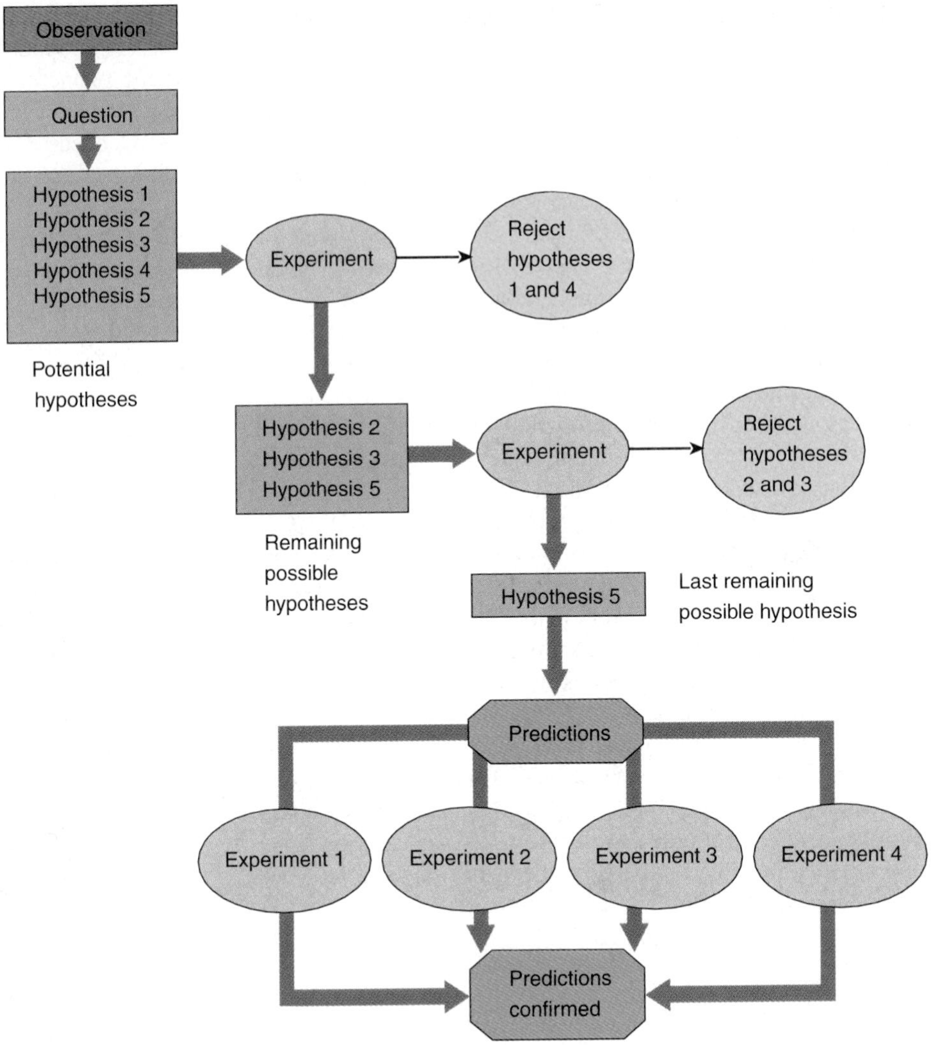

Observation
Question
Hypothesis 1
Hypothesis 2
Hypothesis 3
Hypothesis 4
Hypothesis 5
Potential hypotheses
Experiment
Reject hypotheses 1 and 4
Hypothesis 2
Hypothesis 3
Hypothesis 5
Remaining possible hypotheses
Experiment
Reject hypotheses 2 and 3
Hypothesis 5
Last remaining possible hypothesis
Predictions
Experiment 1
Experiment 2
Experiment 3
Experiment 4
Predictions confirmed

Figure 1.12 The scientific process.

This diagram illustrates the stages of a scientific investigation. First, observations are made. Then a number of potential explanations (hypotheses) are suggested in response to an observation. Experiments are conducted to eliminate any hypotheses. Next, predictions are made based on the remaining hypotheses, and further experiments (including control experiments) are carried out in an attempt to eliminate one or more of the hypotheses. Finally, any hypothesis that is not eliminated is retained. If it is validated by numerous experiments and stands the test of time, a hypothesis may eventually become a theory.

The Scientific Process

Joseph Farman, who first reported the ozone hole, is a practicing scientist, and what he was doing in Antarctica was science. Science is a particular way of investigating the world, of forming general rules about why things happen by observing particular situations. A scientist like Farman is an observer, someone who looks at the world in order to understand how it works.

Scientific investigations can be said to have six stages: (1) observing what is going on, (2) forming a set of hypotheses,

(3) making predictions, (4) testing them, (5) carrying out controls, and (6) forming conclusions after eliminating one or more of the hypotheses (figure 1.12).

1. Observation. The key to any successful scientific investigation is careful **observation.** Farman and other scientists had studied the skies over the Antarctic for many years, noting a thousand details about temperature, light, and levels of chemicals. Had these scientists not kept careful records of what they observed, Farman might not have noticed that ozone levels were dropping.

2. Hypothesis. When the alarming drop in ozone was reported, environmental scientists made a guess about what was destroying the ozone—that perhaps the culprit was CFCs. We call such a guess a hypothesis. A hypothesis is a guess that might be true. What the scientists guessed was that chlorine from CFCs released into the atmosphere was reacting chemically with ozone over the Antarctic, converting ozone (O_3) into oxygen gas (O_2) and in the process removing the ozone shield from our earth's atmosphere. Often, scientists will form **alternative hypotheses** if they have more than one guess about what they observe. In this case, there were several other hypotheses advanced to explain the ozone hole (figure 1.13). One suggestion explained it as the result of convection. A hypothesis was proposed that the seeming depletion of ozone was in fact a normal consequence of the spinning of the earth; the ozone spun away from the polar regions much as water spins away from the center as a clothes washer moves through its spin cycle. Another hypothesis was that the ozone hole was a transient phenomenon, due perhaps to sunspots, and would soon disappear.

3. Predictions. If the CFC hypothesis is correct, then several consequences can reasonably be expected. We call these expected consequences **predictions.** A prediction is what you expect to happen if a hypothesis is true. The CFC hypothesis predicts that if CFCs are responsible for producing the ozone hole, then it should be possible to detect CFCs in the upper Antarctic atmosphere as well as the chlorine released from CFCs that attack the ozone.

4. Testing. Scientists set out to test the CFC hypothesis by attempting to verify some of its predictions. We call the test of a hypothesis an **experiment.** To test the hypothesis, atmospheric samples were collected from the stratosphere over 6 miles up by a high-altitude balloon. Analysis of the samples revealed CFCs, as predicted. Were the CFCs interacting with the ozone? The samples contained free chlorine and fluorine, confirming the breakdown of CFC molecules. The results of the experiment thus support the hypothesis.

5. Controls. Events in the upper atmosphere can be influenced by many factors. We call each factor that might influence a process a **variable.** To evaluate alternative hypotheses about one variable, all the other variables must be kept constant so that we do not get misled or confused by these other influences. This is done by carrying out two experiments in parallel: in the first experimental test, we alter one variable in a known way to test a particular hypothesis; in the second, called a **control experiment,** we do *not* alter that variable. In all other respects, the two experiments are the same. To further test the CFC hypothesis, scientists carried out control experiments in which the key variable was the amount of CFCs in the atmosphere. Working in laboratories, scientists reconstructed the atmospheric conditions, solar bombardment, and extreme temperatures found in the sky far above the Antarctic. If the ozone levels fell without addition of CFCs to the chamber, then CFCs could not be what was attacking the ozone, and the CFC hypothesis must be wrong. Carefully monitoring the chamber, however, scientists detected no drop in ozone levels in the absence of CFCs. The result of the control was thus consistent with the predictions of the hypothesis.

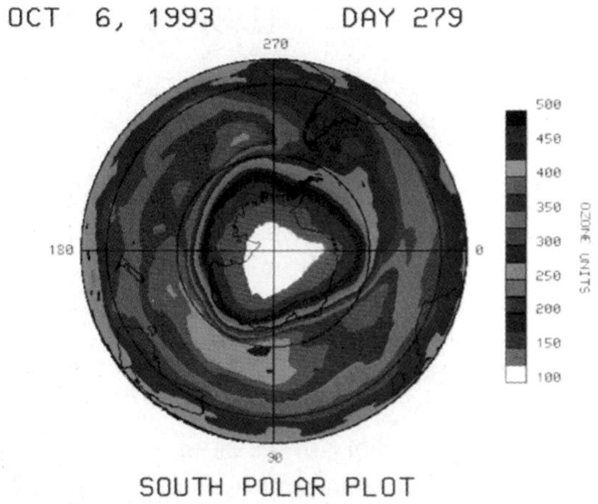

OCT 6, 1993 DAY 279
270
180
0
90
500
450
400
350
300
250
200
150
100
OZONE UNITS
SOUTH POLAR PLOT

Figure 1.13 The ozone hole.

The swirling colors represent different concentrations of ozone over the South Pole as viewed from a satellite. As you can easily see, there is an "ozone hole" over Antarctica, an area about the size of the United States. Careful scientific investigation has eliminated the hypothesis that the "hole" is due to the spinning of the earth.

6. Conclusion. A hypothesis that has been tested and not rejected is tentatively accepted. The hypothesis that CFCs released into the atmosphere are destroying the earth's protective ozone shield is now supported by a great deal of experimental evidence and is widely accepted. While other factors have also been implicated in ozone depletion, destruction by CFCs is clearly the dominant phenomenon. A collection of related hypotheses that have been tested many times is called a **theory.** The theory of the ozone shield, that ozone in the upper atmosphere shields the earth's surface from harmful UV rays by absorbing them, is supported by a wealth of observation and experimentation and is widely accepted.

1.7 Science progresses by systematically eliminating potential hypotheses that are not consistent with observation.

1.8 Theory and Certainty

Hypotheses that stand the test of time—their predictions often tested and never rejected—are sometimes combined into general statements called theories. A theory is a unifying explanation for a broad range of observations. Thus we speak of the theory of gravity, the theory of evolution, and the theory of the atom. Theories are the solid ground of science, that of which we are the most certain. There is no absolute truth in science, however, only varying degrees of uncertainty. The possibility always remains that future evidence will cause a theory to be revised. A scientist's acceptance of a theory is always provisional. For example, in another scientist's experiment, evidence that is inconsistent with a theory may be revealed. As information is shared throughout the scientific community, previous hypotheses and theories may be modified, and scientists may formulate new ideas.

Figure 1.14 Nobel Prize winner.

Robert F. Furchgott is one of the three researchers who won the 1998 Nobel Prize in Physiology or Medicine for the discovery of the physiological role of nitric oxide (see p. 546).

Very active areas of science are often alive with controversy, as scientists grope with new and challenging ideas. This uncertainty is not a sign of poor science but rather of the push and pull that is the heart of the scientific process. The hypothesis that the world's climate is growing warmer due to humanity's excessive production of carbon dioxide (CO_2), for example, has been quite controversial, although the weight of evidence has increasingly supported the hypothesis.

The word theory is thus used very differently by scientists than by the general public. To a scientist, a theory represents that of which he or she is most certain; to the general public, the word theory implies a *lack* of knowledge or a guess. How often have you heard someone say, "It's only a theory!"? As you can imagine, confusion often results. In this text the word theory will always be used in its scientific sense, in reference to a generally accepted scientific principle.

The Scientific "Method"

It was once fashionable to claim that scientific progress is the result of applying a series of steps called the **scientific method;** that is, a series of logical "either/or" predictions tested by experiments to reject one alternative. The assumption was that trial-and-error testing would inevitably lead one through the maze of uncertainty that always slows scientific progress. If this were indeed true, a computer would make a good scientist—but science is not done this way! If you ask successful scientists like Farman how they do their work, you will discover that without exception they design their experiments with a pretty fair idea of how they will come out. Environmental scientists understood the chemistry of chlorine and ozone when they formulated the CFC hypothesis, and they could imagine how the chlorine in CFCs would attack ozone molecules. A hypothesis that a successful scientist tests is not just any hypothesis. Rather, it is a "hunch" or educated guess in which the scientist integrates all that he or she knows. The scientist also allows his or her imagination full play, in an attempt to get a sense of what *might* be true. It is because insight and imagination play such a large role in scientific progress that some scientists are so much better at science than others (figure 1.14)—just as Beethoven and Mozart stand out above most other composers.

The Limitations of Science

Scientific study is limited to organisms and processes that we are able to observe and measure. Supernatural, religious, and unexplained phenomena are beyond the realm of scientific analysis because they cannot be scientifically studied, analyzed, or explained. To some individuals, a nonscientific point of view may have a moral or aesthetic value. However, scientists in their work are limited to objective interpretations of observable phenomena. This does not mean that individuals who are scientists and base their work on the principles of scientific study are less moral. Depending on the society, the culture, and the country, most individuals incorporate many philosophies into their lives.

It is also important to recognize that there are practical limits to what science can accomplish. While scientific study has and continues to revolutionize our world, it cannot be relied upon to solve all problems. For example, we cannot pollute the environment and squander its resources today, in the blind hope that somehow science will make it all right sometime in the future. Nor can science restore an extinct species. Science identifies solutions to problems when solutions exist, but it cannot invent solutions when they don't.

1.8 A scientist does not follow a fixed method to form hypotheses but relies also on judgment and intuition.

CONCEPT REVIEW

1. Metabolism refers to an organism's ability to
   a. reproduce.
   b. use energy.
   c. pass on genes.
   d. move.

2. Key terms for homeostasis are
   a. external environment, stable.
   b. internal environment, unstable.
   c. internal environment, stable.
   d. external environment, unstable.

3. Select the smallest level of organization among the following.
   a. cell
   b. organ
   c. organ system
   d. tissue

4. The change in a species through time is
   a. cooperation.
   b. evolution.
   c. homeostasis.
   d. metabolism.

5. A guess in a scientific process is called a(n)
   a. hypothesis.
   b. observation.
   c. prediction.
   d. theory.

6. A collection of hypotheses that have been repeatedly tested without rejection is called a(n)
   a. control.
   b. observation.
   c. test.
   d. theory.

7. Factors that influence a process in a scientific study are
   a. controls.
   b. tests.
   c. theories.
   d. variables.

8. List the six kingdoms of life.

9. List the five fundamental properties that are shared by all living organisms on earth and that are not exhibited by nonliving things.

10. _________ is the complex linear molecule responsible for heredity.

11. A _________ is a tiny living compartment covered with a membrane.

12. At each level of organization, _________ determines function.

13. Any good scientific investigation begins with careful _________.

14. A _________ is an experiment in which a particular variable is not allowed to change.

CHALLENGE YOURSELF

1. What is the difference between theory and certainty to a scientist? How does the word *hypothesis* fit in with a theory?

2. How does the human heart show all of the general themes of life: levels of organization, homeostasis, etc.?

3. How do you think that the connection between structure and function is the result of evolution?

4. Why is it correct to state that the process of science does not work to discover truth?

5. Imagine that you are a scientist asked to test the following hypothesis: The disappearance of a particular species of fish from a lake in the northeastern United States is due to acid rain resulting from industrial air pollution. What alternative hypotheses could you formulate? What experiments would you conduct to test these hypotheses? How would you use control experiments to isolate the influence of acid rain from that of other variables?

1 eBRIDGE

Reinforcing Key Points

Biology and the Living World

1.1 The Diversity of Life

1.2 Properties of Life

1.3 The Organization of Life

1.4 Biological Themes

The Scientific Process

1.5 The Nature of Science

1.6 Science in Action: A Case Study

1.7 Stages of a Scientific Investigation

Using Science to Make Decisions

1.8 Theory and Certainty

Electronic Learning

Visual Learning

Animations

Levels of Biological Organization

Ecosystem Organization

Homeostasis and Temperature Control

Flow Diagram for Scientific Method

Ozone Layer Depletion

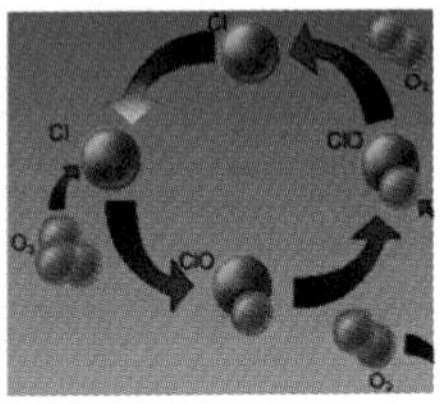

Art Labeling Activities

Levels of Biological Organization (plants)

Levels of Biological Organization (animals)

Author's Corner

Everyday Science. Sometimes the nature of scientific inquiry is most clearly revealed by applying it to everyday matters, like what happens to missing socks, or by contrasting it to fantasy by attempting to evaluate hypotheses about sea monsters, leprechauns, or Santa Claus.

1. What good is science if it can't tell me where all my socks are going?
2. Other attempts to solve the missing sock mystery.
3. The scientific theory of Santa Claus.
4. Further tests of the Santa theory.
5. How to catch a leprechaun.
6. UFOs: Is seeing believing?
7. Yes, Virginia, there is a jackalope.
8. How a scientist came to believe in dragons.
9. Seeking a sea monster named Selma.

Virtual Classroom

Introduction to Aids

The year 2001 marked the twentieth anniversary of the AIDS epidemic. Since the first case was reported on June 5, 1981, the United States has recorded over 770,000 AIDS cases and over 460,000 deaths. Worldwide, the figures are numbing: 60 million cases and 22 million dead. Three million people died of AIDS last year alone. AIDS is caused by a virus called HIV that attacks and destroys the immune system. It is transmitted by sex, needles, and anything else that transfers white blood cells. AIDS is fatal, and there is no cure.

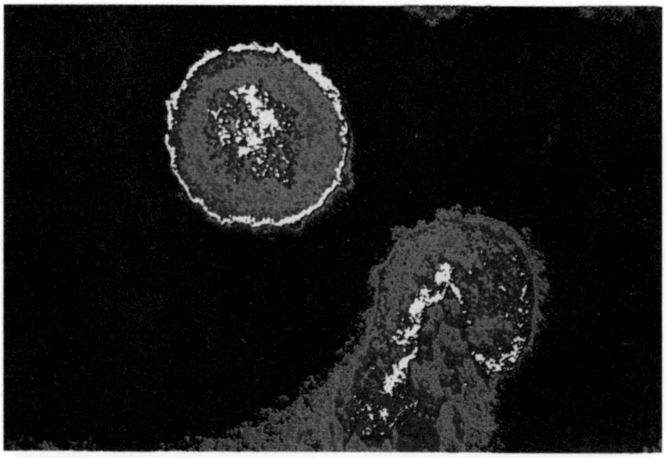

The Challenge of Curing AIDS

Faced with the AIDS plague—no other word will do—scientists all over the world have sought a way to defeat the HIV virus. It has been a discouraging battle. The full nucleotide sequence of the virus is known, and researchers have pieced together a detailed picture of how it infects cells, but all attempts to prepare a vaccine targeted on the HIV coat protein have failed. HIV simply mutates too quickly for any one vaccine to protect many people. New, more promising approaches involve multiple proteins and cell-mediated as well as antibody-based immune defenses.

Virtual Lab

Catching Evolution in Action

To study evolution, biologists have traditionally investigated what has happened in the past by examining fossils. For biologists taking this traditional approach, evolutionary biology is similar to astronomy or history, relying on observation and deduction rather than experimentation. In recent years, however, case studies of natural populations have demonstrated that evolutionary change can occur rapidly, and be studied in action.

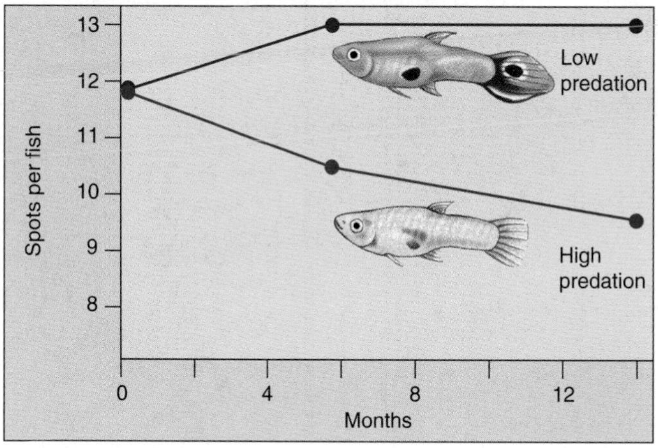
Spots per fish
13
12
11
10
9
8
0
4
8
12
Months
Low predation
High predation

The guppy, *Poecilia reticulata,* offers an excellent experimental opportunity to follow evolutionary change. Guppies inhabit two very different environments in the streams of Venezuela and Trinidad. The streams contain waterfalls that are dispersal barriers to guppies and guppy predators. Guppies above the waterfalls experience little predation and are larger and more colorful than guppies below the waterfalls that share their environment with a voracious predator. Does pressure from predation affect the color and size of guppies? A classic set of laboratory and field experiments initiated by John Endler in the late 1970s demonstrated that natural selection was acting on the Trinidad guppies.

Quizzes | Further Reading | Essential Study Partner | Links | BioCourse.com

CHAPTER 2

EVOLUTION AND ECOLOGY

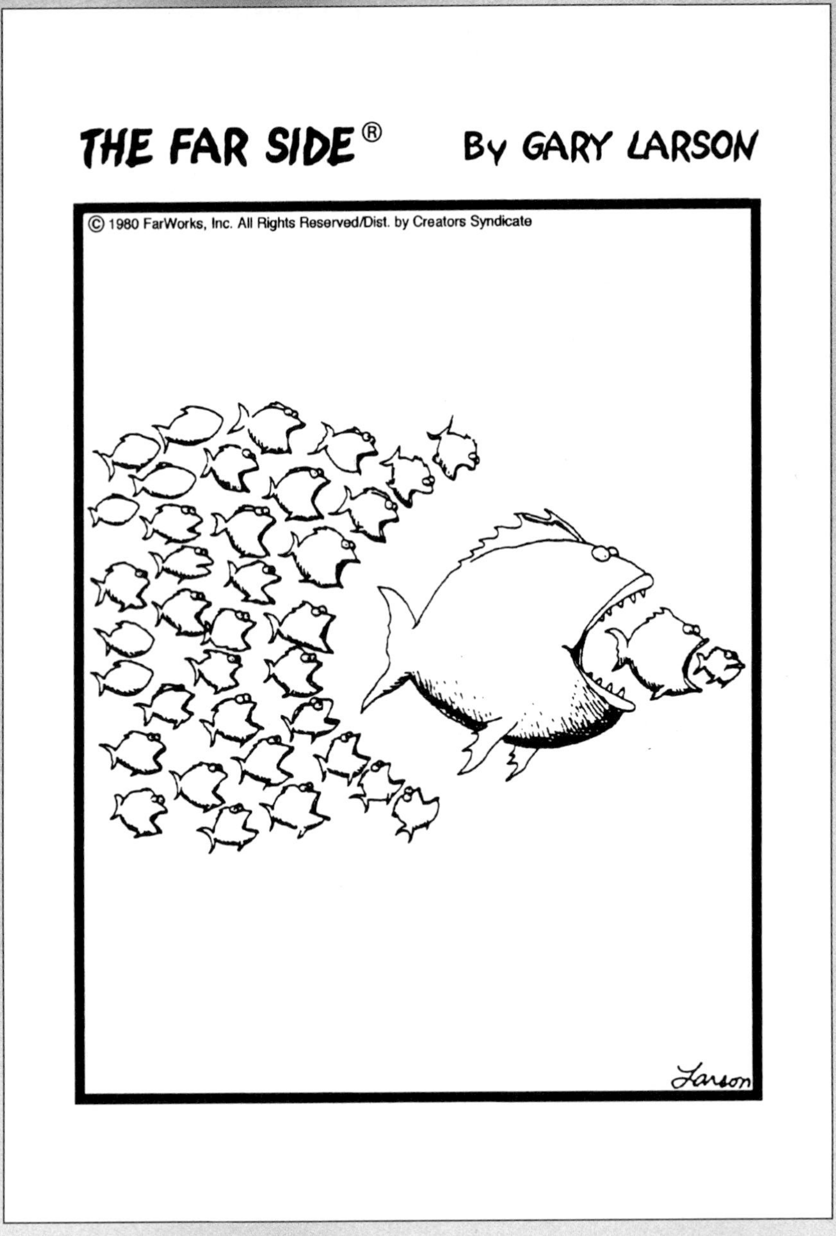
THE FAR SIDE®
By GARY LARSON
© 1980 FarWorks, Inc. All Rights Reserved/Dist. by Creators Syndicate
Larson

CHAPTER 2

Evolution and Ecology

CHAPTER OVERVIEW

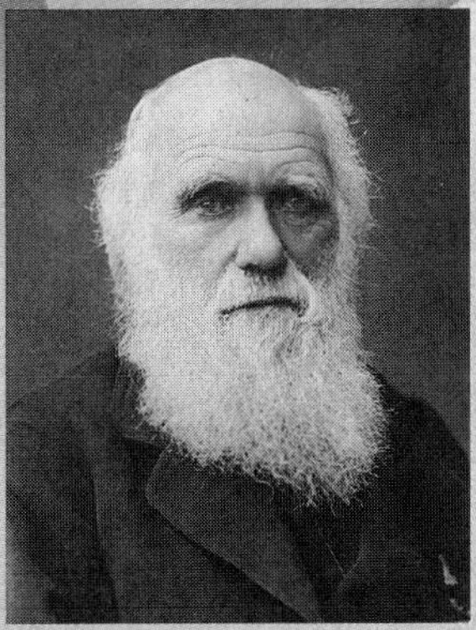

Evolution

- In 1831 Darwin began a trip around the world, closely observing the plants and animals he saw.
- In 1859 Darwin published *On the Origin of Species,* in which he proposed that the mechanism of evolution was natural selection: individuals with characteristics more suitable for survival and reproduction will tend to leave more offspring and so become more common in future generations.

Evolution in Action

- Clusters of species arise when populations differentiate to fill several niches. On islands, differentiation is often rapid because of numerous open habitats. In many continental areas, differentiation is not as rapid; in local situations, as when many different kinds of plants are developing close to one another, differentiation may occur rapidly.

Ecology

- Populations consist of the individuals of a given species that occur together at one place and at one time.
- Populations of different organisms that live together in a particular place are called communities. A community together with the nonliving components of its environment is called an ecosystem.

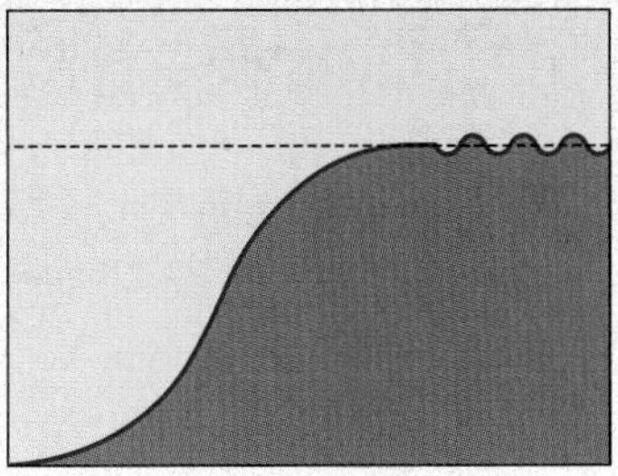

Populations and How They Grow

- Every species has its own life history strategy dictating how often it reproduces and how many offspring it has.
- The size of a population will change if there are unequal rates of birth and death, or if there is net migration into or out of the population.

Biologists believe that the great diversity of life on earth—ranging from bacteria to elephants and roses—is the result of a long process of **evolution,** the change that occurs in organisms' characteristics through time. In 1859, the English naturalist Charles Darwin (1809–82; figure 2.1) first suggested an explanation for why evolution occurs, a process he called **natural selection.** Biologists soon became convinced Darwin was right and now consider evolution one of the central concepts of the science of biology. A second key concept is that of **ecology,** how organisms live in their environment. Ecology is of increasing concern to all of us, as a growing human population places ever-greater stress on our planet. In this chapter, we introduce these two key concepts, evolution and ecology, to provide a foundation as you begin to explore the living world. Both are revisited in more detail later.

Figure 2.1 The theory of evolution was proposed by Charles Darwin.

This newly rediscovered photograph appears to be the last ever taken of the great biologist. It was taken in 1881, the year before Darwin died.

2.1 Darwin's Voyage on HMS *Beagle*

The theory of evolution proposes that a species can gradually evolve, sometimes forming a new species. This famous theory provides a good example of how a scientist develops a hypothesis and how, after much testing, it is eventually accepted as a theory.

Charles Robert Darwin, was an English naturalist who, after 30 years of study and observation, wrote one of the most famous and influential books of all time. This book, *On the Origin of Species by Means of Natural Selection, or The Preservation of Favoured Races in the Struggle for Life*, created a sensation when it was published, and the ideas Darwin expressed in it have played a central role in the development of human thought ever since.

In Darwin's time, most people believed that the various kinds of organisms and their individual structures resulted from direct actions of the Creator (and to this day many people still believe this to be true). Species were thought to be specially created and unchangeable, or immutable, over the course of time. In contrast to these views, a number of earlier philosophers had presented the view that living things must have changed during the history of life on earth. Darwin proposed a concept he called natural selection as a coherent, logical explanation for this process, and he brought his ideas to wide public attention. His book, as its title indicates, presented a conclusion that differed sharply from conventional wisdom. Although his theory did not challenge the existence of a Divine Creator, Darwin argued that this Creator did not simply create things and then leave them forever unchanged. Instead, Darwin's God expressed Himself through the operation of natural laws that produced change over time, or evolution. These views put Darwin at odds with most people of his time, who believed in a literal interpretation of the Bible and accepted the idea of a fixed and constant world.

The story of Darwin and his theory begins in 1831, when he was 22 years old. On the recommendation of one of his professors at Cambridge University, he was selected to serve as naturalist on a five-year navigational mapping expedition around the coasts of South America (figure 2.2), aboard HMS *Beagle* (figure 2.3). During this long voyage, Darwin had the chance to study a wide variety of plants and animals on continents and islands and in distant seas. He was able to explore the biological richness of the tropical forests, examine the extraordinary fossils of huge extinct mammals in Patagonia at the southern tip of South America, and observe the remarkable

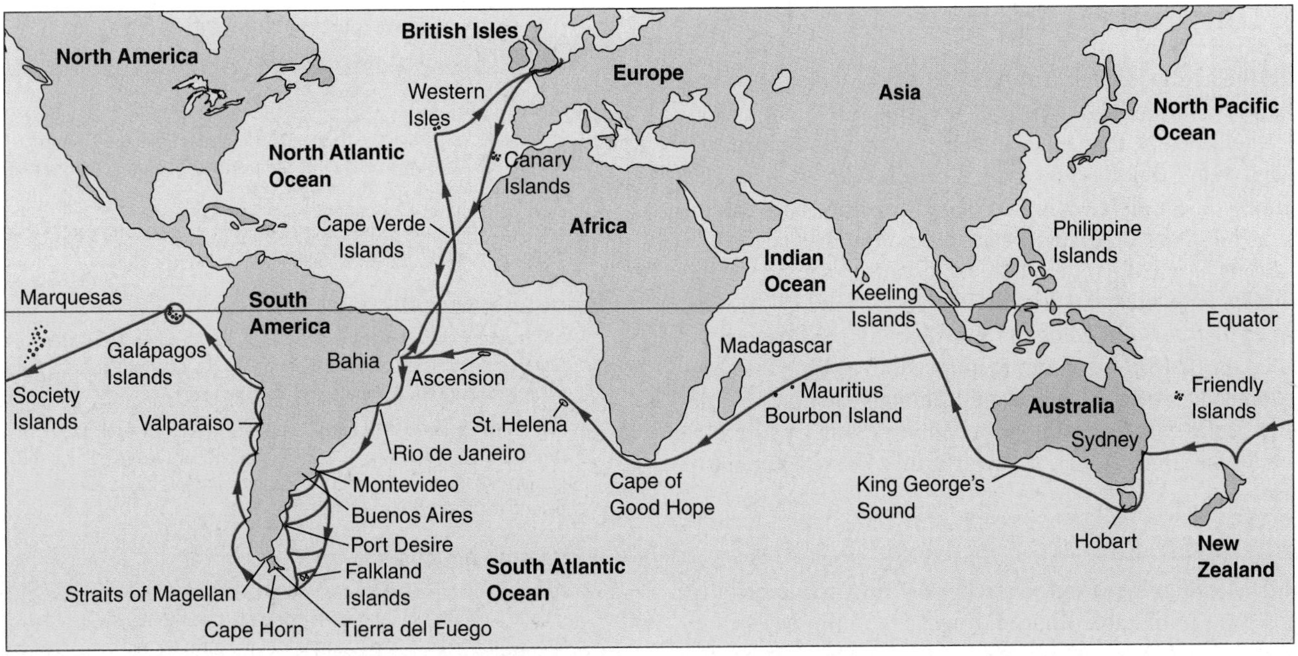
British Isles
North America
Western Isles
Europe
Asia
North Pacific Ocean
North Atlantic Ocean
Canary Islands
Cape Verde Islands
Africa
Philippine Islands
Indian Ocean
Marquesas
South America
Keeling Islands
Equator
Galápagos Islands
Bahia
Madagascar
Ascension
Society Islands
Mauritius
Bourbon Island
Australia
Friendly Islands
Valparaiso
St. Helena
Sydney
Rio de Janeiro
Montevideo
Cape of Good Hope
King George's Sound
Buenos Aires
Port Desire
Hobart
New Zealand
Falkland Islands
South Atlantic Ocean
Straits of Magellan
Cape Horn
Tierra del Fuego

Figure 2.2 The five-year voyage of HMS *Beagle*.

Although the ship sailed around the world, most of the time was spent exploring the coasts and coastal islands of South America, such as the Galápagos Islands. Darwin's studies of the animals of these islands played a key role in the eventual development of his theory of evolution by means of natural selection.

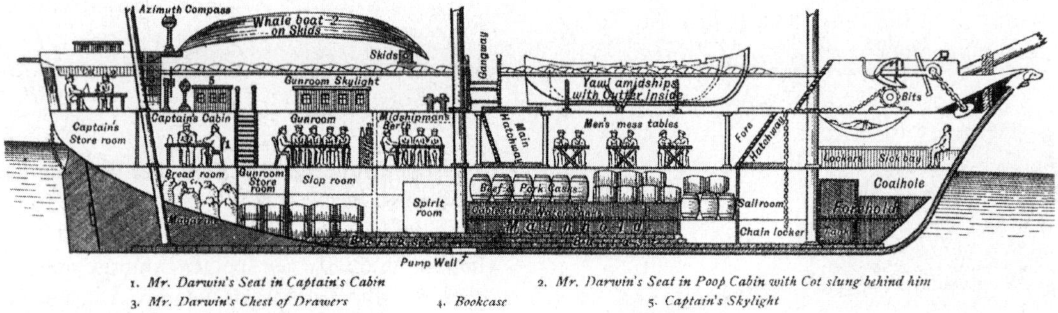
Azimuth Compass
Whale boat on Skids
Skids
Gunroom Skylight
Gangway
Yawl amidships with Cutter inside
Bits
Captain's Store room
Captain's Cabin
Gunroom
Midshipmen's Berth
Main Hatchway
Men's mess tables
Fore Hatchway
Lockers
Sick bay
Coalhole
Bread room
Gunroom Store room
Slop room
Spirit room
Beef & Pork Casks
Sail room
Chain locker
Fore hold
Pump Well
1. Mr. Darwin's Seat in Captain's Cabin
2. Mr. Darwin's Seat in Poop Cabin with Cot slung behind him
3. Mr. Darwin's Chest of Drawers
4. Bookcase
5. Captain's Skylight

Figure 2.3 Cross section of HMS *Beagle*.

HMS *Beagle*, a 10-gun brig of 242 tons, only 90 feet in length, had a crew of 74 people! After he first saw the ship, Darwin wrote to his college professor Henslow: "The absolute want of room is an evil that nothing can surmount."

series of related but distinct forms of life on the **Galápagos Islands,** off the west coast of South America near Ecuador. Such an opportunity clearly played an important role in the development of his thoughts about the nature of life on earth.

When Darwin returned from the voyage at the age of 27, he began a long period of study and contemplation. During the next 10 years, he published important books on several different subjects, including the formation of oceanic islands from coral reefs and the geology of South America. He also devoted eight years of study to barnacles, a group of small marine animals with shells that inhabit rocks and pilings, eventually writing a four-volume work on their classification and natural history. In 1842, Darwin and his family moved out of London to a country home at Down, in the county of Kent. In these pleasant surroundings, Darwin lived, studied, and wrote for the next 40 years.

2.1 Darwin was the first to propose natural selection as the mechanism of evolution that produced the diversity of life on earth.

2.2 Darwin's Evidence

One of the obstacles that had blocked the acceptance of any theory of evolution in Darwin's day was the incorrect notion, widely believed at that time, that the earth was only a few thousand years old. The discovery of thick layers of rocks, evidences of extensive and prolonged erosion, and the increasing numbers of diverse and unfamiliar fossils discovered during Darwin's time made this assertion seem less and less likely. The great geologist Charles Lyell (1797–1875), whose *Principles of Geology* (1830) Darwin read eagerly as he sailed on HMS *Beagle,* outlined for the first time the story of an ancient world of plants and animals in flux. In this world, species were constantly becoming extinct while others were emerging. It was this world that Darwin sought to explain.

What Darwin Saw

When HMS *Beagle* set sail, Darwin was fully convinced that species were immutable. Indeed, it was not until two or three years after his return that he began to seriously consider the possibility that they could change. Nevertheless, during his five years on the ship, Darwin observed a number of phenomena that were of central importance to him in reaching his ultimate conclusion (table 2.1). For example, in the rich fossil beds of southern South America, he observed fossils of extinct armadillos similar in form to the armadillos that still lived in the same area (figure 2.4). Why would similar living and fossil organisms be in the same area unless the earlier form had given rise to the other?

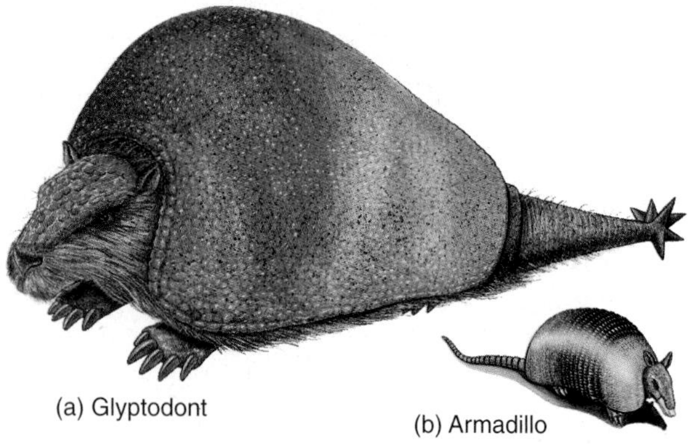
(a) Glyptodont
(b) Armadillo

Figure 2.4 Fossil evidence of evolution.

The now-extinct glyptodont (*a*) was a 2,000-kilogram South American armadillo, much larger than the modern armadillo (*b*), which weighs an average of about 4.5 kilograms. (Drawings are not to scale.) The similarity of fossils such as the glyptodonts to living organisms found in the same regions suggested to Darwin that evolution had taken place.

TABLE 2.1 DARWIN'S EVIDENCE THAT EVOLUTION OCCURS

FOSSILS

1. Extinct species, such as the fossil armadillo shown in figure 2.4, most closely resemble living ones in the same area, suggesting that one had given rise to the other.
2. In rock strata (layers), progressive changes in characteristics can be seen in fossils from earlier and earlier layers.

GEOGRAPHICAL DISTRIBUTION

3. Lands with similar climates, such as Australia, South Africa, California, and Chile, have unrelated plants and animals, indicating that diversity is not entirely influenced by climate and environment.
4. The plants and animals of each continent are distinctive; all South American rodents belong to a single group, structurally similar to the guinea pig, for example, while most of the rodents found elsewhere belong to other groups.

OCEANIC ISLANDS

5. Although oceanic islands have few species, those they do have are often unique (endemic) and show relatedness to one another, such as the Galápagos tortoises. This suggests that the tortoises and other groups of endemic species developed after their mainland ancestors reached the islands and are, therefore, more closely related to one another.
6. Species on oceanic islands show strong affinities to those on the nearest mainland. Thus, the finches of the Galápagos Islands closely resemble a finch seen on the western coast of South America. The Galápagos finches do *not* resemble the birds of the Cape Verde Islands, islands in the Atlantic Ocean off the coast of Africa that are similar to the Galápagos. Darwin personally visited the Cape Verde Islands and many other island groups and was able to make such comparisons on the basis of his own observations.

Repeatedly, Darwin saw that the characteristics of similar species varied somewhat from place to place. These geographical patterns suggested to him that organismal lineages change gradually as species migrate from one area to another. On the Galápagos Islands, Darwin encountered giant land tortoises (figure 2.5). Surprisingly, these tortoises were not all identical. In fact, local residents and the sailors who captured the tortoises for food could tell which island a particular tortoise had come from just by looking at its shell. This distribution of physical variation suggested that all of the tortoises were related, but that they had changed slightly in appearance after becoming isolated on different islands.

Figure 2.5 Galápagos tortoise.
A view of the Galápagos Islands showing a giant land tortoise similar to the ones Darwin saw.

Figure 2.6 Darwin's finches.
(*a*) One of Darwin's Galápagos finches, the medium ground finch. (*b*) The blue-black grassquit, found in grasslands along the Pacific Coast from Mexico to Chile. This species may be the ancestor of Darwin's finches.

In a more general sense, Darwin was struck by the fact that the plants and animals on these relatively young volcanic islands resembled those on the nearby coast of South America (figure 2.6). If each one of these plants and animals had been created independently and simply placed on the Galápagos Islands, why didn't they resemble the plants and animals of islands with similar climates, such as those off the coast of Africa, for example? Why did they resemble those of the adjacent South American coast instead?

2.2 The fossils and patterns of life that Darwin observed on the voyage of HMS *Beagle* eventually convinced him that evolution had taken place.

2.3 Inventing the Theory of Natural Selection

It is one thing to observe the results of evolution but quite another to understand how it happens. Darwin's great achievement lies in his formulation of the hypothesis that evolution occurs because of natural selection.

Darwin and Malthus

Of key importance to the development of Darwin's insight was his study of Thomas Malthus's *Essay on the Principle of Population* (1798). In his book, Malthus pointed out that populations of plants and animals (including human beings) tend to increase geometrically, while the ability of humans to increase their food supply increases only arithmetically. A geometric progression is one in which the elements increase by a constant factor; for example, in the progression 2, 6, 18, 54, . . . each number is three times the preceding one. An arithmetic progression, in contrast, is one in which the elements increase by a constant difference; in the progression 2, 4, 6, 8, . . . each number is two greater than the preceding one (figure 2.7).

Because populations increase geometrically, virtually any kind of animal or plant, if it could reproduce unchecked, would cover the entire surface of the world within a surprisingly short time. Instead, populations of species remain fairly constant year after year, because death limits population numbers. Malthus's conclusion provided the key ingredient that was necessary for Darwin to develop the hypothesis that evolution occurs by natural selection.

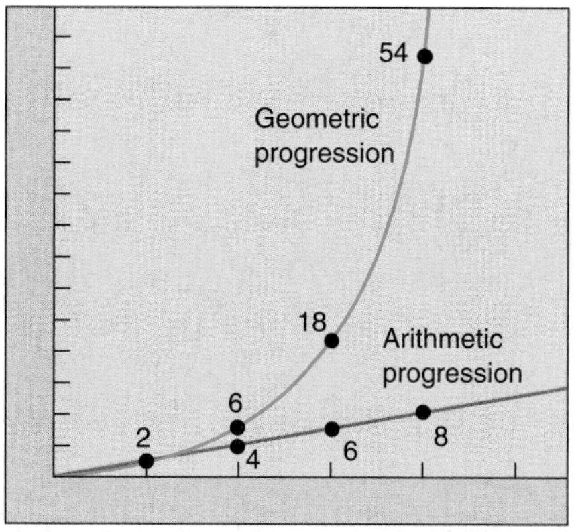

54
Geometric progression
18
Arithmetic progression
6
2
4
6
8

Figure 2.7 Geometric and arithmetic progressions.
An arithmetic progression increases by a constant difference (for example, units of 1 or 2 or 3), while a geometric progression increases by a constant factor (for example, by 2 or by 3 or by 4). Malthus contended that the human growth curve was geometric, but the human food production curve was only arithmetic. Can you see the problems this difference would cause?

Natural Selection

Sparked by Malthus's ideas, Darwin saw that although every organism has the potential to produce more offspring than can survive, only a limited number actually do survive and produce further offspring. Combining this observation with what he had seen on the voyage of HMS *Beagle,* as well as with his own experiences in breeding domestic animals, Darwin made an important association (figure 2.8): Those individuals that possess superior physical, behavioral, or other attributes are more likely to survive than those that are not so well endowed. By surviving, they gain the opportunity to pass on their favorable characteristics to their offspring. As the frequency of these characteristics increases in the population, the nature of the population as a whole will gradually change. Darwin called this process selection. The driving force he identified has often been referred to as survival of the fittest.

"Can we doubt . . . that individuals having any advantage, however slight, over others, would have the best chance of surviving and procreating their kind? On the other hand, we may feel sure that any variation in the least degree injurious would be rigidly destroyed. This preservation of favorable variations, I call Natural Selection."

Figure 2.8 An excerpt from Charles Darwin's *On the Origin of Species.*

Darwin was thoroughly familiar with variation in domesticated animals and began *On the Origin of Species* with a detailed discussion of pigeon breeding. He knew that breeders selected certain varieties of pigeons and other animals, such as dogs, to produce certain characteristics, a process Darwin called artificial selection. Once this had been done, the animals would breed true for the characteristics that had been selected. Darwin had also observed that the differences purposely developed between domesticated races or breeds were often greater than those that separated wild species. Domestic pigeon breeds, for example, show much greater variety than all of the hundreds of wild species of pigeons found throughout the world. Such relationships suggested to Darwin that evolutionary change could occur in nature too. Surely if pigeon breeders could foster such variation by "artificial selection," nature through environmental pressures could do the same, playing the breeder's role in selecting the next generation—a process Darwin called **natural selection.**

Darwin's theory provides a simple and direct explanation of biological diversity, or why animals are different in different places: because habitats differ in their requirements and opportunities, the organisms with characteristics favored locally by natural selection will tend to vary in different places.

Darwin Drafts His Argument

Darwin drafted the overall argument for evolution by natural selection in a preliminary manuscript in 1842. After showing the manuscript to a few of his closest scientific friends, however, Darwin put it in a drawer and for 16 years turned to other research. No one knows for sure why Darwin did not publish his initial manuscript—it is very thorough and outlines his ideas in detail. Some historians have suggested that Darwin was wary of igniting public criticism of his evolutionary ideas—there could have been little doubt in his mind that his theory of evolution by natural selection would spark controversy. Others have proposed that Darwin was simply refining his theory, although there is little evidence he altered his initial manuscript in all that time.

Wallace Has the Same Idea

The stimulus that finally brought Darwin's theory into print was an essay he received in 1858. A young English naturalist named Alfred Russel Wallace (1823–1913) sent the essay to Darwin from Malaysia; it concisely set forth the theory of evolution by means of natural selection, a theory Wallace had developed independently of Darwin. Like Darwin, Wallace had been greatly influenced by Malthus's 1798 book. Colleagues of Wallace, knowing of Darwin's work, encouraged him to communicate with Darwin. After receiving Wallace's essay, Darwin arranged for a joint presentation of their ideas at a seminar in London. Darwin then completed his own book, expanding the 1842 manuscript that he had written so long ago, and submitted it for publication.

Figure 2.9 Darwin greets his monkey ancestor.
In his time, Darwin was often portrayed unsympathetically, as in this drawing from an 1874 publication.

Publication of Darwin's Theory

Darwin's book appeared in November 1859 and caused an immediate sensation. Many people were deeply disturbed by the suggestion that human beings were descended from the same ancestor as apes (figure 2.9). Although people had long accepted that humans closely resembled apes in many characteristics, the possibility that there might be a direct evolutionary relationship was unacceptable to many. Darwin did not actually discuss this idea in his book, but it followed directly from the principles he outlined. In a subsequent book, *The Descent of Man,* Darwin presented the argument directly, building a powerful case that humans and living apes have common ancestors. Darwin's arguments for the theory of evolution by natural selection were so compelling, however, that his views were almost completely accepted within the intellectual community of Great Britain after the 1860s.

2.3 The fact that populations do not really expand geometrically implies that nature acts to limit population numbers. The traits of organisms that survive to produce more offspring will be more common in future generations—a process Darwin called natural selection.

2.4 The Beaks of Darwin's Finches

Darwin's finches are a classic example of evolution by natural selection. He collected 31 specimens of finch from three islands when he visited the Galápagos Islands in 1835. Darwin, not an expert on birds, had trouble identifying the specimens. He believed by examining their bills that his collection contained wrens, "gross-beaks," and blackbirds. You can see Darwin's sketches of four of these birds in figure 2.10.

1
2
3
4

Figure 2.10 Darwin's own sketches of Galápagos finches.
From Darwin's Journal of Researches: (*1*) large ground finch, *Geospiza magnirostris;* (*2*) medium ground finch, *Geospiza fortis;* (*3*) small tree finch, *Camarhynchus parvulus;* (*4*) warbler finch, *Certhidea olivacea.*

The Importance of the Beak

Upon Darwin's return to England, ornithologist John Gould examined the finches. Gould recognized that Darwin's collection was in fact a closely related group of distinct species, all similar to one another except for their bills. In all, there were 13 species. The two ground finches with the larger bills in figure 2.10 feed on seeds, which they crush in their beaks, while the two with narrower bills eat insects. Still another species is a fruit eater, another a cactus eater, and yet another a "vampire" that creeps up on seabirds and uses its sharp beak to drink their blood. Perhaps most remarkable are the tool users, woodpecker finches that pick up a twig, cactus thorn, or leaf stalk, trim it into shape with their bills, and then poke it into dead branches to pry out grubs.

The correspondence between the beaks of the 13 finch species and their food source immediately suggested to Darwin that evolution had shaped them:

> "Seeing this gradation and diversity of structure in one small, intimately related group of birds, one might really fancy that from an original paucity of birds in this archipelago, one species has been taken and modified for different ends."

Was Darwin Wrong?

If Darwin's suggestion that the beak of an ancestral finch had been "modified for different ends" is correct, then it ought to be possible to see the different species of finches acting out their evolutionary roles, each using its bill to acquire its particular food specialty. The four species that crush seeds within their bills, for example, should feed on different seeds, with those with stouter beaks specializing on harder-to-crush seeds.

Many biologists visited the Galápagos after Darwin, but it was 100 years before any tried this key test of his hypothesis. When the great naturalist David Lack finally set out to do this in 1938, observing the birds closely for a full five months, his observations seemed to contradict Darwin's proposal! Lack often observed many different species of finch feeding together on the same seeds. His data indicated that the stout-beaked species and the slender-beaked species were feeding on the very same array of seeds.

We now know that it was Lack's misfortune to study the birds during a wet year, when food was plentiful. The size of the finch's beak is of little importance in such flush times; slender and stout beaks work equally well to gather the abundant tender small seeds. Later work revealed a very different picture during dry years, when few seeds are available.

A Closer Look

Starting in 1973, Peter and Rosemary Grant of Princeton University and generations of their students have studied the medium ground finch, *Geospiza fortis* (figure 2.11), on a tiny island in the center of the Galápagos called Daphne Major. These finches feed preferentially on small tender seeds, abundantly available in wet years. The birds resort to larger, drier seeds that are harder to crush when small seeds are hard to find. Such lean times come during periods of dry weather, when plants produce few seeds, large or small.

The Grants quantified beak shape among the medium ground finches of Daphne Major by carefully measuring beak depth (width of beak, from top to bottom, at its base) on individual birds. Measuring many birds every year, they were able to assemble for the first time a detailed portrait of evolution in action. The Grants found that beak depth changed from one year to the next in a predictable fashion. During droughts, plants produced few seeds, and all available small seeds quickly were eaten, leaving large seeds as the major remaining source of food. As a result, birds with large beaks survived better, because they were better able to break open these large seeds. Consequently, the average beak depth of birds in the population increased the next year, only to decrease again when wet seasons returned (figure 2.12).

Could these changes in beak dimension reflect the action of natural selection? An alternative possibility might be that the changes in beak depth do not reflect changes in gene frequencies but rather are simply a response to diet, with poorly fed birds having stouter beaks. To rule out this possibility, the Grants measured the relation of parent bill size to offspring bill size, examining many broods over several years. The depth of the bill was passed down faithfully from one generation to the next, suggesting the differences in bill size indeed reflected gene differences.

Figure 2.11 The subject of the Grants' study.
The medium ground finch, *Geospiza fortis,* feeds on seeds that it crushes in its bill.

Darwin Was Right After All

If the year-to-year changes in beak depth can be predicted by the pattern of dry years, then Darwin was right after all—natural selection does seem to adjust the beak to its food supply. Birds with stout beaks have an advantage during dry periods, for they can break the large, dry seeds that are the only food available. When small seeds become plentiful once again with the return of wet weather, a smaller beak proves a more efficient tool for harvesting smaller seeds.

2.4 In Darwin's finches, natural selection adjusts the shape of the beak in response to the nature of the food supply, adjustments that are occurring even today.

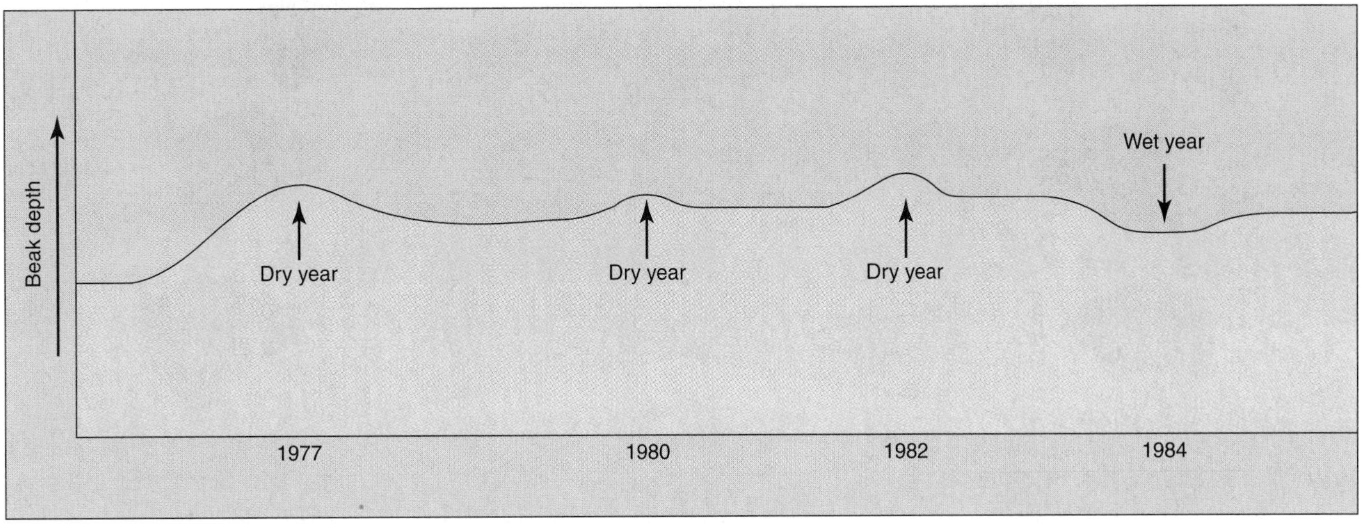
Beak depth
Wet year
Dry year
Dry year
Dry year
1977
1980
1982
1984

Figure 2.12 Evidence that natural selection alters beak size in *Geospiza fortis.*
In dry years, when only large, tough seeds are available, the mean beak size increases. In wet years, when many small seeds are available, smaller beaks become more common.

2.5 Clusters of Species

One of the most visible manifestations of evolution is the existence of clusters of closely related species. These species often have evolved relatively recently from a common ancestor. The phenomenon by which they change, coming to occupy a series of different habitats within a region, is called **adaptive radiation.** Such clusters are often particularly impressive on islands, in series of lakes, or in other sharply discontinuous habitats. One example of a cluster of species that has undergone adaptive radiation to fill a wide range of habitats is Darwin's finches (figure 2.13). Adaptive radiation occurred among the 13 species of Darwin's finches on the Galápagos Islands; 10 of these species are shown in figure 2.13. Presumably, the ancestor of Darwin's finches reached these islands before other land birds, so that when it arrived, all of the niches where birds occur on the mainland were unoccupied. As the new arrivals moved into these vacant niches and adopted new lifestyles, they were subjected to diverse sets of selective pressures. Under these circumstances, the ancestral finches rapidly split into a series of populations, some of which evolved into separate species.

The descendants of the original finches that reached the Galápagos Islands now occupy many different kinds of habitats on the islands. These habitats encompass a variety of niches comparable to those that several distinct groups of birds occupy on the mainland. The 13 species that inhabit the Galápagos comprise four groups:

1. **Ground finches.** There are six species of *Geospiza* ground finches. Most of the ground finches feed on seeds. The size of their bills is related to the size of the seeds they eat. Some of the ground finches feed primarily on cactus flowers and fruits and have longer, larger, more pointed bills than the others.
2. **Tree finches.** There are five species of insect-eating tree finches. Four species have bills that are suitable for feeding on insects. The woodpecker finch has a chisel-like beak. This unique bird carries around a twig or a cactus spine, which it uses to probe for insects in deep crevices.
3. **Warbler finch.** This unusual bird plays the same ecological role in the Galápagos woods that warblers play on the mainland, searching continually over the leaves and branches for insects. It has a slender, warbler-like beak.
4. **Vegetarian finch.** The very heavy bill of this bud-eating bird is used to wrench buds from branches.

2.5 Darwin's finches, all derived from one similar mainland species, have radiated widely on the Galápagos Islands in the absence of competition.

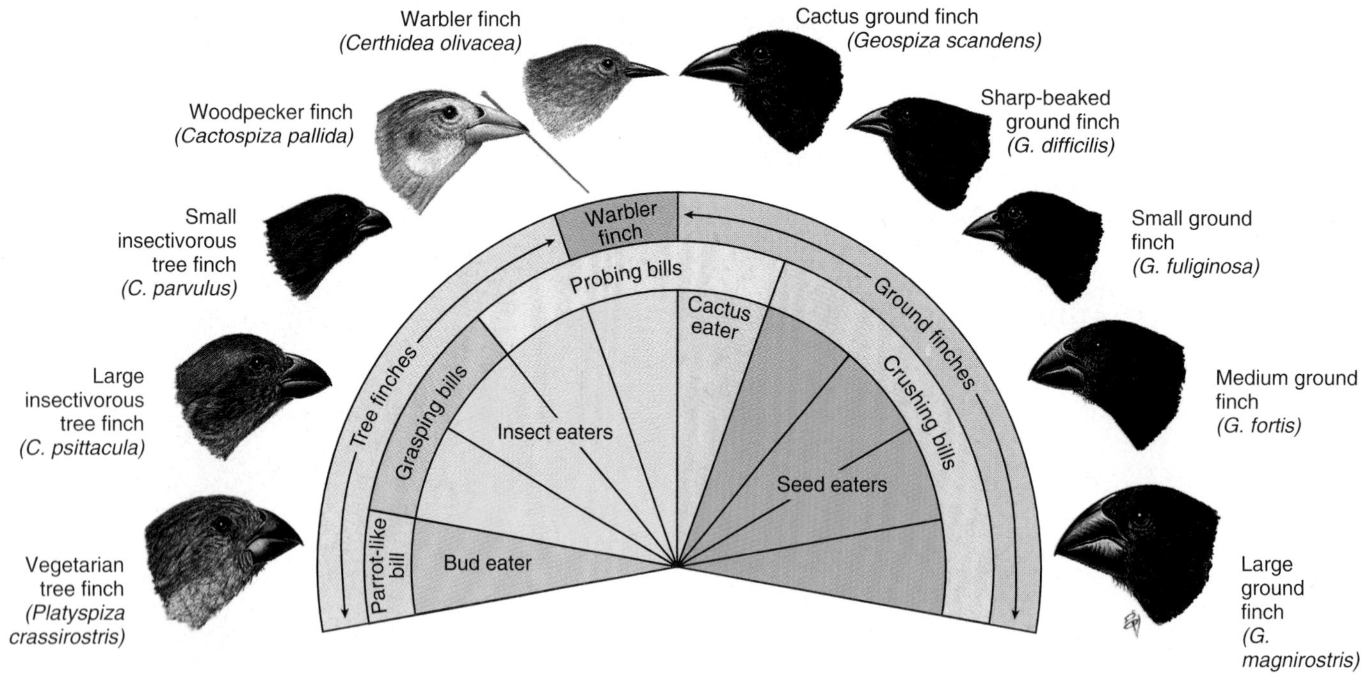

Warbler finch
(Certhidea olivacea)
Cactus ground finch
(Geospiza scandens)
Woodpecker finch
(Cactospiza pallida)
Sharp-beaked ground finch
(G. difficilis)
Small insectivorous tree finch
(C. parvulus)
Small ground finch
(G. fuliginosa)
Large insectivorous tree finch
(C. psittacula)
Medium ground finch
(G. fortis)
Vegetarian tree finch
(Platyspiza crassirostris)
Large ground finch
(G. magnirostris)
Warbler finch
Probing bills
Tree finches
Ground finches
Grasping bills
Crushing bills
Cactus eater
Insect eaters
Seed eaters
Parrot-like bill
Bud eater

Figure 2.13 Darwin's finches.

Ten species of Darwin's finches from Isla Santa Cruz, one of the Galápagos Islands, show differences in bills and feeding habits. The bills of several of these species resemble those of distinct families of birds on the mainland. This condition presumably arose when the finches evolved new species in habitats lacking small birds. The woodpecker finch uses cactus spines to probe in crevices of bark and rotten wood for food. Scientists believe all of these birds derived from a single common ancestor, a finch like the one in figure 2.6*a*.

2.6 Hawaiian *Drosophila*

A classic example of evolutionary diversification is the fly genus *Drosophila* on the Hawaiian Islands. There are at least 1,250 species of this genus throughout the world, and more than a quarter are found only in the Hawaiian Islands (figure 2.14). New species of *Drosophila* are still being discovered in Hawaii, although the rapid destruction of the native vegetation is making the search more difficult. Aside from their sheer number, Hawaiian *Drosophila* species have unique morphological and behavioral traits, some of which will be discussed later in this text. No comparable species of *Drosophila* are found anywhere else in the world.

A second, closely related genus of flies, *Scaptomyza,* also forms a species cluster in Hawaii, where it is represented by as many as 300 species. A few species of *Scaptomyza* are found outside of Hawaii, but the genus is better represented there than elsewhere. In addition, species intermediate between *Scaptomyza* and *Drosophila* exist in Hawaii, but nowhere else. The genera are so closely related that scientists have suggested that all of the estimated 800 species of these two genera that occur in Hawaii may have derived from a single common ancestor.

The native Hawaiian flies are closely associated with the remarkable native plants of the islands and are often abundant in the native vegetation. Evidently, when their ancestors first reached these islands, they encountered many "empty" niches that other kinds of insects and other animals occupied elsewhere. The evolutionary opportunities the ancestral *Drosophila* flies found were similar to those the ancestors of Darwin's finches in the Galápagos Islands encountered, and both groups evolved in a similar way. Many of the Hawaiian *Drosophila* species are highly selective in their choice of host plants for their larvae and in the part of the plant they use. The larvae of various species live in rotting stems, fruits, bark, leaves, or roots, and feed on sap.

New islands have continually arisen from the sea in the region of the Hawaiian Islands. As they have done so, they appear to have been invaded successively by the various *Drosophila* groups present on the older islands. New species have evolved as new islands have been colonized. The Hawaiian species of *Drosophila* have had even greater evolutionary opportunities than Darwin's finches because of their restricted ecological niches and the variable ages of the islands. They clearly tell one of the most unusual evolutionary stories found anywhere in the world.

2.6 The adaptive radiation of about 800 species of the flies *Drosophila* and *Scaptomyza* on the Hawaiian Islands, probably from a single common ancestor, is one of the most remarkable examples of intensive species formation found anywhere on earth.

(a) *Drosophila mulli*

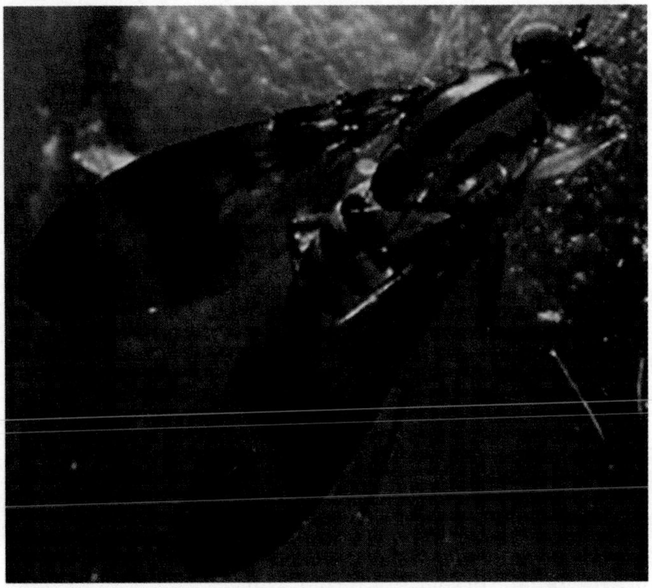

(b) *Drosophila primaeva*

(c) *Drosophila digressa*

Figure 2.14 Hawaiian *Drosophila*.
The hundreds of species that have evolved on the Hawaiian Islands are extremely variable in appearance, although genetically almost identical.

2.7 Lake Victoria Cichlid Fishes

Lake Victoria is an immense, shallow freshwater sea about the size of Switzerland in the heart of equatorial East Africa, until recently home to an incredibly diverse collection of over 200 species of cichlid fishes.

Recent Radiation

This cluster of species appears to have evolved recently and quite rapidly. Researchers like Axel Meyer at the State University of New York, Stony Brook, have been able to estimate that the first cichlids entered Lake Victoria only 200,000 years ago, colonizing from the Nile. Dramatic changes in water level encouraged species formation. As the lake rose, it flooded new areas and opened up new habitat. Many of the species may have originated after the lake dried down 14,000 years ago, isolating local populations in small lakes until the water level rose again.

Figure 2.15 Cichlid fishes of Lake Victoria.

Cichlid fishes are extremely diverse and occupy different niches. Some species feed on arthropods, others on dense stands of plants; there are fish eaters, and still other species feed on fish eggs and larvae. The Nile perch (not shown), a commercial fish introduced into Lake Victoria as a potential food source, is responsible for the virtual extinction of hundreds of species of cichlid fishes.

Cichlid Diversity

These small, perchlike fishes range from 2 to 10 inches in length, and the males come in endless varieties of colors. In initial surveys, far from complete, over 300 closely related species were described! The Lake Victoria cichlids, the most diverse assembly of vertebrates known to science, defy simple description. We can gain some sense of the vast range of types by looking at how different species eat (figure 2.15). There are mud biters, algae scrapers, leaf chewers, snail crushers, snail shellers (who pounce on slow-crawling snails and spear their soft parts with long-curved teeth before the snail can retreat into its shell), zooplankton eaters, insect eaters, prawn eaters, and fish eaters. Scale-scraping cichlids rasp slices of scales off of other fish.

There are even cichlid species that are "pedophages," eating the young of other cichlids. All Lake Victoria cichlids are mouthbrooders, the females keeping their young inside their mouths to protect them. Some pedophage species operate as suckers, others as rammers. Some rammers shoot toward the mother from below and behind, ramming into her throat and then eating the ejected brood before the surprised mother can recover. Another rammer crashes down from above, kamikaze-like, onto the nose of the mother.

Abrupt Extinction

Much of this diversity is gone. In the 1950s, the Nile perch, a commercial fish with a voracious appetite, was introduced on the Ugandan shore of Lake Victoria. Since then it has spread through the lake, eating its way through the cichlids. By 1990 all the open-water cichlid species were extinct, as well as many living in rocky shallow regions. Over 70% of all the named Lake Victoria cichlid species had disappeared, as well as untold numbers of species that had yet to be described. The Nile perch, in the meantime, has become a superb source of food for people living around the lake. The isolation of Lake Victoria from other kinds of fishes played a primary role in the explosive radiation of cichlid fishes, and when that isolation broke down with the introduction of the Nile perch, the bloom of speciation ended.

2.7 Very rapid speciation occurred among cichlid fishes isolated in Lake Victoria, but widespread extinction followed when the isolation ended.

2.8 New Zealand Alpine Buttercups

Adaptive radiations as we have described in Galápagos finches, Hawaiian *Drosophila*, and cichlid fishes seem to be favored by periodic isolation. Another clear example of the role periodic isolation plays in species formation can be seen in the alpine buttercups (genus *Ranunculus*) of New Zealand (figure 2.16). More species of alpine buttercups grow on the two islands of New Zealand than in all of North and South America combined. Detailed studies by the Canadian taxonomist Fulton Fisher revealed that the evolutionary mechanism responsible for inducing this diversity is recurrent isolation associated with the recession of glaciers. The 14 species of alpine *Ranunculus* occupy five distinctive habitats: snowfields (rocky crevices among outcrops in permanent snowfields at 7,000 to 9,000 ft elevation); snowline fringe (rocks at lower margins of snowfields between 4,000 and 7,000 ft); stony debris (scree slopes of exposed loose rocks at 2,000 to 6,000 ft); sheltered situations (shaded by rock or shrubs at 1,000 to 6,000 ft); and boggy habitats (sheltered slopes and hollows, poorly drained tussocks at elevations between 2,500 and 5,000 ft).

Ranunculus species have repeatedly invaded these five habitats as glaciers have formed that join mountains together, receded to isolate mountain habitats, and then reformed to link the habitats together once again (figure 2.17). Parallel sets of "ecospecies" have evolved independently each time in each habitat, so that distantly related species now occur in close proximity on glacial snow, scree slopes, and bogs.

2.8 Recurrent isolation promotes species formation.

Figure 2.16 A New Zealand alpine buttercup.
Fourteen species of alpine *Ranunculus* grow among the glaciers and mountains of New Zealand, including this giant buttercup, *R. lyallii.*

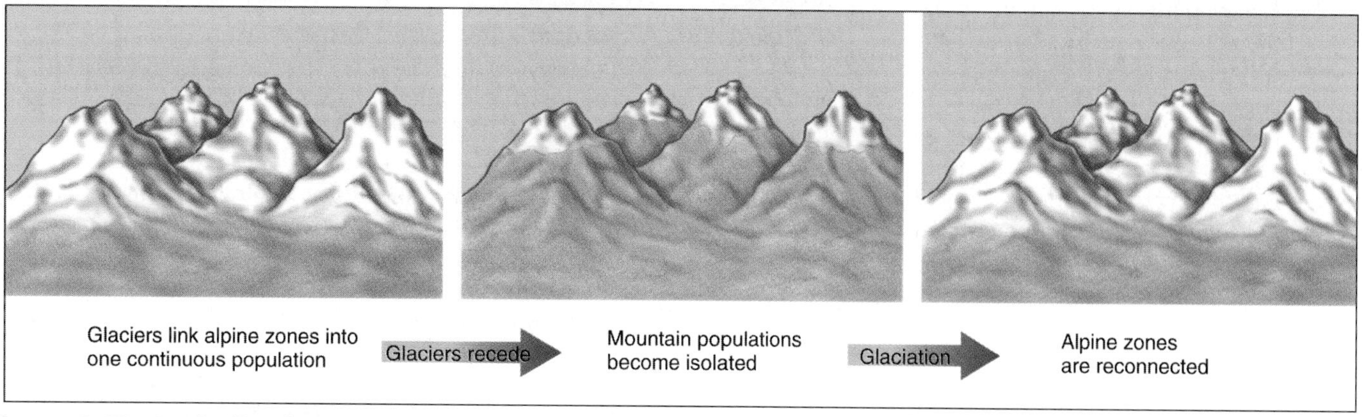
Glaciers link alpine zones into one continuous population
Glaciers recede
Mountain populations become isolated
Glaciation
Alpine zones are reconnected

Figure 2.17 Periodic glaciation encouraged species formation among alpine buttercups in New Zealand.
The formation of extensive glaciers during the Pleistocene linked the alpine zones of many mountains together. When the glaciers receded, these alpine zones were isolated from one another, only to become reconnected with the advent of the next glacial period. During periods of isolation, populations of alpine buttercups diverged in the isolated habitats.

2.9 What Is Ecology?

The word **ecology** was coined in 1866 by the great German biologist Ernst Haekel to describe the study of how organisms interact with their environment. It comes from the Greek words *oikos* (house, place where one lives) and *logos* (study of). Our study of ecology, then, is a study of the house in which we live. Do not forget this simple analogy built into the word *ecology*—most of our environmental problems could be avoided if we treated the world in which we live the same way we treat our own homes. Would you pollute your own house?

Levels of Ecological Organization

Ecologists consider groups of organisms at four progressively more encompassing levels of organization.

1. **Populations.** Individuals of the same species that live together are members of a population. They potentially interbreed with one another, share the same habitat, and use the same pool of resources the habitat provides.
2. **Communities.** Populations of different species that live together in the same place are called communities. Different species typically use different resources within the habitat they share.
3. **Ecosystems.** A community and the nonliving factors with which it interacts is called an ecosystem (figure 2.18). An ecosystem regulates the flow of energy, ultimately derived from the sun, and the cycling of the essential elements on which the lives of its constituent organisms depend.
4. **Biomes.** Biomes are major terrestrial assemblages of plants, animals, and microorganisms that occur over wide geographical areas and that have distinct characteristics. Examples include deserts, tropical forests, and grasslands; similar groupings occur in marine and freshwater habitats.

Figure 2.18 A tropical rain forest ecosystem.
This forest community at 2,120 meters (7,000 ft) in Cibodja, Java, is rich with plant and animal species. Sharing common resources, they possess many adaptations that promote their mutual survival. Complex interaction is the very essence of species-rich communities within ecosystems.

Some ecologists, called **population ecologists,** focus on a particular species and how its populations grow. Other ecologists, called **community ecologists,** study how the different species living in a place interact with one another. Still other ecologists, called **systems ecologists,** are interested in how biological communities interact with their physical environment.

2.9 Ecology is the study of how the organisms that live in a place interact with each other and with their physical habitat.

2.10 Ecosystems

Ecological systems, or **ecosystems,** are the fundamental units of ecology. An ecosystem is basically a biological community and the physical environment in which it lives. It is the most complex biological system that a biologist can study.

Energy Flows Through Ecosystems

All of the organisms within a community require energy to grow, reproduce, and carry out all of the many other activities of living. Almost all of this energy comes, ultimately, from the sun, captured by photosynthesis. The plants, algae, and microbes that capture it are in turn consumed by plant-eating animals called herbivores, and some of the energy captured from sunlight is passed on to the herbivore. The herbivore may then be eaten by a meat-eating animal called a carnivore, which captures some of the energy of the herbivore. Energy thus flows through the ecosystem, from plant to herbivore to carnivore.

Unfortunately, much of the energy is lost at each step of this **food chain.** Thus, because only so much energy arrives from the sun, food chains can be only so long—typically three or four steps. Can you see why there is no lion-eating top carnivore on the African savanna?

Materials Cycle Within Ecosystems

The raw materials that make up organisms—the carbon, nitrogen, phosphorus, and other atoms—are not used up when the organisms die. Instead, as the organisms decompose, the materials of their bodies pass back into the ecosystem, where they can be used to make other organisms. The materials thus cycle between organisms and the physical environment.

Major Ecosystems

While many aspects of the environment act to limit the distribution of particular species, the two most important are rainfall and temperature. Particular organisms are adapted to particular combinations of rainfall and temperature, and every place with that combination of rainfall and temperature will tend to have organisms with similar adaptations living there. On land, these sets of similar plants and animals are referred to as **biomes.** The United States contains a variety of quite different biomes (figure 2.19).

2.10 An ecosystem is a dynamic ecological system that consumes energy and cycles materials.

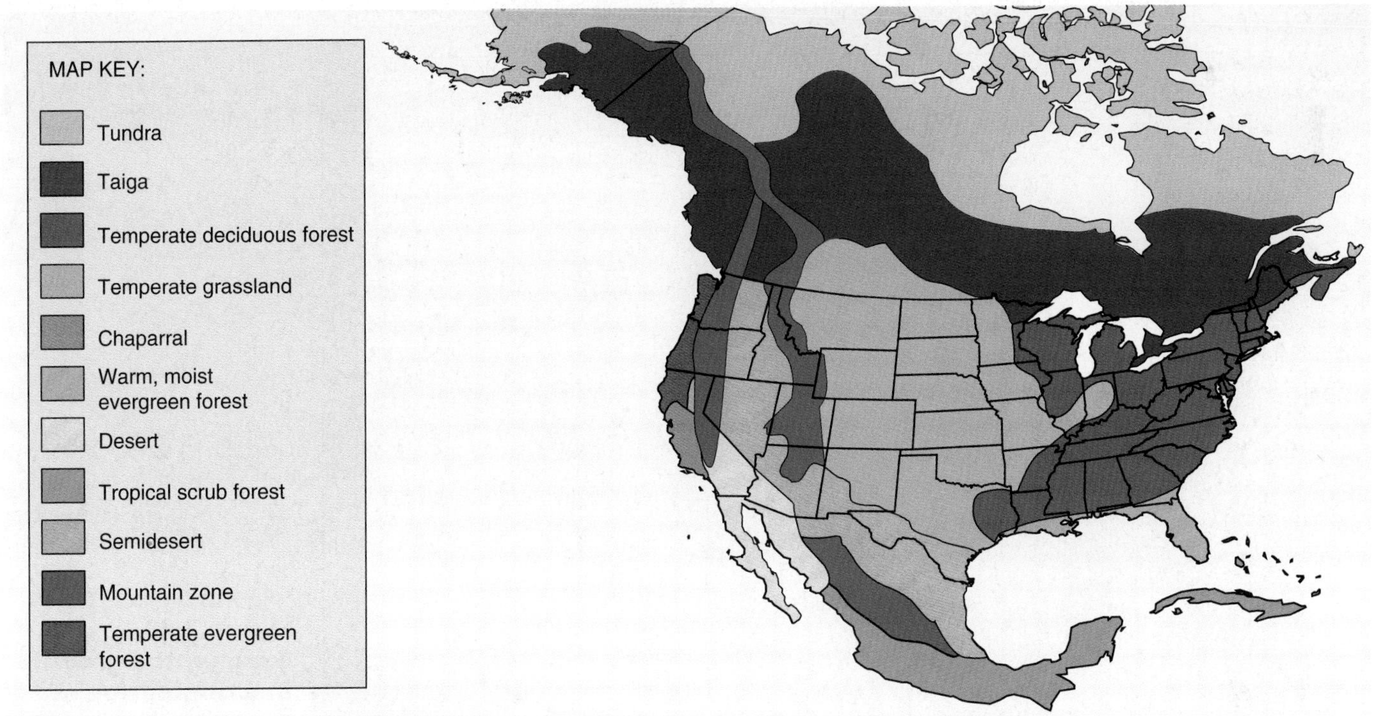
MAP KEY:
Tundra
Taiga
Temperate deciduous forest
Temperate grassland
Chaparral
Warm, moist evergreen forest
Desert
Tropical scrub forest
Semidesert
Mountain zone
Temperate evergreen forest

Figure 2.19 Biomes of North America.

2.11 Patterns of Population Growth

A key characteristic of any population is its capacity to grow. Most populations tend to remain relatively constant in number, regardless of how many offspring the individuals produce. Under certain circumstances, population size can increase rapidly for a time. The rate at which a population will increase when there are no limits on its rate of growth is called the **innate capacity for increase,** or **biotic potential.** This theoretical rate is almost impossible to calculate, however, because there are usually limits to growth. What biologists in fact calculate is the **realized rate of population increase** (abbreviated ***r***). This parameter is defined as the number of individuals added to the population minus the number lost from it. The number added to it equals the birthrate plus the number of **immigrants** (new individuals entering and residing with the population), while the number lost from it equals the death rate plus the number of **emigrants** (individuals leaving the population). Thus:

$$r = \textbf{(birth + immigration)} - \textbf{(death + emigration)}$$

Exponential Growth

A population's innate capacity for growth is constant, determined largely by the organism's physiology. Its actual growth, on the other hand, is not a constant, because r depends on both the birthrate and the death rate, and both of these factors change as the population increases in size. Thus to get the population growth rate, r must be adjusted for population size:

$$\textbf{population growth rate} = rN$$

where r is the realized rate of population increase and N is the number of individuals in the population. In general, the number of individuals grows rapidly at first, and this type of growth is called **exponential growth** (figure 2.20). As a population increases and begins to exhaust its resources, the death rate rises, and the rate of increase slows. Eventually, just as many individuals are dying as are being born. The early rapid phase of population growth lasts only for a short period, usually when an organism reaches a new habitat where resources are abundant. Examples of this pattern include algae colonizing a newly formed pond and the first terrestrial organisms arriving on a recently formed island.

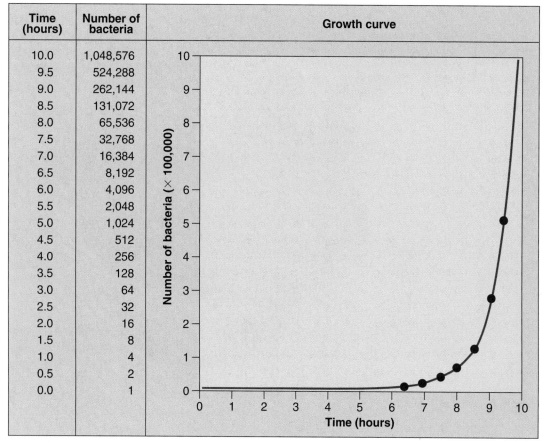

Time (hours)	Number of bacteria
10.0	1,048,576
9.5	524,288
9.0	262,144
8.5	131,072
8.0	65,536
7.5	32,768
7.0	16,384
6.5	8,192
6.0	4,096
5.5	2,048
5.0	1,024
4.5	512
4.0	256
3.5	128
3.0	64
2.5	32
2.0	16
1.5	8
1.0	4
0.5	2
0.0	1

Growth curve

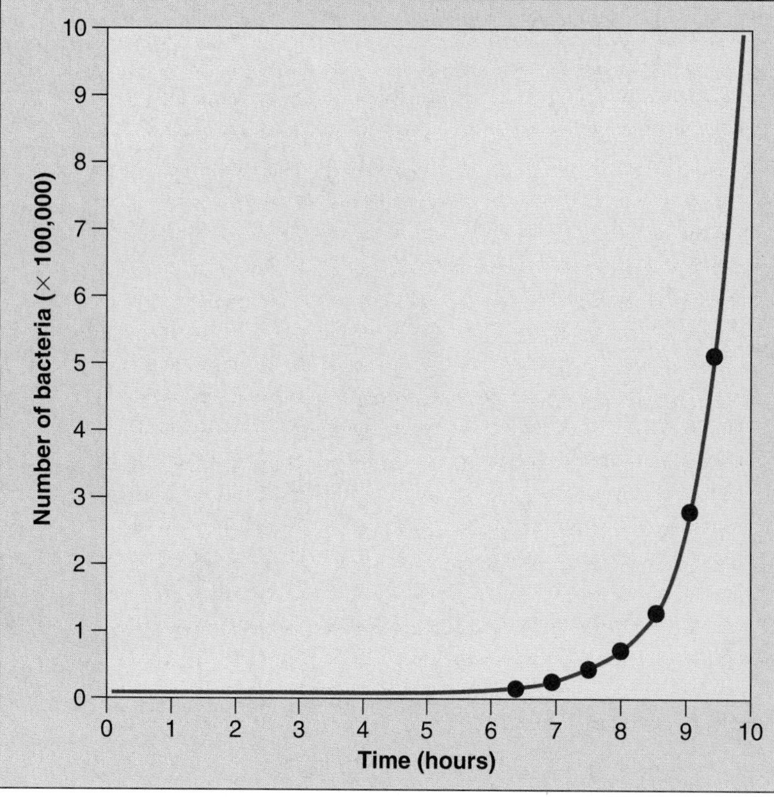
Number of bacteria (× 100,000)
10
9
8
7
6
5
4
3
2
1
0
0
1
2
3
4
5
6
7
8
9
10
Time (hours)

Figure 2.20 Exponential growth in a population of bacteria.

In just 10 hours, this population grew from one individual to over 1 million!

Carrying Capacity

No matter how rapidly new populations grow, they eventually reach an environmental limit imposed by shortages of some important factor, such as space, light, water, or nutrients. A population ultimately stabilizes at a certain size, called the **carrying capacity** of the particular place where the population lives. The carrying capacity, symbolized by *K*, is the number of individuals that can be supported at that place indefinitely. As carrying capacity is approached, the population's rate of growth slows greatly. The growth of a specific population, which is always limited by one or more factors in the environment, can be approximated by the following *logistic growth equation:*

$$\textbf{population growth rate} = rN\left(\frac{K-N}{K}\right)$$

In other words, the growth of the population under consideration equals the ideal rate of increase (*r* multiplied by *N*, the number of individuals present at any one time), adjusted for the amount of resources still available. The adjustment is made by multiplying *rN* by the fraction of *K* still unused (*K* minus *N*, divided by *K*). As *N* increases, the fraction by which *r* is multiplied becomes smaller and smaller, and the rate of increase of the population declines. Graphically, this relationship is the S-shaped **sigmoid growth curve,** characteristic of biological populations (figure 2.21).

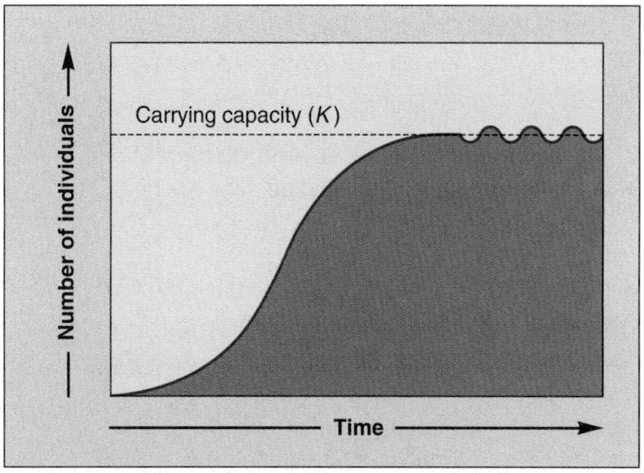
Carrying capacity (K)
Number of individuals
Time

Figure 2.21 The sigmoid growth curve.
The sigmoid growth curve begins with a period of exponential growth like that shown in figure 2.20. When the population approaches its environmental limits, growth slows and finally stabilizes, fluctuating around the carrying capacity (*K*) of the environment.

Life History Strategies

Most populations exhibit characteristics that lie on a continuous spectrum between traits that favor exponential growth by *r* and traits that keep population size under *K*. The particular set of adaptations that adapts an organism's reproductive rate to its environment is called its **life history strategy.** Life history strategies will sometimes favor rapid growth, particularly those of organisms occupying rapidly changing environments. Most mosquitoes, for example, reproduce in temporary puddles and exhibit *r*-selected life histories (figure 2.22). Other life history strategies favor slower growth and are often exhibited by organisms competing for limited resources. Elephants reproduce in a crowded African savanna community and exhibit *K*-selected life histories (figure 2.23).

2.11 Population growth is limited by the ability of the environment to support the population. Organisms in transient environments are often adapted to reproduce rapidly, while those in stable environments tend to reproduce more slowly.

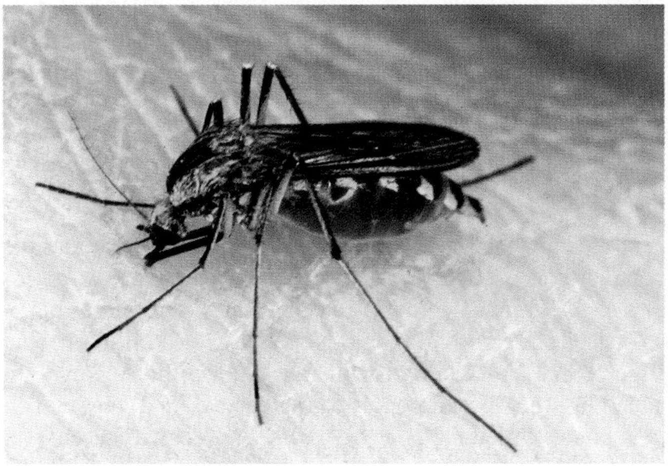

Figure 2.22 An animal with an *r*-selected life history.
Mosquitoes exhibit mainly *r*-selected adaptations, including hundreds of offspring per brood, no investment in parental care (eggs are laid on water and abandoned), rapid growth (young mature within hours), and a short life span numbered in weeks.

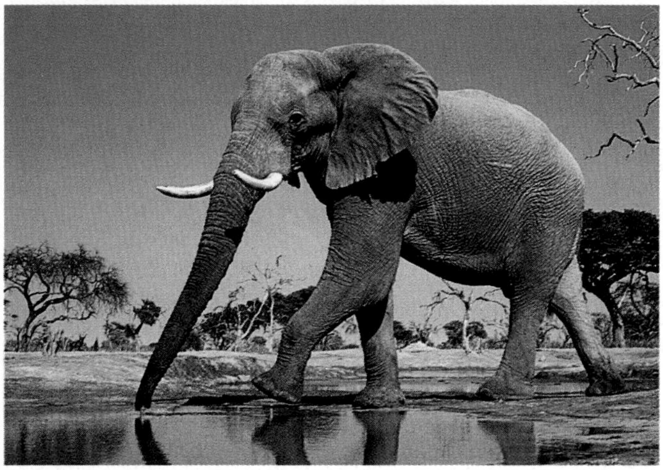

Figure 2.23 An animal with a *K*-selected life history.
Elephants exhibit mainly *K*-selected adaptations, including only a few offspring per brood (typically one), prolonged parental care extending many years, slow growth and maturation, and life spans as long as a hundred years.

2.12 Human Populations

Humans exhibit many *K*-selected life history traits, including small brood size, late reproduction, and a high degree of parental care. These life history traits evolved during the early history of human ancestors, when the limited resources available from the environment controlled population size. Throughout most of human history, our populations have been regulated by food availability, disease, and predators. While unusual disturbances, including floods, plagues, and droughts, no doubt affected the pattern of human population growth, the overall size of the human population grew only slowly during our early history. Two thousand years ago, perhaps 130 million people populated the earth. It took a thousand years for that number to double, and it was 1650 before it had doubled again, to about 500 million. For over 16 centuries, the human population was characterized by very slow growth. In this respect, human populations resembled many other species with predominantly *K*-selected life history adaptations.

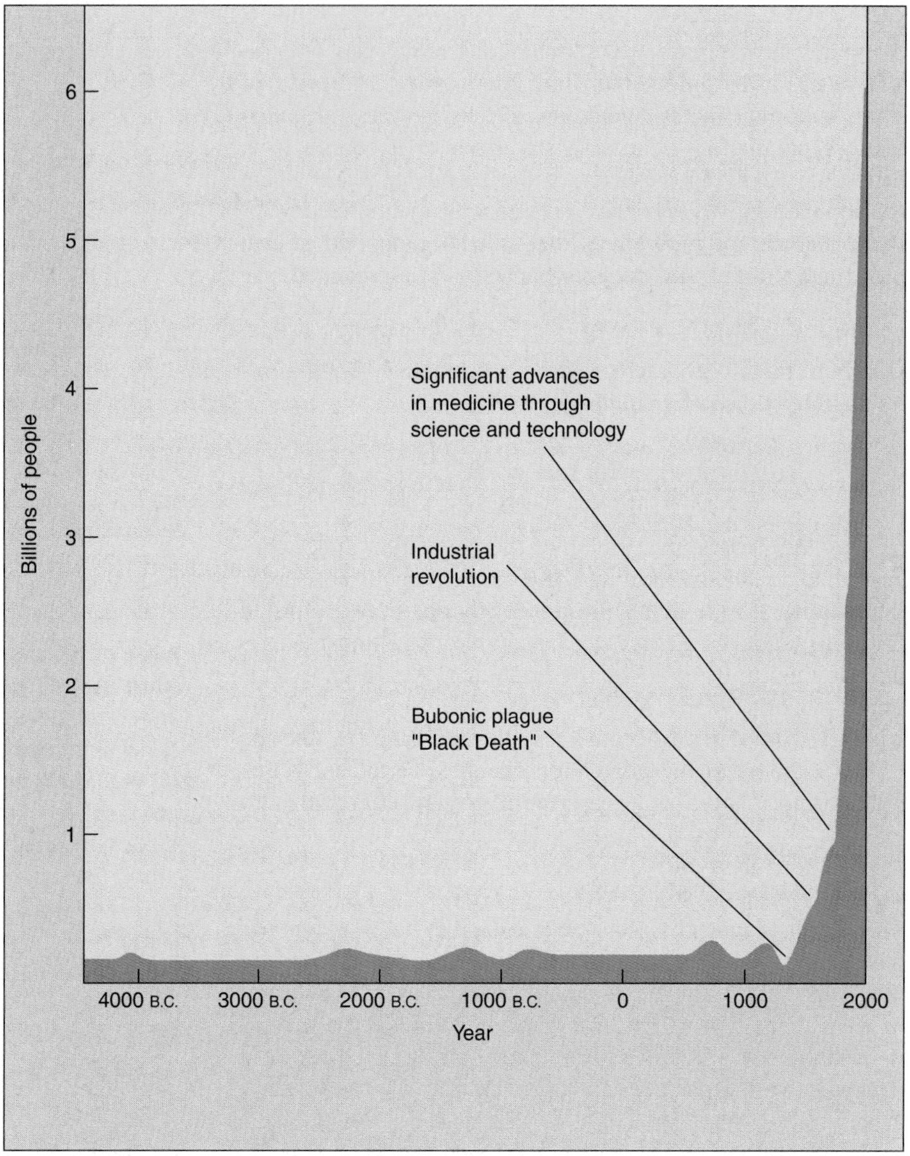

Billions of people
6
5
4
3
2
1
Significant advances in medicine through science and technology
Industrial revolution
Bubonic plague "Black Death"
4000 B.C.
3000 B.C.
2000 B.C.
1000 B.C.
0
1000
2000
Year

Figure 2.24 History of human population size.

Temporary increases in death rate, even severe ones like the Black Death of the 1400s, have little lasting impact. Explosive growth began with the industrial revolution in the 1700s, which produced a significant long-term lowering of the death rate. The current population is about 6 billion, and at the current rate will double in 40 years.

The Advent of Exponential Growth

Starting in the early 1700s, changes in technology have given humans more control over their food supply, enabled them to develop superior weapons to ward off predators, and led to the development of cures for many diseases. At the same time, improvements in shelter and storage capabilities have made humans less vulnerable to climatic uncertainties. These changes allowed humans to expand the carrying capacity of the habitats in which they lived and thus to escape the confines of logistic growth and reenter the exponential phase of the sigmoidal growth curve.

Responding to the lack of environmental constraint, the human population has grown explosively over the last 300 years. While the birthrate has remained essentially unchanged at about 30 per 1,000 per year over this period, the death rate has fallen dramatically, from 29 per 1,000 per year to its present level of about 13 per 1,000 per year. This difference between birth and death rates (17 per 1,000) means that the population is growing at the rate of 1.7% per year.

A 1.7% annual growth rate may not seem large, but it has produced a current human population of 6 billion people (figure 2.24)! At this growth rate, 100 million people are added to the world population annually, and the human population will double in 40 years. As we will discuss in chapter 31, both the current human population level and the projected growth rate have potential consequences for our future that are extremely grave.

Population Pyramids

While the human population as a whole continues to grow rapidly, this growth is not occurring uniformly over the planet. Some countries, like Mexico, are growing rapidly; its birthrate greatly exceeds its death rate (figure 2.25). Other countries are growing much more slowly. The rate at which a population can be expected to grow in the future can be assessed graphically by means of a population pyramid—a bar graph displaying the numbers of people in each age category. Males are conventionally shown to the left of the vertical age axis and females to the right. In most human population pyramids, the number of older females is disproportionately large compared with the number of older males, because females in most regions have a longer life expectancy than males.

Viewing such a pyramid, one can predict demographic trends in births and deaths. In general, rectangular "pyramids" are characteristic of countries whose populations are stable; their numbers are neither growing nor shrinking. A triangular pyramid is characteristic of a country that will exhibit rapid future growth, as most of its population has not yet entered the child-bearing years. Inverted triangles are characteristic of populations that are shrinking.

Examples of population pyramids for the United States and Kenya in 1990 are shown in figure 2.26. In the nearly rectangular population pyramid for the United States, the cohort (group of individuals) 55 to 59 years old represents people born during the Depression and is smaller in size than the cohorts in the preceding and following years. The cohorts 25 to 44 years old represent the "baby boom." The rectangular shape of the population pyramid indicates that the population of the United States is not expanding rapidly. The very triangular pyramid of Kenya, by contrast, predicts explosive future growth. The population of Kenya is predicted to double in less than 20 years.

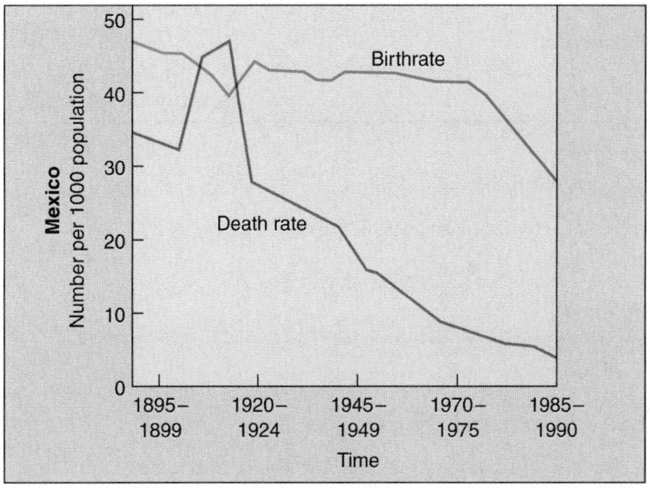
Mexico
Number per 1000 population
50
40
30
20
10
0
Birthrate
Death rate
1895–1899
1920–1924
1945–1949
1970–1975
1985–1990
Time

Figure 2.25 Why Mexico's population is growing.
The death rate (*red line*) in Mexico has been falling, while the birthrate (*blue line*) remained fairly steady until 1970. The difference between birth and death rates has fueled a high growth rate. Efforts begun in 1970 to reduce the birthrate have been quite successful, although the growth rate remains rapid.

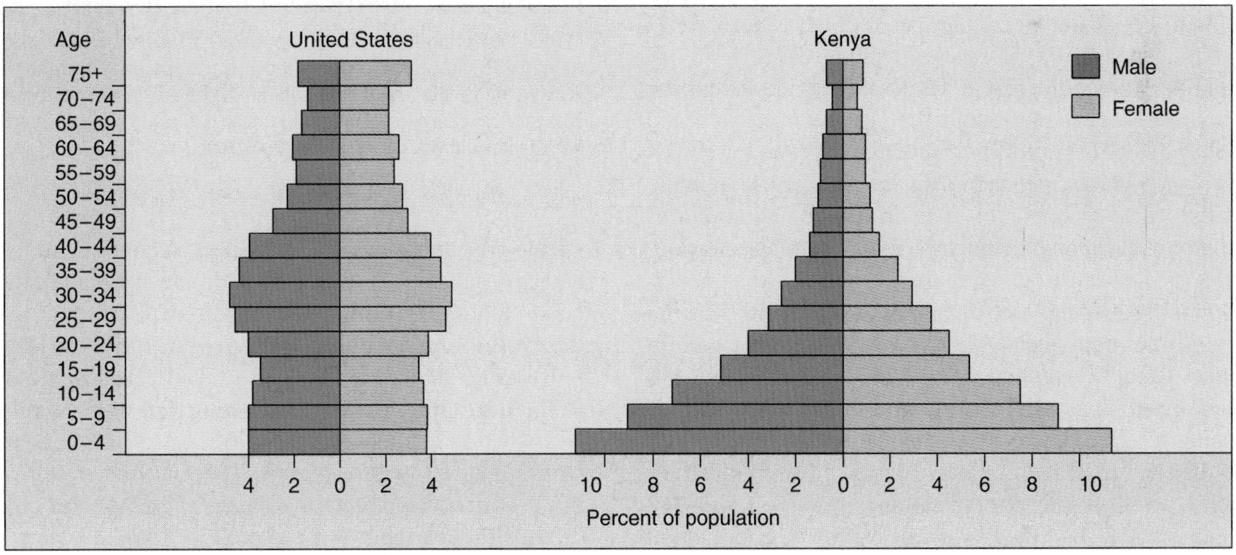
Age
United States
Kenya
Male
Female
75+
70–74
65–69
60–64
55–59
50–54
45–49
40–44
35–39
30–34
25–29
20–24
15–19
10–14
5–9
0–4
4 2 0 2 4
10 8 6 4 2 0 2 4 6 8 10
Percent of population

Figure 2.26 Population pyramids from 1990.
Population pyramids are graphed according to a population's age distribution. Kenya's pyramid has a broad base because of the great number of individuals below child-bearing age. When all of the young people begin to bear children, the population will experience rapid growth. The U.S. pyramid demonstrates a larger number of individuals in the "baby boom" cohort—the pyramid bulges because of an increase in births between 1945 and 1964.

TABLE 2.2 A COMPARISON OF 2000 POPULATION DATA IN DEVELOPED AND DEVELOPING COUNTRIES

	United States (highly developed)	Brazil (moderately developed)	Ethiopia (poorly developed)
Fertility rate	2	2.4	6.7
Doubling time at current rate (yr)	120	45	29
Infant mortality rate (infant deaths/1,000 births)	7	37.99	116
Life expectancy at birth (yr)	77	68	46
Per capita GNP (U.S. $; 1998)	$29,240	$4,630	$100

Source: Population Reference Bureau

An Uncertain Future

The earth's rapidly growing human population constitutes perhaps the greatest challenge to the future of the biosphere, the world's interacting community of living things. Humanity is adding 100 million people a year to the earth's population—a million people every three days, 250 every minute! In more rapidly growing countries, the resulting population increase is staggering (table 2.2). India, for example, had a population of just over 1 billion in 2000; by 2025 its population will reach nearly 1.4 billion!

A key element in the world's population growth is its uneven distribution among countries. Of the billion people added to the world's population in the 1990s, 90% live in developing countries (figure 2.27). This is leading to a major reduction in the fraction of the world's population that lives in industrialized countries. In 1950, fully one-third of the world's population lived in industrialized countries; by 2000 that proportion had fallen to less than one-quarter; in 2025 the proportion will have fallen to less than one-sixth. Thus the world's population growth will be centered in the parts of the world least equipped to deal with the pressures of rapid growth.

Rapid population growth in developing countries has the harsh consequence of increasing the gap between rich and poor. Today 20% of the world's population lives in the industrialized world with a per capita income of $19,480, while 80% of the world's population lives in developing countries with a per capita income of only $1,260. The disproportionate wealth of the industrialized quarter of the world's population is evidenced by the fact that 85% of the world's capital wealth is in the industrial world, with only 15% in developing countries. Eighty percent of all the energy used today is consumed by the industrial world, with only 20% by developing countries. Perhaps most worrisome for the future, fully 94% of all scientists and engineers reside in the industrialized world, with only 6% in developing countries. Thus the problems created by the future's explosive population growth will be faced by countries with little of the world's scientific or technological expertise.

No one knows whether the world can sustain today's population of over 6 billion people, much less the far greater populations expected in the future. As we discuss in later chapters, the world ecosystem is already under considerable stress. We cannot reasonably expect to continue to expand its carrying capacity indefinitely, and indeed we already seem to be stretching the limits. It seems unavoidable that to restrain the world's future population growth, birth and death rates must be equalized. If we are to avoid catastrophic increases in the death rate, the birthrates must fall dramatically. Faced with this grim dichotomy, significant efforts are underway worldwide to lower birthrates. These efforts are already having a noticeable effect in countries like Mexico (see figure 2.25), although there is far to go.

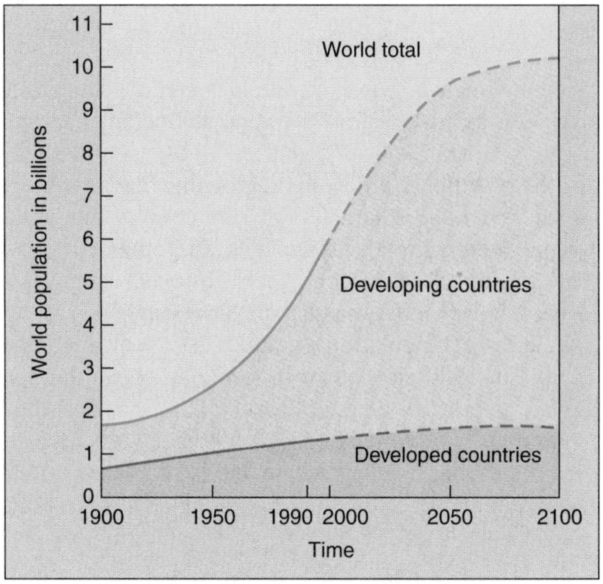

World population in billions
0
1
2
3
4
5
6
7
8
9
10
11
World total
Developing countries
Developed countries
1900
1950
1990
2000
2050
2100
Time

Figure 2.27 World population growth.
Most of the worldwide increase in population has occurred in developing countries. The age structures of developing countries indicate that this trend will increase in the near future. The stabilizing of the world's population at about 10 billion is an optimistic World Bank / United Nations prediction that assumes significant worldwide reductions in growth rate.

2.12 The human population has been growing rapidly for 300 years, since technological innovations dramatically reduced the death rate.

CONCEPT REVIEW

1. Which of the following is *not* one of Darwin's pieces of evidence that evolution occurs?
   a. Species on oceanic islands show strong similarities to species living on the nearest mainland.
   b. The earth was created in 4004 B.C.
   c. Progressive changes in characteristics of plants and animals can be seen in successive rock layers.
   d. The plants and animals of each continent are distinctive.

2. Darwin believed that the major driving force in evolution was
   a. natural selection.
   b. scientific creation.
   c. uniformitarianism.
   d. molecular biology.

3. Adaptive radiation of finches on the Galápagos Islands led to
   a. a variety of bill specializations among the finches.
   b. mutation of the finches' bills.
   c. cooperation among Galápagos tortoises and finches.
   d. none of these.

4. Clusters of species are likely to be extensive in _______ habitats, such as islands, lakes, and mountaintops.
   a. connected
   b. aquatic
   c. discontinuous
   d. continuous

5. An ecosystem consists of a biological community and the _________ components of the environment.
   a. biome
   b. terrestrial
   c. nonliving
   d. aquatic

6. Which of the following factors can affect the population growth rate?
   a. immigration
   b. population size
   c. carrying capacity
   d. all of these

7. Which of the following organisms is not likely to exhibit exponential growth?
   a. gorillas
   b. bacteria
   c. rats
   d. cockroaches

8. Human population growth is greatest in developing countries because
   a. the death rate is higher in developing countries.
   b. the birthrate is higher in developing countries.
   c. much of the population in developing countries has already reached the child-bearing age.
   d. most of the world's population lives in industrialized countries.

9. Darwin's theory of natural selection states that organisms with superior traits will be more likely to survive and pass on their favorable traits to their _________.

10. Almost all of the energy in an ecosystem comes ultimately from ____________.

11. The _________ of a population is the number of individuals that a particular place can support indefinitely.

12. Organisms that reproduce infrequently, produce few offspring, and grow slowly exhibit a ___________ life history strategy.

CHALLENGE YOURSELF

1. On the Galápagos Islands, Darwin saw a variety of finches but few varieties of other small birds. Imagine that you are visiting another group of islands about as far away from the South American mainland as are the Galápagos, but far enough from the Galápagos that no birds travel between the two island groups. Would you expect to find a variety of finches on this second island group as well? Explain your reasoning.

2. The sigmoid growth curve described by the logistic growth equation assumes that the growing population does not destroy the ability of the environment to support the population. Some environmentalists argue that the rapidly growing human population is doing just that. Would you expect this to produce a crash in population numbers or a gradual fall to a new equilibrium?

2 eBRIDGE

Reinforcing Key Points

Evolution

2.1 Darwin's Voyage on HMS *Beagle*

2.2 Darwin's Evidence

2.3 Inventing the Theory of Natural Selection

Evolution in Action

2.4 The Beaks of Darwin's Finches

2.5 Clusters of Species

2.6 Hawaiian *Drosophila*

2.7 Lake Victoria Cichlid Fishes

2.8 New Zealand Alpine Buttercups

Ecology

2.9 What Is Ecology?

2.10 Ecosystems

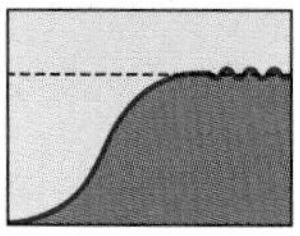

Populations and How They Grow

2.11 Patterns of Population Growth

2.12 Human Populations

Electronic Learning

Visual Learning

Animations

Evolutionary History of *Equus*

Microevolution

Exponential Population Growth

Stages of Population Growth

U.S. Age Distributions, 1910, 1960 and 2010 (projected)

Age Distributions for MDCs and LDCs

World Population Growth, 1750–2150

Resource Consumption for MDCs and LDCs

Helping You Learn

Evolution I: Microevolution and Selection

Author's Corner

Darwin's Critics. While Darwin's theory of evolution by natural selection is almost universally accepted by scientists, it has proven quite a bit more controversial among the general public. Objections to the teaching of evolution in American public schools have focused on a family of arguments with which every student of biology should become familiar. Many of the criticisms of evolution were raised in the years soon after Darwin's publication of *On the Origin of Species.* More recent criticisms doubt that evolution could produce the complexity we see at the cellular and molecular level.

1. 140 Years without Darwin are enough.
2. Answering evolution's critics.
3. Keeping Darwin out of schoolrooms.
4. Darwinism at the cellular level.
5. Darwin and charter schools.

Virtual Classroom

Darwin and Evolution

Evolution is the core of the science of biology. The theory that evolution occurs because of natural selection was advanced by two British naturalists, Charles Darwin and Alfred Wallace, in 1859. The story of how Darwin and Wallace independently came to this key insight says a great deal about the role of observation in science, and the important role played by careful reasoning. Darwin, just out of college at the age of 22, spent five years as a naturalist on a ship that explored the coast of South America, including a group of volcanic islands, the Galápagos Islands. He spent little time on the ship, instead trekking for two years in Argentina and a year and a half in Chile, before sailing around the world to return to England in 1836. It was more than 20 years before Darwin, stimulated by Wallace's similar conclusions after years exploring southeastern Asia, published his book *On the Origin of Species.* There he presents an enormous body of evidence supporting his theory of evolution by natural selection.

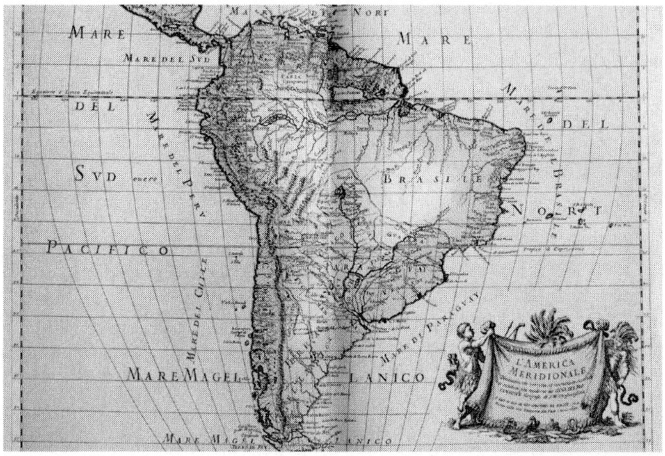
MARE
MARE
DEL
SVD
PACIFICO
BRASILE
MARE MAGELLANICO

Perhaps Darwin's greatest contribution to biology is that he converted evolution from a hypothesis to a documented observation. However controversial his theory among the general public, it is supported by overwhelming evidence, and is universally accepted by biologists.

Virtual Lab

Why Do Tropical Songbirds Lay Fewer Eggs?

British ornithologist Reginald Moreau made an interesting observation in 1944, that songbirds in the tropics lay fewer eggs than their counterparts at higher latitudes. This was particularly interesting because one would expect that natural selection would maximize evolutionary fitness—that is, songbirds the world over should have evolved to produce as many eggs as possible. Why don't the birds living in the tropics produce a maximum number of eggs?

David Lack, a colleague of Moreau's, put forth an argument that natural selection will indeed tend to maximize reproduction rates but only to the greatest level possible within the limits of resources. If there are fewer resources available, birds will lay fewer eggs. This idea was tested by Mark Boyce (now at the University of Alberta, Edmonton) using data collected on the nesting habits and clutch sizes (the number of eggs laid in a nest) of the Great Tit in Oxford, England. A very complete field study was conducted over a 22-year period. The clutch sizes of 4,489 nests were counted, and the number of offspring surviving to the next year determined, to look for the predicted correlation.

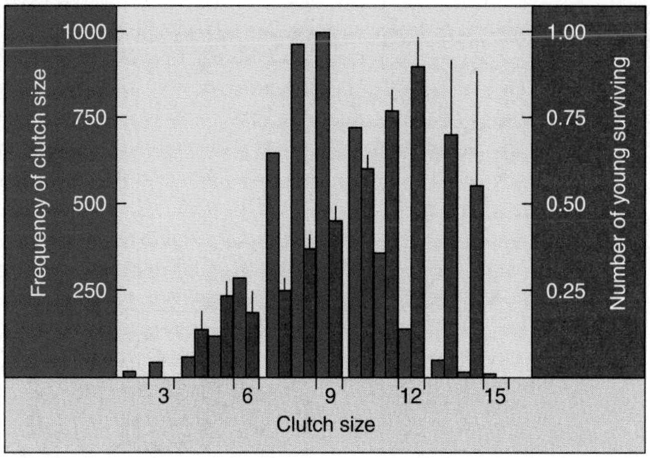
Frequency of clutch size
1000
750
500
250
Number of young surviving
1.00
0.75
0.50
0.25
3
6
9
12
15
Clutch size

Quizzes | Further Reading | Essential Study Partner | Links | BioCourse.com

CHAPTER 3

THE CHEMISTRY OF LIFE

THE FAR SIDE® By GARY LARSON
Larson
My God! You've chosen the seven of CLUBS!.. A superstition will begin among certain Asian cultures that your horn, when ground up into a fine powder, is the source of a powerful aphrodisiac and that belief will in turn cause a great demand which will give rise to a depraved, merciless group of professional killers who will hunt you down, cut off your horn, and leave your carcass for the buzzards!.. You ARE DOOMED! DOOMED! Ha ha ha ha ha ha ha..
I want my $2.00 back.
Madame Carlotta
© 1991 FarWorks, Inc. All Rights Reserved/Dist. by Creators Syndicate

CHAPTER 3

The Chemistry of Life

CHAPTER OVERVIEW

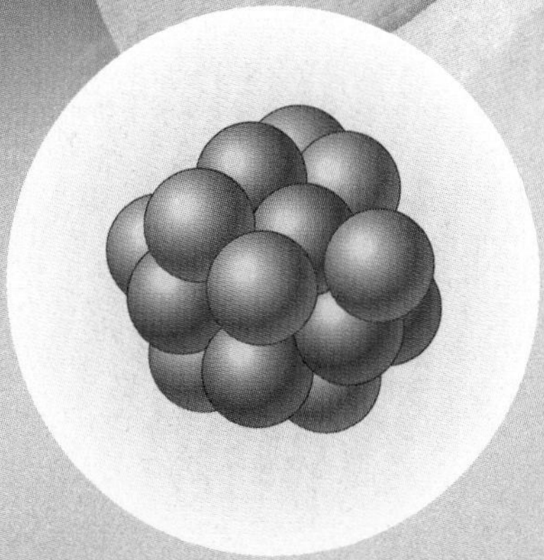

Some Simple Chemistry

3.1 Atoms
3.2 Electrons Determine What Atoms Are Like
3.3 Isotopes
3.4 Molecules

- All matter is composed of atoms, made of protons, neutrons, and electrons.
- Electrons determine the chemical behavior of atoms.
- Molecules are collections of atoms held together by covalent bonds.
- Covalent bonds form when two atoms share electrons.
- Energy can pass from one molecule to another by the transfer of electrons.

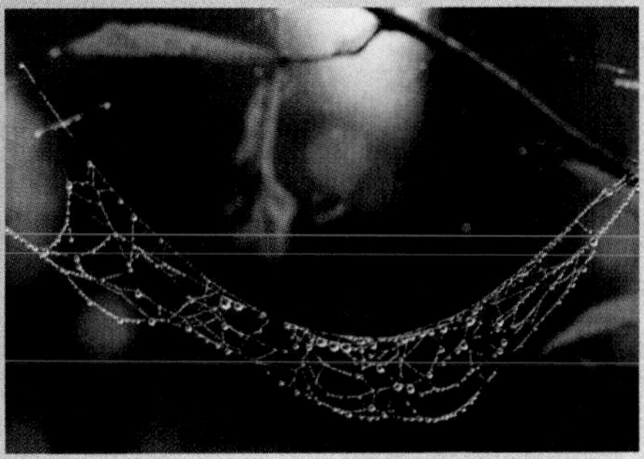

Water: Cradle of Life

3.5 Hydrogen Bonds Give Water Unique Properties
3.6 Water Ionizes

- Water is a highly polar molecule, a characteristic responsible for many of its properties.
- Polar molecules form hydrogen bonds with water, making it an excellent solvent. Nonpolar molecules, called hydrophobic ("water fearing"), tend to aggregate together in water.
- A few molecules of water spontaneously ionize, producing hydrogen ions whose concentration is measured by pH.

Macromolecules

3.7 Forming Macromolecules
3.8 Carbohydrates
3.9 Lipids
3.10 Proteins
3.11 Nucleic Acids

- Cells are built of very large macromolecules, principally carbohydrates, lipids, proteins, and nucleic acids.
- Most macromolecules are assembled by forming chains of subunits.
- The function of a particular macromolecule is critically dependent upon its shape.

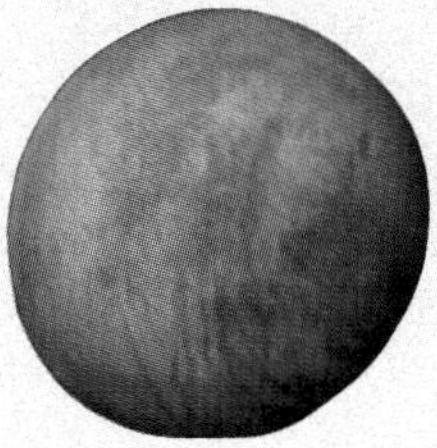

Origin of the First Cells

3.12 Origin of Life
3.13 How Cells Arose
3.14 Has Life Evolved Elsewhere?
3.15 Evolution's Critics

- When conditions resembling those of the early earth are recreated in the laboratory, biological molecules such as amino acids are formed.

3.1 Atoms

Organisms are chemical machines, and to understand them we must learn a little chemistry. Any substance in the universe that has mass and occupies space is defined as matter. All matter is composed of extremely small particles called **atoms.** An atom is the smallest particle into which a substance can be divided and still retain its chemical properties. The terms *mass* and *weight* are often used interchangeably, but they have slightly different meanings. Mass refers to the amount of a substance, while weight refers to the force gravity exerts on a substance. Hence, an object has the same mass whether it is on the earth or the moon, but its weight will be greater on the earth, because the earth's gravitational force is greater than the moon's.

Every atom has the same basic structure: a core nucleus of protons and neutrons surrounded by a cloud of electrons (figure 3.1). At the center of every atom is a small, very dense **nucleus** formed of two subatomic particles, **protons** and **neutrons** (each of which is made of even smaller bits, but we won't go into that!). Whizzing around the core is an orbiting cloud of a third kind of subatomic particle, the **electron.** Neutrons have no electrical charge, while protons have a positive charge and electrons a negative one.

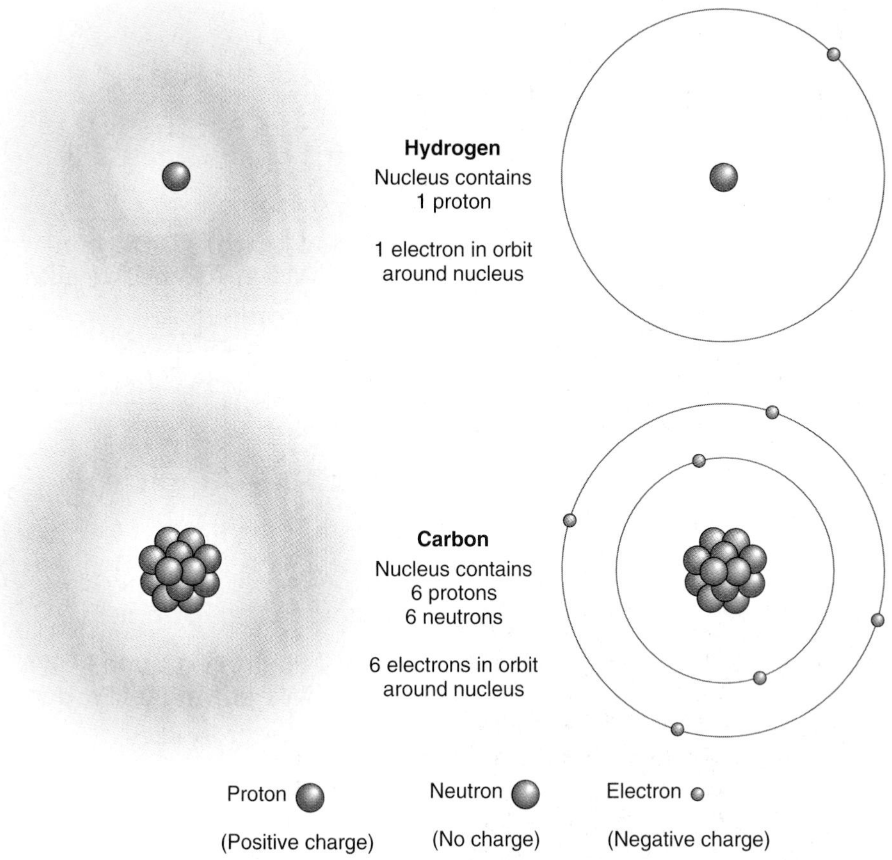
Hydrogen
Nucleus contains
1 proton
1 electron in orbit
around nucleus
Carbon
Nucleus contains
6 protons
6 neutrons
6 electrons in orbit
around nucleus
Proton
(Positive charge)
Neutron
(No charge)
Electron
(Negative charge)

Figure 3.1 Basic structure of atoms.

All atoms have a nucleus consisting of protons and neutrons, except hydrogen, the smallest atom, which has only one proton and no neutrons in its nucleus. Carbon, for example, has six protons and six neutrons in its nucleus. Electrons spin around the nucleus in orbitals a far distance away from the nucleus. The electrons determine how atoms react with each other.

Ions

In a neutral atom (that is, one that carries no electrical charge), there is an orbiting electron for every proton in the nucleus. The electron's negative charge balances the proton's positive charge. Atoms in which the number of electrons does not equal the number of protons are called **ions.** All ions are electrically charged. An atom of sodium, for example, becomes a positively charged sodium ion when it loses an electron, because one proton in the nucleus is left with an unbalanced charge (figure 3.2).

3.1 Atoms, the smallest particles into which a substance can be divided, are composed of electrons orbiting a nucleus composed of protons and neutrons.

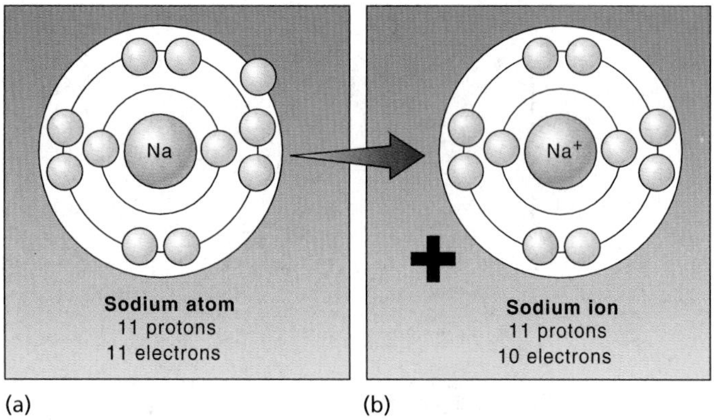
Na
Sodium atom
11 protons
11 electrons
Na^+
Sodium ion
11 protons
10 electrons
(a)
(b)

Figure 3.2 Making a sodium ion.

(*a*) An electrically neutral sodium atom with 11 protons and 11 electrons. (*b*) A sodium ion. Sodium ions bear one positive charge when they ionize and lose one electron. This sodium ion has 11 protons and only 10 electrons.

3.2 Electrons Determine What Atoms Are Like

Electrons have very little mass (only 1/1,840 the mass of a proton). Of all the mass contributing to your weight, the portion that is contributed by electrons is less than the mass of your eyelashes. And yet electrons determine the chemical behavior of atoms because they are the parts of atoms that come close enough to each other in nature to interact. Almost all the volume of an atom is empty space. Protons and neutrons lie at the core of this space, while orbiting electrons are very far from the nucleus. If the nucleus of an atom were the size of an apple, the orbit of the nearest electron would be more than a mile out!

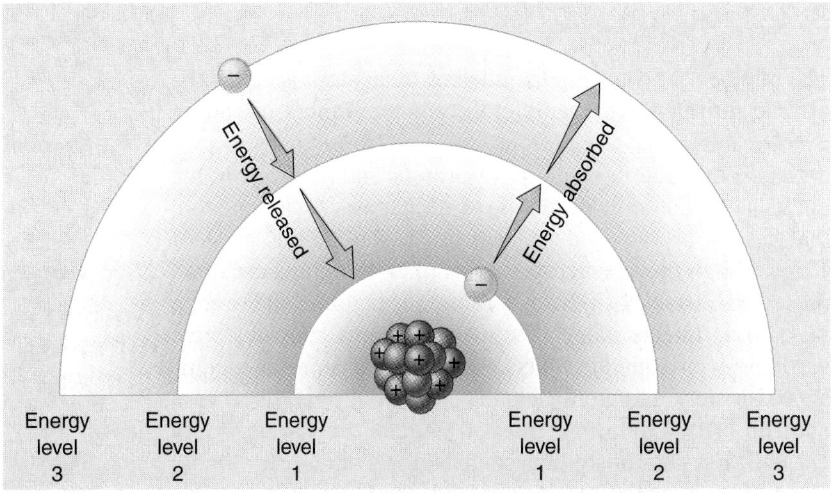
Energy released
Energy absorbed
Energy level 3
Energy level 2
Energy level 1
Energy level 1
Energy level 2
Energy level 3

Figure 3.3 The electrons of atoms possess potential energy.
Electrons circulate rapidly around the nucleus in paths called orbitals. Energy level 1 is the lowest potential energy level because it is closest to the nucleus. When an electron absorbs energy, it moves from level 1 to the next higher energy level (level 2). When an electron loses energy, it falls to a lower energy level closer to the nucleus.

Electrons Carry Energy

Because electrons are negatively charged, they are attracted to the positively charged nucleus. It takes work to keep them in orbit, just as it takes work to hold an apple in your hand when gravity is pulling the apple down toward the ground. The apple in your hand is said to possess **energy,** the ability to do work, because of its position—if you were to release it, the apple would fall. Similarly, electrons have energy of position, called potential energy (figure 3.3). It takes work to oppose the attraction of the nucleus, so moving the electron farther out to a more distant orbit requires an input of energy and results in an electron with greater potential energy. Moving an electron in toward the nucleus has the opposite effect; energy is released, and the electron has less potential energy.

Sometimes an electron is transferred from one atom to another. When an electron is transferred in this way, it keeps its energy of position. In living organisms, chemical energy is stored by using it to move electrons to more distant "high-energy" orbits, and these energetic electrons are frequently transferred from one atom to another. The loss of an electron in such a transfer is called **oxidation;** the gain of an electron is called **reduction** (figure 3.4).

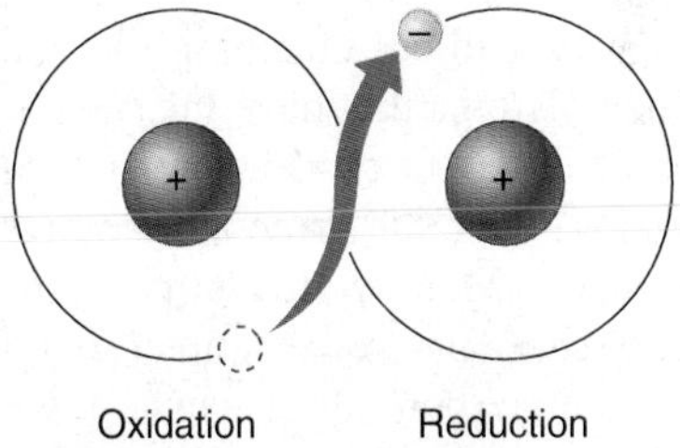
Oxidation
Reduction

Figure 3.4 Oxidation-reduction.
Oxidation is the loss of an electron; reduction is the gain of one.

While the energy levels of an atom are often visualized as well-defined circular orbits around a central nucleus, such a simple picture is not realistic. These energy levels often consist of complex three-dimensional shapes, and the exact location of an individual electron at any given time is impossible to specify. However, some locations are more probable than others, and it is often possible to say where an electron is *most likely* to be located. The volume of space around a nucleus where an electron is most likely to be found is called the **orbital** of that electron.

Each energy level has a specific number of orbitals, and each orbital can hold up to two electrons. The first energy level in any atom contains one orbital. Helium, for example, has one energy level with one orbital that contains two electrons. In atoms with more than one energy level, the second energy level contains four orbitals and holds up to eight electrons. Nitrogen has two energy levels; the first one is completely filled with two electrons, but the four orbitals in the second energy level are not filled because nitrogen's second energy level contains only five electrons. In atoms with more than two energy levels, subsequent energy levels also contain up to four orbitals and a maximum of eight electrons. Atoms with incomplete electron orbitals tend to be more reactive.

3.2 Electrons determine the chemical behavior of atoms. An electron possesses potential energy, which is carried with the electron from one atom to another.

3.3 Isotopes

The number of protons in the nucleus of an atom is called the **atomic number.** For example, the atomic number of carbon is 6 because it has six protons. Neutrons are similar to protons in mass, and the number of protons and neutrons in the nucleus of an atom is called the **atomic mass.** A carbon atom that has six protons and six neutrons has an atomic mass of 12. Although precise measurements of atomic mass are often presented in tables, it is also common to round the atomic mass to an integer value. The mass of atoms and subatomic particles is measured in units called *daltons.* A proton weighs approximately 1 dalton (actually 1.009 daltons), as does a neutron (1.007 daltons). In contrast, electrons weigh only 1/1,840 of a dalton, so their contribution to the overall mass of an atom is negligible. Atoms with the same atomic number (that is, the same number of protons) have the same chemical properties and are said to belong to the same **element.** Formally speaking, an element is any substance that cannot be broken down into any other substance by ordinary chemical means. The atomic numbers and mass numbers of the most common elements on earth are shown in table 3.1.

The number of neutrons that atoms of a particular element have can vary without changing the chemical properties of the element. Atoms that have the same number of protons but different numbers of neutrons are called **isotopes.** Isotopes of an atom have the same atomic number but differ in their atomic mass. Most elements in nature exist as mixtures of different isotopes. For example, there are three isotopes of the element carbon, all of which possess six protons (figure 3.5). The most common isotope of carbon has six neutrons. Because its total mass is 12 (six protons plus six neutrons), it is referred to as carbon-12. The isotope carbon-14 is unstable, and its nucleus tends to break up into particles with lower atomic numbers, a process called **radioactive decay.** Radioactive isotopes are used in dating fossils.

TABLE 3.1 COMMON ELEMENTS ON EARTH

Element	Symbol	Atomic Number	Atomic Mass	Approximate Percent of Earth's Crust by Weight
Oxygen	O	8	15.9994	46.6
Silicon	Si	14	28.086	27.7
Aluminum	Al	13	26.9815	6.5
Iron	Fe	26	55.847	5.0
Calcium	Ca	20	40.08	3.6
Sodium	Na	11	22.989	2.8
Potassium	K	19	39.098	2.6
Magnesium	Mg	12	24.305	2.1
Hydrogen	H	1	1.0079	0.14
Manganese	Mn	25	54.938	0.1
Fluorine	F	9	18.9984	0.07
Phosphorus	P	15	30.9738	0.07
Carbon	C	6	12.0112	0.03
Sulfur	S	16	32.064	0.03
Chlorine	Cl	17	35.453	0.01
Copper	Cu	29	63.546	0.01
Nitrogen	N	7	14.0067	Trace
Boron	B	5	10.811	Trace
Cobalt	Co	27	58.933	Trace
Zinc	Zn	30	65.38	Trace
Selenium	Se	34	78.96	Trace
Molybdenum	Mo	42	95.94	Trace
Tin	Sn	50	118.69	Trace
Iodine	I	53	126.904	Trace

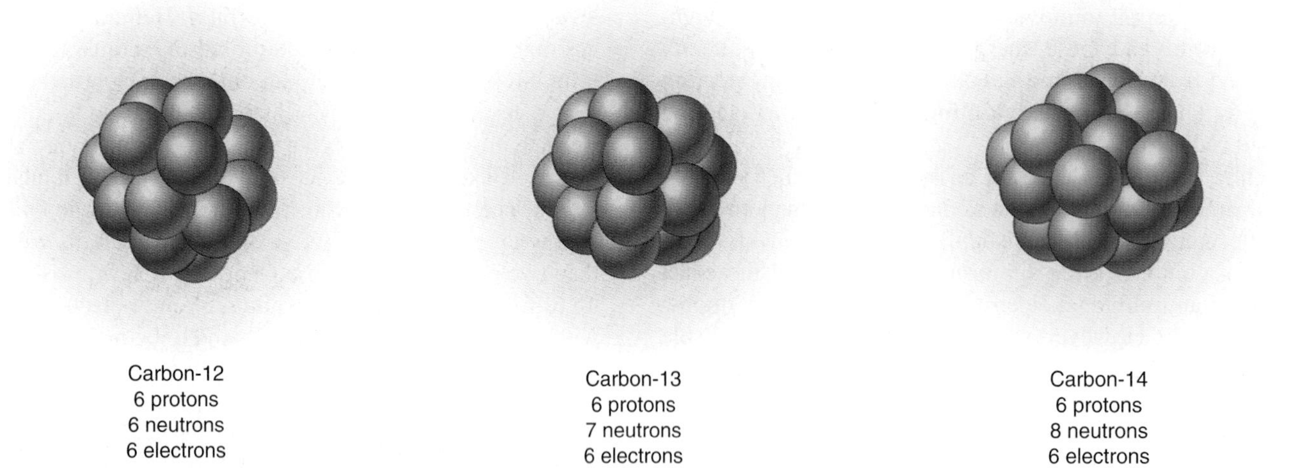
Carbon-12
6 protons
6 neutrons
6 electrons
Carbon-13
6 protons
7 neutrons
6 electrons
Carbon-14
6 protons
8 neutrons
6 electrons

Figure 3.5 Isotopes of the element carbon.

The three most abundant isotopes of carbon are carbon-12, carbon-13, and carbon-14. The yellow "clouds" in the diagrams represent the orbiting electrons, whose numbers are the same for all three isotopes. Protons are shown in purple, and neutrons are shown in pink.

Dating Fossils

Fossils are created when the remains, footprints, or other traces of organisms become buried in sand or sediment. Over time, the calcium in bone and other hard tissues becomes mineralized as the sediment is converted to rock. A fossil is any record of prehistoric life—generally taken to mean older than 10,000 years (figure 3.6). By dating the rocks in which fossils occur, biologists can get a very good idea of how old the fossils are. Rocks are usually dated by measuring the degree of radioactive decay of certain radioactive atoms among rock-forming minerals. A radioactive atom is one whose nucleus contains so many neutrons and protons that it is unstable and eventually flies apart, creating smaller, more stable atoms of another element. Because the rate of decay of a radioactive element (how many of its atoms undergo decay in a minute) is constant, scientists can use the amount of radioactive decay to date fossils. The older the fossil, the greater the fraction of its radioactive atoms that have decayed.

A widely employed method of dating fossils less than 50,000 years old is the carbon-14 (^{14}C) **radioisotopic dating** method (figure 3.7). Most carbon atoms have an atomic weight of 12 (^{12}C), but a fixed proportion of the carbon atoms in the atmosphere consists of carbon atoms with an atomic weight of 14 (^{14}C). This proportion is captured by plants in photosynthesis, and it is present also in the carbon molecules of animal bodies, all of which come ultimately from plants. After a plant or animal dies, its ^{14}C gradually decays over time, losing neutrons to form nitrogen-14 (^{14}N). It takes 5,600 years for half of the ^{14}C present in a sample to be converted to ^{14}N by this process. This length of time is called the **half-life** of the isotope. Because the half-life of an isotope is a constant that never changes, the extent of radioactive decay allows you to date a sample. Thus a sample that had a quarter of its original proportion of ^{14}C remaining would be approximately 11,200 years old (two half-lives).

For fossils older than 50,000 years, there is too little ^{14}C remaining to measure precisely, and scientists instead examine the decay of potassium-40 (^{40}K) into argon-40 (^{40}Ar), which has a half-life of 1.3 billion years.

3.3 Isotopes of an element differ in the number of neutrons they contain, but all have the same chemical properties.

Figure 3.6 Fossil of a transitional bird, *Archaeopteryx*.

This well-preserved fossil of *Archaeopteryx*, a fossil link between reptiles and birds, is about 150 million years old. It was discovered within two years of the publication of *On the Origin of Species*.

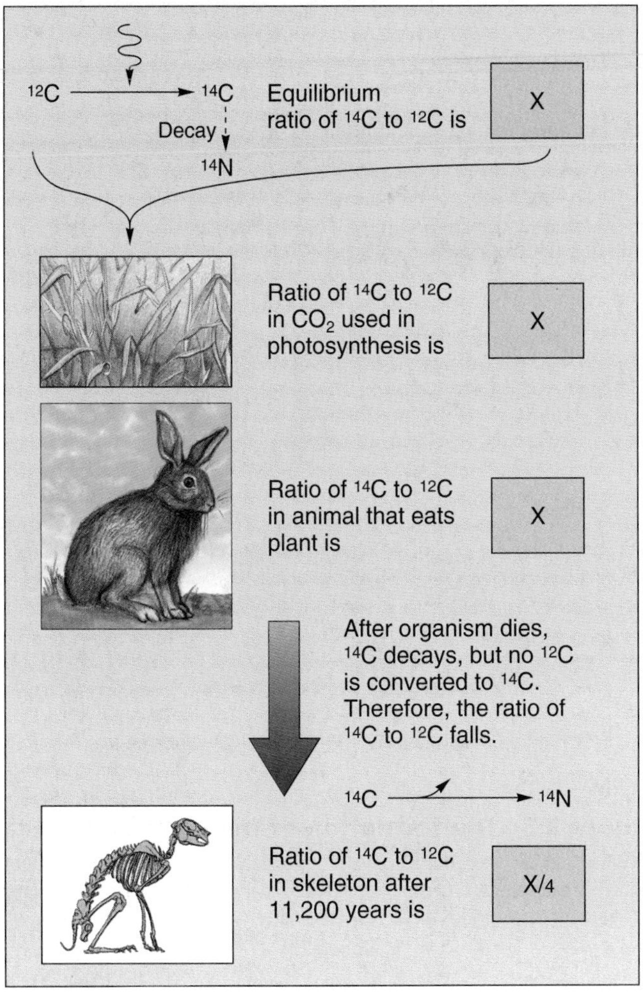

12C → 14C
Decay
14N
Equilibrium ratio of 14C to 12C is
X
Ratio of 14C to 12C in CO2 used in photosynthesis is
X
Ratio of 14C to 12C in animal that eats plant is
X
After organism dies, 14C decays, but no 12C is converted to 14C. Therefore, the ratio of 14C to 12C falls.
14C → 14N
Ratio of 14C to 12C in skeleton after 11,200 years is
X/4

Figure 3.7 Radioactive isotope dating.

This diagram illustrates radioactive dating using carbon-14, a short-lived isotope.

3.4 Molecules

A **molecule** is a group of atoms held together by energy. The energy acts as "glue," ensuring that the various atoms stick to one another. The force holding two atoms together is called a **chemical bond.** There are three principal kinds of chemical bonds: covalent bonds, where the force results from the sharing of electrons, and ionic and hydrogen bonds, where the force is generated by the attraction of opposite electrical charges.

Ionic Bonds

Chemical bonds called **ionic bonds** form when atoms are attracted to each other by opposite electrical charges. Just as the positive pole of a magnet is attracted to the negative pole of another, so an atom can form a strong link with another atom if they have opposite electrical charges. Because an atom with an electrical charge is an ion, these bonds are called ionic bonds.

Everyday table salt is built of ionic bonds. The sodium and chloride atoms of table salt are ions, sodium having given up the sole electron in its outermost sphere (the sphere underneath has eight) and chloride having gained an electron to complete its outermost sphere (figure 3.8). As a result of this electron hopping, sodium atoms in table salt are positive ions and chloride atoms are negative ions. Because each ion is attracted electrically to surrounding ions of opposite charge, this causes the formation of an elaborate matrix of sodium and chloride ionic bonds—a crystal. That is why table salt is composed of tiny crystals and is not a powder.

The two key properties of ionic bonds that make them form crystals are that they are strong (although not as strong as covalent bonds) and that they are *not* directional. A charged atom is attracted to the electrical field contributed by all nearby atoms of opposite charge. Ionic bonds do not play an important part in most biological molecules because of this lack of directionality. Complex, stable shapes require the more specific associations made possible by directional bonds.

Covalent Bonds

Strong chemical bonds called **covalent bonds** form between two atoms when they share electrons. Most of the atoms in your body are linked to other atoms by covalent bonds. Why do atoms in molecules share electrons? All atoms seek to fill up their outermost "sphere" of orbiting electrons, which in all atoms (except tiny hydrogen and helium) takes eight electrons. If an atom has only seven electrons in its outer orbit, it

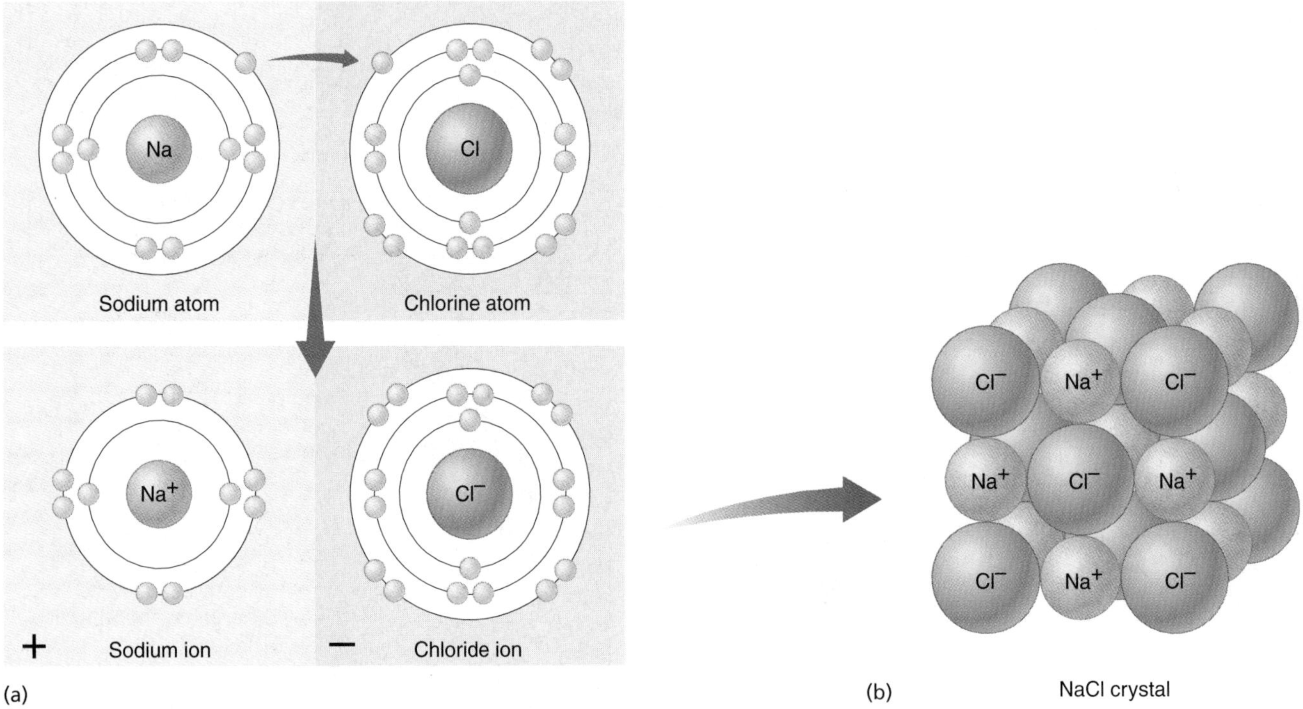
Na
Cl
Sodium atom
Chlorine atom
Na+
Cl−
+
Sodium ion
−
Chloride ion
(a)
Cl−
Na+
Cl−
Na+
Cl−
Na+
Cl−
Na+
Cl−
(b)
NaCl crystal

Figure 3.8 The formation of the ionic bond in table salt.

(*a*) When a sodium atom donates an electron to a chlorine atom, the sodium atom, lacking that electron, becomes a positively charged sodium ion. The chlorine atom, having gained an extra electron, becomes a negatively charged chloride ion. (*b*) Sodium chloride forms a highly regular lattice of alternating sodium ions and chloride ions. You are familiar with these crystals as everyday table salt.

seeks to share them with an atom that has a single electron in its outer orbit, so that the outer sphere of each atom now has eight electrons at least some of the time.

In the same way, an atom with six outer-orbit electrons seeks to share them with an atom that has two outer electrons or with two atoms that have single outer electrons. Water (H_2O), for example, is a molecule in which oxygen (six outer electrons) forms covalent bonds with two hydrogens (one outer electron each) (figure 3.9*a*). Because the atom carbon has four electrons in its outermost sphere, carbon can form as many as four covalent bonds in its attempt to fully populate its outermost sphere of electrons. Because there are many ways four covalent bonds can form, carbon atoms participate in many different kinds of molecules.

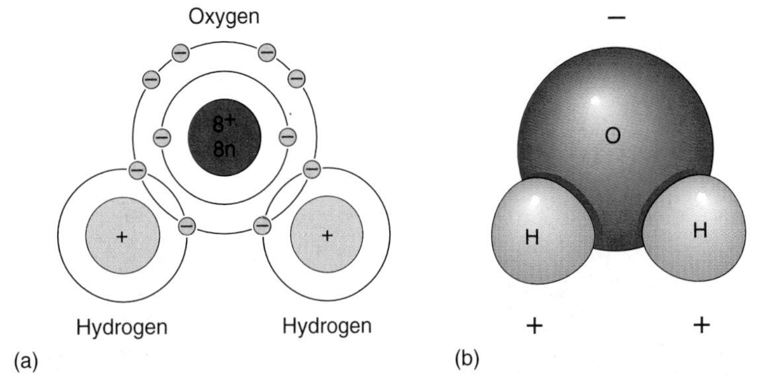
Oxygen
8+
8n
+
+
Hydrogen
Hydrogen
(a)
−
O
H
H
+
+
(b)

Figure 3.9 Water molecules contain two covalent bonds.

Each water molecule is composed of one oxygen atom and two hydrogen atoms. (*a*) The oxygen atom shares one electron with each participating hydrogen atom. (*b*) A three-dimensional representation of a water molecule.

The two key properties of covalent bonds that make them ideal for their molecule-building role in living systems are that (1) they are strong, involving the sharing of lots of energy, and (2) they are very directional—bonds form between two specific atoms, rather than a generalized attraction of one atom for its neighbors.

Hydrogen Bonds

Weak chemical bonds of a very special sort called **hydrogen bonds** play a key role in biology. To understand them we need to look at covalent bonds again briefly. When a covalent bond forms between two atoms, one nucleus may be much better at attracting the shared electrons than the other—in water, for example, the shared electrons are much more strongly attracted to the oxygen atom than to the hydrogen atoms. When this happens, shared electrons spend more time in the vicinity of the more strongly attracting atom, which as a result becomes somewhat negative in charge; they spend less time in the vicinity of the less strongly attracting atom or atoms, and these become somewhat positive in charge (figure 3.9*b*). The charges are not full electrical charges like ions possess but rather tiny partial charges. What you end up with is a sort of molecular magnet, with positive and negative ends, or "poles." Molecules like this are said to be *polar.* A hydrogen bond occurs when the positive end of one **polar molecule** is attracted to the negative end of another, like two magnets drawn to each other (figure 3.10).

Two key properties of hydrogen bonds cause them to play an important role in biological molecules. First, they are weak and so are not effective over long distances like more powerful ionic bonds. Second, as a result of their weakness, hydrogen bonds are highly directional—polar molecules must be very close for the weak attraction to be effective. Hydrogen bonds are too weak to actually form stable molecules by themselves. Instead, they act like Velcro, forming a tight bond by the additive effects of *many* weak interactions.

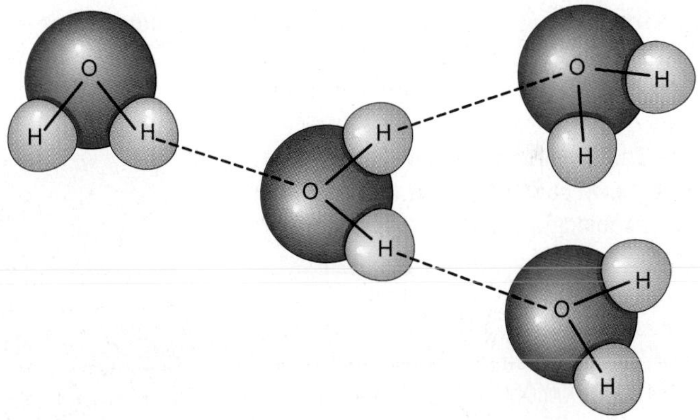
O
H
H
O
H
H
O
H
H
O
H
H

Figure 3.10 Hydrogen bonding in water molecules.

Hydrogen bonds play a key role in determining what shape a protein assumes in the cell. The functional groups of a protein's polar amino acids form many hydrogen bonds with surrounding water molecules, while the nonpolar portions of the polypeptide aggregate in the protein interior.

Hydrogen bonds stabilize the shapes of many important biological molecules by causing certain parts of a molecule to be attracted to other parts. Proteins, which are huge molecules that play many important roles in organisms, each fold into a particular complex shape because of the formation of hydrogen bonds between parts of the protein molecule.

3.4 Molecules are atoms linked together usually by covalent bonds, which are strong linkages created by the sharing of electrons. Hydrogen bonds are important but weaker bonds.

3.5 Hydrogen Bonds Give Water Unique Properties

Three-fourths of the earth's surface is covered by liquid water. About two-thirds of your body is water, and you cannot exist long without it. All other organisms also require water. It is no accident that tropical rain forests are bursting with life, whereas dry deserts seem almost lifeless except after rain. The chemistry of life, then, is water chemistry.

Water has a simple atomic structure, an oxygen atom linked to two hydrogen atoms by single covalent bonds. The chemical formula for water is thus H_2O. It is because the oxygen atom attracts the shared electrons more strongly than the hydrogen atoms that water is a *polar molecule* and so can form *hydrogen bonds*. Water's ability to form hydrogen bonds is responsible for much of the organization of living chemistry, from membrane structure to how proteins fold.

The weak hydrogen bonds that form between a hydrogen atom of one water molecule and the oxygen atom of another produce a lattice of hydrogen bonds within liquid water. Each of these bonds is individually very weak and short-lived—a single bond lasts only 1/100,000,000,000 of a second. However, like the grains of sand on a beach, the cumulative effect of large numbers of these bonds is enormous and is responsible for many of the important physical properties of water (table 3.2).

TABLE 3.2 THE PROPERTIES OF WATER

Property	Explanation
Heat storage	Hydrogen bonds absorb heat when they break and release heat when they form, minimizing temperature changes.
Ice formation	Water molecules in an ice crystal are spaced relatively far apart because of hydrogen bonding.
High heat of vaporization	Many hydrogen bonds must be broken for water to evaporate.
Cohesion	Hydrogen bonds hold molecules of water together.
High polarity	Water molecules are attracted to ions and polar compounds.

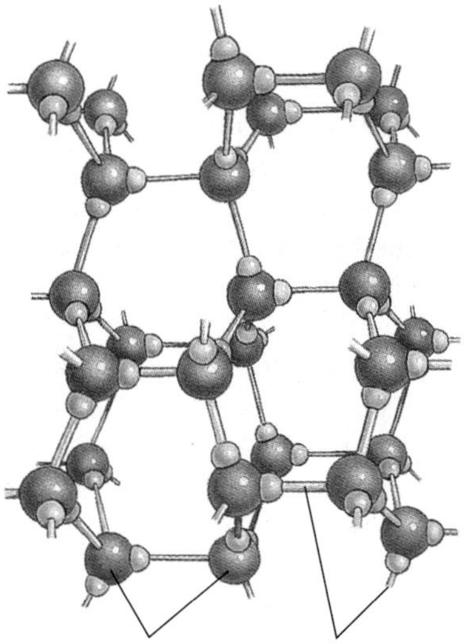
Water molecules
Hydrogen bonds

Figure 3.11 Ice formation. When water cools below 0°C, it forms a regular crystal structure that floats. The individual water molecules are spaced apart and held into position by hydrogen bonds.

Heat Storage

The temperature of any substance is a measure of how rapidly its individual molecules are moving. Because of the many hydrogen bonds that water molecules form with one another, a large input of thermal energy is required to disrupt the organization of liquid water and raise its temperature. Because of this, water heats up more slowly than almost any other compound and holds its temperature longer. That is why your body is able to maintain a relatively constant internal temperature.

Ice Formation

If the temperature is low enough, very few hydrogen bonds break in water so that the lattice of these bonds assumes a crystal-like structure, forming a solid we call ice (figure 3.11). Interestingly, ice is less dense than water—that is why icebergs and ice cubes float. Why is ice less dense? Because the hydrogen bonds space the water molecules apart, preventing them from approaching each other more closely.

High Heat of Vaporization

If the temperature is high enough, many hydrogen bonds break in water so that the liquid is changed into vapor. A considerable amount of heat energy is required to do this—every gram of water that evaporates from your skin removes 586 calories of heat from your body, which is equal to the energy released by lowering the temperature of 586 grams of water 1°C. That is why sweating cools you off.

Cohesion

Because water molecules are very polar, they are attracted to other polar molecules—like glue, hydrogen bonds bind polar molecules to each other. When the other polar molecule is another water molecule, the attraction is called **cohesion.** The surface tension of water is created by cohesion (figure 3.12). When the other polar molecule is a different substance, the attraction is called **adhesion.** Capillary action—water rising up a narrow glass tube—is created by adhesion. Water clings to any substance, such as glass, with which it can form hydrogen bonds. Adhesion is why things get "wet" when they are dipped in water and why waxy substances do not—they are composed of nonpolar molecules.

High Polarity

Water molecules in solution always tend to form the maximum number of hydrogen bonds possible. Polar molecules form hydrogen bonds and are welcomed by water molecules. Polar molecules are called hydrophilic (*hydros*, water, and *philic*, loving), or water-loving, molecules. Water molecules gather closely around any molecule that exhibits an electrical charge, whether a full charge (ion) or partial charge (polar molecule). When a sugar crystal dissolves in water, what really happens is that individual sugar molecules break off from the crystal and become surrounded by water molecules attracted to its slightly polar hydroxyl (OH^-) groups. Water molecules orient around each sugar molecule like a cloud of bees attracted to honey, and this shell of water molecules prevents the sugar molecule from reassociating with the crystal. Similar shells of water form around all polar molecules, and polar molecules that dissolve in water in this way are said to be **soluble** in water (figure 3.13).

Nonpolar molecules like oil do not form hydrogen bonds and are not water-soluble. When nonpolar molecules are placed in water, the water molecules shy away, instead forming hydrogen bonds with other water molecules. The nonpolar molecules are forced into association with one another, crowded together to minimize their disruption of the hydrogen bonding of water. It seems almost as if the nonpolar compounds shrink from contact with water, and for this reason they are called **hydrophobic** (from the Greek *hydros*, water, and *phobos*, fearing). Many biological structures are shaped by such hydrophobic forces. For example, some of the portions of many proteins are hydrophobic, and by forcing these portions into proximity to one another, water causes these proteins to assume particular shapes in solution.

3.5 Water molecules form a network of hydrogen bonds in liquid and dissolve other polar molecules. Many of the key properties of water arise because it takes considerable energy to break liquid water's many hydrogen bonds.

(a)
(b)

Figure 3.12 Cohesion.
(*a*) Cohesion allows water molecules to stick together and form droplets. (*b*) Surface tension is a property derived from cohesion—that is, water has a "strong" surface due to the force of its hydrogen bonds. Some insects, such as this water strider, literally walk on water.

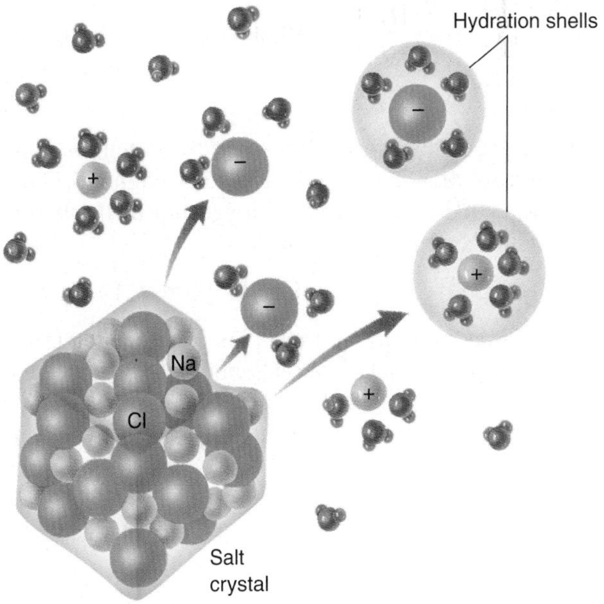

Hydration shells
Na
Cl
Salt crystal

Figure 3.13 How salt dissolves in water.
Salt is soluble in water because the partial charges on water molecules are attracted to the charged sodium and chloride ions. The water molecules surround the ions, forming what are called hydration shells. When all of the ions have been separated from the crystal, the salt is said to be dissolved.

3.6 Water Ionizes

The covalent bonds within a water molecule sometimes break spontaneously. In pure water at 25°C, only 1 out of every 550 million water molecules undergoes this process. When it happens, one of the protons (hydrogen atom nuclei) dissociates from the molecule. Because the dissociated proton lacks the negatively charged electron it was sharing in the covalent bond with oxygen, its own positive charge is no longer counterbalanced, and it becomes a positively charged ion, H^+. The rest of the dissociated water molecule, which has retained the shared electron from the covalent bond, is negatively charged and forms a **hydroxide ion** (OH^-). This process of spontaneous ion formation is called ionization. It can be represented by a simple chemical equation, in which the chemical formulae for water and the two ions are written down, with an arrow showing the direction of the dissociation:

H_2O	$\rightarrow$	OH^-	+	H^+
water		hydroxide ion		hydrogen ion

Because covalent bonds are strong, spontaneous ionization is not common. In a liter of water, only roughly 1 molecule out of each 550 million is ionized at any instant in time, corresponding to 1/10,000,000 (that is, 10^{-7}) of a mole of hydrogen ions (a mole is defined as the weight in grams that corresponds to the molecular mass of a molecule, which in this case equals 1 gram). The concentration of H^+ in water can be written more easily by simply counting the number of decimal places after the digit "1" in the denominator:

$$[H^+] = \frac{1}{10{,}000{,}000}$$

pH

A more convenient way to express the hydrogen ion concentration of a solution is to use the **pH scale** (figure 3.14). This scale defines pH as the negative logarithm of the hydrogen ion concentration in the solution:

$$pH = -\log [H^+]$$

Since the logarithm of the hydrogen ion concentration is simply the exponent of the molar concentration of H^+, the pH equals the exponent times −1. Thus, pure water, with an $[H^+]$ of 10^{-7} mole/liter, has a pH of 7. Recall that for every hydrogen ion formed when water dissociates, a hydroxide ion is also formed, meaning that the dissociation of water produces H^+ and OH^- in equal amounts. Therefore, a pH value of 7 indicates neutrality—a balance between H^+ and OH^-—on the pH scale.

Note that the pH scale is *logarithmic,* which means that a difference of 1 on the pH scale represents a 10-fold change in hydrogen ion concentration. This means that a solution with a pH of 4 has *10 times* the concentration of H^+ present in one with a pH of 5.

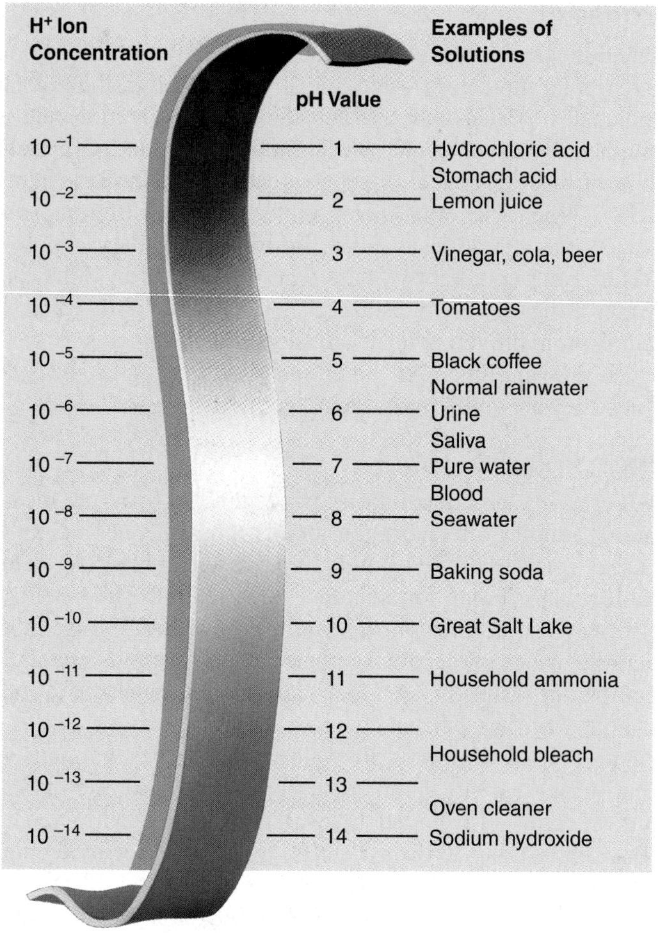

H+ Ion Concentration
Examples of Solutions
pH Value
10^-1 — 1 — Hydrochloric acid
Stomach acid
10^-2 — 2 — Lemon juice
10^-3 — 3 — Vinegar, cola, beer
10^-4 — 4 — Tomatoes
10^-5 — 5 — Black coffee
Normal rainwater
10^-6 — 6 — Urine
Saliva
10^-7 — 7 — Pure water
Blood
10^-8 — 8 — Seawater
10^-9 — 9 — Baking soda
10^-10 — 10 — Great Salt Lake
10^-11 — 11 — Household ammonia
10^-12 — 12
Household bleach
10^-13 — 13
Oven cleaner
10^-14 — 14 — Sodium hydroxide

Figure 3.14 The pH scale.
A fluid is assigned a value according to the number of hydrogen ions present in a liter of that fluid. The scale is logarithmic, so that a change of only 1 means a 10-fold change in the concentration of hydrogen ions; thus lemon juice is 100 times more acidic than tomatoes, and seawater is 10 times more basic than pure water.

Acids. Any substance that dissociates in water to increase the concentration of H^+ is called an acid. Acidic solutions have pH values below 7. The stronger an acid is, the more H^+ it produces and the lower its pH. For example, hydrochloric acid (HCl), which is abundant in your stomach, ionizes completely in water. This means that 10^{-1} mole/liter of HCl will dissociate to form 10^{-1} mole/liter of H^+, giving the solution a pH of 1. The pH of champagne, which bubbles because of the carbonic acid dissolved in it, is about 2.

Bases. A substance that combines with H^+ when dissolved in water is called a base. By combining with H^+, a base lowers the H^+ concentration in the solution. Basic (or alkaline) solutions, therefore, have pH values above 7. Very strong bases, such as sodium hydroxide (NaOH), have pH values of 12 or more.

Buffers

The pH inside almost all living cells, and in the fluid surrounding cells in multicellular organisms, is fairly close to 7. The many proteins that govern metabolism are all extremely sensitive to pH, and slight alterations in pH change their shape and so disrupt their activities. For this reason it is important that a cell maintain a constant pH level. The pH of your blood, for example, is 7.4, and you would survive only a few minutes if it were to fall to 7.0 or rise to 7.8.

Yet the chemical reactions of life constantly produce acids and bases within cells. Furthermore, many animals eat substances that are acidic or basic; Coca Cola, for example, is acidic. What keeps an organism's pH constant? Cells contain chemical substances called buffers that minimize changes in concentrations of H^+ and OH^- (figure 3.15).

A **buffer** is a substance that acts as a reservoir for hydrogen ions, donating them to the solution when their concentration falls and taking them from the solution when their concentration rises. What sort of substance will act in this way? Within organisms, most buffers consist of pairs of substances, one an acid and the other a base. The key buffer in human blood is an acid-base pair consisting of *carbonic acid* (acid) and *bicarbonate* (base). These two substances interact in a pair of reversible reactions. First, carbon dioxide (CO_2) and H_2O join to form carbonic acid (H_2CO_3), which in a second reaction dissociates to yield bicarbonate ion (HCO_3^-) and H^+ (figure 3.16). If some acid or other substance adds H^+ to the blood, the HCO_3^- acts as a base and removes the excess H^+ by forming H_2CO_3. Similarly, if a basic substance removes H^+ from the blood, H_2CO_3 dissociates, releasing more H^+ into the blood. The forward and reverse reactions that interconvert H_2CO_3 and HCO_3^- thus stabilize the blood's pH.

The interaction of carbon dioxide and water has the important consequence that significant amounts of carbon enter into water solution from air in the form of carbonic acid. As we discuss later in this chapter, biologists believe that life first evolved in the early oceans, which were rich in carbon because of this reaction.

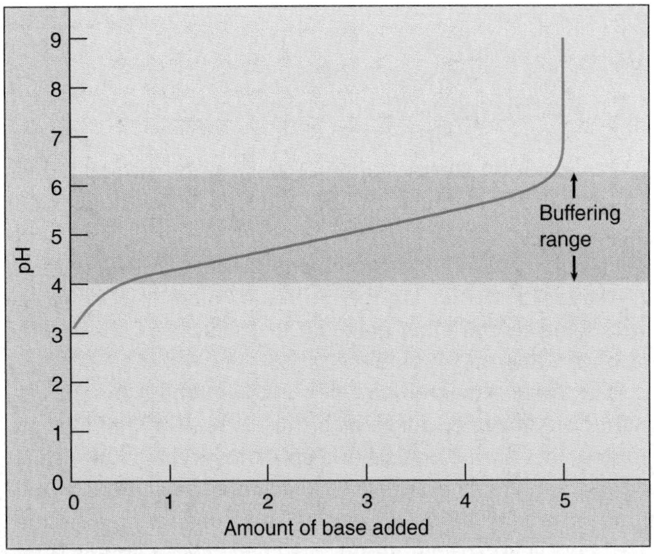
pH
0
1
2
3
4
5
6
7
8
9
Buffering range
0
1
2
3
4
5
Amount of base added

Figure 3.15 Buffers minimize changes in pH.
Adding a base to a solution neutralizes some of the acid present and so raises the pH. Thus, as the curve moves to the right, reflecting more and more base, it also rises to higher pH values. What a buffer does is to make the curve rise or fall very slowly over a portion of the pH scale, called the "buffering range" of that buffer.

3.6 A tiny fraction of water molecules spontaneously ionize at any moment, forming H^+ and OH^-. The pH of a solution is the negative logarithm of the H^+ concentration in the solution. Thus, low pH values indicate high H^+ concentrations (acidic solutions), and high pH values indicate low H^+ concentrations (basic solutions).

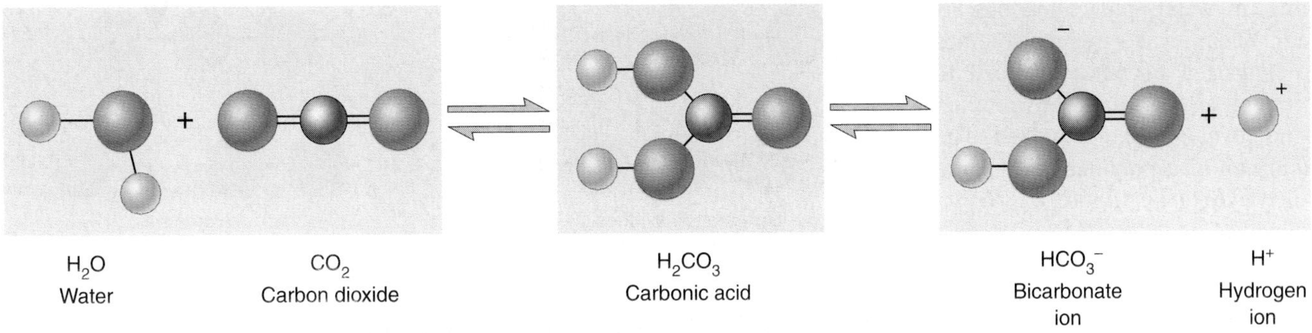
+
+
+
–
H_2O
Water
CO_2
Carbon dioxide
H_2CO_3
Carbonic acid
HCO_3^-
Bicarbonate ion
H^+
Hydrogen ion

Figure 3.16 The buffering action of carbonic acid and bicarbonate.
Carbon dioxide and water combine chemically to form carbonic acid (H_2CO_3), which dissociates in water, freeing H^+. Now if some other source adds H^+ to your blood, the bicarbonate ion acts as a base and removes the excess H^+ from the solution, forming H_2CO_3. Similarly, if some process removes H^+ from your blood, the carbonic acid dissociates, releasing more hydrogen ions into the solution.

3.7 Forming Macromolecules

The bodies of organisms contain thousands of different kinds of molecules, but much of the body is made of just four kinds: *carbohydrates, lipids, proteins,* and *nucleic acids* (table 3.3). Called **macromolecules** because they can be very large, these four are the building materials of cells, the "bricks and mortar" that make up the bodies of cells and the machinery that runs within them.

The body's macromolecules are assembled by sticking smaller bits together, much as a train is built by linking railcars together. A molecule built up of long chains of similar subunits is called a **polymer.** An **organic molecule** is a molecule formed by living organisms that consists of a carbon-based core with special groups attached. These groups of atoms have special chemical properties and are referred to as **functional groups.** Functional groups tend to act as units during chemical reactions and to confer specific chemical properties on the molecules that possess them. The six principal functional groups are listed in figure 3.17.

Group	Structural Formula	Ball-and-Stick Model	Found In:
Hydroxyl	—OH		Alcohols
Carbonyl	—C— ‖ O		Formaldehyde
Carboxyl	—C(=O)OH		Vinegar
Amino	—N(H)H		Ammonia
Sulfhydryl	—S—H		Rubber
Phosphate	—O—P(O⁻)(=O)—O⁻		ATP

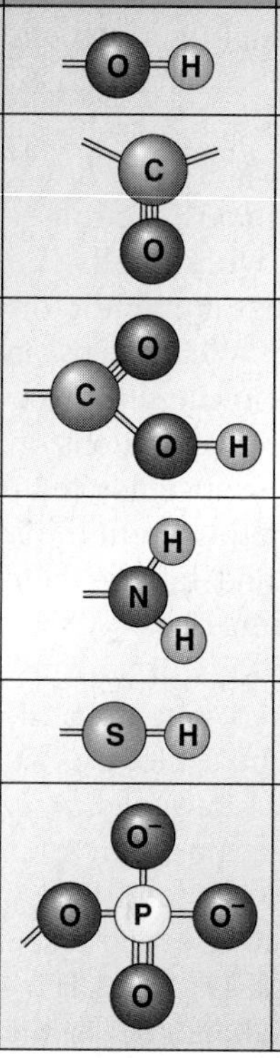

O H
C O
O C O H
H N H
S H
O⁻ O P O⁻ O

Figure 3.17 The six principal functional groups.
Most chemical reactions that occur within organisms involve transferring a functional group from one molecule to another or breaking a carbon–carbon bond.

Making (and Breaking) Macromolecules

The four different kinds of macromolecules put their subunits together in the same way: a covalent bond is formed between two subunits in which a hydroxyl group (OH) is removed from one subunit and a hydrogen (H) is removed from the other. This process is called **dehydration synthesis** because, in effect, the removal of the OH and H groups constitutes removal of a molecule of water—the word *dehydration* means taking away water (figure 3.18*a*). This process requires the help of a special class of proteins called **enzymes** to facilitate the positioning of the molecules so that the correct chemical bonds are stressed and broken. The process of tearing down a molecule, such as the protein or fat contained in food that is consumed, is essentially the reverse of dehydration synthesis: instead of removing a water molecule, one is added. When a water molecule comes in, a hydrogen becomes attached to one subunit and a hydroxyl to another, and the covalent bond is broken. The breaking up of a polymer in this way is called **hydrolysis** (figure 3.18*b*).

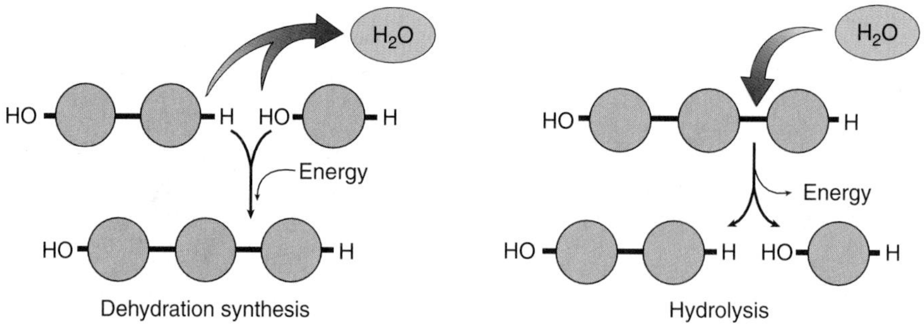

H₂O
HO
H
HO
H
Energy
HO
H
Dehydration synthesis
H₂O
HO
H
Energy
HO
H
HO
H
Hydrolysis

(a) (b)

Figure 3.18 Dehydration and hydrolysis.
(*a*) Biological molecules are formed by linking subunits. The covalent bond between subunits is formed in dehydration synthesis, a process during which a water molecule is eliminated. (*b*) Breaking such a bond requires the addition of a water molecule, a reaction called hydrolysis.

3.7 Macromolecules are formed by linking subunits together into long chains, removing a water molecule as each link is formed.

TABLE 3.3 MACROMOLECULES

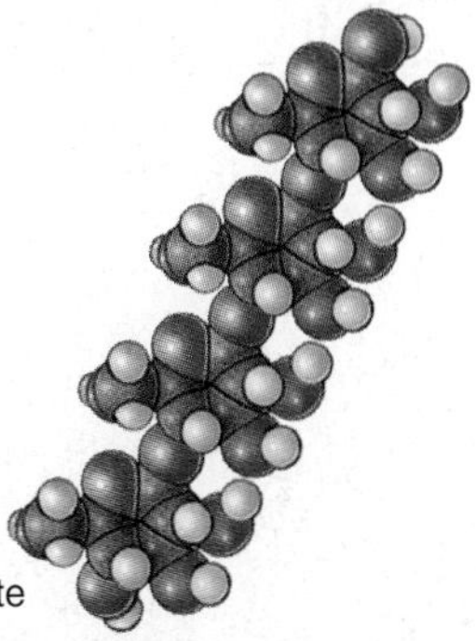
Carbohydrate (starch)

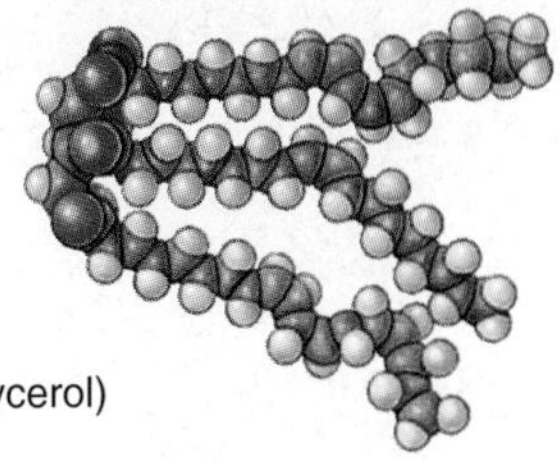
Lipid (triacylglycerol)

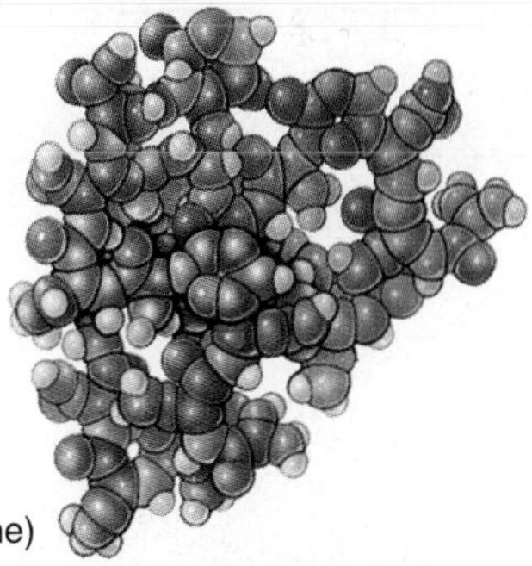
Protein (enzyme)

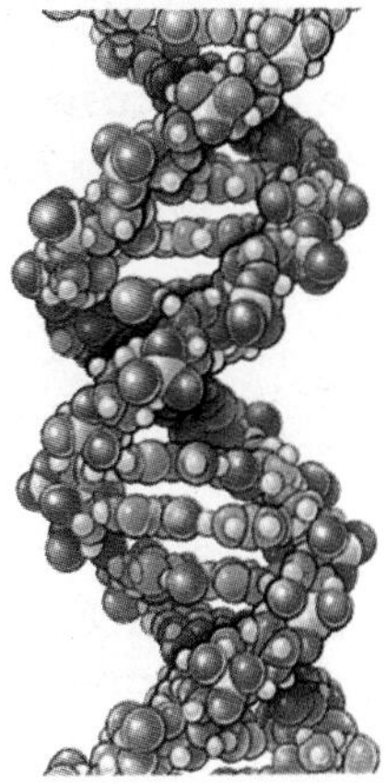
Nucleic acid (DNA)

Macromolecule	Subunit	Function
Carbohydrates		
Starch, glycogen	Glucose	Energy storage
Cellulose	Glucose	Component of plant cell walls
Chitin	Modified glucose	Cell walls of fungi; outer skeleton of insects and related groups
Lipids		
Fats	Glycerol + three fatty acids	Energy storage
Phospholipids	Glycerol + two fatty acids + phosphate	Component of cell membranes
Steroids	Four carbon rings	Message transmission (hormones)
Terpenes	Long carbon chains	Pigments in photosynthesis
Proteins		
Globular	Amino acids	Catalysis
Structural	Amino acids	Support and structure
Nucleic Acids		
DNA	Nucleotides	Encoding of hereditary information
RNA	Nucleotides	Blueprint of hereditary information

3.8 Carbohydrates

Polymers called **carbohydrates** make up the structural framework of cells and play a critical role in energy storage. A carbohydrate is any molecule that contains carbon, hydrogen, and oxygen in the ratio 1:2:1. Some carbohydrates are simple, small monomers or dimers and are called **simple carbohydrates.** Others are long polymers and are called **complex carbohydrates** (figure 3.19). Because they contain many carbon–hydrogen (C–H) bonds, carbohydrates are well-suited for energy storage. Such C–H bonds are the ones most often broken by organisms to obtain energy.

Simple Carbohydrates

The simplest carbohydrates are the *simple sugars* or **monosaccharides** (from the Greek *monos*, single, and *saccharon*, sweet). These molecules consist of one subunit. For example, glucose, the sugar that carries energy to the cells of your body, is made of six carbons and has the chemical formula $C_6H_{12}O_6$ (figure 3.20). Another type of simple carbohydrate is a **disaccharide,** which forms when two monosaccharides link. Sucrose (table sugar) is a disaccharide made of two six-carbon sugars linked together (figure 3.21).

Complex Carbohydrates

Organisms store their metabolic energy by converting sugars, which are soluble, into insoluble forms that can be deposited in specific storage areas in the body. This trick is achieved by linking the sugars together into long polymer chains called **polysaccharides.** Plants and animals store energy in polysaccharides formed from glucose. The glucose polysaccharide that plants use to store energy is called **starch**—that is why potatoes are "starchy" food. In animals, energy is stored in **glycogen,** a highly insoluble thicket of glucose polysaccharides that are very long and highly branched.

Plants and animals also use glucose chains as building materials, linking the subunits together in different orientations not recognized by most enzymes. Chitin (see figure 3.19) and cellulose (a component of plant cell walls) are both polymers composed of long-chain sugar subunits.

3.8 Carbohydrates are molecules made of C, H, and O atoms. As sugars they store energy in C–H bonds.

Figure 3.19 This lobster's shell is made of a complex carbohydrate.

A complex carbohydrate called chitin is the principal structural element in the external skeletons of many invertebrates, including crustaceans and insects, and in the cell walls of fungi.

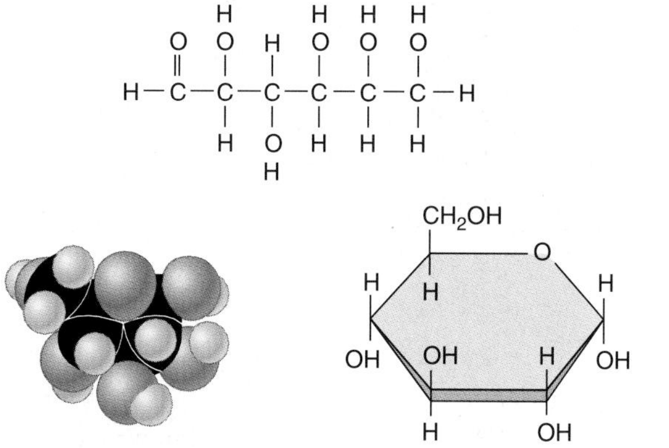

CH_2OH
O
H
OH

Figure 3.20 The structure of glucose.

Glucose is a monosaccharide and consists of a linear six-carbon molecule that forms a ring when added to water. This illustration shows three ways glucose can be represented diagrammatically.

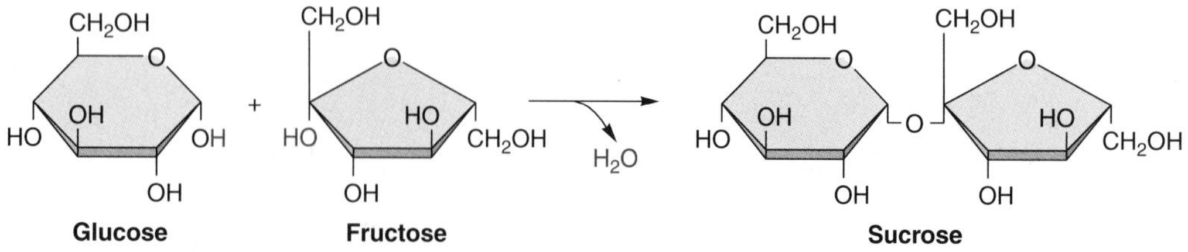

CH_2OH
OH
HO
H_2O
Glucose
Fructose
Sucrose

Figure 3.21 Formation of sucrose.

The disaccharide sucrose is formed from glucose and fructose in a dehydration reaction.

3.9 Lipids

For long-term storage, organisms usually convert glucose into fats, another kind of storage molecule that contains more C–H bonds. Fats and all other biological molecules that are not soluble in water but soluble in oil are called **lipids.** Lipids are insoluble in water not because they are long chains like starches but rather because they are nonpolar. In water, fat molecules cluster together because they cannot form hydrogen bonds with water molecules.

Fats

Fat molecules are composed of two subunits: fatty acids and glycerol. A **fatty acid** is a long hydrocarbon chain ending in a carboxyl (–COOH) group. The three carbons of glycerol form the backbone to which three fatty acids are attached in the dehydration reaction that forms the fat molecule (figure 3.22). Because there are three fatty acids, the resulting fat molecule is called a **triacylglycerol,** or **triglyceride.**

Fatty acids with all internal carbon atoms having two hydrogen side groups contain the maximum number of hydrogen atoms. Fats composed of these fatty acids are said to be **saturated** (figure 3.23*a*). On the other hand, fats composed of fatty acids that have double bonds between one or more pairs of carbon atoms contain fewer than the maximum number of hydrogen atoms and are called **unsaturated** (figure 3.23*b*). Many plant fatty acids are unsaturated. Animal fats, in contrast, are often saturated and occur as hard fats.

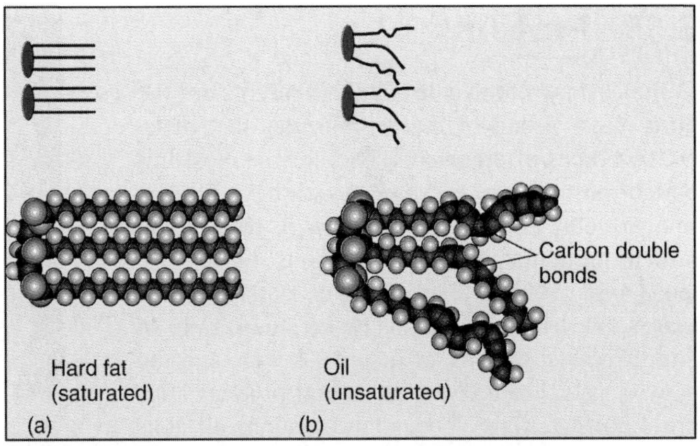

Carbon double bonds
Hard fat (saturated)
Oil (unsaturated)
(a)
(b)

Figure 3.23 Saturated and unsaturated fats.

(a) Most animal fats are "saturated" (every carbon atom carries the maximum load of hydrogens). Their fatty acid chains fit closely together, and these triacylglycerols form immobile arrays called hard fats. (*b*) Most plant fats are unsaturated, which prevents close association between triacylglycerols and produces oils.

Other Types of Lipids

Your body also contains other types of lipids that play many roles in cells in addition to energy storage. The membranes of cells are made of a modified fat called a **phospholipid.** Phospholipids have a polar group at one end and two long tails that are strongly nonpolar. In water, the nonpolar ends of phospholipids aggregate, forming two layers of molecules with the nonpolar tails pointed inside—a lipid bilayer. Membranes also contain a quite different kind of lipid called a **steroid,** composed of four carbon rings. Most animal cell membranes contain the steroid cholesterol. Excess saturated fat intake can cause plugs of cholesterol to form in the blood vessels, which may lead to blockage, high blood pressure, stroke, or heart attack. Male and female sex hormones are also steroids. Other important biological lipids include rubber, waxes, and pigments, such as the chlorophyll that makes plants green and the retinal that your eyes use to detect light.

3.9 Lipids are not water-soluble. Fats contain chains of fatty acid subunits and can store energy.

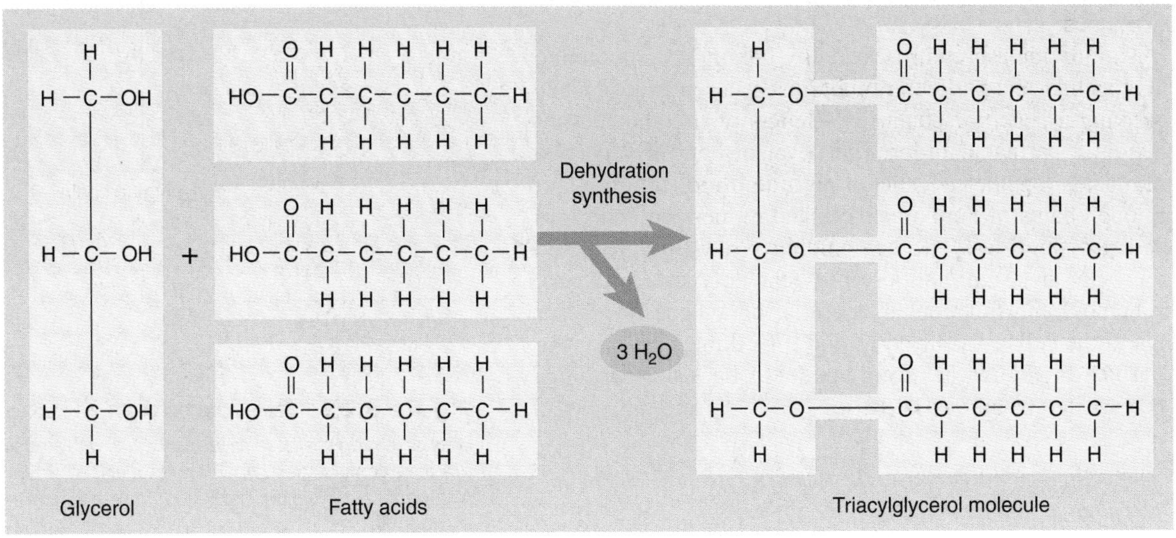

Dehydration synthesis
3 H_2O
Glycerol
Fatty acids
Triacylglycerol molecule

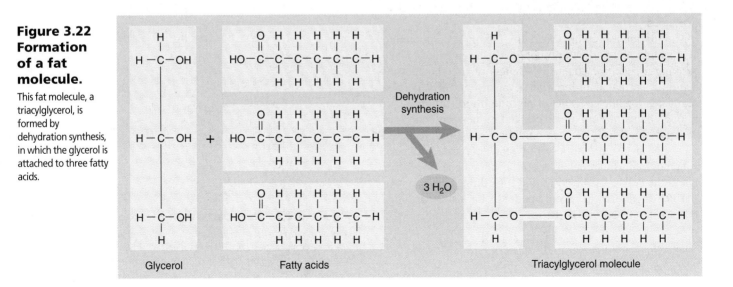

Figure 3.22 Formation of a fat molecule.

This fat molecule, a triacylglycerol, is formed by dehydration synthesis, in which the glycerol is attached to three fatty acids.

3.10 Proteins

Complex macromolecules called **proteins** are the third major group of macromolecules that make up the bodies of organisms. Perhaps the most important proteins are *enzymes,* which have the key role in cells of lowering the energy required to initiate particular chemical reactions. Other proteins play structural roles—the collagen that makes the strings of a tennis racket, the keratin of a bird feather, the silk of a spider's web, and the hair on your head all are structural proteins (figure 3.24). Cartilage, bones, and tendons all contain a structural protein called collagen. Keratin, another structural protein, forms the horns of a rhinoceros and the feathers of a bird. Still other proteins act as chemical messengers within the brain and throughout the body.

Despite their diverse functions, all proteins have the same basic structure: a long polymer chain made of subunits called amino acids. **Amino acids** are small molecules with a simple basic structure: a central carbon atom to which an amino group (–NH_2), a carboxyl group (–COOH), a hydrogen atom (H), and a functional group, designated "R," are bonded. There are 20 common kinds of amino acids.

Each amino acid has the same chemical backbone but can be differentiated from other amino acids by its functional group. Six of the amino acid functional groups are nonpolar, differing chiefly in size—the most bulky contain ring structures, and amino acids containing them are called aromatic. Another six are polar but uncharged, and these differ from one another in the strength of their polarity. Five more are polar and are capable of ionizing to a charged form. The remaining three possess special chemical groups that are important in forming links between protein chains or in forming kinks in their shapes.

An individual protein is made by linking specific amino acids together in a particular order, just as a sentence is made by linking a specific sequence of letters of the alphabet together in a particular order. The covalent bond linking two amino acids together is called a **peptide bond** (figure 3.25), and long chains of amino acids linked by peptide bonds are called **polypeptides.** The hemoglobin proteins in your red blood cells are each composed of four polypeptide chains. The hemoglobin functions as a carrier of oxygen from your lungs to the cells of your body, and it facilitates the carrying of carbon dioxide wastes from your cells back to your lungs to be expelled from your body.

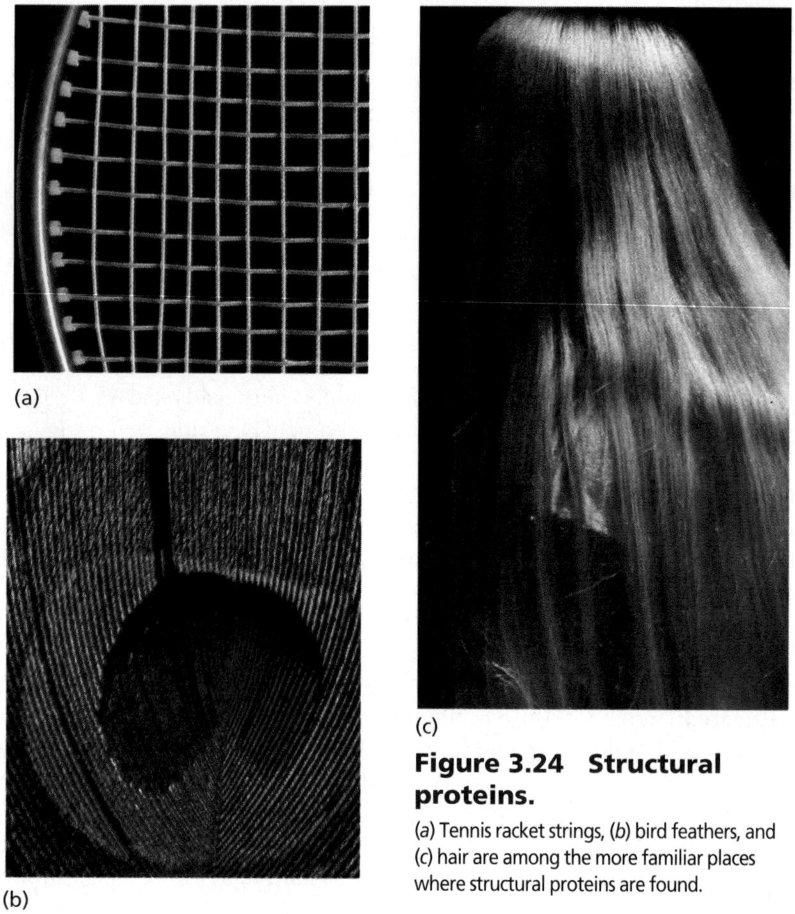
(a)
(b)
(c)

Figure 3.24 Structural proteins.

(*a*) Tennis racket strings, (*b*) bird feathers, and (*c*) hair are among the more familiar places where structural proteins are found.

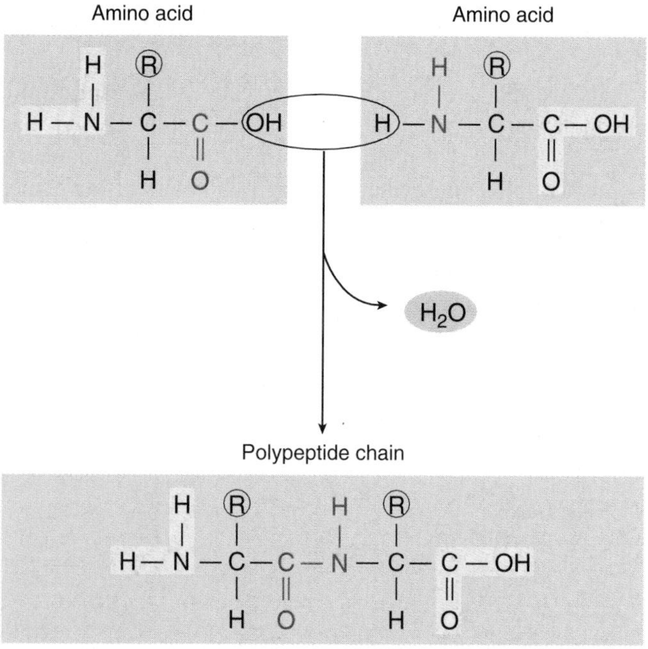
Amino acid
Amino acid
H R
H—N—C—C—OH
H O
H R
H—N—C—C—OH
H O
H2O
Polypeptide chain
H R H R
H—N—C—C—N—C—C—OH
H O H O

Figure 3.25 The formation of a peptide bond.

Every amino acid has the same basic structure, with an amino group at one end and a carboxyl group at the other. The only variable is the functional, or "R," group. Amino acids are linked by dehydration synthesis to form peptide bonds. Chains of amino acids linked in this way are called polypeptides and are the basic structural components of proteins.

Protein Structure

The sequence of amino acids of a polypeptide chain is termed the polypeptide's *primary structure* (figure 3.26). Because some of the amino acids are nonpolar and others are not, a protein chain folds up in solution as the nonpolar regions are forced together. This initial folding is called the *secondary structure* of a protein. The final three-dimensional shape, or *tertiary structure,* of the protein, usually folded and twisted into a globular molecule, is determined by exactly where in a protein chain the nonpolar amino acids occur. When a protein is composed of more than one polypeptide chain, the spatial arrangement of the several component chains is called the *quaternary structure* of the protein.

The shape of the protein is largely the result of the interaction of the amino acid functional groups with water, which tends to shove nonpolar portions of the polypeptide into the protein's interior. If the polar nature of the protein's environment changes, the protein may unfold in a process called **denaturation.** When the polar nature of the solvent is reestablished, proteins may spontaneously refold.

Many structural proteins form long cables that have architectural roles in cells, providing strength and determining shape. The proteins called enzymes have three-dimensional shapes with grooves or depressions that precisely fit a particular sugar or other chemical; once in the groove, the chemical is encouraged to undergo a reaction—often, one of its chemical bonds is stressed as the chemical is bent by the enzyme, like a foot in a flexing shoe. This process of enhancing chemical reactions is called **catalysis,** and proteins are the catalytic agents of cells, determining what chemical processes take place and where and when.

3.10 Proteins are chains of amino acids that fold into complex shapes. A protein's shape depends on its amino acid sequence and determines its function.

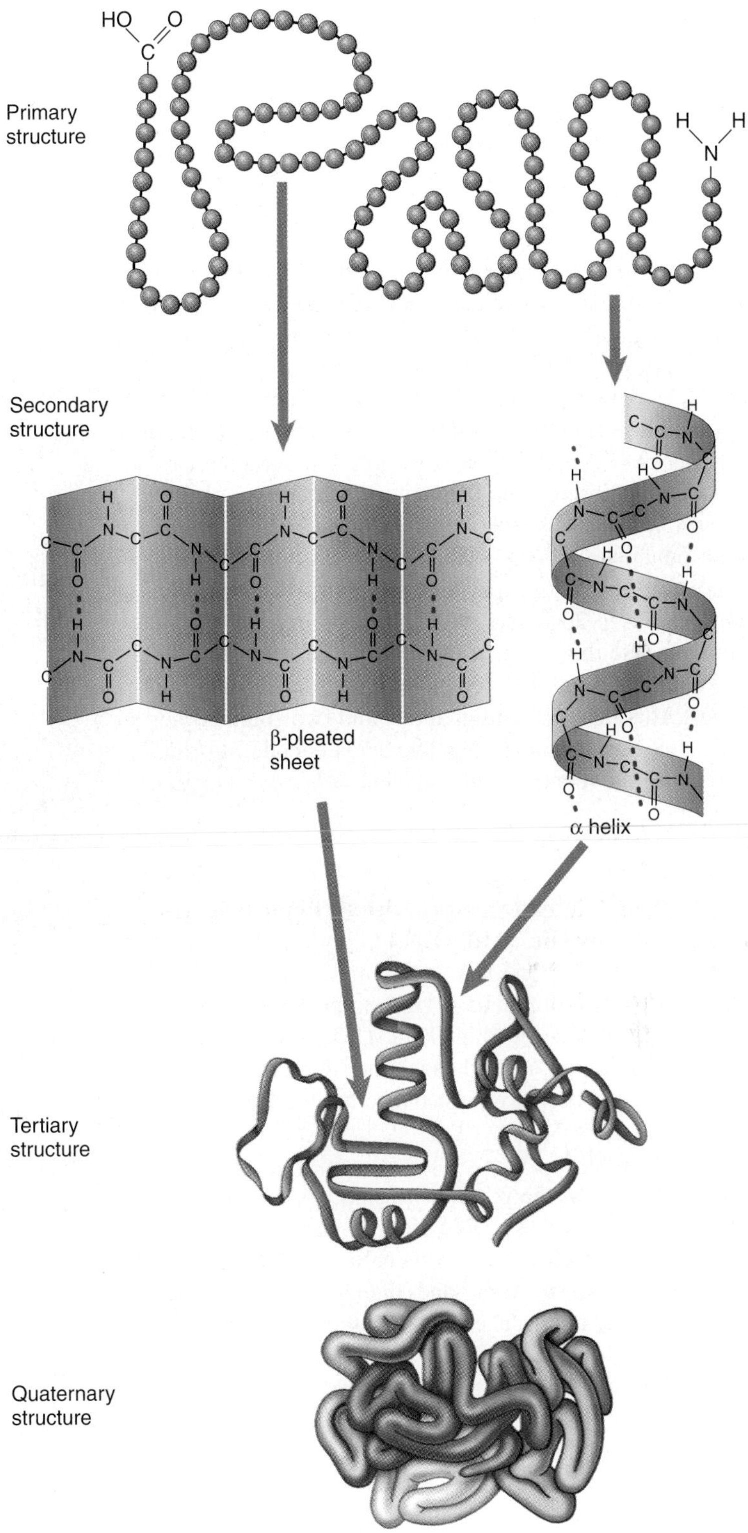
Primary structure
Secondary structure
β-pleated sheet
α helix
Tertiary structure
Quaternary structure

Figure 3.26 Levels of protein structure.
The *primary structure* of a protein is its sequence of amino acids. Twisting or pleating of the chain of amino acids, called *secondary structure*, is due to the formation of localized hydrogen bonds (*red*) within the chain. More complex folding of the chain is referred to as *tertiary structure*. Two or more protein chains associated together form a *quaternary structure*.

3.11 Nucleic Acids

Very long polymers called **nucleic acids** serve as the information storage devices of cells, just as disks or hard drives store the information that computers use. Nucleic acids are long polymers of repeating subunits called **nucleotides.** Each nucleotide is composed of three parts: a five-carbon sugar, a phosphate group (PO_4), and an organic nitrogen-containing (nitrogenous) base (figure 3.27). In the formation of a nucleic acid, the individual sugars are linked in a line by the phosphate groups: —[SUGAR]—phosphate—[SUGAR]—phosphate—, in very long **polynucleotide chains.**

How does the long, chain-like structure of a nucleic acid permit it to store the information necessary to specify what a human being is like? If DNA were simply a monotonous repeating polymer, it could not encode the message of life. Imagine trying to write a story using only the letter *E* and no spaces or punctuation. All you could ever say is "EEEEE EE. . . ." You need more than one letter to write—the English alphabet uses 26 letters. Nucleic acids can encode information because they contain more than one kind of nucleotide. There are four different kinds of nucleotides: two larger ones called adenine and guanine, and two smaller ones called cytosine and thymine. Nucleic acids encode information by varying the identity of the nucleotide at each position in the polymer.

DNA and RNA

Nucleic acids come in two varieties, **deoxyribonucleic acid (DNA)** and **ribonucleic acid (RNA),** both polymers of nucleotides (figure 3.28). RNA is a long, single strand of nucleotides and is used by cells in making proteins using genetic instructions encoded within DNA. DNA consists of *two* nucleotide strands wound around each other in a **double helix,** like strands of a pearl necklace twisted together (figure 3.29).

The Double Helix

Why is DNA a *double* helix? When scientists looked carefully at the structure of the DNA double helix, they found that the nitrogenous bases of the two chains projected inward from the sugar–phosphate backbone, the bases of each chain pointed toward the other. The bases of the two chains are linked in the middle of the molecule by hydrogen bonds, like two columns of people holding hands across. The key to understanding why DNA is a double helix is revealed by looking at the bases: *only two base pairs are possible.* Two big bases cannot pair together—the combination is simply too bulky to fit; similarly, two little ones cannot, as they pinch the helix inward too much. To form a double helix, it is necessary to pair a big base with a little one. *In every DNA double helix, adenine (A) pairs with thymine (T) and guanine (G) pairs with cytosine (C)* (figure 3.30). In case you're wondering, the reason A doesn't pair with C and G doesn't pair with T is that these base pairs cannot form proper hydrogen bonds—the electron-sharing atoms are not pointed at each other.

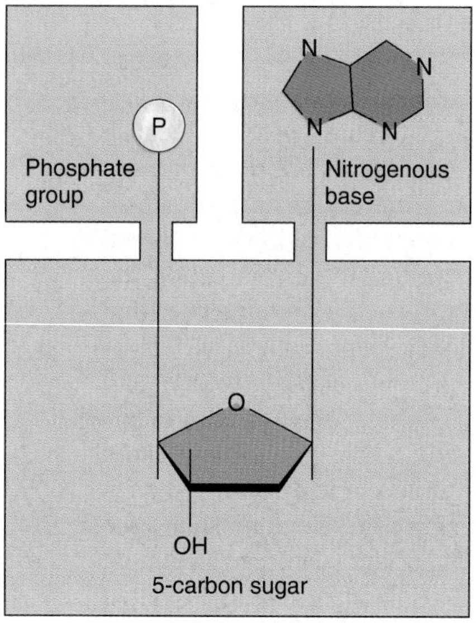

P
Phosphate group
N
N
N
N
Nitrogenous base
O
OH
5-carbon sugar

Figure 3.27 The structure of a nucleotide.
Nucleotides are composed of three parts: a five-carbon sugar, a phosphate group, and an organic nitrogenous base.

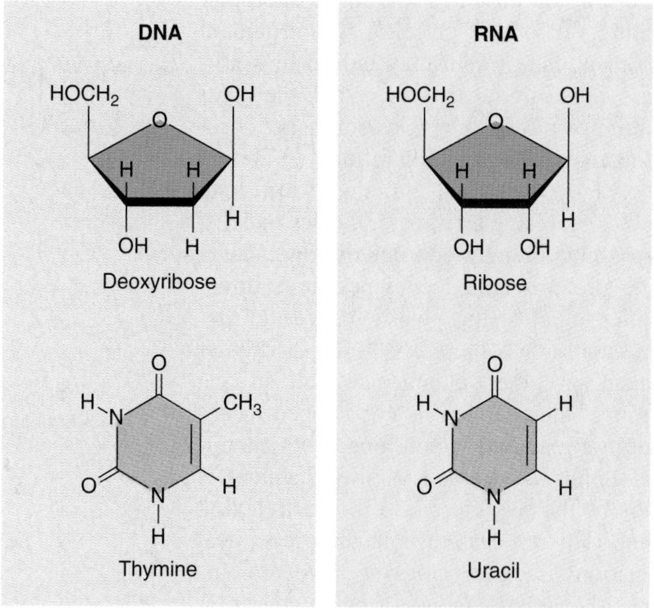

DNA
HOCH2
OH
O
H
H
H
OH
H
Deoxyribose
O
H
N
CH3
O
N
H
H
Thymine
RNA
HOCH2
OH
O
H
H
H
OH
OH
Ribose
O
H
N
H
O
N
H
H
Uracil

Figure 3.28 How DNA differs from RNA.
DNA is similar in structure to RNA but with two major chemical differences: (1) Both contain ribose (five-carbon) sugars, but in DNA one of the sugar's hydroxyl (–OH) groups is replaced by a hydrogen. (That is why DNA is called *deoxyribo*nucleic acid.) (2) One of the four organic bases of DNA, thymine, is changed slightly in RNA by the removal of a $-CH_3$ group and is called uracil.

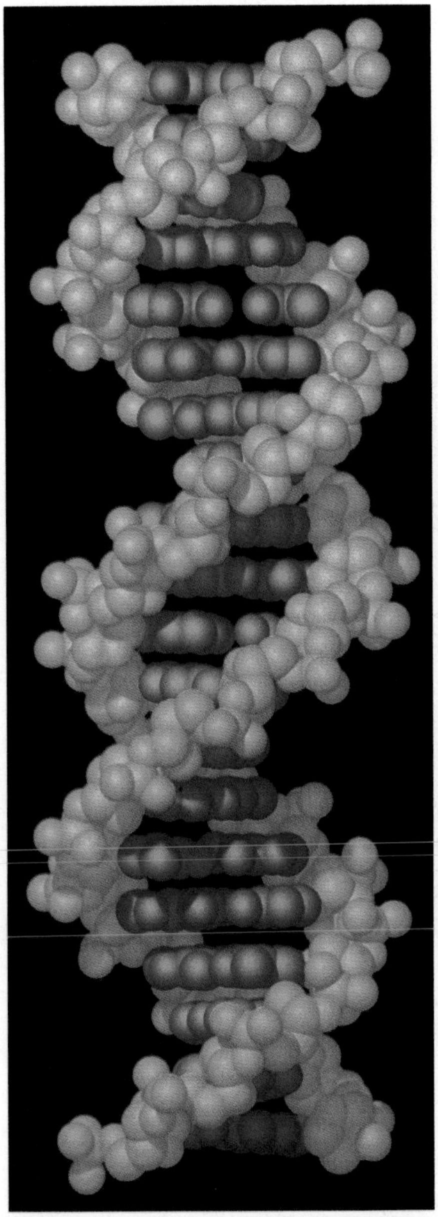

Figure 3.29 The DNA double helix.
The DNA molecule is composed of two nucleotide chains twisted together to form a double helix.

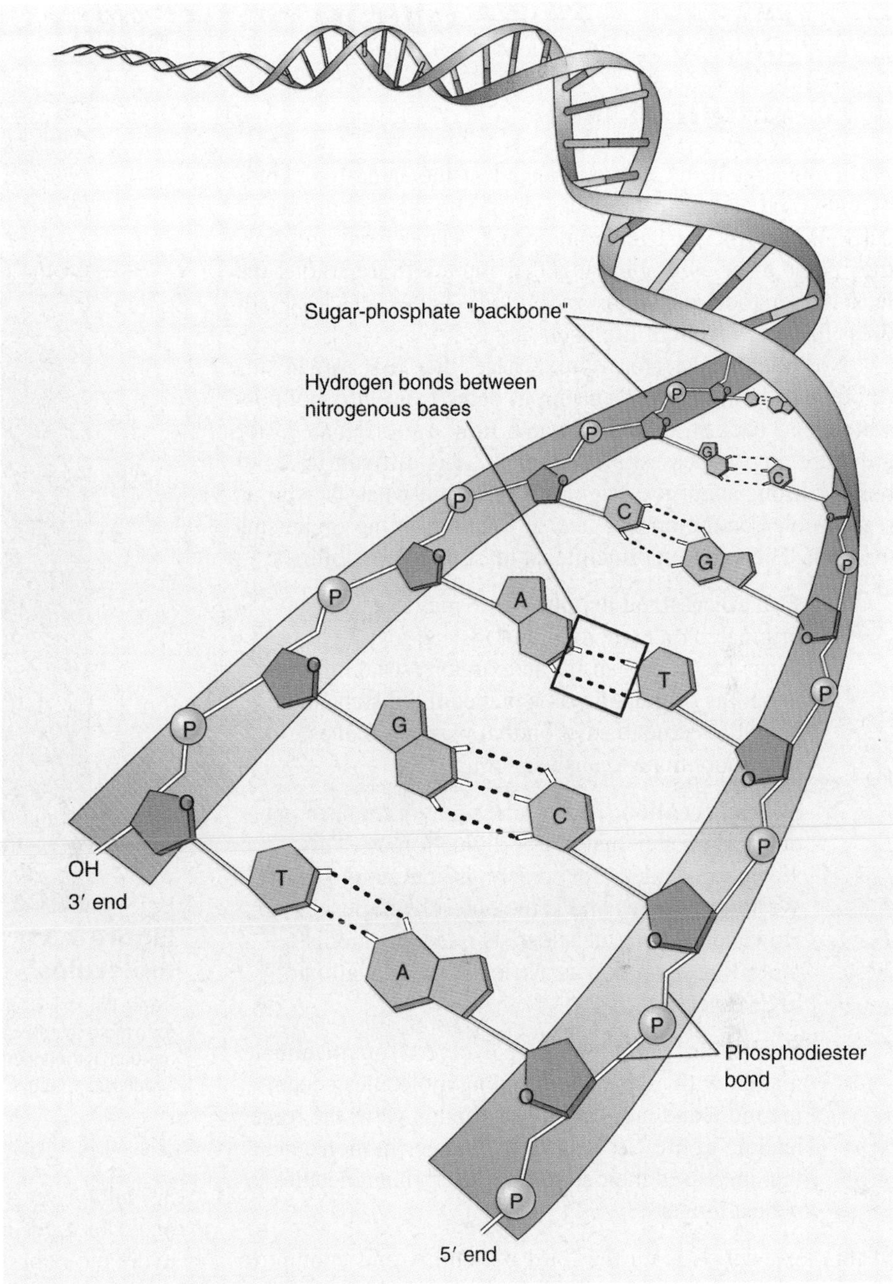
Sugar-phosphate "backbone"
Hydrogen bonds between nitrogenous bases
P
G
C
A
T
OH
3′ end
5′ end
Phosphodiester bond

Figure 3.30 The structure of DNA.
The two chains of the double helix are joined by hydrogen bonds between A–T and G–C base pairs.

The simple A–T, G–C pairs within the DNA double helix allow the cell to copy the information in a very simple way. It just unzips the helix and adds the matching bases to each strand! That is the great advantage of a double helix—it actually contains two copies of the information, one the mirror image of the other. If the sequence of one chain is ATTGCAT, the sequence of its partner in the double helix *must* be TAACGTA. The fidelity with which hereditary information is passed from one generation to the next is a direct result of this simple double-entry bookkeeping, which makes accurate copying of the genetic message possible.

3.11 Nucleic acids like DNA are long chains of the nucleotides A, T, G, and C. The sequence of the nucleotides specifies the amino acid sequence of proteins.

3.12 Origin of Life

All living organisms are constructed of the same four kinds of macromolecules just discussed, the bricks and mortar of cells. Where the first macromolecules came from and how they came to be assembled together into cells are among the least understood questions in biology—questions that address the very origin of life itself.

No one knows for sure where the first organisms (thought to be like today's bacteria) came from. It is not possible to go back in time and watch how life originated, nor are there any witnesses. Nevertheless, it is difficult to avoid being curious about the origin of life, about what, or who, is responsible for the appearance of the first living organisms on earth. There are, in principle, at least three possibilities:

1. **Extraterrestrial origin.** Life may not have originated on earth at all but instead may have been carried to it, perhaps as an extraterrestrial infection of spores originating on a planet of a distant star. How life came to exist on that planet is a question we cannot hope to answer soon.
2. **Special creation.** Life-forms may have been put on earth by supernatural or divine forces. This viewpoint, called creationism, is common to most Western religions and is the oldest hypothesis. However, almost all scientists reject creationism, preferring evolution as a scientific explanation of life's diversity.
3. **Evolution.** Life may have evolved from inanimate matter, with associations among molecules becoming more and more complex. In this view, the force leading to life was selection; changes in molecules that increased their stability caused the molecules to persist longer.

In this text we focus on the third possibility and attempt to understand whether the forces of evolution could have led to the origin of life and, if so, how the process might have occurred. This is not to say that the third possibility, evolution, is definitely the correct one. Any one of the three possibilities might be true. Nor does the third possibility preclude religion: a divine agency might have acted via evolution. Rather, we are limiting the scope of our inquiry to scientific matters. Of the three possibilities, only the third permits testable hypotheses to be constructed and so provides the only scientific explanation—that is, one that could potentially be disproved by experiment.

Forming Life's Building Blocks

How can we learn about the origin of the first cells? One way is to try to reconstruct what the earth was like when life originated

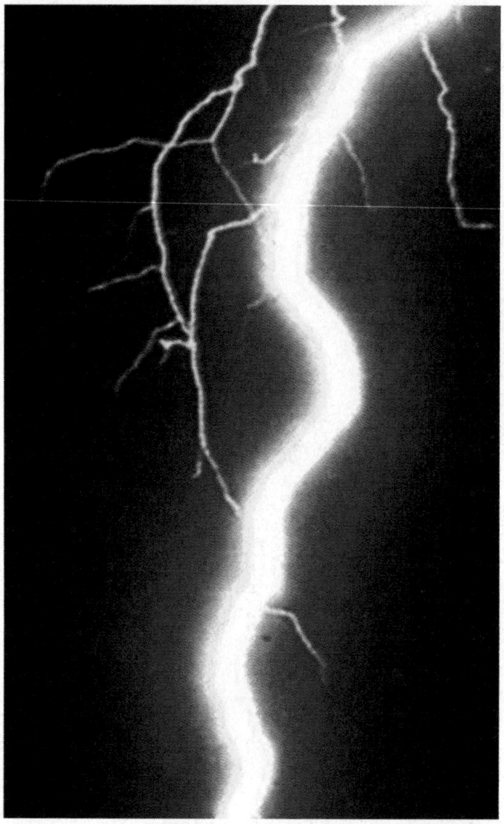

Figure 3.31 Lightning provides energy to form molecules.

Before life evolved, the simple molecules in the earth's atmosphere combined to form more complex molecules. The energy that drove some of these chemical reactions is thought to have come from UV radiation, lightning, and other forms of geothermal energy.

3.5 billion years ago. We know from rocks that there was little or no oxygen in the earth's atmosphere then and more of the hydrogen-rich gases hydrogen sulfide (SH_2), ammonia (NH_3), and methane (CH_4). Electrons in these gases would have been frequently pushed to higher energy levels by photons crashing into them from the sun or by electrical energy in lightning (figure 3.31). Today, high-energy electrons are quickly soaked up by the oxygen in earth's atmosphere (air is 21% oxygen, all of it contributed by photosynthesis) because oxygen atoms have a great "thirst" for such electrons. But in the absence of oxygen, high-energy electrons would have been free to help form biological molecules.

When the scientists Stanley Miller and Harold Urey reconstructed the oxygen-free atmosphere of the early earth in their laboratory and subjected it to the lightning and UV radiation it would have experienced then, they found that many of the building blocks of organisms, such as amino acids and nucleotides, formed spontaneously. They concluded that life

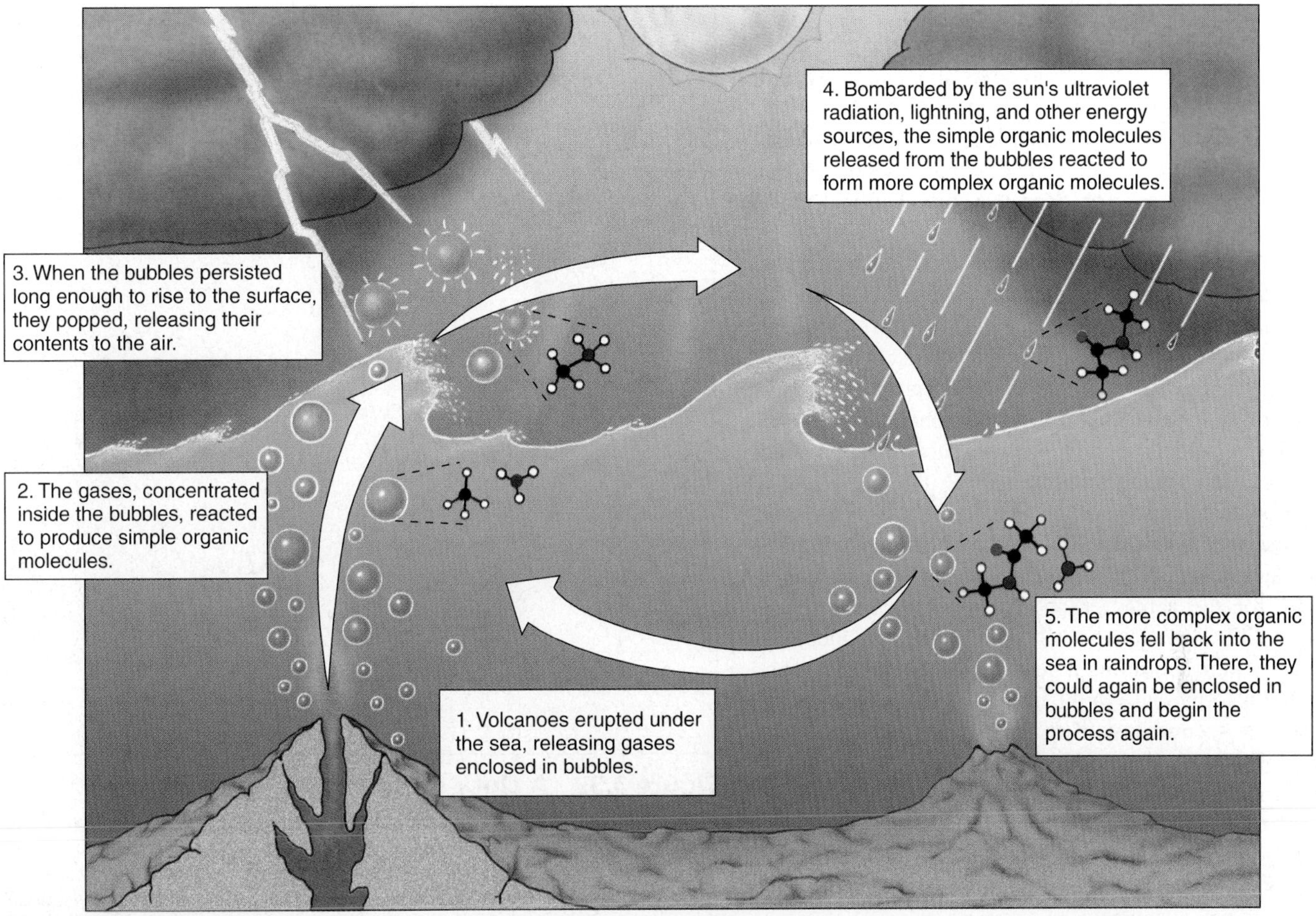
4. Bombarded by the sun's ultraviolet radiation, lightning, and other energy sources, the simple organic molecules released from the bubbles reacted to form more complex organic molecules.
3. When the bubbles persisted long enough to rise to the surface, they popped, releasing their contents to the air.
2. The gases, concentrated inside the bubbles, reacted to produce simple organic molecules.
5. The more complex organic molecules fell back into the sea in raindrops. There, they could again be enclosed in bubbles and begin the process again.
1. Volcanoes erupted under the sea, releasing gases enclosed in bubbles.

Figure 3.32 A chemical process involving bubbles may have preceded the origin of life.
In 1986, geophysicist Louis Lerman proposed that the chemical processes leading to the evolution of life took place within bubbles on the ocean's surface.

may have evolved in a "primordial soup" of biological molecules formed in the ancient earth's oceans.

Recent discoveries of 3.5-billion-year-old fossils have caused scientists to reevaluate the primordial soup hypothesis. This allows less than a half-billion years for life to evolve after the molten earth cooled enough to possess oceans. Also, if the earth's atmosphere had no oxygen 4 billion years ago, as Miller and Urey assumed (and most evidence supports this assumption), then there would have been no protective layer of ozone to shield the earth's surface from the sun's damaging UV radiation. Without an ozone layer, scientists think UV radiation would have destroyed any ammonia and methane present in the atmosphere. When these gases are missing, the **Miller-Urey experiment** does not produce key biological molecules such as amino acids. If the necessary ammonia and methane were not in the atmosphere, where were they?

In the last decade, support has grown among scientists for what has been called the **bubble model.** This model suggests that the problems with the primordial soup hypothesis disappear if the model is "stirred up" a bit. The bubble model proposes that the key chemical processes generating the building blocks of life took place not in a primordial soup but rather within bubbles on the ocean's surface (figure 3.32). Bubbles produced by wind, wave action, the impact of raindrops, and the eruption of volcanoes cover about 5% of the ocean's surface at any given time. Because water molecules are polar, water bubbles tend to attract other polar molecules, in effect concentrating them within the bubbles. This solves two key problems with the primordial soup hypothesis. First, chemical reactions would proceed much faster in bubbles, where polar reactants would be concentrated, and so life could have originated in a much shorter period of time. Second, inside the bubbles, the methane and ammonia required to produce amino acids would have been protected from destruction by UV radiation.

3.12 Life appeared on earth 3.5 billion years ago. It may have arisen spontaneously, although the nature of the process is not clearly understood.

3.13 How Cells Arose

It is one thing to make amino acids spontaneously and quite another to link them together into proteins. Recall from figure 3.25 that making a peptide bond involves producing a molecule of water as one of the products of the reaction. Because this chemical reaction is freely reversible, it should not occur spontaneously in water (an excess of water would push it in the opposite direction). Scientists now suspect that the first macromolecules to form were not proteins but RNA molecules. When "primed" with high-energy phosphate groups (available in many minerals), RNA nucleotides spontaneously form polynucleotide chains that might, folded up, have been capable of catalyzing the formation of the first proteins.

Not everyone accepts the hypothesis that life evolved spontaneously. Those who object say that proteins and RNA could never have assembled spontaneously, for the same reason that shaking a bunch of empty soft drink cans in a box doesn't spontaneously cause them to jump into a neat stack—disorder, not order, tends to increase in the universe. This general rule, called the **second law of thermodynamics,** is a basic principle of chemistry and physics. Does the theory of spontaneous origin violate the second law of thermodynamics? Not at all. The second law of thermodynamics applies only to closed systems (ones in which no energy enters or leaves), while earth and its organisms are open systems.

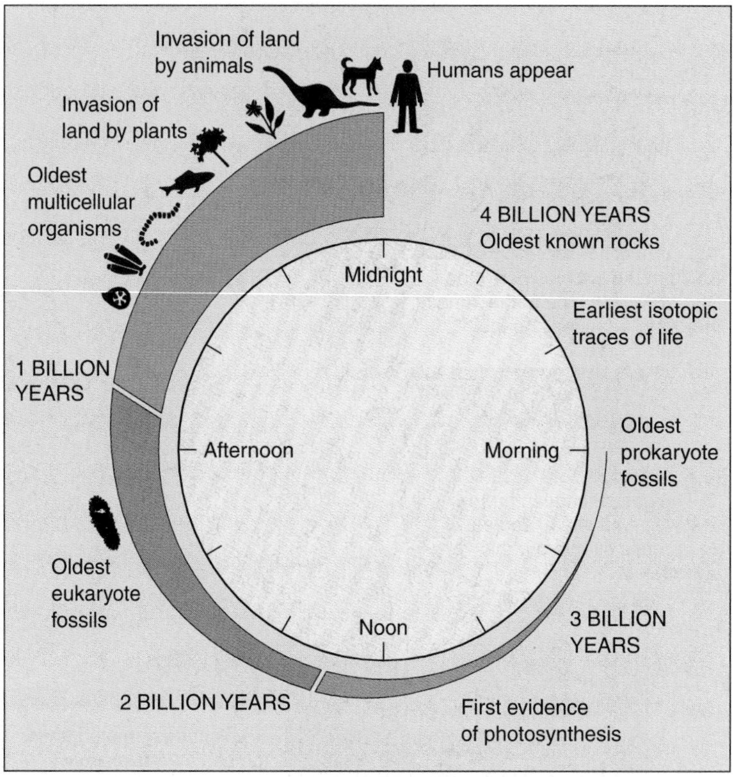
Invasion of land by animals
Humans appear
Invasion of land by plants
Oldest multicellular organisms
4 BILLION YEARS
Oldest known rocks
Midnight
Earliest isotopic traces of life
1 BILLION YEARS
Oldest prokaryote fossils
Afternoon
Morning
Oldest eukaryote fossils
Noon
3 BILLION YEARS
2 BILLION YEARS
First evidence of photosynthesis

Figure 3.33 A clock of biological time.

A billion seconds ago, most students using this text had not yet been born. A billion minutes ago, Jesus was alive and walking in Galilee. A billion hours ago, the first human had not been born. A billion days ago, no biped walked on earth. A billion months ago, the last dinosaurs had not yet been hatched. A billion years ago, no creature had ever walked on the surface of the earth.

The First Cells

We don't know how the first cells formed, but most scientists suspect they aggregated spontaneously. When complex carbon-containing macromolecules are present in water, they tend to gather together, much as the people from the same foreign country tend to aggregate within a large city. Sometimes the aggregations form a cluster big enough to see. Try vigorously shaking a bottle of oil-and-vinegar salad dressing—tiny microspheres form spontaneously, suspended in the vinegar. Similar microspheres might have represented the first step in the evolution of cellular organization. Such microspheres have many cell-like properties—their outer boundary resembles the skin of a cell in that it has two layers, and the microspheres can grow and divide. Over millions of years, those **microdrops** better able to incorporate molecules and energy would have tended to persist longer than others, and when a means occurred to transfer these improvements from parent microdrop to offspring, heredity—and life—began.

When we speak of it having taken millions of years for a cell to develop, it is hard to believe there would be enough time for an organism as complicated as a human to develop. But in the scheme of things, human beings are recent additions. If we look at the development of living organisms as a 24-hour clock of biological time (figure 3.33), with the formation of the earth 4.5 billion years ago being midnight, humans do not appear until the day is almost all over, only minutes before its end.

As you can see, the scientific vision of life's origin is at best a hazy outline. While scientists cannot disprove the hypothesis that life originated naturally and spontaneously, little is known about what actually happened. Many different scenarios seem possible, and some have solid support from experiments. Deep-sea hydrothermal vents are an interesting possibility; the bacteria populating these vents are among the most primitive of living organisms. Other researchers have proposed that life originated deep in the earth's crust.

Because we know so little about how DNA, RNA, and hereditary mechanisms first developed, science is currently unable to resolve disputes concerning the origin of life. How life might have originated naturally and spontaneously remains a subject of intense interest, research, and discussion among scientists.

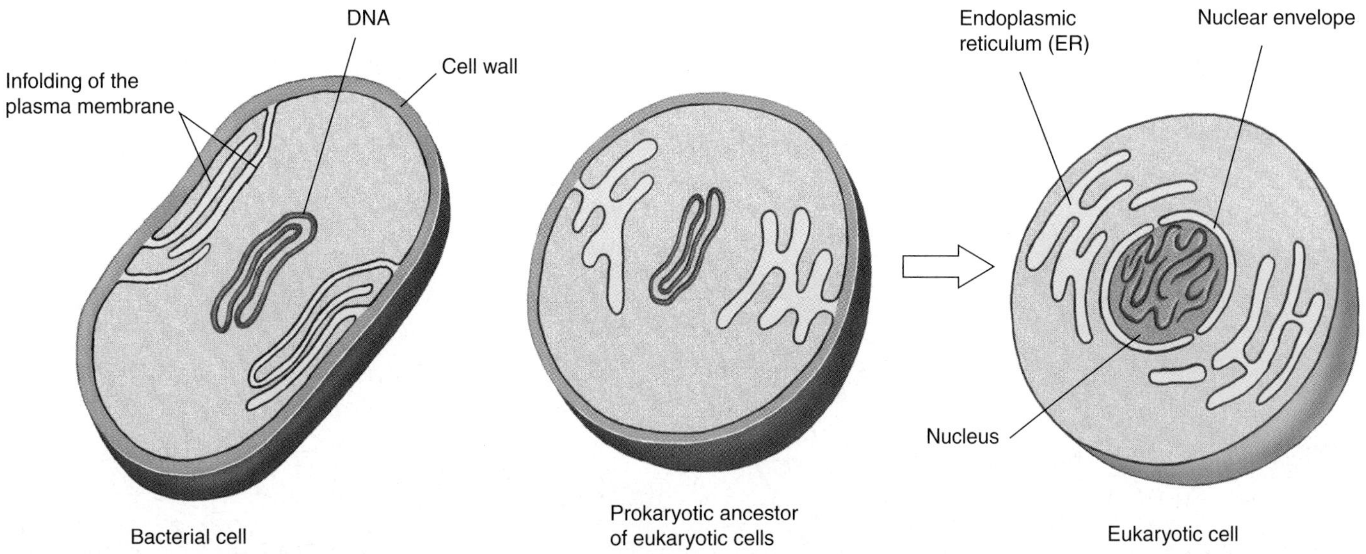
DNA
Cell wall
Infolding of the plasma membrane
Endoplasmic reticulum (ER)
Nuclear envelope
Nucleus
Bacterial cell
Prokaryotic ancestor of eukaryotic cells
Eukaryotic cell

Figure 3.34 Origin of the nucleus and endoplasmic reticulum.
Many bacteria today have infoldings of the plasma membrane. The eukaryotic internal membrane system called the endoplasmic reticulum (ER) and the nuclear envelope may have evolved from such infoldings of the plasma membrane encasing prokaryotic cells that gave rise to eukaryotic cells.

Origin of Eukaryotic Cells

All fossils more than 1.7 billion years old are small, simple cells, similar to the bacteria of today. In rocks about 1.7 billion years old, we begin to see the first microfossils that are noticeably larger than bacteria and have internal membranes and thicker walls. A new kind of organism had appeared, called a **eukaryote,** from the Greek words for "true" and "nucleus," because eukaryotic cells possess an internal structure called a nucleus. Bacteria, by contrast, are called **prokaryotes** ("before the nucleus"). All organisms other than bacteria are eukaryotes.

Many bacteria have infoldings of their outer membranes extending into the interior that serve as passageways to the surface. The network of internal membranes in eukaryotes called the endoplasmic reticulum (ER) is thought to have evolved from such infoldings, as is the nuclear envelope (figure 3.34).

The **endosymbiotic theory,** now widely accepted, suggests that at a critical stage in the evolution of eukaryotic cells, energy-producing bacteria came to reside symbiotically (that is, cooperatively) within larger early eukaryotic cells, eventually evolving into the cell organelles we now know as mitochondria. Similarly, photosynthetic bacteria came to live within other early eukaryotic cells, leading to the evolution of chloroplasts, the photosynthetic organelles of plants and algae. Present-day mitochondria and chloroplasts contain their own DNA, which is remarkably similar to the DNA of bacteria in size and character.

Sexual Reproduction

Eukaryotic cells also possess the ability to reproduce sexually, something prokaryotes cannot do effectively. **Sexual reproduction** is the process of producing offspring, with two copies of each chromosome, by fertilization, the union of two cells that each have one copy of each chromosome. The great advantage of sexual reproduction is that it allows for frequent genetic recombination, which generates the variation that is the raw material for evolution. Not all eukaryotes reproduce sexually, but most have the capacity to do so. The evolution of sexual reproduction led to the tremendous explosion of diversity among the eukaryotes.

Multicellularity

Diversity was also promoted by the development of **multicellularity.** Some single eukaryotic cells began living in association with others, in colonies. Eventually individual members of the colony began to assume different duties, and the colony began to take on the characteristics of a single individual. Multicellularity has arisen many times among the eukaryotes. Practically every organism big enough to see with the unaided eye is multicellular, including all animals and plants. The great advantage of multicellularity is that it fosters specialization; some cells devote all of their energies to one task, other cells to another. Few innovations have had as great an impact on the history of life as the specialization made possible by multicellularity.

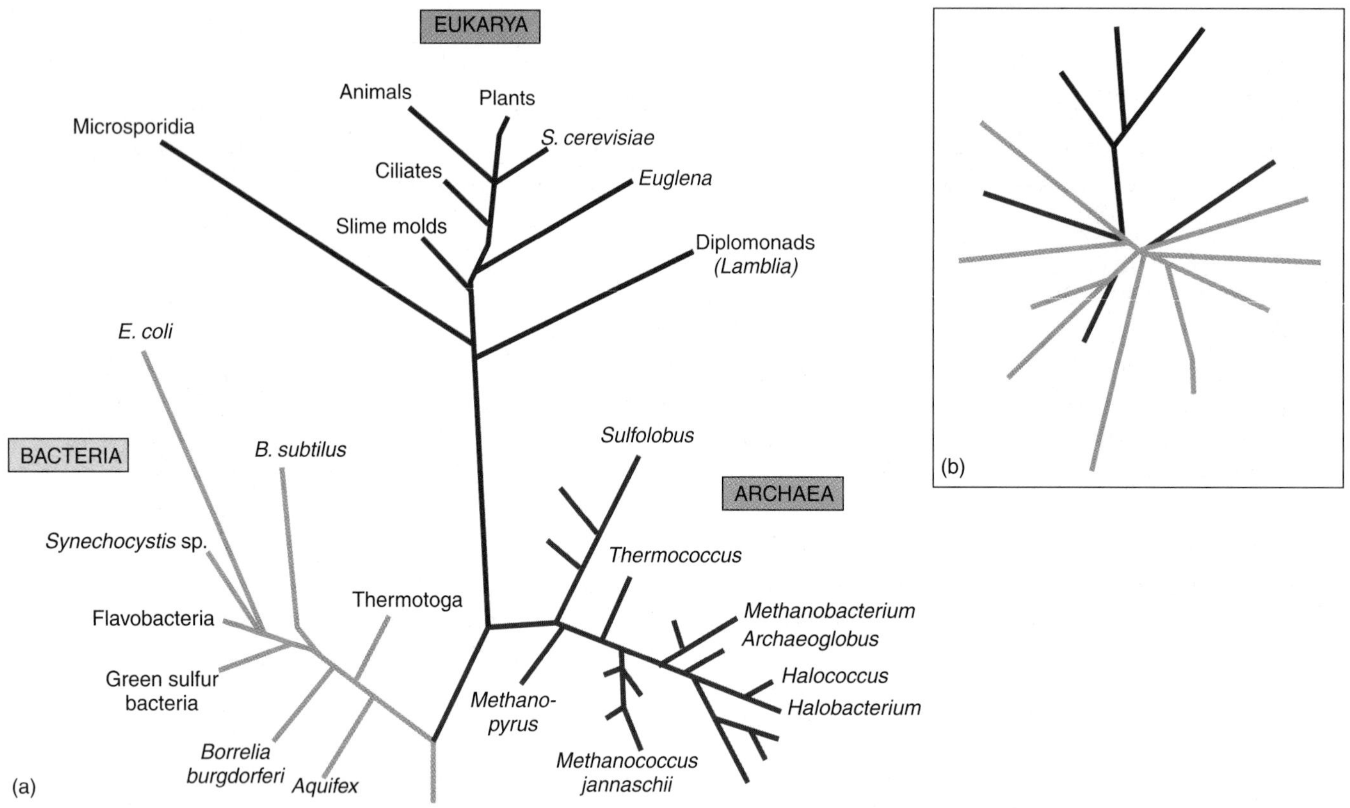
EUKARYA
Animals
Plants
Microsporidia
S. cerevisiae
Ciliates
Euglena
Slime molds
Diplomonads
(Lamblia)
E. coli
BACTERIA
B. subtilus
Sulfolobus
ARCHAEA
Synechocystis sp.
Thermococcus
Thermotoga
Methanobacterium
Flavobacteria
Archaeoglobus
Halococcus
Green sulfur
bacteria
Methano-
pyrus
Halobacterium
Borrelia
burgdorferi
Aquifex
Methanococcus
jannaschii
(a)
(b)

Figure 3.35 The three domains of life.

The kingdoms Archaebacteria and Eubacteria are as different from each other as from eukaryotes, so biologists have assigned them a higher category, a "domain." (*a*) A three-domain tree of life based on ribosomal RNA consists of the Eukarya, Bacteria, and Archaea. (*b*) New analyses of complete genome sequences contradict the rRNA tree, and suggest other arrangements such as this one, which splits the Archaea. Apparently genes hopped from branch to branch as early organisms either stole genes from their food or swapped DNA with their neighbors, even distantly related ones.

The Kingdoms of Life

Confronted with the great diversity of life on earth today, biologists have attempted to categorize similar organisms to better understand them, giving rise to the science of taxonomy. In later chapters, we will discuss taxonomy and classification in detail, but for now we can generalize that all living things fall into one of six kingdoms:

Kingdom Archaebacteria: Prokaryotes that lack a peptidoglycan cell wall, including the methanogens and extreme halophiles and thermophiles.

Kingdom Eubacteria: Prokaryotic organisms with a peptidoglycan cell wall, including cyanobacteria, soil bacteria, nitrogen-fixing bacteria, and pathogenic (disease-causing) bacteria.

Kingdom Protista: Eukaryotic, primarily unicellular (although algae are multicellular), photosynthetic or heterotrophic organisms, such as amoebas and paramecia.

Kingdom Fungi: Eukaryotic, mostly multicellular (although yeasts are unicellular), heterotrophic, usually nonmotile organisms, with cell walls of chitin, such as mushrooms.

Kingdom Plantae: Eukaryotic, multicellular, nonmotile, usually terrestrial, photosynthetic organisms, such as trees, grasses, and mosses.

Kingdom Animalia: Eukaryotic, multicellular, motile, heterotrophic organisms, such as sponges, spiders, newts, penguins, and humans.

As more is learned about living things, particularly from the newer evidence that DNA studies provide, scientists will continue to reevaluate the relationships among the kingdoms of life (figure 3.35).

3.13 For at least the first 2 billion years of life on earth, all organisms were bacteria. About 1.7 billion years ago, the first eukaryotes appeared. Biologists place living organisms into six general categories called kingdoms.

3.14 Has Life Evolved Elsewhere?

We should not overlook the possibility that life processes might have evolved in different ways on other planets. A functional genetic system, capable of accumulating and replicating changes and thus of adaptation and evolution, could theoretically evolve from molecules other than carbon, hydrogen, nitrogen, and oxygen in a different environment. Silicon, like carbon, needs four electrons to fill its outer energy level, and ammonia is even more polar than water. Perhaps under radically different temperatures and pressures, these elements might form molecules as diverse and flexible as those carbon has formed on earth.

The universe has 10^{20} (100,000,000,000,000,000,000) stars similar to our sun. We don't know how many of these stars have planets, but it seems increasingly likely that many do. Since 1996, astronomers have been detecting planets orbiting distant stars. At least 10% of stars are thought to have planetary systems. If only 1 in 10,000 of these planets is the right size and at the right distance from its star to duplicate the conditions in which life originated on earth, the "life experiment" will have been repeated 10^{15} times (that is, a million billion times). It does not seem likely that we are alone.

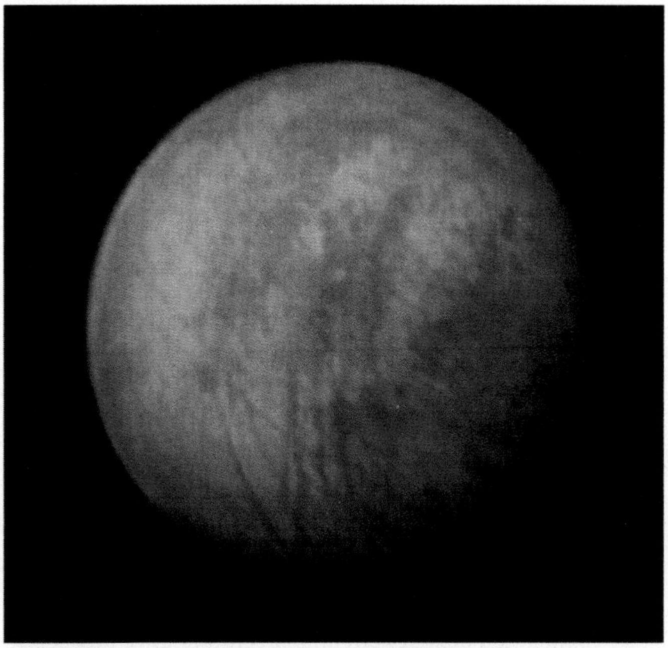

Figure 3.36 Is there life elsewhere?
Currently the most likely candidate for life elsewhere within the solar system is Europa, one of the many moons of the large planet Jupiter.

Ancient Bacteria on Mars?

A dull gray chunk of rock collected in 1984 in Antarctica ignited an uproar about ancient life on Mars with the report that the rock contains evidence of possible life. Analysis of gases trapped within small pockets of the rock indicate it is a meteorite from Mars. It is, in fact, the oldest rock known to science—fully 4.5 billion years old. Back then, when this rock formed on Mars, that cold, arid planet was much warmer, flowed with water, and had a carbon dioxide atmosphere—conditions not too different from those that spawned life on earth.

When examined with powerful electron microscopes, carbonate patches within the meteorite exhibit what look like microfossils, some 20 to 100 nanometers in length. One hundred times smaller than any known bacteria, it is not clear they actually are fossils, but the resemblance to bacteria is striking.

Viewed as a whole, the evidence of bacterial life associated with the Mars meteorite is not compelling. Clearly, more painstaking research remains to be done before the discovery can claim a scientific consensus. However, while there is no conclusive evidence of bacterial life associated with this meteorite, it seems very possible that life has evolved on other worlds in addition to our own.

Deep Sea Vents

The possibility that life on earth actually originated in the vicinity of deep-sea hydrothermal vents is gaining popularity. At the bottom of the ocean, where these vents spewed out a rich froth of molecules, the geological turbulence and radioactive energy battering the land was absent, and things were comparatively calm. The thermophilic archaebacteria found near these vents today are the most ancient group of organisms living on earth. Perhaps the gentler environment of the ocean depths was the actual cradle of life.

Other Planets

Has life evolved on other worlds within our solar system? There are planets other than ancient Mars with conditions not unlike those on earth. Europa, a large moon of Jupiter, is a promising candidate (figure 3.36). Europa is covered with ice, and photos taken in close orbit in the winter of 1998 reveal seas of liquid water beneath a thin skin of ice. Additional satellite photos taken in 1999 suggest that a few miles under the ice lies a liquid ocean of water larger than earth's, warmed by the push and pull of the gravitational attraction of Jupiter's many large satellite moons. The conditions on Europa now are far less hostile to life than the conditions that existed in the oceans of the primitive earth. In coming decades, satellite missions are scheduled to explore this ocean for life.

3.14 There are so many stars that life may have evolved many times. Although evidence for life on Mars is not compelling, the seas of Europa offer a promising candidate that scientists are eager to investigate.

3.15 Evolution's Critics

The theory that life on earth arose spontaneously and evolved into the forms living today is accepted by most, but not all, biologists. Some biologists, and many nonscientists, prefer a religious explanation of life's origin. Critics of evolution raise seven principal objections to evolution:

1. **Evolution is not solidly demonstrated.** *"Evolution is just a theory,"* critics point out, as if theory meant lack of knowledge, some kind of guess. Scientists, however, use the word theory in a very different sense than the general public does. Theories are the solid ground of science, that of which we are most certain. Few of us doubt the theory of gravity because it is "just a theory."
2. **There are no fossil intermediates.** *"No one ever saw a fin on the way to becoming a leg,"* critics claim, pointing to the many gaps in the fossil record in Darwin's day. Since then, however, most fossil intermediates in vertebrate evolution have indeed been found. A clear line of fossils now traces the transition between whales and hoofed mammals, between reptiles and mammals, and between apes and humans. The fossil evidence of evolution between major forms is compelling.
3. **The intelligent design argument.** *"The organs of living creatures are too complex for a random process to have produced—the existence of a clock is evidence of the existence of a clockmaker."* Biologists do not agree. The intermediates in the evolution of the mammalian ear can be seen in fossils, and many intermediate "eyes" are known in various invertebrates. These intermediate forms arose because they have value—being able to detect light a little is better than not being able to detect it at all. Complex structures like eyes evolved as a progression of slight improvements.
4. **Evolution violates the second law of thermodynamics.** *"A jumble of soda cans doesn't by itself jump neatly into a stack—things become more disorganized due to random events, not more organized."* Biologists point out that this argument ignores what the second law really says: disorder increases in a closed system, which the earth most certainly is not. Energy continually enters the biosphere from the sun, fueling life and all the processes that organize it.
5. **Proteins are too improbable.** *"Hemoglobin has 141 amino acids. The probability that the first one would be leucine is 1/20, and that all 141 would be the ones they are by chance is $(1/20)^{141}$, an impossibly rare event."* This is statistical foolishness—you cannot use probability to argue backward. The probability that a student in a classroom has a particular birthday is 1/365; arguing this way, the probability that everyone in a class of 50 would have the birthdays they do is $(1/365)^{50}$, and yet there the class sits.
6. **Natural selection does not imply evolution.** *"No scientist has come up with an experiment where fish evolve into frogs and leap away from predators."* Is microevolution (evolution within a species) the mechanism that has produced macroevolution (evolution among species)? Most biologists that have studied the problem think so. Some kinds of animals produced by man-made selection are remarkably distinctive, such as chihuahuas, dachshunds, and greyhounds. While all dogs are in fact the same species and can interbreed, laboratory selection experiments easily create forms that cannot interbreed and thus would in nature be considered different species. Thus, production of radically different forms has indeed been observed, repeatedly. To object that evolution still does not explain really major differences, like between fish and amphibians, simply takes us back to point 2—these changes take millions of years, and are seen clearly in the fossil record.
7. **The irreducible complexity argument.** *"The intricate molecular machinery of the cell cannot be explained by evolution from simpler stages. Because each part of a complex cellular process like blood clotting is essential to the overall process, how can natural selection fashion any one part?"* What's wrong with this argument is that each part of a complex molecular machine evolves as part of the system. Natural selection can act on a complex system because at every stage of its evolution, the system functions. Parts that improve function are added, and, because of later changes, become essential. The mammalian blood clotting system, for example, evolved from a much simpler system 600 million years ago, and is found today in lampreys, the most primitive fish. One hundred million years later, as vertebrates evolved, proteins were added to the clotting system making it sensitive to substances released from damaged tissues. Fifty million years later a third component was added, triggering clotting by contact with the jagged surfaces produced by injury. At each stage, the clotting system came to depend on the added elements, and thus has become "irreducibly complex."

3.15 The theory of evolution has proven controversial among the general public, although the commonly raised objections are without scientific merit.

CONCEPT REVIEW

1. Select the largest chemical structure.
   a. atom
   b. electron
   c. nucleus
   d. proton

2. Select the correct association.
   a. oxidation—gain of an electron
   b. oxidation—loss of an electron
   c. reduction—gain of a neutron
   d. reduction—loss of a neutron

3. An atom has five electrons in its outer orbit. To complete its outer orbit, it needs _________ electrons.
   a. two
   b. three
   c. four
   d. six

4. Which statement about the hydrogen bond is not true?
   a. It occurs with polar molecules.
   b. It is a weak bond.
   c. It is absent in water.
   d. It is found in proteins.

5. Capillary action is due to
   a. adhesion.
   b. cohesion.

6. Adding an acid to water _________ its pH.
   a. lowers
   b. raises

7. Select the smallest molecule.
   a. glycogen
   b. sucrose
   c. starch
   d. glucose

8. Amino acids are the subunits of
   a. carbohydrates.
   b. lipids.
   c. nucleic acids.
   d. proteins.

9. Each of the following was a molecule of the earth's early atmosphere except
   a. ammonia.
   b. hydrogen sulfide.
   c. methane.
   d. oxygen.

10. Table salt is built from _________ bonds.

11. About _________ of the earth's surface is covered by water.

12. About _________ of the human body consists of water.

13. Polar molecules attract by _________.

14. A _________ is a substance that has a pH greater than 7.

15. Oxygen is supplied to the earth by the process of _______.

CHALLENGE YOURSELF

1. Carbon (atomic number 6) and silicon (atomic number 14) both have vacancies for four electrons in their outer energy levels. Ammonia (NH_3) is even more polar than water. Why do you suppose life evolved into organisms composed of carbon chains in water solution rather than ones composed of silicon in ammonia?

2. Champagne, a carbonic acid buffer, has a pH of about 2. How can we drink such a strong acid?

3. Carbon atoms can share four electron pairs when forming molecules. Why do you suppose that carbon does not form a bimolecular gas, as hydrogen (one pair of shared electrons), oxygen (two pairs of shared electrons), and nitrogen (three pairs of shared electrons) do?

4. Why do long-distance runners eat complex carbohydrates (that is, starches) in preparation for athletic events?

5. Why do you suppose humans circulate the monosaccharide glucose in their blood, rather than employing a disaccharide such as sucrose as a transport sugar, as do plants?

3 eBRIDGE

Reinforcing Key Points

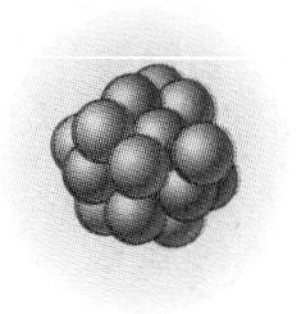

Some Simple Chemistry

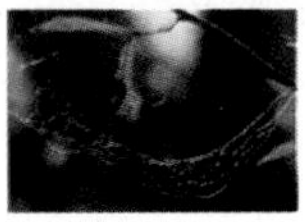

Water: Cradle of Life

Macromolecules

Origin of the First Cells

Electronic Learning

Visual Learning

Animations

- Atomic Structure
- Covalent Bond
- Ionic Bond
- DNA Structure

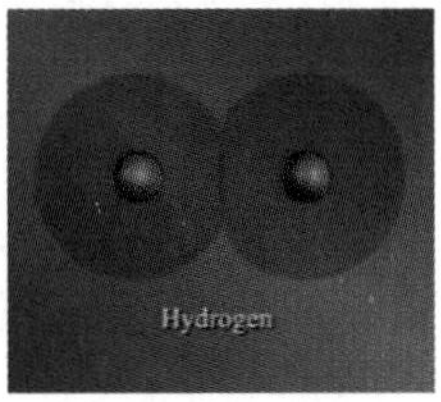
Hydrogen

Helping You Learn

- The Special Properties of Water
- Building Life: 4 Biologically Important Molecules

Explorations

Cell Chemistry: Thermodynamics

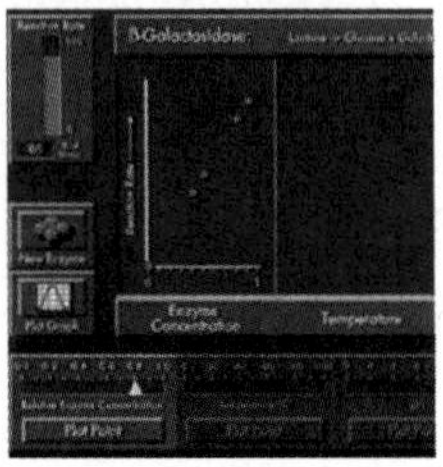

This interactive exercise allows you to explore the way in which reaction conditions affect how an enzyme catalyzes a chemical reaction, focusing on the key roles of enzyme concentration, temperature, and pH.

Author's Corner

Mad Cow Disease. We are accustomed to thinking of proteins as enzymes and structural macromolecules. It has come as something of a shock to discover that a protein can be responsible for infectious disease. Called "prions," these very stable proteins are misfolded versions of normal brain proteins that have the unfortunate ability to induce others of their kind to similarly misfold. A chain reaction of misfolding results, eventually destroying the brain. An epidemic spread among cows by feeding them protein supplements prepared from infected animals, and the prions responsible for mad cow disease have now spread to humans who have eaten infected cows.

1. Mad cows and prions.
2. Do prions threaten our blood supply?
3. The growing epidemic of mad cow disease.
4. Mad deer disease.
5. The mad cow disease epidemic spreads to Europe.

Virtual Classroom

The Search for Extraterrestrial Life

The universe has 10^{20} (100,000,000,000,000,000,000) stars similar to our sun. We don't know how many of these stars have planets, but it seems increasingly likely that many do. Astronomers now estimate that as many as 10% of stars have planetary systems. If only 1 in 10,000 of these planets is the right size and at the right distance from its star as earth, the "life experiment" will have been repeated a million billion times. It does not seem likely that we are alone. It is even possible that other planets orbiting our own sun have harbored life. Mars once was covered by oceans, although it is dry now. Mars meteorites have been claimed to exhibit fossil microbacteria, although the evidence is sketchy at best. Europa, one of the many moons of the large planet Jupiter, is covered with vast seas of liquid water beneath a thin skin of ice. The conditions on Europa now are far less hostile to life than the conditions that existed in the oceans of the primitive earth. In coming decades, satellite missions are scheduled to land on Europa, drill through the ice, and look for life.

Virtual Lab

Unraveling the Mystery of How Geckos Defy Gravity

Geckos exhibit amazing climbing abilities, strolling up walls and across ceilings. How do geckos perform this gripping feat? What force prevents gravity from dropping the gecko to the ground? Tracking down the answer, Kellar Autumn of Lewis & Clark College in Portland, Oregon, and Robert Full of the University of California, Berkeley, took a closer look at gecko feet. Geckos have rows of tiny hairs called setae on the bottoms of their feet. When you look at these hairs under the microscope (each only one-tenth the diameter of a human hair), the end of each seta is divided into 400 to 1,000 fine projections called spatulae. There are about half a million setae on each foot.

Autumn and Full put together an interdisciplinary team of scientists and set out to see if the setae were the adhesive structures responsible for the gecko's gripping abilities, by measuring the adhesive force produced by a single seta. To do this, they had to overcome two significant experimental challenges: isolating a single seta (which had never been done before) and accurately measuring such a very small force.

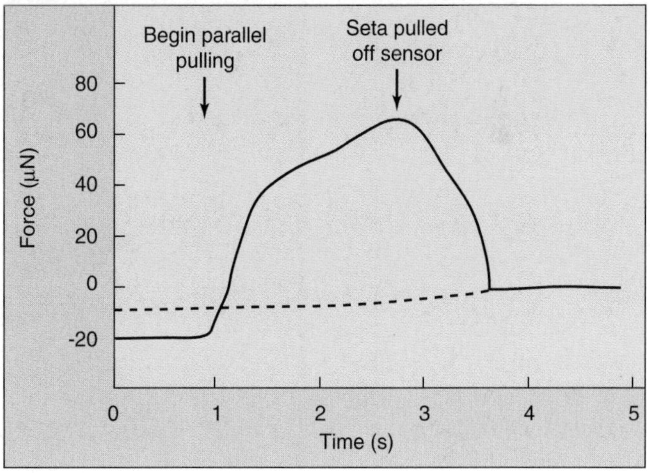
Begin parallel pulling
Seta pulled off sensor
Force (μN)
80
60
40
20
0
-20
0
1
2
3
4
5
Time (s)

Quizzes

Further Reading

Essential Study Partner

Links

BioCourse.com

CHAPTER 4

CELLS

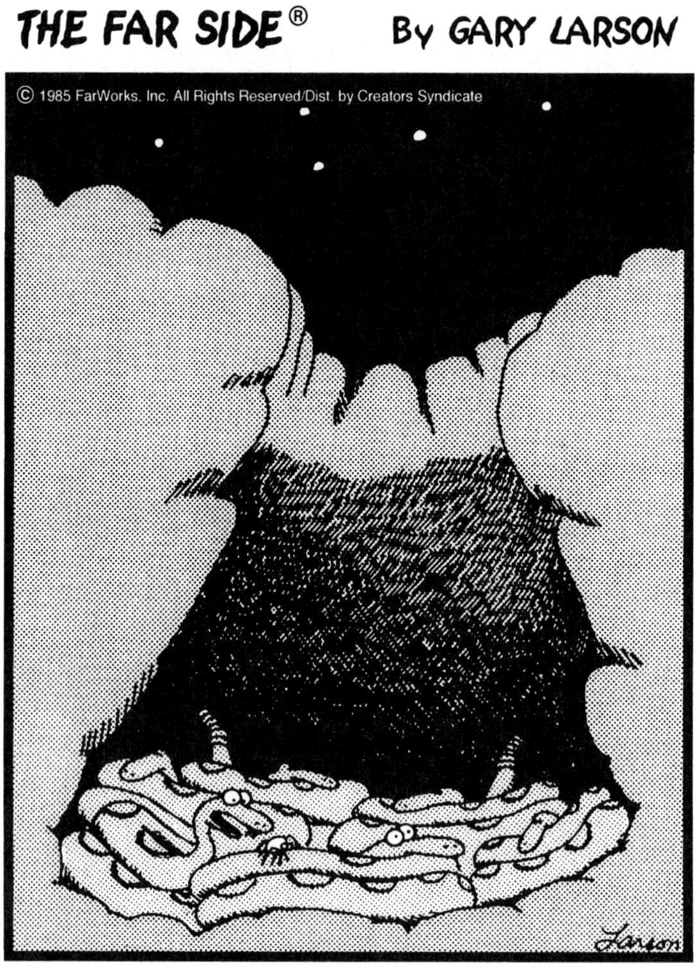
THE FAR SIDE® By GARY LARSON
© 1985 FarWorks, Inc. All Rights Reserved/Dist. by Creators Syndicate
Larson

"Doreen! There's a spider on you! One of those big, hairy, brown ones with the long legs that can move like the wind itself!"

CHAPTER 4

Cells

CHAPTER OVERVIEW

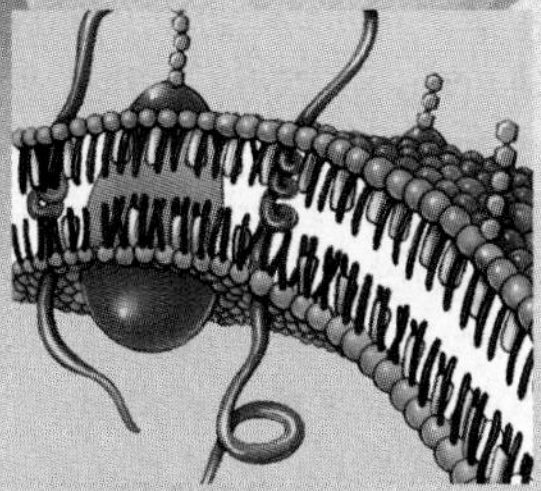

The World of Cells

- Small cells have relatively more surface with which to interact with the environment.
- The cell cytoplasm is bounded by the plasma membrane, a lipid bilayer. Protein-lined passageways through the bilayer allow chemicals to enter and leave the cell.

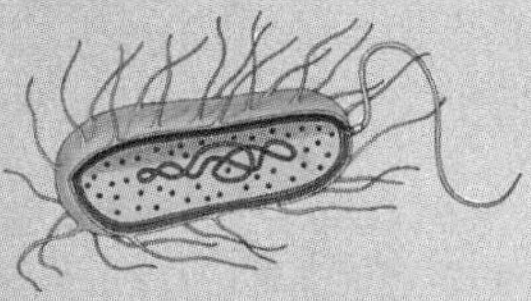

Kinds of Cells

- The simplest cells are bacteria, also called prokaryotes; they lack a nucleus and internal compartments.

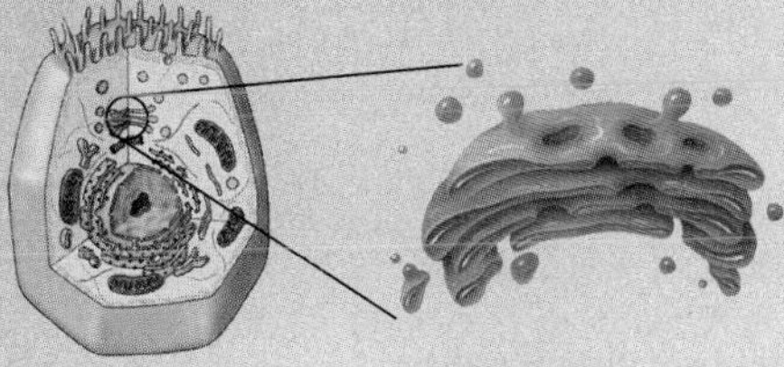

Tour of a Eukaryotic Cell

- The interior of eukaryotic cells is occupied by an extensive membrane system that creates many subcompartments within the cytoplasm.
- The cell's hereditary information, contained within chromosomes, is isolated within a membrane-bounded organelle called the nucleus.
- Cells carry out oxidative metabolism within bacteria-like organelles called mitochondria.

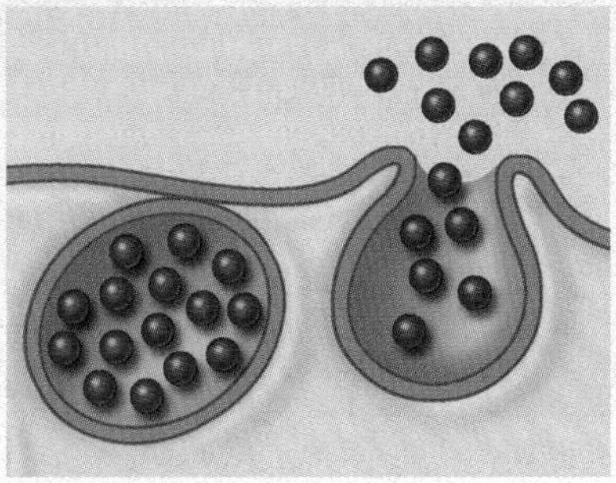

Transport Across Cell Membranes

- Molecules move randomly through the cell by diffusion.
- Osmosis occurs when water molecules are able to move through a membrane and other polar molecules are blocked.
- Selective transport of molecules can occur via facilitated diffusion or via active transport with the expenditure of metabolic energy.

4.1 Cells

Hold your hand up and look at it closely. What do you see? Skin. It looks solid and smooth, creased with lines and flexible to the touch. But if you were able to remove a bit and examine it under a microscope (figure 4.1), it would look very different—a sheet of tiny, irregularly shaped bodies crammed together like shingles on a roof. What you would see are epithelial cells; in fact, every tissue of your body is made of cells, as are the bodies of all organisms. Some organisms are composed of a single cell, and some are composed of many cells. All cells, however, are very small. In this chapter we look more closely at cells and learn something of their internal structure and how they communicate with their environment.

The Cell Theory

Because cells are so small, no one observed them until microscopes were invented in the mid-seventeenth century. Robert Hooke first described cells in 1665, when he used a microscope he had built to examine a thin slice of nonliving plant tissue called cork. Hooke observed a honeycomb of tiny, empty (because the cells were dead) compartments. He called the compartments in the cork *cellulae* (Latin, small rooms), and the term has come down to us as **cells.** For another century and a half, however, biologists failed to

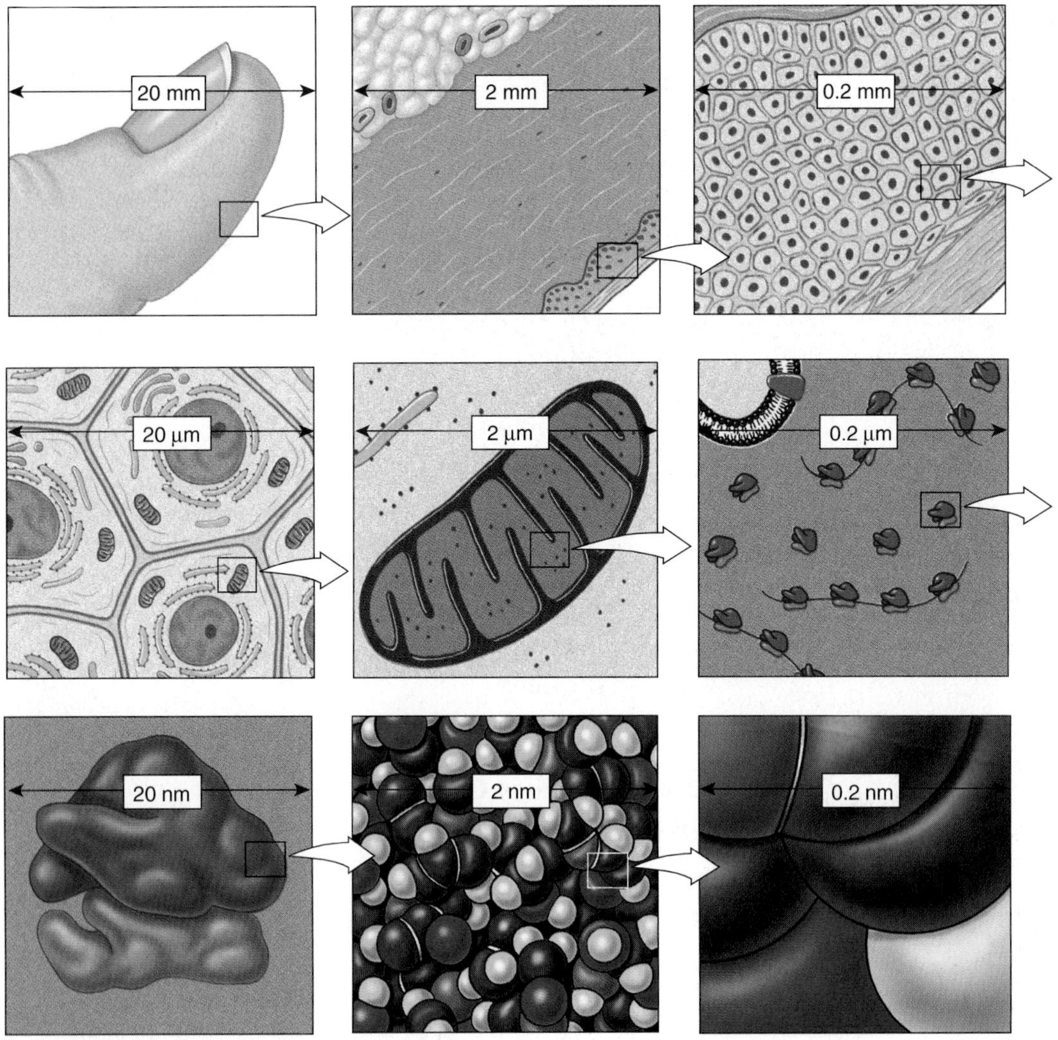
20 mm
2 mm
0.2 mm
20 μm
2 μm
0.2 μm
20 nm
2 nm
0.2 nm

Figure 4.1 The size of cells and their contents.
This diagram shows the size of human skin cells, organelles, and molecules. In general, the diameter of a human skin cell is 20 micrometers (μm), of a mitochondrion is 2 μm, of a ribosome is 20 nanometers (nm), of a protein molecule is 2 nm, and of an atom is 0.2 nm.

recognize the importance of cells. In 1838, botanist Matthias Schleiden made a careful study of plant tissues and developed the first statement of the cell theory. He stated that all plants "are aggregates of fully individualized, independent, separate beings, namely the cells themselves." In 1839, Theodor Schwann reported that all animal tissues also consist of individual cells.

The idea that all organisms are composed of cells is called the **cell theory.** In its modern form, the cell theory includes four principles:

1. All organisms are composed of one or more cells, within which the processes of life occur.
2. Cells are the smallest living things. Nothing smaller than a cell is considered alive.
3. Life evolved only once, 3.5 billion years ago. All organisms living today represent a continuous line of descent from those early cells.
4. Cells arise only by division of a previously existing cell.

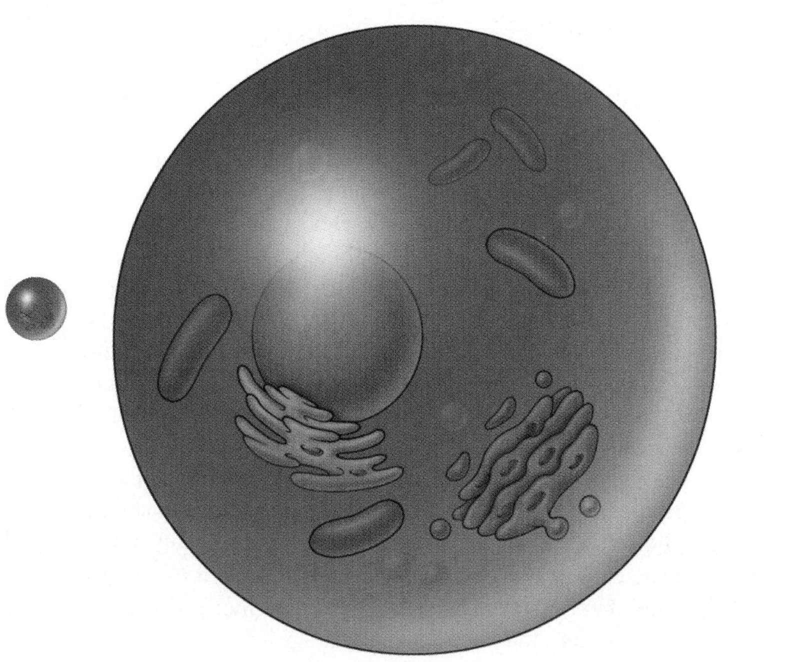

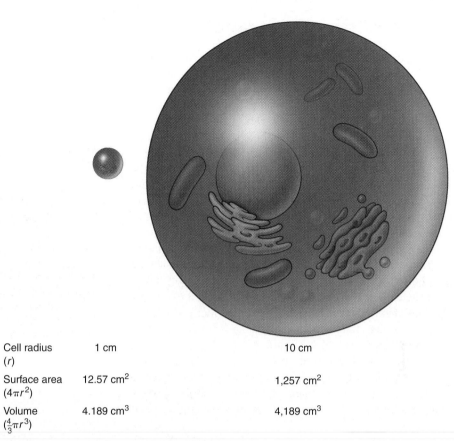

Cell radius (r)	1 cm	10 cm
Surface area ($4\pi r^2$)	12.57 cm^2	1,257 cm^2
Volume ($\frac{4}{3}\pi r^3$)	4.189 cm^3	4,189 cm^3

Figure 4.2 Surface-to-volume ratio.

As a cell gets larger, its volume increases at a faster rate than its surface area. If the cell radius increases by 10 times, the surface area increases by 100 times, but the volume increases by 1,000 times. A cell's surface area must be large enough to meet the needs of its volume.

Most Cells Are Very Small

Cells are not all the same size. Individual cells of the marine alga *Acetabularia,* for example, are up to 5 centimeters long—as long as your little finger. In contrast, the cells of your body are typically from 5 to 20 micrometers in diameter, too small to see. If a typical cell in your body were the size of a shoe box, an *Acetabularia* cell to the same scale would be about 2 kilometers high! The cells of bacteria are even smaller than yours, only a few micrometers thick.

Why Aren't Cells Larger?

Why are most cells so tiny? Most cells are small because larger cells do not function as efficiently. In the center of every cell is a command center that must issue orders to all parts of the cell, directing the synthesis of certain enzymes, the entry of ions and molecules from the exterior, and the assembly of new cell parts. These orders must pass from the core to all parts of the cell, and it takes them a very long time to reach the periphery of a large cell. For this reason, an organism made up of relatively small cells is at an advantage over one composed of larger cells.

Another reason cells are not larger is the advantage of having a greater **surface-to-volume ratio.** As cell size increases, volume grows much more rapidly than surface area (figure 4.2). For a round cell, surface area increases as the square of diameter, whereas volume increases as the cube. Thus a cell with 10 times greater diameter would have 100 (10^2) times the surface area but 1,000 (10^3) times the volume. A cell's surface provides the interior's only opportunity to interact with the environment, and large cells have far less surface for each unit of volume than do small ones.

An Overview of Cell Structure

All cells are surrounded by a delicate membrane that controls the permeability of the cell to water and dissolved substances. A semifluid matrix called **cytoplasm** fills the interior of the cell. It used to be thought that the cytoplasm was uniform, like jello, but we now know that it is highly organized. Your cells, for example, have an internal framework that both gives the cell its shape and positions components and materials within its interior. In the following sections, we first explore the membranes that encase all living cells and then examine in detail their interiors.

4.1 All living things are composed of one or more cells, each a small volume of cytoplasm surrounded by a cell membrane.

4.2 The Plasma Membrane

Encasing all living cells is a delicate sheet only a few molecules thick called the **plasma membrane.** It would take more than 10,000 of these sheets, which are about 7 nanometers thick, piled on top of one another to equal the thickness of this sheet of paper. However, the sheets are not simple in structure, like a soap bubble's skin. Rather, they are made up of a diverse collection of proteins floating within a lipid framework like small boats bobbing on the surface of a pond. Regardless of the kind of cell they enclose, all plasma membranes have the same basic structure of proteins embedded in a sheet of lipid, called the **fluid mosaic model.**

The lipid layer that forms the foundation of a plasma membrane is composed of modified fat molecules called **phospholipids.** One end of a phospholipid molecule has a phosphate chemical group attached to it, making it extremely polar (and thus water-soluble), whereas the other end is composed of two long fatty acid chains that are strongly nonpolar (and thus water-insoluble) (figure 4.3*a*). Because of this structure, with the polar group pointing in one direction and the two nonpolar fatty acids extending in the other direction roughly parallel to each other, phospholipid molecules are usually diagrammed as a (polar) ball with two dangling (nonpolar) tails (figure 4.3*b*).

Imagine what happens when a collection of phospholipid molecules is placed in water. A structure called a **lipid bilayer** forms spontaneously (figure 4.4). How can this happen? The long nonpolar tails of the phospholipid molecules are pushed away by the water molecules that surround them, shouldered aside as the water molecules seek partners that can form hydrogen bonds. After much shoving

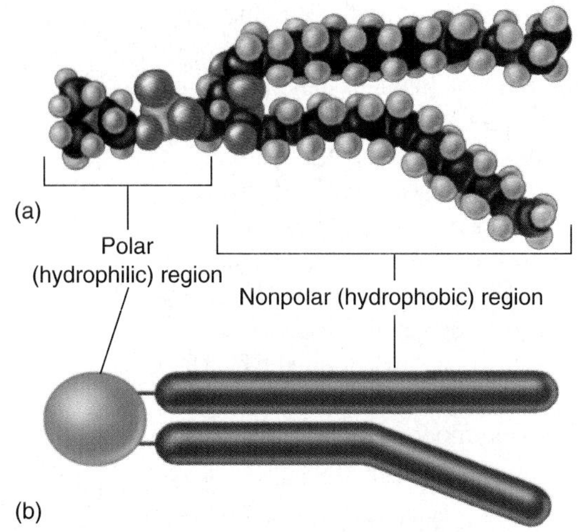
(a)
Polar
(hydrophilic) region
Nonpolar (hydrophobic) region
(b)

Figure 4.3 Phospholipid structure.
One end of a phospholipid molecule is polar and the other is nonpolar. (*a*) The molecular structure is shown by colored spheres representing individual atoms (*black* for carbon, *blue* for hydrogen, *red* for oxygen, and *yellow* for phosphorus). (*b*) The phospholipid is often depicted diagrammatically as a ball with two tails.

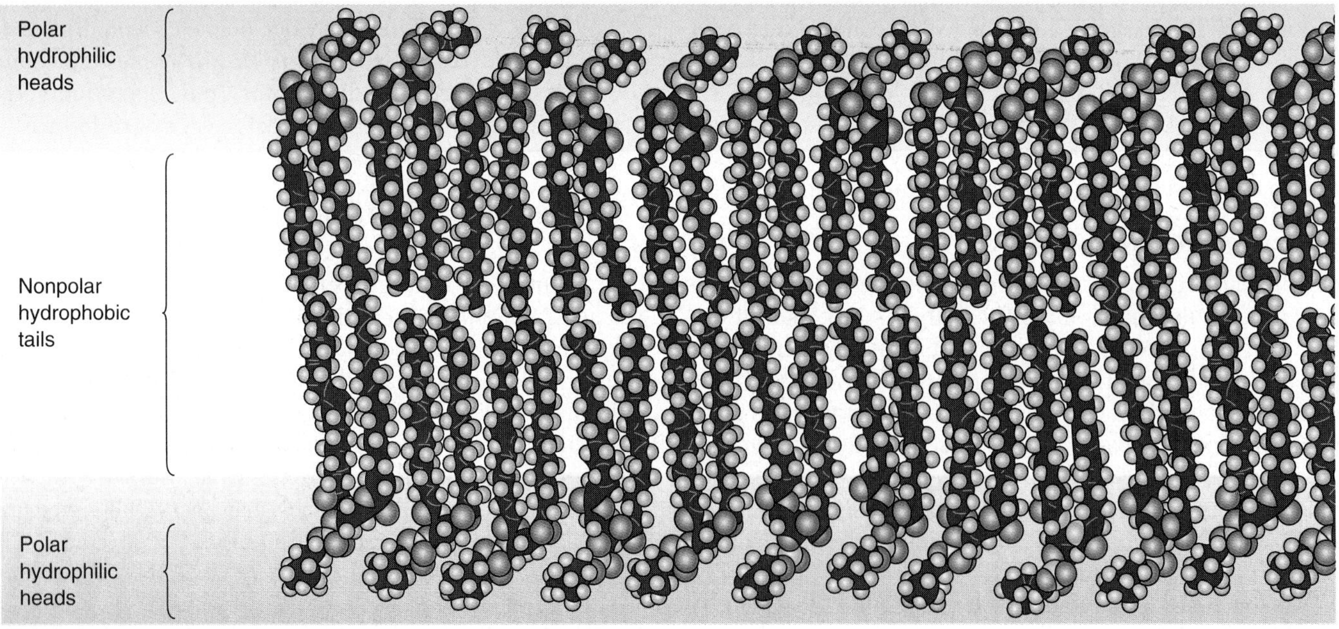
Polar
hydrophilic
heads
Nonpolar
hydrophobic
tails
Polar
hydrophilic
heads

Figure 4.4 The lipid bilayer.
The basic structure of every plasma membrane is a double layer of lipid. This diagram illustrates how phospholipids aggregate to form a bilayer with a nonpolar interior.

and jostling, every phospholipid molecule ends up with its polar head facing water and its nonpolar tail facing away from water. The phospholipid molecules form a *double* layer. Because there are two layers with the tails facing each other, no tails are ever in contact with water.

Because the interior of a lipid bilayer is completely nonpolar, it repels any water-soluble molecules that attempt to pass through it, just as a layer of oil stops the passage of a drop of water (that's why ducks do not get wet). As we will see, it is only because of protein-lined passageways through the bilayer that polar chemicals are able to enter and leave cells.

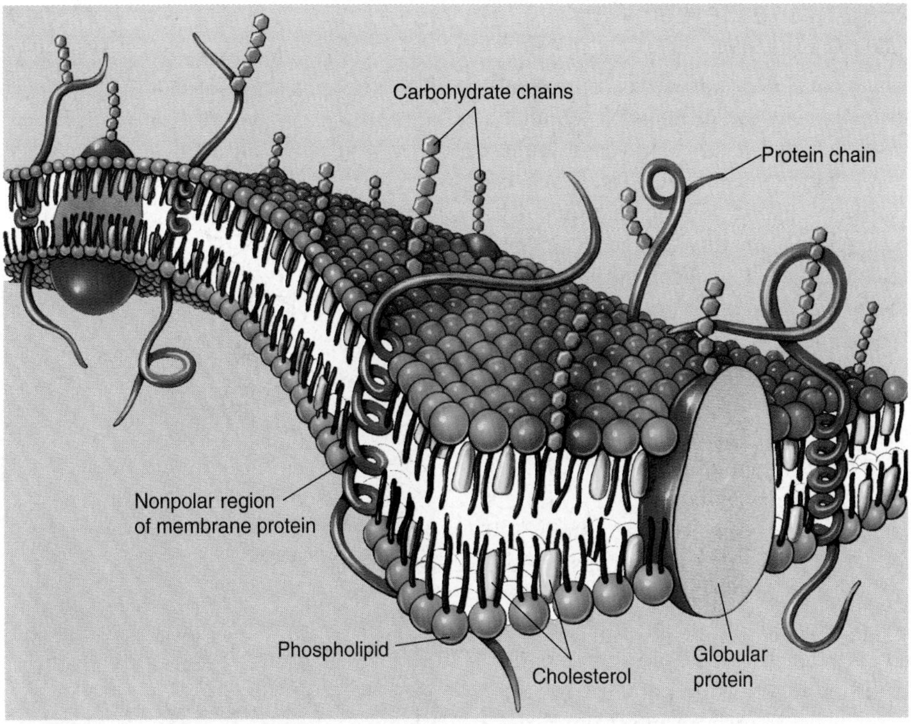

Carbohydrate chains
Protein chain
Nonpolar region of membrane protein
Phospholipid
Cholesterol
Globular protein

Figure 4.5 Proteins are embedded within the lipid bilayer.
A variety of proteins protrude through the lipid bilayer of animal cells. Membrane proteins function as channels, receptors, and cell surface markers. Carbohydrate chains are often bound to these proteins and to phospholipids in the membrane itself as well. These chains serve as distinctive identification tags, unique to particular types of cells.

Proteins Within the Membrane

The second major component of every biological membrane is a collection of **membrane proteins** that float within the lipid bilayer (figure 4.5). In plasma membranes, these proteins provide channels through which molecules and information pass. Membrane proteins are not fixed into position; instead, they move freely. Some membranes are crowded with proteins, packed tightly side by side. In other membranes the proteins are sparsely distributed.

Some membrane proteins project up from the surface of the plasma membrane like buoys, often with carbohydrate chains or lipids attached to their tips like flags. These **cell surface proteins** act as markers to identify particular types of cells or as beacons to bind specific hormones or proteins to the cell. The CD4 protein by which AIDS viruses dock onto human white blood cells is such a beacon.

Other proteins extend all the way across the bilayer, providing channels across which polar ions and molecules can pass into and out of the cell. How do these **transmembrane proteins** manage to span the membrane, rather than just floating on the surface in the way that a drop of water floats on oil? The part of the protein that actually traverses the lipid bilayer is a specially constructed spiral helix of nonpolar amino acids (figure 4.6). Water responds to nonpolar amino acids much as it does to nonpolar lipid chains, and the helical spiral is held within the lipid interior of the bilayer, anchored there by the strong tendency of water to avoid contact with these nonpolar amino acids. Many transmembrane proteins lock this arrangement in place by positioning amino acids with electrical charges (which are very polar) at the two ends of the helical region.

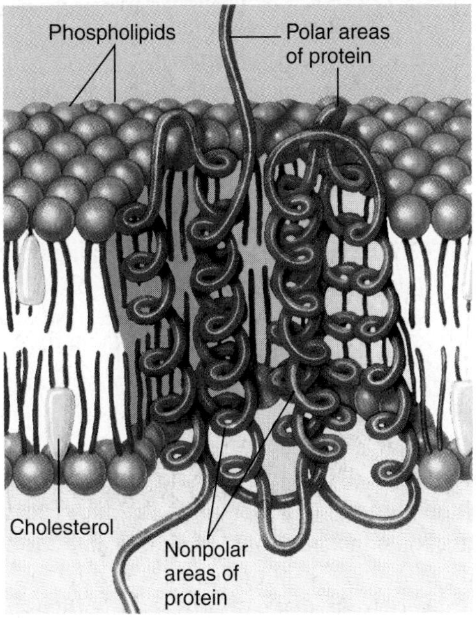

Phospholipids
Polar areas of protein
Cholesterol
Nonpolar areas of protein

Figure 4.6 Nonpolar regions lock proteins into membranes.
A spiral helix of nonpolar amino acids (*red*) extends across the nonpolar lipid interior, while polar (*purple*) portions of the protein protrude out from the bilayer. The protein cannot move in or out because such a movement would drag polar segments of the protein into the nonpolar interior of the membrane.

Membrane Defects Can Cause Disease

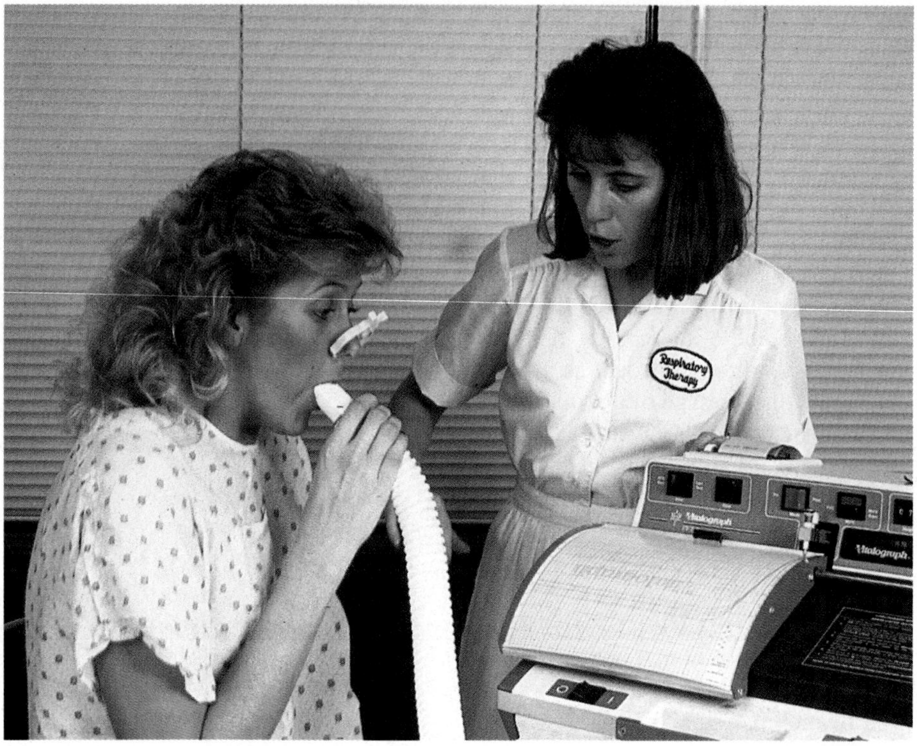

Figure 4.7 Cystic fibrosis.

Cystic fibrosis is a genetic disorder that results from a defective chloride channel. It leads to a buildup of chloride ions within lung cells, causing water to move into the cells by osmosis. Removing water from the surrounding mucus causes it to thicken, leading to the symptoms of the disorder. This cystic fibrosis patient is breathing into a vitalograph, a device that measures lung function.

The year 1993 marked an important milestone in the treatment of human disease. That year the first attempt was made to cure **cystic fibrosis (CF),** a deadly genetic disorder, by transferring healthy genes into sick individuals. Cystic fibrosis is a fatal disease in which the body cells of affected individuals secrete a thick mucus that clogs the airways of the lungs. These same secretions block the ducts of the pancreas and liver so that the few patients who do not die of lung disease die of liver failure. Cystic fibrosis is usually thought of as a children's disease because until recently few affected individuals lived long enough to become adults. Even today half die before their mid-twenties. There is no known cure.

Cystic fibrosis results from a defect in a single gene that is passed down from parent to child. It is the most common fatal genetic disease of Caucasians. One in 20 individuals possesses at least one copy of the defective gene. Most of these individuals are not afflicted with the disease; only those children who inherit a copy of the defective gene from each parent succumb to cystic fibrosis—about 1 in 2,500 infants (figure 4.7).

Cystic fibrosis has proven difficult to study. Many organs are affected, and until recently it was impossible to identify the nature of the defective gene responsible for the disease. In 1985 the first clear clue was obtained. An investigator, Paul Quinton, seized on a commonly observed characteristic of cystic fibrosis patients, that their sweat is abnormally salty, and performed the following experiment. He isolated a sweat duct from a small piece of skin and placed it in a solution of salt (NaCl) that was three times as concentrated as the NaCl inside the duct. He then monitored the movement of ions. Diffusion tends to drive both the sodium (Na^+) and the chloride (Cl^-) ions into the duct because of the higher outer ion concentrations. In skin isolated from normal individuals, Na^+ and Cl^- both entered the duct, as expected. In skin isolated from cystic fibrosis individuals, however, only Na^+ entered the duct—no Cl^- entered. For the first time, the molecular nature of cystic fibrosis became clear. Water accompanies chloride, and was not entering the ducts because chloride was not, creating thick mucus. Cystic fibrosis is a defect in a plasma membrane protein called CFTR (*cystic fibrosis transmembrane conductance regulator*) that normally regulates passage of Cl^- into and out of the body's cells.

The defective *cf* gene was isolated in 1987, and its position on a particular human chromosome (chromosome 7) was pinpointed in 1989. In 1990 a working *cf* gene was successfully transferred via adenovirus into human lung cells growing in tissue culture. The defective cells were "cured," becoming able to transport chloride ions across their plasma membranes. Then in 1991 a team of researchers successfully transferred a normal human *cf* gene into the lung cells of a living animal—a rat. The *cf* gene was first inserted into a cold virus that easily infects lung cells, and the virus was inhaled by the rat. Carried piggyback, the *cf* gene entered the rat lung cells and began producing the normal human CFTR protein within these cells! Tests of gene transfer into cystic fibrosis patients were begun in 1993, and while a great deal of work remains to be done (the initial experiments were not successful), the future for these patients for the first time seems bright.

> **4.2** All cells are encased within a delicate lipid bilayer sheet, the plasma membrane, within which are embedded a variety of proteins that act as markers or channels through the membrane.

4.3 Prokaryotic Cells

There are two major kinds of cells: prokaryotes and eukaryotes. **Prokaryotes** have a relatively uniform cytoplasm that is not subdivided by interior membranes into separate compartments. They do not, for example, have special membrane-bounded compartments, called *organelles,* or a *nucleus* (a membrane-bounded compartment that holds the hereditary information). All bacteria are prokaryotes; all nonbacterial organisms are eukaryotes.

Bacteria are the simplest cellular organisms. Over 2,500 species are recognized, but doubtless many times that number actually exist and have not yet been described. Although these species are diverse in form, their organization is fundamentally similar (figure 4.8): small cells typically about 1 to 10 micrometers thick; enclosed like all cells by a plasma membrane, but with no distinct interior compartments. Outside of almost all bacteria is a *cell wall,* a framework of carbohydrates cross-linked into a rigid structure. In some bacteria another layer called the *capsule* encloses the cell wall. Bacterial cells assume many shapes or can adhere in chains and masses (figure 4.9), but individual cells are separate from one another.

If you were able to magnify your vision and peer into a bacterial cell, you would be struck by its simple organization. The entire interior of the cell, the cytoplasm, is one unit, with no internal support structure (the rigid wall supports the cell's shape) and no internal compartments bounded by membranes. Scattered throughout the cytoplasm of prokaryotic cells are small structures called *ribosomes,* the site where proteins are made. Ribosomes are not considered organelles because they lack a membrane boundary.

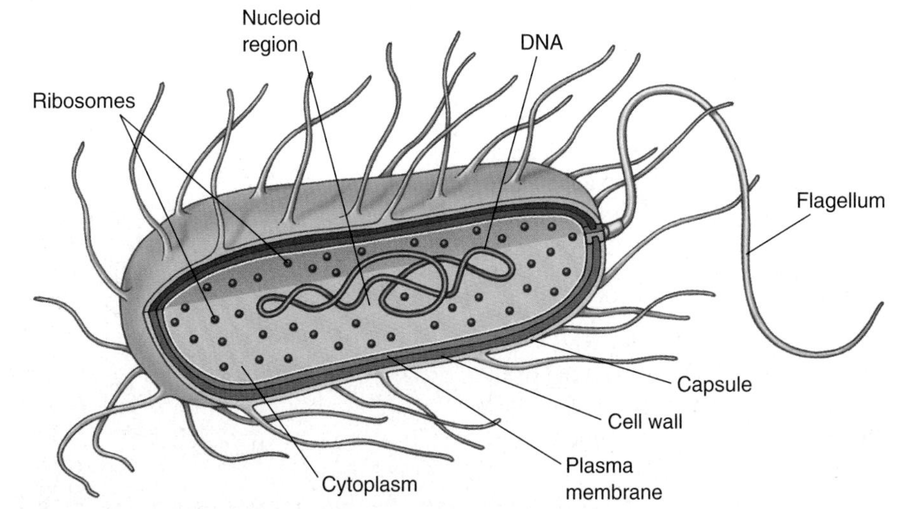
Nucleoid region
DNA
Ribosomes
Flagellum
Capsule
Cell wall
Plasma membrane
Cytoplasm

Figure 4.8 Organization of a bacterial cell.

Bacterial cells lack internal compartments. Not all bacterial cells have a flagellum like the one illustrated here, but all do have a nucleoid region, ribosomes, a plasma membrane, cytoplasm, and a cell wall.

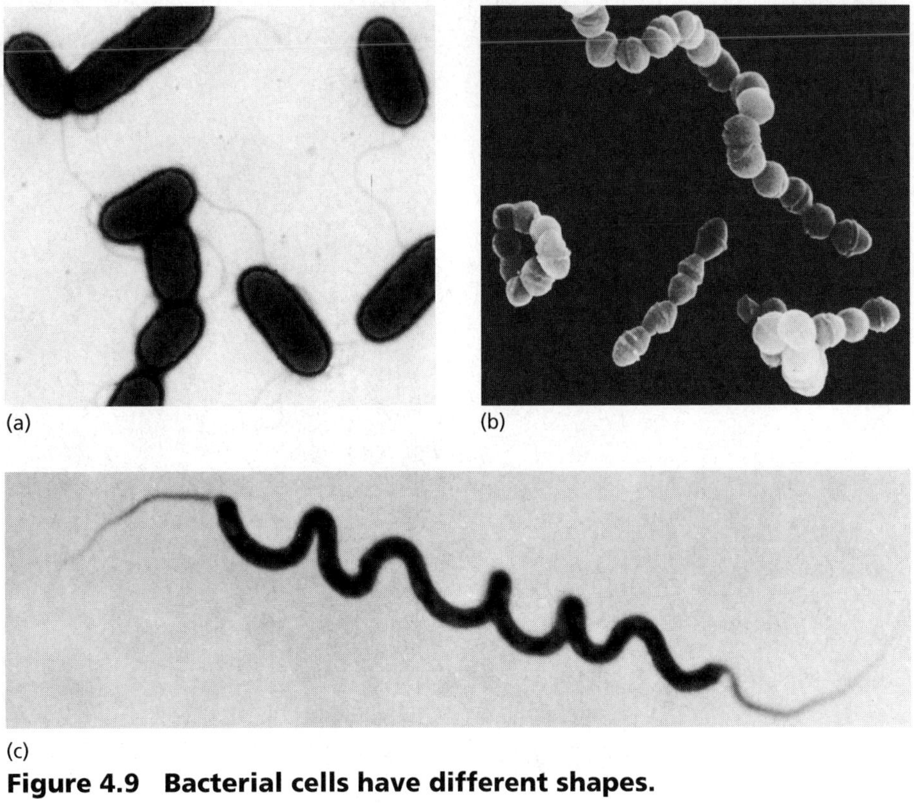

Figure 4.9 Bacterial cells have different shapes.

(a) A rod-shaped bacterium, *Pseudomonas,* is a type associated with many plant diseases. (b) *Streptococcus* is a more or less spherical bacterium in which the individuals adhere in chains. (c) *Spirillum* is a spiral bacterium; this large bacterium has a tuft of flagella at each end.

4.3 Prokaryotic cells lack a nucleus and do not have an extensive system of interior membranes.

4.4 Eukaryotic Cells

For the first 2 billion years of life on earth, all organisms were bacteria, cells with very simple interiors. Then, about 1.5 billion years ago, a new kind of cell appeared for the first time, much larger and with a complex interior organization. All cells alive today except bacteria are of this new kind. Unlike bacteria, these big cells have many membrane-bounded interior compartments and a variety of **organelles** (specialized structures within which particular cell processes occur). One of the organelles is very visible when these cells are examined with a microscope,

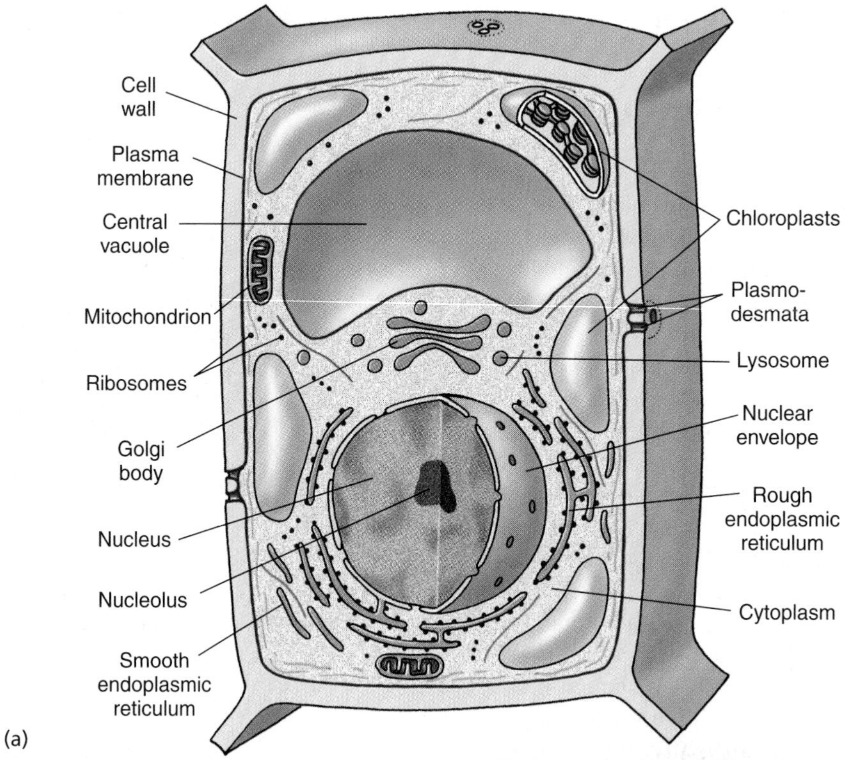

Cell wall
Plasma membrane
Central vacuole
Mitochondrion
Ribosomes
Golgi body
Nucleus
Nucleolus
Smooth endoplasmic reticulum
Chloroplasts
Plasmodesmata
Lysosome
Nuclear envelope
Rough endoplasmic reticulum
Cytoplasm
(a)

Figure 4.10 Structure of a plant cell.

(*a*) An idealized plant cell. The large central vacuole of a plant cell occupies most of the cell's space, providing an expanded surface area. (*b*) Micrograph of a plant cell with drawings detailing organelles.

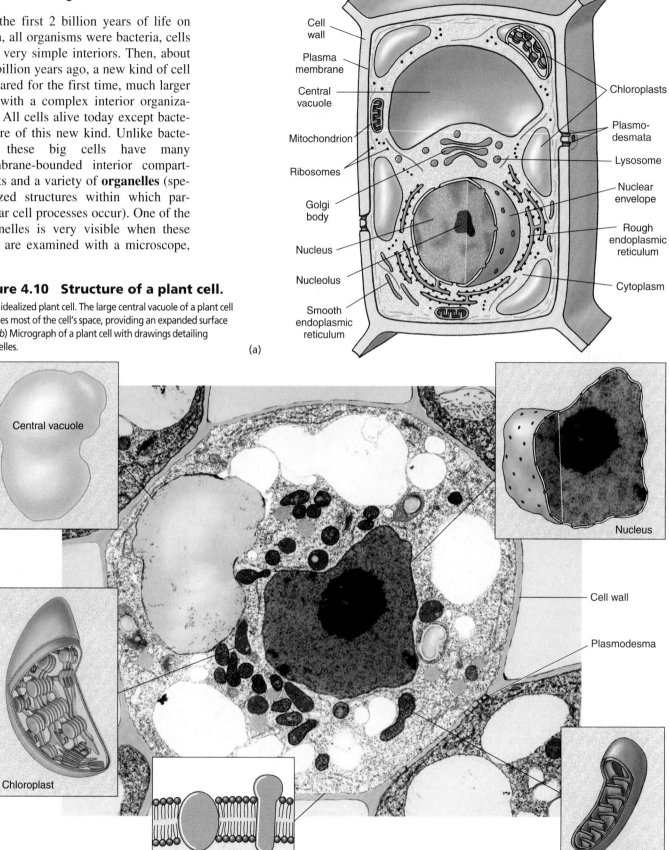

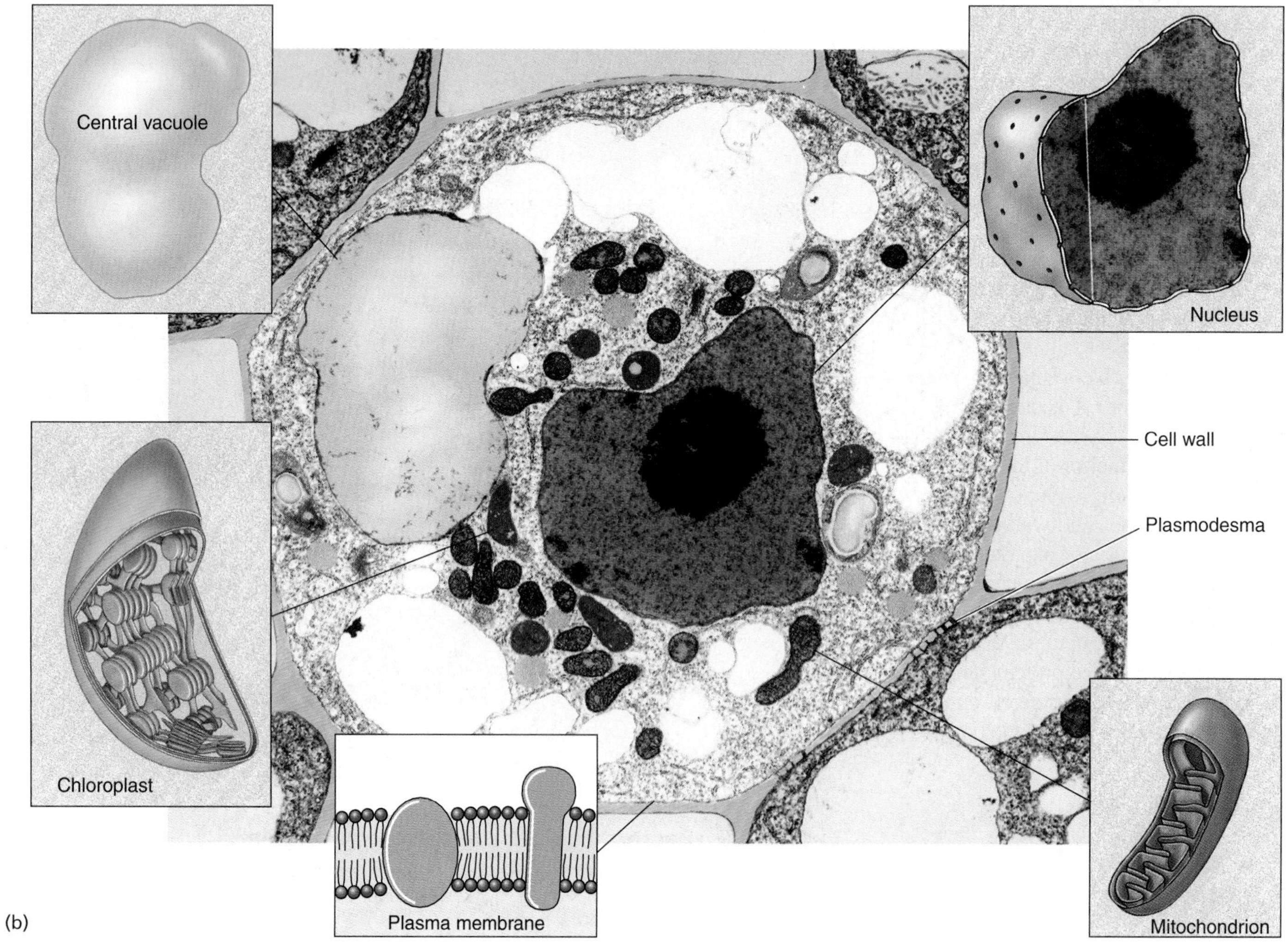

Central vacuole
Nucleus
Cell wall
Plasmodesma
Chloroplast
Plasma membrane
Mitochondrion
(b)

filling the center of the cell like the pit of a peach. Seeing it, the English botanist Robert Brown in 1831 called it the *nucleus,* from the Latin word for "kernel." All cells with nuclei are called **eukaryotes** (from the Greek words *eu,* true, and *karyon,* nut) while bacteria are called *prokaryotes* ("before the nut"). Both plant cells (figure 4.10) and animal cells (figure 4.11) are eukaryotic. We now journey into the interior of a typical eukaryotic cell and explore its complex organization.

4.4 Eukaryotic cells have a system of interior membranes and membrane-bounded organelles that subdivide the interior into functional compartments.

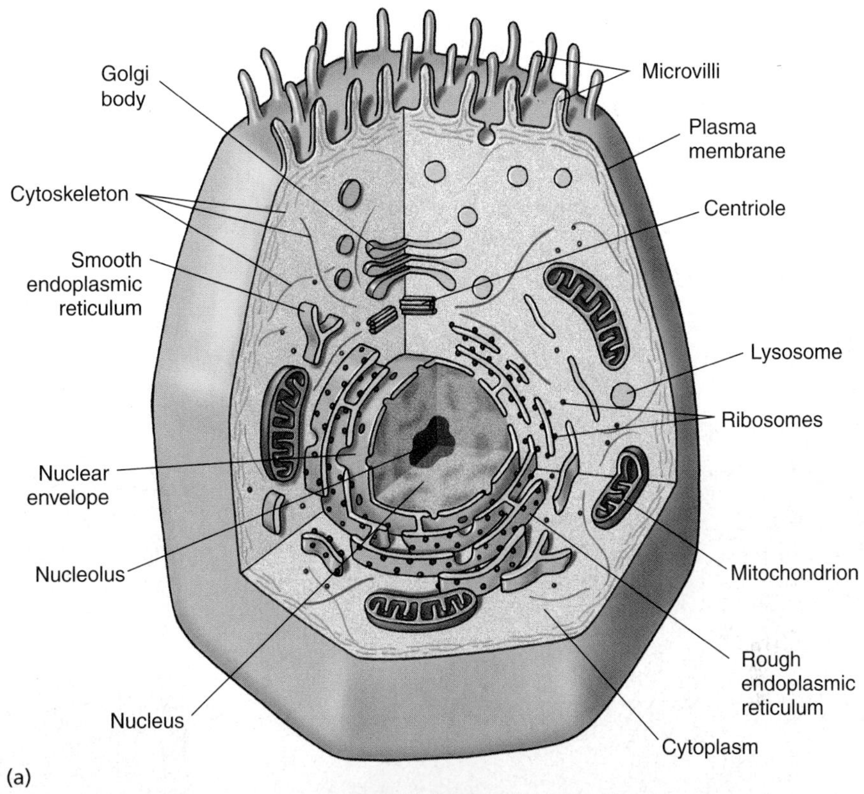

Golgi body
Microvilli
Plasma membrane
Cytoskeleton
Centriole
Smooth endoplasmic reticulum
Lysosome
Ribosomes
Nuclear envelope
Nucleolus
Mitochondrion
Rough endoplasmic reticulum
Nucleus
Cytoplasm
(a)

Figure 4.11 Structure of an animal cell.
(*a*) An idealized animal cell. (*b*) Micrograph of an animal cell with drawings detailing organelles.

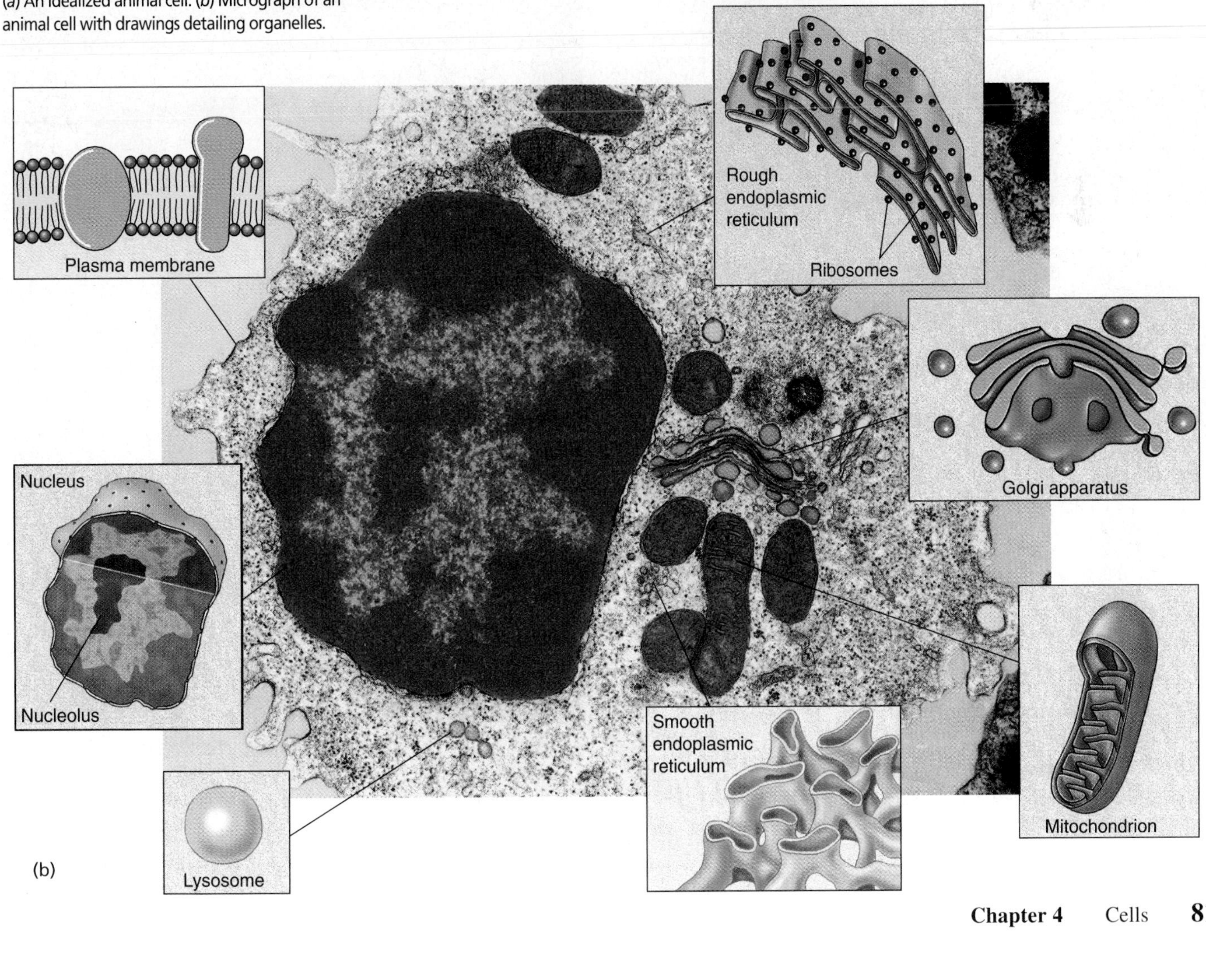

Rough endoplasmic reticulum
Ribosomes
Plasma membrane
Golgi apparatus
Nucleus
Nucleolus
Mitochondrion
Smooth endoplasmic reticulum
Lysosome
(b)

4.5 The Cytoskeleton: Interior Framework of the Cell

If you were to shrink down and enter into the interior of a eukaryotic cell, the first thing you would encounter is a dense network of protein fibers called the **cytoskeleton,** which supports the shape of the cell and anchors organelles such as the nucleus to fixed locations (figure 4.12). This network cannot be seen with a light microscope because the individual fibers are single chains of protein, much too fine for microscopes to resolve. To "see" the cytoskeleton, scientists attach fluorescent antibodies to the protein fibers and then photograph them under fluorescent light (figure 4.13).

The protein fibers of the cytoskeleton are a dynamic system, constantly being formed and disassembled. There are three different kinds of protein fibers (figure 4.14): long, slender **microfilaments** made of the protein actin, hollow tubes called **microtubules** made of the protein tubulin, and thick ropes called **intermediate fibers.** The filaments, microtubules, and fibers are anchored to membrane proteins embedded within the plasma membrane.

The cytoskeleton plays a major role in determining the shape of animal cells, which lack rigid cell walls. Because filaments can form and dissolve readily, the shape of an animal

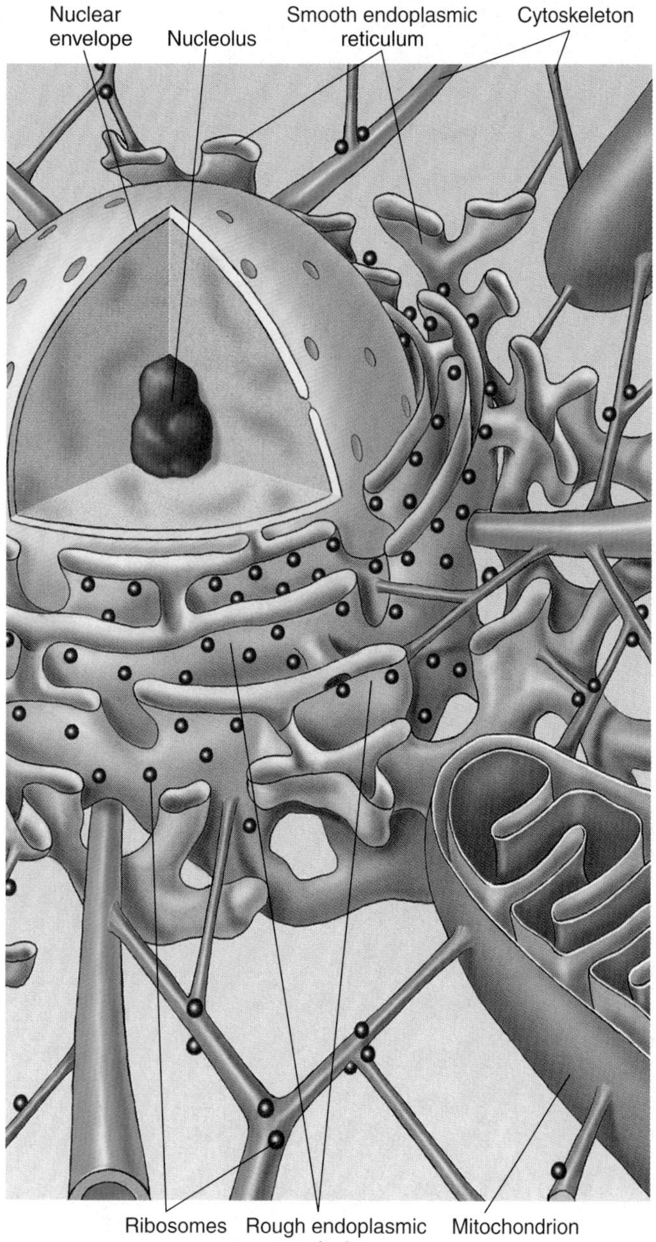
Nuclear envelope
Nucleolus
Smooth endoplasmic reticulum
Cytoskeleton
Ribosomes
Rough endoplasmic reticulum
Mitochondrion

Figure 4.12 The cytoskeleton.
Organelles such as mitochondria are anchored to fixed locations in the cytoplasm by the cytoskeleton, a network of protein fibers.

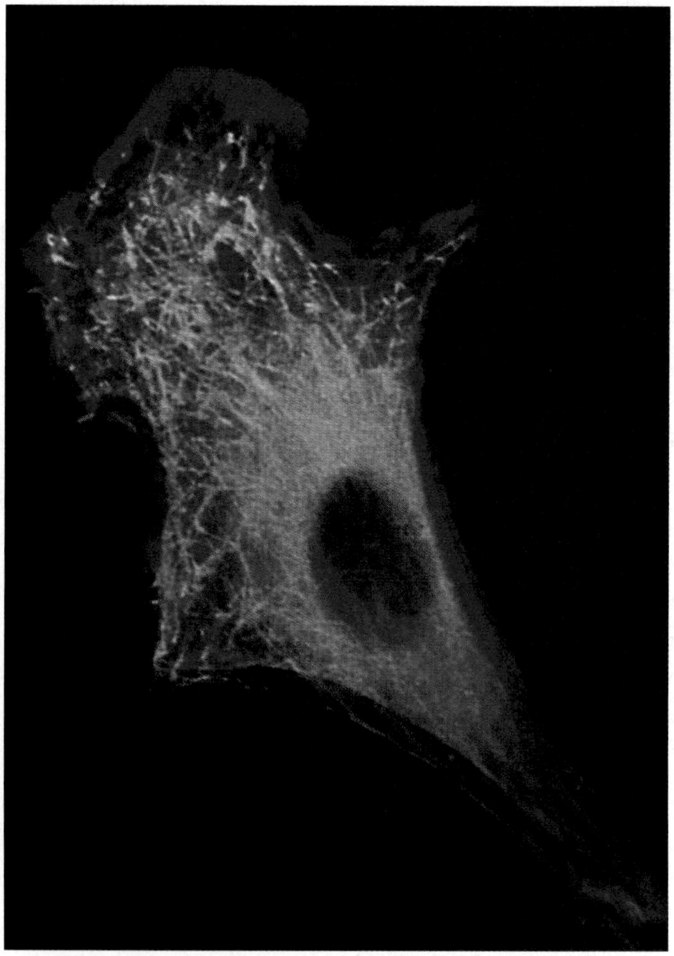

Figure 4.13 Visualizing the cytoskeleton.
A complex web of protein fibers runs through the cytoplasm of the eukaryotic cell. The different proteins are made visible by fluorescence microscopy. Actin appears blue here, microtubules appear green, and intermediate fibers appear red.

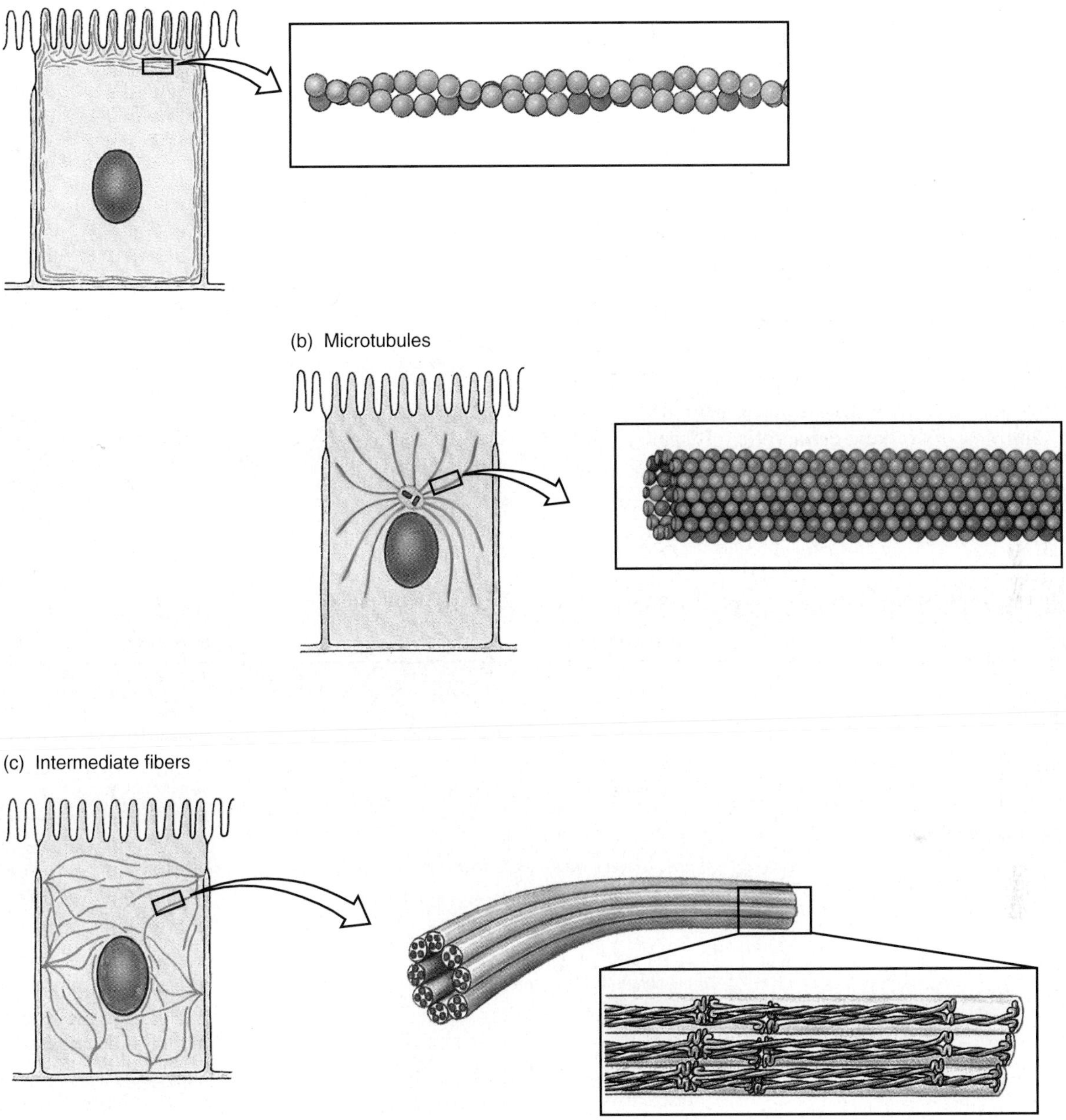
(a) Microfilaments
(b) Microtubules
(c) Intermediate fibers

Figure 4.14 The three protein fibers of the cytoskeleton.
(*a*) Microfilaments parallel the cell surface membrane in bundles of actin known as stress fibers, which may have a contractile function. (*b*) Each microtubule is a hollow tube composed of stacks of a protein called tubulin. There are 13 stacks of tubulin, arrayed side by side in a circle. (*c*) It is not known how the individual subunits are arranged in an intermediate fiber, but the best evidence suggests that three subunits are wound together in a coil, interrupted by uncoiled regions. Intermediate fibers, like microtubules, extend throughout the cytoplasm and probably provide structural reinforcement.

cell can change rapidly. If you examine the surface of an animal cell with a microscope, you will often find it alive with motion, projections shooting out from the surface and then retracting, only to shoot out elsewhere moments later.

The cytoskeleton is not only responsible for the cell's shape, but it also provides a scaffold both for ribosomes to carry out protein synthesis and for enzymes to be localized within defined areas of the cytoplasm. By anchoring particular enzymes near one another, the cytoplasm participates with organelles in organizing the cell's activities.

4.5 The cytoskeleton is a latticework of protein fibers that determines a cell's shape and anchors organelles to particular locations within the cytoplasm.

4.6 An Overview of Cell Organelles

The many parts of the cell (table 4.1) are remarkably similar from plants to animals, or from paramecia to primates. Organelles look similar and carry out similar functions in all organisms. These shared properties must have been derived from common ancestral cells that successfully developed the initial organelles several billion years ago.

How did cells with organelles evolve? Today's eukaryotic cells are thought to have evolved from an endosymbiosis between different species of prokaryotes, in which one species was engulfed by and lived inside the other (figure 4.15). Symbiosis is a close relationship between organisms of different species that live together. According to the theory of endosymbiosis, the engulfed prokaryotes provided their hosts with certain advantages associated with their special metabolic abilities. Two key eukaryotic organelles, mitochondria and chloroplasts, are believed to be the descendants of these endosymbiotic prokaryotes.

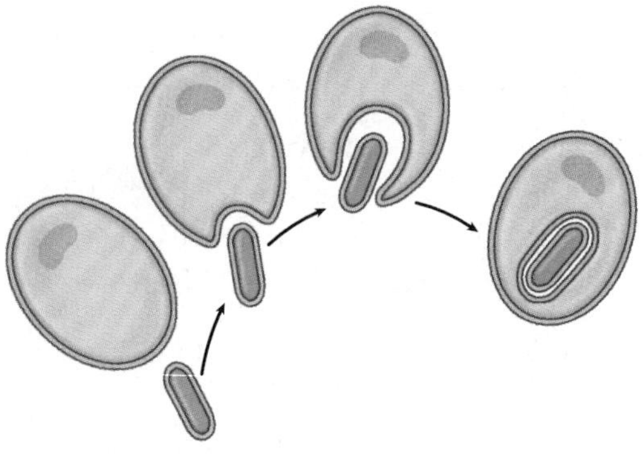

Figure 4.15 Endosymbiosis.
This figure shows how a double membrane may have been created during the symbiotic origin of mitochondria or chloroplasts.

4.6 Eukaryotic cells are thought to have arisen by endosymbiosis.

TABLE 4.1 EUKARYOTIC CELL STRUCTURES AND THEIR FUNCTIONS

Structure	Description	Function
STRUCTURAL ELEMENTS		
Cytoskeleton	Network of protein filaments	Structural support; cell movement
Flagella and cilia	Cellular extensions with 9 + 2 arrangement of pairs of microtubules	Motility or moving fluids over surfaces
ENDOMEMBRANE SYSTEM		
Plasma membrane	Lipid bilayer in which proteins are embedded	Regulates what passes into and out of cell; cell-to-cell recognition
Endoplasmic reticulum	Network of internal membranes	Forms compartments and vesicles
Nucleus	Spherical structure bounded by double membrane; contains chromosomes	Control center of cell; directs protein synthesis and cell reproduction
Golgi complex	Stacks of flattened vesicles	Modifies and packages proteins for export from the cell; forms secretory vesicles
Lysosomes	Vesicles derived from Golgi complex that contain hydrolytic digestive enzymes	Digest worn-out organelles and cell debris; play role in cell death
ENERGY-PRODUCING ORGANELLES		
Mitochondria	Bacteria-like elements with inner membrane	Powerhouse of the cell; site of oxidative metabolism
Chloroplast	Bacteria-like organelle found in plants and algae; complex inner membrane consists of stacked vesicles	Site of photosynthesis
ELEMENTS OF GENE EXPRESSION		
Chromosomes	Long threads of DNA that form a complex with protein	Contain hereditary information
Nucleolus	Site of genes for rRNA synthesis	Assembles ribosomes
Ribosomes	Small, complex assemblies of protein and RNA, often bound to endoplasmic reticulum	Sites of protein synthesis

4.7 The Nucleus: The Cell's Control Center

If you were to continue your journey into the cell's interior, you would eventually reach the center of the cell. There you would find, cradled within a network of fine filaments like a ball in a basket, the **nucleus** (figure 4.16). The nucleus is the command and control center of the cell, directing all of its activities. It is also the genetic library where the hereditary information is stored.

Nuclear Membrane

The surface of the nucleus is bounded by a special kind of membrane called the **nuclear envelope.** The nuclear envelope is actually *two* membranes, one outside the other, like a sweater over a shirt. Scattered over the surface of this envelope are shallow depressions called **nuclear pores.** Nuclear pores form when the two membrane layers of the nuclear envelope pinch together. A nuclear pore is not an empty opening like the hole in a doughnut; rather, it has many proteins embedded within it that permit proteins and RNA to pass into and out of the nucleus.

Chromosomes

In both bacteria and eukaryotes, all hereditary information specifying cell structure and function is encoded in DNA. However, unlike bacterial DNA, the DNA of eukaryotes is divided into several segments and associated with protein, forming **chromosomes.** The proteins in the chromosome permit the DNA to wind tightly and condense during cell division. Under a light microscope, these condensed chromosomes are readily seen in dividing cells as densely staining rods. After cell division, eukaryotic chromosomes uncoil and fully extend into threadlike strands called **chromatin** that can no longer be distinguished individually with a light microscope.

Nucleolus

One region of the nucleus appears darker than the rest; this darker region is called the **nucleolus.** There a cluster of several hundred genes encode rRNA where the ribosome subunits assemble. These subunits leave the nucleus through the nuclear pores and enter the cytoplasm, where final assembly of ribosomes takes place.

To make its many proteins, the cell employs a special structure called a **ribosome,** which reads the RNA copy of a gene and uses that information to direct the construction of a protein. Ribosomes are made up of several special forms of RNA called ribosomal RNA, or rRNA, bound up within a complex of several dozen different proteins.

4.7 The nucleus is the command center of the cell, issuing instructions that control cell activities. It also stores the cell's hereditary information.

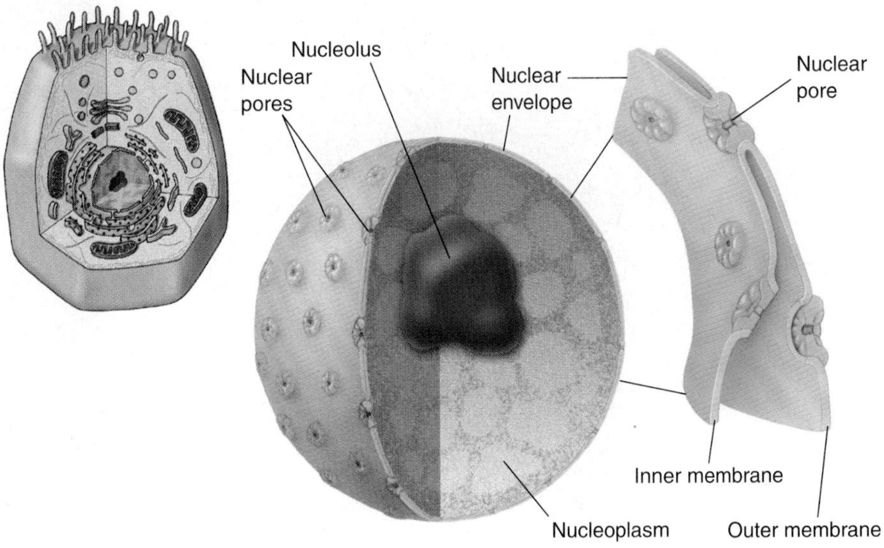
Nucleolus
Nuclear pores
Nuclear envelope
Nuclear pore
Inner membrane
Nucleoplasm
Outer membrane

Figure 4.16 The nucleus.
The nucleus is composed of a double membrane, called a nuclear envelope, enclosing a fluid-filled interior containing the chromosomes. In cross section, the individual nuclear pores are seen to extend through the two membrane layers of the envelope; the dark material within the pore is protein, which acts to control access through the pore.

4.8 The Endomembrane System

Surrounding the nucleus within the interior of the eukaryotic cell is a tightly packed mass of membranes. They fill the cell, dividing it into compartments, channeling the transport of molecules through the interior of the cell and providing the surfaces on which enzymes act. The system of internal compartments created by these membranes in eukaryotic cells constitutes the most fundamental distinction between the cells of eukaryotes and prokaryotes.

Endoplasmic Reticulum: The Transportation System

The extensive system of internal membranes is called the **endoplasmic reticulum,** often abbreviated **ER** (figure 4.17). The term *endoplasmic* means "within the cytoplasm," and the term *reticulum* is a Latin word meaning "little net." The ER, weaving in sheets through the interior of the cell, creates a series of channels and interconnections, and it also isolates some spaces as membrane-enclosed sacs called **vesicles.**

The surface of the ER is the place where the cell manufactures many carbohydrates and lipids. It is also where the cell makes proteins intended for export (such as enzymes secreted from the cell surface). The surface of those regions of the ER devoted to the synthesis of such transported proteins is heavily studded with ribosomes and appears pebbly, like the surface of sandpaper, when seen through an electron microscope. For this reason, these regions are called **rough ER.** Regions in which ER-bounded ribosomes are relatively scarce are correspondingly called **smooth ER.**

The Golgi Complex: The Delivery System

As new molecules are made on the surface of the ER, they are passed out across the ER membrane into flattened stacks of membranes called **Golgi bodies.** These structures are named for Camillo Golgi, the nineteenth-century Italian physician who first called attention to the them. The number of Golgi bodies a cell contains ranges from 1 or a few in protists, to 20 or more in animal cells and several hundred in plant cells. Golgi bodies function in the collection, packaging, and distribution of molecules manufactured in the cell. Scattered through the cytoplasm, Golgi bodies are collectively referred to as the **Golgi complex** (figure 4.18).

The proteins and lipids that are manufactured on the ER membranes are transported through the channels of the ER, or as vesicles budded off from it, into the Golgi bodies. Within the Golgi bodies, many of these molecules become tagged with carbohydrates. The molecules collect at the ends of the membranous folds of the Golgi bodies; these folds are given the special name *cisternae* (Latin, collecting vessels). Vesicles that pinch off from the cisternae carry the molecules to the different compartments of the cell and to the inner surface of the plasma membrane (figure 4.19), where molecules to be secreted are released to the outside.

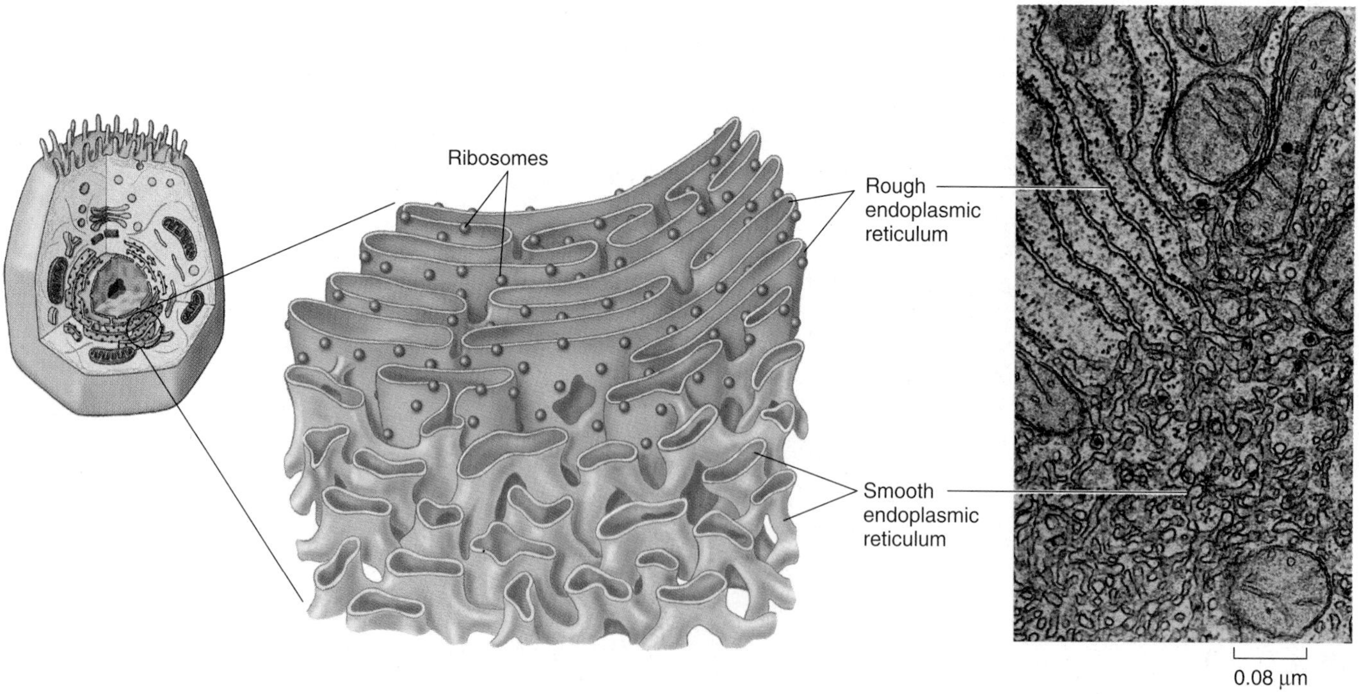
Ribosomes
Rough endoplasmic reticulum
Smooth endoplasmic reticulum
0.08 μm

Figure 4.17 The endoplasmic reticulum (ER).

The endoplasmic reticulum provides the cell with an extensive system of internal membranes for the synthesis and transport of materials. Ribosomes are associated with only one side of the rough ER; the other side is the boundary of a separate compartment within the cell into which the ribosomes extrude newly made proteins destined for secretion. Smooth endoplasmic reticulum has few to no bound ribosomes.

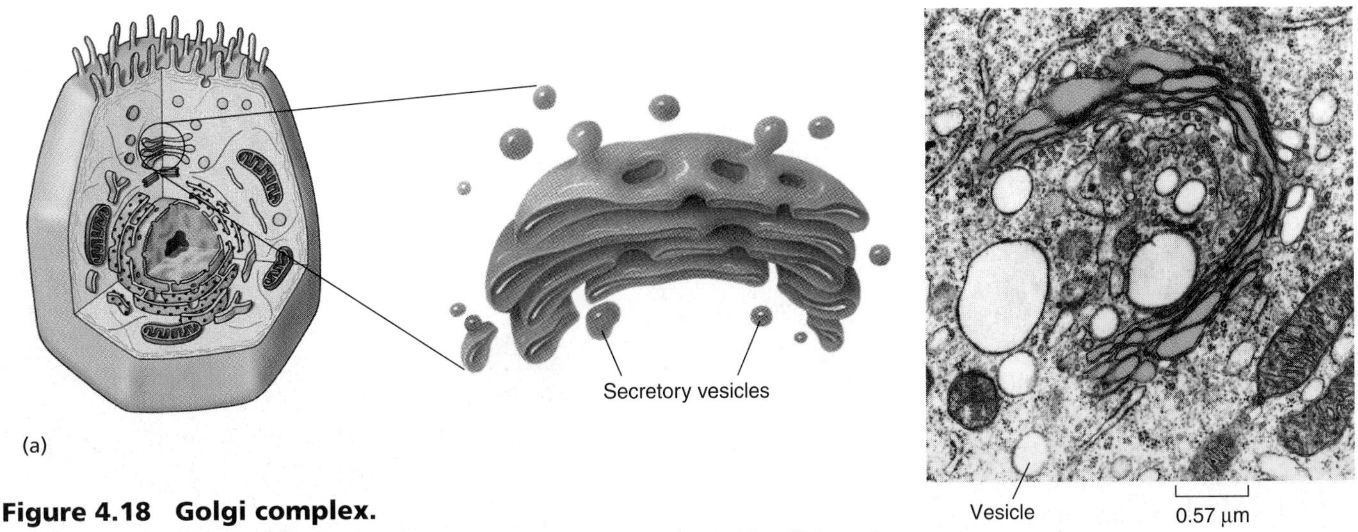
Secretory vesicles
(a)
Vesicle
0.57 μm
(b)

Figure 4.18 Golgi complex.
This vesicle-forming system, called the Golgi complex after its discoverer, is an integral part of the cell's internal membrane system. The Golgi complex processes and packages materials for transport and possible export from the cell. (*a*) Diagram of a Golgi body. (*b*) Micrograph of a Golgi body showing vesicles.

Lysosomes: Recycling Centers

Other organelles called **lysosomes** arise from the Golgi complex and contain a concentrated mix of the powerful enzymes that break down macromolecules. Lysosomes are also the recycling centers of the cell, digesting worn-out cell components to make way for newly formed ones while recycling the proteins and other materials of the old parts. Large organelles called mitochondria are replaced in some human tissues every 10 days, with lysosomes digesting the old ones as the new ones are produced. In addition to breaking down organelles and other structures within cells, lysosomes also eliminate particles (including other cells) that the cell has engulfed.

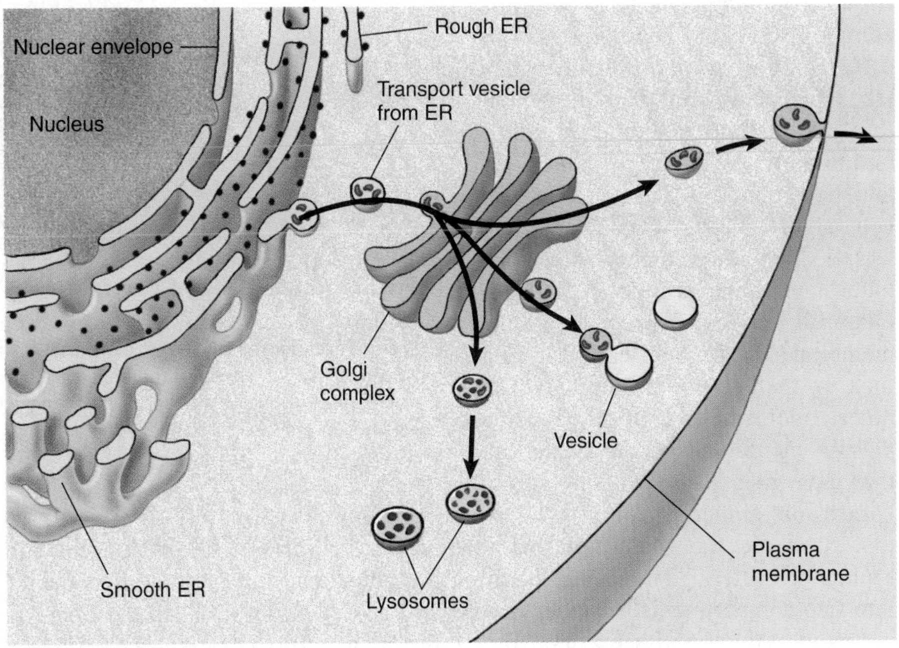
Rough ER
Nuclear envelope
Nucleus
Transport vesicle from ER
Golgi complex
Vesicle
Plasma membrane
Smooth ER
Lysosomes

Figure 4.19 How the endomembrane system works.
A highly efficient highway system within the cell, the endomembrane system transports material from the ER to the Golgi and from there to other destinations.

Peroxisomes: Chemical Specialty Shops

The interior of the eukaryotic cell contains a variety of membrane-bounded spherical organelles derived from the ER that carry out particular chemical functions. Almost all eukaryotic cells, for example, contain **peroxisomes,** vesicles that contain two sets of enzymes and that are about the same size as lysosomes. One set found in plant seeds converts fats to carbohydrates, and the other set found in all eukaryotes detoxifies various potentially harmful molecules—strong oxidants—that form in cells. They do this by using molecular oxygen to remove hydrogen atoms from specific molecules. These chemical reactions would be very destructive to the cell if not confined to the peroxisomes.

4.8 An extensive system of interior membranes organizes the interior of the cell into functional compartments that manufacture and deliver proteins and carry out a variety of specialized chemical processes.

4.9 Organelles That Contain DNA

Eukaryotic cells contain several kinds of complex, cell-like organelles that appear to have been derived from ancient bacteria assimilated by ancestral eukaryotes in the distant past. The three principal kinds are mitochondria (which occur in the cells of all but a very few eukaryotes), chloroplasts (which do not occur in animal cells—they occur only in algae and plants), and centrioles (relict organelles that are not membrane-bounded and occur in the cells of all animals and most protists, but not in plants and fungi).

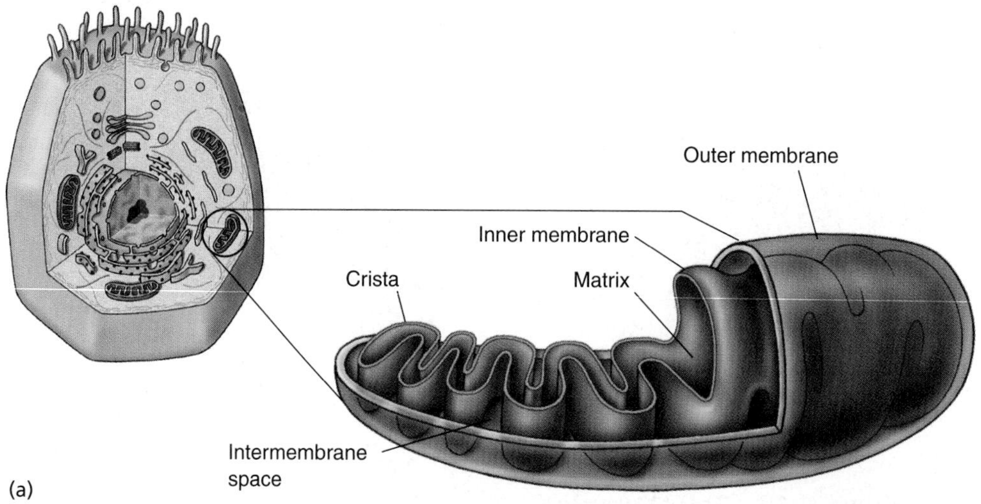

Outer membrane
Inner membrane
Crista
Matrix
Intermembrane space
(a)

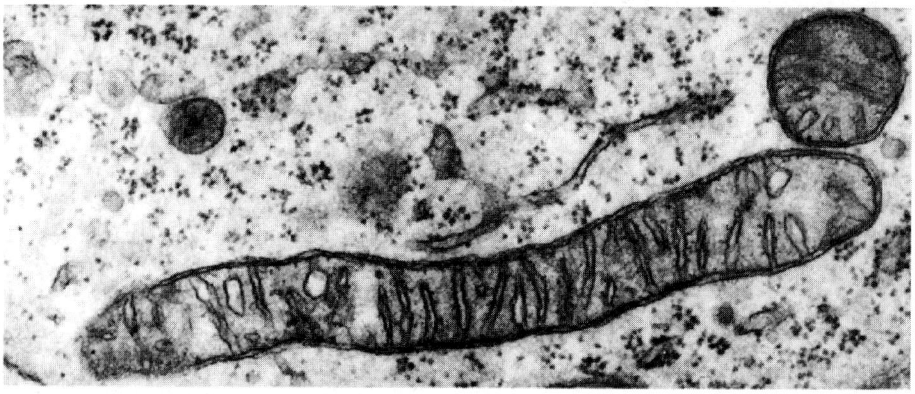

(b)

Figure 4.20 Mitochondria.

The mitochondria of a cell are sausage-shaped organelles within which oxidative metabolism takes place, and energy is extracted from food using oxygen. (*a*) A mitochondrion has a double membrane. The inner membrane is shaped into folds called cristae. The space within the cristae is called the matrix. The cristae greatly increase the surface area for oxidative metabolism. (*b*) Micrograph of two mitochondria, one in cross section, the other cut lengthwise.

Mitochondria: Powerhouses of the Cell

Eukaryotic organisms extract energy from organic molecules ("food") in a complex series of chemical reactions called **oxidative metabolism,** which takes place only in their mitochondria. **Mitochondria** (singular, **mitochondrion**) are sausage-shaped organelles about the size of a bacterial cell (figure 4.20). Mitochondria are bounded by two membranes. The outer membrane is smooth and apparently derives from the plasma membrane of the host cell that first took up the bacterium long ago. The inner membrane, apparently the plasma membrane of the bacterium that gave rise to the mitochondrion, is bent into numerous folds called **cristae** (singular, **crista**) that resemble the folded plasma membranes in various groups of bacteria. The cristae partition the mitochondrion into two compartments, an inner **matrix** and an outer compartment. As you will learn in chapter 5, this architecture is critical to successfully carrying out oxidative metabolism.

During the 1.5 billion years in which mitochondria have existed in eukaryotic cells, most of their genes have been transferred to the chromosomes of the host cells. But mitochondria still have some of their original genes, contained in a circular, closed, naked molecule of DNA (called mitochondrial DNA, or mtDNA) that closely resembles the circular DNA molecule of a bacterium. On this mtDNA are several genes that produce some of the proteins essential for oxidative metabolism. In both mitochondria and bacteria, the circular DNA molecule is replicated during the process of division. When a mitochondrion divides, it splits into two by simple fission, dividing much as bacteria do.

Chloroplasts: Energy-Capturing Centers

All photosynthesis in plants and algae takes place within another bacteria-like organelle, the **chloroplast** (figure 4.21). There is strong evidence that chloroplasts, like mitochondria, were derived by symbiosis from bacteria. A chloroplast is bounded, like a mitochondrion, by two membranes, the inner derived from the original bacterium and the outer from the host cell's ER. Chloroplasts are larger than mitochondria, and their inner membranes have a more complex organization. The inner membranes are fused to form stacks of closed vesicles called **thylakoids.** The light-powered reactions of photosynthesis take place within the thylakoids. The thylakoids are stacked on top of one another to form a column called a **granum** (plural, **grana**). The interior of a chloroplast is bathed with a semiliquid substance called the **stroma.**

Like mitochondria, chloroplasts have a circular DNA molecule. On this DNA are located many of the genes coding

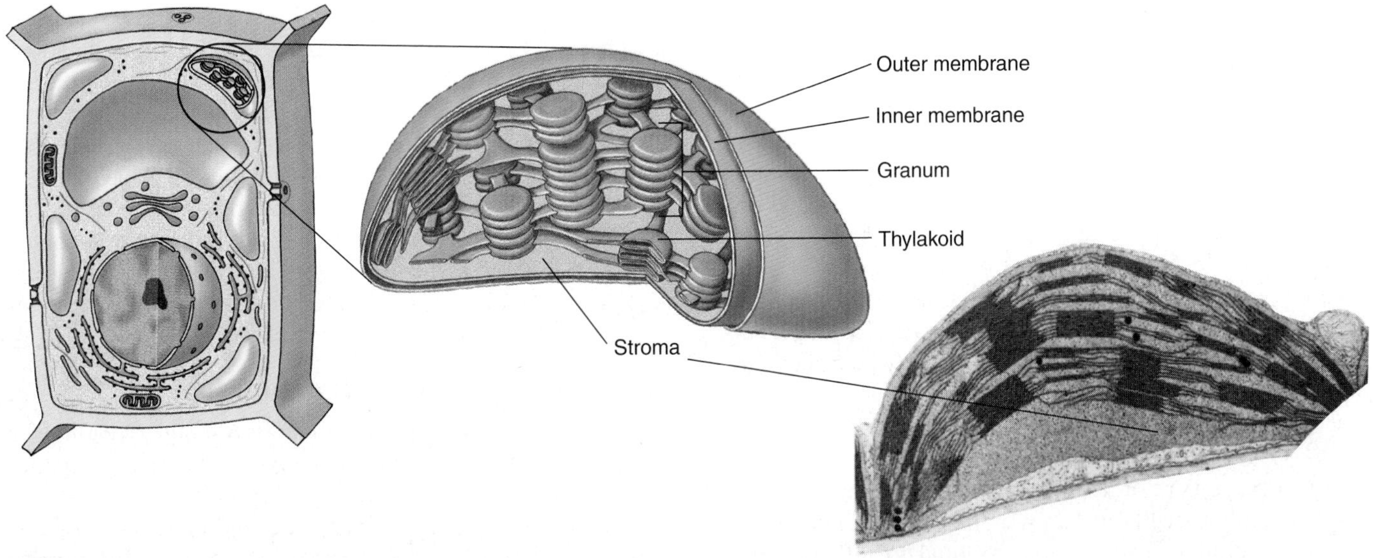
Outer membrane
Inner membrane
Granum
Thylakoid
Stroma

Figure 4.21 A chloroplast.

Bacteria-like organelles called chloroplasts are the sites of photosynthesis in photosynthetic eukaryotes. Like mitochondria, they have a complex system of internal membranes on which chemical reactions take place. The inner membrane of a chloroplast is fused to form stacks of closed vesicles called thylakoids. Photosynthesis occurs within these thylakoids. Thylakoids are stacked one on top of the other in columns called grana. The interior of the chloroplasts is bathed in a semiliquid substance called the stroma.

for the proteins necessary to carry out photosynthesis. Plant cells can contain from one to several hundred chloroplasts, depending on the species. Neither mitochondria nor chloroplasts can be grown in a cell-free culture; they are totally dependent on the cells within which they occur.

Centrioles

Complex structures called **centrioles** (figure 4.22) assemble microtubules from tubulin subunits in the cells of animals and most protists. Centrioles occur in pairs within the cytoplasm, usually located at right angles to one another near the nuclear envelope. They are among the most structurally complex microtubular assemblies of the cell. In cells that contain flagella or cilia, each cilium or flagellum is anchored by a form of centriole called a basal body. Most animal and protist cells have both centrioles and basal bodies; higher plants and fungi lack them, instead organizing microtubules without such structures. Although they lack a membrane, centrioles resemble spirochete bacteria in many other respects, and they contain a circular DNA molecule that is involved in the production of their structural proteins. Some biologists believe that centrioles, like mitochondria and chloroplasts, originated as symbiotic bacteria.

4.9 Eukaryotic cells contain several complex organelles that have their own DNA and are thought to have arisen by endosymbiosis from ancient bacteria.

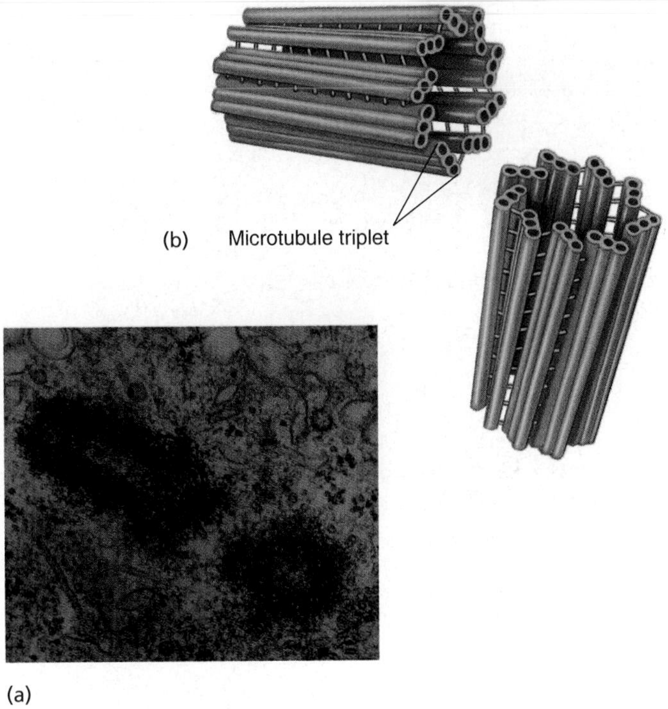
(b) Microtubule triplet
(a)

Figure 4.22 Centrioles.

Centrioles anchor and assemble microtubules. In their anchoring capacity, they can be seen as the basal bodies of eukaryotic flagella. In their organizing capacity, they function during cell division to indicate the plane along which the cell separates. (*a*) A pair of centrioles in a cell are about to divide. Centrioles usually occur in pairs and are located in the cell in characteristic planes. One centriole typically lies parallel to the cell surface, while another lies perpendicular to the surface. (*b*) Centrioles are composed of nine triplets of microtubules.

4.10 Cell Movement

Essentially, all cell motion is tied to the movement of actin filaments, microtubules, or both. Intermediate filaments act as intracellular tendons, preventing excessive stretching of cells, and actin filaments play a major role in determining the shape of cells. Because actin filaments can form and dissolve so readily, they enable some cells to change shape quickly. If you look at the surfaces of such cells under a microscope, you will find them alive with motion. Projections, called **microvilli** in animal cells, shoot outward from the surface and then retract, only to shoot out elsewhere moments later.

Some Cells Crawl

It is the arrangement of actin filaments within the cell cytoplasm that allows cells to "crawl," *literally!* Crawling is a significant cellular phenomenon, essential to inflammation, clotting, wound healing, and the spread of cancer. White blood cells in particular exhibit this ability. Produced in the bone marrow, these cells are released into the circulatory system and then eventually crawl out of capillaries and into the tissues to destroy potential pathogens. The crawling mechanism is an exquisite example of cellular coordination.

Actin filaments play a role in other types of cell movement. For example, during animal cell reproduction (see chapter 6), chromosomes move to opposite sides of a dividing cell because they are attached to shortening microtubules. The cell then pinches in two when a belt of actin filaments contracts like a purse string. Muscle cells also use actin filaments to contract their cytoskeletons. The fluttering of an eyelash, the flight of an eagle, and the awkward crawling of a baby all depend on these cytoskeletal movements within muscle cells.

Swimming with Flagella and Cilia

Flagella (singular, **flagellum**) are fine, long, threadlike organelles protruding from the cell surface. Each flagellum arises from a structure called a **basal body** and consists of a circle of nine microtubule pairs surrounding two central ones (figure 4.23*a*). This **9 + 2 arrangement** is a fundamental feature of eukaryotes and apparently evolved early in their history. Even in cells that lack flagella, derived structures with the same 9 + 2 arrangement often occur, like in the sensory hairs of the human ear. If flagella are numerous and organized in dense rows, they are called **cilia** (figure 4.23*b*). Cilia do not differ from flagella in their structure, but cilia are usually short. In humans, we find a single long flagellum on each sperm cell, and dense mats of cilia line our breathing tube, the trachea, to move mucus and dust particles out of the respiratory tract into the throat (where we can expel these unneeded contaminants by spitting or swallowing).

4.10 Cells can move by changing their shape so as to crawl over a surface, or they can swim along by waving flagella.

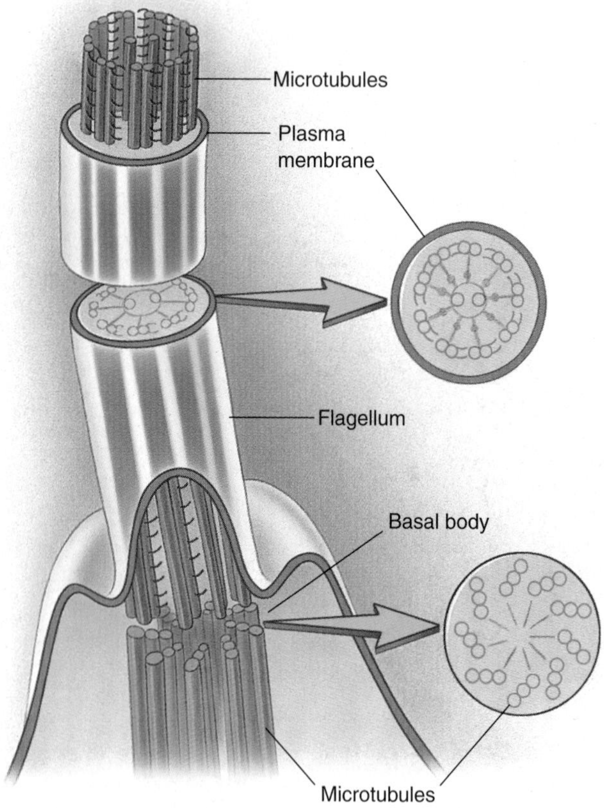

Microtubules
Plasma membrane
Flagellum
Basal body
Microtubules

(a)

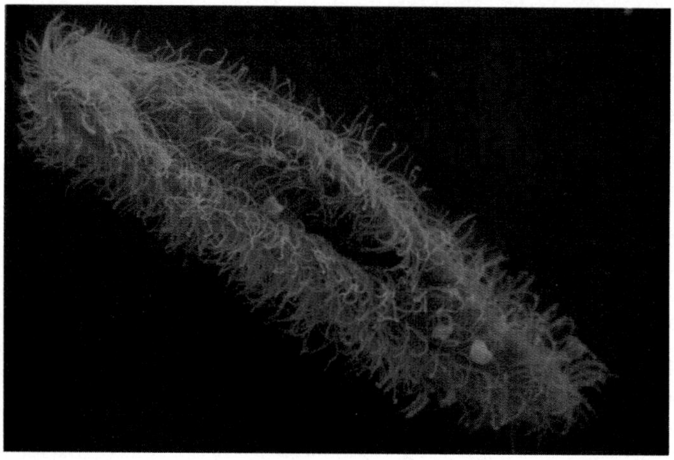

(b)

Figure 4.23 Flagella and cilia.

(*a*) A eukaryotic flagellum springs directly from a basal body and is composed of a ring of nine pairs of microtubules with two microtubules in its core. (*b*) The surface of this paramecium is covered with a dense forest of cilia.

4.11 Special Things About Plant Cells

Vacuoles: A Central Storage Compartment

The center of a plant cell usually contains a large, apparently empty space, called the **central vacuole.** This vacuole is not really empty; it contains large amounts of water and other materials, such as sugars, ions, and pigments. The central vacuole functions as a storage center for these important substances and also helps to increase the surface-to-volume ratio of the plant cell outside the vacuole by applying pressure to the plasma membrane. The plasma membrane expands outward under this pressure, thereby increasing its surface area (figure 4.24).

Cell Walls: Protection and Support

Plant cells share a characteristic with bacteria that is not shared with animal cells—that is, plants have **cell walls,** which protect and support the plant cell. Plant cell walls are chemically and structurally different from bacterial cell walls. Cell walls are also present in fungi and some protists. In plants, cell walls are composed of fibers of the polysaccharide cellulose. **Primary walls** are laid down when the cell is still growing, and between the walls of adjacent cells is a sticky substance called the **middle lamella,** which glues the cells together (figure 4.25). Some plant cells produce strong **secondary walls,** which are deposited inside the primary walls of fully expanded cells.

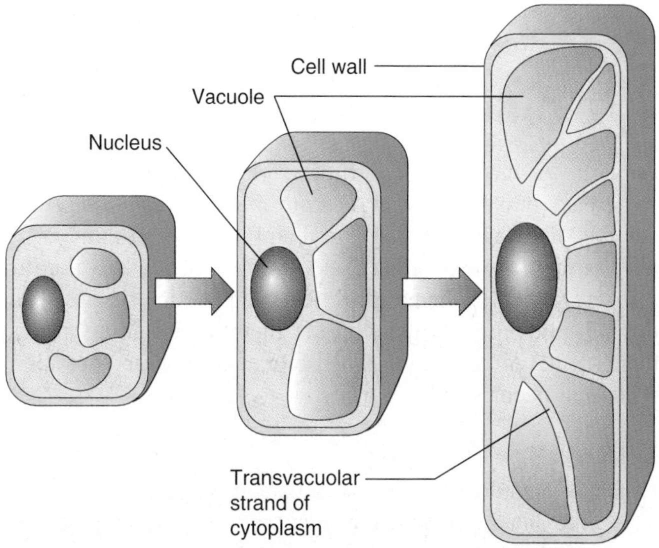
Cell wall
Vacuole
Nucleus
Transvacuolar strand of cytoplasm

Figure 4.24 Vacuole expansion.

A plant's central vacuole can increase in size to increase the surface area of a plant cell. Cytoplasm may be interconnected through the vacuole by cytoplasmic strands.

> **4.11** Plant cells store substances in a large central vacuole and encase themselves within a strong cellulose cell wall.

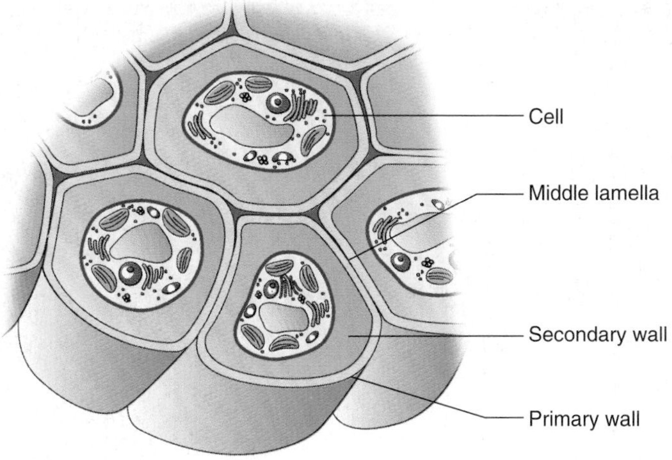
Cell
Middle lamella
Secondary wall
Primary wall

(a)

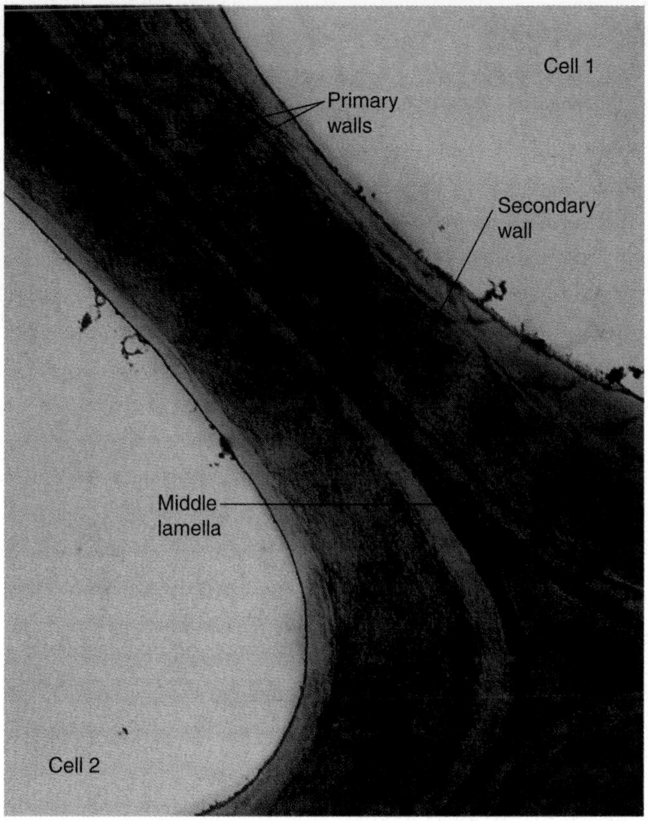
Cell 1
Primary walls
Secondary wall
Middle lamella
Cell 2

(b)

Figure 4.25 Cell walls in plants.

As shown in this drawing (*a*) and transmission electron micrograph (*b*), plant cell walls are fundamentally different from those of bacteria—thicker, stronger, and more rigid. Primary cell walls are laid down when the cell is young. Thicker secondary cell walls may be added later when the cell is fully grown.

4.12 Diffusion and Osmosis

For cells to survive, food particles, water, and other materials must pass into the cell, and waste materials must be eliminated. All of this moving back and forth across the cell's plasma membrane occurs in one of three ways: (1) water diffuses through the membrane; (2) food particles and sometimes liquids are engulfed by the membrane folding around them; or (3) proteins in the membrane act as doors that admit certain molecules only.

Diffusion

How a molecule moves—just where it goes—is totally random, like shaking marbles in a cup, so if two kinds of molecules are added together, they soon mix. The random motion of molecules always tends to produce uniform mixtures, causing a net movement of molecules toward zones where that kind of molecule is scarce (that is, *down* the **concentration gradient**). How does a molecule "know" in what direction to move? It doesn't—molecules don't "know" anything. A molecule is equally likely to move in any direction and is constantly changing course in random ways. There are simply more molecules able to move from where they are common than from where they are scarce. This mixing process is called **diffusion** (figure 4.26). Diffusion is the net movement of molecules down a concentration gradient toward regions of lower concentration (that is, where there are relatively fewer of them) as a result of random motion.

Osmosis

Diffusion allows molecules like oxygen, carbon dioxide, and nonpolar lipids to cross the plasma membrane, but many molecules cannot pass across. This is particularly true of food molecules like sugars and proteins, which are polar and thus unable to freely cross the lipid bilayer of the plasma membrane. Because the plasma membrane holds them in, a cell is able to accumulate and store such food molecules within its cytoplasm. The movement of water molecules is not blocked, however, because they are small enough to pass freely through the membrane. This free movement of water across plasma membranes has a very important consequence—it causes cells to absorb water or to lose water!

To understand this puzzling fact, focus on the water molecules already present inside a cell. What are they doing? Many of them are busily engaged in interacting with the sugars, proteins, and other polar molecules inside. Remember, water is very polar itself and loves to "rub noses" with other polar molecules. These social water molecules are not randomly moving about as they were outside; instead, they remain in place,

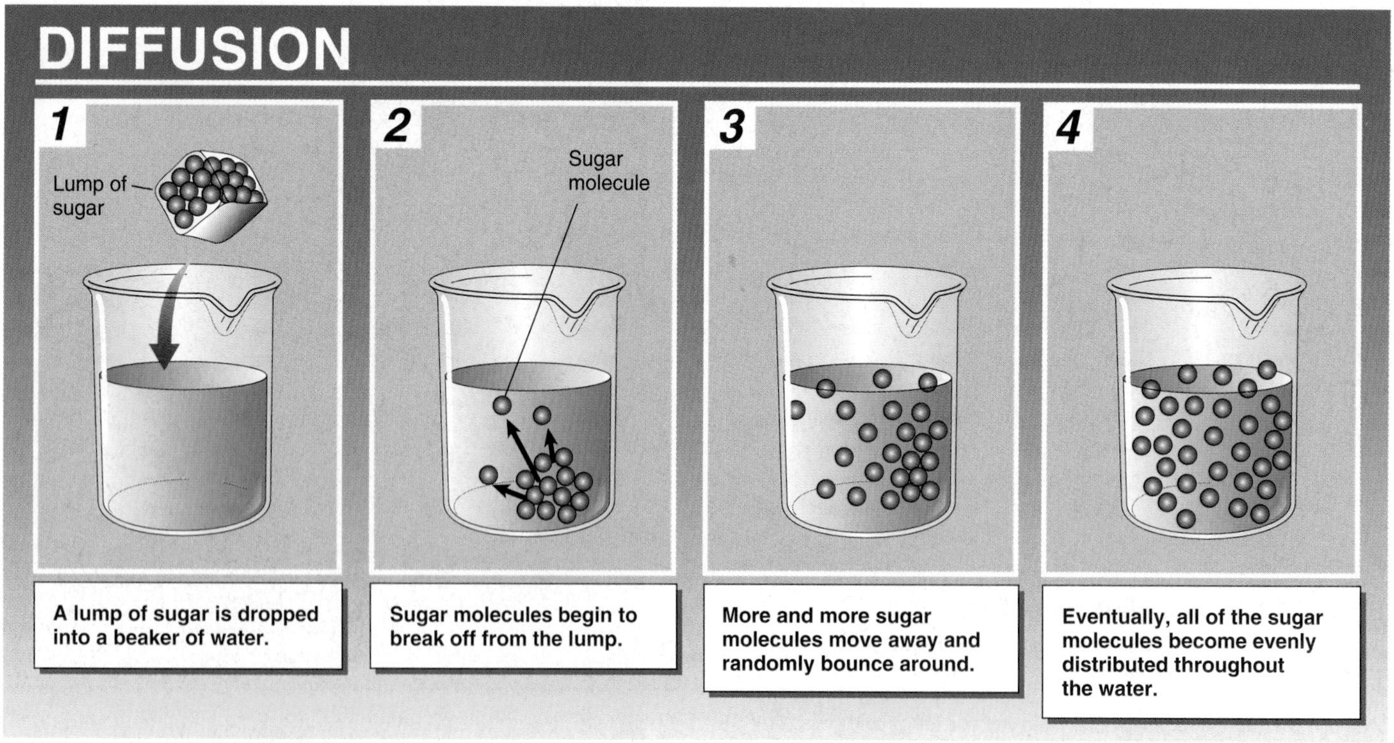
DIFFUSION
1
Lump of sugar
2
Sugar molecule
3
4
A lump of sugar is dropped into a beaker of water.
Sugar molecules begin to break off from the lump.
More and more sugar molecules move away and randomly bounce around.
Eventually, all of the sugar molecules become evenly distributed throughout the water.

Figure 4.26 How diffusion works.

Diffusion is the mixing process that spreads molecules through the cell interior. To see how diffusion works, visualize a simple experiment in which a lump of sugar is dropped into a beaker of water.

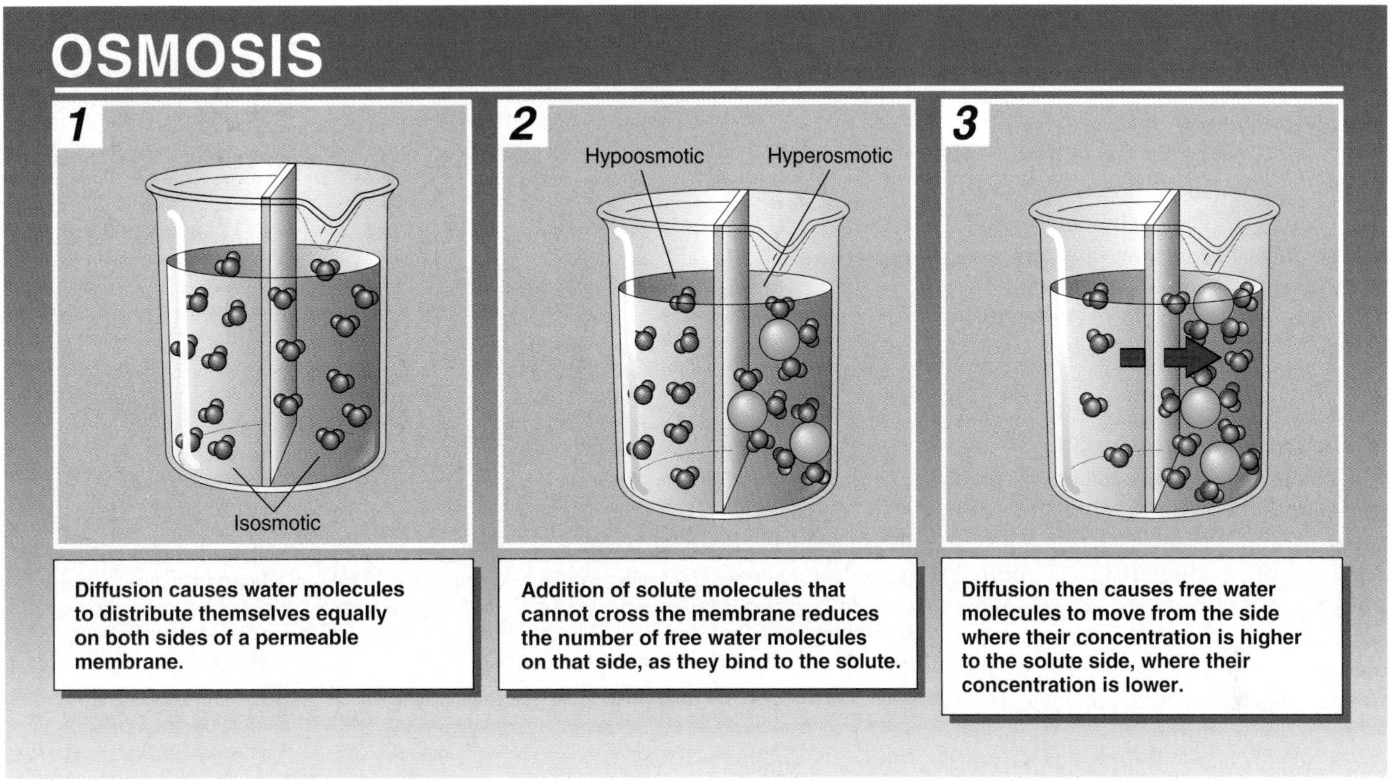
OSMOSIS
1
Isosmotic
Diffusion causes water molecules to distribute themselves equally on both sides of a permeable membrane.
2
Hypoosmotic
Hyperosmotic
Addition of solute molecules that cannot cross the membrane reduces the number of free water molecules on that side, as they bind to the solute.
3
Diffusion then causes free water molecules to move from the side where their concentration is higher to the solute side, where their concentration is lower.

Figure 4.27 How osmosis works.

Osmosis is the net movement of water across a membrane toward the side with less "free" water. To visualize this, imagine adding polar urea molecules to a vessel of water divided in the middle by a membrane. When such a polar solute is added, the water molecules that gather around each urea molecule are no longer free to diffuse across the membrane—in effect, the polar solute has reduced the number of free water molecules. Now imagine you added more urea molecules to one side of the membrane than the other. Because the side of the membrane with less solute (on the left) has more unbound water molecules than the side on the right with more solute, water moves by diffusion from the left to the right.

clustered around the polar molecules they are interacting with. As a result, while water molecules keep coming into the cell by random motion, they don't randomly come out again. Because more water molecules come in than go out, there is a net movement of water into the cell. Diffusion of water across a membrane toward the side with polar molecules that cannot traverse the membrane is called **osmosis** (figure 4.27).

The concentration of *all* molecules dissolved in a solution (called **solutes**) is called the osmotic concentration of the solution. If two solutions have unequal osmotic concentrations, the solution with the higher concentration is said to be **hyperosmotic** (Greek *hyper,* more than), and the solution with the lower concentration is **hypoosmotic** (Greek *hypo,* less than). If the osmotic concentrations of the two solutions are equal, the solutions are **isosmotic** (Greek *iso,* the same).

Movement of water into a cell by osmosis creates pressure, called **osmotic pressure,** which can cause a cell to swell and burst. Most cells cannot withstand osmotic pressure unless their plasma membranes are braced to resist the swelling. If placed in pure water, they soon burst like over-inflated balloons (figure 4.28). That is why the cells of so many kinds of organisms have cell walls to stiffen their exteriors. In animals, the fluids bathing the cells have as many polar molecules dissolved in them as the cells do, so the problem doesn't arise.

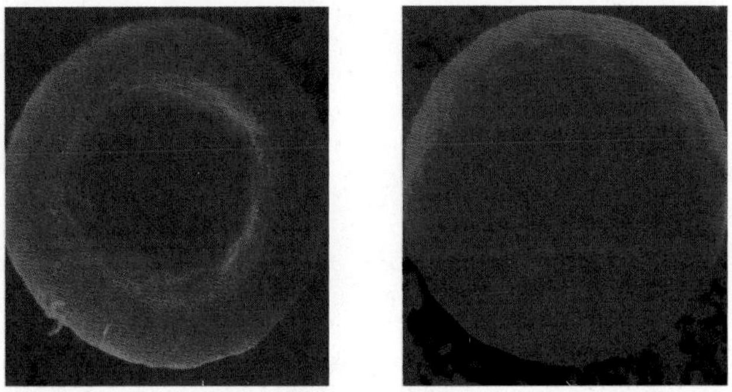

(a) (b)

Figure 4.28 Osmotic pressure in a red blood cell.

(*a*) Normally, a red blood cell has a flattened, pillowlike appearance, the concentrations of solutes inside the cell and in the surrounding fluid are the same, and there is no net movement of water across the membrane. (*b*) If placed in pure water, a red blood cell swells and will ultimately burst because the concentration of solutes is higher inside the cell, and there is a net movement of water into the cell.

4.12 Random movements of molecules cause them to mix uniformly in solution, a process called diffusion. Water associated with polar solutes is not free to diffuse, and there is a net movement of water across a membrane toward the side with less "free" water, a process called osmosis.

4.13 Bulk Passage into and out of the Cell

The cells of many eukaryotes take in food and liquids by extending their plasma membranes outward toward food particles. The membrane engulfs the particle and forms a vesicle—a membrane-bordered sac—around it. This process is called **endocytosis** (figure 4.29). The reverse process, ridding a cell of material by discharging it from vesicles at the cell surface, is called **exocytosis** (figure 4.30).

When the vesicle contains a food particle (or sometimes an entire cell such as a bacterium), that particular version of endocytosis is called **phagocytosis** (from the Greek word *phagein,* to eat). If the material brought in is a liquid containing dissolved molecules, the endocytosis is referred to as **pinocytosis** (from the Greek word *pinein,* to drink). Pinocytosis is common among human cells. Human egg cells, for example, are "nursed" by surrounding cells, which secrete nutrients the maturing egg takes up by pinocytosis. Some types of white blood cells ingest 25% of their volume by pinocytosis each hour!

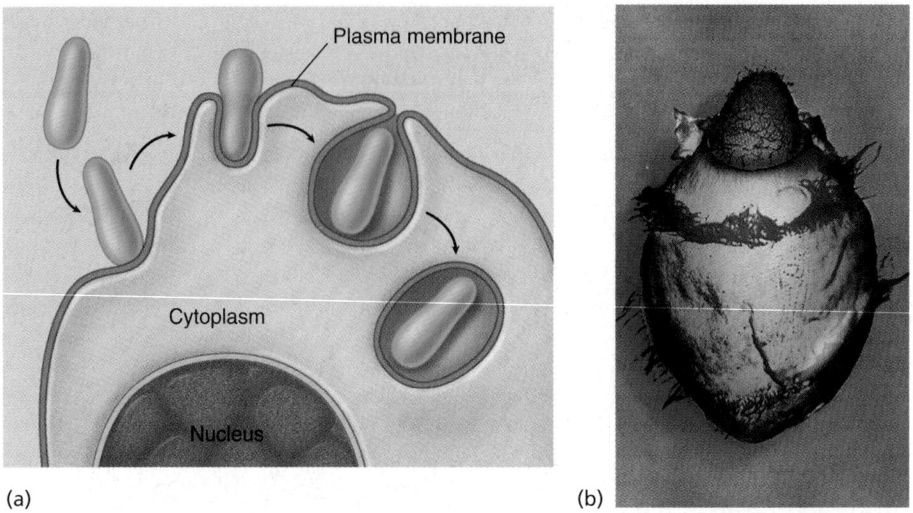

Plasma membrane
Cytoplasm
Nucleus

(a) (b)

Figure 4.29 Endocytosis.

Endocytosis is the process of engulfing material by folding the plasma membrane around it, forming a vesicle. (*a*) When the material is an organism or some other relatively large fragment of organic matter, the process is called phagocytosis. This form of endocytosis can be quite dramatic. In (*b*), the egg-shaped protist *Didinium nasutum* is ingesting the smaller protist, *Paramecium*.

4.13 The plasma membrane can engulf materials by endocytosis, folding the membrane around the material to encase it within a vesicle.

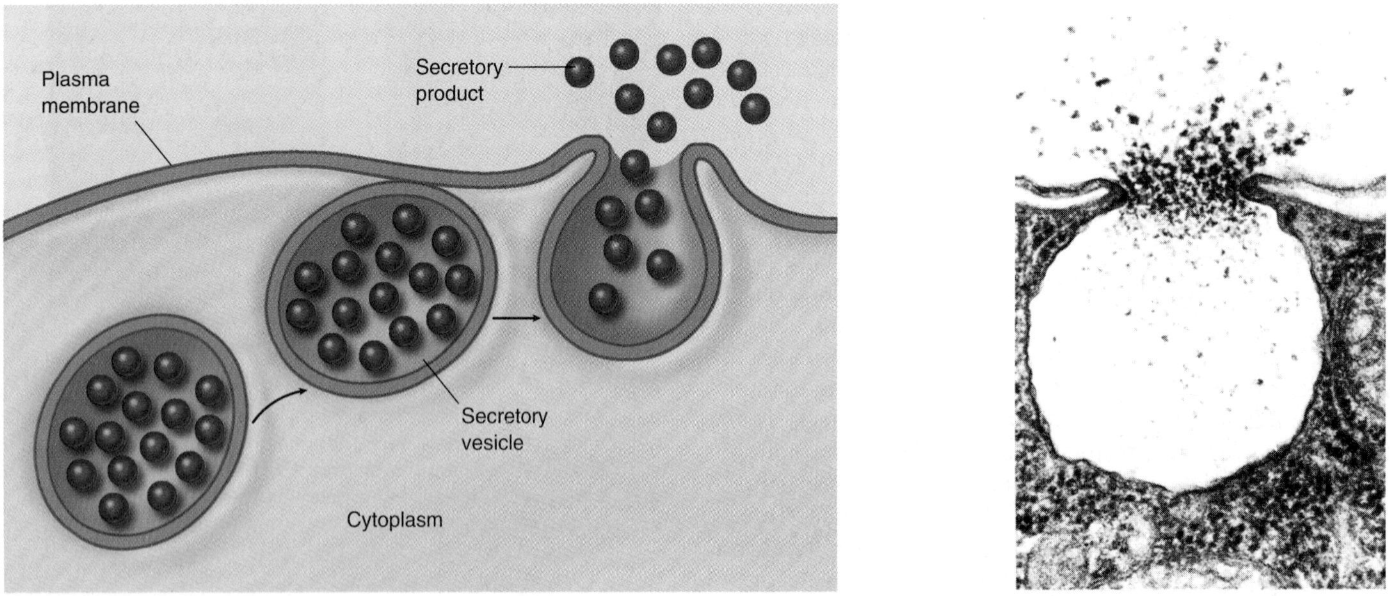

Plasma membrane
Secretory product
Secretory vesicle
Cytoplasm

(a) (b)

Figure 4.30 Exocytosis.

Exocytosis is the discharge of material from vesicles at the cell surface. (*a*) Proteins and other molecules are secreted from cells in small pockets called secretory vesicles, whose membranes fuse with the plasma membrane, thereby allowing the secretory vesicles to release their contents to the cell surface. (*b*) In the photomicrograph, you can see exocytosis taking place explosively.

4.14 Selective Permeability

From the point of view of efficiency, the problem with endocytosis is that it is expensive to carry out—the cell must make and move a lot of membrane. Also, endocytosis is not picky—in pinocytosis particularly, engulfing liquid does not allow the cell to choose which molecules come in. Cells solve this problem by using proteins in the plasma membrane as channels to pass molecules into and out of the cell. Because each kind of channel passes only a certain kind of molecule, the cell can control what enters and leaves, an ability called **selective permeability.**

Selective Diffusion

Some channels act like open doors. As long as a molecule fits the channel, it is free to pass through in either direction. Diffusion tends to equalize the concentration of such molecules on both sides of the membrane, with the molecules moving toward the side where they are scarcest. This mechanism of transport is called **selective diffusion.** One class of selectively open channels consists of ion channels, which have a water-filled pore that spans the membrane. Ions that fit the pore can diffuse through it in either direction. Such ion channels play an essential role in signaling by the nervous system.

Facilitated Diffusion

Most diffusion channels use a special carrier protein. This protein binds only certain kinds of molecules, such as a particular sugar, amino acid, or ion, physically binding them on one side of the membrane and releasing them on the other. The direction of the molecule's net movement depends on its concentration gradient across the membrane. If the concentration is greater in the cytoplasm, the molecule is more likely to bind to the carrier on the cytoplasmic side of the membrane and be released on the extracellular side. If the concentration of the molecule is greater outside in the fluid surrounding the cell, the net movement will be from outside to inside. Thus the net movement always occurs from high concentration to low, just as it does in simple diffusion, but the process is facilitated by the carriers. For this reason, this mechanism of transport is given a special name, **facilitated diffusion** (figure 4.31).

A characteristic feature of transport by carrier proteins is that its rate can be saturated. If the concentration of a substance is progressively increased, the rate of transport of the substance increases up to a certain point and then levels off. There are a limited number of carrier proteins in the membrane, and when the concentration of the transported substance is raised high enough, all the carriers will be in use. The transport system is then said to be "saturated."

4.14 Cells are selectively permeable, admitting only certain molecules. Facilitated diffusion is the transport of molecules and ions across a selectively permeable membrane by specific carrier proteins in the direction of lower concentration of those molecules or ions.

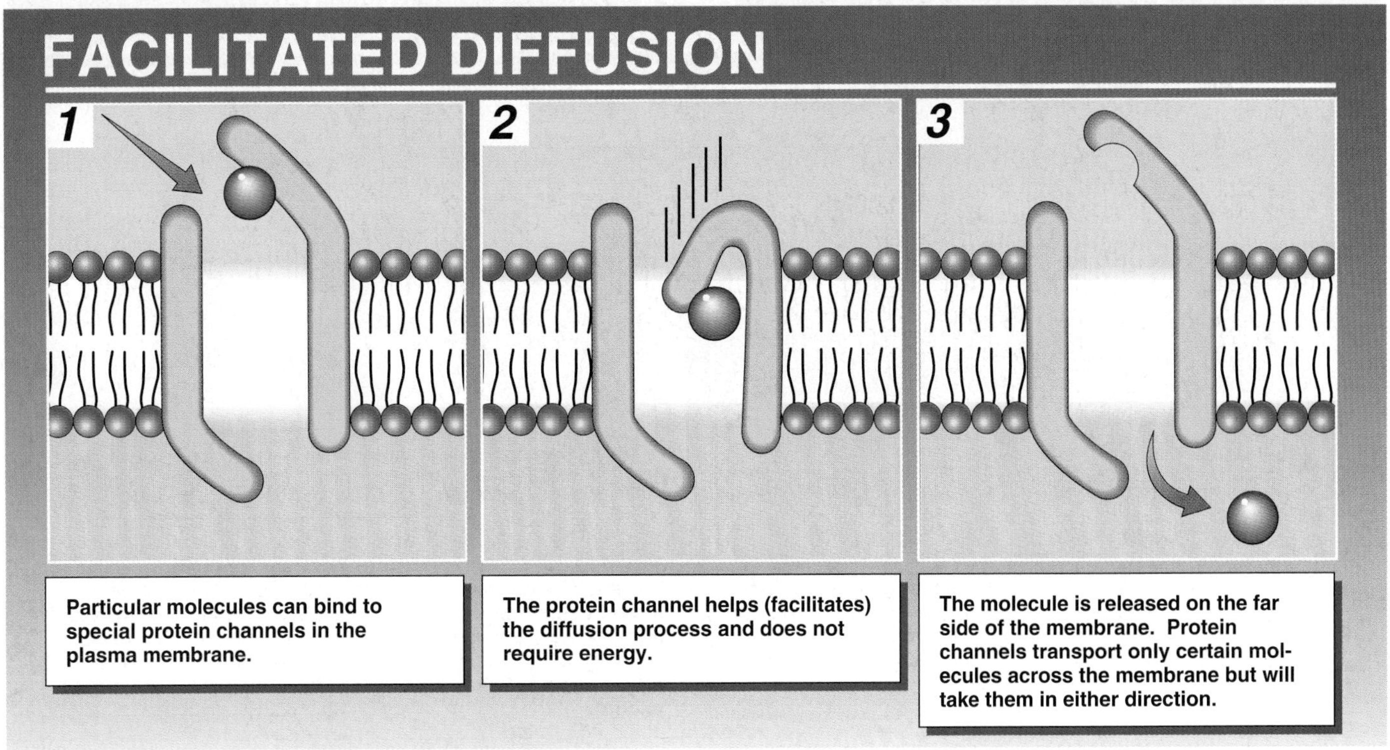
FACILITATED DIFFUSION
1
2
3
Particular molecules can bind to special protein channels in the plasma membrane.
The protein channel helps (facilitates) the diffusion process and does not require energy.
The molecule is released on the far side of the membrane. Protein channels transport only certain molecules across the membrane but will take them in either direction.

Figure 4.31 How facilitated diffusion works.

4.15 Active Transport

Other channels through the plasma membrane are closed doors. These channels open only when energy is provided. They are designed to enable the cell to maintain high or low concentrations of certain molecules, much more or less than exists outside the cell. If the doors were open, the molecules would simply flood in or out by facilitated diffusion. Instead, like motordriven turnstiles, the channels operate only when energy is provided, and they operate only in one direction. The operation of these one-way, energy-requiring channels results in **active transport,** the movement of molecules across a membrane to a region of higher concentration by the expenditure of energy. Active transport is one of the most important activities of any cell, permitting the cell to accumulate food molecules from outside even when the cell has more of them inside its cytoplasm than outside.

You might think that the plasma membrane possesses all sorts of active transport channels for the transport of important sugars, amino acids, and other molecules. You would be wrong. In fact, almost all of the active transport in cells is carried out by only two kinds of channels, the sodium-potassium pump and the proton pump.

The Sodium-Potassium Pump

The first of these, the **sodium-potassium (Na^+-K^+) pump,** expends metabolic energy to actively pump sodium ions (Na^+) out of cells and potassium ions (K^+) in (figure 4.32). More than one-third of all the energy expended by your body's cells is spent driving Na^+-K^+ pump channels. This energy is derived from *adenosine triphosphate (ATP),* a molecule we will learn about in the next chapter. Each channel can move over 300 sodium ions per second when working full tilt. As a result of all this pumping, there are far fewer sodium ions in the cell. This concentration gradient, paid for by the expenditure of considerable metabolic energy in the form of ATP molecules, is exploited by your cells in many ways. Two of the most important are (1) the conduction of signals along nerve cells (discussed in detail in chapter 26) and (2) the pulling into the cell of valuable molecules such as sugars and amino acids *against* their concentration gradient!

We will focus for a moment on this second process. The plasma membranes of many cells are studded with facilitated diffusion channels, which offer a path for sodium ions that have been pumped out by the Na^+-K^+ pump to diffuse back in. There is a catch, however: these channels require that the sodium ions have a partner in order to pass through—like a dancing party where only couples are admitted through the door. These special channels won't let sodium ions across unless another molecule tags along, crossing hand in hand with the sodium ion. In some cases the partner molecule is a sugar, in others an amino acid or other molecule. Because so many sodium ions are trying to get back in, this diffusion pressure drags in the partner molecules as well, even if they

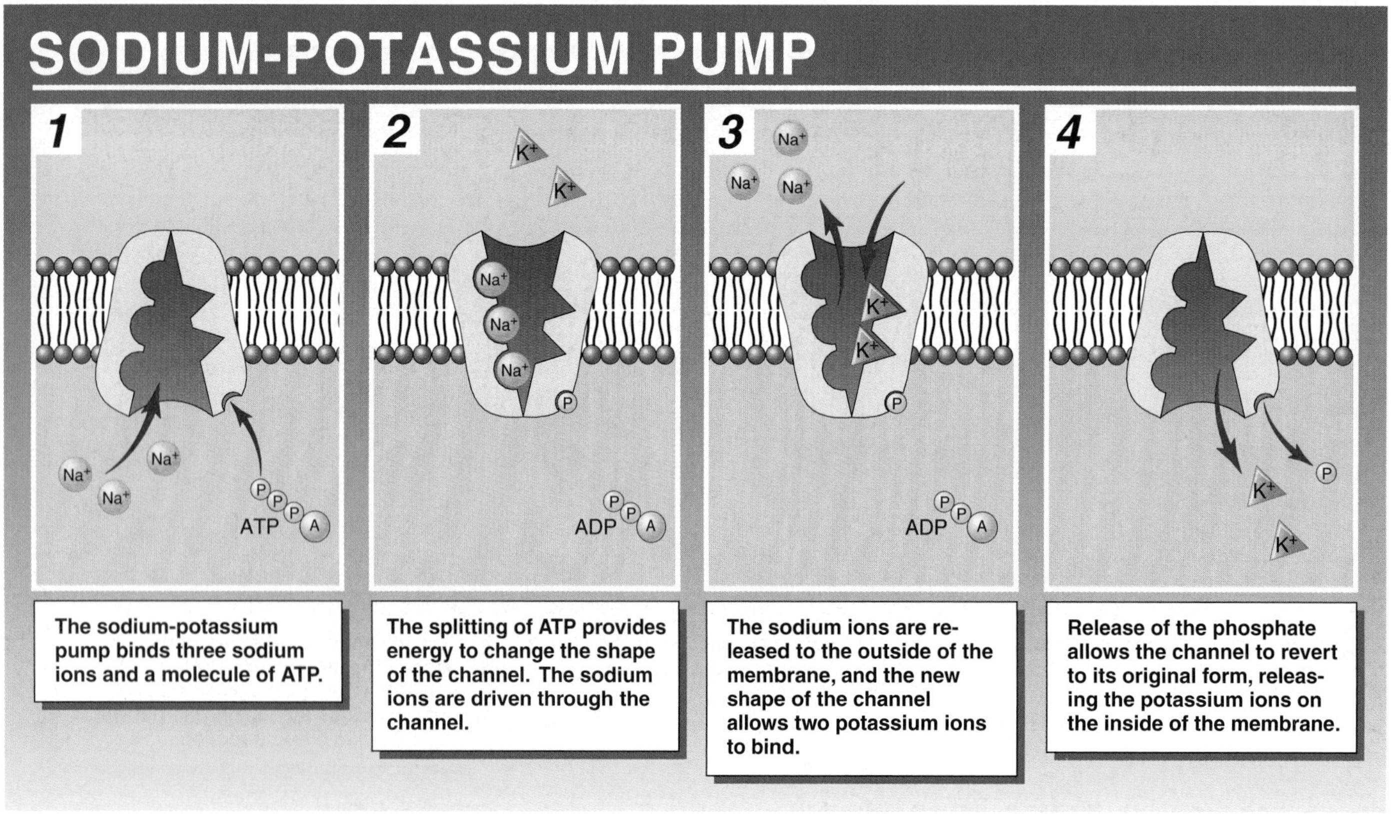

SODIUM-POTASSIUM PUMP
1
Na+
Na+
Na+
P
P
P
ATP
A
2
K+
K+
Na+
Na+
Na+
P
P
P
ADP
A
3
Na+
Na+
Na+
K+
K+
P
P
P
ADP
A
4
K+
K+
P
The sodium-potassium pump binds three sodium ions and a molecule of ATP.
The splitting of ATP provides energy to change the shape of the channel. The sodium ions are driven through the channel.
The sodium ions are released to the outside of the membrane, and the new shape of the channel allows two potassium ions to bind.
Release of the phosphate allows the channel to revert to its original form, releasing the potassium ions on the inside of the membrane.

Figure 4.32 How the sodium-potassium pump works.

are already in high concentration within the cell. In this way, sugars and other actively transported molecules enter the cell—via special **coupled channels** (figure 4.33). Their movement is in fact a form of facilitated diffusion driven by the active transport of sodium ions.

The Proton Pump

The second major active transport channel is the **proton pump,** a complex channel that expends metabolic energy to pump protons across membranes (figure 4.34). Just as in the sodium-potassium pump, this creates a diffusion pressure that tends to drive protons back across again, but in this case the only channels open to them are not coupled channels but rather channels that make ATP, the energy currency of the cell. This pump is the key to cell metabolism, which is the way cells convert photosynthetic energy or chemical energy from food into ATP. Its activity is referred to as **chemiosmosis.** We discuss chemiosmosis at greater length in the next chapter.

4.15 Active transport is the energy-driven transport of molecules across a membrane toward a region of higher concentration. Typically the transport of molecules is coupled to the transport of sodium or hydrogen ions.

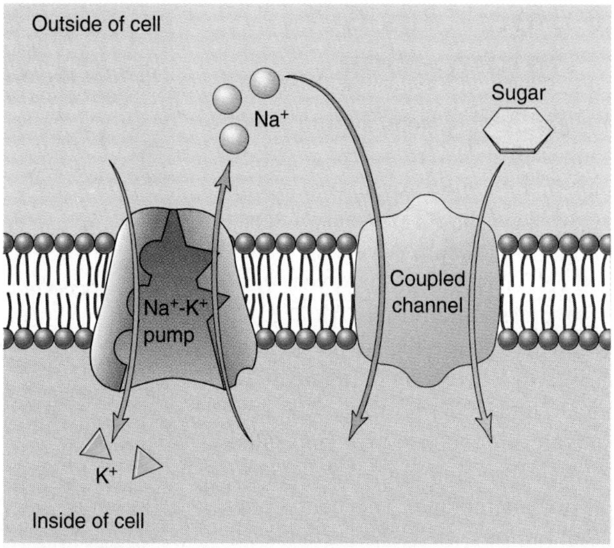

Outside of cell
Na^+
Sugar
Na^+-K^+ pump
Coupled channel
K^+
Inside of cell

Figure 4.33 A coupled channel.

The active transport of a sugar molecule into a cell typically takes place in two stages—facilitated diffusion of the sugar coupled to active transport of sodium ions. The sodium-potassium pump keeps the Na^+ concentration higher outside the cell than inside. For Na^+ to diffuse back in through the coupled channel requires the simultaneous transport of a sugar molecule as well. Because the concentration gradient for Na^+ is steeper than the opposing gradient for sugar, Na^+ and sugar move into the cell.

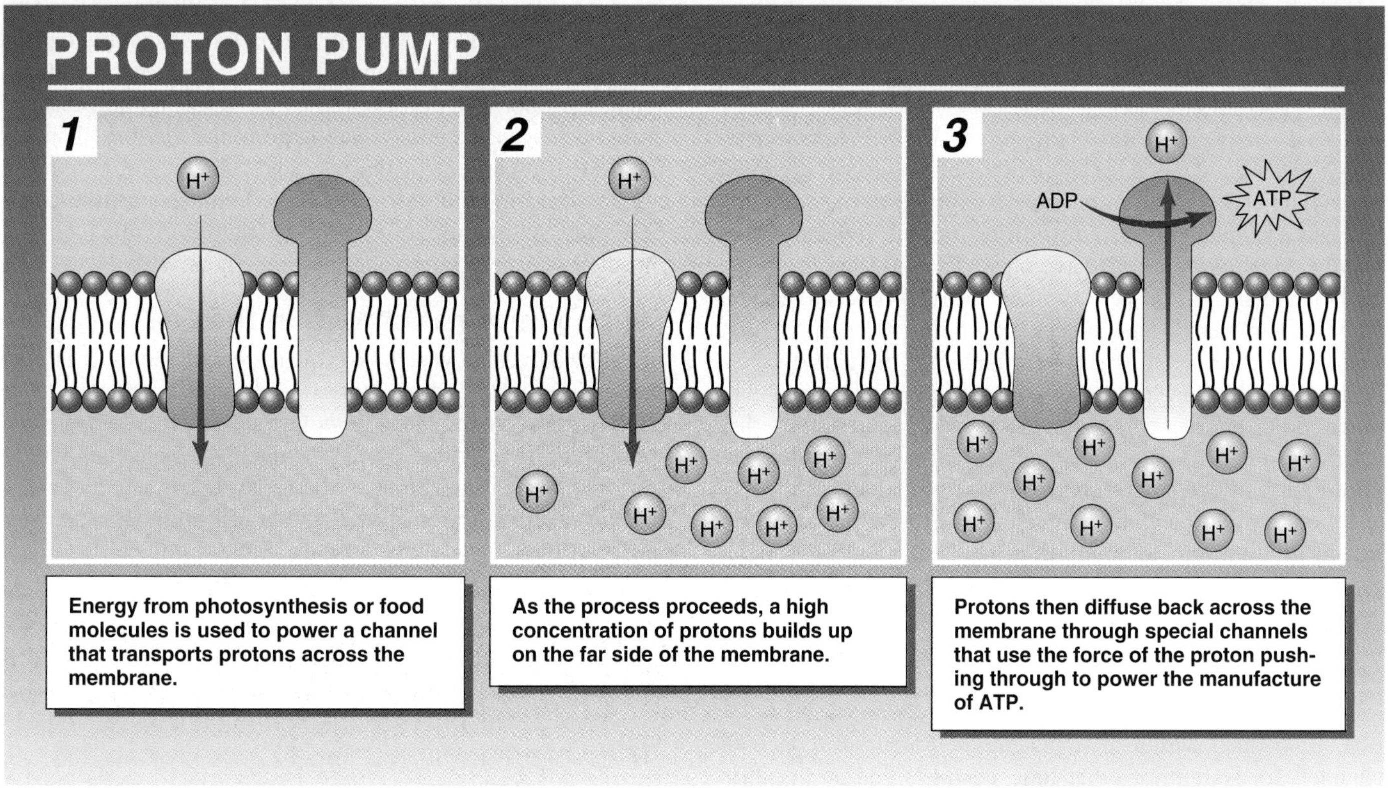

PROTON PUMP
1
H^+
Energy from photosynthesis or food molecules is used to power a channel that transports protons across the membrane.
2
H^+
As the process proceeds, a high concentration of protons builds up on the far side of the membrane.
3
ADP
ATP
H^+
Protons then diffuse back across the membrane through special channels that use the force of the proton pushing through to power the manufacture of ATP.

Figure 4.34 How the proton pump works.

4.16 How Cells Get Information

A cell's ability to respond appropriately to changes in its environment is a key element in its ability to survive. Cells have evolved a variety of ways of sensing things about them. A very few specialized cells have "ears" or "eyes" sensitive to pressure or light. Almost all cells, however, sense their environment primarily by detecting chemical or electrical signals. To do this, cells employ a battery of special proteins called cell surface proteins embedded within the plasma membrane. The many proteins that protrude from the surface of a cell are the cell's only contact with the outside world—its only avenue of communication with its environment.

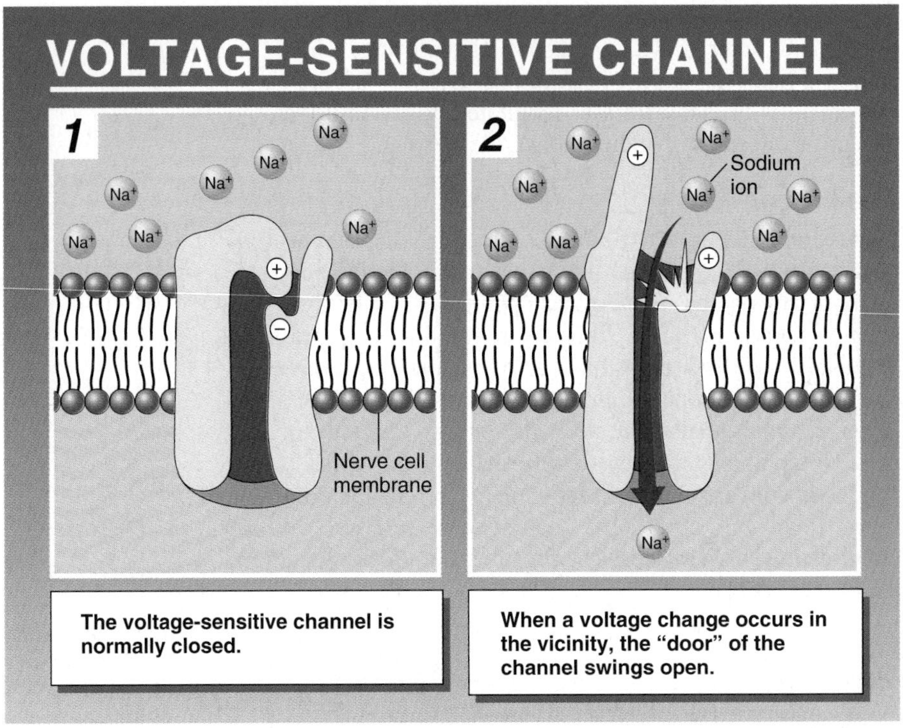

VOLTAGE-SENSITIVE CHANNEL
1
Na+
Nerve cell membrane
The voltage-sensitive channel is normally closed.
2
Sodium ion
When a voltage change occurs in the vicinity, the "door" of the channel swings open.

Figure 4.35 How voltage-sensitive channels work.

Sensing Chemical Information

Cells sense chemical information by means of cell surface proteins called **receptor proteins** projecting from their plasma membranes. These proteins bind a particular kind of molecule, but they do not provide a channel for the molecule to enter the cell. What the receptors do transmit into the cell is information. Often the information is about the presence of other cells in the vicinity—sensing cellular identity is the basis of the immune system, which defends your body from infection. In other instances the information may be a chemical signal sent from other cells.

Your body uses chemical signals called **hormones,** which provide a good example of how receptor proteins work. The end of a receptor protein exposed to the cell surface has a shape that fits to a specific hormone molecule, like insulin. Most of your cells have only a few insulin receptors, but your liver cells possess as many as 100,000 each! When an insulin molecule encounters an insulin receptor on the surface of a liver cell, the insulin molecule binds to the receptor. This binding produces a change in the shape of the other end of the receptor protein (the end protruding into the interior of the cell), just as stamping hard on your foot causes your mouth to open. This change in receptor shape at the interior end initiates a change in cell activity—in this case, the end protruding into the cytoplasm begins to add phosphate groups to proteins, and so it activates a variety of cell processes involved with regulating glucose levels in the blood.

Sensing Voltage

Many cells can sense electrical as well as chemical information. Embedded within their plasma membranes are special channels for sodium or other ions, channels that are usually closed. Unlike the sodium-potassium pump, these closed channels do not open in response to chemical energy. What does open these channels is voltage. Like little magnets, they flip open or shut in response to electrical signals.

It is not difficult to understand how **voltage-sensitive channels** work. The center of the protein that provides the channel through the membrane is occupied by a voltage-sensitive "door"—a portion of the protein containing charged amino acids. When a voltage charge in the vicinity changes, the door flips up out of the way and the channel is open to the passage of sodium or other ions. Voltage-sensitive channels play many important roles within excitable cells of muscle tissue and the nervous system (figure 4.35).

Sensing Information Within the Cell

In eukaryotic cells, which have many compartments, it is very important that the different parts of the cell be able to sense what is going on elsewhere in the cytoplasm. This sort of communication is provided by the diffusion of molecules within the cytoplasm. Some of the chemical signals are molecules used in metabolism; others are within-cell hormones; and still others are ions, particularly those that indicate cell pH.

> **4.16** Cells obtain information about their surroundings from a battery of proteins protruding from the cell membrane. Some are sensitive to chemical information and others to voltage.

CONCEPT REVIEW

1. Select the incorrect statement.
a. All organisms consist of one or more cells.
b. Cells are the smallest living things.
c. Life evolved 2 billion years ago.
d. Cells arise only by the division of other cells.

2. Select the molecule not normally found in the plasma membrane.
a. phospholipid
b. freely moving proteins
c. transmembrane proteins
d. DNA

3. The DNA of eukaryotic cells is fragmented into several pieces called _________, unlike the DNA of bacteria, which is mostly contained in a single circular molecule.
a. ribosomes
b. peroxisomes
c. chromosomes
d. chloroplasts

4. Proteins are made at the
a. ribosome.
b. mitochondrion.
c. lysosome.
d. Golgi complex.

5. Arrange in order the sequence of membrane-bounded structures through which a secreted protein must pass.
a. Golgi body
b. secretory vesicle
c. rough ER
d. smooth ER

6. Which of the following statements does *not* describe osmosis?
a. Water molecules diffuse.
b. Water molecules pass through a membrane.
c. Water enters a hypotonic solution.
d. Movement of water solution creates osmotic pressure.

7. Which of the following processes is distinctly different from the other three?
a. endocytosis
b. exocytosis
c. phagocytosis
d. pinocytosis

8. Which of the following does *not* describe active transport?
a. Active transport requires energy.
b. Molecules move from a lower concentration to a higher concentration.
c. The sodium-potassium pump is one example of active transport.
d. Active transport involves water diffusing through a membrane.

9. _________ are organisms with the simplest cell structure.

10. _________ are long, threadlike organelles projecting from the cell that function in cell movement.

11. _________ are one type of human cell organelle known to have their own DNA.

CHALLENGE YOURSELF

1. Mitochondria and chloroplasts are thought to be the evolutionary descendants of living cells that were engulfed by other cells. Are mitochondria and chloroplasts alive?

2. Some cells are very much larger than others. What would you expect the relationship to be between cell size and level of cell activity?

3. How does a Golgi body "know" where to send what vesicle?

4. Trace a synthesizing protein through the endomembrane system, culminating in release of the protein outside the cell.

5. What mechanism could account for the appearance of the same 9 + 2 structure in three different cellular components (centrioles, flagella, and cilia)?

4 eBRIDGE

Reinforcing Key Points

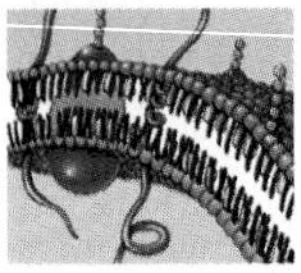

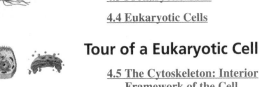

The World of Cells

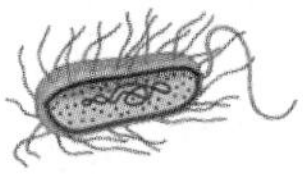

Kinds of Cells

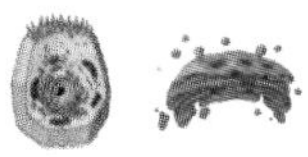

Tour of a Eukaryotic Cell

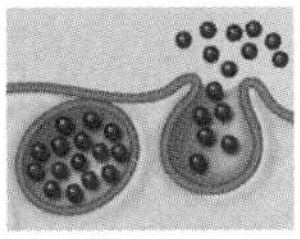

Transport Across Cell Membranes

Electronic Learning

Visual Learning

Animations

Four Animations

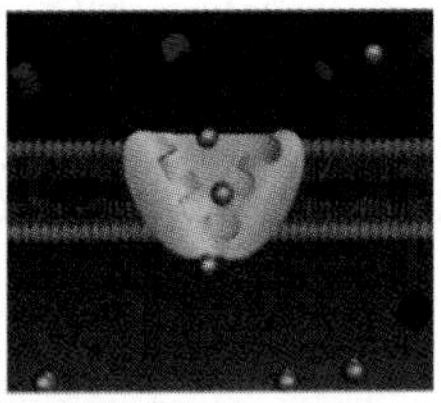

Art Labeling Activities

Thirteen Art Labeling Activities

Helping You Learn

Five Exercises

Explorations

Cell Size

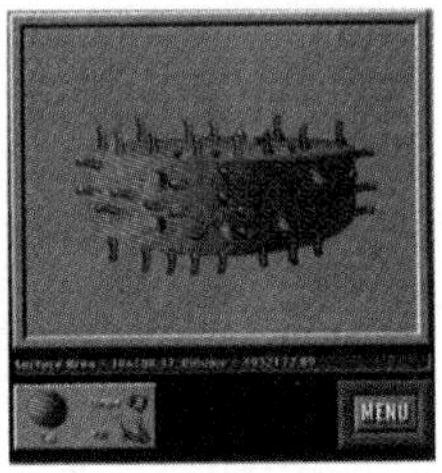

MENU

This interactive exercise allows you to explore the architecture of a cell, observing how changes in cell size and shape affect ease of access to the cell interior.

Active Transport

This interactive exercise allows you to alter ATP concentrations and relative concentrations of substances and then explore transport across a membrane.

Author's Corner

Cytoskeleton Interactions. One of the most exciting developments in cell biology toward the close of the last century was the revelation that a cell's interior is highly organized by a complex of fibers and tubes called the cytoskeleton. Some of the fibers serve structural roles, like the steel girders of a building, while others guide the movement of materials through the cell, much as a system of rails guides the movements of railway trains around the country. This transport function of the cytoplasm plays a key role in the development of a fruit fly, as it allows molecular signals to pass from the nucleus across the cytoplasm to the back end of the egg. The positioning of these molecular signals there determines the future axis of the body. How is this positioning achieved?

1. How does a fruit fly label its rear end, and why should we care?
2. Nobel Prize winner figured out how a cell's proteins decide where they will go.

The Living World

Virtual Classroom

Cystic Fibrosis: A Membrane Disorder

Cystic fibrosis is the most common fatal gene disorder of Caucasians. Every year about 1 in 2,500 American infants is born with the disorder. Their body cells secrete a thick mucus that clogs the airways of the lungs and blocks the ducts of the pancreas and liver. Cystic fibrosis is usually thought of as a children's disease because until recently few affected individuals lived long enough to become adults. Even today half die before their mid-twenties. There is no known cure.

Cystic fibrosis results from a defect in a gene encoding a plasma membrane protein. Called CFTR (*c*ystic *f*ibrosis *t*ransmembrane conductance *r*egulator), this protein channel through the membrane regulates passage of chloride ions into and out of the body's cells. A defective CFTR channel leads to a buildup of chloride ions within lung, pancreas, and liver cells, causing water to move into the cells by osmosis. Removing water from the surrounding mucus causes it to thicken, clogging the passageways. Attempts are underway to cure cystic fibrosis with gene therapy, using a virus to ferry healthy CFTR genes into patients lacking them. While much remains to be done, the future is bright.

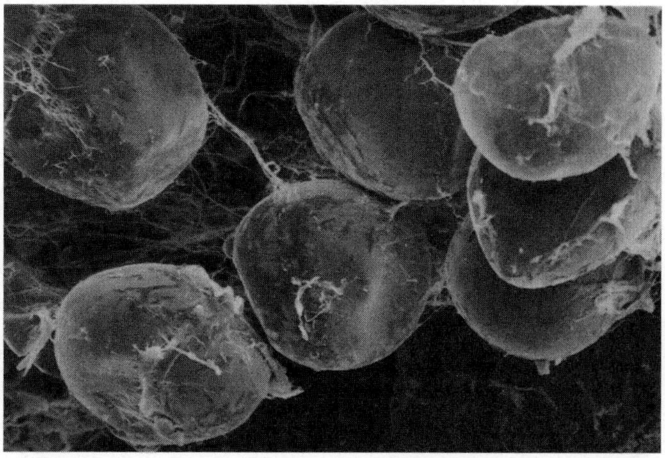

Virtual Lab

How Do the Cells of a Growing Plant Know in Which Direction to Elongate?

Sometimes questions that seem simple can be devilishly difficult to answer. For example, did you ever wonder how the individual cells within a blade of grass know in what direction to grow to keep the blade growing up and not out?

This question has been addressed experimentally in a simple and direct way in the laboratory of Richard Cyr at Pennsylvania State University. Plant cells conduct mechanical force well from one cell to another, and Carol Wymer (then a graduate student in the Cyr lab) suspected some sort of mechanical force was the signal guiding cortical microtubule alignment. Cortical microtubules lie just beneath the cell membrane and control the direction of cell elongation by forming a template for building the cell walls. Wymer set out to test this hypothesis using centrifugation. If cortical microtubules are obtaining their positional information from an applied force, then their realignment should be affected by centrifugal force. Also, chemicals that disrupt microtubule formation would interfere with the realignment of microtubules by the centrifugal force.

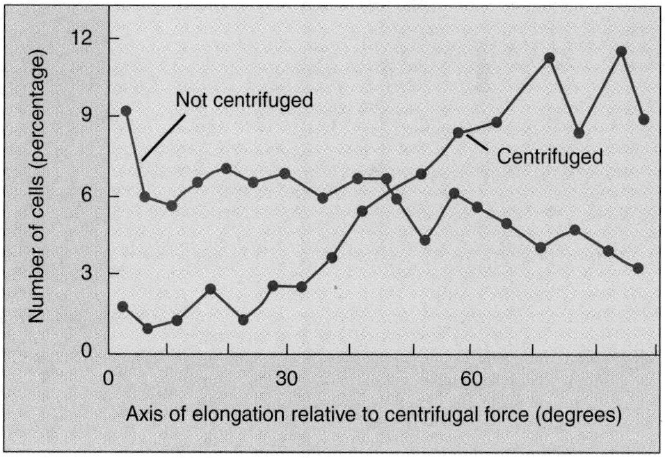

Number of cells (percentage)
12
9
6
3
0
Not centrifuged
Centrifuged
0
30
60
Axis of elongation relative to centrifugal force (degrees)

Quizzes | Further Reading | Essential Study Partner | Links | BioCourse.com

CHAPTER 5

ENERGY AND LIFE

THE FAR SIDE® By GARY LARSON
© 1985 FarWorks, Inc. All Rights Reserved/Dist. by Creators Syndicate
Oh please, oh please...
CAT FUD
CAT FUD
CAT FUD
Larson

CHAPTER 5

Energy and Life

CHAPTER OVERVIEW

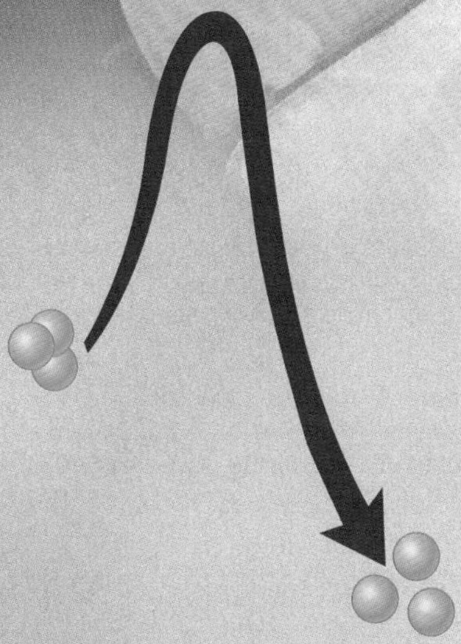

Cells and Energy

- Reactions that release energy are called exergonic, and those where the products contain more energy than the reactants are called endergonic.
- Almost all chemical reactions require an input of energy, called activation energy, to start them off.
- Activation energies of cellular reactions are lowered during catalysis by an enzyme.
- ATP serves as the molecular energy currency for virtually all of the cell's activities. Energy is released when ATP is cleaved into ADP + P_i.

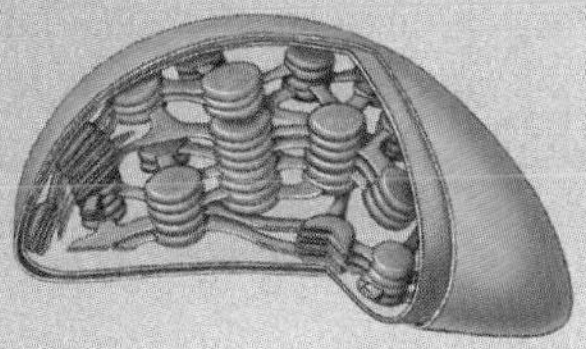

Photosynthesis

- Energy ultimately reaches organisms via photosynthesis, which captures energy from sunlight and uses it to make ATP and NADPH.
- In the Calvin cycle, chloroplasts use the ATP and NADPH to convert CO_2 in the air into organic molecules.
- Ultimately, photosynthesis consumes CO_2 and releases O_2.

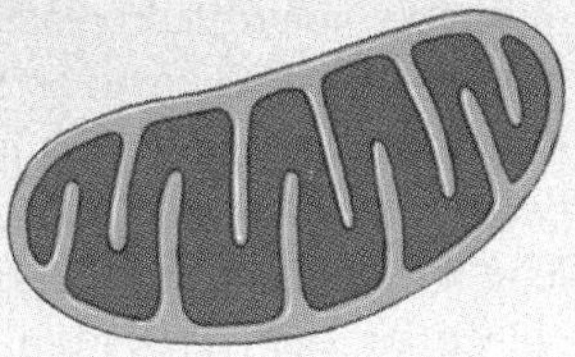

Cellular Respiration

- The first stage of cellular respiration is glycolysis, which does not require oxygen.
- The next stages involve oxidation of the product of glycolysis, first to acetyl-CoA and then, via the Krebs cycle, to CO_2.
- Ultimately, oxidative cellular respiration consumes O_2 and releases CO_2.

5.1 The Flow of Energy in Living Things

Studying the chemistry of the cell may seem to you an uninteresting way to study biology. Indeed, few subjects put off biology students more quickly than chemistry. The prospect of encountering chemical equations seems at the same time both difficult and boring. Here we will try to keep the pain to a minimum, to avoid equations where possible, and to stress ideas rather than formulas. However, there is no avoiding some study of cell chemistry, for the same reason that a successful race car driver must learn how the engine of a car works. We are chemical machines, powered by chemical energy, and if we are to understand ourselves, we must "look under the hood" at the chemical machinery of our cells and see how it operates.

Energy is defined as the capacity to do work. It can be considered to exist in two states. *Kinetic energy* is the energy of motion. Objects that are not actively moving but have the capacity to do so are said to possess *potential energy,* or stored energy. Hence, a boulder perched on a hilltop has potential energy; as it begins to roll downhill, some of its potential energy is converted into kinetic energy. Much of the work carried out by living organisms involves the transformation of potential energy to kinetic energy.

Energy exists in many forms: mechanical energy, heat, sound, electric current, light, or radioactive radiation. Because it can exist in so many forms, there are many ways to measure energy. The most convenient is in terms of heat, because all other forms of energy can be converted into heat. In fact, the study of energy is called **thermodynamics,** meaning heat changes. The unit of heat most commonly employed in biology is the kilocalorie (kcal). One kilocalorie is equal to 1,000 calories (cal), and 1 calorie is the heat required to raise the temperature of 1 gram of water 1 degree Celsius (°C). (It is important not to confuse calories with a term related to diets and nutrition, the Calorie with a capital C, which is actually another term for kilocalorie.)

Oxidation-Reduction

Energy flows into the biological world from the sun, which shines a constant beam of light on the earth. It is estimated that the sun provides the earth with more than 13×10^{23} calories per year, or 40 million billion calories per second! Plants, algae, and certain kinds of bacteria capture a fraction of this energy through photosynthesis. In photosynthesis, energy garnered from sunlight is used to combine small molecules (water and carbon dioxide) into more complex molecules (sugars). The energy is stored in the covalent bonds between atoms in the sugar molecules. Recall from chapter 3 that an atom consists of a central nucleus surrounded by one or more orbiting electrons, and a covalent bond forms when two atomic nuclei share electrons. Breaking such a bond requires energy to pull the nuclei apart. Indeed, the strength of a covalent bond is measured by the amount of energy required to break it. For example, it takes 98.8 kcal to break 1 mole (6.023×10^{23}) of carbon–hydrogen (C–H) bonds.

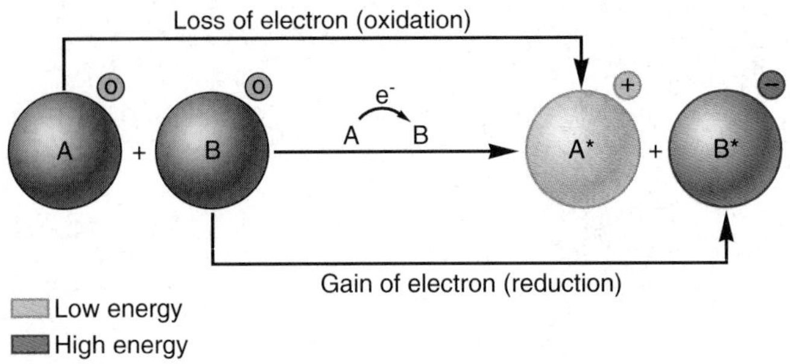
Loss of electron (oxidation)
A
B
e⁻
A B
A*
B*
Gain of electron (reduction)
Low energy
High energy

Figure 5.1 Redox reactions.
Oxidation is the loss of an electron; reduction is the gain of an electron. Here the charges of molecules A and B are shown in small circles to the upper right of each molecule. Molecule A loses energy as it loses an electron, while molecule B gains energy as it gains an electron.

All the chemical activities within cells can be viewed as a series of chemical reactions between molecules. A **chemical reaction** is the making or breaking of chemical bonds—gluing atoms together to form new molecules or tearing molecules apart and sometimes sticking the pieces onto other molecules. During a chemical reaction, the energy stored in chemical bonds may transfer to new bonds. In some of these reactions, electrons actually pass from one atom or molecule to another. When an atom or molecule loses an electron, it is said to be oxidized, and the process by which this occurs is called **oxidation.** The name reflects the fact that in biological systems, oxygen, which attracts electrons strongly, is the most common electron acceptor. Conversely, when an atom or molecule gains an electron, it is said to be reduced, and the process is called **reduction.** Oxidation and reduction always take place together, because every electron that is lost by an atom through oxidation is gained by some other atom through reduction. Therefore, chemical reactions of this sort are called oxidation-reduction (redox) reactions (figure 5.1). Energy is transferred from one molecule to another via redox reactions. The reduced form of a molecule thus has a higher level of energy than the oxidized form. Oxidation-reduction reactions play a key role in the flow of energy through biological systems because the electrons that pass from one atom to another carry energy with them.

> **5.1** Energy is the capacity to do work, either actively (kinetic energy) or stored for later use (potential energy). Energy is often transferred with electrons. Oxidation is the loss of an electron; reduction is the gain of one.

5.2 The Laws of Thermodynamics

Running, thinking, singing, reading these words—all activities of living organisms involve changes in energy. A set of universal laws we call the laws of thermodynamics govern all energy changes in the universe, from nuclear reactions to the buzzing of a bee.

The First Law of Thermodynamics

The first of these universal laws, the **first law of thermodynamics,** concerns the amount of energy in the universe. It states that energy can change from one form to another (from potential to kinetic, for example) but it can never be destroyed, nor can new energy be made. The total amount of energy in the universe remains constant.

A lion eating a giraffe is in the process of acquiring energy. Rather than creating new energy or capturing the energy in sunlight, the lion is merely transferring some of the potential energy stored in the giraffe's tissues to its own body (just as the giraffe obtained the potential energy stored in the plants it ate while it was alive). Within any living organism, this chemical potential energy can be shifted to other molecules and stored in different chemical bonds, or it can be converted into other forms, such as kinetic energy, light, or electricity. During each conversion, some of the energy dissipates into the environment as **heat,** a measure of the random motions of molecules (and, hence, a measure of one form of kinetic energy). Energy continuously flows through the biological world in one direction, with new energy from the sun constantly entering the system to replace the energy dissipated as heat.

Heat can be harnessed to do work only when there is a heat gradient, that is, a temperature difference between two areas (this is how a steam engine functions). Cells are too small to maintain significant internal temperature differences, so heat energy is incapable of doing the work of cells. Thus, although the total amount of energy in the universe remains constant, the energy available to do work decreases, as progressively more of it dissipates as heat.

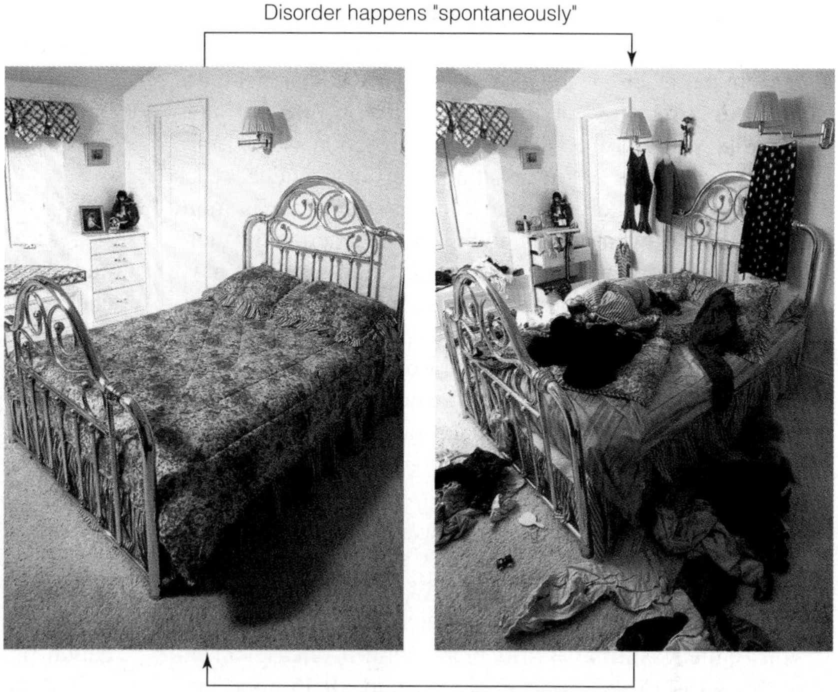
Disorder happens "spontaneously"
Organization requires energy

Figure 5.2 Entropy in action.
As time elapses, a child's room becomes more disorganized. It takes effort to clean it up.

The Second Law of Thermodynamics

The **second law of thermodynamics** concerns this transformation of potential energy into heat, or random molecular motion. It states that the disorder (more formally called entropy) in the universe is continuously increasing. Put simply, disorder is more likely than order. For example, it is much more likely that a column of bricks will tumble over than that a pile of bricks will arrange themselves spontaneously to form a column. In general, energy transformations proceed spontaneously to convert matter from a more ordered, less stable form, to a less ordered, more stable form (figure 5.2).

Entropy

Entropy is a measure of the disorder of a system, so the second law of thermodynamics can also be stated simply as "entropy increases." When the universe formed 10 to 20 billion years ago, it held all the potential energy it will ever have. It has become progressively more disordered ever since, with every energy exchange increasing the amount of entropy in the universe.

5.2 The first law of thermodynamics states that energy cannot be created or destroyed; it can only undergo conversion from one form to another. The second law of thermodynamics states that disorder (entropy) in the universe is increasing. Life converts energy from the sun to other forms of energy that drive life processes; the energy is never lost, but as it is used, more and more of it is converted to heat, the energy of random molecular motion.

5.3 Chemical Reactions

In a chemical reaction, the molecules that you start with are called **reactants** or sometimes **substrates,** while the molecules that you end up with, after the reaction is over, are called the **products** of the reaction. Not all chemical reactions are equally likely to occur. Just as a boulder is more likely to roll downhill than uphill, so a reaction is more likely to occur if it releases energy than if it needs to have energy supplied. Reactions that release energy, ones in which the products contain less energy than the reactants, tend to occur by themselves and are called **exergonic.** By contrast, reactions in which the products contain more energy than the reactants, called **endergonic,** do not occur unless energy is supplied from an outside source.

Activation Energy

If all chemical reactions that release energy tend to occur spontaneously, it is fair to ask, "Why haven't all exergonic reactions occurred already?" Clearly they have not. If you ignite gasoline, it still burns with a release of energy. So why doesn't all the gasoline in all the automobiles in the world just burn up right now? It doesn't because the burning of gasoline, and almost all other chemical reactions, requires an input of energy to get it started—a kick in the pants such as a match or spark plug. Even if the new bonds of the product contain or store less energy than the existing bonds of the reactants, it is first necessary to break those existing bonds, and this takes energy. The extra energy required to destabilize existing chemical bonds and so initiate a chemical reaction is called **activation energy** (figure 5.3*a*). Thus, to roll a boulder downhill, you must first nudge it out of the hole it sits in. Activation energy is simply a chemical nudge.

Catalysis

Just as all the gasoline in the world doesn't burn up right now because the combustion reaction has a sizable activation energy, so all the exergonic reactions in your cells don't spontaneously happen either. Each reaction waits for something to nudge it along, to supply it with sufficient activation energy to get going. One way to make an exergonic reaction more likely to happen is to lower the necessary activation energy. Like digging away the ground below your boulder, lowering activation energy reduces the nudge needed to get things started. The process of lowering the activation energy of a reaction is called **catalysis.** Catalysis cannot make an endergonic reaction occur spontaneously—you cannot avoid the need to supply energy—but it can make an exergonic reaction proceed much faster, because almost all the time required for chemical reactions is tied up in overcoming the activation energy barrier (figure 5.3*b*).

5.3 Chemical reactions occur when the covalent bonds linking atoms together are formed or broken. It takes energy to initiate chemical reactions.

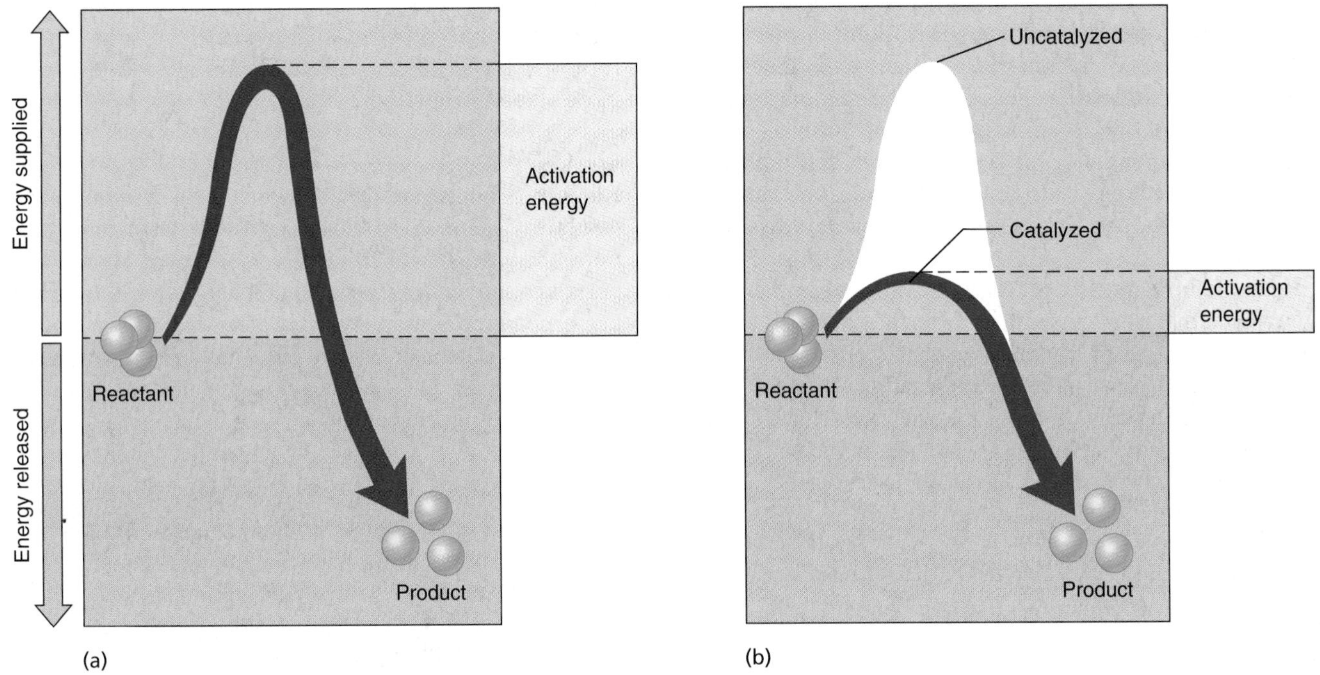

Energy supplied
Energy released
Activation energy
Reactant
Product
(a)
Uncatalyzed
Catalyzed
Activation energy
Reactant
Product
(b)

Figure 5.3 Activation energy and catalysis.

Exergonic reactions do not necessarily proceed rapidly because it takes energy to get them going. (*a*) The "hill" in this energy diagram represents energy that must be supplied to destabilize existing chemical bonds. (*b*) Catalyzed reactions occur faster because the amount of activation energy required to initiate the reaction—the height of the energy hill that must be overcome—is lowered.

5.4 Enzymes

Proteins called **enzymes** are the catalysts used by cells to touch off particular chemical reactions. By controlling which enzymes are present, and when they are active, cells are able to control what happens within themselves, just as a conductor controls the music an orchestra produces by dictating which instruments play when.

How Enzymes Work

An enzyme works by binding to a specific molecule and stressing the bonds of that molecule in such a way as to make a particular reaction more likely. The key to this activity is the shape of the enzyme (figure 5.4). An enzyme is specific for a particular reactant because the enzyme surface provides a mold that fits the desired reactant exactly. Other molecules that fit less perfectly simply don't adhere to the enzyme's surface. The site on the enzyme surface where the reactant fits is called the **active site.** The site on the reactant that binds to an enzyme is called the **binding site.**

An enzyme lowers the activation energy of a particular reaction. It may encourage the breaking of a particular chemical bond in the reactant molecule by weakening it, often by drawing away some of its electrons. Alternatively, an enzyme may encourage the formation of a link between two reactants by holding them near each other (figure 5.5).

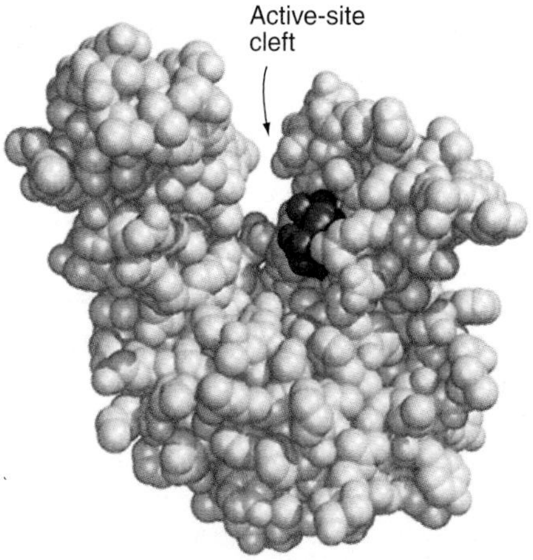

Active-site
cleft

Figure 5.4 Enzyme shape determines its activity.

The adenylate kinase enzyme (*blue* in this diagram) has a deep groove running across it, and at the base of the groove is a pair of substrate binding sites, facing each other. One binds the substrate ATP, the other a second substrate, AMP. By positioning these two molecules so closely together, the enzyme aids the transfer of the terminal phosphate from ATP (three phosphates) to AMP (one phosphate), forming two molecules of ADP (two phosphates). It is the close positioning of these two substrate binding sites that makes the catalysis possible.

From *Biochemistry,* 4th ed. by Stryer. Copyright © 1995 by Lubert Stryer. Used with permission of W.H. Freeman and Company.

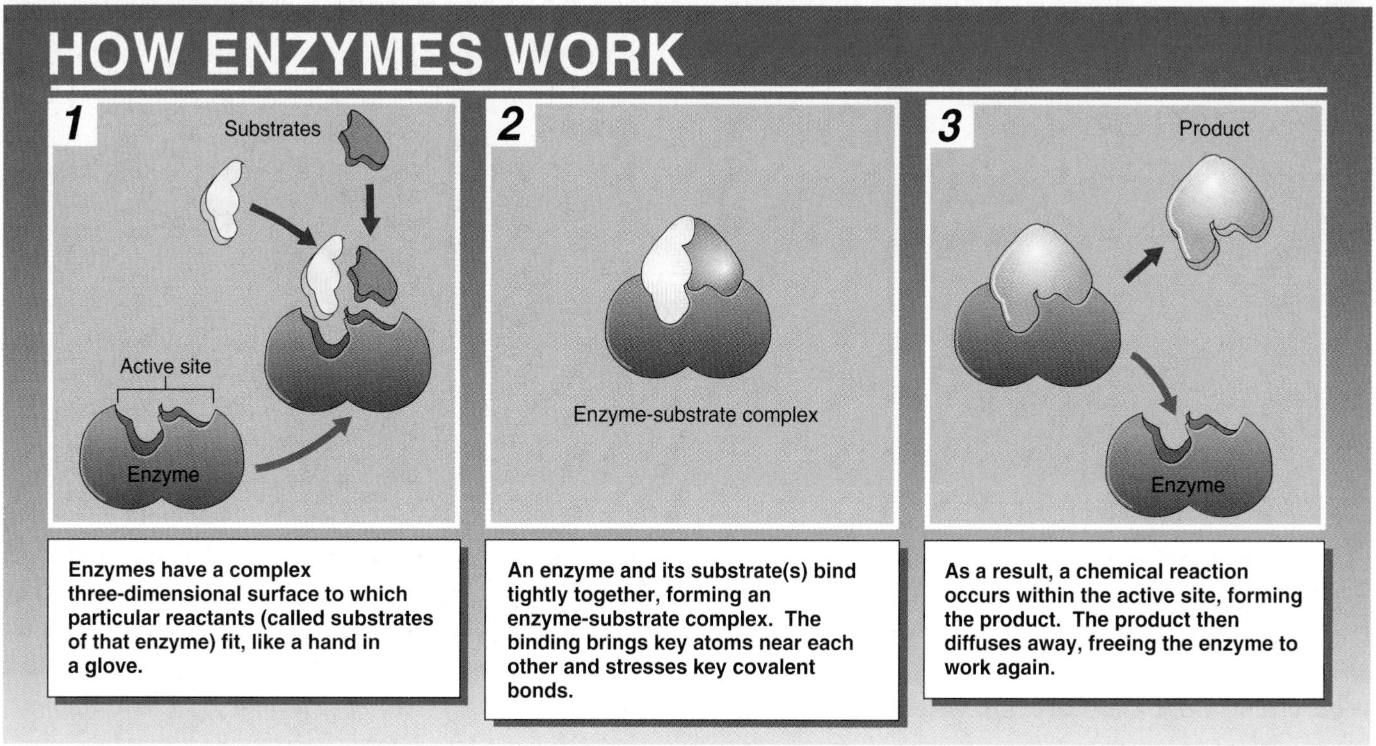

HOW ENZYMES WORK
1
Substrates
Active site
Enzyme
2
Enzyme-substrate complex
3
Product
Enzyme
Enzymes have a complex three-dimensional surface to which particular reactants (called substrates of that enzyme) fit, like a hand in a glove.
An enzyme and its substrate(s) bind tightly together, forming an enzyme-substrate complex. The binding brings key atoms near each other and stresses key covalent bonds.
As a result, a chemical reaction occurs within the active site, forming the product. The product then diffuses away, freeing the enzyme to work again.

Figure 5.5 How enzymes work.

Regulating Enzymes

Because an enzyme must have a precise shape to work correctly, it is possible for the cell to control when an enzyme is active by altering its shape. Many enzymes have shapes that can be altered by the binding to their surfaces of "signal" molecules. Such enzymes are called *allosteric* (Latin, other shape). If the new shape produced by binding the signal molecule is no longer able to fit the reactant, the signal acts as an inhibitor of the enzyme's activity. If the enzyme is unable to bind its reactant unless the signal molecule is bound to it, the signal acts as an activator of the enzyme's activity (figure 5.6). The site where the signal molecule binds to the enzyme surface is called the **allosteric site.**

Factors Affecting Enzyme Activity

Coenzymes. Enzymes often use additional chemical components called **cofactors** as tools to aid catalysis. Metal ions and coenzymes can both act as cofactors. A **coenzyme** is a nonprotein organic molecule that acts as a cofactor. One of the most important coenzymes is the electron acceptor NAD^+. In many enzyme reactions, energy-bearing electrons are passed from the active site of the enzyme to NAD^+. Electrons are usually transferred along with protons as hydrogen atoms. NAD^+ thus becomes NADH and can then carry the electrons to a different enzyme (see figure 5.30).

pH and Temperature. In addition to coenzymes, temperature and pH can also influence the action of enzymes. Enzyme activity is affected by any change in condition that alters the enzyme's three-dimensional shape. When the temperature becomes too low, the bonds that determine enzyme shape are not flexible enough to permit the induced-fit change sometimes necessary for catalysis; at higher temperatures, the bonds are too weak to hold the enzyme's peptide chains in the proper position. As a result, enzymes function best within an optimum temperature range. This range is relatively narrow for most human enzymes. In addition, most enzymes also function within an optimal pH range, because the structural bonds of enzymes are also sensitive to hydrogen ion (H^+) concentration. Most human enzymes, such as the protein-degrading enzyme trypsin, work best within the range of pH 6 to 8. However, some enzymes, such as the digestive enzyme pepsin, are able to function in very acidic environments, such as the stomach.

> **5.4** Enzymes are proteins that catalyze chemical reactions within cells. Their activity can be affected by signal molecules that bind to them, changing their shape. They are also sensitive to temperature and pH, because both of these variables also influence protein shape.

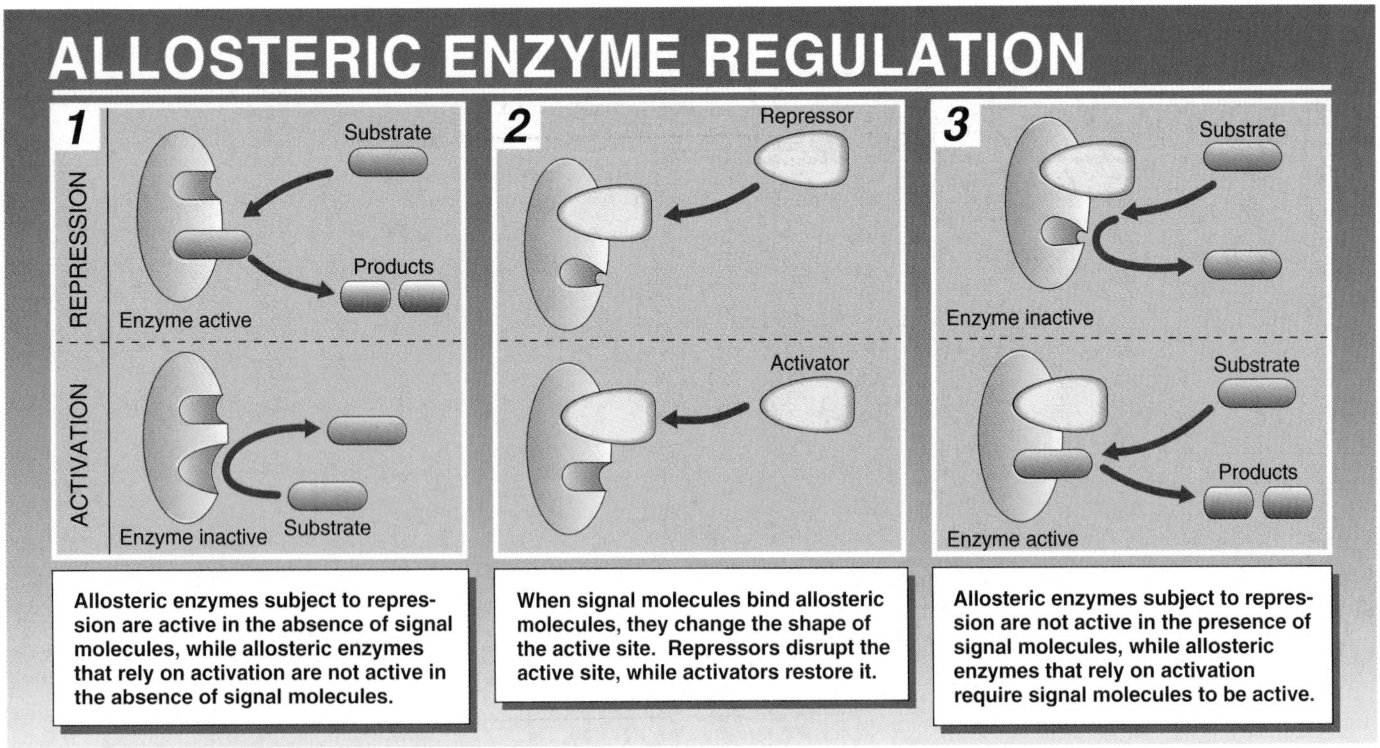
ALLOSTERIC ENZYME REGULATION
1
REPRESSION
ACTIVATION
Substrate
Products
Enzyme active
Enzyme inactive
Substrate
2
Repressor
Activator
3
Substrate
Enzyme inactive
Substrate
Products
Enzyme active
Allosteric enzymes subject to repression are active in the absence of signal molecules, while allosteric enzymes that rely on activation are not active in the absence of signal molecules.
When signal molecules bind allosteric molecules, they change the shape of the active site. Repressors disrupt the active site, while activators restore it.
Allosteric enzymes subject to repression are not active in the presence of signal molecules, while allosteric enzymes that rely on activation require signal molecules to be active.

Figure 5.6 How allosteric enzymes work.
Allosteric enzymes have two binding sites, one for the substrate and another for a signal molecule. When a signal molecule binds to its site, the shape of the enzyme changes. This disrupts some active sites (repression) and restores others to a proper configuration (activation).

5.5 How Cells Use Energy

Cells use energy to do all those things that require work. One of the most obvious of these is *movement*. Some cells swim about by rapidly spinning a long, tail-like flagellum. Other types of cells move about, crawling over one another to reach new positions. Much movement occurs within cells. Mitochondria and cellular materials are passed a meter or more along the narrow nerve cells that connect your feet with your spine. Chromosomes are pulled by microtubules during cell division. All of these movements by cells require the expenditure of energy. The molecule in the body that supplies that energy is **adenosine triphosphate (ATP).**

A second major way cells use energy is to *drive endergonic reactions*. Many of the synthetic activities of the cell are endergonic because building molecules takes energy. Recall that endergonic reactions do not take place spontaneously. The chemical bonds of the products contain more energy than what you started with, so nothing can happen until that extra energy is supplied from somewhere. The somewhere is, as you might guess, the cell's supply of ATP.

Structure of the ATP Molecule

As you have seen, ATP is the energy currency of the cell, the source of the force used to power the cell's activities. Each ATP molecule is composed of three parts: (1) a sugar that serves as the backbone to which the other two parts are attached; (2) adenine, one of the four nucleotide bases in DNA; and (3) a chain of three phosphates (figure 5.7).

Because phosphates have negative electrical charges, it takes considerable chemical energy to hold the line of three phosphates next to one another at the end of ATP. Like a coiled spring, the phosphates are poised to push apart. It is for this reason that the chemical bonds linking the phosphates are such chemically reactive bonds.

When the endmost phosphate is broken off an ATP molecule, a sizable packet of energy is released. The reaction converts ATP to adenosine diphosphate, ADP. The second phosphate group can also be removed, yielding additional energy and leaving adenosine monophosphate (AMP). Most energy exchanges in cells involve cleavage of only the outermost bond, converting ATP into ADP and P_i, inorganic phosphate (figure 5.8).

$$\text{ATP} \rightarrow \text{ADP} + P_i + \text{energy}$$

Because almost all endergonic reactions in cells require less energy than is released by this reaction, ATP is able to power many of the cell's activities.

5.5 Cells store chemical energy in ATP and use ATP to drive endergonic reactions.

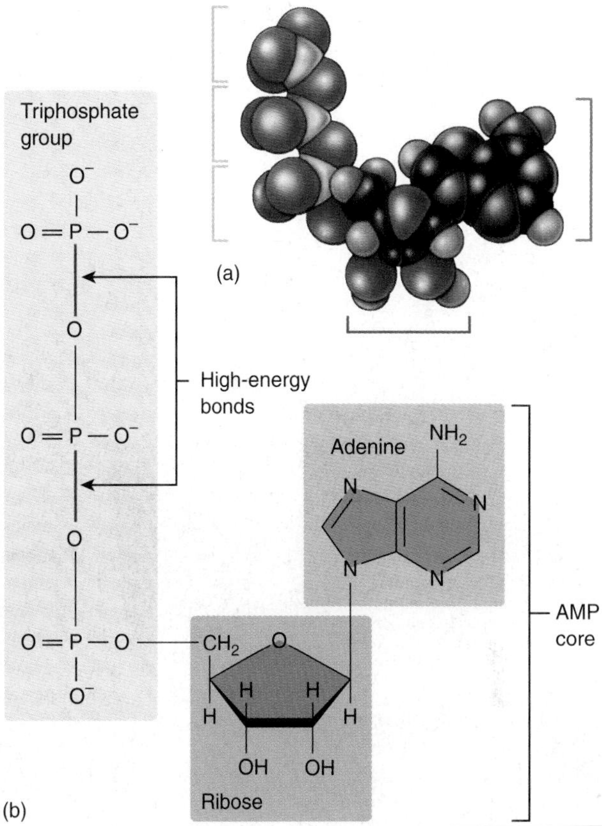

(a)
Triphosphate group
High-energy bonds
Adenine
NH_2
AMP core
CH_2
OH OH
Ribose
(b)

Figure 5.7 The parts of an ATP molecule.
The model (*a*) and structural diagram (*b*) both show that ATP consists of three phosphate groups attached to a ribose (five-carbon sugar) molecule. The ribose molecule is also attached to an adenine molecule (also in one of the nucleotides of DNA). When the outermost phosphate group is split off from the ATP molecule, a great deal of energy is released. The resulting molecule is ADP, or adenosine diphosphate. The second phosphate group can also be removed, yielding additional energy and leaving AMP (adenosine monophosphate).

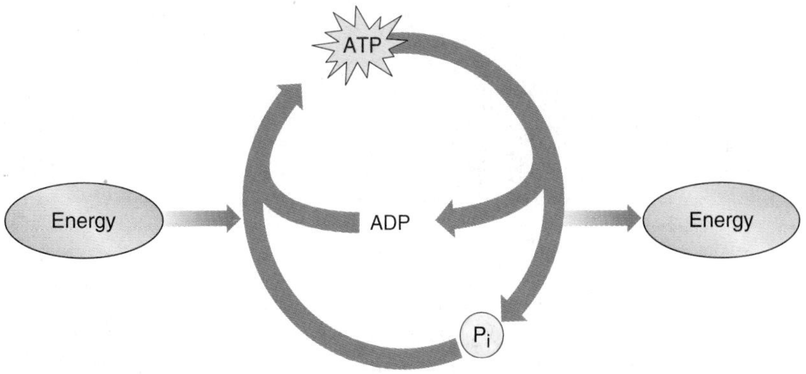

ATP
Energy
ADP
Energy
P_i

Figure 5.8 The ATP-ADP cycle.
In mitochondria and chloroplasts, chemical or photosynthetic energy is harnessed to form ATP from ADP and inorganic phosphate. When ATP is used to drive the living activities of cells, the molecule is cleaved back to ADP and inorganic phosphate, which are then available to form new ATP molecules.

5.6 An Overview of Photosynthesis

Life is powered by sunshine. All of the energy used by living cells comes ultimately from the sun, captured by plants and algae through the process of **photosynthesis.** Thus, life is only possible because our earth is awash in energy streaming inward from the sun. Each day, the radiant energy that reaches the earth is equal to that of about 1 million Hiroshima-sized atomic bombs. About 1% of it is captured by photosynthesis and provides the energy that drives us all.

Photosynthesis occurs in plants within leaves. Recall from chapter 4 that the cells of plant leaves contain organelles called chloroplasts that actually carry out the photosynthesis (figure 5.9). No other structure in a plant other than chloroplasts is able to carry out photosynthesis. Photosynthesis takes place in three stages: (1) capturing energy from sunlight; (2) using the energy to make ATP; and (3) using the ATP to power the synthesis of plant molecules from CO_2 in the air.

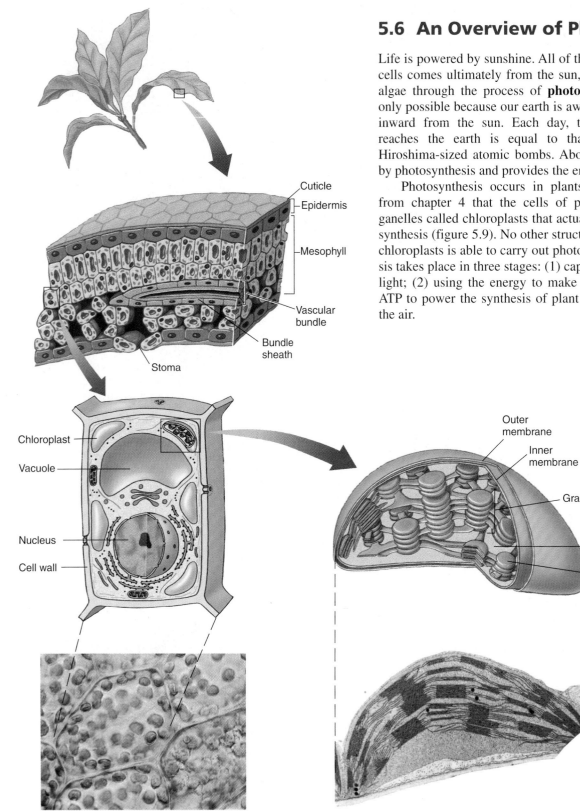

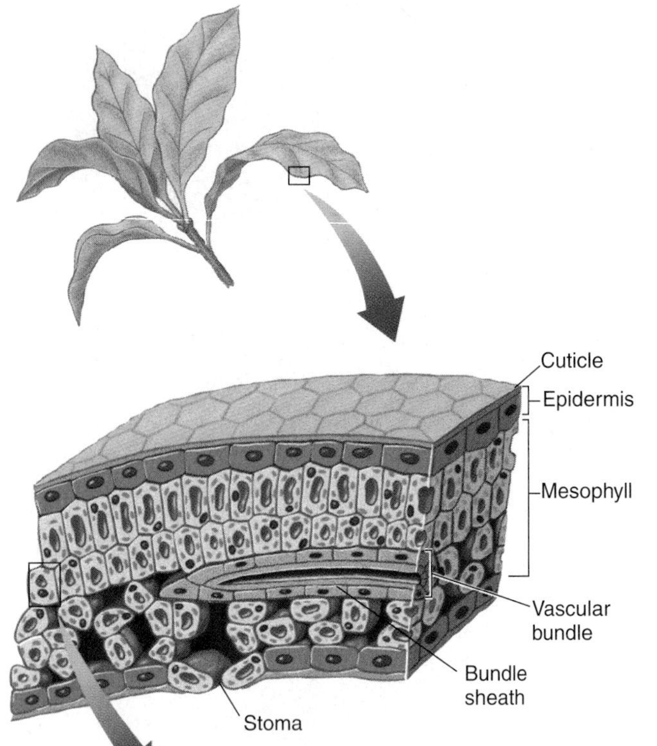
Cuticle
Epidermis
Mesophyll
Vascular bundle
Bundle sheath
Stoma

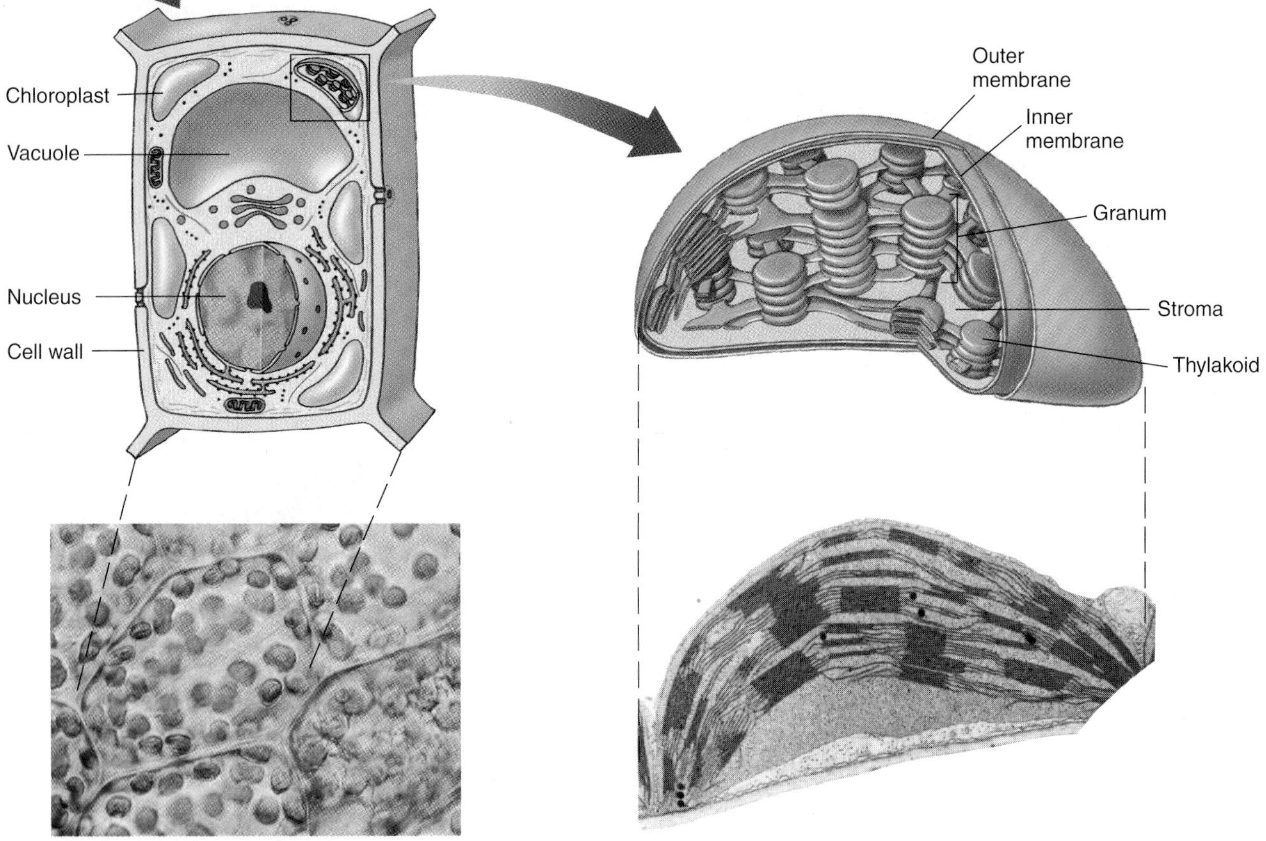
Chloroplast
Vacuole
Nucleus
Cell wall
Outer membrane
Inner membrane
Granum
Stroma
Thylakoid

Figure 5.9 Journey into a leaf.
Plant cells within leaves contain chloroplasts, in which thylakoid membranes are stacked. Within the thylakoid, chlorophyll pigments grouped in photosystems drive the reactions of photosynthesis.

The first two stages take place only in the presence of light and are commonly called the **light reactions.** The third stage, the formation of organic molecules from atmospheric CO_2, is called the **Calvin cycle.** As long as ATP is available, the Calvin cycle may occur in the absence of light.

The overall process of photosynthesis may be summarized by the following simple equation:

$$\mathbf{6\,CO_2 + 12\,H_2O + light\ energy \rightarrow C_6H_{12}O_6 + 6\,H_2O + 6\,O_2}$$

carbon dioxide, water → glucose, water, oxygen

Inside the Chloroplast

The internal membranes of chloroplasts are organized into flattened sacs called *thylakoids,* and often numerous thylakoids are stacked on top of one another in columns called *grana.* Surrounding the thylakoid membrane system is a semiliquid substance called *stroma.* In the membranes of thylakoids, chlorophyll pigments are grouped together in a network called a **photosystem.**

Each chlorophyll molecule within the photosystem network is capable of capturing photons (units of light energy). A lattice of proteins holds the pigments of the photosystem in close contact with one another. When light of the proper wavelength strikes any chlorophyll molecule in the photosystem, the resulting excitation passes from one chlorophyll molecule to another. The excited electron does not transfer physically—it is the *energy* that is passed from one chlorophyll to another. A crude analogy to this form of energy transfer is the initial "break" in a game of pool. If the cue ball squarely hits the point of the triangular array of 15 pool balls, the two balls at the far corners of the triangle fly off, and none of the central balls move at all. The energy is transferred through the central balls to the most distant ones.

Eventually the energy arrives at a key chlorophyll molecule touching a membrane-bound protein. It is transferred to that protein, which passes it in turn to a series of proteins that put the energy to work making ATP and building organic molecules, in a way we will discuss. The photosystem thus acts as a large antenna, amplifying the light-gathering powers of individual chlorophyll molecules.

5.6 Photosynthesis uses energy from sunlight to power the synthesis of organic molecules from CO_2 in the air. In plants, photosynthesis takes place in specialized compartments within chloroplasts.

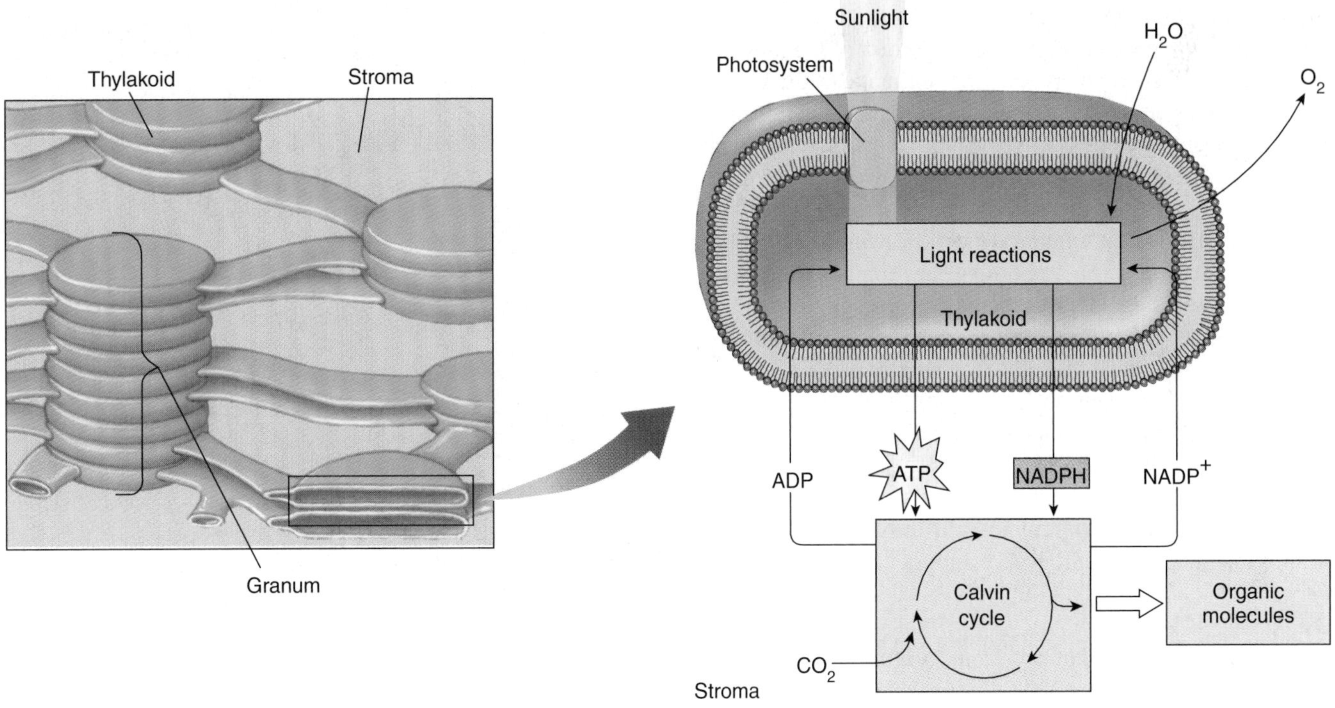
Thylakoid
Stroma
Granum
Sunlight
Photosystem
H_2O
O_2
Light reactions
Thylakoid
ADP
ATP
NADPH
$NADP^+$
Calvin cycle
Organic molecules
CO_2
Stroma

5.7 How Plants Capture Energy from Sunlight

Where is the energy in light? What is there about sunlight that a plant can use to create chemical bonds? The revolution in physics in the twentieth century taught us that light actually consists of tiny packets of energy called **photons.** When light shines on your hand, your skin is being bombarded by a stream of these photons smashing onto its surface.

Sunlight contains photons of many energy levels, only some of which we "see." Some of the photons in sunlight carry a great deal of energy—for example, X rays and ultraviolet light. Others such as radio waves carry very little. In general, high-energy photons have shorter wavelengths than low-energy photons. Our eyes perceive photons carrying intermediate amounts of energy as **visible light** (figure 5.10), because our eyes absorb only that kind of photon. Plants are even more picky, absorbing mainly blue and red light and reflecting back what is left of the visible light, which is why they appear green (figure 5.11).

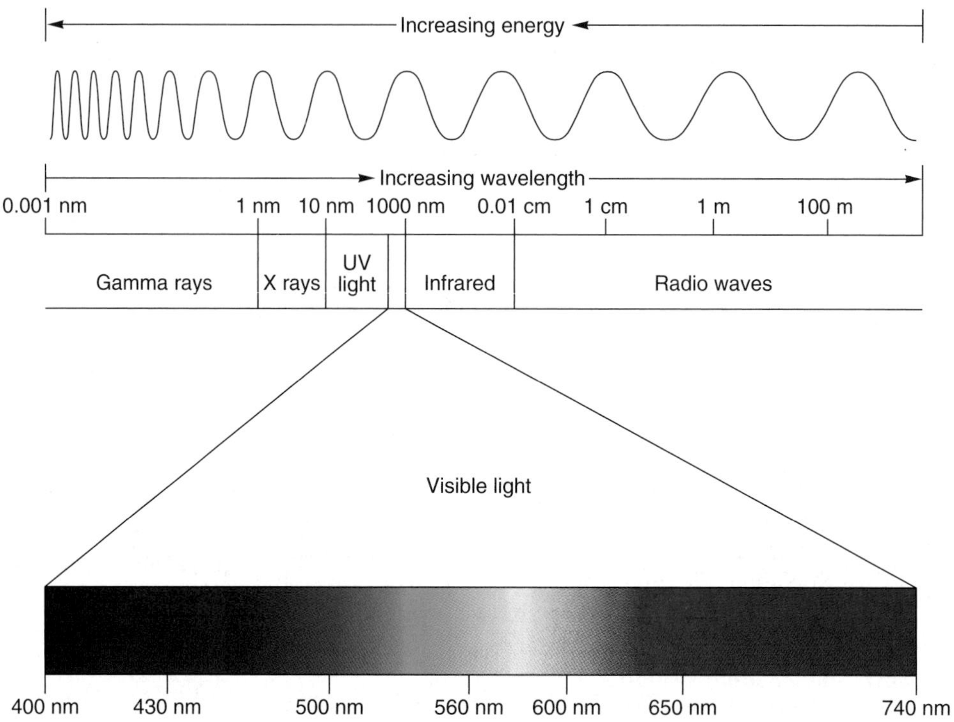

Increasing energy
Increasing wavelength
0.001 nm
1 nm
10 nm
1000 nm
0.01 cm
1 cm
1 m
100 m
Gamma rays
X rays
UV light
Infrared
Radio waves
Visible light
400 nm
430 nm
500 nm
560 nm
600 nm
650 nm
740 nm

Figure 5.10 Photons of different energy: the electromagnetic spectrum.

Light is composed of packets of energy called photons. Some of the photons in light carry more energy than others. Light, a form of electromagnetic energy, is conveniently thought of as a wave. The shorter the wavelength of light, the greater the energy of its photons. Visible light represents only a small part of the electromagnetic spectrum, that between about 400 and 740 nanometers.

Visible light represents only a small portion of the range of photon energies in sunlight. We call the full range of these photons the **electromagnetic spectrum.** The highest-energy photons, which have the shortest wavelengths, are gamma rays with wavelengths of less than 1 nanometer; the lowest-energy photons, with wavelengths of thousands of meters, are radio waves.

How can a leaf or a human eye choose which photons to absorb? The answer to this important question has to do with the nature of atoms. Remember that electrons spin in particular orbits around the atomic nucleus, each at a different energy level. Atoms absorb light by boosting electrons to higher energy levels, using the energy in the photon to power the move. Boosting the electron requires just the right amount of energy, no more and no less, just as when climbing a ladder you must raise your foot just so far to climb a rung. A particular kind of atom absorbs only certain photons of light, those with the appropriate amount of energy.

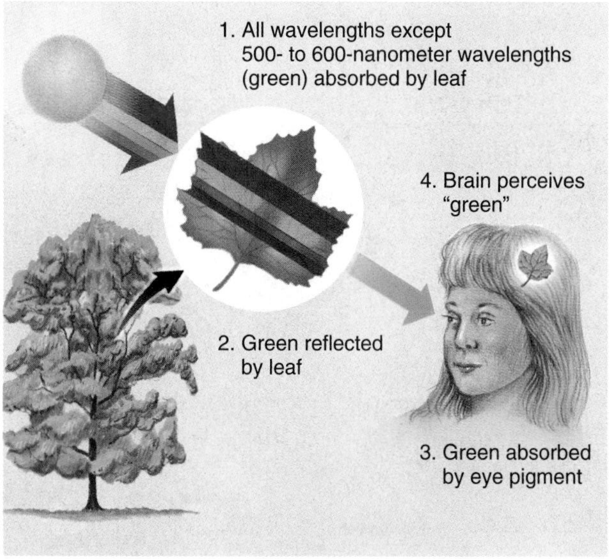

1. All wavelengths except 500- to 600-nanometer wavelengths (green) absorbed by leaf
2. Green reflected by leaf
3. Green absorbed by eye pigment
4. Brain perceives "green"

Figure 5.11 Why are plants green?

A leaf containing chlorophyll absorbs a broad range of photons—all the colors in the spectrum except for the photons around 500 to 600 nanometers. The leaf reflects these colors. These reflected wavelengths are absorbed by the visual pigments in our eyes, and our brains perceive the reflected wavelengths as "green."

Pigments

Molecules that absorb light are called **pigments.** When we speak of visible light, we refer to those wavelengths that the pigment within human eyes, called *retinal,* can absorb—roughly from 380 nanometers (violet) to 750 nanometers (red). Other animals use different pigments for vision and thus "see" a different portion of the electromagnetic spectrum. For example, the pigment in insect eyes absorbs at shorter wavelengths than retinal. That is why bees can see ultraviolet light, which we cannot see, but are blind to red light, which we can see.

The main pigment in plants that absorbs light is **chlorophyll,** which occurs in two forms, chlorophyll *a* and chlorophyll *b* (figure 5.12). While it absorbs fewer kinds of photons than retinal, it is very much more efficient at capturing them. Chlorophylls capture photons with a metal ion (magnesium) that lies at the center of a complex carbon ring. Photons excite electrons of the magnesium ion, which are then channeled away by the carbon atoms. The two kinds of chlorophyll differ in small chemical "side groups" attached to the outside of the ring, which alter slightly the wavelength of the photons they absorb. Another group of pigments, the **carotenoids,** capture light of wavelengths not efficiently absorbed by chlorophylls (figure 5.13).

5.7 Plants use pigments like chlorophyll to capture photons of blue and red light, reflecting photons of green wavelengths.

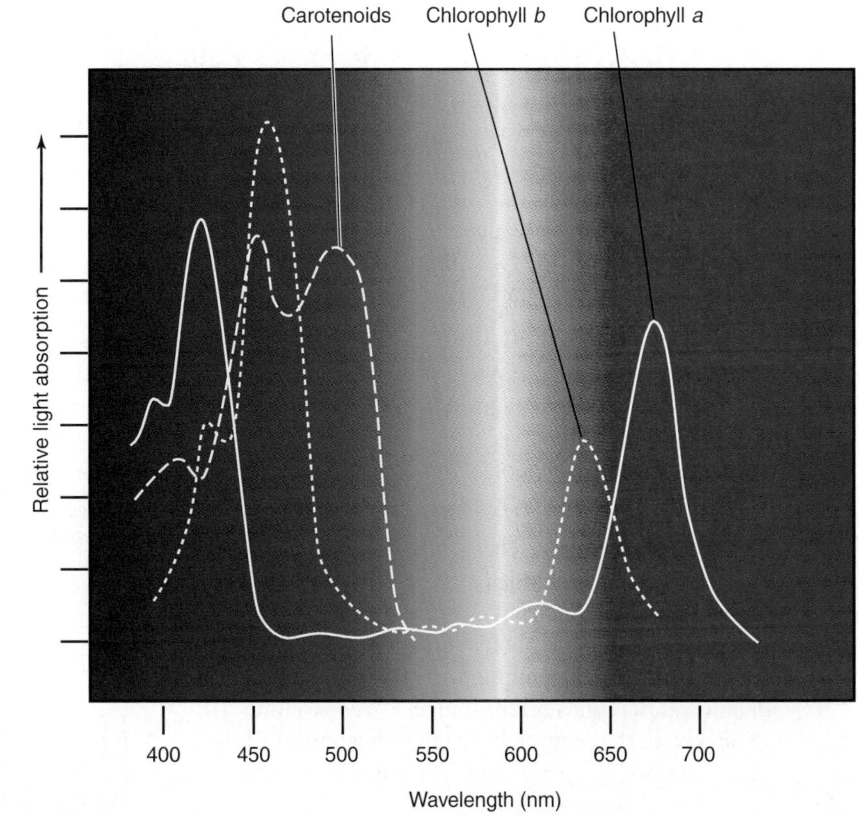
Carotenoids
Chlorophyll *b*
Chlorophyll *a*
Relative light absorption
400
450
500
550
600
650
700
Wavelength (nm)

Figure 5.12 Absorption spectra of chlorophylls and carotenoids.
The peaks represent wavelengths of sunlight strongly absorbed by the two common forms of photosynthetic pigment, chlorophyll *a* and chlorophyll *b*, and by carotenoids. Chlorophylls absorb predominantly violet-blue and red light, in two narrow bands of the spectrum, while they reflect the green light in the middle of the spectrum. Carotenoids absorb mostly blue and green light and reflect orange and yellow light.

Oak leaf in summer
Oak leaf in autumn

Figure 5.13 Fall colors are produced by pigments such as carotenoids.
During the spring and summer, chlorophyll masks the presence in leaves of other pigments called carotenoids. Cool temperatures in the fall cause leaves to cease manufacturing chlorophyll. With chlorophyll no longer present to reflect green light, orange and yellow light reflected by carotenoids give bright colors to the autumn leaves.

5.8 Organizing Pigments into Photosystems

The light reactions of photosynthesis occur on membranes (figure 5.14). In plants, photosynthesis takes place on membranes within organelles that are the evolutionary descendants of photosynthetic bacteria, chloroplasts. The light reactions take place in three stages:

1. **Primary photoevent.** A photon of light is captured by a pigment. The result of this primary photoevent is the excitation of an electron within the pigment.
2. **Electron transport.** The excited electron is shuttled along a series of electron-carrier molecules embedded within the photosynthetic membrane until it arrives at a transmembrane proton-pumping channel (sometimes referred to as a "redox pump"). Its arrival at the pump induces the transport of a proton across the membrane. The electron is then passed to an acceptor.
3. **Chemiosmosis.** The transport of protons drives the chemiosmotic synthesis of ATP, just as it does in aerobic respiration.

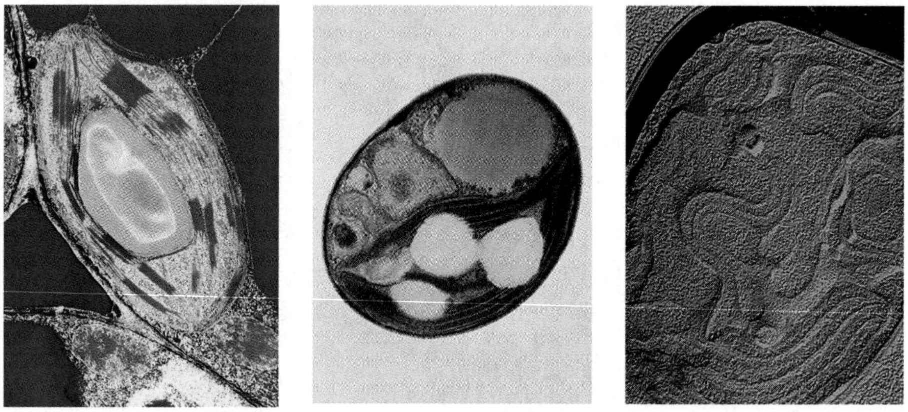
(a) (b) (c)

Figure 5.14 A sampling of photosynthetic membranes.
(*a*) Chloroplast of the green plant *Coleus blumei* (5,000×), (*b*) unicellular green algae *Chlorella,* and (*c*) nitrogen-fixing cyanobacterium *Nostoc museorum* (60,000×).

Discovery of Photosystems

One way to study how pigments absorb light is to measure the dependence of the output of photosynthesis on the intensity of illumination—that is, how much photosynthesis is produced by how much light. When experiments of this sort are done on plants, they show that the output of photosynthesis increases linearly at low intensities but lessens at higher intensities, finally saturating at high-intensity light. Saturation occurs because all the light-absorbing capacity of the plant is in use; adding more light doesn't increase the output because there is nothing to absorb the added photons.

It is tempting to think that at saturation, all of a plant's pigment molecules are in use. In 1932 plant physiologists Robert Emerson and William Arnold set out to test this hypothesis in an organism where they could measure both the number of chlorophyll molecules and the output of photosynthesis. In their experiment, they measure the oxygen yield of photosynthesis when *Chlorella* (unicellular green algae) were exposed to very brief light flashes lasting only a few microseconds. Assuming the hypothesis of pigment saturation to be correct, they expected to find that as they increased the intensity of the flashes, the yield per flash would increase, until each chlorophyll molecule absorbed a photon, which would then be used in the light reactions, producing a molecule of O_2.

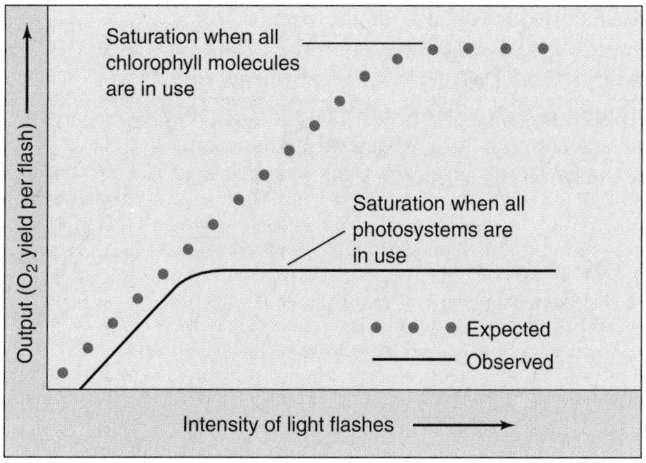
Saturation when all chlorophyll molecules are in use
Saturation when all photosystems are in use
Expected
Observed
Output (O_2 yield per flash)
Intensity of light flashes

Figure 5.15 Emerson and Arnold's experiment.
When photosynthetic saturation is achieved, further increases in intensity cause no increase in output.

Unexpectedly, this is not what happened (figure 5.15). Instead, saturation was achieved much earlier, with only one molecule of O_2 per 2,500 chlorophyll molecules! This led Emerson and Arnold to conclude that light is absorbed not by independent pigment molecules, but rather by clusters of chlorophyll and accessory pigment molecules, which have come to be called *photosystems.* In a photosystem such as those of *Chlorella*, light is absorbed by any one of the hundreds of pigment molecules in a photosystem, which transfer their excitation energy to one with a lower energy level than the others. This **reaction center** of the photosystem acts as an energy sink, trapping the excitation energy. It was the saturation of these reaction centers, not individual molecules, that was observed by Emerson and Arnold.

Architecture of a Photosystem

In chloroplasts and all but the most primitive bacteria, light is captured by such photosystems. Each photosystem is a network of chlorophyll *a* molecules and accessory pigments held within a protein matrix on the surface of the photosynthetic membrane. Like a magnifying glass focusing light on a precise point, a photosystem channels the excitation energy gathered by any one of its pigment molecules to a specific molecule, the reaction center chlorophyll (figure 5.16). This molecule then passes the energy out of the photosystem to drive the synthesis of ATP and organic molecules.

Plants Use Two Photosystems

Green plants and algae use two photosystems (figure 5.17). Energy collected by the first is used to manufacture ATP, which provides the energy to build molecules. Energy gathered by the second drives the production of NADPH from $NADP^+$, which provides the "reducing power" (hydrogen atoms) needed to build sugars and other organic molecules with many C–H bonds.

5.8 Photon energy is captured by pigments that employ it to excite electrons that are channeled away to do the chemical work of producing ATP and NADPH.

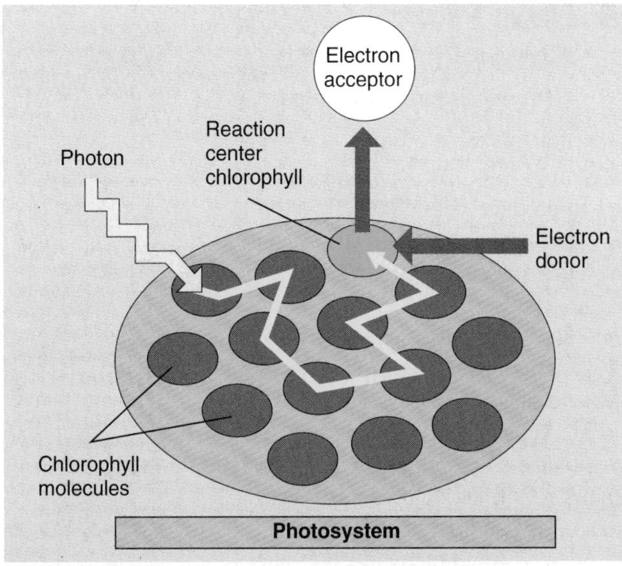

Electron acceptor
Reaction center chlorophyll
Photon
Electron donor
Chlorophyll molecules
Photosystem

Figure 5.16 How a photosystem works.

When light of the proper wavelength strikes any pigment molecule within a photosystem, the light is absorbed and its excitation energy is then transferred from one molecule to another within the cluster of pigment molecules until it encounters the reaction center, which exports the energy as high-energy electrons to an acceptor molecule.

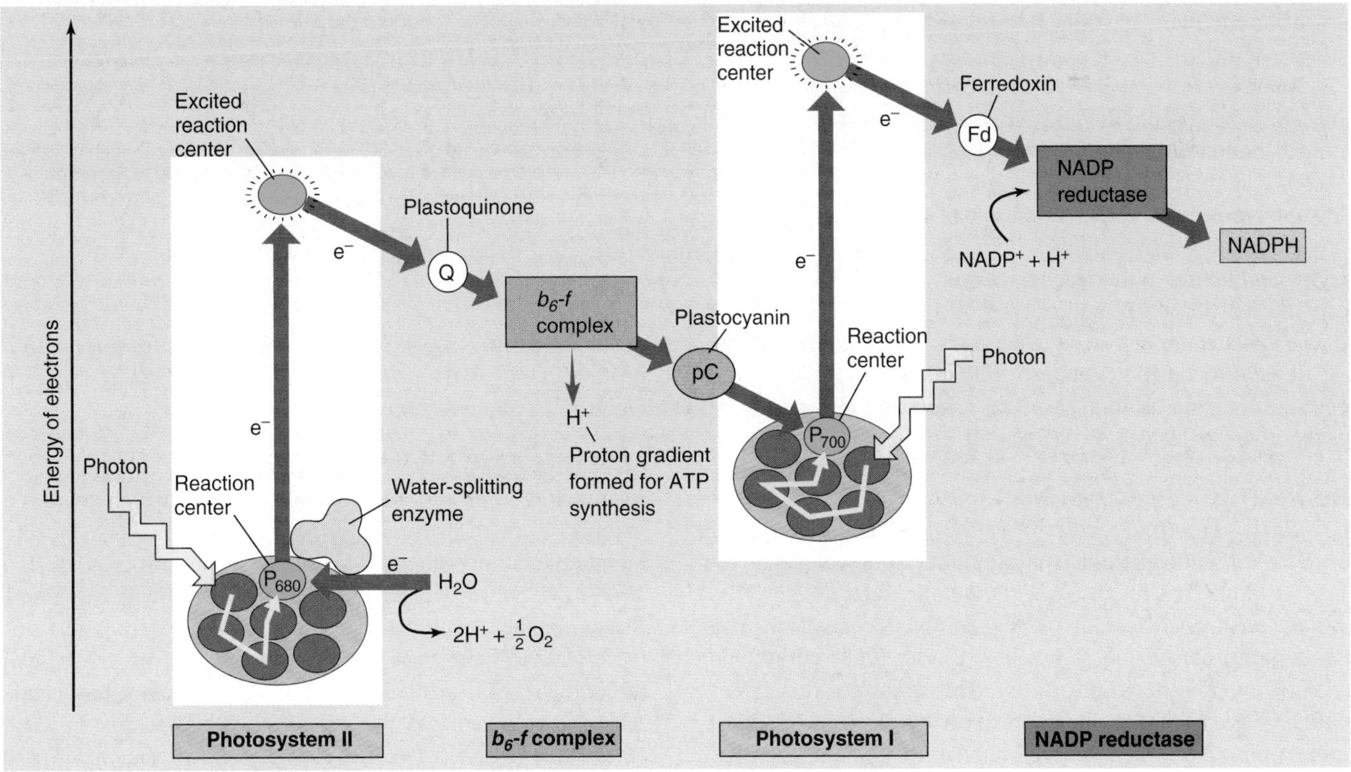

Energy of electrons
Excited reaction center
Plastoquinone
Q
e⁻
b_6-f complex
H⁺
Proton gradient formed for ATP synthesis
Photon
Reaction center
Water-splitting enzyme
P680
H₂O
2H⁺ + ½O₂
Plastocyanin
pC
Excited reaction center
Ferredoxin
Fd
NADP reductase
NADP⁺ + H⁺
NADPH
Reaction center
Photon
P700
Photosystem II
b_6-f complex
Photosystem I
NADP reductase

Figure 5.17 Plants use two photosystems.

First, a high-energy electron from photosystem II is used to pump a proton across the membrane, contributing chemiosmotically to the production of a molecule of ATP. The ejected electron then passes along a chain of cytochromes to photosystem I, which uses it, with a photon of light energy, to drive the formation of NADPH.

5.9 How Photosystems Convert Light to Chemical Energy

Plants use the two photosystems discussed in series, first one and then the other, to produce both ATP and NADPH. This two-stage process is called **noncyclic photophosphorylation,** because the path of the electrons is not a circle—the electrons ejected from the photosystems do not return to them, but rather end up in NADPH. The photosystems are replenished instead with electrons obtained by splitting water. Photosystem II acts first. High-energy electrons generated by photosystem II are used to synthesize ATP and then passed to photosystem I to drive the production of NADPH. For every pair of electrons obtained from water, one molecule of NADPH and slightly more than one molecule of ATP are produced.

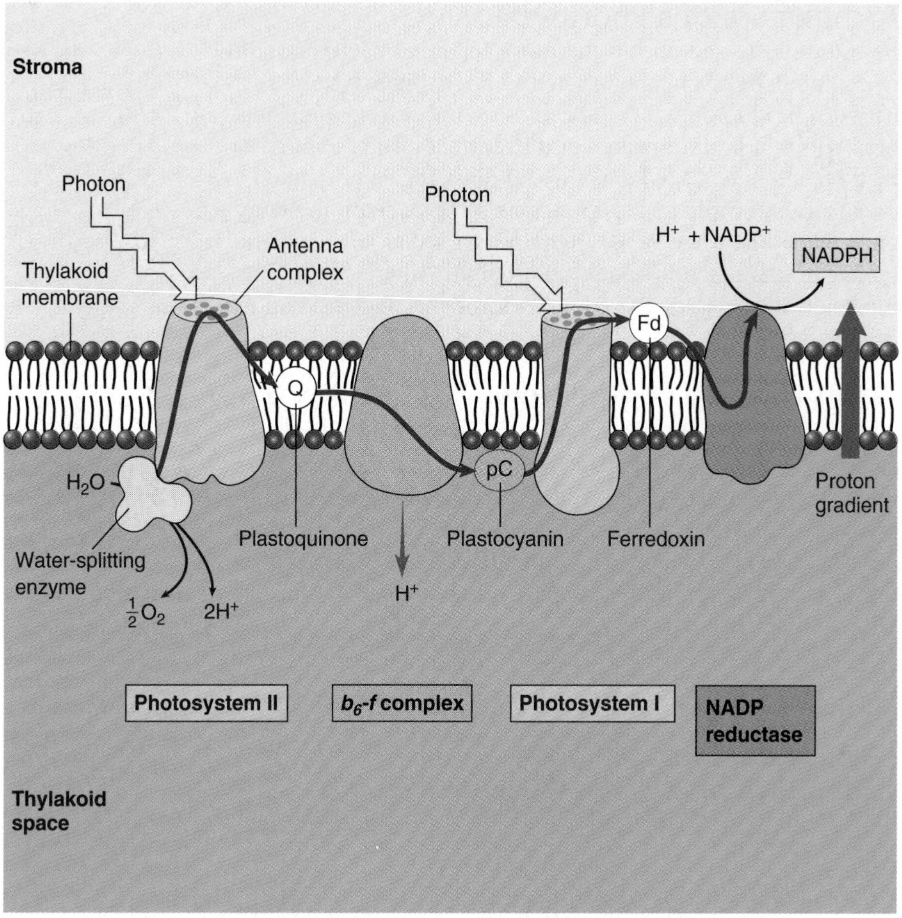

Stroma
Photon
Thylakoid membrane
Antenna complex
Photon
H^+ + $NADP^+$
NADPH
Fd
Q
pC
H_2O
Proton gradient
Water-splitting enzyme
$\frac{1}{2}O_2$
$2H^+$
Plastoquinone
H^+
Plastocyanin
Ferredoxin
Photosystem II
b_6-f complex
Photosystem I
NADP reductase
Thylakoid space

Figure 5.18 The photosynthetic electron transport system.
When a photon of light strikes a pigment molecule in photosystem II, it excites an electron. This electron is coupled to a proton stripped from water by an enzyme and is passed along a chain of membrane-bound cytochrome electron carriers (*red arrow*). When water is split, oxygen is released from the cell, and the hydrogen ions remain in the thylakoid space. At the proton pump (b_6-f complex), the energy supplied by the photon is used to transport a proton across the membrane into the thylakoid. The concentration of hydrogen ions within the thylakoid thus increases further. When photosystem I absorbs another photon of light, its pigment passes a second high-energy electron to a reduction complex, which generates NADPH.

Photosystem II

The reaction center of photosystem II, called P_{680}, closely resembles the reaction center of purple bacteria. It consists of more than 10 transmembrane protein subunits. The *antenna complex* of the photosystem consists of some 250 molecules of chlorophyll *a* and accessory pigments bound to several protein chains. In photosystem II the oxygen atoms of two water molecules bind to a cluster of manganese atoms embedded within an enzyme and bound to the reaction center. In a way that is poorly understood, this enzyme splits water, removing electrons one at a time to fill the holes left in the reaction center by departure of light-energized electrons. As soon as four electrons have been removed from the two water molecules, O_2 is released.

The Path to Photosystem I

The primary electron acceptor for the light-energized electrons leaving photosystem II is a quinone molecule. The reduced quinone that results (*plastoquinone,* symbolized Q) is a strong electron donor; it passes the excited electron to a proton pump called the *b_6-f complex* embedded within the thylakoid membrane (figure 5.18). This complex closely resembles the *bc_1* complex in the respiratory electron transport chain of mitochondria discussed later in this chapter. Arrival of the energetic electron causes the b_6-f complex to pump a proton into the thylakoid space. A small copper-containing protein called *plastocyanin* (symbolized pC) then carries the electron to photosystem I.

Making ATP: Chemiosmosis

Each thylakoid is a closed compartment into which protons are pumped from the stroma by the b_6-f complex. The thylakoid membrane is impermeable to protons, so protons cross back out almost exclusively via the channels provided by *ATP synthases.* These channels protrude like knobs on the external surface of the thylakoid membrane. As protons pass out of the thylakoid through the ATP synthase channel, ADP is phosphorylated to ATP and released into the stroma, the fluid matrix inside the chloroplast (figure 5.19). The stroma contains the enzymes that catalyze the reactions of carbon fixation.

Photosystem I

The reaction center of photosystem I, called P_{700}, is a transmembrane complex consisting of at least 13 protein subunits. Energy is fed to it by an antenna complex consisting of 130 chlorophyll *a* and accessory pigment molecules. Photosystem I accepts an electron from plastocyanin into the hole created by the exit of a light-energized electron. This arriving electron has by no means lost all of its light-excited energy; almost half remains. Thus, the absorption of a photon of light energy by photosystem I boosts the electron leaving the reaction center to a very high energy level. Unlike photosystem II, photosystem I does not rely on quinone as an electron acceptor. Instead, it passes electrons to an iron-sulfur protein called *ferredoxin* (Fd).

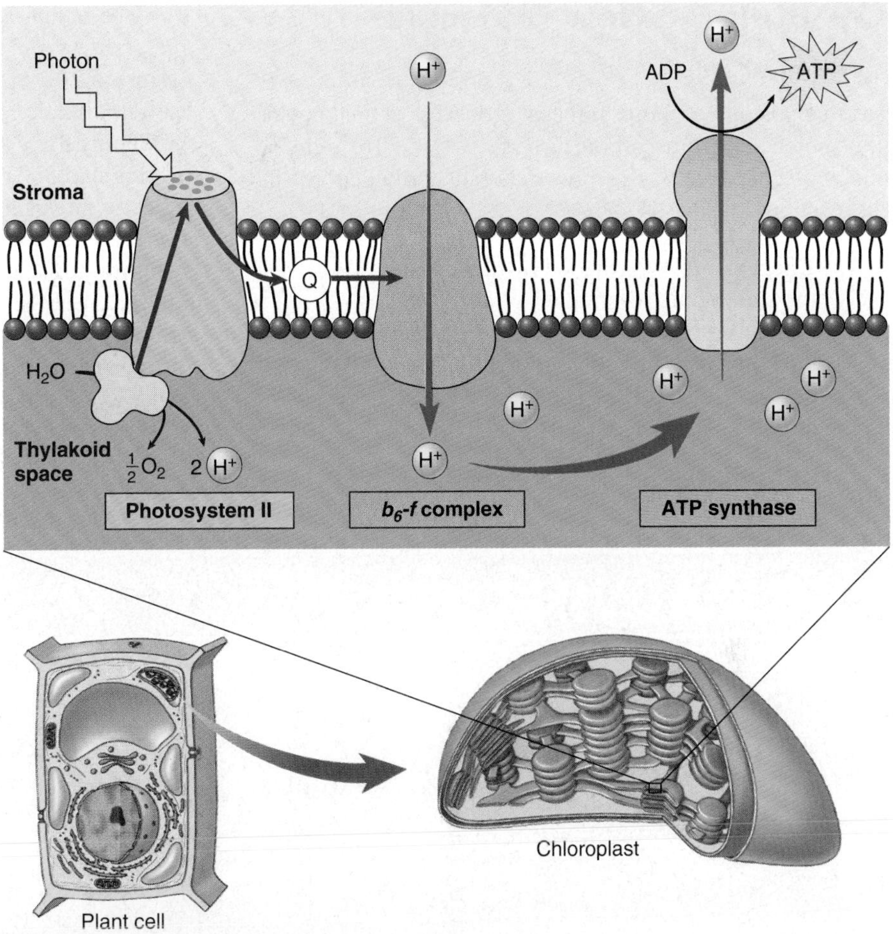

Photon
Stroma
Q
H^+
ADP
ATP
H_2O
Thylakoid space
$\frac{1}{2}O_2$ $2H^+$
Photosystem II
b_6-f complex
ATP synthase
Chloroplast
Plant cell

Figure 5.19 Chemiosmosis in a chloroplast.

The b_6-f complex embedded in the thylakoid membrane pumps protons into the interior of the thylakoid. ATP is produced on the outside surface of the membrane (stroma side), as protons diffuse back out of the thylakoid through ATP synthase channels.

Making NADPH

Photosystem I passes electrons to ferredoxin on the stromal side of the membrane (outside the thylakoid). The reduced ferredoxin carries a very high-potential electron. Two of them, from two molecules of reduced ferredoxin, are then donated to a molecule of $NADP^+$ to form NADPH. The reaction is catalyzed by the membrane-bound enzyme *NADP reductase.* Because the reaction occurs on the stromal side of the membrane and involves the uptake of a proton in forming NADPH, it contributes further to the proton gradient established during photosynthetic electron transport.

Making More ATP

The passage of an electron from water to NADPH in the noncyclic photophosphorylation described generates one molecule of NADPH and slightly more than one molecule of ATP. However, as you will learn later in this chapter, building organic molecules takes more energy than that—it takes one-and-a-half ATP molecules per NADPH molecule to fix carbon. To produce the extra ATP, many plant species are capable of short-circuiting photosystem I, switching photosynthesis into a *cyclic photophosphorylation* mode, so that the light-excited electron leaving photosystem I is used to make ATP instead of NADPH. The energetic electron is simply passed back to the b_6-f complex rather than passing on to $NADP^+$. The b_6-f complex pumps out a proton, adding to the proton gradient driving the chemiosmotic synthesis of ATP. The relative proportions of cyclic and noncyclic photophosphorylation in these plants determine the relative amounts of ATP and NADPH available for building organic molecules.

5.9 The electrons that photosynthesis employs to form energy-rich reduced organic molecules are obtained from water; the residual oxygen atoms of the water molecules combine to form oxygen gas.

5.10 Building New Molecules

The Calvin Cycle

Stated very simply, photosynthesis is a way of making organic molecules from carbon dioxide (CO_2). To build organic molecules, cells use raw materials provided by the light reactions:

1. **Energy.** ATP (provided by photosystem II) drives the endergonic reactions.
2. **Reducing power.** NADPH (provided by photosystem I) provides a source of hydrogens and the energetic electrons needed to bind them to carbon atoms.

The actual assembly of new molecules employs a complex battery of enzymes in what is called the **Calvin cycle** (figure 5.20), or **C_3 photosynthesis.** A carbon atom from a carbon dioxide molecule is first added to a five-carbon sugar, producing two three-carbon sugars. This process is called "fixing carbon" because it attaches a carbon atom that was in a gas to an organic molecule. Then, in a long series of reactions, the carbons are shuffled about. Eventually some of the resulting molecules are channeled off to make sugars, while others are used to re-form the original five-carbon sugar, which is then available to restart the cycle. The cycle has to "turn" six times in order to form a new glucose molecule, because each turn of the cycle adds only one carbon atom from CO_2, and glucose is a six-carbon sugar.

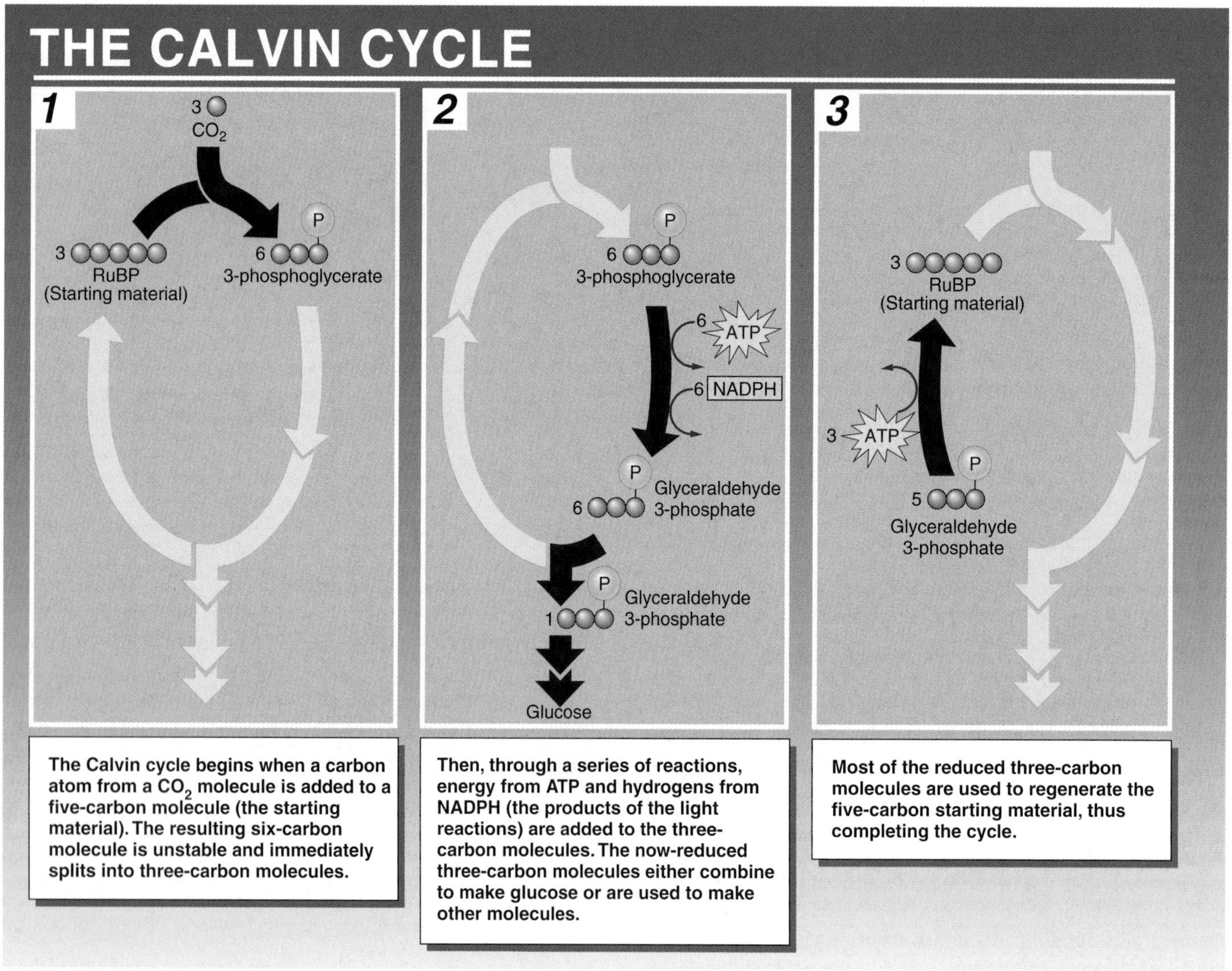

THE CALVIN CYCLE
1
3
CO2
P
3
6
RuBP
(Starting material)
3-phosphoglycerate
The Calvin cycle begins when a carbon atom from a CO2 molecule is added to a five-carbon molecule (the starting material). The resulting six-carbon molecule is unstable and immediately splits into three-carbon molecules.
2
P
6
3-phosphoglycerate
6
ATP
6
NADPH
P
Glyceraldehyde
6
3-phosphate
P
Glyceraldehyde
1
3-phosphate
Glucose
Then, through a series of reactions, energy from ATP and hydrogens from NADPH (the products of the light reactions) are added to the three-carbon molecules. The now-reduced three-carbon molecules either combine to make glucose or are used to make other molecules.
3
3
RuBP
(Starting material)
3
ATP
P
5
Glyceraldehyde
3-phosphate
Most of the reduced three-carbon molecules are used to regenerate the five-carbon starting material, thus completing the cycle.

Figure 5.20 How the Calvin cycle works.

The Calvin cycle takes place in the stroma of the chloroplasts. The NADPH and the ATP that were generated by the light reactions are used in the Calvin cycle to build carbon molecules. The number of carbon atoms at each stage is indicated by the number of balls. It takes six turns of the cycle to make one molecule of glucose.

C_4 Photosynthesis

Many plants have trouble carrying out C_3 photosynthesis when the weather is hot. As temperature increases, plants partially close their leaf openings, called **stomata,** to conserve water. As a result, CO_2 and O_2 are not able to enter and exit the leaves through these openings. The concentration of CO_2 in the leaves falls, while the concentration of O_2 in the leaves rises. These conditions cause RuBP carboxylase, an enzyme that carries out the first step of the Calvin cycle, to engage in **photorespiration,** removing CO_2 from the product of the reaction rather than adding it to the substrate. Photorespiration thus short-circuits the successful performance of the Calvin cycle.

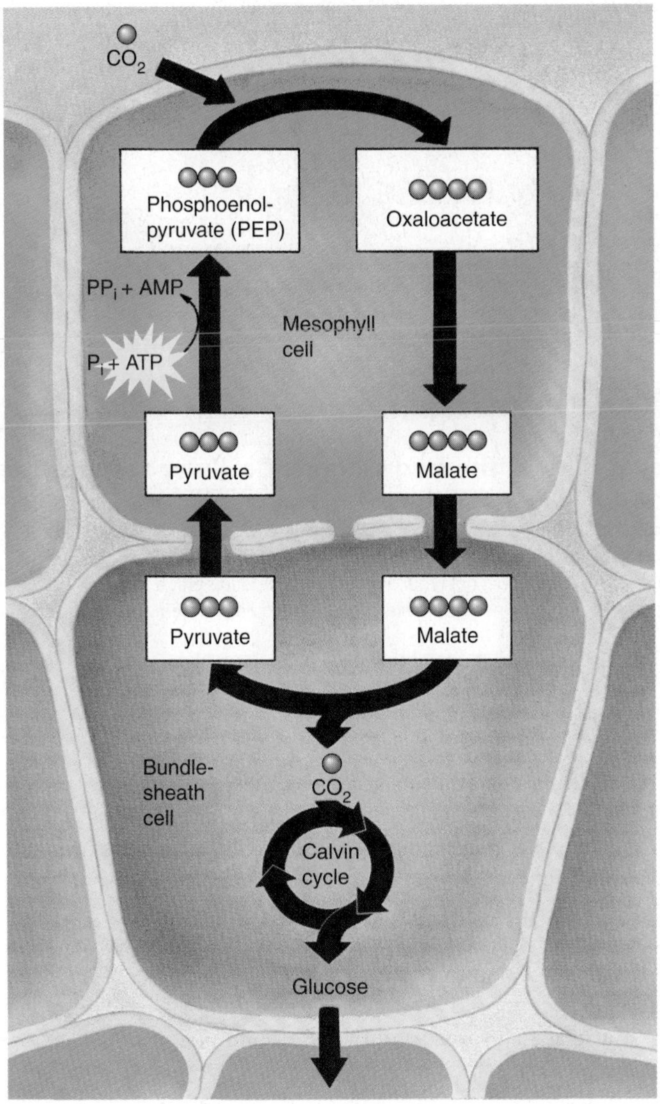

CO_2
Phosphoenol-pyruvate (PEP)
Oxaloacetate
PP_i + AMP
P_i + ATP
Mesophyll cell
Pyruvate
Malate
Pyruvate
Malate
Bundle-sheath cell
CO_2
Calvin cycle
Glucose

Figure 5.21 The path of carbon fixation in C_4 plants.

C_4 plants shuttle a four-carbon molecule (malate) to bundle-sheath cells, where carbon dioxide can be concentrated. In this way, C_4 cells conserve carbon dioxide.

Some plants are able to adapt to climates with higher temperatures by performing **C_4 photosynthesis.** In this process, plants such as sugarcane, corn, and many grasses are able to fix carbon using different types of cells within their leaves and thereby avoiding a reduced yield in photosynthesis due to higher temperatures.

C_4 plants fix CO_2 first as the four-carbon molecule oxaloacetate (hence the name, C_4 photosynthesis), rather than as the three-carbon molecule phosphoglycerate of normal everyday C_3 photosynthesis, such as described previously. C_4 plants carry out this process in the mesophyll cells of their leaves. The oxaloacetate that this process produces is then transferred out to the bundle-sheath cells of the leaf and there broken down to regenerate CO_2, which enters the Calvin cycle (figure 5.21). The bundle-sheath cells are impermeable to CO_2 and therefore hold it within them. The concentration of CO_2 increases and thus decreases the occurrence of photorespiration.

In C_4 photosynthesis, the energetic cost is almost doubled compared to that of C_3 photosynthesis. However, in a hot climate in which photorespiration decreases the growth of C_3 plants, C_4 photosynthesis is the best alternative. For this reason, C_4 plants are more abundant in warmer regions than in cooler ones (figure 5.22).

5.10 In a series of reactions that do not require light, cells use ATP and NADPH provided by photosystems I and II to assemble new organic molecules.

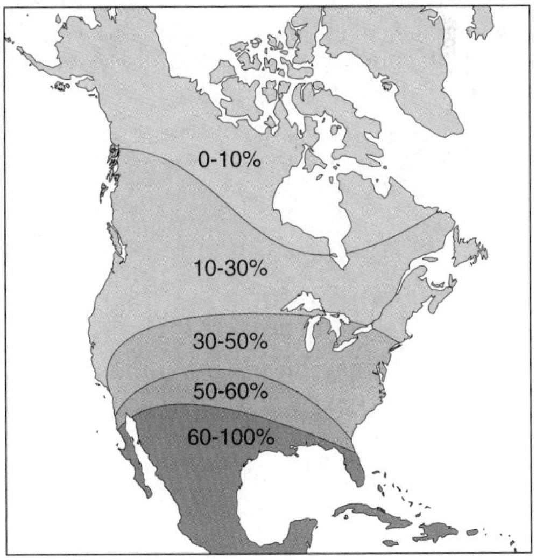

0-10%
10-30%
30-50%
50-60%
60-100%

Figure 5.22 Prevalence of C_4 plants in North America.

Many more C_4 grasses occur in the south, where the average temperatures during the growing season are higher. At higher temperatures, photorespiration wastes more of the products of photosynthesis, and the ability of C_4 plants to counteract photorespiration is more of an advantage than it is in cooler regions, where C_3 grasses predominate. This map shows the percentage of species of C_4 grasses among all grass species.

5.11 An Overview of Cellular Respiration

The cells of plants fuel their activities with sugars and other fuel molecules, just as yours do; only the chloroplasts carry out photosynthesis. No light shines on roots below the ground, and yet root cells are just as alive as the cells in the stem and leaves. In both plants and animals, and in fact in almost all organisms, the energy for living is obtained by breaking down the organic molecules assembled by chloroplasts. The ATP energy and reducing power invested in building the organic molecules are retrieved by stripping away the energetic electrons and using them to make ATP. This is possible because the electrons are still far from the nucleus, still carrying the energy contributed by their original encounter with the photon of light.

When electrons are stripped away from chemical bonds, the food molecules are being oxidized (remember, oxidation is the loss of electrons). The oxidation of foodstuffs to obtain energy is called **cellular respiration.** Do not confuse this with the breathing of oxygen gas that your lungs carry out, which is called simply respiration.

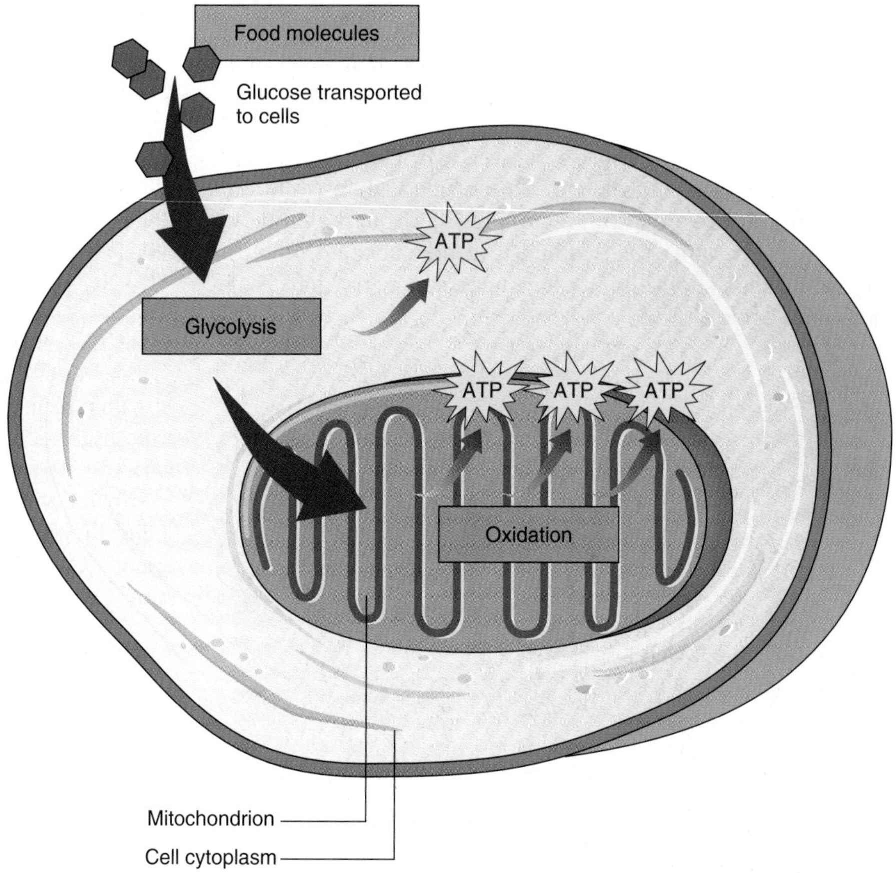
Food molecules
Glucose transported to cells
ATP
Glycolysis
ATP
ATP
ATP
Oxidation
Mitochondrion
Cell cytoplasm

Figure 5.23 An overview of the two stages of cellular respiration. When you take in food, the food is broken down into its various components by your digestive system. Fuel for your cells, in the form of glucose, is delivered to them by your circulatory system. Once inside your cells, the energy stored in glucose is liberated and transferred to ATP molecules by a complex process called *glycolysis*, which takes place in the cytoplasm of the cell, does not require oxygen, and produces only a few ATP molecules. The second phase, *oxidation*, occurs in the cell's mitochondrion, requires oxygen, and produces much larger quantities of ATP.

Aerobic Respiration

In **aerobic respiration,** ATP forms as electrons are harvested, transferred along the electron transport chain, and eventually donated to oxygen gas. Eukaryotes produce the majority of their ATP from glucose in this way. Chemically, there is little difference between this oxidation of carbohydrates in a cell and the burning of wood in a fireplace. In both instances, the reactants are carbohydrates and oxygen, and the products are carbon dioxide, water, and energy:

$$C_6H_{12}O_6 + 6\,O_2 \longrightarrow 6\,CO_2 + 6\,H_2O + \textbf{energy (heat or ATP)}$$

Cellular respiration is carried out in two stages (figure 5.23). The first stage uses coupled reactions to make ATP. This stage, **glycolysis,** takes place in the cell's cytoplasm and does not require oxygen. This ancient energy-extracting process is thought to have evolved over 3 billion years ago, when there was no oxygen in the earth's atmosphere. The second stage harvests electrons from chemical bonds and uses their energy to power the production of ATP. This stage, **oxidation,** takes place only within mitochondria. It is far more powerful than glycolysis at recovering energy from food molecules and is where the bulk of the energy used by animal cells is extracted.

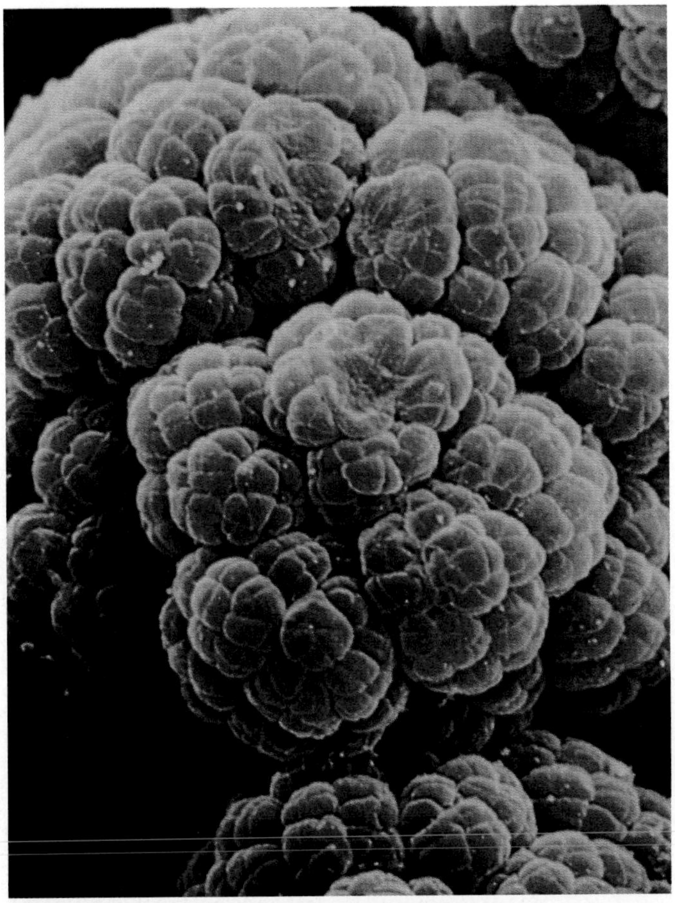

Figure 5.24 ***Methanosarcina barkeri.***
These primitive bacteria are called methanogens. They carry out a form of anaerobic respiration in which carbon dioxide (CO_2) is reduced to methane gas (CH_4).

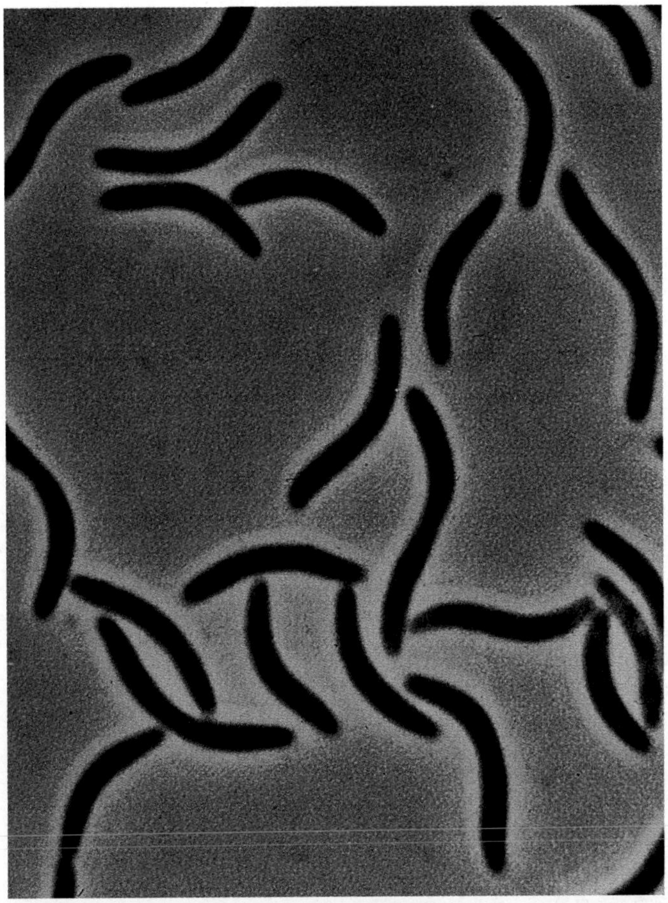

Figure 5.25 ***Desulfovibrio gigas.***
These bacteria carry out a form of anaerobic respiration in which sulfate (SO_4^{2-}) is reduced to hydrogen sulfide (H_2S).

Anaerobic Respiration

In the presence of oxygen, cells can respire aerobically. In the absence of oxygen, some organisms respire anaerobically, using different inorganic electron acceptors than oxygen. For example, many bacteria use sulfur, nitrate, or other inorganic compounds as the electron acceptor in place of oxygen.

Methanogens. Among the heterotrophs that practice anaerobic respiration are primitive archaebacteria such as the thermophilic, or "heat-loving," bacteria. Some of these, called methanogens (figure 5.24), use CO_2 as the electron acceptor, reducing CO_2 to CH_4 (methane) with the hydrogens derived from organic molecules produced by other organisms.

Sulfur Bacteria. Evidence of a second anaerobic respiratory process among primitive bacteria is seen in a group of rocks about 2.7 billion years old, known as the Woman River iron formation. Organic material in these rocks is enriched for the light isotope of sulfur, ^{32}S, relative to the heavier isotope ^{34}S. No known geochemical process produces such enrichment, but biological sulfur reduction does, in a process still carried out today by certain primitive bacteria. In this sulfate respiration, the bacteria derive energy from the reduction of inorganic sulfates (SO_4) to hydrogen sulfide (H_2S) (figure 5.25). The hydrogen atoms are obtained from organic molecules other organisms produce. These bacteria thus do the same thing methanogens do, but they use SO_4 as the oxidizing (that is, electron-accepting) agent in place of CO_2.

The sulfate reducers set the stage for the evolution of photosynthesis, creating an environment rich in H_2S. The first form of photosynthesis obtained hydrogens from H_2S using the energy of sunlight.

5.11 Cellular respiration is the dismantling of food molecules to obtain energy. In aerobic respiration, the cell harvests energy from glucose molecules in two stages, glycolysis and oxidation. Oxygen is the final electron acceptor. Anaerobic respiration donates the harvested electrons to other inorganic compounds.

5.12 Using Coupled Reactions to Make ATP

Glycolysis

The first stage in cellular respiration, glycolysis (figure 5.26), is a series of sequential biochemical reactions, a *biochemical pathway*. In 10 enzyme-catalyzed reactions (figure 5.27), the six-carbon sugar glucose is cleaved into two three-carbon molecules called pyruvate. The biochemical pathway does not employ oxidation to extract energy; it simply shuffles chemical bonds so that two coupled reactions can occur. In each reaction the breaking of a chemical bond contributes enough energy to force the formation of an ATP molecule from ADP. This transfer of a phosphate group from a substrate to ADP is called **substrate-level phosphorylation.** In the process, two energetic electrons are extracted and donated to a carrier molecule called NAD^+. The NAD^+ carries the electrons as NADH to join the other electrons extracted during oxidative respiration, discussed in the following section. Only a small number of ATP molecules are made in glycolysis, two for each molecule of glucose attacked, but in the absence of oxygen it is the only way organisms can get energy from food.

Glycolysis is thought to have been one of the earliest of all biochemical processes to evolve. Every living creature is capable of carrying out glycolysis. Glycolysis, although inefficient, was not discarded during the course of evolution but rather was used as the starting point for the further extraction of energy by oxidation. Nature did not, so to speak, go back to the drawing board and design metabolism from scratch. Rather, new reactions, which make up what is called the *Krebs cycle,* were added onto the old, just as successive layers of paint can be found in an old apartment.

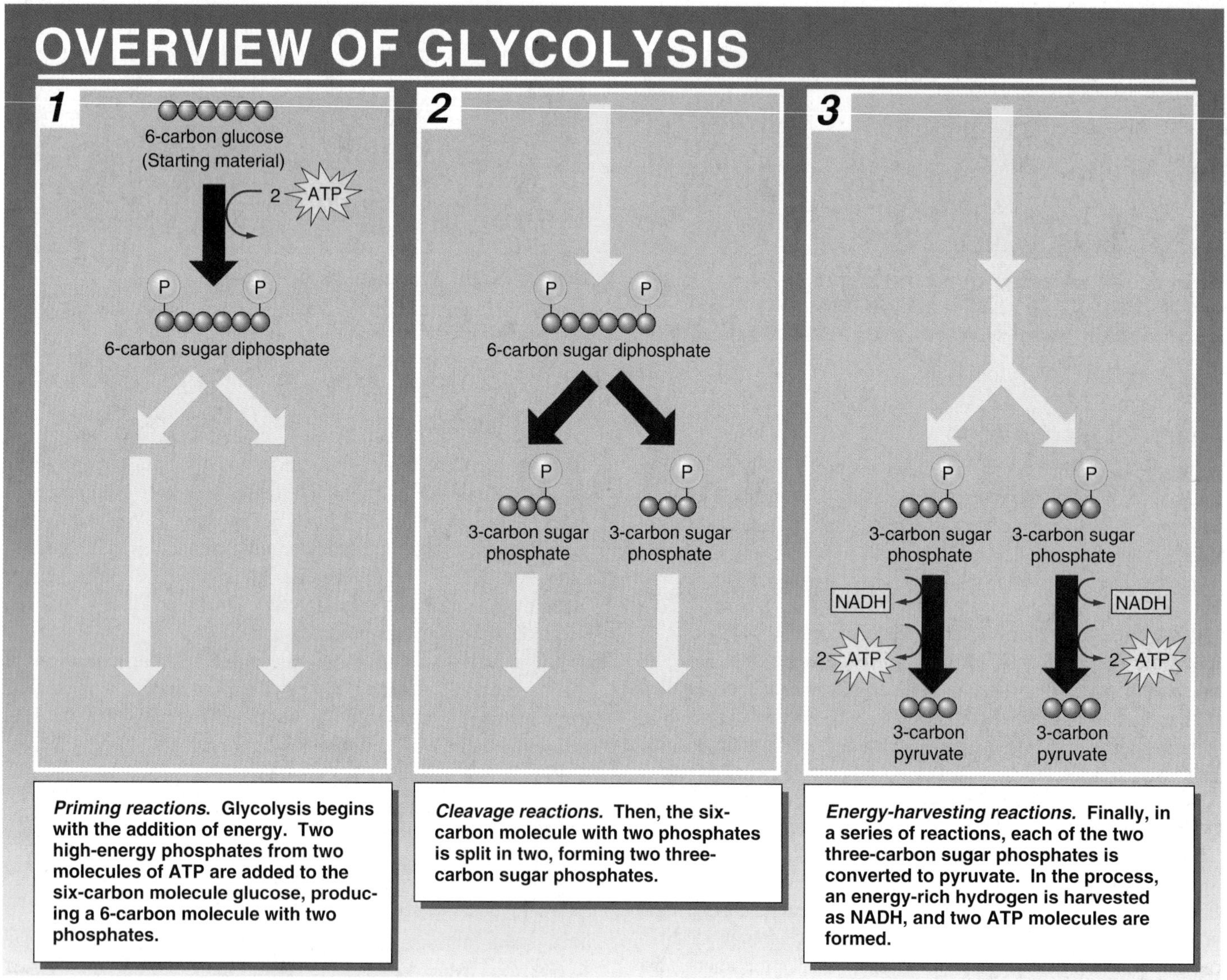

OVERVIEW OF GLYCOLYSIS
1
6-carbon glucose
(Starting material)
2 ATP
P
P
6-carbon sugar diphosphate
2
P
P
6-carbon sugar diphosphate
P
P
3-carbon sugar phosphate
3-carbon sugar phosphate
3
P
P
3-carbon sugar phosphate
3-carbon sugar phosphate
NADH
NADH
2 ATP
2 ATP
3-carbon pyruvate
3-carbon pyruvate
Priming reactions. Glycolysis begins with the addition of energy. Two high-energy phosphates from two molecules of ATP are added to the six-carbon molecule glucose, producing a 6-carbon molecule with two phosphates.
Cleavage reactions. Then, the six-carbon molecule with two phosphates is split in two, forming two three-carbon sugar phosphates.
Energy-harvesting reactions. Finally, in a series of reactions, each of the two three-carbon sugar phosphates is converted to pyruvate. In the process, an energy-rich hydrogen is harvested as NADH, and two ATP molecules are formed.

Figure 5.26 How glycolysis works.

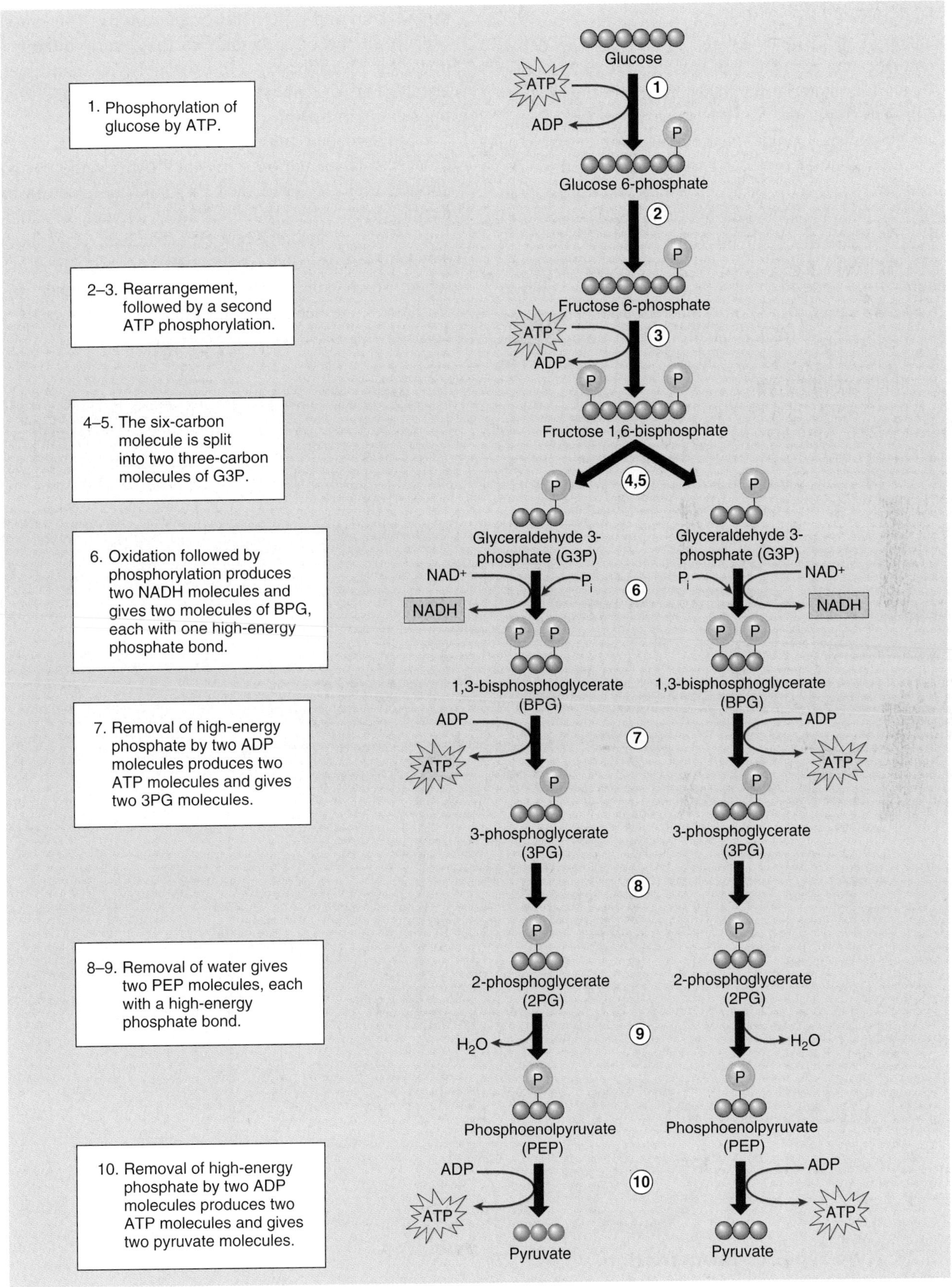
1. Phosphorylation of glucose by ATP.
2–3. Rearrangement, followed by a second ATP phosphorylation.
4–5. The six-carbon molecule is split into two three-carbon molecules of G3P.
6. Oxidation followed by phosphorylation produces two NADH molecules and gives two molecules of BPG, each with one high-energy phosphate bond.
7. Removal of high-energy phosphate by two ADP molecules produces two ATP molecules and gives two 3PG molecules.
8–9. Removal of water gives two PEP molecules, each with a high-energy phosphate bond.
10. Removal of high-energy phosphate by two ADP molecules produces two ATP molecules and gives two pyruvate molecules.
Glucose
ATP
ADP
1
P
Glucose 6-phosphate
2
Fructose 6-phosphate
3
Fructose 1,6-bisphosphate
4,5
Glyceraldehyde 3-phosphate (G3P)
NAD+
P_i
NADH
6
1,3-bisphosphoglycerate (BPG)
7
3-phosphoglycerate (3PG)
8
2-phosphoglycerate (2PG)
9
H_2O
Phosphoenolpyruvate (PEP)
10
Pyruvate

Figure 5.27 Glycolysis.

The process of glycolysis involves ten enzyme-catalyzed reactions.

Fermentation

What happens if there is no oxygen to carry out oxidative respiration? Does the pyruvate that is the product of glycolysis and the starting material for oxidative respiration just accumulate in the cytoplasm? No. It has a different fate. Recall that during glycolysis a single energetic electron is extracted at one step, carried away by a carrier molecule called NAD^+. In the absence of oxygen, these electrons are not used in chemiosmosis, and so soon all the cell's NAD^+ becomes saturated with electrons. With no more NAD^+ available to carry away electrons, glycolysis cannot proceed. Clearly, to obtain energy from food in the absence of oxygen, a solution to this problem is needed. A home must be found for these electrons. Adding the extracted electron to an organic molecule, as animals and plants do when they have no oxygen to take it, is called **fermentation.**

Two types of fermentation are common among eukaryotes (figure 5.28). Animals such as ourselves simply add the extracted electrons to pyruvate, forming lactate. Later, when oxygen becomes available, the process can be reversed and the electrons used for energy production. This is why your arm muscles would feel tired if you were to lift this text up and down 100 times rapidly. The muscle cells use up all the oxygen, and so they start running on ATP made by glycolysis, storing the pyruvate and electrons as lactate. This so-called oxygen debt produces the tired, burning feeling in the muscle.

Single-celled fungi called yeasts adopt a different approach to fermentation. First they convert the pyruvate into another molecule, and then they add the electron extracted during glycolysis, producing ethyl alcohol, or ethanol. For centuries, humans have consumed such ethyl alcohol in wine and beer. Yeasts only conduct this fermentation in the absence of oxygen. That is why wine is made in closed containers—to keep oxygen in the air away from the crushed grapes.

5.12 In the first stage of respiration, glycolysis, cells shuffle chemical bonds so that two coupled reactions can occur, producing ATP by substrate-level phosphorylation. When electrons that are also a product of glycolysis are not donated to oxygen, they are added to organic molecules, a process called fermentation.

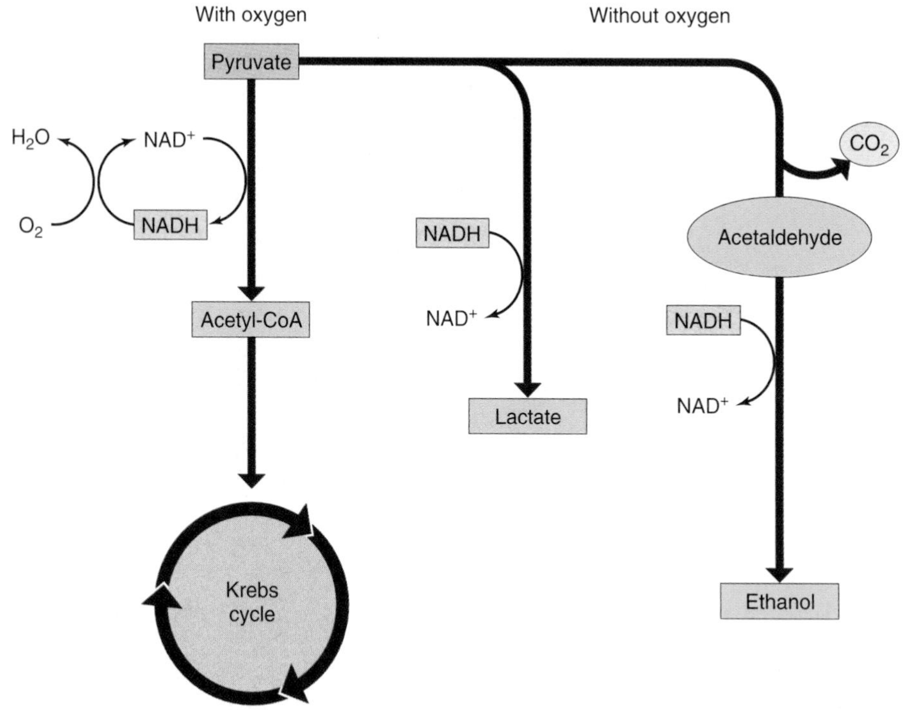

With oxygen
Without oxygen
Pyruvate
H_2O
NAD^+
O_2
NADH
Acetyl-CoA
Krebs cycle
NADH
NAD^+
Lactate
CO_2
Acetaldehyde
NADH
NAD^+
Ethanol

Figure 5.28 Two types of fermentation.

In the presence of oxygen, pyruvate is oxidized to acetyl-CoA and enters the Krebs cycle. In the absence of oxygen, pyruvate instead is reduced, accepting the electrons extracted during glycolysis and carried by NADH. This process is called fermentation. When pyruvate is reduced directly, as it is in your muscles, the product is lactate; when CO_2 is first removed from pyruvate and the remainder is reduced, as it is in yeasts, the product is ethanol.

5.13 Harvesting Electrons from Chemical Bonds

When oxygen is available, the cell need not stop extracting energy after glycolysis. Instead, a second oxidative stage of cellular respiration takes place. The first step of oxidative respiration is to oxidize the three-carbon molecule called pyruvate, which is the end product of glycolysis. Pyruvate molecules produced in the cytoplasm by glycolysis enter the mitochondria, where they are oxidized. The cell harvests pyruvate's considerable energy in two steps: first, by oxidizing pyruvate to form acetyl-CoA, and then by oxidizing acetyl-CoA in the Krebs cycle.

Producing Acetyl-CoA

Pyruvate is oxidized in a single "decarboxylation" reaction that cleaves off one of pyruvate's three carbons. This carbon then departs as CO_2 (figure 5.29). This reaction produces a two-carbon fragment called an acetyl group, as well as a pair of electrons and their associated hydrogen, which reduces NAD^+ to NADH (figure 5.30). In the course of the reaction, the acetyl group removed from pyruvate combines with a cofactor called coenzyme A (CoA), forming a compound known as **acetyl-CoA.** Acetyl-CoA is also generated by the breakdown of proteins and lipids. If the cell has plentiful supplies of ATP, acetyl-CoA is funneled into fat synthesis, with its energetic electrons preserved for later needs. If the cell needs ATP now, the fragment is directed into ATP production through the Krebs cycle.

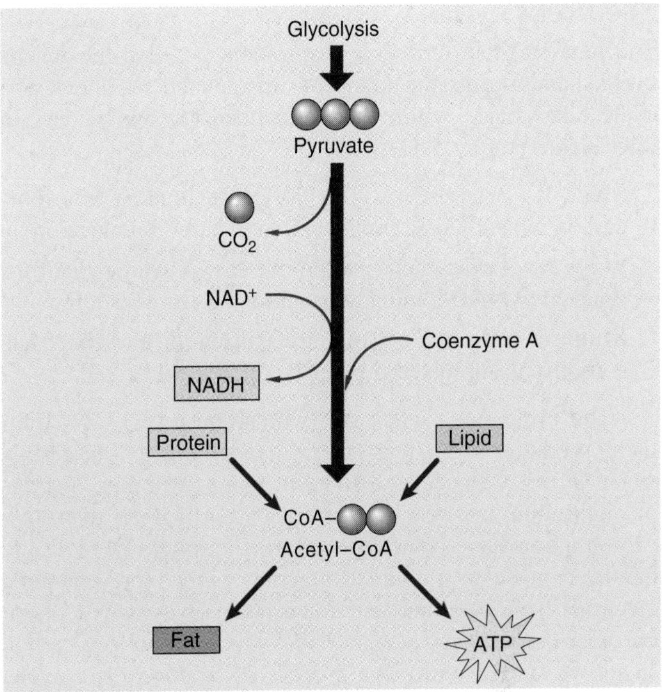

Glycolysis
Pyruvate
CO_2
NAD^+
Coenzyme A
NADH
Protein
Lipid
CoA–
Acetyl–CoA
Fat
ATP

Figure 5.29 Producing acetyl-CoA.
Pyruvate, the three-carbon product of glycolysis, is oxidized to the two-carbon molecule acetyl-CoA, in the process losing one carbon atom as CO_2 and an electron (donated to NAD^+ to form NADH). Almost all the molecules you use as foodstuffs are converted to acetyl-CoA; the acetyl-CoA is then channeled into fat synthesis or into ATP production, depending on your body's needs.

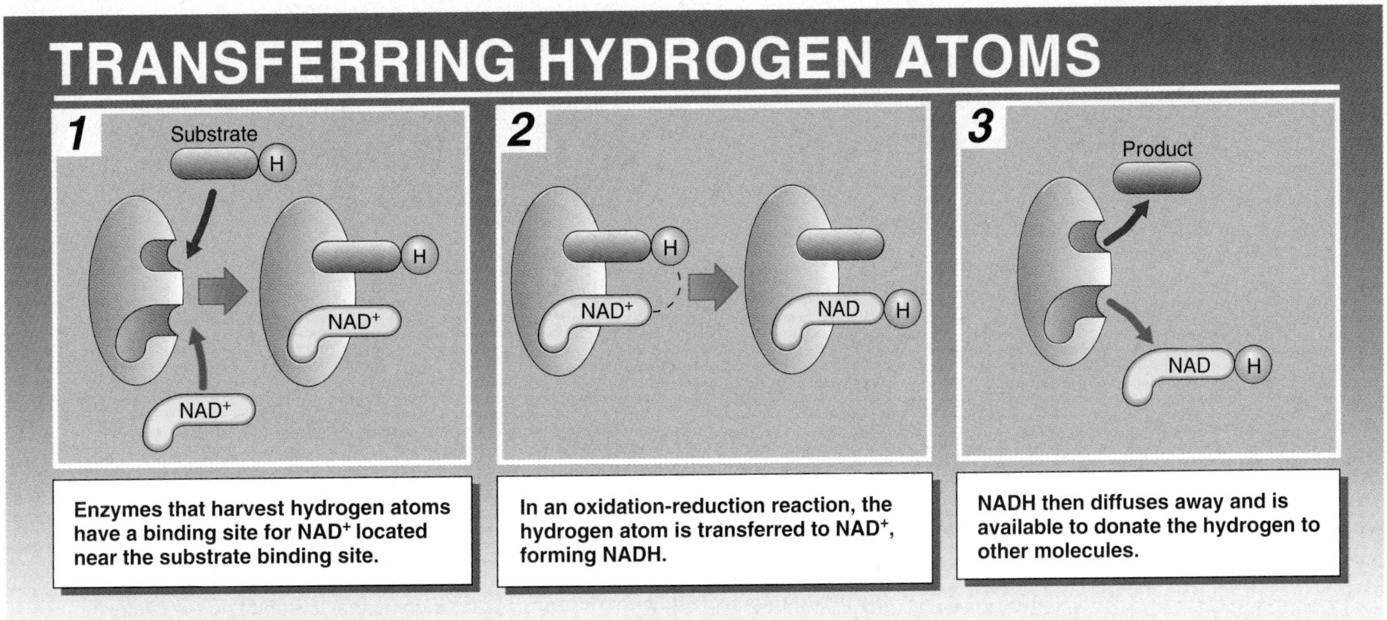

TRANSFERRING HYDROGEN ATOMS
1
Substrate
H
NAD^+
2
NAD
3
Product
Enzymes that harvest hydrogen atoms have a binding site for NAD^+ located near the substrate binding site.
In an oxidation-reduction reaction, the hydrogen atom is transferred to NAD^+, forming NADH.
NADH then diffuses away and is available to donate the hydrogen to other molecules.

Figure 5.30 How NAD^+ works.
Cells use NAD^+ to carry energetic electrons in hydrogen atoms from one molecule to another. Molecules that gain energetic electrons are said to be reduced, while ones that lose energetic electrons are said to be oxidized. NAD^+ oxidizes energy-rich molecules by acquiring their hydrogens (in the figure, this proceeds 1 → 2 → 3) and then reduces other molecules by giving the hydrogens to them (in the figure, this proceeds 3 → 2 → 1).

The Krebs Cycle

The next stage in oxidative respiration is called the **Krebs cycle,** named after the man who discovered it. The Krebs cycle takes place within the mitochondrion and occurs in three stages (figure 5.31):

Stage 1. Acetyl-CoA joins the cycle, binding to a four-carbon molecule and producing a six-carbon molecule.

Stage 2. Two carbons are removed as CO_2, their electrons donated to NAD^+, and a four-carbon molecule is left.

Stage 3. More electrons are extracted, and the four-carbon starting material is regenerated.

The cycle starts when the two-carbon acetyl-CoA fragment produced from pyruvate is stuck onto a four-carbon sugar called oxaloacetate. Then, in rapid-fire order, a series of eight additional reactions occur. When it is all over, two carbon atoms have been expelled as CO_2, one ATP molecule has been made in a coupled reaction, eight more energetic electrons have been harvested and taken away as NADH or on other carriers, and we are left with the same four-carbon sugar we started with. The process of reactions is a cycle, that is, a circle of reactions (figure 5.32). In each turn of the cycle, a new acetyl group replaces the two CO_2 molecules lost, and more electrons are extracted. Note that a single glucose molecule produces *two* turns of the cycle, one for each of the two pyruvate molecules generated by glycolysis.

In the process of cellular respiration, glucose is entirely consumed. The six-carbon glucose molecule is first cleaved into a pair of three-carbon pyruvate molecules during glycolysis. One of the carbons of each pyruvate is then lost as CO_2 in the conversion of pyruvate to acetyl-CoA, and the other two carbons are lost as CO_2 during the oxidations of the Krebs cycle. All that is left to mark the passing of the glucose molecule into six CO_2 molecules is its energy, preserved in four ATP molecules and electrons carried by 10 NADH and two $FADH_2$ carriers.

5.13 The end product of glycolysis, pyruvate, is oxidized to the two-carbon acetyl-CoA, yielding a pair of electrons plus CO_2. Acetyl-CoA then enters the Krebs cycle, yielding ATP, many energized electrons, and two CO_2 molecules.

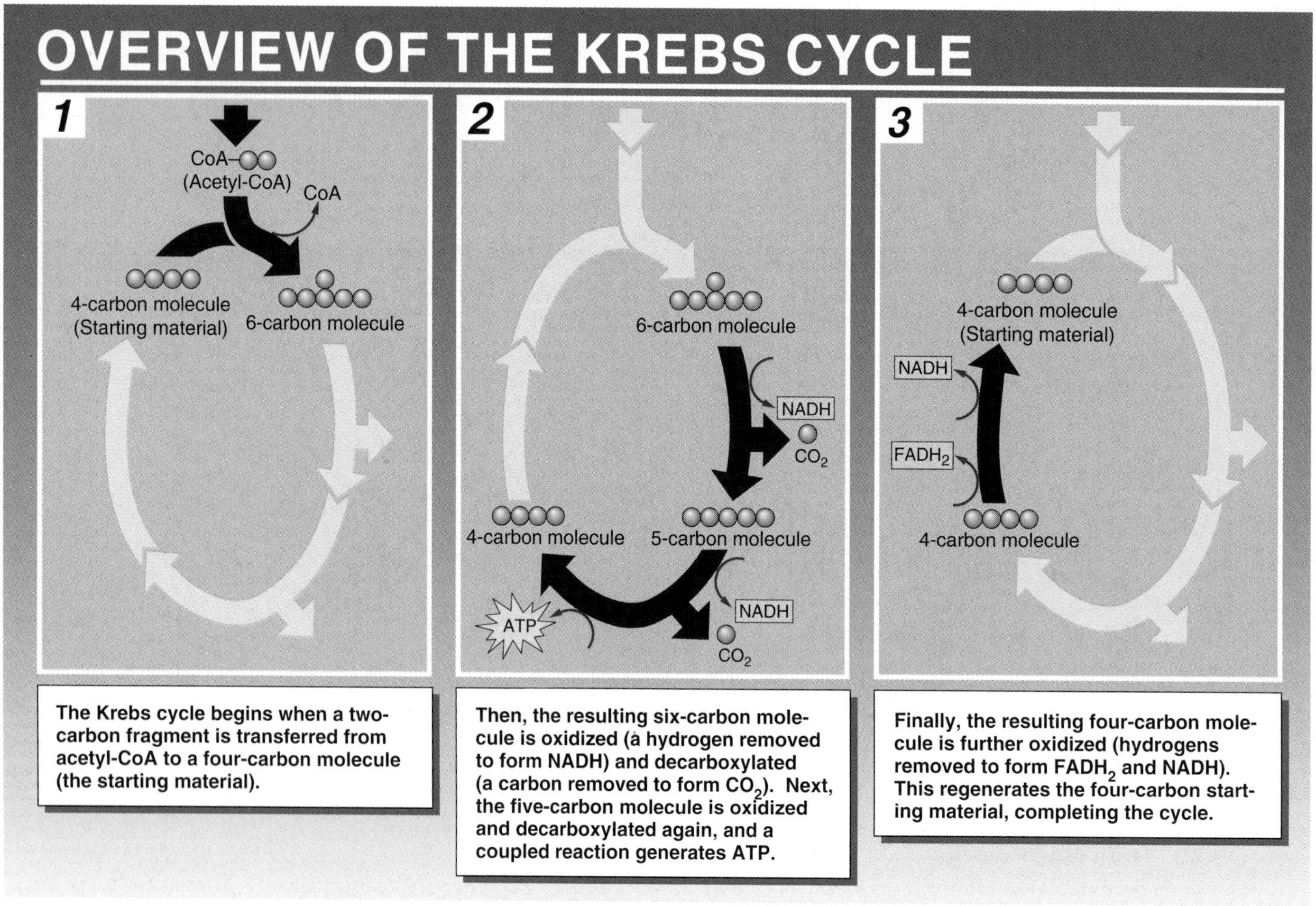
OVERVIEW OF THE KREBS CYCLE
1
CoA–
(Acetyl-CoA)
CoA
4-carbon molecule
(Starting material)
6-carbon molecule
The Krebs cycle begins when a two-carbon fragment is transferred from acetyl-CoA to a four-carbon molecule (the starting material).
2
6-carbon molecule
NADH
CO2
4-carbon molecule
5-carbon molecule
ATP
NADH
CO2
Then, the resulting six-carbon molecule is oxidized (a hydrogen removed to form NADH) and decarboxylated (a carbon removed to form CO2). Next, the five-carbon molecule is oxidized and decarboxylated again, and a coupled reaction generates ATP.
3
4-carbon molecule
(Starting material)
NADH
FADH2
4-carbon molecule
Finally, the resulting four-carbon molecule is further oxidized (hydrogens removed to form FADH2 and NADH). This regenerates the four-carbon starting material, completing the cycle.

Figure 5.31 How the Krebs cycle works.

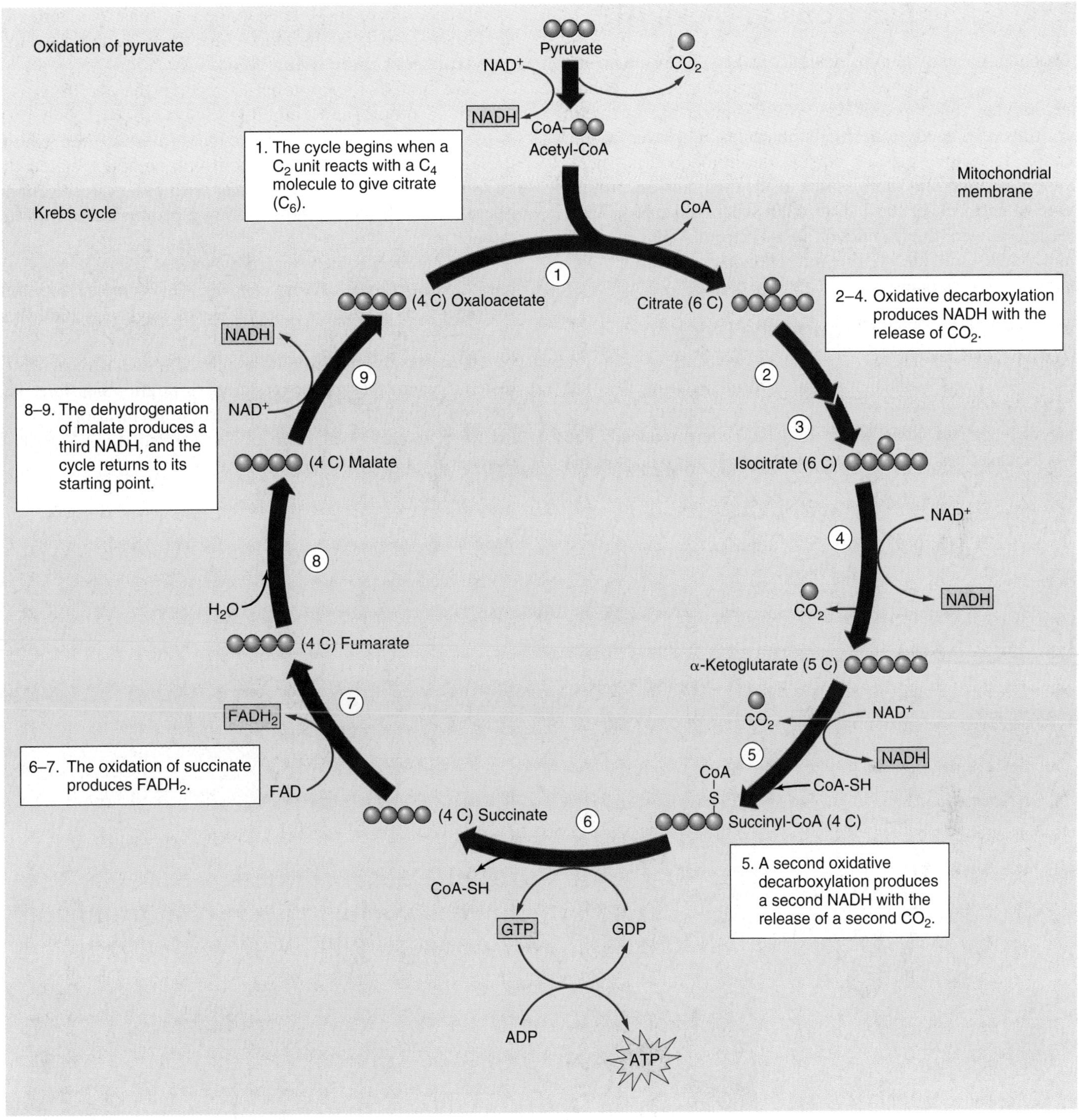

Oxidation of pyruvate
Pyruvate
NAD+
CO2
NADH
CoA
Acetyl-CoA
1. The cycle begins when a C2 unit reacts with a C4 molecule to give citrate (C6).
Mitochondrial membrane
Krebs cycle
CoA
1
(4 C) Oxaloacetate
Citrate (6 C)
2–4. Oxidative decarboxylation produces NADH with the release of CO2.
NADH
9
2
8–9. The dehydrogenation of malate produces a third NADH, and the cycle returns to its starting point.
NAD+
3
(4 C) Malate
Isocitrate (6 C)
NAD+
4
8
NADH
CO2
H2O
(4 C) Fumarate
α-Ketoglutarate (5 C)
7
CO2
NAD+
FADH2
5
NADH
6–7. The oxidation of succinate produces FADH2.
CoA
S
CoA-SH
FAD
(4 C) Succinate
6
Succinyl-CoA (4 C)
5. A second oxidative decarboxylation produces a second NADH with the release of a second CO2.
CoA-SH
GTP
GDP
ADP
ATP

Figure 5.32 The Krebs cycle.

This series of nine enzyme-catalyzed reactions takes place within the mitochondrion.

5.14 Using Electrons to Make ATP

Mitochondria use chemiosmosis to make ATP in much the same way that chloroplasts do, although the proton pumps are oriented the opposite way. Mitochondria use energetic electrons extracted from food molecules to power proton pumps that shove protons outside the mitochondrial membrane. Becoming far more scarce inside than outside, protons drive to get back in and jam through special channels. Their passage powers the production of ATP from ADP. The ATP then passes outside of the mitochondrion through ATP-passing open channels.

Moving Electrons Through the Electron Transport Chain

The NADH and $FADH_2$ molecules formed during the first stages of aerobic respiration each contain a pair of electrons that were gained when NAD^+ and FAD were reduced. The NADH molecules carry their electrons to the inner mitochondrial membrane, where they transfer the electrons to a series of membrane-associated proteins collectively called the **electron transport chain** (figure 5.33).

The first of the proteins to receive the electrons is a complex, membrane-embedded enzyme called *NADH dehydrogenase.* A carrier called *ubiquinone* then passes the electrons to a protein-cytochrome complex called the *bc_1 complex.* This complex, along with others in the chain, operates as a proton pump, driving a proton out across the membrane.

The electron is then carried by another carrier, *cytochrome c,* to the *cytochrome oxidase complex.* This complex uses four such electrons to reduce a molecule of oxygen, which then combines with two hydrogen ions to form water.

It is the availability of a plentiful electron acceptor (often oxygen) that makes oxidative respiration possible. The electron transport chain used in aerobic respiration is similar to, and may well have evolved from, the chain employed in photosynthesis.

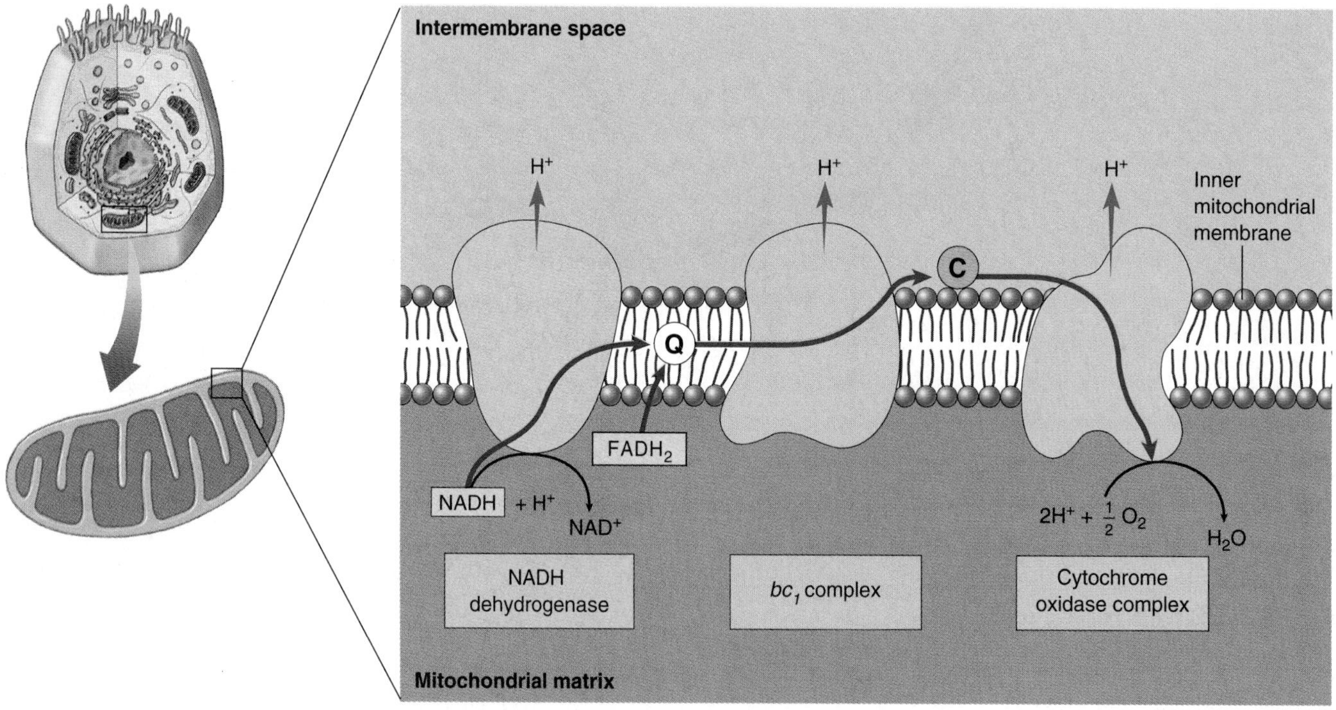

Intermembrane space
H+
H+
H+
Inner mitochondrial membrane
C
Q
FADH2
NADH + H+
NAD+
2H+ + 1/2 O2
H2O
NADH dehydrogenase
bc1 complex
Cytochrome oxidase complex
Mitochondrial matrix

Figure 5.33 The electron transport chain.

High-energy electrons are transported (*red arrows*) by mobile electron carriers (ubiquinone, marked Q, and cytochrome *c*, marked C) along a chain of membrane proteins. Three proteins use portions of the electrons' energy to pump protons (*blue arrows*) out of the matrix and into the intermembrane space. The electrons are finally donated to oxygen to form water.

Building an Electrochemical Gradient

In eukaryotes, aerobic metabolism takes place within the mitochondria present in virtually all cells. The internal compartment, or **matrix,** of a mitochondrion contains the enzymes that carry out the reactions of the Krebs cycle. As the electrons harvested by oxidative respiration are passed along the electron transport chain, the energy they release transports protons out of the matrix and into the outer compartment, sometimes called the **intermembrane space.** Proton pumps, transmembrane proteins in the inner mitochondrial membrane (see chapter 4), actually accomplish the transport. The flow of excited electrons induces a change in the shape of the pump proteins, which causes them to transport protons across the membrane. The electrons contributed by NADH activate three of these proton pumps, and those contributed by $FADH_2$ activate two.

Producing ATP: Chemiosmosis

As the proton concentration in the outer compartment rises above that in the matrix, the matrix becomes slightly negative in charge. This internal negativity attracts the positively charged protons and induces them to reenter the matrix. The higher outer concentration tends to drive protons back in by diffusion; because membranes are relatively impermeable to ions, most of the protons that reenter the matrix pass through special proton channels in the inner mitochondrial membrane. When the protons pass through, these channels synthesize ATP from ADP + P_i within the matrix. The ATP is then transported by facilitated diffusion out of the mitochondrion and into the cell's cytoplasm. It is because the chemical formation of ATP is driven by a diffusion force similar to osmosis that this process is referred to as **chemiosmosis** (figure 5.34).

Thus, the electron transport chain uses electrons harvested in aerobic respiration to pump a large number of protons across the inner mitochondrial membrane. Their subsequent reentry into the mitochondrial matrix drives the synthesis of ATP by chemiosmosis. Figure 5.35 summarizes the overall process.

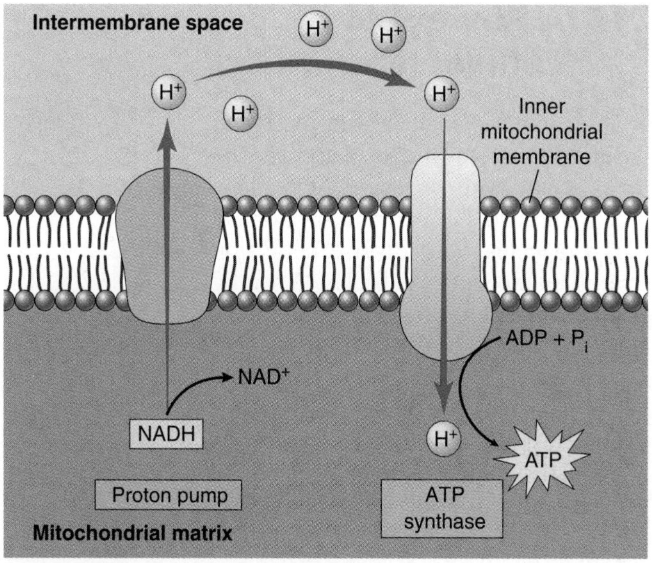

Intermembrane space
H+
H+
H+
H+
H+
Inner mitochondrial membrane
ADP + Pi
NAD+
NADH
H+
ATP
Proton pump
ATP synthase
Mitochondrial matrix

Figure 5.34 Chemiosmosis.

NADH transports high-energy electrons harvested from macromolecules to "proton pumps" that use the energy to pump protons out of the mitochondrial matrix. As a result, the concentration of protons outside the inner mitochondrial membrane rises, inducing protons to diffuse back into the matrix. Many of the protons pass through special channels that couple the reentry of protons to the production of ATP.

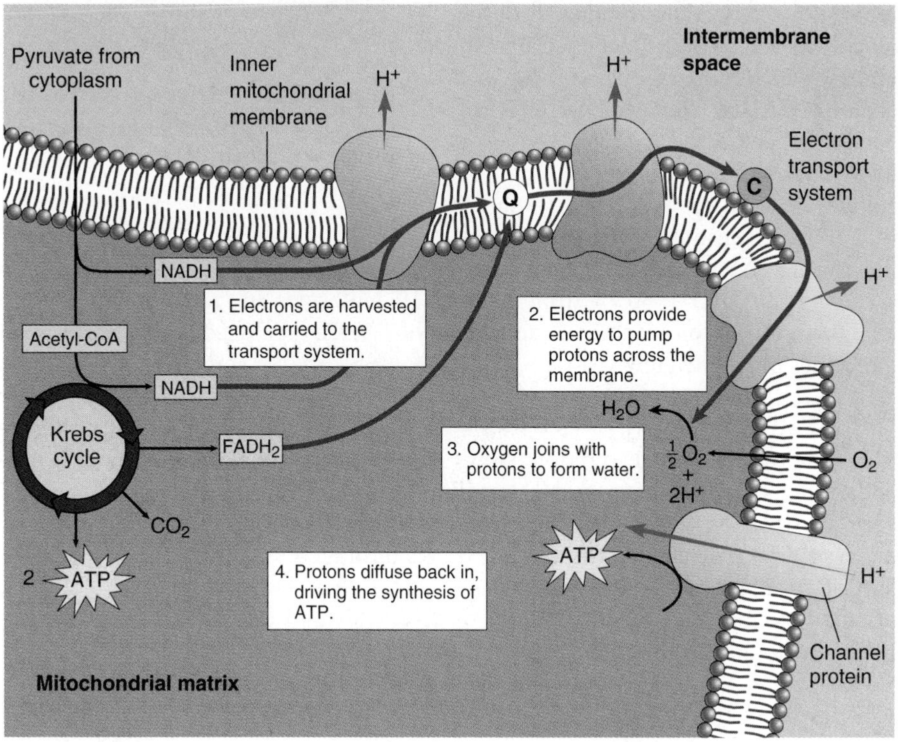

Pyruvate from cytoplasm
Inner mitochondrial membrane
H+
H+
Intermembrane space
Electron transport system
Q
C
NADH
H+
1. Electrons are harvested and carried to the transport system.
2. Electrons provide energy to pump protons across the membrane.
Acetyl-CoA
NADH
H2O
Krebs cycle
FADH2
3. Oxygen joins with protons to form water.
1/2 O2 + 2H+
O2
CO2
2 ATP
4. Protons diffuse back in, driving the synthesis of ATP.
ATP
H+
Channel protein
Mitochondrial matrix

Figure 5.35 An overview of the electron transport chain and chemiosmosis.

> **5.14** The electrons harvested by oxidizing food molecules are used to power proton pumps that chemiosmotically drive the production of ATP.

5.15 A Review of Cellular Respiration

As an example of metabolism in action, consider the fate of a chocolate bar. The bar is composed of sugar, chocolate, other lipids and fats, protein, and many other molecules. This diverse collection of complex molecules is broken down by the process of digestion into simpler molecules such as glucose, amino acids, and fatty acids. These breakdowns produce little or no energy but prepare the way for cellular respiration—that is, glycolysis and oxidative metabolism.

The chemiosmotic model suggests that in oxidative metabolism one ATP molecule is generated for each proton pump activated by the electron transport chain. The electrons from NADH activate three pumps and those from $FADH_2$ activate two. Thus, we would expect each molecule of NADH and $FADH_2$ to generate three and two ATP molecules, respectively. However, because eukaryotic cells carry out glycolysis in their cytoplasm and the Krebs cycle within their mitochondria, they must transport the two molecules of NADH produced during glycolysis across the mitochondrial membranes, which requires one ATP per molecule of NADH. Thus, the net ATP production is decreased by two. Therefore, the overall ATP production resulting from cellular respiration *theoretically* should be 2 (4 from substrate-level phosphorylation during glycolysis − 2 for transport of glycolytic NADH) + 30 (3 from each of 10 molecules of NADH) + 4 (2 from each of 2 molecules of $FADH_2$) = 36 molecules of ATP (figure 5.36).

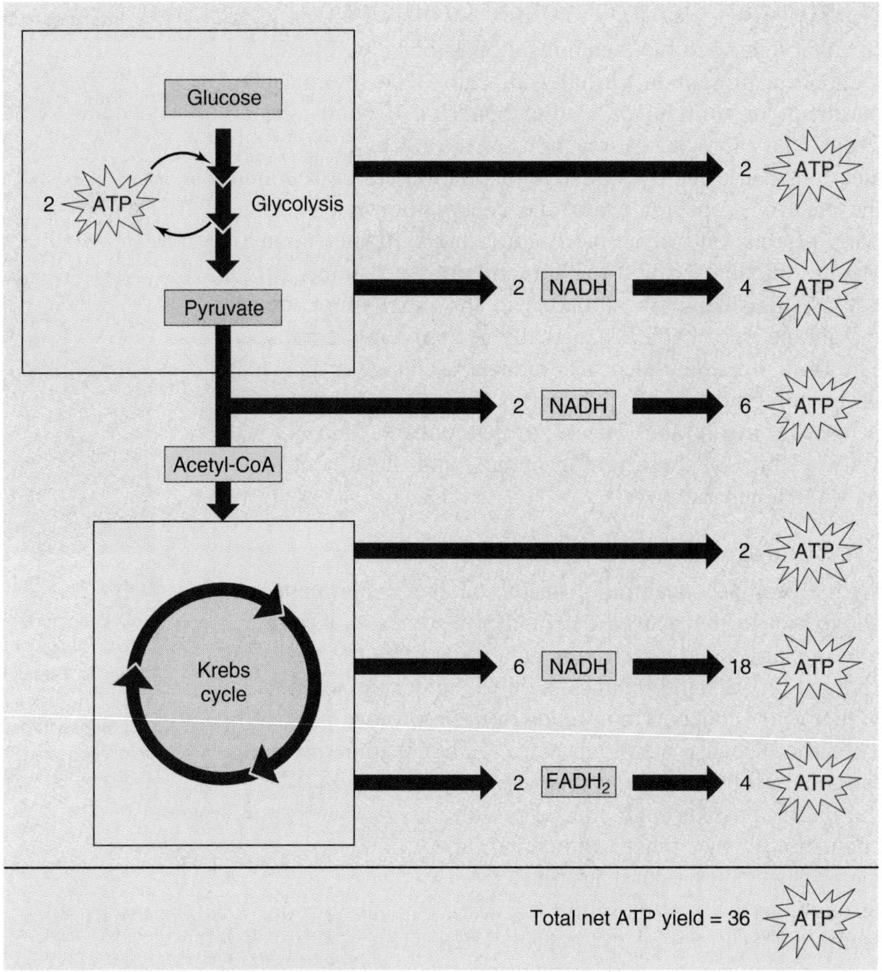

Glucose
2 ATP
Glycolysis
Pyruvate
2 ATP
2 NADH
4 ATP
2 NADH
6 ATP
Acetyl-CoA
Krebs cycle
2 ATP
6 NADH
18 ATP
2 $FADH_2$
4 ATP
Total net ATP yield = 36 ATP

Figure 5.36 Energy extracted during the breakdown of glucose. Most of the energy is extracted during oxidation (oxidation is the removal of electrons). The energetic electrons are carried by NADH and $FADH_2$ and used to produce ATP via chemiosmosis.

Regulating Cellular Respiration

The rate of cellular respiration slows down when your body's cells already have ample supplies of ATP. This is very sensible, but how does each mitochondrion arrive at the appropriate decision? The control works through a system of feedback inhibition in which excess product shuts off the reaction. Key reactions early in glycolysis and the Krebs cycle are catalyzed by enzymes that have a second, allosteric site. This site is the same shape as ATP, and when ATP levels in the cell are high it is very likely that ATP molecules will become stuck to it. The binding of ATP to the allosteric site goads the protein into changing shape to better accommodate the fit—and the new shape is not active as an enzyme! High levels of ATP thus shut down the processes the cell uses to make ATP. Like a well-designed automobile, your energy-producing machinery operates only when you step on the gas.

5.15 Oxidation is very efficient. The breakdown of glucose to pyruvate in glycolysis yields a net of only 2 ATP, while the oxidation of glucose yields an additional 34 ATP molecules.

CONCEPT REVIEW

1. Select the largest molecule.
   a. ADP
   b. ATP
   c. P
   d. K

2. Carbon fixation requires the expenditure of ATP molecules. This ATP is generated by
   a. the Calvin cycle.
   b. replenishment of the photosynthetic pigment.
   c. the light reactions.
   d. none of these.

3. When P_{700} receives photon-induced excitation energy, what is the form of this energy?
   a. light energy
   b. protons
   c. neutrons
   d. electrons

4. When plants and algae employ both photosystem I and II, _________ and _________ are generated, both of which are required to form organic molecules from atmospheric carbon dioxide.
   a. water
   b. ATP
   c. NADH
   d. NADPH

5. Select the first stage of cellular respiration.
   a. chemiosmosis
   b. glycolysis
   c. Krebs cycle
   d. electron transport

6. Entrance of pyruvate into the second stage of cellular respiration requires the presence of
   a. carbon dioxide.
   b. ATP.
   c. ADP.
   d. oxygen.

7. By the end of oxidative metabolism, all of the carbons from glucose are gone. Where did they go?
   a. carbon dioxide
   b. ATP
   c. to make pyruvate
   d. water

8. An _________ reaction requires energy from an outside source.

9. The _________ site is where a reactant attaches to an enzyme.

10. NAD^+ carries an electron as _________.

11. Light energy is "captured" and used to boost an electron to a higher energy level in molecules called _________.

12. All of the oxygen that you breathe has been produced by the splitting of water during ________.

13. The basic photosynthetic unit of a chloroplast is the _________, a flattened membranous sac that occurs in stacked columns called grana.

14. The second stage of cellular respiration is called the _________ cycle.

15. During fermentation, pyruvate is converted to _________ in yeast cells.

CHALLENGE YOURSELF

1. Almost no sunlight penetrates into the deep ocean. However, many fish that live there attract prey and potential mates by producing their own light. Where does that light come from?

2. If you poke a hole in a mitochondrion, can it still perform oxidative respiration? Can fragments of mitochondria perform oxidative respiration?

3. In theory, a plant kept in total darkness could still manufacture glucose—if it were supplied with *which* molecules?

4. Why are plants that consume 30 ATP molecules to produce one molecule of glucose (rather than the usual 18 molecules of ATP per glucose molecule) favored in hot climates but not in cold climates?

5 eBRIDGE

Reinforcing Key Points

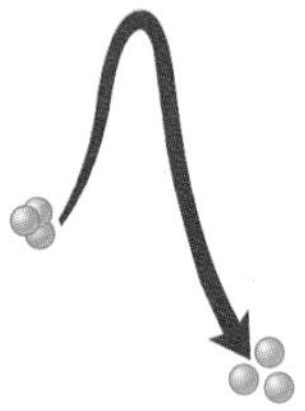

Cells and Energy

- 5.1 The Flow of Energy in Living Things
- 5.2 The Laws of Thermodynamics
- 5.3 Chemical Reactions
- 5.4 Enzymes
- 5.5 How Cells Use Energy

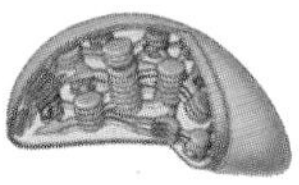

Photosynthesis

- 5.6 An Overview of Photosynthesis
- 5.7 How Plants Capture Energy from Sunlight
- 5.8 Organizing Pigments into Photosystems
- 5.9 How Photosystems Convert Light to Chemical Energy
- 5.10 Building New Molecules

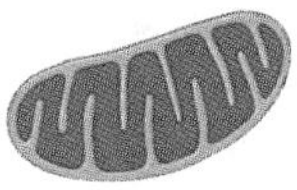

Cellular Respiration

- 5.11 An Overview of Cellular Respiration
- 5.12 Using Coupled Reactions to Make ATP
- 5.13 Harvesting Electrons from Chemical Bonds
- 5.14 Using Electrons to Make ATP
- 5.15 A Review of Cellular Respiration

Electronic Learning

Visual Learning

Animations

Eight Animations

Art Labeling Activities

Five Art Labeling Activities

Helping You Learn

Six Exercises

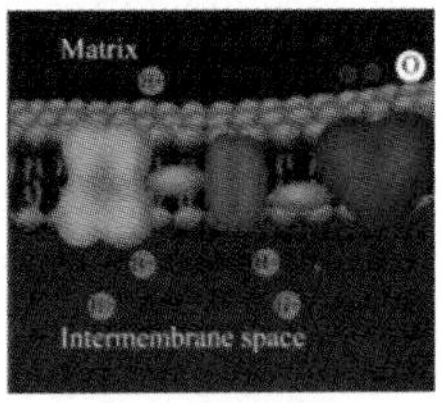
Matrix
Intermembrane space

Explorations

Enzymes in Action: Kinetics

In this exercise, you can compare catalysis ability and the effectiveness of binding a substrate among ten different enzymes.

MENU

Oxidative Respiration

In this exercise, you can vary oxygen levels, food supply, and ATP levels and explore the effects on the mitochondrial membrane.

Author's Corner

Aging. Given enough food to live on, and protection from infectious disease, humans live quite a long time, often for 80 years or more. But they do eventually die. Is this merely a matter of our bodies wearing out, or is our eventual death somehow programmed into the human blueprint? Theories abound. Many involve the progressive accumulation of damage to DNA, as genes that prolong life often affect DNA repair processes. Other theories involve the progressive loss of telomeric DNA from the ends of chromosomes with successive cell divisions. Still other theories focus on caloric restriction, arguing for prolonging life by reducing the efficiency with which energy is gleaned from food.

1. Aging may be the body's way of preventing the development of cancer.
2. Unraveling the mystery of aging.
3. A gene mutation called "I'm not dead yet" may hold the secret of longer life.

Virtual Classroom

Aging: Does Metabolism Limit Life Span?

All the activities of life—growth, communication, reproduction—require energy. It thus should come as no surprise that researchers now suggest aging is related to changes in the way we process metabolic energy. All humans die. After puberty, the rate of death increases exponentially with age. A variety of theories have been advanced to explain why. The oldest theory of aging is simply that cells accumulate mutations as they age. Other related theories focus on the idea that cells wear out over time, accumulating damage until they are no longer able to function. Free radicals produced as a byproduct of oxidative metabolism can be quite destructive in a cell. Also, every time a cell divides, it loses material from the tips of its chromosomes; eventually so much is lost that the chromosome can no longer divide. Some investigators argue that a gene clock controls aging. Single gene mutations can double the life span of fruit flies. When researchers isolated the gene involved, it proved to encode a protein involved in moving preliminary products of food metabolism across membranes to where the food's processing takes place. Surveys of very-long-lived humans also point to a single gene, whose function is being eagerly sought.

Virtual Lab

How Do Proteins Help Chlorophyll Carry Out Photosynthesis?

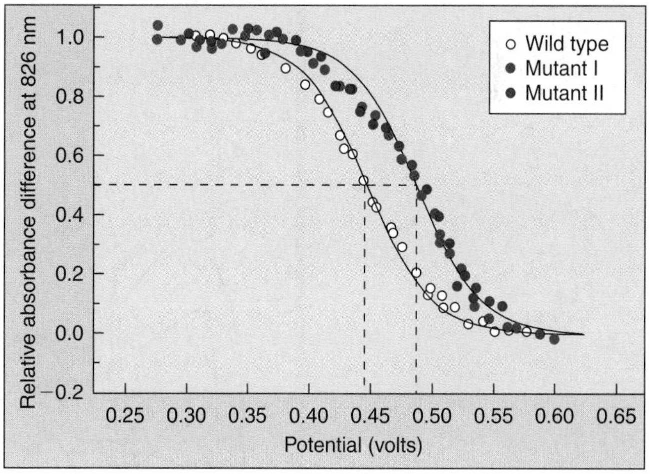
Relative absorbance difference at 826 nm
1.0
0.8
0.6
0.4
0.2
0.0
−0.2
0.25 0.30 0.35 0.40 0.45 0.50 0.55 0.60 0.65
Potential (volts)
Wild type
Mutant I
Mutant II

Great advances in biology have been made in recent years, some more quietly than others. Among these has been unmasking the underlying mechanism of photosynthesis. Plants possess two kinds of photosystems (I and II) that work together to harvest light energy. In the reaction center of photosystem I, a pair of chlorophyll molecules act as the trap for photon energy, passing an excited electron onto an acceptor molecule outside the reaction center. Two proteins (PsaA and PsaB) act as scaffolds to hold the chlorophyll molecules in place. A single amino acid of the PsaB protein, dubbed His-656, has become the focus of efforts to clarify how proteins help chlorophyll carry out photosynthesis. To determine the importance of His-656, Andrew Webber of Arizona State University, working with an international research team, set out to change the animo acid located at position 656 of PsaB in order to see what effects the change might have on photosynthesis. If His-656 indeed plays a critical role in modifying the chlorophylls, then a different amino acid at that position should have profound effects.

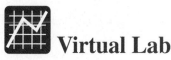

Quizzes | Further Reading | Essential Study Partner | Links | BioCourse.com

PART 3 THE CONTINUITY OF LIFE

CHAPTER 6

How Cells Divide

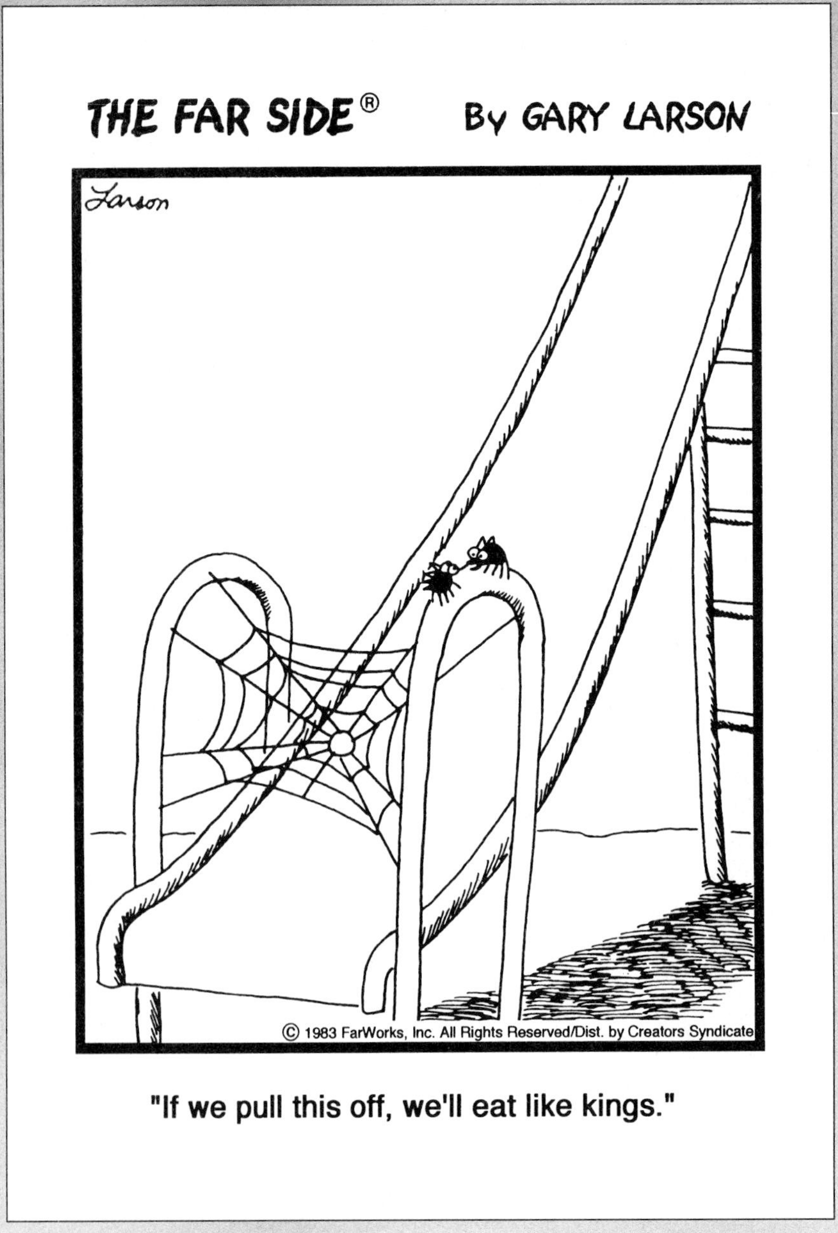
THE FAR SIDE®   By GARY LARSON
Larson
© 1983 FarWorks, Inc. All Rights Reserved/Dist. by Creators Syndicate
"If we pull this off, we'll eat like kings."

CHAPTER 6

How Cells Divide

CHAPTER OVERVIEW

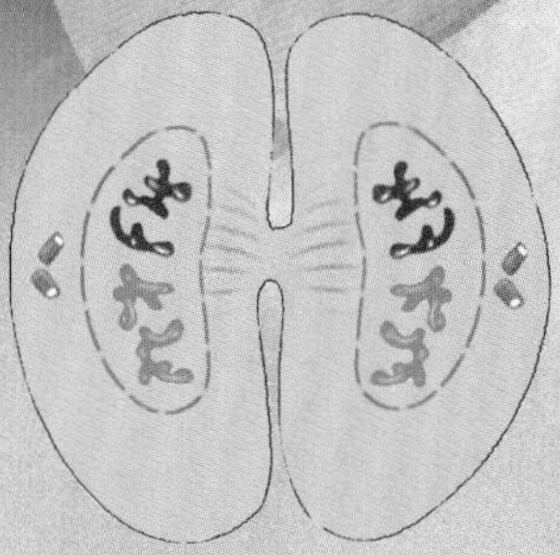

Cell Division

- 6.1 Prokaryotes Have a Simple Cell Cycle
- 6.2 Eukaryotes Have a Complex Cell Cycle
- 6.3 Chromosomes
- 6.4 Cell Division
- 6.5 Controlling the Cell Cycle

- Bacterial cells divide by simply splitting into two halves.
- Eukaryotic DNA forms a complex with histones and other proteins and is packaged into chromosomes.
- In eukaryotic cells, DNA replication is completed during the S phase of the cell cycle, and during the G_2 phase, the cell makes its final preparations for mitosis.
- Eukaryotic cells divide by mitosis, a complex process that delivers one replica of each chromosome to each of the two daughter cells.
- The cel! cycle is regulated at key checkpoints.

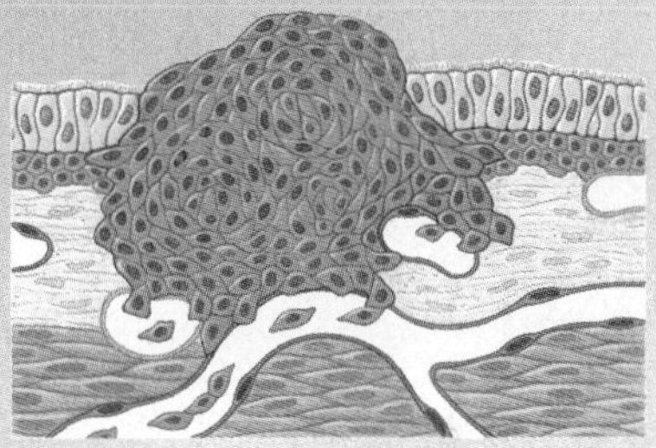

Cancer and the Cell Cycle

- 6.6 What Is Cancer?
- 6.7 Cancer and Control of the Cell Cycle
- 6.8 Curing Cancer

- Cancer is a disease in which the regulatory controls that normally restrain cell division are disrupted.
- A variety of environmental factors, have been implicated in causing cancer.
- New molecular therapies for cancer attempt to counteract the effects of deficiencies in proteins like p53 that normally act to check uncontrolled cell division.

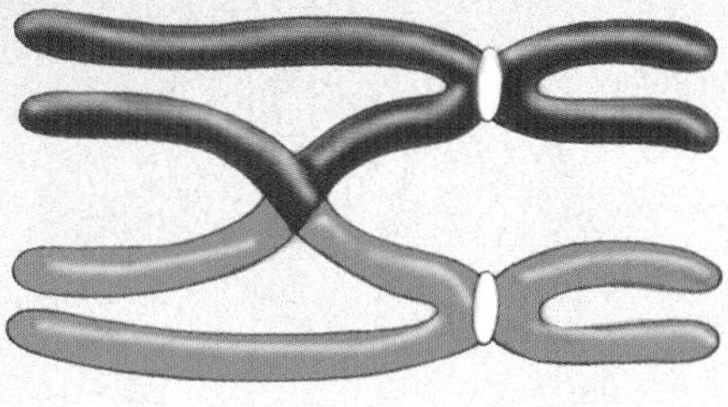

Meiosis

- 6.9 Discovery of Meiosis
- 6.10 The Sexual Life Cycle
- 6.11 The Stages of Meiosis
- 6.12 Unique Features of Meiosis
- 6.13 The Evolutionary Consequences of Sex

- Haploid gametes are produced from diploid cells by a special form of reduction division called meiosis.
- The reduction in chromosome number from diploid to haploid occurs because the chromosomes do not replicate between the two meiotic divisions.
- Early in meiosis, homologous chromosomes pair up closely. At this time, they often exchange segments, a process called crossing over.

6.1 Prokaryotes Have a Simple Cell Cycle

All species reproduce, passing their hereditary information on to their offspring. In this chapter, we begin our consideration of heredity with a look at how cells reproduce. Cell division in bacteria (figure 6.1) takes place in two stages, which together make up a **simple cell cycle.** First the DNA is copied, and then the cell splits.

In bacteria, the hereditary information—that is, the genes that specify the bacteria—is encoded in a single circle of DNA, attached at one point to the inner surface of the plasma membrane like a rope attached to the inner wall of a tent. Before the cell itself divides, the DNA circle makes a copy of itself. Starting at one point, the double helix of DNA begins to unzip, exposing the two strands. A new double helix is then formed on each naked strand by placing on each exposed nucleotide its complementary nucleotide (that is, A with T, G with C as discussed in chapter 3). DNA replication is discussed in more detail in chapter 8. When the unzipping has gone all the way around the circle, the cell possesses two copies of its hereditary information, attached side by side to the interior plasma membrane (figure 6.2).

When the DNA has been copied and the cell reaches an appropriate size, the bacterial cell splits into two equal halves, a process called **binary fission.** New plasma membrane and cell wall are added at a point between where the two DNA copies are attached to the membrane. As the growing plasma membrane pushes inward, the cell is constricted in two, eventually forming two **daughter cells.** Each contains one of the circles of DNA and is a complete living cell in its own right.

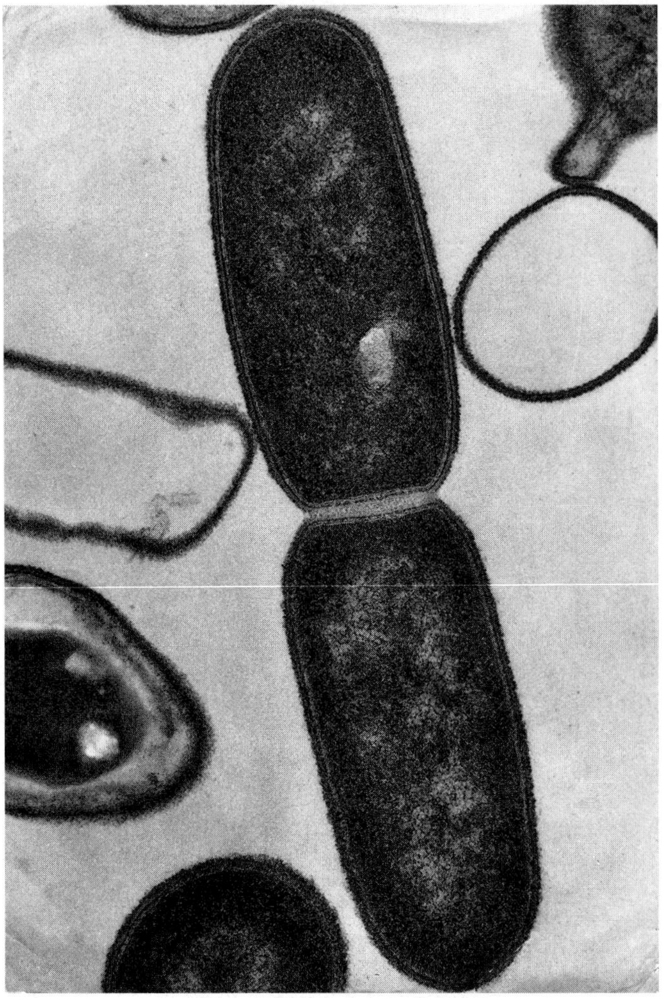

Figure 6.1 Cell division in bacteria.
Bacteria divide by a process of simple cell fission. Here, a cell has divided in two and is about to be pinched apart by the growing plasma membrane.

6.1 Bacteria divide by binary fission after the DNA has replicated. Fission begins between the points where the two copies of DNA are bound to the plasma membrane.

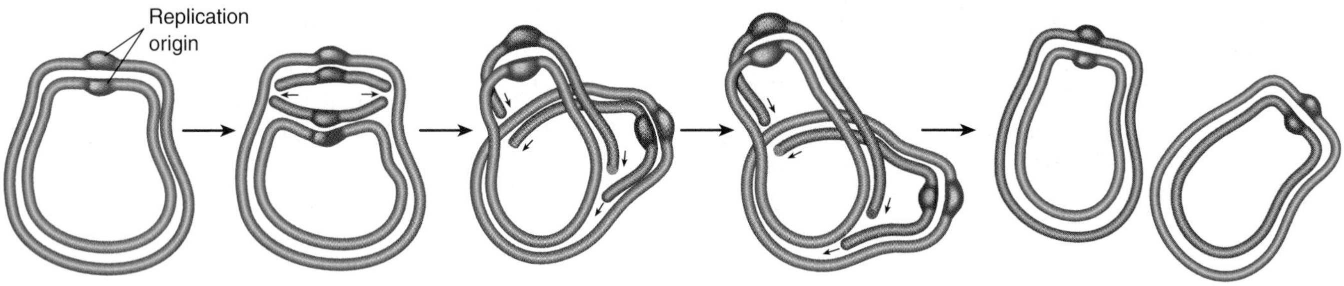
Replication origin

Figure 6.2 How bacterial DNA replicates.
The circular DNA molecule of a bacterium initiates replication at a single site, moving out in both directions. When the two moving replication points meet on the far side of the molecule, its replication is complete.

6.2 Eukaryotes Have a Complex Cell Cycle

The evolution of the eukaryotes introduced several additional factors into the process of cell division. Eukaryotic cells are much larger than bacteria, and they contain much more DNA. Eukaryotic DNA is contained in a number of linear chromosomes, whose organization is much more complex than that of the single, circular DNA molecules in bacteria. A **chromosome** is a single, long DNA molecule packaged with proteins into a compact shape. In chromosomes, DNA forms a complex with packaging proteins called histones and is wound into tightly condensed coils.

Cell division in eukaryotes is more complex than in bacteria, both because eukaryotes contain far more DNA and because it is packaged differently. The cells of eukaryotic organisms either undergo mitosis or meiosis to divide up the DNA. **Mitosis** is the mechanism of cell division that occurs in an organism's nonreproductive cells, or **somatic cells.** A second process, called **meiosis,** divides the DNA in cells that participate in sexual reproduction, or **germ cells.** Meiosis results in the production of gametes, such as sperm and eggs, and is discussed later in this chapter.

The events that prepare the eukaryotic cell for division and the division process itself constitute a **complex cell cycle** (figure 6.3):

G_1 phase. This "first growth" phase is the cell's primary growth phase. For most organisms, this phase occupies the major portion of the cell's life span.

S phase. In this "synthesis" phase, the DNA replicates, producing two copies of each chromosome.

G_2 phase. Cell division preparation continues with the replication of mitochondria, chromosome condensation, and the synthesis of microtubules. G_1, S, and G_2 are collectively referred to as **interphase,** or the period between cell divisions.

M phase. In mitosis, a microtubular apparatus binds to the chromosomes and moves them apart.

C phase. In cytokinesis the cytoplasm divides, creating two daughter cells.

6.2 Eukaryotic cells divide by separating duplicate copies of their chromosomes into daughter cells.

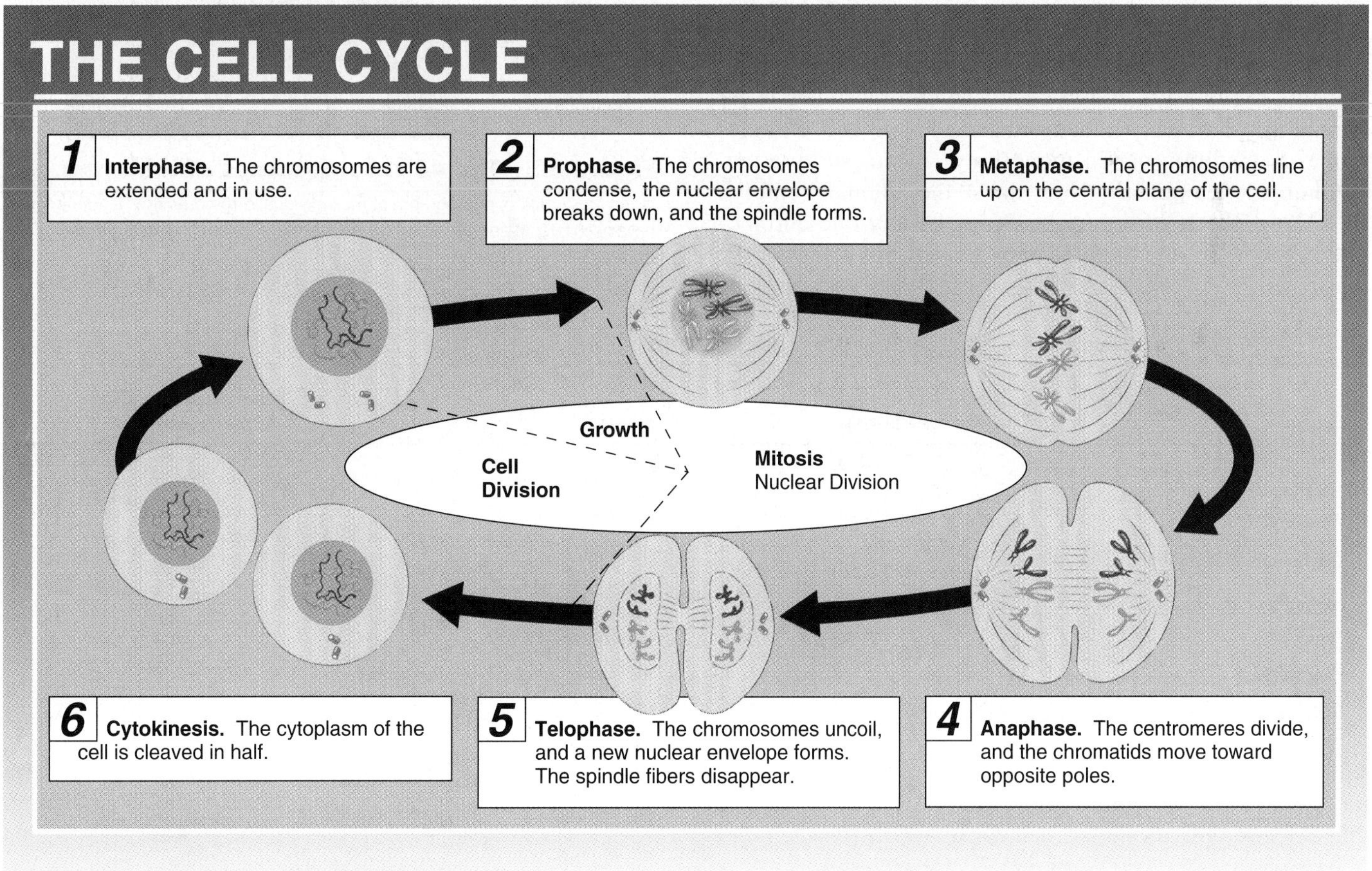
THE CELL CYCLE
1 Interphase. The chromosomes are extended and in use.
2 Prophase. The chromosomes condense, the nuclear envelope breaks down, and the spindle forms.
3 Metaphase. The chromosomes line up on the central plane of the cell.
Growth
Cell Division
Mitosis Nuclear Division
6 Cytokinesis. The cytoplasm of the cell is cleaved in half.
5 Telophase. The chromosomes uncoil, and a new nuclear envelope forms. The spindle fibers disappear.
4 Anaphase. The centromeres divide, and the chromatids move toward opposite poles.

Figure 6.3 How the cell cycle works.

The G_1, S, and G_2 phases occur during interphase, while the cell is growing and preparing to divide. Then, during mitosis, the cell's nucleus divides. Finally, the cell's cytoplasm divides (cytokinesis) with the formation of two daughter cells.

6.3 Chromosomes

Discovery of Chromosomes

Chromosomes were first observed by the German embryologist Walther Fleming in 1882, while he was examining the rapidly dividing cells of salamander larvae. When Fleming looked at the cells through what would now be a rather primitive light microscope, he saw minute threads within their nuclei that appeared to be dividing lengthwise. Fleming called their division *mitosis,* based on the Greek word *mitos,* meaning "thread."

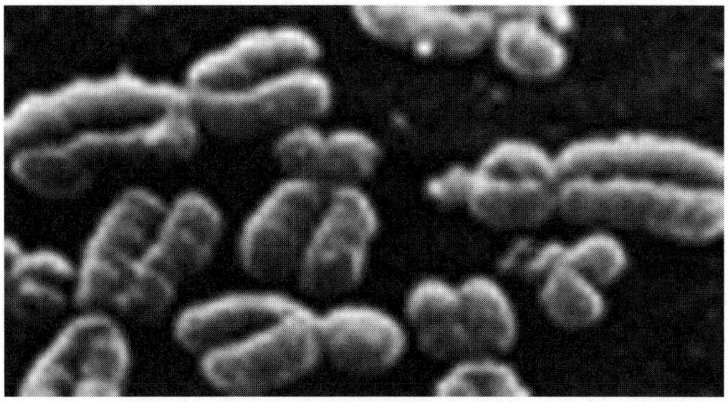

Figure 6.4 Human chromosomes.
The photograph (×950) shows human chromosomes as they appear immediately before nuclear division. Each DNA strand has already replicated.

Chromosome Number

Since their initial discovery, chromosomes have been found in the cells of all eukaryotes examined (figure 6.4). Their number may vary enormously from one species to another. A few kinds of organisms—such as the Australian ant *Myrmecia* spp.; the plant *Haplopappus gracilis,* a relative of the sunflower that grows in North American deserts; and the fungus *Penicillium*—have only 1 pair of chromosomes, while some ferns have more than 500 pairs (table 6.1). Most eukaryotes have between 10 and 50 chromosomes in their body cells.

Human cells each have 46 chromosomes, consisting of 23 nearly identical pairs (figure 6.5). Each chromosome contains thousands of genes that play important roles in determining how a person's body develops and functions. For this reason, possession of all the chromosomes is essential to survival. Humans missing even one chromosome, a condition called monosomy, do not survive embryonic development. Nor does the human embryo develop properly with an extra copy of any one chromosome, a condition called trisomy. For all but a few of the

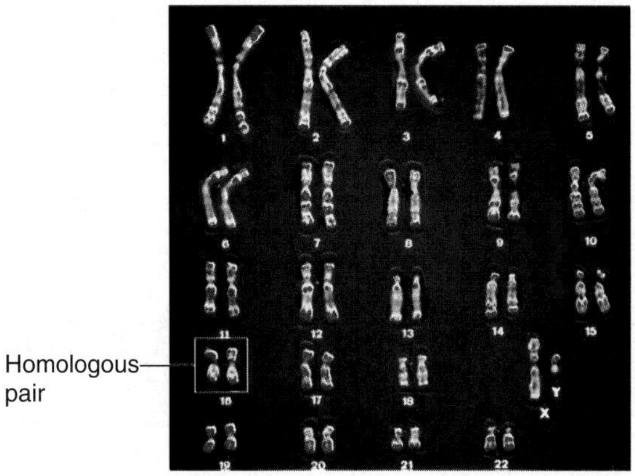
Homologous pair

Figure 6.5 The 46 chromosomes of a human.
In this presentation, photographs of the individual chromosomes of a human male have been cut out and paired with their homologues, creating an organized display called a *karyotype.*

TABLE 6.1 CHROMOSOME NUMBER IN SELECTED EUKARYOTES

Group	Total Number of Chromosomes	Group	Total Number of Chromosomes
		Bread wheat	42
FUNGI			
Neurospora	7	Sugarcane	80
Saccharomyces (a yeast)	16	Horsetail	216
		Adder's tongue fern	1,262
INSECTS			
Mosquito	6	**VERTEBRATES**	
Drosophila	8	Opossum	22
Honeybee	32	Frog	26
Silkworm	56	Mouse	40
		Human	46
PLANTS			
Haplopappus gracilis	2	Chimpanzee	48
Garden pea	14	Horse	64
Corn	20	Chicken	78
		Dog	78

smallest chromosomes, trisomy is fatal; even in those cases, serious problems result.

Chromosome Structure

A eukaryotic cell has far more DNA in it than a bacterium does. The DNA of eukaryotes is divided into several chromosomes. Because the phosphate groups of DNA molecules have negative charges, it is impossible to just tightly wind up DNA because all the negative charges would simply repel one another. Instead, the DNA helix is wrapped around proteins with positive charges called **histones,** the positive (histone) and negative (DNA) charges counteracting each other so that the complex has no net charge. This complex of DNA and histone proteins is then coiled tightly forming a compact chromosome (figure 6.6). For example, a human chromosome is about 40% DNA and 60% protein and typically contains about half a billion nucleotides in one long, unbroken DNA helix that would be about 5 centimeters (2 in.) long if laid out in a straight line. The amount of information in one human chromosome would fill about 2,000 printed books of 1,000 pages each!

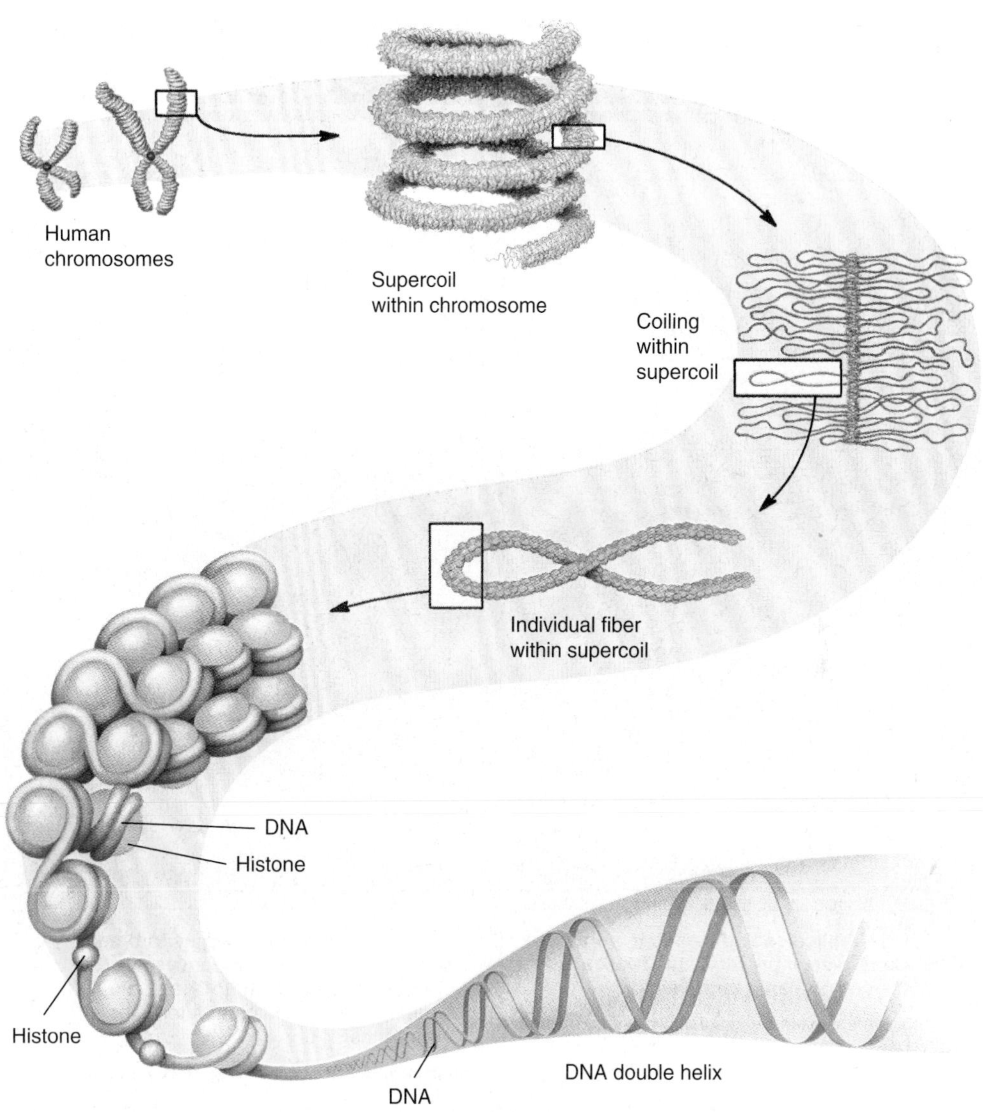
Human
chromosomes
Supercoil
within chromosome
Coiling
within
supercoil
Individual fiber
within supercoil
DNA
Histone
Histone
DNA
DNA double helix

Figure 6.6 Levels of chromosomal organization.
Compact, rod-shaped chromosomes are in fact highly wound-up molecules of DNA.

Chromosomes exist in somatic cells as two nearly identical copies of each other, called **homologous chromosomes,** or **homologues.** Cells that have two of each type of chromosome are called **diploid cells.** Before cell division, each of the two homologues replicates, resulting in two identical copies, called **sister chromatids,** that remain joined together at a special linkage site called the **centromere** (figure 6.7). For example, human body cells have 46 chromosomes. In their unduplicated state, before replication during interphase, you have 23 pairs of homologous chromosomes, one set from your mother and one set from your father. In their duplicated state, there are 23 pairs of duplicated chromosomes, which consist of two sister chromatids each, for a total of 92 chromatids.

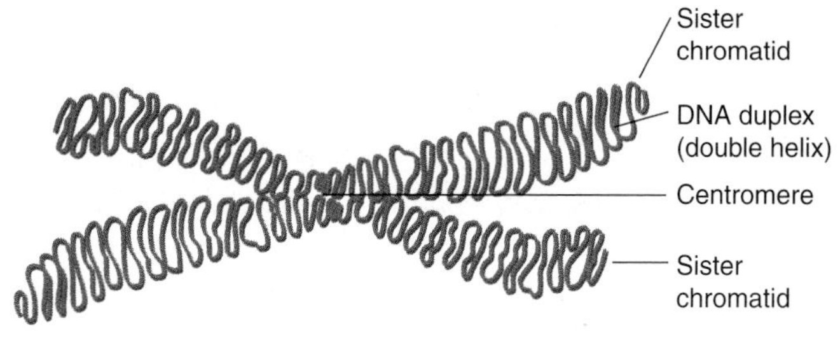
Sister
chromatid
DNA duplex
(double helix)
Centromere
Sister
chromatid

Figure 6.7 Sister chromatids.
A duplicated chromosome looks somewhat like an X and is composed of two sister chromatids held together by a centromere.

6.3 All eukaryotic cells store their hereditary information in chromosomes, but different kinds of organisms use very different numbers of chromosomes to store this information.

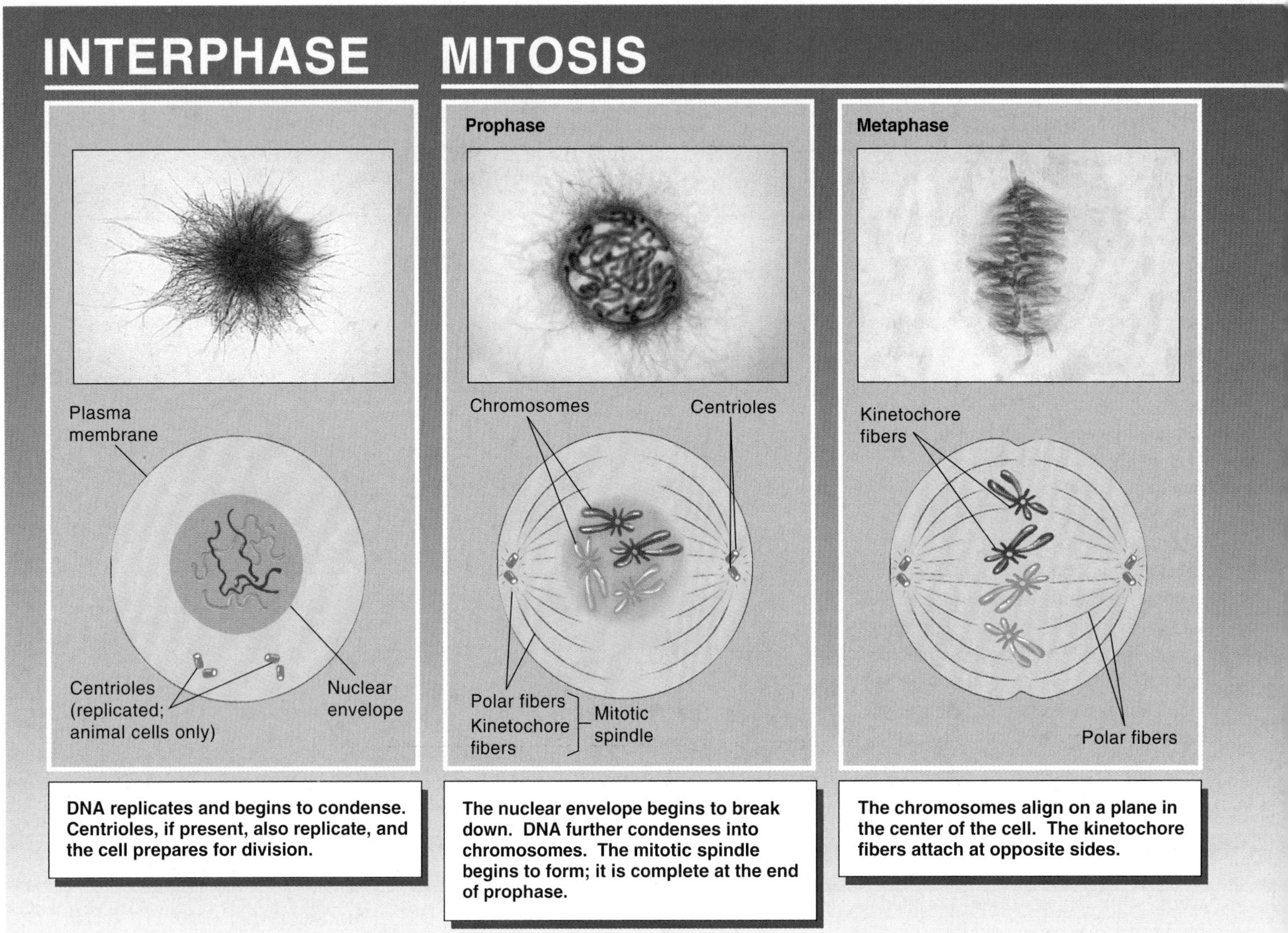
INTERPHASE
MITOSIS
Plasma membrane
Centrioles (replicated; animal cells only)
Nuclear envelope
DNA replicates and begins to condense. Centrioles, if present, also replicate, and the cell prepares for division.
Prophase
Chromosomes
Centrioles
Polar fibers
Kinetochore fibers
Mitotic spindle
The nuclear envelope begins to break down. DNA further condenses into chromosomes. The mitotic spindle begins to form; it is complete at the end of prophase.
Metaphase
Kinetochore fibers
Polar fibers
The chromosomes align on a plane in the center of the cell. The kinetochore fibers attach at opposite sides.

Figure 6.8 How cell division works.

Cell division in eukaryotes begins in interphase, carries through the four stages of mitosis, and ends with cytokinesis. The chromosomes, stained *blue* in the photos, are being drawn to the poles of the dividing cell by microtubules, stained *red* in the photos.

6.4 Cell Division

Interphase

When cell division begins, in the interphase portion of the cell cycle, chromosomes replicate and then begin to wind up tightly, a process called **condensation.** Daughter chromosomes are held together by a complex of proteins called cohesin.

Mitosis

Interphase sets the stage for cell division. It is followed by nuclear division, called *mitosis.* Although the process of mitosis is continuous, with the stages flowing smoothly one into another, for ease of study, mitosis is traditionally subdivided into four stages: prophase, metaphase, anaphase, and telophase (figure 6.8).

Prophase: Mitosis Begins. In **prophase,** the individual condensed chromosomes first become visible with a light microscope.

As the replicated chromosomes condense, the cell dismantles the nuclear membrane and begins to assemble the apparatus it will use to pull the replicated daughter chromosomes to opposite ends ("poles") of the cell. In the center of an animal cell, the pair of centrioles starts to separate; the two centrioles move apart toward opposite poles of the cell, forming between them as they move apart a network of protein cables called the **spindle.** Each cable is called a spindle fiber and is made of microtubules, which are long, hollow tubes of protein. Plant cells lack centrioles and instead brace the ends of the spindle with a support structure called an aster.

As condensation of the chromosomes continues, a second group of microtubules extends out from the centromere of each

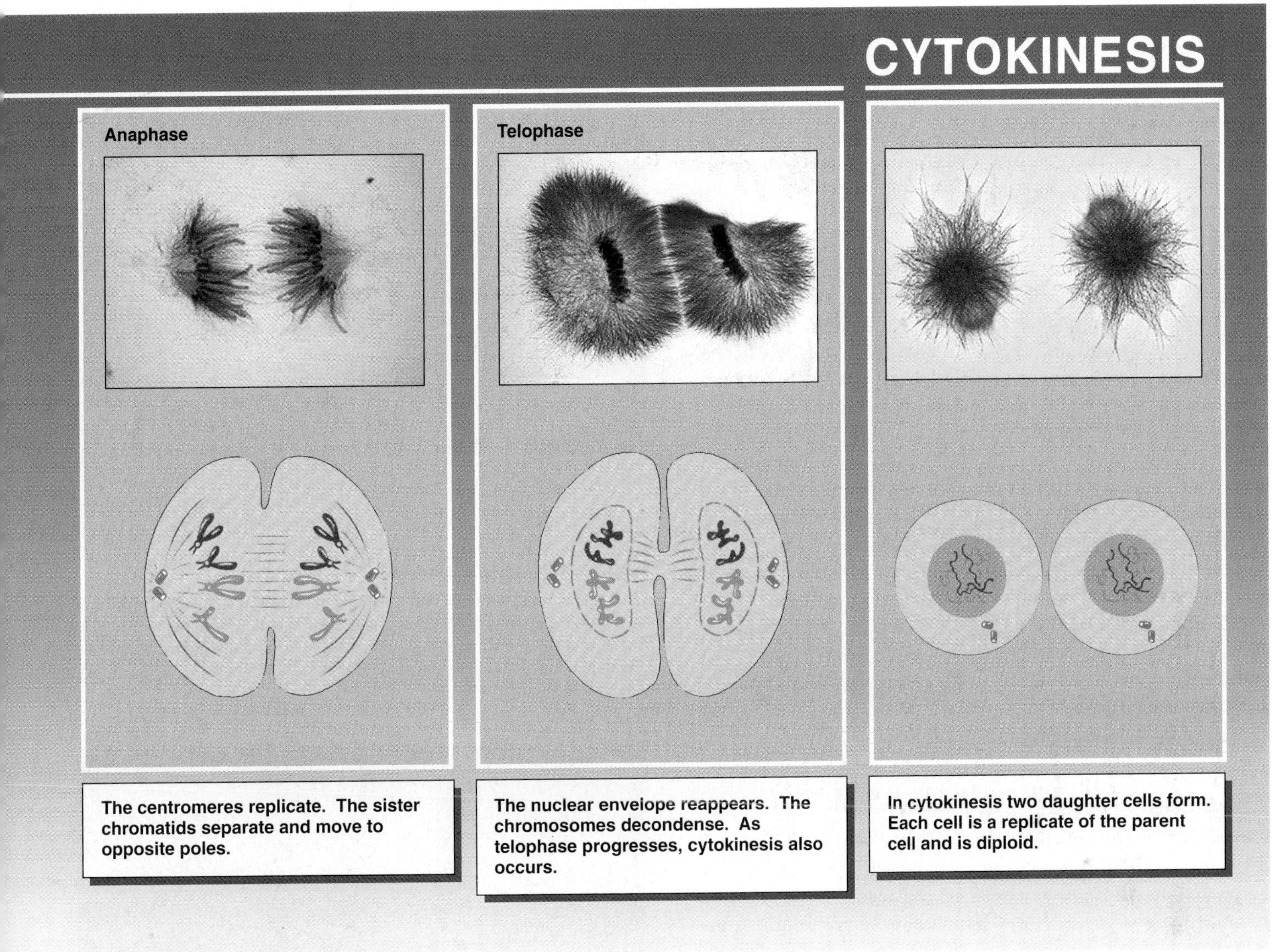
CYTOKINESIS
Anaphase
Telophase
The centromeres replicate. The sister chromatids separate and move to opposite poles.
The nuclear envelope reappears. The chromosomes decondense. As telophase progresses, cytokinesis also occurs.
In cytokinesis two daughter cells form. Each cell is a replicate of the parent cell and is diploid.

chromosome from a disk of protein called a **kinetochore.** The two sets of microtubules extend out from opposite sides of the kinetochore toward opposite poles of the cell. Each set of microtubules continues to grow longer until it makes contact with the pole toward which it is growing. When the process is complete, one daughter chromosome of each pair is attached by microtubules to one pole and the other daughter to the other pole. Also in this stage, the nuclear envelope begins to break down and the mitotic spindle that will be used to separate the sister chromatids is assembled. In animal cells and those of most protists, the spindle fibers are associated with the centrioles. After the centrioles are replicated, the pairs separate and move to opposite poles of the cell.

Metaphase: Alignment of the Chromosomes. The second phase of mitosis, **metaphase,** begins when the chromosomes, each consisting of a pair of chromatids, align in the center of the cell. Microtubules attached to the kinetochores of the centromeres are fully extended back toward the opposite poles of the cell.

Anaphase: Separation of the Chromatids. In **anaphase,** enzymes cleave the cohesin link holding sister chromatids together, the kinetochores split, and the daughter chromosomes are freed from each other. Cell division is now simply a matter of reeling in the microtubules, dragging to the poles the daughter chromosomes like fish on the end of a line—at the poles, the ends of the microtubules are dismantled, one bit after another, making the tubes shorter and shorter and so drawing the chromosome attached to the far end closer and closer to the pole. They move rapidly toward opposite poles of the cell. When they finally arrive, each pole has one complete set of chromosomes. This is the shortest stage of mitosis.

Telophase: Re-formation of the Nuclei. The only tasks that remain in **telophase** are the dismantling of the stage and the removal of the props. The mitotic spindle is disassembled, and a nuclear envelope forms around each set of chromosomes while they begin to uncoil.

Cytokinesis

At the end of telophase, mitosis is complete. The cell has divided its replicated chromosomes into two nuclei, which are positioned at opposite ends of the cell. Following mitosis, **cytokinesis,** the division of the cytoplasm, occurs, and the cell is cleaved into roughly equal halves. Cytoplasmic organelles have already been replicated and reassorted to the areas that will separate and become the daughter cells.

In animal cells, which lack cell walls, cytokinesis is achieved by pinching the cell in two with a contracting belt of microtubules. As contraction proceeds, a **cleavage furrow** becomes evident around the cell's circumference, where the cytoplasm is being progressively pinched inward by the decreasing diameter of the microtubule belt (figure 6.9*a*). The cleavage furrow deepens until the cell is literally pinched in two.

Plant cells have rigid walls that are far too strong to be deformed by microtubule contraction. A different approach to cytokinesis has therefore evolved in plants. Plant cells assemble membrane components in their interior, at right angles to the mitotic spindle. This expanding partition, called a **cell plate,** grows outward until it reaches the interior surface of the plasma membrane and fuses with it, at which point it has effectively divided the cell in two (figure 6.9*b*). Cellulose, the major component of cell walls, is then laid down on the new membranes, creating two new cells.

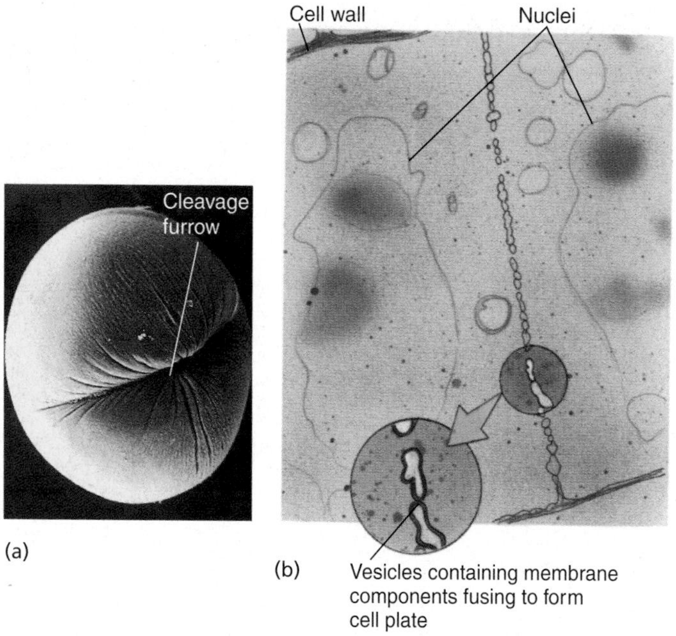
Cleavage furrow
(a)
Cell wall
Nuclei
(b)
Vesicles containing membrane components fusing to form cell plate

Figure 6.9 Cytokinesis.
The division of cytoplasm that occurs after mitosis is called cytokinesis and cleaves the cell into roughly equal halves. (*a*) In an animal cell, such as this sea urchin egg, a cleavage furrow forms around the dividing cell. (*b*) In this dividing plant cell, a cell plate is forming between the two newly forming daughter cells.

Cell Death

Despite the ability to divide, no cell lives forever. The ravages of living slowly tear away at a cell's machinery. To some degree damaged parts can be replaced, but no replacement process is perfect. And sometimes the environment intervenes. If food supplies are cut off, for example, animal cells cannot obtain the energy necessary to maintain their lysosome membranes. The cells die, digested from within by their own enzymes.

During fetal development, many cells are programmed to die. In human embryos, hands and feet appear first as "paddles," but the skin cells between bones die on schedule to form the separated toes and fingers (figure 6.10). In ducks, this cell death is not part of the developmental program, which is why ducks have webbed feet and you don't.

Human cells appear to be programmed to undergo only so many cell divisions and then die, following a plan written into the genes. In tissue culture, cell lines divide about 50 times, and then the entire population of cells dies off. Even if some of the cells are frozen for years, when they are thawed they simply resume where they left off and die on schedule. Only cancer cells appear to thwart these instructions, dividing endlessly. All other cells in your body contain a hidden clock that keeps time by counting cell divisions, and when the alarm goes off the cells die.

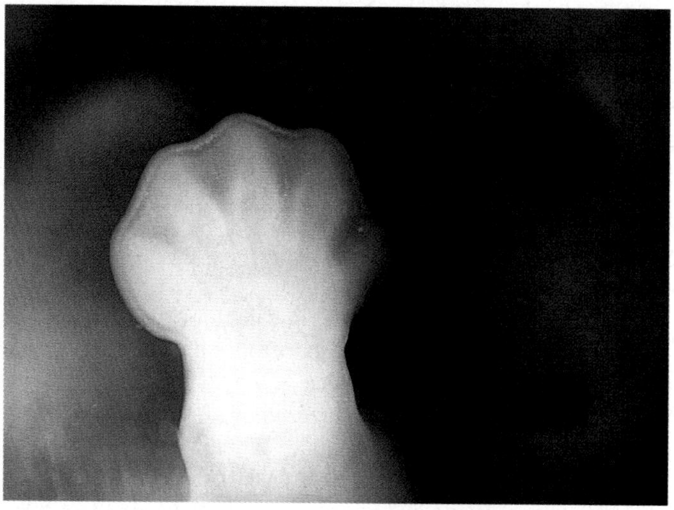

Figure 6.10 Programmed cell death.
In the human embryo, programmed cell death results in the formation of fingers and toes from paddlelike hands and feet.

6.4 The eukaryotic cell cycle starts in interphase with the condensation of replicated chromosomes; in mitosis, these chromosomes are drawn by microtubules to opposite ends of the cell; in cytokinesis the cell is split into two daughter cells.

6.5 Controlling the Cell Cycle

The events of the cell cycle are coordinated in much the same way in all eukaryotes. The control system human cells use first evolved among the protists over a billion years ago; today, it operates in essentially the same way in fungi as it does in humans.

The goal of controlling any cyclic process is to adjust the duration of the cycle to allow sufficient time for all events to occur. In principle, a variety of methods can achieve this goal. For example, an internal clock can be employed to allow adequate time for each phase of the cycle to be completed. This is how many organisms control their daily activity cycles. The disadvantage of using such a clock to control the cell cycle is that it is not very flexible. One way to achieve a more flexible and sensitive regulation of a cycle is simply to let the completion of each phase of the cycle trigger the beginning of the next phase, as a runner passing a baton starts the next leg in a relay race. Until recently, biologists thought this type of mechanism controlled the cell division cycle. However, we now know that eukaryotic cells employ a separate, centralized controller to regulate the process: at critical points in the cell cycle, further progress depends upon a central set of "go/no-go" switches that are regulated by feedback from the cell.

This mechanism is the same one engineers use to control many processes. For example, the furnace that heats a home in the winter typically goes through a daily heating cycle. When the daily cycle reaches the morning "turn on" checkpoint, sensors report whether the house temperature is below the set point (for example, 70°F). If it is, the thermostat triggers the furnace, which warms the house. If the house is already at least that warm, the thermostat does not start the furnace. Similarly, the cell cycle has key checkpoints where feedback signals from the cell about its size and the condition of its chromosomes can either trigger subsequent phases of the cycle or delay them to allow more time for the current phase to be completed.

Three principal checkpoints control the cell cycle in eukaryotes (figure 6.11):

1. **Cell growth is assessed at the G_1 checkpoint.** Located near the end of G_1, just before entry into S phase, this checkpoint makes the key decision of whether the cell should divide, delay division, or enter a resting stage (figure 6.12). In yeasts, where researchers first studied this checkpoint, it is called START. If conditions are favorable for division, the cell begins to copy its DNA, initiating S phase. The G_1 checkpoint is where the more complex eukaryotes typically arrest the cell cycle if environmental conditions make cell division impossible or if the cell passes into G_0 for an extended period.
2. **DNA replication is assessed at the G_2 checkpoint.** The second checkpoint, which occurs at the end of G_2, triggers the start of M phase. If this checkpoint is passed, the cell initiates the many molecular processes that signal the beginning of mitosis.
3. **Mitosis is assessed at the M checkpoint.** Occurring at metaphase, the third checkpoint triggers the exit from mitosis and cytokinesis and the beginning of G_1.

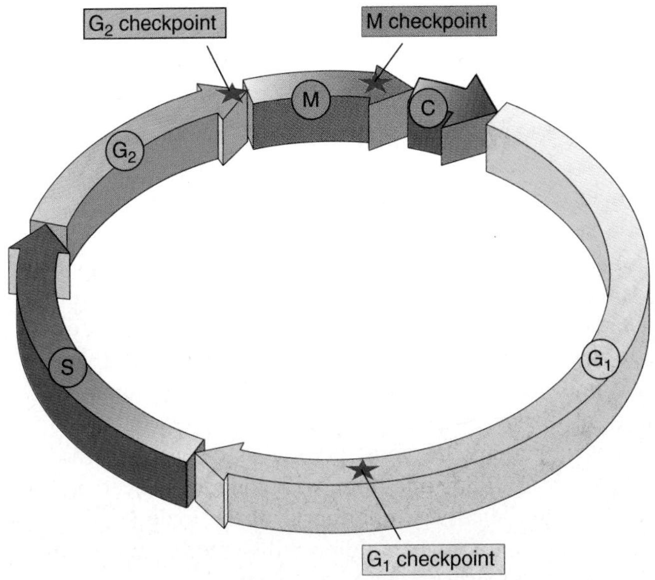

G2 checkpoint
M checkpoint
M
C
G2
S
G1
G1 checkpoint

Figure 6.11 Control of the cell cycle.
Cells use a centralized control system to check whether proper conditions have been achieved before passing three key checkpoints in the cell cycle.

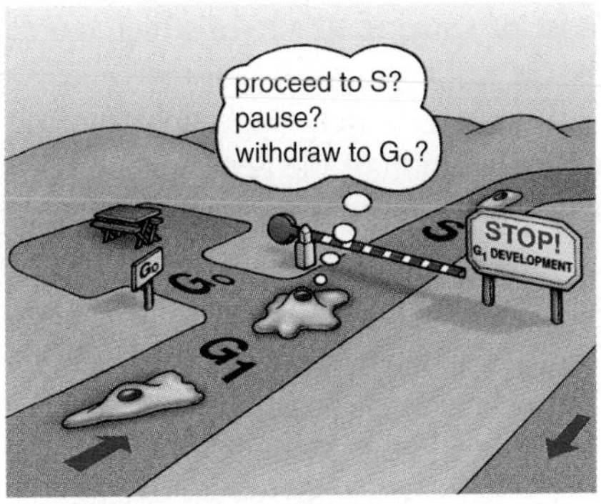

proceed to S?
pause?
withdraw to G0?
STOP!
G1 DEVELOPMENT
G0
G0
G1
S

Figure 6.12 The G_1 checkpoint.
Feedback from the cell determines whether the cell cycle will proceed to the S phase, pause, or withdraw into G_0 for an extended rest period.

6.5 The complex cell cycle of eukaryotes is controlled by feedback at three checkpoints.

6.6 What Is Cancer?

Cancer is a growth disorder of cells. It starts when an apparently normal cell begins to grow in an uncontrolled and invasive way (figure 6.13). The result is a cluster of cells, called a **tumor,** that constantly expands in size. Benign tumors are encapsulated and noninvasive. Malignant tumors are invasive and not encapsulated, shedding cells. Cells that leave the tumor and spread throughout the body, forming new tumors at distant sites, are called **metastases** (figure 6.14). Cancer is perhaps the most pernicious disease. Of the children born in 1985, one-third will contract cancer at some time during their lives; one-fourth of the male children and one-third of the female children will someday die of cancer. Most of us have had family or friends affected by the disease. In 2000, 552,200 Americans died of cancer.

Not surprisingly, researchers are expending a great deal of effort to learn the cause of this disease. Scientists have made considerable progress in the last 30 years using molecular biological techniques, and the rough outlines of understanding are now emerging. We now know that cancer is a gene disorder of somatic tissue, in which damaged genes fail to properly control cell proliferation. The cell division cycle is regulated by a sophisticated group of proteins. Cancer results from the mutation of the genes encoding these proteins.

Cancer can be caused by chemicals that mutate DNA, or in some instances by viruses that circumvent the cell's normal proliferation controls. Whatever the immediate cause, however, all cancers are characterized by unrestrained cell growth and division. The cell cycle never stops in a cancerous line of cells. Cancer cells are virtually immortal—until the body in which they reside dies.

Figure 6.13 Lung cancer cells (×530).
These cells are from a tumor located in the alveolus (air sac) of a lung.

6.6 Cancer is unrestrained cell proliferation caused by damage to genes regulating the cell division cycle.

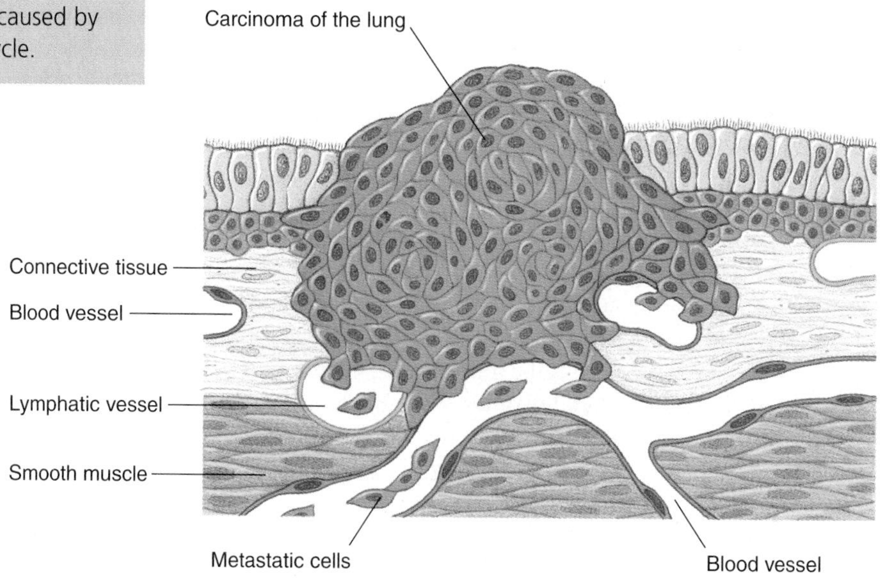
Carcinoma of the lung
Connective tissue
Blood vessel
Lymphatic vessel
Smooth muscle
Metastatic cells
Blood vessel

Figure 6.14 Portrait of a cancer.
This ball of cells is a carcinoma (cancer tumor) developing from epithelial cells that line the interior surface of a human lung. As the mass of cells grows, it invades surrounding tissues, eventually penetrating lymphatic and blood vessels, both of which are plentiful within the lung. These vessels carry metastatic cancer cells throughout the body, where they lodge and grow, forming new masses of cancerous tissue.

6.7 Cancer and Control of the Cell Cycle

As already mentioned, cancer results from damaged genes failing to control cell division. Recent work has identified one of these genes. Working independently, scientists researching such diverse fields as genetics, molecular biology, cell biology, and cancer have repeatedly identified what has proven to be the same gene! Officially dubbed *p53* (researchers italicize the gene symbol to differentiate it from the protein) and popularly referred to as the "guardian angel gene," this gene plays a key role in the G_1 checkpoint of cell division. The gene's product, the p53 protein, monitors the integrity of DNA, checking that it has been successfully replicated and is undamaged. If the p53 protein detects damaged DNA, it halts cell division and stimulates the activity of special enzymes to repair the damage. Once the DNA has been repaired, p53 allows cell division to continue. In cases where the DNA is irreparable, p53 then directs the cell to kill itself, activating an apoptosis (cell suicide) program.

By halting division in damaged cells, *p53* prevents the development of many mutated cells, and it is therefore considered a tumor-suppressor gene (even though its activities are not limited to cancer prevention). Scientists have found that *p53* is entirely absent or damaged beyond use in the majority of cancerous cells they have examined! It is precisely because *p53* is nonfunctional that these cancer cells are able to repeatedly undergo cell division without being halted at the G_1 checkpoint (figure 6.15). To test this, scientists administered healthy p53 protein to rapidly dividing cancer cells in a petri dish: the cells soon ceased dividing and died. Scientists have further reported that cigarette smoke causes mutations in the *p53* gene. This study reinforced the strong link between smoking and cancer and is described in chapter 23.

6.7 Mutations disabling key elements of the G_1 checkpoint are associated with many cancers.

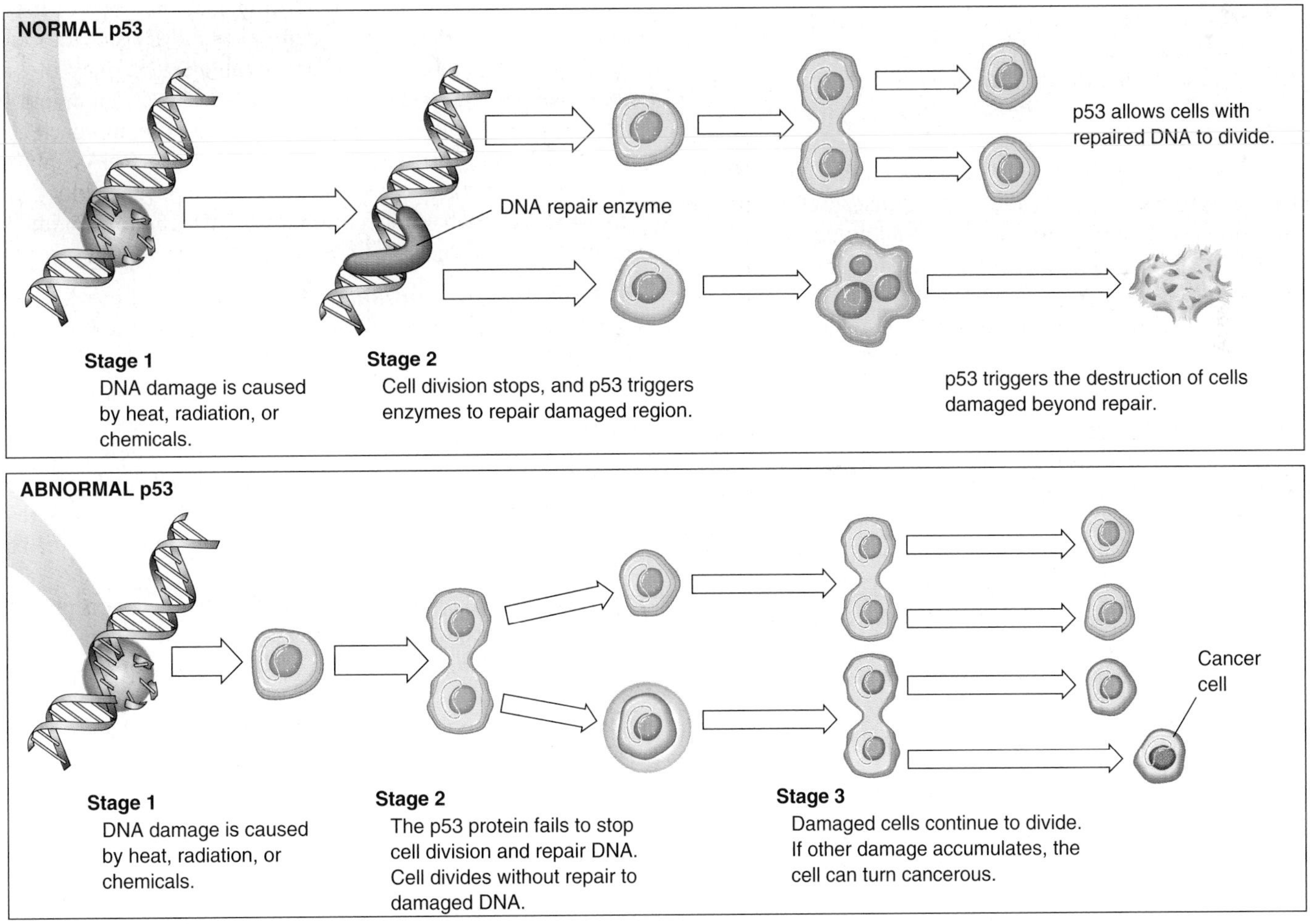

NORMAL p53
DNA repair enzyme
p53 allows cells with repaired DNA to divide.
Stage 1
DNA damage is caused by heat, radiation, or chemicals.
Stage 2
Cell division stops, and p53 triggers enzymes to repair damaged region.
p53 triggers the destruction of cells damaged beyond repair.
ABNORMAL p53
Cancer cell
Stage 1
DNA damage is caused by heat, radiation, or chemicals.
Stage 2
The p53 protein fails to stop cell division and repair DNA. Cell divides without repair to damaged DNA.
Stage 3
Damaged cells continue to divide. If other damage accumulates, the cell can turn cancerous.

Figure 6.15 Cell division and p53 protein.
Normal p53 protein monitors DNA, destroying cells with irreparable damage to their DNA. Abnormal p53 protein fails to stop cell division and repair DNA. As damaged cells proliferate, cancer develops.

6.8 Curing Cancer

Potential cancer therapies are being developed on many fronts (figure 6.16). Some act to prevent the start of cancer within cells. Others act outside cancer cells, preventing tumors from growing and spreading.

Preventing the Start of Cancer

Many promising cancer therapies act within potential cancer cells, focusing on different stages of the cell's "Shall I divide?" decision-making process.

1. Receiving the Signal to Divide. The first step in the decision process is the reception of a "divide" signal, usually a small protein called a growth factor released from a neighboring cell. The growth factor is received by a protein receptor on the cell surface. Mutations that increase the number of receptors on the cell surface amplify the division signal and so lead to cancer. Over 20% of breast cancer tumors prove to overproduce a protein called HER2 associated with the receptor for epidermal growth factor.

Therapies directed at this stage of the decision process utilize the human immune system to attack cancer cells. Special protein molecules called monoclonal antibodies, created by genetic engineering, are the therapeutic agents. These monoclonal antibodies are designed to seek out and stick to HER2. Like waving a red flag, the presence of the monoclonal antibody calls down attack by the immune system on the HER2 cell. Because breast cancer cells overproduce HER2, they are killed preferentially. Genentech's recently approved monoclonal antibody, called herceptin, has given promising results in clinical tests. In other tests, the monoclonal antibody C225, directed against epidermal growth factor receptors, has succeeded in curing advanced colon cancer. Clinical trials of C225 have begun.

2. The Relay Switch. The second step in the decision process is the passage of the signal into the cell's interior, the cytoplasm. This is carried out in normal cells by a protein called Ras that acts as a relay switch. When growth factor binds to a receptor like EGF, the adjacent Ras protein acts like it has been "goosed," contorting into a new shape. This new shape is chemically active, and initiates a chain of reactions that passes the "divide" signal inward toward the nucleus. Mutated forms of the Ras protein behave like a relay switch stuck in the "ON" position, continually instructing the cell to divide when it should not. Thirty percent of all cancers have a mutant form of Ras.

Therapies directed at this stage of the decision process take advantage of the fact that normal Ras proteins are inactive when made. Only after it has been modified by the special enzyme *farnesyl transferase* does Ras protein become able to function as a relay switch. In tests on animals, farnesyl transferase inhibitors induce the regression of tumors and prevent the formation of new ones.

3. Amplifying the Signal. The third step in the decision process is the amplification of the signal within the cytoplasm. Just as a TV signal needs to be amplified in order to be received at a distance, so a "divide" signal must be amplified if it is to reach the nucleus at the interior of the cell, a very long journey at a molecular scale. Cells use an ingenious trick to amplify the signal. Ras, when "ON," activates an enzyme, a protein kinase. This protein kinase activates other protein kinases that in their turn activate still others. The trick is that once a protein kinase enzyme is activated, it goes to work like a demon, activating hoards of others every second! And each and every one it activates behaves the same way too, activating still more, in a cascade of ever-widening effect. At each stage of the relay, the signal is amplified a thousandfold. Mutations stimulating any of the protein kinases can dangerously increase the already amplified signal and lead to cancer. Five percent of all cancers, for example, have a mutant hyperactive form of the protein kinase Src.

Therapies directed at this stage of the decision process employ so-called "anti-sense RNA" directed specifically against Src or other cancer-inducing kinase mutations. The idea is that the *src* gene uses a complementary copy of itself to manufacture the Src protein (the "sense" RNA or messenger RNA), and a mirror image complementary copy of the sense RNA ("anti-sense" RNA) will stick to it, gumming it up so it can't be used to make Src protein. The approach appears promising. In tissue culture, anti-sense RNAs inhibit the growth of cancer cells, and some also appear to block the growth of human tumors implanted in laboratory animals. Human clinical trials are underway.

4. Releasing the Brake. The fourth step in the decision process is the removal of the "brake" the cell uses to restrain cell division. In healthy cells this brake, a tumor suppressor protein called Rb, blocks the activity of a transcription factor protein called E2F. When free, E2F enables the cell to copy its DNA. Normal cell division is triggered to begin when Rb is inhibited, unleashing E2F. Mutations which destroy Rb release E2F from its control completely, leading to ceaseless cell division. Forty percent of all cancers have a defective form of Rb.

Therapies directed at this stage of the decision process are only now being attempted. They focus on drugs able to inhibit E2F, which should halt the growth of tumors arising from inactive Rb. Experiments in mice in which the E2F genes have been destroyed provide a model system to study such drugs, which are being actively investigated.

5. Checking That Everything Is Ready. The fifth step in the decision process is the mechanism used by the cell to ensure that its DNA is undamaged and ready to divide. This job is carried out in healthy cells by the tumor-suppressor protein p53, which inspects the integrity of the DNA. When it detects damaged or foreign DNA, p53 stops cell division and activates the cell's DNA repair systems. If the damage doesn't get repaired in a reasonable time, p53 pulls the plug, triggering events that kill the cell. In this way, mutations such as those that cause cancer are either repaired or the cells containing them eliminated. If p53 is itself destroyed by mutation, future damage accumulates unrepaired. Among this damage are mutations that lead to cancer. Fifty percent of all cancers have a disabled p53. Fully 70% to 80% of lung cancers have a mutant inactive p53—the chemical benzo[*a*]pyrene in cigarette smoke is a potent mutagen of p53.

A promising new therapy using adenovirus (responsible for mild colds) is being targeted at cancers with a mutant p53. To grow in a host cell, adenovirus must use the product of its gene *E1B* to block the host cell's p53, thereby enabling replication of the adenovirus DNA. This means that while mutant adenovirus without *E1B* cannot grow in healthy cells, the mutants should be able to grow in, and destroy, cancer cells with defective p53. When human colon and lung cancer cells are introduced into mice lacking an immune system and allowed to produce substantial tumors, 60% of the tumors simply disappear when treated with E1B-deficient adenovirus, and do not reappear later. Initial clinical trials are less encouraging, as many people possess antibodies to adenovirus.

6. Stepping on the Gas. Cell division starts with replication of the DNA. In healthy cells, another tumor suppressor "keeps the gas tank nearly empty" for the DNA replication process by inhibiting production of an enzyme called telomerase. Without this enzyme, a cell's chromosomes lose material from their tips, called telomeres. Every time a chromosome is copied, more tip material is lost. After some 30 divisions, so much is lost that copying is no longer possible. Cells in the tissues of an adult human have typically undergone 25 or more divisions. Cancer can't get very far with only the five remaining cell divisions, so inhibiting telomerase is a very effective natural break on the cancer process. It is thought that almost all cancers involve a mutation that destroys the telomerase inhibitor, releasing this break and making cancer possible. It should be possible to block cancer by reapplying this inhibition. Cancer therapies that inhibit telomerase are just beginning clinical trials.

Preventing the Spread of Cancer

7. Tumor Growth. Once a cell begins cancerous growth, it forms an expanding tumor. As the tumor grows ever-larger, it requires an increasing supply of food and nutrients, obtained from the body's blood supply. To facilitate this necessary grocery shopping, tumors leak out substances into the surrounding tissues that encourage angiogenesis, the formation of small blood vessels. Chemicals that inhibit this process are called angiogenesis inhibitors. In mice, two such angiogenesis inhibitors, angiostatin and endostatin, caused tumors to regress to microscopic size. This very exciting result has proven controversial, but initial human trials seem promising.

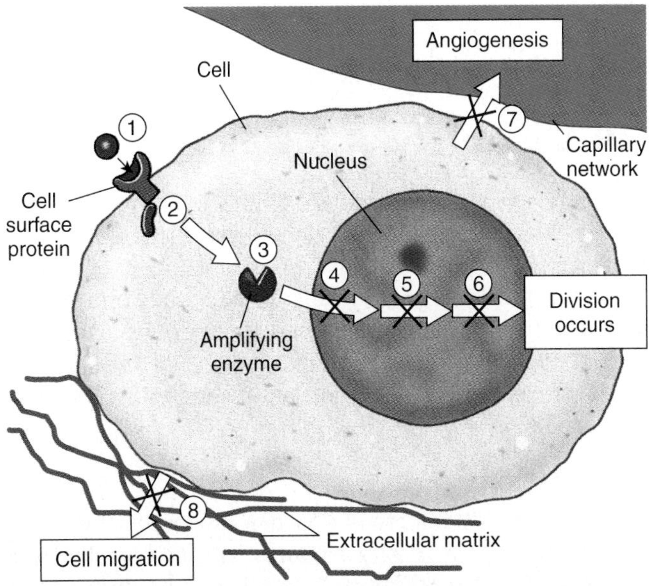

Angiogenesis
Cell
Capillary network
Nucleus
Cell surface protein
Amplifying enzyme
Division occurs
Cell migration
Extracellular matrix
1
2
3
4
5
6
7
8

Figure 6.16 New molecular therapies for cancer target eight different stages in the cancer process.

(*1*) On the cell surface, a growth factor signals the cell to divide. (*2*) Just inside the cell, a protein relay switch passes on the divide signal. (*3*) In the cytoplasm, enzymes amplify the signal. In the nucleus, (*4*) a "brake" preventing DNA replication is released, (*5*) proteins check that the replicated DNA is not damaged, and (*6*) other proteins rebuild chromosome tips so DNA can replicate. (*7*) The new tumor promotes angiogenesis, the formation of growth-promoting blood vessels. (*8*) Some cancer cells break away from the extracellular matrix and invade other parts of the body.

8. Metastasis. If cancerous tumors simply continued to grow where they form, many could be surgically removed, and far fewer would prove fatal. Unfortunately, many cancerous tumors eventually metastasize, individual cancer cells breaking their moorings to the extracellular matrix and spreading to other locations in the body where they initiate formation of secondary tumors. This process involves metal-requiring protease enzymes that cleave the cell-matrix linkage, components of the extracellular matrix such as fibronectin that also promote the migration of several noncancerous cell types, and RhoC, a GTP-hydrolyzing enzyme that promotes cell migration by providing needed GTP. All of these components offer promising targets for future anti-cancer therapy.

Therapies such as those described here are only part of a wave of potential treatments under development and clinical trial. The clinical trials will take years to complete, but in the coming decade we can expect cancer to become a curable disease.

6.8 Understanding of how mutations produce cancer has progressed to the point where promising potential therapies can be tested.

6.9 Discovery of Meiosis

Only a few years after Walther Fleming's discovery of chromosomes in 1882, Belgian cytologist Pierre-Joseph van Beneden was surprised to find different numbers of chromosomes in different types of cells in the roundworm *Ascaris.* Specifically, he observed that the **gametes** (eggs and sperm) each contained two chromosomes, while the *somatic* (nonreproductive) cells of embryos and mature individuals each contained four.

Fertilization

From his observations, van Beneden proposed in 1887 that an egg and a sperm, each containing half the complement of chromosomes found in other cells, fuse to produce a single cell called a **zygote.** The zygote, like all of the somatic cells ultimately derived from it, contains two copies of each chromosome. The fusion of gametes to form a new cell is called **fertilization** or **syngamy.**

Figure 6.17 Sexual and asexual reproduction.
Not all organisms reproduce exclusively asexually or sexually; some do both. The strawberry reproduces both asexually (runners) and sexually (flowers).

Meiosis

It was clear even to early investigators that gamete formation must involve some mechanism that reduces the number of chromosomes to half the number found in other cells. If it did not, the chromosome number would double with each fertilization, and after only a few generations, the number of chromosomes in each cell would become impossibly large. For example, in just 10 generations, the 46 chromosomes present in human cells would increase to over 47,000 (46×2^{10}).

The number of chromosomes does not explode in this way because of a special reduction division that occurs during gamete formation, producing cells with half the normal number of chromosomes. The subsequent fusion of two of these cells ensures a consistent chromosome number from one generation to the next. This reduction division process, known as *meiosis,* is the subject of this section.

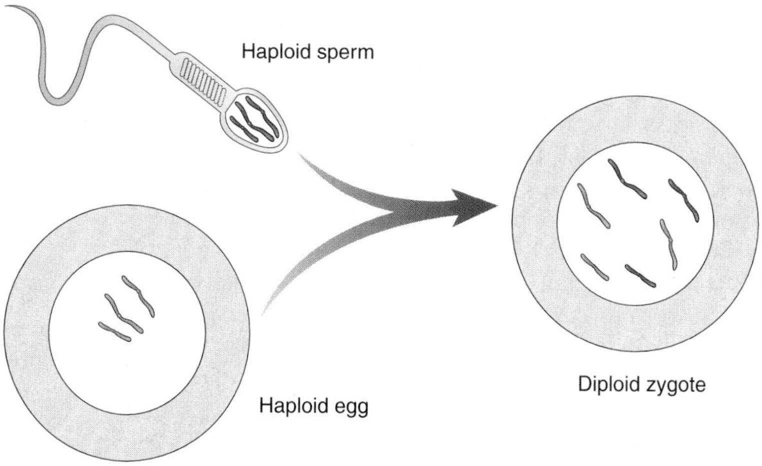
Haploid sperm
Haploid egg
Diploid zygote

Figure 6.18 Diploid cells carry chromosomes from two parents.
A diploid cell contains two versions of each chromosome, one contributed by the haploid egg of the mother, the other by the haploid sperm of the father.

The Sexual Life Cycle

Meiosis and fertilization together constitute a cycle of reproduction. Two sets of chromosomes are present in the somatic cells of adult individuals, making them **diploid** cells (Greek, *di,* two), but only one set is present in the gametes, which are thus **haploid** (Greek, *haploos,* one). Reproduction that involves this alternation of meiosis and fertilization is called **sexual reproduction** (figure 6.17). Its outstanding characteristic is that offspring inherit chromosomes from two parents (figure 6.18). You, for example, inherited 23 chromosomes from your mother, contributed by the egg fertilized at your conception, and 23 from your father, contributed by the sperm that fertilized that egg.

6.9 Meiosis is a process of cell division in which the number of chromosomes in certain cells is halved during gamete formation.

6.10 The Sexual Life Cycle

Somatic Tissues

The life cycles of all sexually reproducing organisms follow the same basic pattern of alternation between the diploid and haploid chromosome numbers (figures 6.19 and 6.20). After fertilization, the resulting zygote begins to divide by mitosis. This single diploid cell eventually gives rise to all of the cells in the adult. These cells are called **somatic** cells, from the Latin word for "body." Except when rare accidents occur, or in special variation-creating situations such as occur in the immune system, every one of the adult's somatic cells is genetically identical to the zygote.

In unicellular eukaryotic organisms, including most protists, individual cells function as gametes, fusing with other gamete cells. The zygote may undergo mitosis, or it may divide immediately by meiosis to give rise to haploid individuals. In plants, the haploid cells that meiosis produces divide by mitosis, forming a multicellular haploid phase. Certain cells of this haploid phase eventually differentiate into eggs or sperm.

Germ-Line Tissues

In animals, the cells that will eventually undergo meiosis to produce gametes are set aside from somatic cells early in the course of development. These cells are often referred to as **germ-line** cells. Both the somatic cells and the gamete-producing germ-line cells are diploid, but while somatic cells undergo mitosis to form genetically identical, diploid daughter cells, gamete-producing germ-line cells undergo meiosis, producing haploid gametes.

6.10 In the sexual life cycle, there is an alternation of diploid and haploid generations.

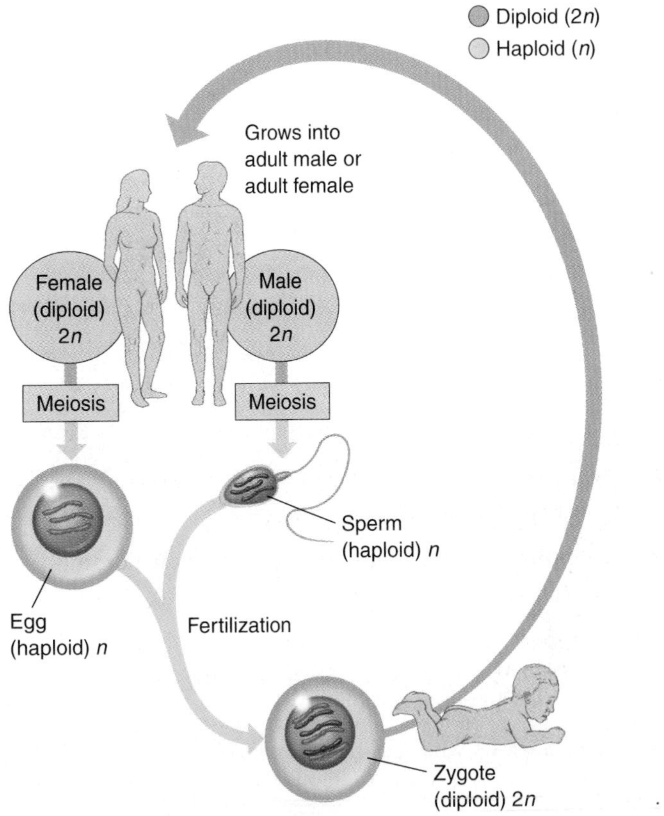

Diploid (2n)
Haploid (n)
Grows into
adult male or
adult female
Female
(diploid)
2n
Male
(diploid)
2n
Meiosis
Meiosis
Sperm
(haploid) n
Egg
(haploid) n
Fertilization
Zygote
(diploid) 2n

Figure 6.19 The sexual life cycle.
In animals, the completion of meiosis is followed soon by fertilization. Thus, the vast majority of the life cycle is spent in the diploid stage. In this text, *n* stands for haploid and 2*n* stands for diploid.

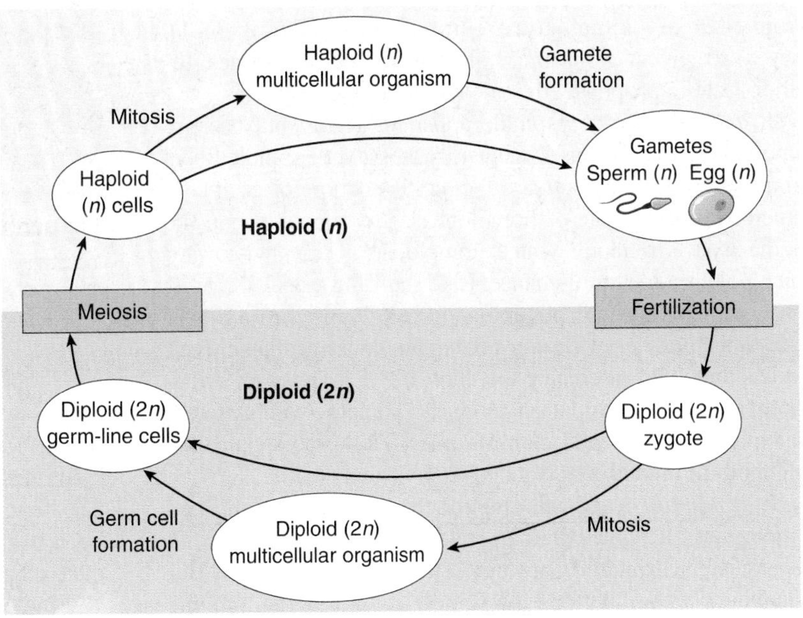

Haploid (n)
multicellular organism
Gamete
formation
Mitosis
Gametes
Sperm (n) Egg (n)
Haploid
(n) cells
Haploid (n)
Meiosis
Fertilization
Diploid (2n)
Diploid (2n)
germ-line cells
Diploid (2n)
zygote
Germ cell
formation
Diploid (2n)
multicellular organism
Mitosis

Figure 6.20 Alternation of generations.
In sexual reproduction, haploid cells or organisms alternate with diploid cells or organisms.

6.11 The Stages of Meiosis

Just as in mitosis, the chromosomes have replicated before meiosis begins, during a period called interphase. The first of the two divisions of meiosis, called **meiosis I,** serves to separate the two versions of each chromosome; the second, **meiosis II,** serves to separate the two replicas of each version. Thus when meiosis is complete, what started out as one diploid cell ends up as four haploid cells. Because there was one replication of DNA but *two* cell divisions, the process halves the number of chromosomes.

Meiosis I

Meiosis I is traditionally divided into four stages:

1. **Prophase I.** The two versions of each chromosome pair up and exchange segments.
2. **Metaphase I.** The chromosomes align on a central plane.
3. **Anaphase I.** One version of each chromosome moves to a pole of the cell, and the other version moves to the opposite pole.
4. **Telophase I.** Individual chromosomes gather together at each of the two poles.

In *prophase I,* individual chromosomes first become visible, as viewed with a light microscope, as their DNA coils more and more tightly. Because the chromosomes (DNA) have replicated before the onset of meiosis, each of the threadlike chromosomes actually consists of two sister chromatids joined at their centromeres. The two homologous chromosomes then line up side by side, and **crossing over** is initiated, in which DNA is exchanged between the two nonsister chromatids of homologous chromosomes (figure 6.21). These crossovers hold the homologous chromosomes together. Late in prophase, the nuclear envelope disperses.

In *metaphase I,* the spindle apparatus forms, but because homologues are held close together by crossovers, spindle fibers can attach to only the outward-facing kinetochore of each centromere. For each pair of homologues, the orientation on the spindle axis is random; which homologue is oriented toward which pole is a matter of chance. Like shuffling a deck of cards, many combinations are possible—in fact, 2 raised to a power equal to the number of chromosome pairs. In a hypothetical cell that has three chromosome pairs, there are eight possible orientations (2^3). Each orientation results in gametes with different combinations of parental chromosomes. This process is called **independent assortment** (figure 6.22).

In *anaphase I,* the spindle attachment is complete, and homologues are pulled apart and move toward opposite poles. Sister chromatids are not separated at this stage. Because the orientation along the spindle equator is random, the chromosome that a pole receives from each pair of homologues is also random with respect to all chromosome pairs. At the end of anaphase I, each pole has half as many chromosomes as were present in the cell in which meiosis began. Remember that the chromosomes replicated and thus contained two sister chromatids before the start of meiosis. The function of the meiotic stages up to this point has not been to reduce the number of chromosomes but to allow for the exchange of genetic material in crossing over.

In *telophase I,* the chromosomes gather at their respective poles to form two chromosome clusters. After an interval of variable length, meiosis II occurs, in which the number of chromosomes is reduced (figure 6.23).

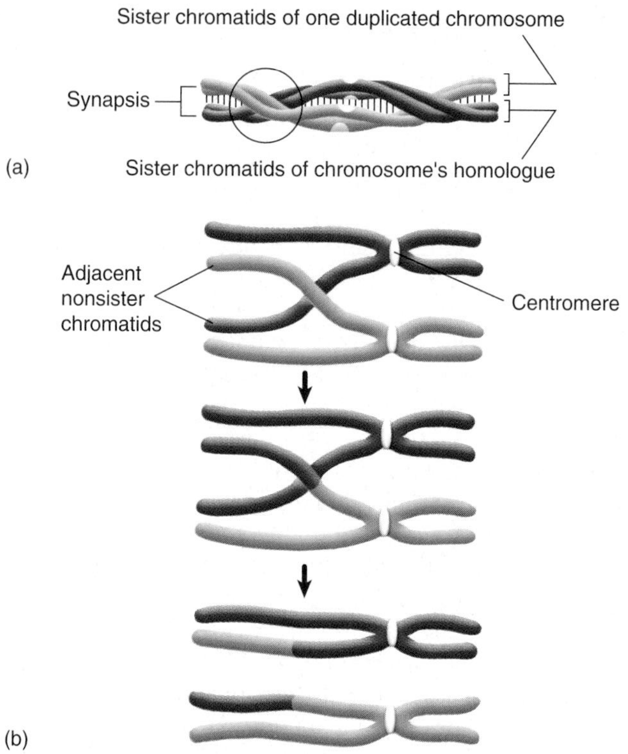

Sister chromatids of one duplicated chromosome
Synapsis
(a)
Sister chromatids of chromosome's homologue
Adjacent nonsister chromatids
Centromere
(b)

Figure 6.21 Crossing over.
In crossing over, the two copies of each chromosome exchange portions. (*a*) The *open circle* highlights the complex series of events that occur during close pairing of the two copies, a process called synapsis. (*b*) During the crossing over process, nonsister chromatids that are next to each other exchange genetic information.

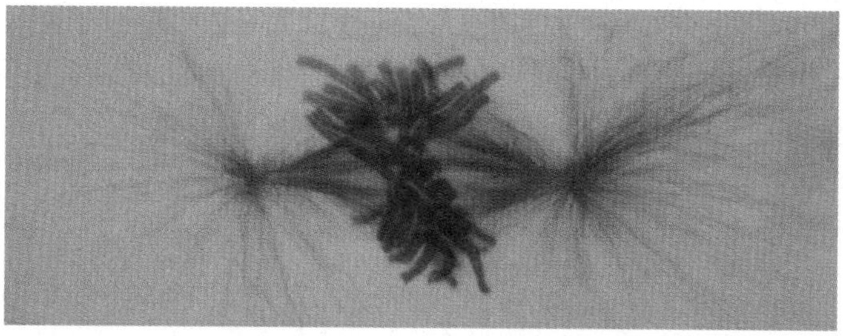

Figure 6.22 Independent assortment.
Independent assortment occurs because the orientation of chromosomes (*blue* in this micrograph) on the metaphase plate is random. Each of the many possible orientations results in gametes with different combinations of parental chromosomes.

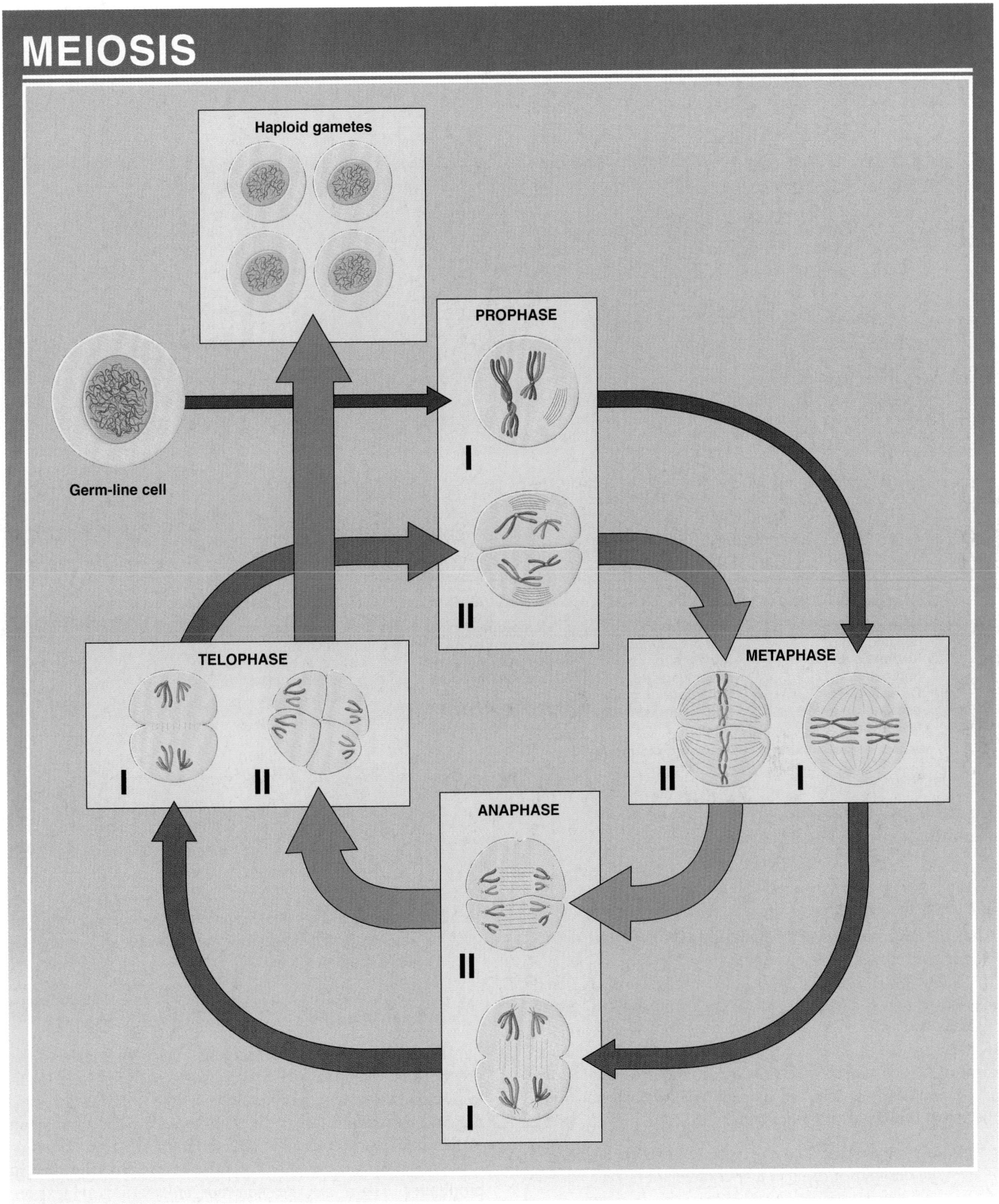
MEIOSIS
Haploid gametes
PROPHASE
I
II
Germ-line cell
TELOPHASE
I
II
METAPHASE
II
I
ANAPHASE
II
I

Figure 6.23 How meiosis works.

Meiosis consists of two rounds of cell division similar to mitosis, and produces four haploid cells.

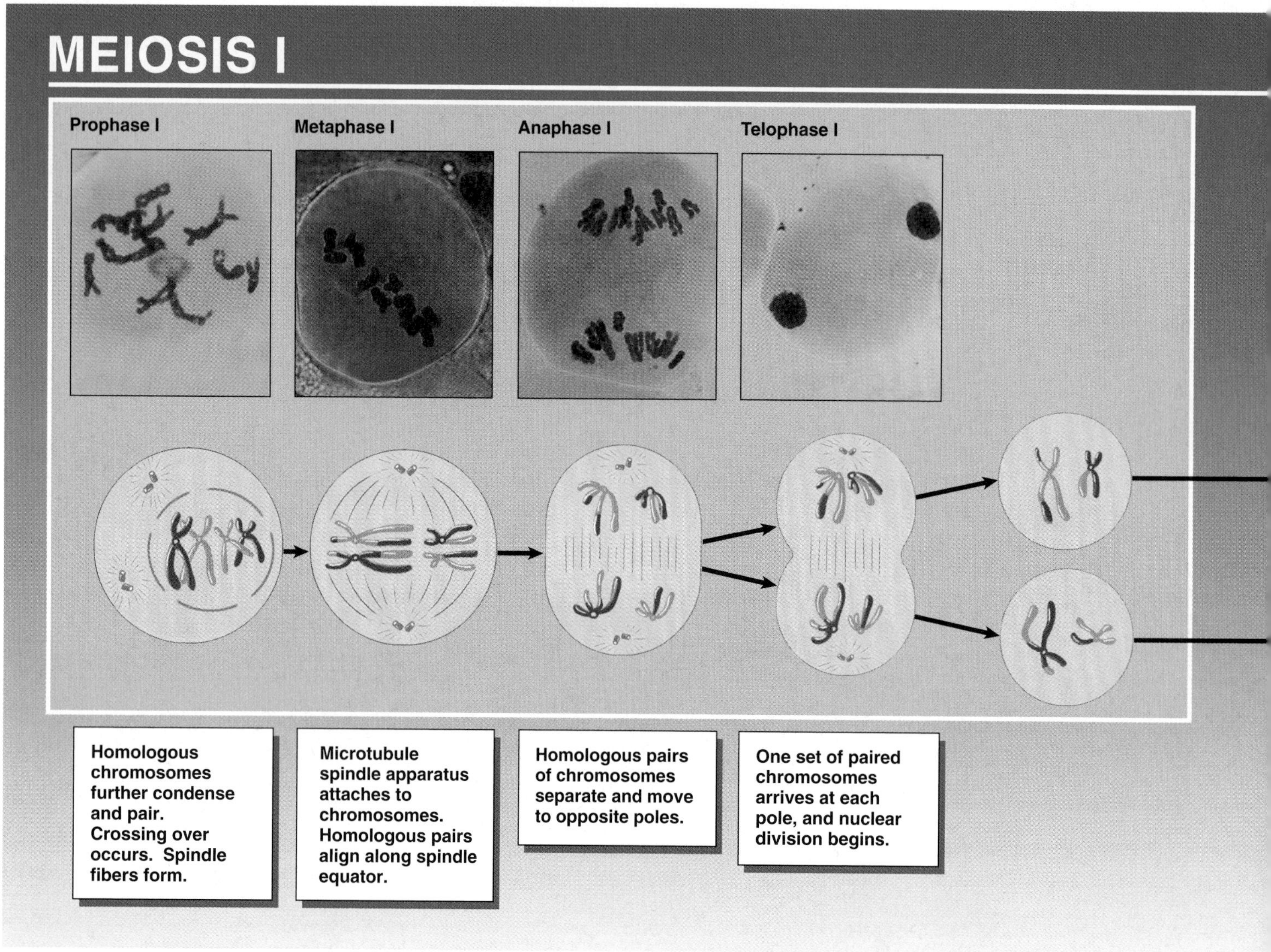
MEIOSIS I
Prophase I
Metaphase I
Anaphase I
Telophase I
Homologous chromosomes further condense and pair. Crossing over occurs. Spindle fibers form.
Microtubule spindle apparatus attaches to chromosomes. Homologous pairs align along spindle equator.
Homologous pairs of chromosomes separate and move to opposite poles.
One set of paired chromosomes arrives at each pole, and nuclear division begins.

Figure 6.24 Meiosis.
Meiosis results in four haploid daughter cells.

Meiosis II

After a brief interphase, in which no DNA synthesis occurs, the second meiotic division begins.

Meiosis II is simply a mitotic division involving the products of meiosis I, except that the sister chromatids are not genetically identical, as they are in mitosis, because of crossing over. At the end of anaphase I, each pole has a haploid complement of chromosomes, each of which is still composed of two sister chromatids attached at the centromere. Like meiosis I, meiosis II is divided into four stages:

1. **Prophase II.** At the two poles of the cell, the clusters of chromosomes enter a brief prophase II, each nuclear envelope breaking down as a new spindle forms.
2. **Metaphase II.** In metaphase II, spindle fibers bind to both sides of the centromeres.
3. **Anaphase II.** The spindle fibers contract, splitting the centromeres and moving the sister chromatids to opposite poles.
4. **Telophase II.** Finally, the nuclear envelope reforms around the four sets of daughter chromosomes.

The main outcome of the four stages of meiosis II—prophase II, metaphase II, anaphase II, and telophase II—is to separate these sister chromatids. At both poles of the original cell, the chromosomes divide mitotically. The final result of this division is four cells containing haploid sets of chromosomes. No two are alike, because of the crossing over in prophase I. The nuclei are then reorganized, and nuclear envelopes form around each haploid set of chromosomes. The cells that contain these haploid nuclei may develop directly into gametes, as they do in animals. Alternatively, they may

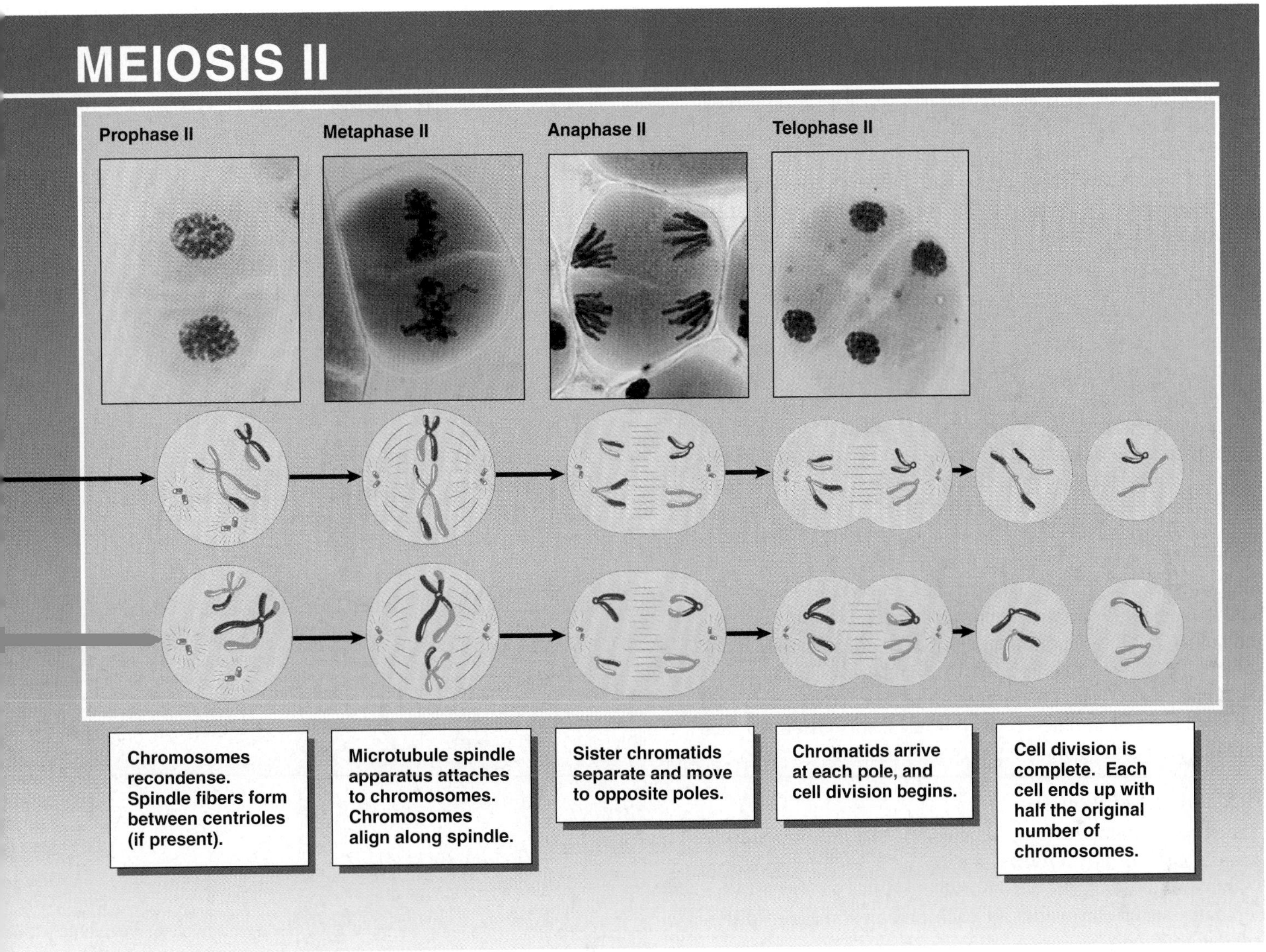
MEIOSIS II
Prophase II
Metaphase II
Anaphase II
Telophase II
Chromosomes recondense. Spindle fibers form between centrioles (if present).
Microtubule spindle apparatus attaches to chromosomes. Chromosomes align along spindle.
Sister chromatids separate and move to opposite poles.
Chromatids arrive at each pole, and cell division begins.
Cell division is complete. Each cell ends up with half the original number of chromosomes.

themselves divide mitotically, as they do in plants, fungi, and many protists, eventually producing greater numbers of gametes or, as in the case of some plants and insects, adult individuals of varying ploidy. Figure 6.24 summarizes the various stages of meiosis.

The Important Role of Crossing Over

If you think about it, the key to meiosis is that the replicas of each chromosome are not separated from each other in the first division. Why not? What prevents microtubules from attaching to them and pulling them to opposite poles of the cell, just as eventually happens later in the second meiotic division? The answer is the crossing over that occurred early in the first division. By exchanging segments, the two versions of each chromosome are tied together by strands of DNA, like two people sharing a belt. It is because microtubules can gain access to only one side of each replicated chromosome that they cannot pull the two replicas apart! Imagine two people dancing closely—you can tie a rope to the back of each person's belt, but you cannot tie a second rope to their belt buckles because the two dancers are facing each other and are very close. In just the same way, microtubules cannot attach to the inside replica of chromosomes because crossing over holds chromosome replicas together like dancing partners.

6.11 During meiosis I, homologous chromosomes move toward opposite poles in anaphase I, and individual chromosomes cluster at the two poles in telophase I. At the end of meiosis II, each of the four haploid cells contains one copy of every chromosome in the set, rather than two. Because of crossing over, no two cells are the same. These haploid cells may develop directly into gametes, as in animals, or they may divide by mitosis, as in plants, fungi, and many protists.

6.12 Unique Features of Meiosis

The mechanism of meiosis varies in important details in different organisms. This is particularly true of chromosomal separation mechanisms, which differ substantially in protists and fungi from the process in plants and animals that we describe here. Although meiosis and mitosis have much in common, meiosis has two unique features: synapsis and reduction division.

Synapsis

The first of these two features happens early during the first nuclear division. Following chromosome replication, homologous chromosomes or homologues *pair all along their length, and genetic exchange occurs between them* while they are thus physically joined (figure 6.25*a*). The process of forming these complexes of homologous chromosomes is called *synapsis,* and the exchange process called crossing over occurs between paired chromosomes. Chromosomes are then drawn together along the equatorial plane of the dividing cell; subsequently, homologues are pulled by microtubules toward opposite poles of the cell. When this process is complete, the cluster of chromosomes at each pole contains one of the two homologues of each chromosome. Each pole is haploid, containing half the number of chromosomes present in the original diploid cell. Sister chromatids do not separate from each other in the first nuclear division, so each homologue is still composed of two chromatids.

Reduction Division

The second unique feature of meiosis is that *the chromosome homologues do not replicate between the two nuclear divisions,* so that chromosome assortment in the second division separates sister chromatids of each chromosome into different daughter cells (figure 6.25*b*).

In most respects, the second meiotic division is identical to a normal mitotic division. However, because of the crossing over that occurred during the first division, the sister chromatids in meiosis II are not identical to each other.

Meiosis is a continuous process, but it is most easily studied when we divide it into arbitrary stages, just as we did for mitosis. Like mitosis, the two stages of meiosis, meiosis I and II, are subdivided into prophase, metaphase, anaphase, and telophase. In meiosis, however, prophase I is more complex than in mitosis. Figure 6.26 compares and contrasts meiosis and mitosis.

6.12 In meiosis, homologous chromosomes become intimately associated and do not replicate between the two nuclear divisions.

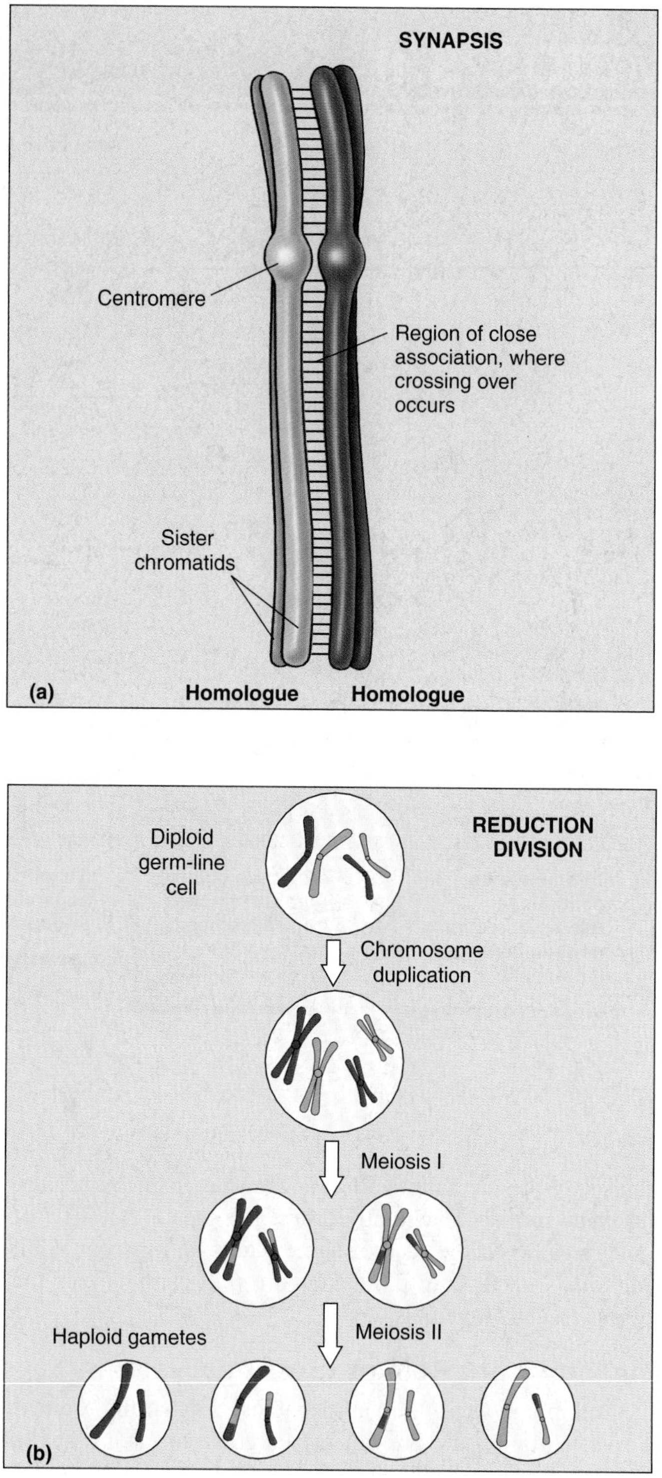
SYNAPSIS
Centromere
Region of close association, where crossing over occurs
Sister chromatids
(a)
Homologue
Homologue
REDUCTION DIVISION
Diploid germ-line cell
Chromosome duplication
Meiosis I
Haploid gametes
Meiosis II
(b)

Figure 6.25 Unique features of meiosis.

(*a*) Synapsis draws homologous chromosomes together, creating a situation where the two chromosomes can physically exchange arms, a process called crossing over. (*b*) Reduction division, omitting a chromosome duplication before meiosis II, produces haploid gametes, thus ensuring that the chromosome number remains stable during the reproduction cycle.

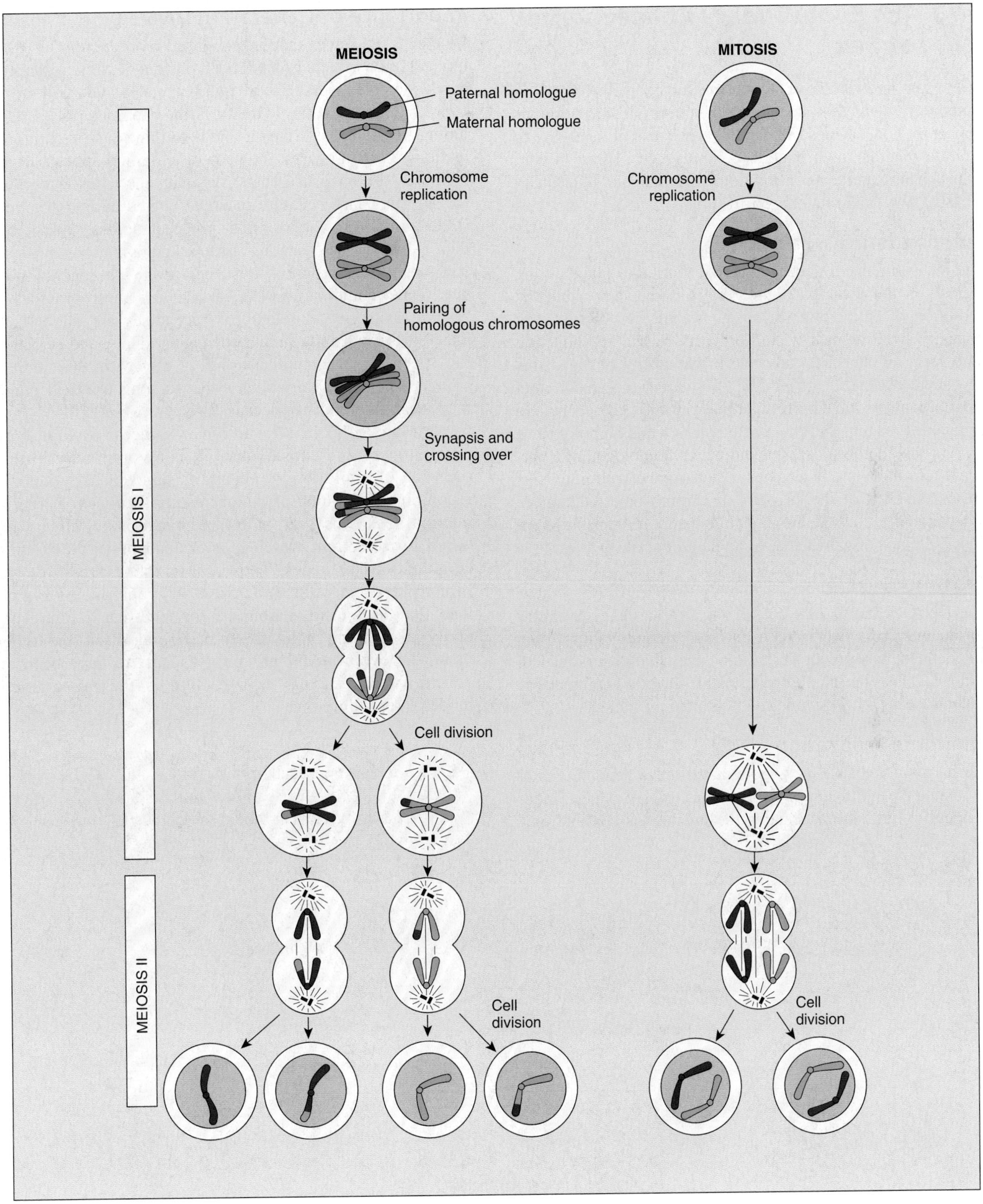
MEIOSIS
MITOSIS
Paternal homologue
Maternal homologue
Chromosome replication
Chromosome replication
Pairing of homologous chromosomes
Synapsis and crossing over
MEIOSIS I
Cell division
MEIOSIS II
Cell division
Cell division

Figure 6.26 A comparison of meiosis and mitosis.

Meiosis involves two nuclear divisions with no DNA replication between them. It thus produces four daughter cells, each with half the original number of chromosomes. Crossing over occurs in prophase I of meiosis. Mitosis involves a single nuclear division after DNA replication. It thus produces two daughter cells, each containing the original number of chromosomes.

6.13 The Evolutionary Consequences of Sex

While our knowledge of how sex evolved is sketchy, it is abundantly clear that sexual reproduction has an enormous impact on how species evolve today, because of its ability to rapidly generate new genetic combinations. Three mechanisms each make key contributions: independent assortment, crossing over, and random fertilization.

Independent Assortment

The reassortment of genetic material that takes place during meiosis is the principal factor that has made possible the evolution of eukaryotic organisms, in all their bewildering diversity, over the past 1.5 billion years. Sexual reproduction represents an enormous advance in the ability of organisms to generate genetic variability. To understand, recall that most organisms have more than one chromosome. In human beings, for example, each gamete receives one homologue of each of the 23 chromosomes, but which homologue of a particular chromosome it receives is determined randomly (figure 6.27). Each of the 23 pairs of chromosomes segregates independently, so there are 2^{23} (more than 8 million) different possible kinds of gametes that can be produced.

Crossing Over

The DNA exchange that occurs when the arms of nonsister chromatids cross over adds even more recombination to the independent assortment of chromosomes that occurs later in meiosis. Thus, the number of possible genetic combinations that can occur among gametes is virtually unlimited.

Random Fertilization

Furthermore, because the zygote that forms a new individual is created by the fusion of two gametes, each produced independently, fertilization squares the number of possible outcomes ($2^{23} \times 2^{23}$ = 70 trillion).

Importance of Generating Diversity

Whatever the forces that led to sexual reproduction, its evolutionary consequences have been profound. No genetic process generates diversity more quickly; and, as you will see in later chapters, genetic diversity is the raw material of evolution, the fuel that drives it and determines its potential directions. In many cases, the pace of evolution appears to increase as the level of genetic diversity increases. Programs for selecting larger stature in domesticated animals such as cattle and sheep, for example, proceed rapidly at first, but then slow as the existing genetic combinations are exhausted; further progress must then await the generation of new gene combinations. Racehorse breeding provides a graphic example: thoroughbred racehorses are all descendants of a small initial number of individuals, and selection for speed has accomplished all it can with this limited amount of genetic variability—the winning times in major races ceased to improve decades ago.

Paradoxically, the evolutionary process is thus both revolutionary and conservative. It is revolutionary in that the pace of evolutionary change is quickened by genetic recombination, much of which results from sexual reproduction. It is conservative in that change is not always favored by selection, which may instead preserve existing combinations of genes. These conservative pressures appear to be greatest in some asexually reproducing organisms that do not move around freely and that live in especially demanding habitats. In vertebrates, on the other hand, the evolutionary premium appears to have been on versatility, and sexual reproduction is the predominant mode of reproduction.

6.13 Sexual reproduction increases genetic variability through independent assortment in metaphase I of meiosis, crossing over in prophase I of meiosis, and random fertilization.

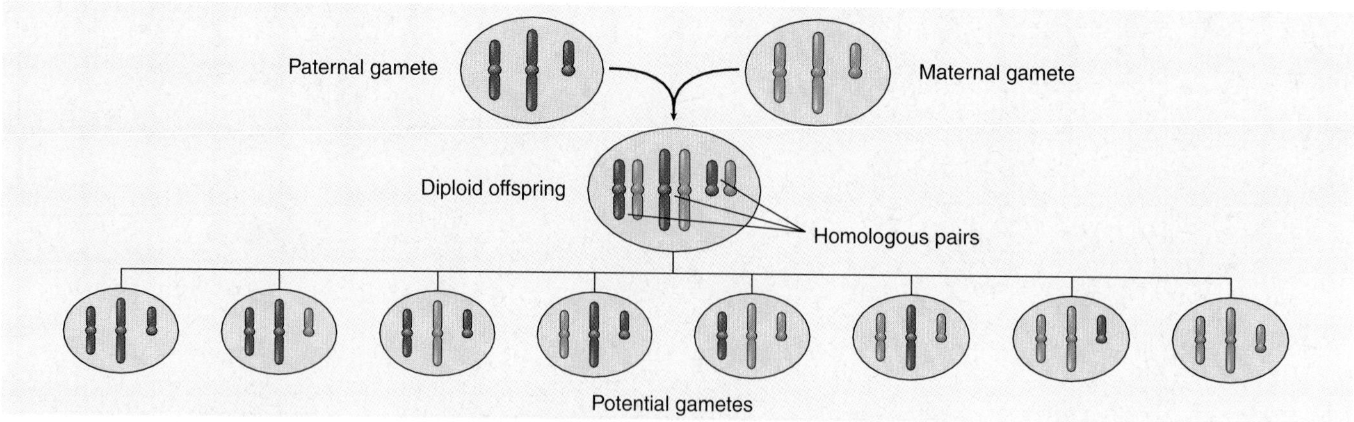

Paternal gamete
Maternal gamete
Diploid offspring
Homologous pairs
Potential gametes

Figure 6.27 Independent assortment increases genetic variability.
Independent assortment contributes new gene combinations to the next generation because the orientation of chromosomes on the metaphase plate is random. In the cell shown above with three chromosome pairs, there are eight different gametes that can result, each with different combinations of parental chromosomes.

CONCEPT REVIEW

1. An organism's nonreproductive cells are called
   a. somatic cells.
   b. germ cells.
   c. gametes.
   d. homologues.

2. During the S phase of the cell cycle,
   a. the cell splits.
   b. DNA is replicated.
   c. the microtubular apparatus is assembled.
   d. cytokinesis occurs.

3. During which phase of mitosis do the sister chromatids move toward opposite poles of the dividing cell?
   a. prophase
   b. metaphase
   c. anaphase
   d. telophase

4. Which of the following does *not* cause cancer?
   a. an increased number of growth factor receptors
   b. viruses
   c. a mutated form of Rb protein
   d. the presence of p53 protein

5. Sexual reproduction in plants and animals involves the production of gametes through a special type of reduction division called _________, followed by the union of haploid gametes to form a _________.
   a. mitosis
   b. cytokinesis
   c. zygote
   d. chromatin
   e. meiosis

6. In diploid somatic cells that have *not* undergone DNA replication, there are ________ copies of each chromosome.
   a. one
   b. two
   c. three
   d. four

7. In diploid somatic cells that have undergone DNA replication, there are _________ sister chromatids per homologous pair of chromosomes.
   a. one
   b. two
   c. three
   d. four

8. Which of the following is not true of crossing over?
   a. It occurs during prophase I.
   b. It occurs during prophase II.
   c. It occurs between homologues.
   d. It holds homologues together.

9. Bacteria split by a process called _________ _________.

10. The _________ phase of the cell cycle is when the microtubular apparatus is assembled.

11. _________ protein inspects DNA at the G_1 checkpoint in the cell cycle.

12. Meiotic recombination greatly enhances the ability of organisms to generate genetic _________.

CHALLENGE YOURSELF

1. If you could construct an artificial chromosome, what elements would you introduce into it so that it could function normally?

2. What problems could develop in organisms whose cells divided too rapidly by mitosis?

3. How can crossing over be advantageous for natural selection?

4. Humans have 23 pairs of chromosomes. Ignoring the effects of crossing over, what proportion of a woman's eggs contain only chromosomes she received from her mother?

5. Many sexually reproducing lizard species are able to generate local populations that reproduce asexually by parthenogenesis. What do you imagine the sex of these local parthenogenetic populations would be: male, female, or neuter? Why?

6 eBRIDGE

Reinforcing Key Points

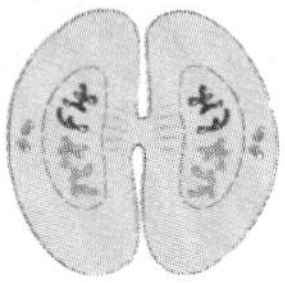

Cell Division

6.1 Prokaryotes Have a Simple Cell Cycle

6.2 Eukaryotes Have a Complex Cell Cycle

6.3 Chromosomes

6.4 Cell Division

6.5 Controlling the Cell Cycle

Cancer and the Cell Cycle

6.6 What Is Cancer?

6.7 Cancer and Control of the Cell Cycle

6.8 Curing Cancer

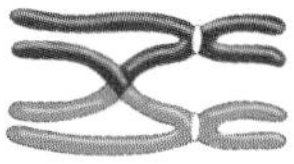

Meiosis

6.9 Discovery of Meiosis

6.10 The Sexual Life Cycle

6.11 The Stages of Meiosis

6.12 Unique Features of Meiosis

6.13 The Evolutionary Consequences of Sex

Electronic Learning

Visual Learning

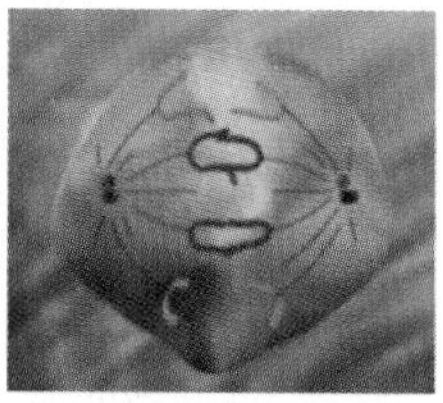

Animations

DNA Packaging

Mitosis

Art Labeling Activities

Plant Cell Mitosis

Helping You Learn

Meiosis

Explorations

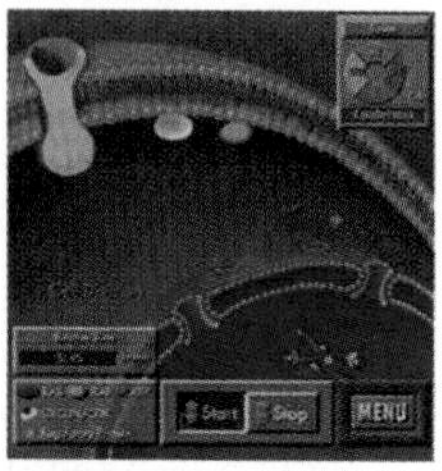

Mitosis: Cell Cycle Regulation

In this exercise, you can explore the ways a cell decides if and when it will divide, a decision that, if made inappropriately, may lead to cancer.

Exploring Meiosis

In this exercise, you can explore how the failure of chromosomes to separate during meiosis can result in Down syndrome.

Author's Corner

Curing Cancer. In 2000, over half a million Americans died of cancer, a growth disorder of cells caused by damage to genes regulating the cell division cycle. In some cases cancerous growth is initiated by the inappropriate activation of proteins that regulate the cell cycle. In others, it results from inactivation of proteins that normally suppress cell division.

1. Can scientists starve cancer tumors?
2. Understanding the processes that lead to cancer.
3. Why is deadly lymphoma becoming more common?
4. High-fiber diets don't protect against colon cancer.
5. Curing colon cancer by blocking cell receptors.
6. The cancer war is yielding promising results.
7. Oncogene blockers—the newest weapons against cancer.

Virtual Classroom

Progress in Curing Cancer

Cancer is a cell growth defect, a breakdown in the cell's control of its proliferation. Progress in curing cancer is focused on the elements of this breakdown. Two sorts of damage to genes must occur to produce cancer. As a crude analogy, imagine a car parked on the side of a road. To get it moving you must step on the accelerator and release the brake. *Accelerators:* Chemical signals called growth factors pass a "divide" signal from the cell surface to the nucleus, where the divide instructions are to be carried out. Any change that facilitates any stage of this information pathway will promote cancer. *Releasing the Brakes:* Cells have three safeguards against uncontrolled cell division, all of which must be removed before cancer can occur. First, cells have proteins called tumor suppressors that normally block cell division which are temporarily inactivated during normal cell division. In cancer cells they are permanently inactivated. Second, cells have error-correcting proteins such as p53 that detect damage to genes that might initiate cancer. In most cancers this error-detection has been disabled. Third, cancer cells rebuild the tips of their chromosomes, removing the normal limit on the cell's life span.

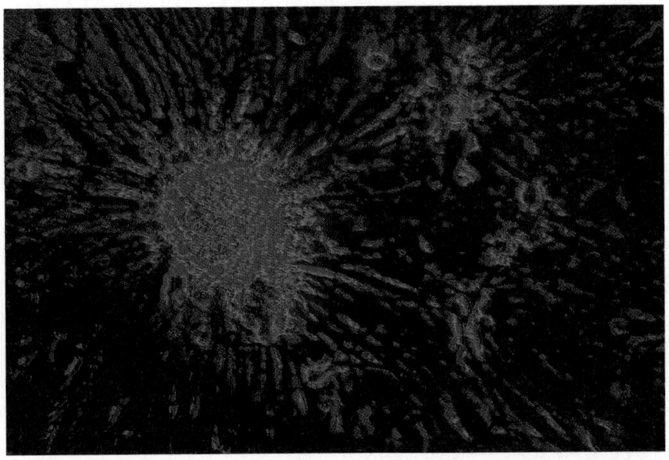

Virtual Lab

Can Cancer Tumors Be Starved to Death?

One of the most exciting recent developments in the war against cancer is the report that it might be possible to starve cancer tumors to death. In order for a tumor to grow, it needs an ample supply of blood. A tumor increases its blood supply by secreting substances into the surrounding tissues that encourage angiogenesis, the formation of small blood vessels. As a solid tumor grows and outstrips its blood supply, its interior becomes hypoxic (oxygen depleted). In response to hypoxia, it appears that genes are turned on that promote survival under low oxygen pressure, including ones that promote angiogenesis. How does hypoxia within a tumor promote angiogenesis?

Randall Johnson of the University of California, San Diego, is studying one possible way. A tumor responding to hypoxia induces a gene-specific transcription factor (a protein that activates the transcription of a particular gene) that promotes angiogenesis. Called HIF-1, for hypoxia inducible factor-1, this transcription factor appears to play a critical role in inducing the transcription of genes necessary for blood vessel formation.

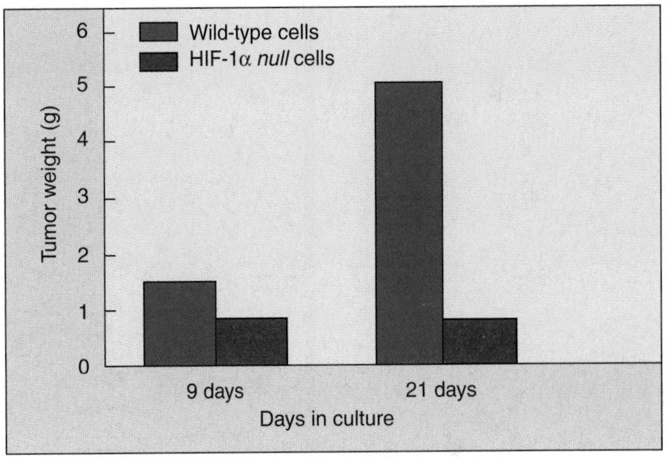
Wild-type cells
HIF-1α *null* cells
Tumor weight (g)
6
5
4
3
2
1
0
9 days
21 days
Days in culture

Quizzes
Further Reading
Essential Study Partner
Links
BioCourse.com

CHAPTER 7

FOUNDATIONS OF GENETICS

THE FAR SIDE®
By GARY LARSON
© 1982 FarWorks, Inc. All Rights Reserved/Dist. by Creators Syndicate
Larson
"This is your side of the family, you realize."

CHAPTER

Foundations of Genetics

CHAPTER OVERVIEW

Mendel

7.1 Mendel and the Garden Pea
7.2 What Mendel Found
7.3 Mendel Proposes a Theory
7.4 Analyzing Mendel's Results
7.5 Mendel's Laws

- In the 1860s, Gregor Mendel conducted genetic crosses of varieties of pea plants.
- Carefully counting the numbers of each kind of offspring, Mendel observed that in crosses of heterozygous parents, one-fourth of the offspring always appear recessive.
- From this simple result, Mendel formulated his theory of heredity.

From Genotype to Phenotype

7.6 Epistasis
7.7 Multiple Alleles
7.8 Other Modifications of the Genotype

- The essence of Mendel's theory is that traits are determined by *information,* which was soon shown to be stored in genes on chromosomes.
- When the pattern of a trait's heredity reflects chromosome segregation, that trait is said to be *Mendelian.*
- Many factors can obscure the underlying pattern of chromosomal segregation, such as when one gene modifies the phenotypic expression of another.

Chromosomes and Heredity

7.9 Chromosomes Are the Vehicles of Mendelian Inheritance
7.10 Human Chromosomes

- The segregation of the white-eye trait in *Drosophila* is associated with the segregation of the X chromosome.
- Nondisjunction results when chromosomes do not separate during meiosis, leading to gametes with missing or extra chromosomes.

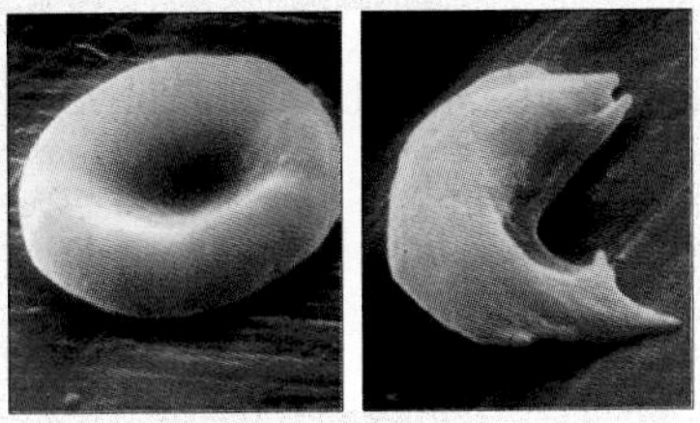

Human Hereditary Disorders

7.11 The Role of Mutations in Human Heredity
7.12 Hemophilia
7.13 Sickle-Cell Anemia
7.14 Other Disorders
7.15 Genetic Counseling and Therapy

- Some hereditary disorders are relatively common in human populations; others are rare.
- Genetic counseling techniques like amniocentesis are important aids in predicting the likelihood of producing children expressing hereditary disorders.

7.1 Mendel and the Garden Pea

When you were born, many things about you resembled your mother or father. This tendency for traits to be passed from parent to offspring is called **heredity.** How does heredity happen? Before DNA and chromosomes were discovered, this puzzle was one of the greatest mysteries of science. The key to understanding the puzzle of heredity was found in the garden of an Austrian monastery over a century ago by a monk named Gregor Mendel (figure 7.1). Crossing pea plants with one another, Mendel developed a series of simple rules that accurately predicted patterns of heredity—that is, how many offspring would be like one parent and how many like the other. When Mendel's rules became widely known, investigators all over the world set out to discover the physical mechanism responsible for them. They learned that hereditary traits are written in *genes,* the instructions carefully laid out in the DNA a child receives from each parent. Mendel's solution to the puzzle of heredity was the first step on this journey of understanding and one of the greatest intellectual accomplishments in the history of science.

(a) (b)

Figure 7.1 Gregor Mendel and the garden pea.

(*a*) The key to understanding the puzzle of heredity was solved by Mendel by cultivating pea plants in the garden of his monastery in Brunn, Austria. (*b*) Because it is easy to cultivate and because there are many distinctive varieties, the garden pea, *Pisum sativum,* had been a popular choice as an experimental subject in investigations of heredity for as long as a century before Mendel's studies.

Early Ideas About Heredity

Mendel was not the first person to try to understand heredity by crossing pea plants. Over 200 years earlier British farmers had performed similar crosses and obtained results similar to Mendel's. They observed that in crosses between two types—tall and short plants, say—the less frequent kind (in this case, short plants) would disappear in one generation, only to reappear in the next. In the 1790s, for example, the British farmer T. A. Knight crossed a variety of the garden pea that had purple flowers with one that had white flowers. All the offspring of the cross had purple flowers. If two of these offspring were crossed, however, some of *their* offspring were purple and some were white. Knight noted that the purple had a "stronger tendency" to appear than white, but he did not count the numbers of each kind of offspring.

Mendel's Experiments

Gregor Mendel was born in 1822 to peasant parents and was educated in a monastery. He became a monk and was sent to the University of Vienna to study science and mathematics. Although he aspired to become a scientist and teacher, he failed his university exams for a teaching certificate and returned to the monastery, where he spent the rest of his life, eventually becoming abbot. Upon his return, Mendel joined an informal neighborhood science club, a group of local farmers and others interested in science. Under the patronage of a local nobleman, each member set out to undertake scientific investigations, which were then discussed at meetings and published in the club's own journal. Mendel undertook to repeat the classic series of crosses with pea plants done by Knight and others, but this time he intended to count the numbers of each kind of offspring in the hope that the numbers would give some hint of what was going on. Quantitative approaches to science—measuring and counting—were just becoming fashionable in Europe.

The Garden Pea

Mendel chose to study the garden pea because several of its characteristics made it easy to work with:

1. Many varieties were available. Mendel selected seven pairs of lines that differed in easily distinguished traits (including the white versus purple flowers that Knight had studied 60 years earlier).
2. Mendel knew from the work of Knight and others that he could expect the infrequent version of a trait to disappear in one generation and reappear in the next. He knew, in other words, that he would have something to count.
3. Pea plants are small, easy to grow, produce large numbers of offspring, and mature quickly.
4. The reproductive organs of peas are enclosed within their flowers (figure 7.2), and, left alone, the flowers

do not open. They simply fertilize themselves with their own pollen (male gametes). To carry out a cross, Mendel had only to pry the petals apart, reach in with a scissors, and snip off the male organs (anthers); he could then dust the female organs with pollen from another plant to make the cross.

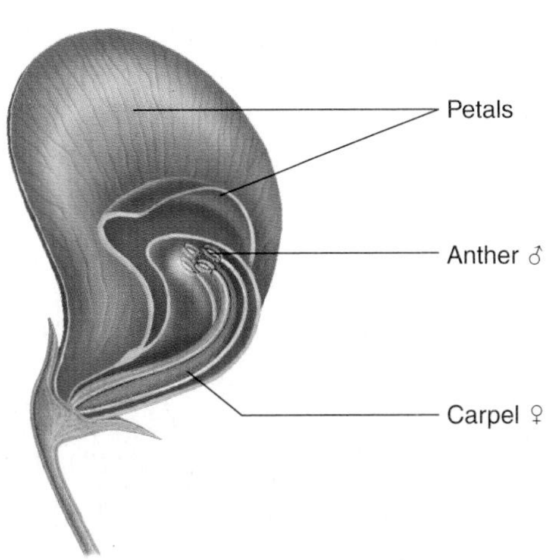

Petals
Anther ♂
Carpel ♀

Figure 7.2 The reproductive organs of a pea flower. In a pea flower, the petals enclose the male (anther) and female (carpel) parts, ensuring that self-fertilization will take place unless the flower is disturbed.

Mendel's Experimental Design

Mendel's experimental design was the same as Knight's, only Mendel counted his plants. The crosses were carried out in three steps:

1. Mendel began by letting each variety self-fertilize for several generations. This ensured that each variety was **true-breeding** and contained no other varieties. The white flower variety, for example, produced only white flowers and no purple ones in each generation. Mendel called these lines the **P generation** (P for parental).
2. Mendel then conducted his experiment: He crossed two pea varieties exhibiting alternative traits, such as white versus purple flowers (figure 7.3). The offspring that resulted he called the **F_1 generation** (F_1 for "first filial" generation, from the Latin word for "son" or "daughter").
3. Finally, Mendel allowed the plants produced in the crosses of step 2 to self-fertilize, and he counted the numbers of each kind of offspring that resulted in this **F_2** ("second filial") **generation.**

7.1 Mendel studied heredity by crossing pure-breeding garden peas that differed in easily scored alternative traits and then allowing the offspring to self-fertilize.

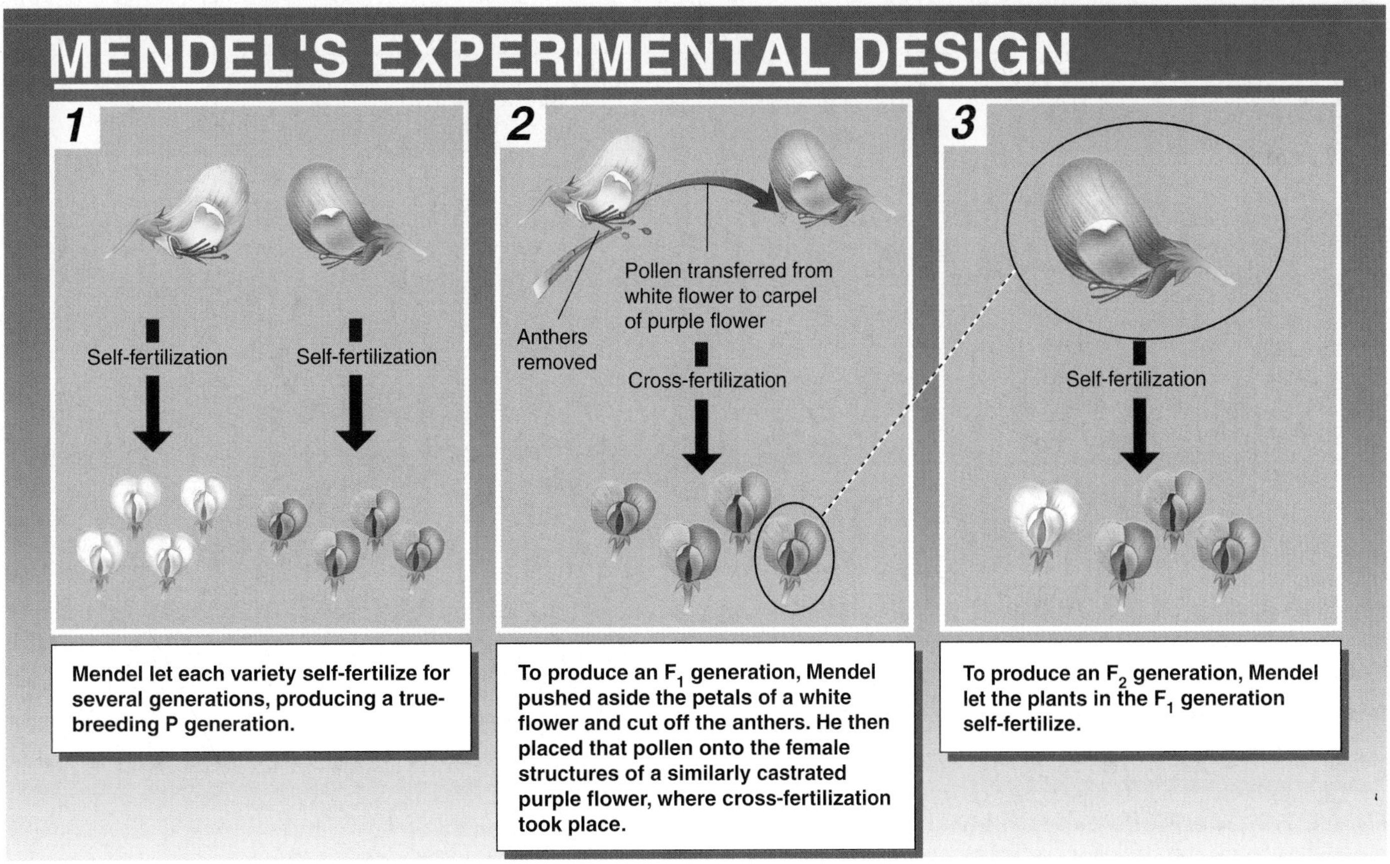

MENDEL'S EXPERIMENTAL DESIGN
1
Self-fertilization
Self-fertilization
2
Pollen transferred from white flower to carpel of purple flower
Anthers removed
Cross-fertilization
3
Self-fertilization
Mendel let each variety self-fertilize for several generations, producing a true-breeding P generation.
To produce an F_1 generation, Mendel pushed aside the petals of a white flower and cut off the anthers. He then placed that pollen onto the female structures of a similarly castrated purple flower, where cross-fertilization took place.
To produce an F_2 generation, Mendel let the plants in the F_1 generation self-fertilize.

Figure 7.3 How Mendel conducted his experiments.

7.2 What Mendel Found

Mendel experimented with a variety of traits in the garden pea and repeatedly came up with similar results. In all, Mendel examined seven pairs of contrasting traits. For each pair of contrasting varieties that Mendel crossed (figure 7.4), he obtained the same result. We will examine in detail Mendel's crosses with flower color.

The F_1 Generation

In the case of flower color, when Mendel crossed purple and white flowers, all the F_1 generation plants were purple, and the contrasting trait, white flowers, was not seen. Mendel called the trait expressed in the F_1 plants **dominant** and the trait not expressed **recessive.** In this case, purple flower color was dominant and white flower color recessive. Mendel studied several other traits in addition to flower color, and for every pair of contrasting traits Mendel examined, one proved to be dominant and the other recessive.

The F_2 Generation

After allowing individual F_1 plants to mature and self-fertilize, Mendel collected and planted the seeds from each plant to see what the offspring in the F_2 generation would

Trait	Dominant vs. recessive	F_2 generation: Dominant form	F_2 generation: Recessive form	Ratio
Flower color	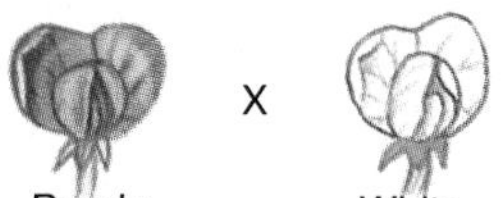 Purple X White	705	224	3.15:1
Seed color	 Yellow X Green	6,022	2,001	3.01:1
Seed shape	 Round X Wrinkled	5,474	1,850	2.96:1
Pod color	 Green X Yellow	428	152	2.82:1
Pod shape	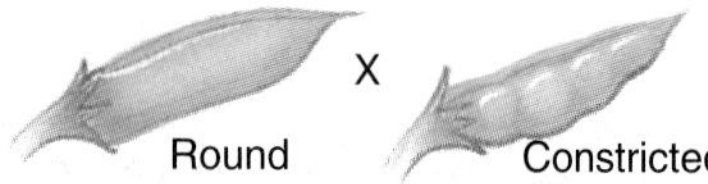 Round X Constricted	882	299	2.95:1
Flower position	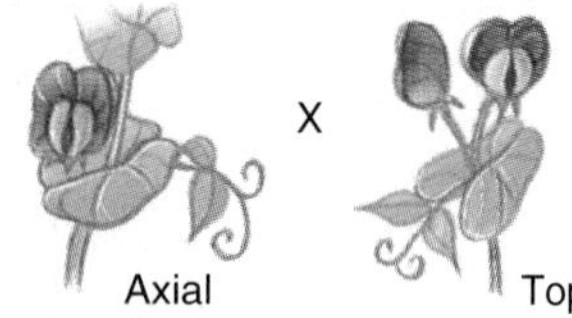 Axial X Top	651	207	3.14:1
Plant height	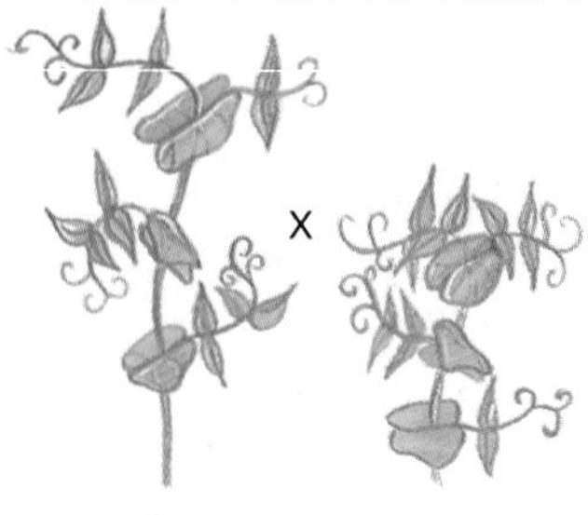 Tall X Dwarf	787	277	2.84:1

Figure 7.4 Mendel's experimental results.

The seven pairs of contrasting traits studied by Mendel in the garden pea are illustrated, along with the data he obtained in crosses of these traits. Every one of the seven traits that Mendel studied yielded results very close to a theoretical 3:1 ratio.

Figure 7.5 Round versus wrinkled seeds.

One of the differences among varieties of pea plants that Mendel studied was the shape of the seed. In some varieties the seeds were round, whereas in others they were wrinkled.

look like. Mendel found (as Knight had earlier) that some F_2 plants exhibited white flowers, the recessive trait. The recessive trait had disappeared in the F_1 generation, only to reappear in the F_2 generation. It must somehow have been present in the F_1 individuals but unexpressed!

At this stage Mendel instituted his radical change in experimental design. He *counted* the number of each type among the F_2 offspring. He believed the proportions of the F_2 types would provide some clue about the mechanism of heredity. In the cross between the purple-flowered F_1 plants, he counted a total of 929 F_2 individuals (see figure 7.4). Of these, 705 (75.9%) had purple flowers and 224 (24.1%) had white flowers. Approximately one-fourth of the F_2 individuals exhibited the recessive form of the trait. Mendel obtained the same numerical result with the other six traits he examined: three-fourths of the F_2 individuals exhibited the dominant form of the trait, and one-fourth displayed the recessive form. In other words, the dominant:recessive ratio among the F_2 plants was always close to 3:1. Mendel carried out similar experiments with other traits, such as round versus wrinkled seeds (figure 7.5) and obtained the same result.

A Disguised 1:2:1 Ratio

Mendel let the F_2 plants self-fertilize for another generation and found that the one-fourth that were recessive were true-breeding—future generations showed nothing but the recessive trait. Thus, the white F_2 individuals described previously showed only white flowers in the F_3 generation. Among the three-fourths of the F_2 plants that had shown the dominant trait, only one-third of the individuals were true-breeding. The others showed both traits in the F_3 generation—and when Mendel counted their numbers, he found the ratio of dominant to recessive to again be 3:1! From these results Mendel concluded that the 3:1 ratio he had observed in the F_2 generation was in fact a disguised 1:2:1 ratio (figure 7.6):

1		2		1
true-breeding dominant	**:**	**not-true-breeding dominant**	**:**	**true-breeding recessive**

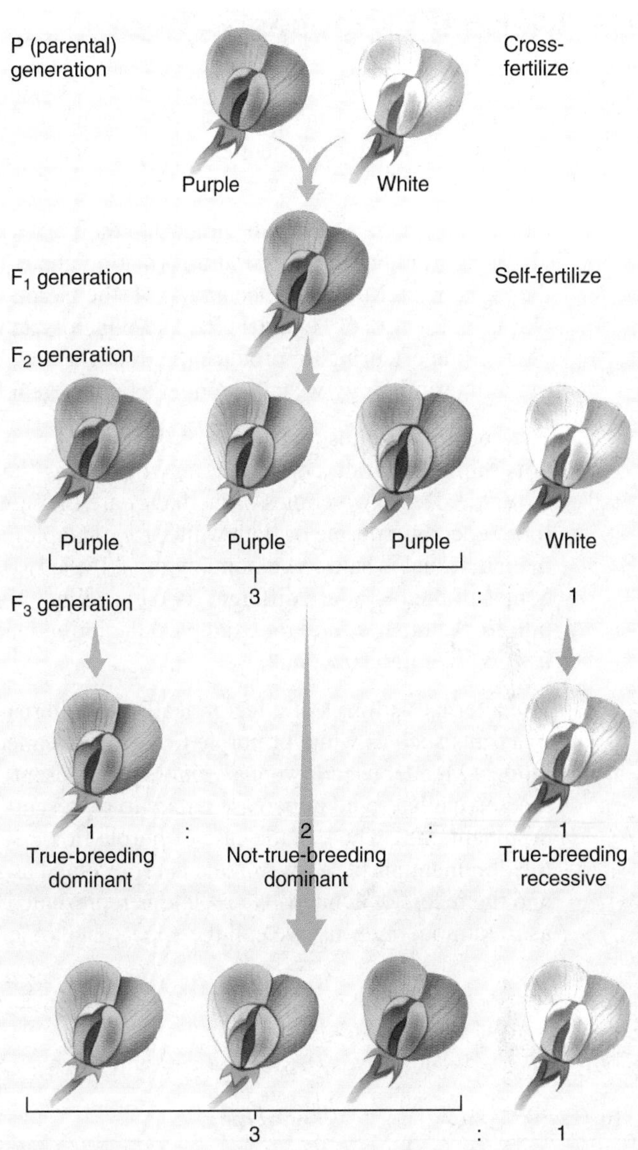

P (parental) generation
Cross-fertilize
Purple
White
F1 generation
Self-fertilize
F2 generation
Purple
Purple
Purple
White
3
:
1
F3 generation
1
:
2
:
1
True-breeding dominant
Not-true-breeding dominant
True-breeding recessive
3
:
1

Figure 7.6 The F_2 generation is a disguised 1:2:1 ratio.

By allowing the F_2 generation to self-fertilize, Mendel found from the offspring (F_3) that the ratio of F_2 plants was one true-breeding dominant, two not-true-breeding dominant, and one true-breeding recessive.

7.2 When Mendel crossed two contrasting varieties and counted the offspring in the subsequent generations, he found all of the offspring in the first generation exhibited one (dominant) trait, and none exhibited the other (recessive) trait. In the following generation, 25% were pure-breeding for the dominant trait, 50% were hybrid for the two traits and appeared dominant, and 25% were pure-breeding for the recessive trait.

7.3 Mendel Proposes a Theory

To explain his results, Mendel proposed a simple set of rules that would faithfully predict the results he had found. It has become one of the most famous theories in the history of science. The theory that Mendel proposed has five elements:

1. Parents do not transmit traits directly to their offspring. Rather, they transmit information about the traits, what Mendel called *merkmal* (the German word for "factor" or "character"). These factors act later, in the offspring, to produce the trait. In modern terminology we call Mendel's factors **genes.**
2. Each parent contains *two* copies of the factor governing each trait, copies that may or may not be the same. If the two copies of the factor are the same (both encoding purple or both white, for example) the individual is said to be **homozygous.** If the two copies of the factor are different (one encoding purple, the other white, for example), the individual is said to be **heterozygous.**
3. The alternative forms of a factor, leading to alternative traits such as white or purple flowers, are called **alleles.** Mendel used lowercase letters to represent recessive alleles and uppercase letters to represent dominant ones. Thus, in the case of purple flowers, the dominant purple flower allele is represented as *P* and the recessive white flower allele is represented as *p*. In modern terms, we call the appearance of an individual, such as possessing white flowers, its **phenotype.** Appearance is determined by which alleles of the flower-color gene the plant receives from its parents, and we call those particular alleles its **genotype.** Thus a pea plant might have the phenotype "white flower" and the genotype *pp*.
4. The two alleles that an individual possesses, one contributed by the male parent and the other by the female parent, do not affect each other, any more than two letters in a mailbox alter each other's contents. Each allele is passed on unchanged when the individual matures and produces its own gametes.
5. The presence of an allele does not ensure that a trait will be expressed in the individual that carries it. In heterozygous individuals, only the dominant allele achieves expression; the recessive allele is present but unexpressed.

These five elements, taken together, constitute Mendel's model of the hereditary process. Many traits in humans also exhibit dominant or recessive inheritance, similar to the traits Mendel studied in peas (table 7.1).

7.3 The genes that an individual has are referred to as its genotype; the outward appearance of the individual is referred to as its phenotype.

TABLE 7.1 SOME DOMINANT AND RECESSIVE TRAITS IN HUMANS

Recessive Traits	Phenotypes	Dominant Traits	Phenotypes
Common baldness	M-shaped hairline receding with age	Mid-digital hair	Presence of hair on middle segment of fingers
Albinism	Lack of melanin pigmentation	Brachydactyly	Short fingers
Alkaptonuria	Inability to metabolize homogenistic acid	Huntington's disease	Degeneration of nervous system, starting in middle age
Red-green color blindness	Inability to distinguish red or green wavelengths of light	Phenylthiocarbamide (PTC) sensitivity	Ability to taste PTC as bitter
Cystic fibrosis	Abnormal gland secretion, leading to liver degeneration and lung failure	Camptodactyly	Inability to straighten the little finger
Duchenne muscular dystrophy	Wasting away of muscles during childhood	Hypercholesterolemia (the most common human Mendelian disorder—1:500)	Elevated levels of blood cholesterol and risk of heart attack
Hemophilia	Inability to form blood clots	Polydactyly	Extra fingers and toes
Sickle-cell anemia	Defective hemoglobin that causes red blood cells to collapse		

7.4 Analyzing Mendel's Results

Consider again Mendel's cross of purple-flowered with white-flowered plants. We will assign the symbol *P* to the dominant allele, associated with the production of purple flowers, and the symbol *p* to the recessive allele, associated with the production of white flowers. By convention, genetic traits are usually assigned a letter symbol referring to their more common forms, in this case "P" for purple flower color. The dominant allele is written in uppercase, as *P;* the recessive allele (white flower color) is assigned the same symbol in lowercase, *p.*

In this system, the genotype of an individual that is true-breeding for the recessive white-flowered trait would be designated *pp.* In such an individual, both copies of the allele specify the white-flowered phenotype. Similarly, the genotype of a true-breeding purple-flowered individual would be designated *PP,* and a heterozygote would be designated *Pp* (dominant allele first). Using these conventions, and denoting a cross between two strains with ×, we can symbolize Mendel's original cross as *pp* × *PP.*

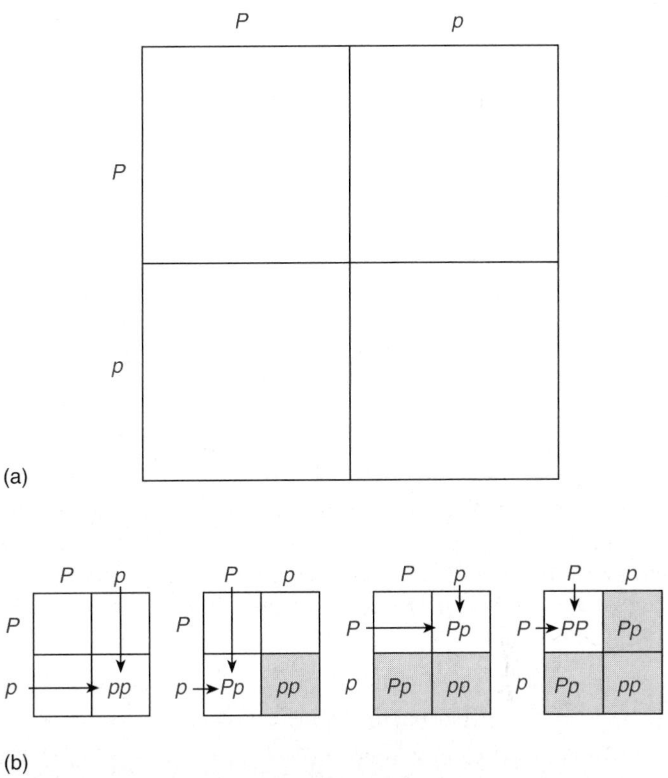
P
p
(a)
P p
Pp
pp
PP
(b)

Figure 7.7 A Punnett square.

A Punnett square analysis is an easy way to determine all the possible genotypes of a particular cross. (*a*) The possible gametes for one parent are placed along one side of the square, and the possible gametes for the other parent are placed along the other side of the square. (*b*) The Punnett square can be used like a mathematical table to visualize the genotypes of all potential offspring.

Punnett Squares

The possible results from a cross between a true-breeding, white-flowered plant (*pp*) and a true-breeding, purple-flowered plant (*PP*) can be visualized with a **Punnett square.** In a Punnett square, the possible gametes of one individual are listed along the horizontal side of the square, while the possible gametes of the other individual are listed along the vertical side. The genotypes of potential offspring are represented by the cells within the square (figure 7.7).

The frequency that these genotypes occur in the offspring is usually expressed by a **probability.** For example, in a cross between a homozygous white-flowered plant (*pp*) and a homozygous purple-flowered plant (*PP*) such as that performed by Mendel, *Pp* is the only possible genotype for all individuals in the F_1 generation (figure 7.8). Because *P* is dominant to *p,* all individuals in the F_1 generation have purple flowers. When individuals from the F_1 generation are crossed, the probability of obtaining a homozygous dominant (*PP*) individual in the F_2 is 25% because one-fourth of the possible genotypes are *PP*. Similarly, the probability of an individual in the F_2 generation being homozygous recessive (*pp*) is 25%. Because the heterozygous genotype (*Pp*) occurs in half of the cells within the square, the probability of obtaining a heterozygous (*Pp*) individual in the F_2 is 50%.

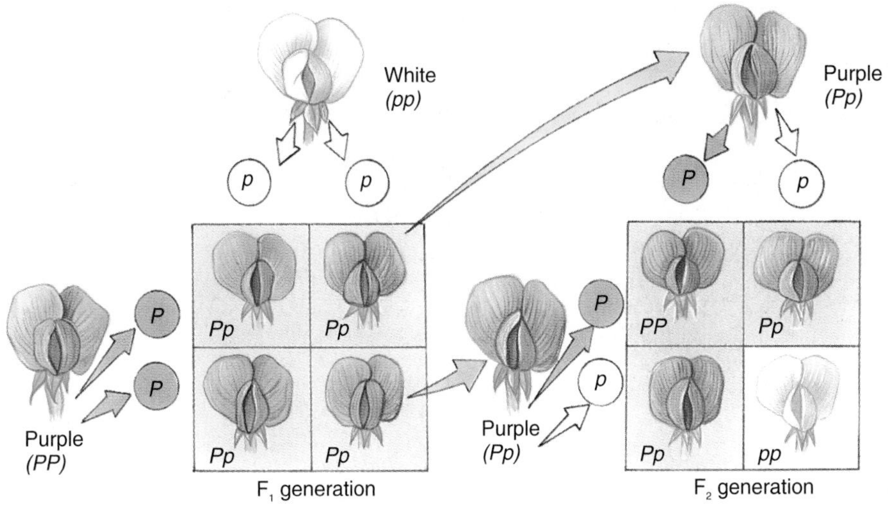
White
(pp)
Purple
(Pp)
Purple
(PP)
Purple
(Pp)
P
p
Pp
PP
pp
F1 generation
F2 generation

Figure 7.8 How Mendel analyzed flower color.

The only possible offspring of the first cross are *Pp* heterozygotes, purple in color. These individuals are known as the F_1 generation. When two heterozygous F_1 individuals cross, three kinds of offspring are possible: *PP* homozygotes (purple flowers); *Pp* heterozygotes (also purple flowers), which may form two ways; and *pp* homozygotes (white flowers). Among these individuals, known as the F_2 generation, the ratio of dominant phenotype to recessive phenotype is 3:1.

The Testcross

How did Mendel know which of the purple-flowered individuals in the F_2 generation (or the P generation) were homozygous (*PP*) and which were heterozygous (*Pp*)? It is not possible to tell simply by looking at them. For this reason, Mendel devised a simple and powerful procedure called the **testcross** to determine an individual's actual genetic composition. Consider a purple-flowered plant. It is impossible to tell whether such a plant is homozygous or heterozygous simply by looking at its phenotype. To learn its genotype, you must cross it with some other plant. What kind of cross would provide the answer? If you cross it with a homozygous dominant individual, all of the progeny will show the dominant phenotype whether the test plant is homozygous or heterozygous. It is also difficult (but not impossible) to distinguish between the two possible test plant genotypes by crossing with a heterozygous individual. However, if you cross the test plant with a homozygous recessive individual, the two possible test plant genotypes will give totally different results (figure 7.9):

Alternative 1: unknown individual homozygous (*PP*). *PP* × *pp:* all offspring have purple flowers (*Pp*).

Alternative 2: unknown individual heterozygous (*Pp*). *Pp* × *pp:* one-half of offspring have white flowers (*pp*) and one-half have purple flowers (*Pp*).

To perform his testcross, Mendel crossed heterozygous F_1 individuals back to the parent homozygous for the recessive trait. He predicted that the dominant and recessive traits would appear in a 1:1 ratio, and that is what he observed.

For each pair of alleles he investigated, Mendel observed phenotypic F_2 ratios of 3:1 (see figure 7.4) and testcross ratios very close to 1:1, just as his model predicted.

Testcrosses can also be used to determine the genotype of an individual when two genes are involved. Mendel carried out many two-gene crosses, some of which we will soon discuss. He often used testcrosses to verify the genotypes of particular dominant-appearing F_2 individuals. Thus an F_2 individual showing both dominant traits (*A_ B_*) might have any of the following genotypes: *AABB, AaBB, AABb,* or *AaBb.* By crossing dominant-appearing F_2 individuals with homozygous recessive individuals (that is, *A_ B_* × *aabb*), Mendel was able to determine if either or both of the traits bred true among the progeny and so determine the genotype of the F_2 parent.

AABB	trait A breeds true	trait B breeds true
AaBB		trait B breeds true
AABb	trait A breeds true	
AaBb		

7.4 Analyses using Punnett squares demonstrate that Mendel's results reflect independent segregation of gametes.

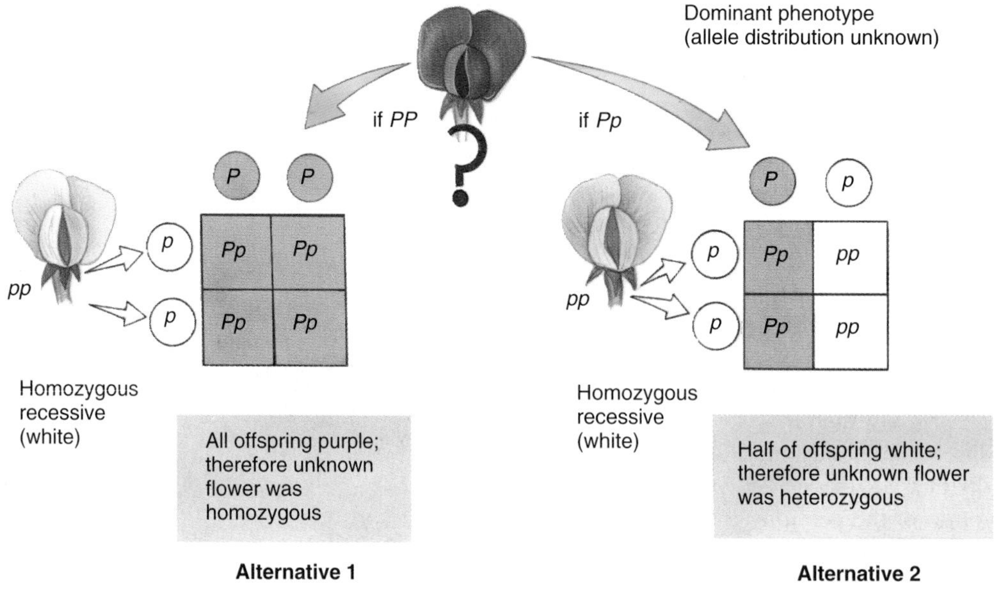

Dominant phenotype
(allele distribution unknown)
if PP
if Pp
?
P
P
p
p
Pp
Pp
Pp
Pp
pp
Homozygous
recessive
(white)
All offspring purple;
therefore unknown
flower was
homozygous
Alternative 1
P
p
p
p
Pp
pp
Pp
pp
pp
Homozygous
recessive
(white)
Half of offspring white;
therefore unknown flower
was heterozygous
Alternative 2

Figure 7.9 How Mendel used the testcross to detect heterozygotes.
To determine whether an individual exhibiting a dominant phenotype, such as purple flowers, is homozygous (*PP*) or heterozygous (*Pp*) for the dominant allele, Mendel devised the testcross. He crossed the individual in question with a known homozygous recessive (*pp*)—in this case, a plant with white flowers.

7.5 Mendel's Laws

Mendel's First Law: Segregation

Mendel's model brilliantly predicts the results of his crosses, accounting in a neat and satisfying way for the ratios he observed. Similar patterns of heredity have since been observed in countless other organisms. Traits exhibiting this pattern of heredity are called *Mendelian traits.* Because of its overwhelming importance, Mendel's theory is often referred to as Mendel's first law, or the **law of segregation.** In modern terms, Mendel's first law states that *only one allele specifying an alternative trait can be carried in a particular gamete, and gametes combine randomly in forming offspring.*

Mendel's Second Law: Independent Assortment

Mendel went on to study how pairs of genes are inherited, such as flower color and plant height. He first established a series of true-breeding lines of peas that differed from one another with respect to two of the seven pairs of characteristics he had studied. Second, he crossed contrasting pairs of true-breeding lines. For example, homozygous individuals with round, yellow seeds crossed with individuals that are homozygous for wrinkled, green seeds produce dihybrid offspring that have round, yellow seeds and are heterozygous for both of these traits. Such F_1 individuals are said to be **dihybrid** (figure 7.10).

Mendel then allowed the dihybrid individuals to self-fertilize. If the segregation of alleles affecting seed shape and seed color were independent, the probability that a particular pair of seed-shape alleles would occur together with a particular pair of seed-color alleles would simply be a product of the two individual probabilities that each pair would occur separately. For example, the probability of an individual with wrinkled, green seeds appearing in the F_2 generation would be equal to the probability of an individual with wrinkled seeds (1 in 4) multiplied by the probability of an individual with green seeds (1 in 4), or 1 in 16.

In his dihybrid crosses, Mendel found that the frequency of phenotypes in the F_2 offspring closely matched the 9:3:3:1 ratio predicted by a Punnett square analysis. He concluded that for the pairs of traits he studied, the inheritance of one trait does not influence the inheritance of any other trait, a result often referred to as Mendel's second law, or the **law of independent assortment.** We now know that this result is only valid for genes not located near one another on the same chromosome. Thus in modern terms Mendel's second law is often stated as follows: *Genes located on different chromosomes are inherited independently of one another.*

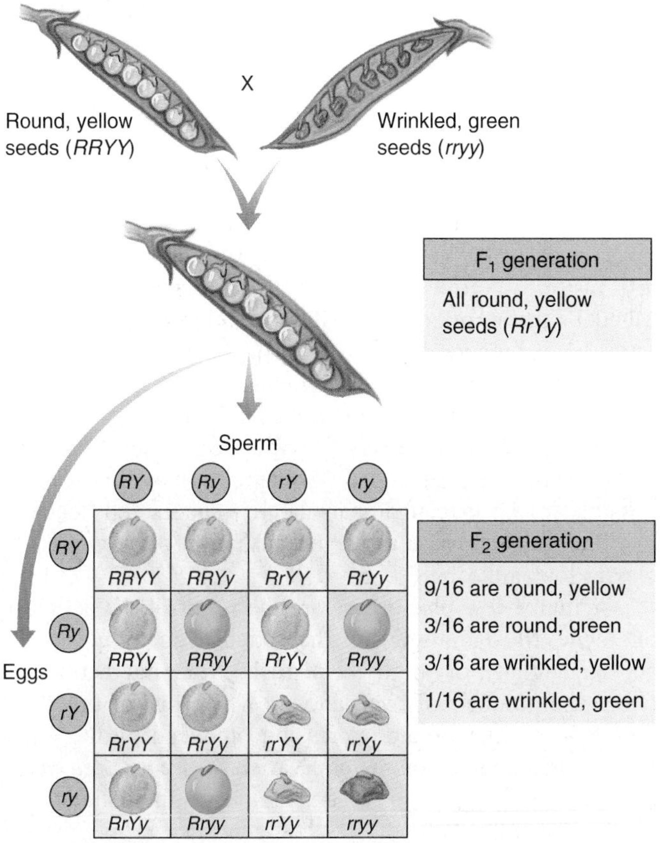
Round, yellow seeds (RRYY)
X
Wrinkled, green seeds (rryy)
F1 generation
All round, yellow seeds (RrYy)
Sperm
RY
Ry
rY
ry
Eggs
RY
Ry
rY
ry
RRYY
RRYy
RrYY
RrYy
RRYy
RRyy
RrYy
Rryy
RrYY
RrYy
rrYY
rrYy
RrYy
Rryy
rrYy
rryy
F2 generation
9/16 are round, yellow
3/16 are round, green
3/16 are wrinkled, yellow
1/16 are wrinkled, green

Figure 7.10 Analysis of a dihybrid cross.
This dihybrid cross shows round (*R*) versus wrinkled (*r*) seeds and yellow (*Y*) versus green (*y*) seeds. The ratio of the four possible combinations of phenotypes is predicted to be 9:3:3:1, the ratio that Mendel found.

Mendel's paper describing his results was published in the journal of his local scientific society in 1866. Unfortunately, his paper failed to arouse much interest, and his work was forgotten. Sixteen years after his death, in 1900, several investigators independently rediscovered Mendel's pioneering paper. They came across it while they were searching the literature in preparation for publishing their own findings, which were similar to those Mendel had quietly presented more than three decades earlier.

7.5 Mendel's theories of segregation and independent assortment are so well supported by experimental results that they are considered "laws."

7.6 Epistasis

Not all traits have the simple correlation between genotype and phenotype that Mendel saw. In the decades following the rediscovery of Mendel in 1900, many investigators set out to test his ideas. However, scientists attempting to confirm Mendel's theory often had trouble obtaining the same simple ratios he had reported. This was particularly true for dihybrid crosses. Recall that when individuals heterozygous for two different genes mate (a dihybrid cross), offspring may display the dominant phenotype for both genes, either one of the genes, or for neither gene. Sometimes, however, it is not possible for an investigator to identify successfully each of the four phenotypic classes, because two or more of the classes look alike.

Actually, few phenotypes are the result of the action of one gene. Most traits reflect the action of many genes that act sequentially or jointly. **Epistasis** is an interaction between the products of two genes in which one of the genes modifies the phenotypic expression produced by the other. For example, some commercial varieties of corn, *Zea mays,* exhibit a purple pigment called anthocyanin in their seed coats, while others do not. In 1918, geneticist R. A. Emerson crossed two pure-breeding corn varieties, neither exhibiting anthocyanin pigment. Surprisingly, all of the F_1 plants produced purple seeds.

When two of these pigment-producing F_1 plants were crossed to produce an F_2 generation, 56% were pigment producers and 44% were not. What was happening? Emerson correctly deduced that two genes were involved in producing pigment, and that the second cross had thus been a dihybrid cross like those performed by Mendel. Mendel had predicted 16 equally possible ways gametes could combine with each other, resulting in genotypes with a phenotypic ratio of 9:3:3:1 (9 + 3 + 3 + 1 = 16). How many of these were in each of the two types Emerson obtained? He multiplied the fraction that were pigment producers (0.56) by 16 to obtain 9 and multiplied the fraction that were not (0.44) by 16 to obtain 7. Thus, Emerson had a **modified ratio** of 9:7 instead of the usual 9:3:3:1 ratio (figure 7.11).

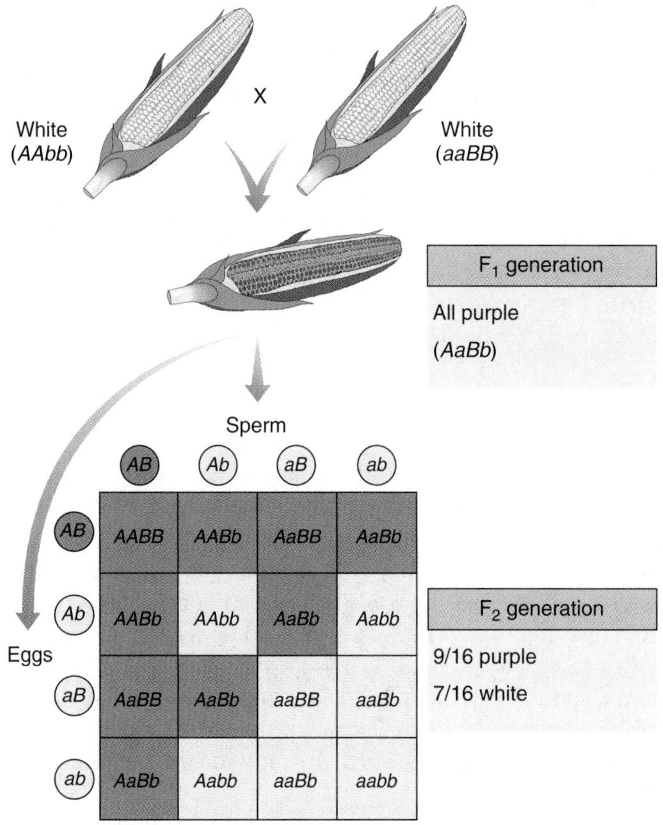
White
(AAbb)
X
White
(aaBB)
F1 generation
All purple
(AaBb)
Sperm
AB
Ab
aB
ab
Eggs
AABB
AABb
AaBB
AaBb
AAbb
Aabb
aaBB
aaBb
aabb
F2 generation
9/16 purple
7/16 white

Figure 7.11 How epistasis affects kernel color.
The purple pigment found in some varieties of corn is the result of two genes. Unless a dominant allele is present at each of the two genes, no pigment is expressed.

Why Was Emerson's Ratio Modified?

It turns out that either one of the two genes that contribute to kernel color can block the expression of the other. One of the genes (B) produces an enzyme that permits colored pigment to be produced only if a dominant allele (*BB* or *Bb*) is present. The other gene (A) produces an enzyme that in its dominant form (*AA* or *Aa*) allows the pigment to be deposited on the seed coat color. Thus, an individual with two recessive alleles for gene A (no pigment deposition) will have white seed coats even though it is able to manufacture the pigment because it possesses dominant alleles for gene B (purple pigment production). Similarly, an individual with dominant alleles for gene A (pigment can be deposited) will also have white seed coats if it has only recessive alleles for gene B (pigment production) and cannot manufacture the pigment.

To produce pigment, a plant must possess at least one functional copy of each enzyme gene. Of the 16 genotypes predicted by random assortment, 9 contain at least one dominant allele of both genes; they produce purple progeny. The remaining 7 genotypes lack dominant alleles at either or both loci (3 + 3 + 1 = 7) and so are phenotypically the same (nonpigmented), giving the phenotypic ratio of 9:7 that Emerson observed.

7.6 Sometimes an allele of one gene makes it difficult to detect the effect of another gene on the phenotype.

7.7 Multiple Alleles

A gene may have more than two alleles in a population, and most genes possess several different alleles. Often, no single allele is dominant; instead, each allele has its own effect, and the alleles are considered **codominant.**

A human gene that exhibits more than one codominant allele is the gene that determines ABO blood type. This gene encodes an enzyme that adds sugar molecules to lipids on the surface of red blood cells. These sugars act as recognition markers for cells in the immune system and are called cell surface antigens. The gene that encodes the enzyme, designated *I*, has three common alleles: I^B, whose product adds the sugar galactose; I^A, whose product adds galactosamine; and *i*, which codes for a protein that does not add a sugar.

Different combinations of the three *I* gene alleles occur in different individuals because each person possesses two copies of the chromosome bearing the *I* gene and may be homozygous for any allele or heterozygous for any two. An individual heterozygous for the I^A and I^B alleles produces both forms of the enzyme and adds both galactose and galactosamine to the surfaces of red blood cells. Because both alleles are expressed simultaneously in heterozygotes, the I^A and I^B alleles are codominant. Both I^A and I^B are dominant over the *i* allele because both I^A or I^B alleles lead to sugar addition and the *i* allele does not. The different combinations of the three alleles produce four different phenotypes (figure 7.12):

1. Type A individuals add only galactosamine. They are either I^AI^A homozygotes or *IAi* heterozygotes.
2. Type B individuals add only galactose. They are either I^BI^B homozygotes or I^Bi heterozygotes.
3. Type AB individuals add both sugars and are I^AI^B heterozygotes.
4. Type O individuals add neither sugar and are *ii* homozygotes.

These four different cell surface phenotypes are called the **ABO blood groups** or, less commonly, the Landsteiner blood groups, after the man who first described them. As Karl Landsteiner noted, a person's immune system can distinguish between these four phenotypes. If a type A individual receives a transfusion of type B blood, the recipient's immune system recognizes that the type B blood cells possess a "foreign" antigen (galactose) and attacks the donated blood cells, causing the cells to clump or agglutinate. This also happens if the donated blood is type AB. However, if the donated blood is type O, no immune attack will occur, as there are no galactose antigens on the surfaces of blood cells produced by the type O donor. In general, any individual's immune system will tolerate a transfusion of type O blood. Because neither galactose nor galactosamine is foreign to type AB individuals (whose red blood cells have both sugars), those individuals may receive any type of blood.

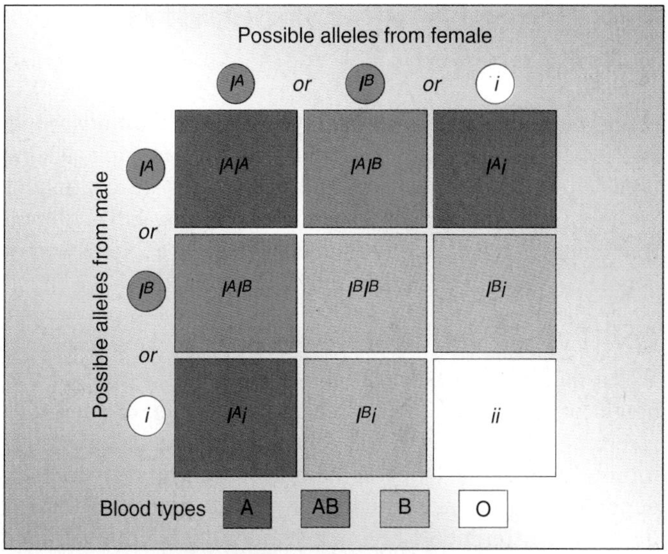
Possible alleles from female
I^A or I^B or i
Possible alleles from male
I^A or I^B or i
I^AI^A
I^AI^B
I^Ai
I^AI^B
I^BI^B
I^Bi
I^Ai
I^Bi
ii
Blood types
A
AB
B
O

Figure 7.12 Multiple alleles controlling the ABO blood groups.

Three common alleles control the ABO blood groups. Different combinations of the three so-called *I* gene alleles occur in different individuals. An individual may be homozygous for any allele or heterozygous for any two. The different combinations of the three alleles result in four different blood type phenotypes: type A (either I^AI^A homozygotes or I^Ai heterozygotes), type B (either I^BI^B homozygotes or I^Bi heterozygotes), type AB (I^AI^B heterozygotes), and type O (*ii* homozygotes).

The Rh Blood Group

Another set of cell surface markers on human red blood cells is the **Rh blood group** antigens, named for the rhesus monkey in which they were first described. About 85% of adult humans have the Rh cell surface marker on their red blood cells and are called Rh-positive. Rh-negative persons lack this cell surface marker because they are homozygous recessive for the gene encoding it.

If an Rh-negative person is exposed to Rh-positive blood, the Rh surface antigens of that blood are treated like foreign invaders by the Rh-negative person's immune system, which proceeds to make antibodies directed against the Rh antigens. This most commonly happens when an Rh-negative woman gives birth to an Rh-positive child (whose father is Rh-positive). Some fetal red blood cells cross the placental barrier and enter the mother's bloodstream, where they induce the production of "anti-Rh" antibodies. In subsequent pregnancies, the mother's antibodies can cross back to the new fetus and cause its red blood cells to clump, leading to a potentially fatal condition called erythroblastosis fetalis.

7.7 Many genes, like those encoding vertebrate blood groups, exhibit several alleles among members of a population.

7.8 Other Modifications of the Genotype

Often, the expression of the genotype is not straightforward. In addition to epistasis and codominance with multiple alleles, other situations that can complicate matters are continuous variation, pleiotropy, lack of complete dominance, and environmental effects (the environment affects the expression of an allele). We will examine each of these briefly.

Continuous Variation

When multiple genes act jointly to influence a trait such as height or weight, the trait often shows a range of small differences. Because all of the genes that play a role in determining phenotypes such as height or weight segregate independently of each other, we see a gradation in the degree of difference when many individuals are examined (figure 7.13). We call this gradation **continuous variation.** The greater the number of genes that influence a trait, the more continuous the expected distribution of the versions of that trait.

How can we describe the variation in a trait such as the height of the individuals in figure 7.13*a?* Individuals range from quite short to very tall, with average heights more common than either extreme. What we often do is to group the variation into categories. Each height, in inches, is a separate phenotypic category. Plotting the numbers in each height category produces a histogram, such as that in figure 7.13*b*. The histogram approximates an idealized bell-shaped curve, and the variation can be characterized by the mean and spread of that curve.

Pleiotropic Effects

Often, an individual allele has more than one effect on the phenotype. Such an allele is said to be **pleiotropic.** When the pioneering French geneticist Lucien Cuenot studied yellow fur in mice, a dominant trait, he was unable to obtain a true-breeding yellow strain by crossing individual yellow mice with one another. Individuals homozygous for the yellow allele died, because the yellow allele was pleiotropic: one effect was yellow color, but another was a lethal developmental defect. A pleiotropic gene alteration may be dominant with respect to one phenotypic consequence (yellow fur) and recessive with respect to another (lethal developmental defect). In pleiotropy, one gene affects many traits, in marked contrast to polygeny, where many genes affect one trait. Pleiotropic effects are difficult to predict, because the genes that affect a trait often perform other functions we may know nothing about.

Pleiotropic effects are characteristic of many inherited disorders, such as cystic fibrosis and sickle-cell anemia, both discussed later in this chapter. In these disorders, multiple symptoms can be traced back to a single gene defect. In cystic fibrosis, patients exhibit clogged blood vessels, overly sticky mucus, salty sweat, liver and pancreas failure, and a battery of other symptoms. All are pleiotropic effects of a single defect, a mutation in a gene that encodes a chloride ion transmembrane channel. In sickle-cell anemia, a defect in the oxygen-carrying hemoglobin molecule causes anemia, heart failure, increased susceptibility to pneumonia, kidney failure, enlargement of the spleen, and many other symptoms. It is usually difficult to deduce the nature of the primary defect from the range of its pleiotropic effects.

(a)

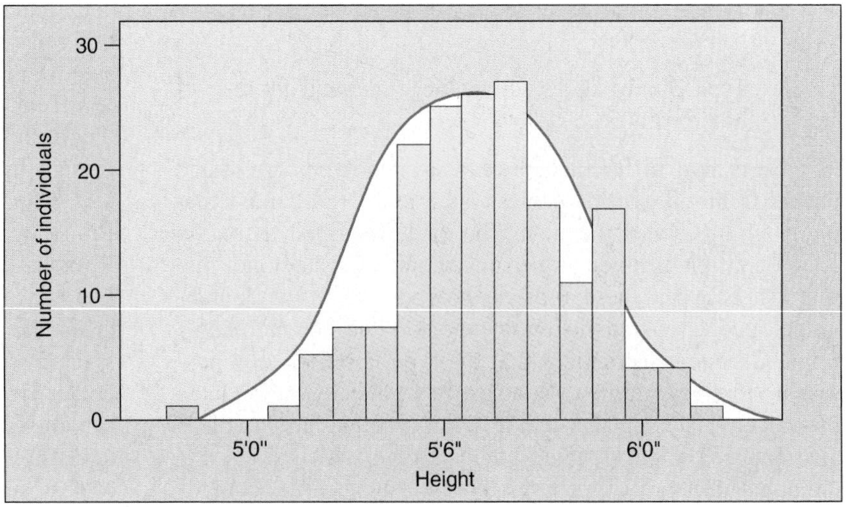
30
20
10
0
Number of individuals
5'0"
5'6"
6'0"
Height

(b)

Figure 7.13 Height is a continuously varying trait.

(*a*) This photograph shows the variation in height among students of the 1914 class of the Connecticut Agricultural College. Because many genes contribute to height and tend to segregate independently of each other, there are many possible combinations of those genes. (*b*) The cumulative contribution of different combinations of alleles to height forms a *continuous* spectrum of possible heights—a random distribution, in which the extremes are much rarer than the intermediate values.

Incomplete Dominance

Not all alternative alleles are fully dominant or fully recessive in heterozygotes. Some pairs of alleles exhibit **incomplete dominance** and produce a heterozygous phenotype that is intermediate between those of the parents. For example, in the cross of red- and white-flowered Japanese four o'clocks described in figure 7.14, all the F_1 offspring had pink flowers—indicating that neither red nor white flower color was dominant. Does this example of incomplete dominance argue that Mendel was wrong? Not at all. When two of the F_1 pink flowers were crossed, they produced red-, pink-, and white-flowered plants in a 1:2:1 ratio. Heterozygotes are simply intermediate in color.

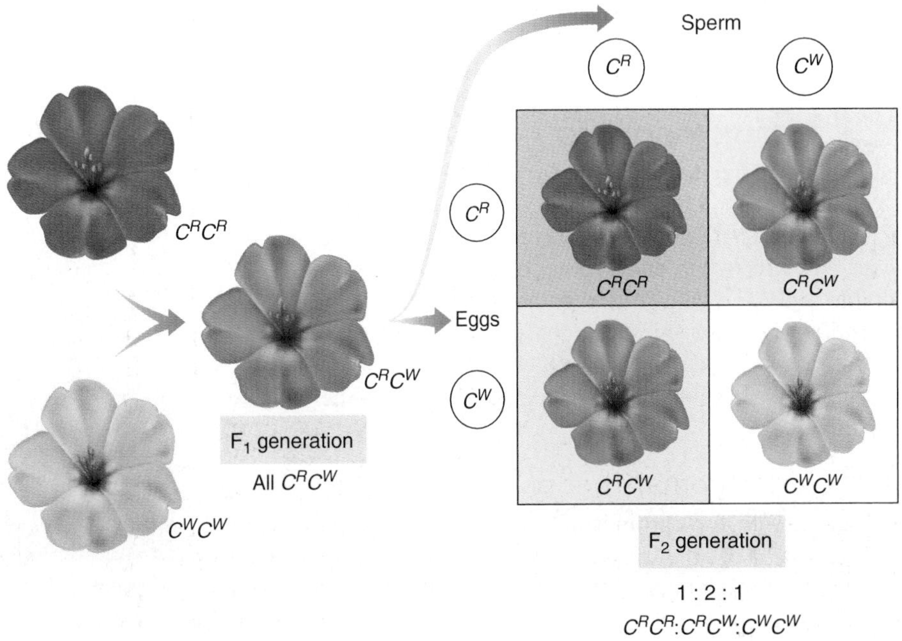

Sperm
C^R
C^W
C^RC^R
C^R
C^RC^R
C^RC^W
Eggs
C^RC^W
C^W
F_1 generation
All C^RC^W
C^RC^W
C^WC^W
C^WC^W
F_2 generation
1 : 2 : 1
C^RC^R:C^RC^W:C^WC^W

Figure 7.14 Incomplete dominance.

In a cross between a red-flowered Japanese four o'clock, genotype C^RC^R, and a white-flowered one (C^WC^W), neither allele is dominant. The heterozygous progeny have pink flowers and the genotype C^RC^W. If two of these heterozygotes are crossed, the phenotypes of their progeny occur in a ratio of 1:2:1 (red:pink:white).

Environmental Effects

The degree to which many alleles are expressed depends on the environment. Some alleles are heat-sensitive, for example. Traits influenced by such alleles are more sensitive to temperature or light than are the products of other alleles. The arctic foxes in figure 7.15, for example, make fur pigment only when the weather is warm. Similarly, the *ch* allele in Himalayan rabbits and Siamese cats encodes a heat-sensitive version of tyrosinase, one of the enzymes mediating the production of melanin, a dark pigment. The ch version of the enzyme is inactivated at temperatures above about 33°C. At the surface of the main body and head, the temperature is above 33°C and the tyrosinase enzyme is inactive, while it is more active at body extremities such as the tips of the ears and tail, where the temperature is below 33°C. The dark melanin pigment this enzyme produces causes the ears, snout, feet, and tail to be black.

7.8 A variety of factors can disguise the Mendelian segregation of alleles. Among them are the continuous variation that results when many genes contribute to a trait; incomplete dominance, which produces heterozygotes unlike either parent; and environmental influences on the expression of phenotypes.

(a)

(b)

Figure 7.15 Environmental effects on an allele.

(*a*) An arctic fox in winter has a coat that is almost white, so it is difficult to see the fox against a snowy background. (*b*) In summer, the same fox's fur darkens to a reddish brown, so that it resembles the color of the surrounding tundra. Heat-sensitive alleles control this color change.

7.9 Chromosomes Are the Vehicles of Mendelian Inheritance

Chromosomes are not the only kinds of structures that segregate regularly when eukaryotic cells divide. Centrioles also divide and segregate in a regular fashion, as do the mitochondria and chloroplasts (when present) in the cytoplasm. Therefore, in the early twentieth century it was by no means obvious that chromosomes were the vehicles of hereditary information.

The Chromosomal Theory of Inheritance

A central role for chromosomes in heredity was first suggested in 1900 by the German geneticist Karl Correns, in one of the papers announcing the rediscovery of Mendel's work. Soon after, observations that similar chromosomes paired with one another during meiosis led directly to the *chromosomal theory of inheritance,* first formulated by the American Walter Sutton in 1902.

Several pieces of evidence supported Sutton's theory. One was that reproduction involves the initial union of only two cells, egg and sperm. If Mendel's model were correct, then these two gametes must make equal hereditary contributions. Sperm, however, contain little cytoplasm, suggesting that the hereditary material must reside within the nuclei of the gametes. Furthermore, while diploid individuals have two copies of each pair of homologous chromosomes, gametes have only one. This observation was consistent with Mendel's model, in which diploid individuals have two copies of each heritable gene and gametes have one. Finally, chromosomes segregate during meiosis, and each pair of homologues orients on the metaphase plate independently of every other pair. Segregation and independent assortment were two characteristics of the genes in Mendel's model.

Problems with the Chromosomal Theory

Investigators soon pointed out one problem with this theory, however. If Mendelian traits are determined by genes located on the chromosomes, and if the independent assortment of Mendelian traits reflects the independent assortment of chromosomes in meiosis, why does the number of traits that assort independently in a given kind of organism often greatly exceed the number of chromosome pairs the organism possesses? This seemed a fatal objection, and it led many early researchers to have serious reservations about Sutton's theory.

Morgan's White-Eyed Fly

The essential correctness of the chromosomal theory of heredity was demonstrated long before this paradox was resolved. A single small fly provided the confirmation. In 1910 Thomas Hunt Morgan, studying the fruit fly *Drosophila melanogaster,* detected a mutant male fly, one that differed strikingly from normal flies of the same species: its eyes were white instead of red (figure 7.16).

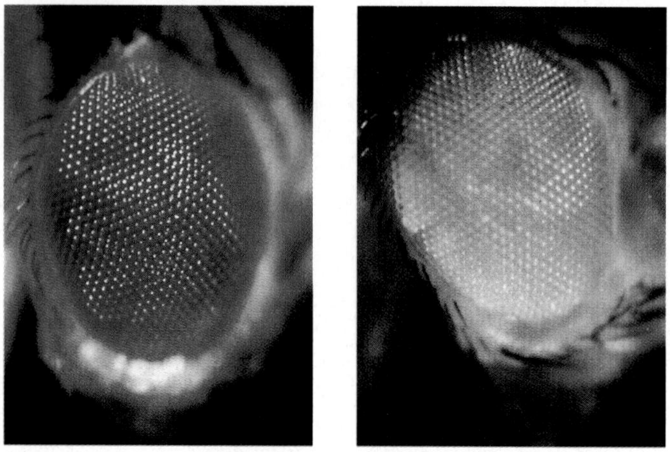

Figure 7.16 Red-eyed (wild type) and white-eyed (mutant) *Drosophila*.
The white-eyed defect is hereditary, the result of a mutation in a gene located on the X chromosome. By studying this mutation, Morgan first demonstrated that genes are on chromosomes.

Morgan immediately set out to determine if this new trait would be inherited in a Mendelian fashion. He first crossed the mutant male with a normal female to see if red or white eyes were dominant. All of the F_1 progeny had red eyes, so Morgan concluded that red eye color was dominant over white. Following the experimental procedure that Mendel had established long ago, Morgan then crossed the red-eyed flies from the F_1 generation with each other. Of the 4,252 F_2 progeny Morgan examined, 782 (18%) had white eyes. Although the ratio of red eyes to white eyes in the F_2 progeny was greater than 3:1, the results of the cross nevertheless provided clear evidence that eye color segregates. However, there was something about the outcome that was strange and totally unpredicted by Mendel's theory—*all of the white-eyed F_2 flies were males!*

How could this result be explained? Perhaps it was impossible for a white-eyed female fly to exist; such individuals might not be viable for some unknown reason. To test this idea, Morgan testcrossed the female F_1 progeny with the original white-eyed male. He obtained white-eyed and red-eyed males and females in a 1:1:1:1 ratio, just as Mendelian theory predicted. Hence, a female could have white eyes. Why, then, were there no white-eyed females among the progeny of the original cross?

Sex Linkage Confirms the Chromosomal Theory

The solution to this puzzle involved sex. In *Drosophila,* the sex of an individual is determined by the number of copies of a particular chromosome, the **X chromosome,** that an individual possesses. A fly with two X chromosomes is a female, and a fly with only one X chromosome is a male. In males, the single X chromosome pairs in meiosis with a large, dissimilar partner called the **Y chromosome.** The female thus produces only X gametes, while the male produces both X and Y gametes. When fertilization involves an X sperm, the result is an XX zygote, which develops into a female; when fertilization involves a Y sperm, the result is an XY zygote, which develops into a male.

The solution to Morgan's puzzle is that the gene causing the white-eye trait in *Drosophila* resides only on the X chromosome—it is absent from the Y chromosome. (We now know that the Y chromosome in flies carries almost no functional genes.) A trait determined by a gene on the sex chromosome is said to be **sex-linked.** Knowing the white-eye trait is recessive to the red-eye trait, we can now see that Morgan's result was a natural consequence of the Mendelian assortment of chromosomes (figure 7.17).

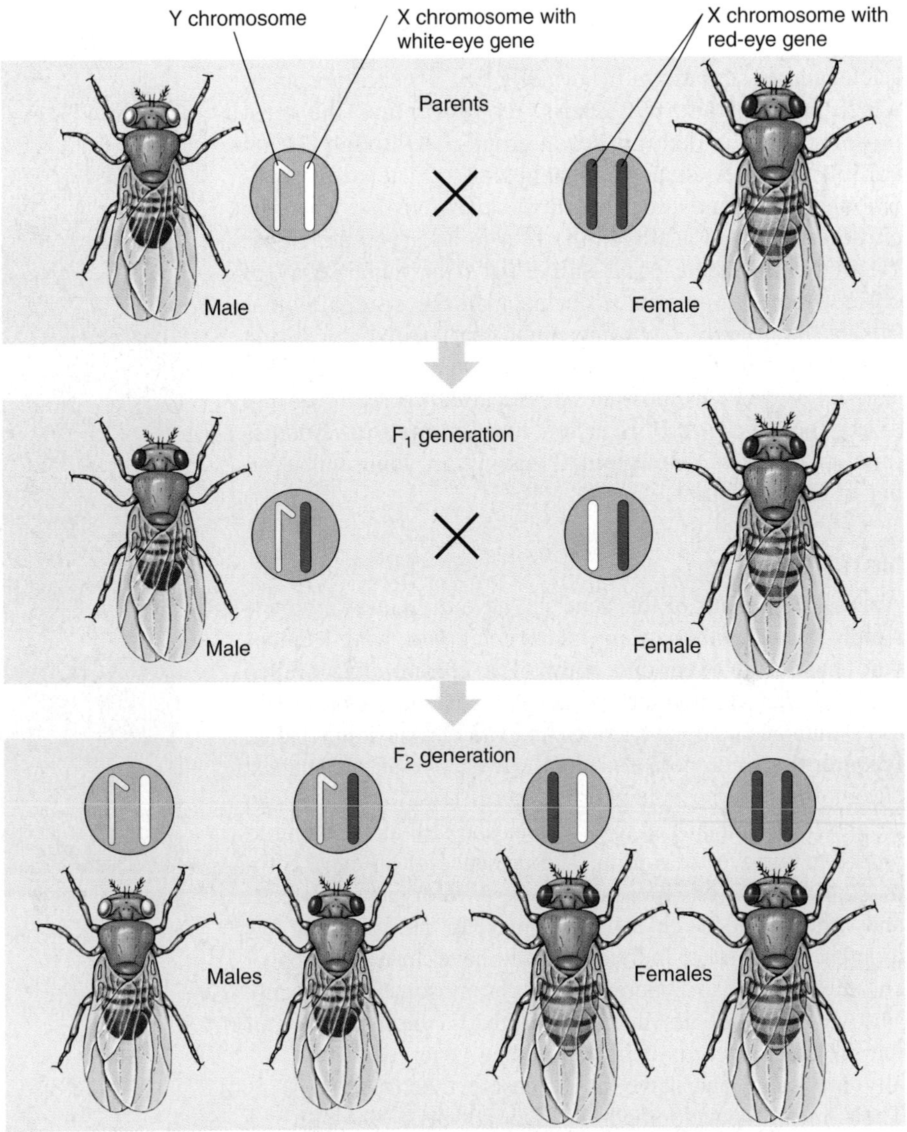
Y chromosome
X chromosome with white-eye gene
X chromosome with red-eye gene
Parents
Male
Female
F_1 generation
Male
Female
F_2 generation
Males
Females

Figure 7.17 Morgan's experiment demonstrating the chromosomal basis of sex linkage in *Drosophila.*

The white-eyed mutant male fly was crossed with a normal female. The F_1 generation flies all exhibited red eyes, as expected for flies heterozygous for a recessive white-eye allele. In the F_2 generation, all of the white-eyed flies were male.

Morgan's experiment was one of the most important in the history of genetics because it presented the first clear evidence that the genes determining Mendelian traits do indeed reside on the chromosomes, as Sutton had proposed. The segregation of the white-eye trait has a one-to-one correspondence with the segregation of the X chromosome. In other words, Mendelian traits such as eye color in *Drosophila* assort independently because chromosomes do. When Mendel observed the segregation of alternative traits in pea plants, he was observing a reflection of the meiotic segregation of chromosomes.

7.9 Mendelian traits assort independently because they are determined by genes located on chromosomes that assort independently in meiosis.

7.10 Human Chromosomes

Each human somatic cell normally has 46 chromosomes, which in meiosis form 23 pairs. By convention, the chromosomes are divided into seven groups (A through G), according to size, shape, and appearance. The differences among the chromosomes are most clearly visible when the chromosomes are arranged in order in a karyotype (figure 7.18). A **karyotype** is an individual's particular array of chromosomes. Sometimes during meiosis, sister chromatids or homologous chromosomes that paired up during metaphase remain stuck together instead of separating. The failure of chromosomes to separate correctly during either meiosis I or II is called **nondisjunction.** Nondisjunction leads to **aneuploidy,** having an abnormal number of chromosomes.

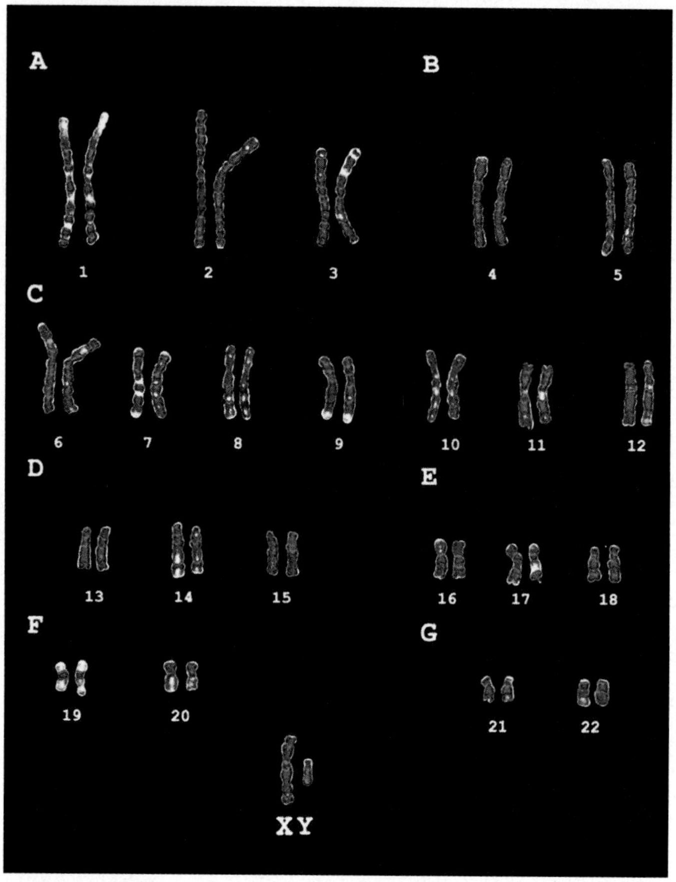
A
B
1
2
3
4
5
C
6
7
8
9
10
11
12
D
13
14
15
E
16
17
18
F
19
20
G
21
22
XY

Figure 7.18 A human karyotype.
This karyotype is arranged by class A to G.

Nondisjunction

Almost all humans of the same sex have the same karyotype simply because other arrangements don't work well. Humans who have lost even one copy of a chromosome (called **monosomics**) do not survive development. In all but a few cases, humans who have gained an extra chromosome (called **trisomics**) also do not survive. However, five of the smallest chromosomes—those numbered 13, 15, 18, 21, and 22—can be present in humans as three copies and still allow the individual to survive for a time. The presence of an extra chromosome 13, 15, or 18 causes severe developmental defects, and infants with such a genetic makeup die within a few months. In contrast, individuals who have an extra copy of chromosome 21 or, more rarely, chromosome 22, usually survive to adulthood. In such individuals, the maturation of the skeletal system is delayed, so they generally are short and have poor muscle tone. Their mental development is also affected, and children with trisomy 21 or trisomy 22 are always mentally impaired.

Down Syndrome. The developmental defect produced by trisomy 21 (figure 7.19) was first described in 1866 by J. Langdon Down; for this reason, it is called **Down syndrome.** About 1 in every 750 children exhibits Down syndrome, and the frequency is similar in all racial groups. It is much more common in children of older mothers—in mothers under 30 years old, the incidence is only about 1 in 1,500 births, while in mothers 30 to 35 years old, the incidence doubles, to 1 in 750 births. In mothers over 45, the risk is as high as 1 in 16 births (figure 7.20). The reason that older mothers are more prone to Down syndrome babies is that all the eggs that a woman will ever produce are present in her ovaries by the time she is born, and as she gets older they accumulate damage.

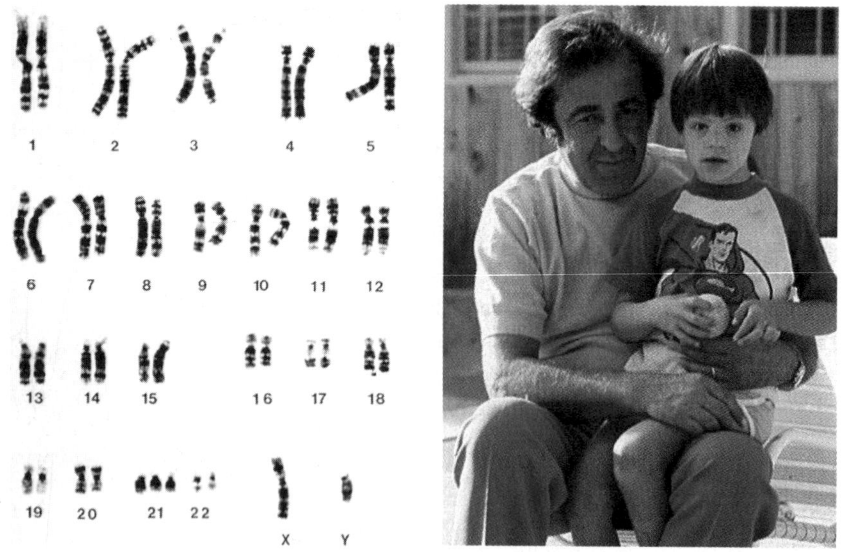
1
2
3
4
5
6
7
8
9
10
11
12
13
14
15
16
17
18
19
20
21
22
X
Y

(a) (b)

Figure 7.19 Down syndrome.
(a) In this karyotype of an individual with Down syndrome, the trisomy at position 21 can be clearly seen. *(b)* A child with Down syndrome sits on his father's knee.

Nondisjunction Involving the Sex Chromosomes

Of the 23 pairs of human chromosomes, 22 are perfectly matched in both males and females and are called autosomes. The remaining pair are the sex chromosomes, X and Y. In humans, as in *Drosophila* (but by no means in all diploid species), females are XX and males XY; any individual with at least one Y chromosome is male. The Y chromosome is highly condensed and bears few functional genes in most organisms. Some of the active genes the Y chromosome does possess are responsible for the features associated with "maleness." Individuals who gain or lose a sex chromosome do not generally experience the severe developmental abnormalities caused by changes in autosomes. Such individuals may reach maturity, but with somewhat abnormal features.

The X Chromosome. When X chromosomes fail to separate during meiosis, some of the gametes that are produced possess both X chromosomes and so are XX gametes; the other gametes that result from such an event have no sex chromosome and are designated "O" (figure 7.21).

If an XX gamete combines with an X gamete, the resulting XXX zygote develops into a female who is sterile but usually normal in other respects. If an XX gamete combines with a Y gamete, the XXY zygote develops into a sterile male who has many female body characteristics and, in some cases, diminished mental capacity. This condition, called *Klinefelter syndrome,* occurs in about 1 in 500 male births.

If an O gamete fuses with a Y gamete, the OY zygote is nonviable and fails to develop further because humans cannot survive when they lack the genes on the X chromosome. If an O gamete fuses with an X gamete, the XO zygote develops into a sterile female of short stature, with a webbed neck and immature sex organs that do not undergo changes during puberty. The mental abilities of an XO individual are in the low to normal range. This condition, called *Turner syndrome,* occurs roughly once in every 5,000 female births.

The Y Chromosome. The Y chromosome can also fail to separate in meiosis, leading to the formation of YY gametes. When these gametes combine with X gametes, the XYY zygotes develop into fertile males of normal appearance. The frequency of the XYY genotype is about 1 per 1,000 newborn males, but it is approximately 20 times higher among males in penal and mental institutions. This observation has led to the highly controversial suggestion that XYY males are inherently antisocial, a suggestion supported by some studies but not by others. In any case, most XYY males do not develop patterns of antisocial behavior.

7.10 Autosome loss is always lethal, and an extra autosome is with few exceptions lethal too. Additional sex chromosomes have less serious consequences, although they can lead to sterility.

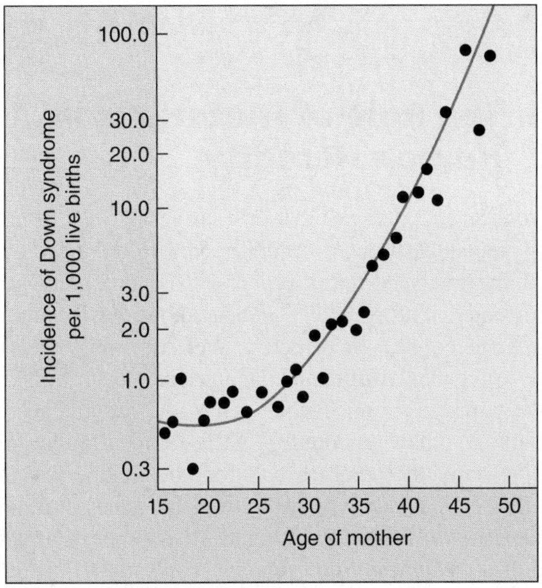
100.0
30.0
20.0
10.0
3.0
2.0
1.0
0.3
Incidence of Down syndrome per 1,000 live births
15
20
25
30
35
40
45
50
Age of mother

Figure 7.20 Correlation between maternal age and the incidence of Down syndrome.

As women age, the chances they will bear a child with Down syndrome increase. After a woman reaches age 35, the frequency of Down syndrome increases rapidly.

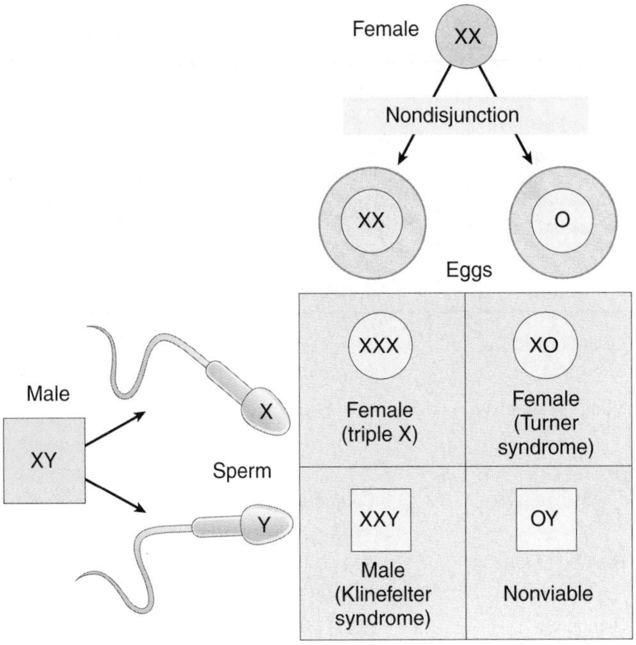
Female
XX
Nondisjunction
XX
O
Eggs
Male
XY
X
Sperm
Y
XXX
Female (triple X)
XO
Female (Turner syndrome)
XXY
Male (Klinefelter syndrome)
OY
Nonviable

Figure 7.21 Nondisjunction of the X chromosome.

Nondisjunction of the X chromosome can produce sex chromosome aneuploidy—that is, abnormalities in the number of sex chromosomes.

7.11 The Role of Mutations in Human Heredity

The proteins encoded by most of your genes must function in a very precise fashion in order for you to develop properly and for the many complex processes of your body to function correctly. Unfortunately, genes sometimes sustain damage or are copied incorrectly. We call these accidental changes in genes **mutations.** Mutations occur only rarely, because your cells police your genes and attempt to correct any damage they encounter. Still, some mutations get through. Many of them are bad for you in one way or another. It is easy to see why. Mutations hit genes at random—imagine that you randomly changed the number of a part on a design of a jet fighter. Sometimes it won't matter critically—a seatbelt becomes a radio, say. But what if a key rivet in the wing becomes a roll of toilet paper? The chance of a random mutation in a gene improving the performance of its protein is about the same as that of a randomly selected part making the jet fly faster.

Most mutations are rare in human populations. Almost all result in recessive alleles, and so they are not eliminated from the population by evolutionary forces—because they are not expressed in most individuals (heterozygotes) in which they occur. Do you see why they occur mostly in heterozygotes? Because mutant alleles are rare, it is unlikely that a person carrying a copy of the mutant allele will marry someone who also carries it. Instead, he or she will marry someone homozygous normal, so their children could not be homozygous for the mutant allele.

In some cases, particular mutant alleles have become more common in human populations. In these cases the harmful effects that they produce are called genetic disorders. Some of the most common genetic disorders are listed in table 7.2. To study human heredity, scientists look at the results of crosses that have already been made. They study family trees, or **pedigrees,** to identify which relatives exhibit a trait. Then they can often determine whether the gene producing the trait is sex-linked or autosomal and whether the trait's phenotype is dominant or recessive. Frequently, they can infer which individuals are homozygous and which are heterozygous for the allele specifying the trait.

> **7.11** Many human hereditary disorders reflect the presence of rare (and sometimes not so rare) mutations within human populations.

TABLE 7.2 SOME IMPORTANT GENETIC DISORDERS

Disorder	Symptom	Defect	Dominant/ Recessive	Frequency Among Human Births
Cystic fibrosis	Mucus clogs lungs, liver, and pancreas	Failure of chloride ion transport mechanism	Recessive	1/2,500 (Caucasians)
Sickle-cell anemia	Poor blood circulation	Abnormal hemoglobin molecules	Recessive	1/625 (African Americans)
Tay-Sachs disease	Deterioration of central nervous system in infancy	Defective enzyme (hexosaminidase A)	Recessive	1/3,500 (Ashkenazi Jews)
Phenylketonuria	Brain fails to develop in infancy	Defective enzyme (phenylalanine hydroxylase)	Recessive	1/12,000
Hemophilia	Blood fails to clot	Defective blood-clotting factor VIII	Sex-linked recessive	1/10,000 (Caucasian males)
Huntington's disease	Brain tissue gradually deteriorates in middle age	Production of an inhibitor of brain cell metabolism	Dominant	1/24,000
Muscular dystrophy (Duchenne)	Muscles waste away	Degradation of myelin coating of nerves stimulating muscles	Sex-linked recessive	1/3,700 (males)
Hypercholesterolemia	Excessive cholesterol levels in blood, leading to heart disease	Abnormal form of cholesterol cell surface receptor	Dominant	1/500

7.12 Hemophilia

When a blood vessel ruptures, the blood in the immediate area of the rupture forms a solid gel called a clot. The clot forms as a result of the polymerization of protein fibers circulating in the blood. A dozen proteins are involved in this process, and all must function properly for a blood clot to form. A mutation causing any of these proteins to lose their activity leads to a form of **hemophilia,** a hereditary condition in which the blood is slow to clot or does not clot at all.

Hemophilias are recessive disorders, expressed only when an individual does not possess any copy of the normal allele and so cannot produce one of the proteins necessary for clotting. Most of the genes that encode the blood-clotting proteins are on autosomes, but two (designated VIII and IX) are on the X chromosome. These two genes are sex-linked: any male who inherits a mutant allele of either of the two will develop hemophilia because his other sex chromosome is a Y chromosome that lacks any alleles of those genes.

The most famous instance of hemophilia, often called the Royal hemophilia, is a sex-linked form that arose in the royal family of England. This hemophilia was caused by a mutation in gene IX that occurred in one of the parents of Queen Victoria of England (1819–1901; figure 7.22). In the five generations since Queen Victoria, 10 of her male descendants have had hemophilia. The present British royal family has escaped the disorder because Queen Victoria's son, King Edward VII, did not inherit the defective allele, and all the subsequent rulers of England are his descendants. Three of Victoria's nine children did receive the defective allele, however, and they carried it by marriage into many of the other royal families of Europe (figure 7.23).

7.12 Hemophilia results from recessive mutations of genes encoding blood-clotting proteins.

Figure 7.22 Queen Victoria of England, surrounded by some of her descendants in 1894.
Of Victoria's four daughters who lived to bear children, two, Alice and Beatrice, were carriers of Royal hemophilia. Two of Alice's daughters are standing behind Victoria (wearing feathered boas): Princess Irene of Prussia (*right*), and Alexandra (*left*), who would soon become Czarina of Russia. Both Irene and Alexandra were also carriers of hemophilia.

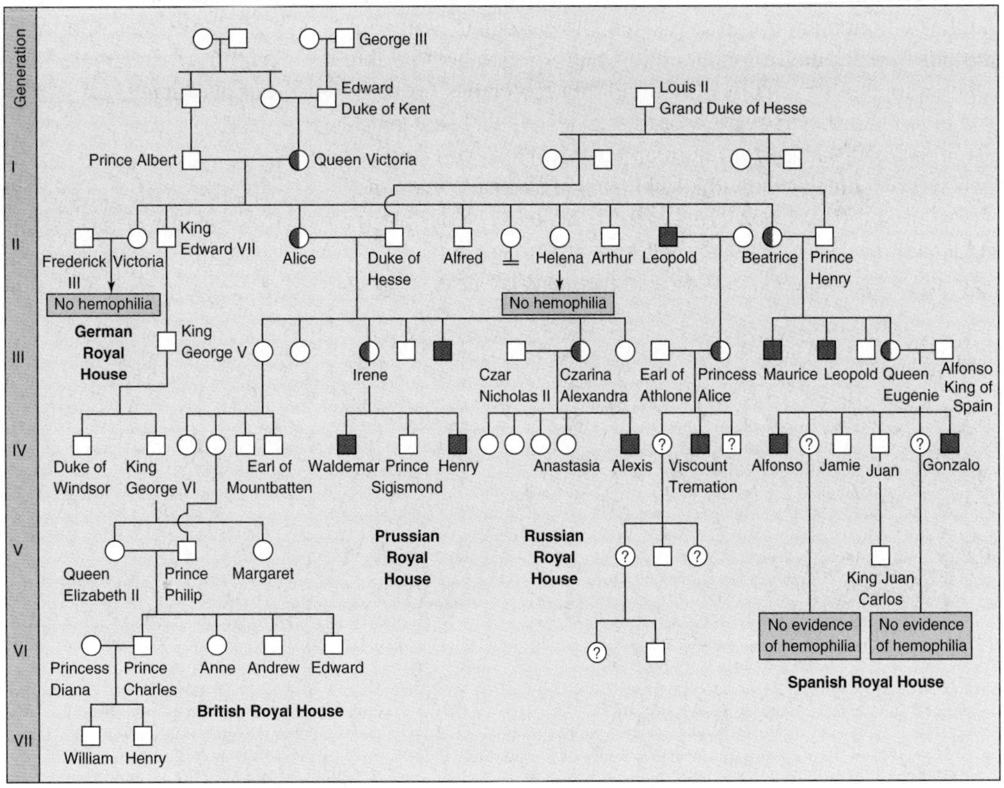

Generation
I
II
III
IV
V
VI
VII
George III
Edward
Duke of Kent
Louis II
Grand Duke of Hesse
Prince Albert
Queen Victoria
Frederick III
Victoria
King Edward VII
Alice
Duke of Hesse
Alfred
Helena
Arthur
Leopold
Beatrice
Prince Henry
No hemophilia
No hemophilia
German Royal House
King George V
Irene
Czar Nicholas II
Czarina Alexandra
Earl of Athlone
Princess Alice
Maurice
Leopold
Queen Eugenie
Alfonso King of Spain
Duke of Windsor
King George VI
Earl of Mountbatten
Waldemar
Prince Sigismond
Henry
Anastasia
Alexis
Viscount Trematon
Alfonso
Jamie
Juan
Gonzalo
Queen Elizabeth II
Prince Philip
Margaret
Prussian Royal House
Russian Royal House
King Juan Carlos
No evidence of hemophilia
No evidence of hemophilia
Spanish Royal House
Princess Diana
Prince Charles
Anne
Andrew
Edward
British Royal House
William
Henry

Figure 7.23 The Royal hemophilia pedigree.
Queen Victoria's daughter Alice introduced hemophilia into the Russian and Austrian royal houses, and her daughter Beatrice introduced it into the Spanish royal house. Victoria's son Leopold, himself a victim, also transmitted the disorder in a third line of descent. Half-shaded symbols represent carriers with one normal allele and one defective allele; fully shaded symbols represent affected individuals. Squares represent males; circles represent females.

7.13 Sickle-Cell Anemia

Sickle-cell anemia is a recessive hereditary disorder in which afflicted individuals have defective molecules of hemoglobin, the protein within red blood cells that carries oxygen. Consequently, these individuals are unable to properly transport oxygen to their tissues. The defective hemoglobin molecules stick to one another, forming stiff, rodlike structures and resulting in the formation of sickle-shaped red blood cells (figure 7.24). As a result of their stiffness and irregular shape, these cells have difficulty moving through the smallest blood vessels; they tend to accumulate in those vessels and form clots. People who have large proportions of sickle-shaped red blood cells tend to have intermittent illness and a shortened life span.

The hemoglobin in the defective red blood cells differs from that in normal red blood cells in only one of hemoglobin's 574 amino acid subunits. In the defective hemoglobin, the amino acid valine replaces a glutamic acid at a single position in the protein. Interestingly, the position of the change is far from the active site of hemoglobin where the iron-bearing heme group binds oxygen. Instead, the change occurs on the outer edge of the protein. Why then is the result so catastrophic? The sickle-cell mutation puts a very nonpolar amino acid on the surface of the hemoglobin protein, creating a "sticky patch" that sticks to other such patches—nonpolar amino acids tend to associate with one another in polar environments like water. As one hemoglobin adheres to another, chains of hemoglobin molecules form.

Individuals heterozygous for the sickle-cell allele are generally indistinguishable from normal persons. However, some of their red blood cells show the sickling characteristic when they are exposed to low levels of oxygen. The allele responsible for sickle-cell anemia is particularly common among people of African descent; about 9% of African Americans are heterozygous for this allele, and about 0.2% are homozygous and therefore have the disorder. In some groups of people in Africa, up to 45% of all individuals are heterozygous for this allele, and fully 6% are homozygous and express the disorder. What factors determine the high frequency of sickle-cell anemia in Africa? It turns out that heterozygosity for the sickle-cell anemia allele increases resistance to malaria, a common and serious disease in central Africa (figure 7.25). Sickle-cell anemia is discussed further in chapter 11.

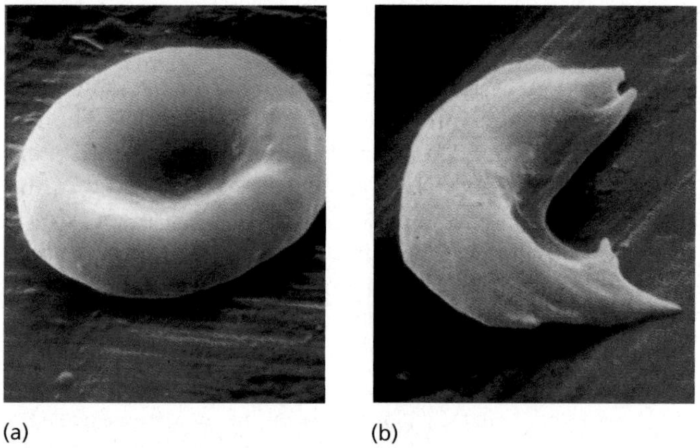

(a)
(b)

Figure 7.24 Normal and sickled red blood cells.
(*a*) A normal red blood cell is shaped like a flattened sphere. Its smooth edges pass easily through the body's tiny capillaries. In individuals homozygous for the sickle-cell trait, many of the red blood cells have sickle shapes (*b*). Their pointed edges get stuck while passing through capillaries, leading to internal bleeding and anemia.

7.13 Sickle-cell anemia is due to a recessive mutation of hemoglobin that is common in Africa because heterozygotes are resistant to malaria.

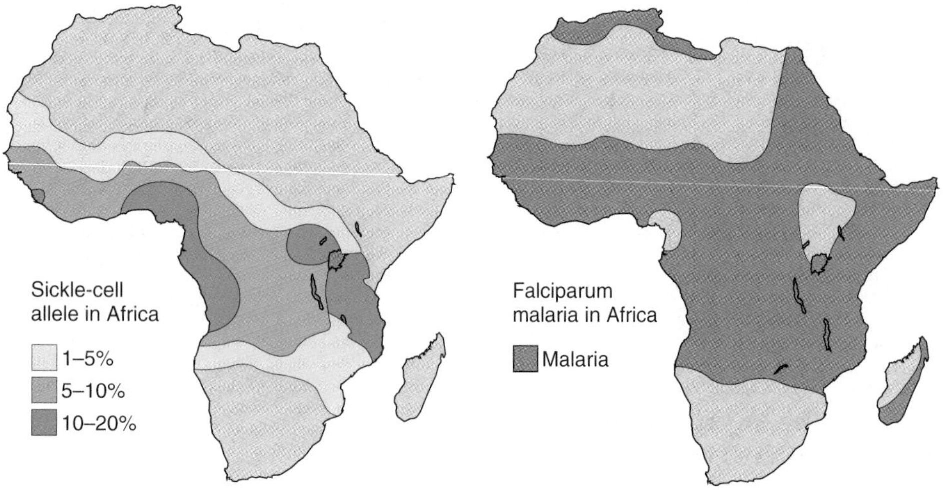

Sickle-cell allele in Africa
1–5%
5–10%
10–20%
Falciparum malaria in Africa
Malaria

Figure 7.25 The sickle-cell allele confers resistance to malaria.
The distribution of sickle-cell anemia closely matches the occurrence of malaria in central Africa. This is not a coincidence. The sickle-cell allele, when heterozygous, confers resistance to malaria, a very serious disease.

7.14 Other Disorders

Tay-Sachs Disease

Tay-Sachs disease is an incurable hereditary disorder in which the brain deteriorates. Affected children appear normal at birth and usually do not develop symptoms until about the eighth month, when signs of mental deterioration appear. The children are blind within a year after birth, and they rarely live past five years of age.

Tay-Sachs disease is rare in most human populations, occurring in only 1 in 300,000 births in the United States. However, the disease has a high incidence among Jews of Eastern and Central Europe (Ashkenazi) and among American Jews, 90% of whom trace their ancestry to Eastern and Central Europe. In these populations, it is estimated that 1 in 28 individuals is a heterozygous carrier of the disease, and approximately 1 in 3,500 infants has the disease. Because the disease is caused by a recessive allele, most of the people who carry the defective allele do not themselves develop symptoms of the disease.

The Tay-Sachs allele produces the disease by encoding a nonfunctional form of the enzyme hexosaminidase A. This enzyme breaks down *gangliosides,* a class of lipids occurring within the lysosomes of brain cells (figure 7.26). As a result, the lysosomes fill with gangliosides, swell, and eventually burst, releasing oxidative enzymes that kill the cells. There is no known cure for this disorder.

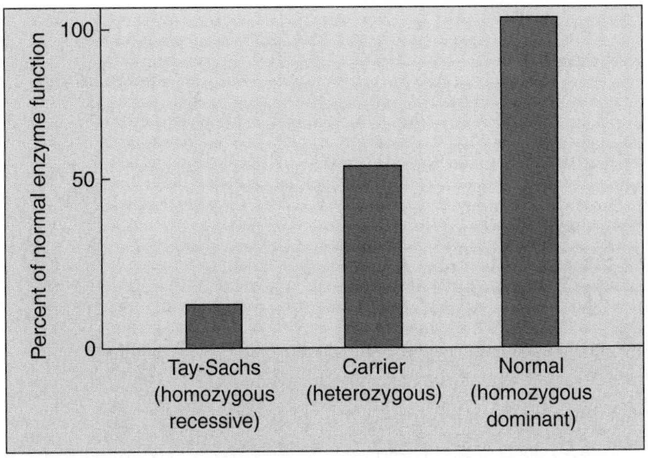

Percent of normal enzyme function
100
50
0
Tay-Sachs (homozygous recessive)
Carrier (heterozygous)
Normal (homozygous dominant)

Figure 7.26 Tay-Sachs disease.

Homozygous individuals (*left bar*) typically have less than 10% of the normal level of hexosaminidase A (*right bar*), while heterozygous individuals (*middle bar*) have about 50% of the normal level—enough to prevent deterioration of the central nervous system.

Huntington's Disease

Not all hereditary disorders are recessive. **Huntington's disease** is a hereditary condition caused by a dominant allele that causes the progressive deterioration of brain cells (figure 7.27). Perhaps 1 in 24,000 individuals develops the disorder. Because the allele is dominant, every individual who carries the allele expresses the disorder. Nevertheless, the disorder persists in human populations because its symptoms usually do not develop until the affected individuals are more than 30 years old, and by that time most of those individuals have already had children. Consequently, the allele is often transmitted before the lethal condition develops. A person who is heterozygous for Huntington's disease has a 50% chance of passing the disease to his or her children (even though the other parent does not have the disorder). In contrast, the carrier of a recessive disorder such as cystic fibrosis has a 50% chance of passing the allele to offspring but must mate with another carrier to risk bearing a child with the disease.

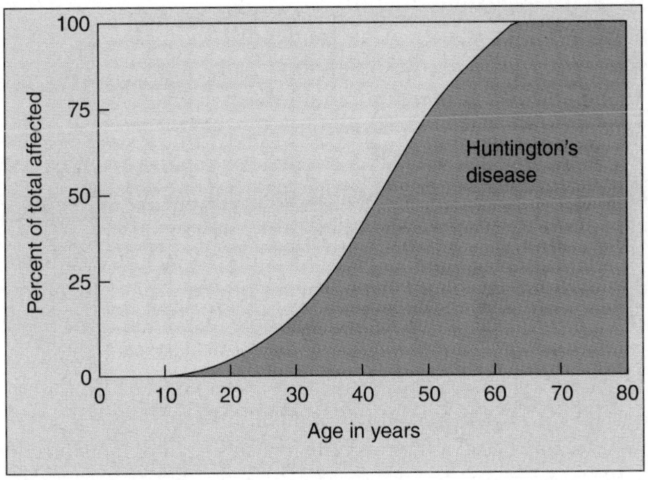

Percent of total affected
100
75
50
25
0
Huntington's disease
0
10
20
30
40
50
60
70
80
Age in years

Figure 7.27 Huntington's disease is a dominant genetic disorder.

It is because of the late age of onset of this disease that it persists despite the fact that it is dominant and fatal.

7.14 While many hereditary disorders are recessive, some are dominant, persisting because they exhibit delayed expression.

7.15 Genetic Counseling and Therapy

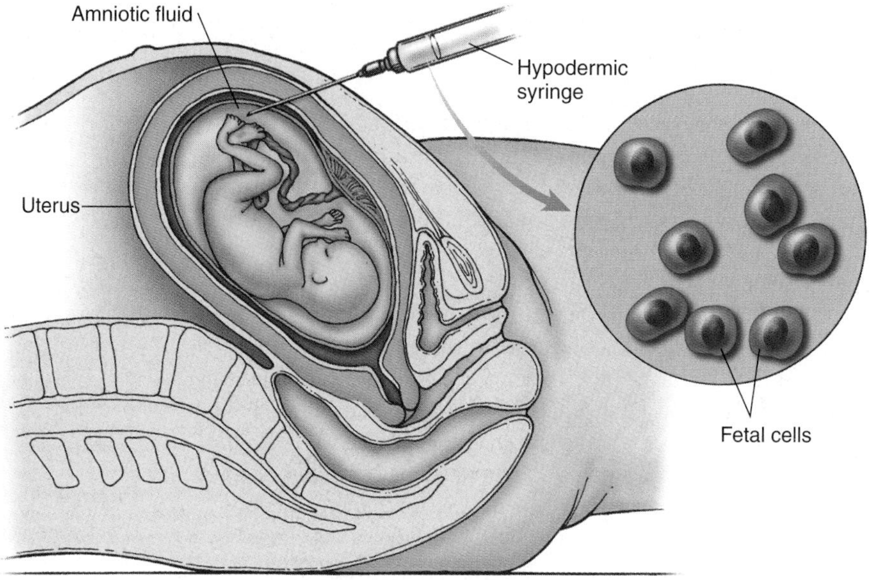

Amniotic fluid
Hypodermic syringe
Uterus
Fetal cells

Figure 7.28 Amniocentesis.

A needle is inserted into the amniotic cavity, and a sample of amniotic fluid, containing some free cells derived from the fetus, is withdrawn into a syringe. The fetal cells are then grown in culture and their karyotype and many of their metabolic functions are examined.

Although most genetic disorders cannot yet be cured, we are learning a great deal about them, and progress toward successful therapy is being made in many cases. In the absence of a cure, however, the only recourse is to try to avoid producing children with these conditions. The process of identifying parents at risk of producing children with genetic defects and of assessing the genetic state of early embryos is called **genetic counseling.**

If a genetic defect is caused by a recessive allele, how can potential parents determine the likelihood that they carry the allele? One way is through pedigree analysis, often employed as an aid in genetic counseling. By analyzing a person's pedigree, it is sometimes possible to estimate the likelihood that the person is a carrier for certain disorders. For example, if one of your relatives has been afflicted with a recessive genetic disorder such as cystic fibrosis, it is possible that you are a heterozygous carrier of the recessive allele for that disorder. When a couple is expecting a child, and pedigree analysis indicates that both of them have a significant probability of being heterozygous carriers of a recessive allele responsible for a serious genetic disorder, the pregnancy is said to be a *high-risk pregnancy.* In such cases, there is a significant probability that the child will exhibit the clinical disorder.

Another class of high-risk pregnancies are those in which the mothers are more than 35 years old. As we have seen, the frequency of birth of infants with Down syndrome increases dramatically in the pregnancies of older women (see figure 7.20).

When a pregnancy is diagnosed as being high risk, many women elect to undergo **amniocentesis,** a procedure that permits the prenatal diagnosis of many genetic disorders (figure 7.28). In the fourth month of pregnancy, a sterile hypodermic needle is inserted into the expanded uterus of the mother, and a small sample of the amniotic fluid bathing the fetus is removed. Within the fluid are free-floating cells derived from the fetus; once removed, these cells can be grown in cultures in the laboratory. During amniocentesis, the position of the needle and that of the fetus are usually observed by means of **ultrasound** (figure 7.29). The sound waves used in ultrasound are not harmful to mother or fetus, and they permit the person withdrawing the amniotic fluid to do so without damaging the fetus. In addition, ultrasound can be used to examine the fetus for signs of major abnormalities.

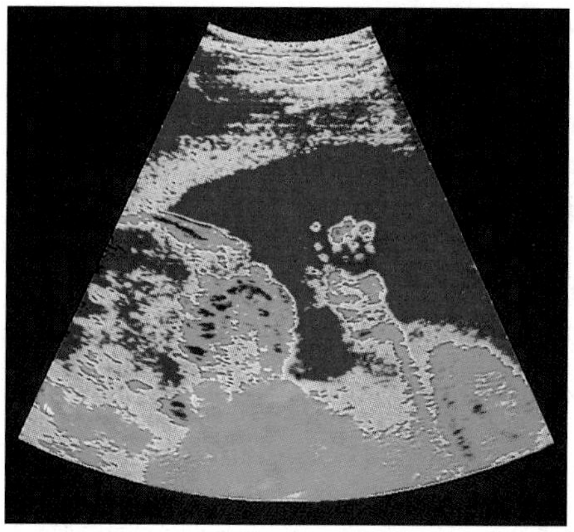

Figure 7.29 An ultrasound view of a fetus.

During the fourth month of pregnancy, when amniocentesis is normally performed, the fetus usually moves about actively. The head of the fetus here is to the *left.*

In recent years, physicians have increasingly turned to a new, less invasive procedure for genetic screening called **chorionic villi sampling.** In this procedure, the physician removes cells from the chorion, a membranous part of the placenta that nourishes the fetus. This procedure can be used earlier in pregnancy (by the eighth week) and yields results much more rapidly than does amniocentesis.

7.15 It has recently become possible to detect genetic defects early in pregnancy.

CONCEPT REVIEW

1. Several people had carried out plant crosses before Mendel, but none are credited with being the founder of the science of genetics. What did Mendel do that was different?
   a. He used garden peas.
   b. He did specific crosses.
   c. He counted the numbers of different types of offspring.
   d. He removed the male parts from flowers.

2. Of Mendel's F_2 plants that showed the dominant character, what proportion was true-breeding?
   a. one-fourth
   b. one-third
   c. one-half
   d. all

3. The white color of a homozygous recessive plant is its
   a. phenotype.
   b. genotype.

4. Of the following crosses, which is a testcross?
   a. $WW \times WW$
   b. $WW \times Ww$
   c. $Ww \times ww$
   d. $Ww \times W$

5. A normally pigmented man mates with an albino woman. They have three children, one of whom is an albino. What is the genotype of the father?
   a. *AA*
   b. *Aa*
   c. *aa*

6. Characteristics that exhibit continuous variation are generally controlled by
   a. a single dominant gene.
   b. pleiotropy.
   c. epistatic interactions.
   d. multiple genes.

7. How many chromosomes would you expect to find in the karyotype of a person with Turner syndrome?
   a. 22
   b. 23
   c. 45
   d. 46

8. Which of the following genetic diseases can be sex-linked?
   a. hemophilia
   b. Tay-Sachs disease
   c. cystic fibrosis
   d. hypercholesterolemia

9. If two alleles for a trait are the same, the individual is _________.

10. When two heterozygous individuals are crossed, the percentage of the progeny that exhibits the recessive trait is _________.

11. _________ is the observation that different genes on different chromosomes segregate independently of one another in genetic crosses.

12. The ability of one gene to modify the phenotypic expression of another is called _________.

CHALLENGE YOURSELF

1. Why did Mendel observe only two alleles of any given trait in the crosses that he carried out?

2. How might Mendel's results and the model he formulated have been different if the traits he chose to study were governed by alleles exhibiting incomplete dominance or codominance?

7 eBRIDGE

Reinforcing Key Points

Mendel

From Genotype to Phenotype

Chromosomes and Heredity

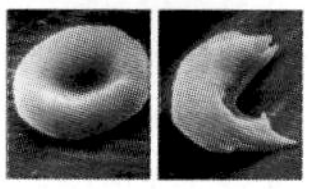

Human Hereditary Disorders

Electronic Learning

Explorations

Constructing a Genetic Map

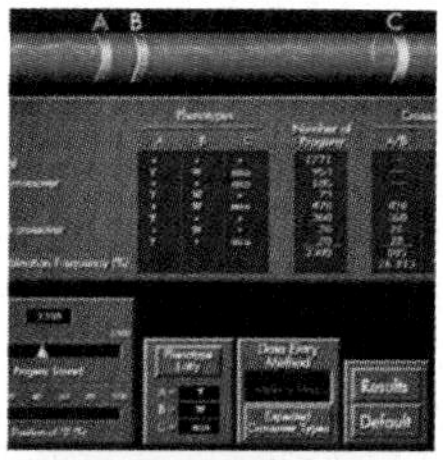

This exercise explores the recombination of three alleles in a dihybrid cross. You can either move the position of genes on a chromosome or enter recombination frequencies and see what genetic map you get.

Gene Segregation Within Families

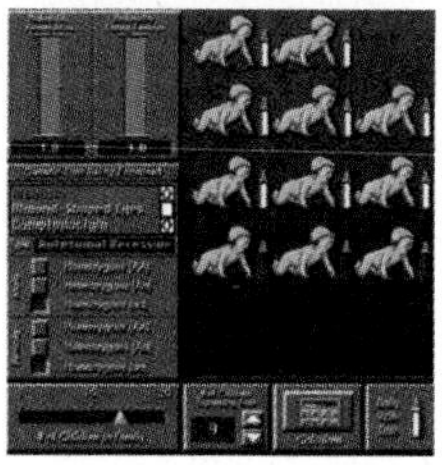

This interactive exercise allows you to explore the potential makeup of families. Specifically, you can assess the likelihood that a family of a given size will have all girls (or boys) or that offspring will be homozygous for a recessive trait.

Heredity in Families

In this exercise, you can explore a bank of pedigrees and analyze the dominance/recessiveness and X-linkage versus autosomal location of alleles within families.

Author's Corner

Gene Therapy. Scientists have long sought a way to introduce "healthy" genes into humans that lack them. Such a therapy was actually achieved in 1990 for a rare blood disorder, but it has been difficult to advance the research further, as the adenovirus used to carry the healthy version of the gene proved inappropriate (it's a cold virus, and most people have antibodies directed against it). New virus vectors avoid these problems, and offer hope of cures for many hereditary disorders. Cystic fibrosis, for example, results from a defect in a single gene encoding a chloride ion transport protein, and gene therapy may finally offer a way to cure it.

1. Improvements in gene vectors renew hope for gene therapy.
2. Altering ANDi: Inserting DNA into a primate.
3. Gene therapy may allow us to combat debilitating Parkinson's disease.

Virtual Classroom

Mendel and Genetics

Few scientists have made as great an impact on biology as Gregor Mendel, a Bavarian monk who over a hundred years ago worked out the basic pattern of heredity. He carried out experimental crosses with common pea plants in his monastery garden, and was able to explain the pattern of their inheritance with a simple "model" in which factors are contributed by each parent to an offspring. We now know Mendel's factors to be chromosomes.

Human Hereditary Disorders

Many inherited disorders reflect a mutation in a specific gene. Cystic fibrosis, for example, is a fatal disorder caused by a defective chloride channel protein—overly thick mucus blocks the passages of lungs and pancreas. The *cf* mutation is carried by 1 in 20 Caucasians. Sickle-cell anemia is caused by a mutation that alters one amino acid in the protein hemoglobin. Unfortunately, the change causes hemoglobin molecules to stick together, clumping up uselessly. The disorder is common in Central Africa because it confers resistance to malaria in heterozygous individuals.

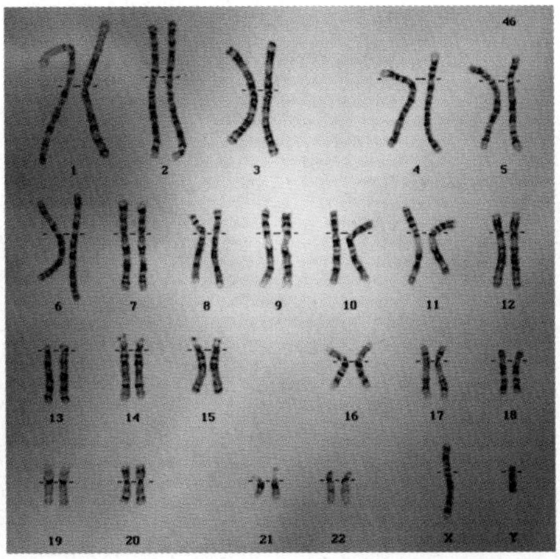
46
1
2
3
4
5
6
7
8
9
10
11
12
13
14
15
16
17
18
19
20
21
22
X
Y

Virtual Lab

How Regulatory Genes Direct Vertebrate Development

Most genetic disorders are the result of chemical modification of DNA. While many mutations have little or no effect on an organism, the mutations causing genetic disorders often have profound effects. Consider the activity of the anterior pituitary gland in humans. The anterior pituitary releases hormones from specialized cells that regulate growth, metabolism, lactation, reproduction, and response to stress. Early in development, the cells of the anterior pituitary differentiate into many specialized hormone-producing cell types, triggered by the activation of specific transcription factors. Many pituitary transcription factors have been identified, and mutations in these transcription factor genes are sometimes associated with inherited pituitary disorders. For example, mutations in the *LHX3* gene, encoding a pituitary transcription factor, have been identified in patients with retarded growth, pituitary hormone deficiency, and spinal deformities. Simon Rhodes, along with colleagues at Indiana University and Purdue University, is investigating the way in which mutations in the *LHX3* gene alter pituitary function to produce these profound effects.

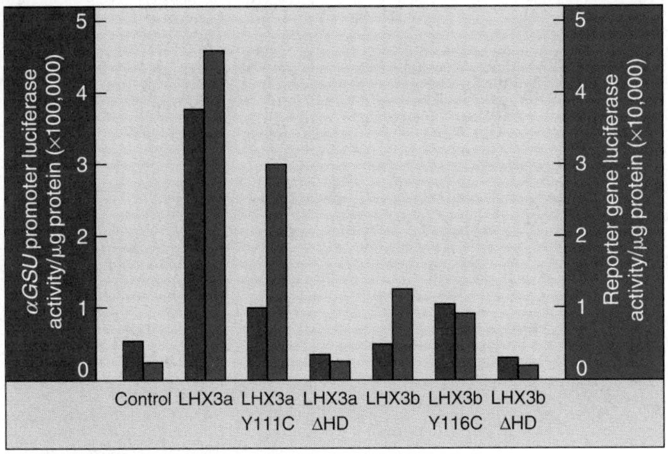
αGSU promoter luciferase activity/μg protein (×100,000)
Reporter gene luciferase activity/μg protein (×10,000)
0
1
2
3
4
5
Control
LHX3a
LHX3a Y111C
LHX3a ΔHD
LHX3b
LHX3b Y116C
LHX3b ΔHD

Quizzes | Further Reading | Essential Study Partner | Links | BioCourse.com

CHAPTER 8

How Genes Work

THE FAR SIDE®
By GARY LARSON
© 1986 FarWorks, Inc. All Rights Reserved/Dist. by Creators Syndicate
Larson
"Bummer of a birthmark, Hal."

CHAPTER 8

How Genes Work

CHAPTER OVERVIEW

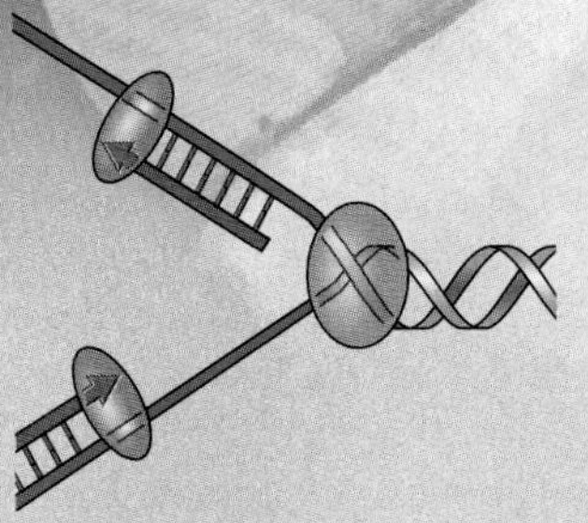

Genes Are Made of DNA

8.1 The Griffith Experiment
8.2 The Avery Experiments
8.3 The Hershey-Chase Experiment
8.4 The Fraenkel-Conrat Experiment
8.5 Discovering the Structure of DNA
8.6 How the DNA Molecule Replicates

- A series of critical experiments demonstrated that genes are composed of DNA.
- A key advance was the demonstration by Avery that the active ingredient in Griffith's transforming extract was DNA.
- DNA is a double helix, in which A on one strand pairs with T on the other, and G similarly pairs with C.

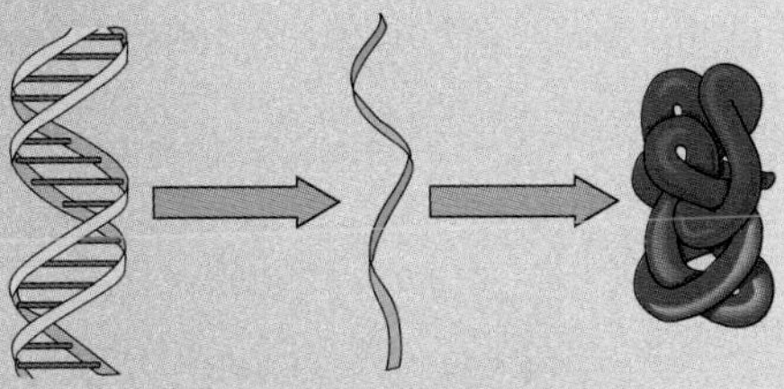

From Gene to Protein

8.7 Transcription
8.8 The Genetic Code
8.9 Translation
8.10 Architecture of the Gene

- DNA serves as a template on which mRNA is assembled in a process called transcription.
- The genetic information is encoded in DNA in three-nucleotide units called codons, which correspond to particular amino acids.
- A ribosome moves along mRNA, adding an amino acid to the end of a growing protein chain as it passes each codon.

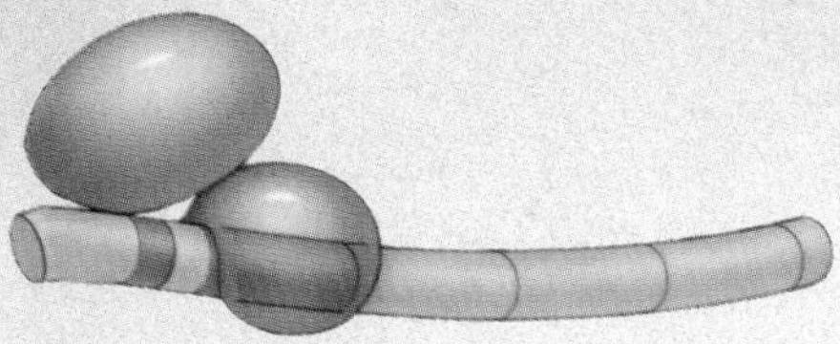

Regulating Gene Expression

8.11 Turning Genes Off and On

- Most genes are regulated by controlling the rate at which they are transcribed.
- Bacteria shut off genes by attaching a repressor protein on or near the promoter, so the polymerase is unable to bind.
- Eukaryotes can use many regulatory proteins simultaneously.

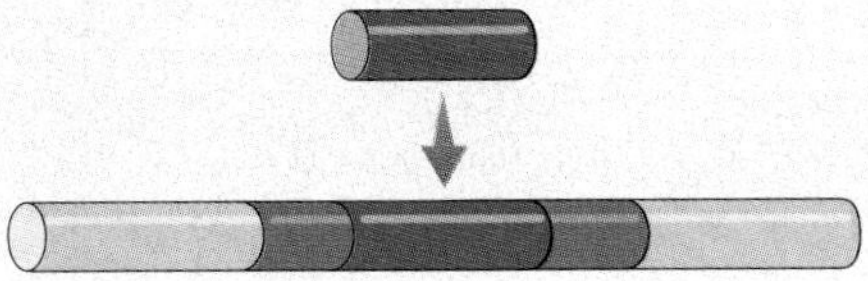

Altering the Genetic Message

8.12 Mutation
8.13 Kinds of Mutation
8.14 Cancer and Mutation

- A mutation is a change in the hereditary message.
- Mutations that change one or only a few nucleotides are called point mutations. They may arise as the result of errors in pairing during DNA replication, ultraviolet radiation, or chemical mutagens.

8.1 The Griffith Experiment

As we learned in chapters 6 and 7, chromosomes contain genes, which, in turn, contain hereditary information. However, Mendel's work left a key question unanswered: What *is* a gene? When biologists began to examine chromosomes in their search for genes, they soon learned that chromosomes are made of two kinds of macromolecules, both of which you encountered in chapter 3: **proteins** (long chains of *amino acid subunits* linked together in a string) and **DNA** (deoxyribonucleic acid) (long chains of *nucleotide* subunits linked together in a string). It was possible to imagine that either of the two was the stuff that genes are made of—information might be stored in a sequence of different amino acids, or of different nucleotides. But which one is the stuff of genes, protein or DNA? This question was answered clearly in a variety of different experiments, all of which shared the same basic design: If you separate the DNA in an individual's chromosomes from the protein, which of the two materials is able to change another individual's genes?

In 1928, British microbiologist Frederick Griffith made a series of unexpected observations while experimenting with pathogenic (disease-causing) bacteria. When he infected mice with a virulent strain of *Streptococcus pneumoniae* bacteria (then known as *Pneumococcus*), the mice died of blood poisoning. However, when he infected similar mice with a mutant strain of *S. pneumoniae* that lacked the virulent strain's polysaccharide coat, the mice showed no ill effects. The coat was apparently necessary for infection. The normal pathogenic form of this bacterium is referred to as the S form because it forms smooth colonies in a culture dish. The mutant form, which lacks an enzyme needed to manufacture the polysaccharide coat, is called the R form because it forms rough colonies.

To determine whether the polysaccharide coat itself had a toxic effect, Griffith injected dead bacteria of the virulent S strain into mice; the mice remained perfectly healthy. Finally, he injected mice with a mixture containing dead S bacteria of the virulent strain and live, coatless R bacteria, each of which by itself did not harm the mice (figure 8.1). Unexpectedly, the mice developed disease symptoms and many of them died. The blood of the dead mice was found to contain high levels of live, virulent *Streptococcus* type S bacteria, which had surface proteins characteristic of the live (previously R) strain. Somehow, the information specifying the polysaccharide coat had passed from the dead, virulent S bacteria to the live, coatless R bacteria in the mixture, permanently transforming the coatless R bacteria into the virulent S variety.

8.1 Hereditary information can pass from dead cells to living ones and transform them.

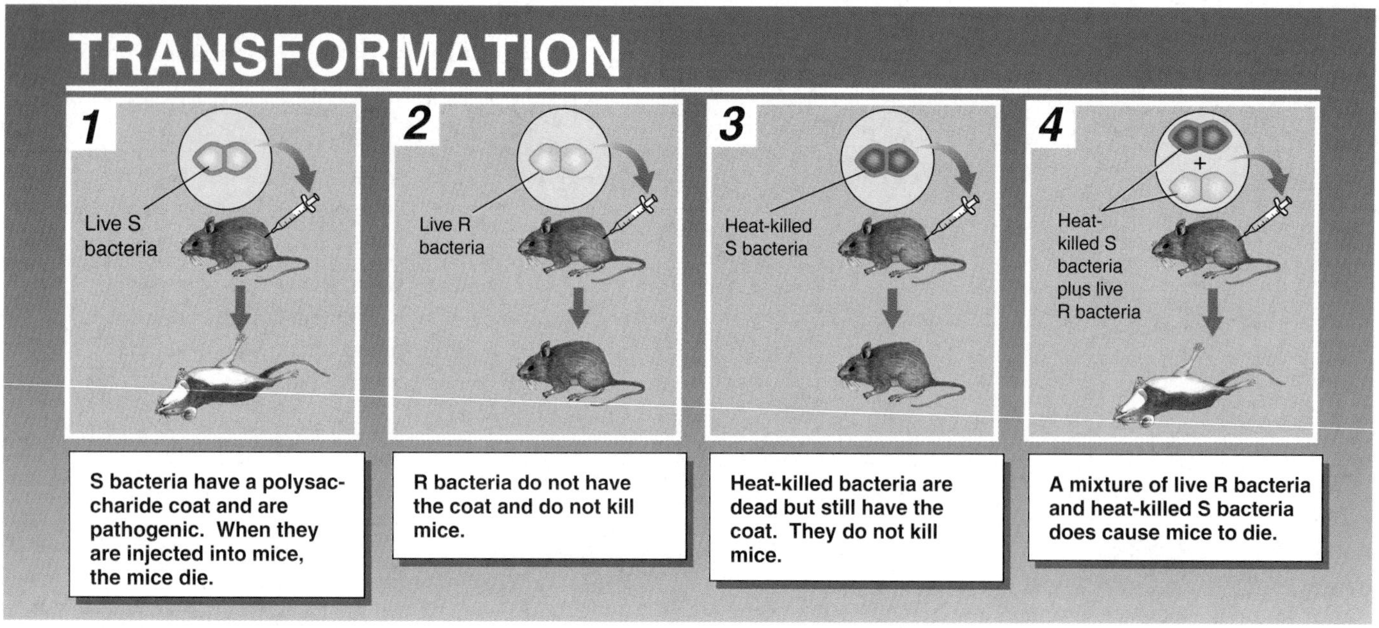

TRANSFORMATION
1
Live S bacteria
S bacteria have a polysaccharide coat and are pathogenic. When they are injected into mice, the mice die.
2
Live R bacteria
R bacteria do not have the coat and do not kill mice.
3
Heat-killed S bacteria
Heat-killed bacteria are dead but still have the coat. They do not kill mice.
4
+
Heat-killed S bacteria plus live R bacteria
A mixture of live R bacteria and heat-killed S bacteria does cause mice to die.

Figure 8.1 How Griffith discovered transformation.

Transformation, the movement of a gene from one organism to another, provided some of the key evidence that DNA is the genetic material. Griffith found that extracts of dead pathogenic strains of the bacterium *Streptococcus pneumoniae* can "transform" live harmless strains into live pathogenic strains. Later, Avery analyzed the extract and demonstrated that the material that had passed from dead to living bacteria was DNA.

8.2 The Avery Experiments

The agent responsible for transforming *Streptococcus* went undiscovered until 1944. In a classic series of experiments, Oswald Avery and his coworkers Colin MacLeod and Maclyn McCarty characterized what they referred to as the "transforming principle." They first prepared the mixture of dead S *Streptococcus* and live R *Streptococcus* that Griffith had used. Then Avery and his colleagues removed as much of the protein as they could from their preparation, eventually achieving 99.98% purity. Despite the removal of nearly all protein, the transforming activity was not reduced (figure 8.2). Moreover, the properties of the transforming principle resembled those of DNA in several ways:

Same chemistry as DNA. When the purified principle was analyzed chemically, the array of elements agreed closely with DNA.

Same behavior as DNA. In an ultracentrifuge, the transforming principle migrated like DNA; in electrophoresis and other chemical and physical procedures, it also acted like DNA.

Not affected by lipid and protein extraction. Extracting the lipid and protein from the purified transforming principle did not reduce its activity.

Not destroyed by protein- or RNA-digesting enzymes. Protein-digesting enzymes did not affect the principle's activity; nor did RNA-digesting enzymes.

Destroyed by DNA-digesting enzymes. The DNA-digesting enzyme destroyed all transforming activity.

The evidence was overwhelming. They concluded that "a nucleic acid of the deoxyribose type is the fundamental unit of the transforming principle of *Pneumococcus* Type III"—in essence, that DNA is the hereditary material.

8.2 Avery's experiments demonstrate conclusively that DNA is the hereditary material.

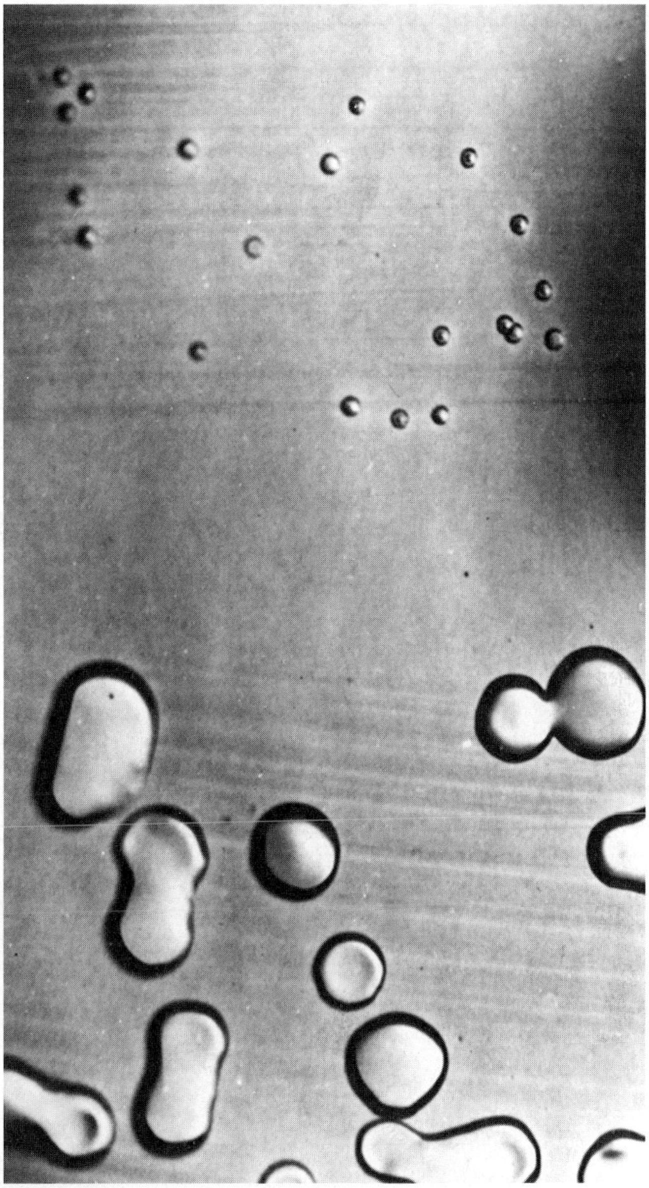

Figure 8.2 Transformation of *Streptococcus*.
This photo, from the original publication of Avery and his coworkers, shows the transformation of nonpathogenic R *Streptococcus* (the small colonies) to pathogenic S *Streptococcus* (the large colonies) in an extract prepared from heat-killed S *Streptococcus*.

8.3 The Hershey-Chase Experiment

Avery's result was not widely appreciated at first, because most biologists still preferred to think that genes were made of proteins. In 1952, however, a simple experiment carried out by Alfred Hershey and Martha Chase was impossible to ignore. The team studied the genes of viruses that infect bacteria. These viruses attach themselves to the surface of bacterial cells and inject their genes into the interior; once inside, the genes take over the genetic machinery of the cell and order the manufacture of hundreds of new viruses. When mature, the progeny viruses burst out to infect other cells. These bacteria-infecting viruses have a very simple structure: a core of DNA surrounded by a coat of protein.

In this experiment, Hershey and Chase used radioactive isotopes to "label" the DNA and protein of the viruses. In one preparation, the viruses were grown so that their DNA contained radioactive phosphorus (^{32}P); in another preparation, the viruses were grown so that their protein coats contained radioactive sulfur (^{35}S). The two preparations were then mixed together, and the mixture was allowed to infect bacteria. After a few minutes Hershey and Chase shook the suspension forcefully to dislodge attacking viruses from the surface of bacteria, used a rapidly spinning centrifuge to isolate the bacteria, and then asked a very simple question: What did the viruses inject into the bacterial cells, protein or DNA? They found that the interiors of the bacterial cells contained the ^{32}P label but not the ^{35}S label. The conclusion is clear: the genes that viruses use to specify new generations of viruses are made of DNA and not protein (figure 8.3).

8.3 The hereditary material of bacteriophages is DNA and not protein.

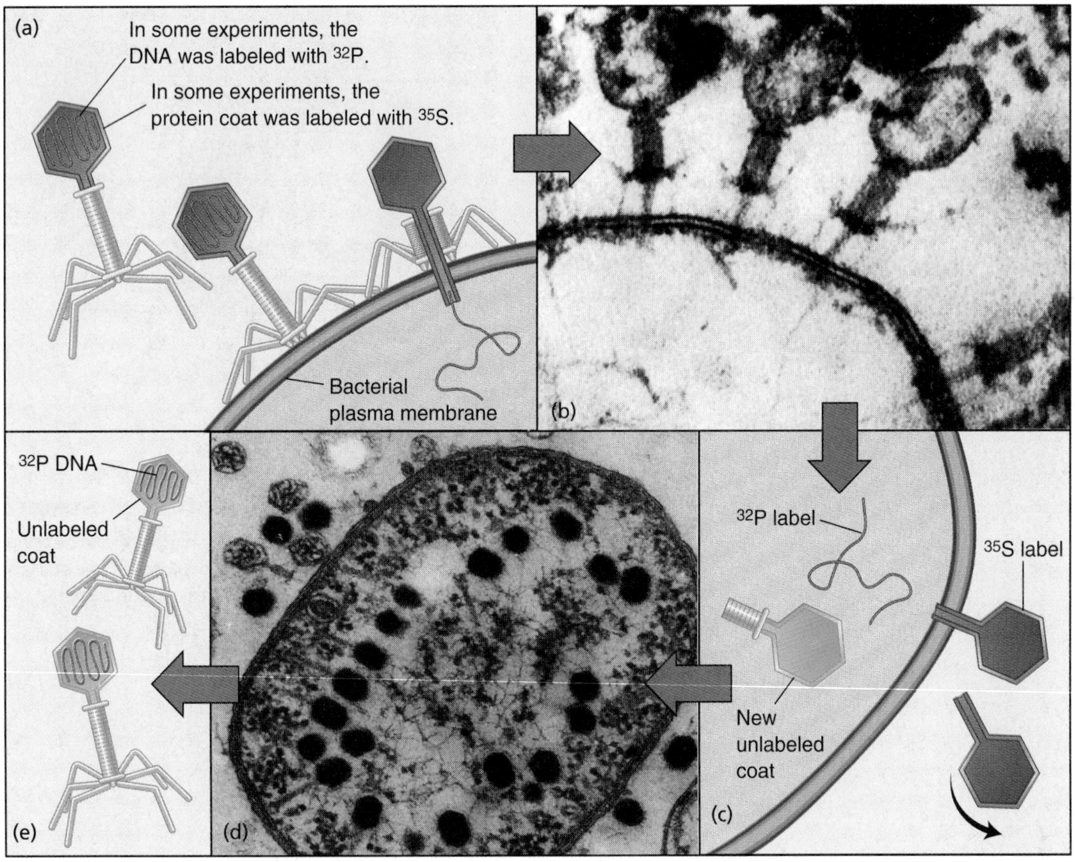
(a)
In some experiments, the DNA was labeled with 32P.
In some experiments, the protein coat was labeled with 35S.
Bacterial plasma membrane
(b)
32P DNA
Unlabeled coat
32P label
35S label
New unlabeled coat
(c)
(d)
(e)

Figure 8.3 The Hershey-Chase experiment.
The experiment that convinced most biologists that DNA is the genetic material was carried out soon after World War II, when radioactive isotopes were first becoming commonly available to researchers. Hershey and Chase used different radioactive labels to "tag" and track protein and DNA. They found that when bacterial viruses inserted their genes into bacteria to guide the production of new viruses, it was DNA and not protein that was inserted. Clearly the virus DNA, not the virus protein, was responsible for directing the production of new viruses.

8.4 The Fraenkel-Conrat Experiment

Some viruses contain RNA instead of DNA, and yet they manage to reproduce quite satisfactorily. What genetic material do *they* use?

In 1957, Heinz Fraenkel-Conrat and his coworkers answered this question for two RNA-containing viruses: tobacco mosaic virus (TMV), which infects the leaves of tobacco plants, and Holmes ribgrass virus (HRV), which infects grass. TMV, the better studied, consists of a single strand of RNA 6,390 nucleotides long, surrounded by a protein coat of 2,130 identical subunits. The protein can be separated from the RNA by a simple chemical treatment. When this is done, the isolated RNA is infective, while the protein is not, suggesting that RNA is the hereditary material of these viruses. If the dissociated RNA and protein subunits are mixed together in solution, they recombine to form fully active virus particles.

Fraenkel-Conrat and his coworkers further investigated this conclusion with a simple but compelling exchange experiment. First they chemically dissociated each virus, separating its protein coat from its RNA. They then manufactured hybrid viruses by combining the protein of one with the RNA of the other. When they infected healthy tobacco plants with a hybrid virus composed of HRV RNA and TMV protein, the tobacco leaves developed lesions characteristic of HRV (figure 8.4). Clearly, the hereditary properties of the virus were determined by the nucleic acid in its core, not the protein in its coat.

Retroviruses

Later studies have shown that many other viruses contain RNA rather than DNA. When DNA viruses infect a cell, their DNA is often inserted into the host cell's DNA as if they were the cell's own genes. Viruses containing RNA use a more indirect method. They first make an intermediate double-stranded form of DNA from the RNA, using a special kind of polymerase enzyme called reverse transcriptase. This DNA copy may then insert into the cell's DNA. Because the path of information flows from RNA to DNA rather than from DNA to RNA, these RNA viruses are called **retroviruses.** The HIV virus that causes acquired immunodeficiency syndrome (AIDS) is a retrovirus, as are many tumor-forming viruses. Transcription of the retrovirus RNA, necessary to produce new virus particles, takes place only after its DNA copy has been integrated into the host DNA. Thus, integration is an obligatory step in the life cycle of a retrovirus.

8.4 DNA is the genetic material for all cellular organisms and most viruses, although some viruses use RNA.

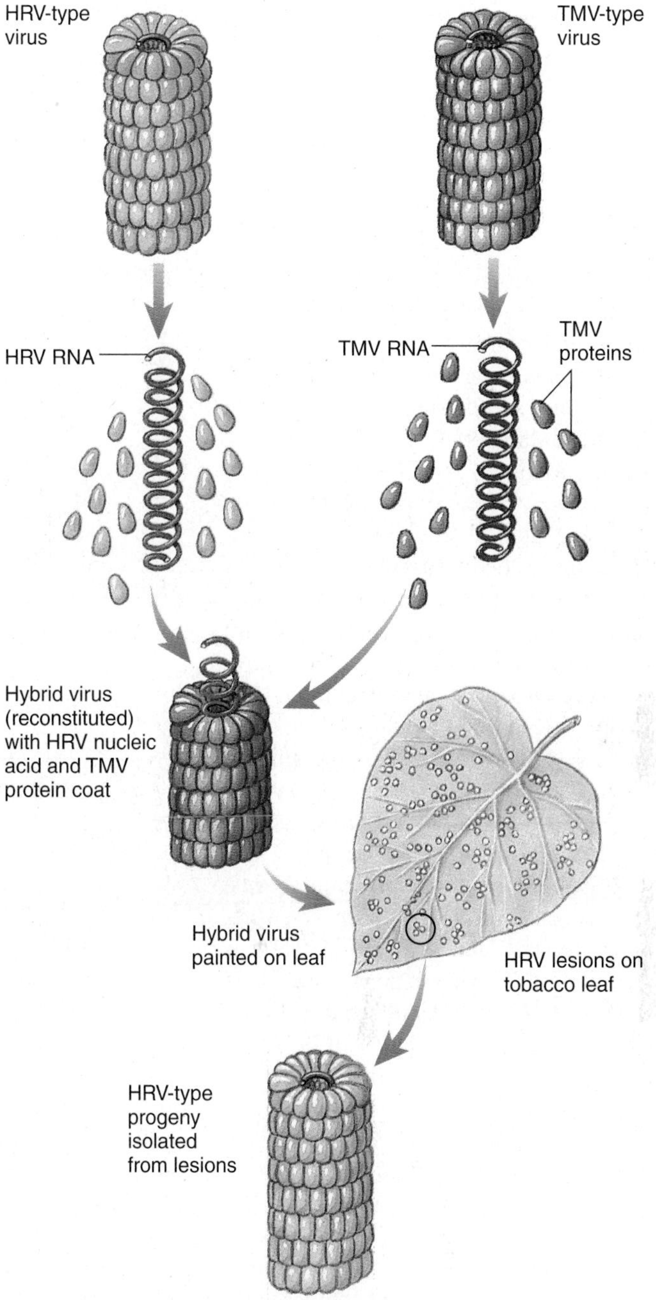

HRV-type virus
TMV-type virus
HRV RNA
TMV RNA
TMV proteins
Hybrid virus (reconstituted) with HRV nucleic acid and TMV protein coat
Hybrid virus painted on leaf
HRV lesions on tobacco leaf
HRV-type progeny isolated from lesions

Figure 8.4 Fraenkel-Conrat's virus reconstitution experiment.

Both TMV and HRV are plant RNA viruses that infect tobacco plants, causing lesions on the leaves. In this experiment, TMV and HRV were both dissociated into protein and RNA. Then hybrid virus particles were produced by mixing the HRV RNA and the TMV protein. When the reconstituted virus particles were painted onto tobacco leaves, HRV-type lesions developed. From these lesions, normal HRV virus particles could be isolated in great numbers; no TMV viruses could be isolated from the lesions. Thus the RNA (HRV) and not the protein (TMV) contains the information necessary to specify the production of the viruses.

8.5 Discovering the Structure of DNA

DNA is a long, chainlike molecule made up of subunits called **nucleotides.** Each nucleotide has three parts: a central sugar, a phosphate (PO_4) group, and an organic base. The sugar and the phosphate group are the same in every nucleotide of DNA, but there are four different kinds of bases, two large ones with double-ring structures and two small ones with single rings (figure 8.5). The large bases, called **purines,** are **A** (adenine) and **G** (guanine). The small bases, called **pyrimidines,** are **C** (cytosine) and **T** (thymine). A key observation, made by Erwin Chargaff, was that DNA molecules always had equal amounts of purines and pyrimidines. In fact, with slight variations due to imprecision of measurement, the amount of A always equals the amount of T, and the amount of G always equals the amount of C. This observation (A = T, G = C), known as **Chargaff's rule,** suggested that DNA had a regular structure.

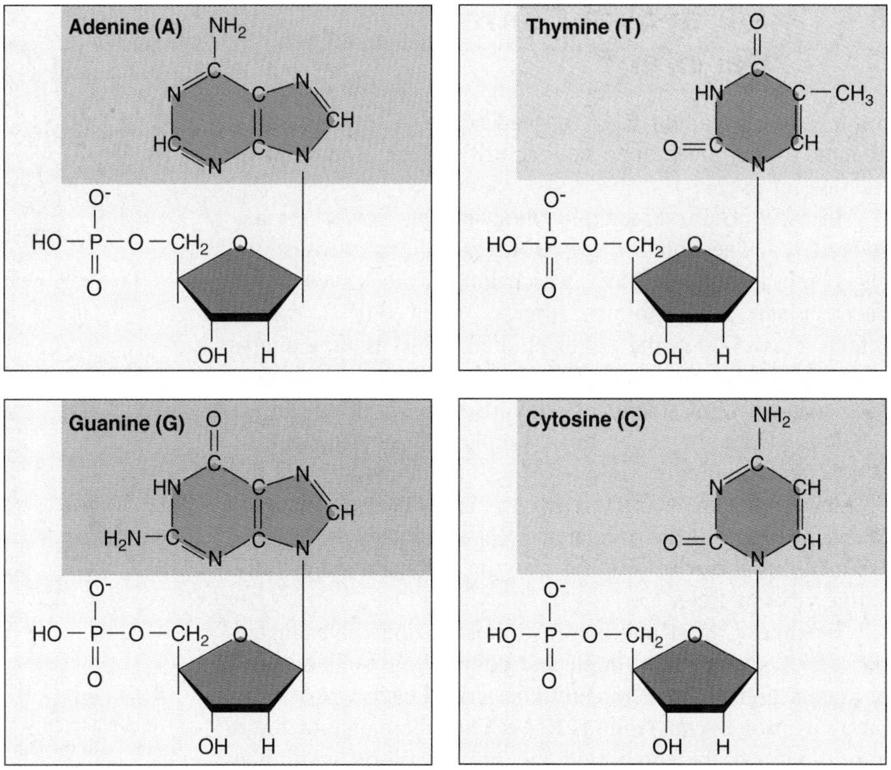
Adenine (A)
Thymine (T)
Guanine (G)
Cytosine (C)
HO – P – O – CH2
OH H

Figure 8.5 The four nucleotide subunits that make up DNA.
The nucleotide subunits of DNA are composed of three elements: a central five-carbon sugar, a phosphate group, and an organic, nitrogen-containing base.

The significance of Chargaff's rule became clear in 1953, when scientists began to study the structure of DNA using X-ray diffraction techniques. In these experiments, DNA molecules are bombarded with X-ray beams, and when individual rays encounter atoms, their paths are bent or diffracted; each atomic encounter creates a pattern on photographic film like the ripples created by tossing a rock into a smooth lake. The first of these studies, carried out by Rosalind Franklin at Kings College, London, suggested that the DNA molecule had the shape of a coiled spring, a form called a **helix.**

Two workers at Cambridge University, Francis Crick and James Watson, learned informally of Franklin's results and, using Tinkertoy models of the bases, deduced the true structure of DNA (figure 8.6): The DNA molecule is a **double helix,** a winding staircase of two strands whose bases face one another. Chargaff's rule is a direct reflection of this structure—every bulky purine on one strand is paired with a slender pyrimidine on the other strand. Specifically, A pairs with T, and G pairs with C (figure 8.7). Because hydrogen bonds can form between the **base pairs,** the molecule keeps a constant thickness.

Figure 8.6 Watson and Crick.
In 1953 Watson and Crick deduced the structure of DNA. James Watson (peering up at their homemade model of the DNA molecule) was a young American postdoctoral student, and Francis Crick (pointing) was an English scientist.

8.5 The DNA molecule has two strands of nucleotides that form hydrogen bonds with each other.

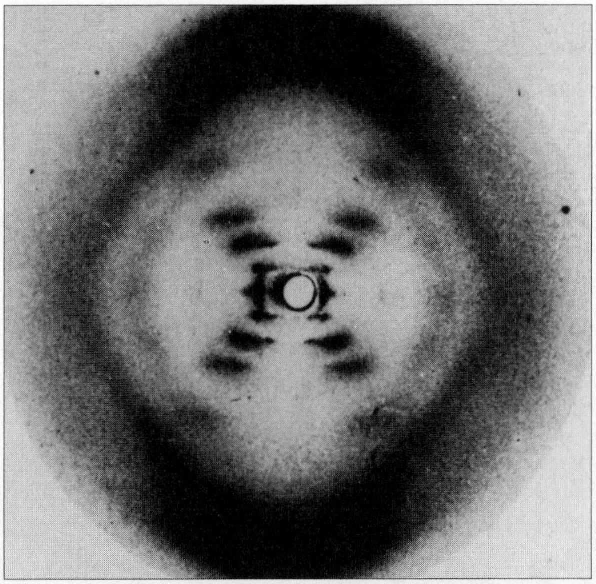

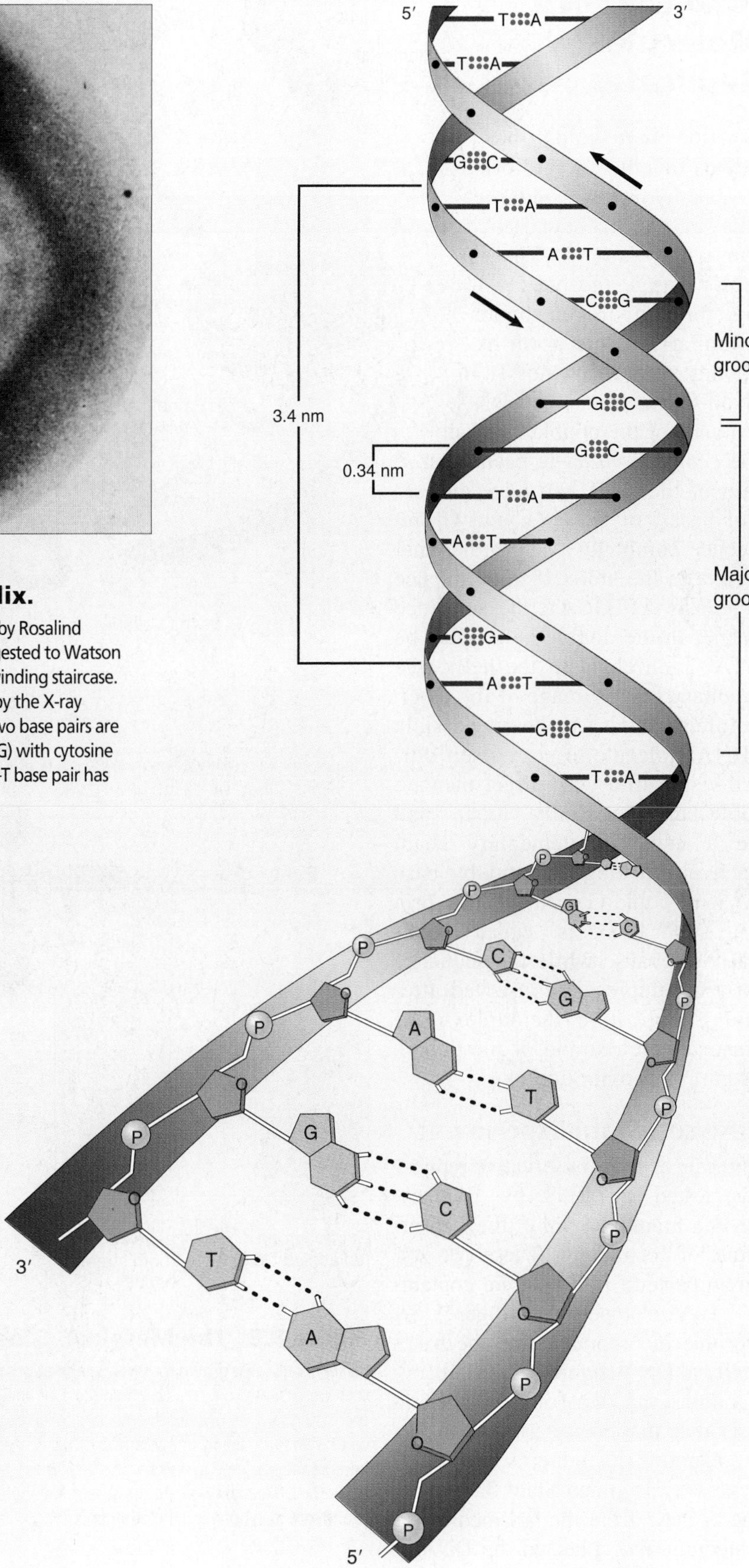
5′
3′
T:::A
T:::A
G:::C
T:::A
A:::T
C:::G
Minor groove
G:::C
3.4 nm
G:::C
0.34 nm
T:::A
A:::T
Major groove
C:::G
A:::T
G:::C
T:::A
P
G
C
A
T
3′
5′

Figure 8.7 The DNA double helix.

This X-ray diffraction photograph was made in 1953 by Rosalind Franklin in the laboratory of Maurice Wilkins. It suggested to Watson and Crick that the DNA molecule was a helix, like a winding staircase. The dimensions of the double helix were suggested by the X-ray diffraction studies. In a DNA duplex molecule, only two base pairs are possible: adenine (A) with thymine (T) and guanine (G) with cytosine (C). A G–C base pair has three hydrogen bonds; an A–T base pair has only two.

8.6 How the DNA Molecule Replicates

The attraction that holds the two DNA strands together is the formation of weak hydrogen bonds between the bases that face each other from the two strands. That is why A pairs with T and not C—it can only form hydrogen bonds with T. Similarly, G can form hydrogen bonds with C but not T. In the Watson-Crick model of DNA, the two strands of the double helix are said to be *complementary* to each other. One chain of the helix can have any sequence of bases, of A, T, G, and C, but this sequence completely determines that of its partner in the helix. If the sequence of one chain is ATTGCAT, the sequence of its partner in the double helix must be TAACGTA. Each chain in the helix is a complementary mirror image of the other. This **complementarity** makes it possible for the DNA molecule to copy itself during cell division in a very direct manner. The double helix need only "unzip" and assemble a new complementary chain along each naked single strand by base pairing A with T and G with C. This form of DNA replication is called **semiconservative,** because while the sequence of the original duplex is conserved after one round of replication, the duplex itself is not. Instead, each strand of the duplex becomes part of another duplex.

The Meselson-Stahl Experiment

The hypothesis of semiconservative replication was tested in 1958 by Matthew Meselson and Franklin Stahl of the California Institute of Technology. These two scientists grew bacteria in a medium containing the heavy isotope of nitrogen, ^{15}N, which became incorporated into the bases of the bacterial DNA (figure 8.8). After several generations, the DNA of these bacteria was denser than that of bacteria grown in a medium containing the lighter isotope of nitrogen, ^{14}N. Meselson and Stahl then transferred the bacteria from the ^{15}N medium to the ^{14}N medium and collected the DNA at various intervals.

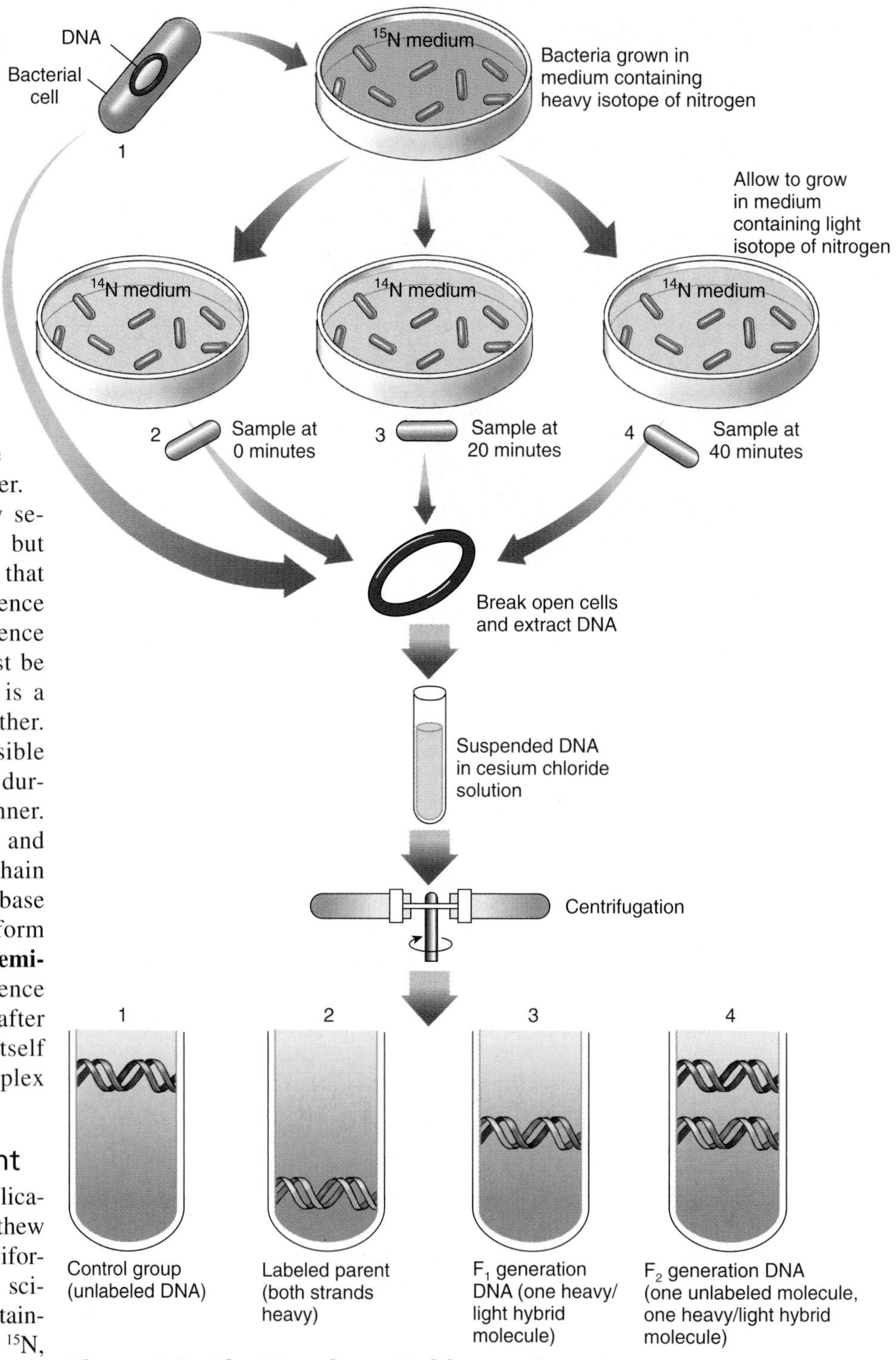

DNA
Bacterial cell
1
^{15}N medium
Bacteria grown in medium containing heavy isotope of nitrogen
Allow to grow in medium containing light isotope of nitrogen
^{14}N medium
^{14}N medium
^{14}N medium
2
Sample at 0 minutes
3
Sample at 20 minutes
4
Sample at 40 minutes
Break open cells and extract DNA
Suspended DNA in cesium chloride solution
Centrifugation
1
2
3
4
Control group (unlabeled DNA)
Labeled parent (both strands heavy)
F$_1$ generation DNA (one heavy/light hybrid molecule)
F$_2$ generation DNA (one unlabeled molecule, one heavy/light hybrid molecule)

Figure 8.8 The Meselson-Stahl experiment.

Bacterial cells were grown for several generations in a medium containing a heavy isotope of nitrogen (^{15}N) and then were transferred to a new medium containing the normal lighter isotope (^{14}N). At various times thereafter, samples of the bacteria were collected, and their DNA was dissolved in a solution of cesium chloride, which was spun rapidly in a centrifuge. The labeled and unlabeled DNA settled in different areas of the tube because they differed in weight. The DNA with two heavy strands settled down toward the bottom of the tube. The DNA with two light strands settled higher up in the tube. The DNA with one heavy and one light strand settled in between the other two.

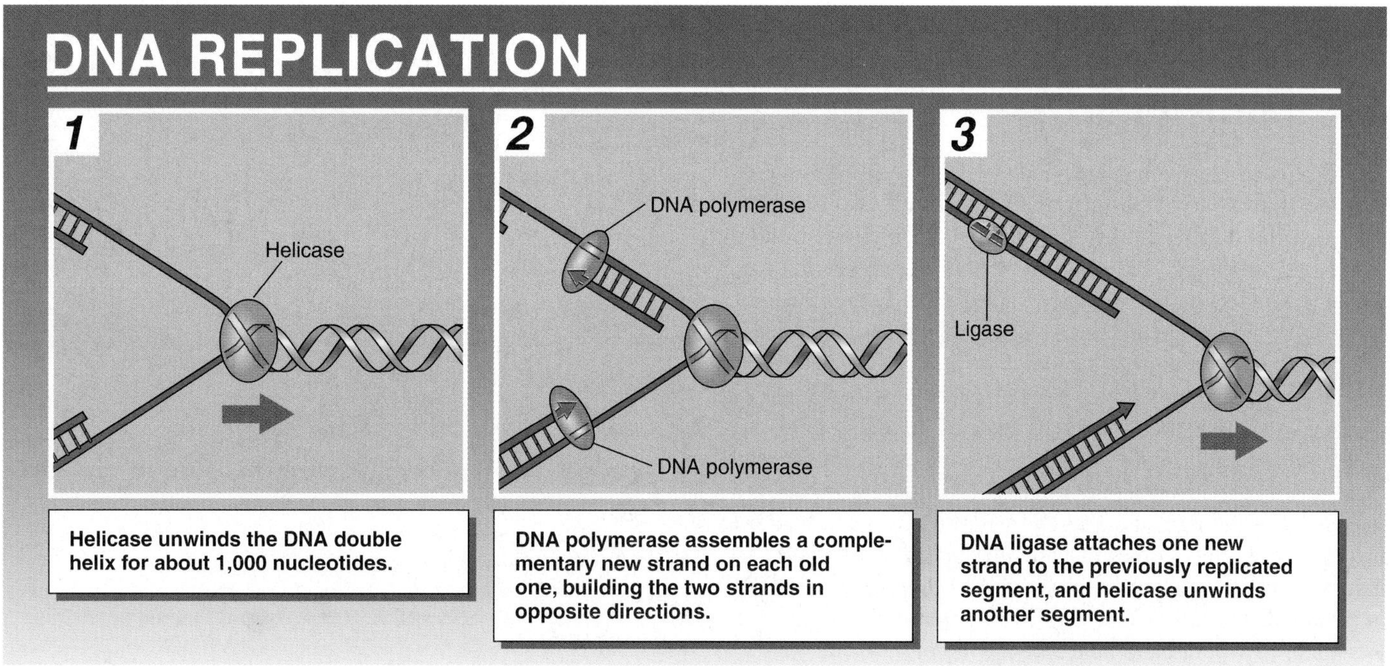
DNA REPLICATION
1
Helicase
Helicase unwinds the DNA double helix for about 1,000 nucleotides.
2
DNA polymerase
DNA polymerase
DNA polymerase assembles a complementary new strand on each old one, building the two strands in opposite directions.
3
Ligase
DNA ligase attaches one new strand to the previously replicated segment, and helicase unwinds another segment.

Figure 8.9 How DNA replication works.

By dissolving the DNA they had collected in a heavy salt called cesium chloride and then spinning the solution at very high speeds in an ultracentrifuge, Meselson and Stahl were able to separate DNA strands of different densities. The centrifugal forces caused the cesium ions to migrate toward the bottom of the centrifuge tube, creating a gradient of cesium concentration, and thus of density. Each DNA strand floats or sinks in the gradient until it reaches the position where its density exactly matches the density of the cesium there. Because ^{15}N strands are denser than ^{14}N strands, they migrate farther down the tube to a denser region of cesium.

The DNA collected immediately after the transfer was all dense. However, after the bacteria completed their first round of DNA replication in the ^{14}N medium, the density of their DNA had decreased to a value intermediate between ^{14}N-DNA and ^{15}N-DNA. After the second round of replication, two density classes of DNA were observed, one intermediate and one equal to that of ^{14}N-DNA.

Meselson and Stahl interpreted their results as follows: after the first round of replication, each daughter DNA duplex was a hybrid possessing one of the heavy strands of the parent molecule and one light strand; when this hybrid duplex replicated, it contributed one heavy strand to form another hybrid duplex and one light strand to form a light duplex. Thus, this experiment clearly confirmed the prediction of the Watson-Crick model that DNA replicates in a semiconservative manner.

How DNA Copies Itself

The copying of DNA before cell division is called **DNA replication** and is overseen by an enzyme called *DNA polymerase.* After an enzyme called *helicase* unwinds the DNA double helix, DNA polymerase reads along each naked single strand and adds the correct complementary nucleotide (A with T, G with C) at each position as it moves, creating a complementary strand. *DNA ligase* joins the ends of newly synthesized segments of DNA (figure 8.9). The place where the parent DNA molecule becomes unzipped is called a **replication fork.** At replication forks, the polymerase very actively shuttles up one strand and down the other. Chromosomes each contain a single, very long molecule of DNA, but it is too long to copy conveniently all the way from one end to the other with a single replication fork. Each chromosome is instead copied in segments; each zone of about 100,000 nucleotides has its own replication fork.

The enormous amount of DNA that resides within the cells of your body represents a long series of DNA replications, starting with the DNA of a single cell—the fertilized egg. Living cells have evolved many mechanisms to avoid errors during DNA replication and to preserve the DNA from damage. These mechanisms of **DNA repair** proofread the strands of each daughter cell against one another for accuracy and correct any mistakes. But the proofreading is not perfect. If it were, no mistakes would occur, no variation in gene sequence would result, and evolution would come to a halt.

8.6 The basis for the great accuracy of DNA replication is complementarity. DNA's two strands are complementary mirror images of each other, so either one can be used as a template to reconstruct the other.

8.7 Transcription

The discovery that genes are made of DNA left unanswered the question of how the information in DNA is used. How does a string of nucleotides in a spiral molecule determine if you have red hair? We now know that the information in DNA is arrayed in little blocks, like entries in a dictionary, each block a gene specifying a protein. These proteins determine what a particular cell will be like.

Just as an architect protects building plans from loss or damage by keeping them safe in a central place and issuing only blueprint copies to on-site workers, so your cells protect their DNA instructions by keeping them safe within a central DNA storage area, the nucleus. The DNA never leaves the nucleus. Instead, "blueprint" copies of particular genes within the DNA instructions are sent out into the cell to direct the assembly of proteins. These working copies of genes are made of ribonucleic acid (RNA) rather than DNA. Recall that RNA is the same as DNA except that the sugars in RNA have an extra oxygen atom and T is replaced by a similar pyrimidine base called U, uracil (see figure 3.28). The path of information is thus:

DNA → RNA → protein

This information path is often called the *central dogma,* because it describes the key organization used by your cells to express their genes (figure 8.10). A cell uses three kinds of RNA in the synthesis of proteins: messenger RNA (mRNA), ribosomal RNA (rRNA), and transfer RNA (tRNA).

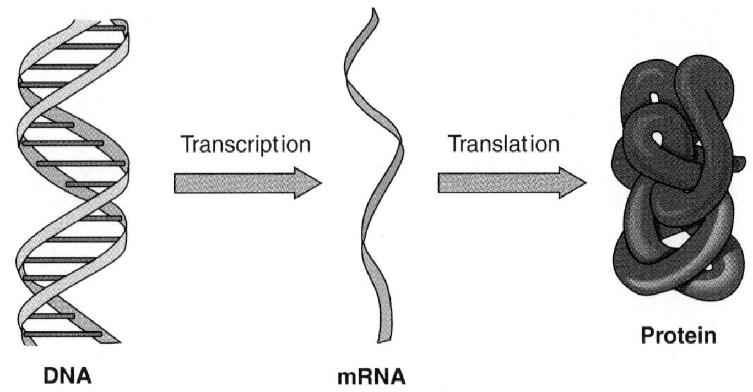
Transcription
Translation
Protein
DNA
mRNA

Figure 8.10 The central dogma of gene expression.
Through the production of mRNA (transcription) and the synthesis of proteins (translation), the information contained in DNA is expressed.

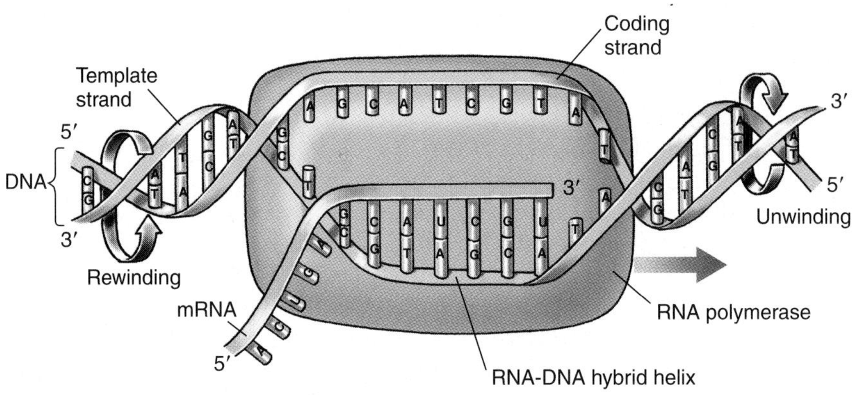
Coding strand
Template strand
5′
3′
DNA
3′
5′
Unwinding
Rewinding
mRNA
5′
3′
RNA polymerase
RNA-DNA hybrid helix

Figure 8.11 Transcription.
One of the strands of DNA functions as a template on which nucleotide building blocks are assembled into mRNA by RNA polymerase as it moves down the DNA strand.

The use of information in DNA to direct the production of particular proteins is called **gene expression.** Gene expression occurs in two stages: in the first stage, called **transcription,** an mRNA molecule is synthesized from a gene within the DNA; in the second stage, called **translation,** this mRNA is used to direct the production of a protein.

The Transcription Process

The RNA copy of a gene used in the cell to produce a protein is called **messenger RNA (mRNA)**—it is the messenger that conveys the information from the nucleus to the cytoplasm. The copying process that makes the mRNA is called transcription—just as monks in monasteries used to make copies of manuscripts by faithfully transcribing each letter, so enzymes within the nuclei of your cells make mRNA copies of your genes by faithfully complementing each nucleotide.

In your cells, the transcriber is a large and very sophisticated protein called **RNA polymerase.** It binds to one strand of a DNA double helix at a particular site called a *promoter* and then moves down the DNA strand like a train on a track. As it goes along the DNA "track," the polymerase pairs each nucleotide with its complementary RNA version (A with U, G with C), building an mRNA chain behind it as it moves down the DNA strand (figure 8.11).

8.7 Transcription is the production of an mRNA copy of a gene by the enzyme RNA polymerase.

8.8 The Genetic Code

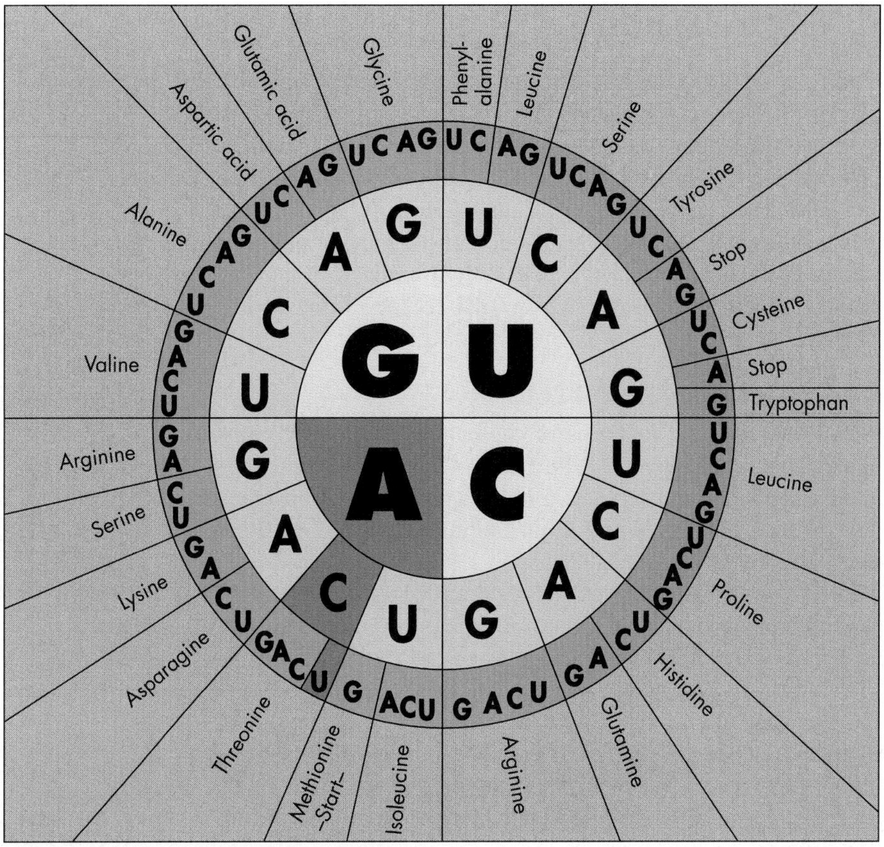

Glutamic acid
Glycine
Phenyl-
alanine
Leucine
Serine
Tyrosine
Stop
Cysteine
Stop
Tryptophan
Leucine
Proline
Histidine
Glutamine
Arginine
Isoleucine
Methionine
–Start–
Threonine
Asparagine
Lysine
Serine
Arginine
Valine
Alanine
Aspartic acid
G
U
A
C

Figure 8.12 The genetic code.

A codon consists of three nucleotides read from the middle of the circle and moving outward. For example, ACU codes threonine.

The essence of Mendelian genetics is that information determining hereditary traits, traits passed from parent to child, is encoded information. The information is written within the chromosomes in blocks called genes. Genes affect Mendelian traits by directing the production of particular proteins. The essence of gene expression, of using your genes, is reading the information encoded within DNA and using that information to direct the production of the correct protein.

To correctly read a gene, a cell must translate the information encoded in DNA into the language of proteins—that is, it must convert the *order* of the gene's nucleotides into the order of amino acids in a protein. The rules that govern this translation are called the **genetic code.** They are very simple:

1. Each gene is read from a fixed starting position, a nucleotide sequence at its beginning called a promoter site where the RNA polymerase first binds to the DNA.
2. RNA polymerase moves down the DNA in steps that are three nucleotides long.
3. Each three-nucleotide block in the gene corresponds to a particular amino acid.

Special three-nucleotide sequences located at the end of genes say "stop." A three-nucleotide sequence on mRNA that corresponds to an amino acid is called a **codon.** Biologists worked out which codons correspond to which amino acids by trial-and-error experiments carried out in test tubes. In these experiments, investigators used artificial mRNAs to direct the synthesis of proteins in the tube and then looked to see the sequence of amino acids in the proteins. An mRNA that was a string of UUUUUU . . . , for example, produced a protein that was a string of phenylalanine (PHE) amino acids, telling investigators that the codon UUU corresponded to the amino acid PHE. The entire **genetic code dictionary** is presented in figure 8.12. Because at each position of a three-letter codon any of the four different nucleotides (A, U, G, C) may be used, there are 64 different possible three-letter codons ($4 \times 4 \times 4 = 64$) in the genetic code.

The genetic code is universal, the same in practically all organisms. GUC codes for valine in bacteria, in fruit flies, in eagles, and in your own cells. The only exceptions biologists have ever found to this rule is in the way in which cell organelles that contain DNA (mitochondria and chloroplasts) and a few microscopic protists read the "stop" codons. In every other instance, the same genetic code is employed by all living things.

> **8.8** The genetic code dictates how a particular nucleotide sequence specifies a particular amino acid sequence. The nucleotide sequence is read in three-base increments called codons. The genetic code determines which amino acid is associated with which codon.

8.9 Translation

The final result of the transcription process is the production of an mRNA copy of a gene. Like a photocopy, the mRNA can be used without damage or wear and tear on the original. After transcription of a gene is finished, the mRNA passes out of the nucleus into the cytoplasm through pores in the nuclear membrane. There, translation of the genetic message occurs. In **translation,** organelles called **ribosomes** use the mRNA produced by transcription to direct the synthesis of a protein.

The Protein-Making Factory

Ribosomes are the factories of the cell. Each is very complex, containing over 50 different proteins and several segments of **ribosomal RNA** (rRNA) (figure 8.13). Ribosomes use mRNA, the "blueprint" copies of nuclear genes, to direct the assembly of a protein.

Ribosomes are composed of two pieces, or subunits, one nested into the other like a fist in the palm of your hand (figure 8.14). The "fist" is the smaller of the two subunits. Its rRNA has a short nucleotide sequence exposed on the surface of the subunit. This exposed sequence is identical to a sequence called the leader region that occurs at the beginning of all genes. Because of this, an mRNA molecule binds to the exposed rRNA of the small subunit like a fly sticking to flypaper.

The Key Role of tRNA

Directly adjacent to the exposed rRNA sequence are three small pockets or dents, called the A, P, and E sites (discussed shortly), in the surface of the ribosome that have just the right shape to bind yet a third kind of RNA molecule, **transfer RNA (tRNA).** It is tRNA molecules that bring amino acids to the ribosome to use in making proteins. tRNA molecules are chains about 80 nucleotides long, folded into a compact shape, with a three-nucleotide sequence jutting out at one end and an amino acid attachment site to the other.

The three-nucleotide sequence, called the **anticodon** is very important: it is the complementary sequence to 1 of the 64 codons of the genetic code! Special enzymes match amino acids with their proper tRNAs, with the anticodon determining which amino acid will attach to a particular tRNA.

Because the first dent in the ribosome (the A site) is directly adjacent to where the mRNA binds to the rRNA, three nucleotides of the mRNA are positioned directly facing the jutting anticodon of the tRNA. Like the address on a letter, the anticodon ensures that an amino acid is delivered to its correct "address" on the mRNA where the ribosome is assembling the protein.

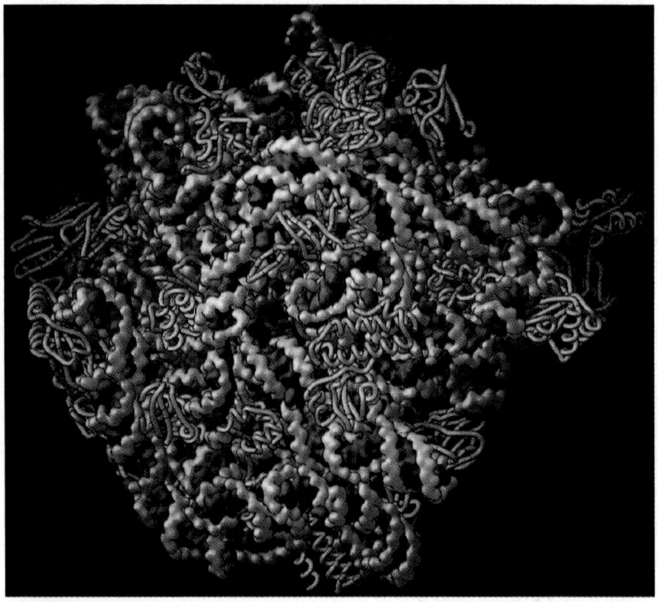

Figure 8.13 Ribosomes are very complex machines.

The complete atomic structure of a bacterial large ribosomal subunit has recently been determined at 2.4 Å resolution. The RNA of the subunit is shown in *gray* and the proteins in *gold*. The subunit's RNA is twisted into irregular shapes that fit together like a three-dimensional jigsaw puzzle. The chemical reactions that form the peptide bond in protein synthesis are carried out deep in the interior by ribosomal RNA. The ribosome is thus a "ribozyme." Proteins are absent from the active site but abundant everywhere on the surface. The proteins stabilize the structure by interacting with adjacent RNA strands.

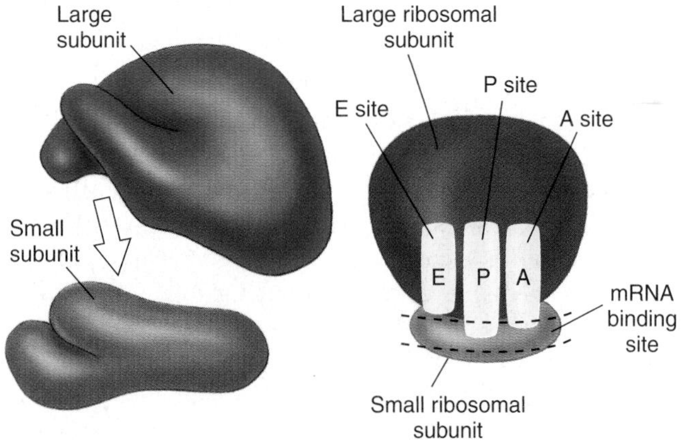

Large subunit
Small subunit
Large ribosomal subunit
E site
P site
A site
E
P
A
mRNA binding site
Small ribosomal subunit

Figure 8.14 A ribosome is composed of two subunits.

The smaller subunit fits into a depression on the surface of the larger one. The A, P, and E sites on the ribosome play key roles in protein synthesis.

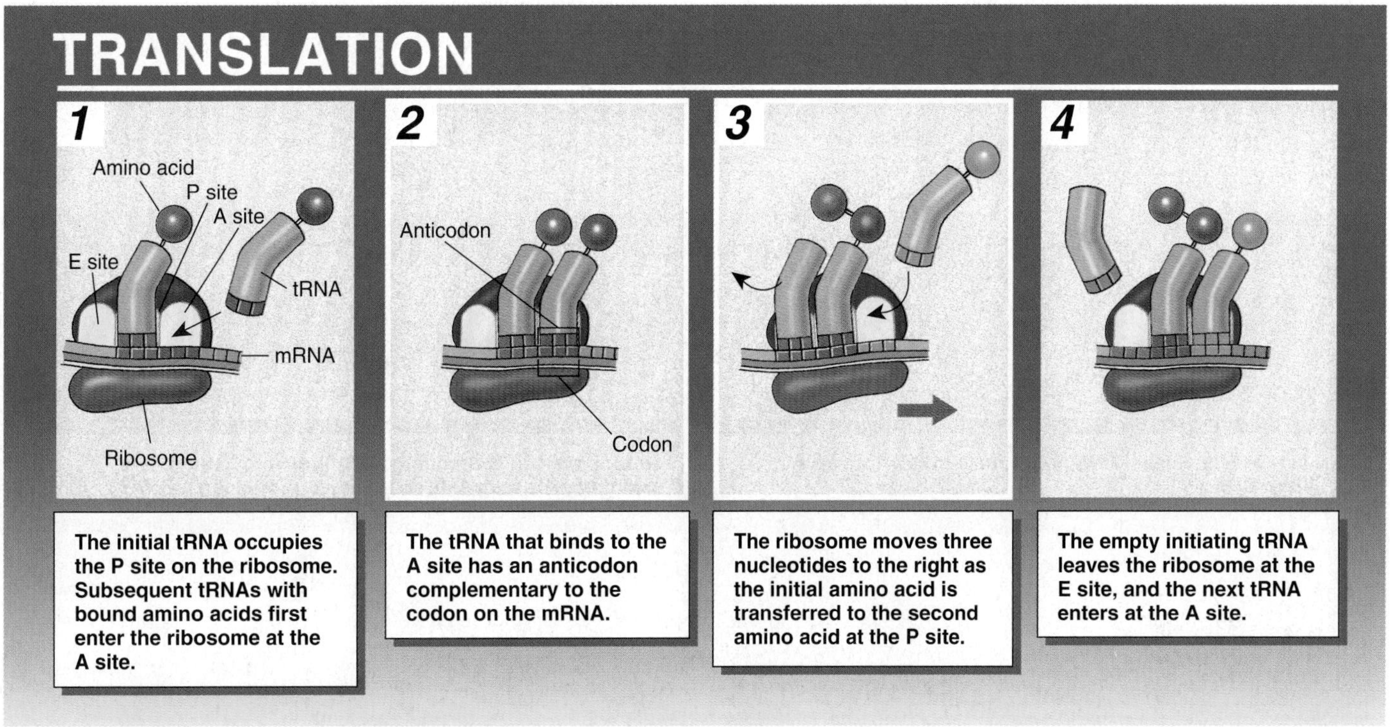
TRANSLATION
1
Amino acid
P site
A site
E site
tRNA
mRNA
Ribosome
2
Anticodon
Codon
3
4
The initial tRNA occupies the P site on the ribosome. Subsequent tRNAs with bound amino acids first enter the ribosome at the A site.
The tRNA that binds to the A site has an anticodon complementary to the codon on the mRNA.
The ribosome moves three nucleotides to the right as the initial amino acid is transferred to the second amino acid at the P site.
The empty initiating tRNA leaves the ribosome at the E site, and the next tRNA enters at the A site.

Figure 8.15 How translation works.

The mRNA strand acts as a template for tRNA molecules. The appropriate tRNA is selected and positioned by the ribosome, which moves along the mRNA in three-nucleotide steps. tRNAs bring amino acids into the ribosome at the A site. A peptide bond is formed between the incoming amino acid and the growing polypeptide chain at the P site, and the empty tRNAs leave the ribosome at the E site.

Making the Protein

Once an mRNA molecule has bound to the small ribosomal subunit, the other larger ribosomal subunit binds as well, forming a complete ribosome. The ribosome then begins the process of translation (figure 8.15). The mRNA begins to thread through the ribosome like a string passing through the hole in a donut. The mRNA passes through in short spurts, three nucleotides at a time, and at each burst of movement a new three-nucleotide codon is positioned opposite the A site in the ribosome, where tRNA molecules first bind (figure 8.16).

As each new tRNA brings in an amino acid to each new codon presented at the A site, the old tRNA paired with the previous codon is passed over to the P site and eventually to the E site, as the amino acid it carried is attached to the end of a growing amino acid chain. So as the ribosome proceeds down the mRNA, one tRNA

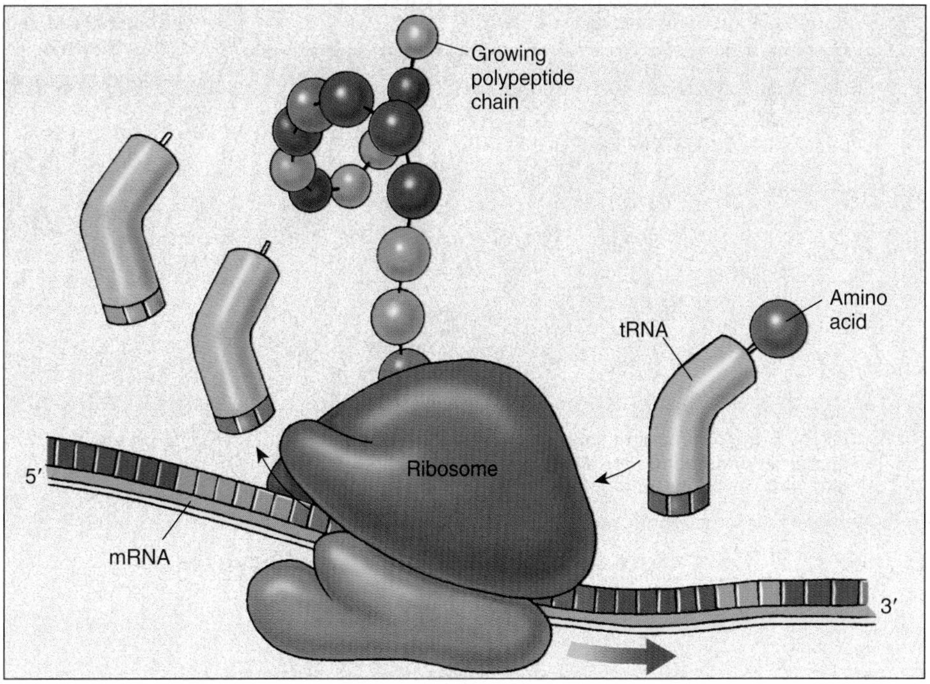
Growing polypeptide chain
Amino acid
tRNA
Ribosome
5′
mRNA
3′

Figure 8.16 Ribosomes guide the translation process.

tRNA binds to an amino acid as determined by the anticodon sequence. Ribosomes bind the loaded tRNAs to their complementary sequences on the strand of mRNA. tRNA adds its amino acid to the growing peptide chain, which is released as the completed protein.

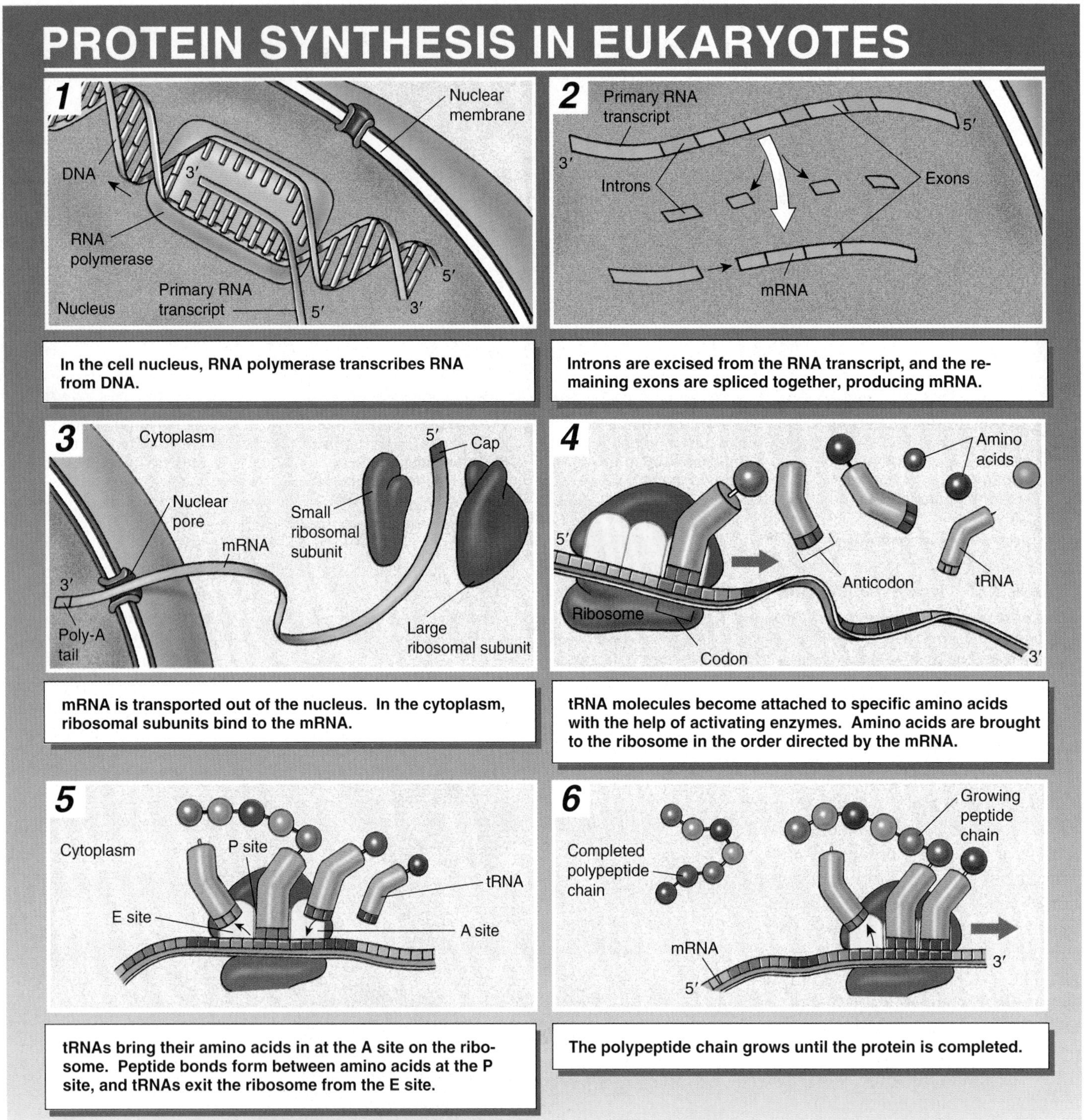
PROTEIN SYNTHESIS IN EUKARYOTES
1
Nuclear membrane
DNA
3′
RNA polymerase
5′
Primary RNA transcript
Nucleus
5′
3′
In the cell nucleus, RNA polymerase transcribes RNA from DNA.
2
Primary RNA transcript
5′
3′
Introns
Exons
mRNA
Introns are excised from the RNA transcript, and the remaining exons are spliced together, producing mRNA.
3
Cytoplasm
5′
Cap
Nuclear pore
Small ribosomal subunit
mRNA
3′
Poly-A tail
Large ribosomal subunit
mRNA is transported out of the nucleus. In the cytoplasm, ribosomal subunits bind to the mRNA.
4
Amino acids
5′
Anticodon
tRNA
Ribosome
Codon
3′
tRNA molecules become attached to specific amino acids with the help of activating enzymes. Amino acids are brought to the ribosome in the order directed by the mRNA.
5
Cytoplasm
P site
tRNA
E site
A site
tRNAs bring their amino acids in at the A site on the ribosome. Peptide bonds form between amino acids at the P site, and tRNAs exit the ribosome from the E site.
6
Growing peptide chain
Completed polypeptide chain
mRNA
5′
3′
The polypeptide chain grows until the protein is completed.

Figure 8.17 How protein synthesis works in eukaryotes.

after another is selected to match the sequence of mRNA codons, until the end of the mRNA sequence is reached. At this point, a codon is encountered for which there is no anticodon on any tRNA molecule. With nothing to fit into the tRNA site, the ribosome complex falls apart, and the newly made protein is released into the cell (figure 8.17).

8.9 Translation is the reading of mRNA by a ribosome, the sequence of mRNA codons dictating the assembly of a corresponding sequence of amino acids in a growing protein chain.

8.10 Architecture of the Gene

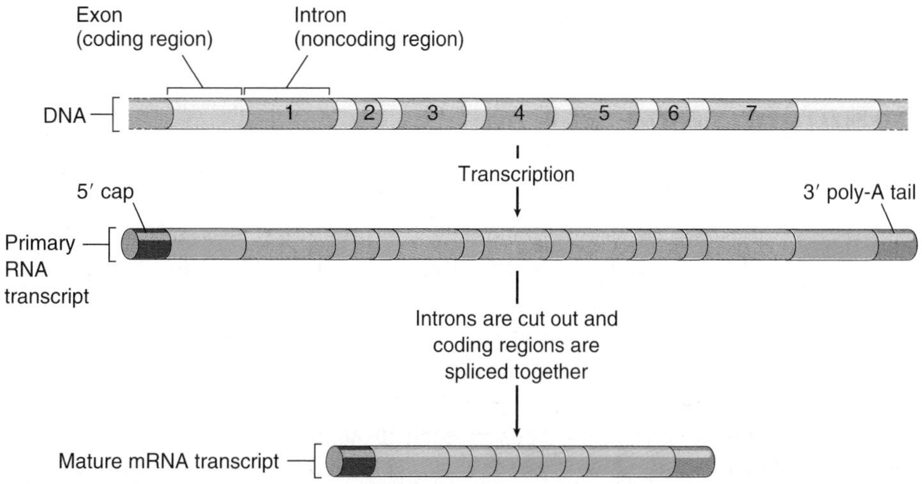

Exon
(coding region)
Intron
(noncoding region)
DNA
1
2
3
4
5
6
7
Transcription
5′ cap
3′ poly-A tail
Primary RNA transcript
Introns are cut out and coding regions are spliced together
Mature mRNA transcript

Figure 8.18 Processing a eukaryotic gene.
The gene shown here codes for a protein called ovalbumin. The ovalbumin gene and its primary transcript contain seven segments not present in the mRNA used by the ribosomes to direct the synthesis of the protein. These segments (introns) are removed by enzymes that cut them out and splice together the exons.

Introns

While it is tempting to think of a gene as simply the nucleotide version of a protein—an uninterrupted stretch of nucleotides that is read three at a time to make a chain of amino acids—this actually occurs only in bacteria. In eukaryotes, genes are fragmented. In these more complex genes, the DNA nucleotide sequences encoding the amino acid sequence of the protein (called **exons**) are interrupted frequently by extraneous "extra stuff" called **introns.** Imagine looking at an interstate highway from a satellite. Scattered randomly along the thread of concrete would be cars, some moving in clusters, others individually; most of the road would be bare. That is what a eukaryotic gene is like: scattered exons embedded within much longer sequences of introns. In humans, only 1% to 1.5% of the genome is devoted to the exons that encode proteins, while 24% is devoted to the noncoding introns within which these exons are embedded.

When a eukaryotic cell transcribes a gene, it first produces a **primary RNA transcript** of the entire gene. The primary transcript is then processed. First, enzyme-RNA complexes excise out the introns and join together the exons to form the shorter mature mRNA transcript that is actually translated into protein (figure 8.18). Because introns are excised from the RNA transcript before it is translated into protein, they do not affect the structure of the protein encoded by the gene in which they occur, despite the fact that introns represent over 90% of the nucleotide sequence of a typical human gene. In addition, enzymes add a *5′ cap* to the mRNA transcript which protects the 5′ end of mRNA from being degraded during its long journey through the cytoplasm. Another enzyme adds about 250 A nucleotides to the 3′ end of the transcript. Called a *3′ poly-A tail,* this long string of A nucleotides also protects the transcript from degradation and appears to make the transcript a better template for protein synthesis.

Why this crazy organization? It appears that many human genes can be spliced together in more than one way. In many instances, exons are not just random fragments, but rather functional modules. One exon encodes a straight stretch of protein, another a curve, yet another a flat place. Like mixing Tinkertoy parts, you can construct quite different assemblies by employing the same exons in different combinations and orders. With this sort of **alternative splicing,** the 30,000 genes of the human genome seem to encode as many as 120,000 different expressed messenger RNAs. It seems that added complexity in humans has been achieved not by gaining more gene parts (we have only twice as many genes as a fruit fly), but rather by learning new ways to put them together.

Gene Families

Eukaryotic genes have other unique characteristics. For example, everything we have said in this chapter assumes that chromosomes each carry one copy of each kind of gene—one to make the enzyme that breaks down lactose, for instance, and one to encode the protein hemoglobin, which carries oxygen in your blood from lungs to tissues. However, most eukaryotic genes exist in multiple copies, clusters of almost identical sequences called **multigene families.** Multigene families may contain as few as three or as many as several hundred versions of a gene.

Transposons: Jumping Genes

Other genes are very unusual in that they are repeated hundreds of thousands of times, scattered randomly about on the chromosomes. These genes, called **transposable sequences** or **transposons,** have the remarkable ability to move about from one chromosomal location to another. Once every few thousand cell divisions, a transposon simply picks up and moves elsewhere, jumping at random to a new location on the chromosome. Transposons appear to be molecular parasites. Fully 45% of the human genome is composed of transposable sequences.

8.10 The coding portions of most eukaryotic genes are embedded as exons within long sequences of noncoding introns. Many eukaryotic genes exist in multiple copies, some of which appear to have moved from one chromosomal location to another.

8.11 Turning Genes Off and On

Being able to translate a gene into a protein is only part of gene expression. Every cell must also be able to regulate when particular genes are used. Imagine if every instrument in a symphony played at full volume all the time, all the horns blowing full blast and each drum beating as fast and loudly as it could! No symphony plays that way, because music is more than noise—it is the controlled expression of sound. In the same way, growth and development is the controlled expression of genes, each brought into play at the proper moment to achieve precise and delicate effects.

Cells control the expression of their genes by saying *when* individual genes are to be transcribed. At the beginning of each gene are special regulatory sites that act as points of control. Specific regulatory proteins within the cell bind to these sites, turning transcription of the gene off or on.

Repressors

Many genes are "negatively" controlled: they are turned off except when needed. In these genes, the regulatory site is located between the place where the RNA polymerase binds to the DNA (called the **promoter** site) and the beginning edge of the gene. When a regulatory protein called a **repressor** is bound to its regulatory site, the **operator,** its presence blocks the movement of the polymerase toward the gene. Imagine if you sat down to eat dinner and someone placed a brick wall between your chair and the table—you could not begin your meal until the wall was removed, any more than the polymerase can begin transcribing the gene until the repressor protein is removed.

To turn on a gene whose transcription is blocked by a repressor, all that is required is to remove the repressor. Cells do this by binding special "signal" molecules to the repressor protein; the binding causes the repressor protein to contort into a shape that doesn't fit DNA, and it falls off, removing the barrier to transcription. A specific example demonstrating how repressor proteins work is the set of genes called the *lac* operon in the bacterium *Escherichia coli.* An **operon** is a segment of DNA containing a cluster of genes that are transcribed as a unit. The operon consists of both protein-encoding (structural) genes and associated regulatory elements—the operator and promoter. When an *E. coli* encounters the sugar lactose, the lactose binds to the repressor protein and induces a twist in its shape that causes it to fall from the DNA. RNA polymerase is no longer blocked, so it starts to transcribe the genes needed to break down the lactose to get energy (figure 8.19).

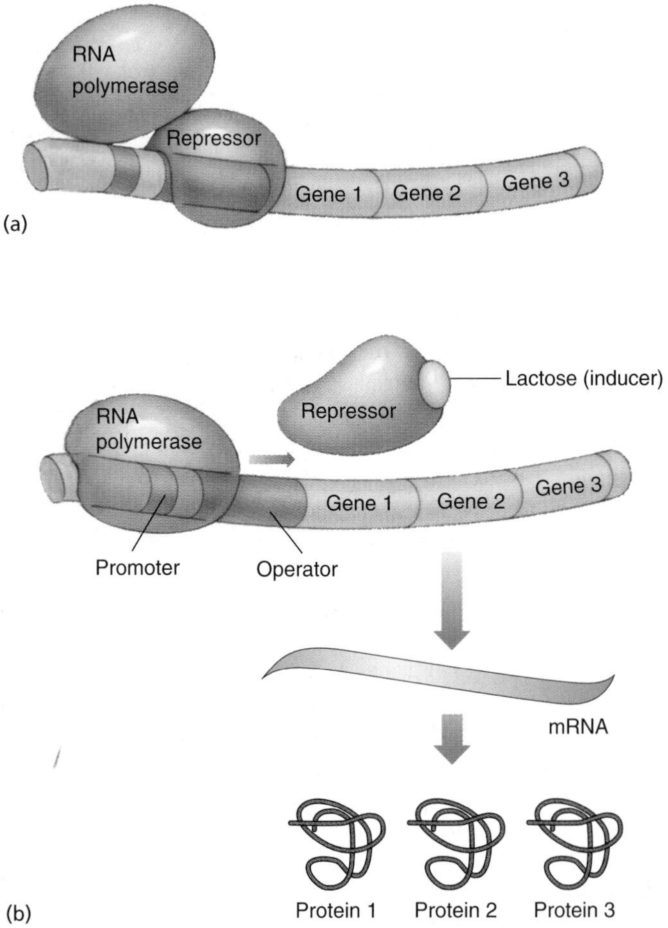
RNA polymerase
Repressor
Gene 1
Gene 2
Gene 3
(a)
Lactose (inducer)
Repressor
RNA polymerase
Gene 1
Gene 2
Gene 3
Promoter
Operator
mRNA
(b)
Protein 1
Protein 2
Protein 3

Figure 8.19 How the *lac* operon works.

(*a*) The *lac* operon is shut down ("repressed") when the repressor protein is bound to the operator site. Because promoter and operator sites overlap, polymerase and repressor cannot bind at the same time, any more than two people can sit in one chair. (*b*) The *lac* operon is transcribed ("induced") when lactose binding to the repressor protein changes its shape so that it can no longer sit on the operator site and block polymerase binding.

Activators

Because RNA polymerase binds to a specific promoter site on one strand of the DNA double helix, it is necessary that the DNA double helix unzip in the vicinity of this site in order for the polymerase protein to be able to sit down properly. In many genes, this unzipping cannot take place without the assistance of a regulatory protein called an **activator** that binds to the DNA in this region and helps it unwind. Just as in the case of the repressor protein described previously, cells can turn genes on and off by binding "signal" molecules to the activator protein. These molecules prevent it from binding to the DNA or enable it to do so.

Activators enable a cell to carry out a second level of control. When a bacterium encounters the sugar lactose, it may not be low on energy. Imagine if you had to eat every time you encountered food! When a bacterial cell already has lots of energy, levels of a special "I'm hungry" signal molecule fall. Without being prodded by this signal molecule, the activator protein, called CAP, cannot twist into the proper shape to fit the DNA unwinding site in front of the lactose-using genes; as a result, the genes are not transcribed, even though the repressor protein does not block the polymerase (figure 8.20).

Enhancers

A third level of control is exercised by expanding access to the gene. To make the promoters of complexly controlled genes accessible to many regulatory proteins simultaneously, many eukaryotic genes possess special associated sequences called **enhancers.** These enhancer sequences bind specific regulatory proteins that interact with the protein transcription factors that help RNA polymerase find and attach to its binding site at the beginning of the structural gene. Unlike promoters and operators, which butt right up to the start of a gene, enhancers are usually located far away from the start of the gene, often thousands of nucleotides distant. Although enhancers occur in exceptional instances in bacteria, they are the rule rather than the exception in eukaryotes.

How can regulatory proteins affect a promoter when they bind to the DNA at enhancer sites located far from the promoter? Apparently the DNA loops around so that the enhancer is positioned near the promoter. This brings the regulatory protein attached to the enhancer into direct contact with the transcription factor complex attached to the promoter (figure 8.21).

The enhancer mode of transcriptional control that has evolved in eukaryotes adds a great deal of flexibility to the control process. The positioning of regulatory sites at a distance permits a large number of different regulatory sequences scattered about the DNA to influence that particular gene.

8.11 Cells control the expression of genes by saying when they are transcribed, not how fast. Some regulatory proteins block the binding of the polymerase, and others facilitate it. In eukaryotes these regulatory proteins are often associated with control genes located on the chromosome far from the gene being regulated.

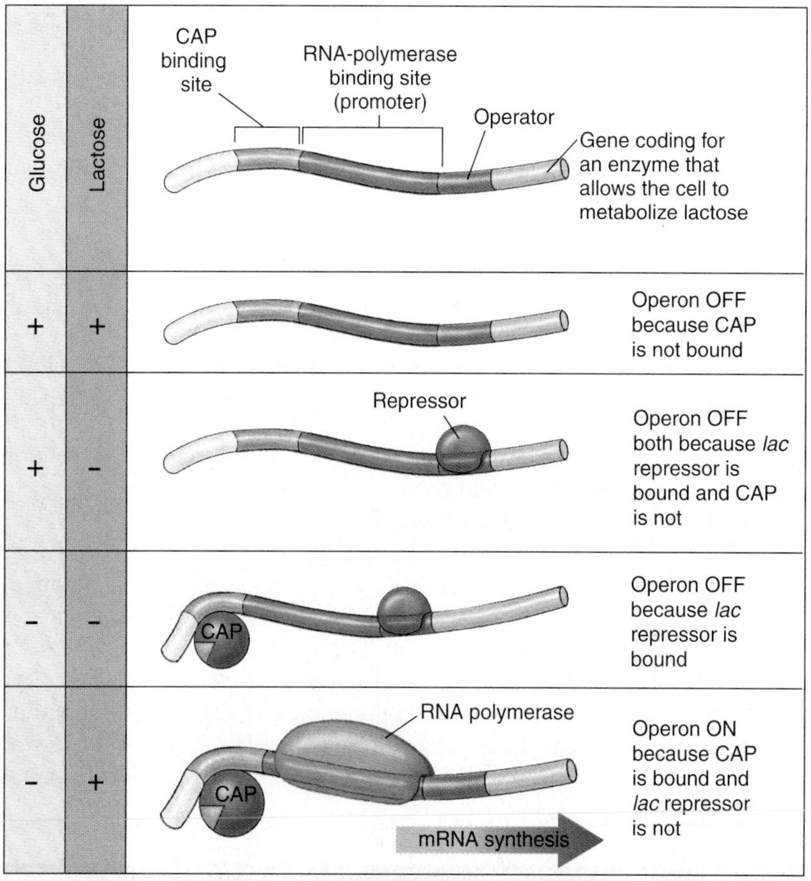

Glucose
Lactose
CAP binding site
RNA-polymerase binding site (promoter)
Operator
Gene coding for an enzyme that allows the cell to metabolize lactose
+
+
Operon OFF because CAP is not bound
+
-
Repressor
Operon OFF both because lac repressor is bound and CAP is not
-
-
CAP
Operon OFF because lac repressor is bound
-
+
RNA polymerase
CAP
mRNA synthesis
Operon ON because CAP is bound and lac repressor is not

Figure 8.20 Activators and repressors at the *lac* operon.
Together, the *lac* repressor and the activator called CAP provide a very sensitive response to the cell's need for and ability to use lactose-metabolizing enzymes.

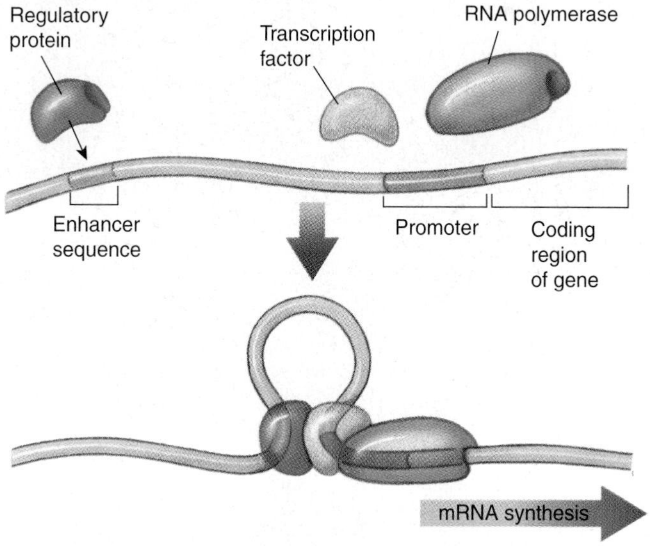

Regulatory protein
Transcription factor
RNA polymerase
Enhancer sequence
Promoter
Coding region of gene
mRNA synthesis

Figure 8.21 How enhancers work.
The enhancer site is located far away from the gene being regulated. Binding of a regulatory protein (*red*) to the enhancer allows the protein to interact with the transcription factors (*green*) associated with RNA polymerase, activating transcription.

8.12 Mutation

There are two general ways in which the genetic message is altered: mutation and recombination. A change in the content of the genetic message—the base sequence of one or more genes—is referred to as a **mutation.** Some mutations alter the identity of a particular nucleotide, while others remove or add nucleotides to a gene. A change in the position of a portion of the genetic message is referred to as **recombination.** Some recombination events move a gene to a different chromosome; others alter the location of only part of a gene. The cells of eukaryotes contain an enormous amount of DNA, and the mechanisms that protect and proofread the DNA are not perfect. If they were, no variation would be generated.

Mistakes Happen

In fact, cells do make mistakes during replication, often causing a change in a cell's genetic message, or a mutation (figure 8.22). However, change is rare. Typically, a particular gene is altered in only one of a million gametes. If changes were common, the genetic instructions encoded in DNA would soon degrade into meaningless gibberish. Limited as it might seem, the steady trickle of change that does occur is the very stuff of evolution. Every difference in the genetic messages that specify different organisms arose as the result of genetic change.

The Importance of Genetic Change

All evolution begins with alterations in the genetic message: mutation creates new alleles, gene transfer and transposition alter gene location, reciprocal recombination shuffles and sorts these changes, and chromosomal rearrangement alters the organization of entire chromosomes. Some changes in germ-line tissue produce alterations that enable an organism to leave more offspring, and those changes tend to be preserved as the genetic endowment of future generations. Other changes reduce the ability of an organism to leave offspring. Those changes tend to be lost, as the organisms that carry them contribute fewer members to future generations.

Evolution can be viewed as the selection of particular combinations of alleles from a pool of alternatives. The rate of evolution is ultimately limited by the rate at which these alternatives are generated. Genetic change through mutation and recombination provides the raw material for evolution.

Genetic changes in somatic cells do not pass on to offspring, and so they have less evolutionary consequence than germ-line change. However, changes in the genes of somatic cells can have an important and immediate impact particularly if the gene affects development or is involved with regulation of cell proliferation.

8.12 Rare changes in genes, called mutations, can have significant effects on the individual when they occur in somatic tissue, but they are inherited only if they occur in germ-line tissue. Inherited changes provide the raw material for evolution.

Figure 8.22 Mutation.
Fruit flies normally have one pair of wings, extending from the thorax. This fly is a *bithorax* mutant. Because of a mutation in a gene regulating a critical stage of development, it possesses *two* thorax segments and thus two sets of wings.

8.13 Kinds of Mutation

Because mutations can occur randomly in a cell's DNA, most mutations are detrimental, just as making a random change in a computer program usually worsens performance. The consequences of a detrimental mutation may be minor or catastrophic, depending on the function of the altered gene.

Mutations in Germ-Line Tissues

The effect of a mutation depends critically on the identity of the cell in which the mutation occurs. During the embryonic development of all multicellular organisms, there comes a point when cells destined to form gametes (germ-line cells) are segregated from those that will form the other cells of the body (somatic cells). Only when a mutation occurs within a germ-line cell is it passed to subsequent generations as part of the hereditary endowment of the gametes derived from that cell.

Mutations in Somatic Tissues

Mutations in germ-line tissue are of enormous biological importance because they provide the raw material from which natural selection produces evolutionary change. Change can occur only if there are new, different allele combinations available to replace the old. Mutation produces new alleles, and recombination puts the alleles together in different combinations. In animals, it is the occurrence of these two processes in germ-line tissue that is important to evolution, because mutations in somatic cells (somatic mutations) are not passed from one generation to the next. However, a somatic mutation may have drastic effects on the individual organism in which it occurs, because it is passed on to all of the cells that are descended from the original mutant cell. Thus, if a mutant lung cell divides, all cells derived from it will carry the mutation. Somatic mutations of lung cells are, as we shall see, the principal cause of lung cancer in humans.

Point Mutations

One category of mutational changes affects the message itself, producing alterations in the sequence of DNA nucleotides. If alterations involve only one or a few base pairs in the coding sequence, they are called **point mutations.** Sometimes the identity of a base changes (**base substitution**), while other times one or a few bases are added (**insertion**) or lost (**deletion**). If an insertion or deletion throws the reading of the gene message out of register, a **frame-shift mutation** results. While some point mutations arise due to spontaneous pairing errors that occur during DNA replication, others result from damage to the DNA caused by **mutagens,** usually radiation or chemicals. The latter class of mutations is of particular importance because modern industrial societies often release many chemical mutagens into the environment.

Changes in Gene Position

Another category of mutations affects the way the genetic message is organized. In both bacteria and eukaryotes, individual genes may move from one place in the genome to another by **transposition.** When a particular gene moves to a different location, its expression or the expression of neighboring genes may be altered. In addition, large segments of chromosomes in eukaryotes may change their relative locations or undergo duplication. Such **chromosomal rearrangements** often have drastic effects on the expression of the genetic message. Table 8.1 reviews the effects of some categories of mutations.

TABLE 8.1 SOME CATEGORIES OF MUTATIONS

Mutation	Result
No Mutation  A B C	A normal B protein is produced by the *B* gene.
Point Mutation	
Base substitution 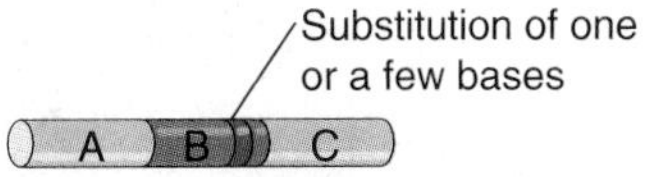 Substitution of one or a few bases A B C	B protein is inactive because changed amino acid disrupts function.
Insertion 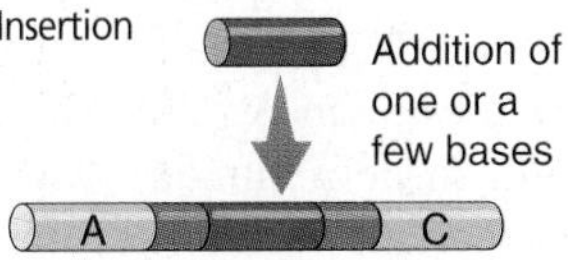 Addition of one or a few bases A C	B protein is inactive because inserted material disrupts proper shape.
Deletion 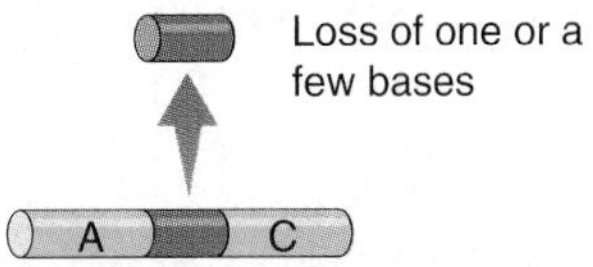 Loss of one or a few bases A C	B protein is inactive because portion of protein is missing.
Changes in Gene Position	
Transposition  A C B	*B* gene or B protein may be regulated differently because of change in gene position.
Chromosomal rearrangement 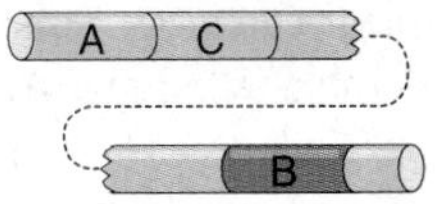 A C B	*B* gene may be inactivated or regulated differently in its new location on chromosome.

8.13 Point mutations are changes in the hereditary message of an organism. They may result from spontaneous errors during DNA replication or from damage to the DNA due to radiation or chemicals.

8.14 Cancer and Mutation

The search for the cause of cancer has focused in part on environmental factors, including ionizing radiation such as X rays and a variety of chemicals. Agents thought to cause cancer are called **carcinogens.** The association of particular chemicals with cancer, particularly chemicals that are potent mutagens, led researchers early on to the suspicion that cancer might be caused, at least in part, by chemicals, the so-called **chemical carcinogenesis theory.**

Early Ideas

The chemical carcinogenesis theory was first advanced over 200 years ago in 1761 by Dr. John Hill, an English physician. Hill noted unusual tumors of the nose in heavy snuff users and suggested tobacco had produced these cancers. In 1775, a London surgeon, Sir Percivall Pott, made a similar observation, noting that men who had been chimney sweeps exhibited frequent cancer of the scrotum. He suggested that soot and tars might be responsible. These and many other observations led to the hypothesis that cancer results from the action of chemicals on the body.

Demonstrating That Chemicals Can Cause Cancer

It was over a century before this hypothesis was directly tested. In 1915, Japanese doctor Katsusaburo Yamagiwa applied extracts of coal tar to the skin of 137 rabbits every two or three days for three months. Then he waited to see what would happen. After a year, cancers appeared at the site of application in seven of the rabbits. Yamagiwa had induced cancer with the coal tar, the first direct demonstration of chemical carcinogenesis. In the decades that followed, this approach demonstrated that many chemicals were capable of causing cancer.

These were lab studies, and many did not accept that they applied to real people. Do tars in fact induce cancer in humans? In 1949, the American physician Ernst Winder and the British epidemiologist Richard Doll independently reported that lung cancer showed a strong link to the smoking of cigarettes, which introduces tars into the lungs. Winder interviewed 684 lung cancer patients and 600 normal controls, asking whether each had ever smoked. Cancer rates were 40 times higher in heavy smokers than in nonsmokers. From these studies, it seemed likely as long as 50 years ago that tars and other chemicals in cigarette smoke induce cancer in the lungs of persistent smokers. While this suggestion was (and is) resisted by the tobacco industry, the evidence that has accumulated since these pioneering studies makes a clear case, and there is no longer any real doubt. Chemicals in cigarette smoke cause cancer.

Carcinogens Are Common

In ongoing investigations over the last 50 years, many hundreds of synthetic chemicals have been shown capable of causing cancer in laboratory animals. Among them are trichloroethylene, asbestos, benzene, vinyl chloride, arsenic, acrylamide, and a host of complex petroleum products with chemical structures resembling chicken wire (table 8.2).

TABLE 8.2 CHEMICAL CARCINOGENS IN THE WORKPLACE

Chemical	Cancer	Workers at Risk for Exposure
Common Exposure		
Benzene	Myelogenous leukemia	Painters; dye users; furniture finishers
Diesel exhaust	Lung	Railroad and bus-garage workers; truckers; miners
Mineral oils	Skin	Metal machining
Pesticides	Lung	Sprayers
Cigarette tar	Lung	Smokers
Uncommon Exposure		
Asbestos	Mesothelioma, lung	Brake-lining, insulation workers
Synthetic mineral fibers	Lung	Wall and pipe insulation; duct wrapping
Hair dyes	Bladder	Hairdressers and barbers
Paint	Lung	Painters
Polychlorinated biphenyls	Liver, skin	Hydraulic fluids and lubricants; inks; adhesives; insecticides
Soot	Skin	Chimney sweeps; brick layers; firefighters; heating-unit service workers
Rare Exposure		
Arsenic	Lung, skin	Insecticide/herbicide sprayers; tanners; oil refiners
Formaldehyde	Nose	Hospital and lab workers; wood product, paper, textiles, and metal product workers

In addition to identifying potentially dangerous substances, what have the studies of potential carcinogens told us about the nature of cancer? What do these cancer-causing chemicals have in common? *They are all mutagens, each capable of inducing changes in DNA.*

8.14 Chemicals that produce mutations in DNA, such as tars in cigarette smoke, are often potent carcinogens.

CONCEPT REVIEW

1. The Hershey-Chase experiment demonstrated that
   a. virus DNA injected into bacterial cells is apparently the factor involved in directing the production of new virus particles.
   b. RNA is the genetic material of some viruses.
   c. ^{32}P-labeled protein is injected into bacterial cells by viruses.
   d. the transforming principle is the DNA, not the polysaccharide coat.

2. According to Chargaff's rule and the Watson-Crick model of DNA, if you know the base sequence of one strand is AATTCG, then the sequence of the complementary strand must be
   a. AATTCG.
   b. TTGGAC.
   c. TTAACG.
   d. TTAAGC.

3. DNA makes RNA by
   a. translation.
   b. replication.
   c. transcription.
   d. repair.

4. RNA makes proteins by
   a. translation.
   b. replication.
   c. transcription.
   d. repair.

5. In the *lac* region of the bacterium *Escherichia coli*'s chromosome, the repressor protein binds at the
   a. activator protein site.
   b. promoter site.
   c. operator site.
   d. transcription unit.

6. Transposons can jump from one chromosome to another.
   a. true
   b. false

7. The transforming principle in the experiments by Griffith and Avery proved to be ________.

8. Erwin Chargaff observed that the amount of adenine in all DNA molecules is equal to the amount of ________ and that the amount of guanine is equal to the amount of ________.

9. The ________ is the cell site of protein synthesis.

10. ________ is the molecule that carries amino acids during translation.

11. Special sequences of eukaryotic genes called ________ make the promoters of complexly controlled genes accessible to many regulatory proteins simultaneously.

12. A regulatory protein responsible for blockage of DNA transcription is called a ________.

13. Exons alternate with ________ on chromosomes.

14. A chemical capable of damaging DNA is called a ________.

CHALLENGE YOURSELF

1. From an extract of human cells growing in tissue culture, you obtain a white, fibrous substance. How would you distinguish whether it was DNA, RNA, or protein?

2. Why is hydrogen bonding important to the structure of DNA?

3. Why must the DNA molecule become unzipped in order to function?

4. The nucleotide sequence of a hypothetical eukaryotic gene is shown below:

   TAC ATA CTA GTT AC<u>G</u> TCG CCC GGA AAT ATC

   If a mutation in this gene changed the fifteenth nucleotide (underlined) from guanine to thymine, what effect do you think it would have on the expression of this gene?

8 eBRIDGE

Reinforcing Key Points

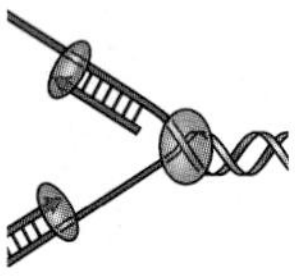

Genes Are Made of DNA

8.1 The Griffith Experiment

8.2 The Avery Experiments

8.3 The Hershey-Chase Experiment

8.4 The Fraenkel-Conrat Experiment

8.5 Discovering the Structure of DNA

8.6 How the DNA Molecule Replicates

From Gene to Protein

8.7 Transcription

8.8 The Genetic Code

8.9 Translation

8.10 Architecture of the Gene

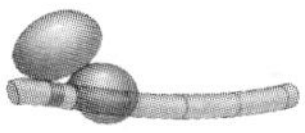

Regulating Gene Expression

8.11 Turning Genes Off and On

Altering the Genetic Message

8.12 Mutation

8.13 Kinds of Mutation

8.14 Cancer and Mutation

Electronic Learning

Visual Learning

Animations

Nine Animations

Including:
Regulation of *E. coli lac* Operon
DNA Structure
Translation

Explorations

Reading DNA

In this exercise, you can explore how regulatory proteins "read" DNA, design proteins with different structural motifs, and test these hypothetical proteins against particular DNA sequences.

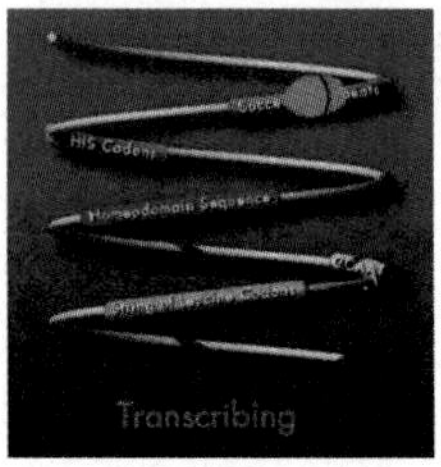
Transcribing

Gene Regulation

This exercise explores the various strategies employed by organisms to regulate the transcription of genes. You can examine bacterial gene regulation and eukaryotic gene regulation.

Author's Corner

Ribosomes Are Ribozymes. The machine within the cell that manufactures proteins is a complex of proteins and RNA molecules called a ribosome. Because the enzymes of the cell are proteins, it had been commonly assumed that the proteins of the ribosomes did the actual chemistry of protein synthesis, the RNA molecules carrying out secondary roles. Recently scientists completed the first detailed analysis of the structure of a ribosome at atomic resolution, and they had quite a surprise. The researchers found that the many proteins of a ribosome are scattered over its surface like decorations on a Christmas tree, linking RNA strands together at key positions like spot-welds. Inside, where the protein-building takes place, there are no proteins, just twists of RNA. It is this RNA that catalyzes the linking together of amino acids to make a new protein. The RNA of the ribosome is acting like an enzyme—a "ribozyme."

1. The fact that ribosomes carry out catalysis may shed light on the origin of life.

Virtual Classroom

Unraveling the Mystery of DNA

The realization that Mendel's patterns of heredity can be explained by the segregation of chromosomes in meiosis raised a question that occupied biologists for 50 years: What exactly is the nature of the connection between hereditary traits and chromosomes? In this lecture we recount the chain of experiments that led to our current understanding of the molecular mechanisms of hereditary. The experiments are among the most elegant in science. Just as in a good detective story, each conclusion has led to new questions. The intellectual path has not always been a straight one, the best questions not always obvious. But however erratic and lurching the course of the experimental journey, our picture of heredity has become progressively clearer, the image more sharply defined. As we have mastered the details of what a gene is, and of how genes do their job of dictating what we are like, DNA has become a household word, and its study the core of the new science of molecular biology. Little of biology, from taxonomy and botany to genetics and cell biology, can be properly understood except in the context of DNA.

Virtual Lab

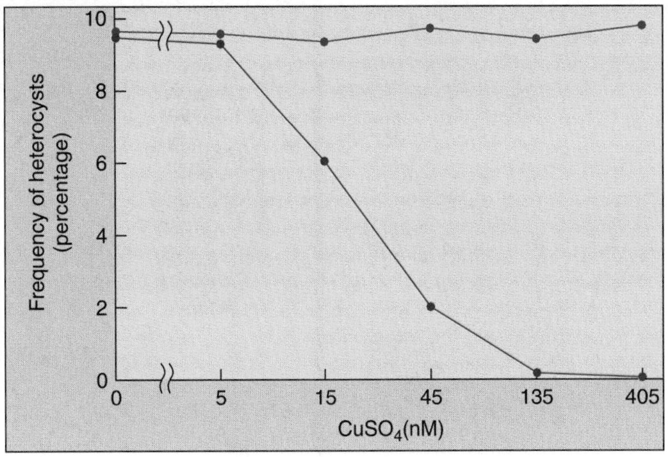

Frequency of heterocysts (percentage)
10
8
6
4
2
0
0
5
15
45
135
405
$CuSO_4$(nM)

Cyanobacteria Control Heterocyst Pattern Formation Through Intracellular Signaling

The regulation of pattern formation in an organism is a fundamental aspect of its development. Pattern formation is under strict genetic regulation, whether in large multicellular organisms or simple bacteria. Studying simple systems is a good way to uncover basic mechanisms. Cyanobacteria, for example, are photosynthetic prokaryotic cells that grow together in filaments. These cyanobacterial filaments exhibit a simple developmental pattern: single heterocysts (specialized nitrogen-fixing cells) are separated by approximately ten photosynthetic vegetative cells.

James Golden and Ho-Sung Yoon at Texas A&M University have identified a small gene, called *patS,* that appears to be crucial for the formation and maintenance of proper pattern formation of heterocysts and vegetative cells. To investigate how *patS* controls heterocyst pattern, Golden and Yoon examined the effects of different levels of *patS* transcription on heterocyst formation. They found that the *patS* gene makes a diffusible protein that inhibits the formation of adjacent heterocysts, in this way maintaining a minimum number of ten vegetative cells in between heterocysts.

Quizzes | Further Reading | Essential Study Partner | Links | BioCourse.com

CHAPTER 9

Gene Technology

THE FAR SIDE® By GARY LARSON
Larson
© 1984 FarWorks, Inc. All Rights Reserved/Dist. by Creators Syndicate

CHAPTER 9

Gene Technology

CHAPTER OVERVIEW

Genetic Engineering

9.1 A Scientific Revolution
9.2 Restriction Enzymes
9.3 The Four Stages of a Genetic Engineering Experiment
9.4 Other Genetic Techniques

- DNA is cut by restriction enzymes into fragments with "sticky ends."
- Any two fragments of DNA cut by the same restriction enzyme can be joined, whatever the source of the DNA.
- DNA fragments can be introduced into bacteria, carried on plasmids or viruses. A bacterial colony that grows from such a cell is called a clone.
- Probes such as cDNA can be used to screen bacterial colonies for the presence of a gene you are attempting to transfer.

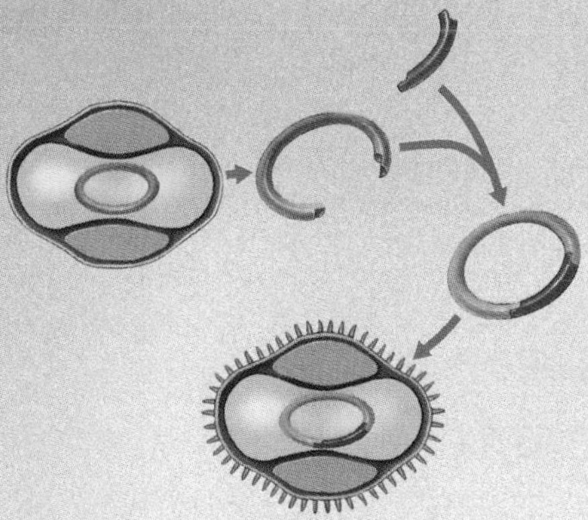

Advances in Medicine

9.5 Genetic Engineering and Medicine

- By transferring human genes into bacteria, it is possible to produce commercial quantities of rare proteins, making many new drugs practical.
- Effective new vaccines are being made by introducing genes specifying surface proteins of pathogenic microbes into harmless infective agents.
- Transfer of healthy genes into cystic fibrosis patients may soon cure this hereditary disorder.

Transforming Agriculture

9.6 Genetic Engineering of Crop Plants
9.7 Genetic Engineering of Farm Animals
9.8 Cloning
9.9 Ethical Issues

- It is difficult to transfer genes into plants because of their tough cell walls—few viruses can penetrate them.
- Today genes are inserted into plant cells using plasmids or DNA particle guns.
- Major advances have been made in cloning, and in transferring genes resistant to herbicides and pests, as well as genes that increase milk production and farm animal size.
- The potential risks of bioengineered crops and animals are controversial.

9.1 A Scientific Revolution

Moving genes from one organism to another is often called **genetic engineering.** Many of the gene transfers have placed eukaryotic genes into bacteria, converting the bacteria into tiny factories that produce prodigious amounts of the protein encoded by the eukaryotic gene. Other gene transfers have moved genes from one animal or plant to another.

Genetic engineering is having a major impact on medicine and agriculture (figure 9.1). Most of the insulin used to treat diabetes is now obtained from bacteria that contain a human insulin gene. In late 1990, the first transfers of genes from one human to another were carried out in attempts to correct the effects of defective genes in a rare genetic disorder called *severe combined immune deficiency.* In addition, cultivated plants and animals can be genetically engineered to resist pests, grow bigger, or grow faster.

9.1 In recent years the ability to manipulate genes and move them from one organism to another has led to great advances in medicine and agriculture.

Producing growth hormone. These two mice are genetically identical, except that the large one has one extra gene, encoding a potent human growth hormone not normally present in mice. Courtesy Ralph L. Brinster, University of Pennsylvania, School of Veterinary Medicine.

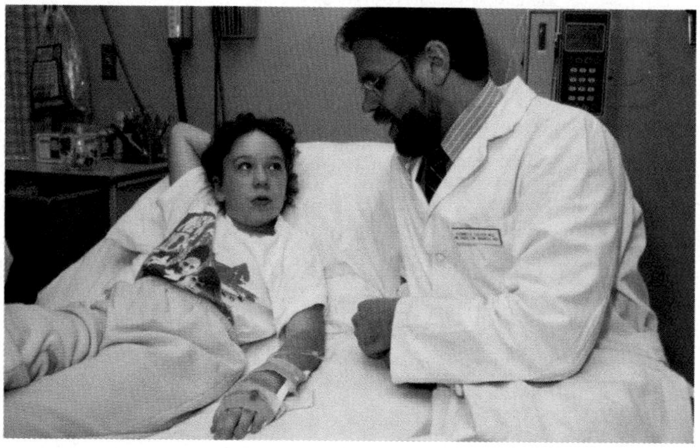

Curing disease. One of two young girls who were the first humans "cured" of a hereditary disorder by transferring into their bodies healthy versions of the gene they lacked. The transfer was successfully carried out in 1990, and the girls remain healthy.

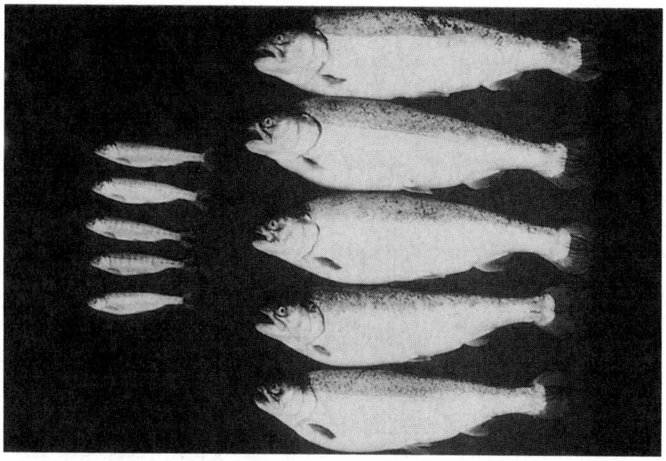

Increasing yields. The genetically engineered salmon on the *right* have shortened production cycles and are heavier than the nontransgenic salmon on the *left.*

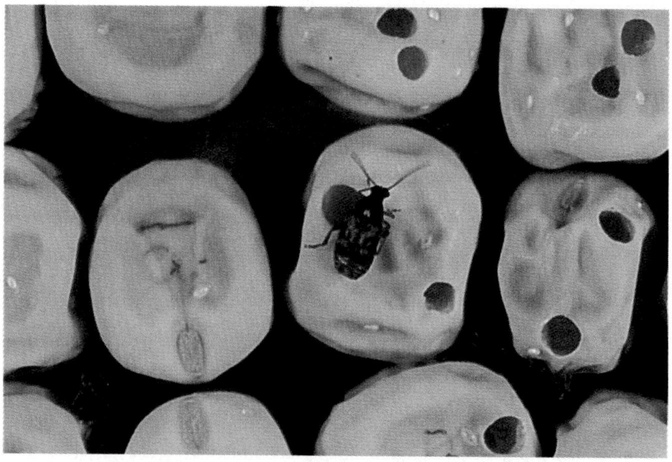

Pest-proofing plants. The genetically engineered peas on the *left* have a gene that inhibits feeding by weevils, one of the most notorious pests of stored grains.

Figure 9.1 Examples of genetic engineering.

9.2 Restriction Enzymes

The first stage in any genetic engineering experiment is to chop up the "source" DNA in order to get a copy of the gene you wish to transfer. This first stage is the key to successful transfer of the gene, and learning how to do it is what has led to the genetic revolution. The trick is in how the DNA molecules are cut. The cutting must be done in such a way that the resulting DNA fragments have "sticky ends" that can later be inserted into another chromosome.

This special form of molecular surgery is carried out by **restriction enzymes,** which are special enzymes that bind to specific short sequences (typically four to six nucleotides long) on the DNA. These sequences are very unusual in that they are symmetrical—the two strands of the DNA duplex have the same nucleotide sequence, running in opposite directions! One of these sequences, for example, is GAATTC. Try writing down the sequence of the opposite strand: it is CTTAAG—the same sequence, written backwards.

What makes the DNA fragments "sticky" is that most restriction enzymes do not make their incision in the center of the sequence; rather, the cut is made to one side. In the sequence written above, for example, the cut is made after the first nucleotide, G/AATTC. This produces a break, with short, single strands of DNA dangling from each end. Because the two single-stranded ends are complementary in sequence, they could pair up and heal the break, with the aid of a sealing enzyme—*or* they could pair with *any other DNA fragment cut by the same enzyme,* because all would have the same single-stranded sticky ends (figure 9.2)! Any gene in any organism cut by the enzyme that attacks CTTAAG sequences can be joined to any other with the aid of a sealing enzyme called a **ligase,** which re-forms the bonds between the sugars and phosphates of DNA.

Hundreds of different restriction enzymes are known, recognizing a wide variety of four- to six-nucleotide sequences. Each kind of enzyme attacks only one sequence and always cuts at the same place. By trying one after another, biologists can almost always find a sequence that cuts out the gene they seek, one present by chance on both ends of the gene but not within it. The restriction enzymes are the basic tools of genetic engineering.

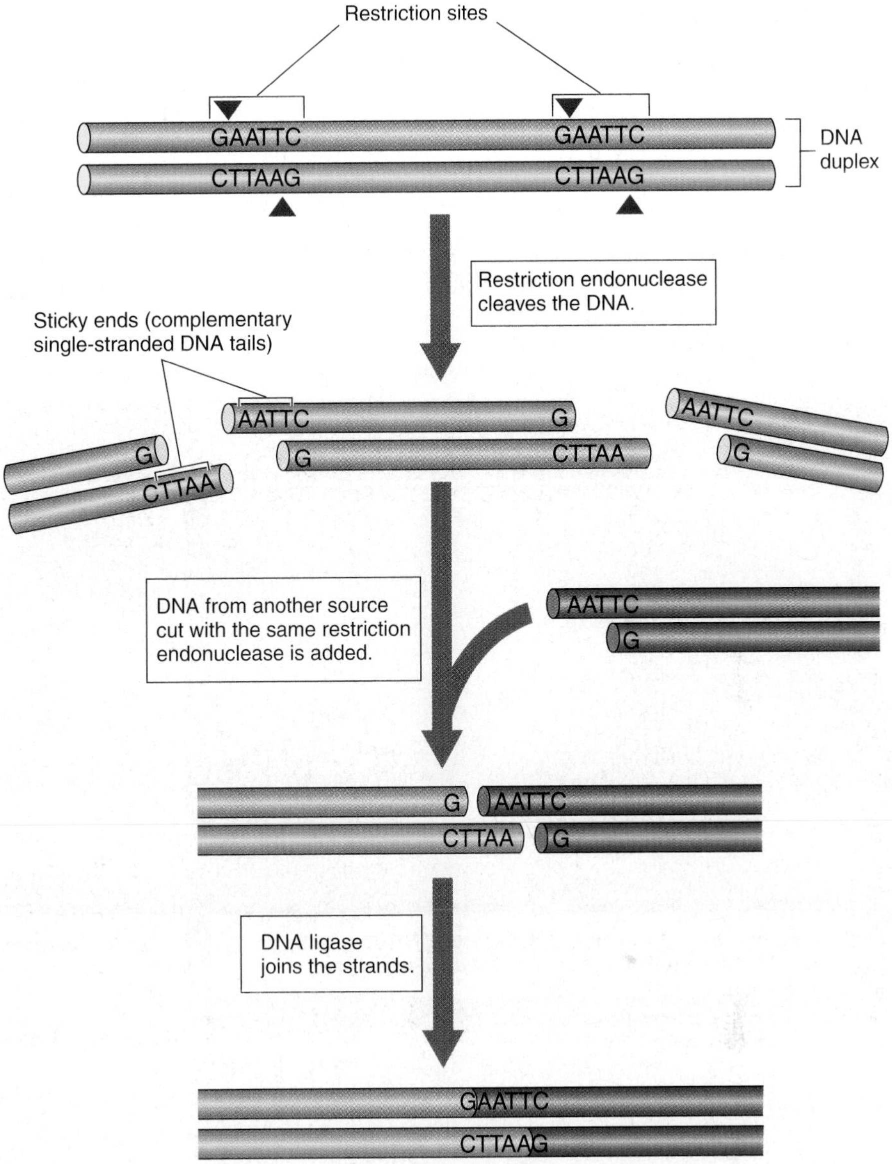

Restriction sites
GAATTC
GAATTC
CTTAAG
CTTAAG
DNA duplex
Restriction endonuclease cleaves the DNA.
Sticky ends (complementary single-stranded DNA tails)
AATTC
G
G
CTTAA
G
CTTAA
AATTC
G
DNA from another source cut with the same restriction endonuclease is added.
AATTC
G
G
AATTC
CTTAA
G
DNA ligase joins the strands.
GAATTC
CTTAAG
Recombinant DNA molecule

Figure 9.2 How restriction enzymes produce DNA fragments with sticky ends.

The restriction enzyme *Eco*RI always cleaves the sequence GAATTC between G and A. Because the same sequence occurs on both strands, both are cut. However, the two sequences run in opposite directions on the two strands. As a result, single-stranded tails are produced that are complementary to each other, or "sticky."

9.2 Restriction enzymes, the key tools that make genetic engineering possible, bind to specific short sequences of DNA and cut the DNA where they bind. This produces fragments with "sticky ends," which can be rejoined in different combinations.

9.3 The Four Stages of a Genetic Engineering Experiment

Every gene transfer presents unique problems, but all share four distinct stages (figure 9.3):

1. Cleaving DNA—cutting up the source chromosome.
2. Producing recombinant DNA—placing the DNA fragments into vehicles that can infect the target cells.
3. Cloning—infecting target cells with DNA-bearing vehicles and allowing infected cells to reproduce.
4. Screening—selecting the particular infected cells that have received the gene of interest.

Stage 1: Cleaving DNA

A restriction enzyme is used to cleave the source DNA into fragments. Because the endonuclease's recognition sequence is likely to occur many times within the source DNA, cleavage

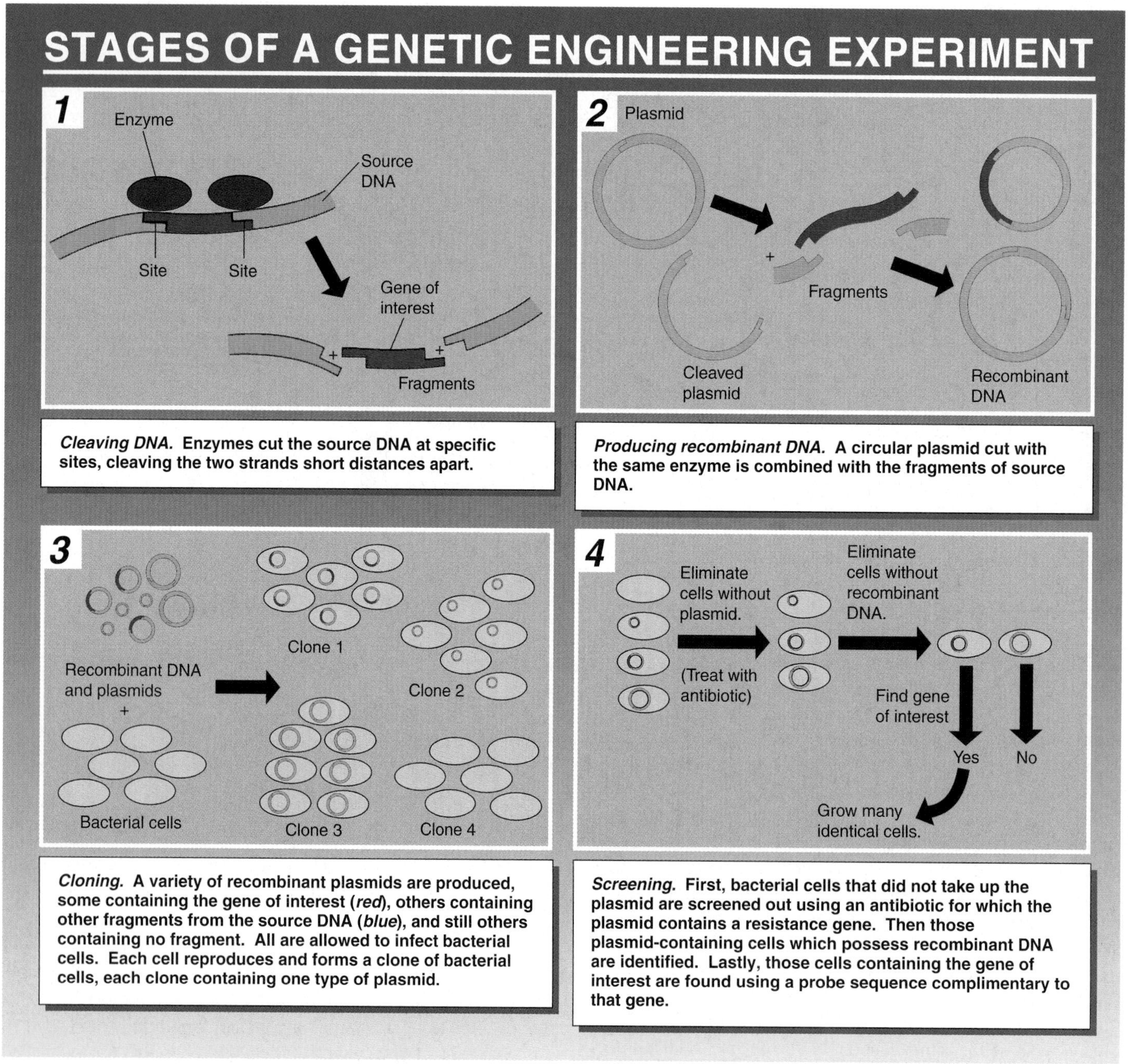

STAGES OF A GENETIC ENGINEERING EXPERIMENT
1
Enzyme
Source DNA
Site
Site
Gene of interest
Fragments
Cleaving DNA. Enzymes cut the source DNA at specific sites, cleaving the two strands short distances apart.
2
Plasmid
Fragments
Cleaved plasmid
Recombinant DNA
Producing recombinant DNA. A circular plasmid cut with the same enzyme is combined with the fragments of source DNA.
3
Recombinant DNA and plasmids
Bacterial cells
Clone 1
Clone 2
Clone 3
Clone 4
Cloning. A variety of recombinant plasmids are produced, some containing the gene of interest (red), others containing other fragments from the source DNA (blue), and still others containing no fragment. All are allowed to infect bacterial cells. Each cell reproduces and forms a clone of bacterial cells, each clone containing one type of plasmid.
4
Eliminate cells without plasmid.
(Treat with antibiotic)
Eliminate cells without recombinant DNA.
Find gene of interest
Yes
No
Grow many identical cells.
Screening. First, bacterial cells that did not take up the plasmid are screened out using an antibiotic for which the plasmid contains a resistance gene. Then those plasmid-containing cells which possess recombinant DNA are identified. Lastly, those cells containing the gene of interest are found using a probe sequence complimentary to that gene.

Figure 9.3 How a genetic engineering experiment works.

will produce a large number of different fragments. Different fragments may be obtained by employing endonucleases that recognize different sequences. The fragments can be separated from each other according to their size by **electrophoresis** (figure 9.4).

Stage 2: Producing Recombinant DNA

The fragments of DNA are inserted into plasmids or viral DNA, which have been cleaved with the same restriction endonuclease as the source DNA. A **plasmid** is a tiny circle of DNA that is able to replicate outside of the main bacterial chromosome.

Stage 3: Cloning

The plasmids or viruses serve as **vectors,** a genome that carries the foreign DNA into the host cell. Host cells are usually, but not always, bacteria. As each cell reproduces, it forms a **clone** of cells that all contain the fragment-bearing vector. Each clone is maintained separately, and all of them together constitute a **clone library** of the original source DNA.

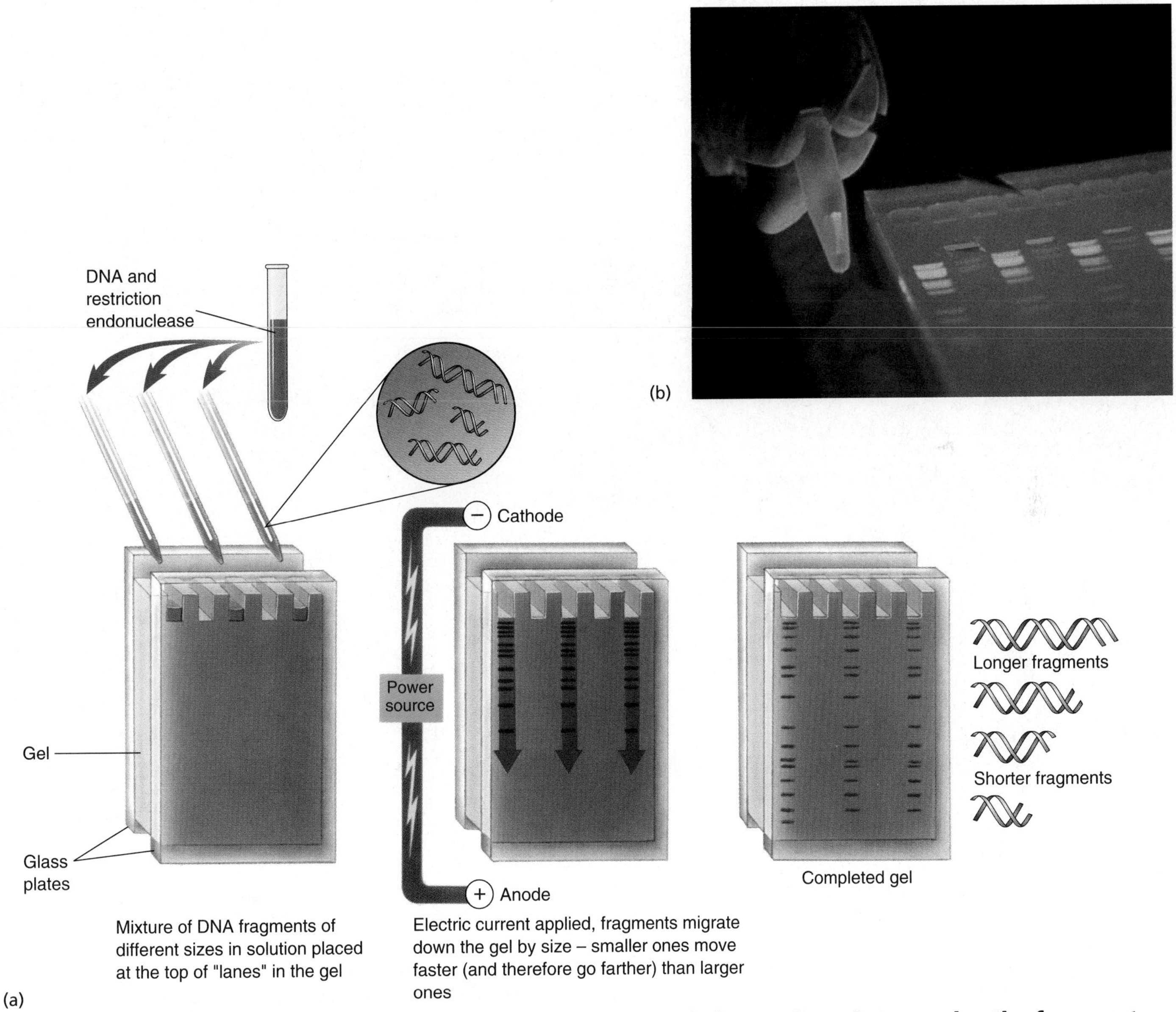
DNA and
restriction
endonuclease
Cathode
Power
source
Gel
Glass
plates
Anode
Completed gel
Longer fragments
Shorter fragments
Mixture of DNA fragments of
different sizes in solution placed
at the top of "lanes" in the gel
Electric current applied, fragments migrate
down the gel by size – smaller ones move
faster (and therefore go farther) than larger
ones
(a)
(b)

Figure 9.4 Using restriction endonucleases to cleave DNA and electrophoresis to resolve the fragment.

Gel electrophoresis allows investigators to separate a complex mixture of DNA fragments according to how long the individual fragments are. (*a*) When placed on one side of a gel, the fragments will migrate different distances when an electric current is applied. The distance a fragment moves relates to its size, larger fragments moving slower and thus traveling shorter distances during the period of time the current is applied. (*b*) The fragments show up as separate bands when viewed under fluorescent light.

Stage 4: Screening

The clones containing a specific DNA fragment of interest, often a fragment that includes a particular gene, are identified from the clone library. Let's examine this stage in more detail, as it is generally the most challenging in any genetic engineering experiment.

The Preliminary Screening of Clones. To make the screening of clones easier, investigators initially try to eliminate from the library any clones that do not contain vectors, as well as clones whose vectors do not contain fragments of the source DNA. The first category of clones can be eliminated by employing a vector with a gene that confers resistance to a specific antibiotic, such as tetracycline, penicillin, or ampicillin. In figure 9.5*a*, the gene *amp*^{r} is incorporated into the plasmid and confers resistance to the antibiotic ampicillin. When the clones are exposed to a medium containing that antibiotic, only clones that contain the vector will be resistant to the antibiotic and able to grow.

One way to eliminate clones with vectors that do not have an inserted DNA fragment is to use a vector that, in addition to containing antibiotic resistance genes, contains the *lacZ′* gene, which is required to produce β-galactosidase, an enzyme that enables the cells to metabolize the sugar X-gal. Metabolism of X-gal results in the formation of a blue reaction product, so any cells whose vectors contain a functional version of this gene will turn blue in the presence of X-gal (figure 9.5*b*). However, if a restriction enzyme whose recognition sequence lies within the *lacZ′* gene is used, the gene will be cleaved when recombinants are formed. If a fragment of the source DNA inserts into the vector at the cleavage site, the gene will be inactivated and the cell will be unable to metabolize X-gal. Therefore, cells with vectors that contain a fragment of source DNA should remain colorless in X-gal media.

Any cells that are able to grow in a medium containing the antibiotic but don't turn blue in the medium with X-gal must have incorporated a vector with a fragment of source DNA. Identifying cells that have a *specific* fragment of the source DNA is the next step in screening clones.

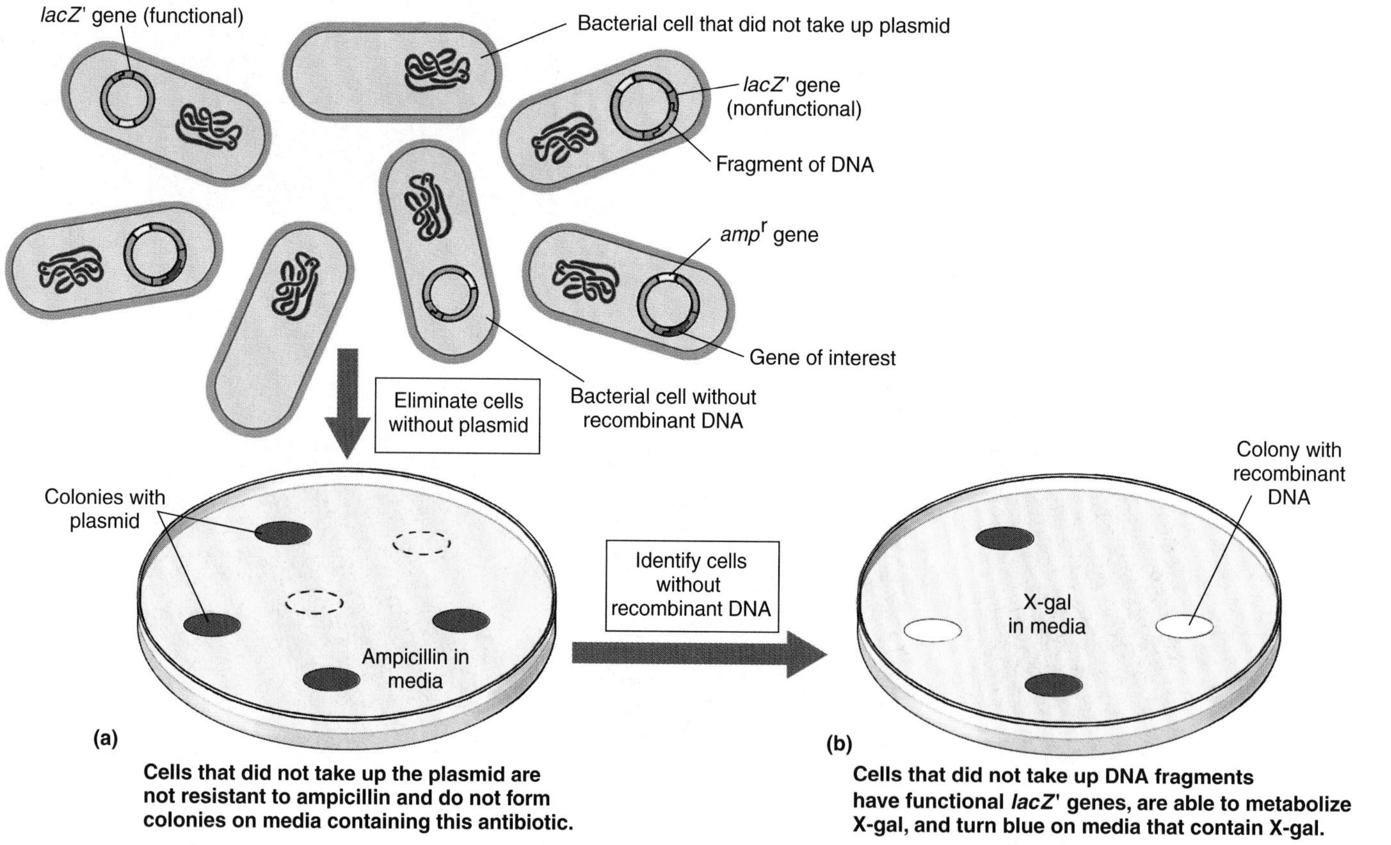

lacZ' gene (functional)
Bacterial cell that did not take up plasmid
lacZ' gene (nonfunctional)
Fragment of DNA
amp^r gene
Gene of interest
Eliminate cells without plasmid
Bacterial cell without recombinant DNA
Colonies with plasmid
Ampicillin in media
Identify cells without recombinant DNA
Colony with recombinant DNA
X-gal in media
(a)
Cells that did not take up the plasmid are not resistant to ampicillin and do not form colonies on media containing this antibiotic.
(b)
Cells that did not take up DNA fragments have functional lacZ' genes, are able to metabolize X-gal, and turn blue on media that contain X-gal.

Figure 9.5 Using antibiotic resistance and X-gal as preliminary screens of restriction fragment clones.
Bacteria are transformed with recombinant plasmids that contain a gene (*amp*^{r}) that confers resistance to the antibiotic ampicillin and a gene (*lacZ'*) that is required to produce β-galactosidase, the enzyme that enables the cells to metabolize the sugar X-gal. (*a*) Only those bacteria that have incorporated a plasmid will be resistant to ampicillin and will grow on a medium that contains the antibiotic. (*b*) Ampicillin-resistant bacteria will be able to metabolize X-gal if their plasmid does *not* contain a DNA fragment inserted in the *lacZ'* gene; such bacteria will turn blue when grown on a medium containing X-gal. Bacteria with a plasmid that has a DNA fragment inserted within the *lacZ'* gene will not be able to metabolize X-gal and, therefore, will remain colorless in the presence of X-gal.

Finding the Gene of Interest. A clone library may contain anywhere from a few dozen to many thousand individual fragments of source DNA. Many of those fragments will be identical, so to assemble a complete library of the entire source genome, several hundred thousand clones could be required. A complete *Drosophila* (fruit fly) library, for example, contains more than 40,000 different clones; a complete human library consisting of fragments 20 kilobases long would require close to a million clones. To search such an immense library for a clone that contains a fragment corresponding to a particular gene requires ingenuity, but many different approaches have been successful.

The most general procedure for screening clone libraries to find a particular gene is hybridization (figure 9.6). In this method, the cloned genes form base pairs with complementary sequences on another nucleic acid. The complementary nucleic acid is called a **probe** because it is used to probe for the presence of the gene of interest. At least part of the nucleotide sequence of the gene of interest must be known to be able to construct the probe.

In this method of screening, bacterial colonies containing an inserted gene are grown on agar. Some cells are transferred to a filter that is pressed onto the colonies, forming a replica plate. The filter is then treated with a solution that denatures the bacterial DNA and that contains a radioactively labeled probe. The probe hybridizes with complementary single-stranded sequences on the bacterial DNA.

When the filter is laid over photographic film, areas that contain radioactivity will expose the film (autoradiography). Only colonies that contain the gene of interest hybridize with the radioactive probe and emit radioactivity onto the film. The pattern on the film is then compared to the original master plate, and the gene-containing colonies may be identified.

9.3 A typical genetic engineering experiment consists of four stages: cleaving DNA into fragments (using restriction enzymes), producing recombinant DNA (adding restriction fragments to plasmids), cloning (growing the plasmids in bacterial cells), and screening (identifying cells containing plasmids with a gene of interest).

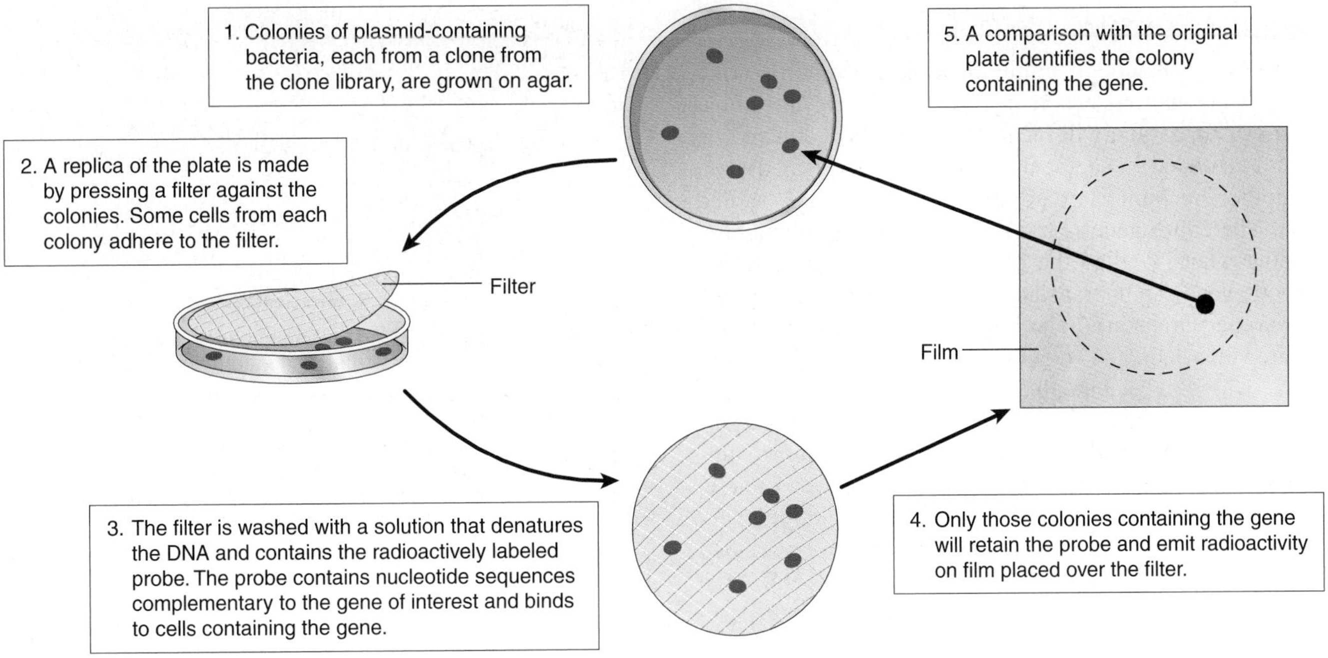

1. Colonies of plasmid-containing bacteria, each from a clone from the clone library, are grown on agar.
2. A replica of the plate is made by pressing a filter against the colonies. Some cells from each colony adhere to the filter.
Filter
3. The filter is washed with a solution that denatures the DNA and contains the radioactively labeled probe. The probe contains nucleotide sequences complementary to the gene of interest and binds to cells containing the gene.
4. Only those colonies containing the gene will retain the probe and emit radioactivity on film placed over the filter.
Film
5. A comparison with the original plate identifies the colony containing the gene.

Figure 9.6 Using hybridization to identify the gene of interest.

(*1*) Each of the colonies on these bacterial culture plates represents millions of clones descended from a single cell. To test whether a certain gene is present in any particular clone, it is necessary to identify colonies whose cells contain DNA that hybridizes with a probe consisting of DNA sequences complementary to the gene. (*2*) Pressing a filter against the master plate causes some cells from each colony to adhere to the filter. (*3*) The filter is then washed with a solution that denatures the DNA and contains the radioactively labeled probe. (*4*) Only those colonies that contain DNA that hybridizes with the probe, and thus contain the gene of interest, will expose film in autoradiography. (*5*) The film is then compared to the master plate to identify the gene-containing colony.

9.4 Other Genetic Techniques

PCR Amplification

Often, instead of inserting a gene into bacteria in order to generate many copies, and so produce large amounts of the protein the gene encodes, researchers use a technique called **PCR,** or the **polymerase chain reaction** (figure 9.7).

To carry out this gene-multiplying procedure, researchers first identify and synthesize short, single-stranded sequences of nucleotides called **primers,** which occur on either side of the DNA region to be amplified. A solution containing DNA and the primers is then heated to about 95°C. This high temperature disrupts the hydrogen bonds holding the DNA duplex together, causing it to denature into single strands. As the solution cools, the primers, which are present in excess, bind to their complementary sequence near the region of interest on the single strands of DNA. Next a heat-stable type of DNA polymerase, called *Taq polymerase,* is added along with a supply of all four nucleotides. Using the primer as a starting point, the polymerase then proceeds down the strand of DNA, adding nucleotides. When it is done, what used to be the primer is now lengthened into a complementary copy of the entire single-stranded fragment. Because both strands behave this way, there are now two copies of the original fragment. This process is repeated many times, resulting in the desired amplification of the DNA region of interest.

Formation of cDNA

Virtually every nucleotide within the transcribed portion of a bacterial gene participates in a code specifying an amino acid. However, as you will recall from chapter 8, eukaryotic genes are encoded in segments called *exons* separated from one another by numerous nontranslated sequences called *introns*. The entire gene is transcribed by RNA polymerase, producing what is called the **primary transcript.** Before a eukaryotic gene can be translated, the introns must be cut out of this primary transcript. The fragments that remain (the exons) are then stitched together to form the **processed mRNA,** which is eventually translated in the cytoplasm. When transferring eukaryotic genes into bacteria, it is desirable to transfer DNA that has already been processed this way, instead of the raw eukaryotic DNA, because bacteria lack the enzymes to carry out the processing. To do this, genetic engineers first isolate from the cytoplasm the processed mRNA corresponding to a particular gene (figure 9.8). The cytoplasmic mRNA has *only* exons, properly stitched together. An enzyme called *reverse transcriptase* is then used to make a DNA version of the mRNA that has been isolated. Such a version of a gene is called copy DNA, or **cDNA.** The

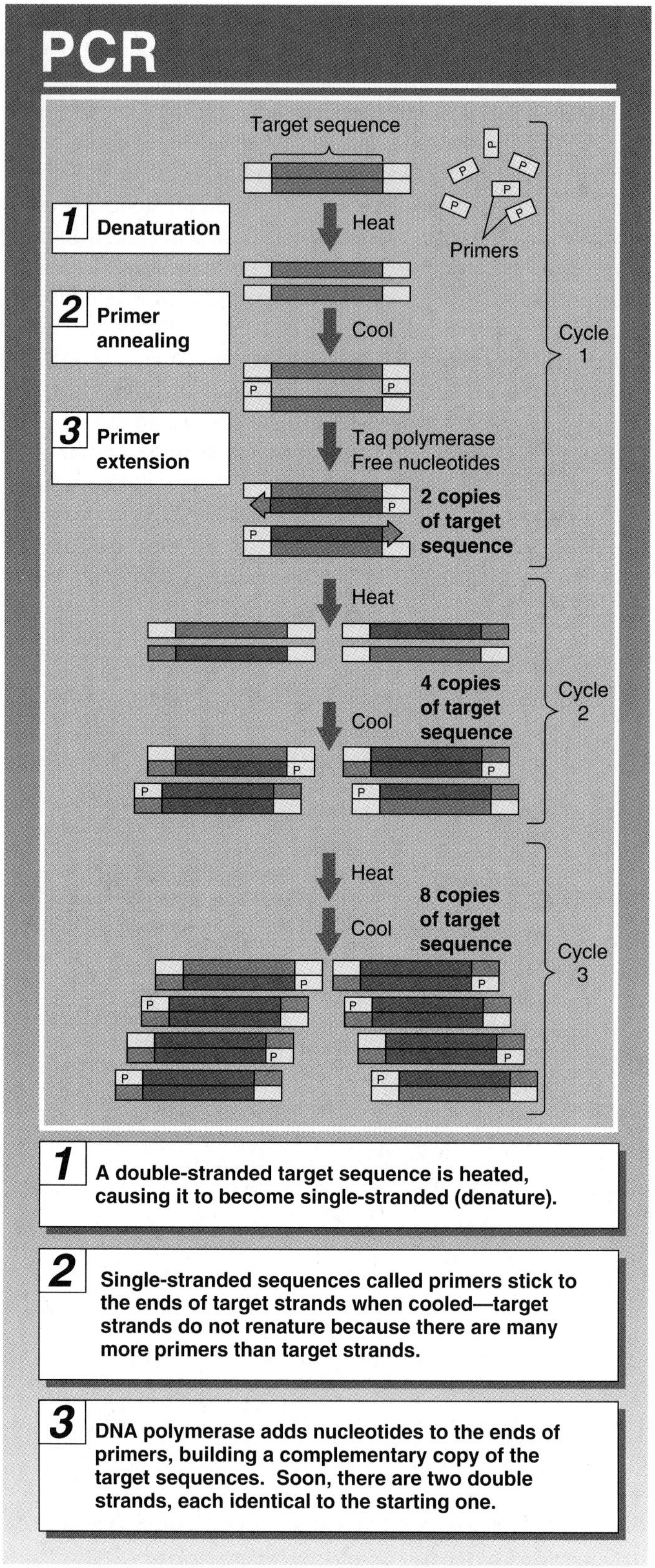
PCR
Target sequence
1 Denaturation
Heat
Primers
2 Primer annealing
Cool
3 Primer extension
Taq polymerase
Free nucleotides
2 copies of target sequence
Cycle 1
Heat
4 copies of target sequence
Cool
Cycle 2
Heat
8 copies of target sequence
Cool
Cycle 3
1 A double-stranded target sequence is heated, causing it to become single-stranded (denature).
2 Single-stranded sequences called primers stick to the ends of target strands when cooled—target strands do not renature because there are many more primers than target strands.
3 DNA polymerase adds nucleotides to the ends of primers, building a complementary copy of the target sequences. Soon, there are two double strands, each identical to the starting one.

Figure 9.7 How the polymerase chain reaction works.

Steps 1 to 3 are repeated many times, each time doubling the number of copies, until enough copies of the DNA fragment exist for analysis.

single strand of cDNA can then be used as a template for the synthesis of a complementary strand. In this way, a gene lacking introns is synthesized that can be used by bacteria to produce proteins.

DNA Fingerprinting

Figure 9.9 shows the DNA fingerprints a prosecuting attorney presented in a rape trial in 1987. They consisted of autoradiographs, parallel bars on X-ray film resembling the line patterns of the universal price code found on groceries. Each bar represents the position of a DNA restriction endonuclease fragment produced by techniques similar to that described in figure 9.4. The lane with many bars represents a standardized control. Two different probes were used to identify the restriction fragments. A vaginal swab had been taken from the victim within hours of her attack; from it semen was collected and the semen DNA analyzed for its restriction endonuclease patterns.

Compare the restriction endonuclease patterns of the semen to that of the suspect, Tommie Lee Andrews. You can see that the suspect's two patterns match that of the rapist (and are not at all like those of the victim). Clearly the semen collected from the rape victim and the blood sample from the suspect came from the same person. On November 6, 1987, the jury returned a verdict of guilty. Andrews became the first person in the United States to be convicted of a crime based on DNA evidence.

Since the Andrews verdict, DNA fingerprinting has been admitted as evidence in more than 2,000 court cases. While some probes highlight profiles shared by many people, others are quite rare. Using several probes, identity can be clearly established or ruled out.

Just as fingerprinting revolutionized forensic evidence in the early 1900s, so DNA fingerprinting is revolutionizing it today. A hair, a minute speck of blood, a drop of semen, all can serve as sources of DNA to damn or clear a suspect. As the man who analyzed Andrews's DNA says: "It's like leaving your name, address, and social security number at the scene of the crime. It's that precise." Of course, laboratory analyses of DNA samples must be carried out properly—sloppy procedures could lead to a wrongful conviction. After widely publicized instances of questionable lab procedures, national standards are being developed.

9.4 Among key techniques used by today's genetic engineers are PCR (used to amplify amounts of DNA samples), cDNA (used to build genes from their mRNA), and fingerprinting (used to identify particular individuals).

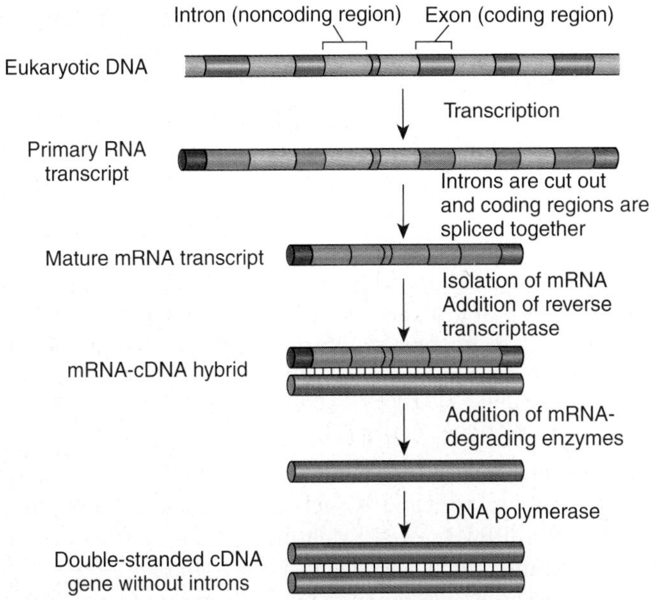
Intron (noncoding region)
Exon (coding region)
Eukaryotic DNA
Transcription
Primary RNA transcript
Introns are cut out and coding regions are spliced together
Mature mRNA transcript
Isolation of mRNA
Addition of reverse transcriptase
mRNA-cDNA hybrid
Addition of mRNA-degrading enzymes
DNA polymerase
Double-stranded cDNA gene without introns

Figure 9.8 Producing an intron-free version of a eukaryotic gene.

The primary transcript of the gene contains numerous introns that are cut out during processing, producing the processed mRNA found in the cytoplasm. The enzyme reverse transcriptase is then used to make a DNA strand complementary to the processed mRNA. Finally, that newly made strand of DNA is used as a template for the enzyme DNA polymerase, which assembles a complementary DNA strand along it, producing a double-stranded DNA version of the gene that is free of introns.

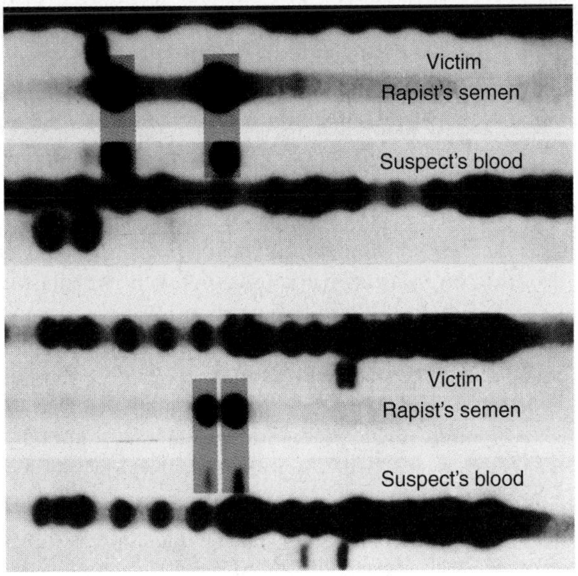
Victim
Rapist's semen
Suspect's blood
Victim
Rapist's semen
Suspect's blood

Figure 9.9 Two of the DNA profiles that led to conviction.

The two DNA probes seen here were used to characterize DNA isolated from the victim, the semen left by the rapist, and the suspect. The dark channels are multiband controls. There is a clear match between the suspect's DNA and the DNA of the rapist's semen in these.

9.5 Genetic Engineering and Medicine

Much of the excitement about genetic engineering has focused on its potential to improve medicine—to aid in curing and preventing illness. Major advances have been made in the production of proteins used to treat illness, in the creation of new vaccines to combat infections, and in the replacement of defective genes.

Making "Magic Bullets"

Many genetic defects occur because our bodies fail to make critical proteins. *Diabetes* is such an illness. The body is unable to control levels of sugar in the blood because a critical protein, **insulin,** cannot be made. These failures can be overcome if the body can be supplied with the protein it lacks. The donated protein is in a very real sense a "magic bullet" to combat the body's inability to regulate itself.

Until recently, the principal problem with using regulatory proteins as drugs was in manufacturing the protein. Proteins that regulate the body's functions are typically present in the body in very low amounts, and this makes them difficult and expensive to obtain in quantity. With genetic engineering techniques, the problem of obtaining large amounts of rare proteins has been largely overcome. The genes encoding the medically important proteins are now introduced into bacteria (table 9.1). Because the host bacteria can be grown cheaply, large amounts of the desired protein can be easily isolated. In 1982, the U.S. Food and Drug Administration approved the use of human insulin produced from genetically engineered bacteria, the first commercial product of genetic engineering.

Today hundreds of pharmaceutical companies around the world are busy producing other medically important proteins using these genetic engineering techniques. These products include **anticoagulants** (proteins involved in dissolving blood clots), which are effective in treating heart attack patients, and **factor VIII,** a protein that promotes blood clotting. A deficiency in factor VIII leads to hemophilia, an inherited disorder discussed in chapter 7, which is characterized by prolonged bleeding. For a long time, hemophiliacs received blood factor VIII that had been isolated from donated blood. Unfortunately, some of the donated blood had been infected with viruses such as HIV and hepatitis B, which were then unknowingly transmitted to those people who received blood transfusions. Today the use of genetically engineered factor VIII eliminates the risks associated with blood products obtained from other individuals.

TABLE 9.1 GENETICALLY ENGINEERED DRUGS

Product	Effects and Uses
Colony-stimulating factors	Stimulate white blood cell production; used to treat infections and immune system deficiencies
Erythropoietin	Stimulates red blood cell production; used to treat anemia in individuals with kidney disorders
Growth factors	Stimulate differentiation and growth of various cell types; used to aid wound healing
Human growth factors	Used to treat dwarfism
Interferons	Disrupt the reproduction of viruses; used to treat some cancers
Interleukins	Activate and stimulate white blood cells; used to treat wounds, HIV infections, cancer, immune deficiencies

Piggyback Vaccines

Vaccines are a way to teach our bodies how to defend against disease agents they have not yet met. A harmless version of a disease-causing microbe is injected into the body so that the immune system will develop defenses against the disease. The injected version serves as a model for the body, which carefully examines its surface and uses what it learns to craft defensive proteins called antibodies.

Traditionally, vaccines have been prepared by killing the disease agent (or sometimes by rendering it weak and unable to grow). That way, injecting it into your body doesn't make you sick. This "weakening" process was used in preparing the first Salk polio vaccines. The problem with this approach is that any failure in the weakening or killing process results in introducing the disease into the very patients seeking protection. This danger is one of the reasons rabies vaccines are administered only when an animal suspected of carrying rabies has actually bitten someone.

Now, with the advent of genetic engineering techniques, a much safer approach is possible: inserting the gene encoding the microbe's surface protein into the DNA of a harmless virus. As illustrated in figure 9.10, the harmless virus now carries the added gene "piggyback" wherever it goes. The modified but still harmless virus becomes an effective and safe **piggyback vaccine.** Injected into a human body, the surfaces of the modified virus display the piggyback surface protein like battle flags. Responding to this challenge, the immune system makes antibodies that attack any cells displaying that protein. As a result, the body is thereafter protected against infection by the microbe from which the piggyback gene was obtained.

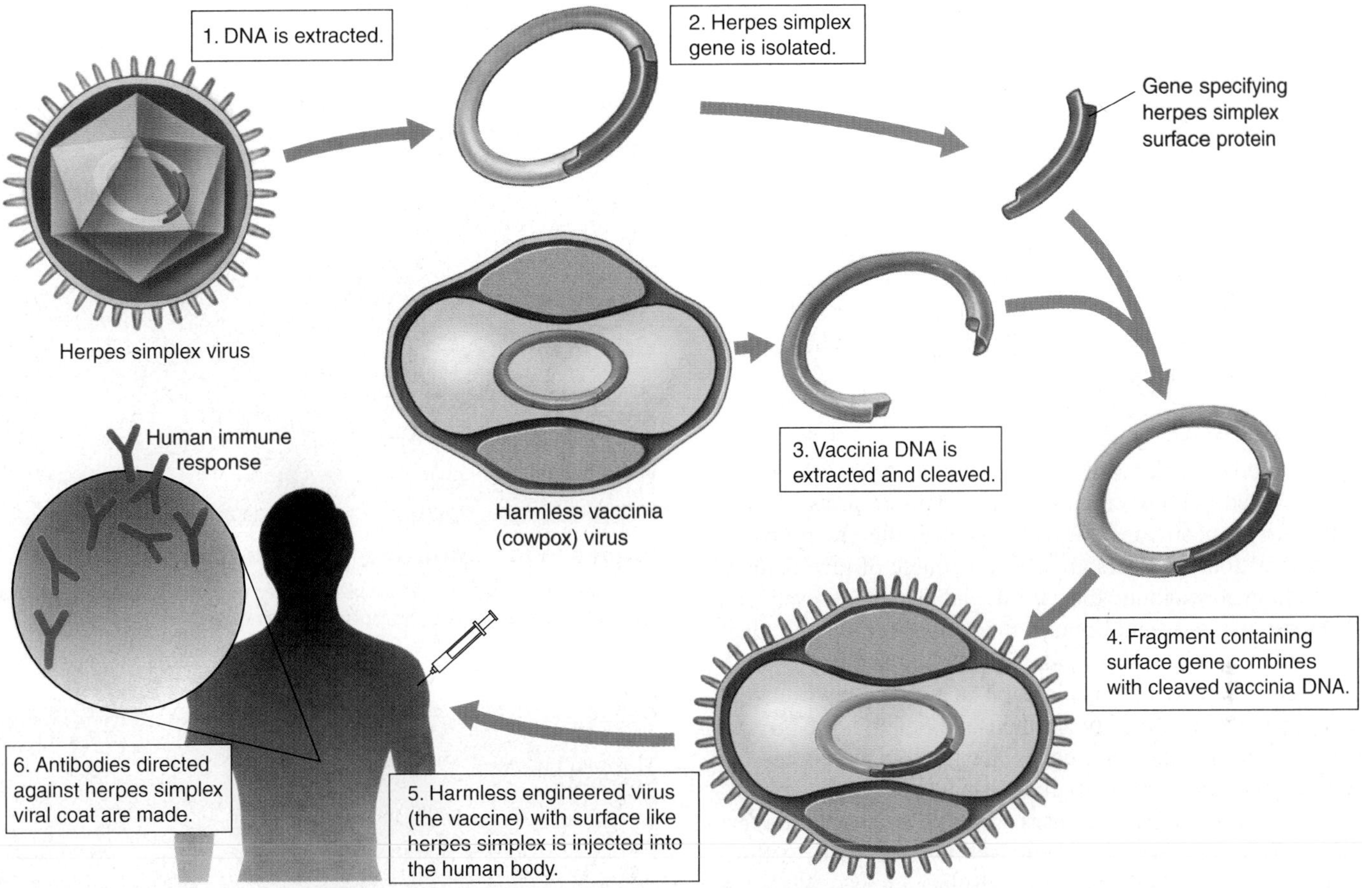
1. DNA is extracted.
2. Herpes simplex gene is isolated.
Gene specifying herpes simplex surface protein
Herpes simplex virus
Harmless vaccinia (cowpox) virus
3. Vaccinia DNA is extracted and cleaved.
Human immune response
4. Fragment containing surface gene combines with cleaved vaccinia DNA.
6. Antibodies directed against herpes simplex viral coat are made.
5. Harmless engineered virus (the vaccine) with surface like herpes simplex is injected into the human body.

Figure 9.10 Constructing a piggyback vaccine for the herpes simplex virus.

Among the many other vaccines now being manufactured in this way are ones directed against the herpes simplex virus and hepatitis B virus, the latter the cause of a highly infectious and very dangerous disease.

Human Gene Therapy

In 1990 researchers first attempted to combat genetic defects by the transfer of human genes. Many so-called genetic diseases occur when a person lacks a particular gene because both chromosomes' copies are either defective or missing. One obvious way to cure such disorders is to give the person a working copy of the gene. With the techniques of genetic engineering, **gene therapy** is being attempted as a way of combating a variety of genetic disorders (table 9.2) such as cystic fibrosis and muscular dystrophy. Among the first successful attempts has been the transfer of an enzyme-encoding gene into the bone marrow of two girls suffering from a rare blood disorder caused by the lack of this enzyme.

TABLE 9.2 DISEASES BEING TREATED IN CLINICAL TRIALS OF GENE THERAPY

Disease	Disease
Cystic fibrosis	Fanconi's anemia
Familial hypercholesterolemia	Hunter's syndrome
Hemophilia	Rheumatoid arthritis
Alpha-1 antitrypsin deficiency	

9.5 Genetic engineering has facilitated the production of medically important proteins, has led to novel vaccines, and offers the promise of curing hereditary disorders.

9.6 Genetic Engineering of Crop Plants

One of the greatest impacts of genetic engineering on society has been the successful manipulation of the genes of crop plants to make them more resistant to disease, frost, and other forms of stress; to improve their nutritional balance and protein content; and to make them resistant to herbicides (chemicals that kill plants).

Pest Resistance

Other important efforts of genetic engineers in agriculture have involved making crops resistant to insect pests without spraying with pesticides, a great saving to the environment. Consider cotton. Its fibers are a major source of raw material for clothing throughout the world, yet the plant itself can hardly survive in a field because many insects attack it. Over 40% of the chemical insecticides used today are employed to kill insects that eat cotton plants. The world's environment would greatly benefit if these thousands of tons of insecticide were not needed. Biologists are now in the process of producing cotton plants that are resistant to attack by insects.

One successful approach uses a kind of soil bacterium that secretes Bt, a protein toxic when eaten by crop pests, such as larvae (caterpillars) of butterflies and other insects. When the gene producing the Bt protein was inserted into the chromosomes of tomatoes, the plants began to manufacture Bt protein, which made them highly toxic to tomato hornworms, one of the most serious pests of commercial tomato crops.

Many important plant pests also attack roots. To combat these pests, genetic engineers are introducing the *Bt* gene into different kinds of bacteria, ones that colonize the roots of crop plants. Any insect eating such roots consume the bacteria and so are lethally attacked by the enzyme.

Herbicide Resistance

A major advance has been the creation of crop plants that are resistant to the herbicide **glyphosate,** the active ingredient in Roundup, a powerful biodegradable herbicide. Roundup is used in orchards and agricultural fields to control weeds. Growing plants need to make a lot of protein, and glyphosate stops them from making protein by destroying an enzyme that manufactures a critical amino acid. Humans are unaffected by glyphosate because we don't make this amino acid anyway—we obtain it from plants we eat! To make crop plants resistant to this powerful plant killer, genetic engineers screened thousands of organisms until they found a species of bacteria that could make the amino acid in the presence of glyphosate. They then isolated the gene encoding the resistant enzyme and, using plasmids or DNA particle guns (figure 9.11), successfully introduced the gene into plants (figure 9.12).

Figure 9.11 Shooting genes into cells.
This DNA particle gun fires tungsten pellets coated with DNA into plant cells such as the ones on the culture plate this experimenter is holding.

Figure 9.12 Genetically engineered herbicide resistance.
All four of these petunia plants were exposed to equal doses of an herbicide. The two on *top* were genetically engineered to be resistant to glyphosate, the active ingredient in the herbicide, whereas the two dead ones on the *bottom* were not.

Agriculture is being revolutionized by herbicide-resistant crops for two reasons. First, it lowers the cost of producing a crop, because a crop resistant to Roundup does not need to be weeded. Also, the creation of glyphosate-resistant crops is of major benefit to the environment. Glyphosate is quickly broken down in the environment, which makes its use a great improvement over most commercial herbicides. Perhaps even more important, not having to plow to remove weeds reduces the loss of fertile topsoil to erosion, one of the greatest environmental challenges facing our country today.

More Nutritious Crops

In the last 10 years the cultivation of genetically modified crops of corn, cotton, and soybeans has become commonplace in the United States—in 1999, over half of the 72 million acres planted with soybeans in the United States were planted with seeds genetically modified to be herbicide resistant, with the result that less tillage has been needed, and as a consequence soil erosion has been greatly lessened. These benefits, while significant, have been largely confined to farmers, making their cultivation of crops cheaper and more efficient. The food that the public gets is the same, it just costs less to get it to the table.

Like the first act of a play, these developments have served mainly to set the stage for the real action, which is only now beginning to happen. The real promise of plant genetic engineering is to produce genetically modified plants with desirable traits that directly benefit the consumer.

One recent advance, nutritionally improved "golden" rice, gives us a hint of what is to come. In developing countries large numbers of people live on simple diets that are poor sources of vitamins and minerals (what botanists called "micronutrients"). Worldwide, the two major micronutrient deficiencies are iron, which affects 1.4 billion women, 24% of the world population, and vitamin A, affecting 40 million children, 7% of the world population. The deficiencies are especially severe in developing countries where the major staple food is rice. In recent research, Swiss bioengineer Ingo Potrykus and his team at the Institute of Plant Sciences, Zurich, have gone a long way towards solving this problem. Supported by the Rockefeller Foundation and with results to be made free to developing countries, the work is a model of what plant genetic engineering can achieve.

To solve the problem of dietary iron deficiency among rice eaters, Potrykus first asked why rice is such a poor source of dietary iron. The problem, and the answer, proved to have three parts:

1. *Too little iron.* The proteins of rice endosperm have unusually low amounts of iron. To solve this problem, a ferritin gene was transferred into rice from beans (figure 9.13). Ferritin is a protein with an extraordinarily high iron content, and so greatly increased the iron content of the rice.
2. *Inhibition of iron absorption by the intestine.* Rice contains an unusually high concentration of a chemical called phytate, which inhibits iron reabsorption in the intestine—it stops your body from taking up the iron in the rice. To solve this problem, a gene encoding an enzyme that destroys phytate was transferred into rice from a fungus.
3. *Too little sulfur for efficient iron absorption.* Sulfur is required for iron uptake, and rice has very little of it. To solve this problem, a gene encoding a particularly sulfur-rich metallothionin protein was transferred into rice from wild rice.

To solve the problem of vitamin A deficiency, the same approach was taken. First, the problem was identified. It turns out rice only goes part way toward making betacarotene (provitamin A); there are no enzymes in rice to catalyze the last four steps. To solve the problem, genes encoding these four enzymes were added to rice from a familiar flower, the daffodil.

Potrykus's development of transgenic rice to combat dietary deficiencies will directly improve the lives of millions of people. His work is representative of the very real promise of genetic engineering to help meet the challenges of the coming new millennium.

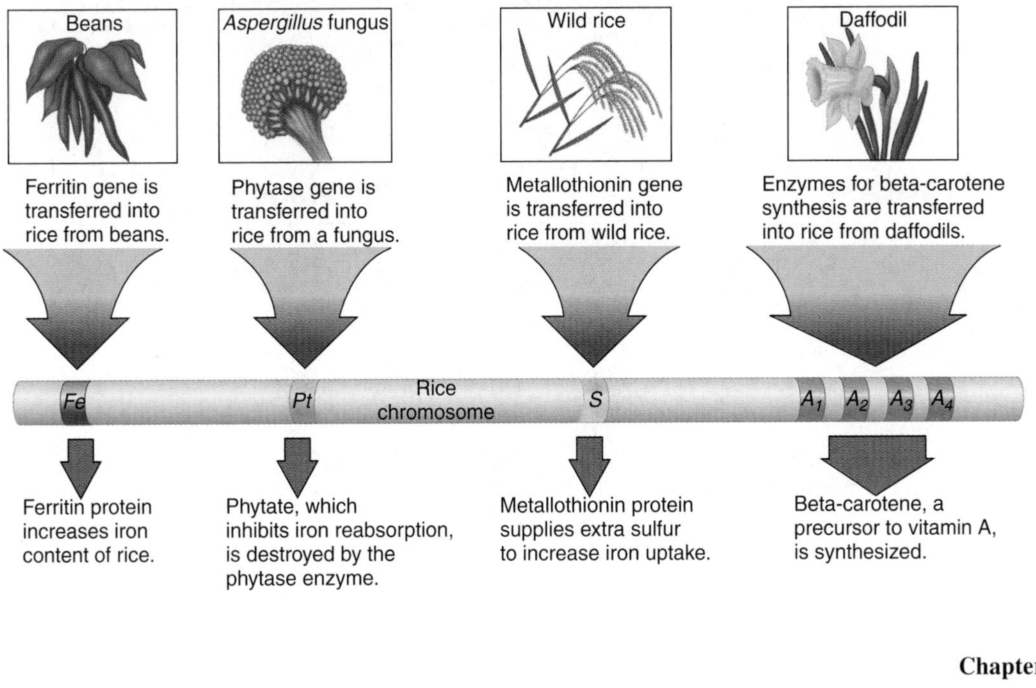

Beans
Aspergillus fungus
Wild rice
Daffodil
Ferritin gene is transferred into rice from beans.
Phytase gene is transferred into rice from a fungus.
Metallothionin gene is transferred into rice from wild rice.
Enzymes for beta-carotene synthesis are transferred into rice from daffodils.
Fe
Pt
Rice chromosome
S
A_1 A_2 A_3 A_4
Ferritin protein increases iron content of rice.
Phytate, which inhibits iron reabsorption, is destroyed by the phytase enzyme.
Metallothionin protein supplies extra sulfur to increase iron uptake.
Beta-carotene, a precursor to vitamin A, is synthesized.

Figure 9.13 Transgenic "golden" rice.

Developed by Swiss bioengineer Ingo Potrykus, transgenic rice offers the promise of improving the diets of people in rice-consuming developing countries, where iron and vitamin A deficiencies are a serious problem.

How Do We Measure the Potential Risks of Genetically Modified Crops?

The advantages afforded by genetic engineering are revolutionizing our lives. But many people, including influential members of the scientific community, have expressed concern. Could genetically engineered products administered to plants or animals turn out to be dangerous after several generations? What kind of unforeseen impact on the ecosystem might "improved" crops have? Two sets of risks need to be considered. The first stems from eating genetically modified foods; the other concerns potential ecological effects.

Is Eating Genetically Modified Food Dangerous? Protesters worry that genetically modified (GM) food may have been rendered somehow dangerous. To sort this out, it is useful to bear in mind that bioengineers modify crops in two quite different ways. One class of gene modification makes the crop easier to grow; a second class of modification is intended to improve the food itself.

The introduction of Roundup-resistant soybeans to Europe is an example of the first class of modification. Is the soybean that results nutritionally different? No. The gene that confers Roundup resistance in soybeans does so by protecting the plant's ability to manufacture so-called "aromatic" amino acids. In unprotected weeds, by contrast, Roundup blocks this manufacturing process, killing the weed. Because humans don't make any aromatic amino acids anyway (we get them in our diets), Roundup doesn't hurt us. The GM soybean we eat is nutritionally the same as an "organic" one, just cheaper to produce.

In the second class of modification, where a gene is added to improve the nutritional character of some food, the food will be nutritionally different. It is necessary to examine the possibility that consumers may prove allergic to the product of the introduced gene. For this reason, screening for allergy problems is now routine.

On both scores, then, the risk of bioengineering to the food supply seems to be very slight. GM foods to date seem completely safe.

Are GM Crops Harmful to the Environment? What are we to make of the much publicized report that Monarch butterflies might be killed by eating pollen blowing out of fields planted with GM corn? First, it should come as no surprise. The GM corn (so-called Bt corn) was engineered to contain an insect-killing toxin (harmless to people) in order to combat corn borer pests. Of course it will kill any butterflies or other insects in the immediate vicinity of the field. However, focus on the fact that the GM cornfields do not need to be sprayed with pesticide to control the corn borer. An estimated $9 billion in damage is caused annually by the application of pesticides in the United States, and billions of insects and other animals, including an estimated 67 million birds, are killed each year. This pesticide-induced murder of wildlife is far more damaging ecologically than any possible effects of GM crops on butterflies.

Will pests become resistant to the GM toxin? Not nearly as fast as they now have become resistant to the far higher levels of chemical pesticide we sprayed on crops.

How about the possibility that introduced genes will pass from GM crops to their wild or weedy relatives? This sort of gene flow happens naturally all the time, and so this is a legitimate question. But so what if genes for resistance to Roundup herbicide spread from cultivated sugar beets to wild populations of sugar beets in Europe? Why would that be a problem? Besides, there is almost never a potential relative around to receive the modified gene from the GM crop. There are no wild relatives of soybeans in Europe, for example. Thus there can be no gene escape from GM soybeans in Europe, any more than genes can flow from you to other kinds of animals.

9.6 Genetic engineering affords great opportunities for progress in food production, although many are concerned about possible risks. On balance, the risks appear slight, and the potential benefits substantial.

Calvin and Hobbes
by Bill Watterson
WHEN I GROW UP, I'M GOING TO BE A SCIENTIST.
I'LL DEDICATE MY CAREER TO THE PROPOSITION THAT MAN CAN RESHAPE THE UNIVERSE ACCORDING TO HIS OWN WHIMS.
I'LL PROBABLY GO INTO GENETIC ENGINEERING AND CREATE NEW LIFE FORMS.
YOU WANT TO PLAY GOD?
NOT EXACTLY. GOD NEVER BOTHERED TO PATENT HIS STUFF.
© 1995 Watterson/Dist. by Universal Press Syndicate

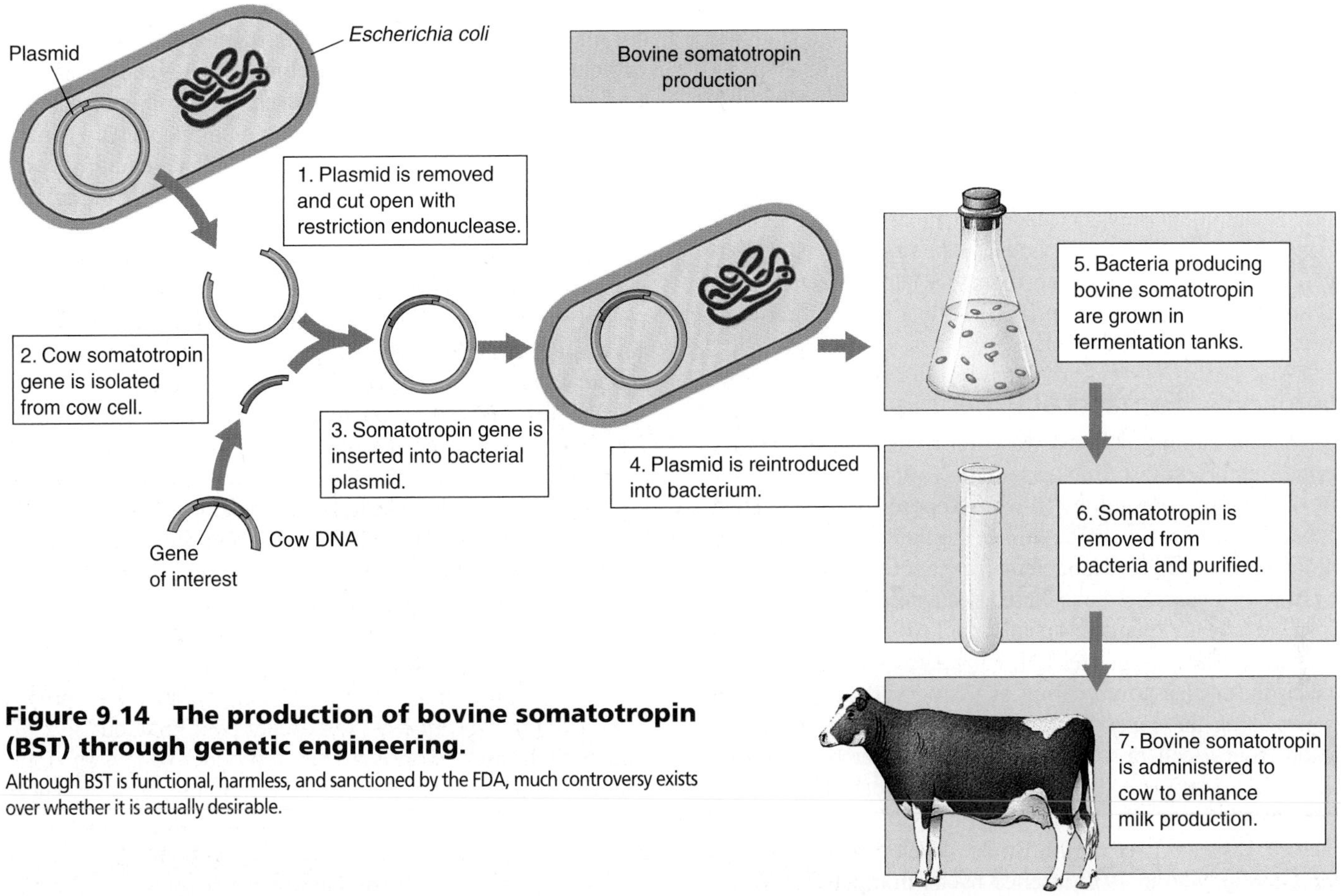
Plasmid
Escherichia coli
Bovine somatotropin production
1. Plasmid is removed and cut open with restriction endonuclease.
2. Cow somatotropin gene is isolated from cow cell.
3. Somatotropin gene is inserted into bacterial plasmid.
4. Plasmid is reintroduced into bacterium.
5. Bacteria producing bovine somatotropin are grown in fermentation tanks.
6. Somatotropin is removed from bacteria and purified.
7. Bovine somatotropin is administered to cow to enhance milk production.
Gene of interest
Cow DNA

Figure 9.14 The production of bovine somatotropin (BST) through genetic engineering.

Although BST is functional, harmless, and sanctioned by the FDA, much controversy exists over whether it is actually desirable.

9.7 Genetic Engineering of Farm Animals

The gene encoding the growth hormone somatotropin was one of the first to be cloned successfully. In 1994 Monsanto received federal approval to make its recombinant bovine somatotropin (BST) commercially available, and dairy farmers worldwide began to add the hormone as a supplement to their cows' diets, increasing the animals' milk production (figure 9.14). Genetically engineered somatotropin is also being tested to see if it increases the muscle weight of cattle and pigs (figure 9.15), and as a treatment for human disorders in which the pituitary gland fails to make adequate levels of somatotropin, producing dwarfism.

BST technology has proven controversial. Consumers worry that somehow the hormone will find its way into a cow's milk. In fact, this sort of transfer through a cow's milk-producing glands does not occur. Even if it did, the hormone ingested in milk would be expected to have little or no effect on humans, because it would be digested in the stomach. Consumers also worry, with some reason, that artificially enhanced levels of milk production may lead to a greater risk of udder infection among dairy cows.

Figure 9.15 A meatier pig.

Researchers are trying to increase the ratio of muscle mass to fat in pigs by introducing pig growth hormone.

9.7 Gene technology can result in the production of animals with desirable traits.

9.8 Cloning

The difficulty in using transgenic animals to improve livestock is in getting enough of them. Breeding produces offspring only slowly, and recombination acts to undo the painstaking work of the genetic engineer. Ideally, we would be able to "Xerox" many exact genetic copies of the transgenic strain, but until recently it was commonly accepted that animals can't be cloned. In 1997, however, scientists announced the first successful cloning of differentiated vertebrate tissue—a lamb grown from a cell taken from an adult sheep. This startling result promises to revolutionize agricultural science.

Spemann's "Fantastical Experiment"

The idea of cloning animals was first suggested in 1938 by German embryologist Hans Spemann (often called the "father of modern embryology"), who proposed what he called a "fantastical experiment": remove the nucleus from an egg cell, and put in its place a nucleus from another cell.

It was 14 years before technology advanced far enough for anyone to take up Spemann's challenge. In 1952, two American scientists, Robert Briggs and T. J. King, used very fine pipettes to suck the nucleus from a frog egg (frog eggs are unusually large, making the experiment feasible) and transfer a nucleus sucked from a cell of a more advanced frog embryo into its place. The experiment did not work when done this way, or when using early frog embryos, but partial success was achieved 18 years later by the British developmental biologist John Gurdon, who in 1970 inserted nuclei from advanced toad embryos and adult small intestine into toad eggs. The toad eggs developed into tadpoles, but almost all died before becoming adults.

The Path to Success

For 14 years, nuclear transplant experiments were attempted without success. Technology continued to advance, however, until finally in 1984, Steen Willadsen, a Danish embryologist working in Texas, succeeded in cloning a sheep using a nucleus from a cell of an early embryo. This exciting result was soon replicated by others in a host of other organisms, including cattle, pigs, and monkeys.

Only early embryo cells seemed to work, however. Researchers became convinced that animal cells become irreversibly "committed" after the first few cell divisions of the developing embryo, and after that, nuclei from differentiated animal cells cannot be used to clone entire organisms.

We now know this conclusion to have been unwarranted. The key advance for unraveling this puzzle was made in Scotland by geneticist Keith Campbell, a specialist in studying the cell cycle of agricultural animals. By the early 1990s, knowledge of how the cell cycle is controlled led to an understanding that cells don't divide until conditions are appropriate, just as a washing machine checks that the water has completely emptied before initiating the spin cycle. Campbell reasoned, "Maybe the egg and the donated nucleus need to be at the same stage in the cell cycle." This proved to be a key insight. In 1994 researcher Neil First, and in 1995 Campbell himself working with reproductive biologist Ian Wilmut, succeeded in cloning farm animals from advanced embryos by first starving the cells, so that they paused at the beginning of the cell cycle at the G_1 checkpoint. Two starved cells are thus synchronized at the same point in the cell cycle.

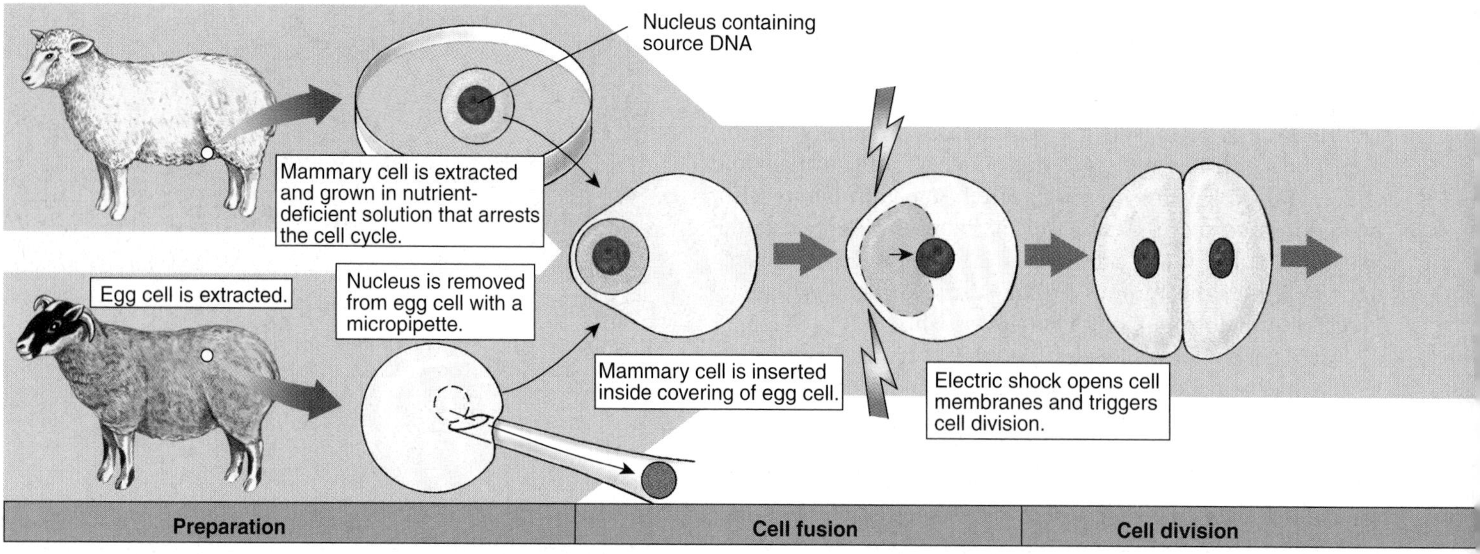

Nucleus containing source DNA
Mammary cell is extracted and grown in nutrient-deficient solution that arrests the cell cycle.
Egg cell is extracted.
Nucleus is removed from egg cell with a micropipette.
Mammary cell is inserted inside covering of egg cell.
Electric shock opens cell membranes and triggers cell division.
Preparation
Cell fusion
Cell division

Figure 9.16 Wilmut's animal cloning experiment.
Wilmut combined a nucleus from a mammary cell and an egg cell (with its nucleus removed) to successfully clone a sheep.

Wilmut's Lamb

Wilmut then set out to attempt the key breakthrough, the experiment that had eluded researchers since Spemann proposed it fifty-nine years ago: to transfer the nucleus from an adult differentiated cell into an enucleated egg, and allow the resulting embryo to grow and develop in a surrogate mother, hopefully producing a healthy animal.

Wilmut removed mammary cells from the udder of a six-year-old sheep (figure 9.16). The origin of these cells gave the clone its name, "Dolly," after the country singer Dolly Parton. The cells were grown in tissue culture, and some were frozen so that in the future it would be possible with genetic fingerprinting to prove that a clone was indeed genetically identical with the cells from the six-year-old sheep.

In preparation for cloning, Wilmut's team greatly reduced for five days the concentration of serum on which the sheep mammary cells were subsisting. In parallel preparation, eggs obtained from a ewe were enucleated, the nucleus of each egg carefully removed with a micropipette.

Mammary cells and egg cells were then surgically combined in January 1996; the mammary cells were inserted inside the covering around the egg cell. Wilmut then applied a brief electrical shock. This causes the plasma membranes surrounding the two cells to become leaky, so that the contents of the mammary cell pass into the egg cell. The shock also kick-starts the cell cycle, causing the cell to begin to divide.

After six days, in 30 of 277 tries, the dividing embryo reached the hollow-ball blastula stage, and 29 of these were transplanted into surrogate mother sheep. A little over five months later, on July 5, 1997, one sheep gave birth to a lamb. This lamb, Dolly, was the first successful clone generated from a differentiated animal cell.

The Future of Cloning

Wilmut's successful cloning of fully differentiated sheep cells is a milestone event in gene technology. Even though his procedure proved inefficient (only 1 of 277 trials succeeded), it established the point beyond all doubt that cloning of adult animal cells *can* be done. In the following four years researchers succeeded in greatly improving the efficiency of cloning. Seizing upon the key idea in Wilmut's experiment, to clone a resting-stage cell, they returned to the nuclear transplant procedure pioneered by Briggs and King. It works well. Many different mammals have been successfully cloned, including mice, pigs, and cattle.

Transgenic cloning can be expected to have a major impact on medicine as well as agriculture. Animals with human genes can be used to produce rare hormones. For example, sheep that have recently been genetically engineered to secrete a protein called alpha-1 antitrypsin (helpful in relieving the symptoms of cystic fibrosis) into their milk may be cloned, greatly cheapening the production of this expensive drug.

It is hard not to speculate on the possibility of cloning a human. There is no reason to believe such an experiment would not work, but there are many reasons to question whether it should be done. Because much of Western thought is based on the concept of human individuality, we can expect the possibility of human cloning to engender considerable controversy.

9.8 Recent experiments have demonstrated the possibility of cloning differentiated mammalian tissue, opening the door for the first time to practical transgenic cloning of farm animals.

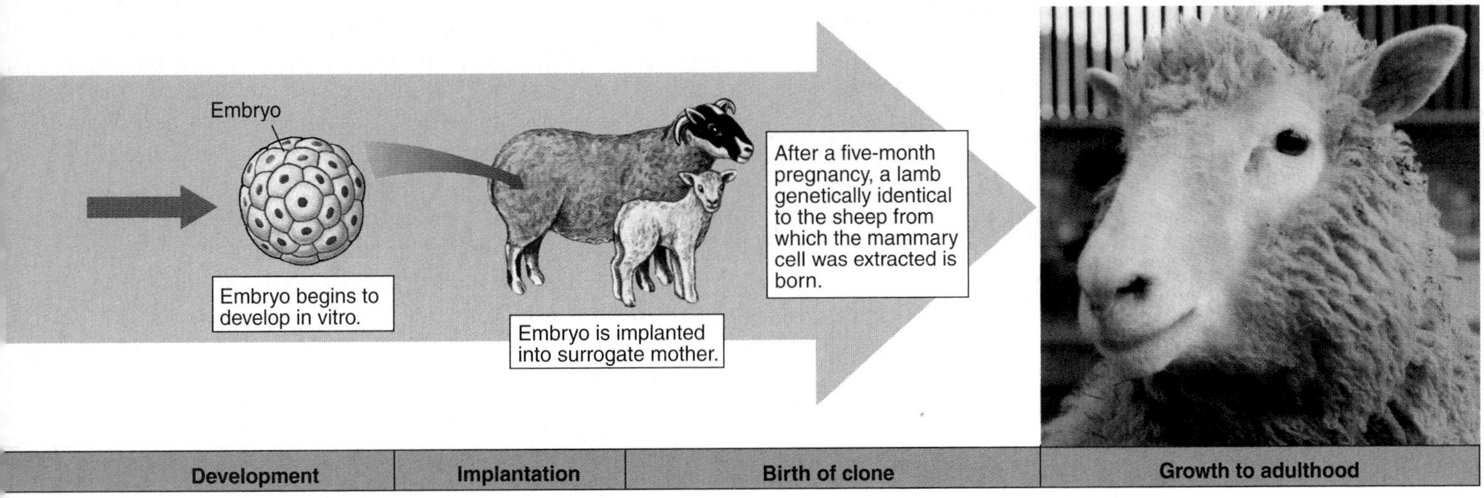
Embryo
Embryo begins to develop in vitro.
Embryo is implanted into surrogate mother.
After a five-month pregnancy, a lamb genetically identical to the sheep from which the mammary cell was extracted is born.
Development
Implantation
Birth of clone
Growth to adulthood

9.9 Ethical Issues

Few scientific advances have created a media uproar like that seen over Dolly, the lamb cloned in Scotland from an adult sheep. While the advance is of undoubted scientific importance, and may aid agriculture and medicine substantially, the same might be said of any number of other scientific advances in recent years. What is it about Dolly that creates such fear and unrest among the general public?

The fear of human cloning is of course at the heart of the public unease. While cloning procedures are still inefficient, there seems to be no technical reason why it would not be possible to clone a human. Indeed, Italian scientists announced in 2001 that they were proceeding to do just that. Their plan to clone children for infertile couples has been met with a storm of protest. Because cloning procedures are so inefficient, most embryos in such a procedure would not be expected to survive. Is it ethical to form so many human embryos condemned to die, in order to garner one that will live? The prospect of cloning humans raises a host of such issues.

Even if the technology is perfected, the cloning of humans raises profound issues of personal identity. If you were to clone a child, the initial cell of the clone, the cell that would go on to form a child in your image, would be in every way the same as the cell that made YOU. What is there to worry about in this? What is so wrong about this cell, that was so right about you?

An obvious answer is that we suppose individuality in every person. Deeply ingrained in our culture is the supposition that we are all different from one another, and that these differences are an essential part of the human condition. When Jefferson wrote that "We hold these truths to be self-evident, that all men are created equal..." he didn't mean that we are all clones, genetically equal to one another, but rather that, despite all our differences, every person ought to be equal under the law. Why treat every person equally? Because, when we were conceived, the genetic dice were rolled for all of us equally. How we turn out as people, Jefferson imagined, depended importantly on opportunity, which should be the same for all. This sense of individual self-worth lies at the heart of Jeffersonian democracy, and is an integral part of our laws and how we look at each other. It is what we mean by the word "individual." Yet, under law, a clone would have every right you do.

You see, it never occurred to Jefferson that it might be possible to load the genetic dice. The possibility of controlling what people will be like, cloning identical offspring of specific individuals, would probably have horrified Jefferson, for the same reason it disturbs so many people today: deeply and fundamentally, we each KNOW ourselves to be unique, special, unlike any other person. It is deeply disquieting to imagine otherwise.

Careful studies of identical twins over the last decade have produced very strong evidence that much of the variation in personality and intelligence between humans is highly heritable—that is, a substantial portion of the variation among individuals in such traits is the result of genetic differences between them. Only some 30% of the variation among individuals in these traits reflects differences in experience, in how they were raised and educated. A bunch of clones would differ from one another by just that much. Is 30% of the variation among us enough to sustain our view of individual uniqueness?

Looming within the blizzard of ethical issues surrounding human cloning is the temptation to yield to the understandable urge to "better" humankind, to achieve in the short run what evolution strives for in the long run. The danger, of course, lies in assuming we know the answer to the challenge of the future.

But the temptation may be difficult to resist. With the human genome sequenced, it will not be long before the genes responsible for most hereditary disorders are identified. Advances in gene therapy are increasingly allowing the transfer of healthy genes to replace defective ones. As cloning technology is perfected, it becomes possible to contemplate creating a child by cloning one of the parents, in the process "correcting" any gene defects. A perfect child is the hope of every parent. It is difficult to imagine a more cloudy ethical issue.

An even more profound problem arises from another direction, one that does not cause public unrest because few of us think in its broad-brush terms: because cloning promotes genetic uniformity, making our genes more like each other's, cloning increases the danger that at some future time a disease might arise against which the "common" cloned form has no resistance. Genetic variation is the chief defense our species has against an uncertain future. To strip ourselves of it, even partially, is to endanger our species.

Asexual reproduction, in which all offspring are genetically identical clones, is common in nature in both plants (dandelions are a common example) and animals (some lizard species have only females), but usually only in extreme or high-risk environments, where survival is uncertain. Nature has not favored asexual reproduction in any mammal because the 30% of variation due to nurture is just not enough protection against an uncertain future if you are going to make a major investment in each offspring. It is thus the very nature of our species that places such value on variation among individuals, perhaps the deepest and most compelling reason to carefully consider the implications of human cloning before proceeding.

9.9 Most biologists are deeply concerned about the ethical issues raised by the potential cloning of humans. It will be important to carefully consider this step before proceeding.

CONCEPT REVIEW

1. In creating recombinant DNA, the sticky ends of source DNA fragments are joined to the sticky ends of _________ produced by the cutting up of a circular bacterial chromosome.

a. clones

b. antibiotics

c. plasmids

d. restriction enzymes

2. In gel electrophoresis, larger DNA fragments move _________ distances when an electric current is applied.

a. longer

b. shorter

3. A clone library is

a. a collection of different lines of bacterial cells, each containing a plasmid carrying one series of different DNA fragments.

b. a collection of books about clones.

c. the particular bacterial line that contains the DNA fragment of interest.

d. a series of restriction enzymes that can be used to produce different clones.

4. PCR is a technique used to

a. create clone libraries.

b. transfer genes.

c. chop up source DNA.

d. amplify a region of DNA.

5. A version of DNA that is made from processed RNA is called

a. a plasmid.

b. cDNA.

c. a probe.

d. mDNA.

6. In piggyback vaccines, a plasmid containing DNA from the disease-causing microbe and from a harmless virus is inserted into a human host. Why is the human body able to form antibodies against the vaccine versus being infected with the disease?

a. The plasmid contains the gene that codes for the microbe's surface proteins but is otherwise harmless.

b. A virulent form of the disease is introduced into the body.

c. PCR was used to create many copies of the gene that encodes the microbe's surface protein.

d. The plasmid acts as a "magic bullet."

7. The transfer of working copies of genes from normal, healthy individuals to afflicted individuals is known as

a. a piggyback vaccine.

b. cloning.

c. human gene therapy.

d. DNA fingerprinting.

8. ________ is the last of the four stages of genetic engineering.

9. The enzyme used to create cDNA from mRNA is called _________.

CHALLENGE YOURSELF

1. A major focus of genetic engineering has been the attempt to produce large quantities of scarce human metabolites by placing the appropriate human genes into bacteria. Human insulin is now manufactured this way. However, if we attempt to use this approach to produce human hemoglobin (β-globin), the experiment does not work. Even if the proper clone from a clone library is identified, the fragment containing the β-globin gene is successfully incorporated into a plasmid, and *E. coli* are successfully infected with the chimeric plasmid, no β-globin is produced by the infected cells. Why doesn't the experiment work?

2. In Michael Crichton's 1990 novel, *Jurassic Park,* genetic engineers re-create dinosaurs from samples of fossilized dinosaur DNA. The story relies in part on discoveries and technology that already exist. For example, mosquitoes and other blood-sucking insects dating from the age of the dinosaurs have been found preserved in amber, researchers have been successful in obtaining and analyzing DNA contained in 18-million-year-old fossils of leaves, and transgenic animals have been produced after foreign genes were inserted into the eggs from which they developed. Given these successes, why hasn't the *Jurassic Park* experiment been completed in reality?

9 eBRIDGE

Reinforcing Key Points

Genetic Engineering

9.1 A Scientific Revolution

9.2 Restriction Enzymes

9.3 The Four Stages of a Genetic Engineering Experiment

9.4 Other Genetic Techniques

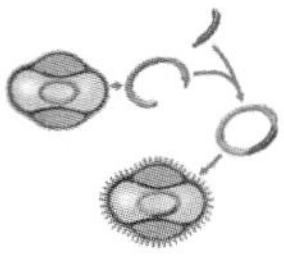

Advances in Medicine

9.5 Genetic Engineering and Medicine

Transforming Agriculture

9.6 Genetic Engineering of Crop Plants

9.7 Genetic Engineering of Farm Animals

9.8 Cloning

9.9 Ethical Issues

Electronic Learning

Enhancement Chapter

Stem Cells: The Revolution in Cell Technology

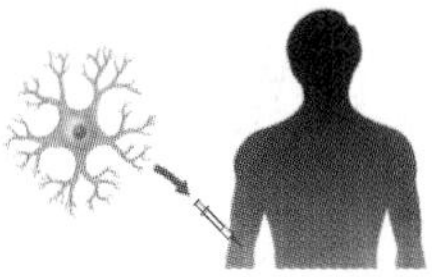

The issue of stem cells has recently burst upon the public awareness. This enhancement chapter introduces you to this revolutionary cell technology, which promises to open a new field of regenerative medicine but whose application to humans is highly controversial.

Explorations

DNA Fingerprinting: You Be the Judge

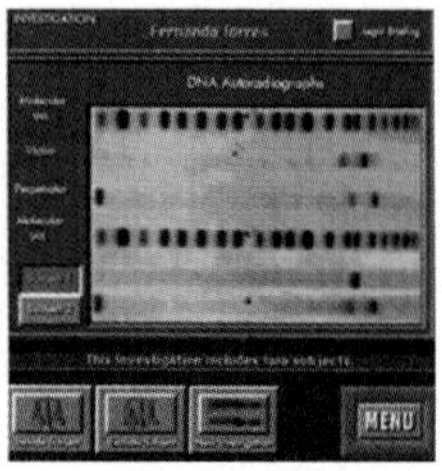
DNA Autoradiographs
MENU

In this interactive exercise, you analyze the DNA evidence presented in real courtroom trials, attempting to ascertain the guilt or innocence of the suspect in each instance. The DNA samples from the victim, perpetrator, and suspect are presented from a library of real cases, and you choose the DNA probes to use and then compare the evidence.

Author's Corner

Genetic Engineering. The ability to move specific genes from one organism to another is revolutionizing biology. But while the potential to improve the human condition is riveting the scientific community, many people are scared that genetic engineers are altering nature with consequences we can neither predict nor control. Much of the controversy has centered on gene-modified foods.

1. How genetic engineering is done.
2. The real promise of genetic engineering.
3. Finding useful genes: Cassava's relatives.
4. Measuring the risks of bioengineering.
5. Should we label genetically modified crops?
6. Gene engineers should renounce the terminator.
7. Frankenstein grass is poised to invade our yards.

Virtual Classroom

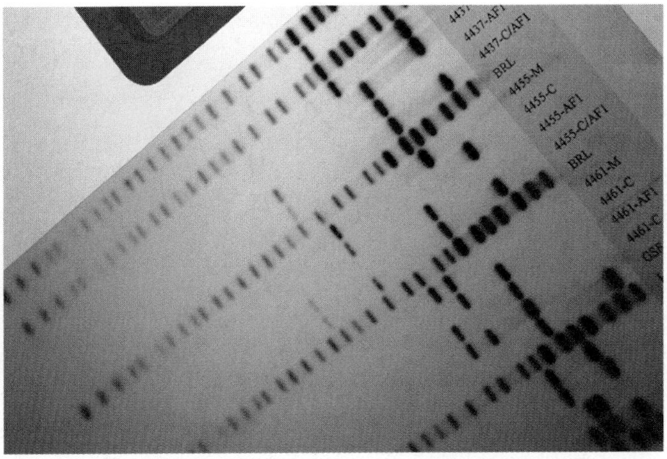

How Genetic Engineering Is Done

All of genetic engineering relies on the surprising discovery of enzymes that make offset cuts in double-stranded DNA—one strand is cut several bases from the other. This creates complementary single-stranded "tails" on the fragments, so that any two genes excised by the same enzyme can be stuck together! The revolution in genetic engineering that is changing modern biology all derives from this one surprising discovery.

Assessing the Risks of GM Foods

There has been considerable public discussion about the potential risks of genetically modified crops, so-called GM foods. The intense feelings generated by this dispute point to the need to understand how we measure the risks associated with gene manipulation of plants. Two sorts of risks need to be considered. First, is eating genetically modified food dangerous? Second, are GM crops potentially harmful to the environment? On both questions, there is profound public distrust on the one hand, and practically universal scientific consensus that GM crops are safe, on the other.

Virtual Lab

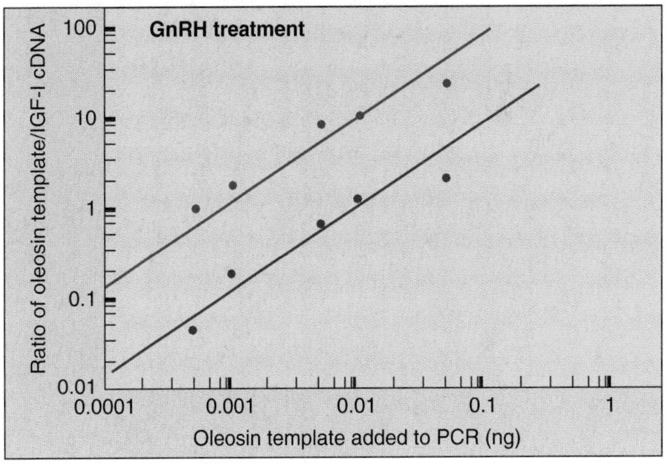
GnRH treatment
Ratio of oleosin template/IGF-I cDNA
100
10
1
0.1
0.01
0.0001
0.001
0.01
0.1
1
Oleosin template added to PCR (ng)

Trading Hormones Among Fishes: Gene Technology Lets Us Watch What Happens

The polymerase chain reaction (PCR) is a technique that allows a researcher to quickly and easily generate many copies of a segment of DNA. Prior to the development of this procedure, researchers had amplified DNA sequences by the laborious procedure of inserting them into bacterial DNA, growing large cultures of the sequence-carrying cells, and then harvesting the desired DNA. The PCR method is both faster and far easier.

Researchers are now expanding the uses of PCR beyond the simple amplification of DNA. Hamid Habibi, Maurice Moloney, and colleagues at the University of Calgary have developed a new laboratory method that uses PCR to quantify the relative levels of DNA in a sample, a method they call Quantitative PCR (Q-PCR). The group had set out to isolate an insulin-like growth factor (IGF-I) from goldfish liver, following a treatment with growth hormone or gonadotropin-releasing hormone. The researchers developed their competitive quantitative PCR technique to provide a highly accurate way of measuring IGF-I mRNA levels in treated goldfish livers.

Quizzes | Further Reading | Essential Study Partner | Links | BioCourse.com

CHAPTER 10

GENOMICS

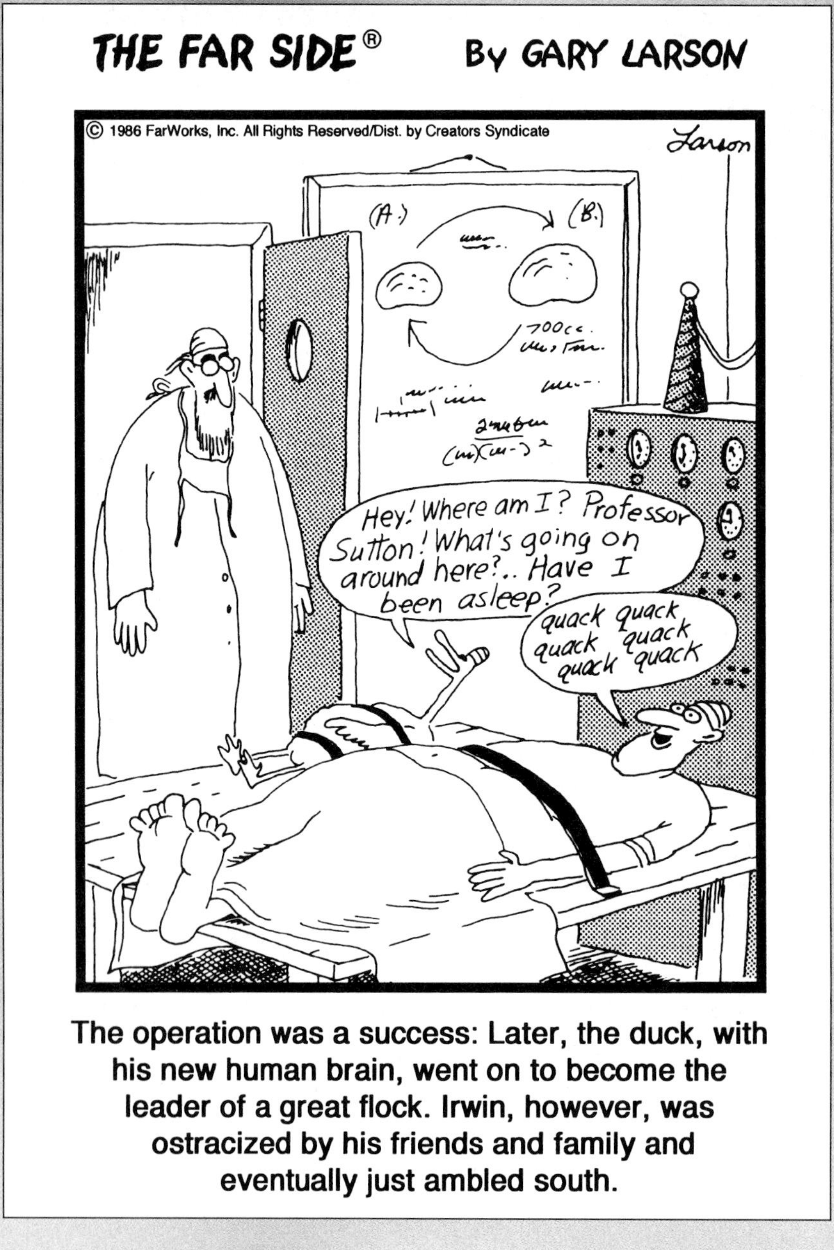
THE FAR SIDE® By GARY LARSON
© 1986 FarWorks, Inc. All Rights Reserved/Dist. by Creators Syndicate
Larson
Hey! Where am I? Professor Sutton! What's going on around here?.. Have I been asleep?
quack quack quack quack quack quack
The operation was a success: Later, the duck, with his new human brain, went on to become the leader of a great flock. Irwin, however, was ostracized by his friends and family and eventually just ambled south.

CHAPTER 10

Genomics

CHAPTER OVERVIEW

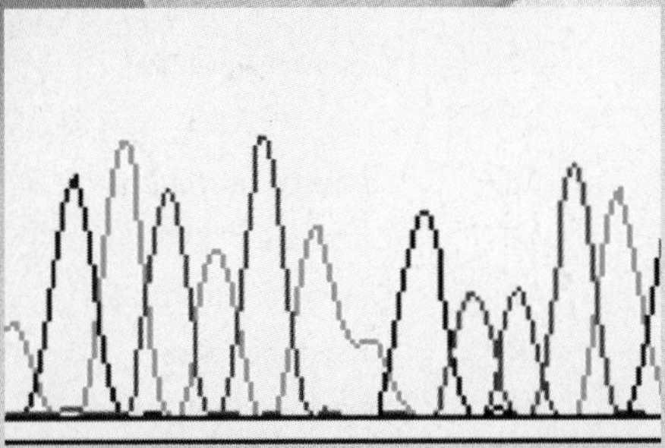

The Challenge of Sequencing Entire Genomes

10.1 Genomics

10.2 The Human Genome Project

10.3 What the Human Genome Is Like

- The full complement of genetic information of an organism is called its genome.
- The human genome contains about 30,000 genes scattered randomly about the chromosomes.
- Only about 1% of human DNA is devoted to protein-encoding genes.
- Much of the rest of the human genome is composed of transposable elements.

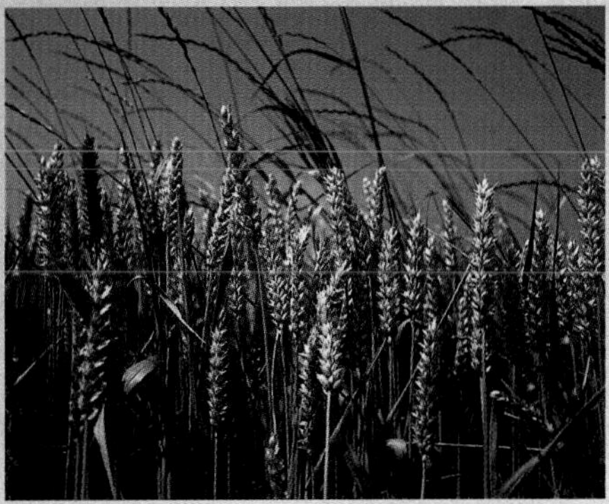

Genome Evolution

10.4 Variation Within the Genome

- An average person differs from the consensus human genome sequence in about 1 out of every 1,200 nucleotides, a total of some 300,000 differences.
- A single nucleotide difference is called a SNP (single nucleotide polymorphism).
- The human genome has acquired numerous genes from bacteria.
- All animal genomes have the same basic set of development-controlling genes, but use them differently.

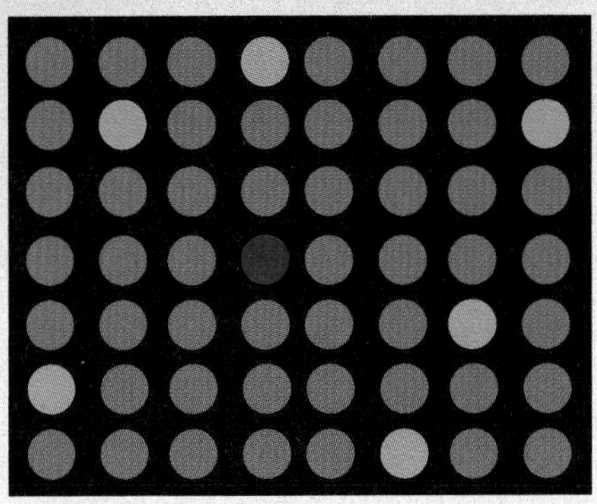

Putting Genomic Information to Work

10.5 Gene Microarrays

10.6 Functional Genomics

10.7 Proteomics

- Gene microarrays are stamp-sized glass chips containing many thousands of gene fragments, allowing rapid screening of gene profiles.
- Functional genomics is the identification of the functions of the genome's many genes.
- Proteomics is the analysis of all the proteins encoded by a genome.

THE CHALLENGE OF SEQUENCING ENTIRE GENOMES

10.1 Genomics

Recent years have seen an explosion of interest in comparing the entire DNA content of different organisms, a new field of biology called **genomics.** While initial successes focused on organisms with a relatively small number of genes, researchers have recently completed the sequencing of several large eukaryotic genomes, including our own.

The full complement of genetic information of an organism—all of its genes and other DNA—is called its **genome.** The first genome to be sequenced was a very simple one: a small bacterial virus called X174. Frederick Sanger, inventor of the first practical ways to sequence DNA, obtained the sequence of this 5,375-nucleotide genome in 1977.

This first genome sequence was a milestone, because it demonstrated that we can locate the exact coding regions for each and every gene from the sequence. Said simply, we now knew we would be able to read the information in the hereditary "library." How do we recognize the presence of a gene within a long sequence of DNA nucleotides? It starts with a "start" codon like ATG, and when translated in three-nucleotide codon units contains no "stop" codons (UAA, UGA, or UAG) for a distance long enough to encode a protein. This "coding region" is referred to as an **open reading frame.** Open reading frames are the translated genes of genomes, the sequences that encode proteins.

A Genomic Time Line

The advent of automated DNA sequencing machines in recent years has made the DNA sequencing of entire genomes practical (table 10.1). Researchers first focused on microorganisms with relatively small genomes, on the order of a few million nucleotide base pairs (Mb). The first free-living organism was sequenced in 1995, the bacterium *Haemophilus influenzae* with 1,830,137 nucleotide base pairs. About half of its 800 genes prove to have a known function. What the other half of the genes are doing is a complete mystery.

The first eukaryotic genome to be sequenced in its entirety was the baker's yeast *Saccharomyces cerevisiae,* 13 million base pairs (Mb) encoding about 6,000 genes. This was followed in 1997 by *E. coli,* a bacterium that has been the focus of much research over the last 50 years.

The first animal to have its DNA completely read was the nematode *C. elegans* (100 Mb) in December 1998, followed by the fruit fly *Drosophila* (120 Mb), and the mouse (300 Mb). The plants *Arabidopsis* (120 Mb) (figure 10.1) and rice (430 Mb) were completed in the year 2001.

10.1 Powerful automated DNA sequencing technology has begun to reveal the DNA sequences of entire organisms.

TABLE 10.1 GENOME SEQUENCING PROJECTS

Organism	Genome Size (Mb)	Description
Archaebacteria		
Methanococcus jannaschii	1.7	Extreme thermophile
Eubacteria		
Mycoplasma genitalium	0.6	Smallest known organism
Helicobacter pylori	1.7	Causes ulcers
Vibrio cholerae	4.0	Causes often fatal disease
Mycobacterium tuberculosis	4.4	Causes tuberculosis
Escherichia coli	4.6	Used as a laboratory standard
Fungi		
Saccharomyces cerevisiae	13	Baker's yeast
Protist		
Plasmodium	30	Malarial parasite
Plant		
Arabidopsis thaliana	120	Relative of mustard plant
Oryza sativa	430	Commercial rice
Animal		
Caenorhabditis elegans	100	Nematode
Drosophila melanogaster	120	Fruit fly
Mus musculus	300	Mouse
Homo sapiens	3,200	Human

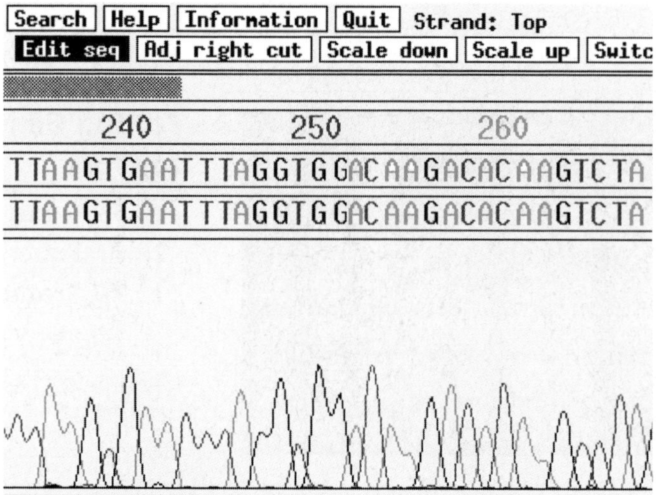
Search
Help
Information
Quit
Strand: Top
Edit seq
Adj right cut
Scale down
Scale up
240
250
260
TTAAGTGAATTTAGGTGGACAAGACACAAGTCTA
TTAAGTGAATTTAGGTGGACAAGACACAAGTCTA

Figure 10.1 Part of the genome sequence of the plant *Arabidopsis*.

Data from an automated DNA-sequencing run show the nucleotide sequence for a small section of the *Arabidopsis* genome.

10.2 The Human Genome Project

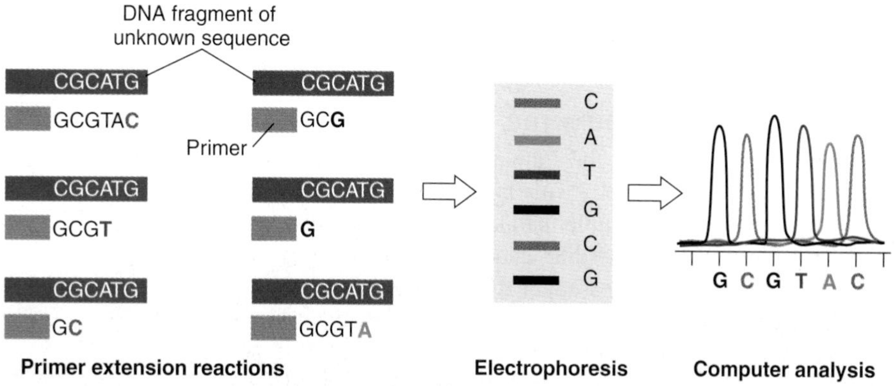
DNA fragment of unknown sequence
CGCATG
GCGTAC
CGCATG
GCG
Primer
CGCATG
GCGT
CGCATG
G
CGCATG
GC
CGCATG
GCGTA
C
A
T
G
C
G
G C G T A C
Primer extension reactions
Electrophoresis
Computer analysis

Figure 10.2 How to sequence DNA.

A DNA strand is sequenced by adding complementary bases to it, and looking to see which base is added at each stage in the process of assembling the new chain (see text for a more detailed explanation of this process).

In 1990 American geneticists embarked on an ambitious attempt to map and ultimately sequence the entire human genome. This effort, which quickly became an international program, presented no small challenge, as the human genome is huge—more than 3 billion base pairs. To get an idea of the magnitude of the task, consider that if all 3.2 billion base pairs were written down on the pages of this book, the book would be 500,000 pages long and it would take you about 60 years, working eight hours a day, every day, at 5 bases a second, to read it all.

In sequencing DNA, a DNA fragment of unknown sequence is first amplified, so there are thousands of copies of the fragment. The fragments are then mixed with a primer, DNA polymerase, a supply of the four nucleotide bases, and a supply of four different chain-terminating chemical tags that each can act as one of the four nucleotide bases in DNA synthesis. After heat is applied to denature the double-stranded DNA, the primer binds to one strand of the DNA, and synthesis of the complementary strand proceeds. Whenever a tag is added instead of a nucleotide base, the synthesis stops. However, because of the relatively low concentration of the chemical tags compared to the nucleotides, a tag that binds to A on the DNA fragment, for example, will not necessarily be added to the first A site. Thus, the mixture will contain a series of fragments of different lengths, corresponding to the different distances the polymerase traveled from the primer before a chain-terminating tag was incorporated.

The series of fragments are then separated according to size by gel electrophoresis. The fragments become arrayed like the rungs of a ladder, each rung one base longer than the one below it. In the manual method of sequencing DNA (the Sanger method), the synthesized fragments are radioactively labeled and are visualized on X-ray film in four different columns that correspond to the four different nucleotide bases. The DNA sequence can then be read directly from the film by researchers. This time-consuming procedure was improved with the development of automated DNA sequencing. In this method, the chemical tags used are fluorescently colored, one color corresponding to each nucleotide. Computers read off the colors on the gel in order to determine the DNA sequence and display this sequence as a series of colored peaks (figure 10.2). What has made the attempt to sequence 3.2 billion bases practical was the development in the mid-1990s of automated sequencers that perform electrophoresis of DNA fragments in capillary tubes instead of the traditional gel slabs. These systems can handle about 1,000 samples a day, with only 15 minutes of human attention. An institute with several hundred such instruments can produce about 100 Mb (million base pairs) every day.

Two Strategies for Sequencing Such a Large Genome

The original plan for the publicly financed Human Genome Project was systematic and conservative. First, detailed genetic maps of each of the 23 human chromosomes would be prepared. For each segment of the map, fragments of DNA would be isolated and cloned into bacterial plasmids. Cloning allows investigators to get enough material to sequence the fragments. The map would then allow the sequenced fragments to be pieced together in the proper order.

Then, in May of 1998, the researcher who had sequenced the first bacterial genome, Craig Venter, announced he had established a private company to sequence the human genome. Venter proposed an astonishing schedule, proposing to finish a rough draft of the entire human genome in only two years. Instead of relying on a map, with the time-consuming ordering of clones used to build it, Venter proposed a **shotgun sequencing** approach in which the mapping step would be skipped altogether. Instead, the entire human genome would be chopped up, the fragments cloned, and the DNA sequence of each clone determined. Finally, the sequences would be pieced together using powerful computer programs that looked for overlaps between fragments.

The publicly financed venture rose to the challenge, and the two ventures raced to see who could complete the human genome first. The upshot of this race was a tie of sorts. On June 26, 2000, the two research groups jointly announced success.

10.2 The entire 3.2 billion base pair human genome has been sequenced, using automated machines to sequence random "shotgun cloned" fragments and powerful computers to order the sequenced fragments.

10.3 What the Human Genome Is Like

Geography of the Genome

The number of genes in the human genome is only about 30,000–40,000 (figure 10.3). This is barely a third more than in nematodes, scarcely double the number in *Drosophila,* and but a quarter of the number that had been anticipated by scientists counting unique messenger RNA (mRNA) molecules.

How can human cells contain four times as many kinds of mRNA as there are genes? A typical human gene is not simply a straight sequence of DNA, with the order of its units corresponding to the sequence of amino acids in a protein. Instead, a human gene is fragmented. The sequence of DNA nucleotides that specifies a protein is broken into many bits called exons, scattered among much longer segments of nontranslated DNA called introns. Imagine this paragraph was a human gene; all the occurrences of the letter "e" could be considered exons, while the rest would be noncoding introns.

When a cell uses a human gene to make a protein, it first manufactures mRNA copies of all the exons (protein-specifying fragments) of the gene, then splices the exons together. Now here's the turn-of-events researchers had not anticipated: the transcripts of human genes are often spliced together in different ways, called alternative splicing. Each exon is actually a module; one exon may code for one part of a protein, another for a different part of a protein. When the exon transcripts are mixed in different ways, very different protein shapes can be built.

With alternative mRNA splicing, it is easy to see how 30,000 genes can encode four times as many proteins. It seems that the added complexity of human proteins has been achieved not by gaining more gene parts, but rather by learning new ways to put them together. Great music is made from simple tunes in much the same way.

Genes are not distributed evenly over the genome. The small chromosome number 19 is packed densely with genes, transcription factors, and other functional elements. The much larger chromosomes numbers 4 and 8, by contrast, have few genes. On most chromosomes, vast stretches of seemingly barren DNA fill the chromosomes between scattered clusters rich in genes.

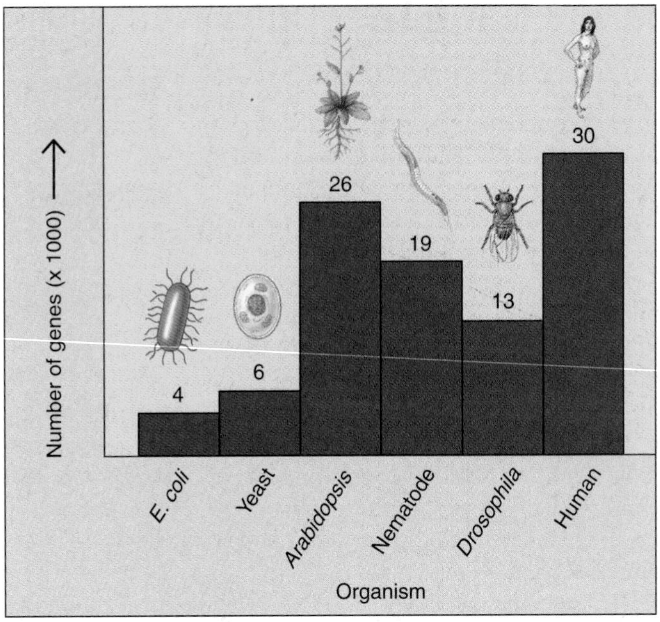

Number of genes (x 1000)
4
6
26
19
13
30
E. coli
Yeast
Arabidopsis
Nematode
Drosophila
Human
Organism

Figure 10.3 What the human genome is like.
The human genome has an unexpectedly small number of genes, some 30,000. This is not many more than the plant *Arabidopsis,* and only a third more than nematode worms.

DNA That Codes for Proteins

Four different classes of protein-encoding genes are found in the human genome, differing largely in gene copy number.

Single-copy genes. Many eukaryotic genes exist as single copies at a particular location on a chromosome. Mutations in these genes produce recessive Mendelian inheritance. Silent copies, inactivated by mutation, are called pseudogenes.

Segmental duplications. Sequencing of the human genome has revealed that human chromosomes contain many segmental duplications, where whole blocks of genes have been copied over from one chromosome to another. Blocks of similar genes in the same order are found throughout the genome, indicating segmental duplication has been a major factor in determining the human genome's present architecture. Chromosome 19 seems to have been the biggest borrower, with blocks of genes shared with 16 other chromosomes.

Multigene families. As we have learned more about the nucleotide sequences of eukaryotic genomes, it has become apparent that many genes exist as parts of multigene families, groups of related but distinctly different genes that often occur together in a cluster. Multigene families differ from tandem clusters (discussed next) in that they contain far fewer genes (from three to several hundred), and those genes differ much more from one another than the genes in tandem clusters. Despite their differences, the genes in a multigene family are clearly related in their sequences, making it likely that they arose from a single ancestral sequence through a series of unequal crossing over events.

Tandem clusters. A second class of repeated gene consists of DNA sequences that are repeated many times, one copy following another in tandem array. By transcribing all of the copies in these tandem clusters simultaneously, a cell can rapidly obtain large amounts of the product they encode. For example, the genes encoding rRNA are typically present in clusters of several hundred copies.

TABLE 10.2 CLASSES OF DNA SEQUENCES FOUND IN THE HUMAN GENOME

Class	Frequency	Description
Protein-encoding genes	1%	Translated exons, within some 30,000 genes scattered about the chromosomes
Introns	24%	Noncoding DNA comprising the great majority of most genes
Structural DNA	20%	Constitutive heterochromatin, localized near centromeres and telomeres
Repeated sequences	3%	Simple sequence repeats (SSRs) of a few nucleotides repeated millions of times
Transposable elements	45%	20% long interspersed elements (LINEs), active transposons 15% other transposable elements, including long terminal repeats (LTRs) 10% the parasite sequence (ALU), present in half a million copies

Noncoding DNA

One of the most notable characteristics of the human genome is the startling amount of noncoding DNA they possess. Only 1% to 1.5% of the human genome is coding DNA, devoted to genes encoding proteins. Each of your cells has about six feet of DNA stuffed into it, but of that, less than one inch is devoted to genes! Nearly 99% of the DNA in your cells seems to have little or nothing to do with the instructions that make you you. True genes are scattered about the human genome in clumps among the much larger amount of noncoding DNA, like isolated hamlets in a desert.

There are four major sorts of noncoding human DNA (table 10.2).

Noncoding DNA within genes. As we discussed on page 236, a human gene is made up of numerous fragments of protein-encoding information (exons) embedded within a much larger matrix of noncoding DNA (introns). Together, introns make up about 24% of the human genome, and exons about 1.5%.

Structural DNA. Some regions of the chromosomes remain highly condensed, tightly coiled, and untranscribed throughout the cell cycle. Called constitutive heterochromatin, these portions tend to be localized around the centromere, or located near the ends of the chromosome.

Repeated sequences. Scattered about chromosomes are simple sequence repeats (SSRs). An SSR is a two- or three-nucleotide sequence like CA or CGG, repeated like a broken record thousands and thousands of times. SSRs make up about 3% of the human genome. An additional 7% is devoted to other sorts of duplicated sequences. Repetitive sequences with excess C and G tend to be found in the neighborhood of genes, while A- and T-rich repeats dominate the non-gene deserts. The light bands on chromosome karyotypes now have an explanation—they are regions rich in GC and genes. Dark bands signal neighborhoods rich in AT and thin on genes. Chromosome 19, dense with genes, has few dark bands. Roughly 25% of the human genome has no genes at all.

Transposable elements. Fully 45% of the human genome consists of mobile bits of DNA called transposable elements. Discovered by Barbara McClintock in 1950 (she won the Nobel Prize for her discovery in 1983), transposable elements are bits of DNA that are able to jump from one location on a chromosome to another, tiny molecular versions of Mexican jumping beans.

Human chromosomes contain five sorts of transposable elements. Fully 20% of the genome consists of long interspersed elements (LINEs). An ancient and very successful element, LINEs are about 6 kb (6,000 DNA bases) long, and contain all the equipment needed for transposition, including genes for a DNA-loop-nicking enzyme and a reverse transcriptase.

Nested within the genome's LINEs are over half a million copies of a parasitic element called ALU, composing 10% of the human genome. ALU is only about 300 bases long, and has no transposition machinery of its own; like a flea on a dog, ALU moves with the LINE it resides within. Just as a flea sometimes jumps to a different dog, so ALU sometimes uses the enzymes of its LINE to move to a new chromosome location. Often jumping right into genes, ALU transpositions cause many harmful mutations.

Three other sorts of transposable elements are also present in the human genome. Eight percent of the genome is devoted to long terminal repeats (LTRs), also called "retroposons." Three percent is devoted to DNA transposons, which copy themselves as DNA rather than RNA. And, some 4% is devoted to dead transposons, elements that have lost the signals for replication and so can no longer jump.

10.3 Gene sequences in humans vary greatly in copy number, some occurring many thousands of times, others only once. Only about 1% of the human genome is devoted to protein-encoding genes. Much of the rest is comprised of transposable elements.

10.4 Variation Within the Genome

One of the most surprising results revealed by these genome sequences has been the discovery of just how similar living things are to one another at the genetic level. Forty-two percent of the genes discovered in *C. elegans* had some sort of match to genes in other organisms only distantly related. Fully 83% of *Drosophila* genes match those of other species. The matches are not perfect, however—the DNA sequences of genes that do the same job have drifted apart over millions of years. But functionality is maintained. For example, when a gene involved in eye development in mice was substituted for its homologue in *Drosophila,* the flies were born with normal, functional eyes.

Perhaps the most striking lesson learned from the sequence of the human genome is how very like other organisms humans are. More than half of the genes of *Drosophila* have human counterparts. Among mammals the differences are even fewer. Humans have only 300 genes that have no counterpart in the mouse genome. This suggests that when an ape genome is sequenced, it will possess practically all of the genes that humans do.

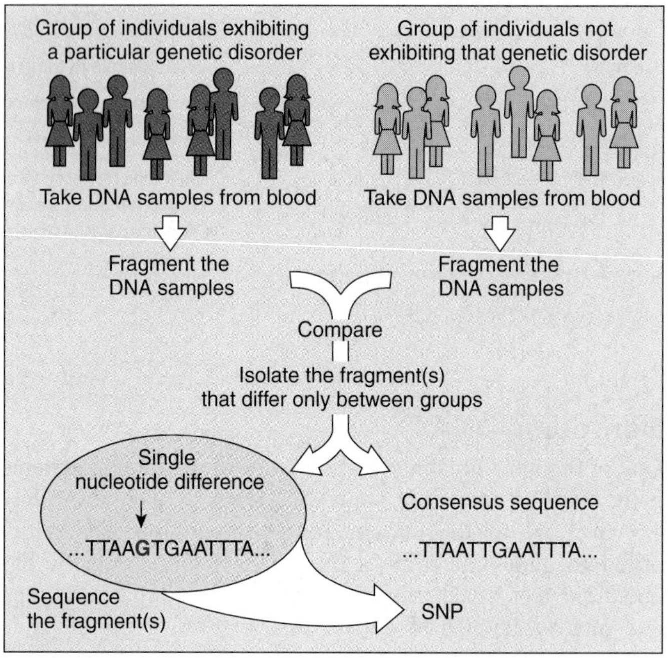
Group of individuals exhibiting a particular genetic disorder
Group of individuals not exhibiting that genetic disorder
Take DNA samples from blood
Take DNA samples from blood
Fragment the DNA samples
Fragment the DNA samples
Compare
Isolate the fragment(s) that differ only between groups
Single nucleotide difference
...TTAAGTGAATTTA...
Consensus sequence
...TTAATTGAATTTA...
Sequence the fragment(s)
SNP

Figure 10.4 Using SNPs to screen for genetic disorders.

Many genetic disorders result from point mutations in genes. Because SNPs are scattered randomly throughout the genome, there is a good chance that one or more SNPs will be located near any particular mutation. Once a SNP has been shown to be associated with a mutation causing a genetic disorder, it can be used as an indicator of the disease—just test individuals for the presence of that SNP.

Single Nucleotide Polymorphisms (SNPs)

There seems to be tremendous variation among individual human genomes. So far we have just two genomes belonging to anonymous individuals. The private company used a male with the code name Celer (a first-century architect who broke with Roman tradition and created a new architectural form). Twelve men and women contributed DNA to Mosaic Man's sequence, the public genome. However, most of the sequence was based on a single male.

SNPs (pronounced snips) appear to be the most effective way to analyze variation at the whole genome level. SNPs, or single nucleotide polymorphisms, are variations (polymorphisms) in single base pairs (figure 10.4). For example, you might have the sequence ACGCTCA, while a friend has the sequence AGGCTCA in the same gene. Two of every three SNPs are the result of replacement of C with T. About 1.42 million SNPs are distributed across our genome. That is, about one in every 1,200 bases that are mapped onto the human genome are SNPs. SNPs can serve as markers of particular human alleles, such as those that cause inherited diseases. About 600,000 SNPs have been identified in exons, and 85% of all exons are less than 5 kb from an identified SNP.

SNPs have a low rate of recurrent mutation, so they are excellent markers of human history. Iceland provides a clear example. Almost no immigration occurred in Iceland for more than a thousand years, so the gene pool among Icelanders today reflects Northern Europe about A.D. 800. The *BRCA2* breast cancer gene in Iceland has only one mutant allele, dating back to a mutation in a sixteenth-century cleric named Einar. A collective genetic and genealogical database being created in Iceland will provide a unique opportunity to study the evolution of SNPs within the human genome.

Genomic Exchanges Between Species

Comparisons of the genomes of different organisms suggests that genes have also moved laterally between organisms. For example, humans have several hundred genes similar to those of bacteria. Because these genes have no known counterpart in roundworms and fruit flies, researchers surmise that at least some of the genes may have been acquired from bacteria by lateral transfer. An example is the gene encoding monoamine oxidase, an important degradative enzyme for the central nervous system.

The many transposons of the human genome also provide a paleontological record of the human genome over several hundred million years. Comparisons among versions of a transposon that has duplicated many times allow researchers to construct a "family tree" to identify the ancestral form of the transposon. The percent sequence divergence of the duplicates allows the researcher to estimate the time when that particular transposon originally invaded the human genome millions of years ago.

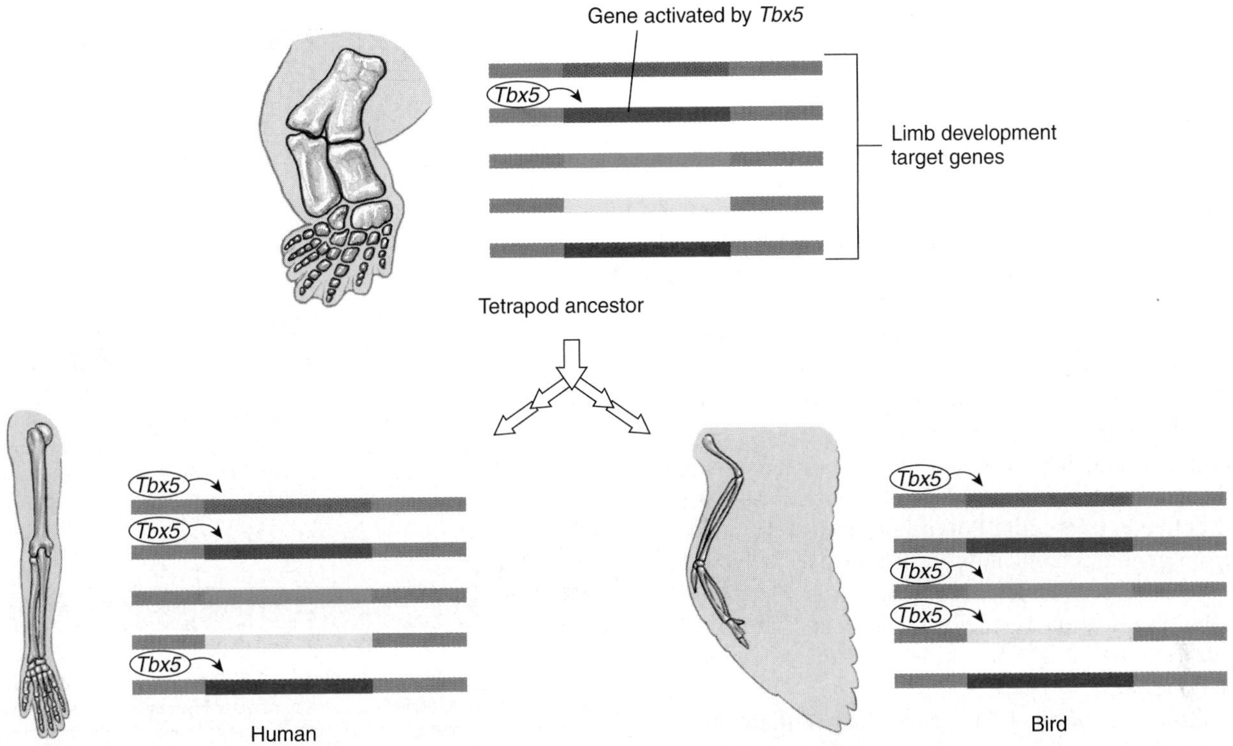

Gene activated by Tbx5
Tbx5
Limb development target genes
Tetrapod ancestor
Tbx5
Tbx5
Tbx5
Human
Tbx5
Tbx5
Tbx5
Bird

Figure 10.5 How can the same gene regulate the production of a wing and an arm?
Tbx5 is a gene with a key region called a T-box. Genes with T-boxes produce proteins that act as transcription factors, turning on genes. Tbx5 proteins turn on the genes that regulate an intricate limb-assembly developmental program that can result in the wing of a bird or a human arm. This same basic *Tbx5* tool gets used in different ways in different organisms, turning on different target genes (represented by the colored bars) in order to produce wing or arm development.

In humans most of the DNA hitchhiking seems to have occurred millions of years ago. Our genome carries many ancient transposons, making it quite different from other genomes that have been studied, like those of fruit flies, worms, and mustard plants. Our genome has hung onto hitchhiking DNA since then, but acquired few new invaders. Mice, by contrast, are continuing to acquire new transposable elements.

Evolution of Development

The use of genomics to investigate how organismal development has evolved is a newly emerging field that combines genetics, genomics, evolution, and development to ask how the diversity of life has arisen. There are some two dozen conserved gene families that regulate development in animals. Some of these gene families have ancestral genes that predate the origins of animals. These genes serve in a sense as a commonly shared tool kit to build an animal.

The paradox that puzzles evolutionary developmental biologists is how this same tool kit can be used to build an insect, a bird, a bat, a whale, or a human. One explanation is that these genes turn on different genes or combinations of genes in different animals. For example, all tetrapods have four limbs, two hind limbs and two forelimbs. The forelimb in a bird is actually the wing. Our forelimb is an arm. Clearly these are two very different structures, but they have a common evolutionary origin. That is, they are homologous structures (see chapter 11).

At the genetic level, humans and birds both express the *Tbx5* gene in developing limb buds. *Tbx5* is a member of a gene family with a specific domain. *Tbx5* encodes a protein domain called the T-box, which is a transcription factor. So *Tbx5* turns on a gene or genes that are needed to make a limb. What seems to have changed as birds and humans evolved are the genes that are transcribed because of the Tbx5 protein (figure 10.5). In the ancestral tetrapod, perhaps Tbx5 protein bound to only one gene and triggered transcription. In humans and birds, a few genes are expressed in response to Tbx5 protein, but they are different genes.

The story of the evolution of limb development includes far more complex gene interactions than just the effects of *Tbx5,* which gets limb development started. The genes that *Tbx5* regulates in turn may affect the expression of other genes. Protein products serve as enzymes in biochemical pathways and provide the building blocks for other protein structures. Mining genome sequences for different organisms will be essential in identifying all the genes involved. Also, development occurs in four dimensions, three-dimensional space over time. That means that the timing of gene expression, as well as the genes that are expressed, can be altered and result in dramatic changes in shape.

Genome Rearrangements

DNA transposons can cause major chromosomal rearrangements in the host within which the transposon resides. Chromosomal rearrangements often have profound developmental effects, because changes in gene location can easily upset the delicate balance of gene regulation. Rearrangements can also have a serious evolutionary impact, leading in many instances in plants to the formation of entire new species.

Some rearrangements cause whole chromosome duplication, and even duplication of entire genomes. Genome duplication has been common in the evolution of plants, and is referred to as a change in **ploidy level.** Plants vary widely in ploidy level, but successful reproduction requires an even number of chromosome sets (figure 10.6).

Meiosis would be a disaster in a $3n$ organism like a banana because three sets of chromosomes cannot be evenly divided between two cells. Polyploidy ($3n$ or greater) can result from genome duplication in one species or hybridization of two different species. Hybridization is often followed by genome duplication so meiosis can occur. Bread wheat arose from two hybridization and whole genome duplication events (figure 10.7).

The wheat genome is 40 times larger than the rice genome (table 10.3), but gene content, order and orientation is conserved between wheat and rice. Insertion of repeated DNA and rearrangements within the genome have resulted in these conserved blocks of DNA moving to different chromosomes. The conservation of these blocks of genes is called conserved synteny. Synteny comes from the Greek *syn* (together) and *taenia* (ribbon). Rice, maize, barley, and wheat diverged more than 50 million years ago, but their chromosomes have synteny at the level of large blocks of DNA. Rice has been sequenced, so the degree of synteny within these DNA blocks can be explored. The sequenced rice genome is incredibly valuable to humans. Rice is the major food source for humans worldwide. Analysis of rice and syntenous cereal genomes will help us identify genes associated with disease resistance, crop yield, nutritional quality, and growth capacity. Comparisons of the genomes of these species should provide clues about the genome of their common ancestor.

10.4 Lateral gene transfer has led to significant mixing of genes among organisms. Animal evolution has involved significant changes in the way a common genetic tool kit is used by different organisms. Gene rearrangements have also had profound effects on genomes, even while preserving considerable synteny.

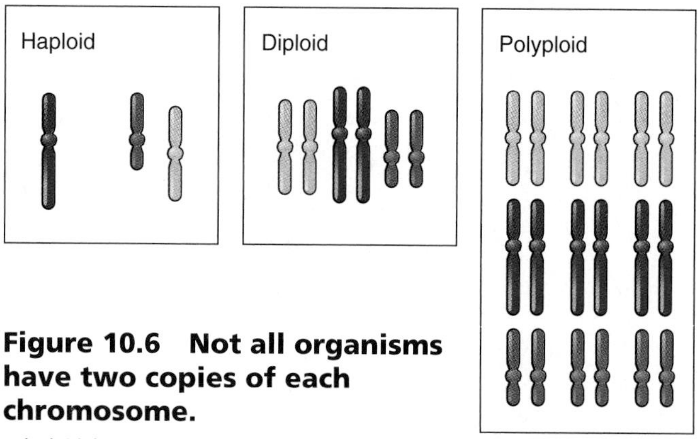
Haploid
Diploid
Polyploid

Figure 10.6 Not all organisms have two copies of each chromosome.

Polyploids have multiple sets of chromosome pairs; for example, bananas have a triple set of chromosomes. Polyploids with an odd number of copies of each chromosome cannot undergo meiosis successfully.

TABLE 10.3 GENOME SIZES OF SOME PLANTS

Scientific Name	Common Name	Genome Size (Millions of Base Pairs)
Arabidopsis thaliana	Arabidopsis	120
Prunus persica	Peach	262
Ricinus communis	Castor bean	323
Citrus sinensis	Orange	367
Oryza sativa spp. *javanica*	Rice	430
Petunia parodii	Petunia	1,221
Pisum sativum	Garden pea	3,947
Avena sativa	Oats	11,315
Triticum aestivum	Wheat	16,960
Tulipa spp.	Garden tulip	24,704

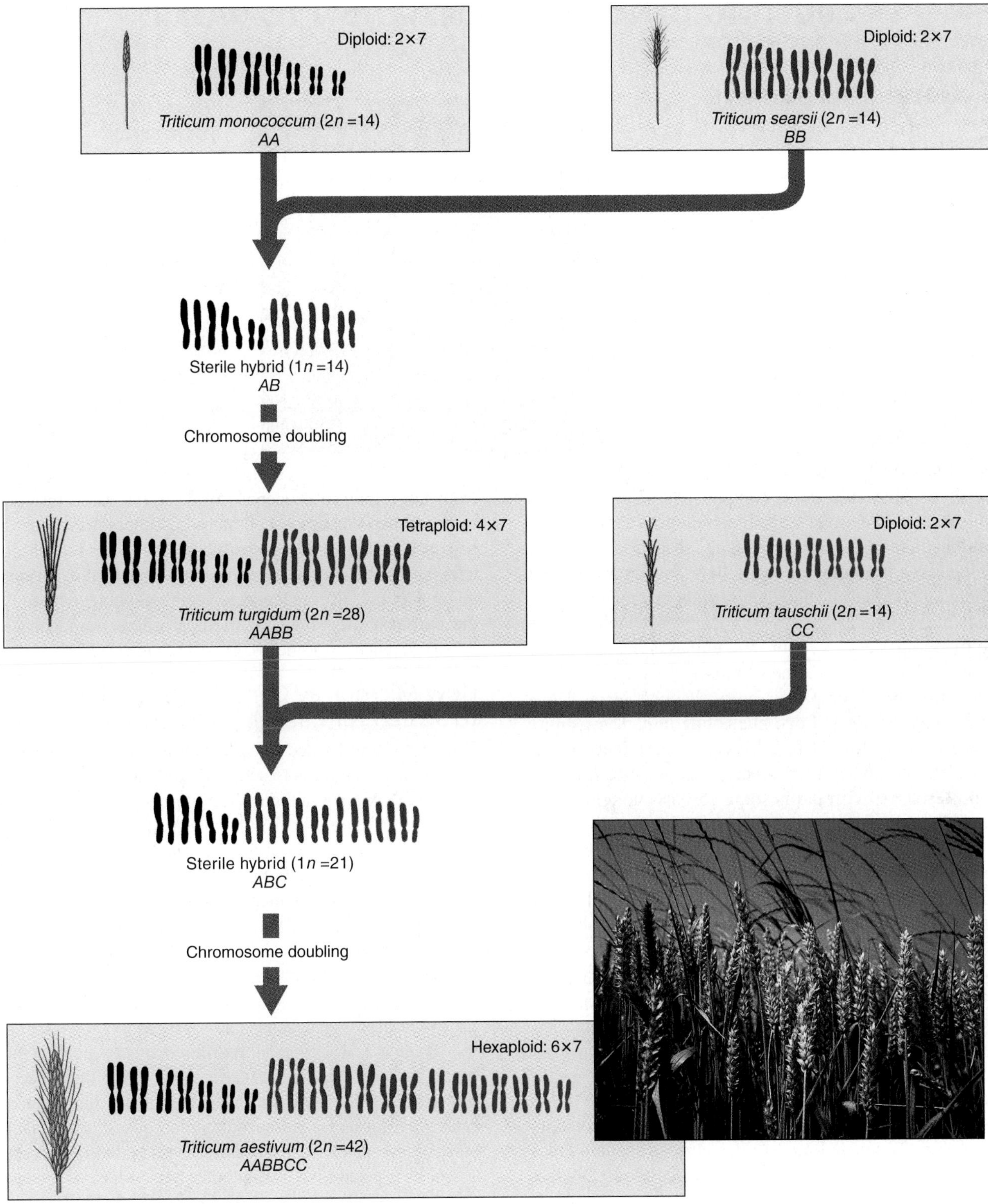

Diploid: 2×7
Triticum monococcum (2n =14)
AA
Diploid: 2×7
Triticum searsii (2n =14)
BB
Sterile hybrid (1n =14)
AB
Chromosome doubling
Tetraploid: 4×7
Triticum turgidum (2n =28)
AABB
Diploid: 2×7
Triticum tauschii (2n =14)
CC
Sterile hybrid (1n =21)
ABC
Chromosome doubling
Hexaploid: 6×7
Triticum aestivum (2n =42)
AABBCC

Figure 10.7 Evolutionary history of wheat.

Domestic wheat arose in southwestern Asia in the hilly country of what is now Iraq. In this region there is a rich assembly of grasses of the genus *Triticum.* Domestic wheat (*T. aestivum*) is a polyploid species of *Triticum* that arose through two so-called "allopolyploid" events. (*1*) Two different diploid species, symbolized here as *AA* and *BB,* hybridized to form an *AB* polyploid; the species were so different that A and B chromosomes could not pair in meiosis, so the *AB* polyploid was sterile. However, in some plants the chromosome number spontaneously doubled due to a failure of chromosomes to separate in meiosis, producing a fertile tetraploid species *AABB.* This wheat is used in the production of pasta. (*2*) In a similar fashion, the tetraploid species *AABB* hybridized with another diploid species *CC* to produce, after another doubling event, the hexaploid *T. aestivum, AABBCC.* This bread wheat is commonly used throughout the world.

10.5 Gene Microarrays

Microarrays

A gene microarray is a square of glass smaller than a postage stamp, covered with millions of strands of DNA arrayed like blades of grass. Microarrays were invented in 1991 by gene scientist Stephen Fodor. In a flash of insight, he saw that photolithography, the process used to etch semiconductor circuits into silicon, could also be used to assemble particular DNA molecules on a chip—a gene microarray.

Think of the chip surface as a field of assembly sites, much as a TV screen is a field of colored dots. Just as a scanning beam moves over each individual TV dot instructing it to be red, green, or blue (the three components of color), so a scanning beam moves over each biochip spot, commanding the addition there of a base to growing strands of DNA. A computer uses a scanning beam of light to unmask all the sites where an "A" is to be added, and adds that nucleotide to thousands of growing DNA strands anchored at each of these spots. It then remasks these spots and unmasks instead all the sites scheduled to receive G, then similarly C and T.

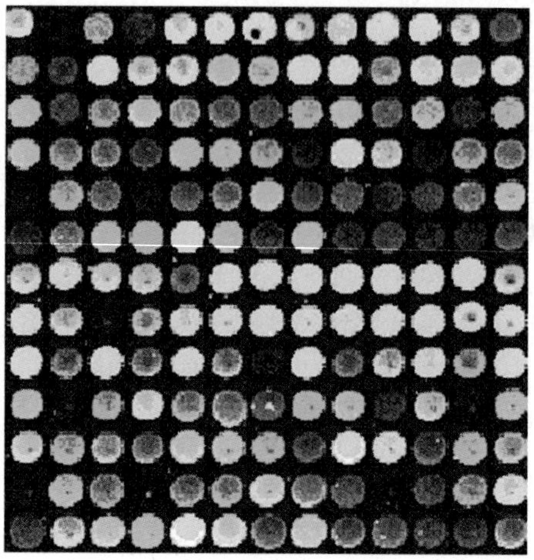

Figure 10.8 A gene microarray.

In gene microarrays, gene expression patterns from two different samples, such as from two different organisms, individuals, or tissues, can be compared. One sample's DNA (actually cDNA that is synthesized from mRNA) is labeled with a green fluorescent dye, and the other sample's with a red fluorescent dye. Spots that fluoresce are places where the samples bind to DNA on the microarray. The intensity and color of the fluorescence correlates with the amount of gene expression and the source of the cDNA (sample 1 or sample 2).

When the entire chip has been scanned, each DNA strand has been lengthened one nucleotide unit. The identity of the added nucleotide is in each case determined by the computer. The computer then repeats the process, layer by layer, until each DNA strand is some 25 bases long, enough to unambiguously identify one specific gene. One gene microarray chip made in this way can contain hundreds of thousands of specific gene recognition sequences.

How could you use such a microarray chip to delve into a person's genes? All you would have to do is to obtain a little of the person's DNA, say from a blood sample. Flush fluid containing the DNA over the chip surface. Every place that the DNA has a gene matching one of the microarray strands, it will stick to it in a way the computer can detect.

Patterns of gene expression can be directly determined with gene microarrays. To do this, mRNA isolated from the cells being studied is reverse transcribed to make copy DNA (cDNA; see chapter 9). Because the reaction mixture contains fluorescently labeled nucleotides, any DNA the cDNA binds to is easily recognized. When this labeled cDNA is mixed with a gene microarray representing many thousands of genes, spots light up corresponding to those genes being transcribed in the cells (figure 10.8).

Researchers are busily comparing the "reference sequence" of the human genome to the DNA of individual people, and noting any differences they detect. In this way, they are finding SNPs, which we discussed earlier. These SNPs, or spot differences in the identity of particular nucleotides, record every way in which a particular individual differs from the reference sequence. Some SNPs cause diseases like cystic fibrosis or sickle-cell anemia. Others may give you red hair or elevated cholesterol in your blood. The human genome tells us that SNPs can be expected to occur at a frequency of about 1 per 1,000 nucleotides, scattered about randomly over the chromosomes. Each of us thus differs from the standard "type sequence" in some 30,000 nucleotide SNPs.

How Microarray Gene Chips Can Be Used to Screen for Cancer

One of the biggest decisions facing an oncologist (cancer doctor) treating a tumor is to select the proper treatment. Most cancer cells look alike, although the tumors may in fact be caused by quite different forms of cancer. If the oncologist could clearly identify the cancer, very targeted therapies might be possible. Unable to tell the difference for sure, however, oncologists take no chances. Tumors are treated with therapy that attacks all cancers, usually with severe side effects.

Researchers Todd Golub and Eric Lander took a vital step toward treating cancer, using gene microarrays to sniff out the differences between different forms of a deadly cancer of the immune system.

The ideal way to tell the difference between two kinds of cancer is to compare the mutations that led to the cancer in the first place. The mutations that cause many lung cancers are caused by a tobacco-induced alteration of a single DNA nucleotide in one gene. Such spot differences between the version of a gene one person has and another person has, or a cancer patient has, are examples of medically important SNPs.

Golub and Lander obtained bone marrow cells from patients with two types of leukemia (cancer of white blood cells), and exposed DNA from each to microarrays containing 6,817 human genes. Using high-speed computer programs, Golub and Lander examined each of the 6,817 positions on the chip. The two forms of leukemia each showed gene changes from normal, but, importantly, the changes were different in each case! Each had their own characteristic SNP profile (figure 10.9).

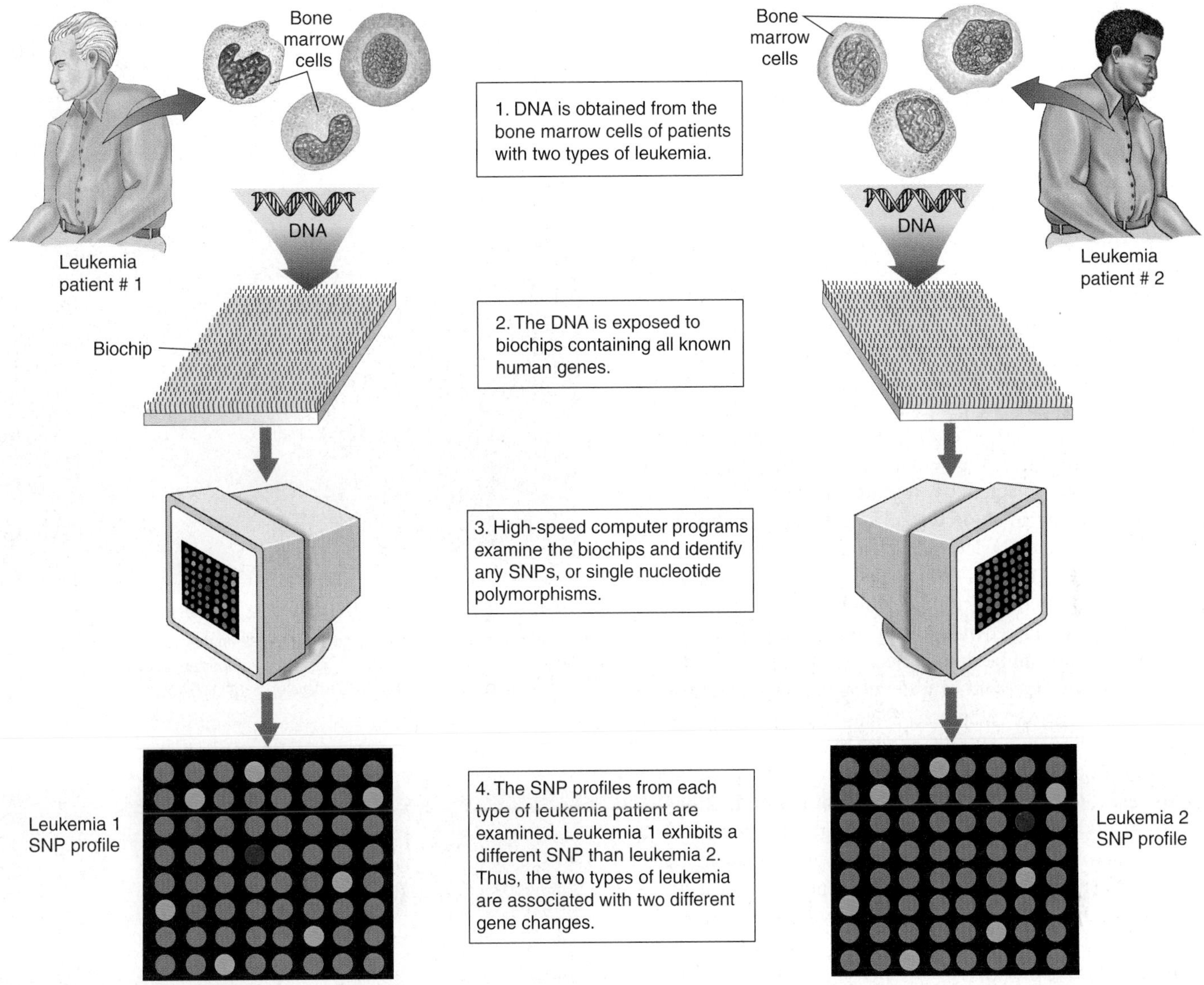
Bone
marrow
cells
Leukemia
patient # 1
DNA
Biochip
1. DNA is obtained from the bone marrow cells of patients with two types of leukemia.
2. The DNA is exposed to biochips containing all known human genes.
3. High-speed computer programs examine the biochips and identify any SNPs, or single nucleotide polymorphisms.
4. The SNP profiles from each type of leukemia patient are examined. Leukemia 1 exhibits a different SNP than leukemia 2. Thus, the two types of leukemia are associated with two different gene changes.
Leukemia 1
SNP profile
Bone
marrow
cells
Leukemia
patient # 2
Leukemia 2
SNP profile

Figure 10.9 Biochips can help in identifying precise forms of cancer.

Screening with the microarray thus provided a rapid and reliable way to determine which form of leukemia a patient possessed, greatly facilitating their treatment.

Researchers have announced plans to compile a database of hundreds of thousands of SNPs over the next two years. Screening SNPs and comparing them to known SNP databases will soon allow doctors to screen each of us for copies of genes leading to genetic diseases.

Biochips Raise Critical Issues of Personal Privacy

Researchers had identified some 300,000 different SNPs by 2001, all of which could reside on a single biochip. When your DNA is flushed over a SNP biochip, the sequences that light up will instantly reveal your SNP profile, the genetic characteristics that make you you. Genes that might affect your health, your behavior, your future potential—all are there to be read. Your SNP profile will reflect all of this variation: a table of contents of your chromosomes, a molecular window to who you are.

When millions of such SNP profiles have been gathered over the coming years, computers will be able to identify other individuals with profiles like yours, and, by examining health records, standard personality tests, and the like, correlate parts of your profile with particular traits. Even behavioral characteristics involving many genes, which until now have been thought too complex to ever analyze, cannot resist a determined assault by a computer comparing SNP profiles.

10.5 A gene microarray is a discrete collection of gene fragments on a stamp-sized chip that can be used to screen for the presence of particular gene variants. Microarrays allow rapid screening of gene profiles, a tool that promises to have a revolutionary impact on medicine and society.

10.6 Functional Genomics

Functional genomics is the identification of the function of genes in a genome. One of the first steps in analyzing a map of a sequenced genome is called annotation, or labeling a stretch of DNA according to its function. Annotation is often speeded by comparing sequences with a library of previously studied sequences of known function. However, while the function of a gene can usually be inferred from its sequence by comparing it to other genes with similar sequences, experiments are necessary to demonstrate that the inference is accurate.

Tools for functional analysis exist in certain model systems like fruit flies, but need to be developed for many other organisms if we are going to succeed in piecing together a comprehensive view of evolutionary history. Model systems like yeast, *Arabidopsis,* the nematode worm, the fruit fly, and the mouse were selected by researchers because they are easy to manipulate in some ways. These organisms all share three characteristics: (1) They have easily studied genetic systems so that mutant genes can be easily analyzed by making crosses; (2) They have fairly short life cycles (imagine trying to study genetics in redwood trees or elephants!); (3) They reproduce in, and can be maintained easily in, the laboratory.

Let's explore the role of a development-regulating gene called *Pax6* in eye development as an example of functional genomics approaches. *Pax6* encodes a transcription factor (that is, a protein that enables a gene to be "read" by RNA polymerase). *Pax6* is important in lens formation in a phylogenetically diverse collection of organisms including fruit flies (compound eyes), fish, frogs, mice, and humans. One way to study the function of this gene is to put it into a different species and see what it does. To do this, the mouse homologue of the *Pax6* gene was inserted into the fly genome, creating a transgenic fly. *Pax6* was expressed and an eye formed on the antenna of the fly (figure 10.10)!

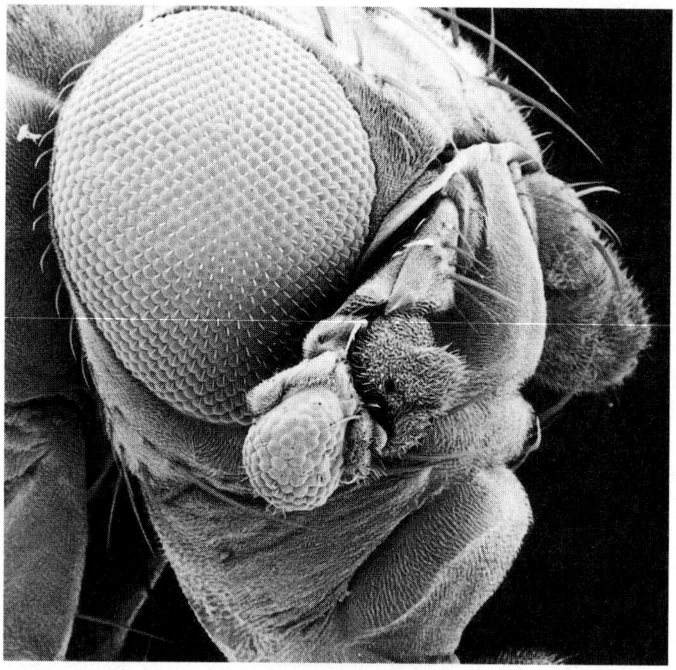

Figure 10.10 An eye on the antenna of a fly.
Expression of the mouse *Pax6* gene in a fly results in the production of an eye on the antenna of the fruit fly. Thus, a gene that is important in mouse eye development also plays a role in eye development in an insect.

The *Pax6* story extends to eyeless fish found in caves (figure 10.11). Fish that live in dark caves need to rely on senses other than sight. In cavefish, *Pax6* gene expression is greatly reduced. Eyes start to develop, but then degenerate. Changes in *Pax6* expression were visualized using gene microarrays to see where and when the gene is being transcribed.

Figure 10.11 Cavefish have lost their sight.
Mexican tetras (*Astyanax mexicanus*) have surface-dwelling members (*a*) and cave-dwelling members (*b*) of the same species. Cavefish that live in caves have very tiny eyes, partly because of reduced expression of *Pax6*.

Another way to determine where a gene is being expressed is to use recombinant DNA technology to attach the promoter of the gene to another gene with a protein product that can be visualized (often after staining). This new gene is then inserted into an organism and its expression is followed as the organism develops (figure 10.12). This is a good way to look at the regulation of gene expression, because you can see *where* the protein is first made.

Because a whole genome contains so many genes, it must be screened before an individual gene can be analyzed by time-consuming methods like creating transgenic organisms. Studying the expression of each of the 26,000

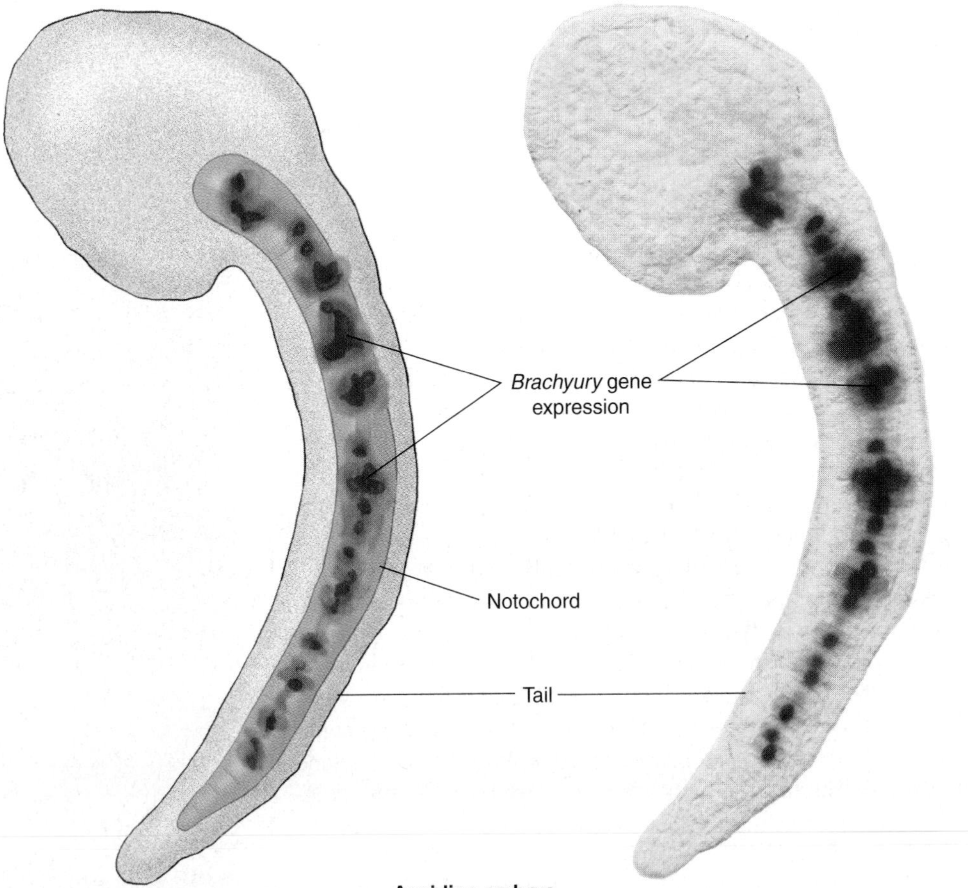
Brachyury gene expression
Notochord
Tail
Ascidian embryo

Figure 10.12 Visualizing gene expression.
Ascidians are basal chordates that have a notochord but no vertebrae. *Brachyury* is a T-box gene expressed in the developing notochord. By attaching the *Brachyury* promoter to a gene with a protein product that stains deep blue and inserting this modified gene into the ascidian, it is possible to see where *Brachyury* is expressed. This is an example of an ancestral gene that has been co-opted by an organism for a new function (making a notochord). *Brachyury* is also expressed in invertebrates, but has a very different function.

Arabidopsis genes in different parts of the plant at different stages in its life cycle, under different growth conditions, on a gene-by-gene basis would take a very very long time. And that would only be the first step in asking what a gene actually does. Plant genomicists involved in an effort called Project 2010 plan to identify the function of all the *Arabidopsis* genes by 2010 using more "high-throughput" approaches. One of the technological advances that will make this possible is the microscaling and automation of gene microarrays. As we have explained, once a genome has been cloned, fragments of the genome can be placed on a microarray. RNA probes are then labeled with a fluorescent dye and hybridized to the DNA. The microarray is scanned, and a computer translates the fluorescing spots into a visual image. It is possible to use two different RNA pools and two dyes on the same microarray. Let's go back to the *Pax6* experiment with flies. Suppose you want to know if different genes are expressed in the fly when the mouse *Pax6* gene is used versus the fly *Pax6* homologue. You would make a microarray using the fly genome (fortunately it has been sequenced already). You would then isolate RNA from a fly that had been exposed to the mouse *Pax6* and label it with one dye and repeat this with RNA from a fly exposed to the fly *Pax6* that is labeled with a different dye. You would hybridize both pools of RNA to one microarray (see figure 10.8). Visually you could determine which genes were expressed in both cases, which were expressed with only the mouse *Pax6*, which only with the fly, and which were not expressed at all. Because you already know the sequence of each spot of DNA on your microarray, you could go back and choose which sequences you wanted to study further.

10.6 To unravel the meaning of a genome's DNA sequence, functional genomics tools are being developed that will allow rapid analysis of huge numbers of genes.

10.7 Proteomics

Bioinformatics

Ideally, a researcher would like to be able to examine a nucleotide sequence and know what sort of functional protein the gene specifies. However, efforts to calculate what shape a protein will assume from knowledge of its amino acid sequence have proven difficult, even with the aid of large computers. However, by also looking at the proteins that are produced by the genes of the human genome, researchers are beginning to get a clearer picture of how gene sequence relates to protein shape and function.

Powerful computer programs are now having considerable success in screening the human genome for particular sorts of sequences, and increasing success in predicting the structure of a protein from the nucleotide sequence of the gene encoding it. This fast-growing area of genomics is loosely called **bioinformatics.** It combines molecular genetics and computational analysis in an attempt to predict what sort of protein a particular sequence encodes.

Proteins are three-dimensional structures that often interact with other proteins and molecules to function. Thus, although the computer model provides a good starting point, a lot of protein biochemistry is still necessary to understand how the protein is actually working. For example, the structure of the Pax6 protein has been deduced from the gene's nucleotide sequence, allowing researchers to predict how it might interact with one of the DNA sequences it regulates (figure 10.13). Determining the structures of all the genes Pax6 regulates can now be attempted using this approach.

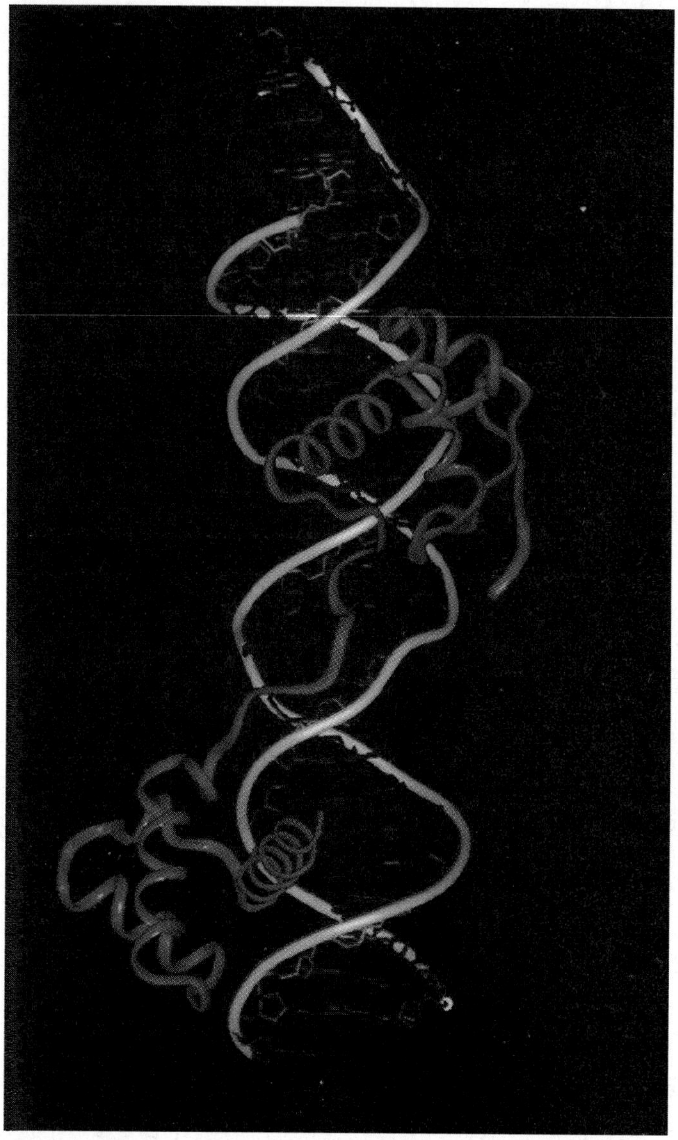

Figure 10.13 Pax6 protein interacting with DNA. The red represents a ribbon running through the main carbon backbone of the Pax6 protein. The phosphates in the DNA backbone are colored blue. Two different domains in this protein allow it to bind to DNA where it initiates transcription.

Proteomics: The Next Frontier

With the sequencing of the human genome now essentially complete, researchers have begun an even more challenging task: the cataloging and analysis of every protein in the human body, an endeavor called *proteomics.* Each gene's nucleotide sequence specifies an amino acid sequence that folds in a certain way, producing a protein whose shape gives it a particular function. Only by understanding the protein shapes that genes produce can we begin to make sense of the human genome.

Protein arrays, just like DNA microarrays, are now being developed to study all the proteins an organism possesses, its **proteome.** These arrays are screened using antibodies to specific proteins. Antibodies are fluorescently labeled so they can be detected, and the patterns on the protein array can then be determined by computer analysis. Technological advances are underway that will allow many proteins to be characterized on a mass scale in much less time than it took to uncover the structure of individual proteins like Pax6 in the past.

Fortunately, while there may be as many as a million different proteins, most are just variations on a handful of themes. The same shared structural motifs—barrels, helices, molecular zippers—are found in the proteins of plants, insects, and humans. The maximum number of distinct motifs has been estimated as fewer than 5,000. About 1,000 of these motifs have already been cataloged. Both public and privately financed efforts are now underway to detail the shapes of all the common motifs.

10.7 Like functional genomics, proteomics is a new approach that will enable analysis of proteins and comparisons at the protein level.

CONCEPT REVIEW

1. When looking at a DNA sequence, we find a stretch of DNA that begins with the "start" codon ATG and continues for 900 base pairs before ending with a "stop" codon. This "coding region" is referred to as
   a. a transposon.
   b. an intron.
   c. an open reading frame.
   d. structural DNA.

2. The number of genes in the human genome is ________ the number of genes in *Arabidopsis*, the first plant genome sequenced.
   a. slightly less than
   b. slightly more than
   c. 10 times greater than
   d. 100 times greater than

3. Which of the following is NOT one of the four classes of protein coding genes?
   a. structural DNA
   b. multigene families
   c. tandem clusters
   d. single-copy genes

4. What percent of the human genome is noncoding DNA?
   a. 50%
   b. 76%
   c. 88%
   d. 99%

5. Why is meiosis impossible in the bananas we buy in the grocery store?
   a. Bananas have been genetically engineered to stop meiosis.
   b. They are polyploid and have three sets of chromosomes.
   c. They have too many transposable elements.
   d. Their genome is too large to undergo meiosis.

6. Domestic wheat has a genome that is 40 times larger than rice. What is one reason for this large difference?
   a. Wheat is far more complex than rice and needs many more genes.
   b. Wheat has the same number of chromosomes as rice, but they are larger, with more noncoding DNA.
   c. Wheat resulted from the hybridization of a tetraploid with a diploid, giving it many more chromosomes than rice.
   d. Rice lost its transposable elements.

7. Gene microarrays can be used in all of the following ways except to
   a. identify different patterns of gene expression.
   b. characterize the proteins of an organism.
   c. screen for copies of genes leading to diseases.
   d. find SNPs.

8. Craig Ventner, the founder of a private company that set out to sequence the human genome, used an approach called ____________, which does not rely on mapping genes.

9. DNA that is highly condensed, untranscribed through the cell cycle, and tends to be near the centromere or the ends of the chromosome is called ____________.

10. There are about ____________ genes in the human genome.

11. When sequences of DNA differ by one base-pair, this variation is called ____________.

12. ____________ is the identification of the functions of genes in a genome, while ____________ is the cataloging and identification of all the proteins an organism possesses.

CHALLENGE YOURSELF

1. Counting the number of protein-encoding genes in the human genome is a difficult process, as the "start" and "stop" boundaries of transcribed genes are only imperfectly understood. Some researchers have taken a different approach, capturing mRNAs from the cytoplasm and matching them with the human genome DNA sequence. They report finding upward of 65,000 discrete mRNA-transcribing sequences in the genome. Does this mean that there are actually 65,000 human genes, rather than the 30,000 originally reported?

2. How would you go about being sure that a gene humans share with bacteria was actually acquired by the human genome via lateral exchange, rather than having simply persisted in our linage since bacterial times?

10 eBRIDGE

Reinforcing Key Points

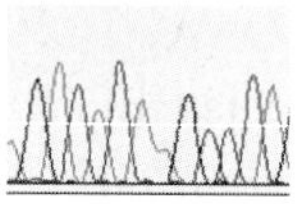

The Challenge of Sequencing Entire Genomes

Genome Evolution

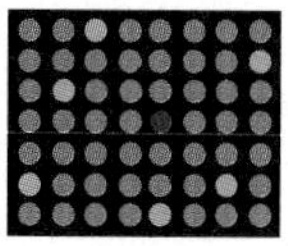

Putting Genomic Information to Work

Electronic Learning

Explorations

Making a Restriction Map

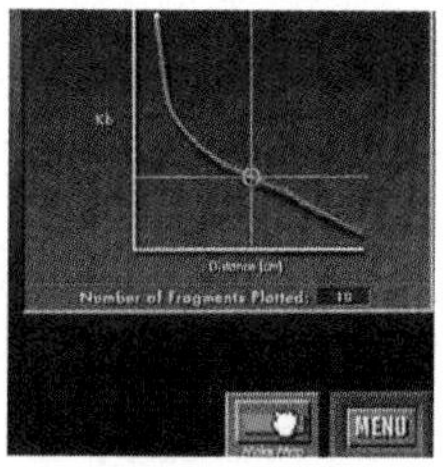
MENU

In this exercise, you construct a restriction map by entering measured band position data from a set of electrophoresis gels. The data may be supplied by the user from real lab experiments, or the user may choose to analyze one of several data sets provided by the interactive exercise.

Cystic Fibrosis

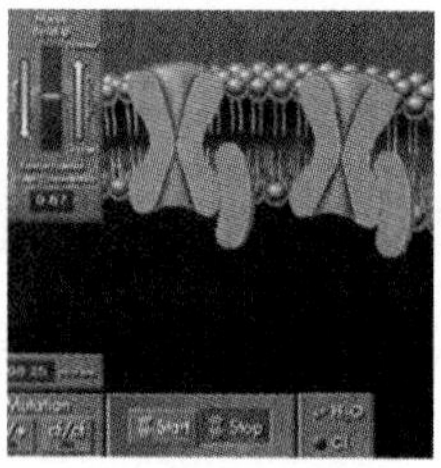

This interactive exercise allows you to explore the way in which transport proteins influence the passage of water in and out of cells by examining the effects of a mutation that disables a particular transport protein, the one responsible for pumping chloride ions out of cells. The water transport into lung cells that results from this mutation is the direct cause of the symptoms of cystic fibrosis, a disease which scientists hope to be able to cure with gene therapy.

Author's Corner

Genomics. The sequencing of the human genome was announced to the public on Darwin's birthday, June 26, 2000. Deciphering all of the DNA in our chromosomes was a major scientific accomplishment, as it involved sequencing some 3.2 billion bases. The genomes of many microbes have been completed, as well as those of solid worms, fruit flies, mice, and flowering plants. The new discipline of genomics promises to have a major impact on biology in the coming years.

1. We humans don't have as many genes as we thought.
2. Very little of the human genome is devoted to being human.
3. Who should own the human genome?
4. The secrets of your genes on a microchip.
5. Gene microarrays and personal privacy

Virtual Classroom

The Human Genome

The human genome sequence was announced on June 26, 2000, Darwin's birthday. It contains some 3.2 billion nucleotide bases, so the sequencing presented no small challenge. There were two quite unexpected results: First, we don't have nearly as many genes as we thought. Each cell contains about six feet of DNA, of which only about one inch is devoted to encoding proteins. Second, practically half the entire human genome is occupied by so-called "transposable elements," bits of DNA that hop around the genome like fleas on a dog.

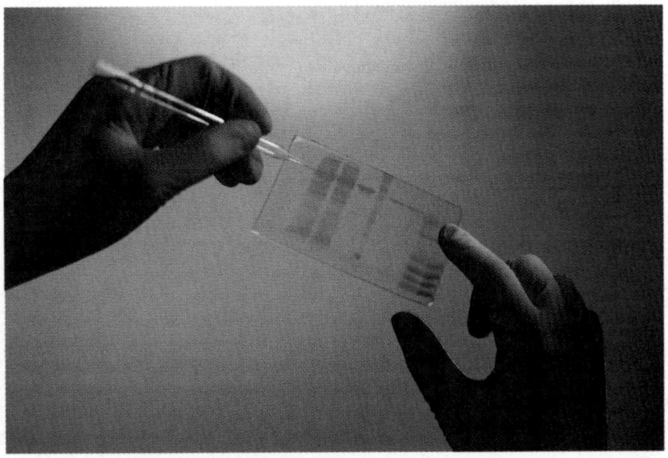

Gene Therapy

While genetically modified foods remain controversial, gene-modified humans are much in demand. The search for a way to introduce "healthy" genes into people suffering from cystic fibrosis, muscular dystrophy, or other gene disorders has finally begun to pay off. Discovery of workable virus "vectors" to carry the genes has researchers pursuing gene therapy cures like hounds to a hot scent.

Virtual Lab

The Control of Patterning in Plant Root Development

With the sequencing of the *Arabidopsis* genome completed, scientists are now poised to learn even more about the genes that control plant development and physiological functions. For example, we are now in a position to explore one of the "deep" questions in botany—what mechanism controls the development of central pattern formation in plants? Using *Arabidopsis,* scientists have been identifying mutations that affect development, and then examining the genes subsequently affected by those mutations. *Arabidopsis* is an ideal model for studying plant pattern formation during development. Individual plants grow quickly in laboratory test tubes, growing no taller than your thumb. Its genome, now fully sequenced, aids in studying the molecular events that underlie pattern formation.

John Schiefelbein and colleagues at the University of Michigan have focused on one sharply defined aspect of plant root pattern formation in *Arabidopsis,* the formation of root hairs on the epidermis. The positioning of root hairs is under tight central control, and proves to be a balancing act between cell division and cell differentiation.

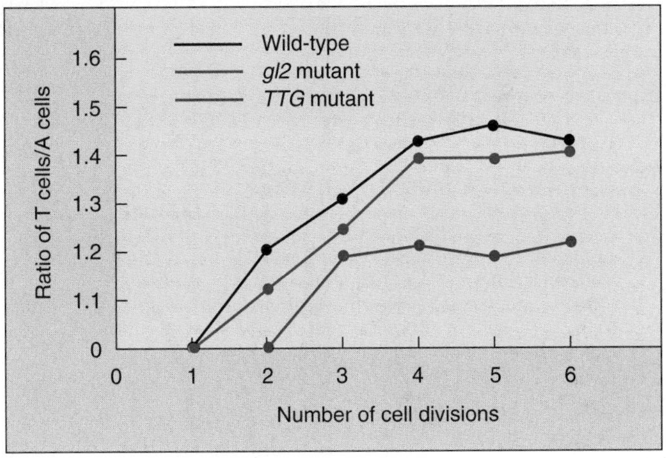
Wild-type
gl2 mutant
TTG mutant
Ratio of T cells/A cells
1.6
1.5
1.4
1.3
1.2
1.1
0
0
1
2
3
4
5
6
Number of cell divisions

Quizzes

Further Reading

Essential Study Partner

Links

BioCourse.com

PART 4 THE EVOLUTION AND DIVERSITY OF LIFE

CHAPTER 11

Evolution and Natural Selection

THE FAR SIDE® By GARY LARSON
© 1991 FarWorks, Inc. All Rights Reserved/Dist. by Creators Syndicate
Okay, who's it gonna be?.. Whooooooooo's it gonna be?..
Larson

CHAPTER 11

Evolution and Natural Selection

CHAPTER OVERVIEW

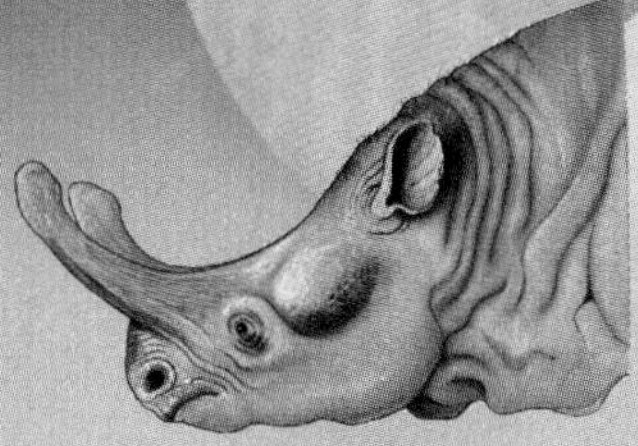

The Evidence for Evolution

11.1 Gene Variation: The Raw Material of Evolution
11.2 The Pace of Evolution
11.3 The Fossil Record
11.4 The Molecular Record
11.5 The Anatomical Record

- If we date fossils, and order them by age, progressive changes are seen. This is direct evidence that evolution has occurred.
- The progressive accumulation of molecular differences and comparisons of living organisms provide additional strong evidence that evolution has occurred.

How Populations Evolve

11.6 The Hardy-Weinberg Rule
11.7 Why Do Allele Frequencies Change?
11.8 Forms of Selection

- In a population not undergoing significant evolutionary change, two alleles present in frequencies p and q will be distributed among the genotypes in the proportions $p^2 + 2pq + q^2$, the Hardy-Weinberg rule.
- Allele frequencies change in nature due to mutation, migration, drift, nonrandom mating, and selection.

Adaptation: Evolution in Action

11.9 Sickle-Cell Anemia
11.10 Peppered Moths and Industrial Melanism

- A mutation in hemoglobin causes a condition known as sickle-cell anemia. This recessive mutation is common in central Africa because it renders heterozygous individuals resistant to malaria.
- Vegetation darkened by industrial soot has favored the evolution of darker moths, better concealed from their predators.

How Species Form

11.11 The Species Concept
11.12 Prezygotic Isolating Mechanisms
11.13 Postzygotic Isolating Mechanisms

- Microevolution leads to macroevolution. Adaptation to local habitats leads to divergence and the evolution of ecological races.
- Isolating mechanisms then reinforce the differences, leading to reproductive isolation and species formation.

11.1 Gene Variation: The Raw Material of Evolution

Microevolution Leads to Macroevolution

When the word *evolution* is mentioned, it is difficult not to conjure up images of dinosaurs, woolly mammoths frozen in blocks of ice, or Darwin observing a monkey. Traces of ancient life-forms, now extinct, survive as fossils that help us piece together the evolutionary story. With such a background, we usually think of evolution in terms of changes that take place over long periods of time, changes in the kinds of animals and plants on earth as new forms replace old ones. This kind of evolution, **macroevolution,** is evolutionary change on a grand scale, encompassing the origins of novel designs, evolutionary trends, new kinds of organisms penetrating new habitats, and major episodes of extinction.

Much of the focus of Darwin's theory of natural selection, however, is directed not at the way in which new species are formed from old ones but rather at the way that changes occur within species. Natural selection is the process whereby some individuals in a population, those that possess certain inherited characteristics, produce more surviving offspring than individuals lacking these characteristics. As a result, the population will gradually come to include more and more individuals with the advantageous characteristics. In this way the population evolves. Changes of this sort within populations—changes in gene frequencies—represent **microevolution.**

Adaptation results from microevolutionary changes that increase the likelihood of survival and reproduction of particular genetic traits in a population. In essence, Darwin's explanation of evolution is that adaptation by natural selection is responsible for evolutionary changes *within* a species (microevolution), and that the accumulation of these changes leads to the development of new species (macroevolution).

The Key Is the Source of the Variation

Darwin agreed with many earlier philosophers and naturalists who deduced that the many kinds of organisms around us were produced by a process of evolution. Unlike his predecessors, however, Darwin proposed natural selection as the mechanism of evolution. He proposed that new kinds of organisms evolve from existing ones because some individuals have traits that allow them to produce more offspring that, in turn, carry those traits.

Natural selection was by no means the only mechanism proposed. A rival theory, championed by the prominent biologist Jean-Baptiste Lamarck, was that evolution occurred by the inheritance of acquired characteristics. According to Lamarck, individuals passed on to offspring body and behavior changes acquired during their lives. Thus, Lamarck proposed that ancestral giraffes with short necks tended to stretch their necks to feed on tree leaves, and this extension of the neck was passed on to subsequent generations, leading to the long-necked giraffe (figure 11.1). In Darwin's theory, by contrast, the variation is not created by experience but already exists when selection acts on it.

Figure 11.1 How did giraffes evolve a long neck?

The giraffe's near relative, the okapi, has a short neck. Lamarck believed the giraffe lengthened its neck by stretching to reach tree leaves and then passed the change to offspring. Darwin proposed instead that giraffes who happened to be born with longer necks were more successful. The key difference in the two proposals is whether the variation is acquired or genetic.

> **11.1** Darwin proposed that natural selection on variants within populations leads to the evolution of different species.

11.2 The Pace of Evolution

Different kinds of organisms evolve at different rates. Mammals, for example, evolve relatively slowly. On the basis of a relatively complete fossil record, it has been estimated that a good average value for the duration of a "typical" mammal species, from formation of the species to its extinction, might be about 200,000 years. American paleontologist George Gaylord Simpson has pointed out that certain groups of animals, such as lungfishes, are apparently evolving even more slowly than mammals. In fact, Simpson estimated that there has been little evolutionary change among lungfishes over the past 150 million years, and even slower rates of evolution occur in other groups.

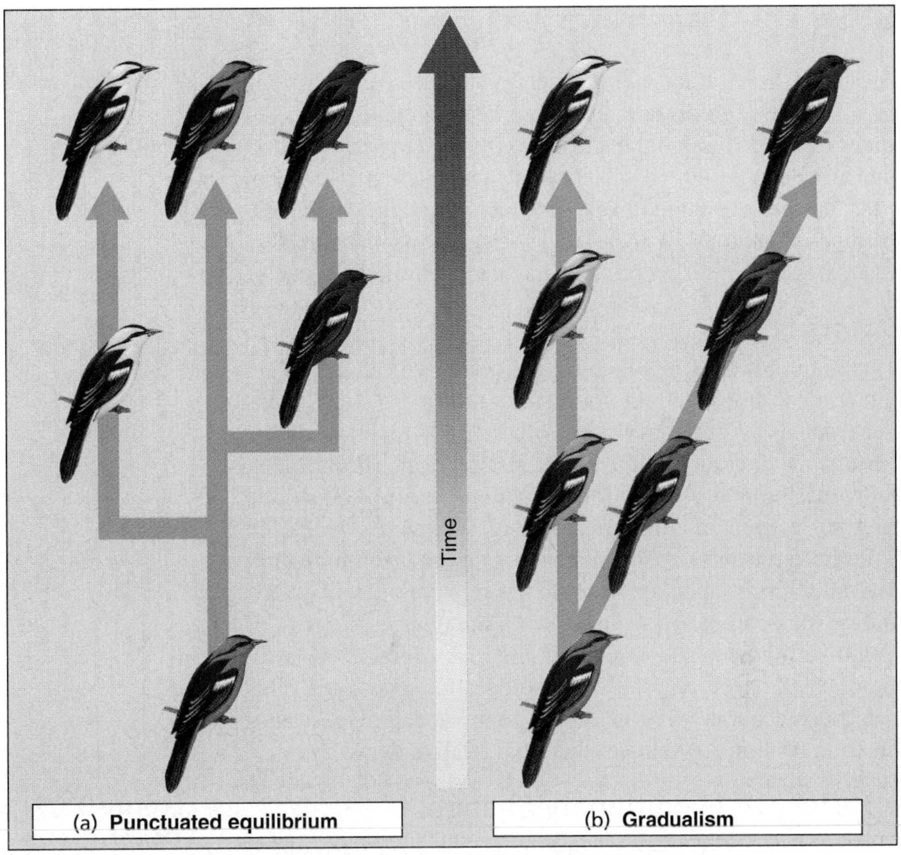

Time
(a) Punctuated equilibrium
(b) Gradualism

Figure 11.2 Two views of the pace of macroevolution.
(*a*) Punctuated equilibrium surmises that species formation occurs in bursts, separated by long periods of quiet, while (*b*) gradualism surmises that species formation is constantly occurring.

Evolution in Spurts?

Not only does the rate of evolution differ greatly from group to group, but evolution within a group apparently proceeds rapidly during some periods and relatively slowly during others. The fossil record provides evidence for such variability in evolutionary rates, and evolutionists are very interested in understanding the factors that account for it. In 1972, paleontologists Niles Eldredge of the American Museum of Natural History in New York and Stephen Jay Gould of Harvard University proposed that evolution normally proceeds in spurts. They claimed that the evolutionary process is a series of **punctuated equilibria.** Evolutionary innovations would occur and give rise to new lines; then these lines might persist unchanged for a long time, in "equilibrium." Eventually there would be a new spurt of evolution, creating a "punctuation" in the fossil record. Eldredge and Gould contrast their theory of punctuated equilibrium with that of **gradualism,** or gradual evolutionary change, which they claimed was what Darwin and most earlier students of evolution had considered normal (figure 11.2).

Eldredge and Gould proposed that stasis, or lack of evolutionary change, would be expected in large populations under diverse and conflicting selective pressures. In contrast, rapid evolution of new species would usually occur when populations were small, isolated, and possibly already differing from their parental population as a result of a "founder effect." This, combined with selective pressures from a new environment, could bring about rapid change.

Unfortunately, the distinctions are not as clear-cut as implied by this discussion. Some well-documented groups such as African mammals clearly have evolved gradually and not in spurts. Other groups, like marine bryozoa, seem to show the irregular pattern of evolutionary change the punctuated equilibrium model predicts. One could argue, however, that the fossil record in these instances is incomplete because of changes in the conditions under which fossils are deposited, thus making the interpretation of many of the "gaps" problematic. Despite these difficulties, the punctuated equilibrium model has provided a useful perspective for considering the mode and pace of evolution.

11.2 The punctuated equilibrium model assumes that evolution occurs in spurts, between which there are long periods in which there is little evolutionary change. The gradualism model assumes that evolution proceeds gradually, with successive change in a given evolutionary line.

11.3 The Fossil Record

The most direct evidence of macroevolution is found in the fossil record. **Fossils** are the preserved remains, tracks, or traces of once-living organisms. Fossils are created when organisms become buried in sediment, the calcium in bone or other hard tissue mineralizes, and the surrounding sediment eventually hardens to form rock. The fossils contained in layers of sedimentary rock reveal a history of life on earth.

Dating Fossils

By dating the rocks in which fossils occur, we can get an accurate idea of how old the fossils are. In Darwin's day, rocks were dated by their position with respect to one another (*relative dating*); rocks in deeper strata are generally older. Knowing the relative positions of sedimentary rocks and the rates of erosion of different kinds of sedimentary rocks in different environments, geologists of the nineteenth century derived a fairly accurate idea of the relative ages of rocks. Today, rocks are dated by measuring the degree of decay of certain radioisotopes contained in the rock (*absolute dating*); the older the rock, the more its isotopes have decayed. This is a more accurate way of dating rocks and provides dates stated in millions of years, rather than relative dates.

A History of Evolutionary Change

When fossils are arrayed according to their age, from oldest to youngest, they often provide evidence of successive evolutionary change. Among the hoofed mammals illustrated in figure 11.3, small, bony bumps on the nose can be seen to change continuously, until they become large, blunt horns. About 200 million years ago, oysters underwent a change from small curved shells to larger, flatter ones, with progressively flatter fossils being seen in the fossil record over a period of 12 million years (figure 11.4). Many other examples illustrate a record of successive change and are one of the strongest lines of evidence that evolution has occurred.

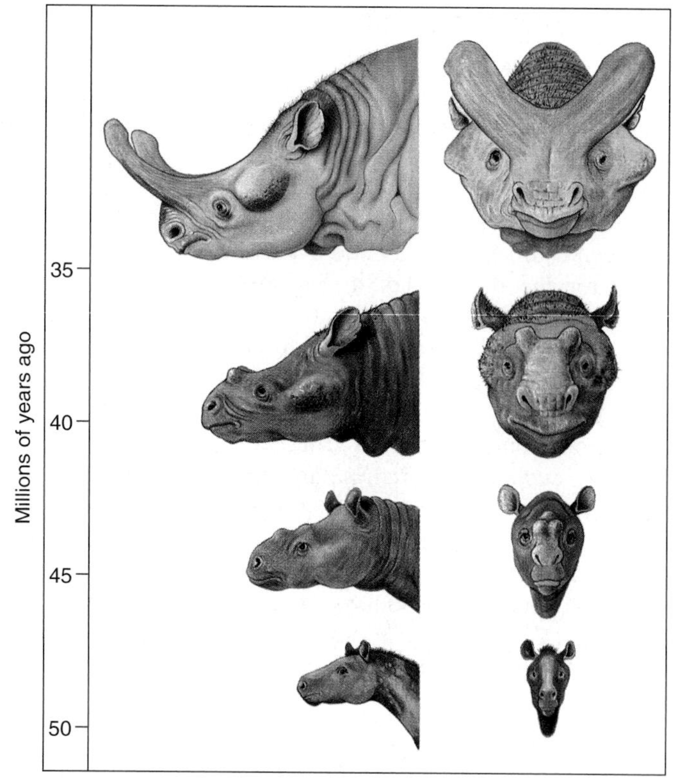

Millions of years ago
35
40
45
50

Figure 11.3 Successive evolution in the titanotheres.

Here you see illustrated changes in a group of hoofed mammals known as titanotheres between about 50 million and 35 million years ago. During this time, the small, bony protuberance located above the nose 50 million years ago evolved into relatively large, blunt horns.

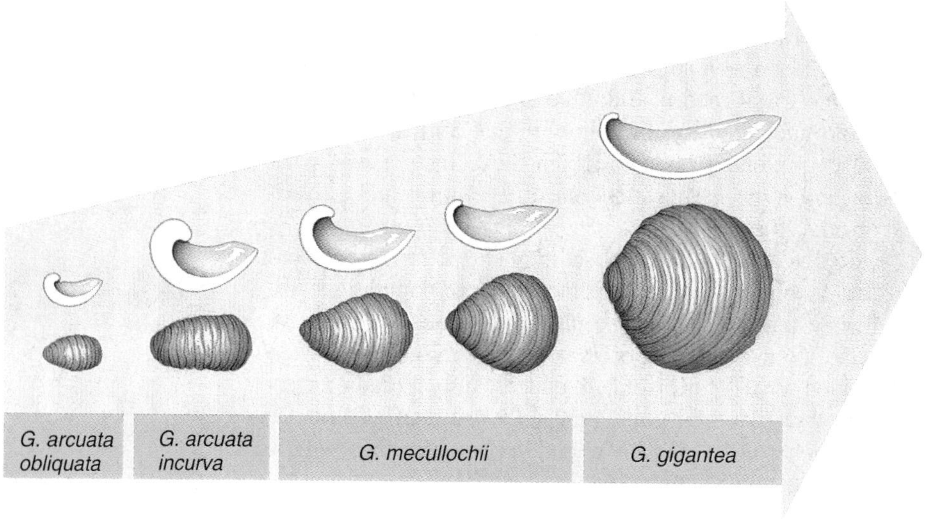

G. arcuata obliquata
G. arcuata incurva
G. mecullochii
G. gigantea

Figure 11.4 Evolution of shell shape in oysters.

During a 12-million-year portion of the early Jurassic period, the shells of a group of coiled oysters became larger, thinner, and flatter. These animals rested on the ocean floor, and the larger, flatter shells may have proven more stable against potentially disruptive water movements.

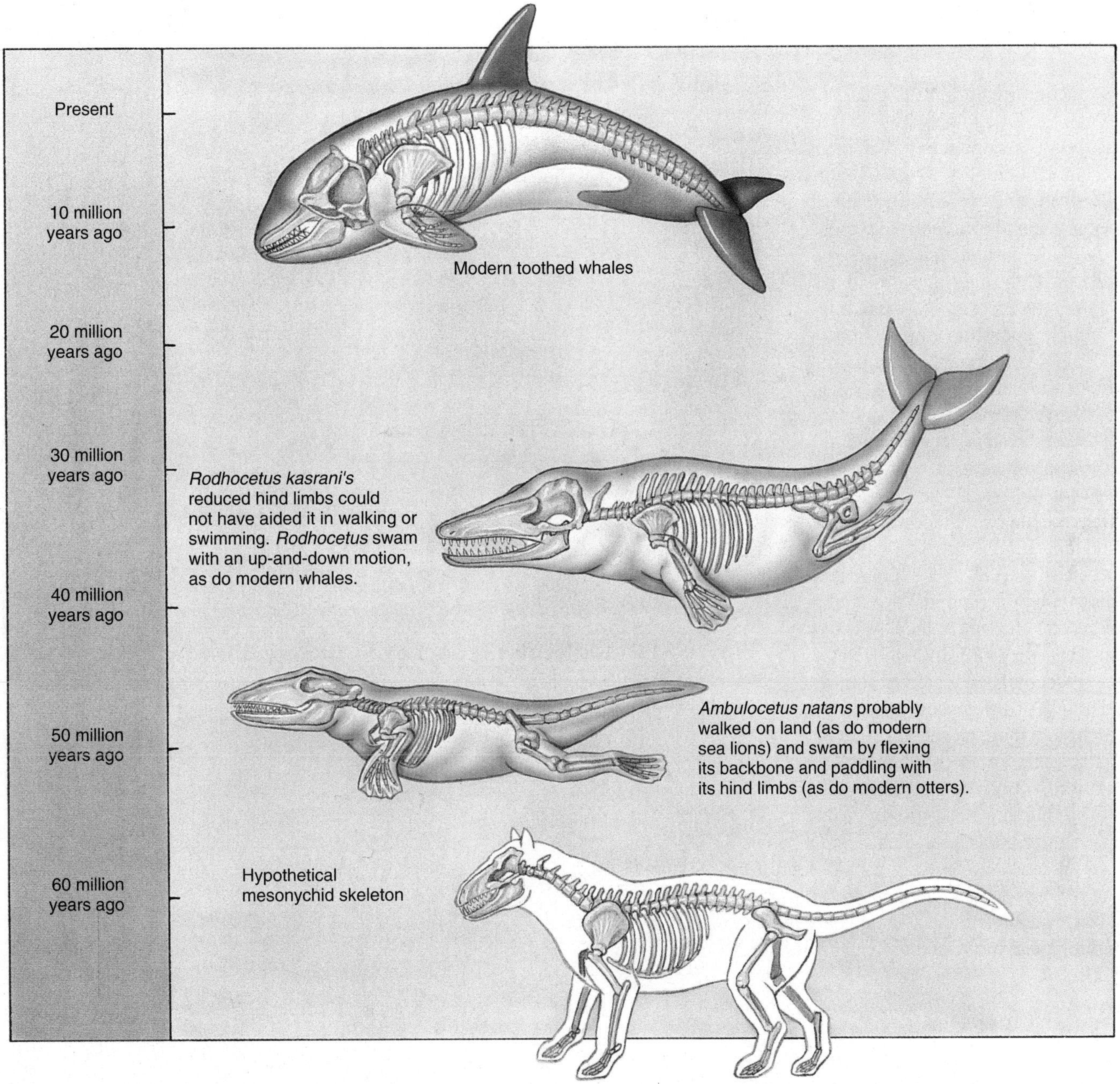
Present
10 million
years ago
20 million
years ago
30 million
years ago
40 million
years ago
50 million
years ago
60 million
years ago
Modern toothed whales
Rodhocetus kasrani's reduced hind limbs could not have aided it in walking or swimming. Rodhocetus swam with an up-and-down motion, as do modern whales.
Ambulocetus natans probably walked on land (as do modern sea lions) and swam by flexing its backbone and paddling with its hind limbs (as do modern otters).
Hypothetical mesonychid skeleton

Figure 11.5 Whale "missing links."
The recent discoveries of *Ambulocetus* and *Rodhocetus* have filled in the gaps between the mesonychids, the hypothetical ancestral link between the whales and the hoofed mammals, and present-day whales.

Gaps in the Fossil Record

While many gaps interrupted the fossil record in Darwin's era, even then, scientists knew of the *Archaeopteryx* fossil transitional between reptiles and birds. Today, the fossil record is far more complete, particularly among the vertebrates; fossils have been found linking all the major groups. The forms linking mammals to reptiles are particularly well known. A series of extinct marine mammals linking whales to terrestrial, four-legged hoofed ancestors filled in one of the last significant gaps (figure 11.5).

11.3 The fossil record provides a clear record of continuous evolutionary change.

11.4 The Molecular Record

Traces of our evolutionary past are also evident at the molecular level. We possess color vision genes that have become more complex as vertebrates have evolved, and we employ pattern formation genes during early development that all animals share. If you think about it, the fact that organisms have evolved successively from relatively simple ancestors implies that a record of evolutionary change is present in the cells of each of us, in our DNA. According to evolutionary theory, every evolutionary change involves the substitution of new versions of genes for old ones. The new alleles arise from the old by mutation and come to predominance through favorable selection. Thus, a series of evolutionary changes involves a continual accumulation of genetic changes in the DNA. Organisms that are more distantly related will have accumulated a greater number of evolutionary differences, while two species that are more closely related will share a greater portion of their DNA. This pattern of divergence is clearly seen in a human hemoglobin polypeptide (figure 11.6). Chimpanzees, gorillas, orangutans, and gibbons, the vertebrates most closely related to humans, have fewer differences from humans in the 146-amino-acid hemoglobin β chain than do more distantly related mammals, like mice. Nonmammalian vertebrates differ even more, and nonvertebrates are the most different of all.

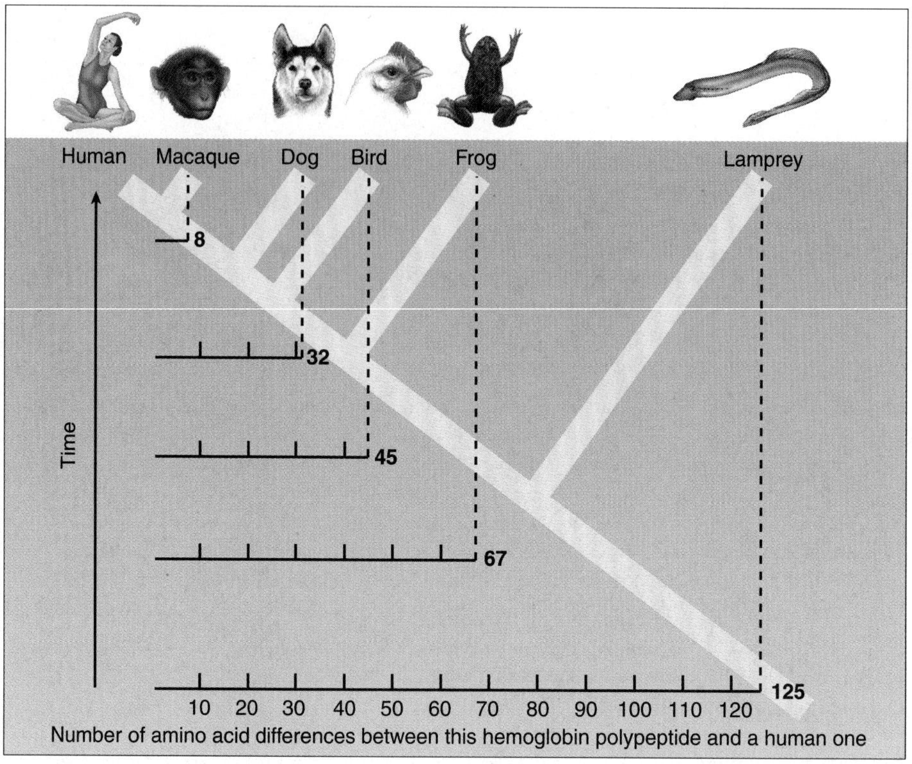

Human
Macaque
Dog
Bird
Frog
Lamprey
Time
8
32
45
67
125
10 20 30 40 50 60 70 80 90 100 110 120
Number of amino acid differences between this hemoglobin polypeptide and a human one

Figure 11.6 Molecules reflect evolutionary divergence.
You can see that the greater the evolutionary distance from humans, the greater the number of amino acid differences in the vertebrate hemoglobin polypeptide.

Molecular Clocks

This same pattern is seen when DNA sequences from various organisms are compared. For example, the longer the time since the organisms diverged, the greater the number of differences in the nucleotide sequence of the cytochrome *c* gene (figure 11.7), which codes for a protein that plays a key role in oxidative metabolism, as we learned in chapter 5. The changes appear to accumulate in cytochrome *c* at a constant rate, a phenomenon sometimes referred to as a **molecular clock.**

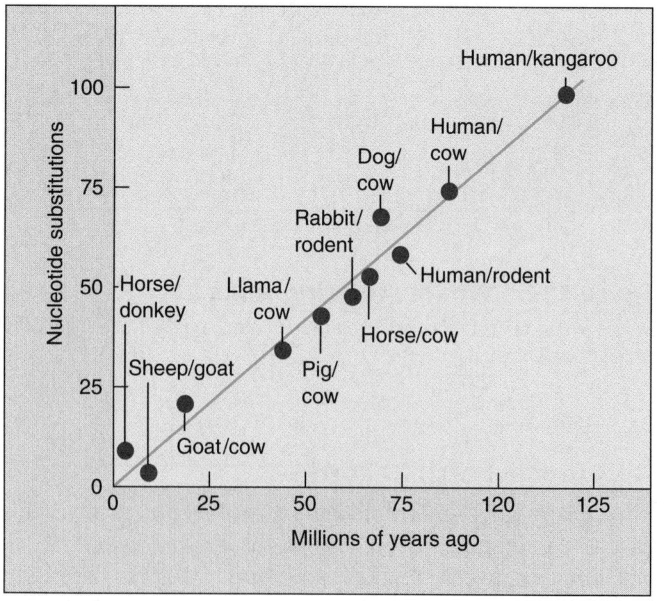

Human/kangaroo
Human/cow
Dog/cow
Rabbit/rodent
Human/rodent
Horse/donkey
Llama/cow
Horse/cow
Sheep/goat
Pig/cow
Goat/cow
Nucleotide substitutions
0
25
50
75
100
0 25 50 75 120 125
Millions of years ago

Figure 11.7 The molecular clock of cytochrome *c*.
When the time since each pair of organisms presumably diverged is plotted against the number of nucleotide differences, the result is a straight line, suggesting that the cytochrome c gene is evolving at a constant rate.

Proteins Evolve at Different Rates

The constant rate at which cytochrome *c* has evolved raises the interesting question of whether other proteins also evolve at constant rates. As a rule of thumb, highly conserved proteins like hemoglobin or cytochrome *c* provide the best molecular clocks, but all proteins for which data are available appear to accumulate changes over time. However, different proteins evolve at very different rates, as can be seen in figure 11.8, which presents estimates of the number of "acceptable" point mutations (that is, changes successfully incorporated into the protein) per 100 amino acids per million years. The fastest rate of change appears to be in fibrinopeptides, while the most highly conserved protein is histone H4. Even faster rates of change are seen for pseudogenes, which are not transcribed, suggesting that molecular evolution proceeds more quickly when less constrained by selection.

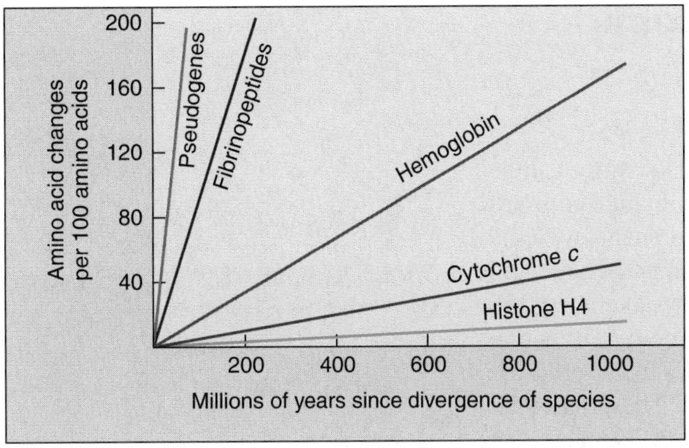

Amino acid changes per 100 amino acids
200
160
120
80
40
Pseudogenes
Fibrinopeptides
Hemoglobin
Cytochrome c
Histone H4
200
400
600
800
1000
Millions of years since divergence of species

Figure 11.8 Different molecular clocks tick at different speeds.

Pseudogenes and fibrinopeptides evolve more rapidly than hemoglobin, which evolves more rapidly than cytochrome *c* or histone H4.

Phylogenetic Trees

The same regular pattern of change is seen in many other proteins. Some genes, such as the ones specifying the hemoglobin protein, have been well studied, and the entire time course of their evolution can be laid out with confidence by tracing the origin of particular substitutions in their nucleotide sequences. The pattern of descent obtained is called a **phylogenetic tree.** It represents the evolutionary history of the gene. Note that the successive changes in the hemoglobin molecule produce a tree that closely reflects the evolutionary relationships predicted by a study of anatomy (figure 11.9). Whales, dolphins, and porpoises cluster together, as do the primates and the hoofed animals. The pattern of accumulating changes seen in the molecular record constitutes strong direct evidence for macroevolution.

11.4 The genes encoding proteins have undergone continual evolution, accumulating increasing numbers of changes over time.

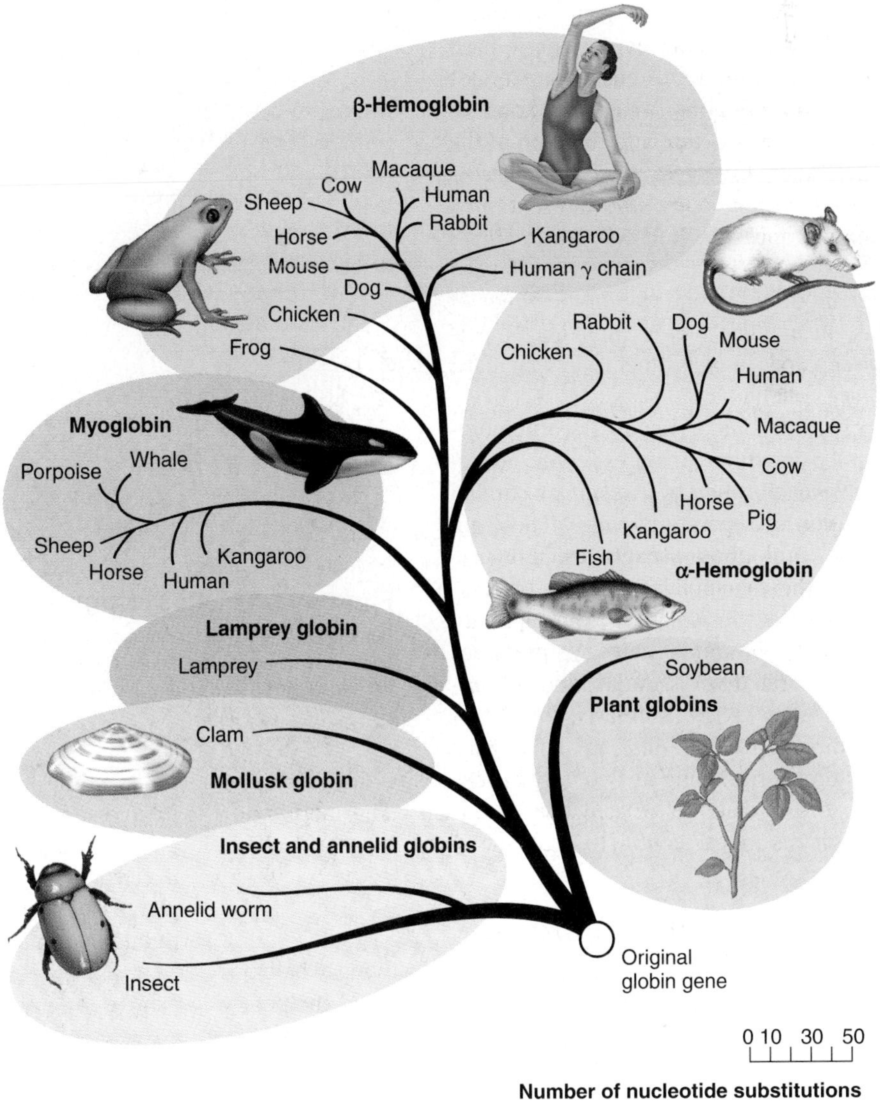

β-Hemoglobin
Macaque
Cow
Sheep
Human
Rabbit
Horse
Kangaroo
Mouse
Human γ chain
Dog
Chicken
Frog
Rabbit
Dog
Mouse
Chicken
Human
Macaque
Cow
Horse
Pig
Kangaroo
Fish
α-Hemoglobin
Myoglobin
Whale
Porpoise
Sheep
Horse
Human
Kangaroo
Lamprey globin
Lamprey
Soybean
Plant globins
Clam
Mollusk globin
Insect and annelid globins
Annelid worm
Insect
Original globin gene
0 10 30 50
Number of nucleotide substitutions

Figure 11.9 An evolutionary tree of the globin gene.

The length of the various lines reflects the number of nucleotide substitutions in the gene. In humans, two of the proteins encoded by versions of the globin gene are α-hemoglobin and β-hemoglobin.

11.5 The Anatomical Record

Development

Much of our evolutionary history can be seen in the way in which human embryos develop. Early in development, we (and all other vertebrate embryos) have gill slits like a fish; at a later stage every human embryo has a long bony tail; and at the fifth month of development, each develops a coat of fine fur! These relict developmental forms strongly suggest that our development has evolved, with new instructions being layered on top of old ones (figure 11.10).

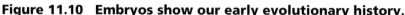

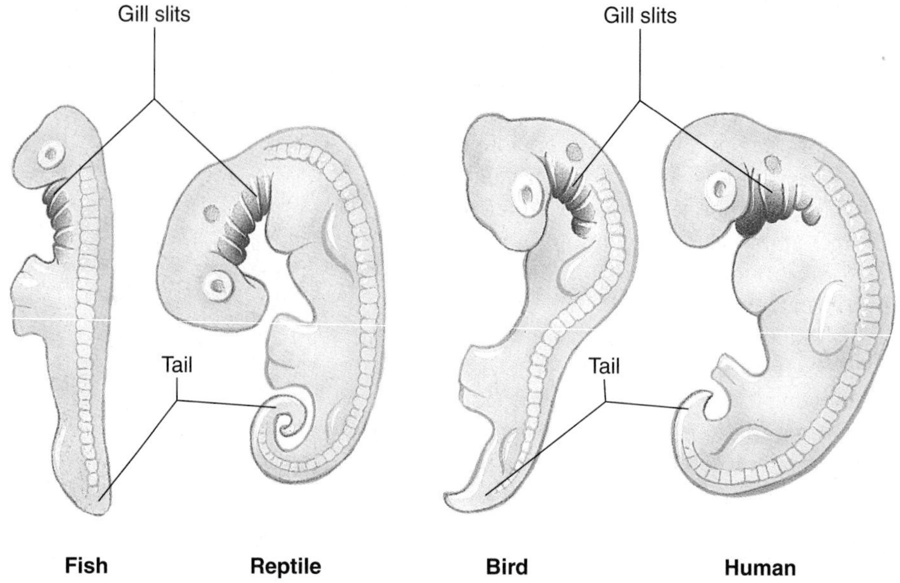
Gill slits
Gill slits
Tail
Tail
Fish
Reptile
Bird
Human

Figure 11.10 Embryos show our early evolutionary history.
These embryos, representing various vertebrate animals, show the primitive features that all share early in their development, such as gill slits and a tail.

Sharing the Same Parts

As vertebrates have evolved, the same bones are sometimes put to different uses, yet they can still be seen, their presence betraying their evolutionary past. For example, the forelimbs of vertebrates are all **homologous structures;** that is, although the structure and function of the bones have diverged, they are derived from the same body part present in a common ancestor. You can see in figure 11.11 how the bones of the forelimb have been modified in one way in the wings of bats, in another way in the fins of porpoises, and in yet other ways in cats, horses, and humans.

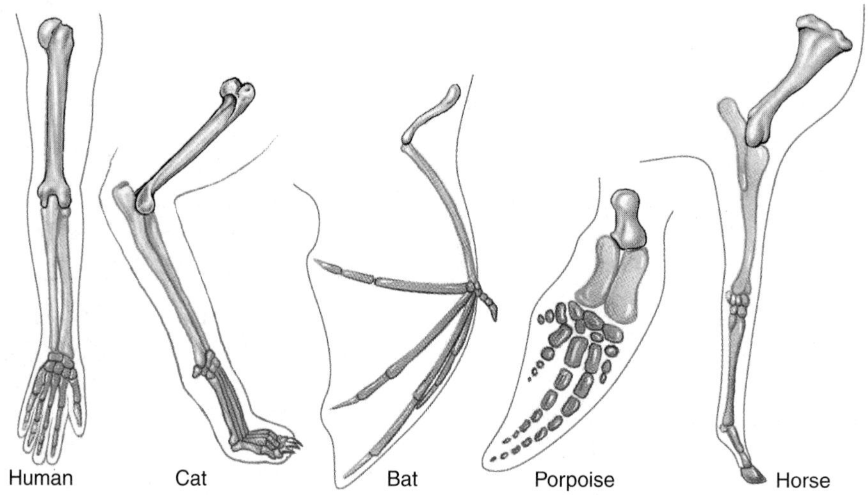
Human
Cat
Bat
Porpoise
Horse

Figure 11.11 Homology among vertebrate limbs.
Homologies among the forelimbs of five mammals show the ways in which the proportions of the bones have changed in relation to the particular way of life of each organism. Although considerable differences can be seen in form and function, the same basic bones are present in each forelimb.

Not all similar features are homologous. Sometimes features come to resemble each other as a result of parallel evolution of separate lineages. These are called **analogous structures.** For example, the marsupial mammals of Australia evolved in isolation from placental mammals, but similar selective pressures have generated very similar kinds of animals (figure 11.12). Similarly, flippers of penguins and dolphins are analogous structures, originating as two very different structures in two ancestors of distantly related lines and then being modified through natural selection to look the same and serve the same function.

"Leftover" (Vestigial) Organs

Sometimes bones are put to no use at all! In living whales, which evolved from hoofed mammals, the bones of the pelvis that formerly anchored the two hind limbs are all that remain of the rear legs, unattached to any other bones and serving no apparent purpose. Another example of a vestigial organ is the human appendix. In the great apes, our closest relatives, we find an appendix much larger than ours, attached to the gut tube, which functions in digestion, holding bacteria used in digesting the cellulose cell walls of the plants eaten by these primates. The human appendix is a vestigial version of this structure that now serves no function.

11.5 Comparative anatomy offers evidence that evolution has occurred, including relict developmental forms, vestigial organs, and homologous structures.

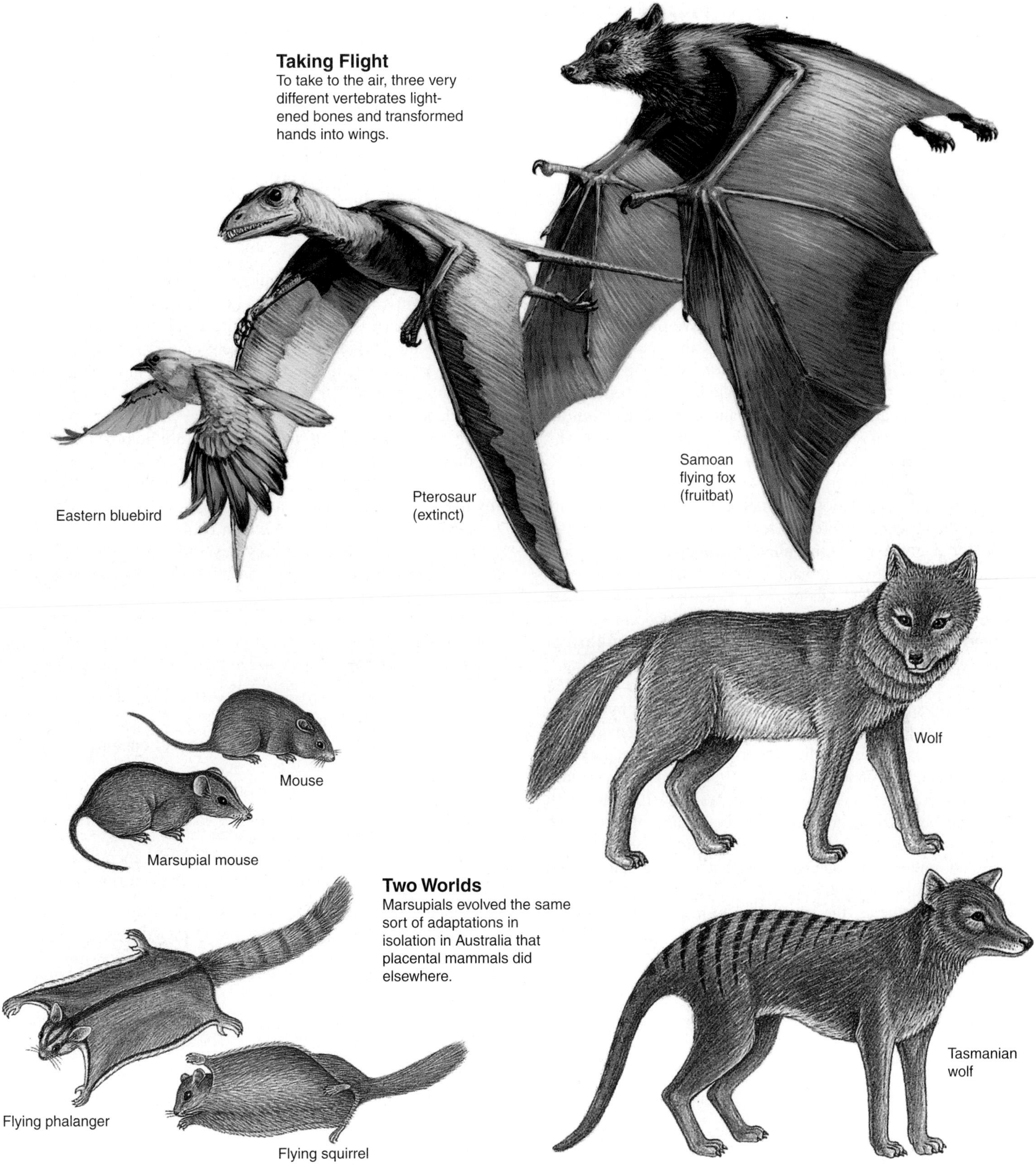
Taking Flight
To take to the air, three very different vertebrates lightened bones and transformed hands into wings.
Eastern bluebird
Pterosaur (extinct)
Samoan flying fox (fruitbat)
Wolf
Mouse
Marsupial mouse
Two Worlds
Marsupials evolved the same sort of adaptations in isolation in Australia that placental mammals did elsewhere.
Tasmanian wolf
Flying phalanger
Flying squirrel

Figure 11.12 Convergent evolution: many paths to one goal.

Over the course of evolution, form often follows function. Members of very different animal groups often adapt in similar fashions when challenged by similar opportunities. These are but a few of many examples of such convergent evolution. The flying vertebrates represent mammals (bat), reptiles (pterosaur), and birds (bluebird). The three pairs of terrestrial vertebrates each contrast a North American placental mammal with an Australian marsupial one.

11.6 The Hardy-Weinberg Rule

Population genetics is the study of the properties of genes in populations. Genetic variation within natural populations was a puzzle to Darwin and his contemporaries. The way in which meiosis produces genetic segregation among the progeny of a hybrid had not yet been discovered. Selection, scientists then thought, should always favor an optimal form.

Genes Within Populations

From the 1920s onward, scientists began to formulate a comprehensive theory of how **alleles,** alternative forms of a gene, behave in populations and how changes in gene frequencies lead to evolutionary change. Variation within populations puzzled many scientists; dominant alleles were believed to drive recessive alleles out of populations, with selection favoring an optimal form. The solution to the puzzle of why genetic variation persists was developed in 1908 by G. H. Hardy and G. Weinberg. Hardy and Weinberg pointed out that in a large population in which there is random mating, and in the absence of forces that change the proportion of alleles of a given gene, the **allele frequencies,** the original genotype proportions, remain constant from generation to generation. Dominant alleles do not, in fact, replace recessive ones. Because their proportions do not change, the genotypes are said to be in **Hardy-Weinberg equilibrium** (figure 11.13). A population that is in Hardy-Weinberg equilibrium is not evolving.

Hardy and Weinberg came to this conclusion by analyzing the frequencies of alleles in successive generations. **Frequency** is defined as the proportion of individuals falling within a category in relation to the total number of individuals being considered. Thus, in a population of 100 cats, with 84 black and 16 white cats, the frequencies of black and white are .84 and .16. By convention, the frequency of the more common of two alleles is designated by the letter p and that of the less common allele by the letter q. Because there are only two alleles, the sum of p and q must always equal 1 ($p + q = 1$).

In algebraic terms, the Hardy-Weinberg equilibrium is written as an equation. For a gene with two alternative alleles B (frequency p) and b (frequency q), the equation looks like this:

$(p + q)^2$	=	p^2	+	$2pq$	+	q^2
		Individuals homozygous for allele *B*		**Individuals heterozygous for alleles *B* and *b***		**Individuals homozygous for allele *b***

This equation lets us calculate allele frequencies (values of p and q) in a very simple way. For example, in the population of 100 cats, if we assume white color to be a recessive trait, then the 16 white cats represent double-recessive individuals, q^2 in the equation. If $q^2 = .16$, then q must equal .40 (the square root of .16 is .4). Now recall that $p + q = 1$. p must thus equal $1 - .40$, or .60. Then, $p^2 = (.60)^2$ or .36, which means that 36 out of the 100 cats are homozygous for the dominant allele B and are black. What is the frequency of the heterozygous genotype? The equation tells us $2pq$ is the frequency for individuals heterozygous for the alleles B and b, and $2pq = (2 \times .60 \times .40)$, which is .48. So, 48 out of the 100 cats are also black, but they have a heterozygous genotype. This result can be easily visualized in a Punnett square diagram (figure 11.14).

The Hardy-Weinberg rule is based on certain assumptions. The previous equation is true only if the following five assumptions are met:

1. The size of the population is very large or effectively infinite.
2. Individuals mate with one another at random.
3. There is no mutation.
4. There is no input of new copies of any allele from any extraneous source (such as from a nearby population or from mutation).
5. All alleles are replaced equally from generation to generation (natural selection is not occurring).

How valid are the predictions made by the Hardy-Weinberg equation? For many genes, they prove to be very accurate. As an example, consider the recessive allele responsible for the serious human disease cystic fibrosis. This allele is

Phenotypes			
Genotypes	*BB*	*Bb*	*bb*
Frequency of genotype in population	0.36	0.48	0.16
Frequency of gametes	0.36 + 0.24 = 0.6*B*		0.24 + 0.16 = 0.4*b*

Figure 11.13 Hardy-Weinberg equilibrium.

In the absence of factors that alter them, the frequencies of gametes, genotypes, and phenotypes remain constant generation after generation. The example shown here involves a population of 100 cats, in which 16 are white and 84 are black. White cats are *bb*, and black cats are *BB* or *Bb*.

present in Caucasians in North America at a frequency of about 22 per 1,000 individuals, or .022. What proportion of Caucasian North Americans, therefore, is expected to express this trait? The frequency of double-recessive individuals (q^2) is expected to be .022 × .022, or about 1 in every 2,000 individuals. What proportion is expected to be heterozygous carriers? If the frequency of the recessive allele q is .022, then the frequency of the dominant allele p must be 1 − .022, or .978. The frequency of heterozygous individuals ($2pq$) is thus expected to be 2 × .978 × .022, or 43 in every 1,000 individuals, very close to real estimates.

Most human populations are large and randomly mating and thus are similar to the ideal population envisioned by Hardy and Weinberg. For some genes, however, the calculated predictions do *not* match the actual values. The reasons why explain a great deal about evolution.

11.6 Mendelian inheritance does not alter allele frequencies. In a large, random-mating population, alternative alleles *B* (frequency *p*) and *b* (frequency *q*) are expected to be present in the proportions $p^2 + 2pq + q^2$.

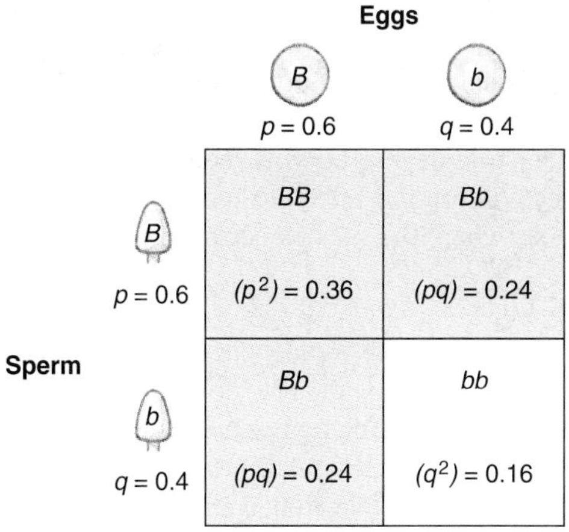
Eggs
B
p = 0.6
b
q = 0.4
Sperm
B
p = 0.6
b
q = 0.4
BB
(p²) = 0.36
Bb
(pq) = 0.24
Bb
(pq) = 0.24
bb
(q²) = 0.16

Figure 11.14 A Punnett square analysis.

The potential crosses in this cat population can be determined using a Punnett square analysis.

11.7 Why Do Allele Frequencies Change?

Many factors can alter allele frequencies. But only five alter the proportions of homozygotes and heterozygotes enough to produce significant deviations from the proportions predicted by the Hardy-Weinberg rule: (1) mutation, (2) migration (including both immigration into and emigration out of a given population), (3) genetic drift (random loss of alleles, which is more likely to occur in small populations), (4) nonrandom mating, and (5) selection (table 11.1).

Mutation

A **mutation** is an error in replication of a nucleotide sequence in DNA. Mutation from one allele to another obviously can change the proportions of particular alleles in a population. But mutation rates are generally too low to significantly alter Hardy-Weinberg proportions of common alleles. Many genes mutate 1 to 10 times per 100,000 cell divisions. Some of these mutations are harmful, while others are neutral or, even rarer, beneficial. This rate is so slow that few populations are around long enough to accumulate significant numbers of mutations.

Migration

Migration, defined in genetic terms as the movement of individuals from one population into another, can be a powerful force upsetting the genetic stability of natural populations. Sometimes, migration is obvious, as when an animal moves from one place to another. If the characteristics of the newly arrived animal differ from those already there, and if the newly arrived individual or individuals can adapt to survive in the new area and mate successfully, then the genetic composition of the receiving population may be altered.

Other important kinds of migration are not as obvious. These subtler movements include the drifting of gametes of plants or immature stages of marine animals or plants from one place to another. However it occurs, migration can alter the genetic characteristics of populations and prevent the maintenance of Hardy-Weinberg equilibrium. However, the evolutionary role of migration is more difficult to assess and depends heavily on the selective forces prevailing at the different places where the species occurs.

Genetic Drift

In small populations, the frequencies of particular alleles may be changed drastically by chance alone. The individual alleles of a given gene may all be represented in few individuals, and some of them may be accidentally lost if those individuals fail to reproduce or die. Allele frequencies appear to change randomly, as if the frequencies were drifting; thus, random loss of alleles is known as **genetic drift.** A series of small populations that are isolated from one another may come to differ strongly as a result of genetic drift.

TABLE 11.1 AGENTS OF EVOLUTIONARY CHANGE

Factor		Description
Mutation	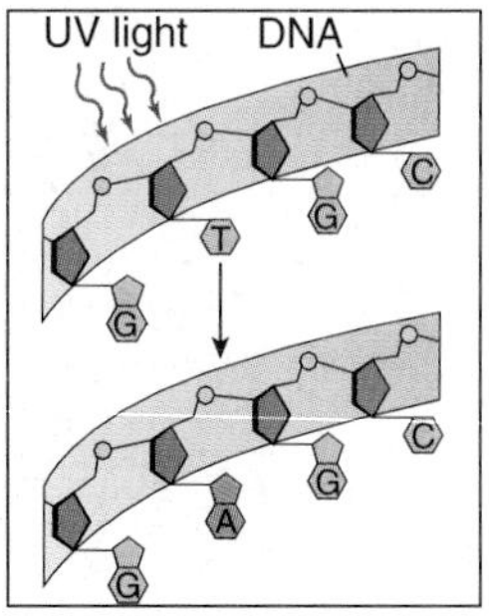 UV light DNA T G C G A G C G	The ultimate source of variation. Individual mutations occur so rarely that mutation alone does not change allele frequency much.
Migration	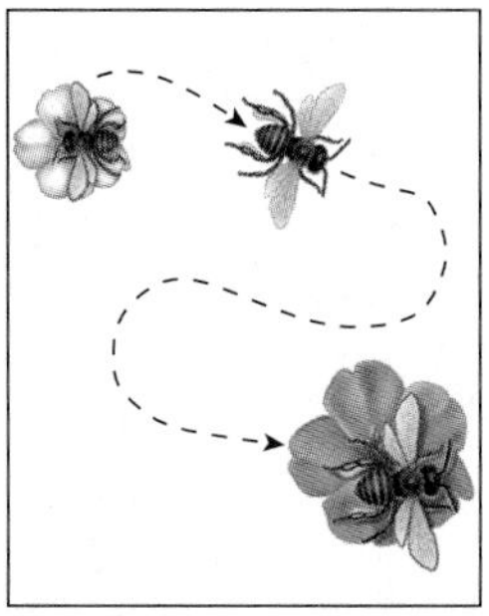	A very potent agent of change. Migration acts to promote evolutionary change by enabling populations that exchange members to converge toward one another.
Genetic drift	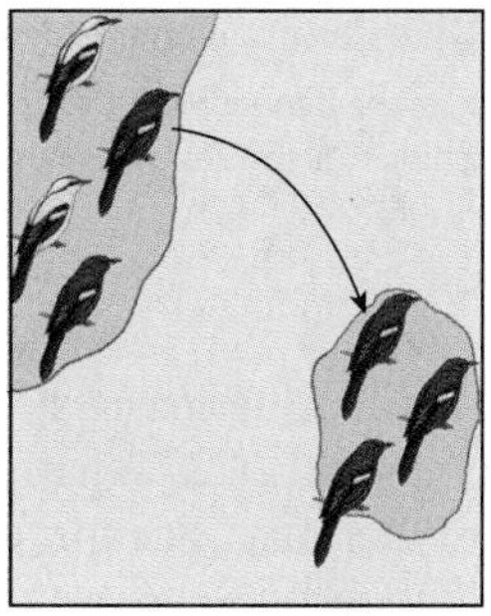	Statistical accidents. Usually occurs only in very small populations.
Nonrandom mating	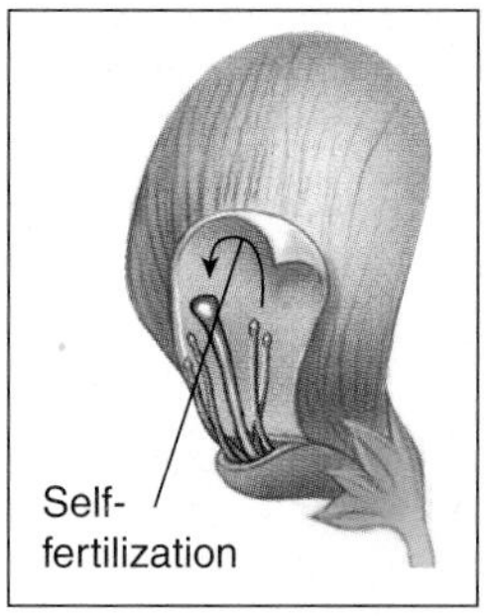 Self-fertilization	Inbreeding is the most common form. It does not alter allele frequency but decreases the proportion of heterozygotes ($2pq$).
Selection	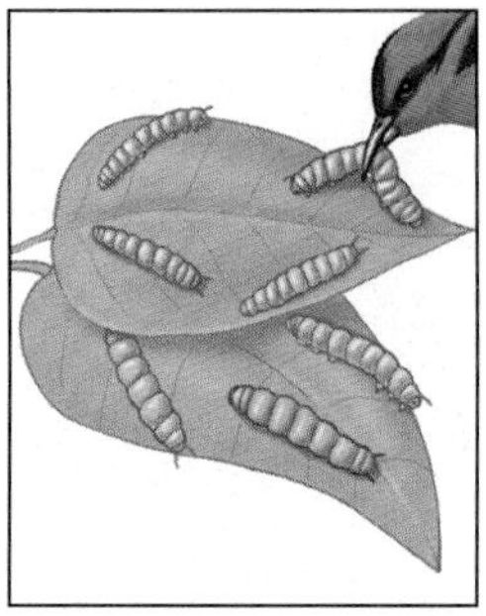	The only form that produces *adaptive* evolutionary changes. Only rapid for allele frequency greater than .01.

When one or a few individuals migrate and become the founders of a new, isolated population at some distance from their place of origin, the alleles that they carry are of special significance. Even if these alleles are rare in the source population, they will be a significant fraction of the new population's genetic endowment. This is called the **founder effect.** As a result of the founder effect, rare alleles and combinations often become more common in new, isolated populations. The founder effect is particularly important in the evolution of organisms that occur on oceanic islands, such as the Galápagos Islands, which Darwin visited. Most of the kinds of organisms that occur in such areas were probably derived from one or a few initial founders. In a similar way, isolated human populations are often dominated by the genetic features that were characteristic of their founders, if only a few individuals were involved initially (figure 11.15).

Even if organisms do not move from place to place, occasionally their populations may be drastically reduced in size. This may result from flooding, drought, earthquakes, and other natural forces or from progressive changes in the environment. The surviving individuals constitute a random genetic sample of the original population. Such a restriction in genetic variability has been termed the **bottleneck effect.**

Some living species appear to be severely depleted genetically and have probably suffered from a bottleneck effect in the past. As shown by Steve O'Brien of the National Institutes of Health, all living cheetahs are practically identical genetically. Lacking the genetic diversity that might lend some of them resistance, they are therefore very susceptible to disease. The simplest explanation for this similarity is that there was a drastic reduction in the size of cheetah populations in the recent past and that all cheetahs living today have descended from a very few individuals; they have passed through a genetic bottleneck.

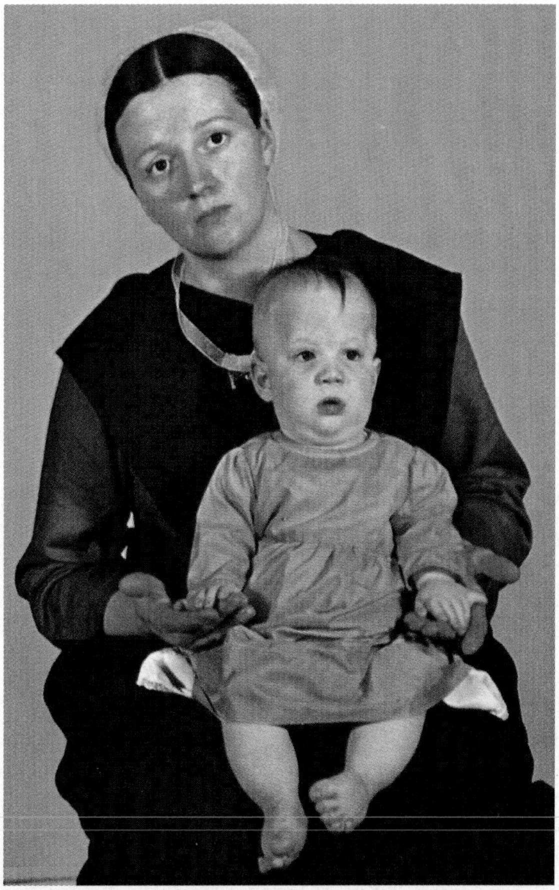

Figure 11.15 A consequence of the founder effect.

This Amish woman is holding her child, who has Ellis-van Creveld syndrome. The characteristic symptoms are short limbs, dwarfed stature, and extra fingers. This disorder was introduced in the Amish community by one of its founders in the eighteenth century and persists to this day because of the reproductive isolation of the Amish.

Nonrandom Mating

Individuals with certain genotypes sometimes mate with one another either more or less commonly than would be expected on a random basis, a phenomenon known as **nonrandom mating.** One type of nonrandom mating characteristic of many groups of organisms is **inbreeding,** mating with relatives. Inbreeding increases the proportions of individuals that are homozygous, and as a result inbred populations contain more homozygous individuals than predicted by the Hardy-Weinberg rule. For this reason, populations of self-fertilizing plants consist primarily of homozygous individuals, whereas outcrossing plants, which interbreed with individuals different from themselves, have a higher proportion of heterozygous individuals.

Selection

As Darwin pointed out, some individuals leave behind more progeny than others, and the likelihood they will do so is affected by their inherited characteristics. The result of this process is called **selection** and was familiar even in Darwin's day to breeders of horses and farm animals. In so-called **artificial selection,** the breeder selects for the desired characteristics. In **natural selection,** Darwin suggested the environment plays this role, with conditions in nature determining which kinds of individuals in a population are the most fit and so affecting the proportions of genes among individuals of future populations. This is the key point in Darwin's proposal that evolution occurs because of natural selection: the environment imposes the conditions that determine the results of selection and, thus, the direction of evolution.

11.7 Five evolutionary forces have the potential to significantly alter allele frequencies in populations: mutation, migration, genetic drift, nonrandom mating, and selection.

11.8 Forms of Selection

Selection operates in natural populations of a species as skill does in a football game. In any individual game, it can be difficult to predict the winner, because chance can play an important role in the outcome; but over a long season, the teams with the most skillful players usually win the most games. In nature, those individuals best suited to their environments tend to win the evolutionary game by leaving the most offspring, although chance can play a major role in the life of any one individual. Selection is a statistical concept, just as betting is. Although you cannot predict the fate of any one individual, or any one coin toss, it is possible to predict which kind of individual will tend to become more common in populations of a species, as it is possible to predict the proportion of heads after many coin tosses.

In nature, many traits, perhaps most, are affected by more than one gene. The interactions between genes are typically complex, as you saw in chapter 7. For example, alleles of many different genes play a role in determining human height (see figure 7.13). In such cases, selection operates on all the genes, influencing most strongly those that make the greatest contribution to the phenotype. How selection changes the population depends on which genotypes are favored. Three types of natural selection have been identified: disruptive selection, stabilizing selection, and directional selection (figure 11.16).

Disruptive Selection

In some situations, selection acts to eliminate, rather than favor, the intermediate type. For example, in different parts of Africa, the color pattern of the butterfly *Papilio dardanus* is dramatically different, although in each instance it closely resembles the coloring of some other butterfly species that birds do not like to eat. Birds quickly detect and eat butterflies that do not resemble distasteful butterflies, so any intermediate color patterns are eliminated. In this case, selection is acting to eliminate the intermediate phenotypes. This form of selection is called **disruptive selection.** Disruptive selection is far less common than the other two types of selection.

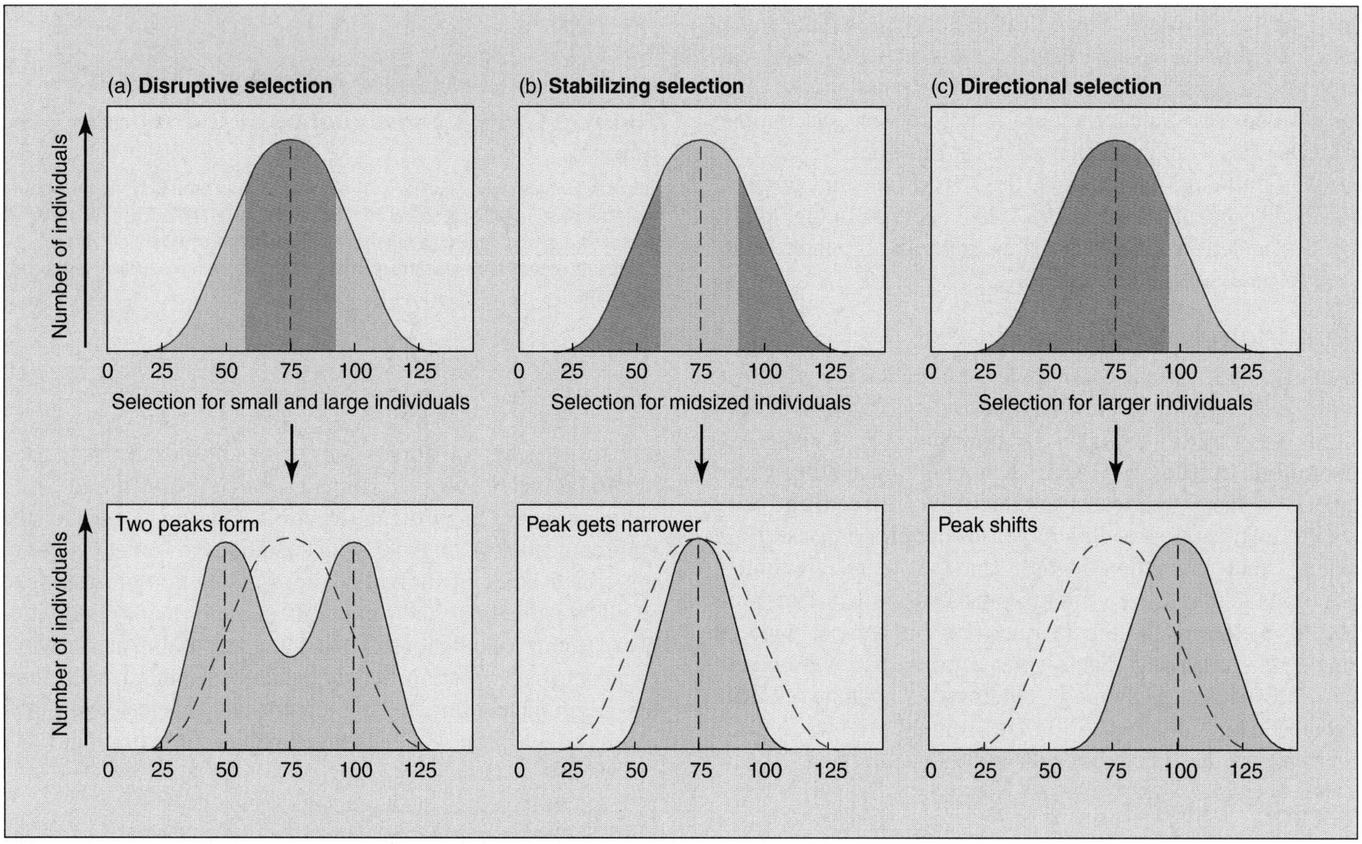
(a) Disruptive selection
(b) Stabilizing selection
(c) Directional selection
Number of individuals
0 25 50 75 100 125
Selection for small and large individuals
Selection for midsized individuals
Selection for larger individuals
Two peaks form
Peak gets narrower
Peak shifts

Figure 11.16 Three kinds of natural selection.

(*a*) In *disruptive selection,* individuals in the middle of the range of phenotypes of a certain trait are selected against, and the extreme forms of the trait are favored. (*b*) In *stabilizing selection,* individuals with midrange phenotypes are favored, with selection acting against both ends of the range of phenotypes. (*c*) In *directional selection,* individuals concentrated toward one extreme of the array of phenotypes are favored.

Stabilizing Selection

When selection acts to eliminate both extremes from an array of phenotypes, the result is to increase the frequency of the already common intermediate type. In effect, selection is operating to prevent change away from this middle range of values. In a classic study carried out after an "uncommonly severe storm of snow, rain, and sleet" on February 1, 1898, 136 starving English sparrows were collected and brought to the laboratory of H. C. Bumpus at Brown University in Providence, Rhode Island. Of these, 64 died and 72 survived. Bumpus took standard measurements on all the birds. He found that among males, the surviving birds tended to be bigger, as one might expect from the action of directional selection. However, among females, the birds that perished were not smaller, on the average, than those that survived. But among them were many more individuals that had extreme measurements—measurements unusual for the population as a whole. Selection had acted most strongly against these individuals. When selection acts in this way, the population contains fewer individuals with alleles promoting extreme types. Selection has not changed the most common phenotype of the population but rather made it even more common by eliminating extremes. Many examples similar to Bumpus's female sparrows are known. In humans, infants with intermediate weight at birth have the highest survival rate (figure 11.17). In ducks and chickens, eggs of intermediate weight have the highest hatching success. This form of selection is called **stabilizing selection.**

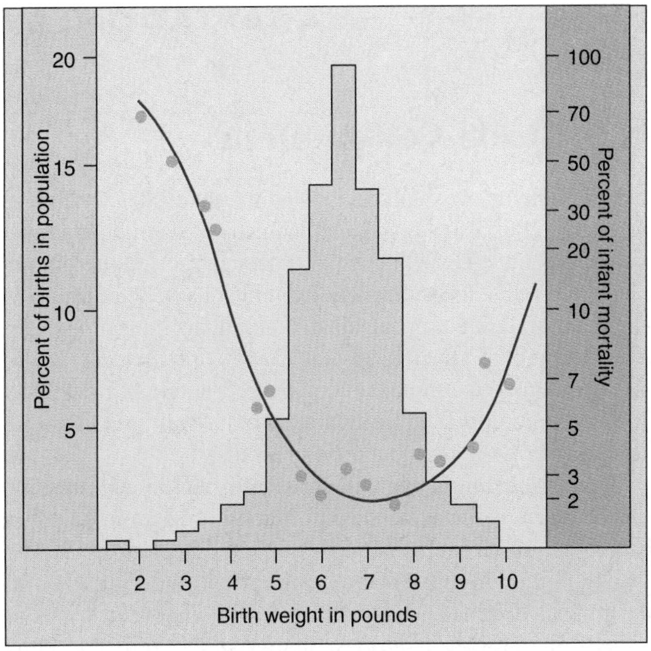

Percent of births in population
20
15
10
5
Percent of infant mortality
100
70
50
30
20
10
7
5
3
2
2 3 4 5 6 7 8 9 10
Birth weight in pounds

Figure 11.17 Stabilizing selection for birth weight.

The death rate among human babies is lowest at an intermediate birth weight between 7 and 8 pounds. Both larger and smaller babies have a greater tendency to die at or near birth. Stabilizing selection acts to eliminate both extremes from an array of phenotypes and to increase the frequency of the intermediate type.

Directional Selection

When selection acts to eliminate one extreme from an array of phenotypes, the genes determining this extreme become less frequent in the population. The environment dictates which extreme will not be favored. Individuals exhibiting the other extreme of an array of phenotypes will leave more offspring than individuals with the unfavored form of the trait and individuals in the middle of the array. Thus, individuals exhibiting forms of a trait at one extreme become more common (figure 11.18). This form of selection is called **directional selection.**

11.8 Selection on traits affected by many genes can favor both extremes of the trait, or intermediate values, or only one extreme.

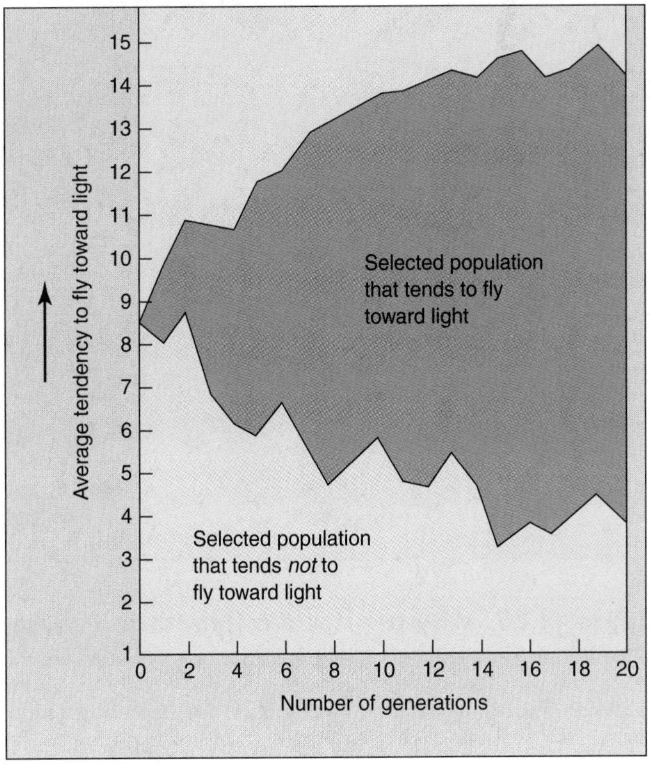

Average tendency to fly toward light
15
14
13
12
11
10
9
8
7
6
5
4
3
2
1
Selected population that tends to fly toward light
Selected population that tends *not* to fly toward light
0 2 4 6 8 10 12 14 16 18 20
Number of generations

Figure 11.18 Directional selection in action.

In generation after generation, individuals of the fly *Drosophila* were selectively bred to obtain two populations. When flies with a strong tendency to fly toward light were used as parents for the next generation, their offspring had a greater tendency to fly toward light (*top curve*). When flies that tended not to fly toward light were used as parents for the next generation, their offspring had an even greater tendency not to fly toward light (*bottom curve*).

11.9 Sickle-Cell Anemia

Darwin's theory of evolution not only states that evolution has occurred, it also suggests a mechanism, natural selection. In the time since Darwin (and Wallace) suggested the pivotal role of natural selection, many examples have been found in which natural selection is clearly acting to change the genetic makeup of species, just as Darwin predicted. Two of the best-studied examples are sickle-cell anemia (a defect in human hemoglobin proteins) and industrial melanism (a darkening of wing pigmentation in moths).

The first example we will consider is **sickle-cell anemia,** a hereditary disease affecting hemoglobin molecules in the blood. Sickle-cell anemia was first detected in 1904 in Chicago in a blood examination of an individual complaining of tiredness (figure 11.19). The disorder arises as a result of a single nucleotide change in the gene encoding β-hemoglobin, which causes hemoglobin molecules to clump together (figure 11.20).

Persons homozygous for the sickle-cell genetic mutation in the β-hemoglobin gene frequently have a reduced life span. This is because the sickled form of hemoglobin does not carry oxygen atoms well and red blood cells that are sickled do not flow smoothly through the tiny capillaries but instead jam up and block blood flow. Heterozygous individuals, who have both a defective and a normal form of the gene, make enough functional hemoglobin to keep their red blood cells healthy.

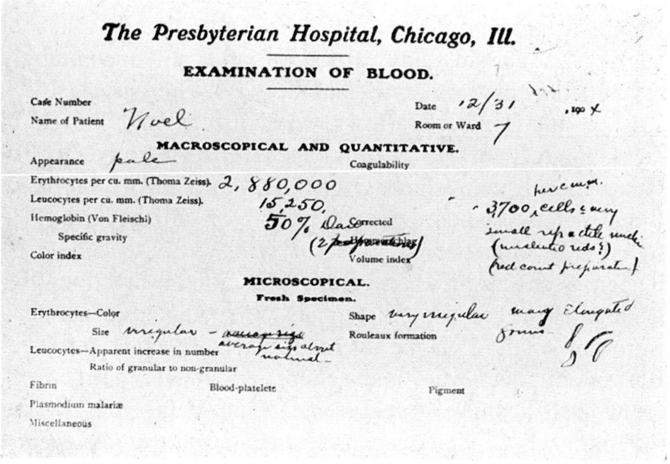
The Presbyterian Hospital, Chicago, Ill.

EXAMINATION OF BLOOD.

Case Number

Name of Patient Noel

Date 12/31, 1904

Room or Ward 7

MACROSCOPICAL AND QUANTITATIVE.

Appearance pale

Coagulability

Erythrocytes per cu. mm. (Thoma Zeiss). 2,880,000

Leucocytes per cu. mm. (Thoma Zeiss). 15,250,

Hemoglobin (Von Fleischl) 50%

Corrected

Specific gravity

Hematokrit

Color index

Volume index

3700 cells

MICROSCOPICAL.

Fresh Specimen.

Erythrocytes—Color

Size irregular

Shape very irregular

Rouleaux formation

Leucocytes—Apparent increase in number

Ratio of granular to non-granular

Fibrin

Blood-platelets

Pigment

Plasmodium malariæ

Miscellaneous

Figure 11.19 The first known sickle-cell anemia patient.

Dr. Ernest Irons's blood examination report on his patient Walter Clement Noel, December 31, 1904, described his oddly shaped red blood cells.

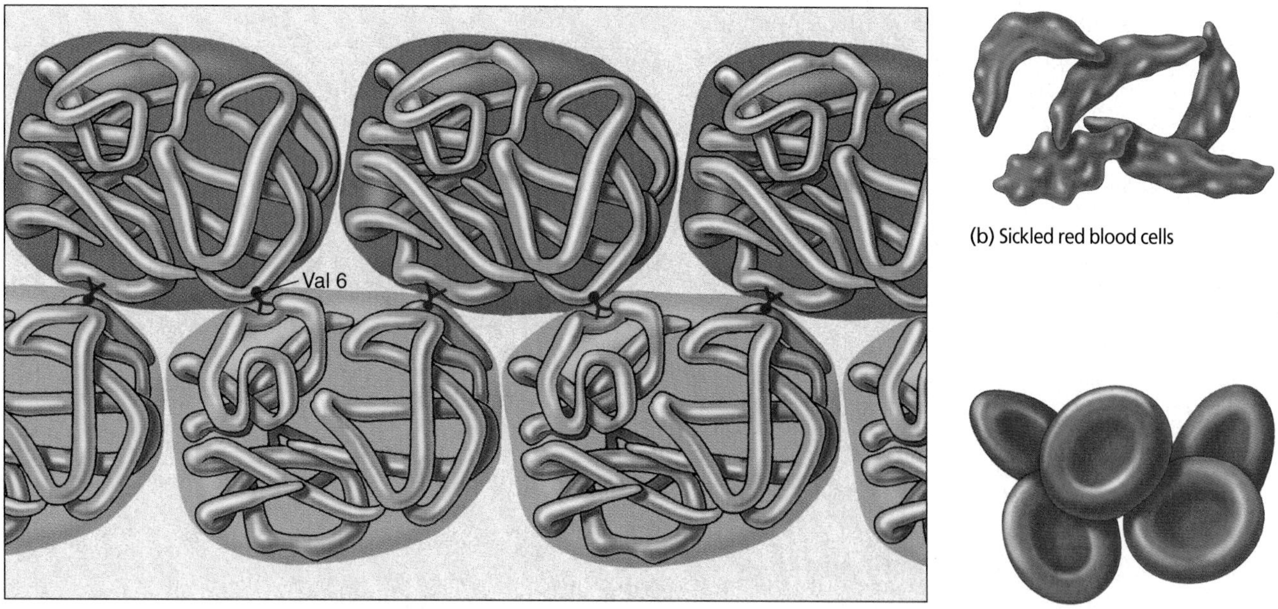
Val 6

(a)

(b) Sickled red blood cells

(c) Normal red blood cells

Figure 11.20 Why the sickle-cell mutation causes hemoglobin to clump.

(*a*) The sickle-cell mutation changes the sixth amino acid in the β-hemoglobin chain (position B6) from glutamic acid (very polar) to valine (nonpolar). The unhappy result of this change is that the nonpolar valine at position B6, protruding from a corner of the hemoglobin molecule, fits nicely into a nonpolar pocket on the opposite side of another hemoglobin molecule. This causes the two molecules to clump together. As each molecule has both a B6 valine and an opposite nonpolar pocket, long chains form. (*b*) The result is a deformed "sickle-shaped" red blood cell. (*c*) When polar glutamic acid (the normal allele) occurs at position B6, it is not attracted to the nonpolar pocket, and no clumping occurs.

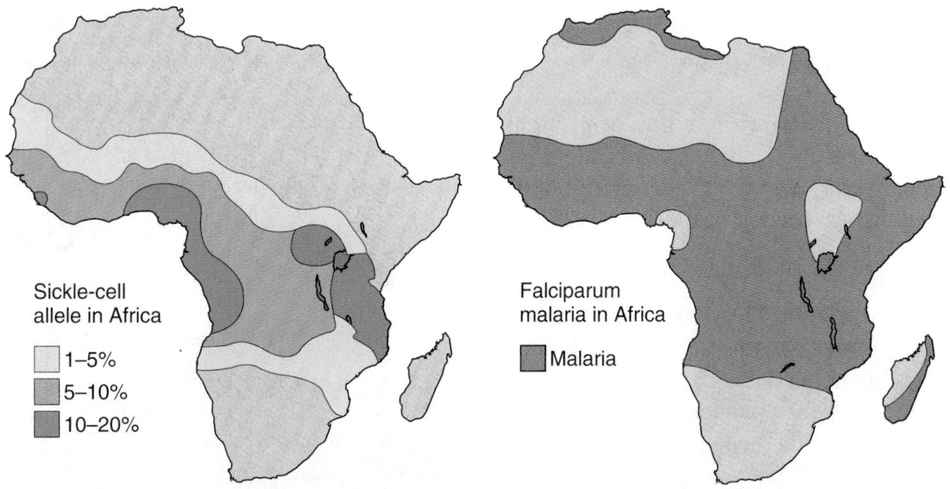
Sickle-cell
allele in Africa
1–5%
5–10%
10–20%
Falciparum
malaria in Africa
Malaria

Figure 11.21 How stabilizing selection maintains sickle-cell anemia.
The diagrams show the frequency of the sickle-cell allele (*left*) and the distribution of falciparum malaria (*right*). Falciparum malaria is one of the most devastating forms of the often fatal disease. As you can see, its distribution in Africa is closely correlated with that of the allele of the sickle-cell characteristic.

The Puzzle: Why So Common?

The disorder is now known to have originated in central Africa, where the frequency of the sickle-cell allele is about .12. One in 100 people is homozygous for the defective allele and develops the fatal disorder. Sickle-cell anemia affects roughly two African Americans out of every thousand but is almost unknown among other racial groups.

If Darwin is right, and natural selection drives evolution, then why has natural selection not acted against the defective allele in Africa and eliminated it from the human population there? Why is this potentially fatal allele instead very common there?

The Answer: Stabilizing Selection

The defective allele has not been eliminated from central Africa because people who are heterozygous for the sickle-cell allele are much less susceptible to malaria, one of the leading causes of death in central Africa (figure 11.21). Even though the population pays a high price—the many individuals in each generation who are homozygous for the sickle-cell allele die—the deaths are far fewer than would occur due to malaria if the heterozygous individuals were not malaria resistant. One in 5 individuals are heterozygous and survive malaria, while only 1 in 100 are homozygous and die of anemia. Natural selection has favored the sickle-cell allele in central Africa because the payoff in survival of heterozygotes there more than makes up for the price in death of homozygotes.

Stabilizing selection (also called "balancing selection") is thus acting on the sickle-cell allele: (1) selection tends to eliminate the sickle-cell allele because of its lethal effects on homozygous individuals; (2) selection tends to favor the sickle-cell allele because it protects heterozygotes from malaria. Like an economist balancing a budget, natural selection increases the frequency of an allele in a species as long as there is something to be gained by it, until the cost balances the benefit.

Stabilizing selection occurs because malarial resistance counterbalances lethal anemia. Malaria is a tropical disease that has not occurred in the United States since the early 1900s, and stabilizing selection has not favored the sickle-cell allele here. Blacks brought to America several centuries ago from Africa have not gained any evolutionary advantage in all that time from being heterozygous for the sickle-cell allele. There is no benefit to being resistant to malaria if there is no danger of getting malaria anyway. As a result, the selection against the sickle-cell allele in America is not counterbalanced by any advantage, and the allele has become far less common among American blacks than among blacks in central Africa.

Stabilizing selection is thought to have influenced many other human genes in a similar fashion. The recessive *cf* allele causing cystic fibrosis is unusually common in northwestern Europeans. Apparently, the bacterium causing typhoid fever uses the healthy version of the CFTR protein (see page 78) to enter the cells it infects, but it cannot use the cystic fibrosis version of the protein. As with sickle-cell anemia, heterozygotes are protected.

11.9 The prevalence of sickle-cell anemia in African populations is thought to reflect the action of natural selection favoring individuals carrying one copy of the sickle-cell allele, because they are resistant to malaria, common in Africa.

11.10 Peppered Moths and Industrial Melanism

The peppered moth, *Biston betularia,* is a European moth that rests on tree trunks during the day. Until the mid-nineteenth century, almost every captured individual of this species had light-colored wings. From that time on, individuals with dark-colored wings increased in frequency in the moth populations near industrialized centers until they made up almost 100% of these populations. Dark individuals had a dominant allele that was present but very rare in populations before 1850. Biologists soon noticed that in industrialized regions where the dark moths were common, the tree trunks were darkened almost black by the soot of pollution. Dark moths were much less conspicuous resting on them than were light moths. In addition, the air pollution that was spreading in the industrialized regions had killed many of the light-colored lichens on tree trunks, making the trunks darker.

Selection for Melanism

Can Darwin's theory explain the increase in the frequency of the dark allele? Why did dark moths gain a survival advantage around 1850? An amateur moth collector named J. W. Tutt proposed what became the most commonly accepted hypothesis explaining the decline of the light-colored moths. He suggested that light forms were more visible to predators on sooty trees that had lost their lichens. Consequently, birds ate the peppered moths resting on the trunks of trees during the day. The dark forms, in contrast, were at an advantage because they were camouflaged (figure 11.22). Although Tutt initially had no evidence, British ecologist Bernard Kettlewell tested the hypothesis in the 1950s by rearing populations of peppered moths with equal numbers of dark and light individuals. Kettlewell then released these populations into two sets of woods: one, near heavily polluted Birmingham, and the other, in unpolluted Dorset. Kettlewell set up rings of traps around the woods to see how many of both kinds of moths survived. To evaluate his results, he had marked the released moths with a dot of paint on the underside of their wings, where birds could not see it.

In the polluted area near Birmingham, Kettlewell trapped 19% of the light moths but 40% of the dark ones. This indicated that dark moths had a far better chance of surviving in these polluted woods, where the tree trunks were dark. In the relatively unpolluted Dorset woods, Kettlewell recovered 12.5% of the light moths but only 6% of the dark ones. This indicated that where the tree trunks were still light-colored, light moths had a much better chance of survival. Kettlewell later solidified his argument by placing hidden blinds in the woods and actually filming birds eating the moths. Sometimes the birds Kettlewell observed actually passed right over a moth that was the same color as its background.

Figure 11.22 Tutt's hypothesis explaining industrial melanism.

These photographs show color variants of the peppered moth, *Biston betularia.* Tutt proposed that the dark moth is more visible to predators on unpolluted trees (*top*), while the light moth is more visible to predators on bark blackened by industrial pollution (*bottom*).

Industrial Melanism

Industrial melanism is a term used to describe the evolutionary process in which darker individuals come to predominate over lighter individuals since the industrial revolution as a result of natural selection. The process is widely believed to have taken place because the dark organisms are better concealed from their predators in habitats that have been darkened by soot and other forms of industrial pollution, as suggested by Kettlewell's research.

Dozens of other species of moths have changed in the same way as the peppered moth in industrialized areas throughout Eurasia and North America, with dark forms becoming more common from the mid-nineteenth century onward as industrialization spread.

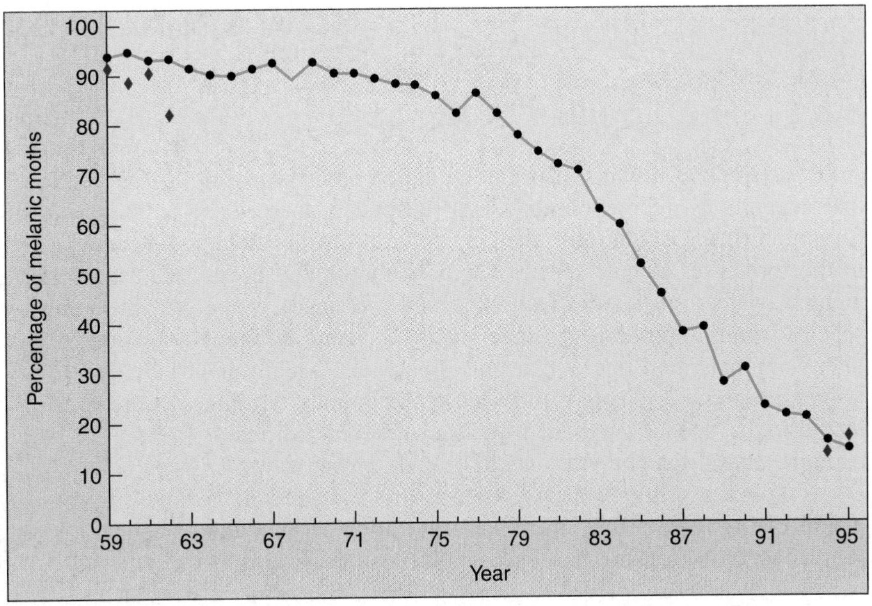
Percentage of melanic moths
100
90
80
70
60
50
40
30
20
10
0
59
63
67
71
75
79
83
87
91
95
Year

Figure 11.23 Selection against melanism.

The circles indicate the frequency of melanic *Biston* moths at Caldy Common in England, sampled continuously from 1959 to 1995. Diamonds indicate frequencies in Michigan from 1959 to 1962 and from 1994 to 1995.

Selection Against Melanism

As of the second half of the twentieth century, with the widespread implementation of pollution controls, these trends are reversing, not only for the peppered moth in many areas in England but also for many other species of moths throughout the northern continents. These examples provide some of the best-documented instances of changes in allelic frequencies of natural populations as a result of natural selection due to specific factors in the environment.

In England, the pollution promoting industrial melanism began to reverse following enactment of Clean Air legislation in 1956. Beginning in 1959, the *Biston* population at Caldy Common outside Liverpool has been sampled each year. The frequency of the melanic (dark) form dropped from a high of 94% in 1960 to a low of 19% in 1995 (figure 11.23). Similar reversals have been documented at numerous other locations throughout England. The drop correlates well with a drop in air pollution, particularly with tree-darkening sulfur dioxide and suspended particulates.

Interestingly, the same reversal of industrial melanism appears to have occurred in America during the same time that it was happening in England. Industrial melanism in the American subspecies of the peppered moth was not as widespread as in England, but it has been well documented at a rural field station near Detroit. Of 576 peppered moths collected there from 1959 to 1961, 515 were melanic, a frequency of 89%. The American Clean Air Act, passed in 1963, led to significant reductions in air pollution. Resampled in 1994, the Detroit field station peppered moth population had only 15% melanic moths (see figure 11.23)! The moths in Liverpool and Detroit, both part of the same natural experiment, exhibit strong evidence of natural selection.

Reconsidering the Target of Natural Selection

Tutt's hypothesis, widely accepted in the light of Kettlewell's studies, is currently being reevaluated. The problem is that the recent selection against melanism does not appear to correlate with changes in tree lichens. At Caldy Common, the light form of the peppered moth began its increase in frequency long before lichens began to reappear on the trees. At the Detroit field station, the lichens never changed significantly as the dark moths first became dominant and then declined over 30 years. In fact, investigators have not been able to find peppered moths on Detroit trees at all, whether covered with lichens or not. Wherever the moths rest during the day, it does not appear to be on tree bark. Some evidence suggests they rest on leaves on the treetops, but no one is sure.

The action of selection may depend on other differences between light and dark forms of the peppered moth besides their wing coloration. Researchers report, for example, a clear difference in their ability to survive as caterpillars under a variety of conditions. Perhaps natural selection is targeting the caterpillars rather than the adults. While we can't yet say exactly what is going on, researchers are actively investigating one of the best-documented instances of natural selection in action.

11.10 Natural selection has favored the dark form of the peppered moth in areas subject to severe air pollution, perhaps because on darkened trees they are less easily seen by moth-eating birds. Selection has in turn favored the light form as pollution has abated.

11.11 The Species Concept

A key aspect of Darwin's theory of evolution is his proposal that adaptation (microevolution) leads ultimately to species formation (macroevolution). The way natural selection leads to the formation of new species has been thoroughly documented by biologists, who have observed the stages of the species-forming process, or speciation, in many different plants, animals, and microorganisms. Speciation usually involves progressive change: first, local populations become increasingly specialized; then, if they become different enough, natural selection may act to keep them that way.

A **species** (figure 11.24) is generally defined as a group of organisms that is unlike other such groups and that does not integrate extensively with other groups in nature. Species formation is generally considered to be the final stage of a long evolutionary process: (1) First, local populations become adapted to the particular set of circumstances that they face. (2) When they become different enough, they are considered to be ecological races. (3) Then, natural selection may act to reinforce the differences between two races by favoring changes that discourage hybrids. These sorts of changes are called **isolating mechanisms.** (4) Finally, the two races become incapable of interbreeding and are considered separate species.

11.11 A species is generally defined as a group of similar organisms that does not exchange genes extensively with other groups in nature.

Figure 11.24 Great genetic variation can occur within a single species.
"Looking different" does not mean two animals are members of different species. All of the different breeds of dogs are members of the same species, *Canis familiaris,* and all can breed successfully with each other. Dog breeders engaging in artificial selection are responsible for the great differences among dog breeds.

11.12 Prezygotic Isolating Mechanisms

Often ecological races continue to diverge, becoming more and more different from each other as natural selection favors different survival strategies in different environments. Eventually a point is reached when the races are so different that biologists consider them separate species. Very often, selection favors changes called *isolating mechanisms,* which prevent the new species from breeding with each other, particularly when the hybrids produced by such matings are ill-suited to either environment. At this point, two species are said to be **reproductively isolated.**

The reproductive isolation between species is created and maintained by two basic kinds of isolating mechanisms: **prezygotic** isolating mechanisms, which prevent the formation of a zygote, and **postzygotic** isolating mechanisms, which prevent the proper development or functioning of zygotes after they are formed.

There are six main types of prezygotic isolating mechanisms. In *geographic isolation,* species are physically separated from one another and have no opportunity to hybridize (figure 11.25). In *ecological isolation,* two species may occur in the same area, but they may use different portions of the habitat and thus not hybridize because they do not encounter each other. In *behavioral isolation,* the often elaborate courtship and mating rituals differ among related species and tend to keep these species distinct in nature even if they inhabit the same places. *Temporal isolation* occurs when species have different breeding seasons, preventing hybridization between the species. In *mechanical isolation,* structural differences prevent mating between related species of animals. Aside from such obvious features as size, the structure and location of the male and female copulatory organs may be incompatible. Lastly, in *prevention of gamete fusion,* the union of gametes may be prevented even following successful mating if the gametes fail to attract one another or function poorly.

11.12 Prezygotic isolating mechanisms lead to reproductive isolation by preventing the formation of hybrid zygotes.

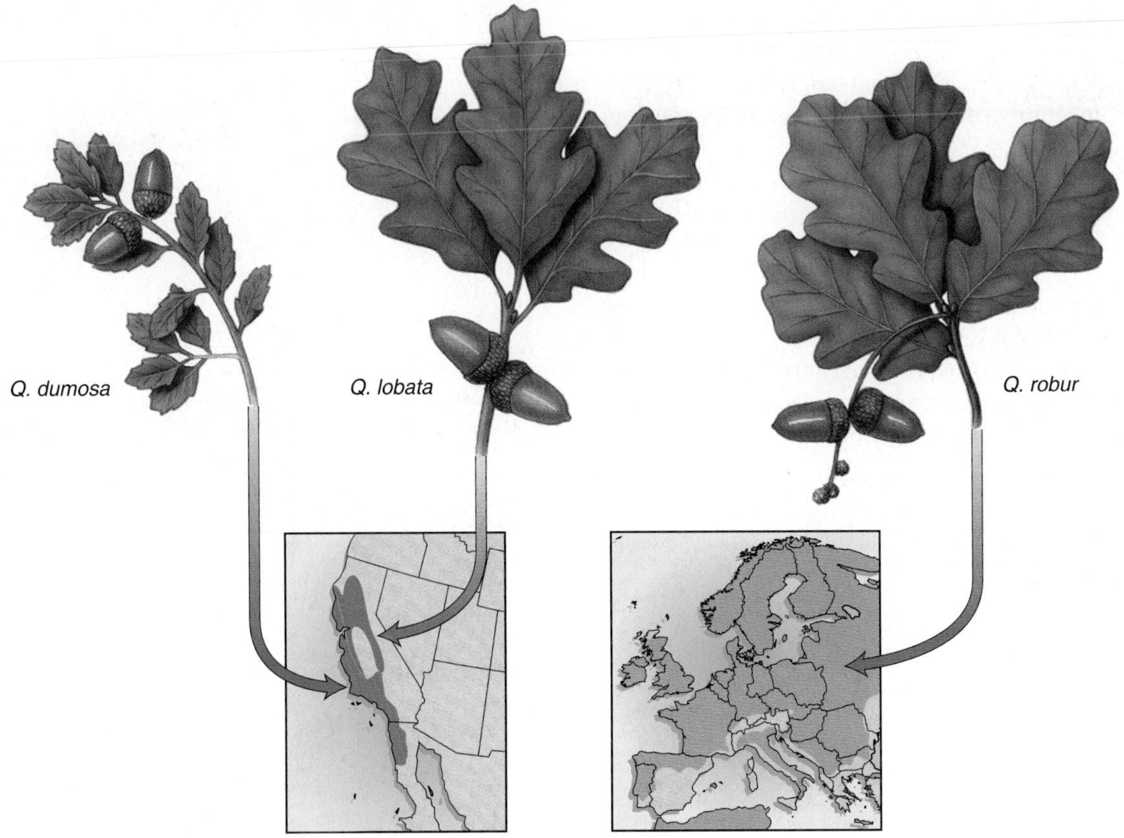
Q. dumosa
Q. lobata
Q. robur

Figure 11.25 Geographical and ecological isolation in oaks.

Three species of oaks—*Quercus robur, Q. lobata,* and *Q. dumosa*—have different leaf and acorn characteristics, yet all of them could hybridize with one another. However, *Q. robur* occurs in Europe, geographically isolating it from the other two species, which occur in California. Ecological isolation keeps *Q. lobata* and *Q. dumosa* separate from each other because they are found in different habitats in California.

11.13 Postzygotic Isolating Mechanisms

All of the factors we have discussed up to this point tend to prevent hybridization (table 11.2). If hybridization does occur and zygotes are produced, there are still many factors that may prevent those zygotes from developing into normal, functional, fertile F_1 individuals. Development in any species is a complex process. In hybrids, the genetic complements of two species may be so different that they cannot function together normally in embryonic development. For example, hybridization between sheep and goats usually produces embryos that die in the earliest developmental stages.

Leopard frogs (*Rana pipiens* complex) of the eastern United States are a group of similar species, assumed for a long time to constitute a single species. Previously, it was assumed that hybrids between some of these frogs were scarce because they came from different regions. But careful examination revealed that although the species appear similar, successful mating between them is rare. Many of the hybrid combinations cannot be produced even in the laboratory.

Many examples of this kind, in which similar species have been distinguished only as a result of hybridization experiments, are known in plants. Sometimes the hybrid embryos can be removed at an early stage and grown in an artificial medium. When these hybrids are supplied with extra nutrients or other supplements that compensate for their weakness or inviability, they may complete their development normally.

Even if the hybrids survive the embryo stage, they may not develop normally. If the hybrids are weaker than their parents, they will almost certainly be eliminated in nature. Even if they are vigorous and strong, as in the case of the mule, a hybrid between a horse and a donkey, they may still be sterile and thus incapable of contributing to succeeding generations. The development of sex organs in hybrids may be abnormal, the chromosomes derived from the respective parents may not pair properly, or their fertility may simply be lower than normal for other reasons.

11.13 Postzygotic isolating mechanisms include improper development of hybrids and failure of hybrids to become established in nature.

TABLE 11.2 ISOLATING MECHANISMS

Mechanism		Description
Prezygotic Isolating Mechanisms		
Geographic Isolation		Species occur in different areas, which are often separated by a physical barrier such as a river or mountain range.
Ecological Isolation		Species occur in the same area but they occupy different habitats. Survival of hybrids is low because they are not adapted to either environment of their parents.
Temporal Isolation		Species reproduce in different seasons or at different times of the day.
Behavioral Isolation		Species differ in their mating rituals.
Mechanical Isolation	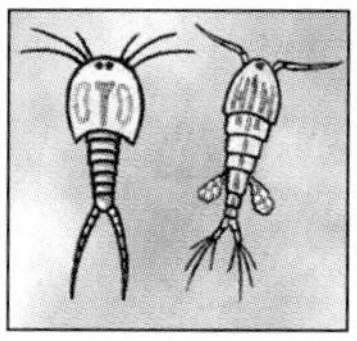	Structural differences between species prevent mating.
Prevention of Gamete Fusion	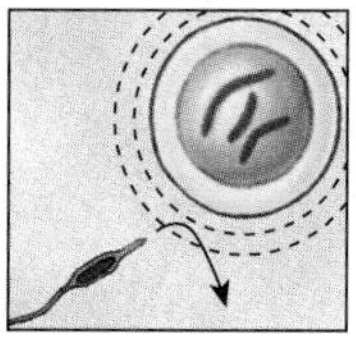	Gametes of one species function poorly with the gametes of another species or within the reproductive tract of another species.
Postzygotic Isolating Mechanisms		
Hybrid Inviability or Infertility		Hybrid embryos do not develop properly, hybrid adults do not survive in nature, or hybrid adults are sterile or have reduced fertility.

CONCEPT REVIEW

1. Microevolution is associated with
   a. extinction episodes.
   b. adaptation.
   c. the evolution of species.
   d. change over long periods of time.

2. Evidence of macroevolution is seen in ____________, while ________ is evidence of microevolution.
   a. the gradualism theory
   b. inbreeding
   c. industrial melanism
   d. fossils

3. Which of the following provides an anatomical record of macroevolution?
   a. vestigial organs
   b. the dinosaurs
   c. DNA
   d. sickle-cell anemia

4. If the frequency of a homozygous recessive genotype is 0.01, what is the expected frequency of the dominant allele in the population?
   a. 0.99
   b. 0.09
   c. 0.90
   d. 0.81

5. If you came across a population of plants and discovered a surprisingly high level of homozygosity, what would you predict about their mating system?
   a. Their pollen is dispersed by the wind.
   b. They probably reproduce asexually.
   c. They are predominantly outcrossing.
   d. They are predominantly self-fertilizing.

6. Which type of selection acts to eliminate both extremes from an array of phenotypes?
   a. directional
   b. stabilizing
   c. disruptive
   d. intensive

7. Heterozygotes for the sickle-cell allele have less resistance to malaria than individuals who do not have the allele.
   a. true
   b. false

8. The arm of a human and the fin of a porpoise are considered to be _________ structures, made of the same bones, which have been modified in size and shape for different uses.

9. By studying changing genes through time, biologists have studied _________ clocks.

10. Hardy-Weinberg equilibrium assumes that population size is _________.

CHALLENGE YOURSELF

1. Imagine that you sit on the Supreme Court and are hearing a case in which it is argued that creation science should be taught in public schools alongside evolution as a legitimate alternative scientific explanation of biological diversity. What is the best case that lawyers might make for and against this proposition? How would you vote and why?

2. Will a dominant allele that is lethal be removed from a large population as a result of natural selection? What factors might prevent this from happening? What if the lethal allele is recessive?

3. In a large, randomly mating population with no forces acting to change gene frequencies, the frequency of homozygous recessive individuals for the characteristic of extra long eyelashes is 90 per 1,000 or .09. What percent of the population carries this desirable trait but displays the dominant phenotype, short eyelashes? Would the frequency of the extra long eyelash allele increase, decrease, or remain the same if long-lashed individuals preferentially mated with each other and no one else?

11 eBRIDGE

Reinforcing Key Points

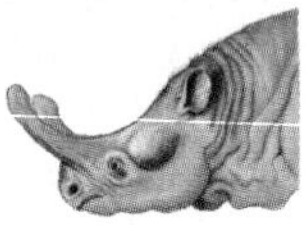

The Evidence for Evolution

11.1 Gene Variation: The Raw Material of Evolution

11.2 The Pace of Evolution

11.3 The Fossil Record

11.4 The Molecular Record

11.5 The Anatomical Record

How Populations Evolve

11.6 The Hardy-Weinberg Rule

11.7 Why Do Allele Frequencies Change?

11.8 Forms of Selection

Adaptation: Evolution in Action

11.9 Sickle-Cell Anemia

11.10 Peppered Moths and Industrial Melanism

How Species Form

11.11 The Species Concept

11.12 Prezygotic Isolating Mechanisms

11.13 Postzygotic Isolating Mechanisms

Electronic Learning

Visual Learning

Animations

Molecular Clock

Hardy-Weinberg Equilibrium

Helping You Learn

Evolution I: Microevolution and Selection

Explorations

How Proteins Function: Hemoglobin

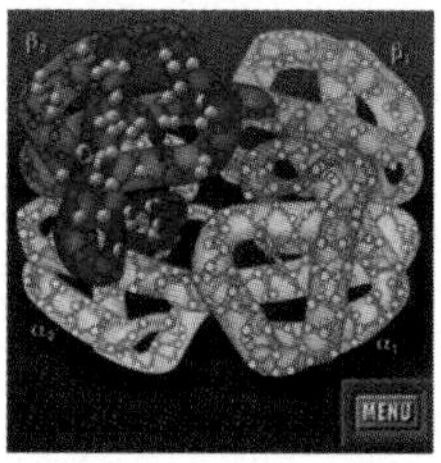
MENU

In this exercise, you can explore how the hemoglobin molecule changes when particular amino acid substitutions are made. The substitutions examined are real alleles that actually occur in human populations (including the sickle-cell allele). Oxygen-carrying ability, stability, and tendency to stick together are characterized for each version of the hemoglobin molecule.

Author's Corner

Evolution in Action. Perhaps the best way to understand evolution is to watch it in action. Biology abounds with interesting studies of species undergoing evolutionary change. Few natural populations, when carefully examined, are not responding in one way or another to natural selection. These six, concerning bacteria, lizards, birds, dogs, and cats, are but a few of the many examples that you might explore.

1. Did Darwin get it wrong about Galápagos finches?
2. Why do tropical songbirds lay fewer eggs?
3. Evolution of the family dog.
4. DNA and Darwin: Lizard evolution repeats itself.
5. How *E. coli* bacteria became deadly: Making a monster.
6. Bird-killing cats may be nature's way of making better birds.

Virtual Classroom

The Evidence for Evolution

At its core, the case for Darwin's theory of evolution by natural selection is built upon three pillars: first, evidence that natural selection can produce evolutionary change; second, evidence from the fossil record that evolution has occurred; third, evidence of evolutionary descent in the DNA sequences of the genomes of different species. In addition, information from many different areas of biology—including areas as divergent as embryology, anatomy, and ecology—can only be interpreted sensibly as the outcome of evolution.

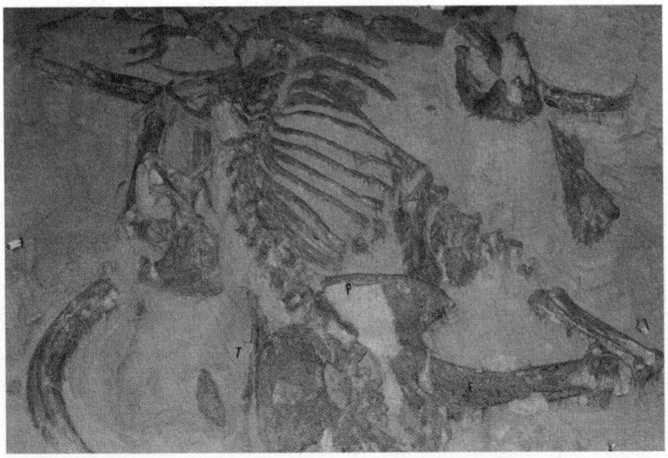

Arguments Against Evolution

Of all the major ideas of biology, the theory that today's organisms evolved from now-extinct ancestors is perhaps the best known to the general public. This is not because the average person truly understands the basic facts of evolution, but rather because many people mistakenly believe that evolution represents a challenge to their religious beliefs. Although highly publicized criticisms of evolution have occurred ever since Darwin's time, the scientific community has found the objections to be without merit.

Virtual Lab

Do Some Genes Maintain More Than One Common Allele in a Population?

A population of bacteria contains variation among its members that arises as a result from random mutations. It is expected that variants will arise in a population, but the fact that they are maintained in the population in significant numbers is *not* expected. Theory predicts that whenever a new variant appears in a population, it will be weighed in the balance by natural selection and either win or lose. One version of the gene should become universal in the population, and the other be eliminated. But contrary to simple theory, natural populations of most species, including bacteria, appear to have lots of common variants—they are said to be "polymorphic." How does a polymorphic condition arise in a bacterial population, and how can it be maintained? To investigate this experimental question, Julian Adams and co-workers at the University of Michigan set out to see if polymorphism for metabolic abilities would develop spontaneously in *Escherichia coli* (*E. coli*) bacteria growing in a uniform environment, and, if polymorphism arises, to investigate how it is maintained in the population.

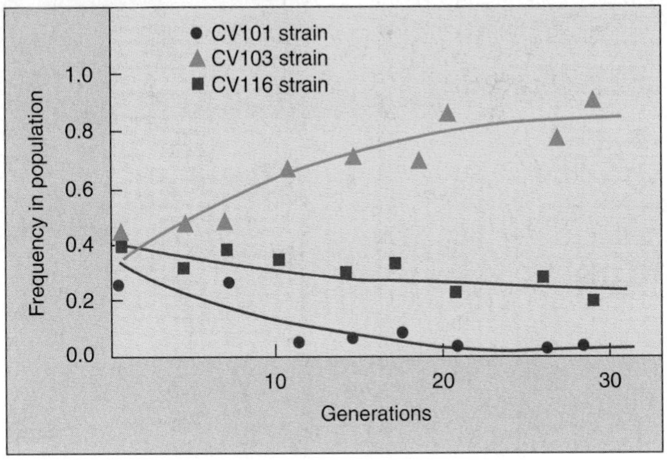
CV101 strain
CV103 strain
CV116 strain
Frequency in population
1.0
0.8
0.6
0.4
0.2
0.0
10
20
30
Generations

Quizzes

Further Reading

Essential Study Partner

Links

BioCourse.com

CHAPTER 12

How We Name Living Things

THE FAR SIDE®
By GARY LARSON
Larson
© 1981 FarWorks, Inc. All Rights Reserved/Dist. by Creators Syndicate

"Hey! What's this *Drosophila melanogaster* doing in my soup?"

CHAPTER 12

How We Name Living Things

CHAPTER OVERVIEW

The Classification of Organisms

12.1 The Invention of the Linnaean System
12.2 Species Names
12.3 Higher Categories
12.4 What Is a Species?

- Linnaeus invented the binomial system for naming species, a vast improvement over complex polynomial names.
- Taxonomists placed organisms into seven hierarchical categories: kingdom, phylum, class, order, family, genus, and species.
- The biological species concept defines species as groups that cannot successfully interbreed. This concept works well for animals and outcrossing plants but poorly for other organisms.

Inferring Phylogeny

12.5 How to Build a Family Tree

- Systematics is the study of the evolutionary relationships among a group of organisms.
- Cladistics builds family trees by clustering together those groups that share an ancestral "derived" character.
- Traditional taxonomy classifies organisms based on a large amount of information available, giving due weight to the evolutionary significance of certain characters.

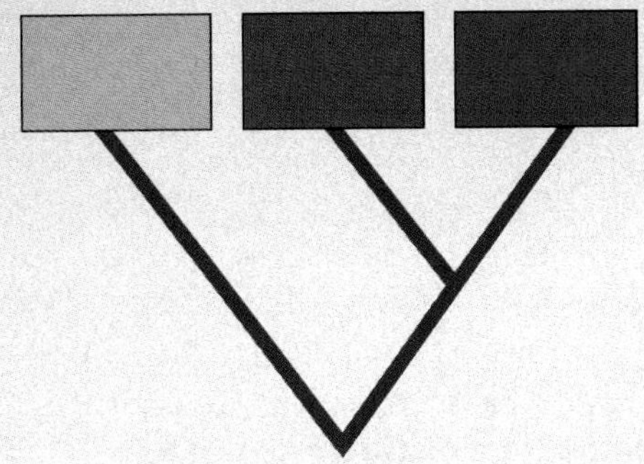

Kingdoms and Domains

12.6 The Kingdoms of Life
12.7 Domain Archaea (Archaebacteria)
12.8 Domain Bacteria (Eubacteria)
12.9 Domain Eukarya (Eukaryotes)
12.10 Viruses: A Special Case

- Prokaryotes, or bacteria, are assigned to two fundamentally different kingdoms, Archaebacteria and Eubacteria.
- Eukaryotes and archaebacteria are more closely related to each other than to eubacteria.
- Many distinctive evolutionary lines that consist mainly of unicellular organisms exist. Most of these are included in a very diverse kingdom, Protista.
- Viruses are not organisms and are not included in the classification of organisms.

12.1 The Invention of the Linnaean System

Our world is populated by some 10 million different kinds of organisms. To talk about them and study them, it is necessary to give them names, just as it is necessary that people have names. Of course, no one can remember the name of every kind of organism, so biologists use a kind of multilevel grouping of individuals called **classification.**

Organisms were first classified more than 2,000 years ago by the Greek philosopher Aristotle, who categorized living things as either plants or animals. He classified animals as either land, water, or air dwellers, and he divided plants into three kinds based on stem differences. This simple classification system was expanded by the Greeks and Romans, who grouped animals and plants into basic units such as cats, horses, and oaks. Eventually, these units began to be called **genera** (singular, **genus**), the Latin word for "group." Starting in the Middle Ages, these names began to be systematically written down, using Latin, the language used by scholars at that time. Thus, cats were assigned to the genus *Felis,* horses to *Equus*, and oaks to *Quercus*—names that the Romans had applied to these groups. For genera that were not known to the Romans, new names were invented.

The classification system of the Middle Ages, called the polynomial system, was used virtually unchanged for hundreds of years, until it was replaced about 250 years ago by the **binomial system** introduced by Linnaeus.

The Polynomial System

Until the mid-1700s, biologists usually added a series of descriptive terms to the name of the genus when they wanted to refer to a particular kind of organism, which they called a **species.** These phrases, starting with the name of the genus, came to be known as **polynomials** (*poly,* many, and *nomial,* name), strings of Latin words and phrases consisting of up to 12 or more words. This would be like the mayor of New York referring to a particular citizen as "Brooklyn resident: Democrat, male, Asian American, middle income, Protestant, elderly, likely voter, short, bald, heavyset, wears glasses, works in the Bronx selling shoes." As you can imagine, these polynomial names were cumbersome. Even more worrisome, the names were altered at will by later authors, so that a given organism really did not have a single name that was its alone.

The Binomial System

A much simpler system of naming animals, plants, and other organisms stems from the work of the Swedish biologist Carolus Linnaeus (1707–78). Linnaeus devoted his life to a challenge that had defeated many biologists before him—cataloging all the different kinds of organisms. In the 1750s he produced several major works that, like his earlier books, employed the polynomial system. But as a kind of shorthand, Linnaeus also included in these books a two-part name for each species. These two-part names, or **binomials** (*bi* is the Latin prefix for "two"), have become our standard way of designating species. For example, he designated the willow oak *Quercus phellos* and the red oak *Quercus rubra* (figure 12.1), even though he also included the polynomial name for these species. This naming is like the mayor of New York referring to the Brooklyn resident as Sylvester Kingston—we also use binomial names for ourselves, our so-called given and family names.

12.1 Two-part (binomial) Latin names, first used by Linnaeus, are now universally employed by biologists to name particular organisms.

(a)
Quercus phellos
(Willow oak)
Quercus rubra
(Red oak)
(b)

Figure 12.1 How Linnaeus named two species of oaks.
(*a*) Willow oak, *Quercus phellos.* (*b*) Red oak, *Quercus rubra.* Although they are clearly oaks (members of the genus *Quercus*), these two species differ sharply in the shapes and sizes of their leaves and in many other features, including their overall geographical distributions.

12.2 Species Names

A group of organisms at a particular level in a classification system is called a **taxon** (plural, **taxa**), and the branch of biology that identifies and names such groups of organisms is called **taxonomy.** Taxonomists are in a real sense detectives, biologists who must use clues of appearance and behavior to identify and assign names to organisms.

By formal agreement among taxonomists throughout the world, no two organisms can have the same name. So that no one country is favored, a language spoken by no country—Latin—is used for the names. Because the scientific name of an organism is the same anywhere in the world, this system provides a standard and precise way of communicating, whether the language of a particular biologist is Chinese, Arabic, Spanish, or English. This is a great improvement over the use of common names, which often vary from one place to the next. As you can see in figure 12.2, corn in Europe refers to the plant Americans call wheat; a bear is a large placental omnivore in the United States but a koala (a vegetarian marsupial) in Australia; and a robin is a very different bird in Europe and North America.

By convention, the first word of the binomial name is the genus to which the organism belongs. This word is always capitalized. The second word refers to the particular species and is not capitalized. The two words together are called the **scientific name** and are written in italics. The system of naming animals, plants, and other organisms established by Linnaeus has served the science of biology well for nearly 250 years.

12.2 By convention, the first part of a binomial species name identifies the genus to which the species belongs, and the second part distinguishes that particular species from other species in the genus.

(a) (b) (c)

Figure 12.2 Common names make poor labels.
The common names corn (*a*), bear (*b*), and robin (*c*) bring clear images to our minds (photos on *left*), but the images would be very different to someone living in Europe or Australia (photos on *right*). There, the same common names are used to label very different species.

12.3 Higher Categories

Like the mayor of New York, a biologist needs more than two categories to classify all the world's living things. Taxonomists group the genera with similar properties into a cluster called a **family,** and place similar families into the same **order** (figure 12.3). Orders with common properties are placed into the same **class,** and classes with similar characteristics into the same **phylum** (plural, **phyla**). Botanists (that is, those who study plants) used to call plant phyla "divisions," but that practice is falling out of fashion. Finally, the phyla are assigned to one of several great groups, the **kingdoms.** Biologists currently recognize six kingdoms: two kinds of bacteria (Archaebacteria and Eubacteria), a largely unicellular group of eukaryotes (Protista), and three multicellular groups (Fungi, Plantae, and Animalia). In order to remember the seven categories in their proper order, it may prove useful to memorize a phrase such as "**k**indly **p**ay **c**ash **o**r **f**urnish **g**ood **s**ecurity" (**k**ingdom–**p**hylum–**c**lass–**o**rder–**f**amily–**g**enus–**s**pecies).

In addition, an eighth level of classification, called *domains,* is sometimes used. Biologists recognize three domains, which are discussed later in this chapter.

Each of the categories in this **Linnaean system of classification** is loaded with information. For example, a honeybee has the species (level 1) name *Apis mellifera.* Its genus name (level 2) *Apis* is a member of the family Apidae (level 3). All members of this family are bees, some solitary, others living in hives as *A. mellifera* does. Knowledge of its order (level 4), Hymenoptera, tells you that *A. mellifera* is likely able to sting and may live in colonies. Its class (level 5), Insecta, indicates that *A. mellifera* has three major body segments, with wings and three pairs of legs attached to the middle segment. Its phylum (level 6), Arthropoda, tells us that the honeybee has a hard cuticle of chitin and jointed appendages. Its kingdom (level 7), Animalia, tells us that *A. mellifera* is a multicellular heterotroph whose cells lack cell walls.

> **12.3** A hierarchical system is used to classify organisms, in which higher categories convey more general information about the group.

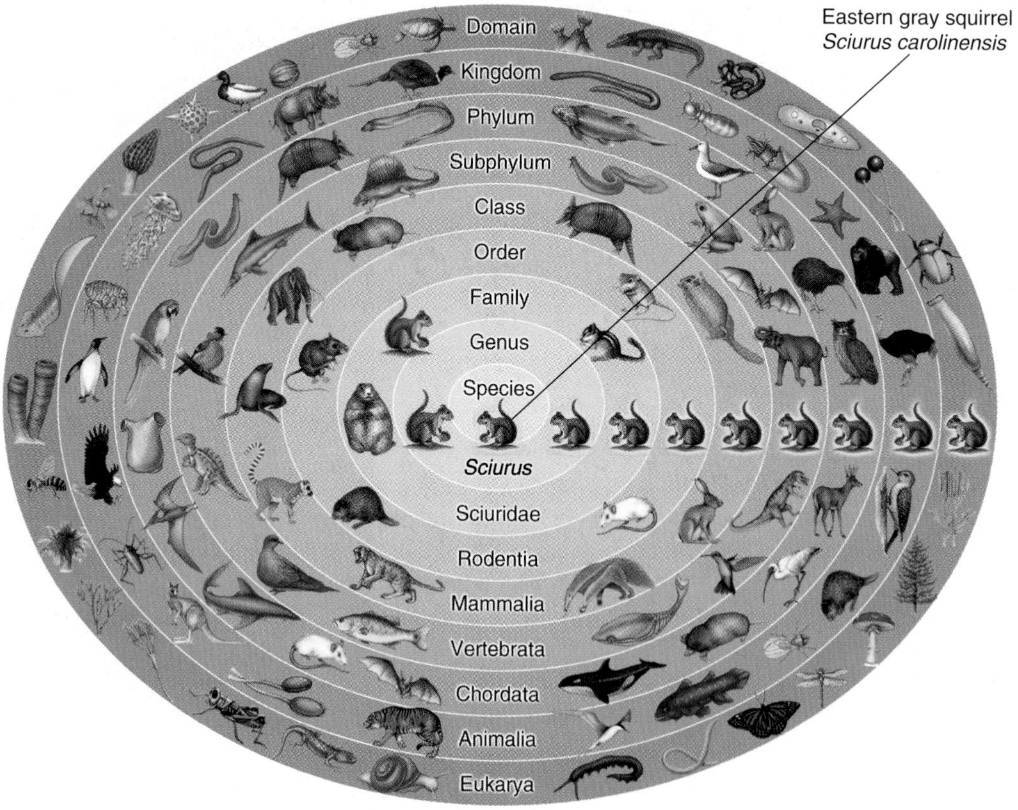

Eastern gray squirrel
Sciurus carolinensis
Domain
Kingdom
Phylum
Subphylum
Class
Order
Family
Genus
Species
Sciurus
Sciuridae
Rodentia
Mammalia
Vertebrata
Chordata
Animalia
Eukarya

Figure 12.3 The hierarchical system used to classify an organism.

The organism is first recognized as a eukaryote (domain: Eukarya). Second, within this domain, it is an animal (kingdom: Animalia). Among the different phyla of animals, it is a vertebrate (phylum: Chordata, subphylum: Vertebrata). The organism's fur characterizes it as a mammal (class: Mammalia). Within this class, it is distinguished by its gnawing teeth (order: Rodentia). Next, because it has four front toes and five back toes, it is a squirrel (family: Sciuridae). Within this family, it is a tree squirrel (genus: *Sciurus*), with gray fur and white-tipped hairs on the tail (species: *Sciurus carolinensis,* the eastern gray squirrel).

12.4 What Is a Species?

The basic biological unit in the Linnaean system of classification is the species. John Ray (1627–1705), an English clergyman and scientist, was one of the first to propose a general definition of species. In about 1700 he suggested a simple way to recognize a species: all the individuals that belong to it can breed with one another and produce fertile offspring. By Ray's definition, the offspring of a single mating were all considered to belong to the same species, even if they contained different-looking individuals, as long as these individuals could interbreed. All domestic cats are one species (they can all interbreed), while carp are not the same species as goldfish (they cannot interbreed). The donkey you see in figure 12.4 is not the same species as the horse, because when they interbreed, the offspring—mules—are sterile.

The Biological Species Concept

With Ray's observation, the species began to be regarded as an important biological unit that could be cataloged and understood, the task that Linnaeus set himself a generation later. Where information was available, Linnaeus used Ray's species concept, and it is still widely used today. When the evolutionary ideas of Darwin were joined to the genetic ideas of Mendel in the 1920s to form the field of population genetics, it became desirable to define the category species more precisely. The definition that emerged, the so-called **biological species concept,** was stated by the American evolutionist Ernst Mayr as follows: Species are "groups of actually or potentially interbreeding natural populations which are reproductively isolated from other such groups." In other words, hybrids between species occur rarely in nature, whereas individuals that belong to the same species are able to interbreed freely.

Problems with the Biological Species Concept

The biological species concept works fairly well for animals, where strong barriers to hybridization between species exist, but very poorly for members of the other kingdoms. The problem is that the biological species concept assumes that organisms regularly **outcross,** that is, interbreed with individuals other than themselves, individuals with a different genetic makeup than their own. Animals regularly outcross, and so the concept works well for animals. However, outcrossing is less common in the other five kingdoms. In bacteria and many protists, fungi, and plants, **asexual reproduction,** reproduction without sex, predominates. These species clearly cannot be characterized in the same way as outcrossing animals and plants—they do not interbreed with one another, much less with the individuals of other species.

Complicating matters further, the reproductive barriers that are the key element in the biological species concept, while common among animal species, are not typical of other kinds of organisms. In fact, there are essentially no barriers to hybridization between the species in many groups of trees and other plants. Even among animals, fish species are able to form fertile hybrids with one another, though they may not do so in nature.

In practice, biologists today recognize species in different groups in much the way they always did, because they differ from one another in their visible features. Within animals, the biological species concept is still widely employed, while among plants and other kingdoms it is not.

How Many Kinds of Species Are There?

Since the time of Linnaeus, about 1.5 million species have been named. But the actual number of species in the world is undoubtedly much greater, judging from the very large numbers that are still being discovered. Some scientists estimate that at least 10 million species exist on earth, and at least two-thirds of these occur in the tropics.

12.4 Among animals, species are generally defined as reproductively isolated groups; among the other kingdoms such a definition is less useful, as their species typically have weaker barriers to hybridization.

Horse Donkey Mule

Figure 12.4 Ray's definition of a species.

According to Ray, donkeys and horses are not the same species. Even though they produce very robust offspring (mules) when they mate, the mules are sterile.

12.5 How to Build a Family Tree

After naming and classifying some 1.5 million organisms, what have biologists learned? One very important advantage of being able to classify particular species of plants, animals, and other organisms is that individuals of species that are useful to humans as sources of food and medicine can be identified. For example, if you cannot tell the fungus *Penicillium* from *Aspergillus,* you have little chance of producing the antibiotic penicillin. In a thousand ways, just having names for organisms is of immense importance in our modern world.

Taxonomy also enables us to glimpse the evolutionary history of life on earth. The more similar two taxa are, the more closely related they are likely to be, for the same reason that you are more like your brothers and sisters than like strangers selected from a crowd. By looking at the differences and similarities between organisms, biologists can attempt to reconstruct the tree of life, inferring which organisms evolved from which other ones, in what order, and when. The reconstruction and study of evolutionary trees, or **phylogenetic trees,** is called **systematics.**

Cladistics

A simple and objective way to construct a phylogenetic tree is to focus on key characters that some organisms share because they have inherited them from a common ancestor. A **clade** is a group of organisms related by descent, and this approach to constructing a phylogeny is called **cladistics.** Cladistics infers phylogeny (that is, builds family trees) according to similarities derived from a common ancestor, so-called **derived characters.** The key to the approach is being able to identify morphological, physiological, or behavioral traits that differ among the organisms being studied and can be attributed to a common ancestor. By examining the distribution of these traits among the organisms, it is possible to construct a **cladogram** (figure 12.5), a branching diagram that represents the phylogeny.

Cladograms are not true family trees. They do not convey direct information about ancestors and descendants—who came from whom. Instead, they convey comparative information about *relative* relationships. Organisms that are closer together on a cladogram simply share a more recent common ancestor than those that are farther apart. Because the analysis is comparative, it is necessary to have something to anchor the comparison to, some solid ground against which the comparisons can be made. To achieve this, each cladogram must contain an **outgroup,** a rather different organism (but not *too* different) to serve as a baseline for comparisons among the other organisms being evaluated, the **ingroup.** For example, in figure 12.5, the lamprey is the outgroup to the clade of animals that have jaws.

Cladistics is a relatively new approach in biology and has become popular among students of evolution. This is because it does a very good job of portraying the *order* in which a series of evolutionary events have occurred. The

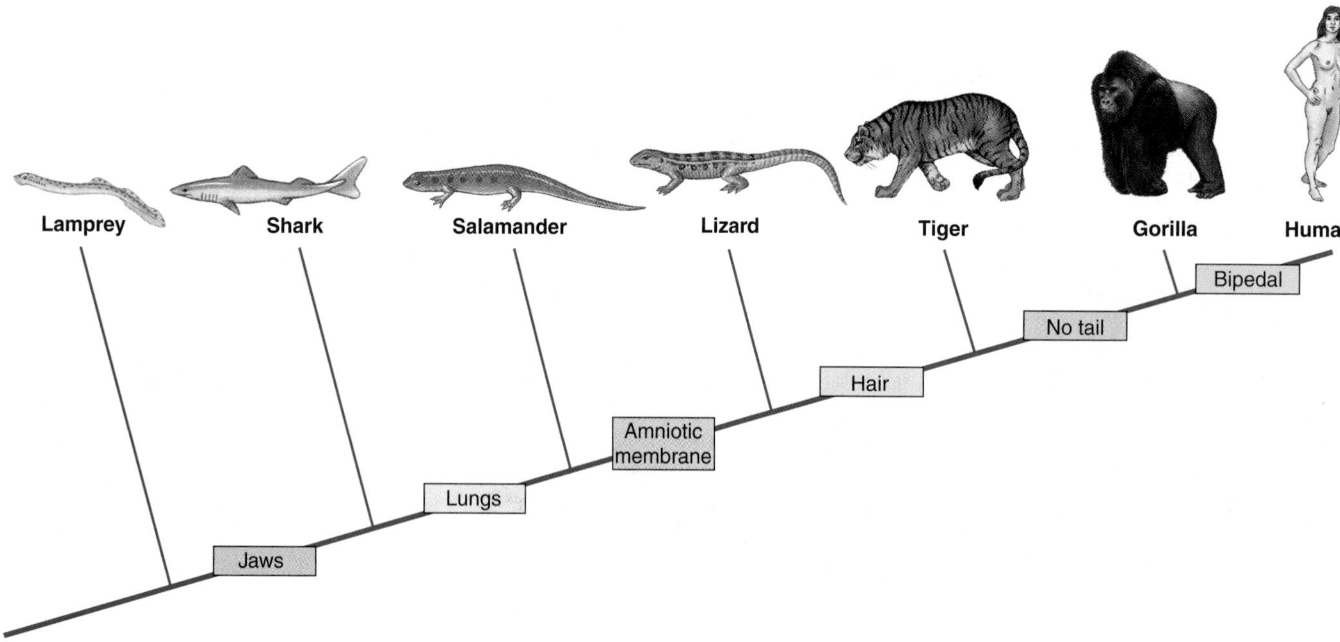
Lamprey
Shark
Salamander
Lizard
Tiger
Gorilla
Human
Jaws
Lungs
Amniotic membrane
Hair
No tail
Bipedal

Figure 12.5 A cladogram of vertebrate animals.
The derived characters between the branch points are shared by all the animals to the right of the characters and are not present in any organisms to the left of it.

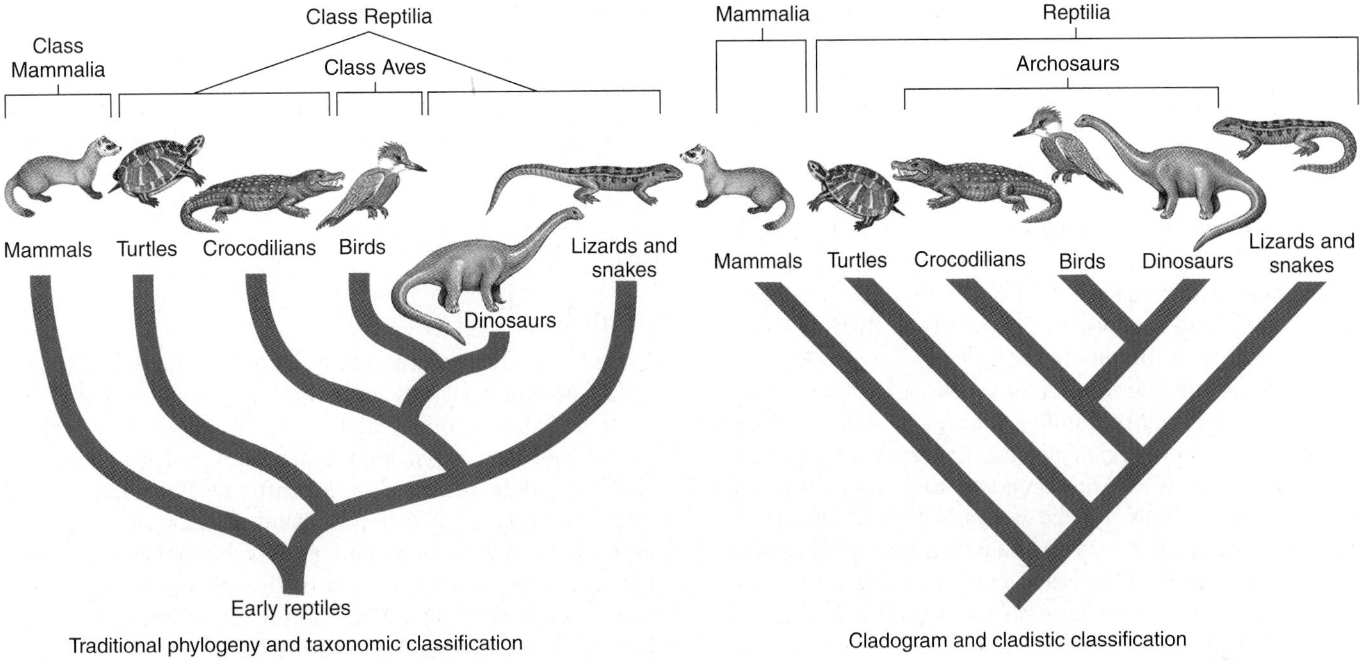

Class Mammalia
Class Reptilia
Class Aves
Mammals
Turtles
Crocodilians
Birds
Lizards and snakes
Dinosaurs
Early reptiles
Traditional phylogeny and taxonomic classification
Mammalia
Reptilia
Archosaurs
Mammals
Turtles
Crocodilians
Birds
Dinosaurs
Lizards and snakes
Cladogram and cladistic classification

Figure 12.6 Two ways to classify terrestrial vertebrates.

Traditional taxonomic analyses place birds in their own class (Aves) because birds have evolved several unique adaptations that separate them from the reptiles. Cladistic analyses, however, place crocodiles, dinosaurs, and birds together (as archaeosaurs) because they share many derived characters.

great strength of a cladogram is that it can be completely objective. A computer fed the data will generate exactly the same cladogram time and again. In fact, most cladistic analyses involve many characters, and computers are required to make the comparisons.

Sometimes it is necessary to "weight" characters, or take into account the variation in the "strength" of a character—the size or location of a fin, the effectiveness of a lung. For example, let's say that the following are five unique events that occurred on November 22, 1963: (1) my cat was declawed, (2) I had a wisdom tooth pulled, (3) I sold my first car, (4) President Kennedy was assassinated, and (5) I passed physics. Without weighting the events, each one is assigned equal importance. In a nonweighted cladistic sense, they are equal (all happened only once, and on that day), but in a practical, real-world sense, they certainly are not. One event, Kennedy's death, had a far greater impact and importance than the others. Because evolutionary success depends so critically on just such high-impact events, modern cladistics attempts to weight the evolutionary significance of the characters being studied.

Traditional Taxonomy

Weighting characters lies at the core of **traditional taxonomy.** In this approach, phylogenies are constructed based on a vast amount of information about the morphology and biology of the organism gathered over a long period of time. Traditional taxonomists use both ancestral and derived characters to construct their trees, while cladists use only derived characters. The large amount of information used by traditional taxonomists permits a knowledgeable weighting of characters according to their biological significance. In traditional taxonomy, the full observational power and judgment of the biologist is brought to bear—and also any biases he or she may have. For example, in classifying the terrestrial vertebrates, traditional taxonomists place birds in their own class (Aves), giving great weight to the characters that made powered flight possible, such as feathers. However, a cladogram of vertebrate evolution lumps birds in among the reptiles with crocodiles and dinosaurs (figure 12.6). This accurately reflects their true ancestry but ignores the immense evolutionary impact of a derived character such as feathers.

Overall, phylogenetic trees based on traditional taxonomy are information-rich, while cladograms do a better job of deciphering evolutionary histories. Thus, traditional taxonomy is the better approach when a great deal of information is available to guide character weighting, while cladistics is the better approach when little information is available about how the character affects the life of the organism. DNA sequence comparisons, for example, lend themselves well to cladistics—you have a great many derived characters (DNA sequence differences) but little or no idea of what impact the sequence differences have on the organism.

12.5 A phylogeny may be represented as a cladogram based on the order in which groups evolved or as a traditional taxonomic tree that weights characters according to assumed importance.

12.6 The Kingdoms of Life

The earliest classification systems recognized only two kingdoms of living things: animals and plants (figure 12.7*a*). But as biologists discovered microorganisms and learned more about other organisms, they added kingdoms in recognition of fundamental differences (figure 12.7*b*). Most biologists now use a six-kingdom system first proposed by Carl Woese of the University of Illinois (figure 12.7*c*).

In this system, four kingdoms consist of eukaryotic organisms. The two most familiar kingdoms, **Animalia** and **Plantae,** contain only organisms that are multicellular during most of their life cycle. The kingdom **Fungi** contains multicellular forms and single-celled yeasts, which are thought to have multicellular ancestors. Fundamental differences divide these three kingdoms. Plants are mainly stationary, but some have motile sperm; fungi have no motile cells; animals are mainly motile. Animals ingest their food, plants manufacture it, and fungi digest it by means of secreted extracellular enzymes. Each of these kingdoms probably evolved from a different single-celled ancestor.

The large number of unicellular eukaryotes are arbitrarily grouped into a single kingdom called **Protista** (see chapter 14). This kingdom includes the algae, all of which are unicellular during important parts of their life cycle.

The remaining two kingdoms, **Archaebacteria** and **Eubacteria,** consist of prokaryotic organisms, which are vastly different from all other living things (see chapter 13). Archaebacteria are a diverse group including the methanogens and extreme thermophiles; they differ from the other bacteria, members of the kingdom Eubacteria.

Domains

As biologists have learned more about the archaebacteria, it has become increasingly clear that this ancient group is very different from all other organisms. When the full genomic DNA sequences of an archaebacterium and a eubacterium were first compared in 1996, the differences proved striking. Archaebacteria are as different from eubacteria as eubacteria are from eukaryotes. Recognizing this, biologists are increasingly adopting a classification of living organisms that recognizes three **domains,** a taxonomic level higher than kingdom (figure 12.7*d*). Archaebacteria are in one domain, eubacteria in a second, and eukaryotes in the third.

12.6 Living organisms are grouped into three categories called domains. One of the domains, Eukarya, is divided into four kingdoms: Protista, Fungi, Plantae, and Animalia.

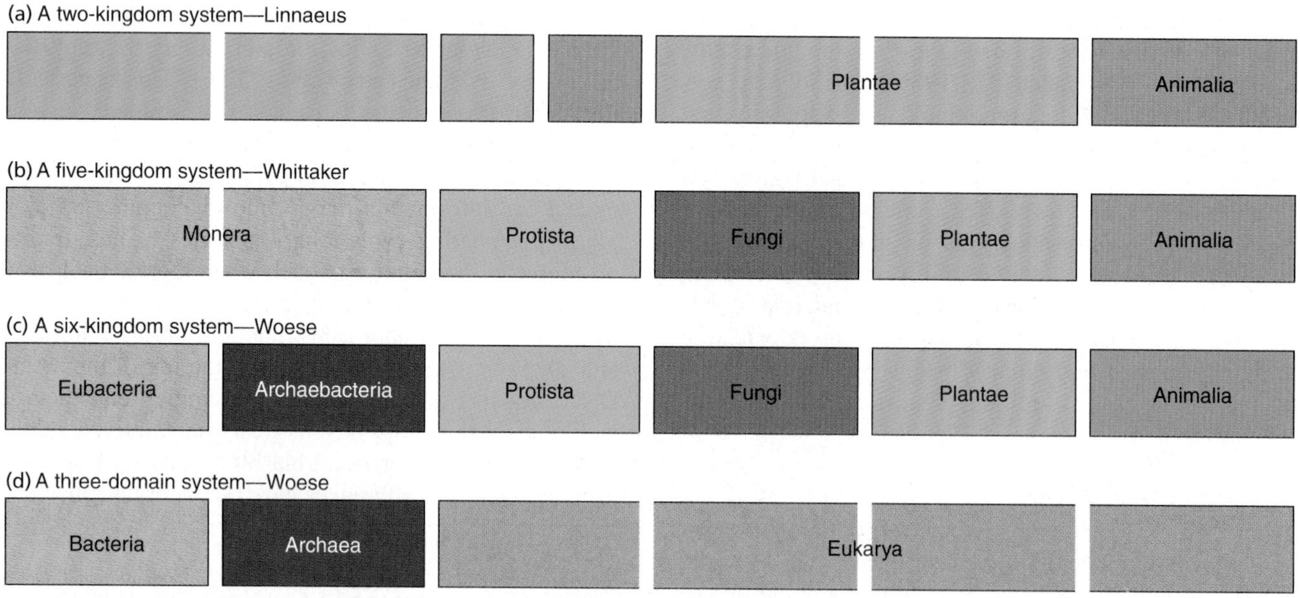
(a) A two-kingdom system—Linnaeus
Plantae
Animalia
(b) A five-kingdom system—Whittaker
Monera
Protista
Fungi
Plantae
Animalia
(c) A six-kingdom system—Woese
Eubacteria
Archaebacteria
Protista
Fungi
Plantae
Animalia
(d) A three-domain system—Woese
Bacteria
Archaea
Eukarya

Figure 12.7 Different approaches to classifying living organisms.
(*a*) Linnaeus popularized a two-kingdom approach, in which the fungi and the photosynthetic protists were classified as plants and the nonphotosynthetic protists as animals; when bacteria were described, they too were considered plants. (*b*) Whittaker in 1969 proposed a five-kingdom system that soon became widely accepted. (*c*) Woese has championed splitting the bacteria into two kingdoms for a total of six kingdoms or even assigning them separate domains (*d*).

12.7 Domain Archaea (Archaebacteria)

The term *archaebacteria* (Greek, *archaio,* ancient) refers to the ancient origin of this group of bacteria, which most likely diverged very early from the eubacteria (figure 12.8). Today, archaebacteria inhabit some of the most extreme environments on earth. Though a diverse group, all archaebacteria share certain key characteristics (table 12.1). Their cell walls lack the peptidoglycan characteristic of other bacteria. They possess very unusual lipids and characteristic ribosomal RNA sequences. Some of their genes possess introns, unlike those of other bacteria.

Archaebacteria are grouped into three general categories: methanogens, extremophiles, and nonextreme archaebacteria.

Methanogens obtain their energy by using hydrogen gas (H_2) to reduce carbon dioxide (CO_2) to methane gas (CH_4). They are strict anaerobes, poisoned by even traces of oxygen. They live in swamps, marshes, and the intestines of mammals. Methanogens release about 2 billion tons of methane gas into the atmosphere each year.

Extremophiles are able to grow under conditions that seem extreme to us.

> *Thermophiles* ("heat lovers") live in very hot places, typically from 60° to 80°C. Many thermophiles have metabolisms based on sulfur. Thus the *Sulfolobus* inhabiting the hot sulfur springs of Yellowstone National Park at 70° to 75°C obtain their energy by oxidizing elemental sulfur to sulfuric acid. The recently described *Pyrolobus fumarii* holds the current record for heat stability, 106°C temperature optimum and 113°C maximum—it is so heat-tolerant that it is not killed by a one-hour treatment in an autoclave (121°C)!
>
> *Halophiles* ("salt lovers") live in very salty places like the Great Salt Lake in Utah, Mono Lake in California, and the Dead Sea in Israel. Whereas the salinity of seawater is around 3%, these bacteria thrive in, and indeed require, water with a salinity of 15% to 20%.
>
> *pH-tolerant* archaebacteria grow in highly acidic (pH = 0.7) and very basic (pH = 11) environments.
>
> *Pressure-tolerant* archaebacteria have been isolated from ocean depths that require at least 300 atmospheres of pressure to survive, and tolerate up to 800 atmospheres!

Nonextreme archaebacteria grow in the same environments eubacteria do. As the genomes of archaebacteria have become better known, microbiologists have been able to identify **signature sequences** of DNA present in all archaebacteria and in no other organisms. When samples from soil or sea water are tested for genes matching these signature sequences, many of the bacteria living there prove to be archaebacteria. Clearly, archaebacteria are not restricted to extreme habitats, as microbiologists used to think.

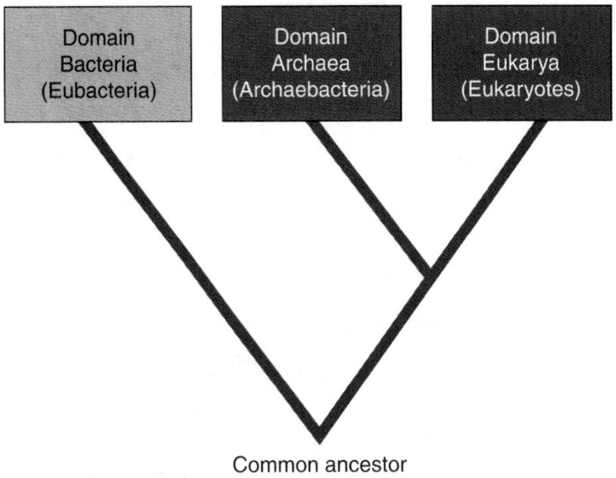
Domain
Bacteria
(Eubacteria)
Domain
Archaea
(Archaebacteria)
Domain
Eukarya
(Eukaryotes)
Common ancestor

Figure 12.8 An evolutionary relationship among the three domains.

Most likely, eubacteria diverged from the evolutionary line that gave rise to the archaebacteria and eukaryotes.

TABLE 12.1 FEATURES OF THE DOMAINS OF LIFE

	Domain		
Feature	**Archaea**	**Bacteria**	**Eukarya**
Amino acid that initiates protein synthesis	Methionine	Formyl-methionine	Methionine
Introns	Present in some genes	Absent	Present
Membrane-bounded organelles	Absent	Absent	Present
Membrane lipid structure	Branched	Unbranched	Unbranched
Nuclear envelope	Absent	Absent	Present
Number of different RNA polymerases	Several	One	Several
Peptidoglycan in cell wall	Absent	Present	Absent
Response to the antibiotics streptomycin and chloramphenical	Growth not inhibited	Growth inhibited	Growth not inhibited

12.7 Archaebacteria are poorly understood bacteria that inhabit diverse environments, some of them extreme.

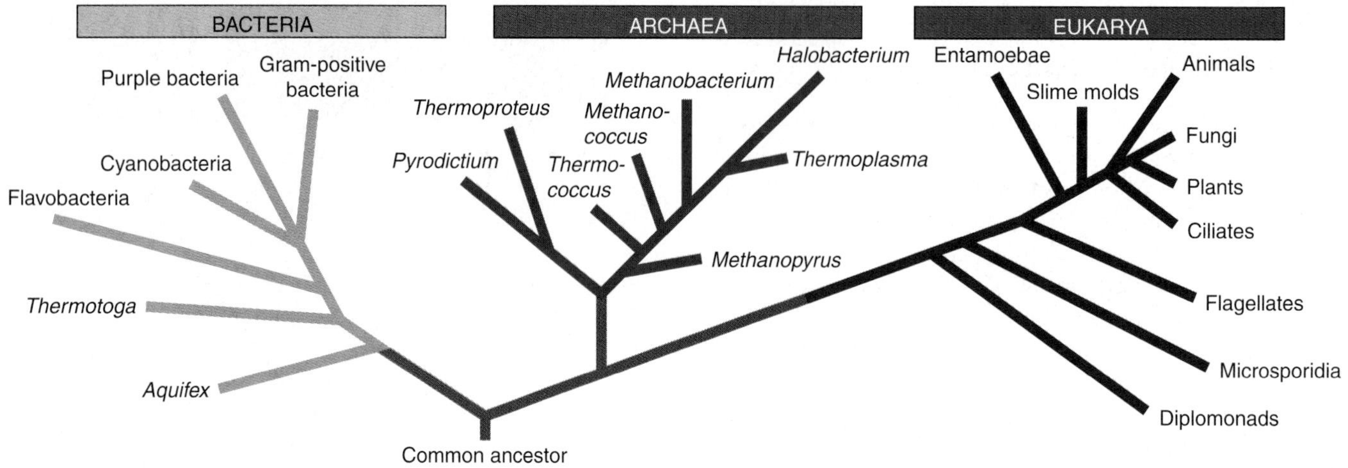

BACTERIA
ARCHAEA
EUKARYA
Purple bacteria
Gram-positive bacteria
Cyanobacteria
Flavobacteria
Thermotoga
Aquifex
Thermoproteus
Pyrodictium
Methanobacterium
Methanococcus
Thermococcus
Halobacterium
Thermoplasma
Methanopyrus
Entamoebae
Slime molds
Animals
Fungi
Plants
Ciliates
Flagellates
Microsporidia
Diplomonads
Common ancestor

Figure 12.9 A tree of life.

This phylogeny, prepared from rRNA analyses, shows the evolutionary relationships among the three domains. The base of the tree was determined by examining genes that are duplicated in all three domains, the duplication presumably having occurred in the common ancestor. When one of the duplicates is used to construct the tree, the other can be used to root it. This approach clearly indicates that the root of the tree is within the eubacterial domain. Archaebacteria and eukaryotes diverged later and are more closely related to each other than either is to eubacteria.

12.8 Domain Bacteria (Eubacteria)

The eubacteria are the most abundant organisms on earth. There are more living eubacteria in your mouth than there are mammals living on earth. Although too tiny to see with the unaided eye, eubacteria play critical roles throughout the biosphere. They extract from the air all the nitrogen used by organisms, and they play key roles in cycling carbon and sulfur.

There are many different kinds of eubacteria, and the evolutionary links between them are not well understood. While there is considerable disagreement among taxonomists about the details of bacterial classification, most recognize 12 to 15 major groups of eubacteria. Comparisons of the nucleotide sequences of ribosomal RNA (rRNA) molecules are beginning to reveal how these groups are related to each other and to the other two domains (figure 12.9). The archaebacteria and eukaryotes are more closely related to each other than to eubacteria and are on a separate evolutionary branch of the tree, even though archaebacteria and eubacteria are both prokaryotes.

12.8 Eubacteria are as different from archaebacteria as from eukaryotes.

TABLE 12.2 CHARACTERISTICS OF THE SIX KINGDOMS

Kingdom		Cell Type	Nuclear Envelope	Mitochondria	Chloroplasts	Cell Wall
Archaebacteria and Eubacteria		Prokaryotic	Absent	Absent	None (photosynthetic membranes in some types)	Noncellulose (polysaccharide plus amino acids)
Protista	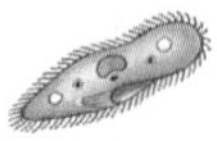	Eukaryotic	Present	Present or absent	Present (some forms)	Present in some forms, various types
Fungi		Eukaryotic	Present	Present or absent	Absent	Chitin and other noncellulose polysaccharides
Plantae		Eukaryotic	Present	Present	Present	Cellulose and other polysaccharides
Animalia		Eukaryotic	Present	Present	Absent	Absent

12.9 Domain Eukarya (Eukaryotes)

For at least 2 billion years, bacteria ruled the earth. No other organisms existed to eat them or compete with them, and their tiny cells formed the world's oldest fossils. The third great domain of life, the eukaryotes, appear in the fossil record much later, only about 1.5 billion years ago. Metabolically, eukaryotes are more uniform than bacteria. Each of the two domains of prokaryotic organisms has far more metabolic diversity than all eukaryotic organisms taken together.

Three Largely Multicellular Kingdoms

Fungi, plants, and animals are well-defined evolutionary groups, each of them clearly stemming from a different single-celled, eukaryotic ancestor. They are largely multicellular, each a distinct evolutionary line from an ancestor that would be classified in the kingdom Protista.

The amount of diversity among the protists, however, is much greater than that within or between the three largely multicellular kingdoms derived from the protists. Because of the size and ecological dominance of plants, animals, and fungi, and because they are predominantly multicellular, we recognize them as kingdoms distinct from Protista.

A Fourth Very Diverse Kingdom

When multicellularity evolved, the diverse kinds of single-celled organisms that existed at that time did not simply become extinct. A wide variety of unicellular eukaryotes and their relatives exists today, grouped together in the kingdom Protista solely because they are not fungi, plants, or animals. Protists are a fascinating group containing many organisms of intense interest and great biological significance.

The characteristics of the six kingdoms are outlined in table 12.2.

Symbiosis and the Origin of Eukaryotes

The hallmark of eukaryotes is complex cellular organization, highlighted by an extensive endomembrane system that subdivides the eukaryotic cell into functional compartments. Not all of these compartments, however, are derived from the endomembrane system. With few exceptions, all modern eukaryotic cells possess energy-producing organelles, the mitochondria. Mitochondria are about the size of bacteria and contain DNA. Comparison of the nucleotide sequence of this DNA with that of a variety of organisms indicates clearly that mitochondria are the descendants of purple bacteria that were incorporated into eukaryotic cells early in the history of the group. Some protist phyla have in addition acquired chloroplasts during the course of their evolution and thus are photosynthetic. These chloroplasts are derived from cyanobacteria that became symbiotic in several groups of protists early in their history.

Mitochondria and chloroplasts are both believed to have entered early eukaryotic cells by a process called endosymbiosis (*endo,* inside). We discussed the theory of the endosymbiotic origin of mitochondria and chloroplasts in chapter 4 and will revisit it in chapter 14.

Some biologists suggest that basal bodies, centrioles, flagella, and cilia may have arisen from endosymbiotic spirochaete-like bacteria. Even today, so many bacteria and unicellular protists form symbiotic alliances that the incorporation of smaller organisms with desirable features into eukaryotic cells appears to be a relatively common process.

12.9 Eukaryotic cells acquired mitochondria and chloroplasts by endosymbiosis, mitochondria being derived from purple bacteria and chloroplasts from cyanobacteria.

TABLE 12.2 (CONTINUED)

Means of Genetic Recombination, if Present	Mode of Nutrition	Motility	Multicellularity
Conjugation, transduction, transformation	Autotrophic (chemosynthetic, photosynthetic) or heterotrophic	Bacterial flagella, gliding or nonmotile	Absent
Fertilization and meiosis	Photosynthetic or heterotrophic, or combination of both	9 + 2 cilia and flagella; ameboid, contractile fibrils	Absent in most forms
Fertilization and meiosis	Absorption	Nonmotile	Present in most forms
Fertilization and meiosis	Photosynthetic; chlorophylls *a* and *b*	None in most forms, 9 + 2 cilia and flagella in gametes of some forms	Present in all forms
Fertilization and meiosis	Digestion	9 + 2 cilia and flagella, contractile fibrils	Present in all forms

12.10 Viruses: A Special Case

Viruses are not living organisms because they do not satisfy the basic criteria of life. Unlike all living organisms, viruses are acellular—that is, they are not cells and do not consist of cells. Unlike all living organisms, viruses do not metabolize energy. Viruses also do not carry out photosynthesis, cellular respiration, or fermentation.

Viruses thus present a special classification problem. Because they are not organisms, we cannot logically place them in any of the kingdoms. Despite their simplicity—merely bits of nucleic acid usually surrounded by a protein coat—viruses are able to invade cells and direct the genetic machinery of these cells to manufacture more virus material (figure 12.10). **Viruses** appear to be fragments of nucleic acids originally derived from the genome of a living cell. Viruses infect organisms at all taxonomic levels (figure 12.11).

12.10 Viruses are not organisms and are not classified in the kingdoms of life.

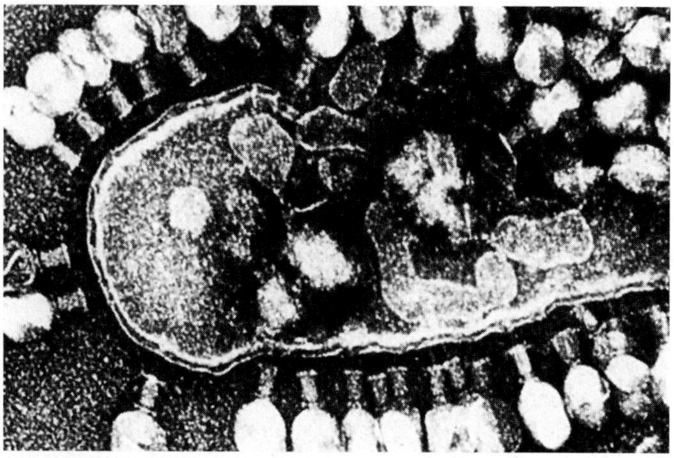

Figure 12.10 Viruses are cell parasites.
In this micrograph, several T4 bacteriophages (viruses) are attacking an *Escherichia coli* bacterium. Some of the viruses have already entered the cell and are reproducing within it.

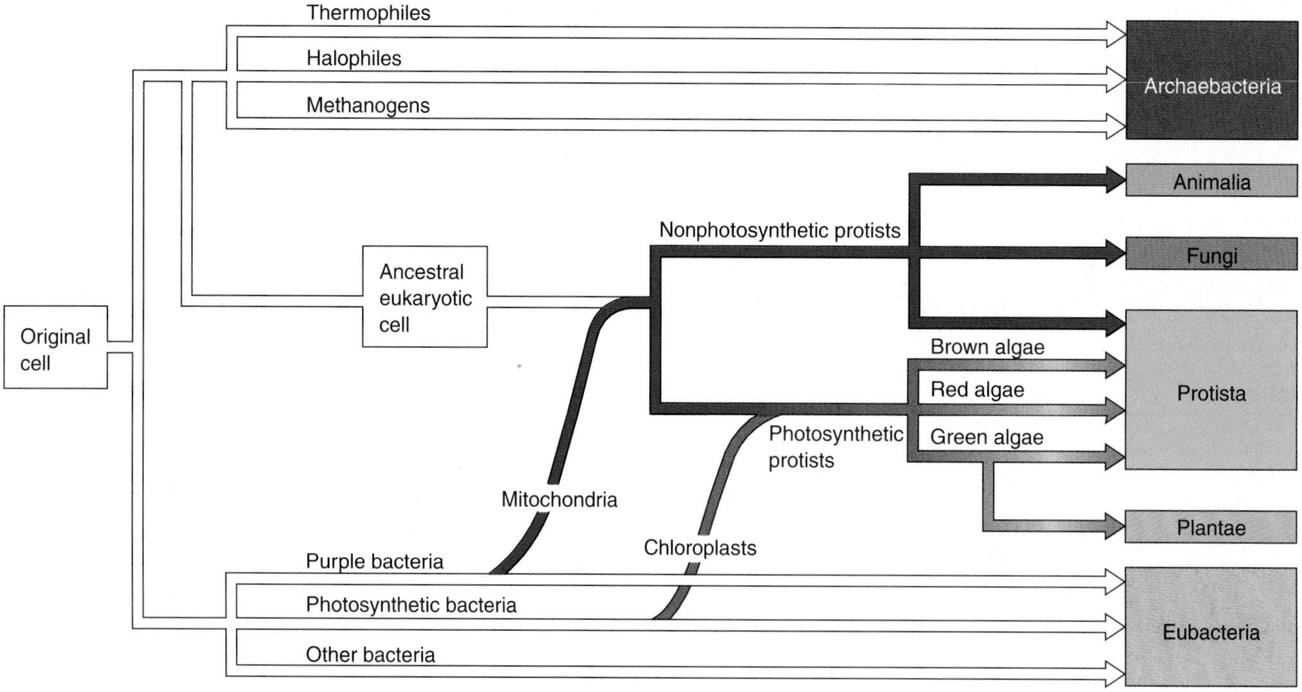
Thermophiles
Halophiles
Methanogens
Archaebacteria
Animalia
Nonphotosynthetic protists
Fungi
Ancestral eukaryotic cell
Original cell
Brown algae
Red algae
Protista
Photosynthetic protists
Green algae
Mitochondria
Chloroplasts
Plantae
Purple bacteria
Photosynthetic bacteria
Eubacteria
Other bacteria

Figure 12.11 Diagram of the evolutionary relationships among the six kingdoms of organisms.
The colored lines indicate symbiotic events. Although not placed in any kingdom, viruses infect organisms in all taxonomic levels.

CONCEPT REVIEW

1. List the taxonomic categories from most inclusive to least inclusive.
   a. order
   b. kingdom
   c. species
   d. phylum
   e. family
   f. genus
   g. class

2. Donkeys and horses are not the same species because the hybrids they produce are
   a. ecological races.
   b. fertile.
   c. sterile.
   d. fertile, but nonreproducing.

3. About _________ species of organisms have been named.
   a. 1.5 million
   b. 10 million
   c. 2 billion
   d. 10,000

4. Organisms within a clade share a more recent _________ than organisms outside the clade.
   a. ingroup
   b. outgroup
   c. common ancestor
   d. species

5. The derived character that is shared by gorillas and humans is
   a. hair.
   b. lungs.
   c. no tail.
   d. jaws.

6. In constructing phylogenies, traditional taxonomists and modern cladists often _________ the characters that have evolutionary significance.
   a. weight
   b. ignore
   c. randomize
   d. create

7. The statement that species are groups of interbreeding natural populations that are reproductively isolated from other such groups is called the _________.

8. _________ is the study of the evolutionary relationships among organisms.

9. A branching tree that reflects the evolutionary relationships among organisms is called a _________.

10. A group of organisms related by descent is called a _________.

11. In a cladogram of terrestrial vertebrates, birds are most closely related to _________ because the two groups share many derived characters.

CHALLENGE YOURSELF

1. What are some of the advantages of the binomial system of naming organisms?

2. Do you think a better system of classification would be to group all photosynthetic eukaryotes, regardless of whether or not they are single-celled, as plants and to group all nonphotosynthetic ones as animals? Present arguments in favor of doing so and other arguments in favor of the system of classification used in this text.

3. How is a species best defined? Would this definition apply to all organisms in all kingdoms? If not, how would you amend your definition to include the other organisms?

12 eBRIDGE

Reinforcing Key Points

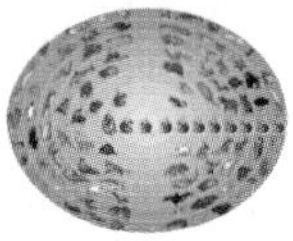

The Classification of Organisms

Inferring Phylogeny

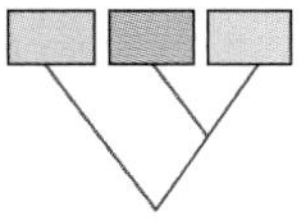

Kingdoms and Domains

Electronic Learning

Visual Learning

Animations

- Classification and Phylogeny
- Allopatric Versus Sympatric Speciation
- Molecular Clock
- Significance of Biochemical Differences
- Genetic Data
- Phylogenetic Tree of the Animal Kingdom
- Phylogenetic Tree of Chordates
- Traditional Five-Kingdom System of Classification
- Reproduction of Retrovirus HIV-1

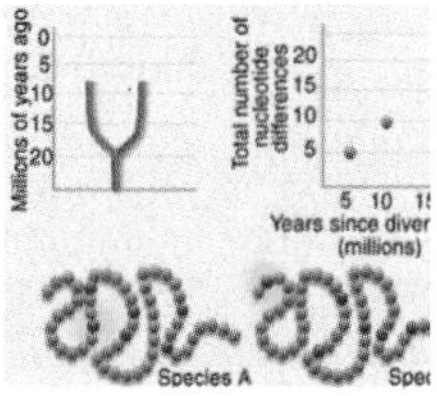
Millions of years ago
Total number of nucleotide differences
Years since diver
(millions)
Species A

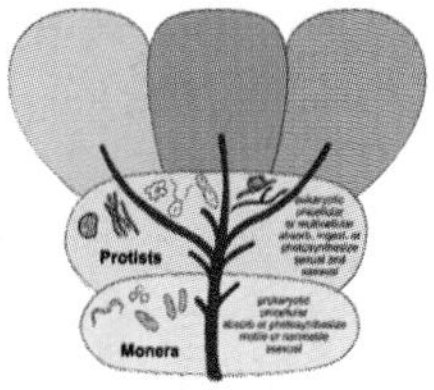
Protists
Monera

Author's Corner

Biodiversity. At its heart, biology is about critters—the great diversity of organisms which share the earth. Research into biodiversity, often carried out in the field, is a rich and rewarding activity. It can also be fun, funny, and intensely interesting.

1. Biodiversity behind bars: Are zoos justified?
2. A fierce argument has broken out among ecologists over biodiversity.
3. Is the number of men in a female's life written in her genes? Polyandry in Galápagos hawks.
4. Violence in Eden: Is Flipper a senseless killer?
5. The killer bees are coming.
6. How did Saint Patrick get the snakes out of Ireland?
7. Going batty: One of the most successful mammals flies at night.

Virtual Classroom

Prions and Mad Cow Disease

Not all diseases are caused by bacteria or viruses. "Mad cow disease" is a fatal and communicable brain disease of cows that is spread by protein molecules from one individual to another. The proteins, called prions, cause brain proteins to fold up incorrectly, eventually leading to brain lesions and death. The prion proteins are very stable, but because cows normally eat grass rather than each other, you would not expect prion infection to be a problem. However, until recent years it was common practice to supplement cattle feed with extra protein, often from the "rendered" bodies of cows that had died in the field. Unfortunately, humans that eat infected cows can acquire the prions too, leading to fatal brain disease for which there is no cure. Some 100 such fatal cases in humans have been reported in Britain, where the outbreak began, but many thousands more are expected. No cases have been reported in the United States, but caution is warranted. Europe was also free of mad cow disease for 15 years after the disease first broke out in England, only to spread to France, Germany, and the rest of Europe in 2000, apparently via contaminated bone meal.

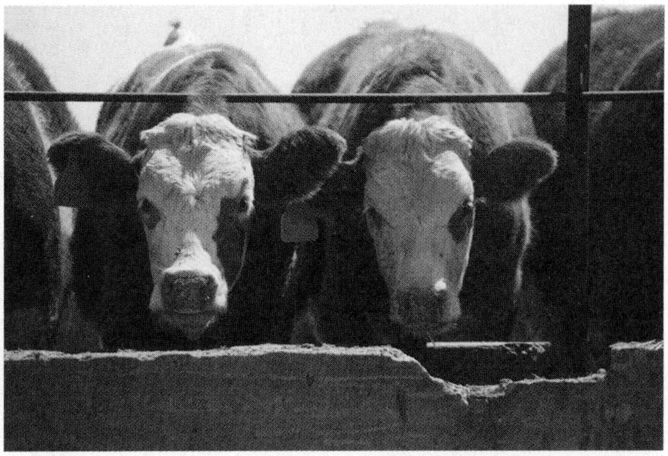

Virtual Lab

Unearthing the Root of Flowering Plant Phylogeny

The flowering plants have dominated the plant world for over 90 million years, with their ancestors appearing about 135 million years ago. What did this ancestral plant look like? Who are its closest living relatives? The Gnetales, a small group of primitive gymnosperms, share some characteristics with angiosperms, but molecular data seems to rule out the Gnetales as the angiosperm's closest ancestor. Fossils provide a glimpse of the morphological characteristics of early angiosperms, but no clear picture. Ancestral gene sequences in DNA may provide the critical information needed to interpret this early morphological evidence. Plant cells carry DNA in three different compartments: the mitochondrion, the chloroplast, and the nucleus genomes. A complete analysis of plant phylogeny needs to consider phylogenetic estimates based on all three of these genomic compartments. Todd Barkman of Western Michigan University and Claude de Pamphilis of Pennsylvania State University have evaluated available plant genomic data. Their study points to an evolutionary family tree that reveals the root of flowering plant phylogeny.

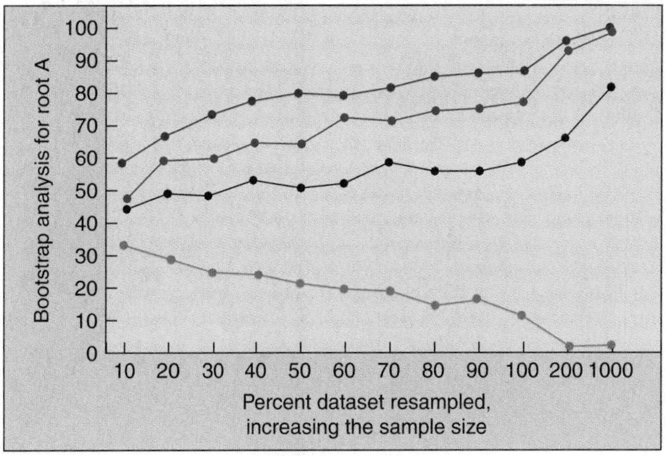
Bootstrap analysis for root A
0
10
20
30
40
50
60
70
80
90
100
10 20 30 40 50 60 70 80 90 100 200 1000
Percent dataset resampled, increasing the sample size

Quizzes | Further Reading | Essential Study Partner | Links | BioCourse.com

CHAPTER 13

The First Single-Celled Creatures

THE FAR SIDE®
By GARY LARSON
© 1985 FarWorks, Inc. All Rights Reserved/Dist. by Creators Syndicate
Larson
Mom! Dad! He followed me home! Can we keep him?

Childhood innocence

CHAPTER 13

The First Single-Celled Creatures

CHAPTER OVERVIEW

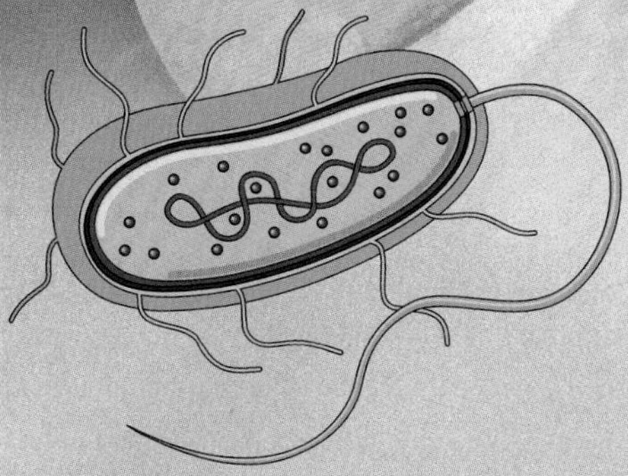

Bacteria

13.1 The Simplest Organisms
13.2 Comparing Bacteria to Eukaryotes
13.3 Importance of Bacteria

- Bacteria are the most ancient form of life on earth.
- Bacteria are small, simply organized, single cells that lack an organized nucleus enclosed within a nuclear membrane.
- Bacteria reproduce by binary fission.
- Some bacteria transfer plasmids from one cell to another in a process called conjugation.

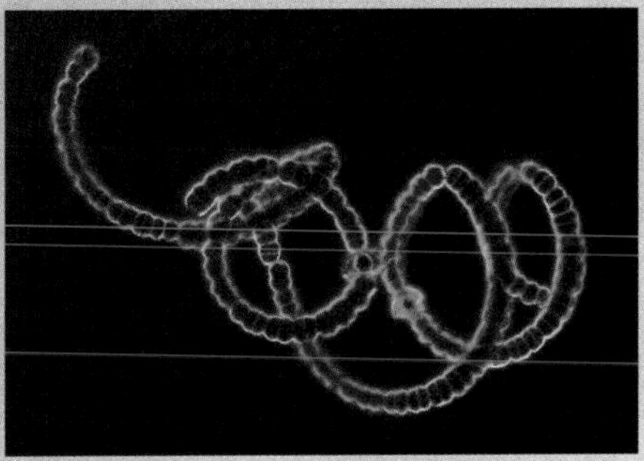

Bacterial Diversity

13.4 Kinds of Bacteria
13.5 Bacterial Lifestyles

- There are two kingdoms of bacteria, archaebacteria and eubacteria.
- The two kingdoms differ in fundamental aspects of their structure and are found living in very different places.
- Bacteria are responsible for much of the world's photosynthesis and all of its nitrogen fixation.

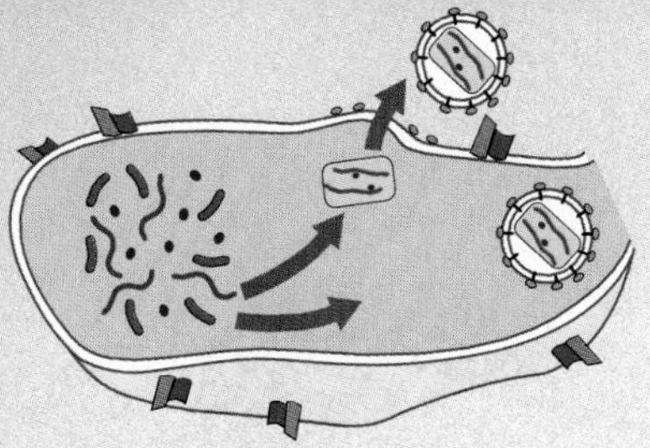

Viruses

13.6 The Discovery of Viruses
13.7 The Structure of Viruses
13.8 How Bacteriophages Enter Cells
13.9 How Animal Viruses Enter Cells
13.10 Disease Viruses

- Viruses are not organisms, as they cannot reproduce outside of cells.
- Every virus has the same basic structure: a core of nucleic acid encased within a sheath of protein.
- Animal viruses enter cells by endocytosis, while bacterial viruses inject their nucleic acid into host bacterial cells.

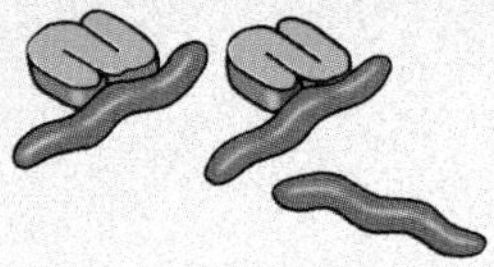

Other Nonliving Infectious Agents

13.11 Prions and Viroids

- Proteins called prions may transmit serious brain diseases from one individual to another, although the theory is controversial.
- RNA molecules can in some instances be infective disease agents.

13.1 The Simplest Organisms

Judging from fossils in ancient rocks, bacteria have been plentiful on earth for over 3.5 billion years. From the diverse array of early living forms, a few became the ancestors of the great majority of organisms alive today. Several forms, including methane-producing bacteria and cyanobacteria, have survived locally or in unusual habitats, but others probably became extinct millions or even billions of years ago. The fossil record indicates that eukaryotic cells, being much larger than bacteria and exhibiting elaborate shapes in some cases, did not appear until about 1.5 billion years ago. Therefore, for at least 2 billion years—nearly half the age of the earth—bacteria were the only organisms that existed.

Today bacteria are the simplest and most abundant form of life on earth. In a spoonful of farmland soil, 2.5 billion bacteria may be present. In 1 hectare (about 2.5 acres) of wheat land in England, the weight of bacteria in the soil is approximately equal to that of 100 sheep!

It is not surprising, then, that bacteria occupy a very important place in the web of life on earth. They play a key role in cycling minerals within the earth's ecosystems. In fact, photosynthetic bacteria were in large measure responsible for the introduction of oxygen into the earth's atmosphere. Bacteria are responsible for some of the most deadly animal and plant diseases, including many that infect humans. Bacteria are our constant companions, present in everything we eat and on everything we touch.

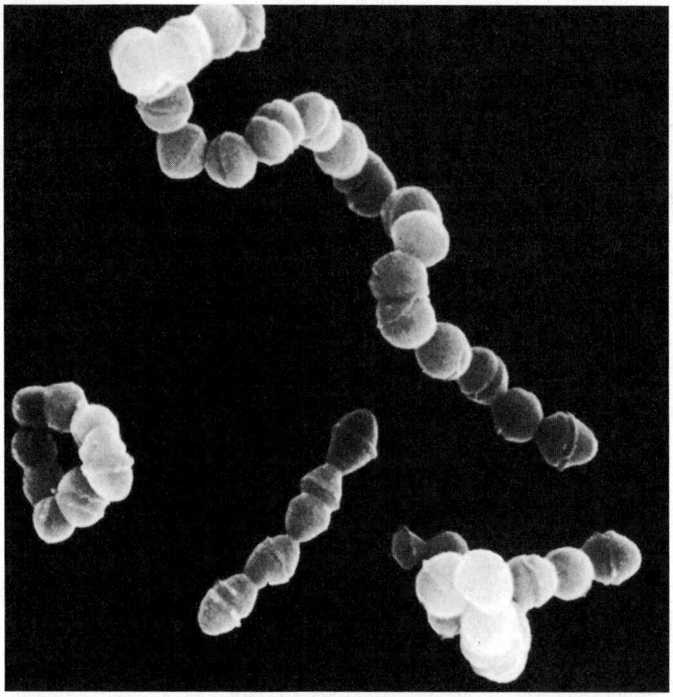

Figure 13.1 Spherical-shaped bacteria.
Bacterial cells are simple in form. Each of the spheres (cocci) adhering in chains in this micrograph is an individual bacterial cell of the genus *Streptococcus*.

The Structure of a Bacterium

The essential character of bacteria can be conveyed in a simple sentence: **Bacteria** are small, simply organized, single cells that lack an organized nucleus. Therefore, they are prokaryotes; their single circle of DNA is not confined by a nuclear membrane in a nucleus, as in the cells of eukaryotes. Too tiny to see with the naked eye, a bacterial cell is usually simple in form, either rod-shaped (bacilli) (see figure 4.9*a*), spherical (cocci) (figure 13.1), or spirally coiled (spirilla) (see figure 4.9*c*). A few kinds of bacteria aggregate into stalked structures or filaments.

The bacterial cell's plasma membrane is encased within a cell wall (figure 13.2). Among members of the more advanced of the two bacterial kingdoms, the cell wall is made of a network of polysaccharide molecules linked together by protein cross-links. Many species of bacteria have only this cell wall, but in others, the cell wall is covered with an outer membrane layer composed of large molecules of lipopolysaccharide (a chain of sugars with lipids attached to it). Bacteria are commonly classified by the presence or absence of this membrane as **gram-positive** (no outer membrane) or **gram-negative** (possess an outer membrane). The name refers to the Danish microbiologist Hans Gram, who

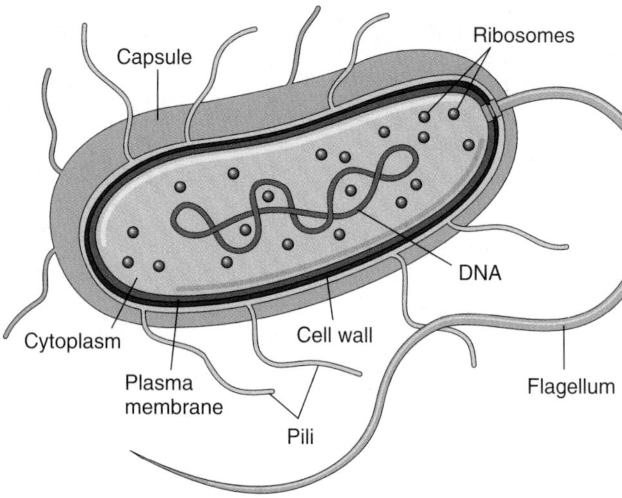

Capsule
Ribosomes
DNA
Cytoplasm
Cell wall
Plasma membrane
Pili
Flagellum

Figure 13.2 The structure of a bacterial cell.
Bacterial cells are small and lack interior organization. The plasma membrane is encased within a rigid cell wall.

developed a staining process that colors the cell wall purple. In bacteria with an outer membrane, the stain cannot reach the cell wall and so the cells do not develop a purple color. The membranes of gram-negative bacteria make them resistant to antibiotics that attack the bacterial cell wall. That is why penicillin, which targets the protein cross-links of the eubacterial cell wall, is effective only against gram-positive bacteria. Outside of the cell wall and membrane, many bacteria possess a gelatinous layer called a **capsule.**

Many kinds of bacteria possess threadlike **flagella,** long strands of protein that may extend out several times the length of the cell body. Bacteria swim by twisting these flagella in a corkscrew motion. Flagella may be distributed all over the cell body or be confined to one or both ends of the cell. Some bacteria also possess shorter outgrowths called **pili** (singular, **pilus**), which act as docking cables, helping the cell to attach to surfaces or other cells.

When exposed to harsh conditions (dryness or high temperature), some bacteria form thick-walled **endospores** around their DNA and a small bit of cytoplasm. These endospores are highly resistant to environmental stress and may germinate to form new active bacteria even after centuries. The formation of endospores is the reason that the bacterium responsible for botulism, *Clostridium botulinum,* sometimes persists in cans and bottles if the containers have not been heated at a high enough temperature to kill the spores.

How Bacteria Reproduce

Bacteria, like all other living cells, grow and divide. Bacteria reproduce using a process called **binary fission,** in which an individual cell simply increases in size and divides in two. Following replication of the bacterial DNA, the plasma membrane and cell wall grow inward and eventually divide the cell by forming a new wall from the outside toward the center of the old cell.

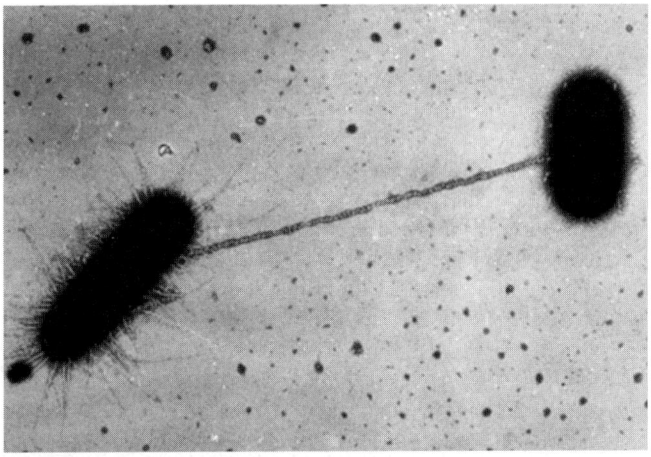

Figure 13.3 Contact by a pilus.

The pilus of a donor cell connects to a recipient cell and draws the two cells close together so that DNA transfer can occur.

Some bacteria can pass plasmids from one cell to another in a process called **conjugation.** Recall from chapter 9 that a plasmid is a small, circular fragment of DNA that replicates outside the main circle of DNA. In bacterial conjugation, the pilus of one cell, called the donor cell, contacts a recipient cell (figure 13.3) and draws the two cells close together. The plasmid within the donor cell is then mobilized for transfer across a conjugation bridge. The plasmid in the donor cell begins to replicate its DNA, passing the replicated copy out across the bridge and into the recipient cell (figure 13.4).

13.1 Bacteria are the smallest and simplest organisms, composed of a single cell with no internal compartments or organelles. They divide by binary fission, simply pinching in two.

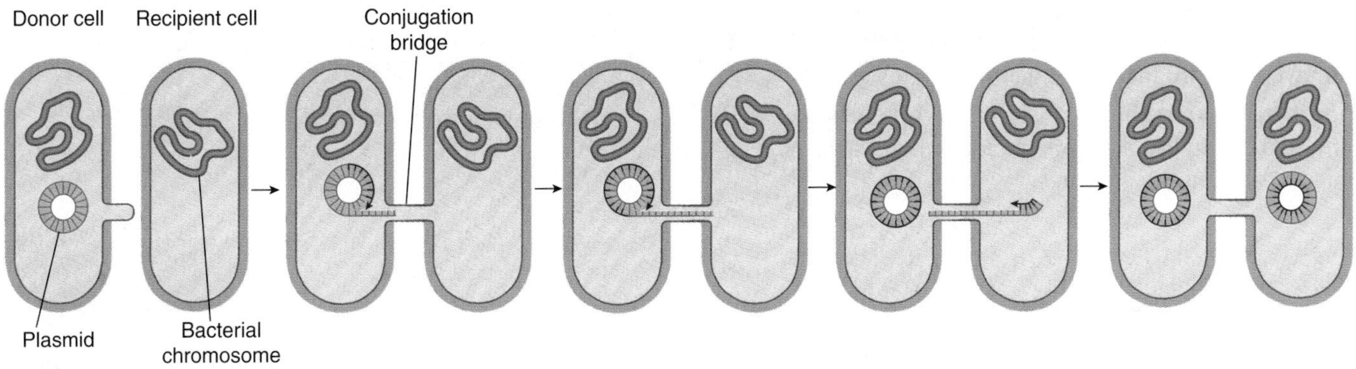
Donor cell
Recipient cell
Conjugation bridge
Plasmid
Bacterial chromosome

Figure 13.4 Bacterial conjugation.

Donor cells contain a plasmid that recipient cells lack. The plasmid replicates itself and transfers the copy across a conjugation bridge. The remaining strand of the plasmid serves as a template to build a replacement. When the single strand enters the recipient cell, it serves as a template to assemble a double-stranded plasmid. When the process is complete, both cells contain a complete copy of the plasmid.

13.2 Comparing Bacteria to Eukaryotes

Bacteria differ from eukaryotes in many respects: the cytoplasm of bacteria has very little internal organization, bacteria are unicellular and much smaller than eukaryotes, the bacterial chromosome is a single circle of DNA, cell division and flagella are simple, and bacteria are far more metabolically diverse than eukaryotes (table 13.1).

Bacterial Metabolism

Bacteria have evolved many more ways than eukaryotes to acquire the carbon atoms and energy necessary for growth and reproduction. Many are **autotrophs,** organisms that obtain their carbon from inorganic CO_2. Autotrophs that obtain their energy from sunlight are called *photoautotrophs,* while those that harvest energy from inorganic chemicals are called *chemoautotrophs.* Other bacteria are **heterotrophs,** organisms that obtain at least some of their carbon from organic molecules like glucose. Heterotrophs that obtain their energy from sunlight are called *photoheterotrophs,* while those that harvest energy from organic molecules are called *chemoheterotrophs.*

Photoautotrophs. Many bacteria carry out photosynthesis, using the energy of sunlight to build organic molecules from carbon dioxide. The cyanobacteria use chlorophyll *a* as the key light-capturing pigment and use H_2O as an electron donor, leaving oxygen gas as a by-product. Other bacteria use bacteriochlorophyll as their pigment and H_2S as an electron donor, leaving elemental sulfur as the by-product.

Chemoautotrophs. Some bacteria obtain their energy by oxidizing inorganic substances. Nitrifiers, for example, oxidize ammonia or nitrite to form the nitrate that is taken up by plants. Other bacteria oxidize sulfur, hydrogen gas, and other inorganic molecules. On the dark ocean floor at depths of 2,500 meters, entire ecosystems subsist on bacteria that oxidize hydrogen sulfide as it escapes from volcanic vents.

Photoheterotrophs. The so-called purple nonsulfur bacteria use light as their source of energy but obtain carbon from organic molecules such as carbohydrates or alcohols that have been produced by other organisms.

Chemoheterotrophs. Most bacteria obtain both carbon atoms and energy from organic molecules. These include decomposers and most pathogens.

13.2 Bacteria differ from eukaryotes in having no nucleus or other interior compartments, in being far more metabolically diverse, and in many other fundamental respects.

TABLE 13.1 BACTERIA COMPARED TO EUKARYOTES

Feature	Example
Internal compartmentalization. Unlike eukaryotic cells, bacterial cells contain no internal compartments, no internal membrane system, and no cell nucleus.	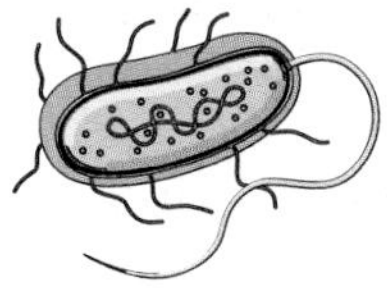 Bacterial cell
Cell size. Most bacterial cells are only about 1 micrometer in diameter, whereas most eukaryotic cells are well over 10 times that size.	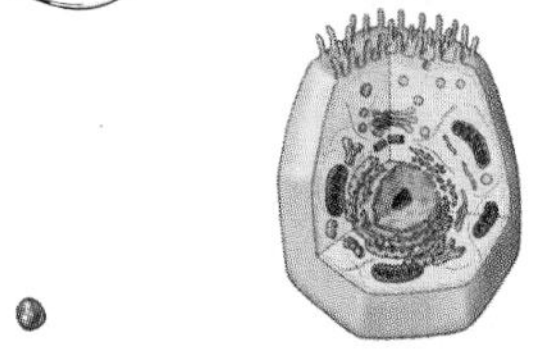 Bacterial cell; Eukaryotic cell
Unicellularity. All bacteria are fundamentally single-celled. Even though some bacteria may adhere together in a matrix or form filaments, their cytoplasm is not directly interconnected, and their activities are not integrated and coordinated, as is the case in multicellular eukaryotes.	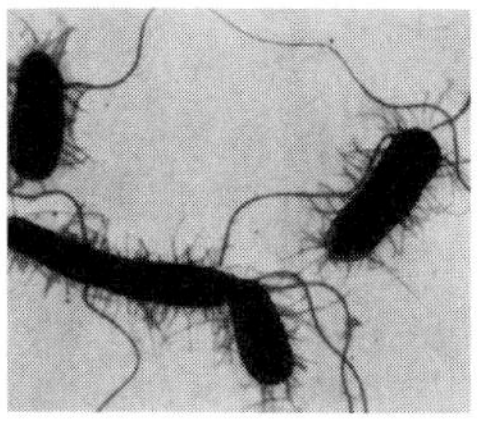 Unicellular bacteria
Chromosomes. Bacteria do not possess chromosomes in which proteins are complexed with the DNA, as eukaryotes do. Instead, their DNA exists as a single circle in the cytoplasm.	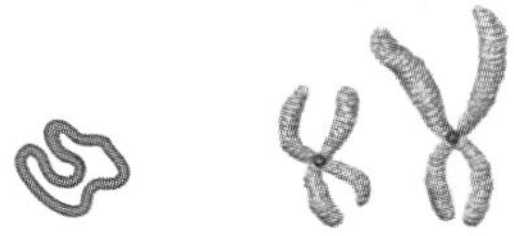 Bacterial chromosome; Eukaryotic chromosomes
Cell division. Cell division in bacteria takes place by binary fission (see chapter 6). The cells simply pinch in two. In eukaryotes, microtubules pull chromosomes to opposite poles during the cell division process, called mitosis.	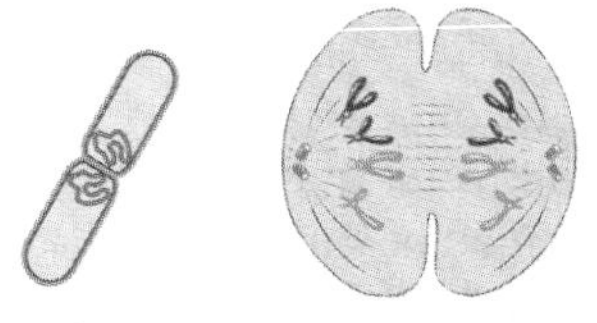 Binary fission in bacteria; Mitosis in eukaryotes
Flagella. Bacterial flagella are simple, composed of a single fiber of protein that is spun like a propeller. Eukaryotic flagella are more complex structures, with a 9 + 2 arrangement of microtubules, that whip back and forth rather than rotate.	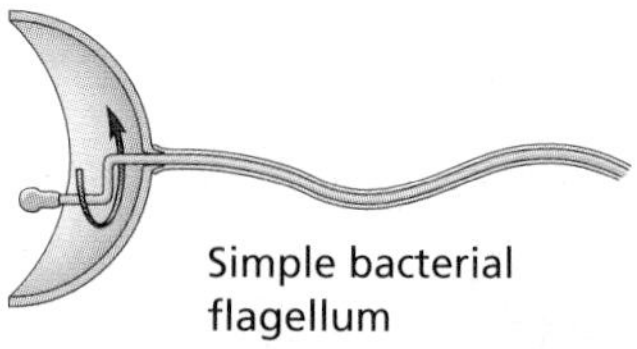 Simple bacterial flagellum
Metabolic diversity. Bacteria possess many metabolic abilities that eukaryotes do not: bacteria perform several different kinds of anaerobic and aerobic photosynthesis; bacteria can obtain their energy from oxidizing inorganic compounds (so-called chemoautotrophs); and bacteria can fix atmospheric nitrogen.	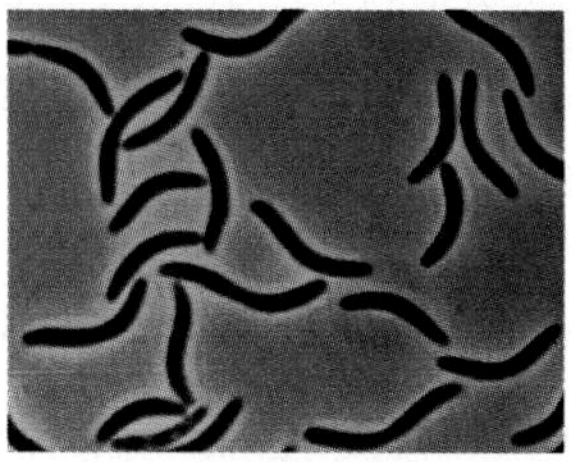 Chemoautotrophs

13.3 Importance of Bacteria

Bacteria, the first organisms to evolve on earth, affect our lives in many important ways.

Bacteria and the Environment

Bacteria were largely responsible for creating the properties of the atmosphere and the soil over billions of years. They are metabolically much more diverse than eukaryotes, which is why they are able to exist in such a wide range of habitats. The many autotrophic bacteria—either photosynthetic or chemoautotrophic—make major contributions to the world carbon balance in terrestrial, freshwater, and marine habitats. Other heterotrophic bacteria play a key role in world ecology by breaking down organic compounds. One of the most important roles of bacteria in the world ecosystem relates to the fact that only a few genera of bacteria—and no other organisms—have the ability to fix atmospheric nitrogen and thus make it available for use by other organisms.

Bacteria and Genetic Engineering

Applying genetic engineering methods to produce improved strains of bacteria for commercial use holds enormous promise for the future. Bacteria are under intense investigation, for example, as nonpolluting insect control agents. *Bacillus thuringiensis* attacks insects in nature, and improved, highly specific strains of *B. thuringiensis* have greatly increased its usefulness as a biological control agent. Genetically modified bacteria have also been extraordinarily useful in producing insulin and other therapeutic proteins. Gene modified bacteria are also playing a part in removing environmental pollutants (figure 13.5).

Bacteria, Disease, and Bioterrorism

Some bacteria cause major diseases in plants and animals, including humans. Among important human bacterial diseases that can be lethal are anthrax, cholera, plague, pneumonia, tuberculosis (TB), and typus. Many pathogenic (disease-causing) bacteria like cholera are dispersed in food and water. Others, like typhus and plague, spread among rodents and humans by fleas. A few, like TB—the leading cause of death from a single infectious agent worldwide in 2001—are spread through the air in water vapor, infecting those who inhale the droplets. Among these inhalation pathogens is anthrax, a disease associated with livestock that rarely kills humans. Most human infections are cutaneous, infecting a cut in the skin, but if a significant number of the spores are inhaled, the pulmonary (lung) infection is almost always fatal. Biological warfare programs in the United States and Russia focused on anthrax as a near-ideal biological weapon, although it has never been used in war. Bioterrorists struck at the United States with anthrax spores in 2001.

13.3 Bacteria make many important contributions to the world ecosystem, including occupying key roles in cycling carbon and nitrogen.

Figure 13.5 Using bacteria to clean up oil spills.
Bacteria can often be used to remove environmental pollutants, such as petroleum hydrocarbons and chlorinated compounds. In areas contaminated by the *Exxon Valdez* oil spill (*rocks on the left*), oil-degrading bacteria produced dramatic results (*rocks on the right*).

13.4 Kinds of Bacteria

Although small and internally simple, all bacteria that live today are diverse in their internal chemistry and in many of the details of how they are assembled (table 13.2). If you were to gauge success by numbers, then bacteria are the most successful of all organisms—there are more bacteria in a spoonful of dirt than all the vertebrates on earth. There are over 4,800 named species of bacteria, and undoubtedly many thousands more awaiting discovery. During the 2 billion years they evolved alone on earth, bacteria adapted to many kinds of environments, including some you might consider harsh. They invaded waters that were very salty, very acidic or alkaline, and very hot or cold—anywhere that would support life, they thrived. Today they are found in hot springs where the temperatures exceed 78°C (180°F) and have been recovered living beneath 430 meters of ice in Antarctica!

Early in the history of life, bacteria split into two kinds, so different in structure and metabolism that biologists assign them to entirely separate kingdoms. One branch produced the **archaebacteria** ("ancient bacteria"), many of its survivors today being confined to extreme environments that may resemble habitats on the early earth. This is the branch from which eukaryotes are thought to have evolved. The other branch produced the **eubacteria** ("true bacteria"), which include nearly all the 4,800 kinds of bacteria that live today.

TABLE 13.2 BACTERIA

Major Group	Typical Examples		Key Characteristics
			Kingdom Archaebacteria
Methanogens, extremophiles, nonextreme archaebacteria	*Methanococcus,* thermophiles, halophiles		Bacteria that are not members of the kingdom Eubacteria. Mostly anaerobic with unusual cell walls. Some produce methane; others reduce sulfur.
			Kingdom Eubacteria
Gram-positive bacteria	*Streptomyces, Actinomyces Bacillus*	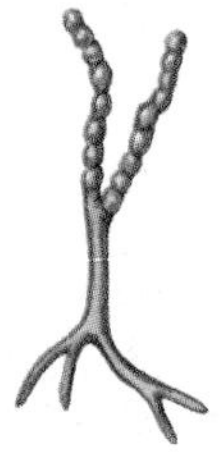	Gram-positive bacteria form branching filaments and produce spores. Sometimes mistaken for fungi. Produce many commonly-used antibiotics. One of the most common types of soil bacteria. Responsible for many important human diseases, including anthrax.
Chemoautotrophs	Sulfur bacteria, *Nitrobacter, Nitrosomonas*	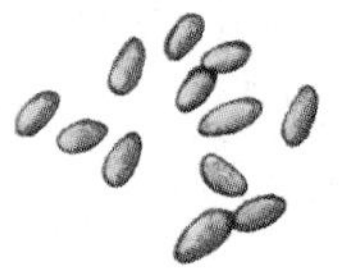	Bacteria able to obtain their energy from inorganic chemicals. Most extract chemical energy from reduced gases such as H_2S (hydrogen sulfide), NH_3 (ammonia), and CH_4 (methane). Play a key role in the nitrogen cycle.
Cyanobacteria	*Anabaena, Nostoc*	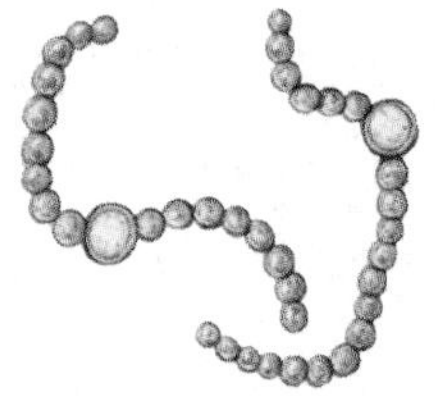	A form of photosynthetic bacteria common in both marine and freshwater environments. Deeply pigmented; often responsible for "blooms" in polluted waters.
Enterobacteria	*Escherichia coli, Salmonella, Vibrio*	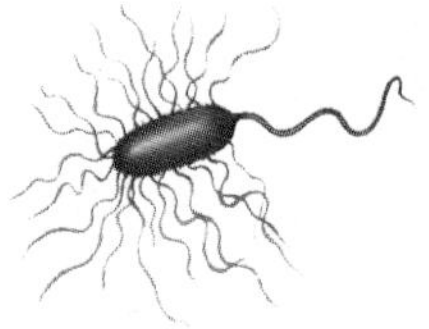	Gram-negative, rod-shaped bacteria. Do not form spores; usually aerobic heterotrophs; cause many important diseases, including bubonic plague and cholera.

Archaebacteria and eubacteria differ in four fundamental ways:

1. **Cell wall.** Both kinds of bacteria have cell walls outside the plasma membrane that strengthen the cell. The cell walls of eubacteria are constructed of carbohydrate-protein complexes called peptidoglycans, which link together to create a strong mesh that gives the eubacterial cell wall great strength. The cell walls of archaebacteria lack peptidoglycans.
2. **Plasma membranes.** All bacteria have plasma membranes with a lipid-bilayer architecture (see chapter 4). The plasma membranes of eubacteria and archaebacteria, however, are made of very different kinds of lipids.
3. **Gene translation machinery.** Eubacteria possess ribosomal proteins and an RNA polymerase that are distinctly different from those of eukaryotes. However, the ribosomal proteins and RNA of archaebacteria are very similar to those of eukaryotes.
4. **Gene architecture.** The genes of eubacteria are not interrupted by introns as are those of eukaryotes, while at least some of the genes of archaebacteria do possess introns.

13.4 There are two major kingdoms of bacteria, archaebacteria and eubacteria, which are as different from each other as you are from a plant.

TABLE 13.2 (CONTINUED)

Major Group	Typical Examples		Key Characteristics
Gliding and budding bacteria	Myxobacteria, *Chondromyces*		Gram-negative bacteria. Exhibit gliding motility by secreting slimy polysaccharides over which masses of cells glide; some groups form upright multicellular structures carrying spores called fruiting bodies.
Pseudomonads	*Pseudomonas*	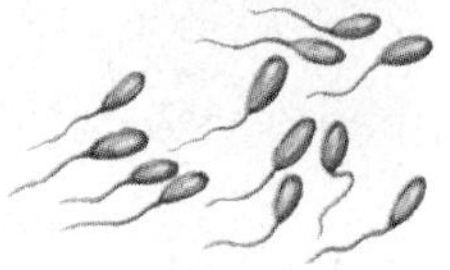	Gram-negative heterotrophic rods with polar flagella. Very common form of soil bacteria; also contain many important plant pathogens.
Rickettsias and chlamydias	*Rickettsia*, *Chlamydia*		Small, gram-negative intracellular parasites. *Rickettsia* life cycle involves both mammals and arthropods such as fleas and ticks; *Rickettsia* are responsible for many fatal human diseases, including typhus (*Rickettsia prowazekii*) and Rocky Mountain spotted fever. Chlamydial infections are one of the most common sexually transmitted diseases.
Spirochaetes	*Treponema*	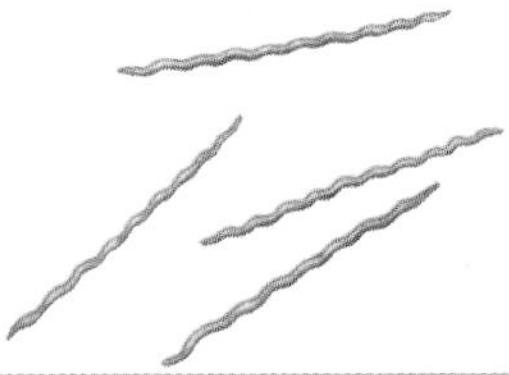	Long, coil-shaped cells. Common in aquatic environments; a parasitic form is responsible for the disease syphilis.

13.5 Bacterial Lifestyles

Archaebacteria

Many of the archaebacteria that survive today are **methanogens,** bacteria that use hydrogen (H_2) gas to reduce carbon dioxide (CO_2) to methane (CH_4). Methanogens are strict anaerobes, poisoned by oxygen gas. They live in swamps and marshes, where other microbes have consumed all the oxygen. The methane that they produce bubbles up as "marsh gas." Methanogens also live in the gut of cows and other herbivores that live on a diet of cellulose, converting the CO_2 produced by that process to methane gas. The best understood archaebacteria are extremophiles that live in unusually harsh environments, such as the very salty Dead Sea and the Great Salt Lake (over 10 times saltier than seawater). **Thermoacidophiles** favor hot, acidic springs such as the sulfur springs of Yellowstone National Park (figure 13.6), where the water is nearly 80°C, with an acidic pH of 2 or 3.

Figure 13.6 Thermoacidophiles live in hot springs.
These archaebacteria growing in Sulfide Spring, Yellowstone National Park, Wyoming, are able to tolerate high acid levels and very hot temperatures.

Eubacteria

Almost all bacteria living today are eubacteria, members of the kingdom Eubacteria. Many are heterotrophs that power their lives by consuming organic molecules, while others are photosynthetic, gaining their energy from the sun. **Cyanobacteria** are among the most prominent of the photosynthetic bacteria. We have already discussed the critical role that the members of this ancient phylum have played in the history of the earth by generating the oxygen in our atmosphere. Cyanobacteria usually have a mucilaginous sheath, which is often deeply pigmented. Nitrogen fixation occurs in almost all cyanobacteria, within specialized cells called **heterocysts,** enlarged cells that occur in many of the filamentous members of this phylum (figure 13.7). These cells begin to form when available nitrogen falls below a certain threshold; when nitrogen is abundant, their formation is inhibited. In **nitrogen fixation,** atmospheric nitrogen is converted to a form in which it can be used by living organisms.

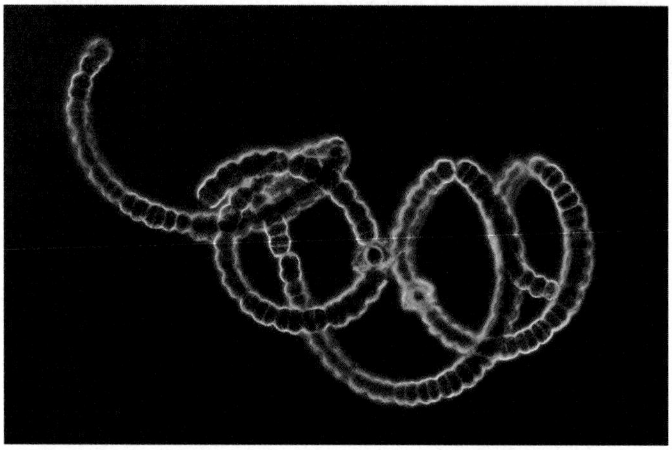

Figure 13.7 The cyanobacterium *Anabaena*.
Individual cells adhere in filaments. The larger cells (areas on the filament that seem to be bulging) are heterocysts, specialized cells in which nitrogen fixation occurs. These organisms exhibit one of the closest approaches to multicellularity among the bacteria.

There are numerous phyla of nonphotosynthetic bacteria. Some are chemoautotrophs, but most are heterotrophs. Some of these heterotrophs are decomposers, breaking down organic material. Bacteria and fungi play the leading role in breaking down organic molecules formed by biological processes, thereby making the nutrients in these molecules available once more for recycling. Decomposition is just as indispensable to the continuation of life on earth as is photosynthesis.

Bacteria cause many diseases in humans (table 13.3), including cholera, diphtheria, and leprosy. Among the most serious of bacterial diseases is **tuberculosis (TB),** a disease of the respiratory tract caused by the bacterium *Mycobacterium tuberculosis.* TB is a leading cause of death throughout the world. Spread through the air, tuberculosis is quite infectious. In most instances, a person becomes infected by inhaling tiny droplets of moisture that contain the bacteria. TB was a major health risk in the United States until the discovery of effective drugs to suppress it in the 1950s. The appearance of drug-resistant strains in the 1990s has raised serious concern within the medical community, as the incidence of TB has increased. The search is on for new types of anti-TB drugs.

13.5 Most commonly encountered bacteria are eubacteria, and some cause significant diseases in humans.

TABLE 13.3 IMPORTANT HUMAN BACTERIAL DISEASES

Disease	Pathogen	Vector/ Reservoir	Epidemiology
Anthrax	*Bacillus anthracis*	Farm animals	Bacterial infection that can be transmitted through inhaled spores, by contact, or ingested. Rare except in sporadic outbreaks. Pulmonary (inhaled) anthrax is usually fatal, while cutaneous anthrax (infection through cuts) is readily treated with antibiotics. Anthrax spores have been used as a biological weapon.
Botulism	*Clostridium botulinum*	Improperly prepared food	Contracted through ingestion or contact with wound. Produces acutely toxic poison; can be fatal.
Chlamydia	*Chlamydia trachomatis*	Humans (STD)	Urogenital infections with possible spread to eyes and respiratory tract. Occurs worldwide; increasingly common over past 20 years.
Cholera	*Vibrio cholerae*	Humans (feces), plankton	Causes severe diarrhea that can lead to death by dehydration; 50% peak mortality if the disease goes untreated. A major killer in times of crowding and poor sanitation; over 100,000 died in Rwanda in 1994 during a cholera outbreak.
Dental caries	*Streptococcus*	Humans	A dense collection of this bacteria on the surface of teeth leads to secretion of acids that destroy minerals in tooth enamel—sugar alone will not cause caries.
Diphtheria	*Corynebacterium diphtheriae*	Humans	Acute inflammation and lesions of mucous membranes. Spread through contact with infected individual. Vaccine available.
Gonorrhea	*Neisseria gonorrhoeae*	Humans only	STD, on the increase worldwide. Usually not fatal.
Hansen's disease (leprosy)	*Mycobacterium leprae*	Humans, feral armadillos	Chronic infection of the skin; worldwide incidence about 10 to 12 million, especially in southeast Asia. Spread through contact with infected individuals.
Lyme disease	*Borrelia burgdorferi*	Ticks, deer, small rodents	Spread through bite of infected tick. Lesion followed by malaise, fever, fatigue, pain, stiff neck, and headache.
Peptic ulcers	*Helicobacter pylori*	Humans	Originally thought to be caused by stress or diet, most peptic ulcers now appear to be caused by this bacterium; good news for ulcer sufferers as it can be treated with antibiotics.
Plague	*Yersinia pestis*	Fleas of wild rodents: rats and squirrels	Killed one-fourth of the population of Europe in the fourteenth century; endemic in wild rodent populations of the western United States today.
Pneumonia	*Streptococcus, Mycoplasma, Chlamydia*	Humans	Acute infection of the lungs, often fatal without treatment.
Tuberculosis	*Mycobacterium tuberculosis*	Humans	An acute bacterial infection of the lungs, lymph, and meninges. Its incidence is on the rise, complicated by the development of new strains of the bacteria that are resistant to antibiotics.
Typhoid fever	*Salmonella typhi*	Humans	A systemic bacterial disease of worldwide incidence. Less than 500 cases a year are reported in the United States. The disease is spread through contaminated water or foods (such as improperly washed fruits and vegetables). Vaccines are available for travelers.
Typhus	*Rickettsia*	Lice, rat fleas, humans	Historically a major killer in times of crowding and poor sanitation; transmitted from human to human through the bite of infected lice and fleas. Typhus has a peak untreated mortality rate of 70%.

13.6 The Discovery of Viruses

The border between the living and the nonliving is very clear to a biologist. Living organisms are cellular and able to grow and reproduce independently, guided by information encoded within DNA. As discussed earlier in this chapter, the simplest creatures living on earth today that satisfy these criteria are bacteria. Viruses, on the other hand, do not satisfy the criteria for "living" because they possess only a portion of the properties of organisms. **Viruses** are literally "parasitic" chemicals, segments of DNA (or sometimes RNA) wrapped in a protein coat. They cannot reproduce on their own, and for this reason they are not considered alive by biologists. They can, however, reproduce within cells, often with disastrous results to the host organism. For this reason viruses have a major impact on the living world.

Viruses are very small (figure 13.8). Biologists first began to suspect the existence of viruses near the end of the nineteenth century. European scientists attempting to isolate the infectious agent responsible for hoof-and-mouth disease in cattle concluded that it was smaller than a bacterium. Investigating the agent further, the scientists found that it could not multiply in solution—it could only reproduce itself within living host cells that it infected. The infecting agents were called viruses.

The true nature of viruses was discovered in 1933, when the biologist Wendell Stanley prepared an extract of a plant virus called *tobacco mosaic virus* (*TMV*) and attempted to purify it. To his great surprise, the purified TMV preparation precipitated (that is, separated from solution) in the form of crystals. This was surprising because precipitation is something that only chemicals do—the TMV virus was acting like a chemical rather than an organism. Stanley concluded that TMV is best regarded as just that—a chemical matter rather than a living organism.

Within a few years, scientists disassembled the TMV virus and found that Stanley was right. TMV was not cellular but rather chemical. Each particle of TMV virus is in fact a mixture of two chemicals: RNA and protein. The TMV virus has the structure of a Twinkie, a tube made of an RNA core surrounded by a coat of protein (figure 13.9). Later workers were able to separate the RNA from the protein and purify and store each chemical. Then, when they reassembled the two components, the reconstructed TMV particles were fully able to infect healthy tobacco plants and so clearly *were* the virus itself, not merely chemicals derived from it. Further experiments carried out on other viruses yielded similar results.

13.6 Viruses are chemical assemblies that can infect cells and replicate within them. They are not alive.

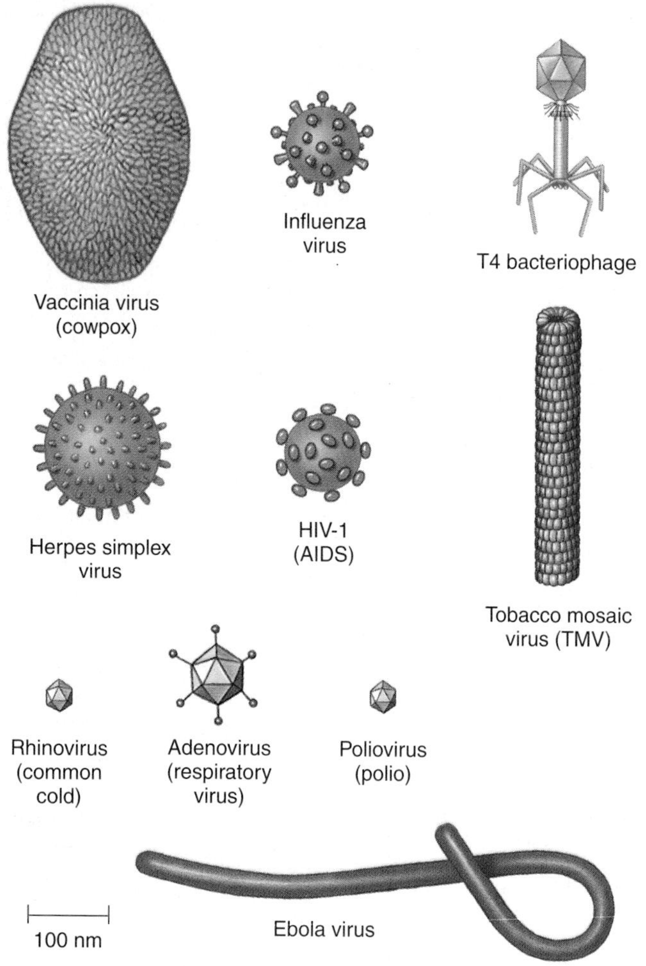

Vaccinia virus (cowpox)
Influenza virus
T4 bacteriophage
Herpes simplex virus
HIV-1 (AIDS)
Tobacco mosaic virus (TMV)
Rhinovirus (common cold)
Adenovirus (respiratory virus)
Poliovirus (polio)
100 nm
Ebola virus

Figure 13.8 Some types of viruses.
A sample of the extensive diversity and small size of viruses is shown. At the scale shown here, a human hair would be 8 meters thick.

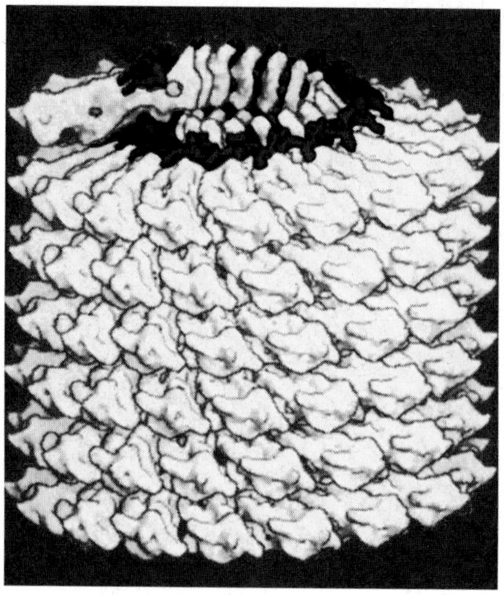

Figure 13.9 Tobacco mosaic virus (TMV).
In this computer-generated model of a portion of tobacco mosaic virus, the *yellow* proteins form a cylindrical coat around the *red* single strand of RNA.

13.7 The Structure of Viruses

All organisms appear to have viruses, and in every case the basic structure is the same, a core of nucleic acid surrounded by protein. There is considerable difference, however, in the details. In figure 13.10 you can compare the structure of bacterial, plant, and animal viruses—they are clearly quite different from one another. Many plant viruses like TMV have a core of RNA, and some animal viruses like HIV do too, but most viruses have DNA in place of RNA. Like TMV, most viruses form a protein sheath, or **capsid,** around their nucleic acid core. The only exceptions are viroids, tiny naked RNA molecules that infect some plant cells. In addition, many viruses form a membranelike **envelope,** rich in proteins, lipids, and glycoprotein molecules, around the capsid.

The smallest viruses are only about 17 nanometers in diameter; the largest ones measure up to 1,000 nanometers, big enough to be barely visible with the light microscope. Viruses are so small that they are smaller than many of the molecules in a cell. Most viruses can be detected only by using the higher resolution of an electron microscope.

Most viruses are more complex than TMV. Several different segments of DNA or RNA may be present in each virus particle, along with many different kinds of protein. Bacterial viruses, called **bacteriophages,** are among the most complex viruses known, built like a lunar lander with a head, neck, tail, base plate, and tail fibers. Most animal and plant viruses are much simpler in appearance than a bacteriophage, being either **helical** (rod-shaped with capsid proteins winding around the core in a helix like a winding staircase) or **isometric** (with the capsid proteins forming a sphere of 20 equal triangular facets like a geodesic dome). The geodesic design of isometric viruses like HIV is the most efficient way to assemble subunits to form an external shell with maximum internal capacity.

13.7 Viruses are not cells and have no true membranes. They are genomes of DNA or RNA, encased in a protein shell.

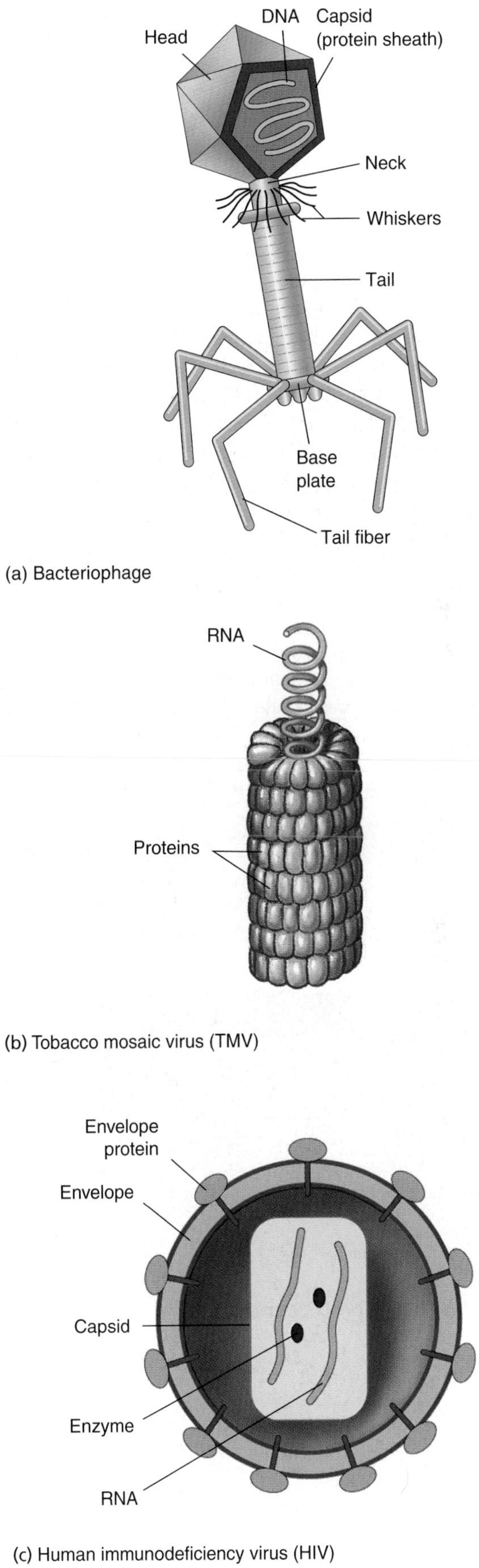
Head
DNA
Capsid (protein sheath)
Neck
Whiskers
Tail
Base plate
Tail fiber
(a) Bacteriophage
RNA
Proteins
(b) Tobacco mosaic virus (TMV)
Envelope protein
Envelope
Capsid
Enzyme
RNA
(c) Human immunodeficiency virus (HIV)

Figure 13.10 The structure of bacterial, plant, and animal viruses.

(*a*) Bacterial viruses, called bacteriophages, often have a complex structure. (*b*) TMV infects plants and consists of 2,130 identical protein molecules (*purple*) that form a cylindrical coat around the single strand of RNA (*green*). The RNA backbone determines the shape of the virus and is protected by the identical protein molecules packed tightly around it. (*c*) In the human immunodeficiency virus (HIV), the RNA core is held within a capsid that is encased by a protein envelope.

13.8 How Bacteriophages Enter Cells

Bacteriophages are viruses that infect bacteria. They are diverse both structurally and functionally and are united solely by their occurrence in bacterial hosts. Double-stranded DNA bacteriophages have played a key role in molecular biology. Many of these bacteriophages are large and complex, with relatively large amounts of DNA and proteins. Some of them have been named as members of a "T" series (T1, T2, and so forth); others have been given different kinds of names. To illustrate the diversity of these viruses, T3 and T7 bacteriophages are icosahedral and have short tails. In contrast, the so-called T-even bacteriophages (T2, T4, and T6) have an icosahedral head, a capsid that consists primarily of three proteins, a connecting neck with a collar and long "whiskers," a long tail, and a complex base plate (figure 13.11).

The Lytic Cycle

During the process of bacterial infection by a T4 bacteriophage, at least one of the tail fibers—they are normally held near the bacteriophage head by the "whiskers"—contacts the lipoproteins of the host bacterial cell wall. The other tail fibers set the bacteriophage perpendicular to the surface of the bacterium and bring the base plate into contact with the cell surface. The tail contracts, and the tail tube passes through an opening that appears in the base plate, piercing the bacterial cell wall. The contents of the head, mostly DNA, are then injected into the host cytoplasm.

The T-series bacteriophages are all virulent viruses, multiplying within infected cells and eventually lysing (rupturing) them. When a virus kills the infected host cell in which it is replicating, the reproductive cycle is referred to as a *lytic cycle* (figure 13.12).

The Lysogenic Cycle

Many bacteriophages do not immediately kill the cells they infect, instead integrating their nucleic acid into the genome of the infected host cell. While residing there, it is called a **prophage.** Among the bacteriophages that do this is the lambda (λ) phage of *Escherichia coli.* We know as much about this bacteriophage as we do about virtually any other biological particle; the complete sequence of its 48,502 bases has been determined. At least 23 proteins are associated with the development and maturation of lambda phage, and many other enzymes are involved in the integration of these viruses into the host genome.

The integration of a virus into a cellular genome is called lysogeny. At a later time, the prophage may exit the genome and initiate virus replication. This sort of reproductive cycle, involving a period of genome integration, is called a *lysogenic cycle.* Viruses that become stably integrated within the genome of their host cells are called lysogenic viruses or temperate viruses.

Lysogenic Conversion

During the integrated portion of a lysogenic reproductive cycle, virus genes are often expressed. RNA polymerase reads them just as if they were host genes, and sometimes expression of these genes has an important effect on the host cell, a process called *lysogenic conversion.* In a more general sense, the genetic alteration of a cell's genome by the introduction of naked DNA is called transformation.

Transforming the Cholera-Causing Bacterium

An important example of this sort of cell transformation directed by viral genes is provided by the bacterium responsible for an often-fatal human disease. The disease-causing

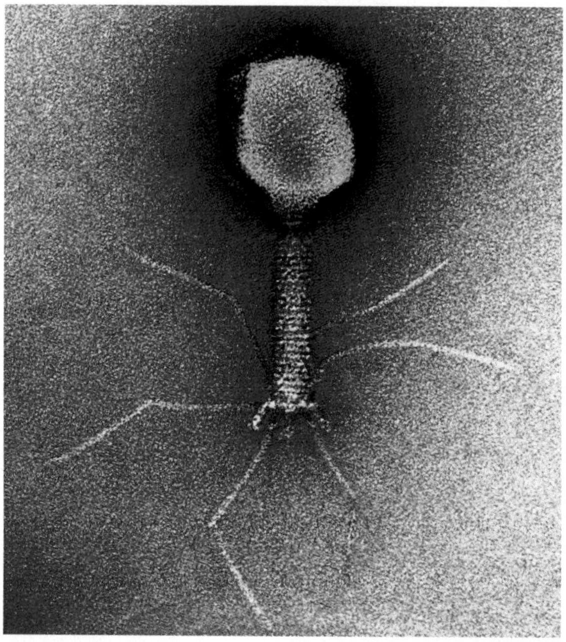

(a)

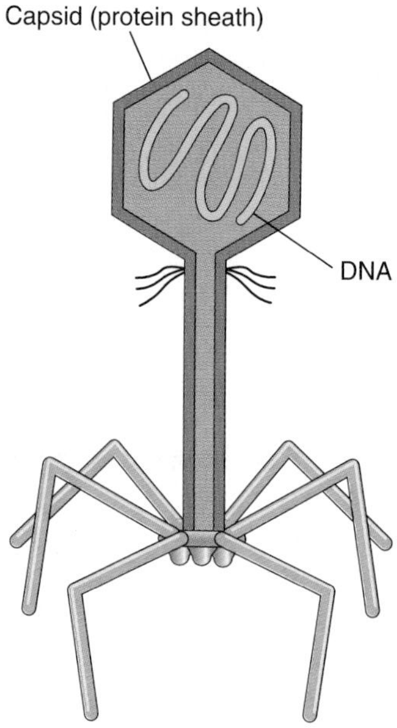
Capsid (protein sheath)
DNA

(b)

Figure 13.11 A bacterial virus.
(*a*) Electron micrograph and (*b*) diagram of the structure of a T4 bacteriophage.

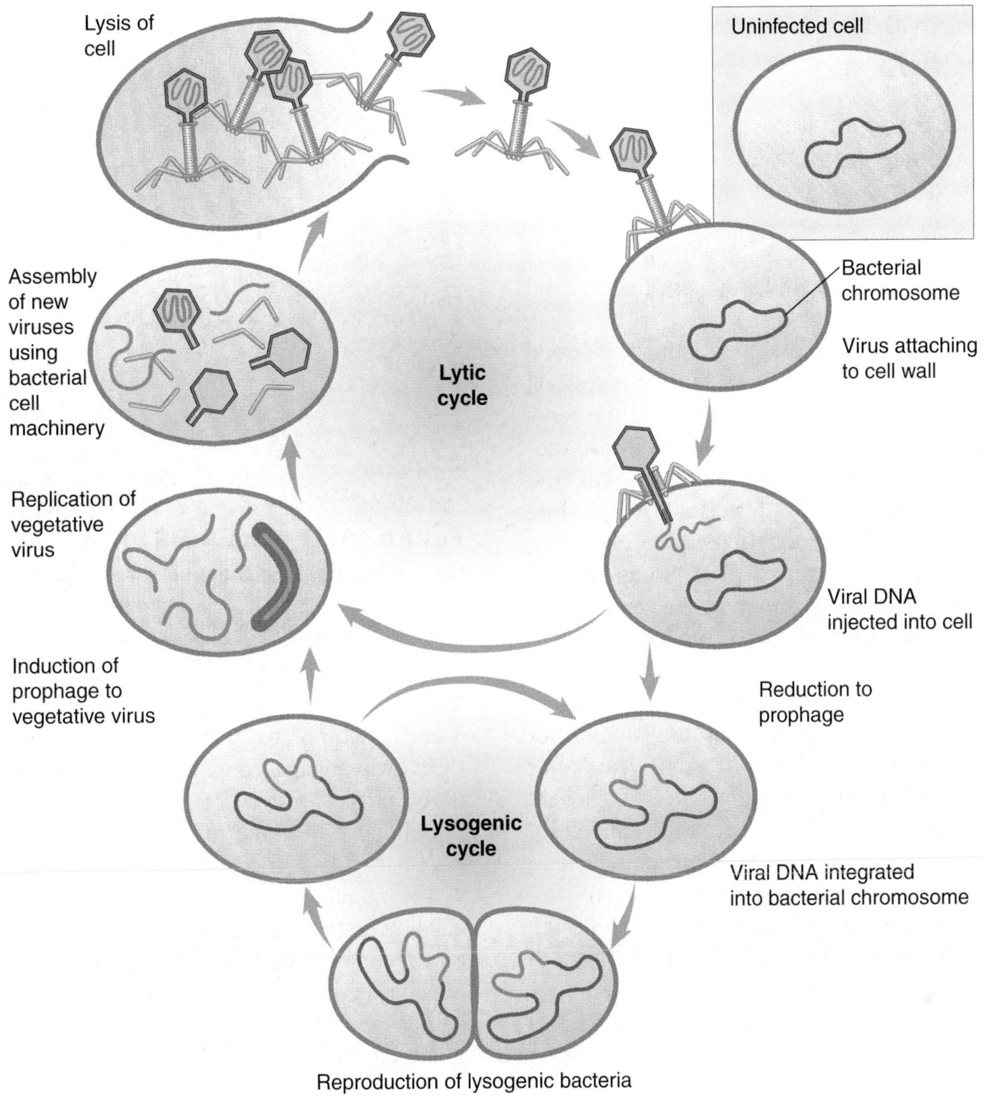
Lysis of cell
Uninfected cell
Assembly of new viruses using bacterial cell machinery
Bacterial chromosome
Virus attaching to cell wall
Lytic cycle
Replication of vegetative virus
Viral DNA injected into cell
Induction of prophage to vegetative virus
Reduction to prophage
Lysogenic cycle
Viral DNA integrated into bacterial chromosome
Reproduction of lysogenic bacteria

Figure 13.12 Lytic and lysogenic cycles of a bacteriophage.

In the lytic cycle, the bacteriophage exists as viral DNA, free in the bacterial host cell's cytoplasm; the viral DNA directs the production of new viral particles by the host cell until the virus kills the cell by lysis. In the lysogenic cycle, the bacteriophage DNA is integrated into the large, circular DNA molecule of the host bacterium and reproduces. It may continue to replicate and produce lysogenic bacteria or enter the lytic cycle and kill the cell. Bacteriophages are much smaller relative to their hosts than illustrated in this diagram.

bacteria *Vibrio cholerae* usually exists in a harmless form, but a second disease-causing, virulent form also occurs. In this latter form, the bacterium causes the deadly disease cholera. Research now shows that a bacteriophage that infects *V. cholerae* introduces into the host bacterial cell a gene that codes for the cholera toxin. This gene becomes incorporated into the bacterial DNA, where it is translated along with the other host genes, thereby transforming the benign bacterium to a disease-causing agent.

Lysogenic conversion is also responsible for the presence of toxin genes in (and much of the virulence of) other pathogens like *C. diphtheria*, which causes diphtheria, *S. pyogenes*, which causes scarlet fever, and *C. botulinum*, which causes botulism.

13.8 Bacteriophages are a diverse group of viruses that attack bacteria. Some kill their host in a lytic cycle; others integrate into the host's genome, initiating a lysogenic cycle. Bacteriophages transform *Vibrio cholerae* bacteria into disease-causing agents.

13.9 How Animal Viruses Enter Cells

As we just discussed, bacterial viruses punch a hole in the bacterial cell wall and inject their DNA inside. Plant viruses like TMV enter plant cells through tiny rips in the cell wall at points of injury. Animal viruses enter their host cells by endocytosis, a process described in chapter 4. In effect, the cell's plasma membrane dimples inward, surrounding and engulfing the virus particle.

A diverse array of viruses occur among animals. A good way to gain a general idea of how they enter cells is to look at one animal virus in detail. Here we will look at the virus responsible for a comparatively new and fatal disease, acquired immunodeficiency syndrome (AIDS). AIDS was first reported in the United States in 1981. It was not long before the infectious agent, human immunodeficiency virus (HIV), was identified by laboratories (figure 13.13). Study of HIV revealed it to be closely related to a chimpanzee virus, suggesting that it might have been introduced to humans in central Africa from chimpanzees.

One of the cruelest aspects of AIDS is that clinical symptoms typically do not begin to develop until long after infection by the HIV virus, generally eight to 10 years after the initial exposure to HIV. During this long interval, carriers of HIV have no clinical symptoms but are apparently fully infectious, which makes the spread of HIV very difficult to control. The reason why HIV remains hidden for so long seems to be that its infection cycle continues throughout the eight- to 10-year latent period without doing serious harm to the infected person. Then, events transpire that allow the virus to quickly overcome the immune defense, starting AIDS.

Attachment

When HIV is introduced into the human bloodstream, the virus particle circulates throughout the entire body but will only infect certain cells, ones called macrophages (Latin, big eaters). Macrophages are the garbage collectors of the body, taking up and recycling fragments of ruptured cells and other bits of organic debris. That HIV specializes in this one kind of cell is not surprising—most other animal viruses are similarly narrow in their requirements. Poliovirus infects only certain spinal nerves, hepatitis virus infects only liver cells, and rabies virus only the brain.

How does a virus such as HIV recognize a specific kind of target cell such as a macrophage? Every kind of cell in the human body has a specific array of cell surface markers that serve to identify them. HIV viruses are able to recognize the macrophage cell surface marker. Studding the surface of each HIV virus are spikes that bang into any cell the virus encounters. Each spike is composed of a protein called **gp120.** Only when gp120 happens onto a cell surface marker that matches its shape does the HIV virus adhere to an animal cell and infect it. It turns out that gp120 precisely fits a cell surface marker called **CD4,** and that CD4 occurs on the surfaces of macrophages (figure 13.14).

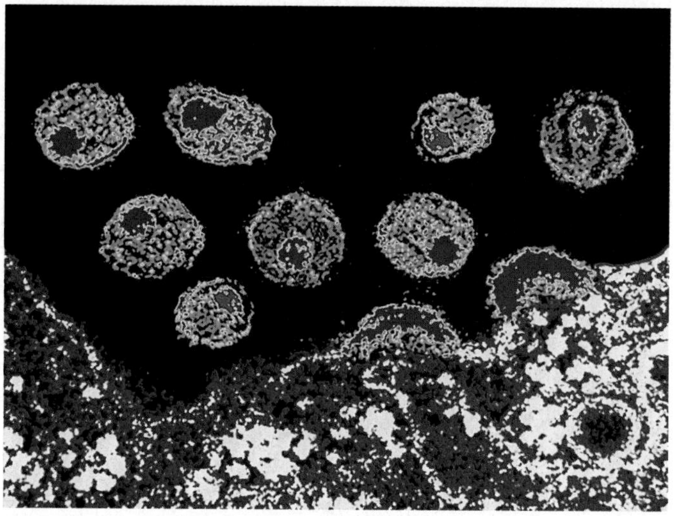

Figure 13.13 The AIDS virus.
HIV particles exit a cell. Soon, they will spread and infect neighboring cells.

Entry into Macrophages

Certain cells of the immune system, called T lymphocytes, or T cells, also possess CD4 markers. Why are they not infected right away, as macrophages are? This is the key question underlying the mystery of the long AIDS latent period. When lymphocytes become infected and killed, AIDS commences. So what holds off lymphocyte infection so long?

Researchers have learned that after docking onto the CD4 receptor of a macrophage, the HIV virus requires a second receptor protein, called **CCR5,** to pull itself across the plasma membrane. After gp120 binds to CD4, its shape becomes twisted (a chemist would say it goes through a conformational change) into a new form that fits the CCR5 coreceptor molecule. Investigators speculate that after the conformational change, the coreceptor CCR5 passes the gp120-CD4 complex through the plasma membrane by triggering endocytosis.

Replication

Once inside the macrophage cell, the HIV virus particle sheds its protective coat. This leaves the virus nucleic acid (RNA in this case) floating in the cell's cytoplasm, along with a virus enzyme that was also within the virus shell. This enzyme, called **reverse transcriptase,** binds to the tip of the virus RNA and slides down it, synthesizing DNA that matches the information contained in the virus RNA. This process translates the RNA language of the virus's genes into the DNA language of the cell, so it can be used by the cell's machinery to direct the production of new viruses.

Importantly, the HIV reverse transcriptase enzyme doesn't do its job very accurately. It often makes mistakes in reading the HIV RNA, and so creates many new mutations. The mistake-ridden double-stranded DNA that it produces then takes over the host cell's machinery, directing it to produce many copies of the virus.

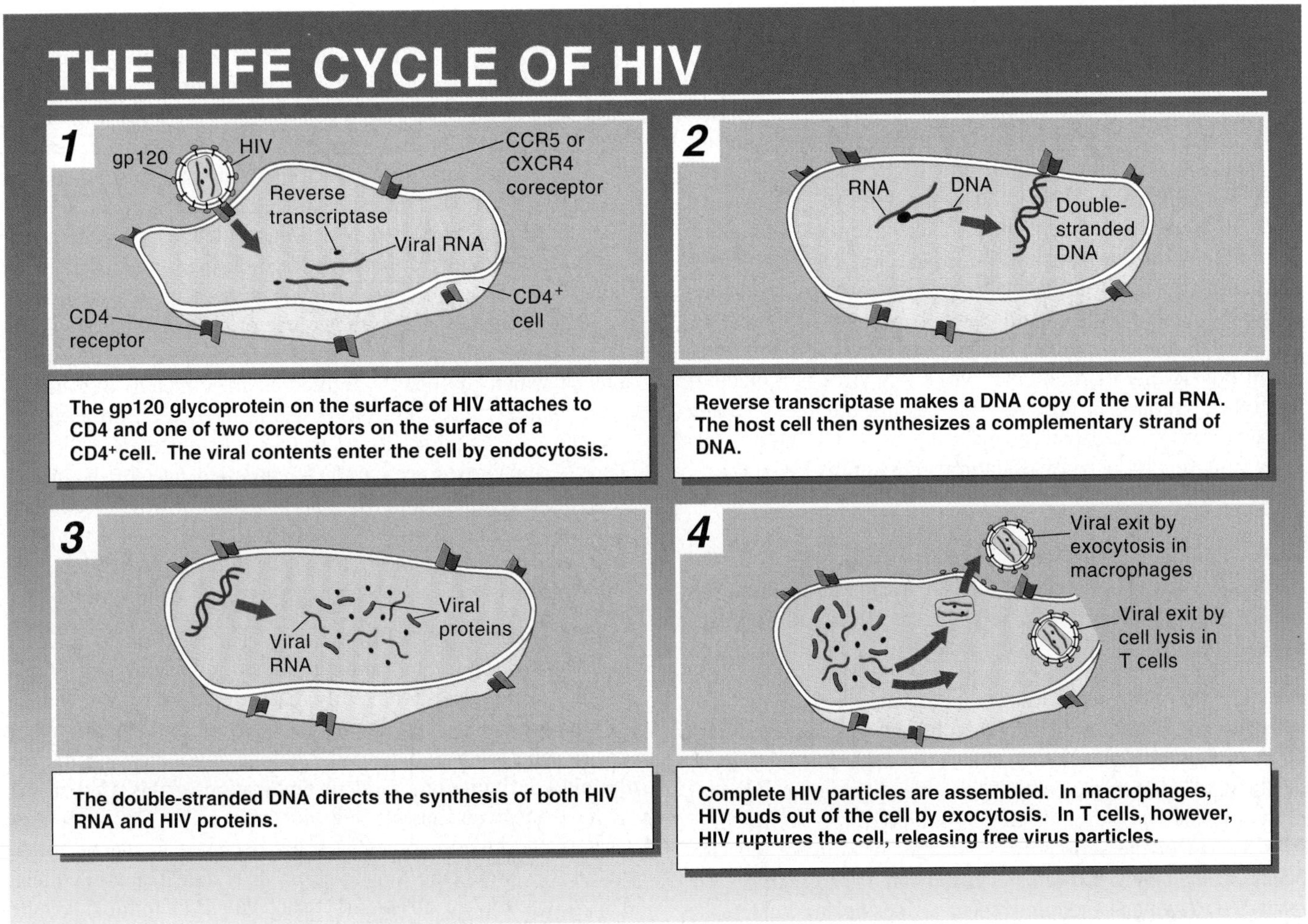
THE LIFE CYCLE OF HIV
1
gp120
HIV
CCR5 or CXCR4 coreceptor
Reverse transcriptase
Viral RNA
CD4 receptor
$CD4^+$ cell
The gp120 glycoprotein on the surface of HIV attaches to CD4 and one of two coreceptors on the surface of a $CD4^+$ cell. The viral contents enter the cell by endocytosis.
2
RNA
DNA
Double-stranded DNA
Reverse transcriptase makes a DNA copy of the viral RNA. The host cell then synthesizes a complementary strand of DNA.
3
Viral RNA
Viral proteins
The double-stranded DNA directs the synthesis of both HIV RNA and HIV proteins.
4
Viral exit by exocytosis in macrophages
Viral exit by cell lysis in T cells
Complete HIV particles are assembled. In macrophages, HIV buds out of the cell by exocytosis. In T cells, however, HIV ruptures the cell, releasing free virus particles.

Figure 13.14 How the HIV infection cycle works.

In all of this process, no lasting damage is done to the host cell. HIV does not rupture and kill the macrophage cells it infects. Instead, the new viruses are released from the cell by exocytosis, folding out in much the same way that HIV initially gained entry into the cell at the start of the infection.

This, then, is the basis of the long latency period characteristic of AIDS. The HIV virus cycles through macrophages over a period of years, multiplying powerfully but doing little apparent damage to the body.

Starting AIDS: Entry into T Cells

During this long latent period, HIV is constantly replicating and mutating as it cycles through successive generations of macrophages. Eventually, by chance, HIV alters the gene for gp120 in a way that causes the gp120 protein to change its coreceptor allegiance. This new form of gp120 protein prefers to bind instead to a different coreceptor, **CXCR4,** a receptor that occurs on the surface of $CD4^+$ T cells (T cells that have the CD4 cell surface marker). Soon the body's T cells become infected with HIV.

This has deadly consequences, as new viruses exit T cells not by harmless exocytosis but by bursting through the plasma membrane. This rupturing destroys the T cell's physical integrity and kills it. As the released viruses infect nearby $CD4^+$ T cells, they in turn are ruptured, in a widening circle of cell death. Soon, the shift to the CXCR4 second receptor produces a steep drop in the number of living T cells. It is this destruction of the body's T cells that blocks the body's immune response and leads directly to the onset of AIDS. Cancers and opportunistic infections are free to invade the defenseless body. Identification of a shift in coreceptor allegiance as the key event triggering AIDS has excited researchers. Any therapy that blocks the CXCR4 coreceptor might prevent the development of full-blown AIDS in HIV-infected individuals.

13.9 Animal viruses enter cells using specific receptor proteins to cross the plasma membrane.

13.10 Disease Viruses

Humans have known and feared diseases caused by viruses for thousands of years. Among the diseases that viruses cause (table 13.4) are influenza, smallpox, infectious hepatitis, yellow fever, polio, rabies, and AIDS, as well as many other diseases not as well known. In addition, viruses have been implicated in some cancers and leukemias. For many autoimmune diseases, such as multiple sclerosis and rheumatoid arthritis, and for diabetes, specific viruses have been found associated with certain cases. In view of their effects, it is easy to see why the late Sir Peter Medawar, Nobel laureate in Physiology or Medicine, wrote, "A virus is a piece of bad news wrapped in protein." Viruses not only cause many human diseases but also cause major losses in agriculture, forestry, and the productivity of natural ecosystems.

Influenza

Perhaps the most lethal virus in human history has been the influenza virus. Some 21 million Americans and Europeans died of flu within 18 months in 1918 and 1919—an astonishing number.

Types. Flu viruses are RNA animal viruses. An individual flu virus resembles a rod studded with spikes composed of two kinds of protein (figure 13.15). There are three general "types" of flu virus, distinguished by their capsid (inner membrane) protein, which is different for each type. Type A flu virus causes most of the serious flu epidemics in humans and also occurs in mammals and birds. Type B and Type C viruses are restricted to humans and rarely cause serious health problems.

Subtypes. Different strains of flu virus, called subtypes, differ in their protein spikes. One of these proteins, **hemagglutinin (H),** aids the virus in gaining access to the cell interior. The other, **neuraminidase (N),** helps the daughter virus break free of the host cell once virus replication has been completed. The structures of both the H and N molecules are known in detail. The H molecule, for example, is made of three parts and stands on the surface of the virus somewhat like a tripod, with clublike projections on top. Each "leg" of the H molecule contains "hot spots" that display an unusual tendency to change as a result of mutation of the virus RNA during imprecise replication. Point mutations cause changes in these spike proteins in 1 of 100,000 viruses during the course of each generation. These very variable segments of the H molecule function as targets against which the body's antibodies are directed. Because of accumulating changes in the H and N molecules, different flu vaccines are required to protect against different subtypes. Type A flu viruses are currently classified into 13 distinct H subtypes and nine distinct N subtypes, each of which requires a different vaccine to protect against infection. Thus the type A virus that caused the Hong Kong flu epidemic of 1968 has type 3 H molecules and type 2 N molecules and is called A(H3N2).

Importance of Recombination. The problem in combating flu viruses arises not through mutation but through recombination. Viral genes are readily reassorted by genetic recombination, sometimes putting together novel combinations of H and N spikes unrecognizable by human antibodies specific for the old configuration. Viral recombination of this kind seems to have been responsible for the three major flu pandemics (that is, worldwide epidemics) that occurred in the twentieth century by producing drastic shifts in H and N combinations. The "killer flu" of 1918, A(H1N1), killed 21 million people. The Asian flu of 1957, A(H2N2), killed over 100,000 Americans. The Hong Kong flu of 1968, A(H3N2), infected 50 million people in the United States alone, of which 70,000 died.

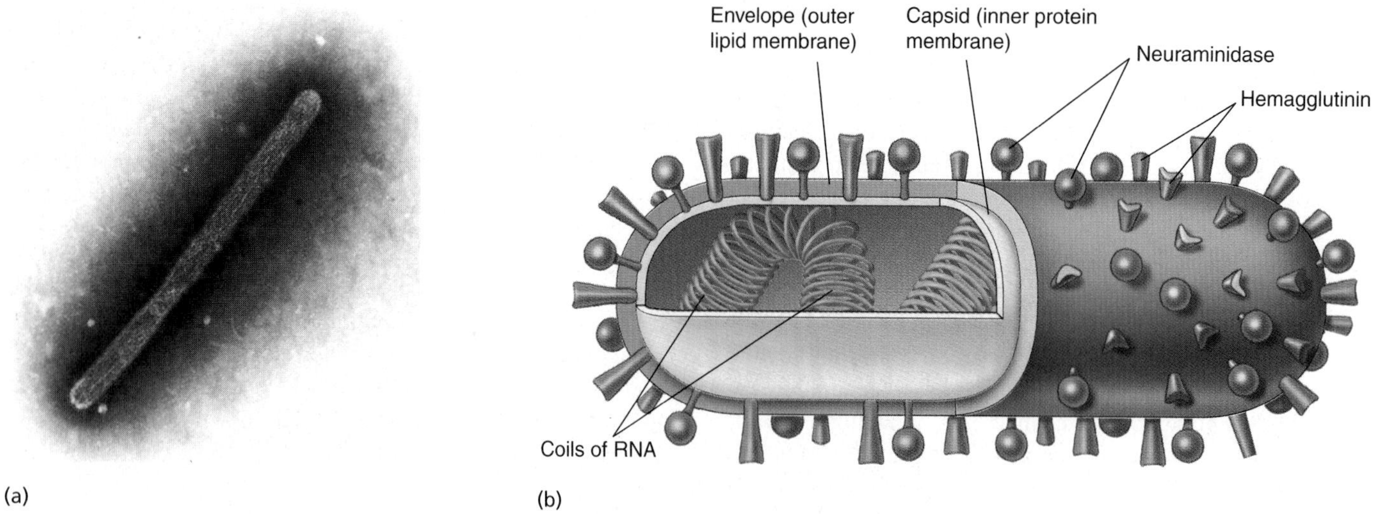
Envelope (outer lipid membrane)
Capsid (inner protein membrane)
Neuraminidase
Hemagglutinin
Coils of RNA
(a)
(b)

Figure 13.15 The influenza virus.

(*a*) TEM of the so-called "bird flu" influenza virus, A(H5N1), which first infected humans in Hong Kong in 1997. (*b*) Diagram of an influenza virus. The coiled RNA has been revealed by cutting through the outer lipid-rich envelope, with its two kinds of projecting spikes, and the inner protein capsid.

TABLE 13.4 IMPORTANT HUMAN VIRAL DISEASES

Disease	Pathogen	Vector/ Reservoir	Epidemiology
AIDS	HIV 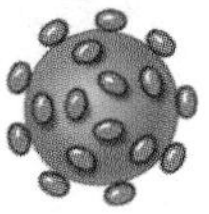	Humans	Destroys immune defenses, resulting in death by infection or cancer. Fourteen million infected worldwide by 1999.
Chicken pox	Human herpes-virus 3 (varicella-zoster)	Humans	Spread through contact with infected individuals. No cure. Rarely fatal. Vaccine approved in United States in early 1995.
Ebola	Filoviruses 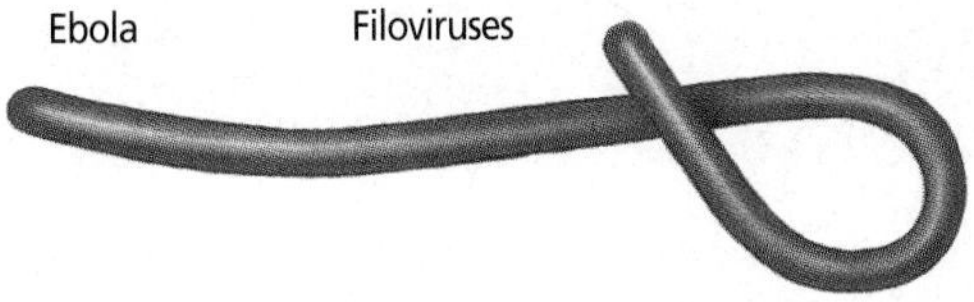	Unknown	Acute hemorrhagic fever; virus attacks connective tissue, leading to massive hemorrhaging and death. Peak mortality is 50% to 90% if the disease goes untreated. Outbreaks confined to local regions of central Africa.
Hepatitis B (viral)	Hepatitis B virus (HBV)	Humans	Highly infectious through contact with infected body fluids. Approximately 1% of U.S. population infected. Vaccine available, no cure. Can be fatal.
Herpes	Herpes simplex virus (HSV) 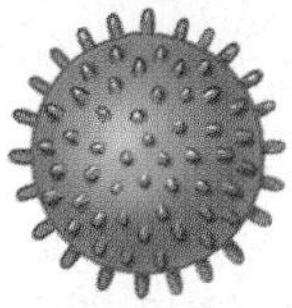	Humans	Fever blisters; spread primarily through contact with infected saliva. Very prevalent worldwide. No cure. Exhibits latency—the disease can be dormant for several years.
Influenza	Influenza viruses 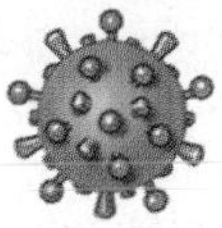	Humans, ducks	Historically a major killer (22 million died in 18 months in 1918–19); wild Asian ducks, chickens, and pigs are major reservoirs. The ducks are not affected by the flu virus, which shuffles its antigen genes while multiplying within them, leading to new flu strains.
Measles	Paramyxoviruses	Humans	Extremely contagious through contact with infected individuals. Vaccine available. Usually contracted in childhood, when it is not serious; more dangerous to adults.
Mononucleosis	Epstein-Barr virus (EBV) 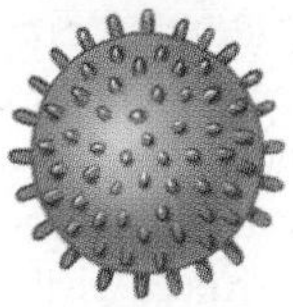	Humans	Spread through contact with infected saliva. May last several weeks; common in young adults. No cure. Rarely fatal.
Mumps	Paramyxovirus 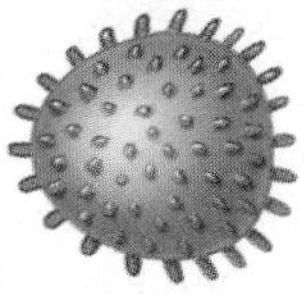	Humans	Spread through contact with infected saliva. Vaccine available; rarely fatal. No cure.
Pneumonia	Influenza virus	Humans	Acute infection of the lungs, often fatal without treatment.
Polio	Poliovirus 	Humans	Acute viral infection of the central nervous system that can lead to paralysis and is often fatal. Prior to the development of Salk's vaccine in 1954, 60,000 people a year contracted the disease in the United States alone.
Rabies	Rhabdovirus 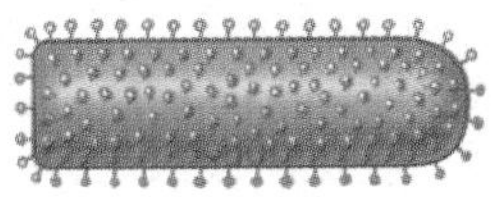	Wild and domestic Canidae (dogs, foxes, wolves, coyotes, etc.)	An acute viral encephalomyelitis transmitted by the bite of an infected animal. Fatal if untreated.
Smallpox	Variola virus	Formerly humans, now thought to exist only in government labs.	Historically a major killer, the last recorded case of smallpox was in 1977. A worldwide vaccination campaign wiped out the disease completely. There is current debate as to whether the virus has been removed from Russian government labs to countries that might make it available to terrorists.
Yellow fever	Flavivirus 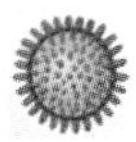	Humans, mosquitoes	Spread from individual to individual by mosquito bites; a notable cause of death during the construction of the Panama Canal. If untreated, this disease has a peak mortality rate of 60%.

(a) (b)

Figure 13.16 The Ebola virus.

(*a*) A virus with a fatality rate that can exceed 90% in humans has reappeared in West Africa after a 20-year hiatus. Health professionals are scrambling to identify natural hosts of the virus and to devise strategies to combat transmission of the disease. (*b*) Two children wait in May 1995 outside the hospital in the Zaire town of Kikwit, where those infected with Ebola were isolated. Seventy-eight percent of the infected died.

Smallpox as a Bioterrorist Threat

One of the greatest triumphs of modern medicine has been the eradication of the highly infectious disease smallpox everywhere in the world. This dreaded disease, caused by inhaling the variola virus, has killed many millions of people. The characteristic fever and skin rash symptoms do not appear for some 12 days, during the last two of which an infected person can infect many others, leading to pandemics. One in three infected people die. When first introduced to the New World by Cortez's troops, smallpox eliminated half or more of the total native population. There were some 15 million cases of smallpox worldwide in 1967. Then an intensive worldwide campaign began to eliminate it. There are no known hosts of the variola virus other than human beings, so that when all the people who were susceptible in areas where smallpox still occurred were innoculated in the 1970s, the disease was eliminated completely. Immunizations for smallpox were discontinued in the United States in 1972.

Unfortunately, the smallpox virus had not been eliminated. It still existed in two high-security laboratories in Atlanta and Moscow. Both the United States and Russia had extensive bioweapons programs in the 1960s. In 1972 the United States and Russia joined 102 other nations in signing a treaty renouncing biological weapons. Unfortunately, the Russians saw the treaty as a strategic opportunity and undertook a program to "weaponize" smallpox. The Russian bioweapons programs were disbanned when the Soviet Union broke up, but many arms control experts fear that Russian bioweapons researchers who took jobs with Iraq and North Korea may have taken smallpox cultures with them. In light of the possibility that samples of the virus might fall into terrorist hands, the United States in 2001 ordered production of 300 million doses of smallpox vaccine, enough to vaccine all Americans.

Emerging Viruses

Sometimes viruses that originate in one organism pass to another, causing a disease in the new host. HIV, for example, arose in chimpanzees and relatively recently passed to humans. Influenza is fundamentally a bird virus. New pathogens arising in this way, called **emerging viruses,** represent a considerable threat in an age when airplane travel potentially allows infected individuals to move about the world quickly, spreading an infection.

Among the most lethal of emerging viruses are a collection of filamentous viruses arising in central Africa that attack human connective tissue. With lethality rates in excess of 50%, these so-called filoviruses are among the most lethal infectious diseases known. One, Ebola virus (figure 13.16), has exhibited lethality rates in excess of 90% in isolated outbreaks in central Africa. The outbreak of Ebola virus in the summer of 1995 in Zaire killed 245 people out of 316 infected—a mortality rate of 78%. A more recent outbreak occurred in Gabon, West Africa, in February 1996. The natural host of Ebola is unknown.

Often carried by monkeys, emerging viruses represent a real potential threat to worldwide human health if they ever achieve widespread dissemination.

13.10 Viruses are responsible for some of the most lethal diseases of humans. Some of the most serious examples are viruses that have transferred to humans from some other host. Influenza, a bird virus, has been responsible for the most devastating epidemics in human history. Newly emerging viruses such as Ebola have received considerable public attention.

13.11 Prions and Viroids

For decades scientists have been fascinated by a peculiar group of fatal brain diseases. These diseases have the unusual property that they are transmissible from one individual to another, but it is years and often decades before the disease is detected in infected individuals. The brains of infected individuals develop numerous small cavities as neurons die, producing a marked spongy appearance. Called **transmissible spongiform encephalopathies (TSEs),** these diseases include scrapie in sheep, "mad cow" disease in cattle, and kuru and Creutzfeldt-Jakob disease in humans.

TSEs can be transmitted between individuals of a species by injecting infected brain tissue into a recipient animal's brain. TSEs can also spread via tissue transplants and, apparently, food. Kuru was common in the Fore people of Papua New Guinea when they practiced ritual cannibalism, literally eating the brains of infected individuals. Mad cow disease spread widely among the cattle herds of England in the 1990s because cows were fed bone meal prepared from cattle carcasses to increase the protein content of their diet. Like the Fore, the British cattle were literally eating the tissue of cattle that had died of the disease.

A Heretical Suggestion

In the 1960s, British researchers T. Alper and J. Griffith noted that infectious TSE preparations remained infectious even after exposed to radiation that would destroy DNA or RNA. They suggested that the infectious agent was a protein. Perhaps, they speculated, the protein usually preferred one folding pattern but could sometimes misfold and then catalyze other proteins to do the same, the misfolding spreading like a chain reaction. This heretical suggestion was not accepted by the scientific community, as it violates a key tenant of molecular biology: only DNA or RNA act as hereditary material, transmitting information from one generation to the next.

Prusiner's Prions

In the early 1970s, physician Stanley Prusiner, moved by the death of a patient from Creutzfeldt-Jakob disease, began to study TSEs. Prusiner became fascinated with Alper and Griffith's hypothesis. Try as he might, Prusiner could find no evidence of nucleic acids or viruses in the infectious TSE preparation. He concluded, as Alper and Griffith had, that the infectious agent was a *protein,* which in a 1982 paper he named a **prion,** for "proteinaceous infectious particle."

Prusiner went on to isolate a distinctive prion protein, and for two decades he continued to amass evidence that prions play a key role in triggering TSEs. The scientific community resisted Prusiner's renegade conclusions, but eventually experiments done in Prusiner's and other laboratories began to convince many. For example, when Prusiner injected prions of different abnormal conformations into several different hosts, these hosts developed prions with the same abnormal conformations as the parent prions (figure 13.17). In another important experiment, Charles Weissmann showed that mice genetically engineered to lack Prusiner's prion protein are immune to TSE infection. However, if brain tissue with the prion protein is grafted into the mice, the grafted tissue—but not the rest of the brain—can then be infected with TSE. In 1997, Prusiner was awarded the Nobel Prize in Physiology or Medicine for his work on prions.

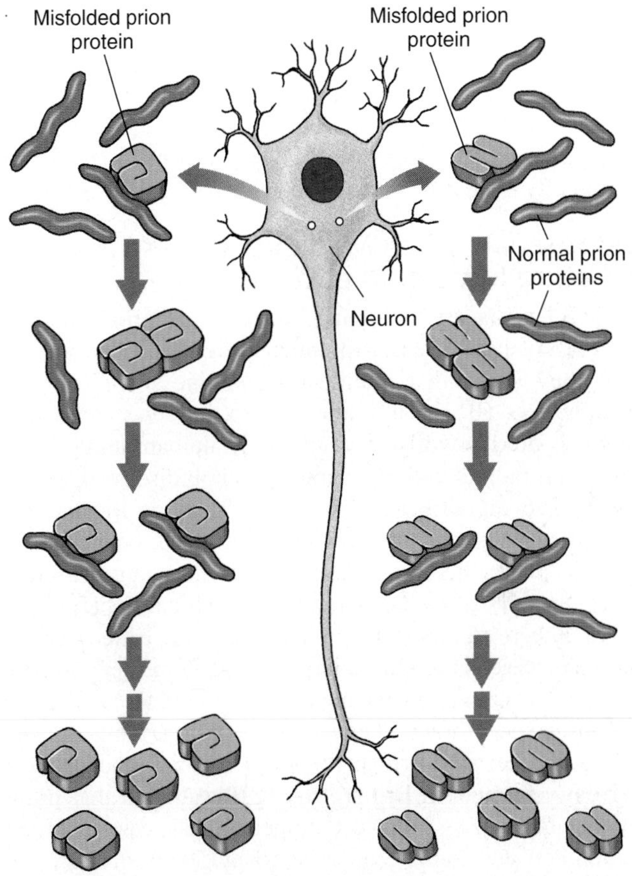
Misfolded prion protein
Misfolded prion protein
Normal prion proteins
Neuron

Figure 13.17 How prions may be maintained. Prions are normal membrane proteins of neurons that serve a critical (though unknown) function. Misfolded prions may cause normal prion protein to also misfold simply by contacting them. When prions misfold in different ways (*blue*) and contact normal prion protein (*purple*), the normal prion protein misfolds in the same way.

Is Prusiner Right?

Despite Prusiner's Nobel Prize, not all scientists are convinced that he is right. So far no one has been able to inject a prion protein synthesized in a test tube (and thus sure to be free of any contaminating virus or nucleic acid) into a healthy animal and make it sick with a TSE. Perhaps the best evidence that prion proteins are the agents of TSE comes from yeast, where a prion-like change has been shown unequivocally to be transmitted by a purified protein.

Can Humans Catch Mad Cow Disease by Eating Infected Meat?

Many scientists are becoming worried that prions may be transmitting such diseases to humans (figure 13.18). Specifically, they worry that prion-caused bovine spongiform encephalopathy (BSE), a brain disease in cows commonly known as **mad cow disease,** may infect humans and produce the similar fatal disorder Creutzfeldt-Jakob disease (CJD). In March 1996, an outbreak of mad cow disease in Britain, with many thousands of cattle apparently affected, created widespread concern. BSE, a degeneration of the brain caused by prions, appears to have entered the British cattle herds from sheep! Sheep are subject to a prion disease called scrapie, and the disease is thought to have passed from sheep to cows through protein-supplemented feed pellets containing ground-up sheep brains. The passage of prions from one species to another has British scientists worried: the death of four dairy farmers in Britain from CJD suggests that prions may be able to pass from cows to people! The case for a connection between eating British beef and CJD is strongly supported by the finding that tissue from the brains of the dead farmers and from BSE cows induces the same brain lesions in mice, while classic CJD produces quite different lesions—clearly their form of CJD was caused by the same agent that caused BSE. Thus, there appears to be legitimate cause for caution. Because the incubation period for CJD can vary from 15 to 45 years, the number of people infected by eating BSE-contaminated meat in Great Britain may not become apparent for some time.

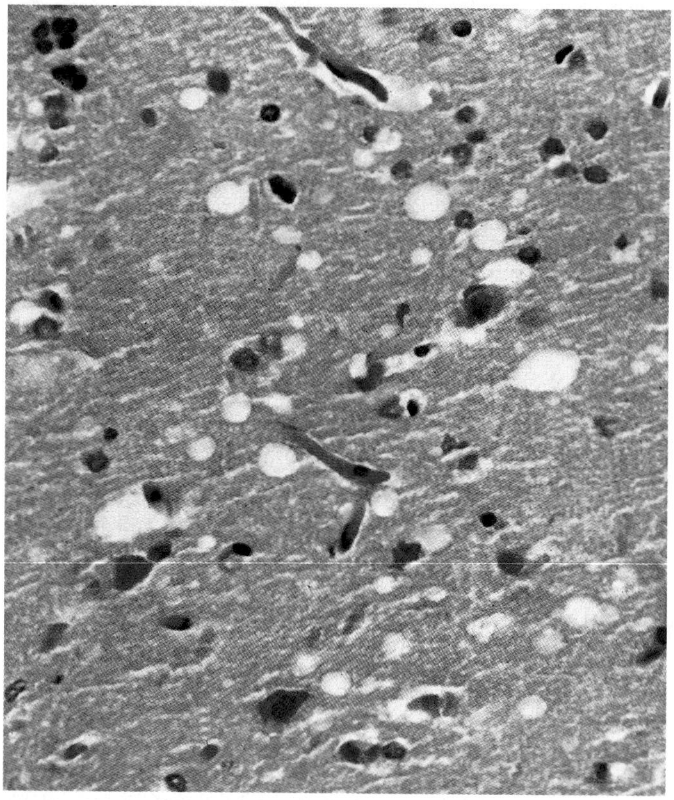

Figure 13.18 Damaged brain tissue.

The lesions or holes in this brain tissue are the result of a TSE called Creutzfeldt-Jakob disease (CJD) that affects humans. Prions play a key role in causing this fatal deterioration of brain tissue.

Viroids

Viroids are tiny, naked molecules of RNA, only a few hundred nucleotides long, that are important infectious disease agents in plants. A recent viroid outbreak killed over 10 million coconut palms in the Philippines. It is not clear how viroids cause disease. One clue is that viroid nucleotide sequences resemble the sequences of introns within ribosomal RNA genes. These sequences are capable of catalyzing excision from DNA—perhaps the viroids are catalyzing the destruction of chromosomal integrity.

13.11 Prions are infectious proteins that many scientists—but not all—believe are responsible for serious brain diseases. In plants, naked RNA molecules called viroids can also transmit disease.

CONCEPT REVIEW

1. Bacteria have been on the earth for over _________ billion years, while eukaryotes did not appear in the fossil record until _________ billion years ago.
   a. 4
   b. 3.5
   c. 2.5
   d. 1.5

2. Prokaryotes, or bacteria, lack
   a. DNA.
   b. cells.
   c. an organized nucleus.
   d. cytoplasm.

3. Cell division in bacteria is achieved by
   a. binary fission.
   b. mitosis.
   c. meiosis.
   d. flagella.

4. Eukaryotes are thought to have evolved from the branch of bacteria called
   a. eubacteria.
   b. archaebacteria.

5. In eubacteria, the genes are not interrupted by
   a. introns.
   b. exons.
   c. restriction enzymes.
   d. "stop" codons.

6. Which three of the following diseases are caused by bacteria?
   a. tetanus
   b. AIDS
   c. tuberculosis
   d. cholera
   e. herpes

7. Stanley's evidence that TMV was chemical rather than cellular was that TMV
   a. did not contain DNA or RNA.
   b. could reproduce on its own.
   c. did not contain any protein.
   d. crystallized in solution.

8. The genetic information of viruses is encoded in a molecule of
   a. RNA.
   b. DNA.
   c. either RNA or DNA, but not both.
   d. either RNA or protein, but not both.

9. Once HIV enters the cell and copies its RNA, it undergoes a _________ period of eight to 10 years before lysing the cell or exiting the cell by exocytosis.
   a. lysogenic
   b. latency
   c. replication
   d. lytic cycle

10. _________ is the branch of bacteria that includes nearly all of the 4,800 kinds of bacteria that live today.

11. In cyanobacteria, nitrogen fixation occurs in specialized cells called _________.

12. _________ are viruses that infect bacterial cells.

CHALLENGE YOURSELF

1. In what ways might the early, self-replicating particles that gave rise to the first organisms have resembled, or differed from, viruses?

2. Justify the assertion that life on earth could not exist without bacteria.

3. Why are there no effective antibiotics against viral infections?

4. Why do we say that bacteria are alive and viruses are not?

13 eBRIDGE

Reinforcing Key Points

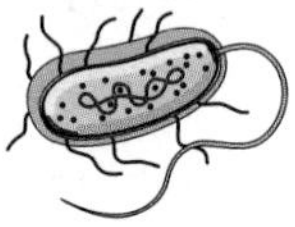

Bacteria

13.1 The Simplest Organisms

13.2 Comparing Bacteria to Eukaryotes

13.3 Importance of Bacteria

Bacterial Diversity

13.4 Kinds of Bacteria

13.5 Bacterial Lifestyles

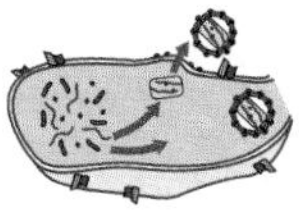

Viruses

13.6 The Discovery of Viruses

13.7 The Structure of Viruses

13.8 How Bacteriophages Enter Cells

13.9 How Animal Viruses Enter Cells

13.10 Disease Viruses

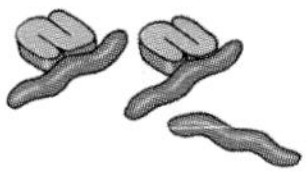

Other Nonliving Infectious Agents

13.11 Prions and Viroids

Electronic Learning

Visual Learning

Animations

- Lytic and Lysogenic Cycles in Prokaryotes
- The Life Cycle of HIV

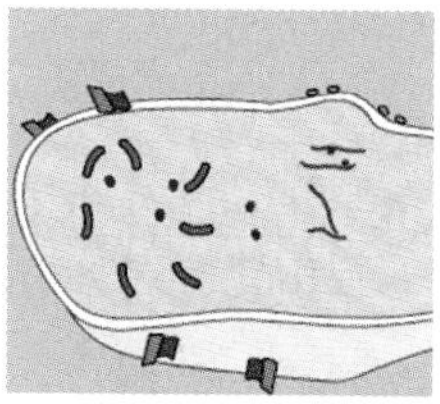

Art Labeling Activities

- Nonphotosynthetic Bacterium
- Cyanobacterium

Enhancement Chapter

Infectious Diease and Bioterrorism

The new century has seen a new and potentially deadly way for disease to spread—by the deliberate actions of terrorists. After a brief discussion of infectious disease, this enhancement chapter reviews how microbes have been "weaponized" in bio-warfare programs, and takes a close look at anthrax and smallpox. It concludes by examining the future threat of gene-modified human and crop-plant pathogens.

Author's Corner

Infectious Disease. One of the biggest impacts of biology on the human condition has been its contribution to the understanding and treatment of infectious disease. Smallpox, one of the great killers of history, remains a potential threat as a bioweapon. Other diseases stalk all of us. Even a Kentucky thoroughbred foal is not immune.

1. Smallpox: Tomorrow's nightmare?
2. Asking the hard questions about bioweapons.
3. The silent epidemic of hepatitis C.
4. Using viruses to combat the rise in antibiotic-resistant pathogens.
5. Foot-and-mouth disease is more about money than health.
6. Science as detective work: The case of the dying racehorse foals.

Virtual Classroom

The Threat of Bioweapons

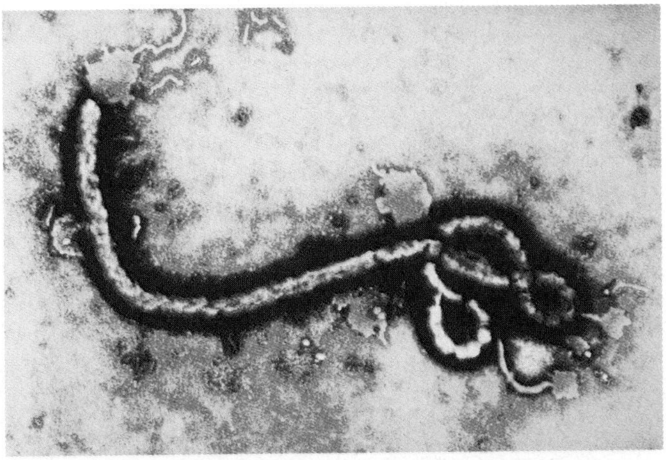

One of the unfortunate nightmares of modern biotechnology is that it makes feasible the production of biological weapons of mass destruction. While plague and most other disease-causing bacteria and viruses are impractical as weapons, two killers are easy to produce, easy to dispense, and deadly: smallpox and anthrax. Anthrax is a deadly disease of cattle caused by a bacterium that forms endospores, making anthrax ideal for aerosol dispersal. Peak mortality for humans that breathe in the anthrax endospores is 89%. The peak mortality of smallpox is 30%, but smallpox can be expected to kill far more people than anthrax, because the virus which causes smallpox is easily transmissible. Infected people will make many others sick before they even know they are infected, causing the disease to spread like a chain reaction. Bioweapons were developed by the former Soviet Union, and there is concern that, with the breakdown of the Russian scientific establishment, strains of weaponized smallpox may reach terrorist nations. It is very important that the American public become informed of this danger, so that steps can be taken to prepare for disaster, should it strike.

Virtual Lab

How *Pseudomonas* "Sugar-Coats" Itself to Cause Chronic Lung Infections

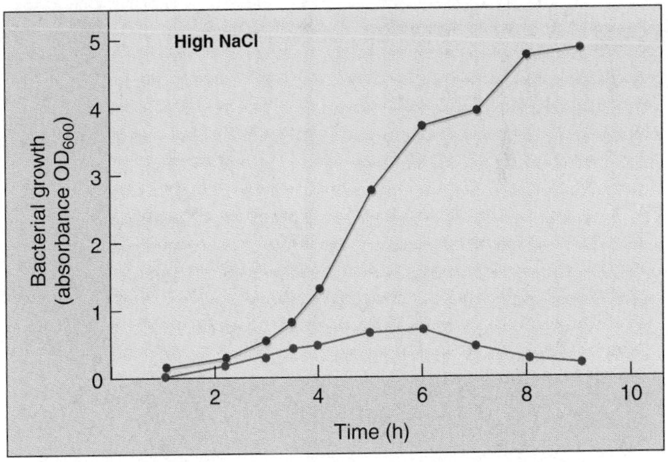
High NaCl
Bacterial growth
(absorbance OD_{600})
0
1
2
3
4
5
2
4
6
8
10
Time (h)

The bacterium *Pseudomonas aeruginosa* infects the respiratory tract of cystic fibrosis patients, and is also found outside of the human body. Bacterial colonies of *P. aeruginosa* associated with cystic fibrosis exhibit a mucoid phenotype caused by the overproduction of a protein called alginate which forms an exopolysaccharide coat around each bacterial cell. This mucoid colony morphology, however, is rarely found in environmental isolates. Several gene mutations convert *P. aeruginosa* to the mucoid phenotype, all located within the so-called *algUmucABCD* gene cluster. What do these genes do? The *mucA, mucB,* and *mucD* genes each appear to have a negative regulatory role, keeping the cell in the nonmucoid phenotype. Very little is known about the function of *mucC.* Vojo Deretic, now at the University of New Mexico, and Donald Rowen, now at the University of Nebraska, Omaha, has been studying the function of the enigmatic *mucC gene.* Some bacterial mutations are "conditional"—their effect on bacterial cells depends on their environment. His studies suggest that *mucC* is such a conditional mutation, acting—but only under adverse growth conditions—to inhibit growth.

Quizzes
Further Reading
Essential Study Partner
Links
BioCourse.com

CHAPTER 14

ADVENT OF THE EUKARYOTES

THE FAR SIDE® By GARY LARSON
© 1981 FarWorks, Inc. All Rights Reserved/Dist. by Creators Syndicate
Larson
"C'mon! Look at these fangs! Look at these claws! ... You think we're supposed to eat just honey and berries?"

CHAPTER 14

Advent of the Eukaryotes

CHAPTER OVERVIEW

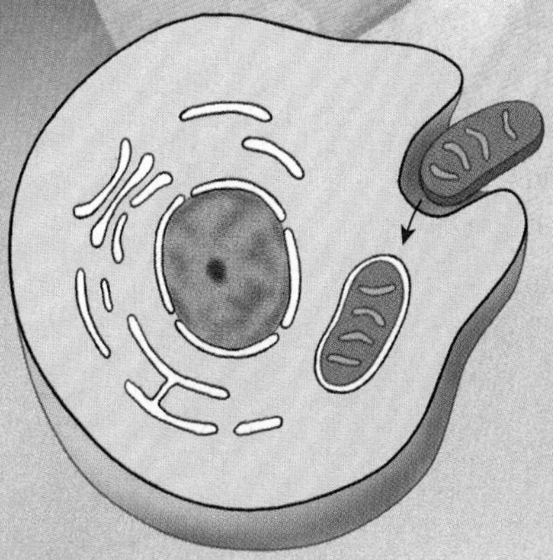

The Evolution of Eukaryotes

14.1 Endosymbiosis

14.2 The Evolution of Sex

14.3 General Biology of the Protists

- The theory of endosymbiosis, accepted by almost all biologists, proposes that mitochondria and chloroplasts were once aerobic eubacteria that were engulfed by ancestral eukaryotes.
- Reproduction without sex is the rule among protists, which typically resort to sexual reproduction only in times of stress.
- Sex appears to have first evolved as a mechanism to repair damage to DNA.
- Colonial organisms are collections of cells that are permanently associated but in which little integration of cell activities occurs.
- Aggregations are transient collections of cells.
- Multicellularity has evolved among the protists many times.
- Multicellular organisms are composed of many cells that integrate their activities.

The Protists

14.4 Classifying the Protists

14.5 Heterotrophs with No Permanent Locomotor Apparatus

14.6 Heterotrophs with Flagella

14.7 Nonmotile Spore-Formers

14.8 Photosynthetic Protists

14.9 Heterotrophs with Restricted Mobility

- Of the six kingdoms, Protista is by far the most diverse.
- Protista is a catchall kingdom containing all eukaryotes that are not fungi, plants, or animals.
- The 15 major phyla of protists can be sorted into five general groups.
- Algae are photosynthetic protists, many of which are multicellular.
- Slime molds and water molds are heterotrophs with restricted mobility.

14.1 Endosymbiosis

What was the first eukaryote like? We cannot be sure, but a good model is *Pelomyxa palustris,* a single-celled, nonphotosynthetic organism that appears to represent an early stage in the evolution of eukaryotic cells (figure 14.1). The cells of ***Pelomyxa*** are much larger than bacterial cells and contain a complex system of internal membranes. Although they resemble some of the largest early fossil eukaryotes, these cells are unlike those of any other eukaryote: *Pelomyxa* lacks mitochondria and does not undergo mitosis. Its nuclei divide somewhat as do those of bacteria, by pinching apart into two daughter nuclei, around which new membranes form. This primitive eukaryote is so distinctive that it is assigned a phylum all its own, Caryoblastea.

Biologists know very little of the origin of *Pelomyxa,* except that in many of its fundamental characteristics it resembles the archaebacteria far more than the eubacteria. Because of this general resemblance, it is widely assumed that the first eukaryotic cells were nonphotosynthetic descendants of archaebacteria.

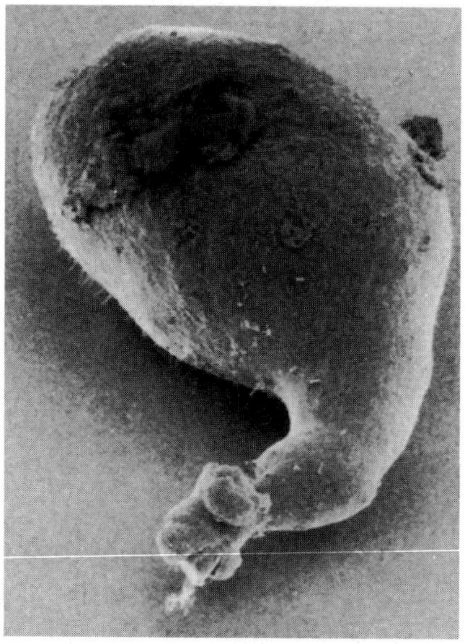

Figure 14.1 *Pelomyxa palustris.* This unique, amoebalike protist lacks mitochondria and may represent a very early stage in the evolution of eukaryotic cells.

Endosymbiosis

What about the wide gap between *Pelomyxa* and all other eukaryotes? Where did mitochondria come from? Most biologists agree with the theory of **endosymbiosis,** which proposes that mitochondria originated as symbiotic, aerobic (oxygen-requiring) eubacteria (figure 14.2). Symbiosis (Greek, *syn,* together with, and *bios,* life) means living together in close association. The aerobic eubacteria that became mitochondria are thought to have been engulfed by ancestral eukaryotic cells much like *Pelomyxa* early in the history of eukaryotes. The most similar eubacteria to mitochondria today are the nonsulfur purple bacteria, which are able to carry out oxidative metabolism (described in chapter 5). Before they had acquired these bacteria, the host cells were unable to carry out the Krebs cycle or other metabolic reactions necessary for living in an atmosphere that contained increasing amounts of oxygen. The engulfed bacteria became the interior portion of the mitochondria we see today.

Mitochondria. Sausage-shaped organelles called **mitochondria** are 1 to 3 micrometers long, about the same size as most eubacteria. Mitochondria are bounded by *two* membranes. The outer membrane is smooth and was apparently derived from the endoplasmic reticulum of the host cell. The inner membrane is folded into numerous layers, resembling the folded membranes of nonsulfur purple bacteria; embedded within this membrane are the proteins that carry out oxidative metabolism.

During the billion-and-a-half years in which mitochondria have existed as endosymbionts within eukaryotic cells, most of their genes have been transferred to the chromosomes of the host cells—but not all. Each mitochondrion still has its own genome, a circular, closed molecule of DNA similar to that found in bacteria, on which is located genes encoding the essential proteins of oxidative metabolism. These genes are transcribed within the mitochondrion, using mitochondrial ribosomes that are smaller than those of eukaryotic cells, very much like bacterial ribosomes in size and structure. Mitochondria divide by simple fission, just as bacteria do, replicating and sorting their DNA much as bacteria do. However, nuclear genes direct the process, and mitochondria cannot be grown outside of the eukaryotic cell, in cell-free culture.

Chloroplasts. Many eukaryotic cells contain other endosymbiotic bacteria in addition to mitochondria. Plants and algae contain chloroplasts, bacterialike organelles that were apparently derived from symbiotic photosynthetic bacteria. Chloroplasts have a complex system of inner membranes and a circle of DNA.

While all mitochondria are thought to have arisen from a single symbiotic event, it is difficult to be sure with chloroplasts. Three biochemically distinct classes of chloroplasts exist, but all appear to have their origin in the cyanobacteria.

Red algae and green algae seem to have acquired cyanobacteria directly as endosymbionts, and may be sister groups. Other algae have chloroplasts of secondary origin, having taken up one of these algae in their past. The chloroplasts of euglenoids are thought to be green algal origin,

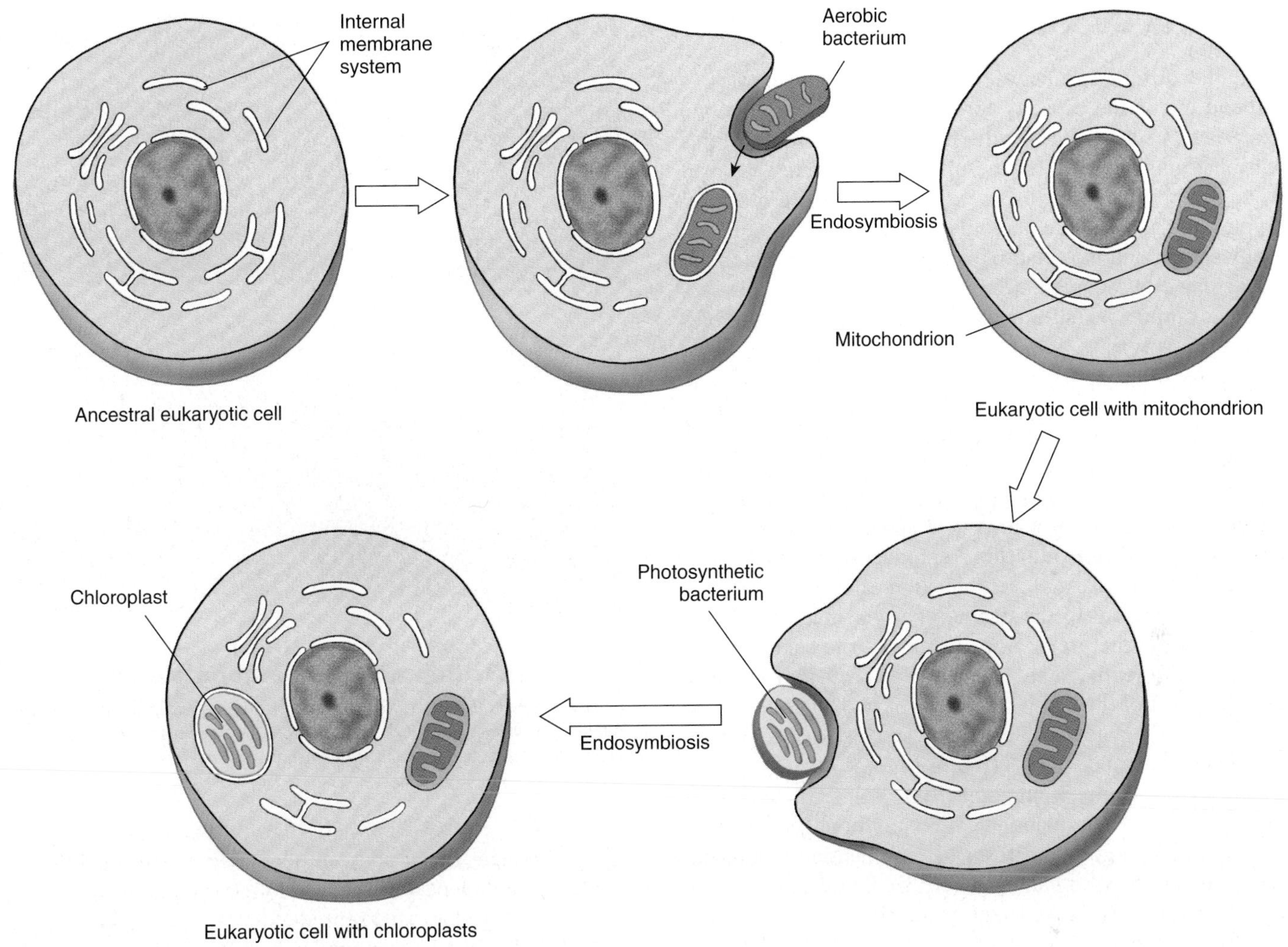
Internal
membrane
system
Aerobic
bacterium
Endosymbiosis
Mitochondrion
Ancestral eukaryotic cell
Eukaryotic cell with mitochondrion
Chloroplast
Photosynthetic
bacterium
Endosymbiosis
Eukaryotic cell with chloroplasts

Figure 14.2 The theory of endosymbiosis.
Scientists propose that ancestral eukaryotic cells engulfed aerobic eubacteria, which then became mitochondria in the eukaryotic cell. Chloroplasts may also have originated this way, with eukaryotic cells engulfing photosynthetic eubacteria.

while those of brown algae and diatoms are likely of red algal origin. The chloroplasts of dinoflagellates seem to be of complex origins, which might include diatoms.

Mitosis

What of mitosis, the other typical eukaryotic process that *Pelomyxa* lacks? The mechanism of mitosis, now so common among eukaryotes, did not evolve all at once. Traces of very different, and possibly intermediate, mechanisms survive today in some of the eukaryotes. In fungi and some groups of protists, for example, the nuclear membrane does not dissolve, and mitosis is confined to the nucleus. When mitosis is complete in these organisms, the nucleus divides into two daughter nuclei, and only then does the rest of the cell divide. This separate nuclear division phase of mitosis does not occur in most protists, or in plants or animals. We do not know if it represents an intermediate step on the evolutionary journey to the form of mitosis that is characteristic of most eukaryotes today or if it is simply a different way of solving the same problem. There are no fossils in which we can see the interiors of dividing cells well enough to be able to trace the history of mitosis.

14.1 The theory of endosymbiosis proposes that mitochondria originated as symbiotic aerobic eubacteria.

14.2 The Evolution of Sex

Of all the differences between bacteria and eukaryotes, a profound one is the capacity for sexual reproduction among eukaryotes. In **sexual reproduction,** two different parents contribute gametes to form the offspring. Gametes are usually formed by meiosis, discussed in chapter 6. In most eukaryotes, the gametes are haploid (have a single copy of each chromosome), and the offspring produced by their fusion are diploid (have two copies of each chromosome). In this section we examine sexual reproduction among the eukaryotes and how it evolved.

Life Without Sex

Sexual reproduction is not the only way that eukaryotes can reproduce. Consider, for example, a sponge. A sponge can reproduce by simply fragmenting its body. Each small portion grows and gives rise to a new sponge. This is an example of **asexual reproduction,** reproduction without forming gametes. In asexual reproduction, the offspring are genetically identical to the parents, barring mutation. Some protists such as the green algae exhibit a true sexual cycle, but only transiently. The majority of protists reproduce asexually most of the time. The fusion of two haploid cells to create a diploid zygote, the essential act of sexual reproduction, occurs only under stress (figure 14.3).

The development of an adult from an unfertilized egg is a form of asexual reproduction called **parthenogenesis.** Parthenogenesis is a common form of reproduction among insects. Among bees, for example, fertilized eggs develop into females, while unfertilized eggs become males. Some lizards, fishes, and amphibians reproduce by parthenogenesis; an unfertilized egg undergoes mitosis without cell cleavage to produce a diploid cell, which then undergoes development as if it had been produced by sexual union of two gametes.

Many plants and marine fishes undergo a form of sexual reproduction that does not involve partners. In **self-fertilization,** one individual provides both male and female gametes. Mendel's peas discussed in chapter 7 produced their F_2 generations by "selfing." Why isn't this asexual reproduction (after all, there is only one parent)? This is considered to be sexual rather than asexual reproduction because the offspring are not genetically identical to the parent. During the production of the gametes, considerable genetic reassortment occurs—that is why Mendel's F_2 plants were not all the same!

Why Sex?

If reproduction without sex is so common among eukaryotes today, it is a fair question to ask why sex occurs at all. Evolution is the result of changes that occur at the level of *individual* survival and reproduction, and it is not immediately obvious what advantage is gained by the progeny of an individual that engages in sexual reproduction. Indeed, the segregation of chromosomes that occurs in meiosis tends to disrupt advantageous combinations of genes more often than it assembles new, better-adapted ones. Because all the progeny could maintain a parent's successful gene combinations if the parent employed asexual reproduction, the widespread use of sexual reproduction among eukaryotes raises a puzzle: Where is the benefit from sex that promoted the evolution of sexual reproduction?

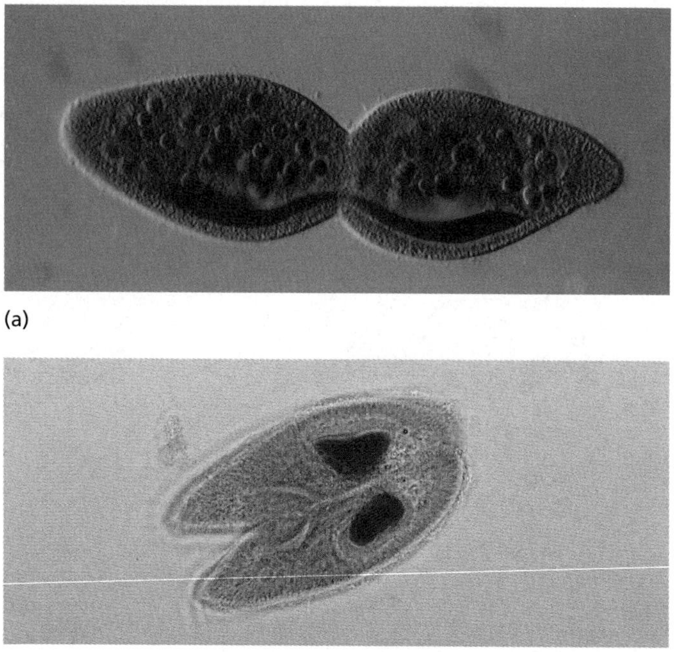

(a)

(b)

Figure 14.3 Reproduction among paramecia.
(*a*) When *Paramecium* reproduces asexually, a mature individual divides, and two complete individuals result. (*b*) In sexual reproduction, two mature cells fuse in a process called conjugation (×100) and exchange haploid nuclei.

How Sex Evolved

In attempting to answer this question, biologists have looked more carefully at where sex first evolved—among the protists. Why do many protists form a diploid cell in response to stress? Biologists think this occurs because only in a diploid cell can certain kinds of chromosome damage be repaired effectively, particularly double-strand breaks in DNA. Such breaks are induced, for example, by desiccation—drying out. The early stages of meiosis, in which the two copies of each chromosome line up and pair with each other, seems to have evolved originally as a mechanism for repairing double-strand damage to DNA by using the undamaged version of the chromosome as a template to guide the fixing of the damaged one. In yeasts, mutations that inactivate the system that repairs double-strand breaks of the chromosomes also prevent crossing over. Thus it seems likely that sexual reproduction and the close association between pairs of chromosomes that occurs during meiosis first evolved as mechanisms to repair chromosomal damage by using the second copy of the chromosome as a template.

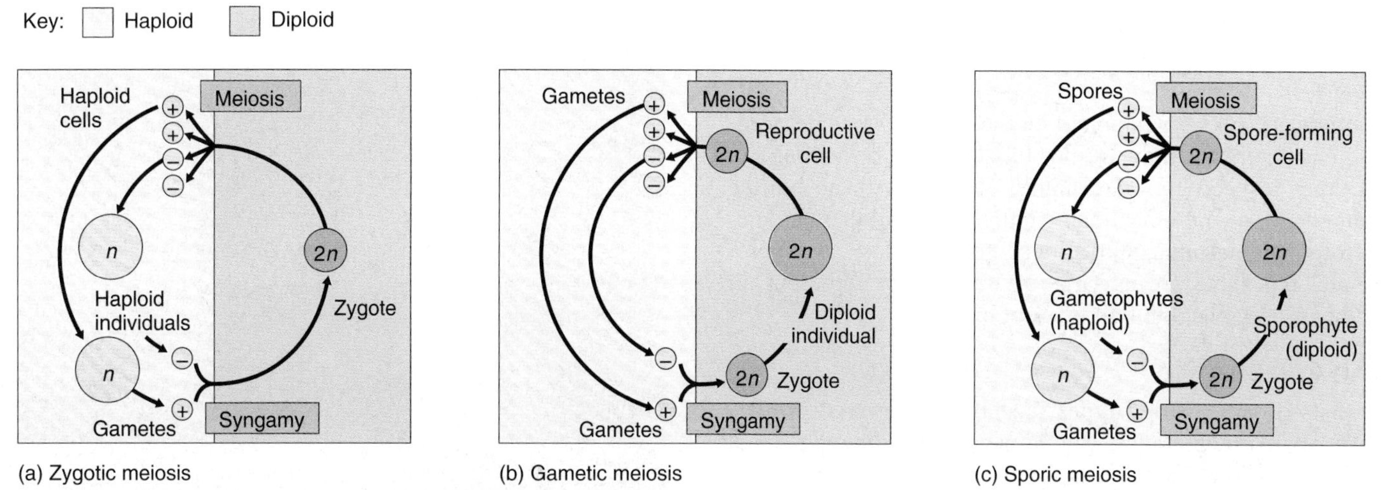
Key: Haploid Diploid
Haploid cells
Meiosis
n
2n
Haploid individuals
Zygote
n
Gametes
Syngamy
(a) Zygotic meiosis
Gametes
Meiosis
Reproductive cell
2n
2n
Diploid individual
2n
Zygote
Gametes
Syngamy
(b) Gametic meiosis
Spores
Meiosis
Spore-forming cell
2n
n
2n
Gametophytes (haploid)
Sporophyte (diploid)
n
2n
Zygote
Gametes
Syngamy
(c) Sporic meiosis

Figure 14.4 Three types of eukaryotic life cycles.

(*a*) Zygotic meiosis, a life cycle found in most protists. (*b*) Gametic meiosis, a life cycle typical of animals. (*c*) Sporic meiosis, a life cycle found in plants and many algae.

Why Sex Is Important

One of the most important evolutionary innovations of eukaryotes was the invention of sex. Sexual reproduction provides a powerful means of shuffling genes, quickly generating different combinations of genes among individuals. Genetic diversity is the raw material for evolution. In many cases the pace of evolution appears to be geared to the level of genetic variation available for selection to act upon—the greater the genetic diversity, the more rapid the evolutionary pace. Programs for selecting larger domestic cattle and sheep, for example, proceed rapidly at first but then slow as all of the existing genetic combinations are exhausted; further progress must then await the generation of new gene combinations. The genetic recombination produced by sexual reproduction has had an enormous evolutionary impact because of its ability to rapidly generate extensive genetic diversity.

Sexual Life Cycles

Many protists are haploid all their lives, but with few exceptions, animals and plants are diploid at some stage of their lives. That is, the body cells of most animals and plants have two sets of chromosomes, one from the male and one from the female parent. The production of haploid gametes by meiosis, followed by the union of two gametes in sexual reproduction, is called the **sexual life cycle.**

Eukaryotes are characterized by three major types of sexual life cycles (figure 14.4):

1. In the simplest of these, found in many algae, the zygote formed by the fusion of gametes is the only diploid cell. This sort of life cycle is said to have **zygotic meiosis,** because in algae the zygote undergoes meiosis. Here the haploid cells occupy the major portion of the life cycle; the zygote undergoes meiosis immediately after it is formed.
2. In most animals, the gametes are the only haploid cells. They exhibit **gametic meiosis,** because in animals meiosis produces the gametes. Here the diploid zygote occupies the major portion of the life cycle.
3. Plants exhibit **sporic meiosis,** because in plants the spore-forming cells undergo meiosis. In plants there is a regular **alternation of generations** between a haploid phase and a diploid phase. The diploid phase produces spores that give rise to the haploid phase, and the haploid phase produces gametes that fuse to give rise to the diploid phase.

The genesis of sex, then, involved meiosis and fertilization with the participation of two parents. We have previously said that bacteria lack true sexual reproduction, although in some groups, two bacteria do pair up in conjugation and exchange parts of their genome. The invention of true sexual reproduction among the protists has no doubt contributed importantly to their tremendous diversification and adaptation to an extraordinary range of ways of life, as we shall see in the following section.

14.2 Sex evolved among eukaryotes as a mechanism to repair chromosomal damage, but its importance is as a means of generating diversity.

14.3 General Biology of the Protists

Protists are eukaryotes united on the basis of a single negative characteristic: they are not fungi, plants, or animals. In all other respects they are highly variable with no uniting features. Many are unicellular (figure 14.5), but there are numerous colonial and multicellular groups. Most are microscopic, but some are as large as trees. They represent all symmetries, and exhibit all types of nutrition.

The Cell Surface

Protists possess varied types of cell surfaces. All protists have plasma membranes. But, some protists, like algae and molds, are additionally encased within strong cell walls. Still others, like diatoms and forams, secrete glassy shells of silica.

Locomotor Organelles

Movement in protists is also accomplished by diverse mechanisms. Protists move chiefly by either flagellar rotation or pseudopodial movement. Many protists wave one or more flagella to propel themselves through the water, while others use banks of short, flagella-like structures called cilia to create water currents for their feeding or propulsion. Pseudopodia are the chief means of locomotion among amoeba, whose pseudopods are large, blunt extensions of the cell body called lobopodia. Other related protists extend thin, branching protrusions called filopodia. Still other protists extend long, thin pseudopodia called axopodia supported by axial rods of microtubules. Axopodia can be extended or retracted. Because the tips can adhere to adjacent surfaces, the cell can move by a rolling motion, shortening the axopodia in front and extending those in the rear.

Cyst Formation

Many protists with delicate surfaces are successful in quite harsh habitats. How do they manage to survive so well? They survive inhospitable conditions by forming **cysts.** A cyst is a dormant form of a cell with a resistant outer covering in which cell metabolism is more or less completely shut down. Not all cysts are so sturdy. Vertebrate parasitic amoebae, for example, form cysts that are quite resistant to gastric acidity, but will not tolerate desiccation or high temperature.

Nutrition

Protists employ every form of nutritional acquisition except chemoautotrophy, which has so far been observed only in bacteria. Some protists are photosynthetic autotrophs and are called **phototrophs.** Others are heterotrophs that obtain energy from organic molecules synthesized by other organisms. Among heterotrophic protists, those that ingest visible particles of food are called **phagotrophs,** or **holozoic feeders.** Those ingesting food in soluble form are called **osmotrophs,** or **saprozoic feeders.**

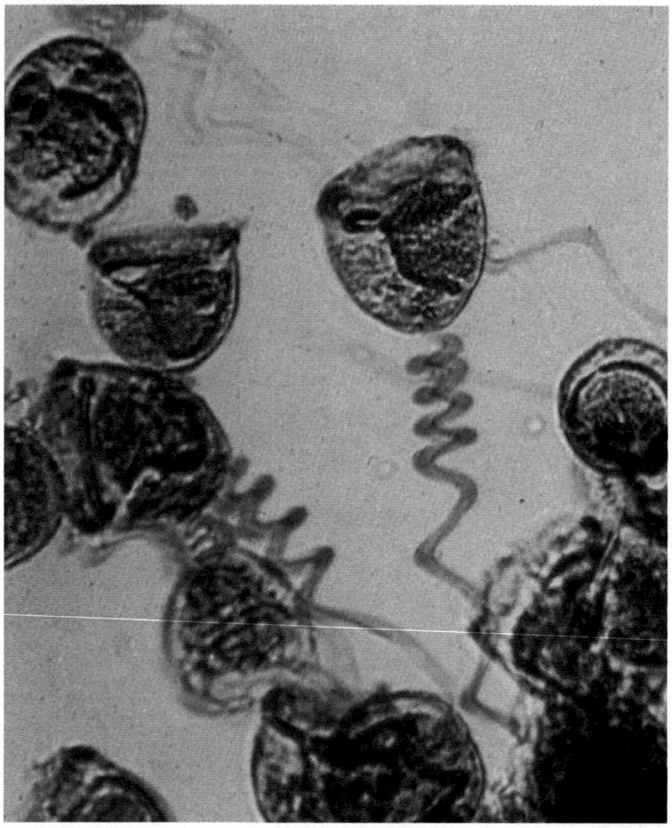

Figure 14.5 A unicellular protist.
The protist kingdom is a catch-all kingdom for many different groups of unicellular organisms, such as this *Vorticella* (phylum Ciliophora), which is heterotrophic, feeds on bacteria, and has a retractable stalk.

Phagotrophs ingest food particles into intracellular vesicles called **food vacuoles** or **phagosomes.** Lysosomes fuse with the food vacuoles, introducing enzymes that digest the food particles within. As the digested molecules are absorbed across the vacuolar membrane, the food vacuole becomes progressively smaller.

Reproduction

Protists typically reproduce asexually, reproducing sexually only in times of stress. Asexual reproduction involves mitosis, but the process is often somewhat different from the mitosis that occurs in multicellular animals. The nuclear membrane, for example, often persists throughout mitosis, with the microtubular spindle forming within it. In some groups, asexual reproduction involves spore formation, in others fission. The most common type of fission is **binary,** in which a cell simply splits into nearly equal halves. When the progeny cell is considerably smaller than its parent, and then grows to adult size, the fission is called **budding.** In multiple fission, or **schizogony,** common among some protists, fission is preceded by several nuclear divisions, so that fission produces several individuals almost simultaneously.

Sexual reproduction also takes place in many forms among the protists. In ciliates and some flagellates, **gametic meiosis** occurs just before gamete formation, as it does in metazoans. In the sporozoans, **zygotic meiosis** occurs directly *after* fertilization, and all the individuals that are produced are haploid until the next zygote is formed. In algae, there is **intermediary meiosis,** producing an alternation of generations similar to that seen in plants, with significant portions of the life cycle spent as haploid as well as diploid.

Multicellularity

A single cell has limits. It can only be so big without encountering serious surface-to-volume problems. Said simply, as a cell becomes larger, there is too little surface area for so much volume. The evolution of multicellular individuals composed of many cells solved this problem. **Multicellularity** is a condition in which an organism is composed of many cells, permanently associated with one another, that integrate their activities. The key advantage of multicellularity is that it allows specialization—distinct types of cells, tissues, and organs can be differentiated within an individual's body, each with a different function. With such functional "division of labor" within its body, a multicellular organism can possess cells devoted specifically to protecting the body, others to moving it about, still others to seeking mates and prey, and yet others to carry on a host of other activities. This allows the organism to function on a scale and with a complexity that would have been impossible for its unicellular ancestors. In just this way, a small city of 50,000 inhabitants is vastly more complex and capable than a crowd of 50,000 people in a football stadium—each city dweller is specialized in a particular activity that is interrelated to everyone else's, rather than just being another body in a crowd.

Colonies. A **colonial organism** is a collection of cells that are permanently associated but in which little or no integration of cell activities occurs. Many protists form colonial assemblies, consisting of many cells with little differentiation or integration. In some protists, the distinction between colonial and multicellular is blurred. For example, in the green algae *Volvox* (figure 14.6), individual motile cells aggregate into a hollow ball of cells that moves by a coordinated beating of the flagella of the individual cells—like scores of rowers all pulling their oars in concert. A few cells near the rear of the moving colony are reproductive cells, but most are relatively undifferentiated.

Aggregates. An **aggregation** is a more transient collection of cells that come together for a period of time and then separate. Cellular slime molds, for example, are unicellular organisms that spend most of their lives moving about and feeding as single-celled amoebas. They are common in damp soil and on rotting logs, where they move about, ingesting bacteria and other small organisms. When the individual amoebas exhaust the supply of bacteria in a given area and are near starvation, all of the individual organisms in that immediate area aggregate into a large moving mass of cells called a slug. By moving to a different location, the aggregation increases the chance that food will be found.

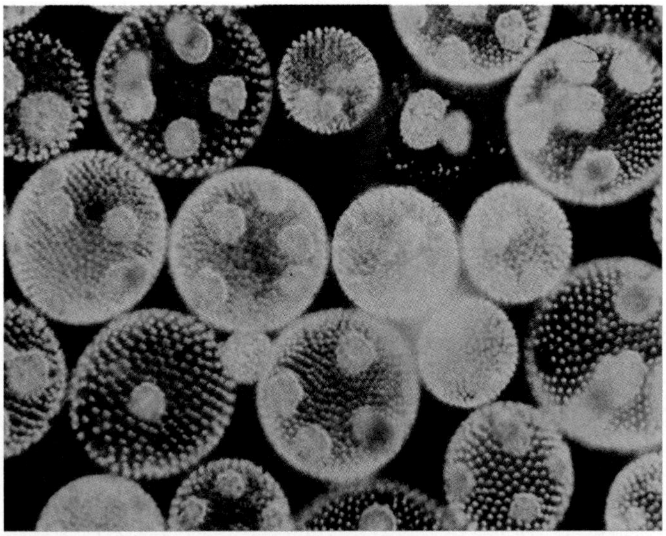

Figure 14.6 A colonial protist.
Individual, motile, unicellular green algae are united in the protist *Volvox* as a hollow colony of cells that moves by the beating of the flagella of its individual cells. Some species of *Volvox* have cytoplasmic connections between the cells that help coordinate colony activities. *Volvox* is a highly complex form of colony that has many of the properties of multicellular life.

Multicellularity. True **multicellularity,** in which the activities of the individual cells are coordinated and the cells themselves are in contact, occurs only in eukaryotes and is one of their major characteristics. Three groups of protists have independently attained true but simple multicellularity—the brown algae (phylum Phaeophyta), green algae (phylum Chlorophyta), and red algae (phylum Rhodophyta). In **multicellular organisms,** individuals are composed of many cells that interact with one another and coordinate their activities.

Simple multicellularity does not imply small size or limited adaptability. Some marine algae grow to be enormous. An individual kelp, one of the brown algae, may grow to tens of meters in length—some taller than a redwood! Red algae grow at great depths in the sea, far below where kelp or other algae are found. Not all algae are multicellular. Green algae, for example, include many kinds of multicellular organisms but an even larger number of unicellular ones.

14.3 Protists exhibit a wide range of forms, locomotion, nutrition, and reproduction. Their cells form clusters with varying degrees of specialization, from transient aggregations to more persistent colonies to permanently multicellular organisms.

14.4 Classifying the Protists

Protists are the most diverse of the four kingdoms in the dormain Eukarya. The kingdom Protista contains many unicellular, colonial and multicellular groups. Probably the most important statement we can make about the kingdom Protista is that it is an artificial group; as a matter of convenience, single-celled eukaryotic organisms have typically been grouped together into this kingdom. This lumps many very different and only distantly related forms together. The "single-kingdom" classification of the Protista is not representative of any evolutionary relationships. The phyla of protists are, with very few exceptions, only distantly related to one another.

New applications of a wide variety of molecular methods are providing important insights into the relationships among the protists. Of all the groups of organisms biologists study, protists are probably in the greatest state of flux when it comes to classification (figure 14.7). There is little consensus, even among experts, as to how the different kinds of protists should be classified. Are they a single, very diverse kingdom, or are they better considered as several different kingdoms, each of equal rank with animals, plants, and fungi?

Figure 14.7 Protist or plant?
Photosynthetic protists called green algae, such as the green alga *Acetabularia* shown above, are on the same evolutionary branch with plants. Most biologists consider them to be protists, however, restricting the plant kingdom to terrestrial organisms.

Because the Protista are still predominantly considered part of one diverse, nonunified group, that is how we will treat them in this chapter, bearing in mind that biologists are rapidly gaining a better understanding of the evolutionary relationships among members of the kingdom Protista. It seems likely that within a few years, the traditional kingdom Protista will be replaced by another more illuminating arrangement.

Five Groups of Protists

There are some 15 distinct phyla of protists. It is difficult to encompass their great diversity with any simple scheme. Traditionally, texts have grouped them artificially (as was done in the nineteenth century) into photosynthesizers (algae), heterotrophs (protozoa), and absorbers (fungus-like protists).

In this text, we group the protists into five general groups according to some of the major shared characteristics. These are characteristics that taxonomists are using today in broad attempts to classify the kingdom Protista. These include (1) the presence or absence and type of cilia or flagella, (2) the presence and kinds of pigments, (3) the type of mitosis, (4) the kinds of cristae present in the mitochondria, (5) the molecular genetics of the ribosomal "S" subunit, (6) the kind of inclusions the protist may have, (7) overall body form (amoeboid, coccoid, and so forth), (8) whether the protist has any kind of shell or other body "armor," and (9) modes of nutrition and movement. These represent only some of the characters used to define phylogenetic relationships.

The criteria we have chosen to define the five general groups are not the only ones that might be chosen, and there is no broad agreement among biologists as to which set of criteria is preferable. As molecular analysis gives us a clearer picture of the phylogenetic relationships among the protists, more evolutionarily suitable groupings will without a doubt replace the one represented here. Table 14.1 summarizes some of the general characteristics and groupings of the 15 major phyla of protists. It is important to remember that while the phyla of protists discussed here are generally accepted taxa, the larger groupings of phyla presented are functional groupings.

14.4 The 15 protist phyla can be grouped into five categories according to major shared characteristics.

TABLE 14.1 KINDS OF PROTISTS

Group	Phylum	Typical Examples		Key Characteristics
HETEROTROPHS WITH NO PERMANENT LOCOMOTOR APPARATUS				
Amoebas	Rhizopoda	*Amoeba*	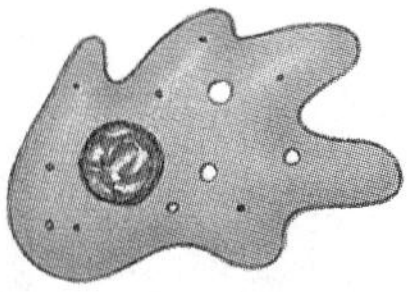	Move by pseudopodia
Forams	Foraminifera	Forams		Rigid shells; move by protoplasmic streaming
Radiolarians	Actinopoda	Radiolarians		Glassy skeletons; needlelike pseudopods
HETEROTROPHS WITH FLAGELLA				
Zoomastigotes	Sarcomastigophora	Trypanosomes		Heterotrophic; unicellular
Ciliates	Ciliophora	*Paramecium*	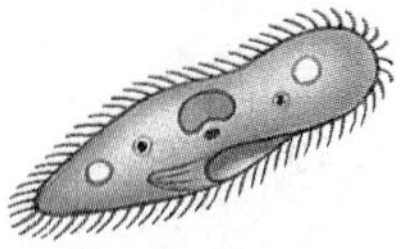	Heterotrophic unicellular protists with cells of fixed shape possessing two nuclei and many cilia; many cells also contain highly complex and specialized organelles
NONMOTILE SPORE-FORMERS				
Sporozoans	Apicomplexa	*Plasmodium*	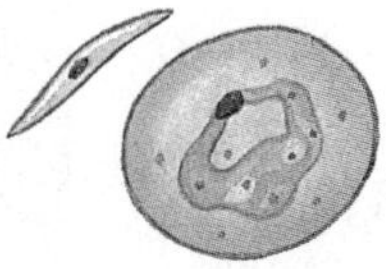	Nonmotile; unicellular; the apical end of the spores contains a complex mass of organelles
PHOTOSYNTHETIC PROTISTS				
Dinoflagellates	Pyrrhophyta	Red tides		Unicellular; two flagella; contain chlorophylls *a* and *b*
Euglenoids	Euglenophyta	*Euglena*	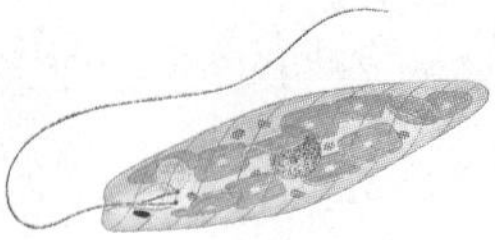	Unicellular; some photosynthetic; others heterotrophic; contain chlorophylls *a* and *b* or none
Diatoms	Chrysophyta	*Diatoma*	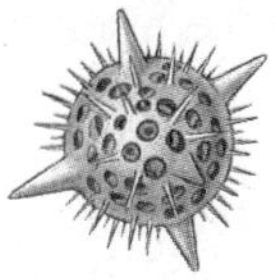	Unicellular; manufacture the carbohydrate chrysolaminarin; unique double shells of silica; contain chlorophylls *a* and *c*
Golden algae	Chrysophyta	*Dinobryon*		Unicellular, but often colonial; manufacture the carbohydrate chrysolaminarin; contain chlorophylls *a* and *c*
Green algae	Chlorophyta	*Chlamydomonas*		Unicellular or multicellular; contain chlorophylls *a* and *b*
Red algae	Rhodophyta	Coralline algae		Most multicellular; contain chlorophyll *a* and a red pigment
Brown algae	Phaeophyta	Kelp		Multicellular; contain chlorophylls *a* and *c*
HETEROTROPHS WITH RESTRICTED MOBILITY				
Cellular slime molds	Acrasiomycota	*Dictyostelium*	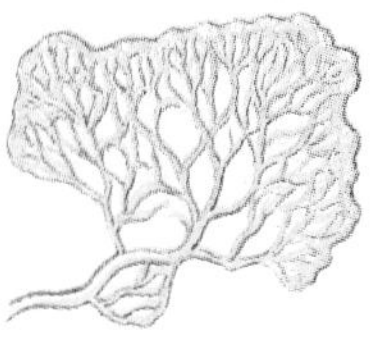	Colonial aggregations of individual cells; most closely related to amoebas
Plasmodial slime molds	Myxomycota	*Fuligo*		Stream along as a multinucleate mass of cytoplasm
Water molds	Oomycota	Water molds and downy mildew		Terrestrial and freshwater

14.5 Heterotrophs with No Permanent Locomotor Apparatus

The largest of the five general groups of protists are distinguished by having no permanent locomotor apparatus. They are all heterotrophs and contain two major phyla, Rhizopoda (amoebas) and Foraminifera (forams), and one minor phylum, Actinopoda (radiolarians).

Amoebas

Amoebas, members of the phylum Rhizopoda, lack flagella and cell walls. There are several hundred species. They move from place to place by **pseudopodia** (Greek, *pseudo,* false, and *podium,* foot), flowing projections of cytoplasm that extend outward (figure 14.8). An amoeba puts a pseudopod forward and then flows into it. Amoebas are abundant in soil, and many are parasites of animals. Reproduction in amoebas occurs by simple *fission* into two daughter cells of equal volume. They do undergo mitosis but lack meiosis and any form of sexuality.

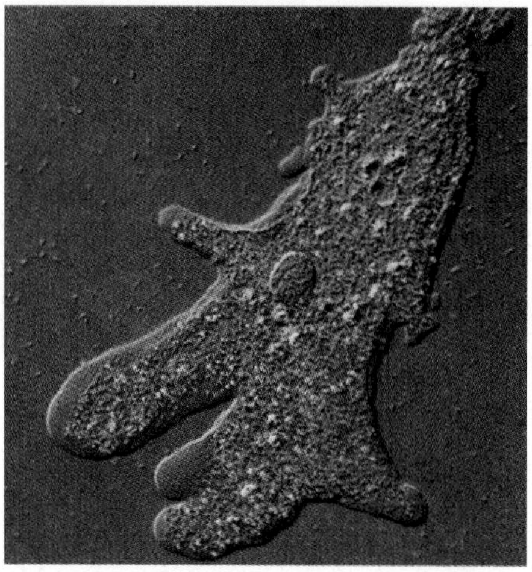

Figure 14.8 An amoeba.
Amoeba proteus is a relatively large amoeba. The projections are pseudopodia; an amoeba moves simply by flowing into them.

Forams

Forams, members of the phylum Foraminifera, possess rigid shells and move by **cytoplasmic streaming.** They are marine protists with pore-studded shells called **tests** that may be as big as several centimeters in diameter. There are several hundred species of forams. Their shells, built largely of calcium carbonate, are often brilliantly colored—vivid yellow, bright red, or salmon pink—and may have many chambers arrayed in a spiral shape resembling a tiny snail. Long, thin cytoplasmic projections called **podia** extend out through the pores in the tests and are used for swimming and capturing prey (figure 14.9). The life cycle of forams is complex, involving alternation between haploid and diploid generations. Limestone is often rich in forams—the White Cliffs of Dover, the famous landmark on the southern England seacoast, is made almost entirely of foram tests.

Figure 14.9 A foram.
In this living foram, a representative of phylum Foraminifera, the podia—thin cytoplasmic projections—extend through pores in the calcareous test, or shell, of the organism.

Radiolarians

Actinosphaerium (figure 14.10) is an unusual kind of amoeba that belongs to another phylum, Actinopoda, whose members have glassy skeletons with many needlelike pseudopods.

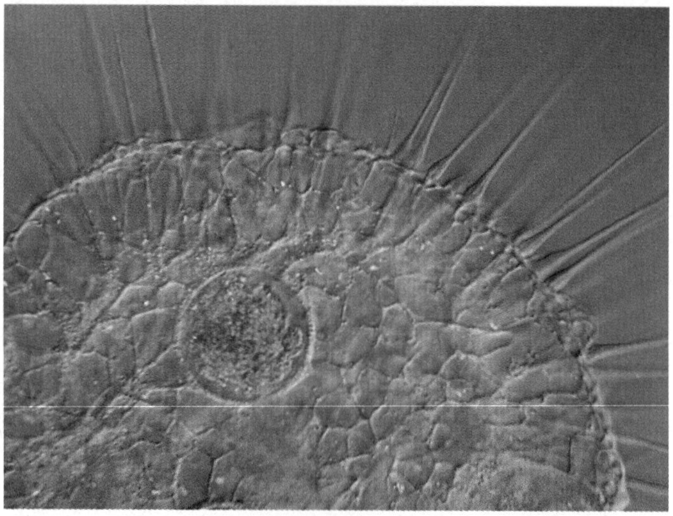

Figure 14.10 A radiolarian.
This amoeba of the phylum Actinopoda, *Actinosphaerium* (×300), has striking needlelike pseudopods.

14.5 Amoebas, forams, and radiolarians have no permanent locomotor apparatus. They use their cytoplasm to aid movement.

14.6 Heterotrophs with Flagella

Zoomastigotes

Some protists move by means of locomotory **flagella.** Sarcomastigophora is one strictly heterotrophic phylum whose members, called zoomastigotes, are flagellated.

Zoomastigotes are of special interest because the ancestor of all animals appears to have been a member of this group. The choanoflagellates are the group from which sponges, and probably all other animals, were derived. Zoomastigotes are unicellular, heterotrophic protists that are highly variable in form. There are several thousand species. All have at least one flagellum, and some species have thousands. Most reproduce only asexually. Members of one group, the trypanosomes, are important pathogens of human beings (figure 14.11). Some zoomastigotes live symbiotically in the guts of termites and provide the enzymes the termites use to digest wood (much as bacteria aid cattle and horses in digesting grass).

Ciliates

The ciliates are very complex and unusual unicellular heterotrophs, with **cilia,** a fixed cell shape, and two nuclei per cell. Ciliates are so different from other eukaryotes (they even use the genetic code differently!) that many taxonomists argue they should be placed in a separate kingdom of their own. There is only one major phylum.

Ciliates, members of the phylum Ciliophora, all possess large numbers of cilia, usually arrayed in long rows down the body or in spirals around it. About 8,000 species have been named. Ciliates have a pellicle, which makes the body wall tough but flexible. The body interior is extremely complex, inspiring some biologists to consider ciliates organisms without cell boundaries rather than unicellular. *Paramecium,* a typical ciliate, has a complex digestive process, with a gullet ("mouth") and intake channel for bacteria and food particles, which are then enclosed in membrane bubbles and digested by enzymes (figure 14.12). Reproduction is usually by fission, with the body splitting in half. Cells divide asexually for about 700 generations and then die if sexual reproduction has not occurred. Sexual reproduction is by conjugation, with duplicate nuclei being exchanged between different mating types across a conjugation bridge.

14.6 Zoomastigotes move by waving flagella, ciliates by beating rows of cilia.

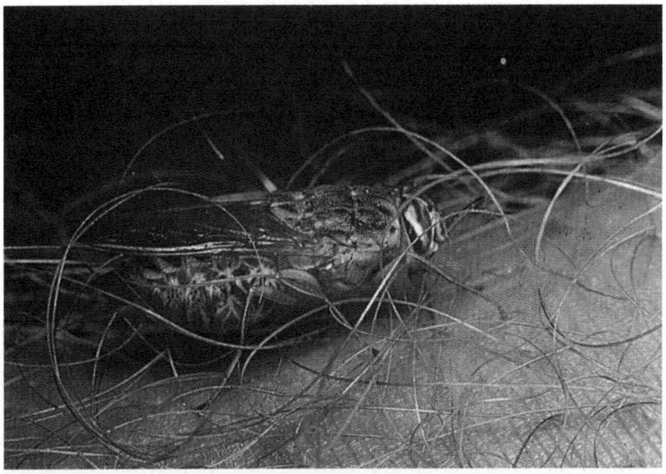

(a)

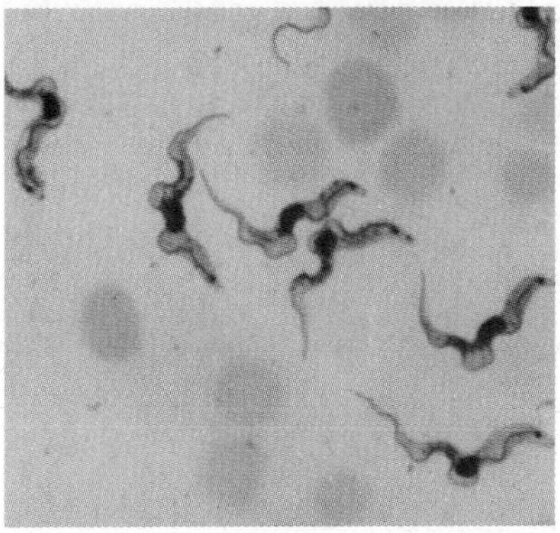

(b)

Figure 14.11
A zoomastigote.
Trypanosomes, a kind of zoomastigote, cause sleeping sickness. (*a*) A tsetse fly, shown here sucking blood from a human arm, in Tanzania, East Africa, can transmit the trypanosomes that cause sleeping sickness. (*b*) The undulating, changeable shapes of *Trypanosoma* are visible among red blood cells in this photograph.

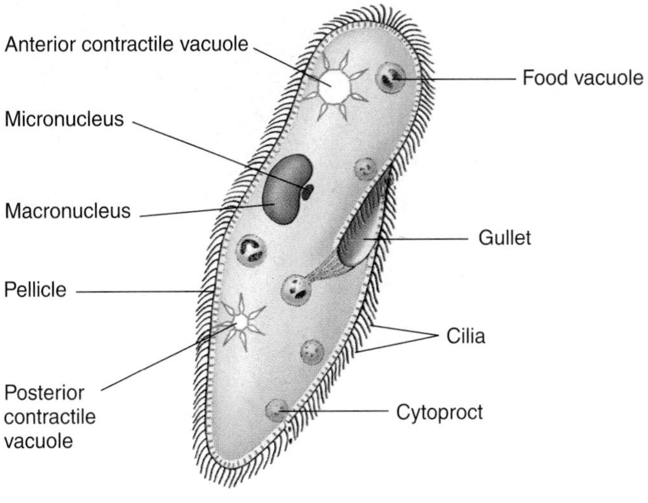
Anterior contractile vacuole
Micronucleus
Macronucleus
Pellicle
Posterior contractile vacuole
Food vacuole
Gullet
Cilia
Cytoproct

Figure 14.12 A ciliate.
In this diagram, the main features of the familiar ciliate *Paramecium* are shown.

14.7 Nonmotile Spore-Formers

Sporozoans are nonmotile, spore-forming, unicellular *parasites* of animals, all members of the phylum Apicomplexa. They are responsible for many diseases in humans and domestic animals. Sporozoans infect animals with small spores that are transmitted from host to host. All sporozoans possess a unique arrangement of microtubules and other organelles clustered at one end of the cell, but adults are nonmotile, having no flagella. There are 3,900 described species of sporozoans, including the malaria-causing parasite *Plasmodium.*

Sporozoans have complex life cycles that involve both asexual and sexual phases. Both haploid and diploid individuals can divide rapidly by mitosis, thus producing a large number of small, infective individuals. Sexual reproduction involves the fertilization of a large female gamete by a small, flagellated male gamete. The zygote that results soon becomes a thick-walled cyst called an oocyst, which is highly resistant to drying out and other unfavorable environmental factors. Within the oocyst, meiotic divisions produce infective haploid spores.

An alternation between different hosts often occurs in the life cycles of sporozoans (figure 14.13). Sporozoans of the genus *Plasmodium* are spread among humans by mosquitoes of the genus *Anopheles;* at least 65 different species of this genus are involved. The sporozoan life cycle stages—called sporozoites, merozoites, and gametocytes—each produce different antigens, and they are sensitive to different antibodies. When a mosquito inserts its proboscis into a human blood vessel, it injects about a thousand sporozoites. They travel to the liver within a few minutes, where they are no longer exposed to antibodies circulating in the blood. If even one sporozoite reaches the liver, it will multiply rapidly there and still cause malaria.

Malaria is one of the most serious diseases in the world. About 500 million people are affected by it at any one time, and approximately 2 million of them, mostly children, die each year. Efforts to eradicate malaria have focused on (1) the elimination of the mosquito vectors, (2) the development of drugs to poison the parasites once they have entered the human body, and (3) the development of vaccines.

14.7 Sporozoans are nonmotile parasites that form spores.

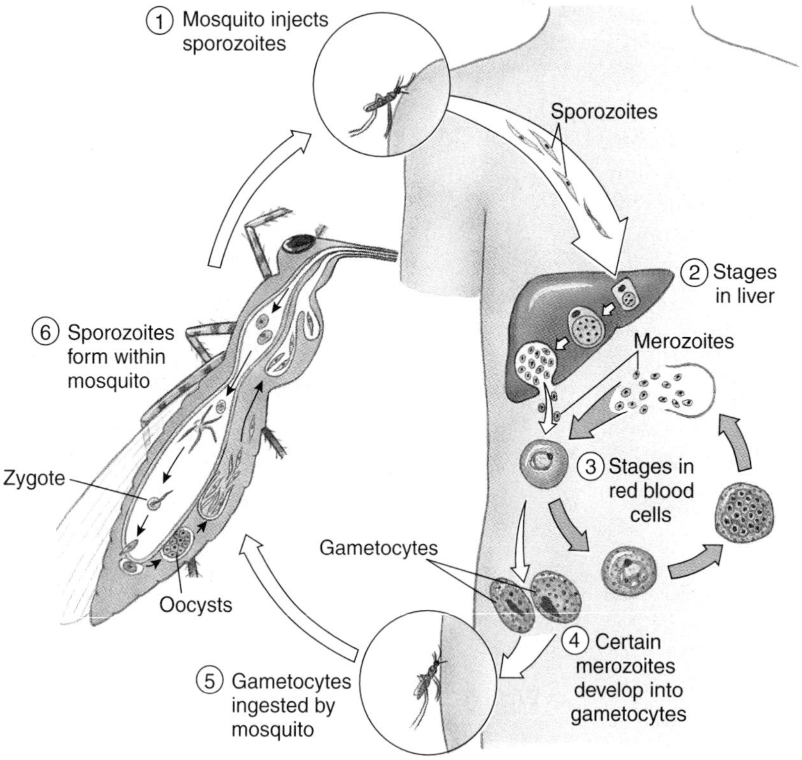

1 Mosquito injects sporozoites
Sporozoites
2 Stages in liver
Merozoites
3 Stages in red blood cells
4 Certain merozoites develop into gametocytes
Gametocytes
5 Gametocytes ingested by mosquito
Oocysts
Zygote
6 Sporozoites form within mosquito

Figure 14.13 A sporozoan life cycle.

Plasmodium is the sporozoan that causes malaria. *Plasmodium* has a complex life cycle that alternates between mosquitoes and mammals.

14.8 Photosynthetic Protists

Dinoflagellates

Dinoflagellates, members of the phylum Pyrrhophyta, are photosynthetic unicellular protists, most with two flagella of unequal length. The dinoflagellates are very unusual protists that do not appear to have any close relatives. There are about 1,000 species. Some occur in freshwater, but most are marine. Luminous dinoflagellates produce the twinkling light sometimes seen in tropical seas at night. Most dinoflagellates have a stiff coat of cellulose, often encrusted with silica, giving them unusual shapes. Their flagella are unique, unlike those of any other phylum: one beats in a groove circling the body like a belt, the other in a groove perpendicular to it (figure 14.14). Their beating rotates the body like a top. A few dinoflagellates produce powerful toxins such as the poisonous "red tides," which are population explosions of such dinoflagellates (figure 14.15). Dinoflagellates reproduce by splitting in half. Their form of mitosis is unique—their chromosomes remain condensed, distributed along the sides of channels containing bundles of microtubules that run through the nucleus.

Euglenoids

Euglenoids, members of the phylum Euglenophyta, are freshwater protists with two flagella that clearly illustrate the folly of attempting to classify protists as tiny animals or plants. About one-third of the 1,000 known species have chloroplasts and are photosynthetic; the others lack chloroplasts, ingest their food, and are heterotrophic. In the dark, many photosynthetic euglenoids reduce the size of their chloroplasts (they may appear to disappear!) and become heterotrophs until put back in light. Euglenoids are very closely related to zoomastigotes, and many taxonomists merge the two phyla.

Euglena, after which the phylum is named, has a protein scaffold called the pellicle inside the plasma membrane and can change shape (figure 14.16). Two flagella are attached at the base of a flask-shaped opening called the reservoir, which is located at the anterior end of the cell. One of the flagella is long and has a row of very fine, short, hairlike projections along one side. A second, shorter flagellum is located within the reservoir but does not emerge from it. Contractile vacuoles collect excess water from all parts of the organism and empty it into the reservoir, which apparently helps regulate the osmotic pressure within the organism. The stigma, an organ that also occurs in the green algae (phylum Chlorophyta), is light-sensitive and aids these photosynthetic organisms to move toward light. Reproduction is by mitotic cell division; no sexual reproduction is known in this group.

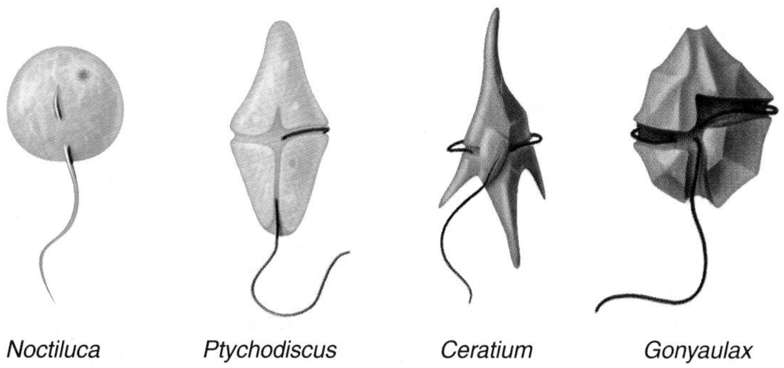

Noctiluca
Ptychodiscus
Ceratium
Gonyaulax

Figure 14.14 Dinoflagellates.
Noctiluca, which lacks the heavy cellulosic armor characteristic of most dinoflagellates, is one of the bioluminescent organisms that causes the waves to sparkle in warm seas at certain times of year. In the other three genera, the shorter encircling flagellum may be seen in its groove, and the longer one projects away.

Figure 14.15 Red tide.
Red tides are caused by population explosions of dinoflagellates. The pigments in the dinoflagellates or, in some cases, other organisms color the water.

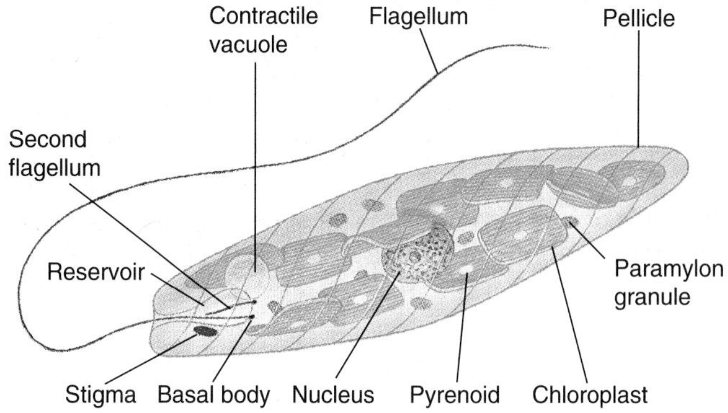

Contractile vacuole
Flagellum
Pellicle
Second flagellum
Reservoir
Paramylon granule
Stigma
Basal body
Nucleus
Pyrenoid
Chloroplast

Figure 14.16 A euglenoid.
In euglenoids such as this *Euglena*, starch forms around pyrenoids; food reserves are stored in paramylon granules.

Diatoms

Diatoms, members of the phylum Chrysophyta, are photosynthetic unicellular protists with a unique double shell of silica. Like tiny oysters, their shells resemble small boxes with lids, one half fitting inside the other. They are sometimes called golden algae, although they do not resemble the three phyla of true algae in any important aspects. There is one major phylum.

Diatoms are abundant in both oceans and lakes. There are over 11,500 species, of two sorts: some with radial symmetry (those in figure 14.17 look like tiny wheels) and others with bilateral (two-sided) symmetry. The shells of fossil diatoms form thick deposits that are mined commercially as "diatomaceous earth," which is used as an abrasive or to make paint sparkle. Diatoms move by protoplasmic streaming along a groove in their shell. Individuals are diploid and usually reproduce asexually by separating the halves of the shell, each half then growing another half to match it. When individuals get too small because of repeated division, they slip out of their shells, grow to full size, and then regenerate a new set of shells. Under stress, diatoms undergo meiosis and reproduce sexually.

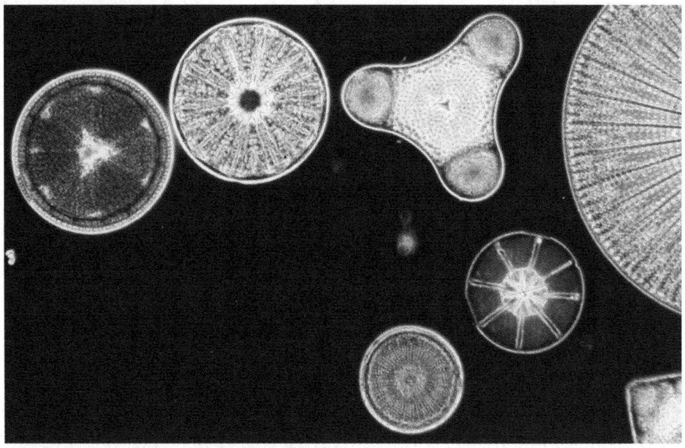

Figure 14.17 Diatoms.
Several different kinds of diatoms with radial symmetry.

The Golden Algae

Also included within the phylum Chrysophyta are the golden algae, named for the yellow and brown carotenoid and xanthophyll accessory pigments in their chloroplasts, which give them a golden color. Unicellular but often colonial, these freshwater protists typically have two flagella, both attached near the same end of the cell. When ponds and lakes dry out in summer, golden algae form resistant cysts. Viable cells emerge from these cysts when wetter conditions recur in the fall.

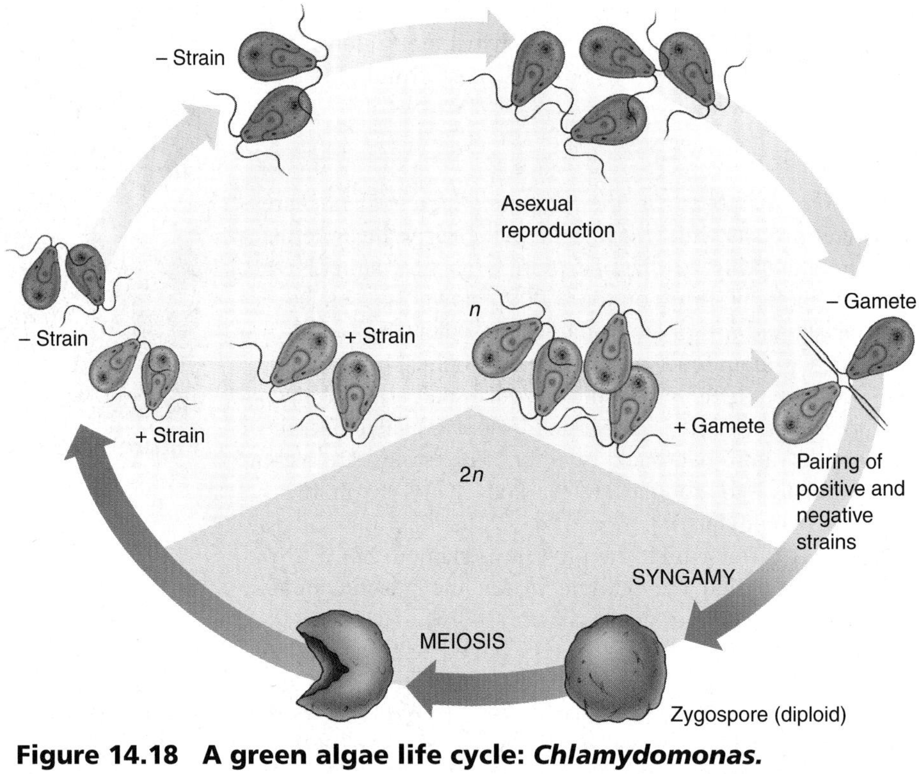
– Strain
Asexual
reproduction
n
– Strain
+ Strain
+ Strain
– Gamete
+ Gamete
Pairing of
positive and
negative
strains
2n
SYNGAMY
MEIOSIS
Zygospore (diploid)

Figure 14.18 A green algae life cycle: *Chlamydomonas.*
Individual cells of *Chlamydomonas* (a microscopic, biflagellated alga, phylum Chlorophyta) are haploid and divide asexually, producing identical copies of themselves. At times, such haploid cells act as gametes, fusing to produce a zygote, as shown in the lower right-hand side of the diagram. The zygote develops a thick, resistant wall, becoming a zygospore. Meiosis takes place, ultimately resulting in the release of four haploid individuals.

Green Algae

Three major phyla of algae contain mostly multicellular organisms. These phyla are distinguished by the kind of chlorophyll pigment their members contain: green, red, or brown. Their chloroplasts appear to have arisen by a single endosymbiotic event, before the differences in chlorophyll evolved.

Green algae, members of the phylum Chlorophyta, are of special interest because the ancestor of true plants was a member of this group. Green algae chloroplasts are similar to plant chloroplasts, and like them they contain chlorophylls *a* and *b*. Green algae are an extremely varied group of more than 7,000 species, mostly mobile and aquatic like *Chlamydomonas* (figure 14.18), but a few (like *Chlorella*) are immobile in moist soil or on tree trunks. Most green algae are microscopic and unicellular, but some, like *Ulva* (sea lettuce), are large and multicellular (figure 14.19). Among the most

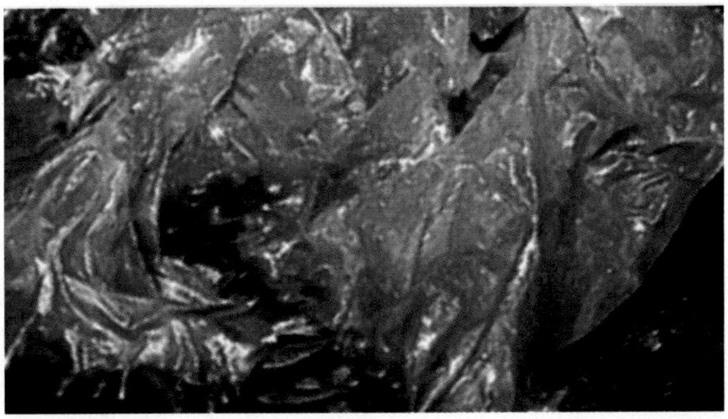

Figure 14.19 Green algae.
Ulva is a large, multicellular green alga.

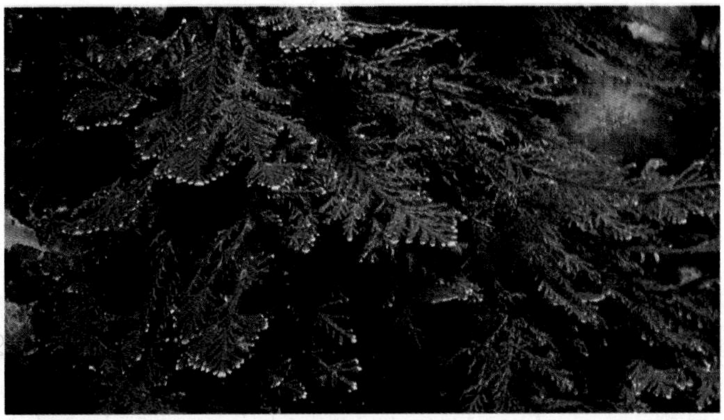

Figure 14.20 Red algae.
These red algae have their cellulose cell walls heavily impregnated with calcium carbonate, the same material of which oyster shells are made. Because they are hard and occur on coral reefs, they are called coralline algae.

elaborate is *Volvox,* a hollow sphere made of tens of thousands of cells. The two flagella of each cell beat in time with all the others to rotate the colony, which has reproductive cells at one end. *Volvox* borders on true multicellularity.

Red Algae

Red algae, members of the phylum Rhodophyta, possess red pigments called phycobilins that give them their characteristic color (figure 14.20). Almost all of the 4,000 species of red algae are multicellular and live in the sea, where they grow more deeply than any other photosynthetic organism. Red algae have complex bodies made of interwoven filaments of cells. The laboratory media *agar* is made from the cell walls of red algae. Their life cycle is complex, usually involving alternation of generations. None of the red algae have flagella or centrioles, suggesting that red algae may be one of the most ancient groups of eukaryotes.

Brown Algae

Brown algae, members of the phylum Phaeophyta, contain the longest, fastest-growing, and most photosynthetically productive living things—giant kelp, with individuals over 100 meters long (figure 14.21). The 1,500 species of brown algae are all multicellular and almost all marine. They are the most conspicuous seaweeds in the ocean. The larger brown algae have flattened blades, stalks, and anchoring bases and often contain complex internal plumbing like that of plants. The life cycle employs an alternation of generations, with the large individuals we see being the sporophyte (diploid) generation.

14.8 Many protists are photosynthetic, including dinoflagellates, euglenoids, diatoms, and four kinds of algae.

Figure 14.21 Brown algae.
Surrounding this diver are massive "groves" of giant kelp, a kind of brown algae. Kelp groves contain some of the largest organisms on earth.

14.9 Heterotrophs with Restricted Mobility

Slime molds and water molds are heterotrophic protists that are sometimes confused with fungi, although they do not resemble them in any significant respect. For example, cellular slime molds and water molds have cell walls made of cellulose, while fungal walls are made of chitin. Also, slime and water molds carry out normal mitosis, while fungal mitosis is unusual. There are three major phyla of protistan molds, each quite different in their structure and life cycles. None of these mold phyla are related.

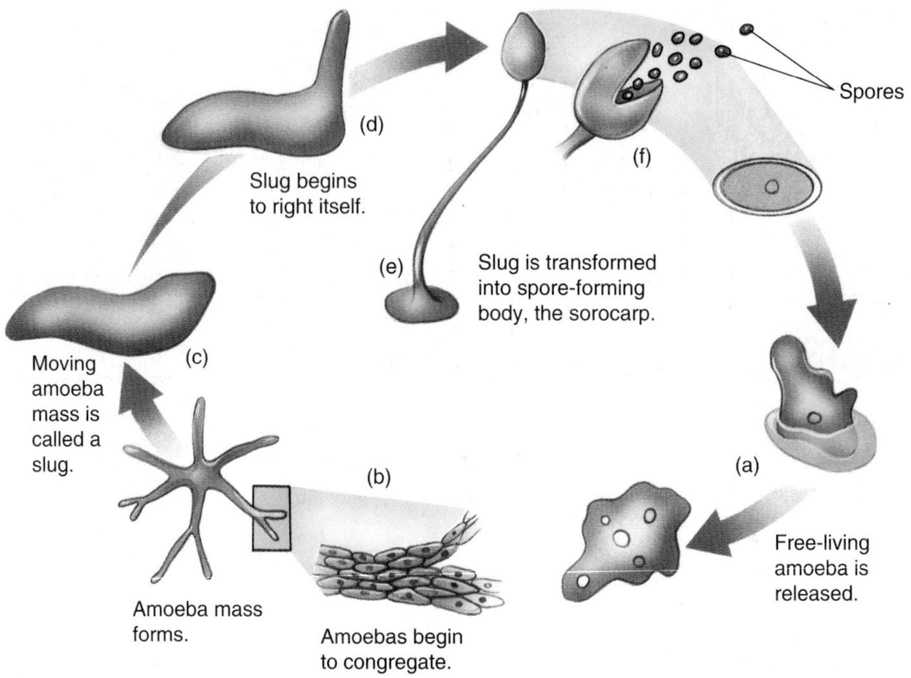

Spores
(d)
Slug begins
to right itself.
(f)
(e)
Slug is transformed
into spore-forming
body, the sorocarp.
Moving
amoeba
mass is
called a
slug.
(c)
(b)
(a)
Free-living
amoeba is
released.
Amoeba mass
forms.
Amoebas begin
to congregate.

Figure 14.22 Life cycle of a cellular slime mold: *Dictyostelium*.

In the cellular slime mold *Dictyostelium discoideum* (phylum Acrasiomycota), (*a*) germinating spores form amoebas. (*b*) The amoebas aggregate and move toward a fixed center. (*c*) They form a multicellular slug 2 to 3 millimeters long that migrates toward light. (*d*) The slug stops moving and begins to differentiate into a spore-forming body, called a sorocarp (*e*). (*f*) Within heads of the sorocarps, the amoebas become encysted as spores.

Cellular Slime Molds

Cellular slime molds, members of the phylum Acrasiomycota, are more closely related to amoebas than any other phylum, but they are able to aggregate in times of stress into mobile, multicellular colonies called **slugs.** There are 70 named species, the best known of which is *Dictyostelium discoideum. Dictyostelium* has been the subject of intensive research by biologists interested in the basic mechanisms of development. The complex developmental cycle of this mold is presented in figure 14.22. *Dictyostelium* is basically a unicellular scavenger. When deprived of food, thousands of individual *Dictyostelium* amoebas come together into a slug that moves to a new habitat. There, the colony differentiates into a base, a stalk, and a swollen tip that develops **spores.** Each of these spores, when released, becomes a new amoeba, which begins to feed and so restarts the life cycle.

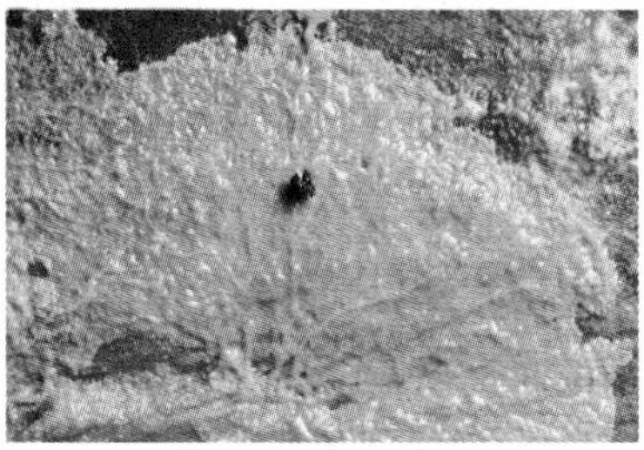

Figure 14.23 A plasmodial slime mold.

This multinucleate plasmodium (phylum Myxomycota) moves about in search of the bacteria and other organic particles that it ingests.

Plasmodial Slime Molds

Plasmodial slime molds, phylum Myxomycota, are a group of about 500 species that stream along as a **plasmodium,** a nonwalled multinucleate mass of cytoplasm (figure 14.23). Plasmodia can flow around obstacles and even pass through the mesh in cloth. Extending pseudopodia as they move, they engulf and digest bacteria and other organic material. A plasmodial slime mold contains many nuclei, but these are not separated by cell walls. All of the nuclei undergo mitosis at the same time, in coordinated fashion. If the plasmodium begins to dry or starve, it migrates away rapidly and then stops and often divides into many small mounds, each of which produces a spore-laden structure. Within the spores, meiosis occurs, producing haploid gametes. Spores germinate when favorable conditions return, the gametes fusing to re-form a diploid plasmodium, in which mitosis continues to occur.

Water Molds

Water molds, members of the phylum Oomycota, are the downy mildews that are often seen in moist environments. There are 580 named species, all of which either parasitize living organisms or feed on dead organic matter. Oomycetes are unusual in that their spores are motile, with two flagella, one pointed forward, the other backward. Many oomycetes are important plant pathogens, including *Phytophthora infestans,* which causes late blight in potatoes. This mold was responsible for the Irish potato famine of 1845–47, during which about 400,000 Irish people starved to death.

14.9 Protistan molds are heterotrophs with restricted mobility, many of which form aggregates.

CONCEPT REVIEW

1. Endosymbiosis of a _________ bacteria most likely gave rise to the chloroplasts of eukaryotic cells of plants and some algae.
   a. chemoautotrophic
   b. *Pelomyxa*
   c. methanogenic
   d. photosynthetic

2. Parthenogenesis is the development of an adult from _________ and is a common form of asexual reproduction among insects.
   a. an unfertilized egg
   b. a fertilized egg
   c. larvae
   d. a zygote

3. In zygotic meiosis, the _________ cells occupy the major portion of the life cycle.
   a. zygotic
   b. diploid
   c. haploid
   d. sporic

4. The kingdom Protista contains all eukaryotes that are not
   a. viruses or bacteria.
   b. animals, fungi, or plants.
   c. photosynthetic.
   d. animals or plants.

5. *Trypanosoma* is a zoomastigote responsible for the disease in humans called
   a. sleeping sickness.
   b. malaria.
   c. red tides.
   d. tuberculosis.

6. Which of the following types of eukaryotes does *not* carry out meiosis?
   a. *Dictyostelium*
   b. *Chlamydomonas*
   c. *Amoeba*
   d. sporozoans

7. Which of the following groups of algae have chloroplasts identical to those found in plants?
   a. euglenoids
   b. red algae
   c. brown algae
   d. green algae

8. The laboratory media agar is made from the cell walls of
   a. diatoms.
   b. red algae.
   c. brown algae.
   d. *Euglena.*

9. Mitochondria are sausage-shaped organelles in eukaryotic cells that are thought to have arisen by _________.

10. In _________ reproduction, no gametes are formed.

11. Sexual reproduction seems to have evolved as a mechanism to repair damaged _________ during the close association between pairs of chromosomes during meiosis.

12. _________ is the group of protists believed to have given rise to animals and certainly gave rise to sponges.

13. Flagella are common among the various phyla of algae except one. Which phyla of algae lacks flagellated cells at any stage of the life cycle? ___________________

14. Red tides are population explosions of protozoans called _________.

CHALLENGE YOURSELF

1. If plants were derived from green algae, why don't we classify green algae as plants in this text?

2. If mitochondria and chloroplasts originated as symbiotic bacteria, what would you suggest were the characteristics of the organism in which they became symbiotic?

3. What are the advantages of sexual reproduction?

4. Coral reefs are one of the most productive communities on earth even though tropical waters are often poor in nutrients. Why? Why do you think coral reefs are only found in tropical and subtropical waters?

14 eBRIDGE

Reinforcing Key Points

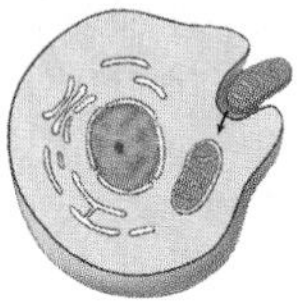

The Evolution of Eukaryotes

14.1 Endosymbiosis

14.2 The Evolution of Sex

14.3 General Biology of the Protists

The Protists

14.4 Classifying the Protists

14.5 Heterotrophs with No Permanent Locomotor Apparatus

14.6 Heterotrophs with Flagella

14.7 Nonmotile Spore-Formers

14.8 Photosynthetic Protists

14.9 Heterotrophs with Restricted Mobility

Electronic Learning

Visual Learning

Animations

Sarcodines

Chlamydomonas

Life Cycle of *Plasmodium vivax*

Deoxygenation of Lakes

Author's Corner

Diversity of Protists. The protists are an extremely diverse group of eukaryotes, and as such, they present a classification challenge. One possible phylogeny for the five groups of protists discussed in this text is a tree in which all branches are equally related to each other. Among the groups of protists, life cycles, structure, and a variety of other characteristics vary greatly.

1. Five groups of protists
2. Heterotrophs with no permanent locomotor apparatus
3. Heterotrophs with flagella
4. Nonmotile spore-formers
5. Photosynthetic protists
6. Heterotrophs with restricted mobility

The Living World

Virtual Classroom

Investigating the Origin of Life

Few questions have fascinated mankind so intensely as the origin of life. There are both religious and scientific views about how life arose on earth. We will limit ourselves to scientific ones here, proposals that are at least in principle subject to test and rejection. There are a great many intriguing ideas advanced by scientists to explain how life may have originated on our planet, but there is very little that we know for sure. It seems clear the earth itself was formed about 4.6 billion years ago. The oldest clear evidence of life—microfossils in ancient rock—are 3.5 billion years old. Thus life arose quickly on earth, within the first billion years, at a time when chemically rich oceans covered much of the earth. One scenario for the origin of life is that it originated spontaneously in this dilute hot chemical soup; another is that it arose in hydrothermal deep-sea vents. Yet another is that life arrived on earth from an extraterrestrial source. Experiments have shown that many of the organic molecules needed to assemble an organism can be produced spontaneously from simple chemicals thought to be present in earth's early oceans, but little is known about how the first cells originated.

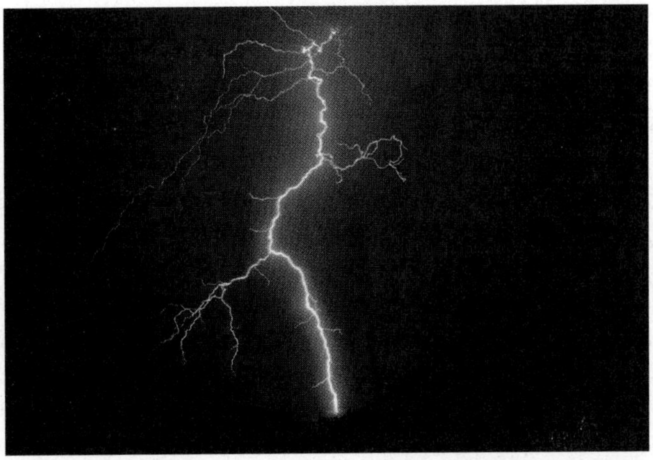

Virtual Lab

Tracking Iron Stress in Diatoms

Algae serve a very important role in many aquatic environments, forming the base of the food chain. Indeed, many of the aquatic animals in such habitats survive solely by eating phytoplankton (small photosynthetic algal organisms). A delicate balance of phytoplankton is needed to maintain a healthy aquatic environment: too much phytoplankton (so-called algal blooms) will deplete the oxygen dissolved in the water; too little phytoplankton creates a food shortage. Diatoms, a type of algae, are important in the aquatic ecosystem of the subarctic Pacific off the western coast of Canada. Scientists monitoring unexpectedly low levels of diatoms and other algae in the North Pacific have discovered that limited iron stores are responsible for the low phytoplankton stocks there. Michael McKay at Bowling Green State University is working on developing biochemical probes that can be used to identify areas of iron deficiencies. Diatoms contain an iron-sulfur protein called ferredoxin. Under low-iron conditions, ferredoxin is replaced by another protein, flavodoxin. McKay uses flavodoxin as a biochemical probe to detect iron deficiency in diatoms.

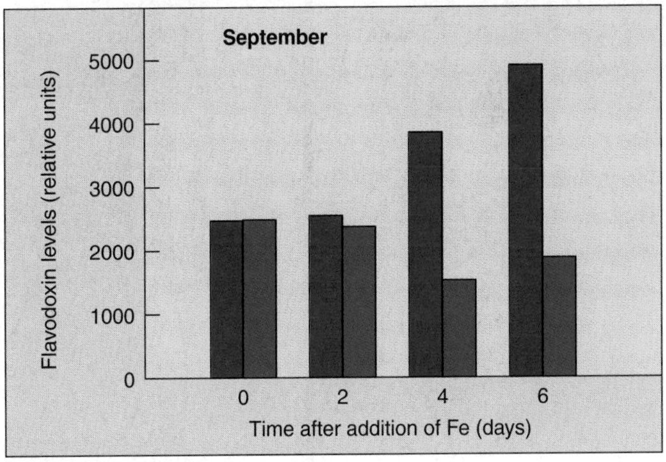
September
Flavodoxin levels (relative units)
5000
4000
3000
2000
1000
0
0
2
4
6
Time after addition of Fe (days)

Quizzes | Further Reading | Essential Study Partner | Links | BioCourse.com

CHAPTER 15

EVOLUTION OF MULTICELLULAR LIFE

THE FAR SIDE®
By GARY LARSON
© 1984 FarWorks, Inc. All Rights Reserved/Dist. by Creators Syndicate
Larson

"Well, of *course* I did it in cold blood, you idiot! ... I'm a reptile!"

CHAPTER 15

Evolution of Multicellular Life

CHAPTER OVERVIEW

Fungi as Multicellular Organisms

15.1 Complex Multicellularity
15.2 A Fungus Is Not a Plant
15.3 Reproduction and Nutrition of Fungi
15.4 Ecology of Fungi

- Complex multicellular organisms are characterized by cell specialization and intercell coordination.
- Fungi are multicellular heterotrophs with filamentous bodies.
- Long strings of fungal cells, called hyphae, form a mass called a mycelium.
- The cells of fungi are interconnected, sharing cytoplasm and nuclei.

Fungal Diversity

15.5 Kinds of Fungi
15.6 Zygomycetes
15.7 Ascomycetes
15.8 Basidiomycetes
15.9 Unicellular and Asexual Fungi

- There are three phyla of fungi: Zygomycota, the zygomycetes; Ascomycota, the ascomycetes; and Basidiomycota, the basidiomycetes.
- Zygomycetes form septa only when gametangia or sporangia are cut off at the ends of their hyphae; otherwise, their hyphae are multinucleate.
- Most hyphae of ascomycetes and basidiomycetes have perforated septa through which the cytoplasm, but not necessarily the nuclei, flows freely.
- The yeasts are a group of unicellular fungi, mostly ascomycetes. They are commercially important because of their roles in baking and fermentation.
- The imperfect fungi are a large artificial group of fungi in which sexual reproduction does not occur or is not known; most are ascomycetes.

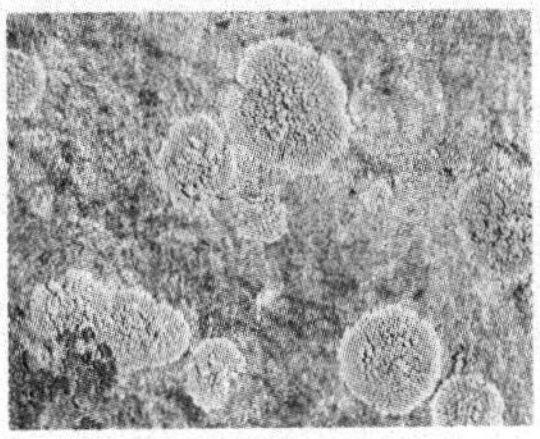

Fungal Associations

15.10 Lichens
15.11 Mycorrhizae

- Lichens are symbiotic associations between a fungus and a photosynthetic partner.
- Mycorrhizae are symbiotic associations between a fungus and the roots of a plant.

15.1 Complex Multicellularity

The algae are structurally simple multicellular organisms that fill the evolutionary gap between unicellular protists and more complexly multicellular organisms (fungi, plants, and animals). In **complex multicellular organisms,** individuals are composed of many highly specialized kinds of cells that coordinate their activities. There are three kingdoms:

1. **Plants.** Multicellular green algae were almost certainly the direct ancestors of the plants (see chapter 16) and were themselves considered plants in the nineteenth century. However, green algae are basically aquatic and much simpler in structure than plants and are considered protists in the six-kingdom system used widely today.
2. **Animals.** Animals arose from a unicellular protist ancestor. Several groups of animal-like protists have been considered to be tiny animals in the past ("protozoa"), including flagellates, ciliates, and amoebas. The simplest (and seemingly most primitive) animals today, the sponges, seem clearly to have evolved from a kind of flagellate.
3. **Fungi.** Fungi also arose from a unicellular protist ancestor, one different from the ancestor of animals. Certain protists, including slime molds and water molds, have been considered fungi ("molds"), although they are usually classified as protists and are not thought to resemble ancestors of fungi. The true protist ancestor of fungi is as yet unknown. This is one of the great unsolved problems of taxonomy.

Two key characteristics of complex multicellular organisms distinguish them from simple multicellular organisms like marine algae: **cell specialization** and **intercell coordination.** In a fungus, plant, or animal, the body of an individual possesses different kinds of cells that have very different structures and are coordinated in complex ways.

Perhaps the most important characteristic of complex multicellular organisms is cell specialization. If you think about it, having a variety of different sorts of cells within the same individual implies something very important about the genes of the individual: *different cells are using different genes!* The process whereby a single cell (in humans, a fertilized egg) becomes a multicellular individual with many different kinds of cells is called **development.** The cell specialization that is the hallmark of complex multicellular life is the direct result of cells developing in different ways by activating different genes.

Figure 15.1 Fungi.
A shelf fungus, *Trametes versicolor.*

A second key characteristic of complex multicellular organisms is intercell coordination, the adjustment of a cell's activity in response to what other cells are doing. The cells of all complex multicellular organisms communicate with one another with chemical signals called hormones. In some organisms like sponges there is relatively little coordination between the cells; in other organisms like humans almost every cell is under complex coordination.

There are three kingdoms of complexly multicellular organisms: Fungi, Plantae, and Animalia. With the exception of yeasts among the fungi, all members of these three great kingdoms are composed of individuals that are complexly multicellular. In the remainder of this chapter we consider the fungi (figure 15.1). In the next three chapters, we examine plants. Then in the following three chapters, we explore in detail how animals have evolved.

15.1 Fungi, plants, and animals are complexly multicellular, with specialized cell types and coordination between cells.

15.2 A Fungus Is Not a Plant

The fungi are a distinct kingdom of organisms, comprising about 73,000 named species. **Mycologists,** scientists who study fungi, believe there may be many more species in existence. Although fungi have traditionally been included in the plant kingdom, they lack chlorophyll and resemble plants only in their general appearance and lack of mobility. Significant differences between fungi and plants include the following:

Fungi are heterotrophs. Perhaps most obviously, a mushroom is not green. Virtually all plants are photosynthesizers, while no fungi carry out photosynthesis. Instead, fungi obtain their food by secreting digestive enzymes onto whatever they are attached to and then absorbing into their bodies the organic molecules that are released by the enzymes.

Fungi have filamentous bodies. Fungi are basically filamentous in their growth form (that is, their body consists of long, slender filaments), even though these filaments may be packed together to form complex structures like the mushroom.

Fungi have nonmotile sperm. Some plants have motile sperm with flagella. No fungi do.

Fungi have cell walls made of chitin. The cell walls of fungi contain chitin, the same tough material that a crab shell is made of. The cell walls of plants are made of cellulose, also a strong building material. Chitin, however, is far more resistant to microbial degradation than is cellulose.

Fungi have nuclear mitosis. Mitosis in fungi is different from plants or any other eukaryote in one key respect: the nuclear envelope does not break down and re-form. Instead, all of mitosis takes place *within* the nucleus. A spindle apparatus forms there, dragging chromosomes to opposite poles of the *nucleus* (not the cell, as in all other eukaryotes).

You could build a much longer list, but already the take-home lesson is clear: Fungi are not like plants at all! Their many unique features are strong evidence that fungi are not closely related to any other group of organisms.

The Body of a Fungus

Fungi exist mainly in the form of slender filaments, barely visible with the naked eye, called **hyphae** (singular, **hypha**). A hypha is basically a long string of cells. The walls dividing one cell from another are called **septa** (singular, **septum**). The presence of septa is another of the ways in which fungi differ fundamentally from all other multicellular organisms, for the septa rarely form a complete barrier! From one fungal cell to the next, cytoplasm flows, streaming freely down the hypha through openings in the septa.

Figure 15.2 A mycelium.

A mycelium is a dense, interwoven mat of fungal hyphae. Most of the body of a fungus is occupied by its mycelium. This fungal mycelium is growing through leaves on the forest floor in Maryland.

The main body of a fungus is not the mushroom, which is a temporary reproductive structure, but rather the extensive network of fine hyphae that penetrate the soil, wood, or flesh in which the fungus is growing. A mass of hyphae is called a **mycelium** (plural, **mycelia**) and may contain many meters of individual hyphae (figure 15.2). This body organization creates a unique relationship between the fungus and its environment. All parts of the fungal body are metabolically active, secreting digestive enzymes and actively attempting to digest and absorb any organic material with which the fungus comes in contact.

Because of cytoplasmic streaming, many nuclei may be connected by the shared cytoplasm of a fungal mycelium. None of them (except for reproductive cells) are isolated in any one cell; all of them are linked cytoplasmically with every cell of the mycelium. Indeed, the entire concept of multicellularity takes on a new meaning among the fungi, the ultimate communal sharers among the multicellular organisms.

15.2 Fungi are not at all like plants. The fungal body is basically a long string of cells, often interconnected.

15.3 Reproduction and Nutrition of Fungi

How Fungi Reproduce

Fungi reproduce both asexually and sexually. All fungal nuclei except for the zygote's are haploid. Often in the sexual reproduction of fungi, individuals of different "mating type" must participate, much as two sexes are required for human reproduction. Sexual reproduction is initiated when two hyphae of genetically different mating types come in contact, and the hyphae fuse. What happens next? In animals and plants, when the two haploid gametes fuse, the two haploid nuclei immediately fuse to form the diploid nucleus of the zygote. As you might by now expect, fungi handle things differently. In most fungi, the two nuclei do not fuse immediately. Instead, they remain unmarried inhabitants of the same house, coexisting in a common cytoplasm for most of the life of the fungus! A fungal hypha that has nuclei within it derived from two genetically different individuals is called a **heterokaryon** (Greek, *heteros,* other, and *karyon,* kernel or nucleus). A fungal hypha in which all the nuclei are genetically similar is said to be a **homokaryon** (Greek, *homo,* one).

When reproductive structures are formed in fungi, complete septa form between cells, the only exception to the free flow of cytoplasm between cells of the fungal body. There are three kinds of reproductive structures: (1) **gametangia,** within which gametes form; (2) **sporangia,** within which spores form; and (3) **conidia,** a form of asexual spore.

Spores are a common means of reproduction among the fungi (figure 15.3). They are well suited to the needs of an organism anchored to one place. They are so small and light that they may remain suspended in the air for long periods of time and may be carried great distances. When a spore lands in a suitable place, it germinates and begins to divide, soon giving rise to a new fungal hypha.

Figure 15.3 Many fungi produce spores.
Spores explode from the surface of a puffball fungus.

How Fungi Obtain Nutrients

All fungi obtain their food by secreting digestive enzymes into their surroundings and then absorbing back into the fungus the organic molecules produced by this **external digestion.** Many fungi are able to break down the cellulose in wood, cleaving the linkages between glucose subunits and then absorbing the glucose molecules as food. That is why fungi are so often seen growing on dead trees.

Just as some plants like the Venus's-flytrap are active carnivores, so some fungi are active predators. For example, the edible oyster fungus *Pleurotus ostreatus* attracts tiny roundworms known as nematodes that feed on it—and secretes a substance that anesthetizes them (figure 15.4). When the worms become sluggish and inactive, the fungal hyphae envelop and penetrate their bodies and absorb their contents, a rich source of nitrogen (always in short supply in natural ecosystems). Other fungi are even more active predators, snaring or trapping prey or firing projectiles into nematodes, rotifers, and other small animals that come near.

Figure 15.4 The oyster mushroom.
This species, *Pleurotus ostreatus,* immobilizes nematodes, which the fungus uses as a source of food.

15.3 Fungi reproduce both asexually and sexually. They obtain their nutrients by secreting digestive enzymes into their surroundings and then absorbing the digested molecules back into the fungal body.

15.4 Ecology of Fungi

Fungi, together with bacteria, are the principal decomposers in the biosphere. They break down organic materials and return the substances locked in those molecules to circulation in the ecosystem. Fungi are virtually the only organisms capable of breaking down lignin, one of the major constituents of wood. By breaking down such substances, fungi release critical building blocks, such as carbon, nitrogen, and phosphorus, from the bodies of dead organisms and make them available to other organisms.

In breaking down organic matter, some fungi attack living plants and animals as a source of organic molecules, while others attack dead ones. Fungi often act as disease-causing organisms for both plants (figure 15.5) and animals, and they are responsible for billions of dollars in agricultural losses every year. Not only are fungi the most harmful pests of living plants, but they also attack food products once they have been harvested and stored. In addition, fungi often secrete substances into the foods that they are attacking that make these foods unpalatable or poisonous.

The same aggressive metabolism that makes fungi ecologically important has been put to commercial use in many ways. The manufacture of both bread and beer depends on the biochemical activities of **yeasts,** single-celled fungi that produce abundant quantities of ethanol and carbon dioxide. Cheese and wine achieve their delicate flavors because of the metabolic processes of certain fungi, and others make possible the manufacture of soy sauce. Vast industries depend on the biochemical manufacture of organic substances such as citric acid by fungi in culture, and yeasts are now used on a large scale to produce protein for the enrichment of animal food. Many antibiotics, including the first one that was used on a wide scale, penicillin, are derived from fungi.

Some fungi are used to convert one complex organic molecule into another, cleaning up toxic substances in the environment. For example, at least three species of fungi have been isolated that combine selenium, accumulated at the San Luis National Wildlife Refuge in California's San Joaquin Valley, with harmless volatile chemicals—thus removing it from the soil.

Two kinds of mutualistic associations between fungi and autotrophic organisms are ecologically important. Lichens are symbiotic associations between fungi and either green algae or cyanobacteria. They are prominent nearly everywhere in the world, especially in unusually harsh habitats such as bare rock. Mycorrhizae, specialized symbiotic associations between the roots of plants and fungi, are characteristic of about 80% of all plants. In each of them, the photosynthetic organisms fix atmospheric carbon dioxide and thus make organic material available to the fungi. The metabolic activities of the fungi, in turn, enhance the overall ability of the symbiotic association to exist in a particular habitat. In the case of mycorrhizae, the fungal partner expedites the plant's absorption of essential nutrients such as phosphorus. Both of these associations are discussed further in this chapter.

15.4 Fungi are key decomposers within almost all terrestrial ecosystems and play many other important ecological and commercial roles.

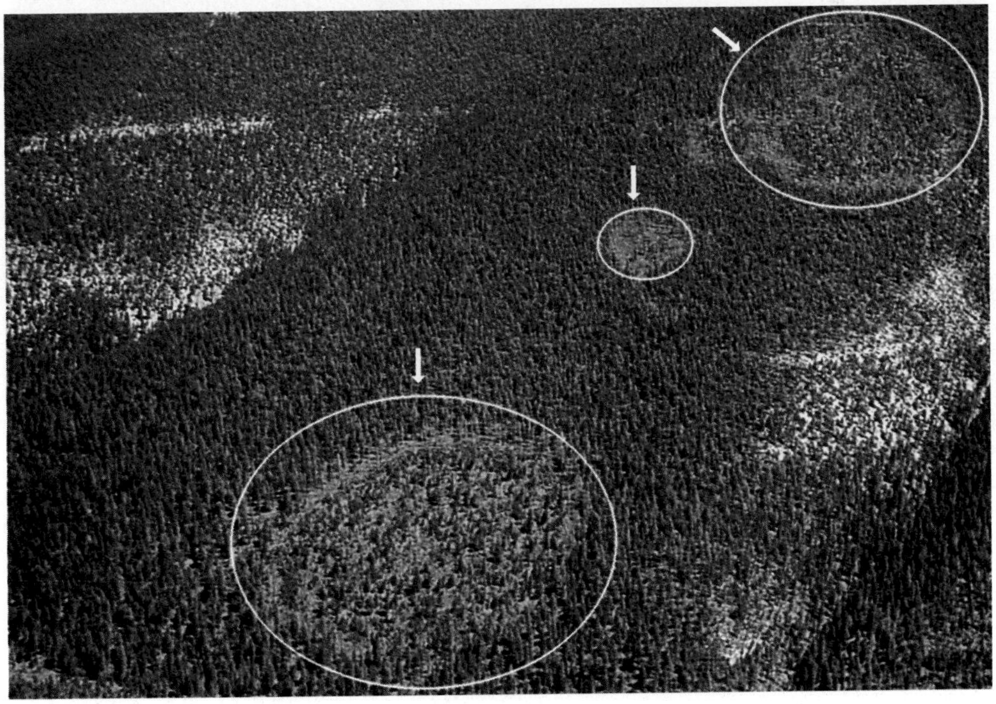

Figure 15.5 World's largest organism?

Armillaria, a pathogenic fungus shown here afflicting three discrete regions of coniferous forest in Montana, grows from a central focus as a single clone. The large patch at the bottom of the picture is almost 8 hectares in diameter. The largest clone measured so far has been 15 hectares in diameter—pretty impressive for a single individual!

15.5 Kinds of Fungi

Fungi are an ancient group of organisms at least 400 million years old. There are nearly 73,000 described species, in four groups (table 15.1), and many more awaiting discovery. Many fungi are harmful because they decay, rot, and spoil many different materials as they obtain food and because they cause serious diseases in animals and particularly in plants. Other fungi, however, are extremely useful. The manufacture of both bread and beer depends on the biochemical activities of yeasts, single-celled fungi that produce abundant quantities of carbon dioxide and ethanol. Fungi are used on a major scale in industry to convert one complex organic molecule into another; many commercially important steroids are synthesized in this way.

The three fungal phyla, distinguished from one another primarily by their mode of sexual reproduction, are the zygomycetes, the ascomycetes, and the basidiomycetes. The evolutionary relationships among these three phyla and the imperfect fungi are not clear, although the zygomycetes, which we examine first, are the simplest.

15.5 The fungal phyla are distinguished primarily by their modes of sexual reproduction. Fungi whose sexual stage has not been observed are lumped into the so-called imperfect fungi.

TABLE 15.1 FUNGI

Phylum	Typical Examples		Key Characteristics	Approximate Number of Living Species
Ascomycota	Yeasts, truffles, morels		Develop by sexual means; ascospores are formed inside a sac called an ascus; asexual reproduction is also common	32,000
Imperfect fungi	*Aspergillus*, *Penicillium*		Sexual reproduction has not been observed; most are thought to be ascomycetes that have lost the ability to reproduce sexually	17,000
Basidiomycota	Mushrooms, toadstools, rusts		Develop by sexual means; basidiospores are borne on club-shaped structures called basidia; the terminal hyphal cell that produces spores is called a basidium; asexual reproduction occurs occasionally	22,000
Zygomycota	*Rhizopus* (black bread mold)		Develop sexually and asexually; multinucleate hyphae lack septa, except for reproductive structures; fusion of hyphae leads directly to formation of a zygote, in which meiosis occurs just before it germinates	1,050

15.6 Zygomycetes

The **zygomycetes,** members of the phylum Zygomycota, are unique among the fungi in that the fusion of hyphae does not produce a heterokaryon. Instead, the two nuclei fuse and form a single diploid nucleus. Just as the fusion of sperm and egg produces a zygote in plants and animals, so this fusion produces a zygote. The name *zygomycetes* means "fungi that make zygotes."

Zygomycetes are the exception to the rule among fungi, and there are not many kinds of them—only about 1,050 named species (about 1% of the named fungi). Included among them are some of the most frequent bread molds (the so-called black molds) and many microscopic fungi found on decaying organic material.

Reproduction among the zygomycetes is typically asexual. A cell at the tip of a hypha becomes walled off by a complete septum, forming an erect stalk tipped by a sporangium within which haploid spores are produced (figure 15.6). These spores are shed into the wind and blown to new locations, where they germinate and attempt to start new mycelia. Sexual reproduction is unusual but may occur in time of stress. It leads to the production of a particularly sturdy and resistant structure called a **zygosporangium** (figure 15.7). The zygosporangium is a very effective survival mechanism, a resting structure that allows the organism to remain dormant for long periods of time when conditions are not favorable.

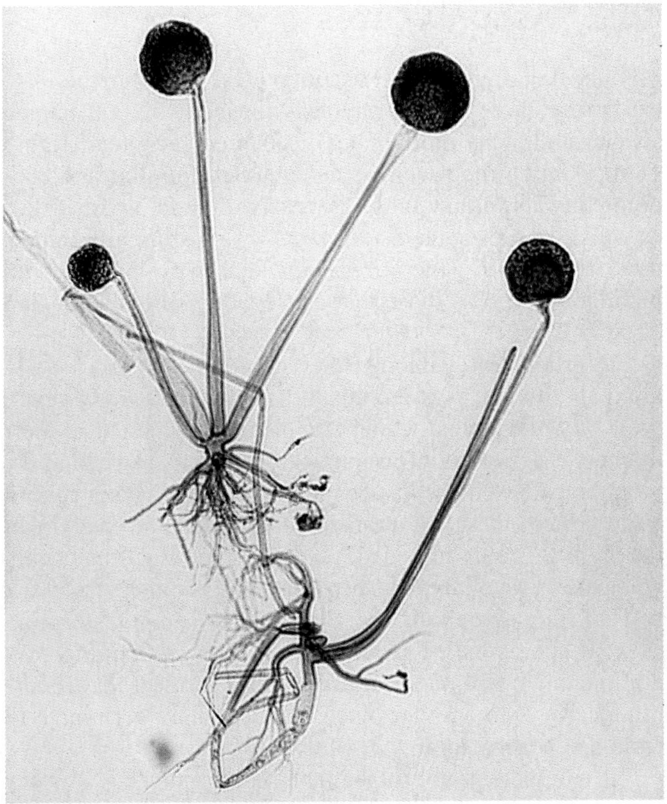

Figure 15.6 A zygomycete.
The spore-bearing structures of *Rhizopus* are about a centimeter tall.

15.6 Zygomycetes are unusual fungi that typically reproduce asexually; when hyphae do fuse, a zygote (one $2n$ nucleus), rather than a heterokaryon (two n nuclei), is produced.

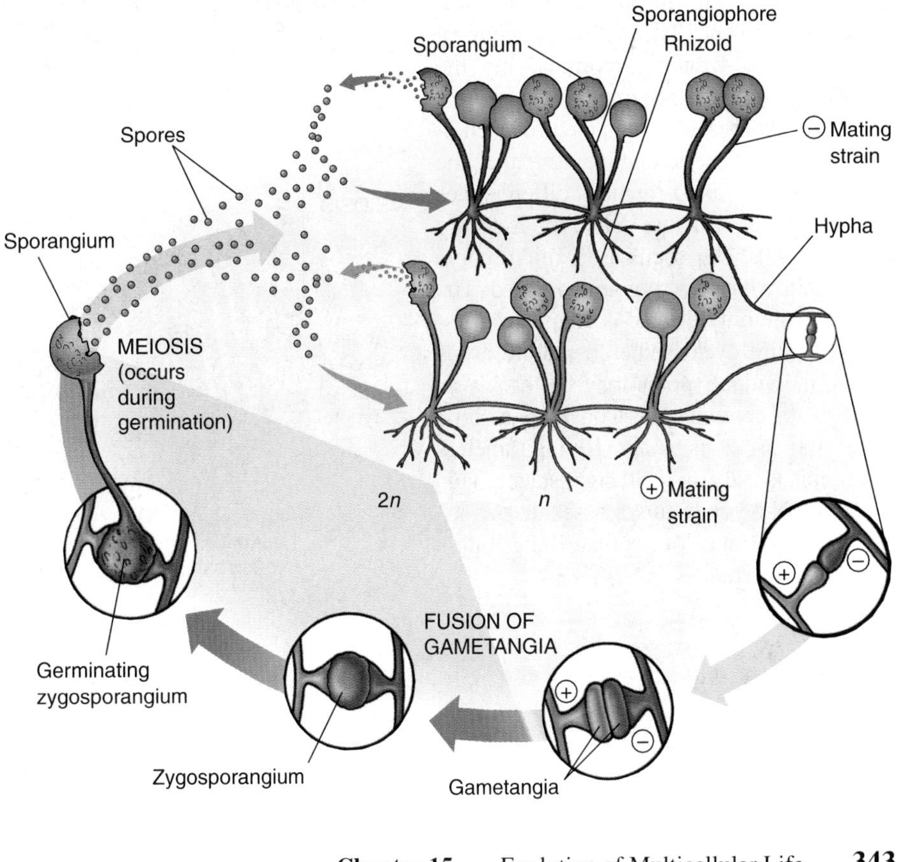
Sporangium
Sporangiophore
Rhizoid
Spores
⊖ Mating strain
Hypha
Sporangium
MEIOSIS (occurs during germination)
2n
n
⊕ Mating strain
+
−
FUSION OF GAMETANGIA
Germinating zygosporangium
Zygosporangium
Gametangia

Figure 15.7 Life cycle of a zygomycete.
The hyphae grow over the surface of the bread or other material on which the fungus feeds, producing erect, sporangium-bearing stalks in clumps. If two hyphae grow together, their nuclei may fuse, producing a zygote. This zygote, which is the only diploid cell of the life cycle, acquires a thick, black coat and is then called a zygosporangium. Meiosis occurs during its germination, and normal, haploid hyphae grow from the haploid cells that result from this process.

15.7 Ascomycetes

Phylum Ascomycota, the **ascomycetes,** is by far the largest of the three fungal phyla, with about 32,000 named species and many more being discovered each year (figure 15.8). Among the ascomycetes are such familiar and economically important fungi as yeasts, morels, and truffles, as well as molds such as *Neurospora* (a historically important organism in genetic research) and many plant fungal pathogens, such as those that cause Dutch elm disease and chestnut blight.

Reproduction among the ascomycetes is usually asexual, just as it is among the zygomycetes. However, unlike zygomycetes, which completely lack septa in their hyphae, the hyphae of ascomycetes possess septa that divide the cells. The septa are incomplete and have a central large pore in them, so the flow of cytoplasm up and down the hypha is not impeded. Asexual reproduction occurs when the tips of hyphae become fully isolated from the rest of the mycelium by a complete septum, forming asexual spores called conidia, each often containing several nuclei. When one of these conidia is released, air currents carry it to another place, where it may germinate to form a new mycelium.

It is important not to get confused by the number of nuclei in conidia. These multinucleate spores are *haploid,* not diploid, because there is only one version of the genome (the set of ascomycete chromosomes) present, while in a diploid cell there are two genetically different sets of chromosomes present. The actual number of nuclei is not what's important—it's the number of different genomes.

The ascomycetes are named for a characteristic sexual reproductive structure, the **ascus** (plural, **asci**), which differentiates within a larger structure called the ascocarp (figure 15.9). The ascus is a microscopic cell within which the zygote is formed. The zygote is the only diploid nucleus of the ascomycete life cycle. When a mature ascus bursts, individual spores may be thrown as far as 30 centimeters. Considering how small the ascus is (only 10 micrometers long), this is truly an amazing distance. This would be like you hitting a baseball 1.25 kilometers, 10 times longer than Babe Ruth's longest home run!

Figure 15.8 An ascomycete.
This cup fungus, phylum Ascomycota, is found in Trinidad.

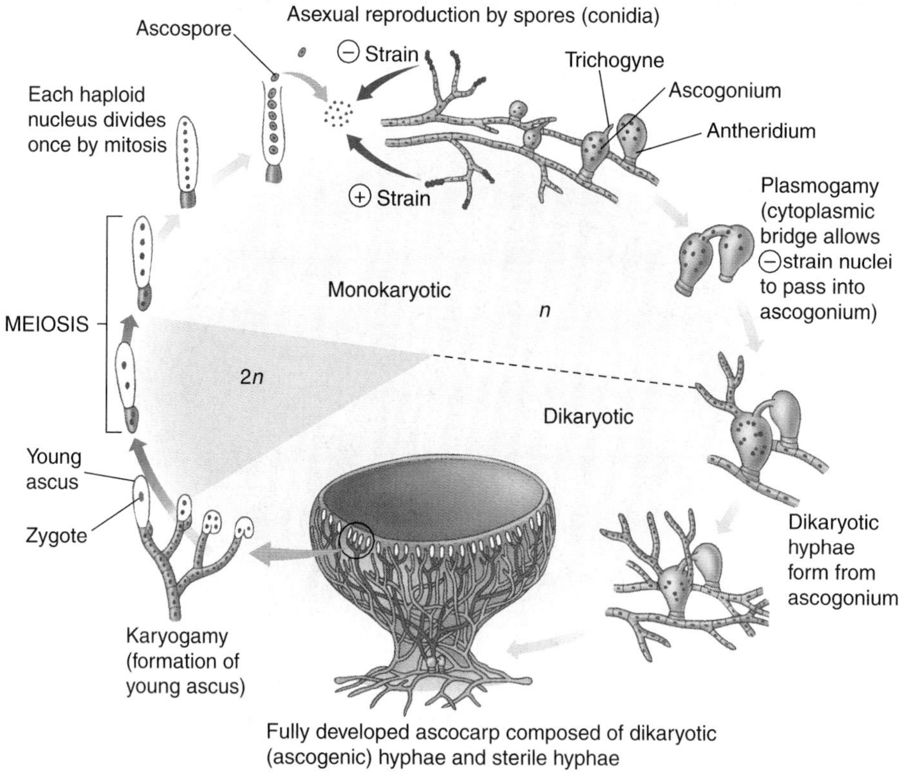
Asexual reproduction by spores (conidia)
Ascospore
⊖ Strain
Trichogyne
Ascogonium
Antheridium
Each haploid nucleus divides once by mitosis
⊕ Strain
Plasmogamy (cytoplasmic bridge allows ⊖strain nuclei to pass into ascogonium)
Monokaryotic
n
MEIOSIS
2n
Dikaryotic
Young ascus
Zygote
Dikaryotic hyphae form from ascogonium
Karyogamy (formation of young ascus)
Fully developed ascocarp composed of dikaryotic (ascogenic) hyphae and sterile hyphae

Figure 15.9 Life cycle of an ascomycete.
Asexual reproduction takes place by means of conidia, spores cut off by septa at the ends of modified hyphae. Sexual reproduction occurs when the female gametangium, or ascogonium, fuses with the male gametangium, or antheridium, through a structure called the trichogyne. Following zygote formation, the ascocarp develops.

15.7 Most fungi are ascomycetes, which form zygotes within reproductive structures called asci.

15.8 Basidiomycetes

The phylum Basidiomycota contains the most familiar of the fungi among their 22,000 named species—the mushrooms, toadstools, puffballs, and shelf fungi. Many mushrooms are used as food, but others are deadly poisonous (figure 15.10). Some species are cultivated as crops—the button mushroom *Agaricus campestris* is grown in more than 70 countries, producing a crop in 1998 with a value of over $15 billion. Also among the **basidiomycetes** are the rusts and smuts, many of them responsible for important plant diseases.

The life cycle of a basidiomycete starts with the production of a hypha from a germinating spore (figure 15.11). These hyphae lack septa at first, just as in zygomycetes. Eventually, however, septa are formed between each of the nuclei—but as in ascomycetes, there are holes in these cell separations, allowing cytoplasm to flow freely between cells. These hyphae grow, forming complex mycelia, and when hyphae of two different mating types fuse, the dikaryon that results goes on to form a dikaryotic mycelium.

The two nuclei in each cell of a dikaryotic hypha can coexist together for a very long time without fusing. Unlike the other two fungal phyla, asexual reproduction is infrequent among the basidiomycetes; they almost always reproduce sexually.

In sexual reproduction, zygotes (the only diploid cells of the life cycle) form when the two nuclei of dikaryotic cells fuse. This occurs within a club-shaped reproductive structure called the **basidium.** Around the many thousands of basidia in which this starts to occur, the mycelium forms a complex structure made of dikaryotic hyphae called the basidiocarp, or **mushroom.** Meiosis occurs in each basidium, forming haploid spores. The basidia occur in a dense layer on the underside of the cap of the mushroom, where the surface is folded like an accordion. It has been estimated that a mushroom with an 8-centimeter cap produces as many as 40 million spores per hour!

15.8 Mushrooms are basidiomycetes, which form club-shaped reproductive structures called basidia.

Figure 15.10 A basidiomycete.
The death cap mushroom, *Amanita phalloides,* is usually fatal when eaten.

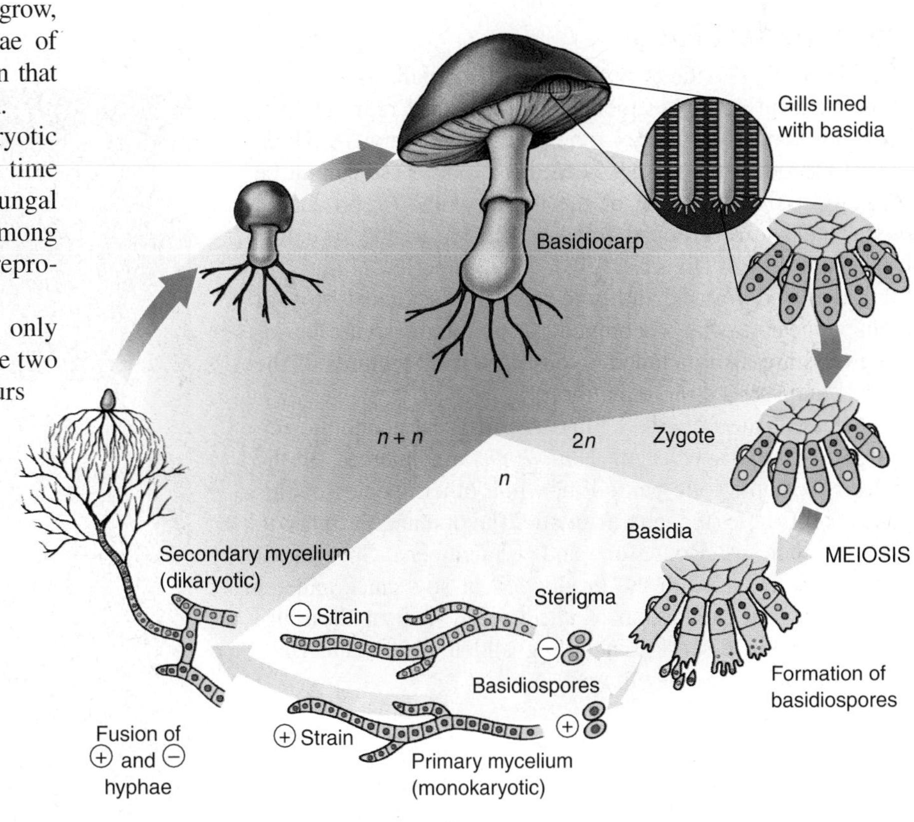

Gills lined with basidia
Basidiocarp
n + n
2n
Zygote
n
Basidia
MEIOSIS
Secondary mycelium (dikaryotic)
⊖ Strain
Sterigma
Basidiospores
Formation of basidiospores
Fusion of ⊕ and ⊖ hyphae
⊕ Strain
Primary mycelium (monokaryotic)

Figure 15.11 Life cycle of a basidiomycete.
Basidiomycetes usually reproduce sexually, with the fusion of nuclei in the basidia to produce a zygote. Meiosis follows syngamy and produces spores that eventually form a basidiocarp.

15.9 Unicellular and Asexual Fungi

Yeasts

Yeast is the generic (general) name given to unicellular fungi. Although single-celled, yeasts appear almost certainly to have been derived from multicellular ancestors. Some yeasts have been derived from each of the three phyla of fungi, most of them from the ascomycetes. There are about 250 named species of yeasts, including *Saccharomyces cerevisiae,* or baker's yeast, used for thousands of years in the production of bread, beer, and wine (figure 15.12). Other yeasts are important pathogens, including *Candida,* a common source of vaginal infection.

Just as in ascomycetes, most of yeast reproduction is asexual and takes place by cell fission or budding (the formation of a small cell from a portion of a larger one). Sexual reproduction among yeasts occurs when two yeast cells fuse. The new cell containing two nuclei functions as an ascus. After the two nuclei fuse, meiosis produces four ascospores, which develop directly into new yeast cells.

Figure 15.12 The use of yeasts.
Baking bread involves the metabolic activities of the yeasts, single-celled ascomycetes. Wine making also uses the metabolic properties of yeasts, and the same species, *Saccharomyces cerevisiae,* is usually employed for both processes. In baking bread, yeasts generate carbon dioxide; in making beer or wine, yeasts are used to produce alcohol.

The Imperfect Fungi

In addition to the three phyla of fungi, which, as we have seen, differ primarily in their mode of sexual reproduction, there are some 17,000 described species of fungi in whom sexual reproduction has not been observed. These cannot be formally assigned to one of the three sexually reproducing phyla and so are grouped for convenience as the so-called **imperfect fungi** (figure 15.13). The imperfect fungi are fungi that have lost the ability to reproduce sexually. Most of them appear to be ascomycetes, although some basidiomycetes are also included—you can tell by features of the hyphae and asexual reproduction.

Many of the imperfect fungi are of great economic importance. Some species of *Penicillium* are sources of the well-known antibiotic penicillin, while other species of this genus contribute the characteristic flavors and aromas to cheeses such as Roquefort and Camembert. Species of *Aspergillus* are used in the production of soy sauce and for the commercial production of citric acid. Most of the fungi that cause skin diseases, including athlete's foot and ringworm, are also imperfect fungi.

15.9 Yeasts are unicellular fungi. Imperfect fungi have lost the ability to reproduce sexually. Many are of great economic importance.

Figure 15.13 Imperfect fungi.
Imperfect fungi are fungi in which sexual reproduction is unknown. *Verticillium alboatrum,* an important pathogen of alfalfa, has whorled conidia. The single-celled conidia of this member of the imperfect fungi are borne at the ends of the conidiophores.

15.10 Lichens

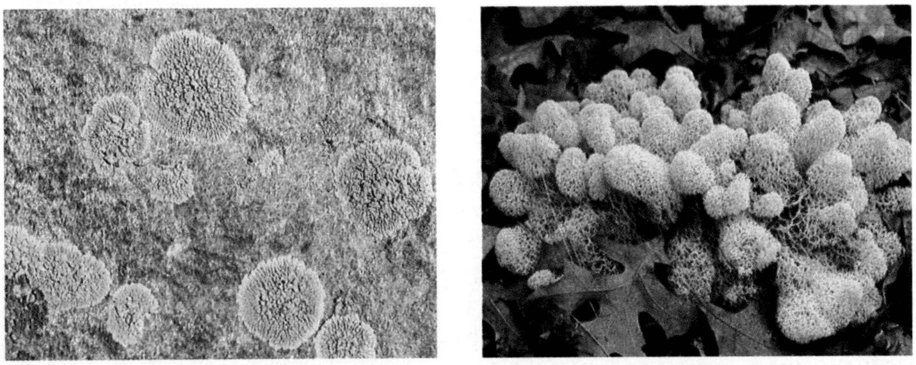

(a) (b)

Figure 15.14 Two types of lichens.

(*a*) Crustose (encrusting) lichens growing on a rock in California. (*b*) A fruticose lichen, *Cladina evansii,* growing on the ground in Florida. Fruticose lichens also grow in deserts because they are efficient in capturing water from moist air.

By now you will be convinced that fungi are very unlike the other multicellular organisms—algae, animals, and plants. But you should not conclude from this that fungi have little to do with other organisms other than to consume their dead bodies. The truth is just the opposite: fungi are involved in a variety of intimate symbiotic associations with algae and plants that play very important roles in the biological world. Mutualism is a form of symbiosis (recall our discussion of endosymbiosis in chapter 14) in which each partner benefits. Among the fungi, these symbiotic associations typically involve a sharing of abilities between a heterotroph (the fungus) and a photosynthesizer (the algae or plant). The fungus contributes the ability to absorb minerals and other nutrients very efficiently from the environment; the photosynthesizer contributes the ability to use sunlight to power the building of organic molecules. Alone, the fungus has no source of food, the photosynthesizer no source of nutrients. Together, each has access to both food and nutrients, a partnership in which both participants benefit.

A **lichen** is a symbiotic association between a fungus and a photosynthetic partner. Ascomycetes are the fungal partners in all but 20 of the 15,000 different species of lichens that have been characterized. Most of the visible body of a lichen consists of its fungus, but interwoven between layers of hyphae within the fungus are cyanobacteria, green algae, or sometimes both. Enough light penetrates the translucent layers of hyphae to make photosynthesis possible. Specialized fungal hyphae envelop and sometimes penetrate the photosynthetic cells, serving as highways to collect and transfer to the fungal body the sugars and other organic molecules manufactured by the photosynthetic cells. The fungus transmits special biochemical signals that direct the cyanobacteria or green algae to produce metabolic substances that they would not if growing independently of the fungus. Indeed, the fungus is not able to grow or survive without its photosynthetic partner. Many biologists characterize this particular symbiotic relationship as one of slavery rather than cooperation, a controlled parasitism of the photosynthetic organism by the fungal host.

The durable construction of the fungus, combined with the photosynthetic abilities of its partner, has enabled lichens to invade the harshest of habitats, from the tops of mountains to dry, bare rock faces in the desert (figure 15.14). In such harsh, exposed areas, lichens are often the first colonists, breaking down the rocks and setting the stage for the invasion of other organisms. A key component of such primary succession, as it is called, are lichens with cyanobacteria that are able to fix atmospheric nitrogen. The activities of these lichens introduce usable nitrogen into the environment in the form of ammonia or organic molecules, where it can be used by other pioneering organisms. Without this nitrogen, the other organisms could not survive.

Lichens are able to survive drying or freezing by converting to a condition we might call suspended animation. When moisture and warmth return, the lichen recovers quickly and resumes its normal metabolic activities, including photosynthesis. In harsh environments, the growth of the lichen may be extremely slow. Some high-mountain lichens covering an area no larger than your fist actually appear to be thousands of years old and therefore are among the oldest living things on earth.

Lichens are extremely sensitive to pollutants in the atmosphere, because they readily absorb substances dissolved in rain and dew. This is why lichens are generally absent in and around cities—they are acutely sensitive to sulfur dioxide produced by automobile traffic and industrial activity. Such pollutants destroy their chlorophyll molecules and thus decrease photosynthesis and upset the physiological balance between the fungus and the algae or cyanobacteria. Degradation or destruction of the lichen results.

15.10 Lichens are symbiotic associations between a fungus and a photosynthetic partner (a cyanobacterium or an alga).

15.11 Mycorrhizae

The roots of about 80% of all kinds of plants are involved in symbiotic associations with certain kinds of fungi. In fact, it has been estimated that fungi account for as much as 15% of the total weight of the world's plant roots! Associations of this kind are called **mycorrhizae** (Greek *myco,* fungus, and *rhizos,* roots) (figure 15.15). In a mycorrhiza, filaments of the fungus act as superefficient root hairs, projecting out from the epidermis, or outermost cell layer, of the terminal portions of the root. The fungal filaments aid in the direct transfer of phosphorus and other minerals from the soil into the roots of the plant, while the plant supplies organic carbon to the symbiotic fungus. Because both partners benefit substantially, the association represents a fine example of mutualism (figure 15.16).

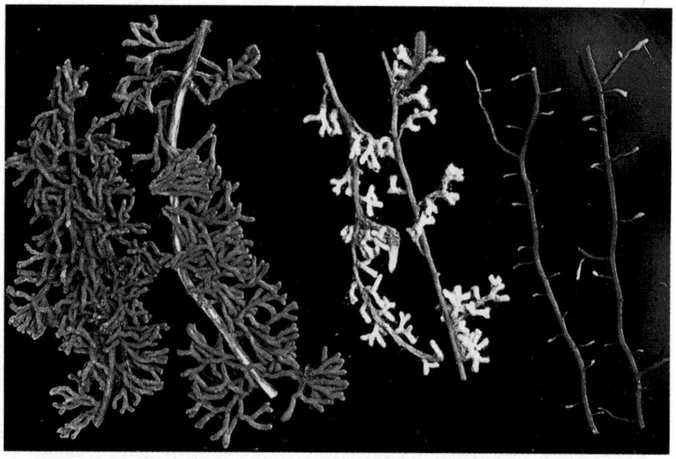

Figure 15.15 Mycorrhizae on the roots of pines.
From left to right are yellow-brown ectomycorrhizae formed by *Pisolithus*, white ectomycorrhizae formed by *Rhizopagon*, and pine roots not associated with a fungus.

Endomycorrhizae

In almost all mycorrhizae, involving perhaps more than 200,000 species of plants, the fungal hyphae of the mycelium actually penetrate the outer cells of the plant root, as well as extending far out into the soil. These are called **endomycorrhizae.** You might be surprised to learn that all of this symbiotic activity is being carried out worldwide by a very small group of fungi—only 30 species of zygomycetes are known to be involved in these hundreds of thousands of associations!

The earliest fossil plants often have endomycorrhizal roots, which are thought to have played an important role in the invasion of land by plants. The soils of that time would have completely lacked in organic matter other than that contributed to beach sands by dead marine life, and mycorrhizal plants are particularly successful in such infertile soils. The most primitive vascular plants surviving today continue to depend strongly on endomycorrhizae.

IS
CK
GM
GE

Figure 15.16 How mutualism aids plant growth.
Mycorrhizae aid plant growth by helping the plant roots absorb nutrients from the soil. Soybeans without mycorrhizae (*far left*) grow far more slowly than soybeans with different strains of mycorrhizae (*center* and *right*).

Ectomycorrhizae

In perhaps 10,000 species of plants, the mycorrhizae surround but do not physically penetrate the plant root cells. These are called **ectomycorrhizae.** In marked contrast to endomycorrhizae, these nonpenetrating ectomycorrhizae represent highly specialized relationships, in which a particular species of plant has become associated with a particular fungus, usually a basidiomycete (although some are ascomycetes). At least 5,000 species of fungi have been identified in different ectomycorrhizae, most restricted to a single species of plant. These sorts of mycorrhizae are important because they involve many commercially significant trees in temperate regions, including pines, firs, oaks, beeches, and willows.

15.11 Mycorrhizae are symbiotic associations between fungi and plant roots.

CONCEPT REVIEW

1. The key advantage of multicellularity is that
   a. it allows the organism to be motile.
   b. it allows the formation of specialized tissues.
   c. photosynthesis is possible.
   d. sexual reproduction is possible.

2. A fungus is not a plant for a variety of reasons. Which of the following characteristics does not distinguish fungi from plants?
   a. The sperm of fungi do not have flagella.
   b. The cell walls of fungi are made of chitin.
   c. Fungi do not carry out photosynthesis.
   d. Fungi are multicellular.

3. A mass of fungal filaments is called a
   a. hyphae.
   b. mycelium.
   c. septum.
   d. colony.

4. Members of the three phyla of fungi reproduce
   a. only asexually.
   b. only sexually.
   c. both asexually and sexually.
   d. either asexually or sexually, but not both.

5. Which of the following statements describing how fungi eat is *not* true?
   a. Fungi secrete a sticky substance that can trap flies.
   b. Fungi can fire projectiles into animals such as rotifers.
   c. Fungi secrete digestive enzymes into their surroundings.
   d. Fungi can feed on dead trees.

6. A mushroom is a complex structure of dikaryotic hyphae called the
   a. basidiocarp.
   b. ascus.
   c. zygote.
   d. mycorrhizae.

7. The slender filaments that make up the body of a fungus are called _________.

8. Fungi are often growing on dead trees because they are able to break down the _________ in the wood and absorb glucose molecules as food.

9. _________ reproduction in the zygomycetes is unusual and frequently occurs in times of _________, when conditions are not favorable.

10. The fungal phylum _________ is the largest of the fungal phyla, with about 32,000 species named.

11. _________ is a member of the imperfect fungi and is a source of a well-known antibiotic.

12. Symbiotic associations between plants and fungi, called _________, aid in the uptake of _________ by plants.

CHALLENGE YOURSELF

1. If fungi have been so successful as to become an entire kingdom, why do you suppose the protist that first gave rise to the fungi has not persisted?

2. What is there about the way fungi live in nature that helps make them particularly valuable in industrial processes?

3. What role did mycorrhizae probably play in plants' invasion of land hundreds of millions of years ago?

4. Many antibiotics, including penicillin, are derived from fungi. Why do you think the fungi produce these substances? Of what use are they to the fungi? In addition, fungi often secrete substances into the food that they are attacking that make these foods unpalatable or even poisonous. What kind of advantage would these substances provide the fungi?

15 eBRIDGE

Reinforcing Key Points

Fungi as Multicellular Organisms

15.1 Complex Multicellularity

15.2 A Fungus Is Not a Plant

15.3 Reproduction and Nutrition of Fungi

15.4 Ecology of Fungi

Fungal Diversity

15.5 Kinds of Fungi

15.6 Zygomycetes

15.7 Ascomycetes

15.8 Basidiomycetes

15.9 Unicellular and Asexual Fungi

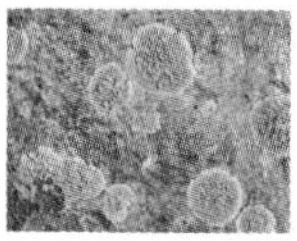

Fungal Associations

15.10 Lichens

15.11 Mycorrhizae

Electronic Learning

Visual Learning

Animations

Black Bread Mold, *Rhizopus stolonifer*

Club Fungi

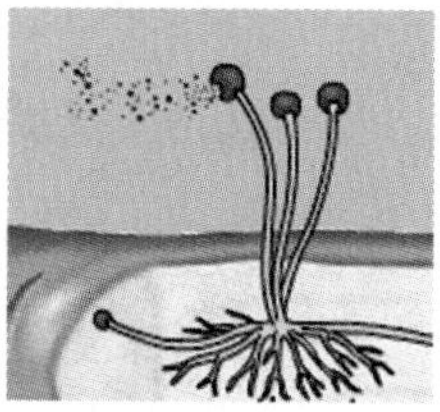

Lichen Morphology

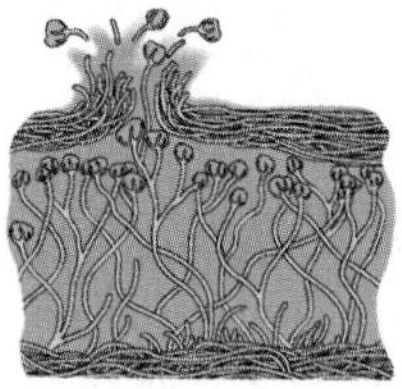

Author's Corner

Characteristics of the Fungi. Many unique features distinguish the groups of fungi from each other and the fungi as a whole from other organisms. It is thought that the ascomycetes and basidiomycetes are more closely related to each other than to zygomycetes, but where exactly the imperfect fungi fit in is under debate.

1. The body of a fungus
2. The three phyla of fungi
3. Phylum Zygomycota
4. Phylum Ascomycota
5. Phylum Basidiomycota
6. The imperfect fungi

Virtual Classroom

Unearthing the Roots of the Tree of Life

Organisms were first classified more than 2,000 years ago by the Greek philosopher Aristotle, who categorized all living things as being either animals or plants. Biologists now call each of these great groups of organisms a "kingdom" and recognize six kingdoms of life: animals, plants, fungi, and protists (all eukaryotes—that is, their cells have a nucleus), and two very different kinds of bacteria (both prokaryotes—their cells lack nuclei). When genomic DNA sequences of the two kingdoms of bacteria were first compared in 1996, the differences were astonishingly great, as great as either kingdom is from any eukaryote. Recognizing this, biologists increasingly group all four eukaryote kingdoms (animals, plants, fungi, and protists) together into one great category called a domain, and place each of the two kingdoms of bacteria into its own domain. How the three domains of life are related to each other is proving difficult to sort out, as there appears to have been considerable exchange of genes early in the history of life between the three groups. At its base, the tree of life seems to be less a trunk than a thicket.

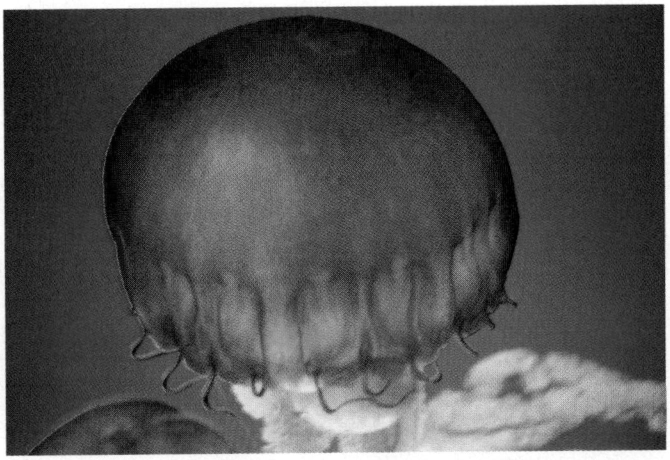

Virtual Lab

How Actin-binding Proteins Interact with the Cytoskeleton to Determine the Morphology of Yeasts

The single-celled yeast *Saccharomyces cerevisiae* is a model eukaryotic cell for scientific research. The ways in which yeasts differ from multicellular eukaryotes can reveal a great deal about the intrinsic nature of the multicellular lifestyle. David Drubin of the University of California, Berkeley, uses yeast to study cytoskeletal proteins involved in intracellular translocation and endocytosis. Actin, a cytoskeletal protein, is among the most highly conserved proteins of the eukaryotic cell. Drubin has been studying a highly conserved actin-binding protein called cofilin. Cofilin binds to actin filaments and stimulates depolymerization. Drubin is investigating the cellular role of cofilin. His studies focus on two temperature-sensitive mutant alleles of cofilin that interfere with actin depolymerization. In his studies, the mutant alleles exhibit a disruption of endocytosis but not of intracellular translocation. This suggests that cofilin's cellular role is limited to the endocytosis process, and that it does not play a broader role in intracellular transport processes.

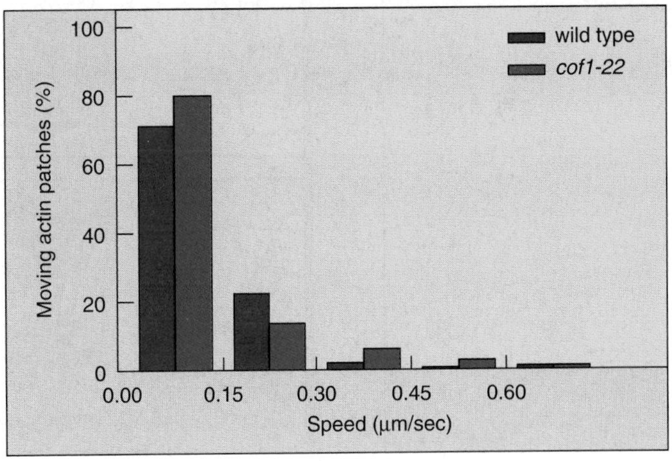
wild type
cof1-22
Moving actin patches (%)
100
80
60
40
20
0
0.00
0.15
0.30
0.45
0.60
Speed (μm/sec)

Quizzes

Further Reading

Essential Study Partner

Links

BioCourse.com

PART 5 PLANT LIFE

CHAPTER 16

EVOLUTION OF PLANTS

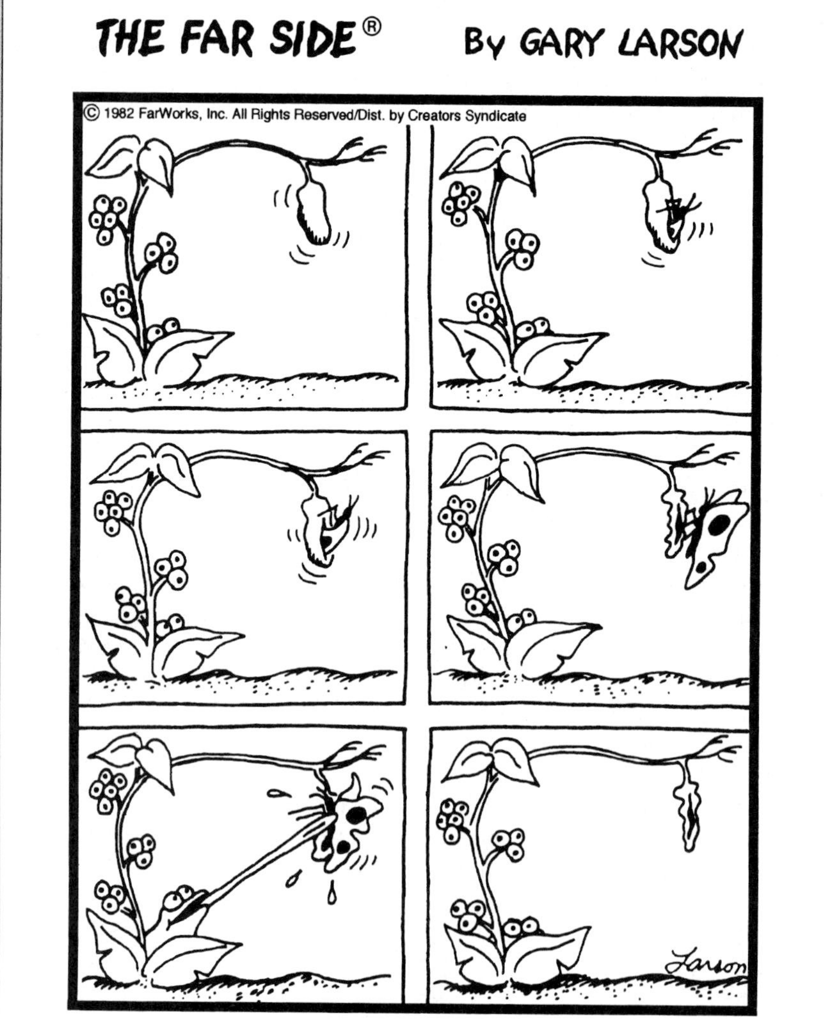
THE FAR SIDE®
By GARY LARSON
© 1982 FarWorks, Inc. All Rights Reserved/Dist. by Creators Syndicate
Larson

Metamorphosis

CHAPTER 16

Evolution of Plants

CHAPTER OVERVIEW

Plants

16.1 Adapting to Terrestrial Living

16.2 Evolution of a Vascular System

- Plants were faced with three key challenges in adapting to life on land: absorbing minerals, conserving water, and transferring gametes during sexual reproduction.
- Plants exhibit alternation of gametophyte (haploid) and sporophyte (diploid) generations.

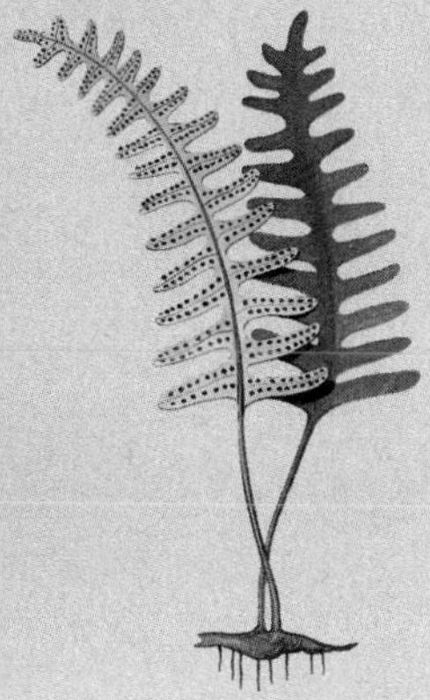

Seedless Plants

16.3 Nonvascular Plants

16.4 The Evolution of Vascular Tissue

16.5 Seedless Vascular Plants

- Sporophytes of mosses, liverworts, and hornworts are usually nutritionally dependent on the more conspicuous gametophytes.
- Vascular plants possess specialized water-conducting tissues.
- The life cycle of vascular plants is dominated by the sporophyte generation.
- Ferns, lycopods, horsetails, and whisk ferns are all seedless vascular plants.

The Advent of Seeds

16.6 Seed Plants

16.7 Gymnosperms

- A seed is a plant embryo with food reserves and a durable, watertight cover.
- In the gymnosperms, the ovules are not completely enclosed within sporophyte tissue when pollinated.
- Seeds often contain nourishment for the germinating seedling.
- Gymnosperms include the conifers, cycads, gnetophytes, and ginkgos.

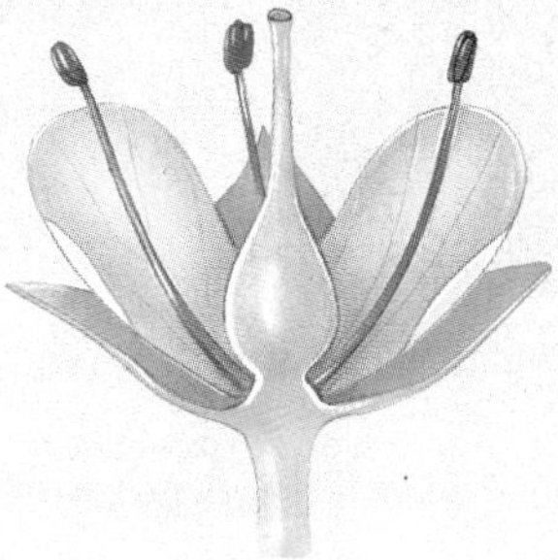

The Evolution of Flowers

16.8 Rise of the Angiosperms

16.9 Why There Are Different Kinds of Flowers

16.10 Improving Seeds: Double Fertilization

16.11 Improving Seed Dispersal: Fruits

- Angiosperm ovules are completely enclosed by the sporophyte carpel when pollinated.
- Angiosperms, or "flowering plants," are the most successful of all plants.
- Angiosperms are characterized by flowers and fruit, both of which aid dispersal of gametes.

16.1 Adapting to Terrestrial Living

Plants are complex multicellular organisms that are terrestrial **autotrophs**—that is, they occur almost exclusively on land and feed themselves by photosynthesis. The name *autotroph* comes from the Greek, *autos,* self, and *trophos,* feeder. Today, plants are the dominant organisms on the surface of the earth. An estimated 288,700 species are now in existence (figure 16.1), covering every part of the terrestrial landscape except the extreme polar regions and the highest mountaintops. Here we examine how plants adapted to life on land.

The green algae that were probably the ancestors of today's plants are aquatic organisms that are not well-adapted to living on land. Before their descendants could live on land, they had to overcome three environmental challenges. First, they had to absorb minerals from the rocky surface. Second, they had to find a means of conserving water. Third, they had to develop a way to reproduce on land.

Figure 16.1 There are many kinds of plants.
The plant kingdom is astonishingly diverse, often displaying remarkable adaptations.

Absorbing Minerals

Plants require relatively large amounts of six inorganic minerals: nitrogen, potassium, calcium, phosphorus, magnesium, and sulfur. Each of these minerals constitutes 1% or more of a plant's dry weight. Algae absorb these minerals from water, but where is a plant on land to get them? The first plants seem to have developed a special relationship with fungi that was a key factor in their ability to absorb minerals in terrestrial habitats. Within the roots of many early fossil plants like *Cooksonia* and *Rhynia* can be seen fungi, living intimately within and among the root cells. As you may recall from chapter 15, these kinds of symbiotic associations are called **mycorrhizae.** In plants with mycorrhizae, the fungi enable the plant to take up phosphorus, zinc, copper, and other nutrients from rocky soil, while the plant supplies organic molecules to the fungus.

Conserving Water

One of the key challenges to living on land is to avoid drying out. To solve this problem, plants have a watertight outer covering called a **cuticle.** The covering is formed from a waxy substance that is impermeable to water. Like the wax on a shiny car, the cuticle prevents water from entering or leaving the stem or leaves. Water enters the plant only from the roots, while the cuticle prevents water loss to the air. Passages do exist through the cuticle, in the form of specialized pores called **stomata** (singular, **stoma**) in the leaves and sometimes the green portions of the stems (figure 16.2). Stomata, which occur on at least some portions of all plants except liverworts, allow carbon dioxide to pass into the plant bodies for photosynthesis and allow water and oxygen gas to pass out of them. The cells that border stomata expand and contract, thus controlling the loss of water while allowing the

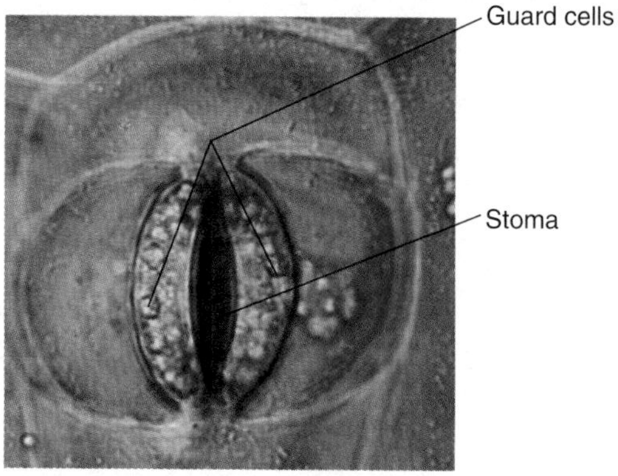
Guard cells
Stoma

Figure 16.2 A stoma.
A stoma is a passage through the cuticle. Water and oxygen pass out through the stoma, and carbon dioxide enters by the same portal. The cells flanking the stoma are called guard cells.

entrance of carbon dioxide. In most plants, water enters through the roots and exits through the underside of the leaves.

Reproducing on Land

To reproduce sexually on land, it is necessary to pass gametes from one individual to another, and because plants cannot move about, it is necessary that the gametes avoid drying out while they are transferred by wind or insects. In the first plants, the eggs were surrounded by a jacket of cells, and a film of water was required for the sperm to swim to the egg and fertilize it. Today, mosses still reproduce this way. However, soon after mosses evolved, changes occurred in the plant life cycle that favored the development of **spores,** which as you will recall from chapter 15 are reproductive cells very resistant to drying out.

Changing the Life Cycle. Among many algae, haploid cells occupy the major portion of the life cycle (see chapter 14). The zygote formed by the fusion of gametes is the only diploid cell, and it immediately undergoes meiosis to form haploid cells again. In early plants, this meiosis became delayed, so that for a significant portion of the life cycle, cells were diploid. This resulted in an **alternation of generations,** in which a diploid generation alternates with a haploid one (figure 16.3). Botanists call the diploid generation the **sporophyte** because it forms haploid spores by meiosis. The haploid generation is called the **gametophyte** because it forms haploid gametes by mitosis. When you look at primitive plants such as mosses and liverworts, you see largely gametophyte tissue—the sporophytes are smaller brown structures attached to or enclosed within the tissues of the larger gametophyte (figure 16.4*a*). When you look at plants that evolved later you see largely sporophyte tissue. The gametophytes are always much smaller than the sporophytes and are often enclosed within sporophyte tissues (figure 16.4*b*).

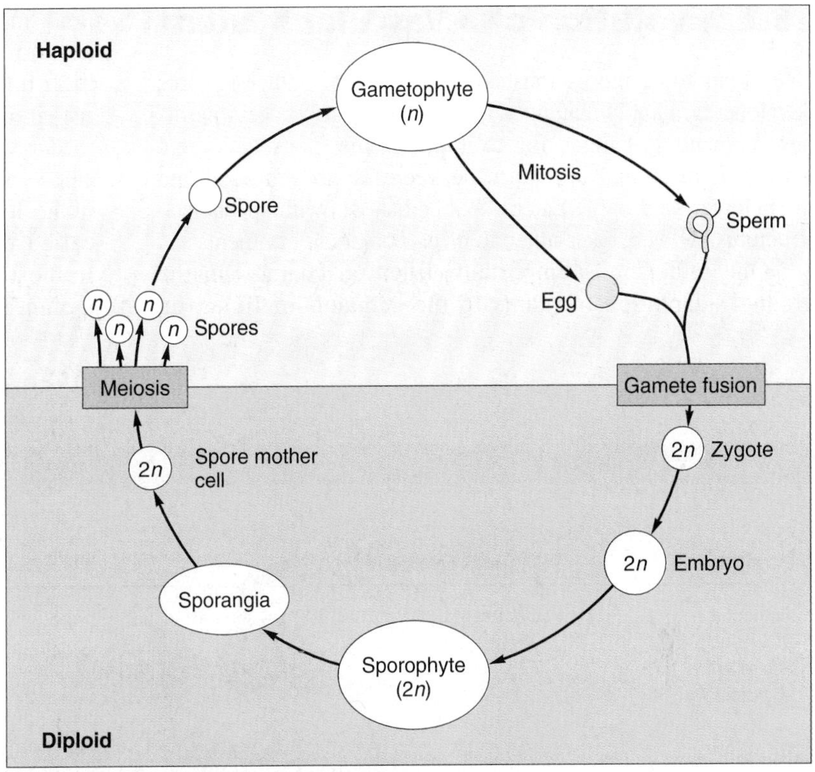

Haploid
Gametophyte
(n)
Mitosis
Spore
Sperm
Egg
n
n
n
n
Spores
Meiosis
Gamete fusion
2n
Zygote
2n
Spore mother
cell
2n
Embryo
Sporangia
Sporophyte
(2n)
Diploid

Figure 16.3 Generalized plant life cycle.
In a plant life cycle, there is alternation of generations, in which a diploid generation alternates with a haploid one. Gametophytes, which are haploid (n), alternate with sporophytes, which are diploid ($2n$). Gametophytes give rise by mitosis to sperm and eggs. The sperm and egg ultimately come together to produce the first diploid cell of the sporophyte generation, the zygote. The zygote undergoes cell division, ultimately forming the sporophyte. Meiosis takes place within the sporangia, the spore-producing organs of the sporophyte, resulting in the production of the spores, which are haploid and are the first cells of gametophyte generations.

16.1 Plants are multicellular terrestrial photosynthesizers, evolved from green algae, that adapted to life on land by developing ways to absorb minerals in partnership with fungi, to conserve water with watertight coverings, and to reproduce with spores and seeds.

(a)

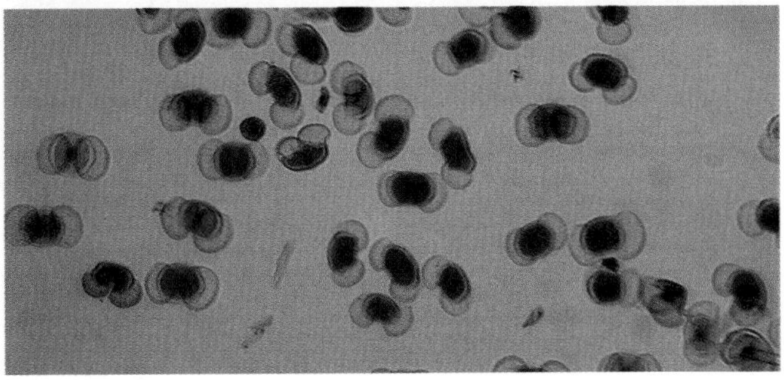
(b)

Figure 16.4 Two types of gametophytes.
(*a*) This gametophyte of a moss (a nonvascular plant) is green and free-living. (*b*) Gametophytes of a pine (a vascular plant) are barely large enough to be visible to the naked eye.

16.2 Evolution of a Vascular System

Once plants became established on land, many other features developed gradually that aided their evolutionary success in this new, demanding habitat. For example, in the first plants there was no fundamental difference between the aboveground and the belowground parts. Later, roots and shoots with specialized structures evolved, each suited to its particular environment.

One of the most important structural changes in the gradual adaptation of plants to the demands of living on land involved developing better ways of moving water around the body of the plant. In order for a plant to grow high into the air, a relatively efficient plumbing system is required to carry water up from the roots to the leaves and to carry carbohydrates down from the leaves to the roots. These plumbing systems, consisting of specialized strands of hollow cells connected end to end like a pipeline, are called **vascular systems** (Latin, *vasculum,* vessel or duct). In most vascular plants today they run from near the tip of a plant's roots all the way up the stem and into the leaves.

TABLE 16.1 PLANTS

Phylum	Typical Examples		Key Characteristics	Approximate Number of Living Species
Anthophyta (flowering plants, also called angiosperms)	Oak trees, corn, wheat, roses		Flowering; characterized by ovules that are fully enclosed by the carpel; fertilization involves two sperm nuclei; one forms the embryo, the other fuses with polar bodies to form endosperm for the seed; after fertilization, carpels and the fertilized ovules (now seeds) mature to become fruit	250,000
Bryophyta (mosses)	*Polytrichum, Sphagnum* (hairy cap and peat moss)	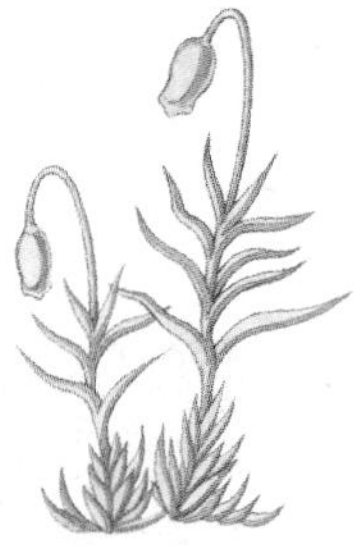	Without true vascular tissues; lack true roots and leaves; live in moist habitats and obtain nutrients by osmosis and diffusion; the three phyla were once grouped together	20,600
Hepaticophyta (liverworts)	*Marchantia*			
Anthocerophyta (hornworts)	*Anthoceros*			
Pterophyta (ferns)	*Azolla, Sphaeropteris* (water and tree ferns)	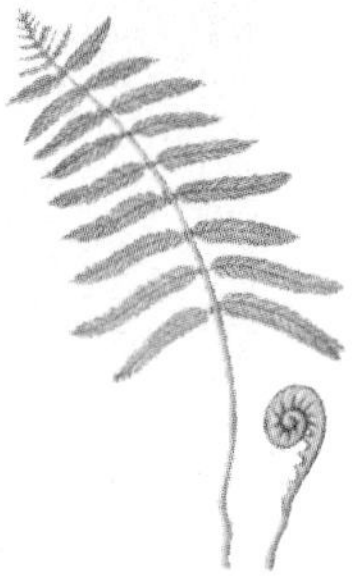	Seedless vascular plants; haploid spores germinate into free-living haploid individuals	12,000
Arthrophyta (horsetails)	*Equisetum*			
Psilophyta (whisk ferns)	*Psilotum*			
Lycophyta (lycopods)	*Lycopodium* (club mosses)	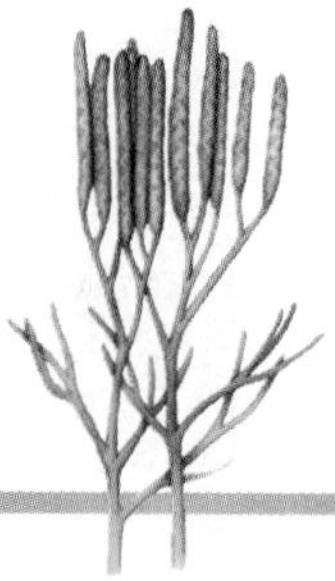	Seedless vascular plants similar in appearance to mosses but diploid; found in moist woodland habitats	1,150

Not all plants have efficient vascular systems. Some plants have none, others simple, inefficient ones, and still others highly sophisticated ones.

Of the 12 phyla of living plants (table 16.1), nine are vascular plants. The first vascular plants of which we have any complete fossils appeared about 410 million years ago. These plants, members of the extinct phylum Rhyniophyta, had no leaves. The plant body was little more than a simple branching axis with spore-bearing sporangia at the tips of the branches. Other ancient vascular plants evolved leaves, which first appeared as simple extensions from the stem; in most modern plants, leaves form on the branches. Vascular plants proved to be phenomenally successful, and they are by far the most common kind of plant alive today.

16.2 Vascular plants have internal plumbing—strands of hollow cells—that permits water and nutrients to move up and down the plant body.

TABLE 16.1 (CONTINUED)

Phylum	Typical Examples	Key Characteristics	Approximate Number of Living Species
Coniferophyta (conifers)	Pines, spruce, fir, redwood, cedar 	Gymnosperms; ovules partially exposed at time of pollination; flowerless; seeds are dispersed by the wind; sperm lack flagella; leaves are needlelike or scalelike; most species are evergreens and live in dense stands; among the most common trees on earth	601
Cycadophyta (cycads)	Cycads, sago palms 	Gymnosperms; very slow growing, palmlike trees; sperm have flagella but reach vicinity of egg by a pollen tube	206
Gnetophyta (shrub teas)	Mormon tea, *Welwitschia* 	Gymnosperms; nonmotile sperm; shrubs and vines	70
Ginkgophyta (ginkgo)	Ginkgo trees 	Gymnosperms; fanlike leaves that are dropped in winter (deciduous); seeds fleshy and ill-scented; motile sperm; trees are either male or female	1

16.3 Nonvascular Plants

Liverworts and Hornworts

The first successful land plants had no vascular system—no tubes or pipes to transport water and nutrients throughout the plant. This greatly limited the maximum size of the plant body because all materials had to be transported by osmosis and diffusion. Only two phyla of living plants, the **liverworts** (phylum Hepaticophyta) and the **hornworts** (phylum Anthocerophyta), completely lack a vascular system. The word *wort* meant *herb* in medieval Anglo-Saxon when these plants were named. Liverworts are the simplest of all living plants. About 8,500 species of liverworts and 100 species of hornworts survive today, usually growing in moist and shady places.

Figure 16.5 Moss sporophytes.

Most of a moss is sporophyte. In this hair-cup moss, *Polytrichum*, the small leaf-like projections at the base belong to the gametophyte, while each of the large, yellowish brown stalks, with the capsule at its summit, is a sporophyte. Although moss sporophytes may be green and carry out a limited amount of photosynthesis when they are immature, they are soon completely dependent, in a nutritional sense, on the gametophyte.

Plants with Simple Vascular Systems: Mosses

Another phylum of plants, the **mosses** (phylum Bryophyta), were the first plants to evolve strands of specialized cells that conduct water and carbohydrates, so-called **vascular tissue.** In many mosses, a central strand of vascular tissue conducts water up the stem of the gametophyte. Simple in design, the moss vascular system is composed of conducting cells without specialized wall thickenings—like soft pipes, they cannot carry water very high. Because their vascular systems are simple, mosses are usually grouped by botanists with the liverworts and hornworts as "nonvascular" plants. Today about 12,000 species of mosses grow in moist places all over the world. In mosses, as in liverworts and hornworts, the sporophytes are borne on the gametophytes, from which the spores derive their food (figure 16.5). Moss spores take 6 to 18 months to develop and are generally elevated on a stalk. The life cycle of a moss is illustrated in figure 16.6.

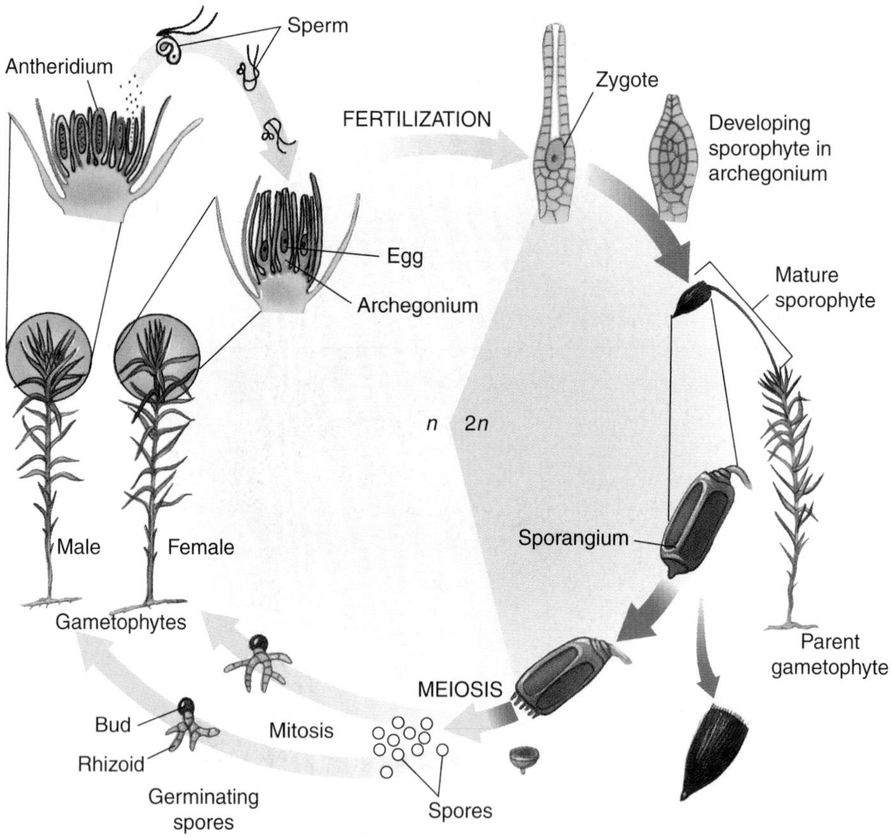
Sperm
Antheridium
FERTILIZATION
Zygote
Developing sporophyte in archegonium
Egg
Archegonium
Mature sporophyte
n 2n
Male
Female
Sporangium
Gametophytes
Parent gametophyte
MEIOSIS
Bud
Rhizoid
Mitosis
Germinating spores
Spores

Figure 16.6 The life cycle of a moss.

On the gametophytes, which are haploid, sperm are released from each antheridium (sperm-producing structure). They then swim through free water to an archegonium (egg-producing structure) and down its neck to the egg. Fertilization takes place there; the resulting zygote develops into a sporophyte, which is diploid. The sporophyte grows out of the archegonium and differentiates into a slender, basal stalk with a swollen capsule, the sporangium, at its apex. The capsule is covered, at least at first, with a cap formed from the swollen archegonium. The sporophyte grows on the gametophyte and eventually produces spores as a result of meiosis. The spores are shed from the capsule. The spores germinate, giving rise to gametophytes. The gametophytes initially are threadlike; they grow along the ground. Ultimately, buds form on them, from which leafy gametophytes arise.

16.3 While liverworts and hornworts totally lack a vascular system, mosses have simple soft strands of conducting tissue.

16.4 The Evolution of Vascular Tissue

The remaining nine phyla of plants, which have efficient vascular systems made of highly specialized cells, are called **vascular plants.** The first vascular plant appeared approximately 430 million years ago, but only incomplete fossils have been found. The first vascular plants for which we have relatively complete fossils, the extinct phylum Rhyniophyta, lived 410 million years ago. Among them is the oldest known vascular plant, *Cooksonia,* with branched, leafless shoots that form spores at their tips (figure 16.7).

Cooksonia and the other early plants that followed it became successful colonizers of the land through the development of efficient water- and food-conducting systems known as **vascular tissues** (figure 16.8). These tissues consist of strands of specialized cylindrical or elongated cells that form a network throughout a plant, extending from near the tips of the roots, through the stems, and into the leaves. The presence of a cuticle and stomata are also characteristic of vascular plants.

Early vascular plants grew by cell division at the tips of the stem and roots. Imagine stacking dishes—the stack can get taller but not wider! This sort of growth is called **primary growth** and was quite successful. During the so-called Age of Coal, when much of the world's fossil fuel was formed, the lowland swamps that covered Europe and North America were dominated by an early form of seedless tree called a lycophyte. Lycophyte trees grew to heights of 10 to 35 meters (33 to 115 ft), and their trunks did not branch until they attained most of their total height. The pace of evolution was rapid during this period, for the world's climate was changing, growing dryer and colder. As the world's swamplands began to dry up, the lycophyte trees vanished, disappearing abruptly from the fossil record. They were replaced by tree-sized ferns, a form of vascular plant that will be described in detail on the following page. Tree ferns grew to heights of more than 20 meters (66 ft) and trunks 30 centimeters (12 in) thick. Like the lycophytes, the trunks of tree ferns were formed entirely by primary growth.

About 380 million years ago, vascular plants developed a new pattern of growth, in which a cylinder of cells beneath the bark divides, producing new cells in regions around the plant's periphery. This growth is called **secondary growth.** Secondary growth makes it possible for a plant to increase in diameter. Only after the evolution of secondary growth could vascular plants become thick-trunked and therefore tall. Redwood trees today reach heights of up to 117 meters (384 ft) and trunk diameters in excess of 11 meters (36 ft). This evolutionary advance made possible the dominance of the tall forests that today cover northern North America. You are familiar with the product of plant secondary growth as **wood.** The growth rings so visible in cross sections of trees are zones of secondary growth (spring–summer) spaced by zones of little growth (fall–winter).

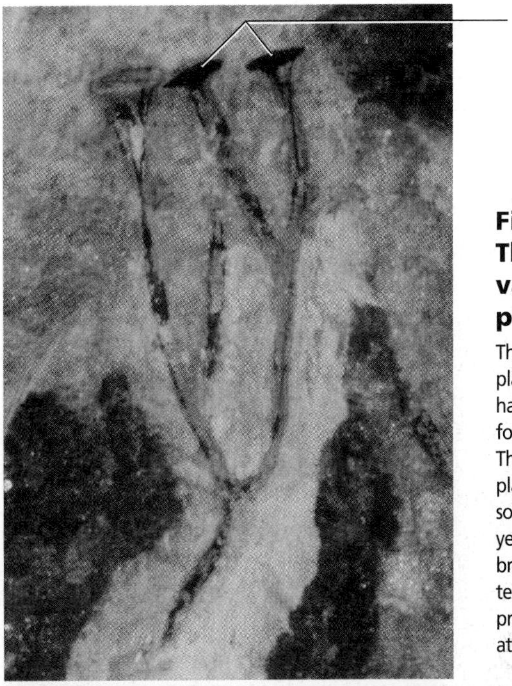
Sporangia

Figure 16.7 The earliest vascular plant.
The earliest vascular plant of which we have complete fossils is *Cooksonia.* This fossil shows a plant that lived some 410 million years ago; its upright branched stems terminated in spore-producing sporangia at the tips.

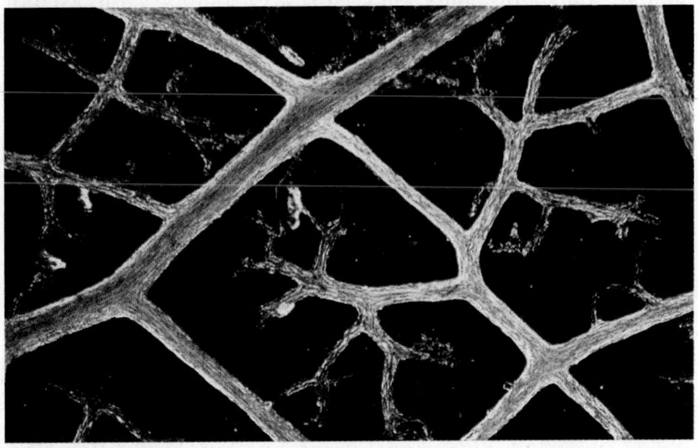

Figure 16.8 The vascular system of a leaf.
The veins of a vascular plant contain strands of specialized cells for conducting carbohydrates, as well as water containing dissolved minerals. The veins run from the tips of the roots to the tips of the shoots and throughout the leaves, as shown in this greatly enlarged photomicrograph of a cleared leaf.

16.4 So-called vascular plants have specialized vascular tissue composed of cylinders of hard hollow cells that conduct water to the leaves and food from them.

16.5 Seedless Vascular Plants

Most vascular plants today are seed plants—that is, the diploid embryos mature into tough, drought-resistant units called seeds. The earliest vascular plants lacked seeds, and four of the nine phyla of modern-day vascular plants do not have seeds. The four phyla of living seedless vascular plants include the ferns, phylum Pterophyta; the club mosses, phylum Lycophyta; the horsetails, phylum Arthrophyta; and the whisk ferns, phylum Psilophyta (figure 16.9). These phyla have free-swimming sperm that require the presence of free water for fertilization.

By far the most abundant of the four phyla of seedless vascular plants contains the **ferns,** with about 12,000 living species. Ferns are found throughout the world, although they are much more abundant in the tropics than elsewhere. Many are small, only a few centimeters in diameter, but some of the largest plants that live today are also ferns. Descendants of ancient tree ferns, they can have trunks more than 24 meters (79 ft) tall and leaves up to 5 meters (16 ft) long!

The Life of a Fern

In ferns, the life cycle of plants begins a revolutionary change that culminates later with seed plants. Nonvascular plants like mosses are made largely of gametophyte (haploid) tissue. Vascular seedless plants like ferns have both gametophyte and sporophyte individuals, each independent and self-sufficient. The gametophyte produces eggs and sperm; when these swim through water and combine in a fertilized egg, the zygote grows into a sporophyte. The sporophyte bears and releases haploid **spores** that float to the ground and germinate, growing into haploid gametophytes. The fern gametophytes are small, thin, heart-shaped photosynthetic plants, usually no more than a centimeter in length, that live in moist places. The fern sporophytes are much larger and more complex, with long vertical leaves called **fronds.** When you see a fern, you are almost always looking at the sporophyte. The fern life cycle is described in figure 16.10.

16.5 Ferns are among the vascular plants that lack seeds, reproducing with spores as nonvascular plants do.

(a)

(b)

(c)

Figure 16.9 Seedless vascular plants.
Three of the four phyla of seedless vascular plants are shown. (*a*) A tree fern in the forests of Malaysia (phylum Pterophyta). The ferns are by far the largest group of spore-producing vascular plants. (*b*) The club moss *Lycopodium lucidulum* (phylum Lycophyta). Although superficially similar to the gametophytes of mosses, the conspicuous club moss plants shown here are sporophytes. (*c*) A horsetail, *Equisetum telmateia,* a representative of the only living genus of the phylum Arthrophyta. This species forms two kinds of erect stems; one is green and photosynthetic, and the other, which terminates in a spore-producing "cone," is mostly light brown.

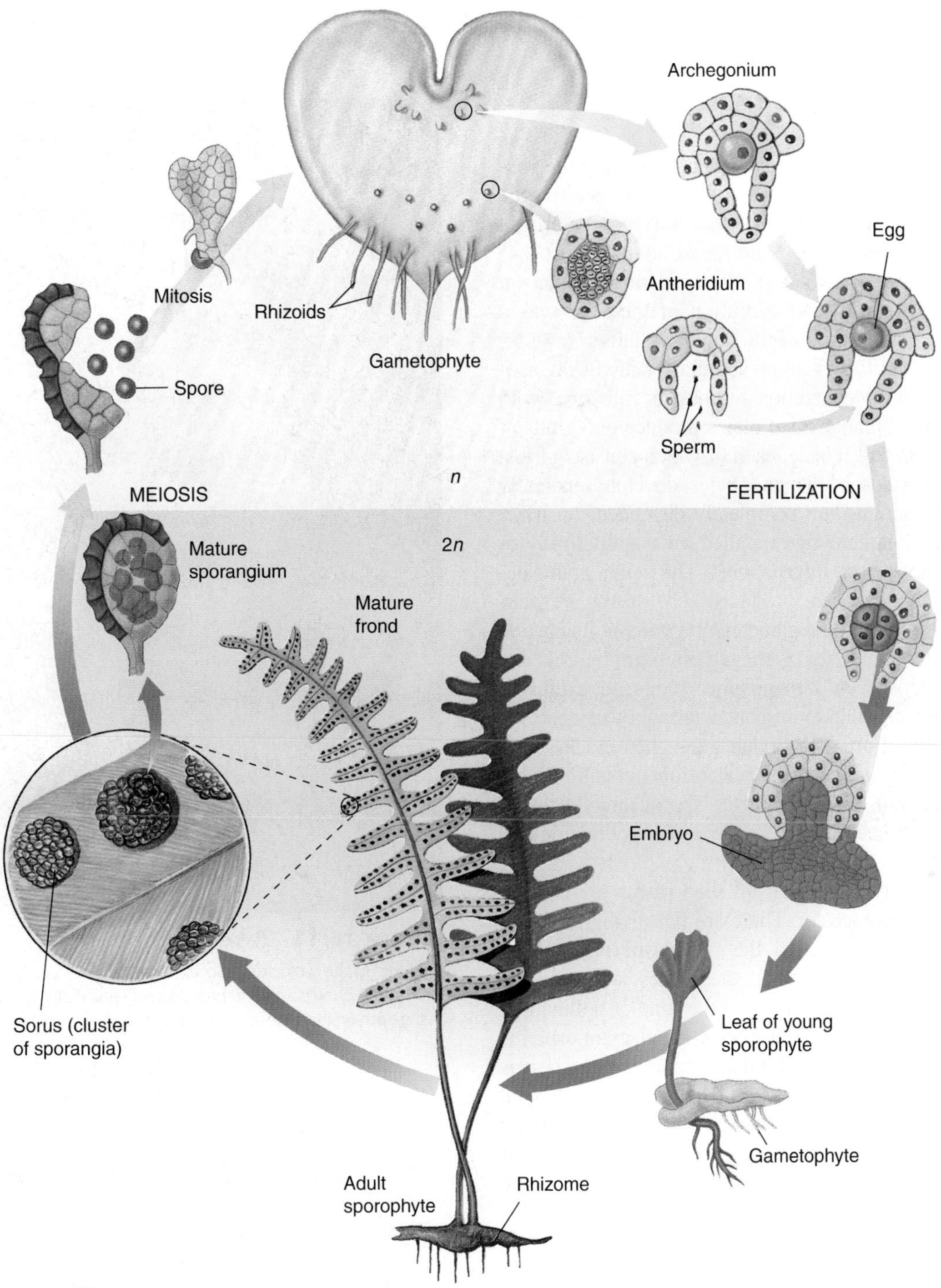
Archegonium
Egg
Antheridium
Sperm
Mitosis
Rhizoids
Gametophyte
Spore
MEIOSIS
n
FERTILIZATION
2n
Mature
sporangium
Mature
frond
Embryo
Sorus (cluster
of sporangia)
Leaf of young
sporophyte
Gametophyte
Adult
sporophyte
Rhizome

Figure 16.10 Fern life cycle.

The gametophytes, which are haploid, grow in moist places. Rhizoids (anchoring structures) project from their lower surface. Eggs develop in an archegonium, and sperm develop in an antheridium, both located on the gametophyte's lower surface. The sperm, when released, swim through free water to the mouth of the archegonium, entering and fertilizing the single egg. Following the fusion of egg and sperm to form a zygote—the first cell of the diploid sporophyte generation—the zygote starts to grow within the archegonium. Eventually, the sporophyte becomes much larger than the gametophyte—it is known as a fern plant. Most ferns have more or less horizontal stems, called rhizomes, that creep along below ground. On the sporophyte's leaves, called fronds, occur clusters of sporangia (called sori; singular, sorus), within which meiosis occurs and spores are formed. The release of these spores, which is explosive in many ferns, and their germination lead to the development of new gametophytes.

16.6 Seed Plants

We now examine a key evolutionary advance among the vascular plants, the development of a protective cover for the embryo. A **seed** is a plant embryo with a durable, watertight cover (figure 16.11). The seed is a crucial adaptation to life on land because it protects the embryonic plant when it is at its most vulnerable stage. The evolution of the seed was a critical step in the domination of the land by plants.

The change in the life cycle of vascular plants in favor of the sporophyte (diploid) generation reaches its full force with the advent of the seed plants. Seed plants produce two kinds of gametophytes—male and female, each of which consists of just a few cells. Both kinds of gametophytes develop separately within the sporophyte and are completely dependent on it for their nutrition. Male gametophytes, called **microgametophytes** or **pollen grains,** arise from **microspores.** The pollen grains become mature when sperm are produced. The sperm are conveyed to the egg in the female gametophyte without using free water. A female gametophyte, or **megagametophyte,** contains the egg and develops from a **megaspore** produced within an **ovule.** The transfer of pollen by insects, wind, or other agents is referred to as **pollination.** The pollen grain then cracks open and sprouts, or germinates, and the pollen tube, containing the sperm cells, grows out, transporting the sperm directly to the egg. Thus there is no need for free water in the pollination and fertilization process.

Botanists generally agree that all seed plants are derived from a single common ancestor. There are five living phyla. In four of them, collectively called the **gymnosperms** (Greek, *gymnos,* naked, and *sperma,* seed), the ovules are not completely enclosed by sporophyte tissue at the time of pollination. Gymnosperms were the first seed plants. From gymnosperms evolved the fifth group of seed plants, called **angiosperms** (Greek, *angion,* vessel, and *sperma,* seed), phylum Anthophyta. Angiosperms are the most recently evolved of all the plant phyla. Angiosperms differ from all gymnosperms in that their ovules are completely enclosed by a vessel of sporophyte tissue called the **carpel** at the time they are pollinated.

The Structure of a Seed

A seed contains a sporophyte embryo surrounded by a protective seed coat (figure 16.12). Seeds are the way in which plants, anchored by their roots to one place in the ground, solve the problem of dispersing their progeny to new locations. The hard cover of the seed (formed from the tissue of the parent plant), protects the seed during its travel through the air. Many seeds have devices to aid in carrying them farther. Most species of pine, for example, have seeds with thin flat wings attached. These wings help catch air currents, which carry the seeds to new areas.

Figure 16.11 A seed plant.
The seeds of this cycad, like all seeds, consist of a plant embryo and a protective covering. A cycad is a gymnosperm (naked-seeded plant), and its seeds develop out in the open on the edges of the cone scales.

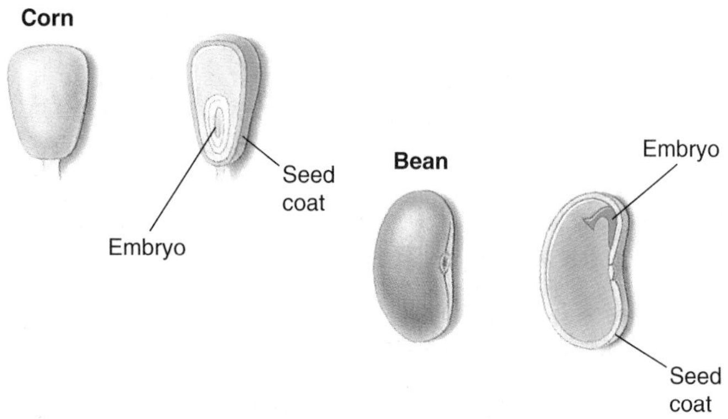
Corn
Seed coat
Embryo
Bean
Embryo
Seed coat

Figure 16.12 Basic structure of seeds.
A seed is a sporophyte (diploid) embryo surrounded by a protective coat. This seed coat, formed of sporophytic tissue from the parent, protects the embryo—the dormant young plant of the next sporophyte generation.

Figure 16.13 Seeds allow plants to bypass the dry season.
When it does rain, seeds can germinate, and plants can grow rapidly to take advantage of the relatively short periods when water is available. This desert tree has tough seeds (*inset*) that germinate only after they are cracked. Rains leach out the chemicals in the seed coats that inhibit germination, and the hard coats of the seeds may be cracked when they are washed down gullies in temporary floods.

Once a seed has fallen to the ground, it may lie there, dormant, for many years. When conditions are favorable, however, and particularly when moisture is present, the seed germinates and begins to grow into a young plant (figure 16.13). Most seeds have abundant food stored in them to provide a ready source of energy for the new plant as it starts its growth.

The advent of seeds had an enormous influence on the evolution of plants. Seeds have greatly improved the adaptation of plants to living on land in at least four respects:

1. **Dispersal.** Most important, seeds facilitate the migration and dispersal of plant offspring into new habitats.
2. **Dormancy.** Seeds permit plants to postpone development when conditions are unfavorable, as during a drought, and to remain dormant until conditions improve.
3. **Germination.** By making the reinitiation of development dependent upon environmental factors such as temperature, seeds permit the course of embryonic development to be synchronized with critical aspects of the plant's habitat, such as the season of the year.
4. **Nourishment.** The seed offers the young plant nourishment during the critical period just after germination, when the seedling must establish itself.

16.6 A seed is a dormant diploid embryo encased with food reserves in a hard protective coat. Seeds play critical roles in improving a plant's chances of successfully reproducing in a varied environment.

16.7 Gymnosperms

Four phyla comprise the gymnosperms (figure 16.14): the conifers (Coniferophyta), the cycads (Cycadophyta), the gnetophytes (Gnetophyta), and the ginkgos (Ginkgophyta). The conifers are the most familiar of the four phyla of gymnosperms and include pine, spruce, hemlock, cedar, redwood, yew, cypress, and fir trees. **Conifers** are trees that produce their seeds in **cones.** The seeds (ovules) of conifers develop on scales within the cones and are exposed at the time of pollination. Most of the conifers have needlelike leaves, an evolutionary adaptation for retarding water loss. Conifers are often found growing in moderately dry regions of the world, including the vast taiga forests of the northern latitudes. Many are very important as sources of timber and pulp.

There are about 600 living species of conifers. The tallest living vascular plant, the coastal sequoia (*Sequoia sempervirens*), found in coastal California and Oregon, is a conifer and reaches 100 meters (328 ft). The biggest redwood, however, is the mountain sequoia redwood species (*Sequoiadendron gigantea*) of the Sierra Nevadas. The largest individual tree is nicknamed after General Sherman of the Civil War, and it stands more than 83 meters (274 ft) tall while measuring 31 meters (102 ft) around its base. Another much smaller type of conifer, the bristlecone pines in Nevada, may be the oldest trees in the world—about 5,000 years old.

The other three gymnosperm phyla are much less widespread. Cycads, the predominant land plant in the early age of dinosaurs, the Jurassic, have short stems and palmlike leaves. They are still widespread throughout the tropics. There is only one living species of ginkgo, the maidenhair tree, which has fan-shaped leaves shed in the autumn. Because ginkgos are resistant to air pollution, they are commonly planted along city streets. The gnetophytes are thought to be the most closely related to angiosperms. This phylum contains only three kinds of plants, all unusual. One of them is perhaps the most bizarre of all plants, *Welwitschia,* which grows on the exposed sands of the harsh Namibian Desert of southwestern Africa. *Welwitschia* acts like a plant standing on its head! Its two beltlike, leathery leaves are generated continuously from their base, splitting as they grow out over the desert sands.

The fossil record indicates that members of the ginkgo phylum were once widely distributed, particularly in the Northern Hemisphere; today only one living species, the maidenhair tree (*Ginkgo biloba),* remains. The reproductive structures of ginkgos are produced on separate trees. The fleshy outer coverings of the seeds of female ginkgo plants exude the foul smell of rancid butter caused by butyric and isobutyric acids. In the Orient, however, the seeds are considered a delicacy. In Western countries, because of the seed odor, male plants vegetatively propagated from shoots are preferred for cultivation.

(a)

(b)

(c)

Figure 16.14 The four kinds of gymnosperms.

(*a*) Slash pines, *Pinus palustris,* in Florida, are representative of the largest phylum of gymnosperms, the Coniferophyta. (*b*) An African cycad, *Encephalartos ferox,* phylum Cycadophyta. The cycads have fernlike leaves and seed-forming cones, like the one shown here. (*c*) Maidenhair tree, *Ginkgo biloba,* the only living representative of the phylum Ginkgophyta, a group of plants that was abundant 200 million years ago. Among living seed plants, only the cycads and ginkgo have swimming sperm. (*d*) *Welwitschia mirabilis,* phylum Gnetophyta, is found in the extremely dry deserts of southwestern Africa. In *Welwitschia*, two enormous, beltlike leaves grow from a circular zone of cell division that surrounds the apex of the carrot-shaped root.

(d)

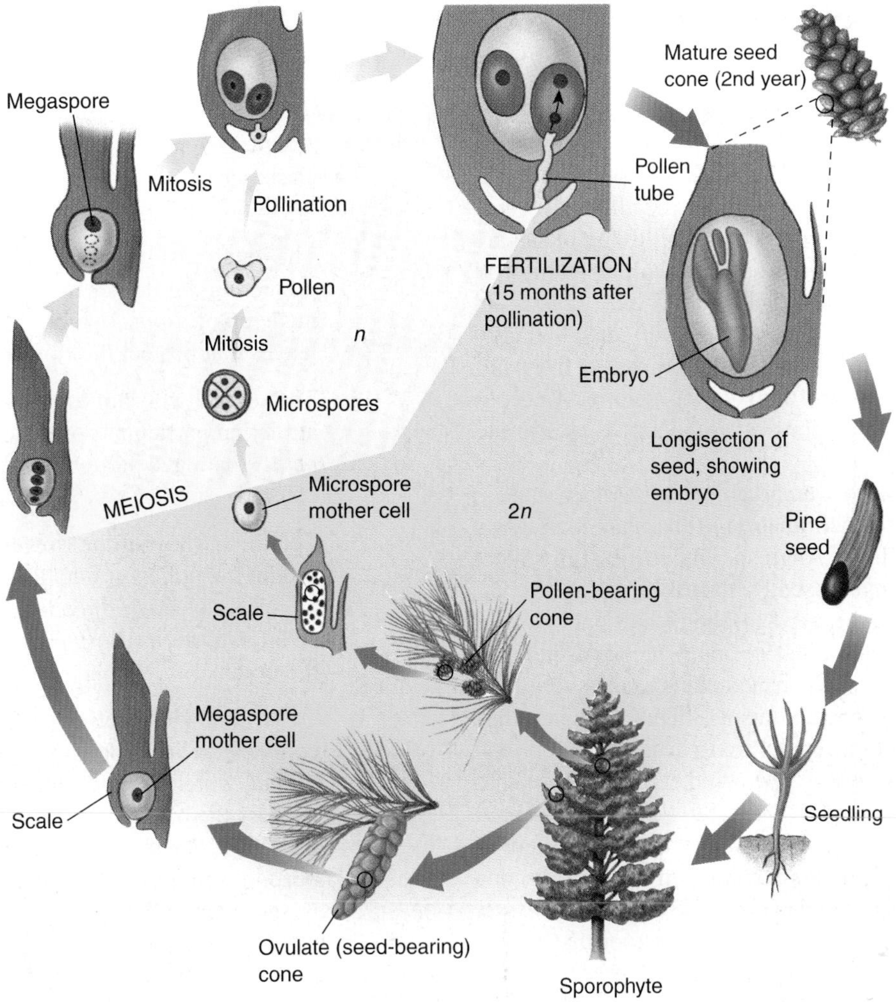

Megaspore
Mitosis
Pollination
Pollen
Mitosis
n
Microspores
MEIOSIS
Microspore
mother cell
Scale
Mature seed
cone (2nd year)
Pollen
tube
FERTILIZATION
(15 months after
pollination)
Embryo
Longisection of
seed, showing
embryo
2n
Pine
seed
Pollen-bearing
cone
Megaspore
mother cell
Scale
Seedling
Ovulate (seed-bearing)
cone
Sporophyte

Figure 16.15 Life cycle of a conifer.

In all seed plants, the gametophyte generation is greatly reduced. In conifers such as pines, the thin scales of the relatively delicate pollen-bearing cones contain microspores, which give rise to pollen grains, the male gametophytes. The familiar seed-bearing cones of pines are much heavier and more substantial structures than the pollen-bearing cones. Two ovules, and ultimately two seeds, are borne on the upper surface of each scale, which contains the megaspores that give rise to the female gametophytes. After a pollen grain has reached a scale, it germinates, and a slender pollen tube grows toward the egg. When the pollen tube grows to the vicinity of the female gametophyte, sperm are released, fertilizing the egg and producing a zygote there. The development of the zygote into an embryo takes place within the ovule, which matures into a seed. Eventually, the seed falls from the cone and germinates, the embryo resuming growth and becoming a new pine tree.

The Life of a Gymnosperm

We will examine conifers as typical gymnosperms. The conifer life cycle is illustrated in figure 16.15. Conifer trees form two kinds of cones. Seed cones contain the female gametophytes, with their egg cells; pollen cones contain pollen grains. Conifer pollen grains are small and light and are carried by the wind to seed cones. Each pollen grain has a pair of air sacs to help carry it in the wind. Because it is very unlikely that any particular pollen grain will succeed in being carried to a seed cone (the wind can take it anywhere), a great many pollen grains are needed to be sure that at least a few succeed in pollinating seed cones. For this reason, pollen grains are shed from their cones in huge quantities, often appearing as a sticky yellow layer on the surfaces of ponds and lakes—and even on windshields.

When a grain of pollen settles down on a scale of a female cone, a slender tube grows out of it down into the scale, delivering the male gamete (the sperm cell) to the female gametophyte containing the egg, or ovum. Fertilization occurs when the sperm cell fuses with the egg, forming a zygote. This zygote is the beginning of the sporophyte generation. What happens next is the essential improvement in reproduction achieved by seed plants. Instead of the zygote simply growing into an adult sporophyte—just as you grow directly into an adult from a fertilized zygote—the fertilized ovule forms a seed.

16.7 Gymnosperms are seed plants in which the ovules are not completely enclosed by diploid tissue at pollination. Gymnosperms do not have flowers.

16.8 Rise of the Angiosperms

Angiosperms, plants in which the ovule is completely enclosed by sporophyte tissue when it is fertilized, are the most successful of all plants. Ninety percent of all living plants are angiosperms, over 250,000 species, including hardwood trees, shrubs, herbs, grasses, vegetables, and grains—in short, nearly all of the plants that we see every day. Virtually all of our food is derived, directly or indirectly, from angiosperms. In fact, more than half of the calories we consume come from just three species: rice, corn (maize), and wheat.

In a very real sense, the remarkable evolutionary success of the angiosperms is the culmination of the plant kingdom's adaptation to life on land. Angiosperms successfully meet the last difficult challenge posed by terrestrial living: the inherent conflict between the need to obtain nutrients (solved by roots, which anchor the plant to one place) and the need to find mates (solved by making the male gametes very tiny, so they can be carried to other plants). This challenge has never really been overcome by gymnosperms, whose pollen grains are carried passively by the wind on the chance that they might by luck encounter a female cone. Think about how inefficient this is! Angiosperms are also able to deliver their pollen directly, as if in an addressed envelope, from one individual of a species to another. How? *By inducing insects and other animals to carry it for them!* The tool that makes this animal-dictated pollination possible, the great advance of the angiosperms, is the flower.

The Flower

Flowers are the reproductive organs of angiosperm plants. A flower is a sophisticated pollination machine. It employs bright colors to attract the attention of insects (or birds or small mammals), nectar to induce the insect to enter the flower, and structures that coat the insect with pollen grains while it is visiting. Then, when the insect visits another flower, it carries the pollen with it into that flower.

The basic structure of a flower consists of four concentric circles (figure 16.16), or **whorls:**

1. The outermost whorl of the flower is concerned with protecting the flower from physical damage. It is made up of *sepals,* which are in effect modified leaves that protect the flower while it is a bud.
2. The second whorl of the flower is concerned with attracting particular pollinators. It is made up of *petals* that have particular pigments, often vividly colored.
3. The third whorl of the flower is concerned with producing pollen grains. It is made up of *stamens,* which are slender, threadlike filaments with a swollen portion (the *anther*) at the tip that contains the pollen.
4. The fourth and innermost whorl of the flower is concerned with producing eggs. It is made of the *carpel,* which is sporophyte tissue that completely encases the ovules within which the egg cell develops. The ovules occur in the swollen lower portion of the carpel, called the *ovary;* usually there is a slender stalk rising from the ovary called the *style,* with a swollen tip called a *stigma* that is receptive to pollen. When the flower is pollinated, a pollen tube grows down from the stigma through the style to the ovary.

16.8 Angiosperms are seed plants in which the ovule is completely enclosed by diploid tissue at pollination. Angiosperms are flowering plants.

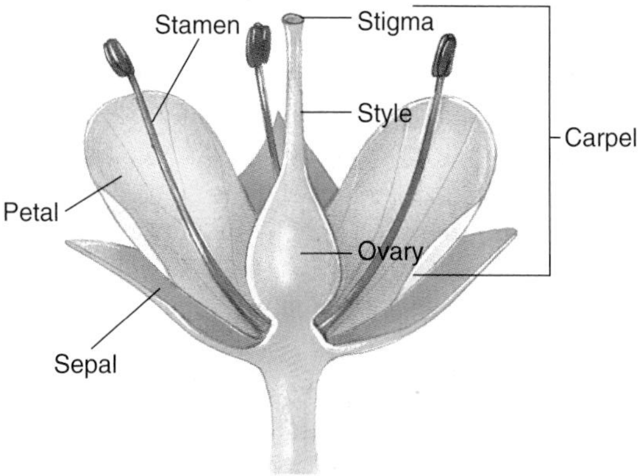
Stamen
Stigma
Style
Carpel
Petal
Ovary
Sepal

(a)

(b)

Figure 16.16 An angiosperm flower.
(*a*) The basic structure of a flower is a series of four concentric circles, or whorls: the sepals, petals, stamens, and the carpel (ovary, style, and stigma). (*b*) This flower of the wild woodland plant *Geranium* shows the five free petals, 10 stamens, and a fused carpel typical of angiosperm flowers.

16.9 Why There Are Different Kinds of Flowers

If you were to watch insects visiting flowers, you would quickly discover that the visits are not random. Instead, certain insects are attracted by particular flowers. Insects recognize a particular color pattern and odor and search for flowers that look similar. Insects and plants have coevolved (see chapter 30) so that certain insects specialize in visiting particular kinds of flowers. As a result, a particular insect carries pollen from one individual to another *of the same species*. It is this keying in on particular species that makes insect pollination so effective.

Of all insect pollinators, the most numerous are bees. Bees evolved soon after flowering plants, some 100 million years ago. Today there are over 20,000 species. Bees locate sources of nectar largely by odor at first (that is why flowers smell sweet) and then focus in on the flower's color and shape. Bee-pollinated flowers are usually yellow or blue, and frequently they have guiding stripes or lines of dots to indicate the position in the flower of the nectar (usually in the throat of the flower). While inside the flower, the bee becomes coated with pollen. This coating is far from accidental. Most of the bees visiting flowers actively seek to acquire pollen, which they use as a rich source of protein to feed their larvae.

Many other insects pollinate flowers. Butterflies tend to visit flowers of plants like phlox that have "landing platforms" on which they can perch. These flowers typically have long, slender floral tubes filled with nectar that a butterfly can reach by uncoiling its long proboscis (a hoselike tube extending out from the mouth). Moths, which visit flowers at night, are attracted to white or very pale-colored flowers, often heavily scented, that are easy to locate in dim light.

Red flowers, interestingly, are not typically visited by insects, most of which cannot "see" red as a distinct color. Who pollinates these flowers? Hummingbirds and sunbirds (figure 16.17)! To these birds, red is a very conspicuous color, just as it is to us. Birds do not have a well-developed sense of smell, and do not orient to odor, which is why red flowers often do not have a strong smell. Mammals such as nocturnal opossums and bats may visit tree species and even giant blooming cacti to eat the pollen, and, in the process, transport pollen grains from one plant to another.

Some angiosperms have reverted to the wind pollination practiced by their ancestors, notably oaks, birches, and, most important, the grasses. The flowers of these plants are small, greenish, and odorless.

16.9 Flowers can be viewed as pollinator-attracting devices, with different kinds of pollinators attracted to different kinds of flowers.

Figure 16.17 Red flowers are pollinated by hummingbirds.
This long-tailed hermit hummingbird is extracting nectar from the red flowers of *Heliconia imbricata* in the forests of Costa Rica. Note the pollen on the bird's beak. Hummingbirds of this group obtain nectar primarily at long, curved flowers that more or less match the length and shape of their beaks.

16.10 Improving Seeds: Double Fertilization

The seeds of gymnosperms often contain food to nourish the developing plant in the critical time immediately after germination, but the seeds of angiosperms have greatly improved on this aspect of seed function. Angiosperms produce a special, highly nutritious tissue called **endosperm** within their seeds. Here is how it happens. In the angiosperm life cycle, the male gametophyte (with haploid gametes) contains *two* sperm, not one (figure 16.18). The first fuses with the egg, as in all sexually reproducing organisms, forming the zygote. The other fuses with two other products of meiosis called polar nuclei to form a triploid (three copies of the chromosomes) endosperm cell. This cell divides much more rapidly than the zygote, giving rise to the nutritive endosperm tissue within the seed. This process of fertilization with two sperm to produce both a zygote and endosperm is called **double fertilization.** Double fertilization occurs only in angiosperms.

In some angiosperms, such as the common bean or pea, the endosperm is fully used up by the time the seed is

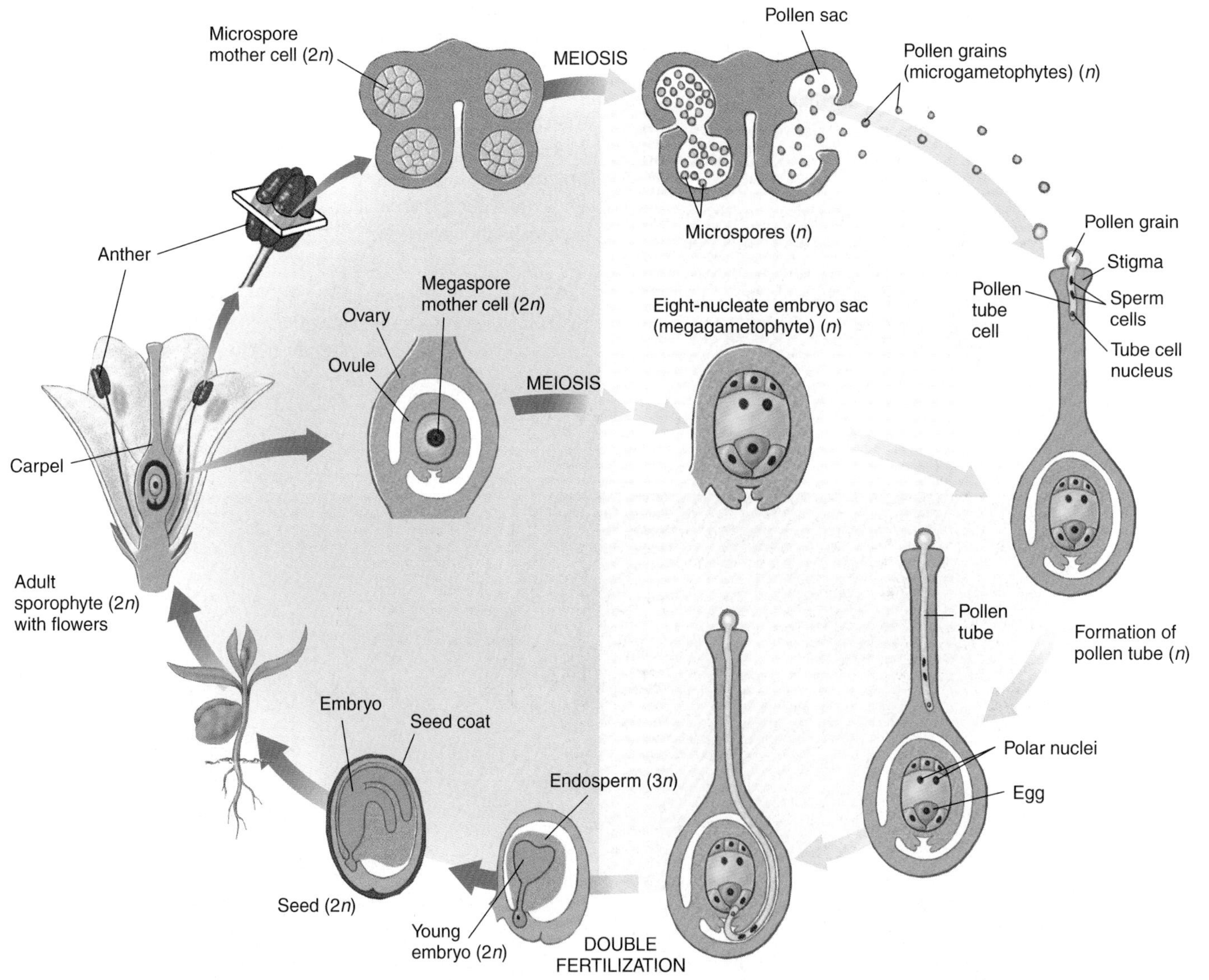

Microspore mother cell (2*n*)
MEIOSIS
Pollen sac
Pollen grains (microgametophytes) (*n*)
Microspores (*n*)
Pollen grain
Stigma
Pollen tube cell
Sperm cells
Tube cell nucleus
Anther
Megaspore mother cell (2*n*)
Ovary
Ovule
MEIOSIS
Eight-nucleate embryo sac (megagametophyte) (*n*)
Carpel
Adult sporophyte (2*n*) with flowers
Pollen tube
Formation of pollen tube (*n*)
Embryo
Seed coat
Polar nuclei
Egg
Endosperm (3*n*)
Seed (2*n*)
Young embryo (2*n*)
DOUBLE FERTILIZATION

Figure 16.18 Life cycle of an angiosperm.

In angiosperms, as in gymnosperms, the sporophyte is the dominant generation. Eggs form within the megagametophyte, or embryo sac, inside the ovules, which, in turn, are enclosed in the carpels. The carpel is differentiated in most angiosperms into a slender portion, or style, ending in a stigma, the surface on which the pollen grains germinate. The microgametophytes, or pollen grains, meanwhile, are formed within the microsporangia of the anthers and complete their differentiation to the mature, three-celled stage either before or after grains are shed. Fertilization is distinctive in angiosperms, being a double process. A sperm and an egg come together, producing a zygote; at the same time, another sperm fuses with the two polar nuclei, producing the primary endosperm nucleus, which is triploid. Both the zygote and the primary endosperm nucleus divide mitotically, giving rise, respectively, to the embryo and the endosperm. The endosperm is the tissue, unique to angiosperms, that nourishes the embryo and young plant.

mature. Food reserves are stored in swollen, fleshy leaves. In other angiosperms, such as corn, the mature seed contains abundant endosperm, which is used after germination.

Some angiosperm embryos have two seed leaves, or cotyledons, and are called dicotyledons, or **dicots.** The first angiosperms were like this. Dicots typically have leaves with netlike branching of veins and flowers with four to five parts per whorl. Oak and maple trees are dicots, as are many shrubs.

The embryos of other angiosperms, which evolved somewhat later, have a single seed leaf and are called monocotyledons, or **monocots.** Monocot leaves typically have parallel veins and flowers with three parts per whorl. Grasses, one of the most abundant of all plants, are wind-pollinated monocots. The basic characteristics of monocots and dicots are outlined in figure 16.19.

> **16.10** Two sperm fertilize each angiosperm ovule. One fuses with the egg, the other with two polar nuclei to form triploid ($3n$) nutritious endosperm.

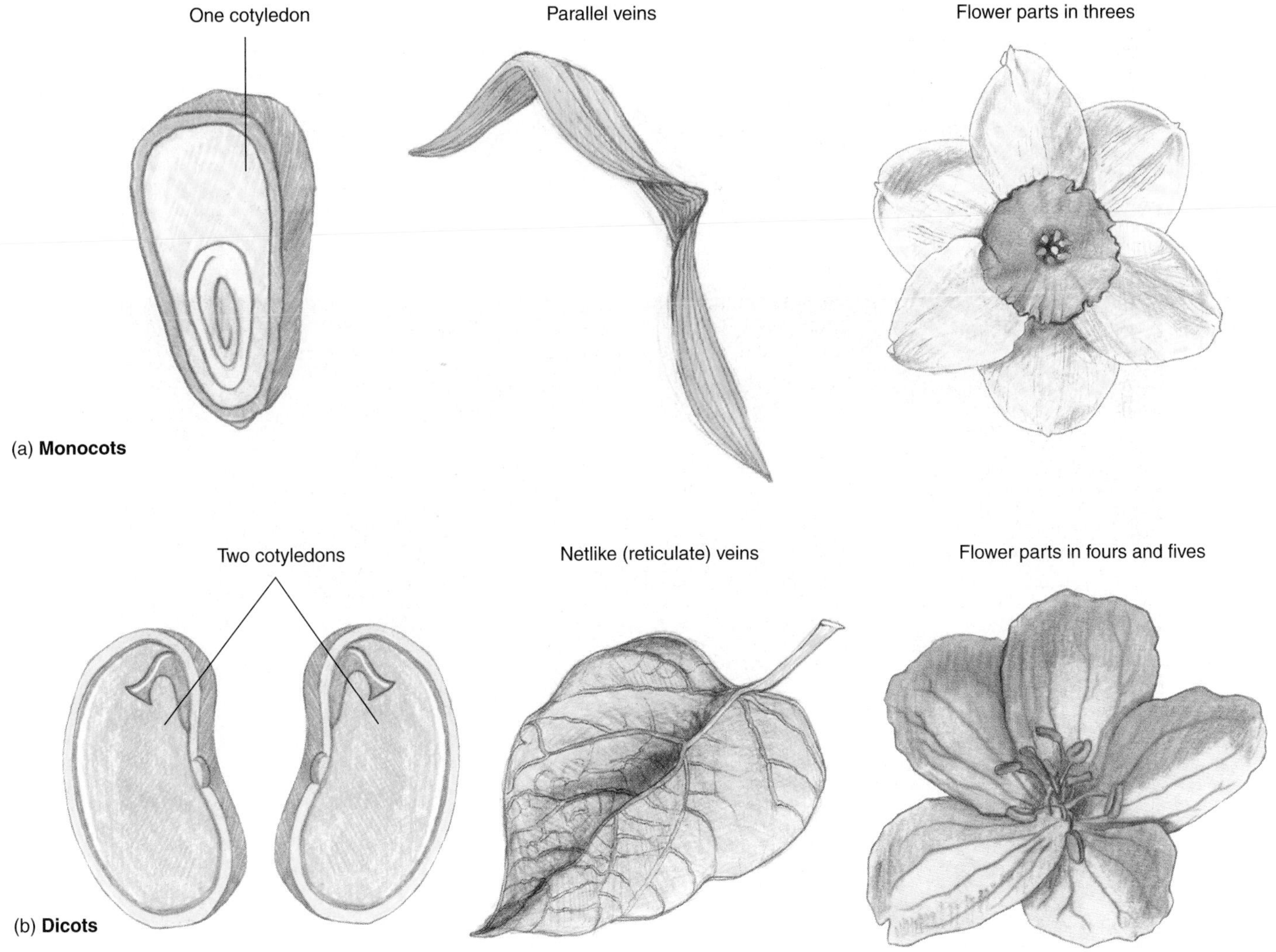
One cotyledon
Parallel veins
Flower parts in threes
(a) Monocots
Two cotyledons
Netlike (reticulate) veins
Flower parts in fours and fives
(b) Dicots

Figure 16.19 Dicots and monocots.

(*a*) Monocots are characterized by one cotyledon, parallel veins, and the occurrence of flower parts in threes. (*b*) Dicots have two cotyledons and netlike (reticulate) veins. Their flower parts occur in fours and fives.

16.11 Improving Seed Dispersal: Fruits

Just as the mature ovules become seeds, so the mature ovary that surrounds the ovules becomes the **fruit.** A fruit is a mature ripened ovary containing fertilized seeds, surrounded by a carpel. Fruits provide angiosperms with a far better way of dispersing their progeny than simply sending their seeds off on the wind. Instead, just as in pollination, they employ animals. By making fruits fleshy and tasty to animals, angiosperms encourage animals to eat them. The seeds within the fruit are resistant to chewing and digestion. They pass out of the animal with the feces, undamaged and ready to germinate at a new location far from the parent plant.

There are three main kinds of fleshy fruits: berries, drupes, and pomes. In *berries*—such as grapes, tomatoes, and dates—which are typically many-seeded, the fleshy portion of the fruit forms on the inner wall of the carpel. In *drupes*—peaches, olives, plums, and cherries—the inner layer of the fruit is stony and adheres tightly to a single seed. In *pomes*—apples and pears—the fleshy portion of the fruit comes from the petals and sepals of the flower, or the receptacle (the swollen end of the flower stem that holds the petals and sepals).

Fleshy fruits usually have distinctively colored coverings, often black, bright blue, or red. These colors help attract animals. Berries, for example, are frequently dispersed by birds and sometimes by other vertebrates. Thus a bear eating blueberries is helping the blueberry plant disperse its seeds.

Other kinds of specialized fruits like burdock lack fleshy carpels and are dispersed by attaching themselves to the fur of animals. Coconuts are dispersed by water, colonizing distant islands (figure 16.20). Coconuts are drupes whose outer layer is fibrous rather than fleshy, playing a protective rather than animal-attracting role.

Many plant fruits are specialized for wind dispersal. The small, nonfleshy fruits of the dandelion, for example, have a plumelike modified calyx that lets them be carried long distances on wind currents. Many grasses have dustlike fruit, so light wind bears them easily. Maples have long wings attached to the fruit. In tumbleweeds, the whole plant breaks off and is blown across open country by the wind, scattering seeds as it moves.

> **16.11** A fruit is a mature ovary containing fertilized seeds, often specialized to aid in seed dispersal.

Figure 16.20 A water-dispersed fruit.

This fruit of the coconut, *Cocos nucifera,* is sprouting on a sandy beach. Coconuts, one of the most useful plants for humans in the tropics, have become established on even the most distant islands by drifting in the waves.

CONCEPT REVIEW

1. The evolution of the cuticle in plants was a key adaptation to terrestrial living because it
   a. enhanced absorption of nutrients from soil.
   b. allowed for an alternation of generations.
   c. helped plants to avoid drying out.
   d. enhanced water loss.

2. In early plants, the gametophyte tissue is usually _________ than the sporophyte tissue.
   a. smaller
   b. larger
   c. more vascularized
   d. less visible

3. Which group of plants completely lacks a vascular system?
   a. liverworts
   b. ferns
   c. mosses
   d. cycads

4. Which of the following are *not* found in all vascular plants?
   a. a waxy cuticle
   b. carbohydrate molecules
   c. seeds
   d. stomata

5. Growth around the plant's periphery, causing it to increase in diameter, is known as ____________ growth.
   a. primary
   b. secondary
   c. tertiary
   d. apical

6. An example of a seedless vascular plant is a
   a. hornwort.
   b. pine tree.
   c. fruit tree.
   d. club moss.

7. Conifers, ginkgos, cycads, and gnetophytes are collectively called
   a. gymnosperms.
   b. angiosperms.
   c. bryophytes.
   d. seedless vascular plants.

8. Which of the following is *not* a part of the innermost whorl of the flower?
   a. stamens
   b. style
   c. carpels
   d. stigma

9. Which of the following is a unique characteristic of angiosperms?
   a. double fertilization
   b. seeds
   c. leaves
   d. cones

10. Unlike the gymnosperms, the ovules of angiosperms are completely enclosed by the _________ at the time of pollination.

11. A sperm is haploid. Endosperm is _________.

12. _________ have one cotyledon, parallel veins, and three flower parts per whorl.

CHALLENGE YOURSELF

1. Compare and contrast the adaptations to a terrestrial existence of fungi versus plants. Which group of organisms has been more successful, and why?

2. Compare and contrast the alternation of gametophyte and sporophyte generations in mosses, ferns, pines, and flowering plants. Which generation is dominant in each case, if any? Is it haploid or diploid? Which plant groups are more advanced? Are there advantages to having the gametophyte or sporophyte generation dominate?

3. Why do mosses and ferns both require free water to complete their life cycles? At what stage of the life cycle is water required? Do angiosperms also require free water to complete their life cycles? What are the reasons for the difference, if any?

4. Why was the development of seeds so important to the success of angiosperms? How did these characteristics help angiosperms become the dominant photosynthetic organisms on land? In what terrestrial communities are angiosperms *not* dominant?

16 eBRIDGE

Reinforcing Key Points

Plants

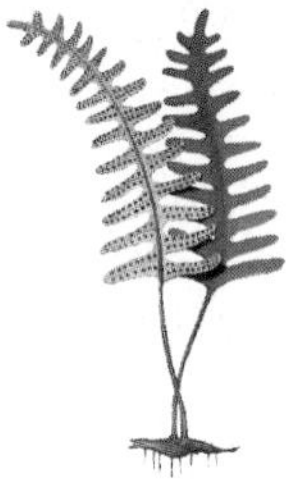

Seedless Plants

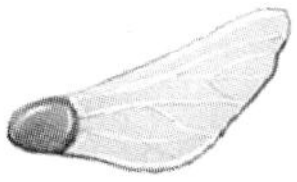

The Advent of Seeds

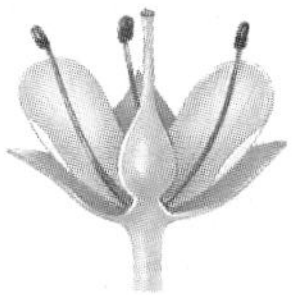

The Evolution of Flowers

Electronic Learning

Visual Learning

Animations

- Effect of Water on Leaves
- Vascular System of Plants
- Moss Life Cycle
- Girth Increase in Woody Dicots
- Fern Life Cycle
- Pine Life Cycle
- Flowering Plant Life Cycle

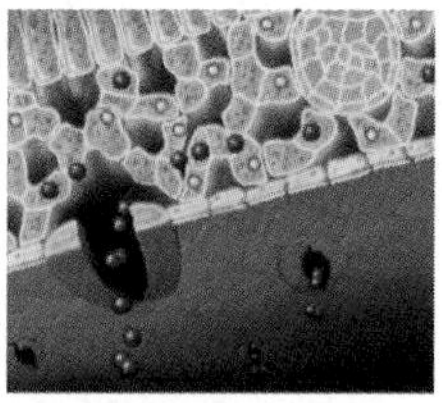

Art Labeling Activities

- Garden Bean Seed Structure
- Corn Grain Structure

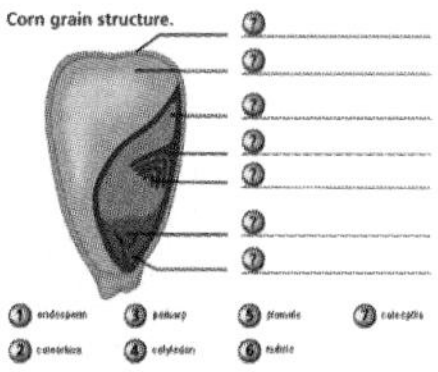

Corn grain structure.

Helping You Learn

- Plant Reproduction

Author's Corner

Four Major Groups of Plants. In this text, four major groups of plants are described: nonvascular plants, seedless vascular plants, gymnosperms, and angiosperms. There are actually five plant lineages, with the seedless vascular plants really consisting of two distinct groups. Nonvascular plants are the most primitive plants, and gymnosperms and angiosperms evolved more recently.

1. The evolutionary origins of plants
2. Nonvascular plants
3. Seedless vascular plants
4. Gymnosperms
5. Angiosperms

The Living World

Virtual Classroom

History of Agriculture

Among plant, fungal, and animal species, only a small minority are edible by humans. Most are useless to us as food because they are indigestible (like wood), poisonous (like some mushrooms), low in nutritional value (like jellyfish), hard to prepare (hard small nuts), difficult to gather (mosquito larvae), or dangerous to hunt (grizzly bears). The few species of plants and animals that we can eat were domesticated early after human agriculture began. The oldest clear evidence of domestication is the dog, which is found in both Asia and North America at about 10,000 B.C. Wheat, sheep, and goats were first domesticated in Southwest Asia around 8500 B.C. Rice and pigs appeared in China by 7500 B.C. Cows arose in India in 6000 B.C., and domestic horses in the Ukraine in 4000 B.C. Corn and beans are first seen in Middle America around 3500 B.C., about the time as potatoes in the Andes.

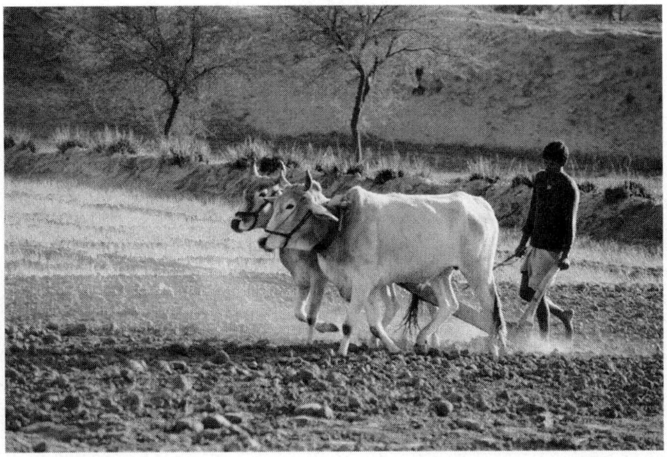

Why these few crops and farm animals, rather than any of the other edible plants and animals? A lot of the answer has to do with ease of domestication—there just aren't many organisms suitable for agriculture, it seems.

Virtual Lab

Why Do Some Plants Accumulate Toxic Levels of Metals?

All plants take up minerals from the soil. A few plant species that grow on soils containing high levels of metals (so-called serpentine soils) are particularly interesting: they have very high levels of metals in their tissues, even more than metal-tolerant plants do. These rare high-metal plants, termed hyperaccumulators, contain in excess of 1,000 micrograms of metal per gram dry weight of tissue! Why have these plant species evolved ways to tolerate high metal levels that would kill an average plant? Several hypotheses have been put forth but an interesting possibility is that the adaptation of the hyperaccumulation of metals makes the plants toxic to herbivores that feed on the plants.

This hypothesis has been examined by Robert Boyd of Auburn University and Scott Martens of the University of California, Davis, using a plant species from the mustard family that hyperaccumulates nickel. By growing the plants on different types of soil that contain various levels of nickel, these researchers were able to vary the level of nickel in the plants, and see what happens to herbivore predators.

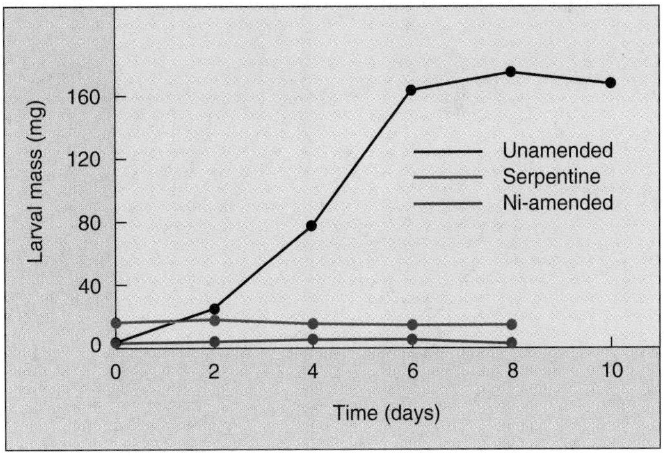
160
120
80
40
0
Larval mass (mg)
0
2
4
6
8
10
Time (days)
Unamended
Serpentine
Ni-amended

Quizzes | Further Reading | Essential Study Partner | Links | BioCourse.com

CHAPTER 17

Plant Form and Function

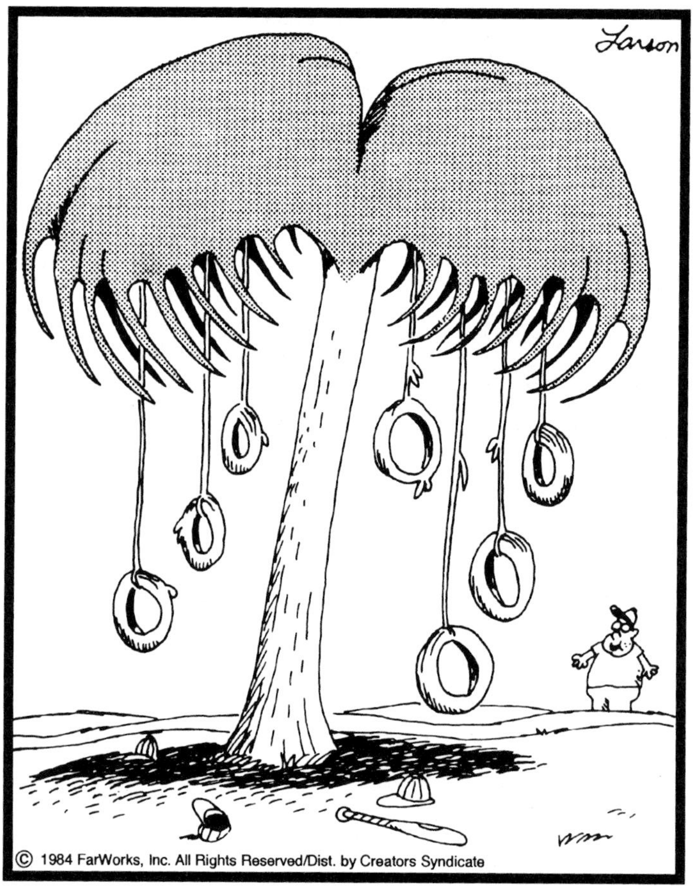

THE FAR SIDE® By GARY LARSON
Larson
© 1984 FarWorks, Inc. All Rights Reserved/Dist. by Creators Syndicate

Her tentacles swaying seductively in the breeze, the Venus Kidtrap was again poised and ready.

CHAPTER 17

Plant Form and Function

CHAPTER OVERVIEW

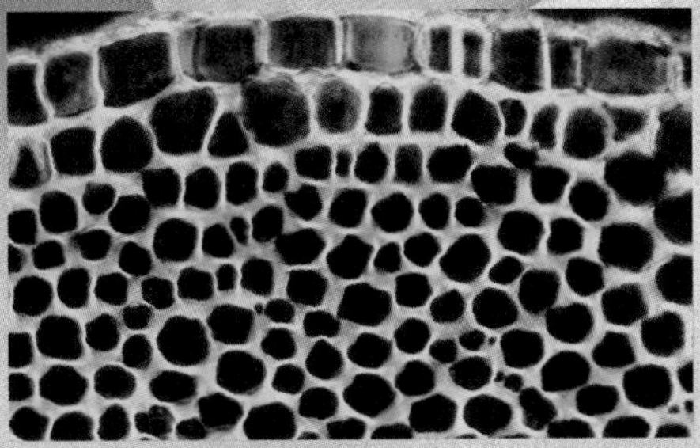

Structure and Function of Plant Tissues

- A vascular plant is organized along a vertical axis.
- The part aboveground, called the shoot, consists of the stem, its branches, and the leaves.
- The part belowground is called the root.
- Plants grow from actively dividing zones called meristems.
- Growth from the tip of the shoot or root is called primary growth.
- Growth in girth is called secondary growth. It takes place in meristems called the vascular or cork cambiums.

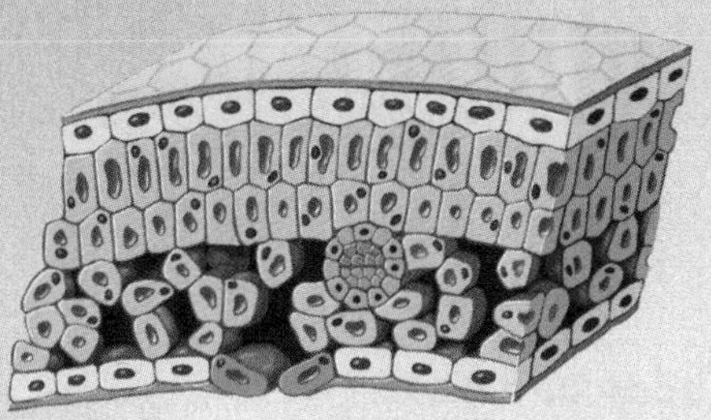

The Plant Body

- Most leaves consist of a flattened blade and a stalk called a petiole. Leaf arrangements on the stem exhibit considerable variation.
- The interior of the leaf, called mesophyll, has veins traversing it.
- Strands of vascular tissue in stems occur as a cylinder toward the edge of the stem in dicots and scattered throughout the stem in monocots.
- The vascular cambium is a cylinder of dividing cells. As a result of their activity, the girth of a plant increases.

Plant Transport and Nutrition

- Water is held up in the conducting vessels of plants by adhesion to the vessel walls and cohesion of the water molecules to one another.
- Water is pulled up the plant by transpiration, which is evaporation from the surface of leaves.
- Carbohydrates are translocated throughout the plant by mass flow.

17.1 Organization of a Vascular Plant

Although the similarities between a cactus, an orchid, and a pine tree may not at first be obvious, most plants possess the same fundamental architecture. All parts of vascular plants have an outer covering of protective tissue and an inner matrix of tissue within which is embedded vascular tissue that conducts water, nutrients, and food throughout the plant. The cells and tissues of vascular plants and how the plant body carries out the functions of living are the focus of this chapter. We discuss the fundamental differences between the belowground portions of the plant—the roots—and the aboveground portions—the shoot—as well as the structural and functional relationships between them.

A vascular plant is organized along a vertical axis, like a pipe. The part belowground is called the **root;** the part aboveground is called the **shoot** (figure 17.1). Although roots and shoots differ in their basic structure, growth at the tips throughout the life of the individual is characteristic of both. The root penetrates the soil and absorbs water and various ions, which are crucial for plant nutrition. It also anchors the plant. The shoot consists of stem and leaves. The **stem** serves as a framework for the positioning of the **leaves,** where most photosynthesis takes place. The arrangement, size, and other characteristics of the leaves are critically important in the plant's production of food. Flowers, and ultimately fruits and seeds are also formed on the shoot.

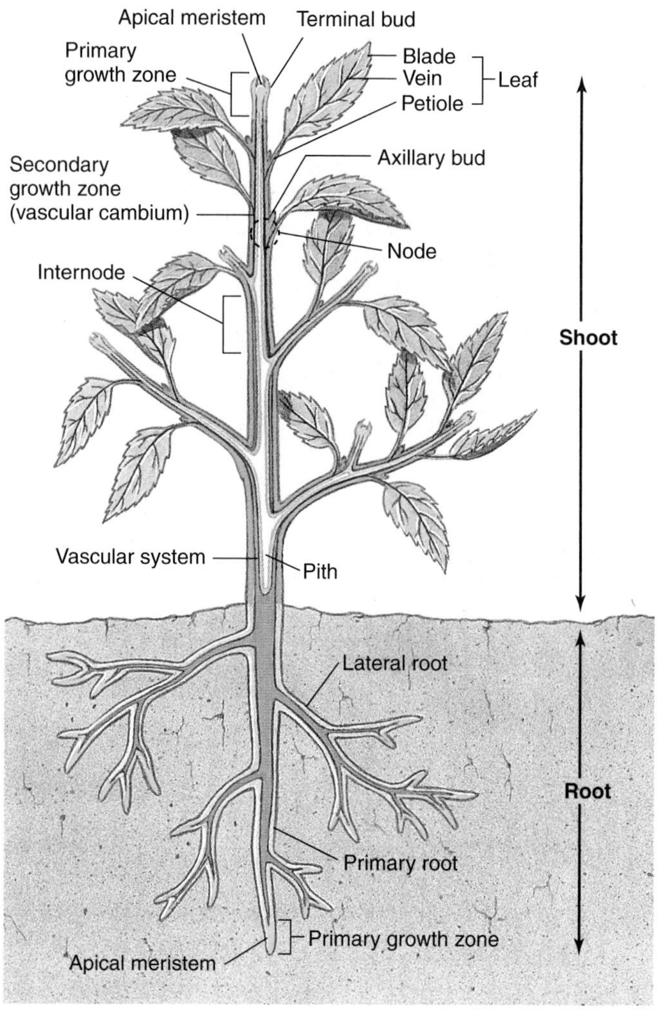
Apical meristem
Terminal bud
Primary growth zone
Blade
Vein
Petiole
Leaf
Secondary growth zone (vascular cambium)
Axillary bud
Node
Internode
Shoot
Vascular system
Pith
Lateral root
Root
Primary root
Primary growth zone
Apical meristem

Figure 17.1 The body of a plant.

The plant body consists of an aboveground portion called the shoot (stems and leaves) and a belowground portion called the root. Elongation of the plant, so-called primary growth, takes place when clusters of cells called the apical meristems (*yellowish green areas*) divide at the ends of the roots and the stems. Thickening of the plant, so-called secondary growth, takes place in a sheath encasing the plant (*lavender area*), allowing the plant to enlarge in girth like letting out a belt.

Meristems

Animals grow all over. As children grow into adults, their torsos grow at the same time their legs do. If, instead, children grew in only one place, with their legs getting longer and longer, they would be growing in a way similar to the way plants grow.

Plants contain growth zones of unspecialized cells called **meristems,** whose only function is to divide. Every time one of these cells divides, one of its two daughter cells remains in the meristem, whereas the other differentiates into one of the three kinds of plant tissue and ultimately becomes part of the plant body.

In plants, **primary growth** is initiated at the tips by the **apical meristems,** regions of active cell division that occur at the tips of roots and shoots. The growth of these meristems results primarily in the extension of the plant body. As it elongates, it forms what is known as the primary plant body, which is made up of the primary tissues.

Growth in thickness, **secondary growth,** involves the activity of the **lateral meristems,** which are cylinders of meristematic tissue. The continued division of their cells results primarily in the thickening of the plant body. There are two kinds of lateral meristems: the **vascular cambium,** which gives rise to ultimately thick accumulations of secondary xylem and phloem, and the **cork cambium,** from which arise the outer layers of bark on both roots and shoots.

17.1 The body of a vascular plant is basically a tube connecting roots to leaves, with growth zones called meristems.

17.2 Plant Tissue Types

The organs of a plant—the leaves, stem, and roots—are composed of different combinations of tissues, just as your legs are composed of bone, muscle, and connective tissue. A tissue is a group of similar cells—cells that are specialized in the same way—organized into a structural and functional unit. Most plants have three major tissue types: (1) *ground tissue,* in which the vascular tissue is embedded; (2) *dermal tissue,* the outer protective covering of the plant; and (3) *vascular tissue,* which conducts water and dissolved minerals up the plant and conducts the products of photosynthesis throughout.

Each major tissue type is composed of distinctive kinds of cells, whose structures are related to the functions of the tissues in which they occur. For example, vascular tissue is composed of *xylem,* which conducts water and dissolved minerals, and *phloem,* which conducts carbohydrates (mostly sucrose), which the plant uses as food.

Ground Tissue

Parenchyma cells are the least specialized and the most common of all plant cell types (figure 17.2); they form masses in leaves, stems, and roots. Parenchyma cells, unlike some other cell types, are characteristically alive at maturity, with fully functional cytoplasm and a nucleus. Most parenchyma cells have only thin cell walls, which are mostly cellulose that is laid down while the cells are still growing.

Collenchyma cells, which are also living at maturity, form strands or continuous cylinders beneath the epidermis of stems or leaf stalks and along veins in leaves. They are usually elongated, with unevenly thickened primary walls, which are their distinguishing feature. Strands of collenchyma provide much of the support for plant organs in which secondary growth has not occurred. These cells also provide the plant with flexibility due to the uneven nature of the walls, the thinner areas becoming flex points (figure 17.3).

In contrast to parenchyma and collenchyma cells, **sclerenchyma cells** have tough, thick cell walls; they usually do not contain living cytoplasm when mature. There are two types of sclerenchyma: **fibers,** which are long, slender cells that usually form strands, and **sclereids,** which are variable in shape but often branched (figure 17.4). Sclereids are sometimes called stone cells because they make up the bulk of the stones of peaches and other "stone" fruits, as well as that of nut shells. Both fibers and sclereids are thick-walled and strengthen the tissues in which they occur.

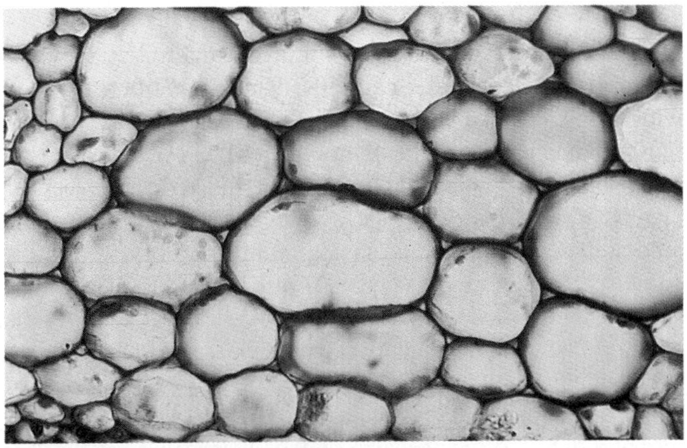

Figure 17.2 Parenchyma cells.

Cross section of parenchyma cells from grass. Only primary cell walls are seen in this living tissue.

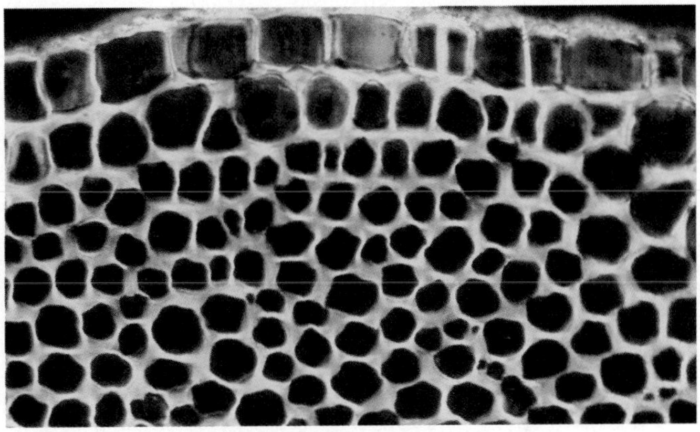

Figure 17.3 Collenchyma cells.

Cross section of collenchyma cells, with thickened side walls, from a young branch of elderberry (*Sambucus*). In other kinds of collenchyma cells, the thickened areas may occur at the corners of the cells or in other kinds of strips.

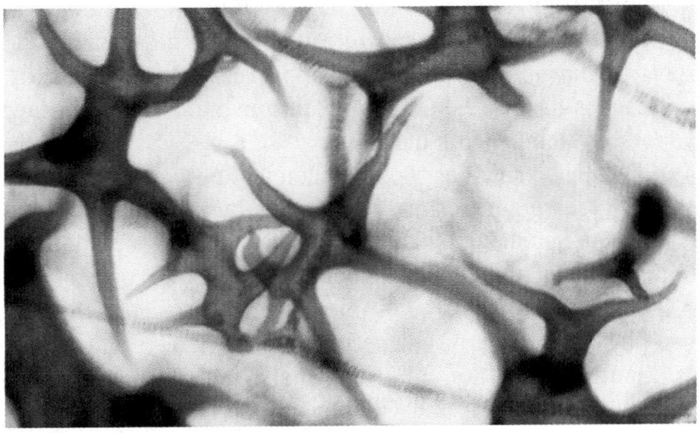

Figure 17.4 Sclereids.

Clusters of sclereids ("stone cells"), stained *blue* in this preparation, in the pulp of a pear. Such clusters of sclereids give pears their gritty texture.

Dermal Tissue

All parts of the outer layer of a primary plant body are covered by flattened epidermal cells, which are often covered with a thick, waxy layer called the **cuticle.** These are the most abundant cells in the plant epidermis, or outer covering. They protect the plant and provide an effective barrier against water loss. One type of specialized cell that occurs among the epidermal cells is the guard cell.

Guard cells are paired cells with openings that lie between them called **stomata** (singular, stoma). Guard cells and stomata occur frequently in the epidermis of leaves and occasionally on other parts of the shoot, such as on stems or fruits (figure 17.5). Oxygen, carbon dioxide, and water pass into and out of the leaves almost exclusively through the stomata, which open and shut in response to such external factors as supply of moisture and light.

Trichomes are outgrowths of the epidermis that occur on the shoot, on the surfaces of stems and leaves. Trichomes vary greatly in form in different kinds of plants. A "fuzzy" or "woolly" leaf is covered with trichomes, which when viewed under the microscope look like a thicket of fibers. Trichomes play an important role in regulating the heat and water balance of the leaf. Much as the hairs of a fur coat provide insulation, so too the trichomes slow heat and water vapor loss from the surface of the leaf. Other kinds of trichomes are glandular, often secreting sticky or toxic substances that may deter potential herbivores.

Other outgrowths of the epidermis occur belowground, on the surface of roots near their tips. Called **root hairs,** these tubular extensions of single epidermal cells keep the root in intimate contact with the particles of soil. Root hairs play an important role in the absorption of water and minerals from the soil, by increasing the surface area of the root.

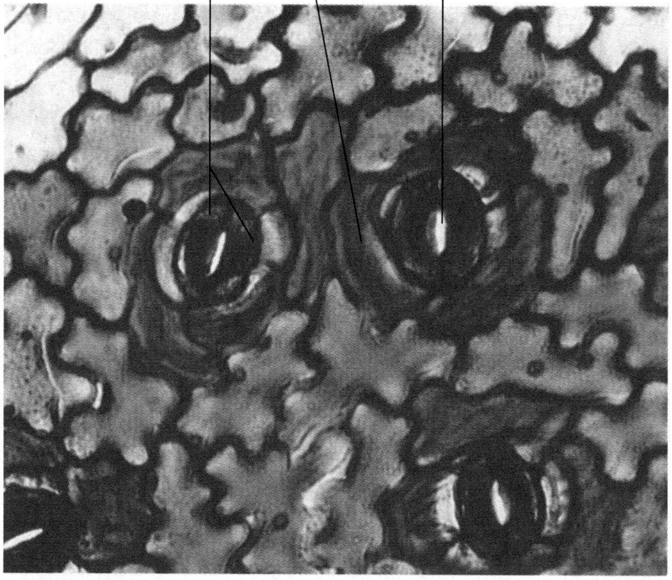
Subsidiary cell
Guard cells
Stomatal opening

Figure 17.5 Guard cells and stomata.
Numerous stomata occur among the leaf epidermal cells of this member of the aralia family (Araliaceae). The epidermis has been peeled off and stained with a red dye.

Vascular Tissue

Vascular plants contain two kinds of conducting, or vascular, tissue: the xylem and the phloem. **Xylem** is the plant's principal water-conducting tissue, forming a continuous system that runs throughout the plant body. Within this system, water (and the minerals dissolved in the water) passes from the roots up through the shoot in an unbroken stream. When water reaches the leaves, much of it passes into the air as water vapor, through the stomata.

The two principal types of conducting elements in the xylem are **tracheids** and **vessel members,** both of which have thick secondary walls, are elongated, and have no living cytoplasm at maturity. In conducting elements composed of tracheids, water flows from tracheid to tracheid through openings called pits in the secondary walls. In contrast, vessel members have not only pits but also definite openings, or perforations, in their end walls by which they are linked together and through which water flows. A linked row of vessel members forms a vessel (figure 17.6). Primitive angiosperms have only tracheids, but the majority of angiosperms have vessels. Vessels conduct water much more efficiently than do strands of tracheids. In addition to conducting cells, xylem includes fibers and parenchyma cells.

Phloem is the principal nutrient-conducting tissue in vascular plants. Food conduction in phloem is carried out through two kinds of elongate cells: **sieve cells** and **sieve-tube members** (figure 17.7). Seedless vascular plants and gymnosperms have only sieve cells; most angiosperms have sieve-tube members, but at least one primitive angiosperm only has sieve cells. Clusters of pores known as *sieve areas* occur on both kinds of cells and connect the cytoplasms of adjoining sieve cells and sieve-tube members. Both cell types are living, but their nuclei are lost during maturation.

In sieve-tube members, some sieve areas have larger pores and are called *sieve plates.* Sieve-tube members occur end to end, forming longitudinal series called **sieve tubes.** Specialized parenchyma cells known as **companion cells** occur regularly in association with sieve-tube members. Sieve cells are less specialized than sieve-tube members, and the pores in all of their sieve areas are roughly of the same diameter. In an evolutionary sense, sieve-tube members are more advanced, more specialized, and, presumably, more efficient.

17.2 Plants contain a variety of ground, dermal, and vascular tissues.

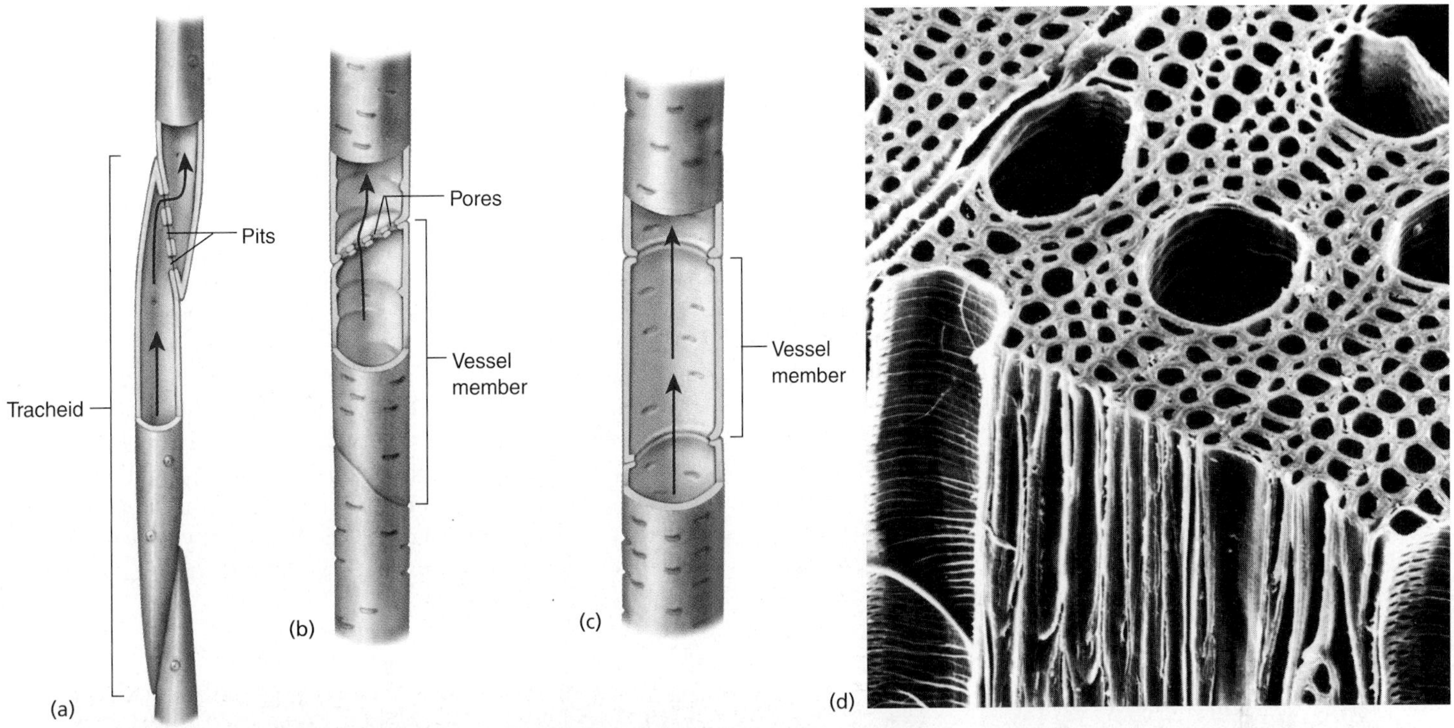
Tracheid
Pits
(a)
Pores
Vessel
member
(b)
Vessel
member
(c)
(d)

Figure 17.6 Comparison of vessel members and tracheids.

(*a*) In tracheids the water passes from cell to cell by means of pits. (*b*) In vessel members, water moves by way of perforation plates, which may be simple or interrupted by bars. (*c*) Open-ended vessel members. (*d*) A scanning electron micrograph of the red maple (*Acer rubrum*), showing the xylem.

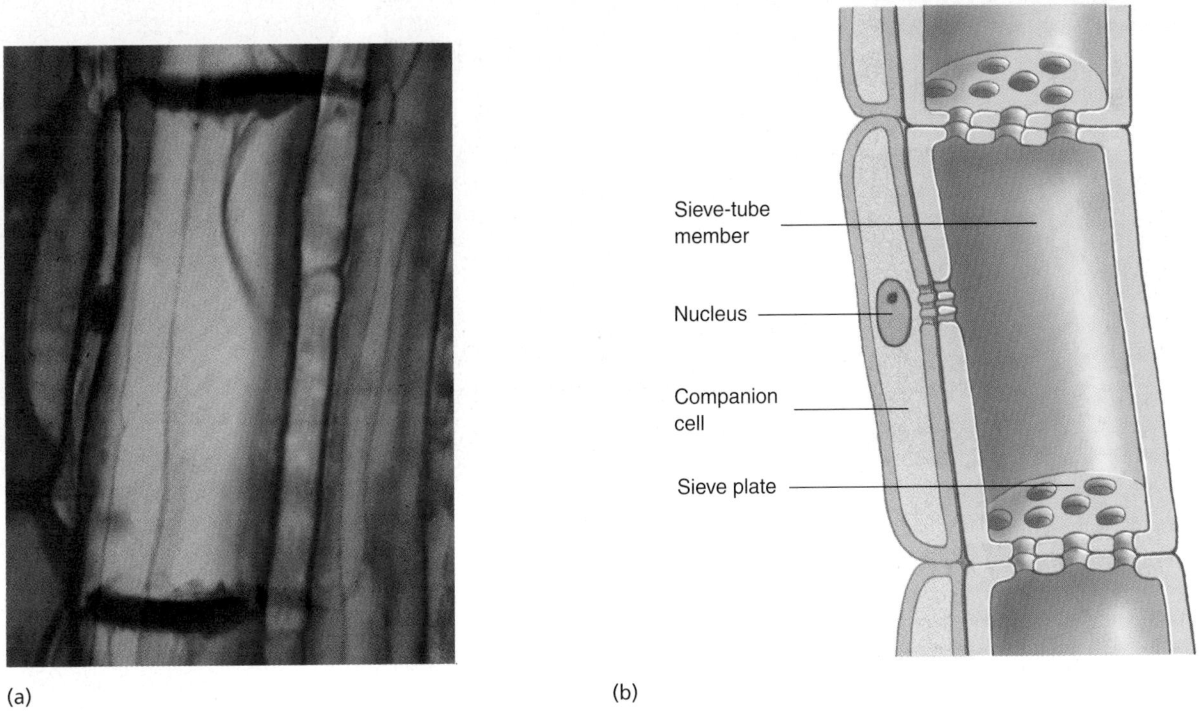
(a)
Sieve-tube
member
Nucleus
Companion
cell
Sieve plate
(b)

Figure 17.7 Sieve tubes.

(*a*) Sieve-tube member from the phloem of squash (*Cucurbita*), connected with the cells above and below to form a sieve tube. (*b*) In this diagram, note the thickened end walls, which are at right angles to the sieve tube. The narrow cell with the nucleus at the left of the sieve-tube member is a companion cell.

17.3 Leaves

We now consider the three kinds of vegetative organs that form the body of a plant: **leaves, stems,** and **roots.** Leaves are usually the most prominent shoot organs and are structurally diverse (figure 17.8). Leaves, outgrowths of the shoot apex, are the major light-capturing organs of most plants. Most of the chloroplast-containing cells of a plant are within its leaves, and it is there where the bulk of photosynthesis occurs. Exceptions to this are found in some plants, such as cacti, whose green stems have largely taken over the function of photosynthesis for the plant. Photosynthesis is conducted mainly by the "greener" parts of plants because they contain more chlorophyll, the most efficient photosynthetic pigment.

The apical meristems of stems and roots are capable of growing indefinitely under appropriate conditions. Leaves, in contrast, grow by means of **marginal meristems,** which flank their thick central portions. These marginal meristems grow outward and ultimately form the **blade** (flattened portion) of the leaf, while the central portion becomes the midrib. Once a leaf is fully expanded, its marginal meristems cease to grow.

In addition to the flattened blade, most leaves have a slender stalk, the **petiole.** Two leaflike organs, the **stipules,** may flank the base of the petiole where it joins the stem. Veins, consisting of both xylem and phloem, run through the leaves. As mentioned in chapter 16, in most dicots, the pattern is net or reticulate venation; in many monocots, the veins are parallel. Leaves may be **alternately** arranged

(a) (b) (c) (d) (e)

Figure 17.8 Leaves.

Angiosperm leaves are stunningly variable. (a) Diverse leaves in the herb layer of a Costa Rican rain forest. (b) A *compound leaf,* marijuana (*Cannabis sativa*). Such a compound leaf is composed of several leaflets. (c) A *simple leaf,* its margin deeply lobed, from the tulip tree (*Liriodendron tulipifera*). A simple leaf has only one leaf element. (*d*) Another compound leaf, from a member of the legume family in the lowland forest of Peru. (e) Many unusual arrangements of leaves occur in different kinds of plants. For example, in this miner's lettuce (*Claytonia perfoliata*), an herb of the Pacific states, two leaves are completely fused below each of the clusters of flowers, which seem, therefore, to arise from the center of a single leaf.

(alternate leaves usually spiral around a shoot) or they may be in **opposite** pairs. Less often, three or more leaves may be in a **whorl,** a circle of leaves at the same level at a node (figure 17.9).

A typical leaf contains masses of parenchyma, called **mesophyll** ("middle leaf"), through which the vascular bundles, or veins, run. Beneath the upper epidermis of a leaf are one or more layers of closely packed, columnlike parenchyma cells called **palisade mesophyll.** The rest of the leaf interior, except for the veins, consists of a tissue called **spongy mesophyll** (figure 17.10). Between the spongy mesophyll cells are large intercellular spaces that function in gas exchange and particularly in the passage of carbon dioxide from the atmosphere to the mesophyll cells. These intercellular spaces are connected, directly or indirectly, with the stomata.

17.3 Leaves, the photosynthetic organs of the plant body, are varied in shape.

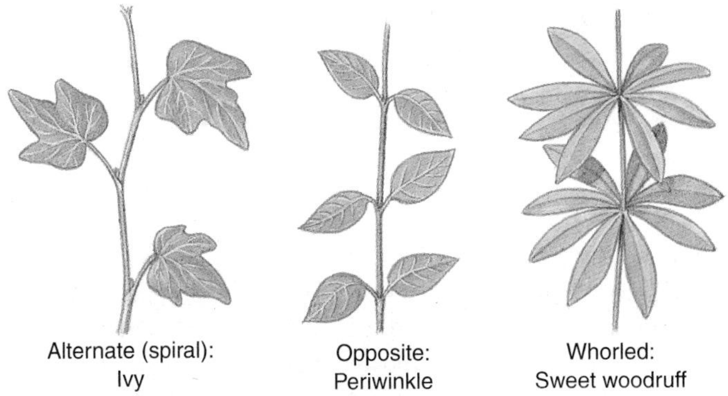
Alternate (spiral):
Ivy
Opposite:
Periwinkle
Whorled:
Sweet woodruff

Figure 17.9 Types of leaf arrangements.
The three common types of leaf arrangements are alternate, opposite, and whorled.

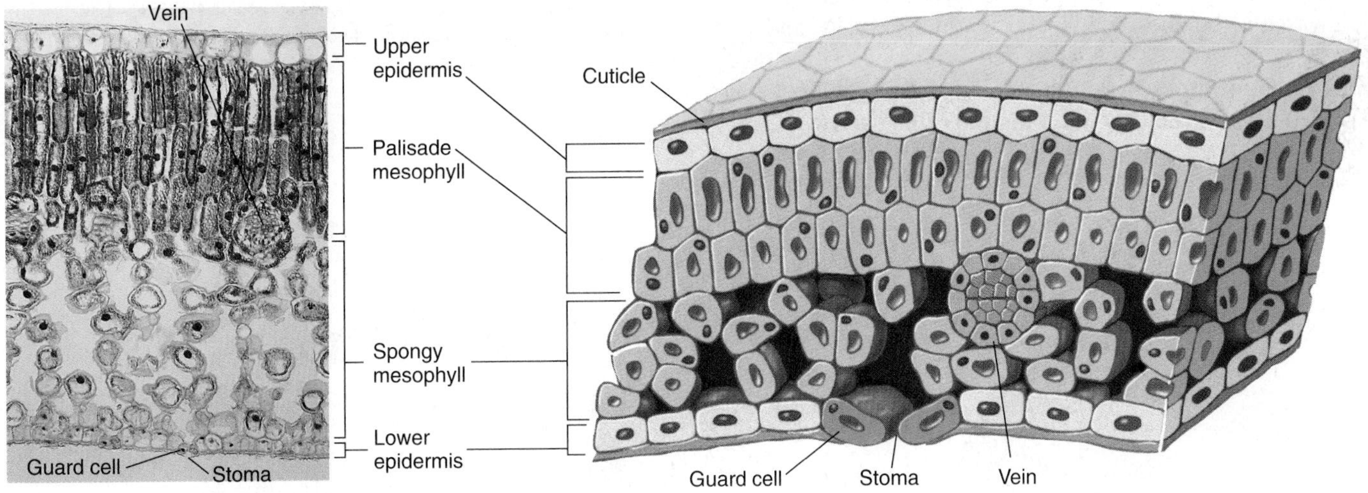
Vein
Upper epidermis
Palisade mesophyll
Spongy mesophyll
Lower epidermis
Guard cell
Stoma
Cuticle
Guard cell
Stoma
Vein

Figure 17.10 A leaf in cross section.
Transection of a leaf, showing the arrangement of palisade and spongy mesophyll, a vascular bundle or vein, and the epidermis, with paired guard cells flanking the stoma.

17.4 Stems

As mentioned earlier, the stem is that part of the shoot that serves as the framework for the positioning of the leaves. Often experiencing both primary and secondary growth, stems are the source of an economically important product—wood.

Primary Growth

In the primary growth of a shoot, leaves first appear as leaf primordia (singular, primordium), or rudimentary young leaves, which cluster around the apical meristem, unfolding and growing as the stem itself elongates (figure 17.11). The places on the stem at which leaves form are called nodes. The portions of the stem between these attachment points are called the internodes. As the leaves expand to maturity, a bud, a tiny, undeveloped side shoot, develops in the **axil** of each leaf, the angle between a leaf and the stem from which it arises. These buds, which have their own immature leaves, may elongate and form lateral branches, or they may remain small and dormant. A hormone moving downward from the terminal bud of the shoot continuously suppresses the expansion of the lateral buds. These buds begin to expand when the terminal bud is removed.

Within the soft, young stems, the strands of vascular tissue, xylem and phloem, either are arranged around the outside of the stem as a cylinder, as is common in dicots, or scattered through it, as is common in monocots (figure 17.12). The vascular bundles contain both primary xylem and primary phloem. At the stage when only primary growth has occurred, the inner portion of the ground tissue of a dicot stem is called the **pith,** and the outer portion is the **cortex.**

Apical meristem
Leaf primordium

Figure 17.11 Where leaves originate.
Scanning electron micrograph of the shoot apex of a silver maple, *Acer saccharinum,* showing a developing shoot during summer, the season of active growth. The apical meristem and leaf primordia are plainly visible.

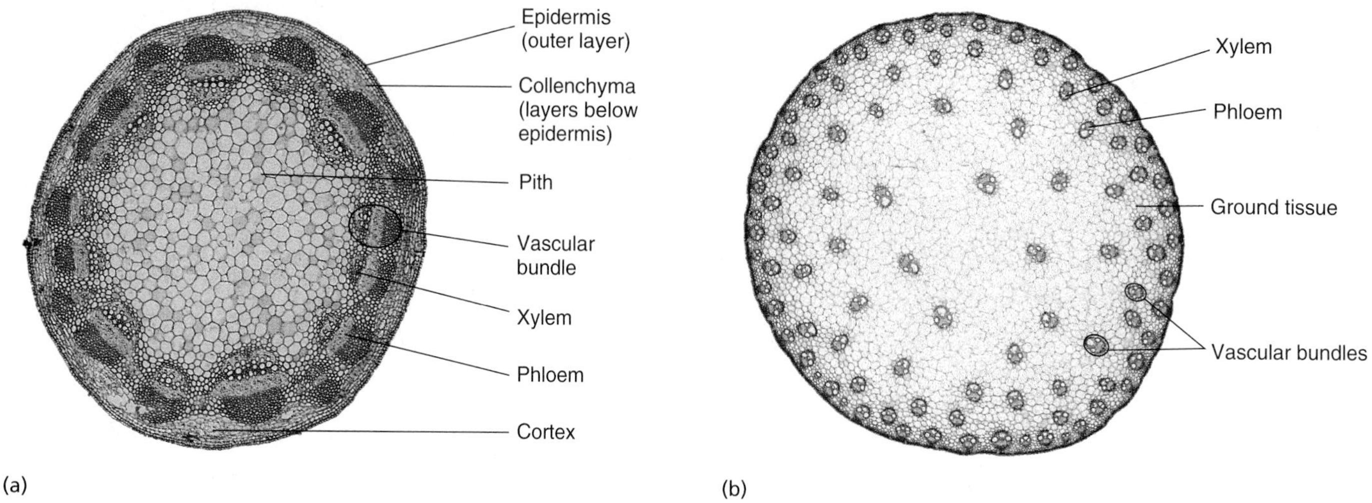
Epidermis (outer layer)
Collenchyma (layers below epidermis)
Pith
Vascular bundle
Xylem
Phloem
Cortex
(a)
Xylem
Phloem
Ground tissue
Vascular bundles
(b)

Figure 17.12 A comparison of dicot and monocot stems.
(*a*) Transection of a young stem of a dicot, the common sunflower, *Helianthus annuus,* in which the vascular bundles are arranged around the outside of the stem.
(*b*) Transection of a monocot stem, corn, *Zea mays,* with the scattered vascular bundles characteristic of the class.

Secondary Growth

In stems, secondary growth is initiated by the differentiation of the **vascular cambium,** which consists of a thin cylinder of actively dividing cells located between the bark and the main stem in woody plants. The vascular cambium develops from cells within the vascular bundles of the stem, between the primary xylem and the primary phloem (figure 17.13). The cylindrical form of the vascular cambium is completed by the differentiation of some of the parenchyma cells that lie between the bundles. Once established, the vascular cambium consists of elongated, somewhat flattened cells with large vacuoles. The cells that divide from the vascular cambium outwardly, toward the bark, become secondary phloem; those that divide from it inwardly become secondary xylem.

While the vascular cambium is becoming established, a second kind of lateral cambium, the cork cambium, develops in the stem's outer layers. The cork cambium usually consists of plates of dividing cells that move deeper and deeper into the stem as they divide. Outwardly, the cork cambium splits off densely packed **cork cells;** they contain a fatty substance and are nearly impermeable to water. Cork cells are dead at maturity. Inwardly, the cork cambium divides to produce a layer of parenchyma cells. The cork, the cork cambium that produces it, and this layer of parenchyma cells make up a layer called the **periderm,** which is the plant's outer protective covering.

Cork, which covers the surfaces of mature stems or roots, takes the place of the epidermis, which performs a similar function in the younger parts of the plant. The term **bark** refers to all of the tissues of a mature stem or root outside of the vascular cambium. Because the vascular cambium has the thinnest-walled cells that occur anywhere in a secondary plant body, it is the layer at which bark breaks away from the accumulated secondary xylem.

Wood is one of the most useful, economically important, and beautiful products obtained from plants. Anatomically, wood is accumulated secondary xylem. As the secondary xylem ages, its cells become infiltrated with gums and resins, and the wood becomes darker. For this reason, the wood located nearer the central regions of a given trunk, called heartwood, is often darker and denser than the wood nearer the vascular cambium, called sapwood, which is still actively involved in transport within the plant. Because of the way it is accumulated, wood often displays rings. In temperate regions, these rings are annual rings (figure 17.14).

17.4 Stems, the aboveground framework of the plant body, grow both at their tips and in circumference.

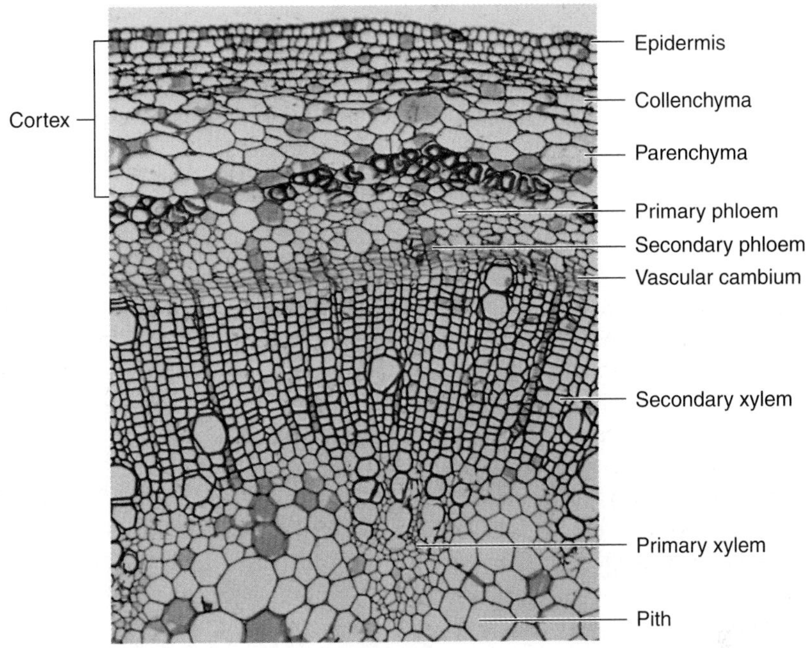
Cortex
Epidermis
Collenchyma
Parenchyma
Primary phloem
Secondary phloem
Vascular cambium
Secondary xylem
Primary xylem
Pith

Figure 17.13 Vascular cambium.
Early stage in vascular cambium differentiation in the castor bean, *Ricirus.*

Figure 17.14 Annual rings in a section of pine.
The rings that you see in this section of pine reflect the fact that the vascular cambium of trees divides more actively in the spring and summer, when water is plentiful and temperatures are suitable for growth, than in the fall and winter, when water is scarce and the weather is cold. As a result, layers of larger, thinner-walled cells formed during the growing season alternate with the smaller, darker layers of thick-walled cells formed during the rest of the year. A count of such annual rings in a tree trunk can be used to calculate the tree's age.

17.5 Roots

Roots have a simpler pattern of organization and development than do stems. Although different patterns exist, the kind of root described here and shown in figure 17.15*a* is found in many dicots. Roots contain xylem and phloem, just as stems do, but there is no pith in the center of the vascular tissue in most dicot roots. Instead, these roots have a central column of xylem with radiating arms. Alternating with the radiating arms of xylem are strands of primary phloem. Surrounding the column of vascular tissue and forming its outer boundary is a cylinder of cells one or more cell layers thick called the **pericycle.** Branch, or lateral, roots are formed from cells of the pericycle. The outer layer of the root, as in the shoot, is the epidermis. The mass of parenchyma in which the root's vascular tissue is located is the cortex. Its innermost layer—the endodermis—consists of specialized cells that regulate the flow of water between the vascular tissues and the root's outer portion (figure 17.16). The endodermis lies just outside of the pericycle. Endodermis cells are encircled by a thickened, waxy band called the **Casparian strip.** By the differential passage of minerals and nutrients through endodermis cells, the plant regulates its supply of minerals.

The apical meristem of the root divides and produces cells both inwardly, back toward the body of the plant, and outwardly. The three primary meristems are the **protoderm,** which becomes the epidermis; the **procambium,** which produces primary vascular tissues (primary xylem and primary phloem); and the **ground meristem,** which differentiates further into ground tissue, which is composed of parenchyma cells. Outward cell division results in the formation of a thimblelike mass of relatively unorganized cells, the **root cap,** which covers and protects the root's apical meristem as it grows through the soil.

The root elongates relatively rapidly just behind its tip. Abundant **root hairs,** extensions of single epidermal cells, form above that zone. Virtually all water and minerals are absorbed from the soil through the root hairs, which greatly increase the root's surface area and absorptive powers. In plants with mycorrhizae (see chapter 15), the root hairs are often greatly reduced in number, and the fungal filaments of the mycorrhizae play a role similar to that of the root hairs, increasing the surface area for absorption.

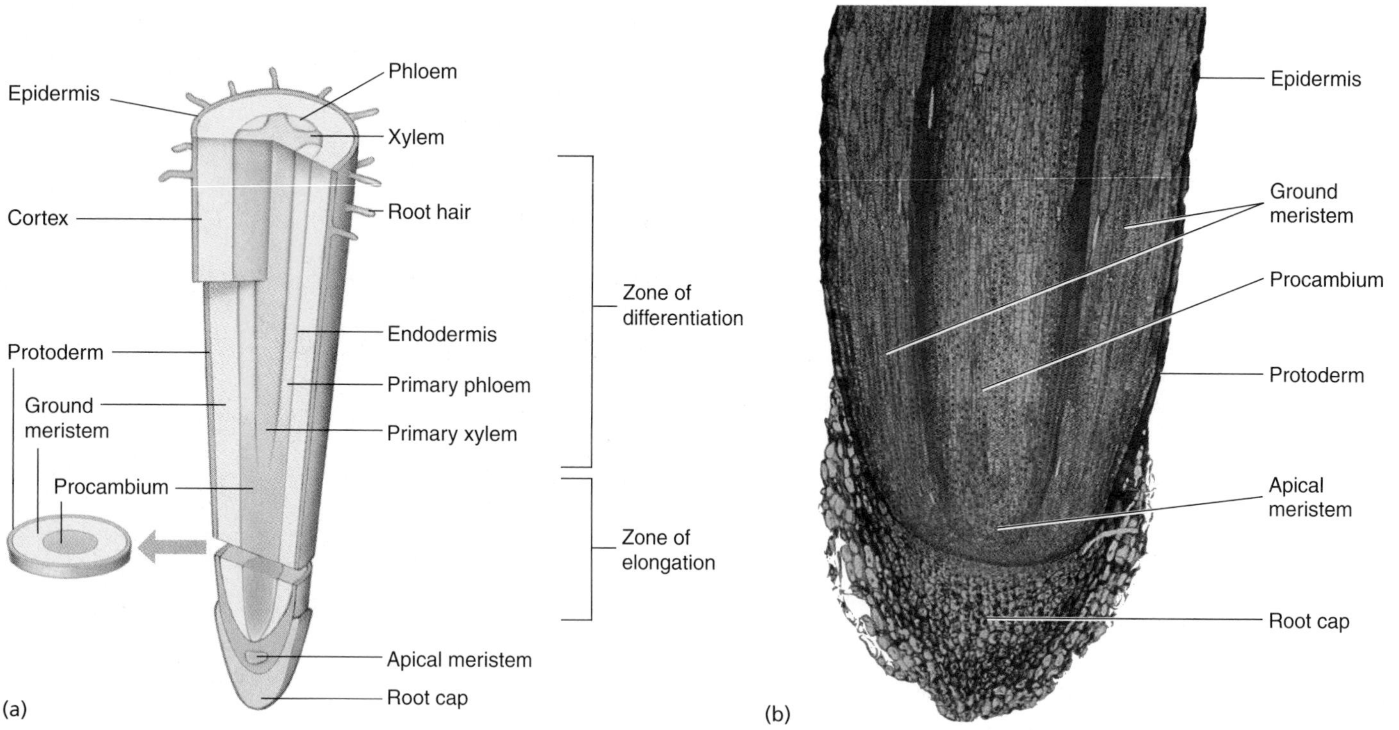

Epidermis
Phloem
Xylem
Cortex
Root hair
Zone of differentiation
Endodermis
Protoderm
Primary phloem
Ground meristem
Primary xylem
Procambium
Zone of elongation
Apical meristem
Root cap
(a)
Epidermis
Ground meristem
Procambium
Protoderm
Apical meristem
Root cap
(b)

Figure 17.15 Root structure.

(*a*) Diagram of primary meristems in a dicot root, showing their relation to the apical meristem. The three primary meristems are the protoderm, which differentiates further into epidermis; the procambium, which differentiates further into primary vascular strands; and the ground meristem, which differentiates further into ground tissue.
(*b*) Median longitudinal section of a monocot root tip in corn, *Zea mays,* showing the differentiation of protoderm, procambium, and ground meristem.

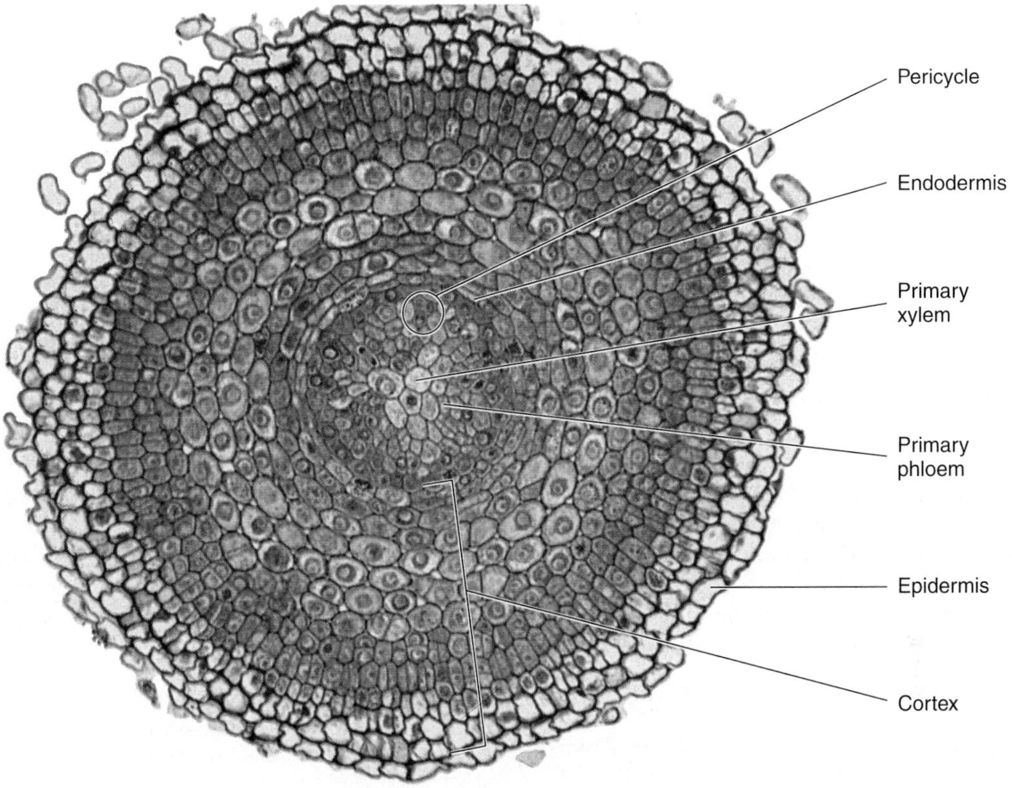

Pericycle
Endodermis
Primary xylem
Primary phloem
Epidermis
Cortex

Figure 17.16 A root cross section.
Cross section through a root of a buttercup, *Ranunculus californicus.*

One of the fundamental differences between roots and shoots has to do with the nature of their branching. In stems, branching occurs from buds on the stem surface; in roots, branching is initiated well back of the root apex as a result of cell divisions in the pericycle. The lateral root primordia grow out through the cortex toward the surface of the root, eventually breaking through and becoming established as lateral roots (figure 17.17).

Secondary growth in roots, both main roots and laterals, is similar to that in stems. In dicots and other plants with secondary growth, part of the pericycle and the parenchyma cells between the phloem patches and the xylem arms become the root vascular cambium, which starts producing secondary xylem to the inside and secondary phloem to the outside. Eventually, the secondary tissues acquire the form of concentric cylinders. The primary phloem, cortex, and epidermis become crushed and are sloughed off as more secondary tissues are added. In the pericycle of woody plants, the cork cambium produces cork cells to the outside and *phelloderm* parenchyma to the inside. The tissue associated with the cork cambium is called the periderm (outer bark), as in stems.

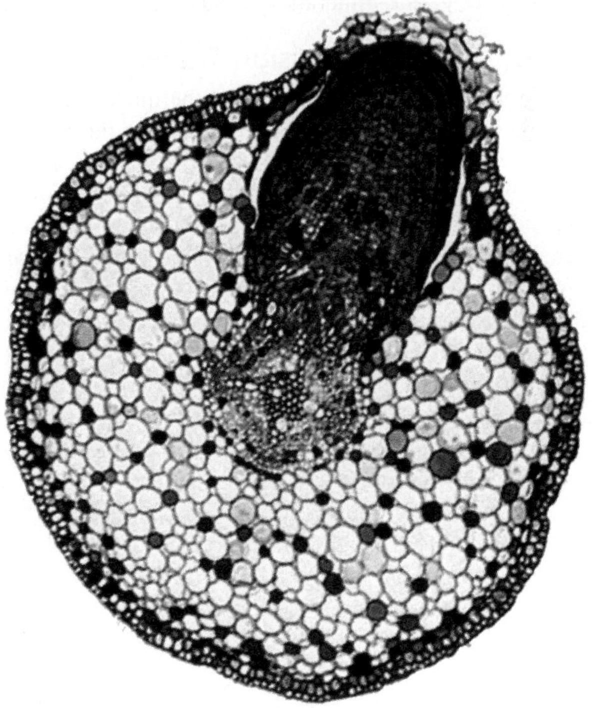

Figure 17.17 Lateral roots.
A lateral root growing out through the cortex of the black willow, *Salix nigra.* Lateral roots originate beneath the surface of the main root, whereas lateral stems originate at the surface.

17.5 Roots, the belowground portion of the plant body, are adapted to absorb water and minerals from the soil.

17.6 Water Movement

Vascular plants have a conducting system, as humans do, for transporting fluids and nutrients from one part to another. Functionally, a plant is essentially a tube with its base embedded in the ground. At the base of the tube are roots, and at its top are leaves. For a plant to function, two kinds of transport processes must occur: first, the carbohydrate molecules produced in the leaves by photosynthesis must be carried to all of the other living plant cells. To accomplish this, liquid, with these carbohydrate molecules dissolved in it, must move both up and down the tube. Second, nutrients and water in the ground must be taken up by the roots and ferried to the leaves and other plant cells. In this process, liquid moves up the tube. Plants accomplish these two processes by using chains of specialized cells: those of the phloem transport photosynthetically produced carbohydrates up and down the tube, and those of the xylem carry water and minerals upward.

Figure 17.18 How does water get to the top of a tree?
We would expect gravity to make such a tall column of water too heavy to be maintained by capillary action. What pulls the water up?

Cohesion-Adhesion-Tension Theory

Many of the leaves of a large tree may be more than 10 stories off the ground. How does a tree manage to raise water so high (figure 17.18)? If a long, hollow tube, closed at one end, is filled with water and placed, open end down, in a full bucket of water, gravity acts (pushes) on the column of air over the bucket. The weight of the air (at sea level) exerts an amount of pressure that is defined as 1 atmosphere downward on the water in the bucket and thus presses the water up into the tube. But gravity also acts to pull the water within the tube down. The interaction of these two forces determines the water level in the tube, about 10.4 meters at sea level.

Opening the tube and blowing air across the upper end demonstrates how water rises higher than 10.4 meters in a plant. The stream of relatively dry air causes water molecules to evaporate from the water surface in the tube. The water level in the tube does not fall because, as water molecules are drawn from the top, they are replenished by new water molecules pulled up from the bottom. This, in essence, is what happens in plants. The passage of air across leaf surfaces results in the loss of water by evaporation, creating a "pull" at the open upper end of the plant. New water molecules that enter the roots are pulled up the plant. *Adhesion* of water molecules to the walls of the narrow vessels in plants also helps to maintain water flow to the tops of plants.

A column of water in a tall tree does not collapse simply because of its weight because water molecules have an inherent strength that arises from their tendency to form hydrogen bonds with one another. These hydrogen bonds cause *cohesion* of the water molecules; in other words, a column of water resists separation. This resistance, called *tensile strength,* varies inversely with the diameter of the column; that is, the smaller the diameter of the column, the greater the tensile strength. Therefore, plants must have very narrow transporting vessels to take advantage of tensile strength.

How the combination of the forces of gravity, tensile strength, and cohesion affect water movement in plants is called the **cohesion-adhesion-tension theory.**

Transpiration

The process by which water leaves a plant is called **transpiration.** More than 90% of the water taken in by plant roots is ultimately lost to the atmosphere, almost all of it from the leaves. It passes out primarily through the stomata in the form of water vapor. On its journey from the plant's interior to the outside, a molecule of water first passes into the pockets of air within the leaf by evaporating from the walls of the spongy mesophyll that lines the intercellular spaces. These intercellular spaces open to the outside of the leaf by way of the stomata. The water that evaporates from these surfaces of the spongy mesophyll cells is continuously replenished from the tips of the veinlets in the leaves. Because the strands of xylem conduct water within the plant in an unbroken stream all the way from the roots to the leaves, when a portion of the water vapor in the intercellular spaces passes out through the stomata, the supply of water vapor in these spaces is continually renewed (figure 17.19).

Structural features such as the stomata, the cuticle, and the intercellular spaces in leaves have evolved in response to one or both of two contradictory requirements: minimizing the loss of water to the atmosphere, on the one hand, and admitting carbon dioxide, which is essential for photosynthesis, on the other. How plants resolve this problem is discussed next.

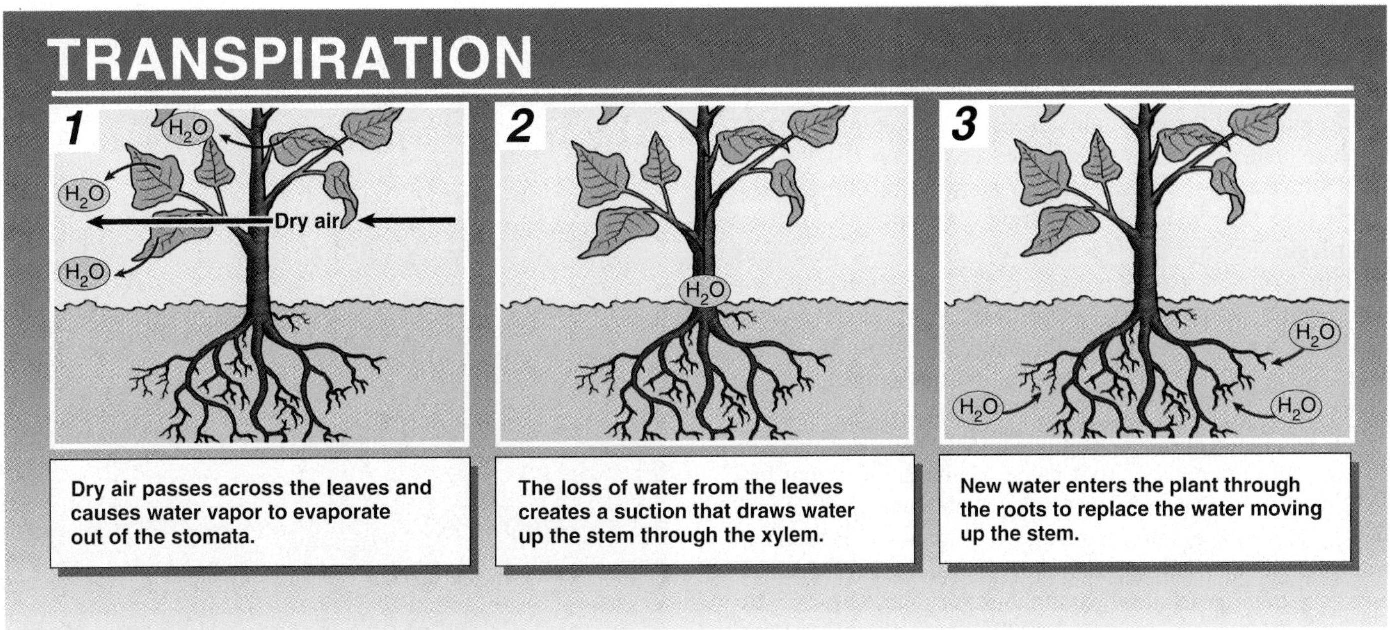
TRANSPIRATION
1
H_2O
H_2O
Dry air
H_2O
2
H_2O
3
H_2O
H_2O
H_2O
Dry air passes across the leaves and causes water vapor to evaporate out of the stomata.
The loss of water from the leaves creates a suction that draws water up the stem through the xylem.
New water enters the plant through the roots to replace the water moving up the stem.

Figure 17.19 How transpiration works.

Regulation of Transpiration: Open and Closed Stomata

The only way plants can control water loss on a short-term basis is to close their stomata. Many plants can do this when subjected to water stress. But the stomata must be open at least part of the time so that carbon dioxide, which is necessary for photosynthesis, can enter the plant. In its pattern of opening or closing its stomata, a plant must respond to both the need to conserve water and the need to admit carbon dioxide.

The stomata open and close because of changes in the water pressure of their guard cells. Stomatal guard cells have a distinctive shape—they are thicker on the side next to the stomatal opening and thinner on their other sides and ends. When the guard cells are turgid (plump and swollen with water), they become bowed in shape, thus opening the stomata as wide as possible (figure 17.20).

A number of environmental factors affect the opening and closing of stomata. The most important is water loss. The stomata of plants that are wilted because of a lack of water tend to close. An increase in carbon dioxide concentration also causes the stomata of most species to close. In most plant species, stomata open in the light and close in the dark. Very high temperatures (above 30° to 35°C) also tend to cause stomata to close.

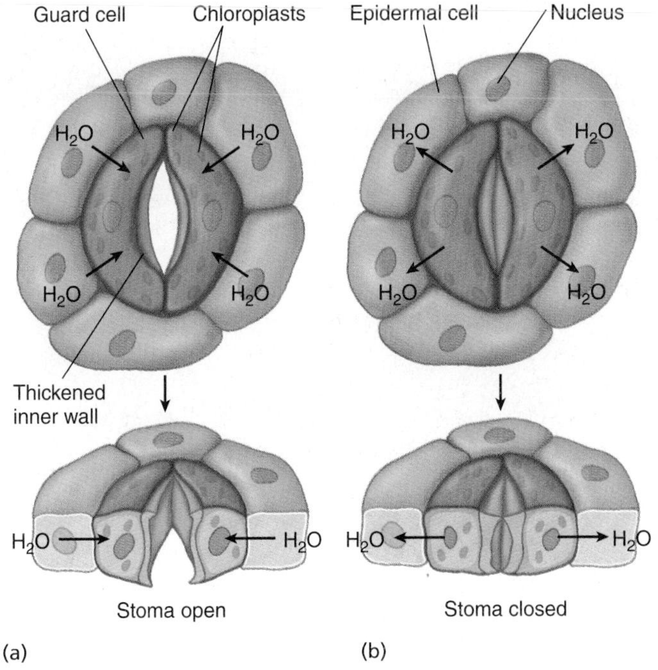
Guard cell
Chloroplasts
Epidermal cell
Nucleus
H_2O
Thickened inner wall
Stoma open
Stoma closed
(a)
(b)

Figure 17.20 How guard cells regulate the opening and closing of stomata.

(a) When guard cells contain a high level of solutes, water enters the guard cells, causing them to bow outward. This bowing opens the stoma. (*b*) When guard cells contain a low level of solutes, water leaves the guard cells, causing them to become flaccid. This flaccidity closes the stoma.

Water Absorption by Roots

Most of the water absorbed by plants comes in through the root hairs (figure 17.21). These root hairs greatly increase the surface area and therefore the absorptive powers of the roots. In plants that have ectomycorrhizae (see chapter 15), the root hairs often are greatly reduced in number; the fungal filaments take their place in promoting absorption. Root hairs are **turgid**—plump and swollen with water—because they contain a higher concentration of dissolved minerals and other solutes than does the water in the soil solution; water, therefore, tends to move into them steadily. Once inside the roots, water passes inward to the conducting elements of the xylem.

Water is not the only substance that enters the roots by passing into the cells of root hairs. Membranes of root hair cells contain a variety of ion transport channels that actively pump specific ions into the plant, even against large concentration gradients. These ions, many of which are plant nutrients, are then transported throughout the plant as a component of the water flowing through the xylem (figure 17.22).

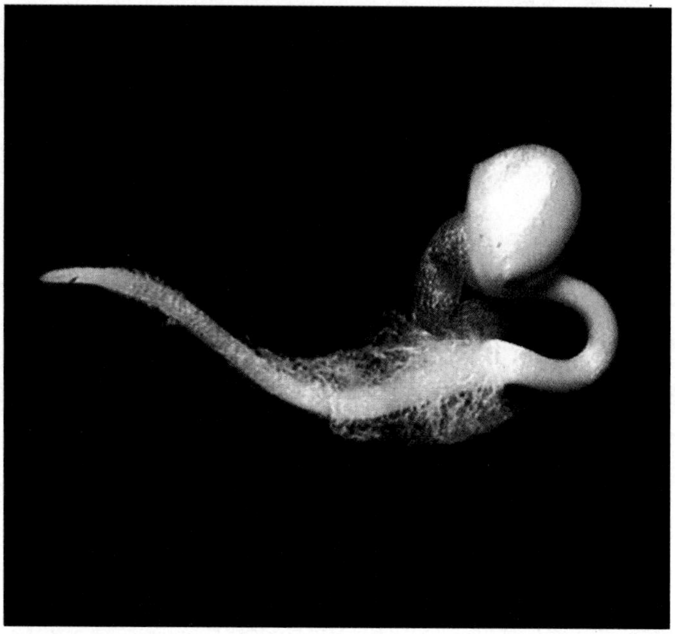

Figure 17.21 Root hairs.

Abundant fine root hairs can be seen in the back of the root apex of this germinating seedling of radish, *Raphanus sativus*.

17.6 Water is drawn up the plant stem from the roots by transpiration from the leaves.

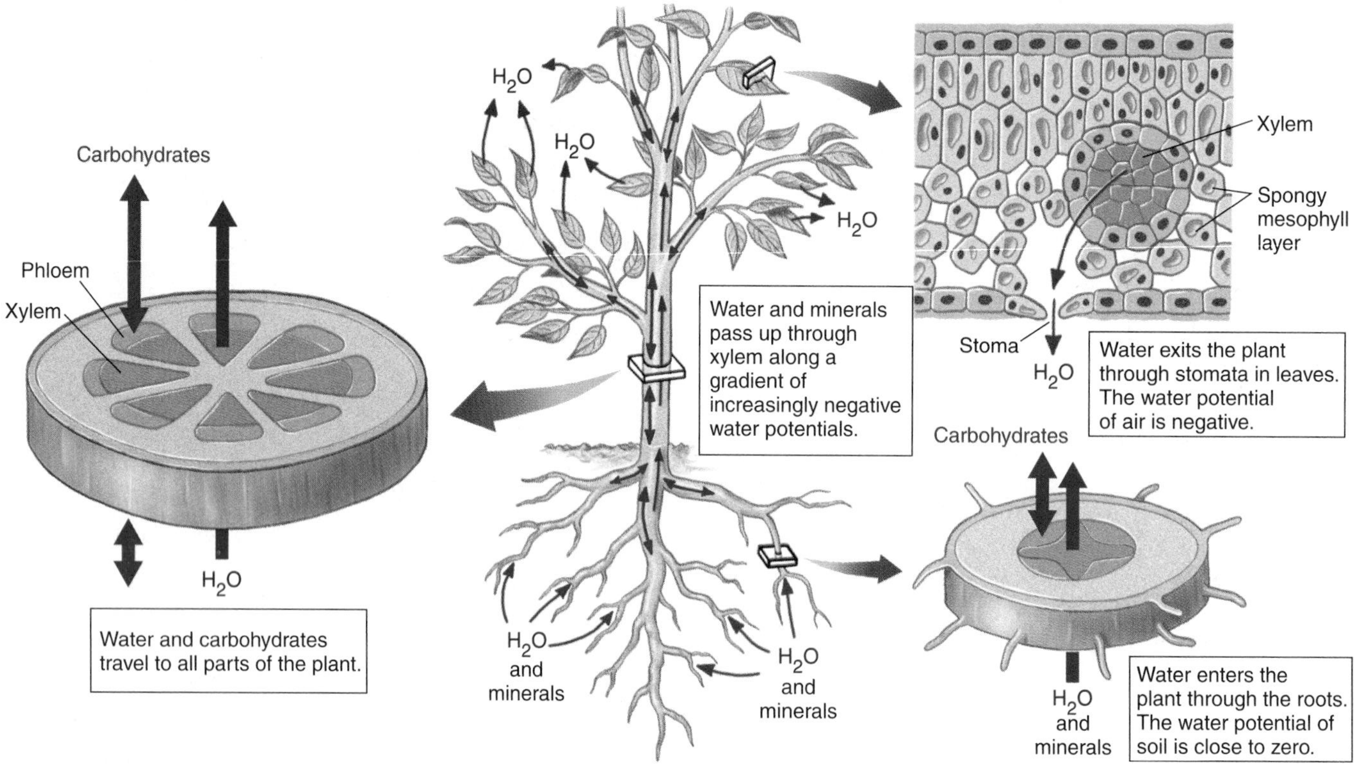

Carbohydrates
Phloem
Xylem
H2O
Water and carbohydrates travel to all parts of the plant.
H2O
H2O
H2O
Water and minerals pass up through xylem along a gradient of increasingly negative water potentials.
H2O and minerals
H2O and minerals
Xylem
Spongy mesophyll layer
Stoma
H2O
Water exits the plant through stomata in leaves. The water potential of air is negative.
Carbohydrates
H2O and minerals
Water enters the plant through the roots. The water potential of soil is close to zero.

Figure 17.22 The flow of materials into, out of, and within a plant.

Water and minerals enter through the roots of a plant and are transported through the xylem to all parts of the plant body (*blue arrows*). Water leaves the plant through the stomata in the leaves. Carbohydrates synthesized in the leaves are circulated throughout the plant by the phloem (*red arrows*).

17.7 Carbohydrate Transport

Most of the carbohydrates manufactured in plant leaves and other green parts are moved through the phloem to other parts of the plant. This process, known as **translocation,** makes suitable carbohydrate building blocks available at the plant's actively growing regions. The carbohydrates concentrated in storage organs such as underground stems, often in the form of starch, are also converted into transportable molecules, such as sucrose, and moved through the phloem.

The pathway that sugars and other substances travel within the plant has been demonstrated precisely by using radioactive isotopes and aphids, a group of insects that suck the sap of plants. Aphids thrust their piercing mouthparts into the phloem cells of leaves and stems to obtain the abundant sugars there. When the aphids are cut off of the leaf, the liquid continues to flow from the detached mouthparts protruding from the plant tissue and is thus available in pure form for analysis. The liquid in the phloem contains 10% to 25% dissolved solid matter, almost all of which is sucrose.

Using aphids to obtain the critical samples and radioactive tracers to mark them, researchers have learned that movement of substances in the phloem can be remarkably fast—rates of 50 to 100 centimeters per hour have been measured. This translocation movement is a passive process that does not require the expenditure of energy. The **mass flow** of materials transported in the phloem occurs because of water pressure, which develops as a result of osmosis. First, sucrose produced as a result of photosynthesis is actively "loaded" into the sieve tubes (or sieve cells) of the vascular bundles. This loading increases the solute concentration of the sieve tubes, so water passes into them by osmosis. An area where the sucrose is made is called a *source;* an area where sucrose is delivered from the sieve tubes is called a *sink.* Sinks include the roots and other regions where the sucrose is being unloaded. There the solute concentration of the sieve tubes is decreased as the sucrose is removed. As a result of these processes, water moves in the sieve tubes from the areas where sucrose is being taken into those areas where it is being withdrawn, and the sucrose moves passively with the water (figure 17.23).

17.7 Carbohydrates move through the plant by the passive osmotic process of translocation.

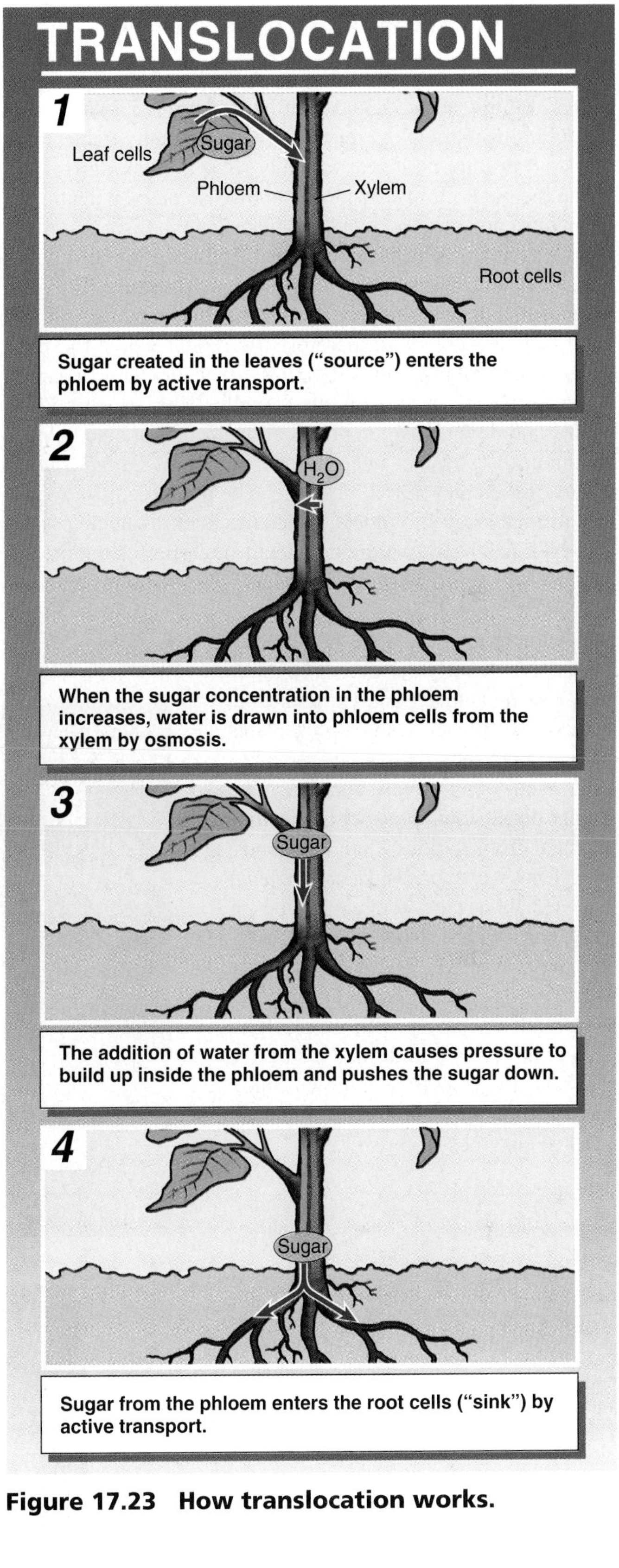
TRANSLOCATION
1
Leaf cells
Sugar
Phloem
Xylem
Root cells
Sugar created in the leaves ("source") enters the phloem by active transport.
2
H_2O
When the sugar concentration in the phloem increases, water is drawn into phloem cells from the xylem by osmosis.
3
Sugar
The addition of water from the xylem causes pressure to build up inside the phloem and pushes the sugar down.
4
Sugar
Sugar from the phloem enters the root cells ("sink") by active transport.

Figure 17.23 How translocation works.

17.8 Essential Plant Nutrients

Just as human beings need certain nutrients, such as carbohydrates, amino acids, and vitamins, to survive, plants also need various nutrients to remain alive and healthy (table 17.1). Lack of an important nutrient may slow a plant's growth or make the plant more susceptible to disease or even death.

Nutrients are involved in plant metabolism in many ways. *Nitrogen* is an essential part of proteins and of nucleic acids. *Potassium* ions regulate the **turgor pressure** (the pressure within a cell that results from water moving into the cell) of guard cells and therefore the rate at which the plant loses water and takes in carbon dioxide. *Calcium* is an essential component of the middle lamellae, the structural elements laid down between plant cell walls, and it also helps to maintain the physical integrity of membranes. *Magnesium* is a part of the chlorophyll molecule. The presence of *phosphorus* in many key biological molecules such as nucleic acids and ATP has been explored in detail in earlier chapters. *Sulfur* is a key component of an amino acid (cysteine) essential in building proteins.

TABLE 17.1 SOME ESSENTIAL PLANT NUTRIENTS

Nutrient	Relative Abundance in Plant Tissue (ppm)
Nitrogen	1,000,000
Potassium	250,000
Calcium	125,000
Magnesium	80,000
Phosphorus	60,000
Sulfur	30,000
Chlorine	3,000
Iron	2,000
Boron	2,000
Manganese	1,000
Zinc	300
Copper	100
Molybdenum	1

Note: ppm = parts per million. Parts per million equals units of an element by weight per million units of oven-dried plant material.

Some plants are able to use other organisms directly as sources of nitrogen, just as animals do. These are the carnivorous plants. Carnivorous plants have adaptations to lure and trap insects and other small animals. The plants digest their prey with enzymes secreted from various kinds of glands. The Venus's-flytrap (*Dionaea muscipula*) has sensitive hairs on each side of each leaf, which, when touched, trigger the two halves of the leaf to snap together (figure 17.24*a,b*). Pitcher plants attract insects with their bright, flowerlike colors, and once inside the pitchers, the insects slide down into a cavity filled with water, sugar, and digestive enzymes (figure 17.24*c*).

17.8 Plants require ample supplies of nitrogen, phosphorus, and potassium and smaller amounts of many other nutrients.

(a)

(b)

(c)

Figure 17.24 Carnivorous plants.

(a) Venus's-flytrap, *Dionaea muscipula,* which inhabits low boggy ground in North and South Carolina. (*b*) A Venus's-flytrap leaf has snapped together, imprisoning a fly. (c) A tropical Asian pitcher plant, *Nepenthes.* Insects seeking nectar enter the pitchers, which are modified leaves, and are trapped and digested. Complex communities of invertebrate animals and protists inhabit the pitchers.

CONCEPT REVIEW

1. Parenchyma cells
   a. are alive at maturity.
   b. have secondary cell walls.
   c. do not contain living cytoplasm when mature.
   d. are a type of dermal tissue in plants.

2. Which of the following plant structures does not consist of dermal tissue?
   a. guard cell
   b. companion cell
   c. root hair
   d. cuticle

3. Some 90% of the water taken up by roots is lost to the atmosphere. Which leaf structure accounts for the greatest portion of this water loss?
   a. cuticle
   b. mesophyll
   c. transpiration
   d. stomata

4. Xylem
   a. is the principal water-conducting tissue of plants.
   b. conducts the products of photosynthesis.
   c. is made of sieve cells.
   d. is not found in roots.

5. What takes place within the intercellular spaces between the spongy mesophyll cells in leaves?
   a. water absorption
   b. gas exchange
   c. photosynthesis
   d. production of the periderm

6. In a young dicot plant, the vascular bundles are arranged in
   a. Casparian strips.
   b. annual rings.
   c. units scattered throughout the pith.
   d. a cylinder around the periphery of the stem.

7. Wood consists of accumulated
   a. primary xylem.
   b. primary phloem.
   c. secondary xylem.
   d. secondary phloem.

8. In dicot roots, xylem and phloem are arranged in
   a. a central column of vascular tissue.
   b. vascular bundles scattered throughout the pericycle.
   c. Casparian strips.
   d. lateral extensions called root hairs.

9. Cells of the _______ regulate the flow of water laterally between the vascular tissues and the cell layers in the outer portions of the root.
   a. periderm
   b. endodermis
   c. pericycle
   d. pith

10. __________ creates a suction that draws water up the stem of a plant.
    a. Diffusion of water into the phloem from the xylem
    b. Water entering the plant from the soil
    c. Evaporation of water vapor from the leaves
    d. The closing of stomata

11. Primary growth is initiated by the _______ meristems, whereas secondary growth involves the activity of the _______ meristems.

12. A special type of _______ cells give pears their gritty texture.

13. Sieve-tube members are found in the nutrient-conducting tissue called _________.

14. In stems, the cylinders that contain both primary xylem and primary phloem are called _______.

15. The process by which water leaves a plant is called _______.

CHALLENGE YOURSELF

1. If you hammer a nail into the trunk of a tree 2 meters above the ground when the tree is 6 meters tall, how far above the ground will the nail be when the tree is 12 meters tall?

2. When plant roots are deprived of oxygen, they lose their ability to absorb ions. Why is this? What does this say about the ion absorption process?

17 eBRIDGE

Reinforcing Key Points

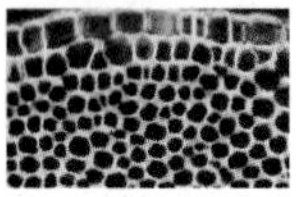

Structure and Function of Plant Tissues

17.1 Organization of a Vascular Plant

17.2 Plant Tissue Types

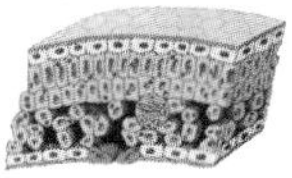

The Plant Body

17.3 Leaves

17.4 Stems

17.5 Roots

Plant Transport and Nutrition

17.6 Water Movement

17.7 Carbohydrate Transport

17.8 Essential Plant Nutrients

Electronic Learning

Visual Learning

Animations

Girth Increase in Woody Dicots

Art Labeling Activities

Leaf Structure

Primary Meristems

Explorations

Photosynthesis

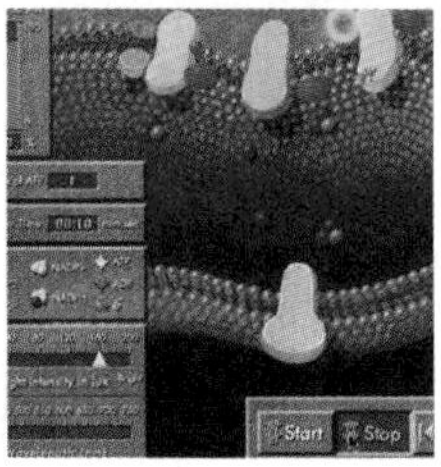

In this exercise, you can explore how the wavelength of light and its intensity affect the output of photosynthesis. A chloroplast membrane containing chlorophyll is examined; chlorophyll, found mostly in the greener portions of plants, like the leaves, is the most efficient photosynthetic pigment. Constructing an action spectrum for chlorophyll and investigating how much light is enough to optimize photosynthetic yield shows how surprisingly little light is required to drive the photosynthetic machinery at full efficiency.

Author's Corner

Special Adaptations in Plants. As plants colonized a wide variety of environments, many adaptations arose that adapted plants to their specific habitats. Modifications of plant organs, such as leaves, stems, and roots, often serve specific purposes and are sometimes quite remarkable. In addition, different mechanisms regulating water loss and transpiration have evolved in plants. Some plants have also adapted to habitats where there is too much water and frequent flooding. These adaptations allow the plants to avoid oxygen-deprivation in both freshwater and saltwater environments.

1. Modified leaves
2. Modified stems
3. Modified roots
4. Adaptations to regulate water loss and survive where there is too much water (flooding)

Virtual Classroom

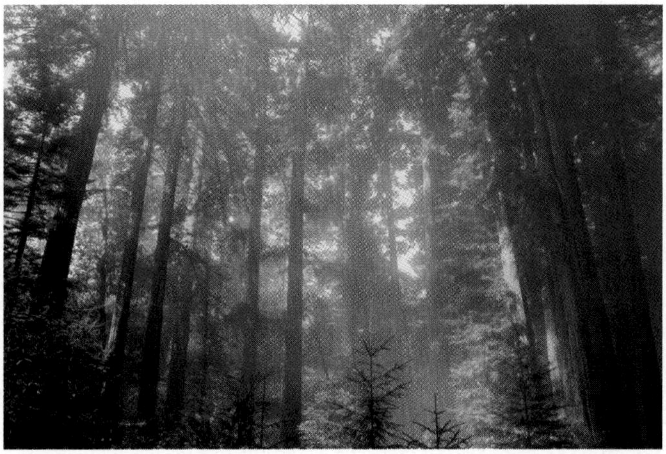

Could Terrorists Declare Biowar on Plants?

The events of the fall of 2001 have made clear the reality of bioterrorism. The anthrax attack on America via the mails, while it harmed relatively few people, underscored the sad fact that terrorists willing to use bioweapons can find the means to do so. The treat of smallpox hangs over us, now no longer a nightmare but distinctly possible. The greatest threat of bioterrorism in the future, however, may lie in another direction. Cereal grains (basically wheat, rice, and corn) supply one-half of the calories consumed by humans. A bioweapon targeted at cereal grains could have a staggering impact.

Many microbes attack plants—one-eighth of crops are lost to disease each year. The two chief plant pathogens are bacteria (chiefly *Pseudomonas*) and fungi. While 600 known plant diseases are caused by viruses, most retard growth rather than killing the plant. The key pathogens attacking cereal grains are "smuts" and "rusts," both fungal infections. Of particular interest is *Puccinia graminis* (wheat stem rust), a highly infectious fungal disease (like smallpox) spread by airborne spores (like anthrax).

Virtual Lab

Which Pest Control Method Is Best for Basil?

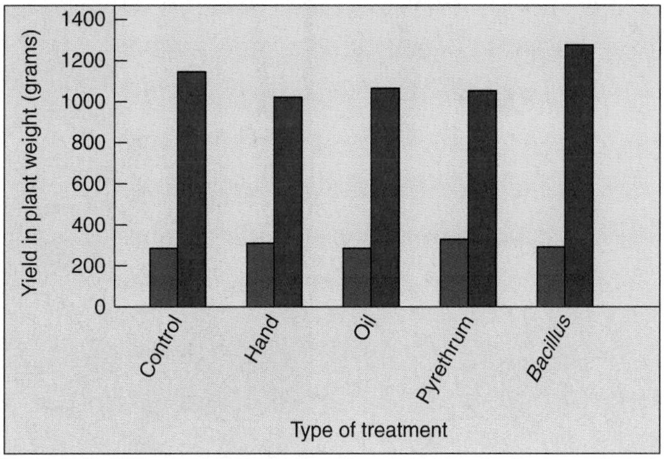
Yield in plant weight (grams)
1400
1200
1000
800
600
400
200
0
Control
Hand
Oil
Pyrethrum
Bacillus
Type of treatment

Sweet basil, *Ocimum basilicum,* is an aromatic plant used as an herb in cooking; some believe it also has medicinal value. Basil is grown and cultivated around the world. The plant is susceptible to damage from persistent insect herbivores, however, which reduce yields. Many forms of pest control are used by basil growers, such as hand removal of insects, the application of horticultural oil spray, treatment with pyrethrum (an insecticidal spray), and treatment with *Bacillus thuringiensis* var. *kurstaki* (a bacterium that releases a toxin after it is ingested by insects feeding on the plant). While many methods of pest control are used, little is known about the effectiveness of these various treatments on maximizing sweet basil yields. James Bidlack of the University of Central Oklahoma set out to evaluate the effectiveness of these methods of pest control on growing basil. He measured the impact of each method on basil crop yields and on harvest time. As another way of evaluating how effectively insect herbivores are being eliminated, he also monitors the activity levels of enzymes synthesized in response to environmental stress, such as predation.

Quizzes | Further Reading | Essential Study Partner | Links | BioCourse.com

CHAPTER 18

Plant Reproduction and Growth

THE FAR SIDE® By GARY LARSON
© 1985 FarWorks, Inc. All Rights Reserved/Dist. by Creators Syndicate
47
Larson

As Harriet turned the page, a scream escaped her lips. There was Donald—his strange disappearance no longer a mystery.

CHAPTER 18

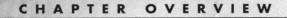

Plant Reproduction and Growth

CHAPTER OVERVIEW

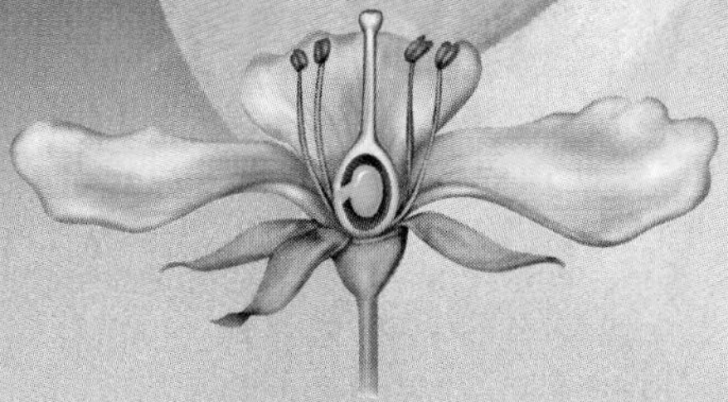

Flowering Plant Reproduction

- 18.1 The Angiosperm Flower
- 18.2 Angiosperm Reproduction
- 18.3 Seeds
- 18.4 Fruit
- 18.5 Germination

- The male and female reproductive structures occupy the third and fourth whorl of a flower.
- If the male and female structures mature at the same time, the flower may self-pollinate.
- If the male and female structures mature at different times, or on different flowers, the pollen grains from the male structures are carried to the female structures of different flowers by insects or by the wind.

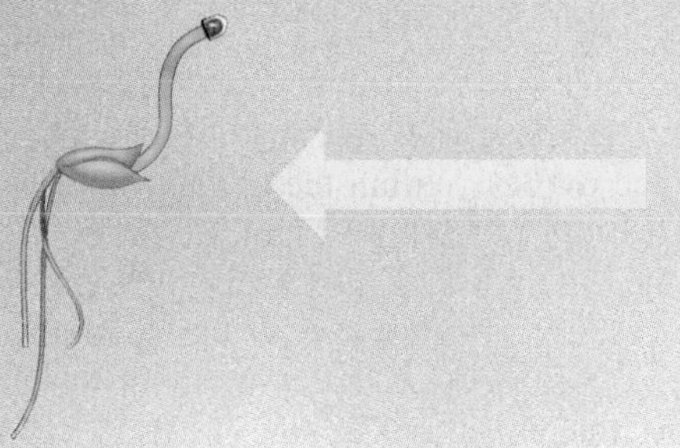

Regulating Plant Growth

- 18.6 Plant Hormones
- 18.7 Auxin
- 18.8 Other Plant Hormones

- Auxin is produced at the tips of shoots and is transported downward. Auxin also moves away from light and promotes the elongation of plant cells on the dark side, causing stems to bend in the direction of light.
- Cytokinins are necessary for mitosis and cell division in plants. They promote growth of lateral buds and inhibit formation of lateral roots.
- Gibberellins, along with auxin, play a major role in stem elongation in most plants.
- Ethylene influences leaf abscission and is widely used in fruit ripening.

Plant Responses to Environmental Stimuli

- 18.9 Photoperiodism and Dormancy
- 18.10 Tropisms

- The period of dark is the critical factor in initiating flowering in many plants.
- A molecule known as phytochrome occurs in two interconvertible forms and plays a role in determining the flowering response.
- Dormancy is a plant adaptation that carries a plant through unfavorable seasons.
- Phototropism is a response to light, gravitropism is a response to gravity, and thigmotropism is a response to contact.

Although reproduction varies greatly among the members of the plant kingdom, we focus in this chapter on sexual reproduction among flowering plants. We then explore how the differentiation of specific tissues in plants is controlled by chemical substances called hormones and how plants grow or die partly in response to their environment.

18.1 The Angiosperm Flower

Plant life cycles are characterized by an alternation of generations, in which a diploid sporophyte generation gives rise to a haploid gametophyte generation. In angiosperms, the developing gametophyte generation is completely enclosed within the tissues of the parent sporophyte. The male gametophytes, or *microgametophytes,* are **pollen grains,** and they develop from *microspores.* The female gametophyte, or *megagametophyte,* is the **embryo sac,** which develops from a *megaspore.* Pollen grains and the embryo sac both are produced in separate, specialized structures of the angiosperm flower.

Like animals, angiosperms have separate structures for producing male and female gametes, but the reproductive organs of angiosperms are different from those of animals in two ways: First, in angiosperms, both male and female structures usually occur together in the same individual flower. Second, angiosperm reproductive structures are not permanent parts of the adult individual. Angiosperm flowers and reproductive organs develop seasonally; these flowering seasons correspond to times of the year most favorable for pollination.

Structure of the Flower

A typical angiosperm flower is composed of whorls, a circle of parts present at a single level along an axis. The outermost whorl consists of structures called **sepals** (figure 18.1). The sepals protect the other whorls and serve as the flower's attachment point to the stalk. The sepals are usually green, although in some angiosperm species, they are brightly colored. All of the sepals together are called the **calyx.**

The second whorl of a flower is composed of **petals.** Most angiosperm petals are vibrantly colored, and in those pollinated by animals (such as insects and birds), the petals may have characteristic shapes that attract the pollinating animal. All of the petals together are called the **corolla.**

The third whorl of a flower is composed of the male reproductive structures. The male reproductive structures are the **stamens.** Each stamen consists of a slender stalk to which is attached the **anther.** The anther consists of two lobes that contain four pollen sacs, or *microsporangia,* within which the microspores form and develop into pollen grains. All of the stamens of an angiosperm flower are called the **androecium,** which means "male household."

The fourth whorl of a flower is composed of female reproductive structures, the **carpel.** Each carpel is composed of a **stigma, style,** and **ovary.** The stigma, the top part of the carpel, is covered with a sticky, sugary liquid to which pollen grains adhere during pollination. The liquid also nourishes the pollen grain as it makes its way to the ovary. The style is the elongated portion of the carpel that leads to the ovary. The ovary contains the **ovules.** Within the ovules, *megasporangia* produce the haploid megaspores, which develop into embryo sacs that will contain the egg. After fertilization, the ovules develop into seeds that give rise to the sporophyte generation. All of the carpels of an angiosperm flower are called the **gynoecium,** meaning "female household."

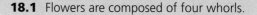

18.1 Flowers are composed of four whorls.

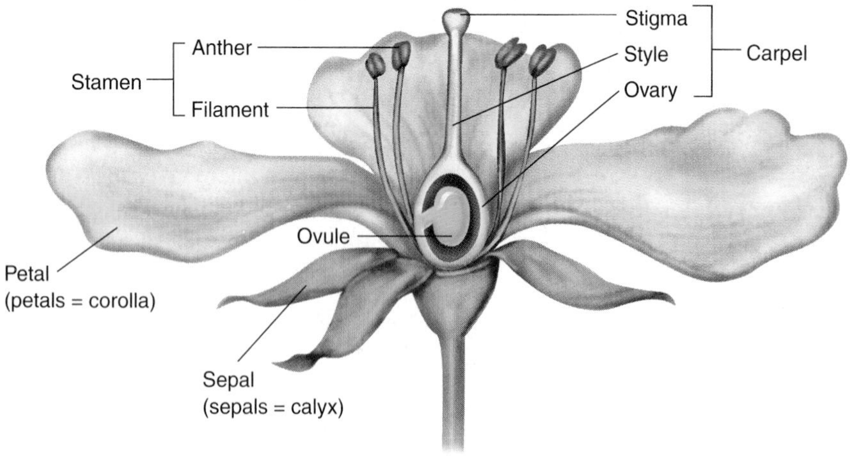
Stigma
Style
Ovary
Carpel
Anther
Filament
Stamen
Ovule
Petal
(petals = corolla)
Sepal
(sepals = calyx)

Figure 18.1 Structure of an angiosperm flower.
The male reproductive structure is called the stamen and consists of the anther and filament. The female reproductive structure is called the carpel and consists of the stigma, style, and ovary. The ovary encloses the ovules, which contain eggs and will develop into seeds after fertilization.

18.2 Angiosperm Reproduction

Pollen Formation

Pollen grains develop from microspores formed in the four pollen sacs located in the anther. Each pollen sac contains specialized chambers in which the *microspore mother cells* are enclosed and protected. Each microspore mother cell undergoes meiosis to form four haploid microspores. Subsequently, mitotic divisions form pollen grains (figure 18.2).

Egg Formation

Eggs develop in the ovules of the angiosperm flower. Within each ovule is a megaspore mother cell. Each megaspore mother cell undergoes meiosis to produce four haploid megaspores. In most plants, only one of these megaspores, however, survives; the rest are absorbed by the ovule. The lone remaining megaspore undergoes repeated mitotic divisions to produce eight haploid nuclei, which are enclosed within a seven-celled embryo sac. Within the embryo sac, the eight nuclei are arranged in precise positions. One nucleus is located near the opening of the embryo sac in the egg cell. Two nuclei are located in a single cell in the middle of the embryo sac and are called polar nuclei. Two nuclei reside in cells that flank the egg cell; and the other three nuclei are located in cells at the end of the embryo sac, opposite the egg cell.

Pollination

Pollination is the process by which pollen is transferred to the stigma. The pollen may be carried to the flower by wind or by animals, or it may originate within the individual flower itself. When pollen from a flower's anther pollinates the same flower's stigma, the self-fertilization process is called *self-pollination.*

In many angiosperms, the pollen grains are carried from flower to flower by insects and other animals that visit the flowers for food or other rewards (figure 18.3) or are deceived into doing so because the flower's characteristics suggest such rewards. A liquid called **nectar,** which is rich in sugar as well as amino acids and other substances, is often the reward sought by animals. Successful pollination depends on the plants attracting insects and other animals regularly enough that the pollen is carried from one flower of that particular species to another.

The relationship between such animals, known as *pollinators,* and the flowering plants has been important to the evolution of both groups. By using insects to transfer pollen, the flowering plants can disperse their gametes on a regular and more or less controlled basis, despite their being anchored to the ground.

For pollination by animals to be effective, a particular insect or other animal must visit plant individuals of the same species. A flower's color and form have been shaped by evolution to promote such specialization. Yellow flowers are particularly attractive to bees, whereas red flowers attract birds but are not particularly noticed by insects. Some flowers have very long floral tubes with the nectar produced deep within them; only the long, slender beaks of hummingbirds or the long, coiled tongues of moths or butterflies can reach such nectar supplies.

In certain angiosperms and all gymnosperms, pollen is blown about by the wind and reaches the stigmas passively. For such a system to operate efficiently, the individuals of a given plant species must grow relatively close together because wind does not carry pollen very far or very precisely, compared to transport by insects or other animals. Because gymnosperms, such as spruces or pines, grow in dense stands, wind pollination is very effective. Wind-pollinated angiosperms, such as birches, grasses, and ragweed, also tend to grow in dense stands. The flowers of wind-pollinated angiosperms are usually small, greenish, and odorless, and their petals are either reduced in size or absent altogether. They typically produce large quantities of pollen.

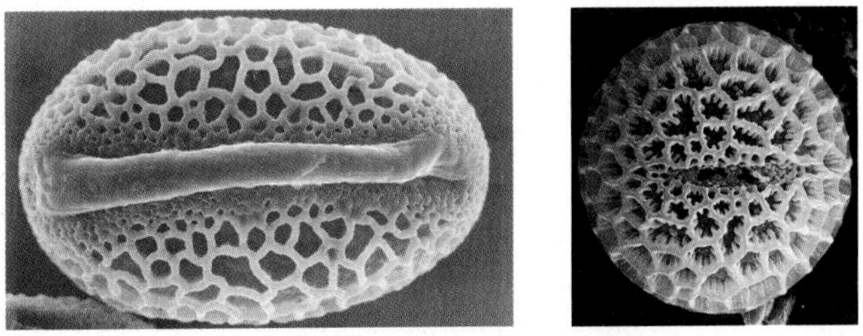

(a) (b)

Figure 18.2 Pollen grains.

(*a*) In the Easter lily, *Lilium candidum,* the pollen tube emerges from the pollen grain through the groove or furrow that occurs on one side of the grain. (*b*) In this plant of the smartweed family, *Polygonum chinense,* three "slits" are hidden among the ornamentation of the pollen grain, only one of which can be seen here. The pollen tube may grow out through any one of them.

Figure 18.3 Insect pollination.

This bumblebee, *Bombus,* has become covered with pollen while visiting a flower. The bee will transfer large quantities of the pollen to the next flower it visits.

Fertilization

Once a pollen grain has been spread by wind, an animal, or self-pollination, it adheres to the sticky, sugary substance that covers the stigma and begins to grow a **pollen tube,** which pierces the style. The pollen tube, nourished by the sugary substance, grows until it reaches the ovule in the ovary. Meanwhile, one of the cells within the pollen grain inside the tube divides to form two sperm cells.

The pollen tube eventually reaches the embryo sac in the ovule. At the entry to the embryo sac, the tip of the pollen tube bursts and releases the two sperm cells. Simultaneously, the two nuclei that flank the egg cell disintegrate, and one of the sperm cells fertilizes the egg cell, forming a zygote. The other sperm cell fuses with the two polar nuclei located at the center of the embryo sac, forming the triploid (3*n*) primary endosperm nucleus (figure 18.4). This unique process of fertilization in angiosperms in which two sperm cells are used is called **double fertilization.** The primary endosperm nucleus eventually develops into the endosperm, which nourishes the embryo.

18.2 In pollination, pollen is transferred to the female stigma. Double fertilization leads to the development of an embryo and endosperm.

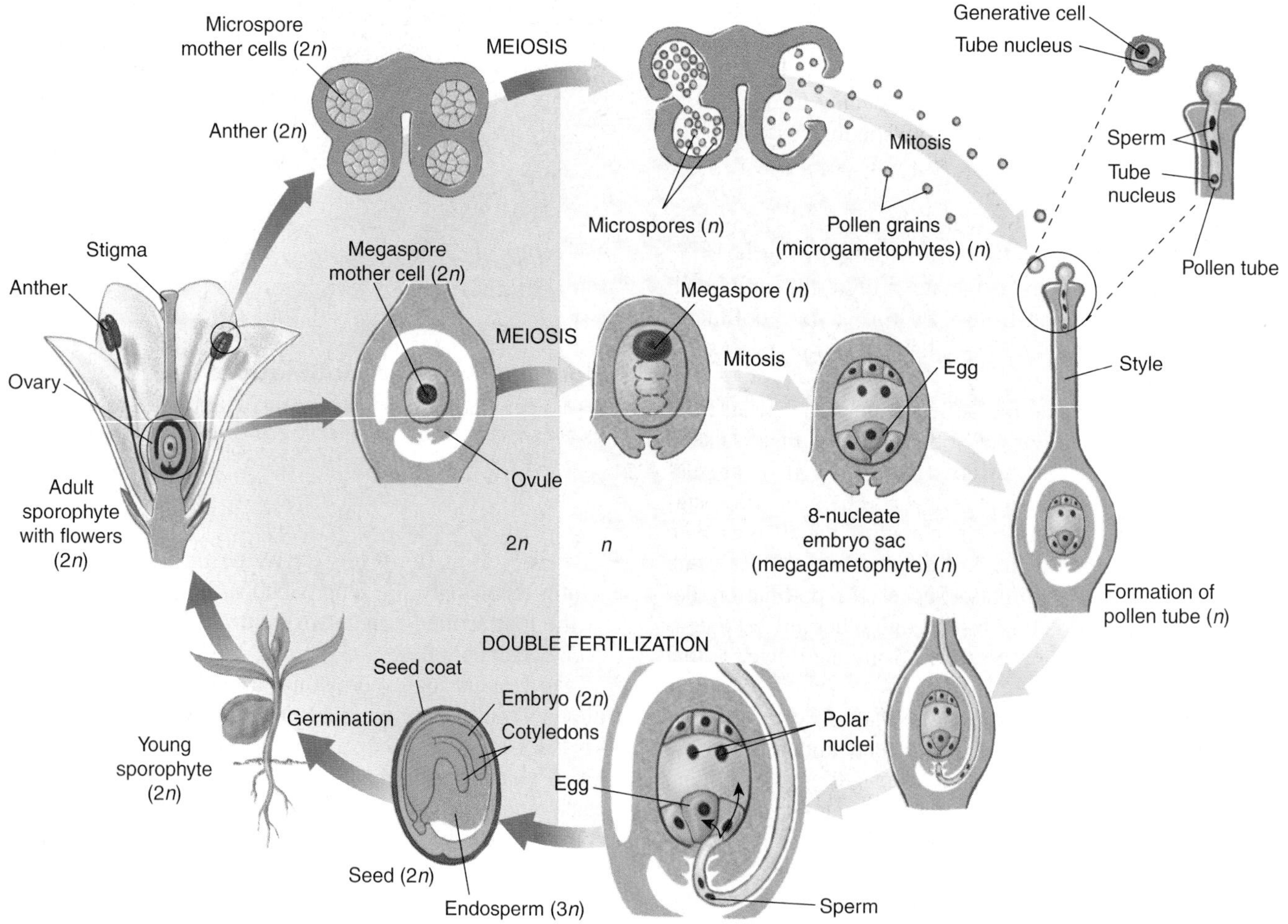

Microspore
mother cells (2n)
MEIOSIS
Anther (2n)
Generative cell
Tube nucleus
Mitosis
Sperm
Tube
nucleus
Microspores (n)
Pollen grains
(microgametophytes) (n)
Pollen tube
Stigma
Anther
Megaspore
mother cell (2n)
Megaspore (n)
MEIOSIS
Mitosis
Egg
Style
Ovary
Ovule
Adult
sporophyte
with flowers
(2n)
2n
n
8-nucleate
embryo sac
(megagametophyte) (n)
Formation of
pollen tube (n)
DOUBLE FERTILIZATION
Seed coat
Embryo (2n)
Cotyledons
Germination
Polar
nuclei
Young
sporophyte
(2n)
Egg
Seed (2n)
Endosperm (3n)
Sperm

Figure 18.4 Angiosperm life cycle.

The egg forms within the embryo sac, inside the ovule. The pollen grains, meanwhile, are formed within the anthers and are shed. Fertilization is a double process. A sperm and an egg come together, producing a zygote; at the same time, another sperm cell fuses with the polar nuclei to produce the endosperm. The endosperm is the tissue, unique to angiosperms, that nourishes the embryo and young plant.

18.3 Seeds

The entire series of events that occurs between fertilization and maturity is called *development.* During development, cells become progressively more specialized, or differentiated. The first stage in the development of a plant zygote is active cell division to form an organized mass of cells, the embryo. In angiosperms, the differentiation of cell types within the embryo begins almost immediately after fertilization (figure 18.5). By the fifth day, the principal tissue systems can be detected within the embryo mass, and within another day, the root and shoot apical meristems can be detected.

Early in the development of an angiosperm embryo, a profoundly significant event occurs: the embryo simply stops developing and becomes dormant as a result of drying. In many plants, embryo development is arrested soon after apical meristems and the first leaves, or **cotyledons,** are differentiated. The integuments—the outermost covering of the ovule—develop into a relatively impermeable seed coat, which encloses the dormant embryo within the seed, together with a source of stored food.

Once the seed coat fully develops around the embryo, most of the embryo's metabolic activities cease; a mature seed contains only about 10% water. Under these conditions, the seed and the young plant within it are very stable.

Germination, or the resumption of metabolic activities that leads to the growth of a mature plant, cannot take place until water and oxygen reach the embryo, a process that sometimes involves cracking the seed. Seeds of some plants have been known to remain viable for hundreds of years. Environmental factors help ensure that the plant germinates only under appropriate conditions.

18.3 A seed contains a dormant embryo and substantial food reserves, encased within a tough drought-resistant coat.

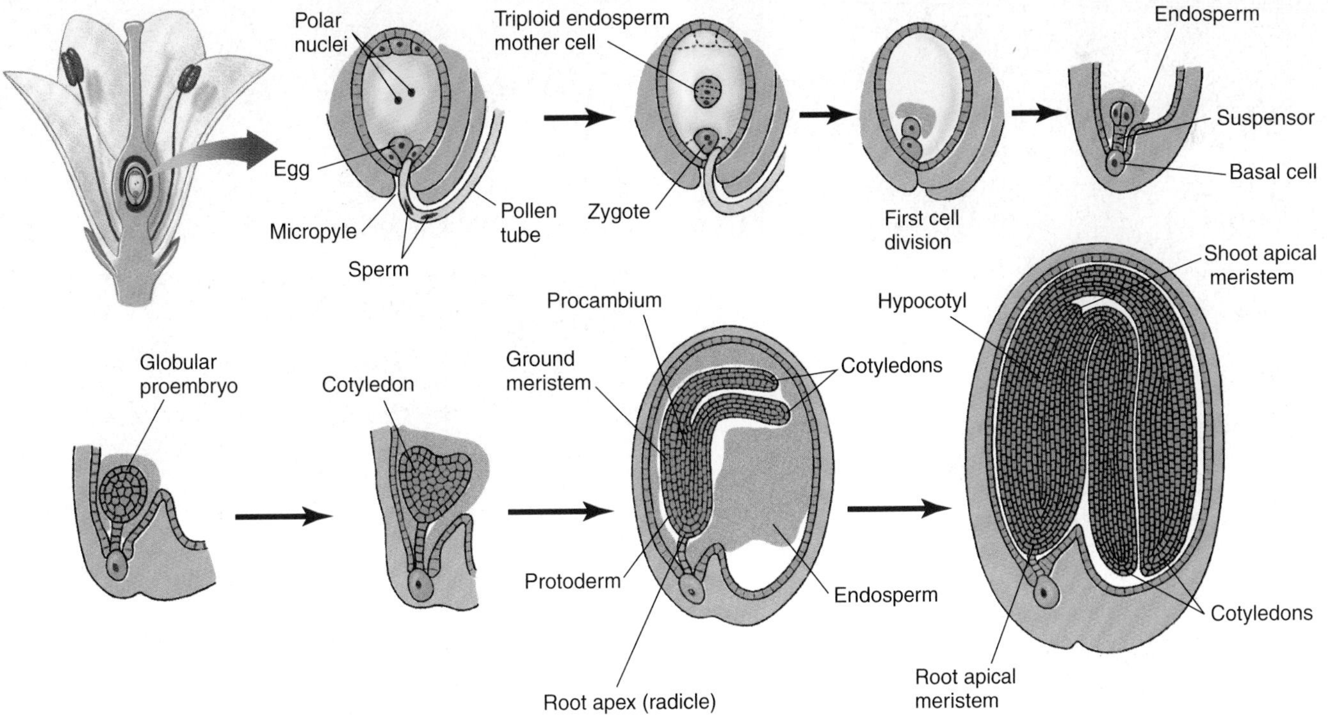

Polar
nuclei
Triploid endosperm
mother cell
Endosperm
Suspensor
Basal cell
Egg
Micropyle
Sperm
Pollen
tube
Zygote
First cell
division
Shoot apical
meristem
Procambium
Hypocotyl
Globular
proembryo
Cotyledon
Ground
meristem
Cotyledons
Protoderm
Endosperm
Cotyledons
Root apex (radicle)
Root apical
meristem

Figure 18.5 Development in an angiosperm embryo.

After the zygote forms, the first cell division is asymmetric. After another division, the basal cell, the one nearest the opening through which the pollen tube entered, undergoes a series of divisions and forms a narrow column of cells called the suspensor. The other three cells continue to divide and form a mass of cells arranged in layers. By about the fifth day of cell division, the principal tissue systems of the developing plant can be detected within this mass.

18.4 Fruit

During seed formation, the flower ovary begins to develop into fruit. Paralleling the evolution of angiosperm flowers, and of equal importance to angiosperm success, has been the evolution of these fruits. Fruits form in many ways and exhibit a wide array of modes of specialization in relation to their dispersal.

Fruits that have fleshy coverings, often black, bright blue, or red, are normally dispersed by birds and other vertebrates. Like the red flowers discussed in relation to pollination by birds, the red fruits signal an abundant food supply (figure 18.6*a*). By feeding on these fruits, the birds and other animals carry seeds from place to place before excreting the seeds as solid waste. The seeds, not harmed by the animal digestive system, thus are transferred from one suitable habitat to another. Other fruits are dispersed by wind, or by attaching themselves to the fur of mammals or the feathers of birds (figure 18.6*b*). Still other fruits, such as those of mangroves, coconuts, and certain other plants that characteristically occur on or near beaches, swamps, or other bodies of water are regularly spread from place to place in the water.

18.4 Fruits are specialized to achieve widespread dispersal by wind, by water, by attachment to animals, or, in the case of fleshy fruits, by being eaten.

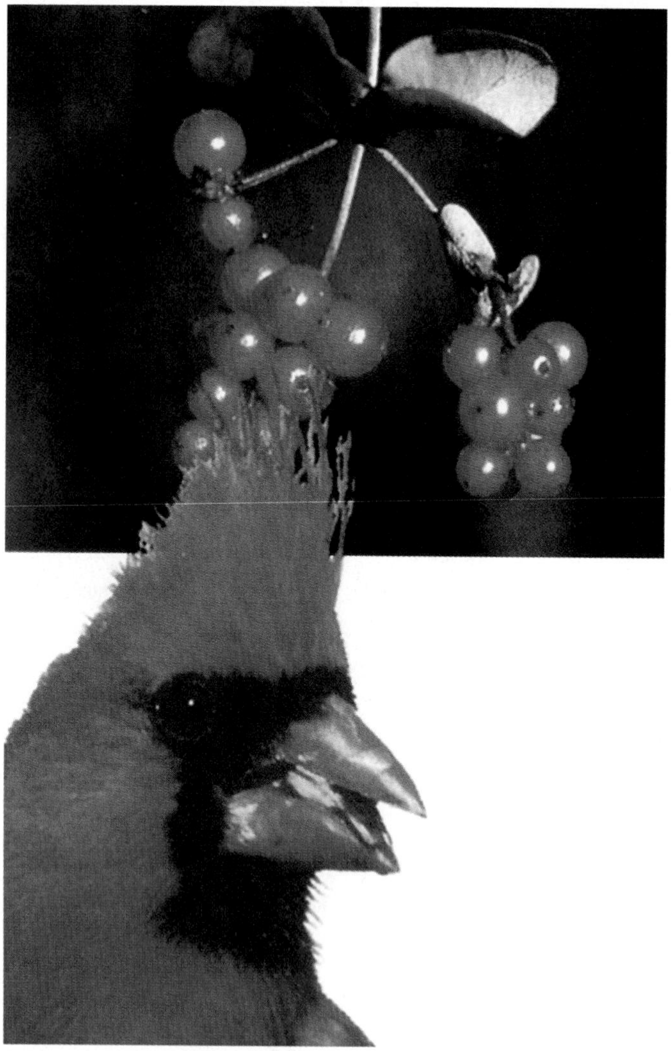

(a)

(b)

Figure 18.6 Animal-dispersed fruits.

(*a*) The bright red berries of this honeysuckle, *Lonicera hispidula,* are highly attractive to birds, just as are red flowers. Birds may carry the berry seeds either internally or stuck to their feet for great distances. (*b*) The spiny fruits of this burgrass, *Cenchrus incertus,* adhere readily to any passing animal, as you will know if you have stepped on them.

18.5 Germination

What happens to a seed when it encounters conditions suitable for its germination? First, it imbibes water. Seed tissues are so dry at the start of germination that the seed takes up water with great force, after which metabolism resumes. Initially, the metabolism may be anaerobic, but when the seed coat ruptures, aerobic metabolism takes over. At this point, oxygen must be available to the developing embryo because plants, which drown for the same reason people do, require oxygen for active growth. Few plants produce seeds that germinate successfully underwater, although some, such as rice, have evolved a tolerance of anaerobic conditions. Figure 18.7 shows the development of a dicot and monocot from germination to maturity.

> **18.5** Germination is the resumption of a seed's growth and reproduction, often triggered by water.

Figure 18.7 Development of angiosperms.

Dicot development in a soybean. The two cotyledons of the dicot are pulled up through the soil along with the hypocotyl (the stem below the cotyledons). The cotyledons are the first leaves that perform photosynthesis. As other leaves develop, they take over photosynthesis entirely, and the cotyledons shrivel and fall off the stem. Flowers develop in buds at the nodes.

Monocot development in corn. Monocots have one single cotyledon, which does not appear in the development of the mature plant. The coleoptile is a tubular sheath; it encloses and protects the shoot and leaves as they push their way up through the soil.

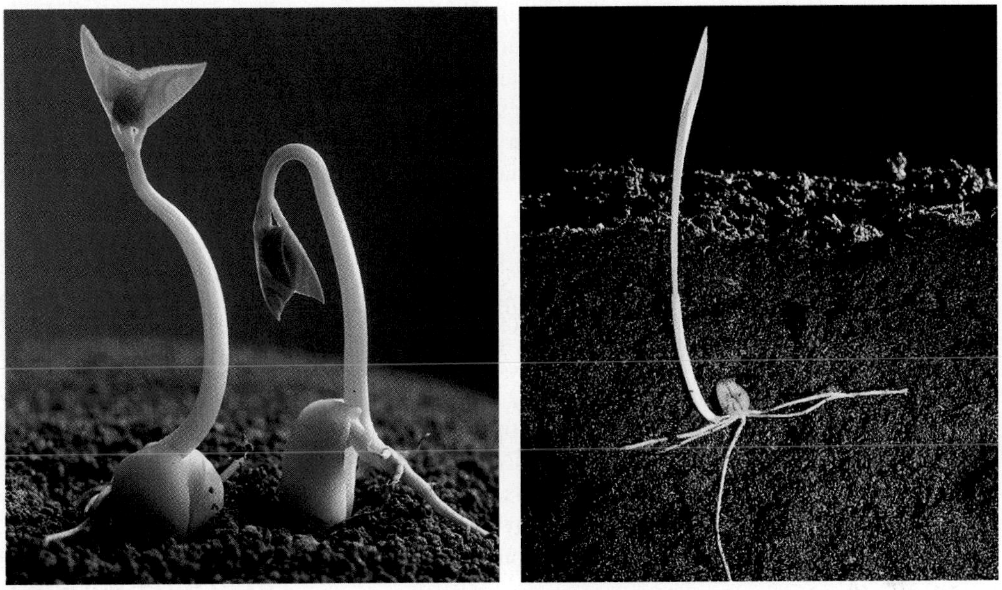

Dicot

Monocot

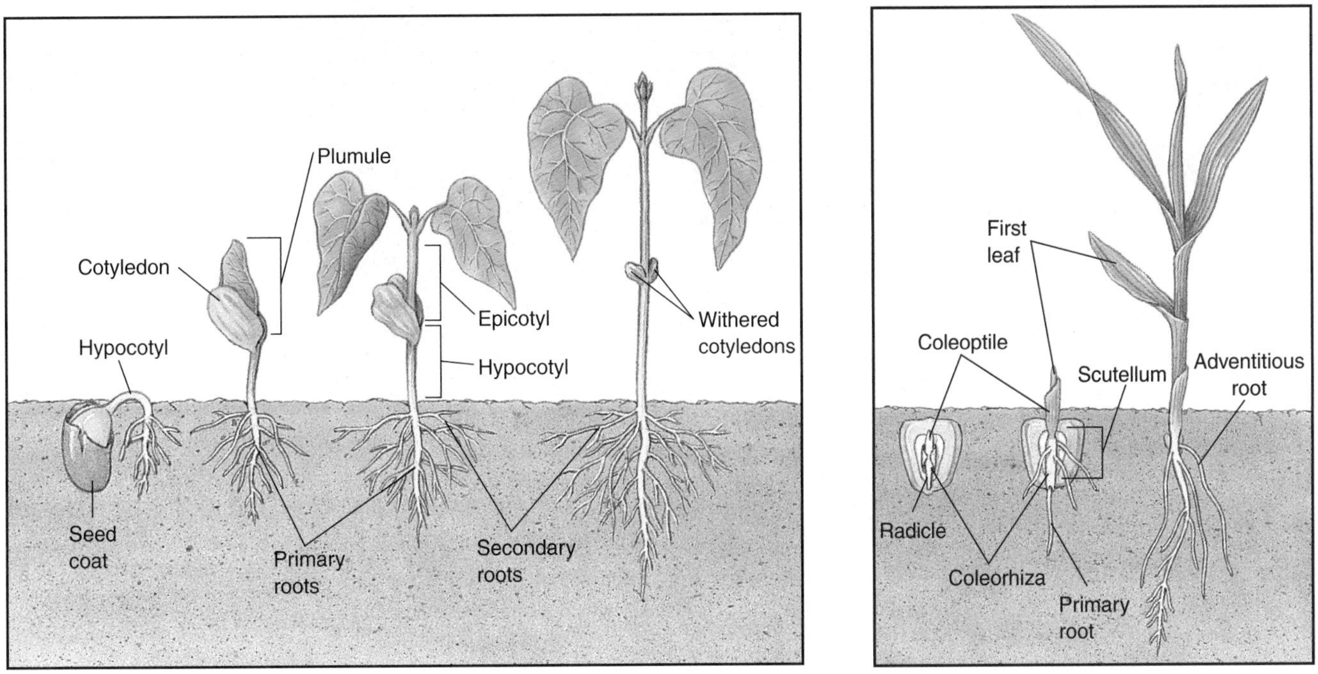
Plumule
Cotyledon
Hypocotyl
Epicotyl
Hypocotyl
Withered
cotyledons
Seed
coat
Primary
roots
Secondary
roots
First
leaf
Coleoptile
Scutellum
Adventitious
root
Radicle
Coleorhiza
Primary
root

18.6 Plant Hormones

After a seed germinates, the pattern of growth and differentiation that was established in the embryo is repeated indefinitely until the plant dies. But differentiation in plants, unlike that in animals, is largely reversible. Botanists first demonstrated in the 1950s that individual differentiated cells isolated from mature individuals could give rise to entire individuals. F. C. Steward was able to induce isolated bits of phloem taken from carrots to form new plants, plants that were normal in appearance and fully fertile (figure 18.8). Regeneration of entire plants from differentiated tissue has since been carried out in many plants, including cotton, tomatoes, and cherries. These experiments clearly demonstrate that the original differentiated phloem tissue still contains all of the genetic potential needed for the differentiation of entire plants. No information is lost during plant tissue differentiation, and no irreversible steps are taken.

Once a seed has germinated, the plant's further development depends on the activities of the meristematic tissues, which interact with the environment. The shoot and root apical meristems give rise to all of the other cells of the adult plant. Differentiation, or the formation of specialized tissues, occurs in five stages in plants (figure 18.9).

The tissue regeneration experiments of Steward and many others have led to the general conclusion that differentiated plant tissue is capable of expressing its hidden genetic complement when provided with suitable environmental signals. What halts the expression of genetic potential when the

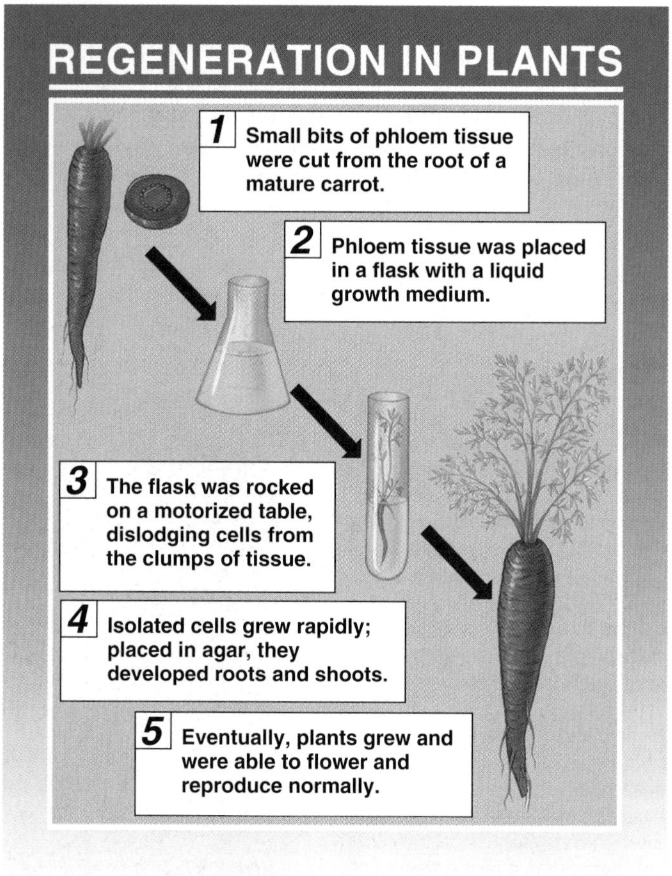
REGENERATION IN PLANTS
1 Small bits of phloem tissue were cut from the root of a mature carrot.
2 Phloem tissue was placed in a flask with a liquid growth medium.
3 The flask was rocked on a motorized table, dislodging cells from the clumps of tissue.
4 Isolated cells grew rapidly; placed in agar, they developed roots and shoots.
5 Eventually, plants grew and were able to flower and reproduce normally.

Figure 18.8 How Steward regenerated a plant from differentiated tissue.

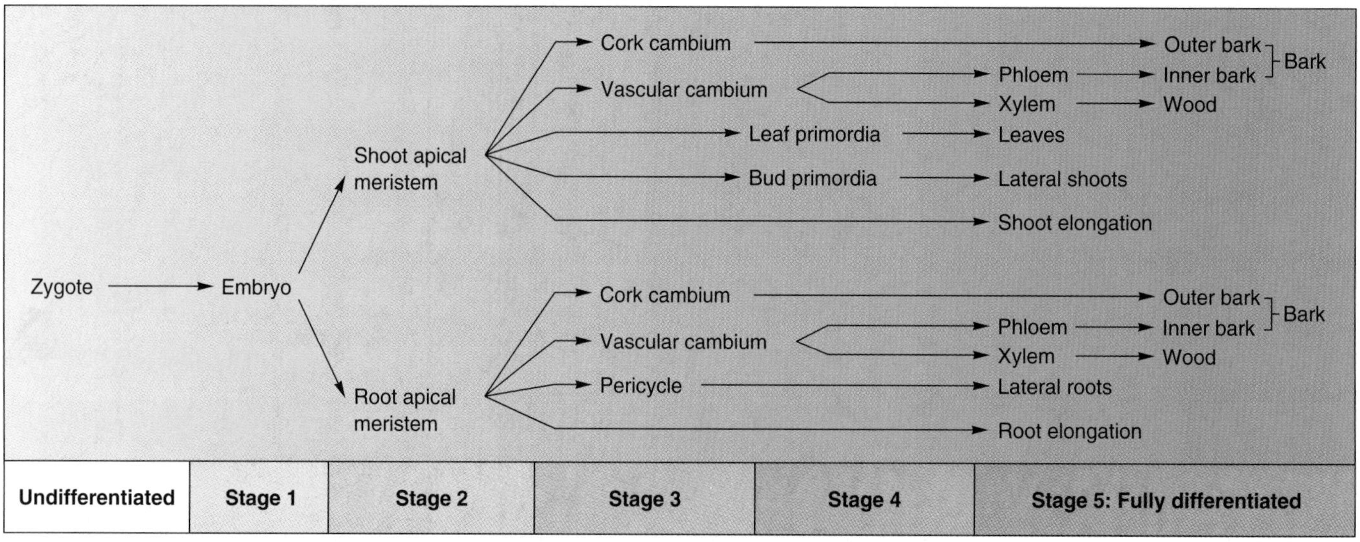
Zygote
Embryo
Shoot apical meristem
Root apical meristem
Cork cambium
Vascular cambium
Leaf primordia
Bud primordia
Pericycle
Phloem
Xylem
Leaves
Lateral shoots
Shoot elongation
Lateral roots
Root elongation
Outer bark
Inner bark
Wood
Bark
Undifferentiated
Stage 1
Stage 2
Stage 3
Stage 4
Stage 5: Fully differentiated

Figure 18.9 Stages of plant differentiation.
As this diagram shows, the different cells and tissues in a plant all originate from the shoot and root apical meristems.

same kinds of cells are incorporated into normal, growing plants? As we will see, the expression of some of these genes is controlled by plant hormones.

Hormones are chemical substances produced in small, often minute quantities in one part of an organism and then transported to another part of the organism, where they stimulate certain physiological processes and inhibit others. How they act in a particular instance is influenced both by what the hormones themselves are and by how they affect the particular tissue that receives their message.

In animals, hormones are usually produced at definite sites, normally in organs that are solely concerned with hormone production. In plants, on the other hand, hormones are produced in tissues that are not specialized for that purpose and instead carry out other functions.

At least five major kinds of hormones are found in plants: auxin, cytokinins, gibberellins, ethylene, and abscisic acid (table 18.1). Other kinds of plant hormones certainly exist but are less well understood. The study of plant hormones, especially how hormones produce their effects, is today an active and important field of research.

18.6 The growth of the plant body does not follow a fixed program, instead adjusting to the surroundings by means of a variety of hormones.

TABLE 18.1 FUNCTIONS OF THE MAJOR PLANT HORMONES

Hormone	Major Functions	Where Produced or Found in Plant
Auxin (IAA)	Promotion of stem elongation and growth; formation of adventitious roots; inhibition of leaf abscission; promotion of cell division (with cytokinins); inducement of ethylene production; promotion of lateral bud dormancy	Apical meristems; other immature parts of plants
Cytokinins	Stimulation of cell division, but only in the presence of auxin; promotion of chloroplast development; delay of leaf aging; promotion of bud formation	Root apical meristems; immature fruits
Gibberellins	Promotion of stem elongation; stimulation of enzyme production in germinating seeds	Root and shoot tips; young leaves; seeds
Ethylene	Control of leaf, flower, and fruit abscission; promotion of fruit ripening	Roots, shoot apical meristems; leaf nodes; aging flowers; ripening fruits
Abscisic acid	Inhibition of bud growth; control of stomatal closure; some control of seed dormancy; inhibition of effects of other hormones	Leaves, fruits, root caps, seeds

18.7 Auxin

In his later years, the great evolutionist Charles Darwin became increasingly devoted to the study of plants. In 1881, he and his son Francis published a book called *The Power of Movement in Plants,* in which they reported their systematic experiments concerning the way in which growing plants bend toward light, a phenomenon known as **phototropism.**

After conducting a series of experiments (figure 18.10), the Darwins hypothesized that when plant shoots were illuminated from one side, an "influence" that arose in the uppermost part of the shoot was then transmitted downward, causing the shoot to bend. Later, several botanists conducted a series of experiments that demonstrated that the substance causing the shoots to bend was a chemical we call **auxin.** Auxin is now known to regulate cell growth in plants.

How auxin controls plant growth was discovered in 1926 by Frits Went, a Dutch plant physiologist, in the course of studies for his doctoral dissertation. From his experiments, described in figure 18.11, Went was able to show that the substance that flowed into the agar from the tips of the light-grown grass seedlings enhanced cell elongation. This chemical messenger caused the tissues on the side of the seedling into which it flowed to grow more than those on the opposite side. He named the substance that he had discovered auxin, from the Greek word *auxein,* meaning "to increase."

Went's experiments provided a basis for understanding the responses the Darwins had obtained some 45 years earlier: Grass seedlings bend toward the light because the auxin contents on the two sides of the shoot differ. The side of the shoot that is in the shade has more auxin; therefore, its cells elongate more than those on the lighted side, bending the plant toward the light. Later experiments by other investigators showed that auxin in normal plants migrates away from the illuminated side toward the dark side in response to light and thus causes the plant to bend toward the light.

Auxin appears to act by increasing the plasticity of the plant cell wall, within minutes of its application. Researchers speculate that the covalent bonds linking the polysaccharides of the cell wall to one another change extensively in response to auxin. This in turn allows auxin-treated cells to take up water and thus enlarge.

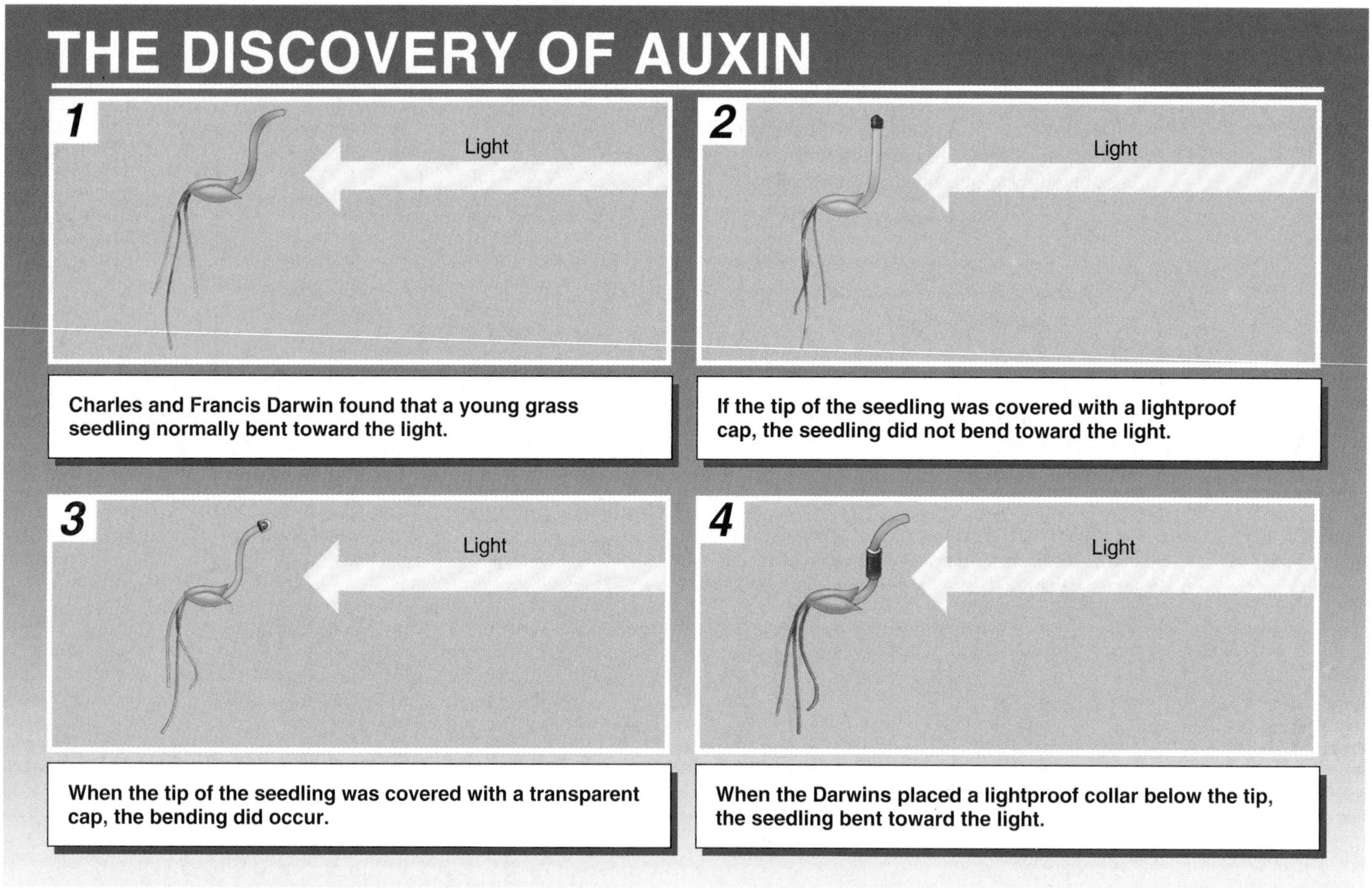
THE DISCOVERY OF AUXIN
1
Light
Charles and Francis Darwin found that a young grass seedling normally bent toward the light.
2
Light
If the tip of the seedling was covered with a lightproof cap, the seedling did not bend toward the light.
3
Light
When the tip of the seedling was covered with a transparent cap, the bending did occur.
4
Light
When the Darwins placed a lightproof collar below the tip, the seedling bent toward the light.

Figure 18.10 How the Darwins discovered auxin.

From these experiments, the Darwins concluded that, in response to light, an "influence" that causes bending was transmitted from the tip of the seedling to the area below the tip, where bending usually occurs.

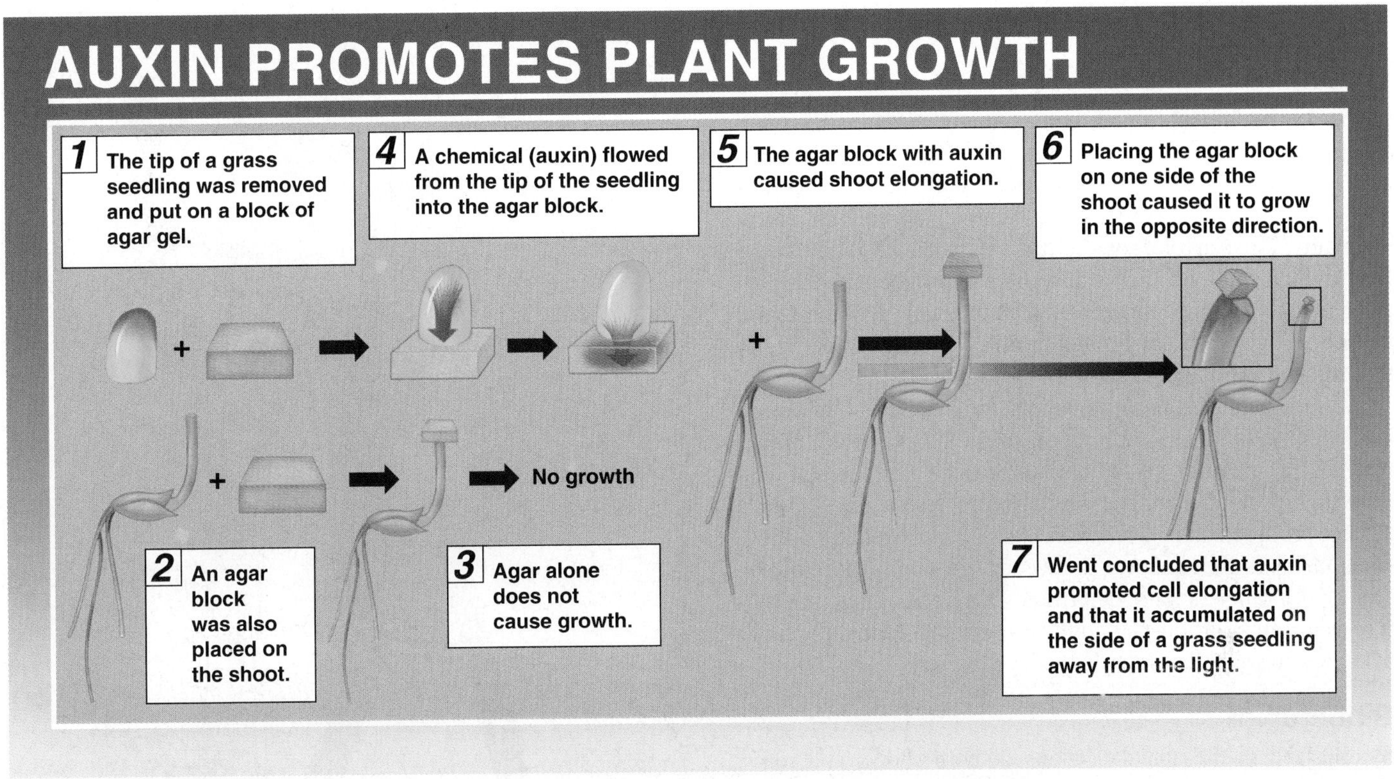
AUXIN PROMOTES PLANT GROWTH
1 The tip of a grass seedling was removed and put on a block of agar gel.
4 A chemical (auxin) flowed from the tip of the seedling into the agar block.
5 The agar block with auxin caused shoot elongation.
6 Placing the agar block on one side of the shoot caused it to grow in the opposite direction.
No growth
2 An agar block was also placed on the shoot.
3 Agar alone does not cause growth.
7 Went concluded that auxin promoted cell elongation and that it accumulated on the side of a grass seedling away from the light.

Figure 18.11 How Went demonstrated the effects of auxin on plant growth.

Synthetic auxins are routinely used to control weeds. When applied as herbicides, they are used in higher concentrations than those at which auxin normally occurs in plants. One of the most important of the synthetic auxins used in this way is 2,4-dichlorophenoxyacetic acid, usually known as 2,4-D (figure 18.12). It kills weeds in lawns without harming the grass because 2,4-D affects only broad-leaved dicots. When treated, the weeds literally "grow to death," rapidly reducing ATP production so that no source of energy remains for transport or other essential functions.

Closely related to 2,4-D is the herbicide 2,4,5-trichlorophenoxyacetic acid (2,4,5-T), which is widely used to kill woody seedlings and weeds. Notorious as the Agent Orange of the Vietnam War, 2,4,5-T is easily contaminated with a by-product of its manufacture, dioxin. Dioxin is harmful to people because it is an **endocrine disrupter,** a chemical that interferes with the course of human development. The growing release of endocrine disrupters as by-products of modern chemical manufacturing is a subject of great environmental concern.

18.7 The primary growth-promoting hormone of plants is auxin, which increases the plasticity of plant cell walls, allowing growth in specific directions.

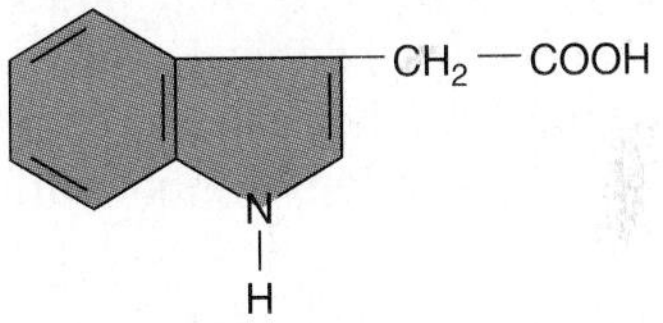
CH2 — COOH
N
H

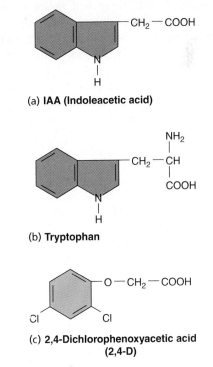

(a) **IAA (Indoleacetic acid)**

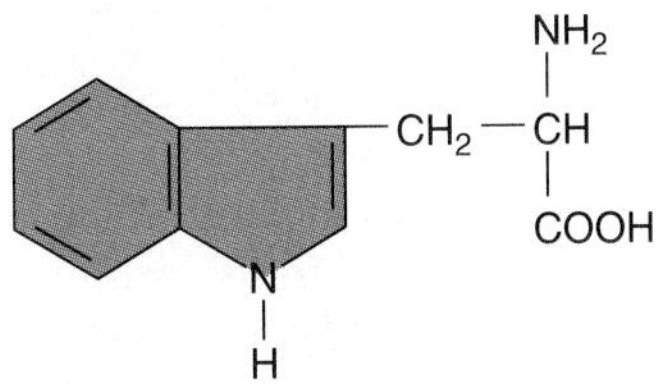
NH2
CH2 — CH
COOH
N
H

(b) **Tryptophan**

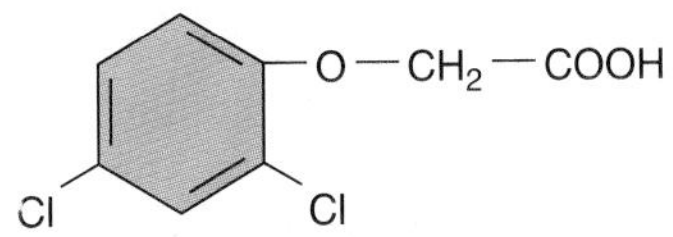
O — CH2 — COOH
Cl
Cl

(c) **2,4-Dichlorophenoxyacetic acid (2,4-D)**

Figure 18.12 Auxins.

(*a*) Indoleacetic acid (IAA) is the only known naturally occurring auxin.
(*b*) Tryptophan is the amino acid from which plants synthesize IAA.
(*c*) 2,4-Dichlorophenoxyacetic acid (2,4-D), a synthetic auxin, is a widely used herbicide.

18.8 Other Plant Hormones

Cytokinins

A **cytokinin** is a plant hormone that, in combination with auxin, stimulates cell division in plants and determines the course of differentiation. Substances with these properties are widespread, both in bacteria and in eukaryotes. In vascular plants, most cytokinins seem to be produced in the roots, from which they are then transported throughout the rest of the plant. Cytokinins apparently stimulate cell division by influencing the synthesis or activation of proteins specifically required for mitosis.

Cytokinins promote growth of lateral buds into branches; thus, along with auxin and ethylene, they play a role in the control of apical dominance and lateral bud growth. Cytokinins inhibit formation of lateral roots, while auxins promote their formation. As a consequence of these relationships, the balance between cytokinins and auxin, along with other factors, determines the appearance of a mature plant. In addition, the application of cytokinins to leaves detached from a plant retards their yellowing.

Gibberellins

Synthesized in the apical portions of both shoots and roots, **gibberellins** have important effects on stem elongation in plants and play the leading role in controlling this process in mature trees and shrubs (figure 18.13). In these plants, the application of gibberellins characteristically promotes elongation within the spaces between leaf nodes on stems, and this effect is enhanced if auxin is also present. Gibberellins also affect a number of other aspects of plant growth and development. The application of gibberellins can often induce biennial plants (plants that live for two years) to flower early during their first year of growth. These hormones also hasten seed germination, apparently because they can substitute for the effects of cold or light requirements in this process.

Figure 18.13 The effect of a gibberellin.
Although more than 60 gibberellins have been isolated from natural sources, apparently only one kind is active in shoot elongation. As shown here, cabbage plants produce tall flowering shoots when treated with this form of gibberellin.

Ethylene

Ethylene is a gas that is produced in relatively large quantities during a certain phase of fruit ripening, when the fruit's respiration is proceeding at its most rapid rate. At this phase, complex carbohydrates are broken down into simple sugars, cell walls become soft, and the volatile compounds associated with flavor and scent in the ripe fruits are produced. When applied to fruits, ethylene hastens their ripening.

One of the first lines of evidence that led to the recognition of ethylene as a plant hormone was the observation that gases from oranges caused premature ripening in bananas. Such relationships have led to major commercial uses. Tomatoes are often picked green and then artificially ripened as desired by the application of ethylene. Ethylene is widely used to speed the color formation of lemons and oranges as well. Carbon dioxide produces effects opposite to those of ethylene in fruits, and fruits that are being shipped are often kept in an atmosphere of carbon dioxide if they are not intended to ripen yet.

Genetic engineers, using techniques described in chapter 9, have placed bacterial genes into tomatoes that slow the ripening process. Until now, commercial tomatoes have had to be picked very early in order to get them to market before they become overripe, and store-bought tomatoes typically lacked the taste of "homegrown" tomatoes. However, in the genetically engineered tomatoes, ripening is delayed, and the tomatoes can be left on the vine longer, greatly improving their taste.

Ethylene also plays an important ecological role. Ethylene production increases rapidly when a plant is exposed to ozone and other toxic chemicals, temperature extremes, drought, attack by pathogens or herbivores, and other stresses. The increased production of ethylene that occurs can accelerate the abscission of leaves (figure 18.14) or fruits that have been damaged by these stresses. It now appears that some of the damage associated with exposure to ozone is due to the ethylene produced by the plants. Some studies suggest that the production of ethylene by plants subjected to attack by herbivores or infected with diseases may be a signal to activate the defense mechanisms of the plants. Such mechanisms may include the production of molecules toxic to the animals or pests attacking them. A full understanding of these relationships is obviously important for agriculture and forestry.

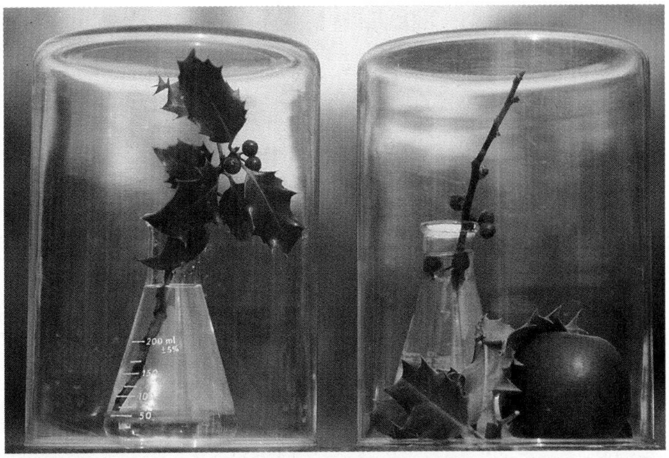

Figure 18.14 The effects of ethylene.

A holly twig was placed under the glass jar on the *left* for a week. Under the jar on the *right*, a holly twig spent a week with a ripe apple. Ethylene produced by the apple caused abscission of the holly leaves.

Abscisic Acid

Abscisic acid is a naturally occurring plant hormone that is synthesized mainly in mature green leaves, fruits, and root caps. The hormone was given its name because it stimulates leaves to age rapidly and fall off (the process of abscission), but evidence that abscisic acid plays an important natural role in this process is scant. In fact, it is believed that abscisic acid may cause ethylene synthesis, and that it is actually the ethylene that promotes senescence and abscision. When abscisic acid is applied to a green leaf, the areas of contact turn yellow. Thus, abscisic acid has the exact opposite effect on a leaf from that of the cytokinins; a yellowing leaf remains green in an area where cytokinins are applied.

Abscisic acid probably induces the formation of winter buds—dormant buds that remain through the winter—by suppressing growth. The conversion of leaf primordia into bud scales follows (figure 18.15*a*). It may also suppress growth of dormant lateral buds. It appears that abscisic acid, by suppressing growth and elongation of buds, can counteract some of the effects of gibberellins (which stimulate growth and elongation of buds). Abscisic acid plays a role in causing the dormancy of many seeds and is also important in controlling the opening and closing of stomata (figure 18.15*b*).

18.8 Plants have four other important hormones: cytokinins, gibberellins, ethylene, and abscisic acid.

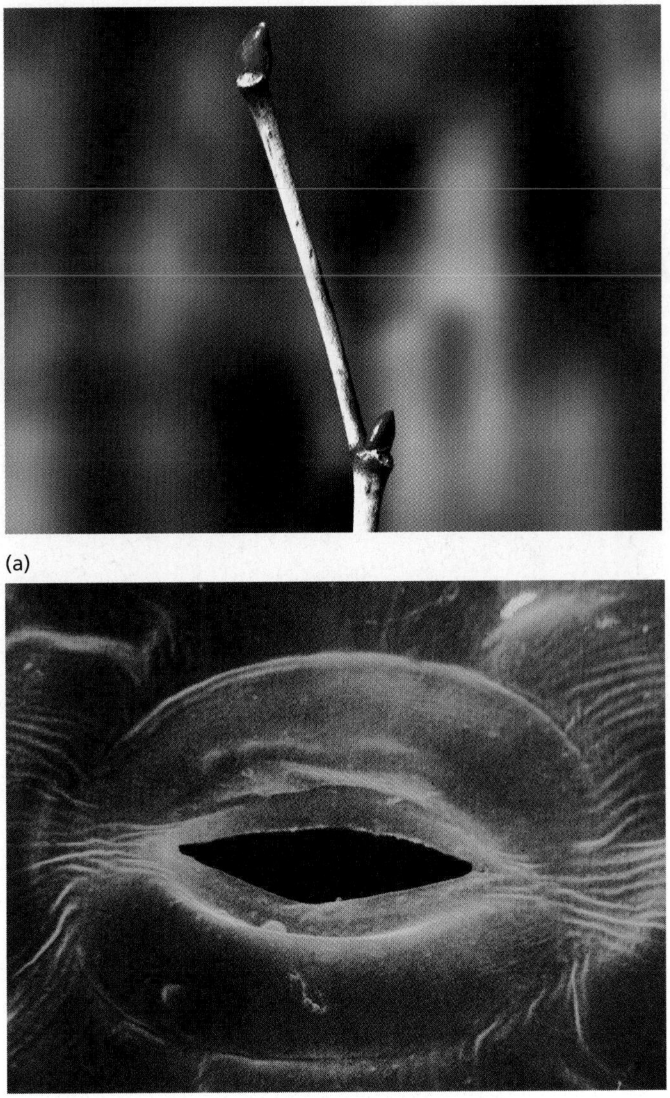

(a)

(b)

Figure 18.15 Effects of abscisic acid.

(*a*) Abscisic acid plays a role in the formation of these winter buds of an American basswood. These buds remain dormant for the winter, and bud scales—modified leaves—protect the buds from desiccation. (*b*) Abscisic acid also affects the closing of stomata by influencing the movement of potassium ions out of guard cells.

18.9 Photoperiodism and Dormancy

Plants respond to different environmental stimuli in a variety of ways. As discussed earlier in the chapter, plants bend toward light as they grow in response to this environmental stimulus. A host of other plant responses, including flowering, dropping of leaves, and yellowing of leaves due to loss of chlorophyll, are also prompted by various environmental stimuli.

Photoperiodism

Essentially all eukaryotic organisms are affected by the cycle of night and day, and many features of plant growth and development are keyed to changes in the proportions of light and dark in the daily 24-hour cycle. Such responses constitute **photoperiodism,** a mechanism by which organisms measure seasonal changes in relative day and night length. One of the most obvious of these photoperiodic reactions concerns angiosperm flower production.

Day length changes with the seasons; the farther from the equator you are, the greater the variation. Plants' flowering responses fall into three basic categories in relation to day length: long-day plants, short-day plants, and day-neutral plants. Long-day plants initiate flowers when nights become shorter than a certain length (and days become longer). Short-day plants, on the other hand, begin to form flowers when nights become longer than a critical length (and days become shorter). Thus, many spring and early summer flowers are long-day plants, and many fall flowers are short-day plants (figure 18.16). The "interrupted night" experiment of figure 18.16 makes it clear that it is the length of uninterrupted dark that is the flowering trigger.

In addition to long-day and short-day plants, a number of plants are described as day-neutral. Day-neutral plants produce flowers whenever environmental conditions are suitable, without regard to day length.

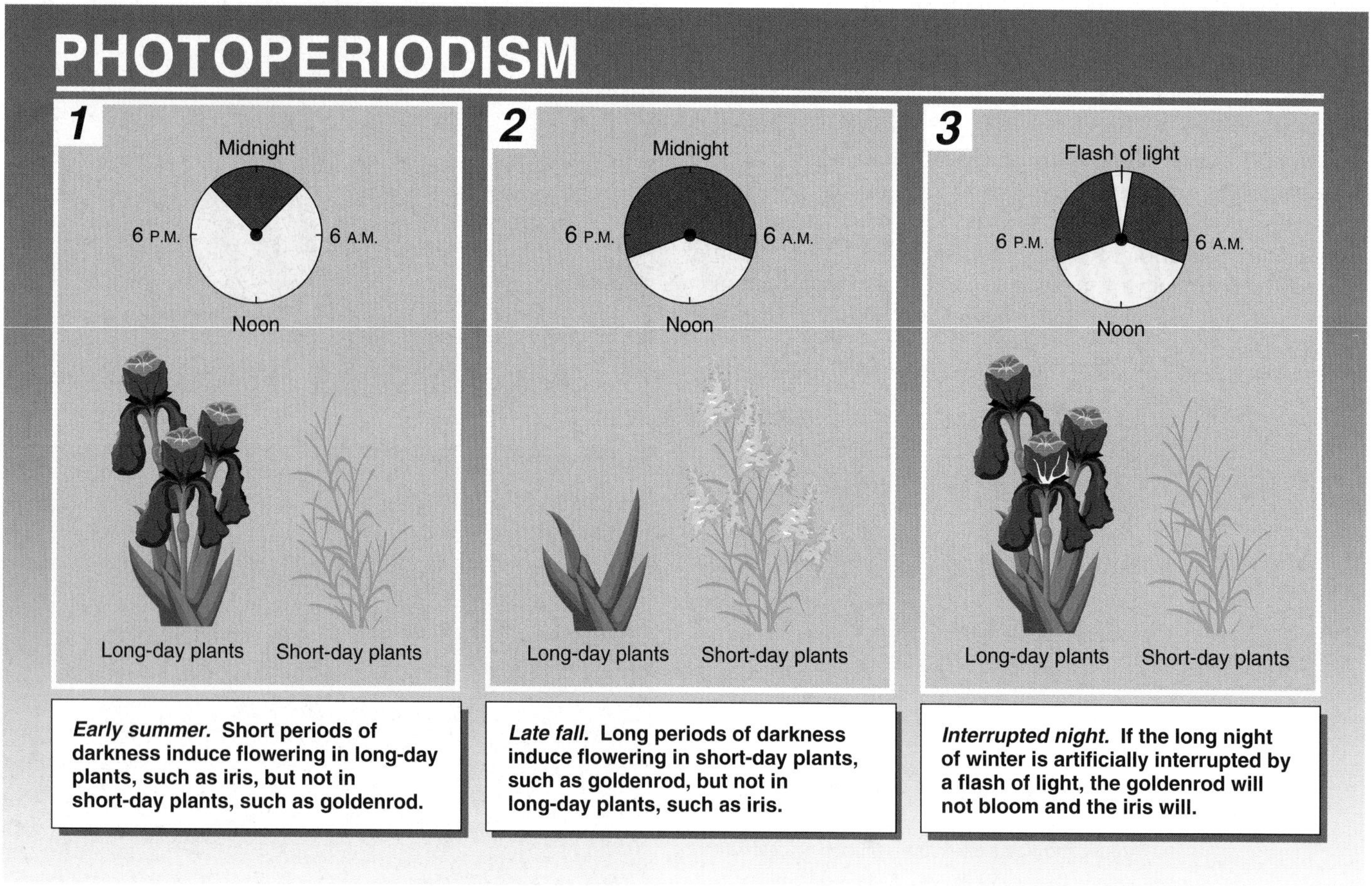
PHOTOPERIODISM
1
Midnight
6 P.M.
6 A.M.
Noon
Long-day plants
Short-day plants
Early summer. Short periods of darkness induce flowering in long-day plants, such as iris, but not in short-day plants, such as goldenrod.
2
Midnight
6 P.M.
6 A.M.
Noon
Long-day plants
Short-day plants
Late fall. Long periods of darkness induce flowering in short-day plants, such as goldenrod, but not in long-day plants, such as iris.
3
Flash of light
6 P.M.
6 A.M.
Noon
Long-day plants
Short-day plants
Interrupted night. If the long night of winter is artificially interrupted by a flash of light, the goldenrod will not bloom and the iris will.

Figure 18.16 How photoperiodism works in plants.
In each case, the duration of uninterrupted darkness determines when flowering occurs.

The Chemical Basis of Photoperiodism

Flowering responses to daylight and darkness are controlled by several chemicals that interact in complex ways. Although the nature of some of these chemicals has been deduced, how the various chemicals work together to promote or inhibit flowering responses is still being debated.

Phytochromes. Plants contain a pigment, **phytochrome,** that exists in two interconvertible forms, P_r and P_{fr}. In the first form, phytochrome absorbs red light; in the second, it absorbs far-red light. When a molecule of P_r absorbs a photon of red light (660 nm), it is instantly converted into a molecule of P_{fr}, and when a molecule of P_{fr} absorbs a photon of far-red light (730 nm), it is instantly converted to P_r. P_{fr} is biologically active and P_r is biologically inactive (figure 18.17). In other words, when P_{fr} is present, a given biological reaction that is affected by phytochrome occurs. When most of the P_{fr} has been replaced by P_r, the reaction does not occur.

Phytochrome is a light receptor, but it does not act directly to bring about reactions to light. In short-day plants, the presence of P_{fr} leads to a biological reaction that suppresses flowering. The amount of P_{fr} steadily declines in darkness, the molecules converting to P_r. When the period of darkness is long enough, the suppression reaction ceases and the flowering response is triggered. However, a single flash of red light at a wavelength of about 660 nanometers converts most of the molecules of P_r to P_{fr}, and the flowering reaction is blocked. Still, because most of the P_{fr} is converted to P_r within the first three to four hours of darkness, the conversion of P_r to P_{fr} cannot fully explain the flowering responses of short-day plants; other factors, still not understood, must also be involved.

Phytochrome is also involved in many other plant growth responses. For example, seed germination is inhibited by far-red light and stimulated by red light in many plants. Because chlorophyll absorbs red light strongly but does not absorb far-red light, light passing through green leaves inhibits seed germination. Consequently, seeds on the ground under deciduous plants that lose their leaves in winter are more apt to germinate in the spring after the leaves have decomposed and the seedlings are exposed to direct sunlight. This greatly improves the chances the seedlings will become established.

The Flowering Hormone. Working with long-day and short-day plants, some investigators have gathered evidence for the existence of a flowering hormone. It has been shown that plants will not flower in response to day-length stimuli if their leaves have been removed before exposure to the light. However, the presence of a single leaf or exposure of a single leaf to the appropriate stimuli usually initiates flowering. If the leaf is removed immediately after exposure, the plant will not produce flowers; but if it is left on the plant for a few hours and then removed, flowering occurs normally. These results indicate that a substance passes from the leaves to the apices of the plant, where it induces flowering. Other experiments have shown that, unlike auxin, the substance cannot be transmitted through agar but actually requires a connection through living plant parts.

Scientists have searched for a flowering hormone for more than 50 years, but their quest has been unsuccessful. A considerable amount of evidence demonstrates the existence of substances that promote flowering and substances that inhibit it. These poorly understood substances appear to interact in complex ways. The complexity of their interactions, as well as the fact that multiple chemical messengers are evidently involved, has made this scientifically and commercially interesting search very difficult, and to this day, the existence of a flowering hormone remains strictly hypothetical.

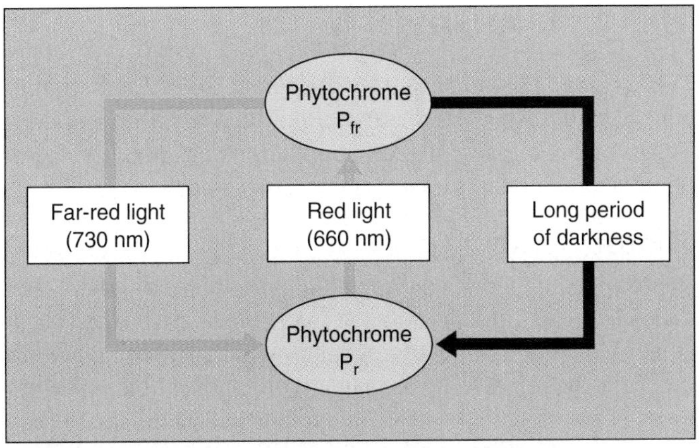

Phytochrome
P_{fr}
Far-red light
(730 nm)
Red light
(660 nm)
Long period
of darkness
Phytochrome
P_r

Figure 18.17 How phytochrome works.
When exposed to red light, the phytochrome molecule P_r is converted to P_{fr}, which is the active form that elicits a response in plants. P_{fr} is converted to P_r when exposed to far-red light, as well as in darkness.

Dormancy

Plants respond to their external environment largely by changes in growth rate. Plants' ability to stop growing altogether when conditions are not favorable—to become dormant—is critical to their survival.

In temperate regions, dormancy is generally associated with winter, when low temperatures and the unavailability of water because of freezing make it impossible for plants to grow. During this season, the buds of deciduous trees and shrubs remain dormant, and the apical meristems remain well protected inside enfolding scales. Perennial herbs spend the winter underground as stout stems or roots packed with stored food. Many other kinds of plants, including most annuals, pass the winter as seeds.

18.9 Plant growth and reproduction are sensitive to photoperiod, using chemicals to link flowering to season.

18.10 Tropisms

Tropisms, or growth responses of plants to external stimuli, control patterns of plant growth and thus plant appearance. Three major classes of plant tropisms are considered here: phototropism, gravitropism, and thigmotropism.

Phototropism

Phototropism, the bending of plants toward directional sources of light, was introduced in the discussion of auxin. In general, stems are positively phototropic, growing toward the light, whereas roots are negatively phototropic, growing away from it. The phototropic reactions of stems are clearly of adaptive value because they allow plants to capture greater amounts of light than would otherwise be possible. Auxin is involved in most, if not all, of plants' phototropic growth responses.

Gravitropism

Another familiar plant response is **gravitropism,** which causes stems to grow upward and roots downward. Both of these responses clearly have adaptive significance: stems that grow upward are apt to receive more light than those that do not; roots that grow downward are more apt to encounter a more favorable environment than those that do not. The phenomenon is called gravitropism because it is clearly a response to gravity. Figure 18.18 illustrates the effect of both phototropism and gravitropism.

Thigmotropism

Still another commonly observed response of plants is **thigmotropism,** a name derived from the Greek root *thigma,* meaning "touch." Thigmotropism is defined as the response of plants to touch. Examples include plant tendrils that rapidly curl around and cling to stems or other objects and twining plants, such as bindweed, that also coil around objects. These behaviors result from rapid growth responses to touch. Specialized groups of cells in the plant epidermis appear to be concerned with thigmotropic reactions, but again, their exact mode of action is not well understood.

18.10 Growth of the plant body is often sensitive to light, gravity, or touch.

Figure 18.18 Tropism guides plant growth.
The branches of this fallen tree are growing straight up because they are gravitropic and also phototropic.

CONCEPT REVIEW

1. Pollen adheres to the sticky top of the carpel called the
   a. ovary.
   b. ovule.
   c. style.
   d. stigma.

2. Which type of cell division produces the gametes in plants?
   a. meiosis
   b. mitosis

3. Pollen grains on the stigma of a flower reach the egg by forming a
   a. cotyledon.
   b. seed.
   c. pollen tube.
   d. carpel.

4. Double fertilization in angiosperms refers to the fact that
   a. two sperm cells fertilize one egg.
   b. two sperm cells fertilize two eggs.
   c. one sperm cell fuses with the polar nuclei and one sperm fertilizes the egg.
   d. one sperm fertilizes the egg and the polar nuclei fuse with each other.

5. Seeds inside fleshy fruits that are brightly colored are often dispersed by
   a. insects.
   b. water.
   c. mammals.
   d. birds.

6. Plants bend toward the light because
   a. cells on the shaded side of the stem elongate more than those on the sunny side.
   b. cells on the sunny side of the stem elongate more than cells on the shaded side.
   c. cells on the sunny side of the stem accumulate auxin.
   d. auxin suppresses lateral bud growth.

7. Which of the following is not a function of the plant hormone ethylene?
   a. Activates the plant's defense mechanisms against pests.
   b. Inhibits the formation of lateral roots.
   c. Controls leaf abscission.
   d. Promotes fruit ripening.

8. Short-day plants flower when the period of darkness
   a. is less than the period of phototropism.
   b. is long.
   c. is interrupted by a flash of light.
   d. equals the period of light.

9. Dormancy in plants is commonly associated with which two of the following?
   a. warm temperatures
   b. cold temperatures
   c. lack of carbon dioxide
   d. lack of available water

10. Microgametophytes are also called ________.

11. _______ is the sweet liquid that insects and other animals seek when they visit flowers and pollinate them.

12. The first step in seed germination occurs when the seed _______.

13. The part of the flower that develops into the fruit is the ________.

14. _________ are light-sensitive pigments in plants that exist in two interconvertible forms.

15. _________ is the response in plants that causes roots to grow downwards.

CHALLENGE YOURSELF

1. Angiosperms usually produce both pollen and ovules within a single flower or at least on a single plant. Why don't all angiosperms simply self-pollinate? What are the advantages of self-pollination?

2. Why do gardeners often remove many of a plant's leaves when transplanting it?

3. If day length in a particular place at a particular time of year were 10 hours, which would produce flowers: a short-day plant, a long-day plant, both, or neither? Why? Do you think there are any short-day plants in the tropics? Why?

4. When poinsettias are kept inside a house after the holidays, they rarely bloom again. Why do you think this might be, and what might you do to get them to produce flowers a second time?

18 eBRIDGE

Reinforcing Key Points

Flowering Plant Reproduction

18.1 The Angiosperm Flower

18.2 Angiosperm Reproduction

18.3 Seeds

18.4 Fruit

18.5 Germination

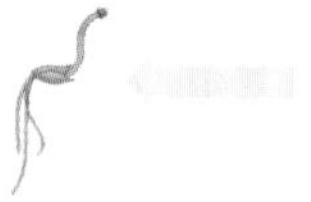

Regulating Plant Growth

18.6 Plant Hormones

18.7 Auxin

18.8 Other Plant Hormones

Plant Responses to Environmental Stimuli

18.9 Photoperiodism and Dormancy

18.10 Tropisms

Electronic Learning

Visual Learning

Animations

- Alternation of Generations Overview
- Fertilization
- Development of a Dicot Embryo
- Common Garden Bean Structure and Germination
- Demonstrating Phototropism
- Auxin Mode of Action
- Gibberellin Mode of Action

Art Labeling Activities

- Garden Bean Seed Structure
- Corn Grain Structure

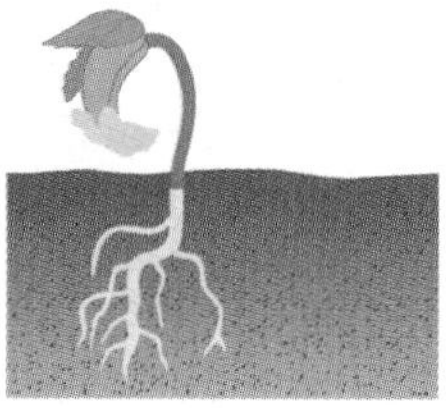

Helping You Learn

- Plant Reproduction

Author's Corner

The Success of Flowering Plants. The remarkable evolutionary success of flowering plants is likely due in part to the unique features of their reproduction. The different types of flowers, mechanisms of gamete formation and transfer, and fruits all contribute to the success of flowering plants, and many of these attributes coevolved with animals to form fascinating symbiotic relationships. Many flowering plants also have the ability to self-pollinate or undergo asexual reproduction, expanding their reproductive opportunities even more.

1. Evolution of the flower
2. Gamete formation and pollination
3. Types of fruit
4. Self-pollination
5. Asexual reproduction

Virtual Classroom

The Promise of GM Crops

In the last decade, the United States has undergone a revolution in agriculture. Genetically modified crops of corn, cotton, and soybeans are now commonplace. In 2000, two-thirds of the 72 million acres planted with soybeans in the United States were planted with seeds genetically modified to be herbicide resistant, with the result that less tillage has been needed, lessening soil erosion. Other crops like corn have been genetically engineered to contain an insecticide called bt, greatly lessening the need to apply chemical pesticides to crops. In another advance, bioengineers have modified rice to improve its nutritional value. Rice is a major food for 3 billion people in the less developed world, yet it is very low in iron, producing iron deficiency in many millions of women. It also lacks vitamin A, affecting the vision of 40 million children. The new "golden" rice contains a gene from beans for a high-iron protein, other genes to promote our body's absorption of the rice iron, and a cluster of genes from daffodils for making beta-carotene (a yellow pigment that is the precursor of vitamin A). The real promise of GM crops lies in its ability to improve the human condition.

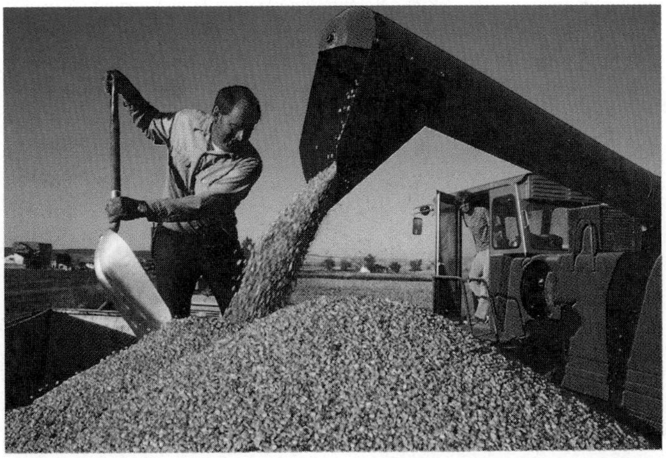

Virtual Lab

How Hormones Protect Seed Development in Peas

The pea plant, *Pisum sativum,* gained everlasting fame as the plant Mendel used to discover the laws of heredity. It continues to be a valuable model organism in studying plant function. In peas, growth of the pericarp (the structure that holds the seeds) requires the physical presence of the seeds. The removal or destruction of the seeds causes a slowing down of pericarp growth, and ultimately results in abscission. However, developing pericarp respond to gibberellin (GAs) and auxin (4-Cl-IAA) hormones. In the absence of seeds, pericarp growth can be induced to continue by the application of GAs or 4-Cl-IAA. It seems that both seeds and 4-Cl-IAA affect a key step in the GA biosynthesis pathway.

Jocelyn Ozga and colleagues at the University of Alberta, Edmonton, investigated the regulation of the GA biosynthesis pathway in pea pericarp at the molecular level. To monitor the genetic expression of proteins in the pathway that produces the active form of gibberellin, the researchers first determined the levels of its mRNA in the pericarp, and then explored how the presence of seeds or the application of auxin or GAs affected these mRNA levels.

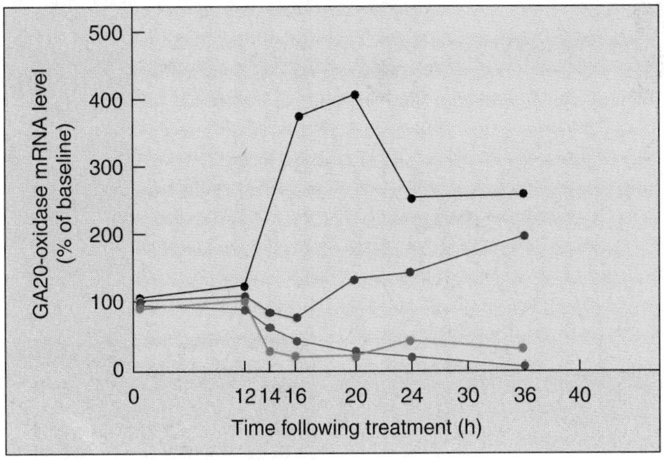
GA20-oxidase mRNA level
(% of baseline)
500
400
300
200
100
0
0
12 14 16
20
24
30
36
40
Time following treatment (h)

Quizzes | Further Reading | Essential Study Partner | Links | BioCourse.com

PART 6 EVOLUTION OF ANIMAL LIFE

CHAPTER 19

EVOLUTION OF THE ANIMAL PHYLA

THE FAR SIDE®
By GARY LARSON
© 1982 FarWorks, Inc. All Rights Reserved/Dist. by Creators Syndicate
Larson

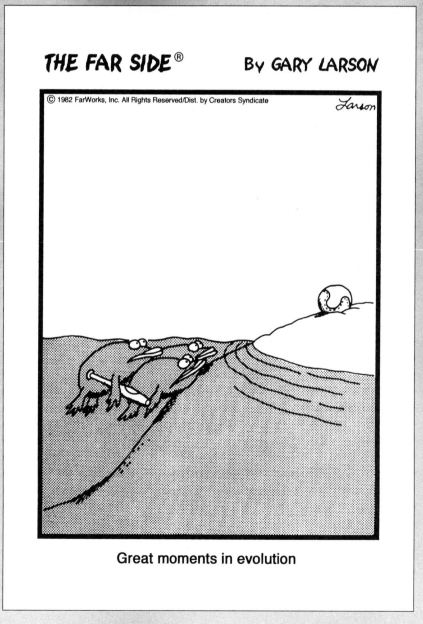

Great moments in evolution

CHAPTER 19

Evolution of the Animal Phyla

CHAPTER OVERVIEW

Introduction to the Animals

19.1 Some General Features of Animals

- There are about 36 phyla of animals.
- Animal cells lack cell walls.
- All animals are multicellular heterotrophs, and most are mobile.

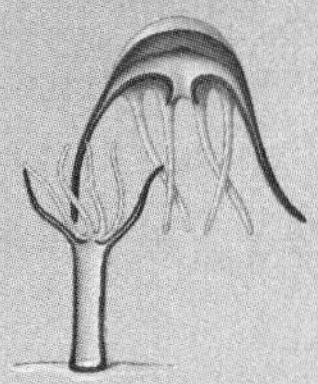

The Simplest Animals

19.2 Sponges: Animals Without Tissues

19.3 Cnidarians: Tissues Lead to Greater Specialization

- Eumetazoans possess tissues and organs, and their bodies have a definite shape and symmetry.
- Protists called choanoflagellates may be the ancestors of all animals.

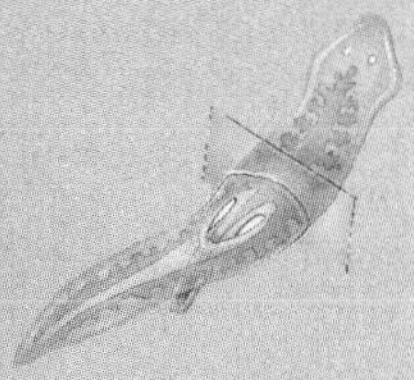

The Advent of Bilateral Symmetry

19.4 Radial Versus Bilateral Symmetry

19.5 Solid Worms: Internal Organs

- All advanced eumetazoans are bilaterally symmetrical at some stage of their life cycle.
- Acoelomate animals have a solid body with no internal cavity; diffusion is the only way they can transport materials.

The Advent of a Body Cavity

19.6 Pseudocoelom Versus Coelom

19.7 Roundworms: Pseudocoelomates

19.8 Mollusks: Coelomates

19.9 Annelids: The Rise of Segmentation

19.10 Arthropods: Advent of Jointed Appendages

- A coelom is a body cavity entirely within the mesoderm. It has the advantage that the primary tissues can interact with one another.
- Segmentation is the building of a body from a series of similar sections; different segments can specialize in different ways.
- A rigid exoskeleton of chitin limits body size because chitin is brittle.

Redesigning the Embryo

19.11 Protostomes and Deuterostomes

19.12 Echinoderms: The First Deuterostomes

19.13 Chordates: Improving the Skeleton

- A notochord is a stiff but flexible rod to which muscles attach, allowing early chordates to swing their bodies back and forth.
- The function of the notochord is taken over by the backbone of vertebrates.

19.1 Some General Features of Animals

We now explore the great diversity of animals, the result of a long evolutionary history. Animals are the eaters of the earth. All are heterotrophs and depend directly or indirectly for nourishment on plants, photosynthetic protists (algae), or autotrophic bacteria. Many animals are able to move from place to place in search of food (figure 19.1). In most, ingestion of food is followed by digestion in an internal cavity.

Figure 19.1 Animals on the move.
These large antelopes, known as wildebeests, move about in large groups consuming grass and finding new areas to feed.

Multicellular Heterotrophs

All animals are multicellular heterotrophs. Several decades ago, a biologist would not have made that statement. In recent years, the generally accepted scientific definition of the kingdom Animalia, which contains the animals, has been changed in one respect. The unicellular, heterotrophic organisms called protozoa, which were at one time regarded as simple animals, are now considered to be members of the kingdom Protista, the large and diverse group we discussed in chapter 14.

Diverse in Form

Almost all animals (99%) are **invertebrates,** animal species that lack a backbone. Of the estimated 10 million living species, only 49,000 have a backbone and are referred to as **vertebrates.** Animals are very diverse in form, ranging in size from ones too small to see with the naked eye to enormous whales and giant squids. The animal kingdom includes about 36 phyla, most of which occur in the sea. Far fewer phyla occur in freshwater and fewer still occur on land. Members of three phyla, Arthropoda (spiders and insects), Mollusca (snails), and Chordata (vertebrates) dominate animal life on land.

No Cell Walls

Animal cells are very diverse in structure and function. They are distinct among multicellular organisms because they lack rigid cell walls and are usually quite flexible. As we shall learn, the cells of all animals except sponges are organized into structural and functional units called **tissues.** A tissue is a collection of cells that have joined together and are specialized to perform a specific function (such as epithelial tissue or muscle tissue).

Active Movement

The ability of animals to move more rapidly and in more complex ways than members of other kingdoms is perhaps their most striking characteristic and one that is directly related to the flexibility of their cells. A remarkable form of movement unique to animals is flying, an ability that is well developed among both insects and vertebrates. Among vertebrates, birds, bats, and pterosaurs (now-extinct flying reptiles) were or are all strong fliers. The only terrestrial vertebrate group never to have had flying representatives are amphibians.

Sexual Reproduction

Most animals reproduce sexually. Animal eggs, which are nonmotile, are much larger than the small, usually flagellated sperm. In animals, cells formed in meiosis function directly as gametes. The haploid cells do not divide by mitosis first, as they do in plants and fungi, but rather fuse directly with one another to form the zygote. Consequently, with a few exceptions, there is no counterpart among animals to the alternation of haploid (gametophyte) and diploid (sporophyte) generations characteristic of plants (see chapter 16).

Embryonic Development

The complex form of a given animal develops from a zygote by a characteristic process of embryonic development. The zygote first undergoes a series of mitotic divisions and becomes a solid ball of cells, the **morula,** and then a hollow ball of cells, the **blastula,** a developmental stage that occurs in all animals. In most, the blastula folds inward at one point to form a hollow sac with an opening at one end called the **blastopore.** An embryo at this stage is called a **gastrula.** The subsequent growth and movement of cells of the gastrula produce the digestive system, also called the gut or intestine. The details of embryonic development differ widely from one phylum of animals to another and often provide important clues to the evolutionary relationships among them. Taken as a whole, the pattern of embryology is characteristic of the animal kingdom.

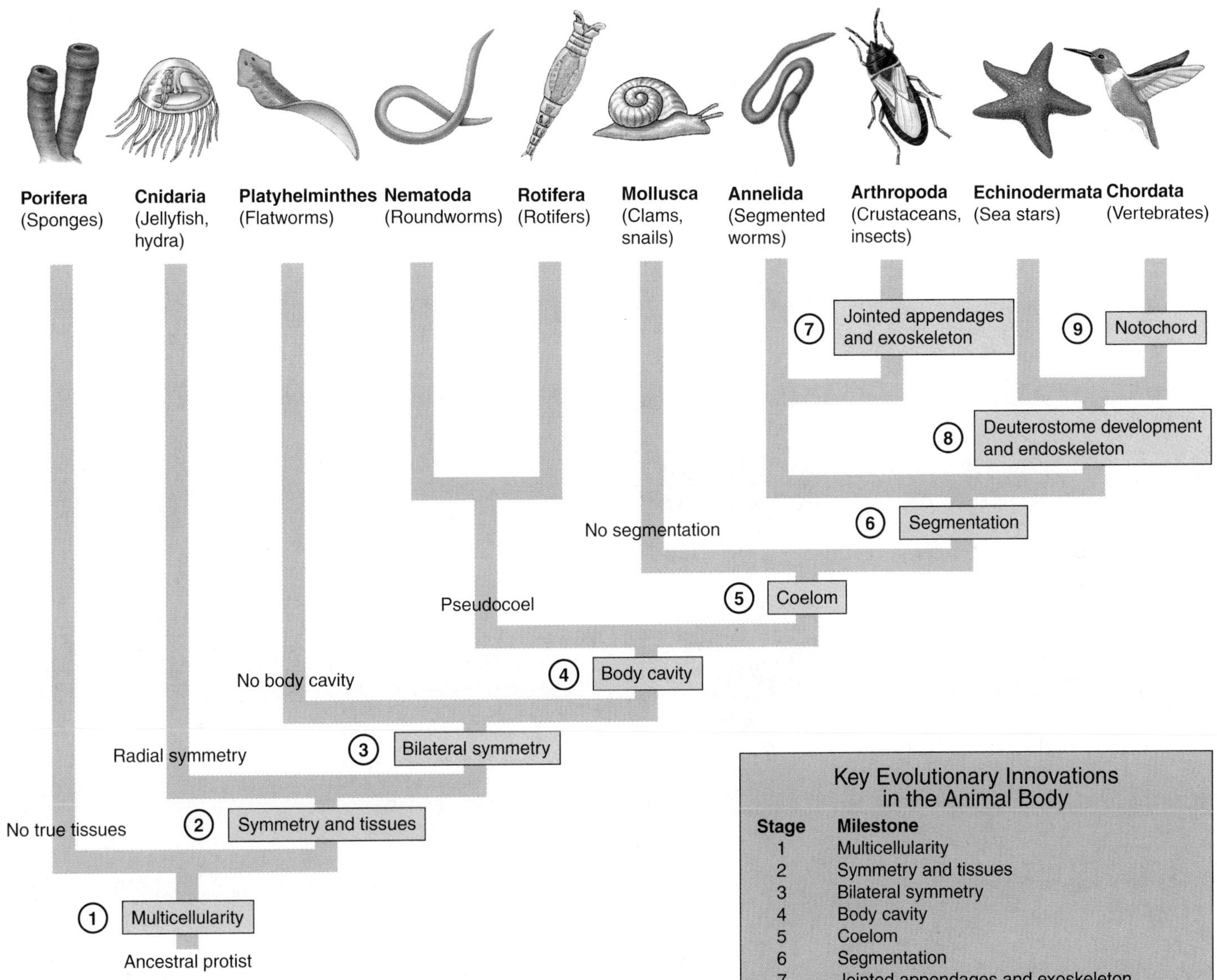

Porifera
(Sponges)
Cnidaria
(Jellyfish, hydra)
Platyhelminthes
(Flatworms)
Nematoda
(Roundworms)
Rotifera
(Rotifers)
Mollusca
(Clams, snails)
Annelida
(Segmented worms)
Arthropoda
(Crustaceans, insects)
Echinodermata
(Sea stars)
Chordata
(Vertebrates)
7 Jointed appendages and exoskeleton
9 Notochord
8 Deuterostome development and endoskeleton
6 Segmentation
No segmentation
5 Coelom
Pseudocoel
4 Body cavity
No body cavity
3 Bilateral symmetry
Radial symmetry
2 Symmetry and tissues
No true tissues
1 Multicellularity
Ancestral protist

Key Evolutionary Innovations in the Animal Body

Stage	Milestone
1	Multicellularity
2	Symmetry and tissues
3	Bilateral symmetry
4	Body cavity
5	Coelom
6	Segmentation
7	Jointed appendages and exoskeleton
8	Deuterostome development and endoskeleton
9	Notochord

Figure 19.2 A possible phylogeny of the major groups of the kingdom Animalia.
Innovations in the body design of animals are noted on the branch stems.

Evolutionary Trends Among the Animals

In this chapter we begin our exploration of animal diversity, examining a series of nine key evolutionary innovations, each exemplified by a major **phylum.** As you will recall from chapter 12, a phylum is the highest level of biological organization within a kingdom. There are about 36 phyla in the animal kingdom. While many other interesting and important innovations also arose, the key innovations discussed in this chapter serve to highlight both the evolutionary trends exhibited by the animal kingdom and the important role of certain key elements of body architecture. The key characteristics of the major animal phyla are shown in table 19.1. The exact order in which the major animal groups evolved is not known with certainty (figure 19.2), but the rough outlines of their evolutionary history can be seen in the successive change brought about by the major adaptations you will now explore. In reviewing these nine stages of animal evolution, you will be tracing the evolutionary history of the animals, an evolutionary journey that started in the sea over 600 million years ago.

19.1 Animals are complex multicellular organisms typically characterized by mobility and heterotrophy. Most animals also exhibit sexual reproduction and tissue-level organization.

TABLE 19.1 THE MAJOR ANIMAL PHYLA

Phylum	Typical Examples		Key Characteristics	Approximate Number of Named Species
Arthropoda (arthropods)	Beetles, other insects, crabs, spiders		Most successful of all animal phyla; chitinous exoskeleton covering segmented bodies with paired, jointed appendages; most insect groups have wings	1,000,000
Mollusca (mollusks)	Snails, oysters, octopuses, nudibranchs		Soft-bodied coelomates whose bodies are divided into three parts: head-foot, visceral mass, and mantle; many have shells; almost all possess a unique rasping tongue called a radula; 35,000 species are terrestrial	110,000
Chordata (chordates)	Mammals, fish, reptiles, birds, amphibians		Segmented coelomates with a notochord; possess a dorsal nerve cord, pharyngeal slits, and a tail at some stage of life; in vertebrates, the notochord is replaced during development by the spinal column; 20,000 species are terrestrial	50,300
Platyhelminthes (flatworms)	*Planaria,* tapeworms, liver flukes	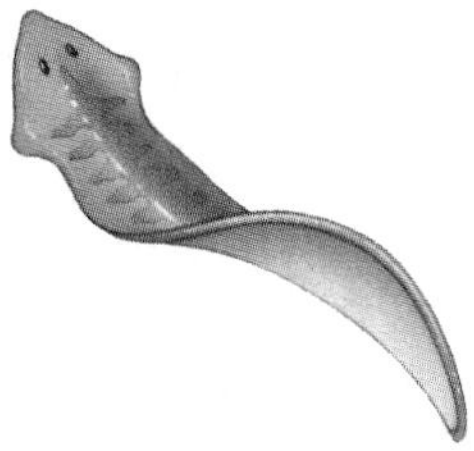	Solid, unsegmented, bilaterally symmetrical worms; no body cavity; digestive cavity, if present, has only one opening	20,000
Nematoda (roundworms)	*Ascaris,* pinworms, hookworms, *Filaria*		Pseudocoelomate, unsegmented, bilaterally symmetrical worms; tubular digestive tract passing from mouth to anus; tiny; without cilia; live in great numbers in soil and aquatic sediments; some are important animal parasites	12,000+
Annelida (segmented worms)	Earthworms, polychaetes, beach tube worms, leeches	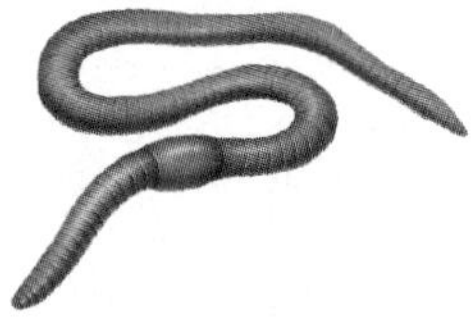	Coelomate, serially segmented, bilaterally symmetrical worms; complete digestive tract; most have bristles called setae on each segment that anchor them during crawling	12,000

TABLE 19.1 (CONTINUED)

Phylum	Typical Examples	Key Characteristics	Approximate Number of Named Species
Cnidaria (cnidarians)	Jellyfish, hydra, corals, sea anemones 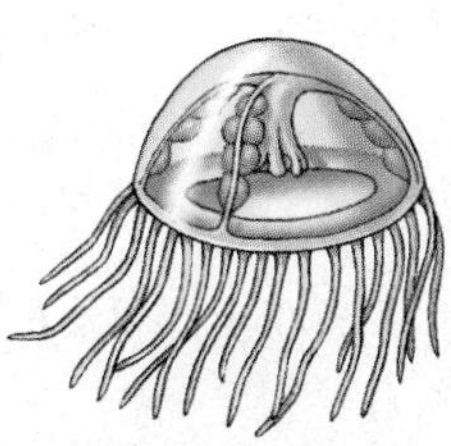	Soft, gelatinous, radially symmetrical bodies whose digestive cavity has a single opening; possess tentacles armed with stinging cells called cnidocytes that shoot sharp harpoons called nematocysts; almost entirely marine	10,000
Echinodermata (echinoderms)	Sea stars, sea urchins, sand dollars, sea cucumbers 	Deuterostomes with radially symmetrical adult bodies; endoskeleton of calcium plates; five-part body plan and unique water vascular system with tube feet; able to regenerate lost body parts; marine	6,000
Porifera (sponges)	Barrel sponges, boring sponges, basket sponges, vase sponges 	Asymmetrical bodies without distinct tissues or organs; saclike body consists of two layers breached by many pores; internal cavity lined with food-filtering cells called choanocytes; most marine (150 species live in freshwater)	5,150
Bryozoa (moss animals)	*Bowerbankia*, *Plumatella*, sea mats, sea moss 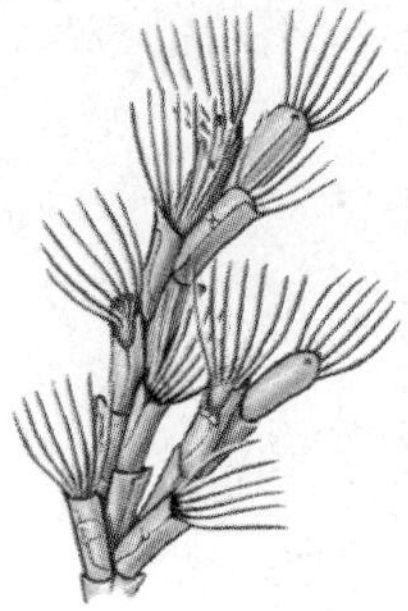	Microscopic, aquatic deuterostomes that form branching colonies, possess circular or U-shaped row of ciliated tentacles for feeding called a lophophore that usually protrudes through pores in a hard exoskeleton; also called Ectoprocta because the anus or proct is external to the lophophore; marine or freshwater	4,000
Rotifera (wheel animals)	Rotifers 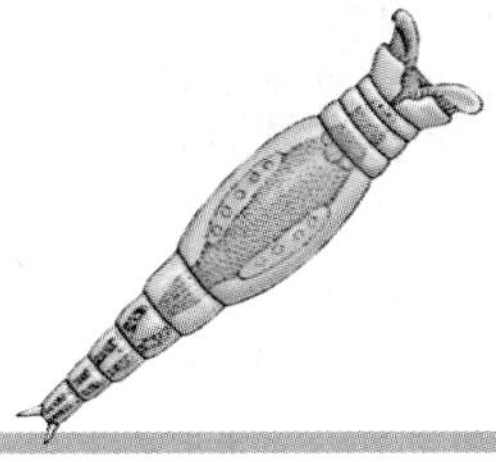	Small, aquatic pseudocoelomates with a crown of cilia around the mouth resembling a wheel; almost all live in freshwater	2,000

19.2 Sponges: Animals Without Tissues

The kingdom Animalia consists of two subkingdoms: (1) *Parazoa,* animals that lack a definite symmetry and possess neither tissues nor organs, and (2) *Eumetazoa,* animals that have a definite shape and symmetry, and in most cases tissues organized into organs. The subkingdom Parazoa consists primarily of the sponges, phylum Porifera. The other animals, comprising about 35 phyla, belong to the subkingdom Eumetazoa.

Sponges, members of the phylum Porifera, are the simplest animals. Most sponges completely lack symmetry, and although some of their cells are highly specialized, they are not organized into tissues. The bodies of sponges consist of little more than masses of specialized cells embedded in a gel-like matrix, like chopped fruit in jello. However, sponge cells do possess a key property of animal cells: cell recognition. For example, when a sponge is passed through a fine silk mesh, individual cells separate and then reaggregate on the other side to re-form the sponge.

About 5,000 species exist, almost all in the sea (a few live in freshwater). Some are tiny, others more than 2 meters in diameter (figure 19.3). The body of an adult sponge is anchored in place on the seafloor and is shaped like a vase. The outside of the sponge is covered with a skin of flattened cells called epithelial cells that protect the sponge.

The body of the sponge is perforated by tiny holes. The name of the phylum, Porifera, refers to this system of pores. Unique flagellated cells called **choanocytes,** or collar cells, line the body cavity of the sponge (figure 19.4). The beating of the flagella of the many choanocytes draws water in through the pores and drives it through the cavity. One cubic centimeter of sponge tissue can propel more than 20 liters of water a day in and out of the sponge body! Why all this moving of water? The sponge is a "filter-feeder." The beating of each choanocyte's flagellum draws water down through its collar, made of small hairlike projections resembling a picket fence. Any food particles in the water, such as protists and tiny animals, are trapped in the fence and later ingested by the choanocyte or other cells of the sponge.

The choanocytes of sponges very closely resemble a kind of protist called choanoflagellates, which seem almost certain to have been the ancestors of sponges. Indeed, they may be the ancestors of *all* animals, although it is difficult to be certain that sponges are the direct ancestors of the other more complex phyla of animals.

19.2 Sponges have a complex multicellular body but lack definite symmetry and organized tissues.

(a)

(b)

Figure 19.3 Diversity in sponges.

These two marine sponges are barrel sponges. They are among the largest of sponges, with well-organized forms. Many are more than 2 meters in diameter (*a*), while others are smaller (*b*).

Phylum Porifera: Sponges

Key Evolutionary Innovation: MULTICELLULARITY

The body of a sponge (phylum Porifera) is complexly multicellular—that is, it contains many cells, of several distinctly different types, whose activities are coordinated with each other. The sponge body is not symmetrical and has no organized tissues.

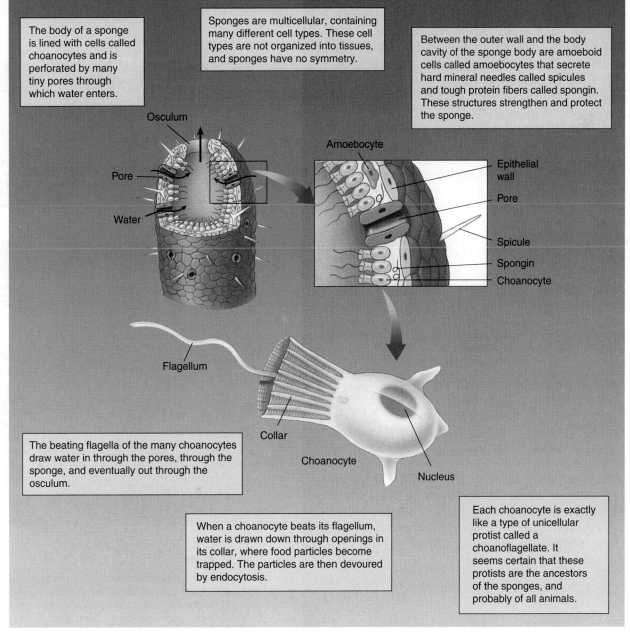

The body of a sponge is lined with cells called choanocytes and is perforated by many tiny pores through which water enters.

Sponges are multicellular, containing many different cell types. These cell types are not organized into tissues, and sponges have no symmetry.

Between the outer wall and the body cavity of the sponge body are amoeboid cells called amoebocytes that secrete hard mineral needles called spicules and tough protein fibers called spongin. These structures strengthen and protect the sponge.

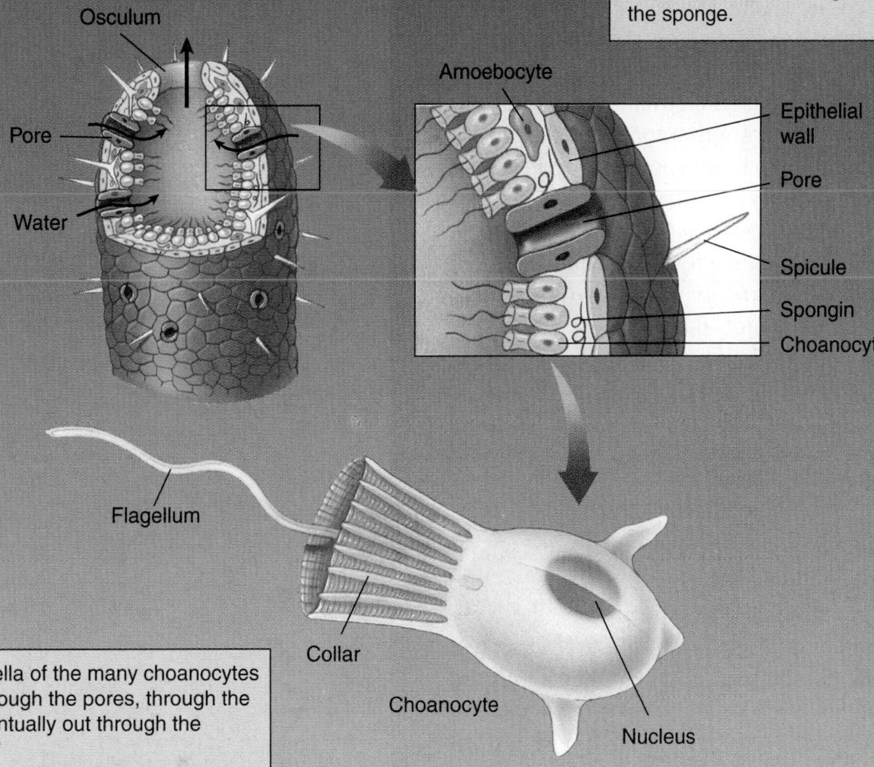

Osculum
Pore
Water
Amoebocyte
Epithelial wall
Pore
Spicule
Spongin
Choanocyte
Flagellum
Collar
Choanocyte
Nucleus

The beating flagella of the many choanocytes draw water in through the pores, through the sponge, and eventually out through the osculum.

When a choanocyte beats its flagellum, water is drawn down through openings in its collar, where food particles become trapped. The particles are then devoured by endocytosis.

Each choanocyte is exactly like a type of unicellular protist called a choanoflagellate. It seems certain that these protists are the ancestors of the sponges, and probably of all animals.

Figure 19.4 Phylum Porifera: sponges.

19.3 Cidarians: Tissues Lead to Greater Specialization

All animals other than sponges have both symmetry and tissues and thus are eumetazoans. The structure of eumetazoans is much more complex than that of sponges. All eumetazoans form distinct embryonic layers. The radially symmetrical eumetazoans have two layers, an outer **ectoderm** and an inner **endoderm.** These layers give rise to the basic body plan, differentiating into the many tissues of the body. No such layers are present in sponges.

The most primitive eumetazoans to exhibit symmetry and tissues are two **radially symmetrical** phyla, with body parts arranged around a central axis like the petals of a daisy. These two phyla are Cnidaria (pronounced ni-DAH-ree-ah), which includes jellyfish, hydra, sea anemones, and corals (figure 19.5), and Ctenophora (pronounced tea-NO-fo-rah), a minor phylum that includes the comb jellies. These two phyla together are called the Radiata. The bodies of all other eumetazoans, the Bilateria, are marked by a fundamental bilateral symmetry. Even sea stars, which are radially symmetrical as adults, are bilaterally symmetrical when young.

A major evolutionary innovation among the radiates is the **extracellular digestion** of food. In sponges, food trapped by a choanocyte is taken directly into that cell, or into a circulating amoeboid cell, by endocytosis. In radiates, food is digested *outside of cells,* in a gut cavity. Extracellular digestion is the same heterotrophic strategy pursued by fungi, except that fungi digest food outside their bodies, while animals digest it within their bodies, in a cavity. This evolutionary advance has been retained by all of the more advanced groups of animals. For the first time it became possible to digest an animal larger than oneself.

Cnidarians

Cnidarians (phylum Cnidaria) are carnivores that capture their prey, such as fishes and shellfish, with tentacles that ring their mouths. These tentacles, and sometimes the body surface cells, bear unique stinging cells called **cnidocytes,** which occur in no other organism and give the phylum its name. Within each cnidocyte is a small but powerful harpoon called a **nematocyst** (figure 19.6), which cnidarians use to spear their prey and then draw the harpooned prey back to the tentacle containing the cnidocyte. The cnidocyte builds up a very high internal osmotic pressure and uses it to push the nematocyst outward so explosively that the barb can penetrate even the hard shell of a crab.

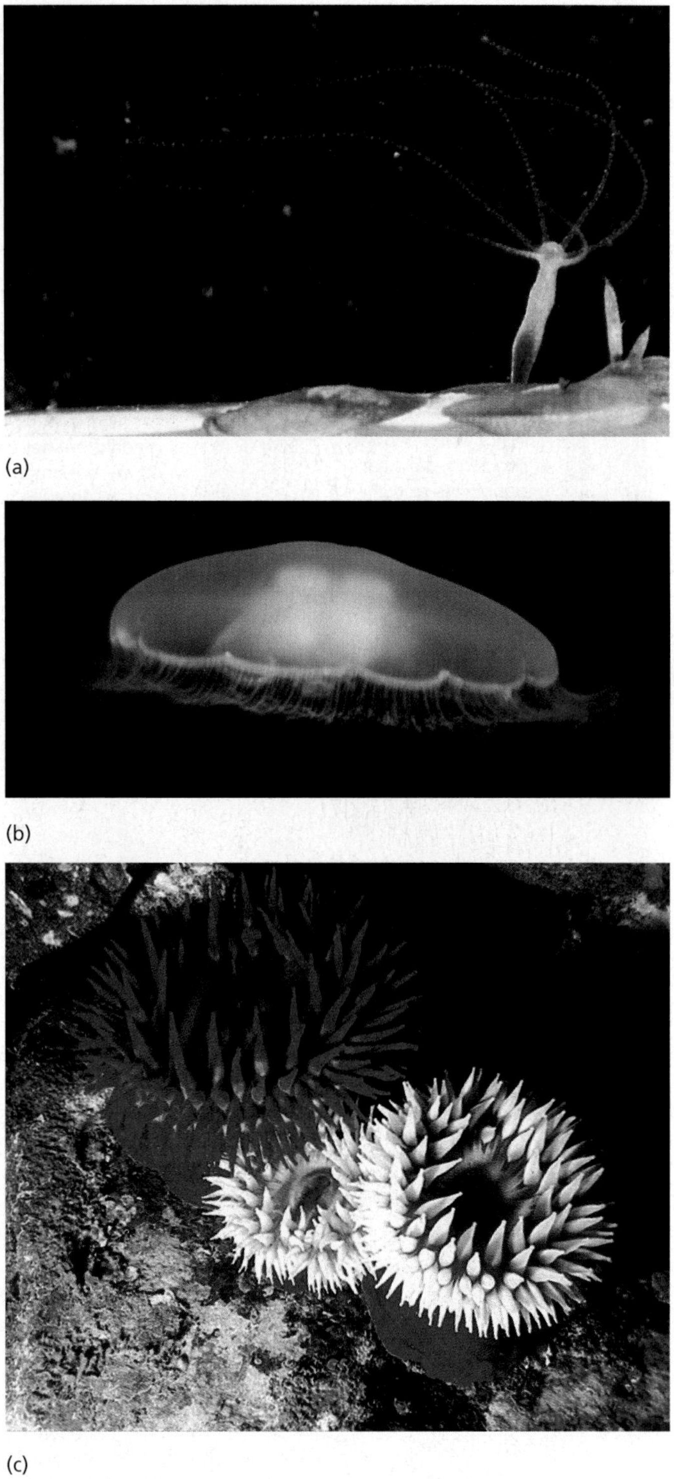

(a)
(b)
(c)

Figure 19.5 Representative cnidarians.

(*a*) Hydroids are a group of cnidarians that are mostly marine and colonial. However, *Hydra,* shown above, is a freshwater genus whose members exist as solitary polyps. (*b*) Jellyfish are translucent, marine cnidarians. (*c*) Together with corals, sea anemones, shown above, comprise the largest group of cnidarians.

Phylum Cnidaria: Cnidarians

Key Evolutionary Innovations: SYMMETRY and TISSUES

The cells of a cnidarian like *Hydra* are organized into specialized tissues. The interior gut cavity is specialized for **extracellular digestion**—that is, digestion within a gut cavity rather than within individual cells. Unlike sponges, cnidarians are **radially symmetrical,** with parts arranged around a central axis like the petals of a daisy.

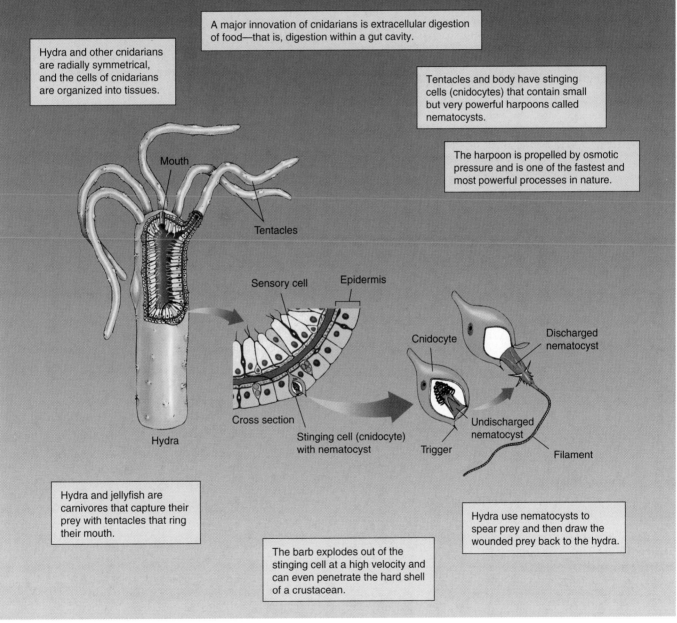

A major innovation of cnidarians is extracellular digestion of food—that is, digestion within a gut cavity.

Hydra and other cnidarians are radially symmetrical, and the cells of cnidarians are organized into tissues.

Tentacles and body have stinging cells (cnidocytes) that contain small but very powerful harpoons called nematocysts.

The harpoon is propelled by osmotic pressure and is one of the fastest and most powerful processes in nature.

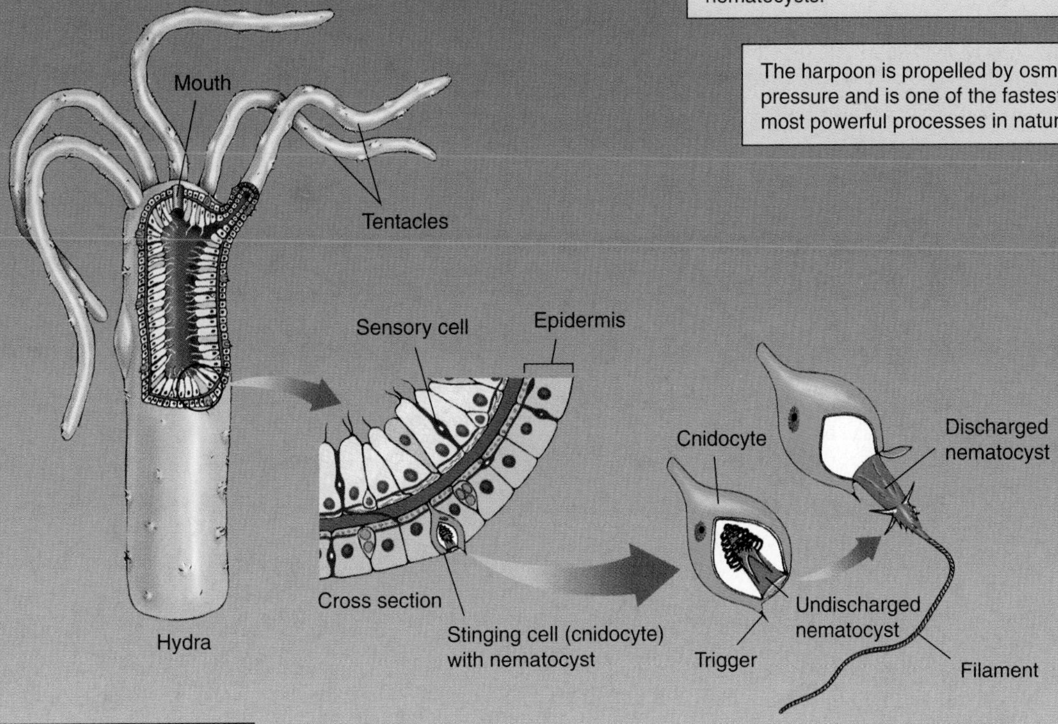
Mouth
Tentacles
Hydra
Sensory cell
Epidermis
Cross section
Stinging cell (cnidocyte) with nematocyst
Cnidocyte
Discharged nematocyst
Undischarged nematocyst
Trigger
Filament

Hydra and jellyfish are carnivores that capture their prey with tentacles that ring their mouth.

Hydra use nematocysts to spear prey and then draw the wounded prey back to the hydra.

The barb explodes out of the stinging cell at a high velocity and can even penetrate the hard shell of a crustacean.

Figure 19.6 Phylum Cnidaria: cnidarians.

Cnidarians have two basic body forms, **medusae** and **polyps** (figure 19.7). Medusae are free-floating, gelatinous, and often umbrella-shaped. Their mouths point downward, with a ring of tentacles hanging down around the edges (hence the radial symmetry). Medusae are commonly called "jellyfish" because of their gelatinous interior or "stinging nettles" because of their nematocysts. Polyps are cylindrical, pipe-shaped animals that are usually attached to a rock. They also exhibit radial symmetry. In polyps, the mouth faces away from the rock and therefore is often directed upward. Many cnidarians exist only as medusae, others only as polyps, and still others alternate between these two phases during the course of their life cycles (figure 19.8).

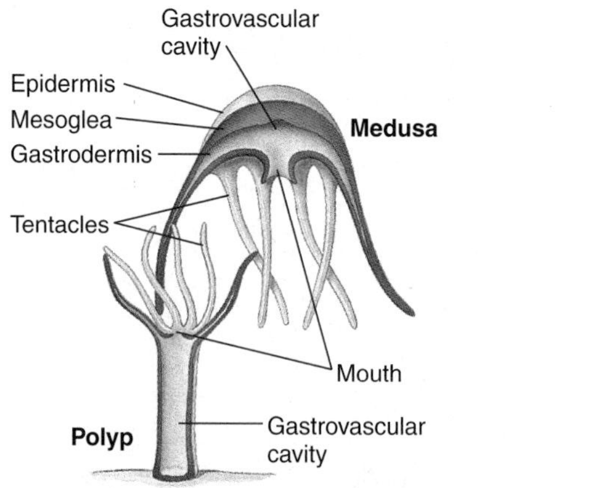

Gastrovascular cavity
Epidermis
Mesoglea
Gastrodermis
Medusa
Tentacles
Mouth
Polyp
Gastrovascular cavity

Figure 19.7 The two basic body forms of cnidarians.

The medusa (*above*) and the polyp (*below*) are the two phases that alternate in the life cycles of many cnidarians, but several species (corals and sea anemones, for example) exist only as polyps.

Ctenophores

The comb jellies (phylum Ctenophora) are transparent relatives of the cnidarians. They propel themselves through the water by beating plates of fused cilia—they are the largest animals that use cilia for locomotion.

19.3 Cnidarians possess radial symmetry and specialized tissues and carry out extracellular digestion.

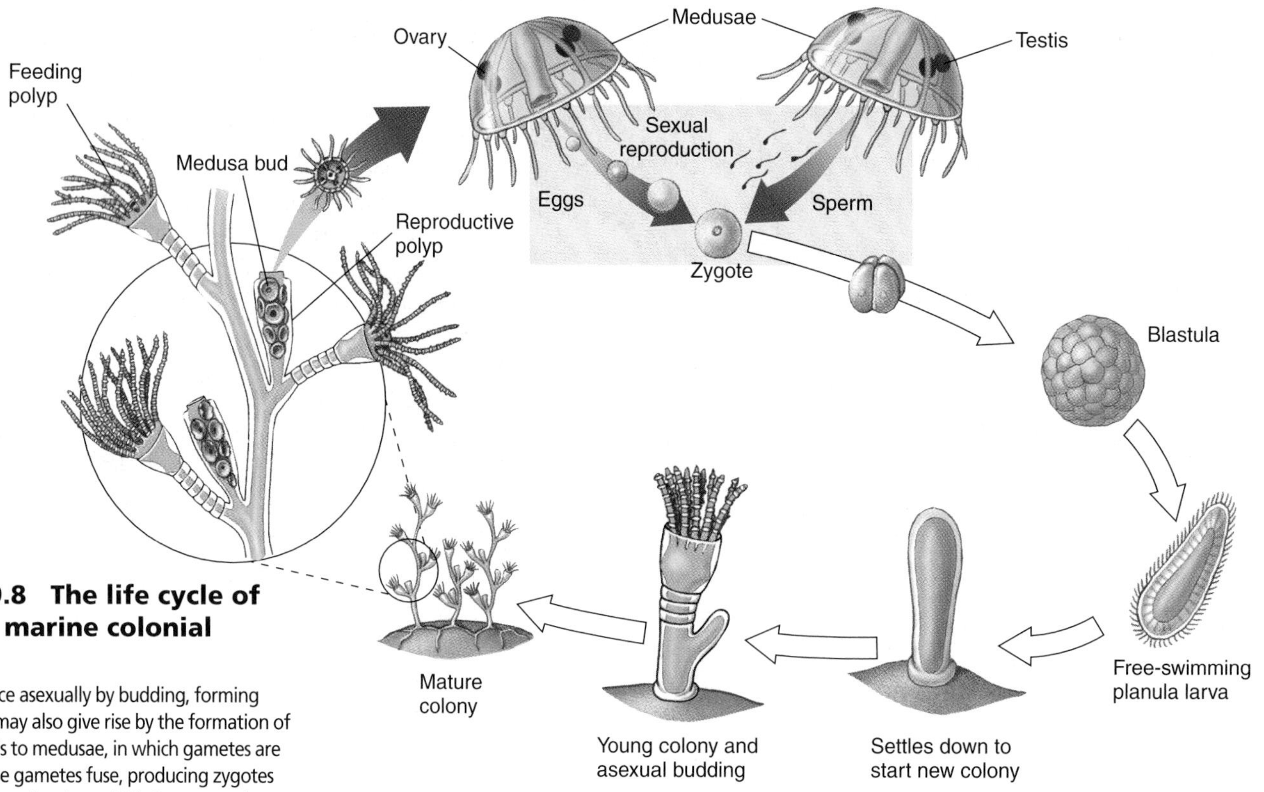

Medusae
Ovary
Testis
Feeding polyp
Medusa bud
Sexual reproduction
Eggs
Sperm
Reproductive polyp
Zygote
Blastula
Free-swimming planula larva
Settles down to start new colony
Young colony and asexual budding
Mature colony

Figure 19.8 The life cycle of *Obelia*, a marine colonial hydroid.

Polyps reproduce asexually by budding, forming colonies. They may also give rise by the formation of specialized buds to medusae, in which gametes are produced. These gametes fuse, producing zygotes that develop into planulae, which, in turn, settle down to produce polyps.

19.4 Radial Versus Bilateral Symmetry

All eumetazoans other than cnidarians and ctenophores are **bilaterally symmetrical**—that is, they have a right half and a left half that are mirror images of each other. In looking at a bilaterally symmetrical animal, you refer to the top half of the animal as **dorsal** and the bottom half as **ventral** (figure 19.9). The front is called **anterior** and the back **posterior.** Bilateral symmetry was a major evolutionary advance among the animals because it allows different parts of the body to become specialized in different ways. For example, most bilaterally symmetrical animals have evolved a definite head end, a process called **cephalization.** Animals that have heads are often active and mobile, moving through their environment headfirst, with sensory organs concentrated in front so the animal can test for food, danger, and mates as it enters new surroundings.

19.4 Bilateral symmetry allows different parts of the body to become specialized in different ways.

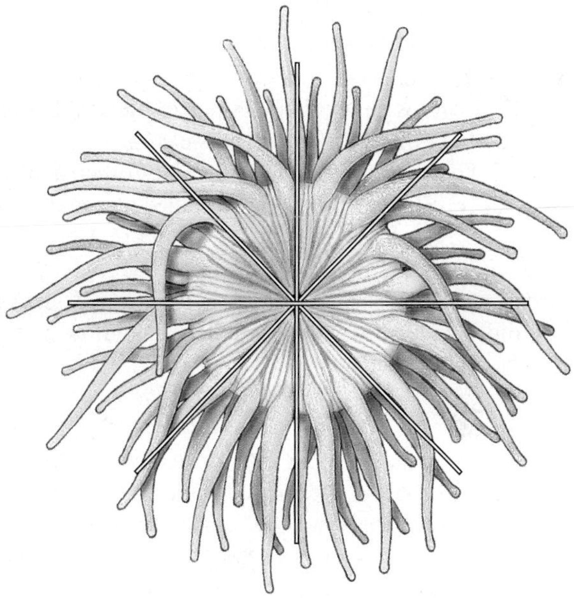

Figure 19.9 How radial and bilateral symmetry differ.

(*a*) Radial symmetry is the regular arrangement of parts around a central axis, so that any plane passing through the central axis divides the organism into halves that are approximate mirror images. (*b*) Bilateral symmetry is reflected in a body form which can only be bisected into halves that are approximate mirror images by one plane (the sagittal plane).

(a)

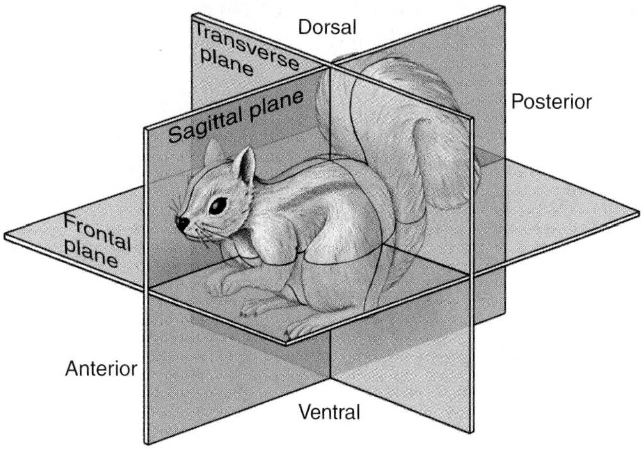
Dorsal
Transverse plane
Sagittal plane
Posterior
Frontal plane
Anterior
Ventral

(b)

19.5 Solid Worms: Internal Organs

The bilaterally symmetrical eumetazoans produce three embryonic layers: an outer ectoderm, an inner endoderm, and a third layer, the **mesoderm,** between the ectoderm and endoderm. In general, the outer coverings of the body and the nervous system develop from the ectoderm, the digestive organs and intestines develop from the endoderm, and the skeleton and muscles develop from the mesoderm (figure 19.10).

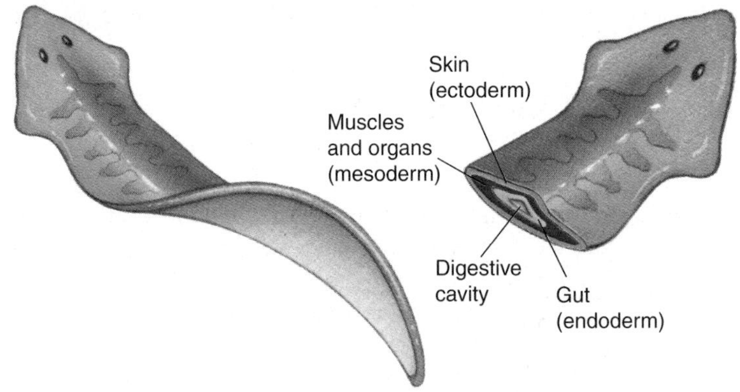

Skin (ectoderm)
Muscles and organs (mesoderm)
Digestive cavity
Gut (endoderm)

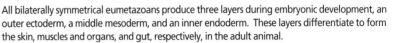

Figure 19.10 Body plan of a solid worm.
All bilaterally symmetrical eumetazoans produce three layers during embryonic development, an outer ectoderm, a middle mesoderm, and an inner endoderm. These layers differentiate to form the skin, muscles and organs, and gut, respectively, in the adult animal.

The simplest of all bilaterally symmetrical animals are the **solid worms.** By far the largest phylum of these, with about 20,000 species, is Platyhelminthes (pronounced plat-ee-hel-MIN-theeze), which includes the flatworms. Flatworms are the simplest animals in which organs occur. An organ is a collection of different tissues that function as a unit. The testes and uterus of flatworms are reproductive organs, for example. The dark spots on the head are eyespots that can detect light, although they cannot focus an image like your eyes can.

Solid worms lack any internal cavity other than the digestive tract. Flatworms are soft-bodied animals flattened from top to bottom, like a piece of tape or ribbon. If you were to cut a flatworm in half across its body, you would see that the gut is completely surrounded by tissues and organs. This solid body construction is called **acoelomate,** meaning without a body cavity.

Flatworms

Although flatworms have a simple body design, they do have a definite head at the anterior end and they do possess organs. Flatworms range in size from a millimeter or less to many meters long, as in some tapeworms. Most species of flatworms are parasitic, occurring within the bodies of many other kinds of animals. Other flatworms are free-living, occurring in a wide variety of marine and freshwater habitats, as well as moist places on land (figure 19.11). Free-living flatworms are carnivores and scavengers; they eat various small animals and bits of organic debris. They move from place to place by means of ciliated epithelial cells, which are particularly concentrated on their ventral surfaces.

There are two classes of parasitic flatworms, which live within the bodies of other animals: flukes and tapeworms. Both groups of worms have epithelial layers resistant to the digestive enzymes and immune defenses produced by their hosts—an important feature in their parasitic way of life. However, they lack certain features of the free-living flatworms, such as cilia in the adult stage, eyespots, and other sensory organs that lack adaptive significance for an organism that lives within the body of another animal (figure 19.12).

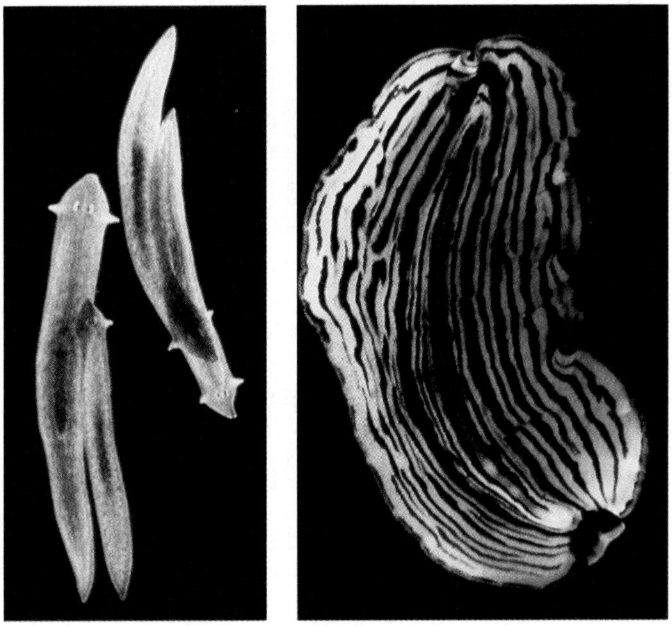

(a) (b)

Figure 19.11 Flatworms.
(*a*) A common flatworm, *Planaria.* (*b*) A marine free-living flatworm.

To human beings, one of the most important parasitic flatworms is the human liver fluke, *Clonorchis sinensis.* It lives in the bile passages of the liver of humans, cats, dogs, and pigs. It is especially common in Asia. The worms are 1 to 2 centimeters long and have a complex life cycle. Other very important flukes are the blood flukes of genus *Schistosoma.* They afflict about 1 in 20 of the world's population, more than 200 million people throughout tropical Asia, Africa, Latin America, and the Middle East. Three species of *Schistosoma* cause the disease called schistosomiasis, or bilharzia. Some 800,000 people die each year from this disease.

Phylum Platyhelminthes: Solid Worms

Key Evolutionary Innovation: BILATERAL SYMMETRY

Acoelomate solid worms such as the flatworms (phylum Platyhelminthes) were the first animals to be **bilaterally symmetrical** and to have a distinct **head.** The evolution of the mesoderm in solid worms allowed the formation of digestive and other organs.

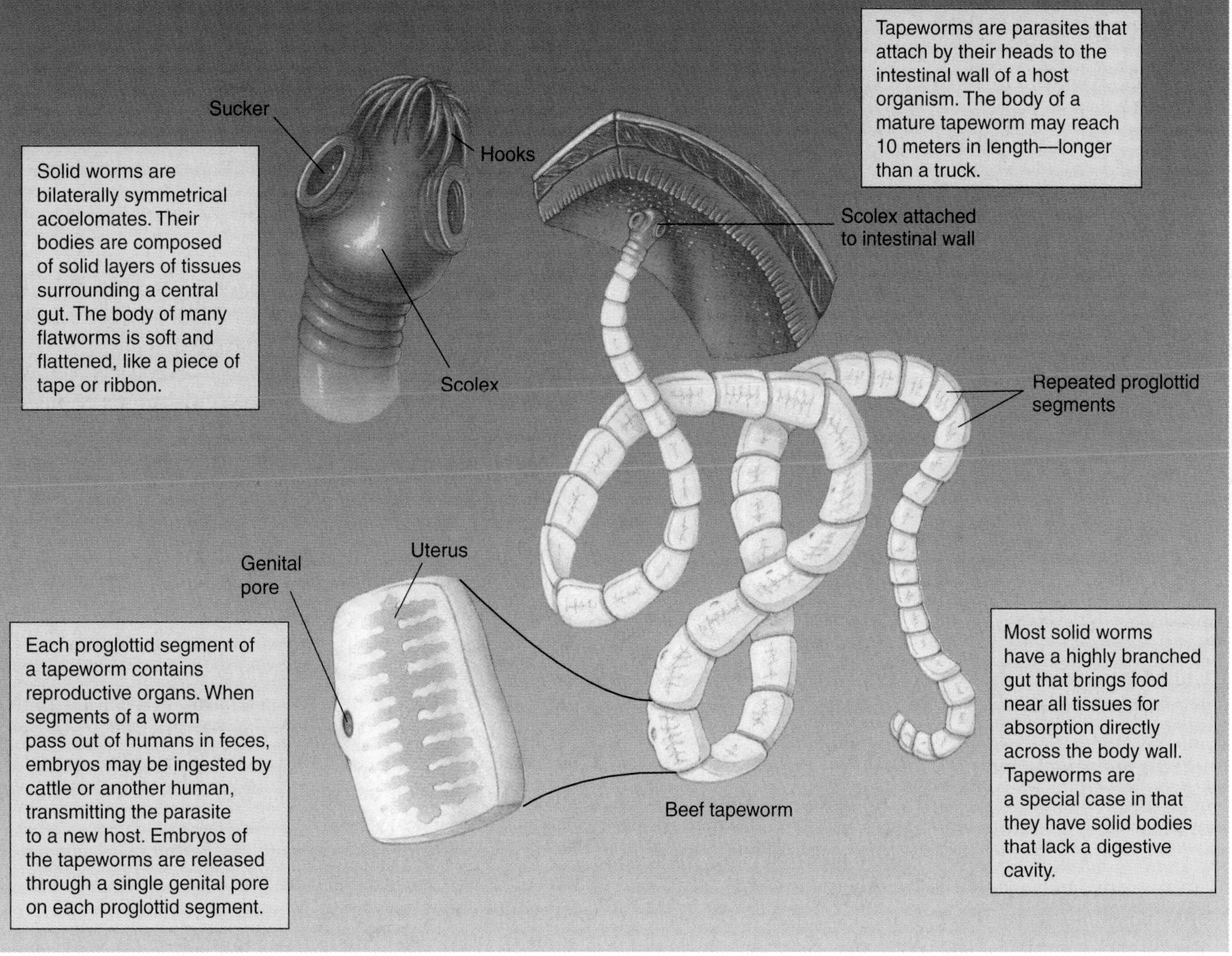

Tapeworms are parasites that attach by their heads to the intestinal wall of a host organism. The body of a mature tapeworm may reach 10 meters in length—longer than a truck.
Sucker
Hooks
Solid worms are bilaterally symmetrical acoelomates. Their bodies are composed of solid layers of tissues surrounding a central gut. The body of many flatworms is soft and flattened, like a piece of tape or ribbon.
Scolex
Scolex attached to intestinal wall
Repeated proglottid segments
Genital pore
Uterus
Each proglottid segment of a tapeworm contains reproductive organs. When segments of a worm pass out of humans in feces, embryos may be ingested by cattle or another human, transmitting the parasite to a new host. Embryos of the tapeworms are released through a single genital pore on each proglottid segment.
Beef tapeworm
Most solid worms have a highly branched gut that brings food near all tissues for absorption directly across the body wall. Tapeworms are a special case in that they have solid bodies that lack a digestive cavity.

Figure 19.12 Phylum Platyhelminthes: solid worms.

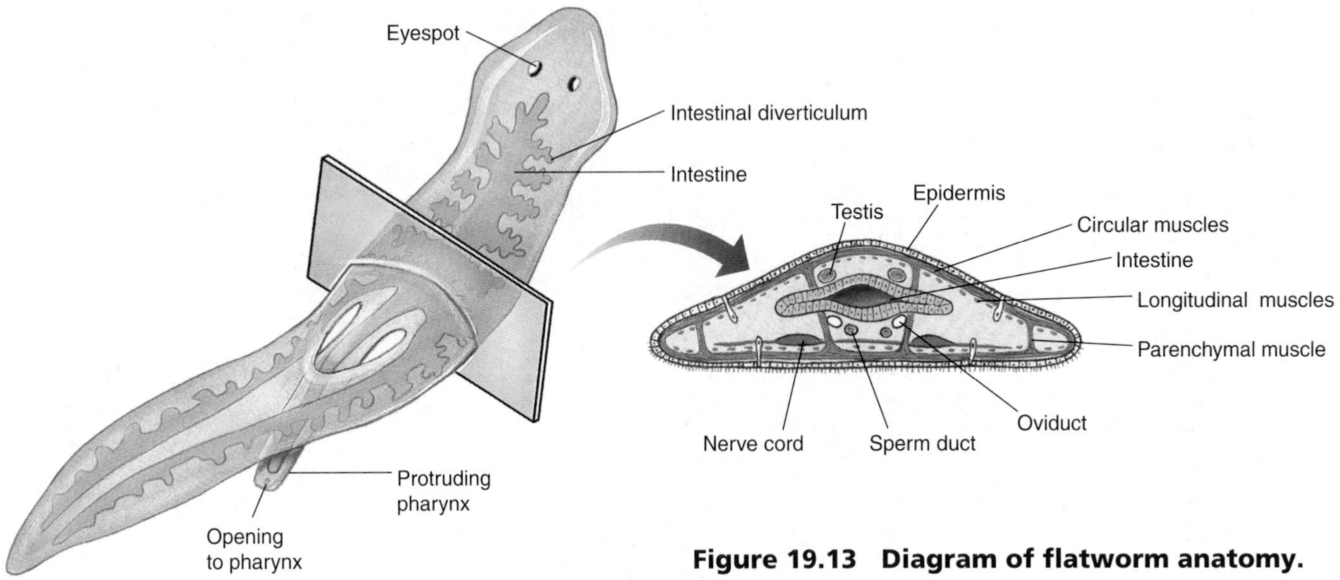

Eyespot
Intestinal diverticulum
Intestine
Testis
Epidermis
Circular muscles
Intestine
Longitudinal muscles
Parenchymal muscle
Oviduct
Nerve cord
Sperm duct
Protruding pharynx
Opening to pharynx

Figure 19.13 Diagram of flatworm anatomy.
The organism shown is *Dugesia,* the familiar freshwater "planaria" of many biology laboratories.

Characteristics of Flatworms

Those flatworms that have a digestive cavity have an incomplete gut, one with only one opening. As a result, they cannot feed, digest, and eliminate undigested particles of food simultaneously. Thus, flatworms cannot feed continuously, as more advanced animals can. The gut is branched and extends throughout the body (figure 19.13), functioning in both digestion and transport of food. Cells that line the gut engulf most of the food particles by phagocytosis and digest them; but, as in the cnidarians, some of these particles are partly digested extracellularly. Tapeworms, which are parasitic flatworms, lack digestive systems. They absorb their food directly through their body walls.

Unlike cnidarians, flatworms have an excretory system, which consists of a network of fine tubules (little tubes) that runs throughout the body. Cilia line the hollow centers of bulblike **flame cells,** which are located on the side branches of the tubules (see figure 24.21). Cilia in the flame cells move water and excretory substances into the tubules and then to exit pores located between the epidermal cells. Flame cells were named because of the flickering movements of the tuft of cilia within them. They primarily regulate the water balance of the organism. The excretory function of flame cells appears to be a secondary one. A large proportion of the metabolic wastes excreted by flatworms probably diffuses directly into the gut and is eliminated through the mouth.

Like sponges, cnidarians, and ctenophorans, flatworms lack a **circulatory system,** a network of vessels that carry fluids, oxygen, and food molecules to parts of the body. Consequently, all flatworm cells must be within diffusion distance of oxygen and food. Flatworms have thin bodies and highly branched digestive cavities that make such a relationship possible.

The nervous system of flatworms is very simple. Like cnidarians, some primitive flatworms have only a nerve net. However, most members of this phylum have longitudinal nerve cords that constitute a simple central nervous system. Between the longitudinal cords are cross-connections, so that the flatworm nervous system resembles a ladder.

Free-living flatworms use sensory pits or tentacles along the sides of their heads to detect food, chemicals, or movements of the fluid in which they are living. Free-living members of this phylum also have eyespots on their heads. These are inverted, pigmented cups containing light-sensitive cells connected to the nervous system. These eyespots enable the worms to distinguish light from dark. Flatworms are far more active than cnidarians or ctenophores. Such activity is characteristic of bilaterally symmetrical animals. In flatworms, this activity seems to be related to the greater concentration of sensory organs and, to some degree, the nervous system elements in the heads of these animals.

The reproductive systems of flatworms are complex. Most flatworms are **hermaphroditic,** with each individual containing both male and female sexual structures. In some parasitic flatworms, there is a complex succession of distinct larval forms. Some genera of flatworms are also capable of asexual regeneration; when a single individual is divided into two or more parts, each part can regenerate an entirely new flatworm.

19.5 Flatworms have internal organs, bilateral symmetry, and a distinct head. They do not have a body cavity.

19.6 Pseudocoelom Versus Coelom

A key transition in the evolution of the animal body plan was the evolution of the body cavity. All bilaterally symmetrical animals other than solid worms have a cavity within their body. The evolution of an internal body cavity was an important improvement in animal body design for three reasons:

1. **Circulation.** Fluids that move within the body cavity can serve as a circulatory system, permitting the rapid passage of materials from one part of the body to another and opening the way to larger bodies.
2. **Movement.** Fluid in the cavity makes the animal's body rigid, permitting resistance to muscle contraction and thus opening the way to muscle-driven body movement.
3. **Organ function.** In a fluid-filled enclosure, body organs can function without being deformed by surrounding muscles. For example, food can pass freely through a gut suspended within a cavity, at a rate not controlled by when the animal moves.

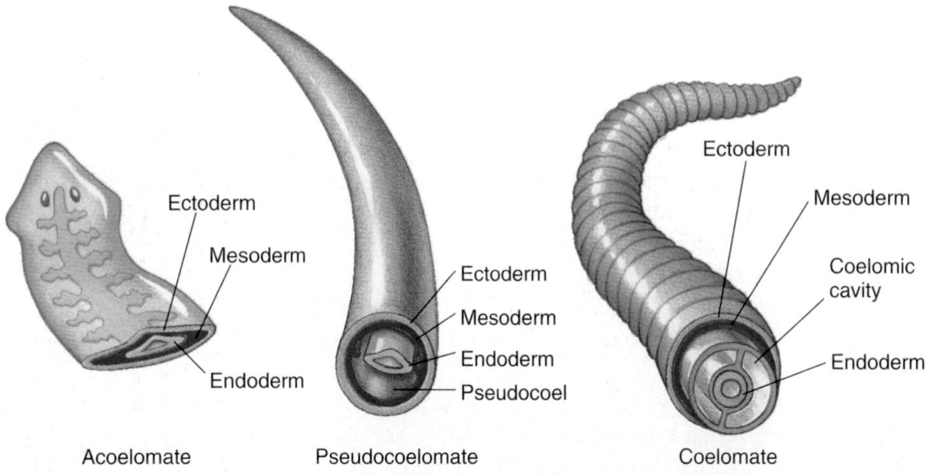
Ectoderm
Mesoderm
Endoderm
Ectoderm
Mesoderm
Endoderm
Pseudocoel
Ectoderm
Mesoderm
Coelomic cavity
Endoderm
Acoelomate
Pseudocoelomate
Coelomate

Figure 19.14 Three body plans for bilaterally symmetrical animals.

Kinds of Body Cavities

There are three basic kinds of body plans found in bilaterally symmetrical animals (figure 19.14). Acoelomates, such as solid worms, have no body cavity. **Pseudocoelomates** have a body cavity called the **pseudocoel** located between the mesoderm and endoderm. A third way of organizing the body is one in which the fluid-filled body cavity develops not between endoderm and mesoderm but rather entirely within the mesoderm. Such a body cavity is called a **coelom,** and animals that possess such a cavity are called **coelomates.** In coelomates, the gut is suspended, along with other organ systems of the animal, within the coelom; the coelom, in turn, is surrounded by a layer of epithelial cells entirely derived from the mesoderm.

The development of the coelom poses a problem—circulation—solved in pseudocoelomates by churning the fluid within the body cavity. In coelomates, the gut is again surrounded by tissue that presents a barrier to diffusion, just as it was in solid worms. This problem is solved among coelomates by the development of a circulatory system. The circulating fluid, or blood, carries nutrients and oxygen to the tissues and removes wastes and carbon dioxide. Blood is usually pushed through the circulatory system by contraction of one or more muscular hearts. In an **open circulatory system,** the blood passes from vessels into sinuses, mixes with body fluid, and then reenters the vessels later in another location. In a **closed circulatory system,** the blood is separate from the body fluid and can be separately controlled. Also, blood moves through a closed circulatory system faster and more efficiently than it does through an open system.

The evolutionary relationship among coelomates, pseudocoelomates, and acoelomates is not clear. Acoelomates, for example, could have given rise to coelomates, but scientists also cannot rule out the possibility that acoelomates were derived from coelomates. The different phyla of pseudocoelomates form two groups that do not appear to be closely related.

19.6 Some body cavities develop between endoderm and mesoderm (pseudocoelomates), others within the mesoderm (coelomates).

19.7 Roundworms: Pseudocoelomates

As we have noted, all bilaterally symmetrical animals except solid worms possess an internal body cavity. Among them, seven phyla are characterized by their possession of a pseudocoel. In all pseudocoelomates, the pseudocoel serves as a hydrostatic skeleton—one that gains its rigidity from being filled with fluid under pressure. The animal's muscles can work against this "skeleton," thus making the movements of pseudocoelomates far more efficient than those of the acoelomates.

Only one of the seven pseudocoelomate phyla includes a large number of species. This phylum, Nematoda, includes some 12,000 recognized species of **nematodes,** eelworms, and other roundworms. Scientists estimate that the actual number might approach 100 times that many. Members of this phylum are found everywhere. Nematodes are abundant and diverse in marine and freshwater habitats, and many members of this phylum are parasites of animals and plants (figure 19.15*a*). Many nematodes are microscopic and live in soil. It has been estimated that a spadeful of fertile soil may contain, on the average, a million nematodes.

A second phylum consisting of animals with a pseudocoelomate body plan is Rotifera, the rotifers. **Rotifers** are common, small, basically aquatic animals that have a crown of cilia at their heads; they range from 0.04 to 2 millimeters long (figure 19.15*b*). About 2,000 species exist throughout the world. Bilaterally symmetrical and covered with chitin, rotifers depend on their cilia for both locomotion and feeding, ingesting bacteria, protists, and small animals.

All pseudocoelomates lack a defined circulatory system; this role is performed by the fluids that move within the pseudocoel. Most pseudocoelomates have a complete, one-way digestive tract that acts like an assembly line. Food is broken down, then absorbed, and then treated and stored.

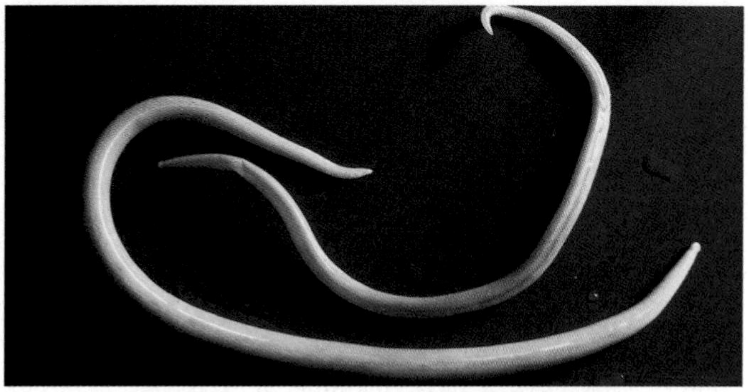

(a)

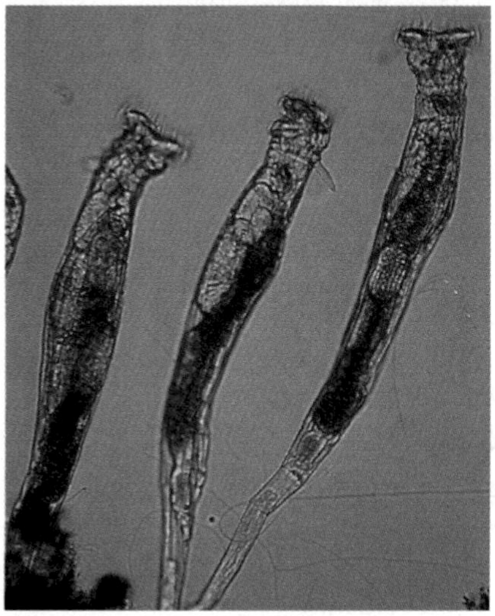

(b)

Figure 19.15 Pseudocoelomates.
(*a*) These nematodes (phylum Nematoda) are intestinal roundworms that infect humans and some other animals. Their fertilized eggs pass out with feces and can remain viable in soil for years. (*b*) Rotifers (phylum Rotifera) are common aquatic animals that depend on their crown of cilia for feeding and locomotion.

Phylum Nematoda: The Roundworms

Nematodes are bilaterally symmetrical, cylindrical, unsegmented worms (figure 19.16). They are covered by a flexible, thick cuticle, which is molted as they grow. Their muscles constitute a layer beneath the epidermis and extend along the length of the worm, rather than encircling its body. These longitudinal muscles pull both against the cuticle and the pseudocoel, which forms a hydrostatic skeleton. When nematodes move, their bodies whip about from side to side.

Near the mouth of a nematode, at its anterior end, are usually 16 raised, hairlike sensory organs. The mouth is often equipped with piercing organs called **stylets.** Food passes through the mouth as a result of the sucking action of a muscular chamber called the **pharynx.** After passing through a short corridor into the pharynx, food continues through the other portions of the digestive tract, where it is broken down and then digested. Some of the water with which the food has been mixed is reabsorbed near the end of the digestive tract, and material that has not been digested is eliminated through the anus.

Nematodes completely lack flagella or cilia, even on sperm cells. Reproduction in nematodes is sexual, with sexes usually separate. Their development is simple, and the adults consist of very few cells. For this reason, nematodes have become extremely important subjects for genetic and developmental studies. The 1-millimeter-long *Caenorhabditis elegans* matures in only three days, its body is transparent, and it has only 959 cells. It is the only animal whose complete developmental cellular anatomy is known, and the first animal whose genome (97 million DNA bases encoding over 19,000 different genes) was fully sequenced.

About 50 species of nematodes, including several that are rather common in the United States, regularly parasitize human beings. The most serious common nematode-caused disease in temperate regions is trichinosis, caused by worms of the genus *Trichinella*. These worms live in the small

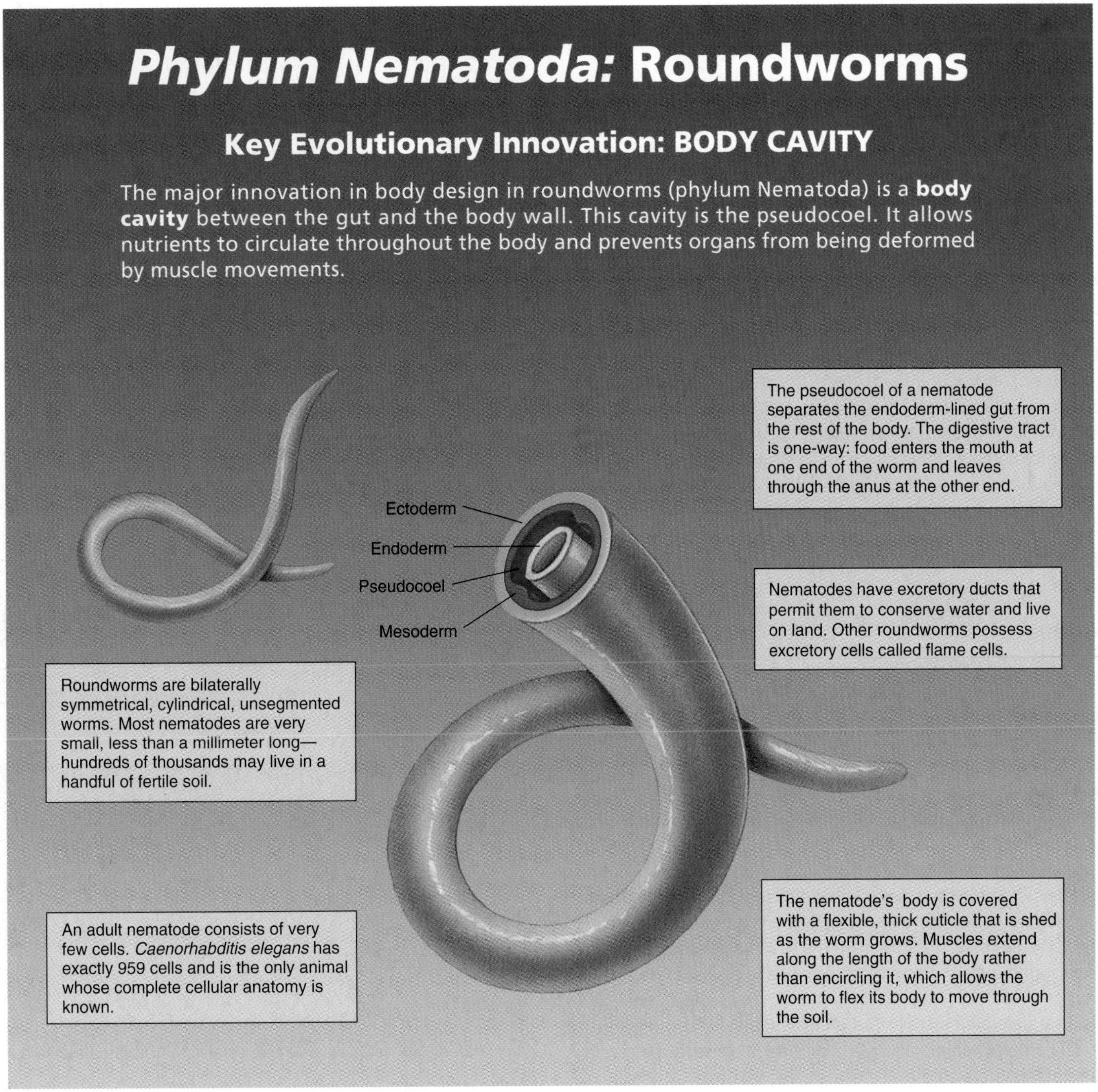
Phylum Nematoda: Roundworms
Key Evolutionary Innovation: BODY CAVITY
The major innovation in body design in roundworms (phylum Nematoda) is a body cavity between the gut and the body wall. This cavity is the pseudocoel. It allows nutrients to circulate throughout the body and prevents organs from being deformed by muscle movements.
The pseudocoel of a nematode separates the endoderm-lined gut from the rest of the body. The digestive tract is one-way: food enters the mouth at one end of the worm and leaves through the anus at the other end.
Ectoderm
Endoderm
Pseudocoel
Mesoderm
Nematodes have excretory ducts that permit them to conserve water and live on land. Other roundworms possess excretory cells called flame cells.
Roundworms are bilaterally symmetrical, cylindrical, unsegmented worms. Most nematodes are very small, less than a millimeter long—hundreds of thousands may live in a handful of fertile soil.
An adult nematode consists of very few cells. Caenorhabditis elegans has exactly 959 cells and is the only animal whose complete cellular anatomy is known.
The nematode's body is covered with a flexible, thick cuticle that is shed as the worm grows. Muscles extend along the length of the body rather than encircling it, which allows the worm to flex its body to move through the soil.

Figure 19.16 Phylum Nematoda: roundworms.

intestine of pigs, where fertilized female worms burrow into the intestinal wall. Once it has penetrated these tissues, each female produces about 1,500 live young. The young enter the lymph channels and travel to muscle tissue throughout the body, where they mature and form highly resistant, calcified cysts. Infection in human beings or other animals arises from eating undercooked or raw pork in which the cysts of *Trichinella* are present. If the worms are abundant, a fatal infection can result, but such infections are rare. It is thought that about 2.4% of the people in the United States are infected with trichinosis, but only about 20 deaths have been attributed to this disease during the past decade. The same parasite occurs in bears, and when humans eat improperly cooked bear meat, there seems to be an even greater chance of infection than with pork.

19.7 Roundworms have a pseudocoel body cavity. Nematodes, a kind of roundworm, are very common in soil.

19.8 Mollusks: Coelomates

Coelomates

Even though acoelomates and pseudocoelomates have proven very successful, the bulk of the animal kingdom consists of coelomates. Coelomates have a new body design that repositions the fluid. What is the functional difference between a pseudocoel and a coelom, and why has the latter kind of body cavity been so overwhelmingly successful? The answer has to do with the nature of animal embryonic development. In animals, development of specialized tissues involves a process called **primary induction,** in which one of the three primary tissues (endoderm, mesoderm, and ectoderm) interacts with another. The interaction requires physical contact. A major advantage of the coelomate body plan is that it allows contact between mesoderm and endoderm, so that primary induction can occur during development. For example, contact between mesoderm and endoderm permits localized portions of the digestive tract to develop into complex, highly specialized regions like the stomach. In pseudocoelomates, mesoderm and endoderm are separated by the body cavity, limiting developmental interactions between these tissues.

Mollusks

The only major phylum of coelomates without segmented bodies are the Mollusca. The **mollusks** are the largest animal phylum, except for the arthropods, with over 110,000 species. Mollusks are mostly marine, but occur almost everywhere.

Mollusks include three classes with outwardly different body plans. However, the seeming differences hide a basically similar body design. The body of mollusks is composed of three distinct parts: a head-foot, a central section called the visceral mass that contains the body's organs, and a mantle. The **mantle** is a heavy fold of tissue wrapped around the visceral mass like a cape, with the gills positioned on its inner surface like the lining of a coat. The **gills** are filamentous projections of tissue, rich in blood vessels, that capture oxygen from the water circulating between the mantle and visceral mass and release carbon dioxide.

The three major classes of mollusks, all different variations upon this same basic design, are gastropods, bivalves, and cephalopods (figure 19.17).

(a)

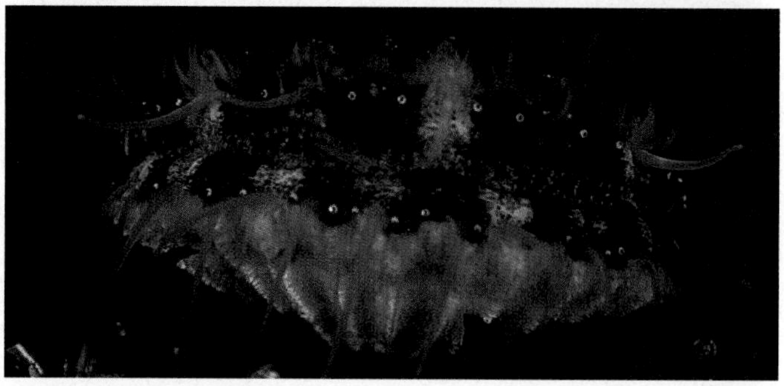

(b)

(c)

Figure 19.17 Three major classes of mollusks.
(*a*) A gastropod. (*b*) A bivalve. (*c*) A cephalopod.

1. **Gastropods** (snails and slugs) use the muscular foot to crawl, and their mantle often secretes a single, hard protective shell. All terrestrial mollusks are gastropods.
2. **Bivalves** (clams, oysters, and scallops) secrete a two-part shell with a hinge, as their name implies. They filter feed by drawing water into their shell.
3. **Cephalopods** (octopuses and squids) have modified the mantle cavity to create a jet propulsion system that can propel them rapidly through the water. Their shell is greatly reduced to an internal structure or is absent.

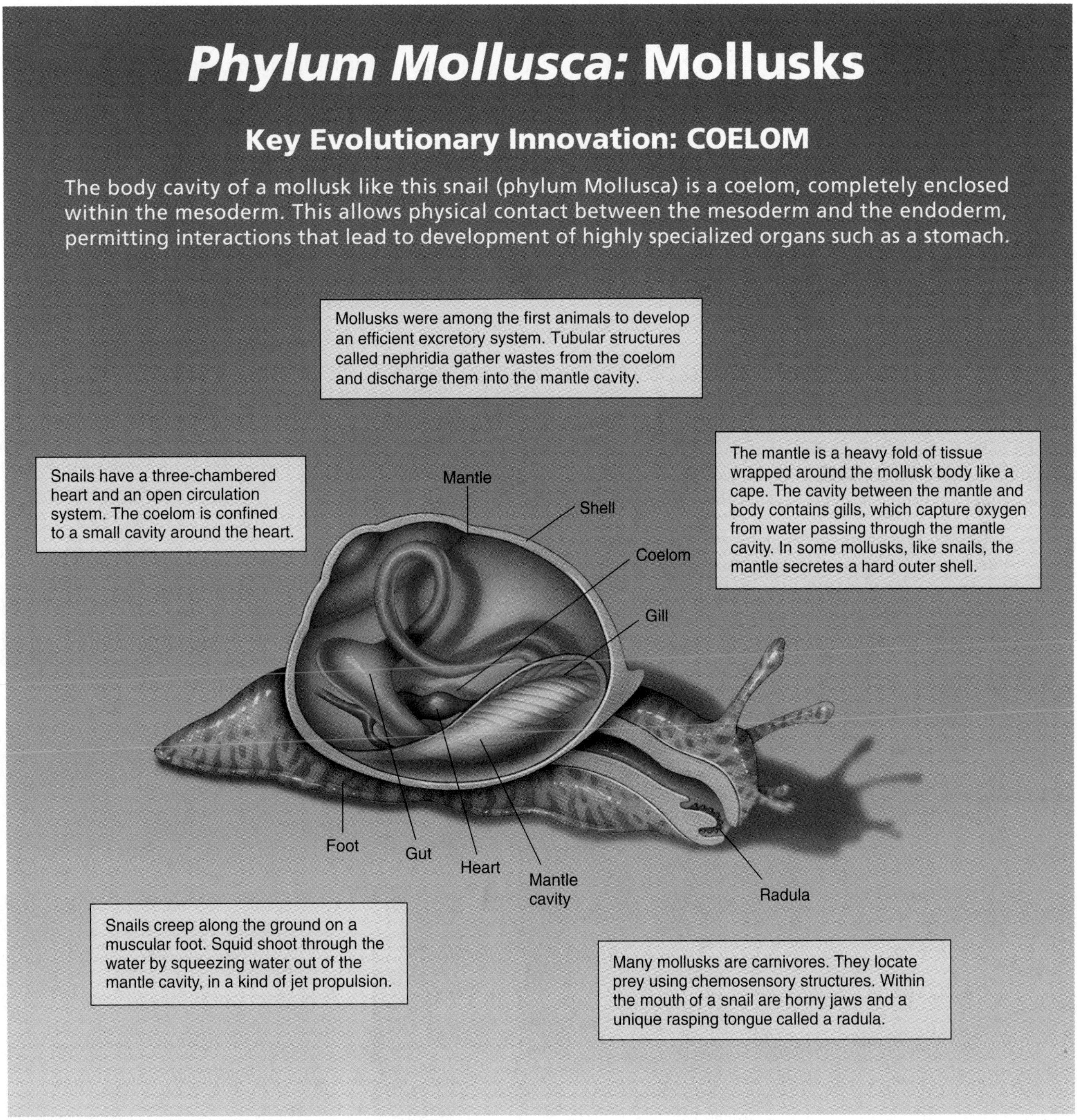
Phylum Mollusca: Mollusks
Key Evolutionary Innovation: COELOM
The body cavity of a mollusk like this snail (phylum Mollusca) is a coelom, completely enclosed within the mesoderm. This allows physical contact between the mesoderm and the endoderm, permitting interactions that lead to development of highly specialized organs such as a stomach.
Mollusks were among the first animals to develop an efficient excretory system. Tubular structures called nephridia gather wastes from the coelom and discharge them into the mantle cavity.
Snails have a three-chambered heart and an open circulation system. The coelom is confined to a small cavity around the heart.
The mantle is a heavy fold of tissue wrapped around the mollusk body like a cape. The cavity between the mantle and body contains gills, which capture oxygen from water passing through the mantle cavity. In some mollusks, like snails, the mantle secretes a hard outer shell.
Mantle
Shell
Coelom
Gill
Foot
Gut
Heart
Mantle cavity
Radula
Snails creep along the ground on a muscular foot. Squid shoot through the water by squeezing water out of the mantle cavity, in a kind of jet propulsion.
Many mollusks are carnivores. They locate prey using chemosensory structures. Within the mouth of a snail are horny jaws and a unique rasping tongue called a radula.

Figure 19.18 Phylum Mollusca: mollusks.

One of the most characteristic features of mollusks is the **radula,** a rasping, tonguelike organ (figure 19.18). With rows of pointed, backward-curving teeth, the radula is used by some snails to scrape algae off rocks. The small holes often seen in oyster shells are produced by gastropods that have bored holes to kill the oyster and extract its body.

19.8 Mollusks have a coelom body cavity but are not segmented. There are more species of terrestrial mollusks than of terrestrial vertebrates.

19.9 Annelids: The Rise of Segmentation

One of the early key innovations in body plan to arise among the coelomates was **segmentation,** the building of a body from a series of similar segments. The first segmented animals to evolve were the **annelid worms,** phylum Annelida. These advanced coelomates are assembled as a chain of nearly identical segments, like the boxcars of a train. The great advantage of such segmentation is the evolutionary flexibility it offers—a small change in an existing segment can produce a new kind of segment with a different function. Thus, some segments are modified for reproduction, some for feeding, and others for eliminating wastes.

Two-thirds of all annelids live in the sea (about 8,000 species), and most of the rest—some 3,100 species—are earthworms (figure 19.19). The basic body plan of an annelid is a tube within a tube: the digestive tract is a tube suspended within the coelom, which is itself a tube running from mouth to anus. There are three characteristics of this organization:

1. **Repeated segments.** The body segments of an annelid are visible as a series of ringlike structures running the length of the body, looking like a stack of donuts (figure 19.20). The segments are divided internally from one another by partitions, just as walls separate the rooms of a building. In each of the cylindrical segments, the digestive, excretory, and locomotor organs are repeated. The body fluid within the coelom of each segment creates a hydrostatic (liquid-supported) skeleton that gives the segment rigidity, like an inflated balloon. Muscles within each segment play against the fluid in the coelom. Because each segment is separate, each is able to expand or contract independently. This lets the worm body move in ways that are quite complex. When an earthworm crawls on a flat surface, for example, it lengthens some parts of its body while shortening others.
2. **Specialized segments.** The anterior (front) segments of annelids contain the sensory organs of the worm. Some of these are organs sensitive to light, and elaborate eyes with lenses and retinas have evolved in some annelids. One anterior segment contains a well-developed cerebral ganglion, or brain.
3. **Connections.** Because partitions separate the segments, it is necessary to provide ways for materials and information to pass between segments. A circulatory system carries blood from one segment to another, while nerve cords connect the nerve centers or ganglia located in each segment with each other and the brain. The brain can then coordinate the worm's activities.

Segmentation underlies the body organization of all complex coelomate animals, not only annelids but also arthropods (crustaceans, spiders, and insects) and chordates (mostly vertebrates). For example, vertebrate muscles develop from repeated blocks of tissue called somites that occur in the embryo. Vertebrate segmentation is also seen in the vertebral column, which is a stack of very similar vertebrae.

> **19.9** Annelids are segmented worms. Most species are marine, but some—about one-third of the species—are terrestrial.

(a)

(b)

Figure 19.19 Representative annelids.

(a) Earthworms are the terrestrial annelids. This night crawler, *Lumbricus terrestris,* is in its burrow. (b) Shiny bristle worm, *Oenone fulgida,* a polychaete.

Phylum Annelida: Annelids

Key Evolutionary Innovation: SEGMENTATION

Marine polychaetes and earthworms (phylum Annelida) were the first organisms to evolve a body plan based on **repeated body segments.** Most segments are identical and are separated from other segments by partitions.

Each segment contains a complete set of digestive and excretory organs and a nerve center.

Earthworms crawl by anchoring bristles called setae to the ground and pulling against them. Polychaete annelids have a flattened body and swim or crawl by flexing it.

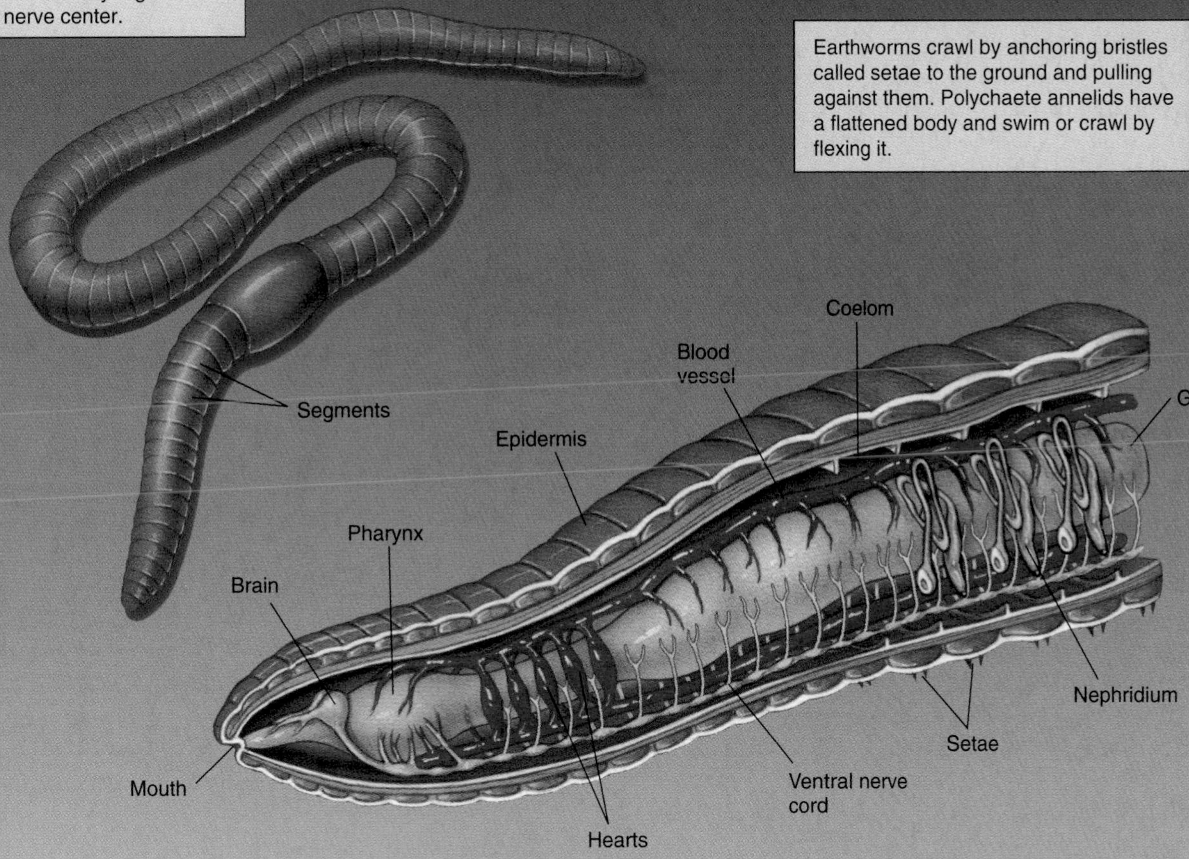

Segments
Coelom
Blood vessel
Gut
Epidermis
Pharynx
Brain
Mouth
Hearts
Ventral nerve cord
Setae
Nephridium

Segments are connected by the circulatory and nervous systems. A series of hearts at the anterior (front) end pump the blood. A well-developed brain located in an anterior segment coordinates the activities of all segments.

Each segment has a coelom. Muscles squeeze the fluid of the coelom, making each segment rigid, like an inflated balloon. Because each segment can contract independently, a worm can crawl by lengthening some segments while shortening others.

Figure 19.20 Phylum Annelida: annelids.

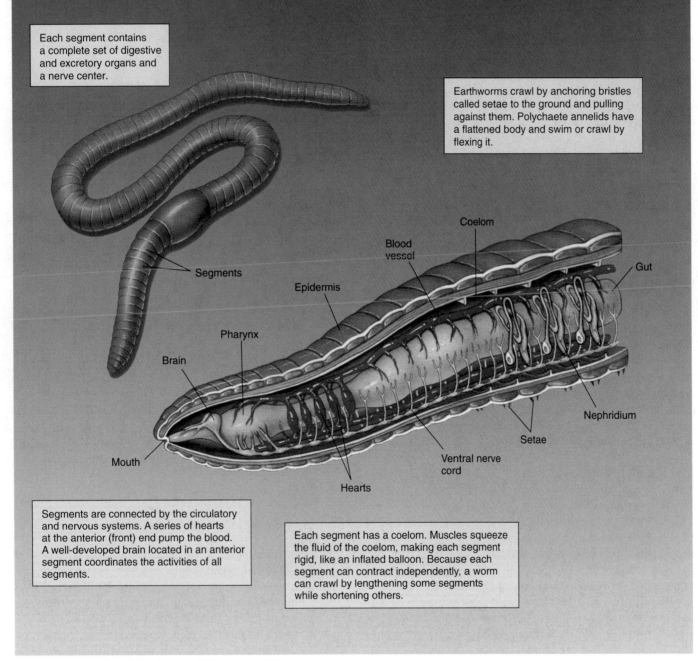

19.10 Arthropods: Advent of Jointed Appendages

The evolution of segmentation among the annelids marked a major innovation in body structure among the coelomates. An even more profound innovation was to come. It marks the origin of the body plan characteristic of the most successful of all animal groups, the **arthropods,** phylum Arthropoda (figure 19.21). This innovation was the development of jointed appendages.

Jointed Appendages

The name *arthropod* comes from two Greek words, *arthros,* jointed, and *podes,* feet. All arthropods have jointed appendages. Some are legs, and others may be modified for other uses. To gain some idea of the importance of jointed appendages, imagine yourself without them—no hips, knees, ankles, shoulders, elbows, wrists, or knuckles. Without jointed appendages, you could not walk or grasp an object. Arthropods use jointed appendages as legs and wings for moving, as antennae to sense their environment, and as mouthparts for sucking, ripping, and chewing prey. A scorpion, for example, seizes and tears apart its prey with mouthpart appendages modified as large pincers.

Rigid Exoskeleton

The arthropod body plan has a second great innovation: Arthropods have a rigid external skeleton, or **exoskeleton,** made of chitin (figure 19.22). In any animal, a key function of the skeleton is to provide places for muscle attachment, and in arthropods the muscles attach to the interior surface of the hard chitin shell, which also protects the animal from predators and impedes water loss.

However, while chitin is hard and tough, it is also brittle and cannot support great weight. As a result, the exoskeleton must be much thicker to bear the pull of the muscles in large insects than in small ones, so there is a limit to how big an arthropod body can be. That is why you don't see beetles as big as birds or crabs the size of a cow—the exoskeleton would be so thick the animal couldn't move its great weight. In fact, the great majority of arthropod species consist of small animals—mostly about a millimeter in length—but members of the phylum range in adult size from about 80 micrometers long (some parasitic mites) to 3.6 meters across (a gigantic crab found in the sea off Japan). Some lobsters are nearly a meter in length. The largest living insects are about 33 centimeters long, but the giant dragonflies that lived 300 million years ago had wingspans of as much as 60 centimeters (2 ft)!

Because this size limitation is inherent in the body design of arthropods, no arthropods have ever grown to great size. As we will see, to overcome this limitation, a strong, flexible endoskeleton is required.

Arthropod bodies are segmented like those of annelids, from which they almost certainly evolved. Individual segments often exist only during early development, however, and fuse into functional groups as adults. For example, caterpillars have many segments, while butterflies have only three functional body units—head, thorax, and abdomen—each composed of several fused segments.

Arthropods have proven very successful—about two-thirds of all named species on earth are arthropods. Scientists estimate that a quintillion (a billion billion) insects are alive at any one time—200 million insects for each living human!

Figure 19.21 Arthropod.
The most common class of arthropods is Insecta, the insects. More than two-thirds of all named animal species are insects. The insect in this photograph is the monarch butterfly, *Danaus plexippus.*

Phylum Arthropoda: Arthropods

Key Evolutionary Innovations: JOINTED APPENDAGES and EXOSKELETON

Insects and other arthropods (phylum Arthropoda) have a coelom, segmented bodies, and **jointed appendages.** The three body regions of an insect (head, thorax, and abdomen) are each actually composed of a number of segments that fuse during development. All arthropods have a strong **exoskeleton** made of chitin. One class of arthropods, the insects, has evolved **wings,** which permit them to fly rapidly through the air.

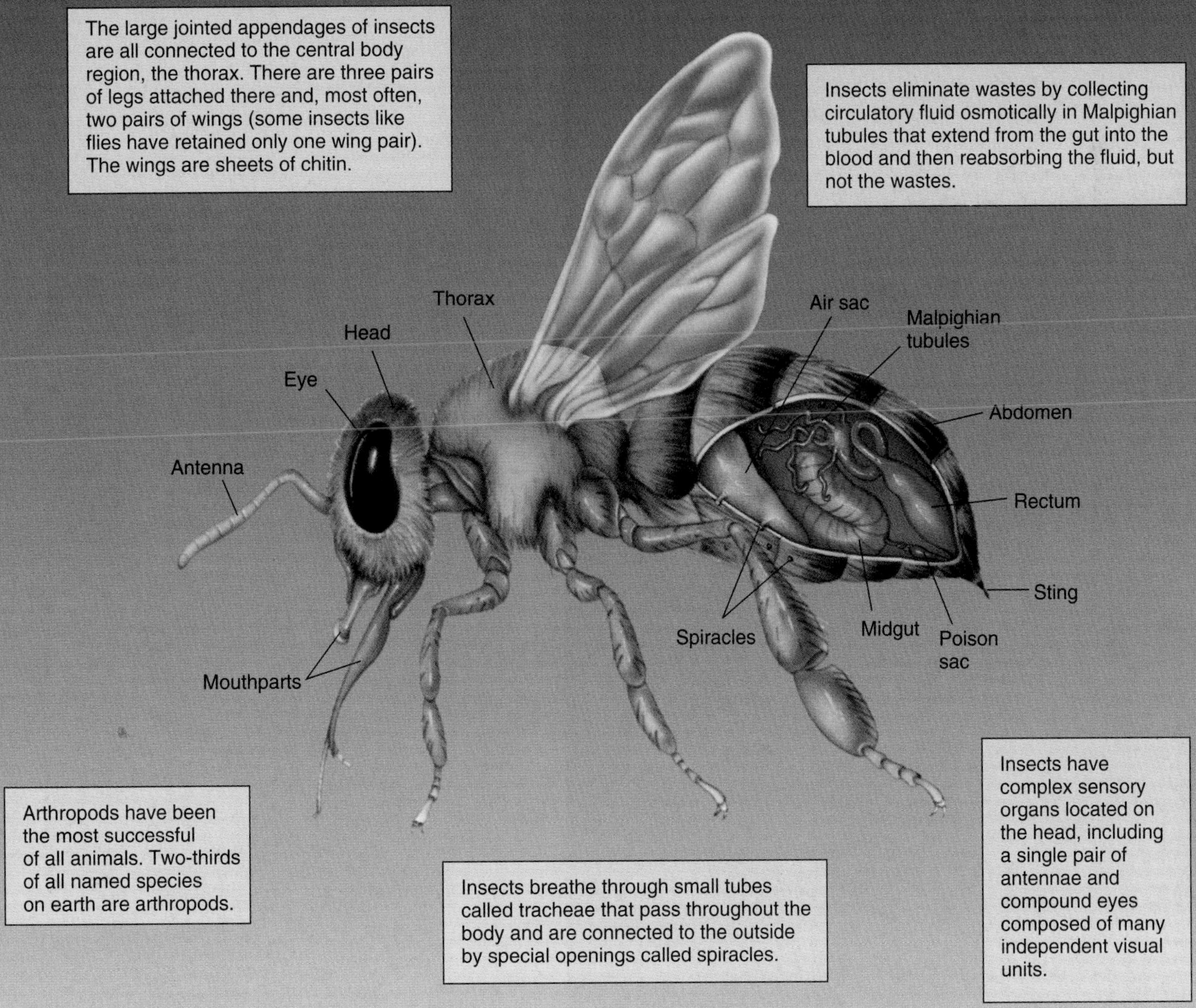

The large jointed appendages of insects are all connected to the central body region, the thorax. There are three pairs of legs attached there and, most often, two pairs of wings (some insects like flies have retained only one wing pair). The wings are sheets of chitin.
Insects eliminate wastes by collecting circulatory fluid osmotically in Malpighian tubules that extend from the gut into the blood and then reabsorbing the fluid, but not the wastes.
Thorax
Head
Eye
Antenna
Mouthparts
Air sac
Malpighian tubules
Abdomen
Rectum
Sting
Spiracles
Midgut
Poison sac
Arthropods have been the most successful of all animals. Two-thirds of all named species on earth are arthropods.
Insects breathe through small tubes called tracheae that pass throughout the body and are connected to the outside by special openings called spiracles.
Insects have complex sensory organs located on the head, including a single pair of antennae and compound eyes composed of many independent visual units.

Figure 19.22 Phylum Arthropoda: arthropods.

Chelicerates

Arthropods such as spiders, mites, scorpions, and a few others lack jaws, or **mandibles,** and are called **chelicerates.** Their mouthparts, known as **chelicerae,** evolved from the appendages nearest the animal's anterior end. The remaining arthropods have mandibles, formed by the modification of one of the pairs of anterior appendages. These arthropods, called **mandibulates,** include the crustaceans, insects, centipedes, millipedes, and a few other groups (figure 19.23).

The chelicerate fossil record goes back as far as that of any multicellular animal, about 630 million years. A major group of arthropods, the now-extinct trilobites, was abundant then, and horseshoe crabs living today seem to be directly descended from them (figure 19.24). Horseshoe crabs feed at night, primarily on mollusks and annelids. They swim on their backs by moving their abdominal plates and walk on their four pairs of legs, which are protected by their shell.

By far the largest of the three classes of chelicerates is the largely terrestrial class Arachnida (figure 19.25), with some 57,000 named species, including the spiders, ticks, mites, scorpions, and daddy longlegs. Arachnids have a pair of chelicerae, a pair of pedipalps, and four pairs of walking legs. The chelicerae consist of a stout basal portion and a movable fang often connected to a poison gland. **Pedipalps,** the next pair of appendages, may resemble legs, but they have one less segment. In scorpions, the pedipalps are large and pinching. Most **arachnids** are carnivorous, although mites are largely herbivorous. Ticks are blood-feeding ectoparasites of vertebrates, and some ticks may carry diseases, such as Rocky Mountain spotted fever and Lyme disease.

Sea spiders are also chelicerates and are relatively common, especially in coastal waters, and more than 1,000 species are in the class.

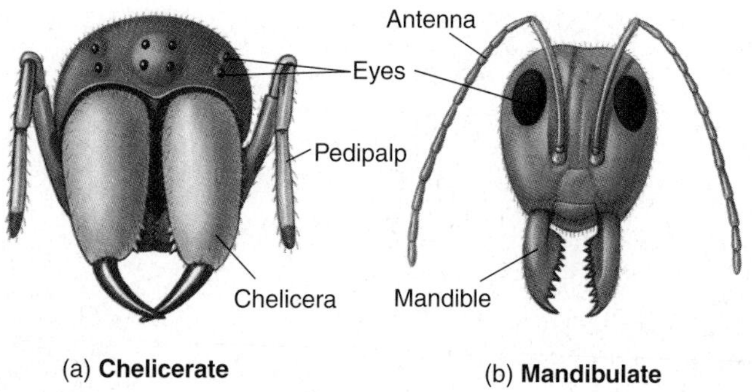
Antenna
Eyes
Pedipalp
Chelicera
Mandible
(a) Chelicerate
(b) Mandibulate

Figure 19.23 Chelicerates and mandibulates.

In the chelicerates, such as the jumping spider (*a*), the chelicerae are the foremost appendages of the body. In contrast, the foremost appendages in the mandibulates, such as the bullfrog ant (*b*), are the antennae, followed by the mandibles.

Figure 19.24 Horseshoe crabs.

These horseshoe crabs, *Limulus,* are emerging from the sea to mate at the edge of Delaware Bay, New Jersey, in early May.

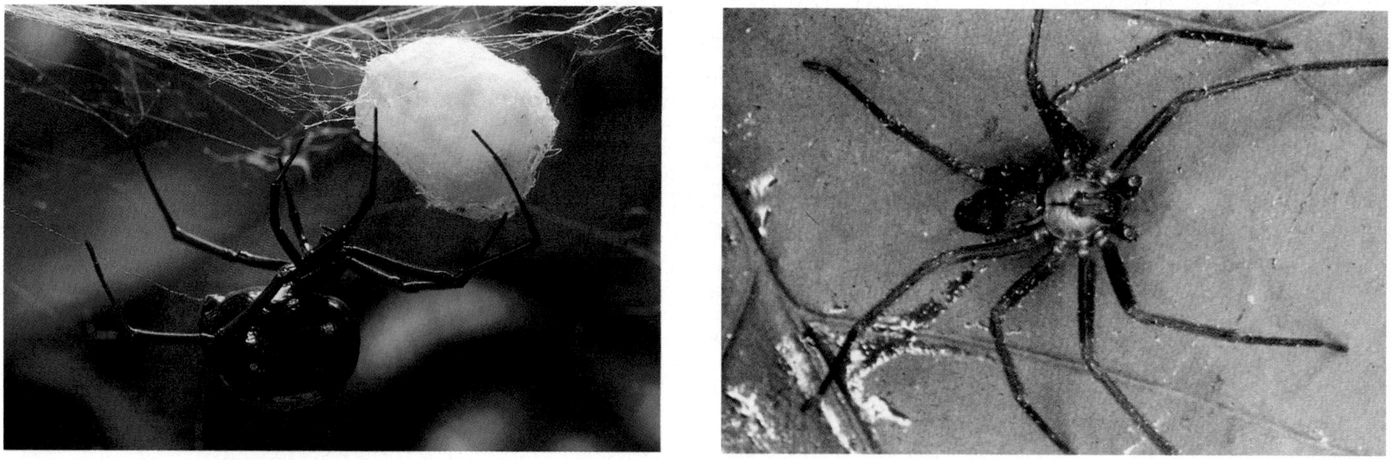

(a) (b)

Figure 19.25 Arachnids.

(*a*) One of the two poisonous spiders in the United States and Canada is the black widow spider, *Latrodectus mactans.* (*b*) Another of the poisonous spiders in this area is the brown recluse, *Loxosceles reclusa.* Both species are common throughout temperate and subtropical North America, but they rarely bite humans.

Mandibulates

Crustaceans. The **crustaceans** (subphylum Crustacea) are a large, diverse group of primarily aquatic organisms, including some 35,000 species of crabs, shrimps, lobsters, crayfish, barnacles, water fleas, pillbugs, sowbugs, and related groups (figure 19.26). Often incredibly abundant in marine and freshwater habitats and playing a role of critical importance in virtually all aquatic ecosystems, crustaceans have been called "the insects of the water." Most crustaceans have two pairs of antennae, three pairs of chewing appendages, and various numbers of pairs of legs (figure 19.27). The nauplius larva stage through which all crustaceans pass provides evidence that all members of this diverse group are descended from a common ancestor. The nauplius hatches with three pairs of appendages and metamorphoses through several stages before reaching maturity. In many groups, this nauplius stage is passed in the egg, and development of the hatchling to the adult form is direct.

Crustaceans differ from the insects—but resemble millipedes and centipedes—in that they have legs on their abdomen as well as on their thorax. Many crustaceans have compound eyes. In addition, they have delicate tactile hairs that project from the cuticle all over the body. Larger crustaceans have feathery gills near the bases of their legs. In smaller members of this class, gas exchange takes place directly through the thinner areas of the cuticle or the entire body. Most crustaceans have separate sexes. Many different kinds of specialized copulation occur among the crustaceans, and the members of some orders carry their eggs with them, either singly or in egg pouches, until they hatch.

(a)
(b)
(c)

Figure 19.26 Crustaceans.

(a) Edible crab, *Cancer pangyrus.* (b) Sowbugs, *Porcellio scaber.* (c) Gooseneck barnacle, *Lepas anatifera,* feeding. These are stalked barnacles; many others lack a stalk.

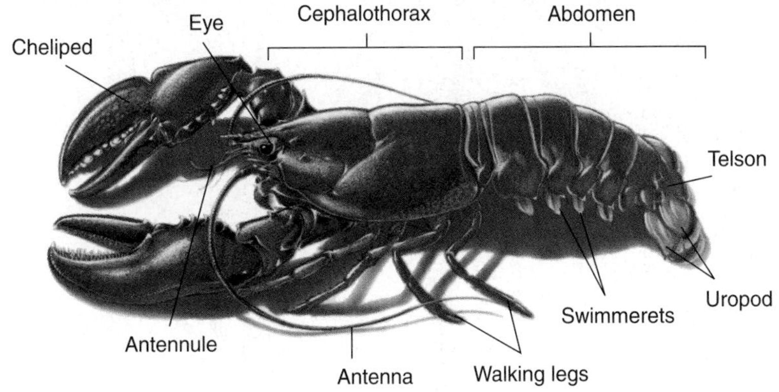

Eye
Cephalothorax
Abdomen
Cheliped
Telson
Uropod
Swimmerets
Antennule
Antenna
Walking legs

Figure 19.27 Body of a lobster, *Homarus americanus.*

Some of the specialized terms used to describe crustaceans are indicated. For example, the head and thorax are fused together into a cephalothorax. Appendages called swimmerets occur in lines along the sides of the abdomen and are used in reproduction and also for swimming. Flattened appendages known as uropods form a kind of compound "paddle" at the end of the abdomen. Lobsters may also have a telson, or tail spine.

Millipedes and Centipedes. Millipedes and centipedes have bodies that consist of a head region followed by numerous similar segments. Centipedes have one pair of legs on each body segment, and millipedes have two (figure 19.28). The centipedes are all carnivorous and feed mainly on insects. The appendages of the first trunk segment are modified into a pair of venomous fangs. In contrast, most millipedes are herbivores, feeding mostly on decaying vegetation. Millipedes live mainly in damp, protected places, such as under leaf litter, in rotting logs, under bark or stones, or in the soil.

Insects. The **insects,** class Insecta, are by far the largest group of arthropods, whether measured in terms of numbers of species or numbers of individuals; as such, they are the most abundant group of eukaryotes on earth. Most insects are relatively small, ranging in size from 0.1 millimeters to about 30 centimeters in length. Insects have three body sections:

1. **Head.** The head of insects is very elaborate, with a single pair of antennae and elaborate mouthparts (figure 19.29). Most insects have compound eyes, which are composed of independent visual units.
2. **Thorax.** The thorax consists of three segments, each of which has a pair of legs. Most insects also have two pairs of wings attached to the thorax. In some insects like flies, one of the pairs of wings has been lost during the course of evolution.
3. **Abdomen.** The abdomen consists of up to 12 segments. Digestion takes place primarily in the stomach, and excretion takes place through organs called Malpighian tubules. Malpighian tubules constitute an efficient mechanism for water conservation and were a key adaptation facilitating invasion of the land by arthropods.

Although primarily a terrestrial group, insects live in every conceivable habitat on land and in freshwater, and a few have even invaded the sea. About 90,000 described species are found in the United States and Canada; many other forms await detection and classification (figure 19.30).

19.10 Arthropods, the most successful animal phylum, have jointed appendages, a rigid exoskeleton, and, in the case of insects, wings.

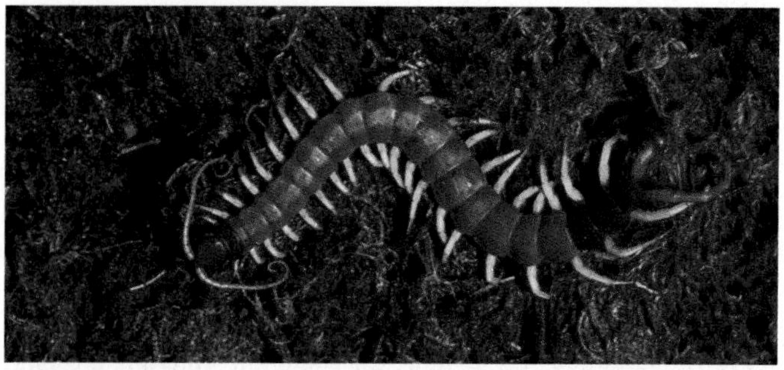

(a)

(b)

Figure 19.28 Centipedes and millipedes.

Centipedes are active predators, whereas millipedes are sedentary herbivores. (*a*) Centipede, *Scolopendra.* (*b*) Millipede, *Sigmoria,* in North Carolina.

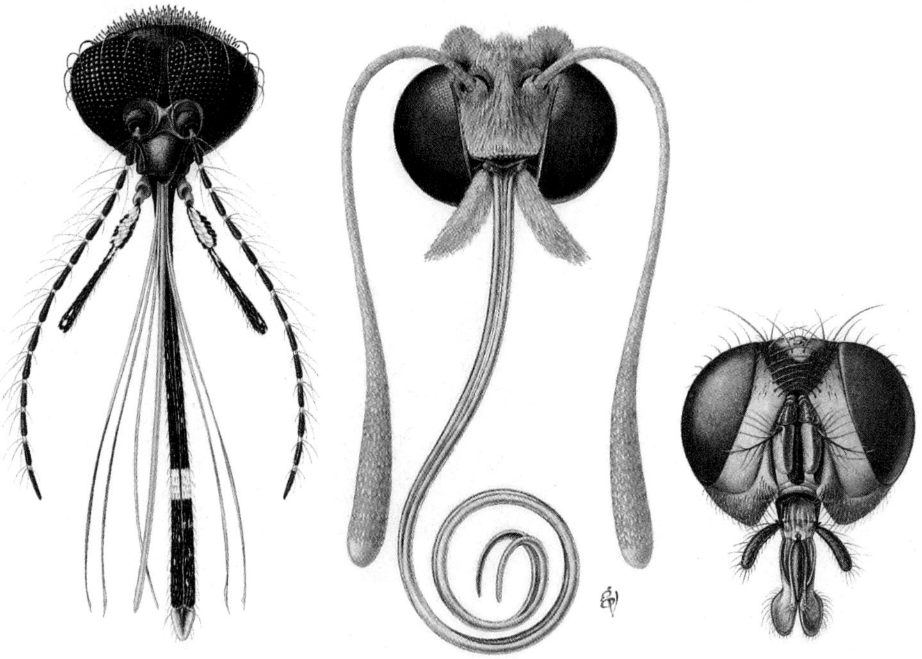

(a) (b) (c)

Figure 19.29 Modified mouthparts in three kinds of insects.

(*a*) Mosquito, *Culex;* mouthparts modified for piercing. (*b*) Alfalfa butterfly, *Colias;* mouthparts modified for sucking nectar from flowers. (*c*) Housefly, *Musca domestica;* mouthparts modified for sopping up liquids.

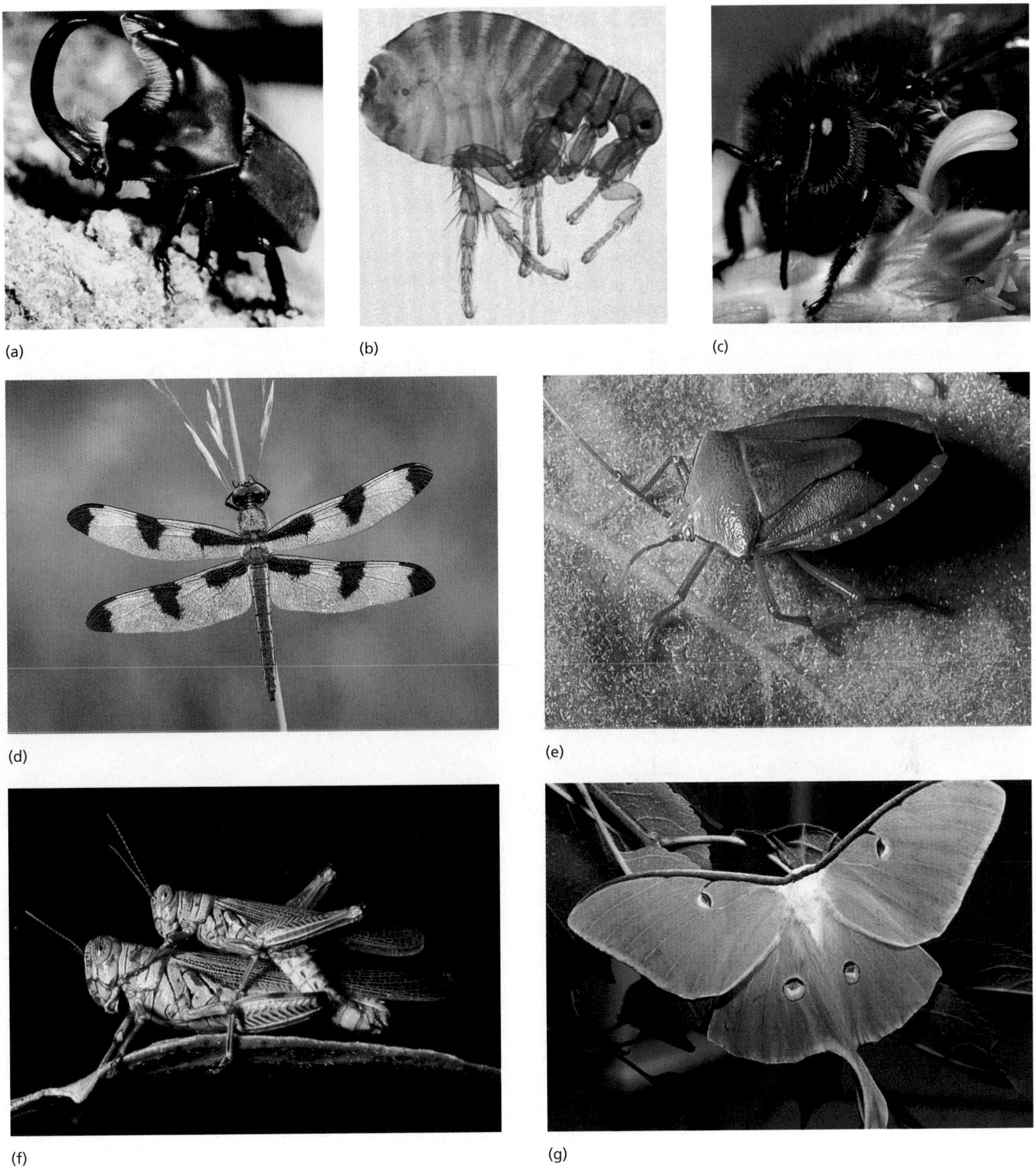
(a)
(b)
(c)
(d)
(e)
(f)
(g)

Figure 19.30 Insect diversity.

(*a*) Some insects have a tough exoskeleton, like this South American scarab beetle, *Dilobderus abderus* (order Coleoptera). (*b*) Human flea, *Pulex irritans* (order Siphonaptera), in California. Fleas are flattened laterally, slipping easily through hair. (*c*) The honeybee, *Apis mellifera* (order Hymenoptera), is a widely domesticated and efficient pollinator of flowering plants. (*d*) This pot dragonfly (order Odonata) has a fragile exoskeleton. (*e*) A true bug, *Edessa rufomarginata* (order Hemiptera), in Panama. (*f*) Copulating grasshoppers (order Orthoptera). (*g*) Luna moth, *Actias luna*, in Virginia. Luna moths and their relatives are among the most spectacular insects (order Lepidoptera).

19.11 Protostomes and Deuterostomes

There are two major kinds of coelomate animals representing two distinct evolutionary lines. All the coelomates we have met so far have essentially the same kind of embryo, starting as a hollow ball of cells, a blastula, which indents to form a two-layer-thick ball with a blastopore opening to the outside. In mollusks, annelids, and arthropods, the mouth (stoma) develops from or near the blastopore. This same pattern of development, in a general sense, is seen in all noncoelomate animals. An animal whose mouth develops in this way is called a **protostome** (from the Greek words *protos,* first, and *stoma,* mouth). If such an animal has a distinct anus or anal pore, it develops later in another region of the embryo. The fact that this kind of developmental pattern is so widespread in diverse phyla makes it likely that it is the original pattern for animals as a whole and that it was characteristic of the common ancestor of all eumetazoan animals.

A second distinct pattern of embryological development occurs in the echinoderms, the chordates, and a few other small, related phyla; it doubtless arose in the common ancestor of this group. In these animals, the anus forms from or near the blastopore, and the mouth forms subsequently on another part of the blastula (figure 19.31). This group of phyla consists of animals that are called the **deuterostomes** (Greek, *deuteros,* second, and *stoma,* mouth). They are clearly related to one another by their shared pattern of embryonic development, which differs radically from the protostome pattern and was most likely derived from it.

Deuterostomes represent a revolution in embryonic development. In addition to the pattern of blastopore formation, deuterostomes differ from protostomes in three other fundamental embryological features:

1. The progressive division of cells during embryonic growth is called *cleavage*. The cleavage pattern relative to the embryo's polar axis determines how the cells array. In nearly all protostomes, each new cell buds off at an angle oblique to the polar axis. As a result, a new cell nestles into the space between the older ones in a closely packed array. This pattern is called **spiral cleavage** because a line drawn through a sequence of dividing cells spirals outward from the polar axis.

   In deuterostomes, the cells divide parallel to and at right angles to the polar axis. As a result, the pairs of cells from each division are positioned directly

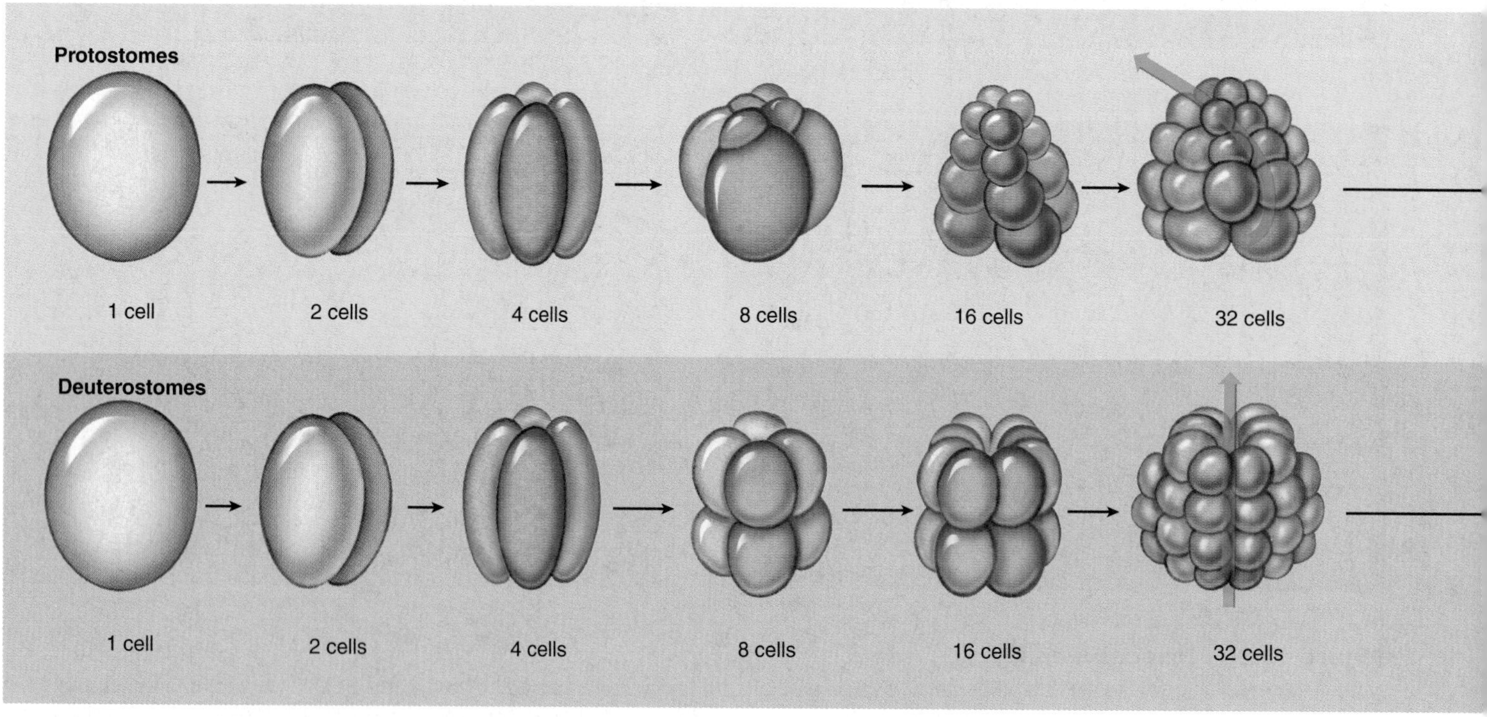
Protostomes
1 cell
2 cells
4 cells
8 cells
16 cells
32 cells
Deuterostomes
1 cell
2 cells
4 cells
8 cells
16 cells
32 cells

Figure 19.31 Embryonic development in protostomes and deuterostomes.
Cleavage of the egg produces a hollow ball of cells called the blastula. Invagination of the blastula produces the blastopore. In protostomes, embryonic cells cleave in a spiral pattern and become tightly packed. The blastopore becomes the animal's mouth, and the coelom originates from a mesodermal split. In deuterostomes, embryonic cells cleave

above and below one another; this process gives rise to a loosely packed array of cells. This pattern is called **radial cleavage** because a line drawn through a sequence of dividing cells describes a radius outward from the polar axis.

2. In protostomes, the developmental fate of each cell in the embryo is fixed when that cell first appears. Even at the four-celled stage, each cell is different, and no one cell, if separated from the others, can develop into a complete animal because the chemicals that act as developmental signals have already been localized in different parts of the egg. Consequently, the cleavage divisions that occur after fertilization separate different signals into different daughter cells. In deuterostomes, on the other hand, the first cleavage divisions of the fertilized embryo produce identical daughter cells, and any single cell, if separated, can develop into a complete organism. The commitment to prescribed developmental pathways occurs later.
3. In all coelomates, the coelom originates from mesoderm. In protostomes, this occurs simply and directly: the cells simply move away from one another as the coelomic cavity expands within the mesoderm. However, in deuterostomes, whole groups of cells usually move around to form new tissue associations. The coelom is normally produced by an evagination of the **archenteron**—the main cavity within the gastrula, also called the primitive gut. This cavity, lined with endoderm, opens to the outside via the blastopore and eventually becomes the gut cavity.

The first abundant and well-preserved animal fossils are about 630 million years old; they occur in the Ediacara series of Australia and similar formations elsewhere. Among these fossils, many represent groups of animals that no longer exist. In addition, these ancient rocks bear evidence of the coelomates, the most advanced evolutionary line of animals, and it is remarkable that their two major subdivisions were differentiated so early. In the coelomates, it is clear that deuterostomes are derived from protostomes. The event, however, occurred very long ago and presumably did not involve groups of organisms that closely resemble any that are living now.

19.11 In deuterostomes, the egg cleaves radially, and the blastopore becomes the animal's anus. In protostomes, the egg cleaves spirally, and the blastopore becomes the mouth.

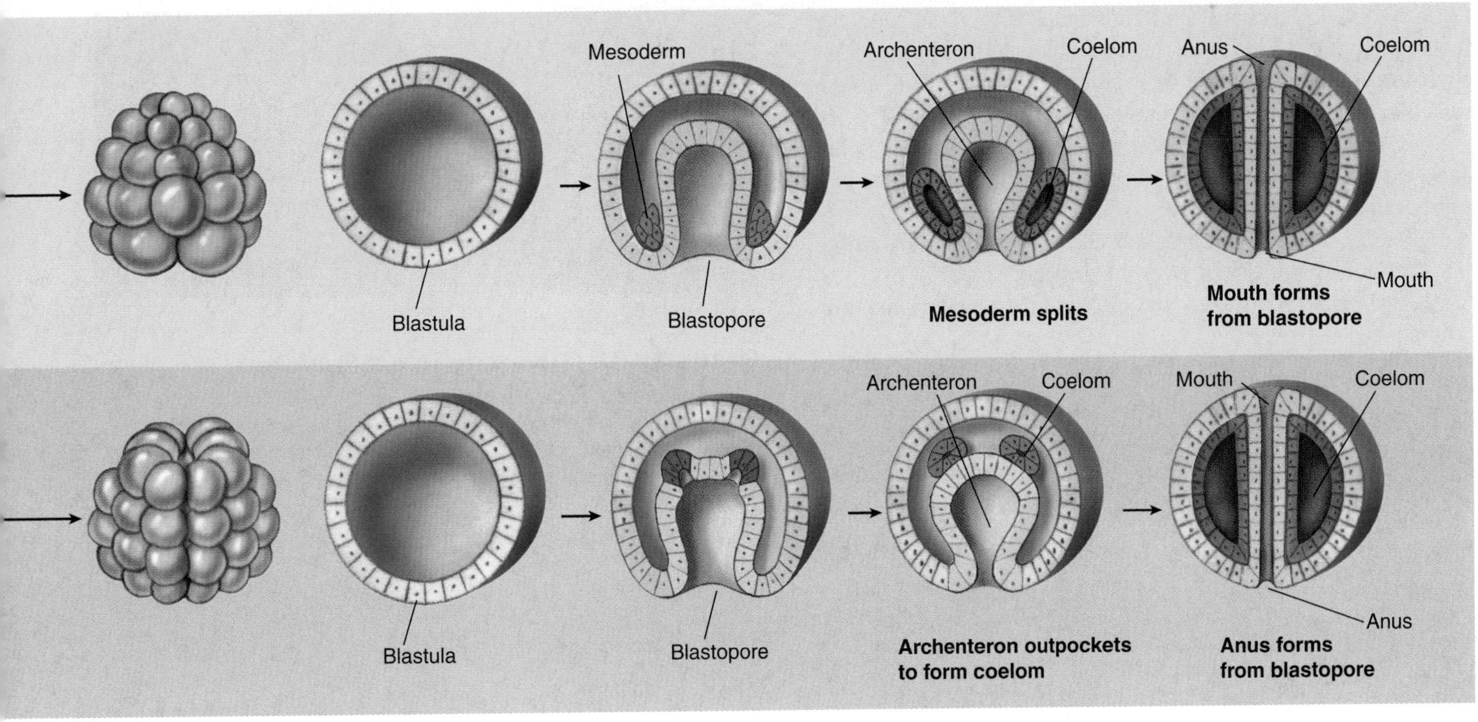

Mesoderm
Archenteron
Coelom
Anus
Coelom
Mouth
Blastula
Blastopore
Mesoderm splits
Mouth forms from blastopore
Archenteron
Coelom
Mouth
Coelom
Anus
Blastula
Blastopore
Archenteron outpockets to form coelom
Anus forms from blastopore

Figure 19.31 (continued)
radially and form a loosely packed array. The blastopore becomes the animal's anus, and the mouth develops at the other end. The coelom originates from an evagination, or outpouching, of the archenteron in deuterostomes.

19.12 Echinoderms: The First Deuterostomes

The first deuterostomes, marine animals called **echinoderms** in the phylum Echinodermata**,** appeared more than 650 million years ago. The term *echinoderm* means "spiny skin" and refers to an **endoskeleton** composed of hard, calcium-rich plates called ossicles just beneath the delicate skin. When they are first formed, the plates are enclosed in living tissue, and so are truly an endoskeleton, although in adults they fuse, forming a hard shell. About 6,000 species of echinoderms are living today, almost all of them on the ocean bottom (figure 19.32). Many of the most familiar animals seen along the seashore are echinoderms, including sea stars (starfish), sea urchins, sand dollars, and sea cucumbers.

The body plan of echinoderms undergoes a fundamental shift during development: All echinoderms are bilaterally symmetrical as larvae but become radially symmetrical as adults. Adult echinoderms have a five-part body plan, easily seen in the five arms of a sea star. Its nervous system consists of a central ring of nerves from which five branches arise—while the animal is capable of complex response patterns, there is no centralization of function, no "brain." Some echinoderms like feather stars have 10 or 15 arms, but always multiples of five.

A key evolutionary innovation of echinoderms is the development of a hydraulic system to aid movement. Called a **water vascular system,** this fluid-filled system is composed of a central ring canal from which five radial canals extend out into the arms. From each radial canal, tiny vessels extend out through short side branches into thousands of tiny, hollow **tube feet.** At the base of each tube foot is a fluid-filled muscular sac that acts as a valve. When a sac contracts, its fluid is prevented from reentering the radial canal and instead is forced into the tube foot, thus extending it. When extended, the tube foot attaches itself to the ocean bottom, often aided by suckers. The sea star can then pull against these tube feet and so haul itself over the seafloor.

Key characteristics of the phylum Echinodermata are summarized in figure 19.33.

19.12 Echinoderms are deuterostomes with an endoskeleton of hard plates, often fused together. Adults are radially symmetrical.

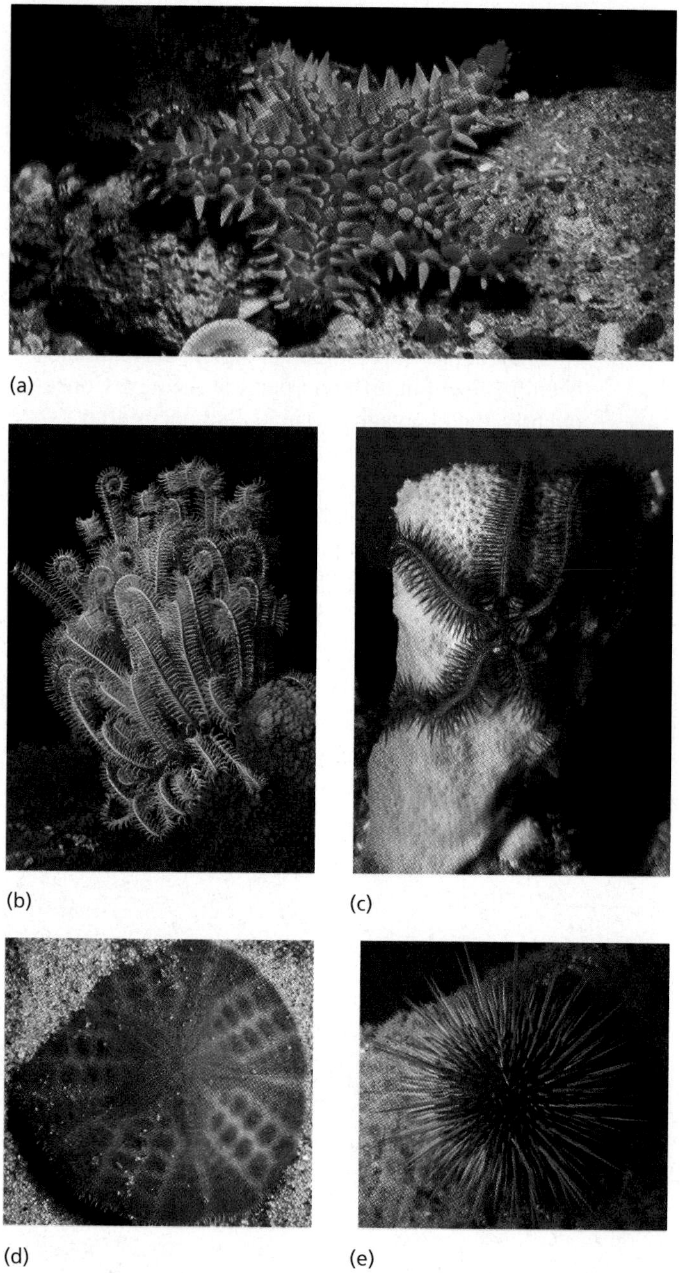
(a)
(b)
(c)
(d)
(e)

Figure 19.32 Diversity in echinoderms.

(a) Sea star, *Oreaster occidentalis* (class Asteroidea), in the Gulf of California, Mexico. (*b*) Feather star (class Crinoidea) on the Great Barrier Reef in Australia. (*c*) Brittle star, *Ophiothrix* (class Ophiuroidea). (*d*) Sand dollar, *Echinarachnius parma.* (e) Giant red sea urchin, *Strongylocentrotus franciscanus.*

Phylum Echinodermata: Echinoderms

Key Evolutionary Innovations: DEUTEROSTOME DEVELOPMENT and ENDOSKELETON

Echinoderms like sea stars (phylum Echinodermata) are coelomates with a **deuterostome** pattern of development. A delicate skin stretches over an **endoskeleton** made of calcium-rich plates, often fused into a continuous, tough spiny layer.

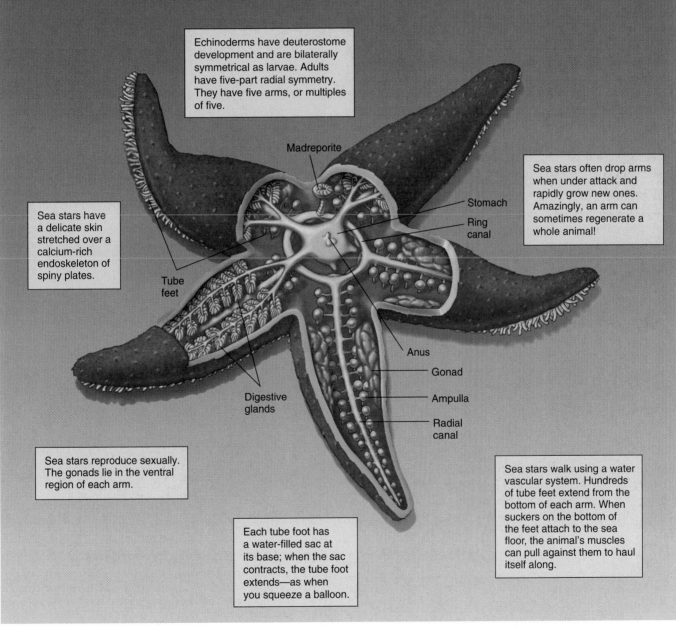

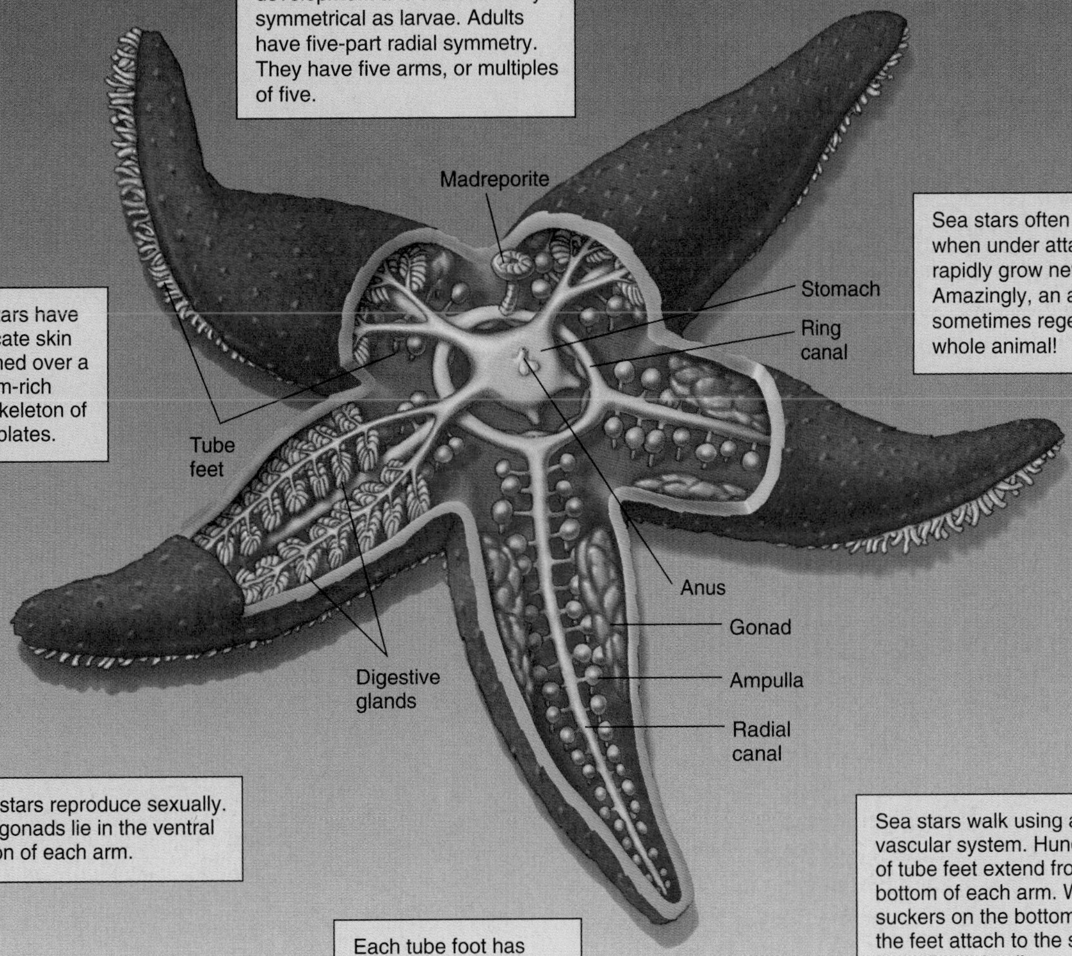
Echinoderms have deuterostome development and are bilaterally symmetrical as larvae. Adults have five-part radial symmetry. They have five arms, or multiples of five.
Madreporite
Stomach
Ring canal
Sea stars often drop arms when under attack and rapidly grow new ones. Amazingly, an arm can sometimes regenerate a whole animal!
Sea stars have a delicate skin stretched over a calcium-rich endoskeleton of spiny plates.
Tube feet
Anus
Gonad
Ampulla
Radial canal
Digestive glands
Sea stars reproduce sexually. The gonads lie in the ventral region of each arm.
Each tube foot has a water-filled sac at its base; when the sac contracts, the tube foot extends—as when you squeeze a balloon.
Sea stars walk using a water vascular system. Hundreds of tube feet extend from the bottom of each arm. When suckers on the bottom of the feet attach to the sea floor, the animal's muscles can pull against them to haul itself along.

Figure 19.33 Phylum Echinodermata: echinoderms.

19.13 Chordates: Improving the Skeleton

General Characteristics of Chordates

Chordates (phylum Chordata) are deuterostome coelomates whose nearest relations in the animal kingdom are the echinoderms, also deuterostomes. Chordates exhibit great improvements in the endoskeleton over what is seen in echinoderms (figure 19.34). The endoskeleton of echinoderms is functionally similar to the exoskeleton of arthropods, in that it is a hard shell that encases the body, with muscles attached to its inner surface. Chordates employ a very different kind of endoskeleton, one that is truly internal. Members of the phylum Chordata are characterized by a flexible rod called a **notochord** that develops along the back of the embryo. Muscles attached to this rod allowed early chordates to swing their backs back and forth, swimming through the water. This key evolutionary innovation, attaching muscles to an internal element, started chordates along an evolutionary path that leads to the vertebrates and for the first time to truly large animals.

The approximately 50,300 species of chordates are distinguished by four principal features:

1. **Notochord.** A stiff, but flexible, rod that forms beneath the nerve cord, between it and the developing gut in the early embryo.
2. **Nerve cord.** A single dorsal (along the back) hollow nerve cord, to which the nerves that reach the different parts of the body are attached.
3. **Pharyngeal slits.** A series of slits behind the mouth into the pharynx, which is a muscular tube that connects the mouth to the digestive tract and windpipe.
4. **Postanal tail.** Chordates have a postanal tail, a tail that extends beyond the anus, at least during their embryonic development. Nearly all other animals have a terminal anus.

All chordates have all four of these characteristics at some time in their lives. For example, human embryos have pharyngeal slits, a nerve cord, and a notochord as embryos.

Key characteristics of the phylum Chordata are summarized in figure 19.35.

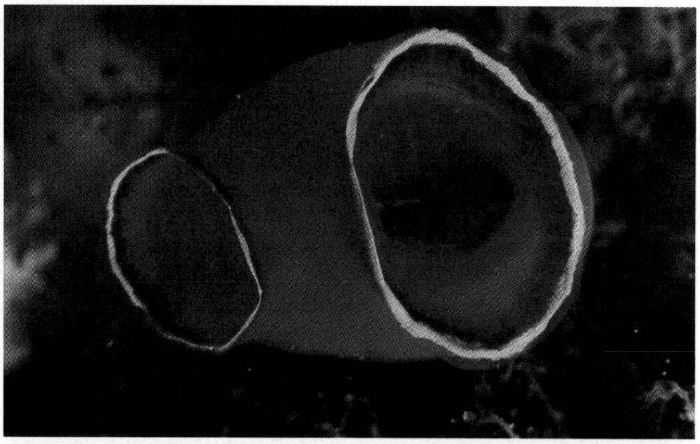

(a)

(b)

(c)

Figure 19.34 Diversity in chordates.

(*a*) A beautiful blue and gold tunicate. (*b*) Two lancelets, *Branchiostoma lanceolatum,* partly buried in shell gravel, with their anterior ends protruding. The muscle segments are clearly visible in this photograph. The numerous square, pale yellow objects along the side of the body are gonads, indicating that these are male lancelets. (*c*) Terrestrial vertebrates are among the most successful animal groups. This lion, stalking a potential dinner, is both large and very mobile, two traits characteristic of many terrestrial vertebrates.

Phylum Chordata: Chordates

Key Evolutionary Innovation: NOTOCHORD

Vertebrates, tunicates, and lancelets are chordates (phylum Chordata), coelomate animals with a stiff, but flexible, rod, the **notochord,** that acts to anchor internal muscles, permitting rapid body movements. Chordates also possess **pharyngeal slits** (relics of their aquatic ancestry) and a dorsal **hollow nerve cord.** In vertebrates, the notochord is replaced during embryonic development by the vertebral column.

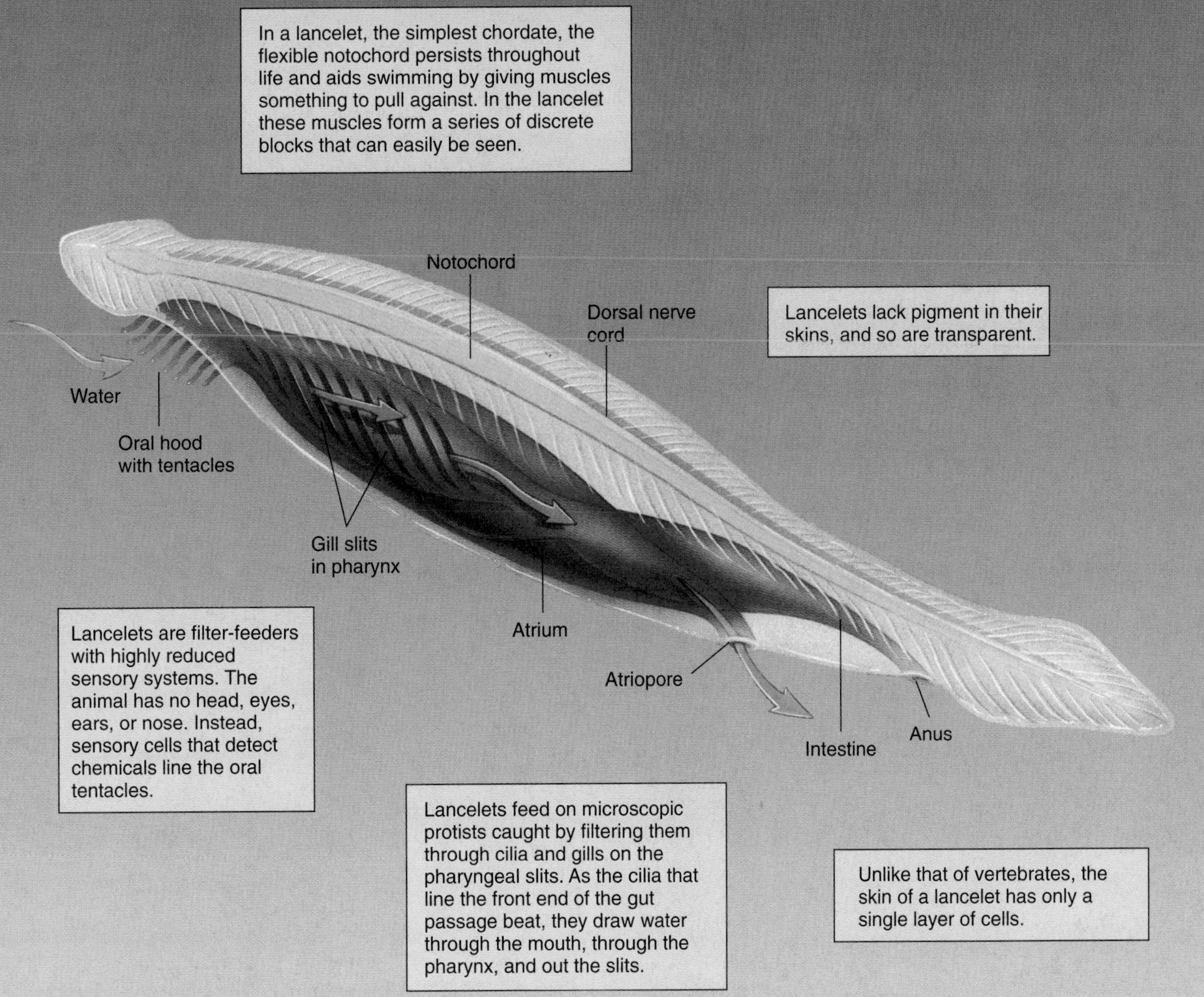
In a lancelet, the simplest chordate, the flexible notochord persists throughout life and aids swimming by giving muscles something to pull against. In the lancelet these muscles form a series of discrete blocks that can easily be seen.
Notochord
Dorsal nerve cord
Lancelets lack pigment in their skins, and so are transparent.
Water
Oral hood with tentacles
Gill slits in pharynx
Atrium
Atriopore
Intestine
Anus
Lancelets are filter-feeders with highly reduced sensory systems. The animal has no head, eyes, ears, or nose. Instead, sensory cells that detect chemicals line the oral tentacles.
Lancelets feed on microscopic protists caught by filtering them through cilia and gills on the pharyngeal slits. As the cilia that line the front end of the gut passage beat, they draw water through the mouth, through the pharynx, and out the slits.
Unlike that of vertebrates, the skin of a lancelet has only a single layer of cells.

Figure 19.35 Phylum Chordata: chordates.

Vertebrates

In their body plan, chordates are segmented, and distinct blocks of muscles can be seen clearly in many forms (figure 19.36). Many chordates have jointed appendages. With the exception of tunicates and a small group of fish-like marine animals, the lancelets, all chordates are **vertebrates.** Vertebrates differ from tunicates and lancelets in two important respects:

1. **Backbone.** In vertebrates, the notochord becomes surrounded and then replaced during the course of the embryo's development by a bony vertebral column, a tube of hollow bones called vertebrae that encloses the dorsal nerve cord like a sleeve and protects it (figure 19.37).
2. **Head.** All vertebrates except the earliest fishes have a distinct and well-differentiated head, with a skull and brain. For this reason, the vertebrates are sometimes called the craniate chordates (Greek, *kranion,* skull).

All vertebrates have an internal skeleton made of bone or cartilage against which the muscles work. This endoskeleton makes possible the great size and extraordinary powers of movement that characterize the vertebrates. An interesting feature of vertebrates and other chordates is that they have a tail that extends beyond the anus, at least during their embryonic development; nearly all other animals have a terminal anus.

The endoskeleton of most vertebrates is made of **bone** (in a few, like sharks, the bone is replaced with more flexible cartilage). Bone is a special form of tissue containing fibers of the protein collagen that are coated with a calcium phosphate salt. Bone is formed in two stages. First, collagen is laid down in a matrix of fibers along lines of stress to provide flexibility, and then calcium minerals impregnate the fibers, providing rigidity. The great advantage of bone over chitin as a structural material is that bone is strong without being brittle.

The evolution and characteristics of the major groups of vertebrates are discussed in detail in the next chapter.

19.13 Chordates have a notochord at some stage of their development. In adult vertebrates, the notochord is replaced by a backbone.

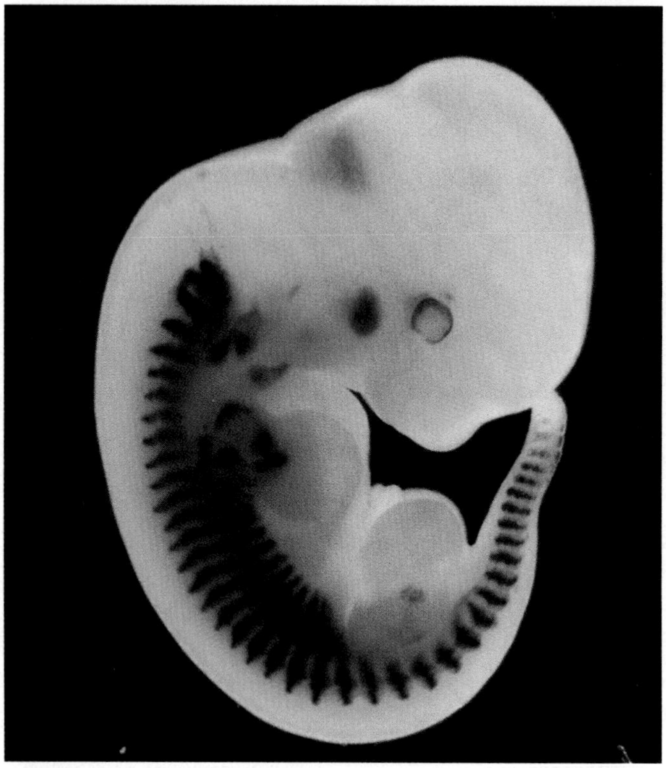

Figure 19.36 A mouse embryo.

At 11.5 days of development, the muscle is already divided into segments called somites (stained *dark* in this photo), reflecting the fundamentally segmented nature of all chordates.

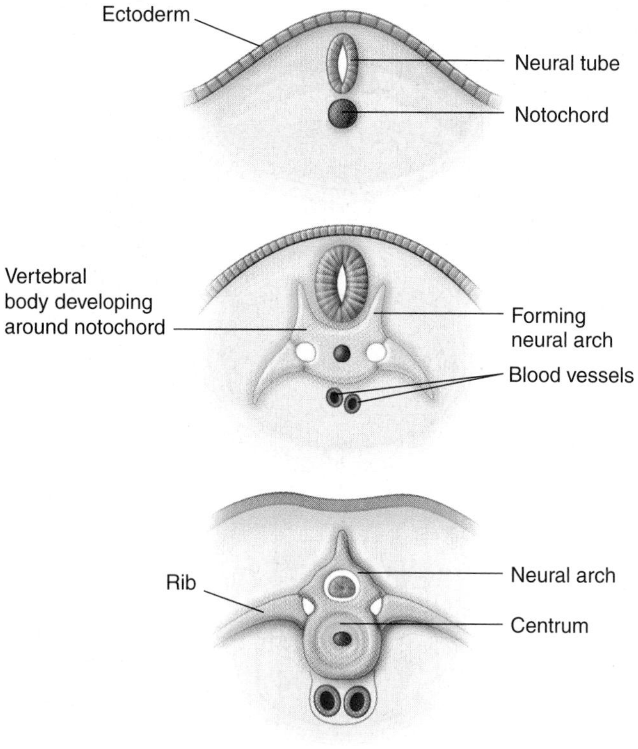
Ectoderm
Neural tube
Notochord
Vertebral body developing around notochord
Forming neural arch
Blood vessels
Rib
Neural arch
Centrum

Figure 19.37 Embryonic development of a vertebrate.

During the course of development, the flexible notochord is surrounded and eventually replaced by a bony covering, the vertebral column, that protects the neural tube and provides a strong, flexible rod against which the muscles pull when the animal swims or moves.

CONCEPT REVIEW

1. The phylum of animals that contains the largest number of named species is
   a. Chordata.
   b. Mammals.
   c. Annelida.
   d. Arthropoda.

2. Choanocytes of sponges bear a striking resemblance to the _________, members of one group of protists.
   a. acoelomates
   b. choanoflagellates
   c. cnidarians
   d. collar cells

3. All animals other than sponges have both _________ and tissues and are called eumetazoans.
   a. multicellularity
   b. choanocytes
   c. symmetry
   d. circulatory systems

4. Which of the following is *not* present in flatworms?
   a. uterus or testes
   b. organs
   c. bilateral symmetry
   d. body cavity

5. Where does a coelom originate?
   a. in the endoderm
   b. between the endoderm and the mesoderm
   c. in the mesoderm
   d. between the mesoderm and the ectoderm

6. Octopuses and squids are in the phylum _________, along with snails and clams.
   a. Echinodermata
   b. Mollusca
   c. Cnidaria
   d. Crustacea

7. Which of the following are *not* arthropods?
   a. earthworms
   b. crayfish
   c. spiders
   d. butterflies

8. Which two phyla listed here are deuterostomes?
   a. Platyhelminthes
   b. Chordata
   c. Echinodermata
   d. Arthropoda

9. Four characteristics that distinguish the chordates from other animals are
   a. a single, hollow dorsal nerve cord.
   b. a notochord at some time in development.
   c. pharyngeal slits at some time in development.
   d. segmentation.
   e. a postanal tail.
   f. a nervous system.

10. Unlike tunicates and lancelets, vertebrates have a(n)
    a. nervous system.
    b. coelom.
    c. exoskeleton.
    d. backbone.

11. The three distinct layers of cells that form in the embryos of all eumetazoans are the _________, _________, and _________.

12. The process of evolving a definite head area is called _________.

13. _________, a type of roundworm, are usually microscopic and are found in large numbers in soil.

14. The key evolutionary advance in annelids is _________.

15. Animals in which the mouth develops from or near the blastopore in the embryo are called _________.

16. Tunicates belong to the phylum _________.

CHALLENGE YOURSELF

1. Why are the most primitive animals encountered in the sea?
2. In what ways is an earthworm more complex than a flatworm?
3. What is the evolutionary advantage of having a shell?
4. What are the key adaptations that facilitated the invasion of the land by arthropods? Why is the phylum so successful?
5. Why is it believed that echinoderms and chordates, which are so dissimilar, are members of the same evolutionary line?

19 eBRIDGE

Reinforcing Key Points

Introduction to the Animals

19.1 Some General Features of Animals

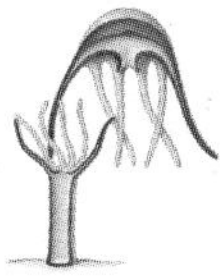

The Simplest Animals

19.2 Sponges: Animals Without Tissues

19.3 Cnidarians: Tissues Lead to Greater Specialization

The Advent of Bilateral Symmetry

19.4 Radial Versus Bilateral Symmetry

19.5 Solid Worms: Internal Organs

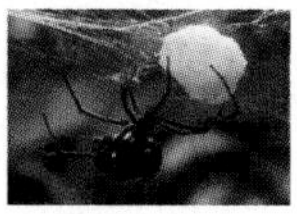

The Advent of a Body Cavity

19.6 Pseudocoelom Versus Coelom

19.7 Roundworms: Pseudocoelomates

19.8 Mollusks: Coelomates

19.9 Annelids: The Rise of Segmentation

19.10 Arthropods: Advent of Jointed Appendages

Redesigning the Embryo

19.11 Protostomes and Deuterostomes

19.12 Echinoderms: The First Deuterostomes

19.13 Chordates: Improving the Skeleton

Electronic Learning

Visual Learning

Animations

Phylogenetic Tree of the Animal Kingdom

Simple Sponge Anatomy

Anatomy of *Hydra*

Obelia Structure and Life Cycle

Symmetry in Nature

Roundworm Anatomy

Bivalve Diversity

Earthworm, *Lumbricus*

Male Crayfish, *Cambarus*

Female Grasshopper, *Romalea*

Echinoderms

Lancelet, *Branchiostoma*

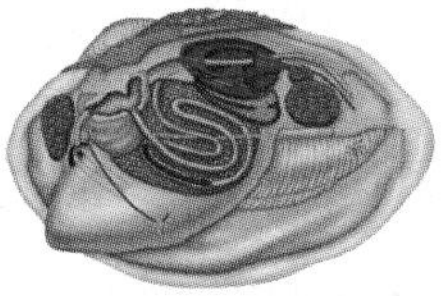

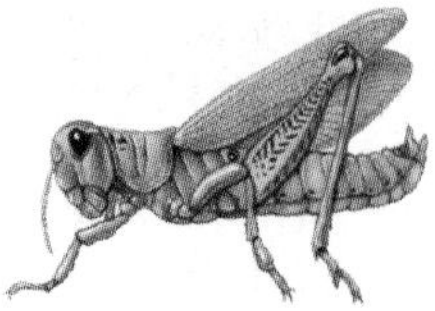

Author's Corner

Urban Wildlife. When one considers the wondrous diversity of the animal phyla, it is easy to slip into thinking of the world's animals as a static collection of different creatures, sort of a great big zoo that we look at but are not part of. Actually, animals interact with urban humans all the time, and always have. Dogs, rats, and cockroaches have evolved in concert with people. In urban areas today, evolutionary interactions abound. The drastic decline of our nation's songbirds due to loss of breeding and overwintering grounds are one example. Another, no longer easy to ignore, is the successful adaptation of deer herds to suburban living.

1. Migrating woodland songbirds are in steep decline.
2. The fastest-growing segment of our suburbs may be the deer population.
3. The problem of exploding deer populations has no easy solution.

The Living World

Virtual Classroom

A Parade of Phyla

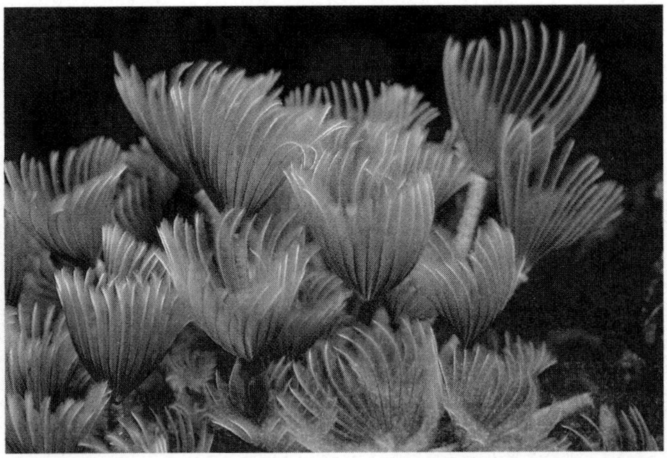

Animals, millions of species of them, are among the most abundant living things. Found in every conceivable habitat, they bewilder us with their diversity. Animals are the eaters of the earth. They are heterotrophs—all animals depend directly or indirectly on photosynthetic plants, algae, and bacteria for their nourishment. Animals are unusual among multicellular organisms in being able to move from place to place in search of food. The evolution of the animal phyla involved five key transitions in body plan: 1. Evolution of Tissues (all animals but sponges have distinct tissues); 2. Evolution of Bilateral Symmetry (jellyfish are radially symmetrical; most other animals with tissues are bilaterally symmetrical); 3. Evolution of a Body Cavity (solid worms lack a body cavity, and nematodes have a simple one, while mollusks have a sophisticated one called a coelom); 4. Evolution of Segmentation (segmentation underlies the body plan of annelid worms, arthropods, and chordates); and 5. Evolution of Deuterostome Development (echinoderms and chordates organize their embryos in a novel fashion).

Virtual Lab

In Pursuit of Preserving Freshwater Mussels

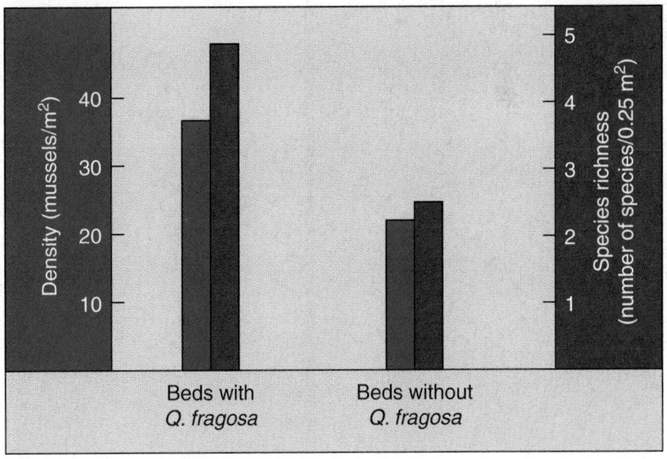
Density (mussels/m²)
10
20
30
40
Species richness (number of species/0.25 m²)
1
2
3
4
5
Beds with Q. fragosa
Beds without Q. fragosa

The lakes and rivers of North America are home to nearly 300 species and subspecies of freshwater mussels, the largest number of species of freshwater mussels in the world. Unfortunately, nearly seventy percent of the taxa are either extinct, endangered, or threatened. To reduce the risk of extinction and maintain freshwater mussel diversity, it is necessary to learn how habitat and community composition affects the distribution of these species.

Nels Troelstrup, Jr., now at South Dakota State University, working in collaborative effort with colleagues from Macalester College and the University of Minnesota, both in St. Paul, examined the physical and biological factors that might affect the distribution of the endangered winged mapleleaf mussel, *Quadrula fragosa*. Their studies reveal that physical factors have some influence on the distribution of the winged mapleleaf mussel, but bed density and species richness significantly influence populations of *Quadrula fragosa*. This suggests that any interventions that improve freshwater mussel habitats in general will also benefit *Q. fragosa.*

Quizzes | Further Reading | Essential Study Partner | Links | BioCourse.com

CHAPTER 20

HISTORY OF THE VERTEBRATES

THE FAR SIDE® By GARY LARSON
© 1983 FarWorks, Inc. All Rights Reserved/Dist. by Creators Syndicate
Larson

CHAPTER 20

History of the Vertebrates

CHAPTER OVERVIEW

Overview of Vertebrate Evolution

- All major groups of living organisms except plants arose in the early Paleozoic era.
- Arthropods and vertebrates invaded land from the sea soon after plants and fungi.
- Periodic mass extinctions have occurred; the greatest, at the end of the Paleozoic era, rendered 90% of all species extinct.
- In the Mesozoic era, reptiles were the dominant terrestrial vertebrate, particularly dinosaurs.
- At the end of the Mesozoic era, dinosaurs disappeared, and mammals took their place.

The Parade of Vertebrates

- Marine and aquatic vertebrates extract oxygen from water very efficiently with gills.
- The first vertebrates were jawless fishes.
- The three key characteristics that led to the success of the bony fishes were the swim bladder, the lateral line system, and the operculum.
- Amphibians, the first vertebrates to live on land, evolved legs, lungs, and the pulmonary vein.
- Reptiles were well adapted to living on land, with dry skin and watertight eggs.
- Birds evolved feathers, hollow bones, and very efficient lungs.
- Most present-day mammals are placental mammals.

20.1 The Paleozoic Era

When scientists first began to study and date fossils, they had to find some way to organize the different time periods from which the fossils came. They divided the earth's past into large blocks of time called **eras.** Eras are further subdivided into smaller blocks of time called **periods,** and some periods, in turn, are subdivided into **epochs,** which can be divided into **ages.**

Virtually all of the major groups of animals that survive at the present time originated in the sea at the beginning of the **Paleozoic era,** during or soon after the Cambrian period (590 to 505 M.Y.A.). Thus, the diversification of animal life on earth is basically a marine record, and the fossils from the Paleozoic era all originated in the sea (figure 20.1).

Many of the animal phyla that appeared in the Cambrian period have no living relatives. Their fossils indicate that this was a period of experimentation with different body forms

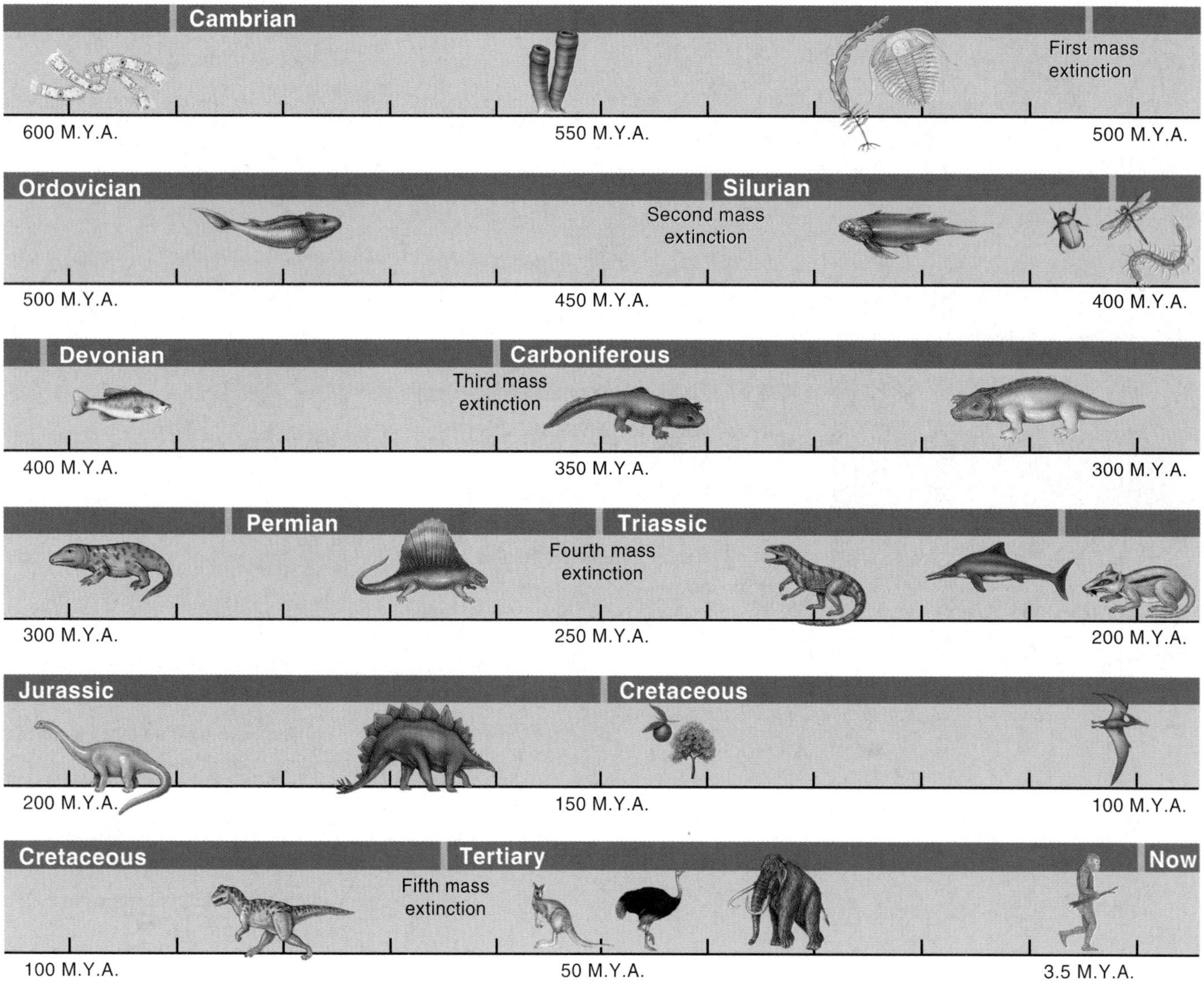

Cambrian
First mass extinction
600 M.Y.A.
550 M.Y.A.
500 M.Y.A.
Ordovician
Silurian
Second mass extinction
500 M.Y.A.
450 M.Y.A.
400 M.Y.A.
Devonian
Carboniferous
Third mass extinction
400 M.Y.A.
350 M.Y.A.
300 M.Y.A.
Permian
Triassic
Fourth mass extinction
300 M.Y.A.
250 M.Y.A.
200 M.Y.A.
Jurassic
Cretaceous
200 M.Y.A.
150 M.Y.A.
100 M.Y.A.
Cretaceous
Tertiary
Now
Fifth mass extinction
100 M.Y.A.
50 M.Y.A.
3.5 M.Y.A.

Figure 20.1 An evolutionary timeline.

Vertebrates evolved in the seas about 470 million years ago (M.Y.A.) and invaded land about 100 million years later. Dinosaurs and mammals evolved in the Triassic period about 220 million years ago. Dinosaurs dominated life on land for over 150 million years, until their sudden extinction 65 million years ago left mammals free to flourish.

and ways of life, some of which ultimately led to the contemporary phyla of animals and others to extinction. For example, the trilobites (figure 20.2) appear to be the ancestors of at least one living group, the horseshoe crabs, whereas the ammonites, which were abundant 100 million years ago, have no surviving descendants.

The first vertebrates evolved about 470 million years ago in the oceans—fishes without jaws. They didn't have paired fins either—many of them looked something like a flat hotdog with a hole at one end and a fin at the other. For 100 million years, a parade of different kinds of fishes were the only vertebrates on earth. They became the dominant creatures in the sea, some bigger than cars.

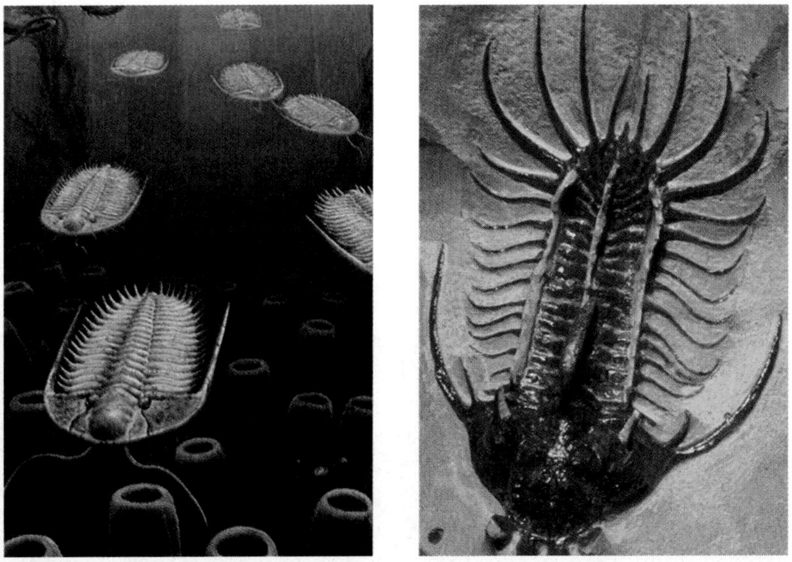

(a) (b)

Figure 20.2 Life in the Cambrian.

(*a*) Trilobites are shown swimming in this reconstruction of a community of marine organisms in the late Cambrian period, 550 to 500 million years ago. Trilobites were early members of the arthropod phylum. On the seafloor is a colony of sponges, members of another ancient animal phylum. (*b*) A fossil trilobite.

Invasion of the Land

Only a few of the animal phyla that evolved in the Cambrian seas have invaded the land successfully; most others have remained exclusively marine. The first organisms to colonize the land were plants, about 410 million years ago. The ancestors of plants were specialized members of a group of photosynthetic protists known as the green algae. It seems probable that plants first occupied the land in symbiotic association with fungi, as discussed in chapter 15.

The second major invasion of the land, and perhaps the most successful, was by the arthropods, a phylum of hard-shelled animals with jointed legs and a segmented body. This invasion of the land occurred soon after the evolution of the plants.

Vertebrates initiated the third major invasion of the land. The first vertebrates to live on land were the amphibians, represented today by frogs, toads, salamanders, and caecilians. The earliest amphibians known are from the Devonian period, and among their descendants are the reptiles which became the ancestors of the dinosaurs, birds, and mammals.

The colonization of the land by all four of the major groups of multicellular organisms—plants, fungi, arthropods, and vertebrates—within a few tens of millions of years of one another is probably related to the development of suitable environmental conditions, such as the formation of a layer of ozone (O_3) in the atmosphere.

Mass Extinctions

The history of life on earth era has been marked by periodic major episodes of extinction, called **mass extinctions.** Four of these mass extinctions occurred during the Paleozoic, the first of them near the end of the Cambrian period about 505 million years ago. At that time, most of the existing families of trilobites (see figure 20.2), a very common type of marine arthropod, became extinct. Additional mass extinctions occurred about 438 and 360 million years ago.

The fourth and most drastic mass extinction in the history of life on earth happened during the last 10 million years of the Permian period, marking the end of the Paleozoic era. It is estimated that 96% of all species of marine animals that were living at that time became extinct! All of the trilobites disappeared forever. Brachiopods, marine animals resembling mollusks but with a different filter-feeding system, were extremely diverse and widespread during the Permian; only a few species survived. Bryozoans, marine filter feeders that formed coral-like colonies in oceans throughout the world in the Permian, became rare afterward.

Mass extinctions left vacant many ecological opportunities, and for this reason they were followed by rapid evolution among the relatively few plants, animals, and other organisms that survived the extinction. Little is known about the causes of major extinctions. Possible causes include volcanic eruptions or a meteor impact. In the case of the Permian mass extinction, some scientists argue that the extinction was brought on by a gradual accumulation of carbon dioxide in ocean waters, the result of large scale volcanism brought on by the collision of the earth's landmasses to form the single large "super-continent" of Pangaea. Such an increase in carbon dioxide would have severely disrupted the ability of animals to carry out metabolism and form their shells.

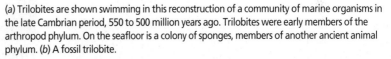

20.1 The diversification of the animal phyla occurred in the sea. Only two animal phyla have invaded the land successfully, arthropods and chordates (the vertebrates).

20.2 The Mesozoic Era

The **Mesozoic era** (248 to 65 M.Y.A.) was a time of intensive evolution of terrestrial plants and animals. The major evolutionary lines on land had been established earlier. However, the evolutionary expansion of these lines, which led to the major groups of organisms living today, occurred just before and during the Mesozoic era, starting with the great evolutionary leap that brought the amphibians onto the land. Frogs and other much larger amphibians that are now extinct were the first vertebrates to live successfully on land—as they still do. Amphibians in turn gave rise to the first reptiles about 300 million years ago. Within 50 million years, the reptiles, better suited than amphibians to living out of water, replaced them as the dominant land animal on earth.

With the success of the reptiles, vertebrates truly came to dominate the surface of the earth. Many kinds of reptiles evolved, from those smaller than a chicken to others bigger than a truck (figure 20.3). Some flew, and others swam. From among them eventually evolved the three great lines of terrestrial vertebrates: dinosaurs, birds, and mammals. Although dinosaurs and mammals appear at about the same time in the fossil record, 220 million years ago, the dinosaurs quickly filled the evolutionary niche for large animals. For over 150 million years, dinosaurs dominated the face of the earth (figure 20.4). (Think of it—over a *million centuries*! If

Figure 20.3 Some dinosaurs were truly enormous.

Here the paleontologist Jim Jensen is standing by a reconstruction of the leg of a sauropod fossil he found. Sauropods were herbivores that had enormous barrel-shaped bodies with heavy columnlike legs and very long necks and tails. Some weighed 55 tons, stood 10 meters tall, and were over 30 meters long.

Carboniferous
Permian
Triassic

Figure 20.4 Dinosaurs.

Some of the remarkable diversity of dinosaurs is shown in this famous reconstruction from the Peabody Museum, Yale University. This painting covers a span of approximately 320 million years during the later Paleozoic and Mesozoic eras, ending 65 million years ago at the end of the Cretaceous period. Throughout this vast period of time, the remarkable you could look back to that distant time and have each century flash by your eye in a brief minute, it would take a thousand days to see it all.) During this period, the largest mammal was no bigger than a cat. Dinosaurs reached the height of their diversification and dominance of the land during the Jurassic and Cretaceous periods.

The Mesozoic era has traditionally been divided into three periods: the Triassic, the Jurassic, and the Cretaceous. Because of the major extinction that ended the Paleozoic, only 4% of species survived into the Mesozoic. These survivors gave rise to new species that radiated to form new genera and families. Both on land and in the sea, the number of species of almost all groups of organisms has been climbing steadily for the past 250 million years and is now at an all-time high.

This extended recovery from the great Permian extinction had one interruption. About 65 million years ago at the end of the Cretaceous period, dinosaurs disappeared, along with the flying reptiles called pterosaurs (figure 20.5) and the great marine reptiles. This extinction marks the end of the Mesozoic era. Mammals quickly took their place, becoming in their turn abundant and diverse—as they are today.

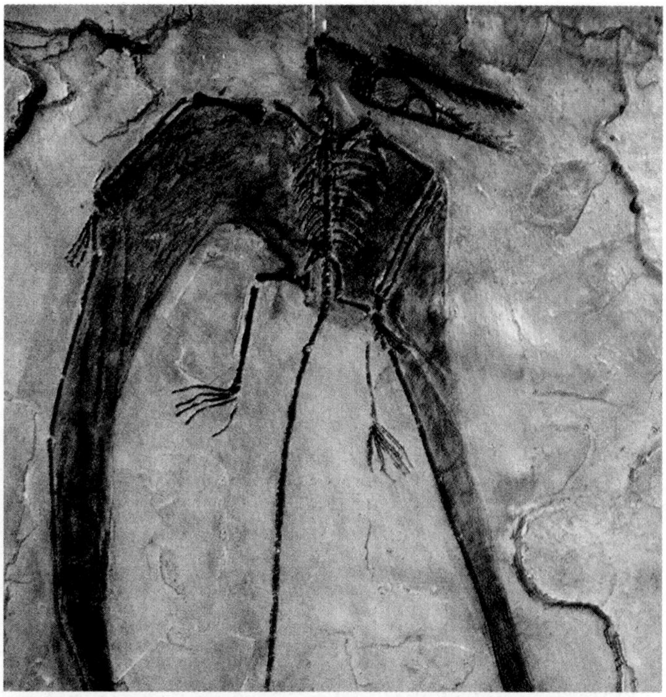

Figure 20.5 An extinct flying reptile.

Pterosaurs, such as the one pictured, became extinct with the dinosaurs about 65 million years ago. Flight has evolved three separate times among vertebrates; however, birds and bats are the only representatives still with us.

Jurassic Cretaceous

Figure 20.4 (continued)

increase in structural complexity and overall diversity of the dinosaurs can be seen, until they abruptly became extinct, giving way to the dominant mammals of the Cenozoic era. Flowering plants can be seen for the first time at the *right-hand side* of the illustration. The names of the geological periods are given along the bottom of the painting.

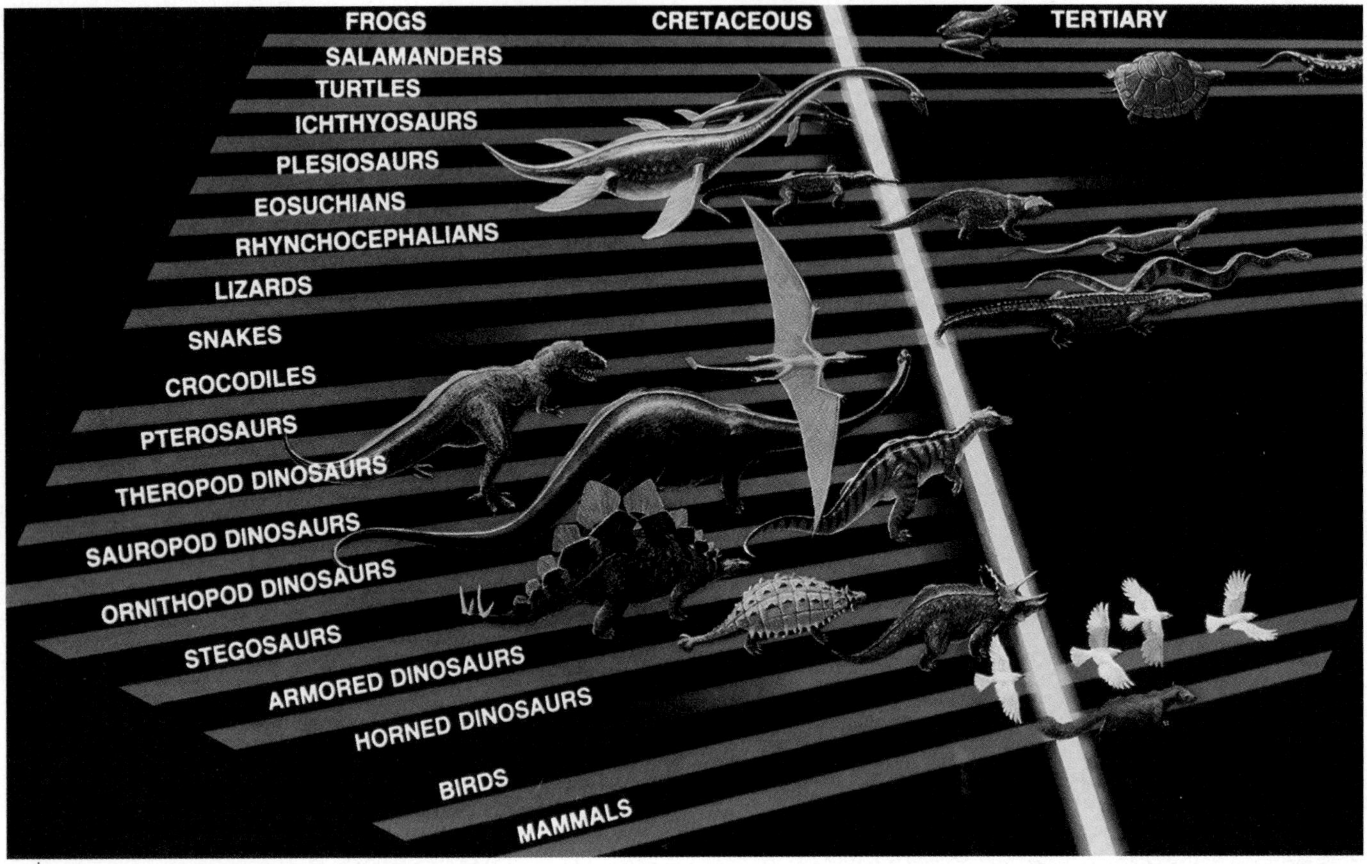

FROGS
CRETACEOUS
TERTIARY
SALAMANDERS
TURTLES
ICHTHYOSAURS
PLESIOSAURS
EOSUCHIANS
RHYNCHOCEPHALIANS
LIZARDS
SNAKES
CROCODILES
PTEROSAURS
THEROPOD DINOSAURS
SAUROPOD DINOSAURS
ORNITHOPOD DINOSAURS
STEGOSAURS
ARMORED DINOSAURS
HORNED DINOSAURS
BIRDS
MAMMALS

Figure 20.6 Extinction of the dinosaurs.

The dinosaurs became extinct 65 million years ago (*yellow line*) in a major extinction event that also eliminated all the great marine reptiles (plesiosaurs and ichthyosaurs) as well as the largest of the primitive land mammals. The birds and smaller mammals survived and went on to occupy the aerial and terrestrial modes of living in the environment left vacant by the dinosaurs. Crocodiles, small lizards, and turtles also survived, but reptiles never again achieved the diversity of the Cretaceous period.

What Happened to the Dinosaurs?

Dinosaurs disappeared abruptly from the fossil record 65 million years ago (figure 20.6), their extinction marking the end of the Mesozoic era. Why? Many explanations have been advanced, including volcanoes and infectious disease. The most widely accepted theory assigns the blame to an asteroid that hit the earth then. Physicist Luis W. Alvarez and his associates discovered that the usually rare element iridium was abundant in many parts of the world in a thin layer of sediment that marked the end of the Cretaceous period. Iridium is rare on earth but common in meteorites. Alvarez and his colleagues have proposed that if a large meteorite 10 kilometers in diameter struck the surface of the earth 65 million years ago, a dense cloud would have been thrown up. The cloud would have been rich in iridium, and as its particles settled, the iridium would have been incorporated into the layers of sedimentary rock being deposited at the time. By darkening the world, the cloud would have greatly slowed or temporarily halted photosynthesis and driven many kinds of organisms to extinction.

The Alvarez hypothesis has become widely accepted among biologists, although it remains controversial among some. These holdouts argue that it is not certain that dinosaurs became extinct suddenly, as they would have by a meteorite collision, and question whether other kinds of animals and plants show the types of patterns that would have been expected as the result of a meteorite collision. The issue was largely settled in the minds of most scientists when an impact crater was discovered in the sea just off the coast of the Yucatán peninsula. For hundreds of kilometers in all directions are signs of the meteor's impact, including quartz crystals with shock patterns that could only have been produced by an enormous impact (they are produced, for example, as a by-product of nuclear tests). Large amounts of soot were deposited in rock worldwide at the time of the extinction, indicating very widespread burning. At what date does radiodating place the Yucatán impact? Sixty-five million years ago.

20.2 The Mesozoic era was the age of dinosaurs. They became extinct abruptly 65 million years ago, probably as the result of a meteor impact.

20.3 The Cenozoic Era

The relatively warm, moist climates of the early **Cenozoic era** (65 M.Y.A. to present) have gradually given way to today's colder and drier climate. The first half of the Cenozoic was very warm, with junglelike forests at the poles. With the extinction of the dinosaurs and many other organisms and this change in climate, new forms of life were able to invade new habitats. Mammals diversified from earlier, small nocturnal forms to many new forms. Most present-day orders of mammals appeared at this time, a period of great diversity.

Then the world's climate began to cool, and ice caps formed at the poles. As glaciation in Antarctica became fully established by about 13 million years ago, regional climates cooled dramatically. A series of so-called ice ages followed, the most recent only a few million years ago. Many very large mammals evolved during the ice ages, including mastodons, mammoths (table 20.1), saber-toothed tigers, and enormous cave bears.

The Antarctic ice mass that formed as a result of these glaciations has made the world's climate cooler near the poles, warmer near the equator, and drier in the middle latitudes than ever before. In general, forests covered most of the land area of continents, except for Antarctica, until about 15 million years ago, when the forests began to recede rapidly and modern plant communities appeared. During the past several million years, the formation of extensive deserts in northern Africa, the Middle East, and India made migration between Africa and Asia very difficult for organisms of tropical forests. Overall, the Cenozoic era has been characterized by sharp differences in habitat, even within small areas, and the regional evolution of distinct groups of plants and animals. These factors have facilitated the rapid formation of many new species.

20.3 We live in the Cenozoic era, the Age of Mammals. Many large mammals common in the ice ages are now extinct.

TABLE 20.1 SOME GROUPS OF EXTINCT MAMMALS

Group	Description
Cave bears 	Numerous in the ice ages, this enormous vegetarian bear slept through the winter in large groups.
Irish elk 	Neither Irish nor an elk (it is a kind of deer), *Megaloceros* was the largest deer that ever lived, with horns spanning nearly 4 meters. Seen in French cave paintings, they became extinct about 2,500 years ago.
Mammoths 	Although only two species of elephants survive today, the elephant family was far more diverse during the late Tertiary. Many were cold-adapted mammoths with fur.
Giant ground sloths 	*Megatherium* was a giant 6-meter ground sloth that weighed 3 tons and was as large as a modern elephant.
Saber-tooth cats 	The jaws of these large, lionlike cats opened an incredible 120 degrees to allow the animal to drive its huge upper pair of saber teeth into prey.

20.4 Fishes Dominate the Sea

A series of key evolutionary advances allowed vertebrates to first conquer the sea and then the land (figure 20.7). About half of all vertebrates are **fishes.** The most diverse and successful vertebrate group, they provided the evolutionary base for invasion of land by amphibians.

Characteristics of Fishes

From whale sharks that are 12 meters long to tiny cichlids no larger than your fingernail, fishes vary considerably in size, shape, color, and appearance. However varied, all fishes have four important characteristics in common:

1. **Gills.** Fish are water-dwelling creatures, and they use gills to extract dissolved oxygen gas from the water around them. Gills are fine filaments of tissue rich in blood vessels. When water passes over the gills in the back of the mouth, oxygen gas diffuses from the water into the fish's blood.
2. **Vertebral column.** All fishes have an internal skeleton with a spine surrounding the dorsal nerve cord, although it may not necessarily be made of bone. The brain is fully encased within a protective box, the skull or cranium, made of bone or cartilage.
3. **Single-loop blood circulation.** Blood is pumped from the heart to the gills. From the gills, the oxygenated blood passes to the rest of the body and then returns to the heart.
4. **Nutritional deficiencies.** Fishes are unable to synthesize the aromatic amino acids and must consume them in their diet. This inability has been inherited by all their vertebrate descendants.

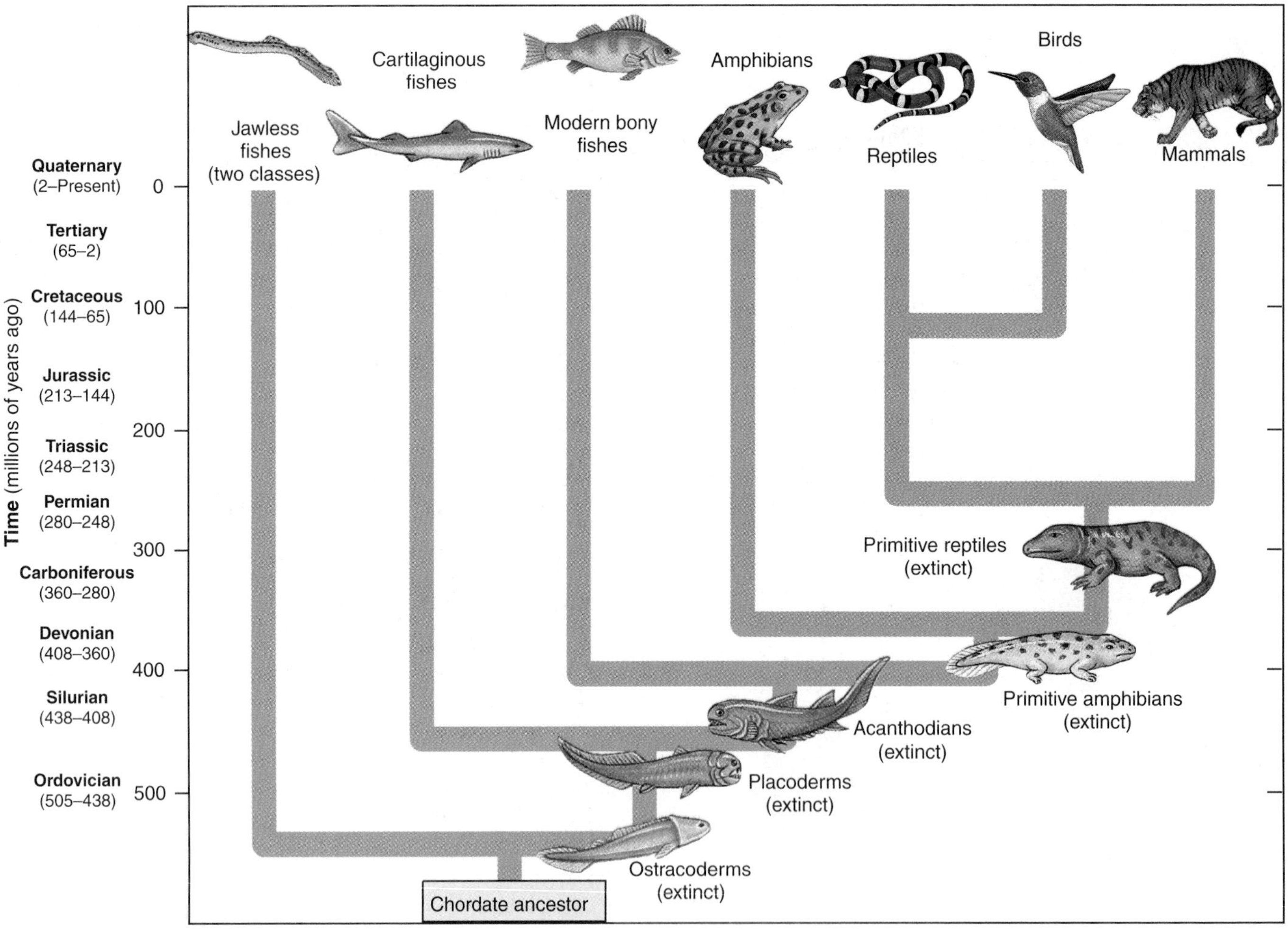
Jawless fishes (two classes)
Cartilaginous fishes
Modern bony fishes
Amphibians
Reptiles
Birds
Mammals
Quaternary (2–Present)
Tertiary (65–2)
Cretaceous (144–65)
Jurassic (213–144)
Triassic (248–213)
Permian (280–248)
Carboniferous (360–280)
Devonian (408–360)
Silurian (438–408)
Ordovician (505–438)
Time (millions of years ago)
0
100
200
300
400
500
Primitive reptiles (extinct)
Primitive amphibians (extinct)
Acanthodians (extinct)
Placoderms (extinct)
Ostracoderms (extinct)
Chordate ancestor

Figure 20.7 Vertebrate family tree.
Primitive amphibians arose from fishes. Primitive reptiles arose from amphibians and gave rise to mammals and to dinosaurs, which survive today as birds.

The First Fishes

The first backboned animals were fishes that appeared in the sea about 470 million years ago. These first fishes were called *ostracoderms,* meaning "shell-skinned." Only their head-shields were made of bone; their elaborate internal skeletons were constructed of cartilage. Wriggling through the water, jawless and toothless, these fishes sucked up small food particles from the ocean floor. Most less than a foot long, they respired with gills but had no fins—just a primitive tail to push them through the water. These first fishes were a great evolutionary success, dominating the world's oceans for about 100 million years. They were eventually replaced by new kinds of fishes that were hunters. One group of jawless fishes, the **agnathans,** survive today as hagfish and parasitic lampreys (figure 20.8).

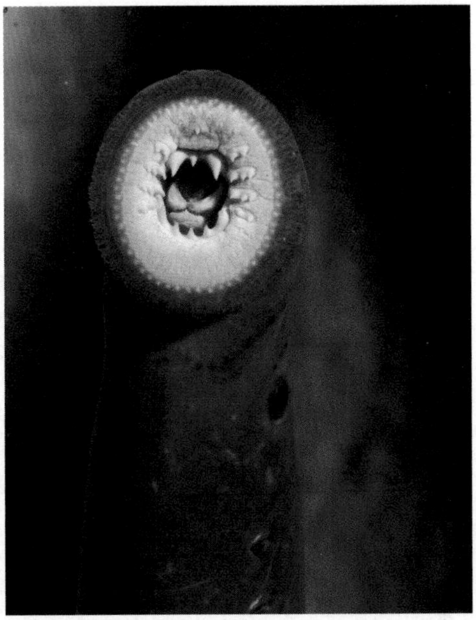

Figure 20.8 Specialized mouth of a lamprey.

Lampreys use their suckerlike mouths to attach themselves to the fish on which they prey. When they have done so, they bore a hole in the fish with their teeth and feed on its blood.

The Evolution of Jaws

The evolution of fishes has been dominated by adaptations to two challenges of surviving as a predator in water:

1. What is the best way to grab hold of potential prey?
2. What is the best way to pursue prey through water?

The fishes that replaced the jawless ones 360 million years ago were powerful predators with much better solutions to both evolutionary challenges. A fundamentally important evolutionary advance was achieved about 410 million years ago—the development of jaws. Jaws seem to have evolved from the front-most of a series of arch supports made of cartilage that were used to reinforce the tissue between gill slits, holding the slits open (figure 20.9).

The earliest jawed fishes had small bodies covered with protective spines and paired fins. Like ostracoderms, these spiny fishes had internal skeletons made of cartilage, but their skin scales contained small plates of bone. Spiny fishes were predators and far better swimmers than ostracoderms, with as many as seven paired fins to aid their swimming.

Later, larger jawed fishes called *placoderms* evolved. These fishes had massive heads armored with heavy bony plates. Many placoderms grew to enormous sizes, some over 9 meters long! Both spiny fishes and placoderms are extinct now, replaced in turn by fishes that evolved even better ways of moving through the water, the sharks and the bony fishes. The replacement did not happen overnight—the earliest sharks and bony fishes first appear in the fossil record soon after spiny fishes and placoderms do. However, after sharing the seas for 150 million years, the long competition finally ended with the complete disappearance of the less maneuverable early jawed fishes. For the last 250 million years, all jawed fishes swimming in the world's oceans and rivers have been either sharks (and their relatives, the rays) or bony fishes.

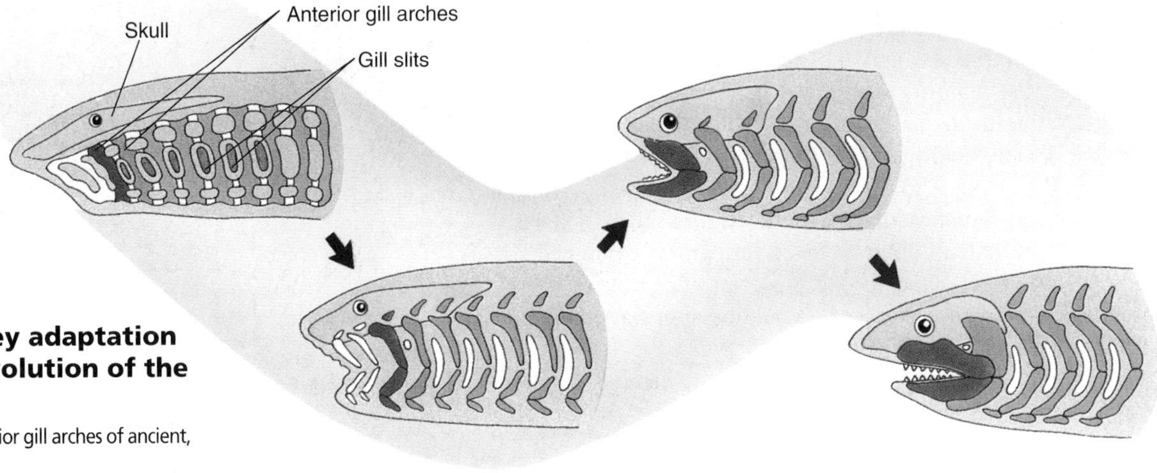
Skull
Anterior gill arches
Gill slits

Figure 20.9 A key adaptation among fishes: evolution of the jaw.

Jaws evolved from the anterior gill arches of ancient, jawless fishes.

Sharks

The problem of fast and maneuverable swimming was solved in sharks by the replacement of the heavy bony skeleton of the early fishes with a far lighter one made of strong, flexible cartilage. Members of this group, the class Chondrichthyes, consist of sharks, skates, and rays. Sharks are very powerful swimmers, with a back fin, a tail fin, and two sets of paired side fins for controlled thrusting through the water (figure 20.10). Skates and rays are flattened sharks that are bottom-dwellers; they evolved some 200 million years after the sharks first appeared. Today there are about 850 species of sharks, skates, and rays.

Some of the largest sharks filter their food from the water like jawless fishes, but most are predators, their mouths armed with rows of hard, sharp teeth. Reproduction among the Chondrichthyes is the most advanced of any fish. Shark eggs are fertilized internally. During mating, the male grasps the female with modified fins called claspers. Sperm run from the male into the female through grooves in the claspers. About 40% of sharks, skates, and rays lay fertilized eggs. The eggs of other species develop within the female's body, and the pups are born alive.

Figure 20.10 Chondrichthyes.
The blue shark is a member of the class Chondrichthyes, which are mainly predators or scavengers and spend most of their time in graceful motion. As they move, they create a flow of water past their gills, from which they extract oxygen.

Bony Fishes

The problem of fast and maneuverable swimming was solved in bony fishes (class Osteichthyes) in a very different way. Instead of gaining speed through lightness, as sharks did, bony fishes adopted a heavy internal skeleton made completely of bone. Such an internal skeleton is very strong, providing a base against which very strong muscles could pull. Bony fishes are still buoyant though because they possess a **swim bladder,** a gas-filled sac that allows them to regulate their buoyant density and so remain effortlessly suspended at any depth in the water (figure 20.11). By adjusting the amount of gas in its swim bladder, a bony fish can rise up and down in the water the same way a submarine does. Sharks, by contrast, increase bouyancy with oil in their liver, but still must move through the water or sink, because their bodies are denser than water. This simple solution to the challenge of swimming has proven to be a great success.

Bony fishes consist of the lobe-finned fishes, ancestors of the first tetrapods, and the ray-finned fishes, which include the vast majority of today's fishes (table 20.2). Bony fishes are the most successful of all fishes, indeed of all vertebrates. Of the nearly 24,500 living species of fishes in the world today, about 23,500 species are bony fishes (class Osteichthyes) with swim bladders (figure 20.12). That's more species than all other kinds of vertebrates combined! In fact, if you could stand in one place and have every vertebrate animal alive today pass by you, one after the other, half of them would be bony fishes.

The remarkable success of the bony fishes has resulted from a series of significant adaptations. These include the swim bladder and gill cover.

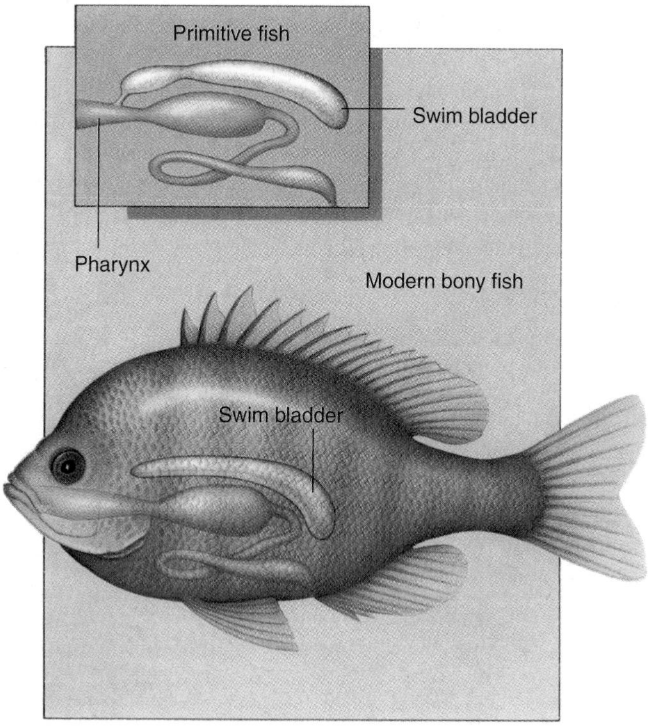

Primitive fish
Swim bladder
Pharynx
Modern bony fish
Swim bladder

Figure 20.11 Diagram of a swim bladder.
The bony fishes use a swim bladder, which evolved as an outpocketing of the pharynx, to control their buoyancy in water.

In addition to the swim bladder, bony fishes have a number of other features that have enabled them to dominate life in the water. They have a highly developed **lateral line system,** a sensory system that enables them to detect changes in water pressure and thus the movement of predators and prey in the water. Also, most bony fishes have a hard plate called the **operculum** that covers the gills on each side of the head. Flexing the operculum permits bony fishes to pump water over their gills. Using the operculum as very efficient bellows, bony fishes can pass water over the gills while stationary in the water. That is what a goldfish is doing when it seems to be gulping in a fish tank.

Figure 20.12 Osteichthyes.

Bony fishes are extremely diverse, containing more species than all other kinds of vertebrates combined. This Korean angelfish, *Pomacanthus semicircularis,* in Fiji, is one of the many striking fishes that live around coral reefs in tropical seas.

20.4 Fishes are characterized by gills, a simple, single-loop circulatory system, and a vertebral column. Sharks are fast swimmers, while the very successful bony fishes have unique characteristics such as swim bladders and lateral line systems.

TABLE 20.2 MAJOR CLASSES OF FISHES

Class	Typical Examples		Key Characteristics	Approximate Number of Living Species
Acanthodii	Spiny fishes	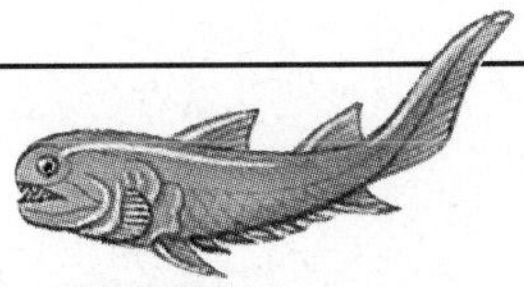	Fishes with jaws; all now extinct; paired fins supported by sharp spines	Extinct
Placodermi	Armored fishes	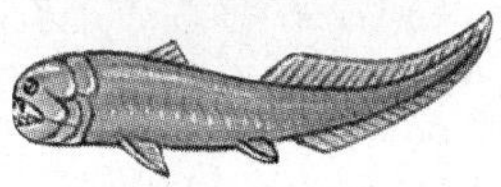	Jawed fishes with heavily armored heads; often quite large	Extinct
Osteichthyes	Ray-finned fishes	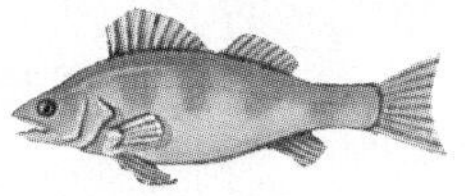	Most diverse group of vertebrates; swim bladders and bony skeletons; paired fins supported by bony rays	23,500
	Lobe-finned fishes		Largely extinct group of bony fishes; ancestral to amphibians; paired lobed fins	7
Chondrichthyes	Sharks, skates, rays		Streamlined hunters; cartilaginous skeletons; no swim bladders; internal fertilization	850
Myxini	Hagfishes		Jawless fishes with no paired appendages; scavengers; mostly blind, but a well-developed sense of smell	43
Cephalaspidomorphi	Lampreys	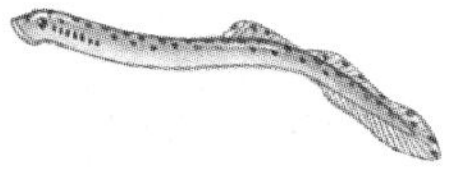	Largely extinct group of jawless fishes with no paired appendages; parasitic and non-parasitic types; all breed in freshwater	17

20.5 Amphibians Invade the Land

Frogs, salamanders, and caecilians, the damp-skinned vertebrates, are direct descendants of fishes. They are the sole survivors of a very successful group, the **amphibians,** the first vertebrates to walk on land (figure 20.13). Amphibians almost certainly evolved from the lobe-finned fishes, fish that have paired fins that consist of a long fleshy muscular lobe, supported by a central core of bones that form fully articulated joints with one another.

Figure 20.13 A representative amphibian.
This red-eyed tree frog, *Agalychnis callidryas*, is a member of the group of amphibians that includes frogs and toads (order Anura).

Characteristics of Amphibians

Amphibians have five key characteristics that allowed them to successfully invade the land.

1. **Legs.** Frogs and salamanders have four legs and can move about on land quite well. Legs, which evolved from fins (figure 20.14), were one of the key adaptations to life on land. Caecilians have lost their legs during the course of adapting to a burrowing existence.

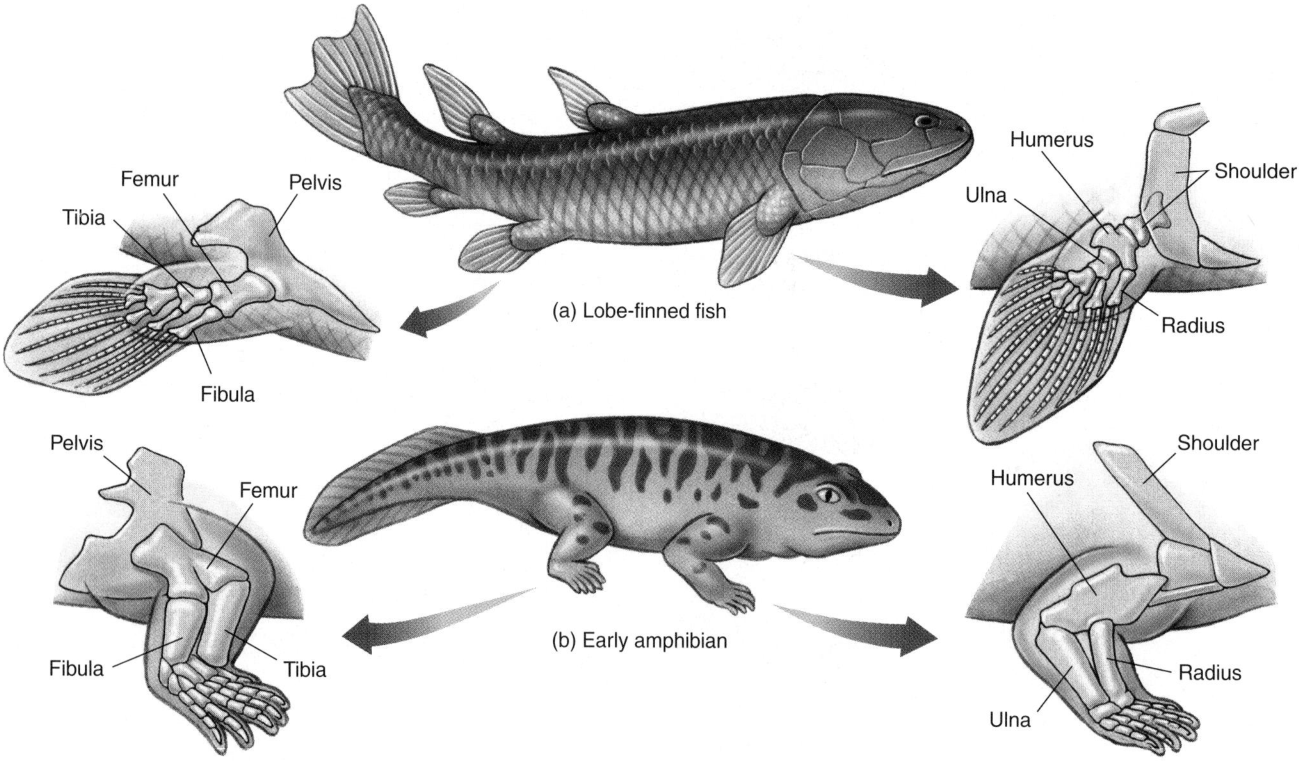

Femur
Pelvis
Tibia
Fibula
(a) Lobe-finned fish
Humerus
Ulna
Shoulder
Radius
Pelvis
Femur
Fibula
Tibia
(b) Early amphibian
Shoulder
Humerus
Radius
Ulna

Figure 20.14 A key adaptation of amphibians: the evolution of legs.
(*a*) The limbs of a lobe-finned fish. Some lobe-finned fishes could move out onto land. (*b*) The limbs of an early amphibian. As illustrated by their skeletal structure, the legs of primitive amphibians could clearly function on land better than could the fins of lobe-finned fishes.

TABLE 20.3 ORDERS OF AMPHIBIANS

Order	Typical Examples		Key Characteristics	Approximate Number of Living Species
Anura	Frogs, toads		Compact tailless body; large head fused to the trunk; rear limbs specialized for jumping	3,680
Urodela (or Caudata)	Salamanders, newts		Slender body; long tail and limbs set out at right angles to the body	369
Apoda (or Gymnophiona)	Caecilians		Tropical group with a snakelike body; no limbs; little or no tail; internal fertilization	160

2. **Cutaneous respiration.** Frogs, salamanders, and caecilians all supplement the use of lungs by respiring directly across their skin, which is kept moist and provides an extensive surface area. This mode of respiration limits the body size of amphibians, because it is only efficient for a high surface-to-volume ratio.
3. **Lungs.** Most amphibians possess a pair of lungs, although the internal surfaces are poorly developed. Lungs were necessary, even though there is far more oxygen in air than water, because the delicate structure of fish gills requires the buoyancy of water to support it.
4. **Pulmonary veins.** After blood is pumped through the lungs, two large veins called pulmonary veins return the aerated blood to the heart for repumping. This allows the aerated blood to be pumped to the tissues at a much higher pressure than when it leaves the lungs.
5. **Partially divided heart.** The heart had to be redesigned to deliver greater amounts of oxygen to the amphibian tissues, because greater amounts of oxygen are required by muscles for walking. The initial chamber of the fish heart is absent in amphibians, and the second and last chambers are separated by a dividing wall that helps prevent aerated blood from the lungs from mixing with nonaerated blood being returned to the heart from the rest of the body. This separates the blood circulation into two separate paths, pulmonary and systemic. The separation is imperfect; the third chamber has no dividing wall.

History of Amphibians

Amphibians were the dominant land vertebrate for 100 million years. They first became common in the late Paleozoic era, when much of North America was covered by lowland tropical swamps. Amphibians reached their greatest diversity during the mid-Permian period, when 40 families existed. Sixty percent of them were fully terrestrial, with bony plates and armor covering their bodies, and many grew to be very large—some as big as a pony! After the great Permian extinction, the terrestrial forms began to decline, and by the time dinosaurs evolved, only 15 families remained, all aquatic. Only two of these families survived the age of the dinosaurs, both aquatic: the anurans (frogs and toads) and the urodeles (salamanders).

Approximately 4,200 species of amphibians exist today, in 37 different families, all aquatic and all descended from the two aquatic families that survived the age of dinosaurs. Three orders comprise the class Amphibia: Anura, frogs and toads; Urodela, salamanders and newts; and Apoda, caecilians (table 20.3). Most of today's amphibians must reproduce in water and live the early part of their lives there, so amphibians are not completely terrestrial. However, in most habitats, particularly in the tropics, they are often today the most abundant and successful vertebrates to be found.

20.5 Amphibians were the first vertebrates to successfully invade land. They developed legs, lungs, and the pulmonary vein, which allowed them to repump oxygenated blood and thus deliver oxygen far more efficiently to the body's muscles.

20.6 Reptiles Conquer the Land

Characteristics of Reptiles

If we think of amphibians as the "first draft" of a manuscript about survival on land, then **reptiles** were the finished book. All living reptiles share certain fundamental characteristics, features they retained from the time when they replaced amphibians as the dominant terrestrial vertebrates. Among the most important are:

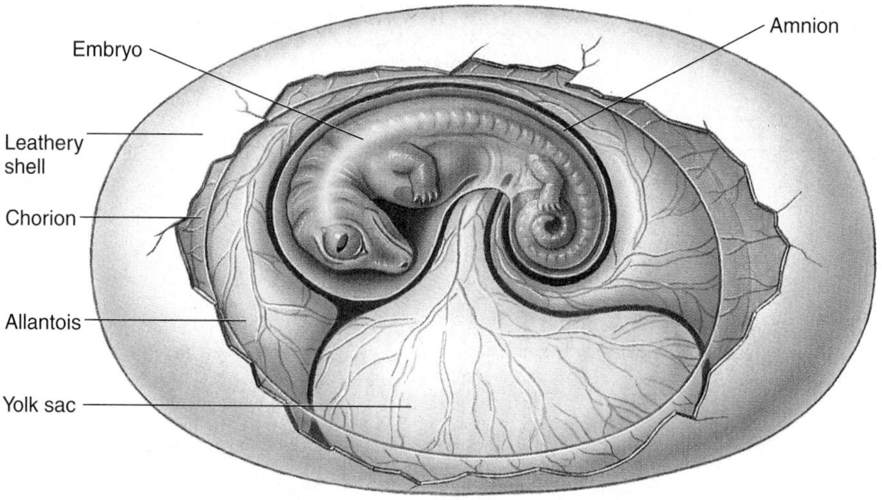

Embryo
Amnion
Leathery shell
Chorion
Allantois
Yolk sac

Figure 20.15 A key adaptation of reptiles: watertight eggs.
The watertight amniotic egg allows reptiles to live in a wide variety of terrestrial habitats.

1. **Amniotic egg.** Amphibians never succeeded in becoming fully terrestrial because amphibian eggs must be laid in water to avoid drying out. Most reptiles lay watertight eggs that contain a food source (the yolk) and a series of four membranes—the chorion, the amnion, the yolk sac, and the allantois (figure 20.15). Each membrane plays a role in making the egg an independent life-support system. The outermost membrane of the egg, the **chorion,** allows oxygen to enter the porous shell but retains water within the egg. The **amnion** encases the developing embryo within a fluid-filled cavity. The **yolk sac** provides food from the yolk for the embryo via blood vessels connecting to the embryo's gut. The **allantois** surrounds a cavity into which waste products from the embryo are excreted.
2. **Dry skin.** Living amphibians have a moist skin and must remain in moist places to avoid drying out. Like earlier amphibians, reptiles have dry skin. A layer of scales or armor covers their bodies, preventing water loss.
3. **Thoracic breathing.** Amphibians breathe by squeezing their throat to pump air into their lungs; this limits their breathing capacity to the volume of their mouth. Reptiles developed pulmonary breathing, expanding and contracting the rib cage to suck air into the lungs and then force it out.

In addition, reptiles improved on the innovations first attempted by amphibians. Legs were arranged to more effectively support the body's weight, allowing reptile bodies to be bigger and to run. Also, the lungs and heart were altered to make them far more efficient.

Today some 7,000 species in the class Reptilia (table 20.4), mostly snakes and lizards, are found in practically every wet and dry habitat on earth. And *today's* reptiles don't begin to tell the tale.

History of Reptiles

Reptiles first evolved about 320 million years ago when the world was entering a long, dry period. At first, "sail-back" **pelycosaurs** rose to prominence. With long, sharp "steak knife" teeth, pelycosaurs were the first land vertebrates able to kill animals their own size. Pelycosaurs were dominant for 50 million years, at their height comprising 70% of all land vertebrates. The **therapsids** replaced the pelycosaurs about 250 million years ago and had a more upright stance than the sprawling pelycosaurs. Therapsids were the immediate ancestors of the mammals, and paleontologists suspect they may have been endothermic, that is, able to maintain a constant high body temperature. This requires considerable fuel to produce body heat, and so therapsids would have had to eat 10 times more frequently than ectothermic pelycosaurs.

The therapsids were in turn replaced by ectothermic reptiles called **thecodonts.** Most thecodonts resembled crocodiles, but later forms were the first reptiles to be bipedal—to stand on two feet. From thecodonts, five great lines of large-bodied reptiles developed: **dinosaurs,** a very diverse group, some of which grew to be larger than houses; **crocodiles,** which have changed little from that time until now; **ichthyosaurs** and **plesiosaurs,** marine reptiles that lived in the oceans; and **pterosaurs,** which were flying reptiles. At about the same time that dinosaurs evolved, about 220 million years ago, mammals evolved, but while dinosaurs flourished there were no large-bodied mammals. That niche was filled.

20.6 Reptiles have three characteristics that suit them well for life on land: a watertight (amniotic) egg, dry skin, and thoracic breathing.

TABLE 20.4 ORDERS OF REPTILES

Order	Typical Examples		Key Characteristics	Approximate Number of Living Species
Ornithischia	Stegosaur		Dinosaurs with two pelvic bones facing backward, like a bird's pelvis; herbivores, with turtlelike upper beak; legs under body	Extinct
Saurischia	Tyrannosaur		Dinosaurs with one pelvic bone facing forward, the other back, like a lizard's pelvis; both plant and flesh eaters; legs under body	Extinct
Pterosauria	Pterosaur		Flying reptiles; wings were made of skin stretched between fourth fingers and body; wingspans of early forms typically 60 centimeters, later forms nearly 8 meters	Extinct
Plesiosauria	Plesiosaur		Barrel-shaped marine reptiles with sharp teeth and large, paddle-shaped fins; some had snakelike necks twice as long as their bodies	Extinct
Ichthyosauria	Ichthyosaur	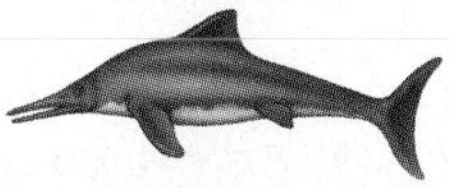	Streamlined marine reptiles with many body similarities to sharks and modern fishes	Extinct
Squamata, suborder Sauria	Lizards		Lizards; limbs set at right angles to body; anus is in transverse (sideways) slit; most are terrestrial	3,800
Squamata, suborder Serpentes	Snakes		Snakes; no legs; move by slithering; scaly skin is shed periodically; most are terrestrial	3,000
Chelonia	Turtles, tortoises, sea turtles		Ancient armored reptiles with shell of bony plates to which vertebrae and ribs are fused; sharp, horny beak without teeth	250
Crocodylia	Crocodiles, alligators, gavials, caimans		Advanced reptiles with four-chambered heart and socketed teeth; anus is a longitudinal (lengthwise) slit; closest living relatives to birds	25
Rhynchocephalia	Tuataras		Sole survivors of a once successful group that largely disappeared before the dinosaurs; fused, wedgelike, socketless teeth; primitive third eye under skin of forehead	2

20.7 Birds Master the Air

Birds evolved from small bipedal dinosaurs about 150 million years ago, but they were not common until the flying reptiles called pterosaurs became extinct along with the dinosaurs. Unlike pterosaurs, birds are insulated with feathers. Birds are so structurally similar to dinosaurs in all other respects that many scientists consider birds to be simply feathered dinosaurs.

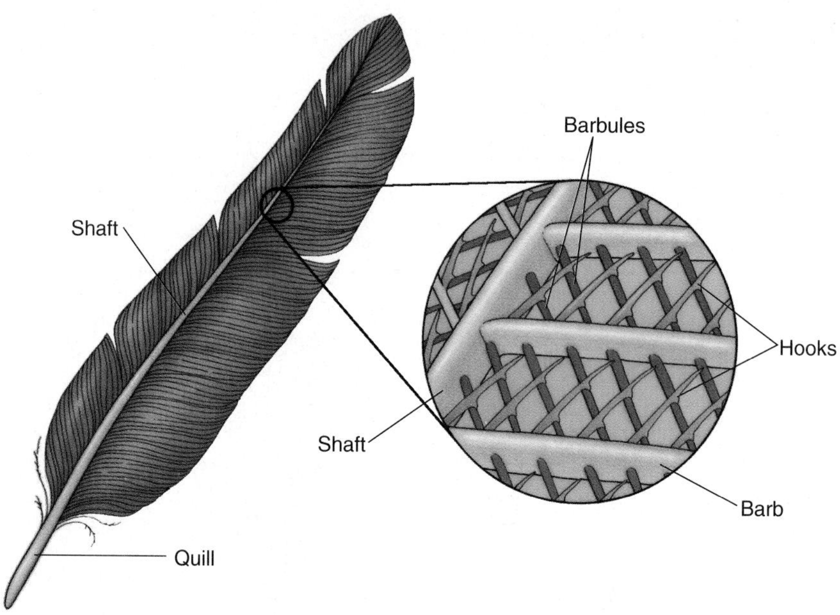

Barbules
Shaft
Hooks
Shaft
Barb
Quill

Figure 20.16 A key adaptation of birds: feathers.
The barbs off the main shaft of a feather have secondary branches called barbules. The barbules of adjacent barbs are attached to one another by microscopic hooks.

Characteristics of Birds

Modern birds lack teeth and have only vestigial tails, but they still retain many reptilian characteristics. For instance, birds lay amniotic eggs, although the shells of bird eggs are hard rather than leathery. Also, reptilian scales are present on the feet and lower legs of birds. What makes birds unique? What distinguishes them from living reptiles?

1. **Feathers.** Derived from reptilian scales, feathers are the ideal adaptation for flight because they are lightweight and easily replaced if damaged (figure 20.16).
2. **Flight skeleton.** The bones of birds are thin and hollow. Many of the bones are fused, making the bird skeleton more rigid than a reptilian skeleton. The fused sections of backbone and of the shoulder and hip girdles form a sturdy frame that anchors muscles during flight. The power for active flight comes from large breast muscles that can make up 30% of a bird's total body weight. They stretch down from the wing and attach to the breastbone, which is greatly enlarged and bears a prominent keel for muscle attachment. They also attach to the fused collarbones that form the so-called wishbone. No other living vertebrates have a fused collarbone or a keeled breastbone.

Birds, like mammals, are endothermic. They generate enough heat through metabolism to maintain a high body temperature. Birds maintain body temperatures significantly higher than most mammals. The high body temperature permits a faster metabolism, necessary to satisfy the large energy requirements of flight.

History of Birds

The oldest bird of which there is a clear fossil is ***Archaeopteryx*** (meaning "ancient wing"), which was about the size of a crow and shared many features with small, bipedal, carnivorous dinosaurs. For example, it had teeth and a long reptilian tail. And unlike the hollow bones of today's birds, its bones were solid. Because of its many dinosaur features, several *Archaeopteryx* fossils were originally classified as *Compsognathus,* a small theropod dinosaur of similar size—until feathers were discovered on the fossils. What makes *Archaeopteryx* distinctly avian is the presence of feathers on its wings and tail. It also has other birdlike features, notably the presence of a wishbone. Dinosaurs lack a wishbone, although thecodonts had them.

By the early Cretaceous, only a few million years after *Archaeopteryx,* a diverse array of birds had evolved, with many of the features of modern birds. Fossils in Mongolia, Spain, and China discovered within the last few years reveal a diverse collection of toothed birds with the hollow bones and breastbones necessary for sustained flight. Other fossils reveal highly specialized, flightless diving birds. The diverse birds of the Cretaceous shared the skies with pterosaurs for 70 million years.

Today nearly 9,000 species of birds (class Aves) occupy a variety of habitats all over the world. The major orders of birds are reviewed in table 20.5. You can tell a great deal by examining the beaks. For example, carnivorous birds such as hawks have a sharp beak for tearing apart meat, the beaks of ducks are flat for shoveling through mud, and the beaks of finches are short and thick for crushing seeds.

20.7 Birds are essentially dinosaurs with feathers. Feathers and a strong, light skeleton make flight possible.

TABLE 20.5 MAJOR ORDERS OF BIRDS

Order	Typical Examples	Key Characteristics	Approximate Number of Living Species
Passeriformes	Crows, mockingbirds, robins, sparrows, starlings, warblers	*Songbirds* Well-developed vocal organs; perching feet; dependent young	5,276 (largest of all bird orders; contains over 60% of all species)
Apodiformes	Hummingbirds, swifts	*Fast fliers* Short legs; small bodies; rapid wing beat	428
Piciformes	Honeyguides, toucans, woodpeckers	*Woodpeckers or toucans* Grasping feet; chisel-like, sharp bills can break down wood	383
Psittaciformes	Cockatoos, parrots	*Parrots* Large, powerful bills for crushing seeds; well-developed vocal organs	340
Charadriiformes	Auks, gulls, plovers, sandpipers, terns	*Shorebirds* Long, stiltlike legs; slender probing bills	331
Columbiformes	Doves, pigeons	*Pigeons* Perching feet; rounded, stout bodies	303
Falconiformes	Eagles, falcons, hawks, vultures	*Birds of prey* Carnivorous; keen vision; sharp, pointed beaks for tearing flesh; active during the day	288
Galliformes	Chickens, grouse, pheasants, quail	*Game birds* Often limited flying ability; rounded bodies	268
Gruiformes	Bitterns, coots, cranes, rails	*Marsh birds* Long, stiltlike legs; diverse body shapes; marsh-dwellers	209
Anseriformes	Ducks, geese, swans	*Waterfowl* Webbed toes; broad bill with filtering ridges	150
Strigiformes	Barn owls, screech owls	*Owls* Nocturnal birds of prey; strong beaks; powerful feet	146
Ciconiiformes	Herons, ibises, storks	*Waders* Long-legged; large bodies	114
Procellariformes	Albatrosses, petrels	*Seabirds* Tube-shaped bills; capable of flying for long periods of time	104
Sphenisciformes	Emperor penguins, crested penguins	*Penguins* Marine; modified wings for swimming; flightless; found only in Southern Hemisphere; thick coats of insulating feathers	18
Dinornithiformes	Kiwis	*Kiwis* Flightless; small; primitive; confined to New Zealand	2
Struthioniformes	Ostriches	*Ostriches* Powerful running legs; flightless; only two toes; very large	1

20.8 Mammals Adapt to Colder Times

Characteristics of Mammals

The **mammals** (class Mammalia) that evolved about 220 million years ago side by side with the dinosaurs would look strange to you, not at all like modern-day lions and tigers and bears. They share three key characteristics with mammals today:

1. **Mammary glands.** Female mammals have mammary glands, which produce milk to nurse the rapidly growing newborns.
2. **Hair.** Among living vertebrates, only mammals have hair (figure 20.17), and all mammals do (even whales have a few sensitive bristles on their snout). A hair is a filament composed of dead cells filled with the protein keratin. The primary function of hair is insulation.
3. **Middle ear.** All mammals have three middle ear bones.

Figure 20.17 A key adaptation of mammals: hair.

These meadow mice are covered with soft fur. There are more species of rodents like these mice than any other kind of mammal. As in the time of the dinosaurs, most mammals living today are smaller than cats.

History of the Mammals

Mammals have been around since the time of the dinosaurs, although they were always small until dinosaurs disappeared. We have learned a lot about the evolutionary history of mammals from their fossils. The first mammals arose from therapsids in the mid-Triassic about 220 million years ago, just as the first dinosaurs evolved from thecodonts. Tiny, shrewlike creatures that lived in trees chasing insects, mammals were only a minor element in a land that quickly came to be dominated by dinosaurs. Fossils reveal that these early mammals had large eye sockets, evidence that they may have been active at night.

For 155 million years, all the time the dinosaurs flourished, mammals were a minor group that changed little. Only five orders of mammals arose in that time, and their fossils are scarce, indicating that mammals were not abundant. However, the two groups to which present-day mammals belong did appear. The most primitive mammals, direct descendents of therapsids, were members of the subclass Prototheria. Most prototherians were small and resembled modern shrews. All prototherians laid shelled eggs, as did their therapsid ancestors. The only prototherians surviving today are the **monotremes**—the duck-bill platypus and the echidnas, or spiny anteaters.

The other major mammalian group is the subclass Theria. Therians are viviparous (that is, their young are born alive). The two major living therian groups are **marsupials,** or pouched mammals, and **placental mammals.** Marsupials, such as kangaroos, opossums, and koalas, do not have shelled eggs. Instead, young are born within the mother's body and within days after birth transferred to pouches, where they can be nursed by the mother's milk in relative protection. About 280 species of marsupials survive today, almost all of them in Australia and New Guinea.

Placental mammals, such as dogs, cats, humans, and many others, invest even more care in nurturing their young than marsupials do. They carry the developing young within the mother's body far longer. During this long period of protected growth and development, the young are fed not by mother's milk but rather by a structure called the *placenta* (that is why they are called "placental mammals").

Today, over 4,000 species of placental mammals occupy all the large-body niches that dinosaurs once claimed, among many others. They range in size from 1.5-gram shrews to 100-ton whales. Almost half of all mammals are rodents—mice and their relatives. Almost one-quarter of all mammals are bats! Mammals have even invaded the seas, as plesiosaurs and ichthyosaurs did so successfully millions of years earlier—79 species of whales and dolphins live in today's oceans. The placental mammals that walked the earth during the ice ages were even larger than today's versions; the world's climate has warmed again in recent times, favoring smaller bodies, which are easier to cool.

Primates, the order to which we belong, are not a major group; there are only 233 known species. Human beings evolved only very recently, less than 2 million years ago. There have been at least three species of humans, but our species, *Homo sapiens,* is the only one that survives today. We are notable among primates for having less hair, walking upright, making tools on a regular basis, and having complicated language. The major orders of mammals are described in table 20.6.

TABLE 20.6 MAJOR ORDERS OF MAMMALS

Order	Typical Examples		Key Characteristics	Approximate Number of Living Species
Rodentia	Beavers, mice, porcupines, rats		*Small plant eaters* Chisel-like incisor teeth	1,814
Chiroptera	Bats		*Flying mammals* Primarily fruit or insect eaters; elongated fingers; thin wing membrane; nocturnal; navigate by sonar	986
Insectivora	Moles, shrews		*Small, burrowing mammals* Insect eaters; most primitive placental mammals; spend most of their time underground	390
Marsupialia	Kangaroos, koalas		*Pouched mammals* Young develop in abdominal pouch	280
Carnivora	Bears, cats, raccoons, weasels, dogs		*Carnivorous predators* Teeth adapted for shearing flesh; no native families in Australia	240
Primates	Apes, humans, lemurs, monkeys		*Tree-dwellers* Large brain size; binocular vision; opposable thumb; end product of a line that branched off early from other mammals	233
Artiodactyla	Cattle, deer, giraffes, pigs		*Hoofed mammals* With two or four toes; mostly herbivores	211
Cetacea	Dolphins, porpoises, whales		*Fully marine mammals* Streamlined bodies; front limbs modified into flippers; no hind limbs; blowholes on top of head; no hair except on muzzle	79
Lagomorpha	Rabbits, hares, pikas		*Rodent-like jumpers* Four upper incisors (rather than the two seen in rodents); hindlegs often longer than forelegs, an adaptation for jumping	69
Pinnipedia	Sea lions, seals, walruses		*Marine carnivores* Feed mainly on fish; limbs modified for swimming	34
Edentata	Anteaters, armadillos, sloths		*Toothless insect eaters* Many are toothless, but some have degenerate, peg-like teeth	30
Perissodactyla	Horses, rhinoceroses, zebras		*Hoofed mammals with one or three toes* Herbivorous teeth adapted for chewing	17
Proboscidea	Elephants		*Long-trunked herbivores* Two upper incisors elongated as tusks; largest living land animal	2

Other Characteristics of Modern Mammals

Endothermy. Mammals, like their therapsid ancestors, are endothermic, a crucial adaptation that has allowed mammals to be active at any time of the day or night and to colonize severe environments, from deserts to ice fields. Many characteristics, such as hair that provides insulation, played important roles in making endothermy possible. Also, the more efficient blood circulation provided by the four-chambered heart and the more efficient breathing provided by the *diaphragm* (a special sheet of muscles below the rib cage that aids breathing) make possible the higher metabolic rate upon which endothermy depends.

Placenta. In most mammal species, females carry their young in the uterus during development, nourishing them by a placenta, and give birth to live young. The placenta is a specialized organ within the womb of the mother that brings the bloodstream of the fetus into close contact with the bloodstream of the mother (figure 20.18). Food, water, and oxygen can pass across from mother to child, and wastes can pass over to the mother's blood and be carried away.

Teeth. Reptiles have homodont dentition: their teeth are all the same. However, mammals have heterodont dentition, with different types of teeth that are highly specialized to match particular eating habits (figure 20.19). It is usually possible to determine a mammal's diet simply by examining its teeth. Compare the skull of a dog (a carnivore) and a deer (an herbivore). The dog's long canine teeth are well suited for biting and holding prey, and some of its premolar and molar teeth are triangular and sharp for ripping off chunks of flesh. In contrast, canine teeth are absent in deer; instead the deer clips off mouthfuls of plants with its flat, chisel-like incisors. The deer's molars are large and covered with ridges to effectively grind and break up tough plant tissues. Rodents, such as a beaver, are gnawers and have long incisors for chewing through branches or stems. These incisors are ever-growing; that is, the ends may become sharp and wear down, but new incisor growth maintains the length.

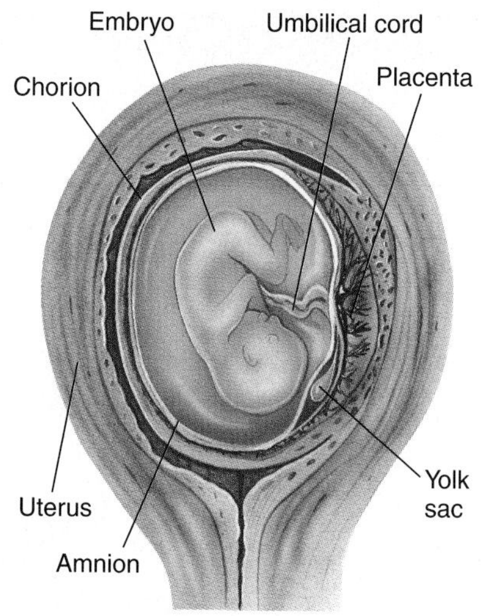
Embryo
Umbilical cord
Chorion
Placenta
Uterus
Amnion
Yolk sac

Figure 20.18 The placenta.

The placenta is characteristic of the largest group of mammals, the placental mammals. It evolved from membranes in the amniotic egg. The umbilical cord evolved from the allantois. The chorion, or outermost part of the amniotic egg, forms most of the placenta itself. The placenta serves as the provisional lungs, intestine, and kidneys of the embryo, without ever mixing maternal and fetal blood.

20.8 Mammals are endotherms that nurse their young with milk and exhibit a variety of different kinds of teeth. All mammals have at least some hair.

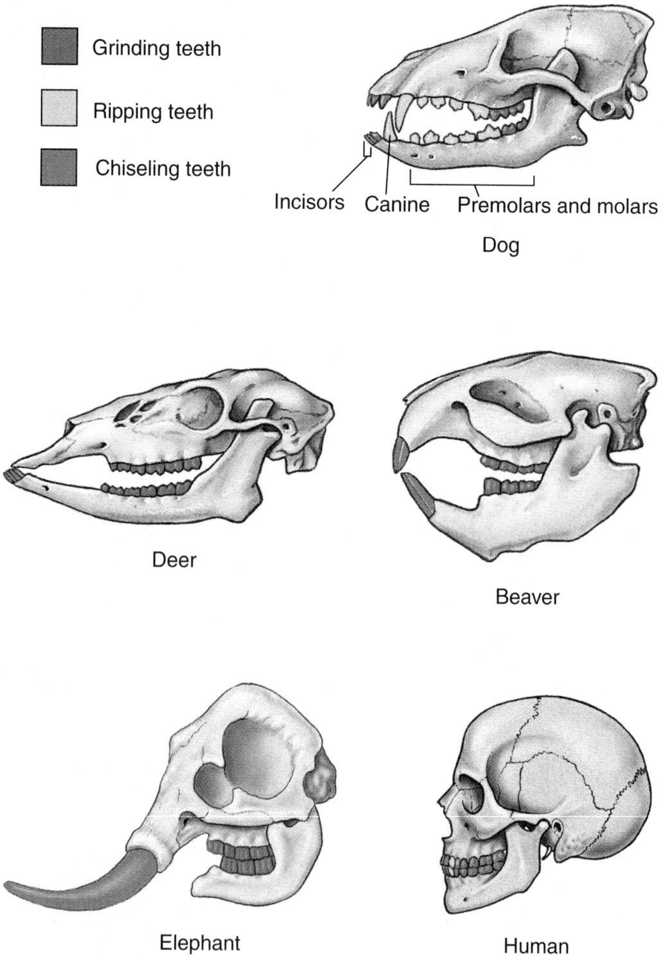
Grinding teeth
Ripping teeth
Chiseling teeth
Incisors
Canine
Premolars and molars
Dog
Deer
Beaver
Elephant
Human

Figure 20.19 Mammals have different types of specialized teeth.

While reptiles have all the same kind of teeth, mammals have different types of teeth specialized for different feeding habits. Carnivores, such as dogs, have *canine* teeth that are able to rip food; some of the *premolars* and *molars* in dogs are also ripping teeth. Herbivores, such as deer, have *incisors* to chisel off vegetation and molars to grind up the plant material. In the beaver, the chiseling incisors dominate. In the elephant, the incisors have become specialized weapons, and molars grind up vegetation. Humans are omnivores; we have ripping, chiseling, and grinding teeth.

CONCEPT REVIEW

1. Which of the following invaded land first?
a. amphibians
b. arthropods
c. plants
d. dinosaurs

2. Match the organism with the period in which it arose.
a. amphibians (1) Ordovician
b. vertebrates (2) Permian
c. dinosaurs (3) Devonian
d. flowering plants (4) Cretaceous

3. During the extinction that marked the end of the Cretaceous period, which of the following did *not* become extinct?
a. ammonites
b. pterosaurs
c. great marine reptiles
d. smaller plankton

4. Which of the following groups of vertebrates do *not* have past or present representatives with flight?
a. amphibians
b. reptiles
c. birds
d. mammals

5. To solve the problem of fast and maneuverable swimming, sharks have a cartilaginous skeleton, while Osteichthyes have
a. heavy bony plates.
b. swim bladders.
c. lungs.
d. suckerlike mouths.

6. Which of the following characteristics are *not* adaptations that contributed to the successful invasion of land by amphibians?
a. legs
b. lungs
c. jaws
d. pulmonary veins

7. Mammals are unique among all vertebrates because they
a. do not lay eggs.
b. have hair.
c. are endothermic.
d. have a cartilaginous skeleton.

8. The first mammals
a. were diurnal.
b. had small eyes.
c. were large, like the mammoths.
d. were shrewlike.

9. _________ are fossil organisms that appear to be the ancestors to horseshoe crabs.

10. The most dramatic extinction event in the history of life on earth happened at the end of the _________ period.

11. More than half of all vertebrate species are _________.

12. One of the pivotal innovations that led to the success of reptiles is the watertight egg, called an _________ egg.

13. The oldest fossil bird, called _________, had feathers and a well-developed collarbone.

CHALLENGE YOURSELF

1. Dinosaurs and mammals both lived throughout the Mesozoic era, a period of more than 150 million years; all this time dinosaurs were the dominant form and mammals a minor group. Both mammals and small reptiles survived the Cretaceous extinction. Why do you suppose reptiles did not go on to become dominant again, rather than mammals?

2. What are the advantages and disadvantages of endothermy compared to ectothermy? How was this thought to have played a role in the rise and fall of some reptile groups?

3. What limits the ability of amphibians to occupy the full range of terrestrial habitats and allows other terrestrial vertebrates to occur in them successfully?

4. Of the 50,300 species of living chordates, twice as many live in the sea as on the surface of the land. Why do you think there are so many more species of chordates in the sea than on land?

20 eBRIDGE

Reinforcing Key Points

Overview of Vertebrate Evolution

20.1 The Paleozoic Era

20.2 The Mesozoic Era

20.3 The Cenozoic Era

The Parade of Vertebrates

20.4 Fishes Dominate the Sea

20.5 Amphibians Invade the Land

20.6 Reptiles Conquer the Land

20.7 Birds Master the Air

20.8 Mammals Adapt to Colder Times

Electronic Learning

Visual Learning

Animations

- Continental Drift
- Phylogenetic Tree of Chordates
- Evolution of Fish Jaws
- Ray-Finned Fishes
- Transition to Land
- Reptilian Diversity

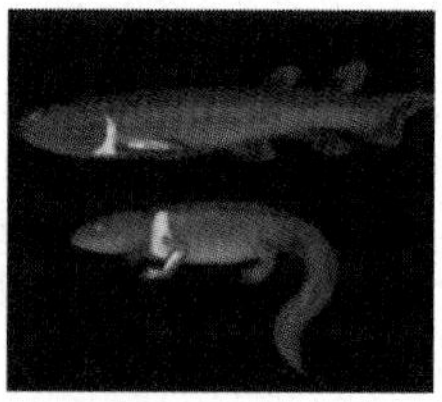

Enhancement Chapter

Dinosaurs

Dinosaurs dominated life on land for 150 million years, an incredibly long time. In this enhancement chapter we first travel back through time and watch how the dinosaurs evolved. We then trace the history of dinosaurs through the Triassic, Jurassic, and Cretaceous, examing in detail each of the major kinds of dinosaurs as they take their turns in this long, very interesting evolutionary story.

Author's Corner

Dinosaurs. Dinosaurs were the dominant large terrestrial vertebrates for 150 million years, until their extinction 65 million years ago, apparently the result of an asteroid's impact. Although classified as reptiles, at least some dinosaurs appear to have been endothermic (warm-blooded). Many of the later dinosaurs sported feathers, and although some ornithologists object to the idea, there seems little doubt that today's birds are the direct descendants of feathered dinosaurs.

1. The isotopes in their bones suggest dinosaurs were warm blooded.
2. The bird you ate for Thanksgiving was a dinosaur.
3. Despite the great "birdosaur" fiasco, birds are still dinosaurs.
4. A small dinosaur with a heart of stone answers an old question.

The Living World

Virtual Classroom

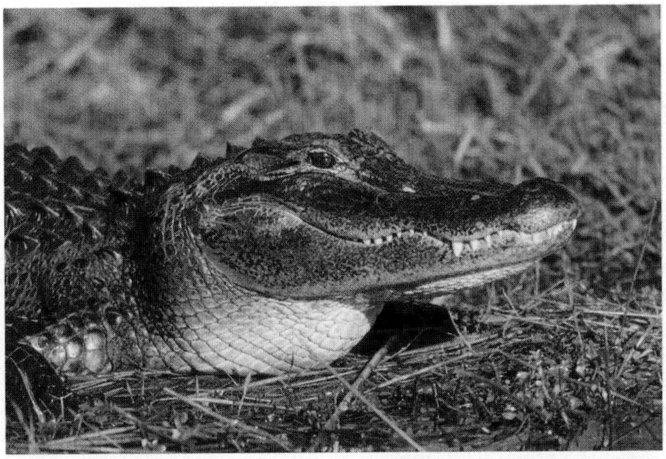

The Challenge of Invading Land

The first vertebrates to invade the land were descendants of bony fishes. To successfully invade land, they had to solve five additional problems: 1. how to get around out of water (legs); 2. how to avoid drying out (watertight skin); 3. how to acquire oxygen from air (lungs); 4. how to get more oxygen to muscles (pulmonary vein, making dual blood circulation possible); and 5. how to reproduce out of water (amniotic egg).

Dinosaurs: An Evolutionary Success Story

Dinosaurs are the most successful of all terrestrial vertebrates. They dominated life on earth for 150 million years, an almost unimaginably long time—for comparison, humans have been on earth only 1 million years. During their long history, dinosaurs changed a great deal, because the world they lived in changed—the world's continents moved, radically altering the earth's climates. Thus we cannot talk about dinosaurs as if they were a particular kind of animal, describing one "type" that is representative of the group. Rather, we have to look at dinosaurs more as a "story," a long parade of change and adaptation.

Virtual Lab

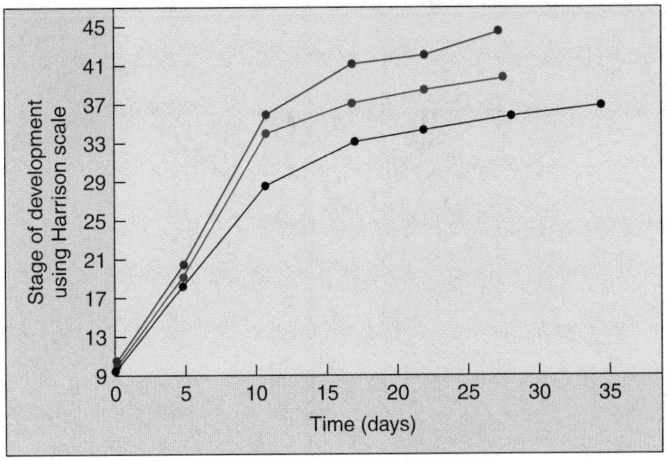
Stage of development using Harrison scale
45
41
37
33
29
25
21
17
13
9
0
5
10
15
20
25
30
35
Time (days)

Amphibian Eggs Hatching in Shallow Ponds
Thirst for Oxygen

Amphibians evolved from aquatic vertebrates, completely dependent upon an aquatic environment for their existence, into semiterrestrial animals less dependent on water. Amphibians, however, must return to water to reproduce. Oxygen availability in the water is an important factor in aquatic ecosystems, because it impacts the development of amphibian embryos. Most amphibians lay their eggs in shallow, temporary ponds or wetlands where the eggs are protected from predatory fish. Such habitats, however, are likely to have low levels of oxygen resulting in hypoxic conditions for the developing amphibian embryos. Few studies have examined the effects of hypoxia on amphibian embryo development, hatching success, and survival. M. Christopher Barnhart of Southwest Missouri State University has undertaken such a study, examining hypoxia in four amphibian species. He removed the eggs of frog and salamander species from their natural environments and placed them into an oxygen-controlled chamber where he could monitor hypoxia-induced changes in the developing embryos under conditions of different oxygen availability.

Quizzes

Further Reading

Essential Study Partner

Links

BioCourse.com

CHAPTER 21

How Humans Evolved

THE FAR SIDE® By GARY LARSON
Larson
© 1982 FarWorks, Inc. All Rights Reserved/Dist. by Creators Syndicate

"Shhh, Zog! ... Here come one now!"

CHAPTER 21

How Humans Evolved

CHAPTER OVERVIEW

The Evolution of Primates

- Primates are the order of mammals that contains humans.
- Prosimians are the oldest primates and are small and nocturnal.
- Anthropoids include the monkeys, apes, and humans and are mostly day-active.
- Hominids are characterized by upright walking and large brains.

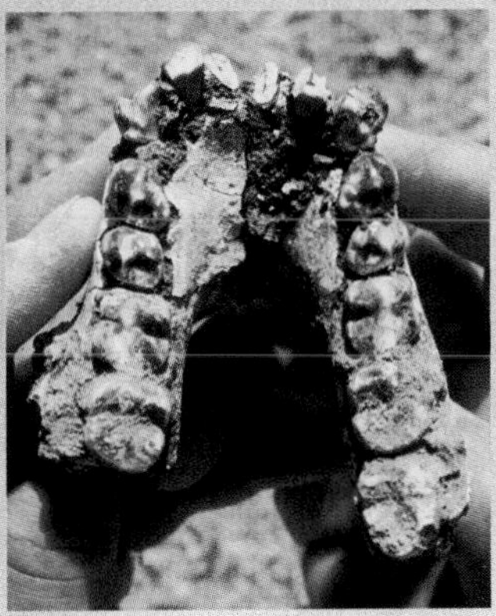

The First Hominids

- Two genera are considered hominid (of the human line): *Australopithecus* and *Homo*.
- Many stocky species of *Australopithecus* appear to be side branches on the evolutionary tree.
- A more slender species, *A. africanus,* appears to be our immediate ancestor.
- The oldest hominids are 4.4 million years old.

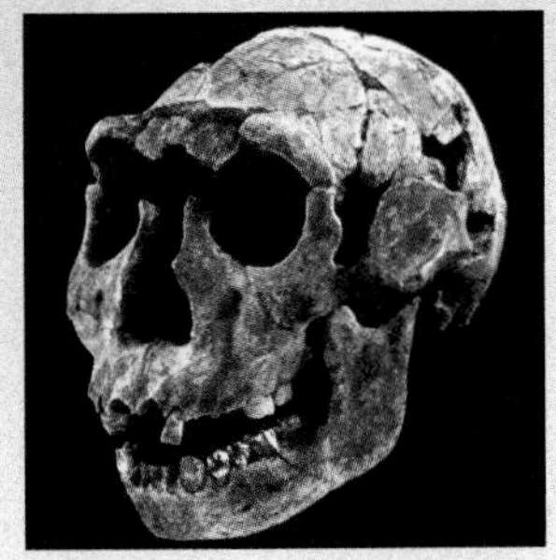

The First Humans

- The first hominid of the genus *Homo* appeared 2 million years ago.
- The most successful species of human, *H. erectus,* appeared in Africa about 1.5 million years ago.
- *H. erectus* migrated out of Africa to Europe and Asia.

Modern Humans

- Modern humans evolved in Africa about half a million years ago.
- Neanderthals, *H. neanderthalensis,* actually had bigger brains than Cro-Magnons, *H. sapiens.*
- Migrating out of Africa, *H. sapiens* retraced the spread of *H. erectus,* eventually supplanting it.

21.1 The Evolutionary Path to Apes

(a)

(b)

Figure 21.1 Prosimians.

(*a*) This tarsier, a prosimian native to tropical Asia, shows the characteristic features of primates: grasping fingers and toes and binocular vision. (*b*) All living lemurs, such as this ringtail lemur, are found in Madagascar.

Humans are new arrivals on the biological scene. Fifty years ago, a visual image was proposed that makes this point in a powerful way. Imagine a motion picture of Earth taken from space, beginning 757 million years ago, with one image being photographed each year. If you project this film at the normal speed of 24 images per second, it would take you a year to view it, with each day representing 2.1 million years. There is no life for the first three months. The first unicellular microbes appear in early April, and protists arrive later that month. In May the first vertebrates are seen, and plants cover the land. Dinosaurs dominate the world for 70 days, from late-September to 1 December, when they disappear abruptly. By late December the modern families of mammals appear, but not until midday on New Year's Eve are direct human ancestors seen. Between 9:30 and 10 p.m., *Homo sapiens* migrates out of Africa to Europe, Asia, and America. At 11:54 p.m., recorded human history begins.

Thus, the story of human evolution begins around 65 million years ago, with the explosive radiation of a group of small, arboreal mammals called the Archonta. These primarily insectivorous mammals had large eyes and were most likely **nocturnal** (active at night). Their radiation gave rise to different types of mammals, including bats, tree shrews, and **primates,** the order of mammals that contains humans.

The Earliest Primates

Primates are mammals with two distinct features that allowed them to succeed in the arboreal, insect-eating environment:

1. **Grasping fingers and toes.** Unlike the clawed feet of tree shrews and squirrels, primates have grasping hands and feet that let them grip limbs, hang from branches, seize food, and, in some primates, use tools. The first digit in many primates is opposable and at least some, if not all, of the digits have nails.
2. **Binocular vision.** Unlike the eyes of shrews and squirrels, which sit on each side of the head so that the two fields of vision do not overlap, the eyes of primates are shifted forward to the front of the face. This produces overlapping binocular vision that lets the brain judge distance precisely—important to an animal moving through the trees.

Other mammals have binocular vision, but only primates have both binocular vision and grasping hands, making them particularly well-adapted to their environment. While early primates were mostly insectivorous, their dentition began to change from the shearing, triangular-shaped molars specialized for insect eating to the more flattened, square-shaped molars and rodentlike incisors specialized for plant eating.

The Evolution of Prosimians

About 40 million years ago, the earliest primates split into two groups: the prosimians and the anthropoids. The **prosimians** ("before monkeys") looked something like a cross between a squirrel and a cat and were common in North America, Europe, Asia, and Africa. Only a few prosimians survive today, representing two lineages that split off and evolved from the earliest primates at roughly the same time as the anthropoids did. One lineage gave rise to lemurs and lorises and another to tarsiers (figure 21.1). In addition to having grasping digits and binocular vision, prosimians have large eyes with increased visual acuity. Most prosimians are nocturnal, feeding on fruits, leaves, and flowers, and many lemurs have long tails for balancing.

Origin of the Anthropoids

The **anthropoids,** or higher primates, include monkeys, apes, and humans. Anthropoids are almost all diurnal—that is, active during the day—feeding mainly on fruits and leaves. Evolution favored many changes in eye design, including color vision, that were adaptations to daytime foraging. An expanded brain governs the improved senses, with the brain case forming a larger portion of the head. Anthropoids, like the relatively few diurnal prosimians, live in groups with complex social interactions. In addition, the anthropoids tend to care for their young for prolonged periods, allowing for a long childhood of learning and brain development.

One of the most contentious issues in primate biology is the identity of the first anthropoid, the common ancestor of monkeys, apes, and humans. One candidate is a 45-million-year-old Chinese fossil called *Eosimias* ("dawn ape") known only from two jaws. It is very old, living just 25 million years after the dinosaurs became extinct. Another candidate, known from several skulls with jaws and teeth, is a 37-million-year-old Egyptian lemurlike primate about the size of a squirrel, called *Catopithecus.* The skulls have accepted anthropoid characteristics such as a complete bony cone around the eye socket, shovel-shaped incisor teeth, and forehead bones that are fused together rather than separate. While no clear determination of the earliest anthropoid is possible without more fossils, current evidence favors *Catopithecus.*

The early anthropoids, now extinct, are thought to have evolved in Africa. Their direct descendants are a very successful group of primates, the monkeys.

New World Monkeys. About 30 million years ago, some anthropoids migrated to South America, where they developed in isolation. Their South American descendants, known as the **New World monkeys,** are easy to identify: all are arboreal, they have flat spreading noses, and many of them grasp objects with long prehensile tails (figure 21.2*a*).

Old World Monkeys. Around 25 million years ago, anthropoids that remained in Africa split into two lineages: one gave rise to the **Old World monkeys** and one gave rise to the hominoids (see page 480). Old World monkeys have been evolving separately from New World monkeys for at least 30 million years and include ground-dwelling as well as arboreal species. None of the Old World monkeys have prehensile tails. Their nostrils are close together, their noses point downward, and some have toughened pads of skin for prolonged sitting (figure 21.2*b*).

21.1 The earliest primates arose from small, tree-dwelling insect eaters and gave rise to prosimians and then anthropoids. Early anthropoids gave rise to New World monkeys and Old World monkeys.

(a) (b)

Figure 21.2 New and Old World monkeys.
(*a*) All New World monkeys, such as this golden lion tamarin, are arboreal, and many have prehensile tails. (*b*) Old World monkeys lack prehensile tails, and many are ground dwellers.

(a) Gibbon

(b) Orangutan

(c) Gorilla

21.2 How the Apes Evolved

From anthropoid ancestors evolved **hominoids,** which are the **apes** and the **hominids** (humans and their direct ancestors). The living apes consist of the gibbon (genus *Hylobates*), orangutan (*Pongo*), gorilla (*Gorilla*), and chimpanzee (*Pan*) (figure 21.3). Apes have larger brains than monkeys, and they lack tails. With the exception of the gibbon, which is small, all living apes are larger than any monkey. Apes exhibit the most adaptable behavior of any mammal except human beings. Once widespread in Africa and Asia, apes are rare today, living in relatively small areas. No apes ever occurred in North or South America.

The First Hominoid

Considerable controversy exists about the identity of the first hominoid. During the 1980s it was commonly believed that the common ancestor of apes and hominids was a late Miocene ape living 5 to 10 million years ago. In 1932, a candidate fossil, an 8-million-year-old jaw with teeth, was unearthed in India. It was called *Ramapithecus* (after the Hindi deity Rama). However, these fossils have never been found in Africa, and more complete fossils discovered in 1981 made it clear that *Ramapithecus* is in fact closely related to the orangutan. Attention has now shifted to an earlier Miocene ape, *Proconsul,* which has many of the characteristics of Old World monkeys but lacks a tail and has apelike hands, feet, and pelvis. However, because very few fossils have been recovered from the period 5 to 10 million years ago, it is not yet possible to identify with certainty the first hominoid ancestor.

(d) Chimpanzee

Figure 21.3 The living apes.

There are four kinds of living apes. (*a*) Mueller gibbon, *Hylobates muelleri.* (*b*) Orangutan, *Pongo pygmaeus.* (*c*) Gorilla, *Gorilla gorilla.* (*d*) Chimpanzee, *Pan troglodytes.*

Which Ape Is Our Closest Relative?

Studies of ape DNA have explained a great deal about how the living apes evolved. The Asian apes evolved first. The line of apes leading to gibbons diverged from other apes about 15 million years ago, while orangutans split off about 10 million years ago (see figure 21.8). Neither are closely related to humans.

The African apes evolved more recently, between 6 and 10 million years ago. These apes are the closest living relatives to humans; some taxonomists have even advocated placing humans and the African apes in the same zoological family, the Hominidae. Chimpanzees are more closely related to humans than gorillas are, diverging from the ape line less than 6 million years ago. Because this split was so recent, the genes of humans and chimpanzees have not had time to evolve many differences—humans and chimpanzees share 98.4% of their nuclear DNA, a level of genetic similarity normally found between sibling species of the same genus! A human hemoglobin molecule differs from its chimpanzee counterpart in only a single amino acid.

Gorilla DNA differs from human DNA by about 2.3%. This somewhat greater genetic difference reflects the greater time since gorillas split off from the ape line, some 8 million years ago. Sometime after the gorilla lineage diverged, the common ancestor of all hominids split off from the ape line to begin the evolutionary journey leading to hominids. Fossils of the earliest hominids, described later in the chapter, suggest that the common ancestor of the hominids was more like a chimpanzee than a gorilla.

Comparing Apes to Hominids

The common ancestor of apes and hominids is thought to have been an arboreal climber. Much of the subsequent evolution of the hominoids reflected different approaches to locomotion. Hominids became **bipedal,** walking upright, while the apes evolved knuckle-walking, supporting their weight on the back sides of their fingers (monkeys, by contrast, use the palms of their hands).

Humans depart from apes in several areas of anatomy related to bipedal locomotion (figure 21.4). Because humans walk on two legs, their vertebral column is more curved than an ape's, and the human spinal cord exits from the bottom rather than the back of the skull. The human pelvis has become broader and more bowl-shaped, with the bones curving forward to center the weight of the body over the legs. The hip, knee, and foot (in which the human big toe no longer splays sideways) have all changed proportions.

Being bipedal, humans carry much of the body's weight on the lower limbs, which comprise 32% to 38% of the body's weight and are longer than the upper limbs; human upper limbs do not bear the body's weight and make up only 7% to 9% of human body weight. African apes walk on all fours, with the upper and lower limbs both bearing the body's weight; in gorillas, the longer upper limbs account for 14% to 16% of body weight, the somewhat shorter lower limbs for about 18%.

21.2 Hominoids, the apes and hominids, arose from Old World monkeys. Among living apes, chimpanzees seem the most closely related to humans.

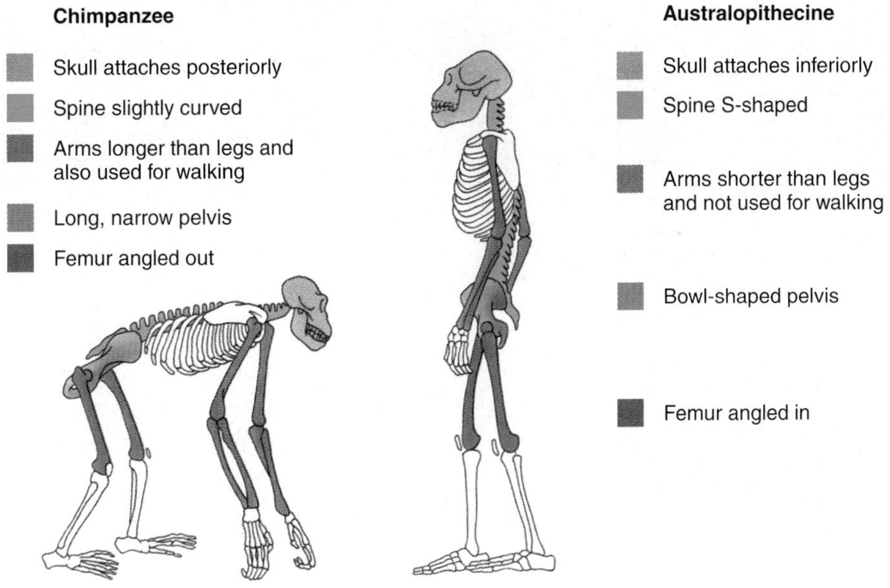
Chimpanzee
Skull attaches posteriorly
Spine slightly curved
Arms longer than legs and also used for walking
Long, narrow pelvis
Femur angled out
Australopithecine
Skull attaches inferiorly
Spine S-shaped
Arms shorter than legs and not used for walking
Bowl-shaped pelvis
Femur angled in

Figure 21.4 A comparison of ape and hominid skeletons.
Early humans, such as australopithecines, were able to walk upright because their arms were shorter, their spinal cord exited from the bottom of the skull, their pelvis was bowl-shaped and centered the body weight over the legs, and their femurs angled inward, directly below the body, to carry its weight.

21.3 An Evolutionary Tree with Many Branches

Five to 10 million years ago, the world's climate began to get cooler, and the great forests of Africa were largely replaced with savannas and open woodland. In response to these changes, a new kind of ape was evolving, one that was bipedal. They are classified as hominids—that is, of the human line.

There are two major groups of hominids: three to seven species of the genus *Homo* (depending how you count them) and seven species of the older, smaller-brained genus ***Australopithecus.*** In every case where the fossils allow a determination to be made, the hominids are bipedal, walking upright. Bipedal locomotion is the hallmark of hominid evolution. We will first discuss *Australopithecus* and then *Homo.*

Discovery of *Australopithecus*

The first hominid was discovered in 1924 by Raymond Dart, an anatomy professor at Johannesburg in South Africa. One day, a mine worker brought him an unusual chunk of sandy rock. Picking away at it, Professor Dart uncovered a skull unlike that of any ape he had ever seen. Beautifully preserved, the skull was of a five-year-old individual, still with its milk teeth. While the skull had many apelike features such as a projecting face and a small brain, it had distinctly human features as well—for example, a rounded jaw unlike the pointed jaw of apes. The ventral position of the foramen magnum (the hole at the base of the skull from which the spinal cord emerges) suggested that the creature had walked upright. Dart concluded it was a human ancestor.

What riveted Dart's attention was that the rock in which the skull was embedded had been collected near other fossils that suggested that the rocks and their fossils were several million years old! At that time, the oldest reported fossils of hominids were less than 500,000 years old, so the ancientness of this skull was unexpected and exciting. Scientists now estimate Dart's skull to be 2.8 million years old. Dart called his find *Australopithecus africanus* (from the Latin *australo,* meaning "southern," and the Greek *pithecus,* meaning "ape"), the ape from the south of Africa.

Today, fossils are dated by the relatively new process of single-crystal laser-fusion dating. A laser beam melts a single potassium feldspar crystal, releasing argon gas, which is measured in a gas mass spectrometer. Because the argon in the crystal has accumulated at a known rate, the amount released reveals the age of the rock and thus of nearby fossils. The margin of error is less than 1%.

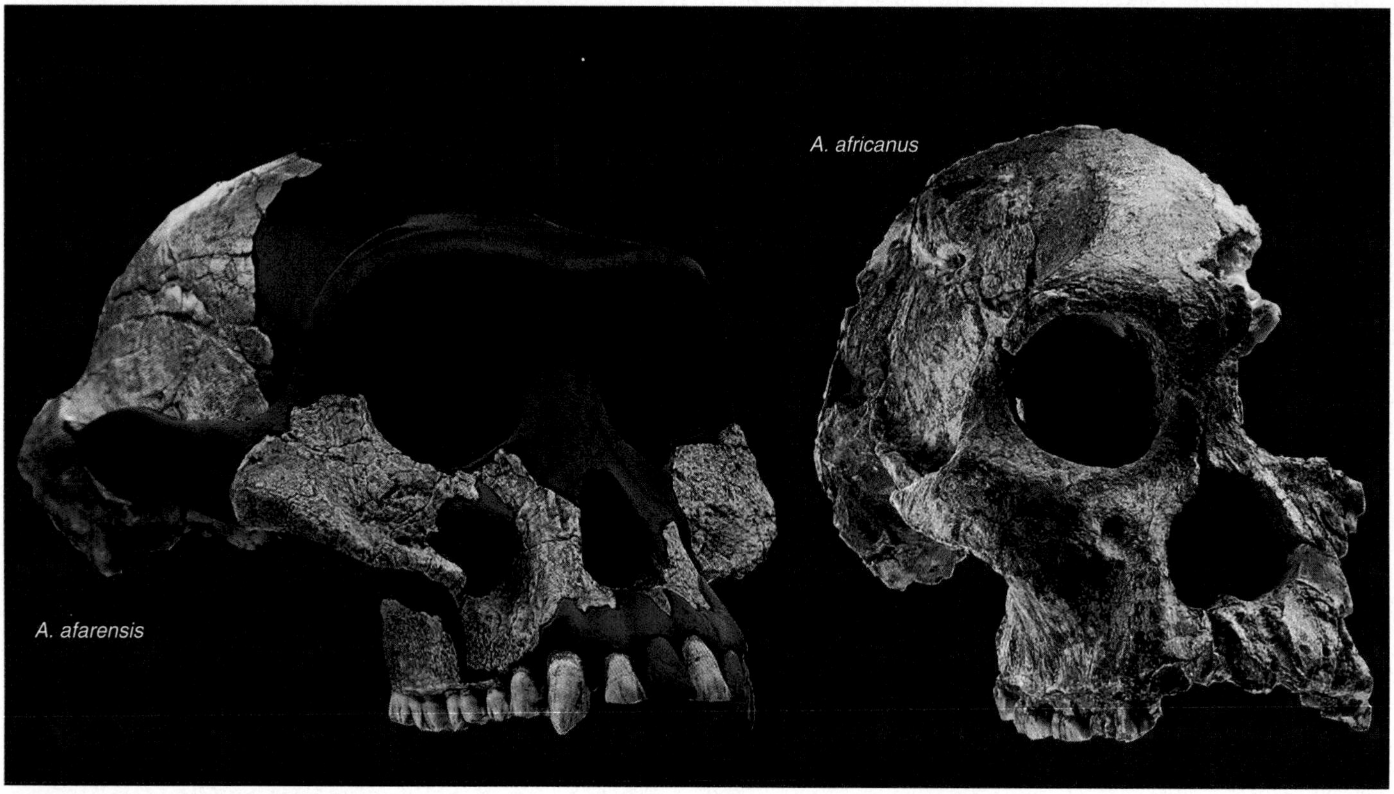
A. africanus
A. afarensis

Figure 21.5 Nearly human.

These four skulls, all photographed from the same angle, are among the best specimens available of the key *Australopithecus* species.

Other Kinds of *Australopithecus*

In 1938, a second, stockier kind of *Australopithecus* was unearthed in South Africa. Called *A. robustus,* it had massive teeth and jaws. In 1959, in East Africa, Mary Leakey discovered a third kind of *Australopithecus*—*A. boisei* (after Charles Boise, an American-born businessman who contributed to the Leakeys' projects)—that was even more stockily built. Like the other australopithecines, *A. boisei* was very old—almost 2 million years. Nicknamed "Nutcracker man," *A. boisei* had a great bony ridge on the crest of the head to anchor its immense jaw muscles; the ridge looked like a Mohawk haircut of bone (figure 21.5).

In 1974, anthropologist Don Johanson went to the remote Afar Desert of Ethiopia in search of early human fossils and hit the jackpot. He found the most complete, best preserved australopithecine skeleton known. Nicknamed "Lucy," the skeleton was 40% complete and over 3 million years old. The skeleton and other similar fossils have been assigned the scientific name *Australopithecus afarensis* (from the Afar Desert). The shape of the pelvis indicated that Lucy was a female, and her leg bones proved she walked upright. Her teeth were distinctly hominid, but her head was shaped like an ape's. Her brain was no larger than that of a chimpanzee, about 400 cubic centimeters, about the size of an orange. (For comparison, the average human brain today is 1,350 cubic centimeters.) More than 300 specimens of *A. afarensis* have since been discovered.

In the last 10 years, three additional kinds of australopithecines have been reported. These seven species provide ample evidence that australopithecines were a diverse group, and additional species will undoubtedly be described by future investigators. The evolution of hominids seems to have begun with an initial radiation of numerous species.

Early Australopithecines Were Bipedal

We now know australopithecines from hundreds of fossils. The structure of these fossils clearly indicate that australopithecines walked upright. These early hominids weighed about 18 kilograms and were about 1 meter tall. Their dentition was distinctly hominid, but their brains were not any larger than those of apes, generally 500 cubic centimeters or less. *Homo* brains, by comparison, are usually larger than 600 cubic centimeters; modern *Homo sapiens* brains average 1,350 cubic centimeters.

Australopithecine fossils have been found only in Africa. Although all the fossils to date come from sites in South and East Africa (except for one specimen from Chad), it is probable that they lived over a much broader area of Africa. Only in South and East Africa, however, are sediments of the proper age exposed to fossil hunters.

21.3 The australopithecines were hominids that walked upright and lived in Africa over 3 million years ago.

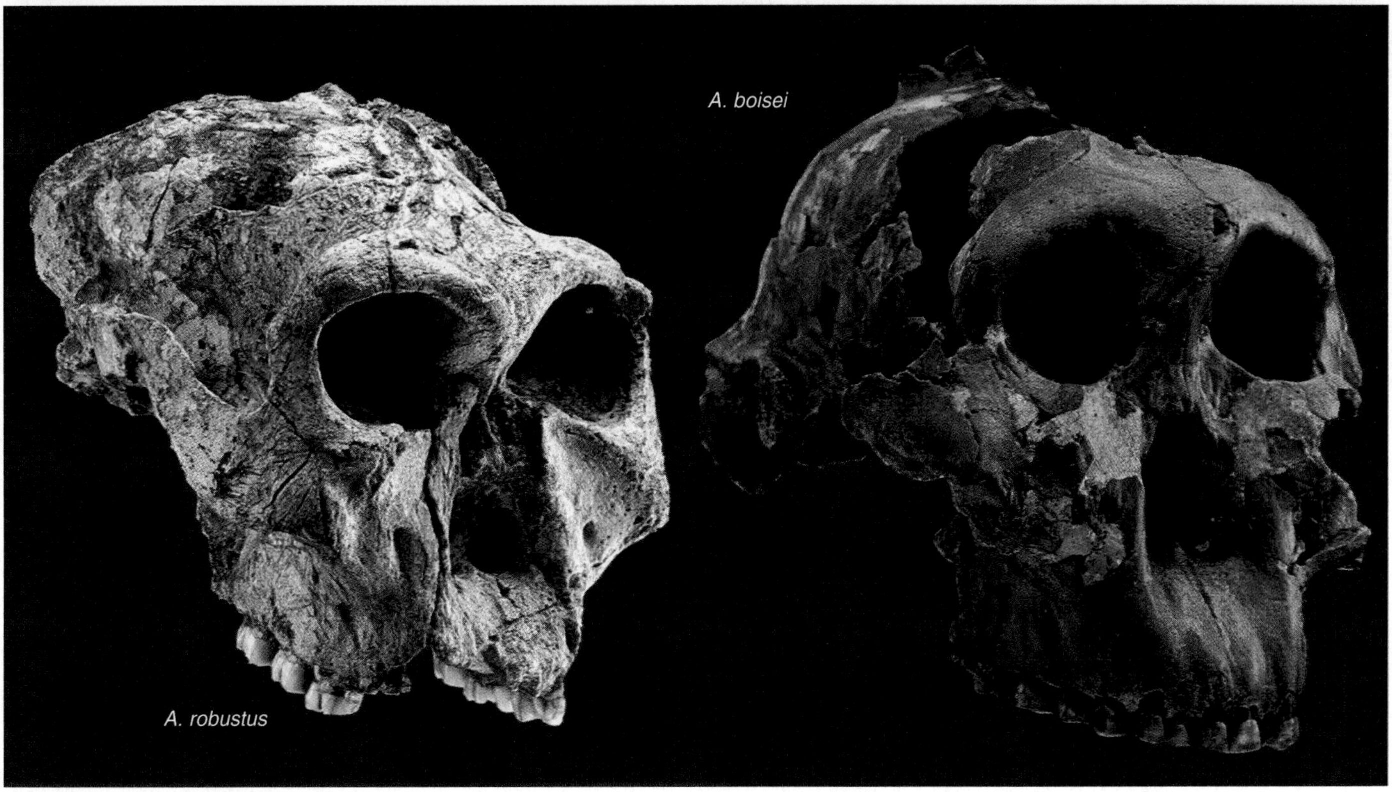
A. boisei
A. robustus

Figure 21.5 (continued)

21.4 The Origins of Bipedalism

For many years, biologists have debated the sequence of events that led to the evolution of hominids. A key element may have been bipedalism. Bipedalism seems to have evolved as our ancestors left dense forests for grasslands and open woodland (figure 21.6). One school of thought proposes that hominid brains enlarged first, and then hominids became bipedal. Another school sees bipedalism as a precursor to bigger brains. Those who favor the brain-first hypothesis speculate that human intelligence was necessary to make the decision to walk upright and move out of the forests and onto the grassland. Those who favor the bipedalism-first hypothesis argue that bipedalism freed the forelimbs to manufacture and use tools, favoring the subsequent evolution of bigger brains.

A treasure trove of fossils unearthed in Africa seems to have settled the debate. These fossils demonstrate that bipedalism extended back 4 million years ago; knee joints, pelvis, and leg bones all exhibit the hallmarks of an upright stance. Substantial brain expansion, on the other hand, did not appear until roughly 2 million years ago. In hominid evolution, upright walking clearly preceded large brains.

Remarkable evidence that early hominids were bipedal is a set of some 69 hominid footprints found at Laetoli, East Africa. Two individuals, one larger than the other, walked side by side for 27 meters, their footprints preserved in 3.7-million-year-old volcanic ash! These footprints came from an early hominid, walking upright. The impression in the ash (figure 21.7) reveals a strong heelstrike and a deep indentation made by the big toe, much as you might make in sand when pushing off to take a step. Importantly, the big toe is not splayed out to the side as in a monkey or ape—the footprints were clearly made by a hominid.

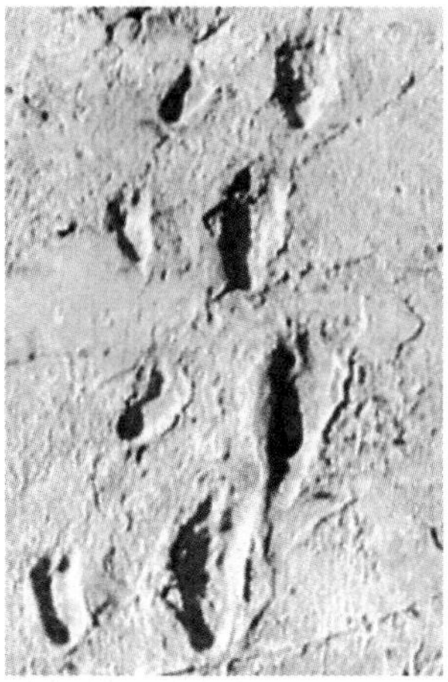

Figure 21.7 The Laetoli footprints.
These *Australopithecus* footprints are 3.7 million years old.

The evolution of bipedalism marks the beginning of the hominids. The reason *why* bipedalism evolved in hominids remains a matter of controversy. No tools appeared until 2.5 million years ago, so toolmaking seems an unlikely cause. Alternative ideas suggest that walking upright is faster and uses less energy than walking on four legs; that an upright posture permits hominids to pick fruit from trees and see over tall grass; that being upright reduces the body surface exposed to the sun's rays; that an upright stance aided the wading of aquatic hominids; and that bipedalism frees the forelimbs of males to carry food back to females, encouraging pair-bonding. All of these suggestions have their proponents, and none is universally accepted. The origin of bipedalism, the key event in the evolution of hominids (figure 21.8), seems destined to remain a mystery for now.

21.4 The evolution of bipedalism—walking upright—marks the beginning of hominid evolution, although no one is quite sure why bipedalism evolved.

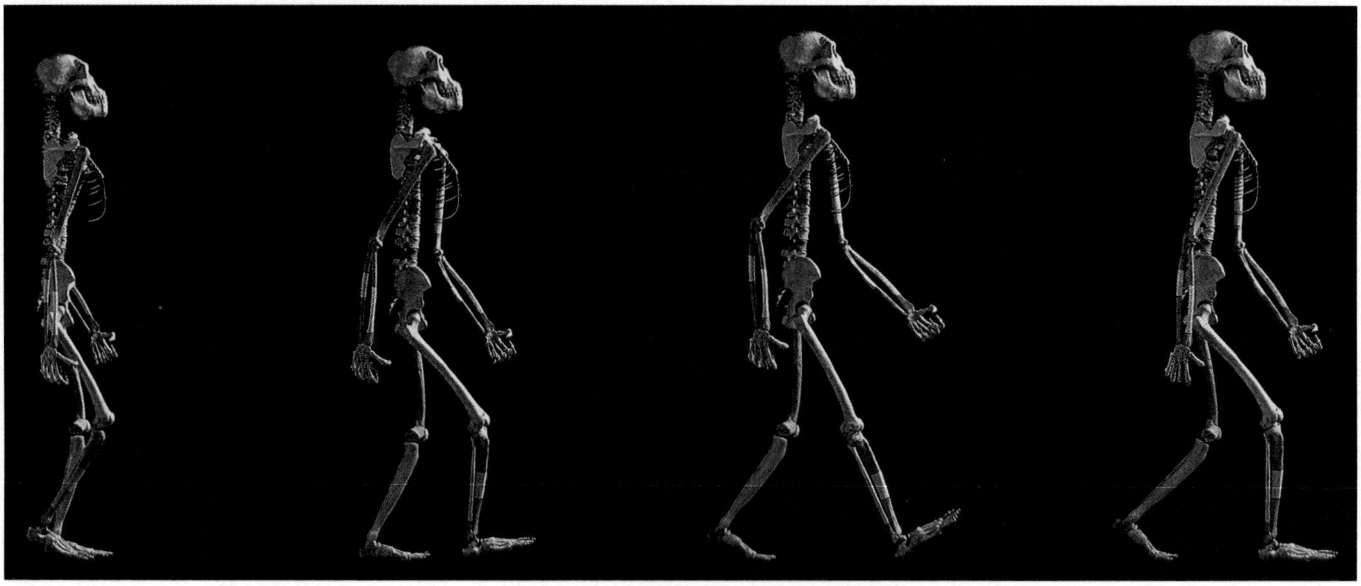

Figure 21.6 A reconstruction of an early hominid walking upright.
These articulated plaster skeletons, made by Owen Lovejoy and his students at Kent State University, depict an early hominid (*Australopithecus afarensis*) walking upright.

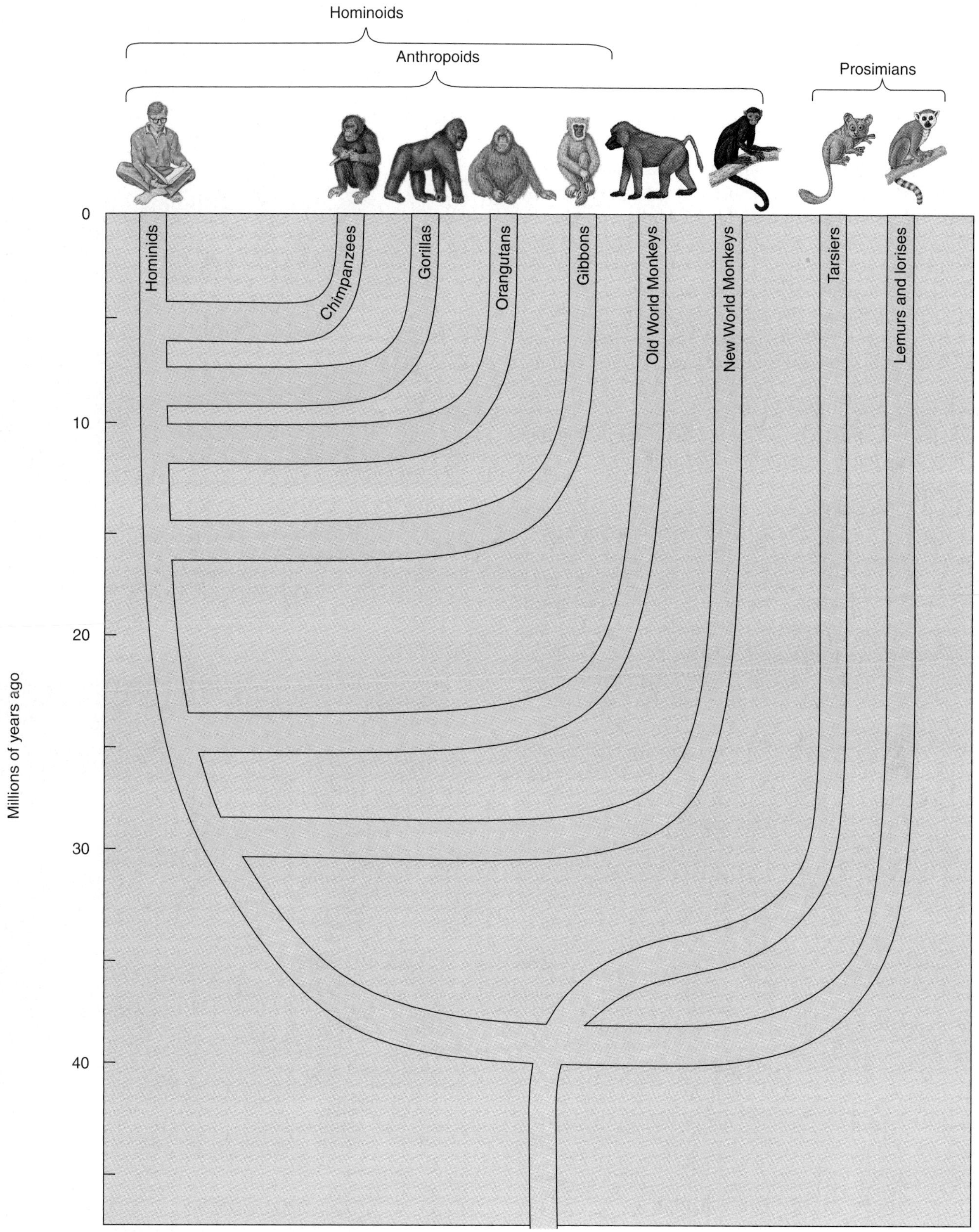
Hominoids
Anthropoids
Prosimians
Hominids
Chimpanzees
Gorillas
Orangutans
Gibbons
Old World Monkeys
New World Monkeys
Tarsiers
Lemurs and lorises
Millions of years ago
0
10
20
30
40

Figure 21.8 A primate evolutionary tree.

The most ancient of the primates are the prosimians, while the hominids were the most recent to evolve.

21.5 The Beginning of Hominid Evolution

The Oldest Known Hominid

In 1994, a remarkable, nearly complete fossil skeleton was unearthed in Ethiopia (the country where Lucy was found). The skeleton is still being painstakingly assembled, but it seems almost certainly to have been bipedal; the foramen magnum, for example, is situated far forward, as in other bipedal hominids. Some 4.4 million years old, it is the most ancient hominid yet discovered. It is significantly more apelike than any australopithecine and so has been assigned to a new genus, *Ardipithecus ramidus* (figure 21.9). In the local Afar language, *ardi* means "ground," while *ramid* means "root"; *pithecus* is the Greek word for "ape."

The First Australopithecine

In 1995, hominid fossils of nearly the same age, 4.2 million years old, were found in the Rift Valley in Kenya. The fossils are fragmentary, but they include complete upper and lower jaws, a piece of the skull, arm bones, and a partial tibia (leg bone). The fossils were assigned to the species *Australopithecus anamensis* (figure 21.10); *anam* is the Turkana word for "lake." They were categorized in the genus *Australopithecus* rather than *Ardipithecus* because the fossils have bipedal characteristics and are much less apelike than *A. ramidus.* While clearly australopithecine, the fossils are intermediate in many ways between apes and *Australopithecus afarensis.* Numerous fragmentary specimens of *Australopithecus anamensis* have since been found.

Most researchers agree that these slightly built *A. anamensis* individuals represent the true base of our family tree, the first members of the genus *Australopithecus,* and thus ancestor to *A. afarensis* and all other australopithecines.

Figure 21.9 Our earliest known ancestor.

A tooth from *Ardipithecus ramidus,* discovered in 1994. The name *ramidus* is from the Latin word for "root," as this is thought to be the root of the hominid family tree. The earliest known hominid, at 4.4 million years old, *A. ramidus* was about the size of a chimpanzee and apparently could walk upright.

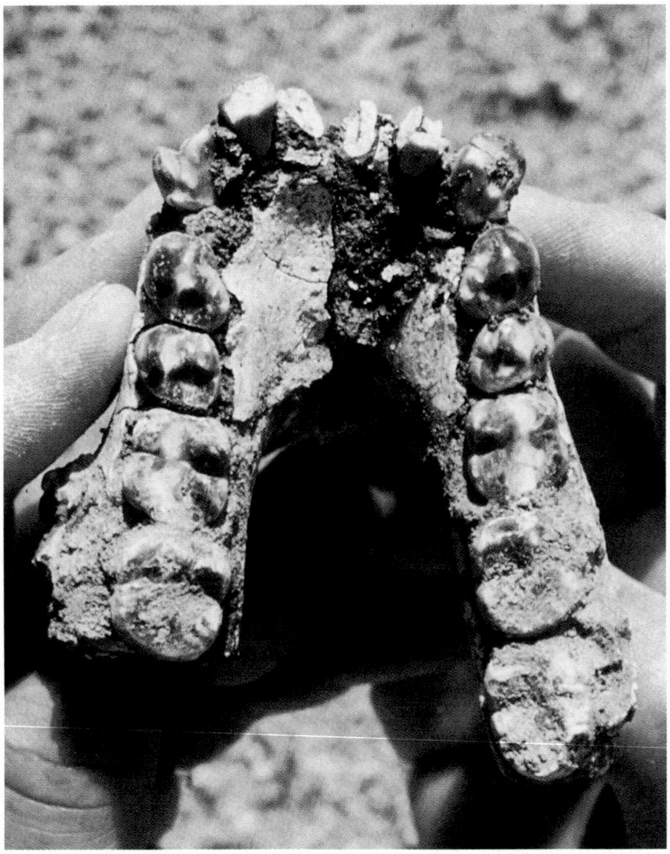

Figure 21.10 The earliest australopithecine.

This fossil jaw of *Australopithecus anamensis* was discovered in 1995 along with a piece of the skull, arm bones, and a leg bone. These fossils are about 4.2 million years old, making *A. anamensis* the oldest known australopithecine.

Differing Views of the Hominid Family Tree

Investigators take two different philosophical approaches to characterizing the diverse group of African hominid fossils. One group focuses on common elements in different fossils and tends to lump together fossils that share key characters. Differences between the fossils are attributed to diversity within the group. The hominid phylogenetic tree in figure 21.11*a* illustrates such a lumper's view.

Other investigators focus more pointedly on the differences between hominid fossils. They are more inclined to assign fossils that exhibit differences to different species. The hominid phylogenetic tree in figure 21.11*b* presents a splitter's view. Where the "lumpers" tree presents three species of *Homo,* for example, the "splitters" tree presents no fewer than seven! The splitter's approach puts more of the data to work in attempting to ferret out evolutionary relationships and more accurately represents the view of most human paleontologists.

21.5 The root of the hominid evolutionary tree is only imperfectly known. The earliest australopithecine yet described is *A. anamensis,* over 4 million years old.

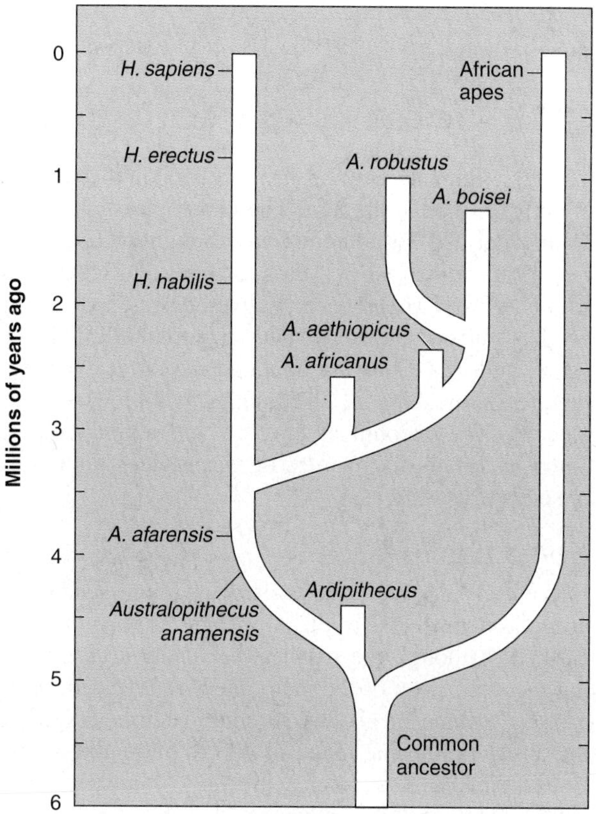

Millions of years ago
0
1
2
3
4
5
6
H. sapiens
H. erectus
H. habilis
A. afarensis
Australopithecus anamensis
African apes
A. robustus
A. boisei
A. aethiopicus
A. africanus
Ardipithecus
Common ancestor

(a)

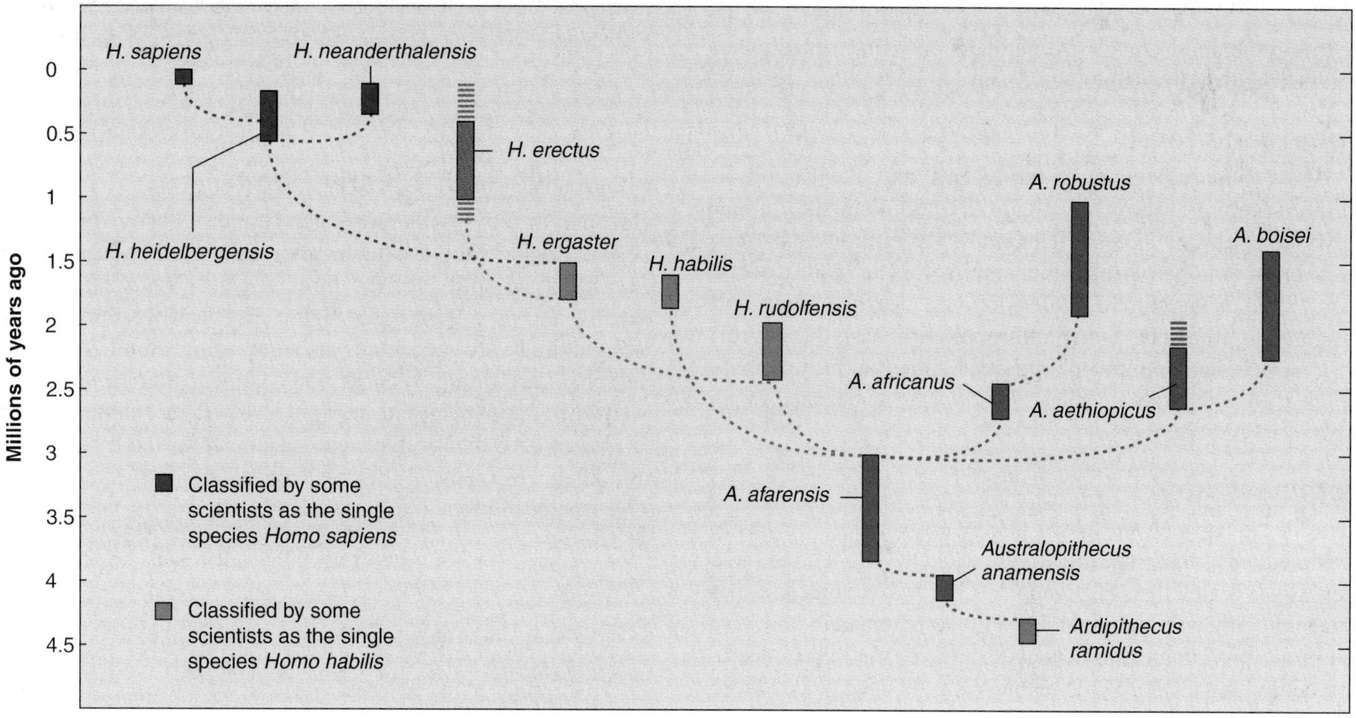

Millions of years ago
0
0.5
1
1.5
2
2.5
3
3.5
4
4.5
H. sapiens
H. neanderthalensis
H. erectus
H. heidelbergensis
H. ergaster
H. habilis
H. rudolfensis
A. robustus
A. boisei
A. africanus
A. aethiopicus
A. afarensis
Australopithecus anamensis
Ardipithecus ramidus
Classified by some scientists as the single species Homo sapiens
Classified by some scientists as the single species Homo habilis

(b)

Figure 21.11 Two views of the hominid phylogenetic tree.

(*a*) A lumper's approach. Six species of *Australopithecus* and only three species of *Homo* are included. (*b*) A splitter's approach. In this tree, the most widely accepted, the vertical bars show the known dates of first and last appearances of proposed species. Six species of *Australopithecus* and seven of *Homo* are included.

21.6 African Origin: Early *Homo*

The first humans evolved from australopithecine ancestors about 2 million years ago. The exact ancestor has not been clearly defined but is commonly thought to be *A. afarensis.* Only within the last 30 years have a significant number of fossils of early *Homo* been uncovered. An explosion of interest has fueled intensive field exploration in the last few years, and new finds are announced yearly; every year, our picture of the base of the human evolutionary tree grows clearer. The account given here will undoubtedly be outdated by future discoveries, but it provides a good example of science at work.

Homo habilis

In the early 1960s, stone tools were found scattered among hominid bones close to the site where *A. boisei* had been unearthed. Although the fossils were badly crushed, painstaking reconstruction of the many pieces suggested a skull with a brain volume of about 680 cubic centimeters, larger than the australopithecine range of 400 to 550 cubic centimeters. Because of its association with tools, this early human was called ***Homo habilis,*** meaning "handy man." Partial skeletons discovered in 1986 indicate that *H. habilis* was small in stature, with arms longer than legs and a skeleton much like *Australopithecus.* Because of its general similarity to australopithecines, many researchers at first questioned whether this fossil was human.

Homo rudolfensis

In 1972, Richard Leakey, working east of Lake Rudolf in northern Kenya, discovered a virtually complete skull about the same age as *H. habilis.* The skull, 1.9 million years old, had a brain volume of 750 cubic centimeters and many of the characteristics of human skulls—it was clearly human and not australopithecine. Some anthropologists assign this skull to *H. habilis,* arguing it is a large male. Other anthropologists assign it to a separate species, *H. rudolfensis,* because of its substantial brain expansion.

Homo ergaster

Some of the early *Homo* fossils being discovered do not easily fit into either of these species (figure 21.12). They tend to have even larger brains than *H. rudolfensis,* with skeletons less like an australopithecine and more like a modern human in both size and proportion. Interestingly, they also have small cheek teeth, as modern humans do. Some anthropologists have placed these specimens in a third species of early *Homo, H. ergaster* (*ergaster* is from the Greek word for "workman").

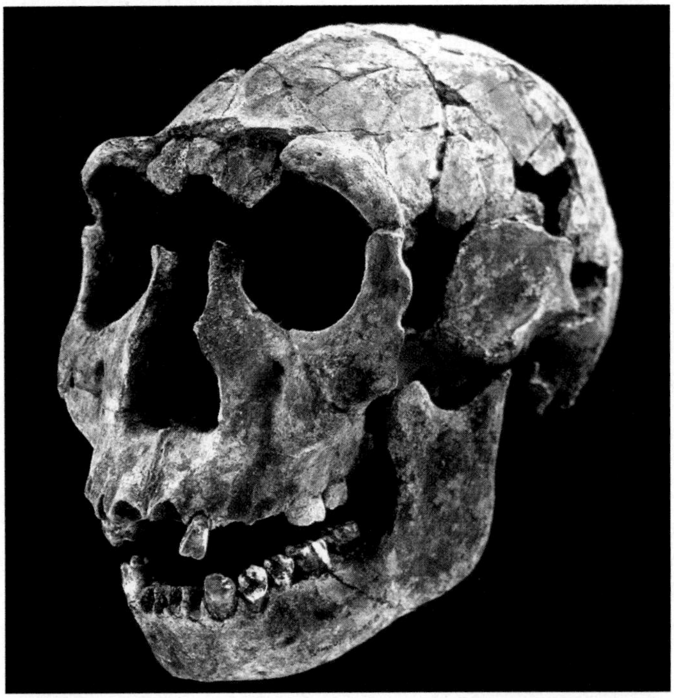

Figure 21.12 Early *Homo.*

This skull of a boy, who apparently died in early adolescence, is 1.6 million years old and has been assigned to the species *Homo ergaster* (a form of *Homo habilis* recognized by some as a separate species). Much larger than earlier hominids, he was about 1.5 meters in height and weighed approximately 47 kilograms.

How Diverse Was Early *Homo?*

Because so few fossils have been found of early *Homo,* there is lively debate about whether they should all be lumped into *H. habilis* or split into the three species *H. rudolfensis, H. habilis,* and *H. ergaster* (see figure 21.11). If the three species designations are accepted, as increasing numbers of researchers are doing, then it would appear that *Homo* underwent an adaptive radiation (as described in chapter 2) with *H. rudolfensis,* the most ancient species, followed by *H. habilis* and then *H. ergaster.* Because of its modern skeleton, *H. ergaster* is thought the most likely ancestor to later species of *Homo.*

21.6 Early species of *Homo,* the oldest members of our genus, had a distinctly larger brain than australopithecines and most likely used tools. There may have been several different species.

21.7 Out of Africa: *Homo erectus*

Our picture of what early *Homo* was like lacks detail because it is based only on a few specimens. Some scientists still dispute *H. habilis*'s qualifications as a true human because it had not moved far from its australopithecine roots. There is no such doubt about the species that replaced it, ***Homo erectus.*** Many specimens have been found (see figure 21.14), and *H. erectus* was without any doubt a true human.

Java Man

After the publication of Darwin's book *On the Origin of Species* in 1859, there was much public discussion about "the missing link," the fossil ancestor common to both humans and apes. Puzzling over the question, a Dutch doctor and anatomist named Eugene Dubois decided to seek fossil evidence of the missing link in the home country of the orangutan, Java. Dubois set up practice in a river village in eastern Java. Digging into a hill that villagers claimed had "dragon bones," he unearthed a skull cap and a thighbone in 1891. He was very excited by his find, informally called **Java man,** for three reasons:

1. The structure of the thigh bone clearly indicated that the individual had long, straight legs and was an excellent walker.
2. The size of the skull cap suggested a *very* large brain, about 1,000 cubic centimeters.
3. Most surprising, the bones seemed as much as 500,000 years old, judged by other fossils Dubois unearthed with them.

The fossil hominid that Dubois had found was far older than any discovered up to that time, and few scientists were willing to accept that it was an ancient species of human.

Peking Man

Another generation passed before scientists were forced to admit that Dubois had been right all along. In the 1920s a skull was discovered near Peking (now Beijing), China, that closely resembled Java man. Continued excavation at the site eventually revealed 14 skulls, many excellently preserved. Crude tools were also found, and most important of all, the ashes of campfires. Casts of these fossils were distributed for study to laboratories around the world. The originals were loaded onto a truck and evacuated from Peking at the beginning of World War II, only to disappear into the confusion of history. No one knows what happened to the truck or its priceless cargo. Fortunately, Chinese scientists have excavated numerous additional skulls of **Peking man** since 1949.

A Very Successful Species

Java man and Peking man are now recognized as belonging to the same species, *H. erectus. H. erectus* was a lot larger than *H. habilis*—about 1.5 meters tall. It had a large brain, about 1,000 cubic centimeters (figure 21.13), and walked erect. Its skull had prominent brow ridges and, like modern humans, a rounded jaw. Most interesting of all, the shape of the skull interior suggests that *H. erectus* was able to talk.

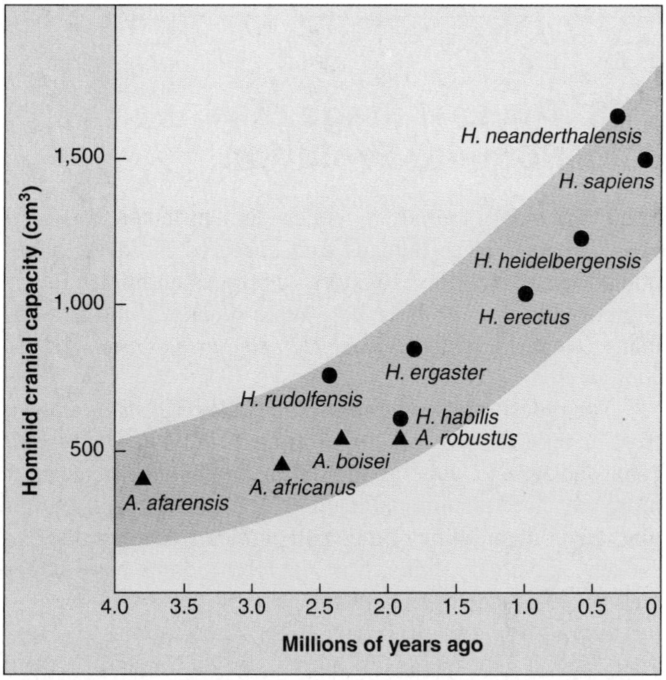

Hominid cranial capacity (cm^3)
1,500
1,000
500
Millions of years ago
4.0
3.5
3.0
2.5
2.0
1.5
1.0
0.5
0
H. neanderthalensis
H. sapiens
H. heidelbergensis
H. erectus
H. ergaster
H. rudolfensis
H. habilis
A. robustus
A. boisei
A. africanus
A. afarensis

Figure 21.13 Brain size increased as hominids evolved.

Homo erectus had a larger brain than early *Homo,* which in turn had larger brains than those of the australopithecines they shared East African grasslands with. Maximum brain size (and apparently body size) was attained by *H. neanderthalensis.* Both brain and body size appear to have declined some 10% in recent millennia.

Where did *H. erectus* come from? It should come as no surprise to you that it came out of Africa. In 1976 a complete *H. erectus* skull was discovered in East Africa. It was 1.5 million years old, a million years older than the Java and Peking finds. Far more successful than *H. habilis, H. erectus* quickly became widespread and abundant in Africa and within 1 million years had migrated into Asia and Europe. A social species, *H. erectus* lived in tribes of 20 to 50 people, often dwelling in caves. They successfully hunted large animals, butchered them using flint and bone tools, and cooked them over fires—the site in China contains the remains of horses, bears, elephants, deer, and rhinoceroses.

H. erectus survived for over a million years, longer than any other species of human. These very adaptable humans disappeared in Africa only about 500,000 years ago, as modern humans were emerging. Interestingly, they survived even longer in Asia, until about 250,000 years ago.

21.7 *Homo erectus* evolved in Africa and migrated from there to Europe and Asia.

21.8 The Last Stage of Hominid Evolution

The evolutionary journey to modern humans entered its final phase when modern humans first appeared in Africa about 600,000 years ago. Investigators who focus on human diversity consider there to have been three species of modern humans: *Homo heidelbergensis, H. neanderthalensis,* and *H. sapiens* (see figure 21.11).

The oldest modern human, *H. heidelbergensis,* is known from a 600,000-year-old fossil from Ethiopia. Although it coexisted with *H. erectus* in Africa, *H. heidelbergensis* has more advanced anatomical features, such as a bony keel running along the midline of the skull, a thick ridge over the eye sockets, and a large brain. Also, its forehead and nasal bones are very like those of *H. sapiens.*

As *H. erectus* was becoming rarer, about 130,000 years ago, a new species of human arrived in Europe from Africa. *H. neanderthalensis* likely branched off from the ancestral line leading to modern humans 500,000 years ago. Compared to modern humans, Neanderthals were short, stocky, and powerfully built. Their skulls were massive, with protruding faces, heavy, bony ridges over the brows (figure 21.14), and larger braincases.

Out of Africa—Again?

The oldest fossil known of ***Homo sapiens,*** our own species, is from Ethiopia and is about 130,000 years old. Outside of Africa and the Middle East, there are no clearly dated *H. sapiens* fossils older than roughly 40,000 years of age. The implication is that *H. sapiens* evolved in Africa and then migrated to Europe and Asia—the Recently-Out-of-Africa model. An opposing view, the Multiregional hypothesis, argues that the human races independently evolved from *H. erectus* in different parts of the world.

Recently, scientists studying human mitochondrial DNA (mDNA) have helped clarify this controversy. Because DNA accumulates mutations over time, the oldest populations should show the greatest genetic diversity. Researchers sequencing the entire mDNA from 53 individuals of differing ethnic backgrounds found that all modern humans shared a common ancestor 170,000 years ago, confirming that *H. sapiens* originated in Africa at about this time. The data reveal a distinct branch on the human family tree 52,000 years ago, separating Africans from non-Africans. This is consistent with the hypothesis that humans originated in Africa, from there spreading to all parts of the world, retracing the path taken by *H. erectus* half a million years before (figure 21.15).

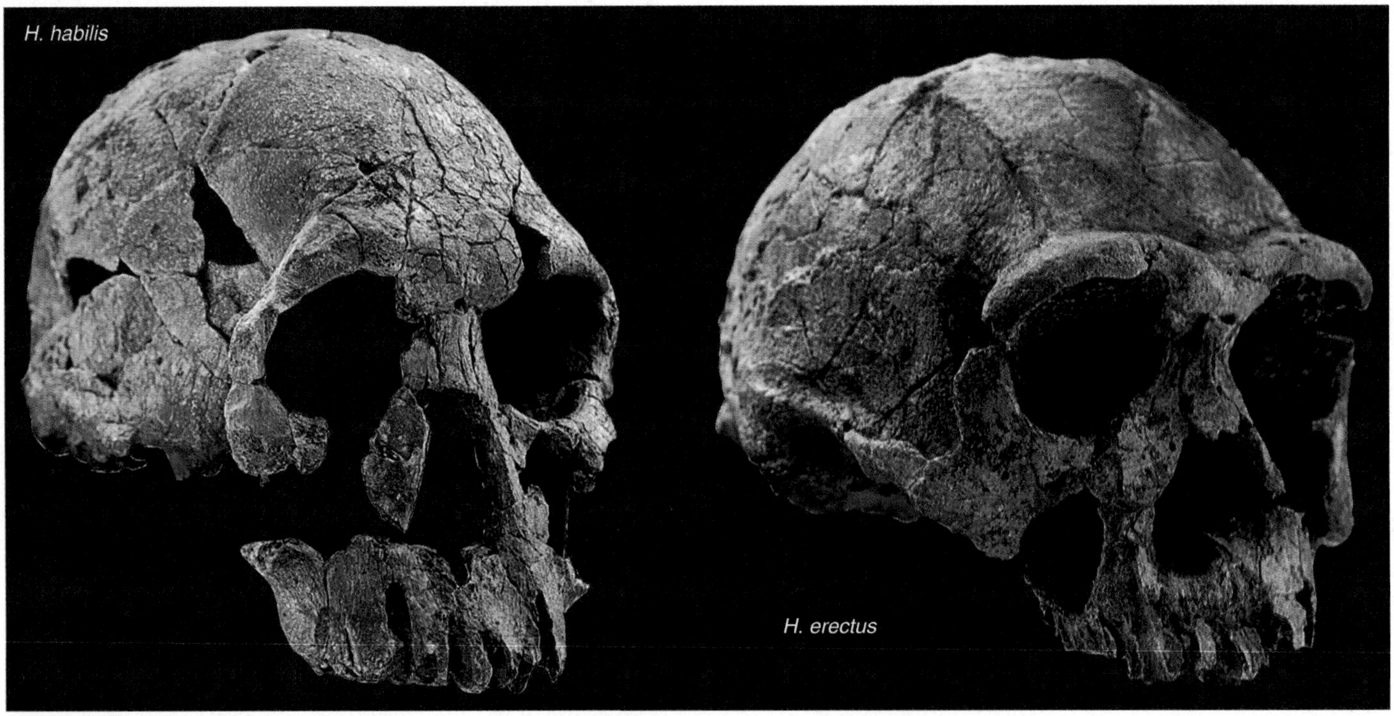
H. habilis
H. erectus

Figure 21.14 Our own genus.
These four skulls illustrate the changes that have occurred during the evolution of the genus *Homo*. The *H. sapiens* skull is essentially the same as human skulls today. The skulls were photographed from the same angle.

Another way to examine the human family tree is to look at genes on the Y chromosome, which does not undergo recombination. Y chromosomes pass down unchanged in males from one generation to the next. Any new changes that arise during evolution are easy to track on the family tree, as they too are passed down unchanged. The researchers in this large study looked at the pattern of gene variation among more than a thousand European males. While they identified many different patterns of variation, fully 80% of European males shared a single pattern, suggesting modern Europeans have a common ancestor. The data indicate the pattern arose some 40,000 to 50,000 years ago. In other words, our species came to Europe recently.

By both these sets of evidence, the Multiregional hypothesis is wrong. Our family tree has a single stem.

21.8 *Homo sapiens*, our species, seems to have evolved in Africa and then, like *H. erectus* before it, migrated to Europe and Asia.

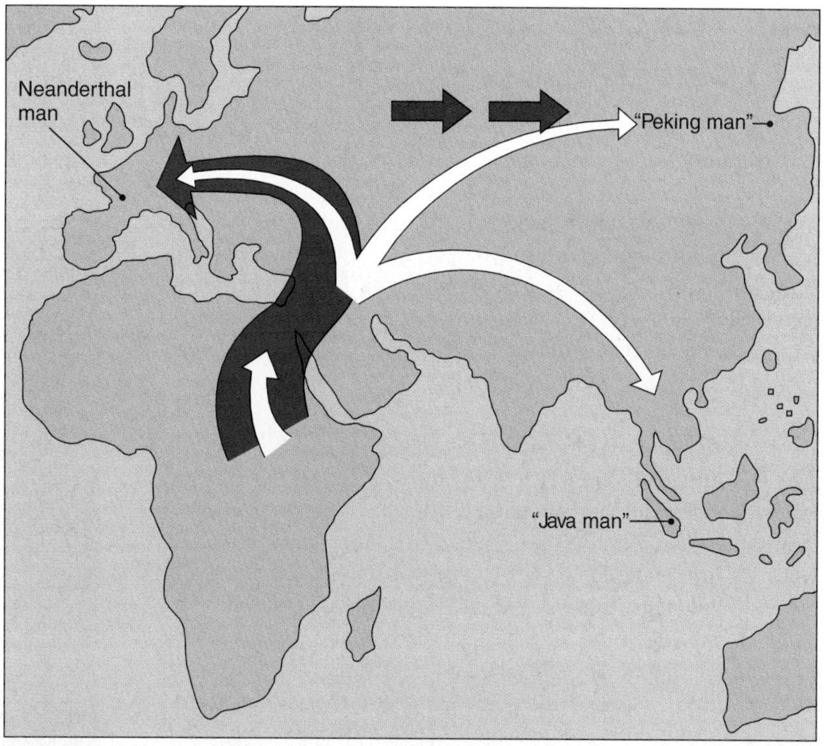

Neanderthal man
"Peking man"
"Java man"

Figure 21.15 Out of Africa—twice.

A still-controversial theory suggests that *Homo* spread from Africa to Europe and Asia twice. First, *H. erectus* (*white arrow*) spread as far as Java and China. Later, *H. erectus* was replaced by *H. sapiens* (*red arrow*) in a second wave of migration.

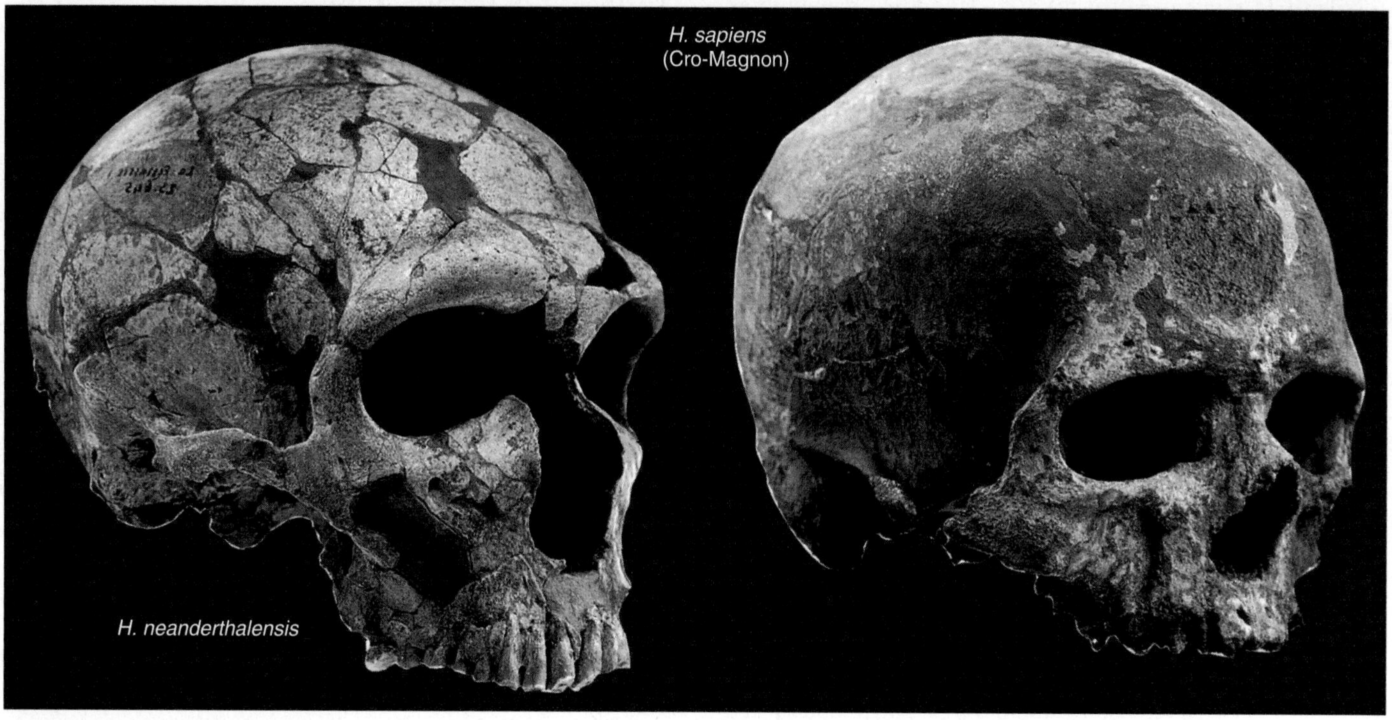

H. sapiens (Cro-Magnon)
H. neanderthalensis

Figure 21.14 (continued)

21.9 Our Own Species: *Homo sapiens*

H. sapiens is the only surviving species of *Homo,* and indeed it is the only surviving hominid. Some of the best fossils of *H. sapiens* are 20 well-preserved skeletons with skulls found in a cave near Nazareth in Israel. Modern dating techniques date these humans between 90,000 and 100,000 years old. The skulls are modern in appearance, with high, short braincases, vertical foreheads with only slight brow ridges, and a cranial capacity of roughly 1,550 cubic centimeters, well within the range of modern humans.

Cro-Magnons Replace the Neanderthals

The **Neanderthals** (classified by some as a separate species, *H. neanderthalensis*) were named after the Neander Valley of Germany, where their fossils were first discovered in 1856. Rare at first outside of Africa, they became progressively more abundant in Europe and Asia and by 70,000 years ago had become common. The Neanderthals made diverse tools, including scrapers, spearheads, and hand axes. They lived in huts or caves. Neanderthals took care of their injured and sick and commonly buried their dead, often placing food, weapons, and even flowers with the bodies. Such attention to the dead strongly suggests that they believed in a life after death. This is the first evidence of the symbolic thinking characteristic of modern humans.

Fossils of *H. neanderthalensis* abruptly disappear from the fossil record about 34,000 years ago and are replaced by fossils of *H. sapiens* called the **Cro-Magnons** (named after the valley in France where their fossils were first discovered). We can only speculate why this sudden replacement occurred, but it was complete all over Europe in a short period. There is some evidence that the Cro-Magnons came from Africa—fossils of essentially modern aspect but as much as 100,000 years old have been found there. Cro-Magnons seem to have completely replaced the Neanderthals in the Middle East by 40,000 years ago and then spread across Europe, coexisting and possibly even interbreeding with the Neanderthals for several thousand years.

The Cro-Magnons used sophisticated stone tools and also made tools out of bones and horns. They had a complex social organization and are thought to have had full language capabilities. They lived by hunting. The world was cooler than it is now—the time of the last great ice age—and Europe was covered with grasslands inhabited by large herds of grazing animals. Pictures of them can be seen in elaborate and often beautiful cave paintings made by Cro-Magnons throughout Europe (figure 21.16).

Humans of modern appearance eventually spread across Siberia to North America, which they reached at least 13,000 years ago, after the ice had begun to retreat and a land bridge still connected Siberia and Alaska. By 10,000 years ago, about 5 million people inhabited the entire world (compared with over 6 *billion* today).

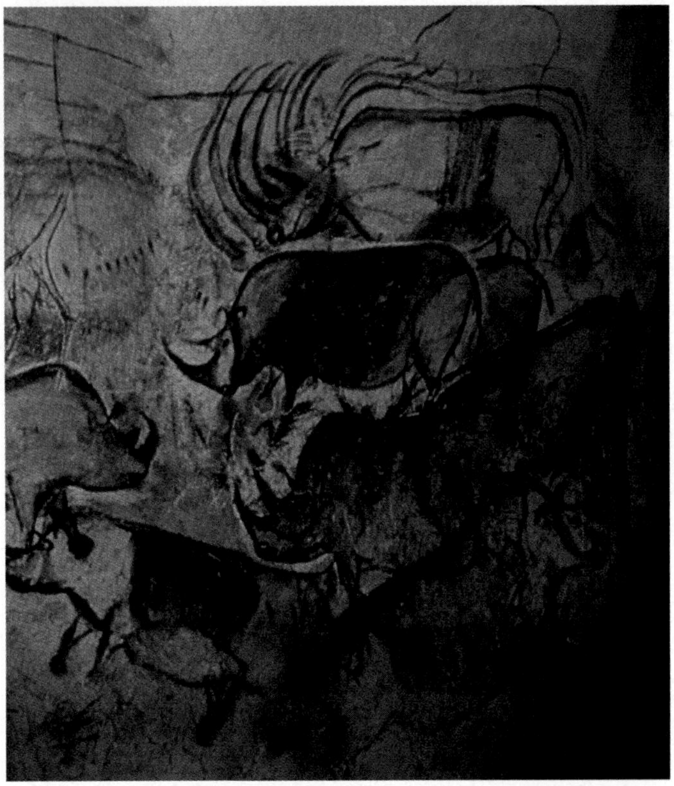

Figure 21.16 Cro-Magnon art.
Rhinoceroses are among the animals depicted in this remarkable cave painting found in 1995 near Vallon-Pont d'Arc, France.

Homo sapiens Are Unique

We humans are animals and the product of evolution. Our evolution has been marked by a progressive increase in brain size, distinguishing us from other animals in several ways. First, humans are able to make and use tools effectively—a capability that, more than any other factor, is responsible for our dominant position in the animal kingdom. Second, while not the only animal capable of conceptual thought, we have refined and extended this ability until it has become the hallmark of our species. Lastly, we use symbolic language and can with words shape concepts out of experience. Our language capability has allowed the accumulation of experience, which can be transmitted from one generation to another. Thus we have what no other animal has ever had: cultural evolution. Through culture, we have found ways to change and mold our environment, rather than changing evolutionarily in response to the demands of the environment. We control our biological future in a way never before possible—an exciting potential and frightening responsibility.

21.9 Our species, *Homo sapiens*, is good at conceptual thought and tool use and is the only animal that uses symbolic language.

CONCEPT REVIEW

1. The ancestors of the first primates were
   a. lemurs. c. mice.
   b. monkeys. d. tree shrews.

2. Unlike African and Asian monkeys, South American monkeys have
   a. color vision. c. prehensile tails.
   b. grasping fingers. d. no tails.

3. Historically, apes occurred in North America.
   a. true
   b. false

4. Arrange the following animals in order of their evolutionary distance from humans, the most distant first.
   a. gibbons d. monkeys
   b. gorillas e. orangutans
   c. prosimians

5. Which of the following characteristics were *not* exhibited by australopithecines?
   a. the use of tools
   b. bipedalism
   c. large brains
   d. omnivorous teeth

6. A fossil named _________ seems to be intermediate between apes and *Australopithecus* and may represent the true base of our family tree.
   a. *Homo habilis*
   b. *Ardipithecus ramidus*
   c. *Homo erectus*
   d. "Lucy"

7. _________ survived longer than any other species of human.
   a. *Homo habilis* c. *Homo erectus*
   b. *Homo sapiens* d. *Homo neanderthalensis*

8. Modern humans first appeared about _________ years ago.
   a. 10 million
   b. 10,000
   c. 100,000
   d. 500,000

9. Arrange the following hominids in the order in which they are thought to have evolved, the most distant first.
   a. *Australopithecus*
   b. Cro-Magnons
   c. *Homo erectus*
   d. *Homo habilis*
   e. *Homo neanderthalensis*

10. Unlike the eyes of squirrels and shrews, the eyes of primates are shifted forward, giving them _________ vision.

11. "Lucy" is the most complete fossil hominid of the genus _________.

12. *Homo* _________ replaced *Homo habilis.*

13. Humans crossed the land bridge across Siberia to North America about _________ years ago.

14. _________ evolution affects modern human races as much as biological evolution.

CHALLENGE YOURSELF

1. Create a hypothesis for why all living lemurs on the earth are on the island of Madagascar.

2. Studies of the DNA of primates have revealed that humans differ from gorillas in only 2.3% of the DNA nucleotide sequences and from chimpanzees in less than 1.6%. This degree of genetic similarity is the same as is usually seen among "sibling" species (that is, species that have only recently evolved from a common ancestor). Yet humans are assigned not only to a different genus but to a different family! Do you think this is legitimate, or are humans just a rather unusual kind of African ape?

3. Why is it incorrect to state that humans evolved from apes or monkeys?

4. What evidence would solve some unanswered questions about human evolution?

5. Modern humans, *Homo sapiens,* evolved from *Homo erectus* fewer than 1 million years ago. Do you think that evolution of the genus *Homo* is over, or might another species of humans evolve within the next million years? Do you think this would involve the extinction of *H. sapiens?*

21 eBRIDGE

Reinforcing Key Points

The Evolution of Primates

21.1 The Evolutionary Path to Apes

21.2 How the Apes Evolved

The First Hominids

21.3 An Evolutionary Tree with Many Branches

21.4 The Origins of Bipedalism

21.5 The Beginning of Hominid Evolution

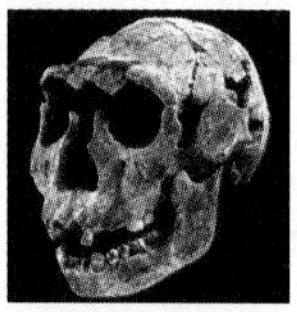

The First Humans

21.6 African Origin: Early *Homo*

21.7 Out of Africa: *Homo erectus*

Modern Humans

21.8 The Last Stage of Hominid Evolution

21.9 Our Own Species: *Homo sapiens*

Electronic Learning

Visual Learning

Animations

Modern Human Skeletal Features and Those of a Gorilla

Binocular Vision

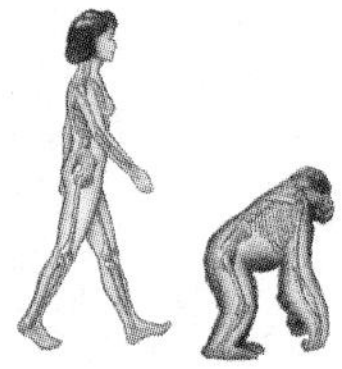

Genetic Data

Origin of Modern Humans

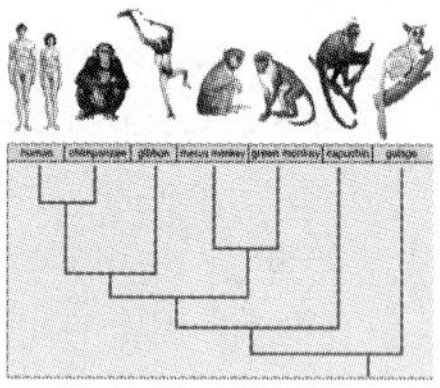

Author's Corner

Human Evolution. Few areas of biology have been as contentious as human evolutionary studies. This is not for lack of data—we know more about how humans evolved than practically any other group—but rather because we are able to ask sophisticated questions, questions for which more than one answer can be defended. Every scientist likes a meaty dispute, human paleontologists more than most. One of the most bitterly contested has been the dispute over the site of origin of our species. Did *Homo sapiens* originate in Africa, then migrate out and replace other earlier species, or did *H. sapiens* evolve in concert at many places?

1. The study of human evolution is alive with controversy.
2. A bitter and angry dispute about the origin of humans has been settled.
3. Ancient campfires tell us that the first American was . . . a Virginian.

Virtual Classroom

The Puzzle of Human Evolution

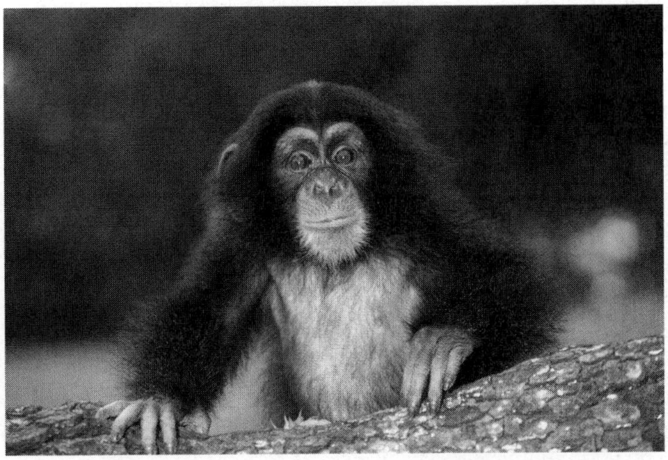

In 1871 Charles Darwin published another ground-breaking book, *The Descent of Man.* In this book, Darwin suggested that humans evolved from the same African ape ancestors that gave rise to the gorilla and the chimpanzee. Numerous fossil discoveries made since then strongly support this hypothesis. Humans and apes are also very similar when compared at the molecular level. For example, a human hemoglobin molecule differs from its chimpanzee counterpart in only a single amino acid. The ability to read genomic DNA reveals very few distinct genes between humans and chimpanzees, the apes most closely related to humans. Human evolution is the part of the evolution story that often interests people most, and it is also the part about which we know the most. The evolutionary journey that has led to humans is best understood by following the story chronologically. It is an exciting story, replete with controversy. Our own species, *Homo sapiens,* appears to have evolved within the last 150,000 years in Africa, and then migrated to Europe and Asia.

Virtual Lab

Plotting an Aerial Attack on Marauding Fire Ants

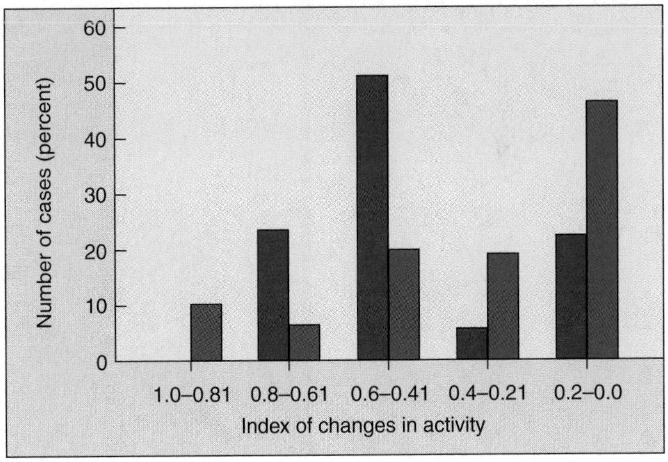

Number of cases (percent)
0
10
20
30
40
50
60
1.0–0.81
0.8–0.61
0.6–0.41
0.4–0.21
0.2–0.0
Index of changes in activity

One of the most significant impacts of human evolution has been on the other elements of the living world. This can be clearly seen in the propensity of modern technological society to introduce animals and plants to new and novel surroundings. Anyone traveling through the American Southwest will have seen kudzu (*Pueraria lobata*), a Chinese vine that was intentionally planted throughout the South in the 1930s to control erosion. It grows as fast as 30 cm *per day,* and now covers nearly 4 million acres, overgrowing everything it encounters. A particularly bothersome new resident of the United States is the red fire ant. Arriving in Mobile, Alabama, on a ship from Brazil in the 1920s, it has spread across America from Florida to California, attacking birds, lizards, and other small animals. In a novel attempt to control their spread, Larry Gilbert of the University of Texas, Austin, is exploring the possibility of releasing phorid flies. A tiny grey insect, *Pseudacteon tricuspis* is a natural predator of fire ants, and appears to keep them well in check in South America. Gilbert has been importing phorid flies from Brazil and Argentina, breeding them in his lab, and releasing them in field experiments to test their ability to keep fire ant populations in check.

Quizzes | Further Reading | Essential Study Partner | Links | BioCourse.com

CHAPTER 22

The Animal Body and How It Moves

THE FAR SIDE®
By GARY LARSON
© 1982 FarWorks, Inc. All Rights Reserved/Dist. by Creators Syndicate
Larson

"Pull out, Betty! Pull out! ... You've hit an artery!"

CHAPTER 22

The Animal Body and How It Moves

CHAPTER OVERVIEW

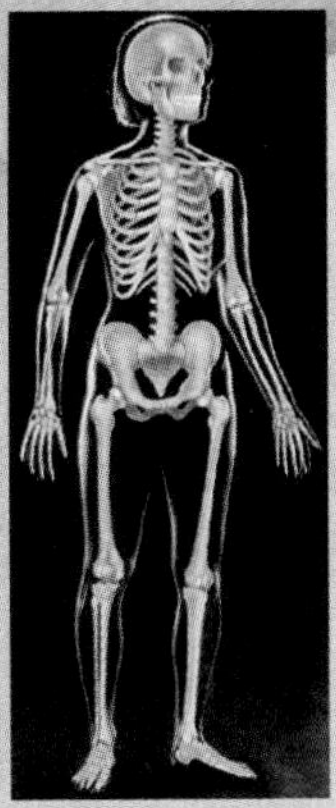

The Animal Body Plan

- Four key transitions in body design among animals include radial to bilateral symmetry, no body cavity to body cavity, nonsegmented to segmented bodies, and protostome to deuterostome development.
- The vertebrate body's 100 trillion cells are organized into tissues, which are the actual structural and functional units of the body.

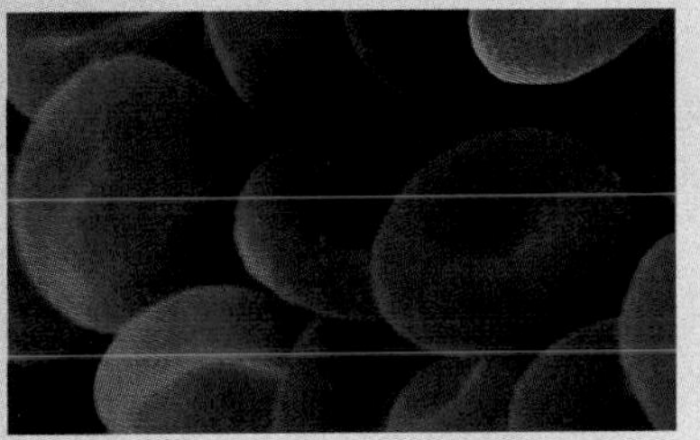

Tissues of the Vertebrate Body

- The outermost of the principal tissues, epithelium, protects the tissues beneath it from dehydration.
- Connective tissue supports the body structurally, defends it with the immune system, and transfers materials via the blood.
- Muscle tissue provides for movement, and nerve tissue provides for regulation.

The Skeletal System

- Animal locomotion is accomplished through the force of muscles acting on the skeletal system of the animal.
- Bone is a dynamic tissue, constantly growing and renewing itself.

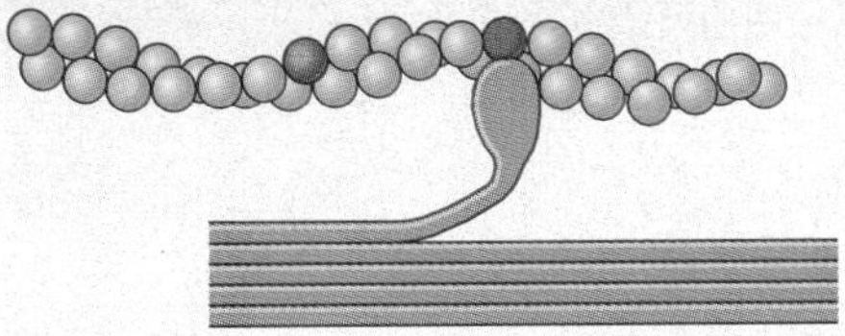

The Muscular System

- Muscle cells do the actual work of movement.
- Muscle cells contain large amounts of the protein filaments actin and myosin.
- Muscle cells contract when myosin "walks" along the actin filament, driven by the cleavage of ATP.

22.1 Radial Versus Bilateral Symmetry

The simplest animals, the sponges, mostly lack any definite symmetry. They grow asymmetrically as irregular masses. Symmetrical bodies first evolved in cnidarians (jellyfish, sea anemones, and corals) and ctenophores (comb jellies). The bodies of these two types of animals exhibit **radial symmetry,** in which the parts of the body are arranged around a central axis in such a way that any plane passing through the central axis divides the organism into halves that are approximate mirror images (figure 22.1).

The bodies of all other animals are marked by a fundamental **bilateral symmetry,** in which the body has a right and a left half that are mirror images of each other (figure 22.2). Bilateral symmetry constitutes a major evolutionary advance. This unique form of organization allows parts of the body to evolve in different ways, permitting different organs to be located in different parts of the body. In some higher animals like echinoderms (sea stars, also called starfish), the adults are radially symmetrical, but even in them, the larvae are bilaterally symmetrical.

Bilaterally symmetrical animals move from place to place more efficiently than radially symmetrical ones, which, in general, lead a sessile or passively floating existence. Bilaterally symmetrical animals are therefore efficient in seeking food and mates and avoiding predators.

During the early evolution of bilaterally symmetrical animals, structures that were important to the organism in monitoring its environment, and thereby in capturing prey or avoiding enemies, came to be grouped at the anterior end. Other functions tended to be located farther back in the body. The number and complexity of sense organs are much greater in bilaterally symmetrical animals than they are in radially symmetrical ones.

Much of the nervous system in bilaterally symmetrical animals is in the form of major longitudinal nerve cords. In a very early evolutionary advance, nerve cells became grouped around the anterior end of the body. These nerve cells probably first functioned mainly to transmit impulses from the anterior sense organs to the rest of the nervous system. This trend ultimately led to the evolution of a definite head and brain area, a process called **cephalization,** as well as to the increasing dominance and specialization of these organs in the more advanced animals.

22.1 The simplest animals lack symmetry. While some marine animals have radial symmetry—notably cnidarians, ctenophores, and adult echinoderms—most animals have bilateral symmetry, which allows parts of the body to specialize in different ways.

Figure 22.1 Radial symmetry.
Cnidarians, like these sea anemones, are radially symmetrical.

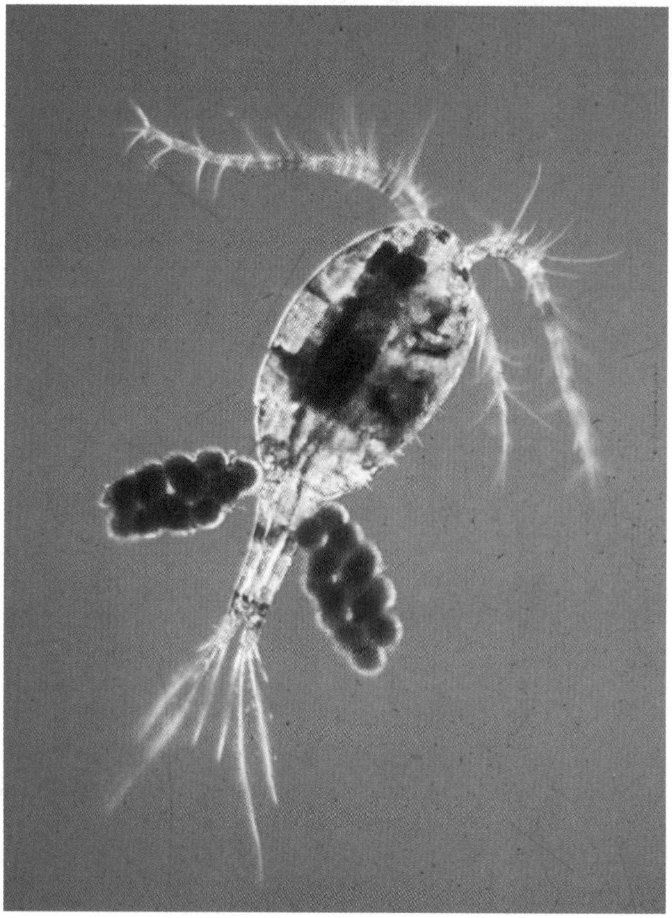

Figure 22.2 Bilateral symmetry.
This copepod, like all animals other than sponges and cnidarians, is bilaterally symmetrical.

22.2 No Body Cavity Versus Body Cavity

A second key transition in the evolution of the animal body plan was the evolution of the body cavity (figures 22.3 and 22.4). The evolution of efficient organ systems within the animal body depended critically upon a body cavity for supporting organs, distributing materials, and fostering complex developmental interactions.

Importance of a Body Cavity

The presence of a body cavity allows the dramatic expansion and sometimes lengthening of portions of the digestive tract, especially in coelomates. This longer passage allows for storage of undigested food, longer exposure to enzymes for more complete digestion, and even storage and final processing of food remnants. Such an arrangement allows an animal to eat a great deal when it is safe to do so and then to hide during the digestive process, thus limiting the animal's exposure to predators. The tube within the body cavity architecture is also more flexible, allowing the animal greater freedom to move.

An internal body cavity also provides space within which the gonads (ovaries and testes) can expand, allowing the accumulation of large numbers of eggs and sperm. Such accumulation helps to make possible all of the diverse modifications of breeding strategy that characterize the more advanced animals. Furthermore, large numbers of gametes can be released when the conditions are as favorable as possible for the survival of the young animals.

Kinds of Body Cavities

In the animal kingdom, we see three body arrangements: Some animals have no body cavity, while others have either of two different types of body cavities, distinguished primarily by where they develop within the three embryonic layers.

1. Animals with no body cavity are called *acoelomates*. Sponges are acoelomates, and so are jellyfish and flatworms.
2. A body cavity that forms between the endoderm and the mesoderm is called a pseudocoel, and the animals in which it occurs are called *pseudocoelomates*. Nematodes are pseudocoelomates.
3. A body cavity that forms entirely within the mesoderm is called a coelom, and animals in which it occurs are called *coelomates*. Mollusks, arthropods, echinoderms, and vertebrates are all coelomates.

22.2 While the simplest animals—sponges, jellyfish, and flatworms—lack a body cavity, most animals possess one, providing support for organs and a ready way to circulate materials.

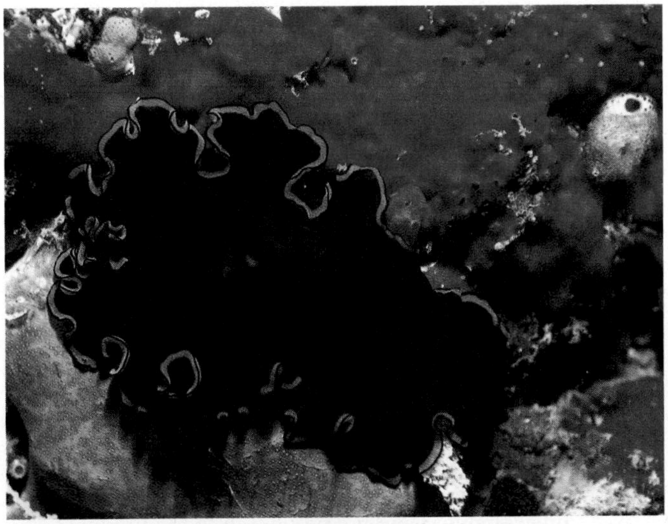

Figure 22.3 No body cavity.
Solid worms, like this marine flatworm, have no body cavity.

Figure 22.4 Body cavity.
This snail, like all mollusks, roundworms, annelids, arthropods, echinoderms, and chordates, has a body cavity.

22.3 Nonsegmented Versus Segmented Bodies

The third key transition in animal body plan involved the subdivision of the body into **segments** (figures 22.5 and 22.6). Just as it is efficient for workers to construct a tunnel from a series of identical prefabricated parts, so segmented animals are "assembled" from a succession of identical segments. During the animal's early development, these segments become most obvious in the mesoderm but later are reflected in the ectoderm and endoderm as well. Two advantages result from early embryonic segmentation:

1. In annelids and other highly segmented animals, each segment may go on to develop a more or less complete set of adult organ systems. Damage to any one segment need not be fatal to the individual, because the other segments duplicate that segment's functions.

2. Locomotion is far more effective when individual segments can move independently because the animal as a whole has more flexibility of movement. Because the separations isolate each segment into an individual hydrostatic skeletal unit, each is able to contract or expand autonomously. Therefore, a long body can move in ways that are often quite complex. When an earthworm crawls on a flat surface, it lengthens some parts of its body while shortening others. The elaborate and obvious segmentation of the annelid body clearly represents an evolutionary specialization. Many scientists believe that segmentation is an adaptation for burrowing; it makes possible the production of strong peristaltic waves along the length of the worm, thus enabling vigorous digging.

Segmentation underlies the organization of all advanced animals. Segmentation can be seen not only in the body of a marine polychaete (see figure 22.6), but also in the bodies of all annelids, arthropods, echinoderms, and chordates. Segmentation is not always obvious. In some adult arthropods, for example, the segments are fused, but segmentation is usually apparent in embryological development. In humans and other vertebrates, the backbone and muscular areas are segmented.

22.3 Segmentation underlies the body organization of all advanced animals, although it is not always evident. Segmentation appears to have provided a very powerful way to efficiently expand the animal body.

Figure 22.5 Nonsegmented body.
Mollusks, like this octopus, do not have segmented bodies.

Figure 22.6 Segmented body.
This polychaete worm, like all annelids, arthropods, echinoderms, and chordates, has a segmented body.

22.4 Protostomes Versus Deuterostomes

Two outwardly dissimilar phyla of animals, the echinoderms (sea stars) and the chordates (vertebrates), together with two smaller phyla of animals, have a series of key embryological features different from those shared by the other animal phyla. Because it is extremely unlikely that these features evolved more than once, it is believed that these four phyla share a common ancestry. They are the members of a group called the **deuterostomes.** All other coelomate animals are called **protostomes** (figure 22.7). Deuterostomes (figure 22.8) evolved from protostomes more than 630 million years ago. Deuterostomes, like protostomes, are coelomates. They differ fundamentally from protostomes, however, in three aspects of embryonic growth.

1. **How cleavage forms a hollow ball of cells.** Deuterostomes differ from protostomes in one of the earliest steps of development, the divisions that determine the plane in which the cells divide. Most protostomes undergo spiral cleavage, in which the plane of cell division is diagonal to the vertical axis of the embryo. Deuterostomes undergo radial cleavage, with the plane of division parallel or perpendicular to the embryo axis.
2. **How the blastopore determines the body axis.** Deuterostomes differ from protostomes in the way in which the embryo grows. Early in a coelomate's embryonic growth, when the embryo is a hollow ball of cells, a portion invaginates inward to form an opening called the blastopore. The blastopore of a protostome becomes the animal's mouth, and the anus develops at the other end. In a deuterostome, by contrast, the blastopore becomes the animal's anus, and the mouth develops at the other end.
3. **How the developmental fate of the embryo is fixed.** Most protostomes undergo determinate cleavage, which rigidly fixes the developmental fate of each cell very early. No one cell isolated at even the four-cell stage can go on to form a normal individual. The cells that make up an embryonic protostome each contain a different portion of the regulatory signals present in the egg, so no one cell of the embryo (or adult) can develop into a complete organism. In marked contrast, deuterostomes undergo indeterminate cleavage, with each cell retaining the capacity to develop into a complete individual.

22.4 Protostomes and deuterostomes differ in several key aspects of early embryonic development.

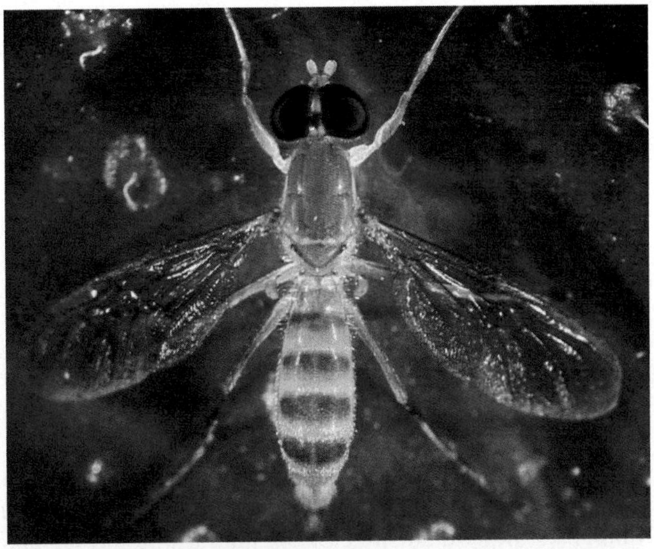

Figure 22.7 Protostome.
Arthropods, like this soldier fly, are protostomes.

Figure 22.8 Deuterostome.
This owl, like all chordates and echinoderms, is a deuterostome.

22.5 Organization of the Vertebrate Body

All vertebrates have the same general architecture: food flows through a long tube from mouth to anus, which is suspended within an internal body cavity called the *coelom.* The coelom of many terrestrial vertebrates is divided into two parts: the *thoracic cavity,* which contains the heart and lungs, and the *abdominal cavity,* which contains the stomach, intestines, and liver. The vertebrate body is supported by an internal scaffold, or skeleton, made up of jointed bones. A bony skull surrounds and protects the brain, while a column of bones, the vertebrae, surrounds the spinal cord.

Like all animals, the vertebrate body is composed of cells—over 100 trillion of them in your body. It's difficult to picture how many 100 trillion actually is. A line with 100 trillion cars in it would stretch from the earth to the sun and back 50 million times! Not all of these 100 trillion cells are the same, of course. If they were, we would not be bodies but amorphous blobs. Vertebrate bodies contain over 100 different kinds of cells.

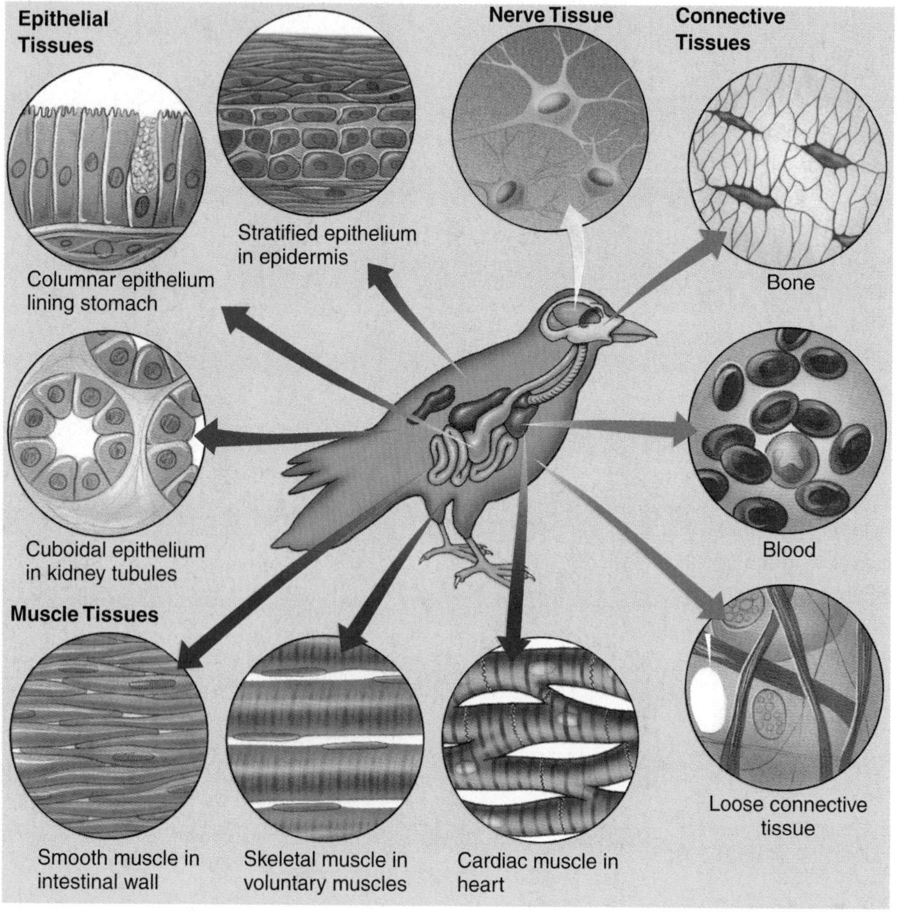

Epithelial Tissues
Nerve Tissue
Connective Tissues
Stratified epithelium in epidermis
Columnar epithelium lining stomach
Bone
Cuboidal epithelium in kidney tubules
Blood
Muscle Tissues
Loose connective tissue
Smooth muscle in intestinal wall
Skeletal muscle in voluntary muscles
Cardiac muscle in heart

Figure 22.9 Vertebrate tissue types.
The four basic classes of tissue are epithelial, connective, muscle, and nerve.

Tissues

Groups of cells of the same type are organized within the body into **tissues,** which are the structural and functional units of the vertebrate body. A tissue is a group of cells of the same type that performs a particular function in the body.

Tissues form as the vertebrate body develops. Early in development, the growing mass of cells that will become a mature animal differentiates into three fundamental layers of cells: endoderm, mesoderm, and ectoderm. These three kinds of embryonic cell layers in turn differentiate into the more than 100 different kinds of cells in the adult body.

It is possible to assemble many different kinds of tissue from 100 cell types, but biologists have traditionally grouped adult tissues into four general classes: *epithelial, connective, muscle,* and *nerve tissue* (figure 22.9). Of these, connective tissues are particularly diverse. The blood cells flowing through veins are connective tissue, and so are the bone cells in the skull.

Organs

Organs are body structures composed of several different tissues grouped together into a larger structural and functional unit, just as a factory is a group of people with different jobs who work together to make something. The heart is an organ. It contains cardiac muscle tissue wrapped in connective tissue and joined to many nerves. All of these tissues work together to pump blood through the body: the cardiac muscles squeeze to push the blood; the connective tissues act as a bag to hold the heart in the proper shape and ensure that the different chambers of the heart contract in the proper order; and the nerves control the rate at which the heart beats. No single tissue can do the job of the heart, any more than one piston can do the job of an automobile engine.

You are probably familiar with many of the major organs of a vertebrate body. Lungs are organs that terrestrial vertebrates use to extract oxygen from the air. Fish use gills to accomplish the same task from seawater. The stomach is an organ that digests food; and the liver an organ that controls the level of sugar and other chemicals in the blood. Organs are the machines of the vertebrate body, each built from several different tissues and each doing a particular job. How many others can you name?

Organ Systems

An **organ system** is a group of organs that work together to carry out an important function. For example, the vertebrate digestive system is an organ system composed of individual organs that break up the food (beaks or teeth), pass the food to the stomach (esophagus), break down the food (stomach), absorb the food (intestine), and expel the solid residue (rectum). If all of these organs do their job right, the body obtains energy and necessary building materials from food. Figure 22.10 illustrates the relationship between cells, tissues, organs, and organ systems.

The vertebrate body contains 11 principal organ systems (figure 22.11):

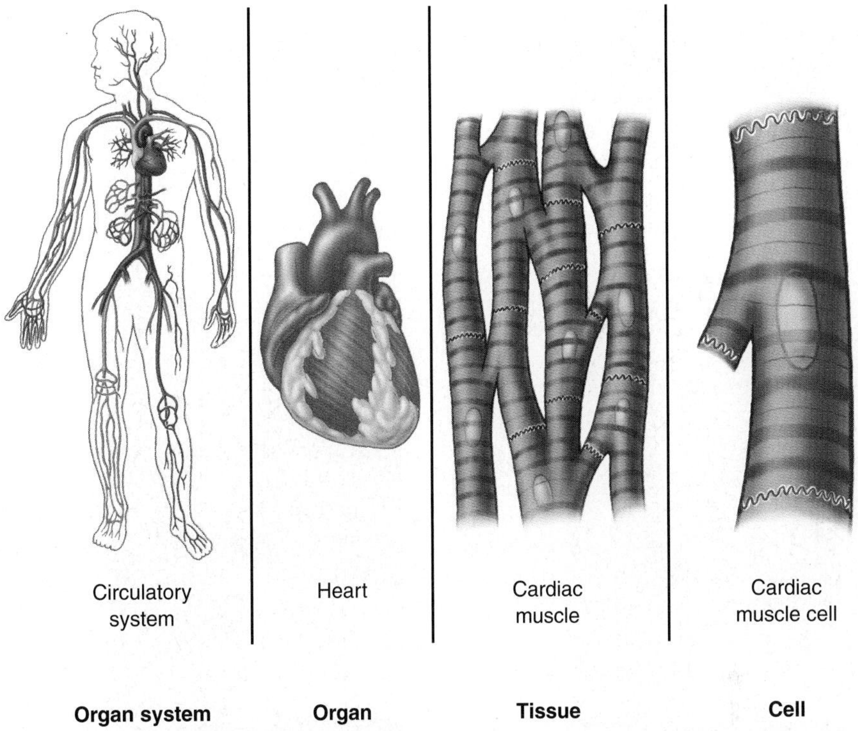

Circulatory system
Heart
Cardiac muscle
Cardiac muscle cell
Organ system
Organ
Tissue
Cell

Figure 22.10 Levels of organization within the vertebrate body.
Similar cell types operate together and form tissues. Tissues functioning together form organs. Several organs working together to carry out a function for the body are called an organ system. The circulatory system is an example of an organ system.

1. **Skeletal.** Perhaps the most important feature of the vertebrate body is its bony internal skeleton. The skeletal system protects the body and provides support for locomotion and movement. Its principal components are bones, skull, cartilage, and ligaments. Like arthropods, vertebrates have jointed appendages—the arms, hands, legs, and feet.
2. **Circulatory.** The circulatory system transports oxygen, nutrients, and chemical signals to the cells of the body and removes carbon dioxide, chemical wastes, and water. Its principal components are the heart, blood vessels, and blood.
3. **Endocrine.** The endocrine system coordinates and integrates the activities of the body. Its principal components are the pituitary, adrenal, thyroid, and other ductless glands.
4. **Nervous.** The activities of the body are coordinated by the nervous system. Its principal components are the nerves, sense organs, brain, and spinal cord.
5. **Respiratory.** The respiratory system captures oxygen and exchanges gases and is composed of the lungs, trachea, and other air passageways.
6. **Immune.** The immune system removes foreign bodies from the bloodstream using special cells, such as lymphocytes, macrophages, and antibodies.
7. **Digestive.** The digestive system captures soluble nutrients from ingested food. Its principal components are the mouth, esophagus, stomach, intestines, liver, and pancreas.
8. **Urinary.** The urinary system removes metabolic wastes from the bloodstream. Its principal components are the kidneys, bladder, and associated ducts.
9. **Muscular.** The muscular system produces movement, both within the body and of its limbs. Its principal components are skeletal muscle, cardiac muscle, and smooth muscle.
10. **Reproductive.** The reproductive system carries out reproduction. Its principal components are the testes in males, ovaries in females, and associated reproductive structures.
11. **Integumentary.** The integumentary system covers and protects the body. Its principal components are the skin, hair, nails, and sweat glands.

22.5 Groups of cells of the same type are organized in the vertebrate body into tissues. Organs are body structures composed of several different tissues. An organ system is a group of organs that work together to carry out an important function.

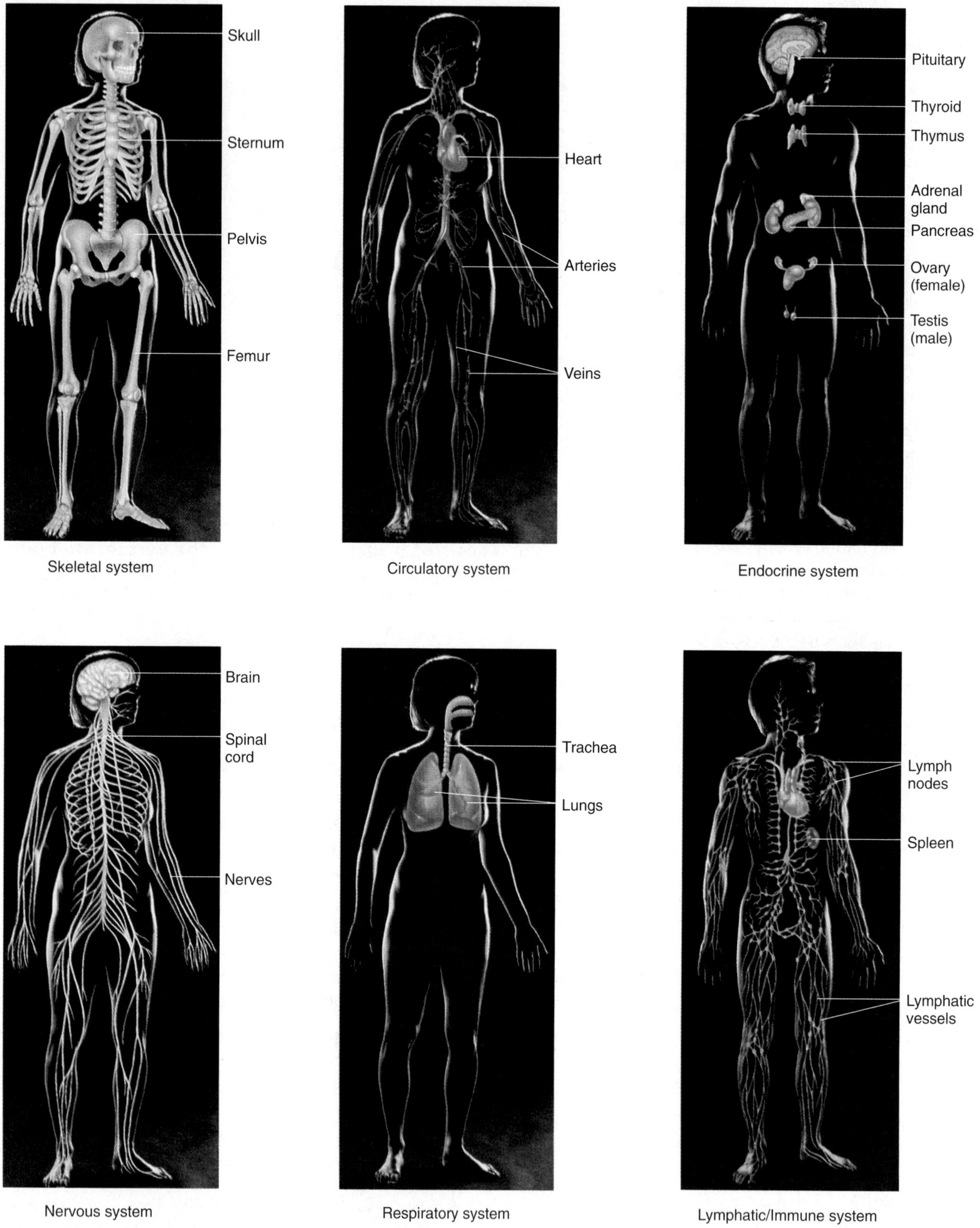
Skull
Sternum
Pelvis
Femur
Skeletal system
Heart
Arteries
Veins
Circulatory system
Pituitary
Thyroid
Thymus
Adrenal gland
Pancreas
Ovary (female)
Testis (male)
Endocrine system
Brain
Spinal cord
Nerves
Nervous system
Trachea
Lungs
Respiratory system
Lymph nodes
Spleen
Lymphatic vessels
Lymphatic/Immune system

Figure 22.11 Vertebrate body organ systems.

The 11 principal organ systems of the human body are shown, including both male and female reproductive systems.

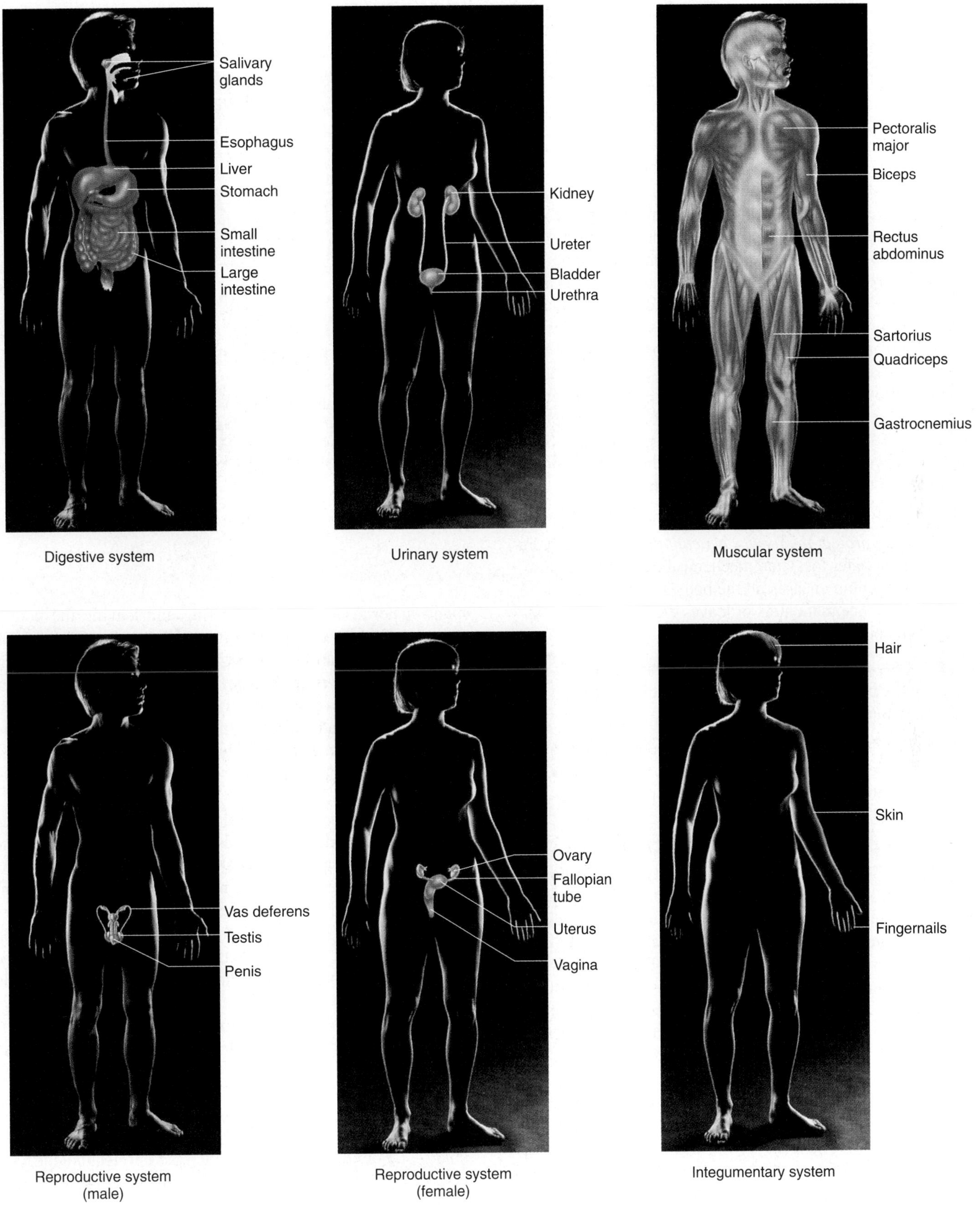
Salivary glands
Esophagus
Liver
Stomach
Small intestine
Large intestine
Digestive system
Kidney
Ureter
Bladder
Urethra
Urinary system
Pectoralis major
Biceps
Rectus abdominus
Sartorius
Quadriceps
Gastrocnemius
Muscular system
Vas deferens
Testis
Penis
Reproductive system (male)
Ovary
Fallopian tube
Uterus
Vagina
Reproductive system (female)
Hair
Skin
Fingernails
Integumentary system

Figure 22.11 (continued)

22.6 Epithelium Is Protective Tissue

As we have stated, the vertebrate body is composed of over 100 kinds of cells, traditionally grouped into four types of tissues: epithelial, connective, muscle, and nerve.

Epithelial cells are the guards and protectors of the body. They cover its surface and determine which substances enter it and which do not. The organization of the vertebrate body is fundamentally tubular, with one tube (the digestive tract) suspended inside another (the body cavity or coelom) like an inner tube inside a tire. The outside of the body is covered with cells (skin) that develop from embryonic *ectoderm* tissue; the body cavity is lined with cells that develop from embryonic *mesoderm* tissue; and the hollow inner core of the digestive tract (the gut) is lined with cells that develop from embryonic *endoderm* tissue. All three kinds of epithelial cells, although different in embryonic origin, are broadly similar in form and function and together are called the **epithelium.**

The body's epithelial layers function in three ways:

1. They *protect the tissues beneath them* from dehydration (water loss) and mechanical damage. Because epithelium encases all the body's surfaces, every substance that enters or leaves the body must cross an epithelial layer (figure 22.12).
2. They *provide sensory surfaces.* Many of a vertebrate's sense organs are in fact modified epithelial cells.
3. They *secrete materials.* Most secretory glands are derived from pockets of epithelial cells that pinch together during embryonic development.

Figure 22.12 The epithelium prevents dehydration.

The tough, scaly skin of this gila monster provides a layer of protection against dehydration and injury. For all land-dwelling vertebrates, the relative impermeability of the surface epithelium (the epidermis) to water offers essential protection from dehydration and from airborne pathogens (disease-causing organisms).

Types of Epithelial Cells and Epithelial Tissues

Epithelial cells are classified into three types according to their shapes: squamous, cuboidal, and columnar. Layers of epithelial tissue are usually only one or a few cells thick. Individual epithelial cells possess only a small amount of cytoplasm and have a relatively low metabolic rate. However, they have remarkable regenerative abilities. The cells of epithelial layers are constantly being replaced throughout the life of the organism. The cells lining the digestive tract, for example, are continuously replaced every few days. The liver, which is a football-sized gland formed of epithelial tissue, can readily regenerate substantial portions of itself removed during surgery.

There are three general kinds of epithelial tissue. First, the membranes that line the lungs and the major cavities of the body are a **simple epithelium** only a single cell layer thick. You can see why—these are surfaces across which many materials must pass, entering and leaving the body's compartments, and it is important that the "road" into and out of the body not be too long. Second, the skin, or epidermis, is a **stratified epithelium** composed of more complex epithelial cells several layers thick. Several layers are necessary to provide adequate cushioning and protection and to enable the skin to continuously replace its cells. Table 22.1 summarizes the characteristics of these two types of epithelial tissues.

The third type of epithelial tissue is found in the **glands** of the body. Endocrine glands secrete hormones into the blood. Exocrine glands (those with ducts that open to the body's outside) secrete sweat, milk, saliva, and digestive enzymes out of the body. Exocrine glands also secrete digestive enzymes into the stomach. If you think about it, the stomach and digestive tract are *outside* the body, because they are the inner canal that passes right through the body. It is possible for a substance to pass all the way through this digestive tract, from mouth to anus, and never enter the body at all. A substance must cross an epithelial layer to truly enter the body.

22.6 Epithelial tissue is the protective tissue of the vertebrate body. In addition to providing protection and support, vertebrate epithelial tissues provide sensory surfaces and secrete key materials.

TABLE 22.1 EPITHELIAL TISSUE

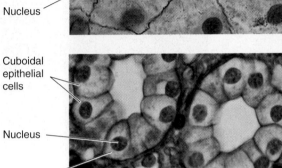

<table>
<tr><th>Tissue</th><th></th><th>Typical Location</th><th>Tissue Function</th></tr>
<tr><td colspan="4">SIMPLE EPITHELIUM</td></tr>
<tr><td>Squamous</td><td>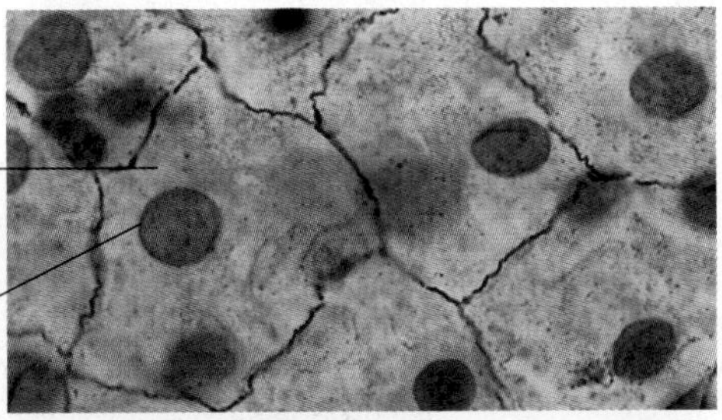
Simple squamous epithelial cell
Nucleus</td><td>Lining of lungs, capillary walls, and blood vessels</td><td>Cells very thin; provides a thin layer across which diffusion can readily occur</td></tr>
<tr><td>Cuboidal</td><td>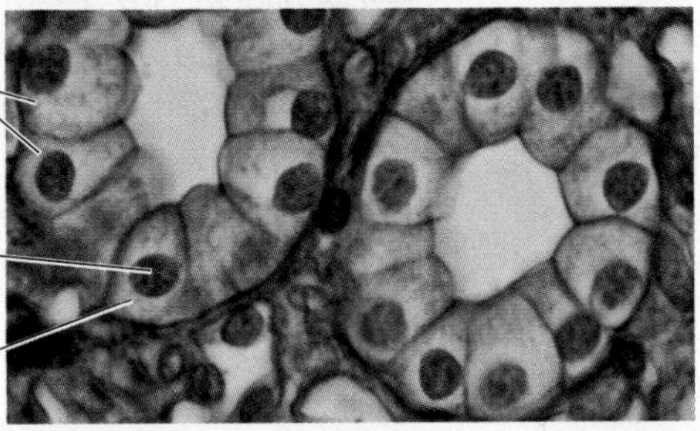
Cuboidal epithelial cells
Nucleus
Cytoplasm</td><td>Lining of some glands and kidney tubules; covering of ovaries</td><td>Cells rich in specific transport channels; functions in secretion and specific absorption</td></tr>
<tr><td>Columnar</td><td>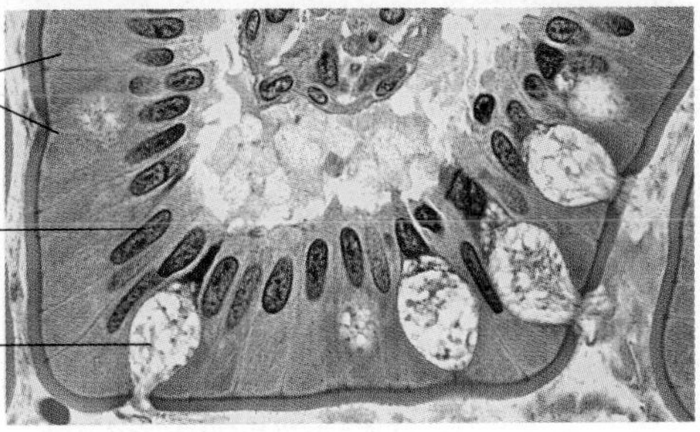
Columnar epithelial cells
Nucleus
Goblet cell</td><td>Surface lining of stomach, intestines, and parts of respiratory tract</td><td>Thicker cell layer; provides protection and functions in secretion and absorption</td></tr>
<tr><td colspan="4">STRATIFIED EPITHELIUM</td></tr>
<tr><td>Squamous</td><td>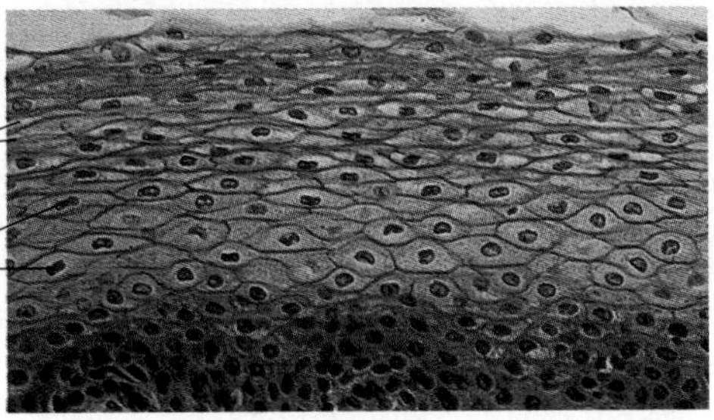
Stratified squamous cells
Nuclei</td><td>Outer layer of skin; lining of mouth</td><td>Tough layer of cells; provides protection</td></tr>
<tr><td>Columnar</td><td>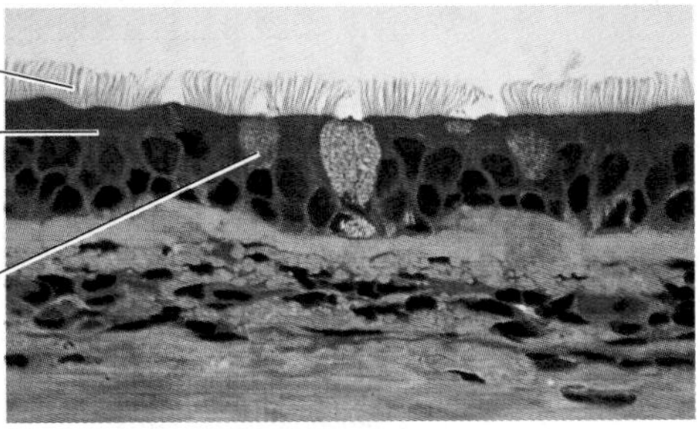
Cilia
Stratified columnar cell
Goblet cell</td><td>Lining of parts of respiratory tract</td><td>Functions in secretion of mucus; dense with cilia (small, hairlike projections) that aid in movement of mucus; provides protection</td></tr>
</table>

22.7 Connective Tissue Supports the Body

The cells of connective tissue provide the vertebrate body with its structural building blocks and also with its most potent defenses. Derived from the mesoderm, these cells are sometimes densely packed together, and sometimes widely dispersed, just as the soldiers of an army are sometimes massed together in a formation and sometimes widely scattered as guerrillas. **Connective tissue** cells fall into three functional categories: (1) the cells of the immune system, which act to defend the body; (2) the cells of the skeletal system, which support the body; and (3) the blood and fat cells, which store and distribute substances throughout the body.

Immune Connective Tissue

The cells of the immune system roam the body within the bloodstream. They are mobile hunters of invading microorganisms and cancer cells. The two principal kinds of immune system cells are **macrophages,** which engulf and digest invading microorganisms, and **lymphocytes** (also called white blood cells), which make antibodies or attack virus-infected cells.

Skeletal Connective Tissue

Three kinds of connective tissue are the principal components of the skeletal system: fibroblasts, cartilage, and bone. Although composed of similar cells, they differ in the nature of the material that is laid down between individual cells.

1. **Fibroblasts.** The most common kind of connective tissue in the vertebrate body consists of flat, irregularly branching cells called fibroblasts that secrete structurally strong proteins into the spaces between the cells. The many types of proteins give tissues different strengths. The most commonly secreted protein, collagen, is the most abundant protein in the human body: in fact, one-quarter of all the protein in your body is collagen! Fibroblasts are active in wound healing; scar tissue, for example, possesses a collagen matrix.
2. **Cartilage.** In cartilage, the collagen matrix between cells forms in long parallel arrays along the lines of mechanical stress. What results is a firm and flexible tissue of great strength, just as strands of nylon molecules laid down in long, parallel arrays produce strong, flexible ropes. Cartilage makes up the entire skeletal system of the modern agnathans and cartilaginous fishes (figure 22.13). In most adult vertebrates, however, cartilage is restricted to the articular (joint) surfaces of bones that form freely movable joints and to other specific locations.
3. **Bone.** Bone is similar to cartilage, except that the collagen fibers are coated with a calcium phosphate salt, making the tissue rigid. The structure of bone and the way it is formed are discussed later in this chapter.

Figure 22.13 This shark's skeleton is made of cartilage.

Cartilage forms tissue of great strength and flexibility. In the jawless fishes and Chondrichthyes (sharks, skates, and rays), cartilage replaces bone in the skeletons that were characteristic of the ancestors of these vertebrate groups.

Storage and Transport Connective Tissue

The third general class of connective tissue is made up of cells that are specialized to accumulate and transport particular molecules. They include the fat-accumulating cells of **adipose tissue.** They also include red blood cells, called **erythrocytes**. About 5 billion erythrocytes are present in every milliliter of your blood. Erythrocytes transport oxygen and carbon dioxide in blood. They are unusual in that during their maturation they lose most of their organelles, including the nucleus, mitochondria, and endoplasmic reticulum. Instead, occupying the interior of each erythrocyte are about 300 million molecules of hemoglobin, the protein that carries oxygen.

The fluid, or **plasma,** in which erythrocytes move is both the "banquet table" and the "refuse heap" of the vertebrate body. Practically every substance used by cells is dissolved in plasma, including inorganic salts like sodium and calcium, body wastes, and food molecules like sugars, lipids, and amino acids. Plasma also contains a wide variety of proteins, including antibodies and particularly albumin, which gives the blood its viscosity. The various types of connective tissue are summarized in table 22.2.

22.7 Connective tissues support the vertebrate body. They include cells of the immune system, which defend the body, cells of the skeletal system, which support the body, and cells like blood and fat cells, which store and distribute substances.

TABLE 22.2 CONNECTIVE TISSUE

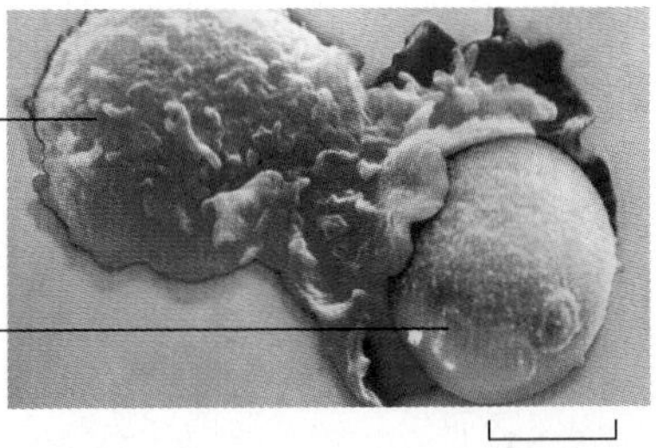
White blood cell
Invading micro- organism
.15 μm

Tissue	Typical Location	Tissue Function	Characteristic Cell Types
IMMUNE			
White blood cells	Circulatory system	Attack invading microorganisms and virus-infected cells	Macrophages; lymphocytes; mast cells

Loose connective tissue (fibroblasts)

Tissue	Typical Location	Tissue Function	Characteristic Cell Types
SKELETAL			
Fibroblasts			
Loose	Beneath skin and other epithelial tissues	Support; provide a fluid reservoir for epithelium	Fibroblasts
Dense	Tendons; sheath around muscles; kidney; liver; dermis of skin	Provide flexible, strong connections	Fibroblasts
Elastic	Ligaments; large arteries; lung tissue; skin	Enable tissues to expand and then return to normal size	Fibroblasts

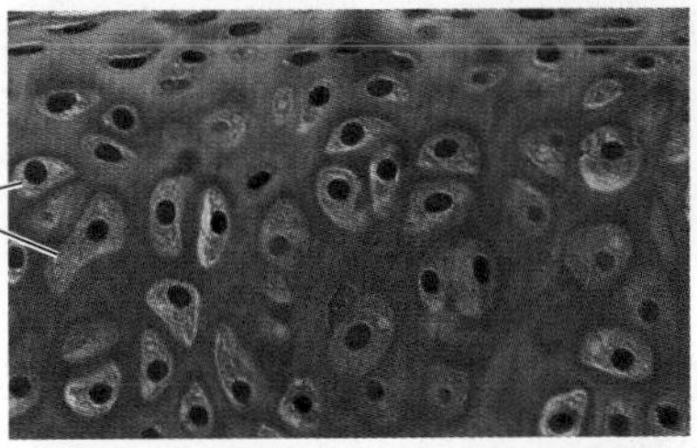
Chondrocytes (cartilage cells)

Tissue	Typical Location	Tissue Function	Characteristic Cell Types
Cartilage	Spinal disks; knees and other joints; ear; nose; tracheal rings	Provides flexible support; functions in shock absorption and reduction of friction on load-bearing surfaces	Chondrocytes (specialized fibroblast-like cells)
Bone	Most of skeleton	Protects internal organs; provides rigid support for muscle attachment	Osteocytes (specialized fibroblast-like cells)

Red blood cells

Tissue	Typical Location	Tissue Function	Characteristic Cell Types
STORAGE AND TRANSPORT			
Red blood cells	In plasma	Transport oxygen	Red blood cells (erythrocytes)

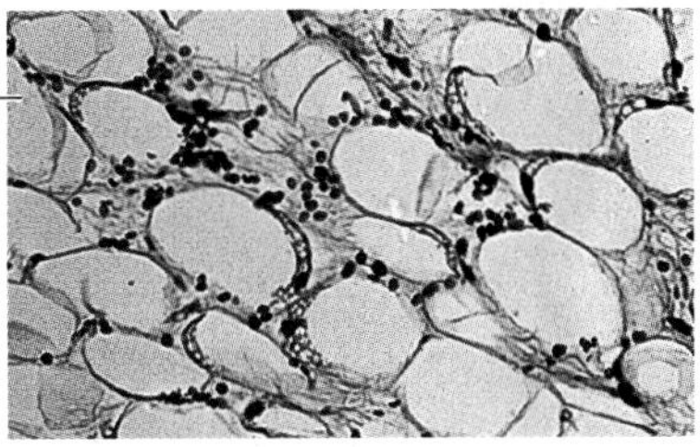
Adipose tissue

Tissue	Typical Location	Tissue Function	Characteristic Cell Types
Adipose tissue	Beneath skin	Stores fat	Specialized fibroblasts (adipocytes)

22.8 Muscle Tissue Lets the Body Move

Muscle cells are the motors of the vertebrate body. The distinguishing characteristic of muscle cells, the thing that makes them unique, is the abundance of contractible protein fibers within them. These fibers, called **microfilaments,** are made of the proteins actin and myosin. Vertebrate cells have a fine network of these microfilaments, but muscle cells have many more than other cells. Crammed in like the fibers of a rope, they take up practically the entire volume of the muscle cell. When actin and myosin slide past each other, shortening the fibers, the muscles contract. Like slamming a spring-loaded door, the shortening of all of these fibers together within a muscle cell can produce considerable force. The vertebrate body possesses three different kinds of muscle cells: **smooth muscle, skeletal muscle,** and **cardiac muscle** (table 22.3). In smooth muscle, the microfilaments are only loosely organized. In cardiac and skeletal muscle, the microfilaments are bunched together into fibers called **myofibrils.** Each myofibril contains many thousands of microfilaments, all aligned to provide maximum force when they simultaneously shorten. Cardiac and skeletal muscle are often called striated muscles because the alignment of so many microfilaments gives the muscle myofibril a banded appearance. The different types of muscle tissue are discussed again later in this chapter.

22.8 Muscle tissue is the tool the vertebrate body uses to move its limbs, contract its organs, and pump the blood through its circulatory system.

TABLE 22.3 MUSCLE TISSUE

Tissue		Typical Location	Tissue Function
Smooth muscle	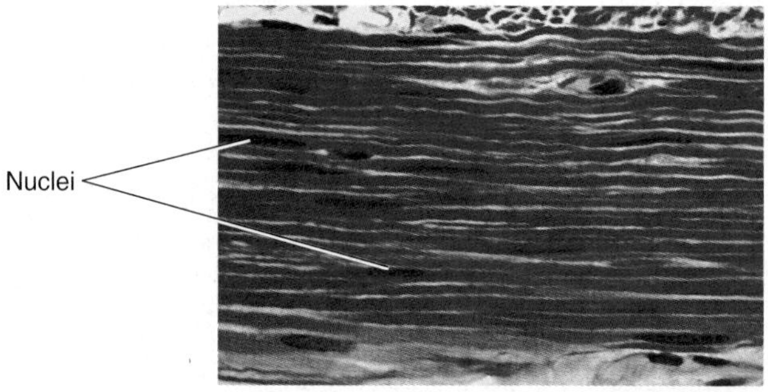 Nuclei	Walls of blood vessels, stomach, and intestines	Powers rhythmic, involuntary contractions commanded by the central nervous system
Skeletal muscle	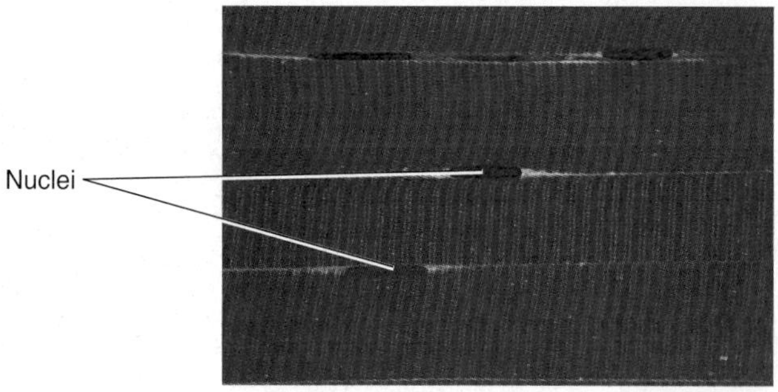 Nuclei	Voluntary muscles	Powers walking, lifting, talking, and all other voluntary movement
Cardiac muscle	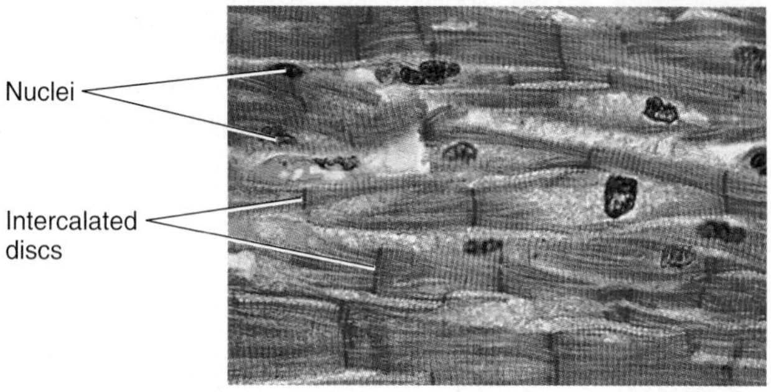 Nuclei Intercalated discs	Walls of heart	Highly interconnected cells; promotes rapid spread of signal initiating contraction

22.9 Nerve Tissue Conducts Signals Rapidly

Nerve cells carry information rapidly from one vertebrate organ to another. Nerve tissue, the fourth major class of vertebrate tissue, is composed of two kinds of cells: (1) **neurons,** which are specialized for the transmission of nerve impulses, and (2) supporting **glial cells,** which supply the neurons with nutrients, support, and insulation.

Neurons have a highly specialized cell architecture that enables them to conduct signals rapidly throughout the body. Their plasma membranes are rich in ion-selective channels that maintain a voltage difference between the interior and the exterior of the cell, the equivalent of a battery. When ion channels in a local area of the membrane open, ions flood in from the exterior, temporarily wiping out the charge difference. This process, called depolarization, tends to open nearby voltage-sensitive channels in the neuron membrane, resulting in a wave of electrical activity that travels down the entire length of the neuron as a nerve impulse.

Each neuron is composed of three parts: (1) a **cell body,** which contains the nucleus; (2) threadlike extensions called **dendrites,** which act as antennae, bringing nerve impulses to the cell body from other cells or sensory systems; and (3) a single, long extension called an **axon,** which carries nerve impulses away from the cell body. Axons often carry nerve impulses for considerable distances: the axons that extend from the skull to the pelvis in a giraffe are about 3 meters long!

TABLE 22.4 NERVE TISSUE

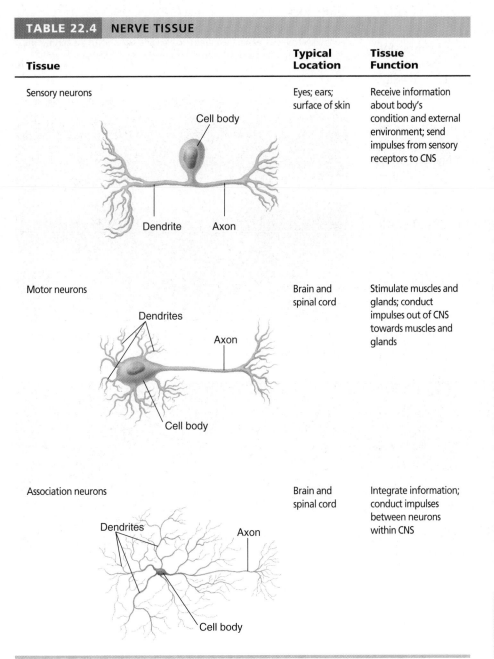

Tissue	Typical Location	Tissue Function
Sensory neurons 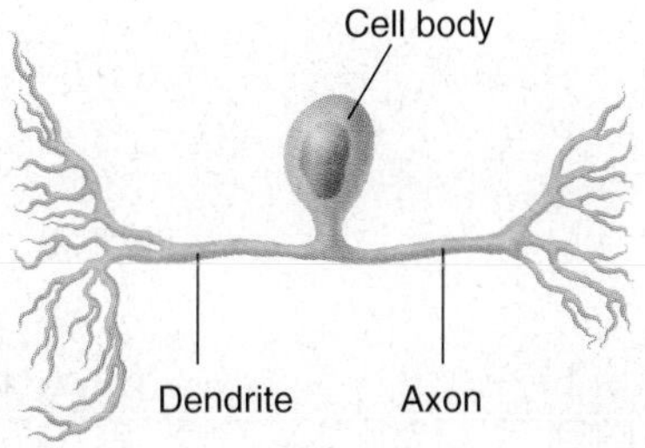 Cell body Dendrite Axon	Eyes; ears; surface of skin	Receive information about body's condition and external environment; send impulses from sensory receptors to CNS
Motor neurons 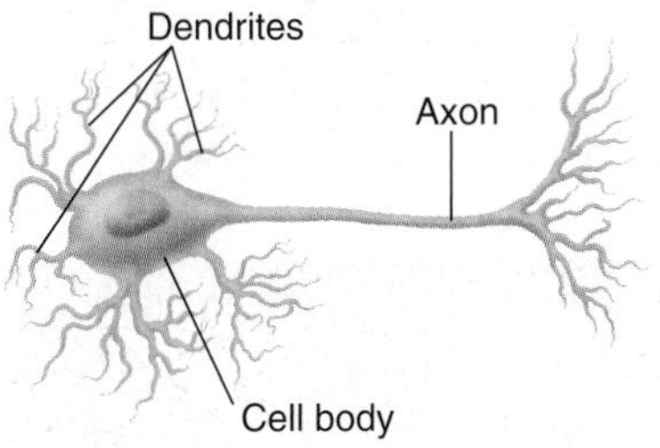 Dendrites Axon Cell body	Brain and spinal cord	Stimulate muscles and glands; conduct impulses out of CNS towards muscles and glands
Association neurons 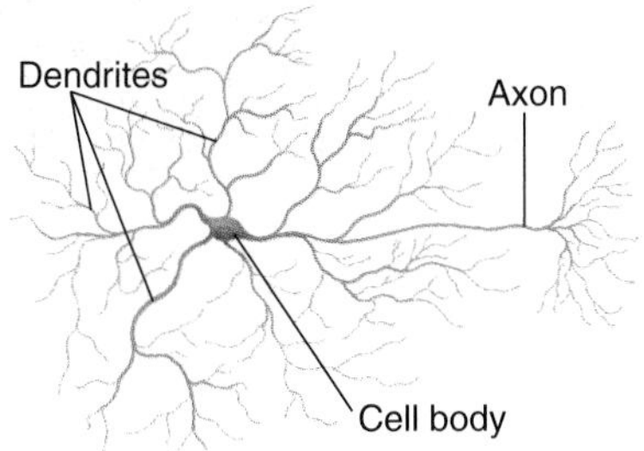 Dendrites Axon Cell body	Brain and spinal cord	Integrate information; conduct impulses between neurons within CNS

The body contains many different kinds of neurons (table 22.4). Some neurons are tiny and have only a few projections, others are bushy and have more projections, and still others have extensions that are meters long. Neurons are not normally in direct contact with one another. Instead, a tiny gap called a **synapse** separates them. Neurons communicate with other neurons by passing chemical signals called **neurotransmitters** across the gap. Special receptor proteins present on the far side of the synapse respond to the arrival of the neurotransmitters by starting a new nerve impulse in the receiving neuron.

Vertebrate nerves appear as fine white threads when viewed with the naked eye, but they are actually composed of bundles of axons. Like a telephone trunk cable, nerves include large numbers of independent communication channels—bundles composed of hundreds of axons, each connecting a nerve cell to a muscle fiber. In addition, the nerve contains numerous supporting glial cells bunched around the axons. It is important not to confuse a nerve with a neuron. A nerve is made up of the axons of many neurons, just as a cable is made of many wires.

22.9 Nerve tissue provides the vertebrate body with a means of communication and coordination.

22.10 Types of Skeletons

With muscles alone, the animal body could not move—it would simply pulsate as its muscles contracted and relaxed in futile cycles. For a muscle to produce movement, it must direct its force against another object. Animals are able to move because the opposite ends of their muscles are attached to a rigid scaffold, or **skeleton,** so that the muscles have something to pull against. There are three types of skeletal systems in the animal kingdom: hydraulic skeletons, exoskeletons, and endoskeletons.

Hydraulic skeletons are found in soft-bodied invertebrates such as earthworms and jellyfish. In this case, a fluid-filled cavity is encircled by muscle fibers that raise the pressure of the fluid when they contract. In an earthworm, for example, a wave of contractions of circular muscles begins anteriorly and compresses the body, so that the fluid pressure pushes it forward (figure 22.14). Contractions of longitudinal muscles then pull the rest of the body.

Exoskeletons surround the body as a rigid hard case in most animals. Arthropods, such as crustaceans (figure 22.15) and insects, have exoskeletons made of the polysaccharide *chitin*. An animal with an exoskeleton cannot get too large because its exoskeleton would have to become thicker and heavier, in order to prevent collapse. If an insect were the size of an elephant, its exoskeleton would have to be so thick and heavy it would hardly be able to move.

Endoskeletons, found in vertebrates and echinoderms, are rigid internal skeletons to which muscles are attached. Vertebrates have a soft, flexible exterior that stretches to accommodate the movements of their skeleton. The endoskeleton of vertebrates is composed of bone (figure 22.16). Unlike chitin, bone is a cellular, living tissue capable of growth, self-repair, and remodeling in response to physical stresses.

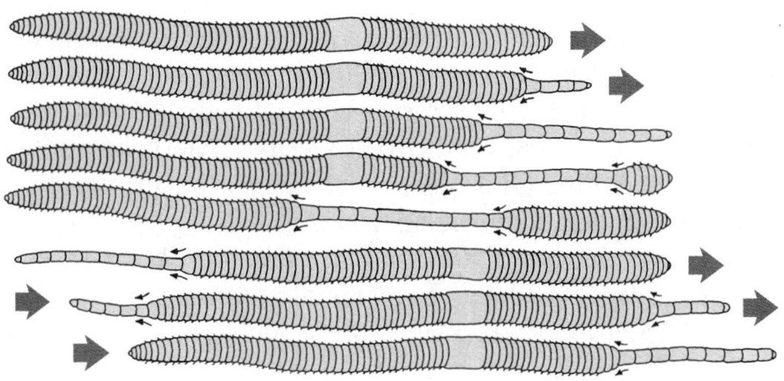

Figure 22.14 Earthworms have a hydraulic skeleton.
When an earthworm's circular muscles contract, the internal fluid presses on the longitudinal muscles, which then stretch to elongate segments of the earthworm. A wave of contractions down the body of the earthworm produces forward movement.

Figure 22.15 Crustaceans have an exoskeleton.
The exoskeleton of this rock crab is bright orange.

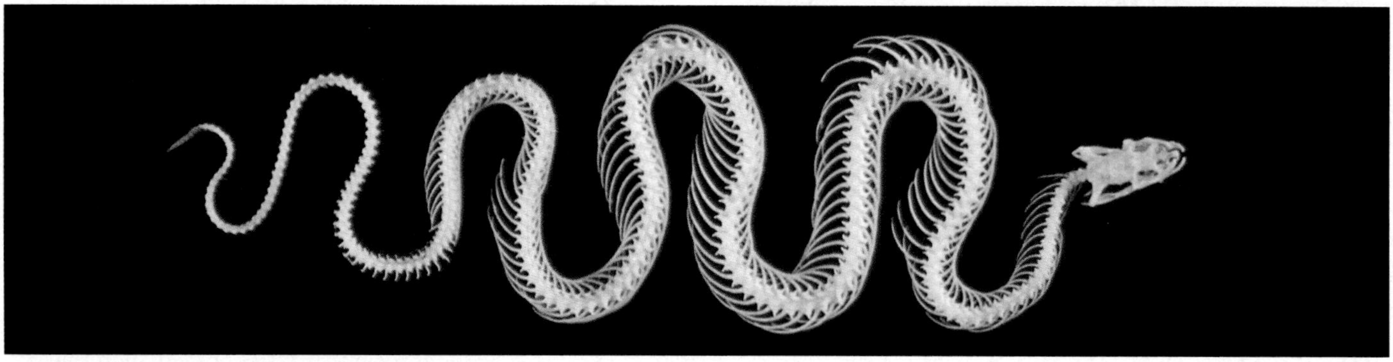

Figure 22.16 The endoskeleton of a snake.
The endoskeleton of most vertebrates is made of bone. A snake's skeleton is specialized for quick lateral movement because it has numerous vertebrae and ribs and no pelvis (pelvic girdle is vestigial in some snakes).

A Vertebrate Endoskeleton: The Human Skeleton

The human skeleton is made up of 206 individual bones. If you saw them as a pile of bones jumbled together, it would be hard to make any sense of them. To understand the skeleton, it is necessary to group the 206 bones according to their function and position in the body. The 80 bones of the **axial skeleton** support the main body axis, while the remaining 126 bones of the **appendicular skeleton** support the arms and legs (figure 22.17). These two skeletons function more or less independently—that is, the muscles controlling the axial skeleton (postural muscles) are managed by the brain separately from those controlling the appendages (manipulatory muscles).

The Axial Skeleton

The axial skeleton is made up of the skull, backbone, and rib cage. Of the skull's 28 bones, only 8 form the cranium, which encases the brain; the rest are facial bones and middle ear bones. An additional bone, the hyoid bone, supports the tongue but is not really part of the skull.

The skull is attached to the upper end of the backbone, which is also called the **spine,** or vertebral column. The spine is made up of 26 vertebrae, stacked one on top of the other to provide a flexible column surrounding and protecting the spinal cord. Curving forward from the vertebrae are 12 pairs of ribs, attached at the front to the breastbone, or sternum, and forming a protective cage around the heart and lungs.

The Appendicular Skeleton

The 126 bones of the appendicular skeleton are attached to the axial skeleton at the shoulders and hips. The shoulder, or **pectoral girdle,** is composed of two large, flat shoulder blades, each connected to the top of the breastbone by a slender, curved collarbone. The arms are attached to the pectoral girdle; each arm and hand contains 30 bones. The collarbone is the most frequently broken bone of the body. Can you guess why? Because if you fall on an outstretched arm, a large component of the force is transmitted to the collarbone.

The **pelvic girdle** forms a bowl that provides strong connections for the legs, which must bear the weight of the body. Each leg and foot contains a total of 30 bones.

22.10 The animal skeletal system provides a framework against which the body's muscles can pull. Many soft-bodied invertebrates employ a hydraulic skeleton, while arthropods have a rigid, hard exoskeleton surrounding their body. Echinoderms and vertebrates have an internal endoskeleton to which muscles attach.

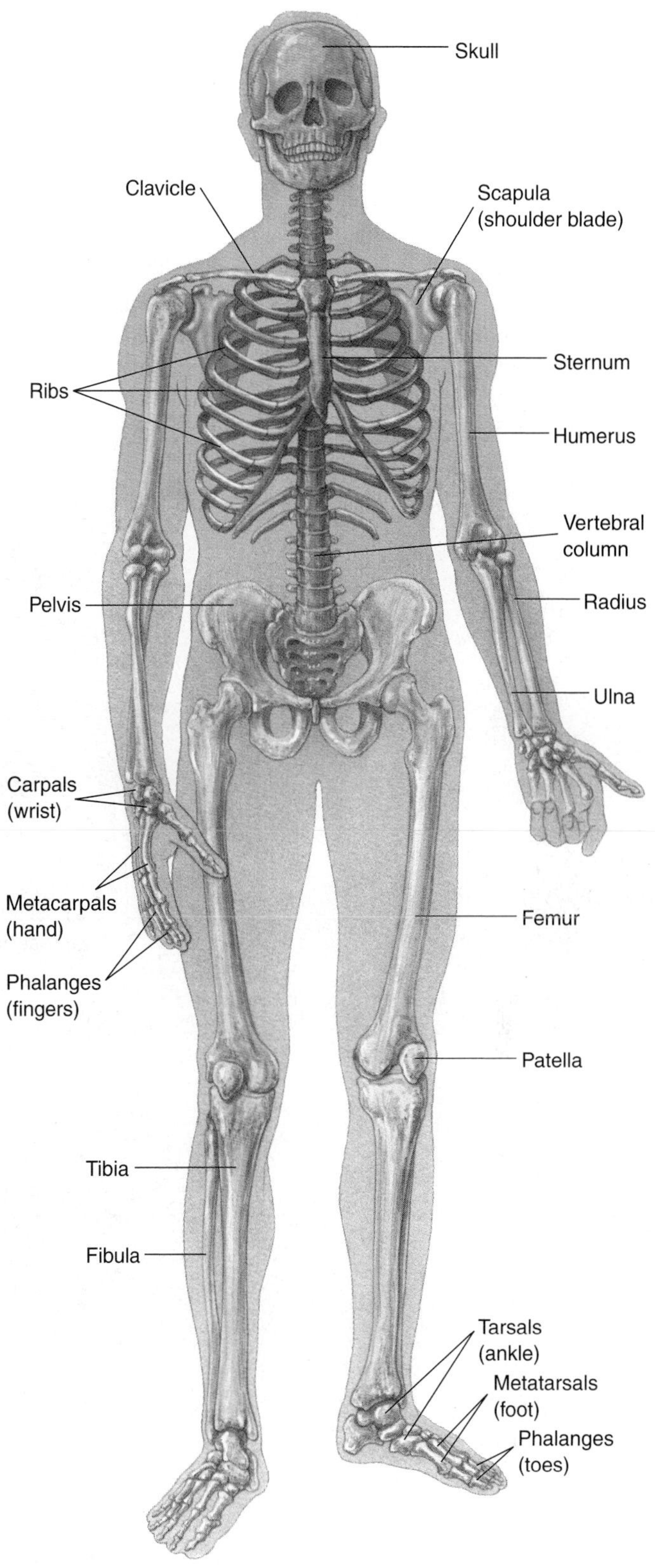

Skull
Clavicle
Scapula (shoulder blade)
Sternum
Ribs
Humerus
Vertebral column
Pelvis
Radius
Ulna
Carpals (wrist)
Metacarpals (hand)
Femur
Phalanges (fingers)
Patella
Tibia
Fibula
Tarsals (ankle)
Metatarsals (foot)
Phalanges (toes)

Figure 22.17 Axial and appendicular skeletons.
The axial skeleton is shown in *purple*, and the appendicular skeleton is shown in *peach*.

22.11 The Structure of Bone

The vertebrate endoskeleton is strong because of the structural nature of bone. Bone is produced by coating collagen fibers with a calcium phosphate salt; the result is a material that is strong without being brittle. To understand how coating collagen fibers with calcium salts makes such an ideal structural material, consider fiberglass. Fiberglass is composed of glass fibers embedded in epoxy glue. The individual fibers are rigid, giving great strength, but they are also brittle. The epoxy glue, on the other hand, is flexible but weak. The composite, fiberglass, is both rigid and strong because when stress causes an individual fiber to break, the crack runs into glue before it reaches another fiber. The glue distorts and reduces the concentration of the stress—in effect, the glue spreads the stress over many fibers.

The construction of bone is similar to that of fiberglass: small, needle-shaped crystals of a calcium phosphate mineral, hydroxyapatite, surround and impregnate collagen fibrils within bone. No crack can penetrate far into bone because any stress that breaks a hard hydroxyapatite crystal passes into the collagenous matrix, which dissipates the stress. The hydroxyapatite mineral provides rigidity, whereas the collagen "glue" provides flexibility.

Most of us think of bones as solid and rocklike. But actually, bone is a dynamic tissue that is constantly being reconstructed (figure 22.18). The outer layer of bone is very dense and compact and so is called **compact bone.**

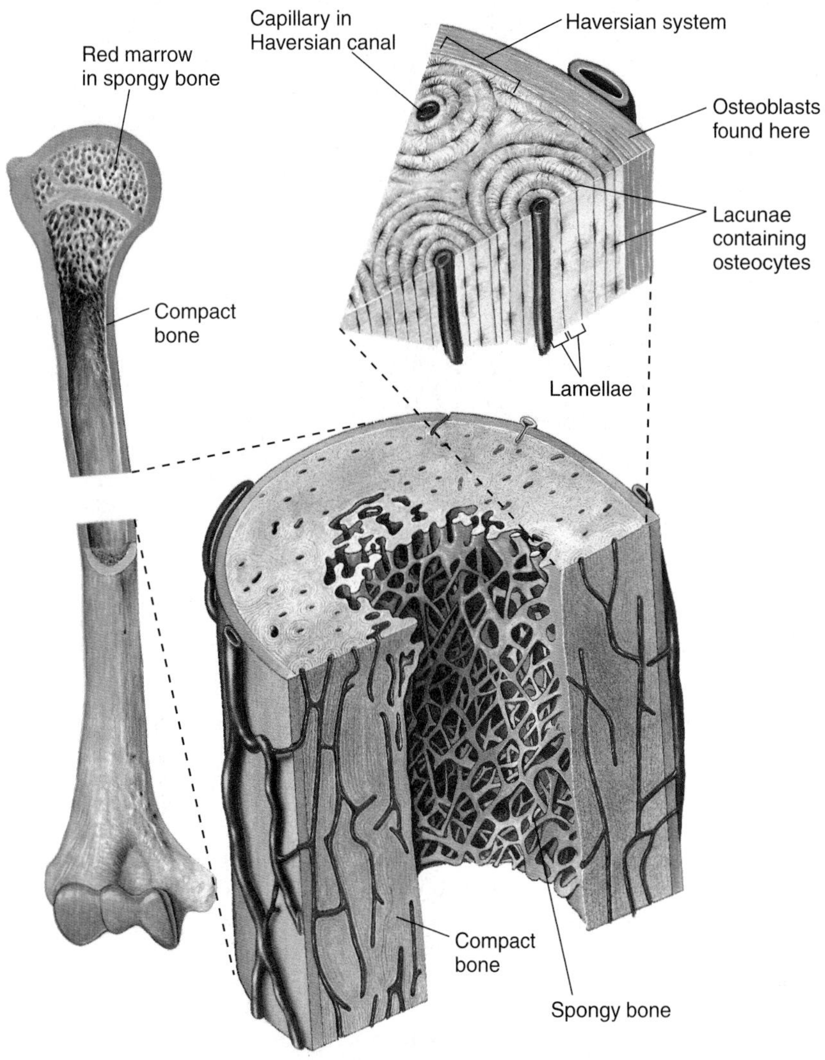
Capillary in
Haversian canal
Haversian system
Red marrow
in spongy bone
Osteoblasts
found here
Lacunae
containing
osteocytes
Compact
bone
Lamellae
Compact
bone
Spongy bone

Figure 22.18 The structure of bone.

Some parts of bones are dense and compact, giving the bone strength. Other parts are spongy, with a more open lattice; it is in the red marrow that most red blood cells are formed.

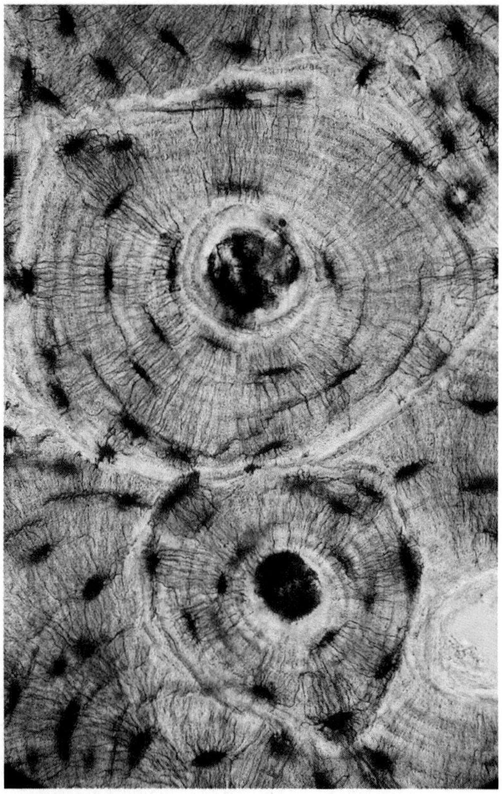

Figure 22.19 Compact bone.

The large, circular structures are Haversian canals containing blood vessels and nerves. Living bone cells occupy the arc-shaped dark spaces.

The interior is less compact, with a more open lattice structure, and is called **spongy bone.** Red blood cells form in the red marrow of spongy bone. New bone is formed in two stages: First, collagen is secreted by cells called **osteoblasts,** which lay down a matrix of fibrils along lines of stress. Then calcium minerals impregnate the fibrils. Bone is laid down in thin, concentric layers, like layers of paint on an old pipe. The layers form as a series of tubes around a narrow central channel called a **Haversian canal** (figure 22.19), which runs parallel to the length of the bone. The many Haversian canals within a bone, all interconnected, contain blood vessels and nerves that provide a lifeline to its living, bone-forming cells.

When bone is first formed in the embryo, osteoblasts use the cartilage skeleton as a template for bone formation. During childhood bones grow actively. The total bone mass in a healthy young adult, by contrast, does not change much from one year to the next. This does not mean change is not occurring. Large amounts of calcium and thousands of osteocytes (bone cells) are constantly being removed and replaced, but total bone mass does not change because deposit and removal take place at about the same rate.

Two cell types are responsible for this dynamic bone "remodeling": *osteoblasts* deposit bone, and *osteoclasts* secrete enzymes that digest the organic matrix of bone, liberating calcium for reabsorption by the bloodstream. The dynamic remodeling of bone adjusts bone strength to work load, new bone being formed along lines of stress. When a bone is subjected to compression, mineral deposition by osteoblasts exceeds withdrawals by osteoclasts. That is why long-distance runners must slowly increase the distances they attempt, to allow their bones to strengthen along lines of stress; otherwise, stress fractures can cripple them.

As a person ages, the backbone and other bones tend to decline in mass. Excessive bone loss is a condition called **osteoporosis** (figure 22.20). After the onset of osteoporosis, the replacement of calcium and other minerals lags behind withdrawal, causing the bone tissue to gradually erode. Eventually the bones become brittle and easily broken. The decreased physical activity typical of older people's lifestyles, a gradual decline in the activity of the body's osteoblasts, excessive protein intake, and sex hormone deficiencies common in older women all contribute to the disorder.

The best defense against osteoporosis is to maximize calcium deposition in bone during the years when active bone remodeling is occurring. High levels of dietary calcium consumed in the first decades of life stimulate osteoblasts to lay down more minerals, producing dense bones. Later in life, when osteoblast activity decreases and mineral withdrawal predominates, there is a much higher mineral reservoir to draw from.

22.11 Bone is a living tissue, strong without being brittle.

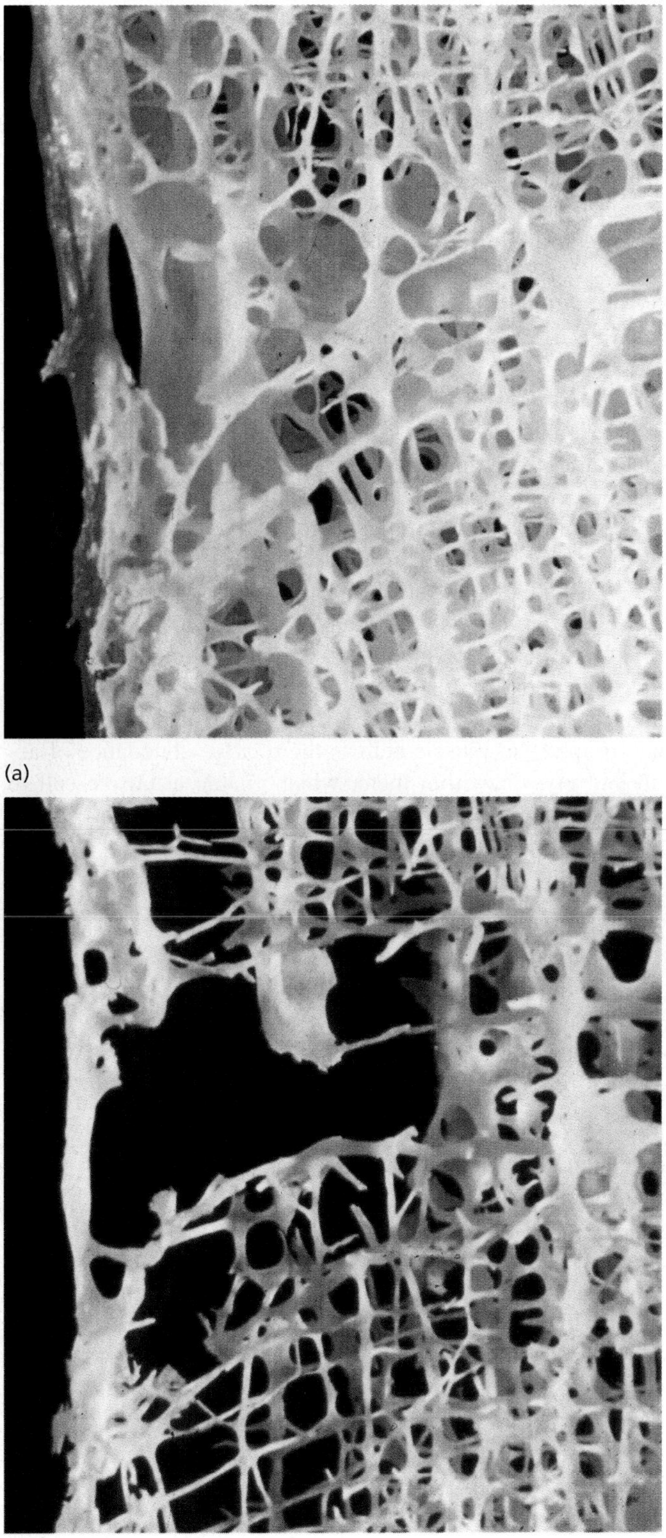

(a)

(b)

Figure 22.20 Osteoporosis.

Common in older women, osteoporosis is a bone disorder in which bones progressively lose minerals. (*a*) Normal bone tissue, in which the rate of bone deposit by osteoblasts equals the rate of bone withdrawal by osteoclasts. (*b*) Advanced osteoporosis, in which an excess of osteoclast activity over a considerable period of time has led to significant bone loss.

22.12 Kinds of Muscle

Three kinds of muscle together form the vertebrate muscular system. The vertebrate body is able to move because *skeletal muscles* pull the bones with considerable force. The heart pumps because of the contraction of *cardiac muscle.* Food moves through the intestines because of the rhythmic contractions of another kind of muscle, *smooth muscle.*

Skeletal Muscle

Skeletal muscles move the bones of the skeleton. Some of the major human muscles are shown in figure 22.21. Skeletal muscle cells are produced during development by the fusion of several cells at their ends to form a very long fiber. Each of these muscle cell fibers still contains all the original nuclei, pushed out to the periphery of the cytoplasm. Each **muscle fiber** consists of many elongated structures called **myofibrils,** and each myofibril is, in turn, composed of many myofilaments (figure 22.22). Furthermore, each myofilament contains the protein filaments actin and myosin. The key property of muscle cells is the relative abundance of actin and myosin within them, which enable a muscle cell to contract. These protein filaments are present as part of the cytoskeleton of all eukaryotic cells, but they are far more abundant in muscle cells.

Cardiac Muscle

The vertebrate heart is composed of striated muscle fibers arranged very differently from the fibers of skeletal muscle. Instead of very long, multinucleate cells running the length of the muscle, heart muscle is composed of chains of single cells, each with its own nucleus. Chains of cells are organized into fibers that branch and interconnect, forming a latticework. This lattice structure is critical to the way heart muscle functions. Each heart muscle cell is coupled to its neighbors electrically by tiny holes called *gap junctions* that pierce the plasma membranes in regions where the cells touch each other. Heart contraction is initiated at one location by the opening of transmembrane channels that depolarize the membrane; a wave of electrical depolarization then passes from cell to cell across the gap junctions, causing the heart to contract in an orderly pulsation.

Smooth Muscle

Smooth muscle cells are long and spindle-shaped, each containing a single nucleus. However, the individual myofilaments are not aligned into orderly assemblies as they are in skeletal and cardiac muscles. Smooth muscle tissue is organized into sheets of cells. In some tissues, smooth muscle cells contract only when they are stimulated by a nerve or hormone. Examples are the muscles that line the walls of many blood vessels and those that make up the iris of the vertebrate eye. In other smooth muscle tissue, such as that found in the wall of the gut, the individual cells contract spontaneously, leading to a slow, steady contraction of the tissue.

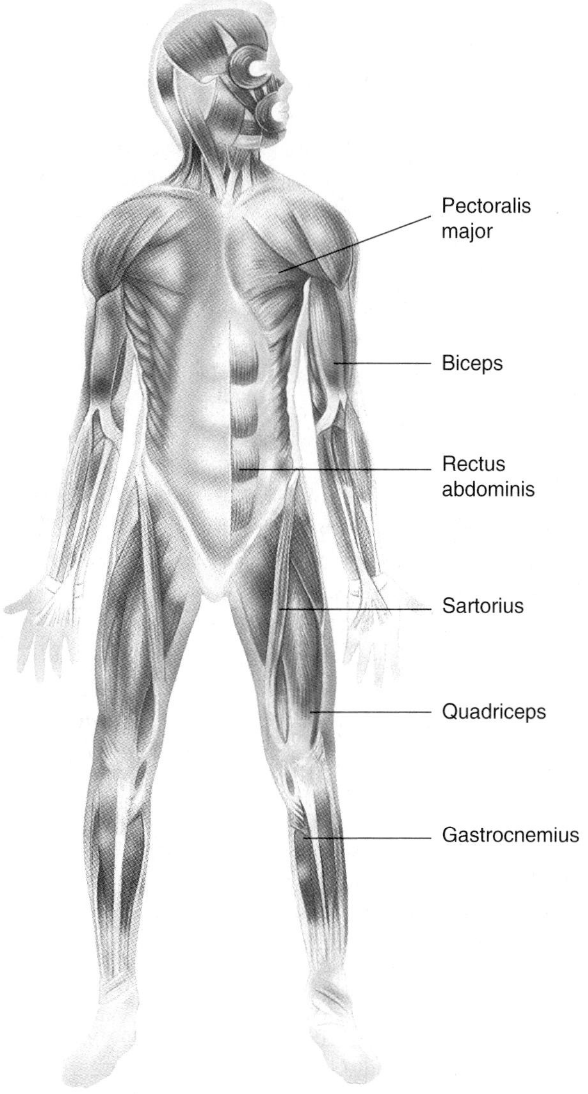
Pectoralis major
Biceps
Rectus abdominis
Sartorius
Quadriceps
Gastrocnemius

Figure 22.21 The muscular system.
The major muscles in the human body are labeled.

Tendons Connect Bone to Muscle

Muscles are attached to bones by straps of dense connective tissue called **tendons.** Bones pivot about flexible connections called **joints,** pulled back and forth by the muscles attached to them. Each muscle pulls on a specific bone. One end of the muscle, the *origin,* is attached by a tendon to a bone that remains stationary during a contraction. This provides an object against which the muscle can pull. The other

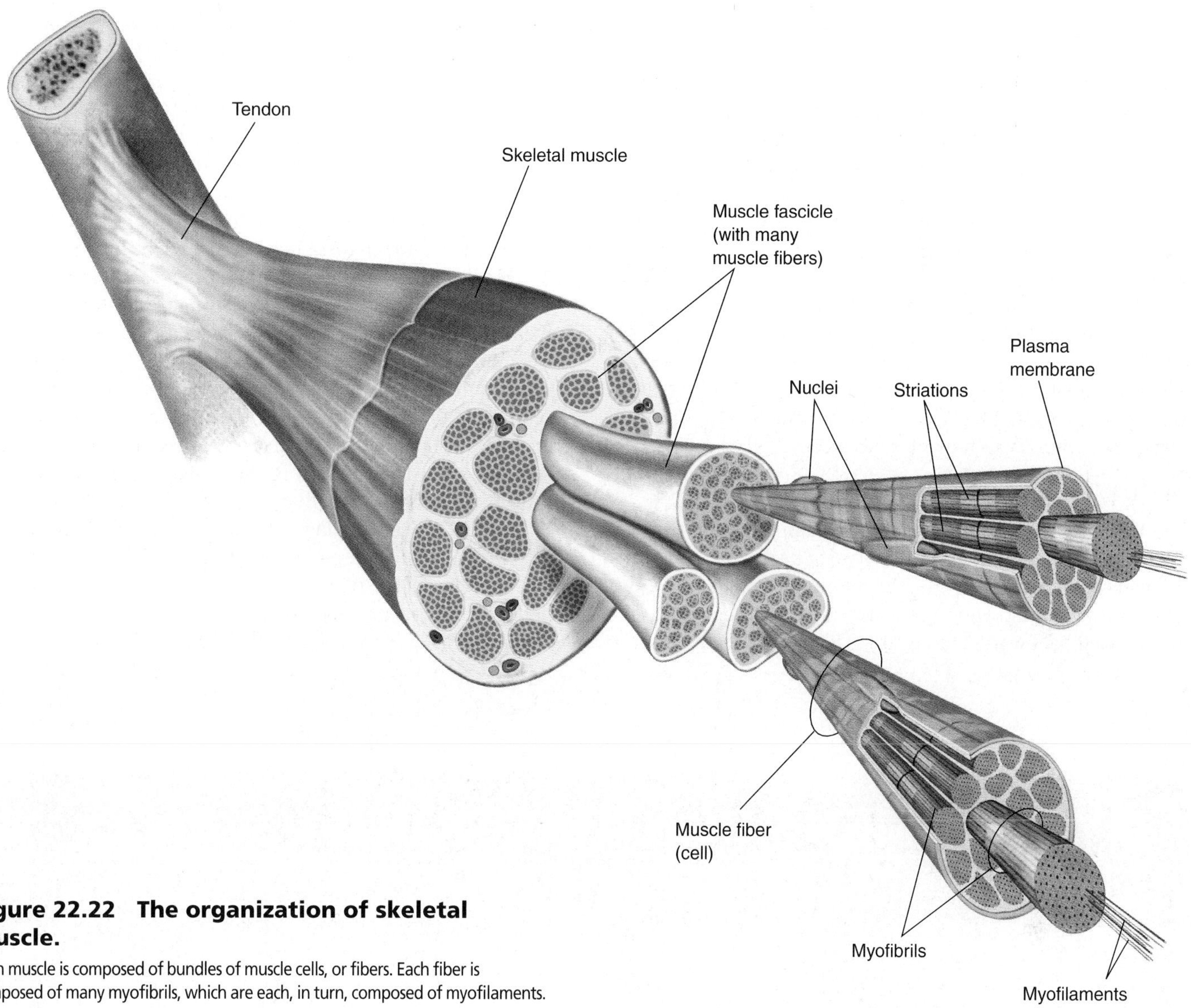
Tendon
Skeletal muscle
Muscle fascicle
(with many
muscle fibers)
Plasma
membrane
Nuclei
Striations
Muscle fiber
(cell)
Myofibrils
Myofilaments

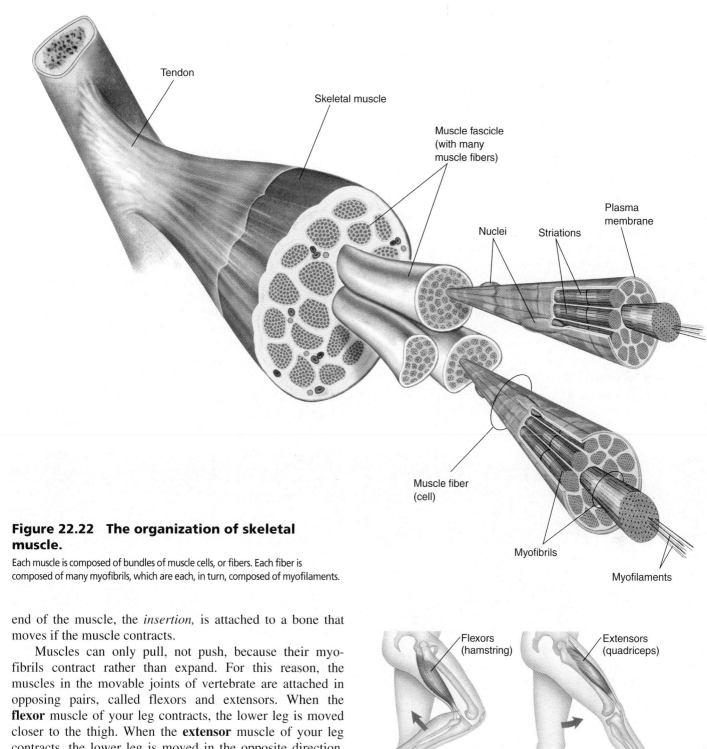

Figure 22.22 The organization of skeletal muscle.

Each muscle is composed of bundles of muscle cells, or fibers. Each fiber is composed of many myofibrils, which are each, in turn, composed of myofilaments.

end of the muscle, the *insertion,* is attached to a bone that moves if the muscle contracts.

Muscles can only pull, not push, because their myofibrils contract rather than expand. For this reason, the muscles in the movable joints of vertebrate are attached in opposing pairs, called flexors and extensors. When the **flexor** muscle of your leg contracts, the lower leg is moved closer to the thigh. When the **extensor** muscle of your leg contracts, the lower leg is moved in the opposite direction, farther away (figure 22.23).

22.12 There are three basic kinds of muscles: skeletal muscle, which moves the bones of the skeleton; the cardiac muscle of the heart; and smooth muscle, which maintains slow contractions around organs. Muscles attach to bones by strong tendons.

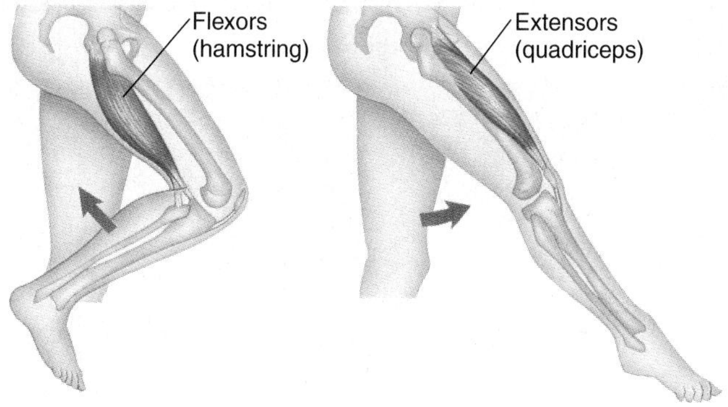
Flexors
(hamstring)
Extensors
(quadriceps)

Figure 22.23 Flexor and extensor muscles.

Limb movement is always the result of muscle contraction, never muscle extension. Muscles that retract limbs are called flexors; those that extend limbs are called extensors. Thus, you bend your leg back by contracting your hamstring muscle, a flexor muscle, and you extend your leg by contracting your quadriceps, an extensor muscle.

22.13 How Muscles Work

Far too fine to see with the naked eye, the individual myofilaments of vertebrate muscles are only 6 nanometers thick. Each is composed of long, threadlike filaments of the proteins actin and myosin. An **actin filament** consists of two strings of actin molecules wrapped around one another, like two strands of pearls loosely wound together. A **myosin filament** is also composed of two strings of protein wound about each other, but a myosin filament is 10 times longer than an actin filament, and the myosin strings have a very unusual shape. One end of a myosin filament consists of a very long rod, while the other end consists of a double-headed globular region, or "head." In electron micrographs, a myosin filament looks like a two-headed snake. This odd structure is the key to how muscles work.

How Myofilaments Contract

Look at the diagram of myosin and actin in a myofilament (figure 22.24), and focus on the myosin heads. When a myofilament contracts, the heads of the myosin filaments move first. Like flexing your hand downward at the wrist, the heads bend backward and inward. This moves them closer to their rodlike backbones and several nanometers in the direction of the flex. In itself, this myosin head-flex accomplishes nothing—but the myosin head is attached to the actin filament! As a result, the actin filament is pulled along with the myosin head as it flexes, causing the actin filament to slide by the myosin filament in the direction of the flex. As one after another myosin head flexes, the myosin in effect "walks" step by step along the actin. Each step uses a molecule of ATP to recock the myosin head before each flex.

How does this sliding of actin past myosin lead to myofilament contraction and muscle cell movement? Within each myofilament, the actin is anchored at one end, at a position in striated muscle called the Z line (figure 22.25). Because it is tethered like this, the actin cannot simply move off. Instead, the actin pulls the anchor with it! As actin moves past myosin, it drags the Z line toward the myosin. The secret of muscle contraction is that each myosin is interposed between two pairs of actin filaments—the myofilament is attached at *both* ends to Z lines by actin. One moving to the left and the other to the right, the two pairs of actin molecules drag the Z lines toward each other as they slide past the myosin core. As the Z lines are pulled closer together, the plasma membranes to which they are attached move toward one another, and the cell contracts.

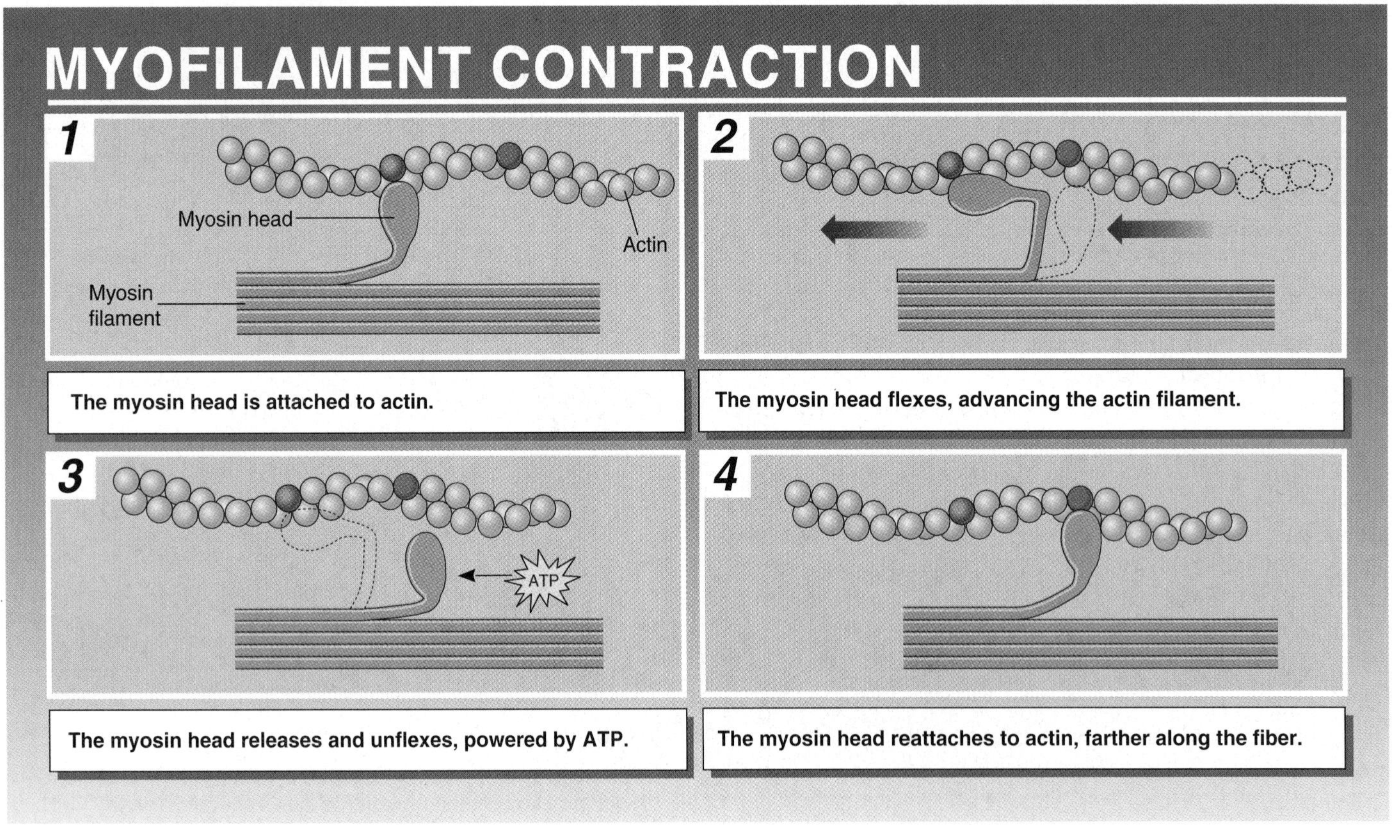
MYOFILAMENT CONTRACTION
1
Myosin head
Actin
Myosin filament
The myosin head is attached to actin.
2
The myosin head flexes, advancing the actin filament.
3
ATP
The myosin head releases and unflexes, powered by ATP.
4
The myosin head reattaches to actin, farther along the fiber.

Figure 22.24 How myofilament contraction works.

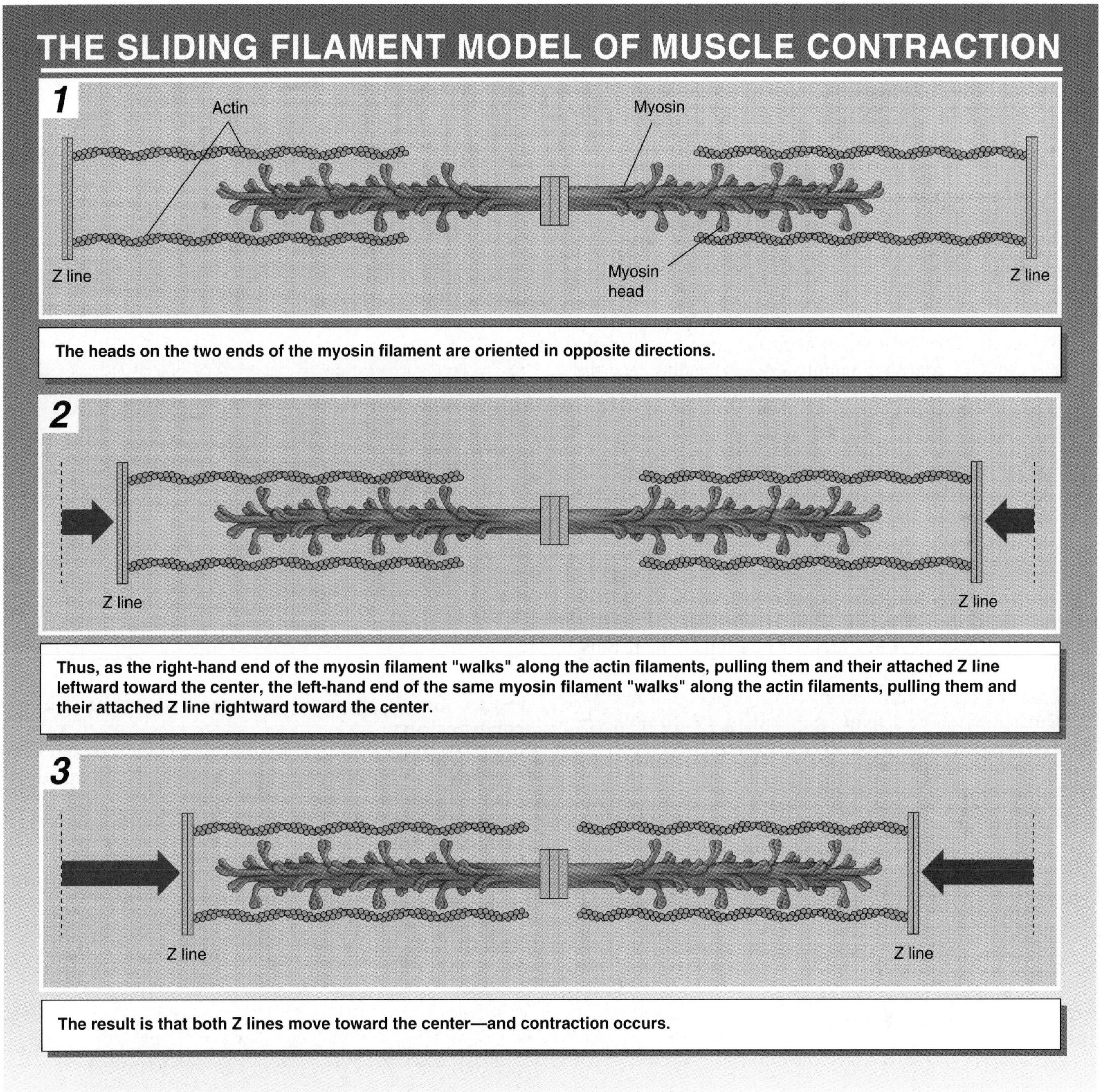
THE SLIDING FILAMENT MODEL OF MUSCLE CONTRACTION
1
Actin
Myosin
Z line
Myosin head
Z line
The heads on the two ends of the myosin filament are oriented in opposite directions.
2
Z line
Z line
Thus, as the right-hand end of the myosin filament "walks" along the actin filaments, pulling them and their attached Z line leftward toward the center, the left-hand end of the same myosin filament "walks" along the actin filaments, pulling them and their attached Z line rightward toward the center.
3
Z line
Z line
The result is that both Z lines move toward the center—and contraction occurs.

Figure 22.25 How actin and myosin filaments interact.

How Nerves Signal Muscles to Contract

In vertebrate skeletal muscle, contraction is initiated by a nerve impulse. The nerve fiber is embedded in the surface of the muscle fiber, forming a **neuromuscular junction.** When a signal reaches the end of a neuron, at the point where the neuron almost touches the muscle cell (the **motor end plate**), the neuron releases the chemical acetylcholine into the tiny gap separating neuron from muscle. Acetylcholine passes across the gap to the muscle plasma membrane, binds to receptor proteins there, and so causes ion channels in the motor end plate to open. The muscle membrane is said to be depolarized, because opening these channels allows ions to move freely in and out, wiping out any differences there might be in ion concentration between inside and outside.

The Role of Calcium Ions in Contraction

When a muscle is relaxed, its myosin heads are "cocked" and ready, through the splitting of ATP, but are unable to bind to actin. This is because the attachment sites for the myosin heads on the actin are physically blocked by another protein, known as **tropomyosin.** Myosin heads therefore cannot bind to actin in the relaxed muscle, and the filaments cannot slide.

In order to contract a muscle, the tropomyosin must be moved out of the way so that the myosin heads can bind to actin. This requires the function of **troponin,** a regulatory protein that binds to the tropomyosin. The troponin and tropomyosin form a complex that is regulated by the calcium ion (Ca^{++}) concentration of the muscle cell cytoplasm.

When the Ca^{++} concentration of the muscle cell cytoplasm is low, tropomyosin inhibits myosin binding, and the muscle is relaxed (figure 22.26). When the Ca^{++} concentration is raised, Ca^{++} binds to troponin. This causes the troponin-tropomyosin complex to be shifted away from the attachment sites for the myosin heads on the actin. When this repositioning has occurred, the myosin heads attach to actin and, using ATP energy, move along the actin in a stepwise fashion to shorten the myofibril.

Where does the Ca^{++} come from? Muscle fibers store Ca^{++} in a modified endoplasmic reticulum called the **sarcoplasmic reticulum** (figure 22.27). When a muscle fiber is stimulated to contract, Ca^{++} is released from the sarcoplasmic reticulum and diffuses into the myofibrils, where it binds to troponin and causes contraction. The contraction of muscles is regulated by nerve activity, and so nerves must influence the distribution of Ca^{++} in the muscle fiber.

22.13 Muscles are made of many tiny threadlike filaments of actin and myosin called myofilaments. Muscles work by using ATP to power the sliding of myosin along actin, causing the myofilaments to contract.

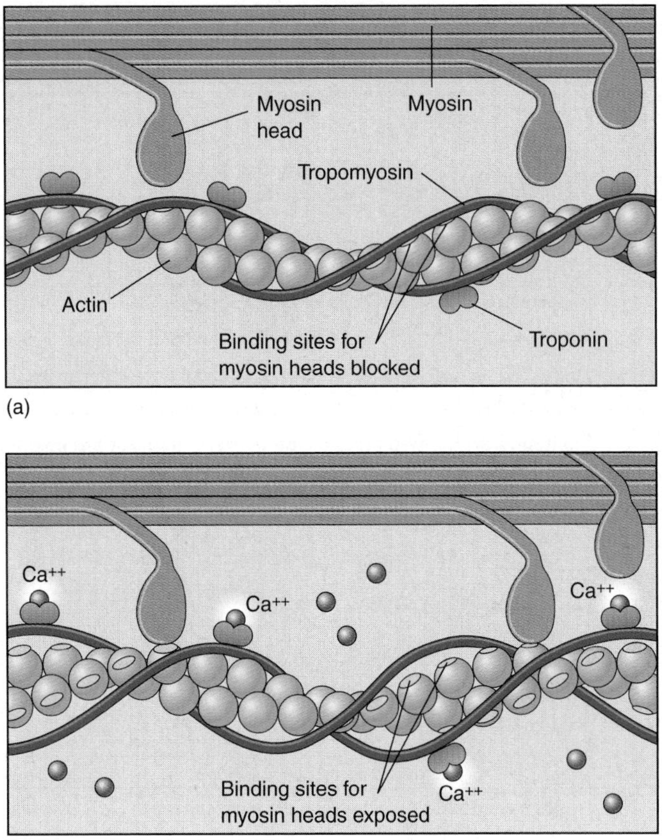

Myosin head
Myosin
Tropomyosin
Actin
Binding sites for myosin heads blocked
Troponin
(a)
Ca++
Ca++
Ca++
Binding sites for myosin heads exposed
Ca++
(b)

Figure 22.26 How calcium controls muscle contraction.

(*a*) When the muscle is at rest, a long filament composed of the molecule tropomyosin blocks the myosin binding sites of the actin molecule. Without actin's ability to form links with myosin at these sites, muscle contraction cannot occur. (*b*) When calcium ions bind to another protein, troponin, the resulting complex displaces the filament of tropomyosin, exposing the myosin binding sites of actin. Myosin heads can bind to the actin, and contraction occurs.

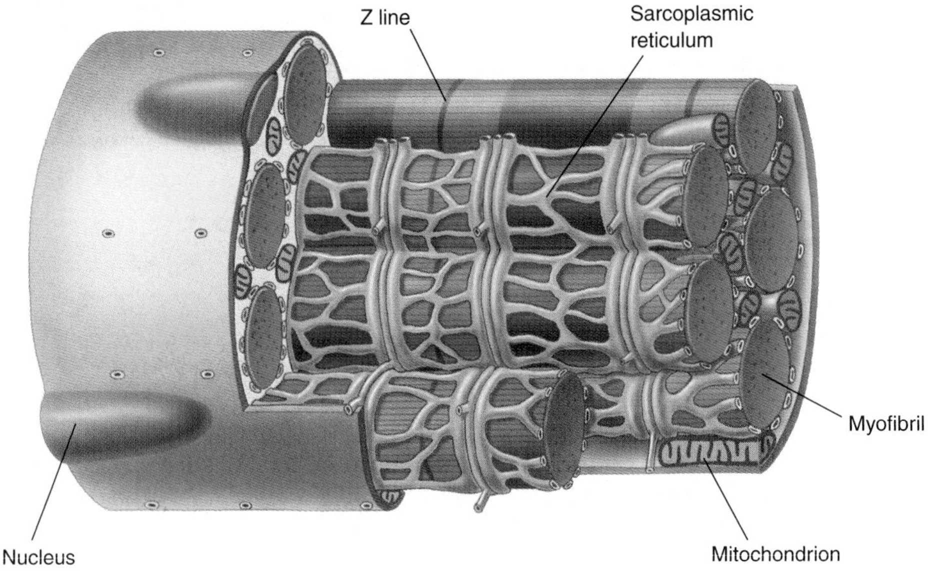

Z line
Sarcoplasmic reticulum
Myofibril
Mitochondrion
Nucleus

Figure 22.27 The sarcoplasmic reticulum.

The sarcoplasmic reticulum is a system of membranes that wraps around the individual myofibrils of a muscle fiber.

CONCEPT REVIEW

1. An organ consists of several types of
   a. organ systems.
   b. organisms.
   c. tissues.
   d. organs.

2. Each of the following is a function of epithelial tissue *except*
   a. movement.
   b. protection.
   c. sensory ability.
   d. secretion.

3. Which of the following is *not* a cell type of the connective tissue?
   a. lymphocytes
   b. fat cells
   c. columnar cells
   d. erythrocytes

4. The erythrocyte
   a. fights infection.
   b. carries oxygen.
   c. coagulates the blood.
   d. dissolves substances.

5. The impulse within a nerve cell travels from
   a. dendrites to axon to cell body.
   b. dendrites to cell body to axon.
   c. cell body to axon to dendrites.
   d. cell body to dendrites to axon.

6. The vertebral column belongs to the _________ skeleton.
   a. appendicular
   b. axial

7. Select the *incorrect* statement about vertebrate bone tissue.
   a. Osteoblasts secrete collagen.
   b. The Haversian canal contains blood vessels.
   c. Bone is a type of connective tissue.
   d. New bone cells are not formed in adults.

8. Select the *incorrect* statement about cardiac muscle.
   a. It is multinucleated.
   b. It comprises the heart.
   c. The fibers are striated.
   d. It is composed of chains of cells.

9. Calcium in the muscle cell attaches to
   a. the neurotransmitter.
   b. actin.
   c. myosin.
   d. troponin.

10. Antibodies are part of the _________ system.

11. _________ tissue covers the surfaces of the body.

12. Neurons have two kinds of projections from the cell body, the _________ and the _________.

13. Among myofilaments, the protein _________ moves past myosin during muscle contraction.

14. When a nerve signal reaches the neuromuscular junction, the chemical _________ is released into the gap between the end of the neuron and the muscle tissue.

15. The source of energy for muscle contraction is _________.

CHALLENGE YOURSELF

1. Chemotherapy is frequently used as a cancer treatment, designed to kill the rapidly dividing malignant cells. Why do you think this treatment typically causes patients to lose their hair and often interferes with the function of their gastrointestinal tract?

2. Is most of bone tissue a living or nonliving substance? Why?

3. In vertebrates, contraction of cardiac muscle in the heart and smooth muscle in the stomach and intestines is initiated by the muscles themselves, independent of the nervous system; in contrast, contraction of the skeletal muscles controlling movements of the jaw and fingers, for example, is initiated by impulses from the nervous system. What do you think are the advantages of having different mechanisms for initiating contractions in muscles such as these?

22 eBRIDGE

Reinforcing Key Points

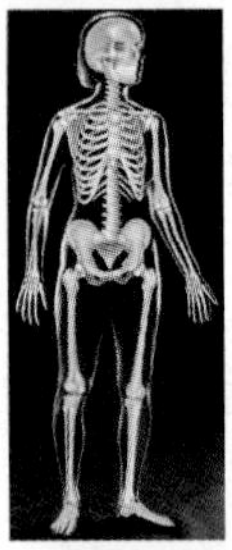

The Animal Body Plan

22.1 Radial Versus Bilateral Symmetry

22.2 No Body Cavity Versus Body Cavity

22.3 Nonsegmented Versus Segmented Bodies

22.4 Protostomes Versus Deuterostomes

22.5 Organization of the Vertebrate Body

Tissues of the Vertebrate Body

22.6 Epithelium Is Protective Tissue

22.7 Connective Tissue Supports the Body

22.8 Muscle Tissue Lets the Body Move

22.9 Nerve Tissue Conducts Signals Rapidly

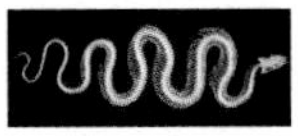

The Skeletal System

22.10 Types of Skeletons

22.11 The Structure of Bone

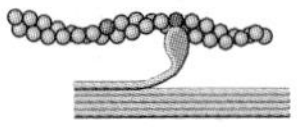

The Muscular System

22.12 Kinds of Muscle

22.13 How Muscles Work

Electronic Learning

Visual Learning

Animations

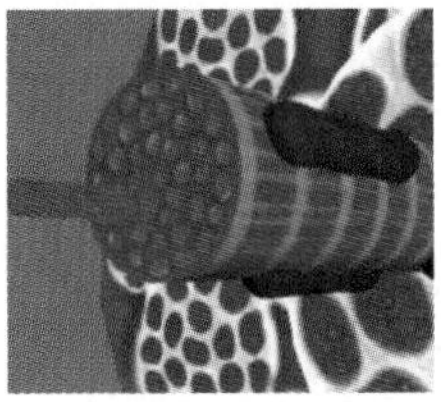

Symmetry in Nature

Striated Muscle Contraction

Walking

Detailed Striated Muscle Contraction

Muscle Contraction Action Potential

Actin-Myosin Crossbridges

Art Labeling Activities

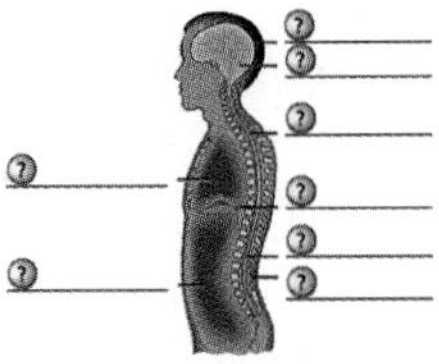

Mammalian Body Cavities

Human Skin Anatomy

Detailed Human Skin Anatomy

Bones of the Pectoral Girdle and Arm

Bones of the Pelvic Girdle and Leg

Attachment of Skeletal Muscles

Skeletal Muscles

Author's Corner

A Closer Look at Jogging. Running and jogging are forms of exercise that many Americans enjoy. However, excessive running can have unfortunate consequences, particularly if you ignore some basic biology about how muscles work and about how bones grow. The fundamental biology of muscle-powered running is the physiology of an impact-driven locomotion device. Successful running depends upon powerful contraction of leg muscles, slamming your feet against the ground, and thrusting your body forward. All this thrusting and slamming places considerable stress on the attachment of the muscle to the leg bones, and of course to the bone itself. Successful running requires a clear understanding of these stresses, and of the precautions necessary to avoid their getting out of hand.

1. Improper running provides a painful lesson in the biology of bones.

Virtual Classroom

The Biology of Jogging

Practically two-thirds of U.S. adults weigh more than they should. An increasing number of Americans are attempting to deal with this dilemma by exercising. You can see them jogging up and down the streets of America every morning and evening. While moderate exercise like jogging is good for you, two important things can go wrong if you get overly enthusiastic too early. The first has to do with how leg and arm muscles are attached to bone. Each major muscle is locked onto a bone by a dense strap of connective tissue called a tendon. Tendons are somewhat elastic, allowing "give-and-take" during muscle contraction, but if you jerk them, you can snap them off the bone—a "pulled muscle." The way to avoid this is to stretch them before running, so they are ready for the added workload. The second thing that can go wrong has to do with how bones grow, which is along lines of stress. If you gradually increase your amount of running, new bone has time to form properly along the lines of stress that shoot with each step up your long leg bones. Push too hard, too fast, and your bones can give way, producing painful tiny cracks called stress fractures.

Virtual Lab

How Honeybees Keep Their Cool

One of the biggest problems terrestrial animals face is that the temperature of their environment keeps changing. Many species of animals have evolved the ability to maintain their body temperature within a narrow range, regardless of the temperature of their surroundings. These animals are called endotherms. You are an endotherm. A honeybee is an endothermic animal which increases its body temperature by shivering its thoracic muscles to allow it to fly in colder temperatures. So, what happens to the honeybee in warmer temperatures? Some animals seek out shade during the heat of the day or rest, reducing their metabolism, but honeybees keep right on flying. Why don't they overheat? A honeybee could in principle reduce its metabolism when it's hot outside, but the act of flying requires a great deal of energy. How could a honeybee reduce metabolism for thermoregulation, and at the same time accomplish flight?

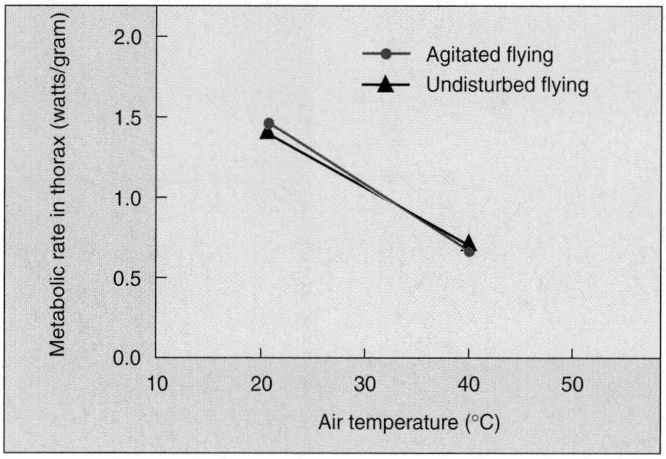
Metabolic rate in thorax (watts/gram)
2.0
1.5
1.0
0.5
0.0
10
20
30
40
50
Air temperature (°C)
Agitated flying
Undisturbed flying

This question has been addressed experimentally by Jon Harrison of Arizona State University, by exposing honeybees to various air temperatures and measuring body temperature, metabolic rate, and wing-beat frequencies.

Quizzes

Further Reading

Essential Study Partner

Links

BioCourse.com

CHAPTER 23

CIRCULATION AND RESPIRATION

THE FAR SIDE® By GARY LARSON
© 1982 FarWorks, Inc. All Rights Reserved/Dist. by Creators Syndicate
Larson
The real reason dinosaurs became extinct

C H A P T E R 23

Circulation and Respiration

CHAPTER OVERVIEW

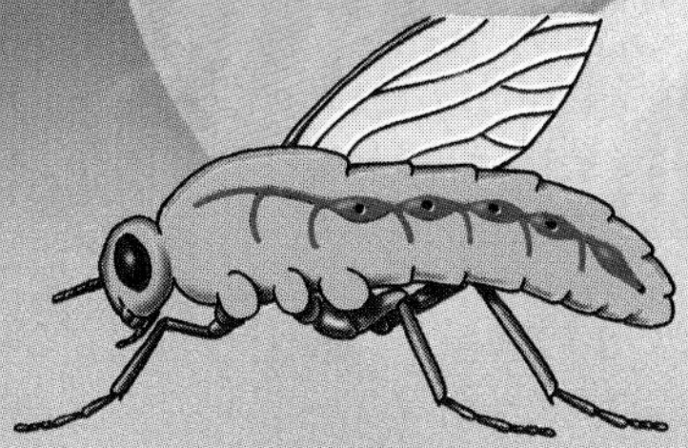

Circulation

23.1 Open and Closed Circulatory Systems

23.2 Architecture of the Vertebrate Circulatory System

23.3 The Lymphatic System: Recovering Lost Fluid

23.4 Blood

- Vertebrates have a closed circulatory system, where the blood stays within vessels as it travels away from and back to the heart.
- Blood is pumped out from the heart through arteries and returns through veins. Materials pass in and out of the blood as it passes through a network of tiny tubes called capillaries, which connects arterial to venous circulation.

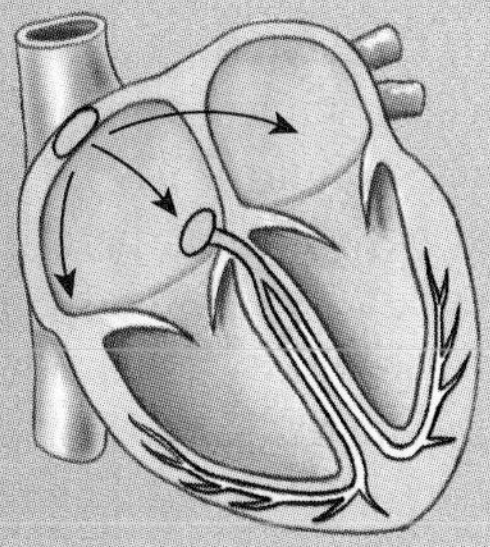

Evolution of Vertebrate Circulatory Systems

23.5 Fish Circulation

23.6 Amphibian and Reptile Circulation

23.7 Mammalian and Bird Circulation

23.8 Monitoring the Heart's Performance

23.9 Cardiovascular Diseases

- In fish, blood from the heart goes to the gills and then to the rest of the body before returning to the heart.
- In terrestrial vertebrates, blood returns from the lungs to the heart before it is pumped to the body.
- The contraction of the heart starts in the upper wall of the right atrium and spreads as a wave across the heart.

Respiration

23.10 Types of Respiratory Systems

23.11 Respiration in Aquatic Vertebrates

23.12 Respiration in Terrestrial Vertebrates

23.13 The Mammalian Respiratory System

23.14 How Respiration Works: Gas Exchange

- Respiration is the uptake of oxygen gas from air and the simultaneous release of carbon dioxide.
- Marine and aquatic vertebrates extract oxygen from water very efficiently with gills.
- Birds have the most efficient respiratory system of terrestrial vertebrates.
- CO_2 is transported largely as bicarbonate ion.

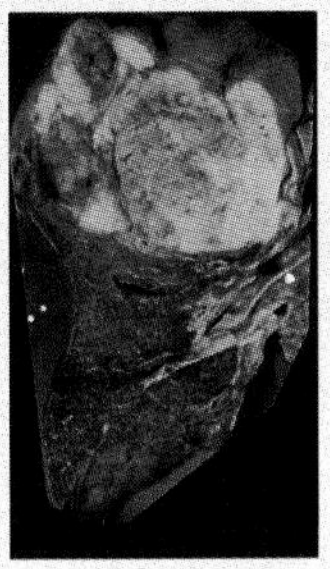

Lung Cancer and Smoking

23.15 The Nature of Lung Cancer

23.16 Smoking Causes Lung Cancer

- Cancer is caused by damage to growth-regulating genes.
- Cigarette smoking is the principal cause of lung cancer.
- Smoking produces lung cancer by introducing carcinogens into the lungs.

Every cell in the vertebrate body must acquire the energy it needs for living from outside organic molecules. Like residents of a city whose food is imported from farms in the countryside, the cells of the body need trucks to carry the food, highways for the trucks to travel on, and fuel to cook the food when it arrives. In the vertebrate body, the trucks are blood, the highways are blood vessels, and the fuel is oxygen molecules. Remember from chapter 5 that cells obtain energy by "burning" sugars like glucose, using up oxygen and generating carbon dioxide. In animals, the organ system that provides the trucks and highways is called the *circulatory system,* while the organ system that acquires the oxygen fuel and disposes of the carbon dioxide waste is called the *respiratory system.* We discuss the functions of these two organ systems in this chapter.

23.1 Open and Closed Circulatory Systems

Among the unicellular protists, oxygen and nutrients are obtained directly by simple diffusion from the aqueous external environment. Cnidarians, such as *Hydra,* and flatworms, such as *Planaria,* have cells that are directly exposed to either the external environment or to a body cavity that functions in both digestion and circulation called the **gastrovascular cavity** (figure 23.1*a*). The gastrovascular cavity of *Hydra* extends even into the tentacles, and that of *Planaria* branches extensively to supply every cell with oxygen and the nourishment obtained by digestion. Larger animals, however, have tissues that are several cell layers thick, so that many cells are too far away from the body surface or digestive cavity to exchange materials directly with the environment. Instead, oxygen and nutrients are transported from the environment and digestive cavity to the body cells by an internal fluid within a **circulatory system.**

There are two main types of circulatory systems: *open* or *closed.* In an **open circulatory system,** such as that found in mollusks and arthropods (figure 23.1*b*), there is no distinction between the circulating fluid (blood) and the extracellular fluid of the body tissues (interstitial fluid or lymph). This fluid is thus called **hemolymph.** Insects have a muscular tube that serves as a heart to pump the hemolymph through a network of channels and cavities in the body. The fluid then drains back into the central cavity.

In a **closed circulatory system,** the circulating fluid, or *blood,* is always enclosed within blood vessels that transport blood away from and back to a pump, the *heart.* Annelids and all vertebrates have a closed circulatory system. In annelids such as an earthworm, a dorsal blood vessel contracts rhythmically to function as a pump. Blood is pumped through five small connecting vessels, which also function as pumps, to a ventral blood vessel, which transports the blood posteriorly until it eventually reenters the dorsal blood vessel. Smaller vessels branch from the ventral blood vessel to supply the tissues of the earthworm with oxygen and nutrients and to transport waste products (figure 23.1*c*).

Blood vessels form a tubular network that permits blood to flow from the heart to all the cells of the body and then back to the heart. *Arteries* carry blood away from the heart, whereas *veins* return blood to the heart. Blood passes from the arterial to the venous system in *capillaries,* which are the thinnest and most numerous of the blood vessels.

As blood plasma passes through capillaries, the pressure of the blood forces some of this fluid out of the capillary walls. Fluid derived this way is called **interstitial fluid.** Some of this fluid returns directly to capillaries, and some enters into **lymph vessels,** located in the connective tissues around the blood vessels. This fluid, now called *lymph,* is returned to the venous blood at specific sites. The lymphatic system is considered a part of the circulatory system and is discussed later in this chapter.

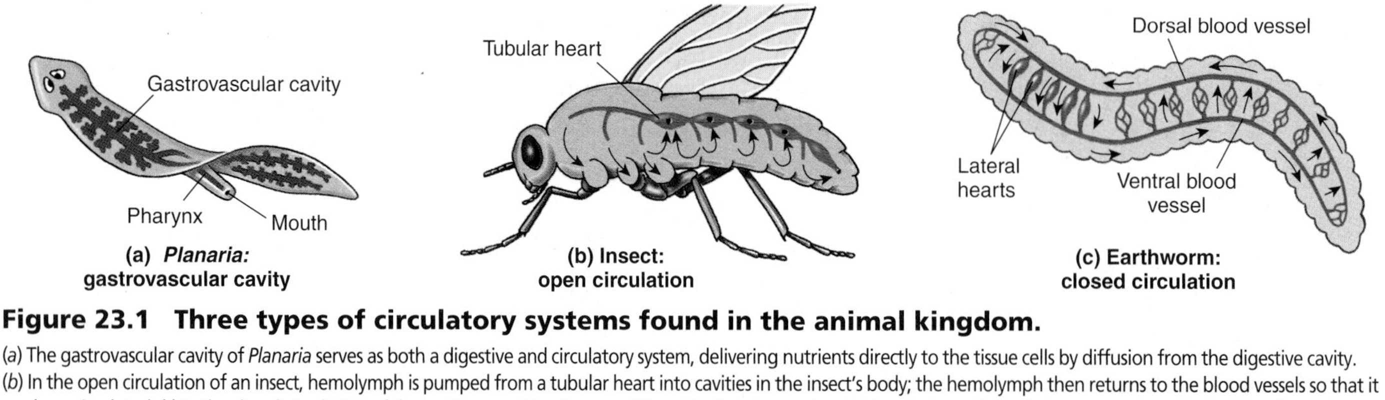
Gastrovascular cavity
Pharynx
Mouth
(a) *Planaria:* gastrovascular cavity
Tubular heart
(b) Insect: open circulation
Dorsal blood vessel
Lateral hearts
Ventral blood vessel
(c) Earthworm: closed circulation

Figure 23.1 Three types of circulatory systems found in the animal kingdom.
(*a*) The gastrovascular cavity of *Planaria* serves as both a digestive and circulatory system, delivering nutrients directly to the tissue cells by diffusion from the digestive cavity. (*b*) In the open circulation of an insect, hemolymph is pumped from a tubular heart into cavities in the insect's body; the hemolymph then returns to the blood vessels so that it can be recirculated. (*c*) In the closed circulation of the earthworm, blood pumped from the hearts remains within a system of vessels that returns it to the hearts. All vertebrates also have closed circulatory systems.

The Functions of Vertebrate Circulatory Systems

The functions of the circulatory system can be divided into three areas: transportation, regulation, and protection.

1. **Transportation.** All of the substances essential for cellular metabolism are transported by the circulatory system. These substances can be categorized as follows:

   Respiratory. Red blood cells, or erythrocytes, transport oxygen to the tissue cells. In the capillaries of the lungs or gills, oxygen attaches to hemoglobin molecules within the erythrocytes and is transported to the cells for aerobic respiration. Carbon dioxide produced by cell respiration is carried by the blood to the lungs or gills for elimination.

   Nutritive. The digestive system is responsible for the breakdown of food so that nutrients can be absorbed through the intestinal wall and into the blood vessels of the circulatory system. The blood then carries these absorbed products of digestion through the liver and to the cells of the body.

   Excretory. Metabolic wastes, excessive water and ions, and other molecules in plasma (the fluid portion of blood) are filtered through the capillaries of the kidneys and excreted in urine.

2. **Regulation.** The cardiovascular system transports hormones and participates in temperature regulation.

   Hormone transport. The blood carries hormones from the endocrine glands, where they are secreted, to the distant target organs they regulate.

   Temperature regulation. In warm-blooded vertebrates, or homeotherms, a constant body temperature is maintained, regardless of the ambient temperature. This is accomplished in part by blood vessels located just under the epidermis. When the ambient temperature is cold, the superficial vessels constrict to divert the warm blood to deeper vessels. When the ambient temperature is warm, the superficial vessels dilate so that the warmth of the blood can be lost by radiation.

   Some vertebrates also retain heat in a cold environment by using a **countercurrent heat exchange.** In this process, a vessel carrying warm blood from deep within the body passes next to a vessel carrying cold blood from the surface of the body (figure 23.2). The warm blood going out heats the cold blood returning from the body surface, so that this blood is no longer cold when it reaches the interior of the body.

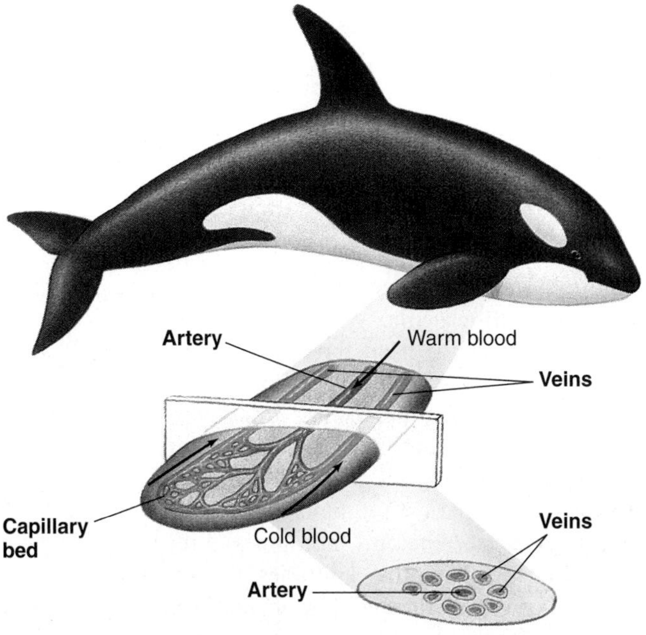

Artery
Warm blood
Veins
Capillary bed
Cold blood
Veins
Artery

Figure 23.2 Countercurrent heat exchange.
Many marine mammals, such as this killer whale, limit heat loss in cold water by countercurrent flow. The warm blood pumped from within the body in arteries warms the cold blood returning from the skin in veins, so that the core body temperature can remain constant in cold water.

3. **Protection.** The circulatory system protects against injury and foreign microbes or toxins introduced into the body.

   Blood clotting. The clotting mechanism protects against blood loss when vessels are damaged. This clotting mechanism involves both proteins from the blood plasma and cell structures called platelets (discussed in the next section).

   Immune defense. The blood contains white blood cells, or leukocytes, that provide immunity against many disease-causing agents. Some white blood cells are phagocytic, some produce antibodies, and some act by other mechanisms to protect the body.

23.1 Circulatory systems may be open or closed. All vertebrates have a closed circulatory system, in which blood circulates away from the heart in arteries and back to the heart in veins. The circulatory system serves a variety of functions, including transportation, regulation, and protection.

23.2 Architecture of the Vertebrate Circulatory System

The vertebrate circulatory system is made up of three elements: (1) the **heart**, a muscular pump that pushes blood through the body; (2) the **blood vessels,** a network of tubes through which the blood moves; and (3) the **blood,** which circulates within these vessels. The plumbing part of the circulatory system, the heart and blood vessels, is sometimes called the **cardiovascular system.**

Blood leaves the heart through vessels known as **arteries.** From the arteries, the blood passes into a network of smaller arteries called **arterioles.** From these, it is eventually forced through the **capillaries,** a fine latticework of very narrow tubes (from the Latin, *capillus*, "a hair"). While passing through the capillaries, the blood exchanges gases and metabolites (glucose, vitamins, hormones) with the cells of the body. After traversing the capillaries, the blood passes into a third kind of vessel, the **venules,** or small veins (figure 23.3). A network of venules empties into larger **veins** that collect the circulating blood and carry it back to the heart.

The capillaries have a much smaller diameter than the other blood vessels of the body. Blood leaves the mammalian heart through a large artery, the aorta, a tube that has a radius of about 1 centimeter (about the same as your thumb), but when it reaches the capillaries it passes through vessels with an average radius of only 8 micrometers, a reduction in radius of some 1,250 times! This decrease in size of blood vessels has a very important consequence. When a fluid such as blood flows through a tube, it meets a frictional resistance as the fluid passes over the walls of the tube—the narrower the tube, the greater the resistance. The capillary network presents a high resistance to flow, a resistance that must be overcome by the strength of the heartbeat.

Because it is a closed loop, every part of the cardiovascular system must have the same overall flow rate (about 5 liters per minute in adult humans). Flow rate (volume per unit time) is not the same as velocity (distance moved per unit time). Even though the velocity of flow in individual capillaries is much less than in arteries, the capillary network has the same overall flow rate as arteries because there are so very many more capillaries.

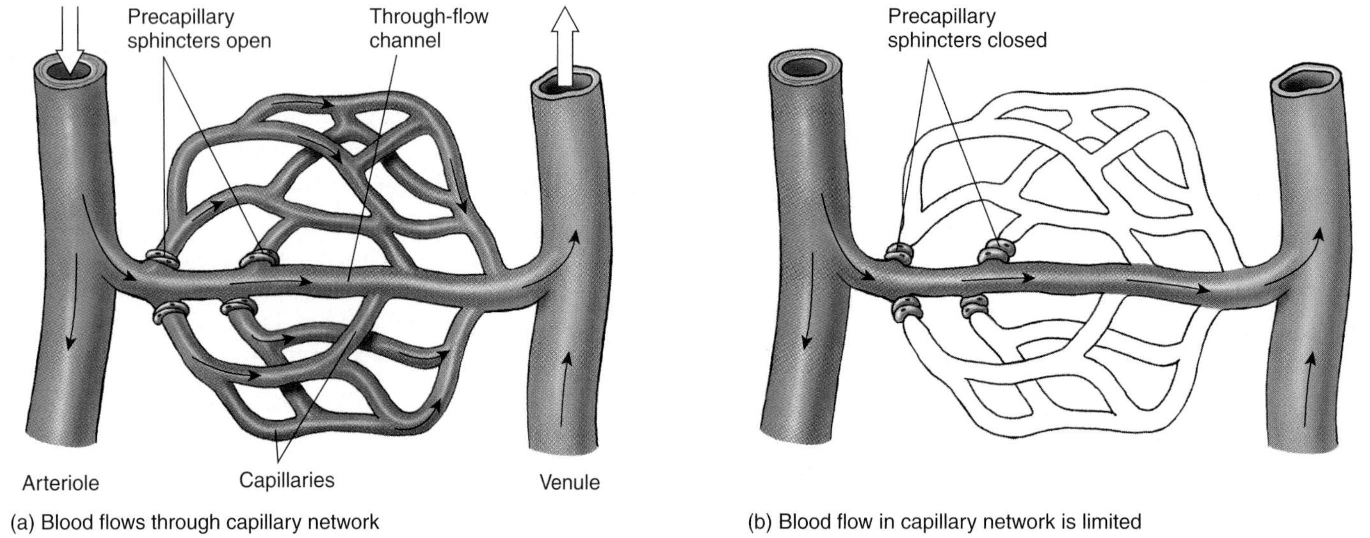

Precapillary sphincters open
Through-flow channel
Arteriole
Capillaries
Venule
(a) Blood flows through capillary network
Precapillary sphincters closed
(b) Blood flow in capillary network is limited

Figure 23.3 The capillary network connects arteries with veins.
Through-flow channels connect arterioles directly to venules. Branching from these through-flow channels is a network of finer channels, the capillaries. Most of the exchange between the body tissues and the red blood cells occurs while they are in this capillary network. Entrance to the capillaries is controlled by bands of muscle called precapillary sphincters located at the entrance to each capillary. (*a*) When a sphincter is open, blood flows through that capillary. (*b*) When a sphincter contracts, it closes off the capillary. By contracting these sphincters, the body can limit the amount of blood in the capillary network of a particular tissue; through-flow channels then allow the blood to bypass the capillary network when it is needed elsewhere.

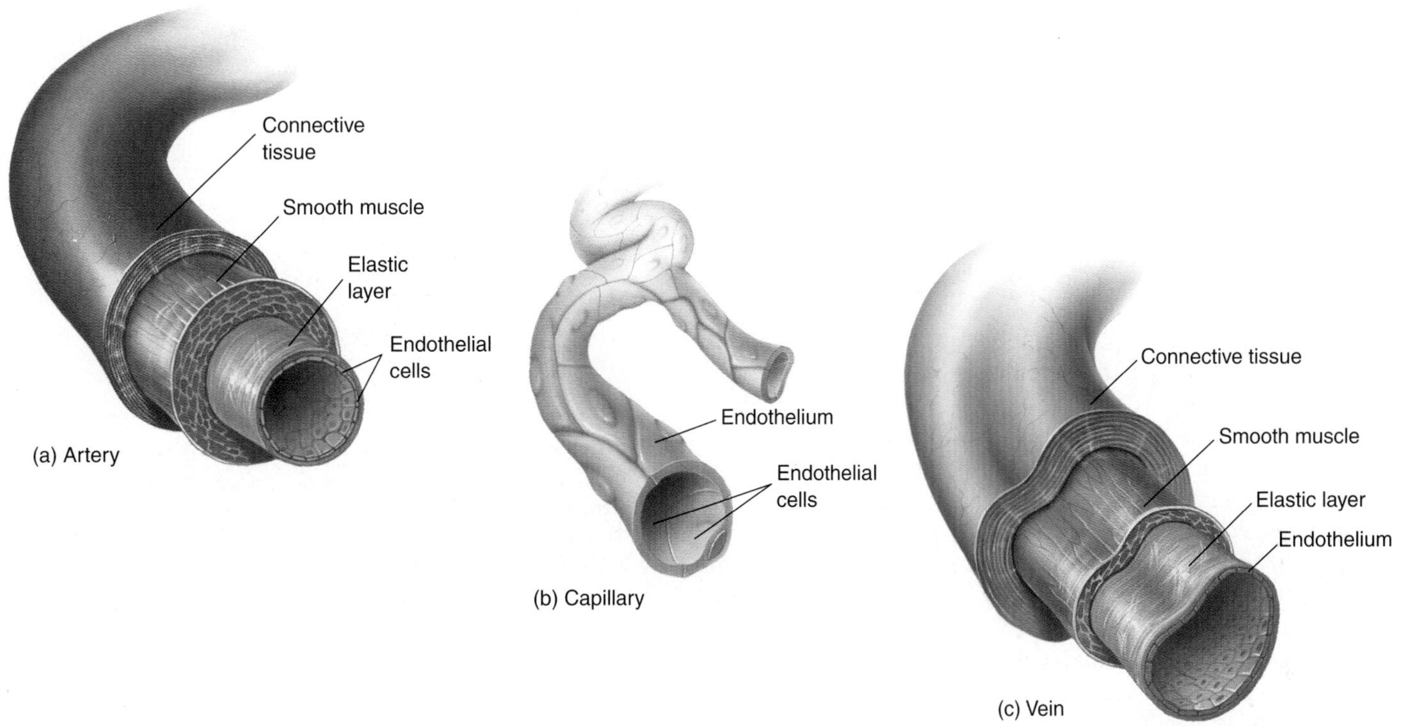
Connective tissue
Smooth muscle
Elastic layer
Endothelial cells
(a) Artery
Endothelium
Endothelial cells
(b) Capillary
Connective tissue
Smooth muscle
Elastic layer
Endothelium
(c) Vein

Figure 23.4 The structure of blood vessels.
(*a*) Arteries, which carry blood away from the heart, are expandable and are composed of layers of tissue. (*b*) Capillaries are simple tubes whose thin walls facilitate the exchange of materials between the blood and the cells of the body. (*c*) Veins, which transport blood back to the heart, do not need to be as sturdy as arteries. The walls of veins have thinner muscle layers than arteries, and they collapse when empty.

Arteries: Highways from the Heart

The arterial system, composed of arteries and arterioles, carries blood away from the heart. An artery is more than simply a pipe. Because blood comes from the heart not in a smooth flow but rather in pulses, slammed into the artery in great big slugs as the heart forcefully ejects its contents with each contraction, the artery has to be able to *expand;* otherwise there would be no place for the added fluid to go. An artery, then, is designed as an expandable tube, with its walls made up of three layers of tissue (figure 23.4). The innermost thin layer is composed of endothelial cells. Surrounding them is a thick layer of smooth muscle and elastic fibers, which in turn is encased within an envelope of protective connective tissue. Because this sheath and envelope are elastic, the artery is able to expand its volume considerably when the heart contracts, shoving a new volume of blood into the artery—just as a tubular balloon expands when you blow more air into it. The steady contraction of the smooth muscle layer strengthens the wall of the vessel against overexpansion.

Arterioles differ from arteries in two ways. They are smaller in diameter, and the muscle layer that surrounds an arteriole can be relaxed under the influence of hormones to enlarge the diameter. When the diameter increases, the blood flow also increases, an advantage during times of high body activity. Most arterioles are also in contact with nerve fibers. When stimulated by these nerves, the muscle lining of the arteriole contracts, constricting the diameter of the vessel. Such contraction limits the flow of blood to the extremities during periods of low temperature or stress. You turn pale when you are scared or cold because the arterioles in your skin are constricting. You blush for just the opposite reason. When you overheat or are embarrassed, the nerve fibers connected to muscles surrounding the arterioles are inhibited, which relaxes the smooth muscle and causes the arterioles in the skin to expand, bringing heat to the surface for escape.

Capillaries: Where Exchange Takes Place

Capillaries are where oxygen and food molecules are transferred from the blood to the body's cells and where waste carbon dioxide is picked up. In order to facilitate this back-and-forth traffic, capillaries have thin walls across which gases and metabolites pass easily. Capillaries have the simplest structure of any element in the cardiovascular system. They are built like a soft-drink straw, simple tubes with walls only one cell thick (see figure 23.4*b*). The average capillary is about 1 millimeter long and connects an arteriole with a venule. All capillaries are very narrow, with an internal diameter of about 8 micrometers, just bigger than the diameter of a red blood cell (5 to 7 micrometers). This design is critical to the function of capillaries. By bumping against the sides of the vessel as they pass through, the red blood cells are forced into close contact with the capillary walls, making exchange easier (figure 23.5).

No cell of the vertebrate body is more than 100 micrometers from a capillary. At any one moment, about 5% of the blood is in capillaries, a network that amounts to several thousand miles in overall length. If all the capillaries in your body were laid end to end, they would extend across the United States! Individual capillaries have high resistance to flow because of their small diameters. However, the total cross-sectional area of the extensive capillary network (that is, the sum of all the diameters of all the capillaries, expressed as area) is greater than that of the arteries leading to it. As a result, the blood pressure is actually far lower in the capillaries than in the arteries. This is important, because the walls of capillaries are not strong, and they would burst if exposed to the pressures that arteries routinely withstand.

Veins: Returning Blood to the Heart

Veins are vessels that return blood to the heart. Veins do not have to accommodate the pulsing pressures that arteries do, because much of the force of the heartbeat is weakened by the high resistance and great cross-sectional area of the capillary network. For this reason, the walls of veins have much thinner layers of muscle and elastic fiber. An empty artery is still a hollow tube, like a pipe, but when a vein is empty, its walls collapse like an empty balloon (figure 23.6).

Because the pressure of the blood flowing within veins is low, it becomes important to avoid any further resistance to flow, lest there not be enough pressure to get the blood back to the heart. Because a wide tube presents much less resistance to flow than a narrow one, the internal passageway of veins is often quite large, requiring only a small pressure difference to return blood to the heart. The diameters of the largest veins in the human body, the venae cavae, which lead into the heart, are fully 3 centimeters, wider than your thumb! Veins also have unidirectional valves that aid the return of blood by preventing it from flowing backward.

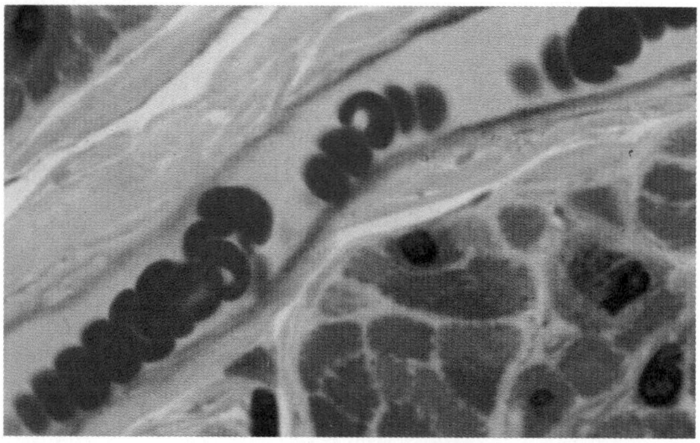

Figure 23.5 Red blood cells within a capillary.

The red blood cells in this capillary pass along in single file. Many capillaries are even narrower than the one shown here. However, red blood cells can pass through capillaries narrower than their own diameter, pushed along by the pressure of the pumping heart.

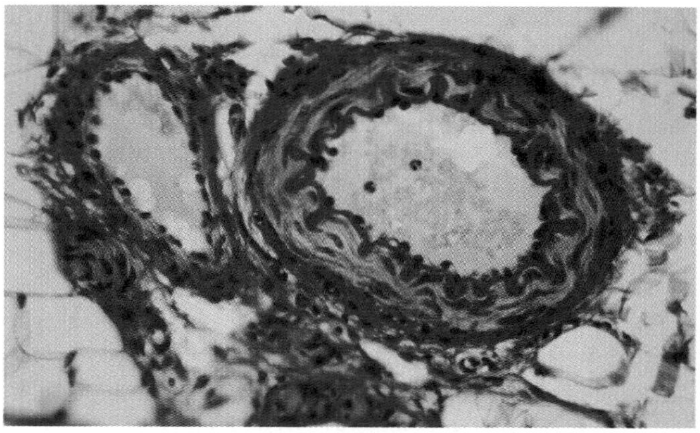

Figure 23.6 Veins and arteries.

The vein (*left*) has the same general structure as the artery (*right*) but much thinner layers of muscle and elastic fiber. An artery retains its shape when empty, but a vein collapses.

> **23.2** The vertebrate circulatory system is composed of arteries and arterioles, which are elastic and carry blood away from the heart; a fine network of capillaries across whose thin walls the exchange of gases and food molecules takes place; and veins, which return blood from the capillaries to the heart.

23.3 The Lymphatic System: Recovering Lost Fluid

The cardiovascular system is very leaky. Fluids are forced out across the thin walls of the capillaries by the pumping pressure of the heart. Although this loss is unavoidable—the circulatory system could not do its job of gas and metabolite exchange without tiny vessels with thin walls—it is important that the loss be made up. In your body, about 4 liters of fluid leave your cardiovascular system in this way each day, more than half the body's total supply of about 5.6 liters of blood! To collect and recycle this fluid, the body uses a second circulatory system called the **lymphatic system** (figure 23.7). Open-ended lymphatic capillaries gather up liquid from the spaces surrounding cells and carry it through a series of progressively larger vessels to two large lymphatic vessels, which drain into veins in the lower part of the neck through one-way valves. Once within the lymphatic system, this fluid is called **lymph.**

Fluid is driven through the lymphatic system when its vessels are squeezed by the movements of the body's muscles. The lymphatic vessels contain a series of one-way valves that permit movement only in the direction of the neck. In some cases, the lymphatic vessels also contract rhythmically. In many fishes, all amphibians and reptiles, bird embryos, and some adult birds, movement of lymph is propelled by **lymph hearts.**

The lymphatic system has three other important functions:

1. **It returns proteins to the circulation.** So much blood flows through the capillaries that there is significant leakage of blood proteins, even though the capillary walls are not very permeable to proteins (proteins are very large molecules). By recapturing the lost fluid, the lymphatic system also returns this lost protein. Whenever this protein is not returned to the blood, the body becomes swollen, a condition known as *edema*.
2. **It transports fats absorbed from the intestine.** Lymph capillaries called *lacteals* penetrate the villi of the small intestine. These lacteals absorb fats from the digestive tract, and by introducing them into the lymph eventually transport them to the circulatory system.
3. **It aids in the body's defense.** Along a lymph vessel are swellings called **lymph nodes.** Inside each node is a honeycomb of spaces filled with white blood cells specialized for defense. The lymphatic system carries bacteria and dead blood cells to the lymph nodes and spleen for destruction.

23.3 The lymphatic system returns to the circulatory system fluids that leak through the capillaries.

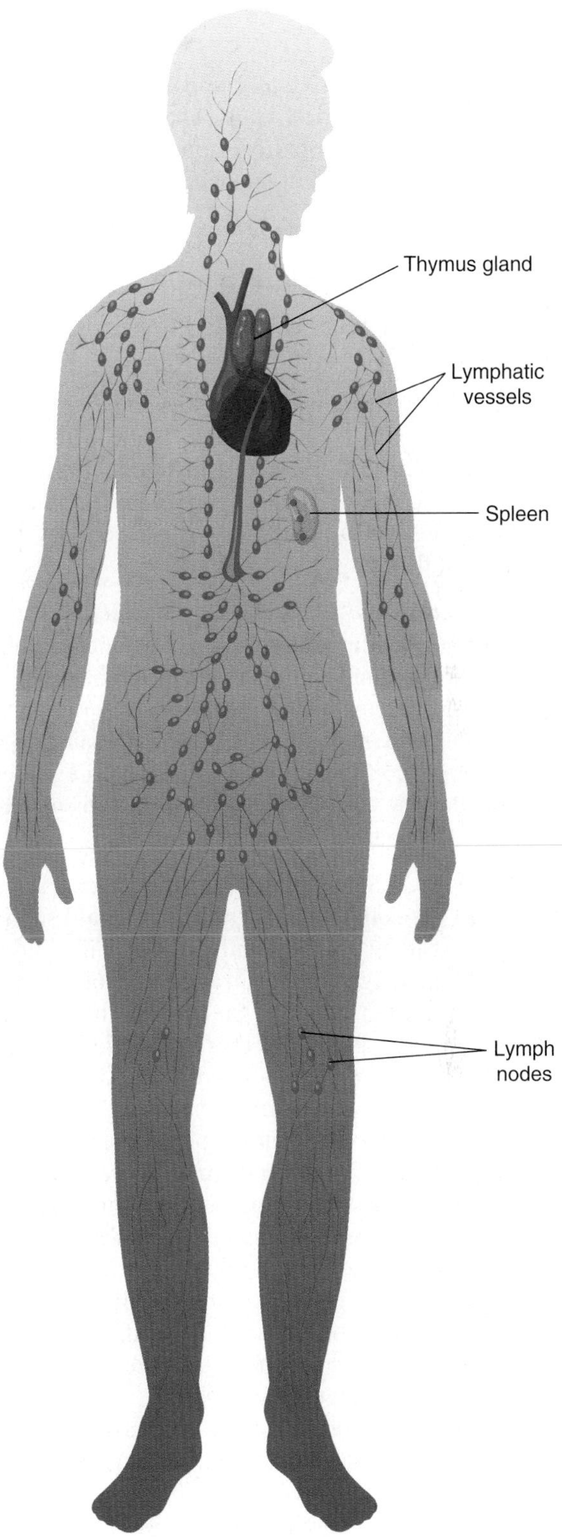

Thymus gland
Lymphatic vessels
Spleen
Lymph nodes

Figure 23.7 The human lymphatic system.
The lymphatic system consists of lymphatic vessels and capillaries, lymph nodes, and lymphatic organs, including the spleen and thymus gland.

23.4 Blood

About 5% of your body mass is composed of the blood circulating through the arteries, veins, and capillaries of your body. This blood is composed of a fluid called **plasma,** together with several different kinds of cells that circulate within that fluid.

Blood Plasma: The Blood's Fluid

Blood plasma is a complex solution of water with three very different sorts of substances dissolved within it:

1. **Metabolites and wastes.** If the circulatory system is the highway of the vertebrate body, the blood contains the traffic traveling on that highway. Dissolved within its plasma are glucose, vitamins, hormones, and wastes that circulate between the cells of the body.
2. **Salts and ions.** Like the seas in which life arose, plasma is a dilute salt solution. The chief plasma ions are sodium, chloride, and bicarbonate. In addition, trace amounts of other salts, such as calcium and magnesium, as well as metallic ions, including copper, potassium, and zinc, are present in plasma. The composition of the plasma is not unlike that of seawater.
3. **Proteins.** Blood plasma is 90% water. Passing by all the cells of the body, blood would soon lose most of its water to them by osmosis if it did not contain as high a concentration of proteins as the cells it passes. Some of the proteins blood plasma contains are antibodies that are active in the immune system. More than half the amount of protein that is necessary to balance the protein content of the cells of the body consists of a single protein, **serum albumin,** which circulates in the blood as an osmotic counterforce. Human blood contains 46 grams of serum albumin per liter—that's over half a pound of it in your body. Starvation and protein deficiency diseases such as kwashiorkor produce swelling of the body because the body's cells take up water from the albumin-deficient blood.

   The liver produces most of the plasma proteins, including albumin, which comprises most of the plasma proteins; the alpha and beta globulins, which serve as carriers of lipids and steroid hormones; and *fibrinogen,* which is required for blood clotting. When blood in a test tube clots, the fibrinogen is converted into insoluble threads of *fibrin* that become part of the clot (figure 23.8). The fluid that's left, which lacks fibrinogen and so cannot clot, is called serum.

Figure 23.8 Threads of fibrin.
This scanning electron micrograph (×1,430) shows fibrin threads among red blood cells. Fibrin is formed from a soluble protein, fibrinogen, in the plasma to produce a blood clot when a blood vessel is damaged.

Blood Cells: Cells That Circulate Through the Body

Although blood is liquid, nearly half of its volume is actually occupied by cells. The fraction of the total volume of the blood that is occupied by cells is referred to as the blood's **hematocrit.** In humans, the hematocrit is usually about 45%. The three principal types of cells in the blood are erythrocytes, leukocytes, and cell fragments called platelets (figure 23.9).

Erythrocytes Carry Hemoglobin. Each milliliter of blood contains about 5 billion **erythrocytes,** also called **red blood cells.** Each erythrocyte is a flat disk with a central depression on both sides, something like a doughnut with a hole that doesn't go all the way through. Erythrocytes carry oxygen to the cells of the body. Almost the entire interior of an erythrocyte is packed with hemoglobin, a protein that binds oxygen in the lungs and delivers it to the cells of the body.

Mature mammalian erythrocytes function like boxcars rather than trucks. Like a vehicle without an engine, erythrocytes contain neither a nucleus nor the machinery to make proteins. Because they lack a nucleus, these cells are unable to repair themselves and therefore have a rather short life; any one erythrocyte lives only about four months. New erythrocytes are constantly being synthesized and released into the blood by cells within the soft interior marrow of bones.

Leukocytes Defend the Body. Less than 1% of the cells in mammalian blood are **leukocytes**, also called **white blood cells.** Leukocytes are larger than red blood cells. They contain no hemoglobin and are essentially colorless. There are several kinds of leukocytes, each with a different function. Neutrophils are the most numerous of the leukocytes, followed in order by lymphocytes, monocytes, eosinophils, and basophils. *Neutrophils* attack like kamikazes, responding to foreign cells by releasing chemicals that kill all the cells in the neighborhood—including the neutrophil. Monocytes give rise to *macrophages,* which attack and kill foreign cells by ingesting them (the name means "large eater"). Lymphocytes include *B cells,* which produce antibodies, and *T cells,* which literally drill holes in invading bacteria.

All of these white blood cell types, and others, help defend the body against invading microorganisms and other foreign substances, as you will see in chapter 25. Unlike other blood cells, leukocytes are not confined to the bloodstream; they are mobile soldiers that also migrate out into the fluid surrounding cells.

Platelets Help Blood to Clot. Certain large cells within the bone marrow, called **megakaryocytes,** regularly pinch off bits of their cytoplasm. These cell fragments, called **platelets,** contain no nuclei. Entering the bloodstream, they play a key role in blood clotting. In a clot, a gluey mesh of fibrin protein fibers sticks platelets together to form a mass that plugs the rupture in the blood vessel. The clot provides a tight, strong seal, much as the inner lining of a tubeless tire seals punctures. The fibrin that forms the clot is made in a series of reactions that start when circulating platelets first encounter the site of an injury. Responding to chemicals released by the damaged blood vessel, platelets release a protein factor into the blood that starts the clotting process.

23.4 Blood is a collection of cells that circulate within a protein-rich salty fluid called plasma. Some of the cells circulating in the blood carry out gas transport; others are engaged in defending the body from infection and cancer.

Blood cell	Life span in blood	Function
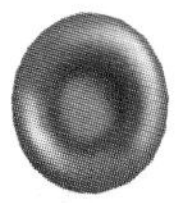 Erythrocyte	120 days	O_2 and CO_2 transport
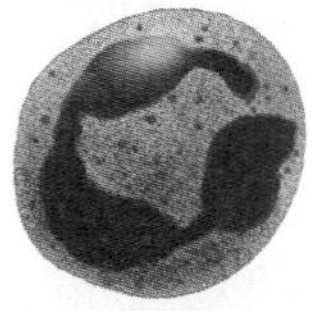 Neutrophil	7 hours	Immune defenses
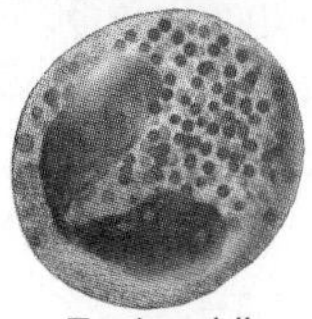 Eosinophil	Unknown	Defense against parasites
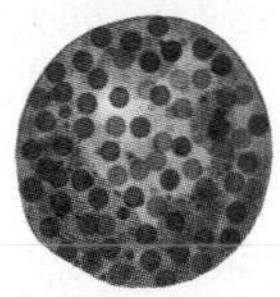 Basophil	Unknown	Inflammatory response
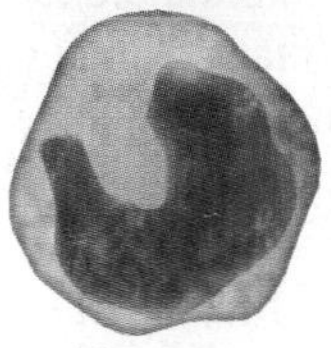 Monocyte	3 days	Immune surveillance (precursor of tissue macrophage)
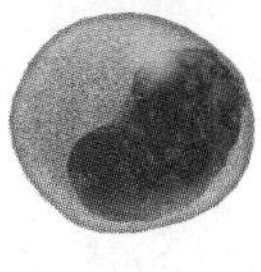 B lymphocyte	Unknown	Antibody production (precursor of plasma cells)
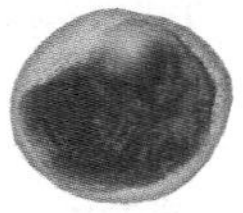 T lymphocyte	Unknown	Cellular immune response
 Platelets	7-8 days	Blood clotting

Figure 23.9 Types of blood cells.
Erythrocytes, leukocytes (neutrophils, eosinophils, basophils, monocytes, and lymphocytes), and platelets are the three principal types of blood cells in vertebrates.

23.5 Fish Circulation

The chordates that were ancestral to the vertebrates are thought to have had simple tubular hearts, similar to those now seen in lancelets (see chapter 19). The heart was little more than a specialized zone of the ventral artery, more heavily muscled than the rest of the arteries, which contracted in simple peristaltic waves. A pumping action results because the uncontracted portions of the vessel have a larger diameter than the contracted portion, and thus present less resistance to blood flow.

The development of gills by fishes required a more efficient pump, and in fishes we see the evolution of a true chamber-pump heart. The fish heart is, in essence, a tube with four chambers arrayed one after the other (figure 23.10*a*). The first two chambers—the **sinus venosus** and **atrium**—are collection chambers, while the second two—the **ventricle** and **conus arteriosus**—are pumping chambers.

As might be expected from the early chordate hearts from which the fish heart evolved, the sequence of the heartbeat in fishes is a peristaltic sequence, starting at the rear and moving to the front. The first of the four chambers to contract is the sinus venosus, followed by the atrium, the ventricle, and finally the conus arteriosus. Despite shifts in the relative positions of the chambers in the vertebrates that evolved later, this heartbeat sequence is maintained in all vertebrates. In fish, the electrical impulse that produces the contraction is initiated in the sinus venosus; in other vertebrates, the electrical impulse is initiated by their equivalent of the sinus venosus.

The fish heart is remarkably well-suited to the gill respiratory apparatus and represents one of the major evolutionary innovations in the vertebrates. Perhaps its greatest advantage is that the blood it delivers to the tissues of the body is fully oxygenated. Blood is pumped first through the gills, where it becomes oxygenated; from the gills, it flows through a network of arteries to the rest of the body; then it returns to the heart through the veins (figure 23.10*b*). This arrangement has one great limitation, however. In passing through the capillaries in the gills, the blood loses much of the pressure developed by the contraction of the heart, so the circulation from the gills through the rest of the body is sluggish. This feature limits the rate of oxygen delivery to the rest of the body.

23.5 The fish heart is a modified tube, consisting of a series of four chambers. Blood first enters the heart at the sinus venosus, where the wavelike contraction of the heart begins.

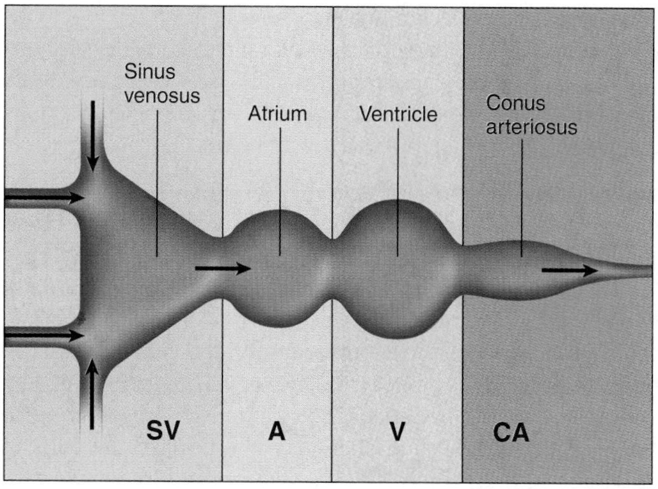
Sinus venosus
Atrium
Ventricle
Conus arteriosus
SV
A
V
CA

(a)

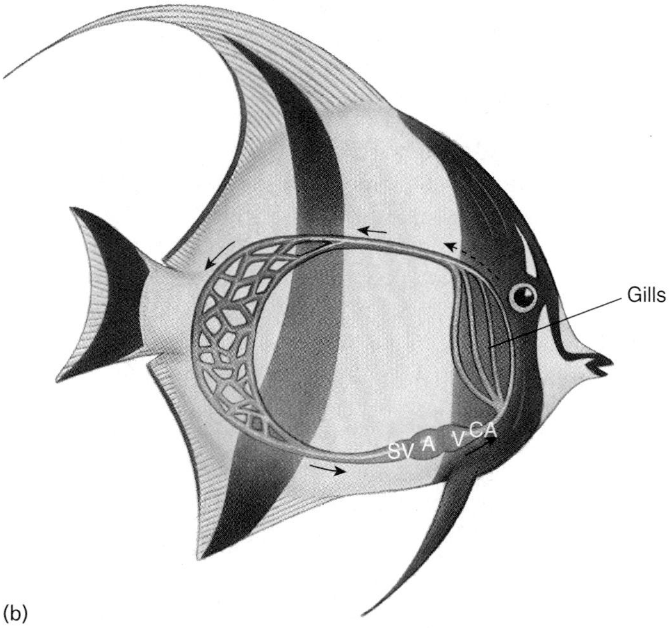
Gills
SV
A
V
CA

(b)

Figure 23.10 The heart and circulation of a fish.
(*a*) Diagram of a fish heart, showing the chambers in series with each other.
(*b*) Diagram of the fish circulation, showing that blood is pumped by the ventricle to the gills, and then blood flows directly to the body, resulting in sluggish circulation. Blood rich in oxygen (oxygenated) is shown in *red;* blood low in oxygen (deoxygenated) is shown in *blue.*

23.6 Amphibian and Reptile Circulation

The advent of lungs involved a major change in the pattern of circulation. After blood is pumped by the heart to the lungs, it does not go directly to the tissues of the body but instead returns to the heart. This results in two circulations: one that goes to and from the heart to the lungs, called the **pulmonary circulation,** and one that goes to and from the heart to the rest of the body, called the **systemic circulation.**

If no changes had occurred in the structure of the heart, the oxygenated blood from the lungs would be mixed in the heart with the deoxygenated blood returning from the rest of the body. Consequently, the heart would pump a mixture of oxygenated and deoxygenated blood rather than fully oxygenated blood. The amphibian heart has two structural features that help reduce this mixing (figure 23.11*a*). First, the atrium is divided into two chambers: the right atrium receives deoxygenated blood from the systemic circulation, and the left atrium receives oxygenated blood from the lungs. These two stores of blood therefore do not mix in the atria, but some mixing might be expected when the contents of each atrium enter the single, common ventricle. Surprisingly, however, little mixing actually occurs. Second, the conus arteriosus is partially separated by a *septum*, or dividing wall, which directs deoxygenated blood into the *pulmonary arteries* to the lungs and oxygenated blood into the *aorta*, the major artery of the systemic circulation to the body.

Because there is only one ventricle in an amphibian heart, the separation of the pulmonary and systemic circulations is incomplete. Amphibians in water, however, can obtain additional oxygen by diffusion through their skin. This process, called **cutaneous respiration,** helps to supplement the oxygenation of the blood in these vertebrates.

Among reptiles, additional modifications have reduced the mixing of blood in the heart still further. In addition to having two separate atria, reptiles have a septum that partially subdivides the ventricle (figure 23.11*b*). This results in an even greater separation of oxygenated and deoxygenated blood within the heart. The separation is complete in one order of reptiles, the crocodiles, which have two separate ventricles divided by a complete septum. Crocodiles therefore have a completely divided pulmonary and systemic circulation. Another change in the circulation of reptiles is that the conus arteriosus has become incorporated into the trunks of the large arteries leaving the heart.

23.6 Amphibians and reptiles have two circulations, pulmonary and systemic, that deliver blood to the lungs and rest of the body, respectively.

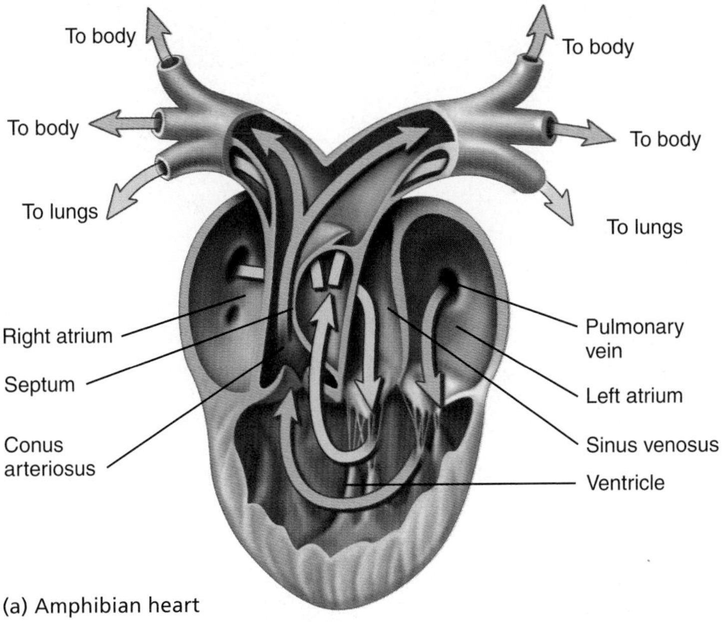

To body
To body
To body
To body
To lungs
To lungs
Right atrium
Septum
Conus arteriosus
Pulmonary vein
Left atrium
Sinus venosus
Ventricle

(a) Amphibian heart

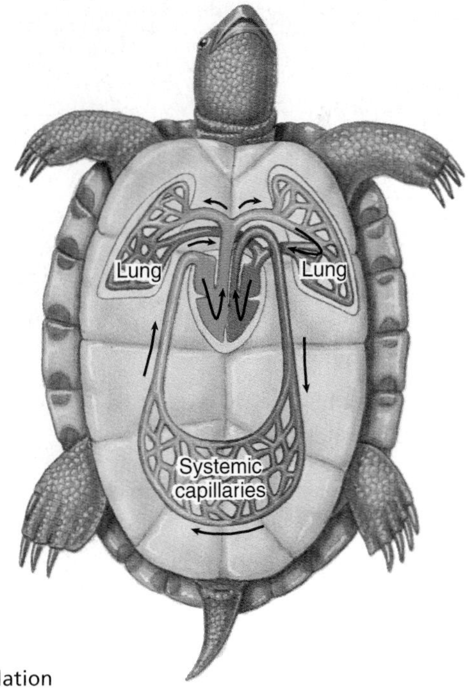

Lung
Lung
Systemic capillaries

(b) Reptilian circulation

Figure 23.11 The amphibian heart and circulation in reptiles.

(*a*) The frog heart has two atria but only one ventricle, which pumps blood both to the lungs and to the body. Despite the potential for mixing, the oxygenated and deoxygenated bloods (*red* and *blue*, respectively) mix very little as they are pumped to the body and lungs. The slight mixing is shown in *purple*. (*b*) In reptiles, not only are there two separate atria, but the ventricle is also partially divided. In amphibians and reptiles, blood is returned to the heart after leaving the lungs and is repumped, resulting in much more vigorous circulation to the body than in fishes.

23.7 Mammalian and Bird Circulation

Mammals, birds, and crocodiles have a four-chambered heart that is really two separate pumping systems operating together within a single unit. One of these pumps blood to the lungs, while the other pumps blood to the rest of the body. The left side has two connected chambers, and so does the right, but the two sides are not connected with one another. The increased efficiency of the double circulatory system in mammals and birds is thought to have been important in the evolution of endothermy (warm-bloodedness), because a more efficient circulation is necessary to support the high metabolic rate required.

Circulation Through the Heart

Let's follow the journey of blood through the mammalian heart, starting with the entry of oxygen-rich blood into the heart from the lungs (figure 23.12). Oxygenated blood from the lungs enters the left side of the heart, emptying directly into the **left atrium** through large vessels called the **pulmonary veins.** From the atrium, blood flows through an opening into the adjoining chamber, the **left ventricle.** Most of this flow, roughly 80%, occurs while the heart is relaxed. When the heart starts to contract, the atrium contracts first, pushing the remaining 20% of its blood into the ventricle.

After a slight delay, the ventricle contracts. The walls of the ventricle are far more muscular than those of the atrium, and thus this contraction is much stronger. It forces most of the blood out of the ventricle in a single strong pulse. The blood is prevented from going back into the atrium by a large, one-way valve, the **mitral valve,** whose flaps are pushed shut as the ventricle contracts.

Prevented from reentering the atrium, the blood within the contracting left ventricle takes the only other passage out. It moves through a second opening that leads into a large blood vessel called the **aorta.** The aorta is separated from the left ventricle by a one-way valve, the **aortic valve.** Unlike the mitral valve, the aortic valve is oriented to permit the

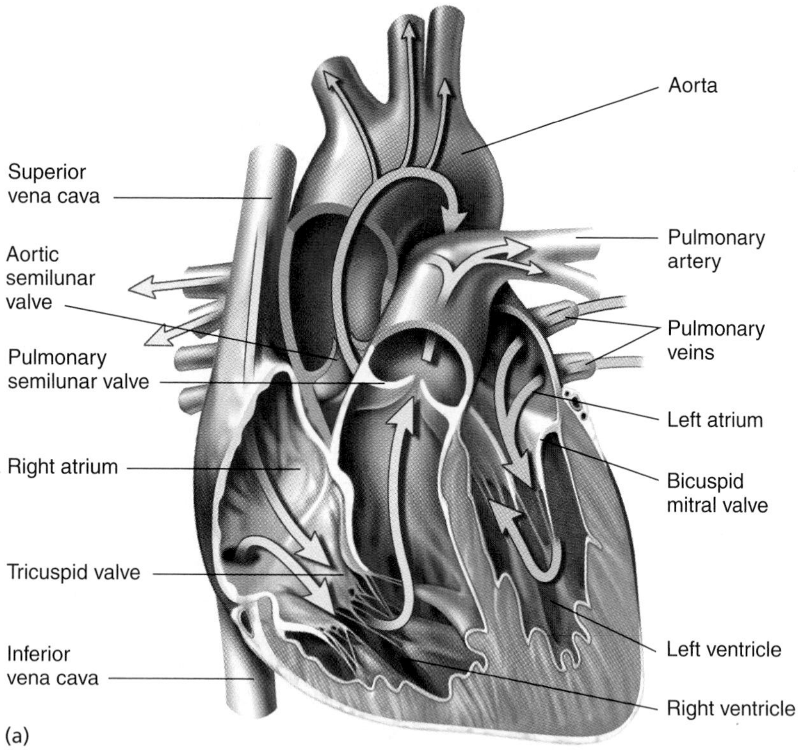
Aorta
Superior vena cava
Pulmonary artery
Aortic semilunar valve
Pulmonary veins
Pulmonary semilunar valve
Left atrium
Right atrium
Bicuspid mitral valve
Tricuspid valve
Left ventricle
Inferior vena cava
Right ventricle
(a)

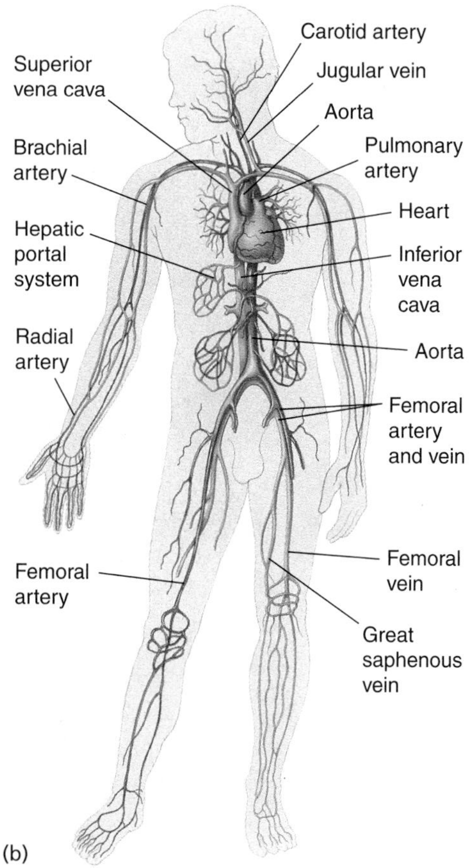
Carotid artery
Superior vena cava
Jugular vein
Aorta
Brachial artery
Pulmonary artery
Heart
Hepatic portal system
Inferior vena cava
Radial artery
Aorta
Femoral artery and vein
Femoral artery
Femoral vein
Great saphenous vein
(b)

Figure 23.12 The heart and circulation of mammals and birds.

(a) Unlike the amphibian heart, this heart has a septum dividing the ventricle into left and right ventricles. Oxygenated blood from the lungs enters the left atrium of the heart by way of the pulmonary veins. This blood then enters the left ventricle, from which it passes into the aorta to circulate throughout the body and deliver oxygen to the tissues. When gas exchange has taken place at the tissues, veins return blood to the heart. After entering the right atrium by way of the superior and inferior venae cavae, deoxygenated blood passes into the right ventricle and then through the pulmonary valve to the lungs by way of the pulmonary artery. (*b*) Some of the major arteries and veins in the human circulatory system are shown.

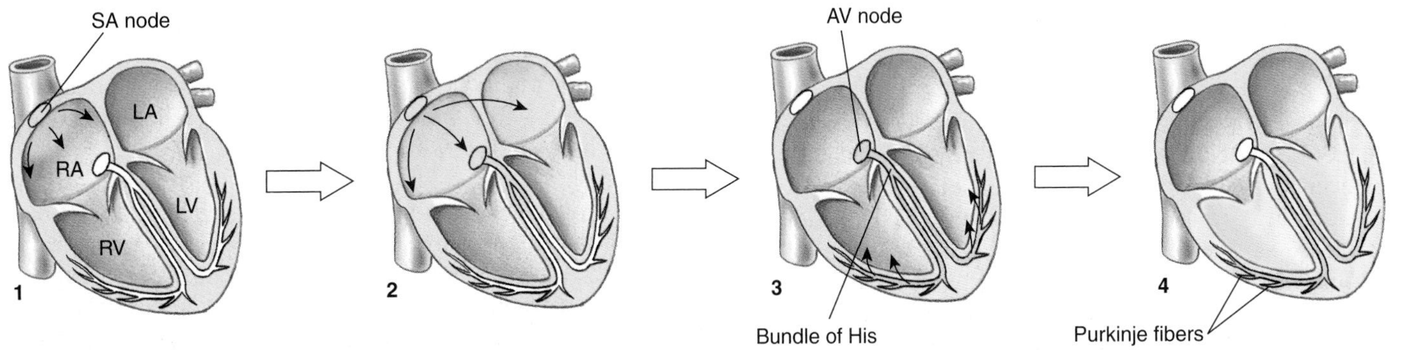
SA node
AV node
LA
RA
LV
RV
1
2
3
4
Bundle of His
Purkinje fibers

Figure 23.13 How the mammalian heart contracts.

Contraction of the mammalian heart is initiated by a wave of depolarization that begins at the SA node. After passing over the right and left atria and causing their contraction, the wave of depolarization reaches the AV node, from which it passes to the ventricles by the bundle of His. The depolarization is then conducted rapidly over the surface of the ventricles by a set of finer fibers called Purkinje fibers, which branch throughout each ventricle.

flow of the blood *out* of the ventricle. Once this outward flow has occurred, the aortic valve closes, preventing the reentry of blood from the aorta into the heart. The aorta and its many branches are systemic arteries and carry oxygen-rich blood to all parts of the body.

Eventually, this blood returns to the heart after delivering its cargo of oxygen to the cells of the body. In returning it passes through a series of progressively larger veins, ending in two large veins that empty into the right atrium of the heart. The **superior vena cava** drains the upper body, and the **inferior vena cava** drains the lower body.

The right side of the heart is similar in organization to the left side. Blood passes from the **right atrium** into the **right ventricle** through a one-way valve, the **tricuspid valve.** It passes out of the contracting right ventricle through a second valve, the **pulmonary valve,** into the **pulmonary arteries,** which carry the deoxygenated blood to the lungs. The blood then returns from the lungs to the left side of the heart with a new cargo of oxygen, which is pumped to the rest of the body.

How the Heart Contracts

Throughout the evolutionary history of the vertebrate heart, the sinus venosus has served as a pacemaker, the site where the impulses that produce the heartbeat originate. Although it constitutes a major chamber in the fish heart, it is reduced in size in amphibians and further reduced in reptiles. In mammals and birds, the sinus venosus is no longer a separate chamber, but some of its tissue remains in the wall of the right atrium, near the point where the systemic veins empty into the atrium. This tissue, which is called the **sinoatrial (SA) node,** is still the site where each heartbeat originates.

The contraction of the heart consists of a carefully orchestrated series of muscle contractions. First, the atria contract together, followed by the ventricles. Contraction is initiated by the sinoatrial (SA) node (figure 23.13). Its membranes spontaneously depolarize with a regular rhythm that determines the rhythm of the heart's beating. Each depolarization initiated within this pacemaker region passes quickly from one heart muscle cell to another in a wave that envelops the left and the right atria almost simultaneously.

But the wave of depolarization does not immediately spread to the ventricles. There is a pause before the lower half of the heart starts to contract. The reason for the delay is that the atria of the heart are separated from the ventricles by connective tissue that does not propagate a depolarization wave. The depolarization would not pass to the ventricles at all except for a slender connection of cardiac muscle cells known as the **atrioventricular (AV) node,** which connects across the gap to a strand of specialized muscle known as the **bundle of His.** Bundle branches divide into fast-conducting **Purkinje fibers,** which initiate the almost simultaneous contraction of all the cells of the right and left ventricles about 0.1 seconds after the atria contract. This delay permits the atria to finish emptying their contents into the corresponding ventricles before those ventricles start to contract.

Because the overall circulatory system is closed, the same volume of blood must move through the pulmonary circulation as through the much larger systemic circulation with each heartbeat. Therefore, the right and left ventricles must pump the same amount of blood each time they contract. If the output of one ventricle did not match that of the other, fluid would accumulate and pressure would increase in one of the circuits. The result would be increased filtration out of the capillaries and edema (as occurs in congestive heart failure, for example). Although the volume of blood pumped by the two ventricles is the same, the pressure they generate is not. The left ventricle, which pumps blood through the higher-resistance systemic pathway, is more muscular and generates more pressure than does the right ventricle.

23.7 The mammalian heart is a two-cycle pump. The left side pumps oxygenated blood to the body's tissues, while the right side pumps O_2-depleted blood to the lungs, where it releases CO_2 and gathers more oxygen.

23.8 Monitoring the Heart's Performance

As you can see, the heartbeat is not simply a squeeze-release, squeeze-release cycle but rather a little play in which a series of events occur in a predictable order. The simplest way to monitor heartbeat is to listen to the heart at work, using a stethoscope. The first sound you hear, a low-pitched *lub,* is the closing of the mitral and tricuspid valves at the start of ventricular contraction. A little later, you hear a higher-pitched *dub,* the closing of the pulmonary and aortic valves at the end of ventricular contraction. If the valves are not closing fully, or if they open incompletely, turbulence is created within the heart. This turbulence can be heard as a **heart murmur.** It often sounds like liquid sloshing.

A second way to examine the events of the heartbeat is to monitor the blood pressure. During the first part of the heartbeat, the atria are filling. At this time the pressure in the arteries leading from the left side of the heart out to the tissues of the body decreases slightly. This low pressure is referred to as the **diastolic** pressure. During the contraction of the left ventricle, a pulse of blood is forced into the systemic arterial system, immediately raising the blood pressure within these vessels. The high blood pressure produced in this pushing period, which ends with the closing of the aortic valve, is referred to as the **systolic** pressure. Normal blood pressure values are 70 to 90 diastolic and 110 to 130 systolic. When the inner walls of the arteries accumulate fats, as they do in the condition known as *atherosclerosis,* the diameters of the passageways are narrowed. If this occurs, the systolic blood pressure is elevated.

A third way to monitor a heartbeat is to measure the waves of depolarization. Because the vertebrate body basically consists of water, it conducts electrical currents rather well. A wave of membrane depolarization passing over the surface of the heart generates an electrical current that passes in a wave throughout the body. The magnitude of this electrical pulse is tiny, but it can be detected by sensors placed on the skin. A recording of these impulses is called an electrocardiogram (figure 23.14).

The evolution of multicellular organisms has depended critically on the ability to circulate materials throughout the body efficiently. Vertebrates carefully regulate the operation of their circulatory systems and are able to integrate their body activities. Indeed, the metabolic demands of different vertebrates have shaped the evolution of circulatory systems.

23.8 The performance of the heart can be monitored by listening to it, by feeling the changes in blood pressure its beating produces, and by measuring the waves of depolarization created by its contraction.

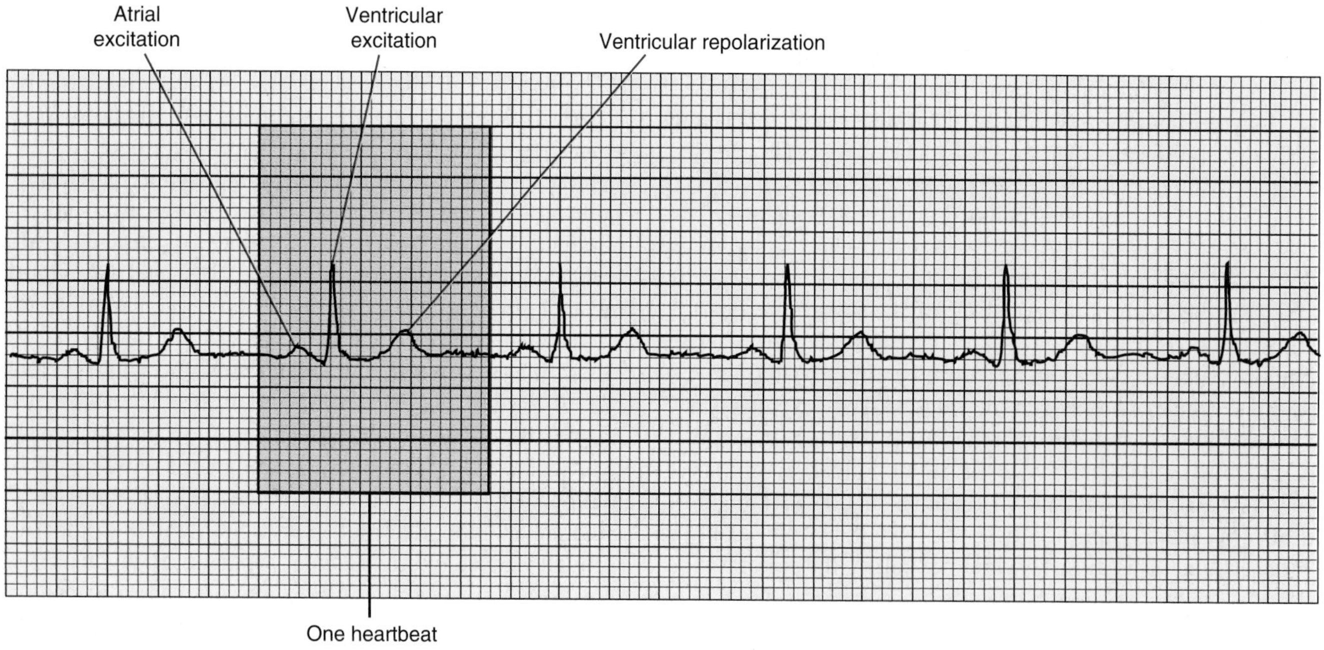

Atrial excitation
Ventricular excitation
Ventricular repolarization
One heartbeat

Figure 23.14 An electrocardiogram.

The chart shows the three successive impulses of a normal heartbeat. The first depolarization represents the atrial excitation. A tenth of a second later, a much stronger wave reflects both the excitation of the ventricles and relaxation of the atria. Finally, perhaps 0.2 seconds later, a third wave results from the relaxation, or repolarization, of the ventricles.

23.9 Cardiovascular Diseases

Cardiovascular diseases are the leading cause of death in the United States; more than 42 million people have some form of cardiovascular disease. Heart attacks are the main cause of cardiovascular deaths in the United States, accounting for about a fifth of all deaths. They result from an insufficient supply of blood reaching one or more parts of the heart muscle, which causes myocardial cells in those parts to die. Heart attacks may be caused by a blood clot forming somewhere in the coronary arteries (the arteries that supply the heart muscle with blood) and blocking the passage of blood through those vessels. They may also result if an artery is blocked by atherosclerosis. Recovery from a heart attack is possible if the portion of the heart that was damaged is small enough that the other blood vessels in the heart can enlarge their capacity and resupply the damaged tissues. **Angina pectoris,** which literally means "chest pain," occurs for reasons similar to those that cause heart attacks, but it is not as severe. The pain may occur in the heart and often also in the left arm and shoulder. Angina pectoris is a warning sign that the blood supply to the heart is inadequate but still sufficient to avoid myocardial cell death.

Strokes are caused by an interference with the blood supply to the brain. They may occur when a blood vessel bursts in the brain or when blood flow in a cerebral artery is blocked by a thrombus (blood clot) or by atherosclerosis. The effects of a stroke depend on how severe the damage is and where in the brain the stroke occurs.

Atherosclerosis is an accumulation within the arteries of fatty materials, abnormal amounts of smooth muscle, deposits of cholesterol or fibrin, or various kinds of cellular debris. These accumulations cause blood flow to be reduced (figure 23.15). The lumen (interior) of the artery may be further reduced in size by a clot that forms as a result of the atherosclerosis. In the severest cases, the artery may be blocked completely. Atherosclerosis is promoted by genetic factors, smoking, hypertension (high blood pressure), and high blood cholesterol levels. Diets low in cholesterol and saturated fats (from which cholesterol can be made) can help lower the level of blood cholesterol, and therapy for hypertension can reduce that risk factor. Stopping smoking, however, is the single most effective action a smoker can take to reduce the risk of atherosclerosis.

Arteriosclerosis, or hardening of the arteries, occurs when calcium is deposited in arterial walls. It tends to occur when atherosclerosis is severe. Not only do such arteries have restricted blood flow, but they also lack the ability to expand as normal arteries do to accommodate the volume of blood pumped out by the heart. This inflexibility forces the heart to work harder.

23.9 Humans are subject to a variety of cardiovascular diseases, many of them associated with the accumulation of fatty materials on the inner surfaces of arteries.

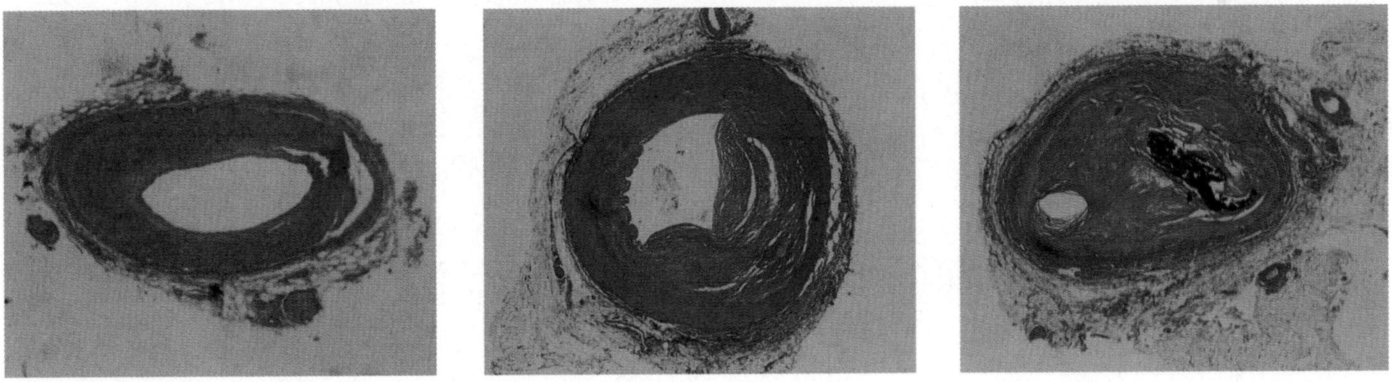

(a) (b) (c)

Figure 23.15 The path to a heart attack.

(*a*) The coronary artery shows only minor blockage. (*b*) The artery exhibits severe atherosclerosis—much of the passage is blocked by buildup on the interior walls of the artery. (*c*) The coronary artery is essentially completely blocked.

23.10 Types of Respiratory Systems

All animals obtain the energy that powers their lives by oxidizing molecules rich in energy-laden carbon–hydrogen bonds. This oxidative metabolism requires a ready supply of oxygen. The uptake of oxygen and the simultaneous release of carbon dioxide together are called **respiration.**

Most of the primitive phyla of organisms obtain oxygen by direct diffusion from seawater, which contains about 10 milliliters of dissolved oxygen per liter (figure 23.16*a*). Sponges, cnidarians, many flatworms and roundworms, and some annelid worms all obtain their oxygen by diffusion from surrounding water. Similarly, members of the vertebrate class Amphibia conduct gas exchange by direct diffusion through their moist skin (figure 23.16*b*).

The more advanced marine invertebrates (mollusks, arthropods, and echinoderms) possess special respiratory organs called gills that increase the surface area available for diffusion of oxygen. A **gill** is basically a thin sheet of tissue that waves through the water. Gills can be simple, as in the papulae of echinoderms (figure 23.16*c*), or complex, as in the highly convoluted gills of fish (figure 23.16*e*). Terrestrial arthropods do not have a single major respiratory organ like a gill. Instead, a network of air ducts called **tracheae,** branching into smaller and smaller tubes, carries air to every part of the body (figure 23.16*d*). The openings of tracheae to the outside are through special structures called **spiracles,** which can be closed and opened. Terrestrial vertebrates, except for some amphibians, do have a single respiratory organ, called the **lung** (figure 23.16*f*).

23.10 Aquatic animals extract oxygen dissolved in water, some by direct diffusion, others with gills. Terrestrial animals use tracheae or lungs.

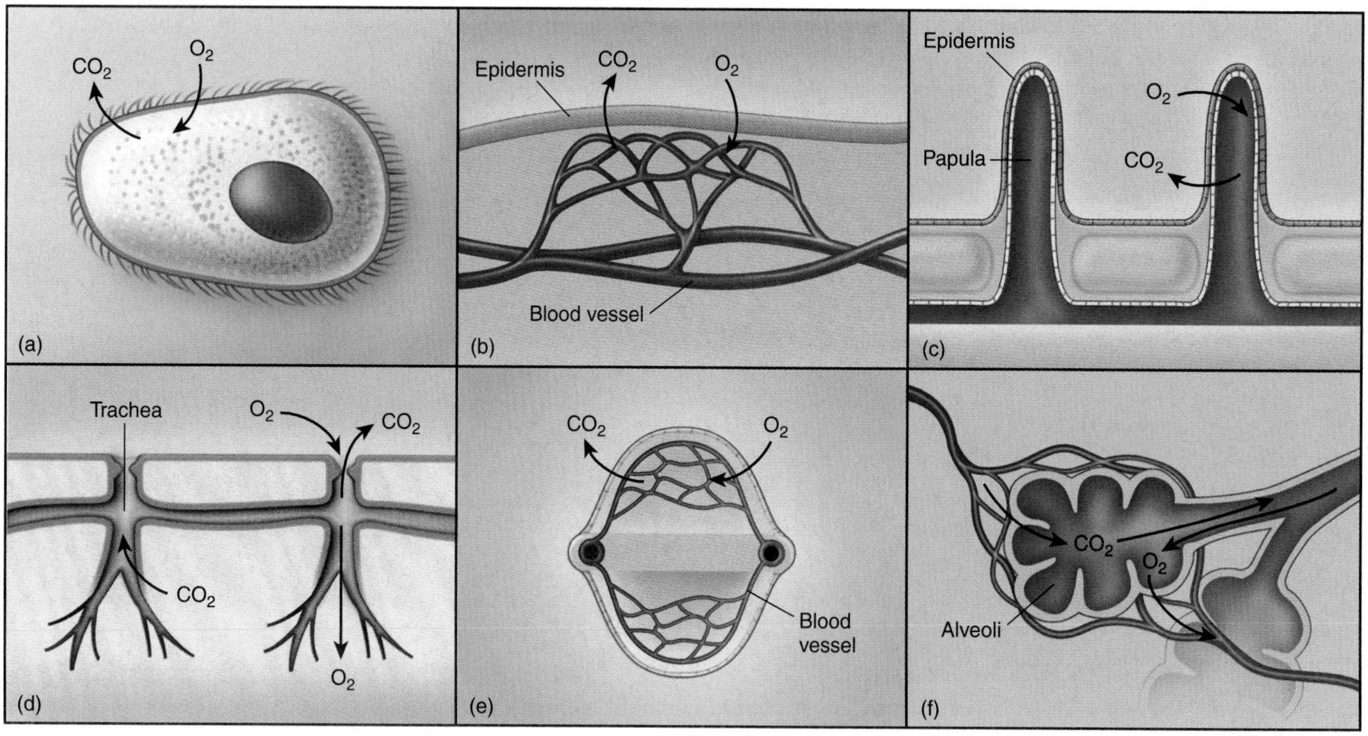
CO2
O2
(a)
Epidermis
CO2
O2
Blood vessel
(b)
Epidermis
O2
Papula
CO2
(c)
Trachea
O2
CO2
CO2
O2
(d)
CO2
O2
Blood
vessel
(e)
CO2
O2
Alveoli
(f)

Figure 23.16 Gas exchange in animals.

(*a*) Gases diffuse directly into single-celled organisms. (*b*) Amphibians and many other multicellular organisms respire through their skin (transcutaneous respiration). (*c*) Echinoderms have protruding papulae, which provide an increased respiratory surface. (*d*) Insects respire through tracheae, which open to the outside. (*e*) Fish gills provide a very large respiratory surface and employ countercurrent flow. (*f*) Mammalian lungs provide a large respiratory surface but do not permit countercurrent flow.

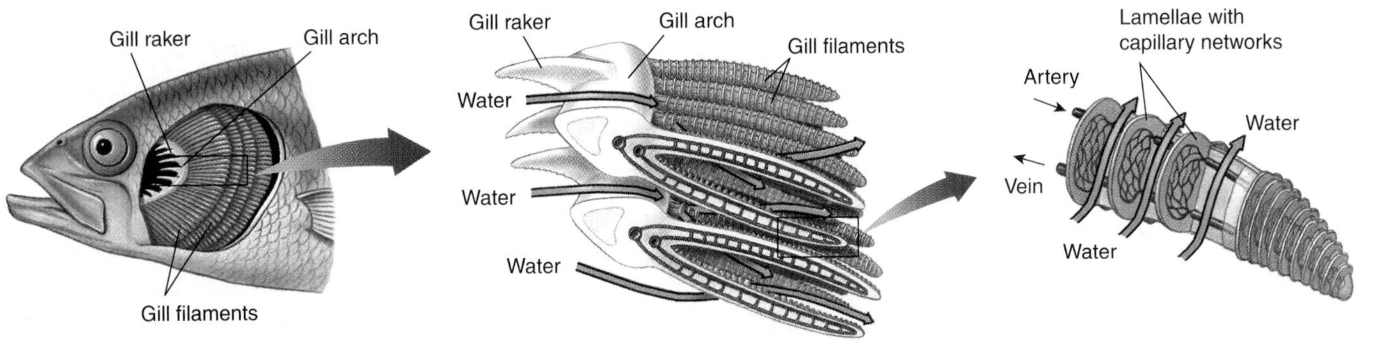
Gill raker
Gill arch
Gill filaments
Gill raker
Gill arch
Gill filaments
Water
Water
Water
Lamellae with capillary networks
Artery
Vein
Water
Water

Figure 23.17 Structure of a fish gill.
Water passes from the gill arch over the filaments (from *left* to *right* in the diagram). Water always passes the lamellae in the same direction, which is opposite to the direction the blood circulates across the lamellae. This opposite orientation of blood and water flow is critical to the success of the gill's operation.

23.11 Respiration in Aquatic Vertebrates

Have you ever seen the face of a swimming fish up close? A fish swimming in water continuously opens and closes its mouth, pushing water through the mouth cavity and out a slit at the rear of the mouth—and (and this is the whole point) past the gills on its one-way journey.

This swallowing process, which seems so awkward, is at the heart of a great advance in gill design achieved by the fishes. What is important about the swallowing is that it causes the water to always move past the fish's gills *in the same direction.* Moving the water past the gills in the same direction permits **countercurrent flow,** which is a supremely efficient way of extracting oxygen. Here is how it works:

Each gill is composed of two rows of gill filaments, thin membranous plates stacked one on top of the other and projecting out into the flow of water (figure 23.17). As water flows past the filaments from front to back, oxygen diffuses from the water into blood circulating within the gill filament. Within each filament the blood circulation is arranged so that the blood is carried in the direction opposite the movement of the water, from the back of the filament to the front.

Because of the countercurrent flow, the blood in the fish's gills can build up oxygen concentrations as high as those of the water entering the gill (figure 23.18). The gills of bony fishes are the most efficient respiratory machines that have ever evolved among organisms. They are able to extract up to 85% of the available oxygen from water.

23.11 Fish gills achieve countercurrent flow, making them very efficient at extracting oxygen.

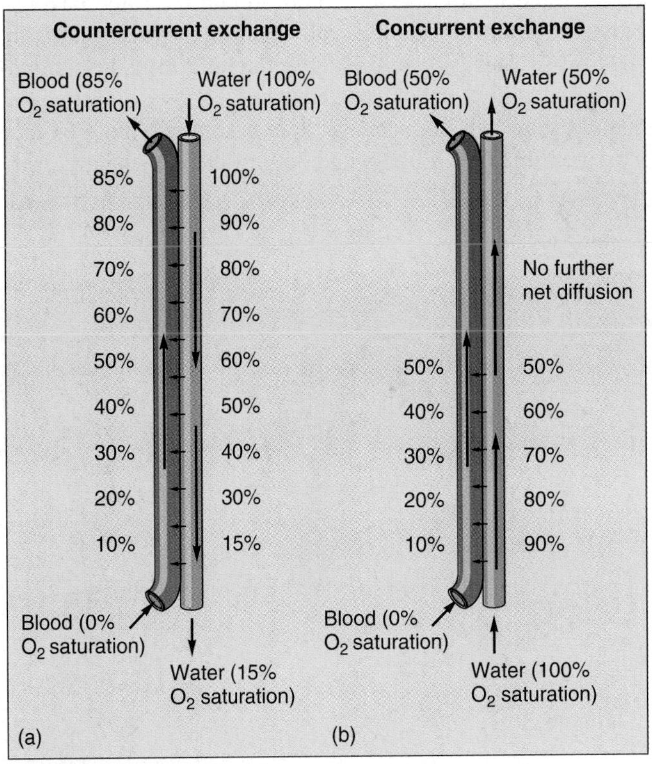
Countercurrent exchange
Concurrent exchange
Blood (85% O_2 saturation)
Water (100% O_2 saturation)
85%
80%
70%
60%
50%
40%
30%
20%
10%
100%
90%
80%
70%
60%
50%
40%
30%
15%
Blood (0% O_2 saturation)
Water (15% O_2 saturation)
(a)
Blood (50% O_2 saturation)
Water (50% O_2 saturation)
No further net diffusion
50%
40%
30%
20%
10%
50%
60%
70%
80%
90%
Blood (0% O_2 saturation)
Water (100% O_2 saturation)
(b)

Figure 23.18 Countercurrent flow.
When blood and water flow in *opposite* directions (*a*), the initial oxygen concentration difference between water and blood is not large, but it is sufficient for oxygen to diffuse from water to blood. As the blood oxygen concentration rises, the blood encounters water with ever higher oxygen concentrations. At every point, the oxygen concentration is higher in the water, so that diffusion continues. In this example, blood attains an oxygen concentration of 85%. When blood and water flow in the *same* direction (*b*), oxygen can diffuse from the water into the blood rapidly at first, but the diffusion rate slows as more oxygen diffuses from the water into the blood, until finally the concentrations of oxygen in water and blood are equal. In this example, blood's oxygen concentration cannot exceed 50%.

23.12 Respiration in Terrestrial Vertebrates

Amphibians Get Oxygen from Air with Lungs

One of the major challenges facing the first land vertebrates was obtaining oxygen from air. Fish gills, which are superb oxygen-gathering machines in water, don't work in air. The gill's system of delicate membranes has no means of support in air, and the membranes collapse on top of one another—that's why a fish dies when kept out of water, literally drowning in air for lack of oxygen.

Unlike a fish, if you lift a frog out of water and place it on dry ground, it doesn't drown. Partly this is because the frog is able to respire through its moist skin, but mainly it is because the frog has lungs. A **lung** is a respiratory organ designed like a bag. The amphibian lung is hardly more than a sac with a convoluted internal membrane (figure 23.19*a*). The air moves into the sac through a tubular passage from the head and then back out again through the same passage. Lungs are not as efficient as gills because new air that is inhaled mixes with old air already in the lung. But, air contains about 210 milliliters of oxygen per liter, over 20 times as much as seawater. So, because there is so much more oxygen *in* air, the lung doesn't have to be as efficient as the gill.

Reptiles and Mammals Increase the Lung Surface

Reptiles are far more active than amphibians, so they need more oxygen. But reptiles cannot rely on their skin for respiration the way amphibians can; their dry scaly skin is "watertight" to avoid water loss. Instead, the lungs of reptiles contain many small air chambers, which greatly increase the surface area of the lung available for diffusion of oxygen (figure 23.19*b*).

Because mammals maintain a constant body temperature by heating their bodies metabolically, they have even greater metabolic demands for oxygen than do reptiles. The problem of harvesting more oxygen is solved by increasing the diffusion surface area within the lung even more. The lungs of mammals possess on their inner surface many small chambers called **alveoli,** which are clustered together like grapes (figure 23.19*c*). Each cluster is connected to the main air sac in the lung by a short passageway called a **bronchiole.** Air within the lung passes through the bronchioles to the alveoli, where all oxygen uptake and carbon dioxide disposal takes place. In more active mammals, the individual alveoli are smaller and more numerous, increasing the diffusion surface area even more. Humans have about 300 million alveoli in each of their lungs, for a total surface area devoted to diffusion of about 80 square meters (about 42 times the surface area of the body)!

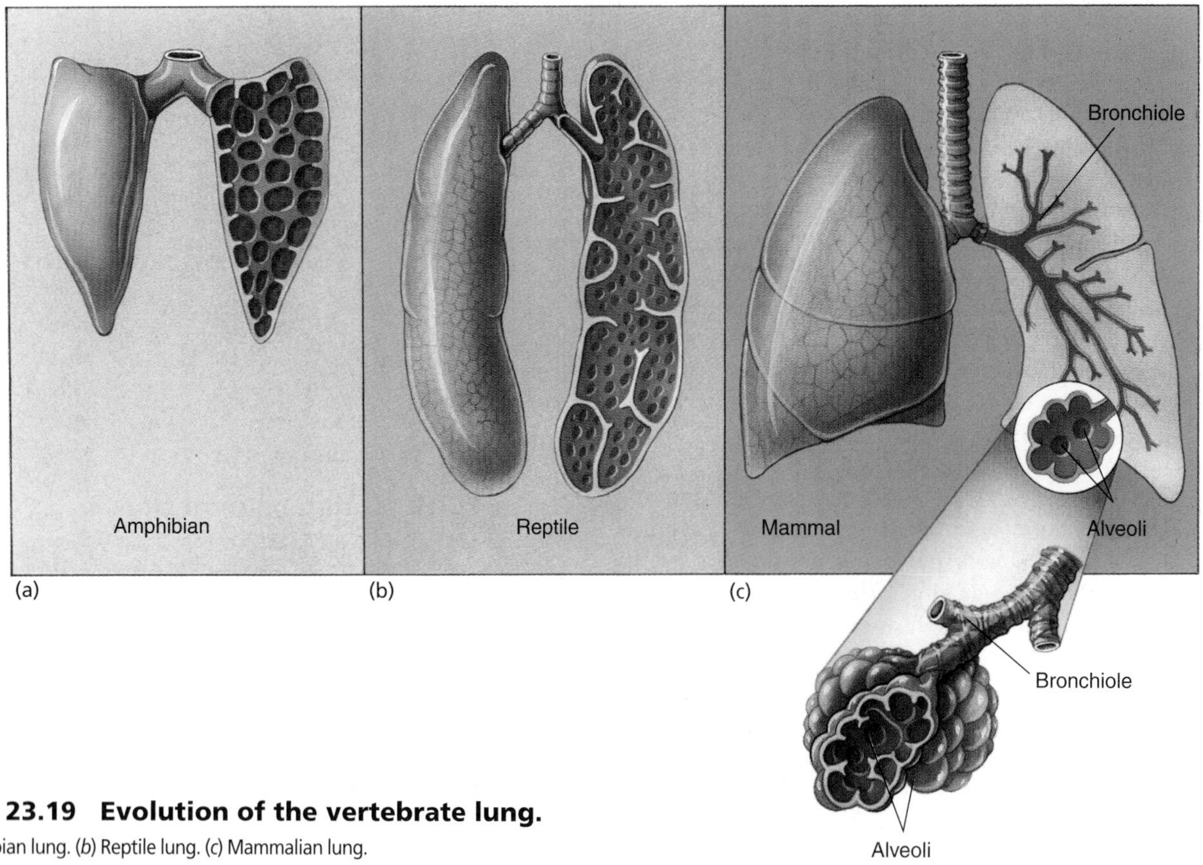
Amphibian
(a)
Reptile
(b)
Mammal
(c)
Bronchiole
Alveoli
Bronchiole
Alveoli

Figure 23.19 Evolution of the vertebrate lung.
(*a*) Amphibian lung. (*b*) Reptile lung. (*c*) Mammalian lung.

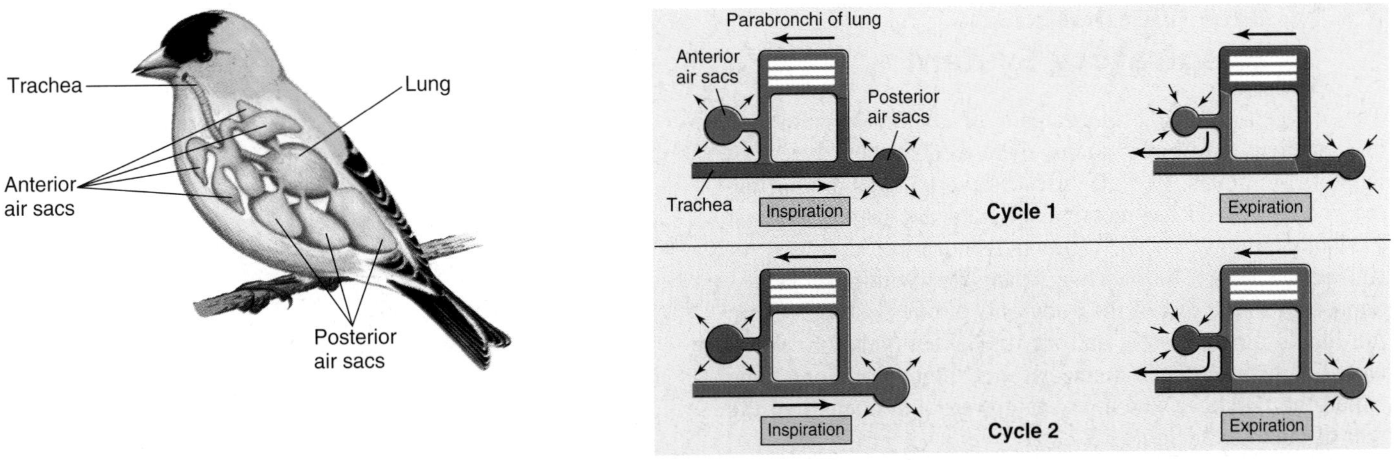
Trachea
Lung
Anterior air sacs
Posterior air sacs
Parabronchi of lung
Anterior air sacs
Posterior air sacs
Trachea
Inspiration
Cycle 1
Expiration
Inspiration
Cycle 2
Expiration

(a) (b)

Figure 23.20 How a bird breathes.

(*a*) A bird's respiratory system is composed of the trachea, anterior air sacs, lungs, and posterior air sacs. (*b*) Breathing occurs in two cycles: in cycle 1, air is drawn from the trachea into the posterior air sacs and then is exhaled into a lung; in cycle 2, the air is drawn from the lung into the anterior air sacs and then is exhaled through the trachea. Passage of air through the lungs is always in the same direction, from posterior to anterior (*right* to *left* here). Because blood circulates in the lung from anterior to posterior, the lung achieves a type of countercurrent flow that is very efficient in picking up oxygen from the air.

Birds Perfect the Lung

There is a limit to how much efficiency can be improved by increasing the surface area of the lung, a limit that has already been reached by the more active mammals. When birds evolved, this efficiency just wasn't good enough. Flying creates a respiratory demand for oxygen that exceeds the capacity of the saclike lungs of even the most active mammal. Unlike bats, whose flight involves considerable gliding, most birds beat their wings rapidly as they fly, often for quite a long time. This intensive wing beating uses up a lot of energy quickly, because the wing muscles must contract very frequently. Flying birds thus must carry out very active oxidative respiration within their cells to replenish the ATP expended by their flight muscles, and this requires a great deal of oxygen.

The bird lung copes with the demands of flight by finding a new way to improve the efficiency of the lung, one that does not involve further increases in its surface area. Can you guess what it is? In effect, they do what fishes do! An avian lung is connected to a series of air sacs behind the lung. When a bird inhales, the air passes directly to these air sacs, which act as holding tanks (figure 23.20). When the bird exhales, the air flows from the air sacs forward into the lung and then on through another set of air sacs in front of the lungs and out of the body. What is the advantage of this complicated passage? It creates a unidirectional flow of air through the lungs.

Air flows through the lungs of birds in one direction only, from back to front. This one-way air flow results in two significant improvements: (1) There is no dead volume, as in the mammalian lung, so the air passing across the diffusion surface of the bird lung is always fully oxygenated. (2) Just as in the gills of fishes, the flow of blood past the lung runs in a direction different from that of the unidirectional air flow. It is not opposite, as in fish; instead, the latticework of capillaries is arranged across the air flow, at a 90-degree angle. This is not as efficient as the 180-degree arrangement of fishes, but the blood leaving the lung can still contain more oxygen than exhaled air, which no mammalian lung can do. That is why a sparrow has no trouble flying at an altitude of 6,000 meters on an Andean mountain peak, while a mouse of the same body mass and with a similar high metabolic rate will stand panting, unable even to walk. The sparrow is simply getting more oxygen than the mouse.

Just as fish gills are the most efficient aquatic respiratory machines, so bird lungs are the most efficient atmospheric ones. Both achieve high efficiency by using forms of countercurrent flow. No one knows what the lungs of dinosaurs were like, but if dinosaurs were indeed endothermic, it is likely that they evolved lungs more efficient than those of today's reptiles. As dinosaurs evolved at about the same time as mammals, they may have adopted the same evolutionary strategy of expanding surface area or adopted the approach seen in the birds.

23.12 Terrestrial vertebrates employ lungs to extract oxygen from air. Bird lungs are the most efficient atmospheric respiratory machines, achieving a form of countercurrent flow.

23.13 The Mammalian Respiratory System

The oxygen-gathering mechanism of mammals, although less efficient than birds, adapts them well to their terrestrial habitat. Mammals, like all terrestrial vertebrates, obtain the oxygen they need for metabolism from air, which is about 21% oxygen gas. A pair of lungs is located in the chest, or **thoracic,** cavity. The two lungs hang free within the cavity, connected to the rest of the body only at one position, where the lung's blood vessels and air tube enter. This air tube is called a **bronchus.** It connects each lung to a long tube called the **trachea,** which passes upward and opens into the rear of the mouth (figure 23.21).

Air normally enters through the nostrils, where it is warmed. In addition, the nostrils are lined with hairs that filter out dust and other particles. As the air passes through the nasal cavity, an extensive array of cilia further filters and moistens the air (figure 23.22). Mucous secretory ciliated cells in the bronchi also trap foreign particles and carry them upward, where they can be swallowed. The air then passes through the back of the mouth, entering first the **larynx** (voice box) and then the trachea. From there, it passes down through the bronchus to the lungs. Because the air crosses the path of food at the back of the throat, a special flap called the epiglottis covers the trachea whenever food is swallowed, to keep it from "going down the wrong pipe."

The mammal respiratory apparatus is simple in structure and functions as a one-cycle pump. The thoracic cavity is bounded on its sides by the ribs and on the bottom by a thick layer of muscle, the **diaphragm,** which separates the thoracic cavity from the abdominal cavity. Each lung is covered by a very thin, smooth membrane called the **pleural membrane.** A second pleural membrane lines the interior of the thoracic cavity, into which the lungs hang. Within the cavity, the weight of the lungs is supported by water, the **interpleural fluid.**

The interpleural fluid not only supports the lungs, but it also plays another important role by permitting an even application of pressure to all parts of the lung. You can visualize the pleural membranes as a system of two balloons of different sizes, one nested inside the other, with the space between them completely filled with a very thin film of water. The inner balloon opens out to the atmosphere, so two forces act upon it: air pressure from the atmosphere pushes it outward, and water pressure from the interpleural fluid pushes it inward.

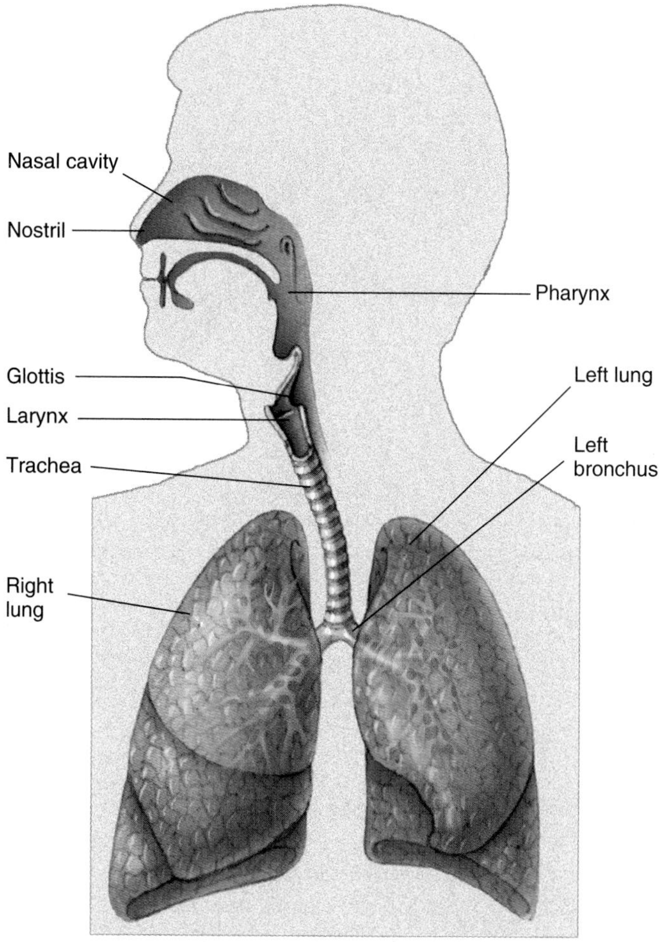
Nasal cavity
Nostril
Pharynx
Glottis
Larynx
Trachea
Left lung
Left bronchus
Right lung

Figure 23.21 The human respiratory system.
The respiratory system consists of the lungs and the passages that lead to them.

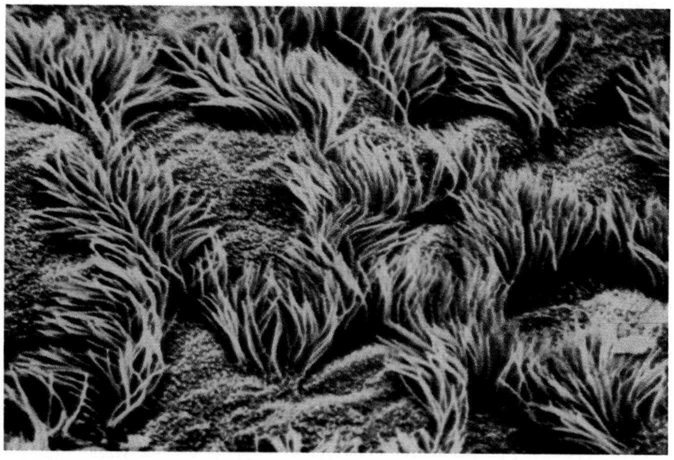

Figure 23.22 Cilia line the respiratory tract.
Cilia, shown here at magnification ×3,500, cover the epithelial lining of the nasal cavity, trachea, and bronchi. These cilia filter and moisten air that is inhaled.

The Mechanics of Breathing

The active pumping of air in and out through the lungs is called **breathing.** During *inhalation,* muscular contraction causes the walls of the chest cavity to expand so that the rib cage moves outward and upward. The diaphragm is dome-shaped when relaxed but moves downward as it flattens during this contraction (figure 23.23). In effect, we have enlarged the outer balloon by pulling it in all directions. This expansion causes the interpleural fluid pressure to decrease to a level less than that of the air pressure within the inner balloon. Because fluids are incompressible, the wall of the inner balloon is pulled out. As the inner balloon expands, its internal air pressure tends to decrease, and because the inner balloon is connected to the atmosphere, air rapidly moves in to equalize pressure.

During *exhalation,* the ribs and diaphragm return to their original resting position. In doing so, they exert pressure on the fluid. This pressure is transmitted uniformly by the fluid over the entire surface of the lung (the inner balloon), forcing air from the inner cavity back out to the atmosphere. In a human, a typical breath at rest moves about 0.5 liters of air, called the **tidal volume.** The extra amount that can be forced into and out of the lung is about 4.5 liters in men and 3.1 liters in women. The air remaining in the lung after such a maximal expiration is the residual volume, typically about 1.2 liters.

When each breath is completed, the lung still contains a volume of air, the **residual volume.** In human lungs this volume is about 1,200 milliliters. Each inhalation adds from 500 milliliters (resting) to 3,000 milliliters (exercising) of additional air. Each exhalation removes approximately the same volume as inhalation added, reducing the air volume in the lungs once more to about 1,200 milliliters. Because the diffusion surfaces of the lungs are not exposed to fully oxygenated air, but rather to a mixture of fresh and partly oxygenated air, the respiratory efficiency of mammalian lungs is far from maximal. A bird, for example, whose lungs do not retain a residual volume, is able to achieve far greater respiratory efficiency.

23.13 In mammals, the lungs are located within a thoracic cavity bounded on the bottom by a muscular diaphragm. By contracting and relaxing, the diaphragm expands or reduces the volume of the cavity, drawing air into the lungs or forcing it out.

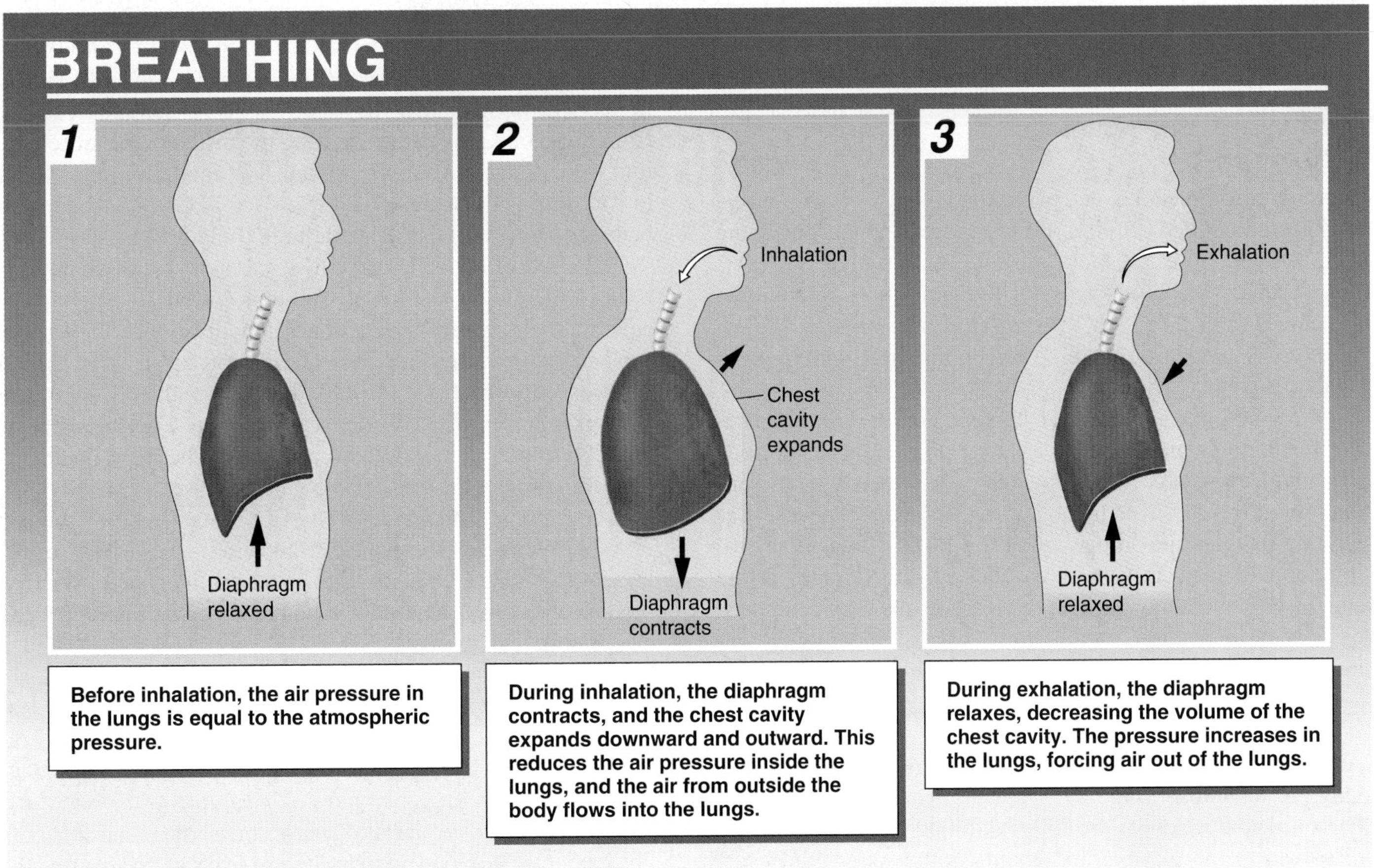

BREATHING
1
Diaphragm relaxed
Before inhalation, the air pressure in the lungs is equal to the atmospheric pressure.
2
Inhalation
Chest cavity expands
Diaphragm contracts
During inhalation, the diaphragm contracts, and the chest cavity expands downward and outward. This reduces the air pressure inside the lungs, and the air from outside the body flows into the lungs.
3
Exhalation
Diaphragm relaxed
During exhalation, the diaphragm relaxes, decreasing the volume of the chest cavity. The pressure increases in the lungs, forcing air out of the lungs.

Figure 23.23 How breathing works.

23.14 How Respiration Works: Gas Exchange

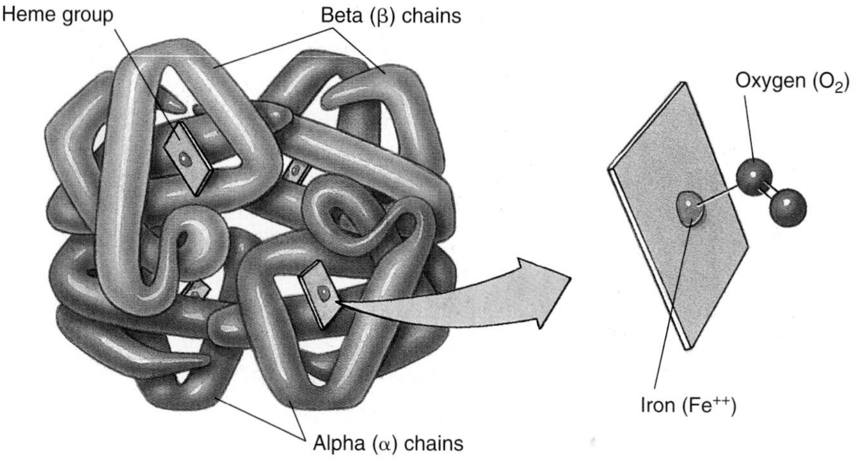

Heme group
Beta (β) chains
Oxygen (O_2)
Iron (Fe^{++})
Alpha (α) chains

Figure 23.24 The hemoglobin molecule.

The hemoglobin molecule is actually composed of four protein chain subunits: two copies of the "alpha chain" and two copies of the "beta chain." Each chain is associated with a heme group, and each heme group has a central iron atom, which can bind to a molecule of oxygen.

When oxygen has diffused from the air into the moist cells lining the inner surface of the lung, its journey has just begun. Passing from these cells into the bloodstream, the oxygen is carried throughout the body by the circulatory system, described earlier in this chapter. It has been estimated that it would take a molecule of oxygen three years to diffuse from your lung to your toe if transport depended only on diffusion, unassisted by a circulatory system.

Oxygen moves within the circulatory system carried piggyback on the protein **hemoglobin.** Hemoglobin molecules contain iron, which combines with oxygen in a reversible way (figure 23.24). Hemoglobin is manufactured within red blood cells and gives them their color. Hemoglobin never leaves these cells, which circulate in the bloodstream like ships bearing cargo. Oxygen binds to hemoglobin within these cells as they pass through the capillaries surrounding the alveoli of the lungs. Carried away within red blood cells, the oxygen-bearing hemoglobin molecules later release their oxygen molecules to metabolizing cells at distant locations in the body.

O_2 Transport

Hemoglobin molecules act like little oxygen (O_2) sponges, soaking up oxygen within red blood cells and causing more to diffuse in from the blood plasma. At the high O_2 levels that occur in the blood supply of the lung, most hemoglobin molecules carry a full load of oxygen atoms. Later, in the tissues, the O_2 levels are much lower, so that hemoglobin gives up its bound oxygen.

In tissue, the presence of carbon dioxide (CO_2) causes the hemoglobin molecule to assume a different shape, one that gives up its oxygen more easily. This speeds up the unloading of oxygen from hemoglobin even more. The effect of CO_2 on oxygen unloading, called the **Bohr effect,** is of real importance, because CO_2 is produced by the tissues at the site of cell metabolism. For this reason the blood unloads oxygen more readily within those tissues undergoing metabolism and generating CO_2.

CO_2 Transport

At the same time the red blood cells are unloading oxygen, they are also absorbing CO_2 from the tissue (figure 23.25). Only a tiny fraction of the CO_2 the blood carries is actually dissolved in the plasma, and about another one-fifth is carried by hemoglobin molecules within red blood cells. All the rest of the CO_2 is carried within the cytoplasm of the red blood cells. How do the red blood cells hold all this CO_2 in? To keep CO_2 from diffusing out of the red blood cells back into the plasma, where CO_2 levels are low, an enzyme, carbonic anhydrase, combines CO_2 molecules with water to form carbonic acid (H_2CO_3). This acid dissociates into **bicarbonate ions,** which do not diffuse outward. This soaking up of CO_2 by red blood cells removes large amounts of CO_2 from the blood plasma, facilitating the diffusion of more CO_2 into it from the surrounding tissue. The facilitation is critical to CO_2 removal, because the difference in CO_2 concentration between blood and tissue is not large (only 5%).

The red blood cells carry their cargo of bicarbonate ions back to the lungs. The lower CO_2 concentration in the air inside the lungs causes the carbonic anhydrase reaction to proceed in the reverse direction, releasing gaseous CO_2, which diffuses outward from the blood into the alveoli. With the next exhalation, this CO_2 leaves the body. The one-fifth of the CO_2 bound to hemoglobin also leaves because hemoglobin has a greater affinity for oxygen than for CO_2 at low CO_2 concentrations. The diffusion of CO_2 outward from the red blood cells causes the hemoglobin within these cells to release its bound CO_2 and take up oxygen instead. The red blood cells, with their newly bound oxygen, then start the next respiratory journey.

NO Transport

Hemoglobin also has the ability to hold and release the gas nitric oxide (NO). Nitric oxide, although a noxious gas in the atmosphere, has an important physiological role in the body, acting on many kinds of cells to change their shape and function. For example, in blood vessels the presence of NO causes the blood vessels to expand because it relaxes the surrounding

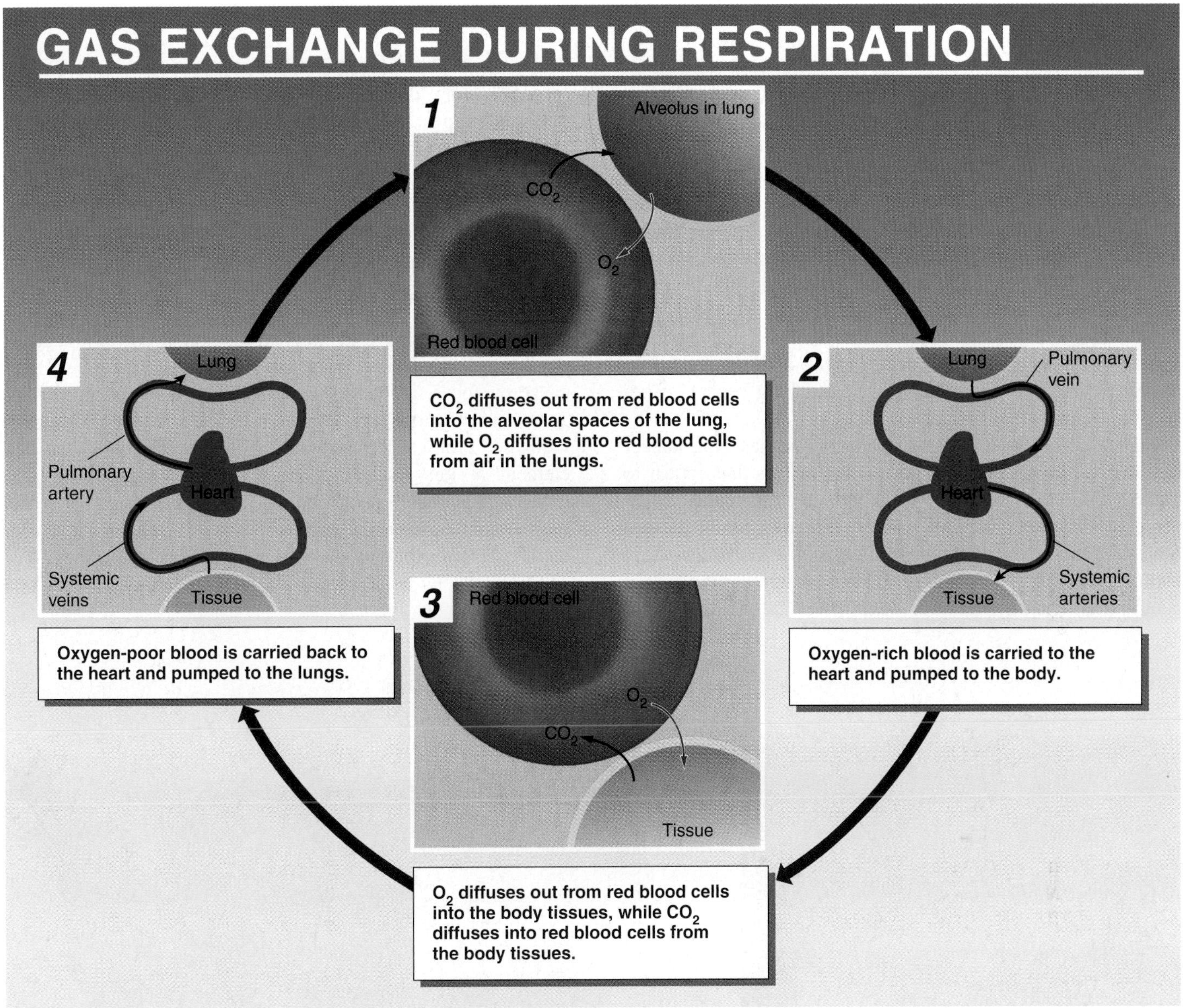
GAS EXCHANGE DURING RESPIRATION
1
Alveolus in lung
CO_2
O_2
Red blood cell
CO_2 diffuses out from red blood cells into the alveolar spaces of the lung, while O_2 diffuses into red blood cells from air in the lungs.
2
Lung
Pulmonary vein
Heart
Tissue
Systemic arteries
Oxygen-rich blood is carried to the heart and pumped to the body.
3
Red blood cell
O_2
CO_2
Tissue
O_2 diffuses out from red blood cells into the body tissues, while CO_2 diffuses into red blood cells from the body tissues.
4
Lung
Pulmonary artery
Heart
Systemic veins
Tissue
Oxygen-poor blood is carried back to the heart and pumped to the lungs.

Figure 23.25 How respiratory gas exchange works.

muscle cells. Thus, blood flow and blood pressure are regulated by the amount of NO released into the bloodstream.

Hemoglobin carries NO in a special form called super nitric oxide. In this form, NO has acquired an extra electron and is able to bind to an amino acid, called cysteine, present in hemoglobin. In the lungs, hemoglobin that is dumping CO_2 and picking up O_2 also picks up NO as super nitric oxide. In blood vessels at the tissues, hemoglobin that is releasing its O_2 and picking up CO_2 can do one of two things with nitric oxide. To increase blood flow, hemoglobin can release the super nitric oxide as NO into the blood, making blood vessels expand because NO acts as a relaxing agent. Or, hemoglobin can trap any excesses of NO on its iron atoms left vacant by the release of oxygen, causing blood vessels to constrict. When the red blood cells return to the lungs, hemoglobin dumps its CO_2 and the regular form of NO bound to the iron atoms. It is then ready to pick up O_2 and super nitric oxide and continue the cycle.

23.14 Oxygen and NO move through the circulatory system carried by the protein hemoglobin within red blood cells. Most CO_2 moves dissolved in the cytoplasm of red blood cells.

23.15 The Nature of Lung Cancer

Of all the diseases to which humans are susceptible, none is more feared than cancer (see chapter 6). One in every four deaths in the United States is caused by cancer. The American Cancer Society estimates that 552,000 people died of cancer in the United States in 2000. About 30% of these—157,000 people—died of **lung cancer.** About 140,000 cases of lung cancer were diagnosed each year in the 1980s, and 90% of these persons died within three years. Lung cancer is one of the leading causes of death among adults in the world today (figure 23.26*a*). What has caused lung cancer to become a major killer of Americans?

The search for a cause of cancers such as lung cancer has uncovered a host of environmental factors that appear to be associated with cancer. For example, the incidence of cancer per 1,000 people is not uniform throughout the United States. Rather, it is centered in cities and in the Mississippi Delta, suggesting that pollution and pesticide runoff may contribute to cancer (figure 23.26*b*). When the many environmental factors associated with cancer are analyzed, a clear pattern emerges: most cancer-causing agents, or carcinogens, share the property of being potent mutagens. Recall from chapter 8 that a mutagen is a chemical or radiation that damages DNA, destroying or changing genes (a change in a gene is called a mutation). The conclusion that cancer is caused by mutation is now supported by an overwhelming body of evidence.

What sort of genes are being mutated? In the last several years, researchers have found that mutation of only a few genes is all that is needed to transform normally dividing cells into cancerous ones. Identifying and isolating these cancer-causing genes, investigators have learned that all are involved with regulating cell proliferation (how fast cells grow and divide). A key element in this regulation are so-called tumor suppressors, genes that actively prevent tumors from forming. Two of the most important tumor-suppressor genes are called *Rb* and *p53,* and the proteins they produce are Rb and p53, respectively (recall from chapter 6 that genes are usually indicated in italics and proteins in regular type).

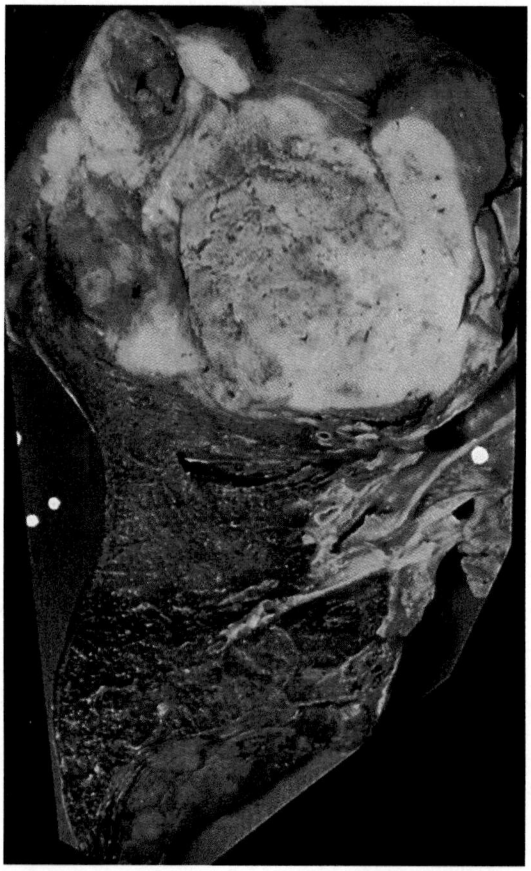

(a)

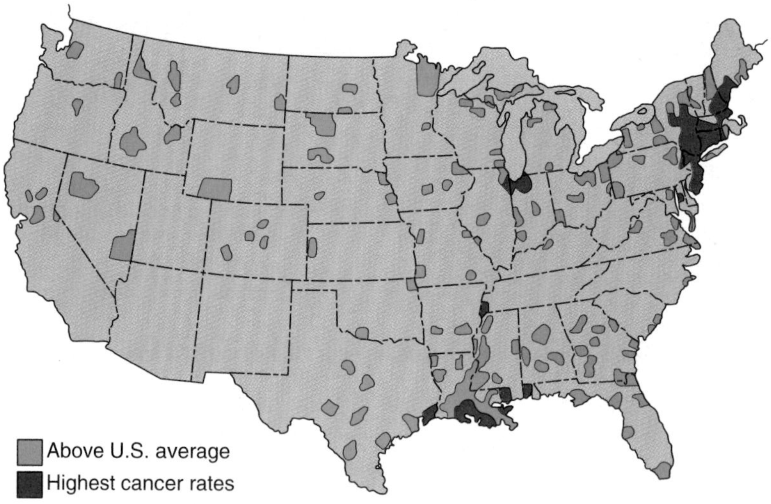
Above U.S. average
Highest cancer rates

(b)

Figure 23.26 Cancer in the United States.

(*a*) Lung cancer is responsible for the most cancer deaths. The bottom half of this lung is normal; the top half has been taken over completely by a cancerous tumor. As you might imagine, a lung in this condition does not function well, but the difficulty in breathing does not cause death. Rather, death usually results when the cancer spreads to other tissues throughout the body. (*b*) The incidence of cancer per 1,000 people is not uniform throughout the United States. It is centered in cities where chemical manufacturing is common and in the Mississippi Delta. This suggests that pollution and pesticide runoff may contribute to the development of cancer.

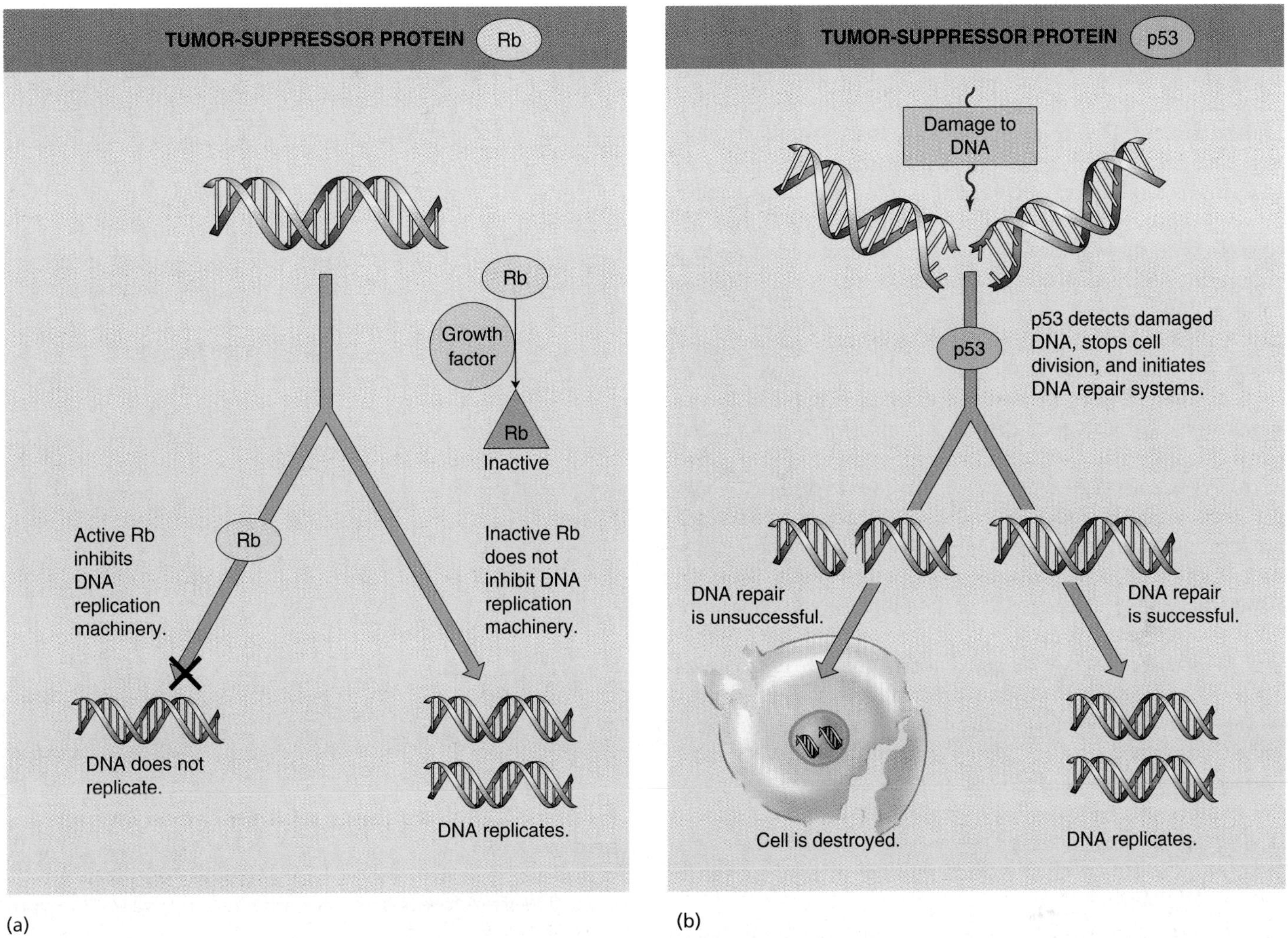
TUMOR-SUPPRESSOR PROTEIN Rb
Rb
Growth factor
Rb
Inactive
Active Rb inhibits DNA replication machinery.
Rb
Inactive Rb does not inhibit DNA replication machinery.
DNA does not replicate.
DNA replicates.
TUMOR-SUPPRESSOR PROTEIN p53
Damage to DNA
p53
p53 detects damaged DNA, stops cell division, and initiates DNA repair systems.
DNA repair is unsuccessful.
DNA repair is successful.
Cell is destroyed.
DNA replicates.
(a)
(b)

Figure 23.27 The roles Rb and p53 play in controlling cell division.

(a) The retinoblastoma protein Rb slows down cell division by inhibiting DNA replication. A normal cell can only reinitiate DNA replication when a growth factor renders Rb inactive. Lung cancer cells often have no Rb protein, and their cells will divide at a rapid rate. (*b*) p53 checks DNA for damage as it gets ready to replicate. If p53 detects damage, cell division is halted, and the cell's DNA repair systems are activated. If the damage does not get repaired, p53 initiates the destruction of the cell.

The Rb Protein

The **Rb protein** (named after retinoblastoma, the rare eye cancer in which it was first discovered) acts as a brake on cell division, attaching itself to the machinery the cell uses to replicate its DNA, and preventing it from doing so. When the cell wants to divide, a growth signal molecule ties up Rb so that it is not available to act as a brake on the division process (figure 23.27*a*). If the gene that produces Rb is disabled, there are no brakes to prevent the cell from replicating its DNA and dividing. The control switch is locked in the "ON" position.

The p53 Protein

The **p53 protein,** a tumor suppressor sometimes called the "guardian angel" of the cell, inspects the DNA to ensure it is ready to divide. When p53 detects damaged or foreign DNA, it stops cell division and activates the cell's DNA repair systems. If the damage doesn't get repaired in a reasonable time, p53 pulls the plug, triggering events that kill the cell (figure 23.27*b*). In this way, mutations such as those that cause cancer are either repaired or the cells containing them eliminated. If the gene that produces p53 is itself destroyed by mutation, future damage accumulates unrepaired. Among this damage are mutations that lead to cancer, mutations that would have been repaired by healthy p53. Fifty percent of all cancers have a disabled *p53* gene.

23.15 Cancer results from destruction by mutation of genes that when healthy enable the cell to regulate cell division.

23.16 Smoking Causes Lung Cancer

If cancer is caused by damage to growth-regulating genes, what then has led to the rapid increase in lung cancer in the United States? Two lines of evidence are particularly telling. The first consists of detailed information about cancer rates among smokers. The annual incidence of lung cancer among nonsmokers is only a few per 100,000 but increases with the number of cigarettes smoked per day to a staggering 300 per 100,000 for those smoking 30 cigarettes a day.

A second line of evidence consists of changes in the incidence of lung cancer that mirror changes in smoking habits. Look carefully at the data presented in figure 23.28. The upper curves are compiled from data on American men and show the incidence of smoking and of lung cancer since 1900. As late as 1920, lung cancer was a rare disease. About 20 years after the incidence of smoking began to increase among men, lung cancer also started to become more common. Now look at the lower curves, which present data on American women. Because of social mores, significant numbers of American women did not smoke until after World War II, when many social conventions changed. As late as 1963, only 6,588 women had died of lung cancer. But as women's frequency of smoking has increased, so has their incidence of lung cancer, again with a lag of about 20 years. American women today have achieved equality with men in the number of cigarettes they smoke, and their lung cancer death rates are now rapidly approaching those for men. This year, an estimated 66,000 women will die of lung cancer in the United States.

How does smoking cause cancer? Cigarette smoke contains many powerful mutagens, among them benzo[*a*]pyrene, and smoking introduces these mutagens to the lung tissues. Benzo[*a*]pyrene, for example, binds to three sites on the *p53* gene and causes mutations at these sites that inactivate the gene. In 1997, scientists studying this tumor-suppressor gene demonstrated a direct link between cigarettes and lung cancer. They found that the *p53* gene is inactivated in 70% of all lung cancers. When these inactivated *p53* genes are examined, they prove to be mutant at just the three sites where benzo[*a*]pyrene binds! Clearly, the chemical in cigarette smoke is responsible for the lung cancer.

In the face of these facts, why do so many people continue to smoke? Because the nicotine in cigarette smoke is an addictive drug. Researchers have identified the receptor on central nervous system neurons that it binds to, and they have demonstrated that smoking leads to a decrease in the brain's population of these receptors, leading to a craving for more cigarettes. This mechanism of nicotine addiction is very similar to that of cocaine addiction; once a person becomes addicted, it is difficult to quit. About half of those who try eventually succeed. Because your life is at stake, it is well worth the effort.

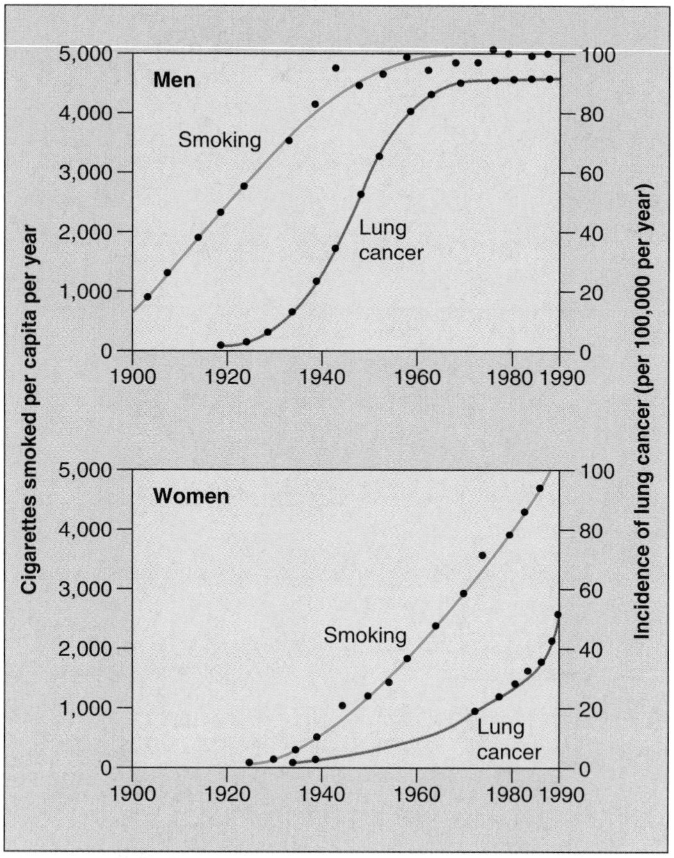
Men
Smoking
Lung cancer
Women
Smoking
Lung cancer
Cigarettes smoked per capita per year
Incidence of lung cancer (per 100,000 per year)
0
1,000
2,000
3,000
4,000
5,000
0
20
40
60
80
100
1900
1920
1940
1960
1980
1990

Figure 23.28 Incidence of lung cancer in men and women.

As men increased smoking in the early 1900s, the incidence of lung cancer also increased. Women followed suit years later, and in 1990, more than 49,000 women died of lung cancer.

Clearly, an effective way to avoid lung cancer is not to smoke. Life insurance companies have computed that, on a statistical basis, smoking a single cigarette lowers your life expectancy 10.7 minutes (more than the time it takes to smoke the cigarette!). Every pack of 20 cigarettes bears an unwritten label: *The price of smoking this pack of cigarettes is 3 1/2 hours of your life.* Smoking a cigarette is very much like going into a totally dark room with a person who has a gun and standing still. The person with the gun cannot see you, does not know where you are, and shoots once in a random direction. A hit is unlikely, and most shots miss. As the person keeps shooting, however, the chance of eventually scoring a hit becomes more likely. Every time an individual smokes a cigarette, mutagens are being shot at his or her genes. Nor do statistics protect any one individual: nothing says the first shot will not hit. Older people are not the only ones who die of lung cancer.

23.16 To avoid lung cancer, don't smoke.

CONCEPT REVIEW

1. Select the smallest type of blood vessel.
   a. artery
   b. capillary
   c. vein
   d. venule

2. Which of the following statements is *not* true about arteries?
   a. They carry blood away from the heart.
   b. They are capable of expanding.
   c. They contain many valves.
   d. They deliver blood to arterioles.

3. Each is a function of the lymphatic system *except*
   a. delivering oxygen to cells.
   b. returning fluid to veins.
   c. absorbing fat.
   d. carrying bacteria to lymph nodes.

4. Blood plasma is about _________ water.
   a. 25%
   b. 50%
   c. 75%
   d. 90%

5. Erythrocytes are packed with
   a. fibrinogen.
   b. serum albumin.
   c. hemoglobin.
   d. platelets.

6. In mammals, birds, and crocodiles, blood pumped out of the left ventricle moves into the
   a. aorta.
   b. pulmonary veins.
   c. right ventricle.
   d. superior vena cava.

7. The lungs of _________ are connected to anterior and posterior air sacs and are the most complex and efficient lungs among vertebrates.
   a. amphibians
   b. reptiles
   c. birds
   d. mammals

8. Arrange the following in the order in which air contacts them during breathing, starting from the mouth and nose.
   a. bronchus
   b. alveoli
   c. larynx
   d. hemoglobin
   e. trachea

9. Which of the following statements is *not* true about the diaphragm?
   a. It separates two body cavities.
   b. It is a thick layer of muscle.
   c. It is located inside the lungs.
   d. It is part of the human respiratory apparatus.

10. _________ capillaries collect fluid that leaks from the cardiovascular system.

11. The function of _________ is to keep the blood plasma in osmotic equilibrium with the cells of the body.

12. The passage of water in the direction opposite to the direction of blood flow in fish gills permits _________, an extremely efficient way of extracting oxygen.

13. A division of the pumping chamber of the heart by a completely extended septum has occurred independently in mammals, birds, and _________.

14. _________ are agents that cause cancer.

CHALLENGE YOURSELF

1. People who appear to have drowned can often be revived, in some cases after being underwater for as long as half an hour. In most of these cases, however, the person has been submerged in very cold water. Try to explain this observation.

2. Why can't humans or other mammals breathe water as well as air? In other words, what is it about the lung as a gas exchange organ that makes it unable to function effectively when filled with water?

3. The evidence associating lung cancer with smoking is overwhelming, and as you have learned in this chapter, we now know in considerable detail the mechanism whereby smoking induces cancer—it introduces powerful mutagens into the lungs, which cause mutations to occur; when a growth-regulating gene is mutated by chance, cancer results. In light of this, why do you suppose that cigarette smoking is still legal?

23 eBRIDGE

Reinforcing Key Points

Circulation

23.1 Open and Closed Circulatory Systems

23.2 Architecture of the Vertebrate Circulatory System

23.3 The Lymphatic System: Recovering Lost Fluid

23.4 Blood

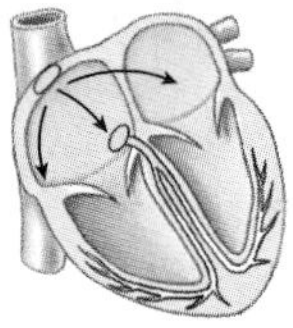

Evolution of Vertebrate Circulatory Systems

23.5 Fish Circulation

23.6 Amphibian and Reptile Circulation

23.7 Mammalian and Bird Circulation

23.8 Monitoring the Heart's Performance

23.9 Cardiovascular Diseases

Respiration

23.10 Types of Respiratory Systems

23.11 Respiration in Aquatic Vertebrates

23.12 Respiration in Terrestrial Vertebrates

23.13 The Mammalian Respiratory System

23.14 How Respiration Works: Gas Exchange

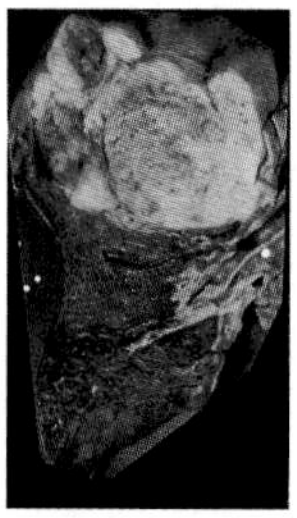

Lung Cancer and Smoking

23.15 The Nature of Lung Cancer

23.16 Smoking Causes Lung Cancer

Electronic Learning

Visual Learning

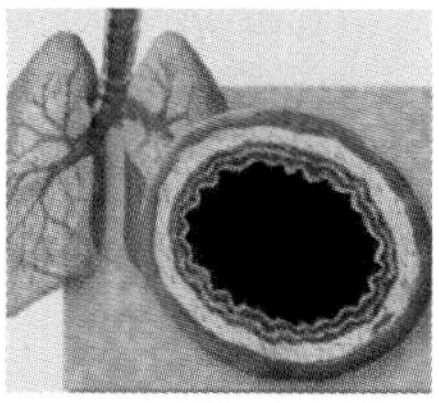

Animations

Nineteen Animations

Art Labeling Activities

Twenty-Seven Art Labeling Activities

Helping You Learn

Two Exercises

Explorations

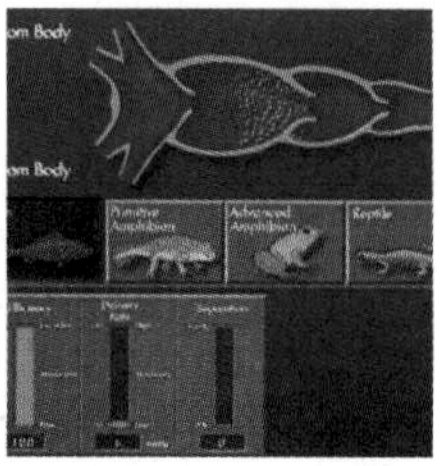

Evolution of the Heart

In this exercise, you can explore how the heart works by examining changes that took place during its evolution. You can watch the heart evolve while monitoring the effects on blood pressure and oxygen delivery to the tissues.

Author's Corner

Cancer and Smoking. One of the great tragedies of modern life is the widespread use of tobacco, particularly smoked in cigarettes. Tobacco contains the chemical nicotine which, unfortunately, proves to be very addictive. Once a person begins to smoke, it is very difficult to quit. Other chemicals in cigarette smoke are powerful mutagens, producing lung cancer with awful regularity. Smokers are thus addicted to a habit that will kill them. The morality of such legalized deadly addiction has in the last decade attracted national attention.

1. Does smoking really cause cancer?
2. Is smoking cigarettes addictive?
3. Nicotine addiction from cigarettes may begin in a few days of smoking.
4. Should the FDA regulate nicotine?
5. Curing lung cancer is a step closer.

Virtual Classroom

Smoking and Cancer

In the year 2000 an estimated 180,000 Americans were diagnosed with lung cancer, and 90% of them will die within 3 years. Ninety-six percent of these cancer victims are cigarette smokers. It speaks to the power of advertising and the addictive nature of nicotine that this grim correlation has done little to reduce smoking. In the United States, 24% of the population smokes, consuming over 450 billion cigarettes in 2000. Few women smoked before World War II, and as late as 1963 lung cancer was thought of as a man's disease, rare among women—only 6,588 American women died of lung cancer that year. In 2000, American women smoke as many cigarettes as men, and more than 55,000 American women died of lung cancer. The problem is that tobacco smoke introduces powerful mutagens to the lungs. One of them, benzo[*a*]pyrene, binds specifically to the tumor suppressor gene *p53*, destroying a key mechanism for preventing uncontrolled cell proliferation. *p53* is mutated to an inactive form in over 70% of lung cancers. When the inactive genes are examined, the mutations are at precisely the spots where benzo[*a*]pyrene binds to the *p53* gene! The cause and effect could not be more clear.

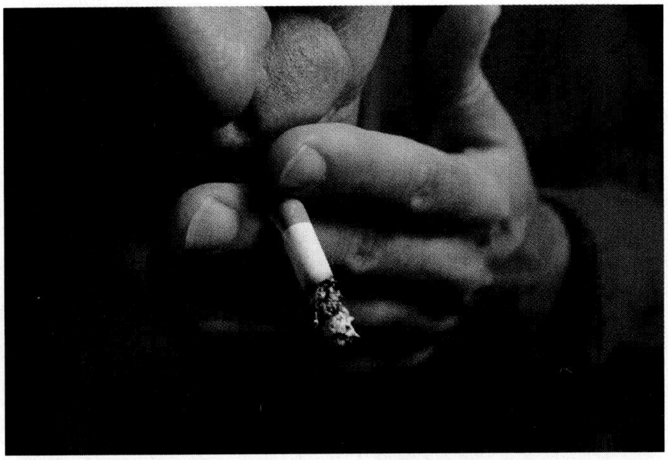

Virtual Lab

Why Some Lizards Take a Deep Breath

A lizard runs with a lateral undulating gait, the body flexing from side to side with each step. The intercostal muscles along the sides of the body contract alternately to move the body from side to side. At rest, lizards breathe by expanding their chest which allows air to rush into the lungs. The intercostal muscles contract simultaneously to expand the chest. Do you see the problem? A running lizard cannot contract its chest muscles on both sides simultaneously for effective breathing and at the same time contract the same muscles alternately for running. This apparent conflict has led to a hypothesis called the "constraint" hypothesis, which states that lizards are subject to a speed-dependent axial constraint that prevents effective lung ventilation while running. However, not all lizards exhibit this speed-dependent axial constraint. The savannah monitor lizard, for example, seems to breathe effectively when running.

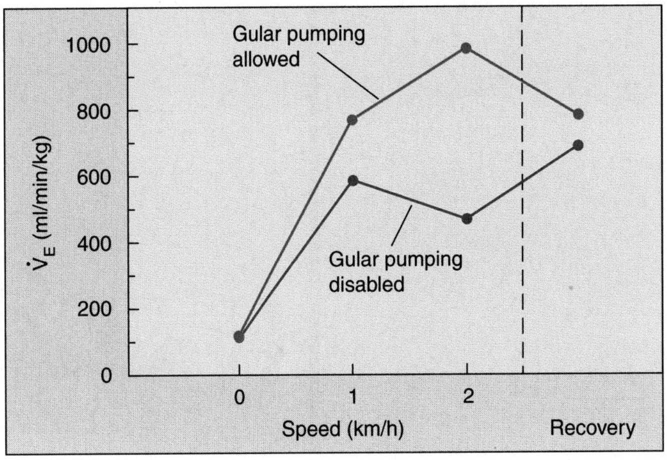
1000
800
600
400
200
0
$\dot{V}_E$ (ml/min/kg)
Gular pumping allowed
Gular pumping disabled
0
1
2
Speed (km/h)
Recovery

Elizabeth Brainerd of the University of Massachusetts, Amherst, and colleagues have investigated breathing in monitor lizards to determine why they do not exhibit speed-dependent axial constraint when running.

Quizzes | Further Reading | Essential Study Partner | Links | BioCourse.com

CHAPTER 24

THE PATH OF FOOD THROUGH THE ANIMAL BODY

THE FAR SIDE® By GARY LARSON
© 1986 FarWorks, Inc. All Rights Reserved/Dist. by Creators Syndicate
Larson
"Just think ... here we are, the afternoon sun beating down on us, a dead, bloated rhino underfoot, and good friends flying in from all over. ... I tell you, Frank, this is the best of times."

CHAPTER 24

The Path of Food Through the Animal Body

CHAPTER OVERVIEW

- Ingested calories are either metabolized by the body or stored as fat.
- Healthy diets contain those necessary substances that animals cannot manufacture, such as essential amino acids, vitamins, and trace elements.

- Digestion is the breaking down of macromolecules into small components that can be metabolized.
- The esophagus contracts in peristaltic waves to drive swallowed food to the stomach.
- Acid predigestion occurs in the stomach; most enzymatic digestion occurs in the duodenum, the initial portion of the small intestine.
- Hormones from the pancreas direct the liver to either store excess glucose as glycogen or break down glycogen into glucose.

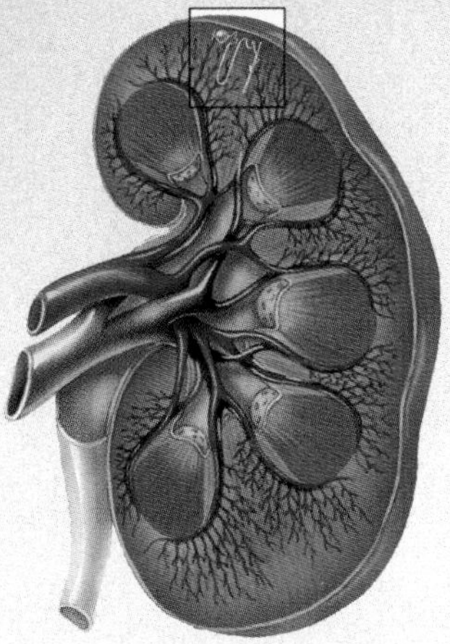

- Regulating body temperature, blood glucose levels, and salt and water balance helps animals achieve homeostasis.
- Freshwater bony fish and marine bony fish have different demands upon their kidneys and other regulatory systems.
- Insects, reptiles, and birds produce uric acid from the amino groups in amino acids; this precipitates so that little water is required for its excretion.
- Mammals concentrate their urine by removing water from it; this allows them to retain the water when the urine is discarded.

24.1 Calories for Energy

The food animals eat provides both a source of energy and a supply of raw materials that the animal body is not able to manufacture for itself. An optimal diet contains more carbohydrates than fats and also a significant amount of protein, as the "pyramid of nutrition" recommended by the federal government indicates (figure 24.1). Fats have a far greater number of energy-rich carbon–hydrogen bonds and thus a much higher energy content per gram than carbohydrates or proteins. That makes fats a very efficient way to store energy. When food is consumed, it is either metabolized by muscles and other cells of the body, or it is converted into fat and stored in fat cells.

Carbohydrates are obtained primarily from cereals and grains, breads, fruits, and vegetables. On the average, carbohydrates contain 4.1 **calories** per gram; fats, by comparison, contain 9.3 calories per gram, over twice as much. Dietary fats are obtained from oils, margarine, and butter and are abundant in fried foods, meats, and processed snack foods, such as potato chips and crackers. Protein, like carbohydrates, has 4.1 calories per gram and can be obtained from dairy products, poultry, fish, meat, and grains.

The body uses carbohydrates for energy and fats to construct cell membranes and other cell structures, to insulate nervous tissue, and to provide energy. Fat-soluble vitamins that are essential for proper health are also absorbed with fats. Proteins are used as building materials for cell struc-

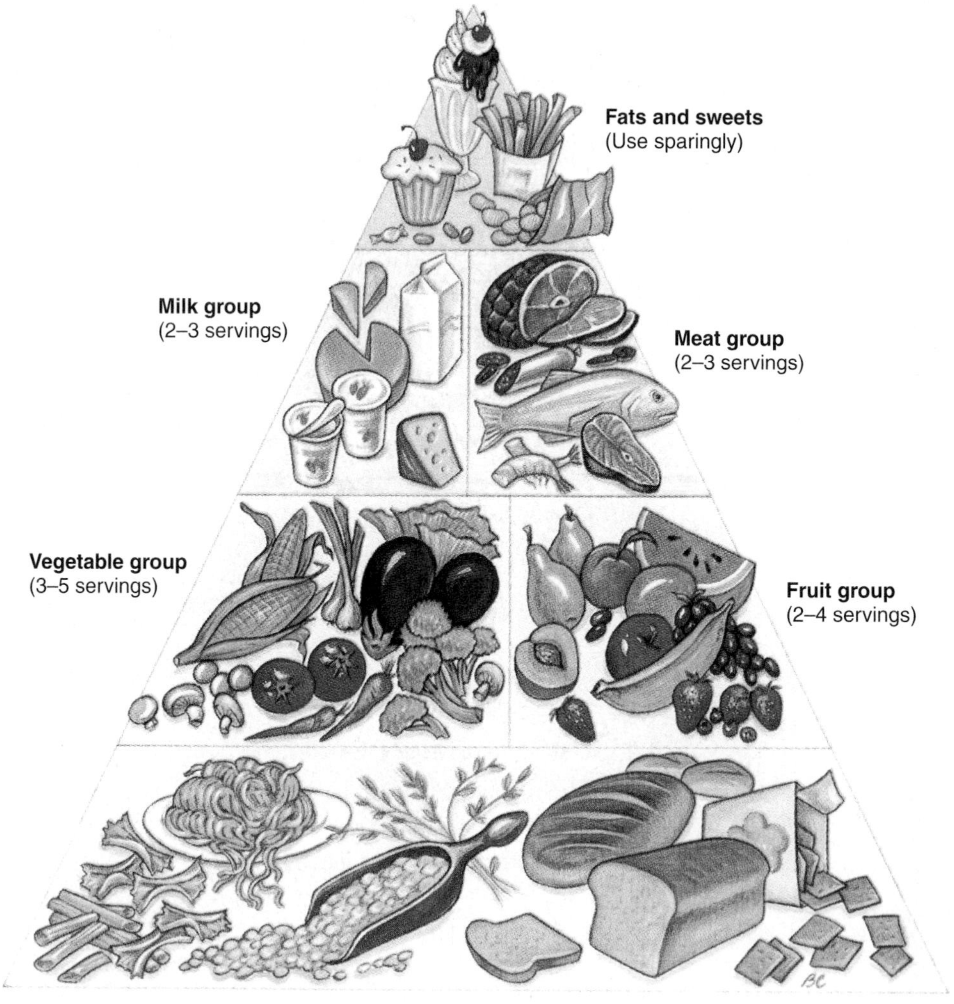
Fats and sweets
(Use sparingly)
Milk group
(2–3 servings)
Meat group
(2–3 servings)
Vegetable group
(3–5 servings)
Fruit group
(2–4 servings)
Bread and cereal group (6–11 servings)

Figure 24.1 The pyramid of nutrition.

A healthy diet uses more foods at the base of the pyramid and fewer at the top. Source: U.S. Department of Agriculture, U.S. Department of Health and Human Services.

25 OVERWEIGHT LIMIT   OVERWEIGHT

WEIGHT / HEIGHT	100	105	110	115	120	125	130	135	140	145	150	155	160	165	170	175	180	185	190	195	200	205
5' 0"	20	21	21	22	23	24	25	26	27	28	29	30	31	32	33	34	35	36	37	38	39	40
5' 1"	19	20	21	22	23	24	25	26	26	27	28	29	30	31	32	33	34	35	36	37	38	39
5' 2"	18	19	20	21	22	23	24	25	26	27	27	28	29	30	31	32	33	34	35	36	37	37
5' 3"	18	19	19	20	21	22	23	24	25	26	27	27	28	29	30	31	32	33	34	35	35	36
5' 4"	17	18	19	20	21	21	22	23	24	25	26	27	27	28	29	30	31	32	33	33	34	35
5' 5"	17	17	18	19	20	21	22	22	23	24	25	26	27	27	28	29	30	31	32	32	33	34
5' 6"	16	17	18	19	19	20	21	22	23	23	24	25	26	27	27	28	29	30	31	31	32	33
5' 7"	16	16	17	18	19	20	20	21	22	23	23	24	25	26	27	27	28	29	30	31	31	32
5' 8"	15	16	17	17	18	19	20	21	21	22	23	24	24	25	26	27	27	28	29	30	30	31
5' 9"	15	16	16	17	18	18	19	20	21	21	22	23	24	24	25	26	27	27	28	29	30	30
5' 10"	14	15	16	17	17	18	19	19	20	21	22	22	23	24	24	25	26	27	27	28	29	29
5' 11"	14	15	15	16	17	17	18	19	20	20	21	22	22	23	24	24	25	26	26	27	28	29
6' 0"	14	14	15	16	16	17	18	18	19	20	20	21	22	22	23	24	24	25	26	26	27	28
6' 1"	13	14	15	15	16	16	17	18	18	19	20	20	21	22	22	23	24	24	25	26	26	27
6' 2"	13	13	14	15	15	16	17	17	18	19	19	20	21	21	22	22	23	24	24	25	26	26
6' 3"	12	13	14	14	15	16	16	17	17	18	19	19	20	21	21	22	22	23	24	24	25	26
6' 4"	12	13	13	14	15	16	16	16	17	18	18	19	19	20	21	21	22	23	23	24	24	25

Figure 24.2 Are you overweight?

This chart presents the body mass index (BMI) values used by federal health authorities to determine who is overweight. Your body mass index is at the intersection of your height and weight. BMI is calculated by dividing body weight in kilograms by height in meters squared. For pounds and feet, first multiply weight in pounds by 703, and then divide by height in inches squared. A BMI value of 25 or over is considered overweight. Source: "Shape Up America" National Institutes of Health.

tures, enzymes, hemoglobin, hormones, and muscle and bone tissue.

In wealthy countries such as those of North America and Europe, being significantly overweight is common, the result of habitual overeating and high-fat diets. The international standard measure of appropriate body weight is the body mass index (BMI), estimated as your body weight in kilograms, divided by your height in meters squared (figure 24.2). In the United States, the National Institutes of Health estimated in 1998 that 55% of middle-aged individuals, 97 million Americans, are overweight, with a body mass index of 25 or more. Those with a body mass index of 30 or more are classified as obese. Being overweight is highly correlated with coronary heart disease and many other disorders.

One essential characteristic of food is its fiber content. Animals have evolved many different ways to process food that has a relatively high fiber content. Diets that are low in fiber, now common in the United States, result in a slower passage of food through the colon. This low dietary fiber content is thought to be associated with levels of colon cancer in the United States, which are among the highest in the world.

24.1 Food is an essential source of calories. It is important to maintain a proper balance of carbohydrate, protein, and fat. Individuals with a body mass index of 25 or more are considered overweight.

24.2 Essential Substances for Growth

Over the course of their evolution, many animals have lost the ability to manufacture certain substances they need, substances that often play critical roles in their metabolism. Mosquitoes and many other blood-sucking insects, for example, cannot manufacture cholesterol, but they obtain it in their diet because human blood is rich in cholesterol. Many vertebrates are unable to manufacture one or more of the 20 amino acids used to make proteins. Humans are unable to synthesize eight amino acids: lysine, tryptophan, threonine, methionine, phenylalanine, leucine, isoleucine, and valine. These amino acids, called **essential amino acids,** must therefore be obtained from proteins in the food we eat (figure 24.3). For this reason, it is important to eat so-called complete proteins, that is, ones containing all the essential amino acids. In addition, all vertebrates have also lost the ability to synthesize certain polyunsaturated fats that provide backbones for the many kinds of fats their bodies manufacture.

Vitamins

Essential organic substances that are used in trace amounts are called **vitamins.** Humans require at least 13 different vitamins (table 24.1). Many vitamins are required cofactors for cellular enzymes. Humans, monkeys, and guinea pigs, for example, have lost the ability to synthesize ascorbic acid (vitamin C) and will develop the potentially fatal disease called scurvy—characterized by weakness, spongy gums, and bleeding of the skin and mucous membranes—if vitamin C is not supplied in their diets. All other mammals are able to synthesize ascorbic acid.

Trace Elements

In addition to supplying energy and essential organic compounds, food that is consumed must also supply the body with *essential minerals* such as calcium and phosphorus, as well as a wide variety of **trace elements,** which are minerals required in very small amounts. Among the trace elements are *iodine* (a component of thyroid hormone), *cobalt* (a component of vitamin B_{12}), *zinc* and *molybdenum* (components of enzymes), *manganese,* and *selenium.* All of these, with the possible exception of selenium, are also essential for plant growth; animals obtain them directly from plants that they eat or indirectly from animals that have eaten plants, or plant eaters.

24.2 Food provides not only calories but also key amino acids that the body cannot manufacture for itself, as well as necessary vitamins and trace elements.

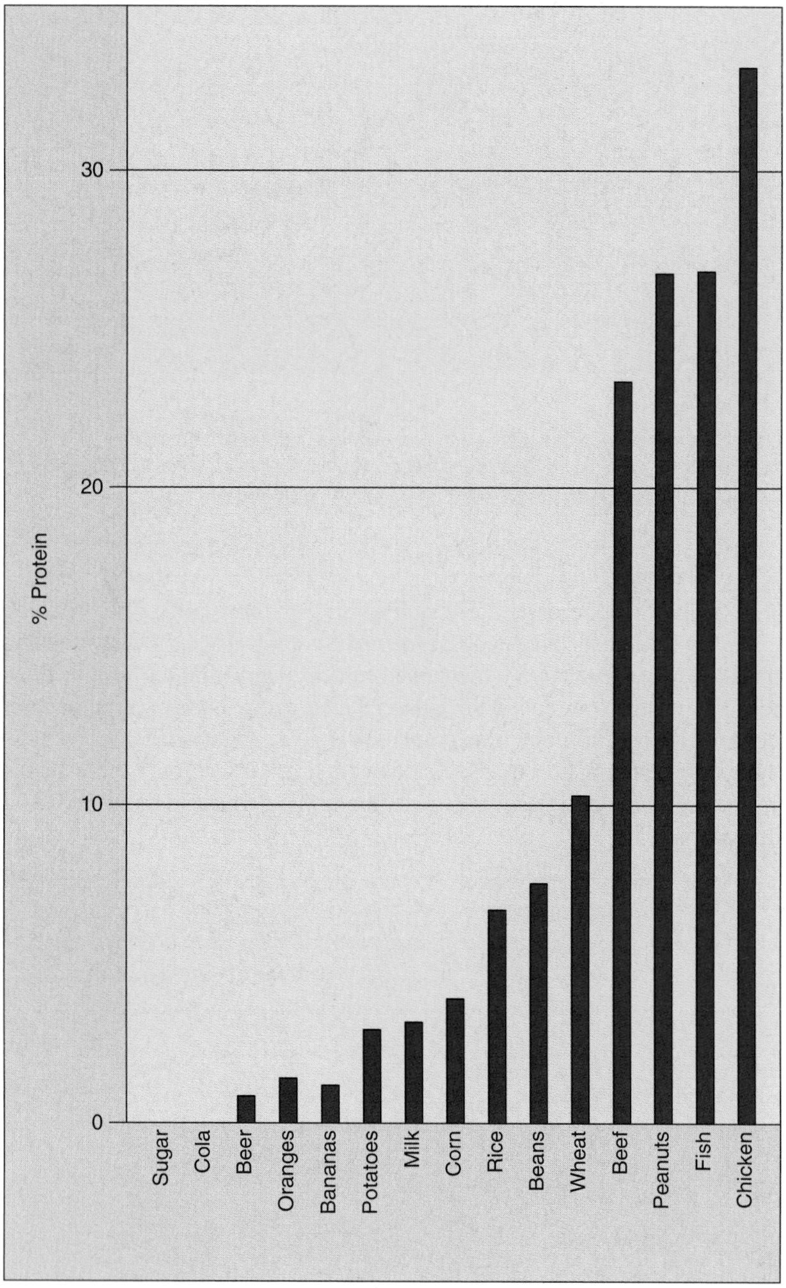

% Protein
30
20
10
0
Sugar
Cola
Beer
Oranges
Bananas
Potatoes
Milk
Corn
Rice
Beans
Wheat
Beef
Peanuts
Fish
Chicken

Figure 24.3 The protein content of a variety of common foods.

Humans and many other vertebrates must eat protein to obtain the amino acids they are unable to synthesize.

TABLE 24.1 MAJOR VITAMINS

Vitamin	Function	Dietary Source	Recommended Daily Allowance (milligrams)	Deficiency Symptoms
Vitamin A	Used in making visual pigments, maintenance of epithelial tissues	Green vegetables, carrots, milk products, liver	1	Night blindness, flaky skin
B-complex vitamins				
B_1	Coenzyme in CO_2 removal during cellular respiration	Meat, grains, legumes	1.5	Beriberi, weakening of heart, edema
B_2 (riboflavin)	Part of coenzymes FAD and FMN, which play metabolic roles	In many different kinds of foods	1.8	Inflammation and breakdown of skin, eye irritation
B_3 (niacin)	Part of coenzymes NAD^+ and $NADP^+$	Liver, lean meats, grains	20	Pellagra, inflammation of nerves, mental disorders
B_5 (pantothenic acid)	Part of coenzyme A, a key connection between carbohydrate and fat metabolism	In many different kinds of foods	5 to 10	Rare: fatigue, loss of coordination
B_6 (pyridoxine)	Coenzyme in many phases of amino acid metabolism	Cereals, vegetables, meats	2	Anemia, convulsions, irritability
B_{12} (cyanocobalamin)	Coenzyme in the production of nucleic acids	Red meat, dairy products	0.003	Pernicious anemia
Biotin	Coenzyme in fat synthesis and amino acid metabolism	Meat, vegetables	Minute	Rare: depression, nausea
Folic acid	Coenzyme in amino acid and nucleic acid metabolism	Green leafy vegetables, whole grain products	0.4	Anemia, diarrhea
Vitamin C	Important in forming collagen, cement of bone, teeth, connective tissue of blood vessels; may help maintain resistance to infection	Fruit, green leafy vegetables	45	Scurvy, breakdown of skin, blood vessels
Vitamin D (calciferol)	Increases absorption of calcium and promotes bone formation	Dairy products, cod liver oil	0.01	Rickets, bone deformities
Vitamin E (tocopherol)	Protects fatty acids and cell membranes from oxidation	Margarine, seeds, green leafy vegetables	15	Rare
Vitamin K	Essential to blood clotting	Green leafy vegetables	0.03	Severe bleeding

24.3 Types of Digestive Systems

Heterotrophs are divided into three groups on the basis of their food sources. Animals that eat plants exclusively are classified as **herbivores;** common examples include cows, horses, rabbits, and sparrows. Animals that are meat eaters, such as cats, eagles, trout, and frogs, are **carnivores. Omnivores** are animals that eat both plants and other animals. We humans are omnivores, as are pigs, bears, and crows.

Single-celled organisms (as well as sponges) digest their food intracellularly. Other animals digest their food extracellularly, within a digestive cavity. In this case, the digestive enzymes are released into a cavity that is continuous with the animal's external environment. In cnidarians and flatworms (such as *Planaria*), the digestive cavity has only one opening that serves as both mouth and anus. There can be no specialization within this type of digestive system, called a gastrovascular cavity (see chapter 23), because every cell is exposed to all stages of food digestion (figure 24.4).

Specialization occurs when the digestive tract, or alimentary canal, has a separate mouth and anus, so that transport of food is one way. The most primitive digestive tract is seen in nematodes (phylum Nematoda), where it is simply a tubular *gut* lined by an epithelial membrane. Earthworms (phylum Annelida) have a digestive tract specialized in different regions for the ingestion, storage, fragmentation, digestion, and absorption of food. All higher animals show similar specializations (figure 24.5).

The ingested food may be stored in a specialized region of the digestive tract or may first be subjected to physical fragmentation through the chewing action of teeth (in the mouth of many vertebrates) or the grinding action of pebbles (in the gizzard of earthworms and birds). Chemical **digestion** then occurs, breaking down the larger food molecules of polysaccharides, fats, and proteins into smaller subunits. Chemical digestion involves hydrolysis reactions that liberate the subunits—primarily monosaccharides, amino acids, and fatty acids—from the food. These products of chemical digestion pass through the epithelial lining of the gut into the blood, in a process known as absorption. Any molecules in the food that are not absorbed cannot be used by the animal. These wastes are excreted from the anus.

24.3 Most animals digest their food extracellularly. A digestive tract with a one-way transport of food allows specialization of regions for different functions.

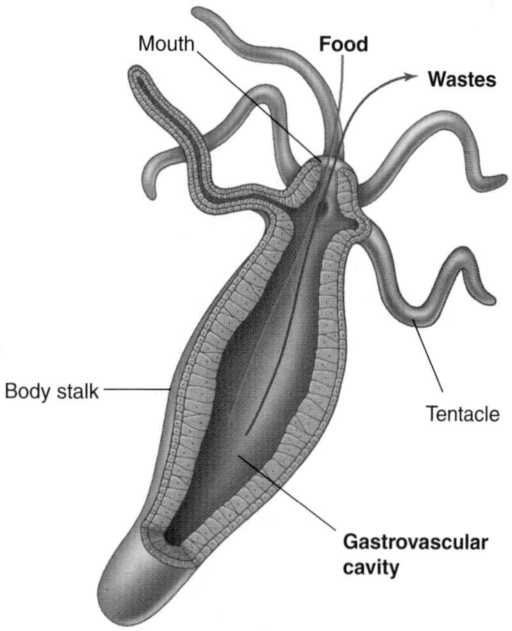
Mouth
Food
Wastes
Body stalk
Tentacle
Gastrovascular cavity

Figure 24.4 The gastrovascular cavity of *Hydra*.
Because there is only one opening, the mouth is also the anus, and no specialization is possible in the different regions that participate in extracellular digestion.

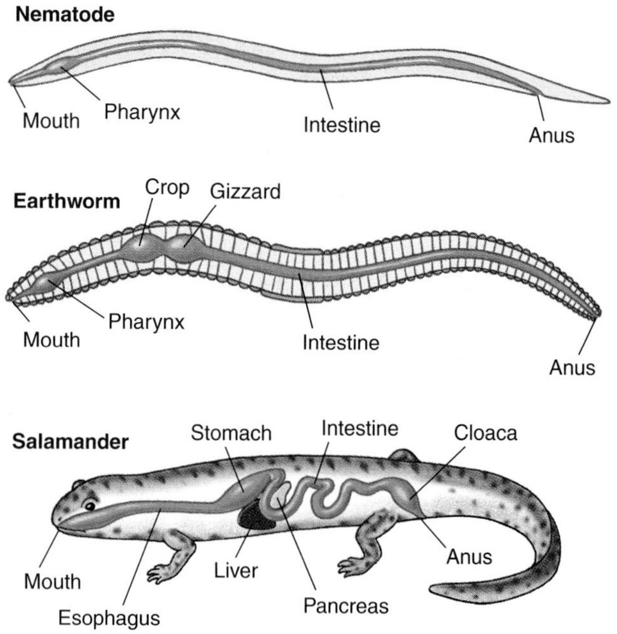
Nematode
Mouth
Pharynx
Intestine
Anus
Earthworm
Crop
Gizzard
Mouth
Pharynx
Intestine
Anus
Salamander
Stomach
Intestine
Cloaca
Mouth
Esophagus
Liver
Pancreas
Anus

Figure 24.5 One-way digestive tracts.
One-way movement through the digestive tract allows different regions of the digestive system to become specialized for different functions.

24.4 Vertebrate Digestive Systems

In humans and other vertebrates, the digestive system consists of a tubular gastrointestinal tract and accessory digestive organs (figure 24.6). The initial components of the gastrointestinal tract are the mouth and the pharynx, which is the common passage of the oral and nasal cavities. The pharynx leads to the esophagus, a muscular tube that delivers food to the stomach, where some preliminary digestion occurs. From the stomach, food passes to the first part of the small intestine, where a battery of digestive enzymes continues the digestive process. The products of digestion then pass across the wall of the small intestine into the bloodstream. The small intestine empties what remains into the large intestine, where water and minerals are absorbed. In most vertebrates other than mammals, the waste products emerge from the large intestine into a cavity called the cloaca (see figure 24.5), which also receives the products of the urinary and reproductive systems. In mammals, the urogenital products are separated from the fecal material in the large intestine; the fecal material enters the rectum and is expelled through the anus.

In general, carnivores have shorter intestines for their size than do herbivores. A short intestine is adequate for a carnivore, but herbivores ingest a large amount of plant cellulose, which resists digestion. These animals have a long, convoluted small intestine. In addition, mammals called *ruminants* (such as cows) that consume grass and other vegetation have stomachs with multiple chambers, where bacteria aid in the digestion of cellulose. Other herbivores, including rabbits and horses, digest cellulose (with the aid of bacteria) in a blind pouch called the **cecum** located at the beginning of the large intestine. Accessory digestive organs described later in this chapter include the liver, the gallbladder, and the pancreas.

The tubular gastrointestinal tract of a vertebrate has a characteristic layered structure (figure 24.7). The innermost layer is the mucosa, an epithelium that lines the interior of the tract (the lumen). The next major tissue layer, composed of connective tissue, is called the submucosa. Just outside the submucosa is the muscularis, which consists of a double layer of smooth muscles. The muscles in the inner layer have a circular orientation, and those in the outer layer are arranged longitudinally. Another connective tissue layer, the serosa, covers the external surface of the tract. Nerves, intertwined in regions called *plexuses,* are located in the submucosa and help regulate the gastrointestinal activities.

24.4 The vertebrate digestive system consists of a tubular gastrointestinal tract, which is modified in different animals, composed of a series of tissue layers.

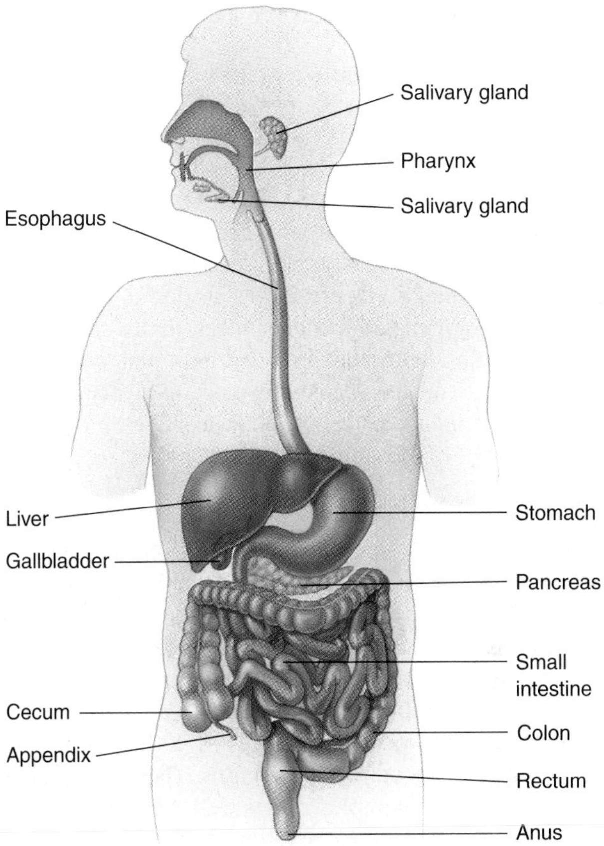

Salivary gland
Pharynx
Salivary gland
Esophagus
Liver
Gallbladder
Stomach
Pancreas
Small intestine
Cecum
Appendix
Colon
Rectum
Anus

Figure 24.6 The human digestive system.
The tubular gastrointestinal tract and accessory digestive organs are shown.

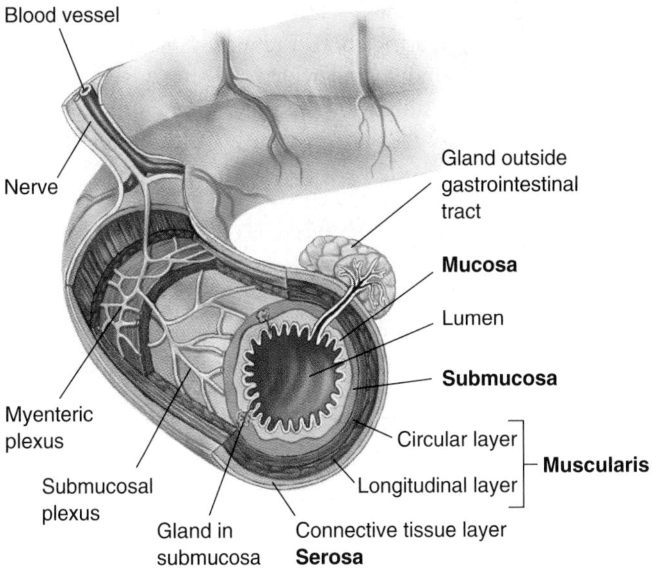

Blood vessel
Nerve
Gland outside gastrointestinal tract
Mucosa
Lumen
Submucosa
Myenteric plexus
Circular layer
Muscularis
Longitudinal layer
Submucosal plexus
Gland in submucosa
Connective tissue layer
Serosa

Figure 24.7 The layers of the gastrointestinal tract.
The mucosa contains a lining epithelium, the submucosa is composed of connective tissue (as is the serosa), and the muscularis consists of smooth muscles.

24.5 The Mouth and Teeth

Specializations of the digestive systems in different kinds of vertebrates reflect differences in the way these animals live. Fishes have a large pharynx with gill slits, while air-breathing vertebrates have a greatly reduced pharynx. Many vertebrates have teeth (figure 24.8), and chewing (*mastication*) breaks up food into small particles and mixes it with fluid secretions. Birds, which lack teeth, break up food in their two-chambered stomachs (figure 24.9). In one of these chambers, the gizzard, small pebbles ingested by the bird are churned together with the food by muscular action. This churning grinds up the seeds and other hard plant material into smaller chunks that can be digested more easily in the second chamber of the stomach.

Vertebrate Teeth

Carnivorous mammals have pointed teeth that lack flat grinding surfaces. Such teeth are adapted for cutting and shearing. Carnivores often tear off pieces of their prey but have little need to chew them, because digestive enzymes can act directly on animal cells. (Recall how a cat or dog gulps down its food.) By contrast, grass-eating herbivores, such as cows and horses, must pulverize the cellulose cell walls of plant tissue before digesting it. These animals have large, flat teeth with complex ridges well suited to grinding.

Humans are omnivores, and human teeth are specialized for eating both plant and animal food. Viewed simply, humans are carnivores in the front of the mouth and herbivores in the back (figure 24.10). The four front teeth in the upper and lower jaws are sharp, chisel-shaped incisors used for biting. On each side of the incisors are sharp, pointed teeth called cuspids (sometimes referred to as "canine" teeth), which are used for tearing food. Behind the canines are two premolars and three molars, all with flattened, ridged surfaces for grinding and crushing food. Children have only 20 teeth, but these deciduous teeth are lost during childhood and are replaced by 32 adult teeth.

Processing Food in the Mouth

Inside the mouth, the tongue mixes food with a mucous solution, **saliva.** In humans, three pairs of salivary glands secrete saliva into the mouth through ducts in the mouth's mucosal lining. Saliva moistens and lubricates the food so that it is easier to swallow and does not abrade the tissue it passes on its way through the esophagus. Saliva also contains the hydrolytic enzyme salivary **amylase,** which initiates the breakdown of the polysaccharide starch into the disaccharide maltose. This digestion is usually minimal in humans, however, because most people don't chew their food very long.

The secretions of the salivary glands are controlled by the nervous system, which in humans maintains a constant flow of about half a milliliter per minute when the mouth is empty of food. This continuous secretion keeps the mouth moist. The presence of food in the mouth triggers an increased rate of secretion, as taste-sensitive neurons in the mouth send impulses to the brain, which responds by stimulating the salivary glands. The most potent stimuli are acidic solutions; lemon juice, for example, can increase the rate of salivation eightfold. The sight, sound, or smell of food can stimulate salivation markedly in dogs, but in humans, these stimuli are much less effective than thinking or talking about food.

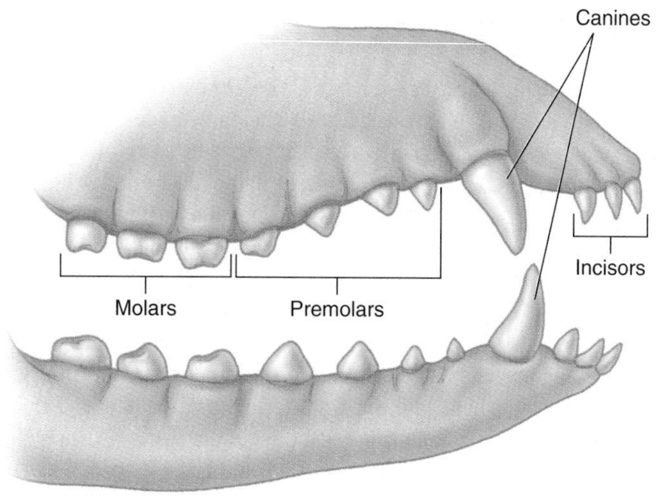
Canines
Incisors
Molars
Premolars

Figure 24.8 Diagram of generalized vertebrate dentition.
Different vertebrates have specific variations from this generalized pattern, depending on whether the vertebrate is an herbivore, carnivore, or omnivore.

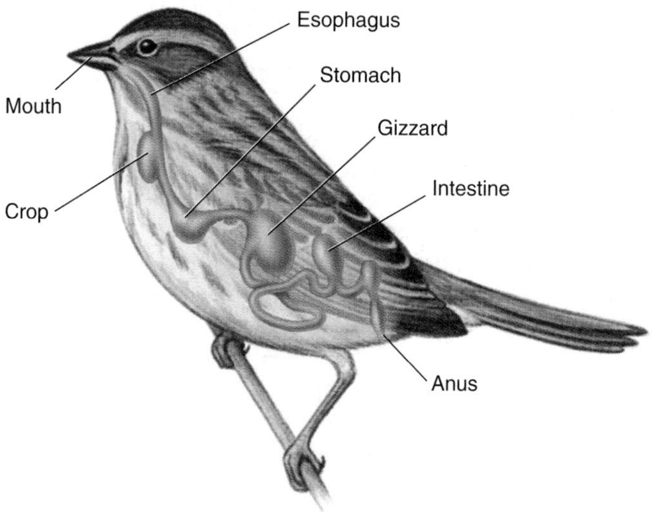
Esophagus
Stomach
Mouth
Gizzard
Intestine
Crop
Anus

Figure 24.9 Birds grind up food in the crop and gizzard.
Birds, which lack teeth, have two chambers, the crop and the gizzard, where food is pulverized and crushed before passing into the small intestine.

Swallowing

When food is ready to be swallowed, the tongue moves it to the back of the mouth. In mammals, the process of swallowing begins when the soft palate elevates, pushing against the back wall of the pharynx (figure 24.11). Elevation of the soft palate seals off the nasal cavity and prevents food from entering it. Pressure against the pharynx stimulates neurons within its walls, which send impulses to the swallowing center in the brain. In response, muscles are stimulated to contract and raise the *larynx* (voice box). This pushes the *glottis,* the opening from the larynx into the trachea (windpipe), against a flap of tissue called the *epiglottis.* These actions keep food out of the respiratory tract, directing it instead into the esophagus.

24.5 In many vertebrates, ingested food is fragmented through the tearing or grinding action of specialized teeth. In birds, this is accomplished through the grinding action of pebbles in the gizzard. Food mixed with saliva is swallowed and enters the esophagus.

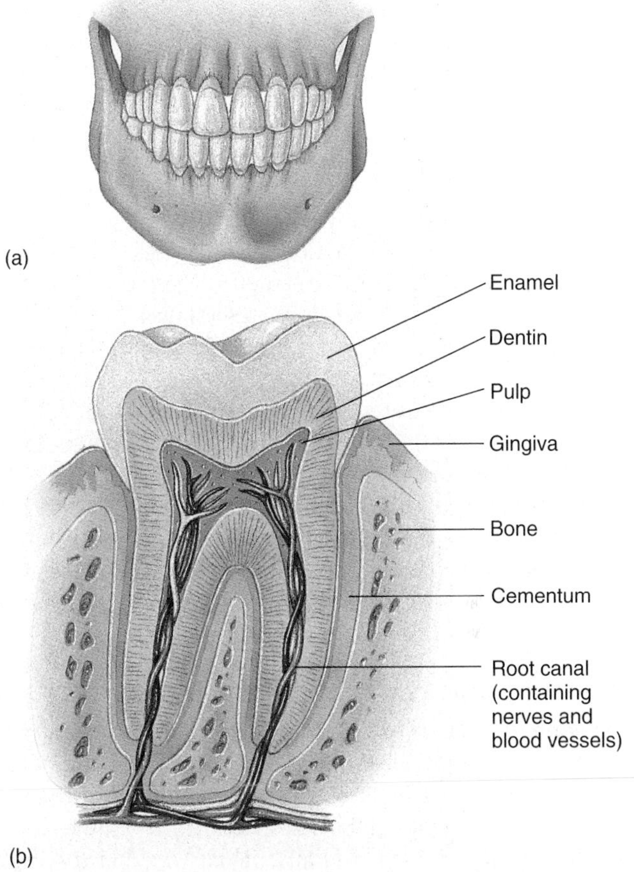

(a)
Enamel
Dentin
Pulp
Gingiva
Bone
Cementum
Root canal (containing nerves and blood vessels)
(b)

Figure 24.10 Human teeth.

(*a*) The front six teeth on the upper and lower jaws are cuspids and incisors. The remaining teeth, running along the sides of the mouth, are grinders called premolars and molars. Hence, humans are carnivores in the front of their mouths and herbivores in the back. (*b*) Each tooth is alive, with a central pulp containing nerves and blood vessels. The actual chewing surface is a hard enamel layered over the softer dentin, which forms the body of the tooth.

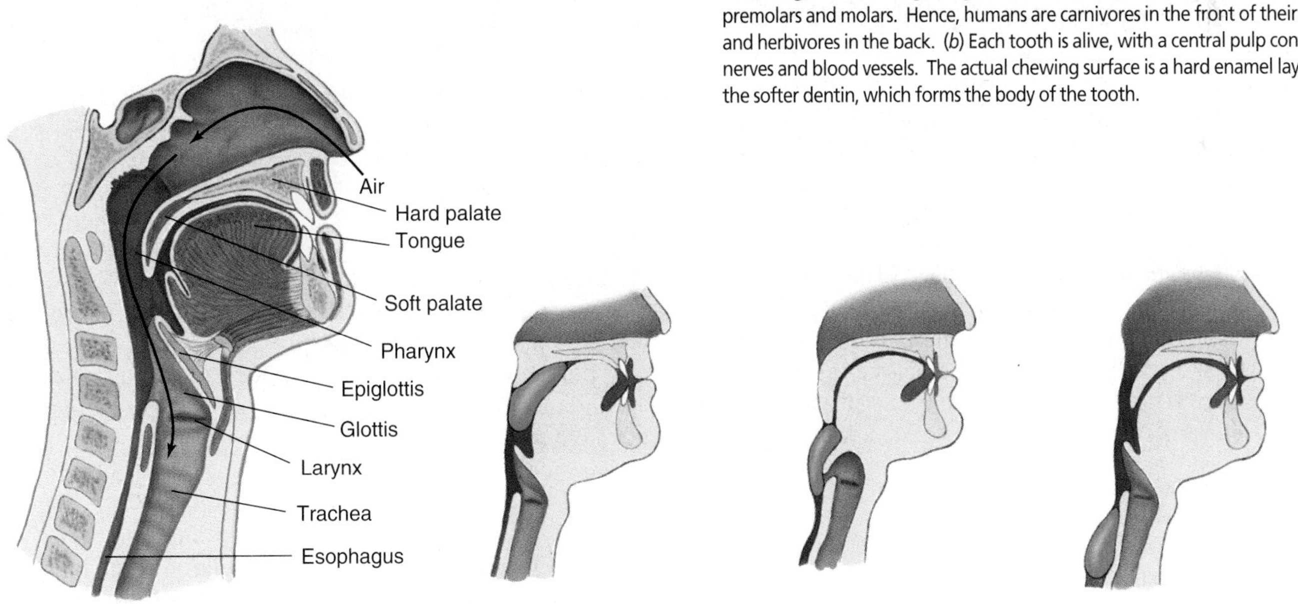

Air
Hard palate
Tongue
Soft palate
Pharynx
Epiglottis
Glottis
Larynx
Trachea
Esophagus

Figure 24.11 The human pharynx, palate, and larynx.

Food that enters the pharynx is prevented from entering the nasal cavity by elevation of the soft palate and from entering the larynx and trachea (the airways of the respiratory system) by elevation of the larynx against the epiglottis.

24.6 The Esophagus and Stomach

Structure and Function of the Esophagus

Swallowed food enters a muscular tube called the **esophagus,** which connects the pharynx to the stomach. In adult humans, the esophagus is about 25 centimeters long; the upper third is enveloped in skeletal muscle, for voluntary control of swallowing, while the lower two-thirds is surrounded by involuntary smooth muscle. The swallowing center stimulates successive waves of contraction in these muscles that move food along the esophagus to the stomach. These rhythmic waves of muscular contraction are called **peristalsis** (figure 24.12); they enable humans and other vertebrates to swallow even if they are upside down.

In many vertebrates, the movement of food from the esophagus into the stomach is controlled by a ring of circular smooth muscle, the **sphincter,** that opens in response to the pressure exerted by the food. Contraction of this sphincter prevents food in the stomach from moving back into the esophagus. Rodents and horses have a true sphincter at this site and thus cannot regurgitate, while humans lack a true sphincter and so are able to regurgitate.

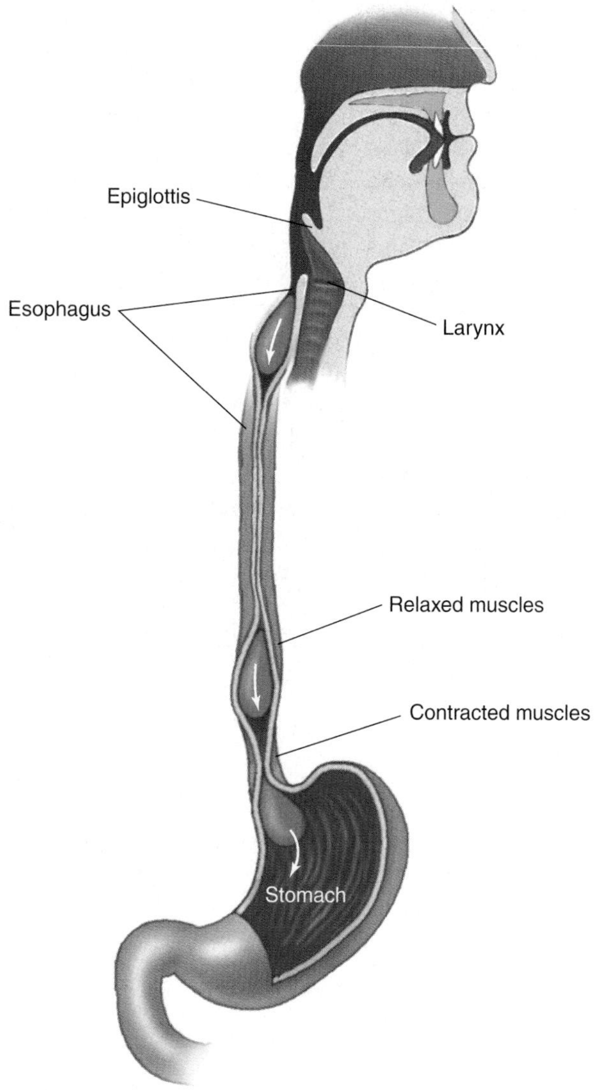
Epiglottis
Esophagus
Larynx
Relaxed muscles
Contracted muscles
Stomach

Figure 24.12 The esophagus and peristalsis.
After food has entered the esophagus, rhythmic waves of muscular contraction, called peristalsis, move the food down to the stomach.

Structure and Function of the Stomach

The **stomach** is a saclike portion of the digestive tract. Its inner surface is highly convoluted, enabling it to fold up when empty and open out like an expanding balloon as it fills with food. Thus, while the human stomach has a volume of only about 50 milliliters when empty, it may expand to contain 2 to 4 liters of food when full. Carnivores that engage in sporadic gorging as an important survival strategy possess stomachs that are able to distend much more than that.

The stomach contains an extra layer of smooth muscle for churning food and mixing it with *gastric juice,* an acidic secretion of the tubular gastric glands of the mucosa (figure 24.13). These exocrine glands contain two kinds of secretory cells: *parietal cells,* which secrete hydrochloric acid (HCl); and *chief cells,* which secrete pepsinogen, a weak protease (protein-digesting enzyme) that requires a very low pH to be active. This low pH is provided by the HCl. Activated pepsinogen molecules then cleave each other at specific sites, producing a much more active protease, pepsin. This process of secreting a relatively inactive enzyme that is then converted into a more active enzyme outside the cell prevents the chief cells from digesting themselves. It should be noted that only proteins are partially digested in the stomach—there is no significant digestion of carbohydrates or fats.

Action of Acid

The human stomach produces about 2 liters of HCl and other gastric secretions every day, creating a very acidic solution inside the stomach. The concentration of HCl in this solution is about 10 millimolar, corresponding to a pH of 2. Thus, gastric juice is about 250,000 times more acidic than blood, whose normal pH is 7.4. The low pH in the stomach helps denature food proteins, making them easier to digest, and keeps pepsin maximally active. Active pepsin hydrolyzes food proteins into shorter chains of polypeptides that are not fully digested until the mixture enters the small intestine. The mixture of partially digested food and gastric juice is called **chyme.**

The acidic solution within the stomach also kills most of the bacteria that are ingested with the food. The few bacteria that survive the stomach and enter the intestine intact are able to grow and multiply there, particularly in the large intestine. In fact, most vertebrates harbor thriving colonies of bacteria within their intestines, and bacteria are a major component of feces. As we discuss later, bacteria that live within the digestive tract of cows and other ruminants play a key role in the ability of these mammals to digest cellulose.

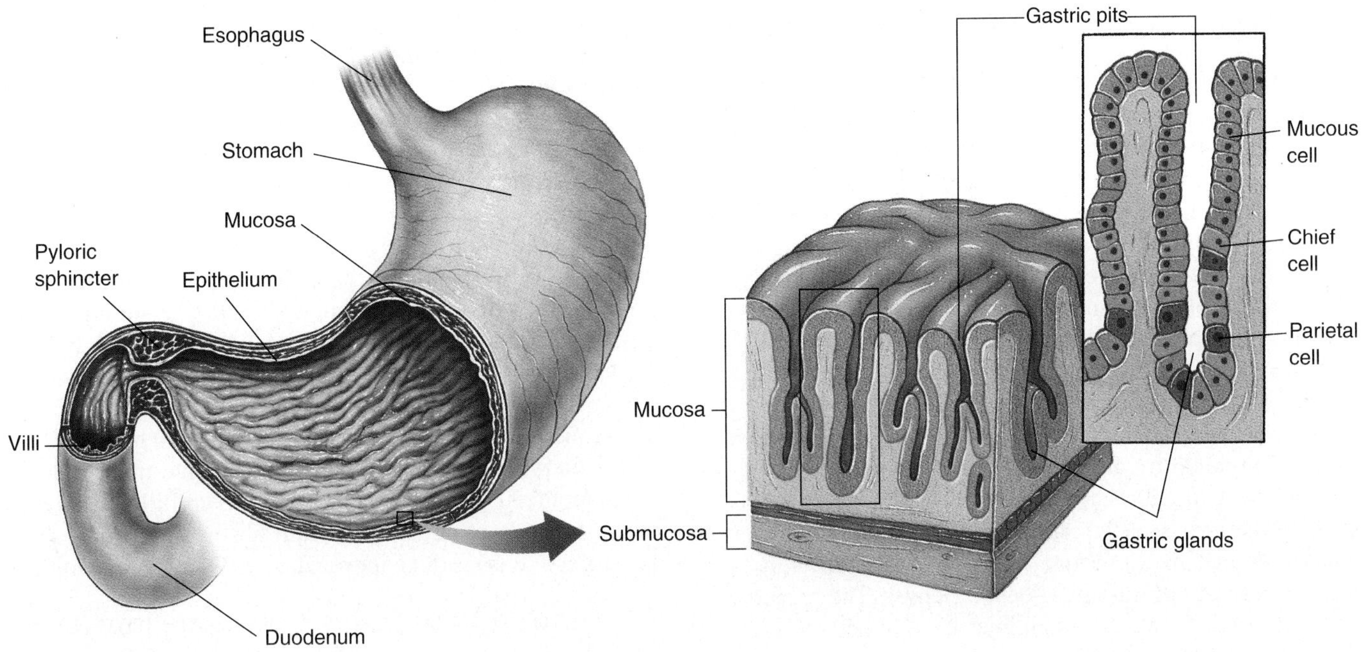

Esophagus
Stomach
Mucosa
Pyloric sphincter
Epithelium
Villi
Duodenum
Gastric pits
Mucous cell
Chief cell
Parietal cell
Mucosa
Submucosa
Gastric glands

Figure 24.13 The stomach and gastric glands.

Food enters the stomach from the esophagus. A band of smooth muscle called the pyloric sphincter controls the entrance to the duodenum, the upper part of the small intestine. The epithelial walls of the stomach are dotted with gastric pits, which contain glands that secrete hydrochloric acid and the enzyme pepsinogen. The gastric glands consist of mucous cells, chief cells that secrete pepsinogen, and parietal cells that secrete HCl. Gastric pits are the openings of the gastric glands.

Ulcers

It is important that the stomach not produce *too* much acid. If it did, the body could not neutralize the acid later in the small intestine, a step essential for the final stage of digestion. Production of acid is controlled by hormones. These hormones are produced by endocrine cells scattered within the walls of the stomach. The hormone **gastrin** regulates the synthesis of HCl by the parietal cells of the gastric pits, permitting HCl to be made only when the pH of the stomach is higher than about 1.5.

Overproduction of gastric acid can occasionally eat a hole through the wall of the stomach. Such gastric **ulcers** are rare, however, because epithelial cells in the mucosa of the stomach are protected somewhat by a layer of alkaline mucus, and because those cells are rapidly replaced by cell division if they become damaged (gastric epithelial cells are replaced every two to three days). Over 90% of gastrointestinal ulcers are duodenal ulcers. These may be produced when excessive amounts of acidic chyme are delivered into the duodenum, so that the acid cannot be properly neutralized through the action of alkaline pancreatic juice (described later). Susceptibility to ulcers is increased when the mucosal barriers to self-digestion are weakened by an infection of the bacterium *Helicobacter pylori.* Indeed, modern antibiotic treatments of this infection can reduce symptoms and often even cure the ulcer.

In addition to producing HCl, the parietal cells of the stomach also secrete intrinsic factor, a polypeptide needed for the intestinal absorption of vitamin B_{12}. Because this vitamin is required for the production of red blood cells, persons who lack sufficient intrinsic factor develop a type of anemia (low red blood cell count) called *pernicious anemia.*

Leaving the Stomach

Chyme leaves the stomach through the *pyloric sphincter* (see figure 24.13) to enter the small intestine. This is where all terminal digestion of carbohydrates, fats, and proteins occurs, and where the products of digestion—amino acids, glucose, and so on—are absorbed into the blood. Only some of the water in chyme and a few substances such as aspirin and alcohol are absorbed through the wall of the stomach.

24.6 Peristaltic waves of contraction propel food along the esophagus to the stomach. Gastric juice contains strong hydrochloric acid and the protein-digesting enzyme pepsin, which begins the digestion of proteins into shorter polypeptides. The acidic chyme is then transferred through the pyloric sphincter to the small intestine.

24.7 The Small and Large Intestines

Digestion and Absorption: The Small Intestine

The digestive tract exits from the stomach into the **small intestine,** where the breaking down of large molecules into small ones occurs. Only relatively small portions of food are introduced into the small intestine at one time, to allow time for acid to be neutralized and enzymes to act. The small intestine is the true digestive vat of the body. Within it, carbohydrates are broken down into sugars, proteins into amino acids, and fats into fatty acids. Once these small molecules have been produced, they pass across the epithelial wall of the small intestine into the bloodstream.

Some of the enzymes necessary for these digestive processes are secreted by the cells of the intestinal wall. Most, however, are made in a large gland called the *pancreas* (discussed in the next section), situated near the junction of the stomach and the small intestine. It is one of the body's major exocrine (secreting through ducts) glands. The pancreas sends its secretions into the small intestine through a duct that empties into its initial segment, the **duodenum.** Your small intestine is approximately 6 meters long—unwound and stood on its end, it would be far taller than you are! Only the first 25 centimeters, about 4% of the total length, is the duodenum. It is within this initial segment, where the pancreatic enzymes enter the small intestine, that digestion occurs.

Much of the food energy the vertebrate body harvests is obtained from fats. Before fats can be digested by enzymes, they must be made soluble in water. This process is carried out by a collection of molecules known as *bile salts* secreted into the duodenum from the *liver* (discussed in the next section). Bile salts solve a key problem of digestion, which is that fat is not soluble in water. How can the body use fats as food if fat is not soluble in the blood, which carries foods to the cells of the body? Bile salts solve this problem by acting as a superdetergent. They combine with fats to form microscopic droplets in a process called emulsification. These tiny droplets remain suspended in water practically indefinitely.

All the rest of the small intestine (96% of its length) is called the **ileum.** The ileum is devoted to absorbing water and the products of digestion into the bloodstream. The lining of the small intestine is covered with fine fingerlike projections called **villi** (singular, **villus**), each too small to see with the naked eye (figure 24.14*a*). In turn, each of the cells covering a villus is covered on its outer surface by a field of cytoplasmic projections called **microvilli** (figure 24.14*b* and *c*). Both kinds of projections greatly increase the absorptive surface of the lining of the small intestine. The average surface area of the small intestine of an adult human is about 300 square meters, more than the surface of many swimming pools!

The amount of material passing through the small intestine is startlingly large. Per day, an average human consumes about 800 grams of solid food, and 1,200 milliliters of water, for a total volume of about 2 liters. To this amount is added about 1.5 liters of fluid from the salivary glands, 2 liters from the gastric secretions of the stomach, 1.5 liters from the pancreas, 0.5 liters from the liver, and 1.5 liters of intestinal secretions. The total adds up to a remarkable 9 liters—more than 10% of the total volume of your body! However, although the flux is great, the *net* passage is small. Almost all these fluids and solids are reabsorbed during their passage through the small intestine—about 8.5 liters across the walls of the small intestine and 0.35 liters across the wall of the large intestine. Of the 800 grams of solid and 9 liters of liquid that enter the digestive tract each day, only about 50 grams of solid and 100 milliliters of liquid leave the body as feces. The fluid absorption efficiency of the digestive tract thus approaches 99%, very high indeed.

Concentration of Solids: The Large Intestine

The **large intestine,** or **colon,** is much shorter than the small intestine, approximately 1 meter long. No digestion takes place within the large intestine, and only about 4% of fluid absorption occurs there. The large intestine is not convoluted, lying instead in three relatively straight segments, and its inner surface does not possess villi. As a consequence, the large intestine has only one-thirtieth the absorptive surface area of the small intestine. Although sodium and vitamin K are absorbed across its walls, the primary function of the large intestine is to act as a refuse dump. Within it, undigested material, including large amounts of plant fiber and cellulose, is compacted and stored. Many bacteria live and actively divide within the large intestine, where they play a role in the processing of undigested material into the final excretory product, **feces.**

The final segment of the digestive tract is a short extension of the large intestine called the **rectum.** Compact solids within the colon pass into the rectum as a result of the peristaltic contractions of the muscles encasing the large intestine. From the rectum, the solid material passes out of the body through the **anus.**

24.7 Most digestion occurs in the initial upper portion of the small intestine, called the duodenum. The rest of the small intestine is devoted to absorption of water and the products of digestion. The large intestine compacts residual solid wastes.

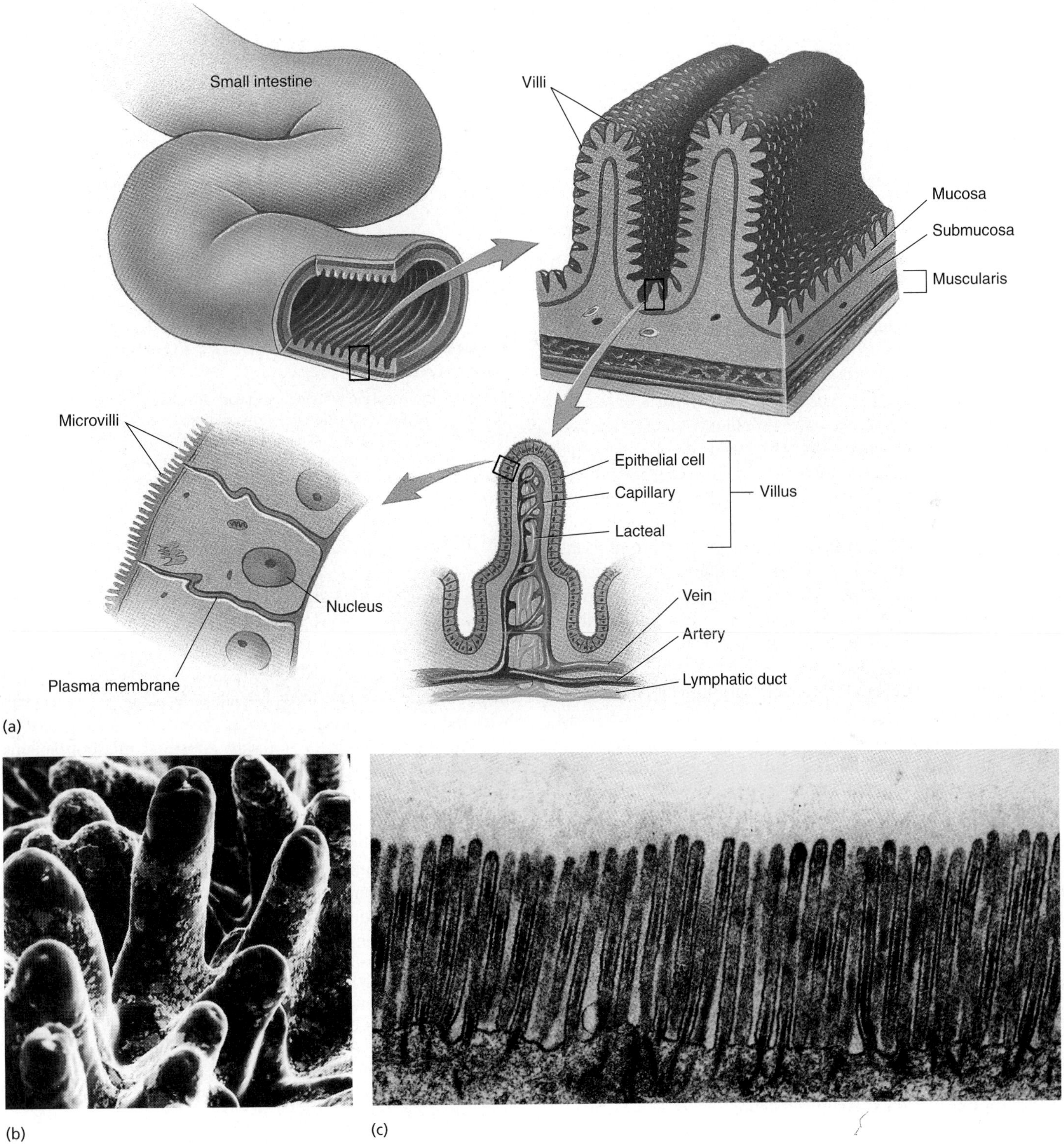
Small intestine
Villi
Mucosa
Submucosa
Muscularis
Microvilli
Epithelial cell
Capillary
Lacteal
Villus
Nucleus
Vein
Artery
Lymphatic duct
Plasma membrane
(a)
(b)
(c)

Figure 24.14 The small intestine.

(*a*) Cross section of the small intestine with details showing villi structure. (*b*) Microvilli, shown in a scanning electron micrograph, are very densely clustered, giving the small intestine an enormous surface area, which is very important for efficient absorption. (*c*) Intestinal microvilli as shown in a transmission electron micrograph.

24.8 Variations in Vertebrate Digestive Systems

Most animals lack the enzymes necessary to digest cellulose, the carbohydrate that functions as the chief structural component of plants. The digestive tracts of some animals, however, contain bacteria and protists that convert cellulose into substances the host can digest. Although digestion by gastrointestinal microorganisms plays a relatively small role in human nutrition, it is an essential element in the nutrition of many other kinds of animals, including insects like termites and cockroaches and a few groups of herbivorous mammals. The relationships between these microorganisms and their animal hosts are mutually beneficial and provide an excellent example of symbiosis.

Cows, deer, and other ruminants have large, divided stomachs (figure 24.15). The first portion consists of the rumen and a smaller chamber, the reticulum; the second portion consists of two additional chambers, the omasum and abomasum. The rumen, which may hold up to 50 gallons, serves as a fermentation vat in which bacteria and protozoa convert cellulose and other molecules into a variety of simpler compounds. The location of the rumen at the front of the four chambers is important because it allows the animal to regurgitate and rechew the contents of the rumen, an activity called *rumination*, or "chewing the cud." The cud is then swallowed and enters the reticulum, from which it passes to the omasum and then the abomasum, where it is finally mixed with gastric juice. Hence, only the abomasum is equivalent to the human stomach in its function. This process leads to a far more efficient digestion of cellulose in ruminants than in mammals that lack a rumen, such as horses.

In horses, rodents, and lagomorphs (rabbits and hares), the digestion of cellulose by microorganisms takes place in the cecum, which is greatly enlarged (figure 24.16). Because the cecum is located beyond the stomach, regurgitation of its contents is impossible. However, rodents and lagomorphs have evolved another way to digest cellulose that achieves a degree of efficiency similar to that of ruminant digestion. They do this by eating their feces, thus passing the food through their digestive tract a second time. The second passage makes it possible for the animal to absorb the nutrients produced by the microorganisms in its cecum. Animals that engage in this practice of **coprophagy** (from the Greek words *copros*, excrement, and *phagein*, eat) cannot remain healthy if they are prevented from eating their feces.

Cellulose is not the only plant product that vertebrates can use as a food source because of the digestive activities of intestinal microorganisms. Wax, a substance indigestible by most terrestrial animals, is digested by symbiotic bacteria living in the gut of honeyguides, African birds that eat the wax in bee nests. In the marine food chain, wax is a major constituent of copepods (crustaceans in the plankton), and many marine fish and birds appear to be able to digest wax with the aid of symbiotic microorganisms.

Another example of the way intestinal microorganisms function in the metabolism of their animal hosts is provided by the synthesis of vitamin K. All mammals rely on intestinal bacteria to synthesize this vitamin, which is necessary for the clotting of blood. Birds, which lack these bacteria, must consume the required quantities of vitamin K in their food. In humans, prolonged treatment with antibiotics greatly reduces the populations of bacteria in the intestine; under such circumstances, it may be necessary to provide supplementary vitamin K.

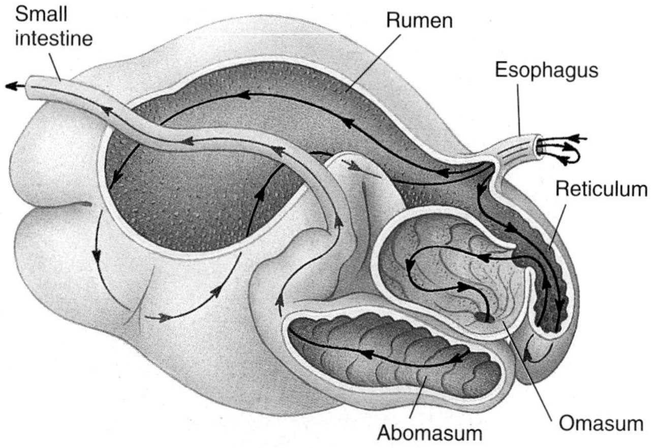
Small intestine
Rumen
Esophagus
Reticulum
Abomasum
Omasum

Figure 24.15 Four-chambered stomach of a ruminant.

The grass and other plants that a ruminant, such as a cow, eats enter the rumen, where they are partially digested. Before moving into a second chamber, the reticulum, the food may be regurgitated and rechewed. The food is then transferred to the rear two chambers, the omasum and abomasum. Only the abomasum is equivalent to the human stomach in its function of secreting gastric juice.

> **24.8** Much of the food value of plants is tied up in cellulose, and the digestive tract of many animals harbors colonies of cellulose-digesting microorganisms. Intestinal microorganisms also produce molecules such as vitamin K that are important to the well-being of their vertebrate hosts.

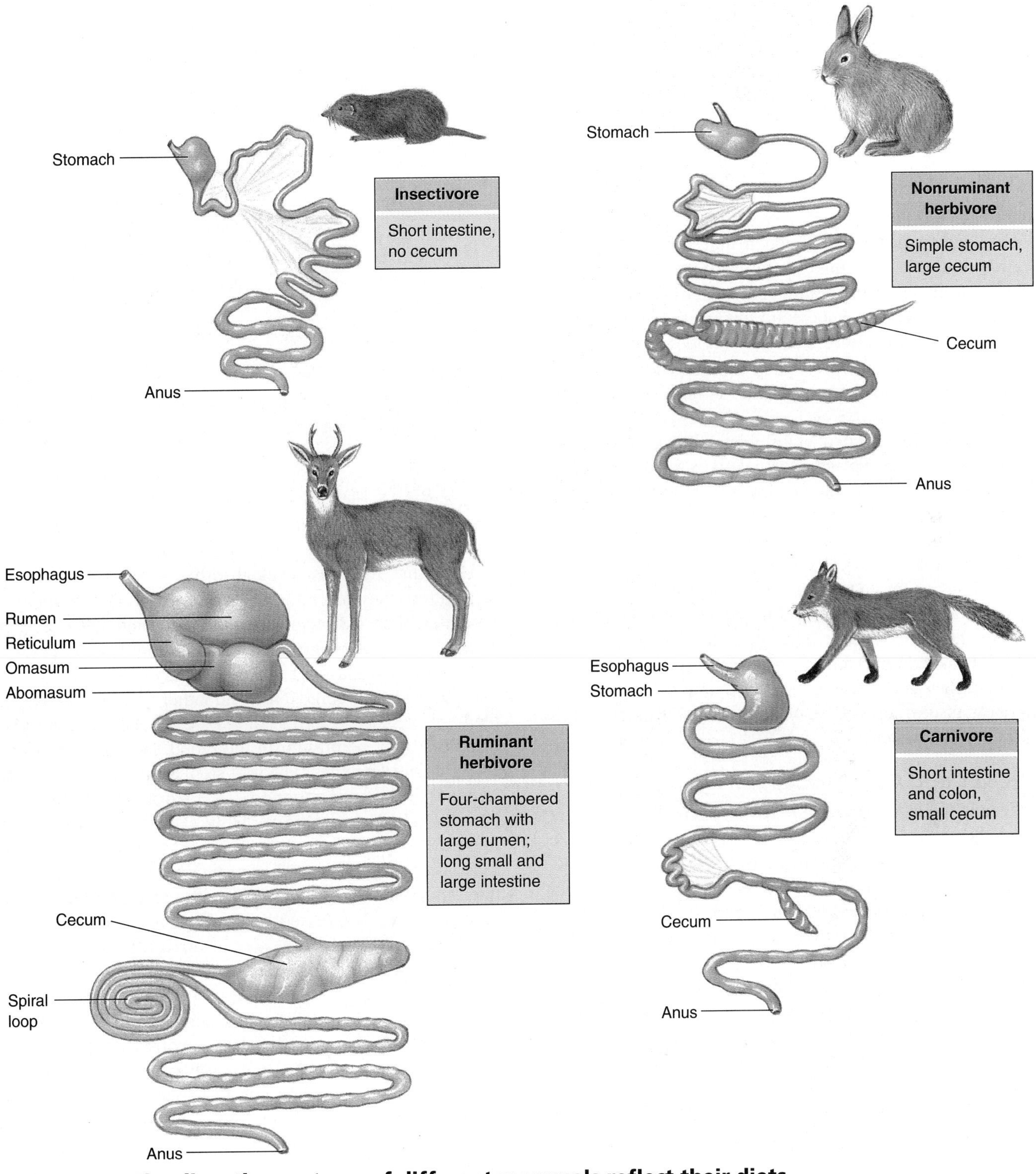
Stomach
Insectivore
Short intestine, no cecum
Anus
Stomach
Nonruminant herbivore
Simple stomach, large cecum
Cecum
Anus
Esophagus
Rumen
Reticulum
Omasum
Abomasum
Ruminant herbivore
Four-chambered stomach with large rumen; long small and large intestine
Cecum
Spiral loop
Anus
Esophagus
Stomach
Carnivore
Short intestine and colon, small cecum
Cecum
Anus

Figure 24.16 The digestive systems of different mammals reflect their diets.
Herbivores require long digestive tracts with specialized compartments for the breakdown of plant matter. Protein diets are more easily digested; thus, insectivorous and carnivorous mammals have short digestive tracts with few specialized pouches.

24.9 Accessory Digestive Organs

The Pancreas

The **pancreas,** a large gland situated near the junction of the stomach and the small intestine, is one of the accessory organs that contribute secretions to the digestive tract. Fluid from the pancreas is secreted into the duodenum through the *pancreatic duct* (figure 24.17). This fluid contains a host of enzymes, including trypsin and chymotrypsin, which digest proteins; pancreatic amylase, which digests starch; and lipase, which digests fats. Inactive forms of these enzymes are released into the duodenum and are then activated by the enzymes of the intestine. Pancreatic enzymes digest proteins into smaller polypeptides, polysaccharides into shorter chains of sugars, and fat into free fatty acids and other products. The digestion of these molecules is then completed by the intestinal enzymes.

Pancreatic fluid also contains bicarbonate, which neutralizes the HCl from the stomach and gives the chyme in the duodenum a slightly alkaline pH. The digestive enzymes and bicarbonate are produced by clusters of secretory cells known as *acini.*

In addition to its exocrine role in digestion, the pancreas also functions as an endocrine gland, secreting several hormones into the blood that control the blood levels of glucose and other nutrients. These hormones are produced in the **islets of Langerhans,** clusters of endocrine cells scattered throughout the pancreas. The two most important pancreatic hormones, insulin and glucagon, are discussed later in this chapter and in chapter 27.

The Liver and Gallbladder

The **liver** is the largest internal organ of the body. In an adult human, the liver weighs about 1.5 kilograms and is the size of a football. The main exocrine secretion of the liver is **bile,** a fluid mixture consisting of *bile pigments* and *bile salts* that is delivered into the duodenum during the digestion of a meal.

The bile salts play a very important role in the digestion of fats. Because fats are insoluble in water, they enter the intestine as drops within the watery chyme. The bile salts, which are partly fat-soluble and partly water-soluble, work like detergents, dispersing the large drops of fat into a fine suspension of smaller droplets. This emulsification process produces a greater surface area of fat upon which the lipase enzymes can act, and thus allows the digestion of fat to proceed more rapidly.

After it is produced in the liver, bile is stored and concentrated in the **gallbladder.** The arrival of fatty food in the duodenum triggers a neural and endocrine reflex that stimulates the gallbladder to contract, causing bile to be transported through the common bile duct and injected into the duodenum. If the bile duct is blocked by a *gallstone* (formed from a hardened precipitate of cholesterol), contraction of the gallbladder causes pain that is generally felt under the right scapula (shoulder blade).

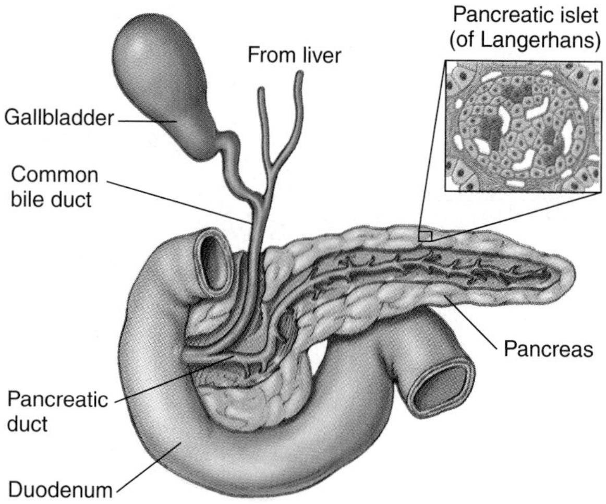

Pancreatic islet (of Langerhans)
From liver
Gallbladder
Common bile duct
Pancreas
Pancreatic duct
Duodenum

Figure 24.17 The pancreatic and bile ducts empty into the duodenum.

Regulatory Functions of the Liver

Because a large vein carries blood from the stomach and intestine directly to the liver, the liver is in a position to chemically modify the substances absorbed in the gastrointestinal tract before they reach the rest of the body. For example, ingested alcohol and other drugs are taken into liver cells and metabolized; this is why the liver is often damaged as a result of alcohol and drug abuse. The liver also removes toxins, pesticides, carcinogens, and other poisons, converting them into less toxic forms (figure 24.18). Also, excess amino acids that may be present in the blood are converted to glucose by liver enzymes. The first step in this conversion is the removal of the amino group ($—NH_2$) from the amino acid, a process called *deamination.* Unlike plants, animals cannot reuse the nitrogen from these amino groups and must excrete it as nitrogenous waste. The product of amino acid deamination, ammonia (NH_3), combines with carbon dioxide to form urea. The urea is released by the liver into the bloodstream, where—as you will learn later in this chapter—the kidneys subsequently remove it.

The liver also produces most of the proteins found in blood plasma. The total concentration of plasma proteins is significant because it must be kept within normal limits in order to maintain osmotic balance between blood and interstitial (tissue) fluid. If the concentration of plasma proteins drops too low, as can happen as a result of liver disease such as cirrhosis, fluid accumulates in the tissues; this condition is called *edema.*

24.9 The pancreas secretes digestive enzymes and bicarbonate into the pancreatic duct. The liver produces bile, which is stored and concentrated in the gallbladder. The liver and the pancreatic hormones regulate blood glucose concentration.

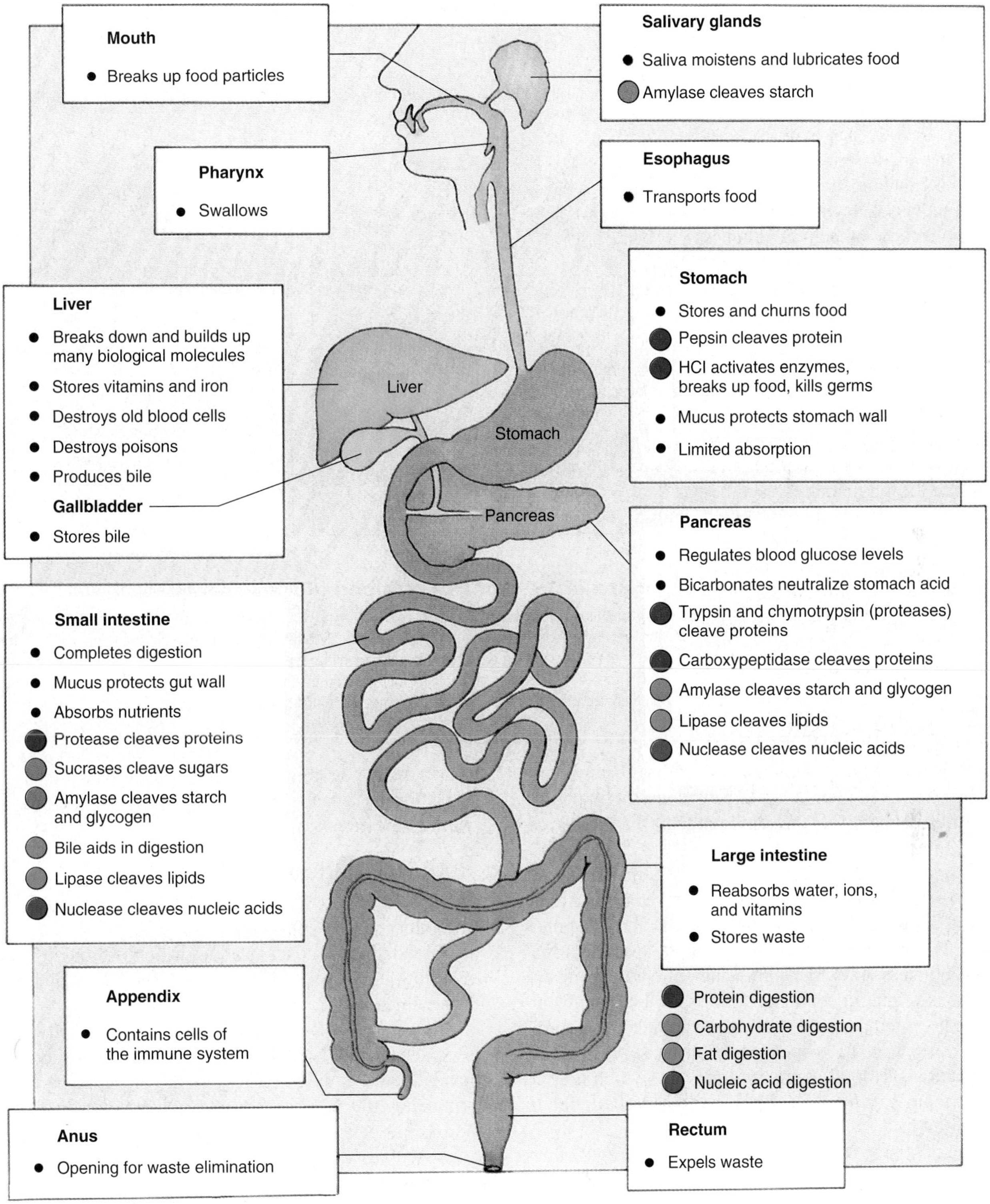

Mouth
• Breaks up food particles
Salivary glands
• Saliva moistens and lubricates food
Amylase cleaves starch
Pharynx
• Swallows
Esophagus
• Transports food
Stomach
• Stores and churns food
Pepsin cleaves protein
HCl activates enzymes, breaks up food, kills germs
• Mucus protects stomach wall
• Limited absorption
Liver
• Breaks down and builds up many biological molecules
• Stores vitamins and iron
• Destroys old blood cells
• Destroys poisons
• Produces bile
Gallbladder
• Stores bile
Liver
Stomach
Pancreas
Pancreas
• Regulates blood glucose levels
• Bicarbonates neutralize stomach acid
Trypsin and chymotrypsin (proteases) cleave proteins
Carboxypeptidase cleaves proteins
Amylase cleaves starch and glycogen
Lipase cleaves lipids
Nuclease cleaves nucleic acids
Small intestine
• Completes digestion
• Mucus protects gut wall
• Absorbs nutrients
Protease cleaves proteins
Sucrases cleave sugars
Amylase cleaves starch and glycogen
Bile aids in digestion
Lipase cleaves lipids
Nuclease cleaves nucleic acids
Large intestine
• Reabsorbs water, ions, and vitamins
• Stores waste
Protein digestion
Carbohydrate digestion
Fat digestion
Nucleic acid digestion
Appendix
• Contains cells of the immune system
Anus
• Opening for waste elimination
Rectum
• Expels waste

Figure 24.18 The organs of the digestive system and their functions.
The digestive system contains some dozen different organs that act on the food that is consumed, starting with the mouth and ending with the anus. All of these organs must work properly in order for the body to effectively obtain nutrition.

24.10 Homeostasis

As the animal body has evolved, specialization has increased. Each cell is a sophisticated machine, finely tuned to carry out a precise role within the body. Such specialization of cell function is possible only when extracellular conditions are kept within narrow limits. Temperature, pH, the concentrations of glucose and oxygen, and many other factors must be held fairly constant for cells to function efficiently and interact properly with one another.

Homeostasis may be defined as the dynamic constancy of the internal environment. The term dynamic is used because conditions are never absolutely constant, but fluctuate continuously within narrow limits. Homeostasis is essential for life, and most of the regulatory mechanisms of the vertebrate body that are not devoted to reproduction are concerned with maintaining homeostasis.

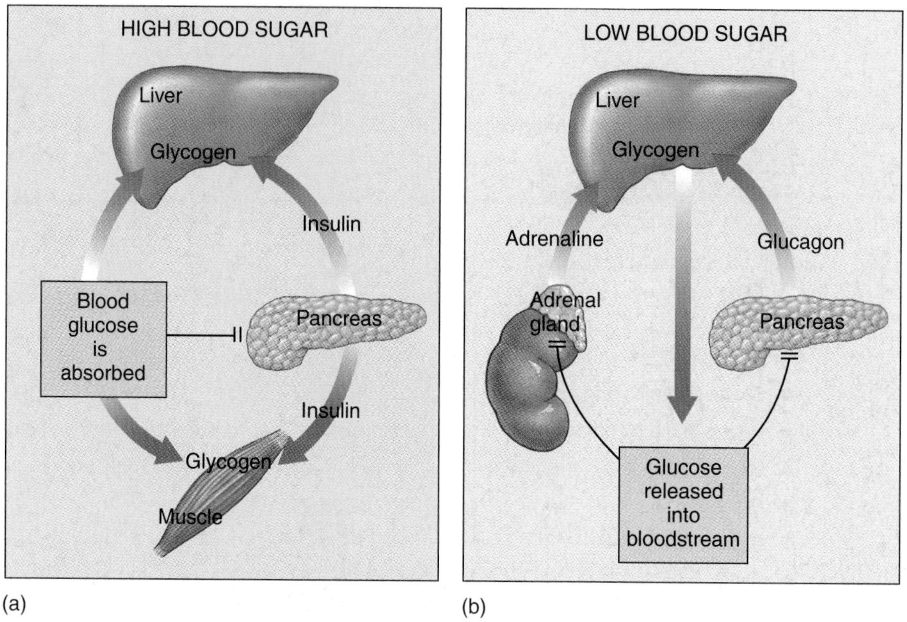
HIGH BLOOD SUGAR
Liver
Glycogen
Insulin
Blood glucose is absorbed
Pancreas
Insulin
Glycogen
Muscle
LOW BLOOD SUGAR
Liver
Glycogen
Adrenaline
Glucagon
Adrenal gland
Pancreas
Glucose released into bloodstream
(a)
(b)

Figure 24.19 Control of blood glucose levels.
(*a*) When blood glucose levels are high, cells within the pancreas produce the hormone insulin, which stimulates the liver and muscles to convert blood glucose into glycogen. (*b*) When blood glucose levels are low, other cells within the pancreas release the hormone glucagon into the bloodstream; in addition, cells within the adrenal gland release the hormone adrenaline into the bloodstream. When they reach the liver, these two hormones act to increase the liver's breakdown of glycogen to glucose.

Regulating Body Temperature

Humans, together with other mammals and with birds, are endothermic; they can maintain relatively constant body temperatures independent of the environmental temperature. When the temperature of your blood exceeds 37°C (98.6°F), neurons in a part of the brain called the hypothalamus (see chapters 26 and 27) detect the temperature change. Acting through the control of neurons, the hypothalamus responds by promoting the dissipation of heat through sweating, dilation of blood vessels in the skin, and other mechanisms. These responses tend to counteract the rise in body temperature. When body temperature falls, the hypothalamus coordinates a different set of responses, such as shivering and the constriction of blood vessels in the skin, which help to raise body temperature and correct the initial challenge to homeostasis.

Vertebrates other than mammals and birds are ectothermic; their body temperatures are more or less dependent on the environmental temperature. However, to the extent that it is possible, many ectothermic vertebrates attempt to maintain some degree of temperature homeostasis. Certain large fish, including tuna, swordfish, and some sharks, for example, can maintain parts of their body at a significantly higher temperature than that of the water. Reptiles attempt to maintain a constant body temperature through behavioral means—by placing themselves in varying locations of sun and shade. That's why you frequently see lizards basking in the sun. Sick lizards even give themselves a "fever" by seeking warmer locations!

Regulating Blood Glucose

When you digest a carbohydrate-containing meal, you absorb glucose into your blood. This causes a temporary rise in the blood glucose concentration, which is brought back down in a few hours. What counteracts the rise in blood glucose following a meal?

Glucose levels within the blood are constantly monitored by a sensor, the islets of Langerhans in the pancreas. When levels increase, the islets secrete the hormone insulin, which stimulates the uptake of blood glucose into muscles, liver, and adipose tissue. The muscles and liver can convert the glucose into the polysaccharide glycogen (figure 24.19); adipose cells can convert glucose into fat. These actions lower the blood glucose and help to store energy in forms that the body can use later. When blood glucose levels decrease, as they do between meals, during periods of fasting, and during exercise, the liver secretes glucose into the blood. This glucose is obtained in part from the breakdown of liver glycogen. This breakdown of liver glycogen is stimulated by another hormone, *glucagon,* which is also secreted by the islets of Langerhans.

Eliminating Nitrogenous Wastes

Amino acids and nucleic acids are nitrogen-containing molecules. When animals catabolize these molecules for energy or convert them into carbohydrates or lipids, they produce nitrogen-containing by-products called **nitrogenous wastes** (figure 24.20) that must be eliminated from the body.

The first step in the metabolism of amino acids and nucleic acids is the removal of the amino (—NH_2) group and its combination with H^+ to form **ammonia** (NH_3) in the liver. Ammonia is quite toxic to cells and therefore is safe only in very dilute concentrations. The excretion of ammonia is not a problem for the bony fish and tadpoles, which eliminate most of it by diffusion through the gills and less by excretion in very dilute urine. In sharks, adult amphibians, and mammals, the nitrogenous wastes are eliminated in the far less toxic form of **urea.** Urea is water-soluble and so can be excreted in large amounts in the urine. It is carried in the bloodstream from its place of synthesis in the liver to the kidneys where it is excreted in the urine.

Reptiles, birds, and insects excrete nitrogenous wastes in the form of **uric acid,** which is only slightly soluble in water. As a result of its low solubility, uric acid precipitates and thus can be excreted using very little water. Uric acid forms the pasty white material in bird droppings called *guano.* The ability to synthesize uric acid in these groups of animals is also important because their eggs are encased within shells, and nitrogenous wastes build up as the embryo grows within the egg. The formation of uric acid, while a lengthy process that requires considerable energy, produces a compound that crystallizes and precipitates. As a precipitate, it is unable to affect the embryo's development even though it is still inside the egg.

Mammals also produce some uric acid, but it is a waste product of the degradation of purine nucleotides (see chapter 3), not of amino acids. Most mammals have an enzyme called *uricase,* which converts uric acid into a more soluble derivative, **allantoin.** Only humans, apes, and the dalmatian dog lack this enzyme and so must excrete the uric acid. In humans, excessive accumulation of uric acid in the joints produces a condition known as *gout.*

24.10 Animals tend to maintain a relatively constant internal environment, a condition called homeostasis. Body temperature, glucose levels, and nitrogenous waste levels are all carefully regulated.

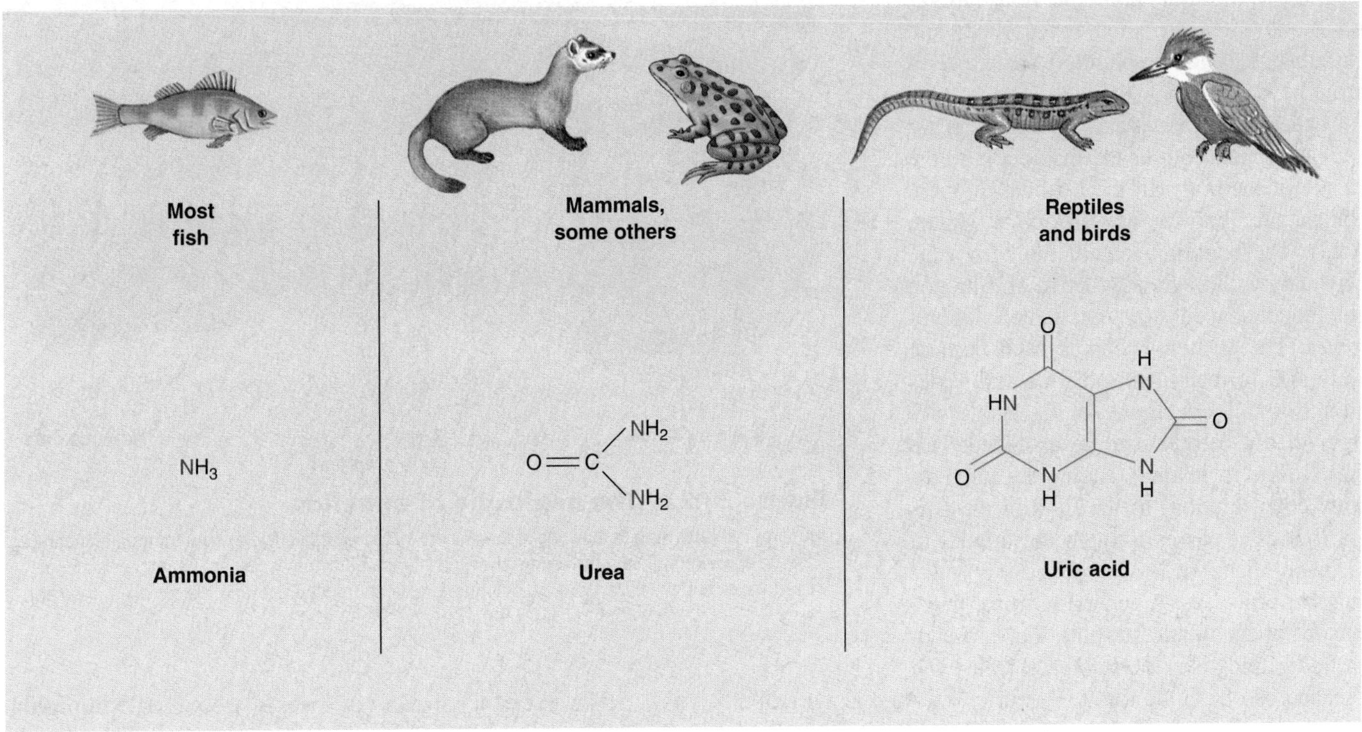
Most fish
Mammals, some others
Reptiles and birds
NH3
Ammonia
O=C NH2 NH2
Urea
HN O N H O N H O N H
Uric acid

Figure 24.20 Nitrogenous wastes.

When amino acids and nucleic acids are metabolized, the immediate by-product is ammonia, which is quite toxic but which can be eliminated through the gills of bony fish. Mammals convert ammonia into urea, which is less toxic. Birds and terrestrial reptiles convert it instead into uric acid, which is insoluble in water.

24.11 Osmoregulatory Organs

Animals must also carefully monitor the water content of their bodies. The first animals evolved in seawater, and the physiology of all animals reflects this origin. Approximately two-thirds of every vertebrate's body is water. If the amount of water in the body of a vertebrate falls much lower than this, the animal dies. Animals use various mechanisms for **osmoregulation,** the regulation of the body's osmotic composition, or how much water and salt it contains. The proper operation of many vertebrate organ systems of the body requires that the osmotic concentration of the blood—the concentration of solutes dissolved within it—be kept within narrow bounds.

Animals have evolved a variety of mechanisms to cope with problems of water balance. In many animals, the removal of water or salts from the body is coupled with the removal of metabolic wastes through the excretory system. Protists employ contractile vacuoles for this purpose, as do sponges. Other multicellular animals have a system of excretory tubules (little tubes) that expel fluid and wastes from the body.

In flatworms, these tubules are called *protonephridia,* and they branch throughout the body into bulblike **flame cells** (figure 24.21). While these simple excretory structures open to the outside of the body, they do not open to the inside of the body. Rather, cilia within the flame cells must draw in fluid from the body. Water and metabolites are then reabsorbed, and the substances to be excreted are expelled through excretory pores.

Other invertebrates have a system of tubules that open both to the inside and to the outside of the body. In the earthworm, these tubules are known as *nephridia* (figure 24.22). The nephridia obtain fluid from the body cavity through a process of filtration into funnel-shaped structures called *nephrostomes*. The term *filtration* is used because the fluid is formed under pressure and passes through small openings, so that molecules larger than a certain size are excluded. This filtered fluid is isotonic (having the same osmotic concentration) to the fluid in the coelom, but as it passes through the tubules of the nephridia, NaCl is removed by active transport processes. A general term for transport out of the tubule and into the surrounding body fluids is *reabsorption*. Because salt is reabsorbed from the filtrate, the urine excreted is more dilute than the body fluids (is hypotonic). The kidneys of mollusks and the excretory organs of crustaceans (called *antennal glands*) also produce urine by filtration and reclaim certain ions by reabsorption.

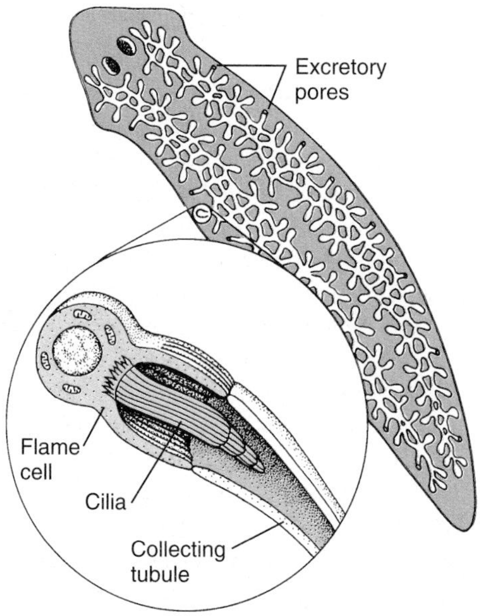
Excretory pores
Flame cell
Cilia
Collecting tubule

Figure 24.21 The protonephridia of flatworms.
A branching system of tubules, bulblike flame cells, and excretory pores make up the protonephridia of flatworms. Cilia inside the flame cells draw in fluids from the body by their beating action. Substances are then expelled through pores that open to the outside of the body.

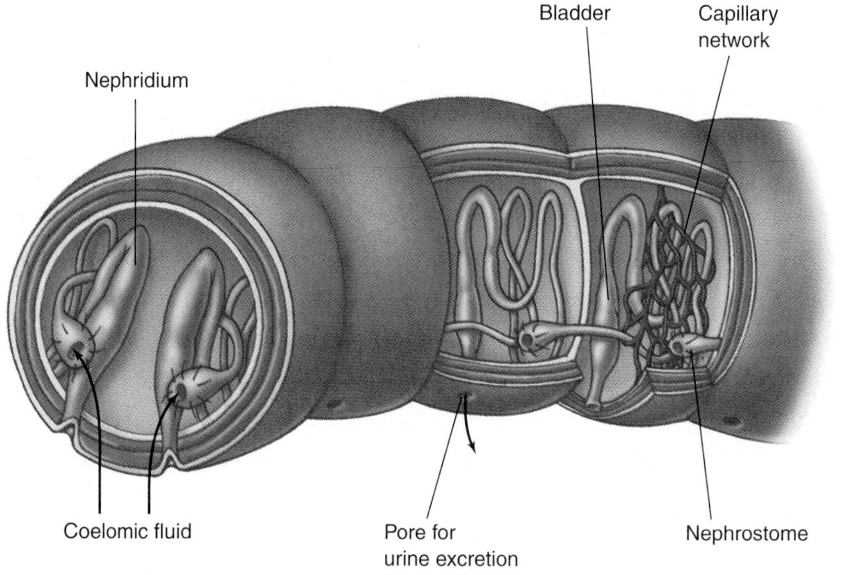
Bladder
Capillary network
Nephridium
Coelomic fluid
Pore for urine excretion
Nephrostome

Figure 24.22 The nephridia of annelids.
Most invertebrates, such as the annelid shown here, have nephridia. These consist of tubules that receive a filtrate of coelomic fluid, which enters the funnel-like nephrostomes. Salt can be reabsorbed from these tubules, and the fluid that remains, urine, is released from pores into the external environment.

The excretory organs in insects are called **Malpighian tubules** (figure 24.23), extensions of the digestive tract that branch off anterior to the hindgut. Urine is not formed by filtration in these tubules, because there is no pressure difference between the blood in the body cavity and the tubule.

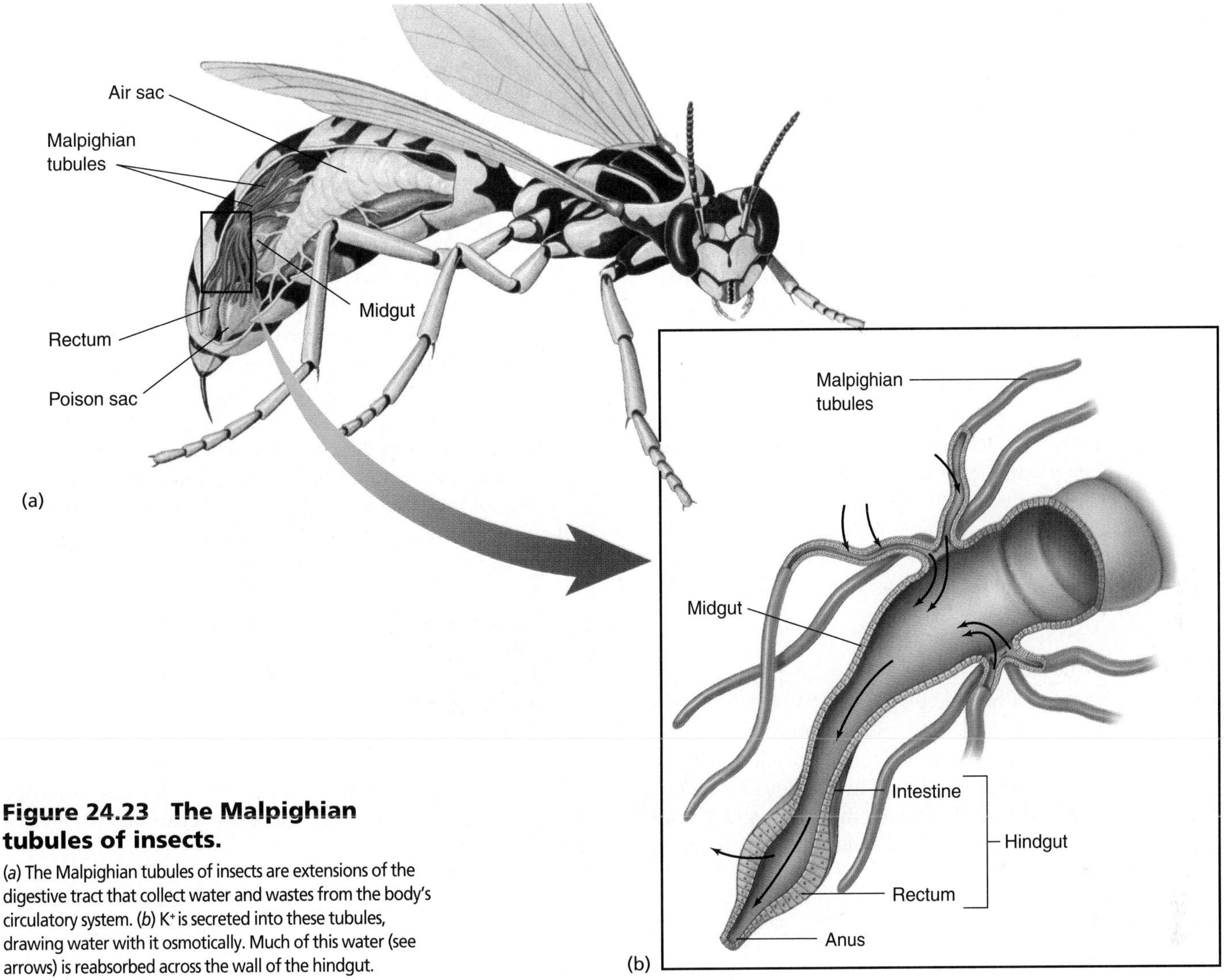
Air sac
Malpighian tubules
Midgut
Rectum
Poison sac
(a)
Malpighian tubules
Midgut
Intestine
Hindgut
Rectum
Anus
(b)

Figure 24.23 The Malpighian tubules of insects.

(*a*) The Malpighian tubules of insects are extensions of the digestive tract that collect water and wastes from the body's circulatory system. (*b*) K^+ is secreted into these tubules, drawing water with it osmotically. Much of this water (see arrows) is reabsorbed across the wall of the hindgut.

Instead, waste molecules and potassium ions (K^+) are secreted into the tubules by active transport. *Secretion* is the opposite of reabsorption—ions or molecules are transported from the body fluid into the tubule. The secretion of K^+ creates an osmotic gradient that causes water to enter the tubules by osmosis from the body's open circulatory system. Most of the water and K^+ is then reabsorbed into the circulatory system through the epithelium of the hindgut, leaving only small molecules and waste products to be excreted from the rectum along with feces. Malpighian tubules thus provide a very efficient means of water conservation.

Kidneys are the excretory organs in vertebrates, and unlike the Malpighian tubules of insects, kidneys create a tubular fluid by filtration of the blood under pressure. In addition to containing waste products and water, the filtrate contains many small molecules that are of value to the animal, including glucose, amino acids, and vitamins. These molecules and most of the water are reabsorbed from the tubules into the blood, while wastes remain in the filtrate. Additional wastes may be secreted by the tubules and added to the filtrate, and the final waste product, urine, is eliminated from the body.

It may seem odd that the vertebrate kidney should filter out almost everything from blood plasma (except proteins, which are too large to be filtered) and then spend energy to take back or reabsorb what the body needs. But selective reabsorption provides great flexibility; various vertebrate groups have evolved the ability to reabsorb different molecules that are especially valuable in particular habitats. This flexibility is a key factor underlying the successful colonization of many diverse environments by the vertebrates.

24.11 Many invertebrates filter fluid into a system of tubules and then reabsorb ions and water, leaving waste products for excretion. Insects create an excretory fluid by secreting K^+ into tubules, which draws water osmotically. The vertebrate kidney produces a filtrate that enters tubules and is modified to become urine.

24.12 Evolution of the Vertebrate Kidney

The kidney is a complex organ made up of thousands of repeating disposal units called **nephrons,** each with the structure of a bent tube (figure 24.24). Blood pressure forces the fluid in blood past a filter, called the *glomerulus,* at the top of each nephron. The glomerulus retains blood cells, proteins, and other useful large molecules in the blood but allows the water, and the small molecules and wastes dissolved in it, to pass through and into the bent tube part of the nephron. As the filtered fluid passes through the nephron tube, useful sugars and ions are recovered from it by active transport, leaving the water and metabolic wastes dissolved in it behind in a fluid urine.

Although the same basic design has been retained in all vertebrate kidneys, there have been a few modifications. Because the original glomerular filtrate is isotonic to blood, all vertebrates can produce a urine that is isotonic to (by reabsorbing ions) or hypotonic to (more dilute than) blood. Only birds and mammals can reabsorb water from their glomerular filtrate to produce a urine that is hypertonic to (more concentrated than) blood.

Freshwater Fish

Kidneys are thought to have evolved first among the freshwater teleosts, or bony fish. Because the body fluids of a freshwater fish have a greater osmotic concentration than the surrounding water, these animals face two serious problems: (1) water tends to enter the body from the environment; and (2) solutes tend to leave the body and enter the environment. Freshwater fish address the first problem by *not* drinking water and by excreting a large volume of dilute urine, which is hypotonic to their body fluids (figure 24.25, *top*). They address the second problem by reabsorbing ions across the nephron tubules, from the glomerular filtrate back into the blood. In addition, they actively transport ions across their gills from the surrounding water into the blood.

Marine Bony Fish

Although most groups of animals seem to have evolved first in the sea, marine bony fish (teleosts) probably evolved from freshwater ancestors. They faced significant new problems in making the transition to the sea because their body fluids are hypotonic to the surrounding seawater. Consequently, water tends to leave their bodies by osmosis across their gills, and they also lose water in their urine. To compensate for this continuous water loss, marine fish drink large amounts of seawater.

Many of the divalent cations (principally Ca^{++} and Mg^{++}) in the seawater that a marine fish drinks remain in the digestive tract and are eliminated through the anus. Some, however, are absorbed into the blood, as are the monovalent ions K^+, Na^+, and Cl^-. Most of the monovalent ions are actively transported out of the blood across the gills, while the divalent ions that enter the blood are secreted into the nephron tubules and excreted in the urine (figure 24.25, *bottom*). In these two ways, marine bony fish eliminate the ions they get from the seawater they drink. The urine they excrete is isotonic to their body fluids. It is more concentrated than the urine of freshwater fish but not as concentrated as that of birds and mammals.

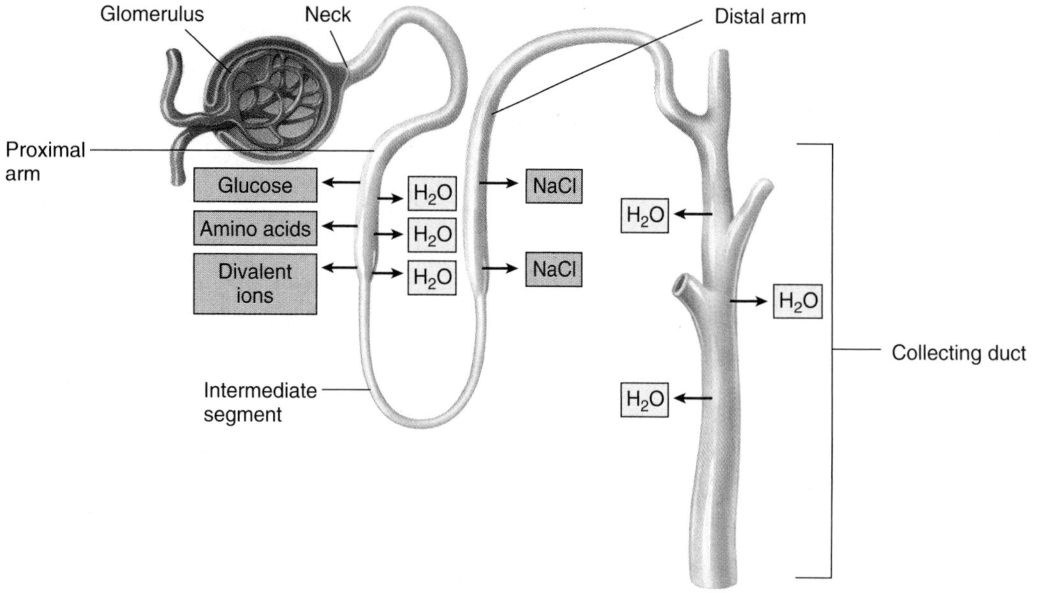

Glomerulus
Neck
Distal arm
Proximal arm
Glucose
Amino acids
Divalent ions
H2O
H2O
H2O
NaCl
NaCl
H2O
H2O
H2O
Collecting duct
Intermediate segment

Figure 24.24 Basic organization of the vertebrate nephron.
The nephron tube of the freshwater fish is a basic design that has been retained in the kidneys of marine fishes and terrestrial vertebrates that evolved later. Sugars, small proteins, and divalent ions such as Ca^{++} are recovered at the beginning of the tube (the so-called proximal arm). Ions such as Na^+ and Cl^- are recovered after the initial bend (the so-called distal arm).

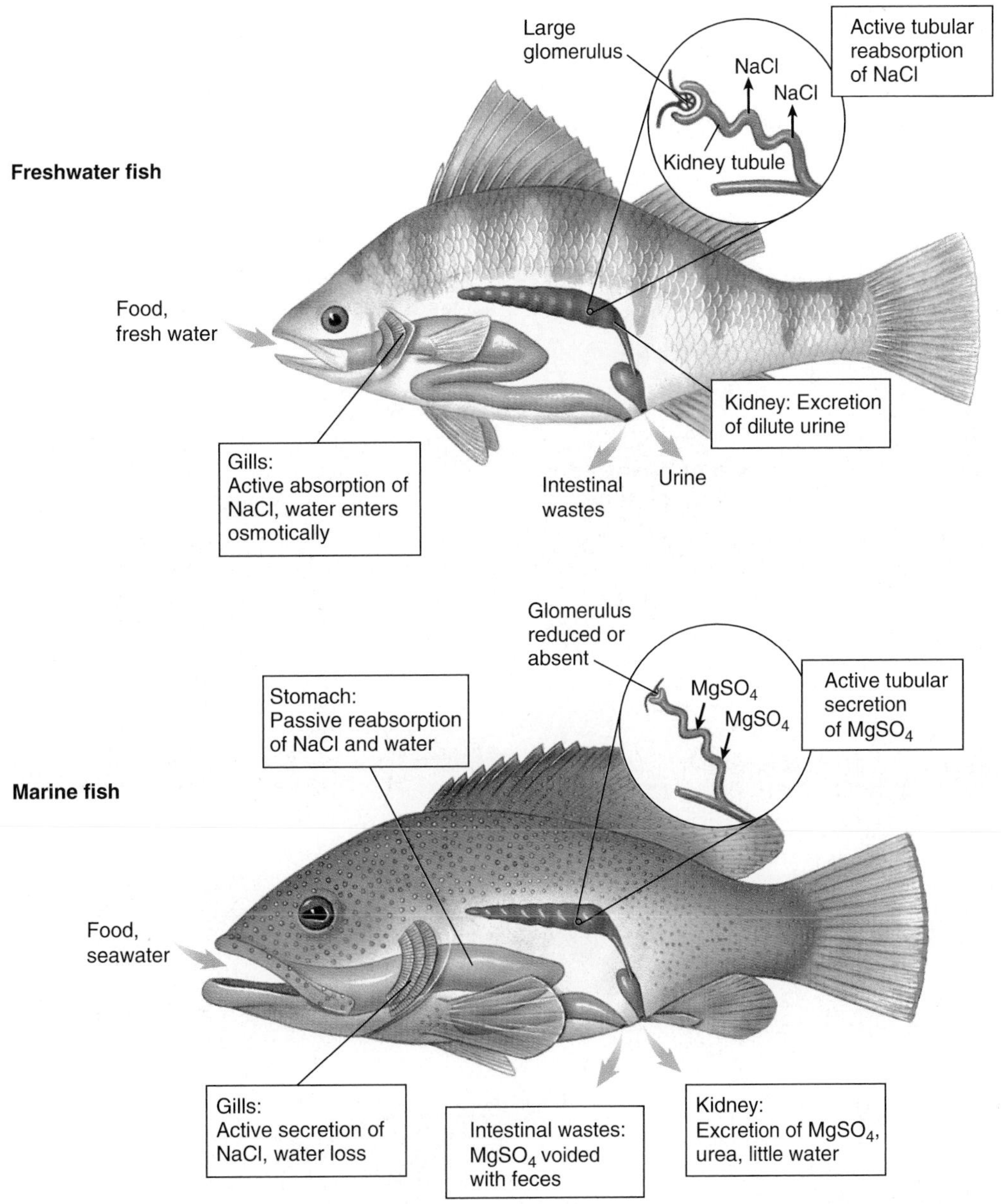

Freshwater fish
Large glomerulus
NaCl
NaCl
Active tubular reabsorption of NaCl
Kidney tubule
Food, fresh water
Kidney: Excretion of dilute urine
Gills: Active absorption of NaCl, water enters osmotically
Intestinal wastes
Urine
Marine fish
Glomerulus reduced or absent
Stomach: Passive reabsorption of NaCl and water
$MgSO_4$
$MgSO_4$
Active tubular secretion of $MgSO_4$
Food, seawater
Gills: Active secretion of NaCl, water loss
Intestinal wastes: $MgSO_4$ voided with feces
Kidney: Excretion of $MgSO_4$, urea, little water

Figure 24.25 Freshwater and marine teleosts (bony fish) face different osmotic problems.
Whereas the freshwater teleost is hypertonic to its environment, the marine teleost is hypotonic to seawater. To compensate for its tendency to take in water and lose ions, a freshwater fish excretes dilute urine, avoids drinking water, and reabsorbs ions across the nephron tubules. To compensate for its osmotic loss of water, the marine teleost drinks seawater and eliminates the excess ions through transport across epithelia.

Cartilaginous Fish

The elasmobranchs—sharks, rays, and skates—are by far the most common subclass in the class Chondrichthyes (cartilaginous fish). Elasmobranchs have solved the osmotic problem posed by their seawater environment in a different way than have the bony fish. Instead of having body fluids that are hypotonic to seawater, so that they have to continuously drink seawater and actively pump out ions, the elasmobranchs reabsorb urea from the nephron tubules and maintain a blood urea concentration that is 100 times higher than that of mammals. This added urea makes their blood approximately isotonic to the surrounding sea. Because there is no net water movement between isotonic solutions, water loss is prevented. Hence, these fish do not need to drink seawater for osmotic balance, and their kidneys and gills do not have to remove large amounts of ions from their bodies. The enzymes and tissues of the cartilaginous fish have evolved to tolerate the high urea concentrations.

Amphibians and Reptiles

The first terrestrial vertebrates were the amphibians, and the amphibian kidney is identical to that of freshwater fish. This is not surprising because amphibians spend a significant portion of their time in freshwater, and when on land, they generally stay in wet places. Amphibians produce a very dilute urine and compensate for their loss of Na^+ by actively transporting Na^+ across their skin from the surrounding water.

Reptiles, on the other hand, live in diverse habitats. Those living mainly in freshwater, like some of the crocodilians, occupy a habitat similar to that of the freshwater fish and amphibians and thus have similar kidneys. Marine reptiles, which consist of other crocodilians, turtles, sea snakes, and one lizard, possess kidneys similar to those of their freshwater relatives but face opposite problems; they tend to lose water and take in salts. Like marine teleosts (bony fish), they drink the seawater and excrete an isotonic urine. Marine teleosts eliminate the excess salt by transport across their gills, while marine reptiles eliminate excess salt through salt glands located near the nose or the eye (figure 24.26).

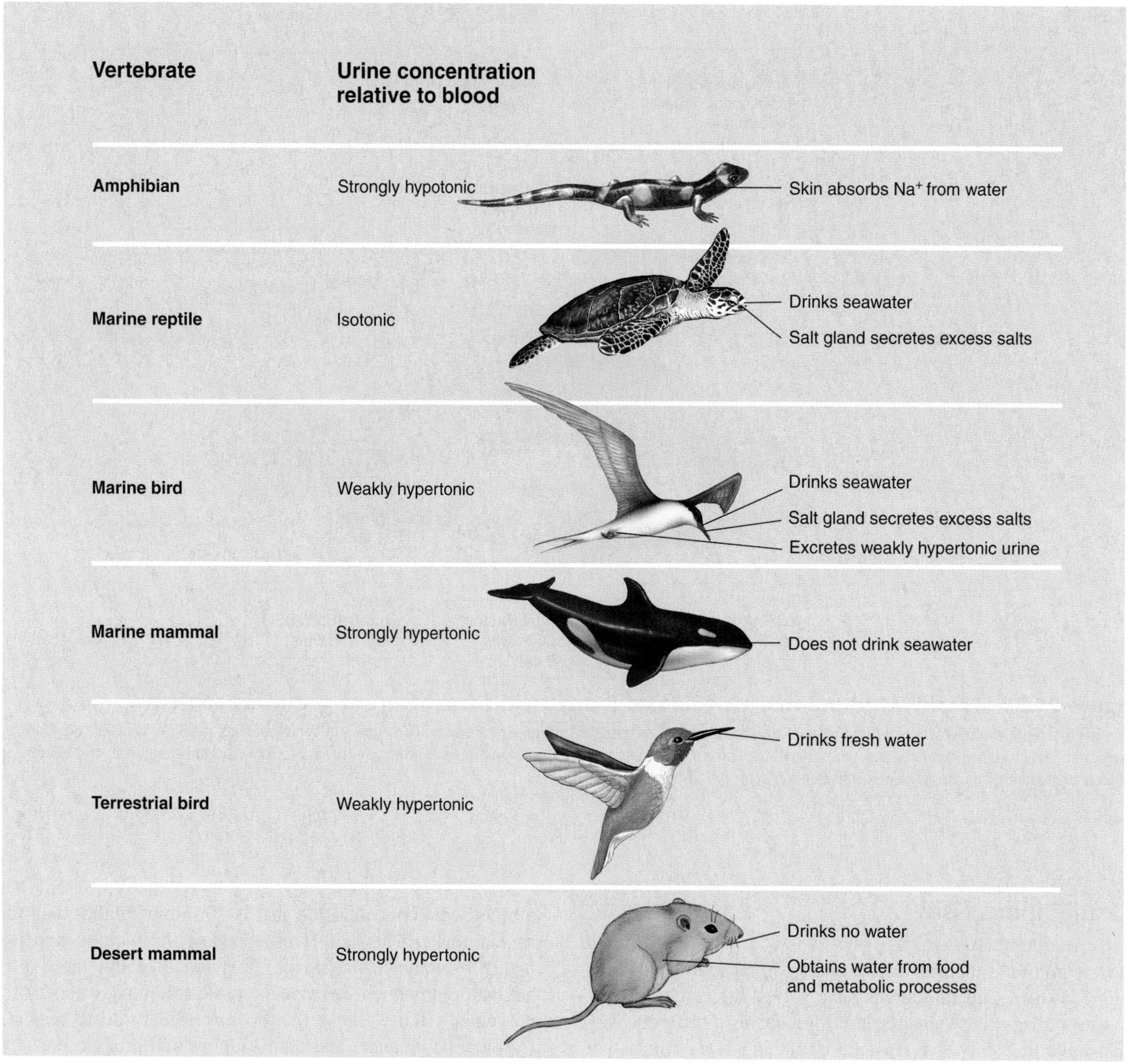

Vertebrate
Urine concentration relative to blood
Amphibian
Strongly hypotonic
Skin absorbs Na+ from water
Marine reptile
Isotonic
Drinks seawater
Salt gland secretes excess salts
Marine bird
Weakly hypertonic
Drinks seawater
Salt gland secretes excess salts
Excretes weakly hypertonic urine
Marine mammal
Strongly hypertonic
Does not drink seawater
Terrestrial bird
Weakly hypertonic
Drinks fresh water
Desert mammal
Strongly hypertonic
Drinks no water
Obtains water from food and metabolic processes

Figure 24.26 Osmoregulation by some vertebrates.
Only birds and mammals can produce a hypertonic urine and thereby retain water efficiently, but marine reptiles and birds can drink seawater and excrete the excess salt through salt glands.

The kidneys of terrestrial reptiles also reabsorb much of the salt and water in the nephron tubules, helping somewhat to conserve blood volume in dry environments. Like fish and amphibians, they cannot produce urine that is more concentrated than the blood plasma. However, when their urine enters their cloaca (the common exit of the digestive and urinary tracts), additional water can be reabsorbed.

Mammals and Birds

Mammals and birds are the only vertebrates able to produce urine with a higher osmotic concentration than their body fluids. This allows these vertebrates to excrete their waste products in a small volume of water, so that more water can be retained in the body. Human kidneys can produce urine that is as much as 4.2 times as concentrated as blood plasma, but the kidneys of some other mammals are even more efficient at conserving water. For example, the camel, gerbil, and pocket mouse, *Perognathus,* can excrete urine 8, 14, and 22 times as concentrated as their blood plasma, respectively. The kidneys of the kangaroo rat (figure 24.27) are so efficient it never has to drink water; it can obtain all the water it needs from its food and from water produced in aerobic cell respiration!

The production of hypertonic urine is accomplished by the *loop of Henle* portion of the nephron (see figure 24.29), found only in mammals and birds. A nephron with a long loop of Henle extends deeper into the renal medulla and can produce more concentrated urine. Most mammals have some nephrons with short loops and other nephrons with loops that are much longer. Birds, however, have relatively few or no nephrons with long loops, so they cannot produce urine that is as concentrated as that of mammals. At most, they can only reabsorb enough water to produce a urine that is about twice the concentration of their blood. Marine birds solve the problem of water loss by drinking seawater and then excreting the excess salt from salt glands near the eyes (figure 24.28).

The moderately hypertonic urine of a bird is delivered to its cloaca, along with the fecal material from its digestive tract. If needed, additional water can be absorbed across the wall of the cloaca to produce a semisolid white paste or pellet, which is excreted.

24.12 The kidneys of freshwater fish must excrete copious amounts of very dilute urine, while marine teleosts drink seawater and excrete an isotonic urine. The basic design and function of the nephron of freshwater fish have been retained in the terrestrial vertebrates. Modifications, particularly the presence of a loop of Henle, allow mammals and birds to reabsorb water and produce a hypertonic urine.

Figure 24.27 The kangaroo rat, *Dipodomys panamintensis.*

This mammal has very efficient kidneys that can concentrate urine to a high degree by reabsorbing water, thereby minimizing water loss from the body. This feature is extremely important to the kangaroo rat's survival in dry or desert habitats.

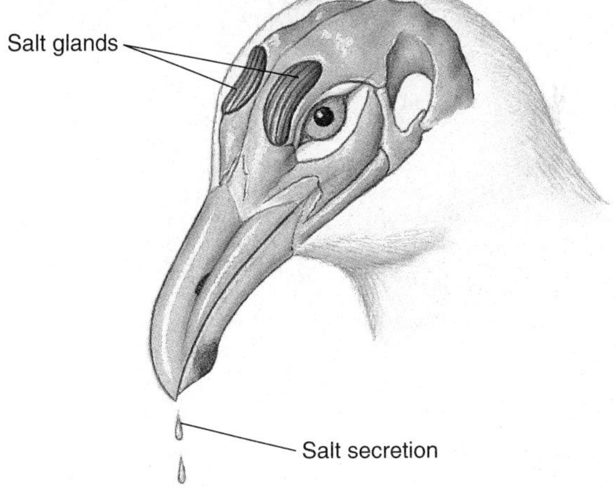
Salt glands
Salt secretion

Figure 24.28 Marine birds drink seawater and then excrete the salt through salt glands.

The salty fluid excreted by these glands can then dribble down the beak.

24.13 The Mammalian Kidney

In humans, the kidneys are fist-sized organs located in the region of the lower back (figure 24.29*a*). Each kidney receives blood from a renal artery, and it is from this blood that urine is produced. Urine drains from each kidney through a **ureter,** which carries the urine to a **urinary bladder.** Within the kidney, the mouth of the ureter flares open to form a funnel-like structure, the *renal pelvis*. The renal pelvis, in turn, has cup-shaped extensions that receive urine from the renal tissue. This tissue is divided into an outer **renal cortex** and an inner **renal medulla** (figure 24.29*b*). Together, these structures perform filtration, reabsorption, secretion, and excretion.

The mammalian kidney is composed of roughly 1 million nephrons (figure 24.29*c*), each of which is composed of three regions:

1. **Filter.** The filtration device at the top of each nephron is called a **Bowman's capsule.** Within each capsule an arteriole enters and splits into a fine network of vessels called a **glomerulus** (figure 24.30). The walls of these capillaries act as a filtration device. Blood pressure forces fluid through the capillary walls. These walls withhold proteins and other large molecules in the blood, while passing water, small molecules, ions, and urea, the primary waste product of metabolism.
2. **Tube.** The Bowman's capsule is connected to a long narrow tube called a renal tubule, which is bent back on itself in its center, called the **loop of Henle.** This long hairpin loop is a reabsorption device.

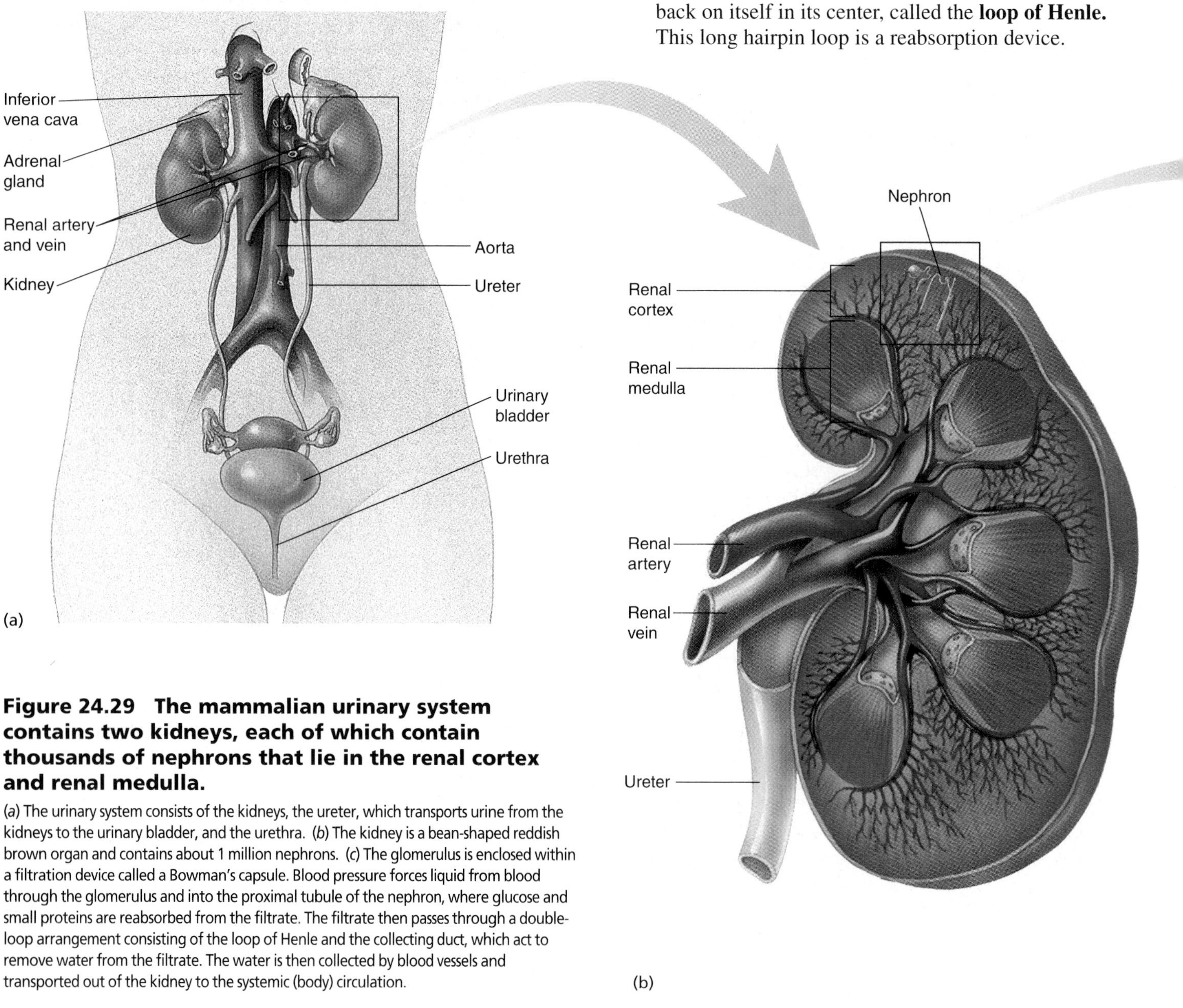
Inferior vena cava
Adrenal gland
Renal artery and vein
Kidney
Aorta
Ureter
Urinary bladder
Urethra
(a)
Nephron
Renal cortex
Renal medulla
Renal artery
Renal vein
Ureter
(b)

Figure 24.29 The mammalian urinary system contains two kidneys, each of which contain thousands of nephrons that lie in the renal cortex and renal medulla.

(*a*) The urinary system consists of the kidneys, the ureter, which transports urine from the kidneys to the urinary bladder, and the urethra. (*b*) The kidney is a bean-shaped reddish brown organ and contains about 1 million nephrons. (*c*) The glomerulus is enclosed within a filtration device called a Bowman's capsule. Blood pressure forces liquid from blood through the glomerulus and into the proximal tubule of the nephron, where glucose and small proteins are reabsorbed from the filtrate. The filtrate then passes through a double-loop arrangement consisting of the loop of Henle and the collecting duct, which act to remove water from the filtrate. The water is then collected by blood vessels and transported out of the kidney to the systemic (body) circulation.

Like the mammalian small intestine, it extracts from the filtrate passing through the tube molecules useful to the body, such as glucose and a variety of ions.

3. **Duct.** The tube empties into a large collection tube called a **collecting duct.** The collecting duct operates as a water conservation device, reclaiming water from the urine so that it is not lost from the body. Human urine is four times as concentrated as blood plasma—that is, the collecting ducts remove much of the water from the filtrate passing through the kidney. Your kidneys achieve this remarkable degree of water conservation by a simple but superbly designed mechanism: they bend the duct back alongside the nephron tube and make the duct permeable to urea. This greatly increases the local salt (urea) concentration in the tissue surrounding the tube, causing water in urine to pass out of the tube by osmosis. The salty tissue sucks up water from the urine like blotting paper, passing it on to blood vessels that carry it out of the kidneys and back to the bloodstream.

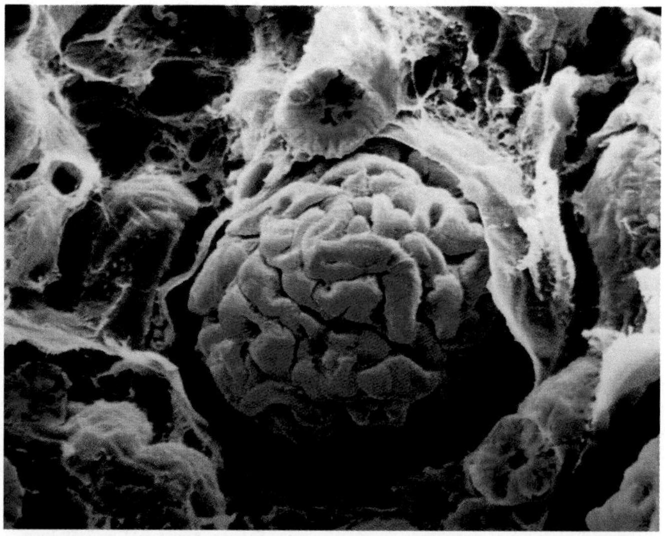

Figure 24.30 A glomerulus in the kidney (×65).
The red spherical structure in this micrograph is a glomerulus, a fine network of capillaries associated with a nephron tubule and connected to a pair of arterioles. From R. G. Kessel and R. H. Kardon, *Tissues and Organs: A Text Atlas of Scanning Electron Microscopy,* 1979, W. H. Freeman Co.

(c)

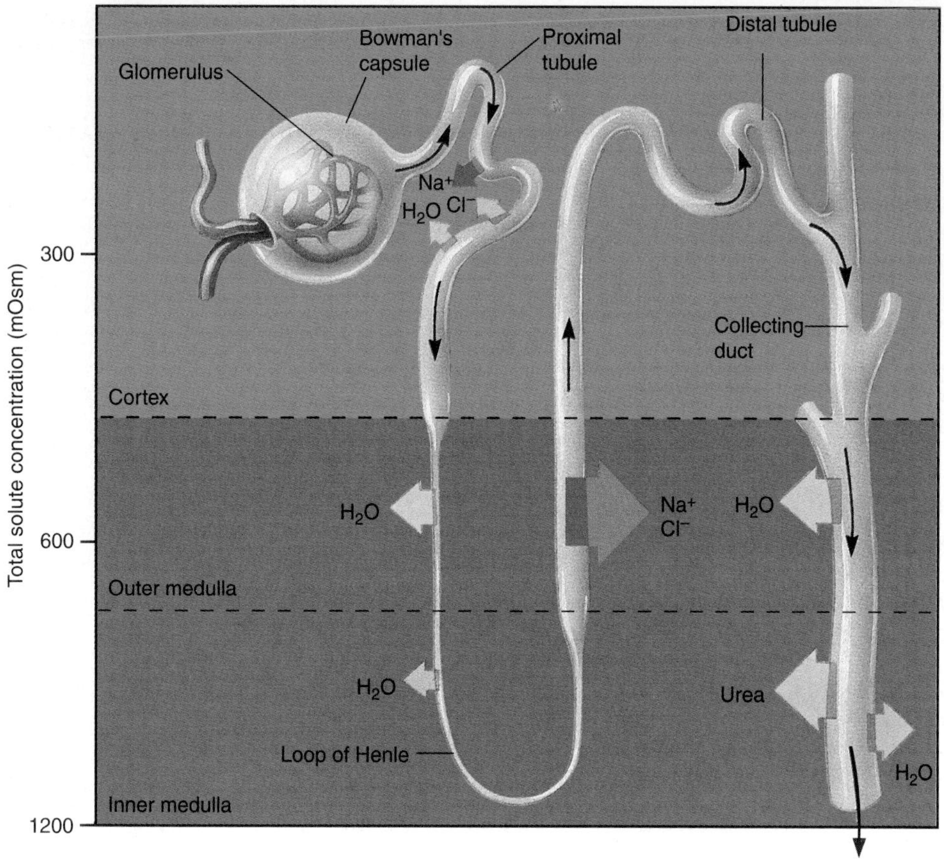

Glomerulus
Bowman's capsule
Proximal tubule
Distal tubule
Na+
H2O Cl−
Collecting duct
Total solute concentration (mOsm)
300
600
1200
Cortex
Outer medulla
Inner medulla
H2O
Na+
Cl−
H2O
H2O
Urea
H2O
Loop of Henle

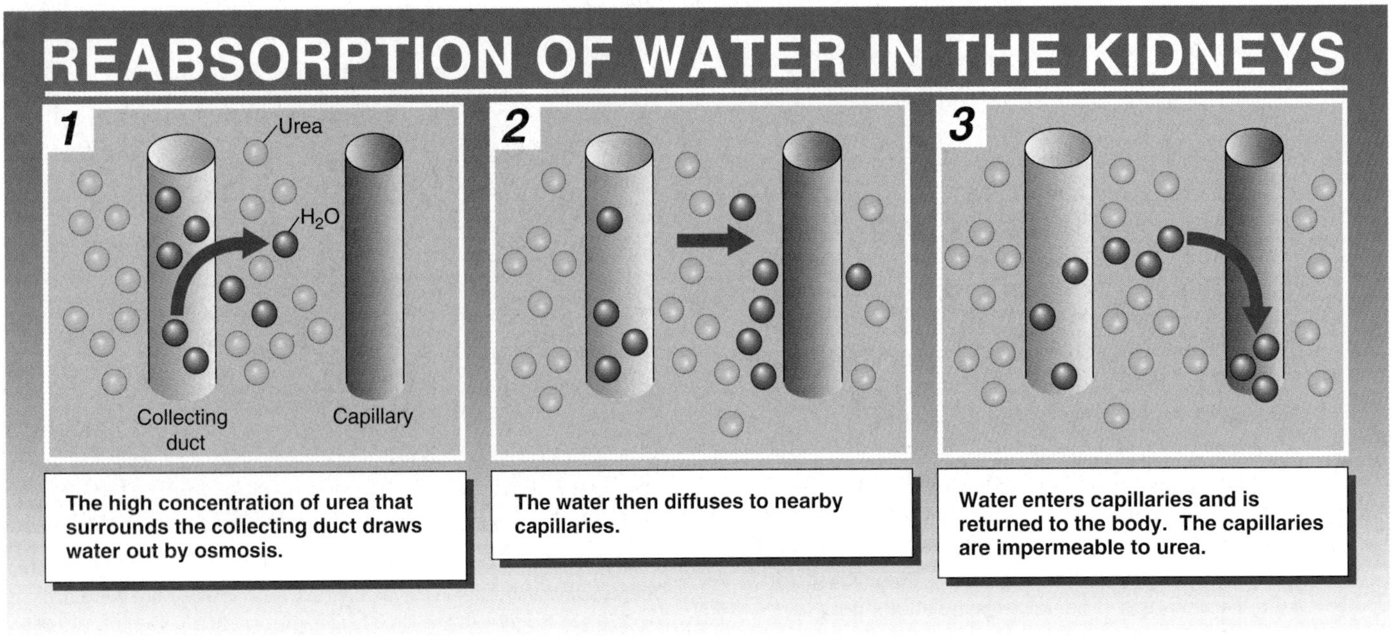
REABSORPTION OF WATER IN THE KIDNEYS
1
Urea
H_2O
Collecting duct
Capillary
2
3
The high concentration of urea that surrounds the collecting duct draws water out by osmosis.
The water then diffuses to nearby capillaries.
Water enters capillaries and is returned to the body. The capillaries are impermeable to urea.

Figure 24.31 How the kidneys reabsorb water.

The Kidney at Work

The formation of urine within the mammalian kidney involves the movement of several kinds of molecules between nephrons and the capillaries that surround them. Five steps are involved: pressure filtration, reabsorption of water, selective reabsorption of ions and nutrients, tubular excretion, and further reabsorption of water.

Pressure Filtration. Driven by the blood pressure, small molecules are pushed across the thin walls of the glomerulus to the inside of the Bowman's capsule. Blood cells and large molecules like proteins cannot pass through, and as a result the blood that enters the glomerulus is divided into two paths: nonfilterable blood components that are retained and leave the glomerulus in the bloodstream and filterable components that pass across and leave the glomerulus in the urine. This filterable stream is called the **glomerular filtrate.** It contains water, nitrogenous wastes (principally urea), nutrients (principally glucose and amino acids), and a variety of ions.

Reabsorption of Water. Filtrate from the glomerulus passes down the descending arm of the loop of Henle. The walls of this portion of the tube are impermeable to either salts or urea but are freely permeable to water. Because (for reasons we discuss later) the surrounding tissue has a high concentration of urea, water passes out of the descending arm by osmosis, leaving behind a more concentrated filtrate.

Selective Reabsorption. At the turn in the loop, the walls of the tubule become permeable to salts and other nutrients, like sugars and amino acids, but much less permeable to water. As the concentrated filtrate passes up this ascending arm, these nutrients pass out into the surrounding tissue, where they are carried away by blood vessels (figure 24.31). In the upper region of the ascending arm are active transport channels that pump out salt (NaCl). Left behind in the filtrate is the urea that initially passed through the glomerulus as nitrogenous waste. The urea concentration is becoming very high within the tubule.

Tubular Excretion. In the ascending loop, substances are also added to the urine by a process called tubular excretion. This active transport process excretes into the urine other nitrogenous wastes such as uric acid and ammonia, as well as excess hydrogen ions.

Further Reabsorption of Water. The tubule then empties into a collecting duct that passes back through the tissue of the kidney. Unlike the tubule, the lower portions of the collecting duct are permeable to urea, some of which diffuses out into the surrounding tissue (that is why the tissue surrounding the descending arm of the loop of Henle has a high urea concentration). A high urea concentration in the tissue results, causing even more water to pass outward from the filtrate by osmosis. The filtrate that is left after salts, nutrients, and water have been removed is urine.

24.13 The mammalian kidney pushes waste molecules through a filter and then reclaims water and useful metabolites and ions from the filtrate before eliminating the residual urine.

CONCEPT REVIEW

1. Starchy foods such as potatoes begin being digested in the
   a. mouth.
   b. esophagus.
   c. stomach.
   d. duodenum.

2. Protein-rich foods, such as steak, are broken down in the stomach by the combined action of (pick two)
   a. low pH.
   b. high pH.
   c. bile.
   d. amylase.
   e. pepsin.

3. Most of the enzymes that complete the breakdown of food items into simple sugars, amino acids, and fatty acids are secreted by which of the following?
   a. large intestine
   b. gastric pits
   c. liver
   d. pancreas

4. The main function of the large intestine is
   a. digestion.
   b. absorption.
   c. compaction.
   d. secretion.

5. Select the *incorrect* statement about the liver.
   a. The liver is the largest internal organ of the human body.
   b. The liver releases glucose into the blood.
   c. The liver stores bile.
   d. The liver modifies excess amino acids.

6. The mammalian body excretes nitrogen mainly as
   a. ammonia.
   b. urea.
   c. uric acid.
   d. water.

7. Because marine fish continuously lose water by osmosis to their salty environment, they must _________ a lot and excrete excess salts.
   a. urinate
   b. swim
   c. eat
   d. drink

8. The glomerulus is a
   a. cup.
   b. network of capillaries.
   c. tubule.
   d. all of these.

9. Which of the following is recovered in the collecting duct of the nephron?
   a. glucose
   b. NaCl
   c. water
   d. proteins

10. A band of muscle called the _________ controls the entrance to the duodenum from the stomach.

11. Food moves down the esophagus by rhythmic muscular contractions called _________.

12. The very numerous _________ that cover the epithelial wall of the small intestine greatly increase the surface area for absorption of digested foods.

13. By the action of the hormone insulin, the liver stores excess glucose as the molecule _________.

14. The _________ ion is secreted into the small intestine by the pancreas to neutralize stomach acid.

15. Pressure _________ occurs inside the Bowman's capsule.

16. The loop of _________ is part of the tube of the nephron.

CHALLENGE YOURSELF

1. While some animals are capable of synthesizing specific vitamins and amino acids, others cannot synthesize these compounds and must obtain them in their diet. What are the relative advantages and disadvantages of an animal that possesses the biochemical machinery to manufacture specific nutrients that other animals cannot manufacture?

2. You are adrift on a life raft on the ocean with no source of freshwater, and you recall that marine bony fishes and some marine birds satisfy their bodies' needs for water by drinking seawater. Should you drink seawater? Note that human kidneys can remove up to about 6 grams of sodium ions from the blood for every liter of urine they produce, and that a liter of seawater contains about 12 grams of sodium ions. Explain your answer.

24 eBRIDGE

Reinforcing Key Points

Food Energy and Essential Nutrients

24.1 Calories for Energy

24.2 Essential Substances for Growth

Digestion

24.3 Types of Digestive Systems

24.4 Vertebrate Digestive Systems

24.5 The Mouth and Teeth

24.6 The Esophagus and Stomach

24.7 The Small and Large Intestines

24.8 Variations in Vertebrate Digestive Systems

24.9 Accessory Digestive Organs

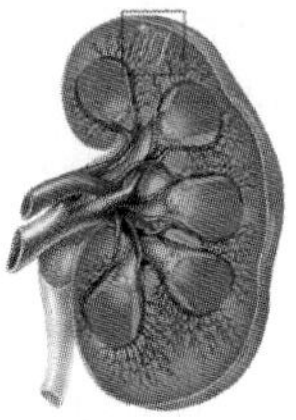

Maintaining the Internal Environment

24.10 Homeostasis

24.11 Osmoregulatory Organs

24.12 Evolution of the Vertebrate Kidney

24.13 The Mammalian Kidney

Electronic Learning

Visual Learning

Animations

Stomach Digestion

Nutritional Mode of a Clam Compared to a Squid

Dental Caries

Digestion

Endoscopy

Ulcers

Small Intestine Digestion

Formation of Gallstones

Kidney Function I

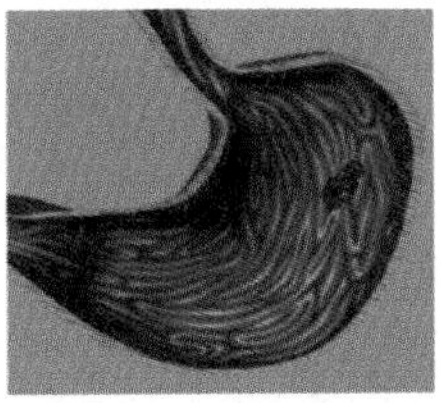

Art Labeling Activities

Wall of Digestive Tract

Anatomy of Kidney and Lobe

Helping You Learn

Human Digestive System: Structure and Function

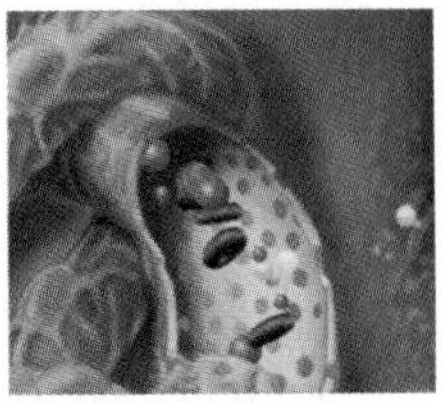

Author's Corner

Dieting. Almost all of us at one time or another worry about our weight. It turns out there's a lot of interesting biology involved.

1. Impossible dreams: Fad diets and our futile search for an easy way to lose weight.
2. Why fat people are hungrier.
3. New drugs reduce cholesterol by inhibiting a key enzyme used in manufacturing it.
4. Diet drugs may be good news for those of us who are overweight.
5. Despite earlier hopes, high-fiber diets don't seem to protect against colon cancer.
6. The turkey's revenge: Why eating Thanksgiving dinner puts on the pounds.
7. In the battle to lose weight, I seem to be losing.

The Living World

Virtual Classroom

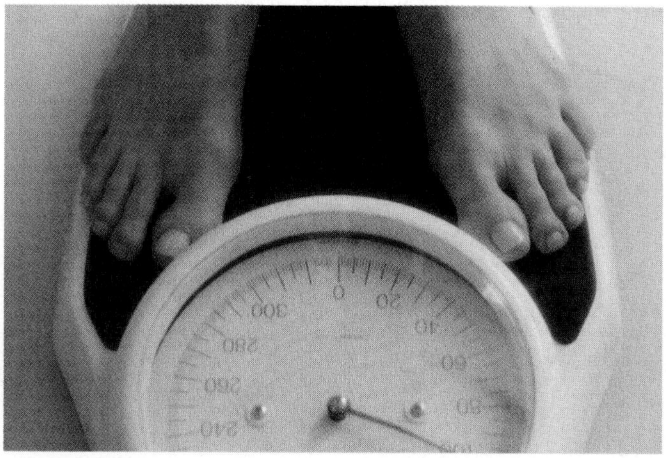

FAD Diets

FAD diets like the Atkins diet are wildly popular, because they promise pain-free weight loss. The combination of hope and hype make this diet, and others like it, perpetual best sellers. The secret of the Atkins diet, put simply, is to avoid carbohydrates. Follow it, and you lose weight rapidly—at first. The temporary weight loss turns out to have a simple explanation: because carbs act as water sponges in your body, depleting your body of carbs causes it to lose water. Government studies have shown that those who achieve longer-lasting benefits from the Atkins diet, or other fad diets like *The Zone,* work not for the bizarre reasons claimed by their promoters, but simply because they are low-calorie diets. There are two basic laws which no diet can successfully violate: 1. all calories are equal, 2. calories in minus calories out equals fat. The fundamental fallacy of all fad diets is the idea that somehow carbohydrate calories are different from fat and protein calories. This is scientific foolishness. Every calorie contributes equally to your weight, whatever its source. The diets work simply because they obey the second law. By reducing calories in, they reduce fat.

Virtual Lab

Discovering the Virus Responsible for Hepatitis C

Hepatitis is a sometimes fatal disease of the liver caused by a virus. There are three distinct forms. Infectious hepatitis or hepatitis A is transmitted by contact with feces from infected individuals. Serum hepatitis or hepatitis B is passed through the blood and other body fluids. A third form called hepatitis C, caused by the hepatitis C virus (HCV), eluded doctors and researchers for many years. It was only isolated in 1990. It too is transmitted through the blood. Hepatitis B and C cause chronic liver damage that can be fatal. Because these viruses are transmitted through the blood, a diagnostic antibody test was necessary to protect the nation's blood supply from contamination. Although a diagnostic test had been developed for the hepatitis B virus, the hepatitis C virus proved challenging because HCV cannot be grown reliably in a laboratory culture and it is strictly a primate virus (infecting only humans, chimpanzees, and tamarins).

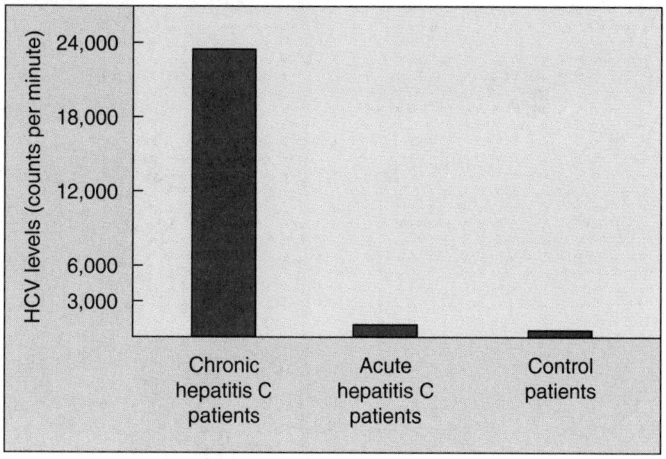
HCV levels (counts per minute)
24,000
18,000
12,000
6,000
3,000
Chronic hepatitis C patients
Acute hepatitis C patients
Control patients

A research team lead by Michael Houghton at Chiron, a California biotechnology company, developed a diagnostic antibody test for HCV. With the HCV diagnostic test in hand, they set out to test its specificity and sensitivity.

Quizzes

Further Reading

Essential Study Partner

Links

BioCourse.com

CHAPTER 25

How the Animal Body Defends Itself

THE FAR SIDE® By GARY LARSON
He's trying to shoot me, all right... Do I know this guy?.. I've got to think!
Larson
© 1984 FarWorks, Inc. All Rights Reserved/Dist. by Creators Syndicate

CHAPTER 25

How the Animal Body Defends Itself

CHAPTER OVERVIEW

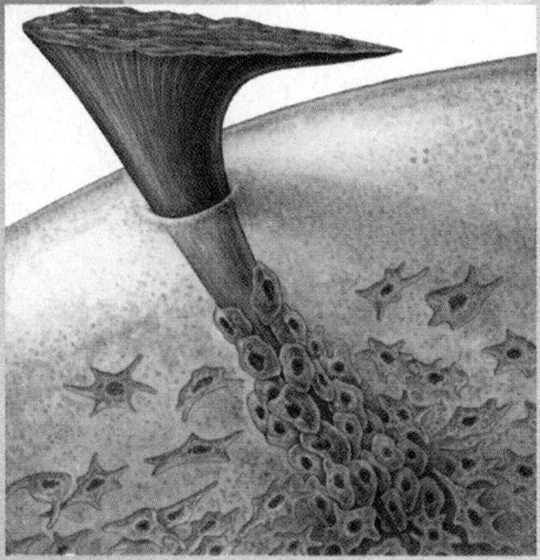

Three Lines of Defense

25.1 Skin: The First Line of Defense
25.2 Cellular Counterattack: The Second Line of Defense
25.3 The Immune System: The Third Line of Defense
25.4 Evolution of the Immune System

- Skin offers an efficient barrier to penetration by microbes.
- Macrophages are white blood cells that patrol the bloodstream, ingesting bacteria by phagocytosis.
- Neutrophils kill all the cells at a site of infection.
- Complement proteins insert into a pathogen's plasma membrane, causing the cell to rupture like a punctured balloon.
- Lymphocytes called T cells are responsible for cell-mediated immunity; lymphocytes called B cells secrete antibodies and produce the humoral response.

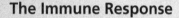

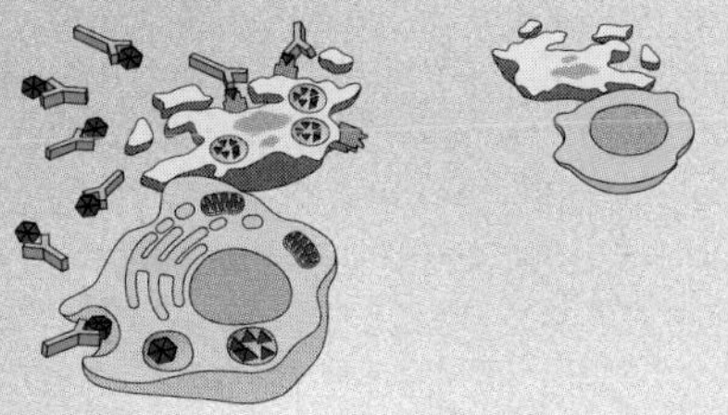

The Immune Response

25.5 Initiating the Immune Response
25.6 T Cells: The Cellular Response
25.7 B Cells: The Humoral Response
25.8 Active Immunity Through Clonal Selection
25.9 Vaccination

- Macrophages release chemicals that initiate the immune response by activating helper T cells.
- Helper T cells in turn activate T cells, which destroy infected body cells, and B cells, which label microbes for destruction by macrophages.
- Memory B cells speed future immune responses such that immunity results.

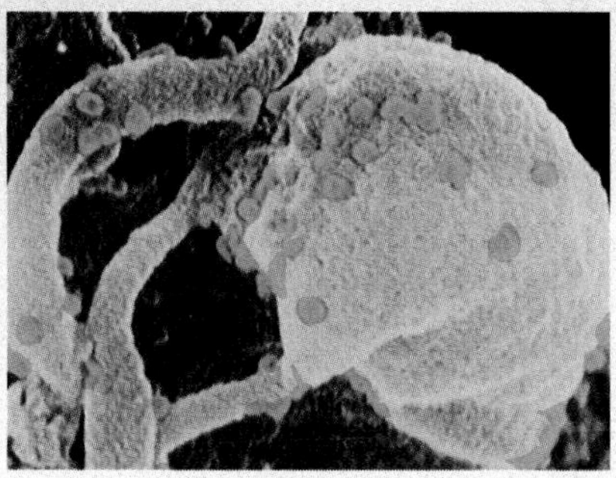

Defeat of the Immune System

25.10 Immune System Failure
25.11 AIDS: Immune System Collapse

- When the body's immune system attacks its own tissues, autoimmune disease results.
- Allergies are inappropriate immune responses to harmless substances.
- AIDS is a disease that destroys the immune response by killing helper T cells and macrophages.
- AIDS is caused by HIV, which is transmitted in body fluids, typically via sexual intercourse or by infected needles during drug use.

Multicellular bodies offer a feast of nutrients for tiny, single-celled creatures, as well as a warm, sheltered environment in which they can grow and reproduce. We live in a world awash with microbes, and no animal can long withstand their onslaught unprotected. Animals survive because they have a variety of very effective defenses against this constant attack.

Figure 25.1 Skin is the vertebrate body's first line of defense.

This young elephant has skin thicker than a belt, allowing it to follow the herd through dense thickets without injury.

25.1 Skin: The First Line of Defense

Overview of the Three Lines of Defense

The vertebrate is defended from infection the same way knights defended medieval cities. "Walls and moats" make entry difficult; "roaming patrols" attack strangers; and "sentries" challenge anyone wandering about and call patrols if a proper "ID" is not presented.

1. **Walls and moats.** The outermost layer of the vertebrate body, the **skin,** is the first barrier to penetration by microbes. Mucous membranes in the respiratory and digestive tracts are also important barriers that protect the body from invasion.
2. **Roaming patrols.** If the first line of defense is penetrated, the response of the body is to mount a **cellular counterattack,** using a battery of cells and chemicals that kill microbes. These defenses act very rapidly after the onset of infection.
3. **Sentries.** Lastly, the body is also guarded by mobile cells that patrol the bloodstream, scanning the surfaces of every cell they encounter. They are part of the **immune system.** One kind of immune cell aggressively attacks and kills any cell identified as foreign, whereas the other type marks the foreign cell or virus for elimination by the roaming patrols.

The Skin

Skin is the outermost layer of the vertebrate body (figure 25.1) and provides the first defense against invasion by microbes. Skin is our largest organ, comprising some 15% of our total weight. One square centimeter of skin from your forearm (about the size of a dime) contains 200 nerve endings, 10 hairs and muscles, 100 sweat glands, 15 oil glands, 3 blood vessels, 12 heat-sensing organs, 2 cold-sensing organs, and 25 pressure-sensing organs. Skin has three distinct layers: an outer **epidermis,** a lower **dermis,** and an underlying **subcutaneous layer.** Cells of the outer epidermis are continually being worn away and replaced by cells moving up from below—in one hour your body loses and replaces approximately 1.5 million skin cells!

The epidermis of skin is from 10 to 30 cells thick, about as thick as this page. The outer layer, called the **stratum corneum,** is the one you see when you look at your arm or face. Cells from this layer are continuously subjected to damage. They are abraded, injured, and worn by friction and stress during the body's many activities. They also lose moisture and dry out. The body deals with this damage not by repairing cells but by replacing them. Cells from the stratum corneum are shed continuously, replaced by new cells produced deep within the epidermis. The cells of the innermost **basal layer** are among the most actively dividing cells of the vertebrate body. New cells formed there migrate upward, and as they move they manufacture keratin protein, which makes them tough. Each cell eventually arrives at the outer surface and takes its turn in the stratum corneum, residing there for about a month before it is shed and replaced by a newer cell. Persistent dandruff (psoriasis) is a chronic skin disorder in which new cells reach the epidermal surface every three or four days, about eight times faster than normal.

The dermis of the skin is from 15 to 40 times thicker than the epidermis. It provides structural support for the epidermis, as well as a matrix for the many specialized cells residing within the skin. The wrinkling that occurs as we grow older occurs here. The leather used to manufacture belts and shoes is derived from thick animal dermis. The layer of subcutaneous tissue below the dermis is composed of fat-rich cells that act as shock absorbers and provide insulation, which conserves body heat.

The skin not only defends the body by providing a nearly impermeable barrier, but it also reinforces this defense with chemical weapons. The oil and sweat glands, for example (figure 25.2), make the skin's surface very acidic, which inhibits the growth of many microbes. Sweat also contains the enzyme lysozyme, which attacks and digests the cell walls of many bacteria.

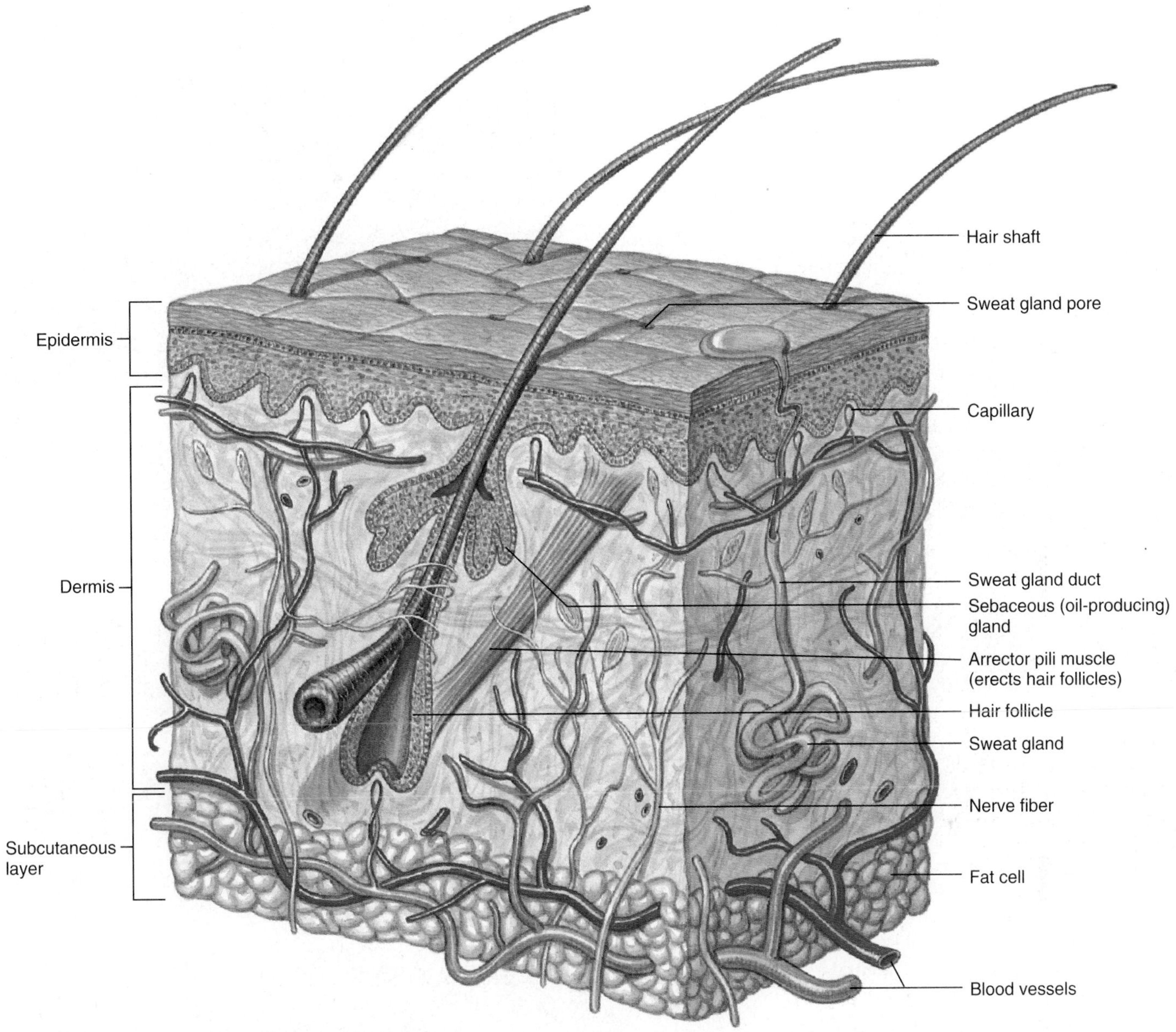
Hair shaft
Sweat gland pore
Epidermis
Capillary
Dermis
Sweat gland duct
Sebaceous (oil-producing) gland
Arrector pili muscle (erects hair follicles)
Hair follicle
Sweat gland
Nerve fiber
Subcutaneous layer
Fat cell
Blood vessels

Figure 25.2 A section of human skin.
The skin defends the body by providing a barrier and sweat and oil glands whose secretions make the skin's surface acidic enough to inhibit the growth of microorganisms.

Other External Surfaces

In addition to the skin, two other potential routes of entry by viruses and microorganisms must be guarded: the *digestive tract* and the *respiratory tract.* Microbes are present in food, but many are killed by saliva (which also contains lysozyme), by the very acidic environment of the stomach, and by digestive enzymes in the intestine. Microorganisms are also present in inhaled air. The cells lining the smaller bronchi and bronchioles secrete a layer of sticky mucus that traps most microorganisms before they can reach the warm, moist lungs, which would provide ideal breeding grounds for them. Other cells lining these passages have cilia that continually sweep the mucus toward the glottis. There it can be swallowed, carrying potential invaders out of the lungs and into the digestive tract.

The surface defenses are very effective, but they are occasionally breached. Through breathing, eating, or cuts and nicks, bacteria and viruses now and then enter our bodies. When these invaders reach deeper tissue, a second line of defense comes into play, a cellular counterattack.

25.1 Skin and the mucous membranes lining the digestive and respiratory tracts are the body's first defenses.

25.2 Cellular Counterattack: The Second Line of Defense

When the body's interior is invaded, a host of cellular and chemical defenses swing into action. Four are of particular importance: (1) cells that kill invading microbes; (2) proteins that kill invading microbes; (3) the inflammatory response, which speeds defending cells to the point of infection; and (4) the temperature response, which elevates body temperature to slow the growth of invading bacteria.

Cells That Kill Invading Microbes

The most important counterattack to infection is mounted by white blood cells, which attack invading microbes. These cells patrol the bloodstream and await invaders within the tissues. The three basic kinds of killing cells are macrophages, neutrophils, and natural killer cells. Each uses a different tactic to kill invading microbes.

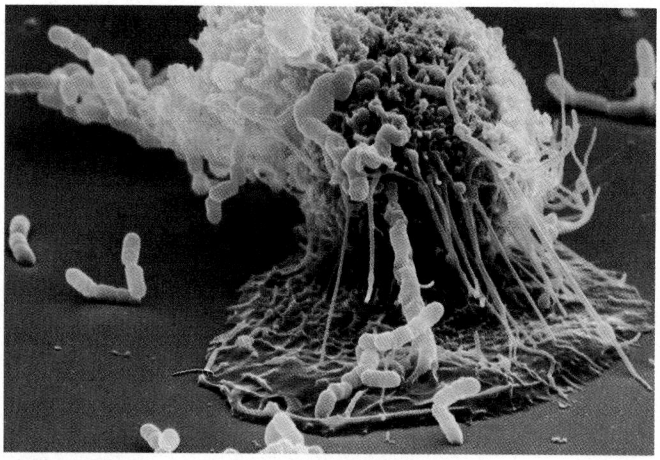

Figure 25.3 A macrophage in action.
In this scanning electron micrograph, a macrophage is "fishing" with long, sticky cytoplasmic extensions. Bacterial cells unfortunate enough to come in contact with the extensions are drawn back to the macrophage and engulfed.

Macrophages. White blood cells called **macrophages** (Greek, "big eaters") kill bacteria by ingesting them (figure 25.3), much as an amoeba ingests a food particle. Although some macrophages are anchored within particular organs, particularly the spleen, most patrol the byways of the body, circulating as precursor cells called **monocytes** in the blood, lymph, and fluid between cells. Macrophages are among the most actively mobile cells of the body.

Neutrophils. Other white blood cells called **neutrophils** act like kamikazes. They release chemicals (identical to household bleach) to "neutralize" the entire area, killing any bacteria in the neighborhood—and themselves in the process. A neutrophil is like a grenade tossed into an infection. It kills everything in the vicinity. Macrophages, by contrast, kill only one invading cell at a time but live to keep on doing it.

Natural Killer Cells. A third kind of white blood cell, called **natural killer cells,** does not attack invading microbes but rather the body cells that are infected by them. Natural killer cells puncture the membrane of the target cell, allowing water to rush in and causing the cell to swell and burst (figure 25.4). Natural killer cells are

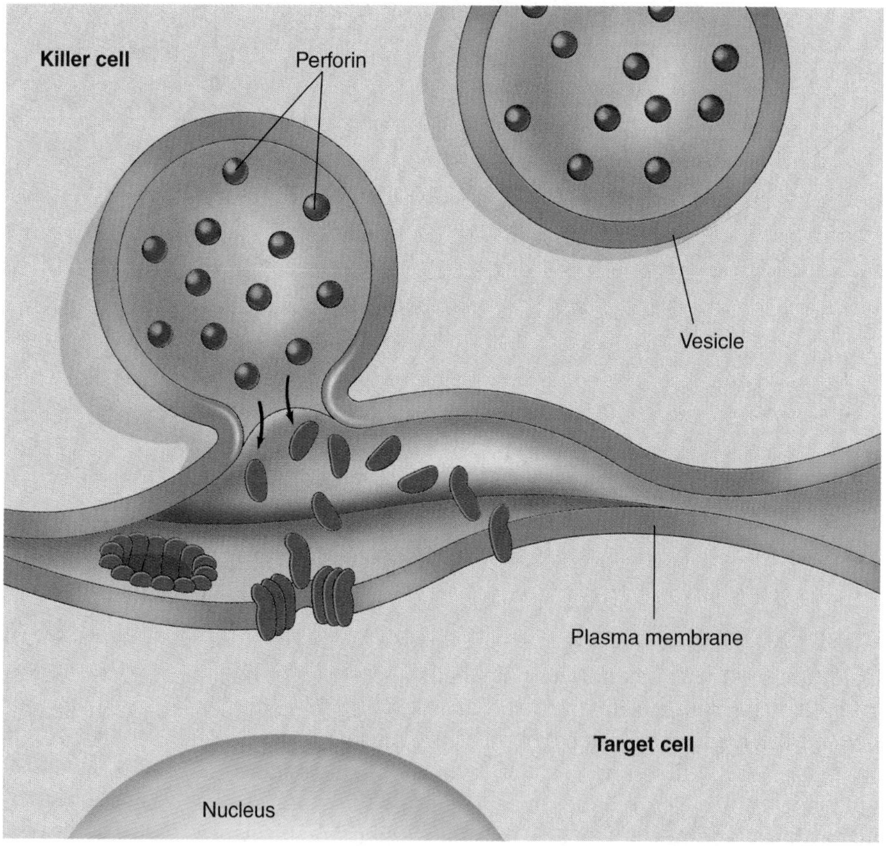

Killer cell
Perforin
Vesicle
Plasma membrane
Target cell
Nucleus

Figure 25.4 How natural killer cells attack target cells.
The initial event is the tight binding of the natural killer cell to the target cell. Binding initiates a chain of events within the killer cell in which vesicles loaded with perforin molecules move to the outer plasma membrane and expel their contents into the intercellular space over the target. The perforin molecules insert into the membrane like boards on a picket fence to form a pore that admits water and ruptures the cell.

particularly effective at detecting and attacking body cells that have been infected with viruses. They are also one of the body's most potent defenses against cancer, which they kill before the cancer cell has a chance to develop into a tumor (figure 25.5).

These three kinds of cells can distinguish the body's cells (self) from foreign cells (nonself) because the body's cells contain self-identifying *MHC proteins* (discussed later in this chapter). When the body's defensive cells fail to make the self versus nonself distinction correctly, they may attack the body's own tissues. Diseases resulting from this failure are known as *autoimmune diseases.*

Proteins That Kill Invading Microbes

The cellular defenses of vertebrates are complemented by a very effective chemical defense called the **complement system.** This system consists of approximately 20 different proteins that circulate freely in the blood plasma in an inactive state. Their defensive activity is triggered when they encounter the cell walls of bacteria or fungi. The complement proteins then aggregate to form a *membrane attack complex* that inserts itself into the foreign cell's plasma membrane, forming a pore like that produced by natural killer cells (figure 25.6). Water enters the foreign cell through this pore, causing the cell to swell and burst. Aggregation of the complement proteins is also triggered by the binding of antibodies to invading microbes, as we'll see in a later section.

The proteins of the complement system can augment the effects of other body defenses. Some amplify the inflammatory response (discussed next) by stimulating histamine release; others attract phagocytes (monocytes and neutrophils) to the area of infection; and still others coat invading microbes, roughening the microbes' surfaces so that phagocytes may attach to them more readily.

Another class of proteins that play a key role in body defense are interferons. There are three major categories of interferons: *alpha, beta,* and *gamma.* Almost all cells in the body make alpha and beta interferons. These polypeptides act as messengers that protect other cells in the vicinity from viral infection. The viruses are still able to penetrate the other cells, but the ability of the viruses to replicate and assemble new virus particles is inhibited. Gamma interferon is produced only by particular lymphocytes and natural killer cells. The secretion of gamma interferon by these cells is part of the immunological defense against infection and cancer.

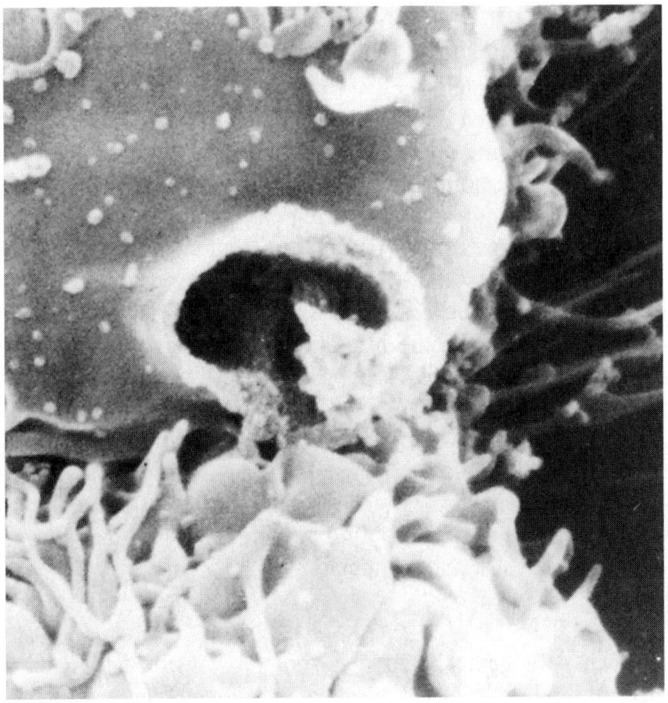

Figure 25.5 Death of a tumor cell.

A natural killer cell has attacked this cancer cell, punching a hole in its plasma membrane. Water has rushed in, making it balloon out. Soon it will burst.

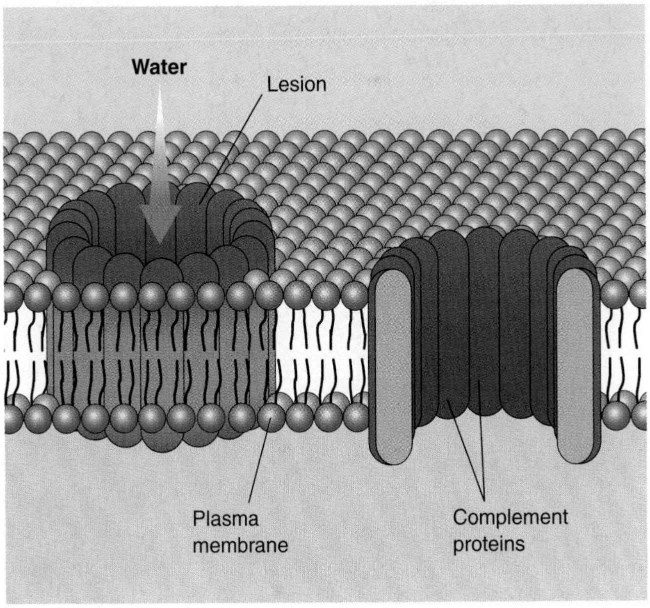
Water
Lesion
Plasma membrane
Complement proteins

Figure 25.6 How complement creates a hole in a cell.

The complement proteins form a transmembrane channel resembling the perforin-lined lesion made by natural killer cells. There are some differences between the two. The complement proteins are free-floating in blood plasma, and they attach to the invading microbe directly. The perforin molecules are produced in killer cells and poke holes in infected body cells, not the actual microbe. But the final results of the two defenses are very similar.

The Inflammatory Response

The aggressive cellular and chemical counterattacks to infection are made more effective by the **inflammatory response.** Infected or injured cells release chemical alarm signals, most notably histamine and prostaglandins, that cause blood vessels to expand, both increasing the flow of blood to the site of infection or injury and, by stretching their thin walls, making the capillaries more permeable (figure 25.7). This produces the redness and swelling so often associated with infection. The increased blood flow through larger, leakier capillaries promotes the migration of phagocytes to the site of infection. Neutrophils arrive first, spilling out chemicals that kill the microbes (as well as tissue cells in the vicinity and themselves). Monocytes follow, become macrophages, and then engulf the pathogens and remains of all the dead cells. This counterattack takes a considerable toll; the pus associated with infections is a mixture of dead or dying neutrophils, tissue cells, and pathogens.

The Temperature Response

Human pathogenic bacteria do not grow well at high temperatures. Thus when macrophages initiate their counterattack, they increase the odds in their favor by sending a message to the brain to raise the body's temperature. The cluster of brain cells that serves as the body's thermostat responds to the chemical signal by boosting the body temperature several degrees above the normal value of 37°C (98.6°F). The higher-than-normal temperature that results is called a **fever.** While fever is quite effective at inhibiting microbial growth, very high fevers are dangerous because excessive heat can inactivate critical cellular enzymes. In general, temperatures greater than 39.4°C (103°F) are considered dangerous; those greater than 40.6°C (105°F) are often fatal.

The second line of defense, with both chemical and cellular weapons, provides a sophisticated defense against microbial infection. Only occasionally do bacteria or viruses overwhelm this defense. When this happens, they face yet a third line of defense, more difficult to evade than any they have encountered. It is the immune system, the most elaborate of the body's defenses. Unlike other defenses, the immune system remembers previous encounters with potential invaders, and if they reappear, the immune system is ready for them.

25.2 Vertebrates respond to infection with a battery of cellular and chemical weapons, including cells and proteins that kill invading microbes and inflammatory and temperature responses.

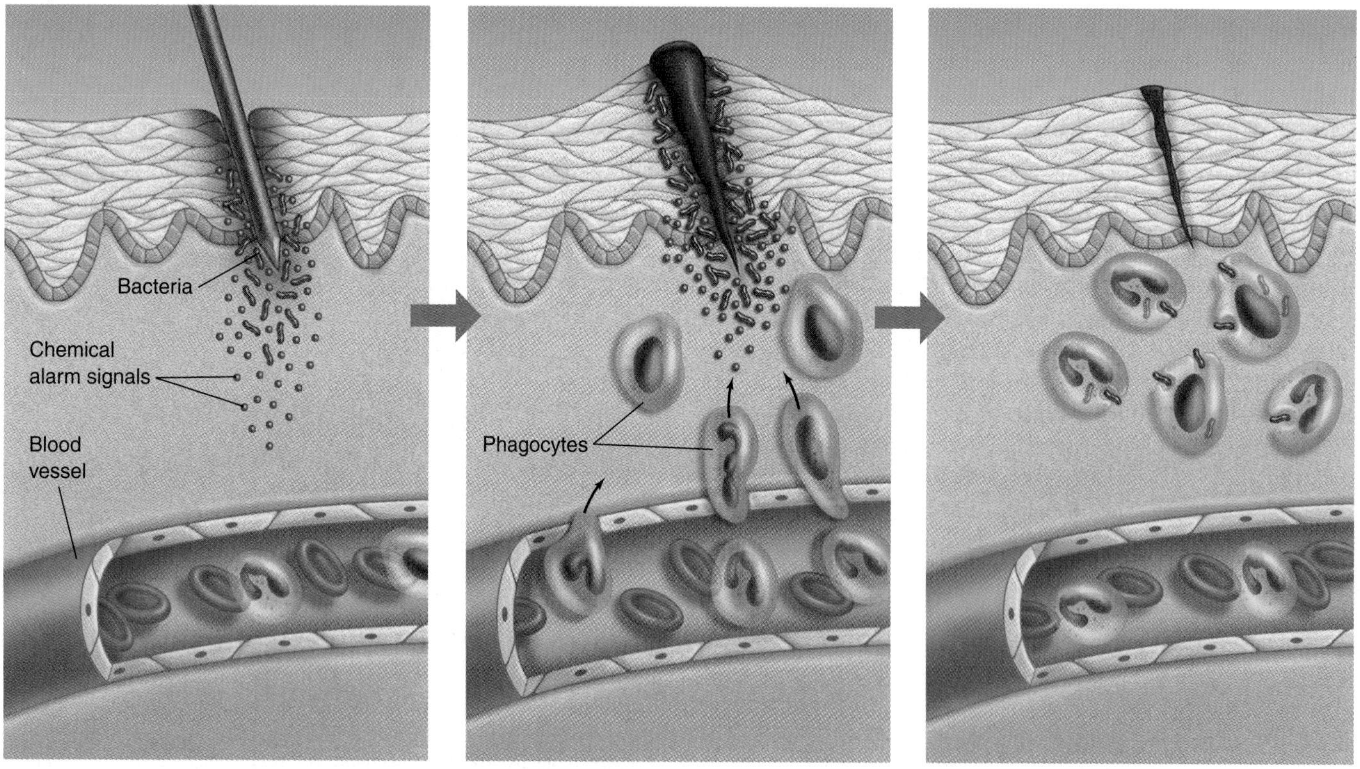

Bacteria
Chemical
alarm signals
Blood
vessel
Phagocytes

Figure 25.7 The events in a local inflammation.
When an invading microbe has penetrated the skin, chemicals, such as histamine and prostaglandins, cause nearby blood vessels to dilate. Increased blood flow brings a wave of phagocytic cells that attack and engulf invading bacteria.

25.3 The Immune System: The Third Line of Defense

The immune defense mechanisms of the body involve the actions of white blood cells, or leukocytes. They are very numerous—of the 100 trillion cells of your body, two in every 100 are white blood cells! Macrophages are white blood cells, as are neutrophils and natural killer cells. In addition, there are T cells, B cells, plasma cells, mast cells, and monocytes (table 25.1). T cells and B cells are called **lymphocytes** and are critical to the specific immune response.

After their origin in the bone marrow, **T cells** migrate to the thymus (hence the designation "T"), a gland just above the heart. There they develop the ability to identify microorganisms and viruses by the antigens exposed on their surfaces. An **antigen** is a molecule that provokes a specific immune response. Antigens are large, complex molecules, such as proteins, and they are generally foreign to the body, usually belonging to bacteria and viruses. Tens of millions of different T cells are made, each specializing in the recognition of one particular antigen. No invader can escape being recognized by at least a few T cells. There are four principal kinds of T cells: inducer T cells oversee the development of T cells in the thymus; helper T cells (often symbolized T_H) initiate the immune response; cytotoxic ("cell-poisoning") T cells (often symbolized T_C) lyse cells that have been infected by viruses; and suppressor T cells terminate the immune response.

Unlike T cells, **B cells** do not travel to the thymus; they complete their maturation in the bone marrow. (B cells are so named because they were originally characterized in a region of chickens called the bursa.) From the bone marrow, B cells are released to circulate in the blood and lymph. Individual B cells, like T cells, are specialized to recognize particular foreign antigens. When a B cell encounters the antigen to which it is targeted, it begins to divide rapidly, and its progeny differentiate into plasma cells and memory cells. Each plasma cell is a miniature factory producing markers called *antibodies*. These antibodies stick like flags to that antigen wherever it occurs in the body, marking any cell bearing the antigen for destruction.

25.3 T cells develop in the thymus, while B cells develop in the bone marrow. When a B cell encounters a specific antigen, it gives rise to plasma cells that produce antibodies.

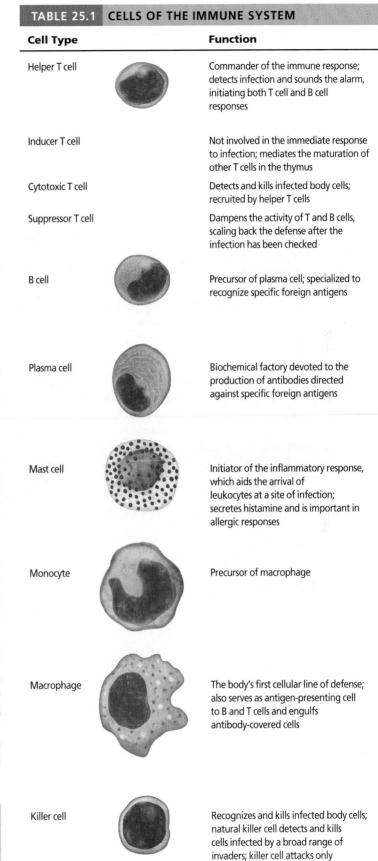

TABLE 25.1 CELLS OF THE IMMUNE SYSTEM

Cell Type		Function
Helper T cell		Commander of the immune response; detects infection and sounds the alarm, initiating both T cell and B cell responses
Inducer T cell		Not involved in the immediate response to infection; mediates the maturation of other T cells in the thymus
Cytotoxic T cell		Detects and kills infected body cells; recruited by helper T cells
Suppressor T cell		Dampens the activity of T and B cells, scaling back the defense after the infection has been checked
B cell	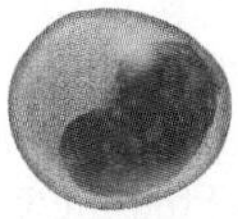	Precursor of plasma cell; specialized to recognize specific foreign antigens
Plasma cell	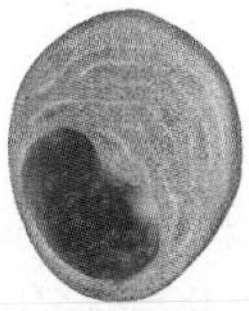	Biochemical factory devoted to the production of antibodies directed against specific foreign antigens
Mast cell	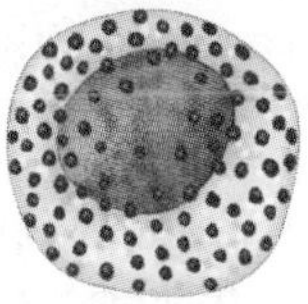	Initiator of the inflammatory response, which aids the arrival of leukocytes at a site of infection; secretes histamine and is important in allergic responses
Monocyte	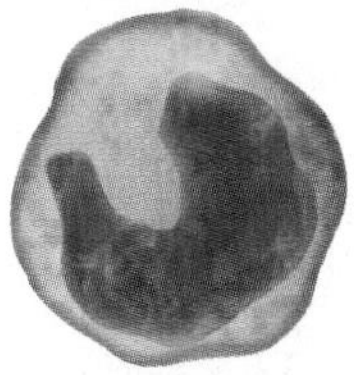	Precursor of macrophage
Macrophage	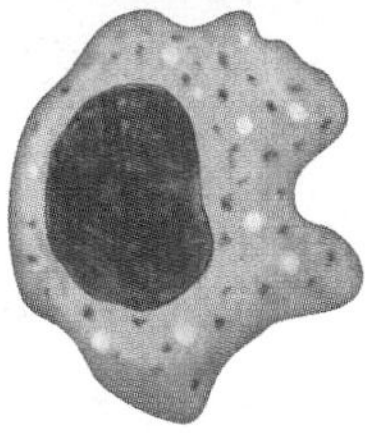	The body's first cellular line of defense; also serves as antigen-presenting cell to B and T cells and engulfs antibody-covered cells
Killer cell	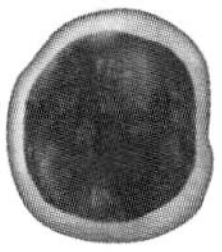	Recognizes and kills infected body cells; natural killer cell detects and kills cells infected by a broad range of invaders; killer cell attacks only antibody-covered cells

25.4 Evolution of the Immune System

All organisms possess mechanisms to protect themselves from the onslaught of smaller organisms and viruses. Bacteria defend against viral invasion by means of *restriction endonucleases*, enzymes that degrade any foreign DNA lacking the specific pattern of DNA methylation characteristic of that bacterium. Multicellular organisms face a more difficult problem in defense because their bodies often take up whole viruses, bacteria, or fungi instead of naked DNA.

Invertebrates

Invertebrate animals solve this problem by marking the surfaces of their cells with proteins that serve as "self " labels. Special amoeboid cells in the invertebrate attack and engulf any invading cells that lack such labels. By looking for the absence of specific markers, invertebrates employ a *negative* test to recognize foreign cells and viruses. This method provides invertebrates with a very effective surveillance system, although it has one great weakness: any microorganism or virus with a surface protein resembling the invertebrate self marker will not be recognized as foreign. An invertebrate has no defense against such a "copycat" invader.

In 1882, Russian zoologist Elie Metchnikoff became the first to recognize that invertebrate animals possess immune defenses. On a beach in Sicily, he collected the tiny transparent larva of a common sea star. Carefully he pierced it with a rose thorn. When he looked at the larva the next morning, he saw a host of tiny cells covering the surface of the thorn as if trying to engulf it (figure 25.8). The cells were attempting to defend the larva by ingesting the invader by phagocytosis (see figure 4.29). For this discovery of what came to be known as the **cellular immune response**, Metchnikoff was awarded the 1908 Nobel Prize in Medicine and Physiology, along with Paul Ehrlich for his work on the other major part of the immune defense, the antibody or **humoral immune response**. The invertebrate immune response shares several elements with the vertebrate immune response.

Phagocytes. All animals possess phagocytic cells that attack invading microbes. These phagocytic cells travel through the animal's circulatory system or circulate within the fluid-filled body cavity. In simple animals like sponges, which lack either a circulatory system or a body cavity, the phagocytic cells circulate among the spaces between cells.

Distinguishing Self from Nonself. The ability to recognize the difference between cells of one's own body and those of another individual appears to have evolved early in the history of life. Sponges, thought to be the oldest animals, attack grafts from other sponges, as do insects and sea stars. None of these invertebrates, however, exhibit any evidence of immunological memory; apparently, the antibody-based humoral immune defense did not evolve until the vertebrates.

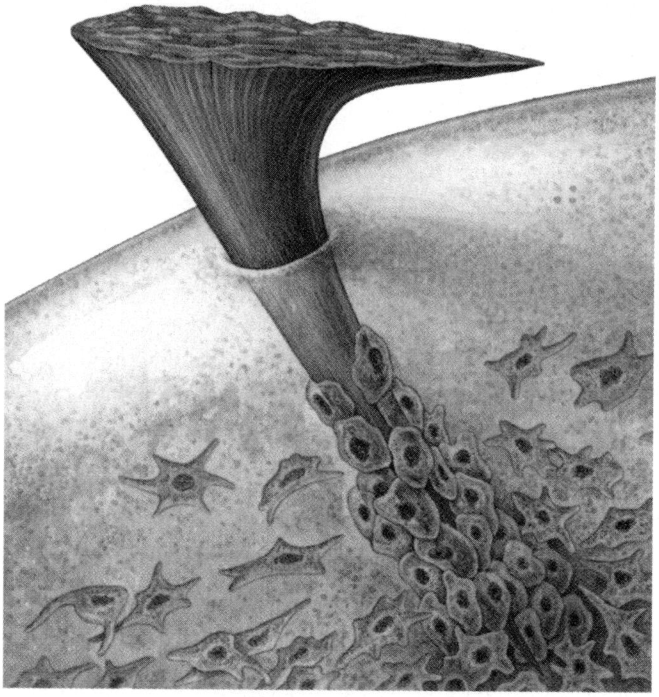

Figure 25.8 Discovering the cellular immune response in invertebrates.
In a Nobel Prize–winning experiment, the Russian zoologist Metchnikoff pierced the larva of a sea star with a rose thorn and the next day found tiny phagocytic cells covering the thorn.

Complement. While invertebrates lack complement, many arthropods (including crabs and a variety of insects) possess an analogous nonspecific defense called the prophenyloxidase (proPO) system. Like the vertebrate complement defense, the proPO defense is activated by a cascade of enzyme reactions, the last of which converts the inactive protein prophenyloxidase into the active enzyme phenyloxidase. Phenyloxidase both kills microbes and aids in encapsulating foreign objects.

Lymphocytes. Invertebrates also lack lymphocytes, but annelid earthworms and other invertebrates do possess lymphocyte-like cells that may be evolutionary precursors of lymphocytes.

Antibodies. All invertebrates possess proteins called lectins that may be the evolutionary forerunners of antibodies. Lectins bind to sugar molecules on cells, making the cells stick to one another. Lectins isolated from sea urchins, mollusks, annelids, and insects appear to tag invading microorganisms, enhancing phagocytosis. The genes encoding vertebrate antibodies are part of a very ancient gene family, the immunoglobulin superfamily. Proteins in this group all have a characteristic recognition structure called the *Ig fold.* The fold probably evolved as a self-recognition molecule in early metazoans. Insect immunoglobulins that bind to microbial surfaces and promote their destruction by phagocytes have

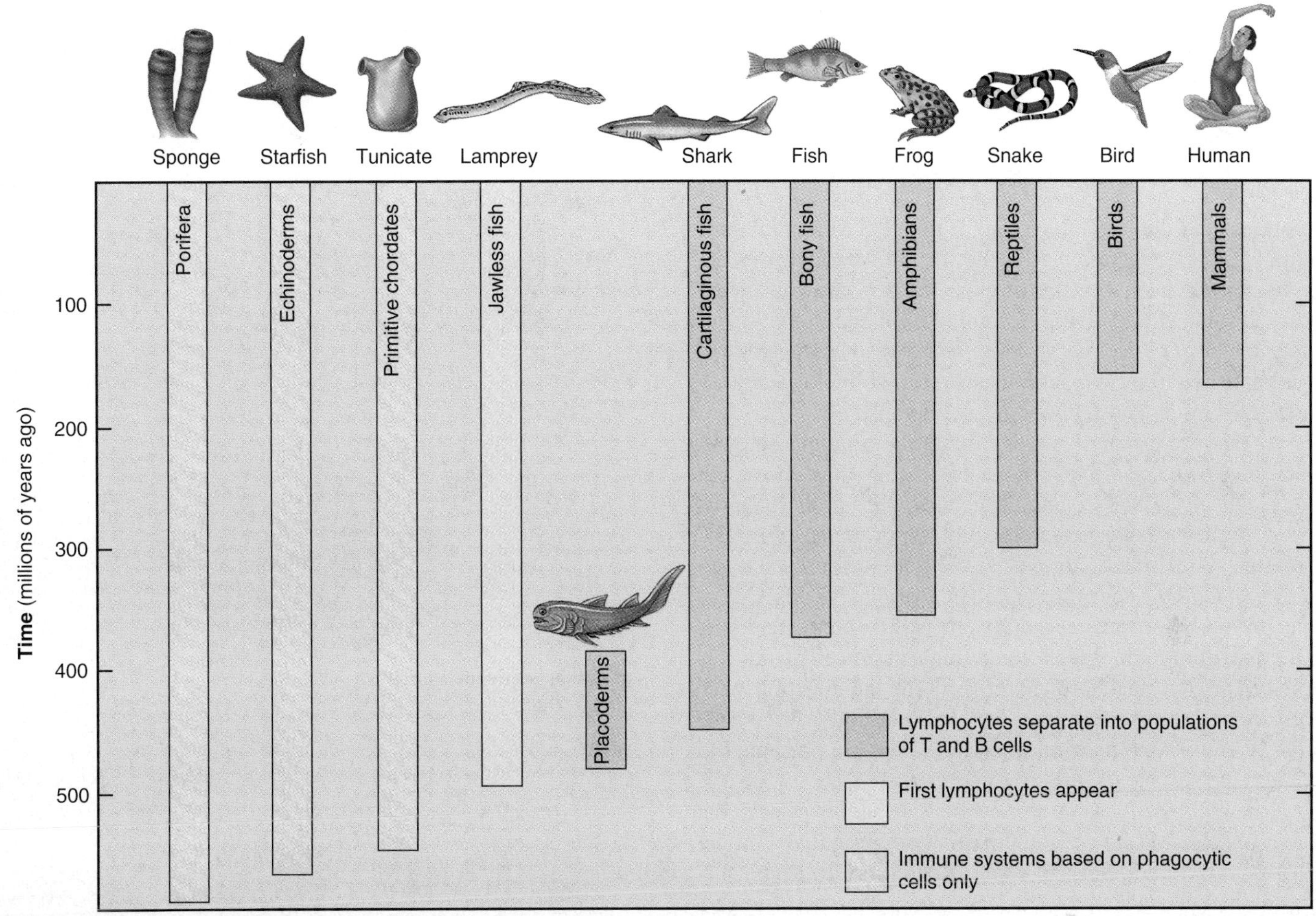
Sponge
Starfish
Tunicate
Lamprey
Shark
Fish
Frog
Snake
Bird
Human
Porifera
Echinoderms
Primitive chordates
Jawless fish
Placoderms
Cartilaginous fish
Bony fish
Amphibians
Reptiles
Birds
Mammals
Time (millions of years ago)
100
200
300
400
500
Lymphocytes separate into populations of T and B cells
First lymphocytes appear
Immune systems based on phagocytic cells only

Figure 25.9 How immune systems evolved in vertebrates.
Lampreys were the first vertebrates to possess an immune system based on lymphocytes, although distinct B cells and T cells did not appear until the jawed fishes evolved. By the time sharks and other cartilaginous fish appeared, the vertebrate immune response was fully formed.

been described in moths, grasshoppers, and flies. The antibody immune response appears to have evolved from these earlier, less complex systems.

Vertebrates

The earliest vertebrates of which we have any clear information, the jawless lampreys that first evolved some 500 million years ago, possess an immune system based on lymphocytes. At this early stage of vertebrate evolution, however, lampreys lack distinct populations of B cells and T cells such as those found in all higher vertebrates (figure 25.9).

With the evolution of fish with jaws, the modern vertebrate immune system first arose. These early fishes, including placoderms and spiny fishes, are now extinct. The oldest surviving group of jawed fishes are the sharks, which evolved some 450 million years ago. By then, the vertebrate immune defense had fully evolved. Sharks have an immune response much like that seen in mammals, with a cellular response carried out by T cell lymphocytes and an antibody-mediated humoral response carried out by B cells. The similarities of the cellular and humoral immune defenses are far more striking than the differences. Both sharks and mammals possess a thymus that produces T cells and a spleen that is a rich source of B cells. Evolution spanning 450 million years did little to change the antibody molecule—the amino acid sequences of shark and human antibody molecules are very similar. The most notable difference between sharks and mammals is that their antibody-encoding genes are arrayed somewhat differently.

25.4 The sophisticated two-part immune defense of mammals evolved about the time jawed fishes appeared. Before then, animals used a simpler immune defense based on mobile phagocytic cells. Many of the elements of today's mammalian immune response can be recognized in analogous systems in invertebrates.

25.5 Initiating the Immune Response

To understand how this third line of defense works, imagine you have just come down with the flu. Influenza viruses enter your body in small water droplets inhaled into your respiratory system. If they avoid becoming ensnared in the mucus lining the respiratory membranes (first line of defense), and avoid consumption by macrophages (second line of defense), the viruses infect and kill mucous membrane cells.

At this point macrophages initiate the immune defense. Macrophages inspect the surfaces of all cells they encounter. Every cell in the body carries special marker proteins on its surface called major histocompatibility proteins, or **MHC proteins.** The MHC proteins are different for each individual, much as fingerprints are. As a result, the MHC proteins on the tissue cells serve as self markers that enable the individual's immune system to distinguish its cells from foreign cells. This distinction is critical for the function of T cells because, unlike B cells, T cells cannot bind to free antigens. T cells can only bind to antigens that are presented to them on the surface of cells (figure 25.10).

When a foreign particle, such as a virus, infects the body, it is taken in by cells and partially digested. Within the cells, the viral antigens are processed and moved to the surface of the plasma membrane. The cells that perform this function are called **antigen-presenting cells.** At the membrane, the processed antigens are complexed with the MHC proteins. T cell receptors can only interact with cells that have this combination of MHC proteins and antigens.

Macrophages that encounter pathogens—either a foreign cell such as a bacterial one, which lacks proper MHC proteins, or a virus-infected body cell with telltale virus proteins stuck to its surface—respond by secreting a chemical alarm signal. The alarm signal is a protein called **interleukin-1** (Latin for "between white blood cells"). This protein stimulates **helper T cells.** The helper T cells respond to the interleukin-1 alarm by simultaneously initiating two different parallel lines of immune system defense: the cellular immune response carried out by T cells and the antibody or humoral response carried out by B cells. The immune response carried out by T cells is called the cellular response because the T lymphocytes attack the cells that carry antigens. The B cell response is called the humoral response because antibodies are secreted into the blood and body fluids (*humor* refers to a body fluid).

25.5 When macrophages encounter cells without the proper MHC proteins, they secrete a chemical alarm that initiates the immune defense.

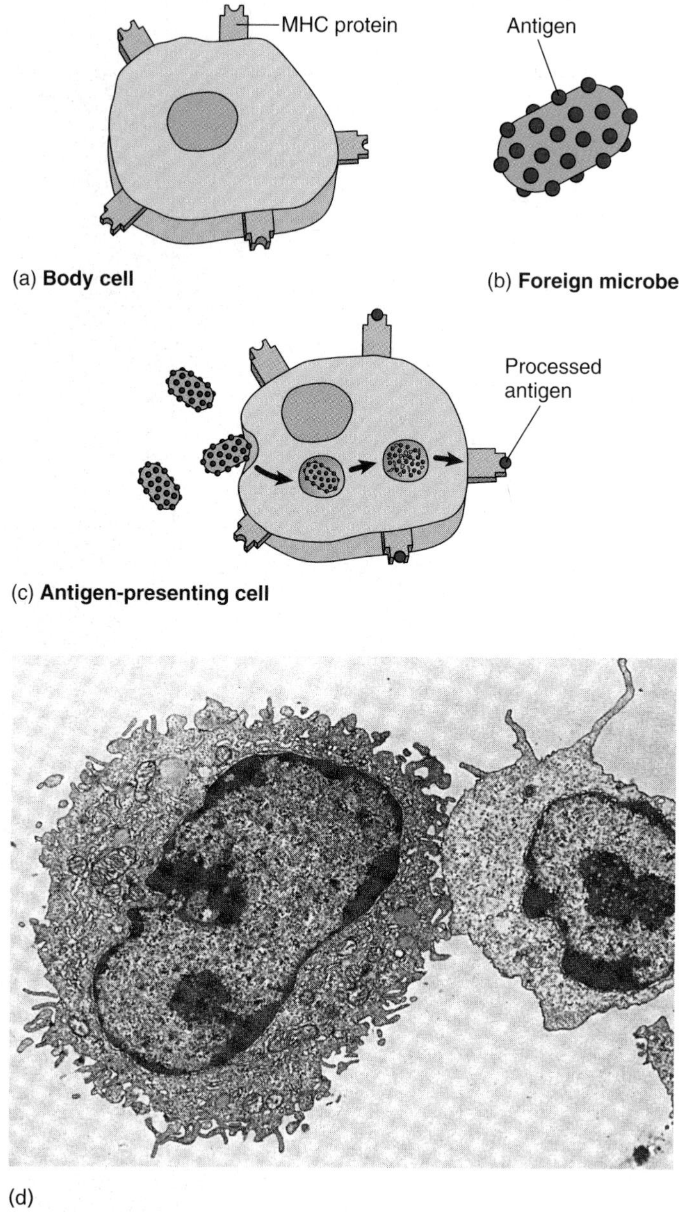
MHC protein
Antigen
(a) **Body cell**
(b) **Foreign microbe**
Processed antigen
(c) **Antigen-presenting cell**
(d)

Figure 25.10 How antigens are presented.
(*a*) Cells of the body have MHC proteins on their surfaces that identify them as "self" cells. Immune system cells do not attack these cells. (*b*) Foreign cells or microbes have antigens on their surfaces. B lymphocytes are able to bind directly to free antigens in the body to initiate an attack on a foreign invader. (*c*) T cells can bind to the antigens to initiate an attack only after the antigens are processed and complexed with MHC proteins on the surface of an antigen-presenting cell. (*d*) In this electron micrograph, a lymphocyte (*right*) contacts a macrophage (*left*), an antigen-presenting cell.

25.6 T Cells: The Cellular Response

When macrophages process the foreign antigens, they secrete interleukin-1, which stimulates cell division and proliferation of T cells. Helper T cells become activated when they bind to the complex of MHC proteins and antigens presented to them by the macrophages. The helper T cells then secrete **interleukin-2,** which stimulates the proliferation of **cytotoxic T cells,** which recognize and destroy body cells infected with the virus (figure 25.11). Cytotoxic T cells can destroy infected cells only if those cells display the foreign antigen together with their MHC proteins.

The body makes millions of different types of T cells. Each type bears a single, unique kind of receptor protein on its membrane, a receptor able to bind to a particular antigen-MHC protein complex on the surface of an antigen-presenting cell. Any cytotoxic T cell whose receptor fits the particular antigen-MHC protein complex present in the body begins to multiply rapidly, soon forming large numbers of T cells capable of recognizing the complex containing the particular foreign protein. Large numbers of infected cells can be quickly eliminated, because the single T cell able to recognize the invading virus is amplified in number to form a large clone of identical T cells, all able to carry out the attack.

Any of the body's cells that bear traces of viral infection are destroyed. The method used by cytotoxic T cells to kill infected body cells is similar to that used by natural killer cells and complement—they puncture the plasma membrane of the infected cell. Cancer cells are also detected and eliminated by this screening.

25.6 The cellular immune response is carried out by T cells, which mount an immediate attack on infecting and infected cells, killing any that present unusual surface antigens.

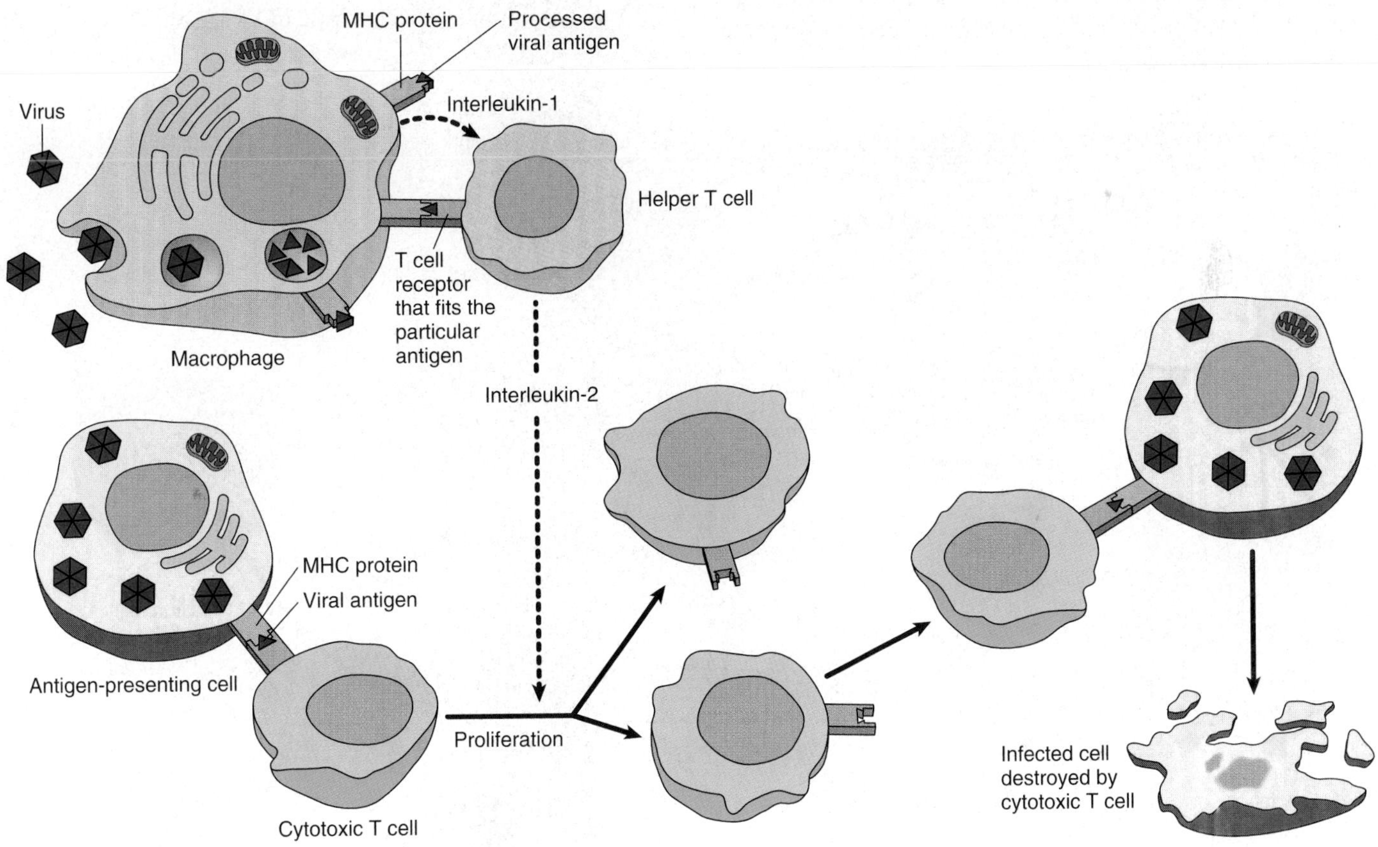

MHC protein
Processed viral antigen
Virus
Interleukin-1
Helper T cell
T cell receptor that fits the particular antigen
Macrophage
Interleukin-2
MHC protein
Viral antigen
Antigen-presenting cell
Proliferation
Cytotoxic T cell
Infected cell destroyed by cytotoxic T cell

Figure 25.11 The T cell immune defense.

After a macrophage has processed an antigen, it releases interleukin-1, signaling helper T cells to bind to the antigen-MHC protein complex. This triggers the helper T cell to release interleukin-2, which stimulates the multiplication of cytotoxic T cells. In addition, proliferation of cytotoxic T cells is stimulated when a T cell with a receptor that fits the antigen displayed by an antigen-presenting cell binds to the antigen-MHC protein complex. Body cells that have been infected by the antigen are destroyed by the cytotoxic T cells. As the infection subsides, suppressor T cells "turn off" the immune response.

25.7 B Cells: The Humoral Response

B cells also respond to helper T cells activated by interleukin-1. Like cytotoxic T cells, B cells have receptor proteins on their surface, one type of receptor for each type of B cell. B cells recognize invading microbes much as cytotoxic T cells recognize infected cells, but unlike cytotoxic T cells, they do not go on the attack themselves. Rather, they mark the pathogen for destruction by mechanisms that have no "ID check" system of their own. Early in the immune response, the markers placed by B cells alert complement proteins to attack the cells carrying them. Later in the immune response, the markers placed by B cells activate macrophages and natural killer cells.

The way B cells do their marking is simple and foolproof. Unlike the receptors on T cells, which bind only to antigen-MHC protein complexes on antigen-presenting cells, B cell receptors can bind to free, unprocessed antigens. When a B cell encounters an antigen, antigen particles enter the B cell by endocytosis and get processed. Helper T cells that are able to recognize the specific antigen bind to the antigen-MHC protein complex on the B cell and release interleukin-2, which stimulates the B cell to divide. In addition, free, unprocessed antigens stick to antibodies on the B cell surface. This antigen exposure triggers even more B cell proliferation. B cells divide to produce long-lived memory B cells and plasma cells that serve as short-lived antibody factories (figure 25.12).

Antibodies are proteins in a class called **immunoglobulins** (abbreviated *Ig*), which is divided into subclasses based on the structures and functions of the antibodies. The five different immunoglobulin subclasses are as follows:

1. **IgM.** This is the first type of antibody to be secreted into the blood during the primary response and to serve as a receptor on the B cell surface. These antibodies also promote agglutination reactions (causing antigen-containing particles to stick together, or agglutinate).
2. **IgG.** This is the major form of antibody secreted in a secondary response and the major one in the blood plasma.
3. **IgD.** These antibodies serve as receptors for antigens on the B cell surface. Their other functions are unknown.
4. **IgA.** This is the form of antibody in external secretions, such as saliva and mother's milk.
5. **IgE.** This form of antibody promotes the release of histamine and other agents that produce allergic symptoms, such as those of hay fever.

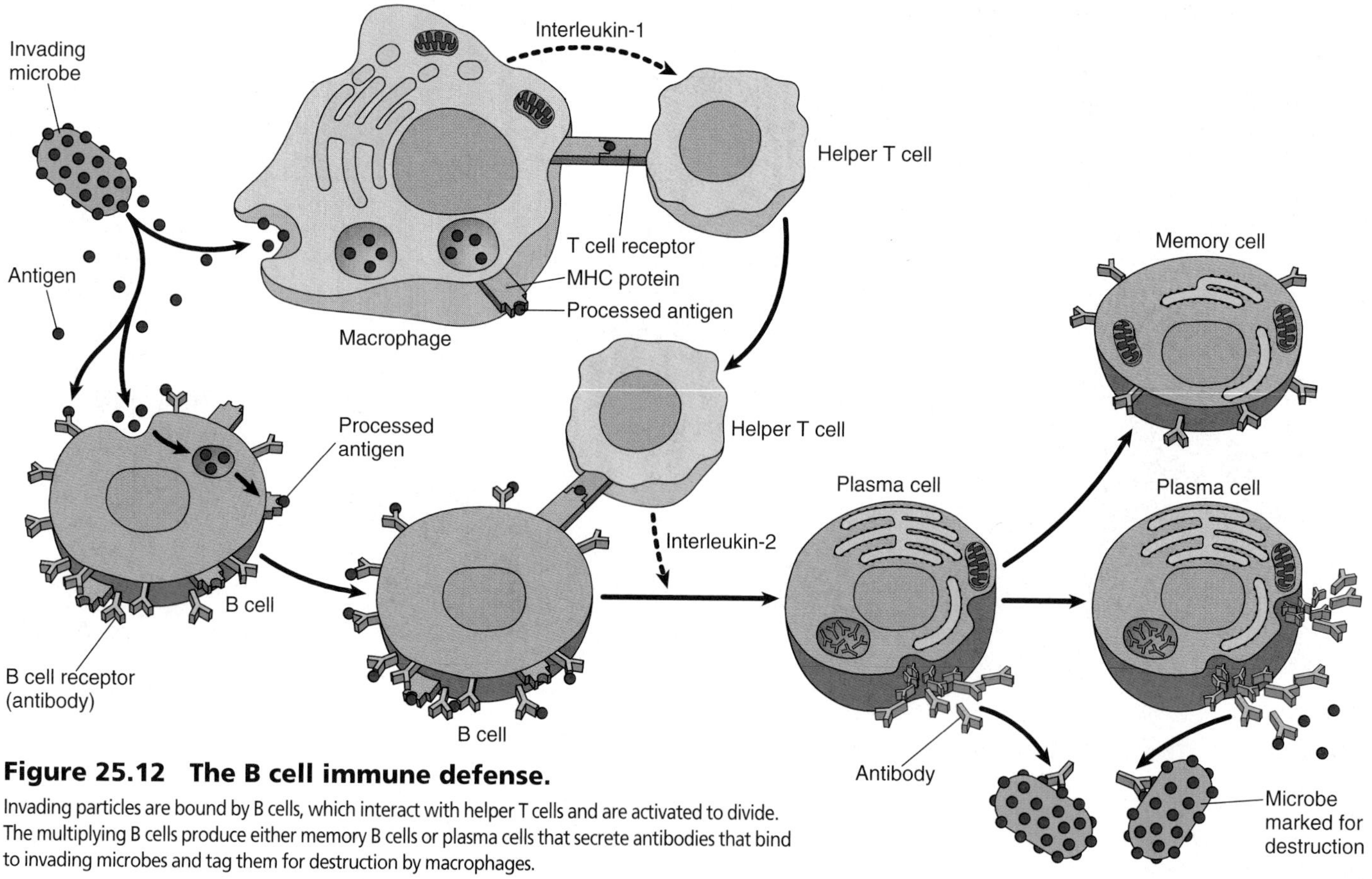

Invading microbe
Interleukin-1
Helper T cell
Antigen
T cell receptor
MHC protein
Processed antigen
Macrophage
Memory cell
Processed antigen
Helper T cell
Plasma cell
Plasma cell
Interleukin-2
B cell
B cell receptor (antibody)
B cell
Antibody
Microbe marked for destruction

Figure 25.12 The B cell immune defense.
Invading particles are bound by B cells, which interact with helper T cells and are activated to divide. The multiplying B cells produce either memory B cells or plasma cells that secrete antibodies that bind to invading microbes and tag them for destruction by macrophages.

Plasma cells produce lots of the same particular antibody that was able to bind to antigen in the initial immune response. Flooding through the bloodstream, these antibody proteins (figure 25.13) are able to stick to antigens on any cells or microbes that present them, flagging those cells and microbes for destruction. Complement proteins, macrophages, or natural killer cells then destroy the antibody-displaying cells and microbes.

The B cell defense is very powerful because it amplifies the reaction to an initial pathogen encounter a millionfold. It is also a very long-lived defense in that a few of the multiplying B cells do not become antibody producers. Instead they become a line of **memory B cells** that continue to patrol your body's tissues, circulating through your blood and lymph for a long time—sometimes for the rest of your life.

Antibody Diversity

The vertebrate immune system is capable of recognizing as foreign practically any nonself molecule presented to it—literally millions of different antigens. Although vertebrate chromosomes contain only a few hundred receptor-encoding genes, it is estimated that human B cells can make between 10^6 and 10^9 different antibody molecules. How do vertebrates generate millions of different antigen receptors when their chromosomes contain only a few hundred versions of the genes encoding those receptors?

The answer to this question is that the millions of immune receptor genes do not exist as single sequences of nucleotides. Rather, they are assembled by stitching together three or four DNA segments that code for different parts of the receptor molecule. When an antibody is assembled, these different sequences of DNA are brought together to form a composite gene (figure 25.14). This process is called **somatic rearrangement.**

Two other processes generate even more sequences. First, the DNA segments are often joined together with one or two nucleotides off-register, shifting the reading frame during gene transcription and so generating a totally different sequence of amino acids in the protein. Second, random mistakes occur during successive DNA replications as the lymphocytes divide during clonal expansion. Both mutational processes produce changes in amino acid sequences, a phenomenon known as **somatic mutation** because it takes place in a somatic cell rather than in a gamete.

Because a cell may end up with any heavy chain gene and any light chain gene during its maturation, the total number of different antibodies possible is staggering: 16,000 heavy chain combinations × 1,200 light chain combinations = 19 million different possible antibodies. If one also takes into account the changes induced by somatic mutation, the total can exceed 200 million! It should be understood that although this discussion has centered on B cells and their receptors, the receptors on T cells are as diverse as those on B cells because they also are subject to similar somatic rearrangements and mutations.

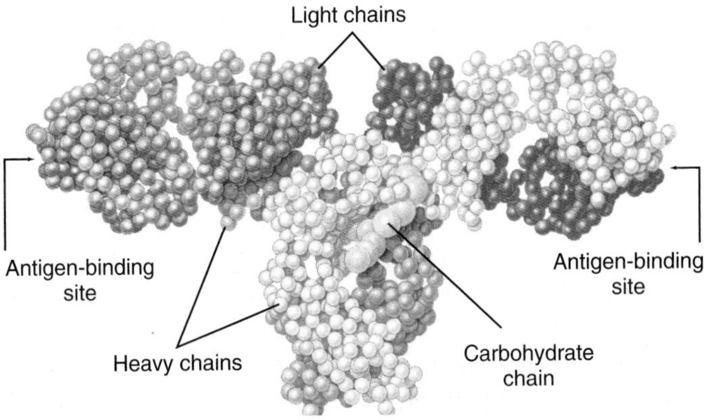
Light chains
Antigen-binding site
Antigen-binding site
Heavy chains
Carbohydrate chain

Figure 25.13 An antibody molecule.

In this molecular model of an antibody molecule, each amino acid is represented by a small sphere. Each molecule consists of four protein chains, two identical "light" (*red*) and two identical "heavy" (*blue*). The four protein chains wind around one another to form a Y shape. Foreign molecules, called antigens, bind to the arms of the Y.

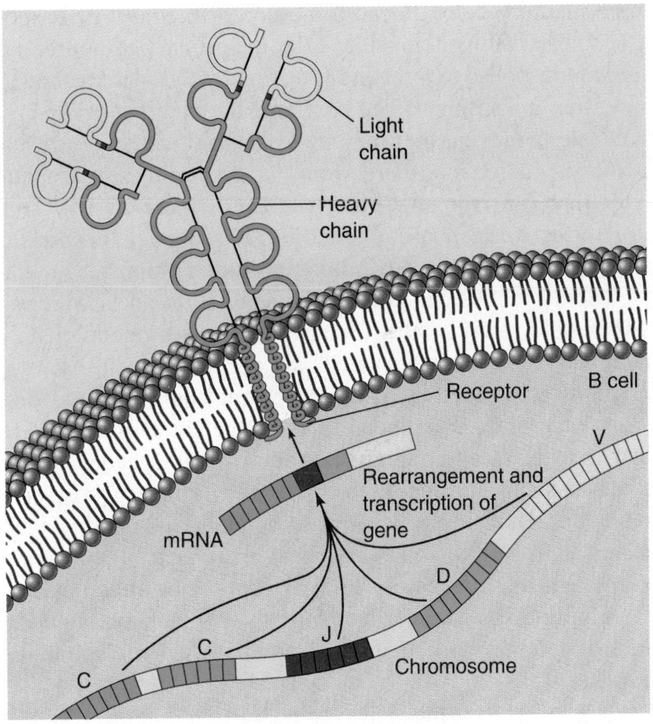
Light chain
Heavy chain
Receptor
B cell
V
Rearrangement and transcription of gene
mRNA
D
C
J
C
Chromosome

Figure 25.14 The lymphocyte receptor molecule is produced by a composite gene.

Different regions of the DNA code for different regions of the receptor structure (*C*, constant regions; *J*, joining regions; *D*, diversity regions; and *V*, variable regions) and are brought together to make a composite gene that codes for the receptor. Through different somatic rearrangements of these DNA segments, an enormous number of different receptor molecules can be produced.

25.7 In the humoral immune response, B cells label infecting and infected cells for destruction by complement proteins, natural killer cells, and macrophages.

25.8 Active Immunity Through Clonal Selection

As we discussed earlier, B cells and T cells have receptors on their cell surfaces that recognize and bind to specific antigens. When a particular antigen enters the body, it must, by chance, encounter the specific lymphocyte with the appropriate receptor in order to provoke an immune response. The first time a pathogen invades the body, there are only a few B cells or T cells that may have the receptors that can recognize the invader's antigens. Binding of the antigen to its receptor on the lymphocyte surface, however, stimulates cell division and produces a *clone* (a population of genetically identical cells). This process is known as **clonal selection.** In this first encounter, there are only a few cells that can mount an immune response and the response is relatively weak. This is called a **primary immune response** (figure 25.15).

If the primary immune response involves B cells, some become plasma cells that secrete antibodies, and some become memory cells. Because a clone of memory cells specific for that antigen develops after the primary response, the immune response to a second infection by the same pathogen is swifter and stronger. The next time the body is invaded by the same pathogen, the immune system is ready. As a result of the first infection, there is now a large clone of lymphocytes that can recognize that pathogen (figure 25.16). This more effective response, elicited by subsequent exposures to an antigen, is called a **secondary immune response.**

Memory cells can survive for several decades, which is why people rarely contract measles a second time after they have had it once. Memory cells are also the reason that vaccinations are effective. The viruses causing childhood diseases have surface antigens that change little from year to year, so the same antibody is effective for decades. Other diseases, such as influenza, are caused by viruses whose protein coat–specifying genes mutate rapidly. This rapid genetic change causes new strains to appear every year or so that are not recognized by memory cells from previous infections.

Figure 25.17 summarizes how the cellular and humoral lines of defense work together to produce the body's immune response.

25.8 A strong immune response is possible because infecting cells stimulate the few responding B cells and T cells to divide repeatedly, forming clones of responding cells.

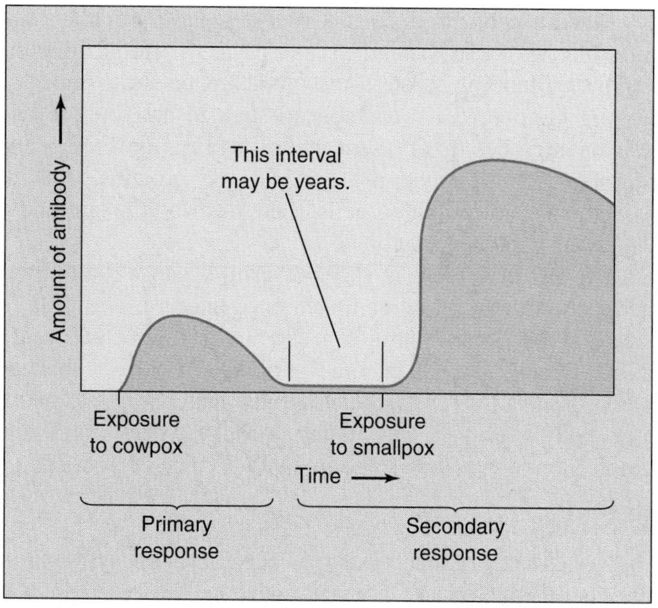

Amount of antibody
This interval may be years.
Exposure to cowpox
Exposure to smallpox
Time
Primary response
Secondary response

Figure 25.15 The development of active immunity.

Immunity to smallpox in Edward Jenner's patients (see next section) occurred because their inoculation with cowpox stimulated the development of lymphocyte clones with receptors that could bind not only to cowpox but also to smallpox antigens. As a result of clonal selection, a second exposure, this time to smallpox, stimulates the immune system to produce large amounts of the antibody more rapidly than before.

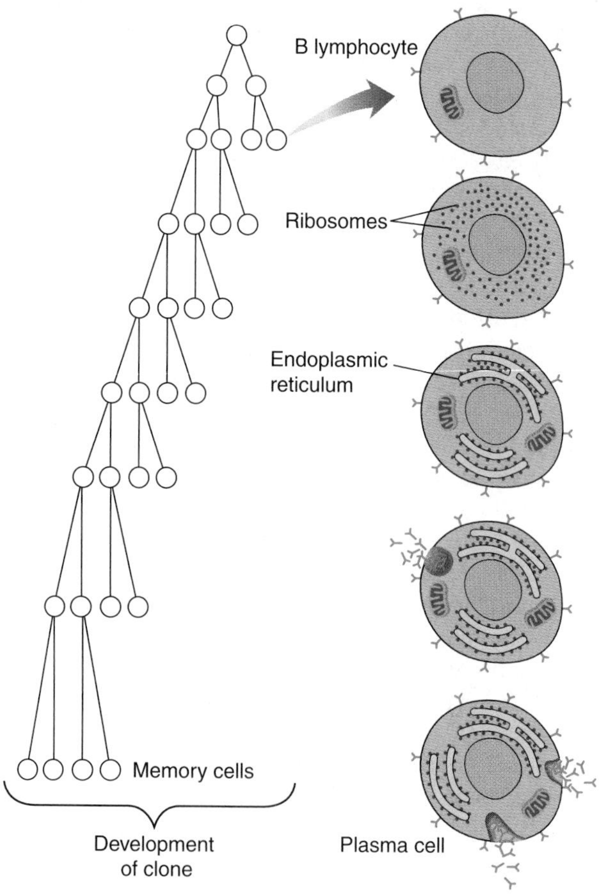

B lymphocyte
Ribosomes
Endoplasmic reticulum
Memory cells
Development of clone
Plasma cell

Figure 25.16 The clonal selection theory of active immunity.

In response to interaction with an antigen that binds specifically to its surface receptors, a B cell divides many times to produce a clone of B cells. Some of these become plasma cells that secrete antibodies for the primary response, while others become memory cells that await subsequent exposures to the antigen for the mounting of a secondary immune response.

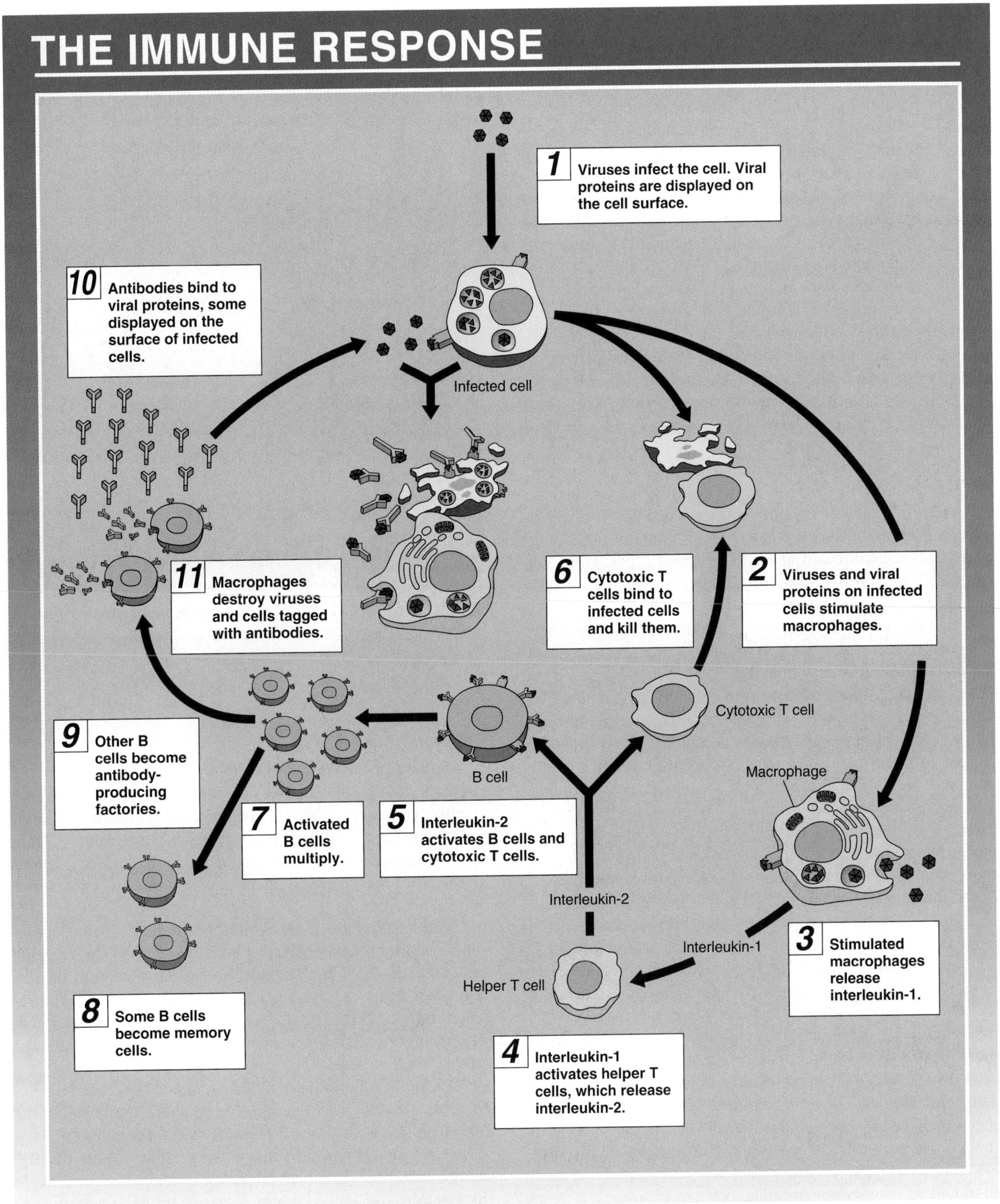
THE IMMUNE RESPONSE
1 Viruses infect the cell. Viral proteins are displayed on the cell surface.
Infected cell
2 Viruses and viral proteins on infected cells stimulate macrophages.
Macrophage
3 Stimulated macrophages release interleukin-1.
Interleukin-1
Helper T cell
4 Interleukin-1 activates helper T cells, which release interleukin-2.
Interleukin-2
5 Interleukin-2 activates B cells and cytotoxic T cells.
B cell
Cytotoxic T cell
6 Cytotoxic T cells bind to infected cells and kill them.
7 Activated B cells multiply.
8 Some B cells become memory cells.
9 Other B cells become antibody-producing factories.
10 Antibodies bind to viral proteins, some displayed on the surface of infected cells.
11 Macrophages destroy viruses and cells tagged with antibodies.

Figure 25.17 How the immune response works.

25.9 Vaccination

In 1796, an English country doctor named Edward Jenner carried out an experiment that marks the beginning of the study of immunology. Smallpox was a common and deadly disease in those days. Jenner observed, however, that milkmaids who had caught a much milder form of "the pox" called cowpox (presumably from cows) rarely caught smallpox. Jenner set out to test the idea that cowpox conferred protection against smallpox. He infected people with mild cowpox (figure 25.18), and as he had predicted, many of them became immune to smallpox.

We now know that smallpox and cowpox are caused by two different but similar viruses. Jenner's patients who were injected with the cowpox virus mounted a defense that was also effective against a later infection of the smallpox virus. Jenner's procedure of injecting a harmless microbe in order to confer resistance to a dangerous one is called vaccination. **Vaccination** is the introduction into your body of a dead or disabled pathogen or, more commonly these days, of a harmless microbe with pathogen proteins displayed on its surface. The vaccination triggers an immune response against the pathogen, without an infection ever occurring. Afterwards, the bloodstream of the vaccinated person contains circulating memory B cells directed against that specific pathogen. The vaccinated person is said to have been "immunized" against the disease.

Through genetic engineering, scientists are now routinely able to produce "piggyback," or subunit, vaccines made of harmless viruses that contain in their DNA a single gene cut out of a pathogen, a gene encoding a protein normally exposed on the pathogen's surface. By splicing the pathogen gene into the DNA of the harmless host, that host is induced to display the protein on its surface. The harmless virus displaying the pathogen protein is like a sheep in wolf's clothing, unable to hurt you but raising alarm as if it could. Your body responds to its presence by making an antibody directed against the pathogen protein, an antibody that acts like an alarm to the immune system, should that pathogen ever visit your body.

If the activities of memory B cells provide such an effective defense against future infection, why can you catch some diseases like flu more than once? The reason you don't stay immune to flu is that the flu virus has evolved a way to evade the immune system—it changes. The genes encoding the surface proteins of the flu virus mutate very rapidly. Thus the shapes of these surface proteins alter swiftly. Your memory B cells do not recognize viruses with altered surface proteins as being the same viruses they have already successfully defeated or been vaccinated against, because the memory B cells' receptors no longer "fit" the new shape of the flu surface proteins. When the new version of flu virus invades your body, you are back where you started and need to mount an entirely new immune defense.

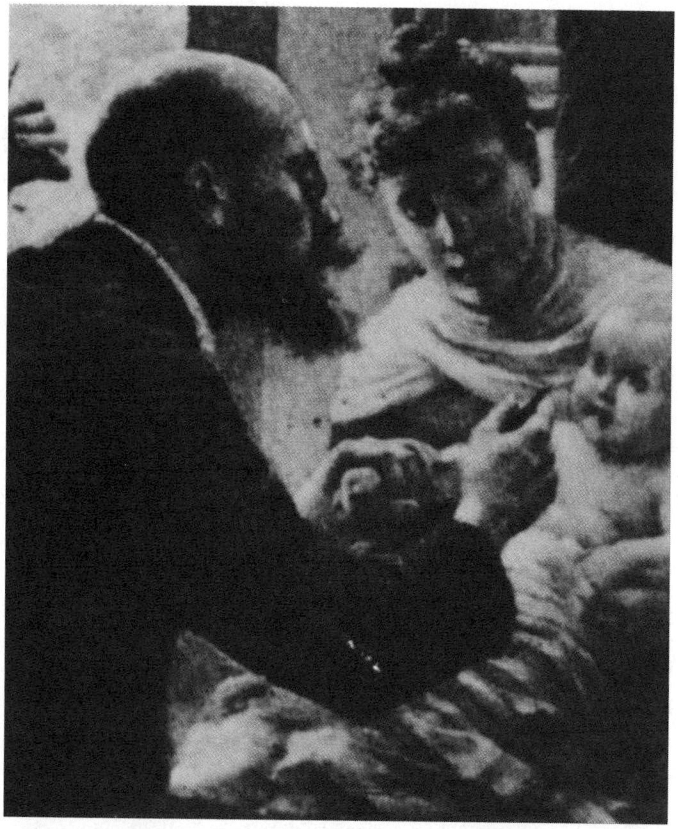

Figure 25.18 The birth of immunology.
This famous painting shows Edward Jenner inoculating patients with cowpox in the 1790s and thus protecting them from smallpox.

Sometimes the new mutations cause the flu virus surface proteins to assume shapes the immune system does not readily recognize. When this happened in 1918, over 22 million Americans and Europeans died in 18 months (figure 25.19). Less profound changes in flu virus surface proteins occur periodically, resulting in new strains of flu for which we are not immune.

One of the most intensive efforts in the history of medicine is currently underway to develop an effective vaccine against HIV, the virus responsible for AIDS (figure 25.20). The HIV virus has nine genes, encoding a variety of proteins. Initial efforts focused on producing a subunit vaccine containing the HIV *env* (envelope) gene, which encodes the protein on the outside at the virus.

Unfortunately for these initial attempts to develop an AIDS vaccine, the HIV virus mutates even more rapidly than the flu virus. Even vaccines that work in the laboratory are not effective outside the laboratory, where new strains of HIV are encountered. This high mutation rate has been the single biggest obstacle to developing a successful AIDS vaccine.

New vaccine approaches look more promising. While mutations of the *env* gene are frequent, they are random events. Because no two virus particles create the same new mutations at the same instant, a vaccine simultaneously directed against three different HIV proteins has a good chance of working—it is very unlikely that a single HIV particle will mutate all three genes simultaneously. To further strengthen new vaccines, multi-protein subunit vaccines are being supplemented with HIV DNA to activate T cell immune defenses. These new AIDS vaccines work well in monkeys, and nineteen different vaccines started human clinical trails in the year 2000.

As you can see, the immune system, often aided by vaccination, can respond in a variety of different ways to different kinds of pathogens. However, as discussed in the next section, its ability to function normally and efficiently is often disrupted.

25.9 Vaccines introduce antigens similar or identical to those of pathogens, eliciting an immune defense that defends against the pathogen too.

Figure 25.19 The flu epidemic of 1918 killed 22 million in 18 months.

With 25 million Americans alone infected during the influenza epidemic, it was hard to provide care for everyone. The Red Cross often worked around the clock.

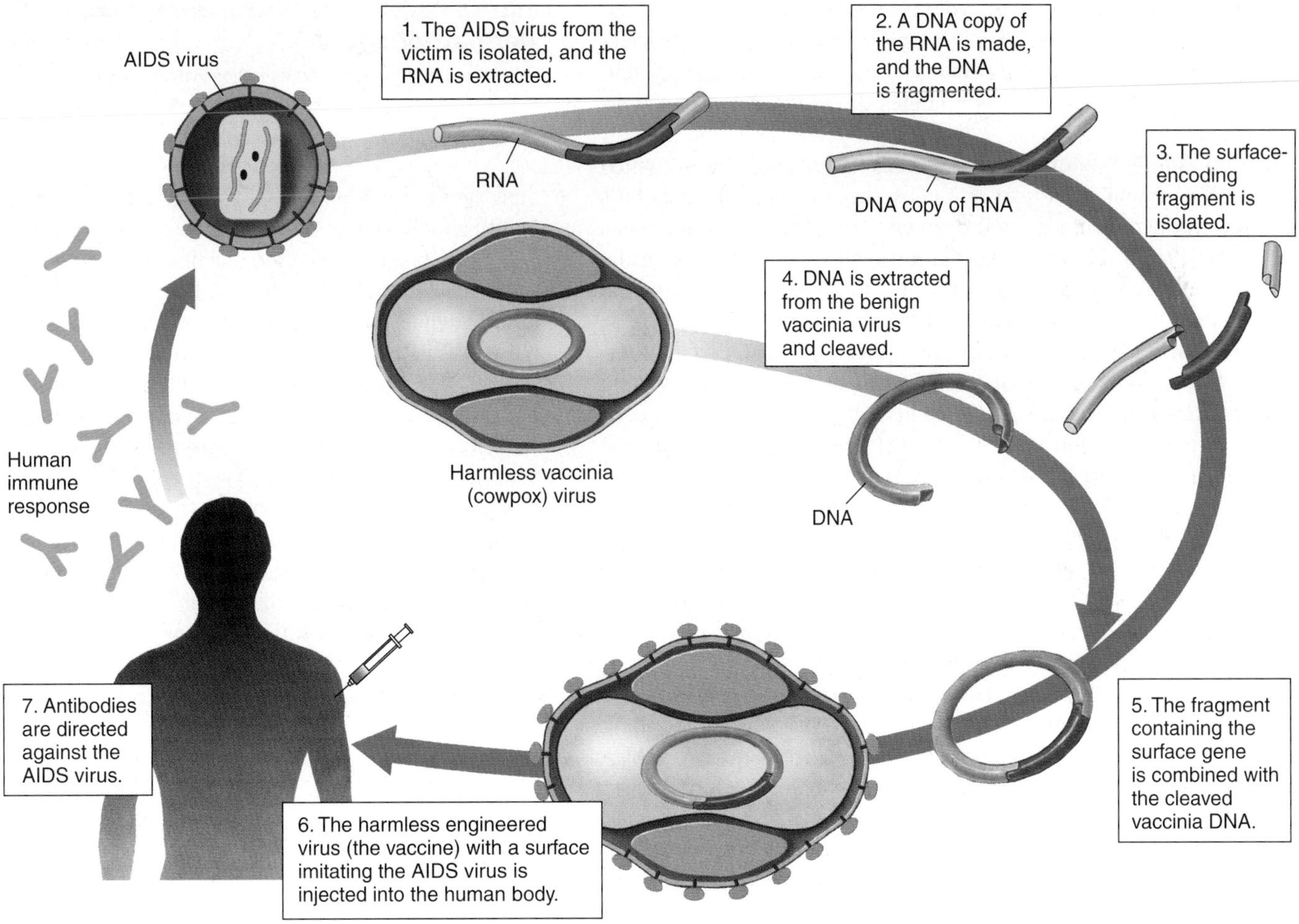

AIDS virus
1. The AIDS virus from the victim is isolated, and the RNA is extracted.
RNA
2. A DNA copy of the RNA is made, and the DNA is fragmented.
DNA copy of RNA
3. The surface-encoding fragment is isolated.
4. DNA is extracted from the benign vaccinia virus and cleaved.
Harmless vaccinia (cowpox) virus
DNA
Human immune response
5. The fragment containing the surface gene is combined with the cleaved vaccinia DNA.
6. The harmless engineered virus (the vaccine) with a surface imitating the AIDS virus is injected into the human body.
7. Antibodies are directed against the AIDS virus.

Figure 25.20 Researchers are attempting to construct an AIDS vaccine.

25.10 Immune System Failure

Although the immune system is one of the most sophisticated systems of the vertebrate body, it is still not perfect. Many of the major diseases we face, and some minor irritations as well, reflect failure of the immune system.

Autoimmune Diseases

The ability of killer T cells and B cells to distinguish cells of your own body—"self" cells—from nonself cells is the key ability of the immune system that makes your body's third line of defense so effective. In certain diseases, this ability breaks down, and the body attacks its own tissues. Such diseases are called **autoimmune diseases.**

Multiple sclerosis is an autoimmune disease that usually strikes people between the ages of 20 and 40. In multiple sclerosis, the immune system attacks and destroys the sheath of myelin that insulates motor nerves (like the rubber covering electrical wires). Degeneration of the myelin sheath interferes with transmission of nerve impulses, until eventually they cannot travel at all. Voluntary functions, such as movement of limbs, and involuntary functions, such as bladder control, are lost, leading finally to paralysis and death. Scientists do not know what stimulates the immune system to attack myelin.

Another autoimmune disease is type I diabetes, in which cells are unable to take in glucose because the pancreas fails to produce insulin (recall from chapter 24 that insulin plays a key role in the liver's regulation of levels of glucose in the blood). Type I diabetes is thought to result from an immune attack on the insulin-manufacturing cells of the pancreas. Again, no one knows why the attack occurs. Other autoimmune diseases are rheumatoid arthritis (an immune system attack on the tissues of the joints), lupus (in which the connective tissue and kidneys are attacked), and Graves' disease (in which the thyroid is attacked).

Allergies

Although your immune system provides very effective protection against fungi, parasites, bacteria, and viruses, sometimes it does its job too well, mounting a major defense against a harmless substance. Such an immune response is called an **allergy.** Hay fever, sensitivity to even tiny amounts of plant pollen, is a familiar example of an allergy. Many people are allergic to proteins released from the feces of a minute mite that lives on grains of house dust (figure 25.21). The dust that the mite calls home is present in mattresses and pillows, and the mite goes out on foraging expeditions and consumes the dead skin cells that many of us shed in large quantities daily. Many people sensitive to feather pillows are in reality allergic to the mites that are residents of the feathers.

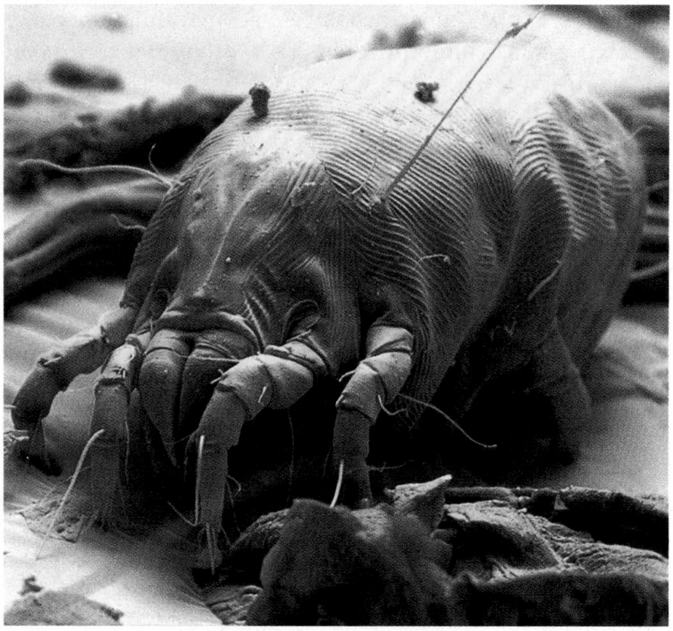

Figure 25.21 The house dust mite, *Dermatophagoides*.
This tiny animal causes an allergic reaction in many people.

What makes an allergic reaction uncomfortable, and sometimes dangerous, is the involvement of antibodies attached to a kind of white blood cell called a **mast cell.** It is the job of the mast cells in an immune response to initiate an inflammatory response. When they encounter something that matches their antibody, mast cells release histamines and other chemicals that cause capillaries to swell. **Histamines** also increase mucus production by cells of the mucous membranes, resulting in runny noses and nasal congestion (all the symptoms of hay fever). Most allergy medicines relieve these symptoms with antihistamines, chemicals that block the action of histamines.

Asthma is a form of allergic response in which histamines cause the narrowing of air passages in the lungs. People who have asthma have trouble breathing when exposed to substances to which they are allergic.

25.10 Allergies are inappropriate immune responses to harmless antigens, while autoimmune diseases are inappropriate responses to "self" cells.

25.11 AIDS: Immune System Collapse

AIDS (acquired immunodeficiency syndrome) was first recognized as a disease in 1981. By the end of 2000, more than 448,060 Americans had died of AIDS, and more than 1.5 million other Americans were thought to be infected with **HIV** (human immunodeficiency virus), the virus that causes the disease (figure 25.22). Worldwide, 36.1 million have become infected, and 21.8 million have died. HIV apparently evolved from a very similar virus that infects chimpanzees in Africa when a mutation arose that allowed the virus to recognize a human cell surface receptor called **CD4.** This receptor is present in the human body on certain immune system cells, notably macrophages and helper T cells. It is the identity of these immune system cells that leads to the devastating nature of the disease.

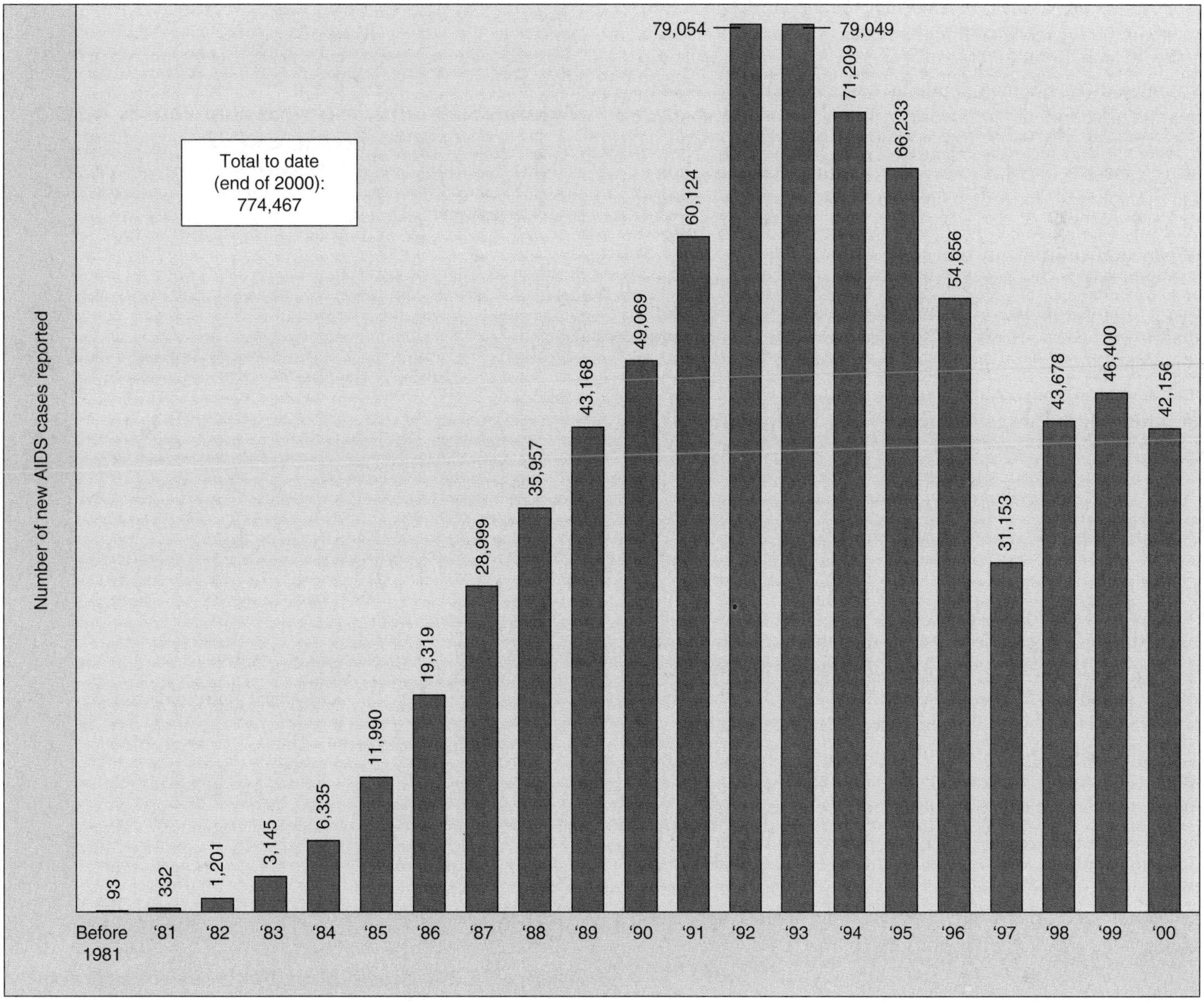

Number of new AIDS cases reported
Total to date (end of 2000): 774,467
93
332
1,201
3,145
6,335
11,990
19,319
28,999
35,957
43,168
49,069
60,124
79,054
79,049
71,209
66,233
54,656
31,153
43,678
46,400
42,156
Before 1981
'81
'82
'83
'84
'85
'86
'87
'88
'89
'90
'91
'92
'93
'94
'95
'96
'97
'98
'99
'00

Figure 25.22 The AIDS epidemic in the United States.

The U.S. Centers for Disease Control (CDC) reports that 46,400 new AIDS cases were reported in 1999 and 42,156 new cases in 2000 in the United States, with a total of 774,467 cases and 448,060 deaths. Over 1.5 million other individuals are thought to be infected with the HIV virus in the United States and 14 million worldwide. The 100,000th AIDS case was reported in August 1989, eight years into the epidemic; the next 100,000 cases took just 26 months; the third 100,000 cases took barely 19 months (May 1993), and the fourth 100,000 took only 13 months (June 1994). The extraordinarily high numbers seen in 1992 reflect an expansion of the definition of what constitutes an AIDS case.

Source: Data from U.S. Centers for Disease Control and Prevention, Atlanta, GA.

How HIV Attacks the Immune System

HIV attacks and cripples the immune system by inactivating $CD4^+$ cells, cells that have CD4 receptors, such as helper T cells. This leaves the immune system unable to mount a response to *any* foreign antigen. AIDS is a deadly disease for just this reason. The AIDS-causing HIV virus mounts a direct attack on $CD4^+$ T cells because it recognizes their CD4 receptors.

HIV's attack on $CD4^+$ T cells progressively cripples the immune system, because HIV-infected cells die only after releasing replicated viruses that proceed to infect other $CD4^+$ T cells, until the body's entire population of $CD4^+$ T cells is destroyed (figure 25.23). In a normal individual, $CD4^+$ T cells make up 60% to 80% of circulating T cells; in AIDS patients, $CD4^+$ T cells often become too rare to detect (figure 25.24), wiping out the human immune defense. With no defense against infection, any of a variety of otherwise commonplace infections proves fatal. With no ability to recognize and destroy cancer cells when they arise, death by cancer becomes far more likely. Indeed, AIDS was first recognized because of a cluster of cases of a rare cancer. More AIDS victims die of cancer than from any other cause.

The fatality rate of AIDS is 100%; no patient exhibiting the symptoms of AIDS has ever been known to survive more than a few years. However, the disease is *not* highly contagious, because it is only transmitted from one individual to another through the transfer of internal body fluids, typically in semen during sexual intercourse and in blood transmitted by needles during drug use.

A variety of drugs inhibit HIV in the test tube. These include AZT and its analogs (which inhibit virus nucleic acid replication) and protease inhibitors (which inhibit the cleavage of the large virus proteins into functional segments). A combination of a protease inhibitor and two AZT analog drugs entirely eliminates the HIV virus from many patients' bloodstreams. Widespread use of this **combination therapy** has cut the U.S. AIDS death rate by almost two-thirds since its introduction in the mid-1990s, from 43,000 AIDS deaths in 1995 to 31,000 in 1996, and just under 17,000 in 1997.

Unfortunately, this sort of combination therapy does not appear to actually succeed in eliminating HIV from the body. While the virus disappears from the bloodstream, traces of it can still be detected in lymph tissue of the patients. When combination therapy is discontinued, virus levels in the bloodstream once again rise. Because of demanding therapy schedules and many side effects, long-term combination therapy does not seem a promising approach.

25.11 HIV cripples the vertebrate immune defense by infecting and killing key lymphocytes.

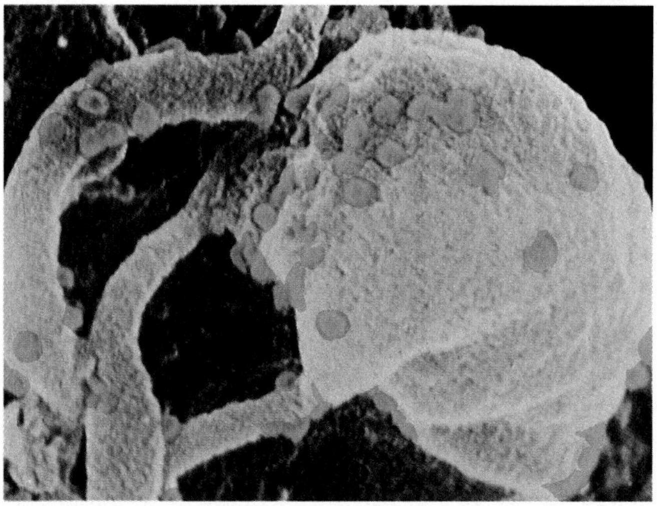

Figure 25.23 HIV, the virus that causes AIDS.

Viruses released from infected $CD4^+$ T cells soon spread to neighboring $CD4^+$ T cells, infecting them in turn. The individual viruses, colored *blue* in this scanning electron micrograph, are extremely small; over 200 million would fit on the period at the end of this sentence.

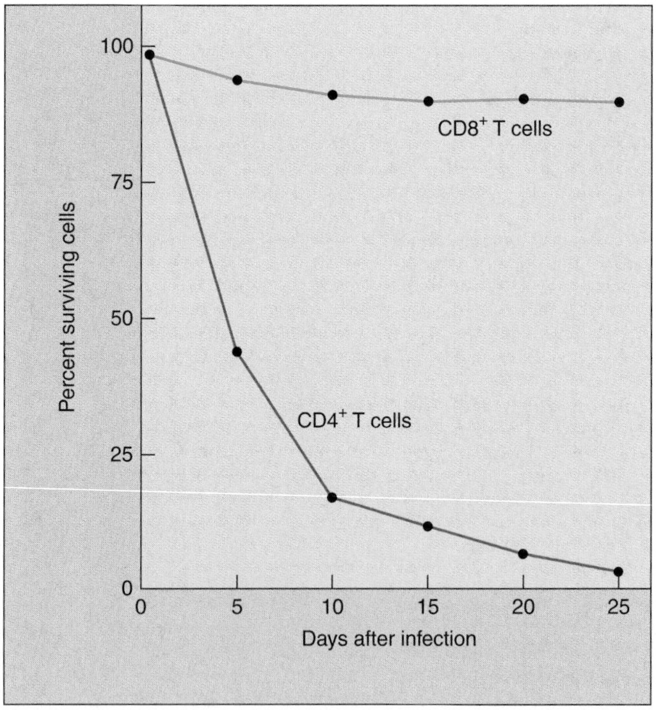
100
75
50
25
0
Percent surviving cells
CD8⁺ T cells
CD4⁺ T cells
0
5
10
15
20
25
Days after infection

Figure 25.24 Survival of T cells in culture after exposure to HIV.

The virus has little effect on the number of $CD8^+$ T cells, T cells with CD8 cell surface receptors. But HIV causes the number of $CD4^+$ T cells (this group includes helper T cells) to decline dramatically.

CONCEPT REVIEW

1. Which of the following statements does *not* describe the skin?
 a. It is the largest body organ.
 b. Its oil and sweat glands make the skin's surface very acidic.
 c. It secretes gastric HCl, which inhibits the growth of many microbes.
 d. It contains sweat glands that secrete an enzyme that attacks the cell walls of many bacteria.

2. White blood cells that kill bacteria by digesting them are called
 a. natural killer cells.
 b. macrophages.
 c. T cells.
 d. lymphocytes.

3. Membrane attack complexes that form holes in the pathogen's membrane are part of the defense called the
 a. complement system.
 b. immune response.
 c. inflammatory response.
 d. temperature response.

4. The inflammatory response results in _________, which promotes the migration of macrophages and neutrophils to the site of infection.
 a. fever
 b. the specific immune response
 c. increased blood flow
 d. the secretion of lysozyme

5. Which of the following cell types in the immune response controls other types of cells?
 a. helper T cell
 b. B cell
 c. cytotoxic T cell
 d. mast cell

6. Unlike B cells, T cells are only able to bind
 a. directly to foreign invaders.
 b. free antigens in the body fluid.
 c. to processed antigens on the surface of an antigen-presenting cell.
 d. to CD4 receptors.

7. The alarm signal protein secreted by macrophages in the immune response is called
 a. a histamine.
 b. an antibody.
 c. MHC.
 d. interleukin-1.

8. Which of the following is *not* an autoimmune disease?
 a. AIDS
 b. type I diabetes
 c. type II diabetes
 d. multiple sclerosis

9. The AIDS virus is remarkably effective at short-circuiting the immune response because it infects
 a. helper T cells.
 b. B cells.
 c. mast cells.
 d. stem cells.

10. _________ is an enzyme in the sweat of the body.

11. B cells secrete protective molecules called _________.

12. _________ is the introduction into your body of a dead pathogen to trigger an immune response against the pathogen.

13. Multiple sclerosis develops from the degeneration of the _________ around neurons.

14. HIV can recognize the surface receptor called ________ on cells of the body.

CHALLENGE YOURSELF

1. Do you think a virus or other pathogen that invades a vertebrate host, avoids the host's immune defenses entirely, reproduces quickly within the host, and leads to the rapid death of the host is well adapted or poorly adapted to that host? Explain.

2. Why might attempting to bring down a slight fever actually be counterproductive?

3. AIDS is a virus that destroys the human immune system by killing helper T cells, which are necessary to activate the immune response. The chimpanzees from which the AIDS virus is thought to have arisen do not suffer from AIDS. How do you imagine they have escaped this?

25 eBRIDGE

Reinforcing Key Points

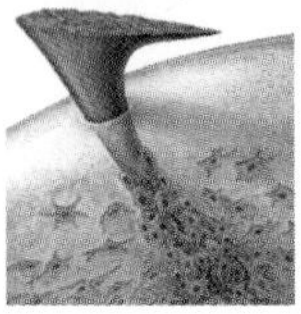

Three Lines of Defense

25.1 Skin: The First Line of Defense

25.2 Cellular Counterattack: The Second Line of Defense

25.3 The Immune System: The Third Line of Defense

25.4 Evolution of the Immune System

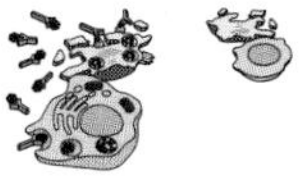

The Immune Response

25.5 Initiating the Immune Response

25.6 T Cells: The Cellular Response

25.7 B Cells: The Humoral Response

25.8 Active Immunity Through Clonal Selection

25.9 Vaccination

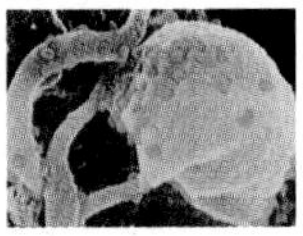

Defeat of the Immune System

25.10 Immune System Failure

25.11 AIDS: Immune System Collapse

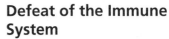

Electronic Learning

Visual Learning

Animations

Phagocytic Cells

Fever

Complement Proteins

Antiviral Defense

T-Cell Function

Clonal Selection

Vaccination

Art Labeling Activities

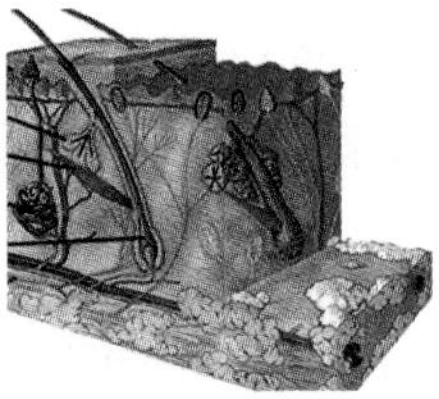

Human Skin Anatomy

Lymphatic System

Bone Marrow

Helping You Learn

The Immune System: The First Line of Defense

The Immune System: The Second Line of Defense

The Immune System: The Third Line of Defense

Author's Corner

AIDS. In the last 20 years, over 22 million people have died of AIDS, a viral infection that destroys the immune system. Researchers have worked feverishly to find a way to halt the epidemic, and after many disappointments seem to be having some success.

1. Combination drug therapy buys time, not a cure.
2. The search for an AIDS vaccine just got harder.
3. Looking for novel ways to hinder the sexual transmission of AIDS.
4. Did a contaminated polio vaccine bring about the AIDS epidemic?
5. The battle against AIDS proceeds along two fronts.
6. Search for AIDS vaccine suddenly looks more promising.
7. AIDS at 20: New approaches to combating HIV

Virtual Classroom

The Continuing Challenge of Infectious Disease

Infectious diseases have had a far greater major impact on human history than most people realize. The introduction of smallpox to the New World by Cortez wiped out 80% of Mexican and Central American Indians in a generation. Flu killed 22 million people in 18 months at the end of World War I, far more people than died of war wounds. In history, typhus and cholera have been among the greatest killers. More troops died of typhus in the Crimean War (104,494) than of combat (63,261). This was before sanitation measures were introduced into armies in the field. Whenever public sanitation breaks down due to war or natural disaster, diseases such as cholera and typhus reappear, killing many. In 1994, some 100,000 people died during the civil unrest in Rwanda of cholera, followed by dysentery and typhus. Nor are these even the greatest killers. Malaria and tuberculosis will each kill some 3 million people THIS YEAR! While many of these diseases have been impacting human history for centuries, others like AIDS and Ebola fever are newly emerging.

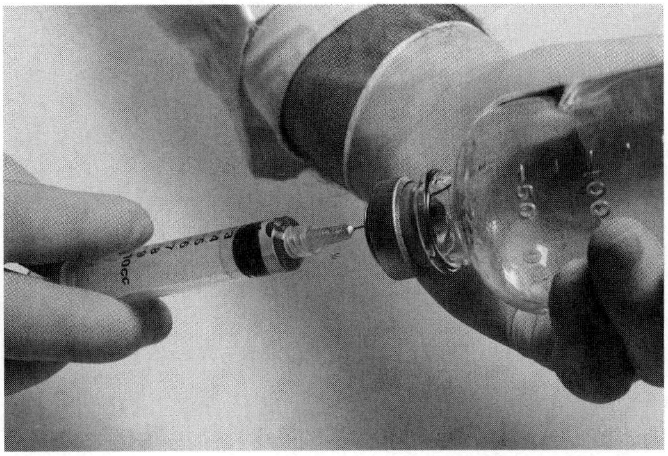

Virtual Lab

In Search of New Antibiotics: How Salamander Skin Secretions Combat Microbial Infections

In response to the growing threat posed by antibiotic-resistant pathogens, researchers are searching for new methods to treat bacterial infections. An interesting opportunity is presented by antimicrobial peptides secreted from the dermal glands of frogs. These small, positively charged peptides kill bacteria by interacting with negatively charged phospholipids and lipopolysaccharides found in bacterial membranes. Bacteria cannot easily develop resistance to the peptides because membrane molecules are less prone to mutations than receptors or enzymes are, the usual targets of antibiotics.

Little is known of salamander production and secretion of antimicrobial peptides. John Dankert of the University of Louisiana, Lafayette, has investigated the antimicrobial peptides secreted from the terrestrial salamander, *Plethodon cinereus*. He collected the secretions from the skin of the tail region, and then tested the secretions for their antimicrobial properties. The tail region is susceptible to injury from predators and competitors, and so might be expected to contain antimicrobial peptides.

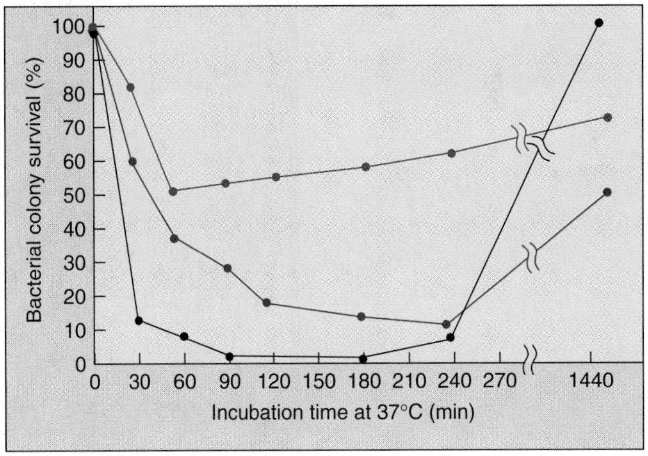
Bacterial colony survival (%)
100
90
80
70
60
50
40
30
20
10
0
0 30 60 90 120 150 180 210 240 270 1440
Incubation time at 37°C (min)

Quizzes

Further Reading

Essential Study Partner

Links

BioCourse.com

CHAPTER 26

THE NERVOUS SYSTEM

THE FAR SIDE® By GARY LARSON
© 1985 FarWorks, Inc. All Rights Reserved/Dist. by Creators Syndicate
Larson
"The picture's pretty bleak, gentlemen. ... The world's climates are changing, the mammals are taking over, and we all have a brain about the size of a walnut."

CHAPTER 26

The Nervous System

CHAPTER OVERVIEW

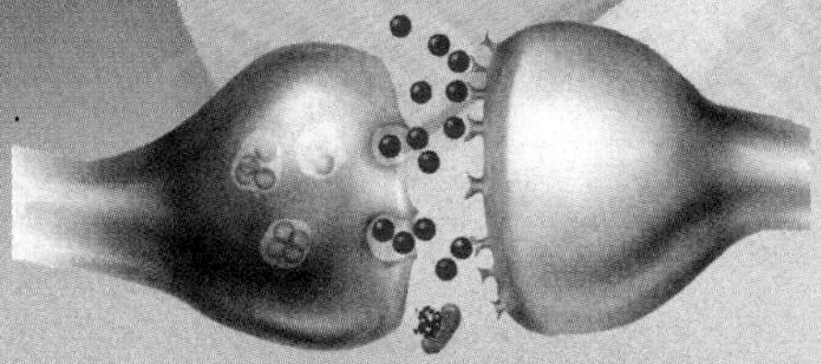

Neurons and How They Work

- The nervous system is composed of sensory, association, and motor neurons.
- A nerve impulse is a propagating depolarization of the plasma membrane.
- Nerve impulses cross synapses via chemicals called neurotransmitters.

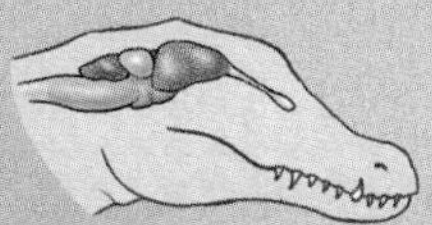

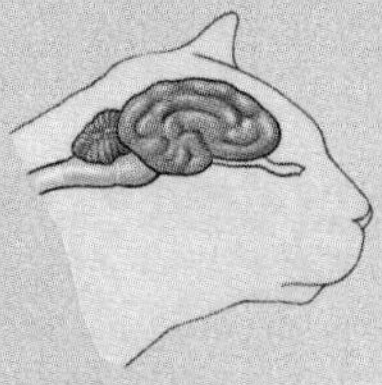

The Central Nervous System

- The central nervous system is composed of the brain and spinal cord.
- Most of the mass of the brain is the cerebrum; its surface is the site of conscious thought.

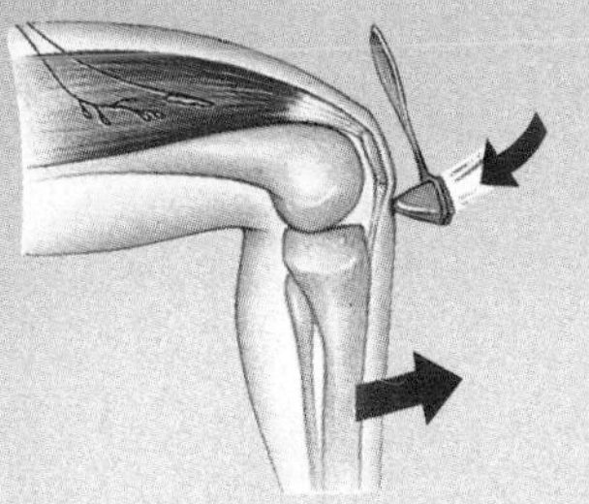

The Peripheral Nervous System

- Voluntary motor neurons send commands from the brain to skeletal muscles.
- Involuntary motor neurons constitute the autonomic nervous system. The autonomic nervous system is composed of two antagonistic elements; the balance of the two determines the degree of stimulation.

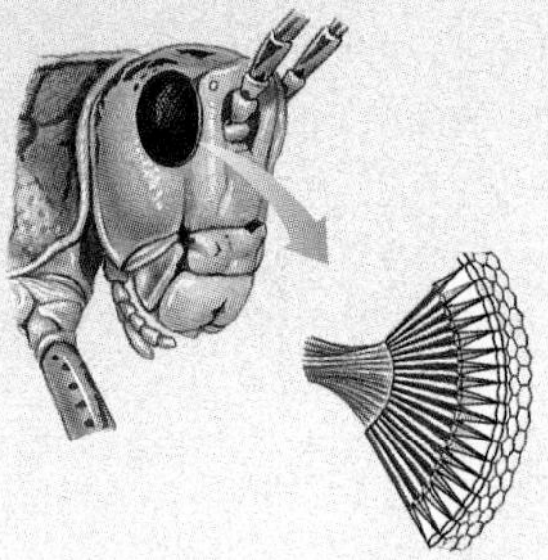

The Sensory Nervous System

- Interoceptors of many kinds monitor the body's internal condition.
- Sensing the exterior world focuses on three sorts of stimuli: chemicals, sound, and light.
- Both sound and light can be used to form three-dimensional images.

26.1 Evolution of the Animal Nervous System

An animal must be able to respond to environmental stimuli. To do this, it must have sensory receptors that can detect the stimulus and motor *effectors* that can respond to it. In most invertebrate phyla and in all vertebrate classes, sensory receptors and motor effectors are linked by way of the **nervous system.** As described in chapter 22, the nervous system consists of neurons and supporting cells. One type of neuron, called **association neurons** (or **interneurons**), is present in the nervous systems of most invertebrates and all vertebrates. These neurons are located in the brain and spinal cord of vertebrates, together called the **central nervous system (CNS)**, where they help provide more complex reflexes and higher associative functions, including learning and memory (figure 26.1).

There are two other types of neurons. **Motor** (or **efferent**) **neurons** carry impulses away from the CNS to effectors—muscles and glands. **Sensory** (or **afferent**) **neurons** carry impulses from sensory receptors to the CNS (figure 26.2). Together, motor and sensory neurons comprise the **peripheral nervous system (PNS)** of vertebrates.

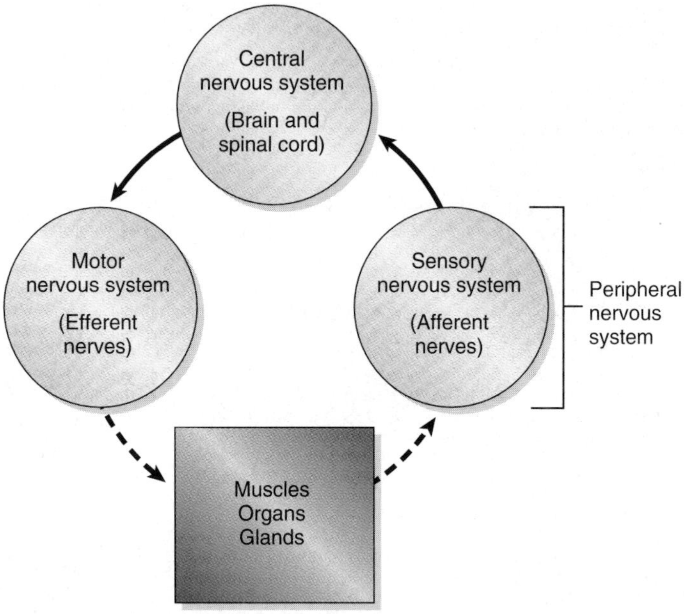
Central nervous system
(Brain and spinal cord)
Motor nervous system
(Efferent nerves)
Sensory nervous system
(Afferent nerves)
Peripheral nervous system
Muscles
Organs
Glands

Figure 26.1 Organization of the vertebrate nervous system.

The central nervous system, consisting of the brain and spinal cord, issues commands via the motor nervous system and receives information from the sensory nervous system. The motor and sensory nervous systems together make up the peripheral nervous system.

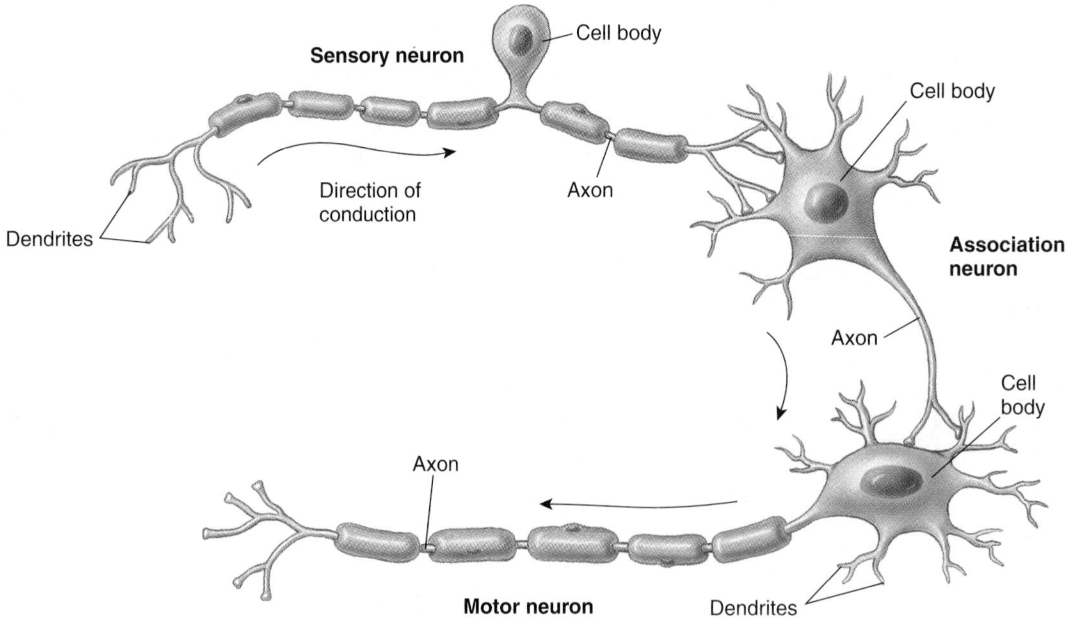
Cell body
Sensory neuron
Cell body
Direction of conduction
Axon
Dendrites
Association neuron
Axon
Cell body
Axon
Motor neuron
Dendrites

Figure 26.2 Three types of neurons.

Sensory neurons carry information about the environment to the brain and spinal cord. *Association neurons* are found in the brain and spinal cord and often provide links between sensory and motor neurons. *Motor neurons* carry impluses to muscles and glands (effectors).

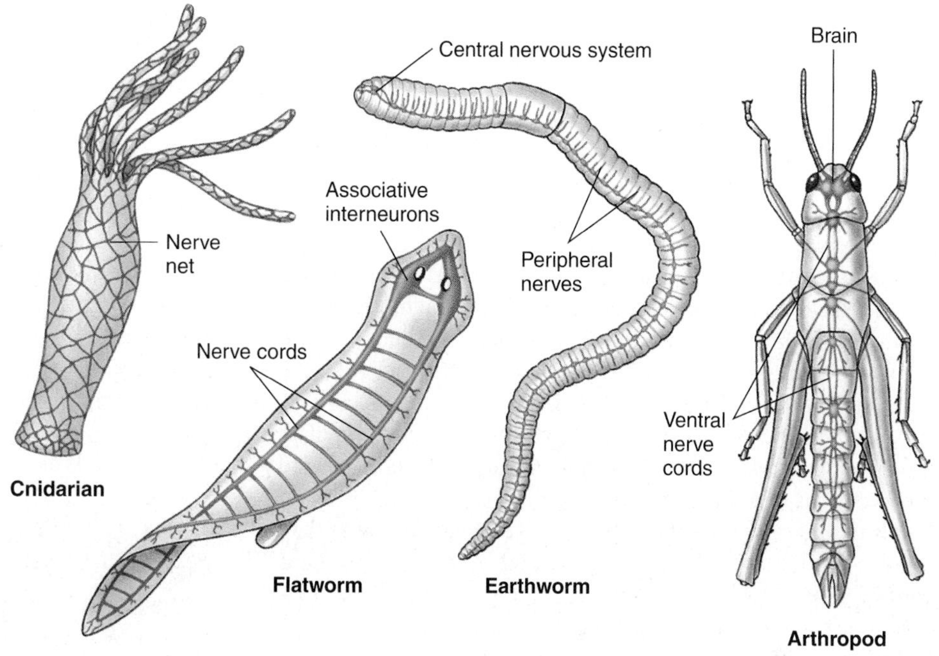
Central nervous system
Brain
Associative
interneurons
Nerve
net
Peripheral
nerves
Nerve cords
Ventral
nerve
cords
Cnidarian
Flatworm
Earthworm
Arthropod

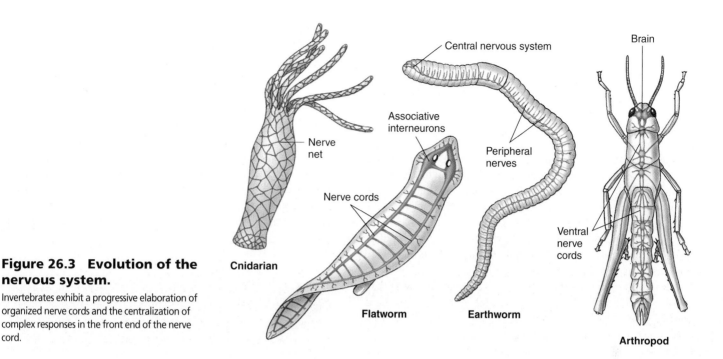

Figure 26.3 Evolution of the nervous system.
Invertebrates exhibit a progressive elaboration of organized nerve cords and the centralization of complex responses in the front end of the nerve cord.

Invertebrate Nervous Systems

Sponges are the only major phylum of multicellular animals that lack nerves. If you prick a sponge, the nearby surface contracts slowly. The cytoplasm of each individual cell conducts an impulse that fades within a few millimeters. No messages dart from one part of the sponge body to another, as they do in all other multicellular animals.

The Simplest Nervous Systems: Reflexes. The simplest nervous systems occur among cnidarians (figure 26.3): all neurons are similar, each having fibers of approximately equal length. Cnidarian neurons are linked to one another in a web, or *nerve net,* dispersed through the body. Although conduction is slow, a stimulus anywhere can eventually spread through the whole net. There is no associative activity, no control of complex actions, and little coordination. Any motion that results is called a **reflex** because it is an automatic consequence of the nerve stimulation.

More Complex Nervous Systems: Associative Activities. The first associative activity in nervous systems is seen in the free-living flatworms, phylum Platyhelminthes. Running down the bodies of these flatworms are two nerve cords; peripheral nerves extend outward to the muscles of the body. The two nerve cords converge at the front end of the body, forming an enlarged mass of nervous tissue that also contains associative neurons that connect neurons to one another. This primitive "brain" is a rudimentary central nervous system and permits a far more complex control of muscular responses than is possible in cnidarians.

The Evolutionary Path to the Vertebrates. All of the subsequent evolutionary changes in nervous systems can be viewed as a series of elaborations on the characteristics already present in flatworms. Five trends can be identified, each becoming progressively more pronounced as nervous systems evolved greater complexity.

1. *More sophisticated sensory mechanisms.* Particularly among the vertebrates, sensory systems become highly complex.
2. *Differentiation into central and peripheral nervous systems.* For example, earthworms exhibit a central nervous system that is connected to all other parts of the body by peripheral nerves.
3. *Differentiation of sensory and motor nerves.* Neurons operating in particular directions (sensory signals traveling to the brain, or motor signals traveling from the brain) become increasingly specialized.
4. *Increased complexity of association.* Central nervous systems with more numerous interneurons evolved, increasing association capabilities dramatically.
5. *Elaboration of the brain.* Coordination of body activities became increasingly localized in arthropods, mollusks, and vertebrates in the front end of the nerve cord.

26.1 As nervous systems became more complex, there was a progressive increase in associative activity, increasingly localized in a brain.

26.2 Neurons

The basic structural unit of the nervous system, whether central, motor, or sensory, is the nerve cell, or **neuron.** All neurons have the same basic structure (figure 26.4*a*). The **cell body** is an enlarged region of the neuron containing the nucleus. Short, slender branches called **dendrites** extend from one end of a neuron's cell body. Dendrites are input channels. Nerve impulses travel inward along them, toward the cell body. Motor and association neurons possess a profusion of highly branched dendrites, enabling those cells to receive information from many different sources simultaneously. Projecting out from the other end of the cell body is a single, long, tubelike extension called an **axon.** Axons are output channels. Nerve impulses travel outward along them, away from the cell body, toward other neurons or to muscles or glands.

Most neurons are unable to survive alone for long; they require the nutritional support provided by companion **neuroglial cells.** More than half the volume of the human nervous system is composed of supporting neuroglial cells. Two of the most important kinds of supporting neurons are the **Schwann cells** and **oligodendrocytes,** which envelop the axon of many neurons with a sheath of fatty material called myelin, which acts as an electrical insulator. Schwann cells produce myelin in the PNS, while oligodendrocytes produce myelin in the CNS. During development, these cells wrap themselves around each axon several times to form a **myelin sheath,** an insulating covering consisting of multiple layers of membrane (figure 26.4*b*). Axons that have myelin sheaths are said to be myelinated, and those that don't are unmyelinated. The myelin sheath is interrupted at intervals, leaving uninsulated gaps called **nodes of Ranvier** where the axon is in direct contact with the surrounding fluid. The nerve impulse jumps from node to node, speeding its travel down the axon.

26.2 Neurons, the basic units of nervous systems, are cells specialized to conduct impulses. Signals typically arrive along any of numerous dendrites, pass over the cell body's surface, and travel outward on a single long axon.

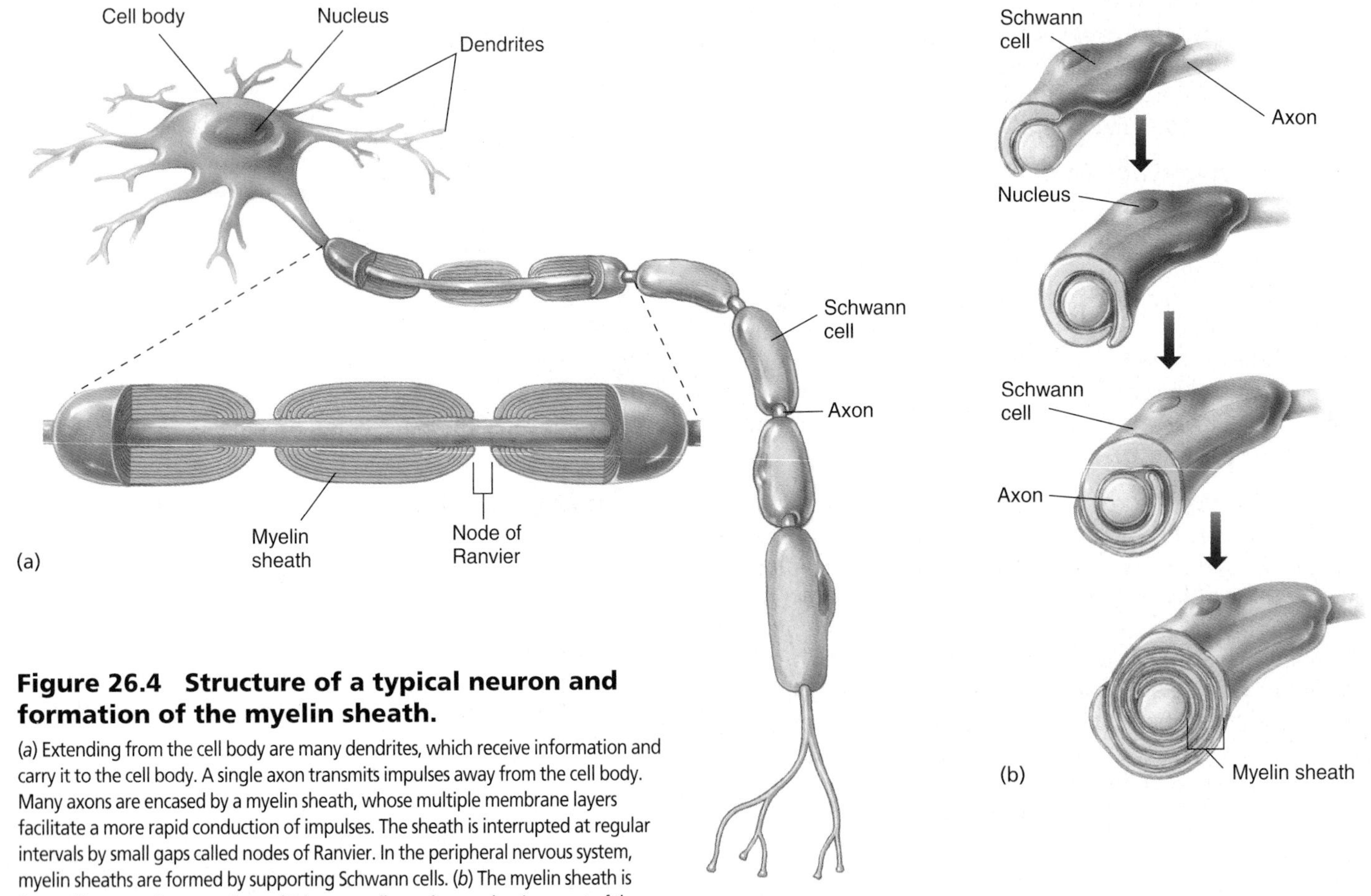

Cell body
Nucleus
Dendrites
Schwann cell
Axon
Myelin sheath
Node of Ranvier
(a)
Schwann cell
Axon
Nucleus
Schwann cell
Axon
(b)
Myelin sheath

Figure 26.4 Structure of a typical neuron and formation of the myelin sheath.

(*a*) Extending from the cell body are many dendrites, which receive information and carry it to the cell body. A single axon transmits impulses away from the cell body. Many axons are encased by a myelin sheath, whose multiple membrane layers facilitate a more rapid conduction of impulses. The sheath is interrupted at regular intervals by small gaps called nodes of Ranvier. In the peripheral nervous system, myelin sheaths are formed by supporting Schwann cells. (*b*) The myelin sheath is formed by successive wrappings of Schwann cell membranes, leaving most of the Schwann cell cytoplasm outside the myelin.

26.3 The Nerve Impulse

When a neuron is "at rest," not carrying an impulse, active transport channels in the neuron's plasma membrane transport sodium ions (Na^+) out of the cell and potassium ions (K^+) in. This sodium-potassium pump was described in chapter 4. Sodium ions cannot easily move back into the cell once they are pumped out, so the concentration of sodium ions builds up outside the cell. Similarly, potassium ions accumulate inside the cell, although not as densely because many potassium ions are able to diffuse out through open channels. The result is to make the outside of the neuron more positive than the inside. The plasma membrane is said to be "polarized."

Neurons are constantly expending energy to pump sodium ions out of the cell, in order to maintain a positive charge on the exterior of the cell and a negative charge on the interior. The net negative charge of most proteins within the cell also adds to this charge difference. This charge separation is called the **resting potential.** Using sophisticated instruments, scientists have been able to measure the voltage difference between the neuron interior and exterior as –70 millivolts (thousandth of a volt). The resting potential is the starting point for a nerve impulse.

A nerve impulse travels along the axon and dendrites as an electrical current gathered by ions moving in and out of the neuron through **voltage-gated channels** (that is, protein channels in the neuron membrane that open and close in response to an electrical voltage). The impulse starts when pressure or other sensory inputs disturb a neuron's plasma membrane, causing sodium channels on a dendrite to open. As a result, sodium ions flood into the neuron from outside, and for a brief moment the inside of the membrane is "depolarized," becoming more positive than the outside in that immediate area of the dendrite.

The sodium channels in the small patch of depolarized membrane remain open for only about a half a millisecond. However, if the change in voltage is big enough, it causes nearby voltage-gated channels to open, which starts a wave of depolarization moving down the neuron, as the opening of the gated channels causes nearby voltage-gated channels to open, like a chain of falling dominoes. This moving local reversal of voltage is called an **action potential.** When the action potential has passed, the voltage-gated sodium channels snap closed again and the resting potential is restored (figure 26.5).

The depolarization and restoration of the resting potential take only about 5 milliseconds. Fully 100 such cycles could occur, one after another, in the time it takes to say the word *nerve.*

26.3 Nerve impulses result from ion movements across the neuron plasma membrane through special protein channels that open and close in response to chemical or electrical stimulation.

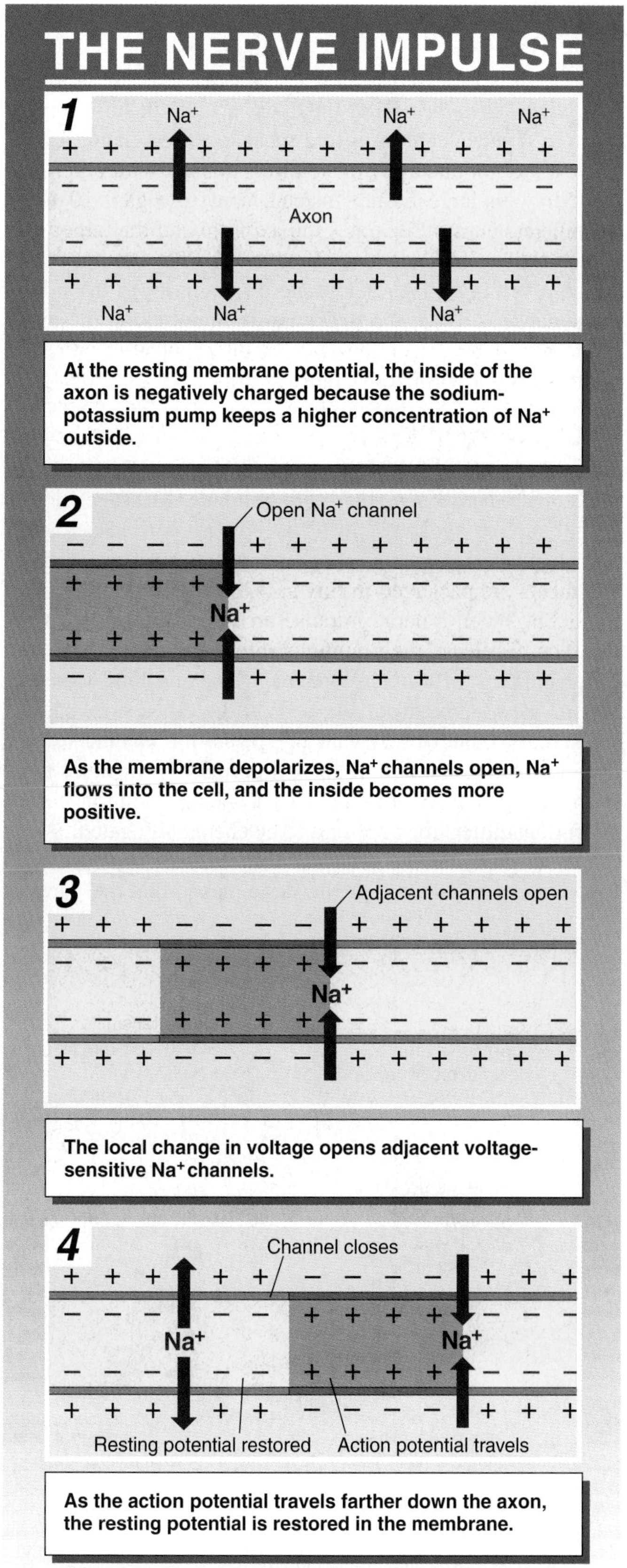
THE NERVE IMPULSE
1
Na+
Axon
At the resting membrane potential, the inside of the axon is negatively charged because the sodium-potassium pump keeps a higher concentration of Na+ outside.
2
Open Na+ channel
Na+
As the membrane depolarizes, Na+ channels open, Na+ flows into the cell, and the inside becomes more positive.
3
Adjacent channels open
Na+
The local change in voltage opens adjacent voltage-sensitive Na+ channels.
4
Channel closes
Na+
Resting potential restored
Action potential travels
As the action potential travels farther down the axon, the resting potential is restored in the membrane.

Figure 26.5 How an action potential works.

26.4 The Synapse

A nerve impulse can travel only so far along a neuron. Eventually it reaches the end of the axon, usually positioned very close to another neuron or to a muscle cell or gland. Axons, however, do not actually make direct contact with other neurons or with large tissue. Instead, a narrow gap, 10 to 20 nanometers across, separates the axon tip and the target neuron or tissue. This junction of an axon with another cell is called a **synapse.** The membrane on the near (axon) side of the synapse is called the **presynaptic membrane;** the membrane on the far (receiving) side of the synapse is called the **postsynaptic membrane** (figure 26.6).

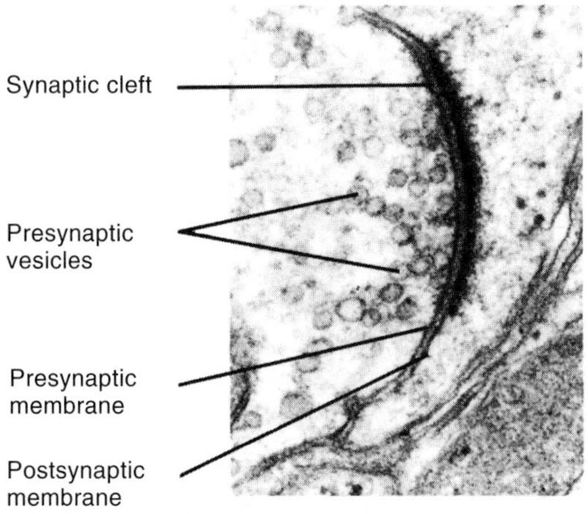
Synaptic cleft
Presynaptic vesicles
Presynaptic membrane
Postsynaptic membrane

Figure 26.6 A synapse between two neurons.
This micrograph clearly shows the space between the presynaptic and postsynaptic membranes, which is called the synaptic cleft.

Neurotransmitters

When a nerve impulse gets to the end of an axon, its message must cross the synapse if it is to continue. Messages do not "jump" across synapses. Instead, they are carried across by chemical messengers called **neurotransmitters.** These chemicals are packaged in tiny sacs, or vesicles, at the tip of the axon. When a nerve impulse arrives at the tip, it causes the sacs to release their contents into the synapse. The neurotransmitters diffuse across the synapse and bind to receptors in the membrane of the cell on the other side, passing the signal to that cell by causing special ion channels in the postsynaptic membrane to open (figure 26.7). Because these channels open when stimulated by a chemical (in this case, a neurotransmitter) they are said to be chemically gated.

Why go to all this trouble? Why not just wire the neurons directly together? For the same reason that the wires of your house are not all connected but instead are separated by a host of switches. When you turn on one light switch, you don't want every light in the house to go on, the toaster to start heating, and the television to come on! If every neuron in your body were connected to every other neuron, it would be impossible to move your hand without moving every other part of your body at the same time. Synapses are the control switches of the nervous system.

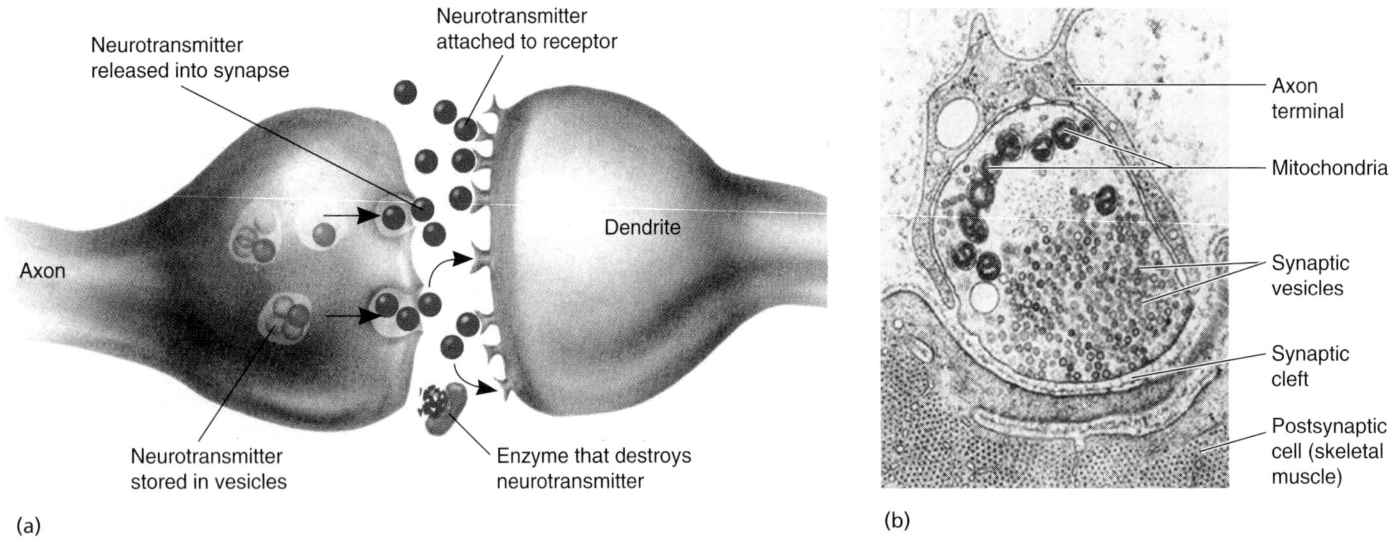
Neurotransmitter released into synapse
Neurotransmitter attached to receptor
Axon
Dendrite
Neurotransmitter stored in vesicles
Enzyme that destroys neurotransmitter
(a)
Axon terminal
Mitochondria
Synaptic vesicles
Synaptic cleft
Postsynaptic cell (skeletal muscle)
(b)

Figure 26.7 Events at the synapse.
(a) When a nerve impulse reaches the end of an axon, it releases a neurotransmitter into the synaptic space. The neurotransmitter molecules diffuse across the synapse and bind to receptors on the postsynaptic cell, a neuron in the case, passing the signal to that cell. Enzymes destroy the neurotransmitter molecules to prevent continuous stimulation of the postsynaptic cell. *(b)* A transmission electron micrograph of the tip of an axon filled with synaptic vesicles.

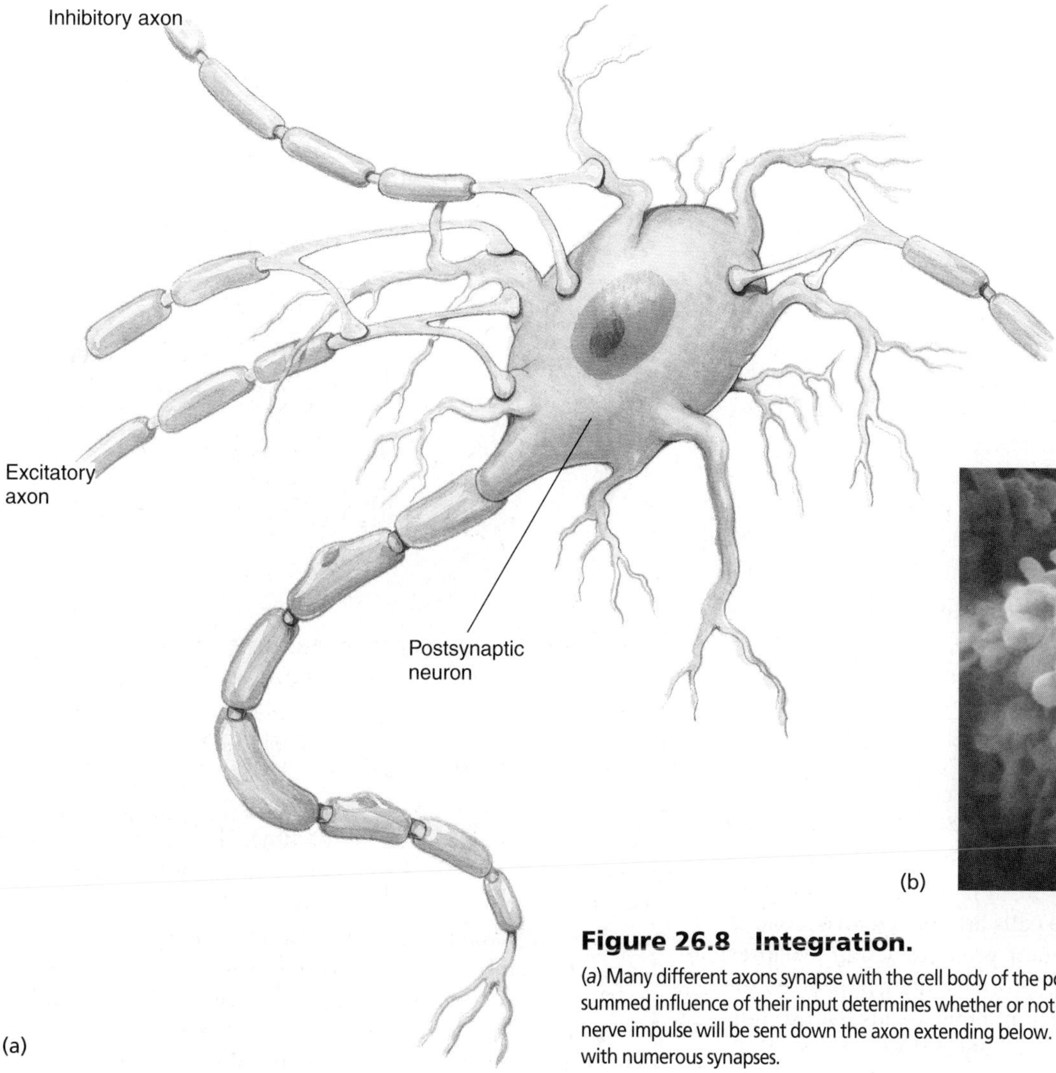
Inhibitory axon
Excitatory axon
Postsynaptic neuron
(a)

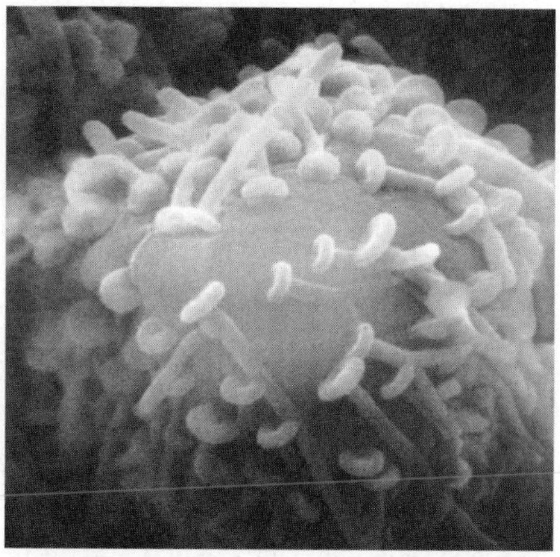
(b)

Figure 26.8 Integration.
(*a*) Many different axons synapse with the cell body of the postsynaptic neuron illustrated here. The summed influence of their input determines whether or not an action potential will be realized and a nerve impulse will be sent down the axon extending below. (*b*) Micrograph of a neuronal cell body with numerous synapses.

Kinds of Synapses

The vertebrate nervous system uses dozens of different kinds of neurotransmitters, each recognized by specific receptors on receiving cells. They fall into two classes, depending on whether they excite or inhibit passage across the synapse.

In an *excitatory synapse,* the receptor protein is a chemically gated sodium channel, meaning that a sodium channel through the membrane is opened by a chemical, the neurotransmitter. On binding with a neurotransmitter whose shape fits it, the sodium channel opens, allowing sodium ions to flood inward. If enough sodium ion channels are opened by neurotransmitters, an action potential begins.

In an *inhibitory synapse,* the receptor protein is a chemically gated potassium channel. Binding with its neurotransmitter opens the potassium channel, leading to the exit of positively charged potassium ions and a more negative interior in the receiving cell. This inhibits the start of an action potential, because the negative voltage change inside means that even more sodium ion channels must be opened to get a domino effect started among voltage-gated sodium channels, and so start an action potential.

An individual nerve cell can possess both kinds of synaptic connections to other nerve cells. When signals from both excitatory and inhibitory synapses reach the body of a neuron, the excitatory effects (which cause less internal negative charge) and the inhibitory effects (which cause more internal negative charge) interact with one another. The result is a process of **integration** in which the various excitatory and inhibitory electrical effects tend to cancel or reinforce one another (figure 26.8). Neurons often receive many inputs. A single motor neuron in the spinal cord may have as many as 50,000 synapses on it!

26.4 A synapse is a junction of an axon with another cell, a gap across which neurotransmitters carry a signal either facilitating or inhibiting transmission of a signal, depending on which ion channels they open.

26.5 Neuromodulators and Drug Addiction

Neuromodulators

The body sometimes deliberately prolongs the transmission of a signal across a synapse by slowing the destruction of neurotransmitters. It does this by releasing into the synapse special long-lasting chemicals called **neuromodulators.** Some neuromodulators aid the release of neurotransmitters into the synapse; others inhibit the reabsorption of neurotransmitters so that they remain in the synapse; still others delay the breakdown of neurotransmitters after their reabsorption, leaving them in the tip to be released back into the synapse when the next signal arrives.

Mood, pleasure, pain, and other mental states are determined by particular groups of neurons in the brain that use special sets of neurotransmitters and neuromodulators. Mood, for example, is strongly influenced by the neurotransmitter serotonin. Many researchers think that depression results from a shortage of serotonin. Prozac, the world's best-selling antidepressant, inhibits the reabsorption of serotonin, thus increasing the amount in the synapse (figure 26.9).

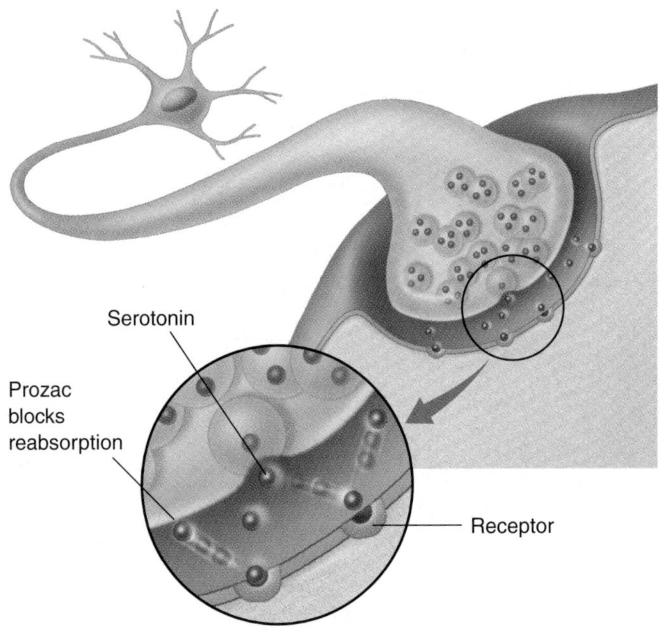

Serotonin
Prozac blocks reabsorption
Receptor

Figure 26.9 Drugs alter transmission of impulses across the synapse.

Depression can result from a shortage of the neurotransmitter serotonin. The antidepressant drug Prozac works by blocking reabsorption of serotonin in the synapse, making up for the shortage.

Drug Addiction

When a cell of the body is exposed to a chemical signal for a prolonged period, it tends to lose its ability to respond to the stimulus with its original intensity. (You are familiar with this loss of sensitivity—when you sit in a chair, how long are you aware of the chair?) Nerve cells are particularly prone to this loss of sensitivity. If receptor proteins within synapses are exposed to high levels of neurotransmitter molecules for prolonged periods, that nerve cell often responds by inserting fewer receptor proteins into the membrane. This feedback is a normal part of the functioning of all neurons, a simple mechanism that has evolved to make the cell more efficient by adjusting the number of "tools" (receptor proteins) in the membrane "workshop" to suit the workload.

Cocaine. The drug cocaine is a neuromodulator that causes abnormally large amounts of neurotransmitters to remain in the synapses for long periods of time. Cocaine affects nerve cells in the brain's pleasure pathways (the so-called limbic system). These cells transmit pleasure messages using the neurotransmitter dopamine. Using radioactively labeled cocaine molecules, investigators found that cocaine binds tightly to the transporter proteins in the gaps between nerves. These proteins normally remove the neurotransmitter dopamine after it has acted. Like a game of musical chairs in which all the chairs become occupied, there are no unoccupied carrier proteins available to the dopamine molecules, so the dopamine stays in the gap, firing the receptors again and again. As new signals arrive, more and more dopamine is added, firing the pleasure pathway more and more often.

When receptor proteins on limbic system nerve cells are exposed to high levels of dopamine neurotransmitter molecules for prolonged periods of time, the nerve cells "turn down the volume" of the signal by lowering the number of receptor proteins on their surfaces. They respond to the greater number of neurotransmitter molecules by simply reducing the number of targets available for these molecules to hit. The cocaine user is now addicted (figure 26.10). **Addiction** occurs when chronic exposure to a drug induces the nervous system to adapt physiologically. With so few receptors, the user needs the drug to maintain even normal levels of limbic activity.

Is Addiction to Smoking Cigarettes Drug Addiction?

Investigators attempting to explore the habit-forming nature of smoking cigarettes used what had been learned about cocaine to carry out what seems a reasonable experiment—they introduced radioactively labeled nicotine from tobacco into the brain and looked to see what sort of carrier protein it attached itself to. To their great surprise, the nicotine ignored proteins in the between-cell gaps and instead bound directly to a specific receptor on the receiving nerve cell surface! This was totally unexpected, as nicotine does not normally occur in the brain—why should it have a receptor there?

Intensive research followed, and researchers soon learned that the "nicotine receptors" normally served to bind the neurotransmitter acetylcholine. It was just an accident of nature that nicotine, an obscure chemical from a tobacco plant, was also able to bind to them. What, then, is the normal function of these receptors? The target of considerable

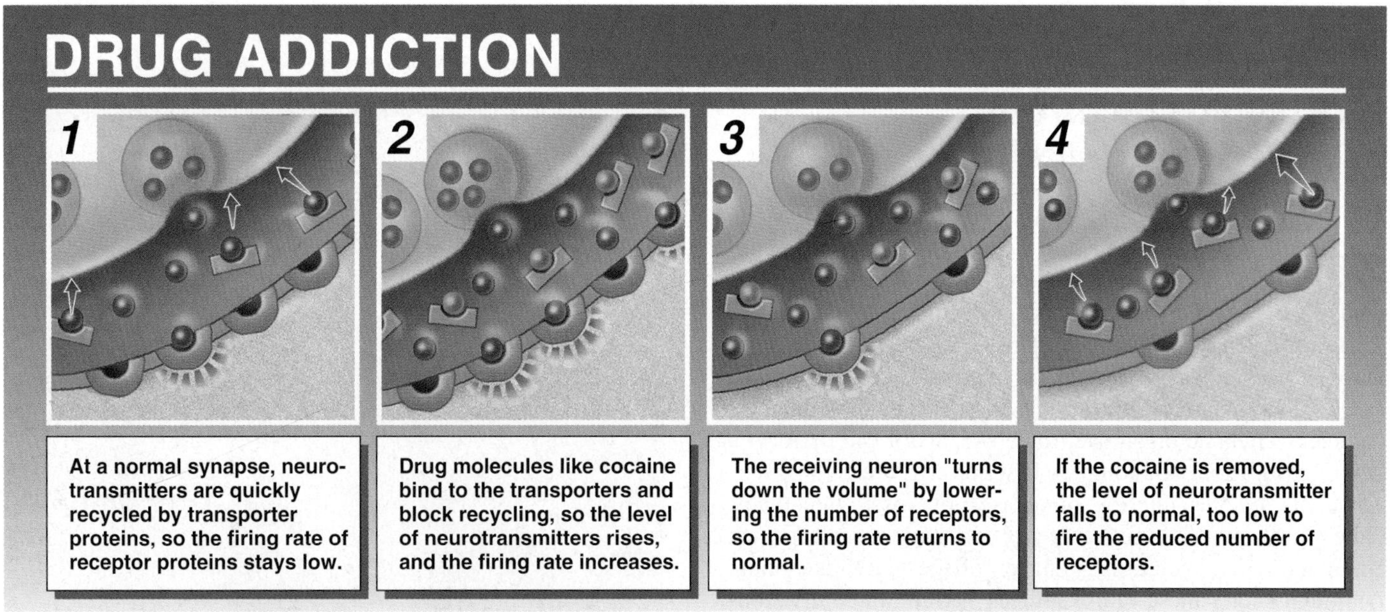
DRUG ADDICTION
1
2
3
4
At a normal synapse, neurotransmitters are quickly recycled by transporter proteins, so the firing rate of receptor proteins stays low.
Drug molecules like cocaine bind to the transporters and block recycling, so the level of neurotransmitters rises, and the firing rate increases.
The receiving neuron "turns down the volume" by lowering the number of receptors, so the firing rate returns to normal.
If the cocaine is removed, the level of neurotransmitter falls to normal, too low to fire the reduced number of receptors.

Figure 26.10 How drug addiction works.

research, these receptors turned out to be one of the brain's most important tools. The brain uses them to coordinate the activities of many other kinds of receptors, acting to "fine-tune" the sensitivity of a wide variety of behaviors.

When neurobiologists compare the limbic system nerve cells of smokers to those of nonsmokers, they find changes in both the number of nicotine receptors and in the levels of RNA used to make the receptors. They have found that the brain adjusts to prolonged exposure to nicotine by "turning down the volume" in two ways: (1) by making fewer receptor proteins to which nicotine can bind; and (2) by altering the pattern of *activation* of the nicotine receptors (that is, their sensitivity to neurotransmitters).

It is this second adjustment that is responsible for the profound effect smoking has on the brain's activities. By overriding the normal system used by the brain to coordinate its many activities, nicotine alters the pattern of release into gaps between nerve cells of many neurotransmitters, including acetylcholine, dopamine, serotonin, and many others. As a result, changes in level of activity occur in a wide variety of nerve pathways within the brain.

Addiction to nicotine occurs because the brain compensates for the many changes nicotine induces by making other changes. Adjustments are made to the numbers and sensitivities of many kinds of receptors within the brain, restoring an appropriate balance of activity.

Now what happens if you stop smoking? Everything is out of whack! The newly coordinated system *requires* nicotine to achieve an appropriate balance of nerve pathway activities. This is addiction in any sensible use of the term. The body's physiological response is profound and unavoidable. There is no way to prevent addiction to nicotine with willpower, any more than willpower can stop a bullet when playing Russian roulette with a loaded gun. If you smoke cigarettes for a prolonged period, you will become addicted.

What do you do if you are addicted to smoking cigarettes and you want to stop? When use of an addictive drug like nicotine is stopped, the level of signaling changes to levels far from normal. If the drug is not reintroduced, the altered level of signaling eventually induces the nerve cells to once again make compensatory changes that restore an appropriate balance of activities within the brain. Over time, receptor numbers, their sensitivity, and patterns of release of neurotransmitters all revert to normal, once again producing normal levels of signaling along the pathways. There is no way to avoid the down side of addiction. The pleasure pathways will not function at normal levels until the number of receptors on the affected nerve cells has time to readjust.

Many people attempt to quit smoking by using patches containing nicotine; the idea is that by providing gradually smaller doses of nicotine, the smoker can be weaned of his or her craving for cigarettes. The patches do reduce the craving for cigarettes—as long as you keep using the patches! Actually, using such patches simply substitutes one (admittedly less dangerous) nicotine source for another. If you are going to quit smoking, there is no way to avoid the necessity of eliminating the drug to which you are addicted. Hard as it is to hear the bad news, there is no easy way out. The only way to quit is to quit.

26.5 Cigarette smokers find it difficult to quit because they have become addicted to nicotine, a powerful neuromodulator.

26.6 Evolution of the Vertebrate Brain

The structure and function of the vertebrate brain have long been the subject of scientific inquiry. Despite ongoing research, scientists are still not sure how the brain performs many of its functions. For instance, scientists continue to look for the mechanism the brain employs to store memories, and they do not understand how some memories can be "locked away," only to surface in times of stress. The brain is the most complex vertebrate organ ever to evolve, and it can perform a bewildering variety of complex functions (figure 26.11).

Casts of the interior braincases of fossil agnathans, fishes that swam 500 million years ago, have revealed much about the early evolutionary stages of the vertebrate brain. Although small, these brains already had the three divisions that characterize the brains of all contemporary vertebrates: (1) the hindbrain, or rhombencephalon; (2) the midbrain, or mesencephalon; and (3) the forebrain, or prosencephalon (figure 26.12).

The hindbrain was the major component of these early brains, as it still is in fishes today. Composed of the *cerebellum*, *pons*, and *medulla oblongata*, the hindbrain may be considered an extension of the spinal cord devoted primarily to coordinating motor reflexes. Tracts containing large numbers of axons run like cables up and down the spinal cord to the hindbrain. The hindbrain, in turn, integrates the many sensory signals coming from the muscles and coordinates the pattern of motor responses.

Figure 26.11 Singing well takes practice.
This baby coyote is greeting the approaching evening. His howling is not as impressive as his dad's—a good performance takes practice. His brain is learning by repetition how to control the vocal cords properly.

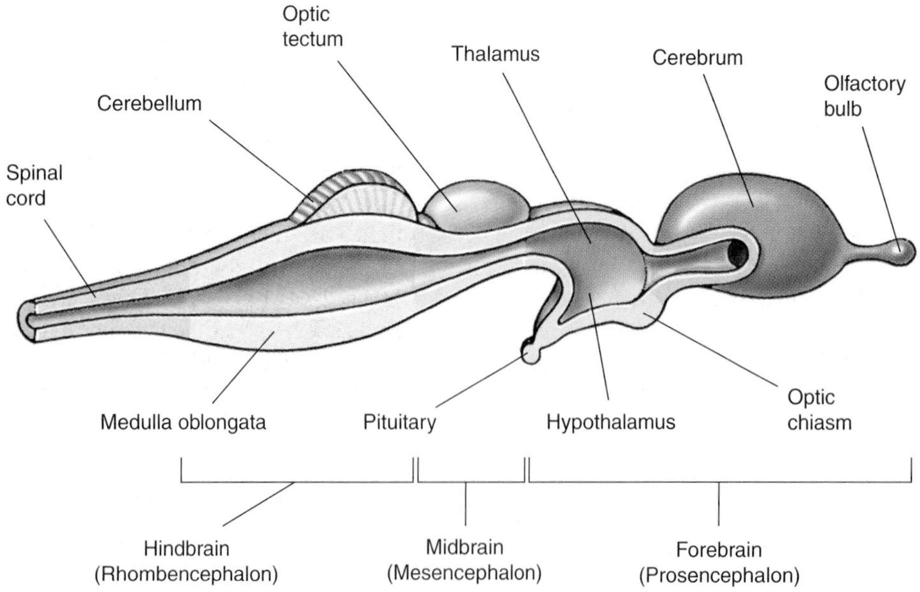
Optic tectum
Thalamus
Cerebrum
Olfactory bulb
Cerebellum
Spinal cord
Medulla oblongata
Pituitary
Hypothalamus
Optic chiasm
Hindbrain (Rhombencephalon)
Midbrain (Mesencephalon)
Forebrain (Prosencephalon)

Figure 26.12 The brain of a primitive fish.
The basic organization of the vertebrate brain can be seen in the brains of primitive fishes. The brain is divided into three regions that are found in differing proportions in all vertebrates: the hindbrain, which is the largest portion of the brain in fishes; the midbrain, which in fishes is devoted primarily to processing visual information; and the forebrain, which is concerned mainly with olfaction (the sense of smell) in fishes. In terrestrial vertebrates, the forebrain plays a far more dominant role in neural processing than it does in fishes.

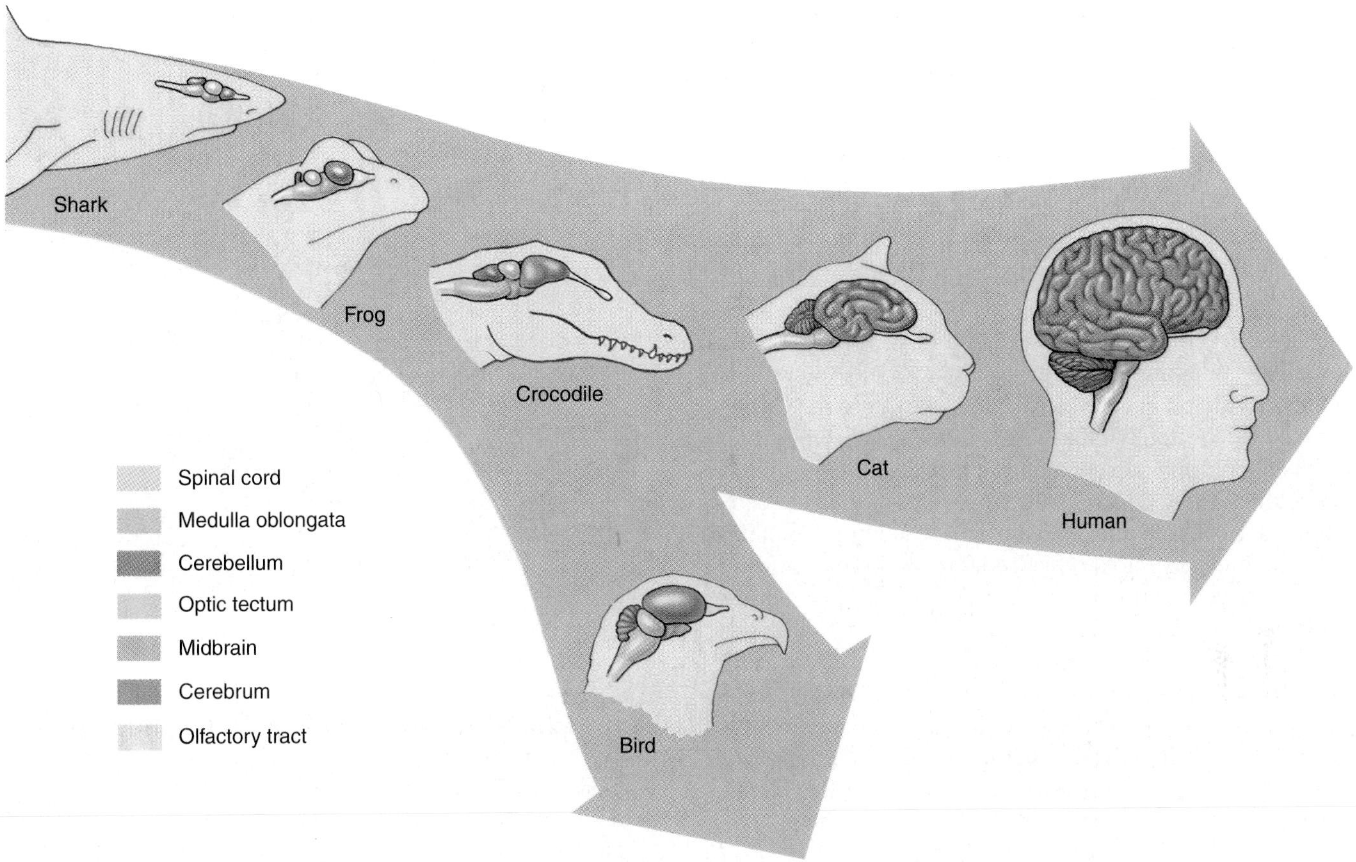
Shark
Frog
Crocodile
Cat
Human
Bird
Spinal cord
Medulla oblongata
Cerebellum
Optic tectum
Midbrain
Cerebrum
Olfactory tract

Figure 26.13 The evolution of the vertebrate brain.
In sharks and other fishes, the hindbrain is predominant, and the rest of the brain serves primarily to process sensory information. In amphibians and reptiles, the forebrain is far larger, and it contains a larger cerebrum devoted to associative activity. In birds, which evolved from reptiles, the cerebrum is even more pronounced. In mammals, the cerebrum covers the optic tectum and is the largest portion of the brain. The dominance of the cerebrum is greatest in humans, where it envelops much of the rest of the brain.

Much of this coordination is carried on within a small extension of the hindbrain called the cerebellum ("little cerebrum"). In more advanced vertebrates, the cerebellum plays an increasingly important role as a coordinating center and is correspondingly larger than it is in the fishes. In all vertebrates, the cerebellum processes data on the current position and movement of each limb, the state of relaxation or contraction of the muscles involved, and the general position of the body and its relation to the outside world. These data are gathered in the cerebellum and synthesized, and the resulting commands are issued to efferent pathways.

In fishes, the remainder of the brain is devoted to the reception and processing of sensory information. The midbrain is composed primarily of the **optic lobes** (tectum), which receive and process visual information, while the forebrain is devoted to the processing of *olfactory* (smell) information. The brains of fishes continue growing throughout their lives. This continued growth is in marked contrast to the brains of other classes of vertebrates, which generally complete their development by infancy. The human brain continues to develop through early childhood, but no new neurons are produced once development has ceased, except in the hippocampus, involved in long-term memory.

The Dominant Forebrain

Starting with the amphibians and continuing more prominently in the reptiles, sensory information is increasingly centered in the forebrain. This pattern was the dominant evolutionary trend in the further development of the vertebrate brain (figure 26.13).

The forebrain in reptiles, amphibians, birds, and mammals is composed of two elements that have distinct functions. The *diencephalon* (Greek, *dia,* between) consists of the thalamus and hypothalamus. The **thalamus** is an integrating and relay center between incoming sensory information and the cerebrum. The **hypothalamus** participates in basic drives and emotions and controls the secretions of the pituitary gland. The *telencephalon*, or "end brain" (Greek, *telos,* end), is located at the front of the forebrain and is devoted largely to associative activity. In mammals, the telencephalon is called the **cerebrum**.

26.6 In fishes, the hindbrain forms much of the brain; as terrestrial vertebrates evolved, the forebrain became increasingly more prominent.

26.7 How the Brain Works

The Cerebrum Is the Control Center of the Brain

Although vertebrate brains differ in the relative importance of different components, the human brain is a good model of how vertebrate brains function. About 85% of the weight of the human brain is made up of the cerebrum (figure 26.14). The cerebrum is the large rounded area of the brain divided by a groove into right and left halves called cerebral hemispheres. It functions in language, conscious thought, memory, personality development, vision, and a host of other activities we call "thinking and feeling." Figure 26.15 shows general areas of the brain and the functions they control. The cerebrum, which looks like a wrinkled mushroom, is positioned over and surrounding the rest of the brain, like a hand holding a fist. Much of the neural activity of the cerebrum occurs within a thin, gray outer layer only a few millimeters thick called the **cerebral cortex** (*cortex* is Latin for "bark of a tree"). This layer is gray because it is densely packed with neuron cell bodies. The human cerebral cortex contains the cell bodies of more than 10 billion nerve cells, roughly 10% of all the neurons in the brain. The wrinkles in the surface of the cerebral cortex increase its surface area (and number of cell bodies) threefold. Underneath the cortex is a solid white region of myelinated nerve fibers that shuttle information between the cortex and the rest of the brain.

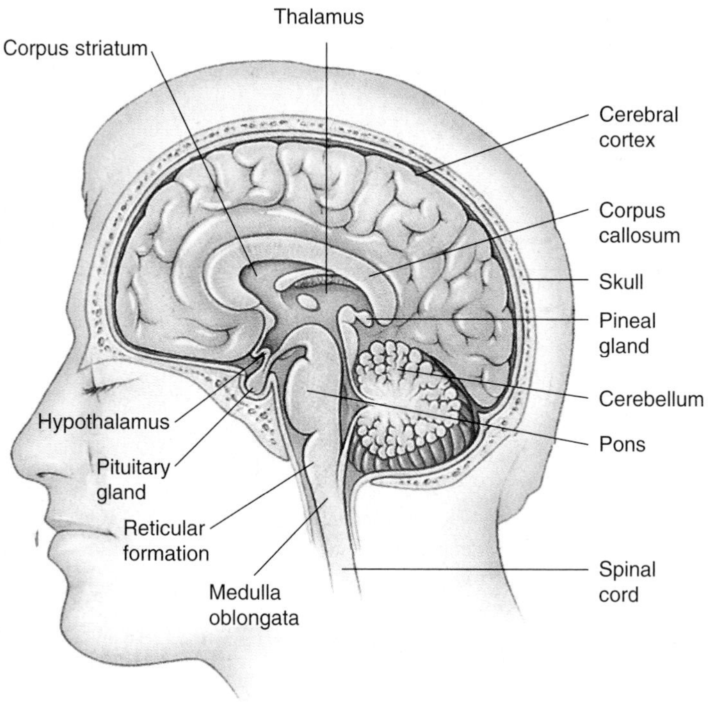

Thalamus
Corpus striatum
Cerebral cortex
Corpus callosum
Skull
Pineal gland
Cerebellum
Pons
Hypothalamus
Pituitary gland
Reticular formation
Medulla oblongata
Spinal cord

Figure 26.14 A section through the human brain.

The cerebrum occupies most of the brain. Only its outer layer, the cerebral cortex, is visible in surface view.

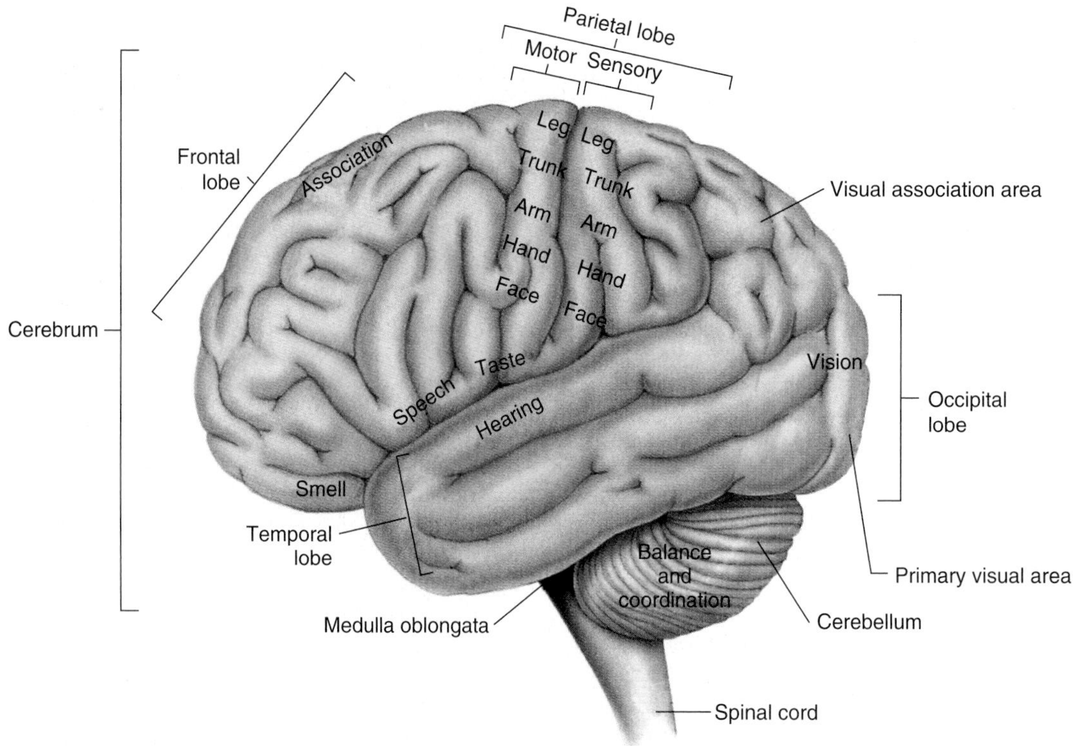

Parietal lobe
Motor
Sensory
Leg
Trunk
Arm
Hand
Face
Leg
Trunk
Arm
Hand
Face
Frontal lobe
Association
Visual association area
Cerebrum
Taste
Speech
Hearing
Vision
Occipital lobe
Smell
Temporal lobe
Balance and coordination
Primary visual area
Medulla oblongata
Cerebellum
Spinal cord

Figure 26.15 The major functional regions of the human brain.

Specific areas of the cerebral cortex are associated with different regions and functions of the body.

The right and left cerebral hemispheres are linked by bundles of neurons called **tracts.** These tracts serve as information highways, telling each half of the brain what the other half is doing. Because these tracts cross over, each half of the brain controls muscles and glands on the opposite side of the body. In general, the left brain is associated with language, speech, and mathematical abilities while the right brain is associated with intuitive, musical, and artistic abilities.

Researchers have found that the two sides of the cerebrum can operate as two different brains. For instance, in some people the tract between the two hemispheres has been cut by accident or surgery. In laboratory experiments, one eye of an individual with such a "split brain" is covered and a stranger is introduced. If the other eye is then covered instead, the person does not recognize the stranger who was just introduced!

Sometimes blood vessels in the brain are blocked by blood clots, causing a disorder called a **stroke.** During a stroke, circulation to an area in the brain is blocked and the brain tissue dies. A severe stroke in one side of the cerebrum may cause paralysis of the other side of the body.

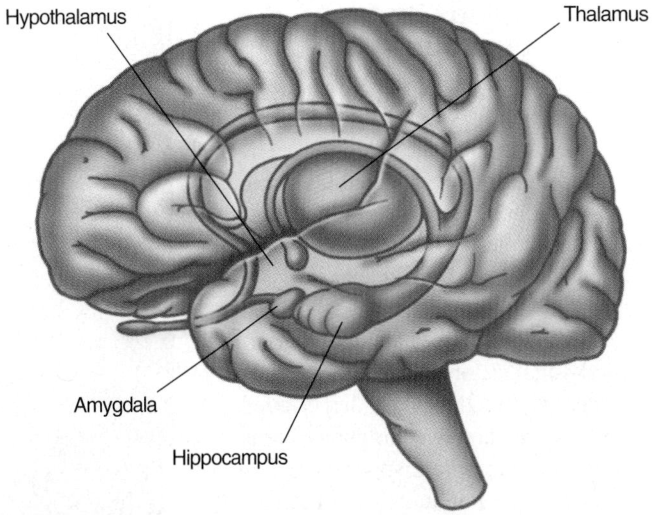
Hypothalamus
Thalamus
Amygdala
Hippocampus

Figure 26.16 The limbic system.
The hippocampus and the amygdala are the major components of the limbic system, which controls our most deep-seated drives and emotions.

The Thalamus and Hypothalamus Process Information

Beneath the cerebrum are the thalamus and hypothalamus, important centers for information processing. The **thalamus** is the major site of sensory processing in the brain. Auditory (sound), visual, and other information from sensory receptors enter the thalamus and then are passed to the sensory areas of the cerebral cortex. The thalamus also controls balance. Information about posture, derived from the muscles, and information about orientation, derived from sensors within the ear, combine with information from the cerebellum and pass to the thalamus. The thalamus processes the information and channels it to the appropriate motor center on the cerebral cortex.

The **hypothalamus** integrates all the internal activities. It controls centers in the brain stem that in turn regulate body temperature, blood pressure, respiration, and heartbeat. It also directs the secretions of the brain's major hormone-producing gland, the pituitary gland. The hypothalamus is linked by an extensive network of neurons to some areas of the cerebral cortex. This network, along with parts of the thalamus and the hypothalamus, is called the **limbic system** (figure 26.16). The operations of the limbic system are responsible for many of the most deep-seated drives and emotions of vertebrates, including pain, anger, sex, hunger, thirst, and pleasure.

The Cerebellum Coordinates Muscle Movements

Extending back from the base of the brain is a structure known as the **cerebellum.** The cerebellum controls balance, posture, and muscular coordination. This small, cauliflower-shaped structure, while well developed in humans and other mammals, is even better developed in birds. Birds perform more complicated feats of balance than we do, because they move through the air in three dimensions. Imagine the kind of balance and coordination needed for a bird to land on a branch, stopping at precisely the right moment without crashing into it.

The Brain Stem Controls Vital Body Processes

The **brain stem,** a term used to collectively refer to the midbrain, pons, and **medulla oblongata,** connects the rest of the brain to the spinal cord. This stalklike structure contains nerves that control your breathing, swallowing, and digestive processes, as well as the beating of your heart and the diameter of your blood vessels. A network of nerves called the **reticular formation** runs through the brain stem and connects to other parts of the brain. Their widespread connections make these nerves essential to consciousness, awareness, and sleep. One part of the reticular formation filters sensory input, enabling you to sleep through repetitive noises such as traffic yet awaken instantly when a telephone rings.

Language and Other Higher Functions

Although the two cerebral hemispheres seem structurally similar, they are responsible for different activities. The most thoroughly investigated example of this lateralization of function is language. The left hemisphere is the "dominant" hemisphere for language—the hemisphere in which most neural processing related to language is performed—in 90% of right-handed people and nearly two-thirds of left-handed people. There are two language areas in the dominant hemisphere: one is important for language comprehension and the formulation of thoughts into speech, and the other is responsible for the generation of motor output needed for language communication (figure 26.17).

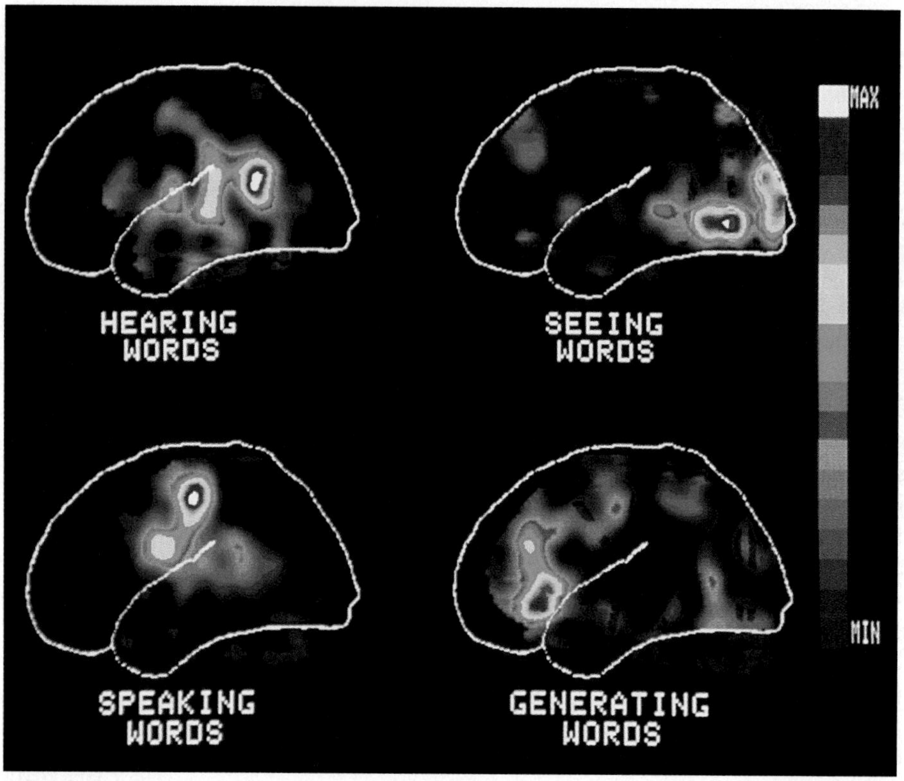
MAX
HEARING WORDS
SEEING WORDS
MIN
SPEAKING WORDS
GENERATING WORDS

Figure 26.17 Different brain regions control various activities. This illustration shows how the brain reacts in human subjects asked to listen to a spoken word, to read that same word silently, to repeat the word out loud, and then to speak a word related to the first. Regions of white, red, and yellow show the greatest activity. Compare this to figure 26.15 to see how regions of the brain are mapped.

While the dominant hemisphere for language is adept at sequential reasoning, like that needed to formulate a sentence, the nondominant hemisphere (the right hemisphere in most people) is adept at spatial reasoning, the type of reasoning needed to assemble a puzzle or draw a picture. It is also the hemisphere primarily involved in musical ability—a person with damage to the speech area in the left hemisphere may not be able to speak but may retain the ability to sing! Damage to the nondominant hemisphere may lead to an inability to appreciate spatial relationships and may impair musical activities such as singing. Reading, writing, and oral comprehension remain normal. The nondominant hemisphere is also important for the consolidation of memories of nonverbal experiences.

One of the great mysteries of the brain is the basis of memory and learning. There is no one part of the brain in which all aspects of a memory appear to reside. Although memory is impaired if portions of the brain, particularly the temporal lobes, are removed, it is not lost entirely. Many memories persist in spite of the damage, and the ability to access them gradually recovers with time. Therefore, investigators who have tried to probe the physical mechanisms underlying memory often have felt that they were grasping at a shadow. Although we still do not have a complete understanding of these mechanisms, we have learned a good deal about the basic processes in which memories are formed.

There appear to be fundamental differences between short-term and long-term memory. Short-term memory is transient, lasting only a few moments. Such memories can readily be erased by the application of an electrical shock, leaving previously stored long-term memories intact. This result suggests that short-term memories are stored electrically in the form of a transient neural excitation. Long-term memory, in contrast, appears to involve structural changes in certain neural connections within the brain. Two parts of the temporal lobes, the hippocampus and the amygdala, are involved in both short-term memory and its consolidation into long-term memory. Damage to these structures impairs the ability to process recent events into long-term memories.

> **26.7** The associative activity of the brain is centered in the wrinkled cerebral cortex, which lies over the cerebrum. Beneath, the thalamus and hypothalamus process information and integrate body activities. At the base of the brain, the cerebellum coordinates muscle movements.

26.8 The Spinal Cord

The **spinal cord** is a cable of neurons extending from the brain down through the backbone (figure 26.18). The gray neuron cell bodies form a column in the center of the cord, surrounded by a sheath of axons and dendrites, which make the outer edges of the cord white because they are coated with myelin. The spinal cord is surrounded and protected by the vertebrae, through which spinal nerves pass out to the body (figure 26.19). Messages from the body and the brain run up and down the spinal cord, an information highway.

In each segment of the spine, motor nerves extend out of the spinal cord to the muscles. Motor nerves from the spine control most of the muscles below the head. This is why injuries to the spinal cord often paralyze the lower part of the body. A muscle is paralyzed and cannot move if its motor neurons are damaged.

Figure 26.18 A view down the human spinal cord.

Pairs of spinal nerves can be seen extending out from the spinal cord. Along these nerves, the brain and spinal cord communicate with the body.

Spinal Cord Regeneration

In the past, scientists have tried to repair severed spinal cords by installing nerves from another part of the body to bridge the gap and act as guides for the spinal cord to regenerate. But most of these experiments have failed because the nerve bridges did not go from white matter to grey matter. Also, there is a factor that inhibits nerve growth in the spinal cord. After discovering that fibroblast growth factor stimulates nerve growth, neurobiologists tried gluing on the nerves, from white to grey matter, with fibrin that had been mixed with the fibroblast growth factor.

Three months later, rats with the nerve bridges began to show movement in their lower bodies. In further analyses of the experimental animals, dye tests indicated that the spinal cord nerves had regrown from both sides of the gap. Many scientists are encouraged by the potential to use a similar treatment in human medicine. However, most spinal cord injuries in humans do not involve a completely severed spinal cord; often, nerves are crushed. Also, while the rats with nerve bridges did regain some locomotory ability, tests indicated that they were barely able to walk or stand.

26.8 The spinal cord, protected in vertebrates by a backbone, extends motor nerves to the muscles below the head.

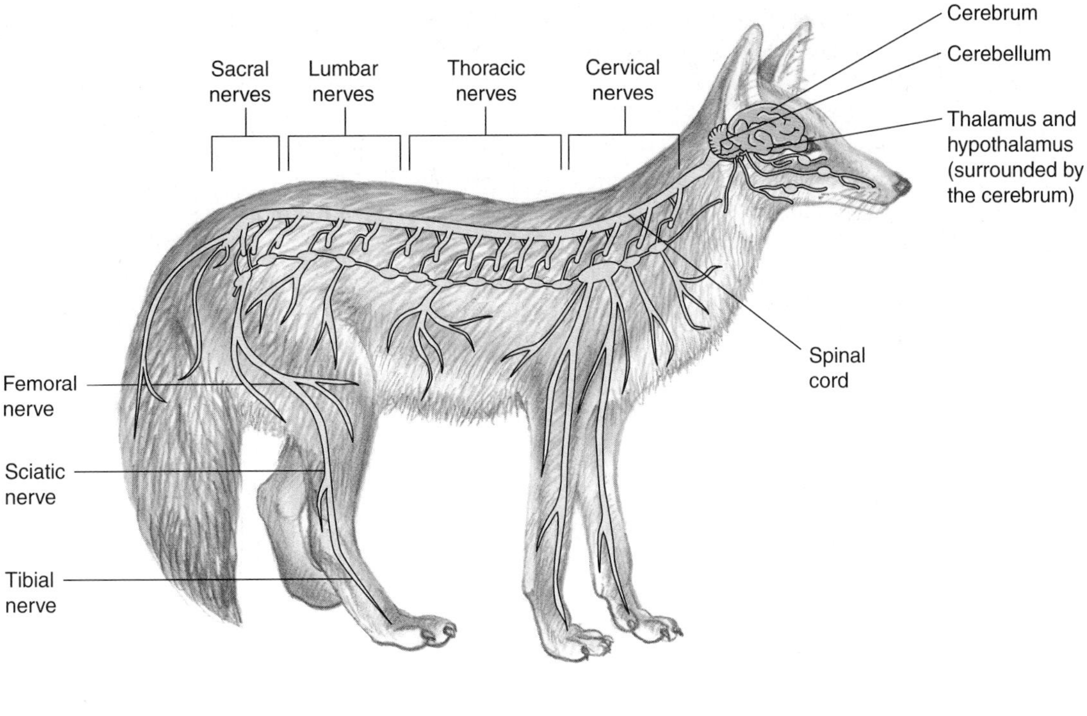

Cerebrum
Cerebellum
Sacral nerves
Lumbar nerves
Thoracic nerves
Cervical nerves
Thalamus and hypothalamus (surrounded by the cerebrum)
Spinal cord
Femoral nerve
Sciatic nerve
Tibial nerve

Figure 26.19 The vertebrate nervous system.

The brain is colored *pink* and the spinal cord and nerves are colored *yellow*.

26.9 Voluntary Nervous System

The motor pathways of a vertebrate can be subdivided into the **voluntary nervous system,** which relays commands to skeletal muscles, and the **autonomic nervous system,** which stimulates glands and relays commands to the smooth muscles of the body. The voluntary nervous system can be controlled by conscious thought. You can, for example, command your hand to move. The autonomic nervous system, by contrast, cannot be controlled by conscious thought. You cannot, for example, tell the smooth muscles in your digestive tract to speed up their action. The central nervous system issues commands over both voluntary and autonomic systems, but you are conscious of only the voluntary commands.

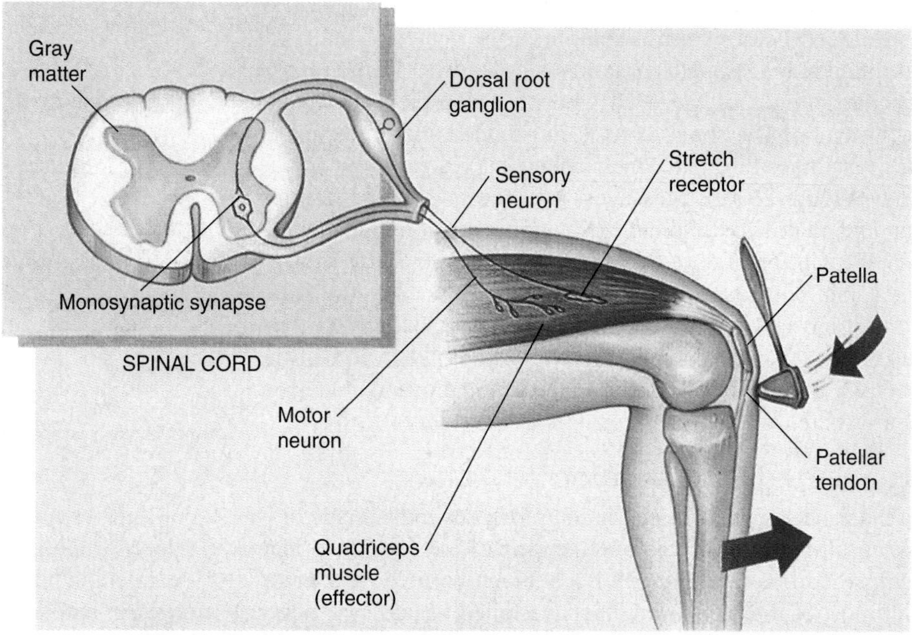
Gray
matter
Dorsal root
ganglion
Sensory
neuron
Stretch
receptor
Patella
Monosynaptic synapse
SPINAL CORD
Motor
neuron
Patellar
tendon
Quadriceps
muscle
(effector)

Figure 26.20 The knee-jerk reflex.

The most famous involuntary response, the knee jerk, is produced by activating stretch receptors in the quadriceps muscle. When a rubber mallet taps the patellar tendon, the muscle and stretch receptors in the muscle are stretched. A signal travels up a sensory neuron to the spine, where the sensory neuron stimulates a motor neuron, which sends a signal to the quadriceps muscle to contract.

Motor neurons carry information from the central nervous system to muscles and glands. For example, if your eyes see a runaway car speeding toward you, the CNS sends messages through motor neurons to glands that secrete the hormone adrenaline. The adrenaline increases your heartbeat and breathing rate. The CNS also sends messages through motor neurons to many muscles, which contract and get your body out of there—fast!

Reflexes Enable Quick Action

The motor neurons of the body have been wired to enable the body to act particularly quickly in time of danger—even before the animal is consciously aware of the threat. These sudden, involuntary movements are called reflexes. A **reflex** produces a rapid motor response to a stimulus because the sensory neuron bringing information about the threat passes the information directly to a motor neuron. The escape reaction of a fly about to be swatted is a reflex. One of the most frequently used reflexes in your body is blinking, a reflex that protects your eyes. If anything, such as an insect or a cloud of dust, approaches your eye, the eyelid blinks closed even before you realize what has happened. The reflex occurs before the cerebrum is aware the eye is in danger.

Because they involve passing information between few neurons, reflexes are very fast. Many reflexes never reach the brain. The "danger" nerve impulse travels only as far as the spinal cord and then comes right back as a motor response. Most reflexes involve a single connecting interneuron between the sensory neuron and the motor neuron. A few, like the knee-jerk reflex (figure 26.20), are monosynaptic reflex arcs. In these, the sensory neuron synapses directly with a motor neuron in the spine—there is no interneuron intermediary between them. If you step on something sharp, your leg jerks away from the danger: the prick causes nerve impulses in sensory neurons, which pass up the spinal cord directly to motor neurons, which cause your leg muscles to contract, jerking your leg up.

26.9 The voluntary nervous system relays commands to skeletal muscles and can be controlled by conscious thought.

26.10 Autonomic Nervous System

Some motor neurons are active all the time, even during sleep. These neurons carry messages from the CNS that keep the body going even when it is not active. These neurons are called the **autonomic nervous system.** The word *autonomic* means involuntary. The autonomic nervous system carries messages to muscles and glands that usually work without the animal noticing.

The autonomic nervous system is the command network the CNS uses to maintain the body's homeostasis. Using it, the CNS regulates heartbeat and controls muscle contractions in the walls of the blood vessels. It directs the muscles that control blood pressure, breathing, and the movement of food through the digestive system. It also carries messages that help stimulate glands to secrete tears, mucus, and digestive enzymes.

The autonomic nervous system is composed of two elements that act in opposition to one another. One division, the **sympathetic nervous system,** dominates in times of stress. It controls the "fight-or-flight" reaction, increasing blood pressure, heart rate, breathing rate, and blood flow to the muscles. It consists of a network of short motor axons extending out from the spine to clusters of neuron cell bodies called **ganglia** located near the spine. It also consists of long motor neurons extending from the ganglia directly to each target organ. Another division, the **parasympathetic nervous system,** has the opposite effect. It conserves energy by slowing the heartbeat and breathing rate and by promoting digestion and elimination. It consists of a network of long axons extending out from motor neurons within the spine; these axons extend to ganglia in the immediate vicinity of an organ. It also consists of short motor neurons extending from the ganglia to the nearby organ.

Most glands, smooth muscles, and cardiac muscles constantly get input from *both* the sympathetic and parasympathetic systems. The CNS controls activity by varying the ratio of the two signals to either stimulate or inhibit the organ (figure 26.21).

26.10 The autonomic nervous system relays commands to muscles and glands that cannot be controlled by conscious thought. It works by varying the ratio of two opposing signals to stimulate or inhibit activity.

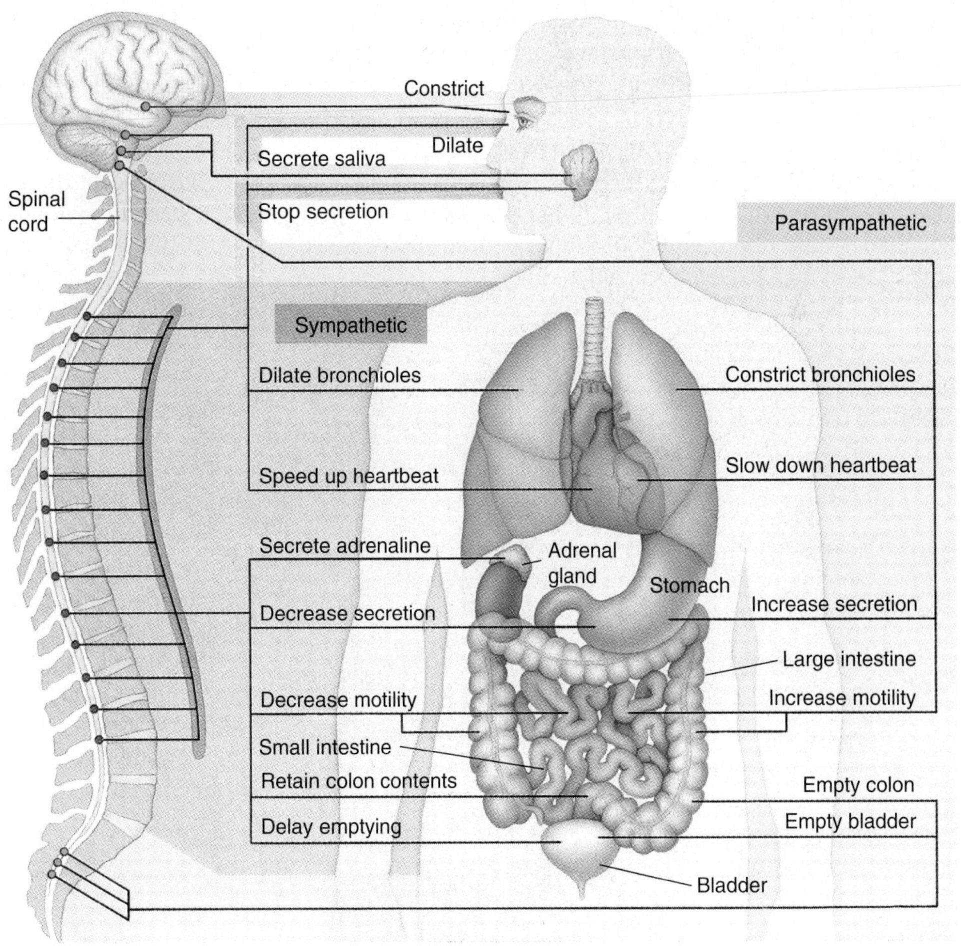

Constrict
Dilate
Secrete saliva
Spinal cord
Stop secretion
Parasympathetic
Sympathetic
Dilate bronchioles
Constrict bronchioles
Speed up heartbeat
Slow down heartbeat
Secrete adrenaline
Adrenal gland
Stomach
Decrease secretion
Increase secretion
Large intestine
Decrease motility
Increase motility
Small intestine
Retain colon contents
Empty colon
Delay emptying
Empty bladder
Bladder

Figure 26.21 How the sympathetic and parasympathetic nervous systems interact.

A nerve path runs from both of the systems to every organ indicated except the adrenal gland.

26.11 Sensory Receptors

Did you ever wonder what it would be like not to know anything about what is going on around you? Imagine if you couldn't hear, or see, or feel, or smell. After a while, a human goes mad if completely deprived of sensory input. The senses are the bridge to experience the way the body relates to everything around it.

The **sensory nervous system** tells the central nervous system what is happening. Sensory neurons carry impulses to the CNS from more than a dozen different types of sensory cells that detect changes outside and inside the body. Called **sensory receptors,** these specialized sensory cells detect many different things, including changes in blood pressure, strain on ligaments, and smells in the air. Particularly complex sensory receptors, made up of many cell and tissue types, are called **sensory organs.** The eyes and ears (figure 26.22) are sensory organs, and so are the taste buds in your mouth.

How does the brain know whether an incoming nerve impulse is light, sound, or pain? This information is built into the "wiring"—into which neurons interact while passing the information to the CNS and into the location in the brain where the information is sent. The brain "knows" it is responding to light because the message from a sensory neuron is wired to light receptor cells. That is why when you press your fingertips gently against the corners of your eyes, you "see stars"—the brain treats any impulse from the eyes as light, even though the eye received no light.

The Path of Sensory Information

The path of sensory information to the CNS is a simple one, composed of three stages:

1. **Stimulation.** A physical stimulus impinges on a sensory receptor.
2. **Transduction.** The sensory receptor initiates the opening or closing of ion channels in a sensory neuron.
3. **Transmission.** The sensory neuron conducts a nerve impulse along an afferent pathway to the CNS.

All sensory receptors are able to initiate nerve impulses by opening or closing **stimulus-gated channels** within sensory neuron membranes. Except for visual photoreceptors, these channels are sodium ion channels that depolarize the membrane and so start an electrical signal. The channels are opened by chemical or mechanical stimulation, often a disturbance such as touch, heat, or cold. The receptors differ from one another in the nature of the environmental input that triggers the opening of the channel. The body contains many sorts of receptors, each sensitive to a different aspect of the body's condition or to a different quality of the external environment.

Figure 26.22 Kangaroo rats have specialized ears.
The ears of kangaroo rats *(Dipodomys)* are adapted to nocturnal life and allow them to hear the low-frequency sounds of their predators, such as an owl's wingbeats or a sidewinder rattlesnake's scales rubbing against the ground. Also, the ears seem to be adapted to the poor sound-carrying quality of dry, desert air.

Exteroceptors are receptors that sense stimuli that arise in the external environment. Almost all of a vertebrate's exterior senses evolved in water before vertebrates invaded the land. Consequently, many senses of terrestrial vertebrates emphasize stimuli that travel well in water, using receptors that have been retained in the transition from the sea to the land. Hearing, for example, converts an airborne stimulus into a waterborne one, using receptors similar to those that originally evolved in the water. A few vertebrate sensory systems that function well in the water, such as the electric organs of fish, cannot function in the air and are not found among terrestrial vertebrates. On the other hand, some land dwellers have sensory systems, such as infrared receptors, that could not function in the sea.

Interoceptors sense stimuli that arise from within the body. These internal receptors detect stimuli related to muscle length and tension, limb position, pain, blood chemistry, blood volume and pressure, and body temperature. Many of these receptors are simpler than those that monitor the external environment and are believed to bear a closer resemblance to primitive sensory receptors.

26.11 Sensory receptors initiate nerve impulses in response to stimulation. All sensory nerve impulses are the same, differing only in the environmental stimulus that fires them and their destination in the brain.

26.12 Sensing the Internal Environment

Sensory receptors inside the body inform the CNS about the condition of the body. Much of this information passes to a coordinating center in the brain, the hypothalamus, the part of the brain responsible for maintaining the body's homeostasis—that is, keeping the body's internal environment constant. The vertebrate body uses a variety of different sensory receptors to respond to different aspects of its internal environment.

Temperature change. Two kinds of nerve endings in the skin are sensitive to changes in temperature, one stimulated by cold, the other by warmth. By comparing information from the two, the CNS can learn what the temperature is and if it is changing.

Blood chemistry. Receptors in the walls of arteries sense CO_2 levels in the blood. The brain uses this information to regulate the body's respiration rate, increasing it when CO_2 levels rise above normal.

Pain. Damage to tissue is detected by special nerve endings within tissues, usually near the surface, where damage is most likely to occur. When these nerve endings are physically damaged or deformed, the CNS responds by reflexively withdrawing the body segment and often by changing heartbeat and blood pressure as well.

Muscle contraction. Buried deep within muscles are sensory receptors called stretch receptors. In each, the end of a sensory neuron is wrapped around a muscle fiber (figure 26.23): when the muscle is stretched, the fiber elongates, stretching the spiral nerve ending and causing repeated signals to be sent to the brain. From these signals the brain can determine the rate of change of muscle length at any given moment. The CNS uses this information to control movements that require the combined action of several muscles, such as those that carry out breathing or locomotion.

Blood pressure. Blood pressure is sensed by neurons called baroreceptors with highly branched nerve endings within the walls of major arteries. When blood pressure increases, the stretching of the arterial wall causes the sensory neuron to increase the rate at which it sends signals to the CNS, while when it decreases, the rate of firing of the sensory neuron goes down (figure 26.24). Thus the frequency of impulses provides the CNS with a continuous measure of blood pressure.

Touch. Touch is sensed by pressure receptors buried below the surface of the skin. There are a variety of different types, some specialized to detect rapid changes in pressure, others to measure the duration and extent to which pressure is applied, and still others sensitive to vibration.

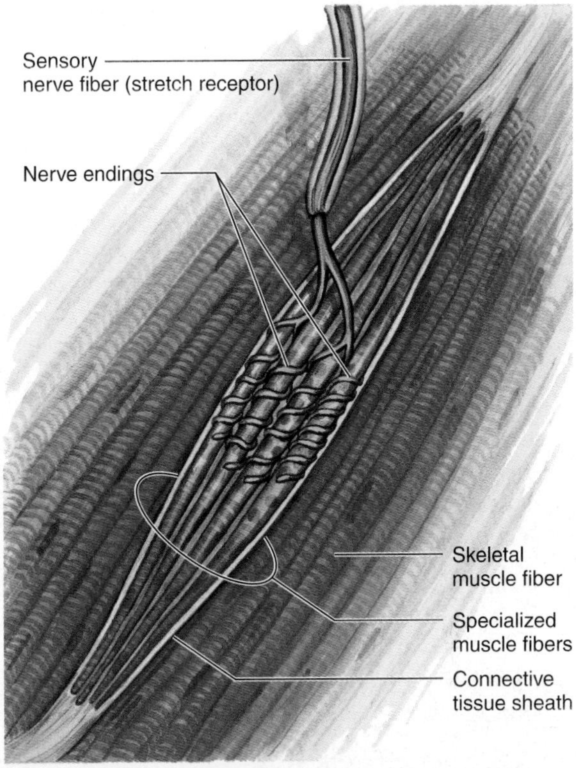
Sensory
nerve fiber (stretch receptor)
Nerve endings
Skeletal
muscle fiber
Specialized
muscle fibers
Connective
tissue sheath

Figure 26.23 A stretch receptor embedded within skeletal muscle.

Stretching the muscle elongates the specialized muscle fibers, which deforms the nerve endings, causing them to send a nerve impulse out along the nerve fiber.

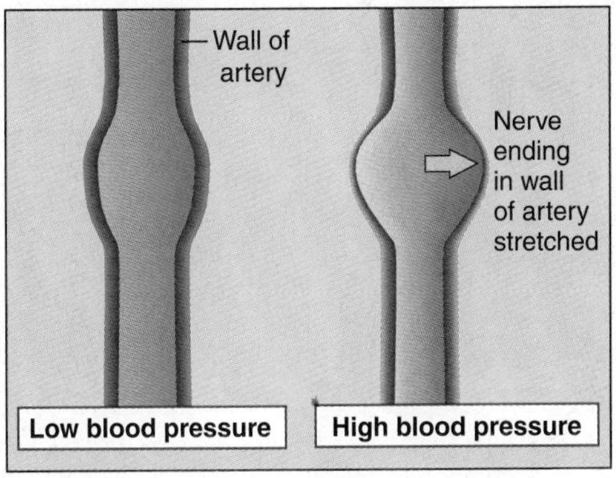
Wall of
artery
Nerve
ending
in wall
of artery
stretched
Low blood pressure
High blood pressure

Figure 26.24 How a baroreceptor works.

A network of nerve endings covers a region where the wall of the artery is thin. High blood pressure causes the wall to balloon out there, stretching the nerve endings and causing them to fire impulses.

26.12 A variety of different sensory receptors inform the hypothalamus about different aspects of the body's internal environment, enabling it to maintain the body's homeostasis.

26.13 Sensing Gravity: Balance

Receptors in the ear inform the brain where the body is in three dimensions. This knowledge enables an animal to move freely and maintain its balance.

Balance. In order to keep the body's balance, the brain needs a frame of reference, and the reference point it uses is gravity. The **otolith** sensory receptors that detect gravity are located in a series of hollow chambers within the inner ear. To illustrate how these receptors work, imagine a pencil standing in a glass. No matter which way you tip the glass, the pencil rolls along the rim, applying pressure to the lip of the glass. If you want to know the direction the glass is tipped, you need only ask where on the rim pressure is being applied. Using gravity receptors, the brain is able to determine its vertical position.

Motion. The brain senses motion in a way similar to that used to determine its vertical position by employing a receptor in which fluid deflects cilia in a direction opposite that of the motion. Within the inner ear are three fluid-filled **semicircular canals,** each oriented in a different plane at right angles to the other two so that motion in any direction can be detected. Protruding into the canal are groups of cilia from sensory cells. The cilia from each cell are arranged in a tentlike assembly called a *cupula* (figure 26.25), which is pushed by moving ear fluid in a direction opposite that of the head's movement. Because the three canals are oriented in all three planes, movement in any plane is sensed by at least one of them, and the brain is able to analyze complex movements by comparing the sensory inputs from each canal.

The brain does not have a speedometer. The semicircular canals do not react if the body moves in a straight line because the fluid in the canals does not move. That is why traveling in a car or airplane at a constant speed in one direction gives no sense of motion.

26.13 The body senses gravity and acceleration by the deflection of cilia by moving objects or fluid. The body cannot sense motion at a constant velocity and direction.

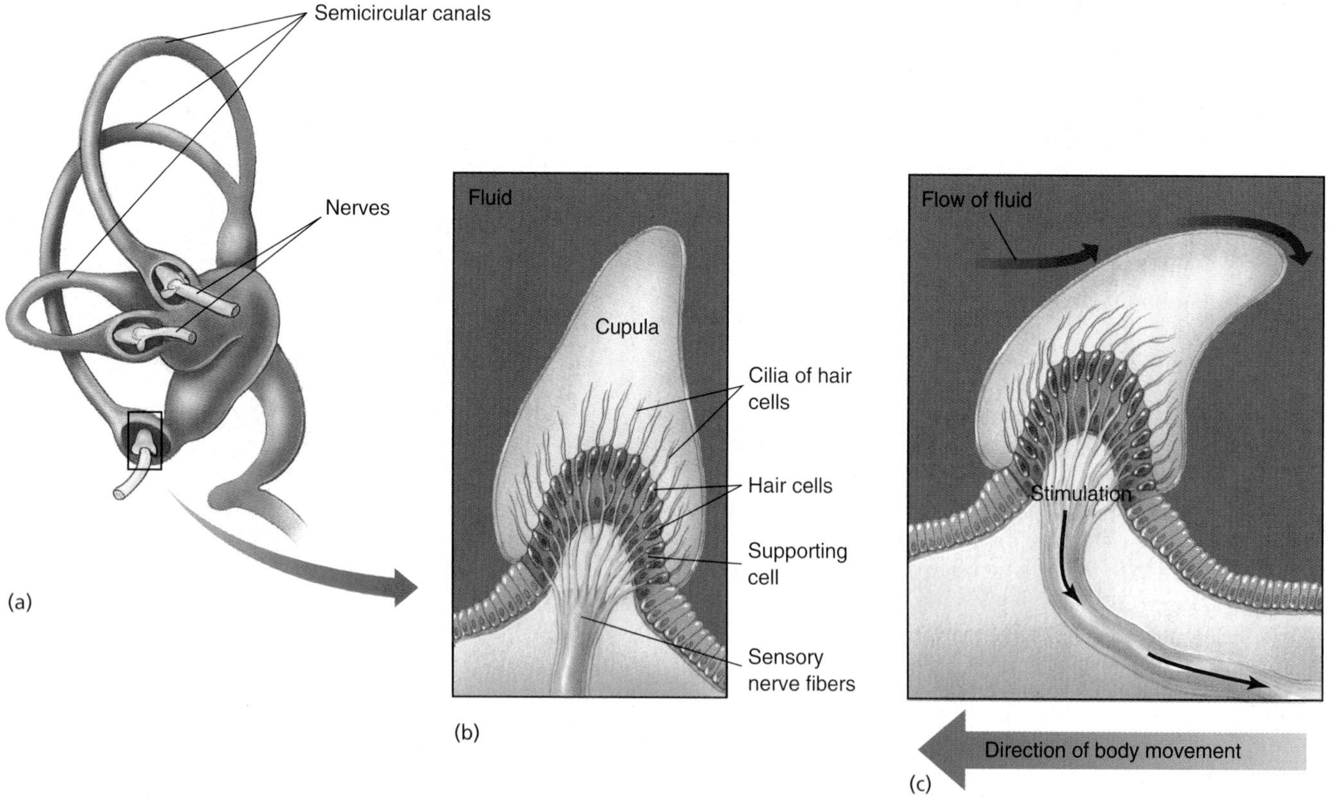

Semicircular canals
Nerves
(a)
Fluid
Cupula
Cilia of hair cells
Hair cells
Supporting cell
Sensory nerve fibers
(b)
Flow of fluid
Stimulation
Direction of body movement
(c)

Figure 26.25 How the inner ear senses motion.
(*a*) The semicircular canals are part of the inner ear. (*b*) The cupula within the semicircular canals are surrounded by fluid and contain hair cells. (*c*) Movement in a particular direction causes fluid in the semicircular canal of that plane to move; the cupula is displaced, thereby stimulating the hair cells.

26.14 Sensing Chemicals: Taste and Smell

Vertebrates are able to detect many of the chemicals in air and in food.

Taste. Embedded within the surface of the tongue are *taste buds,* which are located within raised areas called *papillae* (figure 26.26). Taste buds are onion-shaped structures that contain many taste receptor cells, each of which has fingerlike microvilli that project into an opening called the taste pore. Chemicals from food dissolve in saliva and contact the taste cells through the taste pore. Salty, sour, sweet, and bitter chemicals in food are detected in different ways by taste buds. When the tongue encounters a chemical, information from the taste cells passes to sensory neurons, which transmit the signals to the brain.

Smell. In the nose are chemically sensitive neurons whose cell bodies are embedded within the epithelium of the nasal passage (figure 26.27). When they detect chemicals, these sensory neurons transmit information to a location in the brain where smell information is processed and analyzed. In many vertebrates (dogs are a familiar example), these neurons are far more sensitive than in humans.

Smell as well as taste is very important in telling an animal about its food. That is why when you have a bad cold and your nose is stuffed up, your food has little taste. Other receptors also play a role. Thus the "hot" sensation of foods such as chili peppers is detected by pain receptors, not chemical receptors.

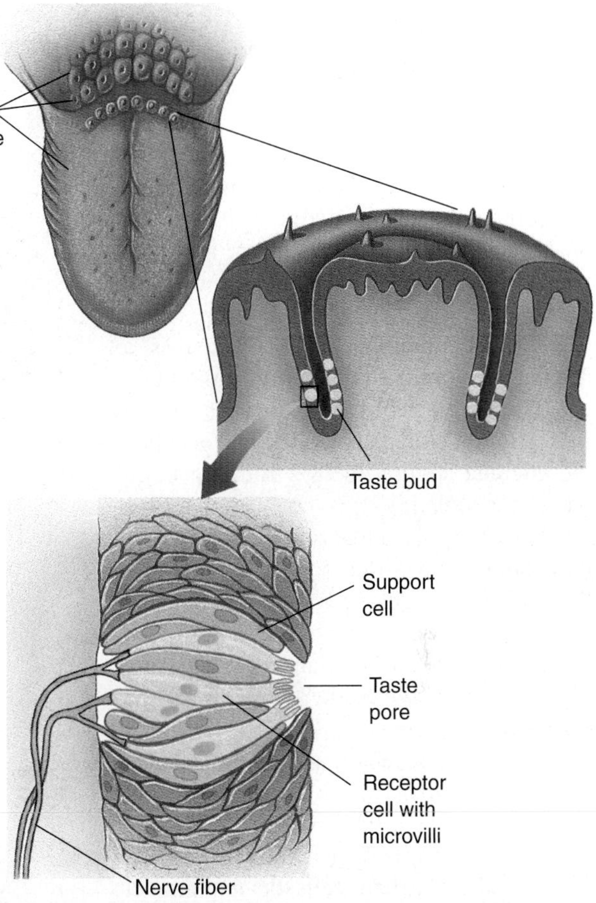
Taste papillae
Taste bud
Support cell
Taste pore
Receptor cell with microvilli
Nerve fiber

Figure 26.26 Taste.

Taste buds on the human tongue are typically grouped into projections called papillae. Individual taste buds are bulb-shaped collections of taste receptor cells that open out into the mouth through a taste pore. Taste buds in humans can detect bitter, sour, salty, and sweet chemicals in food.

26.14 Taste and smell are chemical senses. In many vertebrates, the sense of smell is very well developed.

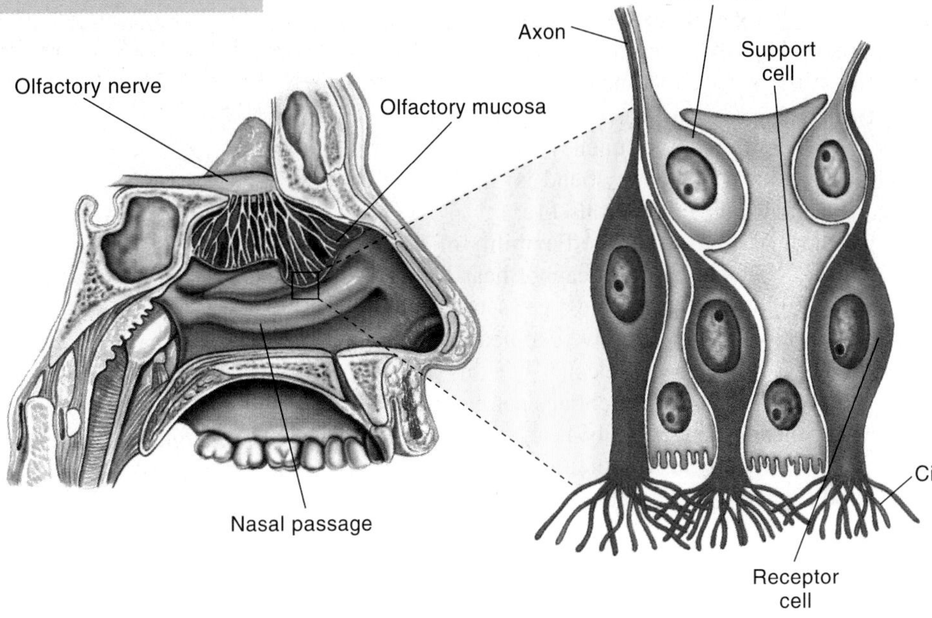
Olfactory nerve
Olfactory mucosa
Nasal passage
Axon
Basal cell
Support cell
Cilia
Receptor cell

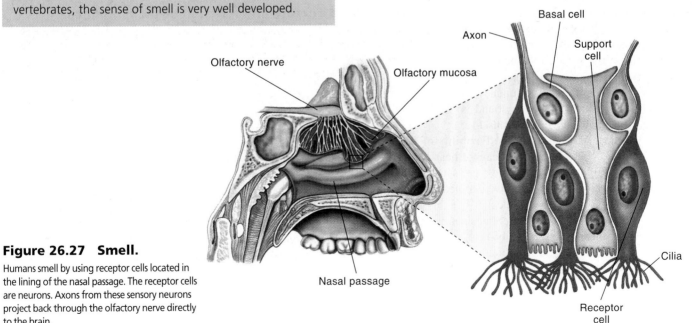

Figure 26.27 Smell.

Humans smell by using receptor cells located in the lining of the nasal passage. The receptor cells are neurons. Axons from these sensory neurons project back through the olfactory nerve directly to the brain.

26.15 Sensing Sounds: Hearing

When you hear a sound, you are detecting the air vibrating—waves of pressure in the air beating against your ear, pushing a membrane called the **eardrum** in and out. On the other side of the eardrum are three small bones that act as a lever system to increase the force of the vibration. They transfer the amplified vibration across a second membrane to fluid within the inner ear. The chamber of the inner ear is shaped like a tightly coiled snail shell and is called the **cochlea** (figure 26.28), from the Latin name for "snail." It is connected to the throat by the eustachian tube in such a way that there is no difference in air pressure between the middle ear and the outside. That is why your ears sometimes "pop" when landing in an airplane—the pressure is equalizing between the two sides of the eardrum.

The sound receptors within the cochlea are hair cells that rest on a membrane that runs up and down the middle of the curving chamber, separating it into two halves like a wall. The hair cells do not project into the fluid filling the cochlea; instead, they are covered by a second membrane. When a sound enters the cochlea, the sound waves cause this membrane "sandwich" to vibrate, bending the hairs pressed against the outer membrane and causing them to send nerve impulses to sensory neurons that travel to the brain.

Sounds of different frequencies cause different parts of the membrane to vibrate, and thus fire different sensory neurons—the identity of the sensory neuron being fired tells the CNS what the frequency of the sound is. The intensity of the sound is determined by how *often* the neurons fire. Our ability to hear depends upon the flexibility of the membranes within the cochlea. Humans cannot hear low-pitched sounds, below 20 vibrations (or cycles) per second, although some vertebrates can. As children, we can hear high-pitched sounds, up to 20,000 cycles per second, but this ability decreases as we get older. Other vertebrates can hear sounds at far higher frequencies. Dogs readily hear sounds of 40,000 cycles per second and so can respond to a high-pitched dog whistle when it seems silent to a human observer.

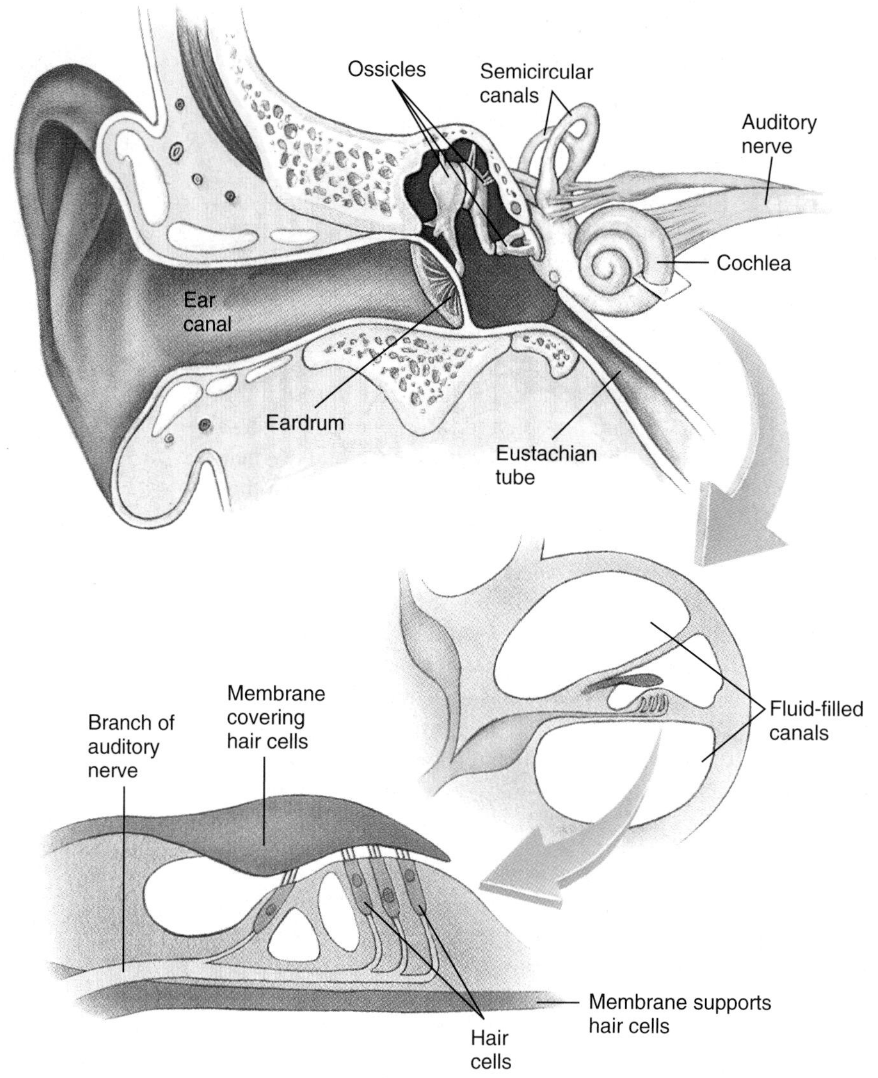

Ossicles
Semicircular canals
Auditory nerve
Cochlea
Ear canal
Eardrum
Eustachian tube
Membrane covering hair cells
Branch of auditory nerve
Fluid-filled canals
Membrane supports hair cells
Hair cells

Figure 26.28 Structure and function of the human ear. Sound waves passing through the ear canal beat on the eardrum, pushing a set of three small bones, or ossicles, against an inner membrane. This sets up a wave motion in the fluid filling the canals within the cochlea. When the sound wave beats against the sides of the canals, the membrane covering the hair cells moves back and forth against the hair cells, which causes associated neurons to fire impulses.

26.15 Sound receptors detect the air vibrating as waves of pressure push against the membrane covering the ear. Inside, these waves are amplified and press down hair cells that send signals to the brain. The location of the stimulated hair cells within the ear tells the brain about the frequency of the sound waves. Intensity is indicated by how often the hair cells fire.

26.16 Sensing Body Position

The Lateral Line System

The lateral line system provides fish with a sense of "distant touch," enabling them to sense objects that reflect pressure waves and low-frequency vibrations. This enables a fish to detect prey, for example, and to swim in synchrony with the rest of its school. It also enables a blind cave fish to sense its environment by monitoring changes in the patterns of water flow past the lateral line receptors. The lateral line system is found in amphibian larvae, but it is lost at metamorphosis and is not present in any terrestrial vertebrate. The lateral line system supplements the fish's sense of hearing, which is performed by a different sensory structure.

The lateral line system consists of sensory structures within a longitudinal canal in the fish's skin that extends along each side of the body and within several canals in the head (figure 26.29). The sensory structures are known as hair cells because they have hairlike processes at their surface that project into a gelatinous membrane called a *cupula* (Latin, "little cup"). The hair cells are innervated by sensory neurons that transmit impulses to the brain. Vibrations carried through the fish's environment produce movements of the cupula, which cause the hairs to bend. When the hair cells bend, the associated sensory neurons are stimulated and generate a receptor potential that is transmitted to the brain.

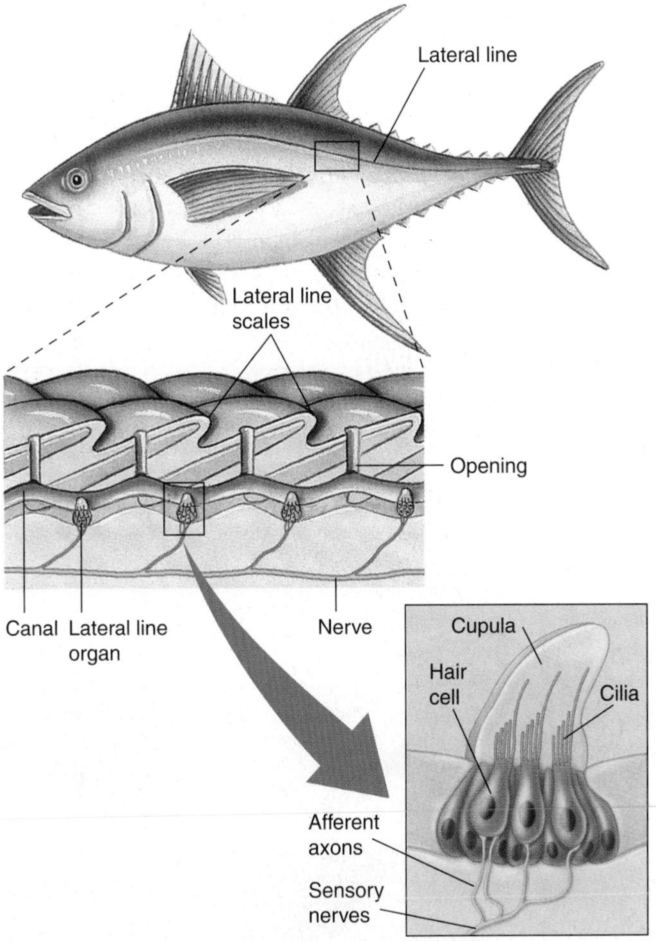

Lateral line
Lateral line scales
Opening
Canal
Lateral line organ
Nerve
Cupula
Hair cell
Cilia
Afferent axons
Sensory nerves

Figure 26.29 The lateral line system.
This system consists of canals running the length of the fish's body beneath the surface of the skin. Within these canals are sensory structures containing hair cells with cilia that project into a gelatinous cupula. Pressure waves traveling through the water in the canals deflect the cilia and depolarize the sensory neurons associated with the hair cells.

Sonar

A few groups of mammals that live and obtain their food in dark environments have circumvented the limitations of darkness. A bat flying in a completely dark room easily avoids objects that are placed in its path—even a wire less than a millimeter in diameter (figure 26.30). Shrews use a similar form of "lightless vision" beneath the ground, as do whales and dolphins beneath the sea. All of these mammals perceive distance by means of sonar. They emit sounds and then determine the time it takes these sounds to reach an object and return to the animal. This process is called **echolocation.** A bat, for example, produces clicks that last 2 to 3 milliseconds and are repeated several hundred times per second. The three-dimensional imaging achieved with such an auditory sonar system can be quite sophisticated.

26.16 Fish sense pressure waves in water much as an ear senses sound. Many vertebrates sense distant objects by bouncing sounds off of them.

Figure 26.30 Using ultrasound to locate a moth.
This bat is emitting high-frequency "chirps" as it flies. It then listens for the sound's reflection against the moth. By timing how long it takes for a sound to return, the bat can "see" the moth even in total darkness.

26.17 Sensing Light: Vision

No other stimulus provides as much detailed information about the environment as light. Vision, the perception of light, is carried out by a special sensory apparatus called an eye. All the sensory receptors described to this point have been chemical or mechanical ones. Eyes contain sensory receptors called rods and cones that respond to photons of light. The light energy is absorbed by pigments in the rods and cones, which respond by triggering nerve impulses in sensory neurons.

Evolution of the Eye

Vision begins with the capture of light energy by photoreceptors. Because light travels in a straight line and arrives virtually instantaneously, visual information can be used to determine both the direction and the distance of an object. No other stimulus provides as much detailed information.

Many invertebrates have simple visual systems with photoreceptors clustered in an eyespot. Although an eyespot can perceive the direction of light, it cannot be used to construct a visual image. The members of four phyla—annelids, mollusks, arthropods, and vertebrates—have evolved well-developed, image-forming eyes. True image-forming eyes in these phyla, though strikingly similar in structure, are believed to have evolved independently (figure 26.31). Interestingly, the photoreceptors in all of them use the same light-capturing molecule, suggesting that not many alternative molecules are able to play this role.

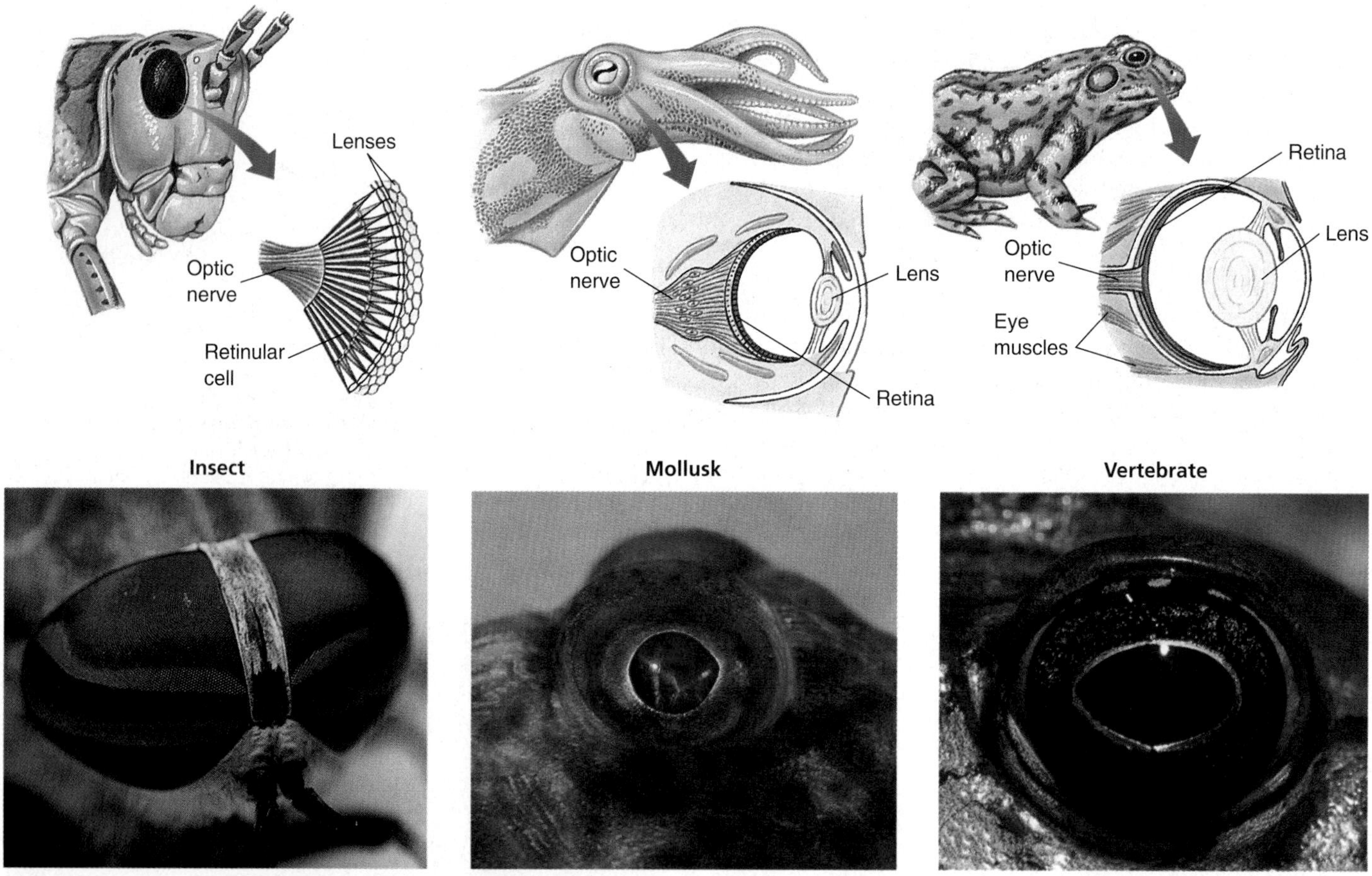

Lenses
Optic nerve
Retinular cell
Optic nerve
Lens
Retina
Retina
Lens
Optic nerve
Eye muscles
Insect
Mollusk
Vertebrate

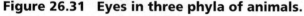

Figure 26.31 Eyes in three phyla of animals.
Although they are superficially similar, these eyes differ greatly in structure and are not homologous. Each has evolved separately and, despite the apparent structural complexity, has done so from simpler structures.

Structure of the Vertebrate Eye

The vertebrate eye works like a lens-focused camera (figure 26.32). Light first passes through a transparent protective covering, the **cornea,** which begins to focus the light onto the rear of the eye. The beam of light then passes through the **lens,** which completes the focusing. The lens is a fat disc, somewhat resembling a flattened balloon. It is attached by suspending ligaments to **ciliary muscles.** When these muscles contract, they change the shape of the lens and thus the point of focus on the rear of the eye. The amount of light entering the eye and reaching the lens is controlled by a shutter, called the **iris,** between the cornea and the lens. The transparent zone in the middle of the iris, the **pupil,** gets larger in dim light and smaller in bright light.

The light that passes through the pupil is focused by the lens onto the back of the eye. An array of light-sensitive receptor cells line the back surface of the eye, called the **retina.** The retina is the light-sensing portion of the eye. The vertebrate retina contains two kinds of photoreceptors, called **rods** and **cones,** which, when stimulated by light, generate nerve impulses that travel to the brain along a short, thick nerve pathway called the optic nerve (figure 26.33). Rods are receptor cells that are extremely sensitive to light, and they can detect various shades of gray even in dim light. However, they cannot distinguish colors, and because they do not detect edges well, they produce poorly defined images. Cones are receptor cells that detect color and are sensitive to edges so that they produce sharp images. The center of the vertebrate retina contains a tiny pit, called the **fovea,** densely packed with some 3 million cones. This area produces the sharpest image, which is why we tend to move our eyes so that the image of an object we want to see clearly falls on this area.

The lens of the vertebrate eye is constructed to filter out short-wavelength light. This solves a difficult optical problem: any uniform lens bends short wavelengths more than it does longer ones, a phenomenon known as chromatic aberration. Consequently, these short wavelengths cannot be brought into focus simultaneously with longer wavelengths. Unable to focus the short wavelengths, the vertebrate eye eliminates them. Insects, whose eyes do not focus light, are able to see these lower, ultraviolet wavelengths quite well and often use them to locate food or mates.

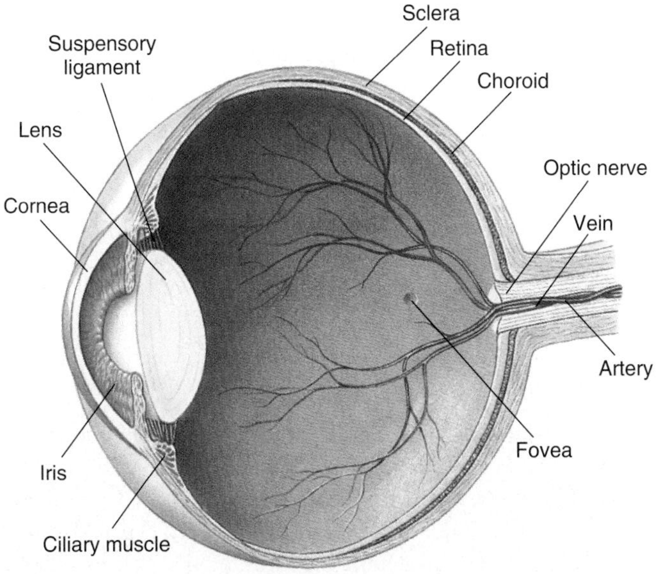
Sclera
Suspensory ligament
Retina
Choroid
Lens
Optic nerve
Cornea
Vein
Artery
Fovea
Iris
Ciliary muscle

Figure 26.32 The structure of the human eye.
Light passes through the transparent cornea and is focused by the lens on the rear surface of the eye, the retina, at a particular location called the fovea. The retina is rich in photoreceptors.

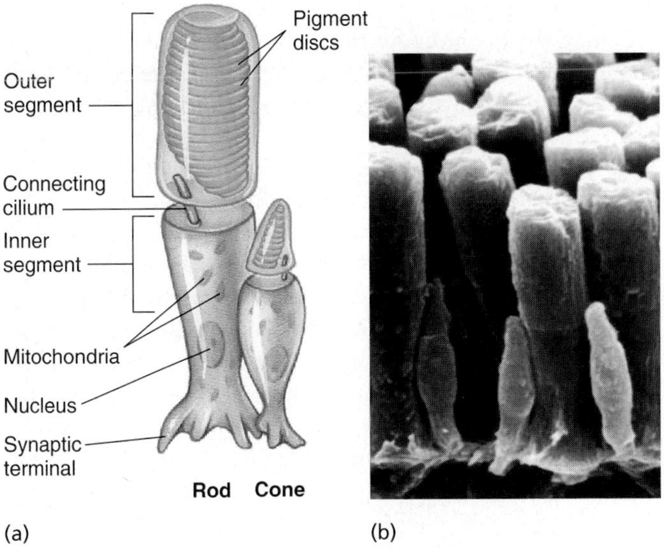
Pigment discs
Outer segment
Connecting cilium
Inner segment
Mitochondria
Nucleus
Synaptic terminal
Rod
Cone
(a)
(b)

Figure 26.33 Rods and cones.
(*a*) The broad tubular cell on the *left* is a rod. The shorter, tapered cell next to it is a cone. (*b*) Electron micrograph of rods and cones.

How Rods and Cones Work

A rod or cone cell in the eye is able to detect a single photon of light. How can it be so sensitive? The primary sensing event of vision is the absorption of a photon of light by a pigment. The pigments in rods and cones are made from plant pigments called carotenoids. That is why eating carrots is said to be good for night vision—the orange color of carrots is due to the presence of carotenoids called carotenes. The visual pigment in the human eye is a fragment of carotene called ***cis*-retinal.** The pigment is attached to a protein called **opsin** to form a light-detecting complex called **rhodopsin.**

When it receives a photon of light, the pigment undergoes a change in shape. This change in shape must be large enough to alter the shape of the opsin protein attached to it. When light is absorbed by the *cis*-retinal pigment, the linear end of the molecule rotates sharply upward, straightening out that end of the molecule (figure 26.34). The new form of the pigment is referred to as ***trans*-retinal.** This radical change in the pigment's shape induces a change in the shape of the protein opsin to which the pigment is bound, initiating a chain of events that leads to the generation of a nerve impulse.

Each rhodopsin activates several hundred molecules of a protein called transducin. Each of these activates several hundred molecules of an enzyme whose product stimulates sodium channels in the photoreceptor membrane at a rate of about 1,000 per second. This cascade of events allows a single photon to have a large effect on the receptor.

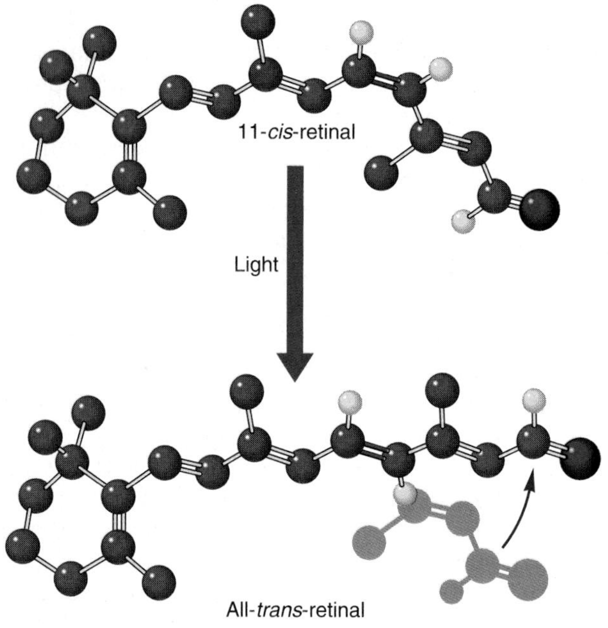

11-*cis*-retinal
Light
All-*trans*-retinal

Figure 26.34 Absorption of light.

When light is absorbed by *cis*-retinal, the pigment undergoes a change in shape and becomes *trans*-retinal. This shape change initiates a chain of events that leads to the generation of a nerve impulse.

Color Vision

Three kinds of **cone cells** provide us with color vision. Each possesses a different version of the opsin protein (that is, one with a distinctive amino acid sequence and thus a different shape). These differences in shape affect the flexibility of the attached retinal pigment, shifting the wavelength at which it absorbs light (figure 26.35). In rods, light is absorbed at 500 nanometers. In cones, the three versions of opsin absorb light at 455 nanometers (blue-absorbing), 530 nanometers (green-absorbing), or 625 nanometers (red-absorbing). By comparing the relative intensities of the signals from the three cones, the brain can calculate the intensity of other colors.

Most vertebrates, particularly those that are diurnal (active during the day), have color vision, as do many insects. Indeed, honeybees can see light in the near-ultraviolet range, which is invisible to the human eye. Color vision requires the presence of more than one photopigment in different receptor cells, but not all animals with color vision have the three-cone system characteristic of humans and other primates. Fish, turtles, and birds, for example, have four or five kinds of cones; the "extra" cones enable these animals to see near-ultraviolet light. Many mammals (such as squirrels), on the other hand, have only two types of cones.

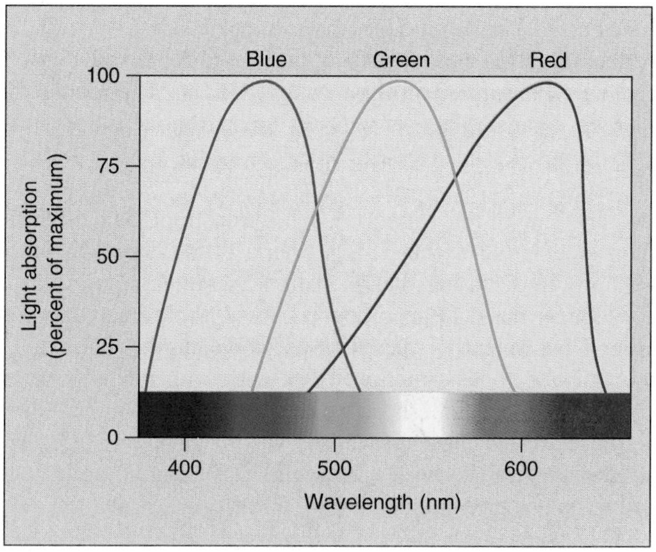

Blue
Green
Red
Light absorption (percent of maximum)
100
75
50
25
0
400
500
600
Wavelength (nm)

Figure 26.35 Color vision.

The absorption spectrum of *cis*-retinal is shifted in cone cells from the 500 nanometers characteristic of rod cells. The amount of the shift determines what color the cone absorbs: a shift down to 455 nanometers yields blue absorption; a shift up to 530 nanometers yields green absorption; and a shift farther up to 625 nanometers yields red absorption.

Conveying the Light Information to the Brain

The path of light through each eye is the reverse of what you might expect. The rods and cones are at the rear of the retina, not the front. Light passes through several layers of ganglion and bipolar cells before it reaches the rods and cones (figure 26.36). Once the photoreceptors are activated, they stimulate bipolar cells, which in turn stimulate ganglion cells. The direction of nerve impulses in the retina is thus opposite to the direction of light.

Action potentials propagated along the axons of ganglion cells are relayed through structures called the *lateral geniculate nuclei* of the thalamus and projected to the occipital lobe of the cerebral cortex. There the brain interprets this information as light in a specific region of the eye's receptive field. The pattern of activity among the ganglion cells across the retina encodes a point-to-point map of the receptive field, allowing the retina and brain to image objects in visual space. In addition, the frequency of impulses in each ganglion cell provides information about the light intensity at each point, while the relative activity of ganglion cells connected (through bipolar cells) with the three types of cones provides color information.

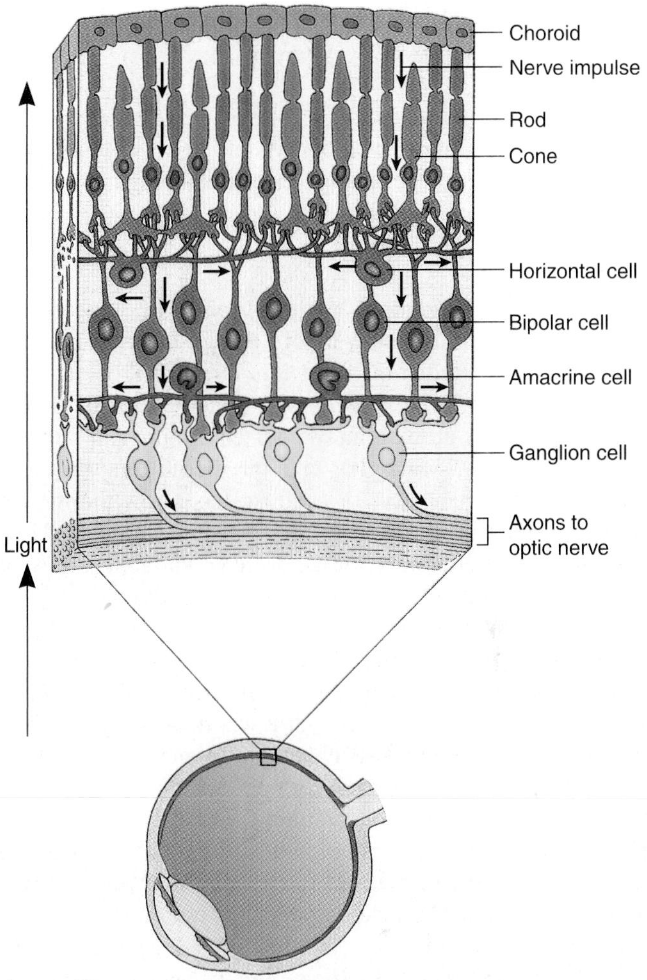

Choroid
Nerve impulse
Rod
Cone
Horizontal cell
Bipolar cell
Amacrine cell
Ganglion cell
Axons to optic nerve
Light

Figure 26.36 Structure of the retina.
The rods and cones are at the rear of the retina, not the front. Light passes over four other types of cells in the retina before it reaches the rods and cones. Nerve impulses then travel through the bipolar cells to the ganglion cells and on to the optic nerve.

Binocular Vision

Primates (including humans) and most predators have two eyes, one located on each side of the face. When both eyes are trained on the same object, the image that each sees is slightly different because each eye views the object from a different angle. This slight displacement of the images permits **binocular vision,** the ability to perceive three-dimensional images and to sense depth. Having their eyes facing forward maximizes the field of overlap in which this stereoscopic vision occurs (figure 26.37).

In contrast, prey animals generally have eyes located to the sides of the head, preventing binocular vision but enlarging the overall receptive field. Depth perception is less important to prey than detection of potential enemies from any quarter. The eyes of the American woodcock, for example, are located at exactly opposite sides of its skull so that it has a 360-degree field of view without turning its head! Most birds have laterally placed eyes and, as an adaptation, have two foveas in each retina. One fovea provides sharp frontal vision, like the single fovea in the retina of mammals, and the other fovea provides sharper lateral vision.

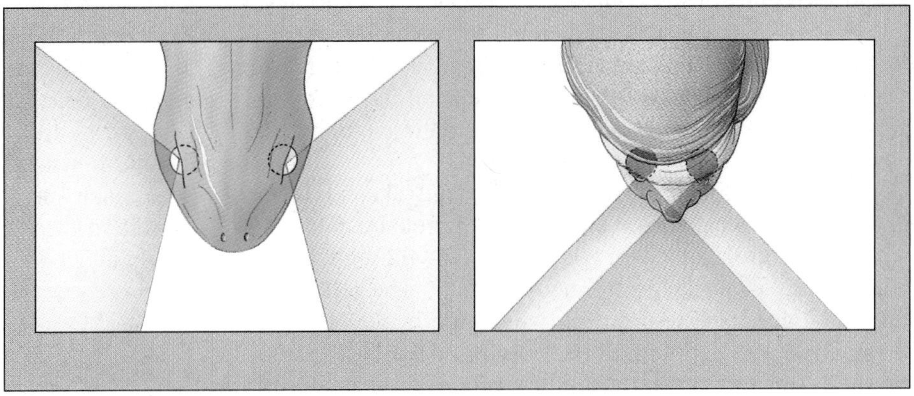

Figure 26.37 Binocular vision.
When the eyes are located on the sides of the head (as on the *left*), the two vision fields do not overlap and binocular vision does not occur. When both eyes are located toward the front of the head (as on the *right*) so that the two fields of vision overlap, depth can be perceived.

26.17 Vision receptors detect reflected light; binocular vision allows the brain to form three-dimensional images of objects.

26.18 Other Types of Sensory Reception

Vision is the primary sense used by all vertebrates that live in a light-filled environment, but visible light is by no means the only part of the electromagnetic spectrum that vertebrates use to sense their environment.

Heat

Electromagnetic radiation with wavelengths longer than those of visible light is too low in energy to be detected by photoreceptors. Radiation from this *infrared* ("below red") portion of the spectrum is what we normally think of as radiant heat. Heat is an extremely poor environmental stimulus in water because water has a high thermal capacity and readily absorbs heat. Air, in contrast, has a low thermal capacity, so heat in air is a potentially useful stimulus. However, the only vertebrates known to have the ability to sense infrared radiation are the snakes known as pit vipers.

The pit vipers possess a pair of heat-detecting pit organs located on either side of the head between the eye and the nostril (figure 26.38). The pit organs permit a blindfolded rattlesnake to accurately strike at a warm, dead rat. Each pit organ is composed of two chambers separated by a membrane, and the organ apparently operates by comparing the temperatures of the two chambers. The nature of the pit organ's thermal receptor is not known; it probably consists of temperature-sensitive neurons innervating the two chambers. The two pit organs appear to provide stereoscopic information, in much the same way that two eyes do. Indeed, the information transmitted from the pit organs is processed by the visual center of the snake brain.

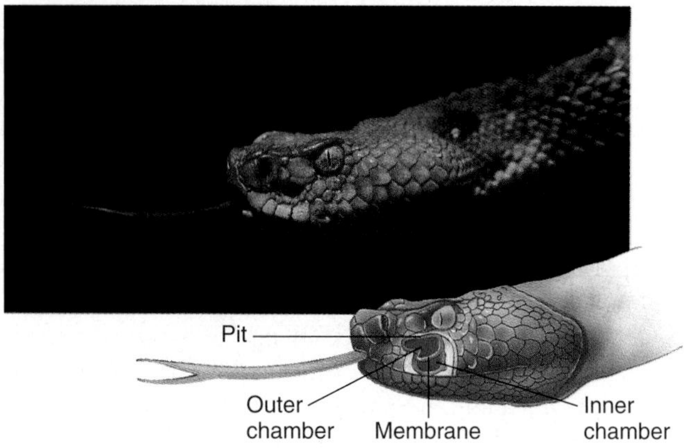

Pit
Outer chamber
Membrane
Inner chamber

Figure 26.38 "Seeing" heat.

The depression between the nostril and the eye of this rattlesnake opens into the pit organ. In the cutaway portion of the diagram, you can see that the organ is composed of two chambers separated by a membrane. Snakes known as pit vipers have this unique ability to sense infrared radiation (heat).

Electricity

While air does not readily conduct an electrical current, water is a good conductor. All aquatic animals generate electrical currents from contractions of their muscles. A number of different groups of fishes can detect these electrical currents. The *electric fish* even have the ability to produce electrical discharges from specialized electric organs. Electric fish use these weak discharges to locate their prey and mates and to construct a three-dimensional image of their environment even in murky water.

The elasmobranchs (sharks, rays, and skates) have electroreceptors called the ampullae of Lorenzini. The receptor cells are located in sacs that open through jelly-filled canals to pores on the body surface. The jelly is a very good conductor, so a negative charge in the opening of the canal can depolarize the receptor at the base, causing the release of neurotransmitters and increased activity of sensory neurons. This allows sharks, for example, to detect the electrical fields generated by the muscle contractions of their prey. Although the ampullae of Lorenzini were lost in the evolution of teleost fish (most of the bony fish), electroreception reappeared in some groups of teleost fish that use sensory structures analogous to the ampullae of Lorenzini. Electroreceptors evolved yet another time, independently, in the duck-billed platypus, an egg-laying mammal. The receptors in its bill can detect the electrical currents created by the contracting muscles of shrimp and fish, enabling the mammal to detect its prey at night and in muddy water.

Magnetism

Eels, sharks, and many birds appear to navigate along the magnetic field lines of the earth. Even some bacteria use such forces to orient themselves. Birds kept in blind cages, with no visual cues to guide them, will peck and attempt to move in the direction in which they would normally migrate at the appropriate time of the year. They will not do so, however, if the cage is shielded from magnetic fields by steel. Indeed, if the magnetic field of a blind cage is deflected 120 degrees clockwise by an artificial magnet, a bird that normally orients to the north will orient toward the east-southeast. There has been much speculation about the nature of the magnetic receptors in these vertebrates, but the mechanism is still very poorly understood.

> **26.18** Pit vipers can locate warm prey by infrared radiation (heat), and many aquatic vertebrates can locate prey and ascertain the contours of their environment by means of electroreceptors. Some vertebrates can even orient themselves with respect to the earth's magnetic field.

CONCEPT REVIEW

1. The sensory nervous system consists of _________ neurons.
   a. afferent
   b. efferent

2. At the synapse, the action potential moves along the neuron in which order?
   a. axon–dendrite–synapse
   b. axon–synapse–dendrite
   c. dendrite–axon–synapse
   d. synapse–axon–dendrite

3. Schwann cells, which envelop the axon at intervals along its length, act as
   a. neurons.
   b. tracts.
   c. electrical insulators.
   d. dendrites.

4. Select the largest region of the human brain.
   a. cerebellum
   b. cerebrum
   c. hypothalamus
   d. thalamus

5. The hindbrain is devoted primarily to
   a. sensory integration.
   b. integrating visceral activities.
   c. processing of visual information.
   d. coordinating motor reflexes.

6. The hypothalamus controls
   a. respiration.
   b. sensory integration.
   c. language.
   d. conscious thought.

7. The parasympathetic branch of the autonomic nervous system _________ the rate of heartbeat.
   a. decreases
   b. increases

8. A reflex does not involve the
   a. sensory neuron.
   b. motor neuron.
   c. cerebrum.
   d. spinal cord.

9. Hair cells in the otolith membrane detect
   a. light.
   b. vibrations.
   c. fluid movement.
   d. gravity.

10. The ability to hear often decreases with age because the
    a. cilia degenerate.
    b. hair cells stiffen.
    c. flexibility of the basilar membrane changes.
    d. tympanic membrane breaks.

11. If light is turned off in a room, the _________ of the retina become more important to vision.
    a. cones
    b. rods

12. The wave of depolarization that moves down a neuron is called an _________.

13. The specialized chemicals that carry out communication across synaptic junctions are called _________.

14. The _________ are three fluid-filled structures that detect motion changes.

CHALLENGE YOURSELF

1. Ouabain is a drug that poisons the sodium-potassium pump. What would be the *immediate* effect of this drug on a neuron's ability to generate action potentials? Would the *long-term* effect be different? Explain.

2. If a nerve impulse can jump from node to node along a myelinated axon, why can't it jump from the presynaptic cell to the postsynaptic cell across a synaptic cleft?

3. Most people at one time or another have sensed an oncoming storm by detecting the increase in humidity in the air. What sort of receptors detect humidity? Why do you suppose that hot days seem so much hotter when the air is humid?

4. Why does traveling in a car offer very little sense of motion?

26 eBRIDGE

Reinforcing Key Points

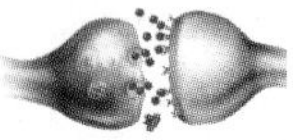

Neurons and How They Work

26.1 Evolution of the Animal Nervous System

26.2 Neurons

26.3 The Nerve Impulse

26.4 The Synapse

26.5 Neuromodulators and Drug Addiction

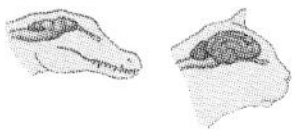

The Central Nervous System

26.6 Evolution of the Vertebrate Brain

26.7 How the Brain Works

26.8 The Spinal Cord

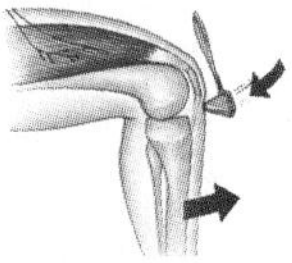

The Peripheral Nervous System

26.9 Voluntary Nervous System

26.10 Autonomic Nervous System

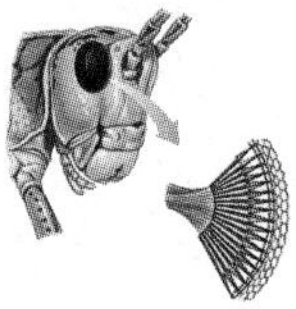

The Sensory Nervous System

26.11 Sensory Receptors

26.12 Sensing the Internal Environment

26.13 Sensing Gravity: Balance

26.14 Sensing Chemicals: Taste and Smell

26.15 Sensing Sounds: Hearing

26.16 Sensing Body Position

26.17 Sensing Light: Vision

26.18 Other Types of Sensory Reception

Electronic Learning

Visual Learning

Animations

Development of Membrane Potential

Action Potential

Stroke

Reflex Arc

Sense of Balance

Sense of Rotational Acceleration

Taste

Smell

Hearing

Vision

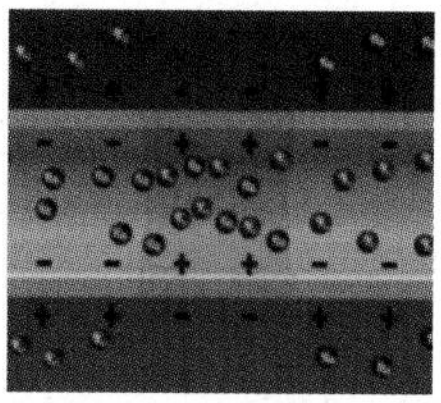

Art Labeling Activities

Central Nervous System

Synapse Structure and Function

Helping You Learn

Nervous Systems

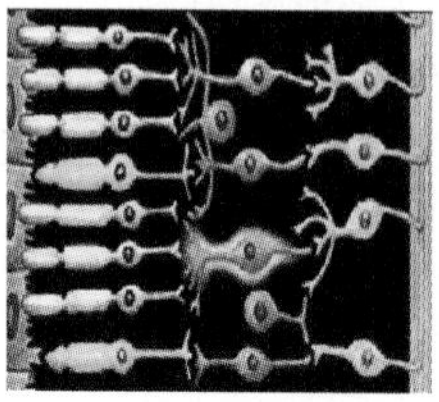

Author's Corner

Learning. Some of the most exciting biology being done today is in neurobiology, where researchers working at the molecular level are beginning to unravel the mystery of memory and learning.

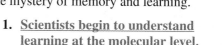

1. Scientists begin to understand learning at the molecular level.
2. Nobel winners show how long-term changes in the brain create mood.
3. Sometimes I wonder if my dog is smarter than I am.
4. One surprisingly small region of the brain is responsible for humans' general intelligence.
5. Simple repetition can have a powerful impact on learning.
6. Pulling an all-nighter doesn't work because learning requires sleep.
7. Lab rats are able to control a robot by thought alone.

Virtual Classroom

Doin' Drugs

Addictive drugs like cocaine or nicotine act at the level of individual synapses within the brain. Typically the drug produces euphoria by stimulating the synapses of pleasure-producing pathways. Cocaine, for example, blocks the reabsorption of the neurotransmitter dopamine in nerve synapses of the limbic system. Cocaine molecules bind tightly to the dopamine-recycling proteins, with the result that more and more dopamine builds up in the synapse, causing the postsynaptic neuron to fire more often, increasing pleasure. Addiction results when the brain attempts to "turn the volume" back down by removing dopamine receptors from the postsynaptic membrane. If you subsequently stop adding cocaine, so that dopamine levels fall to normal, there are not enough "target" postsynaptic receptors to fire the pleasure-producing pathway! To feel even normal, you again take cocaine to re-elevate dopamine levels. You are addicted. The deep lesson of doin' drugs is that addiction is not a matter of will power or moral character—it is just chemistry. You can no more avoid addiction than you can command a bullet speeding toward your head to stop. The way to avoid the bullet is not to pull the trigger.

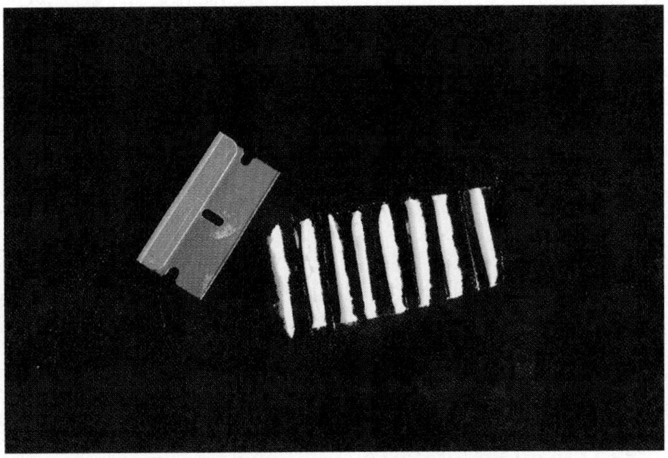

Virtual Lab

How Snails "See" an Invisible Trail

We humans are quite dependent on our senses of sight and sound. Many organisms have evolved other senses, more suited to their particular environments and lifestyles. Gastropod mollusks, for example, deposit mucous trails while crawling over surfaces that other snails of the same species follow. The stimuli in the mucus and the sense organ used by snails to follow the trail are unknown. The hypothesis that originally received the most attention was that the trail contained a chemical concentration gradient which allowed snails to follow the trail in the same direction that it was laid down ("trail polarity"). This hypothesis has been rejected, however, as it conflicts with many experimental observations. Development of alternative hypotheses about how a snail detects the direction of a trail would be greatly aided by a more detailed understanding of the sensory organ involved. Paul Hamilton of the University of Central Arkansas studied the involvement of the cephalic tentacles of the marsh periwinkle snail, *Littoraria irrorata,* on detecting trail polarity. By amputating tentacles, he was able to investigate the role of the tentacles in detecting mucous trail polarity.

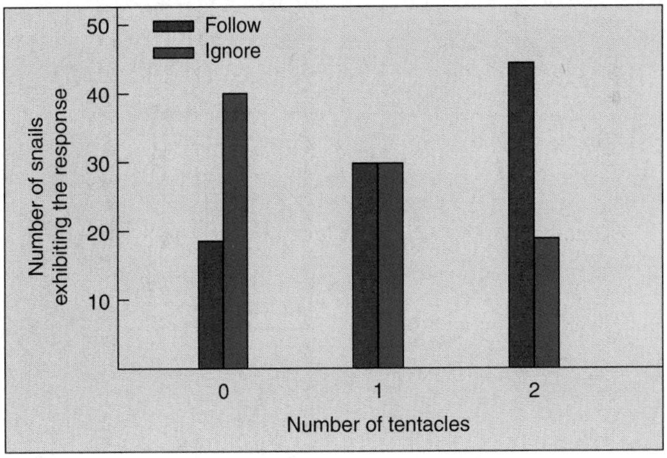
Follow
Ignore
Number of snails exhibiting the response
50
40
30
20
10
0
1
2
Number of tentacles

Quizzes | Further Reading | Essential Study Partner | Links | BioCourse.com

CHAPTER 27

Chemical Signaling Within the Animal Body

THE FAR SIDE® By GARY LARSON
© 1988 FarWorks, Inc. All Rights Reserved/Dist. by Creators Syndicate
Larson
"So! Planning on roaming the neighborhood with some of your buddies today?"

CHAPTER 27

Chemical Signaling Within the Animal Body

CHAPTER OVERVIEW

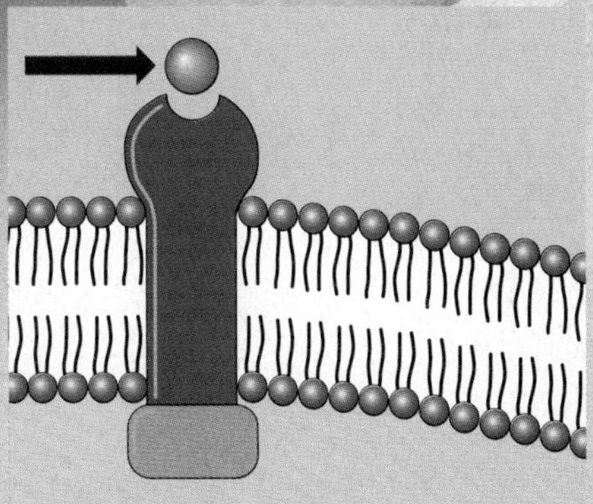

The Neuroendocrine System

- Many of the hormone-producing endocrine glands are under the direct control of the central nervous system.
- The hypothalamus produces releasing hormones that cause the pituitary to release particular hormones that circulate via the bloodstream to their target tissues.
- Steroid hormones enter cells and interact with receptors located in the cytoplasm or nucleus.
- Peptide hormones interact with receptors located on the cell surface.
- Hormones often initiate a cascade of events in the cell that greatly increases the power of the signal.

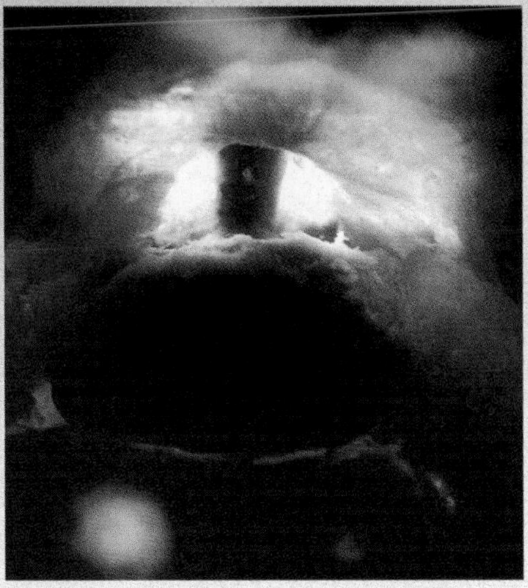

The Major Endocrine Glands

- Axons from the hypothalamus enter the posterior pituitary, carrying ADH and oxytocin; the posterior pituitary stores and secretes these hormones.
- The secretion of hormones produced by the anterior pituitary is controlled by the hypothalamus via releasing and inhibiting hormones.
- The pancreas secretes insulin, which lowers the blood glucose level, and glucagon, which raises the blood glucose level.
- The thyroid secretes thyroxine, which affects the metabolic rate and growth, and calcitonin, which maintains calcium levels in the body.

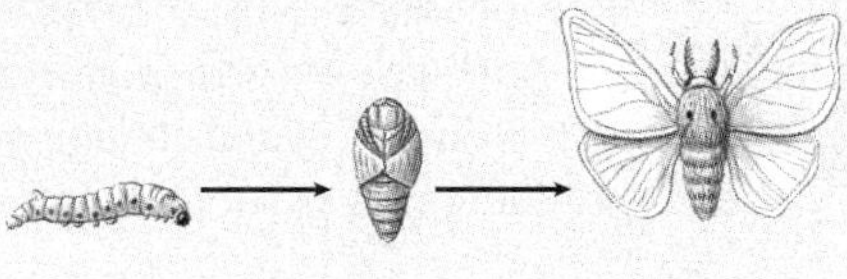

Other Hormone-Producing Glands

- The ovaries and testes produce the sex steroid hormones, including estrogens, progesterone, and testosterone.
- The pineal gland in the brain secretes the hormone melatonin in response to a hypothalamic biological clock.

27.1 Hormones

A **hormone** is a chemical signal produced in one part of the body that is stable enough to be transported in active form far from where it is produced and that typically acts at a distant site. There are three big advantages to using chemical messengers rather than electrical signals (nerves) to control body organs. First, chemical molecules can spread to all tissues via the blood (Imagine trying to wire every cell with its own nerve!). Second, chemical signals can persist much longer than electrical ones, a great advantage for hormones controlling slow processes like growth and development. Third, many different kinds of chemicals can act as hormones, so different hormone molecules can be targeted at different tissues.

Hormones are produced by glands, most of which are controlled by the central nervous system. Because these glands are completely enclosed in tissue rather than having ducts that empty to the outside, they are called **endocrine glands** (from the Greek, *endon,* within). Hormones are secreted from them directly into the bloodstream (this is in contrast to **exocrine glands,** which, like sweat glands, have ducts). Your body has a dozen principal endocrine glands that together make up the endocrine system (figure 27.1).

The *endocrine system* and the *motor nervous system* are the two main routes the central nervous system uses to issue commands to the organs of the body. The two are so closely linked that they are often considered a single system—the **neuroendocrine system.** The hypothalamus can be considered the main switchboard of the neuroendocrine system. The hypothalamus is continually checking conditions inside the body. Is it too hot or too cold? Is it running out of fuel? Is the blood pressure too high? If the internal environment starts to get out of balance, the hypothalamus has several ways to set things right again. For example, if it wants to speed up the heart rate, it can send a nerve signal to the medulla, or it can use a chemical command, causing the adrenal gland to produce the hormone adrenaline, which also speeds up the heart rate. Which command the hypothalamus uses depends on the desired duration of the effect. A chemical message is typically far longer lasting than a nerve signal.

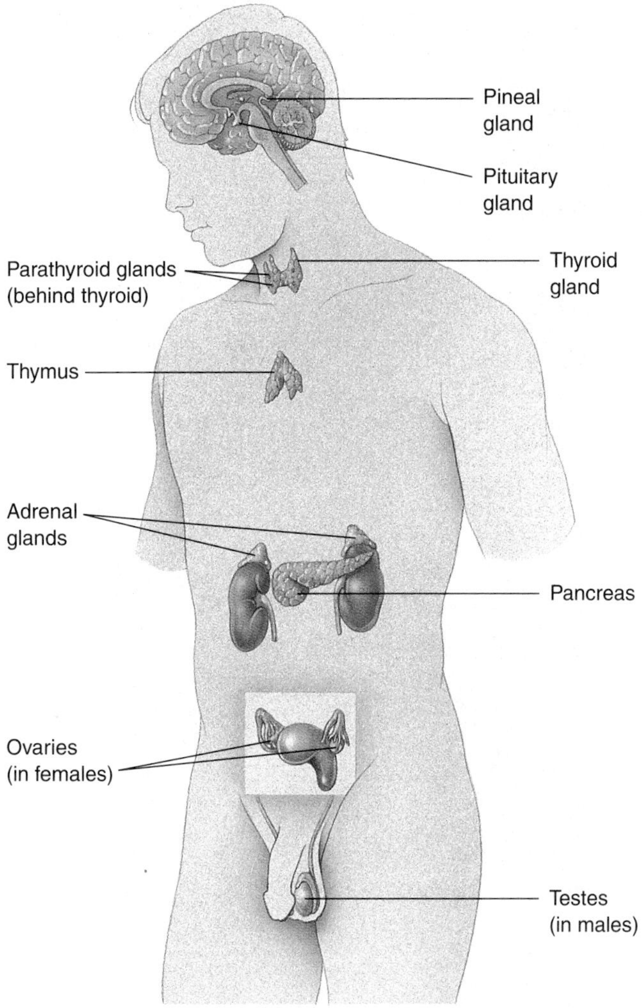
Pineal
gland
Pituitary
gland
Parathyroid glands
(behind thyroid)
Thyroid
gland
Thymus
Adrenal
glands
Pancreas
Ovaries
(in females)
Testes
(in males)

Figure 27.1 Major glands of the human endocrine system.
Hormone-secreting cells are clustered in endocrine glands.

The Chain of Command

The hypothalamus issues commands to a nearby gland, the pituitary, which in turn sends chemical signals to the various hormone-producing glands of the body. The pituitary is suspended from the hypothalamus by a short stalk, across which chemical messages pass from the hypothalamus to the pituitary. The first of these chemical messages to be discovered was a short peptide called thyrotropin-releasing hormone (TRH), which was isolated in 1969. The release of TRH from the hypothalamus triggers the pituitary to release a hormone called thyrotropin, or TSH, which travels to the thyroid and causes the thyroid gland to release thyroid hormones.

Six other hypothalamic hormones have since been isolated, which together govern the pituitary. Thus the CNS regulates the body's hormones through a chain of command. Each of the seven "releasing" hormones made by the hypothalamus causes the pituitary to synthesize a corresponding pituitary hormone, which travels to a distant endocrine gland and causes that gland to begin producing its particular endocrine hormone.

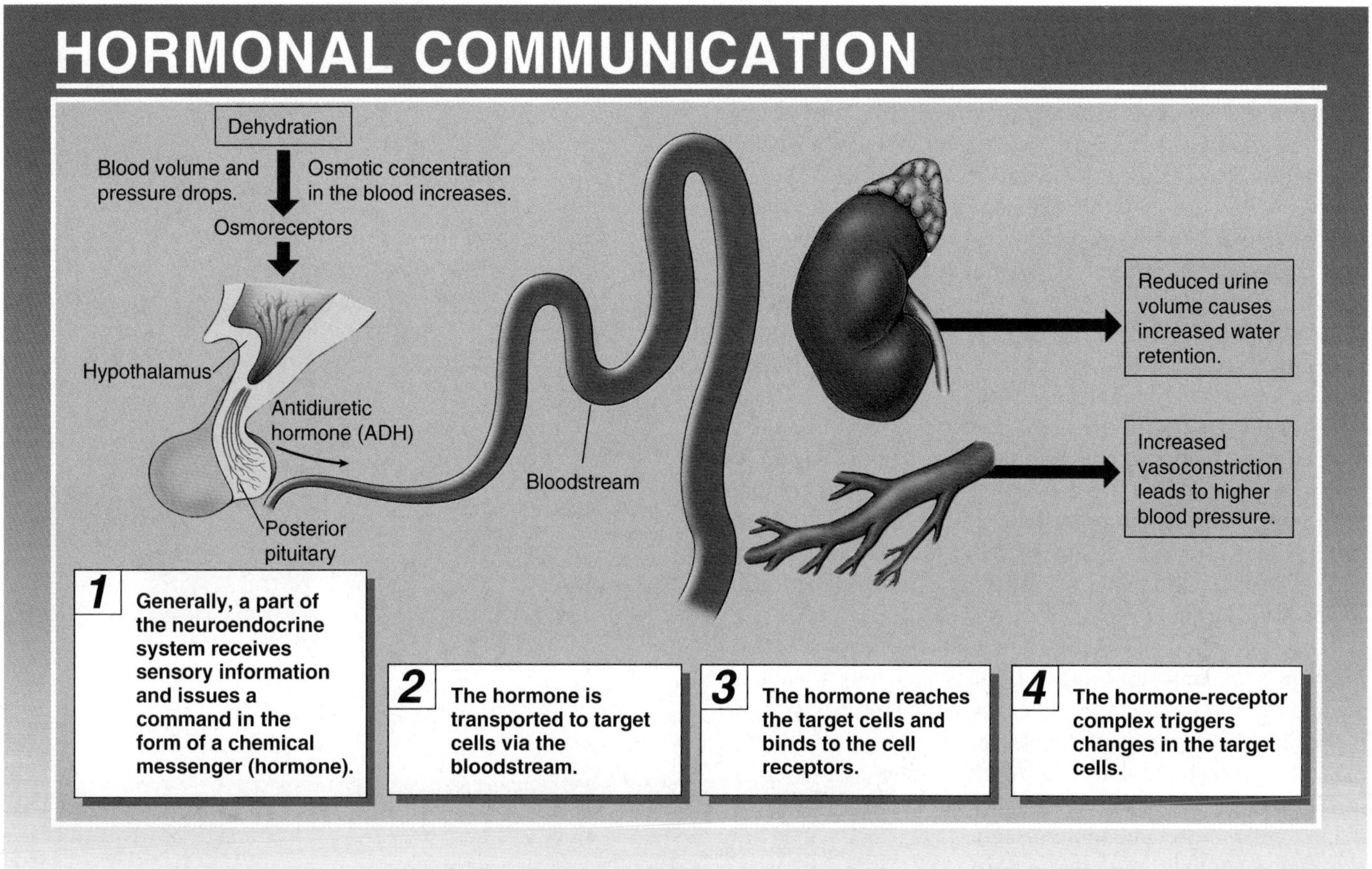
HORMONAL COMMUNICATION
Dehydration
Blood volume and pressure drops.
Osmotic concentration in the blood increases.
Osmoreceptors
Hypothalamus
Antidiuretic hormone (ADH)
Posterior pituitary
Bloodstream
Reduced urine volume causes increased water retention.
Increased vasoconstriction leads to higher blood pressure.
1 Generally, a part of the neuroendocrine system receives sensory information and issues a command in the form of a chemical messenger (hormone).
2 The hormone is transported to target cells via the bloodstream.
3 The hormone reaches the target cells and binds to the cell receptors.
4 The hormone-receptor complex triggers changes in the target cells.

Figure 27.2 How hormonal communication works.

How Hormones Work

The key reason why hormones are effective messengers within the body is because a particular hormone can be directed at a specific target cell. How does the target cell recognize that hormone, ignoring all others? Within the plasma membrane of the target cell are embedded receptor proteins that match the shape of the potential signal hormone like a hand fits a glove. As you recall from chapter 26, nerve cells have highly specific receptors within their synapses, each receptor shaped to "respond" to a different neurotransmitter molecule. Cells that the body has targeted to respond to a particular hormone have receptor proteins shaped to fit that hormone and no other. Thus, chemical communication within the body involves *two* elements: a molecular signal (the hormone) and a protein receptor on target cells. The system is highly specific because each protein receptor has a shape that only a particular hormone fits.

The path of communication taken by a hormonal signal can be visualized as a series of simple steps (figure 27.2):

1. **Issuing the command.** The hypothalamus of the CNS controls the release of many hormones. It chemically signals the pituitary gland, which is located very close to it, to release hormones into the bloodstream.
2. **Transporting the signal.** While hormones can act on an adjacent cell, most are transported throughout the body by the bloodstream.
3. **Hitting the target.** When a hormone encounters a cell with a matching receptor, called a target cell, the hormone binds to that receptor.
4. **Having an effect.** When the hormone binds to the receptor protein, the protein responds by changing shape, which triggers a change in cell activity.

27.1 Hormones are effective because they are recognized by specific receptors. Thus only cells possessing the appropriate receptor will respond to a particular hormone.

27.2 Steroid Hormones Enter Cells

Some protein receptors designed to recognize hormones are located in the cytoplasm of the target cell. The hormones in these cases are lipid-soluble molecules, typically **steroid hormones.** The chemical shape of these molecules resembles chicken wire (figure 27.3). All steroid hormones are manufactured from cholesterol, a complex molecule composed of four rings. The hormones that promote the development of secondary sexual characteristics are steroids. They include cortisol and testosterone, as well as the hormones estrogen and progesterone, which control the female reproductive system and are discussed in the next chapter.

Steroid hormones can pass across the lipid bilayer of the cell plasma membrane and bind to receptors within the cell, often within the nucleus (figure 27.4). This complex of receptor and hormone then binds to the DNA in the nucleus and causes a change in the activity of a particular gene. The change in this gene's activity is responsible for the effect of the hormone.

The steroids that weight lifters and other athletes sometimes use to "bulk up" turn on genes and thus trick their muscle cells into adding growth. **Anabolic steroids** are synthetic compounds that resemble the male sex hormone testosterone. Their injection into muscles causes the muscle cells to produce more protein, resulting in bigger muscles and increased strength. However, anabolic steroids have many dangerous side effects in both men and women, including liver damage, heart disease, and psychological disorders. Some experts also believe that steroid use can be linked to cancer. Anabolic steroids are illegal, and for many sports, athletes are tested for their presence. Fortunately, many athletes have stopped using anabolic steroids because of the side effects and the negative publicity afforded to the abuse of these compounds.

27.2 Steroid hormones pass through the cell's plasma membrane and bind to receptors in the cytoplasm, forming a complex that alters the transcription of specific genes.

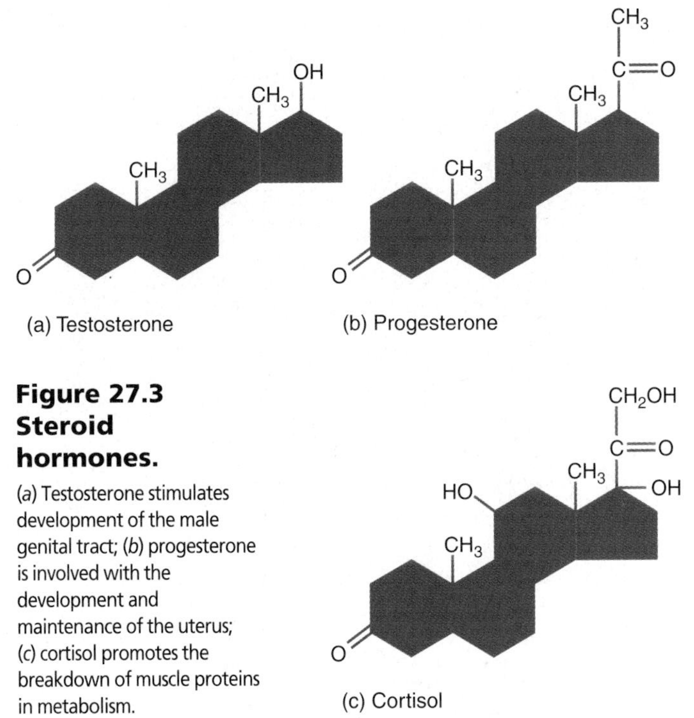

OH
CH3
CH3
O
(a) Testosterone
CH3
C=O
CH3
CH3
O
(b) Progesterone
CH2OH
C=O
CH3
OH
HO
CH3
O
(c) Cortisol

Figure 27.3 Steroid hormones.
(*a*) Testosterone stimulates development of the male genital tract; (*b*) progesterone is involved with the development and maintenance of the uterus; (*c*) cortisol promotes the breakdown of muscle proteins in metabolism.

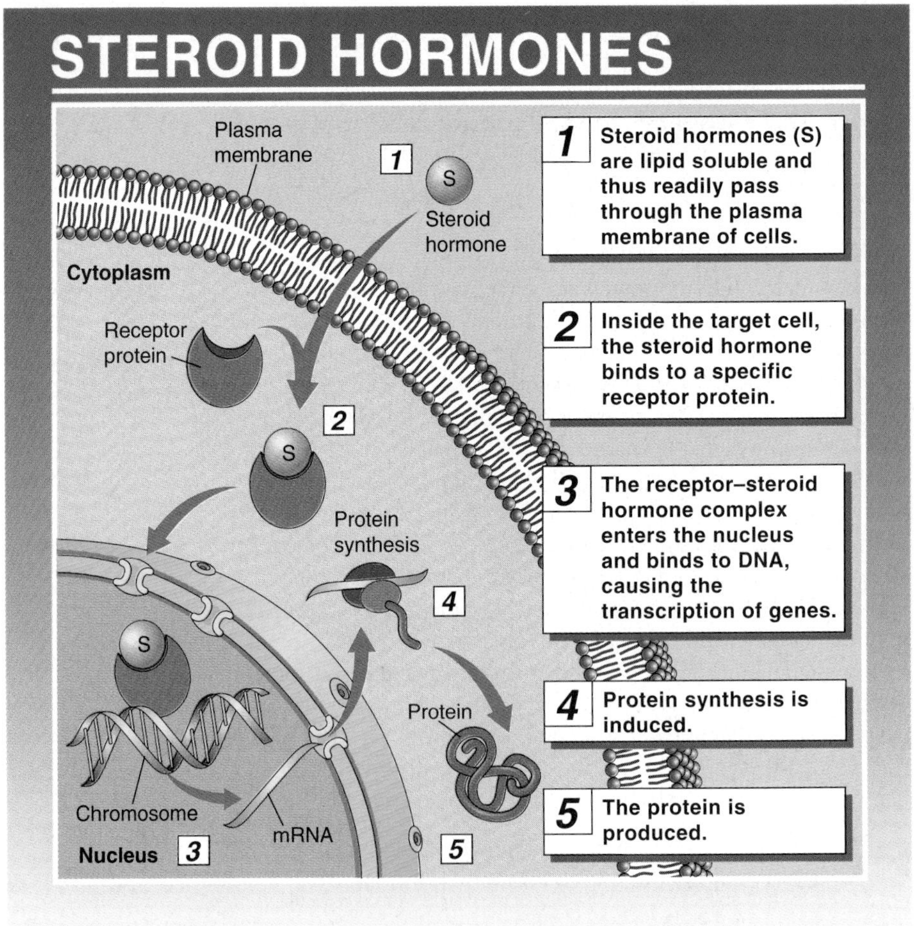

STEROID HORMONES
Plasma membrane
1
S
Steroid hormone
Cytoplasm
Receptor protein
2
S
Protein synthesis
4
S
Protein
Chromosome
mRNA
Nucleus 3
5
1 Steroid hormones (S) are lipid soluble and thus readily pass through the plasma membrane of cells.
2 Inside the target cell, the steroid hormone binds to a specific receptor protein.
3 The receptor–steroid hormone complex enters the nucleus and binds to DNA, causing the transcription of genes.
4 Protein synthesis is induced.
5 The protein is produced.

Figure 27.4 How steroid hormones work.

27.3 Peptide Hormones Act at the Cell Surface

Other hormone receptors are embedded within the plasma membrane, with their recognition region directed outward from the cell surface (figure 27.5). The hormones that bind to these receptors are typically short peptide chains (although some are full-sized proteins). The binding of the **peptide hormone** to the receptor triggers a change in the cytoplasmic end of the receptor protein. This change then triggers events within the cell cytoplasm, usually through intermediate within-cell signals called **second messengers,** which greatly amplify the original signal.

How does a second messenger amplify a hormone's signal? Second messengers activate enzymes (figure 27.6). One of the most common second messengers is cyclic AMP. Cyclic AMP is made from ATP by an enzyme that removes two phosphate units, forming AMP; the ends of the AMP join, forming a circle. A single hormone molecule binding to a receptor in the plasma membrane can result in the formation of many second messengers in the cytoplasm. Each second messenger can, in turn, activate many molecules of a certain enzyme. Sometimes each of these enzymes in turn activates many other enzymes. Thus, second messengers enable each hormone molecule to have a tremendous effect inside the cell, far greater than if the hormone had simply entered the cell and sought out a single target.

The hormone insulin provides a well-studied example of how peptide hormones achieve their effects within target cells. Most human cells have insulin receptors in their membranes—typically only a few hundred but far more in tissues involved in glucose metabolism. A single liver cell, for example, may have 100,000 of them. When insulin binds to one of these insulin receptors, the receptor protein changes its shape, goading an adjacent signal-modulating protein on the cell interior to activate the release of Ca^{++} ions. The Ca^{++} acts as a second messenger, activating a variety of cellular enzymes in a cascading series of events that greatly amplifies the strength of the original signal.

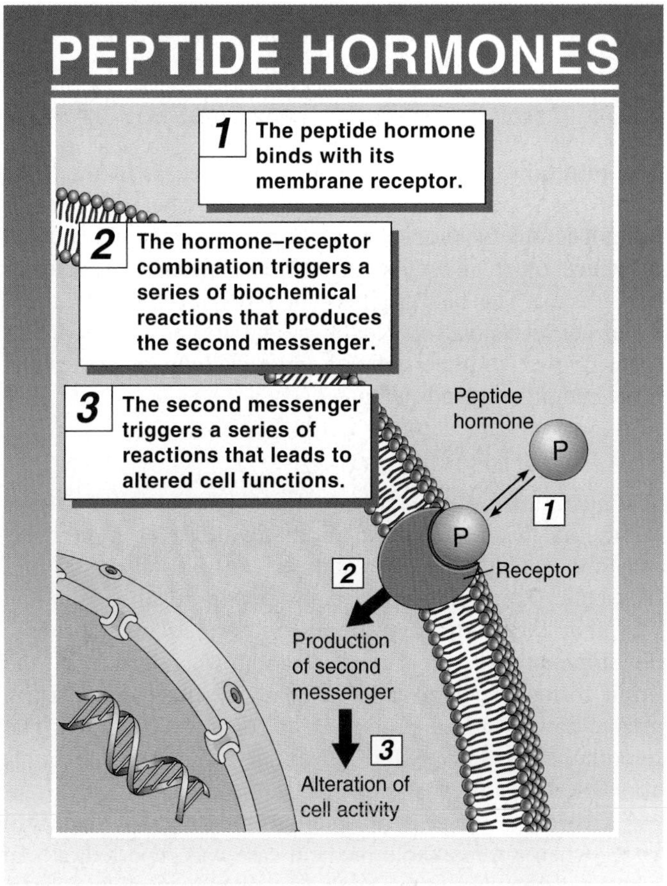
PEPTIDE HORMONES
1 The peptide hormone binds with its membrane receptor.
2 The hormone–receptor combination triggers a series of biochemical reactions that produces the second messenger.
3 The second messenger triggers a series of reactions that leads to altered cell functions.
Peptide hormone
P
1
P
2
Receptor
Production of second messenger
3
Alteration of cell activity

Figure 27.5 How peptide hormones work.

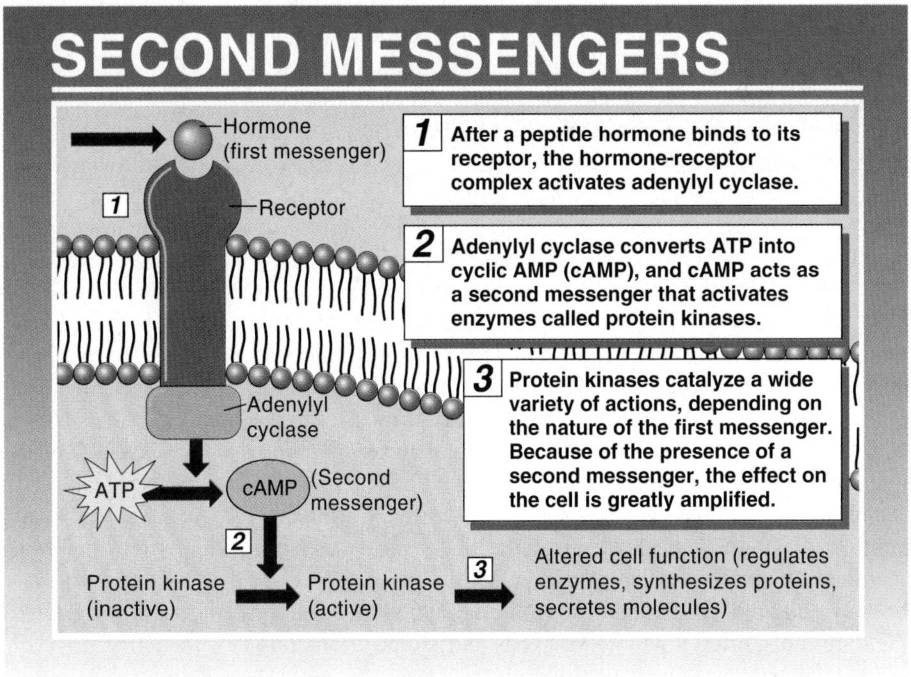
SECOND MESSENGERS
Hormone (first messenger)
Receptor
1
1 After a peptide hormone binds to its receptor, the hormone-receptor complex activates adenylyl cyclase.
2 Adenylyl cyclase converts ATP into cyclic AMP (cAMP), and cAMP acts as a second messenger that activates enzymes called protein kinases.
3 Protein kinases catalyze a wide variety of actions, depending on the nature of the first messenger. Because of the presence of a second messenger, the effect on the cell is greatly amplified.
Adenylyl cyclase
ATP
cAMP
(Second messenger)
2
Protein kinase (inactive)
Protein kinase (active)
3
Altered cell function (regulates enzymes, synthesizes proteins, secretes molecules)

Figure 27.6 How second messengers work.

27.3 Peptide hormones do not enter cells. Instead, they bind to receptors on the target cell surface, triggering a cascade of enzymic activations within the cell.

27.4 The Pituitary: The Master Gland

The **pituitary gland,** located in a bony recess in the brain below the hypothalamus (figure 27.7), is where nine major hormones are produced. These hormones act principally to influence other endocrine glands. The pituitary is actually two glands. The back portion, or *posterior lobe,* regulates water conservation, milk letdown, and uterine contraction in women; the front portion, or *anterior lobe,* regulates the other endocrine glands (figure 27.8).

The Posterior Pituitary

The posterior pituitary appears fibrous because it contains axons that originate in cell bodies within the hypothalamus and extend along the stalk of the pituitary as a tract of fibers. This anatomical relationship results from the way that the posterior pituitary is formed in embryonic development. As the floor of the third ventricle of the brain forms the hypothalamus, part of this neural tissue grows downward to produce the posterior pituitary. The hypothalamus and posterior pituitary thus become interconnected by a tract of axons.

The role of the posterior pituitary first became evident in 1912, when a remarkable medical case was reported: a man who had been shot in the head developed a surprising disorder—he began to urinate every 30 minutes, unceasingly. The bullet had lodged in his pituitary gland, and subsequent research demonstrated that surgical removal of the pituitary also produces these unusual symptoms. Pituitary extracts were shown to contain a substance that makes the kidneys conserve water, and in the early 1950s the peptide hormone **vasopressin** (also called antidiuretic hormone, **ADH**) was isolated. Vasopressin regulates the kidney's retention of water. When vasopressin is missing, the kidneys cannot retain water, which is why the bullet caused excessive urination (and why excessive alcohol, which inhibits vasopressin secretion, has the same effect).

The posterior pituitary also releases a second hormone of very similar structure—both are short peptides composed of nine amino acids—but very different function, called **oxytocin.** Oxytocin initiates uterine contraction during childbirth and milk release in mothers. Here is how milk release works: Sensory receptors in the mother's nipples send messages to the hypothalamus when stimulated by sucking, causing the hypothalamus to make the pituitary release oxytocin. The oxytocin travels in the bloodstream to the breasts, where it stimulates contraction of the muscles around the ducts into which the mammary glands secrete milk. Oxytocin and vasopressin are both synthesized inside neurons within the hypothalamus, and they are transported down nerve axons and stored in the posterior pituitary.

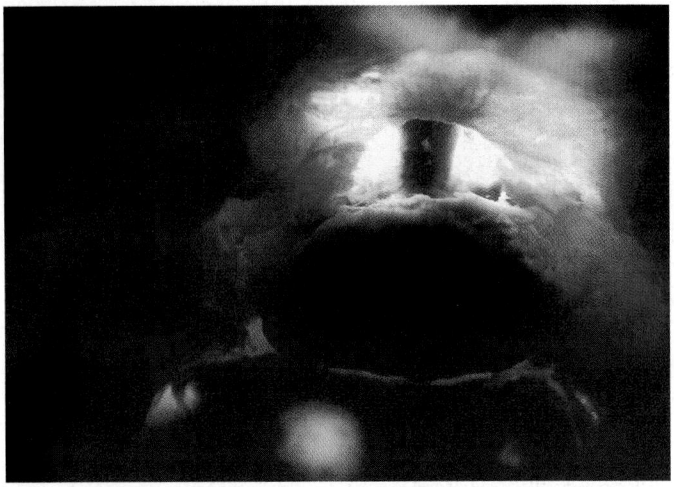

Figure 27.7 The pituitary gland (×7).
The pituitary gland, which hangs by a short stalk from the hypothalamus, regulates the hormone production of many of the body's endocrine glands.

The Anterior Pituitary

The anterior pituitary, unlike the posterior pituitary, does not develop from a downgrowth of the brain; instead, it develops from a pouch of epithelial tissue that pinches off from the roof of the embryo's mouth. Because it is epithelial tissue, the anterior pituitary is a complete gland—it produces the hormones it secretes. The key role of the anterior pituitary first became understood in 1909, when a 38-year-old South Dakota farmer was cured of the growth disorder called acromegaly by the surgical removal of a pituitary tumor. Acromegaly is a form of giantism in which the jaw begins to protrude and the features thicken. It turned out that giantism is almost always associated with pituitary tumors. Robert Wadlow, born in Alton, Illinois, in 1928, grew to a height of 8 feet, 11 inches, and weighed 475 pounds before he died from infection at age 22—the tallest human being ever recorded. Skull X rays showed he had a pituitary tumor. Pituitary hormones have also proven to be the cause of several other well-known cases of giantism, such as the 8-foot, 2-inch Irish giant Charles Byrne, born in 1761; his skeleton, preserved in the Royal College of Surgeons, London, shows the effects of a pituitary tumor.

Why did removal of the pituitary tumor cure the South Dakota farmer? Pituitary tumors produce giants because the tumor cells produce prodigious amounts of a growth-promoting hormone. This **growth hormone (GH),** a long peptide of 191 amino acids, is normally produced in only minute amounts by the anterior pituitary gland and usually only during periods of body growth, such as infancy and puberty.

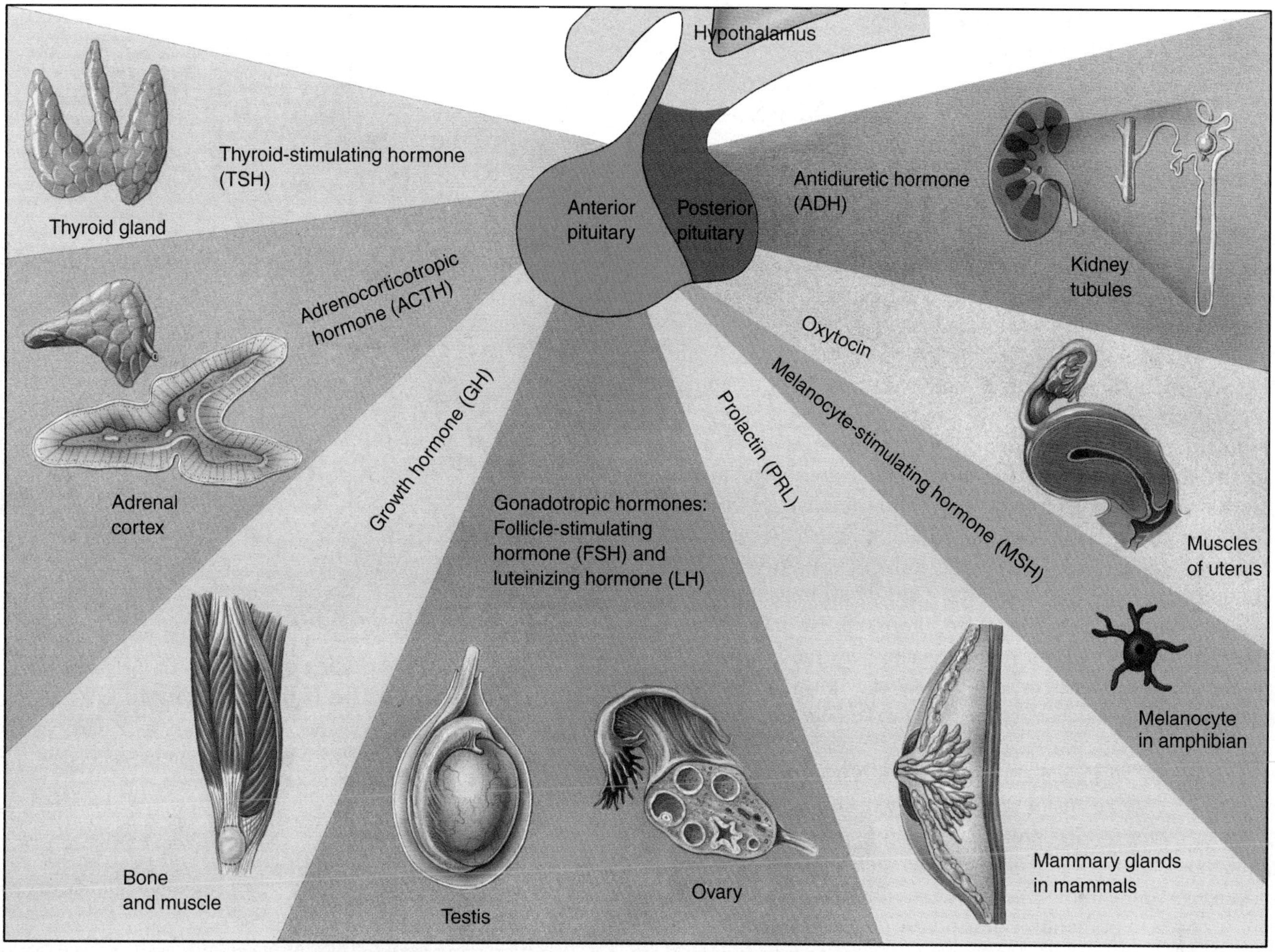

Hypothalamus
Thyroid-stimulating hormone (TSH)
Thyroid gland
Anterior pituitary
Posterior pituitary
Antidiuretic hormone (ADH)
Kidney tubules
Adrenocorticotropic hormone (ACTH)
Oxytocin
Melanocyte-stimulating hormone (MSH)
Prolactin (PRL)
Growth hormone (GH)
Adrenal cortex
Gonadotropic hormones: Follicle-stimulating hormone (FSH) and luteinizing hormone (LH)
Muscles of uterus
Melanocyte in amphibian
Mammary glands in mammals
Bone and muscle
Testis
Ovary

Figure 27.8 The role of the pituitary.
Interactions between the anterior and posterior lobes of the pituitary (two distinct glands) and various organs of the human body are shown in this diagram.

The anterior pituitary gland produces seven major peptide hormones, each controlled by a particular releasing signal secreted from cells in the hypothalamus. All seven of these hormones have major endocrine roles, as well as poorly understood roles in the central nervous system.

1. **Thyroid-stimulating hormone (TSH).** TSH stimulates the thyroid gland to produce the thyroid hormone thyroxine, which in turn stimulates oxidative respiration.
2. **Luteinizing hormone (LH).** LH plays an important role in the female menstrual cycle. It also stimulates the male gonads to produce testosterone, which initiates and maintains the development of male secondary sexual characteristics not involved directly in reproduction.
3. **Follicle-stimulating hormone (FSH).** FSH is significant in the female menstrual cycle, and, in males, it stimulates cells in the testes to produce a hormone that regulates development of the sperm.
4. **Adrenocorticotropic hormone (ACTH).** ACTH stimulates the adrenal gland to produce a variety of steroid hormones. Some regulate the production of glucose from fat; others the balance of sodium and potassium ions in the blood; and still others the development of male secondary sexual characteristics.
5. **Growth hormone (GH).** GH stimulates the growth of muscle and bone throughout the body.
6. **Prolactin.** Prolactin stimulates the breasts to produce milk.
7. **Melanocyte-stimulating hormone (MSH).** In reptiles and amphibians, melatonin stimulates color changes in the epidermis. The function of this hormone in humans is still poorly understood.

How the Hypothalamus Controls the Anterior Pituitary

As noted earlier, the hypothalamus controls production and secretion of the anterior pituitary hormones by means of a family of special hormones. Neurons in the hypothalamus secrete these releasing and inhibiting hormones into blood capillaries at the base of the hypothalamus (figure 27.9). These capillaries drain into small veins that run within the stalk of the pituitary to a second bed of capillaries in the anterior pituitary. This unusual system of vessels is known as the *hypothalamo-hypophyseal portal system*. It is called a portal system because it has a second capillary bed downstream from the first; the only other body location with a similar system is the liver.

Each releasing hormone delivered to the anterior pituitary by this portal system regulates the secretion of a specific hormone. For example, thyrotropin-releasing hormone (TRH) stimulates the release of TSH, corticotropin-releasing hormone (CRH) stimulates the release of ACTH, and gonadotropin-releasing hormone (GnRH) stimulates the release of FSH and LH. A releasing hormone for growth hormone, called growth-hormone-releasing hormone (GHRH) has also been discovered, and a releasing hormone for prolactin has been postulated but has thus far not been identified.

The hypothalamus also secretes hormones that inhibit the release of certain anterior pituitary hormones. To date, three such hormones have been discovered: somatostatin inhibits the secretion of GH; prolactin-inhibiting hormone (PIH) inhibits the secretion of prolactin; and melanotropin-inhibiting hormone (MIH) inhibits the secretion of MSH.

Because hypothalamic hormones control the secretions of the anterior pituitary gland, and the anterior pituitary hormones control the secretions of some other endocrine glands, it may seem that the hypothalamus functions as a "master gland," in charge of hormonal secretion in the body. This idea is not generally valid, however, for two reasons. First, a number of endocrine organs, such as the adrenal medulla and the pancreas, are not directly regulated by this control system. Second, the hypothalamus and the anterior pituitary gland are themselves controlled by the very hormones whose secretion they stimulate! In most cases this is an inhibitory control, where the target gland hormones inhibit the secretions of the hypothalamus and anterior pituitary (figure 27.10). This type of control system is an example of **negative feedback inhibition.**

27.4 The posterior pituitary gland contains axons originating from neurons in the hypothalamus. The anterior pituitary responds to hormonal signals from the hypothalamus by itself producing a family of pituitary hormones that are carried to distant glands and that induce them to produce specific hormones.

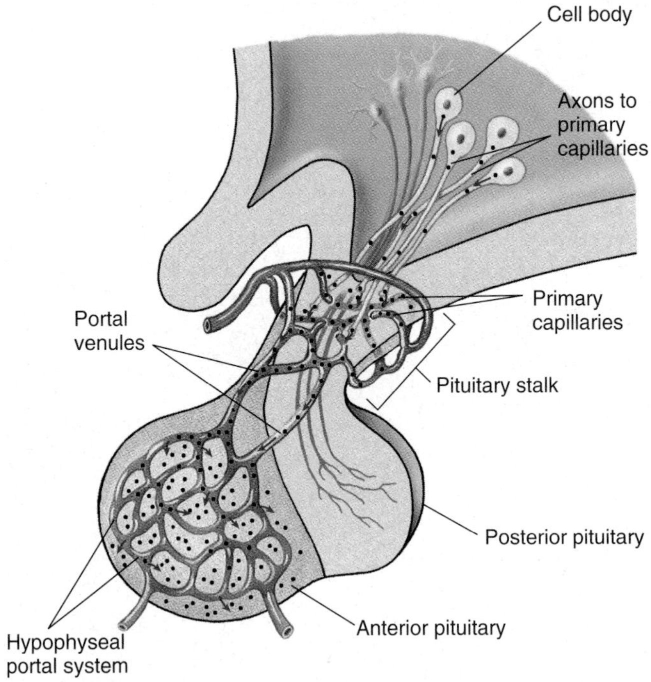

Cell body
Axons to primary capillaries
Primary capillaries
Portal venules
Pituitary stalk
Posterior pituitary
Anterior pituitary
Hypophyseal portal system

Figure 27.9 Hormonal control of the anterior pituitary gland by the hypothalamus.
Neurons in the hypothalamus secrete hormones that are carried by short blood vessels directly to the anterior pituitary gland, where they either stimulate or inhibit the secretion of anterior pituitary hormones.

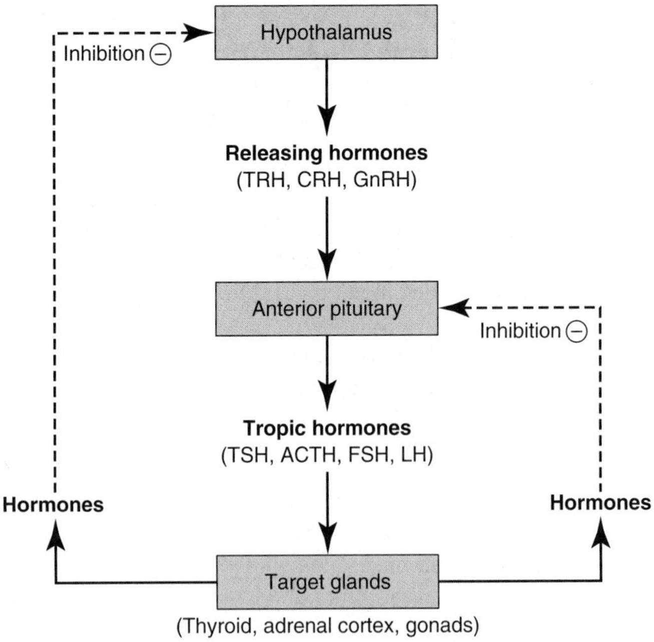

Hypothalamus
Inhibition ⊖
Releasing hormones
(TRH, CRH, GnRH)
Anterior pituitary
Inhibition ⊖
Tropic hormones
(TSH, ACTH, FSH, LH)
Hormones
Hormones
Target glands
(Thyroid, adrenal cortex, gonads)

Figure 27.10 Negative feedback inhibition.
The hormones secreted by some endocrine glands feed back to inhibit the secretion of hypothalamic-releasing hormones and anterior pituitary tropic hormones.

27.5 The Pancreas

The **pancreas** gland is located behind the stomach and is connected to the front end of the small intestine by a narrow tube. It secretes a variety of digestive enzymes into the digestive tract through this tube, and for a long time it was thought to be solely an exocrine gland. In 1869, however, a German medical student named Paul Langerhans described some unusual clusters of cells scattered throughout the pancreas. In 1893, doctors concluded that these clusters of cells, which came to be called islets of Langerhans, produced a substance that prevented diabetes mellitus. **Diabetes mellitus** is a serious disorder in which affected individuals are unable to take up glucose from the blood, even though their levels of blood glucose become very high. Such individuals lose weight and literally starve. Breaking down their fat reserves results in the production of acids called ketones, which lower the pH of the blood. Affected individuals may eventually suffer brain damage and even death. Diabetes is the leading cause of blindness among adults, and it accounts for one-third of all kidney failures. It is the seventh leading cause of death in the United States.

The substance produced by the islets of Langerhans, which we now know to be the peptide hormone *insulin,* was not isolated until 1922. Two young doctors working in a Toronto hospital injected an extract purified from beef pancreas glands into a 13-year-old boy, a diabetic whose weight had fallen to 29 kilograms (65 lb) and who was not expected to survive. The hospital record gives no indication of the historic importance of the trial, only stating, "15 cc of MacLeod's serum. 7-1/2 cc into each buttock." With this single injection, the glucose level in the boy's blood fell 25%—his cells were taking up glucose. A more potent extract soon brought levels down to near normal.

This was the first instance of successful insulin therapy. The islets of Langerhans in the pancreas produce two hormones that interact to govern the levels of glucose in the blood (figure 27.11). These hormones are *insulin* and *glucagon.* Insulin is a storage hormone, designed to put away nutrients for leaner times. It promotes the accumulation of glycogen in the liver and triglycerides in fat cells. When food is consumed, beta cells in the islets of Langerhans secrete insulin, causing the body to store glucose to be used later. When body activity causes the level of glucose in the blood to fall as it is used as fuel, other cells in the islets of Langerhans, called alpha cells, secrete glucagon, which causes liver cells to release stored glucose and fat cells to break down triglycerides. The two hormones work together to keep glucose levels in the blood within narrow bounds.

Twelve million Americans, and over 100 million people worldwide, have **diabetes.** There are *two* kinds of diabetes mellitus. About 10% of affected individuals suffer from type I diabetes, a hereditary autoimmune disease in which the immune system attacks the islets of Langerhans, resulting in abnormally low insulin secretion. Called juvenile onset diabetes, this type usually develops before age 20. Affected individuals can be treated by daily injections of insulin. Active research on the possibility of transplanting islets of Langerhans holds much promise as a lasting treatment for this form of diabetes.

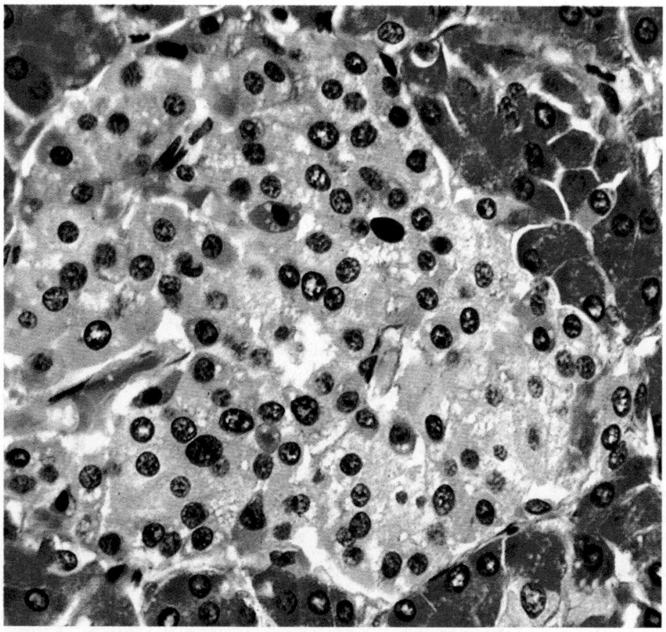

Figure 27.11 Islets of Langerhans.
Glucagon and insulin are produced by clumps of cells within the pancreas called islets of Langerhans. The islet in this photomicrograph is the lighter-colored area.

In type II diabetes, the number of insulin receptors on the target tissue is abnormally low, while the level of insulin in the blood is often higher than normal. This form of diabetes usually develops in people over 40 years of age. It is almost always a consequence of excessive weight; in the United States, 90% of those who develop type II diabetes are obese. Cells of these individuals, sated with food, adjust their appetite for glucose downward by reducing their sensitivity to insulin. As a drug addict's neurons reduce their number of neurotransmitter receptors after continued exposure to a drug, the obese individual's cells reduce their number of insulin receptors. To compensate, the pancreas pumps out ever more insulin, and, in some people, the insulin-producing cells are unable to keep up with the ever-heavier workload and stop functioning. Type II diabetes is usually treatable by diet and exercise, and most affected individuals do not need daily injections of insulin.

27.5 Clusters of cells within the pancreas secrete the hormones insulin and glucagon. Insulin stimulates the storage of glucose as glycogen, while glucagon stimulates glycogen breakdown to glucose. Working together, these hormones keep glucose levels within narrow bounds.

27.6 The Thyroid, Parathyroid, and Adrenal Glands

The Thyroid: A Metabolic Thermostat

The **thyroid gland** is shaped like a shield (its name comes from *thyros,* the Greek word for "shield"). It lies just below the Adam's apple in the front of the neck. The thyroid makes several hormones, the two most important of which are **thyroxine,** which increases metabolic rate and promotes growth, and **calcitonin,** which stimulates calcium uptake by bones.

Thyroxine regulates the level of metabolism in the body in several important ways. Without adequate thyroxine, growth is retarded. For example, children with underactive thyroid glands are not able to carry out carbohydrate breakdown and protein synthesis at normal rates, a condition called cretinism that results in stunted growth. Mental retardation can also result, because thyroxine is needed for normal development of the central nervous system. Thyroxine contains iodine, and if the amount of iodine in the diet is too low, the thyroid cannot make adequate amounts of thyroxine and will grow larger in a futile attempt to manufacture more. The greatly enlarged thyroid gland that results is called a goiter. This need for iodine in the diet is why iodine is added to table salt.

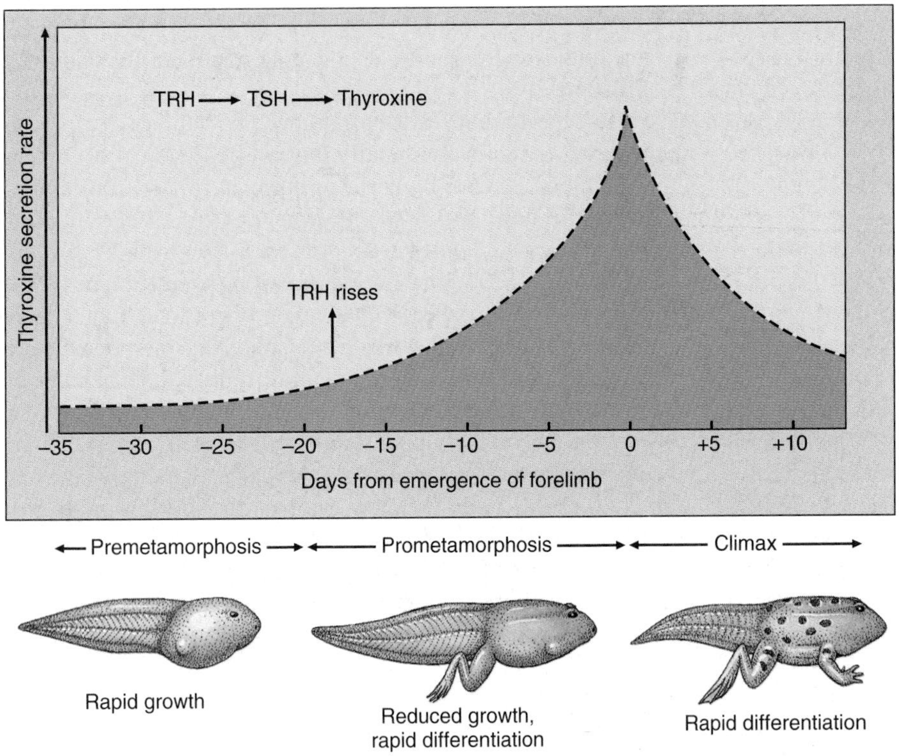

TRH → TSH → Thyroxine
TRH rises
Thyroxine secretion rate
−35 −30 −25 −20 −15 −10 −5 0 +5 +10
Days from emergence of forelimb
Premetamorphosis
Prometamorphosis
Climax
Rapid growth
Reduced growth, rapid differentiation
Rapid differentiation

Figure 27.12 Thyroxine triggers metamorphosis in amphibians.
In tadpoles at the premetamorphic stage, the hypothalamus releases TRH (thyrotropin-releasing hormone), which causes the anterior pituitary to secrete TSH (thyroid-stimulating hormone). TSH then acts on the thyroid gland, which secretes thyroxine. As metamorphosis proceeds, thyroxine reaches its maximal level when the forelimbs begin to form, after the hindlimbs have formed.

There is an additional function of the thyroid gland that is unique to amphibians—thyroid hormones are needed for the metamorphosis of the larvae into adults (figure 27.12). If the thyroid gland is removed from a tadpole, it will not change into a frog. Conversely, if an immature tadpole is fed pieces of a thyroid gland, it will undergo premature metamorphosis and become a miniature frog!

Calcitonin plays a key role in maintaining proper calcium levels in the body. If levels of calcium in the blood become too high, calcitonin stimulates calcium deposition in bone, thus lowering calcium levels in the bloodstream. As you will see, another hormone has an even more critical role in maintaining the body's levels of calcium.

The Parathyroids: Builders of Bone

The **parathyroid glands** are four small glands attached to the thyroid. Small and unobtrusive, they were ignored by researchers until well into the last century. The first suggestion that the parathyroids produce a hormone came from experiments in which they were removed from dogs: the concentration of calcium in the dogs' blood plummeted to less than half the normal level. However, if an extract of the parathyroid gland was administered, calcium levels returned to normal. If an excess was administered, calcium levels in the blood became *too* high, and the bones of the dogs were literally dismantled by the extract. It was clear that the parathyroid glands were producing a hormone that acted on calcium uptake into and out of the bones.

The hormone produced by the parathyroids is **parathyroid hormone (PTH).** It is one of only two hormones in the body that is absolutely essential for survival (the other is aldosterone, a hormone produced by the adrenal glands, discussed on the next page). PTH regulates the level of calcium in blood. Recall that calcium ions are the key actors in muscle contraction—by initiating calcium release, nerve impulses cause muscles to contract. A vertebrate cannot live without the muscles that pump the heart and drive the body, and these muscles cannot function if calcium levels are not kept within narrow limits.

PTH acts as a fail-safe to make sure calcium levels never fall too low (figure 27.13). PTH is released into the bloodstream, where it travels to the bones and acts on the osteoclast cells within bones, stimulating them to dismantle bone tissue and release calcium into the bloodstream. PTH also acts on the kidneys to resorb calcium ions from the urine and leads to

Figure 27.13 Maintenance of proper calcium levels in the blood.

(*a*) When calcium levels in the blood become too low, the parathyroid gland produces additional amounts of PTH, a hormone that stimulates the osteoclast cells to break down calcium in bone. This raises the calcium levels in the blood.
(*b*) Conversely, abnormally high levels of calcium in the blood trigger the thyroid gland to secrete calcitonin, which promotes the activity of osteoblasts to remove calcium from the blood and deposit it in bone.

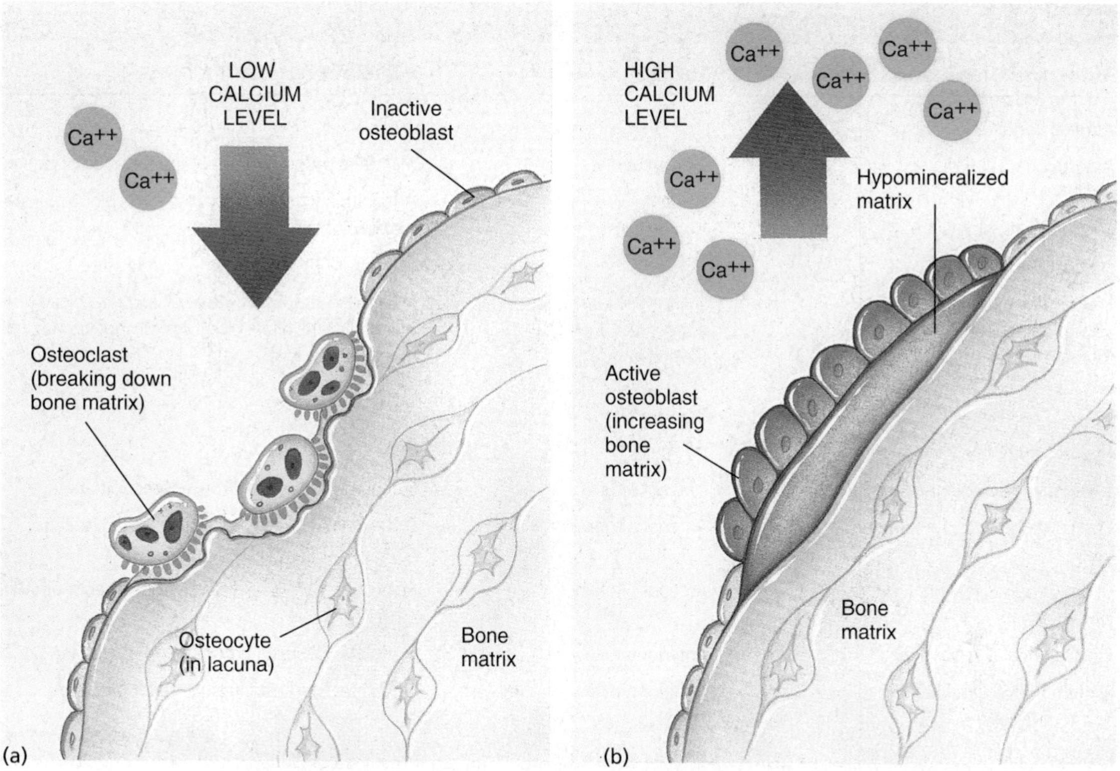

LOW CALCIUM LEVEL
Ca++
Ca++
Inactive osteoblast
Osteoclast (breaking down bone matrix)
Osteocyte (in lacuna)
Bone matrix
(a)
HIGH CALCIUM LEVEL
Ca++
Ca++
Ca++
Ca++
Ca++
Ca++
Ca++
Hypomineralized matrix
Active osteoblast (increasing bone matrix)
Bone matrix
(b)

activation of vitamin D, necessary for calcium absorption by the intestine. A diet deficient in vitamin D leads to poor bone formation, a condition called rickets. When PTH is synthesized by the parathyroids in response to falling levels of calcium in the blood, the body is essentially sacrificing bone to keep calcium levels within the narrow limits necessary for proper functioning of muscle and nerve.

The Adrenals: Two Glands in One

Mammals have two **adrenal glands,** one located just above each kidney. Each adrenal gland is composed of two parts: an inner core called the **medulla,** which produces the hormones adrenaline (also called epinephrine) and norepinephrine, and an outer shell called the **cortex,** which produces the steroid hormones cortisol and aldosterone.

The Adrenal Medulla: Emergency Warning Siren

The medulla releases adrenaline and norepinephrine in times of stress. These hormones act as emergency signals that stimulate rapid deployment of body fuel. The "alarm" response these hormones produce throughout the body is identical to the individual effects achieved by the sympathetic nervous system, but it is much longer lasting. Among the effects of these hormones are an accelerated heartbeat, increased blood pressure, higher levels of blood sugar, and increased blood flow to the heart and lungs. These hormones can thus be thought of as extensions of the sympathetic nervous system.

The Adrenal Cortex: Maintaining the Proper Amount of Salt

The adrenal cortex produces the steroid hormone cortisol. Cortisol (also called hydrocortisone) acts on many different cells in the body to maintain nutritional well-being. It stimulates carbohydrate metabolism and reduces inflammation. Synthetic derivatives of this hormone, such as prednisone, have widespread medical use as anti-inflammatory agents. The ability of many cortisol-derived steroids to stimulate muscle growth has also led to the abuse of anabolic steroids by athletes.

The adrenal cortex also produces aldosterone. Aldosterone acts primarily in the kidney to promote the uptake of sodium and other salts from the urine. Sodium ions play crucial roles in nerve conduction and many other body functions. Aldosterone is, with PTH, one of the two endocrine hormones essential for survival. That is why removal of the adrenal glands is invariably fatal.

Table 27.1 (see next page) summarizes the actions of the principal endocrine glands, including the major ones discussed so far and some others discussed in the next section.

27.6 The thyroid acts as a metabolic thermostat, secreting hormones that adjust metabolic rate. Parathyroid hormone PTH regulates calcium levels in the blood. The adrenal hormone aldosterone promotes the uptake of sodium and other salts from the kidney.

TABLE 27.1 THE PRINCIPAL ENDOCRINE GLANDS

Endocrine Gland and Hormone	Target	Principal Actions
Adrenal Cortex		
Aldosterone	Kidney tubules	Maintains proper balance of sodium and potassium ions
Cortisol	General	Adaptation to long-term stress; raises blood glucose level; mobilizes fat
Adrenal Medulla		
Epinephrine (adrenaline) and norepinephrine (noradrenaline)	Smooth muscle, cardiac muscle, blood vessels, skeletel muscle	Initiate stress responses; increase heart rate, blood pressure, metabolic rate; dilate blood vessels; mobilize fat; raise blood glucose level
Hypothalamus		
Thyrotropin-releasing hormone (TRH)	Anterior pituitary	Stimulates TSH release from anterior pituitary
Corticotropin-releasing hormone (CRH)	Anterior pituitary	Stimulates ACTH release from anterior pituitary
Gonadotropin-releasing hormone (GnRH)	Anterior pituitary	Stimulates FSH and LH release from anterior pituitary
Prolactin-releasing hormone (PRH)	Anterior pituitary	Stimulates PRL release from anterior pituitary
Growth-hormone-releasing hormone (GHRH)	Anterior pituitary	Stimulates GH release from anterior pituitary
Prolactin-inhibiting hormone (PIH)	Anterior pituitary	Inhibits PRL release from anterior pituitary
Growth-hormone-inhibiting hormone (somatostatin)	Anterior pituitary	Inhibits GH release from anterior pituitary
Ovary		
Estrogen	General; female reproductive structures	Stimulates development of secondary sex characteristics in females and growth of sex organs at puberty; prompts monthly preparation of uterus for pregnancy
Progesterone	Uterus, breasts	Completes preparation of uterus for pregnancy; stimulates development of breasts
Pancreas		
Insulin	General	Lowers blood glucose level; increases storage of glycogen in liver
Glucagon	Liver, adipose tissue	Raises blood glucose level; stimulates breakdown of glycogen in liver
Parathyroid Glands		
Parathyroid hormone (PTH)	Bone, kidneys, digestive tract	Increases blood calcium level by stimulating bone breakdown; stimulates calcium reabsorption in kidneys; activates vitamin D

TABLE 27.1 (CONTINUED)

Endocrine Gland and Hormone	Target	Principal Actions
Pineal Gland		
Melatonin	Hypothalamus	Function not well understood; may help control onset of puberty in humans
Posterior Lobe of Pituitary		
Oxytocin (OT)	Uterus Mammary glands	Stimulates contraction of uterus Stimulates ejection of milk
Vasopressin (antidiuretic hormone, ADH)	Kidneys	Conserves water; increases blood pressure
Anterior Lobe of Pituitary		
Growth hormone (GH)	General	Stimulates growth by promoting protein synthesis and breakdown of fatty acids
Prolactin (PRL)	Mammary glands	Sustains milk production after birth
Thyroid-stimulating hormone (TSH)	Thyroid gland	Stimulates secretion of thyroid hormones
Adrenocorticotropic hormone (ACTH)	Adrenal cortex	Stimulates secretion of adrenal cortical hormones
Follicle-stimulating hormone (FSH)	Gonads	In females, stimulates ovarian follicle and secretion of estrogen; in males, stimulates production of sperm cells
Luteinizing hormone (LH)	Ovaries and testes	Stimulates ovulation and corpus luteum formation in females; stimulates secretion of testosterone in males
Melanocyte-stimulating hormone (MSH)	Skin	Stimulates color change in reptiles and amphibians; unknown function in mammals
Testes		
Testosterone	General; male reproductive structures	Stimulates development of secondary sex characteristics in males and growth spurt at puberty; stimulates development of sex organs; stimulates sperm production
Thyroid Gland		
Thyroid hormone (thyroxine, T_4, and triiodothyronine, T_3)	General	Stimulates metabolic rate; essential to normal growth and development
Calcitonin	Bone	Lowers blood calcium level by inhibiting release of calcium from bone
Thymus		
Thymosin	White blood cells	Promotes production and maturation of white blood cells

27.7 A Host of Other Hormones

Sexual Development

The ovaries and testes are important endocrine glands, producing the steroid sex hormones (including estrogens, progesterone, and testosterone), to be described in detail in the next chapter. Estrogens and progesterone regulate the menstrual cycle and are secreted by the placenta during pregnancy as well as by the ovaries.

Biological Clocks

Another endocrine gland is the **pineal gland,** located in the roof of the third ventricle of the brain in most vertebrates (see figure 26.14). It is about the size of a pea and is shaped like a pine cone (hence its name). The pineal gland evolved from a median light-sensitive eye (sometimes called a "third eye," although it could not form images) at the top of the skull in primitive vertebrates. This pineal eye is still present in primitive fish (cyclostomes) and some reptiles. In other vertebrates, however, the pineal gland is buried deep in the brain and functions as an endocrine gland by secreting the hormone melatonin. One of the actions of melatonin is to cause blanching of the skin of lower vertebrates by reducing the dispersal of melanin granules. In vertebrates, the secretion of melatonin by the pineal gland is entrained to cycles of light and dark, decreasing during the day and increasing at night. In some vertebrates, this activity helps to regulate reproductive physiology, but the role of melatonin in human reproduction is controversial.

Nonendocrine Hormones

There are a variety of hormones secreted by nonendocrine organs. The thymus is the site of production of particular lymphocytes called T cells, and it secretes a number of hormones that function in the regulation of the immune system. The right atrium of the heart secretes atrial natriuretic hormone, which stimulates the kidneys to excrete salt and water in the urine. This hormone, therefore, acts antagonistically to aldosterone, which promotes salt and water retention. The kidneys secrete erythropoietin, a hormone that stimulates the bone marrow to produce red blood cells. Even the skin has an endocrine function, because it secretes vitamin D. The gas nitric oxide, made by many different cells, controls blood pressure by dilating arteries. The drug Viagra counters impotency by promoting NO production and so dilating the blood vessels of the penis. Discoverers of the role of nitric oxide were awarded the 1998 Nobel Prize for medicine.

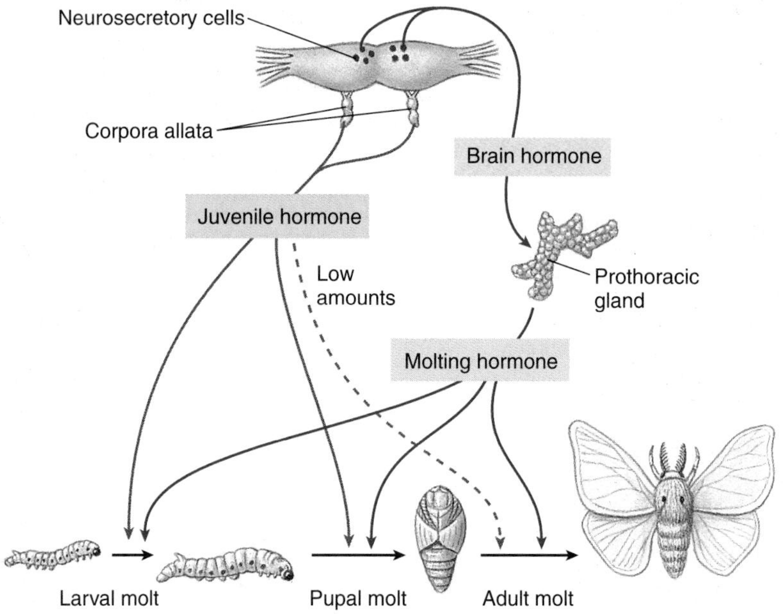
Neurosecretory cells
Corpora allata
Brain hormone
Juvenile hormone
Low amounts
Prothoracic gland
Molting hormone
Larval molt
Pupal molt
Adult molt

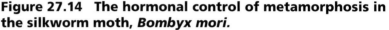

Figure 27.14 The hormonal control of metamorphosis in the silkworm moth, *Bombyx mori.*

While molting hormone (ecdysone), produced by the prothoracic gland, triggers when molting occurs, juvenile hormone, produced by bodies near the brain called the corpora allata, determines the result of a particular molt. High levels of juvenile hormone inhibit the formation of the pupa and adult forms. At the late stages of metamorphosis, therefore, it is important that the corpora allata not produce large amounts of juvenile hormone.

Molting and Metamorphosis in Insects

In insects, hormonal secretions influence both metamorphosis and molting. Prior to molting, neurosecretory cells on the surface of the brain secrete brain hormone, which in turn stimulates a gland in the thorax called the prothoracic gland to produce **molting hormone,** or *ecdysone* (figure 27.14). Another hormone, called juvenile hormone, is produced by bodies near the brain called the corpora allata. For insects to molt, both ecdysone and juvenile hormone must be present, but the level of juvenile hormone determines the result of a particular molt. When juvenile hormone levels are high, the molt produces another larvae. At the late stages of metamorphosis, juvenile hormone levels decrease, and the molt produces a pupa and eventually an adult insect.

27.7 Sex steroid hormones from the gonads regulate reproduction, melatonin secreted by the pineal gland helps regulate circadian rhythms, and thymus hormones help regulate the vertebrate immune system. Molting hormone, or ecdysone, and juvenile hormone regulate metamorphosis and molting in insects.

CONCEPT REVIEW

1. Hormones are
   a. enzymes.
   b. regulatory chemicals.
   c. produced at one place in the body but exert their influence at another.
   d. a–c.
   e. b and c.

2. Which of the following is *not* one of the seven principal pituitary hormones?
   a. thyroid hormone
   b. melanocyte-stimulating hormone
   c. growth hormone
   d. luteinizing hormone

3. Noradrenaline and adrenaline are hormones that produce an "alarm" response. They are produced by the adrenal gland, which is located
   a. in the brain.
   b. just above the kidneys.
   c. in the neck.
   d. in the testes or ovaries.

4. The adrenal cortex controls the uptake of sodium and other salts within the kidney by regulating the production of the hormone
   a. aldosterone.
   b. insulin.
   c. glucagon.
   d. antidiuretic hormone.

5. Type II diabetes
   a. results in abnormally low insulin secretion.
   b. is hereditary.
   c. is most effectively treatable with injections of insulin.
   d. is almost always a consequence of excessive weight.

In questions 6–15, match the hormones in the left-hand column with their effects in the right-hand column.

6. insulin	a. ovulation
7. glucagon	b. secondary sex characteristics
8. ADH	c. reaction to stress
9. TSH	d. sodium balance
10. ACTH	e. reabsorption of water
11. aldosterone	f. milk production
12. epinephrine	g. stimulates adrenal gland
13. prolactin	h. stimulates thyroid gland
14. testosterone	i. increases sugar in the blood
15. LH	j. decreases sugar in the blood

16. Most endocrine glands are under the control of the _________ system.

17. The _________ sends signals to the pituitary gland and thus controls the neuroendocrine system.

18. Hormone molecules are peptides or _________.

19. ADH is secreted by the _________ pituitary gland.

20. The _________ is the outer layer of the adrenal gland.

21. The parathyroids control the level of _________ in the blood.

22. The _________ secretes insulin.

CHALLENGE YOURSELF

1. Why do you suppose the brain goes to the trouble of synthesizing releasing hormones rather than simply directing the production of the pituitary hormones immediately?

2. Why do you suppose steroid hormones do not employ second messengers when so many peptide hormones do?

3. Two different organs, A and B, are sensitive to a particular hormone. The cells in both organs have identical receptors for the hormone, and hormone-receptor binding produces the same intracellular second messenger in both organs. However, the intracellular effects of the hormone on A and B are very different. Explain.

27 eBRIDGE

Reinforcing Key Points

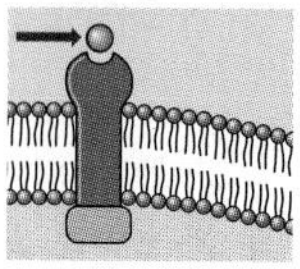

The Neuroendocrine System

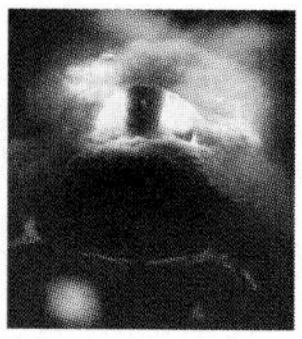

The Major Endocrine Glands

Other Hormone-Producing Glands

Electronic Learning

Visual Learning

Animations

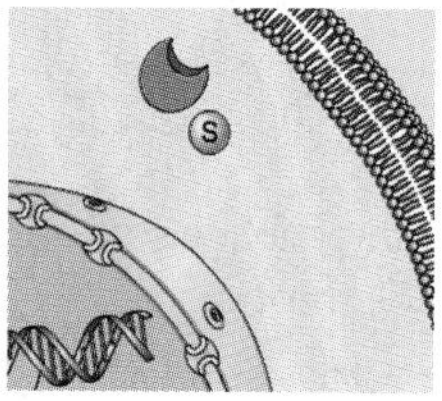

- Hormonal Communication
- Blood Sodium Level and Blood Pressure
- Cellular Activity of Hormones
- Steroid Hormones
- Peptide Hormone Action
- Peptide Hormones
- Second Messengers
- Endocrine System Regulation
- Glucose Regulation
- Regulation of Blood Glucose Level
- Parathyroid Hormone

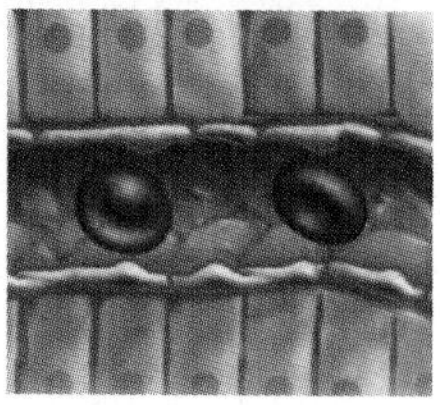

Art Labeling Activities

- The Endocrine System

Author's Corner

Hormones in Action. Much of the long-term signaling that takes place in the body employs chemicals that pass in the bloodstream from one tissue to another. The chemical signals—hormones—may be subtle pheromones that influence mating behavior, or they may help regulate the level of blood sugar. Diabetes, the lack of an effective insulin control of blood sugar levels, is becoming one of the most widespread disorders among adult Americans. Researchers have been actively seeking cures for the two very different kinds of diabetes, with some luck in one instance, much less in the other.

1. Is love some kind of chemical reaction?
2. We may be closing in on the long-sought link between diabetes and obesity.
3. A cure for juvenile diabetes may be near, but late-onset diabetes remains an enigma.

Virtual Classroom

Searching for a Cure for Diabetes

Diabetes is a disorder affecting 15 million Americans and 250 million people worldwide in which the body's cells fail to take up glucose from the blood. Tissues waste away as glucose-starved cells are forced to consume their own protein. Some 15% of diabetics suffer from type I or juvenile diabetes. Their bodies' cells cannot take up glucose because type I diabetics lack the hormone insulin, normally manufactured by islet cells in the pancreas. In juvenile diabetes, the immune system mistakenly attacks and removes these islet cells—and the body's source of insulin. Just as you cannot start a car without inserting the key into the ignition switch, so the cells' glucose transporter channels will not function without insulin to turn them on. Eighty-five percent of diabetics suffer from type II or late-onset diabetes. Typically overweight, these individuals have normal levels of insulin in their blood, and normal insulin receptors on the surfaces of their cells, but have elevated levels of a hormone called resistin produced by fat cells that renders body cells insensitive to insulin. Resistin appears to block the information pathway within the cell between the insulin receptor protein and the cell's glucose transporting machinery.

Virtual Lab

Pheromones Affect Sexual Selection in Cockroaches

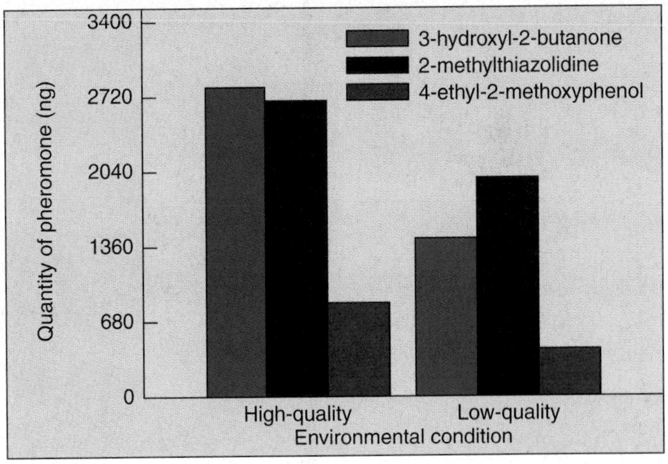
3400
2720
2040
1360
680
0
Quantity of pheromone (ng)
3-hydroxyl-2-butanone
2-methylthiazolidine
4-ethyl-2-methoxyphenol
High-quality
Low-quality
Environmental condition

While hormones act *within* the body as signals for communicating between cells, other chemical signals are released outside the body to influence the behavior or physiological processes of other organisms. These chemicals are called pheromones. Pheromones serve as sex attractants, trail markers, and alarm signals. The actions of pheromones have been studied extensively in insects. Deborah Clark, now at the Middle Tennessee State University, and colleagues have studied the influence of environmental quality on sex pheromones and mating behaviors in the cockroach, *Nauphoeta cineria.* Male cockroaches release a sex pheromone which is used to attract females. The females judge the quality of a suitor as a potential mate by the quality of his pheromone signal. The researchers raised both male and female cockroaches in high- and low-quality environments, and then observed their mating behaviors to determine if the quality of the environment during sexual maturation affects the quality of the sex pheromones in males. They found that differences in male sex pheromones reflect not only genetic differences between males, but also different environmental conditions.

Quizzes | Further Reading | Essential Study Partner | Links | BioCourse.com

CHAPTER 28 Reproduction and Development

THE FAR SIDE® By GARY LARSON
© 1981 FarWorks, Inc. All Rights Reserved/Dist. by Creators Syndicate
Larson
"It's still hungry ... and I've been stuffing worms into it all day."

CHAPTER 28

Reproduction and Development

CHAPTER OVERVIEW

Vertebrate Reproduction

28.1 Asexual and Sexual Reproduction

28.2 Evolution of Reproduction Among the Vertebrates

- Most bony fish and amphibians practice external fertilization, releasing eggs and sperm into water where fertilization occurs.
- Reptiles and birds are oviparous, the young developing in eggs that are deposited externally.
- Most mammals are viviparous, the young developing within the mother.

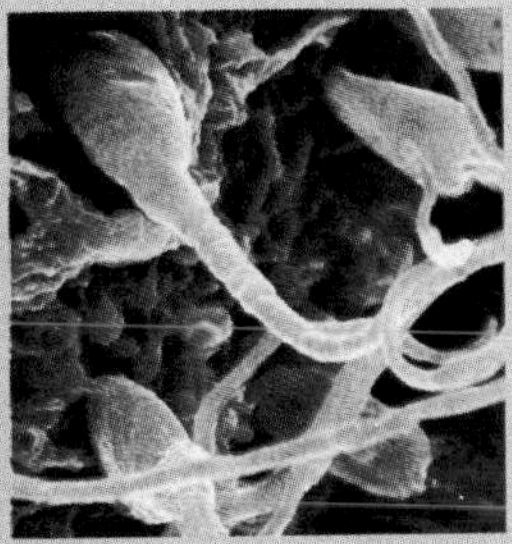

The Human Reproductive System

28.3 Males

28.4 Females

- Male gametes (sperm) are produced in the testes in very large numbers.
- Fertilization occurs in the fallopian tubes a few days after egg release.
- When the developing embryo reaches the uterus, it embeds within the endometrium.

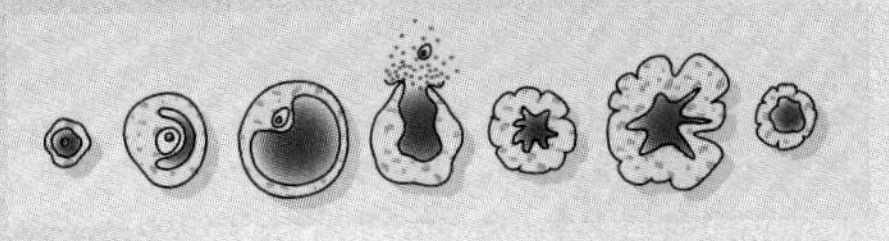

The Female Reproductive Cycle

28.5 Hormones Coordinate the Sexual Cycle

- The hypothalamus regulates the reproductive cycle with a battery of hormones.
- A reproductive cycle takes about 28 days and ends with expulsion of the endometrium.

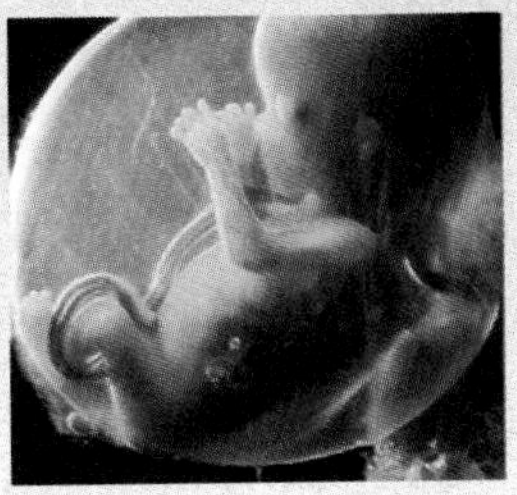

The Course of Development

28.6 Embryonic Development

28.7 Fetal Development

- Cell movements during gastrulation form a hollow embryo with three primary tissues, and by three weeks after fertilization, the body has a neural cord and somites.
- The major organs of the body form in the fourth week, and the limbs of the body assume their adult shape in the second month.

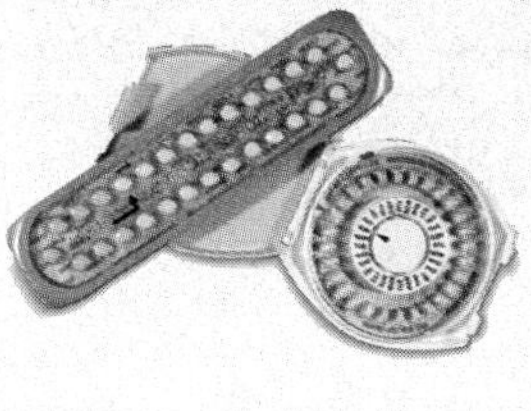

Birth Control

28.8 Approaches to Contraception

- The most effective nonsurgical methods of birth control involve using hormones or other chemicals to prevent egg maturation or embryo implantation.

28.1 Asexual and Sexual Reproduction

Asexual reproduction is the primary means of reproduction among the protists, cnidarians, and tunicates, but it may also occur in some of the more complex animals. Indeed, the formation of identical twins (by the separation of two identical cells of a very early embryo) is a form of asexual reproduction.

Through mitosis, genetically identical cells are produced from a single parent cell. This permits asexual reproduction to occur in protists by division of the organism, or **fission.** Cnidaria commonly reproduce by **budding,** where a part of the parent's body becomes separated from the rest and differentiates into a new individual. The new individual may become an independent animal or may remain attached to the parent, forming a colony.

Sexual reproduction occurs when a new individual is formed by the union of two sex cells, or **gametes,** a term that includes *sperm* and *eggs* (or ova). The union of sperm and egg cells produces a fertilized egg, or **zygote,** that develops by mitotic division into a new multicellular organism. The zygote and the cells it forms by mitosis are diploid; they contain both members of each homologous pair of chromosomes. The gametes, formed by meiosis in the sex organs, or **gonads**—the *testes* and *ovaries*—are haploid (see chapter 6). The processes of spermatogenesis (sperm formation) and oogenesis (egg formation) are described in later sections.

Different Approaches to Sex

Parthenogenesis (virgin birth) is common in many species of arthropods; some species are exclusively parthenogenic (and all female), while others switch between sexual reproduction and parthenogenesis in different generations. In honeybees, for example, a queen bee mates only once and stores the sperm. She then can control the release of sperm. If no sperm are released, the eggs develop parthenogenetically into drones, which are males; if sperm are allowed to fertilize the eggs, the fertilized eggs develop into other queens or worker bees, which are female.

The Russian biologist Ilya Darevsky reported in 1958 one of the first cases of unusual modes of reproduction among vertebrates. He observed that some populations of small lizards of the genus *Lacerta* were exclusively female, and he suggested that these lizards could lay eggs that were viable even if they were not fertilized. In other words, they were capable of asexual reproduction in the absence of sperm, a type of parthenogenesis. Further work has shown that parthenogenesis occurs among populations of other lizard genera.

Another variation in reproductive strategies is **hermaphroditism**, when one individual has both testes and ovaries and so can produce both sperm and eggs (figure 28.1*a*). A tapeworm is hermaphroditic and can fertilize itself, a useful strategy because it is unlikely to encounter another tapeworm. Most hermaphroditic animals, however, require an-

(a)

(b)

Figure 28.1 Hermaphroditism and protogyny.
(*a*) The hamlet bass (genus *Hypoplectrus*) is a deep-sea fish that is a hermaphrodite—both male and female at the same time. In the course of a single pair-mating, one fish may switch sexual roles as many as four times, alternately offering eggs to be fertilized and fertilizing its partner's eggs. Here the fish acting as a male curves around its motionless partner, fertilizing the upward-floating eggs. (*b*) The bluehead wrasse, *Thalassoma bifasciatium,* is protogynous—females sometimes turn into males. Here a large male, or sex-changed female, is seen among females, typically much smaller.

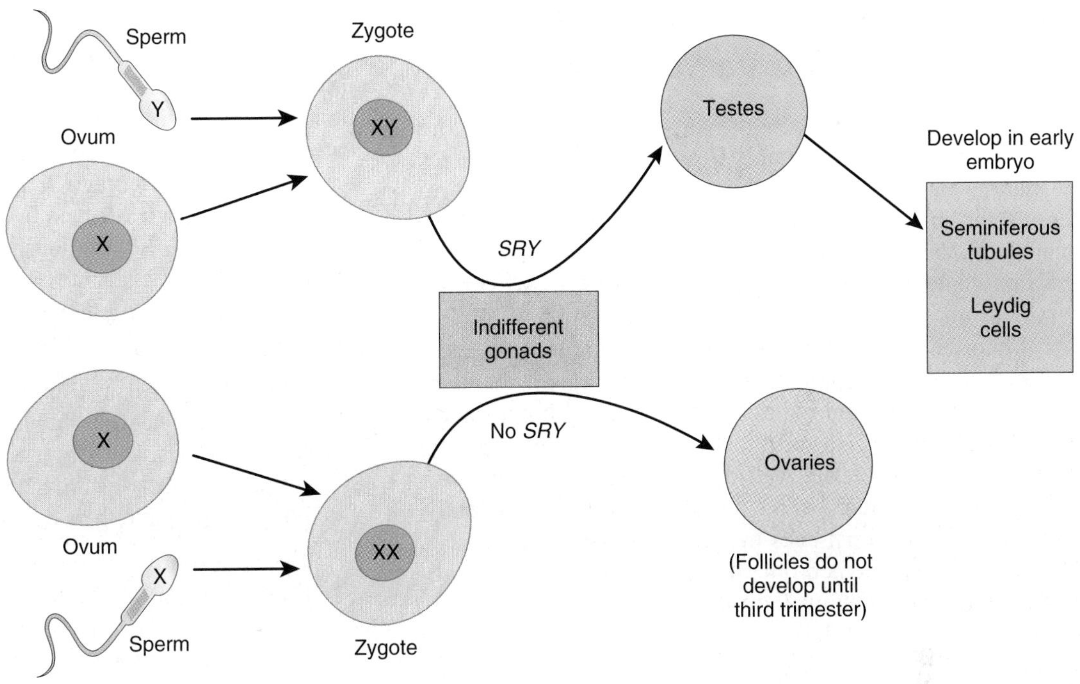
Sperm
Y
Ovum
X
Zygote
XY
SRY
Indifferent gonads
Testes
Develop in early embryo
Seminiferous tubules
Leydig cells
No SRY
Ovaries
(Follicles do not develop until third trimester)
X
Ovum
X
Sperm
XX
Zygote

Figure 28.2 Sex determination.
Sex determination in mammals is made by a region of the Y chromosome designated *SRY*. Testes are formed when the Y chromosome and *SRY* are present; ovaries are formed when they are absent.

other individual to reproduce. Two earthworms, for example, are required for reproduction—each functions as both male and female, and each leaves the encounter with fertilized eggs.

There are some deep-sea fish that are hermaphrodites—both male and female at the same time. Numerous fish genera include species in which individuals can change their sex, a process called *sequential hermaphroditism.* Among coral reef fish, for example, both **protogyny** ("first female," a change from female to male) and **protandry** ("first male," a change from male to female) occur. In fish that practice protogyny (figure 28.1*b*), the sex change appears to be under social control. These fish commonly live in large groups, or schools, where successful reproduction is typically limited to one or a few large, dominant males. If those males are removed, the largest female rapidly changes sex and becomes a dominant male.

Sex Determination

Among the fish just described, and in some species of reptiles, environmental changes can cause changes in the sex of the animal. In mammals, the sex is determined early in embryonic development. The reproductive systems of human males and females appear similar for the first 40 days after conception. During this time, the cells that will give rise to ova or sperm migrate from the yolk sac to the embryonic gonads, which have the potential to become either ovaries in females or testes in males. For this reason, the embryonic gonads are said to be "indifferent." If the embryo is a male, it will have a Y chromosome with a gene whose product converts the indifferent gonads into testes. In females, which lack a Y chromosome, this gene and the protein it encodes are absent, and the gonads become ovaries. Recent evidence suggests that the sex-determining gene may be one known as ***SRY*** (for "sex-determining region of the Y chromosome") (figure 28.2). The *SRY* gene appears to have been highly conserved during the evolution of different vertebrate groups.

Once testes form in the embryo, they secrete testosterone and other hormones that promote the development of the male external genitalia and accessory reproductive organs. If the embryo lacks testes (the ovaries are nonfunctional at this stage), the embryo develops female external genitalia and sex accessory organs. In other words, all mammalian embryos will develop female sex accessory organs and external genitalia unless they are masculinized by the secretions of the testes.

28.1 Sexual reproduction is most common among animals, but many reproduce asexually by fission, budding, or parthenogenesis. Sexual reproduction generally involves the fusion of gametes derived from different individuals of a species, but some species are hermaphroditic.

28.2 Evolution of Reproduction Among the Vertebrates

Vertebrate sexual reproduction evolved in the ocean before vertebrates colonized the land. The females of most species of marine bony fish produce eggs or ova in batches and release them into the water. The males generally release their sperm into the water containing the eggs, where the union of the free gametes occurs. This process is known as **external fertilization.**

Although seawater is not a hostile environment for gametes, it does cause the gametes to disperse rapidly, so their release by females and males must be almost simultaneous. Thus, most marine fish restrict the release of their eggs and sperm to a few brief and well-defined periods. Some reproduce just once a year, while others do so more frequently. There are few seasonal cues in the ocean that organisms can use as signals for synchronizing reproduction, but one all-pervasive signal is the cycle of the moon. Once each month, the moon approaches closer to the earth than usual, and when it does, its increased gravitational attraction causes somewhat higher tides. Many marine organisms sense the tidal changes and entrain the production and release of their gametes to the lunar cycle.

Fertilization is external in most fish but internal in most other vertebrates. The invasion of land posed the new danger of desiccation, a problem that was especially severe for the small and vulnerable gametes. On land, the gametes could not simply be released near each other, because they would soon dry up and perish. Consequently, there was intense selective pressure for terrestrial vertebrates (as well as some groups of fish) to evolve **internal fertilization,** that is, the introduction of male gametes into the female reproductive tract. By this means, fertilization still occurs in a nondesiccating environment, even when the adult animals are fully terrestrial. The vertebrates that practice internal fertilization have three strategies for embryonic and fetal development. Depending upon the relationship of the developing embryo to the mother and egg, those vertebrates with internal fertilization may be classified as oviparous, ovoviviparous, or viviparous.

Oviparity. This is found in some bony fish, most reptiles, some cartilaginous fish, some amphibians, a few mammals, and all birds. The eggs, after being fertilized internally, are deposited outside the mother's body to complete their development.

Ovoviviparity. This is found in some bony fish (including mollies, guppies, and mosquito fish), some cartilaginous fish, and many reptiles. The fertilized eggs are retained within the mother to complete their development, but the embryos still obtain all of their nourishment from the egg yolk. The young are fully developed when they are hatched and released from the mother.

Viviparity. This is found in most cartilaginous fish, some amphibians, a few reptiles, and almost all mammals. The young develop within the mother and obtain nourishment directly from their mother's blood, rather than from the egg yolk (figure 28.3).

Figure 28.3 Viviparous vertebrates carry live, mobile young within their bodies.
The young complete their development within the body of the mother and are then released as small but competent adults. Here a lemon shark has just given birth to a young shark, which is still attached by the umbilical cord.

Fish and Amphibians

Most fish and amphibians, unlike other vertebrates, reproduce by means of external fertilization.

Fish. The eggs of most bony fish and amphibians are fertilized externally. Fertilization in most species of bony fish (teleosts), for example, is external, and the eggs contain only enough yolk to sustain the developing embryo for a short time. After the initial supply of yolk has been exhausted, the young fish must seek its food from the waters around it. Development is speedy, and the young that survive mature rapidly. Although thousands of eggs are fertilized in a single mating, many of the resulting individuals succumb to microbial infection or predation, and few grow to maturity.

In marked contrast to the bony fish, fertilization in most cartilaginous fish is internal! The male introduces sperm into the female through a modified pelvic fin. Development of the young in these vertebrates is generally viviparous.

Amphibians. The amphibians invaded the land without fully adapting to the terrestrial environment, and their life cycle is still tied to the water. Fertilization is external in most amphibians, just as it is in most species of bony fish. Gametes from both males and females are released through the cloaca. Among the frogs and toads, the male grasps the female and discharges fluid containing the sperm onto the eggs as they are released into the water (figure 28.4). Although the eggs of most amphibians develop in the water, there are some interesting exceptions (figure 28.5). In two species of frogs, for example, the eggs develop in the vocal sacs and the stomach, and the young frogs leave through their parent's mouth!

Figure 28.4 The eggs of frogs are fertilized externally.
When frogs mate, as these two are doing, the clasp of the male induces the female to release a large mass of mature eggs, over which the male discharges his sperm.

The time required for development of amphibians is much longer than that for fish, but amphibian eggs do not include a significantly greater amount of yolk. Instead, the process of development in most amphibians is divided into embryonic, larval, and adult stages, in a way reminiscent of the life cycles found in some insects. The embryo develops within the egg, obtaining nutrients from the yolk. After hatching from the egg, the aquatic larva then functions as a free-swimming, food-gathering machine, often for a considerable period of time. The larvae may increase in size rapidly; some tadpoles, which are the larvae of frogs and toads, grow in a matter of weeks from creatures no bigger than the tip of a pencil into individuals as big as a goldfish. When the larva has grown to a sufficient size, it undergoes a developmental transition, or metamorphosis, into the terrestrial adult form.

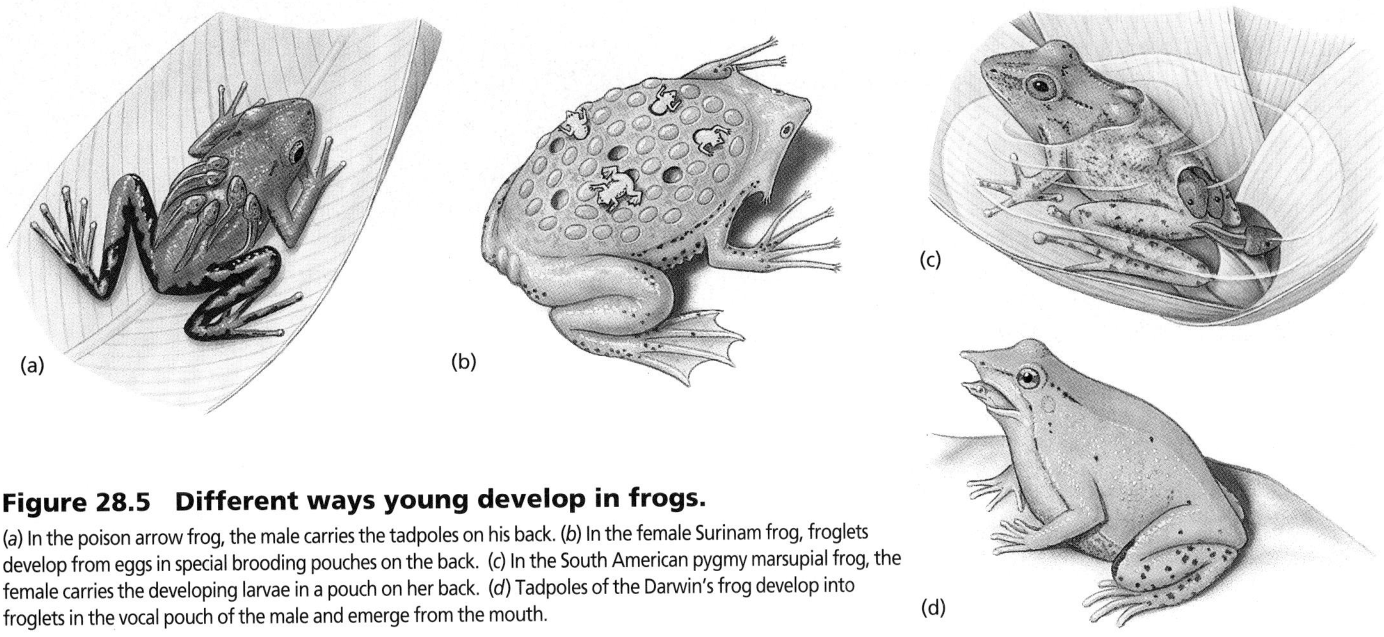
(a)
(b)
(c)
(d)

Figure 28.5 Different ways young develop in frogs.
(*a*) In the poison arrow frog, the male carries the tadpoles on his back. (*b*) In the female Surinam frog, froglets develop from eggs in special brooding pouches on the back. (*c*) In the South American pygmy marsupial frog, the female carries the developing larvae in a pouch on her back. (*d*) Tadpoles of the Darwin's frog develop into froglets in the vocal pouch of the male and emerge from the mouth.

Reptiles and Birds

Most reptiles and all birds are oviparous, laying amniotic eggs that are protected by watertight membranes from desiccation. After the eggs are fertilized internally, they are deposited outside of the mother's body to complete their development. Like most vertebrates that fertilize internally, most male reptiles use a cylindrical organ, the penis, to inject sperm into the female (figure 28.6). The penis, containing erectile tissue, can become quite rigid and penetrate far into the female reproductive tract. Most reptiles are oviparous, laying eggs and then abandoning them. These eggs are surrounded by a leathery shell that is deposited as the egg passes through the oviduct, the part of the female reproductive tract leading from the ovary. Other species of reptiles are ovoviviparous or viviparous, forming eggs that develop into embryos within the body of the mother.

All birds practice internal fertilization, though most male birds lack a penis. In some of the larger birds (including swans, geese, and ostriches), however, the male cloaca extends to form a false penis. As the egg passes along the oviduct, glands secrete albumin proteins (the egg white) and the hard, calcareous shell that distinguishes bird eggs from reptilian eggs. While modern reptiles are poikilotherms (animals whose body temperature varies with the temperature of their environment), birds are homeotherms (animals that maintain a relatively constant body temperature independent of environmental temperatures). Hence, most birds incubate their eggs after laying them to keep them warm (figure 28.7). The young that hatch from the eggs of most bird species are unable to survive unaided because their development is still incomplete. These young birds are fed and nurtured by their parents, and they grow to maturity gradually.

The shelled eggs of reptiles and birds constitute one of the most important adaptations of these vertebrates to life on land, because shelled eggs can be laid in dry places. Such eggs are known as amniotic eggs because the embryo develops within a fluid-filled cavity surrounded by a membrane called the amnion. The amnion is an extraembryonic membrane—that is, a membrane formed from embryonic cells but located outside the body of the embryo. Other extraembryonic membranes in amniotic eggs include the chorion, which lines the inside of the eggshell, the yolk sac, and the allantois. In contrast, the eggs of fish and amphibians contain only one extraembryonic membrane, the yolk sac. The viviparous mammals, including humans, also have extraembryonic membranes that are described later in this chapter.

Figure 28.6 The introduction of sperm by the male into the female's body is called copulation.

Reptiles such as these turtles were the first terrestrial vertebrates to develop this form of reproduction, which is particularly suited to a terrestrial environment.

Figure 28.7 Crested penguins incubating their egg.

This nesting pair is changing the parental guard in a stylized ritual.

Mammals

Some mammals are seasonal breeders, reproducing only once a year, while others have shorter reproductive cycles. Among the latter, the females generally undergo the reproductive cycles, while the males are more constant in their reproductive activity. Cycling in females involves the periodic release of a mature ovum from the ovary in a process known as ovulation. Most female mammals are "in heat," or sexually receptive to males, only around the time of ovulation. This period of sexual receptivity is called **estrus,** and the reproductive cycle is therefore called an **estrous cycle.** The females continue to cycle until they become pregnant.

In the estrous cycle of most mammals, changes in the secretion of follicle-stimulating hormone (FSH) and luteinizing hormone (LH) by the anterior pituitary gland cause changes in egg cell development and hormone secretion in the ovaries. Humans and apes have menstrual cycles that are similar to the estrous cycles of other mammals in their cyclic pattern of hormone secretion and ovulation. Unlike mammals with estrous cycles, however, human and ape females bleed when they shed the inner lining of their uterus, a process called menstruation, and may engage in copulation at any time during the cycle.

Rabbits and cats differ from most other mammals in that they are **induced ovulators.** Instead of ovulating in a cyclic fashion regardless of sexual activity, the females ovulate only after copulation as a result of a reflex stimulation of LH secretion (described later). This makes these animals extremely fertile.

The most primitive mammals, the **monotremes** (consisting solely of the duck-billed platypus and the echidna), are oviparous, like the reptiles from which they evolved. They incubate their eggs in a nest (figure 28.8*a*) or specialized pouch, and the young hatchlings obtain milk from their mother's mammary glands by licking her skin, because monotremes lack nipples. All other mammals are viviparous and are divided into two subcategories based on how they nourish their young. The **marsupials,** a group that includes opossums and kangaroos, give birth to fetuses that are incompletely developed. The fetuses complete their development in a pouch of their mother's skin, where they can obtain nourishment from nipples of the mammary glands (figure 28.8*b*). The **placental mammals** (figure 28.8*c*) retain their young for a much longer period of development within the mother's uterus. The fetuses are nourished by a structure known as the placenta, which is derived from both an extraembryonic membrane (the chorion) and the mother's uterine lining. Because the fetal and maternal blood vessels are in very close proximity in the placenta, the fetus can obtain nutrients by diffusion from the mother's blood. The functioning of the placenta is discussed in more detail later in this chapter.

28.2 Fertilization is external in frogs and most bony fish and internal in other vertebrates. Birds and most reptiles lay watertight eggs, as do monotreme mammals. All other mammals are viviparous, giving birth to live young.

(a) Monotremes (b) Marsupials (c) Placentals

Figure 28.8 Reproduction in mammals.

(*a*) Monotremes, like the duck-billed platypus shown here, lay eggs in a nest. (*b*) Marsupials, such as this kangaroo, give birth to small fetuses that complete their development in a pouch. (*c*) In placental mammals, like this domestic cat, the young remain inside the mother's uterus for a longer period of time and are born relatively more developed.

28.3 Males

The male gamete, or **sperm,** is highly specialized for its role as a carrier of genetic information. Produced after meiosis, the sperm cells have 23 chromosomes instead of the 46 found in all other cells of the male body. Sperm do not complete their development successfully at 37°C (98.6°F), the normal human body temperature. The sperm-producing organs, the **testes,** move during the course of fetal development into a sac called the **scrotum** (figure 28.9), which hangs between the legs of the male maintaining the testes at a temperature about 3°C cooler than the rest of the body.

Male Gametes Are Formed in the Testes

The two testes are composed of several hundred compartments, each packed with large numbers of tightly coiled tubes called **seminiferous tubules.** Within these tubes sperm production takes place (figure 28.10). The number of sperm that is produced is truly incredible. A typical adult male produces several hundred million sperm each day of his life! Those that are not ejaculated from the body are broken down, and their materials are resorbed and recycled.

The testes contain cells that secrete the male sex hormone **testosterone.** Sperm and all the cells of the testes also require a combination of the pituitary hormones FSH and LH for their normal function.

After a sperm cell is manufactured within the testes, it is delivered to a long, coiled tube called the **epididymis,** where it matures. The sperm cell is not motile when it arrives in the epididymis, and it must remain there for at least 18 hours before its motility develops. From there, the sperm is delivered to another long tube, the **vas deferens**. When sperm are delivered during intercourse, they travel through a tube from the vas deferens to the **urethra,** where the reproductive and urinary tracts join, emptying through the penis.

Male Gametes Are Delivered by the Penis

The **penis** is an external tube containing two long cylinders of spongy tissue side by side. Below and between them runs a third cylinder of spongy tissue that contains in its center a tube called the urethra, through which both semen (during ejaculation) and urine (during urination) pass (figure 28.11). Why this unusual design? The penis is designed to inflate! The spongy tissues that make up the three cylinders are riddled with small spaces between the cells, and when nerve impulses from the CNS cause the arterioles leading into this tissue to expand, blood collects within these spaces. Like blowing up a balloon, this causes the penis to become erect and rigid. Continued stimulation by the CNS is required for this erection to be maintained.

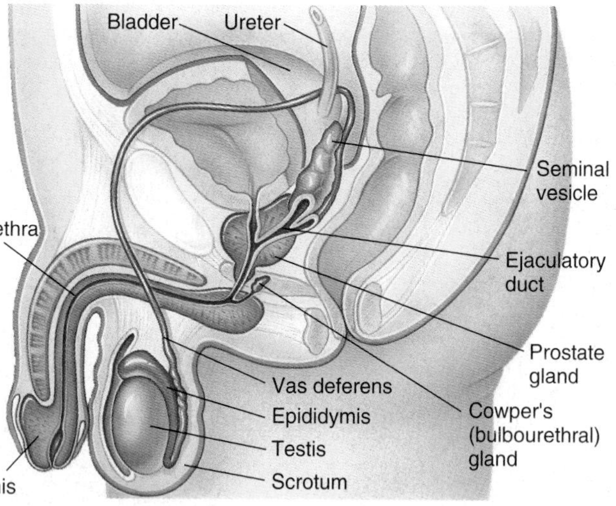
Bladder
Ureter
Seminal vesicle
Urethra
Ejaculatory duct
Prostate gland
Vas deferens
Epididymis
Testis
Scrotum
Cowper's (bulbourethral) gland
Penis

(a)

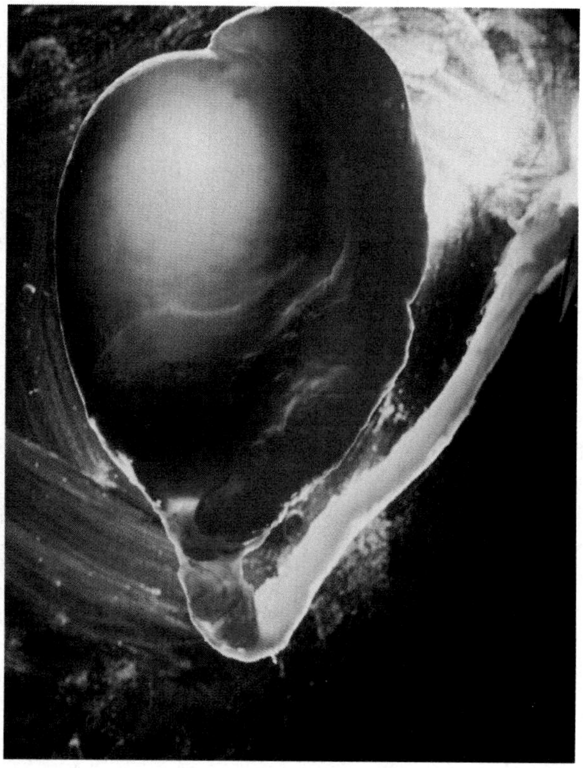

(b)

Figure 28.9 The male reproductive organs.
(*a*) Organization of the male reproductive organs. (*b*) A human testis. The testis is the darker sphere in the center of the photograph; within it, sperm are formed. Cupped above the testicle is the epididymis, a highly coiled passageway within which sperm complete their maturation. Extending away from the epididymis is a long tube, the vas deferens.

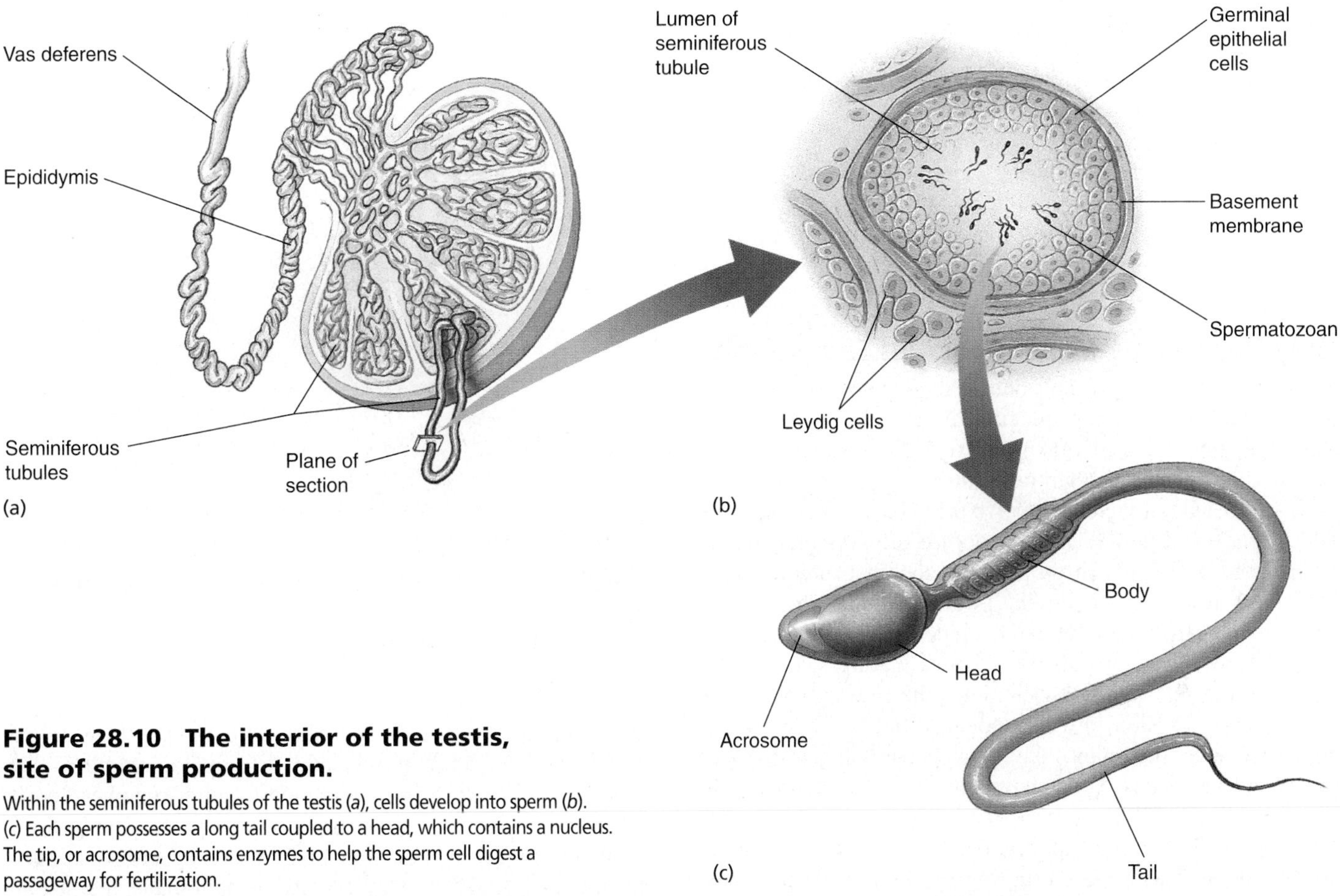
Vas deferens
Epididymis
Seminiferous tubules
Plane of section
(a)
Lumen of seminiferous tubule
Germinal epithelial cells
Basement membrane
Spermatozoan
Leydig cells
(b)
Body
Head
Acrosome
(c)
Tail

Figure 28.10 The interior of the testis, site of sperm production.
Within the seminiferous tubules of the testis (*a*), cells develop into sperm (*b*). (*c*) Each sperm possesses a long tail coupled to a head, which contains a nucleus. The tip, or acrosome, contains enzymes to help the sperm cell digest a passageway for fertilization.

Erection can be achieved without any physical stimulation of the penis, but physical stimulation is required for semen to be delivered. Stimulation of the penis, as by repeated thrusts into the vagina of a female, leads first to the mobilization of the sperm. In this process, muscles encircling the vas deferens contract, moving the sperm along the vas deferens into the urethra. Eventually, the stimulation leads to the violent contraction of the muscles at the base of the penis. The result is **ejaculation,** the forceful ejection of about 5 milliliters of semen. Semen is a collection of secretions from the prostate and other glands that provides metabolic energy sources for the sperm. Within this small 5-milliliter volume are several hundred million sperm. Because the odds against any one individual sperm cell successfully completing the long journey to the egg and fertilizing it are extraordinarily high, successful fertilization requires a high sperm count. Males with fewer than 20 million sperm per milliliter are generally considered sterile.

28.3 Male testes continuously produce large numbers of male gametes, sperm, which mature in the epididymis, are stored in the vas deferens, and are delivered through the penis into the female.

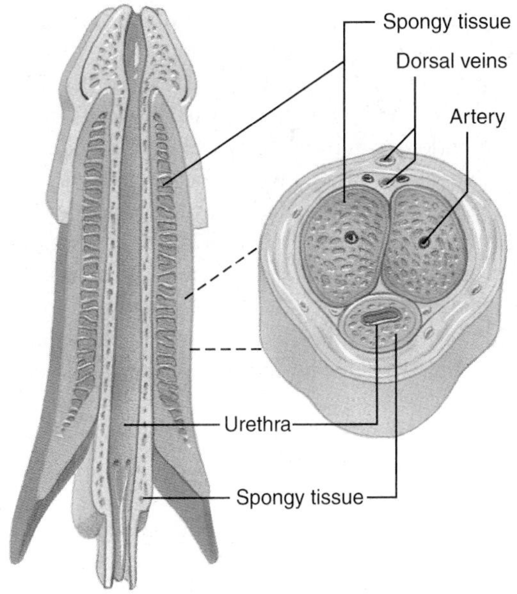
Spongy tissue
Dorsal veins
Artery
Urethra
Spongy tissue

Figure 28.11 Structure of the penis.
(*Left*) Longitudinal section; (*right*) cross section.

28.4 Females

In females, eggs develop from cells called **oocytes,** located in the outer layer of compact masses of cells called **ovaries,** located within the abdominal cavity (figure 28.12). Unlike males, whose gamete-producing cells are constantly dividing, females have at birth all the oocytes that they will ever produce. At each cycle of egg maturation, called **ovulation,** one or a few of these oocytes initiates development; the others remain in a developmental holding pattern.

Only One Female Gamete Matures Each Month

At birth, a female's ovaries contain some 2 million oocytes, all of which have begun the first meiotic division. At this stage they are called **primary oocytes.** Each primary oocyte is poised to develop further, but it does not continue on with meiosis. Instead, it waits to receive the proper developmental "go" signal, and until a primary oocyte receives this signal, its meiosis is arrested in prophase of the first meiotic division. Very few ever receive the awaited signal, which turns out to be the hormone FSH we met in the previous chapter.

With the onset of puberty, females mature sexually. At this time the release of FSH initiates the resumption of the first meiotic division in a few oocytes, but a single oocyte soon becomes dominant, the others regressing. Approximately every 28 days after that, another oocyte matures, although the exact timing may vary from month to month. Only about 400 of the approximately 2 million oocytes a woman is born with mature during her lifetime. When they mature, the egg cells are called **ova** (singular, **ovum**), the Latin word for "egg."

Fertilization Occurs in the Oviducts

The **fallopian tubes** (also called uterine tubes or **oviducts**) transport ova from the ovaries to the **uterus.** In humans, the uterus is a muscular, pear-shaped organ that narrows to form a neck, the cervix, which leads to the vagina (figure 28.13*a*). The uterus is lined with a stratified epithelial membrane called the endometrium. The surface of the endometrium is shed during menstruation, while the underlying portion remains to generate a new surface during the next cycle.

Mammals other than primates have more complex female reproductive tracts, where part of the uterus divides to form uterine "horns," each of which leads to an oviduct (figure 28.13*b, c*). In cats, dogs, and cows, for example, there is one cervix but two uterine horns separated by a septum, or wall. Marsupials, such as opossums, carry the split even further, with two unconnected uterine horns, two cervices, and two vaginas. A male marsupial has a forked penis that can enter both vaginas simultaneously.

Smooth muscles lining the fallopian tubes contract rhythmically, moving the egg down the tube to the uterus in much the same way that food is moved down through your intestines, pushing it along by squeezing the tube behind it. The journey of the egg through the fallopian tube is a slow one, taking from five to seven days to complete. If the egg is unfertilized, it loses its capacity to develop within a few days. Any egg that arrives at the uterus unfertilized can never become so. For this reason the sperm cannot simply lie in wait within the uterus. To fertilize an egg successfully, a sperm must make its way far up the fallopian tube, a long passage that few survive.

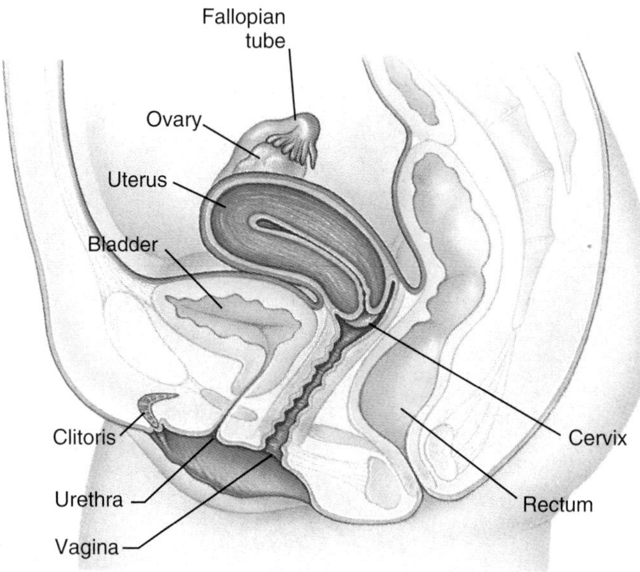

Fallopian tube
Ovary
Uterus
Bladder
Clitoris
Urethra
Vagina
Cervix
Rectum

Figure 28.12 The female reproductive system.
The organs of the female reproductive system are specialized to produce gametes and to provide a site for embryonic development if the gamete is fertilized.

Sperm are deposited within the vagina, a thin-walled muscular tube about 7 centimeters long that leads to the mouth of the uterus. This opening is bounded by a muscular ring called the **cervix.** The uterus is a hollow, pear-shaped organ about the size of a small fist. Its inner wall, the **endometrium,** has two layers. The outer of these layers is shed during menstruation, while the one beneath it generates the next layer. Sperm entering the uterus swim up to and enter the fallopian tube. They swim upward against the current generated by the tube's contractions, which are carrying the ovum downward toward the uterus.

When a sperm succeeds in fertilizing an egg high in the fallopian tube, the fertilized egg—now an embryo—continues on its journey down the fallopian tube. When it reaches the uterus, the new embryo attaches itself to the endometrial lining and starts the long developmental journey that eventually leads to the birth of a child (figure 28.14).

28.4 In human females, hormones trigger the maturation of one or a few oocytes each 28 days. When mature, the egg cells travel down the fallopian tubes and, if fertilized during their journey, implant in the wall of the uterus and initiate embryonic development.

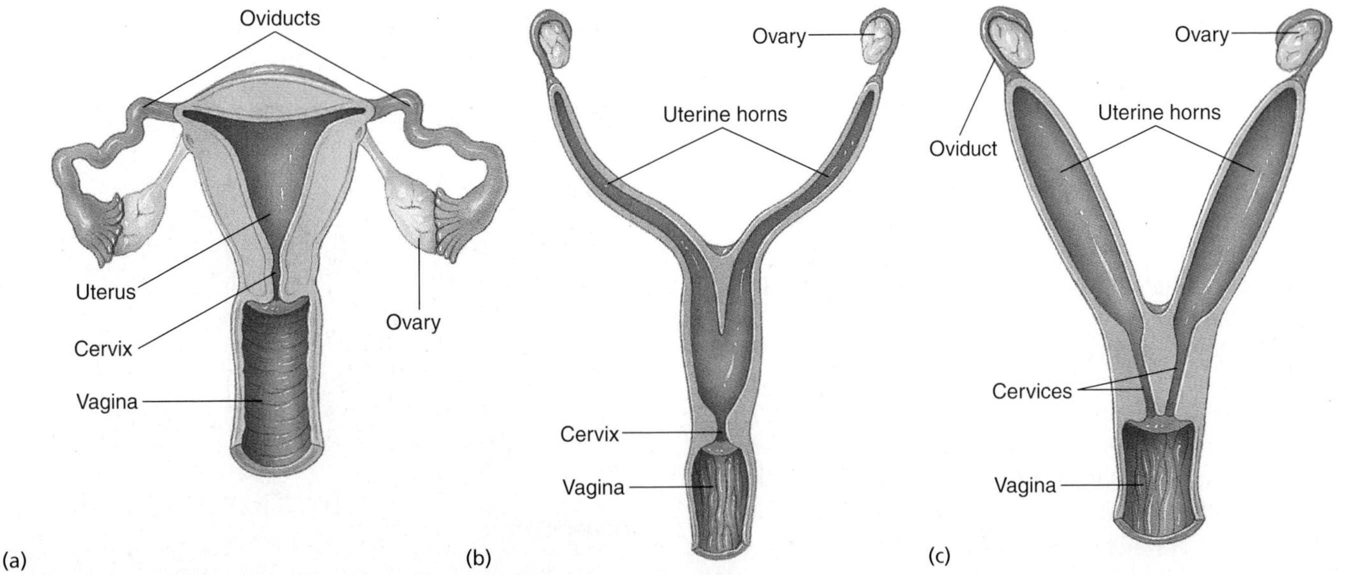
Oviducts
Uterus
Cervix
Vagina
Ovary
(a)
Ovary
Uterine horns
Cervix
Vagina
(b)
Ovary
Oviduct
Uterine horns
Cervices
Vagina
(c)

Figure 28.13 A comparison of mammalian uteruses.

(*a*) Humans and other primates; (*b*) cats, dogs, and cows; and (*c*) rats, mice, and rabbits.

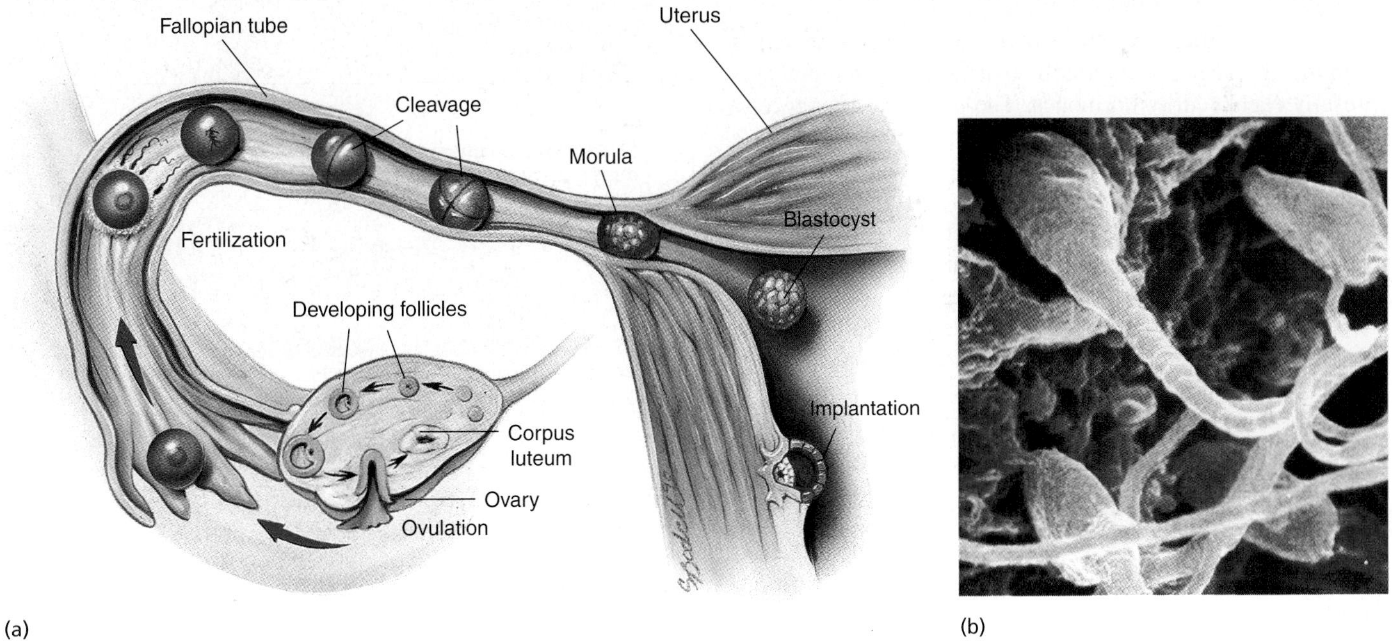
Fallopian tube
Uterus
Cleavage
Morula
Blastocyst
Fertilization
Developing follicles
Implantation
Corpus
luteum
Ovary
Ovulation
(a)
(b)

Figure 28.14 The journey of an ovum.

(*a*) Produced within a follicle and released at ovulation, an ovum is swept up into a fallopian tube and carried down by waves of contraction of the tube walls. Fertilization occurs within the tube, by sperm journeying upward. Several mitotic divisions occur while the fertilized ovum continues its journey down the fallopian tube. The fertilized ovum implants itself within the wall of the uterus, where it continues its development. (*b*) Human sperm fertilizing an ovum. Only the heads and a portion of the long, slender tails of these sperm are shown in this scanning electron micrograph.

28.5 Hormones Coordinate the Sexual Cycle

The female reproductive cycle is composed of two distinct phases, the follicular phase and the luteal phase. These phases are coordinated by a family of hormones. Hormones play many roles in human reproduction. Sexual development, delayed in mammals, is initiated by hormones that coordinate simultaneous sexual development in many kinds of tissues. The production of gametes is another closely orchestrated process, involving a series of carefully timed developmental events. Successful fertilization initiates yet another developmental "program," in which the female body prepares itself for the many changes of pregnancy.

Production of the sex hormones that coordinate all these processes is coordinated by the hypothalamus, which sends releasing hormones to the pituitary, directing it to produce particular sex hormones (figure 28.15). Feedback plays a key role in regulating these activities of the hypothalamus. When target organs receive a pituitary hormone, they begin to produce a hormone of their own, which circulates back to the hypothalamus, shutting down production of the pituitary hormone.

Triggering the Maturation of an Egg

The first, or **follicular, phase** of the reproductive cycle is when the egg develops within the ovary. This development is carefully regulated by hormones. The pituitary, after receiving a chemical signal from the hypothalamus, starts the cycle by secreting **follicle-stimulating hormone (FSH),** which binds to receptors on the surface of cells surrounding the egg (the oocyte and its surrounding mass of tissue is called a **follicle**) and triggers resumption of meiosis. Normally, only a few eggs at any one time have developed far enough to respond immediately to FSH. FSH levels then fall. Because FSH levels are reduced before other eggs ripen to maturity, only a few eggs ripen in every cycle.

The fall of FSH levels is achieved by a feedback command to the pituitary. FSH does not itself carry out this feedback—instead, it sends another hormone as a messenger. FSH not only triggers final egg development, it also causes the ovary to start producing the female sex hormone **estrogen.** Rising levels of estrogen in the bloodstream feed back to the hypothalamus, which responds to the rising estrogen by commanding the pituitary to cut off the further production of FSH. This "shuts the door" on further FSH-induced egg development. The rise in estrogen levels signals the completion of the follicular phase of the reproductive cycle.

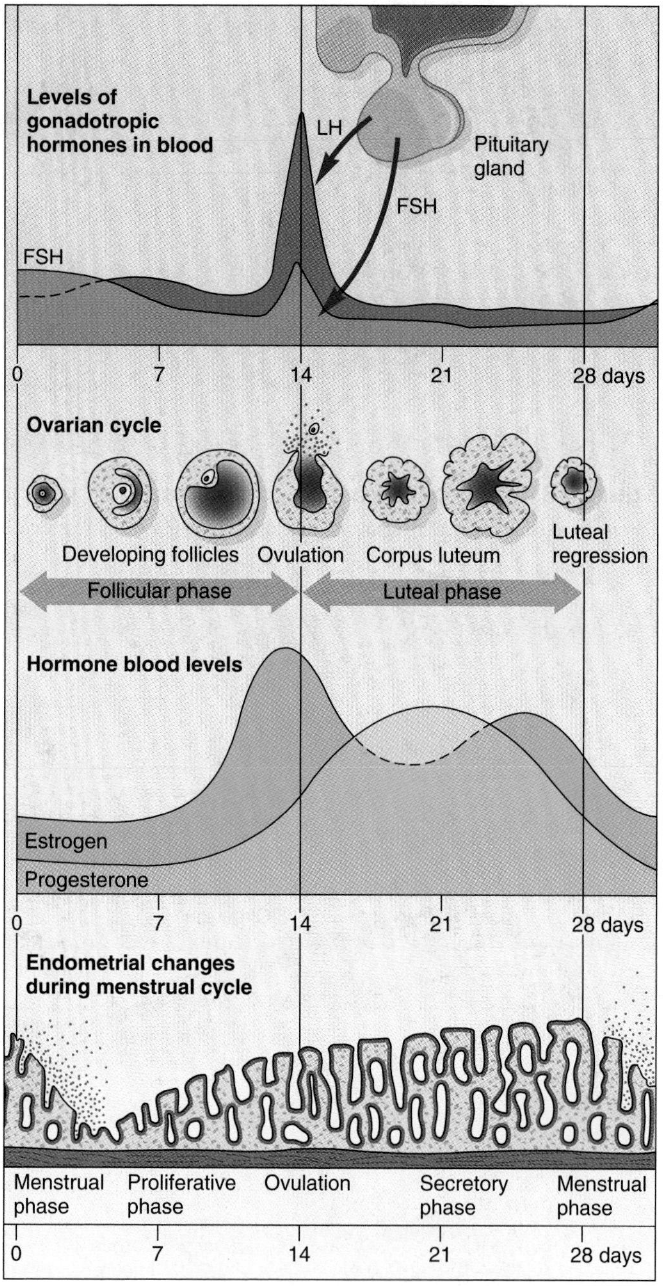
Levels of gonadotropic hormones in blood
LH
Pituitary gland
FSH
FSH
0
7
14
21
28 days
Ovarian cycle
Developing follicles
Ovulation
Corpus luteum
Luteal regression
Follicular phase
Luteal phase
Hormone blood levels
Estrogen
Progesterone
0
7
14
21
28 days
Endometrial changes during menstrual cycle
Menstrual phase
Proliferative phase
Ovulation
Secretory phase
Menstrual phase
0
7
14
21
28 days

Figure 28.15 The human menstrual cycle.
The growth and thickening of the uterine (endometrial) lining is governed by levels of the hormone progesterone; menstruation, the sloughing off of this blood-rich tissue, is initiated by lower levels of progesterone.

Preparing the Body for Fertilization

The second, or **luteal, phase** of the cycle follows smoothly from the first. The hypothalamus responds to estrogen not only by shutting down the pituitary's FSH production but also by causing the pituitary to begin secreting a second hormone, called **luteinizing hormone (LH).** LH is the hormone that causes ovulation, sending the now-mature egg on its journey towards fertilization (figure 28.16). LH is carried in the bloodstream to the developing follicle, where it inhibits further estrogen production and causes the wall of the follicle to burst. The egg within the follicle is released into one of the fallopian tubes extending from the ovary to the uterus.

After the egg's release and departure, the ruptured follicle repairs itself, filling in and becoming yellowish. In this condition it is called the **corpus luteum,** which is simply the Latin phrase for "yellow body." The corpus luteum soon begins to secrete a hormone, **progesterone,** which inhibits FSH (a backup for estrogen in preventing further ovulations). Progesterone is the body's signal to prepare the uterus for fertilization. If fertilization occurs, the corpus luteum continues to produce progesterone for several weeks. The rising levels of progesterone initiate the many physiological changes associated with pregnancy. Among them are thickening of the walls of the uterus in preparation for the implantation of the developing embryo.

If fertilization does *not* occur soon after ovulation, however, production of progesterone slows and eventually ceases, marking the end of the luteal phase. The decreasing levels of progesterone cause the thickened layer of blood-rich tissue to be sloughed off, a process that results in the bleeding associated with menstruation. **Menstruation,** or "having a period," usually occurs about midway between successive ovulations, although its timing varies widely for individual females.

At the end of the luteal phase, neither estrogen nor progesterone is being produced. In their absence, the pituitary can again initiate production of FSH, thus starting another reproductive cycle. Each cycle begins immediately after the preceding one ends. A cycle usually occurs every 28 days, or a little more frequently than once a month, although this varies in individual cases. The Latin word for "month" is *mens,* which is why the reproductive cycle is called the **menstrual cycle,** or monthly cycle.

Two other hormones, both secreted by the pituitary, are important in the female reproductive system. For the first couple of days after childbirth, the mammary glands produce a fluid called colostrum, which contains protein and lactose but little fat. Then milk production is stimulated by the hormone **prolactin,** usually by the third day after delivery. When the infant suckles at the breast, the hormone **oxytocin** is released, initiating milk release. Earlier, in combination with chemicals released from the uterus, oxytocin initiates labor and delivery.

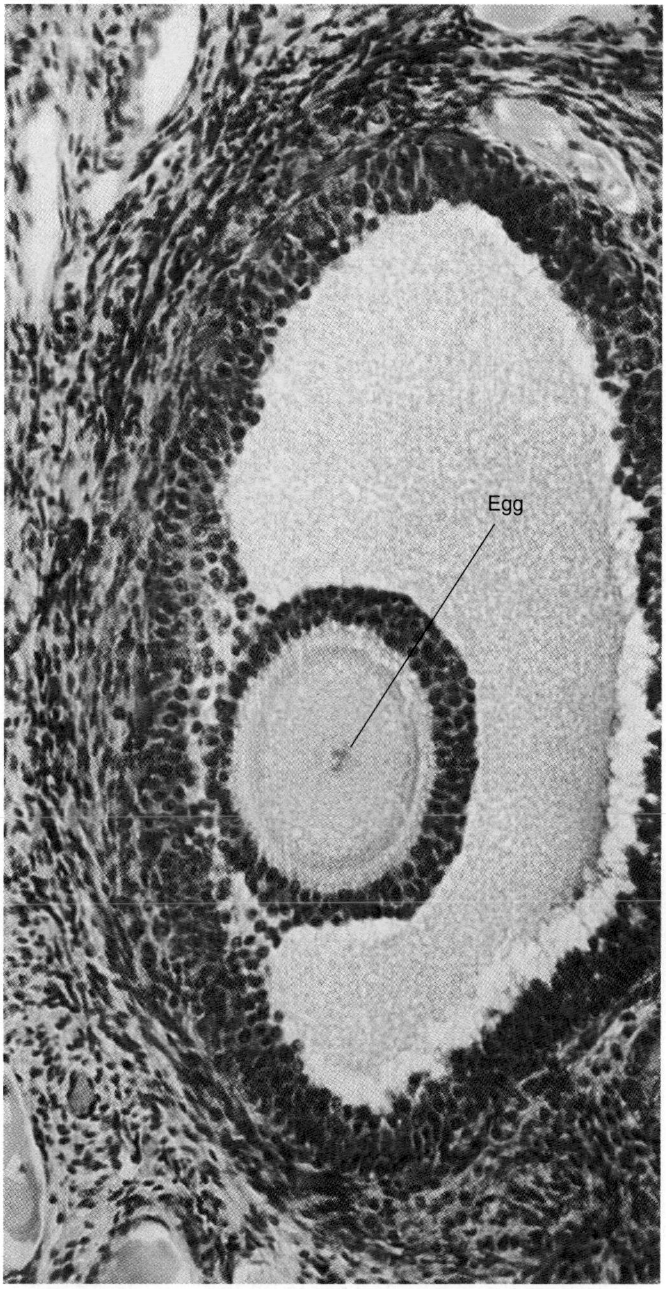
Egg

Figure 28.16 A mature egg within an ovarian follicle.

In each menstrual cycle, a few follicles are stimulated to grow under the influence of FSH, but only one achieves full maturity.

28.5 Humans and apes have menstrual cycles that are similar to the estrus cycles of other mammals in that they are driven by cyclic patterns of hormone secretion and ovulation. The cycle is composed of two distinct phases, follicular and luteal, coordinated by a family of four sex hormones.

28.6 Embryonic Development

Cleavage: Setting the Stage for Development

The first major event in human embryonic development is the rapid division of the zygote into a larger and larger number of smaller and smaller cells, becoming first 2 cells, then 4, then 8, and so on. The first of these divisions occurs about 30 hours after union of the egg and the sperm, and the second, 30 hours later. During this period of division, called **cleavage,** the overall size of the embryo does not increase. The resulting tightly packed mass of about 32 cells is called a **morula,** and each individual cell in the morula is referred to as a **blastomere.**

During this period, the embryo continues its journey down the mother's fallopian tube. On about the sixth day, the embryo reaches the uterus, attaches to the uterine lining, and penetrates into the tissue of the lining. The embryo now begins to grow rapidly, initiating the formation of the membranes that will later surround, protect, and nourish it. One of these membranes, the **amnion,** will enclose the developing embryo, whereas another, the **chorion,** will interact with uterine tissue to form the **placenta,** which will nourish the growing embryo.

The cells of the morula continue to divide without an overall increase in size, each cell secreting a fluid into the center of the cell mass. Eventually, a hollow ball of 500 to 2,000 cells is formed, surrounding a fluid-filled cavity called the **blastocoel.** Within the ball is an inner cell mass concentrated at one pole that goes on to form the developing embryo. This is the **blastocyst.** The outer sphere of cells is called the trophoblast. The trophoblast develops into the membrane surrounding the embryo (the chorion) and a complex series of membranes known as the placenta, which connects the developing embryo to the blood supply of the mother. Fully 61 of the cells at the 64-celled stage develop into the trophoblast and only 3 into the embryo proper.

Gastrulation: The Onset of Developmental Change

Ten to 11 days after fertilization, certain groups of cells move inward from the surface of the cell mass in a carefully orchestrated migration called **gastrulation** (table 28.1). First, the lower cell layer of the blastula cell mass differentiates into **endoderm,** one of the three primary tissues, and the upper layer into **ectoderm.** Just after this differentiation, much of the **mesoderm** and endoderm arise by the invagination of cells that move from the upper layer of the cell mass *inward,* along the edges of a furrow that appears at the midline of the embryo. The site of this invagination appears as a slit in the embryo, called the **primitive streak.**

During gastrulation, about half of the cells of the blastocyst cell mass move into the interior of the human embryo. This movement largely determines the future development of the embryo. By the end of gastrulation, distribution of cells into the three primary germ layers has been completed. The ectoderm is destined to form the epidermis and neural tissue. The mesoderm is destined to form the connective tissue, muscle, and vascular elements. The endoderm forms the lining of the gut and its derivative organs.

Neurulation: Determination of Body Architecture

In the third week of embryonic development, the three primary cell types begin their development into the tissues and organs of the body. As in all vertebrates, this begins with the formation of two characteristic vertebrate features, the notochord and the hollow dorsal nerve cord. This stage in development is called **neurulation.**

The first of these two structures to form is the **notochord,** a flexible rod. Soon after gastrulation is complete, it forms from mesoderm tissue along the midline of the embryo, below its dorsal surface. After the notochord has been formed, the **dorsal nerve cord** forms from the region of the ectoderm that is located above the notochord and later differentiates into the spinal cord and brain.

While the dorsal nerve cord is forming from ectoderm, the rest of the basic architecture of the human body is being rapidly determined by changes in the mesoderm. On either side of the developing notochord, segmented blocks of tissue form. Ultimately, these blocks, or **somites,** give rise to the muscles, vertebrae, and connective tissues. As development continues, more and more somites are formed. Within another strip of mesoderm that runs alongside the somites, many of the significant glands of the body, including the kidneys, adrenal glands, and gonads, develop. The remainder of the mesoderm layer moves out and around the inner endoderm layer of cells and eventually surrounds it entirely. As a result, the mesoderm forms two layers. The outer layer is associated with the body wall and the inner layer is associated with the gut. Between these two layers of mesoderm is the **coelom,** which becomes the body cavity of the adult.

By the end of the third week, over a dozen somites are evident, and the blood vessels and gut have begun to develop. At this point the embryo is about 2 millimeters (less than a tenth of an inch) long.

28.6 The vertebrate embryo develops in three stages. In cleavage, a hollow ball of cells forms. In gastrulation, cells move into the interior, forming the primary tissues. In neurulation, the organs of the body form.

TABLE 28.1 STAGES OF VERTEBRATE DEVELOPMENT

	Stage	Description
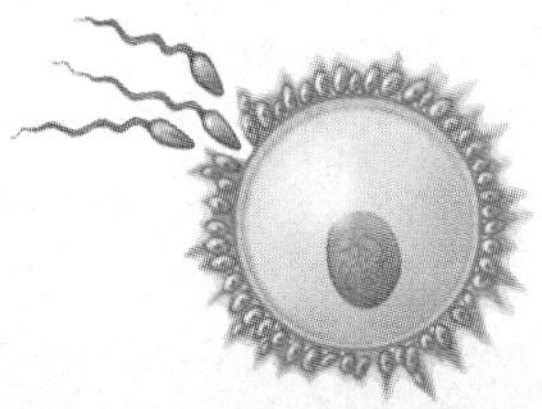	Fertilization	The haploid male and female gametes fuse to form a diploid zygote.
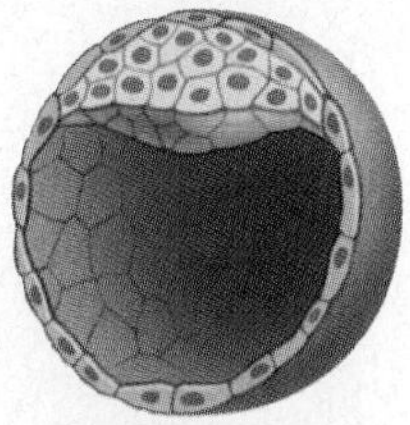	Cleavage	The zygote rapidly divides into many cells, with no overall increase in size. These divisions affect future development, because different cells receive different portions of the egg cytoplasm and, hence, different regulatory signals.
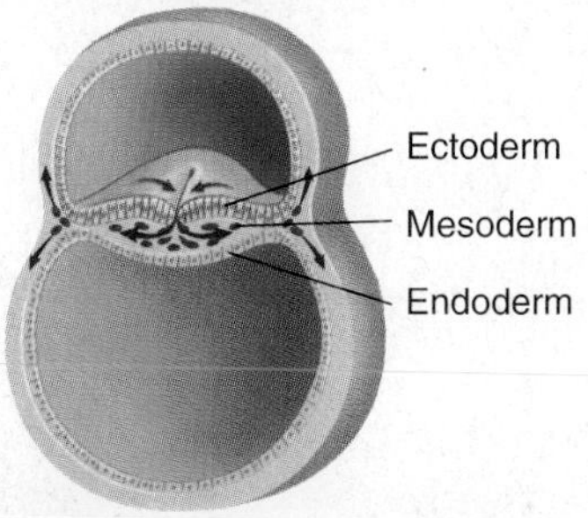 Ectoderm Mesoderm Endoderm	Gastrulation	The cells of the embryo move, forming three primary cell layers: ectoderm, mesoderm, and endoderm.
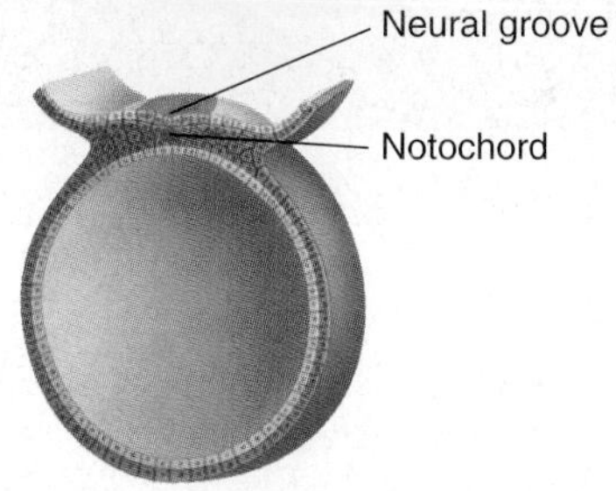 Neural groove Notochord	Neurulation	In all chordates, the first organ to form is the notochord; the second is the dorsal nerve cord.
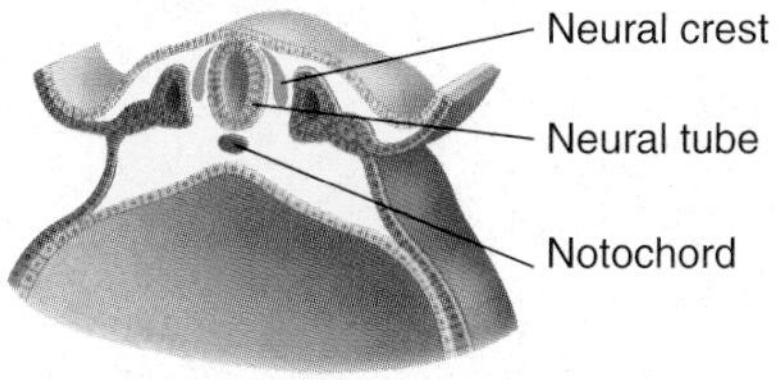 Neural crest Neural tube Notochord	Neural crest	During neurulation, the neural crest is produced as the neural tube is formed. The neural crest gives rise to several uniquely vertebrate structures.
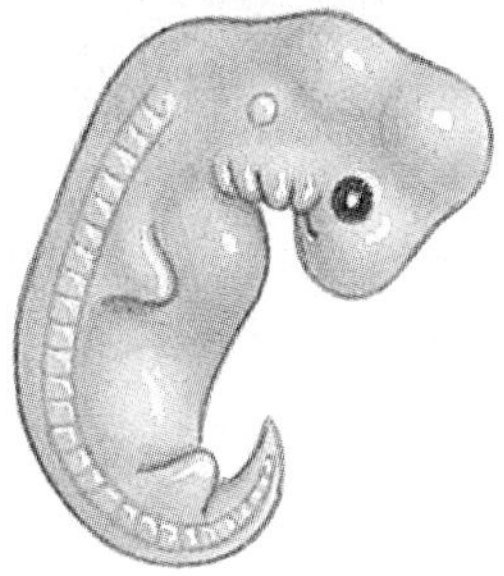	Organogenesis	Cells from the three primary cell layers combine in various ways to produce the organs of the body.

28.7 Fetal Development

The Fourth Week: Organ Formation

In the fourth week of pregnancy, the body organs begin to form, a process called **organogenesis** (figure 28.17*a*). The eyes form, and the heart begins a rhythmic beating and develops four chambers. At 70 beats per minute, the little heart is destined to beat more than 2.5 billion times during a lifetime of about 70 years. More than 30 pairs of somites are visible by the end of the fourth week, and the arm and leg buds have begun to form. The embryo more than doubles in length during this week, reaching about 5 millimeters.

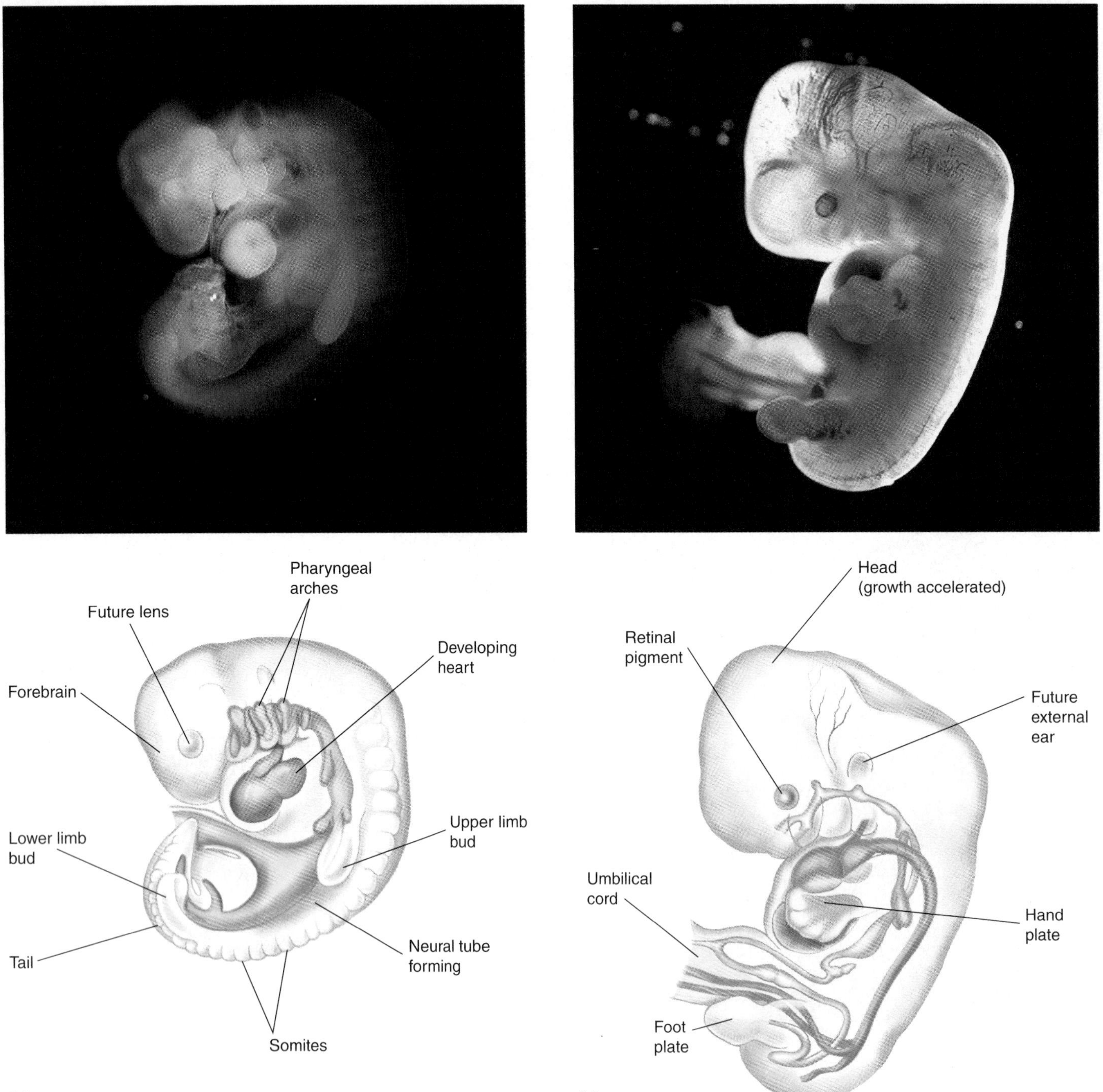
Pharyngeal arches
Future lens
Developing heart
Forebrain
Lower limb bud
Upper limb bud
Tail
Neural tube forming
Somites
Head (growth accelerated)
Retinal pigment
Future external ear
Umbilical cord
Hand plate
Foot plate

(a) (b)

Figure 28.17 The developing human.
(*a*) Four weeks; (*b*) seven weeks; (*c*) three months; and (*d*) four months.

By the end of the fourth week, the developmental scenario is far advanced, although most women are not yet aware that they are pregnant. This is a crucial time in development because the proper course of events can be interrupted easily. For example, alcohol use by pregnant women during the first months of pregnancy is one of the leading causes of birth defects, producing **fetal alcohol syndrome,** in which the baby is born with a deformed face and often severe mental retardation. One in 250 newborns in the United States is affected with fetal alcohol syndrome. Also, most spontaneous abortions (miscarriages) occur during this period.

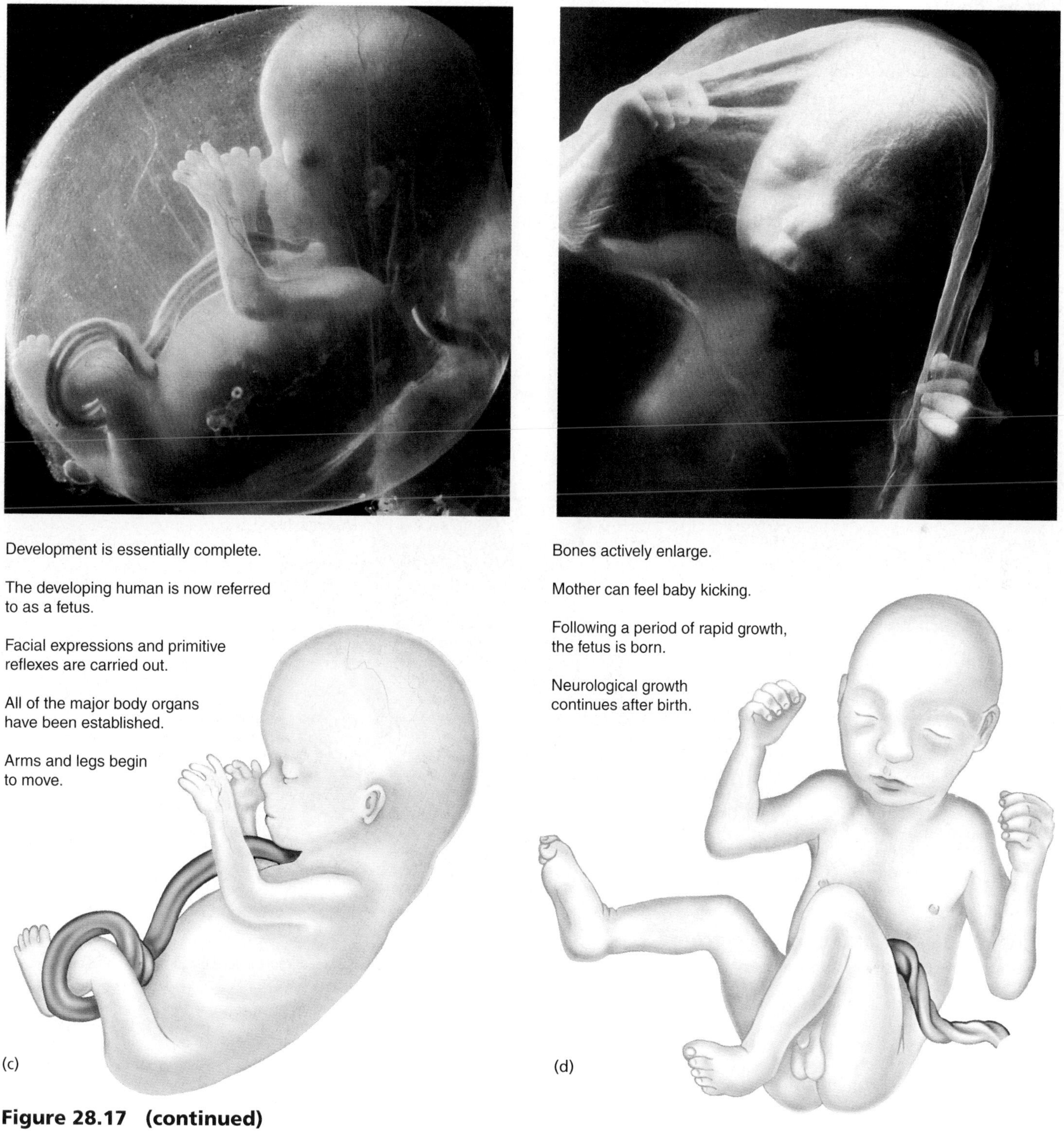

Development is essentially complete.
The developing human is now referred to as a fetus.
Facial expressions and primitive reflexes are carried out.
All of the major body organs have been established.
Arms and legs begin to move.
(c)
Bones actively enlarge.
Mother can feel baby kicking.
Following a period of rapid growth, the fetus is born.
Neurological growth continues after birth.
(d)

Figure 28.17 (continued)

The Second Month: Morphogenesis

During the second month of pregnancy, **morphogenesis** takes place (figure 28.17*b*). The miniature limbs of the embryo assume their adult shapes. The arms, legs, knees, elbows, fingers, and toes can all be seen—as well as a short, bony tail. The bones of the embryonic tail, an evolutionary reminder of our past, later fuse to form the coccyx. Within the body cavity, the major internal organs are evident, including the liver and pancreas. By the end of the second month, the embryo has grown to about 25 millimeters in length—it is 1 inch long. It weighs perhaps a gram and is beginning to look distinctly human.

The Third Month: Completion of Development

Development of the embryo is essentially complete. From this point on, the developing human is referred to as a **fetus** rather than an embryo. What remains is essentially growth. The nervous system and sense organs develop during the third month. The embryo begins to show facial expressions and carries out primitive reflexes such as the startle reflex and sucking. By the end of the third month, all of the major organs of the body have been established and the arms and legs begin to move (figure 28.17*c*).

The Second Trimester: The Fetus Grows in Earnest

The second trimester is a time of growth. In the fourth (figure 28.17*d*) and fifth months of pregnancy, the fetus grows to about 175 millimeters in length (almost 7 in. long), with a body weight of about 225 grams. Bone formation occurs actively during the fourth month. During the fifth month, the head and body become covered with fine hair. This downy body hair, called **lanugo,** is another evolutionary relic and is lost later in development. By the end of the fourth month, the

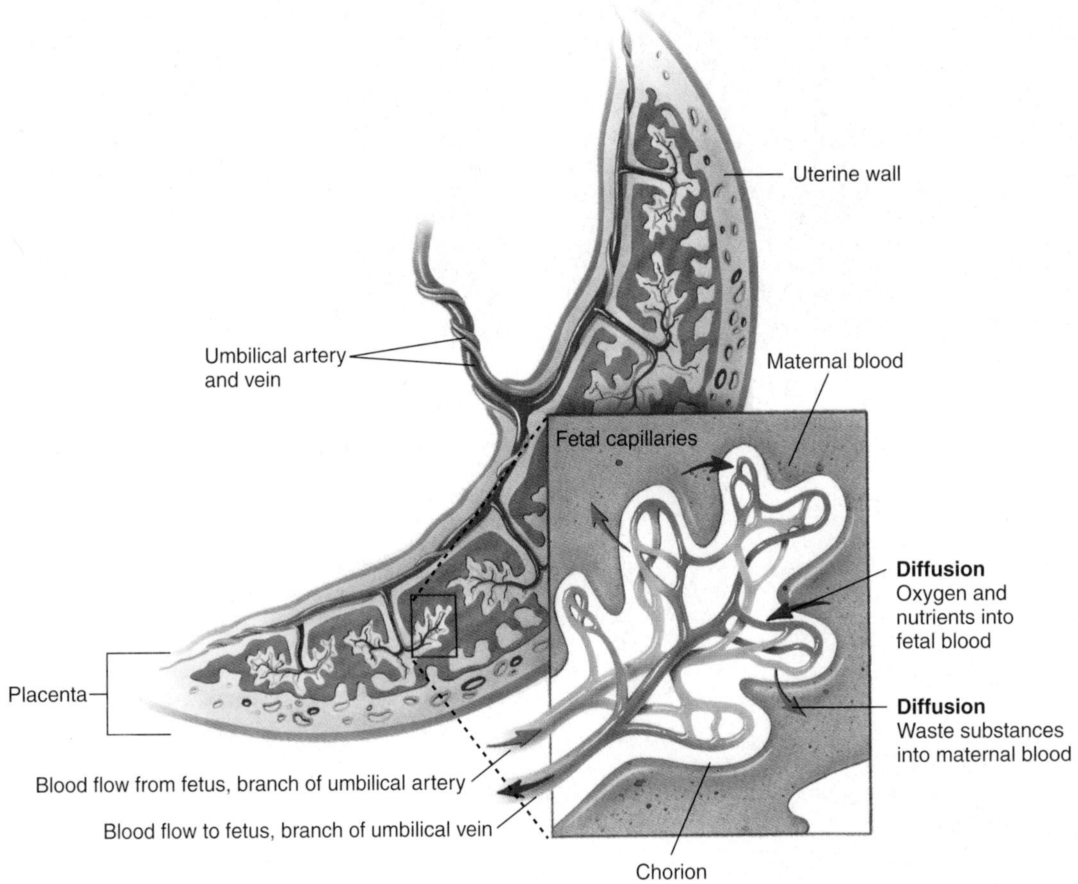

Uterine wall
Umbilical artery and vein
Maternal blood
Fetal capillaries
Diffusion
Oxygen and nutrients into fetal blood
Diffusion
Waste substances into maternal blood
Placenta
Blood flow from fetus, branch of umbilical artery
Blood flow to fetus, branch of umbilical vein
Chorion

Figure 28.18 Structure of the placenta.

Oxygen and nutrients enter the fetal blood from the maternal blood by diffusion. Waste substances enter the maternal blood from the fetal blood, also by diffusion.

mother can feel the baby kicking; by the end of the fifth month, she can hear its rapid heartbeat with a stethoscope.

In the sixth month, growth accelerates. By the end of the sixth month, the baby is over 0.3 meters (1 ft) long and weighs 0.6 kilograms (about 1.5 lb)—and most of its prebirth growth is still to come. At this stage, the fetus cannot yet survive outside the uterus without special medical intervention.

The Third Trimester: The Pace of Growth Accelerates

The third trimester is a period of rapid growth. In the seventh, eighth, and ninth months of pregnancy, the weight of the fetus more than doubles. This increase in bulk is not the only kind of growth that occurs. Most of the major nerve tracts are formed within the brain during this period, as are new brain cells.

All of this growth is fueled by nutrients provided by the mother's bloodstream, passing into the fetal blood supply within the placenta (figure 28.18). The undernourishment of the fetus by a malnourished mother can adversely affect this growth and result in severe retardation of the infant. Retardation resulting from fetal malnourishment is a severe problem in many underdeveloped countries, where poverty and hunger are common.

By the end of the third trimester, the neurological growth of the fetus is far from complete and, in fact, continues long after birth. But by this time the fetus is able to exist on its own. Why doesn't the fetus continue to develop within the uterus until its neurological development is complete? What's the rush to get out and be born? Because physical growth is continuing as well, and the fetus is about as large as it can get and still be delivered through the pelvis without damage to mother or child. As any woman who has had a baby can testify, it is a tight fit. Birth takes place as soon as the probability of survival is high. For better or worse, the infant is then a person.

Postnatal Development

Growth continues rapidly after birth (figure 28.19). Babies typically double their birth weight within a few months. Different organs grow at different rates, however. The reason that adult body proportions are different from infant ones is that different parts of the body grow or cease growing at different times. At birth, the developing nervous system is generating new nerve cells at an average rate of more than 250,000 per minute. Then, about six months after birth, this astonishing production of new neurons ceases permanently. Because both jaw and skull continue to grow at the same rate, the proportions of the head do not change after birth. That is why a young human fetus seems so incredibly adultlike.

Figure 28.19 A mother nursing her infant.
Infants grow very rapidly during the first year after birth. During this time, the nursing mother needs adequate nutrition to enable her to support her growing baby.

28.7 Most of the key events in fetal development occur early. Organs begin to form in the fourth week, and by the end of the second month the developing body looks distinctly human. While many women are not yet aware they are pregnant, development of the embryo is essentially complete. What remains is essentially growth.

28.8 Approaches to Contraception

Not all couples want to initiate a pregnancy every time they have sex, yet sexual intercourse may be a necessary and important part of their emotional lives together. The solution to this dilemma is to find a way to avoid reproduction without avoiding sexual intercourse, an approach that is commonly called **birth control** or contraception. Several different birth-control methods are currently available (figure 28.20). These methods differ from one another in their effectiveness and in their acceptability to different couples.

Abstinence

The simplest and most reliable way to avoid pregnancy is not to have sex at all. Of all methods of birth control, this is the most certain—and the most limiting, because it denies a couple the emotional support of a sexual relationship. A variant of this approach is to avoid sex only on the two days preceding and following ovulation, because this is the period during which successful fertilization is likely to occur. The rest of the sexual cycle is relatively "safe" for intercourse. This approach, called the **rhythm method,** is satisfactory in principle but difficult in application because ovulation is not easy to predict and may occur unexpectedly.

Prevention of Egg Maturation

Since about 1960, a widespread form of birth control in the United States has been the daily ingestion of hormones, or **birth-control pills.** These pills contain estrogen and progesterone. These hormones shut down production of the pituitary hormones FSH and LH, fooling the body into acting as if ovulation had already occurred, when in fact it has not. The ovarian follicles do not ripen in the absence of FSH, and ovulation does not occur in the absence of LH. Taken correctly, birth-control pills provide a very effective means of birth control, with failure rates of less than 10%.

Prevention of Embryo Implantation

The insertion of a coil or other irregularly shaped object into the uterus is an effective means of birth control because the irritation in the uterus prevents the implantation of the descending embryo within the uterine wall. Such **intrauterine devices (IUDs)** are very effective, with a failure rate of less than 4%. Their high degree of effectiveness, like surgically implanted hormones, undoubtedly reflects their being "no-brainers"—once they are inserted, they can be forgotten.

Sperm Blockage

If sperm is not delivered to the uterus, fertilization cannot occur. One way to prevent the delivery of sperm is to encase the penis within a thin rubber bag, or **condom.** In principle, this method is easy to apply and foolproof, but in practice, it proves to be less effective than you might expect, with a failure rate of up to 15%. A second way to prevent the entry of sperm is to cover the cervix with a rubber dome called a **diaphragm,** inserted immediately before intercourse. Because the dimensions of individual cervices vary, diaphragms must be fitted by a physician. Failure rates average 20%.

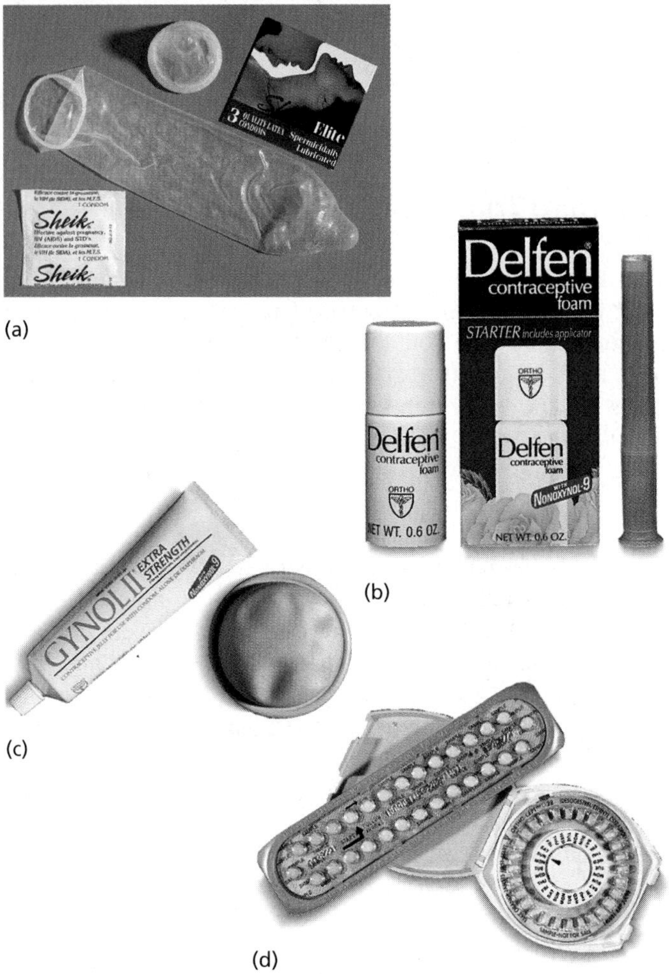
(a)
Sheik
Elite
(b)
Delfen
contraceptive foam
STARTER includes applicator
ORTHO
NET WT. 0.6 OZ
NONOXYNOL-9
(c)
GYNOL II
EXTRA STRENGTH
(d)

Figure 28.20 Four common birth-control methods.

(*a*) Condom; (*b*) foam; (*c*) diaphragm and spermicidal jelly; and (*d*) oral contraceptives.

Sperm Destruction

A third general approach to birth control is to remove or destroy the sperm after ejaculation. Sperm can be destroyed within the vagina with **spermicidal jellies, suppositories,** and **foams.** These require application immediately before intercourse. The failure rate varies widely, from 3% to 22%.

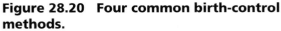

28.8 A variety of birth-control methods are available, many of them quite effective.

CONCEPT REVIEW

1. Select the *incorrect* statement about male gametes.
   a. Sperm begin developing before puberty.
   b. Sperm do not develop successfully at 37°C.
   c. Sperm are made in the seminiferous tubules.
   d. Sperm are capable of movement.

2. The spongy tissue in the penis permits
   a. sperm cell production.
   b. sperm cell storage.
   c. erection.
   d. urination.

3. The most common site of fertilization is the
   a. ovary.
   b. vagina.
   c. fallopian tube.
   d. oocyte.

4. Ovulation is triggered by the increase of
   a. estrogen.
   b. progesterone.
   c. FSH.
   d. LH.

5. After ovulation, the ________ begin(s) to secrete progesterone to prepare the uterus for fertilization.
   a. ovaries
   b. corpus luteum
   c. pituitary
   d. hypothalamus

6. During the period of cleavage, the embryo is a tightly packed mass of cells called a
   a. morula.
   b. blastocyst.
   c. blastomere.
   d. gastrula.

7. Segmented blocks of tissue called ________ on either side of the developing notochord give rise to the muscles, vertebrae, and connective tissue.
   a. mesoderm
   b. somites
   c. neural grooves
   d. spines

8. Which of the following does *not* occur during the first three months of pregnancy?
   a. organogenesis
   b. formation of the placenta
   c. formation of the major nerve tracts within the brain
   d. development of the limb buds

9. By the end of the third trimester, the neurological growth of the fetus is complete.
   a. true
   b. false

10. Which of the following birth-control methods involves sperm blockage?
    a. abstinence
    b. use of a condom
    c. birth-control pill
    d. use of a spermicide

11. The ________ is a muscular tube about 7 centimeters long that leads to the uterus.

12. ________ is the hormone that stimulates milk production after birth, and ________ is the hormone that initiates milk release once the infant suckles at the breast.

13. The ________ is the site of invagination of cells during gastrulation of the embryo.

14. The ________ is the primary layer that forms neural tissue.

15. The ________ is the primary layer that forms the digestive system.

16. The ________ is the primary layer that forms connective tissue.

CHALLENGE YOURSELF

1. Relatively few kinds of animals have both male and female sex organs in the same animal, while most plants do. Propose an explanation for this.

2. How is the location of the testes in the scrotum an advantage for sperm cell production?

3. In mammals, female sexual development is under *negative* control; that is, the absence of the *SRY* gene product and the absence of testosterone result in the development of female structures. Negative control makes it possible for embryos of either sex to develop inside female parents, which have no *SRY* gene product and very low levels of testosterone. If female sexual development were triggered by *positive* controls and male sexual development were under negative controls, how might the development of embryos inside female parents be affected?

28 eBRIDGE

Reinforcing Key Points

Vertebrate Reproduction

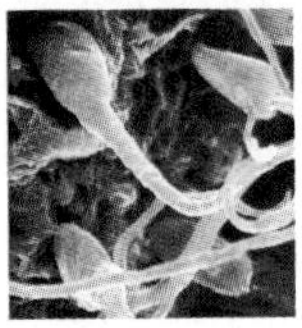

The Human Reproductive System

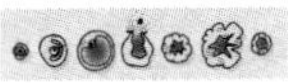

The Female Reproductive Cycle

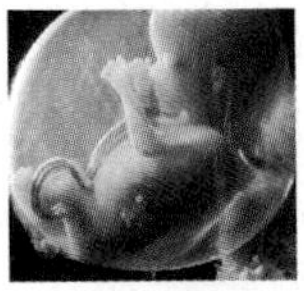

The Course of Development

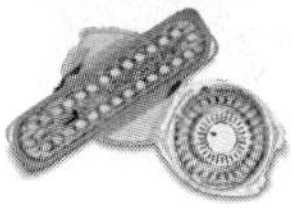

Birth Control

Electronic Learning

Visual Learning

Animations

Nine Animations

Art Labeling Activities

Eight Art Labeling Activities

Helping You Learn

Two Exercises

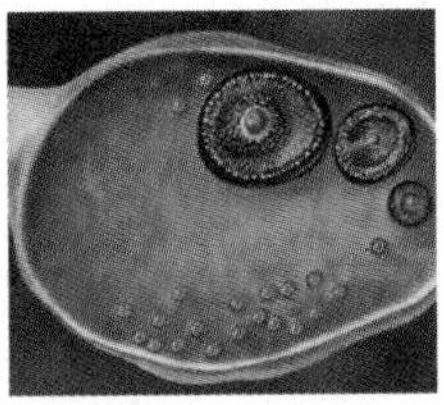

Explorations

Cell-Cell Interactions

In this exercise, you can explore the role of cell-surface receptors and key intracellular cascades in cell-cell communication, a critical aspect of embryonic development. By manipulating the efficiency of various components, you can evaluate how changes in these elements affect communication between cells.

Start
Stop
MENU

Author's Corner

Stem Cells and Cloning. One of the ethically most controversial and scientifically promising areas of research in the new century concerns embryonic stem cells and their potential use to repair damaged tissues. Therapeutic cloning, where the embryonic stem cells are isolated from a clone of the patient's own tissue, is particularly exciting.

1. Stem cell research is a new frontier full of ethical and political questions.
2. Scientists use transplanted stem cells to reverse juvenile diabetes in mice.
3. Ethics and science collide over the therapeutic cloning of human stem cells.
4. Cloning humans is not going to work until scientists solve a key problem.
5. Newspaper reports of genetically modified humans are misleading.

Virtual Classroom

Cloning and Stem Cells

On July 5, 1996, the first mammal, a sheep dubbed Dolly, was cloned from an adult fully differentiated cell. A nucleus was removed from a breast cell and placed in an enucleated egg cell. The resulting cell began to divide, and after a week the early embryo—a cluster of a few hundred embryonic stem cells in a sheath of protective cells—was implanted into a surrogate mother sheep and allowed to come to term. While other mammals have been cloned since, it is difficult to obtain healthy clones because the donated DNA requires time-consuming "programming" from the egg cytoplasm before division starts, programming that a cloned nucleus doesn't get. When this problem is solved, cloning offers a powerful opportunity to cure disorders or conditions involving lost tissue, for which there is now no effective therapy, such as spinal injury, heart muscle loss from repeated heart attacks, juvenile diabetes, and Parkinsons. These conditions have been cured in mice using embryonic stem cells, but the mice had been engineered to lack functional immune systems. Therapeutic cloning using the nucleus from one of a patient's own cells would produce stem cells that could cure the patient, without immunological complications.

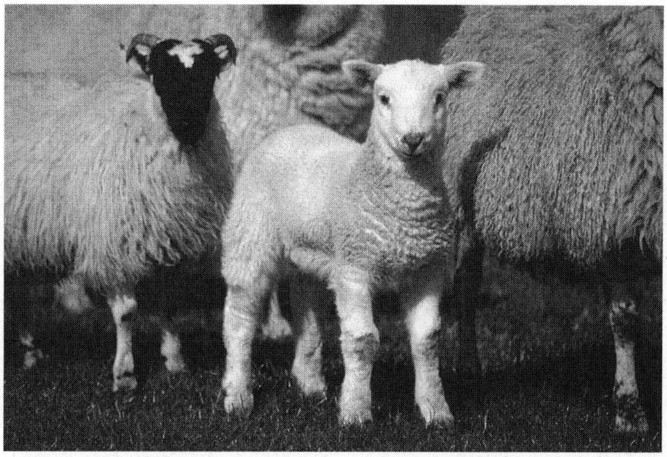

Virtual Lab

Are Pollutants Affecting the Sexual Development of Florida's Alligators?

Alligators are among the most interesting animals for a biologist to study, because their ecology is closely tied to the environment. They spend their lives sloshing around in an aquatic environment. Florida's alligators live in the many lakes that pepper the state, and, living in these lakes, they are exposed all their lives to whatever chemicals happen to enter the lakewater, introduced by chemical spills, industrial wastes, and agricultural runoff. A class of pollutant chemicals called endocrine disrupters have leaked into some of Florida's lakes, and their effects on the wildlife in the lakes could be devastating. The sexual development of male alligators is largely dependent on the androgenic sex hormone testosterone and its derivatives. Endocrine disrupters can block the actions of testosterone, and so interfere with normal sexual development in alligators.

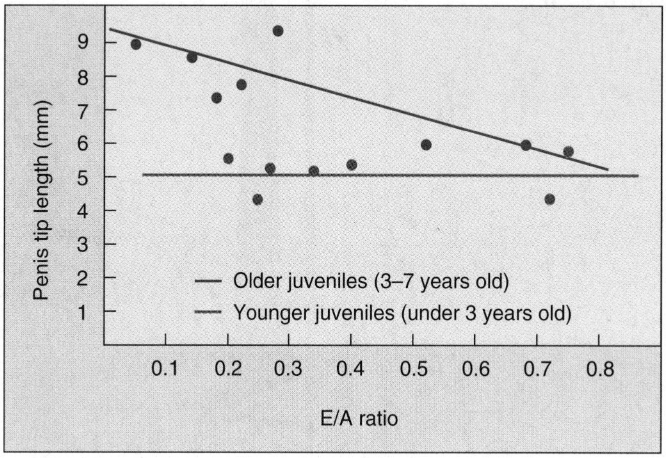
Penis tip length (mm)
1
2
3
4
5
6
7
8
9
Older juveniles (3–7 years old)
Younger juveniles (under 3 years old)
0.1
0.2
0.3
0.4
0.5
0.6
0.7
0.8
E/A ratio

Louis Guillette of the University of Florida, Gainesville, and his students have been studying Florida's alligators to assess the degree to which endocrine disrupters and other chemical pollutants affect the sexual development of male alligators.

Quizzes

Further Reading

Essential Study Partner

Links

BioCourse.com

CHAPTER 29

Ecosystems

THE FAR SIDE® By GARY LARSON
© 1983 FarWorks, Inc. All Rights Reserved/Dist. by Creators Syndicate
Larson
"And see this ring right here, Jimmy? ... That's another time when this old fellow miraculously survived some big forest fire."

CHAPTER 29

Ecosystems

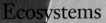

CHAPTER OVERVIEW

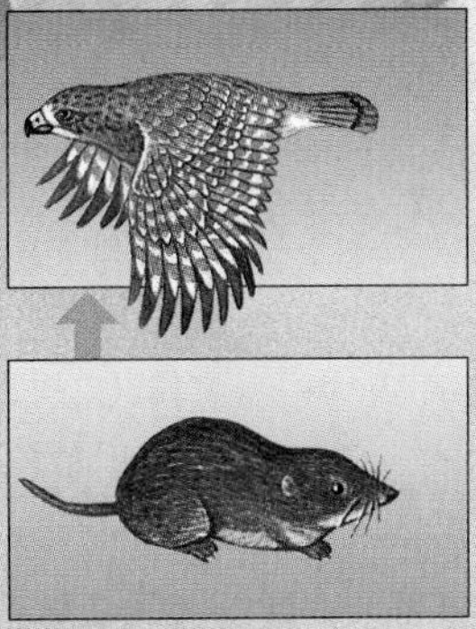

The Energy in Ecosystems

29.1 Trophic Levels

29.2 Energy Flows Through Ecosystems

29.3 Ecological Pyramids

- Ecology is the study of how organisms fit into and interact with their environment.
- An ecosystem is a largely self-sustaining group of organisms and the minerals, water, and weather that make up the habitat.
- Ecologists categorize organisms into trophic or feeding levels based on how they obtain their energy.
- The complex flow of energy among trophic levels is called a food web.

Materials Cycle Within Ecosystems

29.4 The Water Cycle

29.5 The Carbon Cycle

29.6 The Nitrogen Cycle

29.7 The Phosphorus Cycle

- Water cycles either by evaporation and condensation or by absorption and transpiration.
- The breaking up of molecules, the burning of wood and fossil fuels, and erosion are ways in which carbon cycles.
- Although nitrogen is abundant in organisms and the atmosphere, much of life relies on the ability of some bacteria to fix nitrogen.

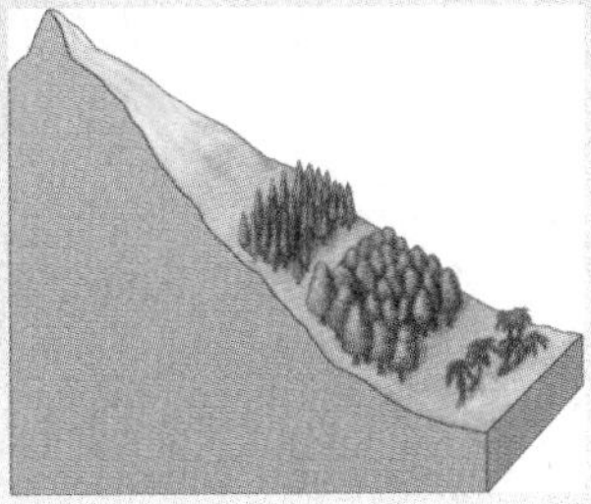

How Weather Shapes Ecosystems

29.8 The Sun and Atmospheric Circulation

29.9 Latitude and Elevation

29.10 Patterns of Circulation in the Ocean

- The climate of a particular area is affected by the intensity of the sun's rays, air currents, elevation, and ocean currents.
- A rain shadow effect is caused by mountains, which force winds upward, causing them to cool and release their moisture on the windward side of mountains.

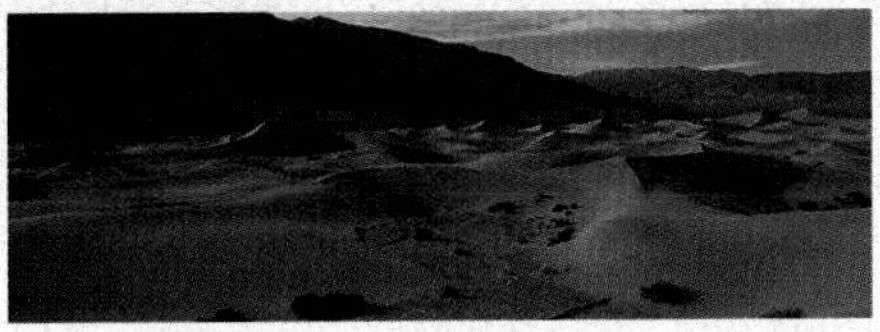

Major Kinds of Ecosystems

29.11 Ocean Ecosystems

29.12 Freshwater Ecosystems

29.13 Land Ecosystems

- Ocean ecosystems consist of highly diverse coastal areas, open waters that contain plankton, and deep waters where little light penetrates.
- A biome is a climatically defined assemblage of organisms that occurs over a wide area.

The ecosystem is the most complex level of biological organization. The biosphere includes all the ecosystems on earth, from the profusion of life in the tropical rain forests to the photosynthetic phytoplankton in the world's oceans. The earth is a closed system with respect to chemicals but an open system in terms of energy. Collectively, the organisms in ecosystems regulate the capture and expenditure of energy and the cycling of chemicals. As we will see in this chapter, all organisms, including humans, depend on the ability of other organisms—plants, algae, and some bacteria—to recycle the basic components of life.

29.1 Trophic Levels

What Is an Ecosystem?

Ecology is the study of the interactions of living organisms with one another and with their physical environment (soil, water, weather, and so on). Ecologists, the scientists who study ecology, view the world as a patchwork quilt of different environments, all bordering on and interacting with one another. Consider for a moment a patch of forest, the sort of place a deer might live. Ecologists call the collection of creatures that live in a particular place a **community**—all the animals, plants, fungi, and microorganisms that live together in a forest, for example, are the forest community. Ecologists call the place where a community lives its **habitat**—the soil, and the water flowing through it, are key components of the forest habitat. The sum of these two, community and habitat, is an ecological system, or **ecosystem.** An ecosystem is a largely self-sustaining collection of organisms and their physical environment.

The Path of Energy: Who Eats Whom in Ecosystems

Energy flows into the biological world from the sun, which shines a constant beam of light on our earth. Life exists on earth because some of that continual flow of light energy can be captured and transformed into chemical energy through the process of photosynthesis and used to make organic molecules. These organic molecules are what we call food. Living organisms use the energy in food to make new materials for growth, to repair damaged tissues, to reproduce, and to do myriad other things that require energy, like turning the pages of this text.

You can think of all the organisms in an ecosystem as chemical machines fueled by energy captured in photosynthesis. The organisms that first capture the energy, the **producers,** are plants (and some bacteria and algae), which produce their own energy-storing molecules by carrying out photosynthesis. All other organisms in an ecosystem are **consumers,** obtaining their energy-storing molecules by consuming plants or other animals. Ecologists assign every organism in an ecosystem to a trophic, or feeding, level, depending on the source of its energy. A **trophic level** is composed of those organisms within an ecosystem whose source of energy is the same number of consumption "steps" away from the sun. Thus, a plant's trophic level is 1, while animals that graze on plants are in trophic level 2, and animals that eat these grazers are in trophic level 3 (figure 29.1). Higher trophic levels exist for animals that eat higher on the food chain. Food energy passes through an ecosystem from one trophic level to another. When the path is a simple linear progression, like the links of a chain, it is called a **food chain.** In most ecosystems, however, the path of energy is not a simple linear one, because individual animals often feed at several trophic levels. This creates a complicated path of energy flow called a **food web** (figure 29.2).

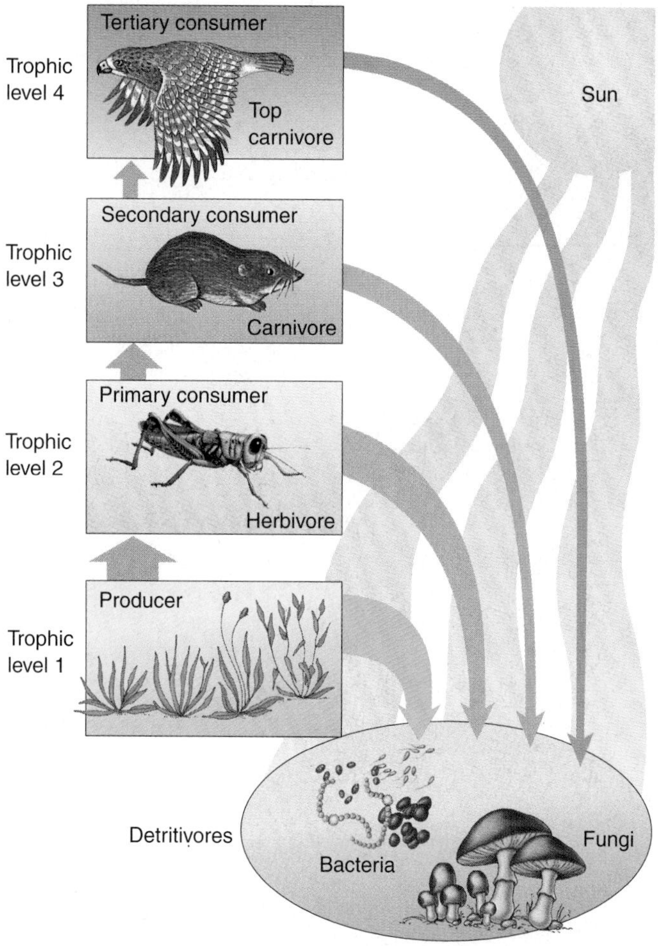
Tertiary consumer
Trophic level 4
Top carnivore
Sun
Secondary consumer
Trophic level 3
Carnivore
Primary consumer
Trophic level 2
Herbivore
Producer
Trophic level 1
Detritivores
Bacteria
Fungi

Figure 29.1 Trophic levels within an ecosystem.
Ecologists assign all the members of a community to various trophic levels based on feeding relationships. Producers, such as photosynthetic plants, obtain their energy directly from the sun and are assigned to trophic level 1. Animals that eat plants (herbivores) are at trophic level 2. Animals that eat plant-eating animals (carnivores) are at trophic level 3 and higher. Detritivores make use of all trophic levels for food.

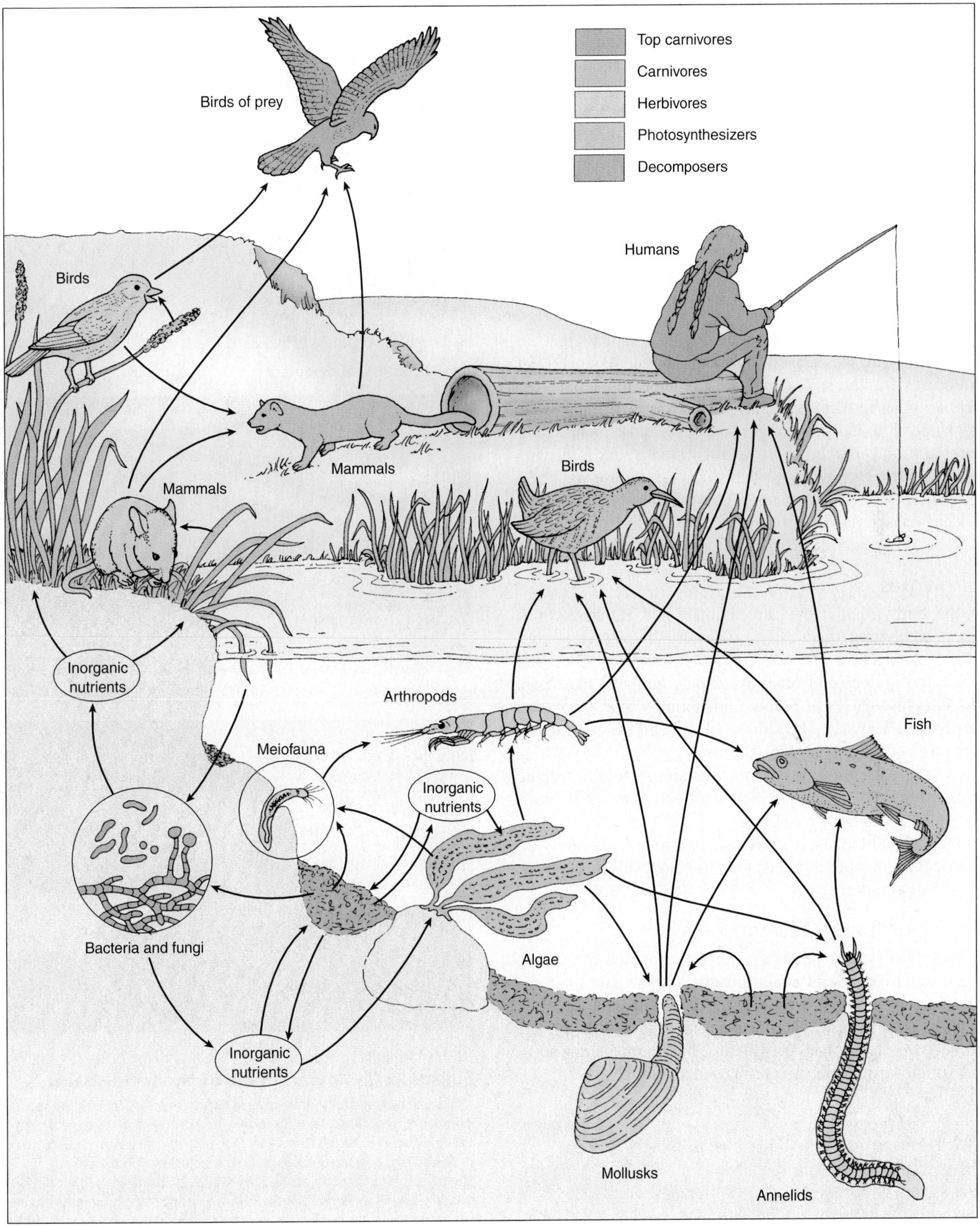
Top carnivores
Carnivores
Herbivores
Photosynthesizers
Decomposers
Birds of prey
Humans
Birds
Mammals
Mammals
Birds
Inorganic nutrients
Arthropods
Fish
Meiofauna
Inorganic nutrients
Bacteria and fungi
Algae
Inorganic nutrients
Mollusks
Annelids

Figure 29.2 A food web.

A food web is much more complicated than a linear food chain. The path of energy passes from one trophic level to another and back again in complex ways.

Producers

The lowest trophic level of any ecosystem is occupied by the producers—green plants in most land ecosystems (and, usually, algae in freshwater). Plants use the energy of the sun to build energy-rich sugar molecules. They also absorb nitrogen and other key substances from the air and soil and build them into biological molecules. It is important to realize that plants consume as well as produce. The roots of a plant, for example, do not carry out photosynthesis—there is no sunlight underground. Roots obtain their energy the same way you do, by using energy-storing molecules produced elsewhere (in this case, in the leaves of the plant).

Herbivores

At the second trophic level are **herbivores,** animals that eat plants. They are the *primary consumers* of ecosystems. Deer and horses are herbivores, and so are rhinoceroses, chickens (primarily herbivores), and caterpillars. Most herbivores rely on "helpers" to aid in the digestion of cellulose, a structural material found in plants. A cow, for instance, has a thriving colony of bacteria in its gut that digests cellulose for it. So does a termite. Humans cannot digest cellulose, because we lack these bacteria—that is why a cow can live on a diet of grass and you cannot.

Carnivores

At the third trophic level are animals that eat herbivores, called **carnivores** (meat eaters). They are the *secondary consumers* of ecosystems. Tigers and wolves are carnivores, and so are mosquitoes and blue jays. Some animals, like bears and humans, eat both plants and animals and are called **omnivores.** They use the simple sugars and starches stored in plants as food and not the cellulose.

Many complex ecosystems contain a fourth trophic level, composed of animals that consume other carnivores. They are called tertiary consumers, or top carnivores. A weasel that eats a blue jay is a tertiary consumer. Only rarely do ecosystems contain more than four trophic levels, for reasons we will discuss later.

Detritivores and Decomposers

In every ecosystem there is a special class of consumers called **detritivores** and **decomposers.** They obtain their energy from the organic wastes and dead bodies that are produced at all trophic levels. Bacteria and fungi are the principal decomposers in land ecosystems, but worms, arthropods, and vultures can be detritivores as well (figure 29.3).

> **29.1** Energy moves through ecological systems from photosynthesizers, which capture it from sunlight, to herbivores, which eat the photosynthesizers, and then to carnivores, which eat the herbivores, and finally to detritivores, which consume the dead bodies of all the others.

(a) Herbivores and producers

(b) Detritivore

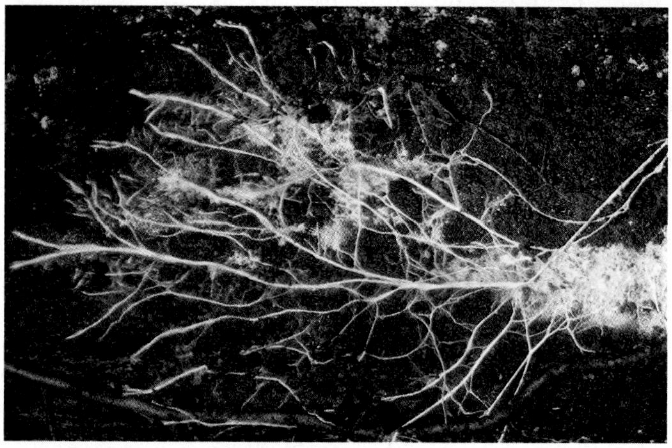

(c) Decomposer

Figure 29.3 Lower levels of the food chain.

(a) The East African grasslands are covered by a dense cover of grasses, with interspersed trees; these plants are primary producers, capturing the energy from the sun. Grazing mammals obtain their food from the plants and may in turn be consumed by predators, such as lions. (*b*) This crab, *Geocarcinus quadratus,* photographed on the beach at Mazatlán, Mexico, is a detritivore, playing the same role that vultures and similar animals do in other ecosystems. (*c*) Fungi, such as the basidiomycete whose mycelium is shown here growing through the soil in Costa Rica, are, together with bacteria, the primary decomposers of terrestrial ecosystems.

29.2 Energy Flows Through Ecosystems

How much energy passes through an ecosystem? **Primary productivity** is the total amount of light energy converted to organic compounds in a given area per unit of time. An ecosystem's **net primary productivity** is the total amount of energy fixed by photosynthesis per unit of time, minus that which is expended by the metabolic activities of ecosystem organisms. The total weight of all ecosystem organisms, called the ecosystem's **biomass,** increases as a result of the ecosystem's net productivity. Some ecosystems, such as cornfields or cattail swamps, have a high net primary productivity. Others, such as tropical rain forests, also have a relatively high net primary productivity, but a rain forest has a much larger biomass than a cornfield. Consequently, a rain forest's net primary productivity is much lower in relation to its biomass.

When a plant uses the energy from sunlight to make structural molecules such as cellulose, it loses a lot of the energy as heat. In fact, only about half of the energy captured by the plant ends up stored in its molecules. The other half of the energy is lost. This is the first of many such losses as the energy passes through the ecosystem. The amount of energy that ends up in the herbivore's body is approximately an order of magnitude less than the energy present in the plant molecules it eats (figure 29.4). Similarly, when a carnivore eats the herbivore, an order of magnitude is lost from the amount of energy present in the herbivore's molecules. Food chains generally consist of only three or four steps. So much energy is lost at each step that very little usable energy remains in the system after it has been incorporated into the bodies of organisms at four successive trophic levels.

Lamont Cole of Cornell University studied the flow of energy in a freshwater ecosystem in Cayuga Lake in upstate New York. He calculated that about 150 of each 1,000 calories of potential energy fixed by algae and cyanobacteria are transferred into the bodies of small heterotrophs. Of these, about 30 calories are incorporated into the bodies of smelt, the principal secondary consumers of the system. If humans eat the smelt, they gain about 6 of the 1,000 calories that originally entered the system. If trout eat the smelt and humans eat the trout, humans gain only about 1.2 calories of the 1,000 (figure 29.5).

29.2 As energy passes through an ecosystem, much is lost as heat at each stage of a food chain. As a result, most food chains are short.

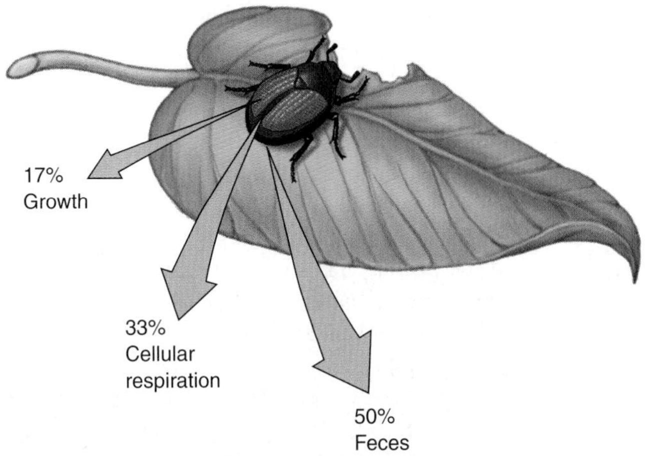
17%
Growth
33%
Cellular
respiration
50%
Feces

Figure 29.4 How heterotrophs use food energy.
A heterotroph assimilates only a fraction of the energy it consumes. For example, if a "bite" is composed of 500 Joules of energy (1 Joule = 0.239 calories), about 50%, 250 J, is lost in feces, about 33%, 165 J, is used to fuel cellular respiration, and about 17%, 85 J, is converted into consumer biomass. Only this 85 J is available to the next trophic level.

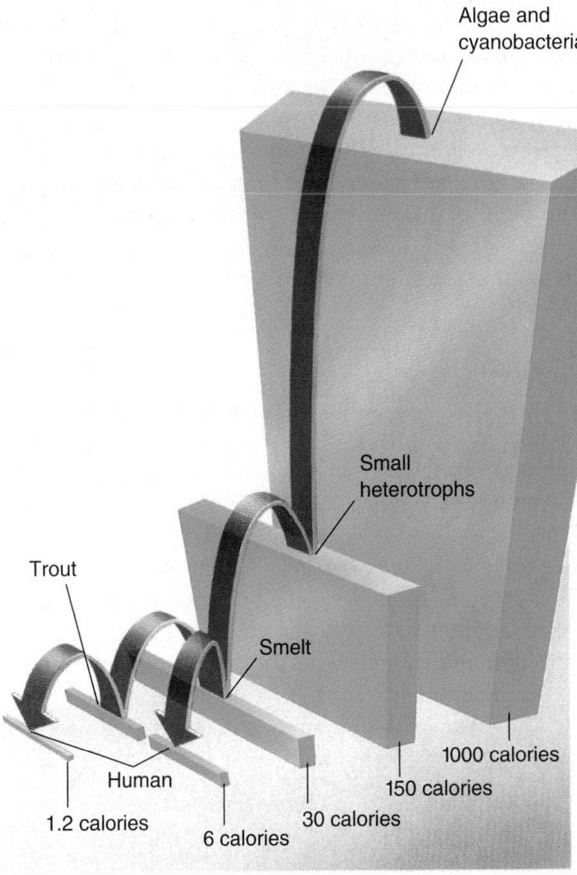
Algae and
cyanobacteria
Small
heterotrophs
Trout
Smelt
Human
1000 calories
150 calories
30 calories
6 calories
1.2 calories

Figure 29.5 Energy loss in an ecosystem.
In a classic study of Cayuga Lake in New York, the path of energy was measured precisely at all points in the food web. This diagram summarizes the results. Photosynthetic algae and cyanobacteria fix the energy of the sun; animal plankton (small heterotrophs) feed on them; and both are consumed by smelt. The smelt are eaten by trout, with about a 10-fold loss in fixed energy.

29.3 Ecological Pyramids

A plant fixes about 1% of the sun's energy that falls on its green parts. The successive members of a food chain, in turn, process into their own bodies about 10% of the energy available in the organisms on which they feed. For this reason, there are generally far more individuals at the lower trophic levels of any ecosystem than at the higher levels. Similarly, the biomass of the primary producers present in a given ecosystem is greater than the biomass of the primary consumers, with successive trophic levels having a lower and lower biomass and correspondingly less potential energy. Larger animals are characteristically members of the higher trophic levels. To some extent, they must be larger to be able to capture enough prey in the lower trophic levels.

These relationships, if shown diagrammatically, appear as pyramids (figure 29.6). We can speak of "pyramids of biomass," "pyramids of energy," "pyramids of number," and so forth, as characteristic of ecosystems.

Inverted Pyramids

Some aquatic ecosystems have inverted biomass pyramids. For example, in a planktonic ecosystem—dominated by small organisms floating in water—the turnover of photosynthetic phytoplankton at the lowest level is very rapid, with zooplankton consuming phytoplankton so quickly that the phytoplankton (the producers at the base of the food chain) can never develop a large population size. Because the phytoplankton reproduce very rapidly, the community can support a population of heterotrophs that is larger in biomass and more numerous than the phytoplankton (see figure 29.6*b*).

Top Carnivores

The loss of energy that occurs at each trophic level places a limit on how many top-level carnivores a community can support. As we have seen, only about one-thousandth of the energy captured by photosynthesis passes all the way through a three-stage food chain to a tertiary consumer such as a snake or hawk. This explains why there are no predators that subsist on lions or eagles—the biomass of these animals is simply insufficient to support another trophic level.

In the pyramid of numbers, top-level predators tend to be fairly large animals. Thus, the small residual biomass available at the top of the pyramid is concentrated in a relatively small number of individuals.

29.3 Because energy is lost at every step of a food chain, the biomass of primary producers (photosynthesizers) tends to be greater than that of the herbivores that consume them, and herbivore biomass tends to be greater than the biomass of the predators that consume them.

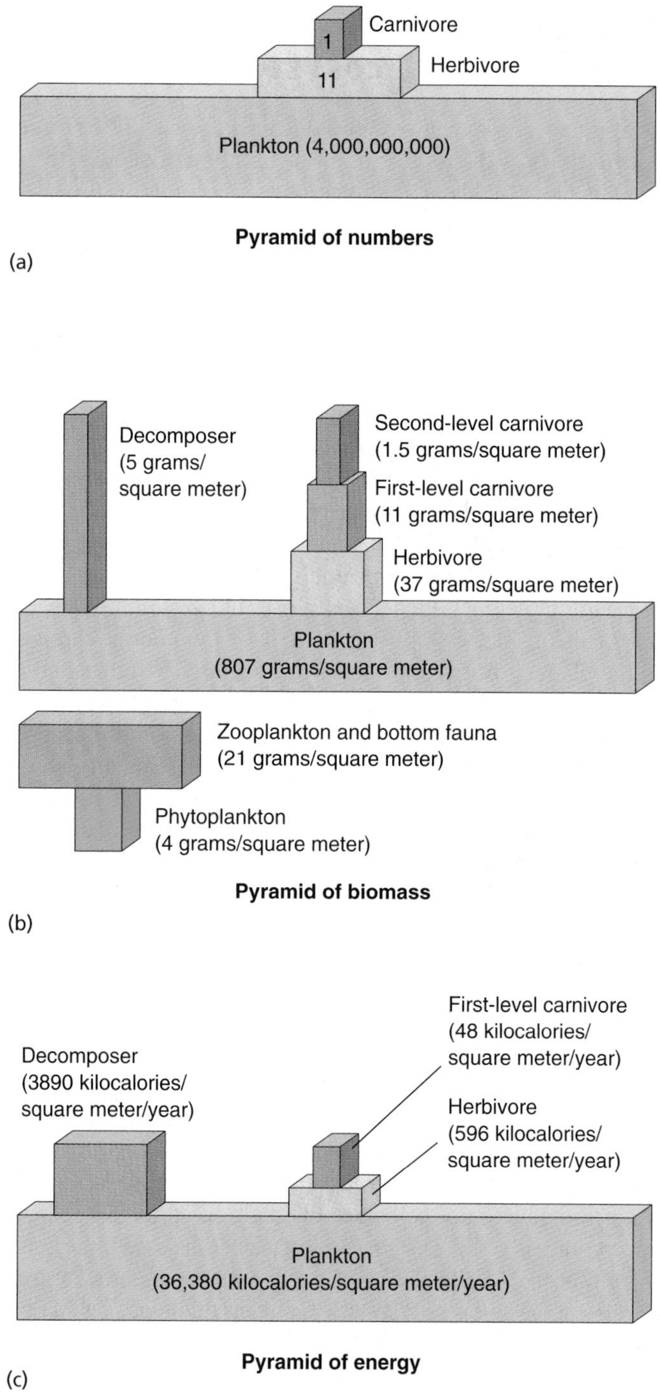

1
Carnivore
11
Herbivore
Plankton (4,000,000,000)
Pyramid of numbers
(a)
Decomposer
(5 grams/
square meter)
Second-level carnivore
(1.5 grams/square meter)
First-level carnivore
(11 grams/square meter)
Herbivore
(37 grams/square meter)
Plankton
(807 grams/square meter)
Zooplankton and bottom fauna
(21 grams/square meter)
Phytoplankton
(4 grams/square meter)
Pyramid of biomass
(b)
First-level carnivore
(48 kilocalories/
square meter/year)
Decomposer
(3890 kilocalories/
square meter/year)
Herbivore
(596 kilocalories/
square meter/year)
Plankton
(36,380 kilocalories/square meter/year)
Pyramid of energy
(c)

Figure 29.6 Ecological pyramids.
Ecological pyramids measure different characteristics of each trophic level. (*a*) Pyramid of numbers. (*b*) Pyramids of biomass, both normal (*top*) and inverted (*bottom*). (*c*) Pyramid of energy.

29.4 The Water Cycle

Unlike energy, which flows through the earth's ecosystems in one direction (from the sun to producers to consumers), the physical components of ecosystems are passed around and reused within ecosystems. Ecologists speak of such constant reuse as recycling or, more commonly, **cycling.** Materials that are constantly recycled include all the inorganic (noncarbon) chemicals that make up the soil, water, and air. While many are important, the proper cycling of three materials is particularly critical to the health of any ecosystem: water, carbon, and soil nutrients (nitrogen plus phosphorus).

The paths of water, carbon, and soil nutrients as they pass from the environment to living organisms and back form closed circles, or cycles. In each cycle, the inorganic substance resides for a time in an organism and then returns to the nonliving environment.

Of all the nonliving components of an ecosystem, water has the greatest influence on the living portion. The availability of water in large measure determines the biological richness of an ecosystem—how many different kinds of creatures live there and how many of each.

Water cycles within ecosystems in two ways: the environmental water cycle and the organismic water cycle.

The Environmental Water Cycle

In the environmental water cycle, water vapor in the atmosphere condenses and falls to the earth's surface as rain or snow. Heated there by the sun, it reenters the atmosphere by **evaporation** from lakes, rivers, and oceans (figure 29.7).

The Organismic Water Cycle

In the organismic water cycle, surface water does not return directly to the atmosphere. Instead, it is taken up by the roots of plants. After passing through the plant, the water reenters the atmosphere through tiny openings in the leaves, evaporating from their surface. This evaporation from leaf surfaces is called **transpiration.** Transpiration is also driven by the sun: the sun's heat creates wind currents that draw moisture from the plant by passing air over the leaves.

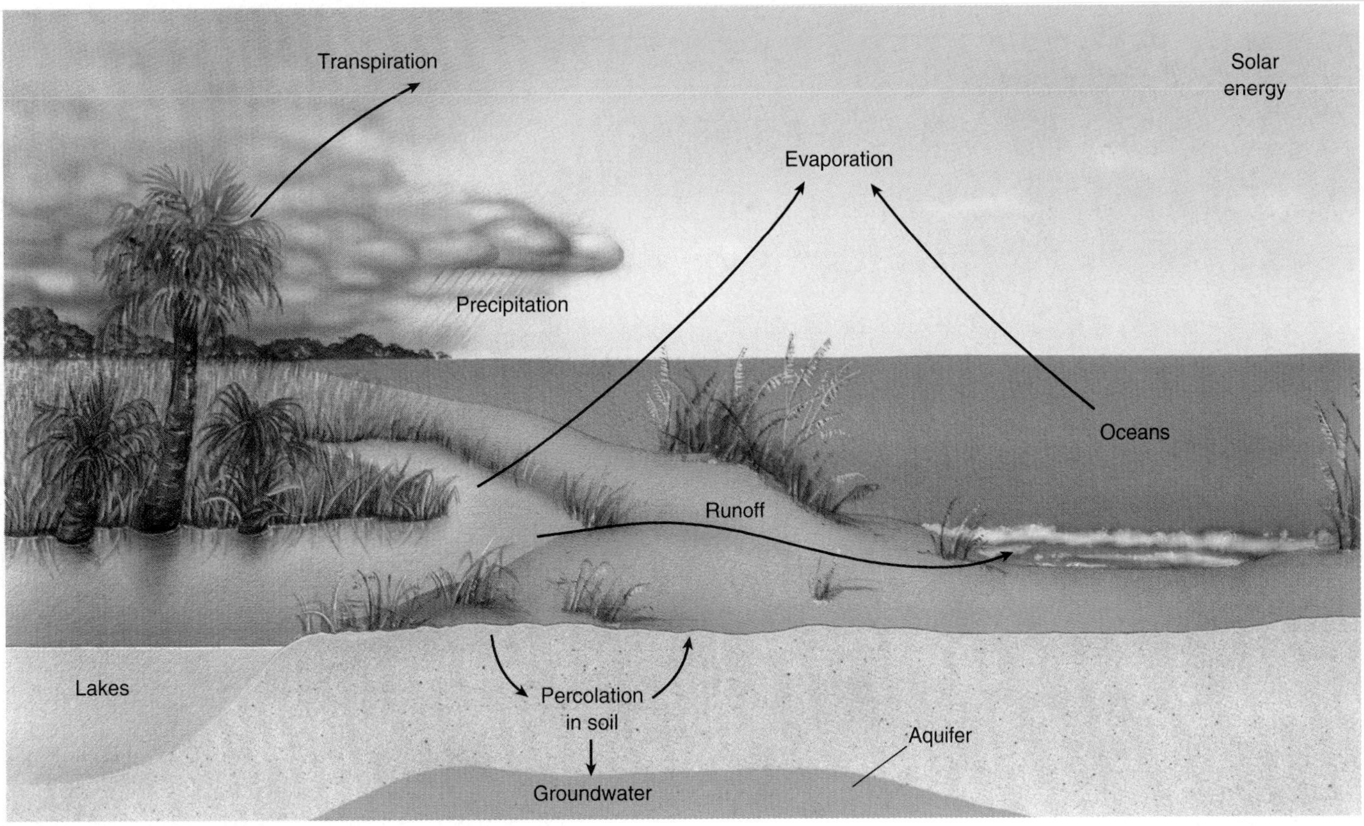
Transpiration
Solar energy
Evaporation
Precipitation
Oceans
Runoff
Lakes
Percolation in soil
Aquifer
Groundwater

Figure 29.7 The environmental water cycle.

Precipitation on land eventually makes its way to the ocean via groundwater, lakes, and finally, rivers. Solar energy causes evaporation, adding water to the sky. Plants give off excess water through transpiration, also adding water to the atmosphere. Atmospheric water falls as rain or snow over land and oceans, completing the water cycle.

Breaking the Cycle

In very dense forest ecosystems, such as tropical rain forests, more than 90% of the moisture in the ecosystem is taken up by plants and then transpired back into the air. Because so many plants in a rain forest are doing this, the vegetation is the primary source of local rainfall. In a very real sense, these plants create their own rain: the moisture that travels up from the plants into the atmosphere falls back to earth as rain.

Where forests are cut down, the organismic water cycle is broken, and moisture is not returned to the atmosphere. Water drains off to the sea instead of rising to the clouds and falling again on the forest. As early as the late 1700s, the great German explorer Alexander von Humbolt reported that stripping the trees from a tropical rain forest in Columbia prevented water from returning to the atmosphere and created a semiarid desert. It is a tragedy of our time that just such a transformation is occurring in many tropical areas, as tropical and temperate rain forests are being clear-cut or burned in the name of "development" (figure 29.8).

Groundwater

Much less obvious than the surface waters seen in streams, lakes, and ponds is the groundwater, which occurs in permeable, saturated, underground layers of rock, sand, and gravel called aquifers. In many areas, groundwater is the most important water reservoir; for example, in the United States, more than 96% of all freshwater is groundwater. Groundwater flows much more slowly than surface water, anywhere from a few millimeters to as much as a meter or so per day. In the United States, groundwater provides about 25% of the water used for all purposes and provides about 50% of the population with drinking water. Rural areas tend to depend on groundwater almost exclusively, and its use is growing at about twice the rate of surface water use.

Because of the greater rate at which groundwater is being used, the increasing chemical pollution of groundwater is a very serious problem. Pesticides, herbicides, and fertilizers are key sources of groundwater pollution. Because of the large volume of water, its slow rate of turnover, and its inaccessibility, removing pollutants from aquifers is virtually impossible.

29.4 Water cycles through ecosystems in the atmosphere via precipitation and evaporation, some of it passing through plants on the way.

Figure 29.8 Burning or clear-cutting forests breaks the water cycle.
The high density and large size of plants in a forest translate into great quantities of water being transpired to the atmosphere, creating rain over the forests. In this way rain forests perpetuate the wet climate that supports them. Tropical deforestation permanently alters the climate in these areas, creating arid zones.

29.5 The Carbon Cycle

The earth's atmosphere contains plentiful carbon, present as carbon dioxide (CO_2) gas. This carbon cycles between the atmosphere and living organisms, often being locked up for long periods of time in fossil organisms. The cycle is begun by plants that use CO_2 in photosynthesis to build organic molecules—in effect, they trap the carbon atoms of CO_2 within the living world. The carbon atoms are returned to the atmosphere's pool of CO_2 through respiration, combustion, and erosion (figure 29.9).

Respiration

Most of the organisms in ecosystems respire—that is, they extract energy from organic food molecules by stripping away the carbon atoms and combining them with oxygen to form CO_2. Plants respire, as do the herbivores, which eat the plants, and the carnivores, which eat the herbivores. All of these organisms use oxygen to extract energy from food, and CO_2 is what is left when they are done.

Combustion

A lot of carbon is tied up in wood, and it may stay trapped there for many years, only returning to the atmosphere when the wood is burned. Sometimes the duration of the carbon's visit to the organic world is long indeed. Plants that become buried in sediment, for example, may be gradually transformed by pressure into coal or oil. The carbon originally trapped by these plants is only released back into the atmosphere when the coal or oil (called **fossil fuels**) is burned.

Erosion

Very large amounts of carbon are present in seawater as dissolved CO_2. Substantial amounts of this carbon are extracted from the water by marine organisms, which use it to build their calcium carbonate shells. When these marine organisms die, their shells sink to the ocean floor, become covered with sediments, and form limestone. Eventually, as the ocean recedes and the limestone becomes exposed to weather and erodes, the carbon is returned to the cycle.

29.5 Carbon captured from the atmosphere by photosynthesis is returned to it through respiration, combustion, and erosion.

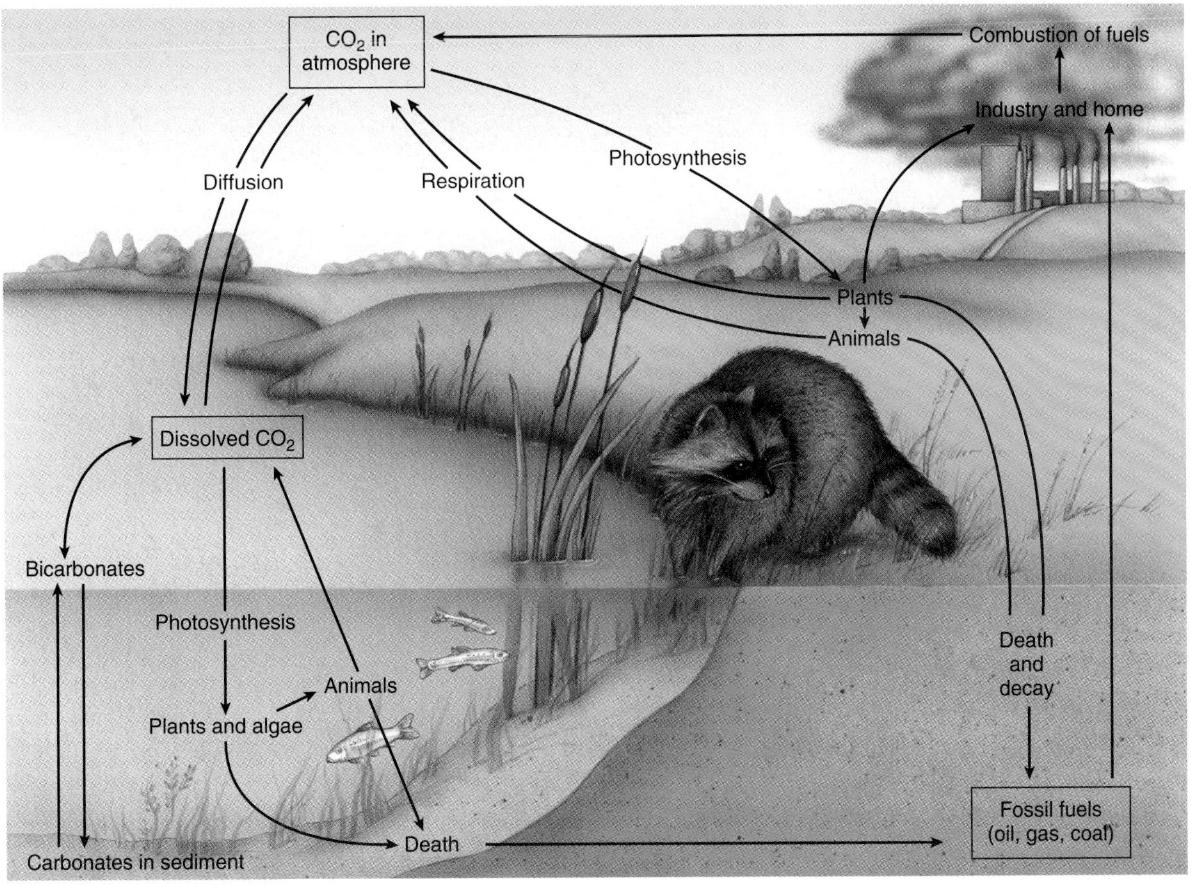

CO_2 in atmosphere
Combustion of fuels
Industry and home
Photosynthesis
Diffusion
Respiration
Plants
Animals
Dissolved CO_2
Bicarbonates
Photosynthesis
Animals
Plants and algae
Death and decay
Fossil fuels (oil, gas, coal)
Death
Carbonates in sediment

Figure 29.9 The carbon cycle. Carbon from the atmosphere and from water is fixed by photosynthetic organisms and returned through respiration, combustion, or erosion.

29.6 The Nitrogen Cycle

Organisms contain a lot of nitrogen (a principal component of protein) and so does the atmosphere, which is 79% nitrogen gas (N_2). However, the chemical connection between these two reservoirs is very delicate, because most living organisms are unable to use the N_2 so plentifully available in the air surrounding them. The two nitrogen atoms of N_2 are bound together by a particularly strong "triple" covalent bond that is very difficult to break. Luckily, a few bacteria can break the nitrogen triple bond and bind its nitrogen atoms to hydrogen (forming "fixed" nitrogen, ammonia [NH_3]) in a process called **nitrogen fixation.** Bacteria evolved the ability to do this early in the history of life, before photosynthesis had introduced oxygen gas into the earth's atmosphere, and that is still the only way the bacteria are able to do it—even a trace of oxygen poisons the process. In today's world, awash with oxygen, these bacteria live encased within bubbles called cysts that admit no oxygen or within special airtight cells in nodules of tissue on the roots of beans, aspen trees, and a few other plants. Figure 29.10 shows how bacteria make needed nitrogen available to other organisms, a process called the nitrogen cycle.

The growth of plants in ecosystems is often severely limited by the availability of "fixed" nitrogen in the soil, which is why farmers fertilize fields. This agricultural practice is a very old one, known even to primitive societies—the American Indians instructed the pilgrims to bury fish, a rich source of fixed nitrogen, with their corn seeds. Today most fixed nitrogen added to soils by farmers is not organic but instead is produced in factories by industrial rather than bacterial nitrogen fixation, a process that accounts for a prodigious 30% of the entire nitrogen cycle.

29.6 Most of the atmosphere is diatomic nitrogen gas, a small fraction of which is cleaved, attaching nitrogen atoms to hydrogen to form ammonia. These nitrogens then cycle through the ecosystem.

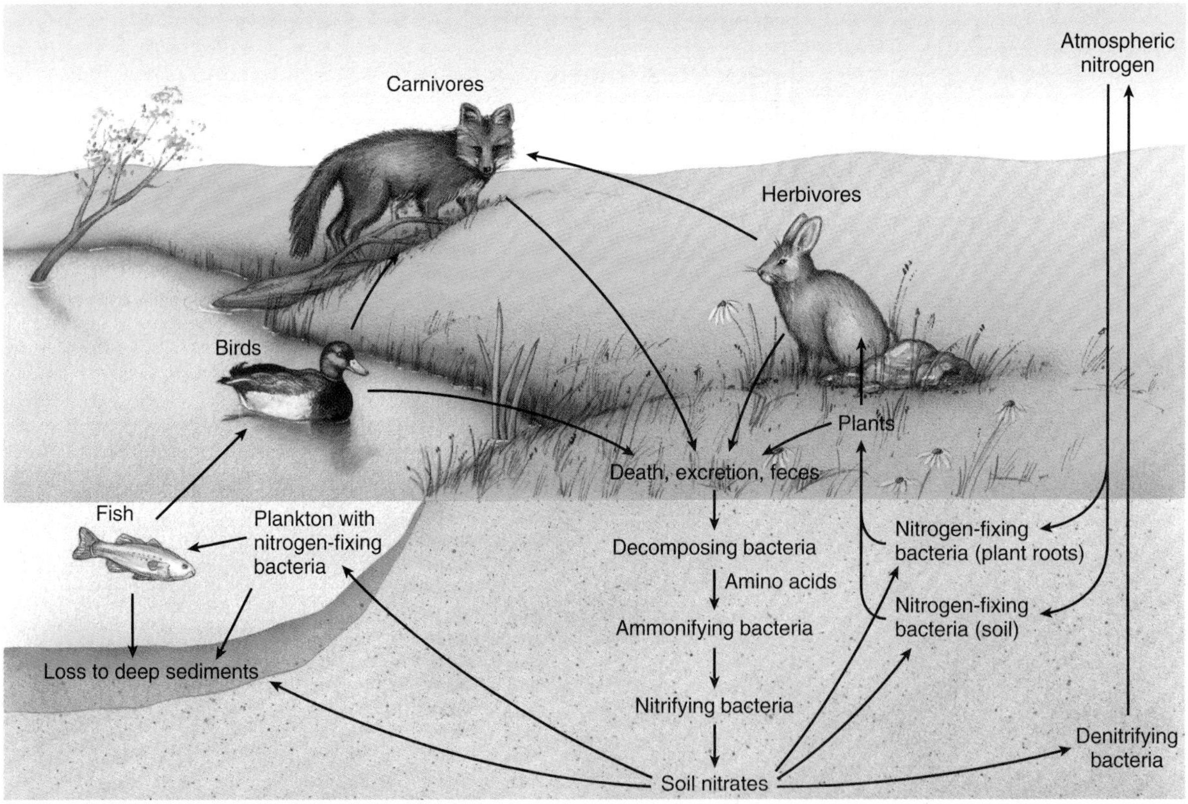
Atmospheric nitrogen
Carnivores
Herbivores
Birds
Plants
Death, excretion, feces
Fish
Plankton with nitrogen-fixing bacteria
Decomposing bacteria
Amino acids
Nitrogen-fixing bacteria (plant roots)
Ammonifying bacteria
Nitrogen-fixing bacteria (soil)
Loss to deep sediments
Nitrifying bacteria
Denitrifying bacteria
Soil nitrates

Figure 29.10 The nitrogen cycle.
Relatively few kinds of organisms—all of them bacteria—can convert atmospheric nitrogen into forms that can be used for biological processes.

29.7 The Phosphorus Cycle

Phosphorus is an essential element in all living organisms, a key part of both ATP and DNA. Phosphorus is often in very limited supply in the soil of particular ecosystems, and because phosphorus does not form a gas, none is available in the atmosphere. Most phosphorus exists in soil and rock as the mineral calcium phosphate, which dissolves in water to form phosphate ions (Coca-Cola is a sweetened solution of phosphate ions). These phosphate ions are absorbed by the roots of plants and used by them to build organic molecules like ATP and DNA. When the plants and animals die and decay, bacteria in the soil convert the organic phosphorus back into phosphorus ions, completing the cycle (figure 29.11).

The phosphorus level in freshwater lake ecosystems is often quite low, preventing much growth of photosynthetic algae in these systems. Such ecosystems are particularly vulnerable to the inadvertent addition of phosphorus by human activity. For example, agricultural fertilizers and many commercial detergents are rich in phosphorus. Pollution of a lake by the addition of phosphorus to its waters first produces a green scum of algal growth on the surface of the lake and then, if the pollution continues, proceeds to "kill" the lake. After the initial bloom of rapid algal growth, aging algae die, and bacteria feeding on the dead algae cells use up so much of the lake's dissolved oxygen that fish and invertebrate animals suffocate. Such rapid, uncontrolled growth caused by excessive nutrients in an aquatic ecosystem is called **eutrophication.**

29.7 Phosphorus, critical to organisms, is available in soil and dissolved in water. It cycles between organisms and the environment and is often the limiting factor in determining what organisms are able to live in an ecosystem.

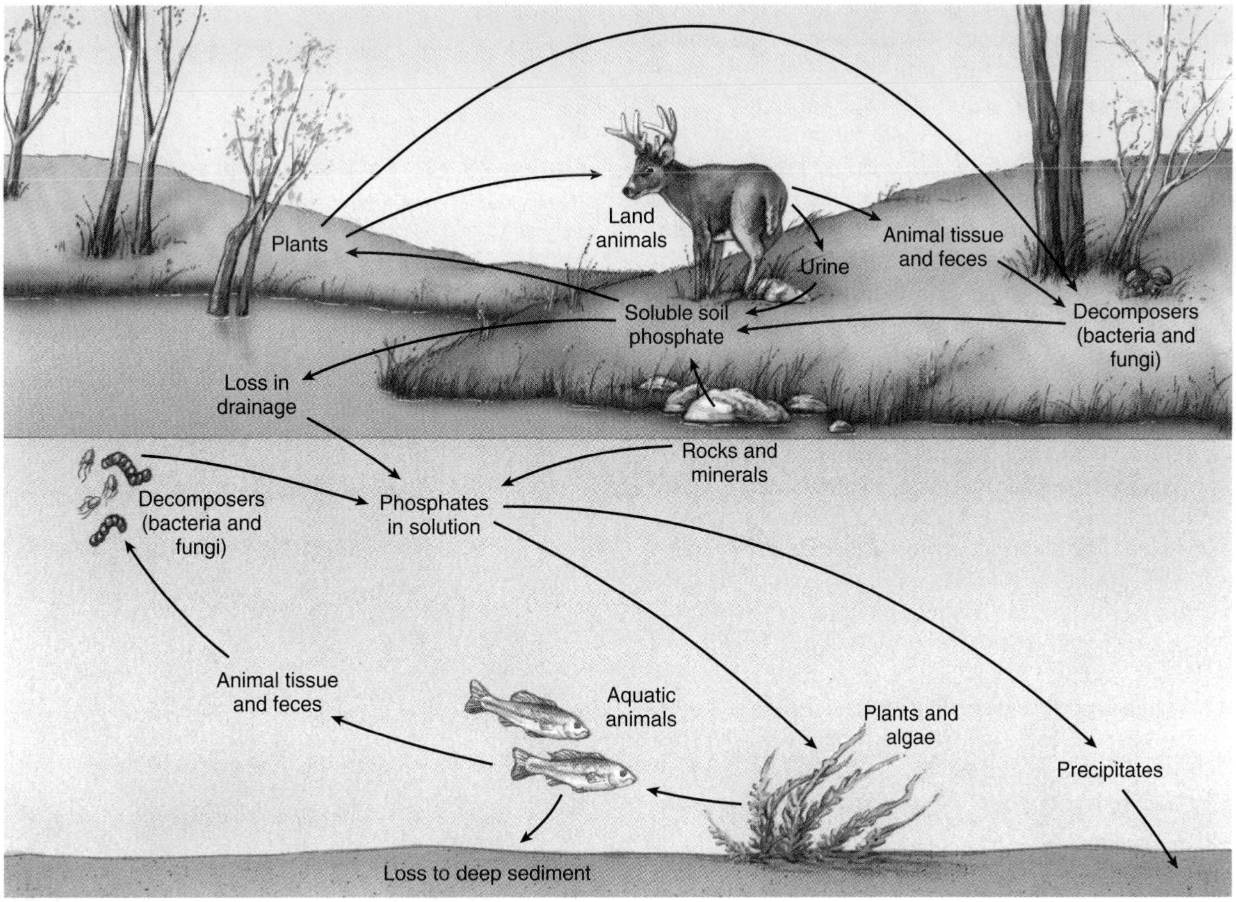

Land animals
Plants
Urine
Animal tissue and feces
Soluble soil phosphate
Decomposers (bacteria and fungi)
Loss in drainage
Rocks and minerals
Decomposers (bacteria and fungi)
Phosphates in solution
Animal tissue and feces
Aquatic animals
Plants and algae
Precipitates
Loss to deep sediment

Figure 29.11 The phosphorus cycle.
Phosphorus plays a critical role in plant nutrition; next to nitrogen, phosphorus is the element most likely to be so scarce that it limits plant growth.

29.8 The Sun and Atmospheric Circulation

The world contains a great diversity of ecosystems because its climate varies a great deal from place to place. On a given day, Miami and Boston often have very different weather. There is no mystery about this. The tropics are warmer than the temperate regions because the sun's rays arrive almost perpendicular (that is, dead on) at regions near the equator. As you move from the equator into temperate latitudes, sunlight strikes the earth at more oblique angles, which spreads it out over a much greater area, thus providing less energy per unit of area (figure 29.12). This simple fact—that because the earth is a sphere some parts of it receive more energy from the sun than others—is responsible for much of the earth's different climates and thus, indirectly, for much of the diversity of its ecosystems.

The earth's annual orbit around the sun and its daily rotation on its own axis are also both important in determining world climate. Because of the daily cycle, the climate at a given latitude is relatively constant. Because of the annual cycle and the inclination of the earth's axis, all parts away from the equator experience a progression of seasons.

The major atmospheric circulation patterns result from the interactions between six large air masses. These great air masses occur in pairs, with one air mass of the pair occurring in the northern latitudes and the other occurring in the southern latitudes. These air masses affect climate because the rising and falling of an air mass influence its temperature, which, in turn, influences its moisture-holding capacity.

Near the equator, warm air rises and flows toward the poles (figure 29.13). As it rises and cools, this air loses most of its moisture because cool air holds less water vapor than warm air. (This explains why it rains so much in the tropics.) When this air has traveled to about 30 degrees north and south latitudes, the cool, dry air sinks and becomes reheated, sucking up water like a sponge as it warms and producing a broad zone of low rainfall. It is no accident that all of the great deserts of the world lie near 30 degrees north or 30 degrees south latitude. Air at these latitudes is still warmer than it is in the polar regions, and thus it continues to flow toward the poles. At about 60 degrees north and south latitudes, air rises and cools and sheds its moisture, and such are the locations of the great temperate forests of the world. Finally, this rising air descends near the poles, producing zones of very low temperatures and precipitation.

29.8 The sun drives circulation of the atmosphere, causing rain in the tropics and a band of deserts at 30 degrees latitude.

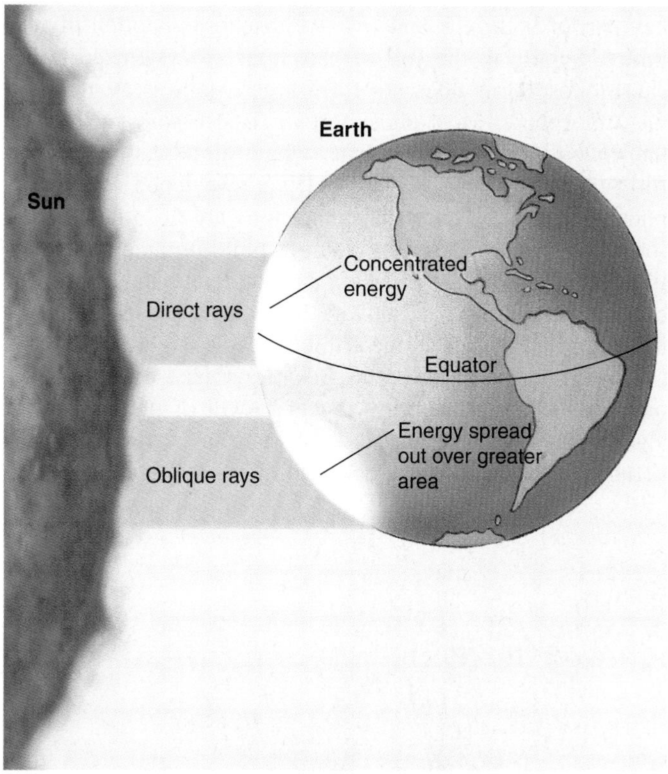
Earth
Sun
Concentrated energy
Direct rays
Equator
Energy spread out over greater area
Oblique rays

Figure 29.12 Latitude affects climate.

The relationship between the earth and sun is critical in determining the nature and distribution of life on earth. The tropics are warmer than the temperate regions because the sun's rays strike at a direct angle, providing more energy per unit of area.

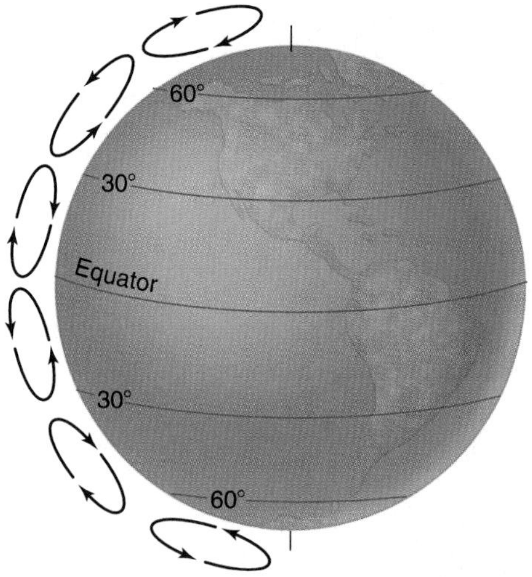
60°
30°
Equator
30°
60°

Figure 29.13 Air rises at the equator and then falls.

The pattern of air movement out from and back to the earth's surface forms three pairs of great cycles. At 30 degrees and at the poles, zones of very dry climate are created by descending dry air. Wet climates occur near the equator and at 60 degrees.

29.9 Latitude and Elevation

Temperatures are higher in tropical ecosystems for a simple reason: more sunlight per unit area falls on tropical latitudes. Solar radiation is most intense when the sun is directly overhead, and this occurs only in the tropics, where sunlight strikes the equator perpendicularly. Temperature also varies with elevation, with higher altitudes becoming progressively colder. At any given latitude, air temperature falls about 6°C for every 1,000-meter increase in elevation. The ecological consequences of temperature varying with elevation are the same as temperature varying with latitude (figure 29.14). Thus, in North America a 1,000-meter increase in elevation results in a temperature drop equal to that of a 880-kilometer increase in latitude. This is why "timberline" (the elevation above which trees do not grow) occurs at progressively lower elevations as one moves farther from the equator.

Rain Shadows

When a moving body of air encounters a mountain, it is forced upward, and as it is cooled at higher elevations the air's moisture-holding capacity decreases, producing rain on the windward side of the mountains—the side from which the wind is blowing. The effect on the other side of the mountain—the leeward side—is quite different. As the air passes the peak and descends on the far side of the mountains, it is warmed, so its moisture-holding capacity increases. Sucking up all available moisture, the air dries the surrounding landscape, often producing a desert. This effect, called a **rain shadow,** is responsible for deserts such as Death Valley, which is in the rain shadow of Mount Whitney, the tallest mountain in the Sierra Nevada (figure 29.15). Mediterranean climates result when winds blow from a cool ocean onto a warm land during the summer. As a result, the air's moisture-holding capacity is increased and precipitation is blocked, similar to what occurs on the leeward side of mountains. This effect accounts for dry, hot summers and cool, moist winters in areas with a Mediterranean climate. Such a climate is unusual on a world scale; in the regions where it occurs, many unusual kinds of plants and animals, often local in distribution, have evolved.

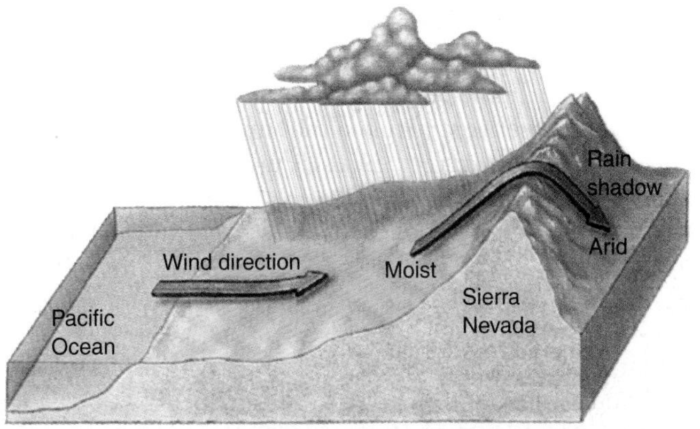

Rain shadow
Arid
Wind direction
Moist
Sierra Nevada
Pacific Ocean

Figure 29.15 The rain shadow effect.
Moisture-laden winds from the Pacific Ocean rise and are cooled when they encounter the Sierra Nevada. As they cool, their moisture-holding capacity decreases and precipitation occurs. As the air descends on the east side of the range, it warms, its moisture-holding capacity increases, and the air picks up moisture from its surroundings. As a result, arid conditions prevail on the east side of these mountains.

29.9 Temperatures fall with increasing latitude and also with increasing altitude. Rainfall is higher on the windward side of mountains, with air losing its moisture as it rises up the mountain; descending on the far side, the dry air warms and sucks up moisture, creating deserts.

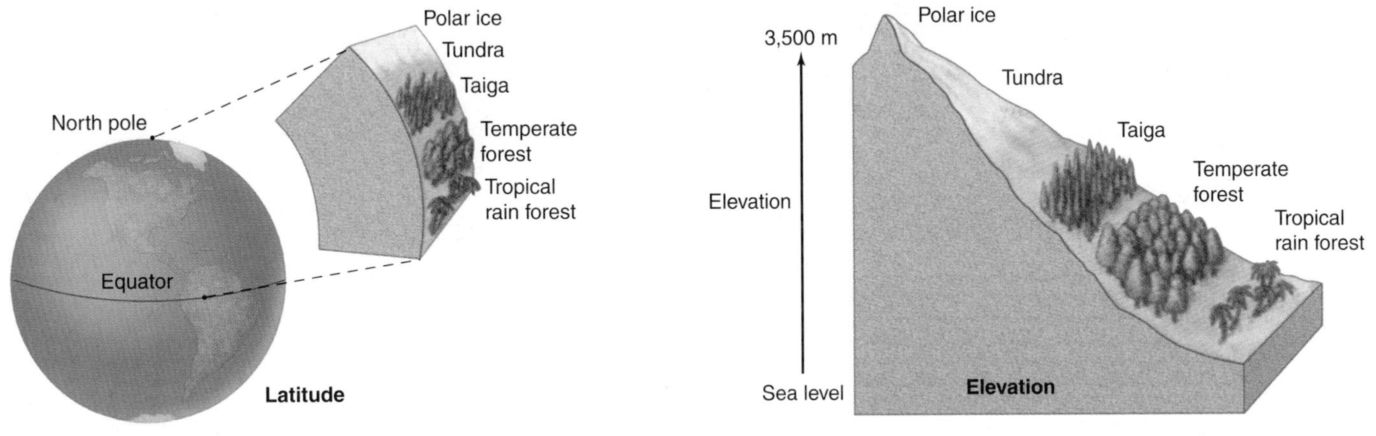

Polar ice
Tundra
Taiga
Temperate forest
Tropical rain forest
North pole
Equator
Latitude
(a)
3,500 m
Polar ice
Tundra
Taiga
Temperate forest
Tropical rain forest
Elevation
Sea level
Elevation
(b)

Figure 29.14 How elevation affects ecosystems.
The same land ecosystems that normally occur north and south of the equator at sea level (*a*) can occur in the tropics as elevation increases (*b*). Thus, on a tall mountain in southern Mexico or Guatemala, you might see a sequence of ecosystems such as is illustrated here.

29.10 Patterns of Circulation in the Ocean

Patterns of ocean circulation are determined by the patterns of atmospheric circulation, but they are modified by the locations of land masses. Oceanic circulation is dominated by huge surface gyres (figure 29.16), which move around the subtropical zones of high pressure between approximately 30 degrees north and south latitudes. These gyres move clockwise in the Northern Hemisphere and counterclockwise in the Southern Hemisphere. The ways they redistribute heat profoundly affects life not only in the oceans but also on coastal lands. For example, the Gulf Stream, in the North Atlantic, swings away from North America near Cape Hatteras, North Carolina, and reaches Europe near the southern British Isles. Because of the Gulf Stream, western Europe is much warmer and more temperate than eastern North America at similar latitudes. As a general principle, western sides of continents in temperate zones of the Northern Hemisphere are warmer than their eastern sides; the opposite is true of the Southern Hemisphere. In addition, winds passing over cold water onto warm land increase their moisture-holding capacity, limiting precipitation.

In South America, the Humboldt current carries phosphorus-rich cold water northward up the west coast. Phosphorus is brought up from the ocean depths by the upwelling of cool water that occurs as offshore winds blow from the mountainous slopes that border the Pacific Ocean. This nutrient-rich current helps make possible the abundance of marine life that supports the fisheries of Peru and northern Chile. Marine birds, which feed on these organisms, are responsible for the commercially important, phosphorus-rich, guano deposits on the seacoasts of these countries.

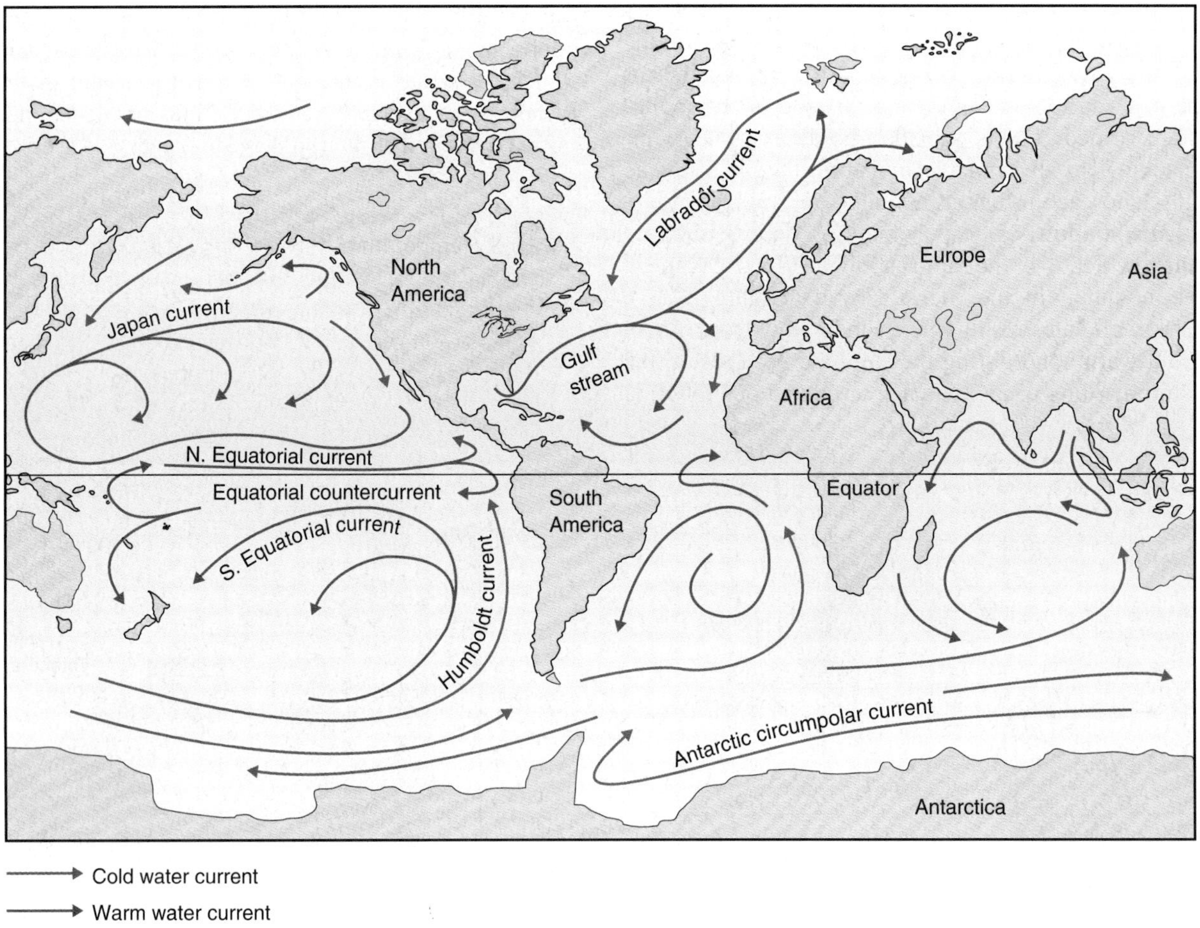

Labrador current
North America
Europe
Asia
Japan current
Gulf stream
Africa
N. Equatorial current
Equatorial countercurrent
Equator
South America
S. Equatorial current
Humboldt current
Antarctic circumpolar current
Antarctica
Cold water current
Warm water current

Figure 29.16 Oceanic circulation.

The circulation in the oceans moves in great surface spiral patterns called gyres; it profoundly affects the climate on adjacent lands.

El Niño Southern Oscillations and Ocean Ecology

Every Christmas a tepid current sweeps down the coast of Peru and Ecuador from the tropics, reducing the fish population slightly and giving local fishers some time off. The local fishers named this Christmas current *El Niño* ("The Christ Child"). Now, though, the term is reserved for a catastrophic version of the same phenomenon, one that occurs every two to seven years and is not only felt locally but on a global scale.

Scientists now have a pretty good idea of what goes on in an El Niño. Normally the Pacific Ocean is fanned by constantly blowing east-to-west trade winds. These winds push warm surface water away from the ocean's eastern side (Peru, Ecuador, and Chile) and allow cold water to well up from the depths in its place, carrying nutrients that feed plankton and hence fish. This surface water piles up in the west, around Australia and the Philippines, making it several degrees warmer and a meter or so higher than the eastern side of the ocean. But if the winds slacken briefly, warm water begins to slosh back across the ocean.

Once this happens, ocean and atmosphere conspire to ensure it keeps happening. The warmer the eastern ocean gets, the warmer and lighter the air above it becomes, and hence more similar to the air on the western side. This reduces the difference in pressure across the ocean. Because a pressure difference is what makes winds blow, the easterly trades weaken further, letting the warm water continue its eastward advance.

The end result is to shift the weather systems of the western Pacific Ocean 6,000 kilometers eastward. The tropical rainstorms that usually drench Indonesia and the Philippines are caused when warm seawater abutting these islands causes the air above it to rise, cool, and condense its moisture into clouds. When the warm water moves east, so do the clouds, leaving the previously rainy areas in drought. Conversely, the western edge of South America, its costal waters usually too cold to trigger much rain, gets a soaking, while the upwelling slows down. During an El Niño, commercial fish stocks virtually disappear from the waters of Peru and northern Chile, and plankton drop to a twentieth of their normal abundance. The commercially valuable anchovy fisheries of Peru were essentially destroyed by the 1972 and 1997 El Niños.

That is just the beginning. El Niño's effects are propagated across the world's weather systems (figure 29.17). Violent winter storms lash the coast of California, accompanied by flooding, and El Niño produces colder and wetter winters than normal in Florida and along the Gulf Coast. The American Midwest experiences heavier-than-normal rains.

Though the effects of an El Niño are now fairly clear, what triggers them still remains a mystery. Models of these weather disturbances suggest that the climatic change that triggers an El Niño is "chaotic." Wind and ocean currents return again and again to the same condition, but never in a regular pattern, and small nudges can send them off in many different directions—including an El Niño.

29.10 The world's oceans circulate in huge gyres deflected by continental land masses. Disturbances in ocean currents like an El Niño can have profound influences on world climate.

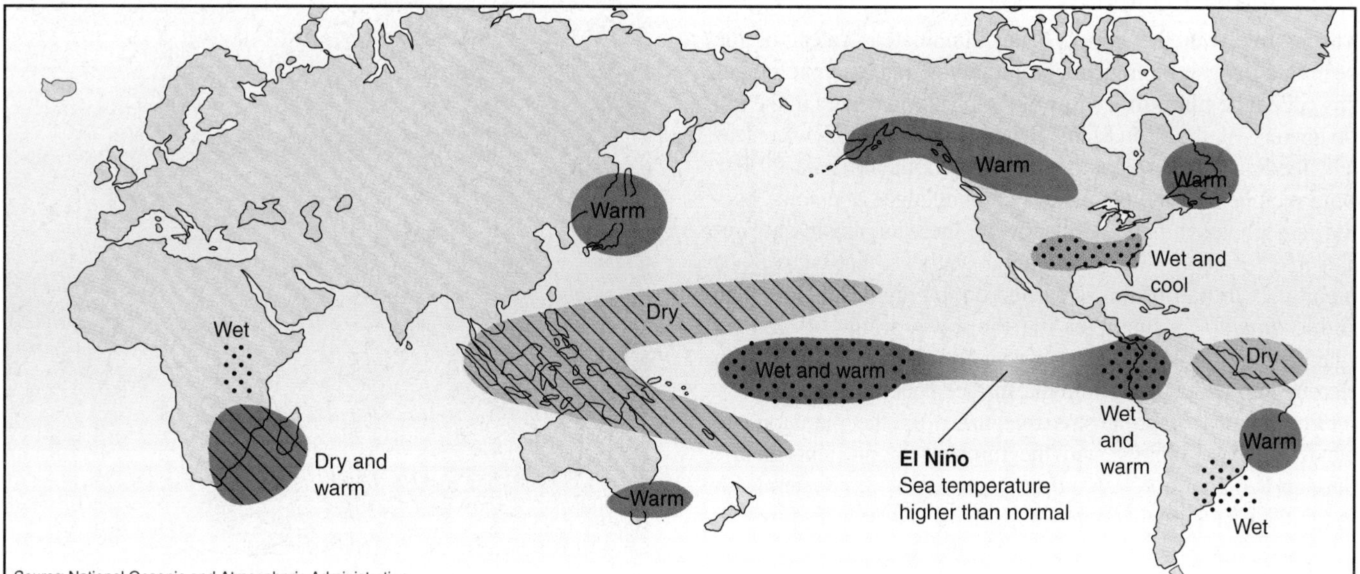

Warm
Warm
Warm
Wet and cool
Dry
Wet
Wet and warm
Dry
Wet and warm
Warm
Dry and warm
Warm
El Niño
Sea temperature higher than normal
Wet

Source: National Oceanic and Atmospheric Administration

Figure 29.17 An El Niño winter.

El Niño currents produce unusual weather patterns all over the world as warm waters from the western Pacific move eastward.

29.11 Ocean Ecosystems

Most of the earth's surface—nearly three-quarters—is covered by water. The seas have an average depth of more than 3 kilometers, and they are, for the most part, cold and dark. Photosynthetic organisms are confined to the upper few hundred meters, because light does not penetrate any deeper. Almost all organisms that live below this level feed on organic debris that rains downward. The three main kinds of ecosystems are shallow waters, open sea surface, and deep-sea waters.

Shallow Waters

Very little of the earth's ocean surface is shallow—mostly that along the shoreline—but this small area contains many more species than other parts of the ocean. The world's great commercial fisheries occur on banks in the coastal zones, where nutrients derived from the land are more abundant than in the open ocean. Part of this zone consists of the **intertidal region,** which is exposed to the air whenever the tides recede (figure 29.18). Partly enclosed bodies of water, such as those that often form at river mouths and in coastal bays, where the salinity is intermediate between that of seawater and freshwater, are called **estuaries.** Estuaries are among the most naturally fertile areas in the world, often containing rich stands of submerged and emergent plants, algae, and microscopic organisms. They provide the breeding grounds for most of the coastal fish and shellfish that are harvested both in the estuaries and in open water.

Figure 29.18 Shallow waters.
Diverse communities occur in intertidal regions. Many different habitats are created by the pounding of the waves and the periodic drying and flooding as the tides move out and in. Organisms that live in tide pools have adaptations that protect them from the dry air.

Open Sea Surface

Drifting freely in the upper, better-illuminated waters of the ocean is a diverse biological community of microscopic organisms called the **plankton.** Most of the plankton occurs in the top 100 meters of the sea. Many fishes swim in these waters as well, feeding on the plankton and one another (figure 29.19). Some members of the plankton, including algae and some bacteria, are photosynthetic. Collectively, these organisms are responsible for about 40% of all photosynthesis that takes place on earth. Over half of this is carried out by organisms less than 10 micrometers in diameter—at the lower limits of size for organisms—and almost all of it near the surface of the sea, in the zone into which light from the surface penetrates freely.

Populations of organisms that make up the plankton are able to increase rapidly, and the turnover of nutrients in the plankton is much more rapid than in most other ecosystems, although the total amounts of nutrients in the sea are very low.

Figure 29.19 Open sea surface.
The upper layers of the open ocean contain plankton and large schools of fish, like these yellow tail grunts.

Deep-Sea Waters

In the deep waters of the sea, below the top 300 meters, little light penetrates. Very few organisms live there, compared to the rest of the ocean, but those that do include some of the most bizarre organisms found anywhere on earth. Many deep-sea inhabitants have bioluminescent (light-producing) body parts that they use to communicate or to attract prey.

The supply of oxygen can often be critical in the deep ocean, and as water temperatures become warmer, the water holds less oxygen. For this reason, the amount of available oxygen becomes an important limiting factor for deep-sea organisms in warmer marine regions of the globe. Carbon dioxide, in contrast, is almost never limited in the deep ocean. The distribution of minerals is much more uniform in the ocean than it is on land, where individual soils reflect the composition of the parent rocks from which they have weathered.

Frigid and bare, the floors of the deep sea have long been considered a biological desert. Recent close-up looks taken by marine biologists, however, paint a different picture (figure 29.20). The ocean floor is teeming with life. Often kilometers deep, thriving in pitch darkness under enormous pressure, crowds of marine invertebrates have been found in hundreds of deep samples from the Atlantic and Pacific. Rough estimates of deep-sea diversity have soared to millions of species. Many appear endemic (local). The diversity of species is so high it may rival that of tropical rain forests! This profusion is unexpected. New species usually require some kind of barrier in order to diverge (see chapter 11), and the ocean floor seems boringly uniform. However, little migration occurs among deep populations, and this lack of movement may encourage local specialization and species formation. A patchy environment may also contribute to species formation there; deep-sea ecologists find evidence that fine but nonetheless formidable resource barriers arise in the deep sea.

From where do deep-sea organisms obtain their energy? While some utilize energy falling to the ocean floor as debris from above, other deep-sea organisms are autotrophic, gaining their energy from **hydrothermal vent systems** and supporting a broad array of other heterotrophic life.

Despite the many new forms of small invertebrates now being discovered on the seafloor, and the huge biomass that occurs in the sea, more than 90% of all *described* species of organisms occur on land. Each of the largest groups of organisms, including insects, mites, nematodes, fungi, and plants, has marine representatives, but they comprise only a very small fraction of the total number of described species. There are two reasons for this. First, barriers between habitats are sharper on land, and variations in elevation, parent rock, degree of exposure, and other factors have been crucial to the evolution of the millions of species of terrestrial organisms. Second, there are simply few deep-sea taxonomists actively classifying the profusion of ocean-floor life being brought to the surface.

Figure 29.20 Deep-sea waters.
Food comes to the ocean floor from above. Looking for all the world like some undersea sunflower, these two sea anemones (actually animals) use a glass-sponge stalk to catch "marine snow," food particles raining down on the ocean floor from the ocean surface kilometers above.

In terms of higher-level diversity, the pattern is quite different. Of the major groups of organisms—phyla—most originated in the sea, and almost every one has representatives in the deep sea. Only a few phyla have been successful on land or in freshwater habitats, although these have given rise to an extraordinarily large number of described species.

29.11 The three principal ocean ecosystems occur in shallow water, in the open-sea surface, and along the deep-sea bottom. Both intertidal shallows and deep-sea communities are very diverse.

29.12 Freshwater Ecosystems

Freshwater ecosystems (lakes, ponds, and rivers) are distinct from both ocean and land ecosystems, and they are very limited in area. Inland lakes cover about 1.8% of the earth's surface and rivers and streams about 0.3%. All freshwater habitats are strongly connected to land ones, with marshes and swamps constituting intermediate habitats. In addition, a large amount of organic and inorganic material continually enters bodies of freshwater from communities growing on the land nearby (figure 29.21). Many kinds of organisms are restricted to freshwater habitats (figure 29.22). When they occur in rivers and streams, they must be able to attach themselves in such a way as to resist or avoid the effects of current or risk being swept away.

Lakes can be divided into two categories, based on their production of organic material. **Eutrophic lakes** have an abundant supply of minerals and organic matter. Oxygen is depleted below the thermocline in the summer because of the abundant organic material and high rate at which aerobic decomposers in the lower layer use oxygen. These stagnant waters again reach the surface after the fall overturn. In **oligotrophic lakes,** on the other hand, organic matter and nutrients are relatively scarce. Such lakes are often deeper than eutrophic ones, and their deep waters are always rich in oxygen. Oligotrophic lakes are highly susceptible to pollution from excess phosphorus from such sources as fertilizer runoff, sewage, and detergents.

Figure 29.21 A nutrient-rich stream.
In this stream in the northern coastal mountains of California, as in all streams, much organic material falls or seeps into the water from communities along the edges. This input is responsible for much of the stream's biological productivity.

(a)

(b)

Figure 29.22 Freshwater organisms.
(*a*) This speckled darter and (*b*) this giant waterbug with eggs on its back can only live in freshwater habitats.

Thermal Stratification

Like the ocean, ponds and lakes have three zones in which organisms live: a shallow "edge" zone, an open-water surface zone, and a deep-water zone where light does not penetrate (figure 29.23). **Thermal stratification,** characteristic of the larger lakes in temperate regions, is the process whereby water at a temperature of 4°C (which is when water is most dense) sinks beneath water that is either warmer or cooler. In winter, water at 4°C sinks beneath cooler water that freezes at the surface at 0°C. Below the ice, the water remains between 0° and 4°C, and plants and animals survive there. In spring, as the ice melts, the surface water is warmed to 4°C and sinks below the cooler water, bringing the cooler water to the top with nutrients from the lake's lower regions. This process is known as the **spring overturn** (figure 29.24).

In summer, warmer water forms a layer over the cooler water (about 4°C) that lies below. In the area between these two layers, called the *thermocline,* temperature changes abruptly. Depending on the climate of the particular area, the warm upper layer may become as much as 20 meters thick during the summer. In autumn, its temperature drops until it reaches that of the cooler layer underneath—4°C. When this occurs, the upper and lower layers mix—a process called the **fall overturn.** Therefore, colder waters reach the surfaces of lakes in the spring and fall, bringing up fresh supplies of dissolved nutrients.

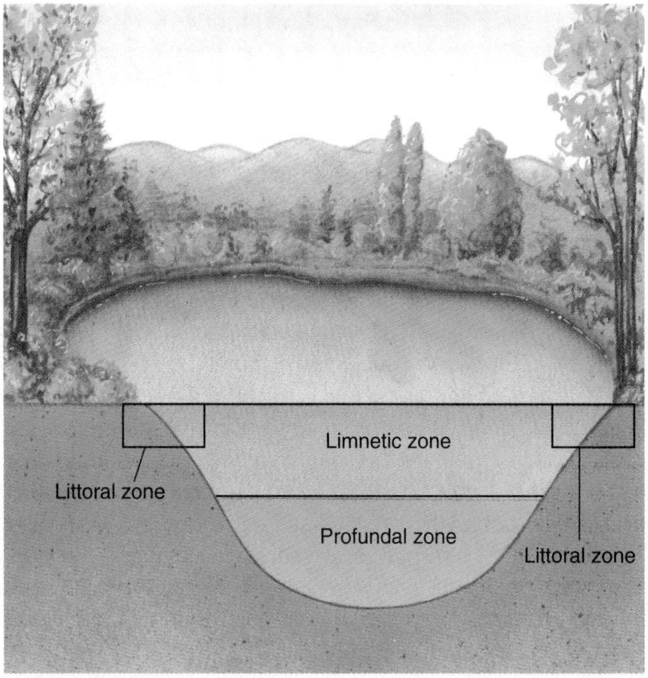

Limnetic zone
Littoral zone
Profundal zone
Littoral zone

Figure 29.23 The three zones in ponds and lakes.

A shallow "edge" (littoral) zone lines the periphery of the lake where attached algae and their insect herbivores live. An open-water surface (limnetic) zone lies across the entire lake and is inhabited by floating algae, zooplankton, and fish. A dark, deep-water (profundal) zone overlies the sediments at the bottom of the lake. The profundal zone contains numerous bacteria and wormlike organisms that consume dead debris settling at the bottom of the lake.

29.12 Freshwater ecosystems cover only about 2% of the earth's surface; all are strongly tied to adjacent terrestrial ecosystems. In some, organic materials are common, and in others, scarce. The temperature zones in lakes overturn twice a year in spring and fall.

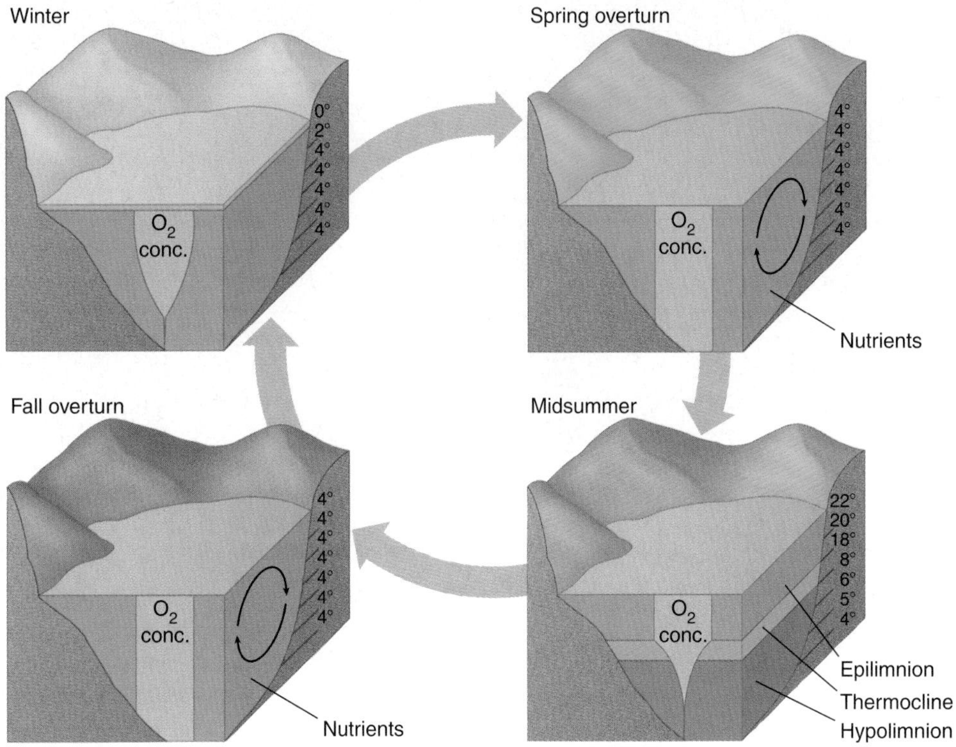

Winter
0°
2°
4°
4°
4°
4°
4°
O2 conc.
Spring overturn
4°
4°
4°
4°
4°
4°
4°
O2 conc.
Nutrients
Fall overturn
4°
4°
4°
4°
4°
4°
4°
O2 conc.
Nutrients
Midsummer
22°
20°
18°
8°
6°
5°
4°
O2 conc.
Epilimnion
Thermocline
Hypolimnion

Figure 29.24 Spring and fall overturns in freshwater ponds or lakes.

The pattern of stratification in a large pond or lake in temperate regions is upset in the spring and fall overturns. Of the three layers of water shown in midsummer (*lower right*), the densest water occurs at 4°C (the hypolimnion). The warmer water at the surface is less dense (the epilimnion). The thermocline is the zone of abrupt change in temperature that lies between them. If you have dived into a pond in temperate regions in the summer, you have experienced the existence of these layers directly.

29.13 Land Ecosystems

Living on land ourselves, we humans tend to focus much of our attention on terrestrial ecosystems. A **biome** is a terrestrial ecosystem that occurs over a broad area. Each biome is characterized by a particular climate and a defined group of organisms.

While biomes can be classified in a number of ways, the seven most widely occurring biomes are (1) tropical rain forest, (2) savanna, (3) desert, (4) temperate grassland, (5) temperate deciduous forest, (6) taiga, and (7) tundra. The reason that there are seven biomes, and not one or 80, is that they have evolved to suit the climate of the region, and the earth has seven principal climates. The seven biomes differ remarkably from one another but are consistent within; a particular biome looks the same, with the same types of creatures living there, wherever it occurs on earth.

There are seven other less widespread biomes: chaparral; polar ice; mountain zone; temperate evergreen forest; warm, moist evergreen forest; tropical monsoon forest; and semidesert. Figure 29.25 shows the geographical distribution of all 14 biomes, both the 7 principal ones and the 7 less common ones.

If there were no mountains and no climatic effects caused by the irregular outlines of the continents and by different sea temperatures, each biome would form an even belt around the globe. In fact, their distribution is greatly affected by these factors, especially by elevation. Thus, the summits of the Rocky Mountains are covered with a vegetation type that resembles tundra, whereas other forest types that resemble taiga occur farther down. It is for reasons such as these that the distributions of the biomes are so irregular. One trend that is apparent is that those biomes that normally occur at high latitudes also follow an altitudinal gradient along mountains. That is, biomes found far north and far south of the equator at sea level also occur in the tropics but at high mountain elevations.

Distinctive features of the seven major biomes—tropical rain forest, savanna, desert, temperate grassland, temperate deciduous forest, taiga, and tundra—are now discussed in more detail.

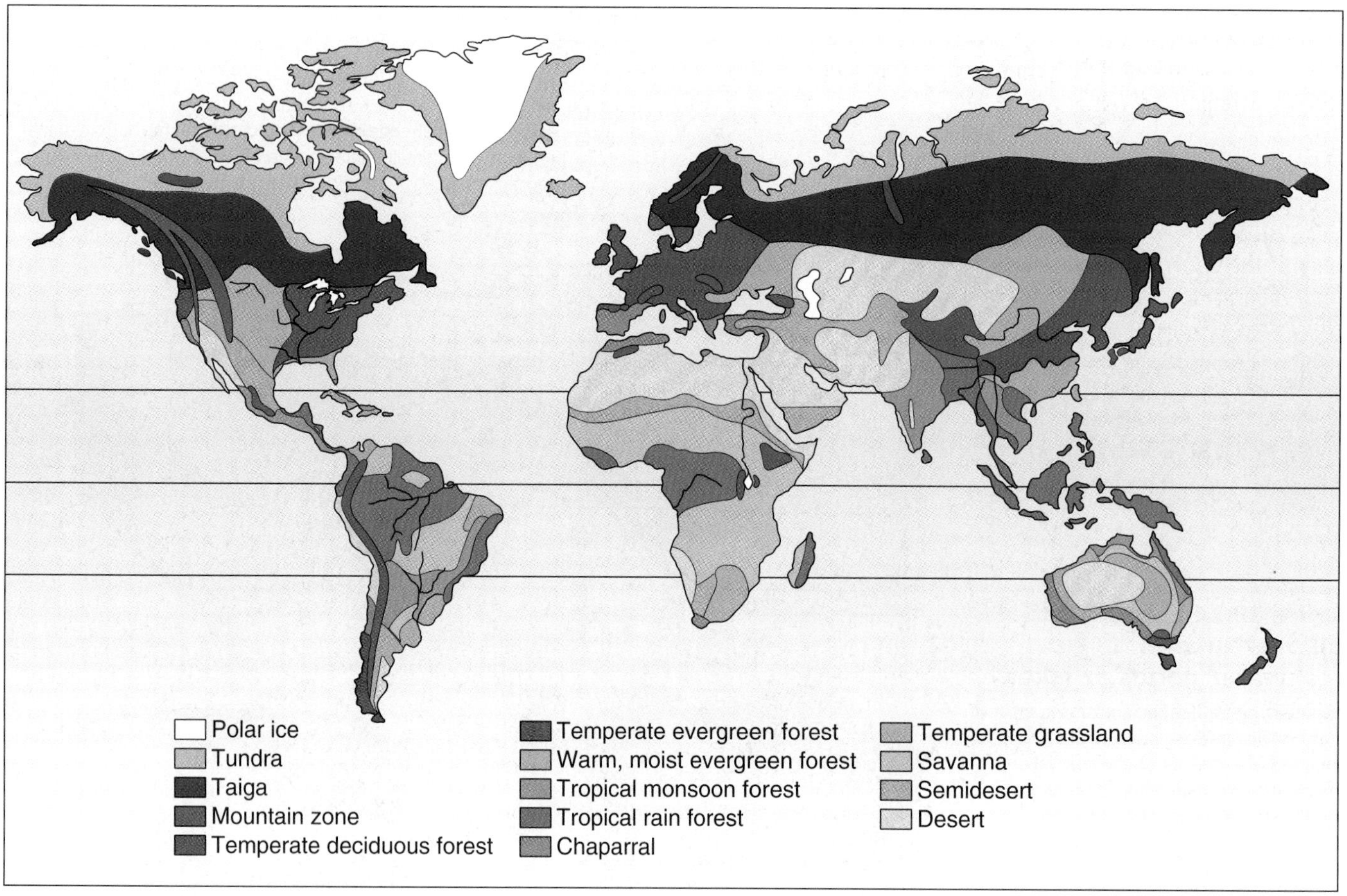

Polar ice
Tundra
Taiga
Mountain zone
Temperate deciduous forest
Temperate evergreen forest
Warm, moist evergreen forest
Tropical monsoon forest
Tropical rain forest
Chaparral
Temperate grassland
Savanna
Semidesert
Desert

Figure 29.25 Distribution of the earth's biomes.

The seven primary types of biomes are tropical rain forest, savanna, desert, temperate grassland, temperate deciduous forest, taiga, and tundra. In addition, seven less widespread biomes are shown.

Lush Tropical Rain Forests

Rain forests, which experience over 250 centimeters of rain a year, are the richest ecosystems on earth (figure 29.26). They contain at least half of the earth's species of terrestrial plants and animals—more than 2 million species! In a single square mile of tropical forest in Rondonia, Brazil, there are 1,200 species of butterflies—twice the total number found in the United States and Canada combined. The communities that make up tropical rain forests are diverse in that each kind of animal, plant, or microorganism is often represented in a given area by very few individuals. There are extensive tropical rain forests in South America, Africa, and Southeast Asia. But the world's tropical rain forests are being destroyed, and with them, countless species, many of them never seen by humans. Perhaps a quarter of the world's species will disappear with the rain forests during the lifetime of many of us.

Figure 29.26 Tropical rain forest.

Savannas: Dry Tropical Grasslands

In the dry climates that border the tropics are found the world's great grasslands, called **savannas.** Landscapes are open, often with widely spaced trees, and rainfall (75 to 125 cm annually) is seasonal. Many of the animals and plants are active only during the rainy season. The huge herds of grazing animals that inhabit the African savanna are familiar to all of us (figure 29.27). Such animal communities occurred in North America during the Pleistocene epoch but have persisted mainly in Africa. On a global scale, the savanna biome is transitional between tropical rain forest and desert. As these savannas are increasingly converted to agricultural use to feed rapidly expanding human populations in subtropical areas, their inhabitants are finding it difficult to survive. The elephant and rhino are now endangered species; lion, giraffe, and cheetah will soon follow them.

Figure 29.27 Savanna.

Deserts: Burning Hot Sands

In the interior of continents are found the world's great deserts, especially in Africa (the Sahara), Asia (the Gobi), and Australia (the Great Sandy Desert). **Deserts** are dry places where fewer than 25 centimeters of rain falls in a year—an amount so low that vegetation is sparse and survival depends on water conservation (figure 29.28). Plants and animals may restrict their activity to favorable times of the year, when water is present. To avoid high temperatures, most desert vertebrates live in deep, cool, and sometimes even somewhat moist burrows. Those that are active over a greater portion of the year emerge only at night, when temperatures are relatively cool. Some, such as camels, can drink large quantities of water when it is available and then survive long, dry periods. Many animals simply migrate to or through the desert, where they exploit food that may be abundant seasonally.

Figure 29.28 Desert.

Grasslands: Seas of Grass

Halfway between the equator and the poles are temperate regions where rich **grasslands** grow. These grasslands once covered much of the interior of North America, and they were widespread in Eurasia and South America as well. Such grasslands are often highly productive when converted to agriculture. Many of the rich agricultural lands in the United States and southern Canada were originally occupied by **prairies,** another name for temperate grasslands. The roots of perennial grasses characteristically penetrate far into the soil, and grassland soils tend to be deep and fertile. Temperate grasslands are often populated by herds of grazing mammals. In North America, the prairies were once inhabited by huge herds of bison and pronghorns (figure 29.29). The herds are almost all gone now, with most of the prairies having been converted to the richest agricultural region on earth.

Figure 29.29 Temperate grassland.

Deciduous Forests: Rich Hardwood Forests

Mild climates (warm summers and cool winters) and plentiful rains promote the growth of **deciduous** ("hardwood") **forests** in Eurasia, the northeastern United States, and eastern Canada (figure 29.30). A deciduous tree is one that drops its leaves in the winter. Deer, bears, beavers, and raccoons are the familiar animals of the temperate regions. Because the temperate deciduous forests represent the remnants of more extensive forests that stretched across North America and Eurasia several million years ago, these remaining areas—especially those in eastern Asia and eastern North America—share animals and plants that were once more widespread. Alligators, for example, are found only in China and in the southeastern United States. The deciduous forest in eastern Asia is rich in species because climatic conditions have remained constant.

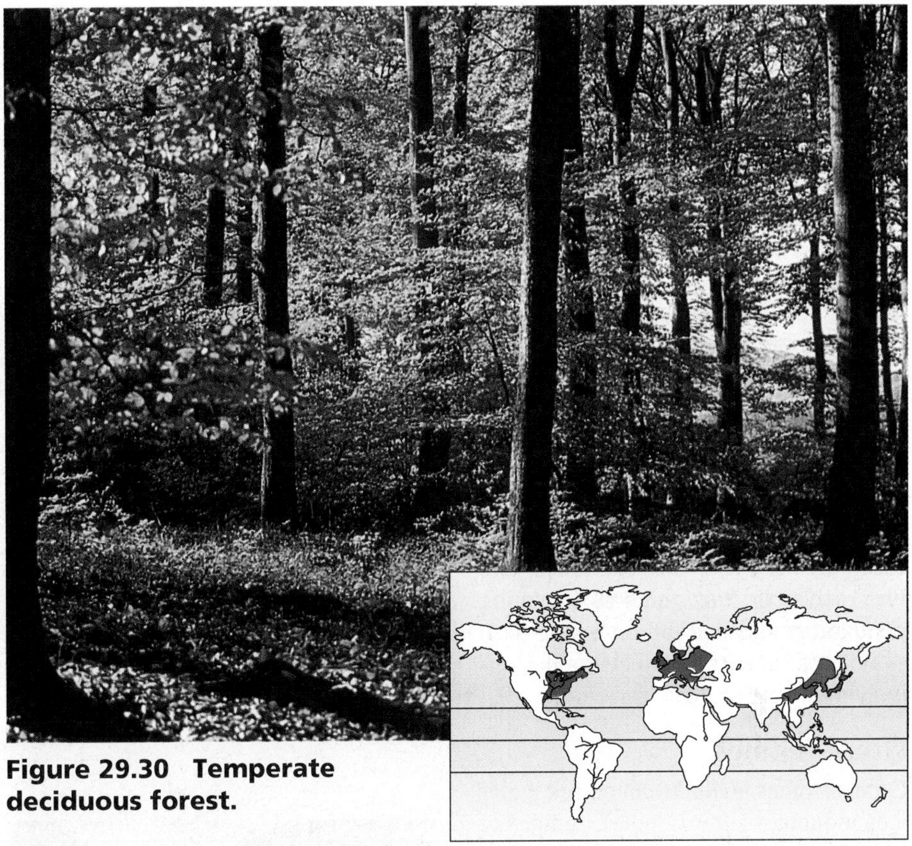

Figure 29.30 Temperate deciduous forest.

Taiga: Trackless Conifer Forests

A great ring of northern forests of coniferous trees (spruce, hemlock, larch, and fir) extends across vast areas of Asia and North America. Coniferous trees are ones with leaves like needles that are kept all year long. This ecosystem, called **taiga,** is one of the largest on earth (figure 29.31). Here, the winters are long and cold, and most of the limited amount of precipitation falls in the summer. Because it has too short a growing season for farming, few people live there. Many large mammals, including elk, moose, deer, and such carnivores as wolves, bears, lynx, and wolverines, live in the taiga. Traditionally, fur trapping has been extensive in this region, which is also important in lumber production. Marshes, lakes, and ponds are common and are often fringed by willows or birches. Most of the trees occur in dense stands of one or a few species.

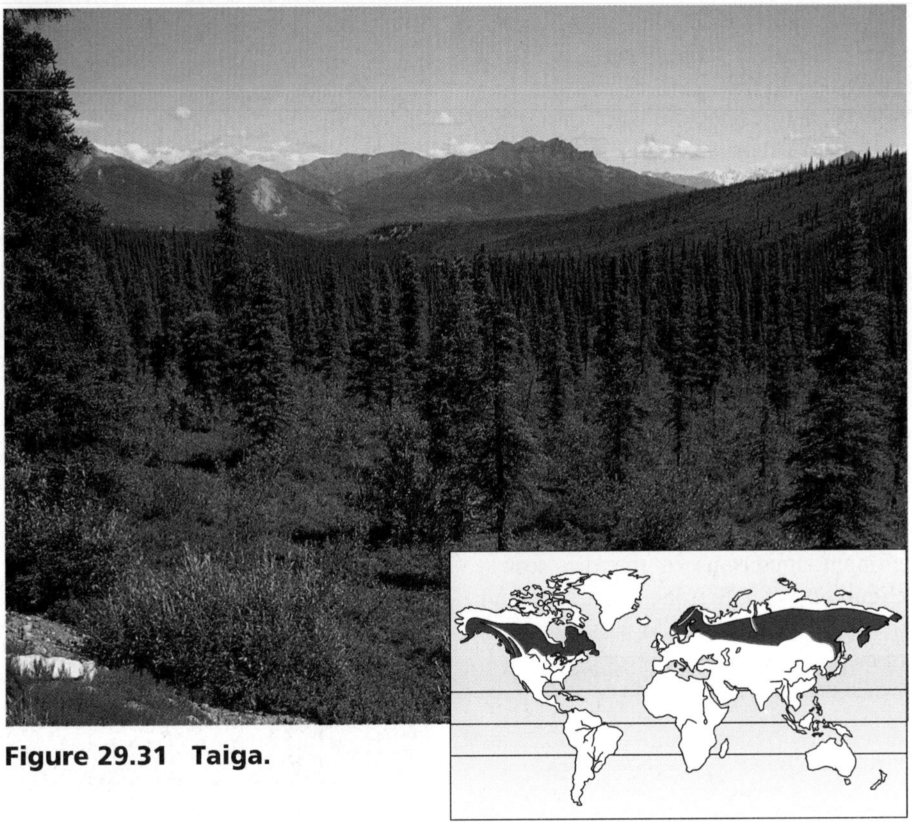

Figure 29.31 Taiga.

Tundra: Cold Boggy Plains

In the far north, above the great coniferous forests and below the polar ice, there are few trees. There the grassland, called **tundra,** is open, windswept, and often boggy (figure 29.32). Enormous in extent, this ecosystem covers one-fifth of the earth's land surface. Very little rain or snow falls. When rain does fall during the brief arctic summer, it sits on frozen ground, creating a sea of boggy ground. **Permafrost,** or permanent ice, usually exists within a meter of the surface. Trees are small and are mostly confined to the margins of streams and lakes. Large grazing mammals, including musk-oxen, caribou, reindeer, and carnivores such as wolves, foxes, and lynx, live in the tundra. Lemming populations rise and fall on a long-term cycle, with important effects on the animals that prey on them.

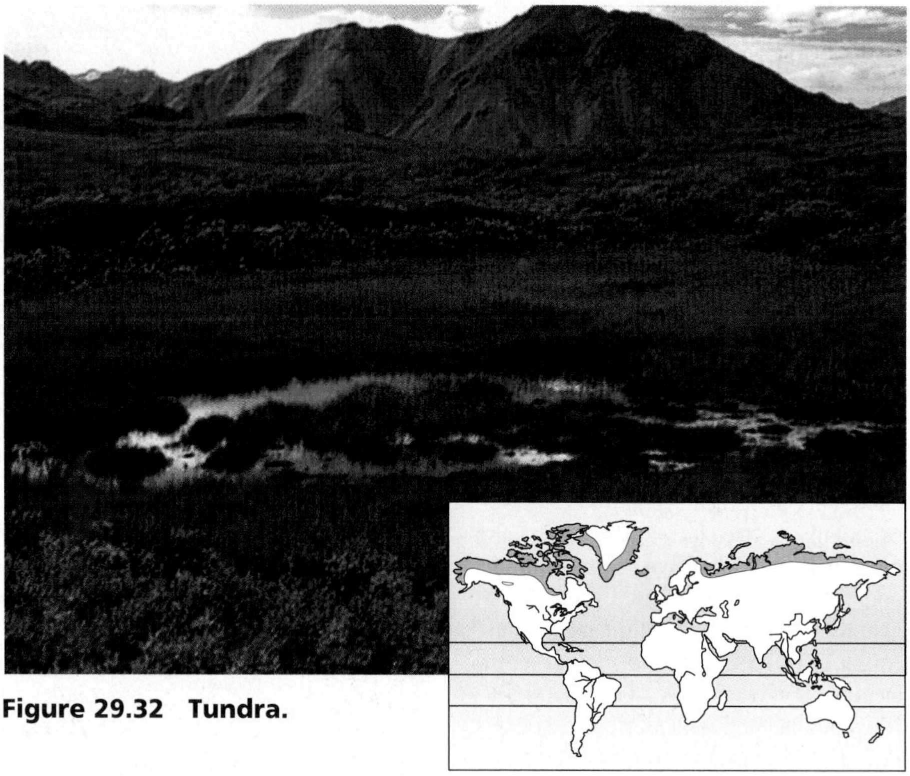

Figure 29.32 Tundra.

Other Biomes

Other biomes include chaparral; polar ice; mountain zone (alpine); temperate evergreen forest; warm, moist evergreen forest; tropical monsoon forest; and semidesert. **Chaparral** consists of evergreen, often spiny shrubs and low trees that form communities in regions with a Mediterranean, dry summer climate (figure 29.33*a*). Many plant species found in chaparral can germinate only when they have been exposed to the hot temperatures generated during a fire. **Polar ice** caps lie over the Arctic Ocean in the north and Antarctica in the south (figure 29.33*b*). The poles receive almost no precipitation, so although ice is abundant, freshwater is scarce. In **tropical monsoon forests,** rainfall is very seasonal, heavy during the monsoon season and approaching drought conditions in the dry season (figure 29.33*c*). **Semidesert** areas occur in tropical regions with less rain than monsoon forests but more rain than savannas (figure 29.33*d*).

29.13 Biomes are major terrestrial communities defined largely by temperature and rainfall patterns.

(a) Chaparral

(b) Polar ice

(c) Tropical monsoon forest

(d) Semidesert

Figure 29.33 Other biomes.

Other less widespread biomes include (*a*) chaparral, (*b*) polar ice, (*c*) tropical monsoon forest, and (*d*) semidesert.

CONCEPT REVIEW

1. Organisms that eat plant eaters are
   a. producers.
   b. in trophic level 3.
   c. in trophic level 2.
   d. detritivores.

2. Herbivores are
   a. primary consumers.
   b. secondary consumers.
   c. decomposers.
   d. photosynthetic.

3. The path of energy in complex ecosystems with animals often feeding at various trophic levels is called a
   a. food chain.
   b. mineral cycle.
   c. community.
   d. food web.

4. In which of the following cycles does the reservoir of the nutrient exist in mineral form?
   a. carbon cycle
   b. phosphorus cycle
   c. nitrogen cycle
   d. two of the above

5. In the environmental water cycle, water is returned to the atmosphere by
   a. evaporation.
   b. condensation.
   c. transpiration.
   d. precipitation.

6. Which of the following does *not* account for the variation in climates in different places on earth?
   a. the interaction between large air masses
   b. spring overturn
   c. the rain shadow effect
   d. the earth's annual orbit around the sun

7. In the rain shadow effect, rain falls on the windward side of mountains because
   a. air is cooled as it is forced upward.
   b. air is warmed as it is forced upward.
   c. warm air collides with cooler air.
   d. the moisture-holding capacity of air is increased.

8. Most of an ocean's diversity occurs in
   a. the open-sea surface.
   b. the shallow waters.
   c. plankton.
   d. the deep-sea waters.

9. In the thermal stratification of lakes, water that is ________ sinks below water that is cooler or warmer.
   a. the coldest
   b. 0°C
   c. 10°C
   d. 4°C

10. One of the largest ecosystems on earth is
    a. tropical rain forests.
    b. polar ice.
    c. taiga.
    d. chaparral.

11. The total amount of energy fixed by photosynthesis per unit of time minus the metabolic expense is called the ecosystem's _________.

12. The total weight of all the organisms living in an ecosystem is called the _________ of that ecosystem.

13. Carbon that is tied up in the wood of plants and is gradually transformed by pressure into coal or oil is called _________.

14. Through a process called _________, bacteria convert the plentifully available nitrogen gas into a form that other organisms are able to use.

15. In _________ lakes, nutrients and organic matter are scarce.

16. The earth's great _________ lie near 30 degrees north or south latitude.

17. The biome that is transitional between tropical rain forest and desert is _________.

CHALLENGE YOURSELF

1. The net carbon dioxide in the atmosphere is increasing. Where is this carbon dioxide coming from? How could you stop this increase?

2. Why does the net productivity of an ecosystem decrease as it becomes mature?

3. What kinds of biological communities would you expect to find on the windward and leeward sides of a mountain range in an area where the annual precipitation ranged between 20 and 100 centimeters per year and was distributed mainly in one rainy season? How would the height of the mountain range affect the situation?

29 eBRIDGE

Reinforcing Key Points

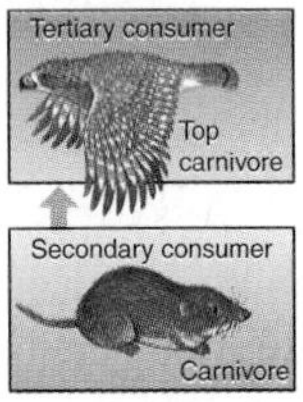
Tertiary consumer
Top carnivore
Secondary consumer
Carnivore

The Energy in Ecosystems

Materials Cycle Within Ecosystems

How Weather Shapes Ecosystems

Major Kinds of Ecosystems

Electronic Learning

Visual Learning

Animations

- Ecosystem Organization
- Hydrologic (Water) Cycle
- Carbon Cycle I
- Carbon Cycle II
- Nitrogen Cycle I
- Nitrogen Cycle II
- Four Seasons
- Distribution of Solar Energy
- Global Air Circulation
- Global Wind Circulation
- Rain Shadow Effect
- Formation of a Rain Shadow
- El Niño Southern Oscillation
- Deoxygenation of Lakes
- Lake Stratification

CO₂

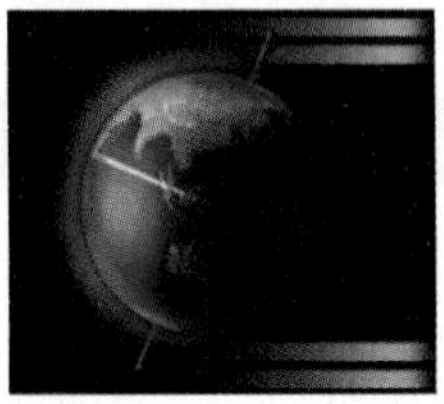

Author's Corner

Ecosystem Destruction. Human impact on the world's ecosystems is harming species in increasingly severe ways. In many cases the exact nature of the damage is not clearly understood. The reasons for amphibian decline have only emerged after a major worldwide research effort. The case for global warming resulting from increased release of carbon dioxide into the atmosphere as a result of human activity, on the other hand, seems clearcut.

1. Research is revealing complex reasons for the global decline in amphibians.
2. Global warming is a problem that is not going to be solved by ignoring it.
3. The problem of exploding urban deer populations has no solutions that please everybody.
4. Migrating woodland songbirds are in steep decline due to habitat loss.

Virtual Classroom

The Global Environmental Challenge

The current world population of more than 6 billion people is placing severe strains on our earth's ability to sustain and support so many people. While the worldwide average birthrate has remained fairly constant for the last 300 years at 25 to 30 births per year per 1,000 people, the death rate has fallen from about 29 per year per 1,000 people to about 19 per year per 1,000 people. This difference between birth and death rates produces an annual worldwide increase of approximately 1.4%. The rate may seem small, but it doubles the world's population every 39 years! The environmental problems that haunt the new century—global warming, ozone depletion, acid rain, chemical pollution—all are a direct result of this explosion of human population growth, the engine which fuels consumption, "development" of third-world resources, and the environmental destruction this economic activity fosters. Perhaps the most serious harm concerns loss of nonrenewable resources, such as topsoil, groundwater, and biodiversity. Current levels of consumption of these resources are not sustainable, and replenishment can take many centuries.

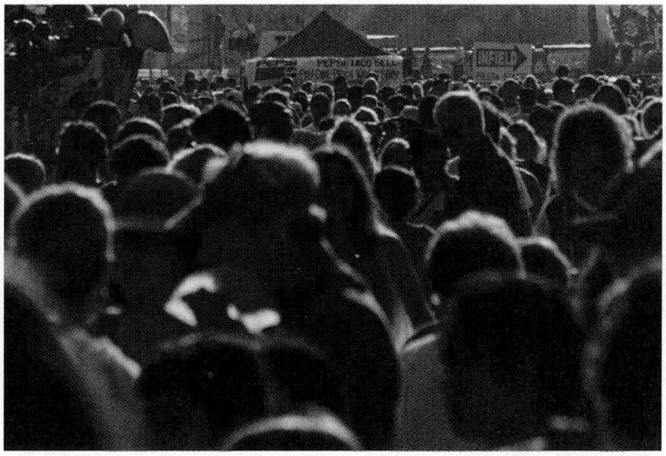

Virtual Lab

Why Does Contamination of a Coastal Salt Marsh with Diesel Fuel Lead to *Increased* Microalgal Biomass?

In shallow coastal waters microscopic algae ("microalgae") are an important food source for the many small animals that live in the sediment. As with any food web, a balance is reached between producers and consumers, but it is not clear what factors directly or indirectly influence that balance. We need to understand these food web relationships to properly assess the potential effects of pollution or environmental changes on the benthic community. Kevin Carman, John Fleeger, and Steven Pomarico, all from Louisiana State University, Baton Rouge, have studied the effects of diesel-fuel-contaminated sediments on the meiofaunal (primarily copepods and nematodes) and microalgae food web of a coastal salt marsh. The researchers set up microcosms in the laboratory using intact natural sediments and salt water. They then exposed the microcosms to various levels of diesel fuel contamination, and observed the effects on the levels of microalgae and meiofauna. Interestingly, the levels of microalgae seem to *increase* with pollution, while meiofauna decrease.

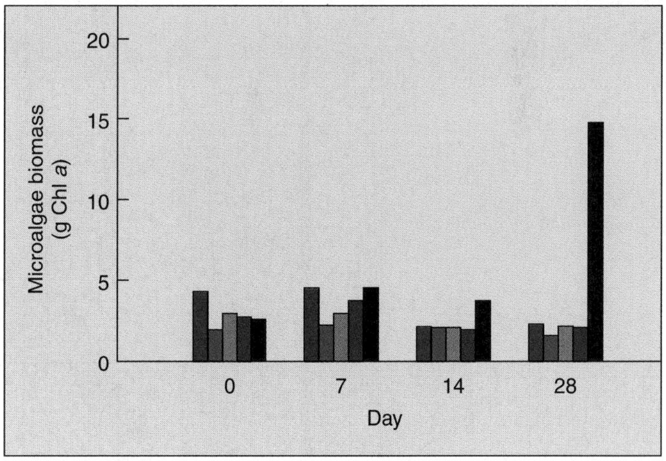
Microalgae biomass
(g Chl *a*)
20
15
10
5
0
0
7
14
28
Day

Quizzes | Further Reading | Essential Study Partner | Links | BioCourse.com

CHAPTER 30

Living in Ecosystems

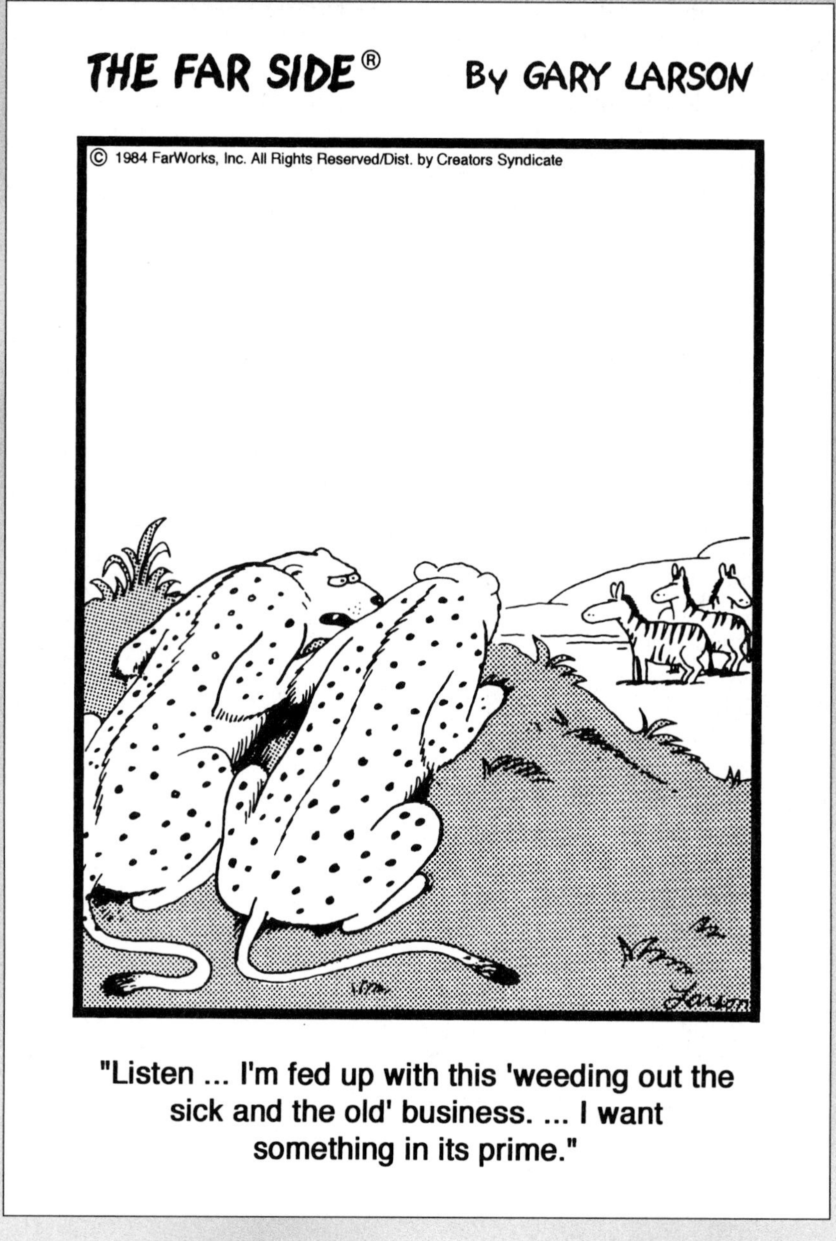
THE FAR SIDE® By GARY LARSON
© 1984 FarWorks, Inc. All Rights Reserved/Dist. by Creators Syndicate
"Listen ... I'm fed up with this 'weeding out the sick and the old' business. ... I want something in its prime."

CHAPTER 30

Living in Ecosystems

CHAPTER OVERVIEW

Population Dynamics

30.1 Population Growth
30.2 Life History Adaptations
30.3 The Influence of Population Density
30.4 Population Demography

- Many populations exhibit a sigmoid growth curve, with a relatively slow start in growth, a rapid increase, and then a leveling off when the carrying capacity of the environment is reached.

How Competition Shapes Communities

30.5 Communities
30.6 The Niche and Competition
30.7 Competitive Exclusion
30.8 Resource Partitioning

- Each species plays a specific role in its ecosystem; this role is called its niche.
- An organism's fundamental niche is the total niche that the organism would occupy in the absence of competition. Its realized niche is the actual niche it occupies in nature.

How Coevolution Shapes Communities

30.9 Coevolution and Symbiosis
30.10 Commensalism
30.11 Mutualism
30.12 Parasitism

- Some symbiotic relationships may be mutually beneficial (mutualism), while in others only one organism benefits while the other is unharmed (commensalism).

Predator and Prey Interactions

30.13 Plant Defenses
30.14 Animal Defenses
30.15 Predator-Prey Cycles
30.16 Mimicry

- Plants are often protected from herbivores by chemicals they manufacture.
- Animals may advertise their toxicity using aposematic coloration or hide their palatability using cryptic coloration.

Community Stability

30.17 Ecological Succession
30.18 Biodiversity
30.19 Island Biogeography

- Primary succession takes place in barren areas.
- The number of different species, or species richness, in an ecosystem determines its biological diversity.

The picture of the biological world painted in the previous chapter is a limited one, in which ecosystems are described as if they were chemical factories fueled by a flow of energy from the sun. In fact, this is only part of the story. The organisms themselves also play a major role in shaping the nature of their ecosystem.

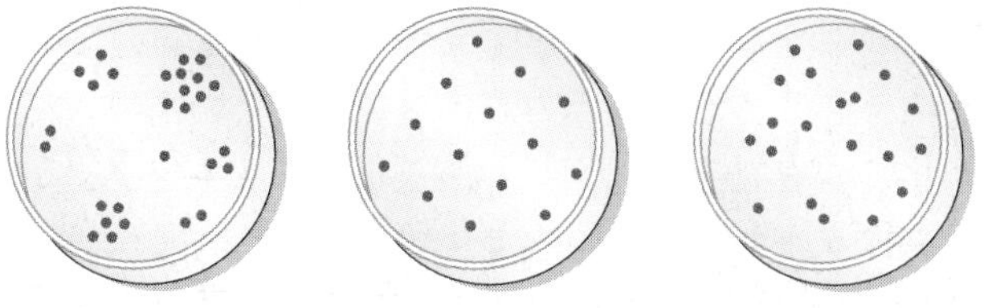

Clumped
Uniform
Random

(a)

(b)

Figure 30.1 Population dispersion.
(a) Different arrangements of bacterial colonies, and (b) starlings distributed uniformly along telephone wires.

30.1 Population Growth

Organisms live as members of **populations,** groups of individuals of a species that live together. One of the critical properties of any population is its **population size**—the number of individuals in the population. For example, if an entire species consists of only one or a few small populations, that species is likely to become extinct, especially if it occurs in areas that have been or are being radically changed. In addition to population size, **population density**—the number of individuals that occur in a unit area, such as per square kilometer—is often an important characteristic. A third significant property is **population dispersion,** the scatter of individual organisms within the population's range. Individuals may be spaced randomly, in clumps, or uniformly (figure 30.1). In addition to size, density, and dispersion, another key characteristic of any population is its capacity to grow. To understand populations, we must consider how they grow and what factors in nature limit **population growth.**

The Exponential Growth Model

The simplest model of population growth assumes a population growing without limits at its maximal rate. This rate, symbolized r and called the **biotic potential,** is the rate at which a population of a given species will increase when no limits are placed on its rate of growth. In mathematical terms, this is defined by the following formula:

$$dN/dt = r_i N$$

where N is the number of individuals in the population, dN/dt is the rate of change in its numbers over time, and r_i is the *intrinsic* rate of natural increase for that population—its innate capacity for growth.

The *actual* rate of population increase, r, is defined as the difference between the birthrate and the death rate corrected for any movement of individuals in or out of the population, whether net emigration (movement out of the area) or net immigration (movement into the area). Thus,

$$r = (b - d) + (i - e)$$

Movements of individuals can have a major impact on population growth rates. For example, the increase in human population in the United States during the closing decades of the twentieth century was mostly due to immigrants. Less than half of the increase came from the reproduction of the people already living there.

The innate capacity for growth of any population is exponential. Even when the *rate* of increase remains constant, the actual increase in the *number* of individuals accelerates rapidly as the size of the population grows. This sort of growth pattern is similar to that obtained by compounding interest on an investment. In practice, such patterns prevail only for short periods, usually when an organism reaches a new habitat with abundant resources. Natural examples include dandelions reaching the fields, lawns, and meadows of North America from Europe for the first time; algae colonizing a newly formed pond; or the first terrestrial immigrants arriving on an island recently thrust up from the sea.

Carrying Capacity

No matter how rapidly populations grow, they eventually reach a limit imposed by shortages of important environmental factors such as space, light, water, or nutrients. A population ultimately stabilizes at a certain size, called the **carrying capacity** of the particular place where it lives. The carrying capacity, symbolized by *K*, is the number of individuals that place can support, a dynamic rather than static measure as the characteristics of the place change.

The Logistic Growth Model

As a population approaches its carrying capacity, its rate of growth slows greatly, because fewer resources remain for each new individual to use. The growth curve of such a population, which is always limited by one or more factors in the environment, can be approximated by the following **logistic growth equation** that adjusts the growth rate to account for the lessening availability of limiting factors:

$$dN/dt = rN \left(\frac{K-N}{K}\right)$$

In this logistic model of population growth, the growth rate of the population (dN/dt) equals its rate of increase (r multiplied by N, the number of individuals present at any one time), adjusted for the amount of resources available. The adjustment is made by multiplying rN by the fraction of K still unused (K minus N, divided by K). As N increases (the population grows in size), the fraction by which r is multiplied (the remaining resources) becomes smaller and smaller, and the rate of increase of the population declines.

In mathematical terms, as N approaches K, the rate of population growth (dN/dt) begins to slow, until it reaches 0 when $N = K$ (figure 30.2). In practical terms, factors such as increasing competition among more individuals for a given set of resources, the buildup of waste, or an increased rate of predation causes the decline in the rate of population growth.

Graphically, if you plot N versus t (time) you obtain an S-shaped **sigmoid growth curve** characteristic of most biological populations. The curve is called "sigmoid" because its shape has a double curve like the letter S. As the size of a population stabilizes at the carrying capacity, its rate of growth slows down, eventually coming to a halt (figure 30.3).

Processes such as competition for resources, emigration, and the accumulation of toxic wastes all tend to increase as a population approaches its carrying capacity for a particular habitat. The resources for which the members of the population are competing may be food, shelter, light, mating sites, mates, or any other factor the species needs to carry out its life cycle and reproduce.

30.1 The size at which a population stabilizes in a particular place is defined as the carrying capacity of that place for that species. Populations grow to the carrying capacity of their environment.

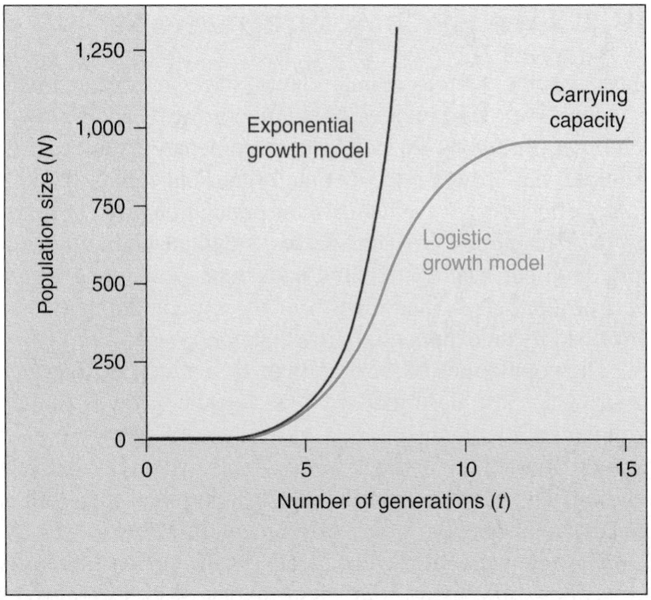

1,250
1,000
750
500
250
0
Population size (N)
Exponential
growth model
Carrying
capacity
Logistic
growth model
0
5
10
15
Number of generations (t)

Figure 30.2 Two models of population growth.
The *red line* illustrates the exponential growth model for a population with an *r* of 1.0. The *blue line* illustrates the logistic growth model in a population with $r = 1.0$ and $K = 1{,}000$ individuals. At first, logistic growth accelerates exponentially, and then, as resources become limiting, the death rate increases and growth slows. Growth ceases when the death rate equals the birthrate. The carrying capacity (K) ultimately depends on the resources available in the environment.

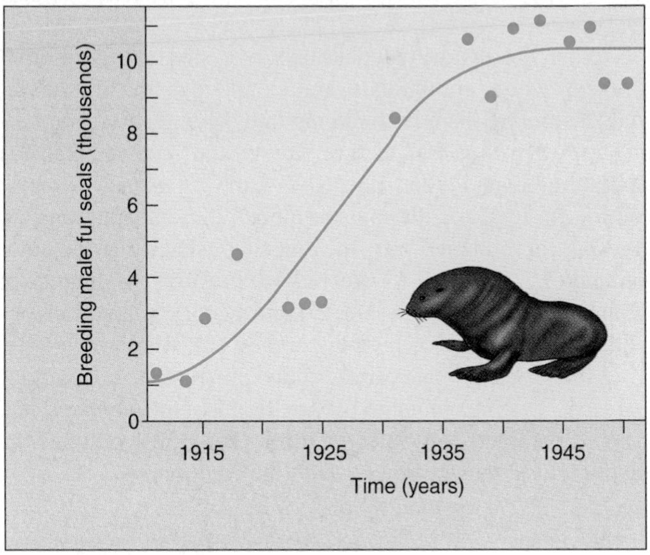

Breeding male fur seals (thousands)
10
8
6
4
2
0
1915
1925
1935
1945
Time (years)

Figure 30.3 Most natural populations exhibit logistic growth.
These data present the history of a fur seal (*Callorhinus ursinus*) population on St. Paul Island, Alaska. Driven almost to extinction by hunting in the late 1800s, the fur seal made a comeback after hunting was banned in 1911. Today the number of breeding males with "harems" oscillates around 10,000 individuals, presumably the carrying capacity of the island for fur seals.

30.2 Life History Adaptations

Many species, such as annual plants, some insects, and bacteria, have very fast rates of population growth. Their growth is not yet limited by dwindling environmental resources. For example, the growth rate of some organisms cannot be controlled effectively by reducing their population sizes or limiting their resources. In such species, small surviving populations soon enter an exponential pattern of growth and regain their original sizes. The growth of these populations is best described by an exponential growth model.

The populations of most animals have much slower rates of growth, best described by the logistic growth model. Populations of organisms that have sigmoid growth curves become limited in number as available resources become limiting. The number of individuals supported at this limit is the carrying capacity of the environment, or K.

The complete life cycle of an organism constitutes its *life history*. Life histories are very diverse, with different organisms making different choices about how to live. Some life history adaptations of a population favor very rapid growth, among them reproducing early, producing many small offspring that mature quickly, and engaging in other aspects of "big bang" reproduction. Using the terms of the exponential model, these adaptations, all favoring a high rate of increase, r, are called ***r*-selected adaptations.** Examples of organisms displaying r-selected life history adaptations include dandelions, aphids, mice, and cockroaches (figure 30.4).

Other life history adaptations favor survival in an environment where individuals are competing for limited resources, among them reproducing late, having small numbers of large offspring that mature slowly and receive intensive parental care, and other aspects of "carrying capacity" reproduction. In terms of the logistic model, these adaptations, all favoring reproduction near the carrying capacity of the environment, K, are called ***K*-selected adaptations.** Examples of organisms displaying K-selected life history adaptations include coconut palms, whooping cranes, and whales.

Most natural populations show life history adaptations that exist along a continuum ranging from completely r-selected traits to completely K-selected traits. Table 30.1 outlines the adaptations at the extreme ends of the continuum.

30.2 Some life history adaptations favor near-exponential growth, others the more competitive logistic growth. Most natural populations exhibit a combination of the two.

Figure 30.4 The consequences of exponential growth.

All organisms have the potential to produce populations larger than those that actually occur in nature. The German cockroach (*Blatella germanica*), a major household pest, produces 80 young every six months. If every cockroach that hatched survived for three generations, kitchens might look like this theoretical culinary nightmare concocted by the Smithsonian Museum of Natural History.

TABLE 30.1 *r*-SELECTED AND *K*-SELECTED LIFE HISTORY ADAPTATIONS

Adaptation	*r*-Selected Populations	*K*-Selected Populations
Age at first reproduction	Early	Late
Homeostatic capability	Limited	Often extensive
Life span	Short	Long
Maturation time	Short	Long
Mortality rate	Often high	Usually low
Number of offspring produced per reproductive episode	Many	Few
Number of reproductions per lifetime	Usually one	Often several
Parental care	None	Often extensive
Size of offspring or eggs	Small	Large

Source: After E. R. Pianka. *Evolutionary Ecology,* 4th ed. New York: Harper & Row, 1987.

30.3 The Influence of Population Density

Many factors act to regulate the growth of populations in nature. Some of these factors act independently of the size of the population; others do not.

Density-Independent Effects

Effects that are independent of the size of a population and act to regulate its growth are called **density-independent effects.** Density-independent effects, such as weather and physical disruption of the habitat, operate regardless of population size.

Density-Dependent Effects

Effects that are dependent on the size of the population and act to regulate its growth are called **density-dependent effects.** Among animals, these effects may be accompanied by hormonal changes that can alter behavior that will directly affect the ultimate size of the population. One striking example occurs in migratory locusts ("short-horned" grasshoppers). When they become crowded, the locusts produce hormones that cause them to enter a migratory phase; the locusts take off as a swarm and fly long distances to new habitats (figure 30.5). Density-dependent effects, in general, have an increasing effect as population size increases. As the population grows, the individuals in the population compete with increasing intensity for limited resources. Charles Darwin proposed that these effects result in natural selection and improved adaptation as individuals compete for the limiting factors.

Figure 30.5 Density-dependent effects.
Migratory locusts, *Locusta migratoria,* are a legendary plague of large areas of Africa and Eurasia. At high population densities, the locusts have different hormonal and physical characteristics and take off as a swarm. The most serious infestation of locusts in 30 years occurred in North Africa in 1988.

Maximizing Population Productivity

In natural systems that are exploited by humans, such as agricultural systems and fisheries, the aim is to maximize productivity by exploiting the population early in the rising portion of its sigmoid growth curve. At such times, populations and individuals are growing rapidly, and net productivity—in terms of the amount of material incorporated into the bodies of these organisms—is highest.

Commercial fisheries attempt to operate so that they are always harvesting populations in the steep, rapidly growing parts of the curve. The point of *maximal sustainable yield* lies partway up the sigmoid curve. Harvesting the population of an economically desirable species near this point will result in the best sustained yields. Overharvesting a population that is smaller than this critical size can destroy its productivity for many years or even drive it to extinction. This evidently happened in the Peruvian anchovy fishery after the populations had been depressed by the 1972 El Niño. It is often difficult to determine population levels of commercially valuable species, and without this information it is hard to determine the yield most suitable for long-term, productive harvesting.

30.3 Density-dependent effects are caused by factors that come into play particularly when the population size is larger; density-independent effects are controlled by factors that operate regardless of population size.

30.4 Population Demography

Demography is the statistical study of populations. The term comes from two Greek words: *demos,* "the people" (the same root we see in the word *democracy*), and *graphos,* "measurement." Demography therefore means measurement of people, or, by extension, of the characteristics of populations. Demography is the science that helps predict how population sizes will change in the future. Populations grow if births outnumber deaths and shrink if deaths outnumber births. Because birth and death rates depend significantly on age and sex, the future size of a population depends on its present age structure and sex ratio.

Age Structure

Many annual plants and insects time their reproduction to particular seasons of the year and then die. All members of these populations are the same age. Perennial plants and longer-lived animals contain individuals of more than one generation, so that in any given year individuals of different ages are reproducing within the population. A group of individuals of the same age is referred to as a **cohort.**

Within a population, every cohort has a characteristic birth rate, or **fecundity,** defined as the number of offspring produced in a standard time (for example, per year), and a characteristic death rate, or **mortality,** the number of individuals that die in that period. The rate of a population's growth depends directly on the difference between these two rates.

The relative number of individuals in each cohort defines a population's age structure. Because individuals of different ages have different fecundity and death rates, age structure has a critical impact on a population's growth rate. Populations with a large proportion of young individuals, for example, tend to grow rapidly because an increasing proportion of their individuals are reproductive.

Sex Ratio

The proportion of males and females in a population is its **sex ratio.** The number of births is usually directly related to the number of females, but it may not be as closely related to the number of males in species where a single male can mate with several females. In deer, elk, lions, and many other animals, a reproductive male guards a "harem" of females with which he mates, while preventing other males from mating with them. In such species, a reduction in the number of males simply changes the identities of the reproductive males without reducing the number of births. Among monogamous species like many birds, by contrast, where pairs form long-lasting reproductive relationships, a reduction in the number of males can directly reduce the number of births.

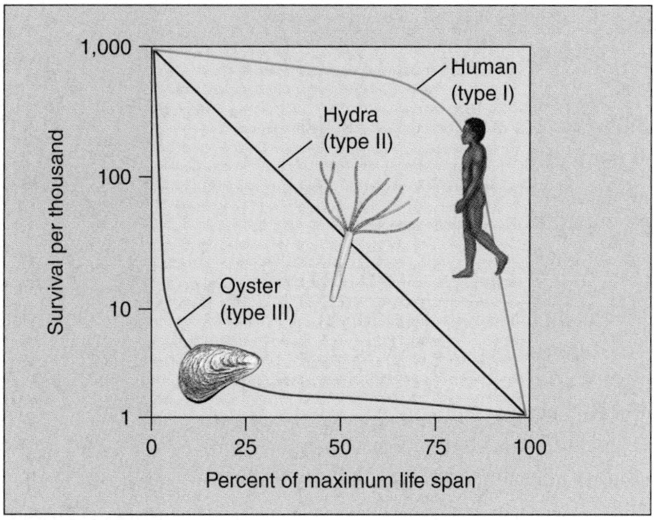

1,000
100
10
1
Survival per thousand
Human
(type I)
Hydra
(type II)
Oyster
(type III)
0
25
50
75
100
Percent of maximum life span

Figure 30.6 Survivorship curves.
By convention, survival (the *vertical axis*) is plotted on a log scale. Humans have a type I life cycle, the hydra (an animal related to jellyfish) type II, and oysters type III.

Mortality and Survivorship Curves

A population's intrinsic rate of increase depends on the ages of the organisms in it and the reproductive performance of the individuals in the various age groups. When a population lives in a constant environment for a few generations, its **age distribution**—the proportion of individuals in different age categories—tends to stabilize. This distribution differs greatly from species to species and even, to some extent, from population to population within a given species. Depending on the mating system of the species, sex ratio and generation time can also have a significant effect on population growth. A population whose size remains fairly constant through time is called a stable population. In such a population, births plus immigration must balance deaths plus emigration.

One way to express the age distribution characteristics of populations is through a **survivorship curve.** Survivorship is defined as the percentage of an original population that survives to a given age. Examples of different kinds of survivorship curves are shown in figure 30.6. In hydra, animals related to jellyfish, individuals are equally likely to die at any age, as indicated by the straight survivorship curve (type II). Oysters, like plants, produce vast numbers of offspring, only a few of which live to reproduce. However, once they become established and grow into reproductive individuals, their mortality is extremely low (type III survivorship curve). Finally, even though human babies are susceptible to death at relatively high rates, mortality in humans, as in many animals and protists, rises in the postreproductive years (type I survivorship curve).

Life Tables

Survivorship curves of natural populations tell us that different age cohorts die at very different rates. To estimate how much longer, on average, an individual of a given age can be expected to live, biologists employ **life tables.** These tables indicate the chance of survival at any given age. Ecologists use them to assess how populations in nature are changing.

Life tables can be constructed by following the fate of a cohort from birth until death, noting the number of individuals that die each year. Constructing such a cohort life table is not an easy task. A cohort must be identified and followed for many years, even though its individuals mingle freely with individuals of other cohorts. A very nice example of such a study was performed by V. Lowe on red deer (*Cervus elaphus*) on the small island of Rhum, Scotland, from 1957 to 1966. The deer live for up to 16 years, and females become capable of breeding when 4 years old. In 1957, Lowe and his coworkers made a careful count of all the deer on the island, including the number of calves (deer less than 1 year old). Lowe chose the female calves as the cohort he would follow in subsequent years. Each year from 1957 to 1966, every deer that had died from natural causes or been shot was examined and its age determined by examining its teeth. This let Lowe determine which dead deer had been calves in 1957. The life table for this cohort of female red deer appears in table 30.2, and the indicated survivorship curve is shown in figure 30.7. The convex shape of the survivorship curve (type I) indicates a fairly consistent increase in the risk of mortality with age.

In table 30.2, the first column indicates the age of the cohort (that is, the number of years since 1957). The second column indicates the proportion of the original 1957 cohort still alive at the beginning of that year. The third column indicates the proportion of the original cohort that died during that year. The fourth column presents the **mortality rate,** the proportion of individuals that started that year alive but died by the end of it. Little mortality occurs among the young; most deer in this female cohort die only after reaching reproductive age (4 years old).

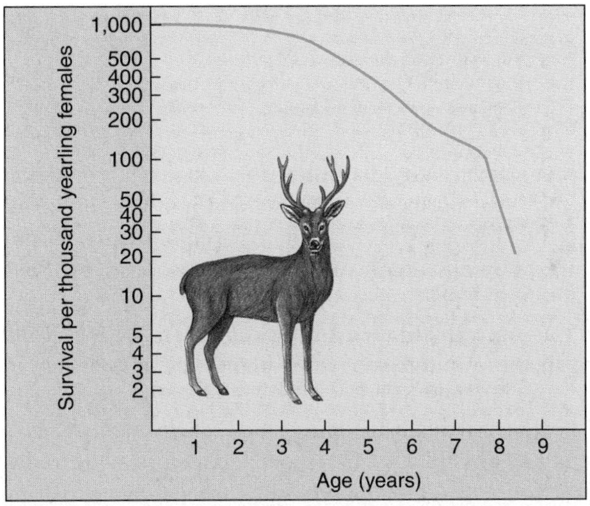

Survival per thousand yearling females
1,000
500
400
300
200
100
50
40
30
20
10
5
4
3
2
1 2 3 4 5 6 7 8 9
Age (years)

Figure 30.7 Survivorship curve for a cohort of female red deer on the island of Rhum.
The cohort studied consisted of calves in the year 1957. Reproductive maturity typically begins at age 4 years. Only then does mortality begin to be significant.

30.4 The growth rate of a population is a sensitive function of its age structure. In some species mortality is focused among the young and in others among the old; in only a few is mortality independent of age.

TABLE 30.2 LIFE TABLE FOR A COHORT OF FEMALE RED DEER

Age (Years) x	Proportion of Original Cohort Surviving to the Beginning of Age-Class X l_x	Proportion of Original Cohort Dying During Age-Class X d_x	Mortality Rate a_x
1	1.000	0	0
2	1.000	0.061	0.061
3	0.939	0.185	0.197
4	0.754	0.249	0.330
5	0.505	0.200	0.396
6	0.305	0.119	0.390
7	0.186	0.054	0.290
8	0.132	0.107	0.810
9	0.025	0.025	1.000

Source: Lowe, V. P. (1969) Population dynamics of the red deer *(Cervus elaphus)* on Rhum. *Journal of Animal Ecology 38*:425–57.

30.5 Communities

The magnificent redwood forest that extends along the coast of central and northern California and into the southwestern corner of Oregon is an example of a **community.** Within it, the most obvious organisms are the redwood trees, *Sequoia sempervirens.* These trees are the sole survivors of a genus that was once distributed throughout much of the Northern Hemisphere. A number of other plants and animals are regularly associated with redwood trees, including the sword fern and beetle illustrated in figure 30.8. Their coexistence is in part made possible by the special conditions the redwood trees themselves create, providing shade, water (dripping from the branches), and relatively cool temperatures. This particular distinctive assemblage of organisms is called the redwood community. The organisms characteristic of this community have each had a complex and unique evolutionary history. They evolved at different times in the past and then came to be associated with the redwoods.

We recognize this community mainly because of the redwood trees, and its boundaries are determined by the redwood's distribution. The distributions of the other organisms in the redwood community may differ a good deal. Some organisms may not be distributed as widely as the redwoods, and some may be distributed over a broader range. In the redwood community or any other community, the ranges of the different organisms overlap; that is why they occur together.

Many communities are very similar in species composition and appearance over wide areas. For example, the open savanna that stretches across much of Africa includes many plant and animal species that coexist over thousands of square kilometers. Interactions between these organisms occur in a similar manner throughout these grassland communities, and some interactions have evolved over millions of years.

30.5 We recognize a community largely because of the presence of its dominant species, but many other kinds of organisms are also characteristic of each community. A community exists in a place because the ranges of its species overlap there.

(a)

(b)

(c)

(d)

Figure 30.8 The redwood community.

(*a*) The redwood forest of coastal California and southwestern Oregon is dominated by the redwoods (*Sequoia sempervirens*) themselves. Other organisms in the redwood community include (*b*) sword ferns (*Polystichum munitum*), (*c*) redwood sorrel (*Oxalis oregana*), and (*d*) ground beetles (*Scaphinotus velutinus*), this one feeding on a slug on a sword fern leaf. The ecological requirements of each of these organisms differ, but they overlap enough that the organisms occur together in the redwood community.

30.6 The Niche and Competition

Each organism in an ecosystem confronts the challenge of survival in a different way. The **niche** an organism occupies is the sum total of all the ways it uses the resources of its environment. The niche of a species may be thought of as its biological role in the community. A niche may be described in terms of space utilization, food consumption, temperature range, appropriate conditions for mating, requirements for moisture, and other factors. *Niche* is not synonymous with **habitat,** the place where an organism lives. *Habitat* is a place, *niche* a pattern of living.

Sometimes organisms are not able to occupy their entire niche because somebody else is using it. We call such situations when two organisms attempt to use the same resource **competition.** Competition is the struggle of two organisms to use the same resource when there is not enough of the resource to satisfy both. Fighting over resources is referred to as **interference competition;** consuming shared resources is called **exploitative competition.**

Interspecific competition refers to the interactions between individuals of different species when both require the same scarce resource. Interspecific competition is often greatest between organisms that obtain their food in similar ways; thus, green plants compete mainly with other green plants, herbivores with other herbivores, and carnivores with other carnivores. In addition, competition is more acute between similar organisms than between those that are less similar. While interspecific competition occurs between members of different species, it is to be distinguished from **intraspecific competition,** which occurs between individuals of a single species.

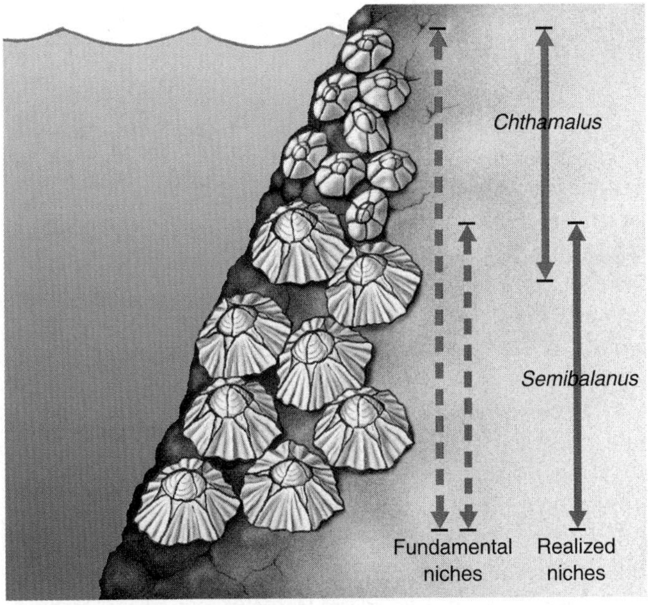

Chthamalus
Semibalanus
Fundamental niches
Realized niches

Figure 30.9 Competition among two species of barnacles limits niche use.
Chthamalus can live in both deep and shallow zones (its fundamental niche), but *Semibalanus* forces *Chthamalus* out of the part of its fundamental niche that overlaps the realized niche of *Semibalanus.*

The Realized Niche

Because of competition, organisms may not be able to occupy the entire niche they are theoretically capable of using, called the **fundamental niche** (or theoretical niche). The actual niche the organism is able to occupy in the presence of competitors is called its **realized niche.**

In a classic study, J. H. Connell of the University of California, Santa Barbara, investigated competitive interactions between two species of barnacles that grow together on rocks along the coast of Scotland. Barnacles are marine animals (crustaceans) that have free-swimming larvae. The larvae eventually settle down, cementing themselves to rocks and remaining attached for the rest of their lives. Of the two species Connell studied, *Chthamalus stellatus* lives in shallower water, where tidal action often exposed it to air, and *Semibalanus balanoides* (called *Balanus balanoides* prior to 1995) lives lower down, where it is rarely exposed to the atmosphere (figure 30.9). In the deeper zone, *Semibalanus* could always outcompete *Chthamalus* by crowding it off the rocks, undercutting it, and replacing it even where it had begun to grow. When Connell removed *Semibalanus* from the area, however, *Chthamalus* was easily able to occupy the deeper zone, indicating that no physiological or other general obstacles prevented it from becoming established there. In contrast, *Semibalanus* could not survive in the shallow-water habitats where *Chthamalus* normally occurs; it evidently does not have the special physiological and morphological adaptations that allow *Chthamalus* to occupy this zone. Thus, the fundamental niche of the barnacle *Chthamalus* in Connell's experiments in Scotland included that of *Semibalanus,* but its realized niche was much narrower because *Chthamalus* was outcompeted by *Semibalanus* in its theoretical niche.

Another interesting example illustrating the properties of a niche involves flour beetles of the genus *Tribolium.* If *Tribolium* is grown in pure flour along with beetles of a second genus, *Oryzaephilus,* it drives *Oryzaephilus* to extinction by means that are only partly understood. If, however, small pieces of glass tubing are placed in the flour, providing refuges for *Oryzaephilus,* then both kinds of flour beetle coexist indefinitely. This experiment suggests why so many kinds of organisms can coexist in a complex ecosystem, such as a tropical rain forest.

30.6 A niche may be defined as the way in which an organism uses its environment.

30.7 Competitive Exclusion

In classic experiments carried out between 1934 and 1935, Russian ecologist G. F. Gause studied competition among three species of *Paramecium,* a tiny protist. All three species grew well alone in culture tubes, preying on bacteria and yeasts that fed on oatmeal suspended in the culture fluid. However, when Gause grew *P. aurelia* together with *P. caudatum* in the same culture tube, the numbers of *P. caudatum* always declined to extinction, leaving *P. aurelia* the only survivor. Why? Gause found *P. aurelia* was able to grow six times faster than its competitor, *P. caudatum,* because it was able to better use the limited available resources.

From experiments such as this, Gause formulated what is now called the *principle of competitive exclusion.* This principle states that if two species are competing for a resource, the species that uses the resource more efficiently will eventually eliminate the other locally—no two species with the same niche can coexist.

Is competitive exclusion the inevitable outcome of competition for limited resources, as Gause's principle states? No. The outcome depends on the fierceness of the competition and on the degree of similarity between the fundamental niches of the competing species. If the species can avoid competing, they may coexist.

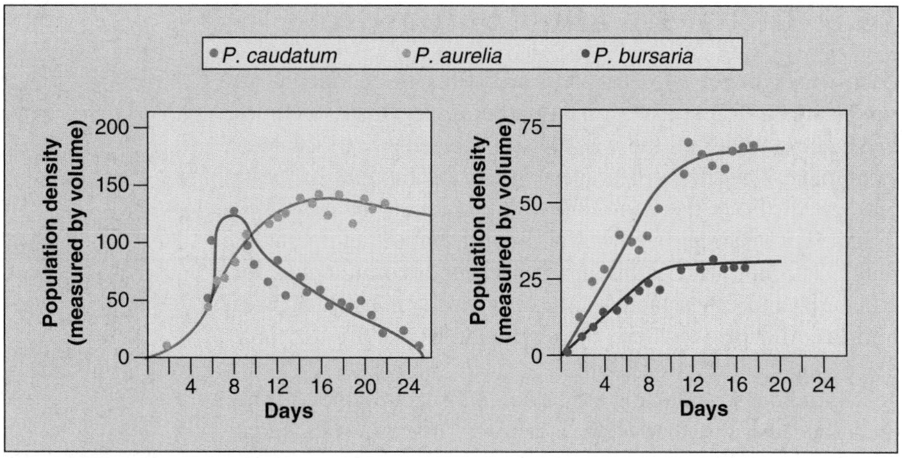

P. caudatum
P. aurelia
P. bursaria
Population density (measured by volume)
200
150
100
50
0
4 8 12 16 20 24
Days
75
50
25
0
4 8 12 16 20 24
Days

Figure 30.10 Competitive exclusion among three species of *Paramecium.*

In the microscopic world, *Paramecium* is a ferocious predator. *Paramecia* eat by ingesting their prey; their plasma membranes surround bacterial or yeast cells, forming a food vacuole containing the prey cell. In his experiments, Gause found that *P. caudatum* would decline to extinction when grown with *P. aurelia* because they shared the same realized niche, and *P. aurelia* outcompeted *P. caudatum* for food resources. However, *P. caudatum* and *P. bursaria* were able to coexist because the two have different realized niches and thus avoid competition.

Niche Overlap

In a revealing experiment, Gause challenged *P. caudatum*—the defeated species in his earlier experiments—with a third species, *P. bursaria.* Because he expected these two species to also compete for the limited bacterial food supply, Gause thought one would win out, as had happened in his previous experiments. But that's not what happened. Instead, both species survived in the culture tubes; the paramecia found a way to divide the food resources. How did they do it? In the upper part of the culture tubes, where the oxygen concentration and bacterial density were high, *P. caudatum* dominated because it was better able to feed on bacteria. However, in the lower part of the tubes, the lower oxygen concentration favored the growth of a different potential food, yeast, and *P. bursaria* was better able to eat this food. The fundamental niche of each species was the whole culture tube, but the realized niche of each species was only a portion of the tube. Because the niches of the two species did not overlap too much, both species were able to survive. Figure 30.10 summarizes Gause's experiments. The graph also demonstrates the negative effect competition had on the participants. Both species reach about twice the density when grown without a competitor.

Competitive Exclusion

Gause's principle of competitive exclusion can be restated to say that *no two species can occupy the same niche indefinitely.* Certainly species can and do coexist while competing for the same resources; we have seen many examples of such relationships. Nevertheless, Gause's theory predicts that when two species coexist on a long-term basis, their niches always differ in one or more features; otherwise, one species outcompetes the other and the extinction of the second species inevitably results, a process referred to as **competitive exclusion.**

Niche is, of course, a complex concept, one that involves all facets of the environment that are important to individual species. In recent years, a vigorous debate has arisen concerning the role of competitive exclusion, not only in determining the structure of communities but also in setting the course of evolution. When one or more resources are obviously limited, as in periods of drought, the role of competition becomes much more obvious than when they are not. On the other hand, especially for plants, the factors important in defining a niche are often difficult to demonstrate, and alternative explanations are being sought for the coexistence of large numbers of species.

30.7 No two species can occupy the same niche indefinitely without competition driving one to extinction.

30.8 Resource Partitioning

Gause's exclusion principle has a very important consequence: persistent competition between two species is rare in natural communities. Either one species drives the other to extinction, or natural selection reduces the competition between them. When the late Princeton ecologist Robert MacArthur studied five species of warblers, small insect-eating forest songbirds, he found that they all appeared to be competing for the same resources. However, when he studied them more carefully, he found that each species actually fed in a different part of spruce trees and so ate different subsets of insects. One species fed on insects near the tips of branches, a second within the dense foliage, a third on the lower branches, a fourth high on the trees, and a fifth at the very apex of the trees. Thus, the species of warblers had *subdivided the niche,* partitioning the available resource so as to avoid direct competition with one another.

Resource partitioning can often be seen in similar species that occupy the same geographical area. Called **sympatric species** (Greek, *syn,* same, and *patria,* country), these species avoid competition by living in different portions of the habitat or by using different food or other resources (figure 30.11). Species that do not live in the same geographical area, called **allopatric species** (Greek, *allos,* other, and *patria,* country), often use the same habitat locations and food resources—because they are not in competition, natural selection does not favor evolutionary changes that subdivide their niche.

When a pair of species occupy the same habitat (that is, when they are sympatric), they tend to exhibit greater differences in morphology and behavior than the same two species do when living in different habitats (that is, when they are allopatric). Called **character displacement,** the differences evident between sympatric species are thought to have been favored by natural selection as a mechanism to facilitate habitat partitioning and thus reduce competition. Thus, the two Darwin's finches in figure 30.12 have bills of similar size where the finches are allopatric, each living on an island where the other does not occur. On islands where they are sympatric, the two species have evolved beaks of different sizes, one adapted to larger seeds, the other to smaller ones.

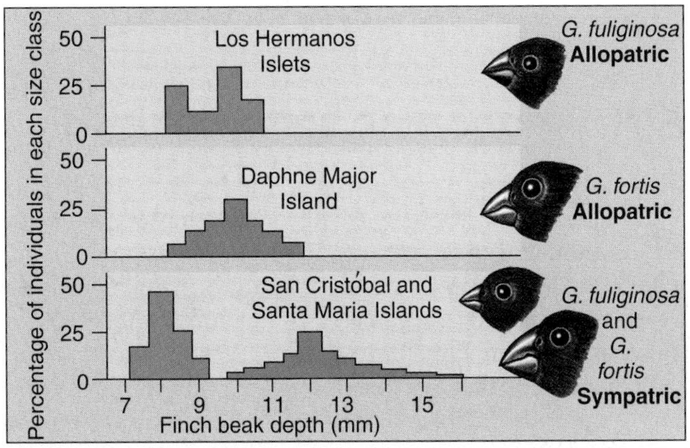
Percentage of individuals in each size class
50
25
0
Los Hermanos Islets
G. fuliginosa
Allopatric
Daphne Major Island
G. fortis
Allopatric
San Cristóbal and Santa Maria Islands
G. fuliginosa and G. fortis
Sympatric
7
9
11
13
15
Finch beak depth (mm)

Figure 30.12 Character displacement in Darwin's finches.

These two species of finches (genus *Geospiza*) have bills of similar size when allopatric but different size when sympatric.

30.8 Sympatric species partition available resources, reducing competition between them.

(a)
(b)
(c)
(d)

Figure 30.11 Resource partitioning among sympatric lizard species.

Species of *Anolis* lizards in the Caribbean studied by Washington University ecologist Jonathan Losos partition their tree habitats in a variety of ways. Some species of anoles occupy the canopy of trees (*a*), others use twigs on the periphery (*b*), and still others are found at the base of the trunk (*c*). In addition, some use grassy areas in the open (*d*). When two species occupy the same part of the tree, they either use different-sized insects as food or partition the thermal microhabitat; for example, one might be found only in the shade, whereas the other would bask only in the sun. Most interestingly, the same pattern of resource partitioning has evolved independently on different Caribbean islands.

30.9 Coevolution and Symbiosis

The plants, animals, protists, fungi, and bacteria that live together in communities have changed and adjusted to one another continually over millions of years. For example, many features of flowering plants have evolved in relation to the dispersal of the plant's gametes by animals (figure 30.13). These animals, in turn, have evolved a number of special traits that enable them to obtain food or other resources efficiently from the plants they visit, often from their flowers. In addition, the seeds of many flowering plants have features that make them more likely to be dispersed to new areas of favorable habitat.

Such interactions, which involve the long-term, mutual evolutionary adjustment of the characteristics of the members of biological communities, are examples of **coevolution.** In this section, we consider some examples of coevolution, including symbiotic relationships and predator-prey interactions.

Symbiosis Is Widespread

In symbiotic relationships, two or more kinds of organisms live together in often elaborate and more or less permanent relationships. All symbiotic relationships carry the potential for coevolution between the organisms involved, and in many instances the results of this coevolution are fascinating. Examples of symbiosis include lichens, which are associations of certain fungi with green algae or cyanobacteria (see chapter 15). Another important example are mycorrhizae, the association between fungi and the roots of most kinds of plants. The fungi expedite the plant's absorption of certain nutrients, and the plants in turn provide the fungi with carbohydrates. Similarly, root nodules that occur in legumes and certain other kinds of plants contain bacteria that fix atmospheric nitrogen and make it available to their host plants.

In the tropics, leaf-cutter ants are often so abundant that they can remove a quarter or more of the total leaf surface of the plants in a given area. They do not eat these leaves directly; rather, they take them to underground nests, where they chew them up and inoculate them with the spores of particular fungi. These fungi are cultivated by the ants and brought from one specially prepared bed to another, where they grow and reproduce. In turn, the fungi constitute the primary food of the ants and their larvae. The relationship between leaf-cutter ants and these fungi is an excellent example of symbiosis.

Figure 30.13 Pollination by bat.
Many flowers have coevolved with other species to facilitate pollen transfer. Insects are widely known as pollinators, but they're not the only ones. Notice the cargo of pollen on the bat's snout.

Kinds of Symbiosis

The major kinds of symbiotic relationships include (1) **commensalism,** in which one species benefits while the other neither benefits nor is harmed; (2) **mutualism,** in which both participating species benefit; and (3) **parasitism,** in which one species benefits but the other is harmed. Parasitism can also be viewed as a form of predation, although the organism that is preyed upon does not necessarily die.

30.9 Coevolution is a term that describes the long-term evolutionary adjustments of species to one another. In symbiosis, two or more species live together.

30.10 Commensalism

Commensalism is a symbiotic relationship that benefits one species and neither hurts nor helps the other. In nature, individuals of one species are often physically attached to members of another. For example, epiphytes are plants that grow on the branches of other plants. In general, the host plant is unharmed, while the epiphyte that grows on it benefits. Similarly, various marine animals, such as barnacles, grow on other, often actively moving sea animals like whales and thus are carried passively from place to place. These "passengers" presumably gain more protection from predation than they would if they were fixed in one place, and they also reach new sources of food. The increased water circulation that such animals receive as their host moves around may be of great importance, particularly if the passengers are filter feeders.

Figure 30.14 Commensalism in the sea.

Clownfishes, such as this *Amphiprion perideraion* in Guam, often form symbiotic associations with sea anemones, gaining protection by remaining among their tentacles and gleaning scraps from their food. Different species of anemones secrete different chemical mediators; these attract particular species of fishes and may be toxic to the fish species that occur symbiotically with other species of anemones in the same habitat. There are 26 species of clownfishes, all found only in association with sea anemones; 10 species of anemones are involved in such associations, so that some of the anemone species are host to more than one species of clownfish.

Examples of Commensalism

The best-known examples of commensalism involve the relationships between certain small tropical fishes and sea anemones, marine animals that have stinging tentacles (see chapter 19). These fish have evolved the ability to live among the tentacles of sea anemones, even though these tentacles would quickly paralyze other fishes that touched them (figure 30.14). The anemone fishes feed on the detritus left from the meals of the host anemone, remaining uninjured under remarkable circumstances.

On land, an analogous relationship exists between birds called oxpeckers and grazing animals such as cattle or rhinoceros. The birds spend most of their time clinging to the animals, picking off parasites and other insects, carrying out their entire life cycles in close association with the host animals. Cattle egrets, which have extended their range greatly during the past few decades, engage in a loosely coupled relationship of this kind (figure 30.15).

Figure 30.15 Commensalism between cattle egrets and an African cape buffalo.

The egrets eat insects off the buffalo.

When Is Commensalism Commensalism?

In each of these instances, it is difficult to be certain whether the second partner receives a benefit or not; there is no clear-cut boundary between commensalism and mutualism. For instance, it may be advantageous to the sea anemone to have particles of food removed from its tentacles; it may then be better able to catch other prey. Similarly, while often thought of as commensalism, the association of grazing mammals and gleaning birds is actually an example of mutualism. The mammal benefits by having parasites and other insects removed from its body, but the birds also benefit by gaining a dependable source of food.

30.10 Commensalism is the benign use of one organism by another.

30.11 Mutualism

Mutualism is a symbiotic relationship among organisms in which both species benefit. Examples of mutualism are of fundamental importance in determining the structure of biological communities. Some of the most spectacular examples of mutualism occur among flowering plants and their animal visitors, including insects, birds, and bats. As we discussed in chapter 18, during the course of their evolution, the characteristics of flowers have evolved in large part in relation to the characteristics of the animals that visit them for food and, in doing so, spread their pollen from individual to individual. At the same time, characteristics of the animals have changed, increasing their specialization for obtaining food or other substances from particular kinds of flowers.

Another example of mutualism involves ants and aphids. Aphids, also called greenflies, are small insects that suck fluids with their piercing mouthparts from the phloem of living plants. They extract a certain amount of the sucrose and other nutrients from this fluid, but they excrete much of it in an altered form through their anus. Certain ants have taken advantage of this—in effect, domesticating the aphids (figure 30.16). The ants carry the aphids to new plants, where they come into contact with new sources of food, and then consume as food the "honeydew" that the aphids excrete.

Figure 30.16 Mutualism: ants and aphids.
These ants are tending to aphids, feeding on the "honeydew" that the aphids excrete continuously, moving the aphids from place to place, and protecting them from potential predators.

Ants and Acacias

A particularly striking example of mutualism involves ants and certain Latin American species of the plant genus *Acacia.* In these species, certain leaf parts, called stipules, are modified as paired, hollow thorns; these particular species are called "bull's horn acacias." The thorns are inhabited by stinging ants of the genus *Pseudomyrmex,* which do not nest anywhere else. Like all thorns that occur on plants, the acacia horns serve to deter herbivores.

At the tip of the leaflets of these acacias are unique, protein-rich bodies called Beltian bodies, named after Thomas Belt, a nineteenth-century British naturalist who first wrote about them after seeing them in Nicaragua. Beltian bodies do not occur in species of *Acacia* that are not inhabited by ants, and their role is clear: they serve as a primary food for the ants. In addition, the plants secrete nectar from glands near the bases of their leaves. The ants consume this nectar as well, feeding it and the Beltian bodies to their larvae.

Obviously, this association is beneficial to the ants, and one can readily see why they inhabit acacias of this group. The ants and their larvae are protected within the swollen thorns, and the trees provide a balanced diet, including the sugar-rich nectar and the protein-rich Beltian bodies. What, if anything, do the ants do for the plants? This question had fascinated observers for nearly a century until it was answered by Daniel Janzen, then a graduate student at the University of California, Berkeley, in a beautifully conceived and executed series of field experiments.

Whenever any herbivore lands on the branches or leaves of an acacia inhabited by ants, the ants immediately attack and devour the herbivore. Thus, the ants protect the acacias from being eaten, and the herbivore also provides additional food for the ants, which continually patrol the acacia's branches. Related species of acacias that do not have the special features of the bull's horn acacias and are not protected by ants have bitter-tasting substances in their leaves that the bull's horn acacias lack. Evidently, these bitter-tasting substances protect the acacias in which they occur from herbivores in a different way.

The ants that live in the bull's horn acacias also help their hosts to compete with other plants. The ants cut away any branches of other plants that touch the bull's horn acacia in which they are living. They create, in effect, a tunnel of light through which the acacia can grow, even in the lush deciduous forests of lowland Central America. Without the ants, as Janzen showed experimentally by poisoning the ant colonies that inhabited individual plants, the acacia is unable to compete successfully in this habitat. Finally, the ants bring organic material into their nests. The parts they do not consume, together with their excretions, provide the acacias with an abundant source of nitrogen.

30.11 Mutualism involves cooperation between species, to the mutual benefit of both.

30.12 Parasitism

Parasitism may be regarded as a special form of symbiosis in which the predator, or parasite, is much smaller than the prey and remains closely associated with it. Parasitism is harmful to the prey organism and beneficial to the parasite. The concept of parasitism seems obvious, but individual instances are often surprisingly difficult to distinguish from predation and from other kinds of symbiosis.

External Parasites

Parasites that feed on the exterior surface of an organism are external parasites, or **ectoparasites.** Many instances of external parasitism are known (figure 30.17*a*). Lice, which live on the bodies of vertebrates—mainly birds and mammals—are normally considered parasites. Mosquitoes are not considered parasites, even though they draw food from birds and mammals in a similar manner to lice, because their interaction with their host is so brief.

Parasitoids are insects that lay eggs on living hosts. This behavior is common among wasps, whose larvae feed on the body of the unfortunate host, often killing it.

Internal Parasites

Vertebrates are parasitized internally by **endoparasites,** members of many different phyla of animals and protists. Invertebrates also have many kinds of parasites that live within their bodies. Bacteria and viruses are not usually considered parasites, even though they fit our definition precisely.

Internal parasitism is generally marked by much more extreme specialization than external parasitism, as shown by the many protist and invertebrate parasites that infect humans. The more closely the life of the parasite is linked with that of its host, the more its morphology and behavior are likely to have been modified during the course of its evolution. The same is true of symbiotic relationships of all sorts. Conditions within the body of an organism are different from those encountered outside and are apt to be much more constant. Consequently, the structure of an internal parasite is often simplified, and unnecessary armaments and structures are lost as it evolves.

Brood Parasitism

Not all parasites consume the body of their host. In brood parasitism, birds like cowbirds and European cuckoos lay their eggs in the nests of other species (figure 30.17*b*). The host parents raise the brood parasite as if it were one of their own clutch, in many cases investing more in feeding the imposter than in feeding their own offspring. The brood parasite reduces the reproductive success of the foster parent hosts, so it is not surprising that evolution has fostered the hosts' ability to detect parasite eggs and reject them.

(a)

(b)

Figure 30.17 Parasitism.

(a) Dodder (*Cuscuta*) is a parasitic plant that has lost its chlorophyll and its leaves in the course of its evolution and is heterotrophic—unable to manufacture its own food. Instead, it obtains its food from the host plants it grows on. (*b*) Brood parasitism. Cuckoos lay their eggs in the nests of other species of birds. Because the young cuckoos (large bird to the *right*) are raised by a different species (like this meadow pipit, smaller bird to the *left*), they have no opportunity to *learn* the cuckoo song; the cuckoo song they later sing is innate.

30.12 In parasitism, one organism serves as a host to another organism, usually to the host's disadvantage.

30.13 Plant Defenses

Predator-prey interactions are interactions between organisms in which one organism uses the other for food. Plants have evolved many mechanisms to defend themselves from herbivores. The most obvious are morphological defenses: thorns, spines, and prickles play an important role in discouraging browsers, and plant hairs, especially those that have a glandular, sticky tip, deter insect herbivores. Some plants, such as grasses, deposit silica in their leaves, both strengthening and protecting themselves. If enough silica is present in their cells, these plants are simply too tough to eat.

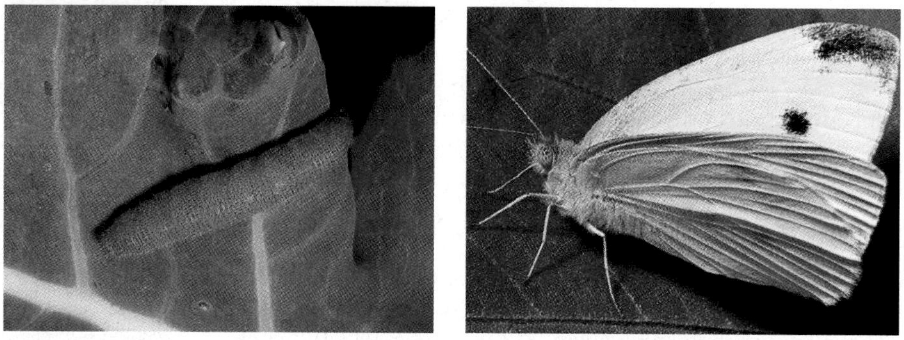

(a) (b)

Figure 30.18 Insect herbivores are well suited to their hosts.

(*a*) The green caterpillars of the cabbage butterfly, *Pieris rapae,* are camouflaged on the leaves of cabbage and other plants on which they feed. Although mustard oils protect these plants against most herbivores, the cabbage butterfly caterpillars are able to break down the mustard oil compounds. (*b*) An adult cabbage butterfly.

Chemical Defenses

Significant as these morphological adaptations are, the chemical defenses that occur so widely in plants are even more crucial. Best known and perhaps most important in the defenses of plants against herbivores are secondary chemical compounds. These are distinguished from primary compounds, which are regular components of the major metabolic pathways, such as respiration. Virtually all plants, and apparently many algae as well, contain very structurally diverse secondary compounds that are either toxic to most herbivores or disturb their metabolism so greatly that they are unable to complete normal development. Consequently, most herbivores tend to avoid the plants that possess these compounds.

The mustard family (Brassicaceae) is characterized by a group of chemicals known as mustard oils. These are the substances that give the pungent aromas and tastes to such plants as mustard, cabbage, watercress, radish, and horseradish. The same tastes we enjoy signal the presence of toxic chemicals to many groups of insects. Similarly, plants of the milkweed family (Asclepiadaceae) and the related dogbane family (Apocynaceae) produce a milky sap that deters herbivores from eating them. In addition, these plants usually contain cardiac glycosides, molecules named for their drastic effect on heart function in vertebrates.

The Evolutionary Response of Herbivores

Certain groups of herbivores are associated with each family or group of plants protected by a particular kind of secondary compound. These herbivores are able to feed on these plants without harm, often as their exclusive food source. For example, cabbage butterfly caterpillars (subfamily Pierinae) feed almost exclusively on plants of the mustard and caper families, as well as on a few other small families of plants that also contain mustard oils (figure 30.18). Similarly, caterpillars of monarch butterflies and their relatives (subfamily Danainae) feed on plants of the milkweed and dogbane families. How do these animals manage to avoid the chemical defenses of the plants, and what are the evolutionary precursors and ecological consequences of such patterns of specialization?

We can offer a potential explanation for the evolution of these particular patterns. Once the ability to manufacture mustard oils evolved in the ancestors of the caper and mustard families, the plants were protected for a time against most or all herbivores that were feeding on other plants in their area. At some point, certain groups of insects—for example, the cabbage butterflies—developed the ability to break down mustard oils and thus feed on these plants without harming themselves. Having evolved this ability, the butterflies were able to use a new resource without competing with other herbivores for it. Often, in groups of insects such as cabbage butterflies, sense organs have evolved that are able to detect the secondary compounds that their food plants produce. Clearly, the relationship that has formed between cabbage butterflies and the plants of the mustard and caper families is an example of coevolution.

30.13 The members of many groups of plants are protected from most herbivores by their secondary compounds. Once the members of a particular herbivore group evolve the ability to feed on them, these herbivores gain access to a new resource, which they can exploit without competition from other herbivores.

30.14 Animal Defenses

Some animals that feed on plants rich in secondary compounds receive an extra benefit. When the caterpillars of monarch butterflies feed on plants of the milkweed family, they do not break down the cardiac glycosides that protect these plants from herbivores. Instead, the caterpillars concentrate and store the cardiac glycosides in fat bodies; they then pass them through the chrysalis stage to the adult and even to the eggs of the next generation. The incorporation of cardiac glycosides thus protects all stages of the monarch life cycle from predators. A bird that eats a monarch butterfly quickly regurgitates it and in the future avoids the conspicuous orange-and-black pattern that characterizes the adult monarch (figure 30.19). Some birds, however, appear to have acquired the ability to tolerate the protective chemicals. These birds eat the monarchs.

(a) (b)

Figure 30.19 A blue jay learns that monarch butterflies taste bad.

(*a*) This cage-reared jay, which had never seen a monarch butterfly before, tried eating one. (*b*) The same jay regurgitated the butterfly a few minutes later. This bird is not likely to attempt to eat an orange-and-black insect again.

Defensive Coloration

Many insects that feed on milkweed plants are brightly colored; they advertise their poisonous nature using an ecological strategy known as **warning coloration** or **aposematic coloration.** Showy coloration is characteristic of animals that use poisons and stings to repel predators, while organisms that lack specific chemical defenses are seldom brightly colored. In fact, many have **cryptic coloration**—color that blends with the surroundings and thus hides the individual from predators (figure 30.20). Camouflaged animals usually do not live together in groups because a predator that discovers one individual gains a valuable clue to the presence of others.

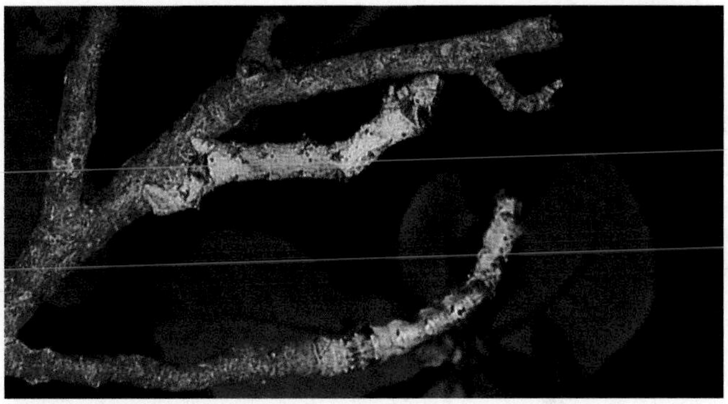

Figure 30.20 Cryptic coloration.

An inchworm caterpillar (*Necophora quernaria*) closely resembles a twig.

Chemical Defenses

Animals also manufacture and use a startling array of substances to perform a variety of defensive functions. Bees, wasps, predatory bugs, scorpions, spiders, and many other arthropods use chemicals to defend themselves and to kill their prey. In addition, various chemical defenses have evolved among marine animals and the vertebrates, including venomous snakes, lizards, fishes, and some birds. The poison-dart frogs of the family Dendrobatidae produce toxic alkaloids in the mucus that covers their brightly colored skin (figure 30.21). Some of these toxins are so powerful that a few micrograms will kill a person if injected into the bloodstream. More than 200 different alkaloids have been isolated from these frogs, and some are playing important roles in neuromuscular research. There is an intensive investigation of marine animals, algae, and flowering plants for new drugs to fight cancer and other diseases, or as sources of antibiotics.

Figure 30.21 Vertebrate chemical defenses.

Dendrobatid frogs advertise their toxicity with aposematic coloration, as shown here.

30.14 Animals defend themselves against predators with warning coloration, camouflage, and chemical defenses such as poisons and stings.

30.15 Predator-Prey Cycles

Figure 30.22 Wolves chasing a moose—what will the outcome be?

On Isle Royale, Michigan, a large pack of wolves pursue a moose. They chased this moose for almost 2 kilometers; it then turned and faced the wolves, who by that time were exhausted from running through chest-deep snow. The wolves lay down, and the moose walked away.

Predation is the consuming of one organism by another. In this sense, predation includes everything from a leopard capturing and eating an antelope to a deer grazing on spring grass.

The relationships between large carnivores and grazing mammals have a major impact on biological communities in many parts of the world. Appearances, however, are sometimes deceiving. On Isle Royale in Lake Superior, moose reached the island by crossing over ice in an unusually cold winter and multiplied freely there in isolation. When wolves later reached the island by crossing over the ice, naturalists widely assumed that the wolves were playing a key role in controlling the moose population. More careful studies have demonstrated that this is not in fact the case. The moose that the wolves eat are, for the most part, old or diseased animals that would not survive long anyway. In general, the moose are controlled by food availability, disease, and other factors rather than by the wolves (figure 30.22).

Refuges Promote Cycles

When experimental populations are set up under simple laboratory conditions, the predator often exterminates its prey and then becomes extinct itself, having nothing left to eat (figure 30.23). However, if refuges are provided for the prey, its population drops to low levels but not to extinction. Low prey population levels then provide inadequate food for the predators, causing the predator population to decrease. When this occurs, the prey population can recover. In this situation the predator and prey populations may continue in this cyclical pattern for some time.

Cycles in Hare Populations: A Case Study

Population cycles are characteristic of some species of small mammals, such as lemmings, and they appear to be stimulated, at least in some situations, by their predators. Ecologists have studied cycles in hare populations since the 1920s. They have found that the North American snowshoe hare *Lepus americanus* follows a "10-year cycle" (in reality, it varies from 8 to 11 years). Its numbers fall 10-fold to 30-fold in a typical cycle, and 100-fold changes can occur. Two

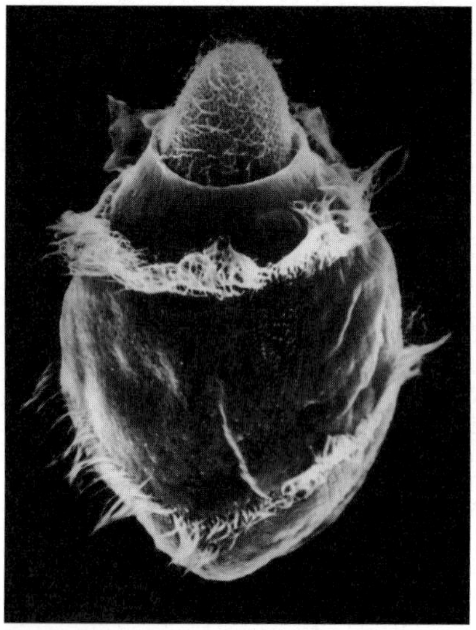

(a)

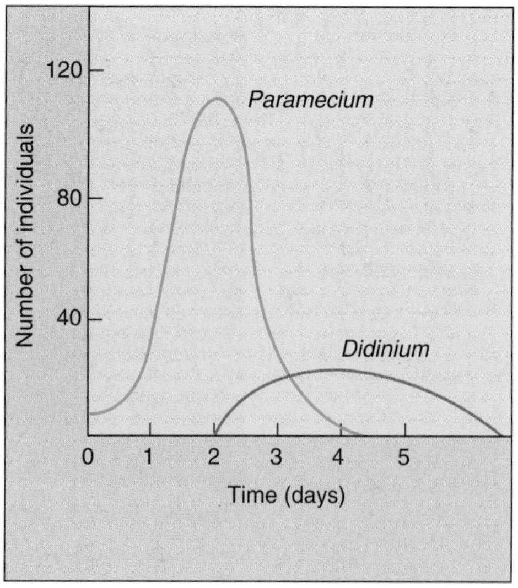

120
80
40
Number of individuals
Paramecium
Didinium
0
1
2
3
4
5
Time (days)

(b)

Figure 30.23 Predator-prey in the microscopic world.

(*a*) Egg-shaped *Didinium,* the predator, has almost completely ingested its prey, the smaller protist *Paramecium.* (*b*) The graph demonstrates that when *Didinium* is added to a *Paramecium* population, the numbers of *Didinium* initially rise, while the numbers of *Paramecium* steadily fall. As the *Paramecium* population is depleted, however, the *Didinium* individuals also die.

(a)

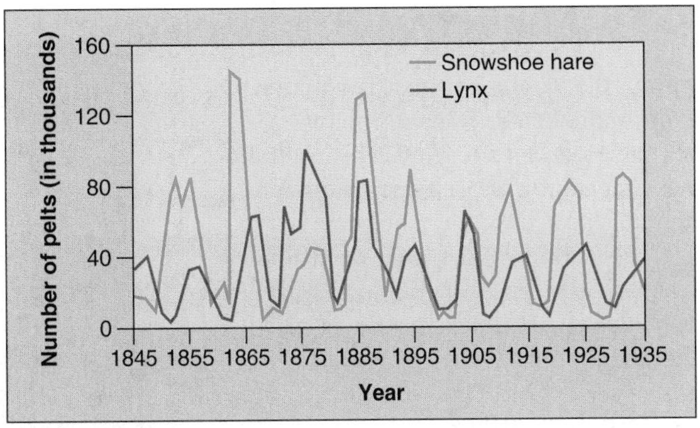
Number of pelts (in thousands)
160
120
80
40
0
Snowshoe hare
Lynx
1845 1855 1865 1875 1885 1895 1905 1915 1925 1935
Year

(b)

Figure 30.24 A predator-prey cycle.

(*a*) A snowshoe hare being chased by a lynx. (*b*) The numbers of lynxes and snowshoe hares oscillate in tune with each other in northern Canada. The data are based on numbers of animal pelts from 1845 to 1935. As the number of hares grows, so does the number of lynxes, with the cycle repeating about every nine years. Both predators (lynxes) and available food resources control the number of hares. The number of lynxes is controlled by the availability of prey (snowshoe hares).

factors appear to be generating the cycle: food plants and predators.

1. **Food plants.** The preferred foods of snowshoe hares are willow and birch twigs. As hare density increases, the quantity of these twigs decreases, forcing the hares to feed on high-fiber (low-quality) food. Lower birth rates, low juvenile survivorship, and low growth rates follow. The result is a precipitous decline in willow and birch twig abundance, and a corresponding fall in hare abundance. It takes two to three years for the quantity of mature twigs to recover.
2. **Predators.** A key predator of the snowshoe hare is the Canada lynx *Lynx canadensis.* The Canada lynx shows a "10-year cycle" of abundance that seems remarkably entrained to the hare abundance cycle (figure 30.24). As hare numbers increase, lynx numbers do, too, rising in response to the increased availability of lynx food. When hare numbers fall, so do lynx numbers, their food supply depleted.

Which factor is responsible for the predator-prey oscillations? Do increasing numbers of hares lead to overharvesting of plants (a hare-plant cycle), or do increasing numbers of lynx lead to overharvesting of hares (a hare-lynx cycle)? Field experiments carried out by C. Krebs and coworkers in 1992 provide an answer. Krebs set up experimental plots in Canada's Yukon containing hare populations. If food is added (no food effect) and predators excluded (no predator effect) from an experimental area, hare numbers increase 10-fold and stay there—the cycle is lost. However, the cycle is retained if either of the factors is allowed to operate alone: exclude predators but don't add food (food effect alone), or add food in presence of predators (predator effect alone). Thus, both factors can affect the cycle, which, in practice, seems to be generated by the interaction between the two factors.

Predation Reduces Competition

Predator-prey interactions are an essential factor in the maintenance of communities that are rich and diverse in species. The predators prevent or greatly reduce competitive exclusion by reducing the numbers of individuals of competing species. For example, in preying selectively on bivalves in marine intertidal habitats, sea stars prevent bivalves from monopolizing such habitats, opening up space for many other organisms. When sea stars are removed, species diversity falls precipitously, the seafloor community coming to be dominated by a few species of bivalves. Because predation tends to reduce competition in natural communities, it is usually a mistake to attempt to eliminate a major predator such as wolves or mountain lions from a community. The result is to decrease rather than increase the biological diversity of the community, the opposite of what is intended.

A given predator may often feed on two, three, or more kinds of plants or animals in a given community. The predator's choice depends partly on the relative abundance of the prey options. A given prey species may be a primary source of food for increasing numbers of species as it becomes more abundant, which tends to limit the size of its population automatically. Such feedback is a key factor in structuring many natural communities.

30.15 Predators and their prey often show similar cyclic oscillations, often promoted by refuges that prevent prey populations from being driven to extinction.

30.16 Mimicry

During the course of their evolution, many unprotected (nonpoisonous) species have come to resemble distasteful ones that exhibit aposematic coloration. The unprotected mimic gains an advantage by looking like the distasteful model. Two types of mimicry have been identified: Batesian mimicry and Müllerian mimicry.

Batesian Mimicry

Batesian mimicry is named for Henry Bates, the nineteenth-century British naturalist who first brought this type of mimicry to general attention in 1857. In his journeys to the Amazon region of South America, Bates discovered many instances of palatable insects that resembled brightly colored, distasteful species. He reasoned that if the unprotected animals (mimics) are present in numbers that are low relative to those of the distasteful species that they resemble (the model), the mimics are avoided by predators, who are fooled by the disguise into thinking the mimic actually is the distasteful model.

It is important that mimics be rare relative to models. If the unprotected mimic animals are too common, or if they do not live among their models, many are eaten by predators that have not yet learned to avoid model individuals with those particular characteristics. As a result, the predator learns that animals with that coloration are good rather than bad to eat.

(a) Model

(b) Batesian Mimic

Figure 30.25 A Batesian mimic.

(*a*) The model. Monarch butterflies (*Danaus plexippus*) are protected from birds and other predators by the cardiac glycosides they incorporate from the milkweeds and dogbanes they feed on as larvae. Adult monarch butterflies advertise their poisonous nature with warning coloration. (*b*) The mimic. Viceroy butterflies, *Limenitis archippus,* are Batesian mimics of the poisonous monarch. Although the viceroy is not related to the monarch, it looks a lot like it, so predators that have learned not to eat distasteful monarchs avoid viceroys, too.

Many of the best-known examples of Batesian mimicry occur among butterflies and moths. Obviously, predators in systems of this kind must use visual cues to hunt for their prey; otherwise, similar color patterns would not matter to potential predators. There is also increasing evidence indicating that Batesian mimicry can also involve nonvisual cues, such as olfaction, although such examples are less obvious to humans.

The kinds of butterflies that provide the models in Batesian mimicry are, not surprisingly, members of groups whose caterpillars feed only on one or a few closely related plant families. The plant families on which they feed are strongly protected by toxic chemicals. The model butterflies incorporate the poisonous molecules from these plants into their bodies. The mimic butterflies, in contrast, belong to groups in which the feeding habits of the caterpillars are not so restricted. As caterpillars, these butterflies feed on a number of different plant families unprotected by toxic chemicals.

One often-studied mimic among North American butterflies is the viceroy, *Limenitis archippus* (figure 30.25). This butterfly, which resembles the poisonous monarch, ranges from central Canada through much of the United States and into Mexico. The caterpillars feed on willows and cottonwoods, and neither caterpillars nor adults were thought to be distasteful to birds, although recent findings may dispute this. Interestingly, the Batesian mimicry seen in the adult viceroy butterfly does not extend to the caterpillars: viceroy caterpillars are camouflaged on leaves, resembling bird droppings, while the monarch's distasteful caterpillars are very conspicuous.

(a)
(b)
(c)
(d)
Müllerian Mimics

Figure 30.26 Müllerian mimics.
Because the color patterns of these insects are very similar, and because they all sting, they are Müllerian mimics. The yellow jacket (*a*), the masarid wasp (*b*), the sand wasp (*c*), and the anthidiine bee (*d*) all act as models for each other, strongly reinforcing the color pattern that they all share.

Müllerian Mimicry

Another kind of mimicry, **Müllerian mimicry,** was named for German biologist Fritz Müller, who first described it in 1878. In Müllerian mimicry, several unrelated but protected animal species come to resemble one another (figure 30.26). Thus, different kinds of stinging wasps have yellow-and-black-striped abdomens, but they may not all be descended from a common yellow-and-black-striped ancestor. In general, yellow-and-black and bright red tend to be common color patterns that warn predators relying on vision. If animals that resemble one another are all poisonous or dangerous, they gain an advantage because a predator learns more quickly to avoid them.

In both Batesian and Müllerian mimicry, mimic and model must not only look alike but also act alike if predators are to be deceived. For example, the members of several families of insects that resemble wasps behave surprisingly like the wasps they mimic, flying often and actively from place to place.

30.16 In Batesian mimicry, unprotected species resemble others that are distasteful. Both species exhibit aposematic coloration. If they are relatively scarce, the unprotected mimics are avoided by predators. In Müllerian mimicry, two or more unrelated but protected species resemble one another, thus achieving a kind of group defense.

30.17 Ecological Succession

Competition and cooperation often produce dramatic changes in communities, with ecosystems tending to change from simple to complex in a process known as **succession.** This process is familiar to anyone who has seen a vacant lot or cleared woods slowly become occupied by an increasing number of plants, or a pond become dry land as it is filled with vegetation encroaching from the sides.

Secondary Succession

If a wooded area is cleared and left alone, plants slowly reclaim the area. Eventually, traces of the clearing disappear and the area is again woods. This kind of succession, which occurs in areas where an existing community has been disturbed, is called **secondary succession.** Humans are often responsible for initiating secondary succession, but it may also take place when a fire has burned off an area or in abandoned agricultural fields.

Primary Succession

In contrast, **primary succession** occurs on bare, lifeless substrate, such as rocks. Primary succession occurs in lakes left behind after the retreat of glaciers, on volcanic islands that rise above the sea, and on land exposed by retreating glaciers (figure 30.27). Primary succession on glacial moraines provides an example (figure 30.28). On bare, mineral-poor soil, lichens grow first, forming small pockets of soil. Acidic secretions from the lichens help to break down the substrate and add to the accumulation of soil. Mosses then colonize these pockets of soil, eventually building up enough nutrients in the soil for alder shrubs to take hold. Over 100 years, the alders build up the soil nitrogen levels until spruce are able to thrive, eventually crowding out the alder and forming a dense spruce forest.

Primary successions often arrive at the same kinds of vegetation—a **climax community** characteristic of the region as a whole. However, because the climate keeps changing, the process of succession is often very slow, and human activities have a major impact, so many successions do not reach climax.

Why Succession Happens

Succession happens because species alter the habitat and the resources available in it, often in ways that favor other species. Three dynamic concepts are of critical importance in the process:

1. **Tolerance.** Early successional stages are characterized by weedy *r*-selected species that do not compete well in established communities but are tolerant of the harsh, abiotic conditions in barren areas.

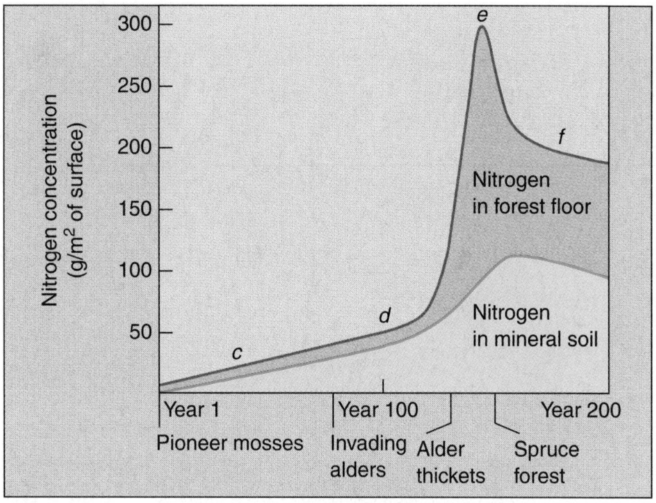
Nitrogen concentration (g/m² of surface)
300
250
200
150
100
50
e
f
d
c
Nitrogen in forest floor
Nitrogen in mineral soil
Year 1
Year 100
Year 200
Pioneer mosses
Invading alders
Alder thickets
Spruce forest

Figure 30.27 Plant succession produces progressive changes in the soil.

Initially the glacial moraine at Glacier Bay, Alaska, portrayed in figure 30.28, had little soil nitrogen, but nitrogen-fixing alders led to a buildup of nitrogen in the soil, encouraging the subsequent growth of the conifer forest. Letters in the graph correspond to photographs in figure 30.28 *c–f*.

2. **Facilitation.** The weedy early successional stages introduce local changes in the habitat that favor other, less weedy species. Thus, the mosses in the Glacier Bay succession of figure 30.28 fix nitrogen, which allows alders to invade. The alders in turn lower soil pH as their fallen leaves decompose, and spruce and hemlock, which require acidic soil, are able to invade.
3. **Inhibition.** Sometimes the changes in the habitat caused by one species, while favoring other species, inhibit the growth of the species that caused them. Alders, for example, do not grow as well in acidic soil as the spruce and hemlock that replace them.

As ecosystems mature, and more *K*-selected species replace *r*-selected ones, species richness (see next section) and total biomass increase but net productivity decreases. Because earlier successional stages are more productive than later ones, agricultural systems are intentionally maintained in early successional stages to keep net productivity high.

30.17 Communities evolve to have greater total biomass and species richness in a process called succession.

(a) Retreating Glacier
(b) Barren Moraine
(c) Pioneering Mosses
(d) Invading Alders
(e) Alder Thickets
(f) Spruce Forest

Figure 30.28 Primary succession at Alaska's Glacier Bay.

The sides of the glacier (*a*) have been retreating at a rate of some 8 meters a year, leaving behind exposed soil (*b*) from which nitrogen and other minerals have been leached out. The first invaders of these exposed sites are pioneer moss species (*c*) with nitrogen-fixing mutualistic microbes. Within 20 years, young alder shrubs take hold (*d*). Rapidly fixing nitrogen, they soon form dense thickets (e). As soil nitrogen levels rise, spruce crowd out the mature alders, forming a forest (*f*).

30.18 Biodiversity

Mature ecosystems in nature persist because they are able to cope with changes in climate and with the many conflicting demands of their inhabitants. Climate, the coevolution of species, and competition—these driving forces of evolutionary change have molded ecosystems over long periods of time to create the world we see today. However, the most important single influence on natural ecosystems today—human activity—is not acting slowly. Human influence now extends to all parts of the globe, to every blade of grass growing anywhere on our world. The fate of every ecosystem on earth in this century will be influenced more by its stability in the face of human disruption than by any other factor. Thus it is essential that we consider how ecosystems respond to disruption and try to gain an understanding of why some ecosystems are more stable than others.

Ecologists have long sought to understand why some ecosystems are more stable than others—better able to avoid permanent change and return to normal after disturbances like land clearing, fire, invasion by plagues of insects, or severe storm damage. Most ecologists now agree that biologically diverse ecosystems are in general more stable than simple ones. Ecosystems with more kinds of different organisms support a more complex web of interactions, and as a result, an alternative niche is more likely to be able to compensate for the effect of a disruption. The number of species in an ecosystem, called **species richness,** is the quantity usually measured by biologists in attempting to characterize an ecosystem's **biodiversity** (figure 30.29).

However, the very complexity of highly diverse ecosystems, while buffering them from everyday insult, also makes them more likely to have particular points of vulnerability that, if damaged, can have far-reaching consequences. A species that interacts with many other elements of an ecosystem in critical ways is called a **keystone species.** Because the loss of a keystone species may affect many other organisms in a diverse community, such species represent points of particular sensitivity in the ecosystem—places where the complex ecological machine can be easily broken. Think of how you would design a computer to protect it from "crashing"—like a complex ecosystem, you would make it with a diverse array of redundant (repeated) circuits that could take over the function of others if the need arose. However, this added complexity has a price: The advanced chip is uniquely vulnerable to damage at those points where functional crossover takes place—a single flaw can destroy the operation of the entire computer, while a simple chip would have lost only one circuit.

Thus, if we wish to preserve natural ecosystems, we should do all we can to preserve their biological diversity, while at the same time realizing that diverse ecosystems can be damaged too.

Figure 30.29 How many species are there?
Scientists are attempting to determine the species richness in the canopy of moist tropical forests, the most diverse biological community on earth. In these experiments, the insects and other animals living in the treetops are sprayed with insecticide and then sampled as they drop onto the sheets below. By such methods, Terry L. Erwin, Jr., of the Smithsonian Institution, has estimated that there may be as many as 30 million kinds of organisms in the world. Taxonomists have thus far recognized only about 1.4 million of these.

No ecosystem, however complex or simple, can be stable in the face of massive disruption. Evolution has provided no mechanism to buffer ecosystems from total destruction by human activity. Unhappily, much of today's disruption of the world's ecosystems is of this drastic sort—replacing forests with grazing land, with "pure stands" of a single kind of tree, with plowed fields for farming, or with asphalt.

What Promotes Biodiversity?

If diversity is such a good thing, why doesn't evolution drive all ecosystems toward high species richness? Why isn't tundra as diverse as tropical rain forest? If we examine all the species-rich ecosystems on earth, and ask what they have in common that species-poor ecosystems lack, two general trends emerge: large size and tropical latitude.

Ecosystem Size

Larger ecosystems, because they contain a more varied array of physical habitats, are able to support a greater number of different species. Because they offer a more diverse array of potential niches, more species can be packed into them.

This dependence of diversity on the physical size of an ecosystem has a very important consequence: If you reduce the size of the ecosystem, you reduce the number of species it can support. In today's world, this happens with increasing frequency. A new road cut across a forest divides the ecosystem in two for the many animals that cannot or do not cross it. In extreme cases, reduction in ecosystem area can produce *faunal collapse,* a situation in which many of the animal species living in the ecosystem become extinct there because the smaller ecosystem is simply unable to divide its resources into that many pieces.

For exactly this reason, in all but the largest of our national parks, a high proportion of the mammals once there have become extinct. Different animals have been lost from different parks, so nothing is entirely lost, but the pattern of species loss is clear. When the parks were formed at the beginning of the twentieth century, their animals were part of much larger communities, but, walled off from their surroundings by society's growing development around the parks, the animals have been restricted to much smaller areas. The loss of species richness has been the result. If we want to maintain animal diversity for future generations, the parks must be managed carefully, and lost species must be reintroduced. In today's world, there is no longer any true wilderness. Our national parks, like gardens, have to be tended.

Latitude

Compared to the arctic region, tropical areas have many more species (figure 30.30), more than 6 million of the estimated 10 million species of organisms that exist on earth. Why should this be so? There are two principal reasons:

1. **Length of growing season.** As a general rule of thumb, ecosystems with more resources are able to subdivide them into more niches, and so maintain a greater number of species. In the tropics, with ample sunlight, warm temperatures, and generous rainfall throughout the year, the growing season never stops. Progressing toward the poles, however, growing seasons shorten, so that polar ecosystems have much less energy to work with in weaving their food webs.
2. **Climatic stability.** The climate in the tropics has not been subjected to glaciers and other major disruptions during the evolutionary past. Thus the unchanging physical conditions have provided a longer evolutionary window in the tropics for specialized relationships to develop.

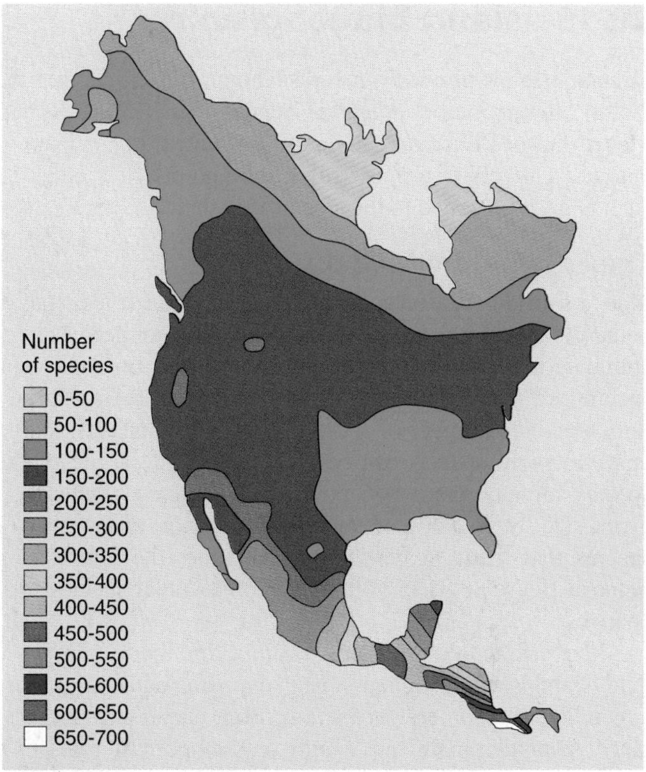
Number
of species
0-50
50-100
100-150
150-200
200-250
250-300
300-350
350-400
400-450
450-500
500-550
550-600
600-650
650-700

Figure 30.30 A latitudinal cline in species richness.

Among North and Central American birds, a marked increase in the number of species occurs as one moves towards the tropics. Fewer than 100 species are found at arctic latitudes, while more than 600 species live in southern Central America.

30.18 Ecosystems with biodiversity are more stable, while, at the same time, they may also have points of vulnerability.

30.19 Island Biogeography

Oceanic islands provide a natural laboratory for studying the factors that promote species richness. In 1967, Robert MacArthur of Princeton University and Edward O. Wilson of Harvard University proposed that the number of species on such islands is related to the size of the island.

The Equilibrium Model

MacArthur and Wilson reasoned that species are constantly being dispersed to islands, so islands have a tendency to accumulate more and more species. At the same time that new species are added, however, other species are lost by extinction. Once the number of species fills the capacity of that island, no more species can be established unless one of the species already there becomes extinct. Every island of a given size, then, has a characteristic equilibrium number of species that tends to persist through time (the intersection point in figure 30.31*a*), although the individual species may change.

MacArthur and Wilson's equilibrium theory proposes that island species richness is a dynamic equilibrium between colonization and extinction. Both island size and distance from the mainland would play important roles. We would expect smaller islands to have higher rates of extinction because their population sizes would, on average, be smaller. Also, we would expect fewer colonizers to reach islands that lie farther from the mainland. Thus, small islands far from the mainland have the fewest species; large islands near the mainland have the most (figure 30.31*b*).

The predictions of this simple model bear out well in field data. Asian Pacific bird species (figure 30.31*c*) exhibit a positive correlation of species richness with island size but a negative correlation of species richness with distance from the mainland (New Guinea).

Figure 30.32 A test of the equilibrium model.
Scientists erected a scaffold to entirely cover this small mangrove island in the Florida Keys. Covered with a plastic sheet, the island became an enclosed experimental system.

Testing the Equilibrium Model

An interesting test of the equilibrium model was carried out on small mangrove islands in the Florida Keys. Islands about 12 meters across were inhabited by 25 to 40 species of arthropods; larger islands had more species, and smaller islands had fewer. After a careful census, six of the islands were covered with plastic sheets and fumigated with an insecticide to kill all arthropods (figure 30.32). These islands were recolonized within a year and rapidly attained a steady-state number of species. For each island, the capacity of that island set the equilibrium number of species, while chance determined the exact identity of the species.

30.19 Species richness on islands is a dynamic equilibrium between colonization and extinction.

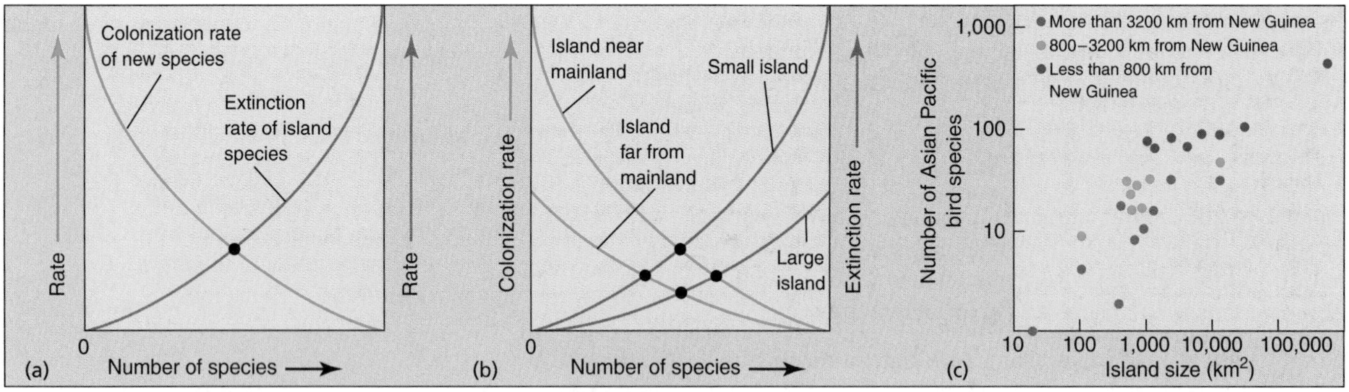

Colonization rate of new species
Extinction rate of island species
Rate
0
Number of species
(a)
Island near mainland
Small island
Island far from mainland
Large island
Colonization rate
Extinction rate
Rate
0
Number of species
(b)
Number of Asian Pacific bird species
1,000
100
10
More than 3200 km from New Guinea
800–3200 km from New Guinea
Less than 800 km from New Guinea
10
100
1,000
10,000
100,000
Island size (km²)
(c)

Figure 30.31 The equilibrium model of island biogeography.
(*a*) Island species richness reaches an equilibrium (*black dot*) when the colonization rate of new species equals the extinction rate of species on the island. (*b*) The equilibrium shifts when the colonization rate is affected by distance from the mainland and when the extinction rate is affected by size of the island. Species richness is positively correlated with island size and inversely correlated with distance from the mainland. (*c*) Small distant Asian islands have fewer bird species, bearing out the principles of the equilibrium model.

CONCEPT REVIEW

1. The hormonal change of locusts to their migratory phase, which is brought on by crowding when populations reach large numbers, is an example of
   a. carrying capacity.
   b. exponential growth.
   c. density-dependent effects.
   d. density-independent effects.

2. What is the difference between a type II and a type III survivorship curve?
   a. In type II, mortality is higher in the early stages than it is in type III.
   b. In type III, mortality is the same in early stages as it is in type II.
   c. In type III, mortality is higher in the early stages than it is in type II.
   d. In type II, a large proportion of individuals survive to the maximum age, whereas in type III they do not.

3. Which of the following is most likely to exhibit character displacement?
   a. allopatric species
   b. sympatric species
   c. isolated species
   d. predators and their prey

4. Which of the following is *not* an example of mutualism?
   a. barnacles growing on the backs of whales
   b. mycorrhizae
   c. leaf-cutter ants and the fungi they grow
   d. ants protecting aphids while using their honeydew as food

5. Animals that are cryptically colored
   a. are distasteful.
   b. tend not to live in groups.
   c. exhibit colors rarely found in the animal's habitat.
   d. often have chemical defenses.

6. Adult monarch butterflies are primarily protected from birds by
   a. Müllerian mimicry.
   b. resembling the poisonous viceroy butterfly.
   c. manufacturing poisons from molecules that are toxic to birds.
   d. obtaining poisons from the plants that their larvae eat and retaining them.

7. A _________ species is one that interacts with many components of an ecosystem in critical ways.
   a. cryptic
   b. diverse
   c. keystone
   d. competitive

8. Which of the following does *not* promote biodiversity?
   a. introduction of exotic species
   b. larger size of an ecosystem
   c. longer growing season
   d. climatic stability

9. Most natural populations exhibit __________ growth.

10. _________ is the use of different parts of a resource by different species as a result of competition among these species for the resource.

11. If two species live together with one benefiting, and the other neither benefiting nor being harmed, their relationship is described as _________.

12. Poisonous animals advertise their toxicity using warning or _________ coloration.

13. _________ succession occurs in areas where there has been previous growth.

14. Biologists measure the _________ of an ecosystem to characterize its overall biodiversity.

CHALLENGE YOURSELF

1. How can the principle of competitive exclusion be applied to the origin of large numbers of species of certain genera in the Hawaiian Islands? What might happen in the future to the hundreds of species of *Drosophila* that occur there, ignoring the possibility of their extinction through habitat destruction?

2. Most chemically protected prey species produce toxins that make them taste bad or make the predator sick. If the prey is "tasted" by an unsuspecting predator, the prey usually dies. If they are dead, how have the chemicals protected them? Would it be an advantage if the prey could produce a toxin that killed its predator?

30 eBRIDGE

Reinforcing Key Points

Population Dynamics

30.1 Population Growth

30.2 Life History Adaptations

30.3 The Influence of Population Density

30.4 Population Demography

How Competition Shapes Communities

30.5 Communities

30.6 The Niche and Competition

30.7 Competitive Exclusion

30.8 Resource Partitioning

How Coevolution Shapes Communities

30.9 Coevolution and Symbiosis

30.10 Commensalism

30.11 Mutualism

30.12 Parasitism

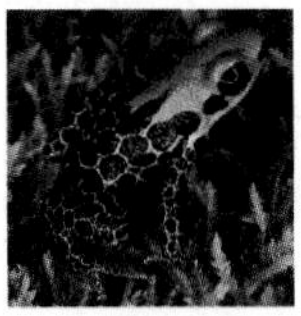

Predator and Prey Interactions

30.13 Plant Defenses

30.14 Animal Defenses

30.15 Predator-Prey Cycles

30.16 Mimicry

Community Stability

30.17 Ecological Succession

30.18 Biodiversity

30.19 Island Biogeography

Electronic Learning

Visual Learning

Animations

Exponential Population Growth

Stages of Population Growth

Feeding Niches for Wading Birds

Predator-Prey Interaction Between a Lynx and a Snowshoe Hare

Eagles and Bears Feed on Spawning Salmon

Secondary Succession in a Forest

Species Richness of Communities

Author's Corner

Natural History. Living in ecosystems is something every organism does, in one way or another. The particular fashion in which a species goes about meeting the challenges of survival defines its niche in the ecosystem. Natural history is, in the broadest sense, the telling of these individual stories. Each story is unique, each approach individual, but taken together such stories weave a rich tapestry of ongoing evolutionary adaptation, the natural history of life in an ecosystem.

1. **Is Flipper a senseless killer?**
2. **The great pigeon race disaster of 1997 suggests the answer to an enduring mystery.**
3. **How Saint Patrick charmed the snakes out of Ireland.**
4. **The saga of Lonesome George the tortoise, a Saint Valentine's Day hero.**

Virtual Classroom

Local Environmental Change

Global environmental change, while of pressing concern to the world's future, is but a small part of a much more intricate pattern of change taking place largely at the local level. The accidental release of aggressive African bees in Brazil in 1956 provides an example. By now the so-called "killer" bees have spread over 5 million square miles, as far as Texas. The salinization of California's Central Valley offers another example, where irrigation has added salts to the groundwater underlying the valley. Salty enough to kill plant roots, the groundwater is rising like water in a bathtub, and is now within 10 feet of the surface of over 186,000 acres of our nation's richest farmland.

Conservation Biology

Among the greatest challenges facing the biosphere is the accelerating pace of species extinctions—not since the dinosaurs have so many species become extinct in so short a period of time. This challenge has led to the emergence in the last decade of the new discipline of conservation biology, seeking to learn how to better preserve species, communities, and ecosystems.

Virtual Lab

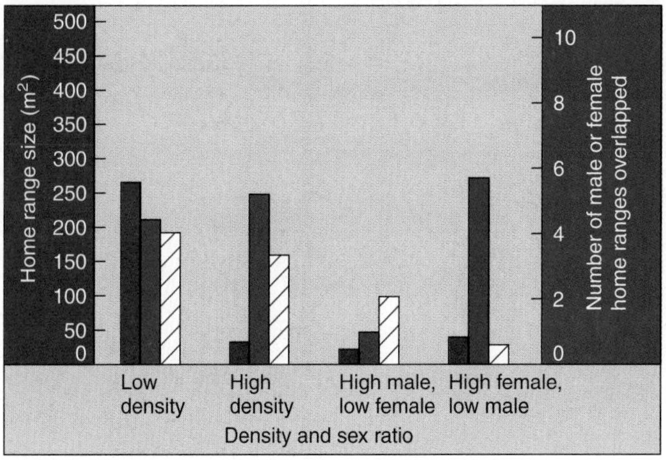
Home range size (m^2)
500
450
400
350
300
250
200
150
100
50
0
10
8
6
4
2
0
Number of male or female home ranges overlapped
Low density
High density
High male, low female
High female, low male
Density and sex ratio

Factors Limiting the Home Range of Male Voles

Male and female mammals utilize space differently because they have different goals in maximizing fitness. Females attempt to maximize offspring survival by providing greater parental investment than males do. Males, by contrast, maximize fitness by attempting to mate with as many females as possible. Thus, there are two factors influencing male reproductive success: (1) the number of females he has access to; and (2) the number of male competitors he has to interact with. Male small mammals typically have larger home range sizes than those of females, and their ranges overlap with females and other males. But, the factors that influence home range size and amount of overlap are not well understood. Jerry Wolff of the University of Memphis has studied the relative influence of male competition versus access to females in establishing home range sizes in the gray-tailed vole. As expected, increased density causes males to establish smaller home ranges with more overlapping ranges. However, regardless of population density or sex ratios, male home ranges are usually limited in their overlap with each other.

Quizzes | Further Reading | Essential Study Partner | Links | BioCourse.com

CHAPTER 31

Planet Under Stress

THE FAR SIDE® By GARY LARSON
© 1984 FarWorks, Inc. All Rights Reserved/Dist. by Creators Syndicate
Larson

"You know what I'm sayin'? ... Me, for example. I couldn't work in some stuffy little office. ... The outdoors just calls to me."

CHAPTER 31

Planet Under Stress

CHAPTER OVERVIEW

Global Change

- Air and water pollution result mainly from increased industrialization, oil spills, the use of agricultural chemicals, and unwise disposal of wastes.
- Acid rain, which is precipitation polluted by sulfuric acid originating from factories, kills trees and lakes by lowering the pH level.
- Chlorofluorocarbons (CFCs) used in coolants, aerosols, and foaming agents are eating the earth's ozone, which exposes life on earth to harmful ultraviolet radiation.
- An increase in carbon dioxide in the atmosphere is responsible for increased global warming, or the greenhouse effect.

Saving Our Environment

- Antipollution laws, pollution taxes, and economic evaluations are underway to reduce pollution.
- We must slow down our use of topsoil and groundwater, resources that took thousands of years to accumulate.
- With the destruction of tropical and temperate rain forests, organisms with potentially vital roles in the ecosystem are being lost.
- The alarmingly high human population growth rate is at the core of many environmental problems.

Solving Environmental Problems

- Environmental problems have been overcome through assessment, analysis, education, and follow-through.
- You can do your part by recycling, educating others about the environment, voting, and writing letters.

31.1 Pollution

Our world is one ecological continent, one highly interactive biosphere, and damage done to any one ecosystem can have ill effects on many others. Burning high-sulfur coal in Illinois kills trees in Vermont, while dumping refrigerator coolants in New York destroys atmospheric ozone over Antarctica and leads to increased skin cancer in Madrid. Biologists call such widespread effects on the worldwide ecosystem **global change.** The pattern of global change that has become evident within recent years, including chemical pollution, acid rain, the ozone hole, and the greenhouse effect, is one of the most serious problems facing humanity's future.

Air Pollution

Air pollution is a major problem in the world's cities. In Mexico City, oxygen is sold routinely on corners for patrons to inhale. Cities such as New York, Boston, and Philadelphia are known as brown-air cities because the pollutants in the air are usually sulfur oxides emitted by industry. Cities such as Los Angeles, however, are called gray-air cities because the pollutants in the air undergo chemical reactions in the sunlight.

Chemical Pollution

The problem posed by chemical pollution has grown very serious in recent years, both because of the growth of heavy industry and because of an overly casual attitude in industrialized countries. In one example, a poorly piloted oil tanker named the *Exxon Valdez* ran aground in Alaska in 1989 and heavily polluted many kilometers of North American coastline and the organisms that live there with oil. If the tanker had been loaded no higher than the waterline, little oil would have been lost, but it was loaded far higher than that, and the weight of the above-waterline oil forced thousands of tons of oil out the hole in the ship's hull. Why do policies permit overloading like this?

Agricultural Chemicals

The spread of "modern" agriculture, and particularly the Green Revolution, which brought high-intensity farming to the Third World, has caused very large amounts of many kinds of new chemicals to be introduced into the global ecosystem, particularly pesticides, herbicides, and fertilizers. Industrialized countries like the United States now attempt to carefully monitor side effects of these chemicals. Unfortunately, large quantities of many toxic chemicals no longer manufactured still circulate in the ecosystem.

For example, the chlorinated hydrocarbons, a class of compounds that includes DDT, chlordane, lindane, and dieldrin, have all been banned for normal use in the United States, where they were once widely used. They are still manufactured in the United States and exported to other countries, where their use continues. Chlorinated hydrocarbon molecules break down slowly and accumulate in animal fat. Furthermore, as they pass through a food chain, they become increasingly concentrated in a process called **biological magnification** (figure 31.1). DDT caused serious problems by leading to the production of thin, fragile eggshells in many predatory bird species in the United States and elsewhere until the late 1960s, when it was banned in time to save the birds from extinction. Chlorinated compounds have other undesirable side effects and exhibit hormonelike activities in the bodies of animals.

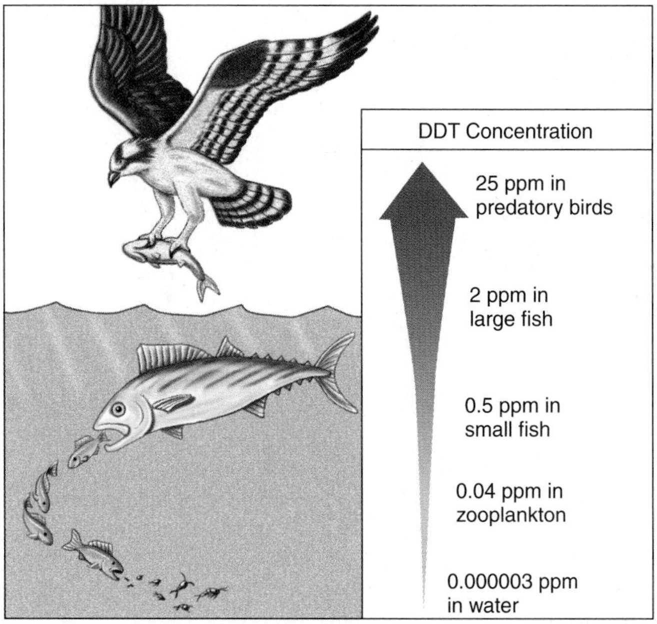
DDT Concentration
25 ppm in predatory birds
2 ppm in large fish
0.5 ppm in small fish
0.04 ppm in zooplankton
0.000003 ppm in water

Figure 31.1 Biological magnification of DDT.
Because DDT accumulates in animal fat, the compound becomes increasingly concentrated in higher levels of the food chain. Before DDT was banned in the United States, predatory bird populations drastically declined because DDT made their eggshells thin and fragile enough to break during incubation.

Water Pollution

Water pollution is a very serious consequence of our casual attitude about pollution. "Flushing it down the sink" doesn't work in today's crowded world. There is simply not enough water available to dilute the many substances that the enormous human population produces continuously. Despite improved methods of sewage treatment, lakes and rivers throughout the world are becoming increasingly polluted.

31.1 All over the globe, increasing industrialization is leading to higher levels of pollution.

31.2 Acid Rain

The smokestacks you see in figure 31.2 are those of the Four Corners power plant in New Mexico. This facility burns coal, sending the smoke high into the atmosphere through these stacks, each of which is over 65 meters tall. The smoke the stacks belch out contains high concentrations of sulfur, because the cheap coal that the plant burns is rich in sulfur. The sulfur-rich smoke is dispersed and diluted by winds and air currents. The first tall stacks were introduced in Britain in the mid-1950s. Such tall stacks rapidly became popular in the United States and Europe—there are now over 800 of them in the United States alone.

In the 1970s, 20 years after the stacks were introduced, ecologists began to report evidence that the tall stacks were not eliminating the problem, just exporting the ill effects elsewhere. Throughout northern Europe, lakes were reported to have suffered drastic drops in biodiversity, some even becoming devoid of life. The trees of the great Black Forest of Germany seemed to be dying—and the damage was not limited to Europe. It turned out that the sulfur introduced into the upper atmosphere by high smokestacks combines with water vapor to produce sulfuric acid. When this water later falls back to earth as rain or snow, it carries the sulfuric acid with it. Drifting to the northeast in the United States, the acid is taken far from its source, but the result is now clear. When schoolchildren measured the pH of natural rainwater as part of a nationwide project in 1989, locations around the United States rarely had a pH lower than 5.6, except in the northeastern United States. Rain and snow in the Northeast now have a pH of about 4.3, about 100 times more acidic than the rest of the country. This pollution-acidified precipitation is called **acid rain.**

Acid rain destroys life. Many of the forests of the northeastern United States and Canada have been seriously damaged. In fact, it is now estimated that at least 1.4 million acres of forests in the Northern Hemisphere have been adversely affected (figure 31.3). Not just trees are affected. Thousands of lakes in Sweden and Norway no longer support fish—these lakes are now eerily clear. In the northeastern United States and Canada, too, tens of thousands of lakes are dying biologically as their pH levels fall to below 5.0.

The solution at first seems obvious: Capture and remove the emissions instead of releasing them into the atmosphere. But there have been serious problems with implementing this solution. First, it is expensive. Estimates of the cost of installing and maintaining the necessary "scrubbers" in the United States are on the order of $5 billion a year. An additional difficulty is that the polluter and the recipient of the pollution are far from one another, and neither wants to pay so much for what they view as someone else's problem. The Clean Air Act revisions of 1990 have begun to address this problem by mandating some cleaning of emissions in the United States, although much still remains to be done worldwide.

Figure 31.2 Tall stacks export pollution.
Tall stacks like those of the Four Corners coal-burning power plant in New Mexico send pollution far up into the atmosphere.

Figure 31.3 Acid rain.
Acid precipitation is killing many of the trees in North American and European forests. Much of the damage is done to the mycorrhizae, fungi growing within the cells of the tree roots. Trees need mycorrhizae in order to extract nutrients from the soil.

31.2 Pollution-acidified precipitation—loosely called acid rain—is destroying forest and lake ecosystems in Europe and North America. The solution is to clean up the emissions.

31.3 The Ozone Hole

Living things were able to leave the oceans and colonize the surface of the earth only after a protective shield of ozone had been added to the atmosphere by photosynthesis. For 3 billion years before that, life was trapped in the oceans because radiation from the sun seared the earth's surface unchecked. Nothing could survive that bath of destructive energy. Imagine if that shield were taken away. Alarmingly, it appears that we are destroying it ourselves. Starting in 1975, the earth's ozone shield began to disintegrate. Over the South Pole in September of that year, satellite photos revealed that the ozone concentration was unexpectedly less than elsewhere in the earth's atmosphere. It was as if some "ozone eater" were chewing it up in the Antarctic sky, leaving a mysterious zone of lower-than-normal ozone concentration, an **ozone hole.** Every year after that, more of the ozone has been depleted, and the hole grows bigger and deeper (figure 31.4).

What is eating the ozone? Scientists soon discovered that the culprit was a class of chemicals that everyone had thought to be harmless: **chlorofluorocarbons (CFCs).** CFCs were invented in the 1920s, a miracle chemical that was stable, harmless, and a near-ideal heat exchanger. Throughout the world, CFCs are used in large amounts as coolants in refrigerators and air conditioners, as the gas in aerosol dispensers, and as the foaming agent in Styrofoam containers. All of these CFCs eventually escape into the atmosphere, but no one worried about this until recently, both because CFCs were thought to be chemically inert and because everyone tends to think of the atmosphere as limitless. But CFCs are very stable chemicals, so over many years they have accumulated in the atmosphere.

It turned out that the CFCs were causing mischief the chemists had not imagined. High over the South and North Poles, nearly 50 kilometers up, where it was very, very cold, the CFCs stuck to frozen water vapor and began to act as catalysts of a chemical reaction. Just as an enzyme carries out a reaction in your cells without being changed itself, so the CFCs began to catalyze the conversion of ozone (O_3) into oxygen (O_2) without being used up themselves. Very stable, the CFCs in the atmosphere just kept at it—little machines that never stop. They are still there, still doing it, today. The drop in ozone worldwide is now over 3%.

Ultraviolet radiation is a serious human health concern. Every 1% drop in the atmospheric ozone content is estimated to lead to a 6% increase in the incidence of skin cancers. At middle latitudes, the drop of approximately 3% that has occurred worldwide is estimated to have led to an increase of perhaps as much as 20% in lethal melanoma skin cancers. Importantly, photosynthetic plankton species, critically important to global productivity, are apparently much more highly susceptible than humans.

International agreements to lower the level of CFC production worldwide over the next several decades have been signed, but no one knows if this gradual phase-out will be adequate, or too little too late. The vast majority of the CFCs that have already been manufactured have not yet reached the upper atmosphere, and an enormous additional amount is scheduled to be produced in the coming decades. We can only guess how much of the earth's ozone will be destroyed.

31.3 CFCs and other chemicals are catalytically destroying the ozone in the upper atmosphere, exposing earth's surface to dangerous radiation.

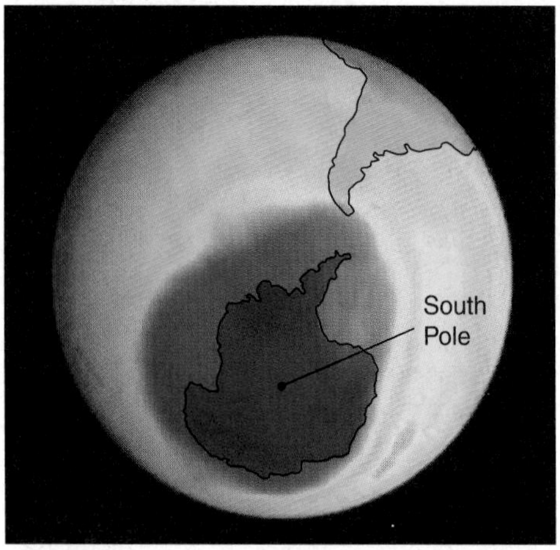
South
Pole

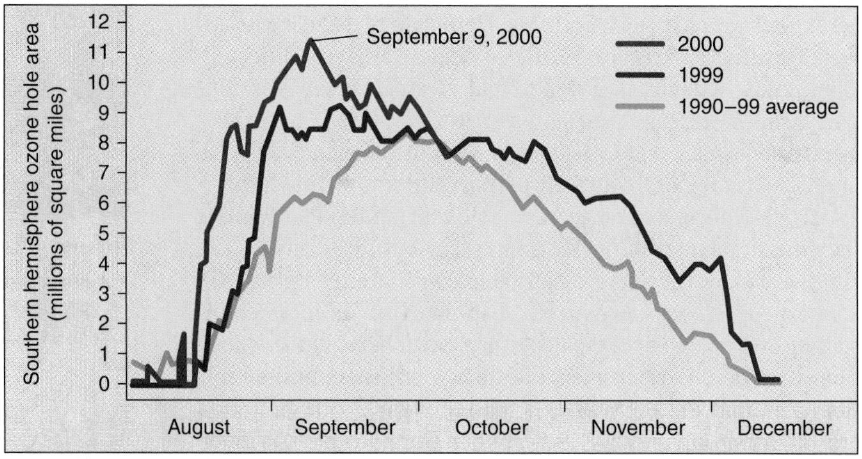
Southern hemisphere ozone hole area
(millions of square miles)
0
1
2
3
4
5
6
7
8
9
10
11
12
September 9, 2000
2000
1999
1990–99 average
August
September
October
November
December

Figure 31.4 The ozone hole over Antarctica is still growing.

For decades NASA satellites have tracked the extent of ozone depletion over Antarctica. Every year since 1979 an ozone "hole" has appeared in August when sunlight triggers chemical reactions in cold air trapped over the South Pole during Antarctic winter. The hole intensifies during September before tailing off as temperatures rise in November–December. In 2000, the 11.4-million-square-mile hole (dark purple in the satellite image) covered an area larger than the United States, Canada, and Mexico combined, the largest hole ever recorded. In September 2000, the hole extended over Punta Arenas, a city of about 120,000 people in southern Chile, exposing residents to very high levels of UV radiation.

31.4 The Greenhouse Effect

For over 150 years, the growth of our industrial society has been fueled by cheap energy, much of it obtained by burning fossil fuels—coal, oil, and gas. Coal, oil, and gas are the remains of ancient plants, transformed by pressure and time into carbon-rich "fossil fuels." When such fossil fuels are burned, this carbon is combined with oxygen atoms, producing carbon dioxide (CO_2). Industrial society's burning of fossil fuels has released huge amounts of carbon dioxide into the atmosphere. As with CFCs, no one paid any attention to this because the carbon dioxide was thought to be harmless and because the atmosphere was thought to be a limitless reservoir, able to absorb and disperse any amount. It turns out neither assumption was true, and in recent decades, the levels of carbon dioxide in the atmosphere have risen sharply and continue to rise.

What is alarming is that the carbon dioxide doesn't just sit in the air doing nothing. The chemical bonds in carbon dioxide molecules transmit radiant energy from the sun but trap the longer wavelengths of infrared light, or heat, and prevent them from radiating into space. This creates what is known as the **greenhouse effect.** Planets that lack this type of "trapping" atmosphere are much colder than those that possess one. If the earth did not have a "trapping" atmosphere, the average earth temperature would be about –20°C, instead of the actual +15°C.

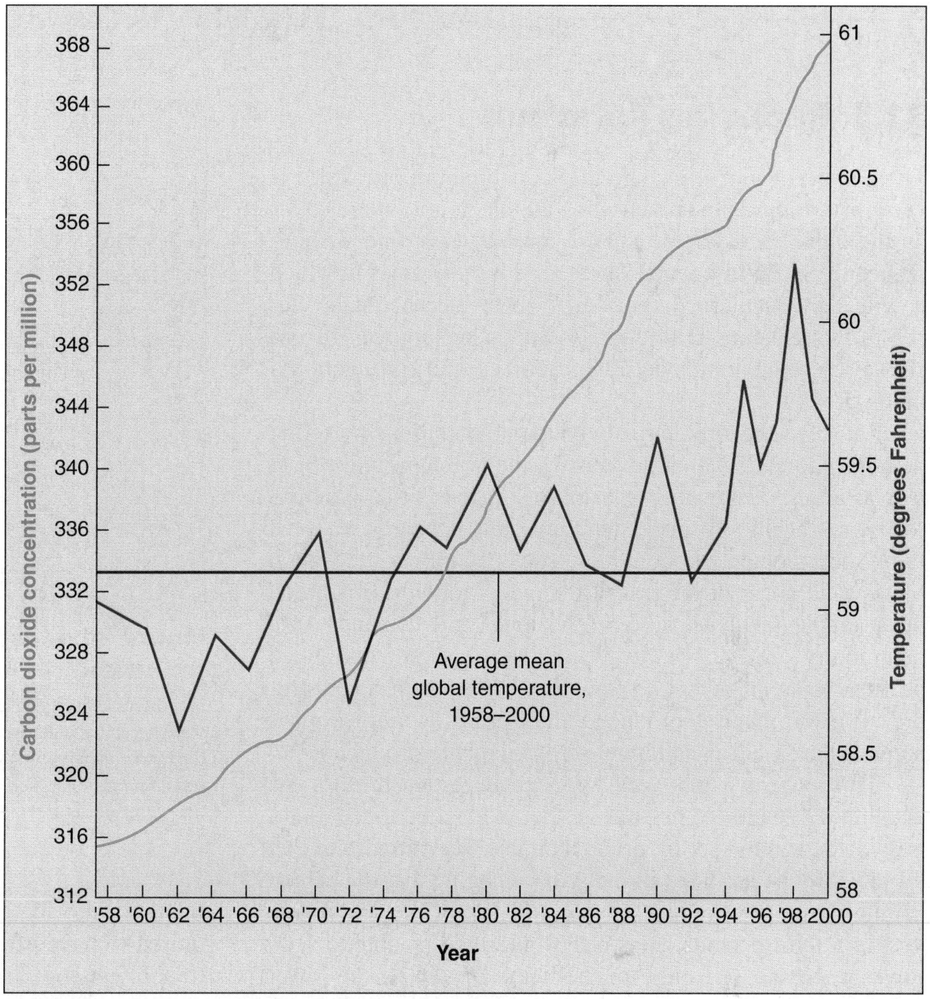

Carbon dioxide concentration (parts per million)
368
364
360
356
352
348
344
340
336
332
328
324
320
316
312
Temperature (degrees Fahrenheit)
61
60.5
60
59.5
59
58.5
58
Average mean
global temperature,
1958–2000
'58 '60 '62 '64 '66 '68 '70 '72 '74 '76 '78 '80 '82 '84 '86 '88 '90 '92 '94 '96 '98 2000
Year

Figure 31.5 The greenhouse effect.
The concentration of carbon dioxide in the atmosphere has shown a steady increase for many years (*blue line*). The *red line* shows the average global temperature for the same period of time. Note the general increase in temperature since the 1950s and, specifically, the sharp rise beginning in the 1980s. Data from Geophysical Monograph, American Geophysical Union, National Academy of Sciences, and National Center for Atmospheric Research.

The earth's greenhouse effect is intensifying with increased fossil fuel combustion and certain types of waste disposal. These activities are increasing the amounts of carbon dioxide, CFCs, nitrogen oxides, and methane—all "greenhouse gases"—in the atmosphere. The rise in average global temperatures during recent decades is consistent with increased carbon dioxide concentrations in the atmosphere (figure 31.5). The idea of **global warming** due to accumulation of greenhouse gases in earth's atmosphere has been controversial, because correlations do not prove a cause-and-effect. However, as more data become available, a growing consensus of scientists accept global warming as an unwelcome reality.

Increases in the amounts of greenhouse gases could increase average global temperatures from 1° to 4°C, which could have serious associated effects in prime agricultural lands, changes in sea levels, and alterations in rain patterns. There is considerable disagreement among governments about what ought to be done about global warming.

31.4 Humanity's burning of fossil fuels has greatly increased atmospheric levels of CO_2, leading to global warming.

31.5 Reducing Pollution

The pattern of global change that is overtaking our world is very disturbing. Human activities are placing a severe stress on the global ecosystem, and we must quickly find ways to reduce the harmful impact. There are four key areas in which it will be particularly important to meet the challenge successfully: reducing pollution, finding other sources of energy, preserving nonreplaceable resources, and curbing population growth.

To solve the problem of industrial pollution, it is first necessary to understand the cause of the problem. In essence, it is a failure of our economy to set a proper price on environmental health. To understand how this happens, we must think for a moment about money. The economy of the United States (and much of the rest of the industrial world) is based on a simple feedback system of supply and demand. As a commodity gets scarce, its price goes up, and this added profit acts as an incentive for more of the item to be produced; if too much is produced, the price falls and less of it is made because it is no longer so profitable to produce it.

This system works very well and is responsible for the economic strength of our nation, but it has one great weakness. If demand is set by price, then it is very important that all the costs be included in the price. Imagine that the person selling the item were able to pass off part of the production cost to a third person. The seller would then be able to set a lower price and sell more of the item! Driven by the lower price, the buyer would purchase more than if all the costs had been added into the price.

Unfortunately, that sort of pricing error is what has driven the pollution of the environment by industry. The true costs of energy and of the many products of industry are composed of direct production costs, such as materials and wages, and of indirect costs, such as pollution of the ecosystem. Economists have identified an "optimum" amount of pollution based on how much it costs to reduce pollution versus the social and environmental cost of allowing pollution (figure 31.6).

The indirect costs of pollution are usually not taken into account. However, the indirect costs do not disappear because we ignore them. They are simply passed on to future generations, which must pay the bill in terms of damage to the ecosystems on which we all depend. Increasingly, the future is now. Our world, unable to support more damage, is demanding that something be done—that we finally pay up.

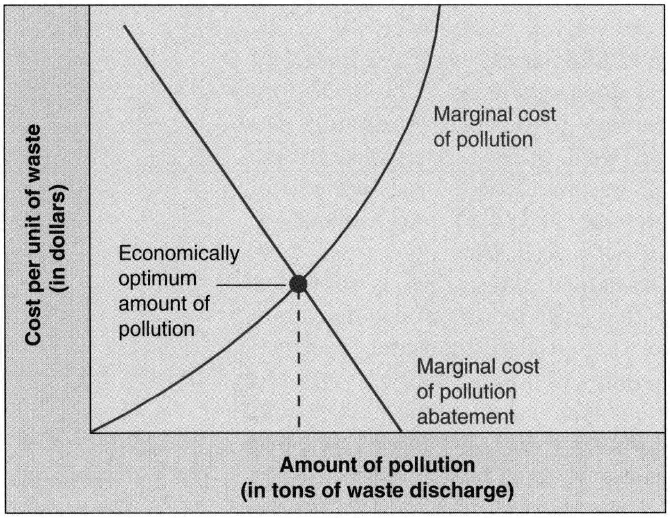
Marginal cost of pollution
Economically optimum amount of pollution
Marginal cost of pollution abatement
Cost per unit of waste (in dollars)
Amount of pollution (in tons of waste discharge)

Figure 31.6 Is there an optimum amount of pollution?
Economists identify the "optimum" amount of pollution as the amount whose marginal cost equals the marginal cost of pollution abatement (the point where the two curves intersect). If more pollution than the optimum is allowed, the social cost is unacceptably high. If less than the optimum amount of pollution is allowed, the economic cost is unacceptably high.

Antipollution Laws

Two effective approaches have been devised to curb pollution in this country. The first is to pass laws forbidding it. In the last 20 years, laws have begun to significantly curb the spread of pollution by setting stiff standards for what can be released into the environment. For example, all cars are required to have effective catalytic converters to eliminate automobile smog. Similarly, the Clean Air Act of 1990 requires that power plants install scrubbers on their smokestacks to eliminate sulfur emissions. The effect is that the consumer pays to avoid polluting the environment. The cost of the converters makes cars more expensive, and the cost of the scrubbers increases the price of the energy. The new, higher costs are closer to the true costs, lowering consumption to more appropriate levels.

Pollution Taxes

A second approach to curbing pollution has been to increase the consumer costs directly by placing a tax on the pollution, in effect an artificial price hike imposed by the government as a tax added to the price of production. This added cost lowers consumption too, but by adjusting the tax, the government can attempt to balance the conflicting demands of environmental safety and economic growth. Such taxes, often imposed as "pollution permits," are becoming an increasingly important part of antipollution laws.

31.5 Free market economies often foster pollution when prices do not include environmental costs. Laws and taxes are being designed in an attempt to compensate.

31.6 Finding Other Sources of Energy

The pollution generated by burning coal and oil, the increasing scarcity of oil, and the potential contributions of carbon dioxide to global warming all make it desirable to find alternative energy sources. Many countries are turning to nuclear power for their growing energy needs. In less than 50 years, nuclear power has become a leading source of energy. In 1995, more than 500 nuclear reactors were producing power worldwide. Over 70% of France's electricity is now produced by nuclear power plants.

Nuclear power plants have not been as popular in this country as in the rest of the world, because we have ample access to cheap coal and because the public fears the consequences of an accident. A reactor partial meltdown at the Three Mile Island nuclear plant in Pennsylvania in 1979 released little radiation into the environment but galvanized these fears. There has been little nuclear power development in this country since then (figure 31.7).

In theory nuclear power can provide plentiful, cheap energy, but the reality is less encouraging. Nuclear power presents several problems that must be overcome if it is to provide a significant portion of the energy that will fuel our future world.

Safe operation. Because of the potential for vast radioactive contamination such as occurred when a reactor exploded at Chernobyl in the Ukraine in 1986 (figure 31.8), it is crucial that nuclear power plants be designed "fail-safe" and be sited away from densely populated areas. While new power plant designs are far safer than those of Chernobyl, there is still much room for improvement.

Waste disposal. Spent nuclear fuel remains very radioactive for thousands of years, and the plant itself quickly wears out—after about 25 years the intense radioactivity makes its metal pipes brittle. As yet, no safe way has been devised to dispose of this growing mountain of highly radioactive material. To date, not a single nuclear power plant in France or the United States has been successfully dismantled and its dangerously radioactive components disposed of safely.

Security. Some of the uranium in nuclear fuel is converted to plutonium during its decay, and this plutonium could be recovered from spent fuel and used to make nuclear weapons. It will be very important to find ways to effectively guard nuclear power plants against terrorism.

31.6 Safety, security, and particularly waste disposal remain serious obstacles to the widespread use of nuclear power.

Figure 31.7 Three Mile Island nuclear power plant.

Since a nuclear accident here in 1979, the building of nuclear power stations in the United States has slowed dramatically.

Figure 31.8 A nuclear disaster.

The Russian scientists in this helicopter measure radiation levels over the Chernobyl reactor days after a 1986 explosion blew it apart.

31.7 Preserving Nonreplaceable Resources

Among the many ways ecosystems are being damaged, one class of problem stands out as more serious than all the rest: consuming or destroying resources that we all share in common (figure 31.9) but cannot replace in the future. While a polluted stream can be cleaned, no one can restore an extinct species. In the United States, three sorts of nonreplaceable resources are being consumed at alarming rates: topsoil, groundwater, and biodiversity.

Topsoil

The United States is one of the most productive agricultural countries on earth, largely because much of it is covered with particularly fertile soils. Our midwestern farm belt sits astride what was once a great prairie. The soil of that ecosystem accumulated bit by bit from countless generations of animals and plants until, by the time humans came to plow, the rich soil extended down several feet.

We can never replace this rich **topsoil,** the capital upon which our country's greatness is built, yet we are allowing it to be lost at a rate of centimeters every decade. Our country has lost one-quarter of its topsoil since 1950! By repeatedly tilling (turning the soil over) to eliminate weeds, we permit rain to wash more and more of the topsoil away, into rivers and eventually out to sea. New approaches are desperately needed to lessen the reliance on intensive cultivation. Some possible solutions include using genetic engineering to make crops resistant to weed-killing herbicides and terracing to recapture lost topsoil.

Groundwater

A second resource that we cannot replace is **groundwater,** water trapped beneath the soil within porous rock reservoirs called aquifers. This water seeped into its underground reservoir very slowly during the last ice age over 12,000 years ago. We should not waste this treasure, for we cannot replace it.

In most areas of the United States, local governments exert relatively little control over the use of groundwater. As a result, a very large portion is wasted watering lawns, washing cars, and running fountains. A great deal more is inadvertently being polluted by poor disposal of chemical wastes—and once pollution enters the groundwater, there is no effective means of removing it.

"Freedom in a Commons Brings Ruin to All"

The essence of Hardin's original essay:

Picture a pasture open to all. It is expected that each herdsman will try to keep as many cattle as possible on [this] commons....What is the utility...of adding one more animal?...Since the herdsman receives all the proceeds from the sale of the additional animal, the positive utility [to the herdsman] is nearly +1.... Since, however, the effects of overgrazing are shared by all the herdsmen, the negative utility for any particular decision-making herdsman is only a fraction of -1. Adding together the...partial utilities, the rational herdsman concludes that the only sensible course for him to pursue is to add another animal to [the] herd. And another; and another.... Therein is the tragedy. Each man is locked into a system that [causes] him to increase his herd without limit—in a world that is limited....Freedom in a commons brings ruin to all.

—G. Hardin, "The Tragedy of the Commons," *Science* **162,** 1243 (1968), p. 1244

Figure 31.9 The tragedy of the commons.
In a now-famous essay, ecologist Garrett Hardin argues that destruction of the environment is driven by freedom without responsibility.

Biodiversity

The number of species in danger of extinction during your lifetime is far greater than the number that became extinct with the dinosaurs. This disastrous loss of **biodiversity** is important to every one of us, because as these species disappear, so does our chance to learn about them and their possible benefits for ourselves. The fact that our entire supply of food is based on 20 kinds of plants, out of the 250,000 available, should give us pause. Like burning a library without reading the books, we don't know what it is we waste. All we can be sure of is that we cannot retrieve it. Extinct is forever.

Over the last 20 years, about half of the world's tropical rain forests have been either burned to make pasture land or cut for timber (figure 31.10). Over 6 million square kilometers have been destroyed. Every year the rate of loss increases as the human population of the tropics grows. About 160,000 square kilometers were cut each year in the 1990s, a rate greater than 0.6 hectares (1.5 acres) per second! At this rate, all the rain forests of the world will be gone in your lifetime. In the process, it is estimated that one-fifth or more of the world's species of animals and plants will become extinct—more than a million species. This would be an extinction event unparalleled for at least 65 million years, since the age of the dinosaurs.

You should not be lulled into thinking that loss of biodiversity is a problem limited to the tropics. The ancient forests of the Pacific Northwest are being cut at a ferocious rate today, largely to supply jobs (the lumber is exported), with much of the cost of cutting it down subsidized by our government (the Forest Service builds the necessary access roads, for example). At the current rate, very little will remain in a decade. Nor is the problem restricted to one area. Throughout our country, natural forests are being "clear-cut," replaced by pure stands of lumber trees planted in rows like so many lines of corn. It is difficult to scold those living in the tropics when we ourselves do such a poor job of preserving our own country's biodiversity.

31.7 Nonreplaceable resources are being consumed at an alarming rate all over the world, key among them topsoil, groundwater, and biodiversity.

(a)
(b)
(c)

Figure 31.10 Destroying the tropical rain forests.

(*a*) These fires are destroying the rain forest in Brazil, which is being cleared for cattle pasture. (*b*) The flames are so widespread and so high that their smoke can be viewed from space. This satellite photo shows a plume of smoke generated from the burning of the rain forest. (*c*) The consequences of deforestation can be seen on these middle-elevation slopes in Ecuador, which now support only low-grade pastures and used to support highly productive forest, which protected the watersheds of the area.

31.8 Curbing Population Growth

If we were to solve all the problems mentioned in this chapter, we would merely buy time to address the fundamental problem: There are getting to be too many of us.

Humans first reached North America at least 12,000 to 13,000 years ago, crossing the narrow straits between Siberia and Alaska and moving swiftly to the southern tip of South America. By 10,000 years ago, when the continental ice sheets withdrew and agriculture first developed, about 5 million people lived on earth, distributed over all the continents except Antarctica. With the new and much more dependable sources of food that became available through agriculture, the human population began to grow more rapidly. By the time of Christ, an estimated 130 million people lived on earth. By the year 1650, the world's population had doubled, and doubled again, reaching 500 million.

Since 1650, and probably for much longer, the average human birthrate has remained nearly constant, at about 30 births per year per 1,000 people worldwide. However, with the spread of better sanitation and improved medical techniques, the death rate has fallen steadily, to an estimated 9 deaths per 1,000 people per year in 1998. The difference between these two figures amounts to an annual worldwide increase in human population of approximately 1.4%. This number may seem small, but don't be deceived—it leads to a doubling of the world's population in only 39 years!

One of the most alarming trends taking place in developing countries is the massive movement to urban centers. For example, Mexico City, the largest city in the world, is plagued by smog, traffic, inadequate waste disposal, and other problems; it has a population of about 26 million people (figure 31.11). The prospects of supplying adequate food, water, and sanitation to this city's people are almost unimaginable. The lot of the rural poor, mainly farmers, in Mexico is even worse.

The world population passed 6 billion people in 1999, and the annual increase now amounts to about 80 million people. Put another way, more than 216,000 people are added to the world population each day, or more than 150 every minute (figure 31.12). At this rate, the world's population will continue to grow over 6 billion, and perhaps stabilize at a figure between 8.5 billion and 20 billion. Such growth cannot continue, because our world cannot support it. Just as a cancer cannot grow unabated in your body without eventually killing you, so humanity cannot continue to grow unchecked in the biosphere without killing it.

In view of the limited resources available to the human population and the need to learn how to manage those resources well, the first and most necessary step toward global prosperity is to stabilize the human population. One of the surest signs of the pressure being placed on the environment is human use of about 40% of the total net global photosynthetic productivity on land. Given that statistic, a doubling of the human population in 39 years poses extraordinarily severe problems. The facts virtually demand restraint in population growth. If and when technology is developed that would allow greater numbers of people to inhabit the earth in a stable condition, the human population can be increased to whatever level might be appropriate.

Figure 31.11 The world's population is centered in mega-cities.

Mexico City, the world's largest city, has about 26 million inhabitants.

About 60% of the people in the world live in countries that are at least partly tropical or subtropical. An additional 20% live in China, and the remaining 20% in the so-called developed, or industrialized, countries: Europe, Russia, Japan, the United States, Canada, Australia, and New Zealand. Whereas the populations of the developed countries are growing at an annual rate of only about 0.3%, those of the less developed, mostly tropical countries (excluding China) are growing at an annual rate estimated to be about 2.2%.

Most countries are devoting considerable attention to slowing the growth rate of their populations, and there are genuine signs of progress. If it continues, the United Nations estimates that the world's population may stabilize by the close of this century at a level of 13 to 15 billion people, nearly three times the number living today. No one knows whether the world can support so many people indefinitely.

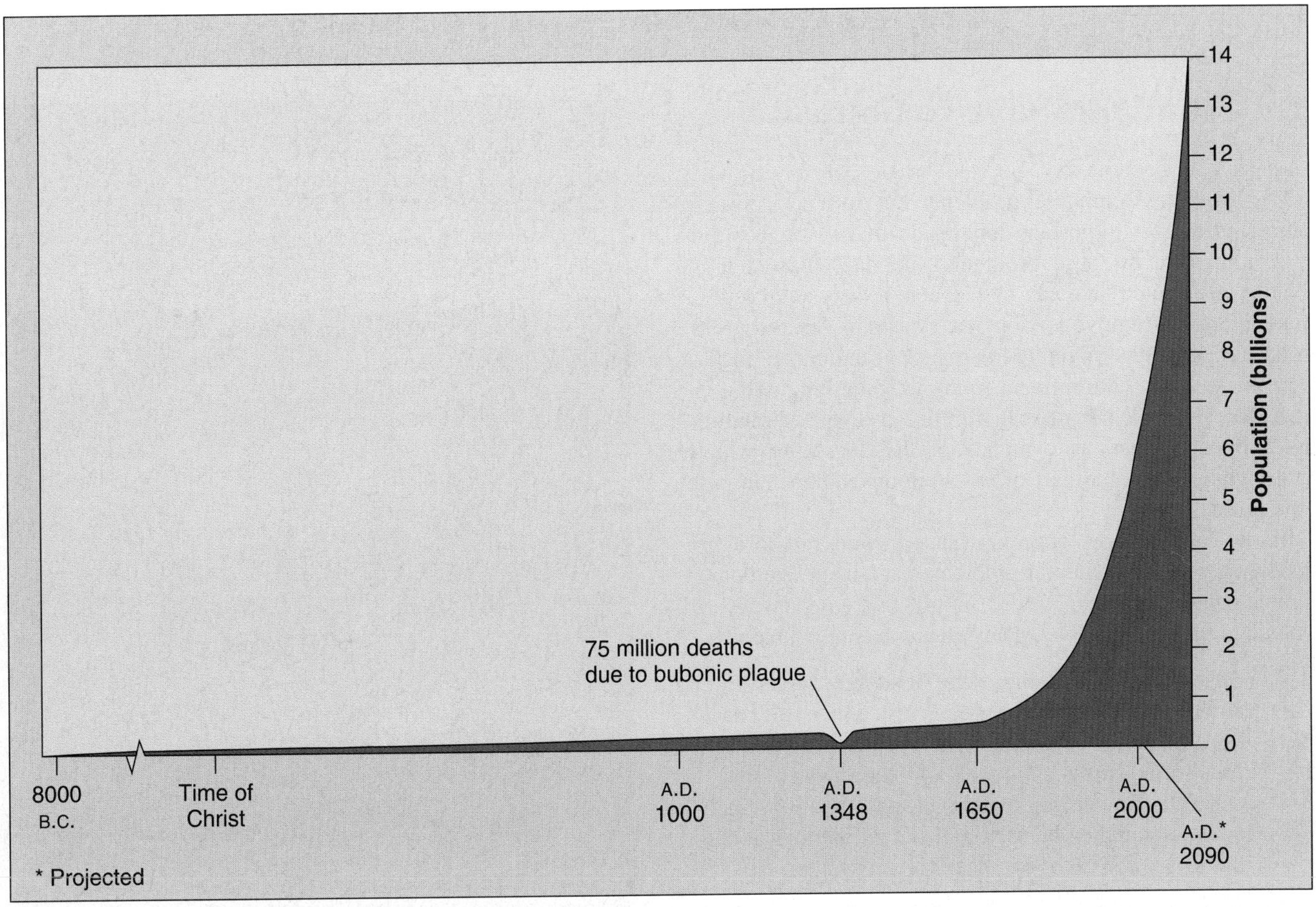
14
13
12
11
10
9
8
7
6
5
4
3
2
1
0
Population (billions)
75 million deaths
due to bubonic plague
8000
B.C.
Time of
Christ
A.D.
1000
A.D.
1348
A.D.
1650
A.D.
2000
A.D.*
2090
* Projected

Figure 31.12 Growth curve of the human population.
Over the past 300 years, the world population has been growing steadily. Currently, there are over 6 billion people on the earth.

Finding a way to do so is the greatest task facing humanity. The quality of life that will be available for your children in the coming century depends to a large extent on our success.

Population Growth Rate Starting to Decline

In 1995, the United Nations announced that the world population growth rate and world fertility were less than expected for that year. In 1994, the world population growth rate was predicted to be 1.57 percent a year, and the actual growth rate from 1990 to 1995 was 1.48 percent. Also, they projected the average number of children per woman would be 3.1, while it was actually 2.96 from 1990 to 1995.

The United Nations attributes the decline to increased family planning efforts and the increased economic power and social status of women. While the United Nations applauds the United States for leading the world in funding family planning programs abroad, some oppose spending money on international family planning. The opposition states that money is better spent on improving education and the economy in other countries, leading to an increased awareness and lowered fertility rates. The United Nations certainly supports the improvement of education programs in developing countries, but, interestingly, it has reported increased education levels *following* a decrease in family size as a result of family planning.

Slowing population growth will help sustain the world's resources, but per capita consumption is also important. Even though the vast majority of the world's population is in developing countries, the vast majority of resource consumption occurs in the developed world—the wealthiest 20% of the world's population accounts for 80% of the world's consumption of resources, whereas the poorest 20% is responsible for only 1.3% of consumption. The developed world must lessen the impact each of us makes.

31.8 The root problem at the core of all other environmental concerns is the rapid growth of the world's human population. Serious efforts are being made to slow its growth.

31.9 What You Have to Contribute

It is easy to become discouraged when considering the world's many environmental problems, but do not lose track of the single most important conclusion that emerges from our examination of these problems—the fact that each is solvable. A polluted lake can be cleaned; a dirty smokestack can be altered to remove noxious gas; waste of key resources can be stopped. What is required are a clear understanding of the problem and a commitment to doing something about it. The extent to which American families **recycle** aluminum cans and newspapers is evidence of the degree to which people want to become part of the solution, rather than part of the problem.

If we look at how success was achieved in those instances where environmental problems have been overcome, a simple pattern emerges. Viewed simply, five components are involved in successfully solving any environmental problem:

1. **Assessment.** The first stage of addressing any environmental problem is scientific analysis. Data must be collected and experiments performed in order to construct a "model" of the ecosystem that describes how it is responding to the situation. Such a model can then be used to make predictions about the future course of events in the ecosystem.
2. **Risk analysis.** Using the results of the scientific investigation as a tool, it is possible to predict the consequences of environmental intervention. It is necessary to evaluate not only the potential for solving the environmental problem but also any adverse effects the action plan might create.
3. **Public education.** When a clear choice can be made among alternative courses of action, the public must be informed. This involves explaining the problem in terms people can understand, presenting the alternative actions available, and describing the probable costs and results of the different choices.
4. **Political education.** The public, through its elected officials, selects a course of action and implements it. Individuals can have a major impact at this stage by exercising their right to vote and be heard. Many voters do not understand the magnitude of what they can achieve by writing letters and supporting special-interest groups.
5. **Follow-through.** The results of any action should be carefully monitored to see if the environmental problem is being solved and, more basically, to evaluate and improve the initial assessment and modeling of the problem. We learn by doing.

Figure 31.13 The future is now.
The girl gazing from this page faces an uncertain future. She is a refugee, an Afghani. The whims of war have destroyed her home, her family, and all that is familiar to her. Her expression carries a message about our own future: The problems humanity faces on an increasingly unstable, overcrowded, and polluted earth are no longer hypothetical. They are with us today and demand solutions.

What You Have to Contribute

You cannot hope to preserve what you do not understand. The world's environmental problems are acute, and a knowledge of biology is an essential tool you will need to contribute to the effort to solve them. It has been said that we do not inherit the earth from our parents—we borrow it from our children. We must preserve for them a world in which they can live. That is our challenge for the future, and it is a challenge that must be met soon. In many parts of the world, the future is happening right now (figure 31.13).

31.9 Biological literacy is no longer a luxury for scientists—it has become a necessity for all of us.

CONCEPT REVIEW

1. Which of the following statements does *not* describe acid rain?
   a. It is sulfuric acid formed from the sulfur-rich smoke emitted by factories and water vapor in the atmosphere.
   b. It has caused dramatic drops in biodiversity throughout the forests and lakes of Europe, Canada, and the northeastern United States.
   c. Wind and air currents dilute acid rain because factories are designed to release it high into the air.
   d. It causes the pH levels of lakes to fall below 5.0, an acidity in which amphibians cannot develop properly.

2. Why is the concentration of ozone less over the North and South Poles?
   a. CFCs accumulate only in areas where the air is cold.
   b. CFCs stick to frozen water vapor and are able to act as catalysts.
   c. CFC use is highest in these areas.
   d. Ultraviolet rays are stronger in these areas.

3. _________ is caused by an accumulation of large amounts of CO_2 and other gases in the atmosphere.
   a. Acid rain
   b. The ozone hole
   c. Skin cancer
   d. The greenhouse effect

4. Which of the following is *not* a problem that has prevented nuclear power from becoming more popular?
   a. securing nuclear power plants against terrorism
   b. decreasing the production of acid rain by nuclear power plants
   c. disposal of nuclear wastes
   d. safely operating nuclear power plants

5. Topsoil must be conserved because
   a. it is fertile soil that has accumulated over thousands of years and prevents erosion.
   b. its underground reservoir is being used at a fast rate.
   c. the agricultural practice of terracing has caused the loss of centimeters per decade.
   d. none of these.

6. The loss of biodiversity is mainly a problem limited to the tropics.
   a. true
   b. false

7. _________ catalyze the conversion of ozone into oxygen without being used.

8. Industrial burning of fossil fuels has released huge amounts of _________ into the atmosphere.

9. Two measures that have been taken to reduce pollution are passing antipollution laws and imposing pollution _________.

10. The three sorts of nonreplaceable resources that are being consumed at an alarming rate in the United States are _________, _________, and _________.

11. Tropical forests are being destroyed for three major reasons: firewood gathering, lumbering, and _________.

12. When agriculture was first developed about 10,000 years ago, there were about _________ people on earth.

13. The present population of the world is about _________.

CHALLENGE YOURSELF

1. Can we ever produce enough food and other materials so that population growth will not be a matter of concern? How?

2. Decreased birthrates have ultimately followed decreased death rates in many countries (for example, Germany and Great Britain). Do you think this will eventually occur worldwide and solve our population problems?

3. Some have argued that attempts by the United States to promote lowering of the birthrate in the underdeveloped countries of the tropics is nothing more than economic imperialism, and that it is in the best interest of these countries that their populations grow as rapidly as possible. Give reasons why you agree or disagree with this assessment.

31 eBRIDGE

Reinforcing Key Points

Global Change

31.1 Pollution

31.2 Acid Rain

31.3 The Ozone Hole

31.4 The Greenhouse Effect

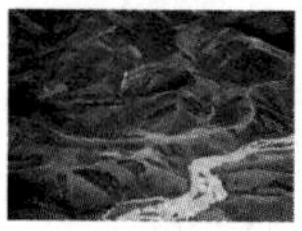

Saving Our Environment

31.5 Reducing Pollution

31.6 Finding Other Sources of Energy

31.7 Preserving Nonreplaceable Resources

31.8 Curbing Population Growth

Solving Environmental Problems

31.9 What You Have to Contribute

Electronic Learning

Visual Learning

Animations

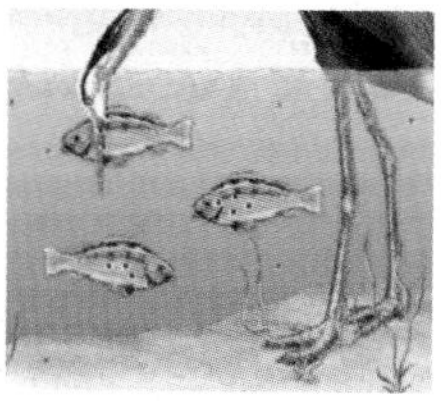

Bioaccumulation

Deoxygenation of Lakes

Acid Rain

Ozone Layer Depletion

Global Warming I

Restoration of the Everglades

Enhancement Chapter

Conservation Biology

Not since the disappearance of the dinosaurs have so many species become extinct in so short a period of time as today. In this enhancement chapter we explore the new discipline of conservation biology. Conservation biology is an applied field that seeks to learn why species go extinct and how it might be possible to preserve species, communities, and ecosystems. Case studies are the tools used to analyze the factors responsible, and the recovery efforts underway.

Author's Corner

Science in the New Millennium. We are entering a new century, one which presents unique challenges. The ability of modern society to sustain itself will depend in important ways on the challenge presented by problems involving biology, challenges that we all must come to understand if we as a society are to meet them.

1. In the new millennium, science must become more than a tool for growth.
2. Was Malthus mistaken? A 150-year-old warning about population growth may have been off only in timing.
3. Seniors graduating from college enter a century of challenging responsibilities.
4. Two minor science stories—both about ice—are likely to profoundly affect the new century.
5. Recent data suggest that, global warming aside, our winters may be getting a lot colder.

Virtual Classroom

Eco-economics and Population Growth

The world faces two serious environmental challenges. The most immediate is degradation of the global environment. This can be seen in acid rain, atmospheric ozone depletion, and global warming due to atmospheric accumulation of carbon dioxide. At least as serious a challenge is posed by overutilization of resources that cannot be replaced, like groundwater, topsoil, and biodiversity. At the core of both challenges is the deeper issue of population growth, and the resulting encouragement of economic "development."

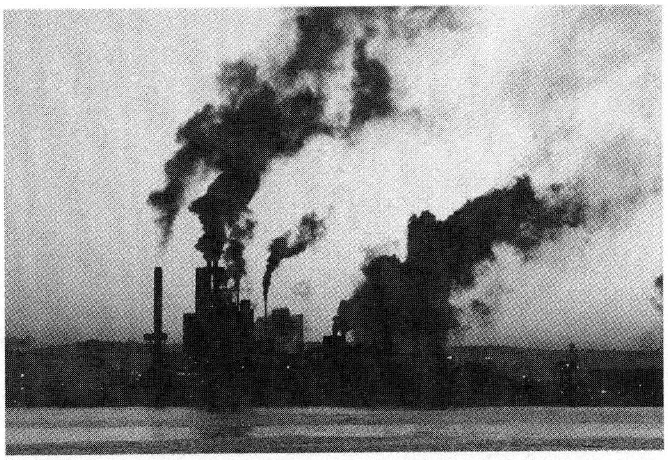

Solving Environmental Problems

Every successful solution of an environmental problem involves the same key elements: assessment, risk analysis, public education, political action, and follow-through. Very often success results from the determined actions of one individual. The cleaning up of Lake Washington, a large freshwater lake near Seattle that had become polluted by treated sewage, and of the Nashua River in New England, polluted by chemical discharges from industrial plants, offer two clear examples.

Virtual Lab

Identifying the Environmental Culprit Harming Amphibians

Scientists are reporting an alarming number of deformed frogs all across the United States and Canada. Some environmental scientists suspect that chemical pollutants in the water might be causing the deformities. Other scientists caution that a different factor might be responsible. Although chemicals such as pesticides certainly *could* produce deformities in localized situations, say near a chemical spill, so too could other environmental factors affecting local habitats, such as parasitic infections.

In a particularly clear example of the kind of investigation that will be needed to sort out this complex issue, Andrew Blaustein of Oregon State University headed a team of scientists that set out to examine the effects of UV-B radiation on amphibians in natural populations. In a series of experiments carried out in the field, they attempted to assess the degree to which UV-B radiation promoted amphibian developmental deformities under natural conditions.

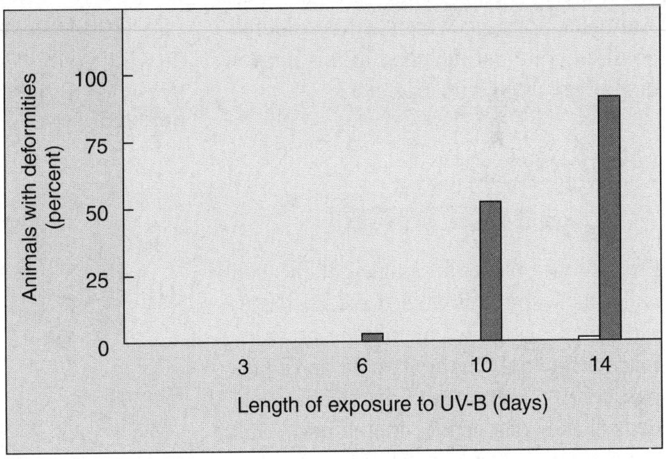
Animals with deformities (percent)
100
75
50
25
0
3
6
10
14
Length of exposure to UV-B (days)

Quizzes | Further Reading | Essential Study Partner | Links | BioCourse.com

APPENDIX A

Classification of Organisms

The classification used in this book is explained in chapter 12. Responding to a wealth of recent molecular data, it recognizes two separate kingdoms for the bacteria (prokaryotes): archaebacteria and eubacteria. It divides the eukaryotes into four kingdoms: the diverse and predominantly unicellular Protista and three large, characteristic multicellular groups derived from them: Fungi, Plantae, and Animalia. Viruses, which are considered nonliving, are not included in this appendix but are treated in chapter 13.

Kingdom Archaebacteria

Prokaryotic bacteria; single-celled, cell walls lack muramic acid. Like all bacteria, they lack a membrane-bounded nucleus, sexual recombination, and internal cell compartments. Archaebacterial cells have distinctive membranes and unique rRNA and metabolic cofactors. Many are capable of living in an anaerobic environment rich in CO_2 and H_2.

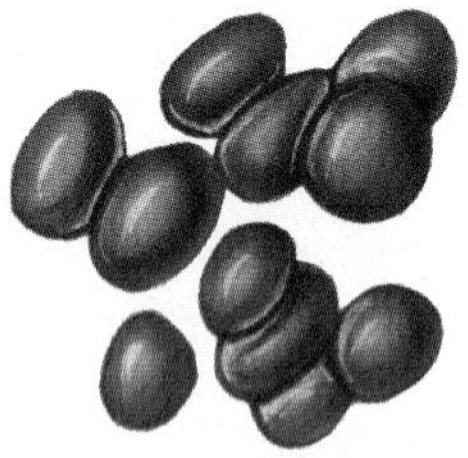

Archaebacteria: a halobacterium.

Kingdom Eubacteria

Single-celled prokaryotic bacteria; cell walls contain muramic acid. Like archaebacteria, they lack a true nucleus and true internal cell compartments. Their flagella are simple, composed of a single fiber of protein. Their reproduction is predominantly asexual. Eubacteria often form filaments or other forms of colonies. A very diverse group metabolically; over 4,800 species have been named and far more undoubtedly exist.

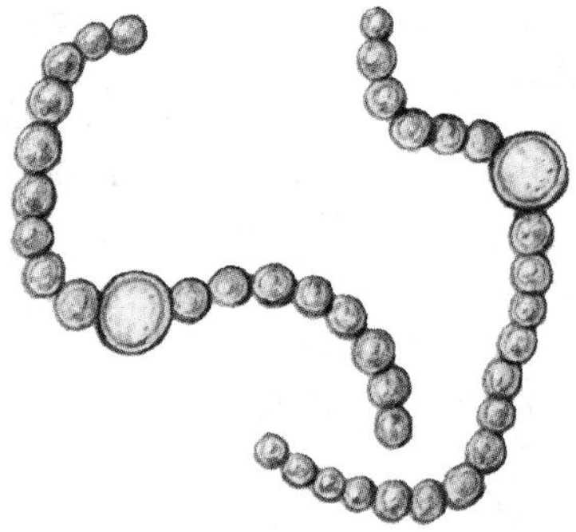

Eubacteria: a cyanobacterium.

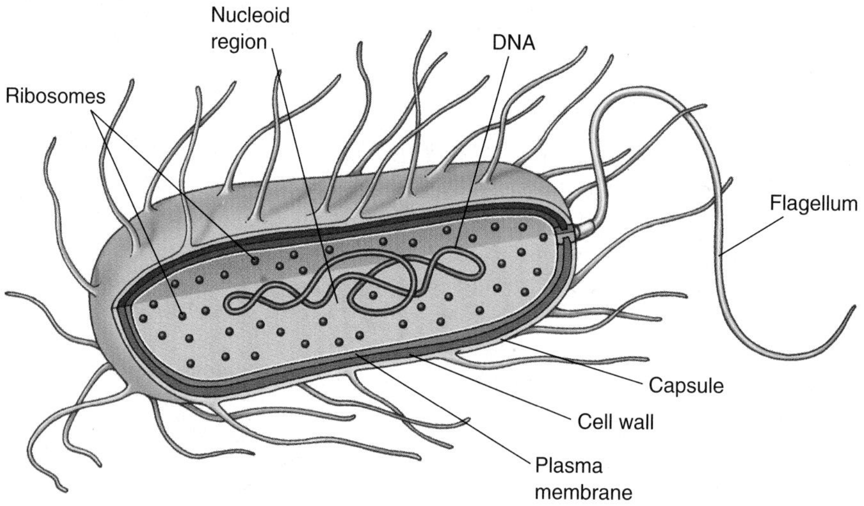
Nucleoid region
DNA
Ribosomes
Flagellum
Capsule
Cell wall
Plasma membrane

Eubacteria: diagram of a typical cell.

Kingdom Protista

Eukaryotic organisms, including many evolutionary lines of primarily single-celled organisms. Eukaryotes have a membrane-bounded nucleus and chromosomes, sexual recombination, and extensive internal compartmentalization of the cells; their flagella are complex, with 9 + 2 internal organization. They are diverse metabolically but much less so than are bacteria; protists are heterotrophic or autotrophic and may capture prey, absorb their food, or photosynthesize. Reproduction in protists is either sexual, involving meiosis and syngamy, or asexual.

Phylum Caryoblastea

One species of primitive amoebalike organism, *Pelomyxa palustris,* which lacks mitosis, mitochondria, and chloroplasts.

Phylum Rhizopoda

Amoebas; heterotrophic, unicellular organisms that move from place to place by cellular extensions called pseudopods and reproduce only asexually, by fission. Hundreds of species.

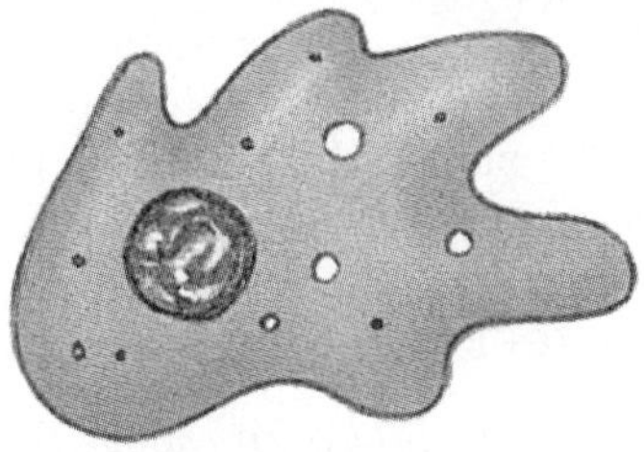

Phylum Rhizopoda: *Amoeba*.

Phylum Foraminifera

Heterotrophic marine organisms; characteristic feature is a pore-studded shell, or test, some of them brilliantly colored.

Phylum Actinopoda

Amoeboid protists with glassy skeletons and needlelike pseudopods.

Phylum Sarcomastigophora

Zoomastigotes; a highly diverse phylum of mostly unicellular, heterotrophic, flagellated, free-living or parasitic protists (flagella: one to thousands). Thousands of species.

Phylum Ciliophora

Ciliates; diverse, mostly unicellular, heterotrophic protists, characteristically with large numbers of cilia. About 8,000 species.

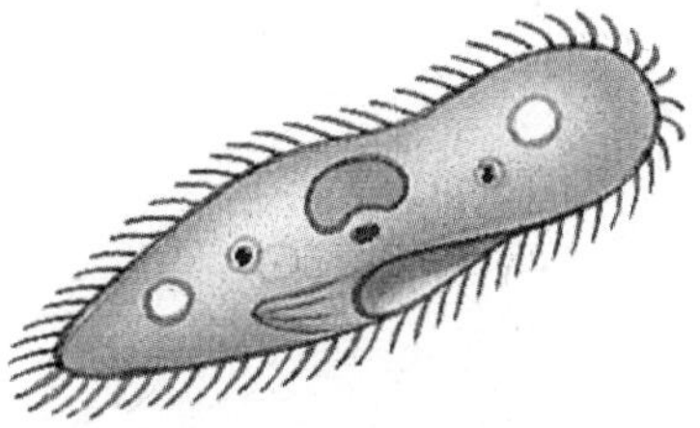

Phylum Ciliophora: *Paramecium*.

Phylum Apicomplexa

Sporozoans; unicellular, heterotrophic, nonmotile, spore-forming parasites of animals. About 3,900 species.

Phylum Pyrrhophyta

Dinoflagellates; unicellular, photosynthetic organisms, most of which are clad in stiff, cellulose plates and have two unequal flagella that beat in grooves encircling the body at right angles. About 1,000 species.

Phylum Euglenophyta

Euglenoids; mostly unicellular, photosynthetic, or heterotrophic protists with two unequal flagella. About 1,000 species.

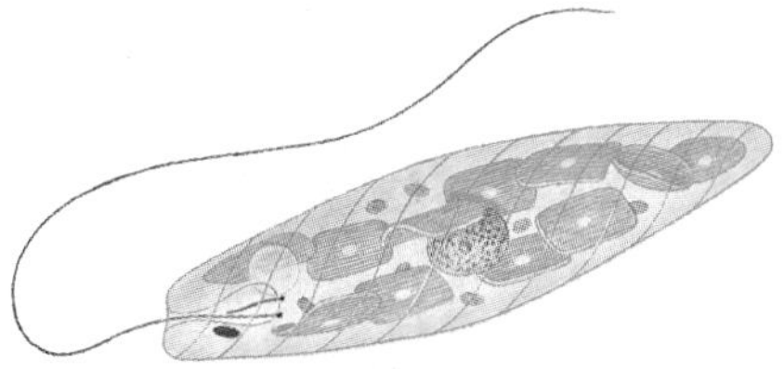

Phylum Euglenophyta: *Euglena*.

Phylum Chrysophyta

Diatoms and related groups; mostly unicellular, photosynthetic organisms with chlorophylls *a* and *c* and fucoxanthin. Diatoms have a unique double shell made of opaline silica. About 11,500 living species.

Phylum Chlorophyta

Green algae; a large and diverse phylum of unicellular or multicellular, mostly aquatic organisms with chlorophylls *a* and *b*, carotenoids, and starch, accumulated within the plastids (as it also is in plants) as the food storage product. About 7,000 species.

Phylum Rhodophyta

Red algae; mostly marine, mostly multicellular protists with chloroplasts containing chlorophyll *a* and phycobilins; no flagellated cells present. About 4,000 species.

Phylum Phaeophyta

Brown algae; multicellular, photosynthetic, mostly marine protists with chlorophylls *a* and *c* and a carotenoid that colors the organisms brownish. About 1,500 species.

Phylum Acrasiomycota

Cellular slime molds; unicellular, amoebalike, heterotrophic organisms that aggregate in masses at certain stages of their life cycle and form compound sporangia. About 70 species.

Phylum Myxomycota

Plasmodial slime molds; heterotrophic organisms that move from place to place as a multicellular, gelatinous mass, forming sporangia at times. About 500 species.

Phylum Oomycota

Oomycetes; water molds, white rusts, and downy mildews. Aquatic or terrestrial unicellular or multicellular parasites or saprobes that feed on dead organic matter. About 580 species.

Kingdom Fungi

Filamentous, multinucleate, heterotrophic eukaryotes with cell walls rich in chitin; no flagellated cells present. Mitosis in fungi takes place within the nuclei; the nuclear envelope never breaks down. The filaments of fungi grow through the substrate, secreting enzymes and digesting the products of their activity. Septa between the nuclei in the hyphae normally complete only when sexual or asexual reproductive structures are being cut off. Asexual reproduction is frequent in some groups. The nuclei of fungi are haploid, with the zygote the only diploid stage in the life cycle. About 73,000 named species. The major groups of fungi used to be called divisions, but currently taxonomists call these groups phyla.

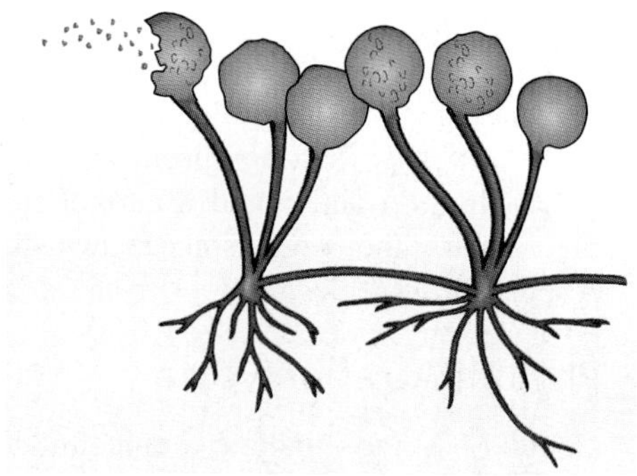

Phylum Zygomycota: haploid spore-bearing stalks.

Phylum Zygomycota

Zygomycetes; bread molds and other microscopic fungi that occur on decaying organic matter. Zygomycetes are the fungal partners in endomycorrhizae. Hyphae aseptate except when forming sporangia or gametangia. Sexual reproduction by equal gametangia containing numerous nuclei; after fusion, zygotes are formed within a zygospore, a structure that often forms a characteristic thick wall. Meiosis occurs during the germination of the zygospore. About 1,050 species.

Phylum Ascomycota

Ascomycetes; yeasts, molds, many important plant pathogens, morels, cup fungi, and truffles. Hyphae divided by incomplete septa except when asci, the structures characteristic of sexual reproduction, are formed. Dikaryotic hyphae form after appropriate fusions of monokaryotic ones, and eventually differentiate asci, which are often club-shaped, within an ascocarp. Meiosis takes place within asci. About 32,000 named species.

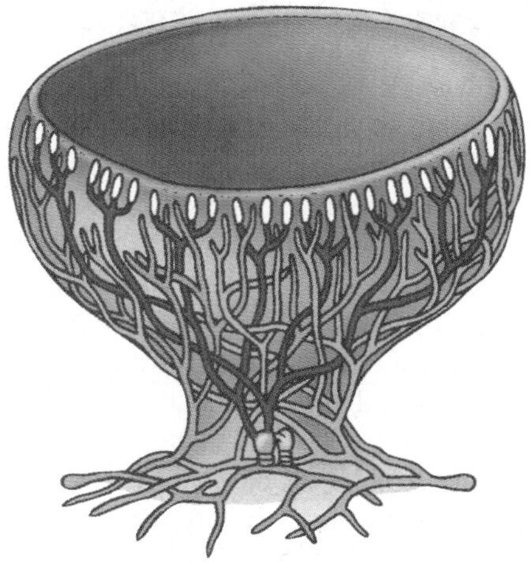

Phylum Ascomycota: ascocarp.

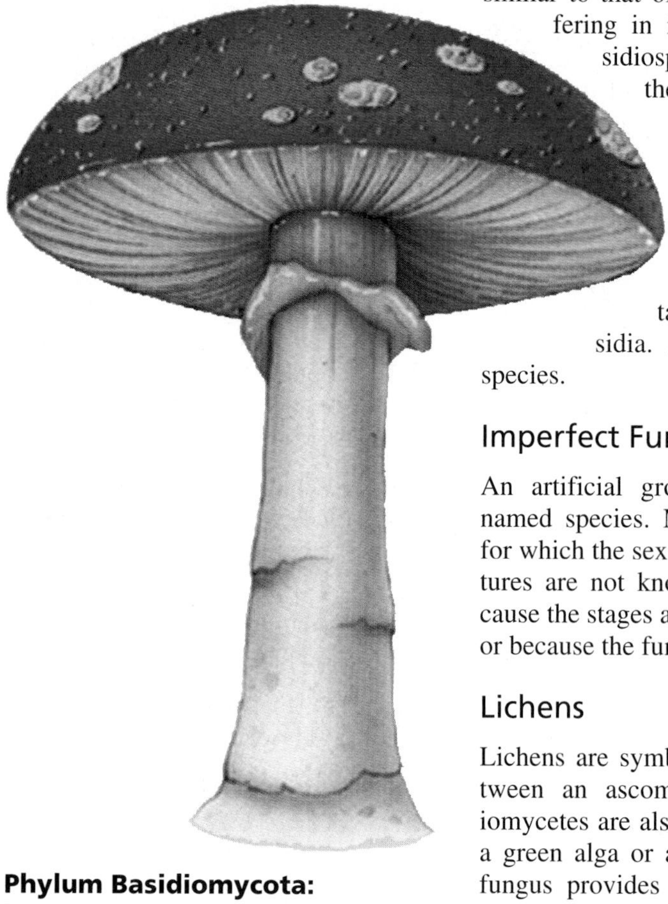

Phylum Basidiomycota: basidiocarp.

Phylum Basidiomycota

Basidiomycetes; mushrooms, toadstools, bracket and shelf fungi, rusts, and smuts. Most ectomycorrhizae involve basidiomycetes; a few involve ascomycetes. Hyphae and life cycle similar to that of ascomycetes but differing in important details. Basidiospores elevated from the basidium on thread-like projections called sterigmata; basidiocarps may be large and elaborate or absent, depending on the group. Meiosis takes place within basidia. About 22,000 named species.

Imperfect Fungi

An artificial group of about 17,000 named species. Most are ascomycetes for which the sexual reproductive structures are not known; this may be because the stages are rare or do not occur or because the fungus is poorly known.

Lichens

Lichens are symbiotic associations between an ascomycete (a few basidiomycetes are also involved) and either a green alga or a cyanobacterium; the fungus provides protection and structure. At least 13,500 species.

Kingdom Plantae

Multicellular, photosynthetic, primarily terrestrial eukaryotes derived from the green algae (phylum Chlorophyta) and, like them, containing chlorophylls *a* and *b*, together with carotenoids, in chloroplasts and storing starch in chloroplasts. The cell walls of plants have a cellulose matrix and sometimes become lignified; cell division is by means of a cell plate that forms across the mitotic spindle. The vascular plants have an elaborate system of conducting cells consisting of xylem (in which water and minerals are transported) and phloem (in which carbohydrates are transported); the mosses have a reduced vascular system, which the liverworts and hornworts (which may not be directly related to the mosses) lack. Plants have a waxy cuticle that helps them to retain water. Most have stomata, flanked by specialized guard cells, which allow water to escape and carbon dioxide to reach the chloroplast-containing cells within their leaves and stems. All plants have an alternation of generations with reduced gametophytes and multicellular gametangia. About 288,700 species. The major groups of plants used to be called divisions, but currently taxonomists call these groups phyla.

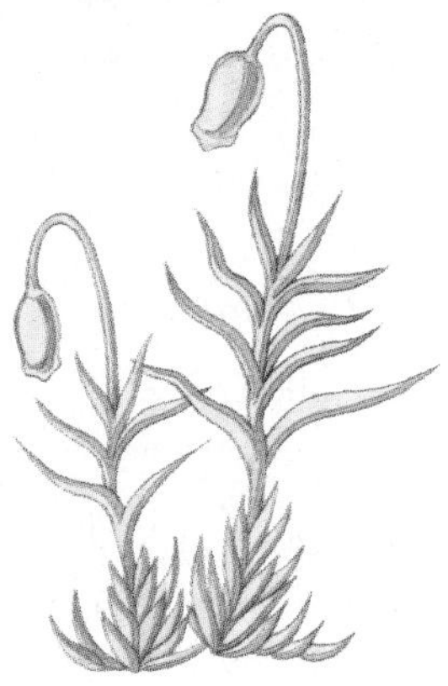

Phylum Bryophyta: a moss.

Phylum Bryophyta

Mosses. Bryophytes have green photosynthetic gametophytes and usually brownish or yellowish sporophytes with little or no chlorophyll. About 20,600 species. Hornworts and liverworts are members of two related phyla.

Phylum Psilophyta

Whisk ferns; a group of vascular plants. Two genera and several species.

Phylum Lycophyta

Lycopods (including clubmosses and quillworts); vascular plants. Five genera and about 1,150 species.

Phylum Arthrophyta

Horsetails; vascular plants. One genus (*Equisetum*) and 15 species.

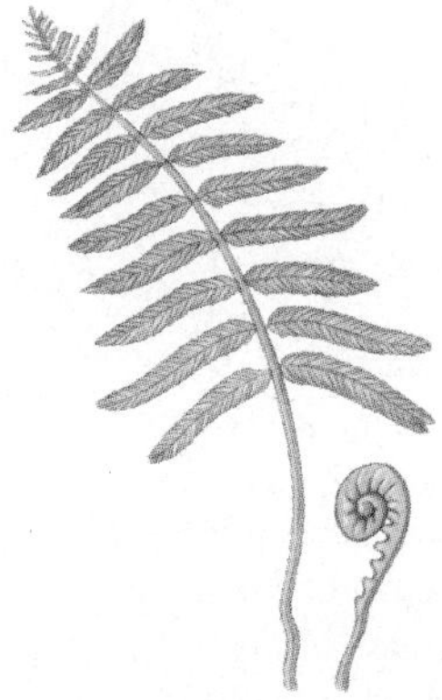

Phylum Pterophyta: a fern.

Phylum Pterophyta

Ferns; vascular plants, often with characteristically divided, feathery leaves (fronds). About 12,000 species.

Phylum Coniferophyta: a conifer.

Phylum Coniferophyta

Conifers; seed-forming vascular plants; mainly trees and shrubs. Archegonia multicellular, antheridia lacking; sperm immotile, carried to the vicinity of the egg by the pollen tube. Gametophytes reduced, held within the ovule or pollen grain. About 50 genera, with about 600 species.

Phylum Cycadophyta

Cycads; tropical and subtropical palm-like gymnosperms. Ten genera and about 206 species.

Phylum Gnetophyta

Gnetophytes, a very diverse group of three genera of gymnosperms. About 70 species.

Phylum Ginkgophyta: a ginkgo.

Phylum Ginkgophyta

One species, the ginkgo or maidenhair tree; a tall deciduous tree with fan-shaped leaves.

Phylum Anthophyta: a fruit tree.

Phylum Anthophyta

Flowering plants, or angiosperms, the dominant group of plants; characterized by a specialized reproductive system involving flowers and fruits. Sperm are immotile, carried to the vicinity of the ovule by the pollen tube. Fertilization is indirect, and the ovules are enclosed at the time of pollination. About 250,000 species.

Kingdom Animalia

Animals are multicellular eukaryotes that characteristically ingest their food. Their cells are usually flexible. In all of the approximately 35 phyla except sponges, these cells are organized into structural and functional units called tissues, which in turn make up organs in most animals. In animals, the cells move extensively during the development of the embryos; the blastula, a hollow ball of cells, forms early in this process and is characteristic of the group. Most animals reproduce sexually; their nonmotile eggs are much larger than their small, flagellated sperm. The gametes fuse directly to produce a zygote and do not divide by mitosis as in plants. More than a million species of animals have been described, and at least several times that many await discovery.

Phylum Porifera

Sponges; animals that mostly lack definite symmetry and possess neither tissues nor organs. About 5,000 species, mostly marine.

Phylum Cnidaria

Corals, jellyfish, hydras; mostly marine, radially symmetrical animals that usually have distinct tissues; two forms, polyps and medusae. About 10,000 species.

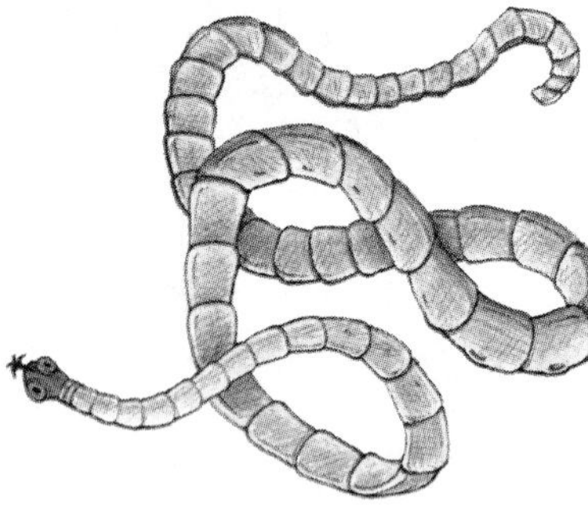

Phylum Platyhelminthes: a tapeworm.

Phylum Platyhelminthes

Flatworms; bilaterally symmetrical acoelomates; the simplest animals that have organs. About 20,000 species.

Phylum Nematoda

Nematodes, eelworms, and roundworms; ubiquitous, bilaterally symmetrical, cylindrical, unsegmented, pseudocoelomate worms, including many important parasites of plants and animals. More than 12,000 described species, but the actual number is probably 500,000 or more species.

Phylum Mollusca: a snail.

Phylum Mollusca

Mollusks; bilaterally symmetrical, protostome coelomate animals that occur in marine, freshwater, and terrestrial habitats. Many mollusks possess a shell. At least 110,000 species.

Phylum Annelida

Annelids; segmented, bilaterally symmetrical, protostome coelomates; the segments are divided internally by septa. About 12,000 species.

Phylum Arthropoda: a beetle.

Phylum Arthropoda

Arthropods; bilaterally symmetrical protostome coelomates with a segmented body, chitinous exoskeleton, complete digestive tract, dorsal brain and paired nerve cord, and jointed appendages. Arthropods are the largest phylum of animals, with nearly a million species described and many more to be found.

Phylum Echinodermata: a sea star.

Phylum Echinodermata

Echinoderms; sea stars, brittle stars, sand dollars, sea cucumbers, and sea urchins. Complex deuterostome, coelomate, marine animals that are more or less radially symmetrical as adults. About 6,000 living species.

Phylum Chordata: a dinosaur.

Phylum Chordata

Chordates; bilaterally symmetrical, deuterostome, coelomate animals that have at some stage of their development a notochord, pharyngeal slits, a hollow nerve cord on their dorsal side, and a tail. The best-known group of animals; about 50,300 species.

APPENDIX B

Answers to Concept Reviews

Chapter 1

1. b. use energy **2. c.** internal environment, stable **3. a.** cell **4. b.** evolution **5. a.** hypothesis **6. d.** theory **7. d.** variables **8.** Archaebacteria, Eubacteria, Protista, Fungi, Plantae, Animalia **9.** cellular organization, metabolism, homeostasis, reproduction, heredity **10.** DNA (deoxyribonucleic acid) **11.** cell **12.** structure **13.** observation **14.** control

Chapter 2

1. b. The earth was created in 4004 B.C. **2. a.** natural selection **3. a.** a variety of bill specializations among the finches **4. c.** discontinuous **5. c.** nonliving **6. d.** all of these **7. a.** gorillas **8. b.** the birthrate is higher in developing countries. **9.** offspring. **10.** the sun. **11.** carrying capacity **12.** *K*-selected

Chapter 3

1. a. atom **2. b.** oxidation—loss of an electron **3. b.** three **4. c.** It is absent in water **5. a.** adhesion **6. a.** lowers **7. d.** glucose **8. d.** proteins **9. d.** oxygen **10.** ionic **11.** three-fourths **12.** two-thirds **13.** cohesion **14.** base **15.** photosynthesis

Chapter 4

1. c. Life evolved 2 billion years ago **2. d.** DNA **3. c.** chromosomes **4. a.** ribosome **5. c, d, a,** then **b** **6. c.** water enters a hypotonic solution **7. b.** exocytosis **8. d.** active transport involves water diffusing through a membrane **9.** bacteria **10.** flagella **11.** mitochondria

Chapter 5

1. b. ATP **2. c.** the light reactions **3. d.** electrons **4. b** ATP, and **d.** NADPH **5. b.** glycolysis **6. d.** oxygen **7. a.** carbon dioxide **8.** endergonic **9.** binding **10.** NADH **11.** pigments **12.** photosynthesis **13.** thylakoid **14.** Krebs **15.** ethyl alcohol

Chapter 6

1. a. somatic cells **2. b.** DNA is replicated **3. c.** anaphase **4. d.** the presence of p53 protein **5. e.** meiosis, then **c.** zygote **6. b.** two **7. d.** four **8. b.** It occurs during prophase II **9.** binary fission **10.** M **11.** p53 **12.** variability

Chapter 7

1. c. He counted the numbers of different types of offspring **2. b.** one-third **3. a.** phenotype **4. c.** *Ww* × *ww* **5. b.** *Aa* **6. d.** multiple genes **7. c.** 45 **8. a.** hemophilia **9.** homozygous **10.** 25% **11.** independent assortment **12.** epistasis

Chapter 8

1. a. virus DNA injected into bacterial cells is apparently the factor involved in directing the production of new virus particles **2. d.** TTAAGC **3. c.** transcription **4. a.** translation **5. c.** operator site **6. a.** true **7.** DNA **8.** thymine, cytosine **9.** ribosome **10.** Transfer RNA **11.** enhancers **12.** repressor protein **13.** introns **14.** mutagen

Chapter 9

1. c. plasmids **2. b.** shorter **3. a.** a collection of different lines of bacterial cells, each containing a plasmid carrying one series of different DNA fragments **4. d.** amplify a region of DNA **5. b.** cDNA **6. a.** The plasmid contains the gene that codes for the microbe's surface proteins but is otherwise harmless **7. c.** human gene therapy **8.** Screening **9.** reverse transcriptase

Chapter 10

1. c. an open reading frame. **2. b.** slightly more than **3. a.** structural DNA **4. d.** 99% **5. b.** They are polyploid and have three sets of chromosomes **6. c.** Wheat resulted from the hybridization of a tetraploid with a diploid, giving it many more chromosomes than rice **7. b.** characterize the proteins of an organism **8.** shotgun sequencing **9.** structural DNA **10.** 30,000 **11.** single nucleotide polymorphisms (SNPs) **12.** Functional genomics, proteomics

Chapter 11

1. b. adaptation **2. d.** fossils, then **c.** industrial melanism **3. a.** vestigial organs **4. c.** 0.90 **5. d.** They are predominantly self-fertilizing **6. b.** stabilizing **7. b.** false **8.** homologous **9.** molecular **10.** effectively infinite

Chapter 12

1. b. kingdom, **d.** phylum, **g.** class, **a.** order, **e.** family, **f.** genus, **c.** species **2. c.** sterile **3. a.** 1.5 million **4. c.** common ancestor **5. c.** no tail **6. a.** weight **7.** biological species concept **8.** Systematics **9.** phylogeny **10.** clade **11.** dinosaurs

Chapter 13

1. b. 3.5, then **d.** 1.5 **2. c.** an organized nucleus **3. a.** binary fission **4. b.** archaebacteria **5. a.** introns **6. a.** tetanus, **c.** tuberculosis, and **d.** cholera **7. d.** crystallized in solution **8. c.** either RNA or DNA, but not both **9. b.** latency **10.** Eubacteria **11.** heterocysts **12.** Bacteriophages

Chapter 14

1. d. photosynthetic **2. a.** an unfertilized egg **3. c.** haploid **4. b.** animals, fungi, or plants **5. a.** sleeping sickness **6. c.** *Amoeba* **7. d.** green algae **8. b.** red algae **9.** endosymbiosis **10.** asexual **11.** DNA **12.** Choanoflagellates **13.** red algae (phylum Rhodophyta) **14.** dinoflagellates

Chapter 15

1. b. it allows the formation of specialized tissues **2. d.** Fungi are multicellular **3. b.** mycelium **4. c.** both asexually and sexually **5. a.** Fungi secrete a sticky substance that can trap flies **6. a.** basidiocarp **7.** hyphae **8.** cellulose **9.** Sexual, stress **10.** Ascomycota **11.** *Penicillium* **12.** mycorrhizae, minerals

Chapter 16

1. c. helped plants to avoid drying out **2. b.** larger **3. a.** liverworts **4. c.** seeds **5. b.** secondary **6. d.** club moss **7. a.** gymnosperms **8. a.** stamens **9. a.** double fertilization **10.** carpel **11.** triploid **12.** Monocots

Chapter 17

1. a. are alive at maturity **2. b.** companion cell **3. d.** stomata **4. a.** is the principal water-conducting tissue of plants **5. b.** gas exchange **6. d.** a cylinder around the periphery of the stem **7. c.** secondary xylem **8. a.** a central column of vascular tissue **9. b.** endodermis **10. c.** Evaporation of water vapor from the leaves **11.** apical, lateral **12.** sclerenchyma **13.** phloem **14.** vascular bundles **15.** transpiration

Chapter 18

1. d. stigma **2. b.** mitosis **3. c.** pollen tube **4. c.** one sperm cell fuses with the polar nuclei and one sperm fertilizes the egg **5. d.** birds **6. a.** cells on the shaded side of the stem elongate more than those on the sunny side **7. b.** inhibit the formation of lateral roots **8. b.** is long **9. b.** cold temperatures and **d.** lack of available water **10.** pollen grains **11.** nectar **12.** imbibes water **13.** ovary **14.** phytochromes **15.** Gravitropism

Chapter 19

1. d. Arthropoda **2. b.** choanoflagellates **3. c.** symmetry **4. d.** body cavity **5. c.** in the mesoderm **6. b.** Mollusca **7. a.** earthworms **8. b.** Chordata and **c.** Echinodermata **9. a.** a single, hollow dorsal nerve cord, **b.** a notochord at some time in development, **c.** pharyngeal slits at some time in development, **e.** a postanal tail **10. d.** backbone **11.** endoderm, mesoderm, ectoderm **12.** cephalization **13.** Nematodes **14.** segmentation **15.** protostomes **16.** Chordata

Chapter 20

1. c. plants **2. a.** (3), **b.** (1), **c.** (2), **d.** (4) **3. d.** smaller plankton **4. a.** amphibians **5. b.** swim bladders **6. c.** jaws **7. b.** have hair **8. d.** were shrewlike **9.** Trilobites **10.** Permian **11.** bony fishes **12.** amniotic **13.** *Archaeopteryx*

Chapter 21

1. d. tree shrews **2. c.** prehensile tails **3. b.** false **4. c.** prosimians, **d.** monkeys, **a.** gibbons, **e.** orangutans, **b.** gorillas **5. a.** the use of tools **6. b.** *Ardipithecus ramidus* **7. c.** *Homo erectus* **8. d.** 500,000 **9. a.** *Australopithecus*, **d.** *Homo habilis*, **c.** *Homo erectus*, **e.** *Homo neanderthalensis*, **b.** Cro-Magnons **10.** binocular **11.** *Australopithecus* **12.** *erectus* **13.** 13,000 **14.** Cultural

Chapter 22

1. c. tissues **2. a.** movement **3. c.** columnar cells **4. b.** carries oxygen **5. b.** dendrites to cell body to axon **6. b.** axial **7. d.** New bone cells are not formed in adults **8. a.** It is multinucleated **9. d.** troponin **10.** immune (lymphatic) **11.** Epithelial **12.** dendrites, axon **13.** actin **14.** acetylcholine **15.** ATP

Chapter 23

1. b. capillary **2. c.** They contain many valves **3. a.** delivering oxygen to cells **4. d.** 90% **5. c.** hemoglobin **6. a.** aorta **7. c.** birds **8. c.** larynx, **e.** trachea, **a.** bronchus, **b.** alveoli, **d.** hemoglobin **9. c.** It is located inside the lungs **10.** lymphatic **11.** serum albumin **12.** countercurrent flow **13.** crocodiles **14.** Carcinogens

Chapter 24

1. a. mouth **2. a.** low pH and **e.** pepsin **3. d.** pancreas **4. c.** compaction **5. c.** The liver stores bile **6. b.** urea **7. d.** drink **8. b.** network of capillaries **9. c.** water **10.** pyloric sphincter **11.** peristalsis **12.** villi **13.** glycogen **14.** bicarbonate **15.** filtration **16.** Henle

Chapter 25

1. c. It secretes gastric HCL, which inhibits the growth of many microbes **2. b.** macrophages **3. a.** complement system **4. c.** increased blood flow **5. a.** helper T cell **6. c.** to processed antigens on the surface of an antigen-presenting cell **7. d.** interleukin-1

8. c. type II diabetes **9. a.** helper T cells **10.** lysozyme **11.** antibodies **12.** Vaccination **13.** myelin sheath **14.** CD4

Chapter 26

1. a. afferent **2. b.** axon—synapse—dendrite **3. c.** electrical insulators **4. b.** cerebrum **5. d.** coordinating motor reflexes **6. a.** respiration **7. a.** decreases **8. c.** cerebrum **9. d.** gravity **10. c.** flexibility of the basilar membrane changes **11. b.** rods **12.** action potential **13.** neurotransmitters **14.** semicircular canals

Chapter 27

1. e. b and c **2. a.** thyroid hormone **3. b.** just above the kidneys **4. a.** aldosterone **5. d.** is almost always a consequence of excessive weight **6.** j **7.** i **8.** e **9.** h **10.** g **11.** d **12.** c **13.** f **14.** b **15.** a **16.** nervous **17.** hypothalamus **18.** steroids **19.** posterior **20.** adrenal cortex **21.** calcium **22.** pancreas

Chapter 28

1. a. Sperm begin developing before puberty **2. c.** erection **3. c.** fallopian tube **4. d.** LH **5. b.** corpus luteum **6. a.** morula **7. b.** somites **8. c.** formation of the major nerve tracks within the brain **9. b.** false **10. b.** use of a condom **11.** vagina **12.** Prolactin, oxytocin **13.** primitive streak **14.** ectoderm **15.** endoderm **16.** mesoderm

Chapter 29

1. b. in trophic level 3 **2. a.** primary consumers **3. d.** food web **4. b.** phosphorus cycle **5. a.** evaporation **6. b.** spring overturn **7. a.** air is cooled as it is forced upwards **8. b.** the shallow waters **9. d.** 4°C **10. c.** taiga **11.** net primary productivity **12.** biomass **13.** fossil fuel **14.** nitrogen fixation **15.** oligotrophic **16.** deserts **17.** savanna

Chapter 30

1. c. density-dependent effects **2. c.** In type III, mortality is higher in the early stages than it is in type II. **3. b.** sympatric species **4. a.** barnacles growing on the backs of whales **5. b.** tend not to live in groups **6. d.** obtaining poisons from the plants that their larvae eat and retaining them **7. c.** keystone **8. a.** introduction of exotic species **9.** logistic **10.** Resource partitioning **11.** commensalism **12.** aposematic **13.** Secondary **14.** species richness

Chapter 31

1. c. Wind and air currents dilute acid rain because factories are designed to release it high into the air **2. b.** CFCs stick to frozen water vapor and are able to act as catalysts **3. d.** The greenhouse effect **4. b.** decreasing the production of acid rain by nuclear power plants **5. a.** it is fertile soil that has accumulated over thousands of years and prevents erosion **6. b.** false **7.** Chlorofluorocarbons (CFCs) **8.** carbon dioxide (CO_2) **9.** taxes **10.** topsoil, groundwater, biodiversity **11.** pastureland (grazing) **12.** 5 million **13.** 6 billion

GLOSSARY

Key Terms and Concepts

A

absorption (L. *absorbere,* to swallow down) The movement of water and of substances dissolved in water into a cell, tissue, or organism.

acid Any substance that dissociates to form H^+ ions when dissolved in water.

acoelomate (Gr. *a,* not + *koiloma,* cavity) A bilaterally symmetrical animal not possessing a body cavity, such as a flatworm.

actin (Gr. *actis,* ray) One of the two major proteins that make up myofilaments (the other is myosin). It provides the cell with mechanical support and plays major roles in determining cell shape and cell movement.

action potential A single nerve impulse. A transient all-or-none reversal of the electrical potential across a neuron membrane. Because it can activate nearby voltage-sensitive channels, an action potential propagates along a nerve cell.

activation energy The energy a molecule must acquire to undergo a specific chemical reaction.

active transport The transport of a solute across a membrane by protein carrier molecules to a region of higher concentration by the expenditure of chemical energy. One of the most important functions of any cell.

adaptation (L. *adaptare,* to fit) Any peculiarity of structure, physiology, or behavior that promotes the likelihood of an organism's survival and reproduction in a particular environment.

adenosine triphosphate (ATP) A molecule composed of ribose, adenine, and a triphosphate group. ATP is the chief energy currency of all cells. Cells focus all of their energy resources on the manufacture of ATP from ADP and phosphate, which requires the cell to supply 7 kilocalories of energy obtained from photosynthesis or from electrons stripped from foodstuffs to form 1 mole of ATP. Cells then use this ATP to drive endergonic reactions.

adhesion (L. *adhaerere,* to stick to) The molecular attraction exerted between the surfaces of unlike bodies in contact, as water molecules to the walls of the narrow tubes that occur in plants.

aerobic (Gr. *aer,* air + *bios,* life) Oxygen-requiring.

allele (Gr. *allelon,* of one another) One of two or more alternative forms of a gene.

allele frequency The relative proportion of a particular allele among individuals of a population. Not equivalent to gene frequency, although the two terms are sometimes confused.

allosteric interaction (Gr. *allos,* other + *stereos,* shape) The change in shape that occurs when an activator or inhibitor binds to an enzyme. These changes result when specific, small molecules bind to the enzyme, molecules that are not substrates of that enzyme.

alternation of generations A reproductive life cycle in which the diploid phase produces spores that give rise to the haploid phase and the haploid phase produces gametes that fuse to give rise to the zygote. The zygote is the first cell of the multicellular diploid phase.

alveolus, *pl.* alveoli (L. *alveus,* a small cavity) One of the many small, thin-walled air sacs within the lungs in which the bronchioles terminate.

amniotic egg An egg that is isolated and protected from the environment by a more or less impervious shell. The shell protects the embryo from drying out, nourishes it, and enables it to develop outside of water.

anaerobic (Gr. *an,* without + *aer,* air + *bios,* life) Any process that can occur without oxygen. Includes glycolysis and fermentation. Anaerobic organisms can live without free oxygen.

anaphase In mitosis and meiosis II, the stage initiated by the separation of sister chromatids, during which the daughter chromosomes move to opposite poles of the cell; in meiosis I, marked by separation of replicated homologous chromosomes.

angiosperms The flowering plants, one of five phyla of seed plants. In angiosperms, the ovules at the time of pollination are completely enclosed by tissues.

anterior (L. *ante,* before) Located before or toward the front. In animals, the head end of an organism.

anther (Gr. *anthos,* flower) The part of the stamen of a flower that bears the pollen.

antibody (Gr. *anti,* against) A protein substance produced in the blood by a B cell lymphocyte in response to a foreign substance (antigen) and released into the bloodstream. Binding to the antigen, antibodies mark them for destruction by other elements of the immune system.

anticodon The three-nucleotide sequence at the end of a tRNA molecule that is complementary to, and base pairs with, an amino acid-specifying codon in mRNA.

antigen (Gr. *anti,* against + *genos,* origin) A foreign substance, usually a protein, that stimulates lymphocytes to proliferate and secrete specific antibodies that bind to the foreign substance, labeling it as foreign and destined for destruction.

apical meristem (L. *apex,* top + Gr. *meristos,* divided) In vascular plants, the growing point at the tip of the root or stem.

aposematic coloration An ecological strategy of some organisms that "advertise" their poisonous nature by the use of bright colors.

appendicular skeleton (L. *appendicula,* a small appendage) The skeleton of the limbs of the human body containing 126 bones.

archaebacteria A group of bacteria that are among the most primitive still in existence, characterized by the absence of peptidoglycan in their cell walls, a feature that distinguishes them from all other bacteria.

asexual Reproducing without forming gametes. Asexual reproduction does not involve sex. Its outstanding characteristic is that an individual offspring is genetically identical to its parent.

atom (Gr. *atomos,* indivisible) A core (nucleus) of protons and neutrons surrounded by an orbiting cloud of electrons. The chemical behavior of an atom is largely determined by the distribution of its electrons, particularly the number of electrons in its outermost level.

atomic mass The atomic mass of an atom consists of the combined weight of all of its protons and neutrons.

atomic number The number of protons in the nucleus of an atom. In an atom that does not bear an electric charge (that is, one that is not an ion), the atomic number is also equal to the number of electrons.

autonomic nervous system (Gr. *autos,* self + *nomos,* law) The motor pathways that carry commands from the central nervous system to regulate the glands and nonskeletal muscles of the body. Also called the involuntary nervous system.

autosome (Gr. *autos,* self + *soma,* body) Any of the 22 pairs of human chromosomes that are similar in size and morphology in both males and females.

autotroph (Gr. *autos,* self + *trophos,* feeder) Self-feeder. An organism that can harvest light energy from the sun or from the oxidation of inorganic compounds to make organic molecules.

axial skeleton The skeleton of the head and trunk of the human body containing 80 bones.

axon (Gr., axle) A process extending out from a neuron that conducts impulses away from the cell body.

B

bacterium, *pl.* bacteria (Gr. *bakterion,* dim. of *baktron,* a staff) The simplest cellular organism. Its cells are smaller and prokaryotic in structure, and they lack internal organization.

Barr body After Murray L. Barr, Canadian anatomist. The inactivated X chromosome in female mammals that can be seen as a deeply staining body, which remains attached to the nuclear membrane.

basal body In cells that contain flagella or cilia, a form of centriole that anchors each flagellum.

base Any substance that combines with H^+ ions. Having a pH value above 7.

Batesian mimicry After Henry W. Bates, English naturalist. A situation in which a palatable or nontoxic organism resembles another kind of organism that is distasteful or toxic. Both species exhibit warning coloration.

B cell A lymphocyte that recognizes invading pathogens much as T cells do, but instead of attacking the pathogens directly, it marks them for destruction by the nonspecific body defenses.

bilateral symmetry (L. *bi,* two + *lateris,* side; Gr. *symmetria,* symmetry) A body form in which the right and left halves of an organism are approximate mirror images of each other.

binary fission (L. *binarius,* consisting of two things or parts + *fissus,* split) Asexual reproduction of a cell by division into two equal, or nearly equal, parts. Bacteria divide by binary fission.

binomial system (L. *bi,* twice, two + Gr. *nomos,* usage, law) A system of nomenclature that uses two words. The first names the genus, and the second designates the species.

biomass (Gr. *bios,* life + *maza,* lump or mass) The total weight of all of the organisms living in an ecosystem.

biome (Gr. *bios,* life + *-oma,* mass, group) A major terrestrial assemblage of plants, animals, and microorganisms that occur over wide geographical areas and have distinct characteristics. The largest ecological unit.

C

calorie (L. *calor,* heat) The amount of energy in the form of heat required to raise the temperature of 1 gram of water 1 degree Celsius.

calyx (Gr. *kalyx,* a husk, cup) The sepals collectively. The outermost flower whorl.

cancer Unrestrained invasive cell growth. A tumor or cell mass resulting from uncontrollable cell division.

capillary (L. *capillaris,* hairlike) A blood vessel with a very slender, hairlike opening. Blood exchanges gases and metabolites within capillaries. Capillaries join the end of an artery to the beginning of a vein.

carbohydrate (L. *carbo,* charcoal + *hydro,* water) An organic compound consisting of a chain or ring of carbon atoms to which hydrogen and oxygen atoms are attached in a ratio of approximately 1:2:1. A compound of carbon, hydrogen, and oxygen having the generalized formula $(CH_2O)_n$, where n is the number of carbon atoms.

carcinogen (Gr. *karkinos,* cancer + -gen) Any cancer-causing agent.

cardiovascular system (Gr. *kardia,* heart + L. *vasculum,* vessel) The blood circulatory system and the heart that pumps it. Collectively, the blood, heart, and blood vessels.

carpel (Gr. *karpos,* fruit) A leaflike organ in angiosperms that encloses one or more ovules.

carrying capacity The maximum population size that a habitat can support.

catabolism (Gr. *katabole,* throwing down) A process in which complex molecules are broken down into simpler ones.

catalysis (Gr. *katalysis,* dissolution + *lyein,* to loosen) The enzyme-mediated process in which the subunits of polymers are held together and their bonds are stressed.

catalyst (Gr. *kata,* down + *lysis,* a loosening) A general term for a substance that speeds up a specific chemical reaction by lowering the energy required to activate or start the reaction. An enzyme is a biological catalyst.

cell (L. *cella,* a chamber or small room) The smallest unit of life. The basic organizational unit of all organisms. Composed of a nuclear region containing the hereditary apparatus within a larger volume called the cytoplasm bounded by a lipid membrane.

cell cycle The repeating sequence of growth and division through which cells pass each generation.

cellular respiration The process in which the energy stored in a glucose molecule is released by oxidation. Hydrogen atoms are lost by glucose and gained by oxygen.

central nervous system The brain and spinal cord, the site of information processing and control within the nervous system.

centromere (Gr. *kentron,* center + *meros,* a part) A constricted region of the chromosome joining two sister chromatids, to which the kinetochore is attached.

chemical bond The force holding two atoms together. The force can result from the attraction of opposite charges (ionic bond) or from the sharing of one or more pairs of electrons (a covalent bond).

chemiosmosis The cellular process responsible for almost all of the adenosine triphosphate (ATP) harvested from eaten food and for all the ATP produced by photosynthesis.

chemoautotroph An autotrophic bacterium that uses chemical energy released by specific inorganic reactions to power its life processes, including the synthesis of organic molecules.

chloroplast (Gr. *chloros,* green + *plastos,* molded) A cell-like organelle present in algae and plants that contains chlorophyll (and usually other pigments) and carries out photosynthesis.

chiasma, *pl.* chiasmata (Gr. a cross) In meiosis, the points of crossing-over where portions of chromosomes have been exchanged during synapsis. A chiasma appears as an X-shaped structure under a light microscope.

chromatid (Gr. *chroma,* color + L. *-id,* daughters of) One of two daughter strands of a duplicated chromosome that is joined by a single centromere.

chromatin (Gr. *chroma,* color) The complex of DNA and proteins of which eukaryotic chromosomes are composed.

chromosome (Gr. *chroma,* color + *soma,* body) The vehicle by which hereditary information is physically transmitted from one generation to the next. In a eukaryotic cell, long threads of DNA that are associated with protein and that contain hereditary information.

cilium, *pl.* cilia (L. eyelash) Refers to flagella, which are numerous and organized in dense rows. Cilia propel cells through water. In human tissue, they move water over the tissue surface.

cladistics A taxonomic technique used for creating hierarchies of organisms that represent true phylogenetic relationship and descent.

class A taxonomic category ranking below a phylum (division) and above an order.

clone (Gr. *klon,* twig) A line of cells, all of which have arisen from the same single cell by mitotic division. One of a population of individuals derived by asexual reproduction from a single ancestor. One of a population of genetically identical individuals.

codominance In genetics, a situation in which the effects of both alleles at a particular locus are apparent in the phenotype of the heterozygote.

codon (L. code) The basic unit of the genetic code. A sequence of three adjacent nucleotides in DNA or mRNA that codes for one amino acid or for polypeptide termination.

coelom (Gr. *koilos,* a hollow) A body cavity formed between layers of mesoderm and in which the digestive tract and other internal organs are suspended.

coenzyme A cofactor that is a nonprotein organic molecule.

coevolution (L. *co-,* together + *e-,* out + *volvere,* to fill) A term that describes the long-term evolutionary adjustment of one group of organisms to another.

commensalism (L. *cum,* together with + *mensa,* table) A symbiotic relationship in which one species benefits while the other neither benefits nor is harmed.

community (L. *communitas,* community, fellowship) The population of different species that live together and interact in a particular place.

competition Interaction between individuals of two or more species for the same scarce resources. Intraspecific competition is interaction for the same scarce resources between individuals of a single species.

competitive exclusion The hypothesis that if two species are competing with one another for the same limited resource in the same place, one will be able to use that resource more efficiently than the other and eventually will drive that second species to extinction locally.

complement system The chemical defense of a vertebrate body that consists of a battery of proteins that become activated by the walls of bacteria and fungi. Complements the cellular defenses.

concentration gradient The concentration difference of a substance as a function of distance. In a cell, a greater concentration of its molecules in one region than in another.

condensation The coiling of the chromosomes into more and more tightly compacted bodies begun during the G_2 phase of the cell cycle.

conjugation (L. *conjugare,* to yoke together) An unusual mode of reproduction that characterizes the ciliates, in which nuclei are exchanged between individuals through tubes connecting them during conjugation.

consumer In ecology, a heterotroph that derives its energy from living or freshly killed organisms or parts thereof. Primary consumers are herbivores; secondary consumers are carnivores or parasites.

corolla (L. *cornea,* crown) The petals of a flower, collectively. Usually, the conspicuously colored flower whorl.

cortex (L. bark) In vascular plants, the primary ground tissue of a stem or root, bounded externally by the epidermis and internally by the central cylinder of vascular tissue. In animals, the outer, as opposed to the inner, part of an organ, as in the adrenal, kidney, and cerebral cortexes.

cotyledon (Gr. *kotyledon,* a cup-shaped hollow) Seed leaf. Monocot embryos have one cotyledon, and dicots have two.

countercurrent flow In organisms, the passage of heat or of molecules (such as oxygen, water, or sodium ions) from one circulation path to another moving in the opposite direction. Because the flow of the two paths is in opposite directions, a concentration difference always exists between the two channels, facilitating transfer.

covalent bond (L. *co-,* together + *valare,* to be strong) A chemical bond formed by the sharing of one or more pairs of electrons.

crossing over An essential element of meiosis occurring during prophase when nonsister chromatids exchange portions of DNA strands.

cuticle (L. *cutis,* skin) A very thin film covering the outer skin of many plants.

cytokinesis (Gr. *kytos,* hollow vessel + *kinesis,* movement) The C phase of cell division in which the cell itself divides, creating two daughter cells.

cytoplasm (Gr. *kytos,* hollow vessel + *plasma,* anything molded) A semifluid matrix that occupies the volume between the nuclear region and the cell membrane. It contains the sugars, amino acids, proteins, and organelles (in eukaryotes) with which the cell carries out its everyday activities of growth and reproduction.

cytoskeleton (Gr. *kytos,* hollow vessel + *skeleton,* a dried body) In the cytoplasm of all eukaryotic cells, a network of protein fibers that supports the shape of the cell and anchors organelles, such as the nucleus, to fixed locations.

D

deciduous (L. *decidere,* to fall off) In vascular plants, shedding all the leaves at a certain season.

dehydration reaction Water-losing. The process in which a hydroxyl (OH) group is removed from one subunit of a polymer and a hydrogen (H) group is removed from the other subunit.

demography (Gr. *demos,* people + *graphein,* to draw) The statistical study of population. The measurement of people or, by extension, of the characteristics of people.

density The number of individuals in a population in a given area.

deoxyribonucleic acid (DNA) The basic storage vehicle or central plan of heredity information. It is stored as a sequence of nucleotides in a linear nucleotide polymer. Two of the polymers wind around each other like the outside and inside rails of a circular staircase.

depolarization The movement of ions across a cell membrane that wipes out locally an electrical potential difference.

determinate Having flowers that arise from terminal buds and thus terminate a stem or branch.

deuterostome (Gr. *deuteros,* second + *stoma,* mouth) An animal in whose embryonic development the anus forms from or near the blastopore, and the mouth forms later on another part of the blastula. Also characterized by radial cleavage.

dicot Short for dicotyledon; a class of flowering plants generally characterized by having two cotyledons, netlike veins, and flower parts in fours or fives.

diffusion (L. *diffundere,* to pour out) The net movement of molecules to regions of lower concentration as a result of random, spontaneous molecular motions. The process tends to distribute molecules uniformly.

dihybrid (Gr. *dis,* twice + L. *hibrida,* mixed offspring) An individual heterozygous for two genes.

dioecious (Gr. *di,* two + *eikos,* house) Having male and female flowers on separate plants of the same species.

diploid (Gr. *diploos,* double + *eidos,* form) A cell, tissue, or individual with a double set of chromosomes.

directional selection A form of selection in which selection acts to eliminate one extreme from an array of phenotypes. Thus, the genes promoting this extreme become less frequent in the population.

disaccharide (Gr. *dis,* twice + *sakcharon,* sugar) A sugar formed by linking two monosaccharide molecules together. Sucrose (table sugar) is a disaccharide formed by linking a molecule of glucose to a molecule of fructose.

disruptive selection A form of selection in which selection acts to eliminate rather than favor the intermediate type.

diurnal (L. *diurnalis,* day) Active during the day.

division Traditionally, a major taxonomic group of the plant kingdom comparable to a phylum of the animal kingdom. Today divisions are called phyla.

dominant allele An allele that dictates the appearance of heterozygotes. One allele is said to be dominant over another if an individual heterozygous for that allele has the same appearance as an individual homozygous for it.

dorsal (L. *dorsum,* the back) Toward the back, or upper surface. Opposite of ventral.

double fertilization A process unique to the angiosperms, in which one sperm nucleus fertilizes the egg and the second one fuses with the polar nuclei. These two events result in the formation of the zygote and the primary endosperm nucleus, respectively.

E

ecdysis (Gr. *ekdysis,* stripping off) The shedding of the outer covering or skin of certain animals. Especially the shedding of the exoskeleton by arthropods.

ecology (Gr. *oikos,* house + *logos,* word) The study of the relationships of organisms with one another and with their environment.

ecosystem (Gr. *oikos,* house + *systema,* that which is put together) A community, together with the nonliving factors with which it interacts.

ecotype (Gr. *oikos,* house + L. *typus,* image) A locally adapted variant of an organism, differing genetically from other ecotypes.

ectothermic Referring to animals whose body temperature is regulated by their behavior or their surroundings.

electron A subatomic particle with a negative electrical charge. The negative charge of one electron exactly balances the positive charge of one proton. Electrons orbit the atom's positively charged nucleus and determine its chemical properties.

electron transport system A collective term describing the series of membrane-associated electron carriers generated by the citric acid cycle. It puts the electrons harvested from the oxidation of glucose to work driving proton-pumping channels.

element A substance that cannot be separated into different substances by ordinary chemical methods.

endergonic (Gr. *endon,* within + *ergon,* work) Describing reactions in which the products contain more energy than the reactants and require an input of usable energy from an outside source before they can proceed. These reactions are not spontaneous.

endocrine gland (Gr. *endon,* within + *krinein,* to separate) A ductless gland producing hormonal secretions that pass directly into the bloodstream or lymph.

endocrine system The dozen or so major endocrine glands of a vertebrate.

endocytosis (Gr. *endon,* within + *kytos,* cell) The process by which the edges of plasma membranes fuse together and form an enclosed chamber called a vesicle. It involves the incorporation of a portion of an exterior medium into the cytoplasm of the cell by capturing it within the vesicle.

endoskeleton (Gr. *endon,* within + *skeletos,* hard) In vertebrates, an internal scaffold of bone to which muscles are attached.

endosperm (Gr. *endon,* within + *sperma,* seed) A nutritive tissue characteristic of the seeds of angiosperms that develops from the union of a male nucleus and the polar nuclei of the embryo sac. The endosperm is either digested by the growing embryo or retained in the mature seed to nourish the germinating seedling.

endosymbiotic (Gr. *endon,* within + *bios,* life) theory Proposes that eukaryotic cells arose from large prokaryotic cells that engulfed smaller ones of a different species, which were not consumed but continued to live and function within the larger host cell. Organelles that are believed to have entered larger cells in this way are mitochondria, chloroplasts, and centrioles.

endothermic Referring to the ability of animals to maintain a constant body temperature.

energy The capacity to bring about change, to do work.

enhancer A site of regulatory protein binding on the DNA molecule distant from the promoter and start site for a gene's transcription.

entropy (Gr. *en,* in + *tropos,* change in manner) A measure of the disorder of a system. A measure of energy that has become so randomized and uniform in a system that the energy is no longer available to do work.

enzyme (Gr. *enzymos,* leavened; from *en,* in + *zyme,* leaven) A protein capable of speeding up specific chemical reactions by lowering the energy required to activate or start the reaction but that remains unaltered in the process.

epidermis (Gr. *epi,* on or over + *derma,* skin) The outermost layer of cells. In vertebrates, the nonvascular external layer of skin of ectodermal origin; in invertebrates, a single layer of ectodermal epithelium; in plants, the flattened, skinlike outer layer of cells.

epistasis (Gr. *epistasis,* a standing still) An interaction between the products of two genes in which one modifies the phenotypic expression produced by the other.

epithelium (Gr. *epi,* on + *thele,* nipple) A thin layer of cells forming a tissue that covers the internal and external surfaces of the body. Simple epithelium consists of the membranes that line the lungs and major body cavities and that are a single cell layer thick. Stratified epithelium (the skin or epidermis) is composed of more complex epithelial cells that are several cell layers thick.

erythrocyte (Gr. *erythros,* red + *kytos,* hollow vessel) A red blood cell, the carrier of hemoglobin. Erythrocytes act as the transporters of oxygen in the vertebrate body. During the process of their maturation in mammals, they lose their nucleus and mitochondria, and their endoplasmic reticulum is reabsorbed.

estrus (L. *oestrus,* frenzy) The period of maximum female sexual receptivity. Associated with ovulation of the egg. Being "in heat."

estuary (L. *aestus,* tide) A partly enclosed body of water, such as those that often form at river mouths and in coastal bays, where the salinity is intermediate between that of saltwater and freshwater.

ethology (Gr. *ethos,* habit or custom + *logos,* discourse) The study of patterns of animal behavior in nature.

euchromatin (Gr. *eu,* good + *chroma,* color) Chromatin that is extended except during cell division, from which RNA is transcribed.

eukaryote (Gr. *eu,* good + *karyon,* kernel) A cell that possesses membrane-bounded organelles, most notably a cell nucleus, and chromosomes whose DNA is associated with proteins; an organism composed of such cells. The appearance of eukaryotes marks a major event in the evolution of life, as all organisms on earth other than bacteria are eukaryotes.

eumetazoan (Gr. *eu,* good + *meta,* with + *zoion,* animal) A "true animal." An animal with a definite shape and symmetry and nearly always distinct tissues.

eutrophic (Gr. *eutrophos,* thriving) Refers to a lake in which an abundant supply of minerals and organic matter exists.

evaporation The escape of water molecules from the liquid to the gas phase at the surface of a body of water.

evolution (L. *evolvere,* to unfold) Genetic change in a population of organisms over time (generations). Darwin proposed that natural selection was the mechanism of evolution.

exergonic (L. *ex,* out + Gr. *ergon,* work) Describes any reaction that produces products that contain less free energy than that possessed by the original reactants and that tends to proceed spontaneously.

exocytosis (Gr. *ex,* out of + *kytos,* cell) The extrusion of material from a cell by discharging it from vesicles at the cell surface. The reverse of endocytosis.

exoskeleton (Gr. *exo,* outside + *skeletos,* hard) An external hard shell that encases a body. In arthropods, comprised mainly of chitin; in vertebrates, comprised of bone.

experiment The test of a hypothesis. A successful experiment is one in which one or more alternative hypotheses are demonstrated to be inconsistent with experimental observation and are thus rejected.

F

facilitated diffusion The transport of molecules across a membrane by a carrier protein in the direction of lowest concentration.

family A taxonomic group ranking below an order and above a genus.

feedback inhibition A regulatory mechanism in which a biochemical pathway is regulated by the amount of the product that the pathway produces.

fermentation (L. *fermentum,* ferment) A catabolic process in which the final electron acceptor is an organic molecule.

fertilization (L. *ferre,* to bear) The union of male and female gametes to form a zygote.

fitness The genetic contribution of an individual to succeeding generations, relative to the contributions of other individuals in the population.

flagellum, *pl.* flagella (L. *flagellum,* whip) A fine, long, threadlike organelle protruding from the surface of a cell. In bacteria, a single protein fiber capable of rotary motion that propels the cell through the water. In eukaryotes, an array of microtubules with a characteristic internal 9 + 2 microtubule structure that is capable of vibratory but not rotary motion. Used in locomotion and feeding. Common in protists and motile gametes. A cilium is a small flagellum. The inward movement of certain cell groups from the surface of the blastula.

food web The food relationships within a community. A diagram of who eats whom.

founder principle The effect by which rare alleles and combinations of alleles may be enhanced in new populations.

frequency In statistics, defined as the proportion of individuals in a certain category, relative to the total number of individuals being considered.

fruit In angiosperms, a mature, ripened ovary (or group of ovaries) containing the seeds. Also applied informally to the reproductive structures of some other kinds of organisms.

G

gamete (Gr. wife) A haploid reproductive cell. Upon fertilization, its nucleus fuses with that of another gamete of the opposite sex. The resulting diploid cell (zygote) may develop into a new diploid individual, or in some protists and fungi, may undergo meiosis to form haploid somatic cells.

gametophyte (Gr. *gamete,* wife + *phyton,* plant) In plants, the haploid (n), gamete-producing generation, which alternates with the diploid ($2n$) sporophyte.

ganglion, *pl.* ganglia (Gr. a swelling) A group of nerve cells forming a nerve center in the peripheral nervous system.

gastrulation The inward movement of certain cell groups from the surface of the blastula.

gene (Gr. *genos,* birth, race) The basic unit of heredity. A sequence of DNA nucleotides on a chromosome that encodes a polypeptide or RNA molecule and so determines the nature of an individual's inherited traits.

gene expression The process in which an RNA copy of each active gene is made, and the RNA copy directs the sequential assembly of a chain of amino acids at a ribosome.

gene frequency The frequency with which individuals in a population possess a particular gene. Often confused with allele frequency.

genetic code The "language" of the genes. The mRNA codons specific for the 20 common amino acids constitute the genetic code.

genetic drift Random fluctuations in allele frequencies in a small population over time.

genetic map A diagram showing the relative positions of genes.

genetics (Gr. *genos,* birth, race) The study of the way in which an individual's traits are transmitted from one generation to the next.

genome (Gr. *genos,* offspring + L. *oma,* abstract group) The genetic information of an organism.

genomics The study of genomes as opposed to individual genes.

genotype (Gr. *genos,* offspring + *typos,* form) The total set of genes present in the cells of an organism. Also used to refer to the set of alleles at a single gene locus.

genus, *pl.* genera (L. race) A taxonomic group that ranks below a family and above a species.

germination (L. *germinare,* to sprout) The resumption of growth and development by a spore or seed.

gland (L. *glandis,* acorn) Any of several organs in the body, such as exocrine or endocrine, that secrete substances for use in the body. Glands are composed of epithelial tissue.

glomerulus (L. a little ball) A network of capillaries in a vertebrate kidney, whose walls act as a filtration device.

glycolysis (Gr. *glykys,* sweet + *lyein,* to loosen) The anaerobic breakdown of glucose; this enzyme-catalyzed process yields two molecules of pyruvate with a net of two molecules of ATP.

gravitropism (L. *gravis,* heavy + *tropes,* turning) The response of a plant to gravity, which generally causes shoots to grow up and roots to grow down.

greenhouse effect The process in which carbon dioxide and certain other gases, such as methane, that occur in the earth's atmosphere transmit radiant energy from the sun but trap the longer wavelengths of infrared light, or heat, and prevent them from radiating into space.

guard cells Pairs of specialized epidermal cells that surround a stoma. When the guard cells are turgid, the stoma is open; when they are flaccid, it is closed.

gymnosperm (Gr. *gymnos,* naked + *sperma,* seed) A seed plant with seeds not enclosed in an ovary. The conifers are the most familiar group.

H

habitat (L. *habitare,* to inhabit) The place where individuals of a species live.

half-life The length of time it takes for half of the carbon-14 present in a sample to be converted to carbon-12.

haploid (Gr. *haploos,* single + *eidos,* form) The gametes of a cell, tissue, or individual with only one set of chromosomes.

Hardy-Weinberg equilibrium After G. H. Hardy, English mathematician, and G. Weinberg, German physician. A mathematical description of the fact that the relative frequencies of two or more alleles in a population do not change because of Mendelian segregation. Allele and genotype frequencies remain constant in a random-mating population in the absence of inbreeding, selection, or other evolutionary forces. Usually stated as: If the frequency of allele A is p and the frequency of allele a is q, then the genotype frequencies after one generation of random mating will always be $(p + q)^2 = p^2 + 2pq + q^2$.

Haversian canal After Clopton Havers, English anatomist. Narrow channels that run parallel to the length of a bone and contain blood vessels and nerve cells.

helper T cell A class of white blood cells that initiates both the cell-mediated immune response and the humoral immune response; helper T cells are the targets of the AIDS virus (HIV).

hemoglobin (Gr. *haima,* blood + L. *globus,* a ball) A globular protein in vertebrate red blood cells and in the plasma of many invertebrates that carries oxygen and carbon dioxide.

herbivore (L. *herba,* grass + *vorare,* to devour) Any organism that eats plants.

heredity (L. *heredis,* heir) The transmission of characteristics from parent to offspring.

heterochromatin (Gr. *heteros,* different + *chroma,* color) That portion of a eukaryotic chromosome that remains permanently condensed and therefore is not transcribed into RNA. Most centromere regions are heterochromatic.

heterokaryon (Gr. *heteros,* other + *karyon,* kernel) A fungal hypha that has two or more genetically distinct types of nuclei.

heterotroph (Gr. *heteros,* other + *trophos,* feeder) An organism that does not have the ability to produce its own food. *See also* autotroph.

heterozygote (Gr. *heteros,* other + *zygotos,* a pair) A diploid individual carrying two different alleles of a gene on its two homologous chromosomes.

hierarchical (Gr. *hieros,* sacred + *archos,* leader) Refers to a system of classification in which successively smaller units of classification are included within one another.

histone (Gr. *histos,* tissue) A complex of small, very basic polypeptides rich in the amino acids arginine and lysine. A basic part of chromosomes, histones form the core around which DNA is wrapped.

homeostasis (Gr. *homeos,* similar + *stasis,* standing) The maintaining of a relatively stable internal physiological environment in an organism or steady-state equilibrium in a population or ecosystem.

homeotherm (Gr. *homeo,* similar + *therme,* heat) An organism, such as a bird or mammal, capable of maintaining a stable body temperature independent of the environmental temperature. "Warm-blooded."

hominid (L. *homo,* man) Human beings and their direct ancestors. A member of the family Hominidae. *Homo sapiens* is the only living member.

homologous chromosome (Gr. *homologia,* agreement) One of the two nearly identical versions of each chromosome. Chromosomes that associate in pairs in the first stage of meiosis. In diploid cells, one chromosome of a pair that carries equivalent genes.

homology (Gr. *homologia,* agreement) A condition in which the similarity between two structures or functions is indicative of a common evolutionary origin.

homozygote (Gr. *homos,* same or similar + *zygotos,* a pair) A diploid individual whose two copies of a gene are the same. An individual carrying identical alleles on both homologous chromosomes is said to be homozygous for that gene.

hormone (Gr. *hormaein,* to excite) A chemical messenger, often a steroid or peptide, produced in a small quantity in one part of an organism and then transported to another part of the organism, where it brings about a physiological response.

hybrid (L. *hybrida,* the offspring of a tame sow and a wild boar) A plant or animal that results from the crossing of dissimilar parents.

hybridization The mating of unlike parents of different taxa.

hydrogen bond A molecular force formed by the attraction of the partial positive charge of one hydrogen atom of a water molecule with the partial negative charge of the oxygen atom of another.

hydrolysis reaction (Gr. *hydro,* water + *lyse,* break) The process of tearing down a polymer by adding a molecule of water. A hydrogen is attached to one subunit and a hydroxyl to the other, which breaks the covalent bond. Essentially the reverse of a dehydration reaction.

hydrophobic (Gr. *hydro,* water + *phobos,* hating) Refers to nonpolar molecules, which do not form hydrogen bonds with water and therefore are not soluble in water.

hydroskeleton (Gr. *hydro,* water + *skeletos,* hard) The skeleton of most soft-bodied invertebrates that have neither an internal nor an external skeleton. They use the relative incompressibility of the water within their bodies as a kind of skeleton.

hyperosmotic The condition in which a (hyperosmotic) solution has a higher osmotic concentration than that of a second solution.

hypertonic (Gr. *hyper,* above + *tonos,* tension) Refers to a cell that contains a higher concentration of solutes than its surrounding solution.

hypha, ***pl.*** **hyphae (Gr. *hyphe,* web)** A filament of a fungus. A mass of hyphae comprises a mycelium.

hypoosmotic The condition in which a (hypoosmotic) solution has a lower osmotic concentration than that of a second solution.

hypothalamus (Gr. *hypo,* under + *thalamos,* inner room) The region of the brain under the thalamus that controls temperature, hunger, and thirst and that produces hormones that influence the pituitary gland.

hypothesis (Gr. *hypo,* under + *tithenai,* to put) A proposal that might be true. No hypothesis is ever proven correct. All hypotheses are provisional—proposals that are retained for the time being as useful but that may be rejected in the future if found to be inconsistent with new information. A hypothesis that stands the test of time—often tested and never rejected—is called a theory.

hypotonic (Gr. *hypo,* under + *tonos,* tension) Refers to the solution surrounding a cell that has a lower concentration of solutes than does the cell.

I

inbreeding The breeding of genetically related plants or animals. In plants, inbreeding results from self-pollination. In animals, inbreeding results from matings between relatives. Inbreeding tends to increase homozygosity.

incomplete dominance The ability of two alleles to produce a heterozygous phenotype that is different from either homozygous phenotype.

independent assortment Mendel's second law: the principle that segregation of alternative alleles at one locus into gametes is independent of the segregation of alleles at other loci. Only true for gene loci located on different chromosomes or those so far apart on one chromosome that crossing over is very frequent between the loci.

industrial melanism (Gr. *melas,* black) Phrase used to describe the evolutionary process in which initially light-colored organisms become dark as a result of natural selection.

inflammatory response (L. *inflammare,* to flame) A generalized nonspecific response to infection that acts to clear an infected area of infecting microbes and dead tissue cells so that tissue repair can begin.

integument (L. *integumentum,* covering) The natural outer covering layers of an animal. Develops from the ectoderm.

interneuron A nerve cell found only in the middle of the spinal cord that acts as a functional link between sensory neurons and motor neurons.

internode The region of a plant stem between nodes where stems and leaves attach.

interoception (L. *interus,* inner + Eng. *[re]ceptive*) The sensing of information that relates to the body itself, its internal condition, and its position.

interphase That portion of the cell cycle preceding mitosis. It includes the G_1 phase, when cells grow, the S phase, when a replica of the genome is synthesized, and a G_2 phase, when preparations are made for genomic separation.

intron (L. *intra,* within) A segment of DNA transcribed into mRNA but removed before translation. These untranslated regions make up the bulk of most eukaryotic genes.

ion An atom in which the number of electrons does not equal the number of protons. An ion does carry an electrical charge.

ionic bond A chemical bond formed between ions as a result of the attraction of opposite electrical charges.

ionizing radiation High-energy radiation, such as X rays and gamma rays.

isolating mechanisms Mechanisms that prevent genetic exchange between individuals of different populations or species. May be behavioral, morphological, or physiological.

isosmotic solutions The conditions in which the osmotic concentrations of two solutions are equal, so that no net water movement occurs between them by osmosis.

isotonic (Gr. *isos,* equal + *tonos,* tension) Refers to a cell with the same concentration of solutes as its environment.

isotope (Gr. *isos,* equal + *topos,* place) An atom that has the same number of protons but different numbers of neutrons.

J

joint The part of a vertebrate where one bone meets and moves on another.

K

karyotype (Gr. *karyon,* kernel + *typos,* stamp or print) The particular array of chromosomes that an individual possesses.

kinetic energy The energy of motion.

kinetochore (Gr. *kinetikos,* putting in motion + *choros,* chorus) A disk of protein bound to the centromere to which microtubules attach during mitosis, linking each chromatid to the spindle.

kingdom The chief taxonomic category. This book recognizes six kingdoms: Archaebacteria, Eubacteria, Protista, Fungi, Animalia, and Plantae.

L

lamella, *pl.* lamellae (L. a little plate) A thin, platelike structure. In chloroplasts, a layer of chlorophyll-containing membranes. In bivalve mollusks, one of the two plates forming a gill. In vertebrates, one of the thin layers of bone laid concentrically around the Haversian canals.

ligament (L. *ligare,* to bind) A band or sheet of connective tissue that links bone to bone.

linkage The patterns of assortment of genes that are located on the same chromosome. Important because if the genes are located relatively far apart, crossing over is more likely to occur between them than if they are close together.

lipid (Gr. *lipos,* fat) A loosely defined group of molecules that are insoluble in water but soluble in oil. Oils such as olive, corn, and coconut are lipids, as well as waxes, such as beeswax and earwax.

lipid bilayer The basic foundation of all biological membranes. In such a layer, the nonpolar tails of phospholipid molecules point inward, forming a nonpolar zone in the interior of the bilayers. Lipid bilayers are selectively permeable and do not permit the diffusion of water-soluble molecules into the cell.

littoral (L. *litus,* shore) Referring to the shoreline zone of a lake or pond or the ocean that is exposed to the air whenever water recedes.

locus, *pl.* loci (L. place) The position on a chromosome where a gene is located.

loop of Henle After F. G. J. Henle, German anatomist. A hairpin loop formed by a urine-conveying tubule when it enters the inner layer of the kidney and then turns around to pass up again into the outer layer of the kidney.

lymph (L. *lympha,* clear water) In animals, a colorless fluid derived from blood by filtration through capillary walls in the tissues.

lymphatic system An open circulatory system composed of a network of vessels that function to collect the water within blood plasma forced out during passage through the capillaries and to return it to the bloodstream. The lymphatic system also returns proteins to the circulation, transports fats absorbed from the intestine, and carries bacteria and dead blood cells to the lymph nodes and spleen for destruction.

lymphocyte (Gr. *lympha,* water + Gr. *kytos,* hollow vessel) A white blood cell. A cell of the immune system that either synthesizes antibodies (B cells) or attacks virus-infected cells (T cells).

lyse (Gr. *lysis,* loosening) To disintegrate a cell by rupturing its cell membrane.

M

macromolecule (Gr. *makros,* large + L. *moliculus,* a little mass) An extremely large molecule. Refers specifically to carbohydrates, lipids, proteins, and nucleic acids.

macrophage (Gr. *makros,* large + *-phage,* eat) A phagocytic cell of the immune system able to engulf and digest invading bacteria, fungi, and other microorganisms, as well as cellular debris.

marrow The soft tissue that fills the cavities of most bones and is the source of red blood cells.

mass In chemistry, the total number of protons and neutrons in the nucleus of an atom. Approximately equal to the atomic weight.

mass flow The overall process by which materials move in the phloem of plants.

meiosis (Gr. *meioun,* to make smaller) A special form of nuclear division that precedes gamete formation in sexually reproducing eukaryotes.

Mendelian ratio After Gregor Mendel, Austrian monk. Refers to the characteristic 3:1 segregation ratio that Mendel observed, in which pairs of alternative traits are expressed in the F_2 generation in the ratio of three-fourths dominant to one-fourth recessive.

menstruation (L. *mens,* month) Periodic sloughing off of the blood-enriched lining of the uterus when pregnancy does not occur.

meristem (Gr. *merizein,* to divide) In plants, a zone of unspecialized cells whose only function is to divide.

mesoderm (Gr. *mesos,* middle + *derma,* skin) One of the three embryonic germ layers that form in the gastrula. Gives rise to muscle, bone, and other connective tissue; the peritoneum; the circulatory system; and most of the excretory and reproductive systems.

mesophyll (Gr. *mesos,* middle + *phyllon,* leaf) The photosynthetic parenchyma of a leaf, located within the epidermis. The vascular strands (veins) run through the mesophyll.

metabolism (Gr. *metabole,* change) The process by which all living things assimilate energy and use it to grow.

metamorphosis (Gr. *meta,* after + *morphe,* form + *osis,* state of) Process in which form changes markedly during postembryonic development—for example, tadpole to frog or larval insect to adult.

metaphase (Gr. *meta,* middle + *phasis,* form) The stage of mitosis characterized by the alignment of the chromosomes on a plane in the center of the cell.

metastasis, *pl.* metastases (Gr. to place in another way) The spread of cancerous cells to other parts of the body, forming new tumors at distant sites.

microevolution (Gr. *mikros,* small + L. *evolvere,* to unfold) Refers to the evolutionary process itself. Evolution within a species. Also called adaptation.

microfilament (Gr. *mikros,* small + L. *filum,* a thread) In cells, a protein thread composed of parallel fibers of actin cross-connected by myosin. Their movement results from an ATP-driven shape change in myosin. The contraction of vertebrate muscles and many other kinds of cell movement in eukaryotes result from the movements of microfilaments within cells.

microtubule (Gr. *mikros,* small + L. *tubulus,* little pipe) In eukaryotic cells, a long, hollow cylinder about 25 nanometers in diameter and composed of the protein tubulin. Microtubules influence cell shape, move the chromosomes in cell division, and provide the functional internal structure of cilia and flagella.

mimicry (Gr. *mimos,* mime) The resemblance in form, color, or behavior of certain organisms (mimics) to other more powerful or more protected ones (models), which results in the mimics being protected in some way.

mitochondrion, *pl.* mitochondria (Gr. *mitos,* thread + *chondrion,* small grain) A tubular or sausage-shaped organelle 1 to 3 micrometers long. Bounded by two membranes, mitochondria closely resemble the aerobic bacteria from which they were originally derived. As chemical furnaces of the cell, they carry out its oxidative metabolism.

mitosis (Gr. *mitos,* thread) The M phase of cell division in which the microtubular apparatus is assembled, binds to the chromosomes, and moves them apart. This phase is the essential step in the separation of the two daughter cell genomes.

mole (L. *moles,* mass) The atomic weight of a substance, expressed in grams. One mole is defined as the mass of 6.0222×10^{23} atoms.

molecule (L. *moliculus,* a small mass) The smallest unit of a compound that displays the properties of that compound.

monocot Short for monocotyledon; flowering plant in which the embryos have only one cotyledon, the flower parts are often in threes, and the leaves typically are parallel-veined.

monosaccharide (Gr. *monos,* one + *sakcharon,* sugar) A simple sugar.

morphogenesis (Gr. *morphe,* form + *genesis,* origin) The formation of shape. The growth and differentiation of cells and tissues during development.

motor endplate The point where a neuron attaches to a muscle. A neuromuscular synapse.

multicellularity A condition in which the activities of the individual cells are coordinated and the cells themselves are in contact. A property of eukaryotes alone and one of their major characteristics.

muscle (L. *musculus,* mouse) The tissue in the body of humans and animals that can be contracted and relaxed to make the body move.

muscle cell A long, cylindrical, multinucleated cell that contains numerous myofibrils and is capable of contraction when stimulated.

muscle spindle A sensory end organ that is attached to a muscle and sensitive to stretching.

mutagen (L. *mutare,* to change) A chemical capable of damaging DNA.

mutation (L. *mutare,* to change) A change in a cell's genetic message.

mutualism (L. *mutuus,* lent, borrowed) A symbiotic relationship in which both participating species benefit.

mycelium, *pl.* mycelia (Gr. *mykes,* fungus) In fungi, a mass of hyphae.

mycology (Gr. *mykes,* fungus) The study of fungi. A person who studies fungi is called a mycologist.

mycorrhiza, *pl.* mycorrhizae (Gr. *mykes,* fungus + *rhiza,* root) A symbiotic association between fungi and plant roots.

myofibril (Gr. *myos,* muscle + L. *fibrilla,* little fiber) A contractile microfilament, composed of myosin and actin, within muscle.

myosin (Gr. *myos,* muscle + *in,* belonging to) One of two protein components of myofilaments. (The other is actin.)

N

natural selection The differential reproduction of genotypes caused by factors in the environment. Leads to evolutionary change.

nematocyst (Gr. *nema,* thread + *kystos,* bladder) A coiled, threadlike stinging process of cnidarians that is discharged to capture prey and for defense.

nephron (Gr. *nephros,* kidney) The functional unit of the vertebrate kidney. A human kidney has more than 1 million nephrons that filter waste matter from the blood. Each nephron consists of a Bowman's capsule, glomerulus, and tubule.

nerve A bundle of axons with accompanying supportive cells, held together by connective tissue.

nerve impulse A rapid, transient, self-propagating reversal in electrical potential that travels along the membrane of a neuron.

neuromodulator A chemical transmitter that mediates effects that are slow and longer lasting and that typically involve second messengers within the cell.

neuromuscular junction The structure formed when the tips of axons contact (innervate) a muscle fiber.

neuron (Gr. nerve) A nerve cell specialized for signal transmission.

neurotransmitter (Gr. *neuron,* nerve + L. *trans,* across + *mitere,* to send) A chemical released at an axon tip that travels across the synapse and binds a specific receptor protein in the membrane on the far side.

neurulation (Gr. *neuron,* nerve) The elaboration of a notochord and a dorsal nerve cord that marks the evolution of the chordates.

neutron (L. *neuter,* neither) A subatomic particle located within the nucleus of an atom. Similar to a proton in mass, but as its name implies, a neutron is neutral and possesses no charge.

neutrophil An abundant type of white blood cell capable of engulfing microorganisms and other foreign particles.

niche (L. *nidus,* nest) The role an organism plays in the environment; actual niche is the niche that an organism occupies under natural circumstances; theoretical niche is the niche an organism would occupy if competitors were not present.

nitrogen fixation The incorporation of atmospheric nitrogen into nitrogen compounds, a process that can be carried out only by certain microorganisms.

nocturnal (L. *nocturnus,* night) Active primarily at night.

node (L. *nodus,* knot) The place on the stem where a leaf is formed.

node of Ranvier After L. A. Ranvier, French histologist. A gap formed at the point where two Schwann cells meet and where the axon is in direct contact with the surrounding intercellular fluid.

nonrandom mating A phenomenon in which individuals with certain genotypes sometimes mate with one another more commonly than would be expected on a random basis.

notochord (Gr. *noto,* back + L. *chorda,* cord) In chordates, a dorsal rod of cartilage that forms between the nerve cord and the developing gut in the early embryo.

nucleic acid A nucleotide polymer. A long chain of nucleotides. Chief types are deoxyribonucleic acid (DNA), which is double-stranded, and ribonucleic acid (RNA), which is typically single-stranded.

nucleosome (L. *nucleus,* kernel + *soma,* body) The basic packaging unit of eukaryotic chromosomes, in which the DNA molecule is wound around a ball of histone proteins. Chromatin is composed of long strings of nucleosomes, like beads on a string.

nucleotide A single unit of nucleic acid, composed of a phosphate, a five-carbon sugar (either ribose or deoxyribose), and a purine or a pyrimidine.

nucleus (L. *a kernel,* dim. Fr. *nux,* nut) A spherical organelle (structure) characteristic of eukaryotic cells. The repository of the genetic information that directs all activities of a living cell. In atoms, the central core, containing positively charged protons and (in all but hydrogen) electrically neutral neutrons.

O

oocyte (Gr. *oion,* egg + *kytos,* vessel) A cell in the outer layer of the ovary that gives rise to an ovum. A primary oocyte is any of the 2 million oocytes a female is born with, all of which have begun the first meiotic division.

operon (L. *operis,* work) A cluster of functionally related genes transcribed onto a single mRNA molecule. A common mode of gene regulation in prokaryotes; it is rare in eukaryotes other than fungi.

order A taxonomic category ranking below a class and above a family.

organ (L. *organon,* tool) A complex body structure composed of several different kinds of tissue grouped together in a structural and functional unit.

organelle (Gr. *organella,* little tool) A specialized compartment of a cell. Mitochondria are organelles.

organism Any individual living creature, either unicellular or multicellular.

organ system A group of organs that function together to carry out the principal activities of the body.

osmoconformer An animal that maintains the osmotic concentration of its body fluids at about the same level as that of the medium in which it is living.

osmoregulation The maintenance of a constant internal solute concentration by an organism, regardless of the environment in which it lives.

osmosis (Gr. *osmos,* act of pushing, thrust) The diffusion of water across a membrane that permits the free passage of water but not that of one or more solutes.

osmotic pressure The increase of hydrostatic water pressure within a cell as a result of water molecules that continue to diffuse inward toward the area of lower water concentration (the water concentration is lower inside than outside the cell because of the dissolved solutes in the cell).

osteoblast (Gr. *osteon,* bone + *blastos,* bud) A bone-forming cell.

osteocyte (Gr. *osteon,* bone + *kytos,* hollow vessel) A mature osteoblast.

outcross A term used to describe species that interbreed with individuals other than those like themselves.

oviparous (L. *ovum,* egg + *parere,* to bring forth) Refers to reproduction in which the eggs are developed after leaving the body of the mother, as in reptiles.

ovulation The successful development and release of an egg by the ovary.

ovule (L. *ovulum,* a little egg) A structure in a seed plant that becomes a seed when mature.

ovum, ***pl.*** **ova (L. egg)** A mature egg cell. A female gamete.

oxidation (Fr. ***oxider,*** **to oxidize)** The loss of an electron during a chemical reaction from one atom to another. Occurs simultaneously with reduction. Is the second stage of the 10 reactions of glycolysis.

oxidative metabolism A collective term for metabolic reactions requiring oxygen.

oxidative respiration Respiration in which the final electron acceptor is molecular oxygen.

P

parasitism (Gr. ***para,*** **beside +** ***sitos,*** **food)** A symbiotic relationship in which one organism benefits and the other is harmed.

parthenogenesis (Gr. ***parthenos,*** **virgin + Eng.** ***genesis,*** **beginning)** The development of an adult from an unfertilized egg. A common form of reproduction in insects.

partial pressures (P) The components of each individual gas—such as nitrogen, oxygen, and carbon dioxide—that together constitute the total air pressure.

pathogen (Gr. ***pathos,*** **suffering + Eng.** ***genesis,*** **beginning)** A disease-causing organism.

pedigree (L. ***pes,*** **foot +** ***grus,*** **crane)** A family tree. The patterns of inheritance observed in family histories. Used to determine the mode of inheritance of a particular trait.

peptide (Gr. ***peptein,*** **to soften, digest)** Two or more amino acids linked by peptide bonds.

peptide bond A covalent bond linking two amino acids. Formed when the positive (amino, or NH_2) group at one end and a negative (carboxyl, or COOH) group at the other end undergo a chemical reaction and lose a molecule of water.

peristalsis (Gr. ***peri,*** **around +** ***stellein,*** **to wrap)** The rhythmic sequences of waves of muscular contraction in the walls of a tube.

pH Refers to the concentration of H^+ ions in a solution. The numerical value of the pH is the negative of the exponent of the molar concentration. Low pH values indicate high concentrations of H^+ ions (acids), and high pH values indicate low concentrations.

phagocyte (Gr. ***phagein,*** **to eat +** ***kytos,*** **hollow vessel)** A cell that kills invading cells by engulfing them. Includes neutrophils and macrophages.

phagocytosis (Gr. ***phagein,*** **to eat +** ***kytos,*** **hollow vessel)** A form of endocytosis in which cells engulf organisms or fragments of organisms.

phenotype (Gr. ***phainein,*** **to show +** ***typos,*** **stamp or print)** The realized expression of the genotype. The observable expression of a trait (affecting an individual's structure, physiology, or behavior) that results from the biological activity of proteins or RNA molecules transcribed from the DNA.

pheromone (Gr. ***pherein,*** **to carry + [hor]mone)** A chemical signal emitted by certain animals that signals their reproductive readiness.

phloem (Gr. ***phloos,*** **bark)** In vascular plants, a food-conducting tissue basically composed of sieve elements, various kinds of parenchyma cells, fibers, and sclereids.

phosphodiester bond The bond that results from the formation of a nucleic acid chain in which individual sugars are linked together in a line by the phosphate groups. The phosphate group of one sugar binds to the hydroxyl group of another, forming an —O—P—O bond.

photon (Gr. ***photos,*** **light)** The unit of light energy.

photoperiodism (Gr. ***photos,*** **light +** ***periodos,*** **a period)** A mechanism that organisms use to measure seasonal changes in relative day and night length.

photorespiration A process in which carbon dioxide is released without the production of ATP or NADPH. Because it produces neither ATP nor NADPH, photorespiration acts to undo the work of photosynthesis.

photosynthesis (Gr. ***photos,*** **light +** ***-syn,*** **together +** ***tithenai,*** **to place)** The process by which plants, algae, and some bacteria use the energy of sunlight to create from carbon dioxide (CO_2) and water (H_2O) the more complicated molecules that make up living organisms.

phototropism (Gr. ***photos,*** **light +** ***trope,*** **turning to light)** A plant's growth response to a unidirectional light source.

phylogeny (Gr. ***phylon,*** **race, tribe)** The evolutionary relationships among any group of organisms.

phylum, ***pl.*** **phyla (Gr.** ***phylon,*** **race, tribe)** A major taxonomic category, ranking above a class.

physiology (Gr. ***physis,*** **nature +** ***logos,*** **a discourse)** The study of the function of cells, tissues, and organs.

pigment (L. ***pigmentum,*** **paint)** A molecule that absorbs light.

pinocytosis (Gr. ***pinein,*** **to drink +** ***kytos,*** **cell)** A form of endocytosis in which the material brought into the cell is a liquid containing dissolved molecules.

pistil (L. ***pistillum,*** **pestle)** Central organ of flowers, typically consisting of ovary, style, and stigma; a pistil may consist of one or more fused carpels and is more technically and better known as the gynoecium.

plankton (Gr. ***planktos,*** **wandering)** The small organisms that float or drift in water, especially at or near the surface.

plasma (Gr. form) The fluid of vertebrate blood. Contains dissolved salts, metabolic wastes, hormones, and a variety of proteins, including antibodies and albumin. Blood minus the blood cells.

plasma membrane A lipid bilayer with embedded proteins that control the cell's permeability to water and dissolved substances.

plasmid (Gr. ***plasma,*** **a form or something molded)** A small fragment of DNA that replicates independently of the bacterial chromosome.

platelet (Gr. dim of ***plattus,*** **flat)** In mammals, a fragment of a white blood cell that circulates in the blood and functions in the formation of blood clots at sites of injury.

pleiotropy (Gr. ***pleros,*** **more +** ***trope,*** **a turning)** Describing a gene that produces more than one phenotypic effect.

polarization The charge difference of a neuron so that the interior of the cell is negative with respect to the exterior.

polar molecule A molecule with positively and negatively charged ends. One portion of a polar molecule attracts electrons more strongly than another portion, with the result that the molecule has electron-rich (–) and electron-poor (+) regions, giving it magnetlike positive and negative poles. Water is one of the most polar molecules known.

pollen (L. fine dust) A fine, yellowish powder consisting of grains or microspores, each of which contains a mature or immature male gametophyte. In flowering plants, pollen is released from the anthers of flowers and fertilizes the pistils.

pollen tube A tube that grows from a pollen grain. Male reproductive cells move through the pollen tube into the ovule.

pollination The transfer of pollen from the anthers to the stigmas of flowers for fertilization, as by insects or the wind.

polygyny (Gr. *poly,* many + *gyne,* woman, wife) A mating choice in which a male mates with more than one female.

polymer (Gr. *polus,* many + *meris,* part) A large molecule formed of long chains of similar molecules.

polymerase chain reaction (PCR) A process by which DNA polymerase is used to copy a sequence of interest repeatedly, making millions of copies of the same DNA.

polymorphism (Gr. *polys,* many + *morphe,* form) The presence in a population of more than one allele of a gene at a frequency greater than that of newly arising mutations.

polynomial system (Gr. *polys,* many + [bi]nomial) Before Linnaeus, naming a genus by use of a cumbersome string of Latin words and phrases.

polyp A cylindrical, pipe-shaped cnidarian usually attached to a rock with the mouth facing away from the rock on which it is growing. Coral is made up of polyps.

polypeptide (Gr. *polys,* many + *peptein,* to digest) A general term for a long chain of amino acids linked end to end by peptide bonds. A protein is a long, complex polypeptide.

polysaccharide (Gr. *polys,* many + *sakcharon,* sugar) A sugar polymer. A carbohydrate composed of many monosaccharide sugar subunits linked together in a long chain.

population (L. *populus,* the people) Any group of individuals, usually of a single species, occupying a given area at the same time.

posterior (L. *post,* after) Situated behind or farther back.

potential difference A difference in electrical charge on two sides of a membrane caused by an unequal distribution of ions.

potential energy Energy with the potential to do work. Stored energy.

predation (L. *praeda,* prey) The eating of other organisms. The one doing the eating is called a predator, and the one being consumed is called the prey.

primary growth In vascular plants, growth originating in the apical meristems of shoots and roots, as contrasted with secondary growth; results in an increase in length.

primary nondisjunction The failure of homologous chromosomes to separate in meiosis I. The cause of Down syndrome.

primary plant body The part of a plant that arises from the apical meristems.

primary producers Photosynthetic organisms, including plants, algae, and photosynthetic bacteria.

primary structure of a protein The sequence of amino acids that makes up a particular polypeptide chain.

primordium, *pl.* primordia (L. *primus,* first + *ordiri,* begin) The first cells in the earliest stages of the development of an organ or structure.

productivity The total amount of energy of an ecosystem fixed by photosynthesis per unit of time. Net productivity is productivity minus that which is expended by the metabolic activity of the organisms in the community.

prokaryote (Gr. *pro,* before + *karyon,* kernel) A simple bacterial organism that is small and single-celled, lacks external appendages, and has little evidence of internal structure.

promoter An RNA polymerase binding site. The nucleotide sequence at the end of a gene to which RNA polymerase attaches to initiate transcription of mRNA.

prophase (Gr. *pro,* before + *phasis,* form) The first stage of mitosis during which the chromosomes become more condensed, the nuclear envelope is reabsorbed, and a network of microtubules (called the spindle) forms between opposite poles of the cell.

protein (Gr. *proteios,* primary) A long chain of amino acids linked end to end by peptide bonds. Because the 20 amino acids that occur in proteins have side groups with very different chemical properties, the function and shape of a protein is critically affected by its particular sequence of amino acids.

protist (Gr. *protos,* first) A member of the kingdom Protista, which includes unicellular eukaryotic organisms and some multicellular lines derived from them.

proton A subatomic particle in the nucleus of an atom that carries a positive charge. The number of protons determines the chemical character of the atom because it dictates the number of electrons orbiting the nucleus and available for chemical activity.

protostome (Gr. *protos,* first + *stoma,* mouth) An animal in whose embryonic development the mouth forms at or near the blastopore. Also characterized by spiral cleavage.

protozoa (Gr. *protos,* first + *zoion,* animal) The traditional name given to heterotrophic protists.

pseudocoel (Gr. *pseudos,* false + *koiloma,* cavity) A body cavity similar to the coelom except that it is unlined.

punctuated equilibrium A hypothesis of the mechanism of evolutionary change that proposes that long periods of little or no change are punctuated by periods of rapid evolution.

Q

quaternary structure of a protein A term to describe the way multiple protein subunits are assembled into a whole.

R

radial symmetry (L. *radius,* a spoke of a wheel + Gr. *summetros,* symmetry) The regular arrangement of parts around a central axis so that any plane passing through the central axis divides the organism into halves that are approximate mirror images.

radioactivity The emission of nuclear particles and rays by unstable atoms as they decay into more stable forms. Measured in curies, with 1 curie equal to 37 billion disintegrations a second.

radula (L. scraper) A rasping, tongue-like organ characteristic of most mollusks.

recessive allele An allele whose phenotype effects are masked in heterozygotes by the presence of a dominant allele.

recombination The formation of new gene combinations. In bacteria, it is accomplished by the transfer of genes into cells, often in association with viruses. In eukaryotes, it is accomplished by reassortment of chromosomes during meiosis and by crossing over.

reducing power The use of light energy to extract hydrogen atoms from water.

reduction (L. *reductio*, a bringing back; originally, "bringing back" a metal from its oxide) The gain of an electron during a chemical reaction from one atom to another. Occurs simultaneously with oxidation.

reflex (L. *reflectere*, to bend back) An automatic consequence of a nerve stimulation. The motion that results from a nerve impulse passing through the system of neurons, eventually reaching the body muscles and causing them to contract.

refractory period The recovery period after membrane depolarization during which the membrane is unable to respond to additional stimulation. The period after ejaculation, lasting 20 minutes or longer, during which males lose their erection, arousal is difficult, and ejaculation is almost impossible.

renal (L. *renes*, kidneys) Pertaining to the kidney.

repression (L. *reprimere*, to press back, keep back) The process of blocking transcription by the placement of the regulatory protein between the polymerase and the gene, thus blocking movement of the polymerase to the gene.

repressor (L. *reprimere*, to press back, keep back) A protein that regulates transcription of mRNA from DNA by binding to the operator and so preventing RNA polymerase from attaching to the promoter.

resolving power The ability of a microscope to distinguish two lines as separate.

respiration (L. *respirare*, to breathe) The utilization of oxygen. In terrestrial vertebrates, the inhalation of oxygen and the exhalation of carbon dioxide.

resting membrane potential The charge difference that exists across a neuron's membrane at rest (about 70 millivolts).

restriction endonuclease A special kind of enzyme that can recognize and cleave DNA molecules into fragments. One of the basic tools of genetic engineering.

restriction fragment-length polymorphism (RFLP) An associated genetic mutation marker detected because the mutation alters the length of DNA segments.

retrovirus (L. *retro*, turning back) A virus whose genetic material is RNA rather than DNA. When a retrovirus infects a cell, it makes a DNA copy of itself, which it can then insert into the cellular DNA as if it were a cellular gene.

ribose A five-carbon sugar.

ribosome An organelle composed of protein and RNA that translates RNA copies of genes into protein.

RNA polymerase The enzyme that transcribes RNA from DNA.

S

saltatory conduction A very fast form of nerve impulse conduction in which the impulses leap from node to node over insulated portions.

sarcoma (Gr. *sarx*, flesh) A cancerous tumor that involves connective or hard tissue, such as muscle.

sarcomere (Gr. *sarx*, flesh + *meris*, part of) The fundamental unit of contraction in skeletal muscle. The repeating bands of actin and myosin that appear between two Z lines.

sarcoplasmic reticulum (Gr. *sarx*, flesh + *plassein*, to form, mold; L. *reticulum*, network) The endoplasmic reticulum of a muscle cell. A sleeve of membrane that wraps around each myofilament.

scientific creationism A view that the biblical account of the origin of the earth is literally true, that the earth is much younger than most scientists believe, and that all species of organisms were individually created just as they are today.

secondary growth In vascular plants, growth that results from the division of a cylinder of cells around the plant's periphery. Secondary growth causes a plant to grow in diameter.

second messenger An intermediary compound that couples extracellular signals to intracellular processes and also amplifies a hormonal signal.

seed A structure that develops from the mature ovule of a seed plant. Contains an embryo surrounded by a protective coat.

selection The process by which some organisms leave more offspring than competing ones and their genetic traits tend to appear in greater proportions among members of succeeding generations than the traits of those individuals that leave fewer offspring.

self-fertilization The transfer of pollen from an anther to a stigma in the same flower or to another flower of the same plant, leading to self-fertilization.

sepal (L. *sepalum*, a covering) A member of the outermost whorl of a flowering plant. Collectively, the sepals constitute the calyx.

septum, *pl.* septa (L. *saeptum*, a fence) A partition or cross-wall, such as those that divide fungal hyphae into cells.

sex chromosomes The X and Y chromosomes, which are different in the two sexes and are involved in sex determination.

sex-linked characteristic A genetic characteristic that is determined by genes located on the sex chromosomes.

sexual reproduction Reproduction that involves the regular alternation between syngamy and meiosis. Its outstanding characteristic is that an individual offspring inherits genes from two parent individuals.

shoot In vascular plants, the aboveground parts, such as the stem and leaves.

sieve cell In the phloem (food-conducting tissue) of vascular plants, a long, slender sieve element with relatively unspecialized sieve areas and with tapering end walls that lack sieve plates. Found in all vascular plants except angiosperms, which have sieve-tube members.

soluble Refers to polar molecules that dissolve in water and are surrounded by a hydration shell.

solute The molecules dissolved in a solution. *See also* solution, solvent.

solution A mixture of molecules, such as sugars, amino acids, and ions, dissolved in water.

solvent The most common of the molecules dissolved in a solution. Usually a liquid, commonly water.

somatic cells (Gr. *soma,* body) All the diploid body cells of an animal that are not involved in gamete formation.

somite A segmented block of tissue on either side of a developing notochord.

species, *pl.* species (L. kind, sort) A level of taxonomic hierarchy; a species ranks next below a genus.

sperm (Gr. *sperma,* sperm, seed) A sperm cell. The male gamete.

spindle The mitotic assembly that carries out the separation of chromosomes during cell division. Composed of microtubules and assembled during prophase at the equator of the dividing cell.

spore (Gr. *spora,* seed) A haploid reproductive cell, usually unicellular, that is capable of developing into an adult without fusion with another cell. Spores result from meiosis, as do gametes, but gametes fuse immediately to produce a new diploid cell.

sporophyte (Gr. *spora,* seed + *phyton,* plant) The spore-producing, diploid ($2n$) phase in the life cycle of a plant having alternation of generations.

stabilizing selection A form of selection in which selection acts to eliminate both extremes from a range of phenotypes.

stamen (L. thread) The part of the flower that contains the pollen. Consists of a slender filament that supports the anther. A flower that produces only pollen is called staminate and is functionally male.

steroid (Gr. *stereos,* solid + L. *ol,* from oleum, oil) A kind of lipid. Many of the molecules that function as messengers and pass across cell membranes are steroids, such as the male and female sex hormones and cholesterol.

steroid hormone A hormone derived from cholesterol. Those that promote the development of the secondary sexual characteristics are steroids.

stigma (Gr. mark) A specialized area of the carpel of a flowering plant that receives the pollen.

stoma, *pl.* stomata (Gr. mouth) A specialized opening in the leaves of some plants that allows carbon dioxide to pass into the plant body and allows water and oxygen to pass out of them.

stratum corneum The outer layer of the epidermis of the skin of the vertebrate body.

substrate (L. *substratus,* strewn under) A molecule on which an enzyme acts.

substrate-level phosphorylation The generation of ATP by coupling its synthesis to a strongly exergonic (energy-yielding) reaction.

succession In ecology, the slow, orderly progression of changes in community composition that takes place through time. Primary succession occurs in nature over long periods of time. Secondary succession occurs when a climax community has been disturbed.

sugar Any monosaccharide or disaccharide.

surface tension A tautness of the surface of a liquid, caused by the cohesion of the liquid molecules. Water has an extremely high surface tension.

surface-to-volume ratio Describes cell size increases. Cell volume grows much more rapidly than surface area.

symbiosis (Gr. *syn,* together with + *bios,* life) The condition in which two or more dissimilar organisms live together in close association; includes parasitism, commensalism, and mutualism.

synapse (Gr. *synapsis,* a union) A junction between a neuron and another neuron or muscle cell. The two cells do not touch. Instead, neurotransmitters cross the narrow space between them.

synapsis (Gr. *synapsis,* contact, union) The close pairing of homologous chromosomes that occurs early in prophase I of meiosis. With the genes of the chromosomes thus aligned, a DNA strand of one homologue can pair with the complementary DNA strand of the other.

syngamy (Gr. *syn,* together with + *gamos,* marriage) Fertilization. The union of male and female gametes.

T

taxonomy (Gr. *taxis,* arrangement + *nomos,* law) The science of the classification of organisms.

T cell A type of lymphocyte involved in cell-mediated immune responses and interactions with B cells. Also called a T lymphocyte.

tendon (Gr. *tenon,* stretch) A strap of cartilage that attaches muscle to bone.

tertiary structure of a protein The three-dimensional shape of a protein. Primarily the result of hydrophobic interactions of amino acid side groups and, to a lesser extent, of hydrogen bonds between them. Forms spontaneously.

test cross A cross between a heterozygote and a recessive homozygote. A procedure Mendel used to further test his hypotheses.

theory (Gr. *theorein,* to look at) A well-tested hypothesis supported by a great deal of evidence.

thigmotropism (Gr. *thigma,* touch + *trope,* a turning) The growth response of a plant to touch.

thorax (Gr. a breastplate) The part of the body between the neck and the abdomen.

thylakoid (Gr. *thylakos,* sac + *-oides,* like) A flattened, saclike membrane in the chloroplast of a eukaryote. Thylakoids are stacked on top of one another in arrangements called grana and are the sites of photosystem reactions.

tissue (L. *texere,* to weave) A group of similar cells organized into a structural and functional unit.

trachea, *pl.* tracheae (L. windpipe) In vertebrates, the windpipe.

tracheid (Gr. *tracheia,* rough) An elongated cell with thick, perforated walls that carries water and dissolved minerals through a plant and provides support. Tracheids form an essential element of the xylem of vascular plants.

transcription (L. *trans,* across + *scribere,* to write) The first stage of gene expression in which the RNA polymerase enzyme synthesizes an mRNA molecule whose sequence is complementary to the DNA.

translation (L. *trans,* across + *latus,* that which is carried) The second stage of gene expression in which a ribosome assembles a polypeptide, using the mRNA to specify the amino acids.

translocation (L. *trans,* across + *locare,* to put or place) In plants, the process in which most of the carbohydrates manufactured in the leaves and other green parts of the plant are moved through the phloem to other parts of the plant.

transpiration (L. *trans,* across + *spirare,* to breathe) The loss of water vapor by plant parts, primarily through the stomata.

transposon (L. *transponere,* to change the position of) A DNA sequence carrying one or more genes and flanked by insertion sequences that confer the ability to move from one DNA molecule to another. An element capable of transposition (the changing of chromosomal location).

trophic level (Gr. *trophos,* feeder) A step in the movement of energy through an ecosystem.

tropism (Gr. *trop,* turning) A plant's response to external stimuli. A positive tropism is one in which the movement or reaction is in the direction of the source of the stimulus. A negative tropism is one in which the movement or growth is in the opposite direction.

turgor pressure (L. *turgor,* a swelling) The pressure within a cell that results from the movement of water into the cell. A cell with high turgor pressure is said to be turgid.

U

unicellular Composed of a single cell.

urea (Gr. *ouron,* urine) An organic molecule formed in the vertebrate liver. The principal form of disposal of nitrogenous wastes by mammals.

urine (Gr. *ouron,* urine) The liquid waste filtered from the blood by the kidneys.

V

vaccination The injection of a harmless microbe into a person or animal to confer resistance to a dangerous microbe.

vacuole (L. *vacuus,* empty) A cavity in the cytoplasm of a cell that is bound by a single membrane and contains water and waste products of cell metabolism. Typically found in plant cells.

variable Any factor that influences a process. In evaluating alternative hypotheses about one variable, all other variables are held constant so that the investigator is not misled or confused by other influences.

vascular bundle In vascular plants, a strand of tissue containing primary xylem and primary phloem. These bundles of elongated cells conduct water with dissolved minerals and carbohydrates throughout the plant body.

vascular cambium In vascular plants, the meristematic layer of cells that gives rise to secondary phloem and secondary xylem. The activity of the vascular cambium increases stem or root diameter.

ventral (L. *venter,* belly) Refers to the bottom portion of an animal.

vertebrate An animal having a backbone made of bony segments called vertebrae.

vesicle (L. *vesicula,* a little bladder) Membrane-enclosed sacs within eukaryotic organisms created by weaving sheets of endoplasmic reticulum through the cell's interior.

vessel element In vascular plants, a typically elongated cell, dead at maturity, that conducts water and solutes in the xylem.

villus, *pl.* villi (L. a tuft of hair) In vertebrates, fine, microscopic, fingerlike projections lining the small intestine that serve to increase the absorptive surface area of the intestine.

vitamin (L. *vita,* life + *amine,* of chemical origin) An organic substance required in minute quantities by an organism for growth and activity but that the organism cannot synthesize.

viviparous (L. *vivus,* alive + *parere,* to bring forth) Refers to reproduction in which eggs develop within the mother's body and young are born free-living.

voltage-gated channel A transmembrane pathway for an ion that is opened or closed by a change in the voltage, or charge difference, across the cell membrane.

W

water vascular system The system of water-filled canals connecting the tube feet of echinoderms.

whorl A circle of leaves or of flower parts present at a single level along an axis.

wood Accumulated secondary xylem. Heartwood is the central, nonliving wood in the trunk of a tree. Hardwood is the wood of dicots, regardless of how hard or soft it actually is. Softwood is the wood of conifers.

X

xylem (Gr. *xylon,* wood) In vascular plants, a specialized tissue, composed primarily of elongate, thick-walled conducting cells, that transports water and solutes through the plant body.

Y

yolk (O.E. *geolu,* yellow) The stored substance in egg cells that provides the embryo's primary food supply.

Z

zygote (Gr. *zygotos,* paired together) The diploid ($2n$) cell resulting from the fusion of male and female gametes (fertilization).

CREDITS

Photographs

Chapter 1

1.1 (top left): © J.J. Cardamone, Jr. & B.K. Pugashetti/BPS/Tom Stack & Associates; (top center): Robert Simpson/Tom Stack & Associates; (top right): © Marty Snyderman; (bottom left): © T.J. Beveridge/Visuals Unlimited; (bottom center): © James L. Castner; (bottom right): © Kjell Sandved/Butterfly Alphabet; 1.2: © T.E. Adams/Visuals Unlimited; 1.3: © Tom J. Ulrich/ Visuals Unlimited; 1.4, page 6(above): Photo Lennart Nilsson/Albert Bonniers Forlag AB, Behold Man, Little, Brown & Co.; (below): From C.P. Morgan & R.A. Jersid, Anatomical Record, 166: 575-586, 1970 © John Wiley & Sons; 1.4, page 7 (top left): © Tom J. Ulrich/Visuals Unlimited; (bottom left): © Ed Reschke; (top right): © Robert & Jean Pollock; (right,2nd fr. top): © Kirtley Perkins/Visuals Unlimited; (right, 3rd fr. top, left): © Arthur Morris/Visuals Unlimited; (right, 3rd fr. top, right): © Alan Nelson/Animals Animals/Earth Scenes; (bottom right): © Maslowski/Visuals Unlimited; 1.5A: © Joe McDonald/Animals Animals /Earth Scenes; 1.5B: © Kenneth Fink/Photo Researchers; 1.5C: © Tom McHugh/Photo Researchers; 1.6: © Michio Hoshino/Minden Pictures; 1.7: © Michio Hoshino/Minden Pictures; 1.8: © Runk/ Schoenberger/Grant Heilman Photography; 1.9: © Frans Lanting/Minden Pictures; 1.13: NASA; 1.14: AP/Wide World Photos; Page 16(top left): © Tom J. Ulrich/Visuals Unlimited; (top right): AP/Wide World Photos; Page 17: © PhotoDisc/ Vol. # 29

Chapter 2

2.1: © Mohr/Darwin Collection; Herbert Rose Barraud, photographer. Reproduced by permission of the Huntington Library, San Marino, California; 2.3: From DARWIN by Adrian Desmond. Copyright © 1991 by Adrian Desmond & James Moore. By permission of Warner Books Inc.; 2.5: © Cleveland Hickman; 2.6A: © Frank B. Gill/ Vireo; 2.6B: © C.H. Greenewalt/Vireo; 2.9: © Mary Evans Picture Library/Photo Researchers; 2.10: Smithsonian Institution Libraries © 2000 Smithsonian Institutions; 2.11: © T.J. Ulrich/ Vireo; 2.11(all): © William P. Mull; 2.16: © G. R. Roberts/G .R. Roberts Photo Library; 2.18: © Kjell Sandved/Butterfly Alphabet; 2.22: © Richard Walters/Visuals Unlimited; 2.23: © Frans Lanting/ Minden Pictures; Page 40(top left): Mohr/Darwin Collection; Herbert Rose Barraud, photographer. Reproduced by permission of the Huntington Library, San Marino, California; (top right): © Kjell Sandved/Butterfly Alphabet; Page 41: © Corbis/Vol. # 175

Chapter 3

3.6: © Mary Evans Picture Library/Photo Researchers; 3.11: © Frans Lanting/Minden Pictures; 3.12A: © John Eastcott/Yva Momatiuk/ The Image Works; 3.12B: © George I. Bernard/ Animals Animals/Earth Scenes; 3.19: ©Scott Johnson/Animals Animals/Earth Scenes; 3.24A: © Michael Pasdzior/The Image Bank; 3.24B: © George Bernard/Animals Animals/Earth Scenes; 3.25C: © Scott Blackman/Tom Stack & Associates; 3.29: © Ken Eward/Biografix/Science Source/Photo Researchers; 3.31: © Bob McKeever/Tom Stack & Associates; 3.36: NASA; Page 70 (left): © John Eastcott/Yva Momatiuk/The Image Works; (top right): NASA; Page 71: © PhotoDisc/Vol. # 1

Chapter 4

4.7: © SIU/Visuals Unlimited; 4.9A: © J.J. Cardamore/BPS/Tom Stack & Associates; 4.9B: © David M. Phillips/Visuals Unlimited; 4.9C: © Ed Reschke; 4.10B: © Biophoto Associates/ Photo Researchers; 4.11B: © K.G. Murti/Visuals Unlimited; 4.13: Courtesy of Drs. J.V. Small & G. Rinnerthaler; 4.17: © R. Bolender & D. Fawcett/Visuals Unlimited; 4.18B: Courtesy of Dr. Charles Flickinger, *Medical Cellular Biology,* W.B. Saunders, 1979; 4.20B: © Don W. Fawcett/Visuals Unlimited; 4.21: Courtesy of Dr. Kenneth Miller, Brown University; 4.22A: © Don W. Fawcett/Visuals Unlimited; 4.23B: © Stanley Flegler/Visuals Unlimited; 4.25: © Biophoto Associates/Photo Researchers; 4.28(both): © David M. Phillips/Visuals Unlimited; 4.29B: © Biophoto Associates/Photo Researchers; 4.30B: Courtesy of Dr. Birgit H. Satir; Page 101: © Corbis/Vol. # 40

Chapter 5

5.2(both): © Spencer Grant/Photo Edit; 5.4: From *Biochemistry* 4th ed. by Stryer. Copyright © 1995 by Lubert Stryer. Used with permission of W.H. Freeman & Company; 5.9(left): © Manfred Kage/ Peter Arnold, Inc.; (right): © Dr. Lewis K. Shumway; 5.13(both): © Eric Soder/Tom Stack & Associates; 5.14A: © SPL/Photo Researchers; 5.14B: Courtesy of Dr. Linda E. Graham & Bruce Iverson; 5.14C: © J. Robert Waaland/Visuals Unlimited; 5.24: © R. Robinson /Visuals Unlimited; 5.25: © F. Widell/Visuals Unlimited; Page 133: © PhotoDisc/Vol. # 46

Chapter 6

6.1: © Lee D. Simon/Photo Researchers; 6.4: © Biophoto Associates/Photo Researchers; 6.5: © SPL/Photo Researchers; 6.8(all): © Andrew S. Bajer; 6.9A: © David M. Phillips/Visuals Unlimited; 6.10: © Cabisco/Visuals Unlimited; 6.13: © Custom Medical Stock Photos; 6.18: © David Cavagnaro/Peter Arnold, Inc.; 6.22: © Andrew S. Bajer; 6.24(all): © C.A. Hasenkampf/Biological Photo Service; Page 159: © PhotoDisc/Vol. # 29

Chapter 7

7.1A: Courtesy of American Museum of Natural History; 7.1B: © Richard Gross/Biological Photography; 7.5: Courtesy R. W. Van Norman; 7.13A: From Albert & Blakeslee “Corn and Man,” *Journal of Heredity* v. 5, pg. 511, 1914, Oxford University Press; 7.15A,B: © Fred Bruemmer; 7.16A(both): © CBS/Phototake; 7.18: © Leonard Lessin/Peter Arnold; 7.19A: © L. Willatt, East Anglian Regional Genetics Service/SPL/Photo Researchers; 7.19B: © Richard Hutchings/Photo Researchers; 7.22: © Corbis; 7.24A(both): © Stanley Flegler/ Visuals Unlimited; 7.29: © Yoav Levy/ Phototake; Page 184(top left): Courtesy of American Museum of Natural History; (top right): © Richard Hutchings/Photo Researchers; (bottom right): © Stanley Flegler/Visuals Unlimited; Page 185: © PhotoDisc/Vol. # 72

Chapter 8

8.2: O.T. Avery, C.M. Macleod & M. McCarty, “Studies on the Chemical Nature of the Substance Inducing Transformation of Pneumococcal Types.” Reproduced from the *Journal of Experimental Medicine* 79 (1944): 137-158. fig.1 by copyright permission of the Rockefeller University Press. Reproduced by permission. Photograph made by Mr. Joseph B. Haulenbeek; 8.3B: © Lee D. Simon/Photo Researchers; 8.3D: Courtesy of Prof. A.K. Kleinschmidt; 8.6: © A.C. Barrington Brown/ Photo Researchers; 8.7: From J.D. Watson, *The Double Helix* Atheneum, New York, 1968. Cold Spring Harbor Lab; 8.13: N. Ban, P. Nissen, J. Hansen, P.B. Moore & T.A. Steitz, “The Complete Atomic Structure of the Large Ribosomal subunit at 2.4A Resolution,” Reprinted with permission from *Science,* v. 289 #5481, p917, © 2000 American Association for the Advancement of Science; 8.22: © David Scharf; Page 209: © PhotoDisc/Vol. # 18

Chapter 9

9.1(top left): R.L. Brinster, U. of Pennsylvania School of Vet. Med.; (top right): Courtesy of Dr. Ken Culver, Photo by John Crawford, National Institutes of Health; (bottom left): Robert H. Devlin/Fisheries & Oceans Canada; (bottom right): Courtesy of Dr. Richard Shade; 9.4: Courtesy of

Bio-Rad Laboratories; 9.9: Courtesy of Lifecodes Corp., Stamford, CT; 9.11: Courtesy of Dr. John Sanford, Cornell University; 9.12: Monsanto Co.; 9.15: © Roger Minkoff/Animals Animals/Earth Scenes; 9.16: AP/Wide World Photos; Page 230 (top left): Courtesy of Bio-Rad Laboratories; (top right): Courtesy of Dr. John Sanford, Cornell University; Page 231: © Corbis/Vol. # 40

Chapter 10

10.1: Nancy Federspiel; 10.7: © Heather Angel; 10.8: Courtesy of Tyson Clark, U. of California, Santa Cruz; 10.10: Courtesy of Prof. Walter J. Gehring; 10.11(both): Courtesy of Dr. William Jeffery; 10.12: Courtesy of Anna Di Gregorio; 10.13: H. Eric Xu, Mark A. Rould, Wenqing Xu, Jonathan A. Epstein, Richard L. Maas, and Carl O. Pabo: "Crystal structure of the human Pax6 paired domain-DNA complex reveals specific roles for the linker region and carboxy-terminal subdomain in DNA binding" *Genes Dev.* 1999 13: 1263-1275; Page 248: © Heather Angel; Page 249: © PhotoDisc/Vol. # 72

Chapter 11

11.1: © M. Philip Kahl, Jr./Photo Researchers; 11.15: Courtesy of Dr. Victor A. McKusick, Johns Hopkins University; 11.19: Courtesy of the University of Chicago Library/Dept. of Special Collections and Todd L. Savitt; 11.22(both): © Breck P. Kent/Animals Animals/Earth Scenes; 11.24(top left): © Hank De Lespinasse/The Image Bank; (center): © Steve Krongard/The Image Bank; (top right): © Chuck Kuhn/The Image Bank; (bottom left): © Dan Coffey/The Image Bank; (bottom right): © G. Casperen/The Image Bank; Page 274: © Breck P. Kent/Animals Animals/Earth Scenes; Page 275: © PhotoDisc/Vol. # 6

Chapter 12

12.2A(left): © Dwight R. Kuhn; (right) © Henry Ausloos/Animals Animals/Earth Scenes; 12.2B(left): © Heather Angel; (right): © John Cancalosi/Peter Arnold Inc.; 12.2C(left): © S. Maslowski/Visuals Unlimited; (right): © Manfred Danegger/Peter Arnold, Inc.; 12.4(left): © Gerard Lacz/Peter Arnold, Inc.; (center): © Ralph Reinhold/Animals Animals/Earth Scenes; (right): © Grant Heilman/Grant Heilman Photography; F12.10: © Lee D. Simon/Photo Researchers; Page 291: © PhotoDisc/Vol. # 19

Chapter 13

13.1: © David M. Phillips/Visuals Unlimited; 13.3: Courtesy of Dr. Charles Brinton; p.296 (above): © Abraham & Beachey/BPS/Tom Stack & Associates; (below): © F. Widell/Visuals Unlimited; 13.5: © Science VU/Visuals Unlimited; 13.6: © Leonard Lessin/Peter Arnold, Inc.; 13.7: © Dwight R. Kuhn; 13.9: Reprinted with permission from Donald L.D. Caspar, Science 227:773-776 Feb. 1985 © American Association for the Advancement of Science; 13.11A: © Dept. of Microbiology, Biozentrum/Science Photo Library/Photo Researchers; 13.13: © Scott Camazine/Photo Researchers; 13.15A: Cynthia Goldsmith & Jackie Katz, Centers for Disease Control & Prevention; 13.16A: © Larry Mulvehill/ Photo Researchers; 13.16B: AP/Wide World Photos; 13.18: © Ralph Eagle Jr./Photo Researchers; Page 314: © Dwight R. Kuhn; Page 315: © Corbis/Vol. # 40

Chapter 14

14.1: Courtesy of Dr. Edward W. Daniels, Argonne National Lab, the University of Illinois College of Medicine at Chicago; 14.3A: © Brian Parker/Tom Stack & Associates; 14.3B: © Manfred Kage/Peter Arnold, Inc.; 14.5,14.6: © John D. Cunningham/Visuals Unlimited; 14.7: © L.L. Sims/Visuals Unlimited; 14.8: © Phil A Harrington/Peter Arnold, Inc.; 14.9: © Manfred Kage/Peter Arnold, Inc.; 14.10: © Phil A. Harrington/Peter Arnold, Inc.; 14.11A: © Edward S. Ross; 14.11B: © Ed Reschke; 14.15: © Kevin Schafer/Peter Arnold, Inc.; 14.17: © D.P. Wilson/ Photo Researchers; 14.19: © John D. Cunningham/ Visuals Unlimited; 14.20: © Rick Harbo; 14.21: © Bob Evans/Peter Arnold, Inc.; 14.23: © Edward S. Ross; Page 334: © Bob Evans/Peter Arnold, Inc.; Page 335: © Photodisc/Vol. # 2

Chapter 15

15.1: © Arthur E. Morris/ Visuals Unlimited; 15.2: © Kjell Sandved/Butterfly Alphabet; 15.3: © Bill Keogh/Visuals Unlimited; 15.4: © L. West/Photo Researchers; 15.5: Ralph Williams, USDA, Forest Service; 15.6: ©Cabisco /Phototake; 15.8: © Ray Coleman/Photo Researchers; 15.10: © Alexander Lowry/The National Audubon Society Collection/ Photo Researchers; 15.12: © Barry L. Runk/Grant Heilman Photography; 15.13: © H.C. Huang/ Visuals Unlimited; 15.14A: © Edward S. Ross; F15.14B: © James L. Castner; 15.15: © D.H. Marx/Visuals Unlimited; 15.16: © R. Roncadori/ Visuals Unlimited; Page 350 (top left): © L. West/ Photo Researchers; (top right): © Edward S. Ross; Page 351: © PhotoDisc/Vol. # 2

Chapter 16

16.1: © Tim Houf/ Visuals Unlimited; 16.2: © Terry Ashley/Tom Stack & Associates; 16.4A: © Edward S. Ross; 16.4B: © Richard Gross/ Biological Photography; 16.5: © Edward S. Ross; 16.7: Courtesy of Hans Steur,The Netherlands; 16.8: © E.J. Cable/Tom Stack & Associates; 16.9A: © Edward S. Ross; 16.9B: © Rod Planck/ Tom Stack & Associates; 16.9C: © Edward S. Ross; 16.11: © Kingsley R. Stern; 16.13: T. Walker; (inset): © R.J. Delorit, Agronomy Publications; 16.14A: © Edward S. Ross; 16.14B: © David Cavagnaro/Peter Arnold, Inc.; 16.14C: © Runk/Schoenberger/Grant Heilman Photography; 16.14D: © Kjell Sandved/Butterfly Alphabet; 16.16B: © Ed Pembleton; 16.17: © Michael & Patricia Fogden; 16.20: © James L. Castner; Page 372: © Tim Houf/Visuals Unlimited; Page 373: © PhotoDisc/Vol. # 60

Chapter 17

17.2: © Biophoto Associates/ Photo Researchers; 17.3: © George Wilder/Visuals Unlimited; 17.4: © Richard H. Gross/Biological Photography; 17.5: © E.J. Cable /Tom Stack & Associates; 17.6D: Courtesy of Wilfred Cote, SUNY College of Environmental Forestry; 17.7A: © Randy Moore/ Visuals Unlimited; 17.8(top left) © Michael & Patricia Fogden; (all others): © Edward S. Ross; 17.10: © Ed Reschke; 17.11: © R.A. Gregory/ Visuals Unlimited; 17.12A,B: © Ed Reschke; 17.13: © John D. Cunningham/Visuals Unlimited; 17.14: © CBS/Phototake; 17.15B: © Terry Ashley/Tom Stack & Associates; 17.16: © Jim Solliday/Biological Photo Service; 17.17: © E.J. Cable/Tom Stack & Associates; 17.18: © John Shaw/Tom Stack & Associates; 17.21: © Doug Wechsler/Animals Animals/Earth Scenes; 17.24A: © J.A.L. Cooke/Animals Animals/Earth Scenes; 17.24B: © Robert Mitchell/Tom Stack & Associates; 17.24C: © Kjell Sandved/Butterfly Alphabet; Page 392: © George Wilder/Visuals Unlimited; Page393: © PhotoDisc/Vol. # 16

Chapter 18

18.2A: Courtesy of William F. Chissoe, Noble Microscopy Lab, University of Oklahoma; 18.2B: Courtesy of Dr. Joan Nowicke; 18.3: © OSF/ Animals Animals/Earth Scenes; 18.6A(above): © Edward S. Ross; (below): © Kirtley-Perkins/ Visuals Unlimited; 18.6B(above): © James L. Castner; (below): © Dwight R. Kuhn; F18.7(left): © Adam Hart-Davis/SPL/Photo Researchers; (right): © Helmut Gritscher/Peter Arnold Inc.; 18.13: © Sylvan H. Wittwer/Visuals Unlimited; 18.14: © Kingsley R. Stern; 18.15A: © John Solden/Visuals Unlimited; 18.15B: © David M. Phillips/Visuals Unlimited; 18.18: © John D. Cunningham/Visuals Unlimited; Page 413: © Corbis/Vol. # 102

Chapter 19

19.1: © Edward S. Ross; 19.3A: © Jim & Cathy Church; 19.3B: © David J. Wrobel/Biological Photo Service; 19.5A: © Gwen Fidler/Tom Stack & Associates; 19.5B: © Gwen Fidler/Tom Stack & Associates; 19.5C: © Daniel Gotshall; 19.11A: © T.E. Adams/Visuals Unlimited; 19.11B: © Stan Elems/Visuals Unlimited; 19.15A: © Larry Jensen/ Visuals Unlimited; 19.15B: © T.E. Adams/Visuals Unlimited; 19.17A: © Milton Rand/Tom Stack & Associates; 19.17B: © Kjell Sandved/Butterfly Alphabet; 19.17C: © Fred Bavendam/Peter Arnold Inc.; 19.19A: © David M. Dennis/Tom Stack & Associates; 19.19B: © Kjell Sandved/Butterfly Alphabet; 19.21: © S. Giordano III; 19.24: © Frans Lanting/Minden Pictures; 19.25A: © George & Kathy Dodge; 19.25B: © Anne Moreton/Tom Stack & Associates; 19.26A: © G. Lacz/Natural History Photographic Agency; 19.26B: © Edward S. Ross; 19.26C: © Kjell Sandved/Butterfly Alphabet; 19.28A: © Alex Kerstitch/Visuals Unilimited; 19.28B: © Edward S. Ross; 19.30A: © Kjell Sandved/Butterfly Alphabet; 19.30B: © Edward S. Ross; 19.30C: © Don Valenti/ Tom Stack & Associates; 19.30D: © John Gerlach/ Visuals Unlimited; 19.30E: © Edward S. Ross; 19.30F: © J.A. Alcock/Visuals Unlimited; 19.30G: © Cleveland P. Hickman; 19.32A: © Alex Kirstitch/ Visuals Unlimited; 19.32B: © Carl Roessler/Tom Stack & Associates; 19.32C: © William C. Ober; 19.32D: © Jeff Rotman; 19.32E: © Daniel W. Gotshall; 19.34A: © Jim & Cathy Church; 19.34B: © Heather Angel; 19.34C: © Stephen J. Krasemann/ DRK Photo; 19.36: © Eric N. Olsen/The University of Texas MD Anderson Cancer Center; Page 450 (above): © George & Kathy Dodge; (below): © Stephen J. Krasemann/DRK Photo; Page 451: © Corbis/Vol. # 64

Chapter 20

20.2A: © John Gurche; 20.2B: © Alex Kerstitch/ Visuals Unlimited; 20.3: ©Louis Psihoyos/Matrix;

20.4: The Age of Reptiles, a mural by Rudolph F. Zallinger. Copyright © 1966, 1975, 1985, 1989, Peabody Museum of Natural History, Yale University, New Haven, CT; 20.5: © E.R. Degginger/Animals Animals/Earth Scenes; 20.8: © Tom McHugh/Photo Researchers; 20.10: © Marty Snyderman; 20.12: © John D. Cunningham/Visuals Unlimited; 20.13: © John Shaw/Tom Stack & Associates; 20.17: © OSF/Animals Animals/Earth Scenes; Page 474: © John D. Cunningham/Visuals Unlimited; Page 475: © PhotoDisc/Vol. # 44

Chapter 21

21.1A: © Alan Nelson/Animals Animals/Earth Scenes; 21.2A: © Denise Tackett/Tom Stack & Associates; 21.2B: © Edward S. Ross; 21.3(all): © Russell A. Mittermeier; 21.5, page 482 (left): © David L. Brill 1985/National Museum of Ethiopia, Addis Ababa; (right): © David L. Brill 1985/Transvaal Museum, Pretoria; 21.5, page 483(left): © David L. Brill 1985/Transvaal Museum, Pretoria; (right): © David L. Brill 1985/National Museum of Tanzania, Dar es Salaam; 21.6(all): © David L. Brill/Dr. Owen Lovejoy & students, Kent State University; 21.7: © John Reader/SPL/Photo Researchers; 21.9: © 1994 Tim D. White/Brill Atlanta, Housed in National Museum of Ethiopia; 21.10: © Bob Campbell/Sygma; 21.12: © 1985 David L. Brill/National Museums of Kenya, Nairobi; 21.14. Page 490 (both): © 1985 David L. Brill/National Museums of Kenya, Nairobi; 21.14, page 491(both): © 1985 David L. Brill/Musee de L'Homme, Paris; 21.14D: © 1985 David l. Brill/Musee de L'Homme, Paris; 21.16: AP/Wide World Photos; Page 494(top left): © Alan Nelson/Animals Animals/Earth Scenes; (bottom left): © Bob Campbell/Corbis Sygma; (top right): © 1985 David L. Brill/National Museums of Kenya, Nairobi; (bottom right): AP/Wide World Photos; Page 495: © PhotoDisc/Vol. # 44

Chapter 22

22.1: © Daniel Gotshall; 22.2: © T.E. Adams/Visuals Unlimited; 22.3: © Denise Tackett/Tom Stack & Associates; 22.4: © O.S.F./Animals Animals/Earth Scenes; 22.5: © Alex Kerstitch/Visuals Unlimited; 22.6: © Kjell Sandved/Visuals Unlimited; 22.7: © Kjell Sandved/Butterfly Alphabet; 22.8: © Alan Nelson/Animals Animals/Earth Scenes; 22.11: © Cynthia Turner Alexander, Terry Cockerhan, Synapse Media Production, from C.P. Anthony & G.A. Thibodeau: Textbook of Anatomy & Physiology, 10/e, 1979, C.V. Mosby; 22.12: © Jim Merli/Visuals Unlimited; p. 507A (top down): © Ed Reschke/Peter Arnold Inc.; B,C: © Ed Reschke; D: © Fred Hossler/Visuals Unlimited; E: © Ed Reschke; 22.13: © Marty Snyderman; p. 509A(top down): © Biology Media/Photo Researchers; B: © Biophoto Associates/Photo Researchers; C: © Chuck Brown/Photo Researchers; D: © Ken Edward/Science Source/Photo Researchers; E: © Biophoto Associates/Photo Researchers; p.510(all): © Ed Reschke; 22.15: © Cleveland P. Hickman, Jr.; 22.16: © David M. Dennis/Tom Stack & Associates; 22.19: © Ed Reschke; 22.20(both): © Dr. Michael Klein/Peter Arnold, Inc.; Page 522(top left): © Cynthia Turner Alexander, Terry Cockerhan, Synapse Media Production, from C.P. Anthony & G.A. Thibodeau: *Textbook of Anatomy & Physiology,* 10/e, 1979, C.V. Mosby; (bottom left): © Ken Edward/Science source/Photo Researchers; (right): © David M. Dennis/Tom Stack & Associates; Page 523: © PhotoDisc/Vol. # 01

Chapter 23

23.5,23.6: © Ed Reschke; 23.8: © Manfred Kage/Peter Arnold Inc.; 23.15(all): Courtesy of Frank P. Sloop, Jr.; 23.22: © Ellen Dirkson/Visuals Unlimited; 23.26: Courtesy of American Cancer Society; Page 552: Courtesy of American Cancer Society; Page 553: © PhotoDisc/Vol. # 40

Chapter 24

24.14B: © David M. Phillips/Visuals Unlimited; 24.14C: © Biophoto Associates/Science Source/Photo Researchers; 24.27: © Larry Brock/Tom Stack & Associates; 24.30: © Prof. P. Motta/Dept. of Anatomoy/University "La Sapienza," Rome/Science Photo Library/Photo Researchers; Page 585: © Corbis/Vol. # 76

Chapter 25

25.1: © Joe McDonald/Visuals Unlimited; 25.3: © Manfred Kage/Peter Arnold, Inc.; 25.5: Courtesy of Dr. Gilla Kaplan; 25.10D: From Alan S. Rosenthal, New England Journal of Medicine 303: 1153, 1980; 25.18: © Visuals Unlimited; 25.19: National Library of Medicine; F25.21: © Oliver Meckes/Photo Researchers; 25.23: © CDC/Science Source/Photo Researchers; Page 608: © CDC/Science source/Photo Researchers; Page 609: © Corbis/Vol. # 106

Chapter 26

26.6: © Springer-Verlag, New York, Inc.; 26.7: © John Heuser, Washington University School of Medicine, St. Louis, MO; F26.8: © E.R. Lewis, YY Zeevi, T.E. Everhart, U. of California/Biological Photo Service; 26.11: © Ernest Wilkinson/Animals Animals/Earth Scenes; 26.17: Dr. Marcus E. Rachle, Washington University, McDonnell Center for High Brain Function/Peter Arnold, Inc.; 26.18: Photo Lennart Nilsson/Albert Bonniers Forlag AB, *Behold Man,* Little Brown & Co.; 26.22: © Wendy Shatil/Bob Rozinski/Tom Stack & Associates; 26.30: © Stephen Dalton/Animals Animals/Earth Scenes; 26.31(left): © David M. Dennis/Tom Stack & Associates; (center, right): Kjell Sandved/Butterfly Alphabet; 26.33: Courtesy of Beckman Vision Center at UCSF School of Medicine; D. Copenhagen, S. Mittman, M. Maglio; 26.38: © Leonard L. Rue, III; Page 641: © PhotoDisc/Vol. # 25

Chapter 27

27.7: Photo Lennart Nilsson/Albert Bonniers Forlag AB, *Behold Man,* Little Brown & Co.; 27.11: © Ed Reschke; Page 658: Photo Lennart Nilsson/Albert Bonniers Forlag AB, *Behold Man,* Little Brown & Co.; Page 659: © PhotoDisc/Vol. # 17

Chapter 28

28.1A: © Chuck Wise/Animals Animals/Earth Scenes; 28.1B: © Fred McConnaughey/The National Audubon Society Collection/Photo Researchers; 28.3: © David Doubilet; 28.4: © Hans Pfletschinger/Peter Arnold Inc.; 28.6: © Cleveland P. Hickman; 28.7: © Frans Lanting/Minden Pictures; 28.8A: © Jean Phillippe Varin/Jacana/Photo Researchers; 28.8B: © Tom McHugh/The National Audubon Society Collection/Photo Researchers; 28.8C: © Fritz Prenzel/Animals Animals/Earth Scenes; 28.9: Photo Lennart Nilsson/Albert Bonniers Forlag AB, *A Child is Born,* Dell Publishing Co.; 28.14: Photo Lennart Nilsson/Albert Bonniers Forlag AB/*Behold Man,* Little Brown & Co.; 28.16: © Ed Reschke; 28.17A,B,C: Photo Lennart Nilsson/Albert Bonniers Forlag AB, *A Child is Born,* Dell Publishing Co.; D:Photo Lennart Nilsson/Albert Bonniers Forlag AB, *Behold Man,* Little Brown & Co.; 28.19: © Photo Researchers; 28.20(all): © McGraw-Hill Higher Education Group, Inc./Bob Coyle, photographer; Page 682(top left): © Hans Pfletschinger/Peter Arnold Inc.; (bottom left): Photo Lennart Nilsson/Albert Bonniers Forlag AB/*Behold Man,* Little Brown & Co.; (top right): Photo Lennart Nilsson/Bonniers Forlag AB, *A Child is Born,* Dell Publishing Co.; (bottom right): © McGraw-Hill Higher Education Group, Inc./Bob Coyle, photographer; Page 683: © Corbis/Vol. # 10

Chapter 29

29.3(all): © Edward S. Ross; 29.8: © Doug Sokell/Tom Stack & Associates; 29.18: © Anne Wertheim/Animals Animals/Earth Scenes; 29.19: © W. Gregory Brown/Animals Animals/Earth Scenes; 29.20: © Kenneth L. Smith; 29.21: © Edward S. Ross; 29.22A: © Fred Rhode/Visuals Unlimited; 29.22B: © Dwight Kuhn; 29.26: © Michael Graybill & Jan Hodder/Biological Photo Service; 29.27: © E.R. Degginger/Photo Researchers; 29.28: © S.J. Krasemann/Peter Arnold, Inc.; 29.29: © J. Weber/Visuals Unlimited; 29.30: © IFA/Peter Arnold, Inc.; 29.31: © Charlie Ott/The National Audubon Society Collection/Photo Researchers; 29.32: © John Shaw/Tom Stack & Associates; 29.33A: © Tom McHugh/Photo Researchers; 29.33B: © Dave Watts/Tom Stack & Associates; 29.33C: © E.R. Degginger/Animals Animals/Earth Scenes; 29.33D: © Gunter Ziesler/Peter Arnold, Inc.; Page 710: © S.J. Krasemann/Peter Arnold Inc.; Page 711: © Corbis/Vol. # 58

Chapter 30

30.1: © Manfred Danegger/Peter Arnold, Inc.; 30.4: Courtesy of National Museum of Natural History, Smithsonian Institution; 30.5(both): © Jean Vie/Gamma Liaison; 30.8A: © Vanessa Vick/Photo Researchers; 30.8B,C,D: © Edward S. Ross; 30.11(all): Courtesy of J.B. Losos; F30.13: © Merlin D. Tuttle, Bat Conservation International; 30.14: © Jim Harvey/Visuals Unlimited; 30.15: © Gunter Ziesler/ Peter Arnold, Inc.; 30.16: © N&C Photography/Peter Arnold, Inc.; 30.17A: © Edward S. Ross; 30.17B: © Roger Wilmshurst/The National Audubon Society Collection/Photo Researchers; 30.18(both): © Edward S. Ross; 30.19(both): © Lincoln P. Brower; 30.20: © James L. Castner; 30.21: © James L. Castner; 30.22: Courtesy of Rolf O. Peterson; 30.23A: © Biophoto Associates/Science Source/Photo Researchers; 30.24: © Tom J. Ulrich/Visuals Unlimited; 30.25A: © Edward S. Ross; 30.25B: Courtesy of Dr. Paul A. Opler; 30.26(all):.© Edward S. Ross; 30.28(all): © Tom Bean; 30.29: © Edward S. Ross; 30.32: Courtesy of E.O. Wilson; Page 740 (top left): © Jean Vie/Gamma Liaison; (bottom left): © Gunter Ziesler/Peter Arnold, Inc.; (top right): © James L. Castner; (bottom right): © Tom Bean; Page 741: © Corbis/Vol. # 86

Chapter 31

31.2: © Grant Heilman/Grant Heilman Photography; 31.3: Courtesy of Richard Klein; 31.4: NASA; 31.7: © Byron Augustine/Tom Stack & Associates; 31.8: Sovfoto/Eastfoto; 31.9: © Gary Griffen/Animals Animals/Earth Scenes; 31.10A: © James Blair/National Geographic Society; 31.10B: NASA; 31.10C: © Frans Lanting/Minden Pictures; 31.11: © Stephanie Maze/Woodfin Camp & Associates; 31.13: © Steve McCurry/National Geographic Society; Page 756 (left): © Frans Lanting/Minden Pictures; (right): © Steve McCurry/National Geographic Society; Page 757: © Corbis/Vol. # 6

Line Art and Text

Chapter 1

Chapter Opener: **The Far Side® by Gary Larson © 1984 FarWorks, Inc. All Rights Reserved. Used with permission.**

Chapter 2

Chapter Opener: **The Far Side® by Gary Larson © 1980 FarWorks, Inc. All Rights Reserved. Used with permission.**

Chapter 3

Chapter Opener: **The Far Side® by Gary Larson © 1991 FarWorks, Inc. All Rights Reserved. Used with permission.**

Chapter 4

Chapter Opener: **The Far Side® by Gary Larson © 1985 FarWorks, Inc. All Rights Reserved. Used with permission.** Figure 4.1: Copyright © 1988 From *Essential Cell Biology: An Introduction to the Molecular Biology of the Cell, 1st Edition* by Bruce Alberts, et al. Reproduced by permission of Routledge, Inc., part of The Taylor & Francis Group.

Chapter 5

Chapter Opener: **The Far Side® by Gary Larson © 1985 FarWorks, Inc. All Rights Reserved. Used with permission.**

Chapter 6

Chapter Opener: **The Far Side® by Gary Larson © 1983 FarWorks, Inc. All Rights Reserved. Used with permission.**

Chapter 7

Chapter Opener: **The Far Side® by Gary Larson © 1982 FarWorks, Inc. All Rights Reserved. Used with permission.** Figure 7.28: From Kent M. Van De Graaff and Stuart I. Fox, *Concepts of Human Anatomy and Physiology,* 4th ed. Copyright © 1995 McGraw-Hill Companies, Inc., Dubuque, IA. All rights reserved. Reprinted by permission.

Chapter 8

Chapter Opener: **The Far Side® by Gary Larson © 1986 FarWorks, Inc. All Rights Reserved. Used with permission.** Figure 8.12: From *Biology: The Living Science* by Kenneth R. Miller and Joseph Levine. Copyright © 1998 by Pearson Education, Inc., publishing as Prentice Hall. Used by permission. p. 224 CALVIN AND HOBBES © Watterson. Reprinted with permission of Universal Press Syndicate. All rights reserved.

Chapter 9

Chapter Opener: **The Far Side® by Gary Larson © 1984 FarWorks, Inc. All Rights Reserved. Used with permission.**

Chapter 10

Chapter Opener: **The Far Side® by Gary Larson © 1986 FarWorks, Inc. All Rights Reserved. Used with permission.** Table 10.3: From Plant Biochemistry and Molecular Biology, by P.J. Lea and R.C. Leegods (eds.). Copyright © 1993 John Wiley & Sons Limited. Reproduced with permission.

Chapter 11

Chapter Opener: **The Far Side® by Gary Larson © 1991 FarWorks, Inc. All Rights Reserved. Used with permission.** Figure 11.12: From The New York Times, December 15, 1998. Copyright © 1998 The New York Times. Reprinted with permission.

Chapter 12

Chapter Opener: **The Far Side® by Gary Larson © 1981 FarWorks, Inc. All Rights Reserved. Used with permission.** Figure 12.3: From Niles Eldredge, "Life in the Balance," in *Natural History,* June 1998. Reprinted by permission of Patricia J. Wynne.

Chapter 13

Chapter Opener: **The Far Side® by Gary Larson © 1985 FarWorks, Inc. All Rights Reserved. Used with permission.**

Chapter 14

Chapter Opener: **The Far Side® by Gary Larson © 1981 FarWorks, Inc. All Rights Reserved. Used with permission.**

Chapter 15

Chapter Opener: **The Far Side® by Gary Larson © 1984 FarWorks, Inc. All Rights Reserved. Used with permission.**

Chapter 16

Chapter Opener: **The Far Side® by Gary Larson © 1982 FarWorks, Inc. All Rights Reserved. Used with permission.**

Chapter 17

Chapter Opener: **The Far Side® by Gary Larson © 1984 FarWorks, Inc. All Rights Reserved. Used with permission.**

Chapter 18

Chapter Opener: **The Far Side® by Gary Larson © 1985 FarWorks, Inc. All Rights Reserved. Used with permission.**

Chapter 19

Chapter Opener: **The Far Side® by Gary Larson © 1982 FarWorks, Inc. All Rights Reserved. Used with permission.**

Chapter 20

Chapter Opener: **The Far Side® by Gary Larson © 1983 FarWorks, Inc. All Rights Reserved. Used with permission.**

Chapter 21

Chapter Opener: **The Far Side® by Gary Larson © 1982 FarWorks, Inc. All Rights Reserved. Used with permission.**

Chapter 22

Chapter Opener: **The Far Side® by Gary Larson © 1982 FarWorks, Inc. All Rights Reserved. Used with permission.**

Chapter 23

Chapter Opener: **The Far Side® by Gary Larson © 1982 FarWorks, Inc. All Rights Reserved. Used with permission.**

Chapter 24

Chapter Opener: **The Far Side® by Gary Larson © 1986 FarWorks, Inc. All Rights Reserved. Used with permission.** Figure 24.16: From Cleveland P. Hickman, Jr., et al., *Integrated Principles of Zoology,* 10th edition. Copyright © 1997 McGraw-Hill Companies, Inc., Dubuque, IA. All rights reserved. Reprinted by permission.

Chapter 25

Chapter Opener: **The Far Side® by Gary Larson © 1984 FarWorks, Inc. All Rights Reserved. Used with permission.** Figure 25.8: From "Immunity and the Invertebrates," Scientific American, November 1996. Copyright Roberto Osti Illustrations. Reprinted with permission.

Chapter 26

Chapter Opener: **The Far Side® by Gary Larson © 1985 FarWorks, Inc. All Rights Reserved. Used with permission.**

Chapter 27

Chapter Opener: **The Far Side® by Gary Larson © 1988 FarWorks, Inc. All Rights Reserved. Used with permission.**

Chapter 28

Chapter Opener: **The Far Side® by Gary Larson © 1981 FarWorks, Inc. All Rights Reserved. Used with permission.**

Chapter 29

Chapter Opener: **The Far Side® by Gary Larson © 1983 FarWorks, Inc. All Rights Reserved. Used with permission.**

Chapter 30

Chapter Opener: **The Far Side® by Gary Larson © 1984 FarWorks, Inc. All Rights Reserved. Used with permission.**

Chapter 31

Chapter Opener: **The Far Side® by Gary Larson © 1984 FarWorks, Inc. All Rights Reserved. Used with permission.**

INDEX

Metric–English Conversions

Length

ENGLISH (USA)	=	METRIC
inch	=	2.54 cm, 25.4 mm
foot	=	0.30 m, 30.48 cm
yard	=	0.91 m, 91.4 cm
mile (statute) (5280 ft)	=	1.61 km, 1609 m
mile (nautical) (6077 ft, 1.15 statute mi)	=	1.85 km, 1850 m

METRIC	=	ENGLISH (USA)
millimeter	=	0.039 in
centimeter	=	0.39 in
meter	=	3.28 ft, 39.37 in
kilometer	=	0.62 mi, 1091 yd, 3273 ft

Weight

ENGLISH (USA)	=	METRIC
grain	=	64.80 mg
ounce	=	28.35 g
pound	=	453.60 g, 0.45 kg
ton (short–2000 lb)	=	0.91 metric ton (907 kg)

METRIC	=	ENGLISH (USA)
milligram	=	0.02 grain (0.000035 oz)
gram	=	0.04 oz
kilogram	=	35.27 oz, 2.20 lb
metric ton (1000 kg)	=	1.10 tons

Volume

ENGLISH (USA)	=	METRIC
cubic inch	=	16.39 cc
cubic foot	=	0.03 m^3
cubic yard	=	0.765 m^3
ounce	=	0.03 liter (3 ml)*
pint	=	0.47 liter
quart	=	0.95 liter
gallon	=	3.79 liters

METRIC	=	ENGLISH (USA)
milliliter	=	0.03 oz
liter	=	2.12 pt
liter	=	1.06 qt
liter	=	0.27 gal

1 liter ÷ 1000 = milliliter or cubic centimeter (10^{-3} liter)
1 liter ÷ 1,000,000 = microliter (10^{-6} liter)
**Note:* 1 ml = 1 cc

Fahrenheit-Celsius Conversion

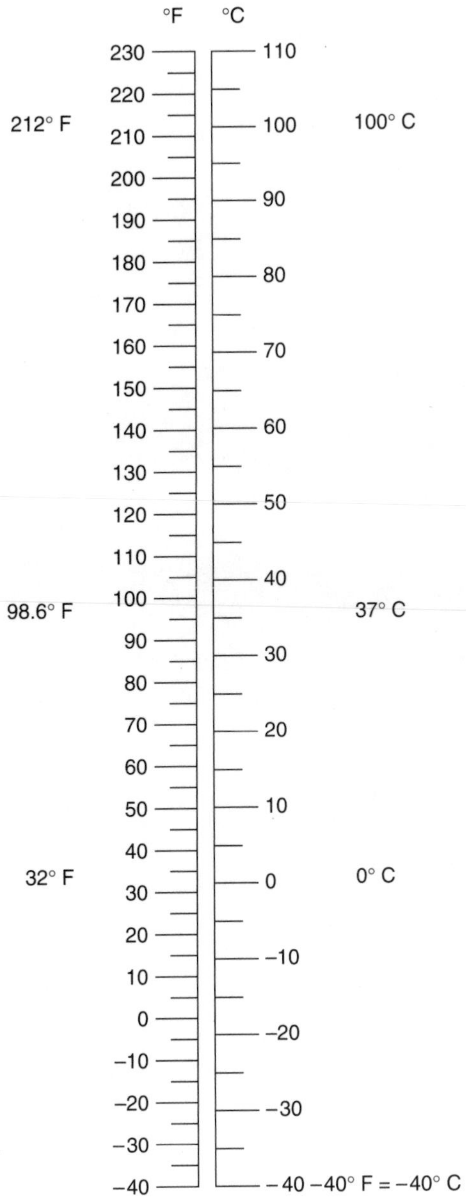

°F
°C
230
220
210
200
190
180
170
160
150
140
130
120
110
100
90
80
70
60
50
40
30
20
10
0
–10
–20
–30
–40
110
100
90
80
70
60
50
40
30
20
10
0
–10
–20
–30
–40
212° F
100° C
98.6° F
37° C
32° F
0° C
–40° F = –40° C

To convert temperature scales:
Fahrenheit to Celsius °C = $\frac{5}{9}$ (°F) –32

Celsius to Fahrenheit °F = $\frac{9}{5}$ (°C) +32